## Table of Problem-Solving Approaches


| | PROBLEM-SOLVING APPROACH | PAGE |
|---|---|---|
| | Problem-solving approach | 51 |
| 2.1 | Motion with constant acceleration | 54 |
| 3.1 | Projectile motion problems | 88 |
| 5.1 | Equilibrium problems | 135 |
| 5.2 | Dynamics problems | 139 |
| 6.1 | Circular dynamics problems | 180 |
| 7.1 | Rotational dynamics problems | 227 |
| 8.1 | Static equilibrium problems | 247 |
| 9.1 | Conservation of momentum problems | 290 |
| 10.1 | Conservation of energy problems | 328 |
| 11.1 | Energy efficiency problems | 357 |
| 12.1 | Calorimetry problems | 419 |
| 16.1 | Standing waves | 561 |
| 20.1 | Electric forces and Coulomb's law | 707 |
| 21.1 | Conservation of energy in charge interactions | 741 |
| 23.1 | Resistor circuits | 812 |
| 24.1 | Magnetic-force problems | 860 |
| 25.1 | Electromagnetic induction | 890 |


## Table of Synthesis Boxes


| | SYNTHESIS | PAGE |
|---|---|---|
| 2.1 | Describing motion in one dimension | 49 |
| 3.1 | Projectile motion | 87 |
| 5.1 | A catalog of forces | 157 |
| 7.1 | Linear and circular motion | 213 |
| 7.2 | Linear and rotational dynamics | 225 |
| 9.1 | Momentum and impulse | 283 |
| 10.1 | Energy and its conservation | 331 |
| 14.1 | Describing simple harmonic motion | 493 |
| 15.1 | Wave motion | 528 |
| 15.2 | Wave power and intensity | 535 |
| 16.1 | Standing wave modes | 561 |
| 17.1 | Double-slit interference and diffraction gratings | 598 |
| 18.1 | Lenses, mirrors, and their sign conventions | 645 |
| 20.1 | Two key electric fields | 718 |
| 21.1 | The parallel-plate capacitor: Potential and electric field | 746 |
| 21.2 | Potential of a point charge and a charged sphere | 750 |
| 24.1 | Fields from currents | 851 |
| 25.1 | Electromagnetic waves | 896 |
| 26.1 | AC circuit elements | 931 |
| 30.1 | Nuclear decay modes | 1066 |




## Table of Math Relationships


| MATH RELATIONSHIP | PAGE |
|---|---|
| Proportional relationships | 39 |
| Quadratic relationships | 50 |
| Inversely proportional relationships | 116 |
| Inverse-square relationships | 191 |
| Square-root relationships | 401 |
| Sinusoidal relationships | 491 |
| Exponential decay | 503 |


*Note for users of the two-volume edition:*
Volume 1 (pp. 1–583)
includes Chapters 1–16.
Volume 2 (pp. 584–1093)
includes Chapters 17–30.

# Brief Contents


## PART I Force and Motion

Chapter 1 Representing Motion 4
2 Motion in One Dimension 32
3 Vectors and Motion in Two Dimensions 71
4 Forces and Newton's Laws of Motion 105
5 Applying Newton's Laws 134
6 Circular Motion, Orbits, and Gravity 175
7 Rotational Motion 205
8 Equilibrium and Elasticity 244

## PART II Conservation Laws

Chapter 9 Momentum 278
10 Energy and Work 309
11 Using Energy 354

## PART III Properties of Matter

Chapter 12 Thermal Properties of Matter 396
13 Fluids 441

## PART IV Oscillations and Waves

Chapter 14 Oscillations 484
15 Traveling Waves and Sound 518
16 Superposition and Standing Waves 549

## PART V Optics

Chapter 17 Wave Optics 586
18 Ray Optics 620
19 Optical Instruments 657

## PART VI Electricity and Magnetism

Chapter 20 Electric Fields and Forces 696
21 Electric Potential 734
22 Current and Resistance 774
23 Circuits 800
24 Magnetic Fields and Forces 838
25 EM Induction and EM Waves 879
26 AC Electricity 916

## PART VII Modern Physics

Chapter 27 Relativity 950
28 Quantum Physics 986
29 Atoms and Molecules 1019
30 Nuclear Physics 1054

# ENGAGE today's students

For the fourth edition of *College Physics: A Strategic Approach*, we expand our focus from HOW students learn physics to WHY students study physics. We now make connections to biology and other sciences throughout the text to keep students engaged, presenting content that is relevant to today's students. This new edition is one of the best college physics book on the market for non-physics majors.

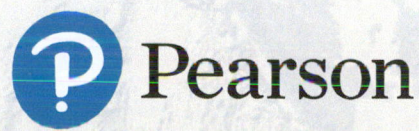

Pearson

# More connections to life science

**Build students' problem-solving skills in a context they care about while using real-life data and examples to keep their interest piqued.**

## 13.7 The Circulatory System BIO

### The Arteries and Capillaries

In the human body, blood pumped from the heart to the body starts its journey in a single large artery, the aorta. The flow then branches into smaller blood vessels, the large arteries that feed the head, the trunk, and the limbs. These branch into still smaller arteries, which then branch into a network of much smaller arterioles, which branch further into the capillaries. FIGURE 13.37 shows a schematic outline of the circulation, with average values for the diameters of the individual vessels, the total cross-section area of all of each type of vessel considered together, and the pressure in these vessels, assuming that the person is lying down so that there is no pressure change due to differences in elevation.

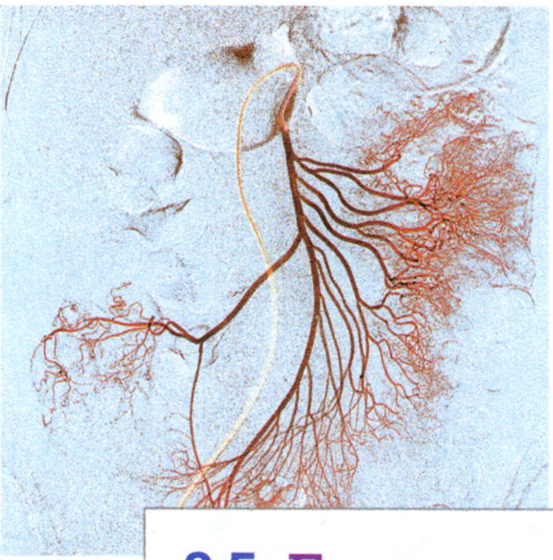

This preserved section of blood vessels shows the tremendous increase in number and in total area as blood vessels branch from large arteries to arterioles. One large artery gives rise to thousands of smaller vessels.

**NEW! Topics of interest to life science students,** such as the nature of the drag force at different scales and qualitative and quantitative descriptions of diffusion, provide current coverage of relevant topics based on the evolving consensus in the introductory physics for the life sciences community.

**NEW! Material stressing the application of physics to life sciences** includes structural color in animals and plants, the electric sense of different animals, the circulatory system (13.7) and on forces and torques in the body (8.5).

## 8.5 Forces and Torques in the Body BIO

Let's take your foot as the object of interest. When you stand on tiptoe, your foot pivots about your ankle. As shown in FIGURE 8.27, the forces on one foot are an upward force on your toes from the floor, a downward force on your ankle from the lower leg bone, and an upward force on the heel of your foot from your Achilles tendon. Suppose a 61 kg woman stands on one foot, on tiptoe, with the sole of her foot making a 25° angle with the floor; the distances are as shown in Figure 8.27. What is the magnitude of the tension force in the tendon? By what fraction does this force exceed the woman's weight? What is the magnitude of the force in the ankle joint?

FIGURE 8.27 Forces on the foot when standing on tiptoe.

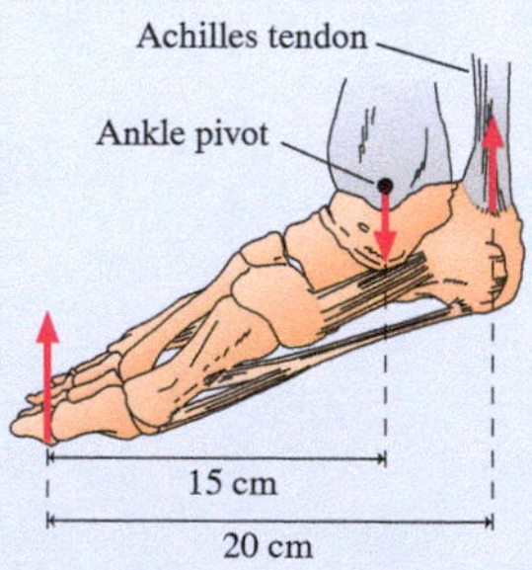

# create relevance to students' lives

**EXAMPLE 2.16** **Finding the height of a leap** BIO

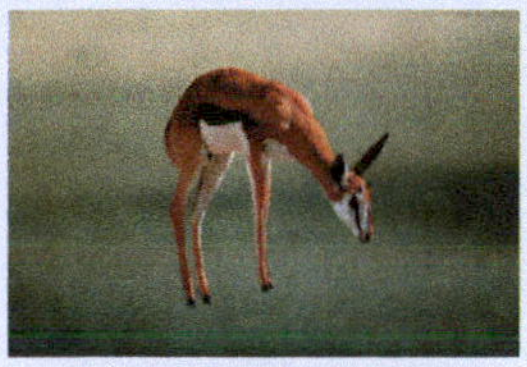

A springbok is an antelope found in southern Africa that gets its name from its remarkable jumping ability. When a springbok is startled, it will leap straight up into the air—a maneuver called a "pronk." A particular springbok goes into a crouch to perform a pronk. It then extends its legs forcefully, accelerating at 35 $m/s^2$ for 0.70 m as its legs straighten. Legs fully extended, it leaves the ground and rises into the air.

a. At what speed does the springbok leave the ground?
b. How high does it go?

***STRATEGIZE*** This is a two-part problem. In the first phase of its motion, the springbok accelerates upward, reaching some maximum speed just as it leaves the ground. As soon as it does so, the springbok is subject to only the force of gravity, so it is in free fall. For both phases, we will use the constant-acceleration equations from Synthesis 2.1.

**NEW! STRATEGIZE step in Examples** shows students the "big picture" view before delving into the details. Classroom testing of this addition has shown it to be popular with students and effective in teaching problem-solving skills.

8. || A hippo's body is 4.0 m long with front and rear feet BIO located as in Figure P8.8. The hippo carries 60% of its weight on its front feet. How far from its tail is the hippo's center of gravity?

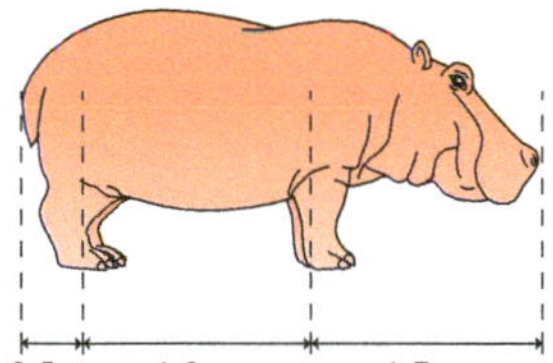


FIGURE P8.8

**NEW! End-of-chapter problem sets** now include real-life data and examples, helping students build transferable skills for their future courses and careers.

**NEW! Learning Objectives,** keyed to relevant end-of-chapter problems, help students check their understanding and guide them in choosing appropriate problems to optimize their study time.

**Learning Objectives** After studying this chapter, you should be able to:

- Use motion diagrams to interpret motion. *Conceptual Question 2.3; Problems 2.1, 2.2, 2.59*
- Use and interpret motion graphs. *Conceptual Questions 2.5, 2.13; Problems 2.4, 2.18, 2.19, 2.22, 2.62*
- Calculate the velocity of an object. *Conceptual Question 2.9; Problems 2.8, 2.15, 2.57*
- Solve problems about an object in uniform motion. *Problems 2.9, 2.10, 2.11, 2.13, 2.58*
- Calculate the acceleration of an object. *Problems 2.25, 2.27, 2.32, 2.33, 2.72*
- Determine and interpret the sign of acceleration. *Conceptual Questions 2.2, 2.8; Problem 2.50*
- Use the problem-solving approach to solve problems of motion with constant acceleration and free fall. *Problems 2.36, 2.40, 2.41, 2.47, 2.52, 2.75*

# Prepare students for engagement

**NEW! What the Physics? Videos** bring relatable content to engage students with what they are learning and promote curiosity for natural phenomena. These short videos present visually stimulating physical phenomena, pause throughout to address misconceptions, and ask conceptual questions about the physics at hand. Quantitative questions follow some of the videos and will be assignable in Mastering™ Physics and embedded in the eText.

**Prelecture Videos,** presented by co-author Brian Jones, expand on the Chapter Previews, giving context, examples, and a chance for students to practice the concepts they are studying via short multiple-choice questions. **NEW! Qualitative and Quantitative prelecture videos** now available with assessment as well!

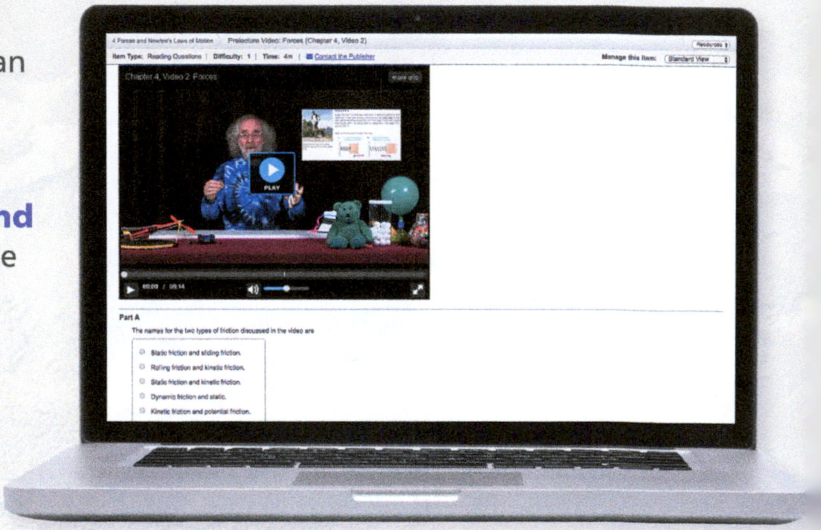


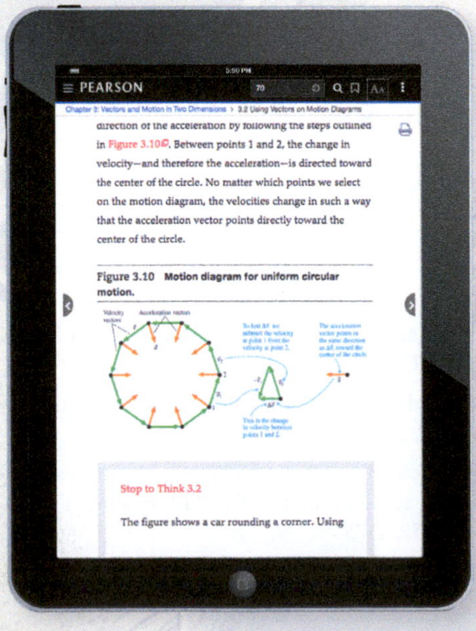


**NEW! eText, optimized for mobile,** seamlessly integrates videos and other rich media with the text and gives students access to their textbook anytime, anywhere. eText is available with Mastering Physics when packaged with new books, or as an upgrade students can purchase online.

# in lecture with interactive media

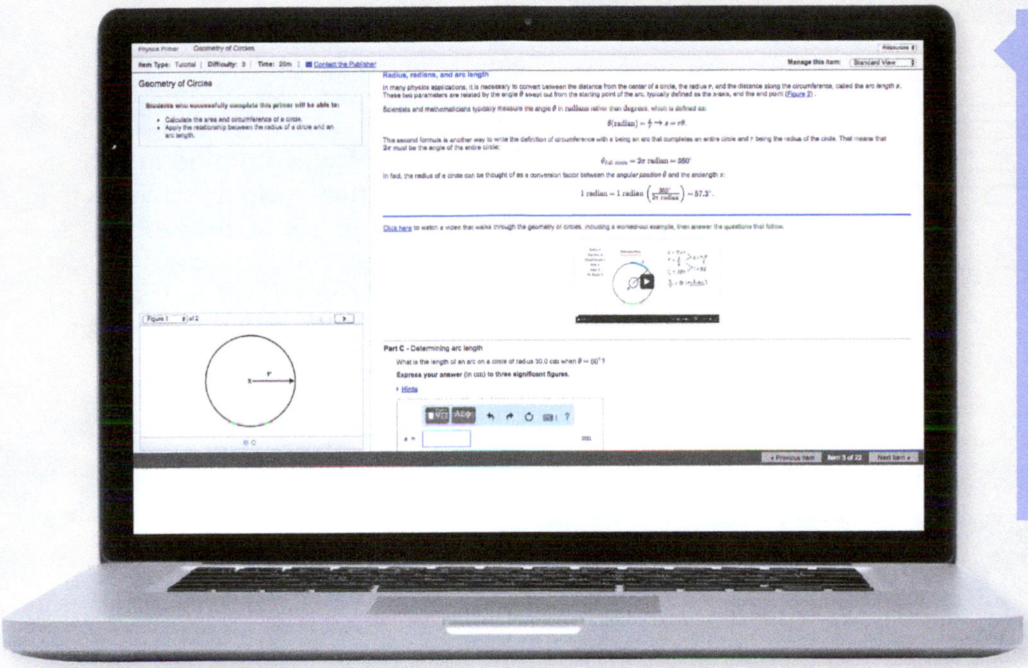


**NEW! The Physics Primer** relies on videos, hints, and feedback to refresh students' math skills in the context of physics and prepare them for success in the course. These tutorials can be assigned before the course begins as well as throughout the course. They ensure students practice and maintain their math skills, while tying together mathematical operations and physics analysis.

**Dynamic Study Modules (DSMs)** help students study effectively on their own by continuously assessing their activity and performance in real time and adapting to their level of understanding. The content focuses on definitions, units, and the key relationships for topics across all of mechanics and electricity and magnetism.

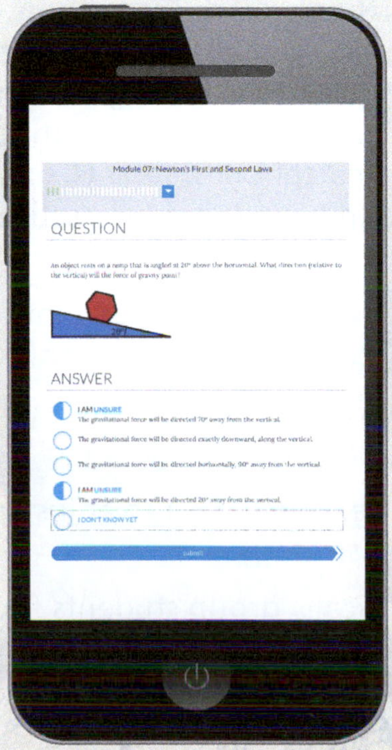

# Enhance students' understanding when

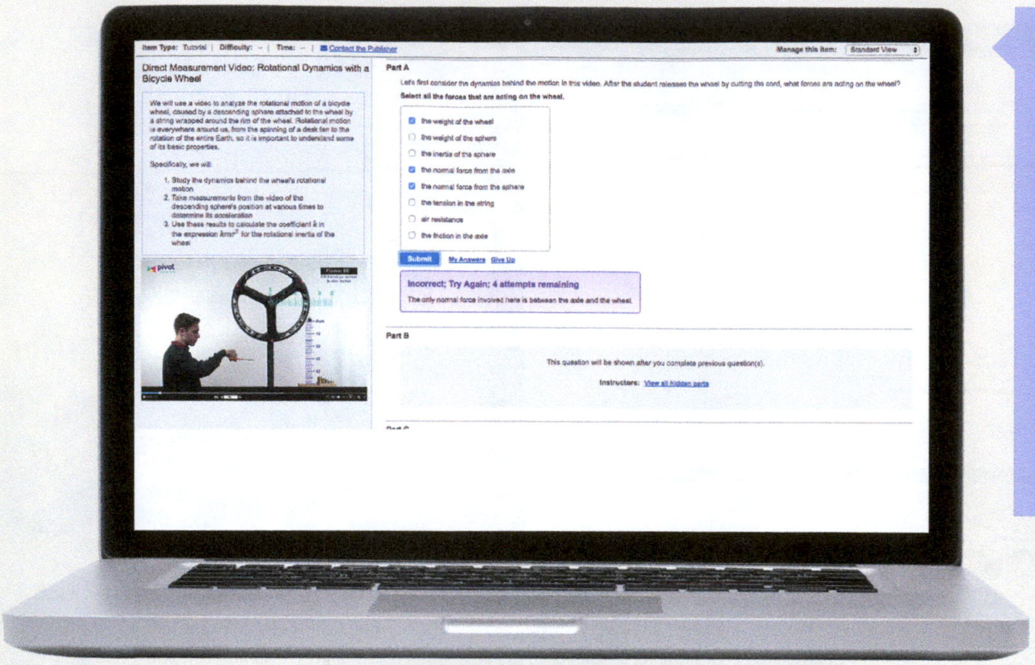

**NEW! Direct Measurement Videos** are short videos that show real situations of physical phenomena. Grids, rulers, and frame counters appear as overlays, helping students to make precise measurements of quantities such as position and time. Students then apply these quantities along with physics concepts to solve problems and answer questions about the motion of the objects in the video.

**Learning Catalytics™** helps generate class discussion, customize lectures, and promote peer-to-peer learning with real-time analytics. Learning Catalytics acts as a student response tool that uses students' smartphones, tablets, or laptops to engage them in more interactive tasks and thinking:

- **NEW!** Upload a full PowerPoint® deck for easy creation of slide questions.
- Monitor responses to find out where your students are struggling.
- Rely on real-time data to adjust your teaching strategy.
- Automatically group students for discussion, teamwork, and peer-to-peer learning.

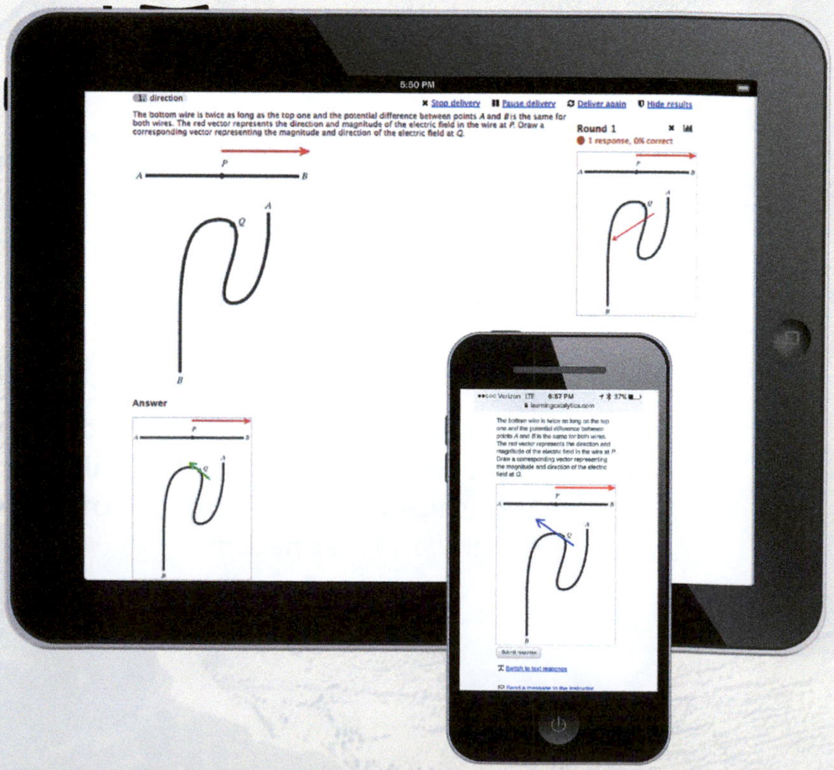

# they apply what they've learned



## 10.6 Potential Energy

17. Below we see a 1 kg object that is initially 1 m above the ground and rises to a height of 2 m. Anjay and Brittany each measure its position but use a different coordinate system to do so. Fill in the table to show the initial and final gravitational potential energies and $\Delta U$ as measured by Anjay and Brittany.

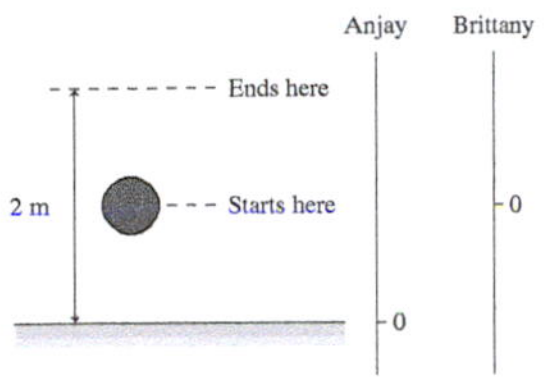


| | $U_i$ | $U_f$ | $\Delta U$ |
|---|---|---|---|
| Anjay | | | |
| Brittany | | | |

18. Three balls of equal mass are fired simultaneously with *equal* speeds from the same height above the ground. Ball 1 is fired straight up, ball 2 is fired straight down, and ball 3 is fired horizontally. Rank in order, from largest to smallest, their speeds $v_1$, $v_2$, and $v_3$ as they hit the ground.

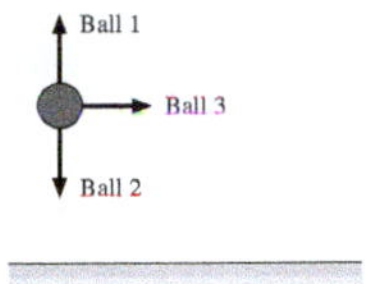


*Order:*

*Explanation:*

19. Below are shown three frictionless tracks. Ablock is released from rest at the position shown on the left. To which point does the block make it on the right before reversing direction and sliding back? Point B is the same height as the starting position.

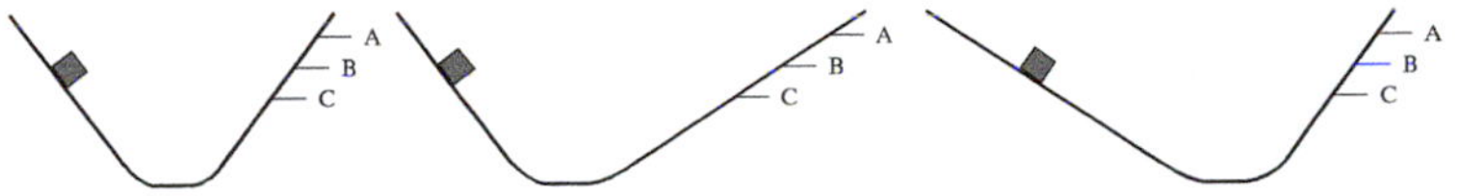


Makes it to ______ Makes it to ______ Makes it to ______

**A key component** of *College Physics: A Strategic Approach* is the accompanying ***Student Workbook.*** The workbook bridges the gap between textbook and homework problems by providing students the opportunity to learn and practice skills prior to using those skills in quantitative end-of-chapter problems, much as a musician practices technique separately from performance pieces. The workbook exercises, which are keyed to each section of the textbook, focus on developing specific skills, ranging from identifying forces and drawing free-body diagrams to interpreting field diagrams.

# Instructor tools help shape your course more efficiently

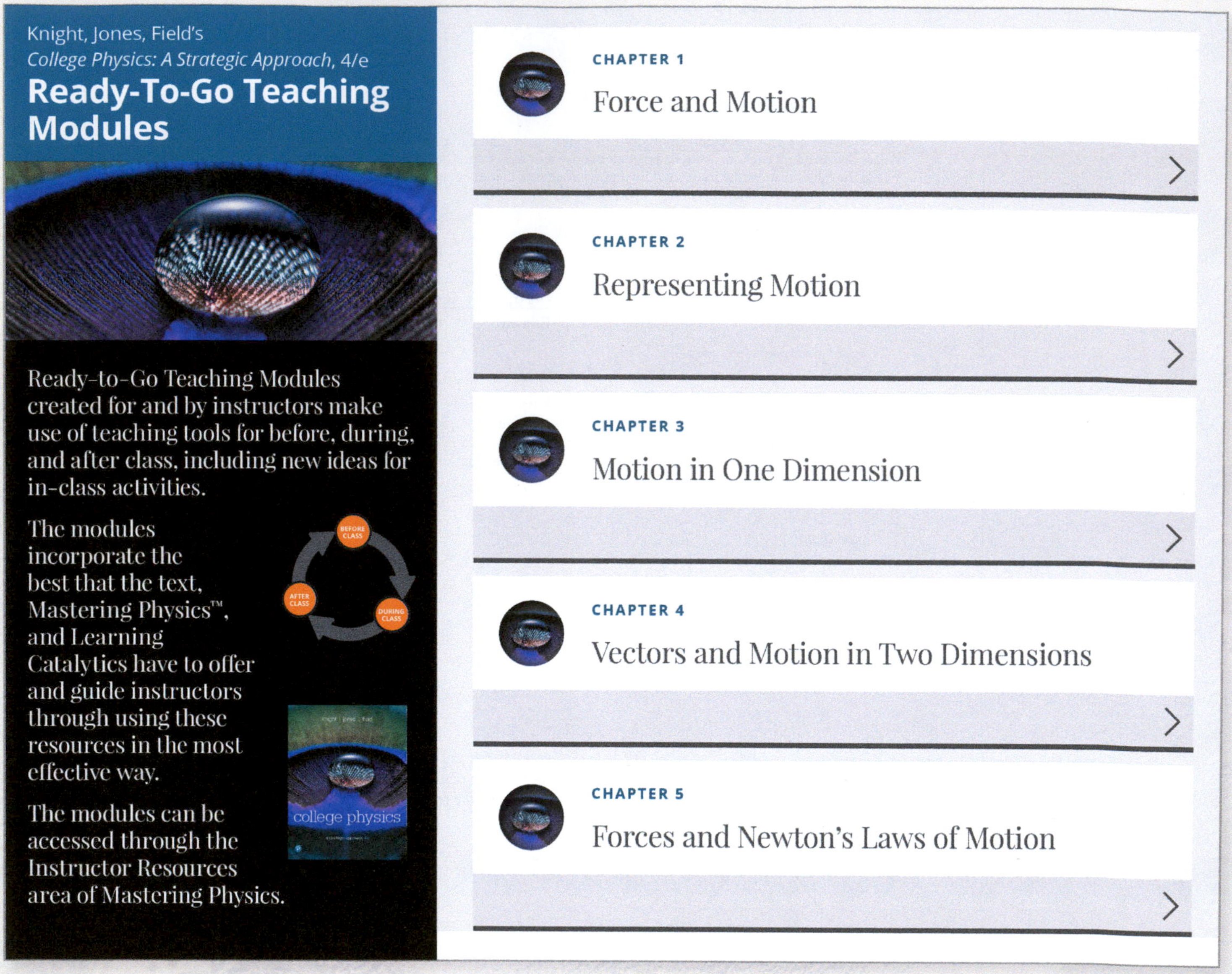

Knight, Jones, Field's
*College Physics: A Strategic Approach*, 4/e
Ready-To-Go Teaching Modules
Ready-to-Go Teaching Modules created for and by instructors make use of teaching tools for before, during, and after class, including new ideas for in-class activities.
The modules incorporate the best that the text, Mastering Physics™, and Learning Catalytics have to offer and guide instructors through using these resources in the most effective way.
The modules can be accessed through the Instructor Resources area of Mastering Physics.
BEFORE CLASS
DURING CLASS
AFTER CLASS
college physics
CHAPTER 1
Force and Motion
CHAPTER 2
Representing Motion
CHAPTER 3
Motion in One Dimension
CHAPTER 4
Vectors and Motion in Two Dimensions
CHAPTER 5
Forces and Newton's Laws of Motion


**NEW! Ready-to-Go Teaching Modules,** created for and by instructors, make use of teaching tools for before, during, and after class, including new ideas for in-class activities. The modules incorporate the best that the text, Mastering Physics, and Learning Catalytics have to offer and guide instructors through using these resources in the most effective way. The modules can be accessed through the Instructor Resources Area of Mastering Physics and as pre-built, customizable assignments.

# college physics

## a strategic approach 4e


randall d. knight | *California Polytechnic State University, San Luis Obispo*

brian jones | *Colorado State University*

stuart field | *Colorado State University*



Pearson

330 Hudson Street, NY NY 10013

*Editor in Chief, Director Physical Science Courseware Portfolio:* Jeanne Zalesky
*Editor, Physics Courseware Portfolio Analyst:* Darien Estes
*Senior Content Producer:* Martha Steele
*Senior Content Developer:* David Hoogewerff
*Managing Producer:* Kristen Flathman
*Courseware Director, Content Development:* Jennifer Hart
*Senior Analyst, Content Development, Science:* Suzanne Olivier
*Courseware Editorial Assistant:* Kristen Stephens
*Rich Media Content Producer:* Dustin Hennessey
*Full-Service Vendor:* Nesbitt Graphics/Cenveo® Publisher Services
*Copyeditor:* Carol Reitz
*Compositor:* Nesbitt Graphics/Cenveo® Publisher Services
*Art and Design Director:* Mark Ong/Side By Side Studios
*Interior/Cover Designer:* tani hasegawa
*Rights & Permissions Project Manager:* Katrina Mohn
*Rights & Permissions Management:* Ben Ferrini
*Photo Researcher:* Dena Digilio Betz
*Manufacturing Buyer:* Stacey J. Weinberger/LSC Communications
*Product Marketing Manager, Physical Sciences:* Elizabeth Bell
*Cover Photo Credit:* MirageC/Getty



**Library of Congress Cataloging in Publication Control Number: 2017044155**

Pearson

ISBN 10: 0-134-60903-4; ISBN 13: 978-0-134-60903-4 (Student edition)
ISBN 10: 0-134-77921-5; ISBN 13: 978-0-134-77921-8 (NASTA)
2 18

# About the Authors

**Randy Knight** taught introductory physics for 32 years at Ohio State University and California Polytechnic State University, where he is Professor Emeritus of Physics. Professor Knight received a Ph.D. in physics from the University of California, Berkeley and was a post-doctoral fellow at the Harvard-Smithsonian Center for Astrophysics before joining the faculty at Ohio State University. It was at Ohio State that he began to learn about the research in physics education that, many years later, led to *Five Easy Lessons: Strategies for Successful Physics Teaching* and this book, as well as *Physics for Scientists and Engineers: A Strategic Approach.* Professor Knight's research interests are in the fields of laser spectroscopy and environmental science. When he's not in front of a computer, you can find Randy hiking, sea kayaking, playing the piano, or spending time with his wife Sally and their five cats.

**Brian Jones** has won several teaching awards at Colorado State University during his 30 years teaching in the Department of Physics. His teaching focus in recent years has been the College Physics class, including writing problems for the MCAT exam and helping students review for this test. In 2011, Brian was awarded the Robert A. Millikan Medal of the American Association of Physics Teachers for his work as director of the Little Shop of Physics, a hands-on science outreach program. He is actively exploring the effectiveness of methods of informal science education and how to extend these lessons to the college classroom. Brian has been invited to give workshops on techniques of science instruction throughout the United States and in Belize, Chile, Ethiopia, Azerbaijan, Mexico, Slovenia, Norway, and Namibia. Brian and his wife Carol have dozens of fruit trees and bushes in their yard, including an apple tree that was propagated from a tree in Isaac Newton's garden.

**Stuart Field** has been interested in science and technology his whole life. While in school he built telescopes, electronic circuits, and computers. After attending Stanford University, he earned a Ph.D. at the University of Chicago, where he studied the properties of materials at ultralow temperatures. After completing a postdoctoral position at the Massachusetts Institute of Technology, he held a faculty position at the University of Michigan. Currently at Colorado State University, Stuart teaches a variety of physics courses, including algebra-based introductory physics, and was an early and enthusiastic adopter of Knight's *Physics for Scientists and Engineers.* Stuart maintains an active research program in the area of superconductivity. Stuart enjoys Colorado's great outdoors, where he is an avid mountain biker; he also plays in local ice hockey leagues.

# Preface to the Instructor

In 2006, we published *College Physics: A Strategic Approach,* a new algebra-based physics textbook for students majoring in the biological and life sciences, architecture, natural resources, and other disciplines. As the first such book built from the ground up on research into how students can more effectively learn physics, it quickly gained widespread critical acclaim from professors and students alike. For this fourth edition, we have continued to build on the research-proven instructional techniques introduced in the first edition while working to make the book more useful for instructors, more relevant to the students who use it, and more connected to the other subjects they study.

## Objectives

Our primary goals in writing *College Physics: A Strategic Approach* are:

- To provide students with a textbook that's a more manageable size, less encyclopedic in its coverage, and better designed for learning.
- To integrate proven techniques from physics education research into the classroom in a way that accommodates a range of teaching and learning styles.
- To help students develop both quantitative reasoning skills and solid conceptual understanding, with special focus on concepts well documented to cause learning difficulties.
- To help students develop problem-solving skills and confidence in a systematic manner using explicit and consistent tactics and strategies.
- To motivate students by integrating real-world examples that are relevant to their majors—especially from biology, sports, medicine, the animal world—and that build upon their everyday experiences.
- To utilize proven techniques of visual instruction and design from educational research and cognitive psychology that improve student learning and retention and address a range of learner styles.

A more complete explanation of these goals and the rationale behind them can be found in Randy Knight's paperback book, *Five Easy Lessons: Strategies for Successful Physics Teaching*. Please request a copy from your local Pearson sales representative if it is of interest to you (ISBN 978-0-805-38702-5).

## What's New to This Edition

In previous editions of the text, we focused on *how* students learn physics. Each chapter was built from the ground up to present concepts and problem-solving strategies in an engaging and effective manner. In this edition, we are focusing on *why* students learn physics. This is a question our students often ask. Why should a biology major take physics? A student planning a career in medicine? This book is for a physics course, but it's a course that will generally be taken by students in other fields.

The central goal of this edition is to make the text more relatable to the students who will use it, to add examples, explanations, and problems that show physics at work in contexts the students will find engaging. We've considered extensive feedback from scores of instructors and thousands of students as we worked to enhance and improve the text, figures, and end-of-chapter problems. Instructors need not be specialists in the life sciences or other fields to appreciate the new material. We've done the work to connect physics to other disciplines so that instructors can use this material to engage their students while keeping their focus on the basic physics.

Making the text more relatable meant making significant changes throughout the book. These edits aren't cosmetic add-ons; they reflect a thorough reworking of each chapter. Changes include:

- Guided by an evolving consensus in the Introductory Physics for the Life Sciences community, we have included **new sections** on the nature of the drag force at different scales, qualitative and quantitative descriptions of diffusion, and other topics of interest to life science students.
- We have added a great deal of **new material** that stresses the application of physics to life science topics. For example, we have expanded our treatment of vision and vision correction, included new material on structural color in animals and plants and the electric sense of different animals, and added new sections on the circulatory system and on forces and torques in the body.
- We have made **new connections** between physics topics and other courses that students are likely to take. For example, a new section connects the concept of the conservation of energy to topics from chemistry, including ionization energy and the role of catalysts in reactions. We have continued this approach when we introduced the concept of electric potential energy.
- Hundreds of **new end-of-chapter questions and problems** show physics at work in realistic, interesting situations. We have replaced problems that are artificial and abstract with problems that use real data from research in life science fields, problems that show the physics behind modern technologies, and problems that use physics to explore everyday phenomena. We have used the wealth of data from Mastering™ Physics to make sure that we have problems of a wide range of difficulties for each topic and problem-solving approach. A rigorous blind-solving and accuracy cross-checking process has been used to check all new problems to be sure that they are clearly worded and correct in all details, that they are accompanied by carefully worked out solutions.
- **New examples** throughout the book use the concepts of the chapters to explore realistic situations of interest to the students—from how bees use electric fields to locate promising flowers to how a study of force and torque in the jaw explains why dogs have long snouts and cats don't.
- We have changed the **photos and captions** at the starts of the chapters and parts of the text to better interest and engage students. The questions that are raised at the starts of the chapters aren't rhetorical; they are questions that will be answered in the flow of the chapter.

We have also made a number of changes to make the text an even more effective tool for students:

- A new **STRATEGIZE** step in examples shows students the "big picture" view before we delve into the details. Classroom testing of this addition has shown it to be quite popular with students, and quite effective in teaching problem-solving skills.
- **Key Concept figures** encourage students to actively engage with key or complex figures by asking them to reason with a related STOP TO THINK question.
- Additional **STOP TO THINK questions** provide students with more crucial practice and concept checks as they go through the chapters. The solutions to these questions have been moved to a more prominent location.
- We now provide **Learning Objectives** keyed to relevant end-of-chapter problems to help students check their understanding and guide them in choosing appropriate problems to optimize their study time.
- **Streamlined text and figures** tighten and focus the presentation to more closely match student needs. We've scrutinized every figure, caption, discussion, and photo in order to enhance their clarity and focus their role.
- Increased emphasis on **critical thinking, modeling, and reasoning,** both in worked examples and in end-of-chapter problems, promotes these key skills. These skills are especially important for students who are taking the MCAT exam.

- Expanded use of **realistic and real-world data** ensures students can make sense of answers that are grounded in the real world. Our examples and problems use real numbers and real data; they test different types of reasoning using equations, ratios, and graphs.

We have made many small changes to the flow of the text throughout, streamlining derivations and discussions, providing more explanation for complex concepts and situations, and reordering and reorganizing material so that each section and each chapter have a clearer focus. We have updated our treatment of entropy and the second law to better match current thinking. We have reordered the presentation of material on motion in two dimensions to be more logical. Every chapter has significant and meaningful changes, making this course especially relevant for today's students.

We know that students increasingly rely on sources of information beyond the text, and instructors are looking for quality resources that prepare students for engagement in lecture. The text will always be the central focus, but we have added additional media elements closely tied to the text that will enhance student understanding. In the Technology Update to the Second Edition, we added Class Videos, Video Tutor Solutions, and Video Tutor Demonstrations. In the Third Edition, we added an exciting new supplement, **Prelecture Videos,** short videos with author Brian Jones that introduce the topics of each chapter with accompanying assessment questions. In the front of this book, you'll find an illustrated walkthrough of the new media available in this technology update for the third edition:

- **NEW! What the Physics? Videos** bring new, relatable content to engage students with what they are learning and promote curiosity for natural phenomena. These short videos present visually stimulating physical phenomena and pause throughout to address misconceptions and ask conceptual questions about the physics at hand. The videos are embedded in the eText as well as assignable in Mastering Physics. Quantitative questions are also available for assignment.
- **NEW! Direct Measurement Videos** are short videos that show real situations of physical phenomena. Grids, rulers, and frame counters appear as overlays, helping students to make precise measurements of quantities such as position and time. Students then apply these quantities along with physics concepts to solve problems and answer questions about the motion of the objects in the video. The problems are assignable in Mastering Physics and can be used to replace or supplement traditional word problems, or as open-ended questions to help develop problem-solving skills.
- **NEW! The Physics Primer** relies on videos, hints, and feedback to refresh students' math skills in the context of physics and prepares them for success in the course. These tutorials can be assigned before the course begins or throughout the course as just-in-time remediation. They ensure students practice and maintain their math skills, while tying together mathematical operations and physics analysis.
- **NEW! Quantitative Prelecture Videos** are assignable, interactive videos that complement the Conceptual Prelecture Videos, giving students exposure to concepts before class and helping them learn how problems for those concepts are worked.
- **NEW! Ready-to-Go Teaching Modules** provide instructors with easy-to-use tools for teaching the toughest topics in physics. These modules demonstrate how your colleagues effectively use all the resources Pearson has to offer to accompany *College Physics: A Strategic Approach,* including, but not limited to, Mastering Physics items. Ready-to-Go Teaching Modules were created for and by instructors to provide easy-to-use assignments for before, during, and after class. Assets also include in-class activities and questions in Learning Catalytics™.
- **Dynamic Study Modules (DSMs)** help students study on their own by continuously assessing their activity and performance in real time. Students complete a set of questions with a unique answer format that repeats each question until students can answer them all correctly and confidently.

- **Dynamic Figure Videos in each chapter** are one-minute videos based on figures from the textbook that depict important, but often challenging, physics principles.
- **Video Tutor Solutions** created by co-author Brian Jones are an engaging and helpful walkthrough of worked examples and select end-of-chapter (EOC) problems designed to help students solve problems for each main topic. Each chapter has seven Video Tutor Solutions.
- **Prep questions aligned with the MCAT exam** are based on the Foundational Concepts and Content Categories outlined by the Association of American Medical Colleges. These 140 new problems are assignable in Mastering Physics and available for self-study in the Study Area.
- **Video Tutor Demonstrations** feature "pause-and-predict" demonstrations of key physics concepts and incorporate assessment with answer-specific feedback.

## Textbook Organization

*College Physics: A Strategic Approach* is a 30-chapter text intended for use in a two-semester course. The textbook is divided into seven parts: Part I: *Force and Motion,* Part II: *Conservation Laws,* Part III: *Properties of Matter,* Part IV: *Oscillations and Waves,* Part V: *Optics,* Part VI: *Electricity and Magnetism,* and Part VII: *Modern Physics*.

Part I covers Newton's laws and their applications. The coverage of two fundamental conserved quantities, momentum and energy, is in Part II, for two reasons. First, the way that problems are solved using conservation laws—comparing an *after* situation to a *before* situation—differs fundamentally from the problem-solving strategies used in Newtonian dynamics. Second, the concept of energy has a significance far beyond mechanical (kinetic and potential) energies. In particular, the key idea in thermodynamics is energy, and moving from the study of energy in Part II into thermal physics in Part III allows the uninterrupted development of this important idea.

Optics (Part V) is covered directly after oscillations and waves (Part IV), but *before* electricity and magnetism (Part VI). Further, we treat wave optics before ray optics. Our motivations for this organization are twofold. First, wave optics is largely just an extension of the general ideas of waves; in a more traditional organization, students will have forgotten much of what they learned about waves by the time they get to wave optics. Second, optics as it is presented in introductory physics makes no use of the properties of electromagnetic fields. The documented difficulties that students have with optics are difficulties with waves, not difficulties with electricity and magnetism. There's little reason other than historical tradition to delay optics. However, the optics chapters are easily deferred until after Part VI for instructors who prefer that ordering of topics.

- **Complete edition,** with Mastering™ Physics and Student Workbook (ISBN 978-0-134-64149-2): Chapters 1–30.
- **Books a la Carte edition,** with Mastering™ Physics and Student Workbook (ISBN 978-0-134-64414-1)
- **Complete edition,** without Mastering™ Physics (ISBN 978-0-134-60903-4): Chapters 1–30.
- **Volume 1** without Mastering™ Physics (ISBN 978-0-134-61045-0): Chapters 1–16.
- **Volume 2** without Mastering™ Physics (ISBN 978-0-134-61046-7): Chapters 17–30.

Split your text the way you split your course! Log on to www.pearsonhighered.com/collections/educator-features.html and create your own splits of *College Physics: A Strategic Approach,* 4e, including *your* choice of chapters.

# Instructional Package

*College Physics: A Strategic Approach,* fourth edition, provides an integrated teaching and learning package of support material for students and instructors.

**NOTE** For convenience, most instructor supplements can be downloaded from the "Instructor Resources" area of Mastering Physics and the Instructor Resource Center (www.pearson.com/us/higher-education/customers/educators.html).

| Supplement | Print | Online | Instructor or Student Supplement | Description |
|---|---|---|---|---|
| Mastering Physics with Pearson eText (ISBN 0134671023) | | ✓ | Instructor and Student Supplement | This product features all of the resources of Mastering Physics in addition to the new Pearson eText 2.0. Now available on smartphones and tablets, Pearson eText 2.0 comprises the full text, including videos and other rich media. |
| Instructor's Solutions Manual (ISBN 0134796829) | | ✓ | Instructor Supplement | This comprehensive solutions manual contains complete solutions to all end-of-chapter questions and problems. |
| TestGen Test Bank (ISBN 0134702409) | | ✓ | Instructor Supplement | The Test Bank contains more than 2,000 high-quality problems, with a range of multiple-choice, true/false, short answer, and regular homework-type questions. Test files are provided in both TestGen® and Word format. |
| Instructor's Resource Materials | ✓ | ✓ | Instructor Supplement | All art, photos, and tables from the book are available in JPEG format and as modifiable PowerPoints™. In addition, instructors can access lecture outlines as well as "clicker" questions in PowerPoint format, editable content for key features, all the instructor's resources listed above, and solutions to the Student Workbook. Materials are accessible to download from the Instructor Resource area of Mastering Physics. |
| Student's Workbook Standard (CH1–30) (ISBN 0134609891X) Volume 1 (CH1–16) (ISBN 0134724828) Volume 2 (CH17–30) (ISBN 0134724801) | ✓ | | Student Supplement | For a more detailed description of the *Student's Workbook,* see page ix. |
| Student's Solutions Manual Volume 1 (CH1–16) (ISBN 0134704193) Volume 2 (CH17–30) (ISBN 0134724798) | ✓ | | Student Supplement | These solutions manuals contain detailed solutions to all of the odd-numbered end-of-chapter problems from the textbook. |
| Ready-to-Go Teaching Modules | | ✓ | Instructor Supplement | Ready-to-Go Teaching Modules provide instructors with easy-to-use tools for teaching the toughest topics in physics. Created by the authors and designed to be used before, during, and after class, these modules demonstrate how to effectively use all the book, media, and assessment resources that accompany *College Physics: A Strategic Approach 4e.* |

## The Student Workbook

A key component of *College Physics: A Strategic Approach* is the accompanying *Student Workbook*. The workbook bridges the gap between textbook and homework problems by providing students the opportunity to learn and practice skills prior to using those skills in quantitative end-of-chapter problems, much as a musician practices technique separately from performance pieces. The workbook exercises, which are keyed to each section of the textbook, focus on developing specific skills, ranging from identifying forces and drawing free-body diagrams to interpreting field diagrams.

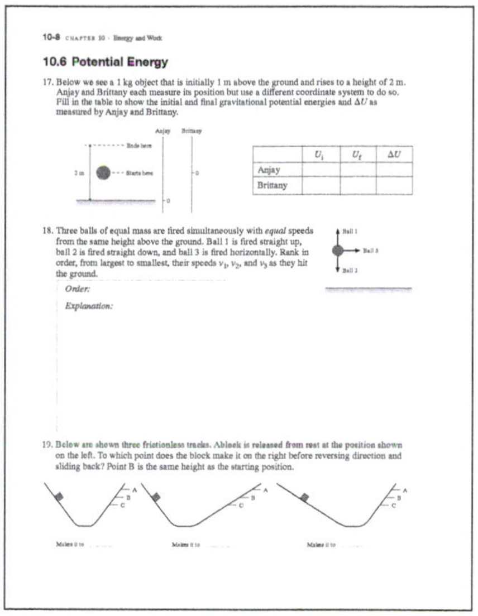
10-8 CHAPTER 10 · Energy and Work

10.6 Potential Energy

17. Below we see a 1 kg object that is initially 1 m above the ground and rises to a height of 2 m. Anjay and Brittany each measure its position but use a different coordinate system to do so. Fill in the table to show the initial and final gravitational potential energies and $\Delta U$ as measured by Anjay and Brittany.

| | $U_i$ | $U_f$ | $\Delta U$ |
|---|---|---|---|
| Anjay | | | |
| Brittany | | | |

18. Three balls of equal mass are fired simultaneously with *equal* speeds from the same height above the ground. Ball 1 is fired straight up, ball 2 is fired straight down, and ball 3 is fired horizontally. Rank in order, from largest to smallest, their speeds $v_1$, $v_2$, and $v_3$ as they hit the ground.

*Order:*

*Explanation:*

19. Below are shown three frictionless tracks. A block is released from rest at the position shown on the left. To which point does the block make it on the right before reversing direction and sliding back? Point B is the same height as the starting position.

Makes it to

Makes it to

Makes it to

The workbook exercises, which are generally qualitative and/or graphical, draw heavily upon the physics education research literature. The exercises deal with issues known to cause student difficulties and employ techniques that have proven to be effective at overcoming those difficulties. Also included are *jeopardy problems* that ask students to work backward from equations to physical situations, enhancing their understanding and critical thinking skills. The workbook exercises can be used in-class as part of an active-learning teaching strategy, in recitation sections, or as assigned homework. More information about effective use of the *Student Workbook* can be found in the *Instructor's Guide* in the Ready-to-Go modules.

Available versions: Standard Edition (ISBN 978-0-134-60989-8): Chapters 1-30, Volume 1 (ISBN 978-0-134-72482-9): Chapters 1–16, and Volume 2 (978-0-134-72480-5): Chapters 17–30.

## Acknowledgments

We have relied upon conversations with and, especially, the written publications of many members of the physics education community. Those who may recognize their influence include the late Arnold Arons, Uri Ganiel, Fred Goldberg, Ibrahim Halloun, Paula Heron, David Hestenes, the late Leonard Jossem, Jill Larkin, Priscilla Laws, John Mallinckrodt, Richard Mayer, Lillian McDermott and members of the Physics Education Research Group at the University of Washington, Edward "Joe" Redish, Fred Reif, Rachel Scherr, Bruce Sherwood, David Sokoloff, Ronald Thornton, Sheila Tobias, Alan Van Heuvelen, Carl Wieman, and Michael Wittman.

We are very grateful to Larry Smith for the difficult task of writing the *Instructor Solutions Manual;* to Scott Nutter for writing out the Student Workbook answers; to Wayne Anderson, Jim Andrews, Nancy Beverly, David Cole, Karim Diff, Jim Dove, Marty Gelfand, Kathy Harper, Charlie Hibbard, Robert Lutz, Matt Moelter, Kandiah Manivannan, Ken Robinson, Cindy Schwarz-Rachmilowitz, and Rachel Jones for their contributions to the end-of-chapter questions and problems; to Wayne again for helping with the Test Bank questions; and to Steven Vogel for his careful review of the biological content of many chapters and for helpful suggestions.

We especially want to thank Editor-in-Chief, Director Physical Science Courseware Portfolio, Jeanne Zalesky; Editor, Physics Courseware Portfolio Analyst, Darien Estes; Courseware Director, Content Development, Jennifer Hart; Senior Analyst, Courseware Development, Suzanne Olivier; Senior Content Producer, Martha Steele; and all the other staff at Pearson for their enthusiasm and hard work on this project. Having a diverse author team is one of the strengths of this book, but it has meant that we rely a great deal on Darien to help us keep to a single focus. Special thanks is due Martha for her keen attention to all details, careful scheduling and shepherding of all the elements of this complex project, and gentle prodding to keep everything moving forward.

Rose Kernan and the team at Nesbitt Graphics/Cenveo, copy editor Carol Reitz, and photo researcher Dena Digilio Betz get much credit for making this complex project all come together. In addition to the reviewers and classroom testers listed below, who gave invaluable feedback, we are particularly grateful to Ansel Foxley for his close scrutiny of every word, symbol, number, and figure.

**Randy Knight:** I would like to thank my Cal Poly colleagues for many valuable conversations and suggestions. I am endlessly grateful to my wife Sally for her love, encouragement, and patience, and to our many cats for nothing in particular other than being cats.

**Brian Jones:** I would like to thank my fellow AAPT and PIRA members for their insight and ideas, the students and colleagues who are my partners in the Little Shop of Physics, the students in my College Physics classes who teach me so much about the world, and the team at Pearson who help me develop as an educator and writer. Most of all, I thank my wife Carol, my best friend and gentlest editor, whose love makes the journey worthwhile.

**Stuart Field:** I would like to thank my wife Julie and my children, Sam and Ellen, for their love, support, and encouragement.

# Reviewers and Classroom Testers

## Reviewers for the Fourth Edition

Eduardo Araujo, *Miami Dade College*
Landon Bellavia, *The University of Findlay*
John Calarco, *University of New Hampshire*
Josh Colwell, *University of Central Florida*
Daniel Constantino, *Pennsylvania State University*
Chad Davies, *Gordon College*
Kimberly Fermoyle, *Elmhurst College*
Lilit Haroyan, *Los Angeles Mission College*
Tracy Hodge, *Berea College*
Kathleen Johnston, *Louisiana Tech University*
Wafaa Khattou, *Valencia College West*
Elena Kuchina, *Thomas Nelson Community College*
Ravi Kumar, *Wayne Community College, Goldsboro*
Jorge Lopez, *University of Texas*
Luiz Manzoni, *Concordia College*
Dario Martinez, *Lone Star College*
John McClain, *Temple College*
Irina Perevalova, *Middle Tennessee State University*
Amy Pope, *Clemson University*
Firouzeh Sabri, *University of Memphis*
Chandralekha Singh, *University of Pittsburgh*
Brenda Skoczelas, *Lake Sumter State College*
Valeriia Starovoitova, *Idaho State University*
Susa Stonedahl, *St. Ambrose University*
Ulrich Zurcher, *Cleveland State University*

## Reviewers of Previous Editions

Special thanks go to our third edition review panel: Taner Edis, Marty Gelfand, Jason Harlow, Charlie Hibbard, Jeff Loats, Amy Pope, and Bruce Schumm.

David Aaron, *South Dakota State University*
Susmita Acharya, *Cardinal Stritch University*
Ugur Akgun, *University of Iowa*
Ralph Alexander, *University of Missouri—Rolla*
Kyle Altmann, *Elon University*
Donald Anderson, *Ivy Tech*
Michael Anderson, *University of California—San Diego*
Steve Anderson, *Montana Tech*
James Andrews, *Youngstown State University*
Charles Ardary, *Edmond Community College*
Charles Bacon, *Ferris State University*
John Barry, *Houston Community College*
David H. Berman, *University of Northern Iowa*
Phillippe Binder, *University of Hawaii—Hilo*
Jeff Bodart, *Chipola College*
James Boger, *Flathead Valley Community College*
Richard Bone, *Florida International University*
James Borgardt, *Juniata College*
Daniela Bortoletto, *Purdue University*
Don Bowen, *Stephen F. Austin State University*
Asa Bradley, *Spokane Falls Community College*
Elena Brewer, *SUNY at Buffalo*
Dieter Brill, *University of Maryland*
Hauke Busch, *Augusta State University*
Kapila Castoldi, *Oakland University*
Raymond Chastain, *Louisiana State University*
Michael Cherney, *Creighton University*
Lee Chow, *University of Central Florida*
Song Chung, *William Paterson University*
Alice Churukian, *Concordia College*
Christopher M. Coffin, *Oregon State University*
John S. Colton, *Brigham Young University*
Kristi Concannon, *Kings College*
Teman Cooke, *Georgia Perimeter College at Lawrenceville*
Daniel J. Costantino, *The Pennsylvania State University*
Jesse Cude, *Hartnell College*
Melissa H. Dancy, *University of North Carolina at Charlotte*
Loretta Dauwe, *University of Michigan—Flint*
Mark Davenport, *San Antonio College*
Chad Davies, *Gordon College*
Lawrence Day, *Utica College*
Carlos Delgado, *Community College of Southern Nevada*
David Donovan, *Northern Michigan University*
James Dove, *Metropolitan State University of Denver*
Archana Dubey, *University of Central Florida*
Andrew Duffy, *Boston University*
Taner Edis, *Truman State University*
Ralph Edwards, *Lurleen B. Wallace Community College*
Steve Ellis, *University of Kentucky*
Paula Engelhardt, *Tennessee Technical University*
Davene Eryes, *North Seattle Community College*
Gerard Fasel, *Pepperdine University*
Luciano Fleischfresser, *OSSM Autry Tech*
Cynthia Galovich, *University of Northern Colorado*
Bertram Gamory, *Monroe Community College*
Sambandamurthy Ganapathy, *SUNY at Buffalo*
Delena Gatch, *Georgia Southern University*
Richard Gelderman, *Western Kentucky University*
Martin Gelfand, *Colorado State University*
Terry Golding, *University of North Texas*
Robert Gramer, *Lake City Community College*
William Gregg, *Louisiana State University*
Paul Gresser, *University of Maryland*
Robert Hagood, *Washtenaw Community College*
Jason Harlow, *University of Toronto*
Heath Hatch, *University of Massachusetts*

Carl Hayn, *Santa Clara University*
James Heath, *Austin Community College*
Zvonko Hlousek, *California State University Long Beach*
Greg Hood, *Tidewater Community College*
Sebastian Hui, *Florence-Darlington Technical College*
Eric Hudson, *The Pennsylvania State University*
Joey Huston, *Michigan State University*
David Iadevaia, *Pima Community College—East Campus*
Fred Jarka, *Stark State College*
Ana Jofre, *University of North Carolina—Charlotte*
Daniel Jones, *Georgia Tech*
Erik Jensen, *Chemeketa Community College*
Todd Kalisik, *Northern Illinois University*
Ju H. Kim, *University of North Dakota*
Armen Kocharian, *California State University Northridge*
J. M. Kowalski, *University of North Texas*
Laird Kramer, *Florida International University*
Christopher Kulp, *Eastern Kentucky University*
Richard Kurtz, *Louisiana State University*
Kenneth Lande, *University of Pennsylvania*
Tiffany Landry, *Folsom Lake College*
Todd Leif, *Cloud County Community College*
John Levin, *University of Tennessee—Knoxville*
John Lindberg, *Seattle Pacific University*
Jeff Loats, *Metropolitan State University of Denver*
Rafael López-Mobilia, *The University of Texas at San Antonio*
Robert W. Lutz, *Drake University*
Lloyd Makorowitz, *SUNY Farmingdale*
Colleen Marlow, *Rhode Island College*
Eric Martell, *Millikin University*
Mark Masters, *Indiana University—Purdue*
John McClain, *Temple College*
Denise Meeks, *Pima Community College*
Henry Merrill, *Fox Valley Technical College*
Mike Meyer, *Michigan Technological University*
Karie Meyers, *Pima Community College*
Tobias Moleski, *Nashville State Tech*
April Moore, *North Harris College*
Gary Morris, *Rice University*
Krishna Mukherjee, *Slippery Rock University*
Charley Myles, *Texas Tech University*
Meredith Newby, *Clemson University*
David Nice, *Bryn Mawr*
Fred Olness, *Southern Methodist University*
Charles Oliver Overstreet, *San Antonio College*
Paige Ouzts, *Lander University*
Russell Palma, *Minnesota State University—Mankato*
Richard Panek, *Florida Gulf Coast University*
Joshua Phiri, *Florence-Darling Technical College*
Iulia Podariu, *University of Nebraska at Omaha*
David Potter, *Austin Community College*
Promod Pratap, *University of North Carolina—Greensboro*
Michael Pravica, *University of Nevada, Las Vegas*
Earl Prohofsky, *Purdue University*
Marilyn Rands, *Lawrence Technological University*
Andrew Rex, *University of Puget Sound*
Andrew Richter, *Valparaiso University*
William Robinson, *North Carolina State University*
Phyliss Salmons, *Embry—Riddle Aeronautical University*
Michael Schaab, *Maine Maritime Academy*
Bruce Schumm, *University of California, Santa Cruz*
Mizuho Schwalm, *University of Minnesota Crookston*
Cindy Schwarz, *Vassar College*
Natalia Semushkhina, *Shippensburg University*
Khazhgery (Jerry) Shakov, *Tulane University*
Kathy Shan, *University of Toledo*
Anwar Sheikh, *Colorado Mesa University*
Bart Sheinberg, *Houston Community College*
Marllin Simon, *Auburn University*
Kenneth Smith, *Pennsylvania State University*
Michael Smutko, *Northwestern University*
Jon Son, *Boston University*
Noel Stanton, *Kansas State University*
Donna Stokes, *University of Houston*
Chuck Stone, *North Carolina A&T*
Chun Fu Su, *Mississippi State University*
Jeffrey Sudol, *West Chester University*
Scott Thompson, *Georgia Gwinnett College*
William Tireman, *Northern Michigan University*
Negussie Tirfessa, *Manchester Community College*
Rajive Tiwari, *Belmont Abbey College*
Herman Trivilino, *College of the Mainland*
Dmitri Tsybychev, *Stony Brook University*
Douglas Tussey, *Pennsylvania State University*
Stephen Van Hook, *Pennsylvania State University*
Manuel Valera, *Slippery Rocky University*
Christos Valiotis, *Antelope Valley College*
James Vesenka, *University of New England*
Stamatis Vokos, *Seattle Pacific University*
James Wanliss, *Embry—Riddle Aeronautical University*
Henry Weigel, *Arapahoe Community College*
Luc T. Wille, *Florida Atlantic University*
Courtney Willis, *University of Northern Colorado*
Katherine Wu, *University of Tampa*
Ali Yazdi, *Jefferson State Community College*
David Young, *Louisiana State University*
Hsiao-Ling Zhou, *Georgia State University*
Todd Zimmerman, *University of Wisconsin—Stout*
Ulrich Zurcher, *Cleveland State University*

## About the Cover

The cover image isn't just a pretty picture. All of the elements—the lensing by the water droplet, the structure of the feather, the mechanism for the feather colors—make an appearance as applications of the physics concepts that students are learning.

# Preface to the Student

*One may say the eternal mystery of the world is its comprehensibility.*
—Albert Einstein

If you are taking a course for which this book is assigned, you probably aren't a physics major or an engineering major. It's likely that you aren't majoring in a physical science. So why are you taking physics?

It's almost certain that you are taking physics because you are majoring in a discipline that requires it. Someone, somewhere, has decided that it's important for you to take this course. And they are right. There is a lot you can learn from physics, even if you don't plan to be a physicist. We regularly hear from doctors, physical therapists, biologists, and others that physics was one of the most interesting and valuable courses they took in college.

So, what can you expect to learn in this course? Let's start by talking about what physics is. Physics is a way of thinking about the physical aspects of nature. Physics is not about "facts." It's far more focused on discovering *relationships* between facts and the *patterns* that exist in nature than on learning facts for their own sake. Our emphasis will be on thinking and reasoning. We are going to look for patterns and relationships in nature, develop the logic that relates different ideas, and search for the reasons *why* things happen as they do.

The concepts and techniques you will learn will have a wide application. In this text we have a special emphasis on applying physics to understanding the living world. You'll use your understanding of charges and electric potential to analyze the electrical signal produced when your heart beats. You'll learn how sharks can detect this signal to locate prey and, further, how and why this electrical sensitivity seems to allow hammerhead sharks to detect magnetic fields, aiding navigation in the open ocean.

Like any subject, physics is best learned by doing. "Doing physics" in this course means solving problems, applying what you have learned to answer questions at the end of the chapter. When you are given a homework assignment, you may find yourself tempted to simply solve the problems by thumbing through the text looking for a formula that seems like it will work. This isn't how to do physics; if it was, whoever required you to take this course wouldn't bother. The folks who designed your major want you to learn to *reason*, not to "plug and chug." Whatever you end up studying or doing for a career, this ability will serve you well.

How do you learn to reason in this way? There's no single strategy for studying physics that will work for all students, but we can make some suggestions that will certainly help:

- **Read each chapter *before* it is discussed in class.** Class attendance is much more effective if you have prepared.
- **Use the other resources that accompany the text.** The text includes many videos and online tools to help you better master new material.
- **Participate actively in class.** Take notes, ask and answer questions, take part in discussion groups. There is ample scientific evidence that *active participation* is far more effective for learning science than is passive listening.
- **After class, go back for a careful rereading of the chapter.** In your second reading, pay close attention to the details and the worked examples. Look for the *logic* behind each example, not just at what formula is being used.
- **Apply what you have learned to the homework problems at the end of each chapter.** By following the techniques of the worked examples, applying the tactics and problem-solving strategies, you'll learn how to apply the knowledge you are gaining.
- **Form a study group with two or three classmates.** There's good evidence that students who study regularly with a group do better than the rugged individualists who try to go it alone.
- **Don't be afraid to ask questions.** The more you engage with your instructor and other students, the more successful you will be.

We have one final suggestion. As you read the book, take part in class, and work through problems, step back every now and then to appreciate the big picture. You are going to study topics that range from motions in the solar system to the electrical signals in the nervous system that let you tell your hand to turn the pages of this book. It's a remarkable breadth of topics and techniques that is based on a very compact set of organizing principles.

Now, let's get down to work.

# Studying for and Taking the MCAT Exam

If you are taking the College Physics course, there's a good chance that you are majoring in the biological sciences. There's also a good chance that you are preparing for a career in the health professions, and so might well be required to take the Medical College Admission Test, the MCAT exam.

The *Chemical and Physical Foundations of Biological Systems* section of the MCAT assesses your understanding of the concepts of this course by testing your ability to apply these concepts to living systems. You will be expected to use what you've learned to analyze situations you've never seen before, making simplified but realistic models of the world. Your reasoning skills will be just as important as your understanding of the universal laws of physics.

## Structure of the MCAT Exam

Most of the test consists of a series of passages of technical information followed by a series of questions based on each passage, much like the passage problems at the end of each chapter in this book. Some details:

- **The passages and the questions are *always* integrated.** Understanding the passage and answering the questions will require you to use knowledge from several different areas of physics.
- **Passages will generally be about topics for which you do not have detailed knowledge.** But, if you read carefully, you'll see that the treatment of the passage is based on information you should know well.
- **The test assumes a basic level of background knowledge.** You'll need to have facility with central themes and major concepts, but you won't need detailed knowledge of any particular topic. Such detailed information, if needed, will be provided in the passage.
- **You can't use calculators on the test, so any math that you do will be reasonably simple.** Quickly estimating an answer with ratio reasoning or a knowledge of the scale of physical quantities will be a useful skill.
- **The answers to the questions are all designed to be plausible.** You can't generally weed out the "bad" answers with a quick inspection.
- **The test is given online.** Practicing with Mastering Physics will help you get used to this format.

## Preparing for the Test

Because you have used this book as a tool for learning physics, you should use it as a tool for reviewing for the MCAT exam. Several of the key features of the book will be useful for this, including some that were explicitly designed with the MCAT exam in mind.

As you review the chapters:

- Start with the *Chapter Previews,* which provide a "big picture" overview of the content. What are the major themes of each chapter?
- Look for the *Synthesis* boxes that bring together key concepts and equations. These show connections and highlight differences that you should understand and be ready to apply.
- Go through each chapter and review the *Stop to Think* exercises. These are a good way to test your understanding of the key concepts and techniques.
- Each chapter closes with a passage problem that is designed to be "MCAT-exam-like." They'll give you good practice with the "read a passage, answer questions" structure of the MCAT exam.

The passage problems are a good tool, but the passages usually don't integrate topics that span several chapters—a key feature of the MCAT exam. For integrated passages and problems, turn to the *Part Summaries:*

- For each Part Summary, read the *One Step Beyond* passage and answer the associated questions.
- After this, read the passages and answer the questions that end each Part Summary section. These passages and associated problems are—by design—very similar to the passages and questions you'll see on the actual MCAT exam.

## Taking the Test: Reading the Passage

As you read each passage, you'll need to interpret the information presented and connect it with concepts you are familiar with, translating it into a form that makes sense based on your background.

The next page shows a passage that was written to very closely match the style and substance of an actual MCAT passage. Blue annotations highlight connections you should make as you read. The passage describes a situation (the mechanics and energetics of sled dogs) that you probably haven't seen before. But the basic physics (friction, energy conversion) are principles that you are familiar with, principles that you have seen applied to related situations. When you read the passage, think about the underlying physics concepts and how they apply to this case.

### Translating the Passage

As you read the passage, do some translation. Connect the scenario to examples you've seen before, translate given information into forms you are familiar with, think about the basic physical principles that apply.

**Passage X**

For travel over snow, a sled with runners that slide on snow is the best way to get around. Snow is slippery, but there is still friction between runners and the ground; the forward force required to pull a sled at a constant speed might be 1/6 of the sled's weight.

As you read this part of the passage, think about the forces involved: For a sled moving at a constant speed, there is no net force. The downward weight force is equal to the upward normal force; the forward pulling force must be equal to the friction force, which is acting opposite the sled's motion. There are many problems like this in Chapter 5.

Part of translating is converting given information into a more usual or more useful form. This is really a statement about the coefficient of kinetic friction.

The pulling force might well come from a dog. In a typical sled, the rope that the dog uses to pull attaches at a slight angle, as in Figure 1. The pulling force is the horizontal component of the tension in the rope.

The force applied to the sled is the tension force in the rope, which is shown at an angle. The horizontal component is the pulling force; you're told this. There is a vertical component of the force as well.

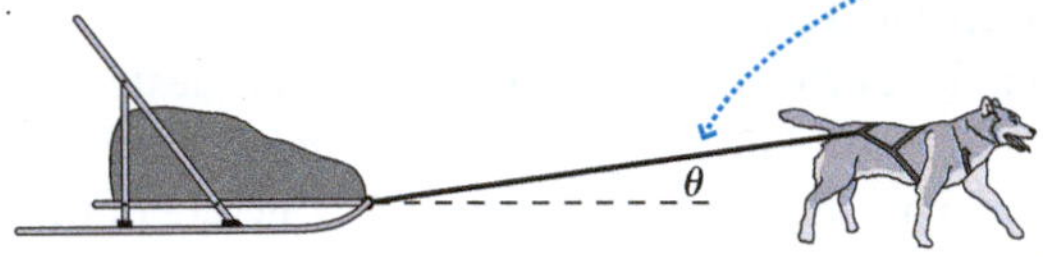


**Figure 1**

In the data given here, and the description given above, the sled moves at a constant speed—there is no mention of acceleration anywhere in this passage. In such cases, the net force is zero and the kinetic energy of the sled isn't changing.

Sled dogs have great aerobic capacity; a 40 kg dog can provide output power to pull with a 60 N force at 2.2 m/s for hours. The output power is related to force and velocity by $P = F \cdot v$, so they can pull lighter loads at higher speeds.

Notice that the key equation relating power, force, and velocity is given to you. That's to be expected. Any specific information, including equations, constants, and other such details, will generally be given in the passage. The MCAT is a test of reasoning, not recall.

Doing 100 J of work means that a dog must expend 400 J of metabolic energy. The difference must be exhausted as heat; given the excellent insulation provided by a dog's fur, this is mostly via evaporation as it pants. At a typical body temperature, the evaporation of 1.0 l of water carries away 240,000 J, so this is an effective means of cooling.

The concepts of metabolic energy and energy output are treated in Chapter 11. The details here match those in the chapter (as they should!); this corresponds to an efficiency of 25%. 400 J of energy is used by the body; 25% of this, 100 J, is the energy output. This means that 300 J is exhausted as heat.

Chapter 12 discusses means of heat transfer: conduction, convection, radiation, evaporation. This paragraph gives biological details about dogs that you can interpret as follows: A dog's fur limits transfer by conduction, convection, and radiation; evaporation of water by a panting dog must take up the slack.

The specific data for energy required to evaporate water are given. If you need such information to answer questions, it will almost certainly be provided. As we noted above, this is a test of reasoning, not recall.

**FIGURE MCAT-EXAM.1** Interpreting a passage.

## Taking the Test: Answering the Questions

The passages on the MCAT exam seem complicated at first, but, as we've seen, they are about basic concepts and central themes that you know well. The same is true of the questions; they aren't as difficult as they may seem at first. As with the passage, you should start by translating the questions, identifying the physical concepts that apply in each case. You then proceed by reasoning, determining the solution to the question, using your understanding of these basic concepts. The practical suggestions below are followed by a detailed overview of the solutions to the questions based on the passage on the previous page.

## You Can Answer the Questions in Any Order

The questions test a range of skills and have a range of difficulties. Many questions will involve simple reading comprehension; these are usually quite straightforward. Some require sophisticated reasoning and (slightly) complex mathematical manipulations. Start with the easy ones, ones that you can quickly solve. Save the more complex ones for later, and skip them if time is short.

## Take Steps to Simplify or Eliminate Calculations

You won't be allowed to use a calculator on the exam, so any math that you do will be reasonably straightforward. To rapidly converge on a correct answer choice, there are some important "shortcuts" that you can take.

- **Use ratio reasoning.** What's the relationship between the variables involved in a question? You can use this to deduce the answer with only a very simple calculation, as we've seen many times in the book. For instance, suppose you are asked the following question:

  *A model rocket is powered by chemical fuel. A student launches a rocket with a small engine containing 1.0 g of combustible fuel. The rocket reaches a speed of 10 m/s. The student then launches the rocket again, using an engine with 4.0 g of fuel. If all other parameters of the launch are kept the same, what final speed would you expect for this second trial?*

  This is an energy conversion problem: Chemical energy of the fuel is converted to kinetic energy of the rocket. Kinetic energy is related to the speed by $K = \frac{1}{2}mv^2$. The chemical energy—and thus the kinetic energy—in the second trial is increased by a factor of 4. Since $K \sim v^2$, the speed must increase by a factor of 2, to 20 m/s.
- **Simplify calculations by liberally rounding numbers.** You can round off numbers to make calculations more straightforward. Your final result will probably be close enough to choose the correct answer from the list given. For instance, suppose you are asked the following question:

  *A ball moving at 2.0 m/s rolls off the edge of a table that's 1.2 m high. How far from the edge of the table does the ball land?*

  *A. 2 m  B. 1.5 m  C. 1 m  D. 0.5 m*

  We know that the vertical motion of the ball is free fall; so the vertical distance fallen by the ball in a time $\Delta t$ is $\Delta y = -\frac{1}{2}gt^2$. The time to fall 1.2 m is $\Delta t = \sqrt{2(1.2\text{ m})/g}$. Rather than complete this calculation, we estimate the results as follows: $\Delta t = \sqrt{2.4/9.8} \approx \sqrt{1/4} = 1/2 = 0.5\text{s}$. During this free-fall time, the horizontal motion is constant at 2.0 m/s, so we expect the ball to land about 1 m away. Our quick calculation shows us that the correct answer is choice C—no other answer is close.
- **For calculations using values in scientific notation, compute either the first digits or the exponents, not both.** In some cases, a quick calculation can tell you the correct leading digit, and that's all you need to figure out the correct answer. In other cases, you'll find possible answers with the same leading digit but very different exponents or decimal places. In this case, all you need is a simple order-of-magnitude estimate to decide on the right result.
- **Where possible, use your knowledge of the expected scale of physical quantities to quickly determine the correct answer.** For instance, suppose a question asks you to find the photon energy for green light of wavelength 550 nm. Visible light has photon energies of about 2 eV, or about $3 \times 10^{-19}$ J, and that might be enough information to allow you to pick out the correct answer with no calculation.
- **Beware of "distractors," answers that you'll get if you make common mistakes.** For example, Question 4 on the next page is about energy conversion. The dog is keeping the sled in motion, so it's common for students to say that the dog is converting chemical energy in its body into kinetic energy. However, the kinetic energy isn't changing. The two answer choices that involve kinetic energy are common, but incorrect, choices. Be aware that the questions are constructed to bring out such misconceptions and that these tempting, but wrong, answer choices will be provided.

## One Final Tip: Look at the Big Picture

The MCAT exam tests your ability to look at a technical passage about which you have some background knowledge and quickly get a sense of what it is saying, enough to answer questions about it. Keep this big picture in mind:

- **Don't get bogged down in technical details of the particular situation.** Focus on the basic physics.
- **Don't spend too much time on any one question.** If one question is taking too much time, make an educated guess and move on.
- **Don't get confused by details of notation or terminology.** For instance, different people use different symbols for physical variables. In this text, we use the symbol $K$ for kinetic energy; others use $E_K$.

Finally, don't forget the most important aspect of success on the MCAT exam: The best way to prepare for this or any test is simply to understand the subject. As you prepare for the test, focus your energy on reviewing and refining your knowledge of central topics and techniques, and practice applying your knowledge by solving problems like you'll see on the actual MCAT.

**Translating**

Look at the questions and think about the physics principles that apply, how they connect to concepts you know and understand.

**Tips**

- Numerical choices are presented in order; that's the usual practice on the test. Estimate the size of the answer, and think about where it falls.
- For questions with sentences as choices, decide on the solution before you look at the choices; this will save time reading.

**Reasoning**

Think about the question and the range of possible answers, and converge to a solution with as few steps as possible—time is limited!

This is a question about the size of the friction force. You are told that it takes a force that's about 1/6 of the sled's weight to pull it forward on snow. You can estimate the friction coefficient from this information.

**1.** What is the approximate coefficient of kinetic friction for a sled on snow?

A. 0.35
B. 0.25
C. 0.15
D. 0.05

For an object on level ground, the normal force equals the weight force. If the sled is moving at a constant speed, the pulling force equals the friction force. This implies that $\mu = f_k/n = f_{pull}/w = 1/6$. Two of the answer choices convert easily to fractions: 0.25 = 1/4; 0.05 = 1/20. 1/6 is between these, so C must be our choice. (Indeed, 1/6 = 0.167, so 0.15 is pretty close.)

If the speed is constant, there is no net force. We are told that the pulling force is the horizontal component of the tension force, not the tension force itself. Because there is no net force, this horizontal component is equal to the friction force, which is directed backward. So this is really a question about the friction force.

**2.** If a rope pulls at an angle, as in Figure 1, how will this affect the pulling force necessary to keep the sled moving at a constant speed?

A. This will reduce the pulling force.
B. This will not change the pulling force.
C. This will increase the pulling force.
D. It will increase or decrease the pulling force, depending on angle.

A vertical component of the tension force will reduce the normal force, reducing the friction force—and thus the pulling force.

We assume that the output power is the same for the two cases—this is implied in the passage.

**3.** A dog pulls a 40 kg sled at a maximum speed of 2 m/s. What is the maximum speed for an 80 kg sled?

A. 2 m/s
B. 1.5 m/s
C. 1.0 m/s
D. 0.5 m/s

Doubling the weight doubles the normal force, which doubles the friction force. This will double the necessary pulling force as well. Given the expression for power given in the passage, this means the maximum speed will be halved.

This is a question about energy transformation. For such questions, think about changes. What forms of energy are *changing*? We know that thermal energy is part of the picture because some of the chemical energy is converted to thermal energy in the dog's body.

**4.** As a dog pulls a sled at constant speed, chemical energy in the dog's body is converted to

A. kinetic energy
B. thermal energy
C. kinetic energy and thermal energy
D. kinetic energy and potential energy

Choice B is correct, but A and C are clever distractors. It's tempting to choose an answer that includes kinetic energy. The sled is in motion, after all! But don't be swayed. The kinetic energy isn't changing, and friction to the sled converts any energy the dog supplies into thermal energy.

Increasing speed increases power, as the passage told us. But the energy to pull the sled is not the *power*, it's the *work*, and we know that the work is $W = F\Delta x$. This is a question about work and energy, not about power.

**5.** A dog pulls a sled for a distance of 1.0 km at a speed of 1 m/s, requiring an energy output of 60,000 J. If the dog pulls the sled at 2 m/s, the necessary energy is

A. 240,000 J
B. 120,000 J
C. 60,000 J
D. 30,000 J

Doubling the speed doubles the power, but it doesn't change the force; that's fixed by friction. The distance is the same as well, and so is the work done, the energy required. Since the speed doubles, it's tempting to think the energy doubles, though. This "obvious" but incorrect solution is one of the choices—expect such situations on the actual MCAT.

The passage tells us that the dog uses 400 J of metabolic energy to do 100 J of work. 300 J, or 75%, must be exhausted to the environment. We can assume the same efficiency here.

**6.** A dog uses 100,000 J of metabolic energy pulling a sled. How much energy must the dog exhaust by panting?

A. 100,000 J
B. 75,000 J
C. 50,000 J
D. 25,000 J

If 75% of the energy must be exhausted to the environment, that's 75,000 J.

**FIGURE MCAT-EXAM.2** Answering the questions for the passage of Figure MCAT-EXAM.1.

# Real-World Applications

Applications of biological or medical interest are marked BIO in the list below, including MCAT-style Passage Problems. Other end-of-chapter problems of biological or medical interest are marked BIO in the chapter.


**Chapter 1**

Depth gauges 8
Accuracy of long jumps 14
Mars Climate Orbiter: unit error 16
BIO Navigating geese 24

**Chapter 2**

Crash cushions 44
BIO Animal acceleration 44
Rocket launch 47–48, 53
BIO Swan's takeoff 50
BIO Chameleon tongues 52
Runway design 54
Braking distance 55
BIO A springbok's pronk 58
BIO Cheetah vs. gazelle 60

**Chapter 3**

Designing speed-ski slopes 80
Dock jumping 87
London Eye Ferris wheel 90
Optimizing javelin throws 93
BIO Albatross's speed 93
BIO Record-breaking frog jumps 95

**Chapter 4**

BIO Snake-necked turtle attack 105, 124
Voyager and Newton's first law 106
Seatbelts and Newton's first law 107
Prosthetics for athletes 110
BIO Scallop propulsion 113
Feel the difference (inertia) 117
Race-car driver mass 118
Bullets and Newton's third law 123
Rocket propulsion 124
A mountain railway 125
Pulling a stuck car from the mud 133

**Chapter 5**

BIO Gliding sifaka 125, 154
Human tower 135
Jumping on the moon 141
Jumping in an elevator 142
Weightless astronauts 144
Antilock brakes 148
BIO Penguin drag coefficient 153
Skydiver terminal speed 153–154
BIO Mass of a pollen grain 155
BIO *Paramecium* stopping distance 156
BIO Traction 161
Stopping distances 164

**Chapter 6**

Scottish heavy hammer throw 178
BIO Runner on a circular track 180–181
Car cornering speed 181–182
Wings on Indy racers 182
Banked racetrack turns 183
BIO Maximum walking speed 168
BIO How you sense "up" 185
Fast-spinning planets 186
BIO Centrifuges 186–187
BIO Human centrifuge 187
BIO Rotating space stations 189
Variable gravity 192
BIO Walking on the moon 193
Hunting with a sling 195–196

**Chapter 7**

Clockwise clocks 211
Starting a bike 216
Designing wheelchair hand-rims 216
Turning a capstan 218
Granite ball and inertia 224
Golf putter moment of inertia 225
Balancing on a high wire 226
Rolling vs. sliding: ancient movers 231
Spinning a gyroscope 232–233
BIO Bunchberry petal release 242

**Chapter 8**

BIO Human stability and balance 244, 251
BIO Muscle forces 245
BIO Finding the body's center of gravity 249
Rollover safety for cars 250
Ladder safety 251
Balancing soda can 251
Elasticity of a golf ball 252
Running shoes and springs 253
BIO Bone strength 257–258
BIO Forces in the ankle joint 259–260
BIO Forces and torque in the elbow 260–261
BIO Carrying a heavy appliance 261
Elevator cable stretch 262
BIO Elasticity of a Ligament

**Part I Summary**

Dark matter 276
BIO Animal athletes 277
BIO Drag on a paramecium 277
BIO Diving falcon 278
Bending beams 278

**Chapter 9**

BIO Ram skull adaptations 278, 284
BIO Optimizing frog jumps 281
Stretchy seatbelts 284
BIO Hedgehog spines 284
BIO Birds taking flight 286–287
BIO Squid propulsion 292
Colliding billiard balls 288
Ice-skating spins 297
Hurricanes 298
Free-falling geckos 298
Acrobatic bats 298
Aerial firefighting 299
BIO Analyzing a dance move 308

**Chapter 10**

Energy around us 318–319
Flywheel energy storage on the ISS 328
Why racing bike wheels are light 328–329
Empire State Building race 330–331
BIO Energy storage in the Achilles tendon 332
BIO Jumping locusts 336–337
Summet Plummet water slide speed 337
Runaway-truck ramps 352

**Chapter 11**

BIO Kangaroo locomotion 354, 390
Lightbulb efficiency 357
BIO Biochemical efficiency 358
BIO Energy in the body: inputs 358–359
BIO Calorie content of foods 359
BIO Energy in the body: outputs 359–360
BIO Daily energy use for mammals and reptiles 361
BIO Energy and locomotion 363
Optical molasses 365
Heat engines 370
Reversible heat pump 373
BIO Entropy in biological systems 378–379
Efficiency of an automobile 380–381
BIO Animal power 389

### Part II Summary

Order out of chaos 391
BIO Squid propulsion 393
Golf club collisions 393

### Chapter 12

BIO Infrared images 396, 426, 431
Temperature in space 400
Frost on Mars 401
BIO Swim-bladder damage to caught fish 403
Tire gauges 404
Elevation and bag expansion 408
Chinook winds 411
Thermal expansion joints 412
Temperate lakes 414
BIO Frogs that survive freezing 416
BIO Keeping cool 418
Sweating glass 418
BIO Penguin feet 424
BIO Penguin feathers 424
Heat transfer on earth 425
BIO Hellbender respiration 429
BIO Photosynthesis and carbon dioxide diffusion 429
BIO Breathing in cold air 430
Ocean temperature 440
BIO The work of breathing 440

### Chapter 13

BIO Albatross and Bernoulli effect, 441, 458
Submarine windows 445
Pressure zones on weather maps 447
Barometers 447
Floating icebergs and boats 451
Hot-air balloons 452–453
Manatees: masters of density 453
Airplane lift 458
BIO Prairie dog burrows 459
BIO Measuring arterial pressure 460
BIO Measuring blood pressure 463
BIO Blood pressure in giraffes 465
BIO Blood pressure and flow 465–466, 477
BIO Capillaries and diffusion 466
BIO Horse hooves and circulation 468
BIO Intravenous transfusions 468–469

### Part III Summary

BIO Scales of living creatures 479
BIO Perspiration 480
Weather balloons 480
Passenger balloons 481

### Chapter 14

BIO Gibbon brachiation 484, 502
BIO Heart rhythms 485
BIO Simple harmonic motion 486
Metronomes 490
Microwaves and SHM 493
Swaying buildings 493
Measuring mass in space 496
BIO Weighing DNA 497–498
Car collision times 499
BIO Animal locomotion 502
Shock absorbers 503
Tidal resonance 504
Musical glasses 505
BIO Hearing (resonance) 506
Springboard diving 507–508
BIO Achilles tendon as a spring 517
BIO Spider-web oscillations 517

### Chapter 15

BIO Echolocation 518, 529, 548
"The Wave" 519
BIO Frog wave-sensors 521
BIO Spider vibration sense 522
Distance to a lightning strike 523
BIO Range of hearing 529
BIO Ultrasound imaging 530
BIO Owl ears 533
BIO Blue whale vocalization 534
BIO Hearing (cochlea hairs) 535
Solar surface waves 536
Red shifts in astronomy 538
BIO Wildlife tracking with weather radar 538
BIO Doppler blood flow meter 539
Earthquake waves 540

### Chapter 16

Digdgeridoos 549, 558, 564
Windows and transmitting light 554
The Tacoma bridge standing wave 557
String musical instruments 557
Microwave cold spots 558
Fiery sound waves 560
BIO Resonances of the ear canal 561–562
Wind musical instruments 562
BIO Speech and hearing 563–565
BIO Vowels and formants 564
Saying "ah" 565
BIO High-frequency hearing loss 565
Active noise reduction 565
Controlling exhaust noise 569
BIO The bat detector 570
BIO Dogs' growls 571
Harmonics and harmony 579

### Part IV Summary

Tsunamis 581
Deep-water waves 582
BIO Attenuation of ultrasound 582
Measuring speed of sound 583

### Chapter 17

BIO Iridescent feathers 586, 603
DVD colors 597
Antireflection coatings 601
Colors of soap bubbles and oil slicks 603
BIO Peacock fern and structural color 603–604
BIO Bluebirds and blue skies 605
Breakwaters and Huygens' principle 606
Laser range finding 611
BIO The Blue Morpho Butterfly 619

### Chapter 18

Glass repair and index of refraction 628
Binoculars 630
Snell's window 630
Optical fibers 631
BIO Arthroscopic surgery 631
BIO Water droplet as converging lens 638
BIO Mirrored eyes (*Gigantocypris*) 641
Supermarket mirrors 643
BIO Optical fiber imaging 647–648
Mirages 656

### Chapter 19

BIO The Anableps "four-eyed" fish 657, 667
BIO A Nautilus eye 658
Cameras 658–660
BIO The human eye 660–667
Prescription lenses 660
BIO Eye's range of accommodation, 663
BIO Vision correction 664, 665–667
BIO Penguin eyes 667
Forced perspective in movies 668
BIO Microscopes 670–673, 681
Telescopes above the atmosphere 673
Rainbows 676
BIO Absorption of chlorophyll 677
Fixing the HST 678
BIO Optical and electron micrographs 680
BIO Visual acuity for a kestrel 682
BIO The blind spot 685
BIO Surgical vision correction 689

### Part V Summary

BIO Scanning confocal microscopy 691
BIO Horse vision 692
BIO Eye shine 692–693
BIO Pupil size 693

### Chapter 20

BIO Bee navigation 696, 711
Amber attraction 697
Dust in the wind 699
Static electricity on a slide 700
BIO Bees picking up pollen 702
BIO Hydrogen bonds in DNA 705

BIO Separating sperm cells 706
Electrostatic precipitators 715–716
BIO Electric field of the heart 718
Static protection 720
Lightning rods 720
BIO Electrolocation 720
Cathode-ray tubes 723–724
BIO Flow cytometry 733

**Chapter 21**

BIO Rays and electric potential 734, 757
Cause of lightning 738, 773
BIO Membrane potential 739
BIO Proton therapy 743
BIO Bumblebee electric-field sensitivity 749
Airplanes and lightning strikes 754
BIO The electrocardiogram 755–756
BIO Shark electroreceptors 757
Random-access memory 759
BIO Plant roots and capacitance 761
Camera flashes 761
BIO Defibrillators 762
Fusion in the sun 764

**Chapter 22**

BIO Percentage body-fat measurement 774, 790–791
BIO The electric torpedo ray 780
Lightbulb filaments 783
Testing drinking water 784
BIO Impedance tomography 784
Photoresistor night lights 785
Cooking hot dogs with electricity 788
Monitoring corrosion in power lines 789
Lightbulb failure 799

**Chapter 23**

BIO Electric fish 800, 837
Christmas-tree lights 807
Headlight wiring 809
Thermistors in measuring devices 811
Cell phone circuitry, 812
Flashing bike light 816
Intermittent windshield wipers 819
BIO Electricity in the nervous system 819–825
BIO Electrical nature of nerve and muscle cells 819
BIO Interpreting brain electric potentials 822
Soil moisture measurement 826
BIO Cardiac defibrillators 837

**Chapter 24**

BIO Magnetic resonance imaging 838, 850, 864
BIO Magnetotactic bacteria 843
BIO Magnetocardiograms 849
Digital compass 853
The aurora 855
BIO Cyclotrons 856
BIO Electromagnetic flowmeters 857
Electric motors 864
Hard disk data storage 867
Rotating magnetic bacteria 868
The velocity selector 877
Ocean potentials 878
The mass spectrometer 878

**Chapter 25**

BIO Color vision in animals 879, 903
BIO Shark navigation 882
Generators 883–884
Wind turbine 884
Electric guitar pickups 889
Magnetic braking 892
BIO Transcranial magnetic stimulation 892
Radio transmission 894
Solar tower of power 895
Polarizers 896–898
BIO Polarizers and blood-glucose tests 898
BIO Polarization analysis 897
BIO Honeybee navigation 898
BIO Wildlife tracking 901
Colors of glowing objects 901
BIO Infrared sensors in snakes 902
BIO Seeing (infra)red 903
Astronomical images 904
Tethered satellite circuits 905
BIO Electromagnetic wave penetration 915
Metal detectors 915

**Chapter 26**

Transformers 919–921
Charging electric toothbrushes 921
Power transmission 922
Household wiring 923–925
BIO Electrical safety 925–927
The lightning crouch 927
GFI circuits 927, 936
Laptop trackpads 928
Under-pavement car detectors 930
Cleaning up computer power 931
BIO Nuclear magnetic resonance 934
BIO Cell membrane capacitance 943
Halogen bulbs 943

**Part VI Summary**

The greenhouse effect and global warming 945
BIO Taking x rays 946
Electric cars 947
Wireless power transmission 947

**Chapter 27**

Global Positioning Systems 950, 960, 963, 976–977
The Stanford Linear Accelerator 968
Nuclear fission 976
BIO Pion therapy 985

**Chapter 28**

BIO Electron microscopy 986
BIO X-ray imaging 987
BIO X-ray diffraction of DNA 989
BIO Biological effects of UV 992
BIO Frequencies for photosynthesis 993
BIO Waves, photons, and vision 996
BIO Transmission electron microscopes 999–1000
Scanning tunneling microscopy 1007
BIO Magnetic resonance imaging 1008–1009
Electron diffraction 1017–1018
Collisional excitation 1018

**Chapter 29**

Spectroscopy 1020
Colors of nebulae 1021
Sodium filters for telescopes 1040
BIO Fluorescence 1041–1042
BIO LASIK surgery 1044
Compact fluorescent lighting 1044–1045
Light-emitting diodes 1053

**Chapter 30**

Body scans 1054
BIO Deuterium and tracking migrating birds 1056
Nuclear fusion in the sun 1059
Nuclear power 1060
Smoke detectors and sensing circuits 1063
BIO Beta decay in the human body 1065
BIO Food irradiation 1068
BIO Graves disease and radioactive iodine 1070
Plutonium “batteries” 1071
Radiocarbon dating 1072–1073
BIO Gamma-ray medical sterilization 1074
BIO Radiation dose from environmental sources 1074
BIO Nuclear medicine 1075–1076
BIO Nuclear imaging, PET scans 1076–1078
Čerenkov radiation 1082
BIO Plutonium-powered exploration 1089
Nuclear fission 1089

**Part VII Summary**

Low-temperature physics 1091
Splitting the atom 1092

# Detailed Contents


Preface to the Instructor iv
Preface to the Student xii
Studying for and Taking the MCAT Exam xiii
Real-World Applications xvii

PART I Force and Motion

OVERVIEW The Science of Physics 3

Chapter 1 **Representing Motion** **4**
1.1 Motion: A First Look 5
1.2 Models and Modeling 7
1.3 Position and Time: Putting Numbers on Nature 8
1.4 Velocity 11
1.5 A Sense of Scale: Significant Figures, Scientific Notation, and Units 13
1.6 Vectors and Motion: A First Look 19
1.7 Where Do We Go from Here? 23
SUMMARY 25
QUESTIONS AND PROBLEMS 26

Chapter 2 **Motion in One Dimension** **32**
2.1 Describing Motion 33
2.2 Uniform Motion 37
2.3 Instantaneous Velocity 41
2.4 Acceleration 44
2.5 Motion with Constant Acceleration 47
2.6 Solving One-Dimensional Motion Problems 51
2.7 Free Fall 55
SUMMARY 61
QUESTIONS AND PROBLEMS 62

Chapter 3 **Vectors and Motion in Two Dimensions** **71**
3.1 Using Vectors 72
3.2 Coordinate Systems and Vector Components 74
3.3 Motion on a Ramp 79
3.4 Motion in Two Dimensions 82
3.5 Projectile Motion 85
3.6 Projectile Motion: Solving Problems 87
3.7 Circular Motion 90
3.8 Relative Motion 93
SUMMARY 96
QUESTIONS AND PROBLEMS 97

Chapter 4 **Forces and Newton's Laws of Motion** **105**
4.1 Motion and Forces 106
4.2 A Short Catalog of Forces 109
4.3 Identifying Forces 113
4.4 What Do Forces Do? 115
4.5 Newton's Second Law 117
4.6 Free-Body Diagrams 120
4.7 Newton's Third Law 122
SUMMARY 126
QUESTIONS AND PROBLEMS 127

Chapter 5 **Applying Newton's Laws** **134**
5.1 Equilibrium 135
5.2 Dynamics and Newton's Second Law 138
5.3 Mass and Weight 141
5.4 Normal Forces 144
5.5 Friction 146
5.6 Drag 151
5.7 Interacting Objects 157
5.8 Ropes and Pulleys 159
SUMMARY 165
QUESTIONS AND PROBLEMS 166

**Chapter 6 Circular Motion, Orbits, and Gravity** 175

6.1 Uniform Circular Motion 176
6.2 Dynamics of Uniform Circular Motion 178
6.3 Apparent Forces in Circular Motion 184
6.4 Circular Orbits and Weightlessness 187
6.5 Newton's Law of Gravity 190
6.6 Gravity and Orbits 193
SUMMARY 197
QUESTIONS AND PROBLEMS 198

**Chapter 7 Rotational Motion** 205

7.1 Describing Circular and Rotational Motion 206
7.2 The Rotation of a Rigid Body 211
7.3 Torque 215
7.4 Gravitational Torque and the Center of Gravity 219
7.5 Rotational Dynamics and Moment of Inertia 223
7.6 Using Newton's Second Law for Rotation 227
7.7 Rolling Motion 231
SUMMARY 234
QUESTIONS AND PROBLEMS 235

**Chapter 8 Equilibrium And Elasticity** 244

8.1 Torque and Static Equilibrium 245
8.2 Stability and Balance 250
8.3 Springs and Hooke's Law 252
8.4 Stretching and Compressing Materials 255
8.5 Forces and Torques in the Body 258
SUMMARY 263
QUESTIONS AND PROBLEMS 264

PART I SUMMARY Force and Motion 272
ONE STEP BEYOND Dark Matter and the Structure of the Universe 273
PART I PROBLEMS 274

**PART II Conservation Laws**

OVERVIEW Why Some Things Stay the Same 277

**Chapter 9 Momentum** 278

9.1 Impulse 279
9.2 Momentum and the Impulse-Momentum Theorem 280
9.3 Solving Impulse and Momentum Problems 285
9.4 Conservation of Momentum 287
9.5 Inelastic Collisions 292
9.6 Momentum and Collisions in Two Dimensions 294
9.7 Angular Momentum 295
SUMMARY 300
QUESTIONS AND PROBLEMS 301

**Chapter 10 Energy and Work** 309

10.1 The Basic Energy Model 310
10.2 Work 314
10.3 Kinetic Energy 318
10.4 Potential Energy 321
10.5 Thermal Energy 325
10.6 Conservation of Energy 327
10.7 Energy Diagrams 331
10.8 Molecular Bonds and Chemical Energy 335
10.9 Energy in Collisions 339
10.10 Power 341
SUMMARY 345
QUESTIONS AND PROBLEMS 346

**Chapter 11 Using Energy 354**
11.1 Transforming Energy 355
11.2 Energy in the Body 358
11.3 Temperature, Thermal Energy, and Heat 364
11.4 The First Law of Thermodynamics 367
11.5 Heat Engines 369
11.6 Heat Pumps 372
11.7 Entropy and the Second Law of Thermodynamics 374
11.8 Systems, Energy, and Entropy 377
SUMMARY 382
QUESTIONS AND PROBLEMS 383

PART II SUMMARY Conservation Laws 390
ONE STEP BEYOND Order Out of Chaos 391
PART II PROBLEMS 392


## PART III Properties of Matter


OVERVIEW Beyond the Particle Model 395


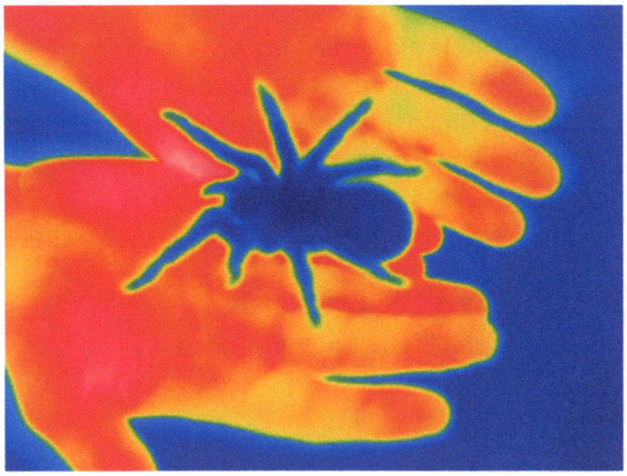


**Chapter 12 Thermal Properties of Matter 396**
12.1 The Atomic Model of Matter 397
12.2 The Atomic Model of an Ideal Gas 399
12.3 Ideal-Gas Processes 406
12.4 Thermal Expansion 412
12.5 Specific Heat and Heat of Transformation 414
12.6 Calorimetry 418
12.7 Specific Heats of Gases 421
12.8 Heat Transfer 422
12.9 Diffusion 426
SUMMARY 431
QUESTIONS AND PROBLEMS 432

**Chapter 13 Fluids 441**
13.1 Fluids and Density 442
13.2 Pressure 443
13.3 Buoyancy 448
13.4 Fluids in Motion 454
13.5 Fluid Dynamics 457
13.6 Viscosity and Poiseuille's Equation 461
13.7 The Circulatory System 463
SUMMARY 470
QUESTIONS AND PROBLEMS 471

PART III SUMMARY Properties of Matter 478
ONE STEP BEYOND Size and Life 479
PART III PROBLEMS 480


## PART IV Oscillations and Waves


OVERVIEW Motion That Repeats Again and Again 483

**Chapter 14 OSCILLATIONS 484**
14.1 Equilibrium and Oscillation 485
14.2 Linear Restoring Forces and SHM 487
14.3 Describing Simple Harmonic Motion 489
14.4 Energy in Simple Harmonic Motion 494
14.5 Pendulum Motion 499
14.6 Damped Oscillations 502
14.7 Driven Oscillations and Resonance 504
SUMMARY 509
QUESTIONS AND PROBLEMS 510

**Chapter 15 Traveling Waves and Sound 518**
15.1 The Wave Model 519
15.2 Traveling Waves 520
15.3 Graphical and Mathematical Descriptions of Waves 524
15.4 Sound and Light Waves 529
15.5 Energy and Intensity 531
15.6 Loudness of Sound 533
15.7 The Doppler Effect and Shock Waves 536
SUMMARY 541
QUESTIONS AND PROBLEMS 542

Chapter 16 **Superposition and Standing Waves** 549
16.1 The Principle of Superposition 550
16.2 Standing Waves 551
16.3 Standing Waves on a String 553
16.4 Standing Sound Waves 558
16.5 Speech and Hearing 563
16.6 The Interference of Waves from Two Sources 565
16.7 Beats 569
SUMMARY 572
QUESTIONS AND PROBLEMS 573

PART IV SUMMARY Oscillations and Waves 580
ONE STEP BEYOND Waves in the Earth and the Ocean 581
PART IV PROBLEMS 582

## PART V Optics

OVERVIEW Light Is a Wave 585

Chapter 17 **Wave Optics** 586
17.1 What Is Light? 587
17.2 The Interference of Light 590
17.3 The Diffraction Grating 594
17.4 Thin-Film Interference 599
17.5 Single-Slit Diffraction 605
17.6 Circular-Aperture Diffraction 609
SUMMARY 612
QUESTIONS AND PROBLEMS 613

Chapter 18 **Ray Optics** 620
18.1 The Ray Model of Light 621
18.2 Reflection 624
18.3 Refraction 627
18.4 Image Formation by Refraction 631
18.5 Thin Lenses: Ray Tracing 633
18.6 Image Formation with Spherical Mirrors 639
18.7 The Thin-Lens Equation 644
SUMMARY 649
QUESTIONS AND PROBLEMS 650

Chapter 19 **Optical Instruments** 657
19.1 The Camera 658
19.2 The Human Eye 660
19.3 The Magnifier 667
19.4 The Microscope 670
19.5 The Telescope 673
19.6 Color and Dispersion 674
19.7 Resolution of Optical Instruments 677
SUMMARY 683
QUESTIONS AND PROBLEMS 684

PART V SUMMARY Optics 690
ONE STEP BEYOND Scanning Confocal Microscopy 691
PART V PROBLEMS 692

## PART VI Electricity and Magnetism

OVERVIEW Charges, Currents, and Fields 695

Chapter 20 **Electric Fields and Forces** 696
20.1 Charges and Forces 697
20.2 Charges, Atoms, and Molecules 703
20.3 Coulomb's Law 705
20.4 The Concept of the Electric Field 710
20.5 The Electric Field of Multiple Charges 713
20.6 Conductors and Electric Fields 719
20.7 Forces and Torques in Electric Fields 721
SUMMARY 725
QUESTIONS AND PROBLEMS 726

Chapter 21 **Electric Potential** 734
21.1 Electric Potential Energy and Electric Potential 735
21.2 Sources of Electric Potential 737
21.3 Electric Potential and Conservation of Energy 740
21.4 Calculating the Electric Potential 743
21.5 Connecting Potential and Field 751
21.6 The Electrocardiogram 755
21.7 Capacitance and Capacitors 757
21.8 Energy and Capacitors 761
SUMMARY 765
QUESTIONS AND PROBLEMS 766

Chapter 22 **Current and Resistance** **774**

22.1 A Model of Current 775
22.2 Defining and Describing Current 777
22.3 Batteries and emf 779
22.4 Connecting Potential and Current 780
22.5 Ohm's Law and Resistor Circuits 784
22.6 Energy and Power 788
SUMMARY 792
QUESTIONS AND PROBLEMS 793

Chapter 23 **Circuits** **800**

23.1 Circuit Elements and Diagrams 801
23.2 Kirchhoff's Laws 802
23.3 Series and Parallel Circuits 805
23.4 Measuring Voltage and Current 810
23.5 More Complex Circuits 811
23.6 Capacitors in Parallel and Series 814
23.7 *RC* Circuits 816
23.8 Electricity in the Nervous System 819
SUMMARY 827
QUESTIONS AND PROBLEMS 828

Chapter 24 **Magnetic Fields and Forces** **838**

24.1 Magnetism 839
24.2 The Magnetic Field 840
24.3 Electric Currents Also Create Magnetic Fields 844
24.4 Calculating the Magnetic Field Due to a Current 847
24.5 Magnetic Fields Exert Forces on Moving Charges 851
24.6 Magnetic Fields Exert Forces on Currents 858
24.7 Magnetic Fields Exert Torques on Dipoles 861
24.8 Magnets and Magnetic Materials 865
SUMMARY 869
QUESTIONS AND PROBLEMS 870

Chapter 25 **EM Induction and EM Waves** **879**

25.1 Induced Currents 880
25.2 Motional emf 881
25.3 Magnetic Flux and Lenz's Law 884
25.4 Faraday's Law 889
25.5 Electromagnetic Waves 893
25.6 The Photon Model of Electromagnetic Waves 899
25.7 The Electromagnetic Spectrum 900
SUMMARY 906
QUESTIONS AND PROBLEMS 907

Chapter 26 **AC Electricity** **916**

26.1 Alternating Current 917
26.2 AC Electricity and Transformers 919
26.3 Household Electricity 923
26.4 Biological Effects and Electrical Safety 925
26.5 Capacitor Circuits 927
26.6 Inductors and Inductor Circuits 929
26.7 Oscillation Circuits 932
SUMMARY 937
QUESTIONS AND PROBLEMS 938

PART VI SUMMARY Electricity and Magnetism 944
ONE STEP BEYOND The Greenhouse Effect and Global Warming 945
PART VI PROBLEMS 946

PART VII Modern Physics

OVERVIEW New Ways of Looking at the World 949

Chapter 27 **Relativity** 950
- 27.1 Relativity: What's It All About? 951
- 27.2 Galilean Relativity 951
- 27.3 Einstein's Principle of Relativity 954
- 27.4 Events and Measurements 957
- 27.5 The Relativity of Simultaneity 960
- 27.6 Time Dilation 962
- 27.7 Length Contraction 968
- 27.8 Velocities of Objects in Special Relativity 970
- 27.9 Relativistic Momentum 971
- 27.10 Relativistic Energy 973
- SUMMARY 978
- QUESTIONS AND PROBLEMS 979

Chapter 28 **Quantum Physics** 986
- 28.1 X Rays and X-Ray Diffraction 987
- 28.2 The Photoelectric Effect 989
- 28.3 Photons 995
- 28.4 Matter Waves 997
- 28.5 Energy Is Quantized 1000
- 28.6 Energy Levels and Quantum Jumps 1002
- 28.7 The Uncertainty Principle 1005
- 28.8 Applications and Implications of Quantum Theory 1007
- SUMMARY 1010
- QUESTIONS AND PROBLEMS 1011

Chapter 29 **Atoms and Molecules** 1019
- 29.1 Spectroscopy 1020
- 29.2 Atoms 1022
- 29.3 Bohr's Model of Atomic Quantization 1025
- 29.4 The Bohr Hydrogen Atom 1027
- 29.5 The Quantum-Mechanical Hydrogen Atom 1032
- 29.6 Multi-electron Atoms 1034
- 29.7 Excited States and Spectra 1037
- 29.8 Molecules 1041
- 29.9 Stimulated Emission and Lasers 1042
- SUMMARY 1046
- QUESTIONS AND PROBLEMS 1047

Chapter 30 **Nuclear Physics** 1054
- 30.1 Nuclear Structure 1055
- 30.2 Nuclear Stability 1057
- 30.3 Forces and Energy in the Nucleus 1060
- 30.4 Radiation and Radioactivity 1062
- 30.5 Nuclear Decay and Half-Lives 1068
- 30.6 Medical Applications of Nuclear Physics 1073
- 30.7 The Ultimate Building Blocks of Matter 1079
- SUMMARY 1083
- QUESTIONS AND PROBLEMS 1084

PART VII SUMMARY Modern Physics 1090
ONE STEP BEYOND The Physics of Very Cold Atoms 1091
PART VII PROBLEMS 1092

Appendix A Mathematics Review A-1
Appendix B Periodic Table of Elements A-3
Appendix C Atomic and Nuclear Data A-4
Answers to Odd-Numbered Problems A-7
Credits C-1
Index I-1

PART I

# Force and Motion

The cheetah is the fastest land animal, able to run at speeds exceeding 60 miles per hour. Nonetheless, the rabbit has an advantage in this chase. It can *change* its motion more quickly and will likely escape. How can you tell, by looking at the picture, that the cheetah is changing its motion?

OVERVIEW

## The Science of Physics

Physics is the foundational science that underlies biology, chemistry, earth sciences, and all other fields that attempt to understand our natural world. Physicists couple careful experimentation with deep theoretical insights to build powerful and predictive models of how the world works. A key aspect of physics is that it is a *unifying* discipline: A relatively small number of key concepts can explain a vast array of natural phenomena. In this text, we have organized the chapters into parts according to seven of these unifying principles. Each of the seven parts of this text opens with an overview that gives you a look ahead, a glimpse of where your journey will take you in the next few chapters. It's easy to lose sight of the big picture while you're busy negotiating the terrain of each chapter. In Part I, the big picture is, in a word, *change*.

### Why Things Change

Simple observations of the world around you show that most things change. Some changes, such as aging, are biological. Others, such as the burning of gasoline in your car, are chemical. We will look at changes that involve *motion* of one form or another—running and jumping, throwing balls, lifting weights.

There are two big questions we must tackle to study how things change by moving:

- **How do we describe motion?** How should we measure or characterize the motion if we want to analyze it mathematically?
- **How do we explain motion?** Why do objects have the particular motion they do? Why, when you toss a ball upward, does it go up and then come back down rather than keep going up? What are the "laws of nature" that allow us to predict an object's motion?

Two key concepts that will help answer these questions are *force* (the "cause") and *acceleration* (the "effect"). Our basic tools will be three laws of motion worked out by Isaac Newton. Newton's laws relate force to acceleration, and we will use them to explain and explore a wide range of problems. As we learn to solve problems dealing with motion, we will learn basic techniques that we can apply in all the parts of this text.

### Simplifying Models

Another key aspect of physics is the importance of models. Suppose we want to analyze a ball moving through the air. Is it necessary to analyze the way the atoms in the ball are connected? Or the details of how the ball is spinning? Or the small drag force it experiences as it moves? These are interesting questions, of course. But if our task is to understand the motion of the ball, we need to simplify!

We can conduct a perfectly fine analysis of the ball's motion if we treat the ball as a single particle moving through the air. This is a *model* of the situation. A model is a simplified description of reality that is used to reduce the complexity of a problem so it can be analyzed and understood. Model building is a major part of the strategy that we will develop for solving problems throughout the text. Learning how to simplify a situation is the essence of successful modeling—and successful problem solving.

# 1 Representing Motion

As this falcon moves in a graceful arc through the air, the direction of its motion and the distance between each of its positions and the next are constantly changing. What language should we use to describe this motion?

## LOOKING AHEAD ▸▸

### Chapter Preview

Each chapter starts with a preview outlining the major topics and what you'll be learning for each topic.

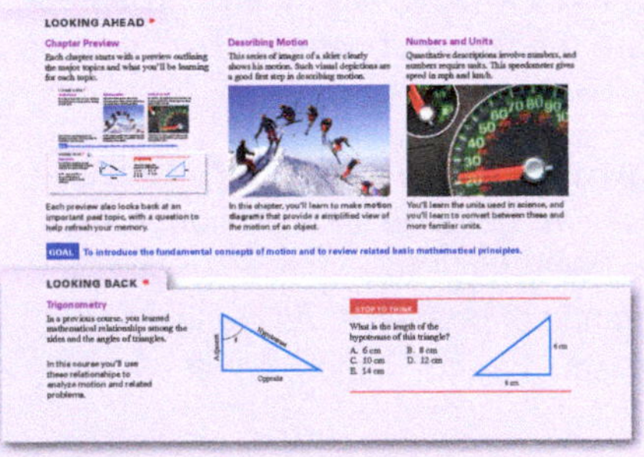

Each preview also looks back at an important past topic, with a question to help refresh your memory.

### Describing Motion

This series of images of a skier clearly shows his motion. Such visual depictions are a good first step in describing motion.

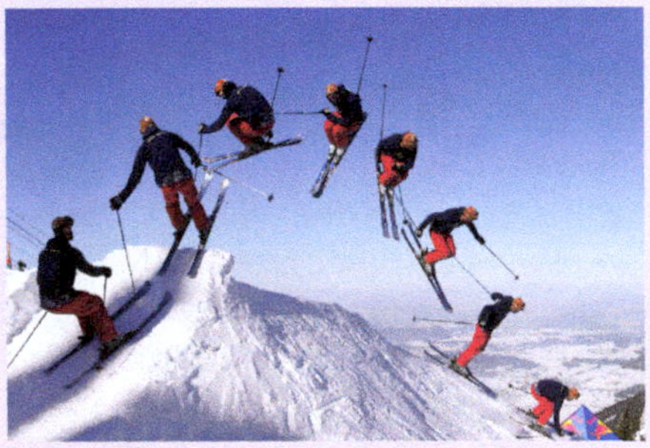

In this chapter, you'll learn to make **motion diagrams** that provide a simplified view of the motion of an object.

### Numbers and Units

Quantitative descriptions involve numbers, and numbers require units. This speedometer gives speed in mph and km/h.

You'll learn the units used in science, and you'll learn to convert between these and more familiar units.

**GOAL** To introduce the fundamental concepts of motion and to review related basic mathematical principles.

## LOOKING BACK ◂◂

### Trigonometry

In a previous course, you learned mathematical relationships among the sides and the angles of triangles.

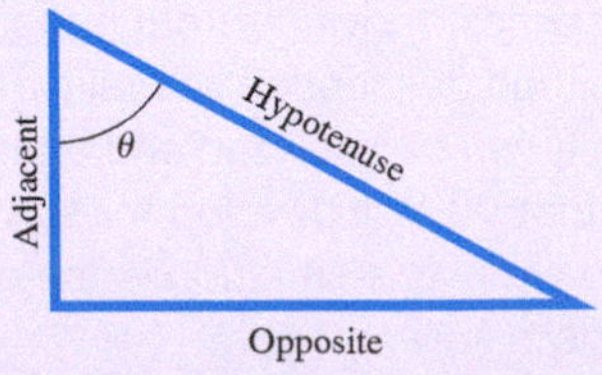


In this course you'll use these relationships to analyze motion and related problems.

**STOP TO THINK**

What is the length of the hypotenuse of this triangle?

A. 6 cm B. 8 cm
C. 10 cm D. 12 cm
E. 14 cm

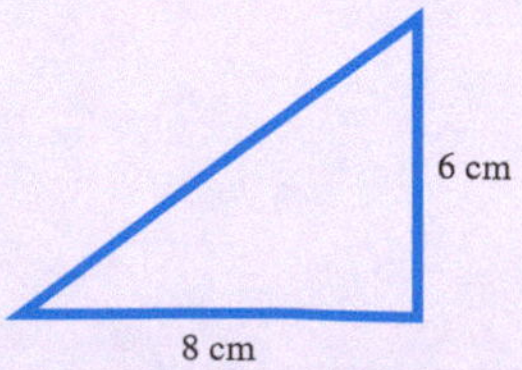

## 1.1 Motion: A First Look

Motion is a theme that will appear in one form or another throughout this entire text. You have a well-developed intuition about motion based on your experiences, but we'll find that some of the most important aspects of motion can be rather subtle. We need to develop some tools to help us explain and understand motion, so rather than jumping immediately into a lot of mathematics and calculations, this first chapter focuses on *visualizing* motion and becoming familiar with the *concepts* needed to describe a moving object.

One key difference between physics and other sciences is how we set up and solve problems. We'll often use a two-step process to solve motion problems. The first step is to develop a simplified *representation* of the motion so that key elements stand out. For example, the photo of the falcon at the start of the chapter allows us to observe its position at many successive times. We will begin our study of motion by considering this sort of picture. The second step is to analyze the motion with the language of mathematics. The process of putting numbers on nature is often the most challenging aspect of the problems you will solve. In this chapter, we will explore the steps in this process as we introduce the basic concepts of motion.

### Types of Motion

As a starting point, let's define **motion** as the change of an object's position or orientation with time. Examples of motion are easy to list. Bicycles, baseballs, cars, airplanes, and rockets are all objects that move. The path along which an object moves, which might be a straight line or might be curved, is called the object's **trajectory.**

FIGURE 1.1 shows four basic types of motion that we will study in this text. In this chapter, we will focus on the first type of motion in the figure, motion along a straight line, or *straight-line motion*. In later chapters, we will learn about *circular motion*, which is the motion of an object along a circular path; *projectile motion*, the motion of an object through the air; and *rotational motion*, the spinning of an object about an axis.

FIGURE 1.1 Four basic types of motion.

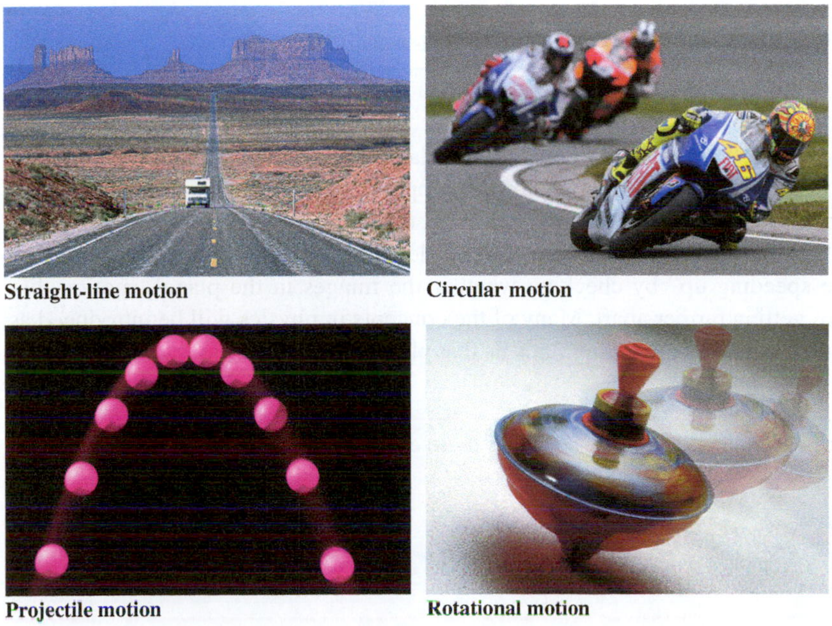


FIGURE 1.2 Several frames from the video of a car.

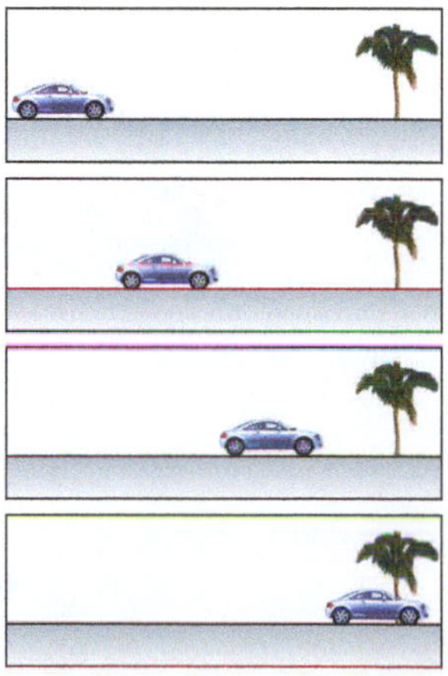

### Making a Motion Diagram

An easy way to study motion is to record a video of a moving object with a stationary camera. A video camera takes images at a fixed rate, typically 30 images every second. Each separate image is called a *frame*. As an example, FIGURE 1.2 shows several frames from a video of a car going past, with the camera in a fixed position. Not surprisingly, the car is in a different position in each frame.

**FIGURE 1.3** A motion diagram of the car shows all the frames simultaneously.

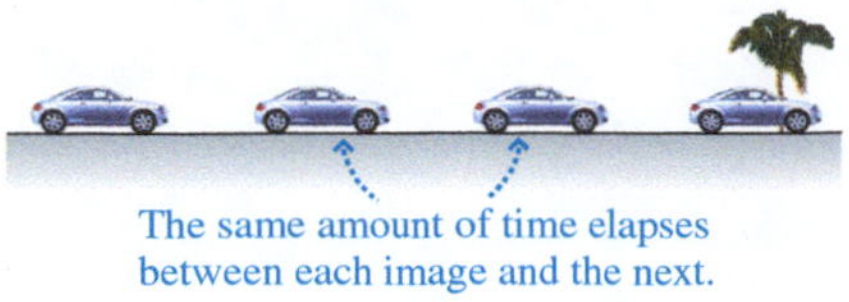

The same amount of time elapses
between each image and the next.

Suppose we now edit the video by layering the frames on top of each other. We end up with the picture in **FIGURE 1.3.** This composite image, showing an object's positions at several *equally spaced instants of time,* is called a **motion diagram.** As simple as motion diagrams seem, they will turn out to be powerful tools for analyzing motion.

Now let's take our camera out into the world and make some motion diagrams. The following table illustrates how a motion diagram shows important features of different kinds of motion.

**Examples of motion diagrams**

Images that are *equally spaced* indicate an object moving with *constant speed.*

**A skateboarder rolling down the sidewalk.**

An *increasing distance* between the images shows that the object is *speeding up.*

**A sprinter starting the 100 meter dash.**

A *decreasing distance* between the images shows that the object is *slowing down.*

**A car stopping for a red light.**

A more complex motion diagram shows changes in speed and direction.

**A basketball free throw.**

We have defined several concepts (constant speed, speeding up, and slowing down) in terms of how the moving object appears in a motion diagram. These are called **operational definitions,** meaning that the concepts are defined in terms of a particular procedure or operation. For example, we could answer the question Is the airplane speeding up? by checking whether the images in the plane's motion diagram are getting farther apart. Many of the concepts in physics will be introduced as operational definitions. This reminds us that physics is an experimental science.

**STOP TO THINK 1.1** Which car is going faster, A or B? Assume there are equal intervals of time between the frames of both videos.

Car A

Car B

**NOTE** ► Each chapter in this text has several *Stop to Think* questions. These questions are designed to see if you've understood the basic ideas that have just been presented. The answers are given at the end of the chapter, but you should make a serious effort to think about these questions before turning to the answers. ◄

## 1.2 Models and Modeling

The real world is messy and complicated. Our goal in studying physics is to brush aside many of the real-world details in order to discern patterns that occur over and over. For example, a swinging pendulum, a vibrating guitar string, a sound wave, and jiggling atoms in a crystal are all very different—yet they share a common core characteristic: Each is an example of an *oscillating system*, something that moves back and forth around an equilibrium position. If we focus on understanding a very simple oscillating system, such as a block (generically, a "mass") attached to a spring, we'll automatically understand quite a bit about the many real-world examples of oscillations.

Stripping away the details to focus on essential features is a process called *modeling*. A **model** is a highly simplified picture of reality, but one that still captures the essence of what we want to study. Thus a mass attached to a spring is a simple but realistic model of many oscillating systems.

Models allow us to make sense of complex situations by providing a framework for thinking about them. One could go so far as to say that developing and testing models is at the heart of the scientific process. Albert Einstein once said, "Physics should be as simple as possible—but not simpler." We want to find the simplest model that allows us to understand the phenomenon we're studying, but we can't make the model so simple that key aspects of the phenomenon get lost.

We'll develop and use many models throughout this text; they'll be one of our most important thinking tools. These models will be of two types:

- *Descriptive models:* What are the essential characteristics and properties of a phenomenon? How do we describe it in the simplest possible terms? For example, the mass-on-a-spring model of an oscillating system is a descriptive model.
- *Explanatory models:* Why do things happen as they do? Explanatory models, based on the laws of physics, have predictive power. They allow us to test—against experimental data—whether a model provides an adequate explanation of our observations. For example, the *charge model* that we will introduce in Chapter 20 helps us explain and predict a wide range of experimental outcomes related to electric forces.

When we solve physics problems, one of the most important steps is choosing an appropriate model for the system we are studying. In the worked examples in this text, in the first "Strategize" step, we'll point out the model being used, when appropriate.

### The Particle Model

For many objects, the motion of the object *as a whole* is not influenced by the details of the object's size and shape. To describe the object's motion, all we really need to keep track of is the motion of a single point: You could imagine looking at the motion of a dot painted on the side of the object.

In fact, for the purposes of analyzing the motion, we can often consider the object *as if* it were just a single point. We can also treat the object *as if* all of its mass were concentrated into this single point. An object that can be represented as a mass at a single point in space is called a **particle**.

If we treat an object as a particle, we can represent the object in each frame of a motion diagram as a simple dot. FIGURE 1.4 shows how much simpler motion diagrams appear when the object is represented as a particle. Note that the dots have been numbered 0, 1, 2, . . . to tell the sequence in which the frames were exposed. These diagrams still convey a complete understanding of the object's motion.

In representing the car in Figure 1.4 as a particle, we have discarded many of the details of the car, such as the shape of its body and the motion of its wheels, which are unimportant in understanding its overall motion. In other words, we have developed a model for moving objects, the **particle model,** that allows us to see

FIGURE 1.4 Simplifying a motion diagram using the particle model.

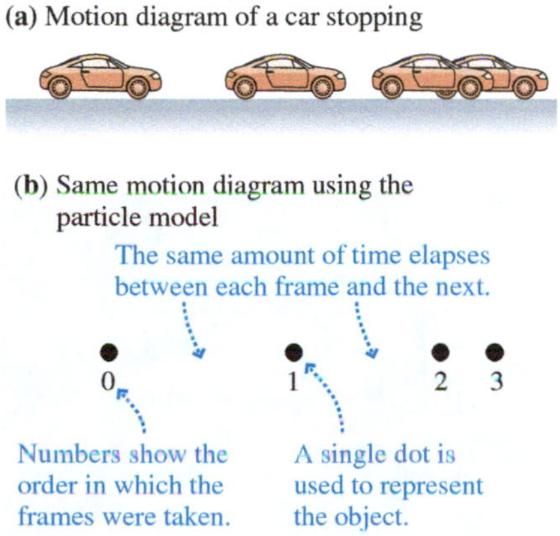


A video to support a section's topic is embedded in the eText.

**Video** Figure 1.4

**FIGURE 1.5** The particle model for two falling objects.

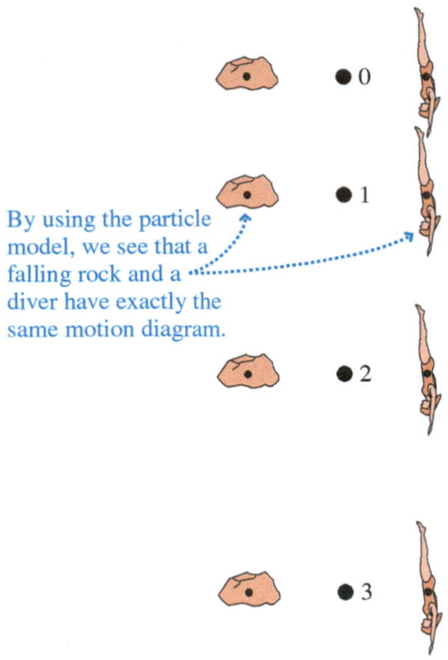


connections that are very important but that are obscured or lost by examining all the parts of an extended, real object. Consider the motion of the rock and the diver shown in **FIGURE 1.5**. These two very different objects have exactly the same motion diagram. As we will see, all objects falling under the influence of gravity move in exactly the same manner if no other forces act. The simplification of the particle model has revealed something about the physics that underlies both of these situations.

**STOP TO THINK 1.2** Three motion diagrams are shown. Which is a dust particle settling to the floor at constant speed, which is a ball dropped from the roof of a building, and which is a descending rocket slowing to make a soft landing on Mars?

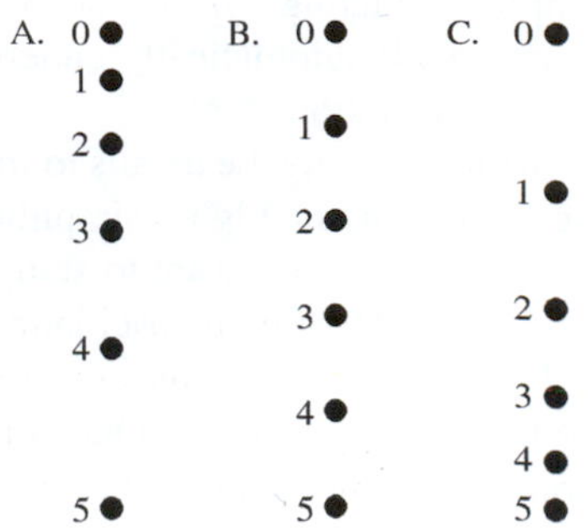


## 1.3 Position and Time: Putting Numbers on Nature

To develop our understanding of motion further, we need to be able to make quantitative measurements: We need to use numbers. As we analyze a motion diagram, it is useful to know where the object is (its *position*) and when the object was at that position (the *time*). We'll start by considering the motion of an object that can move only along a straight line. Examples of this **one-dimensional** or "1-D" motion are a car moving along a long, straight road; an airplane taxiing down a runway; and an elevator moving up and down a shaft.

### Position and Coordinate Systems

**FIGURE 1.6** Describing your position.

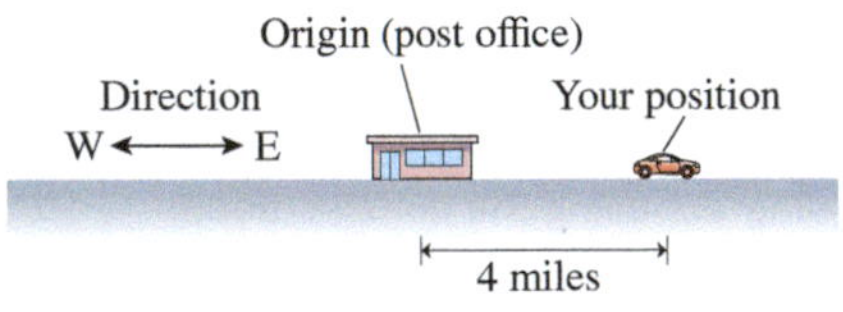


Suppose you are driving along a long, straight country road, as in **FIGURE 1.6**, and your friend calls and asks where you are. You might reply that you are 4 miles east of the post office, and your friend would then know just where you were. Your location at a particular instant in time (when your friend phoned) is called your **position.** Notice that to know your position along the road, your friend needed three pieces of information. First, you had to give her a reference point (the post office) from which all distances are to be measured. We call this fixed reference point the **origin.** Second, she needed to know how far you were from that reference point or origin—in this case, 4 miles. Finally, she needed to know which side of the origin you were on: You could be 4 miles to the west of it or 4 miles to the east.

We will need these same three pieces of information in order to specify any object's position along a line. We first choose our origin, from which we measure the distance to the object. The position of the origin is arbitrary, and we are free to place it where we like. Usually, however, there are certain points (such as the well-known post office) that are more convenient choices than others.

This gauge's vertical scale measures the depth of snow when it falls. It has a natural origin at the level of the road.

In order to specify how far our object is from the origin, we lay down an imaginary axis along the line of the object's motion. Like a ruler, this axis is marked off in equally spaced divisions of distance, perhaps in inches, meters, or miles, depending on the problem at hand. We place the zero mark of this ruler at the origin, allowing us to locate the position of our object by reading the ruler mark where the object is.

Finally, we need to be able to specify which side of the origin our object is on. To do this, we imagine the axis extending from one side of the origin with increasing positive markings; on the other side, the axis is marked with increasing *negative* numbers. By reporting the position as either a positive or a negative number, we know on what side of the origin the object is.

These elements—an origin and an axis marked in both the positive and negative directions—can be used to unambiguously locate the position of an object. We call this a **coordinate system.** We will use coordinate systems throughout this text, and we will soon develop coordinate systems that can be used to describe the positions of objects moving in more complex ways than just along a line. FIGURE 1.7 shows a coordinate system that we can use to locate various objects along the country road discussed earlier.

FIGURE 1.7 The coordinate system used to describe objects along a country road.

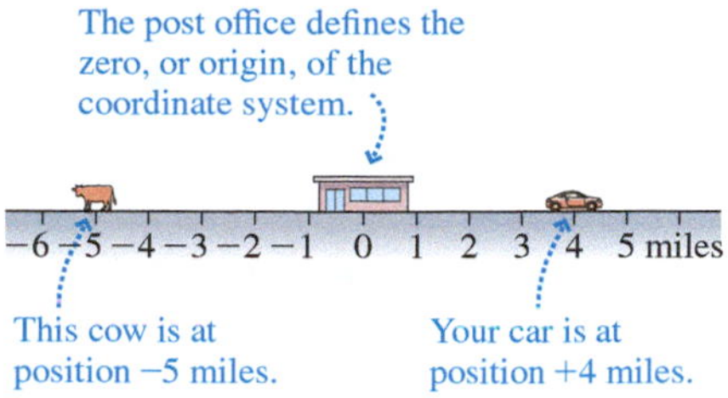


Although our coordinate system works well for describing the positions of objects located along the axis, our notation is somewhat cumbersome because we keep needing to say things like "the car is at position +4 miles." A better notation, and one that will become particularly important when we study motion in two dimensions, is to use a symbol such as $x$ or $y$ to represent the position along the axis. Then we can say "the cow is at $x = -5$ miles." The symbol that represents a position along an axis is called a **coordinate.** The introduction of symbols to represent positions (and, later, velocities and accelerations) also allows us to work with these quantities mathematically.

FIGURE 1.8 shows how we would set up a coordinate system for a sprinter running a 50 meter race (we use the standard abbreviation "m" for meters). For horizontal motion like this we usually use the coordinate $x$ to represent the position.

FIGURE 1.8 A coordinate system for a 50 meter race.

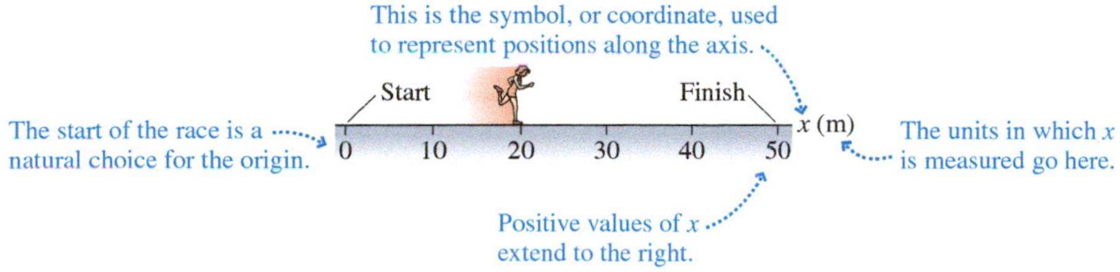


Motion along a straight line need not be horizontal. As shown in FIGURE 1.9, a rock falling vertically downward and a skier skiing down a straight slope are also examples of straight-line or one-dimensional motion.

FIGURE 1.9 Examples of one-dimensional motion.

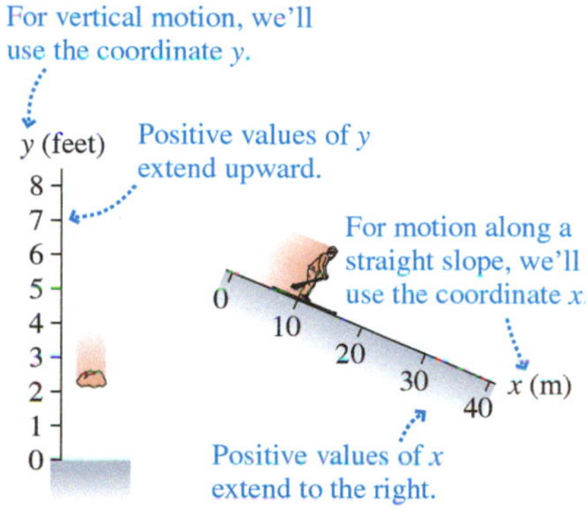


## Time

The pictures in Figure 1.9 show the position of an object at just one instant of time. But a full motion diagram represents how an object moves as time progresses. So far, we have labeled the dots in a motion diagram by the numbers 0, 1, 2, . . . to indicate the order in which the frames were taken. But to fully describe the motion, we need to indicate the *time,* as read off a clock or a stopwatch, at which each frame of a video was made. This is important, as we can see from the motion diagram of a stopping car in FIGURE 1.10. If the frames were taken 1 second apart, this motion diagram shows a leisurely stop; if 1/10 of a second apart, it represents a screeching halt.

For a complete motion diagram, we thus need to label each frame with its corresponding time (symbol $t$) as read off a clock. But when should we start the clock? Which frame should be labeled $t = 0$? This choice is much like choosing the origin $x = 0$ of a coordinate system: You can pick any arbitrary point in the motion and label it "$t = 0$ seconds." This is simply the instant you decide to start your clock or stopwatch, so it is the origin of your time coordinate. A video frame labeled "$t = 4$ seconds" means it was taken 4 seconds after you started your clock. We typically choose $t = 0$ to represent the "beginning" of a problem, but the object may have been moving before then.

FIGURE 1.10 Is this a leisurely stop or a screeching halt?

FIGURE 1.11 The motion diagram of a car that travels at constant speed and then brakes to a halt.

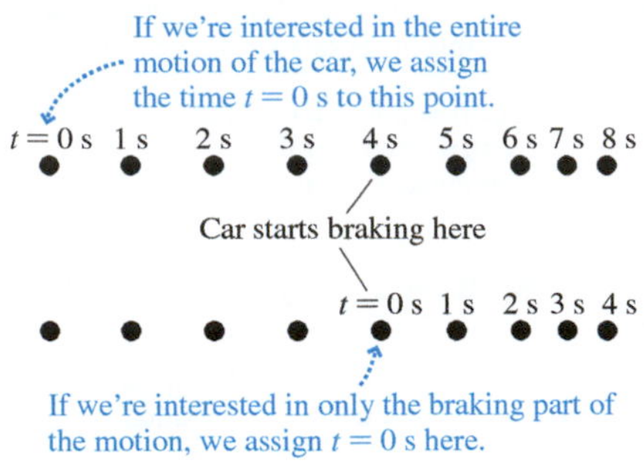


To illustrate, FIGURE 1.11 shows the motion diagram for a car moving at a constant speed and then braking to a halt. Two possible choices for the frame labeled $t = 0$ seconds are shown; our choice depends on what part of the motion we're interested in. Each successive position of the car is then labeled with the clock reading in seconds (abbreviated by "s").

## Changes in Position: Displacement

Now that we've seen how to measure position and time, let's return to the problem of motion. To describe motion we'll need to measure the *changes* in position that occur with time. Consider the following:

> Sam is standing 50 feet (ft) east of the corner of 12th Street and Vine. He then walks to a second point 150 ft east of Vine. What is Sam's change of position?

FIGURE 1.12 Sam undergoes a displacement $\Delta x$ from position $x_i$ to position $x_f$.

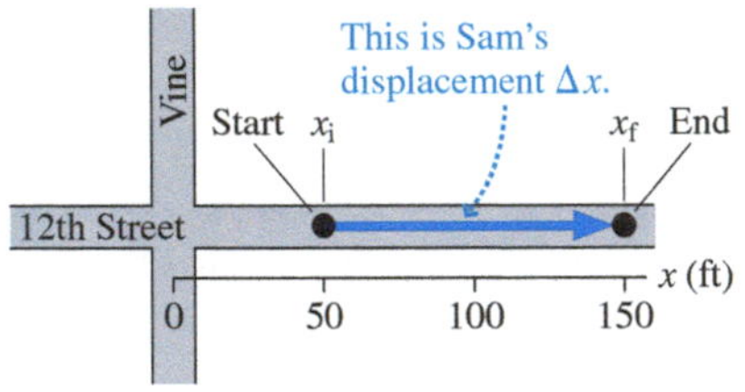


FIGURE 1.12 shows Sam's motion on a map. We've placed a coordinate system on the map, using the coordinate $x$. We are free to place the origin of our coordinate system wherever we wish, so we have placed it at the intersection. Sam's initial position is then at $x_i = 50$ ft. The positive value for $x_i$ tells us that Sam is east of the origin.

> **NOTE** ▸ We will label special values of $x$ or $y$ with subscripts. The value at the start of a problem is usually labeled with a subscript "i," for *initial*, and the value at the end is labeled with a subscript "f," for *final*. For cases having several special values, we will usually use subscripts "1," "2," and so on. ◂

Sam's final position is $x_f = 150$ ft, indicating that he is 150 ft east of the origin. You can see that Sam has changed position, and a *change* of position is called a **displacement.** His displacement is the distance labeled $\Delta x$ in Figure 1.12. The Greek letter delta ($\Delta$) is used in math and science to indicate the *change* in a quantity. Thus $\Delta x$ indicates a change in the position $x$.

> **NOTE** ▸ $\Delta x$ is a *single* symbol. You cannot cancel out or remove the $\Delta$ in algebraic operations. ◂

To get from the 50 ft mark to the 150 ft mark, Sam clearly had to walk 100 ft, so the change in his position—his displacement—is 100 ft. We can think about displacement in a more general way, however. Displacement is the *difference* between a final position $x_f$ and an initial position $x_i$. Thus we can write

$$\Delta x = x_f - x_i = 150 \text{ ft} - 50 \text{ ft} = 100 \text{ ft}$$

> **NOTE** ▸A general principle, used throughout this text, is that the change in any quantity is the final value of the quantity minus its initial value. ◂

FIGURE 1.13 A displacement is a signed quantity. Here $\Delta x$ is a negative number.

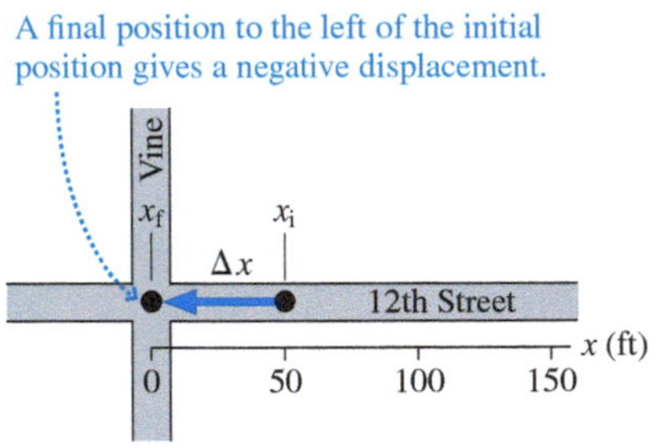


Displacement is a *signed quantity;* that is, it can be either positive or negative. If, as shown in FIGURE 1.13, Sam's final position $x_f$ had been at the origin instead of the 150 ft mark, his displacement would have been

$$\Delta x = x_f - x_i = 0 \text{ ft} - 50 \text{ ft} = -50 \text{ ft}$$

The negative sign tells us that he moved to the *left* along the $x$-axis, or 50 ft *west.*

## Changes in Time

FIGURE 1.14 The motion diagram of a bicycle moving to the right at a constant speed.

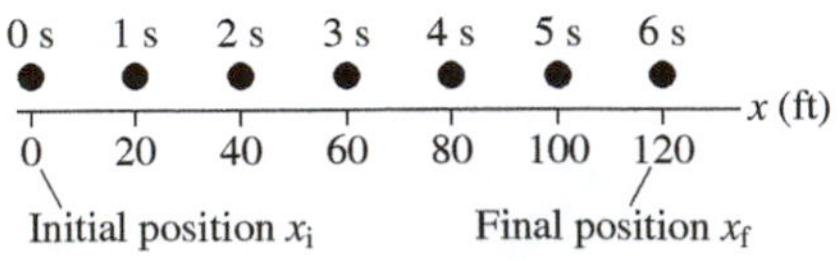


A displacement is a change in position. In order to quantify motion, we'll need to also consider changes in *time,* which we call **time intervals.** We've seen how we can label each frame of a motion diagram with a specific time, as determined by our stopwatch. FIGURE 1.14 shows the motion diagram of a bicycle moving at a constant speed, with the times of the measured points indicated.

The displacement between the initial position $x_i$ and the final position $x_f$ is

$$\Delta x = x_f - x_i = 120 \text{ ft} - 0 \text{ ft} = 120 \text{ ft}$$

Similarly, we define the time interval between these two points to be

$$\Delta t = t_f - t_i = 6\text{ s} - 0\text{ s} = 6\text{ s}$$

**A time interval $\Delta t$ measures the elapsed time as an object moves from an initial position $x_i$ at time $t_i$ to a final position $x_f$ at time $t_f$.** Note that, unlike $\Delta x$, $\Delta t$ is always positive because $t_f$ is always greater than $t_i$.

**EXAMPLE 1.1** **How long a ride?**

Emily is enjoying a bicycle ride on a country road that runs east-west past a water tower. At noon, Emily is 3 miles (mi) east of the water tower. A half-hour later, she is 2 mi west of the water tower. What is her displacement during that half-hour?

**STRATEGIZE** We will use the particle model to represent Emily. She is riding along a straight east-west road, so we will make use of a one-dimensional coordinate system to describe her motion.

**PREPARE** Although it may seem like overkill for such a simple problem, you should start by making a drawing, like the one in FIGURE 1.15, with the $x$-axis along the road. We choose our coordinate system so that increasing $x$ means moving to the east. Distances are measured with respect to the water tower, so it is a natural origin for the coordinate system. Once the coordinate system is established, we can show Emily's initial and final positions and her displacement between the two.

FIGURE 1.15 A drawing of Emily's motion.

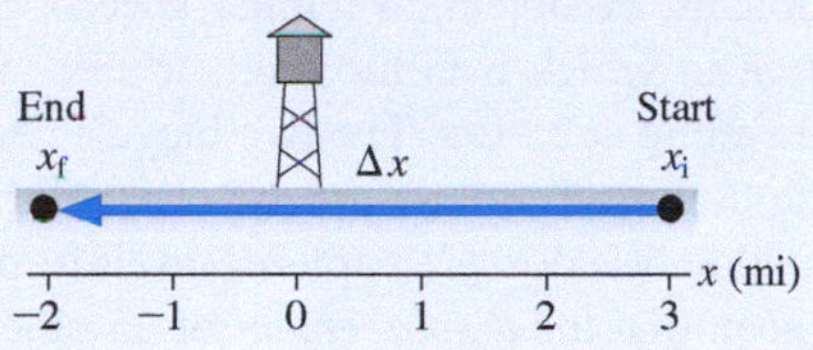


**SOLVE** We've specified values for Emily's initial and final positions in our drawing. We can thus compute her displacement:

$$\Delta x = x_f - x_i = (-2\text{ mi}) - (3\text{ mi}) = -5\text{ mi}$$

**ASSESS** Once we've completed the solution to the problem, we need to go back to see if it makes sense. Emily is moving to the west, so we expect her displacement to be negative—and it is. We can see from our drawing in Figure 1.15 that she has moved 5 miles from her starting position, so our answer seems reasonable. As part of the Assess step, we also check our answers to see if they make physical sense. Emily travels 5 miles in a half-hour, quite a reasonable pace for a cyclist.

**NOTE** ▸ All of the numerical examples in the text are worked out with the same four-step process: Strategize, Prepare, Solve, Assess. It's tempting to cut corners, especially for the simple problems in these early chapters, but you should take the time to do all of these steps now, to practice your problem-solving technique. We'll have more to say about our general problem-solving approach in Chapter 2. ◂

**STOP TO THINK 1.3** Sarah starts at a positive position along the $x$-axis. She then undergoes a negative displacement. Her final position

A. Is positive. B. Is negative. C. Could be either positive or negative.

## 1.4 Velocity

We all have an intuitive sense of whether something is moving very fast or just cruising slowly along. To make this intuitive idea more precise, let's start by examining the motion diagrams of some objects moving along a straight line at a *constant* speed, objects that are neither speeding up nor slowing down. This motion at a constant speed is called **uniform motion.** As we saw for the skateboarder in Section 1.1, for an object in uniform motion, successive frames of the motion diagram are *equally spaced,* so the object's displacement $\Delta x$ is the *same* between successive frames.

To see how an object's displacement between successive frames is related to its speed, consider the motion diagrams of a bicycle and a car, traveling along the same street as shown in FIGURE 1.16. Clearly the car is moving faster than the bicycle: In any 1 second time interval, the car undergoes a displacement $\Delta x = 40$ ft, while the bicycle's displacement is only 20 ft.

FIGURE 1.16 Motion diagrams for a car and a bicycle.

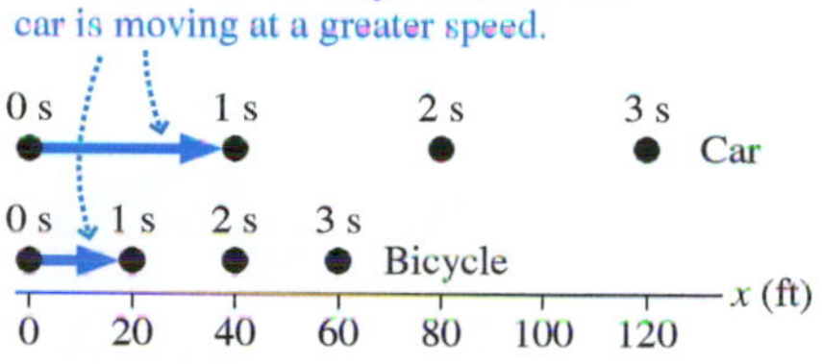

The greater the distance traveled by an object in a given time interval, the greater its speed. This idea leads us to define the speed of an object as

$$\text{speed} = \frac{\text{distance traveled in a given time interval}}{\text{time interval}} \tag{1.1}$$

Speed of an object in uniform motion

For the bicycle, this equation gives

$$\text{speed} = \frac{20\ \text{ft}}{1\ \text{s}} = 20\ \frac{\text{ft}}{\text{s}}$$

while for the car we have

$$\text{speed} = \frac{40\ \text{ft}}{1\ \text{s}} = 40\ \frac{\text{ft}}{\text{s}}$$

The speed of the car is twice that of the bicycle, which seems reasonable.

NOTE ▸ The division gives units that are a fraction: ft/s. This is read as "feet per second," just like the more familiar "miles per hour." ◂

FIGURE 1.17 Two bicycles traveling at the same speed, but with different velocities.

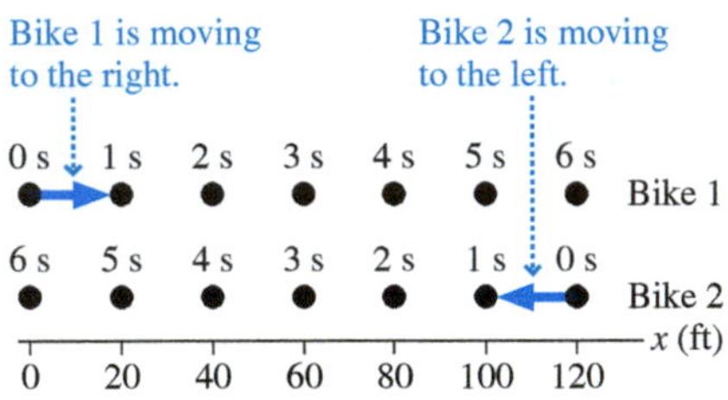


To fully characterize the motion of an object, we must specify not only the object's speed but also the *direction* in which it is moving. FIGURE 1.17 shows the motion diagrams of two bicycles traveling at 20 ft/s. The two bicycles have the same speed, but something about their motion is different—the *direction* of their motion.

The "distance traveled" in Equation 1.1 doesn't capture any information about the direction of travel. But we've seen that the *displacement* of an object does contain this information. We can introduce a new quantity, the **velocity,** as

$$\text{velocity} = \frac{\text{displacement}}{\text{time interval}} = \frac{\Delta x}{\Delta t} \tag{1.2}$$

Velocity of a moving object

The velocity of bicycle 1 in Figure 1.17, computed using the 1 second time interval between the $t = 0$ s and $t = 1$ s positions, is

$$v = \frac{\Delta x}{\Delta t} = \frac{x_1 - x_0}{1\ \text{s} - 0\ \text{s}} = \frac{20\ \text{ft} - 0\ \text{ft}}{1\ \text{s}} = +20\ \frac{\text{ft}}{\text{s}}$$

while the velocity of bicycle 2, during the same time interval, is

$$v = \frac{\Delta x}{\Delta t} = \frac{x_1 - x_0}{1\ \text{s} - 0\ \text{s}} = \frac{100\ \text{ft} - 120\ \text{ft}}{1\ \text{s}} = -20\ \frac{\text{ft}}{\text{s}}$$

NOTE ▸ We have used $x_0$ for the position at time $t = 0$ seconds and $x_1$ for the position at time $t = 1$ second. The subscripts serve the same role as before—identifying particular positions—but in this case the positions are identified by the time at which each position is reached. ◂

The two velocities have opposite signs because the bicycles are traveling in opposite directions. **Speed measures only how fast an object moves, but velocity tells us both an object's speed *and its direction*.** In this text, we'll use a positive velocity to indicate motion to the right or, for vertical motion, upward. We'll use a negative velocity for an object moving to the left, or downward.

NOTE ▸ Learning to distinguish between speed, which is always a positive number, and velocity, which can be either positive or negative, is one of the most important tasks in the analysis of motion. ◂

The velocity as defined by Equation 1.2 is actually what is called the *average* velocity. On average, over each 1 s interval bicycle 1 moves 20 ft, but we don't know if it was moving at exactly the same speed at every moment during this time interval. In Chapter 2, we'll develop the idea of *instantaneous* velocity, the velocity of an object at a particular instant in time. Since our goal in this chapter is to *visualize* motion with motion diagrams, we'll somewhat blur the distinction between average and instantaneous quantities, refining these definitions in Chapter 2.

## The "Per" in Miles Per Hour

The units for speed and velocity are a unit of distance, such as feet, meters, or miles, divided by a unit of time, such as seconds or hours. Thus we could measure velocity in units of mi/h (or mph) or m/s, pronounced "miles *per* hour" or "meters *per* second." The word "per" will often arise in physics when we consider the ratio of two quantities. What do we mean, exactly, by "per"?

If a car moves with a speed of 23 m/s, we mean that it travels 23 meters *for each* second of elapsed time. The word "per" thus associates the number of units in the numerator (23 m) with *one* unit of the denominator (1 s). We'll see many other examples of this idea as the text progresses. You may already know a bit about *density;* you can look up the density of gold and you'll find that it is 19.3 g/cm$^3$ ("grams *per* cubic centimeter"). This means that there are 19.3 grams of gold *for each* cubic centimeter of the metal.

**EXAMPLE 1.2 Finding the speed of a seabird** BIO

Albatrosses are seabirds that spend most of their lives flying over the ocean looking for food. With a stiff tailwind, an albatross can fly at high speeds. Satellite data on one particularly speedy albatross showed it 60 miles east of its roost at 3:00 PM and then, at 3:15 PM, 80 miles east of its roost. What was its velocity?

***STRATEGIZE*** We will assume that the albatross is flying at a constant speed; that is, it is in uniform motion. Using the particle model, we can represent its motion on a coordinate system.

***PREPARE*** The statement of the problem provides us with a natural coordinate system: We can measure distances with respect to the roost, with distances to the east as positive. With this coordinate system, the motion of the albatross appears as in FIGURE 1.18. The motion takes place between 3:00 and 3:15, a time interval of 15 minutes. If we want our final velocity to be in the familiar units of miles per hour, or mph, we need this time interval in hours (abbreviated "h"). So we make the conversion 15 min = 0.25 h.

FIGURE 1.18 The motion of an albatross at sea.

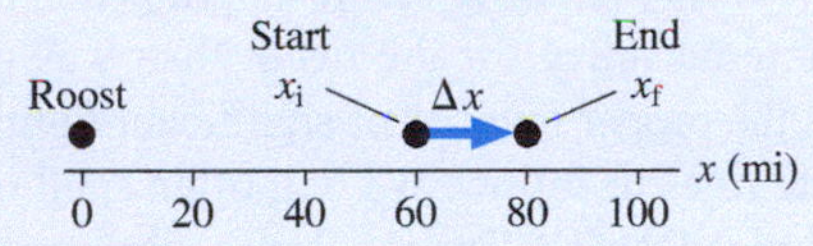


***SOLVE*** We know the initial and final positions, and we know the time interval, so we can calculate the velocity:

$$v = \frac{\Delta x}{\Delta t} = \frac{x_f - x_i}{0.25\ \text{h}} = \frac{20\ \text{mi}}{0.25\ \text{h}} = 80\ \text{mph}$$

***ASSESS*** The velocity is positive, which makes sense because Figure 1.18 shows that the motion is to the right. A speed of 80 mph is certainly fast, but the problem said it was a "particularly speedy" albatross, so our answer seems reasonable. (Indeed, albatrosses have been observed to fly at such speeds in the very fast winds of the Southern Ocean. This problem is based on real observations, as will be our general practice in this text.)

**STOP TO THINK 1.4** Jane starts from her house to take a stroll in her neighborhood. After walking for 2 hours at a steady pace, she has walked 4 miles and is 2 miles from home. For this time interval, what was her speed?

A. 4 mph B. 3 mph C. 2 mph D. 1 mph

# 1.5 A Sense of Scale: Significant Figures, Scientific Notation, and Units

Physics attempts to explain the natural world, from the very small to the exceedingly large. And in order to understand our world, we need to be able to *measure* quantities both minuscule and enormous. A properly reported measurement has three elements. First, we can measure our quantity with only a certain precision. To make this precision clear, we need to make sure that we report our measurement with the correct number of *significant figures.*

Second, writing down the really big and small numbers that often come up in physics can be awkward. To avoid writing all those zeros, scientists use *scientific notation* to express numbers both big and small.

Finally, we need to choose an agreed-upon set of *units* for the quantity. For speed, common units include meters per second and miles per hour. For mass, the kilogram is the most commonly used unit. Every physical quantity that we can measure has an associated set of units.

## Measurements and Significant Figures

**FIGURE 1.19** The precision of a measurement depends on the instrument used to make it.

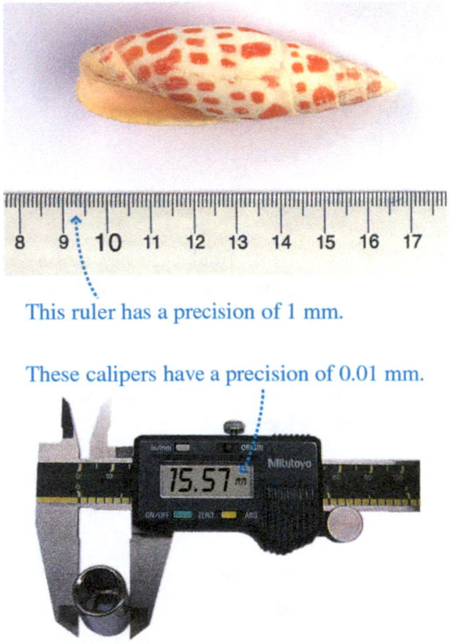


When we measure any quantity, such as the length of a bone or the weight of a specimen, we can do so with only a certain *precision.* If you make a measurement with the ruler shown in **FIGURE 1.19**, you probably can't be more accurate than about $\pm 1$ mm, so the ruler has a precision of 1 mm. The digital calipers shown can make a measurement to within $\pm 0.01$ mm, so it has a precision of 0.01 mm. The precision of a measurement can also be affected by the skill or judgment of the person performing the measurement. A stopwatch might have a precision of 0.001 s, but, due to your reaction time, your measurement of the time of a sprinter would be much less precise.

It is important that your measurement be reported in a way that reflects its actual precision. Suppose you use a ruler to measure the length of a particular frog. You judge that you can make this measurement with a precision of about 1 mm, or 0.1 cm. In this case, the frog's length should be reported as, say, 6.2 cm. We interpret this to mean that the actual value falls between 6.15 cm and 6.25 cm and thus rounds to 6.2 cm. If you reported the frog's length as simply 6 cm, you would be saying less than you know; you would be withholding information. If you reported the number as 6.213 cm, however, anyone reviewing your work would interpret this to mean that the actual length falls between 6.2125 cm and 6.2135 cm, a precision of 0.001 cm. In this case, you would be claiming to have more information than you really possessed.

The way to state your knowledge precisely is through the proper use of **significant figures.** You can think of a significant figure as a digit that is reliably known. A measurement such as 6.2 cm has *two* significant figures, the 6 and the 2. The next decimal place—the hundredths—is not reliably known and is thus not a significant figure. Similarly, a time measurement of 34.62 s has four significant figures, implying that the 2 in the hundredths place is reliably known.

Walter Davis's best long jump on this day was reported as 8.24 m. This implies that the actual length of the jump was between 8.235 m and 8.245 m, a spread of only 0.01 m, which is 1 cm. Does this claimed accuracy seem reasonable?

When we perform a calculation such as adding or multiplying two or more measured numbers, we can't claim more accuracy for the result than was present in the initial measurements. Determining the proper number of significant figures is straightforward, but there are a few definite rules to follow. We will often spell out such technical details in what we call a "Tactics Box." A Tactics Box is designed to teach you particular skills and techniques. Each Tactics Box will include the icon to designate exercises in the *Student Workbook* that you can use to practice these skills.

**TACTICS BOX 1.1 Using significant figures**

❶ When you multiply or divide several numbers, or when you take roots, the number of significant figures in the answer should match the number of significant figures of the *least* precisely known number used in the calculation:

Three significant figures

$$3.73 \times 5.7 = 21$$

Two significant figures

Answer should have the *lower* of the two, or two significant figures.

*Continued*

❷ When you add or subtract several numbers, the number of decimal places in the answer should match the *smallest* number of decimal places of any number used in the calculation:

$$\begin{array}{r} 18.54 \\ +106.6 \\ \hline 125.1 \end{array}$$

18.54 — Two decimal places

+106.6 — One decimal place

125.1 — Answer should have the *lower* of the two, or one decimal place.

❸ **Exact numbers** have no uncertainty and, when used in calculations, do not change the number of significant figures of measured numbers. Examples of exact numbers are $\pi$ and the number 2 in the relation $d = 2r$ between a circle's diameter and radius.

There is one notable exception to these rules:

- It is acceptable to keep one or two extra digits during *intermediate* steps of a calculation to minimize round-off errors in the calculation. But the *final* answer must be reported with the proper number of significant figures.

Exercise 15

**EXAMPLE 1.3** **Using significant figures when measuring the velocity of a car**

To measure the velocity of a car, clocks A and B are set up at two points along the road, as shown in FIGURE 1.20. Clock A is precise to 0.01 s, while clock B is precise to only 0.1 s. The distance between these two clocks is carefully measured to be 124.5 m. The two clocks are automatically started when the car passes a trigger in the road; each clock stops automatically when the car passes that clock. After the car has passed both clocks, clock A is found to read $t_A = 1.22$ s, and clock B to read $t_B = 4.5$ s. The time from the less-precise clock B is correctly reported with fewer significant figures than that from A. What is the velocity of the car, and how should it be reported with the correct number of significant figures?

FIGURE 1.20 Measuring the velocity of a car.

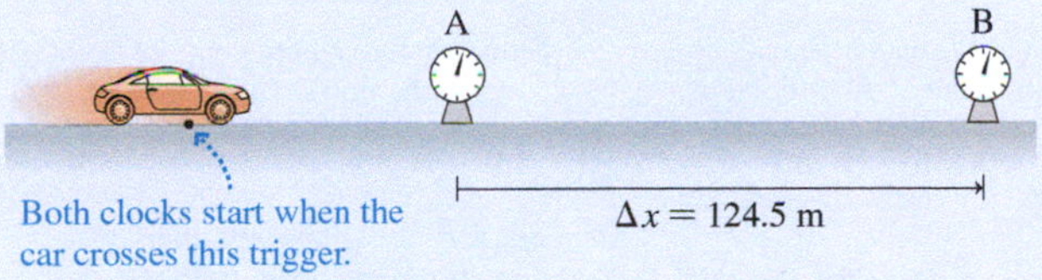


***STRATEGIZE*** To find the car's velocity with the correct precision, we will need to take into account the significant figures in the measured quantities and then apply the rules of Tactics Box 1.1.

***PREPARE*** To calculate the velocity, we need the displacement $\Delta x$ and the time interval $\Delta t$ as the car moves between the two clocks. The displacement is given as $\Delta x = 124.5$ m; we can calculate the time interval as the difference between the two measured times.

***SOLVE*** The time interval is

$$\Delta t = t_B - t_A = (4.5 \text{ s}) - (1.22 \text{ s}) = 3.3 \text{ s}$$

(4.5 s): This number has one decimal place.
(1.22 s): This number has two decimal places.
3.3 s: By rule 2 of Tactics Box 1.1, the result should have *one* decimal place.

We can now calculate the velocity with the displacement and the time interval:

$$v = \frac{\Delta x}{\Delta t} = \frac{124.5 \text{ m}}{3.3 \text{ s}} = 38 \text{ m/s}$$

124.5 m: The displacement has four significant figures.
3.3 s: The time interval has two significant figures.
38 m/s: By rule 1 of Tactics Box 1.1, the result should have *two* significant figures.

***ASSESS*** Our final value has two significant figures. Suppose you had been hired to measure the speed of a car this way, and you reported 37.72 m/s. It would be reasonable for someone looking at your result to assume that the measurements you used to arrive at this value were correct to four significant figures and thus that you had measured time to the nearest 0.001 second. Our correct result of 38 m/s has all of the accuracy that you can claim, but no more!

## Scientific Notation

It's easy to write down measurements of ordinary-sized objects: Your height might be 1.72 meters, the weight of an apple 0.34 pound. But the radius of a hydrogen atom is 0.000 000 000 053 m, and the distance to the moon is 384,000,000 m. Keeping track of all those zeros is quite cumbersome.

Beyond requiring you to deal with all the zeros, writing quantities this way makes it unclear how many significant figures are involved. For the distance to the moon, how many of those digits are significant? Three? Four? All nine?

Writing numbers using **scientific notation** avoids both these problems. A value in scientific notation is a number with one digit to the left of the decimal point and zero or more to the right of it, multiplied by a power of ten. This solves the problem of writing so many zeros and makes the number of significant figures immediately apparent. In scientific notation, writing the distance to the moon as $3.84 \times 10^8$ m indicates that three digits are significant; writing it as $3.8 \times 10^8$ m indicates that only two digits are.

Even for smaller values, scientific notation can clarify the number of significant figures. Suppose a distance is reported as 1200 m. How many significant figures does this measurement have? It's ambiguous, but using scientific notation can remove any ambiguity. If this distance is known to within 1 m, we can write it as $1.200 \times 10^3$ m, showing that all four digits are significant; if it is accurate to only 100 m or so, we can report it as $1.2 \times 10^3$ m, indicating two significant figures.

**TACTICS BOX 1.2** **Using scientific notation**

To convert a number into scientific notation:

1. For a number greater than 10, move the decimal point to the left until only one digit remains to the left of the decimal point. The remaining number is then multiplied by 10 to a power; this power is given by the number of spaces the decimal point was moved. Here we convert the radius of the earth to scientific notation:

   We move the decimal point until there is only one digit to its left, counting the number of steps.

   Since we moved the decimal point 6 steps, the power of ten is 6.

   $$6\,370\,000 \text{ m} = 6.37 \times 10^6 \text{ m}$$

   The number of digits here equals the number of significant figures.

2. For a number less than 1, move the decimal point to the right until it passes the first digit that isn't a zero. The remaining number is then multiplied by 10 to a negative power; the power is given by the number of spaces the decimal point was moved. For the diameter of a red blood cell we have:

   We move the decimal point until it passes the first digit that is not a zero, counting the number of steps.

   Since we moved the decimal point 6 steps, the power of ten is −6.

   $$0.000\,007\,5 \text{ m} = 7.5 \times 10^{-6} \text{ m}$$

   The number of digits here equals the number of significant figures.

Exercise 16

◀ **The importance of units** In 1999, the $125 million Mars Climate Orbiter burned up in the Martian atmosphere instead of entering a safe orbit from which it could perform observations. The problem was faulty units! An engineering team had provided critical data on spacecraft performance in English units, but the navigation team assumed these data were in metric units. As a consequence, the navigation team had the spacecraft fly too close to the planet, and it burned up in the atmosphere.

## Units

As we have seen, in order to measure a quantity we need to give it a numerical value. But a measurement is more than just a number—it requires a *unit* to be given. You can't go to the deli and ask for "three quarters of cheese." You need to use a unit—here, one of weight, such as pounds—in addition to the number.

In your daily life, you probably use the English system of units, in which distances are measured in inches, feet, and miles. These units are well adapted for daily life, but they are rarely used in scientific work. Given that science is an international discipline, it is also important to have a system of units that is recognized around the world. For these reasons, scientists use a system of units called *le Système Internationale d'Unités,* commonly referred to as **SI units.** We often refer to these as *metric units* because the meter is the basic standard of length.

The three basic SI quantities, shown in TABLE 1.1, are time, length (or distance), and mass. The SI units for these quantities are meters, seconds, and kilograms, respectively. Other quantities needed to understand motion can be expressed as combinations of these basic units. For example, speed and velocity are expressed in meters per second or m/s. This combination is a ratio of the length unit (the meter) to the time unit (the second).

TABLE 1.1 Common SI units

| Quantity | Unit | Abbreviation |
|---|---|---|
| Time | second | s |
| Length | meter | m |
| Mass | kilogram | kg |

## Using Prefixes

We will have many occasions to use lengths, times, and masses that are either much less or much greater than the standards of 1 meter, 1 second, and 1 kilogram. We will do so by using *prefixes* to denote various powers of ten. For instance, the prefix "kilo" (abbreviation k) denotes $10^3$, or a factor of 1000. Thus 1 km equals 1000 m, 1 MW (megawatt) equals $10^6$ watts, and 1 $\mu$V (microvolt) equals $10^{-6}$ volts. TABLE 1.2 lists the common prefixes that will be used frequently throughout this text. A more extensive list of prefixes is shown at the front of the text.

Although prefixes make it easier to talk about quantities, those given with prefixed units are usually converted to base SI units of seconds and meters before any calculations are done. Thus 23.0 cm should be converted to 0.230 m before starting calculations. (The exception is the kilogram, which is already the base SI unit.)

TABLE 1.2 Common prefixes for powers of 10

| Prefix | Abbreviation | Power of 10 |
|---|---|---|
| mega- | M | $10^6$ |
| kilo- | k | $10^3$ |
| centi- | c | $10^{-2}$ |
| milli- | m | $10^{-3}$ |
| micro- | $\mu$ | $10^{-6}$ |
| nano- | n | $10^{-9}$ |

## Unit Conversions Between Measurement Systems

Although SI units are our standard, we cannot entirely forget that the United States still uses English units. Even after repeated exposure to metric units in classes, most of us "think" in English units. Thus it remains important to be able to convert back and forth between SI units and English units. TABLE 1.3 shows some frequently used conversions that will come in handy.

TABLE 1.3 Useful unit conversions

| |
|---|
| 1 inch (in) = 2.54 cm |
| 1 foot (ft) = 0.305 m |
| 1 mile (mi) = 1.609 km |
| 1 mile per hour (mph) = 0.447 m/s |
| 1 m = 39.37 in |
| 1 km = 0.621 mi |
| 1 m/s = 2.24 mph |

One effective method of performing unit conversions begins by noticing that since, for example, 1 mi = 1.609 km, the ratio of these two distances—*including their units*—is equal to 1, so that

$$\frac{1\text{ mi}}{1.609\text{ km}} = \frac{1.609\text{ km}}{1\text{ mi}} = 1$$

A ratio of values equal to 1 is called a **conversion factor.** The following Tactics Box shows how to make a unit conversion.

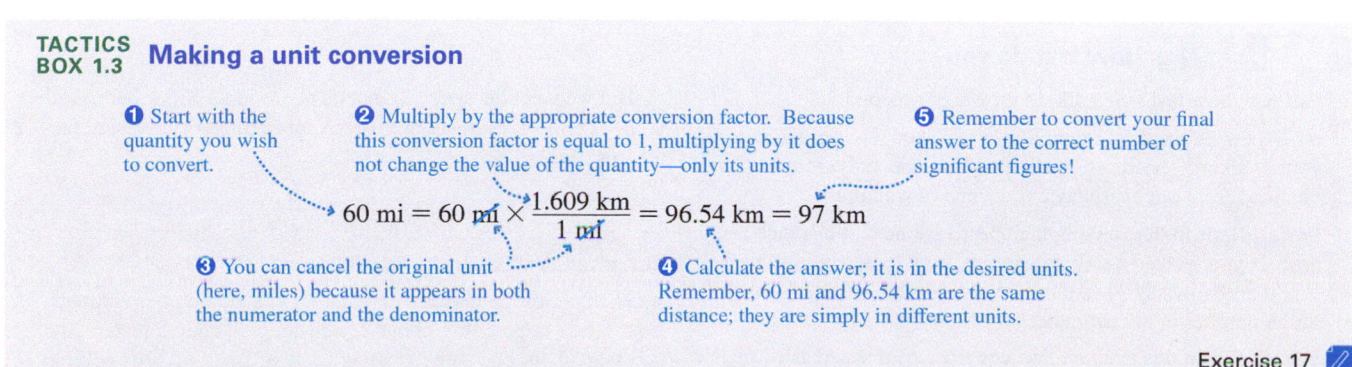

**TACTICS BOX 1.3 Making a unit conversion**

❶ Start with the quantity you wish to convert.

❷ Multiply by the appropriate conversion factor. Because this conversion factor is equal to 1, multiplying by it does not change the value of the quantity—only its units.

❺ Remember to convert your final answer to the correct number of significant figures!

$$60\text{ mi} = 60\text{ mi} \times \frac{1.609\text{ km}}{1\text{ mi}} = 96.54\text{ km} = 97\text{ km}$$

❸ You can cancel the original unit (here, miles) because it appears in both the numerator and the denominator.

❹ Calculate the answer; it is in the desired units. Remember, 60 mi and 96.54 km are the same distance; they are simply in different units.

Exercise 17

More complicated conversions can be done with several successive multiplications of conversion factors, as we see in the example on the next page.

## EXAMPLE 1.4 Can a bicycle go that fast?

In Section 1.4, we calculated the speed of a bicycle to be 20 ft/s. Is this a reasonable speed for a bicycle?

**STRATEGIZE** In order to determine whether or not this speed is reasonable, we will convert it to more familiar units. For speed, the unit you are most familiar with is likely miles per hour.

**PREPARE** We need the following unit conversions:

$$1\text{ mi} = 5280\text{ ft} \qquad 1\text{ hour (1 h)} = 60\text{ min} \qquad 1\text{ min} = 60\text{ s}$$

**SOLVE** We then multiply our original value by successive factors of 1 in order to convert the units:

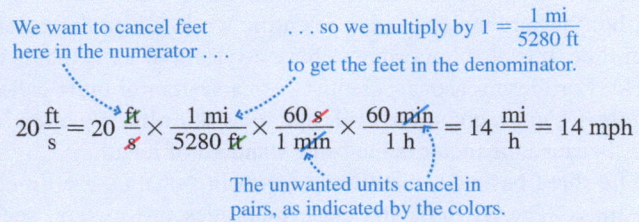

We want to cancel feet here in the numerator . . .

. . . so we multiply by $1 = \frac{1\text{ mi}}{5280\text{ ft}}$ to get the feet in the denominator.

$$20\,\frac{\text{ft}}{\text{s}} = 20\,\frac{\text{ft}}{\text{s}} \times \frac{1\text{ mi}}{5280\text{ ft}} \times \frac{60\text{ s}}{1\text{ min}} \times \frac{60\text{ min}}{1\text{ h}} = 14\,\frac{\text{mi}}{\text{h}} = 14\text{ mph}$$

The unwanted units cancel in pairs, as indicated by the colors.

**ASSESS** Our final result of 14 miles per hour (14 mph) is a very reasonable speed for a bicycle, which gives us confidence in our answer. If we had calculated a speed of 140 miles per hour, we would have suspected that we had made an error because this is quite a bit faster than the average bicyclist can travel!

◄ The man has a mass of 70 kg. What is the mass of the elephant standing next to him? By thinking about the relative dimensions of the two, you can make a reasonable one-significant-figure *estimate*.

## Estimation

Precise calculations are appropriate when we have precise data, but there are many times when just a rough estimate is sufficient. Suppose you saw a rock fall off a cliff and wanted to know how fast it was going when it hit the ground. By doing a mental comparison with the speeds of familiar objects, such as cars and bicycles, you might judge that the rock was traveling at about 20 mph.

This is a one-significant-figure estimate: You can probably distinguish 20 mph from either 10 mph or 30 mph, but you certainly cannot distinguish 20 mph from 21 mph just from an observation. A one-significant-figure estimate or calculation, such as this estimate of speed, is called an **order-of-magnitude estimate.** An order-of-magnitude estimate is indicated by the symbol ~ , which indicates even less precision than the "approximately equal" symbol ≈ . You would report your estimate of the speed of the falling rock as $v \sim 20$ mph.

It's a useful skill to make reliable order-of-magnitude estimates on the basis of known information (or information found on the Internet), simple reasoning, and common sense. It may help to convert from SI units to more familiar units to make such estimates. You can also do this to assess problem solutions given in SI units. TABLE 1.4 lists some approximate conversion factors to apply in such cases.

TABLE 1.4 Some approximate conversion factors

| Quantity | SI unit | Approximate conversion |
|---|---|---|
| Mass | kg | 1 kg ≈ 2 lb |
| Length | m | 1 m ≈ 3 ft |
| | cm | 3 cm ≈ 1 in |
| | km | 5 km ≈ 3 mi |
| Speed | m/s | 1 m/s ≈ 2 mph |
| | km/h | 10 km/h ≈ 6 mph |

## EXAMPLE 1.5 How fast do you walk?

Estimate how fast you walk, in meters per second.

**STRATEGIZE** In this example we're asked for an *estimate* of your walking speed, so we'll need to use only rough values obtained from our everyday experience of walking.

**PREPARE** In order to compute speed, we need a distance and a time. If you walked a mile to campus, how long would this take? You'd probably say 30 minutes or so—half an hour. Let's use this rough number in our estimate.

**SOLVE** Given this estimate, we compute your speed as

$$\text{speed} = \frac{\text{distance}}{\text{time}} \sim \frac{1\text{ mile}}{1/2\text{ hour}} = 2\,\frac{\text{mi}}{\text{h}}$$

But we want the speed in meters per second. Since our calculation is only an estimate, we use an approximate conversion factor from Table 1.4:

$$1\,\frac{\text{mi}}{\text{h}} \sim 0.5\,\frac{\text{m}}{\text{s}}$$

This gives an approximate walking speed of 1 m/s.

**ASSESS** Is this a reasonable value? Let's do another estimate. Your stride is probably about 1 yard long—about 1 meter. And you take about one step per second; next time you are walking, you can count and see. So a walking speed of 1 meter per second sounds pretty reasonable.

**STOP TO THINK 1.5** Rank in order, from the most to the fewest, the number of significant figures in the following numbers. For example, if B has more than C, C has the same number as A, and A has more than D, give your answer as B > C = A > D.

A. 0.43    B. 0.0052    C. 0.430    D. $4.321 \times 10^{-10}$

## 1.6 Vectors and Motion: A First Look

Many physical quantities, such as time, temperature, and mass, can be described completely by a number with a unit. For example, the mass of an object might be 6 kg and its temperature 30°C. When a physical quantity is described by a single number (with a unit), we call it a **scalar quantity.** A scalar can be positive, negative, or zero.

Many other quantities, however, have a directional quality and cannot be described by a single number. To describe the motion of a car, for example, you must specify not only how fast it is moving, but also the *direction* in which it is moving. A **vector quantity** is a quantity that has both a *size* (How far? or How fast?) and a *direction* (Which way?). The size or length of a vector is called its **magnitude.** The magnitude of a vector can be positive or zero, but it cannot be negative.

**We graphically represent a vector as an *arrow*,** as illustrated for the velocity and force vectors. The arrow is drawn to point in the direction of the vector quantity, and the *length* of the arrow is proportional to the magnitude of the vector quantity.

When we want to represent a vector quantity with a *symbol,* we need somehow to indicate that the symbol is for a vector rather than for a scalar. We do this by drawing an arrow over the letter that represents the quantity. Thus $\vec{r}$ and $\vec{A}$ are symbols for vectors, whereas $r$ and $A$, without the arrows, are symbols for scalars. In handwritten work you *must* draw arrows over all symbols that represent vectors. This may seem strange until you get used to it, but it is very important because we will often use both $r$ and $\vec{r}$ or both $A$ and $\vec{A}$ in the same problem, and they mean different things!

**NOTE** ▸ The arrow over the symbol always points to the right, regardless of which direction the actual vector points. Thus we write $\vec{r}$ or $\vec{A}$, never $\overleftarrow{r}$ or $\overleftarrow{A}$. ◂

**Scalars and vectors**

**Scalars**

Time, temperature, and mass are all *scalar* quantities. To specify the current time, the temperature outside, or your mass, we need only a single number.

**Vectors**

The velocity of the race car is a *vector*. To fully specify a velocity, we need to give its magnitude (e.g., 120 mph) *and* its direction (e.g., west).

The force with which the boy pushes on his friend is another example of a vector. To completely specify this force, we must know not only how hard he pushes (the magnitude) but also in which direction.

### Displacement Vectors

For motion along a line, we found in Section 1.3 that the displacement is a quantity that specifies not only how *far* an object moves but also the *direction*—to the left or to the right—that the object moves. Since displacement is a quantity that has both a magnitude (How far?) and a direction, it can be represented by a vector, the **displacement vector.** FIGURE 1.21 shows the displacement vector for Sam's trip that we discussed earlier. We've simply drawn an arrow—the vector—from his initial to his final position and assigned it the symbol $\vec{d}_S$. Because $\vec{d}_S$ has both a magnitude and a direction, it is convenient to write Sam's displacement as $\vec{d}_S = (100 \text{ ft, east})$. The first value in the parentheses is the magnitude of the vector (i.e., the size of the displacement), and the second value specifies its direction.

FIGURE 1.21 Two displacement vectors.

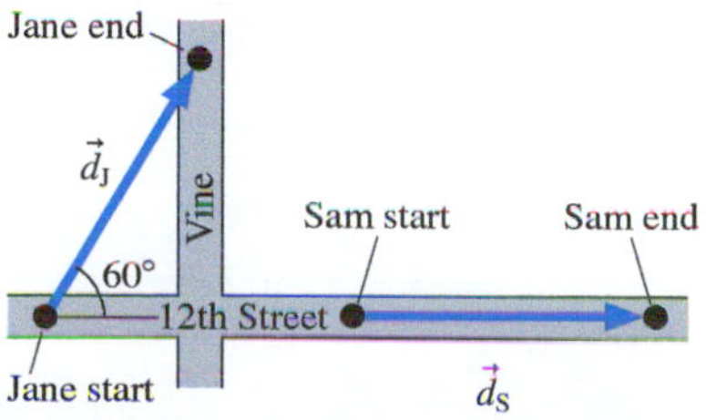


Also shown in Figure 1.21 is the displacement vector $\vec{d}_J$ for Jane, who started on 12th Street and ended up on Vine. As with Sam, we draw her displacement vector as an arrow from her initial to her final position. In this case, $\vec{d}_J = (100 \text{ ft}, 60° \text{ north of east})$.

Jane's trip illustrates an important point about displacement vectors. Jane started her trip on 12th Street and ended up on Vine, leading to the displacement vector

The boat's displacement is the straight-line connection from its initial to its final position.

shown. But to get from her initial to her final position, she didn't have to walk along the straight-line path denoted by $\vec{d}_J$. If she walked east along 12th Street to the intersection and then headed north on Vine, her displacement would still be the vector shown. **An object's displacement vector is drawn from the object's initial position to its final position, regardless of the actual path followed between these two points.**

## Vector Addition

Let's consider one more trip for the peripatetic Sam. In FIGURE 1.22, he starts at the intersection and walks east 50 ft; then he walks 100 ft to the northeast through a vacant lot. His displacement vectors for the two legs of his trip are labeled $\vec{d}_1$ and $\vec{d}_2$ in the figure.

FIGURE 1.22 Sam undergoes two displacements.

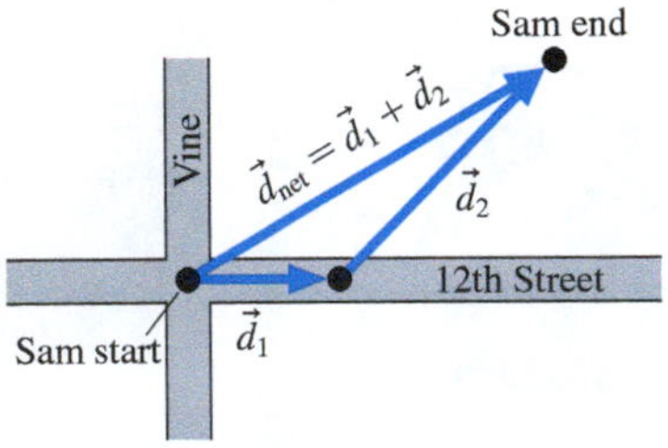


Sam's trip consists of two legs that can be represented by the two vectors $\vec{d}_1$ and $\vec{d}_2$, but we can represent his trip as a whole, from his initial starting position to his overall final position, with the *net* displacement vector labeled $\vec{d}_{net}$. Sam's net displacement is in a sense the *sum* of the two displacements that made it up, so we can write

$$\vec{d}_{net} = \vec{d}_1 + \vec{d}_2$$

Sam's net displacement thus requires the *addition* of two vectors, but vector addition obeys different rules from the addition of two scalar quantities. The directions of the two vectors, as well as their magnitudes, must be taken into account. Sam's trip suggests that we can add vectors together by putting the "tail" of one vector at the tip of the other. This idea, which is reasonable for displacement vectors, in fact is how *any* two vectors are added. Tactics Box 1.4 shows how to add two vectors $\vec{A}$ and $\vec{B}$ to get their **vector sum** $\vec{A} + \vec{B}$.

**TACTICS BOX 1.4 Adding vectors**

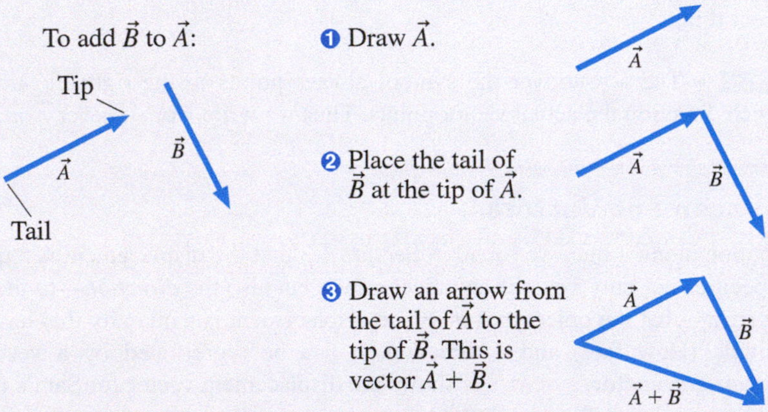


Exercise 21

## Vectors and Trigonometry

When we need to add displacements or other vectors in more than one dimension, we'll end up computing lengths and angles of triangles. This is the job of trigonometry. FIGURE 1.23 reviews the basic ideas of trigonometry.

**KEY CONCEPT** **FIGURE 1.23** Relating sides and angles of right triangles using trigonometry.

We specify the sides of a right triangle in relation to one of the angles.

The longest side, opposite to the right angle, is the **hypotenuse**.

This is the side **opposite** to angle $\theta$.

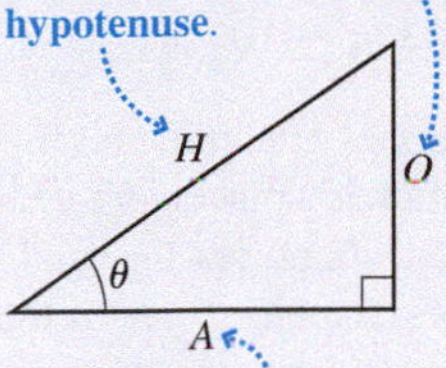


This is the side **adjacent** to angle $\theta$.

The three sides are related by the *Pythagorean theorem*:

$$H = \sqrt{A^2 + O^2}$$

The sine, cosine, and tangent of angle $\theta$ are defined as ratios of the side lengths.

$$\sin\theta = \frac{O}{H}$$

$$\cos\theta = \frac{A}{H}$$

$$\tan\theta = \frac{O}{A}$$

We can rearrange these equations in useful ways:

$$O = H\sin\theta$$

$$A = H\cos\theta$$

Given the length of the hypotenuse and one angle, we can find the side lengths.

$y$ is opposite to the angle; use the sine formula.

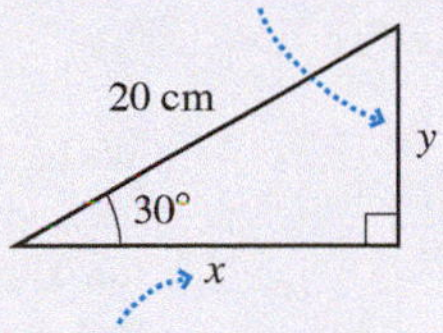


$x$ is adjacent to the angle; use the cosine formula.

$$x = (20\text{ cm})\cos(30°) = 17\text{ cm}$$

$$y = (20\text{ cm})\sin(30°) = 10\text{ cm}$$

Inverse trig functions let us find angles given lengths.

$$\theta = \sin^{-1}\left(\frac{O}{H}\right)$$

$$\theta = \cos^{-1}\left(\frac{A}{H}\right)$$

$$\theta = \tan^{-1}\left(\frac{O}{A}\right)$$

If we are given the lengths of the triangle's sides, we can find angles.

$\theta$ is adjacent to the 10 cm side; use the $\cos^{-1}$ formula.

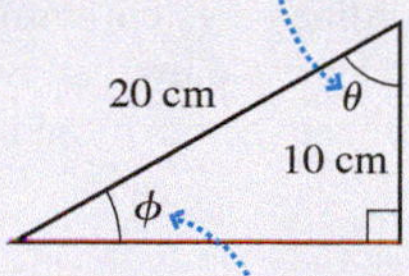


$\phi$ is opposite to the 10 cm side; use the $\sin^{-1}$ formula.

$$\theta = \cos^{-1}\left(\frac{10\text{ cm}}{20\text{ cm}}\right) = 60°$$

$$\phi = \sin^{-1}\left(\frac{10\text{ cm}}{20\text{ cm}}\right) = 30°$$

**STOP TO THINK 1.6** Using the information in Figure 1.23, what is the distance $x$, to the nearest cm, in the triangle at the right?

A. 26 cm B. 20 cm C. 17 cm D. 15 cm

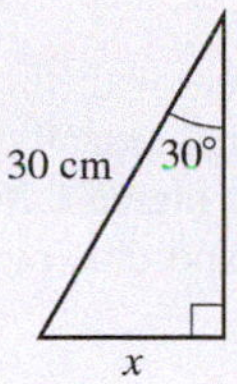


## EXAMPLE 1.6 How far north and east?

Suppose Alex is navigating using a compass. She starts walking at an angle 60° north of east and walks a total of 100 m. How far north is she from her starting point? How far east?

**STRATEGIZE** We'll need to use trigonometry to solve this problem. To do so, we'll need to sketch the situation so that we can identify a right triangle along with its hypotenuse and adjacent and opposite sides.

**PREPARE** A sketch of Alex's motion is shown in **FIGURE 1.24a**. We've shown north and east as they are on a map, and we've noted Alex's displacement as a vector, giving its magnitude and direction. **FIGURE 1.24b** shows a triangle with this displacement as the hypotenuse. Alex's distance north of her starting point is this triangle's opposite side, and her distance east of her starting point is its adjacent side.

**SOLVE** Because we want to find $O$ and $A$ for a triangle with $\theta = 60°$ and $H = 100$ m, we use the equations $O = H\sin\theta$ and $A = H\cos\theta$, giving

$$\text{distance north of start} = (100\text{ m})\sin(60°) = 87\text{ m}$$

$$\text{distance east of start} = (100\text{ m})\cos(60°) = 50\text{ m}$$

**ASSESS** Both of the distances we calculated are less than 100 m, as they must be, and the distance east is less than the distance north, as our diagram in Figure 1.24b shows it should be. Our answers seem reasonable. In finding the solution to this problem, we "broke down" the displacement into two different distances, one north and one east. This hints at the idea of the *components* of a vector, something we'll explore in the next chapter.

**FIGURE 1.24** An analysis of Alex's motion.

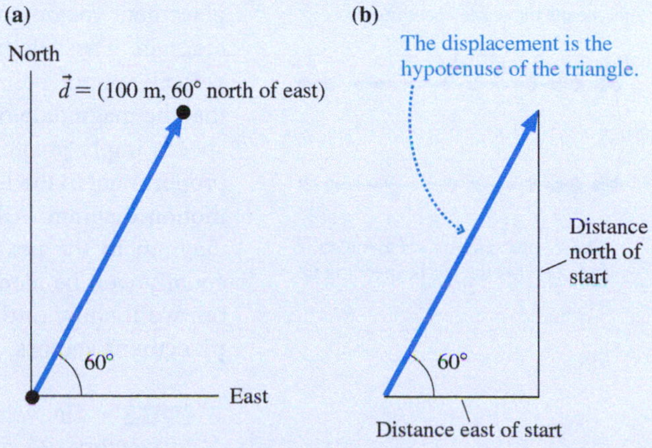

**EXAMPLE 1.7** **How far away is Anna?**

Anna walks 90 m due east and then 50 m due north. What is her displacement from her starting point?

***STRATEGIZE*** Again, we will need to use trigonometry to solve this problem, so we'll draw a right triangle and identify its sides.

***PREPARE*** Let's start with the sketch in FIGURE 1.25a. We set up a coordinate system with Anna's original position as the origin, and then we drew her two subsequent motions as the two displacement vectors $\vec{d}_1$ and $\vec{d}_2$.

FIGURE 1.25 Analyzing Anna's motion.

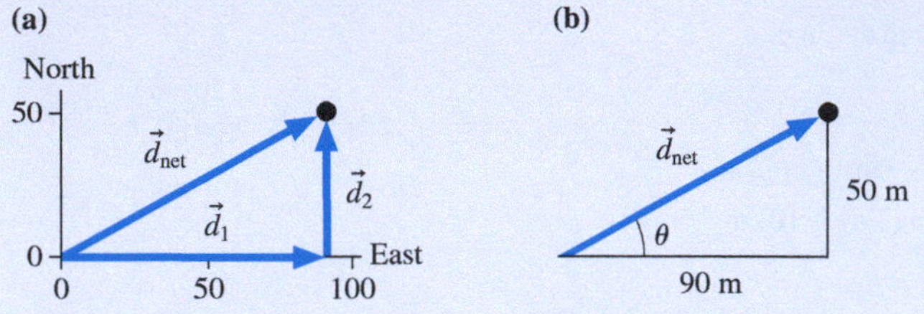


***SOLVE*** We drew the two vector displacements with the tail of one vector starting at the tip of the previous one—exactly what is needed to form a vector sum. The vector $\vec{d}_{net}$ in Figure 1.25a is the vector sum of the successive displacements and thus represents Anna's net displacement from the origin.

Anna's distance from the origin is the length of this vector $\vec{d}_{net}$. FIGURE 1.25b shows that this vector is the hypotenuse of a right triangle with sides 50 m (because Anna walked 50 m north) and 90 m (because she walked 90 m east). We can compute the magnitude of this vector, her net displacement, using the Pythagorean theorem:

$$d_{net} = \sqrt{(50\text{ m})^2 + (90\text{ m})^2} = 103\text{ m} \approx 100\text{ m}$$

We have rounded off to the appropriate number of significant figures, giving us 100 m for the magnitude of the displacement vector. How about the direction? Figure 1.25b identifies the angle that gives the angle north of east of Anna's displacement. In the right triangle, 50 m is the opposite side and 90 m is the adjacent side, so the angle is

$$\theta = \tan^{-1}\left(\frac{50\text{ m}}{90\text{ m}}\right) = \tan^{-1}\left(\frac{5}{9}\right) = 29°$$

Putting it all together, we get a net displacement of

$$\vec{d}_{net} = (100\text{ m}, 29° \text{ north of east})$$

***ASSESS*** We can use our drawing to assess our result. If the two sides of the triangle are 50 m and 90 m, a length of 100 m for the hypotenuse seems about right. The angle is certainly smaller than 45°, but not too much smaller, so 29° seems reasonable.

## Velocity Vectors

We've seen that a basic quantity describing the motion of an object is its velocity. Velocity is a vector quantity because its specification involves how fast an object is moving (its speed) and also the direction in which the object is moving. We thus represent the velocity of an object by a **velocity vector** $\vec{v}$ that points in the direction of the object's motion, and whose magnitude is the object's speed.

FIGURE 1.26 The motion diagram for a car starting from rest.

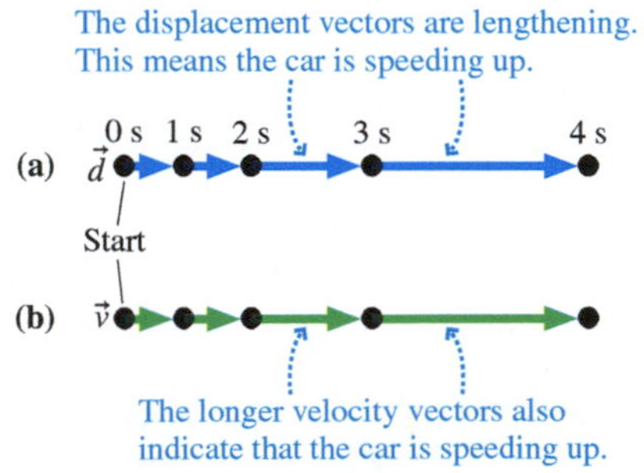


FIGURE 1.26a shows the motion diagram of a car accelerating from rest. We've drawn vectors showing the car's displacement between successive positions in the motion diagram. To draw the velocity vectors, first note that the direction of the displacement vector is the direction of motion between successive points in the motion diagram. The velocity of an object also points in the direction of motion, so the velocity vector points in the same direction as its displacement vector. Next, note that the magnitude of the velocity vector—How fast?—is the object's speed. Higher speeds imply greater displacements, so the length of the velocity vector should be proportional to the length of the displacement vector between successive points on a motion diagram. All this means that the vectors connecting each dot of a motion diagram to the next dot, which we have labeled as displacement vectors, could equally well be identified as velocity vectors, as shown in FIGURE 1.26b. **From now on, we'll show and label velocity vectors on motion diagrams rather than displacement vectors.**

NOTE ► The velocity vectors shown in Figure 1.26b are actually *average* velocity vectors. Because the velocity is steadily increasing, it's a bit less than this average at the start of each time interval, and a bit greater at the end. In Chapter 2 we'll refine these ideas as we develop the idea of instantaneous velocity. ◄

**EXAMPLE 1.8** **Drawing a ball's motion diagram**

Jake hits a ball at a 60° angle from the horizontal. It is caught by Jim. Draw a motion diagram of the ball that shows velocity vectors rather than displacement vectors.

***STRATEGIZE*** This example is typical of how many problems in physics are worded. The problem does not give a clear statement of where the motion begins or ends. Are we interested in the motion of the ball only during the time it is in the air between Jake and Jim? What about the motion *as* Jake hits it (ball rapidly speeding up) or *as* Jim catches it (ball rapidly slowing down)? Should we include Jim dropping the ball after he catches it? The point is that *you* will often be called on to make a *reasonable interpretation* of a problem statement. In this problem, the details of hitting and catching the ball are complex. The motion of the ball through the air is easier to describe, and it's a motion you might expect to learn about in a physics class. So our *interpretation* is that the motion diagram should start as the ball leaves Jake's bat (ball already moving) and should end the instant it touches Jim's hand (ball still moving).

***PREPARE*** We model the ball as a particle, and sketch a motion diagram that represents the motion of a thrown ball along an arc.

***SOLVE*** FIGURE 1.27 shows the motion diagram of the ball. Notice how, in contrast to the car of Figure 1.26, the ball is already moving as the motion diagram begins. As before, the velocity vectors are shown by connecting the dots with arrows. You can see that the velocity vectors get shorter (ball slowing down), get longer (ball speeding up), and change direction. Each $\vec{v}$ is different, so this is *not* constant-velocity motion.

FIGURE 1.27 The motion diagram of a ball traveling from Jake to Jim.

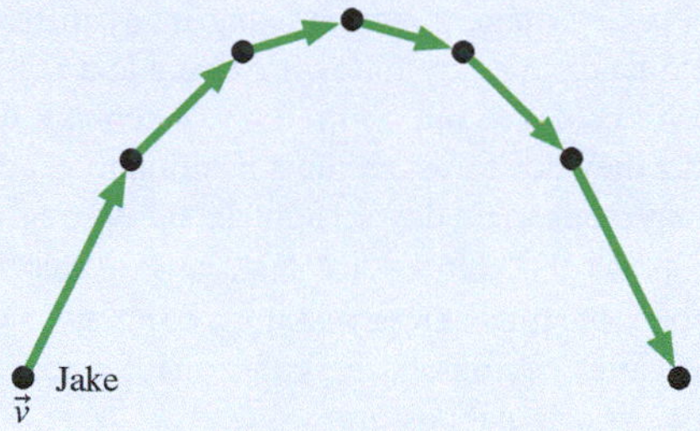


***ASSESS*** We haven't learned enough to make a detailed analysis of the motion of the ball, but it's still worthwhile to do a quick assessment. Does our diagram make sense? Think about the velocity of the ball—we show it moving upward at the start and downward at the end. This does match what happens when you toss a ball back and forth, so our answer seems reasonable.

**STOP TO THINK 1.7** $\vec{P}$ and $\vec{Q}$ are two vectors of equal length but different direction. Which vector shows the sum $\vec{P}+\vec{Q}$?

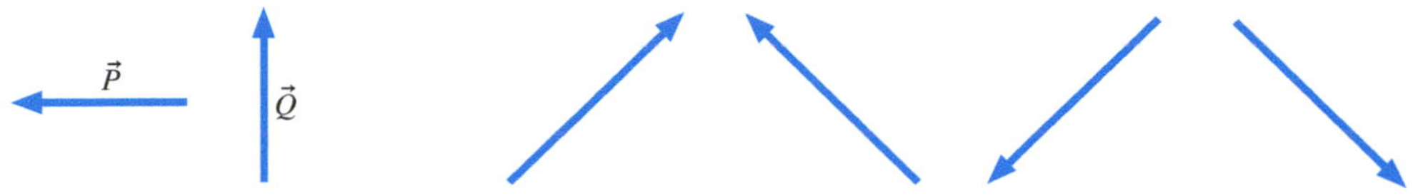


## 1.7 Where Do We Go from Here?

This first chapter has been an introduction to some of the fundamental ideas about motion and some of the basic techniques that you will use in the rest of the course. You have seen some examples of how to make *models* of a physical situation, thereby focusing on the essential elements of the situation. You have learned some practical ideas, such as how to convert quantities from one set of units to another. The rest of this text—and the rest of your course—will extend these themes.

In each chapter of this text, you'll learn both new principles and more tools and techniques. As you proceed, you'll find that each new chapter depends on those that preceded it. The principles and the problem-solving strategies you learned in this chapter will still be needed in Chapter 30.

We'll give you some assistance integrating new ideas with the material of previous chapters. When you start a chapter, the **chapter preview** will let you know which topics are especially important to review. And the last element in each chapter will be an **integrated example** that brings together the principles and techniques you have just learned with those you learned previously. The integrated nature of

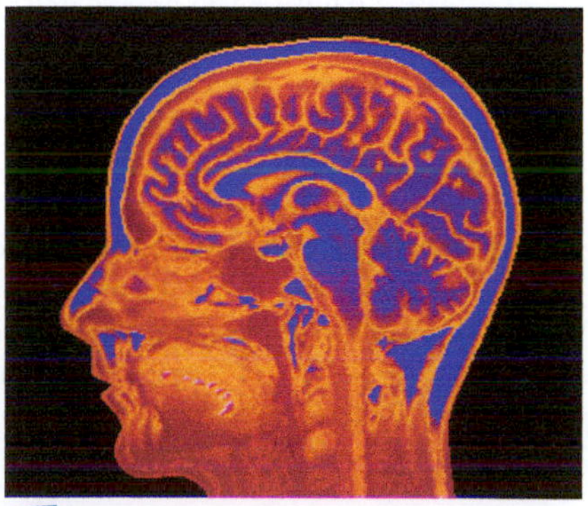

BIO Chapter 28 ends with an integrated example that explores the basic physics of magnetic resonance imaging (MRI), explaining how the interaction of magnetic fields with the nuclei of atoms in the body can be used to create an image of the body's interior.

these examples will also be a helpful reminder that the problems of the real world are similarly complex, and solving such problems requires you to do just this kind of integration.

Our first integrated example is reasonably straightforward because there's not much to integrate yet. The examples in future chapters will be much richer.

## INTEGRATED EXAMPLE 1.9 A goose gets its bearings BIO

Migrating geese determine direction using many different tools: by noting local landmarks, by following rivers and roads, and by using the position of the sun in the sky. When the weather is overcast so that they can't use the sun's position to get their bearings, geese may start their day's flight in the wrong direction. FIGURE 1.28 shows the path of a Canada goose that flew in a straight line for some time before making a corrective right-angle turn. One hour after beginning, the goose made a rest stop on a lake due east of its original position.

FIGURE 1.28 Trajectory of a misdirected goose.

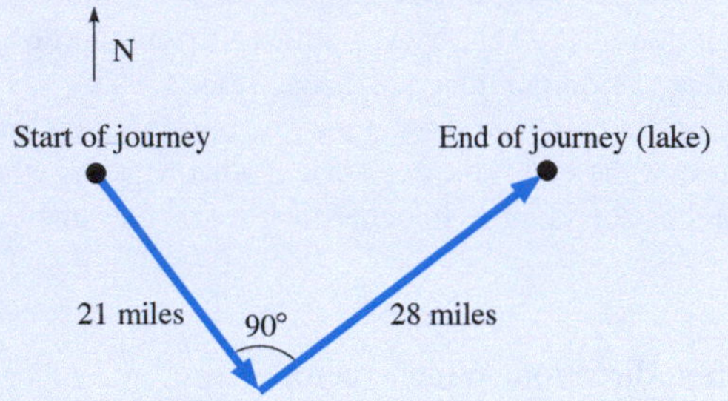


a. How much extra distance did the goose travel due to its initial error in flight direction? That is, how much farther did it fly than if it had simply flown directly to its final position on the lake?
b. What was the flight speed of the goose?
c. A typical flight speed for a migrating goose is 80 km/h. Given this, does your result seem reasonable?

**STRATEGIZE** In this integrated example, we'll need to pull together all our knowledge about right triangles, speed, and significant figures.

**PREPARE** Figure 1.28 shows the trajectory of the goose, but it's worthwhile to redraw Figure 1.28 and note the displacement from the start to the end of the journey, the shortest distance the goose could have flown. (The examples in the chapter to this point have used professionally rendered drawings, but these are much more careful and detailed than you are likely to make. FIGURE 1.29 shows a drawing that is more typical of what you might actually do when working problems yourself.) Drawing and labeling the displacement between the starting and ending points in Figure 1.29 show that it is the hypotenuse of a right triangle, so we can use our rules for triangles as we look for a solution.

FIGURE 1.29 A typical student sketch shows the motion and the displacement of the goose.

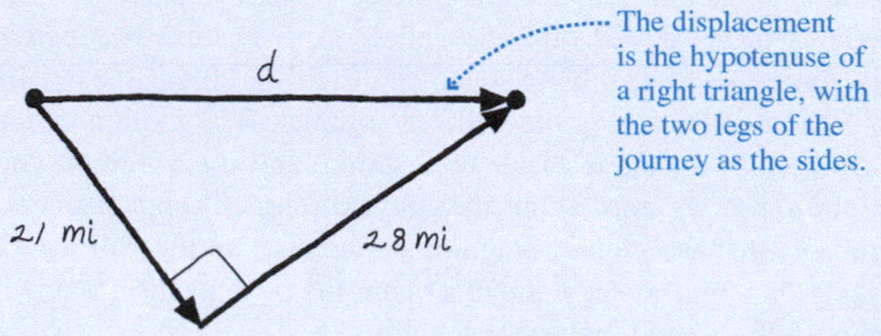


**SOLVE**

a. The minimum distance the goose *could* have flown, if it flew straight to the lake, is the hypotenuse of a triangle with sides 21 mi and 28 mi. This straight-line distance is

$$d = \sqrt{(21\text{ mi})^2 + (28\text{ mi})^2} = 35\text{ mi}$$

The actual distance the goose flew is the sum of the distances traveled for the two legs of the journey:

$$\text{distance traveled} = 21\text{ mi} + 28\text{ mi} = 49\text{ mi}$$

The extra distance flown is the difference between the actual distance flown and the straight-line distance—namely, 14 miles.

b. To compute the flight speed, we need to consider the distance that the bird actually flew. The flight speed is the total distance flown divided by the total time of the flight:

$$v = \frac{49\text{ mi}}{1.0\text{ h}} = 49\text{ mi/h}$$

c. To compare our calculated speed with a typical flight speed, we must convert our solution to km/h, rounding off to the correct number of significant figures:

$$49\,\frac{\text{mi}}{\text{h}} \times \frac{1.61\text{ km}}{1.00\text{ mi}} = 79\,\frac{\text{km}}{\text{h}}$$

A calculator will return many more digits, but the original data had only two significant figures, so we report the final result to this accuracy.

**ASSESS** In this case, an assessment was built into the solution of the problem. The calculated flight speed matches the expected value for a goose, which gives us confidence that our answer is correct. As a further check, our calculated net displacement of 35 mi seems about right for the hypotenuse of the triangle in Figure 1.29.

# SUMMARY

**GOAL** To introduce the fundamental concepts of motion and to review related basic mathematical principles.

## IMPORTANT CONCEPTS

### Motion Diagrams

The **particle model** represents a moving object as if all its mass were concentrated at a single point. Using this model, we can represent motion with a **motion diagram,** where dots indicate the object's positions at successive times. In a motion diagram, the time interval between successive dots is always the same.

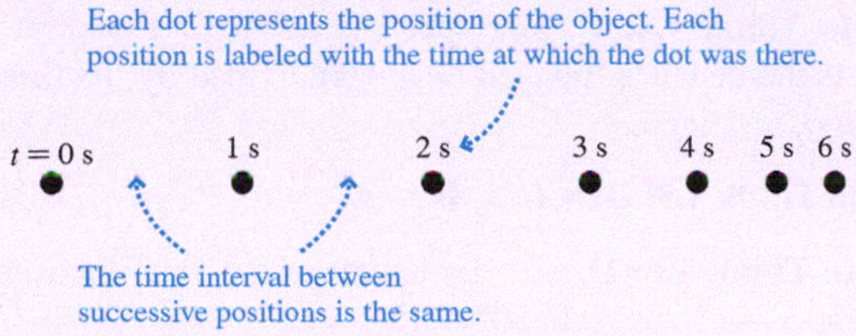

Each dot represents the position of the object. Each position is labeled with the time at which the dot was there.
t = 0 s
1 s
2 s
3 s
4 s
5 s
6 s
The time interval between successive positions is the same.

### Scalars and Vectors

**Scalar** quantities have only a magnitude and can be represented by a single number. Temperature, time, and mass are scalars.

A **vector** is a quantity described by both a magnitude and a direction. Velocity and displacement are vectors.

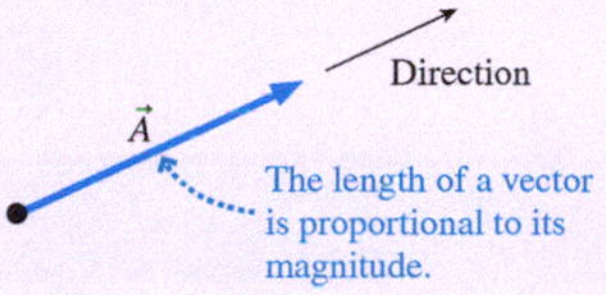


**Velocity vectors** can be drawn on a motion diagram by connecting successive points with a vector.

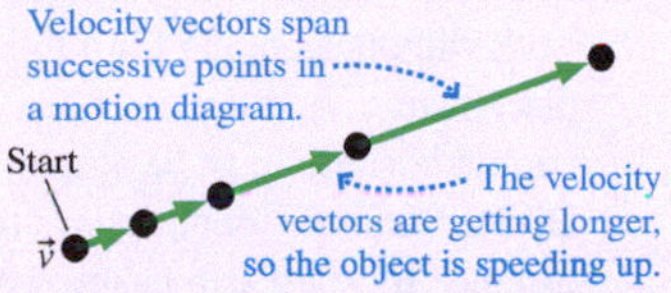


### Describing Motion

**Position** locates an object with respect to a chosen coordinate system. It is described by a **coordinate.**

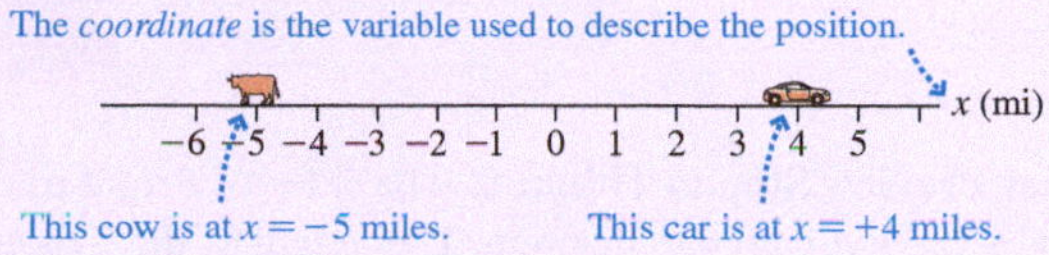


A change in position is called a **displacement**. For motion along a line, a displacement is a signed quantity. The displacement from $x_i$ to $x_f$ is $\Delta x = x_f - x_i$ .

**Time** is measured from a particular instant to which we assign $t = 0$. A **time interval** is the elapsed time between two specific instants $t_i$ and $t_f$. It is given by $\Delta t = t_f - t_i$.

**Velocity** is the ratio of the displacement of an object to the time interval during which this displacement occurs:

$$v = \frac{\Delta x}{\Delta t}$$

### Units

Every measurement of a quantity must include a **unit.**

The standard system of units used in science is the **SI system.** Common SI units include:

- Length: meters (m)
- Time: seconds (s)
- Mass: kilograms (kg)

## APPLICATIONS

### Working with Numbers

In **scientific notation,** a number is expressed as a decimal number between 1 and 10 multiplied by a power of ten. In scientific notation, the diameter of the earth is $1.27 \times 10^7$ m.

A **prefix** can be used before a unit to indicate a multiple of 10 or 1/10. Thus we can write the diameter of the earth as 12,700 km, where the k in km denotes 1000.

We can perform a **unit conversion** to convert the diameter of the earth to a different unit, such as miles. We do so by multiplying by a conversion factor equal to 1, such as 1 = 1 mi/1.61 km.

**Significant figures** are reliably known digits. The number of significant figures for:

- **Multiplication, division, and powers** is set by the value with the fewest significant figures.
- **Addition and subtraction** is set by the value with the smallest number of decimal places.

An **order-of-magnitude estimate** is an estimate that has an accuracy of about one significant figure. Such estimates are usually made using rough numbers from everyday experience.

## Learning Objectives

After studying this chapter, you should be able to:

- Draw and interpret motion diagrams to represent motion. *Conceptual Questions 1.2, 1.13; Problems 1.1, 1.2, 1.3*
- Describe motion in terms of position, velocity, and time. *Conceptual Question 1.8; Problems 1.5, 1.6, 1.7, 1.8, 1.9*
- Calculate the speed and velocity of an object. *Conceptual Question 1.5; Problems 1.11, 1.12, 1.13, 1.15*
- Use scientific notation. *Problems 1.23, 1.24, 1.25*
- Express quantities with the appropriate units and the proper number of significant figures. *Problems 1.19, 1.20, 1.21, 1.22*
- Perform unit conversions. *Conceptual Question 1.17; Problems 1.16, 1.17, 1.18*
- Describe motion using vectors and trigonometry. *Conceptual Question 1.16; Problems 1.27, 1.28, 1.32, 1.35, 1.38*

## STOP TO THINK ANSWERS

**Chapter Preview Stop to Think: C.** The sides of a right triangle are related by the Pythagorean theorem. The length of the hypotenuse is thus $\sqrt{(6\text{ cm})^2 + (8\text{ cm})^2} = 10$ cm. Note that this triangle is a version of a 3-4-5 right triangle; the lengths of the sides are in this ratio.

**Stop to Think 1.1: B.** The images of B are farther apart, so B travels a greater distance than does A during the same intervals of time.

**Stop to Think 1.2: A.** Dropped ball. B. Dust particle. C. Descending rocket.

**Stop to Think 1.3: C.** Depending on her initial positive position and how far she moves in the negative direction, she could end up on either side of the origin.

**Stop to Think 1.4: C.** Her speed is given by Equation 1.1. Her speed is the distance traveled (4 miles) divided by the time interval (2 hours), or 2 mph.

**Stop to Think 1.5: D > C > B = A.**

**Stop to Think 1.6: D.** $x$ is the length of the side opposite the 30° angle, so $x = (30\text{ cm})\sin 30° = 15$ cm.

**Stop to Think 1.7: B.** The vector sum is found by placing the tail of one vector at the tip of the other vector.

Video Tutor Solution Chapter 1

# QUESTIONS

## Conceptual Questions

1. A softball player slides into second base. Use the particle model to draw a motion diagram of the player from the time he begins to slide until he reaches the base. Number the dots in order, starting with zero.
2. A car travels to the left at a steady speed for a few seconds, then brakes for a stop sign. Use the particle model to draw a motion diagram of the car for the entire motion described here. Number the dots in order, starting with zero.
3. BIO The bush baby, a small African mammal, is a remarkable jumper. Although only about 8 inches long, it can jump, from a standing start, straight up to a height of over 7 feet! Use the particle model to draw a motion diagram for a bush baby's jump, from its start until it reaches its highest point.
4. A ball is dropped from the roof of a tall building and students in a physics class are asked to sketch a motion diagram for this situation. A student submits the diagram shown in Figure Q1.4. Is the diagram correct? Explain.

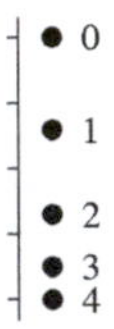


FIGURE Q1.4

5. Mark and Sofia walk together down a long, straight road. They walk without stopping for 4 miles. At this point Sofia says their displacement during the trip must have been 4 miles; Mark says their current position must be 4 miles. Who, if either, is correct? Explain.
6. Give an example of a trip you might take in your car for which the distance traveled as measured on your car's odometer is not equal to the displacement between your initial and final positions.
7. Write a sentence or two describing the difference between speed and velocity. Give one example of each.
8. The motion of a skateboard along a horizontal axis is observed for 5 s. The initial position of the skateboard is negative with respect to a chosen origin, and its velocity throughout the 5 s is also negative. At the end of the observation time, is the skateboard closer to or farther from the origin than initially? Explain.
9. You are standing on a straight stretch of road and watching the motion of a bicycle; you choose your position as the origin. At one instant, the position of the bicycle is negative and its velocity is positive. Is the bicycle getting closer to you or farther away? Explain.
10. Two friends watch a jogger complete a 400 m lap around the track in 100 s. One of the friends states, "The jogger's velocity was 4 m/s during this lap." The second friend objects, saying, "No, the jogger's speed was 4 m/s." Who is correct? Justify your answer.

Problem difficulty is labeled as | (straightforward) to ||||| (challenging). Problems labeled INT integrate significant material from earlier chapters; Problems labeled BIO are of biological or medical interest.

The eText icon indicates when there is a video tutor solution available for the chapter or for a specific problem. To launch these videos, log into your eText through Mastering™ Physics or log into the Study Area.

11. A softball player hits the ball and starts running toward first base. Draw a motion diagram, using the particle model, showing her velocity vectors during the first few seconds of her run.
12. A child is sledding on a smooth, level patch of snow. She encounters a rocky patch and slows to a stop. Draw a motion diagram, using the particle model, showing her velocity vectors.
13. A skydiver jumps out of an airplane. Her speed steadily increases until she deploys her parachute, at which point her speed quickly decreases. She subsequently falls to earth at a constant rate, stopping when she lands on the ground. Draw a motion diagram, using the particle model, that shows her position at successive times and includes velocity vectors.
14. Your roommate drops a tennis ball from a third-story balcony. It hits the sidewalk and bounces as high as the second story. Draw a motion diagram, using the particle model, showing the ball's velocity vectors from the time it is released until it reaches the maximum height on its bounce.
15. A car is driving north at a steady speed. It makes a gradual 90° left turn without losing speed, then continues driving to the west. Draw a motion diagram, using the particle model, showing the car's velocity vectors as seen from a helicopter hovering over the highway.
16. Three displacement vectors have lengths 1 m, 2 m, and 4 m. Could they possibly add together to get a vector of length zero?
17. Your friend Travis claims to have set the new world speed record for riding a unicycle. His top speed, he says, was 55 m/s. Do you believe him? Explain.

## Multiple-Choice Questions

18. | A student walks 1.0 mi west and then 1.0 mi north. Afterward, how far is she from her starting point?
A. 1.0 mi B. 1.4 mi C. 1.6 mi D. 2.0 mi
19. | You throw a rock upward. The rock is moving upward, but it is slowing down. If we define the ground as the origin, the position of the rock is ______ and the velocity of the rock is ______.
A. positive, positive B. positive, negative
C. negative, positive D. negative, negative
20. | Which of the following motions could be described by the motion diagram of Figure Q1.20?
A. A hockey puck sliding across smooth ice.
B. A cyclist braking to a stop.
C. A sprinter starting a race.
D. A ball bouncing off a wall.

FIGURE Q1.20

21. | Which of the following motions is described by the motion diagram of Figure Q1.21?
A. An ice skater gliding across the ice.
B. An airplane braking to a stop after landing.
C. A car pulling away from a stop sign.
D. A pool ball bouncing off a cushion and reversing direction.

FIGURE Q1.21

22. | A bird flies 3.0 km due west and then 2.0 km due north. What is the magnitude of the bird's displacement?
A. 2.0 km B. 3.0 km C. 3.6 km D. 5.0 km
23. || Weddell seals make holes in sea ice so that they can swim down to forage on the ocean floor below. Measurements for one seal showed that it dived straight down from such an opening, reaching a depth of 0.30 km in a time of 5.0 min. What was the speed of the diving seal?
BIO
A. 0.60 m/s B. 1.0 m/s C. 1.6 m/s D. 6.0 m/s
E. 10 m/s
24. || A bird flies 3.0 km due west and then 2.0 km due north. Another bird flies 2.0 km due west and 3.0 km due north. What is the angle between the net displacement vectors for the two birds?
A. 23° B. 34° C. 56° D. 90°
25. || Hicham El Guerrouj of Morocco holds the world record in the 1500 m running race. He ran the final 400 m in a time of 51.9 s. What was his average speed in mph over the last 400 m?
A. 14.2 mph B. 15.5 mph
C. 17.2 mph D. 23.9 mph
26. | Compute 3.24 m + 0.532 m to the correct number of significant figures.
A. 3.7 m B. 3.77 m
C. 3.772 m D. 3.7720 m
27. || An American football field is 109.7 m long and 48.8 m wide. To the correct number of significant figures, what is its area?
A. 5351 $m^2$ B. $5.35 \times 10^3$ $m^2$
C. 5351.17 $m^2$ D. 5400 $m^2$
28. | The earth formed $4.57 \times 10^9$ years ago. What is this time in seconds?
A. $1.67 \times 10^{12}$ s B. $4.01 \times 10^{13}$ s
C. $2.40 \times 10^{15}$ s D. $1.44 \times 10^{17}$ s
29. ||| An object's average density $\rho$ is defined as the ratio of its mass to its volume: $\rho = M/V$. The earth's mass is $5.94 \times 10^{24}$ kg, and its volume is $1.08 \times 10^{12}$ $km^3$. What is the earth's average density?
A. $5.50 \times 10^3$ $kg/m^3$ B. $5.50 \times 10^6$ $kg/m^3$
C. $5.50 \times 10^9$ $kg/m^3$ D. $5.50 \times 10^{12}$ $kg/m^3$

# PROBLEMS

### Section 1.1 Motion: A First Look

1. | A car skids to a halt to avoid hitting an object in the road. Draw a motion diagram of the car from the time the skid begins until the instant the car stops.
2. | A man rides a bike along a straight road for 5 min, then has a flat tire. He stops for 5 min to repair the flat, but can't fix it. He walks the rest of the way, which takes him another 10 min. Use the particle model to draw a motion diagram of the man for the entire motion described here. Number the dots in order, starting with zero.
3. | Amanda has just entered an elevator. The elevator rises and stops at the third floor. Use the particle model to draw a motion diagram of Amanda during her entire ride on the elevator. Number the dots in order, starting from zero. (Be sure to consider how the elevator speeds up and slows down.)

### Section 1.2 Models and Modeling

### Section 1.3 Position and Time: Putting Numbers on Nature

4. | Figure P1.4 shows Sue along the straight-line path between her home and the cinema. What is Sue's position $x$ if
   a. Her home is the origin?
   b. The cinema is the origin?

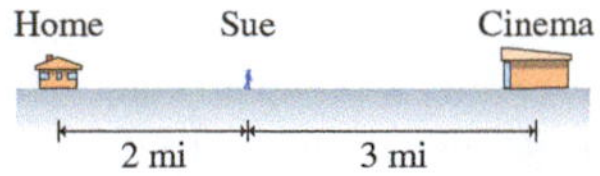


FIGURE P1.4

5. | Figure P1.4 shows Sue along the straight-line path between her home and the cinema. Now Sue walks home. What is Sue's displacement if
   a. Her home is the origin?
   b. The cinema is the origin?
6. | BIO Logan observes a paramecium under a microscope. The eyepiece of the microscope has a horizontal scale marked in mm. The paramecium starts at the 65 mm mark and ends up at the 42 mm mark. What is the paramecium's displacement?
7. | Keira starts at position $x = 23$ m along a coordinate axis. She then undergoes a displacement of $-45$ m. What is her final position?
8. | A car travels along a straight east-west road. A coordinate system is established on the road, with $x$ increasing to the east. The car ends up 14 mi west of the origin, which is defined as the intersection with Mulberry Road. If the car's displacement was $-23$ mi, what side of Mulberry Road did the car start on? How far from the intersection was the car at the start?
9. | BIO Foraging bees often move in straight lines away from and toward their hives. Suppose a bee starts at its hive and flies 500 m due east, then flies 400 m west, then 700 m east. How far is the bee from the hive?

### Section 1.4 Velocity

10. | A security guard walks at a steady pace, traveling 110 m in one trip around the perimeter of a building. It takes him 240 s to make this trip. What is his speed?
11. || List the following items in order of decreasing speed, from greatest to least: (i) A wind-up toy car that moves 0.15 m in 2.5 s. (ii) A soccer ball that rolls 2.3 m in 0.55 s. (iii) A bicycle that travels 0.60 m in 0.075 s. (iv) A cat that runs 8.0 m in 2.0 s.
12. || Figure P1.12 shows the motion diagram for a horse galloping in one direction along a straight path. Not every dot is labeled, but the dots are at equally spaced instants of time. What is the horse's velocity
    a. During the first 10 seconds of its gallop?
    b. During the interval from 30 s to 40 s?
    c. During the interval from 50 s to 70 s?

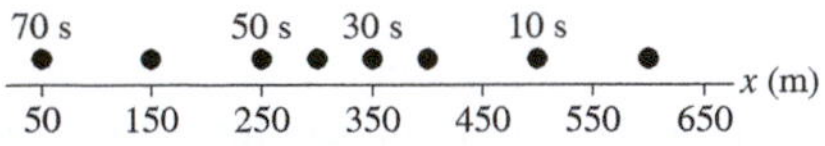


FIGURE P1.12

13. || It takes Harry 35 s to walk from $x = -12$ m to $x = -47$ m. What is his velocity?
14. | A dog trots from $x = -12$ m to $x = 3$ m in 10 s. What is its velocity?

Watch Video Solution Problem 1.25

15. || BIO In Michael Johnson's world-record 400 m sprint, he ran the first 100 m in 11.20 s; then he reached the 200 m mark after a total time of 21.32 s had elapsed, reached the 300 m mark after 31.76 s, and finished in 43.18 s.
    a. During what 100 m segment was his speed the highest?
    b. During this segment, what was his speed in m/s?

### Section 1.5 A Sense of Scale: Significant Figures, Scientific Notation, and Units

16. || Convert the following to SI base units:
    a. 9.12 $\mu$s  b. 3.42 km
    c. 44 cm/ms  d. 80 km/h
17. | Convert the following to SI units:
    a. 8.0 in  b. 66 ft/s  c. 60 mph
18. | Convert the following to SI units:
    a. 1.0 hour  b. 1.0 day  c. 1.0 year
19. || How many significant figures does each of the following numbers have?
    a. 6.21  b. 62.1
    c. 0.620  d. 0.062
20. | How many significant figures does each of the following numbers have?
    a. 0.621  b. 0.006200
    c. 1.0621  d. $6.21 \times 10^3$
21. | Compute the following numbers to three significant figures.
    a. $33.3 \times 25.4$  b. $33.3 - 25.4$
    c. $\sqrt{33.3}$  d. $333.3 \div 25.4$
22. || BIO If you make multiple measurements of your height, you are likely to find that the results vary by nearly half an inch in either direction due to measurement error and actual variations in height. You are slightly shorter in the evening, after gravity has compressed and reshaped your spine over the course of a day. One measurement of a man's height is 6 feet and 1 inch. Express his height in meters, using the appropriate number of significant figures.

23. | Mount Everest has a height of 29,029 ft above sea level. Express this height in meters, giving your result in scientific notation with the correct number of significant figures.
24. ||| BIO Blades of grass grow from the bottom, so, as growth occurs, the top of the blade moves upward. During the summer, when your lawn is growing quickly, estimate this speed, in m/s. Make this estimate from your experience noting, for instance, how often you mow the lawn and what length you trim. Express your result in scientific notation.
25. || BIO Estimate the average speed, in m/s, with which the hair on your head grows. Make this estimate from your own experience noting, for instance, how often you cut your hair and how much you trim. Express your result in scientific notation.

### Section 1.6 Vectors and Motion: A First Look

26. | Loveland, Colorado, is 18 km due south of Fort Collins and 31 km due west of Greeley. What is the distance between Fort Collins and Greeley?

27. || A city has streets laid out in a square grid, with each block 135 m long. If you drive north for three blocks, then west for two blocks, how far are you from your starting point?

28. | Joe and Max shake hands and say goodbye. Joe walks east 0.55 km to a coffee shop, and Max flags a cab and rides north 3.25 km to a bookstore. How far apart are their destinations?

29. || In downtown Chicago, the east-west blocks are 400 ft long while the north-south blocks are 280 ft long. Because of the many one-way streets, it can be challenging to get around. Veronica starts at the corner of Dearborn and Ohio Streets. She drives four blocks north to Superior, two blocks east to Wabash, then a block south to get to her destination at Wabash and Huron. What is the straight-line distance from her starting point?

30. || A butterfly flies from the top of a tree in the center of a garden to rest on top of a red flower at the garden's edge. The tree is 8.0 m taller than the flower, and the garden is 12 m wide. Determine the magnitude of the butterfly's displacement.

31. ||| A garden has a circular path of radius 50 m. John starts at the easternmost point on this path, then walks counterclockwise around the path until he is at its southernmost point. What is John's displacement? Use the (magnitude, direction) notation for your answer.

32. ||| Luis is visiting a public garden that has a large, circular path. When he has walked one-quarter of the distance around the path, the magnitude of his displacement is 180 m. What is the diameter of the path?

33. | BIO Migrating geese tend to travel at approximately constant speed, flying in segments that are straight lines. A goose flies 32 km south, then turns to fly 20 km west. Afterward, how far is the goose from its original position?

34. || A circular test track for cars in England has a circumference of 3.2 km. A car travels around the track from the southernmost point to the northernmost point.
   a. What distance does the car travel?
   b. What is the car's displacement from its original position?

35. || BIO Black vultures excel at gliding flight; they can move long distances through the air without flapping their wings while undergoing only a modest drop in height. A vulture in a typical glide in still air moves along a path tipped 3.5° below the horizontal. If the vulture moves a horizontal distance of 100 m, how much height does it lose?

36. ||| Figure P1.36 shows a map of Olivia's trip to a coffee shop. She gets on her bike at Loomis and then rides south 0.8 mi to Broadway. She turns east onto Broadway, rides 0.8 mi to where Broadway turns, and then continues another 1.2 mi to the shop. What is the total displacement of her trip, in (magnitude, direction) form?

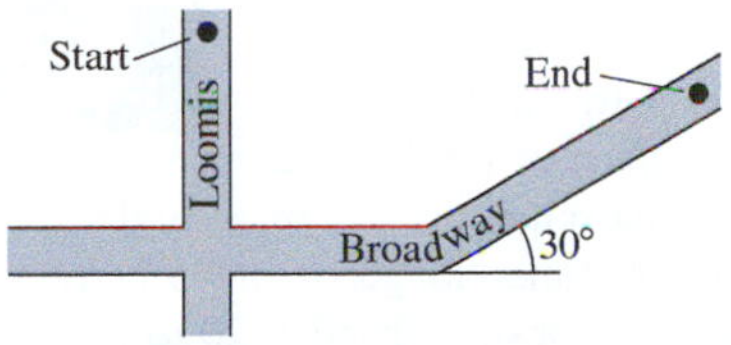

FIGURE P1.36

37. || The Great Pyramid of Giza is 139 m tall, with a slope of 51.8°. If you were to climb the pyramid from base to top (which is forbidden!), what distance along the face of the pyramid would you travel?

38. | A hiker is climbing a steep 10° slope. Her pedometer shows that she has walked 1500 m along the slope. How much elevation has she gained?

39. |||| A ball on a porch rolls 60 cm to the porch's edge, drops 40 cm, continues rolling on the grass, and eventually stops 80 cm from the porch's edge. What is the magnitude of the ball's net displacement, in centimeters?

40. || A kicker punts a football from the very center of the field to the sideline 43 yards downfield. What is the net displacement of the ball? (A football field is 53 yards wide.)

Problems 41 and 42 relate to the gliding flight of flying squirrels. These squirrels glide from tree to tree at a constant speed, moving in a straight line tipped below the vertical and steadily losing altitude as they move forward. Short and long glides have different profiles.

41. || BIO A squirrel completing a short glide travels in a straight line tipped 40° below the horizontal. The squirrel starts 9.0 m above the ground on one tree and glides to a second tree that is a horizontal distance of 3.5 m away.
   a. What is the length of the squirrel's glide path?
   b. What is the squirrel's height above the ground when it lands?

42. || BIO A squirrel in a typical long glide covers a horizontal distance of 16 m while losing 8.0 m of elevation. During this glide,
   a. What is the angle of the squirrel's path below the horizontal?
   b. What is the total distance covered by the squirrel?

## General Problems

Problems 43 through 49 are motion problems similar to those you will learn to solve in Chapter 2. For now, simply *interpret* the problem by drawing a motion diagram showing the object's position and its velocity vectors. **Do *not* solve these problems** or do any mathematics.

43. || BIO In a typical greyhound race, a dog accelerates to a speed of 20 m/s over a distance of 30 m. It then maintains this speed. What would be a greyhound's time in the 100 m dash?

44. || Billy drops a watermelon from the top of a three-story building, 10 m above the sidewalk. How fast is the watermelon going when it hits?

45. || A skateboarder starts from rest at the top of a ramp. He rolls down the ramp and then continues rolling on the smooth, horizontal floor.

46. || A speed skater moving across frictionless ice at 8.0 m/s hits a 5.0-m-wide patch of rough ice. She slows steadily, then continues on at 6.0 m/s. What is her acceleration on the rough ice?

47. || BIO The giant eland, an African antelope, is an exceptional jumper, able to leap 1.5 m off the ground. To jump this high, with what speed must the eland leave the ground?

48. || A ball rolls along a smooth horizontal floor at 10 m/s, then starts up a 20° ramp. How high does it go before rolling back down?

49. || A motorist is traveling at 20 m/s. He is 60 m from a stop light when he sees it turn yellow. His reaction time, before stepping on the brake, is 0.50 s. What steady deceleration while braking will bring him to a stop right at the light?

Watch Video Solution Problem 1.61

Problems 50 through 54 show a motion diagram. For each of these problems, write a one or two sentence "story" about a *real object* that has this motion diagram. Your stories should talk about people or objects by name and say what they are doing. Problems 43 through 49 are examples of motion short stories.

50. |

FIGURE P1.50

51. |

FIGURE P1.51

52. |

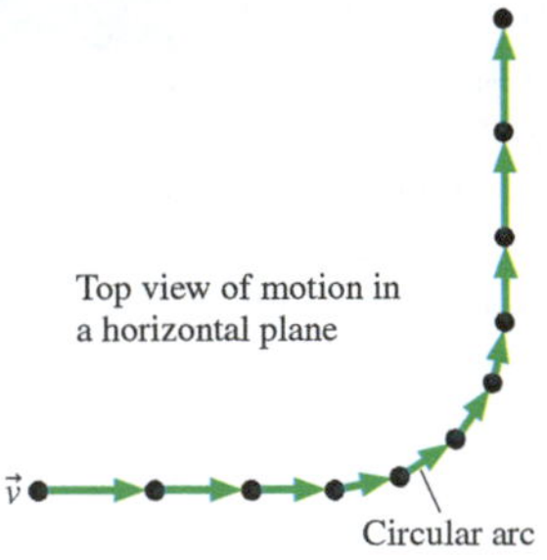


FIGURE P1.52

53. |

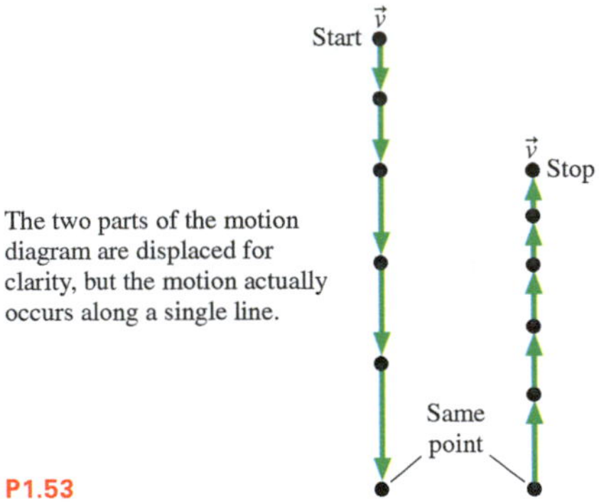

Start
Stop
The two parts of the motion diagram are displaced for clarity, but the motion actually occurs along a single line.
Same point

FIGURE P1.53

54. |

FIGURE P1.54

55. || Estimate the length of a human lifetime, in seconds.

56. ||| On a highway trip, Joseph drives the first 25 miles at 55 mph, and the next 15 miles at 70 mph. What is his average speed for this trip?

57. ||| Evan is just leaving his house to visit his grandmother. Normally, the trip takes him 25 minutes on the freeway, going 55 mph. But tonight he's running 5 minutes late. How fast will he need to drive on the freeway to make up the 5 minutes?

58. ||| Gretchen runs the first 4.0 km of a race at 5.0 m/s. Then a stiff wind comes up, so she runs the last 1.0 km at only 4.0 m/s. If she later ran the same course again, what constant speed would let her finish in the same time as in the first race?

59. |||| If you swim with the current in a river, your speed is increased by the speed of the water; if you swim against the current, your speed is decreased by the water's speed. The current in a river flows at 0.52 m/s. In still water you can swim at 1.78 m/s. If you swim downstream a certain distance, then back again upstream, how much longer, in percent, does it take compared to the same trip in still water?

60. || The end of Hubbard Glacier in Alaska advances by an average of 105 feet per year. What is the speed of advance of the glacier in m/s?

61. | The earth completes a circular orbit around the sun in one year. The orbit has a radius of 93,000,000 miles. What is the speed of the earth around the sun in m/s? Report your result using scientific notation.

62. || The Greenland shark is thought to be the longest-living
BIO vertebrate on earth. Early estimates of its maximum age were based on the fact that such sharks grow in length only about a centimeter per year, and yet an adult shark can reach a length of 15 feet. Estimate how long a 15 foot shark might have lived. (A newborn shark is about 1 foot long.)

63. ||| The winner of the 2016 Keystone (Colorado) Uphill/Downhill mountain bike race finished in a total time of 47 minutes and 25 seconds. The uphill leg was 4.6 miles long, and on this leg his average speed was 8.75 mph. The downhill leg was 6.9 miles. What was his average speed on this leg?

64. ||| Shannon decides to check the accuracy of her speedometer. She adjusts her speed to read exactly 70 mph on her speedometer and holds this steady, measuring the time between successive mile markers separated by exactly 1.00 mile. If she measures a time of 54 s, is her speedometer accurate? If not, is the speed it shows too high or too low?

65. || The Nardo ring is a circular test track for cars. It has a circumference of 12.5 km. Cars travel around the track at a constant speed of 100 km/h. A car starts at the easternmost point of the ring and drives for 15 minutes at this speed.
   a. What distance, in km, does the car travel?
   b. What is the magnitude of the car's displacement, in km, from its initial position?
   c. What is the speed of the car in m/s?

66. || Motor neurons in mammals transmit signals from the brain
BIO to skeletal muscles at approximately 25 m/s. Estimate how much time in ms ($10^{-3}$ s) it will take for a signal to get from your brain to your hand.

67. ||| Satellite data taken several times per hour on a particular
BIO albatross showed travel of 1200 km over a time of 1.4 days.
   a. Given these data, what was the bird's average speed in mph?
   b. Data on the bird's position were recorded only intermittently. Explain how this means that the bird's actual average speed was higher than what you calculated in part a.

68. || The bacterium *Escherichia coli* (or *E. coli*) is a
BIO single-celled organism that lives in the gut of healthy humans and animals. Its body shape can be modeled as a 2-$\mu$m-long cylinder with a 1 $\mu$m diameter, and it has a mass of $1 \times 10^{-12}$ g. Its chromosome consists of a single double-stranded chain of DNA 700 times longer than its body length. The bacterium moves at a constant speed of 20 $\mu$m/s, though not always in the same direction. Answer the following questions about *E. coli* using SI base units (unless specifically requested otherwise) and correct significant figures.

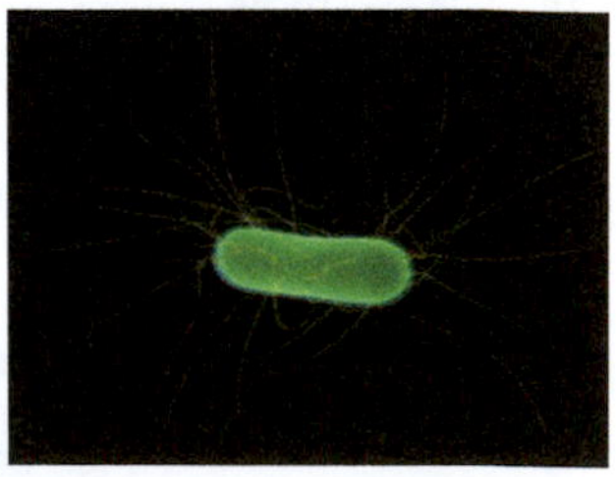

   a. What is its length?
   b. Diameter?
   c. Mass?
   d. What is the length of its DNA, in millimeters?
   e. If the organism were to move along a straight path, how many meters would it travel in one day?

69. | | The bacterium *Escherichia coli* (or *E. coli*) is a single-celled organism that lives in the gut of healthy humans and animals. When grown in a uniform medium rich in salts and amino acids, it swims along zig-zag paths at a constant speed changing direction at varying time intervals. Figure P1.69 shows the positions of an *E. coli* as it moves from point A to point J. Each segment of the motion can be identified by two letters, such as segment BC. During which segments, if any, does the bacterium have the same

BIO

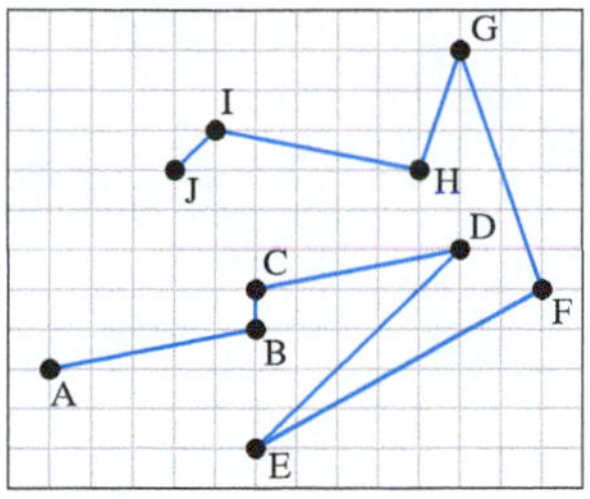


FIGURE P1.69

a. Displacement? b. Speed? c. Velocity?

70. | | | The sun is 30° above the horizon. It makes a 52-m-long shadow of a tall tree. How high is the tree?

71. | | | Weddell seals foraging in open water dive toward the ocean bottom by swimming forward in a straight-line path tipped below the horizontal. The tracking data for one seal showed it taking 4.0 min to descend 360 m below the surface while moving 920 m horizontally.

BIO

a. What was the angle of the seal's path below the horizontal?
b. What distance did the seal cover in making this dive?
c. What was the seal's speed, in m/s?

72. | | | Erica is participating in a road race. The first part of the race is on a 5.2-mile-long straight road oriented at an angle of 25° north of east. The road then turns due north for another 4.0 mi to the finish line. In miles, what is the straight-line distance from the starting point to the end of the race?

73. | | | Whale sharks swim forward while ascending or descending. They swim along a straight-line path at a shallow angle as they move from the surface to deep water or from the depths to the surface. In one recorded dive, a shark started 50 m below the surface and swam at 0.85 m/s along a path tipped at a 13° angle above the horizontal until reaching the surface.

BIO

a. What was the horizontal distance between the shark's starting and ending positions?
b. What was the total distance that the shark swam?
c. How much time did this motion take?

74. | | | Starting from its nest, an eagle flies at constant speed for 3.0 min due east, then 4.0 min due north. From there the eagle flies directly to its nest at the same speed. How long is the eagle in the air?

75. | | | John walks 1.00 km north, then turns right and walks 1.00 km east. His speed is 1.50 m/s during the entire stroll.

a. What is the magnitude of his displacement, from beginning to end?
b. If Jane starts at the same time and place as John, but walks in a straight line to the endpoint of John's stroll, at what speed should she walk to arrive at the endpoint just when John does?

## MCAT-Style Passage Problems

### Growth Speed

The images of trees in Figure P1.76 come from a catalog advertising fast-growing trees. If we mark the position of the top of the tree in the successive years, as shown in the graph in the figure, we obtain a motion diagram much like ones we have seen for other kinds of motion. The motion isn't steady, of course. In some months the tree grows rapidly; in other months, quite slowly. We can see, though, that the average speed of growth is fairly constant for the first few years.

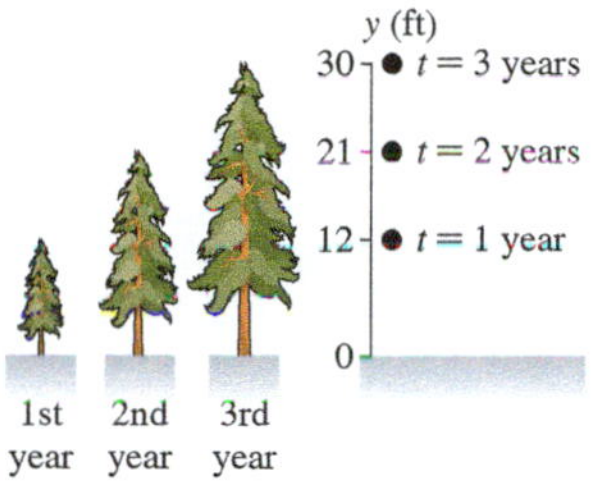


FIGURE P1.76

76. | What is the tree's speed of growth, in feet per year, from $t = 1$ yr to $t = 3$ yr?

A. 12 ft/yr
B. 9 ft/yr
C. 6 ft/yr
D. 3 ft/yr

77. | What is this speed in m/s?

A. $9 \times 10^{-8}$ m/s
B. $3 \times 10^{-9}$ m/s
C. $5 \times 10^{-6}$ m/s
D. $2 \times 10^{-6}$ m/s

78. | At the end of year 3, a rope is tied to the very top of the tree to steady it. This rope is staked into the ground 15 feet away from the tree. What angle does the rope make with the ground?

A. 63°
B. 60°
C. 30°
D. 27°

# 2 Motion in One Dimension

A horse can run at 35 mph, much faster than a human. And yet, surprisingly, a man can win a race against a horse if the length of the course is right. When, and how, can a man outrun a horse?

## LOOKING AHEAD ➧

### Uniform Motion

Successive images of the Segway rider are the same distance apart, so his velocity is constant. This is **uniform motion**.

You'll learn to describe motion in terms of quantities such as distance and velocity, an important first step in analyzing motion.

### Acceleration

A cheetah is capable of running at very high speeds but, more important, it is capable of a rapid *change* in speed—a large **acceleration.**

You'll use the concept of acceleration to solve problems of changing velocity, such as races or predators chasing prey.

### Free Fall

When the diver jumps, his motion—both going up and coming down—is determined by gravity alone. We call this **free fall.**

How long will it take this diver to reach the water? This is the type of free-fall problem you'll learn to solve.

**GOAL** To describe and analyze motion along a line.

## LOOKING BACK ➧

### Motion Diagrams

As you saw in Section 1.6, a good first step in analyzing motion is to draw a motion diagram, marking the position of an object at successive times.

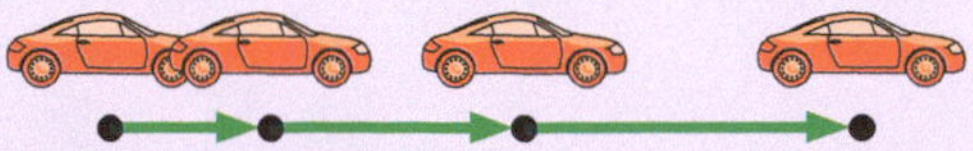

In this chapter, you'll learn to create motion diagrams for different types of motion along a line. Drawing pictures like this is a good starting point for solving problems.

**STOP TO THINK**

A bicycle is moving to the left with increasing speed. Which of the following motion diagrams illustrates this motion?

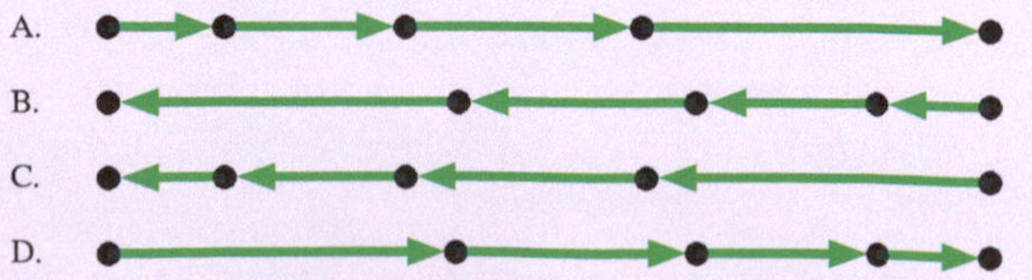

## 2.1 Describing Motion

In this chapter, we'll focus on how to *describe* motion using several different representations, including motion diagrams, graphs, and mathematical equations. We will defer a treatment of *why* objects move as they do until Chapter 4. The branch of physics that deals with the description of motion is **kinematics,** from the Greek word *kinema,* meaning "movement." You know this word through its English variation *cinema*—motion pictures!

### Representing Position

As we saw in Chapter 1, *kinematic variables* such as position and velocity are measured with respect to a coordinate system, an axis that *you* impose on a system. We will use an $x$-axis to analyze both horizontal motion and motion on a ramp; a $y$-axis will be used for vertical motion. We will adopt the convention that the positive end of an $x$-axis is to the right and the positive end of a $y$-axis is up. This convention is illustrated in FIGURE 2.1.

FIGURE 2.1 Sign conventions for position.

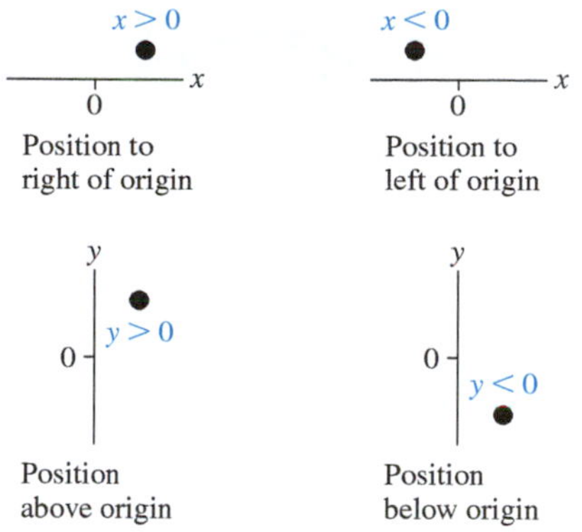


NOTE ▸ The conventions illustrated in Figure 2.1 aren't absolute. In most cases, we are free to define the coordinate system, and doing so in this standardized way makes sense. In some cases, though, we'll want to make a different choice. ◂

Now, let's look at a practical problem. FIGURE 2.2 is a motion diagram of a straightforward situation, a student walking to school. She is moving horizontally, so we use the variable $x$ to describe her motion. We have set the origin of the coordinate system, $x = 0$, at her starting position, and we measure her position in meters. We have included velocity vectors connecting successive positions on the motion diagram, as we saw we could do in Chapter 1.

FIGURE 2.2 The motion diagram of a student walking to school and a coordinate axis for making measurements.

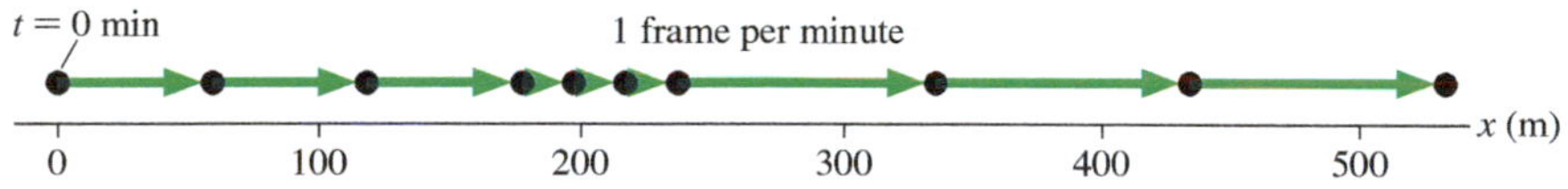


TABLE 2.1 Measured positions of a student walking to school

| Time $t$ (min) | Position $x$ (m) | Time $t$ (min) | Position $x$ (m) |
|---|---|---|---|
| 0 | 0 | 5 | 220 |
| 1 | 60 | 6 | 240 |
| 2 | 120 | 7 | 340 |
| 3 | 180 | 8 | 440 |
| 4 | 200 | 9 | 540 |

The motion diagram shows that she leaves home at a time we choose to call $t = 0$ min, and then makes steady progress for a while. Beginning at $t = 3$ min there is a period in which the distance traveled during each time interval becomes shorter—perhaps she slowed down to speak with a friend. Then, at $t = 6$ min, the distances traveled within each interval are longer—perhaps, realizing she is running late, she begins walking more quickly.

Every dot in the motion diagram of Figure 2.2 represents the student's position at a particular time. For example, the student is at position $x = 120$ m at $t = 2$ min. TABLE 2.1 lists her position for every point in the motion diagram.

The motion diagram of Figure 2.2 is one way to represent the student's motion. Presenting the data as in Table 2.1 is a second way to represent this motion. A third way to represent the motion is to use the data to make a graph. FIGURE 2.3 is a graph of the positions of the student at different times; we say it is a graph of $x$ versus $t$ for the student.

FIGURE 2.3 A graph of the student's motion.

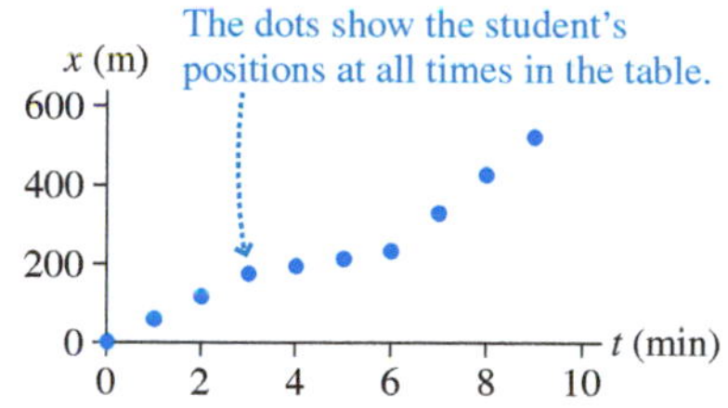


NOTE ▸ A graph of "$a$ versus $b$" means that $a$ is graphed on the vertical axis and $b$ on the horizontal axis. We say that such a graph represents $a$ "as a function of" $b$. ◂

We can flesh out the graph of Figure 2.3, though. We can assume that the student moved *continuously* through all intervening points of space, so we can represent her motion as a continuous curve that passes through the measured points, as shown in FIGURE 2.4. Such a continuous curve that shows an object's position as a function of time is called a **position-versus-time graph** or, sometimes, just a *position graph.*

FIGURE 2.4 Extending the graph of Figure 2.3 to a position-versus-time graph.

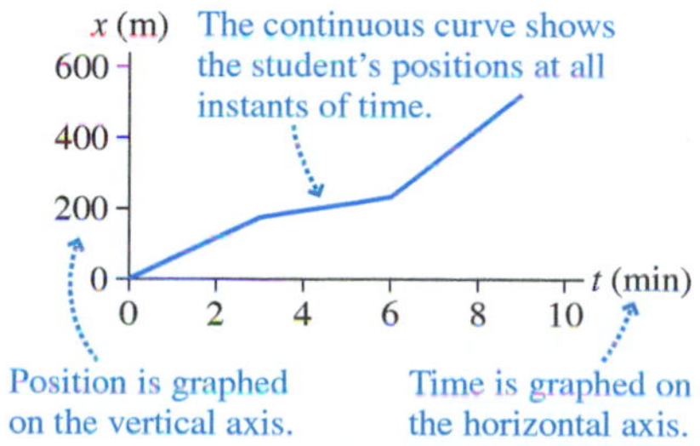

NOTE ▶ A graph is *not* a "picture" of the motion. The student is walking along a straight line, but the graph itself is not a straight line. Further, we've graphed her position on the vertical axis even though her motion is horizontal. A graph is an *abstract representation* of motion. ◀

CONCEPTUAL EXAMPLE 2.1 **Interpreting a car's position-versus-time graph**

The graph in FIGURE 2.5 represents the motion of a car along a straight road. Describe (in words) the motion of the car.

FIGURE 2.5 Position-versus-time graph for the car.

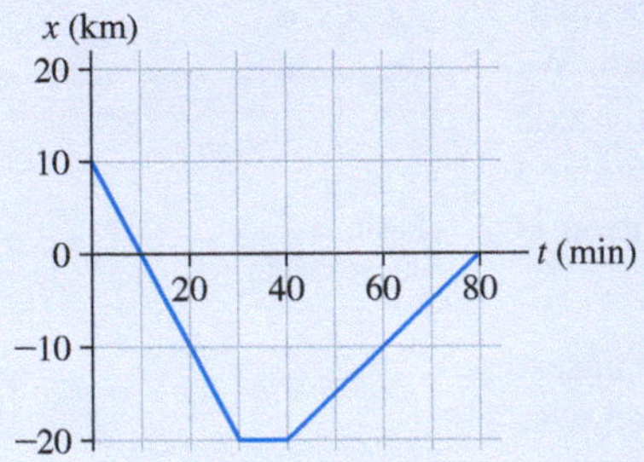


FIGURE 2.6 Looking at the position-versus-time graph in detail.

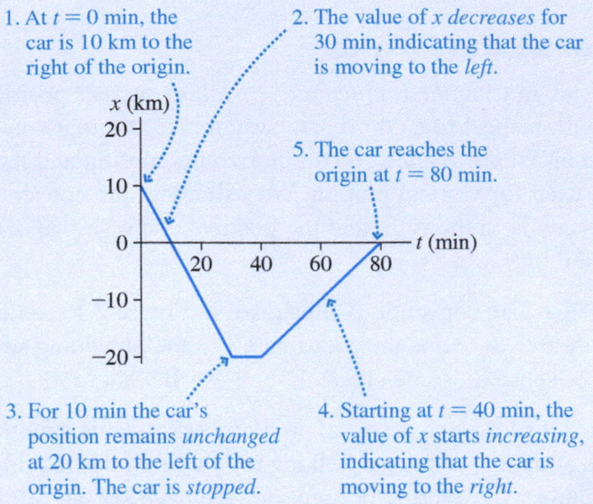

1. At t = 0 min, the car is 10 km to the right of the origin.
2. The value of x decreases for 30 min, indicating that the car is moving to the left.
5. The car reaches the origin at t = 80 min.
x (km)
20
10
0
−10
−20
20
40
60
80
t (min)
3. For 10 min the car's position remains unchanged at 20 km to the left of the origin. The car is stopped.
4. Starting at t = 40 min, the value of x starts increasing, indicating that the car is moving to the right.


***REASON*** The vertical axis in Figure 2.5 is labeled "$x$ (km)," so the car's position is measured in kilometers. Our convention for motion along the $x$-axis given in Figure 2.1 tells us that $x$ increases as the car moves to the right and $x$ decreases as the car moves to the left. As FIGURE 2.6 explains in detail, the graph thus shows that the car travels to the left for 30 minutes, stops for 10 minutes, then travels to the right for 40 minutes. It ends up 10 km to the left of where it began.

***ASSESS*** The car travels to the left for 30 minutes and to the right for 40 minutes. Nonetheless, it ends up to the left of where it started. This means that the car was moving faster when it was moving to the left than when it was moving to the right. We can deduce this fact from the graph as well, as we will see in the next section.

## Representing Velocity

Velocity is a vector, having both a magnitude and a direction. When we draw a general velocity vector on a diagram, we use an arrow labeled with the symbol $\vec{v}$.

For motion in one dimension, however, velocity vectors are restricted to point only forward or backward for horizontal motion, or up or down for vertical motion. This restriction lets us simplify our notation for velocity vectors in one dimension. When we solve problems for motion along an $x$-axis, we represent the velocity with the simple variable $v_x$. As FIGURE 2.7 shows, we adopt the convention that for an object moving to the right, $v_x$ is positive, whereas for motion to the left, $v_x$ is negative.

FIGURE 2.7 Sign conventions for velocity.

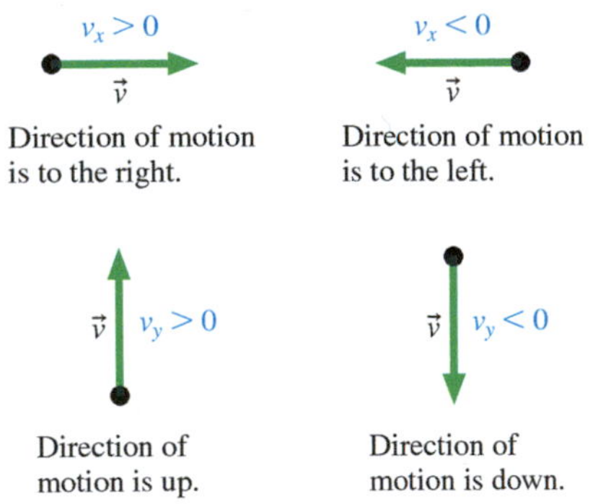


For vertical motion along the $y$-axis, we use the symbol $v_y$ to represent the velocity. The sign conventions for this vertical motion are also shown in Figure 2.7.

We use the symbol $v$, with no subscript, to represent the *speed* of an object. **Speed is the *magnitude* of the velocity vector** and is thus always positive.

In Chapter 1 we defined an object's velocity as $\Delta x/\Delta t$, where $\Delta x = x_f - x_i$ is the *displacement*, or change in position, as the object moves from an initial position $x_i$ to a final position $x_f$, and $\Delta t$ is the interval of time during which the motion occurs. For motion along a horizontal line, we can write

$$v_x = \frac{\Delta x}{\Delta t} \tag{2.1}$$

This agrees with the sign conventions in Figure 2.7. If $\Delta x$ is positive, $x$ is increasing, the object is moving to the right, and Equation 2.1 gives a positive value for velocity. If $\Delta x$ is negative, $x$ is decreasing, the object is moving to the left, and Equation 2.1 gives a negative value for velocity.

Equation 2.1 is the first of many *kinematic equations* we'll see in this chapter. We'll often specify equations in terms of the coordinate $x$, but if the motion is vertical, in which case we use the coordinate $y$, the equations can be easily adapted. For example, Equation 2.1 for motion along a vertical axis becomes

$$v_y = \frac{\Delta y}{\Delta t} \tag{2.2}$$

## From Position to Velocity

How is an object's velocity related to its position-versus-time graph? To find out, let's take another look at the motion diagram of the student walking to school. As we see in FIGURE 2.8, where we have repeated the motion diagram of Figure 2.2, her motion has three clearly defined phases. In each phase her speed is constant (because the velocity vectors have the same length) but the speed varies from phase to phase.

FIGURE 2.8 Revisiting the motion diagram of the student walking to school.

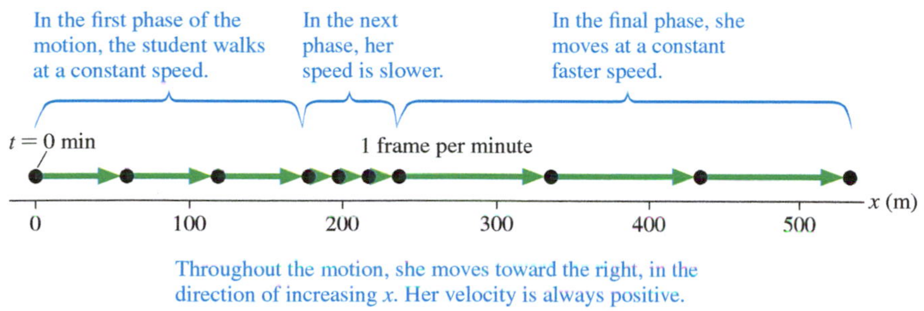

In the first phase of the motion, the student walks at a constant speed.
In the next phase, her speed is slower.
In the final phase, she moves at a constant faster speed.
t = 0 min
1 frame per minute
x (m)
0
100
200
300
400
500
Throughout the motion, she moves toward the right, in the direction of increasing x. Her velocity is always positive.

FIGURE 2.9 Interpreting the slope of the position graph for the student walking to school.

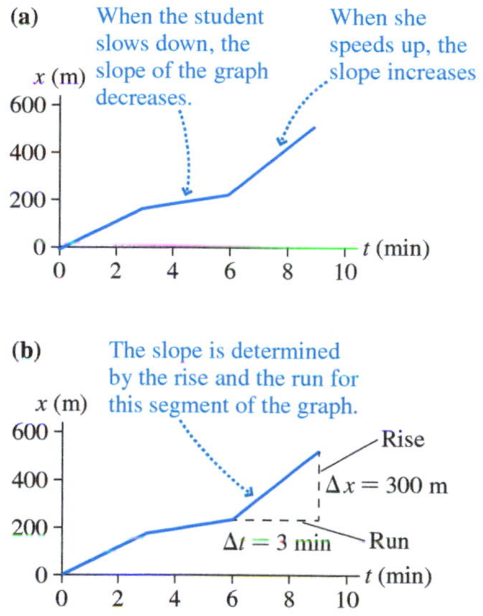


Just as her motion has three different phases, her position-versus-time graph (redrawn in FIGURE 2.9a) has three clearly defined segments with three different *slopes*. We can see that there's a relationship between her speed and the slope of the graph: **A faster speed corresponds to a steeper slope.**

The correspondence is actually deeper than this. Let's look at the slope of the third segment of the position-versus-time graph, as shown in FIGURE 2.9b. The slope of a graph is defined as the ratio of the "rise," the vertical change, to the "run," the horizontal change. For the segment of the graph shown, the slope is

$$\text{slope of graph} = \frac{\text{rise}}{\text{run}} = \frac{\Delta x}{\Delta t}$$

This ratio has a physical meaning—it's the velocity, exactly as we defined it in Equation 2.1. We've shown this correspondence for one particular graph, but it is a general principle: **The slope of an object's position-versus-time graph is the object's velocity at that point in the motion.** This principle also holds for negative slopes, which correspond to negative velocities. We can associate the slope of a position-versus-time graph, a *geometrical* quantity, with velocity, a *physical* quantity.

### TACTICS BOX 2.1 Interpreting position-versus-time graphs

Information about motion can be obtained from position-versus-time graphs as follows:

1. Determine an object's *position* at time $t$ by reading the graph at that instant of time.
2. Determine the object's *velocity* at time $t$ by finding the slope of the position graph at that point. Steeper slopes correspond to faster speeds.
3. Determine the *direction of motion* by noting the sign of the slope. Positive slopes correspond to positive velocities and, hence, to motion to the right (or up). Negative slopes correspond to negative velocities and, hence, to motion to the left (or down).

Exercises 3,4

NOTE ▸ The slope is a ratio of intervals, $\Delta x/\Delta t$, not a ratio of coordinates; that is, the slope is *not* simply $x/t$. ◂

NOTE ▸ We are distinguishing between the actual slope—the slope on the graph—and the *physically meaningful* slope. If you were to use a ruler to measure the rise and the run of the graph, you could compute the actual slope of the line as drawn on the page. That is not the slope we are referring to when we equate the velocity with the slope of the line. Instead, we find the *physically meaningful* slope by measuring the rise and run using the scales along the axes. The "rise" $\Delta x$ is some number of meters; the "run" $\Delta t$ is some number of seconds. The physically meaningful rise and run include units, and the ratio of these units gives the units of the slope. ◂

FIGURE 2.10 Deducing the velocity-versus-time graph from the position-versus-time graph.

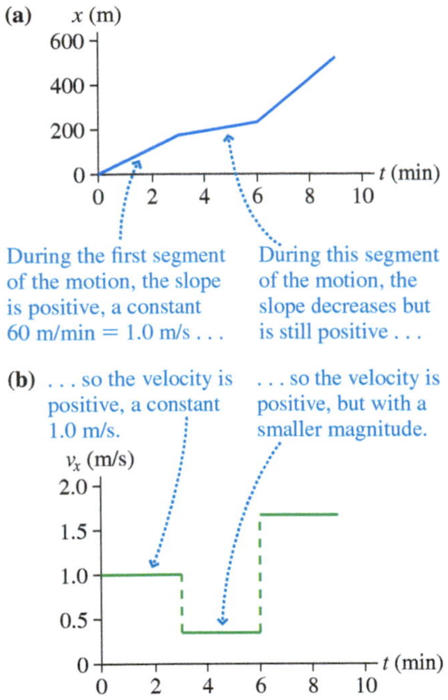


We can now use the approach of Tactics Box 2.1 to analyze the student's position-versus-time graph, redrawn in FIGURE 2.10a. We can determine her velocity during the first phase of her motion by measuring the slope of the line:

$$v_x = \text{slope} = \frac{\Delta x}{\Delta t} = \frac{180\text{ m}}{3.0\text{ min}} = 60\frac{\text{m}}{\text{min}} \times \frac{1\text{ min}}{60\text{ s}} = 1.0\text{ m/s}$$

In completing this calculation, we've converted to more usual units for speed, m/s. During this phase of the motion, her velocity is constant, so a graph of velocity versus time appears as a horizontal line at 1.0 m/s, as shown in FIGURE 2.10b. We can do similar calculations to show that her velocity during the second phase of her motion is +0.33 m/s, and then increases to +1.7 m/s during the final phase. We combine this information to create the **velocity-versus-time graph** shown in Figure 2.10b.

An inspection of the velocity-versus-time graph shows that it matches our understanding of the student's motion: There are three phases of the motion, each with constant speed. In each phase, the velocity is positive because she is always moving to the right. The second phase is slow (low velocity) and the third phase is fast (high velocity). All of this can be clearly seen on the velocity-versus-time graph, which is yet another way to represent her motion.

NOTE ▸ The velocity-versus-time graph in Figure 2.10b includes vertical segments in which the velocity changes instantaneously. Such rapid changes are an idealization; it actually takes at least a small amount of time to change velocity. ◂

## EXAMPLE 2.2 Finding a car's velocity graph from its position graph

FIGURE 2.11 gives the position-versus-time graph of a car.

a. Draw the car's velocity-versus-time graph.
b. Describe the car's motion in words.

FIGURE 2.11 The position-versus-time graph of a car.

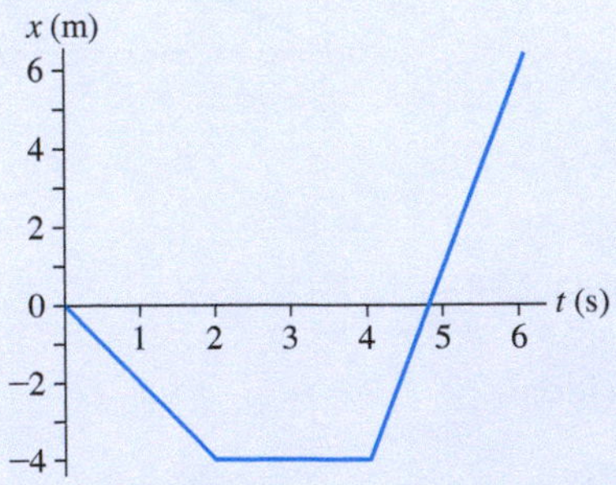


***STRATEGIZE*** We will use the steps from Tactics Box 2.1 to understand the car's motion and to draw its velocity-versus-time graph based on its position graph.

***PREPARE*** Figure 2.11 is a graphical representation of the motion. The car's position-versus-time graph is a sequence of three straight lines. Each of these straight lines represents uniform motion at a constant velocity. We can determine the car's velocity during each interval of time by measuring the slope of the line.

***SOLVE***

a. From $t = 0$ s to $t = 2$ s ($\Delta t = 2$ s) the car's displacement is $\Delta x = -4\text{ m} - 0\text{ m} = -4\text{ m}$. The velocity during this interval is

$$v_x = \frac{\Delta x}{\Delta t} = \frac{-4\text{ m}}{2\text{ s}} = -2\text{ m/s}$$

The car's position does not change from $t = 2$ s to $t = 4$ s ($\Delta x = 0$ m), so $v_x = 0$ m/s. Finally, the displacement between $t = 4$ s and $t = 6$ s ($\Delta t = 2$ s) is $\Delta x = 10$ m. Thus the velocity during this interval is

$$v_x = \frac{10\text{ m}}{2\text{ s}} = 5\text{ m/s}$$

These velocities are represented graphically in FIGURE 2.12.

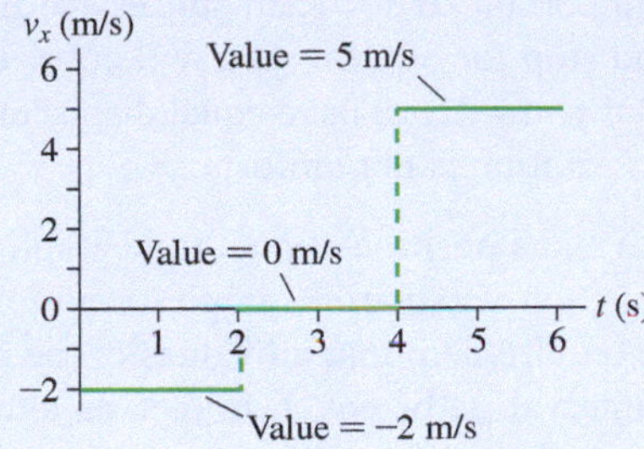


FIGURE 2.12 The velocity-versus-time graph for the car.

b. The velocity-versus-time graph of Figure 2.12 shows the motion in a way that we can describe in a straightforward manner: The car backs up for 2 s at 2 m/s, sits at rest for 2 s, then drives forward at 5 m/s for 2 s.

**ASSESS** Notice that the velocity graph and the position graph look completely different. They should! The value of the velocity graph at any instant of time equals the *slope* of the position graph. Since the position graph is made up of segments of constant slope, the velocity graph should be made up of segments of constant *value*, as it is. This gives us confidence that the graph we have drawn is correct.

## From Velocity to Position

We've now seen how to move between different representations of uniform motion. There's one last issue to address: If you have a graph of velocity versus time, how can you determine the position graph?

Suppose you leave a lecture hall and begin walking toward your next class, which is down the hall to the east. You then realize that you left your textbook at your seat. You turn around and run back to the lecture hall to retrieve it. A velocity-versus-time graph for this motion appears in FIGURE 2.13a. There are two clear phases to the motion: walking away from the lecture hall (velocity +1.0 m/s) and running back (velocity −3.0 m/s). How can we deduce your position-versus-time graph?

As before, we can analyze the graph segment by segment, as shown in Figure 2.13. For the first segment, the velocity graph in Figure 2.13a indicates motion with a constant velocity of +1.0 m/s. This tells us that the corresponding position graph must be a straight line with a positive slope of +1.0 m/s, as shown in the position graph of FIGURE 2.13b. For the second segment, where the velocity is −3.0 m/s, as in Figure 2.13a, the position graph must be a line with a negative slope of −3.0 m/s, also shown in the position graph of Figure 2.13b.

The position graph makes sense: It shows 15 seconds of slowly increasing position (walking away from the lecture hall) and then 5 seconds of rapidly decreasing position (running back). And you end up back where you started.

There's one important detail that we didn't talk about in the preceding paragraph: How did we know that the position graph started at $x = 0$ m? The velocity graph tells us the *slope* of the position graph, but it doesn't tell us where the position graph should start. Although you're free to select any point you choose as the origin of the coordinate system, here it seems reasonable to set $x = 0$ m at your starting point in the lecture hall; as you walk away, your position increases.

FIGURE 2.13 Deducing a position graph from a velocity-versus-time graph.

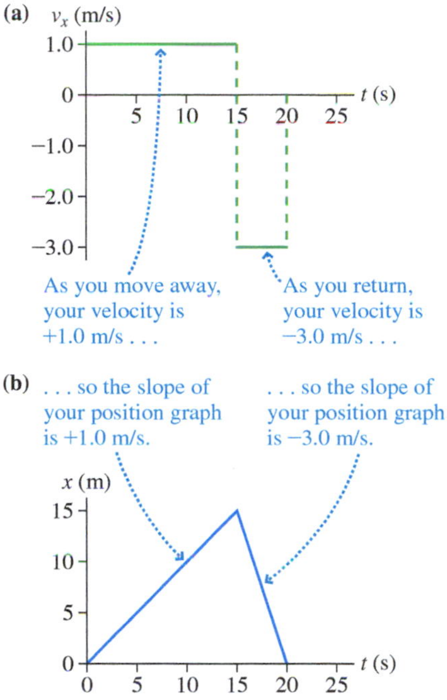


**STOP TO THINK 2.1** Which position-versus-time graph best describes the motion diagram at left?

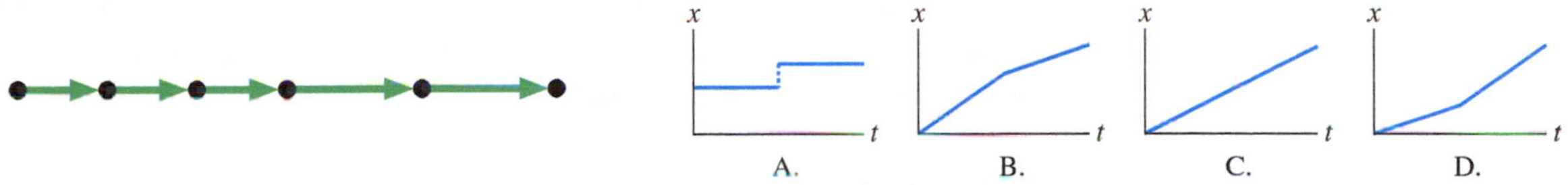


# 2.2 Uniform Motion

If you drive your car on a straight road at a perfectly steady 60 miles per hour (mph), you will cover 60 mi during the first hour, another 60 mi during the second hour, yet another 60 mi during the third hour, and so on. This is an example of what we call *uniform motion.* **Straight-line motion in which equal displacements occur during any successive equal-time intervals is called uniform motion or constant-velocity motion.**

NOTE ► The qualifier "any" is important. If for each successive hour of a trip you drive 120 mph for 30 min and stop for 30 min, you will cover 60 mi during each successive 1 hour interval. But you will *not* have equal displacements during successive 30 min intervals, so this motion is not uniform. ◄

FIGURE 2.14 Motion diagram and position-versus-time graph for uniform motion.

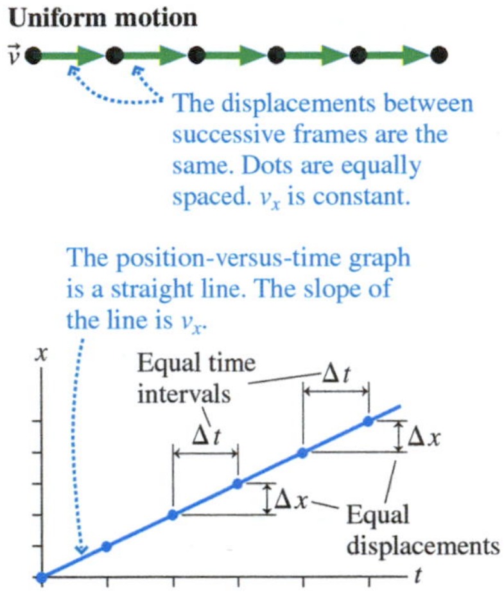


FIGURE 2.14 shows a motion diagram and a position-versus-time graph for an object in uniform motion. Notice that the position-versus-time graph for uniform motion is a straight line. This follows from the requirement that all values of the displacement $\Delta x$ corresponding to the same time interval $\Delta t$ be equal. In fact, an alternative definition of uniform motion is: **An object's motion is uniform if and only if its position-versus-time graph is a straight line.**

## Equations of Uniform Motion

We've just learned that an object in uniform motion along the $x$-axis will have a linear (straight-line) position-versus-time graph like the one shown in FIGURE 2.15. Recall from Chapter 1 that we denote the object's initial position as $x_i$ at time $t_i$. The term "initial" refers to the starting point of our analysis or the starting point in a problem. The object may or may not have been in motion prior to $t_i$. We use the term "final" for the ending point of our analysis or the ending point of a problem, and denote the object's final position $x_f$ at the time $t_f$. As we've seen, the object's velocity $v_x$ along the $x$-axis can be determined by finding the slope of the graph:

$$v_x = \frac{\text{rise}}{\text{run}} = \frac{\Delta x}{\Delta t} = \frac{x_f - x_i}{t_f - t_i} \tag{2.3}$$

FIGURE 2.15 Position-versus-time graph for an object in uniform motion.

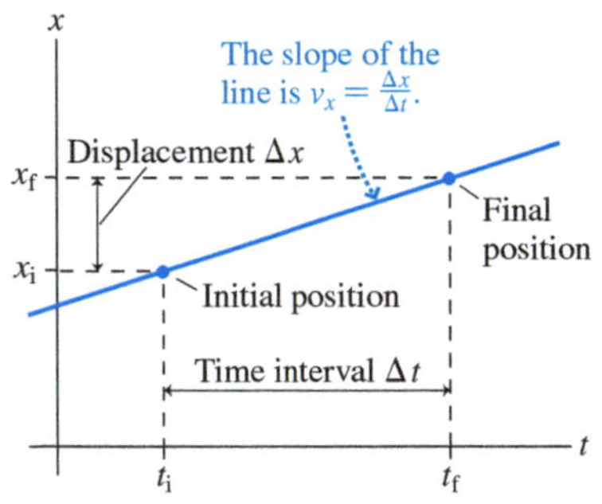


Here, the displacement $\Delta x = x_f - x_i$ is the change in position that occurs during the time interval $\Delta t = t_f - t_i$. Equation 2.3 can be rearranged to give

$$x_f = x_i + v_x \, \Delta t \tag{2.4}$$

Position equation for an object in uniform motion ($v_x$ is constant)

where $\Delta t$ is the interval of time in which the object moves from position $x_i$ to position $x_f$. Equation 2.4 applies to any time interval $\Delta t$ during which the velocity is constant. We can also write this in terms of the object's displacement $\Delta x$:

$$\Delta x = v_x \, \Delta t \tag{2.5}$$

The velocity of an object in uniform motion tells us the amount by which its position changes during each second. An object with a velocity of 20 m/s *changes* its position by 20 m during every second of motion: by 20 m during the first second of its motion, by another 20 m during the next second, and so on. We say that position is changing at the *rate* of 20 m/s. If the object starts at $x_i = 10$ m, it will be at $x = 30$ m after 1 s of motion and at $x = 50$ m after 2 s of motion. Thinking of velocity like this will help you develop an intuitive understanding of the connection between velocity and position.

## Mathematical Relationships

Physics may seem densely populated with equations, but most equations follow a few basic forms. FIGURE 2.16 shows three graphs: a mathematical equation, the kinetic energy of a moving object versus its speed, and the potential energy of a spring versus how far the spring is compressed.

All of these graphs have the same overall appearance. The three expressions differ in their variables, but all three equations have the same **mathematical**

**FIGURE 2.16** Three graphs with the same mathematical relationship.

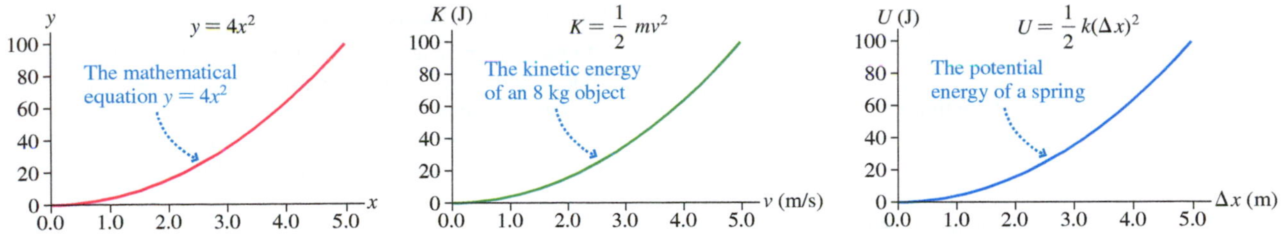

The mathematical equation $y = 4x^2$

$K = \frac{1}{2}mv^2$

The kinetic energy of an 8 kg object

$U = \frac{1}{2}k(\Delta x)^2$

The potential energy of a spring


**relationship.** We'll use only a handful of different mathematical relationships in this text. As we meet each relationship for the first time, we will give an overview of its most important properties. When you see the relationship again in a new equation, we'll insert an icon, such as, that refers back to the overview so that you can remind yourself of the key details.

For instance, the mathematical form of Equation 2.5 is a type that we will see often: The displacement $\Delta x$ is *proportional* to the time interval $\Delta t$. The following proportional relationships overview gives the details.

## Proportional relationships

We say that $y$ is **proportional** to $x$ if they are related by an equation of the form

$$y = Cx$$

$y$ is proportional to $x$

We call $C$ the **proportionality constant.** A graph of $y$ versus $x$ is a straight line that passes through the origin.

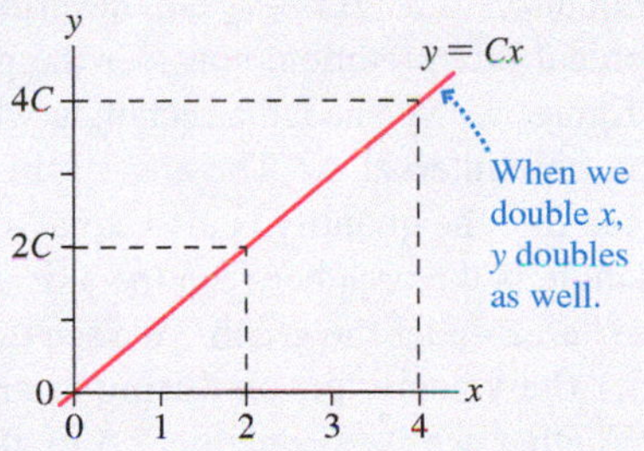

When we double $x$, $y$ doubles as well.


**SCALING** If $x$ has the initial value $x_1$, then $y$ has the initial value $y_1 = Cx_1$. Changing $x$ from $x_1$ to $x_2$ changes $y$ from $y_1$ to $y_2$. The ratio of $y_2$ to $y_1$ is

$$\frac{y_2}{y_1} = \frac{Cx_2}{Cx_1} = \frac{x_2}{x_1}$$

The ratio of $y_2$ to $y_1$ is exactly the same as the ratio of $x_2$ to $x_1$. If $y$ is proportional to $x$, which is often written $y \propto x$, then $x$ and $y$ change by the same factor:

- If you double $x$, you double $y$.
- If you decrease $x$ by a factor of 3, you decrease $y$ by a factor of 3.

If two variables have a proportional relationship, we can draw important conclusions from ratios without knowing the value of the proportionality constant $C$. We can often solve problems in a very straightforward manner by looking at such ratios. This is an important skill called *ratio reasoning*.

Exercise 11

**EXAMPLE 2.3** **If a train leaves Cleveland at 2:00 ...**

A train is moving due west at a constant speed. A passenger notes that it takes 10 minutes to travel 12 km. How long will it take the train to travel 60 km?

**STRATEGIZE** For an object in uniform motion, Equation 2.5 shows that the distance traveled $\Delta x$ is proportional to the time interval $\Delta t$, so this is a good problem to solve using ratio reasoning.

**PREPARE** We are comparing two cases: the time $\Delta t_1 = 10$ min it takes to travel the distance $\Delta x_1 = 12$ km, and the (unknown) time $\Delta t_2$ it will take to travel $\Delta x_2 = 60$ km. Ratio reasoning tells us that $\Delta x_2/\Delta x_1 = \Delta t_2/\Delta t_1$.

*Continued*

**SOLVE** The ratio of the distances is

$$\frac{\Delta x_2}{\Delta x_1} = \frac{60\text{ km}}{12\text{ km}} = 5$$

This is equal to the ratio of the times:

$$\frac{\Delta t_2}{\Delta t_1} = \frac{\Delta t_2}{10\text{ min}} = \frac{\Delta x_2}{\Delta x_1} = 5$$

$$\Delta t_2 = 5 \times (10\text{ min}) = 50\text{ min}$$

It takes 10 minutes to travel 12 km, so it will take 50 minutes—5 times as long—to travel 60 km.

**ASSESS** For an object in steady motion, it makes sense that 5 times the distance requires 5 times the time. We can see that using ratio reasoning is a straightforward way to solve this problem. We don't need to know the proportionality constant (in this case, the velocity); we just used ratios of distances and times.

## A Second Way to Find Position from Velocity

Earlier, we saw that we could deduce an object's position graph from its velocity graph by drawing a position graph in which the slopes everywhere matched the velocity graph. But there's another way to understand the relationship between velocity and position graphs—by looking at what we call the *area under the graph.* Let's look at an example.

FIGURE 2.17 Displacement is the area under a velocity-versus-time graph.

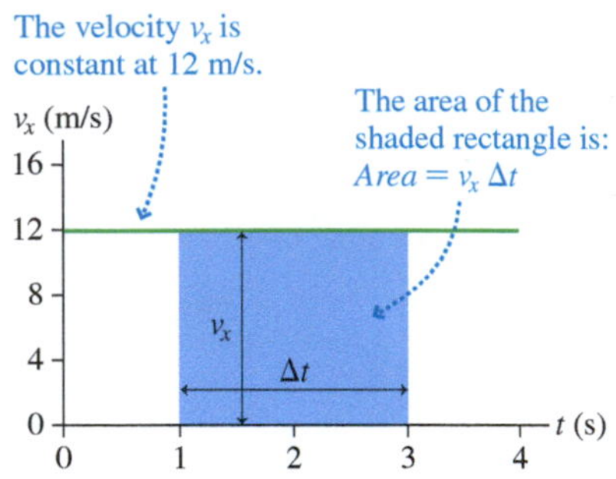


Suppose a car is in uniform motion at 12 m/s. How far does it travel—that is, what is its displacement—during the time interval between $t = 1.0$ s and $t = 3.0$ s?

Equation 2.5, $\Delta x = v_x\,\Delta t$, describes the displacement mathematically; for a graphical interpretation, consider the graph of velocity versus time in FIGURE 2.17. In the figure, we've shaded a rectangle whose height is the velocity $v_x$ and whose base is the time interval $\Delta t$. The area of this rectangle is $v_x\,\Delta t$. Looking at Equation 2.5, we see that the quantity is also equal to the displacement of the car. The area of this rectangle is the area between the axis and the line representing the velocity; we call it the "area under the graph." We see that **the displacement $\Delta x$ is equal to the area under the velocity graph during interval $\Delta t$.**

Whether we use Equation 2.5 or the area under the graph to compute the displacement, we get the same result:

$$\Delta x = v_x\,\Delta t = (12\text{ m/s})\,(2.0\text{ s}) = 24\text{ m}$$

Although we've shown that the displacement is the area under the graph only for uniform motion, where the velocity is constant, we'll soon see that this result applies to any one-dimensional motion.

**NOTE** ▶ Wait a minute! The displacement $\Delta x = x_f - x_i$ is a length. How can a length equal an area? Recall that earlier, when we found that the velocity is the slope of the position graph, we made a distinction between the *actual* slope and the *physically meaningful* slope? The same distinction applies here. The velocity graph does indeed bound a certain area on the page. That is the actual area—say, in square inches of paper—but it is *not* the area to which we are referring. Once again, we need to measure the quantities we are using, $v_x$ and $\Delta t$, by referring to the scales on the axes. $\Delta t$ is some number of seconds, while $v_x$ is some number of meters per second. When these are multiplied, the *physically meaningful* area has units of meters, appropriate for a displacement. ◀

**STOP TO THINK 2.2** Four objects move with the velocity-versus-time graphs shown. Which object has the largest displacement between $t = 0$ s and $t = 2$ s?

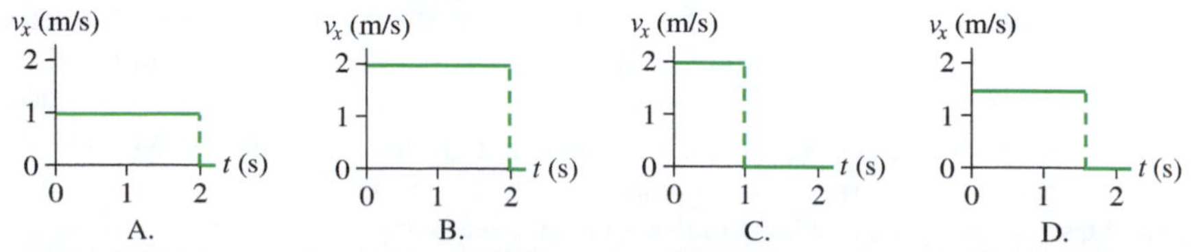

## 2.3 Instantaneous Velocity

The objects we've studied so far have moved with a constant, unchanging velocity or, like the car in Example 2.1, have a velocity that changes abruptly from one constant value to another. This is not very realistic. Real moving objects speed up and slow down, with their velocity changing smoothly. Suppose you're sitting at a red light at the start of a freeway ramp. When the light turns green, you increase your speed steadily from 0 mph to 60 mph until you merge onto the freeway.

A car's speed increases smoothly as it enters the freeway.

Perhaps as you speed down the ramp you glance at your speedometer and notice that, at that specific instant, it reads 40 mph. The speedometer indicates how fast you're moving *at a particular instant* in time. An object's velocity—its speed and direction—at a specific instant of time $t$ is called the object's **instantaneous velocity.**

But what does it mean to have a velocity "at an instant"? An instantaneous velocity with a speed of 40 mph means that the rate at which your car's position is changing—at that exact instant—is such that it would travel a distance of 40 miles in 1 hour if it continued at that rate without change. If you speed up to pass a car that is traveling at a steady 40 mph, then at the very moment that your instantaneous velocity is 40 mph, your speed will match that of the other car—but an instant later, you'll be moving faster than 40 mph.

The velocity we introduced in Section 1.4 is really the *average velocity*; it is the velocity *averaged* over a *finite* time interval, such as 1 s or 1 min. **From now on, though, the word "velocity" will always mean instantaneous velocity**—the velocity at a single instant of time.

For uniform motion, an object's position-versus-time graph is a straight line and the object's velocity is the slope of that line. In contrast, FIGURE 2.18 shows that the position-versus-time graph for a car entering a freeway is a *curved* line. The displacement $\Delta x$ during equal intervals of time gets greater as the car speeds up. Even so, we can use the slope of the position graph to measure the car's velocity. We can say that

$$\text{instantaneous velocity } v_x \text{ at time } t = \text{slope of position graph at time } t \qquad (2.6)$$

But how do we determine the slope of a curved line at a particular point? The following table shows how.

FIGURE 2.18 Position-versus-time graph for a car entering a freeway.

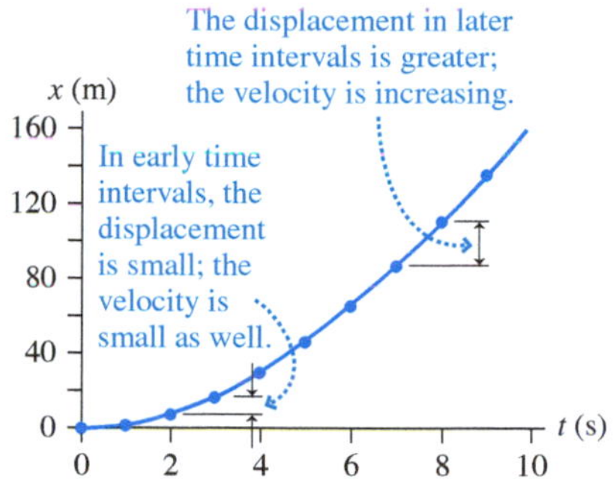


### Finding the instantaneous velocity

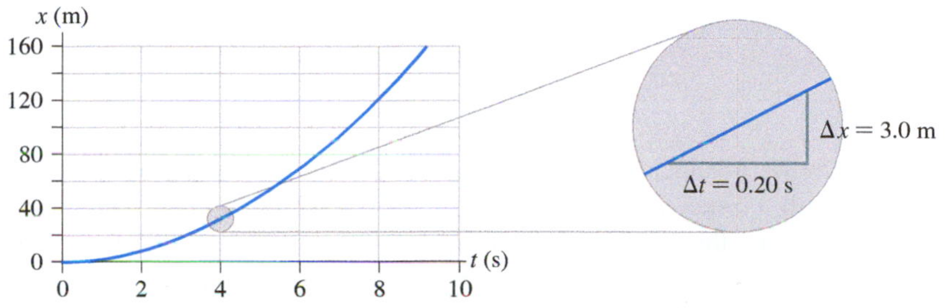


If the velocity changes, the position graph is a curved line. But we can compute a slope at a point by considering a small segment of the graph. Let's look at the motion in a very small time interval right around $t = 4.0$ s. This is highlighted with a circle, and we show a closeup in the next graph at the right.

In this magnified segment of the position graph, the curve isn't apparent. It appears to be a line segment. We can find the slope by calculating the rise over the run, just as before:

$$v_x = (3.0\text{ m})/(0.20\text{ s}) = 15\text{ m/s}$$

This is the slope at $t = 4.0$ s and thus the velocity at this instant of time.

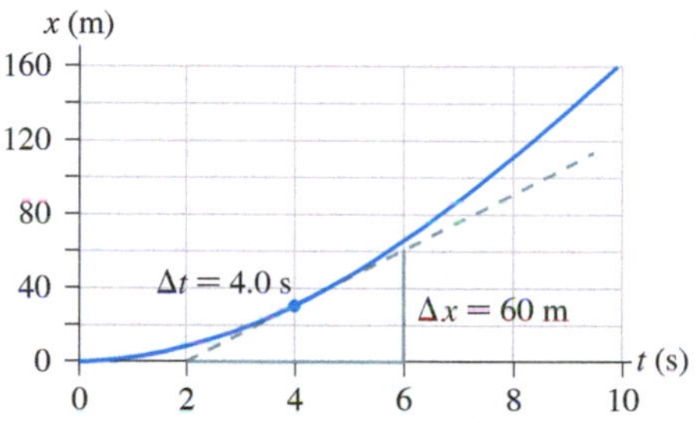


Graphically, the slope of the curve at a point is the same as the slope of a straight line drawn *tangent* to the curve at that point. Calculating rise over run for the tangent line, we get

$$v_x = (60\text{ m})/(4.0\text{ s}) = 15\text{ m/s}$$

This is the same value we obtained from the close-up view. **The slope of the tangent line is the instantaneous velocity at that instant of time.**

## CONCEPTUAL EXAMPLE 2.4 Analyzing a hockey player's position graph

A hockey player moves in a straight line along the length of the ice in a game. We measure position from the center of the rink. FIGURE 2.19 shows a position-versus-time graph for his motion.

a. Sketch an approximate velocity-versus-time graph.
b. At which point or points is the player moving the fastest?
c. Is the player ever at rest? If so, at which point or points?

FIGURE 2.19 The position-versus-time graph for a hockey player.

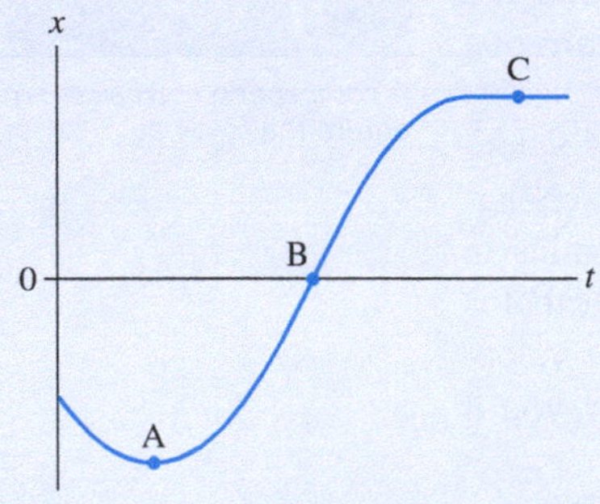


FIGURE 2.20 Finding a velocity graph from a position graph.

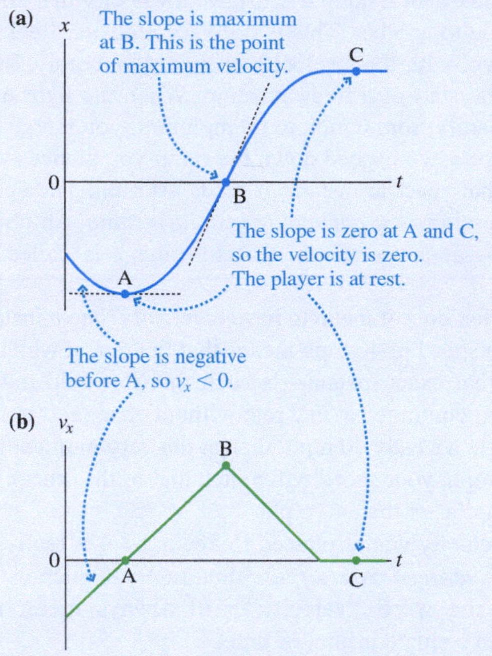


**REASON** a. The velocity at a particular instant of time is the slope of the tangent line to the position-versus-time graph at that time. We can move point-by-point along the position-versus-time graph, noting the slope of the tangent at each point to find the velocity at that point.

Initially, to the left of point A, the slope is negative and thus the velocity is negative (i.e., the player is moving to the left). But the slope decreases as the curve flattens out, and by the time the graph gets to point A, the slope is zero. The slope then increases to a maximum value at point B, decreases back to zero a little before point C, and remains at zero thereafter. This reasoning process is outlined in FIGURE 2.20a, and FIGURE 2.20b shows the approximate velocity-versus-time graph that results.

The other questions were answered during the construction of the graph:

b. The player moves the fastest at point B where the slope of the position graph is the steepest.
c. If the player is at rest, $v_x = 0$. Graphically, this occurs at points where the line tangent to the position-versus-time graph is horizontal and thus has zero slope. Figure 2.20 shows that the slope is zero at point A and for a small range of times near point C. At point A, the velocity is only instantaneously zero—the player is reversing direction, changing from moving to the left to moving to the right. Near point C, he has stopped moving and stays at rest.

**ASSESS** The best way to check our work is to look at different segments of the motion and see if the velocity and position graphs match. Until point A, $x$ is decreasing. The player is moving to the left, so the velocity should be negative, which our graph shows. Between points A and C, $x$ is increasing, so the velocity should be positive, which is also a feature of our graph. The steepest slope is at point B, so this should be the high point of our velocity graph, as it is.

FIGURE 2.21 Velocity-versus-time graph for a lion pursuing prey.

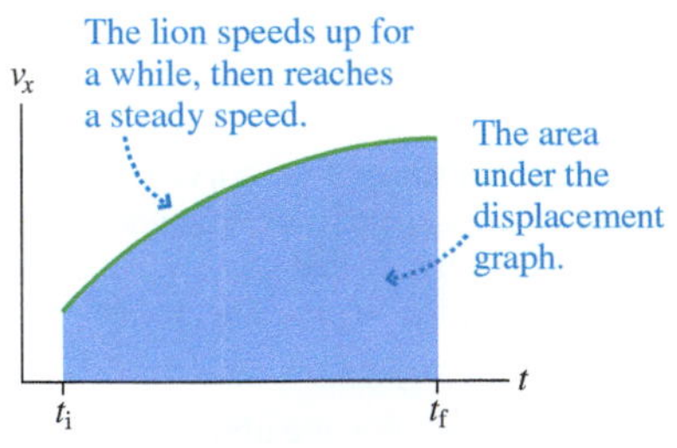


FIGURE 2.21 shows a velocity-versus-time graph for a lion speeding up to pursue prey. Even though the speed varies, we can still use the graph to determine how far the lion moves during the time interval $t_i$ to $t_f$. For uniform motion we showed that the displacement $\Delta x$ is the area under the velocity-versus-time graph during the time interval. But there was nothing special about the type of motion: We can generalize this idea to the case of an object whose velocity varies. If we draw a velocity graph for the motion, the object's displacement is given by

$$x_f - x_i = \text{area under the velocity graph between } t_i \text{ and } t_f \qquad (2.7)$$

The area under the graph in Figure 2.21 tells us how far the lion ran during this segment of the chase.

In many cases, as in the next example, the area under the graph is a simple shape whose area we can easily compute. If the shape is complex, however, we can approximate the area using a number of simpler shapes that closely match it.

**EXAMPLE 2.5 Calculating the displacement of a car during a rapid start**

FIGURE 2.22 shows the velocity-versus-time graph of a car pulling away from a stop. How far does the car move during the first 3.0 s?

***STRATEGIZE*** The question How far? indicates that we need to find a displacement $\Delta x$ rather than a position $x$. Graphically, the displacement is given by the area under the velocity-versus-time graph.

***PREPARE*** In Figure 2.22 we have shaded the area we need to find. It is the area between the straight line of the velocity graph and the $t$-axis, between $t_i = 0$ s and $t_f = 3.0$ s.

FIGURE 2.22 Velocity-versus-time graph for the car of Example 2.5.

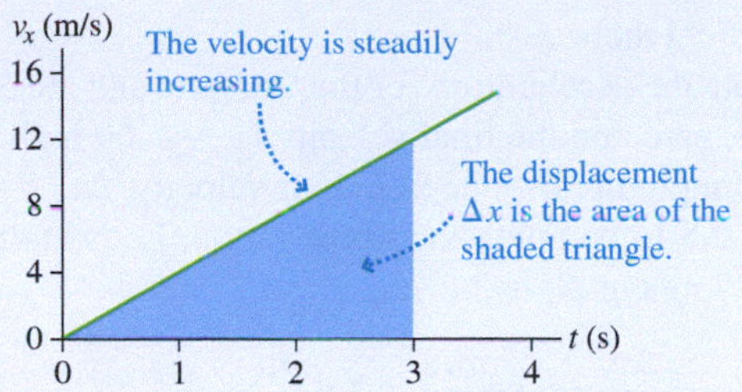


***SOLVE*** The graph in this case is an angled line, so the area is that of a triangle:

$$\Delta x = \text{area of triangle between } t = 0 \text{ s and } t = 3.0 \text{ s}$$
$$= \tfrac{1}{2} \times \text{base} \times \text{height} = \tfrac{1}{2} \times 3.0 \text{ s} \times 12 \text{ m/s} = 18 \text{ m}$$

The car moves 18 m during the first 3 seconds as its velocity changes from 0 to 12 m/s.

***ASSESS*** The physically meaningful area is a product of a time in s and a velocity in m/s, so $\Delta x$ has the proper units of m. Let's check the numbers to see if they make physical sense. The final velocity, 12 m/s, is about 25 mph. Pulling away from a stop, you might expect to reach this speed in about 3 s—at least if you have a reasonably sporty vehicle! Another check is to realize that if the car had moved at a constant 12 m/s (the final velocity) during these 3 s, the distance would be 36 m. The actual distance traveled during the 3 s is 18 m—half of 36 m. This makes sense, as the velocity was 0 m/s at the start of the problem and increased steadily to 12 m/s.

**STOP TO THINK 2.3** Which velocity-versus-time graph goes with the position-versus-time graph on the left?

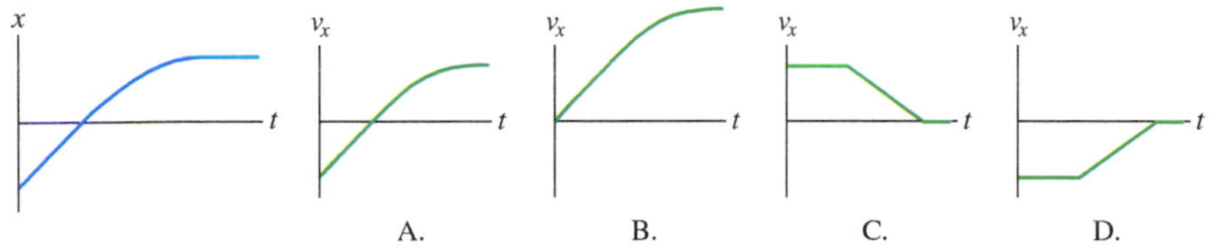


## 2.4 Acceleration

The goal of this chapter is to describe motion. We've seen that velocity describes the rate at which an object changes position. We need one more motion concept to complete the description, one that will describe an object whose velocity is changing.

As an example, let's look at a frequently quoted measurement of car performance, the time it takes the car to go from 0 to 60 mph. TABLE 2.2 shows this time for two different cars, a sporty Corvette and a compact Sonic with a much more modest engine.

TABLE 2.2 Performance data for vehicles

| Vehicle | Time to go from 0 to 60 mph |
|---|---|
| 2016 Chevy Corvette | 3.6 s |
| 2016 Chevy Sonic | 9.0 s |

Let's look at motion diagrams for the Corvette and the Sonic in FIGURE 2.23. We can see two important facts about the motion. First, the lengths of the velocity vectors are increasing, showing that the speeds are increasing. Second, the velocity vectors for the Corvette are increasing in length more rapidly than those of the Sonic. The quantity we seek is one that measures how rapidly an object's velocity vectors change in length.

When we wanted to measure changes in position, the ratio $\Delta x/\Delta t$ was useful. This ratio, which we defined as the velocity, is the *rate of change of position.* Similarly, we

FIGURE 2.23 Motion diagrams for the Corvette and Sonic.

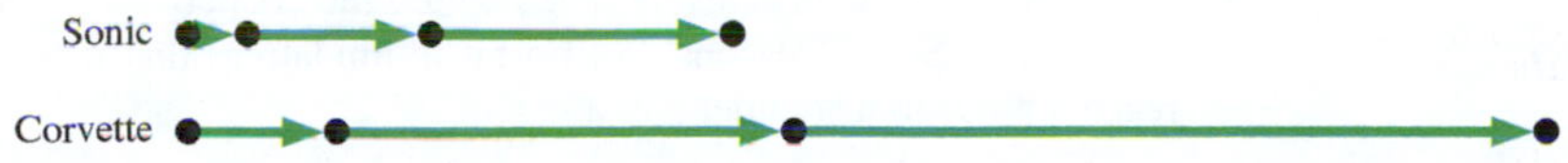

can measure how rapidly an object's velocity changes with the ratio $\Delta v_x / \Delta t$. Given our experience with velocity, we can say a couple of things about this new ratio:

- The ratio $\Delta v_x / \Delta t$ is the *rate of change of velocity.*
- The ratio $\Delta v_x / \Delta t$ is the *slope of a velocity-versus-time graph.*

We will define this ratio as the **acceleration,** for which we use the symbol $a_x$:

$$a_x = \frac{\Delta v_x}{\Delta t} \tag{2.8}$$

Definition of acceleration as the rate of change of velocity

**Cushion kinematics** When a car hits an obstacle head-on, the damage to the car and its occupants can be reduced by making the acceleration as small as possible. As we can see from Equation 2.8, acceleration can be reduced by making the *time* for a change in velocity as long as possible. This is the purpose of the yellow crash cushion barrels you may have seen in work zones on highways—to lengthen the time of a collision with a barrier.

Similarly, $a_y = \Delta v_y / \Delta t$ for vertical motion.

As an example, let's calculate the acceleration for the Corvette and the Sonic. For both, the initial velocity $(v_x)_i$ is zero and the final velocity $(v_x)_f$ is 60 mph. Thus the *change* in velocity is $\Delta v_x = 60$ mph. In m/s, our SI unit of velocity, $\Delta v_x = 27$ m/s.

Now we can use Equation 2.8 to compute the acceleration. Let's start with the Corvette, which speeds up to 27 m/s in $\Delta t = 3.6$ s:

$$a_{\text{Corvette}\,x} = \frac{\Delta v_x}{\Delta t} = \frac{27\text{ m/s}}{3.6\text{ s}} = 7.5\,\frac{\text{m/s}}{\text{s}}$$

Here's the meaning of this final figure: Every second, the Corvette's velocity changes by 7.5 m/s. In the first second of motion, the Corvette's velocity increases by 7.5 m/s; in the next second, it increases by another 7.5 m/s, and so on. Thus after 1 second, the velocity is 7.5 m/s; after 2 seconds, it is 15 m/s. We thus interpret the units as 7.5 meters per second, per second—7.5 (m/s)/s.

The Sonic's acceleration is

$$a_{\text{Sonic}\,x} = \frac{\Delta v_x}{\Delta t} = \frac{27\text{ m/s}}{9.0\text{ s}} = 3.0\,\frac{\text{m/s}}{\text{s}}$$

In each second, the Sonic changes its speed by 3.0 m/s. This is only 2/5 the acceleration of the Corvette! The reason the Corvette is capable of greater acceleration has to do with what causes the motion. We will explore the reasons for acceleration in Chapter 4. For now, we will simply note that the Corvette is capable of much greater acceleration, something you would have suspected.

**NOTE** ▸ It is customary to abbreviate the acceleration units (m/s)/s as $\text{m/s}^2$, which we say as "meters per second squared." For example, the Sonic has an acceleration of $3.0\text{ m/s}^2$. When you use this notation, keep in mind its *meaning* as "(meters per second) per second." ◂

### EXAMPLE 2.6 Animal acceleration BIO

Lions, like most predators, are capable of very rapid starts. From rest, a lion can sustain an acceleration of $9.5\text{ m/s}^2$ for up to one second. How much time does it take a lion to go from rest to a typical recreational runner's top speed of 10 mph?

**STRATEGIZE** The lion's speed increases by $9.5\text{ m/s}^2$ each second. Once we know the runner's speed in m/s, we will calculate the time it would take for the lion to reach that speed.

**PREPARE** We start by converting to SI units. The speed the lion must reach is

$$v_f = 10\text{ mph} \times \frac{0.45\text{ m/s}}{1.0\text{ mph}} = 4.5\text{ m/s}$$

The lion can accelerate at $9.5\text{ m/s}^2$, changing its speed by 9.5 m/s per second, for only 1.0 s—long enough to reach 9.5 m/s. It will take the lion less than 1.0 s to reach 4.5 m/s, so we can use $a_x = 9.5\text{ m/s}^2$ in our solution.

**SOLVE** We know the acceleration and the desired change in velocity, so we can rearrange Equation 2.8 to find the time:

$$\Delta t = \frac{\Delta v_x}{a_x} = \frac{4.5\text{ m/s}}{9.5\text{ m/s}^2} = 0.47\text{ s}$$

**ASSESS** The lion changes its speed by 9.5 meters per second in one second. So it's reasonable (if a bit intimidating) that it will reach 4.5 m/s in just under half a second.

## From Velocity to Acceleration

Let's use the values we have computed for acceleration to make a table of velocities for the Corvette and the Sonic we considered earlier. TABLE 2.3 uses the idea that the Sonic's velocity increases by 3.0 m/s every second while the Corvette's velocity increases by 7.5 m/s every second. The data in Table 2.3 are the basis for the velocity-versus-time graphs in FIGURE 2.24. As you can see, **an object undergoing constant acceleration has a straight-line velocity graph.**

TABLE 2.3 Velocity data for the Sonic and the Corvette

| Time (s) | Velocity of Sonic (m/s) | Velocity of Corvette (m/s) |
|---|---|---|
| 0 | 0 | 0 |
| 1 | 3.0 | 7.5 |
| 2 | 6.0 | 15.0 |
| 3 | 9.0 | 22.5 |
| 4 | 12.0 | 30.0 |

FIGURE 2.24 Velocity-versus-time graphs for the two cars.

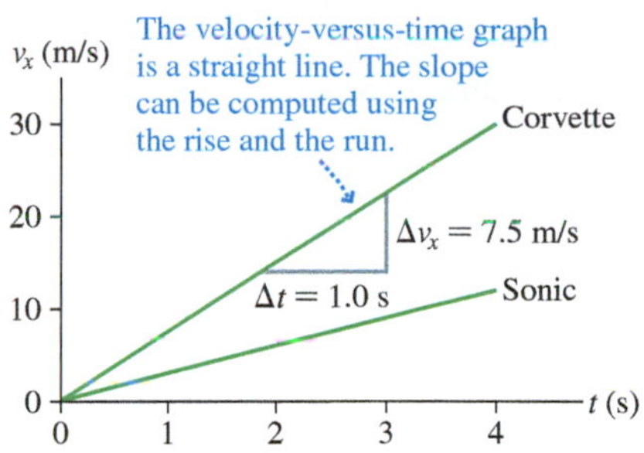


The slope of either of these lines—the rise over the run—is $\Delta v_x / \Delta t$. Comparing this with Equation 2.8, we see that the equation for the slope is the same as that for the acceleration. That is, **an object's acceleration is the slope of its velocity-versus-time graph:**

$$\text{acceleration } a_x \text{ at time } t = \text{slope of velocity graph at time } t \qquad (2.9)$$

The Sonic has a smaller acceleration, so its velocity graph has a smaller slope.

### CONCEPTUAL EXAMPLE 2.7 Analyzing a car's velocity-versus-time graph

FIGURE 2.25a is a graph of velocity versus time for a car. Sketch a graph of the car's acceleration versus time.

***REASON*** The graph can be divided into three sections:

- An initial segment, in which the velocity increases at a steady rate
- A middle segment, in which the velocity is constant
- A final segment, in which the velocity decreases at a steady rate

In each section, the acceleration is the slope of the velocity-versus-time graph. Thus the initial segment has constant, positive acceleration, the middle segment has zero acceleration, and the final segment has constant, *negative* acceleration. The acceleration graph appears in FIGURE 2.25b.

***ASSESS*** This process is analogous to finding a velocity graph from the slope of a position graph. It is important to understand that the zero acceleration in the middle segment does *not* mean that the velocity there is zero. In this segment the velocity is *constant*, which means that it is *not changing* and thus the car is not accelerating.

In the first and last segments, the velocity is changing, and so the car does have a nonzero acceleration. In the first segment, the acceleration is positive; in the last segment, it is negative. What does the *sign* of the acceleration tell us? We will address this issue in the next section.

FIGURE 2.25 Finding an acceleration graph from a velocity graph.

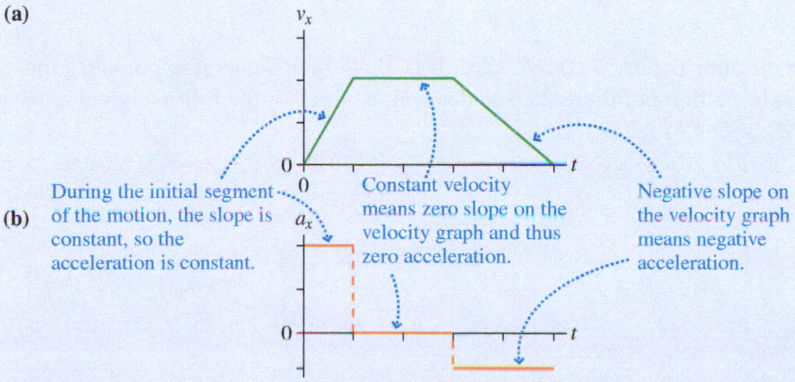

**STOP TO THINK 2.4** A particle moves with the velocity-versus-time graph shown here. At which labeled point is the magnitude of the acceleration the greatest?

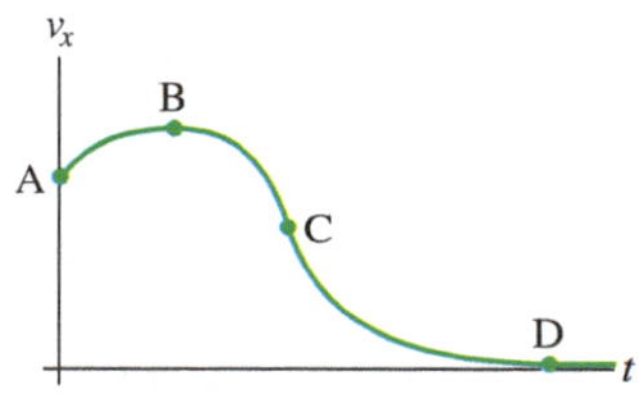


## The Sign of the Acceleration

A video to support a section's topic is embedded in the eText.
**Video** Motion Along a Straight Line

It's a natural tendency to think that a positive value of $a_x$ or $a_y$ describes an object that is speeding up while a negative value describes an object that is slowing down. Unfortunately, this simple interpretation *is not correct.*

Acceleration, like velocity, is a vector. Specifically, **the acceleration vector points in the *same* direction as the velocity vector for an object that is speeding up, and *opposite* to the velocity vector for an object that is slowing down.** Regardless of which way an object moves, an acceleration vector that points in the same direction as the velocity "pulls" the velocity vectors to make them longer and longer—speeding up—while an acceleration vector that points opposite to the velocity "pushes against" the velocity vectors to make them shorter and shorter—slowing down. You can see this in the four situations shown in FIGURE 2.26.

Just as we did with velocity, we can simplify our analysis of one-dimensional motion by using an ordinary variable $a_x$, which can be positive or negative, to represent the one-dimensional acceleration. The sign convention for $a_x$ (and $a_y$) is exactly

KEY CONCEPT **FIGURE 2.26** Determining the sign of the acceleration.

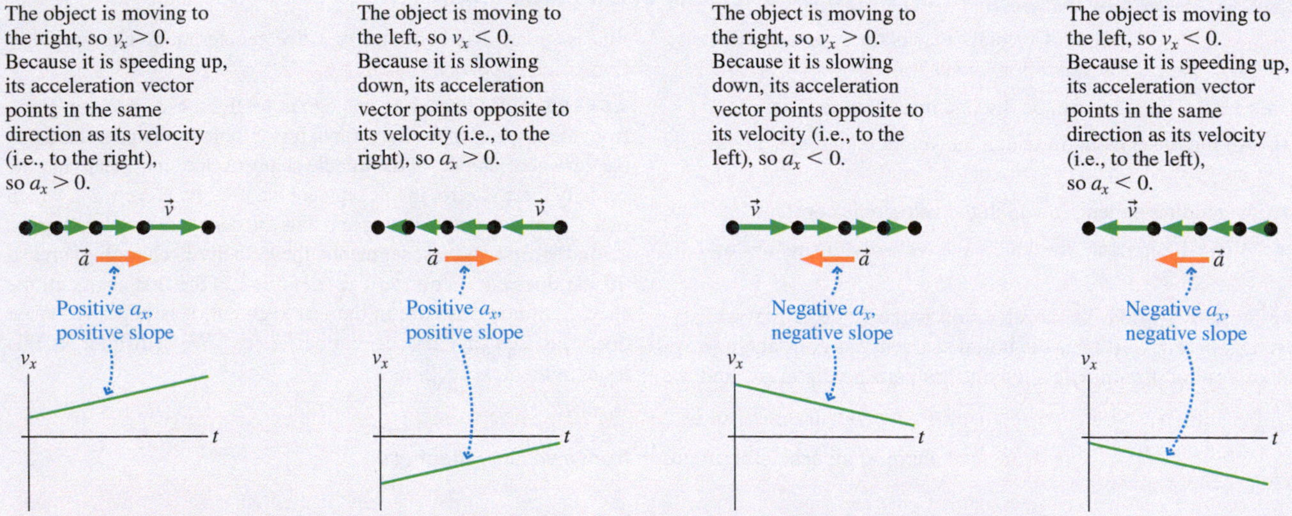


**STOP TO THINK 2.5** An elevator is moving downward. It is slowing down as it approaches the ground floor. Adapt the information in Figure 2.26 to determine which of the following velocity graphs best represents the motion of the elevator.

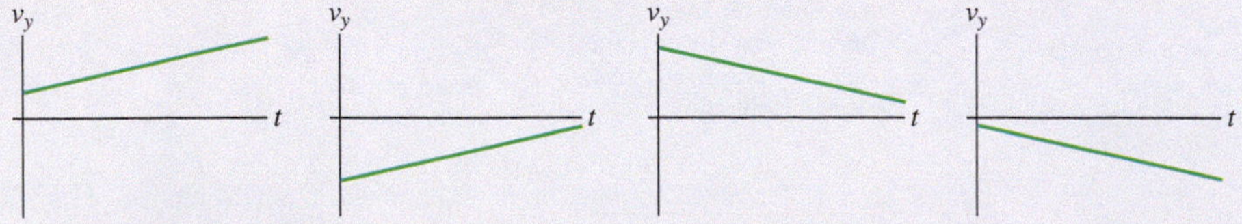


A. B. C. D.

the same as the sign convention for velocity that was shown in Figure 2.7: $a_x$ (or $a_y$) is positive when the acceleration vector points to the right (or up), negative when the acceleration vector points to the left (or down).

Notice that the first two situations in Figure 2.26, where the acceleration vectors point to the right, both have positive values of the acceleration $a_x$ even though one shows an object speeding up and the other an object slowing down. **The sign of $a_x$ is based on the direction the acceleration vector points,** not on whether the object is speeding up or slowing down.

Figure 2.26 illustrates two more ideas: First, our convention for the sign of the acceleration is consistent with what you just saw about an object's acceleration being the slope of its velocity graph. Second, **an object is speeding up if $v_x$ and $a_x$ have the same sign, and slowing down if they have opposite signs.**

## 2.5 Motion with Constant Acceleration

For uniform motion—motion with constant velocity—we found in Equation 2.3 a simple relationship between position and time. It's no surprise that there are also simple relationships that connect the various kinematic variables in constant-acceleration motion. We will start with a concrete example, the launch of a Saturn V rocket like the one that carried the Apollo astronauts to the moon in the 1960s and 1970s. FIGURE 2.27 shows one frame from a video of a rocket lifting off the launch pad. The red dots show the positions of the top of the rocket at equally spaced intervals of time in earlier frames of the video. This is a motion diagram for the rocket, and we can see that the velocity is increasing. The graph of velocity versus time in FIGURE 2.28 shows that the velocity is increasing at a fairly constant rate. We can approximate the rocket's motion as having constant acceleration.

FIGURE 2.27 The red dots show the positions of the top of the Saturn V rocket at equally spaced intervals of time during liftoff.

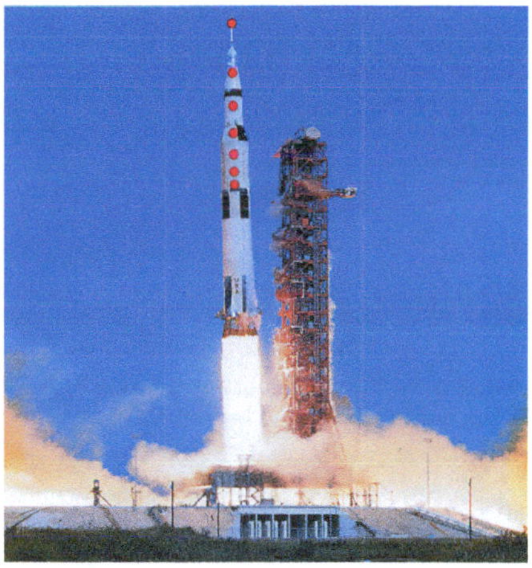

We can use the slope of the graph in Figure 2.28 to determine the acceleration of the rocket:

$$a_y = \frac{\Delta v_y}{\Delta t} = \frac{27\ \text{m/s}}{1.5\ \text{s}} = 18\ \text{m/s}^2$$

This acceleration is more than double the acceleration of the Corvette we discussed earlier, and it goes on for a long time—the first phase of the launch lasts over 2 minutes! How fast is the rocket moving at the end of this acceleration, and how far has it traveled? To answer questions like these, we first need to work out some basic kinematic equations for motion with constant acceleration.

FIGURE 2.28 A graph of the rocket's velocity versus time.

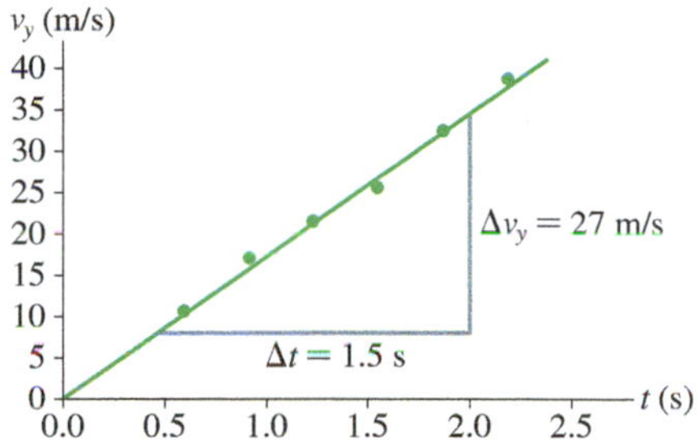


### Constant-Acceleration Equations

Consider an object whose acceleration $a_x$ remains constant during the time interval $\Delta t = t_f - t_i$. At the beginning of this interval, the object has initial velocity $(v_x)_i$ and initial position $x_i$. Note that $t_i$ is often zero, but it need not be. FIGURE 2.29a shows the acceleration-versus-time graph. It is a horizontal line between $t_i$ and $t_f$, indicating a *constant* acceleration.

FIGURE 2.29 Acceleration and velocity graphs for motion with constant acceleration.

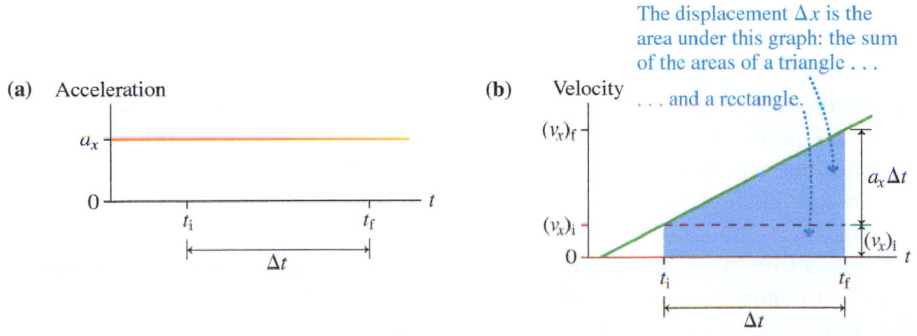

The object's velocity is changing because the object is accelerating. We can use the acceleration to find $(v_x)_f$ at a later time $t_f$. We defined acceleration as

$$a_x = \frac{\Delta v_x}{\Delta t} = \frac{(v_x)_f - (v_x)_i}{\Delta t} \tag{2.10}$$

which is rearranged to give

$$(v_x)_f = (v_x)_i + a_x \Delta t \tag{2.11}$$

Velocity equation for an object with constant acceleration

**NOTE** ► We have expressed this equation for motion along the $x$-axis, but it is a general result that will apply to any axis. ◄

The velocity-versus-time graph for this constant-acceleration motion, shown in FIGURE 2.29b, is a straight line with value $(v_x)_i$ at time $t_i$ and with slope $a_x$.

We would also like to know the object's position $x_f$ at time $t_f$. As you learned earlier, the displacement $\Delta x$ during a time interval $\Delta t$ is the area under the velocity-versus-time graph. This area is shown shaded in Figure 2.29b. The shaded area can be divided into a rectangle of area $(v_x)_i \Delta t$ and a triangle of area $\frac{1}{2}(a_x \Delta t)(\Delta t) = \frac{1}{2}a_x(\Delta t)^2$. Adding these gives

$$x_f = x_i + (v_x)_i \Delta t + \tfrac{1}{2}a_x(\Delta t)^2 \tag{2.12}$$

Position equation for an object with constant acceleration

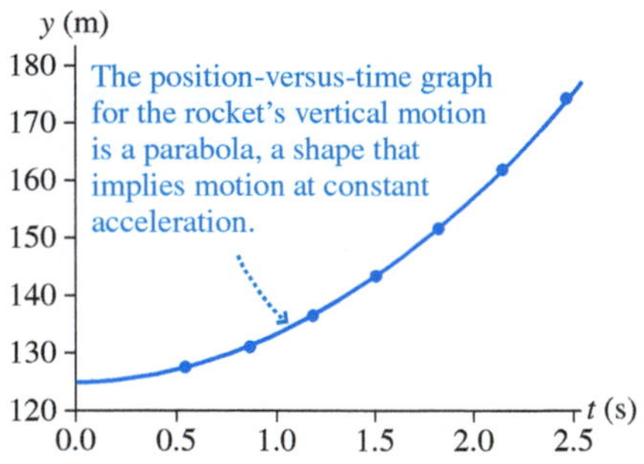


FIGURE 2.30 Position-versus-time graph for the Saturn V rocket launch.

where $\Delta t = t_f - t_i$ is the elapsed time. The fact that the time interval $\Delta t$ appears in the equation as $(\Delta t)^2$ causes the position-versus-time graph for constant-acceleration motion to have a parabolic shape. For the rocket launch of Figure 2.27, a graph of the position of the top of the rocket versus time appears as in FIGURE 2.30.

Equations 2.11 and 2.12 are two of the basic kinematic equations for motion with constant acceleration. They allow us to predict an object's position and velocity at a future instant of time. We need one more equation to complete our set, a direct relationship between displacement and velocity. To derive this relationship, we first use Equation 2.11 to write $\Delta t = ((v_x)_f - (v_x)_i)/a_x$. We can substitute this into Equation 2.12 to obtain

$$(v_x)_f^{\,2} = (v_x)_i^{\,2} + 2a_x \Delta x \tag{2.13}$$

Relating velocity and displacement for constant-acceleration motion

In Equation 2.13, $\Delta x = x_f - x_i$ is the *displacement* (not the distance!), so it can be positive or negative. Notice that Equation 2.13 does not require that we know the time interval $\Delta t$. This is an important equation in problems where we're not given information about times.

At this point, it's worthwhile to summarize the relationships among kinematic variables that we've seen. This will help you solve problems by gathering together the most important information that you'll use in your solutions. But, more important, gathering this information together allows you to compare graphs, equations, and details from different parts of the chapter in one place. This will help you make important connections. The emphasis is on *synthesis*—hence the title of this box. You'll find other such synthesis boxes in most chapters.

## SYNTHESIS 2.1 Describing motion in one dimension

We describe motion in terms of position, velocity, and acceleration.

For **all motion:**

Velocity is the rate of change of position, in m/s. $v_x = \dfrac{\Delta x}{\Delta t}$

Acceleration is the rate of change of velocity, in m/s$^2$. $a_x = \dfrac{\Delta v_x}{\Delta t}$

For **uniform motion:**
- acceleration is zero
- velocity is constant
- position changes steadily

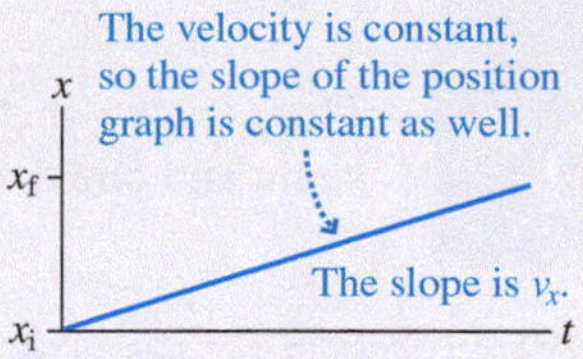


Final and initial position (m)

$$x_f = x_i + v_x \Delta t \qquad (2.4)$$

Velocity (m/s) Time interval (s)

For **motion with constant acceleration:**

- acceleration is steady; it does not change

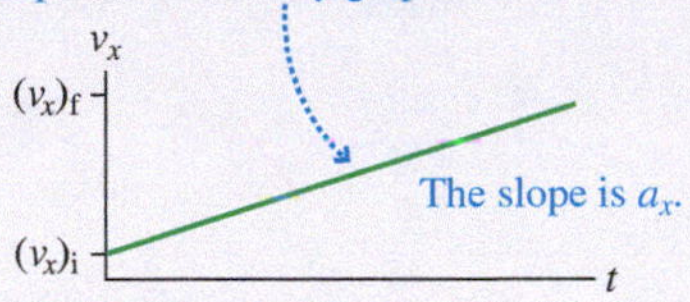


- velocity changes steadily

Final and initial velocity (m/s)

$$(v_x)_f = (v_x)_i + a_x \Delta t \qquad (2.11)$$

Acceleration (m/s$^2$) Time interval (s)

- the position changes as the square of the time interval

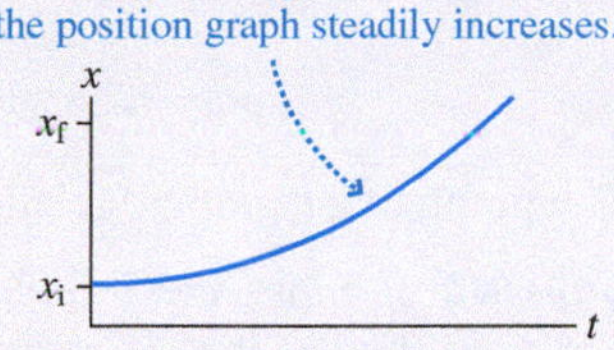


Final and initial position (m) Time interval (s)

$$x_f = x_i + (v_x)_i \,\Delta t + \tfrac{1}{2} a_x (\Delta t)^2 \qquad (2.12)$$

Initial velocity (m/s) Acceleration (m/s$^2$)

- we can also express the change in velocity in terms of **displacement, not time**

This gives us a third equation, which is useful for many kinematics problems.

Final and initial velocity (m/s)

$$(v_x)_f^2 = (v_x)_i^2 + 2a_x \Delta x \qquad (2.13)$$

Acceleration (m/s$^2$) Change in position (m)

## EXAMPLE 2.8 Coming to a stop in a car

As you drive in your car at 15 m/s (just a bit under 35 mph), you see a child's ball roll into the street ahead of you. You hit the brakes and stop as quickly as you can. In this case, you come to rest in 1.5 s. How far does your car travel as you brake to a stop?

**STRATEGIZE** The problem states that your car begins to slow down when you hit the brakes; we'll model this as constant-acceleration motion. We know the initial and final speeds, and we want to find the distance traveled. These observations suggest that we use Equation 2.12 of Synthesis 2.1.

**PREPARE** The problem gives us a description of motion in words. To help us visualize the situation, FIGURE 2.31 illustrates the key features of the motion with a motion diagram and a velocity graph. The graph is based on the car slowing from 15 m/s to 0 m/s in 1.5 s.

FIGURE 2.31 Motion diagram and velocity graph for a car coming to a stop.

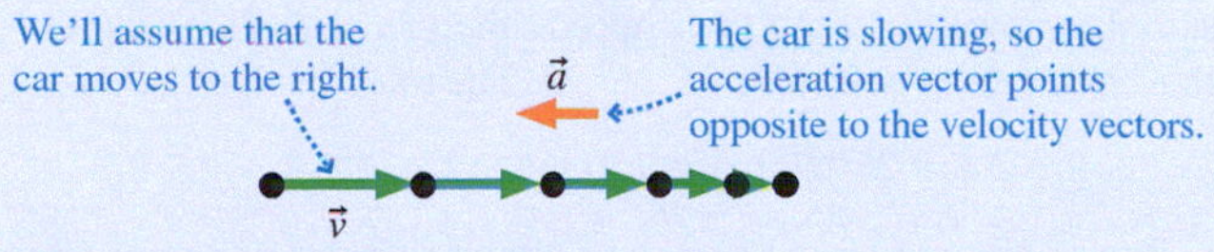


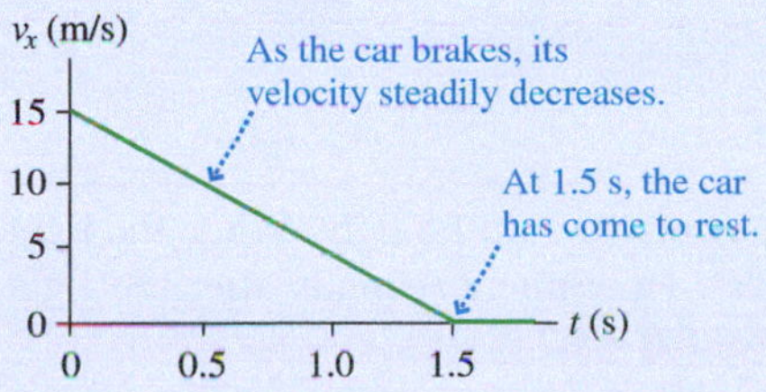


**SOLVE** We've assumed that your car is moving to the right, so its initial velocity is $(v_x)_i = +15\text{ m/s}$. After you come to rest, your final velocity is $(v_x)_f = 0\text{ m/s}$. We use the definition of acceleration from Synthesis 2.1:

$$a_x = \frac{\Delta v_x}{\Delta t} = \frac{(v_x)_f - (v_x)_i}{\Delta t} = \frac{0\text{ m/s} - 15\text{ m/s}}{1.5\text{ s}} = -10\text{ m/s}^2$$

Now that we know the acceleration, we can compute the distance that the car moves as it comes to rest using Equation 2.12:

$$x_f - x_i = (v_x)_i \Delta t + \tfrac{1}{2} a_x (\Delta t)^2$$
$$= (15\text{ m/s})(1.5\text{ s}) + \tfrac{1}{2}(-10\text{ m/s}^2)(1.5\text{ s})^2 = 11\text{ m}$$

**ASSESS** 11 m is a little over 35 feet. That's a reasonable distance for a quick stop while traveling at about 35 mph.

We found that the acceleration $a_x$ is negative. This makes sense from two perspectives. First, as we learned in Figure 2.26, an object moving to the right and slowing down has a negative acceleration. Second, the slope of the velocity graph in Figure 2.31 is negative, again indicating a negative acceleration.

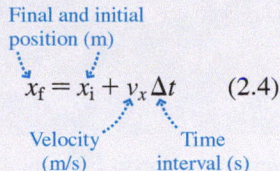

**Getting up to speed** BIO A bird must have a minimum speed to fly. Generally, the larger the bird, the faster the takeoff speed. Small birds can get moving fast enough to fly with a vigorous jump, but larger birds may need a running start. This swan must accelerate for a long distance in order to achieve the high speed it needs to fly, so it makes a frenzied dash across the frozen surface of a pond. Swans require a long, clear stretch of water or land to become airborne.

As we've noted, for motion at constant acceleration, the position changes as the square of the time interval. If the initial velocity $(v_x)_i$ is zero, Equation 2.12 from Synthesis 2.1 can be written as

$$x_f - x_i = \Delta x = \frac{1}{2}a_x(\Delta t)^2$$

This is a new mathematical relationship—a *quadratic relationship*—that we will see again and one that we can use as the basis of reasoning to solve problems.

## Quadratic relationships

Two quantities are said to have a **quadratic relationship** if $y$ is proportional to the square of $x$. We write the mathematical relationship as

$$y = Ax^2$$

$y$ is proportional to $x^2$

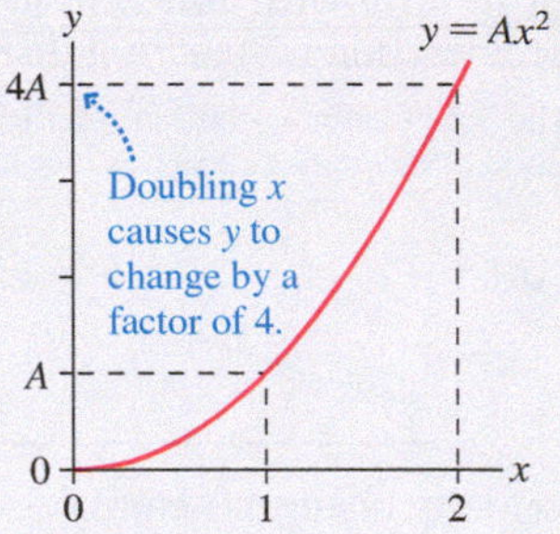


The graph of a quadratic relationship is a parabola.

**SCALING** If $x$ has the initial value $x_1$, then $y$ has the initial value $y_1 = A(x_1)^2$. Changing $x$ from $x_1$ to $x_2$ changes $y$ from $y_1$ to $y_2$. The ratio of $y_2$ to $y_1$ is

$$\frac{y_2}{y_1} = \frac{A(x_2)^2}{A(x_1)^2} = \left(\frac{x_2}{x_1}\right)^2$$

The ratio of $y_2$ to $y_1$ is the square of the ratio of $x_2$ to $x_1$. If $y$ is a quadratic function of $x$, a change in $x$ by some factor changes $y$ by the square of that factor:

- If you increase $x$ by a factor of 2, you increase $y$ by a factor of $2^2 = 4$.
- If you decrease $x$ by a factor of 3, you decrease $y$ by a factor of $3^2 = 9$.

Generally, we can say that:

**Changing $x$ by a factor of $c$ changes $y$ by a factor of $c^2$.**

Exercise 19

### EXAMPLE 2.9 Finding the displacement of a drag racer

A drag racer, starting from rest, travels 6.0 m in 1.0 s. Suppose the car continues this acceleration for an additional 4.0 s. How far from the starting line will the car be?

**STRATEGIZE** We assume that the acceleration is constant. Because the initial position and velocity are zero, the displacement will then scale as the square of the time; we can then use ratio reasoning to solve the problem.

**PREPARE** After 1.0 s, the car has traveled 6.0 m; after another 4.0 s, a total of 5.0 s will have elapsed.

**SOLVE** The initial elapsed time was 1.0 s, so the elapsed time increases by a factor of 5. The displacement thus increases by a factor of $5^2$, or 25. The total displacement is

$$\Delta x = 25(6.0\text{ m}) = 150\text{ m}$$

**ASSESS** This is a big distance in a short time, but drag racing is a fast sport, so our answer makes sense.

**STOP TO THINK 2.6** A cyclist is at rest at a traffic light. When the light turns green, he begins accelerating at 1.2 m/s$^2$. How many seconds after the light turns green does he reach his cruising speed of 6.0 m/s?

A. 1.0 s B. 2.0 s C. 3.0 s D. 4.0 s E. 5.0 s

## 2.6 Solving One-Dimensional Motion Problems

The big challenge when solving a physics problem is to translate the words into symbols that can be manipulated, calculated, and graphed. This translation from words to symbols is the heart of problem solving in physics. Ambiguous words and phrases must be clarified, the imprecise must be made precise, and you must arrive at an understanding of exactly what the question is asking.

**PROBLEM-SOLVING APPROACH**

The first step in solving a seemingly complicated problem is to break it down into a series of smaller steps. In worked examples in the text, we use a problem-solving approach that consists of four steps: *strategize, prepare, solve,* and *assess*. Each of these steps has important elements that you should follow when you solve problems on your own.

***STRATEGIZE*** The Strategize step of the solution is where you address the *big-picture* questions about the problem. Here, you take a step back from the details of the problem to ask:

- **What kind of problem is this?** From reading the problem statement, try to categorize the problem in terms of what you've learned in the chapter. If, for instance, the problem refers to a bicyclist riding at a constant 7.0 m/s, this suggests the problem is about uniform motion.
- **What's the correct general approach?** What principles, strategies, and tactics that you've learned are relevant in solving this problem? For example, if you're given a position-versus-time graph and are asked to find the velocity, the principle that the velocity is related to the slope of the position graph is likely to be important.
- **What should the answer look like?** Is a numerical answer asked for? Do you need a graph or a sketch?

***PREPARE*** The Prepare step of a solution is where you identify important elements of the problem and collect information. It's tempting to jump right to the Solve step, but a skilled problem solver spends the most time on preparation, which includes:

- **Drawing a picture.** This is often the most important part of a problem. The picture lets you model the problem and identify the important elements. As you add information to your picture, the outline of the solution will take shape. For the problems in this chapter, a picture could be a motion diagram or a graph—or perhaps both.
- **Collecting necessary information.** The problem's statement may give you some values of variables. Other information may be implied, or looked up in a table, or estimated or measured.
- **Doing preliminary calculations.** Some calculations, such as unit conversions, are best done in advance.

***SOLVE*** The Solve step of a solution is where you actually do the mathematics or reasoning necessary to arrive at the answer. This is the part of the problem-solving approach that you likely think of as "solving problems." The Strategize and Prepare steps help you be certain you understand the problem before you start putting numbers in equations.

***ASSESS*** The Assess step of your solution is very important. Once you have an answer, you should check to see whether it makes sense. Ask yourself:

- **Does my solution answer the question that was asked?** Make sure you have addressed all parts of the question and clearly written down your solutions.
- **Does my answer have the correct units and number of significant figures?**
- **Does the value I computed make physical sense?** In this book all calculations use physically reasonable numbers. If your answer seems unreasonable, go back and check your work.
- **Can I estimate what the answer should be to check my solution?**
- **Does my final solution make sense in the context of the material I am learning?**

### The Pictorial Representation

Many physics problems, including one-dimensional motion problems, have several variables and other pieces of information to keep track of. The best way to tackle such problems is to draw a picture, as we noted when we introduced a general problem-solving approach. But what kind of picture should you draw?

In this section, we will begin to draw **pictorial representations** as an aid to solving problems. A pictorial representation shows all of the important details that we need to keep track of and will be very important in solving motion problems.

**Dinner at a distance** BIO A chameleon's tongue is a powerful tool for catching prey. Certain species can extend the tongue to a distance of over 1 ft in less than 0.1 s! A study of the kinematics of the motion of the chameleon tongue reveals that the tongue has a period of rapid acceleration followed by a period of constant velocity. This knowledge is a very valuable clue in the analysis of the evolutionary relationships between chameleons and other animals.

### TACTICS BOX 2.2 Drawing a pictorial representation

❶ **Sketch the situation.** Not just any sketch: Show the object at the *beginning* of the motion, at the *end,* and at any point where the character of the motion changes. Very simple drawings are adequate.

❷ **Establish a coordinate system.** Select your axes and origin to match the motion.

❸ **Define symbols.** Use the sketch to define symbols representing quantities such as position, velocity, acceleration, and time. *Every* variable used later in the mathematical solution should be defined on the sketch.

We will generally combine the pictorial representation with a **list of values.** In this list, you should:

❹ *List the known information.* Make a table of the quantities whose values you can determine from the problem statement or that you can find quickly with simple geometry or unit conversions.

❺ *Identify the desired unknowns.* What quantity or quantities will allow you to answer the question?

Exercise 21

### EXAMPLE 2.10 Drawing a pictorial representation

Complete a pictorial representation and a list of values for the following problem: A rocket sled accelerates at 50 m/s$^2$ for 5 s. What are the total distance traveled and the final velocity?

***STRATEGIZE*** We'll prepare the pictorial representation and list of values according to the steps of Tactics Box 2.2.

***PREPARE*** FIGURE 2.32a shows a pictorial representation as drawn by an artist in the style of the figures in this book. This is certainly neater and more artistic than the sketches you will make when solving problems yourself! FIGURE 2.32b shows a sketch like one you might actually draw. It's less formal, but it contains all of the important information you need to solve the problem. The circled numbers in the sketch correspond to the steps in Tactics Box 2.2.

NOTE ▶ Throughout this book we will illustrate select examples with actual hand-drawn figures so that you have them to refer to as you work on your own pictures for homework and practice. ◀

Let's look at how these pictures were constructed. The motion has a clear beginning and end; these are the points in the motion that we've sketched. A coordinate system has been chosen with the origin at the starting point. The quantities $x$, $v_x$, and $t$ are needed at both points, so these have been defined on the sketch and distinguished by subscripts. The acceleration is associated with an interval between these points. Values for two of these quantities are given in the problem statement. Others, such as $x_i = 0$ m and $t_i = 0$ s, are inferred from our choice of coordinate system. The value $(v_x)_i = 0$ m/s is part of our *interpretation* of the problem. Finally, we identify $x_f$ and $(v_x)_f$ as the quantities that will answer the question. We now understand quite a bit about the problem and would be ready to start a quantitative analysis.

***ASSESS*** We didn't *solve* the problem; that was not our purpose. Constructing a pictorial representation and a list of values is part of a systematic approach to interpreting a problem and getting ready for a mathematical solution.

FIGURE 2.32 Constructing a pictorial representation and a list of values.

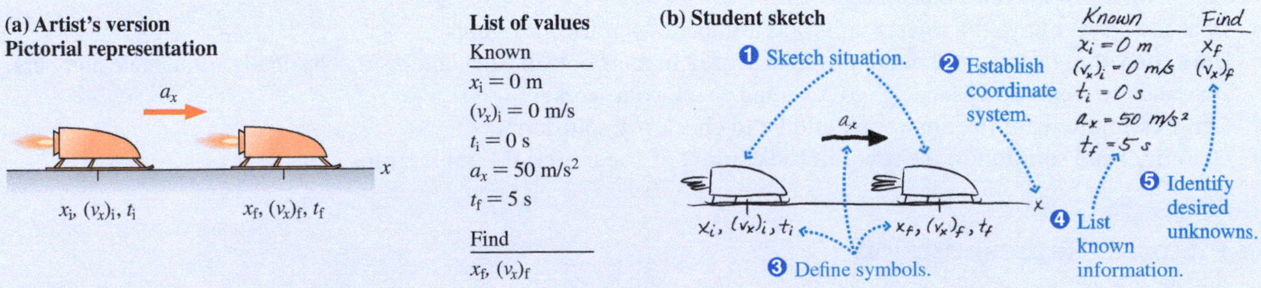


## The Visual Overview

Video Motion with Constant Acceleration

The pictorial representation and the list of values are a very good complement to the motion diagram and other ways of looking at a problem that we have seen. As we translate a problem into a form we can solve, we will combine these elements into what we will term a **visual overview.** The visual overview will consist of some or all of the following elements:

- A *motion diagram.* A good approach for solving a motion problem is to start by drawing a motion diagram.
- A *pictorial representation,* as defined in Tactics Box 2.2.
- A *list of values,* also described in Tactics Box 2.2. This list should sum up all of the important values in the problem.
- A *graphical representation.* For motion problems, it is often quite useful to include a graph of position and/or velocity.

Future chapters will add other elements to this visual overview of the physics.

**EXAMPLE 2.11** **Kinematics of a rocket launch**

A Saturn V rocket is launched straight up with a constant acceleration of 18 m/s$^2$. After 150 s, how fast is the rocket moving and how far has it traveled?

***STRATEGIZE*** We are given the acceleration and the time interval, suggesting that this is a constant-acceleration problem. We will find the velocity from Equation 2.11 and the position from Equation 2.12.

***PREPARE*** FIGURE 2.33 shows a visual overview of the rocket launch that includes a motion diagram, a pictorial representation, and a list of values. The visual overview shows the whole problem in a nutshell. The motion diagram illustrates the motion of the rocket. The pictorial representation (produced according to Tactics Box 2.2) shows the axis, identifies the important points of the motion, and defines the variables. Finally, we include a list of values that gives the known and unknown quantities (again according to Tactics Box 2.2). In the visual overview we have taken the statement of the problem in words and made it much more precise. The overview contains everything we need to know about the problem.

***SOLVE*** Our first task is to find the final velocity. Our list of values includes the initial velocity, the acceleration, and the time interval, so we can use Equation 2.11 of Synthesis 2.1 to find the final velocity:

$$(v_y)_f = (v_y)_i + a_y\,\Delta t = 0\text{ m/s} + (18\text{ m/s}^2)(150\text{ s})$$
$$= 2700\text{ m/s}$$

The distance traveled is found using Equation 2.12 of Synthesis 2.1:

$$y_f = y_i + (v_y)_i\,\Delta t + \tfrac{1}{2}a_y(\Delta t)^2$$
$$= 0\text{ m} + (0\text{ m/s})(150\text{ s}) + \tfrac{1}{2}(18\text{ m/s}^2)(150\text{ s})^2$$
$$= 2.0 \times 10^5\text{ m} = 200\text{ km}$$

***ASSESS*** The acceleration is very large, and it goes on for a long time, so the large final velocity and large distance traveled seem reasonable.

FIGURE 2.33 Visual overview of the rocket launch.

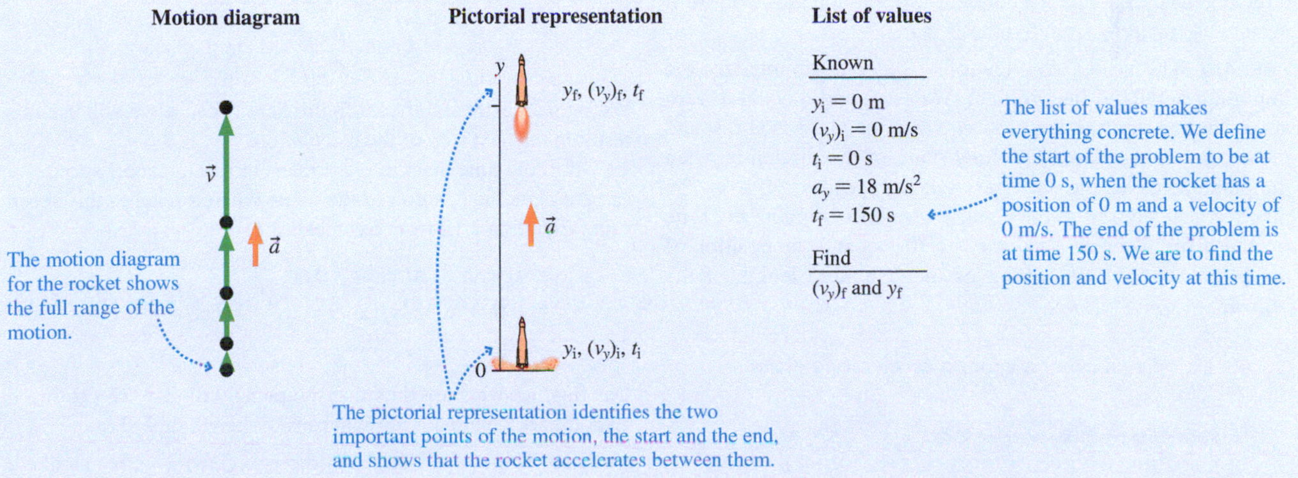


## Problem-Solving Approach for Motion with Constant Acceleration

Earlier in this section, we introduced a general problem-solving approach. In this and future chapters we will adapt this general approach to specific types of problems.

**PROBLEM-SOLVING APPROACH 2.1** **Motion with constant acceleration**

Problems involving constant acceleration—speeding up, slowing down, vertical motion, horizontal motion—can all be treated with the same problem-solving approach.

**STRATEGIZE** Identify the problem as one involving constant-acceleration motion: Look for statements that give the acceleration or indicate that the speed or velocity of an object is changing. Free-fall problems, discussed in the next section, are always constant-acceleration problems. Solve constant-acceleration problems using the ideas in Synthesis 2.1.

**PREPARE** Draw a visual overview of the problem. This should include a motion diagram, a pictorial representation, and a list of values; a graphical representation may be useful for certain problems.

**SOLVE** The mathematical solution is based on the three constant-acceleration equations in Synthesis 2.1.

- Though the equations are phrased in terms of the variable $x$, it's customary to use $y$ for motion in the vertical direction.
- Use the equation that best matches what you know and what you need to find. For example, if you know acceleration and time and are looking for a change in velocity, the first equation is the best one to use.
- Uniform motion with constant velocity has $a = 0$.

**ASSESS** Is your result believable? Does it have proper units? Does it make sense?

Exercise 25

**EXAMPLE 2.12** **Calculating the minimum length of a runway**

A fully loaded Boeing 747 with all engines at full thrust accelerates at $2.6\ \text{m/s}^2$. Its minimum takeoff speed is 70 m/s. How much time will the plane take to reach its takeoff speed? What minimum length of runway does the plane require for takeoff?

**STRATEGIZE** The acceleration of the plane is given, which directly tells us that this is a constant-acceleration problem. We'll need to use material from Synthesis 2.1.

**PREPARE** The visual overview of FIGURE 2.34 summarizes the important details of the problem. We set $x_i$ and $t_i$ equal to zero at the starting point of the motion, when the plane is at rest and the acceleration begins. The final point of the motion is when the plane achieves the necessary takeoff speed of 70 m/s. The plane is accelerating to the right, so we will compute the time for the plane to reach a velocity of 70 m/s and the position of the plane at this time, giving us the minimum length of the runway.

FIGURE 2.34 Visual overview for an accelerating plane.

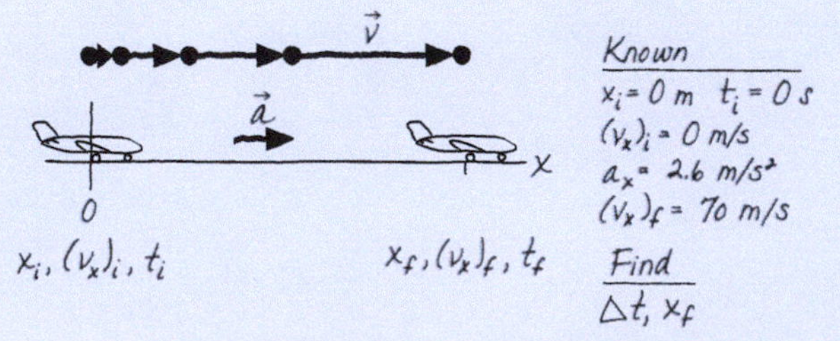


**SOLVE** First we solve for the time required for the plane to reach takeoff speed. We can use Equation 2.11 from Synthesis 2.1 to compute this time:

$$(v_x)_f = (v_x)_i + a_x\,\Delta t$$

$$70\ \text{m/s} = 0\ \text{m/s} + (2.6\ \text{m/s}^2)\,\Delta t$$

$$\Delta t = \frac{70\ \text{m/s}}{2.6\ \text{m/s}^2} = 26.9\ \text{s}$$

We keep an extra significant figure here because we will use this result in the next step of the calculation.

Given the time that the plane takes to reach takeoff speed, we can compute the position of the plane when it reaches this speed using Equation 2.12 from Synthesis 2.1:

$$\begin{aligned} x_f &= x_i + (v_x)_i\,\Delta t + \tfrac{1}{2}a_x(\Delta t)^2 \\ &= 0\ \text{m} + (0\ \text{m/s})(26.9\ \text{s}) + \tfrac{1}{2}(2.6\ \text{m/s}^2)(26.9\ \text{s})^2 \\ &= 940\ \text{m} \end{aligned}$$

Our final answers are thus that the plane will take 27 s to reach takeoff speed, with a minimum runway length of 940 m.

**ASSESS** Think about the last time you flew; 27 s seems like a reasonable time for a plane to accelerate on takeoff. Actual runway lengths at major airports are 3000 m or more, a few times greater than the minimum length, because they have to allow for emergency stops during an aborted takeoff. (If we had calculated a distance far greater than 3000 m, we would know we had done something wrong!)

**EXAMPLE 2.13** **Finding the braking distance of a car on the highway**

A car is traveling at a speed of 30 m/s, a typical highway speed, on wet pavement. The driver sees an obstacle ahead and decides to stop. From this instant, it takes him 0.75 s to begin applying the brakes. Once the brakes are applied, the car experiences an acceleration of $-6.0\ \text{m/s}^2$. How far does the car travel from the instant the driver notices the obstacle until it stops?

**STRATEGIZE** The wording of this problem makes it clear that, while it is braking, the car is experiencing constant acceleration. We'll once more use the information from Synthesis 2.1.

**PREPARE** This problem is more involved than previous problems we have solved, so we will take more care with the visual overview in FIGURE 2.35. In addition to a motion diagram and a pictorial representation, we include a graphical representation. Notice that there are two different phases of the motion: a constant-velocity phase before braking begins, and a steady slowing down once the brakes are applied. We will need to do two different calculations, one for each phase. Consequently, we use numerical subscripts rather than the simple i and f.

**SOLVE** From $t_1$ to $t_2$ the velocity stays constant at 30 m/s. This is uniform motion, so we compute the position at time $t_2$ using Equation 2.4 from Synthesis 2.1:

$$x_2 = x_1 + (v_x)_1(t_2 - t_1) = 0\ \text{m} + (30\ \text{m/s})(0.75\ \text{s})$$
$$= 22.5\ \text{m}$$

Starting at $t_2$, the velocity begins to decrease at a steady $-6.0\ \text{m/s}^2$ until the car comes to rest at $t_3$. We can compute this time interval using Equation 2.11 from Synthesis 2.1, $(v_x)_3 = (v_x)_2 + a_x\,\Delta t$:

$$\Delta t = t_3 - t_2 = \frac{(v_x)_3 - (v_x)_2}{a_x} = \frac{0\ \text{m/s} - 30\ \text{m/s}}{-6.0\ \text{m/s}^2} = 5.0\ \text{s}$$

We can compute the position at time $t_3$ using Equation 2.12 from Synthesis 2.1. We take point 2 as the initial point and point 3 as the final point for this phase of the motion and use $\Delta t = t_3 - t_2$:

$$x_3 = x_2 + (v_x)_2\,\Delta t + \tfrac{1}{2}a_x(\Delta t)^2$$
$$= 22.5\ \text{m} + (30\ \text{m/s})(5.0\ \text{s}) + \tfrac{1}{2}(-6.0\ \text{m/s}^2)(5.0\ \text{s})^2$$
$$= 98\ \text{m}$$

$x_3$ is the position of the car at the end of the problem—and so the car travels 98 m before coming to rest.

**ASSESS** The numbers for the reaction time and the acceleration on wet pavement are reasonable ones for an alert driver in a car with good tires. The final distance is quite large—more than the length of a football field.

FIGURE 2.35 Visual overview for a car braking to a stop.

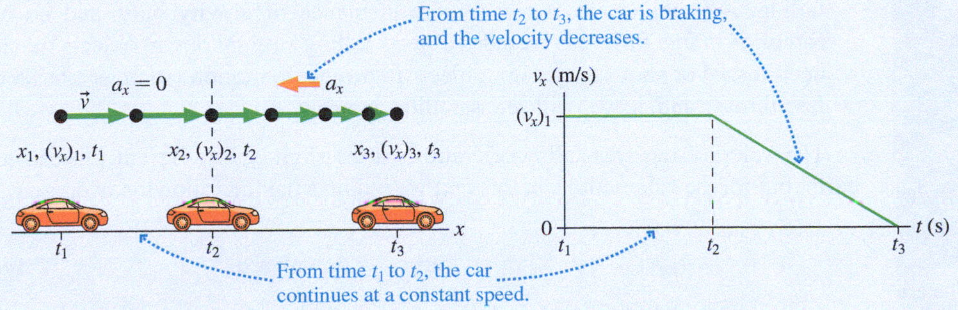


Known
$t_1 = 0$ s
$x_1 = 0$ m
$(v_x)_1 = 30$ m/s
$t_2 = 0.75$ s
$(v_x)_2 = 30$ m/s
$(v_x)_3 = 0$ m/s
Between $t_2$ and $t_3$, $a_x = -6.0\ \text{m/s}^2$

Find
$x_3$

## 2.7 Free Fall

If you drop a hammer and a feather, you know what will happen. The hammer quickly strikes the ground, and the feather drifts slowly downward and lands some time later. But if you do this experiment on the moon, the result is strikingly different: Both the hammer and the feather experience the exact same acceleration, undergo the exact same motion, and strike the ground at the same time.

The moon lacks an atmosphere, and so objects falling to its surface experience no air resistance. They are acted upon by only one force—gravity. If an object moves under the influence of gravity only, and no other forces, we call the resulting motion **free fall.** Many experiments have shown that **all objects in free fall, regardless of their mass, have the same acceleration.** Thus, if you drop two objects and they are both in free fall, they hit the ground at the same time.

On the earth, air resistance is a factor. But when you drop a heavy object like a hammer, air resistance can be ignored, so we make only a slight error in treating the hammer *as if* it were in free fall. Motion with air resistance is a problem we will

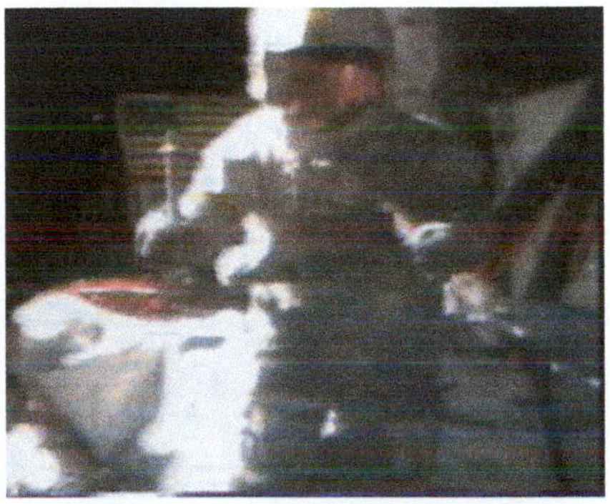

**Free-falling feather** Apollo 15 lunar astronaut David Scott performed a classic experiment on the moon, simultaneously dropping a hammer and a feather from the same height. Both hit the ground at the exact same time—something that would not happen in the atmosphere of the earth!

study in Chapter 5. Until then, we will restrict our attention to situations in which air resistance can be ignored, and we will assume that falling objects are in free fall.

FIGURE 2.36a shows the motion diagram for an object that was released from rest and falls freely. Since the acceleration is the same for all objects, the diagram and graph would be the same for a falling baseball or a falling boulder! FIGURE 2.36b shows the object's velocity graph. The velocity changes at a steady rate. The slope of the velocity-versus-time graph is the free-fall acceleration $a_{\text{free fall}}$.

Video Figure 2.36

FIGURE 2.36 Motion of an object in free fall.

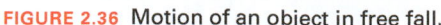

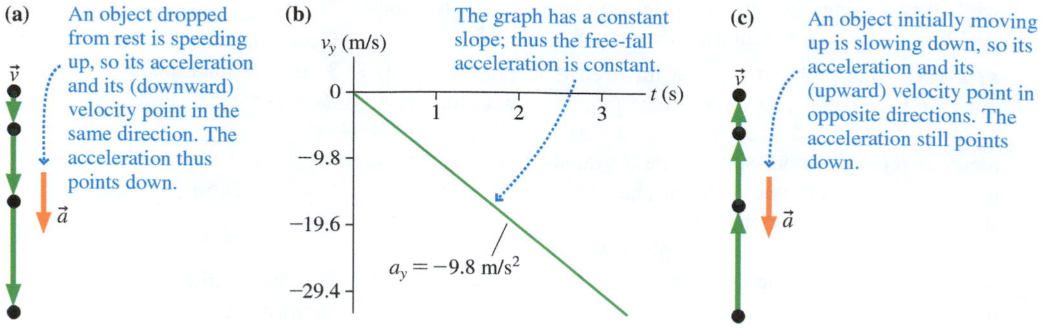

(a)
An object dropped from rest is speeding up, so its acceleration and its (downward) velocity point in the same direction. The acceleration thus points down.
$\vec{v}$
$\vec{a}$
(b)
$v_y$ (m/s)
The graph has a constant slope; thus the free-fall acceleration is constant.
0
−9.8
−19.6
−29.4
1
2
3
t (s)
$a_y = -9.8$ m/s$^2$
(c)
An object initially moving up is slowing down, so its acceleration and its (upward) velocity point in opposite directions. The acceleration still points down.
$\vec{v}$
$\vec{a}$


Instead of dropping the object, suppose we throw it upward. What happens then? You know that the object will move up and that its speed will decrease as it rises. This is illustrated in the motion diagram of FIGURE 2.36c, which shows a surprising result: Even though the object is moving up, its acceleration still points down. In fact, **the free-fall acceleration always points down,** no matter what direction an object is moving.

Video Free Fall

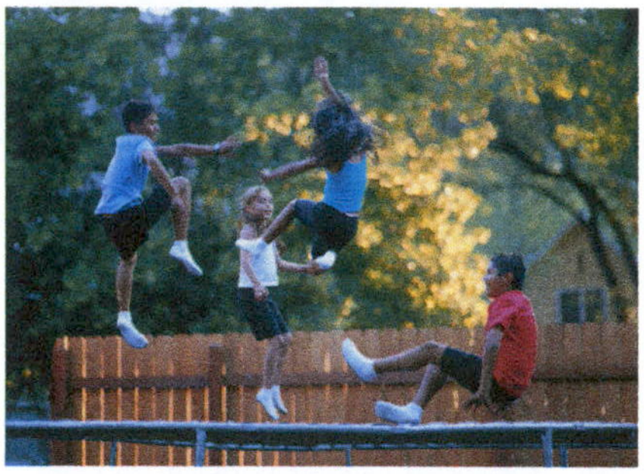

Some of the children on this trampoline are moving up and some are moving down, but all are in free fall—and so are accelerating downward at 9.8 m/s$^2$.

NOTE ▶ Despite the name, free fall is not restricted to objects that are literally falling. Any object moving under the influence of gravity only, and no other forces, is in free fall. This includes objects falling straight down, objects that have been tossed or shot straight up, objects in projectile motion (such as a basketball free throw), and, as we will see, satellites in orbit. ◀

The value of the free-fall acceleration varies slightly at different places on the earth, but for the calculations in this text we will use the the following average value:

$$\vec{a}_{\text{free fall}} = (9.80 \text{ m/s}^2, \text{ vertically downward}) \tag{2.14}$$

Standard value for the acceleration of an object in free fall

The magnitude of the **free-fall acceleration** has the special symbol $g$:

$$g = 9.80 \text{ m/s}^2$$

We will generally work with two significant figures and so will use $g = 9.8 \text{ m/s}^2$.

Several points about free fall are worthy of note:

- $g$, by definition, is *always* positive. **There will never be a problem that uses a negative value for $g$.**
- The velocity graph in Figure 2.36b has a negative slope. Even though a falling object speeds up, it has *negative* acceleration. Thus $g$ is *not* the object's acceleration, simply the *magnitude* of the acceleration. The one-dimensional acceleration is
$$a_y = -g$$
- Because free fall is motion with constant acceleration, we can use the kinematic equations for constant acceleration with $a_y = -g$.
- Once an object is acted upon by only the force of gravity, it is in free fall with $a_y = -9.8 \text{ m/s}^2$. It doesn't matter how the object entered free fall: Once in the air, a football that was punted straight up has the *same* acceleration of $-g$ as a stone dropped off a bridge.

- $g$ is not called "gravity." Gravity is a force, not an acceleration. $g$ is the *free-fall acceleration.*
- $g = 9.80\ \text{m/s}^2$ only on earth. Other planets have different values of $g$. You will learn in Chapter 6 how to determine $g$ for other planets.
- We will sometimes compute acceleration of objects not in free fall in units of $g$. An acceleration of $9.8\ \text{m/s}^2$ is an acceleration of $1g$; an acceleration of $19.6\ \text{m/s}^2$ is $2g$. Generally, we can compute

$$\text{acceleration (in units of } g\text{, or } g\text{'s)} = \frac{\text{acceleration (in units of m/s}^2\text{)}}{9.8\ \text{m/s}^2} \qquad (2.15)$$

This allows us to express accelerations in units that have a definite physical reference.

## EXAMPLE 2.14 Analyzing a rock's fall

A heavy rock is dropped from rest at the top of a cliff and falls 100 m before hitting the ground. How long does the rock take to fall to the ground, and what is its velocity when it hits?

***STRATEGIZE*** This is a free-fall problem, so it is a constant-acceleration problem with $a_y = -g$. We will use the constant-acceleration equations from Synthesis 2.1.

***PREPARE*** FIGURE 2.37 shows a visual overview with all necessary data. We have placed the origin at the ground, so that $y_i = 100$ m.

FIGURE 2.37 Visual overview of a falling rock.

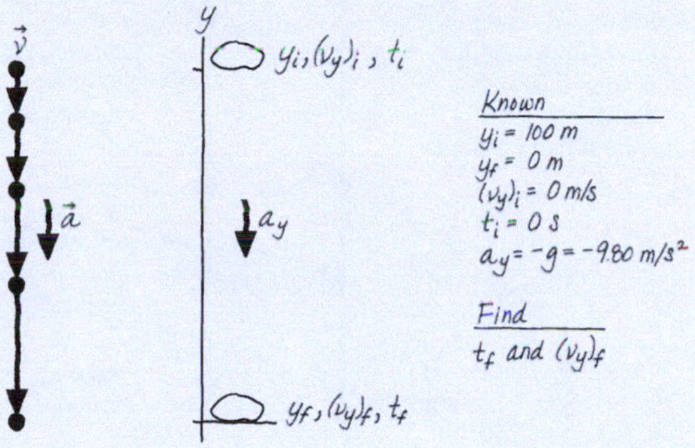


***SOLVE*** The first question in the problem statement involves a relationship between time and distance, a relationship expressed by Equation 2.12 in Synthesis 2.1. Using $(v_y)_i = 0$ m/s and $t_i = 0$ s, we find

$$y_f = y_i + (v_y)_i\,\Delta t + \tfrac{1}{2}a_y(\Delta t)^2 = y_i - \tfrac{1}{2}g(\Delta t)^2 = y_i - \tfrac{1}{2}g t_f^2$$

We can now solve for $t_f$:

$$t_f = \sqrt{\frac{2(y_i - y_f)}{g}} = \sqrt{\frac{2(100\ \text{m} - 0\ \text{m})}{9.80\ \text{m/s}^2}} = 4.52\ \text{s}$$

Now that we know the fall time, we can use Equation 2.11 to find $(v_y)_f$:

$$(v_y)_f = (v_y)_i - g\,\Delta t = -g t_f = -(9.80\ \text{m/s}^2)(4.52\ \text{s})$$
$$= -44.3\ \text{m/s}$$

***ASSESS*** Are the answers reasonable? Well, 100 m is about 300 feet, which is about the height of a 30-floor building. How long does it take an object to fall 30 floors? Four or five seconds seems pretty reasonable. How fast would the object be going at the bottom? Using an approximate version of our conversion factor 1 m/s ≈ 2 mph, we find that 44.3 m/s ≈ 90 mph. That also seems like a pretty reasonable speed for something that has fallen 30 floors. Suppose we had made a mistake. If we had misplaced a decimal point, we could have calculated a speed of 443 m/s, or about 900 mph! This is clearly *not* reasonable. If we had misplaced the decimal point in the other direction, we would have calculated a speed of 4.3 m/s ≈ 9 mph. This is another unreasonable result, because this is slower than a typical bicycling speed.

## CONCEPTUAL EXAMPLE 2.15 Analyzing the motion of a ball tossed upward

Draw a motion diagram and a velocity-versus-time graph for a ball tossed straight up in the air from the point that it leaves the hand until just before it is caught.

***REASON*** You know what the motion of the ball looks like: The ball goes up, and then it comes back down again. This complicates the drawing of a motion diagram a bit because the ball retraces its route as it falls. A literal motion diagram would show the upward motion and downward motion on top of each other, leading to confusion. We can avoid this difficulty by horizontally separating the upward motion and downward motion diagrams. This will not affect our conclusions because it does not change any of the vectors. The motion diagram and velocity-versus-time graph appear as in FIGURE 2.38.

***ASSESS*** The highest point in the ball's motion, where it reverses direction, is called a *turning point.* What are the velocity and the acceleration at this point? We can see from the motion diagram that the velocity vectors are pointing upward but getting shorter as the ball approaches the top. As it starts to fall, the velocity vectors are pointing downward and getting longer. There must be a moment—just an instant as $\vec{v}$ switches from pointing up to pointing down—when the velocity is zero. Indeed, the ball's velocity *is* zero for an instant at the precise top of the motion! We can also see on the velocity graph

*Continued*

that there is one instant of time when $v_y = 0$. This is the turning point.

But what about the acceleration at the top? You might expect the acceleration to be zero at the highest point. But recall that the velocity at the top point is changing—from up to down. If the velocity is changing, there *must* be an acceleration. The slope of the velocity graph at the instant when $v_y = 0$—that is, at the highest point—is no different than at any other point in the motion. The ball is still in free fall with acceleration $a_y = -g$!

Another way to think about this is to note that zero acceleration would mean no change of velocity. When the ball reached zero velocity at the top, it would hang there and not fall if the acceleration were also zero!

**FIGURE 2.38** Motion diagram and velocity graph of a ball tossed straight up in the air.

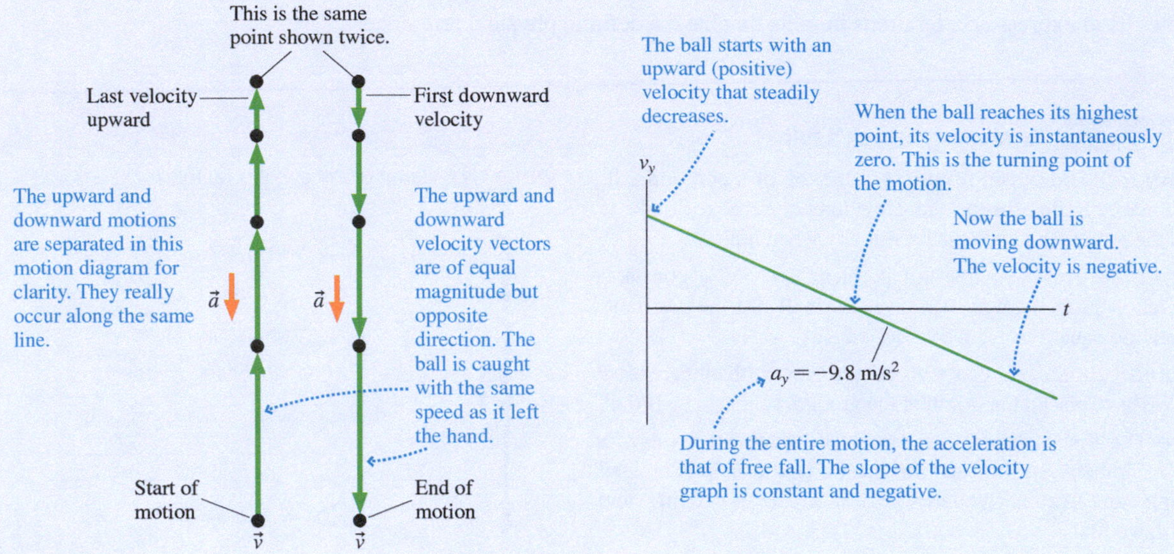


**EXAMPLE 2.16** **Finding the height of a leap** BIO

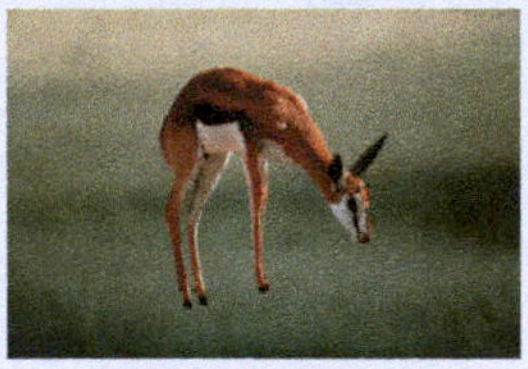

A springbok is an antelope found in southern Africa that gets its name from its remarkable jumping ability. When a springbok is startled, it will leap straight up into the air—a maneuver called a "pronk." A particular springbok goes into a crouch to perform a pronk. It then extends its legs forcefully, accelerating at 35 m/s$^2$ for 0.70 m as its legs straighten. Legs fully extended, it leaves the ground and rises into the air.

a. At what speed does the springbok leave the ground?
b. How high does it go?

**STRATEGIZE** This is a two-part problem. In the first phase of its motion, the springbok accelerates upward, reaching some maximum speed just as it leaves the ground. As soon as it does so, the springbok is subject to only the force of gravity, so it is in free fall. For both phases, we will use the constant-acceleration equations from Synthesis 2.1.

**PREPARE** We begin with the visual overview shown in FIGURE 2.39, where we've identified the two different phases of the motion: the springbok pushing off the ground and the springbok rising into the air. We'll treat these as two separate problems that we solve in turn. We will "re-use" the variables $y_i$, $y_f$, $(v_y)_i$, and $(v_y)_f$ for the two phases of the motion.

For the first part of our solution, in Figure 2.39a we choose the origin of the $y$-axis at the position of the springbok deep in its crouch. The final position is the top extent of the push, at the instant the springbok leaves the ground. We want to find the velocity at this position because that's how fast the springbok is moving as it leaves the ground. Figure 2.39b essentially starts over—we have defined a new vertical axis with its origin at the ground, so the highest point of the springbok's motion is its distance above the ground. The table of values shows the key piece of information for this second part of the problem: The initial velocity for part b is the final velocity from part a.

After the springbok leaves the ground, this is a free-fall problem because the springbok is moving under the influence of gravity only. We want to know the height of the leap, so we are looking for the height at the top point of the motion. This is a turning point of the motion, with the instantaneous velocity equal to zero. Thus $y_f$, the height of the leap, is the springbok's position at the instant $(v_y)_f = 0$.

**SOLVE** a. For the first phase, pushing off the ground, we have information about displacement, initial velocity, and acceleration, but we don't know anything about the time interval. Equation 2.13 from Synthesis 2.1 is perfect for this type of situation. We can use it to solve for the velocity with which the springbok lifts off the ground:

$$(v_y)_f^2 = (v_y)_i^2 + 2a_y\Delta y$$
$$= (0\ \text{m/s})^2 + 2(35\ \text{m/s}^2)(0.70\ \text{m}) = 49\ \text{m}^2/\text{s}^2$$
$$(v_y)_f = \sqrt{49\ \text{m}^2/\text{s}^2} = 7.0\ \text{m/s}$$

The springbok leaves the ground with a speed of 7.0 m/s.

b. Now we are ready for the second phase of the motion, the vertical motion after leaving the ground. Equation 2.13 is again appropriate because again we don't know the time. Because $y_i = 0$, the springbok's displacement is $\Delta y = y_f - y_i = y_f$, the height of the vertical leap. From part a, the initial velocity is $(v_y)_i = 7.0$ m/s, and the final velocity is $(v_y)_f = 0$. This is free-fall motion, with $a_y = -g$; thus

$$(v_y)_f^2 = 0 = (v_y)_i^2 - 2g\Delta y = (v_y)_i^2 - 2gy_f$$

which gives

$$(v_y)_i^2 = 2gy_f$$

Solving for $y_f$, we get a jump height of

$$y_f = \frac{(7.0 \text{ m/s})^2}{2(9.8 \text{ m/s}^2)} = 2.5 \text{ m}$$

**ASSESS** 2.5 m is a remarkable leap—a bit over 8 ft—but these animals are known for their jumping ability, so this seems reasonable.

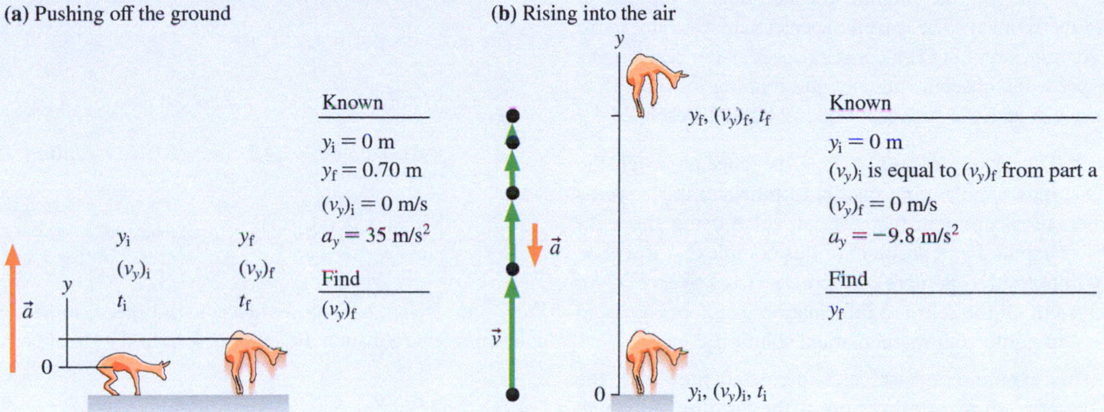


**FIGURE 2.39** A visual overview of the springbok's leap.

The caption accompanying the photo at the start of the chapter suggested a question about animals and their athletic abilities: Who is the winner in a race between a horse and a man? The surprising answer is "It depends." Specifically, the winner depends on the length of the race.

Some animals are capable of high speed; others are capable of great acceleration. Horses can run much faster than humans, but, when starting from rest, humans are capable of much greater initial acceleration. **FIGURE 2.40** shows velocity and position graphs for an elite male sprinter and a thoroughbred racehorse. The horse's maximum velocity is about twice that of the man, but the man's initial acceleration—the slope of the velocity graph at early times—is greater than that of the horse. As the second graph shows, a man could win a short race. For a longer race, the horse's higher maximum velocity will put it in the lead. The men's world-record time for the mile is a bit under 4 min, but a horse can easily run a mile in less than 2 min.

For a race of many miles, another factor comes into play: energy. A very long race is less about velocity and acceleration than about endurance—the ability to continue expending energy for a long time. In such endurance trials, humans often win. We will explore such energy issues in Chapter 11.

**FIGURE 2.40** BIO Velocity-versus-time and position-versus-time graphs for a sprint between a man and a horse.

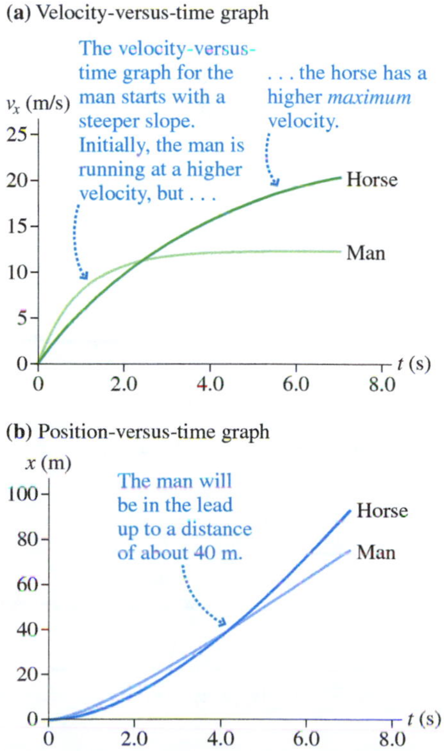

(a) Velocity-versus-time graph
$v_x$ (m/s)
The velocity-versus-time graph for the man starts with a steeper slope. Initially, the man is running at a higher velocity, but . . .
. . . the horse has a higher *maximum* velocity.
Horse
Man
$t$ (s)
(b) Position-versus-time graph
$x$ (m)
The man will be in the lead up to a distance of about 40 m.
Horse
Man
$t$ (s)


**STOP TO THINK 2.7** A volcano ejects a chunk of rock straight up at a velocity of $v_y = 30$ m/s. Ignoring air resistance, what will be the velocity $v_y$ of the rock when it falls back into the volcano's crater?

A. $> 30$ m/s B. 30 m/s C. 0 m/s D. $-30$ m/s E. $< -30$ m/s

## INTEGRATED EXAMPLE 2.17 Speed versus endurance BIO

Cheetahs have the highest top speed of any land animal, but they usually fail in their attempts to catch their prey because their endurance is limited. They can maintain their maximum speed of 30 m/s for only about 15 s before they need to stop.

Thomson's gazelles, their preferred prey, have a lower top speed than cheetahs, but they can maintain this speed for a few minutes. When a cheetah goes after a gazelle, success or failure is a simple matter of kinematics: Is the cheetah's high speed enough to allow it to reach its prey before the cheetah runs out of steam? The following problem uses realistic data for such a chase.

A cheetah has spotted a gazelle. The cheetah leaps into action, reaching its top speed of 30 m/s in a few seconds. At this instant, the gazelle, 160 m from the running cheetah, notices the danger and heads directly away. The gazelle accelerates at 4.5 m/s$^2$ for 6.0 s, then continues running at a constant speed. After reaching its maximum speed, the cheetah can continue running for only 15 s. Does the cheetah catch the gazelle, or does the gazelle escape?

**STRATEGIZE** The example asks, "Does the cheetah catch the gazelle?" Our most challenging task is to translate these words into a mathematical problem that we can solve using the techniques of this chapter. For a problem of this complexity, it will be particularly important to prepare a complete visual overview. An overview lays out all the relevant information in a concise visual form, helping to guide your mathematical solution.

**PREPARE** This example consists of two related problems, the motion of the cheetah and the motion of the gazelle, for which we'll use the subscripts "C" and "G." Let's take our starting time, $t_1 = 0$ s, as the instant that the gazelle notices the cheetah and begins to run. We'll take the position of the cheetah at this instant as the origin of our coordinate system, so $x_{1C} = 0$ m and $x_{1G} = 160$ m—the gazelle is 160 m away when it notices the cheetah. We've used this information to draw the visual overview in FIGURE 2.41, which includes motion diagrams and velocity graphs for the cheetah and the gazelle. The visual overview sums up everything we know about the problem.

With a clear picture of the situation, we can now rephrase the problem this way: Compute the position of the cheetah and the position of the gazelle at $t_3 = 15$ s, the time when the cheetah needs to break off the chase. If $x_{3G} > x_{3C}$, then the gazelle stays out in front and escapes. If $x_{3G} \le x_{3C}$, the cheetah wins the race—and gets its dinner.

**SOLVE** The cheetah is in uniform motion for the entire duration of the problem, so we can use Equation 2.4 of Synthesis 2.1 to solve for its position at $t_3 = 15$ s:

$$x_{3C} = x_{1C} + (v_x)_{1C}\Delta t = 0\text{ m} + (30\text{ m/s})(15\text{ s}) = 450\text{ m}$$

The gazelle's motion has two phases: one of constant acceleration and then one of constant velocity. We can solve for the position and the velocity at $t_2$, the end of the first phase, using Equations 2.11 and 2.12 of Synthesis 2.1. Let's find the velocity first:

$$(v_x)_{2G} = (v_x)_{1G} + (a_x)_G\Delta t = 0\text{ m/s} + (4.5\text{ m/s}^2)(6.0\text{ s}) = 27\text{ m/s}$$

The gazelle's position at $t_2$ is

$$x_{2G} = x_{1G} + (v_x)_{1G}\Delta t + \tfrac{1}{2}(a_x)_G(\Delta t)^2$$

$$= 160\text{ m} + 0 + \tfrac{1}{2}(4.5\text{ m/s}^2)(6.0\text{ s})^2 = 240\text{ m}$$

The gazelle has a head start; it begins at $x_{1G} = 160$ m.

$\Delta t$ is the time for this phase of the motion, $t_2 - t_1 = 6.0$ s.

From $t_2$ to $t_3$ the gazelle moves at a constant speed, so we can use the uniform motion equation, Equation 2.4, to find its final position:

The gazelle begins this phase of the motion at $x_{2G} = 240$ m.

$\Delta t$ for this phase of the motion is $t_3 - t_2 = 9.0$ s.

$$x_{3G} = x_{2G} + (v_x)_{2G}\,\Delta t = 240\text{ m} + (27\text{ m/s})(9.0\text{ s}) = 480\text{ m}$$

$x_{3C}$ is 450 m; $x_{3G}$ is 480 m. The gazelle is 30 m ahead of the cheetah when the cheetah has to break off the chase, so the gazelle escapes.

**ASSESS** Does our solution make sense? Let's look at the final result. The numbers in the problem statement are realistic, so we expect our results to mirror real life. The speed for the gazelle is close to that of the cheetah, which seems reasonable for two animals known for their speed. And the result is the most common occurrence—the chase is close, but the gazelle gets away.

FIGURE 2.41 Visual overview for the cheetah and for the gazelle.

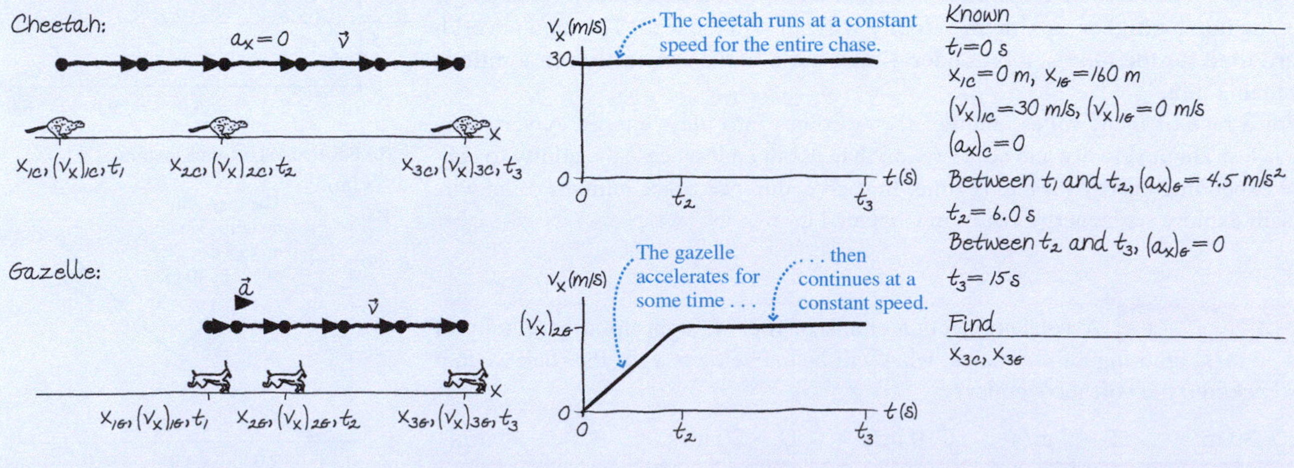

# SUMMARY

**GOAL** To describe and analyze motion along a line.

## GENERAL STRATEGIES

### Problem-Solving Approach

Our general problem-solving approach has four parts:

***STRATEGIZE*** Think about the big picture: What kind of problem is this? What general approach should be used? What should the answer look like?

***PREPARE*** Set up the problem:

- Draw a picture.
- Collect necessary information.
- Do preliminary calculations.

***SOLVE*** Do the necessary mathematics or reasoning.

***ASSESS*** Check your answer to see if it is complete in all details and makes physical sense.

### Visual Overview

A visual overview consists of several parts that completely specify a problem. This may include any or all of the elements below:

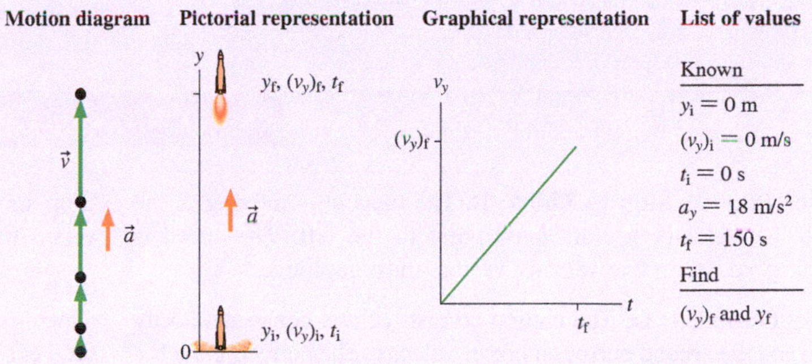


## IMPORTANT CONCEPTS

**Velocity** is the rate of change of position:

$$v_x = \frac{\Delta x}{\Delta t}$$

**Acceleration** is the rate of change of velocity:

$$a_x = \frac{\Delta v_x}{\Delta t}$$

The units of acceleration are m/s$^2$.

An object is speeding up if $v_x$ and $a_x$ have the same sign, slowing down if they have opposite signs.

A **position-versus-time graph** plots position on the vertical axis against time on the horizontal axis.

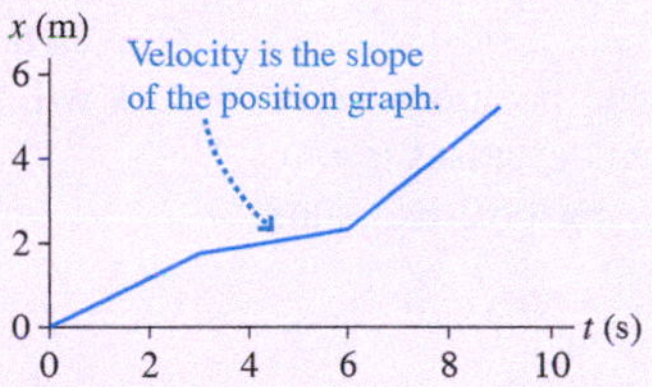


A **velocity-versus-time graph** plots velocity on the vertical axis against time on the horizontal axis.

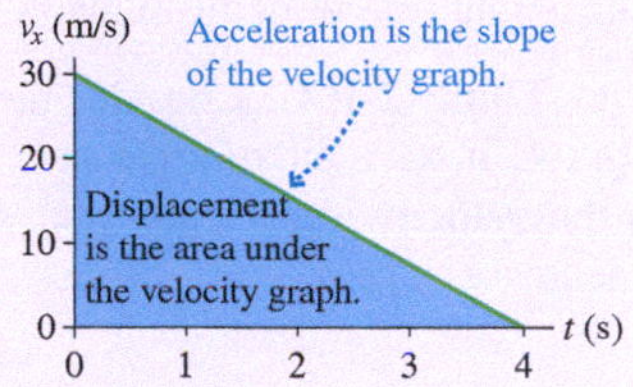


## APPLICATIONS

### Uniform Motion

An object in uniform motion has a constant velocity. Its velocity graph is a horizontal line; its position graph is linear.

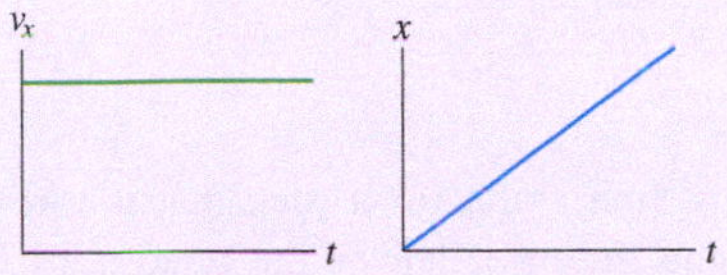


Kinematic equation for uniform motion:

$$x_f = x_i + v_x \Delta t$$

Uniform motion is a special case of constant-acceleration motion, with $a_x = 0$.

### Motion with Constant Acceleration

An object with constant acceleration has a constantly changing velocity. Its velocity graph is linear; its position graph is a parabola.

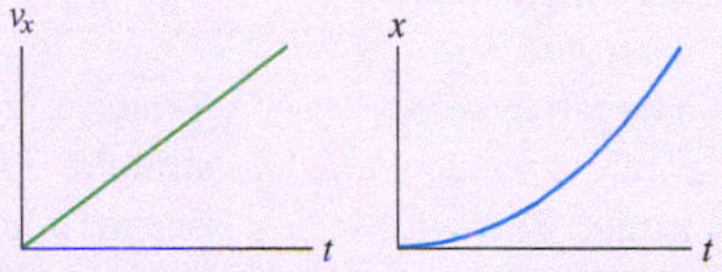


Kinematic equations for motion with constant acceleration:

$$(v_x)_f = (v_x)_i + a_x \Delta t$$

$$x_f = x_i + (v_x)_i \Delta t + \tfrac{1}{2} a_x (\Delta t)^2$$

$$(v_x)_f{}^2 = (v_x)_i{}^2 + 2a_x \Delta x$$

### Free Fall

Free fall is a special case of constant-acceleration motion. The acceleration has magnitude $g = 9.80$ m/s$^2$ and is always directed vertically downward whether an object is moving up or down.

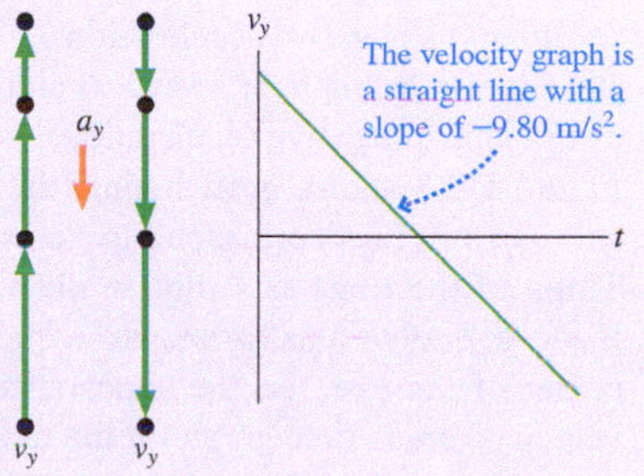

## Learning Objectives After studying this chapter, you should be able to:

- Use motion diagrams to interpret motion. *Conceptual Question 2.3; Problems 2.1, 2.2, 2.59*
- Use and interpret motion graphs. *Conceptual Questions 2.5, 2.13; Problems 2.4, 2.18, 2.19, 2.22, 2.62*
- Calculate the velocity of an object. *Conceptual Question 2.9; Problems 2.8, 2.15, 2.57*
- Solve problems about an object in uniform motion. *Problems 2.9, 2.10, 2.11, 2.13, 2.58*
- Calculate the acceleration of an object. *Problems 2.25, 2.27, 2.32, 2.33, 2.72*
- Determine and interpret the sign of acceleration. *Conceptual Questions 2.2, 2.8; Problem 2.50*
- Use the problem-solving approach to solve problems of motion with constant acceleration and free fall. *Problems 2.36, 2.40, 2.41, 2.47, 2.52, 2.75*

## STOP TO THINK ANSWERS

**Chapter Preview Stop to Think: B.** The bicycle is moving to the left, so the velocity vectors must point to the left. The speed is increasing, so successive velocity vectors must get longer.

**Stop to Think 2.1: D.** The motion consists of two constant-velocity phases, and the second one has a higher velocity. The correct graph has two straight-line segments, with the second one having a steeper slope.

**Stop to Think 2.2: B.** The displacement is the area under a velocity-versus-time graph. In all four cases, the graph is a straight line, so the area under the graph is a rectangle. The area is the product of the length and the height, so the largest displacement belongs to the graph with the largest product of the length (the time interval, in s) and the height (the velocity, in m/s).

**Stop to Think 2.3: C.** Consider the slope of the position-versus-time graph. It starts out positive and constant, then decreases to zero. Thus the velocity graph must start with a constant positive value, then decrease to zero.

**Stop to Think 2.4: C.** Acceleration is the slope of the velocity-versus-time graph. The largest magnitude of the slope is at point C.

**Stop to Think 2.5: B.** The elevator is moving down, so $v_y < 0$. It is slowing down, so the magnitude of the velocity is decreasing. As time goes on, the velocity graph should get closer to the origin. This means that the acceleration is positive, and the slope of the graph is positive.

**Stop to Think 2.6: E.** An acceleration of 1.2 $m/s^2$ corresponds to an increase of 1.2 m/s every second. At this rate, the cruising speed of 6.0 m/s will be reached after 5.0 s.

**Stop to Think 2.7: D.** The final velocity will have the same *magnitude* as the initial velocity, but the velocity is negative because the rock will be moving downward.

Video Tutor Solution Chapter 2

# QUESTIONS

## Conceptual Questions

1. A person gets in an elevator on the ground floor and rides it to the top floor of a building. Sketch a velocity-versus-time graph for this motion.
2. a. Give an example of a vertical motion with a positive velocity and a negative acceleration.
   b. Give an example of a vertical motion with a negative velocity and a negative acceleration.
3. Figure Q2.3 shows growth rings in a tree's trunk. The wide BIO and narrow rings correspond to years of fast and slow growth. Think of the rings as a motion diagram for the tree's growth. If we define an axis as shown, with $x$ measured out from the center of the tree, use the appearance of the rings to sketch a velocity-versus-time graph for the radial growth of the tree.

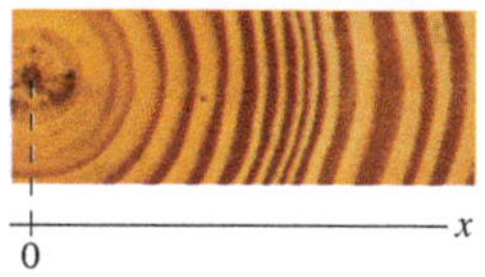


FIGURE Q2.3

4. Sketch a velocity-versus-time graph for a rock that is thrown straight upward, from the instant it leaves the hand until the instant it hits the ground.
5. You are driving down the road at a constant speed. Another car going a bit faster catches up with you and passes you. Draw a position graph for both vehicles on the same set of axes, and note the point on the graph where the other vehicle passes you.

Problem difficulty is labeled as | (straightforward) to ||||| (challenging). Problems labeled INT integrate significant material from earlier chapters; Problems labeled BIO are of biological or medical interest.

The eText icon indicates when there is a video tutor solution available for the chapter or for a specific problem. To launch these videos, log into your eText through Mastering[TM] Physics or log into the Study Area.

6. Figure Q2.6 shows the velocity-versus-time graphs for two objects A and B. Students Zach and Victoria are asked to tell stories that correspond to the motion of the objects. Zach says, "The graph could represent two cars traveling in opposite directions that pass each other." Victoria says, "No, I think they could be two rocks thrown vertically from a bridge; rock A is thrown upward and rock B is thrown downward." Which student, if either, is correct? Explain.

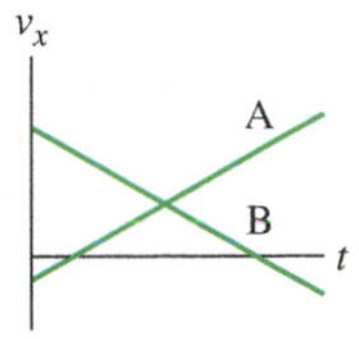


FIGURE Q2.6

7. Certain animals are capable of running at great speeds; other BIO animals are capable of tremendous accelerations. Speculate on which would be more beneficial to a predator—large maximum speed or large acceleration.
8. A ball is thrown straight up into the air. At each of the following instants, is the ball's acceleration $a_y$ equal to $g$, $-g$, 0, $<g$, or $>g$?
   a. Just after leaving your hand?
   b. At the very top (maximum height)?
   c. Just before hitting the ground?
9. Janelle stands on a balcony, two stories above Michael. She throws one ball straight up and one ball straight down, but both with the same initial speed. Eventually each ball passes Michael. Which ball, if either, is moving faster when it passes Michael? Explain.
10. Figure Q2.10 shows an object's position-versus-time graph. The letters A to E correspond to various segments of the motion in which the graph has constant slope.
    a. Write a realistic motion short story for an object that would have this position graph.
    b. In which segment(s) is the object at rest?
    c. In which segment(s) is the object moving to the right?
    d. Is the speed of the object during segment C greater than, equal to, or less than its speed during segment E? Explain.

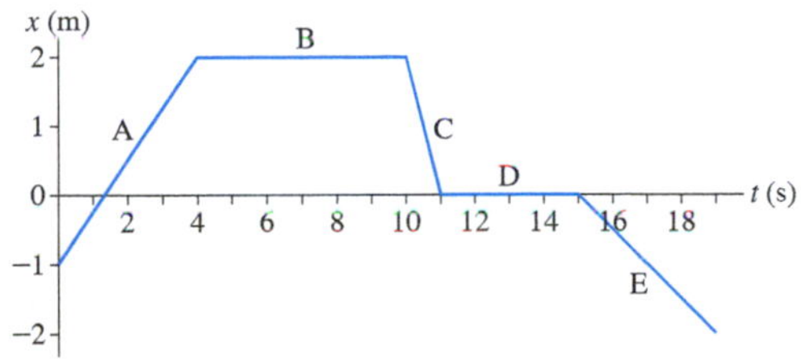


FIGURE Q2.10

11. Figure Q2.11 shows the position graph for an object moving along the horizontal axis.
    a. Write a realistic motion short story for an object that would have this position graph.
    b. Draw the corresponding velocity graph.

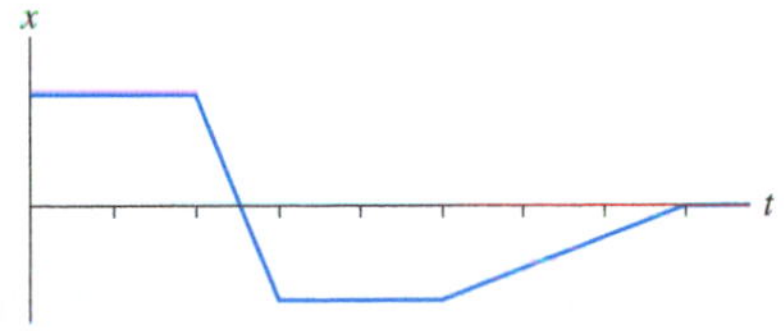


FIGURE Q2.11

12. Figure Q2.12 shows the position-versus-time graphs for two objects, A and B, that are moving along the same axis.
    a. At the instant $t = 1$ s, is the speed of A greater than, less than, or equal to the speed of B? Explain.
    b. Do objects A and B ever have the *same* speed? If so, at what time or times? Explain.

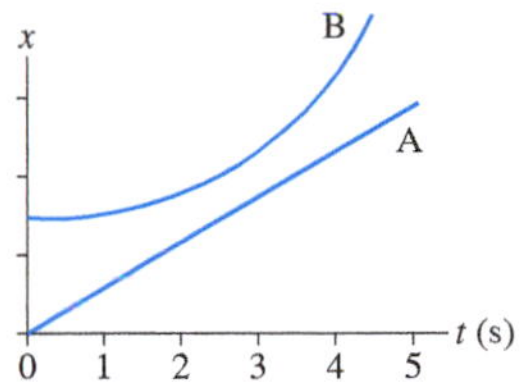


FIGURE Q2.12

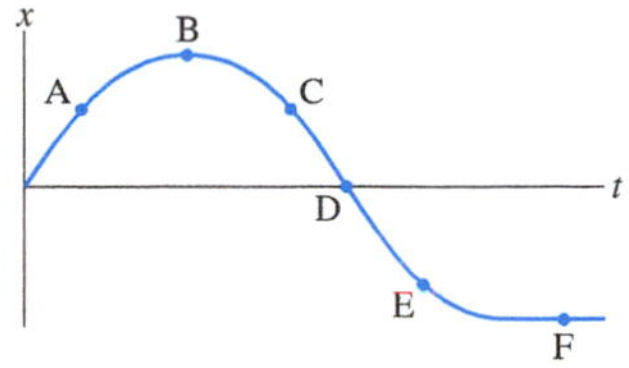


FIGURE Q2.13

13. Figure Q2.13 shows a position-versus-time graph. At which lettered point or points is the object
    a. Moving the fastest?
    b. Moving to the left?
    c. Speeding up?
    d. Slowing down?
    e. Turning around?
14. Figure Q2.14 is the velocity-versus-time graph for an object moving along the $x$-axis.
    a. During which segment(s) is the velocity constant?
    b. During which segment(s) is the object speeding up?
    c. During which segment(s) is the object slowing down?
    d. During which segment(s) is the object standing still?
    e. During which segment(s) is the object moving to the right?

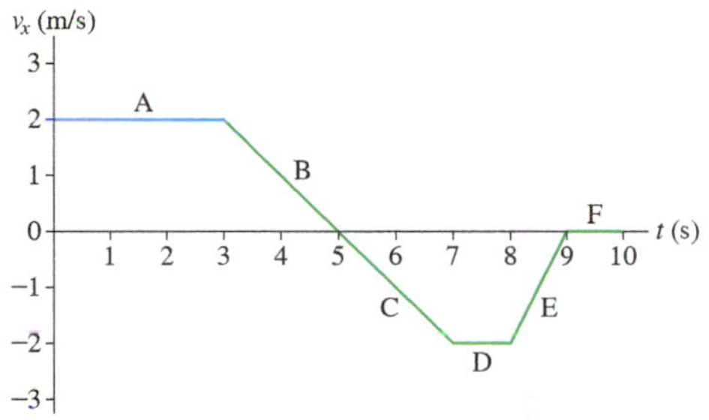


FIGURE Q2.14

## Multiple-Choice Questions

15. | Figure Q2.15 shows the position graph of a car traveling on a straight road. At which labeled instant is the speed of the car greatest?

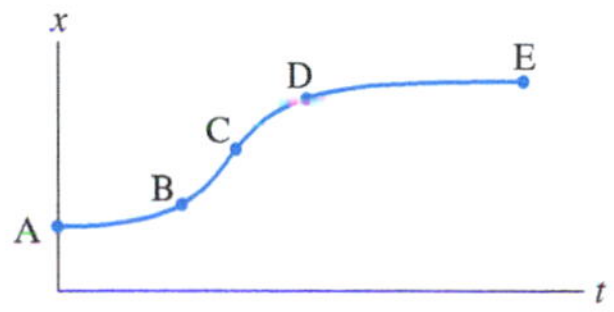


FIGURE Q2.15

16. | Figure Q2.16 shows the position graph of a car traveling on a straight road. The velocity at instant 1 is __________ and the velocity at instant 2 is __________.
    A. positive, negative
    B. positive, positive
    C. negative, negative
    D. negative, zero
    E. positive, zero

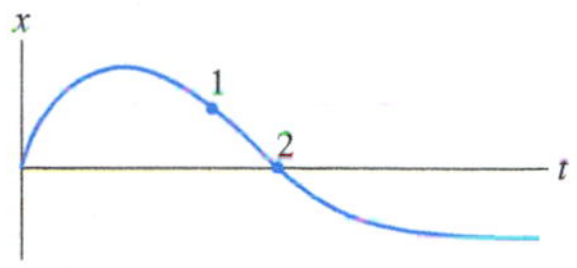


FIGURE Q2.16

17. | Figure Q2.17 shows an object's position-versus-time graph. What is the velocity of the object at $t = 6$ s?
A. 0.67 m/s B. 0.83 m/s
C. 3.3 m/s D. 4.2 m/s
E. 25 m/s

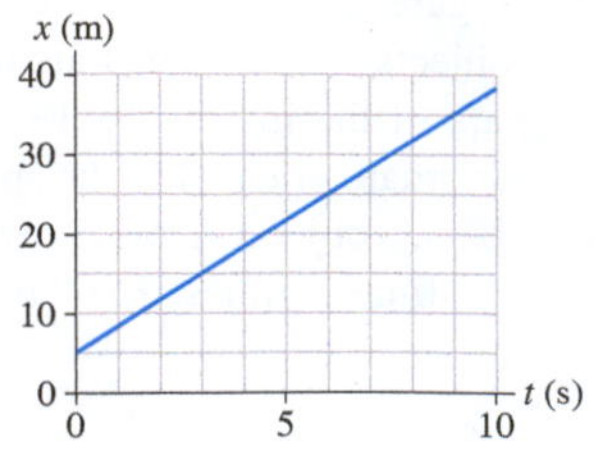

FIGURE Q2.17

18. | The following options describe the motion of four cars A–D. Which car has the largest acceleration?
A. Goes from 0 m/s to 10 m/s in 5.0 s
B. Goes from 0 m/s to 5.0 m/s in 2.0 s
C. Goes from 0 m/s to 20 m/s in 7.0 s
D. Goes from 0 m/s to 3.0 m/s in 1.0 s

19. | A car is traveling at $v_x = 20$ m/s. The driver applies the brakes, and the car slows with $a_x = -4.0\ \text{m/s}^2$. What is the stopping distance?
A. 5.0 m
B. 25 m
C. 40 m
D. 50 m

20. || Velocity-versus-time graphs for three drag racers are shown in Figure Q2.20. At $t = 5.0$ s, which car has traveled the farthest?
A. Andy
B. Rachel
C. Carl
D. All have traveled the same distance

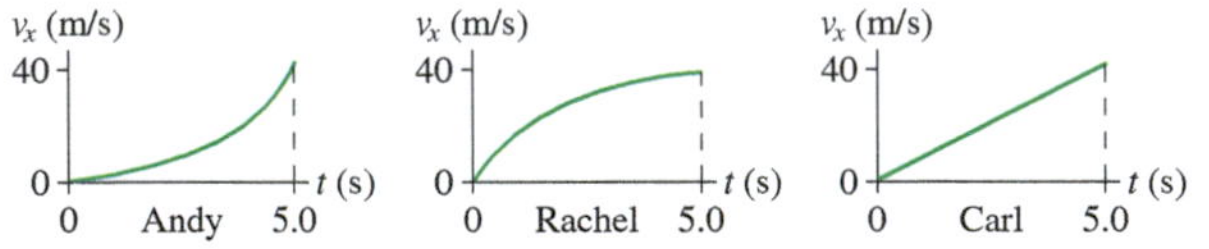

FIGURE Q2.20

21. | Which of the three drag racers in Question 20 had the greatest acceleration at $t = 0$ s?
A. Andy
B. Rachel
C. Carl
D. All had the same acceleration

22. || Chris is holding two softballs while standing on a balcony. She throws ball 1 straight up in the air and, at the same instant, releases her grip on ball 2, letting it drop over the side of the building. Which velocity graph in Figure Q2.22 best represents the motion of the two balls?

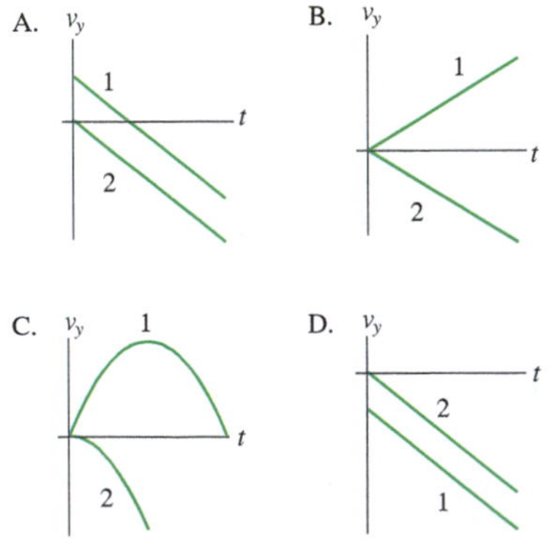

FIGURE Q2.22

23. || Suppose a plane accelerates from rest for 30 s, achieving a takeoff speed of 80 m/s after traveling a distance of 1200 m down the runway. A smaller plane with the same acceleration has a takeoff speed of 40 m/s. Starting from rest, after what distance will this smaller plane reach its takeoff speed?
A. 300 m B. 600 m C. 900 m D. 1200 m

24. || Figure Q2.24 shows a motion diagram with the clock reading (in seconds) shown at each position. From $t = 9$ s to $t = 15$ s the object is at the same position. After that, it returns along the same track. The positions of the dots for $t \geq 16$ s are offset for clarity. Which graph best represents the object's *velocity?*

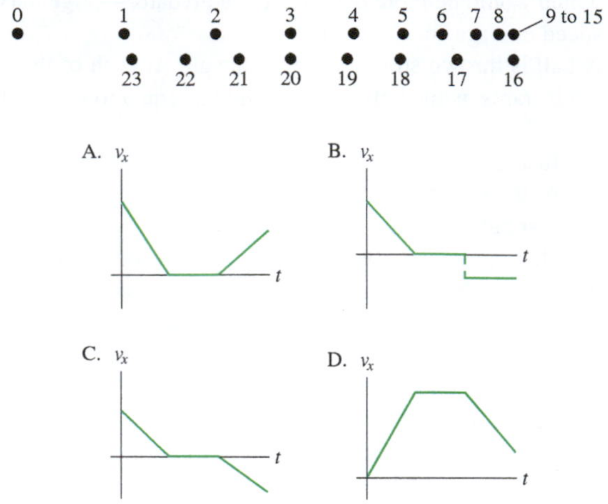

FIGURE Q2.24

25. || Nate throws a ball straight up to Kayla, who is standing on a balcony 3.8 m above Nate. When she catches it, the ball is still moving upward at a speed of 2.8 m/s. With what initial speed did Nate throw the ball?
A. 7.0 m/s B. 12.3 m/s
C. 9.1 m/s D. 10.6 m/s

26. | Starting from rest, a car takes 2.4 s to travel the first 15 m. Assuming a constant acceleration, how long will it take the car to travel the next 15 m?
A. 0.67 s B. 1.0 s C. 1.8 s D. 3.6 s

27. | The velocity-versus-time graph for a car driving down a straight road is shown in Figure Q2.27. What is the acceleration of the car during the period shown?
A. $1.0\ \text{m/s}^2$ B. $2.5\ \text{m/s}^2$
C. $3.8\ \text{m/s}^2$ D. $5.0\ \text{m/s}^2$

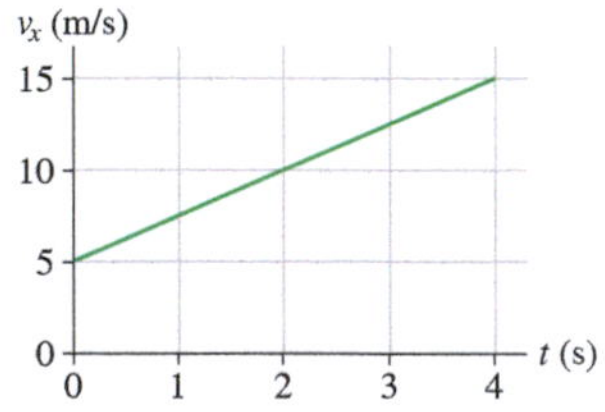

FIGURE Q2.27

28. || The velocity-versus-time graph for a car driving down a straight road is shown in Figure Q2.27. How far does the car travel during the time interval from $t = 0$ s to $t = 4.0$ s?
A. 10 m B. 20 m C. 40 m D. 60 m

# PROBLEMS

## Section 2.1 Describing Motion

1. ||| Figure P2.1 shows a motion diagram of a car traveling down a street. The camera took one frame every second. A distance scale is provided.
   a. Use the scale to determine the $x$-value of the car at each dot. Place your data in a table, similar to Table 2.1, showing each position and the instant of time at which it occurred.
   b. Make a graph of $x$ versus $t$, using the data in your table. Because you have data only at certain instants of time, your graph should consist of dots that are not connected together.

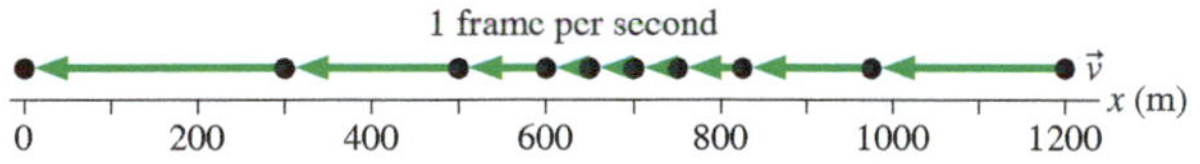


FIGURE P2.1

2. | For each motion diagram in Figure P2.2, determine the sign (positive or negative) of the position and the velocity.

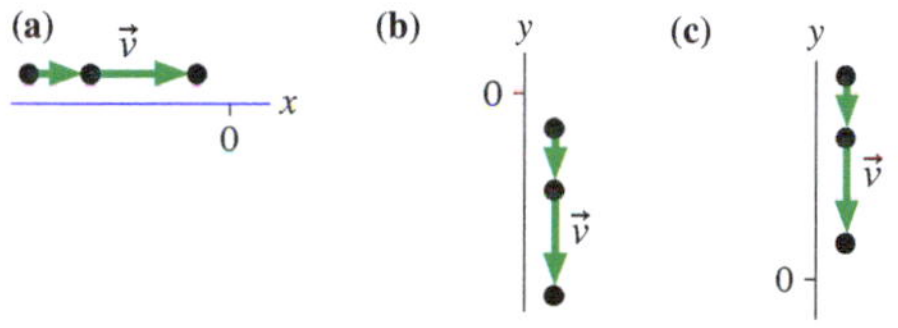


FIGURE P2.2

3. || The position graph of Figure P2.3 shows a dog slowly sneaking up on a squirrel, then putting on a burst of speed.
   a. For how many seconds does the dog move at the slower speed?
   b. Draw the dog's velocity-versus-time graph. Include a numerical scale on both axes.

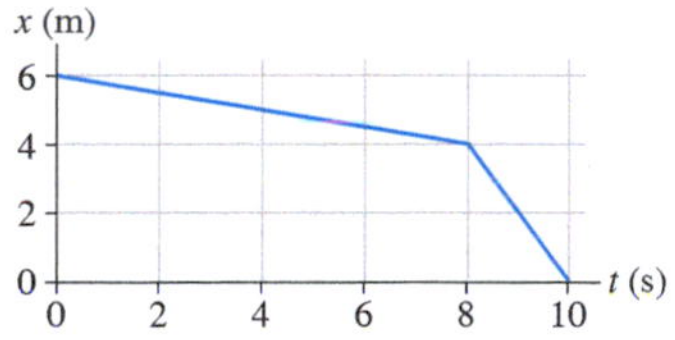


FIGURE P2.3

4. || A rural mail carrier is driving slowly, putting mail in mailboxes near the road. He overshoots one mailbox, stops, shifts into reverse, and then backs up until he is at the right spot. The velocity graph of Figure P2.4 represents his motion.
   a. Draw the mail carrier's position-versus-time graph. Assume that $x = 0$ m at $t = 0$ s.
   b. What is the position of the mailbox?

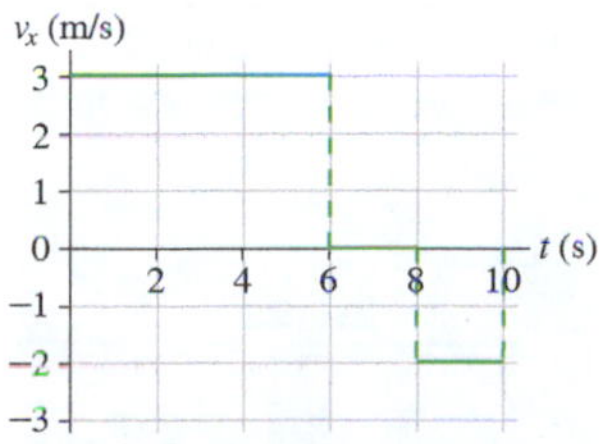


FIGURE P2.4

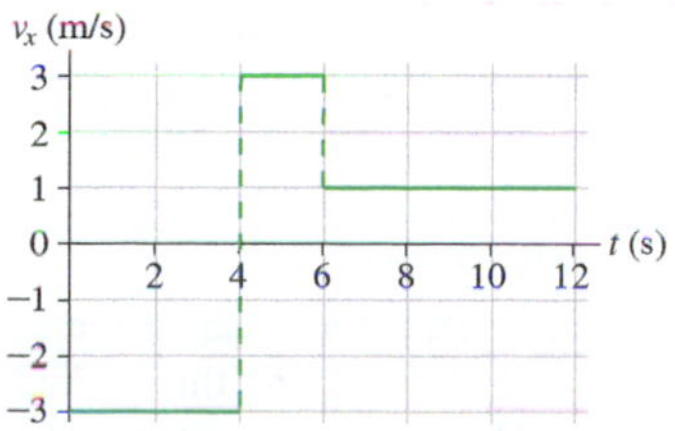


FIGURE P2.5

5. || For the velocity-versus-time graph of Figure P2.5:
   a. Draw the corresponding position-versus-time graph. Assume that $x = 0$ m at $t = 0$ s.
   b. What is the object's position at $t = 12$ s?
   c. Describe a moving object that could have these graphs.
6. || Starting at 48th Street, Dylan rides his bike due east on Meridian Road with the wind at his back. He rides for 20 min at 15 mph. He then stops for 5 min, turns around, and rides back to 48th Street; because of the headwind, his speed is only 10 mph.
   a. How long does his trip take?
   b. Assuming that the origin of his trip is at 48th Street, draw a position-versus-time graph for his trip.
7. ||| An elevator in a high-rise building goes up and down at the same speed. Starting at the ground floor, Rafael and Monica ride up five floors, a vertical rise of 20 m. The elevator stops for 10 s as Monica gets off. Rafael then goes back down two floors. Rafael's entire trip takes 24 s. Taking the origin to be at the ground floor, draw position-versus-time and velocity-versus-time graphs for Rafael's trip.
8. || A bicyclist has the position-versus-time graph shown in Figure P2.8. What is the bicyclist's velocity at $t = 10$ s, at $t = 25$ s, and at $t = 35$ s?

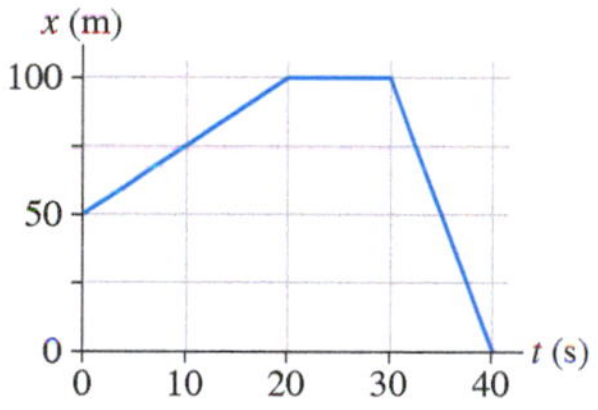


FIGURE P2.8

## Section 2.2 Uniform Motion

9. | In major league baseball, the pitcher's mound is 60 feet from the batter. If a pitcher throws a 95 mph fastball, how much time elapses from when the ball leaves the pitcher's hand until the ball reaches the batter?
10. | In college softball, the distance from the pitcher's mound to the batter is 43 feet. If the ball leaves the bat at 100 mph, how much time elapses between the hit and the ball reaching the pitcher?
11. || Alan leaves Los Angeles at 8:00 AM to drive to San Francisco, 400 mi away. He travels at a steady 50 mph. Beth leaves Los Angeles at 9:00 AM and drives a steady 60 mph.
    a. Who gets to San Francisco first?
    b. How long does the first to arrive have to wait for the second?
12. || Richard is driving home to visit his parents. 125 mi of the trip are on the interstate highway where the speed limit is 65 mph. Normally Richard drives at the speed limit, but today he is running late and decides to take his chances by driving at 70 mph. How many minutes does he save?
13. ||| In a 5.00 km race, one runner runs at a steady 12.0 km/h and another runs at 14.5 km/h. How long does the faster runner have to wait at the finish line to see the slower runner cross?
14. |||| In an 8.00 km race, one runner runs at a steady 11.0 km/h and another runs at 14.0 km/h. How far from the finish line is the slower runner when the faster runner finishes the race?

15. ‖ Figure P2.15 shows actual data from Usain Bolt's 2009 word-record run in the 100 m sprint. From this graph, estimate his top speed in m/s and in mph.

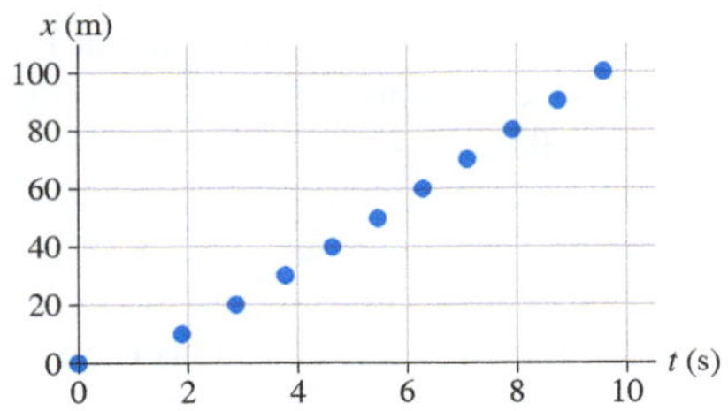


FIGURE P2.15

16. ‖ While running a marathon, a long-distance runner uses a stopwatch to time herself over a distance of 100 m. She finds that she runs this distance in 18 s. Answer the following by considering ratios, without computing her velocity.
   a. If she maintains her speed, how much time will it take her to run the next 400 m?
   b. How long will it take her to run a mile at this speed?

## Section 2.3 Instantaneous Velocity

17. | Figure P2.17 shows the position graph of a particle.
   a. Draw the particle's velocity graph for the interval $0\text{ s} \le t \le 4\text{ s}$.
   b. Does this particle have a turning point or points? If so, at what time or times?

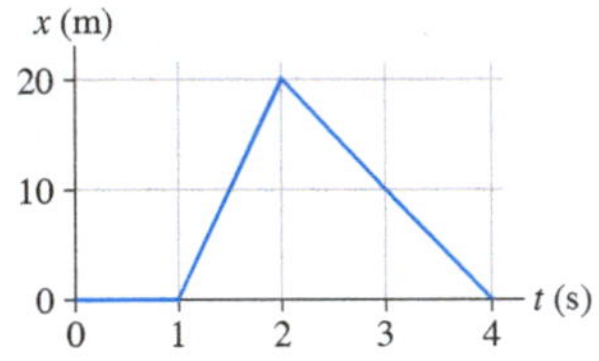


FIGURE P2.17

18. ‖ A somewhat idealized graph of the speed of the blood in
BIO the ascending aorta during one beat of the heart appears as in Figure P2.18.
   a. Approximately how far, in cm, does the blood move during one beat?
   b. Assume similar data for the motion of the blood in your aorta, and make a rough estimate of the distance from your heart to your brain. Estimate how many beats of the heart it takes for blood to travel from your heart to your brain.

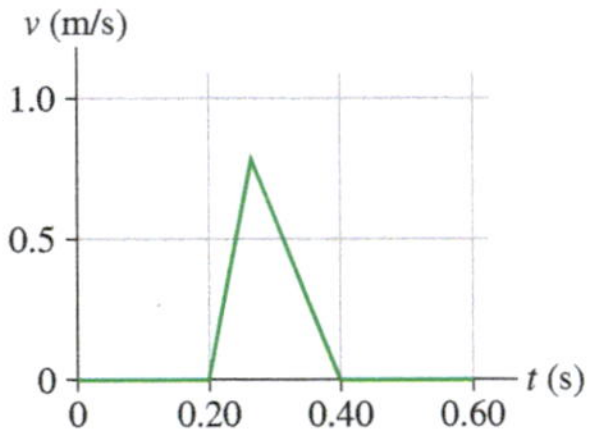


FIGURE P2.18

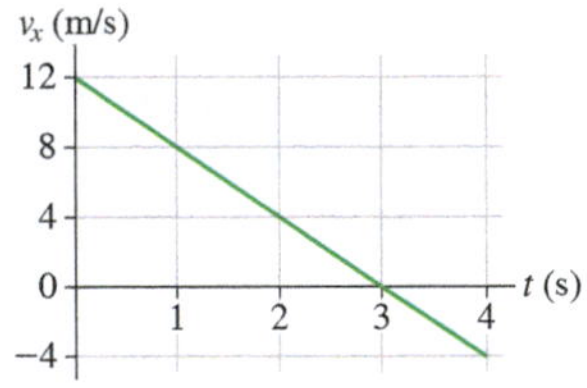


FIGURE P2.19

19. ||| A car starts from $x_i = 10$ m at $t_i = 0$ s and moves with the velocity graph shown in Figure P2.19.
   a. What is the car's position at $t = 2$ s, 3 s, and 4 s?
   b. Does this car ever change direction? If so, at what time?

20. ‖ Figure P2.20 shows a graph of actual position-versus-time data for a particular type of drag racer known as a "funny car."
   a. Estimate the car's velocity at 2.0 s.
   b. Estimate the car's velocity at 4.0 s.

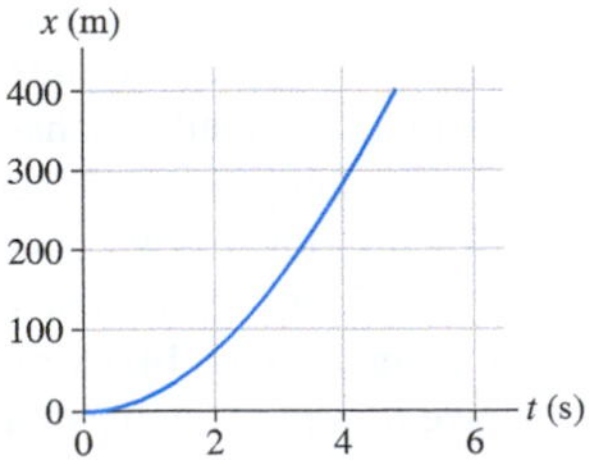


FIGURE P2.20

## Section 2.4 Acceleration

21. ‖ Figure P2.21 shows the velocity graph of a bicycle. Draw the bicycle's acceleration graph for the interval $0\text{ s} \le t \le 4\text{ s}$. Give both axes an appropriate numerical scale.

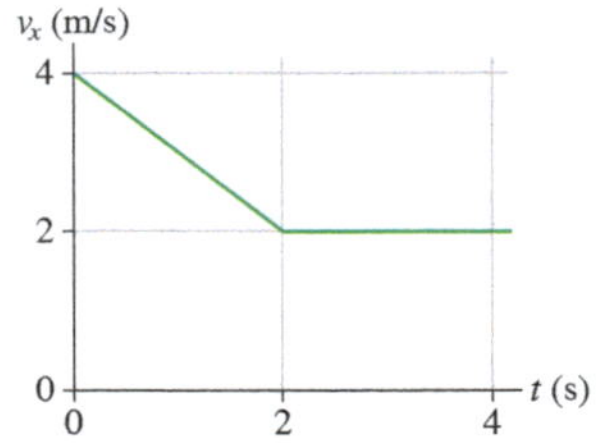


FIGURE P2.21

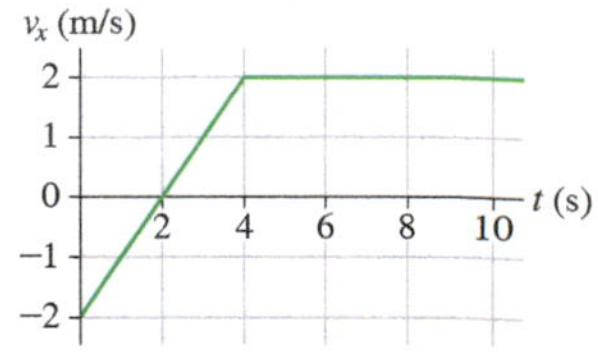


FIGURE P2.22

22. ||| We set the origin of a coordinate system so that the position of a train is $x = 0$ m at $t = 0$ s. Figure P2.22 shows the train's velocity graph.
   a. Draw position and acceleration graphs for the train.
   b. Find the acceleration of the train at $t = 3.0$ s.

23. ||| An object has the acceleration graph shown in Figure P2.23. Its velocity at $t = 0$ s is $v_x = 2.0$ m/s. Draw the object's velocity graph.

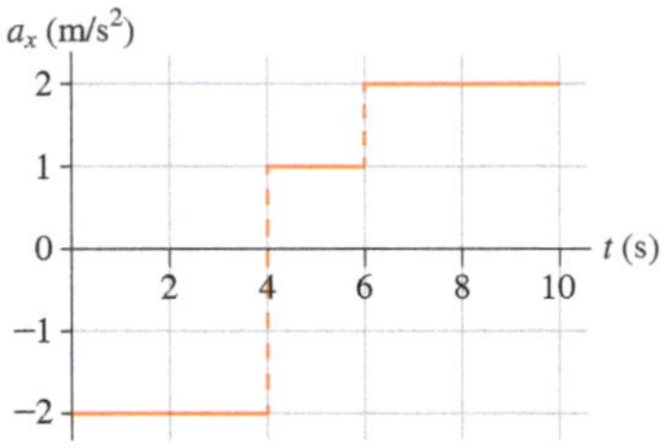


FIGURE P2.23

24. ‖ Figure P2.18 showed data for the speed of blood in the aorta.
BIO Determine the magnitude of the acceleration for both phases, speeding up and slowing down.

25. ‖ Figure P2.25 is a somewhat simplified velocity graph for Olympic sprinter Carl Lewis starting a 100 m dash. Estimate his acceleration during each of the intervals A, B, and C.

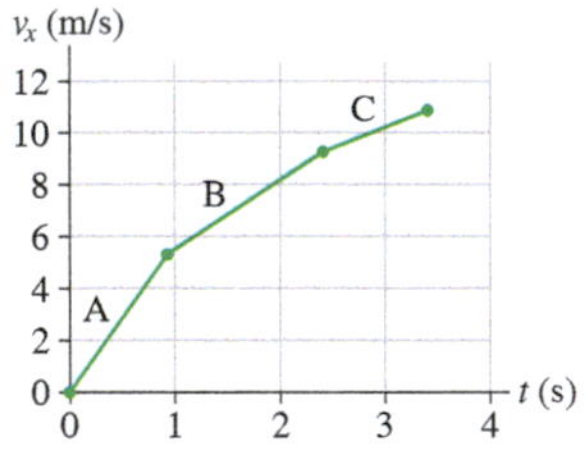


FIGURE P2.25

26. | Small frogs that are good jumpers are capable of remarkable
BIO accelerations. One species reaches a takeoff speed of 3.7 m/s in 60 ms. What is the frog's acceleration during the jump?

27. | A Thomson's gazelle can reach a speed of 13 m/s in 3.0 s. A lion
BIO can reach a speed of 9.5 m/s in 1.0 s. A trout can reach a speed of 2.8 m/s in 0.12 s. Which animal has the largest acceleration?

28. ||| When striking, the pike, a
BIO predatory fish, can accelerate from rest to a speed of 4.0 m/s in 0.11 s.
   a. What is the acceleration of the pike during this strike?
   b. How far does the pike move during this strike?

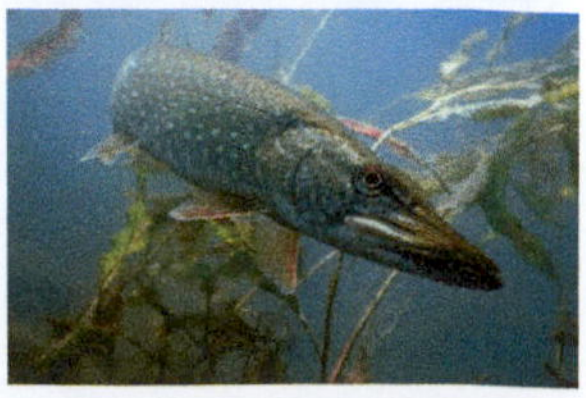

### Section 2.5 Motion with Constant Acceleration

29. || a. What constant acceleration, in SI units, must a car have to go from zero to 60 mph in 10 s?
   b. What fraction of $g$ is this?
   c. How far has the car traveled when it reaches 60 mph? Give your answer both in SI units and in feet.

30. || When jumping, a flea rapidly extends its legs, reaching a takeoff speed of 1.0 m/s over a distance of 0.50 mm.
BIO
   a. What is the flea's acceleration as it extends its legs?
   b. How long does it take the flea to leave the ground after it begins pushing off?

31. ||| In a car crash, large accelerations of the head can lead to severe injuries or even death. A driver can probably survive an acceleration of $50g$ that lasts for less than 30 ms, but in a crash with a $50g$ acceleration lasting longer than 30 ms, a driver is unlikely to survive. Imagine a collision in which a driver's head experienced a $50g$ acceleration.
BIO
   a. What is the highest speed that the car could have had such that the driver survived?
   b. What is the shortest survivable distance over which the driver's head could have come to rest?

32. || Light-rail passenger trains that provide transportation within and between cities speed up and slow down with a nearly constant (and quite modest) acceleration. A train travels through a congested part of town at 5.0 m/s. Once free of this area, it speeds up to 12 m/s in 8.0 s. At the edge of town, the driver again accelerates, with the same acceleration, for another 16 s to reach a higher cruising speed. What is the final speed?

33. ||| A cross-country skier is skiing along at a zippy 8.0 m/s. She stops pushing and simply glides along, slowing to a reduced speed of 6.0 m/s after gliding for 5.0 m. What is the magnitude of her acceleration as she slows?

34. || A small propeller airplane can comfortably achieve a high enough speed to take off on a runway that is 1/4 mile long. A large, fully loaded passenger jet has about the same acceleration from rest, but it needs to achieve twice the speed to take off. What is the minimum runway length that will serve? **Hint:** You can solve this problem using ratios without having any additional information.

35. || Formula One racers speed up much more quickly than normal passenger vehicles, and they also can stop in a much shorter distance. A Formula One racer traveling at 90 m/s can stop in a distance of 110 m. What is the magnitude of the car's acceleration as it slows during braking?

36. || Figure P2.36 shows a velocity-versus-time graph for a particle moving along the $x$-axis. At $t = 0$ s, assume that $x = 0$ m.
   a. What are the particle's position, velocity, and acceleration at $t = 1.0$ s?
   b. What are the particle's position, velocity, and acceleration at $t = 3.0$ s?

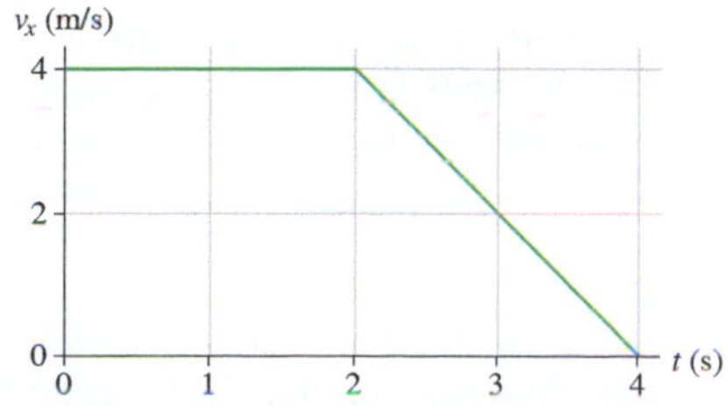


FIGURE P2.36

### Section 2.6 Solving One-Dimensional Motion Problems

37. || A driver has a reaction time of 0.50 s, and the maximum deceleration of her car is 6.0 m/s$^2$. She is driving at 20 m/s when suddenly she sees an obstacle in the road 50 m in front of her. Can she stop the car in time to avoid a collision?

38. || Chameleons catch insects with their tongues, which they can rapidly extend to great lengths. In a typical strike, the chameleon's tongue accelerates at a remarkable 250 m/s$^2$ for 20 ms, then travels at constant speed for another 30 ms. During this total time of 50 ms, 1/20 of a second, how far does the tongue reach?
BIO

39. ||| You're driving down the highway late one night at 20 m/s when a deer steps onto the road 35 m in front of you. Your reaction time before stepping on the brakes is 0.50 s, and the maximum deceleration of your car is 10 m/s$^2$.
   a. How much distance is between you and the deer when you come to a stop?
   b. What is the maximum speed you could have and still not hit the deer?

40. || Upon impact, bicycle helmets compress, thus lowering the potentially dangerous acceleration experienced by the head. A new kind of helmet uses an airbag that deploys from a pouch worn around the rider's neck. In tests, a headform wearing the inflated airbag is dropped onto a rigid platform; the speed just before impact is 6.0 m/s. Upon impact, the bag compresses its full 12.0 cm thickness, slowing the headform to rest. What is the acceleration, in $g$'s, experienced by the headform? (An acceleration greater than $60g$ is considered especially dangerous.)
BIO

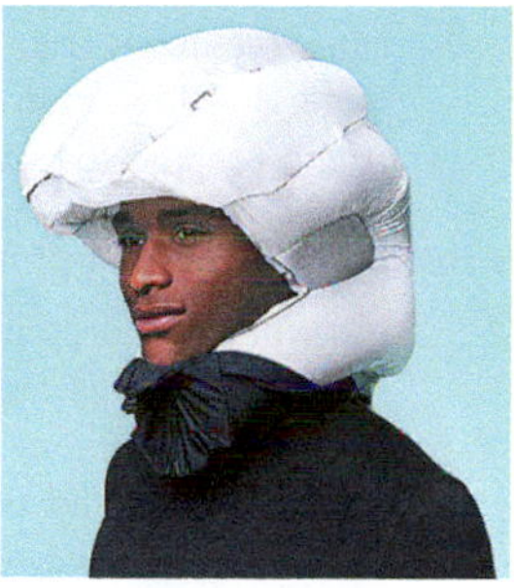

41. || A car is traveling at a steady 80 km/h in a 50 km/h zone. A police motorcycle takes off at the instant the car passes it, accelerating at a steady 8.0 m/s$^2$.
   a. How much time elapses before the motorcycle is moving as fast as the car?
   b. How far is the motorcycle from the car when it reaches this speed?

42. ||| The velocity-versus-time graph for the vertical jump of a green leafhopper, a small insect, is shown in Figure P2.42. This insect is unusual because it jumps with nearly constant acceleration.
BIO
   a. Estimate the leafhopper's acceleration.
   b. About how far does it move during this phase of its jump?

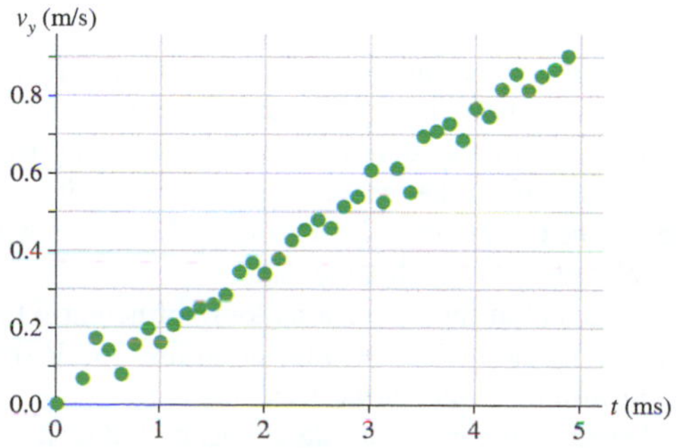


FIGURE P2.42

Watch Video Solution Problems 2.47 and 2.57

43. ||| A simple model for a person running the 100 m dash is to assume the sprinter runs with constant acceleration until reaching top speed, then maintains that speed through the finish line. If a sprinter reaches his top speed of 11.2 m/s in 2.14 s, what will be his total time?

## Section 2.7 Free Fall

44. ||| BIO Scientists have investigated how quickly hoverflies start beating their wings when dropped both in complete darkness and in a lighted environment. Starting from rest, the insects were dropped from the top of a 40-cm-tall box. In the light, those flies that began flying 200 ms after being dropped avoided hitting the bottom of the box 80% of the time, while those in the dark avoided hitting only 22% of the time.
   a. How far would a fly have fallen in the 200 ms before it began to beat its wings?
   b. How long would it take for a fly to hit the bottom if it never began to fly?

45. || BIO Here's an interesting challenge you can give to a friend. Hold a $1 (or larger!) bill by an upper corner. Have a friend prepare to pinch a lower corner, putting her fingers near but not touching the bill. Tell her to try to catch the bill when you drop it by simply closing her fingers. This seems like it should be easy, but it's not. After she sees that you have released the bill, it will take her about 0.25 s to react and close her fingers—which is not fast enough to catch the bill. How much time does it take for the bill to fall beyond her grasp? The length of a bill is 16 cm.

46. ||| BIO In the preceding problem we saw that a person's reaction time is generally not quick enough to allow the person to catch a $1 bill dropped between the fingers. The 16 cm length of the bill passes through a student's fingers before she can grab it if she has a typical 0.25 s reaction time. How long would a bill need to be for her to have a good chance of catching it?

47. | A gannet is a seabird that fishes by diving from a great height. If a gannet hits the water at 32 m/s (which they do), what height did it dive from? Assume that the gannet was motionless before starting its dive.

48. |||| BIO Steelhead trout migrate upriver to spawn. Occasionally they need to leap up small waterfalls to continue their journey. Fortunately, steelhead are remarkable jumpers, capable of leaving the water at a speed of 8.0 m/s.
   a. What is the maximum height that a steelhead can jump?
   b. Leaving the water vertically at 8.0 m/s, a steelhead lands on the top of a waterfall 1.8 m high. How long is it in the air?

49. ||| In a circus act, an acrobat rebounds upward from the surface of a trampoline at the exact moment that another acrobat, perched 9.0 m above him, releases a ball from rest. While still in flight, the acrobat catches the ball just as it reaches him. If he left the trampoline with a speed of 8.0 m/s, how long is he in the air before he catches the ball?

50. | A student at the top of a building of height $h$ throws ball A straight upward with speed $v_0$ and throws ball B straight downward with the same initial speed.
   a. Compare the balls' accelerations, both direction and magnitude, immediately after they leave her hand. Is one acceleration larger than the other? Or are the magnitudes equal?
   b. Compare the final speeds of the balls as they reach the ground. Is one larger than the other? Or are they equal?

51. || BIO Excellent human jumpers can leap straight up to a height of 110 cm off the ground. To reach this height, with what speed would a person need to leave the ground?

52. || A football is kicked straight up into the air; it hits the ground 5.2 s later.
   a. What was the greatest height reached by the ball? Assume it is kicked from ground level.
   b. With what speed did it leave the kicker's foot?

53. ||||| In an action movie, the villain is rescued from the ocean by grabbing onto the ladder hanging from a helicopter. He is so intent on gripping the ladder that he lets go of his briefcase of counterfeit money when he is 130 m above the water. If the briefcase hits the water 6.0 s later, what was the speed at which the helicopter was ascending?

54. ||| BIO Spud Webb was, at 5 ft 8 in, one of the shortest basketball players to play in the NBA. But he had an amazing vertical leap; he could jump to a height of 1.1 m off the ground, so he could easily dunk a basketball. For such a leap, what was his "hang time"—the time spent in the air after leaving the ground and before touching down again?

55. |||| A rock climber stands on top of a 50-m-high cliff overhanging a pool of water. He throws two stones vertically downward 1.0 s apart and observes that they cause a single splash. The initial speed of the first stone was 2.0 m/s.
   a. How long after the release of the first stone does the second stone hit the water?
   b. What was the initial speed of the second stone?
   c. What is the speed of each stone as it hits the water?

## General Problems

56. ||| BIO Actual velocity data for a lion pursuing prey are shown in Figure P2.56. Estimate:
   a. The initial acceleration of the lion.
   b. The acceleration of the lion at 2 s and at 4 s.
   c. The distance traveled by the lion between 0 s and 8 s.

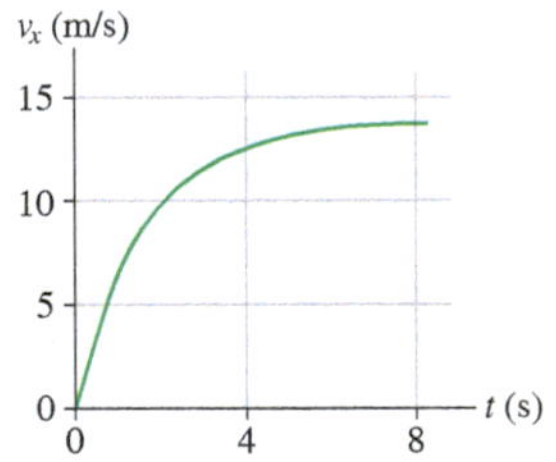


FIGURE P2.56

57. || A truck driver has a shipment of apples to deliver to a destination 440 miles away. The trip usually takes him 8 hours. Today he finds himself daydreaming and realizes 120 miles into his trip that he is running 15 minutes later than his usual pace at this point. At what speed must he drive for the remainder of the trip to complete the trip in the usual amount of time?

58. || Jenny and Alyssa are members of the cross-country team. On a training run, Jenny starts off and runs at a constant 3.8 m/s. Alyssa starts 15 s later and runs at a constant 4.0 m/s. At what time after Jenny's start does Alyssa catch up with Jenny?

59. || Figure P2.59 shows the motion diagram, made at two frames of film per second, of a ball rolling along a track. The track has a 3.0-m-long sticky section.
 a. Use the scale to determine the positions of the center of the ball. Place your data in a table, similar to Table 2.1, showing each position and the instant of time at which it occurred.
 b. Make a graph of $x$ versus $t$ for the ball. Because you have data only at certain instants of time, your graph should consist of dots that are not connected together.
 c. What is the *change* in the ball's position from $t = 0$ s to $t = 1.0$ s?
 d. What is the *change* in the ball's position from $t = 2.0$ s to $t = 4.0$ s?
 e. What is the ball's velocity before reaching the sticky section?
 f. What is the ball's velocity after passing the sticky section?
 g. Determine the ball's acceleration on the sticky section of the track.

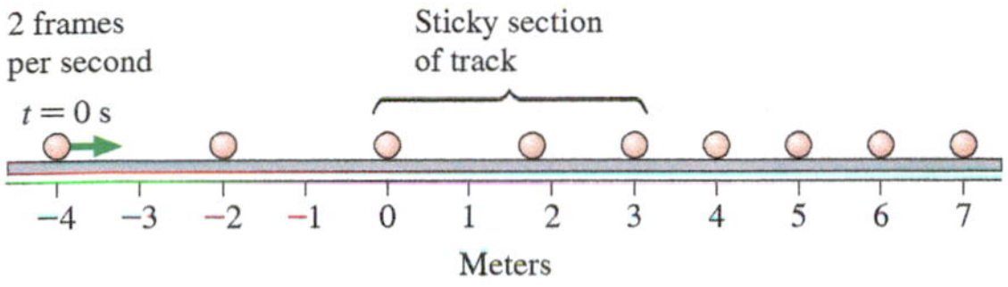

FIGURE P2.59

60. ||| In a 5000 m race, the athletes run $12\frac{1}{2}$ laps; each lap is 400 m. Kara runs the race at a constant pace and finishes in 17.5 min. Hannah runs the race in a blistering 15.3 min, so fast that she actually passes Kara during the race. How many laps has Hannah run when she passes Kara?

61. | The takeoff speed for an Airbus A320 jetliner is 80 m/s. Velocity data measured during takeoff are as shown in the table.
 a. What is the jetliner's acceleration during takeoff, in $m/s^2$ and in $g$'s?
 b. At what time do the wheels leave the ground?
 c. For safety reasons, in case of an aborted takeoff, the length of the runway must be three times the takeoff distance. What is the minimum length runway this aircraft can use?

| $t$ (s) | $v_x$ (m/s) |
|---|---|
| 0 | 0 |
| 10 | 23 |
| 20 | 46 |
| 30 | 69 |

62. |||| Does a real automobile have constant acceleration? Measured data for a Porsche 944 Turbo at maximum acceleration are as shown in the table.
 a. Convert the velocities to m/s, then make a graph of velocity versus time. Based on your graph, is the acceleration constant? Explain.
 b. Estimate how far the car traveled in the first 10 s.
 c. Draw a smooth curve through the points on your graph, then use your graph to *estimate* the car's acceleration at 2.0 s and 8.0 s. Give your answer in SI units. **Hint:** Remember that acceleration is the slope of the velocity graph.

| t (s) | $v_x$ (mph) |
|---|---|
| 0 | 0 |
| 2 | 41 |
| 4 | 66 |
| 6 | 83 |
| 8 | 97 |
| 10 | 110 |

63. || Scientists have studied two species of sand lizards, the Mojave fringe-toed lizard and the western zebra-tailed lizard, to understand the extent to which the different structure of the two species' toes is related to their preferred habitats—fine sand for the Mojave lizard and coarse sand for the zebra-tailed lizard. Figure P2.63 shows a somewhat simplified velocity-versus-time graph for the Mojave fringe-toed lizard.

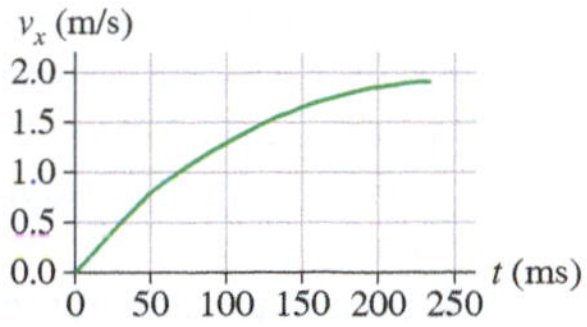

FIGURE P2.63

 a. Estimate the maximum acceleration of the lizard in both $m/s^2$ and $g$'s.
 b. Estimate its acceleration at $t = 150$ ms.
 c. Estimate how far it travels in the first 50 ms.

64. ||| You are driving to the grocery store at 20 m/s. You are 110 m from an intersection when the traffic light turns red. Assume that your reaction time is 0.70 s and that your car brakes with constant acceleration.
 a. How far are you from the intersection when you begin to apply the brakes?
 b. What acceleration will bring you to rest right at the intersection?
 c. How long does it take you to stop?

65. | When you blink your eye, the upper lid goes from rest with your
BIO eye open to completely covering your eye in a time of 0.024 s.
 a. Estimate the distance that the top lid of your eye moves during a blink.
 b. What is the acceleration of your eyelid? Assume it to be constant.
 c. What is your upper eyelid's final speed as it hits the bottom eyelid?

66. ||| A bush baby, an African primate, is
BIO capable of a remarkable vertical leap. The bush baby goes into a crouch and extends its legs, pushing upward for a distance of 0.16 m. After this upward acceleration, the bush baby leaves the ground and travels upward for 2.3 m. What is the acceleration during the pushing-off phase? Give your answer in $m/s^2$ and in $g$'s.

67. |||| When jumping, a flea reaches a takeoff speed of 1.0 m/s
BIO over a distance of 0.50 mm.
 a. What is the flea's acceleration during the jump phase?
 b. How long does the acceleration phase last?
 c. If the flea jumps straight up, how high will it go? (Ignore air resistance for this problem; in reality, air resistance plays a large role, and the flea will not reach this height.)

68. ||| Certain insects can achieve seem-
BIO ingly impossible accelerations while jumping. The click beetle accelerates at an astonishing $400g$ over a distance of 0.60 cm as it rapidly bends its thorax, making the "click" that gives it its name.

 a. Assuming the beetle jumps straight up, at what speed does it leave the ground?
 b. How much time is required for the beetle to reach this speed?
 c. Ignoring air resistance, how high would it go?

Watch Video Solution Problem 2.75

69. |||| A student standing on the ground throws a ball straight up. The ball leaves the student's hand with a speed of 15 m/s when the hand is 2.0 m above the ground. How long is the ball in the air before it hits the ground? (The student moves her hand out of the way.)

70. |||| A rock is tossed straight up with a speed of 20 m/s. When it returns, it falls into a hole 10 m deep.
   a. What is the rock's velocity as it hits the bottom of the hole?
   b. How long is the rock in the air, from the instant it is released until it hits the bottom of the hole?

71. |||| In springboard diving, the diver strides out to the end of the board, takes a jump onto its end, and uses the resultant spring-like nature of the board to help propel him into the air. Assume that the diver's motion is essentially vertical. He leaves the board, which is 3.0 m above the water, with a speed of 6.3 m/s.
   a. How long is the diver in the air, from the moment he leaves the board until he reaches the water?
   b. What is the speed of the diver when he reaches the water?

72. ||| Haley is driving down a straight highway at 75 mph. A construction sign warns that the speed limit will drop to 55 mph in 0.50 mi. What constant acceleration (in m/s) will bring Haley to this lower speed in the distance available?

73. |||| A car starts from rest at a stop sign. It accelerates at 2.0 $m/s^2$ for 6.0 seconds, coasts for 2.0 s, and then slows down at a rate of 1.5 $m/s^2$ for the next stop sign. How far apart are the stop signs?

74. ||| BIO Chameleons can rapidly project their very long tongues to catch nearby insects. The tongue of the tiny Rosette-nosed chameleon has the highest acceleration of a body part of any amniote (reptile, bird, or mammal) ever measured. In a somewhat simplified model of its tongue motion, the tongue, starting from rest, first undergoes a constant-acceleration phase with an astounding magnitude of 2500 $m/s^2$. This acceleration brings the tongue up to a final speed of 5.0 m/s. It continues at this speed for 22 ms until it hits its target.
   a. How long does the acceleration phase last?
   b. What is the total distance traveled by the chameleon's tongue?

75. ||| Heather and Jerry are standing on a bridge 50 m above a river. Heather throws a rock straight down with a speed of 20 m/s. Jerry, at exactly the same instant of time, throws a rock straight up with the same speed. Ignore air resistance.
   a. How much time elapses between the first splash and the second splash?
   b. Which rock has the faster speed as it hits the water?

76. || BIO A Thomson's gazelle can run at very high speeds, but its acceleration is relatively modest. A reasonable model for the sprint of a gazelle assumes an acceleration of 4.2 $m/s^2$ for 6.5 s, after which the gazelle continues at a steady speed.
   a. What is the gazelle's top speed?
   b. A human would win a very short race with a gazelle. The best time for a 30 m sprint for a human runner is 3.6 s. How much time would the gazelle take for a 30 m race?
   c. A gazelle would win a longer race. The best time for a 200 m sprint for a human runner is 19.3 s. How much time would the gazelle take for a 200 m race?

77. ||| BIO We've seen that a man's higher initial acceleration means that he can outrun a horse in very short race. A simple—but plausible—model for a sprint by a man and a horse uses these assumptions: The man accelerates at 6.0 $m/s^2$ for 1.8 s and then runs at a constant speed. A horse accelerates at 5.0 $m/s^2$ but continues accelerating for 4.8 s and then continues at a constant speed. A man and a horse are competing in a 200 m race. The man is given a 100 m head start, so he begins 100 m from the finish line. How much time does the man take to complete the race? How much time does the horse take? Who wins the race?

78. || A pole-vaulter is nearly motionless as he clears the bar, set 4.2 m above the ground. He then falls onto a thick pad. The top of the pad is 80 cm above the ground, and it compresses by 50 cm as he comes to rest. What is his acceleration as he comes to rest on the pad?

79. || A Porsche challenges a Honda to a 400 m race. Because the Porsche's acceleration of 3.5 $m/s^2$ is larger than the Honda's 3.0 $m/s^2$, the Honda gets a 100-m head start—it is only 300 m from the finish line. Assume, somewhat unrealistically, that both cars can maintain these accelerations the entire distance. Who wins, and by how much time?

80. ||||| The minimum stopping distance for a car traveling at a speed of 30 m/s is 60 m, including the distance traveled during the driver's reaction time of 0.50 s.
   a. Draw a position-versus-time graph for the motion of the car. Assume the car is at $x_i = 0$ m when the driver first sees the emergency situation ahead that calls for a rapid halt.
   b. What is the minimum stopping distance for the same car traveling at a speed of 40 m/s?

81. ||||| A rocket is launched straight up with constant acceleration. Four seconds after liftoff, a bolt falls off the side of the rocket. The bolt hits the ground 6.0 s later. What was the rocket's acceleration?

## MCAT-Style Passage Problems

### Free Fall on Different Worlds

Objects in free fall on the earth have acceleration $a_y = -9.8\ m/s^2$. On the moon, free-fall acceleration is approximately 1/6 of the acceleration on earth. This changes the scale of problems involving free fall. For instance, suppose you jump straight upward, leaving the ground with velocity $v_i$ and then steadily slowing until reaching zero velocity at your highest point. Because your initial velocity is determined mostly by the strength of your leg muscles, we can assume your initial velocity would be the same on the moon. But considering the final equation in Synthesis 2.1 we can see that, with a smaller free-fall acceleration, your maximum height would be greater. The following questions ask you to think about how certain athletic feats might be performed in this reduced-gravity environment.

82. | If an astronaut can jump straight up to a height of 0.50 m on earth, how high could he jump on the moon?
   A. 1.2 m B. 3.0 m C. 3.6 m D. 18 m

83. | On the earth, an astronaut can safely jump to the ground from a height of 1.0 m; her velocity when reaching the ground is slow enough to not cause injury. From what height could the astronaut safely jump to the ground on the moon?
   A. 2.4 m B. 6.0 m C. 7.2 m D. 36 m

84. | On the earth, an astronaut throws a ball straight upward; it stays in the air for a total time of 3.0 s before reaching the ground again. If a ball were to be thrown upward with the same initial speed on the moon, how much time would pass before it hit the ground?
   A. 7.3 s B. 18 s C. 44 s D. 108 s

# 3 Vectors and Motion in Two Dimensions

Once the leopard jumps, its trajectory is fixed by the initial speed and angle of the jump. How can we work out where the leopard will land?

## LOOKING AHEAD ▸▸

### Vectors and Components

The dark green vector represents the basketball's velocity. The light green **component vectors** show its horizontal and vertical velocities.

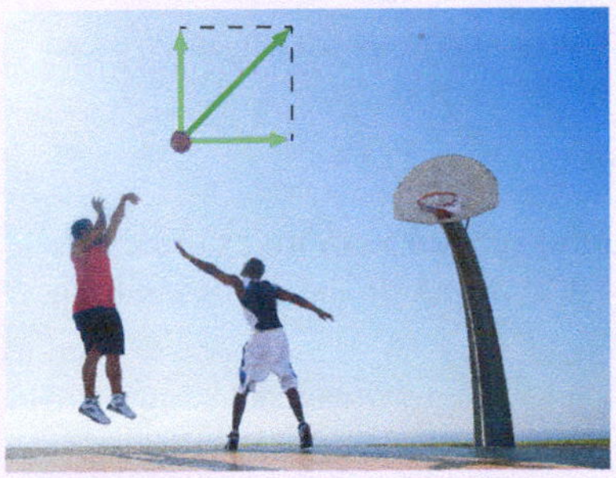

You'll learn how to find components of vectors and how to use these components to solve problems.

### Projectile Motion

A leaping fish's parabolic arc is an example of **projectile motion.** The details are the same for a fish or a basketball.

You'll see how to solve projectile motion problems, determining how long an object is in the air and how far it travels.

### Circular Motion

These riders move in a circle at a constant speed, but they have an acceleration because the direction of their motion is constantly changing.

You'll learn how to determine the magnitude and the direction of the acceleration for an object in circular motion.

**GOAL** **To learn more about vectors and to use vectors as a tool to analyze motion in two dimensions.**

## LOOKING BACK ◂◂

### Free Fall

You learned in Section 2.7 that an object tossed straight up is in free fall. The acceleration is the same whether the object is going up or coming back down.

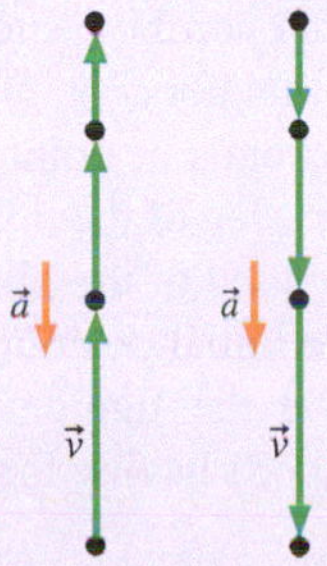


For an object in projectile motion, the vertical component of the motion is also free fall. You'll use your knowledge of free fall to solve projectile motion problems.

**STOP TO THINK**

A player kicks a soccer ball straight up into the air. The ball takes 2.0 s to reach its highest point. Approximately how fast was the ball moving when it left the player's foot?

A. 5 m/s  B. 10 m/s
C. 15 m/s  D. 20 m/s

## 3.1 Using Vectors

In Chapter 2, we solved problems in which an object moved in a straight-line path. In this chapter, objects move in curving paths, so that their motion is in two dimensions. A ball arcing through the air is an example of this two-dimensional motion, because it is moving horizontally and vertically at the same time. Since the direction of motion will be so important, we need to develop an appropriate mathematical language to describe it—the language of vectors.

We introduced the concept of a vector in ◀◀ SECTION 1.6. In the next few sections we will develop techniques for working with vectors as a tool for studying motion in two dimensions. As we learned, a vector is a quantity with both a size (magnitude) and a direction. FIGURE 3.1 shows how to represent a particle's velocity as a vector $\vec{v}$. The particle's speed at this point is 5 m/s *and* it is moving in the direction indicated by the arrow. Recall that the magnitude of a vector is represented by the letter symbol of the vector, but without an arrow. In this case, the particle's speed—the magnitude of the velocity vector $\vec{v}$—is $v = 5$ m/s. The magnitude of a vector, a *scalar* quantity, cannot be a negative number.

FIGURE 3.1 The velocity vector $\vec{v}$ has both a magnitude and a direction.

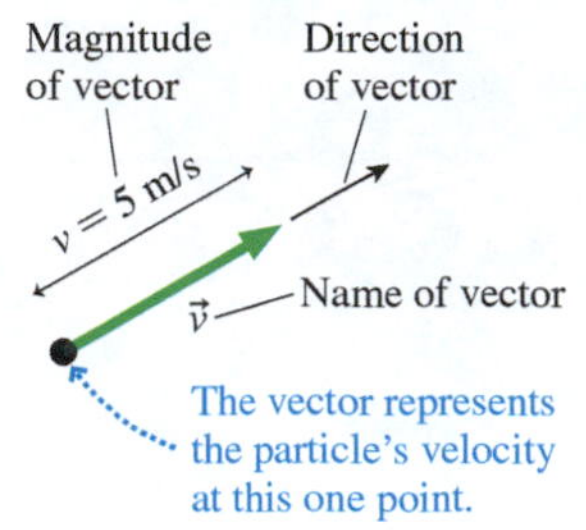


**NOTE** ▶ Although the vector arrow is drawn across the page, from its tail to its tip, this arrow does *not* indicate that the vector "stretches" across this distance. Instead, the arrow tells us the value of the vector quantity only at the one point where the tail of the vector is placed. ◀

We saw in Chapter 1 that the displacement of an object is a vector drawn from its initial position to its position at some later time. Because displacement is an easy concept to think about, we will use it to introduce some of the properties of vectors. However, all the properties we will discuss in this chapter **(addition, subtraction, multiplication, components) apply to all types of vectors, not just to displacement.**

Suppose that Sam, our old friend from Chapter 1, starts from his front door, walks across the street, and ends up 200 ft to the northeast of where he started. Sam's displacement, which we will label $\vec{d}_S$, is shown in FIGURE 3.2a. The displacement vector is a straight-line connection from his initial to his final position, not necessarily his actual path. The dashed line indicates a possible route Sam might have taken, but his displacement is the vector $\vec{d}_S$ from his initial to his final position.

FIGURE 3.2 Displacement vectors.

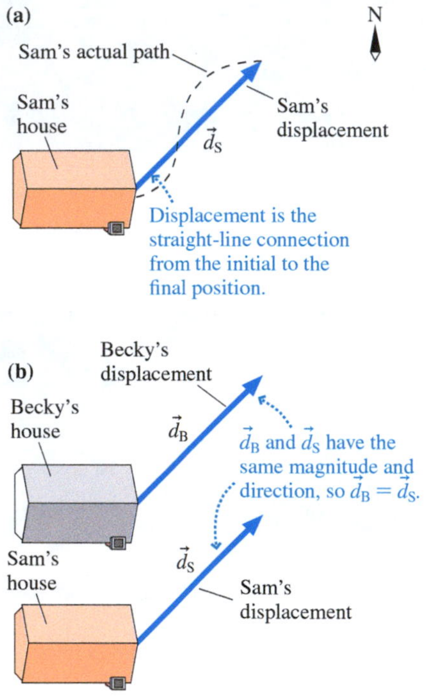


To describe a vector we must specify both its magnitude and its direction. We can write Sam's displacement as

$$\vec{d}_S = (200 \text{ ft, northeast})$$

where the first number specifies the magnitude and the second item gives the direction. The magnitude of Sam's displacement is $d_S = 200$ ft, the distance between his initial and final points.

Sam's next-door neighbor Becky also walks 200 ft to the northeast, starting from her own front door. Becky's displacement $\vec{d}_B = (200 \text{ ft, northeast})$ has the same magnitude and direction as Sam's displacement $\vec{d}_S$. Because vectors are defined by their magnitude and direction, **two vectors are equal if they have the same magnitude and direction.** This is true regardless of the individual starting points of the vectors. Thus the two displacements in FIGURE 3.2b are equal to each other, and we can write $\vec{d}_B = \vec{d}_S$.

### Vector Addition

As we saw in Chapter 1, we can combine successive displacements by vector addition. Let's review and extend this concept. FIGURE 3.3 shows the displacement of a hiker who starts at point P and ends at point S. She first hikes 4 miles to the east, then 3 miles to the north. The first leg of the hike is described by the displacement vector $\vec{A} = (4 \text{ mi, east})$. The second leg of the hike has displacement $\vec{B} = (3 \text{ mi, north})$. By definition, a vector from her initial position P to her final position S is also a displacement. This is vector $\vec{C}$ on the figure. $\vec{C}$ is the *net displacement* because it describes the net result of the hiker's having first displacement $\vec{A}$, then displacement $\vec{B}$.

FIGURE 3.3 The net displacement $\vec{C}$ resulting from two displacements $\vec{A}$ and $\vec{B}$.

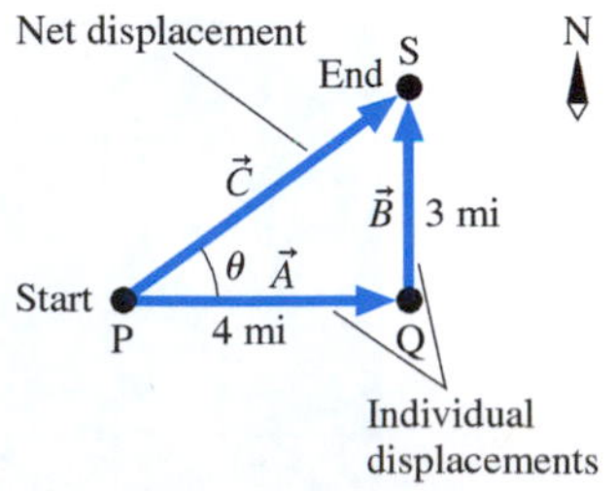

The word "net" implies addition. The net displacement $\vec{C}$ is an initial displacement $\vec{A}$ *plus* a second displacement $\vec{B}$, or

$$\vec{C} = \vec{A} + \vec{B} \tag{3.1}$$

The sum of two vectors is called the **resultant vector.** Vector addition is commutative: $\vec{A} + \vec{B} = \vec{B} + \vec{A}$. You can add vectors in any order you wish.

Look back at ◀◀ TACTICS BOX 1.4 to review the three-step procedure for adding two vectors. This *tip-to-tail* method for adding vectors, which is used to find $\vec{C} = \vec{A} + \vec{B}$ in Figure 3.3, is called *graphical addition.*

There is a second method for adding vectors that is sometimes more convenient than the tip-to-tail method. FIGURE 3.4a shows the addition of two vectors $\vec{D}$ and $\vec{E}$ using the tip-to-tail rule. But recall that a vector such as $\vec{E}$ is not fixed in place on the diagram; it can be "slid" to any other location as long as its magnitude and direction remain unchanged. In particular, it can be moved so that its tail coincides with the tail of $\vec{D}$, as shown in FIGURE 3.4b. In this case, you can see that the vector sum $\vec{D} + \vec{E}$ can be found as the diagonal of the parallelogram defined by $\vec{D}$ and $\vec{E}$. This method is called the *parallelogram rule* of vector addition. It is especially useful when two vectors are already drawn with their tails touching.

FIGURE 3.4 Two vectors can be added using the tip-to-tail rule or the parallelogram rule.

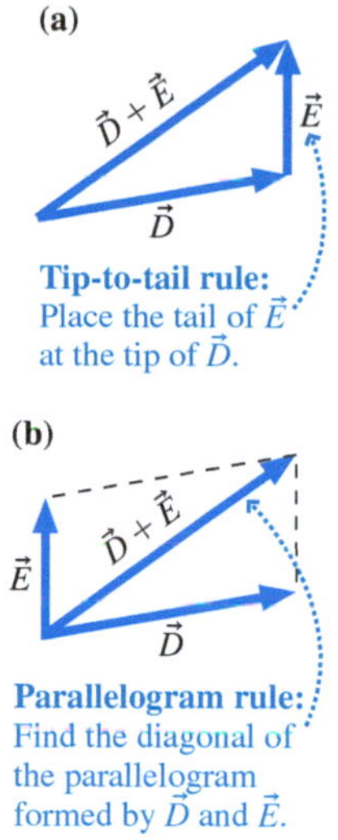


## Multiplication by a Scalar

The hiker in Figure 3.3 started with displacement $\vec{A}_1 = (4 \text{ mi, east})$. Suppose a second hiker walks twice as far to the east. The second hiker's displacement will then certainly be $\vec{A}_2 = (8 \text{ mi, east})$. The words "twice as" indicate a multiplication, so we can say

$$\vec{A}_2 = 2\vec{A}_1$$

**Multiplying a vector by a positive scalar gives another vector of *different magnitude* but pointing in the *same direction.***

Suppose a vector $\vec{A}$ has magnitude $A$ and direction $\theta$. Now we let $\vec{B} = c\vec{A}$, where $c$ is a positive scalar constant. Then vector $\vec{B}$ is stretched or compressed by the factor $c$ (i.e., it has magnitude $B = cA$), but $\vec{B}$ points in the *same* direction as $\vec{A}$. This is illustrated in FIGURE 3.5.

FIGURE 3.5 Multiplication of a vector by a positive scalar.

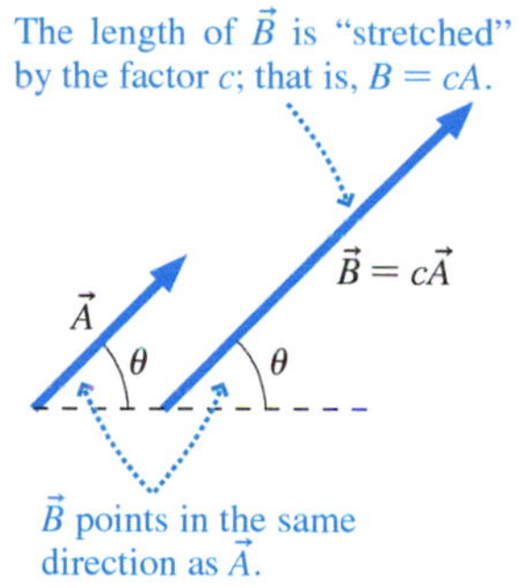


What happens if we multiply $\vec{A}$ by zero? We get a vector that has zero length or magnitude. This vector is known as the **zero vector,** denoted $\vec{0}$. The direction of the zero vector is not defined; you cannot describe the direction of an arrow of zero length!

Finally, what happens if we multiply a vector by a negative number? Consider the vector $-\vec{A}$, which is equivalent to multiplying $\vec{A}$ by $-1$. Because

$$\vec{A} + (-\vec{A}) = \vec{0} \tag{3.2}$$

the vector $-\vec{A}$ must be such that, when it is added to $\vec{A}$, the resultant is the zero vector $\vec{0}$. In other words, the *tip* of $-\vec{A}$ must return to the *tail* of $\vec{A}$, as shown in FIGURE 3.6. This will be true only if $-\vec{A}$ is equal in magnitude to $\vec{A}$ but opposite in direction. Thus we can conclude that

$$-\vec{A} = (A, \text{direction opposite } \vec{A}) \tag{3.3}$$

FIGURE 3.6 Vector $-\vec{A}$.

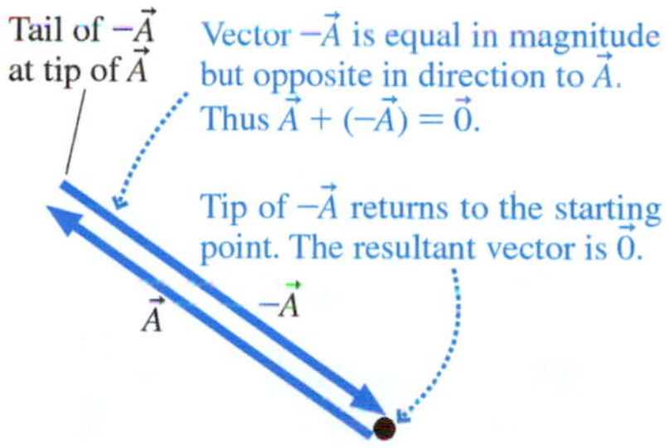


Multiplying a vector by $-1$ reverses its direction without changing its length.

FIGURE 3.7 shows vectors $\vec{A}$, $2\vec{A}$, and $-3\vec{A}$. Multiplication by 2 doubles the length of the vector but does not change its direction. Multiplication by $-3$ stretches the length by a factor of 3 *and* reverses the direction.

FIGURE 3.7 Vectors $\vec{A}$, $2\vec{A}$, and $-3\vec{A}$.

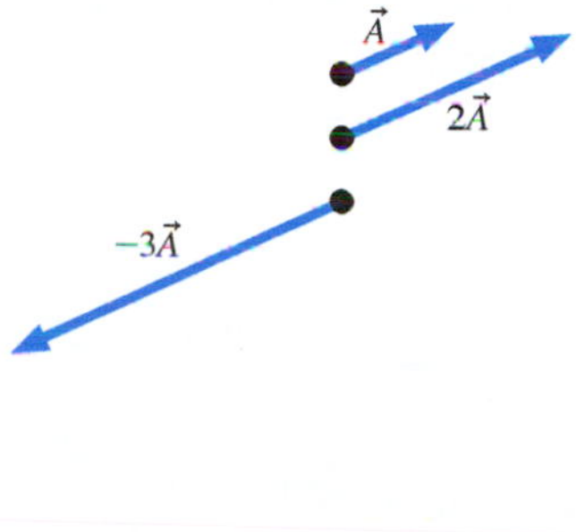


## Vector Subtraction

How might we *subtract* vector $\vec{B}$ from vector $\vec{A}$ to form the vector $\vec{A} - \vec{B}$? With numbers, subtraction is the same as the addition of a negative number. That is, $5 - 3$ is the same as $5 + (-3)$. Similarly, $\vec{A} - \vec{B} = \vec{A} + (-\vec{B})$. We can use the rules for vector addition and the fact that $-\vec{B}$ is a vector opposite in direction to $\vec{B}$ to form rules for vector subtraction.

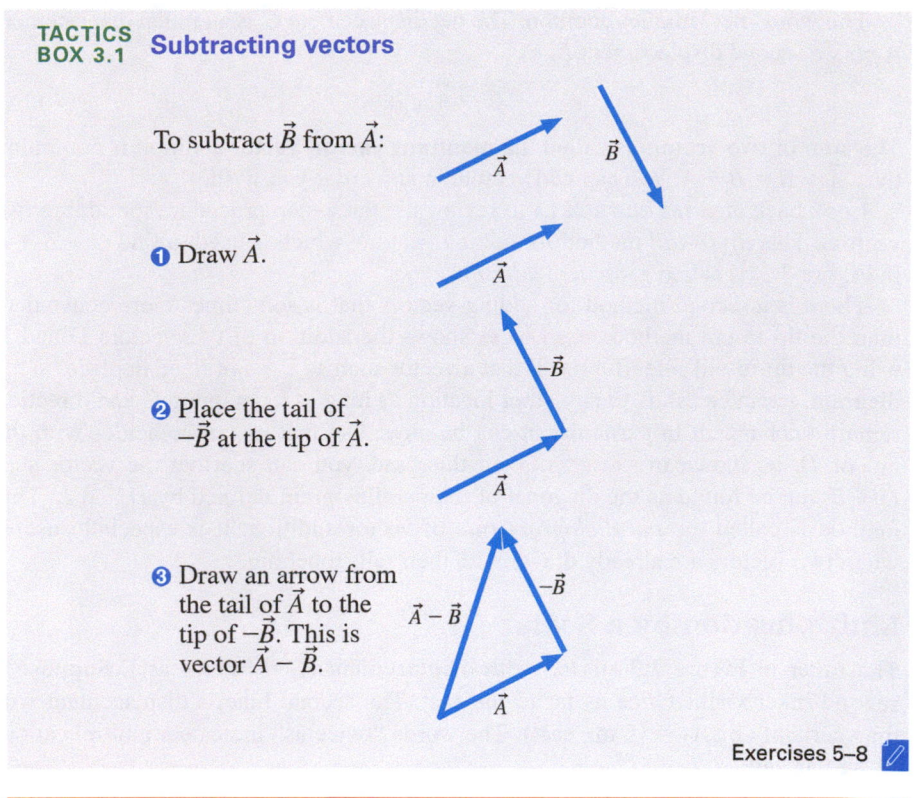

**TACTICS BOX 3.1 Subtracting vectors**

To subtract $\vec{B}$ from $\vec{A}$:

1. Draw $\vec{A}$.
2. Place the tail of $-\vec{B}$ at the tip of $\vec{A}$.
3. Draw an arrow from the tail of $\vec{A}$ to the tip of $-\vec{B}$. This is vector $\vec{A} - \vec{B}$.

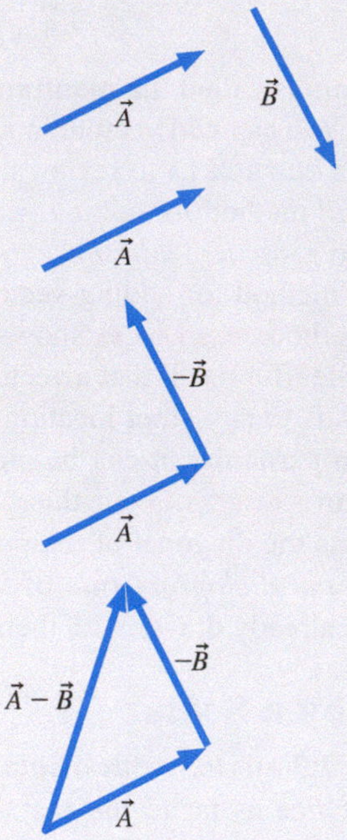


Exercises 5–8

**STOP TO THINK 3.1** Which figure shows $2\vec{P} - \vec{Q}$?

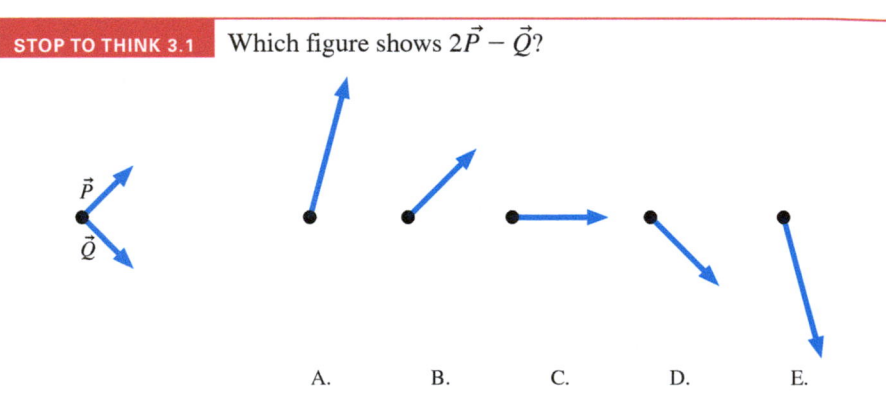

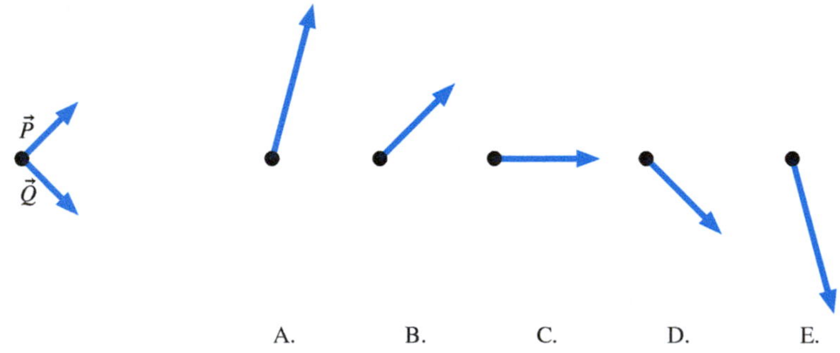

A coordinate system for an archaeological excavation.

## 3.2 Coordinate Systems and Vector Components

We've just learned how to add and subtract vectors graphically. But working with vectors graphically is not an especially good way to find quantitative results. In this section we will introduce a *coordinate description* of vectors that will be the basis for doing vector calculations.

### Coordinate Systems

As we saw in Chapter 1, the world does not come with a coordinate system attached to it. A coordinate system is an artificially imposed grid that you place on a problem in order to make quantitative measurements. The right choice of coordinate system will make a problem easier to solve. We will generally use **Cartesian coordinates,** the familiar rectangular grid with perpendicular axes, as illustrated in FIGURE 3.8.

FIGURE 3.8 A Cartesian coordinate system.

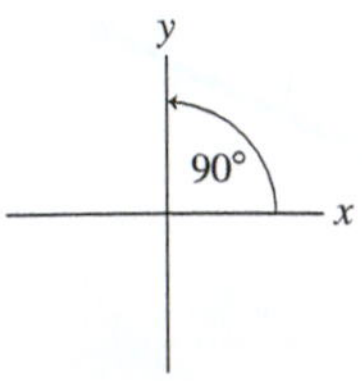


Coordinate axes have a positive end and a negative end, separated by zero at the origin where the two axes cross. When you draw a coordinate system, it is important to label the axes. This is done by placing $x$ and $y$ labels at the *positive* ends of the axes, as in Figure 3.8.

## Component Vectors

FIGURE 3.9 shows a vector $\vec{A}$ and an *xy*-coordinate system that we've chosen. Once the directions of the axes are known, we can define two new vectors *parallel to the axes* that we call the **component vectors** of $\vec{A}$. Vector $\vec{A}_x$, called the *x-component vector,* is the projection of $\vec{A}$ along the *x*-axis. Vector $\vec{A}_y$, the *y-component vector,* is the projection of $\vec{A}$ along the *y*-axis.

You can see, using the parallelogram rule, that $\vec{A}$ is the vector sum of the two component vectors:

$$\vec{A} = \vec{A}_x + \vec{A}_y \tag{3.4}$$

In essence, we have "broken" vector $\vec{A}$ into two perpendicular vectors that are parallel to the coordinate axes. We say that we have **decomposed,** or **resolved,** vector $\vec{A}$ into its component vectors.

FIGURE 3.9 Component vectors $\vec{A}_x$ and $\vec{A}_y$ are drawn parallel to the coordinate axes such that $\vec{A} = \vec{A}_x + \vec{A}_y$.

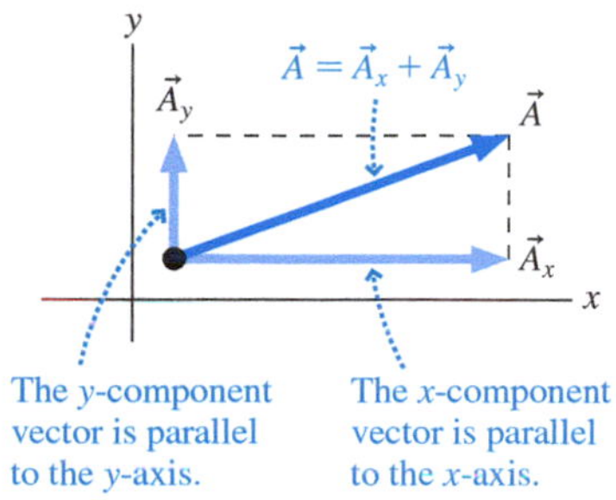


NOTE ▸ It is not necessary for the tail of $\vec{A}$ to be at the origin. All we need to know is the *orientation* of the coordinate system so that we can draw $\vec{A}_x$ and $\vec{A}_y$ parallel to the axes. ◂

## Components

As we learned from studying velocity in one-dimensional motion, we give $v_x$ a positive sign if the velocity vector $\vec{v}$ points toward the positive end of the *x*-axis and a negative sign if $\vec{v}$ points in the negative *x*-direction. Thus the *sign* of $v_x$ indicates the *direction* of the velocity. We can extend this idea to vectors in general.

Suppose we have a vector $\vec{A}$ that has been decomposed into component vectors $\vec{A}_x$ and $\vec{A}_y$ parallel to the coordinate axes. We can describe each component vector with a single number (a scalar) called the **component.** The *x-component* and *y-component* of vector $\vec{A}$, denoted $A_x$ and $A_y$, are determined as follows:

**TACTICS BOX 3.2 Determining the components of a vector**

1. The absolute value $|A_x|$ of the *x*-component $A_x$ is the magnitude of the component vector $\vec{A}_x$.
2. The *sign* of $A_x$ is positive if $\vec{A}_x$ points in the positive *x*-direction, negative if $\vec{A}_x$ points in the negative *x*-direction.
3. The *y*-component $A_y$ is determined similarly.

Exercises 12–14

FIGURE 3.10 Determining the components of a vector.

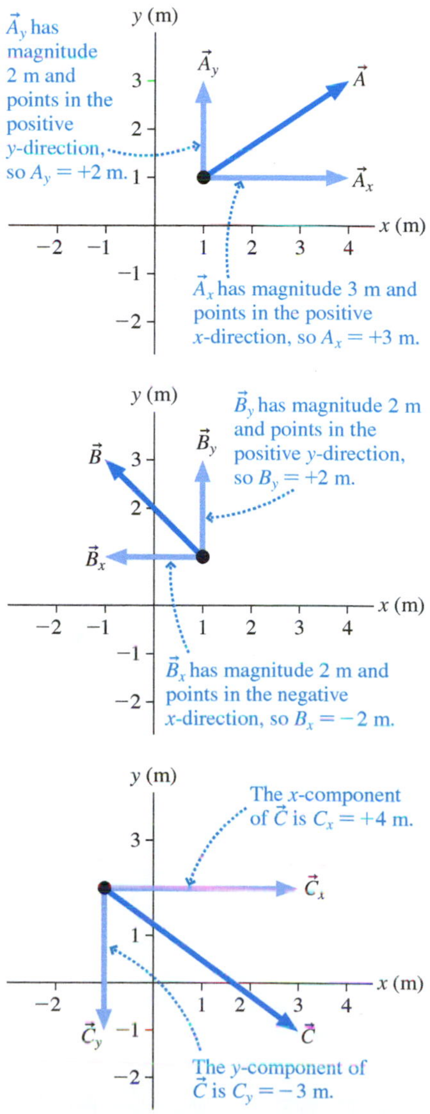


In other words, the component $A_x$ tells us two things: how big $\vec{A}_x$ is and toward which end of the axis $\vec{A}_x$ points. FIGURE 3.10 shows three examples of determining the components of a vector.

NOTE ▸ $\vec{A}_x$ and $\vec{A}_y$ are component *vectors*—they have a magnitude and a direction. $A_x$ and $A_y$ are simply *components.* The components $A_x$ and $A_y$ are scalars—just numbers (with units) that can be positive or negative. ◂

Much of physics is expressed in the language of vectors. We will frequently need to decompose a vector into its components or to "reassemble" a vector from its components, moving back and forth between the graphical and the component representations of a vector.

Let's start with the problem of decomposing a vector into its *x*- and *y*-components. FIGURE 3.11a shows a vector $\vec{A}$ at an angle $\theta$ above horizontal. It is *essential* to use a picture or diagram such as this to define the angle you are using to describe a vector's direction. $\vec{A}$ points to the right and up, so Tactics Box 3.2 tells us that the components $A_x$ and $A_y$ are both positive.

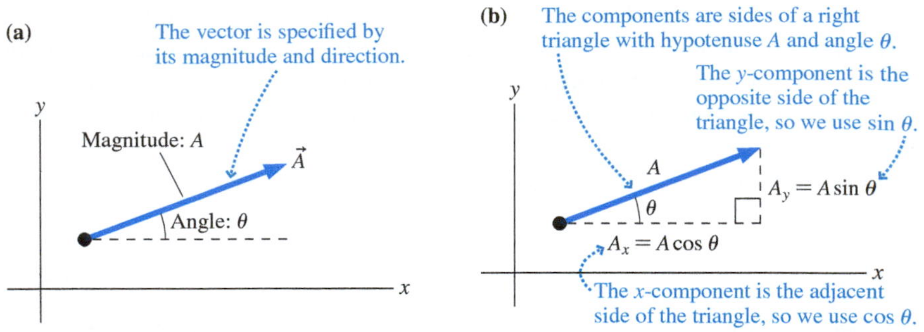


FIGURE 3.11 Finding the components of a vector.

We can find the components using trigonometry, as illustrated in FIGURE 3.11b. For this case, we find that

$$A_x = A\cos\theta$$
$$A_y = A\sin\theta \tag{3.5}$$

where $A$ is the magnitude, or length, of $\vec{A}$. These equations convert the length and angle description of vector $\vec{A}$ into the vector's components, but they are correct *only* if $\theta$ is measured from horizontal.

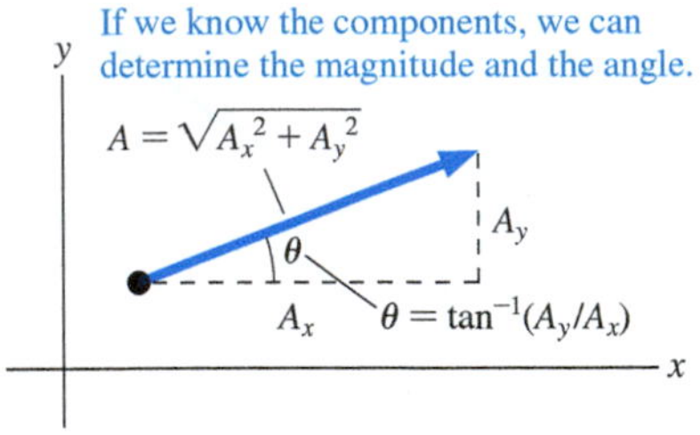


FIGURE 3.12 Specifying a vector from its components.

Alternatively, if we are given the $x$- and $y$-components of a vector, we can determine the length and angle of the vector, as shown in FIGURE 3.12. Because $A$ in Figure 3.12 is the hypotenuse of a right triangle, its length is given by the Pythagorean theorem:

$$A = \sqrt{A_x^2 + A_y^2} \tag{3.6}$$

Similarly, the tangent of angle $\theta$ is the ratio of the opposite side to the adjacent side, so

$$\theta = \tan^{-1}\left(\frac{A_y}{A_x}\right) \tag{3.7}$$

Equations 3.6 and 3.7 can be thought of as the "inverse" of Equations 3.5.

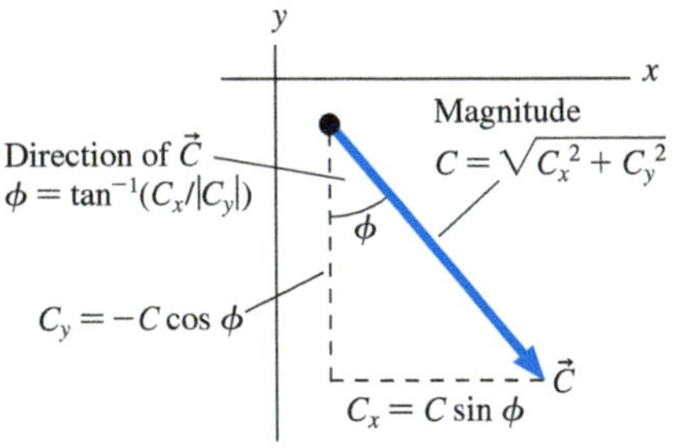


FIGURE 3.13 Relationships for a vector with a negative component.

Vectors may not always point to the right and up as in Figure 3.11, and the angle defining their direction may not always be measured from the $x$-axis. For example, FIGURE 3.13 shows vector $\vec{C}$ pointing down and to the right. In this case, the component vector $\vec{C}_y$ is pointing *down,* in the negative $y$-direction, so the $y$-component $C_y$ is a *negative* number. In addition, the angle $\phi$ is drawn measured from the $y$-axis, so the components of $\vec{C}$ are

$$C_x = C\sin\phi$$
$$C_y = -C\cos\phi \tag{3.8}$$

The roles of sine and cosine are reversed from those in Equations 3.5 because the angle $\phi$ is measured with respect to vertical, not horizontal.

**NOTE** ▶ Whether the $x$- and $y$-components use the sine or cosine depends on how you define the vector's angle. As noted previously, you *must* draw a diagram to define the angle that you use, and you must be sure to refer to the diagram when computing components. Don't use Equations 3.5 or 3.8 as general rules—they aren't! They appear as they do because of how we defined the angles. ◀

Next, let's look at the "inverse" problem for this case: determining the length and direction of the vector given the components. The signs of the components don't matter for determining the length; the Pythagorean theorem always works to find the length or magnitude of a vector because the squares eliminate any concerns over the signs. The length of the vector in Figure 3.13 is simply

$$C = \sqrt{C_x^2 + C_y^2} \tag{3.9}$$

When we determine the direction of the vector from its components, we must consider the signs of the components. Finding the angle of vector $\vec{C}$ in Figure 3.13 requires the length of $C_y$ *without* the minus sign, so vector $\vec{C}$ has direction

$$\phi = \tan^{-1}\left(\frac{C_x}{|C_y|}\right) \tag{3.10}$$

Notice that the roles of $x$ and $y$ differ from those in Equation 3.7.

**EXAMPLE 3.1** **Finding the components of an acceleration vector**

Find the $x$- and $y$-components of the acceleration vector $\vec{a}$ shown in FIGURE 3.14.

FIGURE 3.14 An acceleration vector $\vec{a}$.

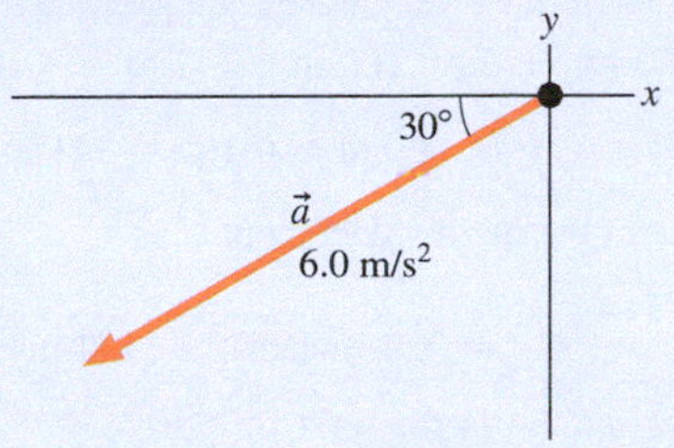


FIGURE 3.15 The components of the acceleration vector.

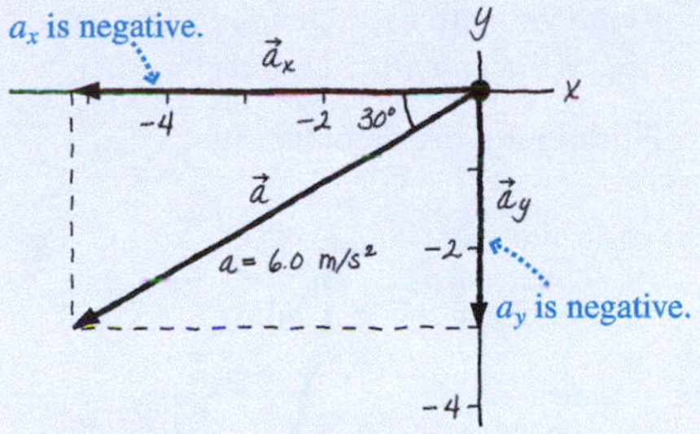


***STRATEGIZE*** We can find the components from the vector's magnitude and by using trigonometry, paying special attention to how the angle is defined in the problem.

***PREPARE*** Making a sketch is crucial to setting up this problem. FIGURE 3.15 shows the original vector $\vec{a}$ decomposed into component vectors parallel to the axes.

***SOLVE*** The acceleration vector $\vec{a} = (6.0\text{ m/s}^2, 30°$ below the negative $x$-axis) points to the left (negative $x$-direction) and down (negative $y$-direction), so the components $a_x$ and $a_y$ are both negative:

$$a_x = -a\cos 30° = -(6.0\text{ m/s}^2)\cos 30° = -5.2\text{ m/s}^2$$

$$a_y = -a\sin 30° = -(6.0\text{ m/s}^2)\sin 30° = -3.0\text{ m/s}^2$$

***ASSESS*** The magnitude of the $y$-component is smaller than that of the $x$-component, as seems to be the case in Figure 3.15, a good check on our work. The units of $a_x$ and $a_y$ are the same as the units of vector $\vec{a}$. Notice that we had to insert the components' minus signs manually after observing that the vector points down and to the left.

**STOP TO THINK 3.2** What are the $x$- and $y$-components $C_x$ and $C_y$ of vector $\vec{C}$?

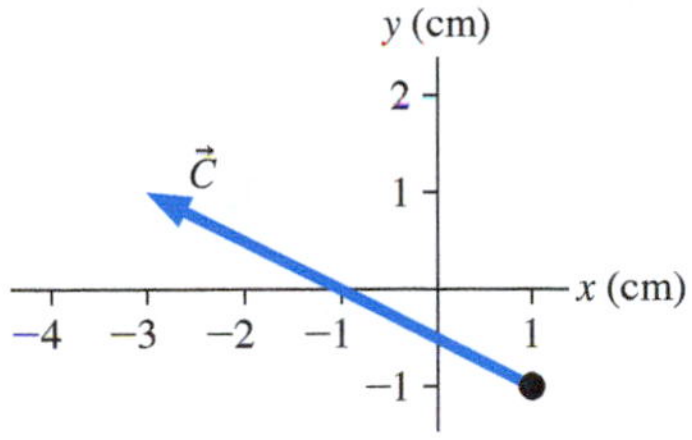


## Working with Components

We've seen how to add vectors graphically, but there's an easier way: using components. To illustrate, let's look at the vector sum $\vec{C} = \vec{A} + \vec{B}$ for the vectors shown in FIGURE 3.16. You can see that the component vectors of $\vec{C}$ are the sums of the component vectors of $\vec{A}$ and $\vec{B}$. The same is true of the components: $C_x = A_x + B_x$ and $C_y = A_y + B_y$.

In general, if $\vec{D} = \vec{A} + \vec{B} + \vec{C} + \cdots$, then the $x$- and $y$-components of the resultant vector $\vec{D}$ are

$$D_x = A_x + B_x + C_x + \cdots$$
$$D_y = A_y + B_y + C_y + \cdots \tag{3.11}$$

This method of vector addition is called *algebraic addition*.

FIGURE 3.16 Using components to add vectors.

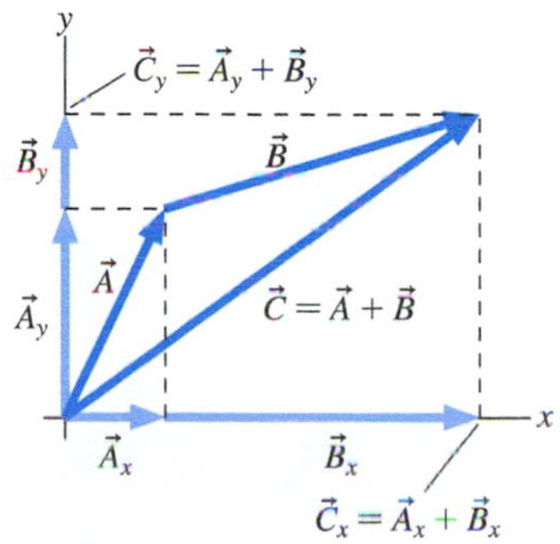

## EXAMPLE 3.2 Using algebraic addition to find a bird's displacement

A bird flies 100 m due east from a tree, then 200 m northwest (that is, 45° north of west). What is the bird's net displacement?

**STRATEGIZE** To find the bird's net displacement using algebraic addition, we will first need to find the components of each of its two individual displacements. Then we will find the components of the net displacement by adding the individual components.

**PREPARE** FIGURE 3.17a shows the displacement vectors $\vec{A} = (100 \text{ m, east})$ and $\vec{B} = (200 \text{ m, northwest})$ and also the net displacement $\vec{C}$. We draw vectors tip-to-tail if we are going to add them graphically, but it's usually easier to draw them all from the origin if we are going to use algebraic addition. FIGURE 3.17b shows the vectors redrawn with their tails together.

FIGURE 3.17 Finding the net displacement.

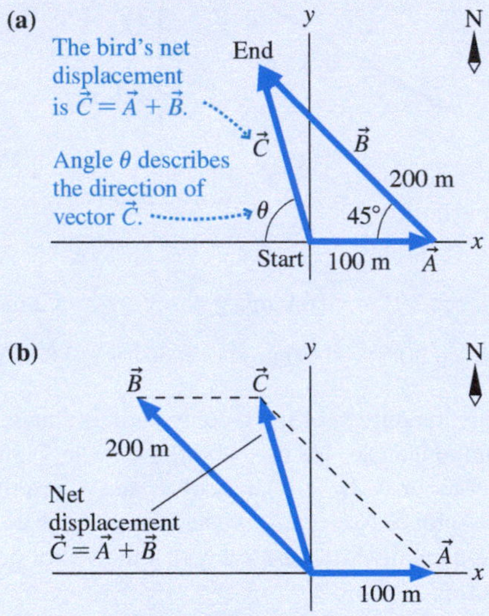


**SOLVE** To add the vectors algebraically we must know their components. From the figure these are seen to be

$$A_x = 100 \text{ m}$$
$$A_y = 0 \text{ m}$$
$$B_x = -(200 \text{ m}) \cos 45° = -141 \text{ m}$$
$$B_y = (200 \text{ m}) \sin 45° = 141 \text{ m}$$

We learned *from the figure* that $\vec{B}$ has a negative $x$-component. Adding $\vec{A}$ and $\vec{B}$ by components gives

$$C_x = A_x + B_x = 100 \text{ m} - 141 \text{ m} = -41 \text{ m}$$
$$C_y = A_y + B_y = 0 \text{ m} + 141 \text{ m} = 141 \text{ m}$$

The magnitude of the net displacement $\vec{C}$ is

$$C = \sqrt{C_x{}^2 + C_y{}^2} = \sqrt{(-41 \text{ m})^2 + (141 \text{ m})^2} = 147 \text{ m}$$

The angle $\theta$, as defined in Figure 3.17, is

$$\theta = \tan^{-1}\left(\frac{C_y}{|C_x|}\right) = \tan^{-1}\left(\frac{141 \text{ m}}{41 \text{ m}}\right) = 74°$$

Thus the bird's net displacement is $\vec{C} = (147 \text{ m}, 74° \text{ north of west})$.

**ASSESS** The final values of $C_x$ and $C_y$ match what we would expect from the sketch in Figure 3.17. The geometric addition was a valuable check on the answer we found by algebraic addition.

Vector subtraction and the multiplication of a vector by a scalar are also easily performed using components. To find $\vec{D} = \vec{P} - \vec{Q}$ we compute

$$D_x = P_x - Q_x$$
$$D_y = P_y - Q_y \tag{3.12}$$

Similarly, $\vec{T} = c\vec{S}$ has components

$$T_x = cS_x$$
$$T_y = cS_y \tag{3.13}$$

The next few chapters will make frequent use of *vector equations*. For example, you will learn that the equation to calculate the net force on a car skidding to a stop is

$$\vec{F} = \vec{n} + \vec{w} + \vec{f} \tag{3.14}$$

Equation 3.14 is really just a shorthand way of writing the two simultaneous equations:

$$F_x = n_x + w_x + f_x$$
$$F_y = n_y + w_y + f_y \tag{3.15}$$

In other words, a vector equation is interpreted as meaning: Equate the $x$-components on both sides of the equals sign, then equate the $y$-components. Vector notation allows us to write these two equations in a more compact form.

## Tilted Axes

Although we are used to having the $x$-axis horizontal, there is no requirement that it has to be that way. In Chapter 1, we saw that for motion on a slope, it is often most convenient to put the $x$-axis along the slope. When we add the $y$-axis, this gives us a tilted coordinate system such as that shown in FIGURE 3.18.

Finding components with tilted axes is no harder than what we have done so far. Vector $\vec{C}$ in Figure 3.18 can be decomposed into component vectors $\vec{C}_x$ and $\vec{C}_y$, with $C_x = C\cos\theta$ and $C_y = C\sin\theta$.

FIGURE 3.18 A coordinate system with tilted axes.

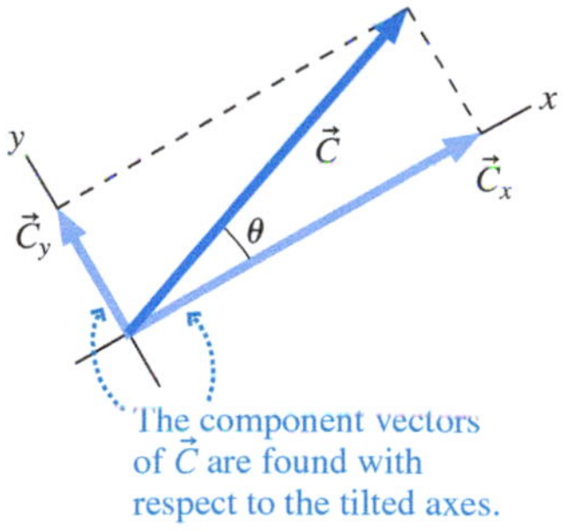


**STOP TO THINK 3.3** Angle $\phi$ that specifies the direction of $\vec{C}$ is computed as

A. $\tan^{-1}(C_x/C_y)$.
B. $\tan^{-1}(C_x/|C_y|)$.
C. $\tan^{-1}(|C_x|/|C_y|)$.
D. $\tan^{-1}(C_y/C_x)$.
E. $\tan^{-1}(C_y/|C_x|)$.
F. $\tan^{-1}(|C_y|/|C_x|)$.

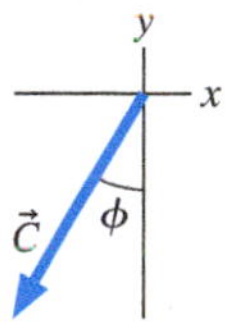


# 3.3 Motion on a Ramp

In this section, we will examine the problem of motion on a ramp, or *incline*. There are three reasons to look at this problem. First, it will provide good practice using vectors to analyze motion. Second, it is a simple problem for which we can find an exact solution. Third, this seemingly abstract problem has real and important applications.

A video to support a section's topic is embedded in the eText.

**Video** Motion on a Ramp

## Constant-Velocity Motion on a Ramp

We begin with a constant-velocity example to give us some practice with vectors and components before moving on to the more general case of accelerated motion.

**EXAMPLE 3.3 Finding the height gained by a car moving up a slope**

A car drives up a steep 10° slope at a constant speed of 15 m/s. After 10 s, how much height has the car gained?

**STRATEGIZE** The vertical motion of the car is determined by the vertical component of the velocity.

**PREPARE** FIGURE 3.19 is a visual overview, with $x$- and $y$-axes defined. The velocity vector $\vec{v}$ points up the slope. We decompose $\vec{v}$ into component vectors $\vec{v}_x$ and $\vec{v}_y$ as shown.

FIGURE 3.19 Visual overview of a car moving up a slope.

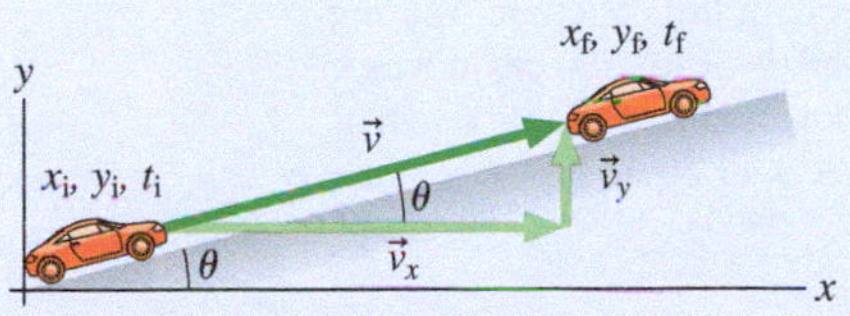


Known
$x_i = y_i = 0$ m
$t_i = 0$ s, $t_f = 10$ s
$v = 15$ m/s
$\theta = 10°$

Find
$\Delta y$

**SOLVE** The velocity component we need is $v_y$, which describes the vertical motion of the car. Using the rules for finding components outlined previously, we find

$$v_y = v\sin\theta = (15\text{ m/s})\sin 10° = 2.6\text{ m/s}$$

Because the velocity is constant, the car's vertical displacement (i.e., the height gained) during 10 s is

$$\Delta y = v_y\,\Delta t = (2.6\text{ m/s})(10\text{ s}) = 26\text{ m}$$

**ASSESS** The car is traveling at a pretty good clip—15 m/s is a bit faster than 30 mph—up a steep slope, so it should climb a respectable height in 10 s. 26 m, or about 80 ft, seems reasonable.

## Accelerated Motion on a Ramp

FIGURE 3.20a shows a crate sliding down a frictionless (i.e., smooth) ramp tilted at angle $\theta$. The crate accelerates due to the action of gravity, but it is *constrained* to accelerate parallel to the surface. What is the acceleration?

A motion diagram for the crate is drawn in FIGURE 3.20b. The velocity vectors point in the direction of motion, downhill and parallel to the ramp's surface. Because the crate is speeding up, its acceleration points in the same direction as its velocity. We can take advantage of the properties of vectors to find the crate's acceleration.

FIGURE 3.20 Acceleration on an incline.

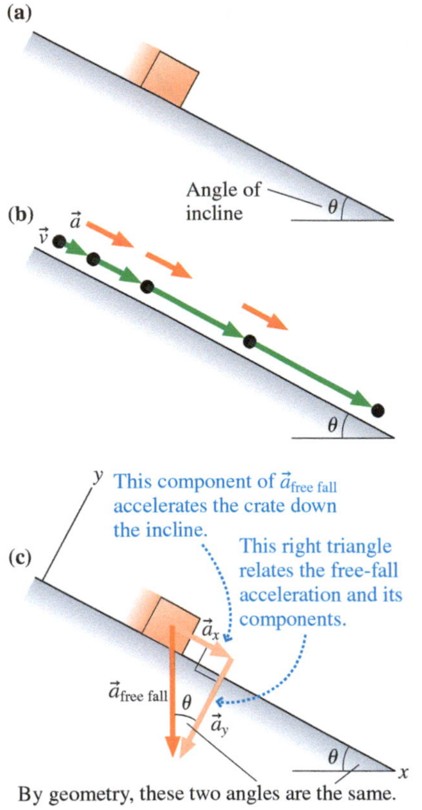


To do so, FIGURE 3.20c sets up a coordinate system with the $x$-axis along the ramp and the $y$-axis perpendicular to the ramp. All motion is along the $x$-axis.

If the incline suddenly vanished, the object would have a free-fall acceleration $\vec{a}_{\text{free fall}}$ straight down. As Figure 3.20c shows, this acceleration vector can be decomposed into two component vectors: a vector $\vec{a}_x$ that is *parallel* to the incline and a vector $\vec{a}_y$ that is *perpendicular* to the incline. The vector addition rules studied earlier in this chapter tell us that $\vec{a}_{\text{free fall}} = \vec{a}_x + \vec{a}_y$.

The motion diagram shows that the object's actual acceleration $\vec{a}_x$ is parallel to the incline. The surface of the incline somehow "blocks" the other component of the acceleration $\vec{a}_y$, through a process we will examine in Chapter 5, but $\vec{a}_x$ is unhindered. It is this component of $\vec{a}_{\text{free fall}}$, parallel to the incline, that accelerates the object.

We can use trigonometry to work out the magnitude of this acceleration. Figure 3.20c shows that the three vectors $\vec{a}_{\text{free fall}}$, $\vec{a}_y$, and $\vec{a}_x$ form a right triangle with angle $\theta$ as shown. This angle is the same as the angle of the incline. By definition, the magnitude of $\vec{a}_{\text{free fall}}$ is $g$. This vector is the hypotenuse of the right triangle. The vector we are interested in, $\vec{a}_x$, is opposite angle $\theta$. Thus the value of the acceleration along a frictionless slope is

$$a_x = \pm g \sin\theta \tag{3.16}$$

NOTE ► The correct sign depends on the direction in which the ramp is tilted. The acceleration in Figure 3.20 is $+g\sin\theta$, but upcoming examples will show situations in which the acceleration is $-g\sin\theta$. ◄

Let's look at Equation 3.16 to verify that it makes sense. A good way to do this is to consider **limiting cases** in which the angle is at one end of its range. In these cases, the physics is clear and we can check our result. Let's look at two such possibilities:

1. Suppose the angle of the ramp is lowered until the ramp is perfectly horizontal, with $\theta = 0°$. An object sliding on a (frictionless) horizontal surface would move with a constant velocity and so have zero acceleration. Equation 3.16 gives $a_x = 0$ when $\theta = 0°$, in agreement with our expectations.
2. Now suppose the ramp is tilted until it becomes vertical, with $\theta = 90°$. You know what happens—the object will be in free fall, parallel to the vertical surface. Equation 3.16 gives $a_x = \pm g$ when $\theta = 90°$, again in agreement with our expectations.

NOTE ► Checking your answer by looking at such limiting cases is a very good way to see if your answer makes sense. We will often do this in the Assess step of a solution. ◄

◄ **Extreme physics** A speed skier, on wide skis with little friction, wearing an aerodynamic helmet and crouched low to minimize air resistance, moves in a straight line down a steep slope—pretty much like an object sliding down a frictionless ramp. There is a maximum speed that a skier could possibly achieve at the end of the slope.

EXAMPLE 3.4 **Maximum possible speed for a skier**

The Willamette Pass ski area in Oregon was the site of the 1993 U.S. National Speed Skiing Competition. The skiers started from rest and then accelerated down a stretch of the mountain with a reasonably constant slope, aiming for the highest possible speed at the end of this run. During this acceleration phase, the skiers traveled 360 m while dropping a vertical distance of 170 m. What is the fastest speed a skier could achieve at the end of this run?

***STRATEGIZE*** The fastest possible run is one with no friction or air resistance, meaning the acceleration down the slope is given by Equation 3.16. The problem is then one-dimensional motion with constant acceleration.

***PREPARE*** We make the visual overview in FIGURE 3.21. The motion diagram shows the acceleration of the skier and the pictorial representation gives an overview of the problem including the dimensions of the slope. As before, we put the $x$-axis along the slope.

FIGURE 3.21 Visual overview of a skier accelerating down a slope.

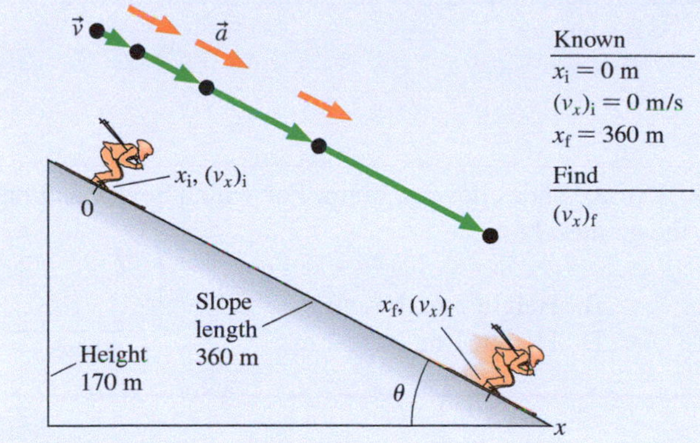


**SOLVE** The acceleration is in the positive $x$-direction, so we use the positive sign in Equation 3.16. What is the angle in Equation 3.16? Figure 3.21 shows that the 360-m-long slope is the hypotenuse of a triangle of height 170 m, so we use trigonometry to find

$$\sin\theta = \frac{170\text{ m}}{360\text{ m}}$$

which gives $\theta = \sin^{-1}(170/360) = 28°$. Equation 3.16 then gives

$$a_x = +g\sin\theta = (9.8\text{ m/s}^2)(\sin 28°) = 4.6\text{ m/s}^2$$

For linear motion with constant acceleration, we can use Equation 2.13 in Synthesis 2.1: $(v_x)_f{}^2 = (v_x)_i{}^2 + 2a_x\,\Delta x$. The initial velocity $(v_x)_i$ is zero; thus

This is the distance along the slope, the length of the run.

$$(v_x)_f = \sqrt{2a_x\Delta x} = \sqrt{2(4.6\text{ m/s}^2)(360\text{ m})} = 58\text{ m/s}$$

This is the fastest that any skier could hope to be moving at the end of the run. Any friction or air resistance would decrease this speed.

**ASSESS** The final speed we calculated is 58 m/s, which is about 130 mph, reasonable because we expect a high speed for this sport. In the competition noted, the actual winning speed was 111 mph, not much slower than the result we calculated. Obviously, efforts to minimize friction and air resistance are working!

Skis on snow have very little friction, but there are other ways to reduce the friction between surfaces. For instance, a roller coaster car rolls along a track on low-friction wheels. No drive force is applied to the cars after they are released at the top of the first hill; the speed changes due to gravity alone. The cars speed up as they go down hills and slow down as they climb.

## EXAMPLE 3.5 Speed of a roller coaster

A classic wooden coaster has cars that go down a big first hill, gaining speed. The cars then ascend a second hill with a slope of 30°. If the cars are going 25 m/s at the bottom and it takes them 2.0 s to climb this hill, how fast are they going at the top?

**STRATEGIZE** The coaster's velocity is directed up the ramp. But because the cars are slowing, the acceleration points opposite to the velocity, or down the hill. The acceleration is constant and given by Equation 3.16; we will use the kinematic equations from Synthesis 2.1 to solve for the speed at the top.

**PREPARE** The visual overview in FIGURE 3.22 includes a motion diagram, a pictorial representation, and a list of values. We've done this with a sketch such as you might draw for your homework. Notice how the motion diagram of Figure 3.22 differs from that of Figure 3.21 in Example 3.4: The motion is along the $x$-axis, as before, but the acceleration vector points in the negative $x$-direction, so the component $a_x$ is negative.

FIGURE 3.22 The coaster's speed decreases as it goes up the hill.

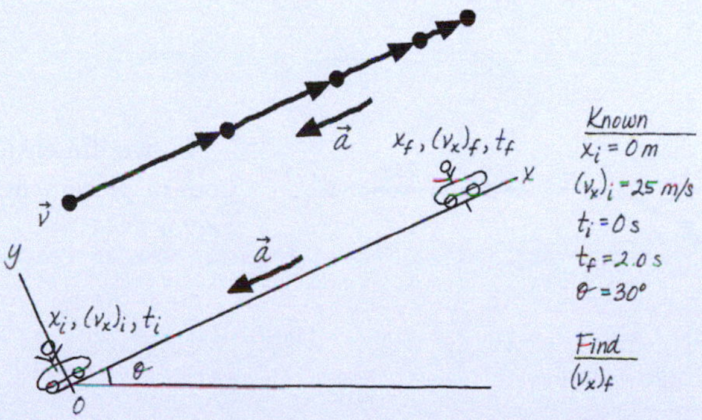


**SOLVE** To determine the final speed, we need to know the acceleration. We will assume that there is no friction or air resistance, so the magnitude of the roller coaster's acceleration is given by Equation 3.16 using the minus sign, as noted:

$$a_x = -g\sin\theta = -(9.8\text{ m/s}^2)\sin 30° = -4.9\text{ m/s}^2$$

*Continued*

The speed at the top of the hill can then be computed using Equation 2.11 of Synthesis 2.1:

$$(v_x)_f = (v_x)_i + a_x\,\Delta t = 25\text{ m/s} + (-4.9\text{ m/s}^2)(2.0\text{ s}) = 15\text{ m/s}$$

**ASSESS** The speed is less at the top of the hill than at the bottom, as it should be, but the coaster is still moving at a pretty good clip at the top—almost 35 mph. This seems reasonable—a fast ride is a fun ride.

**STOP TO THINK 3.4** A block of ice slides down a ramp. For which height and base lengths is the acceleration the greatest?

A. Height 4 m, base 12 m
B. Height 3 m, base 6 m
C. Height 2 m, base 5 m
D. Height 1 m, base 3 m

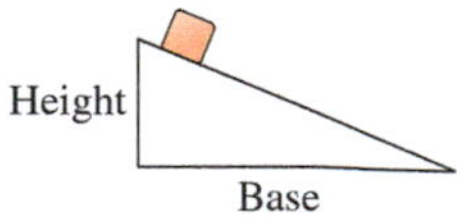


## 3.4 Motion in Two Dimensions

So far, we've studied in detail only one-dimensional motion, such as a sprinter running a 100 m race or a skier accelerating down a slope. For such motion we can represent the object's displacement, velocity, and acceleration by vectors that are directed along the single axis of motion.

Now we'll examine the more complex case of *two-dimensional motion.* In this kind of motion, the object moves not along a single line, but in an entire *plane.* For example, a skater gliding around a rink moves in the plane defined by the horizontal ice surface. A rider on a Ferris wheel moves in a vertical plane. As an object undergoes two-dimensional motion, its displacement, velocity, and acceleration vectors all change, and point in many different directions.

FIGURE 3.23a shows a motion diagram for a car, as seen from above, rounding a curve in the road. This is two-dimensional motion in the plane of the road. The motion diagram shows that the car slows as it approaches the curve, rounds the curve slowly, and then speeds up as it exits the curve. We have drawn displacement vectors $\vec{d}$ connecting the dots that represent the car's positions at successive times.

FIGURE 3.23 Motion diagram for a car going around a curve.

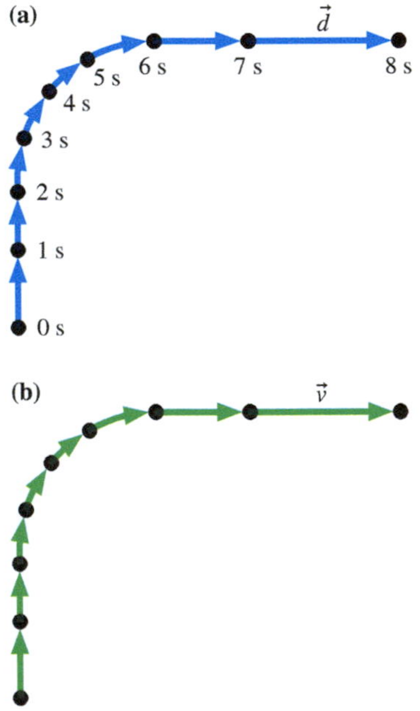


We can also represent the car's velocity vectors on a motion diagram. In Chapter 2, we defined velocity for one-dimensional motion as an object's displacement—its change in position—divided by the time interval in which the change occurs:

$$v_x = \frac{\Delta x}{\Delta t} = \frac{x_f - x_i}{\Delta t}$$

In two dimensions, an object's displacement is a vector. Suppose an object undergoes displacement $\vec{d}$ during the time interval $\Delta t$. Then we define the object's velocity *vector* as

$$\vec{v} = \frac{\vec{d}}{\Delta t} = \left(\frac{d}{\Delta t}, \text{ same direction as } \vec{d}\right) \tag{3.17}$$

Definition of velocity in two or more dimensions

The velocity vector is simply the displacement vector multiplied by the scalar $1/\Delta t$. Consequently, **the velocity vector points in the same direction as the displacement.** As a result, we represent the velocity on a motion diagram by vectors that extend from one dot to the next, just as the displacement vectors do. Using this idea, we draw the velocity vectors for the car's motion as in FIGURE 3.23b.

**EXAMPLE 3.6 A kayaker goes over a waterfall**

A kayaker paddles off the edge of a waterfall, arcing down to the water below. The partial motion diagram in FIGURE 3.24 shows his position at 0.1 s intervals. Estimate the kayaker's velocity during the time interval from $t = 0.5$ s to $t = 0.6$ s.

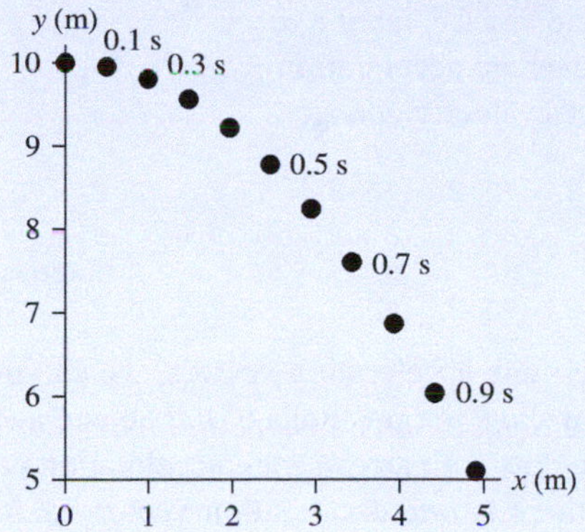


FIGURE 3.24 Partial motion diagram for a kayaker.

**STRATEGIZE** The problem asks for the kayaker's velocity, so we'll need to specify both his speed and his direction. We'll first find his displacement from the motion diagram and then use Equation 3.17 to find his velocity.

**PREPARE** During this time interval, we estimate the distance he traveled to be 0.8 m, and the angle of his motion looks to be about 45° below the horizontal.

**SOLVE** From our estimates, the kayaker's displacement during the stated time interval is

$$\vec{d} = (0.8 \text{ m}, 45^\circ \text{ below horizontal})$$

Then from Equation 3.17 his velocity is

$$\vec{v} = \left(\frac{0.8 \text{ m}}{0.1 \text{ s}}, 45^\circ \text{ below horizontal}\right)$$
$$= (8 \text{ m/s}, 45^\circ \text{ below horizontal})$$

**ASSESS** The kayaker's *speed* of 8 m/s ≈ 18 mph seems reasonable for a daredevil kayaker.

## Acceleration in Two Dimensions

In ◀◀ SECTION 2.4 we defined an object's acceleration in one dimension as the change in its velocity divided by the time interval in which the change occurs, so that $a_x = \Delta v_x/\Delta t$. In two dimensions, we need to use a vector to describe acceleration. The vector definition of acceleration is a straightforward extension of the one-dimensional version:

$$\vec{a} = \frac{\vec{v}_f - \vec{v}_i}{t_f - t_i} = \frac{\Delta\vec{v}}{\Delta t} \tag{3.18}$$

Definition of acceleration in two or more dimensions

From its definition, we see that **$\vec{a}$ points in the same direction as $\Delta\vec{v}$,** the change in velocity. Because velocity in two dimensions is a vector, as an object moves, its velocity vector can change in two possible ways:

1. The magnitude of $\vec{v}$ can change, which indicates that the object's *speed* has changed.
2. The direction of $\vec{v}$ can change, which indicates that the object's *direction of motion* has changed.

In Chapter 2 we saw how to compute an acceleration vector for the first case, in which an object speeds up or slows down while moving in a straight line. Now we will examine the second case, in which an object's direction of motion changes.

Suppose an object has initial velocity $\vec{v}_i$ and, a short time $\Delta t$ later, final velocity $\vec{v}_f$. During this time interval, the velocity changes by an amount $\Delta\vec{v} = \vec{v}_f - \vec{v}_i$. Because the velocity *changes,* the object must undergo an *acceleration* during the time interval $\Delta t$, as Equation 3.18 shows. Equation 3.18 further shows that the acceleration vector points in the *same direction* as the change in velocity $\Delta\vec{v} = \vec{v}_f - \vec{v}_i$. This means that in order to find the acceleration, we have to subtract two velocity vectors.

Tactics Box 3.1 showed how to perform vector subtraction. Tactics Box 3.3 shows how to use vector subtraction to find the acceleration vector.

### TACTICS BOX 3.3 Finding the acceleration vector

To find the acceleration as the velocity changes from $\vec{v}_i$ to $\vec{v}_f$:

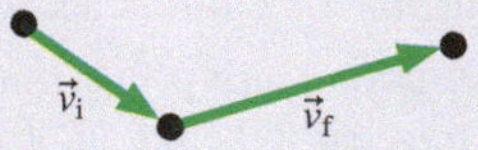


❶ Draw the velocity vector $\vec{v}_f$.

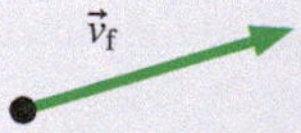


❷ Draw $-\vec{v}_i$ at the tip of $\vec{v}_f$.

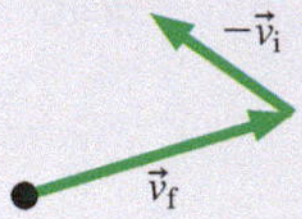


❸ Draw $\Delta\vec{v} = \vec{v}_f - \vec{v}_i$
$= \vec{v}_f + (-\vec{v}_i)$
This is the direction of $\vec{a}$.

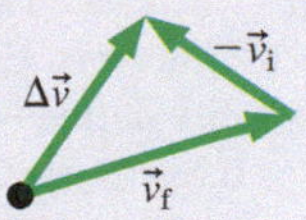


❹ In the original motion diagram, draw a vector at the middle point in the direction of $\Delta\vec{v}$; label it $\vec{a}$. This is the average acceleration as the object's velocity changes from $\vec{v}_i$ to $\vec{v}_f$.

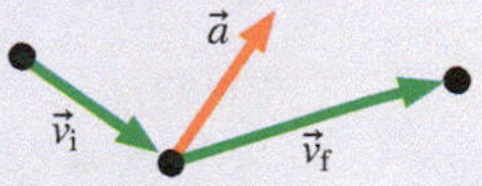


Exercises 27, 28

Now that we know how to determine acceleration vectors, we can make a complete motion diagram with dots that show the positions of the object, average velocity vectors found by connecting the dots with arrows, and acceleration vectors found using Tactics Box 3.3. Note that there is *one* acceleration vector linking each *two* velocity vectors, and $\vec{a}$ is drawn at the dot between the two velocity vectors it links.

### EXAMPLE 3.7 Skiing through a valley

A skier skis down a long, straight hill, through a short valley, and then back up a hill on the other side. Draw a complete motion diagram of her motion.

**STRATEGIZE** This is two-dimensional motion. As the skier skis down through the valley, both her speed and the direction of her velocity change. Both these changes contribute to her acceleration, which we will determine using the steps of Tactics Box 3.3.

**PREPARE** In FIGURE 3.25 we have drawn dots to represent her positions at equally spaced times.

**SOLVE** The velocity vectors of the skier's motion are found by connecting successive dots in the motion diagram with arrows. Her motion down the hill is just one-dimensional motion on a ramp; because she is speeding up, her acceleration vectors point in the same direction as her velocity vectors—that is, downhill. Similarly, for her straight-line motion up the hill on the other side she is slowing, so her acceleration vectors point in the direction opposite to her velocity.

As the skier moves through the valley, where both her speed and the direction of her velocity are changing, we must use the steps of Tactics Box 3.3 to find her acceleration. We need to use vector subtraction to find the direction of $\Delta\vec{v}$ and thus of $\vec{a}$. The procedure is shown at two points in the motion diagram.

**ASSESS** This example illustrates a new and counterintuitive result: For two-dimensional motion, the acceleration need not point in the same (or opposite) direction as the velocity! In particular, at the very bottom of the valley, the skier is neither speeding up nor slowing down, so her speed is, at this instant in time, constant. But she still has an acceleration, directed at right angles to her velocity. We'll see more examples of this interesting behavior of acceleration as we go through this chapter.

FIGURE 3.25 Motion diagram for a skier skiing through a valley.

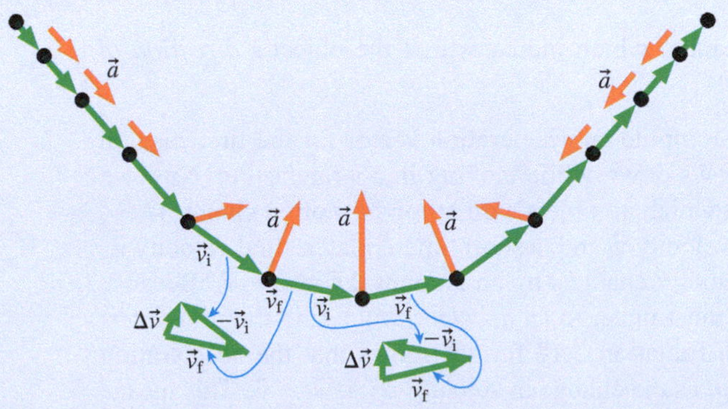

As Example 3.7 shows, the relationship between velocity and acceleration can sometimes be subtle. But the example illustrates some important ideas. First, as shown in Tactics Box 3.3, we can always find the acceleration graphically by taking the difference between the two velocity vectors on either side of one dot in a motion diagram. Second, even if the object has a constant speed (as at the bottom of the valley), it will have an acceleration if its path is curving; the acceleration in this case points at right angles to the velocity, directed toward the "inside" of the curve. Finally, if the object is moving along a straight-line segment of its path, just as in Chapter 2 the acceleration points in the direction of the velocity (object speeding up) or in the direction opposite to the velocity (object slowing down).

**STOP TO THINK 3.5** The positions of four cars on four different roads are indicated by the dots. Which cars could possibly have the acceleration shown?

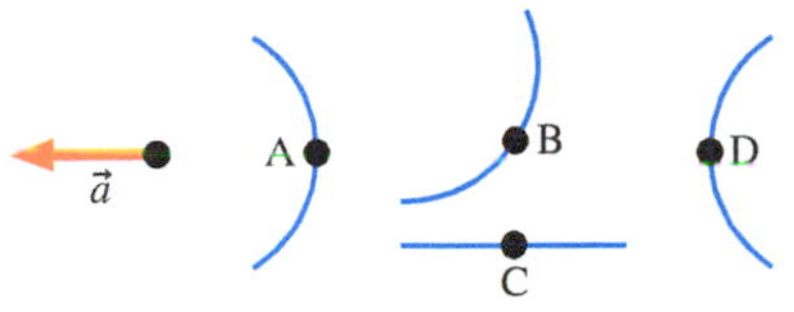


## 3.5 Projectile Motion

Balls flying through the air, long jumpers, and cars doing stunt jumps are all examples of the two-dimensional motion that we call **projectile motion.** Projectile motion is an extension to two dimensions of the free-fall motion we studied in ◀◀ SECTION 2.7. **A projectile is an object that moves in two dimensions under the influence of gravity and nothing else.** Although real objects are also influenced by air resistance, the effect of air resistance is small for reasonably dense objects moving at modest speeds, so we can ignore it for the cases we consider in this chapter. As long as we can neglect air resistance, any projectile will follow the same type of path. Because the form of the motion will always be the same, the strategies we develop to solve one projectile problem can be applied to others as well.

Video Video Dropped and Thrown Balls
Video Figure 3.26

FIGURE 3.26 The motions of two balls launched at the same time.

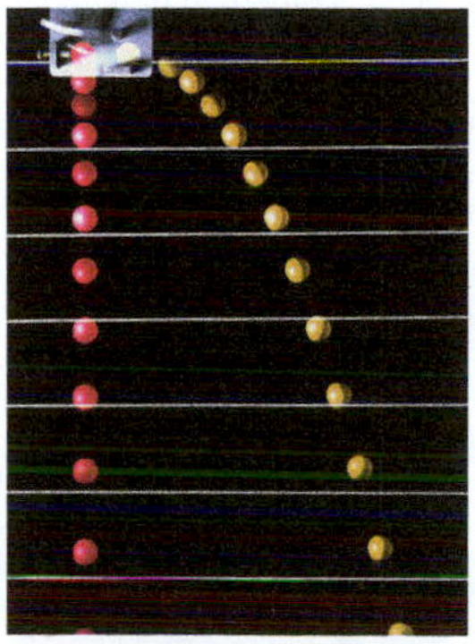

FIGURE 3.26 is a strobe photograph of two falling balls. The red ball was released from rest and undergoes ordinary free fall. At the same instant the red ball was released, the yellow ball was shot horizontally. Note that the *vertical* motions of the two balls are identical—at the same time after release, they are at the same height—and they hit the floor simultaneously. Evidently the vertical motion of the yellow ball is not affected by the fact that it is moving horizontally. Thus the vertical motion of each ball is free fall, which is the same for all objects, as we learned in Chapter 2. Furthermore, a careful look at the *horizontal* motion of the yellow ball shows that it is uniform motion; the horizontal motion continues as if the ball were not falling.

So, for an object in projectile motion, the initial horizontal velocity has *no* influence over the vertical motion, and vice versa. This is a general rule for *any* projectile motion: **The horizontal and vertical components of an object undergoing projectile motion are independent of each other.**

**CONCEPTUAL EXAMPLE 3.8 Time and distance for balls rolled off the table**

Two balls are rolling toward the edge of a table, with ball 1 rolling twice as fast as ball 2. Both balls leave the table edge at the same time. Which ball hits the ground first? Which ball goes farther?

***REASON*** The vertical motion of both balls is the same—free fall—and they both fall from the same height. Both balls are in the air for the same time interval and hit the ground at the exact same time. During this time interval, the two balls continue moving horizontally at the speed with which they left the table. Ball 2 has twice the horizontal speed of ball 1; it will therefore go twice as far.

***ASSESS*** This result makes sense. The vertical and horizontal motions are independent of each other, so we can analyze them separately. If you drop two objects from the same height, they hit the ground at the same time, so both balls should land at the same time. And as they drop, the one that was moving faster horizontally will continue to do so—thus it will go farther.

FIGURE 3.27 The motion of a tossed ball.

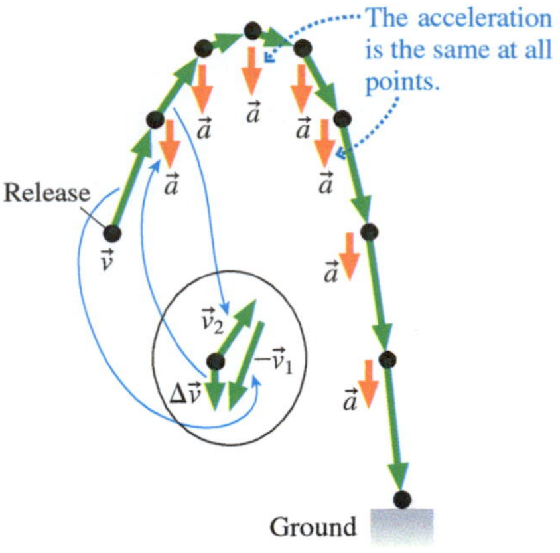


In Chapter 2, you learned that for an object in free fall, the acceleration is always directed vertically downward, with magnitude $g$. FIGURE 3.27 shows that this statement is also true for a ball tossed into the air at an angle—that is, for projectile motion. We can use the procedure of Tactics Box 3.3 to find the ball's acceleration. The inset in the figure shows this procedure at one point in the ball's motion; we can see that the ball's acceleration vector points straight down. In fact, a careful analysis would find that *at every point in the motion* the acceleration vector points straight down and has magnitude $g = 9.80\ \text{m/s}^2$. This should not be surprising. If we ignore the tiny effect of air resistance, a projectile is an object moving under the influence of gravity only, so it is in free fall with acceleration $\vec{a}_{\text{free fall}} = (9.80\ \text{m/s}^2$, vertically downward).

As the projectile moves, the free-fall acceleration will change the *vertical* component of the velocity, but there will be no change to the *horizontal* component of the velocity. **The vertical component of acceleration $a_y$ for all projectile motion is the familiar $-g$ of free fall, while the horizontal component $a_x$ is zero.**

## Analyzing Projectile Motion

FIGURE 3.28 The launch and motion of a projectile.

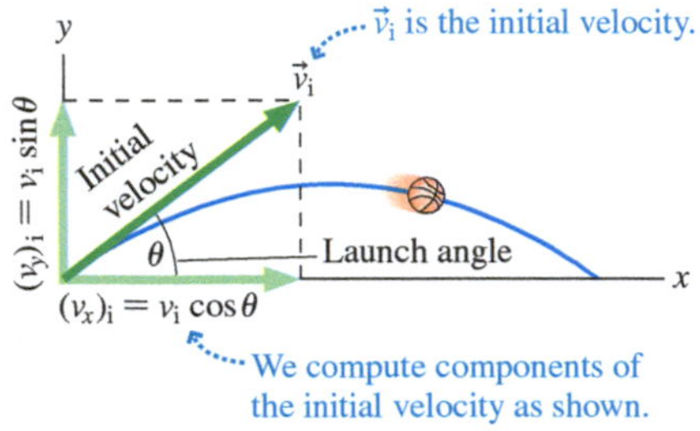


Suppose you toss a basketball down the court, as shown in FIGURE 3.28. To study the resulting projectile motion, we've established a coordinate system with the $x$-axis horizontal and the $y$-axis vertical. In projectile motion, the angle of the initial velocity above the horizontal (i.e., above the $x$-axis) is called the **launch angle.** As you learned in Section 3.2, the initial velocity vector can be expressed in terms of its $x$- and $y$-components, which are shown in the figure.

**NOTE** ▸ The components $(v_x)_i$ and $(v_y)_i$ are not always positive. A projectile launched at an angle *below* the horizontal (such as a ball thrown downward from the roof of a building) has *negative* values for $\theta$ and $(v_y)_i$. However, the *speed* $v_i$ is always positive. ◂

FIGURE 3.29 The velocity and acceleration vectors of a projectile.

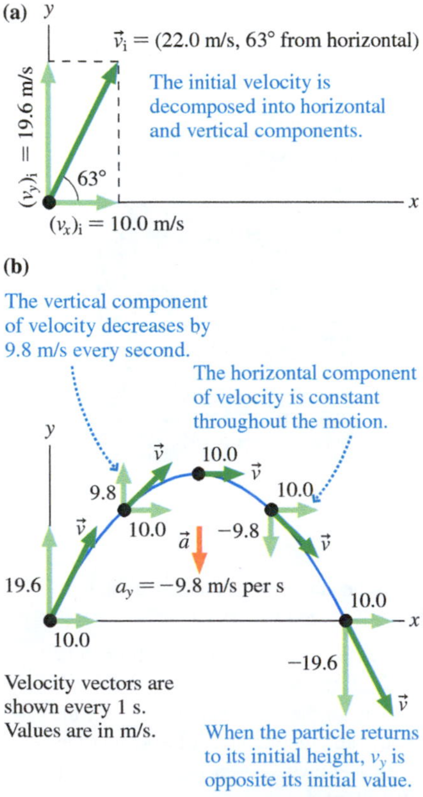


Once the basketball leaves your hand, its subsequent motion is determined by the initial components of the velocity and the acceleration. To see how this plays out in practice, let's look at a specific case with definite numbers. FIGURE 3.29 shows a projectile launched at a speed of 22.0 m/s at an angle of 63° from the horizontal. In Figure 3.29a, the initial velocity vector $v_i$ is broken into its horizontal and vertical components. In Figure 3.29b, the velocity vector and its component vectors are shown every 1.0 s. Because there is no horizontal acceleration ($a_x = 0$) the value of $v_x$ never changes. In contrast, $v_y$ decreases by 9.8 m/s every second. (This is what it *means* to accelerate at $a_y = -9.8\ \text{m/s}^2 = (-9.8\ \text{m/s})$ per second.) There is nothing *pushing* the projectile along the curve. As the ball moves, the downward acceleration changes the velocity vector as shown, causing it to tip downward as the motion proceeds. At the end of the arc, when the ball is at the same height as it started, $v_y$ is $-19.6$ m/s, the negative of its initial value. **The ball finishes its motion moving downward at the same speed as it started moving upward,** just as we saw in the case of one-dimensional free fall in Chapter 2.

You can see from Figure 3.29 that **projectile motion is made up of two independent motions: uniform motion at constant velocity in the horizontal direction and free-fall motion in the vertical direction.** The two motions are independent, but because they occur together they must be analyzed together, as we'll see.

In Chapter 2, we saw kinematic equations for constant-velocity and constant-acceleration motion. We can adapt these equations to this current case. The horizontal motion is constant-velocity motion at $(v_x)_i$, and the vertical motion is constant-acceleration motion with initial velocity $(v_y)_i$ and an acceleration of $a_y = -g$.

Let's summarize and synthesize everything we've learned to this point about projectile motion, including the equations from Chapter 2 for the horizontal and vertical motion. We'll use this information to solve some problems in the next section.

**SYNTHESIS 3.1 Projectile motion**

The horizontal and vertical components of projectile motion are independent, but must be analyzed together.

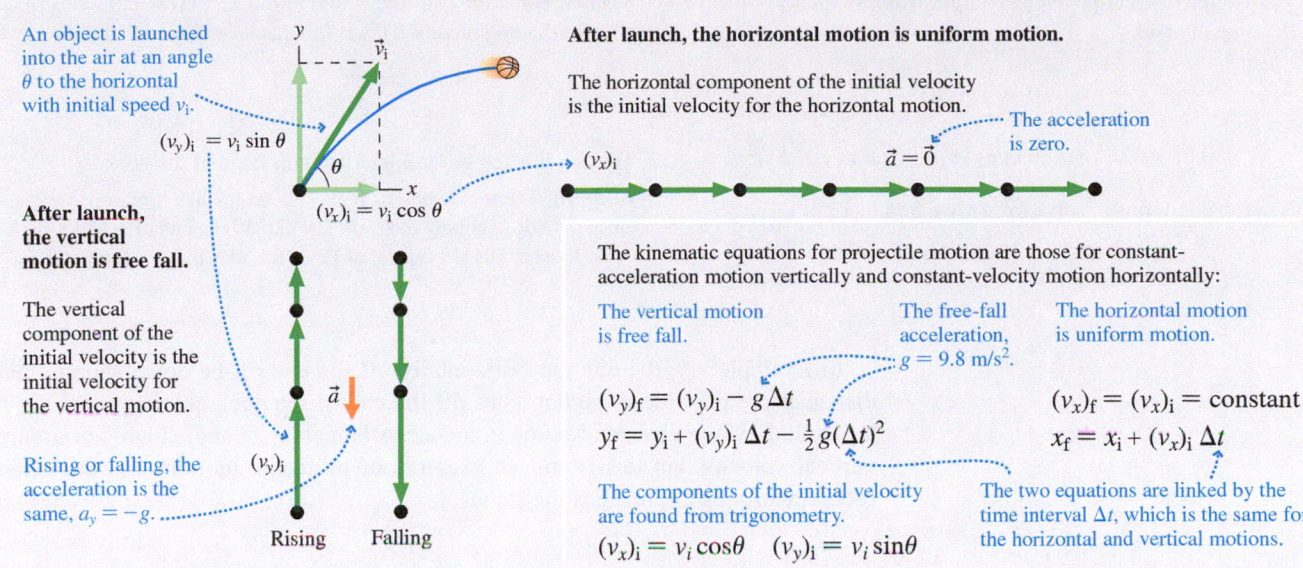

An object is launched into the air at an angle $\theta$ to the horizontal with initial speed $v_i$.

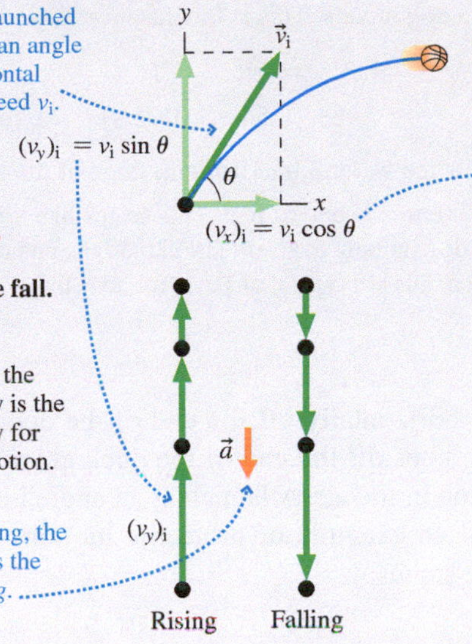


**After launch, the vertical motion is free fall.**

The vertical component of the initial velocity is the initial velocity for the vertical motion.

Rising or falling, the acceleration is the same, $a_y = -g$.

**After launch, the horizontal motion is uniform motion.**

The horizontal component of the initial velocity is the initial velocity for the horizontal motion.

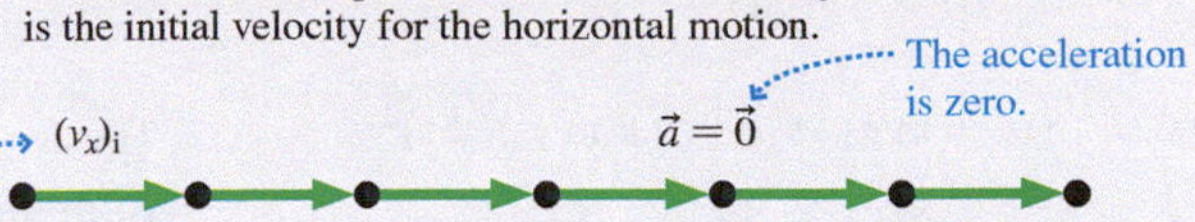


The kinematic equations for projectile motion are those for constant-acceleration motion vertically and constant-velocity motion horizontally:

The vertical motion is free fall. The free-fall acceleration, $g = 9.8\text{ m/s}^2$. The horizontal motion is uniform motion.

$$(v_y)_f = (v_y)_i - g\,\Delta t \qquad (v_x)_f = (v_x)_i = \text{constant}$$

$$y_f = y_i + (v_y)_i\,\Delta t - \tfrac{1}{2}g(\Delta t)^2 \qquad x_f = x_i + (v_x)_i\,\Delta t$$

The components of the initial velocity are found from trigonometry.

$$(v_x)_i = v_i\cos\theta \qquad (v_y)_i = v_i\sin\theta$$

The two equations are linked by the time interval $\Delta t$, which is the same for the horizontal and vertical motions.

**STOP TO THINK 3.6** A 100 g ball rolls off a table and lands 2 m from the base of the table. A 200 g ball rolls off the same table with the same speed. How far does it land from the base of the table?

A. <1 m
B. 1 m
C. Between 1 m and 2 m
D. 2 m
E. Between 2 m and 4 m
F. 4 m

**Video** What the Physics? Projectile Paths
**Video** Ball Fired Upward from Moving Cart

## 3.6 Projectile Motion: Solving Problems

Now that we have a good idea of how projectile motion works, we can use that knowledge to solve some true two-dimensional motion problems.

**EXAMPLE 3.9 Dock jumping**

In the sport of dock jumping, dogs run at full speed off the end of a dock that sits a few feet above a pool of water. The winning dog is the one that lands farthest from the end of the dock. If a dog runs at 8.5 m/s (a pretty typical speed for this event) straight off the end of a dock that is 0.61 m (2 ft, a standard height) above the water, how far will the dog go before splashing down?

**STRATEGIZE** We can treat this as a projectile motion problem, so the dog's vertical motion is free fall and its horizontal motion is uniform motion. The relevant equations are given in Synthesis 3.1.

**PREPARE** We make a visual overview of the situation as in FIGURE 3.30. We have chosen to put the origin of the coordinate system at the base of the dock. The dog runs horizontally off the end of the dock, so the initial components of the velocity are $(v_x)_i = 8.5$ m/s and $(v_y)_i = 0$ m/s.

We know that the horizontal and vertical motions are independent. The fact that the dog is falling toward the water doesn't affect its horizontal motion. When the dog leaves the end of the dock, it will continue to move horizontally at 8.5 m/s. The vertical motion is free fall. The jump ends when the dog hits the water—that is, when it has dropped by 0.61 m. We are ultimately interested in how far the dog goes, but to make this determination we'll need to find the time interval $\Delta t$ that the dog is in the air.

**FIGURE 3.30** Visual overview for Example 3.9.

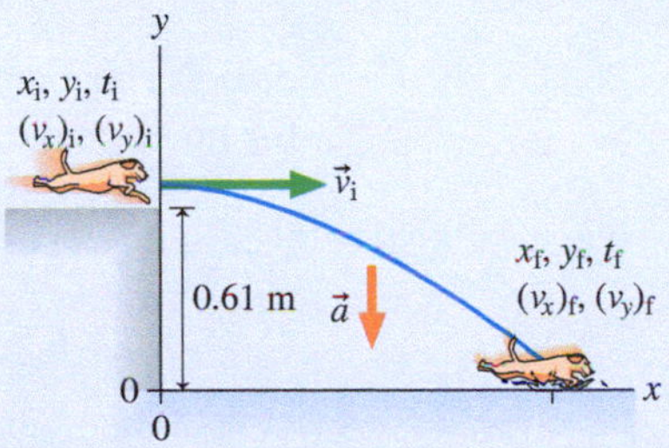

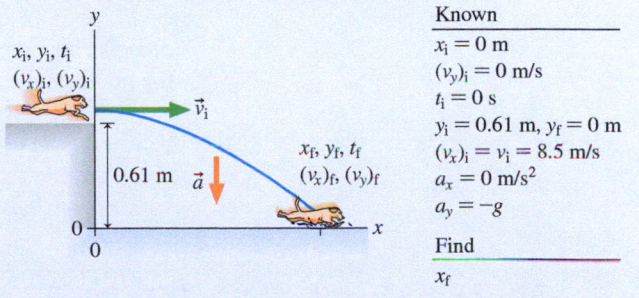

**Known**
$x_i = 0$ m
$(v_y)_i = 0$ m/s
$t_i = 0$ s
$y_i = 0.61$ m, $y_f = 0$ m
$(v_x)_i = v_i = 8.5$ m/s
$a_x = 0\text{ m/s}^2$
$a_y = -g$

**Find**
$x_f$

*Continued*

**SOLVE** We'll start by solving for the time interval $\Delta t$, the time the dog is in the air. This time is determined by the vertical motion, which is free fall with an initial velocity $(v_y)_i = 0$ m/s. We use the vertical-position equation from Synthesis 3.1 to find the time interval:

$$y_f = y_i + (v_y)_i\,\Delta t - \frac{1}{2}g(\Delta t)^2$$

$$0\text{ m} = 0.61\text{ m} + (0\text{ m/s})\Delta t - \frac{1}{2}(9.8\text{ m/s}^2)(\Delta t)^2$$

Rearranging terms to solve for $\Delta t$, we find that

$$\Delta t = 0.35\text{ s}$$

This is how long it takes the dog's vertical motion to reach the water. During this time interval, the dog's horizontal motion is uniform motion at the initial velocity. We can use the horizontal-position equation with the initial speed and $\Delta t = 0.35$ s to find how far the dog travels. This is the distance we are looking for:

$$x_f = x_i + (v_x)_i\,\Delta t$$

$$= 0\text{ m} + (8.5\text{ m/s})(0.35\text{ s}) = 3.0\text{ m}$$

The dog hits the water 3.0 m from the edge of the dock.

**ASSESS** 3.0 m is about 10 feet. This seems like a reasonable distance for a dog running at a very fast clip off the end of a 2-foot-high dock. Indeed, this is a typical distance for dogs in such competitions.

In Example 3.9, the dog ran horizontally off the end of the dock. Much greater distances are possible if the dog goes off the end of the dock at an angle above the horizontal, which gives more time in the air. A launch at an angle involves an initial vertical velocity, but this is still an example of projectile motion, and the general problem-solving approach is the same.

## PROBLEM-SOLVING APPROACH 3.1 Projectile motion problems

**STRATEGIZE** We will solve projectile motion problems by considering the horizontal and vertical motions as separate but related problems.

**PREPARE** There are a number of steps that you should go through in setting up the solution to a projectile motion problem:

- Make simplifying assumptions. Whether the projectile is a car or a basketball, the motion will be the same.
- Draw a visual overview including a pictorial representation showing the beginning and ending points of the motion.
- Establish a coordinate system with the $x$-axis horizontal and the $y$-axis vertical. In this case, you know that the horizontal acceleration will be zero and the vertical acceleration will be free fall: $a_x = 0$ and $a_y = -g$.
- Draw a vector representing the initial velocity, and find its $x$- and $y$-components in terms of the initial speed and the launch angle.
- Define symbols and write down a list of known values. Identify what the problem is trying to find.

**SOLVE** There are two sets of kinematic equations for projectile motion, one for the horizontal component and one for the vertical:

| Horizontal | Vertical |
|---|---|
| $x_f = x_i + (v_x)_i\,\Delta t$ | $y_f = y_i + (v_y)_i\,\Delta t - \frac{1}{2}g(\Delta t)^2$ |
| $(v_x)_f = (v_x)_i = \text{constant}$ | $(v_y)_f = (v_y)_i - g\,\Delta t$ |

**$\Delta t$ is the same for the horizontal and vertical components of the motion.** Find $\Delta t$ by solving for the vertical or the horizontal component of the motion; then use that value to complete the solution for the other component.

**ASSESS** Check that your result has the correct units, is reasonable, and answers the question.

## EXAMPLE 3.10 A free kick

In a soccer free kick, a player kicks a stationary ball toward the goal that is 18 m away. He kicks the ball at an angle of 22° from the horizontal at a speed of 23 m/s. How long does the ball take to reach the goal? And how far off the ground is the ball when it reaches the goal?

**STRATEGIZE** This is a projectile motion problem with the initial velocity at an angle above the horizontal. This means that we will have both vertical and horizontal components of the initial velocity when we apply the equations of Problem-Solving Approach 3.1.

**PREPARE** Following Problem-Solving Approach 3.1, we prepare the visual overview shown in FIGURE 3.31. In choosing our axes, we've placed the origin at the point where the ball is kicked. The initial velocity vector is tilted at 22° above the horizontal, so the components of the initial velocity are

$$(v_x)_i = v_i \cos\theta = (23\text{ m/s})(\cos 22°) = 21.3\text{ m/s}$$

$$(v_y)_i = v_i \sin\theta = (23\text{ m/s})(\sin 22°) = 8.6\text{ m/s}$$

FIGURE 3.31 Visual overview of a kicked soccer ball.

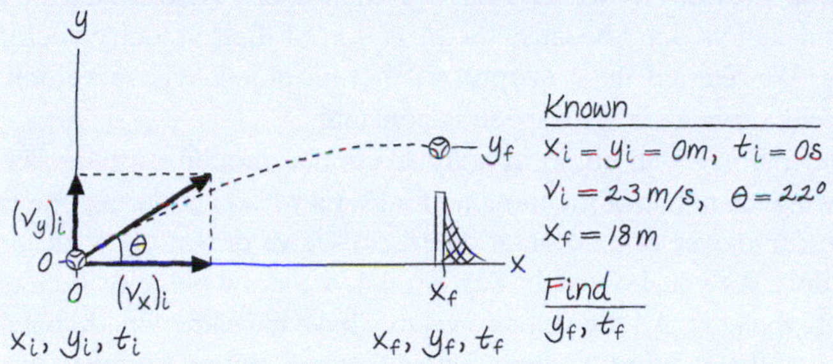


**SOLVE** The problem-solving approach suggests using one component of the motion to solve for $\Delta t$. We can do so using the kinematic equation for the horizontal position,

$$x_f = x_i + (v_x)_i \Delta t$$

We know that $x_i = 0$ m, so we can solve this equation for $\Delta t$ to get

$$\Delta t = \frac{x_f}{(v_x)_i} = \frac{18\text{ m}}{21.3\text{ m/s}} = 0.845\text{ s}$$

where we have kept one extra significant figure for our calculations.

We can now use the kinematic equation for the vertical position to find the height of the ball when it reaches the goal:

$$y_f = y_i + (v_y)_i \Delta t - \tfrac{1}{2} g (\Delta t)^2$$

$$= 0\text{ m} + (8.6\text{ m/s})(0.845\text{ s}) - \tfrac{1}{2}(9.8\text{ m/s}^2)(0.845\text{ s})^2 = 3.8\text{ m}$$

The ball is 3.8 m off the ground when it reaches the goal.

**ASSESS** A height of 3.8 m—about 12 ft—seems reasonable for the height of a kicked ball. Unfortunately, the height of the goal itself is only 2.44 m, so this kick sails easily over the crossbar.

## The Range of a Projectile

When the quarterback throws a football down the field, how far will it go? What will be the **range** for this particular projectile motion—the horizontal distance traveled?

For a projectile, its initial speed and its launch angle are the two variables that determine the range. A higher speed means a greater range, of course. But how does the angle figure in?

FIGURE 3.32 shows the trajectory that a projectile launched at 100 m/s will follow for different launch angles. At very small or very large angles, the range is quite small. If you throw a ball at a 75° angle, it will do a great deal of up-and-down motion, but it won't achieve much horizontal travel. If you throw a ball at a 15° angle, the ball won't be in the air long enough to go very far. These cases both have the same range, as Figure 3.32 shows.

If the angle is too small or too large, the range is shorter than it could be. The "just right" case that gives the maximum range when the landing is at the same elevation as the launch is a launch angle of 45°, as Figure 3.32 shows.

For real-life projectiles, such as golf balls and baseballs, the optimal angle may be less than 45° because of air resistance. Up to this point we've ignored air resistance, but for small objects traveling at high speeds, air resistance is critical. Aerodynamic forces come into play, causing the projectile's trajectory to deviate from a parabola. The maximum range for a golf ball comes at an angle much less than 45°, as you no doubt know if you have ever played golf.

FIGURE 3.32 Trajectories of a projectile for different launch angles, assuming air resistance can be ignored.

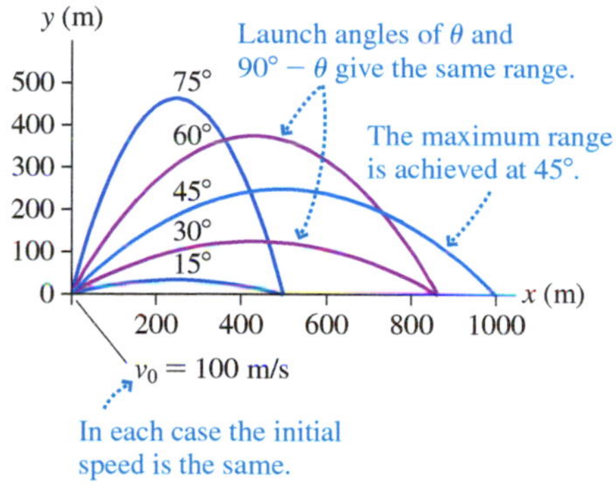


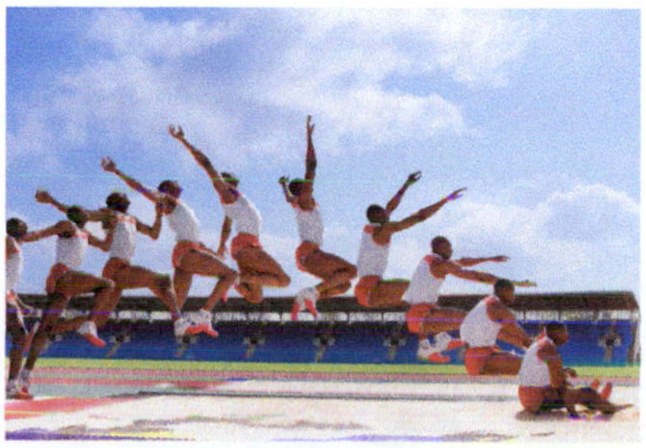

▶ **A long long jump** A 45° angle gives the greatest range for a projectile, so why do long jumpers take off at a much shallower angle? Two of the assumptions that lead to the 45° optimal angle don't apply here. The athlete changes the position of his legs in the air—he doesn't land at the same height as that from which he took off. Also, athletes can't keep the same launch speed for different angles—they can jump faster at smaller angles. The gain from the faster speed outweighs the effect of the smaller angle.

**STOP TO THINK 3.7** A baseball player is taking batting practice. He hits two successive pitches at different angles but at exactly the same speed. Ball 1 and ball 2 follow the paths shown. Which ball is in the air for a longer time? Assume that you can ignore air resistance for this problem.

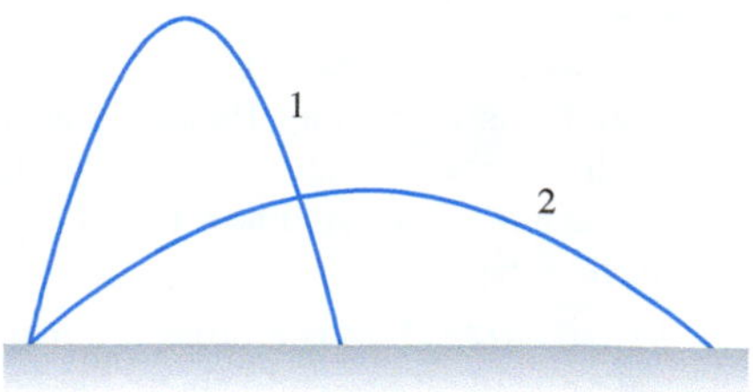


A. Ball 1  B. Ball 2
C. Both balls are in the air for the same amount of time.

## 3.7 Circular Motion

The London Eye Ferris wheel.

The 32 cars on the London Eye Ferris wheel move at about 0.5 m/s in a vertical circle with a 65 m radius. The cars have a *constant speed,* and so we say that the cars are in **uniform circular motion.** However, although their speed is constant, the cars do not move with constant *velocity* because the *direction* of their velocity vectors is constantly changing. We learned in ◀◀ SECTION 3.4 that an object whose velocity is changing has an acceleration, even if its speed is constant.

Let's find the acceleration for an object in uniform circular motion. FIGURE 3.33 is a motion diagram showing ten points of the motion of a Ferris wheel car during one complete revolution. The car moves at a constant speed, so we've drawn equal distances between successive dots. As you learned in Section 3.4, we draw velocity vectors by connecting each dot to the next. All the velocity vectors have the same length, but each has a different direction. We can find the direction of the acceleration by following the steps listed in Tactics Box 3.3, where we found that the acceleration points in the same direction as the *change* in velocity $\Delta\vec{v}$. Figure 3.33 shows how to do this for two successive velocities $\vec{v}_1$ and $\vec{v}_2$. We see that $\Delta\vec{v}$—and therefore the acceleration—is directed

KEY CONCEPT **FIGURE 3.33** Motion diagram for uniform circular motion.

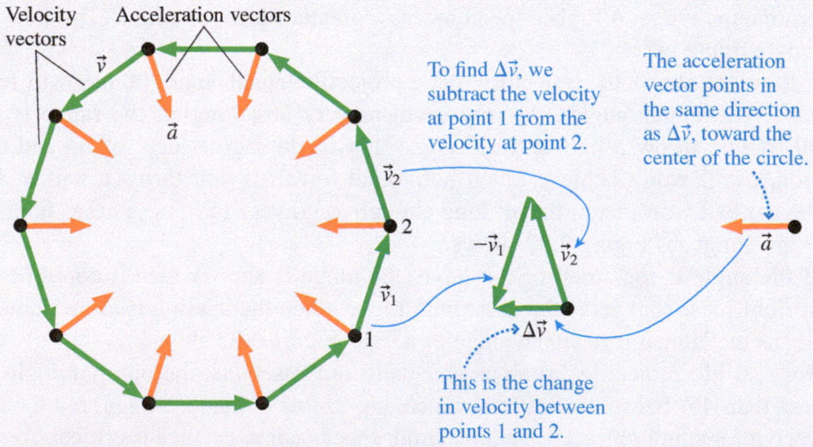

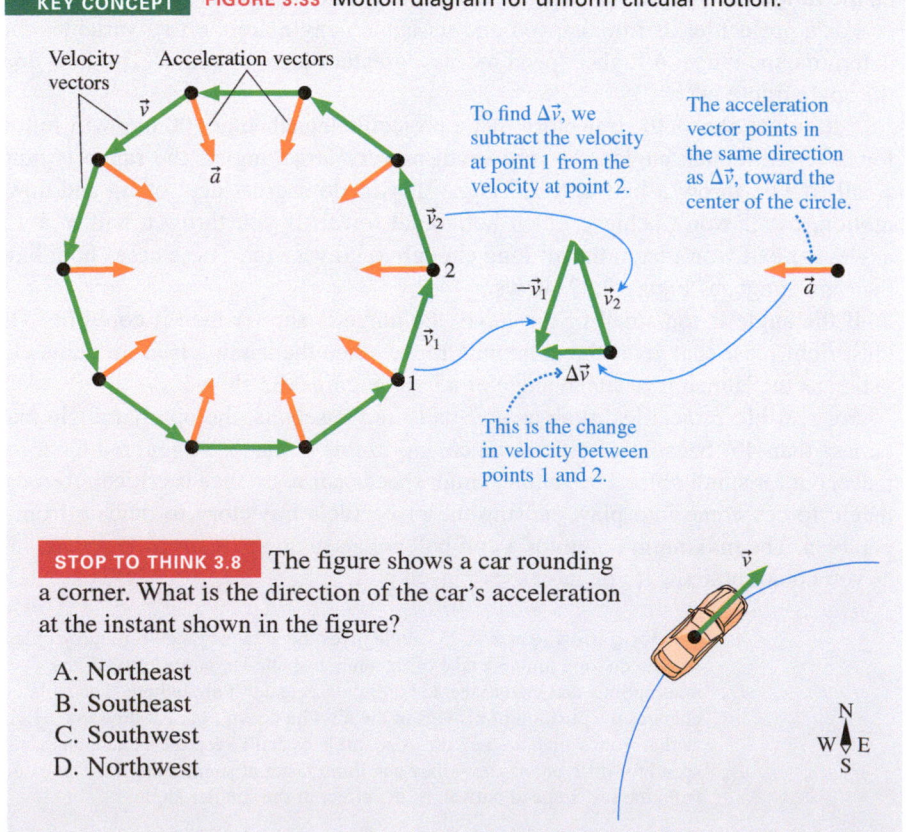

**STOP TO THINK 3.8** The figure shows a car rounding a corner. What is the direction of the car's acceleration at the instant shown in the figure?

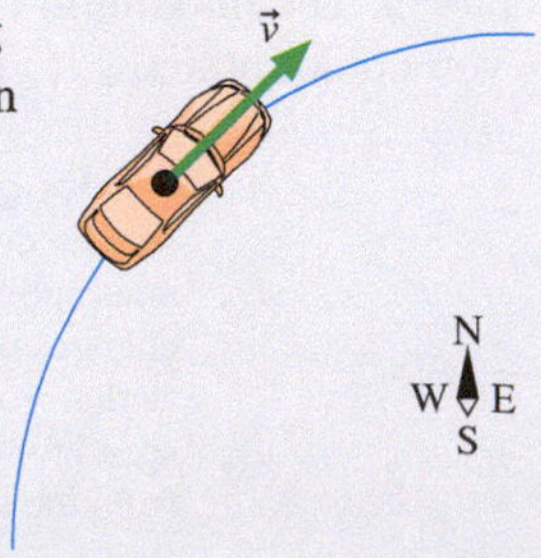


A. Northeast
B. Southeast
C. Southwest
D. Northwest

toward the center of the circle. No matter which points we select on the motion diagram, the velocities change in such a way that **the acceleration vector $\vec{a}$ points toward the center of the circle.** This is a key result for uniform circular motion.

The acceleration that always points directly toward the center of a circle is called **centripetal acceleration.** The word "centripetal" comes from a Greek root meaning "center seeking." FIGURE 3.34 summarizes what we've learned about centripetal acceleration.

FIGURE 3.34 The velocity and acceleration vectors for circular motion.

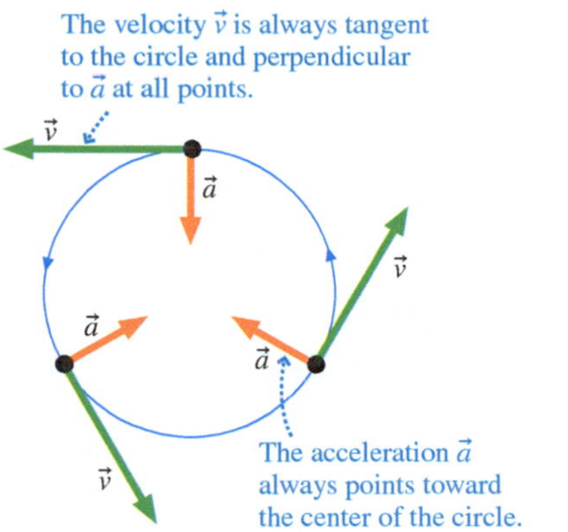
The velocity $\vec{v}$ is always tangent to the circle and perpendicular to $\vec{a}$ at all points.
$\vec{v}$
$\vec{a}$
$\vec{v}$
$\vec{a}$
$\vec{a}$
$\vec{v}$
The acceleration $\vec{a}$ always points toward the center of the circle.

NOTE ▸ Centripetal acceleration is not a new type of acceleration. It is simply the name given to the acceleration of an object moving in a circular path. ◂

To complete our description of circular motion, we need to find a quantitative relationship between the magnitude of the acceleration $a$ and the speed $v$. Let's return to the case of the Ferris wheel. As shown in FIGURE 3.35a, during a time $\Delta t$ in which a car on the Ferris wheel moves along the circle from point 1 to point 2, the car moves through an angle $\theta$ and undergoes a displacement $\vec{d}$. We've chosen a relatively large angle $\theta$ for our drawing so that angular relationships can be clearly seen, but for a small angle the displacement is essentially identical to the actual distance traveled along the circular arc, and we'll make this approximation.

FIGURE 3.35 Changing position and velocity for an object in circular motion.

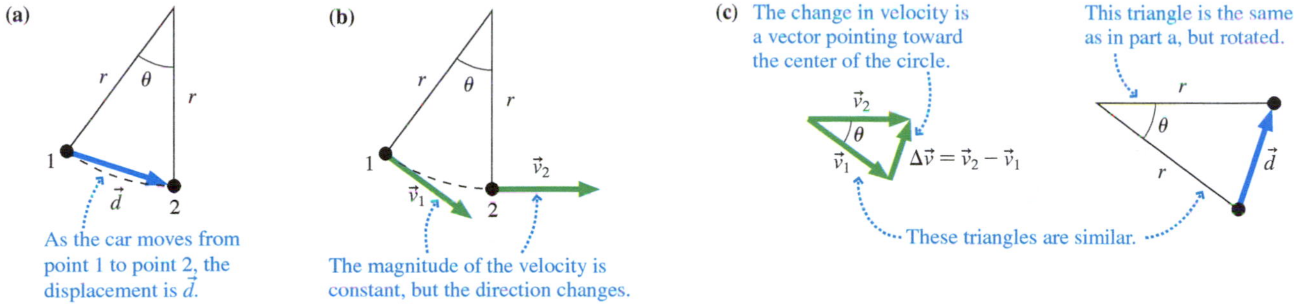


FIGURE 3.35b shows how the velocity changes as the car moves, and FIGURE 3.35c shows the vector calculation of the change in velocity. The triangle we use to make this calculation is geometrically *similar* to the one that shows the displacement, as Figure 3.35c shows. This is a key piece of information: You'll remember from geometry that similar triangles have equal ratios of their sides, so we can write

$$\frac{\Delta v}{v} = \frac{d}{r} \tag{3.19}$$

where $\Delta v$ is the magnitude of the velocity-change vector $\Delta\vec{v}$. We've used the unsubscripted speed $v$ for the length of a side of the first triangle because it is the same for velocities $\vec{v}_1$ and $\vec{v}_2$.

Now we're ready to compute the acceleration. The displacement is just the speed $v$ times the time interval $\Delta t$, so we can write

$$d = v\Delta t$$

We can substitute this for $d$ in Equation 3.19 to obtain

$$\frac{\Delta v}{v} = \frac{v\Delta t}{r}$$

which we can rearrange as

$$\frac{\Delta v}{\Delta t} = \frac{v^2}{r}$$

We recognize the left-hand side of the equation as the acceleration, so this becomes

$$a = \frac{v^2}{r}$$

Combining this magnitude with the direction we noted previously, we can write the centripetal acceleration as

$$\vec{a} = \left(\frac{v^2}{r}, \text{ toward center of circle}\right) \tag{3.20}$$

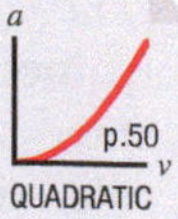


Centripetal acceleration of an object moving in a circle of radius $r$ at speed $v$

**Video** Acceleration Due to Changing Direction

## CONCEPTUAL EXAMPLE 3.11 Acceleration on a swing

A child is riding a playground swing. The swing rotates in a segment of a circle around a central point where the rope or chain for the swing is attached. The speed isn't changing at the lowest point of the motion, but the direction is—this is circular motion, with an acceleration directed upward, as shown in FIGURE 3.36. More acceleration will mean a more exciting ride.

FIGURE 3.36 A child at the lowest point of motion on a swing.

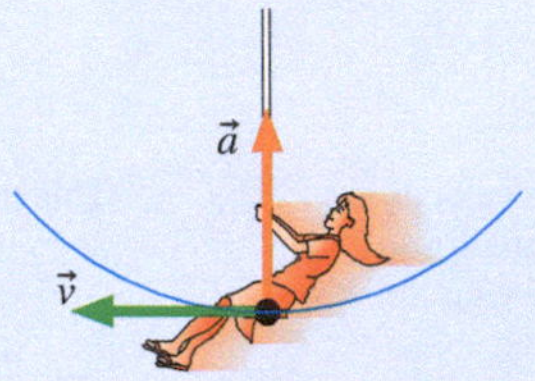


What change could the child make to increase the acceleration she experiences?

**REASON** The acceleration the child experiences is the "changing direction" acceleration of circular motion, given by Equation 3.20. The acceleration depends on the speed and the radius of the circle. The radius of the circle is determined by the length of the chain or rope, so the only easy way to change the acceleration is to change the speed, which she could do by swinging higher. Because the acceleration is proportional to the square of the speed, doubling the speed means a fourfold increase in the acceleration.

**ASSESS** If you have ever ridden a swing, you know that the acceleration you experience is greater the faster you go—so our answer makes sense.

## EXAMPLE 3.12 Finding the acceleration of speed skaters

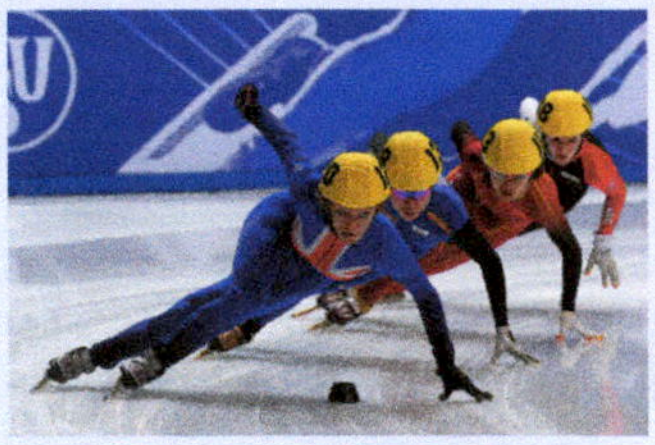

World-class female short-track speed skaters can cover the 500 m of a race in 45 s. The most challenging elements of the race are the turns, which are very tight, with a radius of approximately 11 m. Estimate the magnitude of the skater's centripetal acceleration in a turn.

**STRATEGIZE** The skaters are moving in a circular path, so they have a centripetal acceleration directed inward, toward the center of the curve. We'll need to know their speed, which we can find from the distance around the track and the time to complete the race.

**PREPARE** The centripetal acceleration depends on two quantities: the radius of the turn (given as approximately 11 m) and the speed. The speed varies during the race, but we can make a good estimate of the speed by using the total distance and time:

$$v \simeq \frac{500\text{ m}}{45\text{ s}} = 11\text{ m/s}$$

**SOLVE** We can use these values to estimate the magnitude of the acceleration:

$$a = \frac{v^2}{r} \simeq \frac{(11\text{ m/s})^2}{11\text{ m}} = 11\text{ m/s}^2$$

**ASSESS** This is a large acceleration—a bit more than $g$—but the photo shows the skaters leaning quite hard into the turn, so such a large acceleration seems quite reasonable.

**STOP TO THINK 3.9** Which of the following particles has the greatest centripetal acceleration?

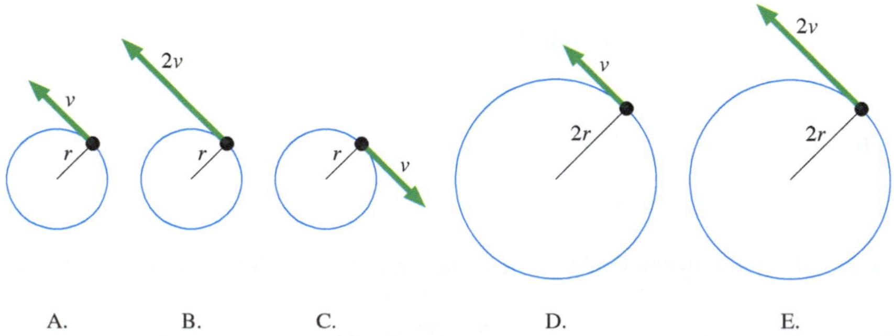

# 3.8 Relative Motion

You've now dealt many times with problems that say something like "A car travels at 30 m/s" or "A plane travels at 300 m/s." But, as we will see, we may need to be a bit more specific.

In FIGURE 3.37, Amy, Bill, and Carlos are watching a runner. According to Amy, the runner's velocity is $v_x = 5$ m/s. But to Bill, who's riding a bicycle alongside, the runner is lifting his legs up and down but going neither forward nor backward relative to Bill. As far as Bill is concerned, the runner's velocity is $v_x = 0$ m/s. Driving his car, Carlos sees the runner receding in his rearview mirror, in the *negative* $x$-direction, getting 10 m farther away from him every second. According to Carlos, the runner's velocity is $v_x = -10$ m/s. Which is the runner's *true* velocity?

As we'll see, there is no single true velocity for an object; we can define an object's velocity only *relative* to the observer who is measuring it. The runner's velocity *relative to Amy* is 5 m/s; that is, his velocity is 5 m/s in a coordinate system attached to Amy and in which Amy is at rest. The runner's velocity relative to Bill is 0 m/s, and the velocity relative to Carlos is −10 m/s. These are all valid descriptions of the runner's motion.

FIGURE 3.37 Amy, Bill, and Carlos each measure the velocity of the runner. The velocities are shown relative to Amy.

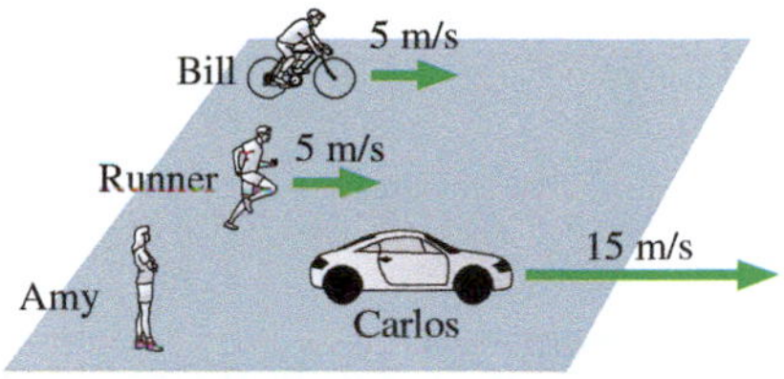


## Relative Velocity

Suppose we know that the runner's velocity relative to Amy is 5 m/s; we will call this velocity $(v_x)_{RA}$. The second subscript "RA" means "**R**unner relative to **A**my." We also know that the velocity of **C**arlos relative to **A**my is 15 m/s; we write this as $(v_x)_{CA} = 15$ m/s. It is equally valid to compute Amy's velocity relative to Carlos. From Carlos's point of view, Amy is moving to the left at 15 m/s. We write **A**my's velocity relative to **C**arlos as $(v_x)_{AC} = -15$ m/s. Note that $(v_x)_{AC} = -(v_x)_{CA}$.

Given the runner's velocity relative to Amy and Amy's velocity relative to Carlos, we can compute the runner's velocity relative to Carlos by combining the two velocities we know. The subscripts as we have defined them are our guide for this combination:

$$(v_x)_{RC} = (v_x)_{RA} + (v_x)_{AC} \tag{3.21}$$

The "A" appears on the right of the first expression and on the left of the second; when we combine these velocities, we "cancel" the A to get $(v_x)_{RC}$.

**Throwing for the gold** An athlete throwing the javelin does so while running. It's harder to throw the javelin on the run, but there's a very good reason to do so. The distance of the throw will be determined by the velocity of the javelin with respect to the ground—which is the sum of the velocity of the throw plus the velocity of the athlete. A faster run means a longer throw.

Generally, we can add two relative velocities in this manner, by "canceling" subscripts as in Equation 3.21. In Chapter 27, when we learn about relativity, we will have a more rigorous scheme for computing relative velocities, but this technique will serve our purposes at present.

**EXAMPLE 3.13 Speed of a seabird**

Researchers doing satellite tracking of albatrosses in the Southern Ocean observed a bird maintaining sustained flight speeds of 35 m/s—nearly 80 mph! This seems surprisingly fast until you realize that this particular bird was flying with the wind, which was moving at 23 m/s. What was the bird's airspeed—its speed relative to the air? This is a truer measure of its flight speed.

**STRATEGIZE** We are asked how fast the bird appears to be moving relative to someone drifting along at the same speed as the wind. We will use Equation 3.21 to find this relative velocity.

**PREPARE** FIGURE 3.38 shows the wind and the albatross moving to the right, so all velocities will be positive. We've shown the velocity $(v_x)_{bw}$ of the **b**ird with respect to the **w**ater, which is the measured flight speed, and the velocity $(v_x)_{aw}$ of the **a**ir with respect to the **w**ater, which is the known wind speed. We want to find the **b**ird's airspeed—the speed of the **b**ird with respect to the **a**ir.

FIGURE 3.38 Relative velocities for the albatross and the wind for Example 3.13.

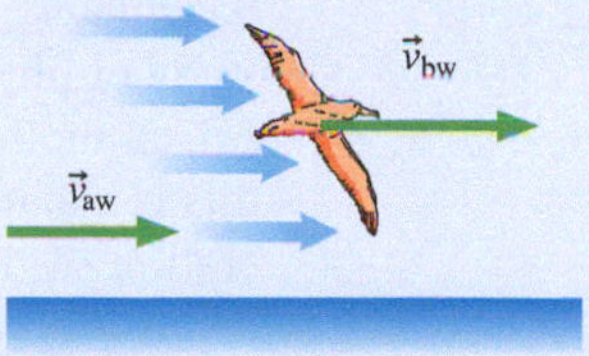


**Known**
$(v_x)_{bw} = 35$ m/s
$(v_x)_{aw} = 23$ m/s
**Find**
$(v_x)_{ba}$

*Continued*

**SOLVE** We've noted three different velocities that are important in the problem: $(v_x)_{bw}$, $(v_x)_{aw}$, and $(v_x)_{ba}$. We can combine these in the usual way:

$$(v_x)_{bw} = (v_x)_{ba} + (v_x)_{aw}$$

Then, to solve for $(v_x)_{ba}$, we can rearrange the terms:

$$(v_x)_{ba} = (v_x)_{bw} - (v_x)_{aw} = 35 \text{ m/s} - 23 \text{ m/s} = 12 \text{ m/s}$$

**ASSESS** 12 m/s—about 25 mph—is a reasonable airspeed for a bird. And it's slower than the observed flight speed, which makes sense because the bird is flying with the wind.

This technique for finding relative velocities also works for two-dimensional situations, another good exercise in working with vectors.

**EXAMPLE 3.14 Finding the ground speed of an airplane**

Cleveland is approximately 300 miles east of Chicago. A plane leaves Chicago flying due east at 500 mph. The pilot forgot to check the weather and doesn't know that the wind is blowing to the south at 100 mph. What is the plane's velocity relative to the ground?

**STRATEGIZE** We will generalize the idea of relative velocity to the case where the velocities are not directed along a single line by using *vectors* to represent the velocities in Equation 3.21.

**PREPARE** FIGURE 3.39 is a visual overview of the situation. We are given the velocity of the **p**lane relative to the **a**ir ($\vec{v}_{pa}$) and the velocity of the **a**ir relative to the **g**round ($\vec{v}_{ag}$); the velocity of the **p**lane relative to the **g**round will be the vector sum of these velocities:

$$\vec{v}_{pg} = \vec{v}_{pa} + \vec{v}_{ag}$$

This vector sum is shown in Figure 3.39.

FIGURE 3.39 The wind causes a plane flying due east in the air to move to the southeast relative to the ground.

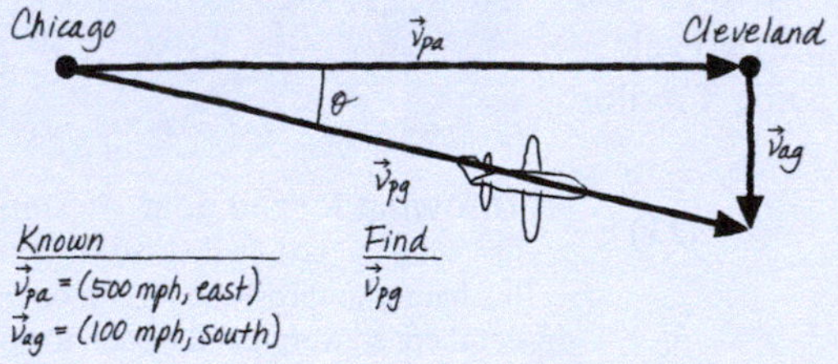


**SOLVE** The plane's speed relative to the ground is the hypotenuse of the right triangle in Figure 3.39; thus:

$$v_{pg} = \sqrt{v_{pa}^2 + v_{ag}^2} = \sqrt{(500 \text{ mph})^2 + (100 \text{ mph})^2} = 510 \text{ mph}$$

The plane's direction can be specified by the angle $\theta$ measured from due east:

$$\theta = \tan^{-1}\left(\frac{100 \text{ mph}}{500 \text{ mph}}\right) = \tan^{-1}(0.20) = 11°$$

The velocity of the plane relative to the ground is thus

$$\vec{v}_{pg} = (510 \text{ mph}, 11° \text{ south of east})$$

**ASSESS** The good news is that the wind is making the plane move a bit faster relative to the ground. The bad news is that the wind is making the plane move in the wrong direction!

**STOP TO THINK 3.10** The water in a river flows downstream at 3.0 m/s. A boat is motoring upstream against the flow at 5.0 m/s relative to the water. What is the boat's speed relative to the riverbank?

A. 8.0 m/s  B. 5.0 m/s  C. 3.0 m/s  D. 2.0 m/s

## What Comes Next: Forces

Kinematics, the mathematical description of motion, is a good place to start our study of physics because motion is very visible and very familiar. But what actually *causes* motion? The skaters in Example 3.12 were "leaning quite hard into the turn"; this is a statement about the forces acting on them, the forces that cause the acceleration of the turn. In the next chapter, we'll explore the nature of forces and the connection to motion, giving us the tools to treat a much broader range of problems.

## INTEGRATED EXAMPLE 3.15 World-record jumpers BIO

Frogs, with their long, strong legs, are excellent jumpers. And thanks to the good folks of Calaveras County, California, who have a jumping frog contest every year in honor of a Mark Twain story, we have very good data as to just how far a determined frog can jump. The current record holder is Rosie the Ribeter, a bullfrog that made a leap of 2.2 m from a standing start. This compares favorably with the world record for a human, which is a mere 3.7 m.

Typical data for a serious leap by a bullfrog look like this: The frog goes into a crouch, then rapidly extends its legs by 15 cm as it pushes off, leaving the ground at an angle of 30° to the horizontal. It's in the air for 0.49 s before landing at the same height from which it took off. Given this leap, what is the acceleration while the frog is pushing off? How far does the frog jump?

**STRATEGIZE** The problem really has two parts: the leap through the air and the acceleration required to produce this leap. We'll need to analyze the leap—the projectile motion—first, which will give us the frog's launch speed and the distance of the jump. Once we know the velocity with which the frog leaves the ground, we can calculate its acceleration while pushing off the ground.

**PREPARE** Let's start with a visual overview of the two parts, as shown in FIGURE 3.40. Notice that the second part of the problem uses a different $x$-axis, tilted as we did earlier for motion on a ramp.

**SOLVE** The "flying through the air" part shown in Figure 3.40a is projectile motion. The frog lifts off at a 30° angle with a speed $v_i$. The $x$- and $y$-components of the initial velocity are

$$(v_x)_i = v_i \cos 30°$$
$$(v_y)_i = v_i \sin 30°$$

The vertical motion can be analyzed as we did in Example 3.10. The kinematic equation is

$$y_f = y_i + (v_y)_i\,\Delta t - \tfrac{1}{2}g(\Delta t)^2$$

We know that $y_f = y_i = 0$, so this reduces to

$$(v_y)_i = \tfrac{1}{2}g\Delta t = \tfrac{1}{2}(9.8\ \text{m/s}^2)(0.49\ \text{s}) = 2.4\ \text{m/s}$$

We know the $y$-component of the velocity and the angle, so we can find the magnitude of the velocity and the $x$-component:

$$v_i = \frac{(v_y)_i}{\sin 30°} = \frac{2.4\ \text{m/s}}{\sin 30°} = 4.8\ \text{m/s}$$
$$(v_x)_i = v_i \cos 30° = (4.8\ \text{m/s})\cos 30° = 4.2\ \text{m/s}$$

The horizontal motion is uniform motion, so the frog's horizontal position when it returns to the ground is

$$x_f = x_i + (v_x)_i\,\Delta t = 0 + (4.2\ \text{m/s})(0.49\ \text{s}) = 2.1\ \text{m}$$

This is the length of the jump.

Now that we know how fast the frog is going when it leaves the ground, we can calculate the acceleration necessary to produce this jump—the "pushing off the ground" part shown in Figure 3.40b. We've drawn the $x$-axis along the direction of motion, as we did for problems of motion on a ramp. We know the displacement $\Delta x$ of the launch but not the time, so we can use Equation 2.13 from Synthesis 2.1:

$$(v_x)_f^2 = (v_x)_i^2 + 2a_x\,\Delta x$$

The initial velocity is zero, the final velocity is $(v_x)_f = 4.8$ m/s, and the displacement is the 15 cm (or 0.15 m) stretch of the legs during the jump. Thus the frog's acceleration while pushing off is

$$a_x = \frac{(v_x)_f^2}{2\,\Delta x} = \frac{(4.8\ \text{m/s})^2}{2(0.15\ \text{m})} = 77\ \text{m/s}^2$$

**ASSESS** A 2.1 m jump is an impressive jump, but it's less than the record for a frog, so the final result for the distance seems reasonable. Such a long jump must require a large acceleration during the pushing-off phase, which is what we found.

FIGURE 3.40 A visual overview for the leap of a frog.

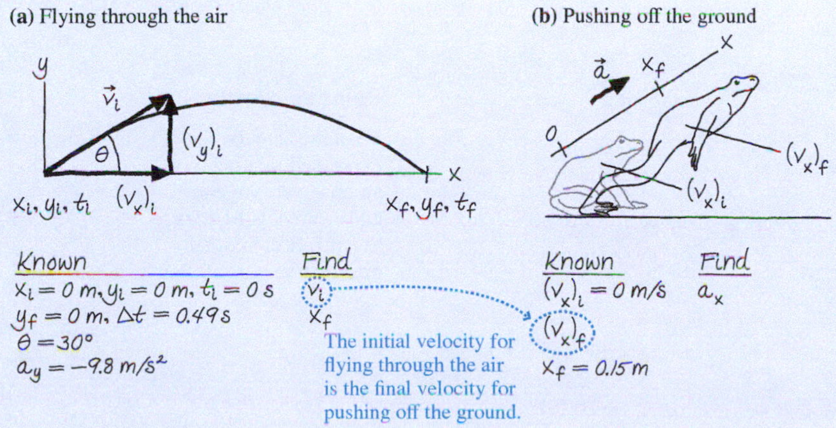

# SUMMARY

**GOAL** To learn more about vectors and to use vectors as a tool to analyze motion in two dimensions.

## GENERAL PRINCIPLES

### Projectile Motion

A projectile is an object that moves through the air under the influence of gravity and nothing else.

The path of the motion is a parabola.

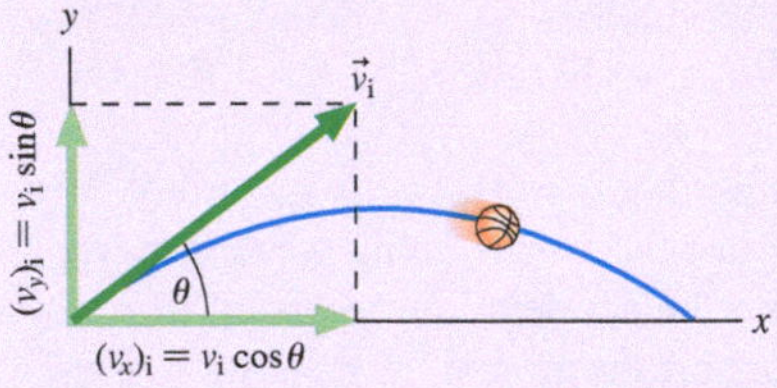


The motion consists of two pieces:

1. Vertical motion with free-fall acceleration, $a_y = -g$
2. Horizontal motion with constant velocity

Kinematic equations:

$$x_f = x_i + (v_x)_i\,\Delta t$$
$$(v_x)_f = (v_x)_i = \text{constant}$$
$$y_f = y_i + (v_y)_i\,\Delta t - \tfrac{1}{2}g(\Delta t)^2$$
$$(v_y)_f = (v_y)_i - g\,\Delta t$$

### Circular Motion

An object moving in a circle at a constant speed has a velocity that is constantly changing direction, and so experiences an acceleration:

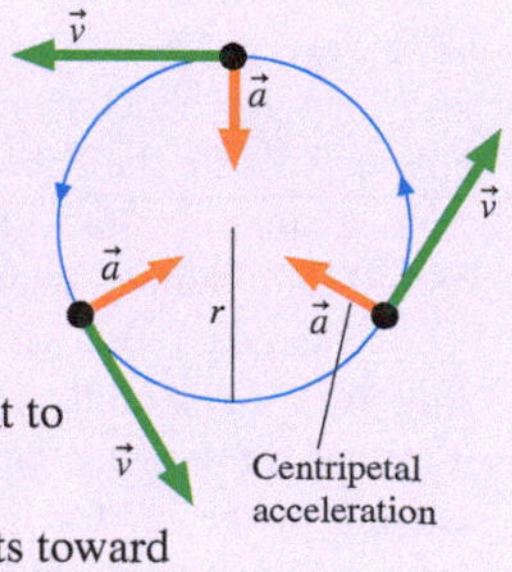

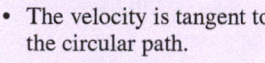

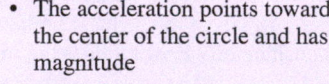

- The velocity is tangent to the circular path.
- The acceleration points toward the center of the circle and has magnitude

$$a = \frac{v^2}{r}$$

## IMPORTANT CONCEPTS

### Vectors and Components

A vector can be decomposed into *x*- and *y*-**components.**

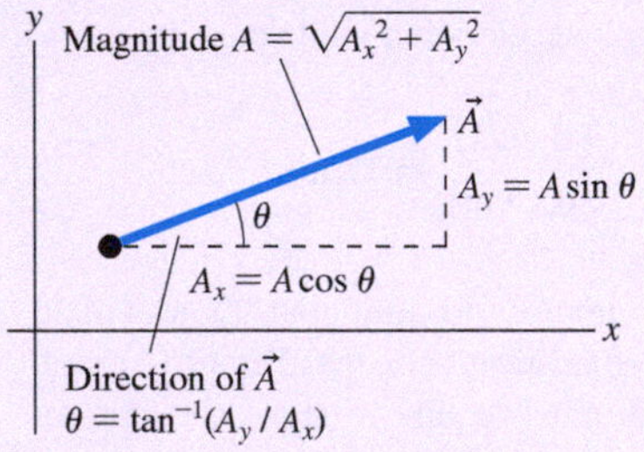


The magnitude and direction of a vector can be expressed in terms of its components.

The sign of the components depends on the direction of the vector:

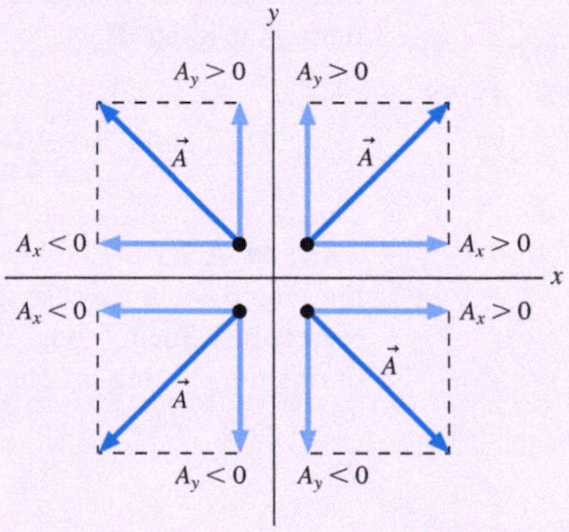


### The Acceleration Vector

We define the acceleration vector as

$$\vec{a} = \frac{\vec{v}_f - \vec{v}_i}{t_f - t_i} = \frac{\Delta\vec{v}}{\Delta t}$$

We find the acceleration vector on a motion diagram as follows:

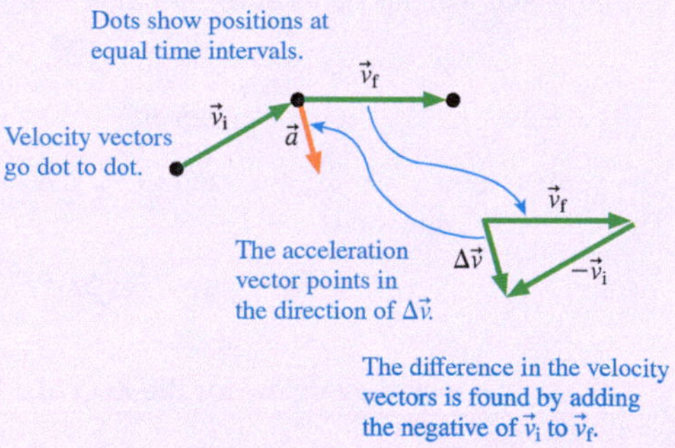


## APPLICATIONS

### Motion on a ramp

An object sliding down a ramp will accelerate parallel to the ramp:

$$a_x = \pm g\sin\theta$$

The correct sign depends on the direction in which the ramp is tilted.

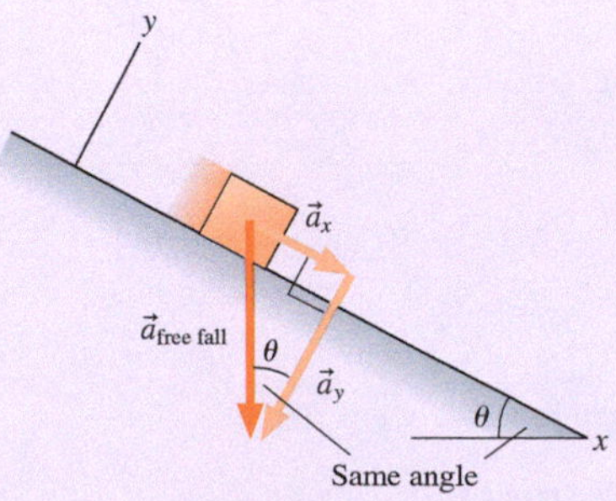


### Relative motion

Velocities can be expressed relative to an observer. We can add relative velocities to convert to another observer's point of view.

c = car, r = runner, g = ground

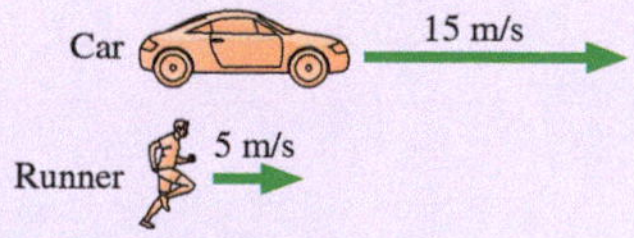


The speed of the car with respect to the runner is

$$(v_x)_{cr} = (v_x)_{cg} + (v_x)_{gr}$$

## Learning Objectives
After studying this chapter, you should be able to:

- Add and subtract vectors graphically. *Conceptual Questions 3.2, 3.3; Problems 3.1, 3.2, 3.3, 3.4*
- Use vectors and motion diagrams to find acceleration. *Conceptual Question 3.12; Problems 3.21, 3.22, 3.23, 3.24, 3.25*
- Perform vector calculations using vector components. *Problems 3.5, 3.6, 3.9, 3.13, 3.52*
- Solve problems about an object moving on a ramp. *Conceptual Question 3.11; Problems 3.16, 3.17, 3.18, 3.19, 3.20, 3.59*
- Understand and use relative velocity. *Conceptual Question 3.19; Problems 3.45, 3.46, 3.47, 3.49, 3.76*
- Solve problems for projectiles that follow parabolic trajectories. *Conceptual Questions 3.6, 3.9; Problems 3.27, 3.28, 3.31, 3.34, 3.66*
- Calculate and use centripetal acceleration. *Conceptual Questions 3.15, 3.17; Problems 3.37, 3.38, 3.40, 3.41, 3.43*

## STOP TO THINK ANSWERS

**Chapter Preview Stop to Think: D.** The ball took 2.0 s to reach the highest point, at which time it is (momentarily) at rest. As it rises, its vertical velocity is changing at $a_y = -g = -9.8 \text{ m/s}^2 \approx -10 \text{ m/s}^2$. The change in speed is $\Delta v_y \approx (-10 \text{ m/s}^2)(2.0 \text{ s}) = -20 \text{ m/s}$, so the initial speed is approximately 20 m/s.

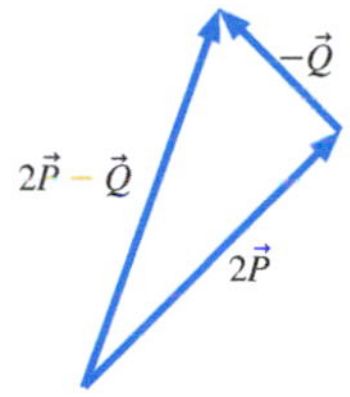


**Stop to Think 3.1: A.** The graphical construction of $2\vec{P} - \vec{Q}$ is shown at right.

**Stop to Think 3.2: B.** From the axes on the graph, we can see that the $x$- and $y$-components are $-4$ cm and $+2$ cm, respectively.

**Stop to Think 3.3: C.** Vector $\vec{C}$ points to the left and down, so both $C_x$ and $C_y$ are negative. $C_x$ is in the numerator because it is the side opposite $\phi$.

**Stop to Think 3.4: B.** The angle of the slope is greatest in this case, leading to the greatest acceleration.

**Stop to Think 3.5: A, B, and C.** If car A has a constant speed, then at the point shown its acceleration will be to the left; this is like the valley of Example 3.7 turned counterclockwise. Car B can also have the acceleration shown; imaging tilting the valley slightly counterclockwise. Car C is undergoing linear motion, so its acceleration will point to the left if it is either moving to the left and speeding up, or moving to the right and slowing down. But car D's acceleration can never point as shown: In the valley example, the acceleration vectors always point to the inside of the curve, never to the outside, no matter whether the skier is speeding up or slowing down.

**Stop to Think 3.6: D.** Mass does not appear in the kinematic equations, so the mass has no effect. The balls follow the same path.

**Stop to Think 3.7: A.** The time in the air is determined by the vertical component of the velocity. Ball 1 has a higher vertical velocity, so it will be in the air for a longer time.

**Stop to Think 3.8: B.** The acceleration is directed toward the center of the circle. At the instant shown, the acceleration is down and to the right—to the southeast.

**Stop to Think 3.9: B.** The magnitude of the acceleration is $v^2/r$. Acceleration is greatest for the combination of highest speed and smallest radius.

**Stop to Think 3.10: D.** Let's define upstream as positive and downstream as negative. Then the speed of the boat with respect to the water is $v_{bw} = +5$ m/s, and the speed of the water with respect to the riverbank is $v_{wr} = -3$ m/s. Thus the speed of the boat with respect to the riverbank is

$$\begin{aligned} v_{br} &= v_{bw} + v_{wr} \\ &= 5 \text{ m/s} - 3 \text{ m/s} = 2 \text{ m/s} \end{aligned}$$

**Video Tutor Solution** Chapter 3

# QUESTIONS

## Conceptual Questions

1. a. Can a vector have nonzero magnitude if a component is zero? If no, why not? If yes, give an example.
   b. Can a vector have zero magnitude and a nonzero component? If no, why not? If yes, give an example.
2. If $\vec{C} = \vec{A} + \vec{B}$ can $C = 0$? Can $C < 0$? For each question, show how or explain why not.
3. Suppose two vectors have unequal magnitudes. Can their sum be $\vec{0}$? Explain.
4. At the instant shown, is the particle in Figure Q3.4 speeding up, slowing down, or traveling at a constant speed?

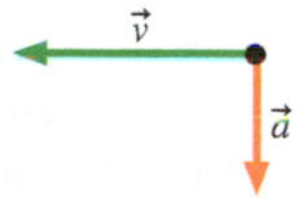


FIGURE Q3.4

5. For a projectile, which of the following quantities are constant during the flight: $x$, $y$, $v_x$, $v_y$, $v$, $a_x$, $a_y$? Which of the quantities are zero throughout the flight?

Problem difficulty is labeled as | (straightforward) to ||||| (challenging). Problems labeled INT integrate significant material from earlier chapters; Problems labeled BIO are of biological or medical interest.

The eText icon indicates when there is a video tutor solution available for the chapter or for a specific problem. To launch these videos, log into your eText through Mastering[TM] Physics or log into the Study Area.

6. A baseball player throws a ball at a 40° angle to the ground. The ball lands on the ground some distance away.
   a. Is there any point on the trajectory where $\vec{v}$ and $\vec{a}$ are parallel to each other? If so, where?
   b. Is there any point where $\vec{v}$ and $\vec{a}$ are perpendicular to each other? If so, where?
7. An athlete performing the long jump tries to achieve the maximum distance from the point of takeoff to the first point of touching the ground. After the jump, rather than land upright, she extends her legs forward as in the photo. How does this affect the time in the air? How does this give the jumper a longer range?

8. If you kick a football, at what angle to the ground should you kick the ball for the maximum range—that is, the greatest distance down the field? At what angle to the ground should you kick the ball for the maximum "hang time"—that is, the maximum time in the air?
9. Carla holds a ball 1.5 m above the ground. Daniel, leaning out of a car window, also holds a ball 1.5 m above the ground. Daniel drives past Carla at 40 mph and, just as he passes her, both release their balls at the same instant. Whose ball hits the ground first? Explain.
10. A cart that is rolling at a constant velocity on a level table fires a ball straight up out of a vertical tube.
    a. When the ball comes back down, will it land in front of the launching tube, behind the launching tube, or directly into the tube? Explain.
    b. Does your answer change if the cart is accelerating in the forward direction? If so, how?
11. If you go to a ski area, you'll likely find that the beginner's slope has the smallest angle. Use the concept of acceleration on a ramp to explain why this is so.
12. In an amusement-park ride, cars rolling along at high speed suddenly head up a long, straight ramp. They roll up the ramp, reverse direction at the highest point, then roll backward back down the ramp. In each of the following segments of the motion, which way does the acceleration vector point?
    a. As the cars roll up the ramp.
    b. At the highest point on the ramp.
    c. As the cars roll back down the ramp.
13. There are competitions in which pilots fly small planes low over the ground and drop weights, trying to hit a target. A pilot flying low and slow drops a weight; it takes 2.0 s to hit the ground, during which it travels a horizontal distance of 100 m. Now the pilot does a run at the same height but twice the speed. How much time does it take the weight to hit the ground? How far does it travel before it lands?
14. Roller coaster loops are rarely perfectly circular. Instead, they are tightly curved at the top, where the cars are moving more slowly, and they have a gentler curve at the bottom, where the cars are moving much faster. Explain why the loops are designed this way.

15. You are cycling around a circular track at a constant speed. Does the magnitude of your acceleration change? The direction?
16. An airplane has been directed to fly in a clockwise circle, as seen from above, at constant speed until another plane has landed. When the plane is going north, is it accelerating? If so, in what direction does the acceleration vector point? If not, why not?
17. When you go around a corner in your car, your car follows a path that is a segment of a circle. To turn safely, you should keep your car's acceleration below some safe upper limit. If you want to make a "tighter" turn—that is, turn in a circle with a smaller radius—how should you adjust your speed? Explain.
18. A person trying to throw a ball as far as possible will run forward during the throw. Explain why this increases the distance of the throw.
19. Anna is running to the right, as shown in Figure Q3.19. Balls 1 and 2 are thrown toward her by friends standing on the ground. According to Anna, both balls are approaching her at the same speed. Which ball was thrown with the faster speed? Or were they thrown with the same speed? Explain.

FIGURE Q3.19

## Multiple-Choice Questions

20. ‖ Which combination of the vectors shown in Figure Q3.20 has the largest magnitude?

A. $\vec{A}+\vec{B}+\vec{C}$ B. $\vec{B}+\vec{A}-\vec{C}$
C. $\vec{A}-\vec{B}+\vec{C}$ D. $\vec{B}-\vec{A}-\vec{C}$

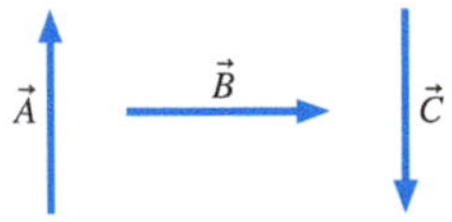


FIGURE Q3.20

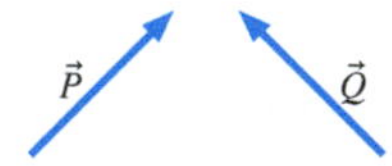


FIGURE Q3.21

21. ‖ Two vectors appear as in Figure Q3.21. Which combination points directly to the left?

A. $\vec{P}+\vec{Q}$ B. $\vec{P}-\vec{Q}$ C. $\vec{Q}-\vec{P}$ D. $-\vec{Q}-\vec{P}$

22. | The gas pedal in a car is sometimes referred to as "the accelerator." Which other controls on the vehicle can be used to produce acceleration?

A. The brakes. B. The steering wheel.
C. The gear shift. D. All of the above.

23. | A car travels at constant speed along the curved path shown from above in Figure Q3.23. Five possible vectors are also shown in the figure; the letter E represents the zero vector. Which vector best represents
    a. The car's velocity at position 1?
    b. The car's acceleration at position 1?
    c. The car's velocity at position 2?
    d. The car's acceleration at position 2?
    e. The car's velocity at position 3?
    f. The car's acceleration at position 3?

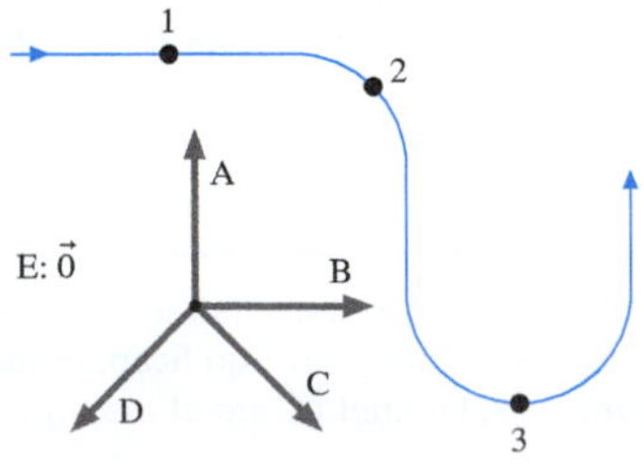


FIGURE Q3.23

24. | A ball is fired from a cannon at point 1 and follows the trajectory shown in Figure Q3.24. Air resistance may be neglected. Five possible vectors are also shown in the figure; the letter E represents the zero vector. Which vector best represents
   a. The ball's velocity at position 2?
   b. The ball's acceleration at position 2?
   c. The ball's velocity at position 3?
   d. The ball's acceleration at position 3?

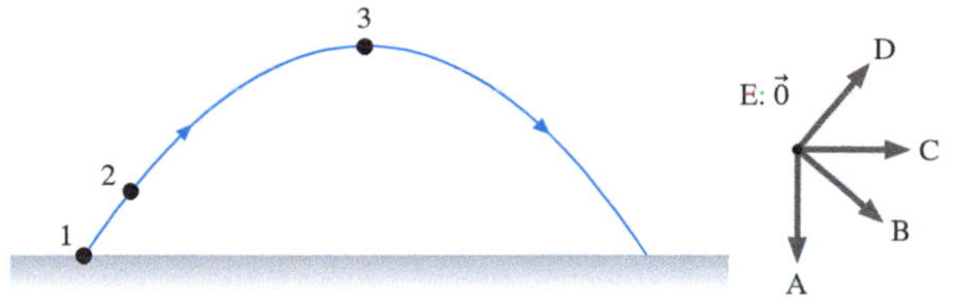


FIGURE Q3.24

25. | A ball thrown at an initial angle of 37.0° and initial velocity of 23.0 m/s reaches a maximum height $h$, as shown in Figure Q3.25. With what initial speed must a ball be thrown *straight up* to reach the same maximum height $h$?
   A. 13.8 m/s B. 17.3 m/s C. 18.4 m/s D. 23.0 m/s

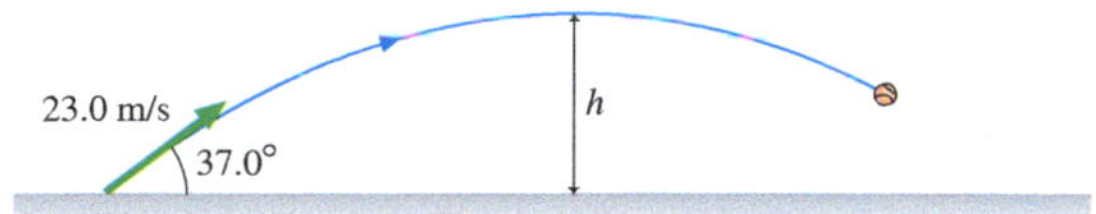


FIGURE Q3.25

26. || At a football game, an air gun fires T-shirts into the crowd. The gun is fired at an angle of 40° from the horizontal with an initial speed of 28 m/s. A fan who is sitting 60 m horizontally from the gun, but high in the stands, catches a T-shirt.
   a. How long does it take for the T-shirt to reach the fan?
      A. 1.3 s B. 2.0 s C. 2.8 s D. 3.4 s E. 3.7 s
   b. At what height $h$ is the fan from the ground?
      A. 3.2 m B. 4.3 m C. 6.6 m D. 9.2 m E. 12.0 m

27. || Liam throws a water balloon horizontally at 8.2 m/s out of a window 10 m from the ground. How far from the base of the building does the balloon land?
   A. 4.2 m B. 8.0 m C. 12 m D. 15 m E. 18 m

28. | A football is kicked at an angle of 30° with a speed of 20 m/s. To the nearest second, how long will the ball stay in the air?
   A. 1 s B. 2 s C. 3 s D. 4 s

29. | A football is kicked at an angle of 30° with a speed of 20 m/s. To the nearest 5 m, how far will the ball travel?
   A. 15 m B. 25 m C. 35 m D. 45 m

30. || An ultracentrifuge spins biological samples at very high speeds
BIO in order to separate their constituent microscopic components. One such instrument spins at 120,000 rpm, and the samples experience a centripetal acceleration of $6.6 \times 10^6$ m/s$^2$ What is the radius of the circle around which the samples travel?
   A. 2.6 cm B. 4.2 cm C. 7.8 cm D. 12 cm E. 17 cm

31. | Formula One race cars are capable of remarkable accelerations when speeding up, slowing down, and turning corners. At one track, cars round a corner that is a segment of a circle of radius 95 m at a speed of 68 m/s. What is the approximate magnitude of the centripetal acceleration, in units of $g$?
   A. $1g$ B. $2g$ C. $3g$ D. $4g$ E. $5g$

# PROBLEMS

## Section 3.1 Using Vectors

1. || Trace the vectors in Figure P3.1 onto your paper. Then use graphical methods to draw the vectors (a) $\vec{A} + \vec{B}$ and (b) $\vec{A} - \vec{B}$.

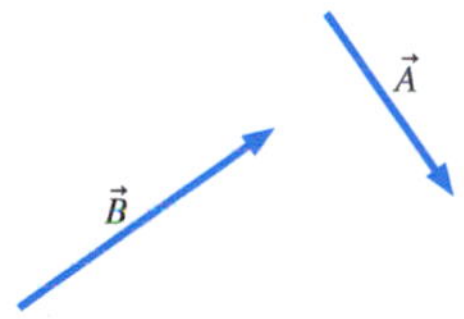


FIGURE P3.1

2. ||| Trace the vectors in Figure P3.2 onto your paper. Then use graphical methods to draw the vectors (a) $\vec{A} + \vec{B}$ and (b) $\vec{A} - \vec{B}$.

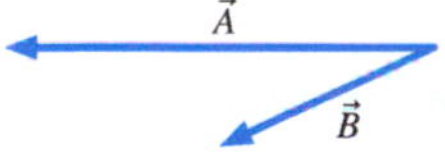


FIGURE P3.2

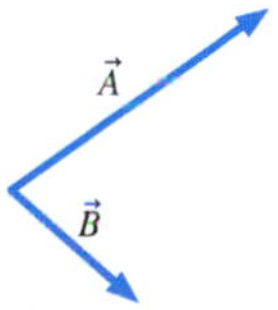


FIGURE P3.3

3. | Trace the vectors in Figure P3.3 onto your paper. Then draw the vector $\vec{C}$ such that $\vec{A} + \vec{B} + \vec{C} = 0$.

4. || Two vectors $\vec{A}$ and $\vec{B}$ are at right angles to each other. The magnitude of $\vec{A}$ is 1. What should be the length of $\vec{B}$ so that the magnitude of their vector sum is 2?

## Section 3.2 Coordinate Systems and Vector Components

5. || A position vector with magnitude 10 m points to the right and up. Its $x$-component is 6.0 m. What is the value of its $y$-component?

6. ||| A velocity vector 40° above the positive x-axis has a $y$-component of 10 m/s. What is the value of its $x$-component?

7. || A cannon tilted upward at 30° fires a cannonball with a speed of 100 m/s. At that instant, what is the component of the cannonball's velocity parallel to the ground?

8. || a. What are the $x$- and $y$-components of vector $\vec{E}$ of Figure P3.8 in terms of the angle $\theta$ and the magnitude $E$?
   b. For the same vector, what are the $x$- and $y$-components in terms of the angle $\phi$ and the magnitude $E$?

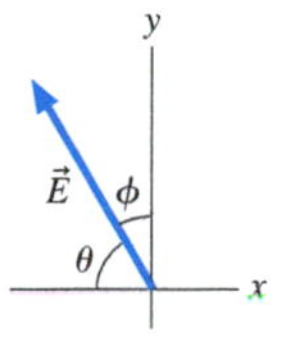


FIGURE P3.8

9. | Draw each of the following vectors, then find its $x$- and $y$-components.
   a. $\vec{d} = (100 \text{ m}, 45° \text{ below } +x\text{-axis})$
   b. $\vec{v} = (300 \text{ m/s}, 20° \text{ above } +x\text{-axis})$
   c. $\vec{a} = (5.0 \text{ m/s}^2, -y\text{-direction})$

10. || Draw each of the following vectors, then find its $x$- and $y$-components.
   a. $\vec{d} = (2.0 \text{ km}, 30° \text{ left of } +y\text{-axis})$
   b. $\vec{v} = (5.0 \text{ cm/s}, -x\text{-direction})$
   c. $\vec{a} = (10 \text{ m/s}^2, 40° \text{ left of } -y\text{-axis})$

11. | Each of the following vectors is given in terms of its $x$- and $y$-components. Draw the vector, label an angle that specifies the vector's direction, then find the vector's magnitude and direction.
   a. $v_x = 20$ m/s, $v_y = 40$ m/s
   b. $a_x = 2.0$ m/s$^2$, $a_y = -6.0$ m/s$^2$

Watch Video Solution Problem 3.19

12. | Each of the following vectors is given in terms of its *x*- and *y*-components. Draw the vector, label an angle that specifies the vector's direction, then find the vector's magnitude and direction.
 a. $v_x = 10$ m/s, $v_y = 30$ m/s
 b. $a_x = 20$ m/s$^2$, $a_y = 10$ m/s$^2$

13. || A wildlife researcher is tracking a flock of geese. The geese fly 4.0 km due west, then turn toward the north by 40° and fly another 4.0 km. How far west are they of their initial position? What is the magnitude of their displacement?
BIO

14. || Jack and Jill ran up the hill at 3.0 m/s. The horizontal component of Jill's velocity vector was 2.5 m/s.
 a. What was the angle of the hill?
 b. What was the vertical component of Jill's velocity?

15. | Josh is climbing up a steep 34° slope, moving at a steady 0.75 m/s along the ground. How many meters of elevation does he gain in one minute of this climb?

## Section 3.3 Motion on a Ramp

16. ||| You begin sliding down a 15° ski slope. Ignoring friction and air resistance, how fast will you be moving after 10 s?

17. |||| A car traveling at 30 m/s runs out of gas while traveling up a 5.0° slope. How far will it coast before starting to roll back down?

18. || In the Soapbox Derby, young participants build non-motorized cars with very low-friction wheels. Cars race by rolling down a hill. The track at Akron's Derby Downs, where the national championship is held, begins with a 55-ft-long section tilted 13° below horizontal.

 a. What is the maximum possible acceleration of a car moving down this stretch of track?
 b. If a car starts from rest and undergoes this acceleration for the full 55 ft, what is its final speed in m/s?

19. ||| A piano has been pushed to the top of the ramp at the back of a moving van. The workers think it is safe, but as they walk away, it begins to roll down the ramp. If the back of the truck is 1.0 m above the ground and the ramp is inclined at 20°, how much time do the workers have to get to the piano before it reaches the bottom of the ramp?

20. || In the winter sport of bobsledding, athletes push their sled along a horizontal ice surface and then hop on the sled as it starts to careen down the steeply sloped track. In one event, the sled reaches a top speed of 9.2 m/s before starting down the initial part of the track, which is sloped downward at an angle of 6.0°. What is the sled's speed after it has traveled the first 100 m?

## Section 3.4 Motion in Two Dimensions

21. | A car goes around a corner in a circular arc at constant speed. Draw a motion diagram including positions, velocity vectors, and acceleration vectors.

22. |||| Figure 3.33 showed the motion diagram for a rider on a Ferris wheel that was turning at a constant speed. The inset to the figure showed how to find the acceleration vector at the rightmost point. Use a similar analysis to find the rider's acceleration vector at the leftmost position of the motion diagram, then at one of the highest positions and at one of the lowest positions. Use a ruler so that your analysis is accurate.

Problems 23 and 24 show partial motion diagrams. For each:
 a. Complete the motion diagram by adding acceleration vectors.
 b. Write a physics *problem* for which this is the correct motion. Be imaginative! Don't forget to include enough information to make the problem complete and to state clearly what is to be found.

23. ||

FIGURE P3.23

24. ||

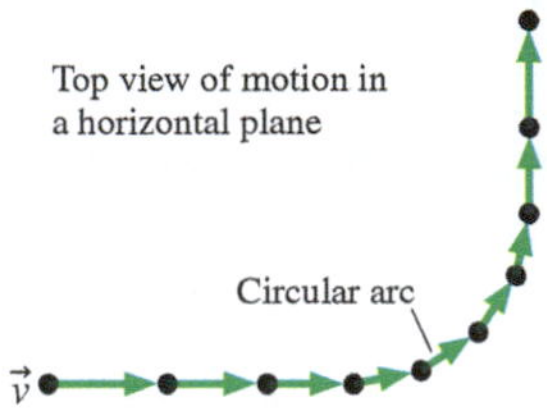


FIGURE P3.24

Answer Problems 25 and 26 by choosing one of the eight labeled acceleration vectors or selecting option I: $\vec{a} = \vec{0}$.

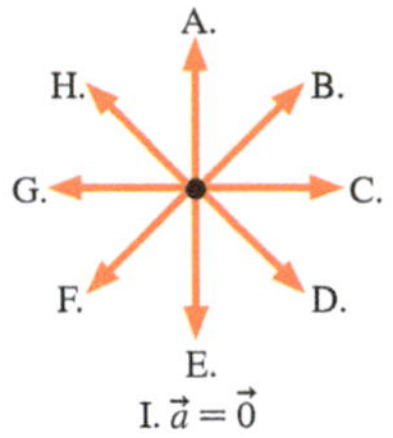


25. || At this instant, the particle has steady speed and is curving to the right. What is the direction of its acceleration?

FIGURE P3.25

26. || At this instant, the particle is speeding up and curving upward. What is the direction of its acceleration?

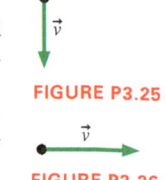

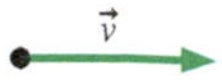


FIGURE P3.26

## Section 3.5 Projectile Motion

## Section 3.6 Projectile Motion: Solving Problems

27. ||| A ball is thrown horizontally from a 20-m-high building with a speed of 5.0 m/s.
 a. Make a sketch of the ball's trajectory.
 b. Draw a graph of $v_x$, the horizontal velocity, as a function of time. Include units on both axes.
 c. Draw a graph of $v_y$, the vertical velocity, as a function of time. Include units on both axes.
 d. How far from the base of the building does the ball hit the ground?

28. || A ball with a horizontal speed of 1.25 m/s rolls off a bench 1.00 m above the floor.
 a. How long will it take the ball to hit the floor?
 b. How far from a point on the floor directly below the edge of the bench will the ball land?

29. || A pipe discharges storm water into a creek. Water flows horizontally out of the pipe at 1.5 m/s, and the end of the pipe is 2.5 m above the creek. How far out from the end of the pipe is the point where the stream of water meets the creek?

30. || On a day when the water is flowing relatively gently, water in the Niagara River is moving horizontally at 4.5 m/s before shooting over Niagara Falls. After moving over the edge, the water drops 53 m to the water below. If we ignore air resistance, how much time does it take for the water to go from the top of the falls to the bottom? How far does the water move horizontally during this time?

31. ||| A running mountain lion can make a leap 10.0 m long, reaching a maximum height of 3.0 m.
BIO
   a. What is the speed of the mountain lion just as it leaves the ground?
   b. At what angle does it leave the ground?

32. ||| A rifle is aimed horizontally at a target 50 m away. The bullet hits the target 2.0 cm below the aim point.
   a. What was the bullet's flight time?
   b. What was the bullet's speed as it left the barrel?

33. ||| A gray kangaroo can bound across a flat stretch of ground with each jump carrying it 10 m from the takeoff point. If the kangaroo leaves the ground at a 20° angle, what are its (a) takeoff speed and (b) horizontal speed?
BIO

34. ||| On the Apollo 14 mission to the moon, astronaut Alan Shepard hit a golf ball with a golf club improvised from a tool. The free-fall acceleration on the moon is 1/6 of its value on earth. Suppose he hit the ball with a speed of 25 m/s at an angle 30° above the horizontal.
   a. How long was the ball in flight?
   b. How far did it travel?
   c. Ignoring air resistance, how much farther would it travel on the moon than on earth?

35. || Emily throws a soccer ball out of her dorm window to Allison, who is waiting below to catch it. If Emily throws the ball at an angle of 30° below horizontal with a speed of 12 m/s, how far from the base of the dorm should Allison stand to catch the ball? Assume the vertical distance between where Emily releases the ball and Allison catches it is 6.0 m.

36. || A soccer player takes a free kick from a spot that is 20 m from the goal. The ball leaves his foot at an angle of 32°, and it eventually hits the crossbar of the goal, which is 2.4 m from the ground. At what speed did the ball leave his foot?

## Section 3.7 Circular Motion

37. | Racing greyhounds are capable of rounding corners at very high speeds. A typical greyhound track has turns that are 45-m-diameter semicircles. A greyhound can run around these turns at a constant speed of 15 m/s. What is its acceleration in $m/s^2$ and in units of $g$?
BIO

38. || To withstand "g-forces" of up to 10 g's, caused by suddenly pulling out of a steep dive, fighter jet pilots train on a "human centrifuge." 10 g's is an acceleration of 98 $m/s^2$. If the length of the centrifuge arm is 12 m, at what speed is the rider moving when she experiences 10 g's?
BIO

39. || The moon completes one (circular) orbit of the earth in 27.3 days. The distance from the earth to the moon is $3.84 \times 10^8$ m. What is the moon's centripetal acceleration?

40. || In a roundabout (or traffic circle), cars go around a 25-m-diameter circle. If a car's tires will skid when the car experiences a centripetal acceleration greater than $0.60g$, what is the maximum speed of the car in this roundabout?

41. || A particle rotates in a circle with centripetal acceleration $a = 8.0$ $m/s^2$. What is $a$ if
   a. The radius is doubled without changing the particle's speed?
   b. The speed is doubled without changing the circle's radius?

42. || Entrance and exit ramps for freeways are often circular stretches of road. As you go around one at a constant speed, you will experience a constant acceleration. Suppose you drive through an entrance ramp at a modest speed and your acceleration is 3.0 $m/s^2$. What will be the acceleration if you double your speed?

43. || A peregrine falcon in a tight, circular turn can attain a centripetal acceleration 1.5 times the free-fall acceleration. If the falcon is flying at 20 m/s, what is the radius of the turn?
BIO

## Section 3.8 Relative Motion

44. || An airplane cruises at 880 km/h relative to the air. It is flying from Denver, Colorado, due west to Reno, Nevada, a distance of 1200 km, and will then return. There is a steady 90 km/h wind blowing to the east. What is the difference in flight time between the two legs of the trip?

45. | Anita is running to the right at 5 m/s, as shown in Figure P3.45. Balls 1 and 2 are thrown toward her at 10 m/s by friends standing on the ground. According to Anita, what is the speed of each ball?

FIGURE P3.45

46. ||| In the 2016 Olympics in Rio, after the 50 m freestyle competition, a problem with the pool was found. In lane 1 there was a gentle 1.2 cm/s current flowing in the direction that the swimmers were going, while in lane 8 there was a current of the same speed but directed opposite to the swimmers' direction. Suppose a swimmer could swim the 50 m in 25.0 s in the absence of any current. What would be her time in lane 1? In lane 8? How does the difference in these times compare to the actual 0.06 s difference in times between the gold medal winner and the fourth-place finisher?

47. | Anita is running to the right at 5 m/s, as shown in Figure P3.47. Balls 1 and 2 are thrown toward her by friends standing on the ground. According to Anita, both balls are approaching her at 10 m/s. According to her friends, with what speeds were the balls thrown?

FIGURE P3.47

48. || Two children who are bored while waiting for their flight at the airport decide to race from one end of the 20-m-long moving sidewalk to the other and back. Phillippe runs on the sidewalk at 2.0 m/s (relative to the sidewalk). Renee runs on the floor at 2.0 m/s. The sidewalk moves at 1.5 m/s relative to the floor. Both make the turn instantly with no loss of speed.
   a. Who wins the race?
   b. By how much time does the winner win?

49. ||| A boat takes 3.0 h to travel 30 km down a river, then 5.0 h to return. How fast is the river flowing?

## General Problems

50. | Suppose $\vec{C} = \vec{A} + \vec{B}$ where vector $\vec{A}$ has components $A_x = 5$, $A_y = 2$ and vector $\vec{B}$ has components $B_x = -3$, $B_y = -5$.
   a. What are the $x$- and $y$-components of vector $\vec{C}$?
   b. Draw a coordinate system and on it show vectors $\vec{A}$, $\vec{B}$, and $\vec{C}$.
   c. What are the magnitude and direction of vector $\vec{C}$?

51. | Suppose $\vec{D} = \vec{A} - \vec{B}$ where vector $\vec{A}$ has components $A_x = 5$, $A_y = 2$ and vector $\vec{B}$ has components $B_x = -3$, $B_y = -5$.
   a. What are the $x$- and $y$-components of vector $\vec{D}$?
   b. Draw a coordinate system and on it show vectors $\vec{A}$, $\vec{B}$, and $\vec{D}$.
   c. What are the magnitude and direction of vector $\vec{D}$?

52. || Suppose $\vec{E} = 2\vec{A} + 3\vec{B}$ where vector $\vec{A}$ has components $A_x = 5$, $A_y = 2$ and vector $\vec{B}$ has components $B_x = -3$, $B_y = -5$.
   a. What are the $x$- and $y$-components of vector $\vec{E}$?
   b. Draw a coordinate system and on it show vectors $\vec{A}$, $\vec{B}$, and $\vec{E}$.
   c. What are the magnitude and direction of vector $\vec{E}$?

53. || For the three vectors shown in Figure P3.53, the vector sum $\vec{D} = \vec{A} + \vec{B} + \vec{C}$ has components $D_x = 2$ and $D_y = 0$.
   a. What are the $x$- and $y$-components of vector $\vec{B}$?
   b. Write $\vec{B}$ as a magnitude and a direction.

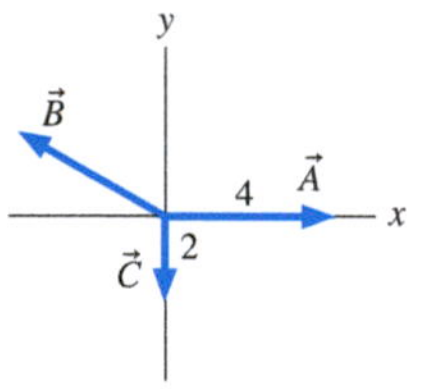

FIGURE P3.53

54. || Let $\vec{A} = (3.0 \text{ m}, 20° \text{ south of east})$, $\vec{B} = (2.0 \text{ m, north})$, and $\vec{C} = (5.0 \text{ m}, 70° \text{ south of west})$.
   a. Draw and label $\vec{A}$, $\vec{B}$, and $\vec{C}$ with their tails at the origin. Use a coordinate system with the $x$-axis to the east.
   b. Write the $x$- and $y$-components of vectors $\vec{A}$, $\vec{B}$, and $\vec{C}$.
   c. Find the magnitude and the direction of $\vec{D} = \vec{A} + \vec{B} + \vec{C}$.

55. || To get to his office from home, Greg walks 5 blocks north and then 3 blocks east. After work he meets some friends at a café; to get there he walks 2 blocks south and 5 blocks west. All blocks are 660 feet long. What is the straight-line distance from the café to his home?

56. ||| A pilot in a small plane encounters shifting winds. He flies 26.0 km northeast, then 45.0 km due north. From this point, he flies an additional distance in an unknown direction, only to find himself at a small airstrip that his map shows to be 70.0 km directly north of his starting point. What were the length and direction of the third leg of his trip?

57. ||| In punting a football, the kicker tries to maximize both the distance of the kick and its "hang time"—the time that the ball is in the air. A kicker gets off a great punt with a hang time of 5.0 s that lands 50 yards from the kicker.
   a. What is the speed of the ball as it leaves the kicker's foot?
   b. What is the angle of the ball's initial velocity?

Watch Video Solution Problem 3.53

58. || BIO The bacterium *Escherichia coli* (or *E. coli*) is a single-celled organism that lives in the gut of healthy humans and animals. When grown in a uniform medium rich in salts and amino acids, these bacteria swim along zig-zag paths at a constant speed of 20 $\mu$m/s. Figure P3.58 shows the trajectory of an *E. coli* as it moves from point A to point E. Each segment of the motion can be identified by two letters, such as segment BC.
   a. For each of the four segments in the bacterium's trajectory, calculate the $x$- and $y$-components of its displacement and of its velocity.
   b. Calculate both the total distance traveled and the magnitude of the net displacement for the entire motion.
   c. What are the magnitude and the direction of the bacterium's average velocity for the entire trip?

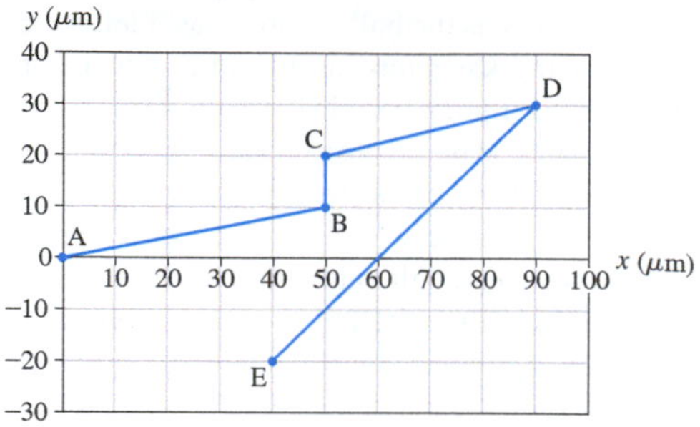

FIGURE P3.58

59. ||| A skier gliding across the snow at 3.0 m/s suddenly starts down a 10° incline, reaching a speed of 15 m/s at the bottom. Friction between the snow and her freshly waxed skis is negligible.
   a. What is the length of the incline?
   b. How long does it take her to reach the bottom?

60. ||| As shown in Figure P3.60, a skier speeds along a flat patch of snow, and then flies horizontally off the edge at 12.0 m/s. He eventually lands on a straight, sloped section that is at an angle of 45° below the horizontal. How long is he in the air?

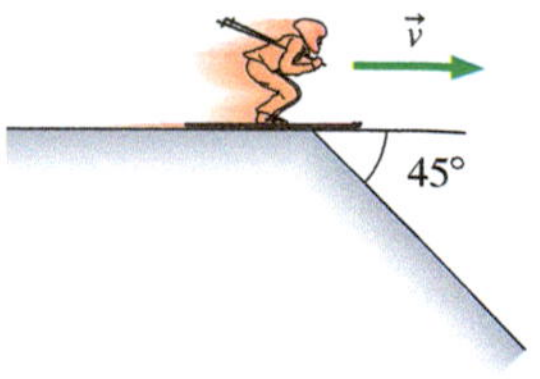

FIGURE P3.60

61. || A physics student on Planet Exidor throws a ball, and it follows the parabolic trajectory shown in Figure P3.61. The ball's position is shown at 1.0 s intervals until $t = 3.0$ s. At $t = 1.0$ s, the ball's velocity has components $v_x = 2.0$ m/s, $v_y = 2.0$ m/s.

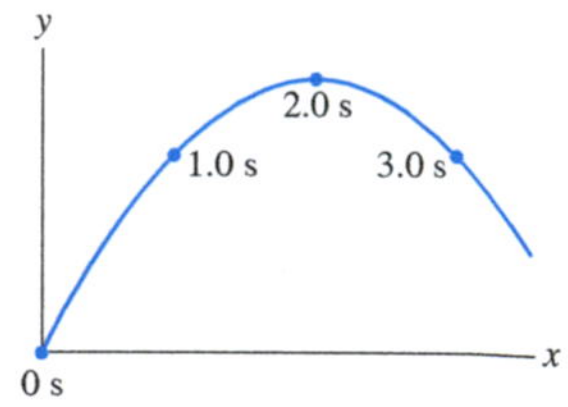

FIGURE P3.61

   a. Determine the $x$- and $y$-components of the ball's velocity at $t = 0.0$ s, 2.0 s, and 3.0 s.
   b. What is the value of $g$ on Planet Exidor?
   c. What was the ball's launch angle?

62. ||| BIO The archerfish uses a remarkable method for catching insects sitting on branches or leaves above the waterline. The fish rises to the surface and then shoots out a stream of water precisely aimed to knock the insect off its perch into the water, where the archerfish gobbles it up. Scientists have measured the speed of the water stream exiting the fish's mouth to be 3.7 m/s. An archerfish spots an insect sitting 19 cm above the waterline and a horizontal distance of 30 cm away. The fish aims its stream at an angle of 39° from the waterline. Does the stream hit its mark?

63. ||| In 1780, in what is now referred to as "Brady's Leap," Captain Sam Brady of the U.S. Continental Army escaped certain death from his enemies by running horizontally off the edge of the cliff above Ohio's Cuyahoga River, which is confined at that spot to a gorge. He landed safely on the far side of the river. It was reported that he leapt 22 ft across while falling 20 ft. Tall tale, or possible?

FIGURE P3.63

a. What is the minimum speed with which he'd need to run off the edge of the cliff to make it safely to the far side of the river?
b. The world-record time for the 100 m dash is approximately 10 s. Given this, is it reasonable to expect Brady to be able to run fast enough to achieve Brady's leap?

64. ||| The longest recorded pass in an NFL game traveled 83 yards in the air from the quarterback to the receiver. Assuming that the pass was thrown at the optimal 45° angle, what was the speed at which the ball left the quarterback's hand?

65. ||| A spring-loaded gun, fired vertically, shoots a marble 6.0 m straight up in the air. What is the marble's range if it is fired horizontally from 1.5 m above the ground?

66. || Small-plane pilots regularly compete in "message drop" competitions, dropping heavy weights (for which air resistance can be ignored) from their low-flying planes and scoring points for having the weights land close to a target. A plane 60 m above the ground is flying directly toward a target at 45 m/s.
a. At what distance from the target should the pilot drop the weight?
b. The pilot looks down at the weight after she drops it. Where is the plane located at the instant the weight hits the ground—not yet over the target, directly over the target, or past the target?

67. ||| Paintball guns were originally developed to mark trees for logging. A forester aims his gun directly at a knothole in a tree that is 4.0 m above the gun. The base of the tree is 20 m away. The speed of the paintball as it leaves the gun is 50 m/s. How far below the knothole does the paintball strike the tree?

68. ||| Trained dolphins are capable of a vertical leap of 7.0 m
BIO straight up from the surface of the water—an impressive feat.
INT Suppose you could train a dolphin to launch itself out of the water at this same speed but at an angle. What maximum horizontal range could the dolphin achieve?

69. ||| A tennis player hits a ball 2.0 m above the ground. The ball leaves his racquet with a speed of 20 m/s at an angle 5.0° above the horizontal. The horizontal distance to the net is 7.0 m, and the net is 1.0 m high. Does the ball clear the net? If so, by how much? If not, by how much does it miss?

70. ||| The shot put is a track-
INT and-field event in which athletes throw a heavy ball—the shot—as far as possible. The best athletes can throw the shot as far as 23 m. Athletes who use the "glide" technique push the shot outward in a reasonably straight line, accelerating it over a distance of about 2.0 m. What acceleration do they provide to the shot as they push on it? Assume that the shot is launched at an angle of 37°, a reasonable value for an excellent throw. You can assume that the shot lands at the same height from which it is thrown; this simplifies the calculation considerably, and makes only a small difference in the final result.

71. ||| Water at the top of Horseshoe Falls (part of Niagara Falls) is moving horizontally at 9.0 m/s as it goes off the edge and plunges 53 m to the pool below. If you ignore air resistance, at what angle is the falling water moving as it enters the pool?

72. || A supply plane needs to drop a package of food to scientists working on a glacier in Greenland. The plane flies 100 m above the glacier at a speed of 150 m/s. How far short of the target should it drop the package?

73. |||| A BMX bicycle rider takes off from a ramp at a point 1.8 m above the ground. The ramp is angled at 40° from the horizontal, and the rider's speed is 6.7 m/s when he leaves the ramp. How far from the end of the ramp does he land?

74. ||||| Ships A and B leave port together. For the next two hours, ship A travels at 20 mph in a direction 30° west of north while ship B travels 20° east of north at 25 mph.
a. What is the distance between the two ships two hours after they depart?
b. What is the speed of ship A as seen by ship B?

75. ||| A flock of ducks is trying to migrate south for the winter, but they keep being blown off course by a wind blowing from the west at 12 m/s. A wise elder duck finally realizes that the solution is to fly at an angle to the wind. If the ducks can fly at 16 m/s relative to the air, in what direction should they head in order to move directly south?

76. ||| A kayaker needs to paddle north across a 100-m-wide harbor. The tide is going out, creating a tidal current flowing east at 2.0 m/s. The kayaker can paddle with a speed of 3.0 m/s.
a. In which direction should he paddle in order to travel straight across the harbor?
b. How long will it take him to cross?

77. |||| A plane has an airspeed of 200 mph. The pilot wishes to reach a destination 600 mi due east, but a wind is blowing at 50 mph in the direction 30° north of east.
a. In what direction must the pilot head the plane in order to reach her destination?
b. How long will the trip take?

78. ||| The Gulf Stream off the east coast of the United States can flow at a rapid 3.6 m/s to the north. A ship in this current has a cruising speed of 10 m/s. The captain would like to reach land at a point due west from the current position.
a. In what direction with respect to the water should the ship sail?
b. At this heading, what is the ship's speed with respect to land?

79. ||| A ball thrown horizontally at 25 m/s travels a horizontal distance of 50 m before hitting the ground. From what height was the ball thrown?

80. ||| A sports car is advertised as capable of "reaching 60 mph in
INT 5 seconds flat, cornering at $0.85g$, and stopping from 70 mph in only 168 feet." In which of those three situations is the magnitude of the car's acceleration the largest? In which is it the smallest?

81. || A Ford Mustang can accelerate from 0 to 60 mph in a time
INT of 5.6 s. A Mini Cooper isn't capable of such a rapid start, but it can turn in a very small circle 34 ft in diameter. How fast would you need to drive the Mini Cooper in this tight circle to match the magnitude of the Mustang's acceleration?

82. ‖ The "Screaming Swing" is a carnival ride that is—not surprisingly—a giant swing. It's actually two swings moving in opposite directions. At the bottom of its arc, a rider in one swing is moving at 30 m/s with respect to the ground in a 50-m-diameter circle. The rider in the other swing is moving in a similar circle at the same speed, but in the exact opposite direction.
   a. What is the acceleration, in $m/s^2$ and in units of $g$, that riders experience?
   b. At the bottom of the ride, as they pass each other, how fast do the riders move with respect to each other?

83. ‖ On an otherwise straight stretch of road near Moffat, Colorado, the road suddenly turns. This bend in the road is a segment of a circle with radius 110 m. Drivers are cautioned to slow down to 40 mph as they navigate the curve.
   a. If you heed the sign and slow to 40 mph, what will be your acceleration going around the curve at this constant speed? Give your answer in $m/s^2$ and in units of $g$.
   b. At what speed would your acceleration be double that at the recommended speed?

## MCAT-Style Passage Problems

### Riding the Water Slide

A rider on a water slide goes through three different kinds of motion, as illustrated in Figure P3.84. Use the data and details from the figure to answer the following questions.

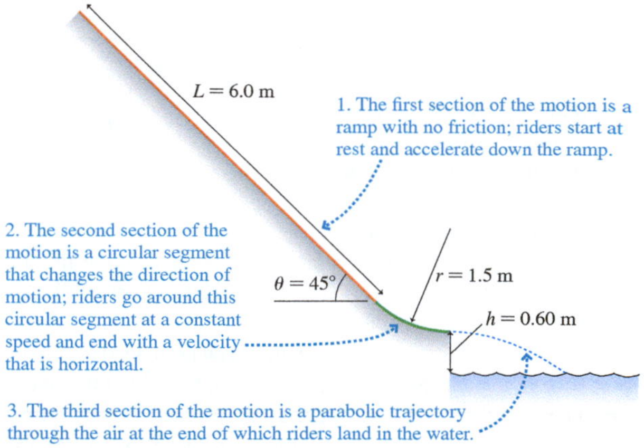

$L = 6.0$ m
1. The first section of the motion is a ramp with no friction; riders start at rest and accelerate down the ramp.
2. The second section of the motion is a circular segment that changes the direction of motion; riders go around this circular segment at a constant speed and end with a velocity that is horizontal.
$\theta = 45°$
$r = 1.5$ m
$h = 0.60$ m
3. The third section of the motion is a parabolic trajectory through the air at the end of which riders land in the water.


FIGURE P3.84

84. | At the end of the first section of the motion, riders are moving at what approximate speed?
   A. 3 m/s B. 6 m/s C. 9 m/s D. 12 m/s

85. | Suppose the acceleration during the second section of the motion is too large to be comfortable for riders. What change could be made to decrease the acceleration during this section?
   A. Reduce the radius of the circular segment.
   B. Increase the radius of the circular segment.
   C. Increase the angle of the ramp.
   D. Increase the length of the ramp.

86. | What is the vertical component of the velocity of a rider as he or she hits the water?
   A. 2.4 m/s B. 3.4 m/s C. 5.2 m/s D. 9.1 m/s

87. | Suppose the designers of the water slide want to adjust the height $h$ above the water so that riders land twice as far away from the bottom of the slide. What would be the necessary height above the water?
   A. 1.2 m B. 1.8 m C. 2.4 m D. 3.0 m

88. | During which section of the motion is the magnitude of the acceleration experienced by a rider the greatest?
   A. The first. B. The second.
   C. The third. D. It is the same in all sections.

# 4 Forces and Newton's Laws of Motion

We don't normally think of turtles as speedy, but the snake-necked turtle is an ambush predator, catching prey by surprise with a very rapid acceleration of its head. How is it able to achieve this feat?

## LOOKING AHEAD ➧

### Forces

A force is a push or a pull. It is an interaction between two objects, the **agent** (here, the woman) and the **object** (the car).

In this chapter, you'll learn how to identify different forces, and you'll learn their properties.

### Forces and Motion

Forces cause objects to *accelerate*. A forward acceleration of the sled requires a forward force.

A larger acceleration requires a larger force. You'll learn this connection between force and motion, part of *Newton's second law*.

### Reaction Forces

The hammer exerts a downward force on the nail. Surprisingly, the nail exerts an equal force on the hammer, directed upward.

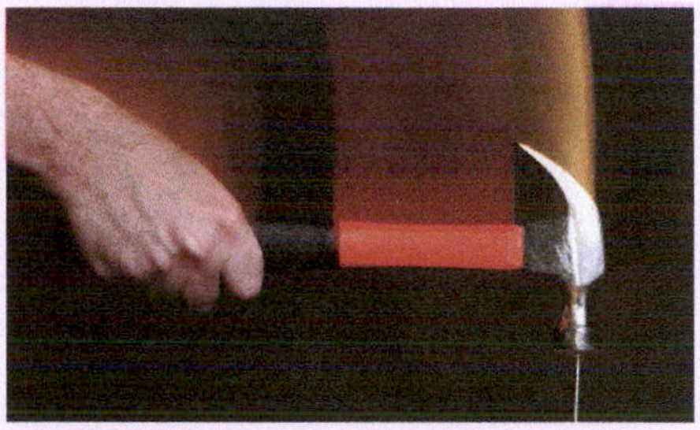

You'll learn how to identify and reason with **action/reaction pairs** of forces according to *Newton's third law*.

**GOAL** **To establish the connection between force and motion.**

## LOOKING BACK ➧

### Acceleration

You learned in Chapters 2 and 3 that acceleration is a vector that points in the direction of the change in velocity.

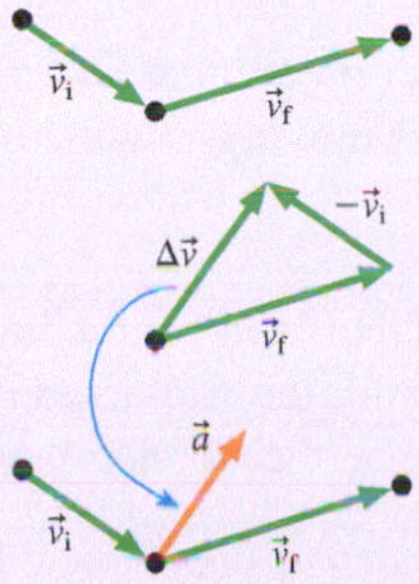


If the velocity is changing, there is an acceleration. And so, as you'll learn in this chapter, there must be a net force.

**STOP TO THINK**

A swan is landing on an icy lake, sliding across the ice and gradually coming to a stop. As the swan slides, the direction of the acceleration is

A. To the left.
B. To the right.
C. Upward.
D. Downward.

**Interstellar coasting** A nearly perfect example of Newton's first law is the pair of Voyager space probes launched in 1977. Both spacecraft long ago ran out of fuel and are now coasting through the frictionless vacuum of space. Although not entirely free of influence from the sun's gravity, they are now so far from the sun and other stars that gravitational influences are very nearly zero, and the probes will continue their motion for billions of years.

## 4.1 Motion and Forces

The snake-necked turtle in the photo at the beginning of the chapter can accelerate its head forward at 40 m/s$^2$ to capture prey. In Chapters 1 through 3, we've learned how to *describe* this and other types of motion with pictures, graphs, and equations, and you know enough about the scale of things to see that this acceleration—which is about $4g$—is quite impressive. But until now we've said nothing to *explain* the motion, to say how the turtle is able to achieve this feat. In this chapter, we will turn our attention to the *cause* of motion—**forces.** This topic is called **dynamics,** which joins with kinematics to form **mechanics,** the general science of motion. We'll begin our study of dynamics qualitatively in this chapter and then add quantitative detail over the next four chapters.

### What Causes Motion?

Let's start with a basic question: Do you need to keep pushing on something—to keep applying a force—to keep it going? Your daily experience might suggest that the answer is Yes. If you slide your textbook across the desk and then stop pushing, the book will quickly come to rest. Other objects will continue to move for a longer time: When a hockey puck is sliding across the ice, it keeps going for a long time, but it, too, comes to rest at some point. Now let's take a closer look to see whether this idea holds up to scrutiny.

FIGURE 4.1 shows a series of motion experiments. Tyler slides down a hill on his sled and then out onto a horizontal patch of smooth snow, as shown in Figure 4.1a. Even if the snow is smooth, the friction between the sled and the snow will soon cause the sled to come to rest. What if Tyler slides down the hill onto some very slick ice, as in Figure 4.1b? The friction is much less, so the sled could slide for quite a distance before stopping. Now, *imagine* the situation in Figure 4.1c, where the sled slides on idealized *frictionless* ice. In this case, the sled, once started in its motion, would continue in motion forever, moving in a straight line with no loss of speed.

FIGURE 4.1 A sled sliding on increasingly smooth surfaces.

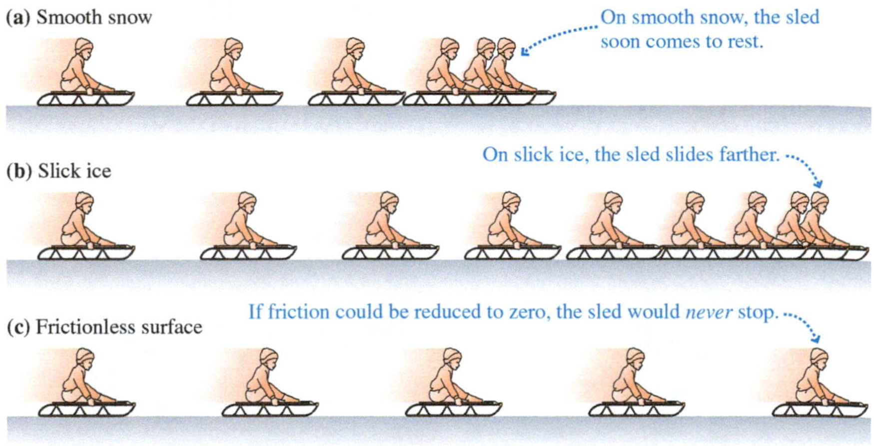

(a) Smooth snow
On smooth snow, the sled soon comes to rest.
(b) Slick ice
On slick ice, the sled slides farther.
(c) Frictionless surface
If friction could be reduced to zero, the sled would *never* stop.

In the absence of friction, **if the sled is moving, it will stay in motion.** It's also true that if the sled were sitting still, it wouldn't start moving on its own; if the sled is at rest, it will stay at rest. Careful experiments done over the past few centuries, notably by Galileo and then by Isaac Newton, verify that this is in fact the way the world works. What we have concluded for the sled is actually a general rule that applies to other similar situations. We call the generalization **Newton's first law** of motion:

> **Newton's first law** Consider an object that has no forces acting on it. If it is at rest, it will remain at rest. If it is moving, it will continue to move in a straight line at a constant speed.

As an important application of Newton's first law, consider the crash test of FIGURE 4.2. As the car contacts the wall, the wall exerts a force on the car and the car begins to slow. But the wall is a force on the *car,* not on the dummy. In accordance with

FIGURE 4.2 Newton's first law tells us: Wear your seatbelts!

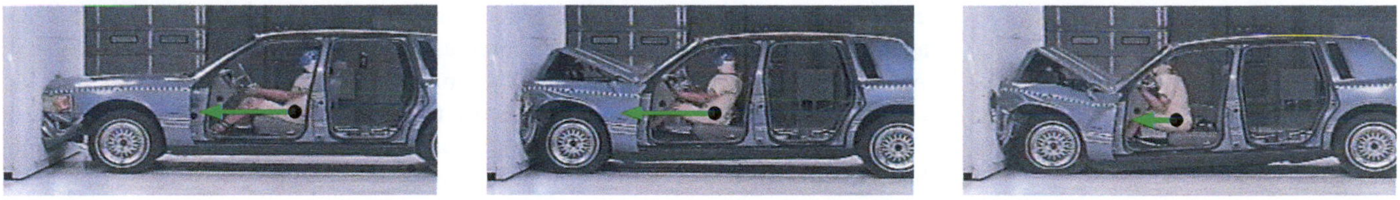
At the instant of impact, the car and the dummy are moving at the same speed.
The car slows as it hits, but the dummy continues at the same speed . . .
. . . until it hits the now-stationary dashboard. Ouch!

Newton's first law, the unbelted dummy continues to move straight ahead at its original speed. Sooner or later, a force will act to bring the dummy to rest. The only questions are when and how large the force will be. In the case shown, the dummy comes to rest in a short, violent collision with the dashboard of the stopped car. Seatbelts and air bags slow a dummy—or a driver—at a much lower rate and provide a much gentler stop.

## Forces

We started with the question What causes motion? Our answer is Newton's first law, which tells us that an object in motion needs *no* cause—or *force*—to keep moving in a straight line forever. But this law does not explain in any detail exactly what a force *is*. The concept of force is best introduced by looking at examples of some common forces and considering the basic properties shared by all forces. Let's begin by examining the properties that all forces have in common, as presented in the table below.

### What is a force?

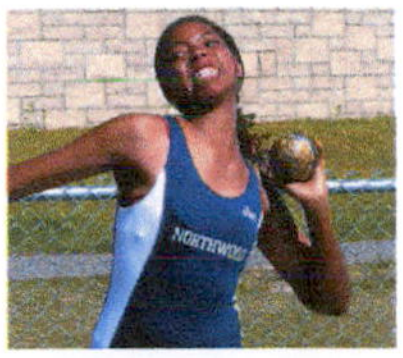

**A force is a push or a pull.**

Our commonsense idea of a **force** is that it is a *push* or a *pull*. We will refine this idea as we go along, but it is an adequate starting point. Notice our careful choice of words: We refer to "*a* force" rather than simply "force." We want to think of a force as a very specific *action*, so that we can talk about a single force or perhaps about two or three individual forces that we can clearly distinguish—hence the concrete idea of "a force" acting on an object.

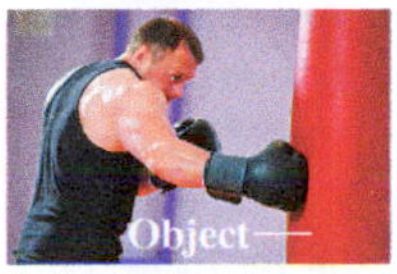


**A force acts on an object.**

Implicit in our concept of force is that **a force acts on an object.** In other words, pushes and pulls are applied *to* something—an object. From the object's perspective, it has a force *exerted* on it. Forces do not exist in isolation from the object that experiences them.

**A force requires an agent.**

Every force has an **agent,** something that acts or pushes or pulls; that is, a force has a specific, identifiable *cause*. As you throw a ball, it is your hand, while in contact with the ball, that is the agent or the cause of the force exerted on the ball. *If* a force is being exerted on an object, you must be able to identify a specific cause (i.e., the agent) of that force. Conversely, a force is not exerted on an object *unless* you can identify a specific cause or agent. Note that an agent can be an inert, inanimate object such as a tabletop or a wall. Such agents are the cause of many common forces.

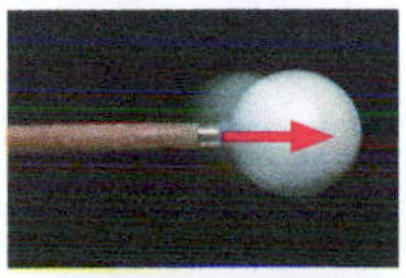

**A force is a vector.**

If you push an object, you can push either gently or very hard. Similarly, you can push either left or right, up or down. To quantify a push, we need to specify both a magnitude *and* a direction. It should thus come as no surprise that a force is a vector quantity. The general symbol for a force is the vector symbol $\vec{F}$. The size or strength of such a force is its magnitude $F$.

**A force can be either a contact force . . .**

There are two basic classes of forces, depending on whether the agent touches the object or not. **Contact forces** are forces that act on an object by touching it at a point of contact. The bat must touch the ball to hit it. A string must be tied to an object to pull it. The majority of forces that we will examine are contact forces.

**. . . or a long-range force.**

**Long-range forces** are forces that act on an object without physical contact. Magnetism is an example of a long-range force. You have undoubtedly held a magnet over a paper clip and seen the paper clip leap up to the magnet. A coffee cup released from your hand is pulled to the earth by the long-range force of gravity.

There's one more important aspect of forces. If you push against a door (the object) to close it, the door pushes back against your hand (the agent). If a tow rope pulls on a car (the object), the car pulls back on the rope (the agent). In general, if an agent exerts a force on an object, the object exerts a force on the agent. We really need to think of a force as an *interaction* between two objects. Although the interaction perspective is a more exact way to view forces, it adds complications that we would like to avoid for now. Our approach will be to start by focusing on how a single object responds to forces exerted on it. Later, in Section 4.7, we'll return to the larger issue of how two or more objects interact with each other.

## Force Vectors

We can use a simple diagram to visualize how forces are exerted on objects. Because we are using the particle model, in which objects are treated as points, the process of drawing a force vector is straightforward:

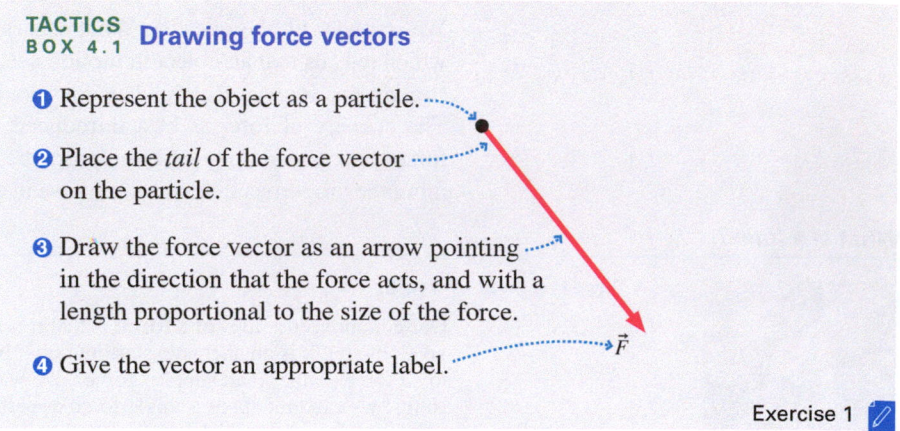

**TACTICS BOX 4.1 Drawing force vectors**

1. Represent the object as a particle.
2. Place the *tail* of the force vector on the particle.
3. Draw the force vector as an arrow pointing in the direction that the force acts, and with a length proportional to the size of the force.
4. Give the vector an appropriate label.

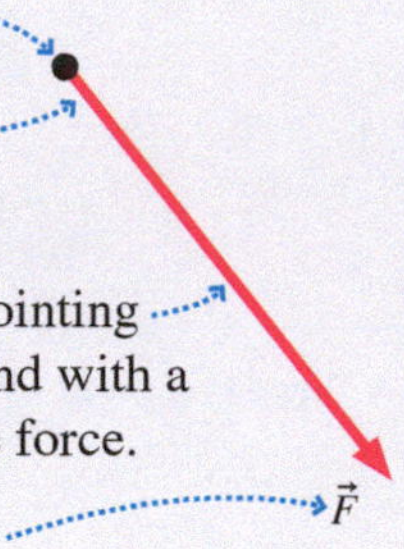


Exercise 1

Step 2 may seem contrary to what a "push" should do (it may look as if the force arrow is *pulling* the object rather than *pushing* it), but recall that moving a vector does not change it as long as the length and angle do not change. The vector $\vec{F}$ is the same regardless of whether the tail or the tip is placed on the particle. Our reason for using the tail will become clear when we consider how to combine several forces.

FIGURE 4.3 shows three examples of force vectors. One is a pull, one a push, and one a long-range force, but in all three the *tail* of the force vector is placed on the particle that represents the object. Although the generic symbol for a force is $\vec{F}$, as the figure shows we often use special symbols for certain forces that arise frequently, such as $\vec{T}$ for the tension force in a rope, $\vec{F}_{sp}$ for the force of a spring, and $\vec{w}$ for the force of gravity (an object's *weight*).

FIGURE 4.3 Three force vectors.

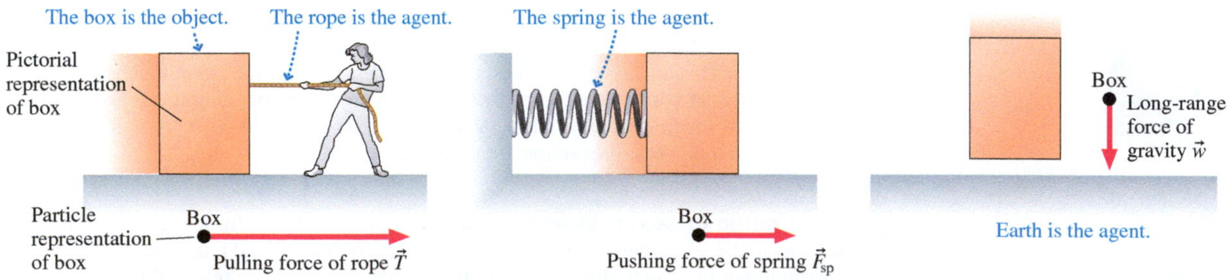

## Combining Forces

Force is a vector quantity. We saw in ◀◀ SECTION 3.1 how to find the *vector sum* of two vectors. FIGURE 4.4a shows a top view of a box being pulled by two ropes, each exerting a force on the box. How will the box respond? Experiments show that when several forces $\vec{F}_1, \vec{F}_2, \vec{F}_3, \ldots$ are exerted on an object, they combine to form a **net force** that is the vector sum of all the forces:

$$\vec{F}_{\text{net}} = \vec{F}_1 + \vec{F}_2 + \vec{F}_3 + \cdots \tag{4.1}$$

That is, the single force $\vec{F}_{\text{net}}$ causes the exact same motion of the object as this combination of original forces $\vec{F}_1, \vec{F}_2, \vec{F}_3, \ldots$. FIGURE 4.4b shows the net force on the box.

NOTE ▶ It is important to realize that the net force $\vec{F}_{\text{net}}$ is not a new force acting *in addition* to the original forces $\vec{F}_1, \vec{F}_2, \vec{F}_3, \ldots$. Instead, we should think of the original forces being *replaced* by $\vec{F}_{\text{net}}$. ◀

FIGURE 4.4 Two forces applied to a box.

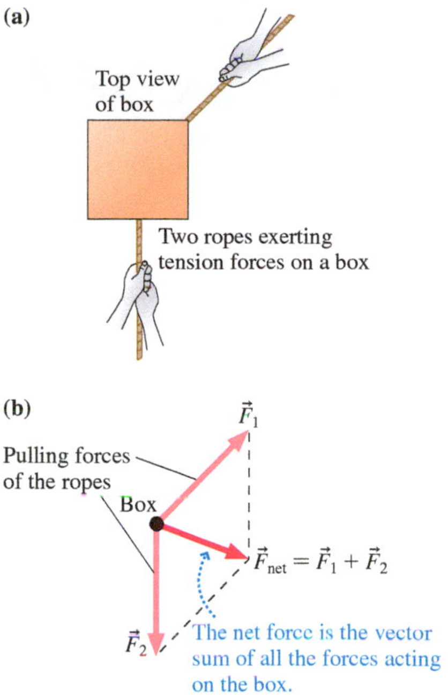


**STOP TO THINK 4.1** Two of the three forces exerted on an object are shown. The net force points directly to the left. Which is the missing third force?

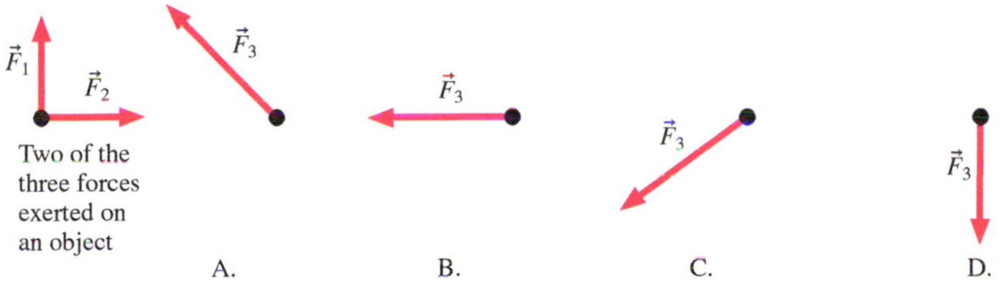


## 4.2 A Short Catalog of Forces

There are many forces we will deal with over and over. This section will introduce you to some of them and to the symbols we use to represent them.

### Weight

A falling rock is pulled toward the earth by the long-range force of gravity. Gravity is what keeps you in your chair, keeps the planets in their orbits around the sun, and shapes the large-scale structure of the universe. We'll have a thorough look at gravity in Chapter 6. For now we'll concentrate on objects on or near the surface of the earth (or other planet).

The gravitational pull of the earth on an object on or near the surface of the earth is called **weight.** The symbol for weight is $\vec{w}$. Weight is the only long-range force we will encounter in the next few chapters. The agent for the weight force is the *entire earth* pulling on an object. The weight force is in some ways the simplest force we'll study. As FIGURE 4.5 shows, **an object's weight vector always points vertically downward,** no matter how the object is moving.

NOTE ▶ We often refer to "the weight" of an object. This is an informal expression for $w$, the magnitude of the weight force exerted on the object. Note that **weight is not the same thing as mass.** We will briefly examine mass later in the chapter, and we'll explore the connection between weight and mass in Chapter 5. ◀

FIGURE 4.5 Weight always points vertically downward.

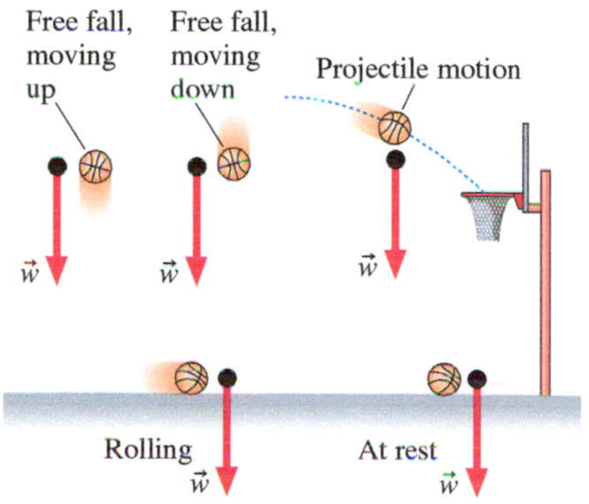

## Spring Force

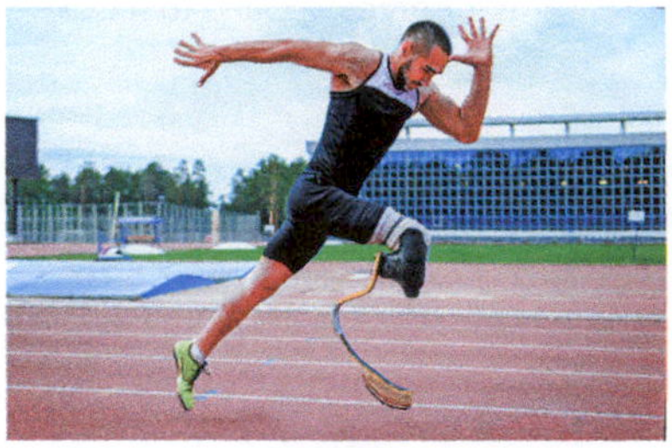

When the flexible blade of this athlete's prosthesis hits the ground, it compresses just like an ordinary spring.

Springs exert one of the most basic contact forces. A spring can either push (when compressed) or pull (when stretched). FIGURE 4.6 shows the **spring force.** In both cases, pushing and pulling, the tail of the force vector is placed on the particle in the force diagram. There is no special symbol for a spring force, so we simply use a subscript label: $\vec{F}_{sp}$.

FIGURE 4.6 The spring force is parallel to the spring.

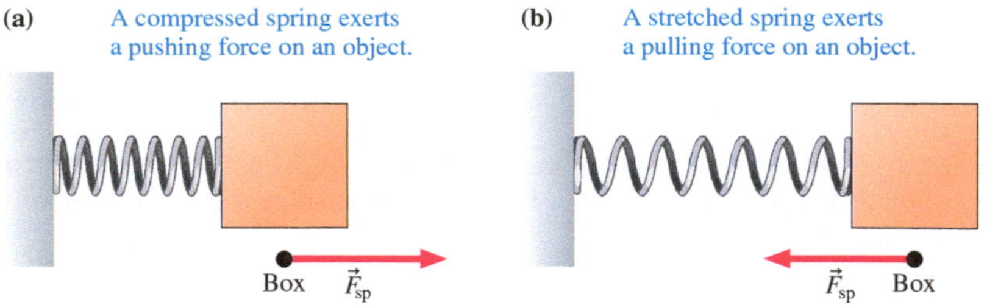


Although you may think of a spring as a metal coil that can be stretched or compressed, this is only one type of spring. Hold a ruler, or any other thin piece of wood or metal, by the ends and bend it slightly. It flexes. When you let go, it "springs" back to its original shape. This is just as much a spring as is a metal coil.

## Tension Force

FIGURE 4.7 Tension is parallel to the rope.

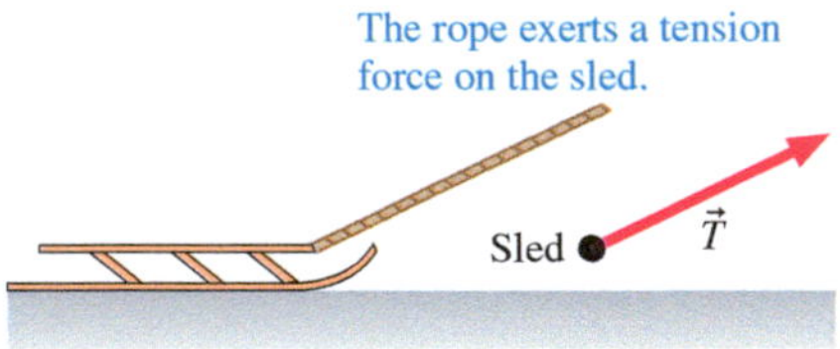


When a string or rope or wire pulls on an object, it exerts a contact force that we call the **tension force,** represented by $\vec{T}$. **The direction of the tension force is always in the direction of the string or rope,** as you can see in FIGURE 4.7. When we speak of "the tension" in a string, this is an informal expression for $T$, the size or magnitude of the tension force. Note that the tension force can only *pull* in the direction of the string; if you try to *push* with a string, it will go slack and be unable to exert a force.

We can think about the tension force using a microscopic picture. If you were to use a very powerful microscope to look inside a rope, you would "see" that it is made of *atoms* joined together by *molecular bonds.* Molecular bonds are not rigid connections between the atoms. They are more accurately thought of as tiny *springs* holding the atoms together, as in FIGURE 4.8. Pulling on the ends of a string or rope stretches the spring-like molecular bonds ever so slightly. The tension within a rope and the tension force experienced by an object at the end of the rope are really the net spring force exerted by billions and billions of microscopic springs.

FIGURE 4.8 An atomic model of tension.

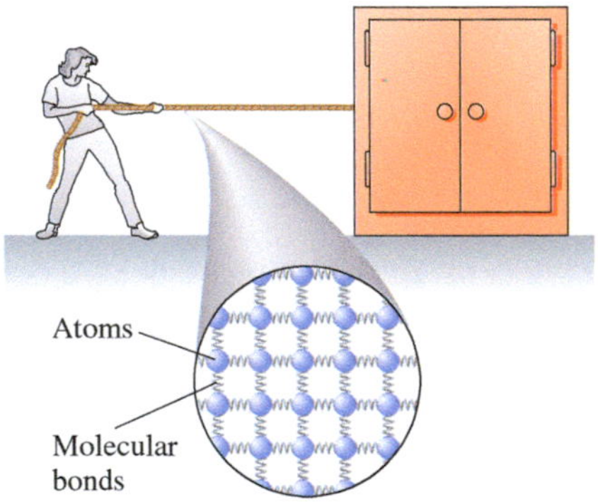


This atomic-level view of tension introduces a new idea: a microscopic **atomic model** for understanding the behavior and properties of **macroscopic** (i.e., containing many atoms) objects. We will frequently use atomic models to obtain a deeper understanding of our observations.

The atomic model of tension also helps to explain one of the basic properties of ropes and strings. When you pull on a rope tied to a heavy box, the rope in turn exerts a tension force on the box. If you pull harder, the tension force on the box becomes greater. How does the box "know" that you are pulling harder on the other end of the rope? According to our atomic model, when you pull harder on the rope, its microscopic springs stretch a bit more, increasing the spring force they exert on each other—and on the box they're attached to.

## Normal Force

If you sit on a bed, the springs in the mattress compress and, as a consequence of the compression, exert an upward force on you. Stiffer springs would show less compression but would still exert an upward force. The compression of extremely stiff springs might be measurable only by sensitive instruments. Nonetheless, the springs would compress ever so slightly and exert an upward spring force on you.

FIGURE 4.9 shows a book resting on top of a sturdy table. The table may not visibly flex or sag, but—just as you do to the bed—the book compresses the molecular "springs" in the table. The compression is very small, but it is not zero. As a consequence, the compressed molecular springs *push upward* on the book. We say that "the table" exerts the upward force, but it is important to understand that the pushing is *really* done by molecular springs. Similarly, an object resting on the ground compresses the molecular springs holding the ground together and, as a consequence, the ground pushes up on the object.

We can extend this idea. Suppose you place your hand on a wall and lean against it, as shown in FIGURE 4.10. Does the wall exert a force on your hand? As you lean, you compress the molecular springs in the wall and, as a consequence, they push outward *against* your hand. So the answer is Yes, the wall does exert a force on you. It's not hard to see this if you examine your hand as you lean: You can see that your hand is slightly deformed, and becomes more so the harder you lean. This deformation is direct evidence of the force that the wall exerts on your hand. Consider also what would happen if the wall suddenly vanished. Without the wall there to push against you, you would topple forward.

The force the table surface exerts is vertical, while the force the wall exerts is horizontal. In all cases, the force exerted on an object that is pressing against a surface is in a direction *perpendicular* to the surface. Mathematicians refer to a line that is perpendicular to a surface as being *normal* to the surface. In keeping with this terminology, we define the **normal force** as the force exerted by a surface (the agent) against an object that is pressing against the surface. The symbol for the normal force is $\vec{n}$.

We're not using the word "normal" to imply that the force is an "ordinary" force or to distinguish it from an "abnormal force." A surface exerts a force *perpendicular* (i.e., normal) to itself as the molecular springs press *outward*. FIGURE 4.11 shows an object on an inclined surface, a common situation. Notice how the normal force $\vec{n}$ is perpendicular to the surface.

The normal force is a very real force arising from the very real compression of molecular bonds. It is in essence just a spring force, but one exerted by a vast number of microscopic springs acting at once. The normal force is responsible for the "solidness" of solids. It is what prevents you from passing right through the chair you are sitting in and what causes the pain and the lump if you bang your head into a door. Your head can then tell you that the force exerted on it by the door was very real!

## Friction

You've certainly observed that a rolling or sliding object, if not pushed or propelled, slows down and eventually stops. You've probably discovered that you can slide better across a sheet of ice than across asphalt. And you also know that most objects stay in place on a table without sliding off even if the table is tilted a bit. The force responsible for these sorts of behavior is **friction.** The symbol for friction is $\vec{f}$.

Friction, like the normal force, is exerted by a surface. Unlike the normal force, however, **the frictional force is always *parallel* to the surface,** not perpendicular to it. (In many cases, a surface will exert *both* a normal and a frictional force.) On a microscopic level, friction arises as atoms from the object and atoms on the surface run into each other. The rougher the surface is, the more these atoms are forced into close proximity and, as a result, the larger the friction force. We will develop a simple model of friction in the next chapter that will be sufficient for our needs. For now, it is useful to distinguish between two kinds of friction:

- *Kinetic friction,* denoted $\vec{f}_k$, acts as an object *slides* across a surface. Kinetic friction is a force that always "opposes the motion," meaning that the friction force $\vec{f}_k$ on a sliding object points in the direction opposite to the direction of the object's motion.

FIGURE 4.9 An atomic model of the force exerted by a table.

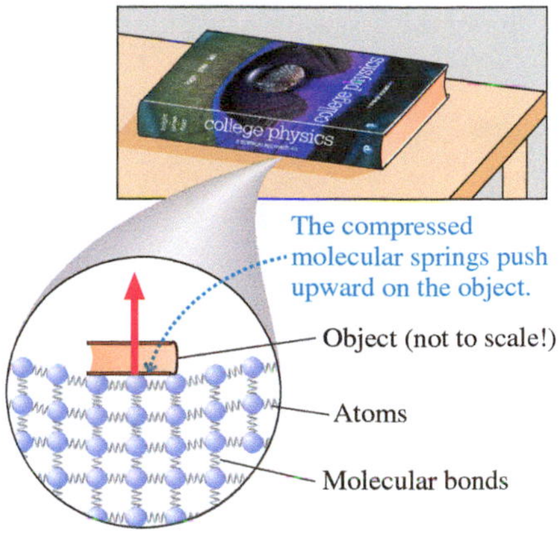


A video to support a section's topic is embedded in the eText.

**Video** Figure 4.9

**Video** What the Physics? Pushing Down to Move Upward

FIGURE 4.10 The wall pushes outward against your hand.

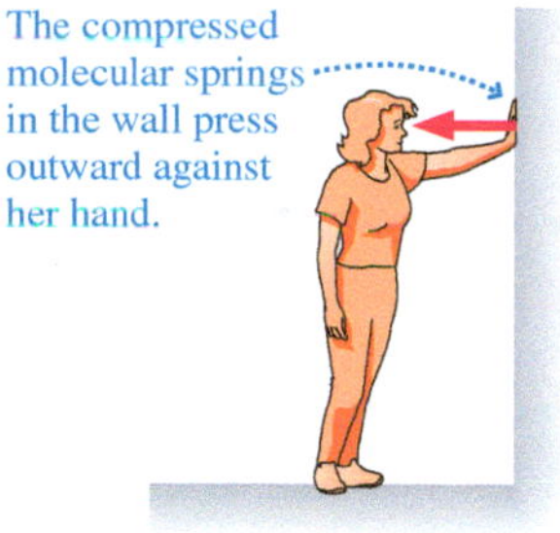


FIGURE 4.11 The normal force is perpendicular to the surface.

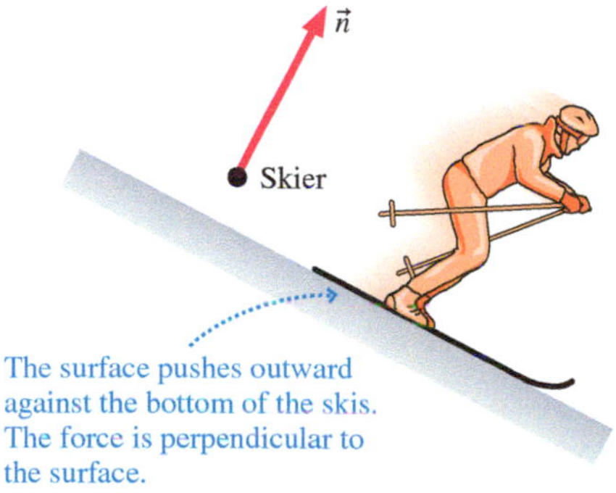

- *Static friction,* denoted $\vec{f}_s$, is the force that keeps an object "stuck" on a surface and prevents its motion relative to the surface. Finding the direction of $\vec{f}_s$ is a little trickier than finding the direction of $\vec{f}_k$. Static friction points opposite the direction in which the object *would* move if there were no friction; that is, it points in the direction necessary to *prevent* motion.

Examples of kinetic and static friction are shown in FIGURE 4.12.

KEY CONCEPT **FIGURE 4.12** Kinetic and static friction are parallel to the surface.

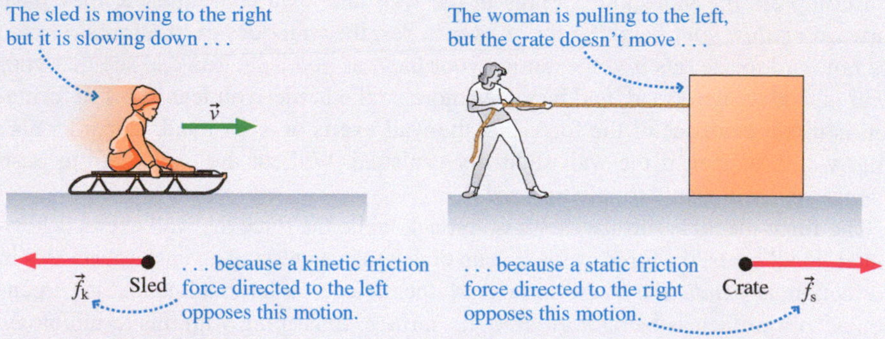
The sled is moving to the right but it is slowing down . . .
$\vec{v}$
. . . because a kinetic friction force directed to the left opposes this motion.
$\vec{f}_k$
Sled
The woman is pulling to the left, but the crate doesn't move . . .
. . . because a static friction force directed to the right opposes this motion.
Crate
$\vec{f}_s$

**STOP TO THINK 4.2** A frog is resting on a slope. Given what you learned in Figure 4.12, what can you say about the friction force acting on the frog?

A. There is no friction force.
B. There is a kinetic friction force directed up the slope.
C. There is a static friction force directed up the slope.
D. There is a kinetic friction force directed down the slope.
E. There is a static friction force directed down the slope.

**FIGURE 4.13** Air resistance is an example of drag.

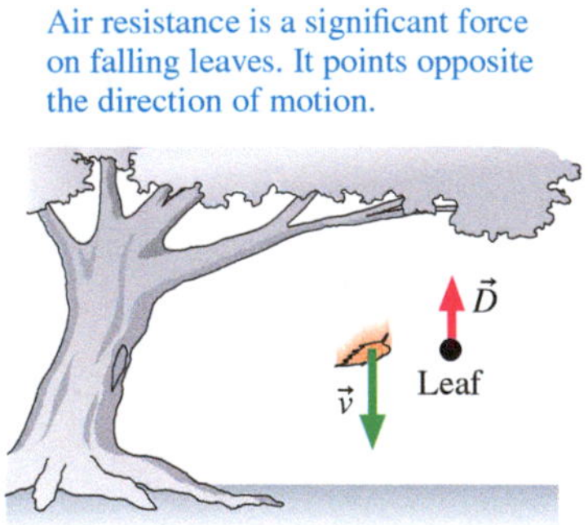
Air resistance is a significant force on falling leaves. It points opposite the direction of motion.
$\vec{D}$
$\vec{v}$
Leaf

**FIGURE 4.14** The thrust force on a rocket is opposite the direction of the expelled gases.

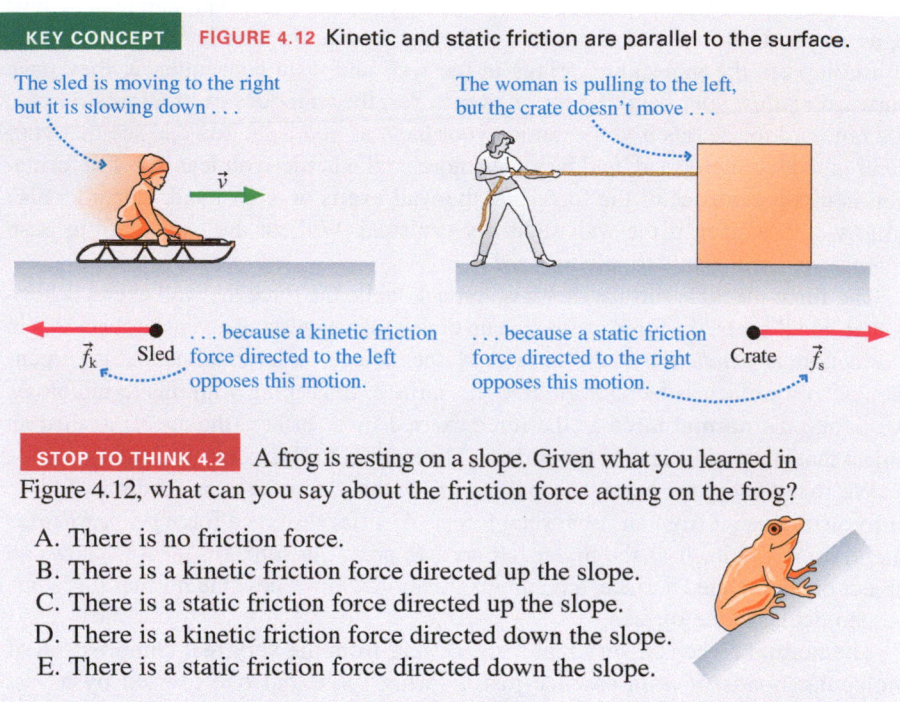

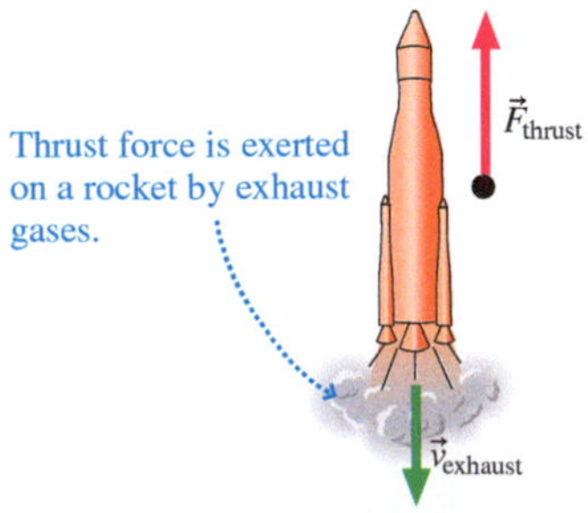
Thrust force is exerted on a rocket by exhaust gases.
$\vec{F}_{thrust}$
$\vec{v}_{exhaust}$

## Drag

Friction at a surface is one example of a *resistive force,* a force that opposes or resists motion. Resistive forces are also experienced by objects moving through *fluids*—gases (like air) and liquids (like water). This kind of resistive force—the force of a fluid on a moving object—is called **drag** and is symbolized as $\vec{D}$. Like kinetic friction, **drag points opposite the direction of motion.** FIGURE 4.13 shows an example of drag.

Drag can be a large force for objects moving at high speeds or in dense fluids. Hold your arm out the window as you ride in a car and feel how hard the air pushes against your arm. Note also how the air resistance against your arm increases rapidly as the car's speed increases. For a small particle moving in water, such as a swimming *Paramecium,* drag can be the dominant force.

On the other hand, for objects that are heavy and compact, moving in air, and with a speed that is not too great, the drag force of air resistance is fairly small. To keep things as simple as possible, **you can neglect air resistance in all problems unless a problem explicitly asks you to include it.** The error introduced into calculations by this approximation is generally pretty small.

## Thrust

A jet airplane obviously has a force that propels it forward; likewise for the rocket in FIGURE 4.14. This force, called **thrust,** occurs when a jet or rocket engine expels gas molecules at high speed. Thrust is a contact force, with the exhaust gas being the agent that pushes on the engine. The process by which thrust is generated is rather subtle and requires an appreciation of Newton's third law, introduced later in this chapter. For now, we need only consider that **thrust is a force opposite the direction in which the exhaust gas is expelled.** There's no special symbol for thrust, so we call it $\vec{F}_{thrust}$.

**It's not just rocket science** BIO Rockets are propelled by thrust, but many animals are as well. Scallops are shellfish with no feet and no fins, but they can escape from predators or move to new territory by using a form of jet propulsion. A scallop forcibly ejects water from the rear of its shell, resulting in a thrust force that moves it forward.

### Electric and Magnetic Forces

Electricity and magnetism, like gravity, exert long-range forces. The forces of electricity and magnetism act on charged particles. We will study electric and magnetic forces in detail in Part VI of this text. These forces—and the forces inside the nucleus, which we will also see later in the text—won't be important for the dynamics problems we consider in the next several chapters.

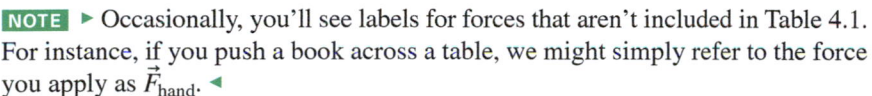

**STOP TO THINK 4.3** A boy is using a rope to pull a sled to the right. What are the directions of the tension force and the friction force on the sled, respectively?

A. Right, right  B. Right, left
C. Left, right  D. Left, left

## 4.3 Identifying Forces

A typical physics problem describes an object that is being pushed and pulled in various directions. Some forces are given explicitly, while others are only implied. In order to proceed, it is necessary to determine all the forces that act on the object. It is also necessary to avoid including forces that do not really exist. Now that you have learned the properties of forces and seen a catalog of typical forces, we can develop a step-by-step method for identifying each force in a problem. A list of the most common forces we'll come across in the next few chapters is given in TABLE 4.1.

**Video** Identifying Forces

**NOTE** ▶ Occasionally, you'll see labels for forces that aren't included in Table 4.1. For instance, if you push a book across a table, we might simply refer to the force you apply as $\vec{F}_{hand}$. ◀

Tactics Box 4.2 will help you correctly identify the forces acting on an object. It's followed by two examples that explicitly illustrate the steps of the Tactics Box.

**TACTICS BOX 4.2 Identifying forces**

❶ **Identify the object of interest.** This is the object whose motion you wish to study.

❷ **Draw a picture of the situation.** Show the object of interest and all other objects—such as ropes, springs, and surfaces—that touch it.

❸ **Draw a closed curve around the object.** Only the object of interest is inside the curve; everything else is outside.

❹ **Locate every point on the boundary of this curve where other objects touch the object of interest.** These are the points where *contact forces* are exerted on the object.

❺ **Name and label each contact force acting on the object.** There is at least one force at each point of contact; there may be more than one. When necessary, use subscripts to distinguish forces of the same type.

❻ **Name and label each long-range force acting on the object.** For now, the only long-range force we'll consider is weight.

Exercises 4–8

**TABLE 4.1** Common forces and their notation

| Force | Notation |
|---|---|
| General force | $\vec{F}$ |
| Weight | $\vec{w}$ |
| Spring force | $\vec{F}_{sp}$ |
| Tension | $\vec{T}$ |
| Normal force | $\vec{n}$ |
| Static friction | $\vec{f}_s$ |
| Kinetic friction | $\vec{f}_k$ |
| Drag | $\vec{D}$ |
| Thrust | $\vec{F}_{thrust}$ |

## CONCEPTUAL EXAMPLE 4.1 Identifying forces on a bungee jumper

A bungee jumper has leapt off a bridge and is nearing the bottom of her fall. What forces are being exerted on the bungee jumper?

**REASON** FIGURE 4.15 Forces on a bungee jumper.

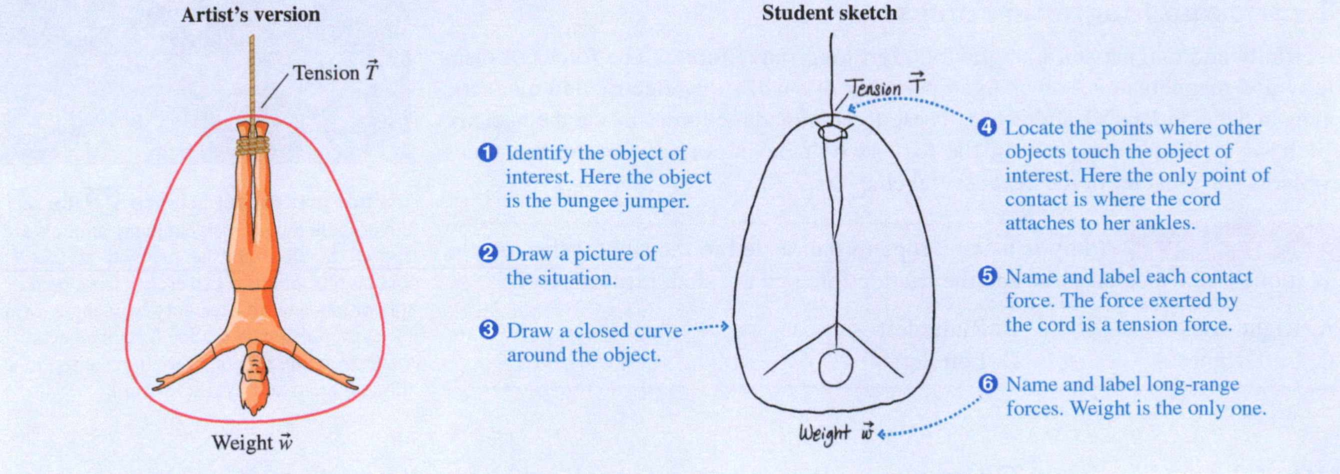

Artist's version
Tension $\vec{T}$
Weight $\vec{w}$
Student sketch
Tension $\vec{T}$
Weight $\vec{w}$
❶ Identify the object of interest. Here the object is the bungee jumper.
❷ Draw a picture of the situation.
❸ Draw a closed curve around the object.
❹ Locate the points where other objects touch the object of interest. Here the only point of contact is where the cord attaches to her ankles.
❺ Name and label each contact force. The force exerted by the cord is a tension force.
❻ Name and label long-range forces. Weight is the only one.


## CONCEPTUAL EXAMPLE 4.2 Identifying forces on a skier

A skier is being towed up a snow-covered hill by a tow rope. What forces are being exerted on the skier?

**REASON** FIGURE 4.16 Forces on a skier.

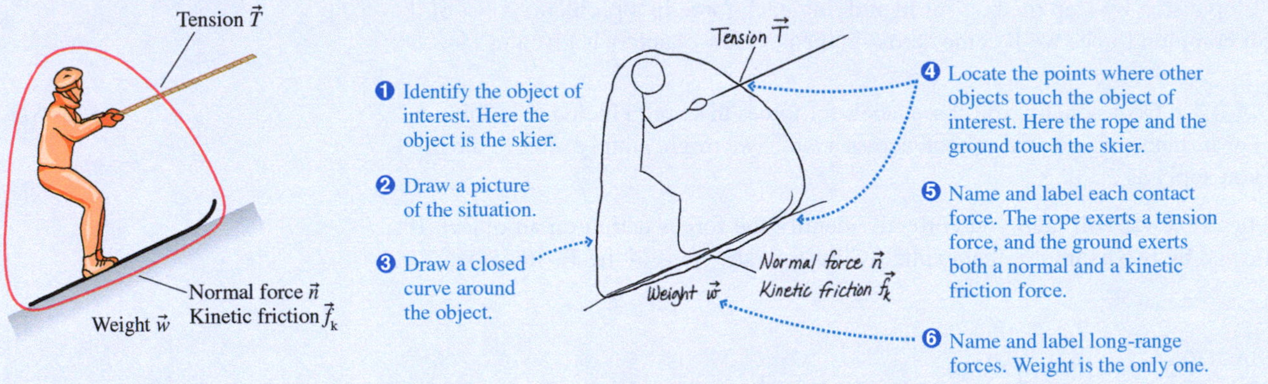

Tension $\vec{T}$
Normal force $\vec{n}$
Kinetic friction $\vec{f}_k$
Weight $\vec{w}$
❶ Identify the object of interest. Here the object is the skier.
❷ Draw a picture of the situation.
❸ Draw a closed curve around the object.
Tension $\vec{T}$
Normal force $\vec{n}$
Kinetic friction $\vec{f}_k$
Weight $\vec{w}$
❹ Locate the points where other objects touch the object of interest. Here the rope and the ground touch the skier.
❺ Name and label each contact force. The rope exerts a tension force, and the ground exerts both a normal and a kinetic friction force.
❻ Name and label long-range forces. Weight is the only one.


**NOTE** ▸ You might have expected two friction forces and two normal forces in Example 4.2, one on each ski. Keep in mind, however, that we're working within the particle model, which represents the skier by a single point. A particle has only one contact with the ground, so there is a single normal force and a single friction force. The particle model is valid if we want to analyze the motion of the skier as a whole, but we would have to go beyond the particle model to find out what happens to each ski. ◂

## CONCEPTUAL EXAMPLE 4.3 Identifying forces on a rocket

A rocket is flying upward through the air, high above the ground. Air resistance is not negligible. What forces are being exerted on the rocket?

**REASON**

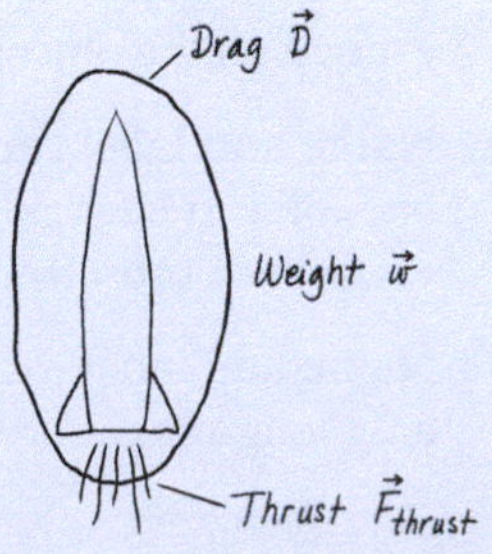

Drag $\vec{D}$
Weight $\vec{w}$
Thrust $\vec{F}_{thrust}$


FIGURE 4.17 Forces on a rocket.

**STOP TO THINK 4.4** You've just kicked a rock, and it is now sliding across the ground about 2 meters in front of you. Which of these are forces acting on the rock? Include all that apply.

A. Gravity, acting downward
B. The normal force, acting upward
C. The force of the kick, acting in the direction of motion
D. Friction, acting opposite the direction of motion
E. Air resistance, acting opposite the direction of motion

## 4.4 What Do Forces Do?

The fundamental question is: How does an object move when a force is exerted on it? The only way to answer this question is to do experiments. To do experiments, however, we need a way to reproduce the same force again and again, and we need a standard object so that our experiments are repeatable.

FIGURE 4.18 shows how you can use your fingers to stretch a rubber band to a certain length—say, 10 centimeters—that you can measure with a ruler. We'll call this the *standard length.* You know that a stretched rubber band exerts a force because your fingers *feel* the pull. Furthermore, this is a reproducible force. If you use the same rubber band stretched to the standard length, it will exert the same force. We'll call the magnitude of this force the *standard force F*. Not surprisingly, two identical rubber bands, each stretched to the standard length, exert twice the force of one rubber band; three rubber bands exert three times the force; and so on.

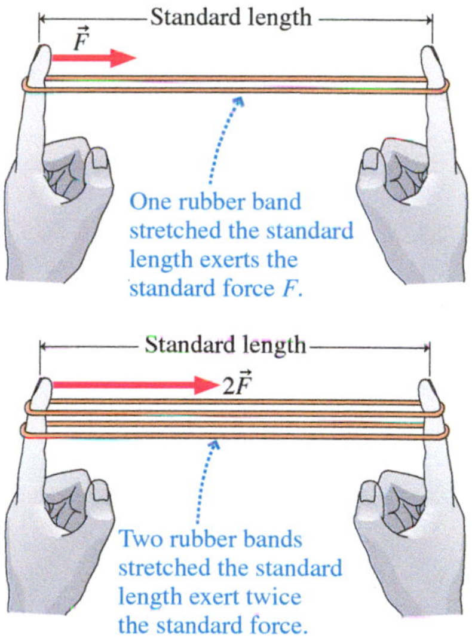

FIGURE 4.18 A reproducible force.

We'll also need several identical standard objects to which the force will be applied. As we learned in Chapter 1, the SI unit of mass is the kilogram (kg). For our standard objects, we will make ourselves several identical blocks, each with a mass of 1 kg.

Now we're ready to start a virtual experiment. First, place one of the 1 kg blocks on a frictionless surface. (In a real experiment, we can nearly eliminate friction by floating the block on a cushion of air.) Second, attach a rubber band to the block and stretch the band to the standard length. Then the block experiences the same force $F$ as your finger did. As the block starts to move, in order to keep the pulling force constant you must *move your hand* in just the right way to keep the length of the rubber band—and thus the force—*constant.* FIGURE 4.19 shows the experiment being carried out. Once the motion is complete, you can use motion diagrams and kinematics to analyze the block's motion.

FIGURE 4.19 Measuring the motion of a 1 kg block that is pulled with a constant force.

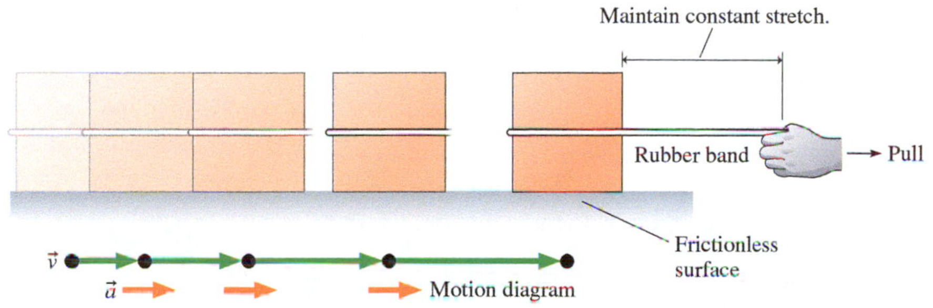


The motion diagram in Figure 4.19 shows that the velocity vectors are getting longer, so the velocity is increasing: The block is *accelerating.* Furthermore, a close inspection of the motion diagram shows that the acceleration vectors are all the same length. This is the first important finding of this experiment: **An object pulled with a constant force moves with a constant acceleration.** This finding could not have been anticipated in advance. It's conceivable that the object would speed up for a

FIGURE 4.20 Graph of acceleration versus force.

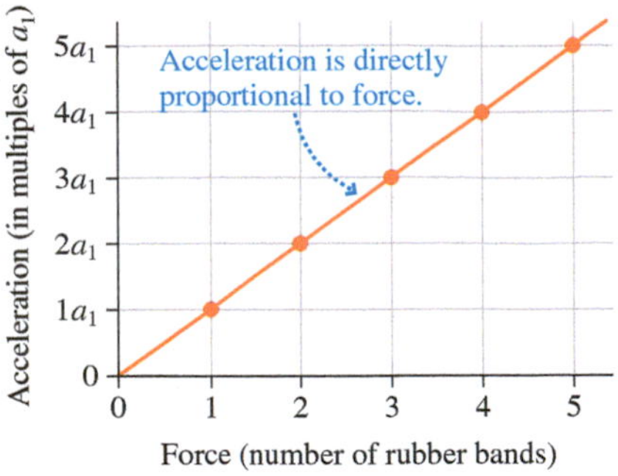


FIGURE 4.21 Graph of acceleration versus number of blocks.

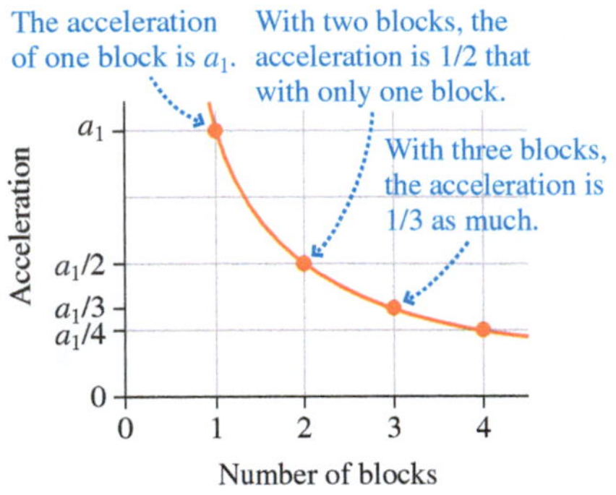


while and then move with a steady speed. Or that it would continue to speed up, but the *rate* of increase, the acceleration, would steadily decline. But these descriptions do not match what happens. Instead, the object continues *with a constant acceleration* for as long as you pull it with a constant force. We'll call this constant acceleration of *one* block pulled by *one* band $a_1$.

What happens if you increase the force by using several rubber bands? To find out, use two rubber bands. Stretch both to the standard length to double the force to $2F$, then measure the acceleration. Measure the acceleration due to three rubber bands, then four, and so on. FIGURE 4.20 is a graph of the results. Force is the independent variable, the one you can control, so we've placed force on the horizontal axis to make an acceleration-versus-force graph. The graph reveals our second important finding: **Acceleration is directly proportional to force.**

The final question for our virtual experiment is: How does the acceleration of an object depend on the mass of the object? To find out, glue two of the 1 kg blocks together, so that we have a block with twice as much matter as a 1 kg block—that is, a 2 kg block. Now apply the same force—a single rubber band—as you applied to the single 1 kg block. FIGURE 4.21 shows that the acceleration is *one-half* as great as that of the single block. If we glue three blocks together, making a 3 kg object, we find that the acceleration is only *one-third* of the 1 kg block's acceleration. In general, we find that the acceleration is proportional to the *inverse* of the mass of the object. So our third important result is: **Acceleration is *inversely proportional* to an object's mass.**

## Inversely proportional relationships

Two quantities are said to be **inversely proportional** to each other if one quantity is proportional to the *inverse* of the other. Mathematically, this means that

$$y = \frac{A}{x}$$

*y* is inversely proportional to *x*

Here, $A$ is a proportionality constant. This relationship is sometimes written as $y \propto 1/x$.

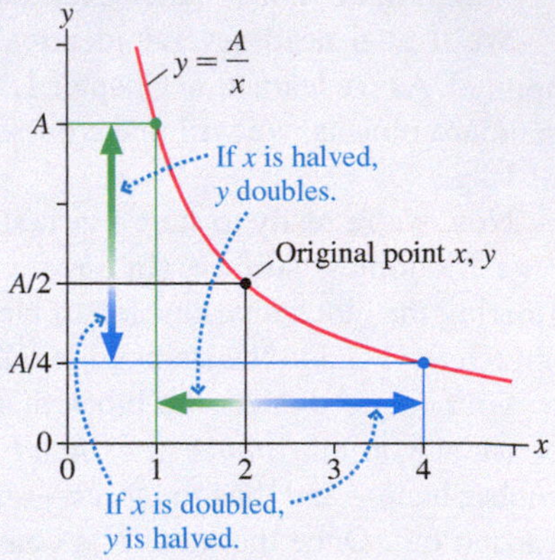


**SCALING**
- If you double $x$, you halve $y$.
- If you triple $x$, $y$ is reduced by a factor of 3.
- If you halve $x$, $y$ doubles.
- If you reduce $x$ by a factor of 3, $y$ becomes 3 times as large.

**RATIOS** For any two values of $x$—say, $x_1$ and $x_2$—we have

$$y_1 = \frac{A}{x_1} \quad \text{and} \quad y_2 = \frac{A}{x_2}$$

Dividing the $y_1$ equation by the $y_2$ equation, we find

$$\frac{y_1}{y_2} = \frac{A/x_1}{A/x_2} = \frac{A}{x_1}\frac{x_2}{A} = \frac{x_2}{x_1}$$

That is, the ratio of $y$-values is the inverse of the ratio of the corresponding values of $x$.

**LIMITS**
- As $x$ gets very large, $y$ approaches zero.
- As $x$ approaches zero, $y$ gets very large.

Exercises 10, 11

You're familiar with this idea: It's much harder to get your car rolling by pushing it than to get your bicycle rolling, and it's harder to stop a heavily loaded grocery cart than to stop a skateboard. This tendency to resist a change in velocity (i.e., to resist speeding up or slowing down) is called **inertia.** Thus we can say that more massive objects have more inertia.

**Feel the difference** Because of its high sugar content, a can of regular soda has a mass about 4% greater than that of a can of diet soda. If you try to judge which can is more massive by simply holding one in each hand, this small difference is almost impossible to detect. If you *move* the cans up and down, however, the difference becomes subtly but noticeably apparent: People evidently are more sensitive to how the mass of each can resists acceleration than they are to the cans' weights alone.

**EXAMPLE 4.4 Finding the mass of an unknown block**

When a rubber band is stretched to pull on a 1.0 kg block with a constant force, the acceleration of the block is measured to be 3.0 m/s$^2$. When a block with an unknown mass is pulled with the same rubber band, using the same force, its acceleration is 5.0 m/s$^2$. What is the mass of the unknown block?

**STRATEGIZE** Because acceleration is inversely proportional to mass, we will use ratio reasoning to solve this problem.

**PREPARE** We denote the mass of the unknown block by $m$.

**SOLVE** We can use the result of the Inversely proportional relationships box to write

$$\frac{3.0\ \text{m/s}^2}{5.0\ \text{m/s}^2} = \frac{m}{1.0\ \text{kg}}$$

or

$$m = \frac{3.0\ \text{m/s}^2}{5.0\ \text{m/s}^2} \times (1.0\ \text{kg}) = 0.60\ \text{kg}$$

**ASSESS** With the same force applied, the unknown block had a *larger* acceleration than the 1.0 kg block. It makes sense, then, that its mass—its resistance to acceleration—is *less* than 1.0 kg.

**Video** Newton's Second Law

**STOP TO THINK 4.5** Two rubber bands stretched to the standard length cause an object to accelerate at 2 m/s$^2$. Suppose another object with twice the mass is pulled by four rubber bands stretched to the standard length. What is the acceleration of this second object?

A. 1 m/s$^2$ B. 2 m/s$^2$ C. 4 m/s$^2$ D. 8 m/s$^2$ E. 16 m/s$^2$

## 4.5 Newton's Second Law

**Video** Newton's Second Law Application

We can now summarize the results of our experiments. We've seen that **a force causes an object to accelerate. The acceleration $a$ is directly proportional to the force $F$ and inversely proportional to the mass $m$.** We can express both these relationships in equation form as

$$a = \frac{F}{m} \tag{4.2}$$

Note that if we double the size of the force $F$, the acceleration $a$ will double, as we found experimentally. And if we triple the mass $m$, the acceleration will be only one-third as great, again agreeing with our experiments.

Equation 4.2 tells us the magnitude of an object's acceleration in terms of its mass and the force applied. But our experiments also had another important finding: The *direction* of the acceleration was the same as the direction of the force. We can express this fact by writing Equation 4.2 in *vector* form as

$$\vec{a} = \frac{\vec{F}}{m} \tag{4.3}$$

Finally, our experiment was limited to looking at an object's response to a *single* applied force acting in a single direction. Realistically, an object is likely to be subjected to several distinct forces $\vec{F}_1, \vec{F}_2, \vec{F}_3, \ldots$ that may point in different directions. What happens then? Experiments show that the acceleration of the object is determined by the *net force* acting on it. Recall from Figure 4.4 and Equation 4.1 that the net force is the *vector sum* of all forces acting on the object. So if several forces are acting, we use the net force in Equation 4.4.

Newton was the first to recognize these connections between force and motion. This relationship is known today as **Newton's second law.**

**Newton's second law** An object of mass $m$ subjected to forces $\vec{F}_1, \vec{F}_2, \vec{F}_3, \ldots$ will undergo an acceleration $\vec{a}$ given by

$$\vec{a} = \frac{\vec{F}_{\text{net}}}{m} \tag{4.4}$$

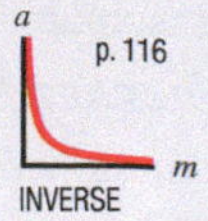

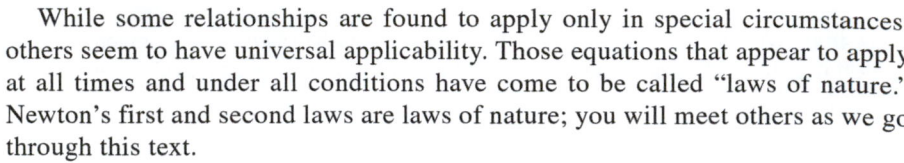

where the net force $\vec{F}_{\text{net}} = \vec{F}_1 + \vec{F}_2 + \vec{F}_3 + \cdots$ is the vector sum of all forces acting on the object. **The acceleration vector $\vec{a}$ points in the same direction as the net force vector $\vec{F}_{\text{net}}$.**

While some relationships are found to apply only in special circumstances, others seem to have universal applicability. Those equations that appear to apply at all times and under all conditions have come to be called "laws of nature." Newton's first and second laws are laws of nature; you will meet others as we go through this text.

We can rewrite Newton's second law in the form

$$\vec{F}_{\text{net}} = m\vec{a} \tag{4.5}$$

which is how you'll see it presented in many textbooks and how, in practice, we'll often use the second law. Equations 4.4 and 4.5 are mathematically equivalent, but Equation 4.4 better describes the central idea of Newtonian mechanics: A force applied to an object causes the object to accelerate and the acceleration is in the direction of the net force.

**NOTE** ► When several forces act on an object, be careful not to think that the strongest force "overcomes" the others to determine the motion on its own. It is $\vec{F}_{\text{net}}$, the sum of *all* the forces, that determines the acceleration $\vec{a}$. ◄

**Size matters?** Race car driver Danica Patrick was the subject of controversial comments by drivers who thought her relatively small mass of 45 kg gave her an unfair advantage. Because every driver's car must have the same mass, Patrick's overall racing mass was lower than any other driver's, so her car could be expected to have a slightly greater acceleration.

**CONCEPTUAL EXAMPLE 4.5** **Acceleration of a wind-blown basketball**

You drop a basketball while a stiff breeze is blowing to the right. In what direction does the ball accelerate?

***REASON*** Wind is just air in motion. If the air is moving to the *right* with respect to the ball, then the ball is moving to the *left* with respect to the air. There will be a drag force opposite the velocity of the ball relative to the air, to the right. So, as FIGURE 4.22a shows, two forces are acting on the ball: its weight $\vec{w}$ directed downward and the drag force $\vec{D}$ directed to the right. Newton's second law tells us that the direction of the acceleration is the same as the direction of the net force $\vec{F}_{\text{net}}$. In FIGURE 4.22b we find $\vec{F}_{\text{net}}$ by graphical vector addition of $\vec{w}$ and $\vec{D}$. We see that $\vec{F}_{\text{net}}$ and therefore $\vec{a}$ point downward and to the right.

FIGURE 4.22 A basketball falling in a strong breeze.

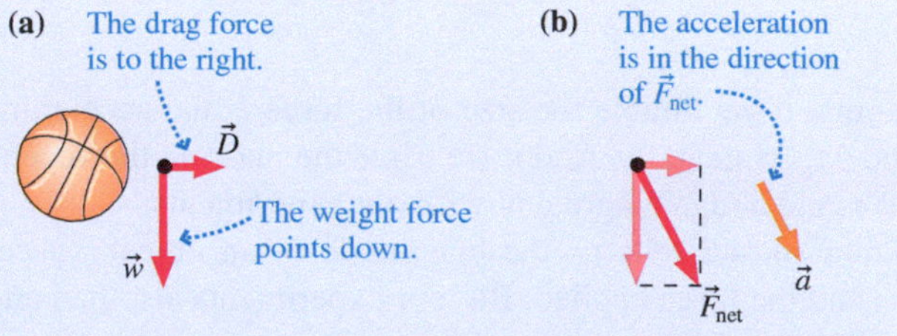


***ASSESS*** This makes sense on the basis of your experience. Weight pulls the ball down, and the wind pushes the ball to the right. The net result is an acceleration down and to the right.

## The Unit of Force

Because $\vec{F}_{net} = m\vec{a}$, the units of force must be the unit of mass (kg) multiplied by the unit of acceleration (m/s$^2$); thus the units of force are kg · m/s$^2$. This unit of force is called the **newton:**

$$1 \text{ newton} = 1 \text{ N} = 1 \frac{\text{kg} \cdot \text{m}}{\text{s}^2}$$

The abbreviation for the newton is N. TABLE 4.2 lists some typical forces in newtons.

The newton is a *secondary unit,* meaning that it is defined in terms of the *primary units* of kilograms, meters, and seconds.

The unit of force in the English system is the *pound* (abbreviated lb). Although the definition of the pound has varied, it is now defined in terms of the newton:

$$1 \text{ pound} = 1 \text{ lb} = 4.45 \text{ N}$$

You very likely associate pounds with kilograms rather than with newtons. Everyday language often confuses the ideas of mass and weight, but we're going to need to make a clear distinction between them. We'll have more to say about this in the next chapter.

TABLE 4.2 Approximate magnitude of some typical forces

| Force | Approximate magnitude (newtons) |
|---|---|
| Weight of a U.S. nickel | 0.05 |
| Weight of ¼ cup of sugar | 0.5 |
| Weight of a 1 pound object | 5 |
| Weight of a typical house cat | 50 |
| Weight of a 110 pound person | 500 |
| Propulsion force of a car | 5000 |
| Thrust force of a small jet engine | 50,000 |
| Pulling force of a locomotive | 500,000 |

### EXAMPLE 4.6 Racing down the runway

A Boeing 737—a small, short-range jet with a mass of 51,000 kg—sits at rest at the start of a runway. The pilot turns the pair of jet engines to full throttle, and the thrust accelerates the plane down the runway. After traveling 940 m, the plane reaches its takeoff speed of 70 m/s and leaves the ground. What is the thrust of each engine?

**STRATEGIZE** If we assume that the plane undergoes a constant acceleration (a reasonable assumption), we can use kinematics to find the magnitude of that acceleration. Then we will use Newton's second law to find the force—the thrust—that produced this acceleration.

**PREPARE** FIGURE 4.23 is a visual overview of the airplane's motion.

FIGURE 4.23 Visual overview of the accelerating airplane.

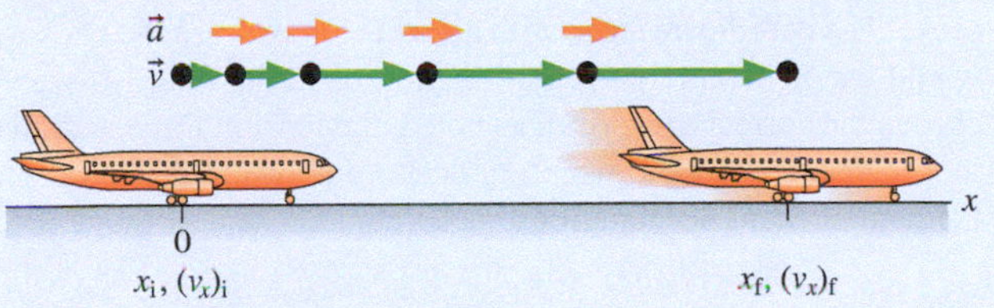

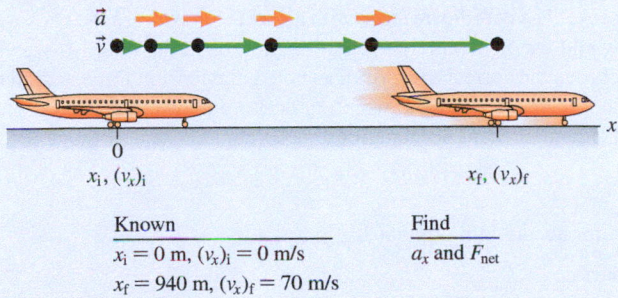

Known
$x_i = 0$ m, $(v_x)_i = 0$ m/s
$x_f = 940$ m, $(v_x)_f = 70$ m/s

Find
$a_x$ and $F_{net}$

**SOLVE** We don't know how much time it took the plane to reach its takeoff speed, but we do know that it traveled a distance of 940 m. We can solve for the acceleration by using Equation 2.13 of Synthesis 2.1 :

$$(v_x)_f^2 = (v_x)_i^2 + 2a_x\Delta x$$

The displacement is $\Delta x = x_f - x_i = 940$ m, and the initial velocity is 0. We can rearrange the equation to solve for the acceleration:

$$a_x = \frac{(v_x)_f^2}{2\Delta x} = \frac{(70 \text{ m/s})^2}{2(940 \text{ m})} = 2.61 \text{ m/s}^2$$

We've kept an extra significant figure because this isn't our final result—we are asked to solve for the thrust. We complete the solution by using Newton's second law:

$$F = ma_x = (51{,}000 \text{ kg})(2.61 \text{ m/s}^2) = 133{,}000 \text{ N}$$

The thrust of each engine is half of this total force:

$$\text{Thrust of one engine} = 67{,}000 \text{ N} = 67 \text{ kN}$$

**ASSESS** An acceleration of about ¼g seems reasonable for an airplane: It's zippy, but it's not a thrill ride. And the final value we find for the thrust of each engine is close to the value given in Table 4.2. This gives us confidence that our final result makes good physical sense.

**STOP TO THINK 4.6** Three forces act on an object. In which direction does the object accelerate?

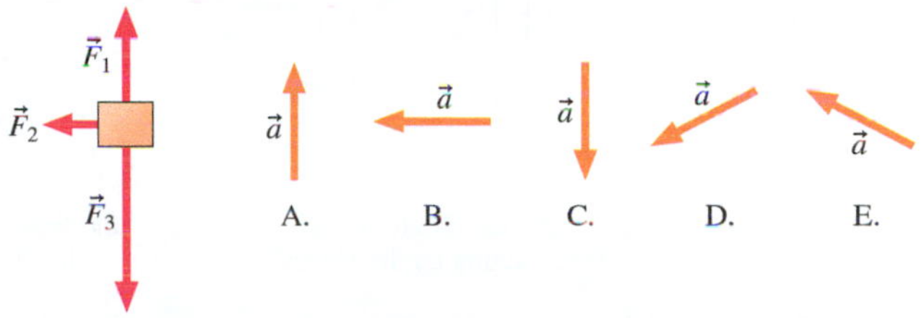

## 4.6 Free-Body Diagrams

When we solve a dynamics problem, it is useful to assemble all of the information about the forces that act on an object (or "body") into a single diagram called a **free-body diagram.** A free-body diagram represents the object as a particle and shows *all* of the forces that act on the object. Learning how to draw a correct free-body diagram is a very important skill, one that in the next chapter will become a critical part of our approach to solving motion problems.

**TACTICS BOX 4.3 Drawing a free-body diagram**

1. **Identify all forces acting on the object.** This step was described in Tactics Box 4.2.
2. **Draw a coordinate system.** Use the axes defined in your pictorial representation (see Tactics Box 2.2). If those axes are tilted, for motion along an incline, then the axes of the free-body diagram should be similarly tilted.
3. **Represent the object as a dot at the origin of the coordinate axes.** This is the particle model.
4. **Draw vectors representing each of the identified forces.** This was described in Tactics Box 4.1. Be sure to label each force vector.
5. **Draw and label the *net force* vector $\vec{F}_{net}$.** Draw this vector beside the diagram, not on the particle. Then check that $\vec{F}_{net}$ points in the same direction as the acceleration vector $\vec{a}$ on your motion diagram. Or, if appropriate, write $\vec{F}_{net} = \vec{0}$.

Exercises 17–22

**EXAMPLE 4.7 Forces on an elevator**

An elevator, suspended by a cable, speeds up as it moves upward from the ground floor. Draw a free-body diagram of the elevator.

***STRATEGIZE*** We'll follow the steps of Tactics Box 4.3. We note that the elevator is moving upward and its speed is increasing. That means the acceleration is directed upward, so that $\vec{F}_{net}$ must be directed upward as well.

***PREPARE*** In FIGURE 4.24 we illustrate the steps listed in Tactics Box 4.3. Because the net force is directed upward, the magnitude of the (upward) tension force $\vec{T}$ must be greater than that of the (downward) weight force $\vec{w}$. We have shown this in the free-body diagram by drawing the tension vector longer than the weight vector.

***ASSESS*** Let's take a look at our picture and see if it makes sense. The coordinate axes, with a vertical $y$-axis, are the ones we would use in a pictorial representation of the motion, so we've chosen the correct axes. And, as noted, the tension force is drawn longer than the weight, which indicates an upward net force and hence an upward acceleration.

FIGURE 4.24 Free-body diagram of an elevator accelerating upward.

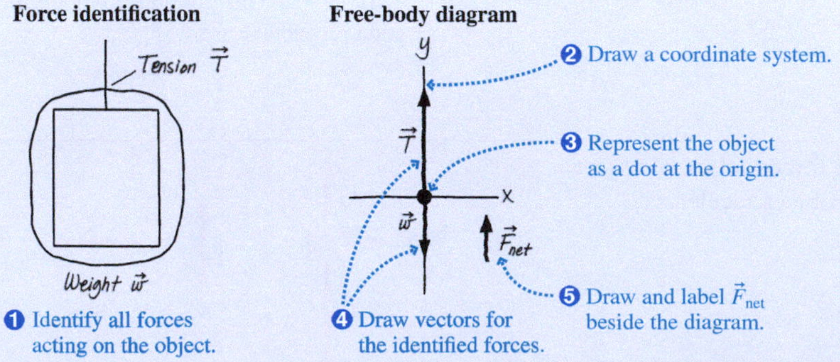

## EXAMPLE 4.8 Forces on the world's fastest car

The Thrust SSC is the world's fastest car.

The speed record for the world's fastest car is held by the jet-powered Thrust SSC, whose top speed exceeds the speed of sound. At high speeds, air drag is significant, but when the car is just starting to accelerate from rest, friction and drag are negligible compared to the jet thrust. Draw a visual overview—motion diagram, force identification picture, and free-body diagram—for the car as it starts to move forward from rest.

**STRATEGIZE** To draw the visual overview, we'll use the steps listed in Tactics Box 4.3.

**PREPARE** We treat the car as a particle. The visual overview is shown in FIGURE 4.25.

**ASSESS** The motion diagram tells us that the acceleration is in the positive $x$-direction. According to the rules of vector addition, this can be true only if the upward-pointing $\vec{n}$ and the downward-pointing $\vec{w}$ are equal in magnitude and thus cancel each other. The vectors have been drawn accordingly, and this leaves the net force vector pointing toward the right, in agreement with $\vec{a}$ from the motion diagram.

FIGURE 4.25 Visual overview for a carbon dioxide racer.

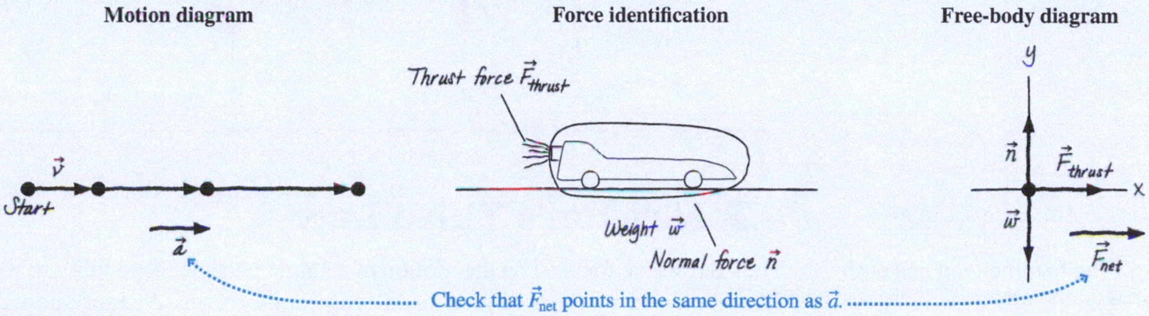


## EXAMPLE 4.9 Forces on a towed skier

A tow rope pulls a skier up a snow-covered hill at a constant speed. Draw a full visual overview of the skier.

**STRATEGIZE** This is example 4.2 again, with the additional information that the skier is moving at a constant speed. Because his acceleration is zero, the net force acting on him must be zero as well.

**PREPARE** If we were doing a kinematics problem, the pictorial representation would use a tilted coordinate system with the $x$-axis parallel to the slope, so we use these same tilted coordinate axes for the free-body diagram. The full visual overview is shown in FIGURE 4.26.

**ASSESS** We have shown $\vec{T}$ pulling parallel to the slope and $\vec{f}_k$, which opposes the direction of motion, pointing down the slope. The normal force $\vec{n}$ is perpendicular to the surface and thus along the $y$-axis. Finally, and this is important, the weight $\vec{w}$ is *vertically* downward, *not* along the negative $y$-axis.

The skier moves in a straight line with constant speed, so $\vec{a} = \vec{0}$. Newton's second law then tells us that $\vec{F}_{net} = m\vec{a} = \vec{0}$. Thus we have drawn the vectors such that the forces add to zero. We'll learn more about how to do this in Chapter 5.

FIGURE 4.26 Visual overview for a skier being towed at a constant speed.

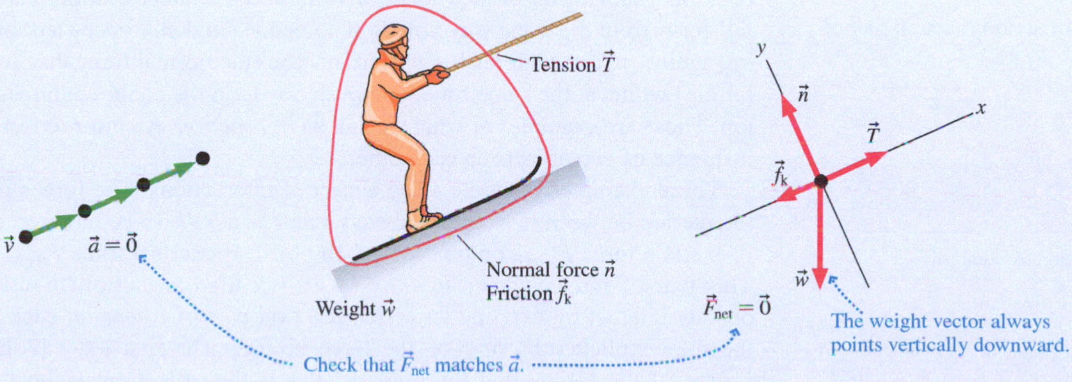

Free-body diagrams will be our major tool for the next several chapters. Careful practice with the workbook exercises and homework in this chapter will pay immediate benefits in the next chapter. Indeed, it is fair to say that a problem is more than half solved when you correctly complete the free-body diagram.

**STOP TO THINK 4.7** An elevator suspended by a cable is moving upward and slowing to a stop. Which free-body diagram is correct?

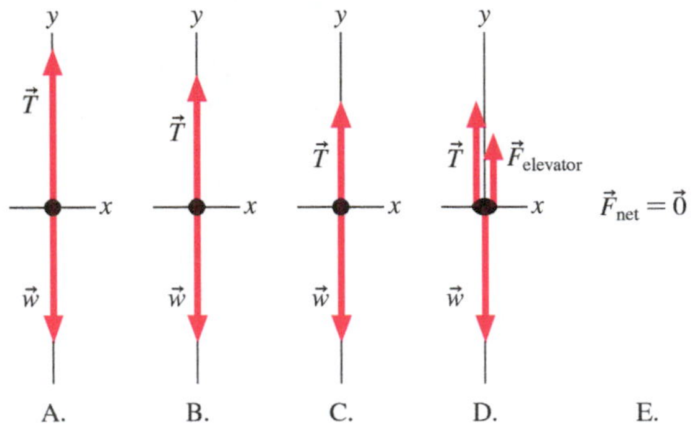


## 4.7 Newton's Third Law

FIGURE 4.27 The hammer and nail each exert a force on the other.

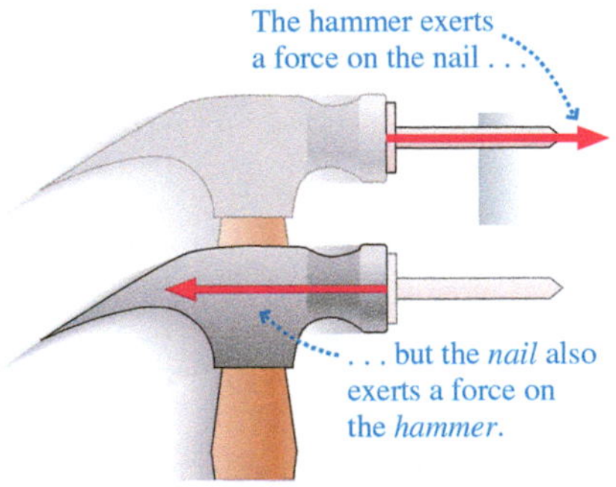


Thus far, we've focused on the motion of a single particle responding to well-defined forces exerted by other objects or to long-range forces. A skier sliding downhill, for instance, is subject to frictional and normal forces from the slope and the pull of gravity on his body. Once we have identified these forces, we can use Newton's second law to calculate the acceleration, and hence the overall motion, of the skier.

But motion in the real world often involves two or more objects *interacting* with each other. Consider the hammer and nail in FIGURE 4.27. As the hammer hits the nail, the nail pushes back on the hammer. A bat and a ball, your foot and a soccer ball, and the earth–moon system are other examples of interacting objects. We need to consider how the forces on these interacting objects are related to each other.

### Interacting Objects

Think about the hammer and nail in Figure 4.27 again. The hammer certainly exerts a force on the nail as it drives the nail forward. At the same time, the nail exerts a force on the hammer. If you are not sure that it does, imagine hitting the nail with a glass hammer. It's the force of the nail on the hammer that would cause the glass to shatter.

Indeed, any time that object A pushes or pulls on object B, object B pushes or pulls back on object A. As you push on a filing cabinet to move it, the cabinet pushes back on you. (If you pushed forward without the cabinet pushing back, you would fall forward in the same way you do if someone suddenly opens a door you're leaning against.) Your chair pushes upward on you (the normal force that keeps you from falling) while, at the same time, you push down on the chair, compressing the cushion. These are examples of what we call an *interaction*. An **interaction** is the mutual influence of two objects on each other.

FIGURE 4.28 An action/reaction pair of forces.

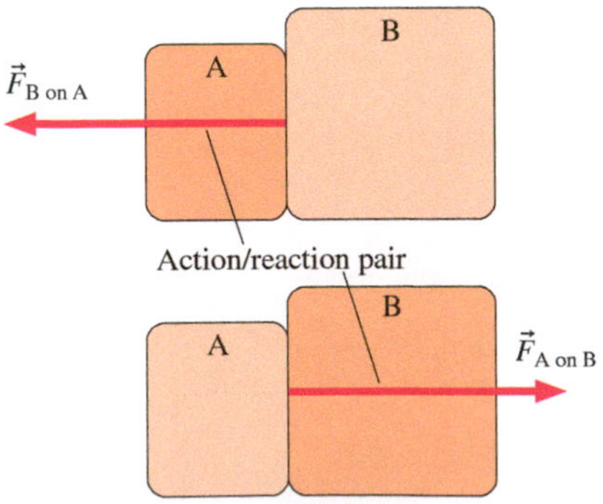


These examples illustrate a key aspect of interactions: The forces involved in an interaction between two objects always occur as a *pair*. To be more specific, if object A exerts a force $\vec{F}_{A\,on\,B}$ on object B, then object B exerts a force $\vec{F}_{B\,on\,A}$ on object A. This pair of forces, shown in FIGURE 4.28, is called an **action/reaction pair.** Two objects interact by exerting an action/reaction pair of forces on each other. Notice the very explicit subscripts on the force vectors. The first letter is the *agent*—the source of the force—and the second letter is the *object* on which the force acts. $\vec{F}_{A\,on\,B}$ is thus the force exerted *by* A *on* B.

**NOTE** ▸ The name "action/reaction pair" is somewhat misleading. The forces occur simultaneously, and we cannot say which is the "action" and which the "reaction." Neither is there any implication about cause and effect: The action does not cause the reaction. **An action/reaction pair of forces exists as a pair, or not at all.** For action/reaction pairs, the labels are the key: Force $\vec{F}_{\text{A on B}}$ is paired with force $\vec{F}_{\text{B on A}}$. ◂

## Reasoning with Newton's Third Law

Two objects always interact via an action/reaction pair of forces. Newton was the first to recognize how the two members of an action/reaction pair of forces are related to each other. Today we know this as **Newton's third law:**

**Newton's third law** Every force occurs as one member of an action/reaction pair of forces.

- The two members of an action/reaction pair act on two *different* objects.
- The two members of an action/reaction pair point in *opposite* directions and are *equal in magnitude.*

Newton's third law is often stated: "For every action there is an equal but opposite reaction." While this is a catchy phrase, it lacks the preciseness of our preferred version. In particular, it fails to capture an essential feature of the two members of an action/reaction pair—that each acts on a *different* object. This is shown in FIGURE 4.29, where a hammer hitting a nail exerts a force $\vec{F}_{\text{hammer on nail}}$ on the nail, and by the third law, the nail must exert a force $\vec{F}_{\text{nail on hammer}}$ to complete the action/reaction pair.

Figure 4.29 also illustrates that these two forces point in *opposite directions.* This feature of the third law is also in accord with our experience. If the hammer hits the nail with a force directed to the right, then the force of the nail on the hammer is directed to the left. If the force of me on my chair pushes down, the force of the chair on me pushes up.

Finally, Figure 4.29 shows that, according to Newton's third law, the two members of an action/reaction pair have *equal* magnitudes, so that $F_{\text{hammer on nail}} = F_{\text{nail on hammer}}$. This is something new, and it is by no means obvious.

FIGURE 4.29 Newton's third law.

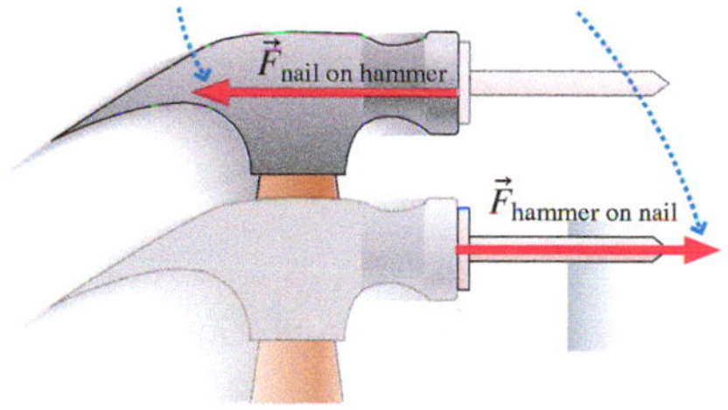


**Video** Weighing a Hovering Magnet

### CONCEPTUAL EXAMPLE 4.10 Forces in a collision

A 10,000 kg truck has a head-on collision with a 1000 kg compact car. During the collision, is the force of the truck on the car greater than, less than, or equal to the force of the car on the truck?

***REASON*** Newton's third law tells us that the magnitude of the force of the car on the truck must be equal to that of the truck on the car! How can this be, when the car is so small compared to the truck? The source of puzzlement in problems like this is that Newton's third law equates the sizes of the *forces* acting on the two objects, not their *accelerations*. The acceleration of each object depends not only on the force applied to it but also, according to Newton's second law, on its mass. The car and the truck do in fact feel forces of equal strength from the other, but the car, with its smaller mass, undergoes a much greater acceleration than the more massive truck.

***ASSESS*** This is a type of question for which your intuition may need to be refined. When you think about questions of this sort, be sure to separate the *effects* (the accelerations) from the *causes* (the forces themselves). Because two interacting objects can have very different masses, their accelerations can be very different. Don't let this dissuade you from realizing that the interaction forces are of the same strength.

**Revenge of the target** We normally think of the damage that the force of a bullet inflicts on its target. But according to Newton's third law, the target exerts an equal force on the bullet. The bullet on the left has not been fired. The bullets on the right have been fired into a test target at increasing speeds and were clearly damaged by this interaction.

A runner starts from rest and then begins to move down the track. Because he's accelerating, there must be a force on him in the forward direction. The *energy* to put his body into motion comes from inside his body (we'll consider this type of problem in detail in Chapter 11). But where does the *force* come from?

If you tried to walk across a frictionless floor, your foot would slip and slide *backward*. In order for you to walk, the floor needs to have friction so that your foot *sticks* to the floor as you straighten your leg, moving your body forward. The friction that prevents slipping is *static* friction. Static friction, you will recall, acts in the direction that prevents slipping, so the static friction force $\vec{f}_{\text{surface on person}}$ has to point in the *forward* direction to prevent your foot from slipping backward. As shown in FIGURE 4.30a, it is this forward-directed static friction force that propels you forward! The force of your foot on the floor, $\vec{f}_{\text{person on surface}}$, is the other half of the action/reaction pair, and it points in the opposite direction as you push backward against the floor. So, when the runner starts down the track—or when you start walking across the floor—it is the static friction force between the ground and the runner that provides the acceleration. This may seem surprising, but imagine that the race was held on an icy pond. The runners would have much more trouble getting started.

FIGURE 4.30 Pushing backward to move forward.

(a)

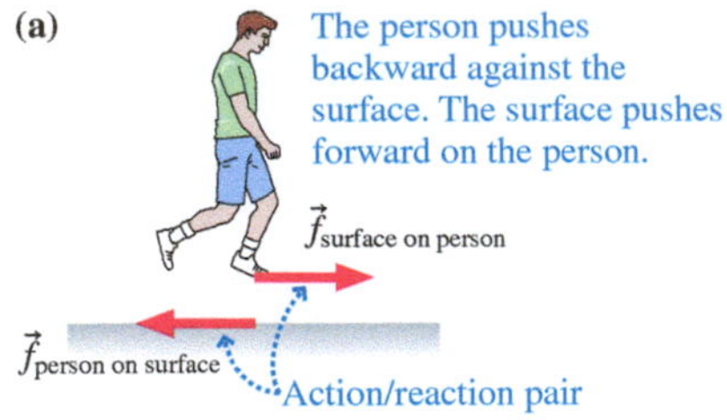


(b)

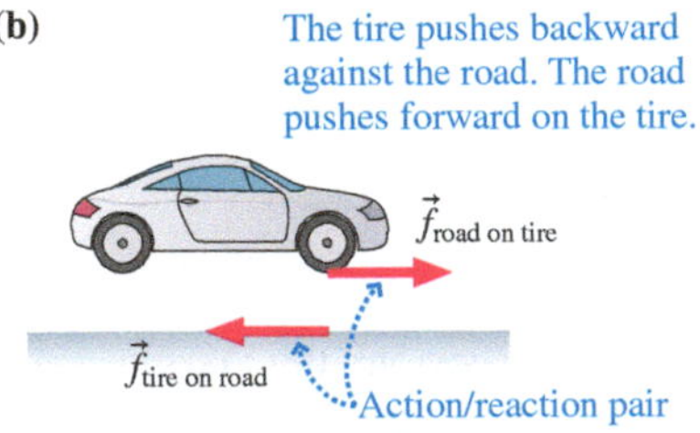


**NOTE** ▸ A counterintuitive notion for many students is that it is *static* friction that pushes you forward. You are moving, so how can this be static friction? It's true that your body is in motion, but your feet are not sliding, so this is not kinetic friction, but static. ◂

Similarly, the car in FIGURE 4.30b uses static friction to propel itself. The car uses its motor to turn the tires, causing the tires to push backward against the road ($\vec{f}_{\text{tire on road}}$). The road surface responds by pushing the car forward ($\vec{f}_{\text{road on tire}}$). Again, the forces involved are *static* friction forces. The tire is rolling, but the bottom of the tire, where it contacts the road, is instantaneously at rest. If it weren't, you would leave one giant skid mark as you drove and would burn off the tread within a few miles.

FIGURE 4.31 Rocket propulsion.

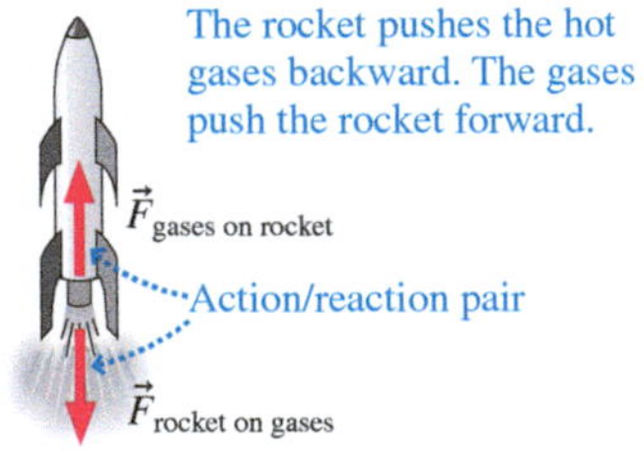


Rocket motors provide propulsion as well, but there is a difference from the earlier cases: A rocket doesn't push against the ground or even the atmosphere. Instead, the rocket engine pushes hot, expanding gases out of the back of the rocket, as shown in FIGURE 4.31. In response, the exhaust gases push the rocket forward with the force we've called *thrust*. That's why rocket propulsion works in the vacuum of space.

Now let's consider the opposite case—a pull *in* rather than a push *out*. The chapter opened with a photo of a snake-necked turtle. The muscles in the turtle's neck aren't strong enough to snap its head forward with the observed acceleration. Instead, the turtle uses a different approach. The turtle opens its mouth and forcefully pulls water into its throat. The turtle's head and the water form an action/reaction pair: The water is pulled backward, and this results in a forward force on the turtle's head. This is just the reverse of what happens in a rocket, and it is a surprisingly effective technique, enabling the turtle to strike more rapidly than many predatory fish.

Now we've assembled all the pieces we need in order to start solving problems in dynamics. We have seen what forces are and how to identify them, and we've learned how forces cause objects to accelerate according to Newton's second law. We've also found how Newton's third law governs the interaction forces between two objects. Our goal in the next several chapters is to apply Newton's laws to a variety of problems involving straight-line and circular motion.

**STOP TO THINK 4.8** A small car is pushing a larger truck that has a dead battery. The mass of the truck is greater than the mass of the car. The car and the truck are moving to the right and are speeding up. Which of the following statements is true?

A. The car exerts a force on the truck, but the truck doesn't exert a force on the car.
B. The car exerts a larger force on the truck than the truck exerts on the car.
C. The car exerts the same amount of force on the truck as the truck exerts on the car.
D. The truck exerts a larger force on the car than the car exerts on the truck.
E. The truck exerts a force on the car, but the car doesn't exert a force on the truck.

## INTEGRATED EXAMPLE 4.11 Pulling an excursion train

An engine slows as it pulls two cars of an excursion train up a mountain. Draw a visual overview (motion diagram, force identification diagram, and free-body diagram) for the car just behind the engine. Ignore friction.

**STRATEGIZE** We will treat the train car as a particle. We'll prepare the visual overview using the steps of Tactics Box 4.2 to identify the forces, and Tactics Box 4.3 to draw the free-body diagram.

**PREPARE** Because the train is slowing down, the motion diagram consists of a series of particle positions that become closer together at successive times; the corresponding velocity vectors become shorter and shorter. To identify the forces acting on the car we use the steps of Tactics Box 4.2. Finally, we can draw a free-body diagram using Tactics Box 4.3.

**SOLVE** Finding the forces acting on car 1 can be tricky. The engine exerts a forward force $\vec{F}_{\text{engine on 1}}$ on car 1 where the engine touches the front of car 1. At its back, car 1 touches car 2, so car 2 must also exert a force on car 1. The direction of this force can be understood from Newton's third law. Car 1 exerts an uphill force on car 2 in order to pull it up the mountain. Thus, by Newton's third law, car 2 must exert an oppositely directed *downhill* force on car 1. This is the force we label $\vec{F}_{\text{2 on 1}}$. The three diagrams that make up the full visual overview are shown in FIGURE 4.32.

**ASSESS** Correctly preparing the three diagrams illustrated in this example is critical for solving problems using Newton's laws. The motion diagram allows you to determine the direction of the acceleration and hence of $\vec{F}_{\text{net}}$. Using the force identification diagram, you will correctly identify all the forces acting on the object and, just as important, not add any extraneous forces. And by properly drawing these force vectors in a free-body diagram, you'll be ready for the quantitative application of Newton's laws that is the focus of Chapter 5.

FIGURE 4.32 Visual overview for a slowing train car being pulled up a mountain.

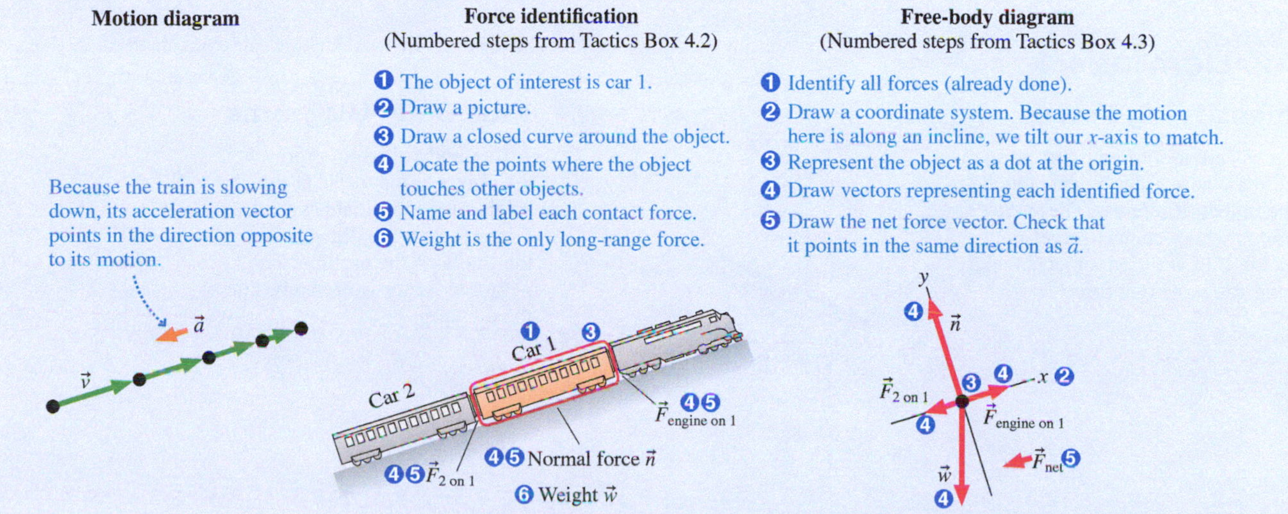

Motion diagram
Because the train is slowing down, its acceleration vector points in the direction opposite to its motion.
$\vec{a}$
$\vec{v}$
Force identification
(Numbered steps from Tactics Box 4.2)
1 The object of interest is car 1.
2 Draw a picture.
3 Draw a closed curve around the object.
4 Locate the points where the object touches other objects.
5 Name and label each contact force.
6 Weight is the only long-range force.
1 Car 1 3
Car 2
4 5 $\vec{F}_{\text{engine on 1}}$
4 5 $\vec{F}_{\text{2 on 1}}$
4 5 Normal force $\vec{n}$
6 Weight $\vec{w}$
Free-body diagram
(Numbered steps from Tactics Box 4.3)
1 Identify all forces (already done).
2 Draw a coordinate system. Because the motion here is along an incline, we tilt our x-axis to match.
3 Represent the object as a dot at the origin.
4 Draw vectors representing each identified force.
5 Draw the net force vector. Check that it points in the same direction as $\vec{a}$.
y
4 $\vec{n}$
3 4 x 2
$\vec{F}_{\text{2 on 1}}$ 4
$\vec{F}_{\text{engine on 1}}$
$\vec{w}$ 4
$\vec{F}_{\text{net}}$ 5

# SUMMARY

**GOAL** To establish the connection between force and motion.

## GENERAL PRINCIPLES

### Newton's First Law

Consider an object with no force acting on it. If it is at rest, it will remain at rest. If it is in motion, then it will continue to move in a straight line at a constant speed.

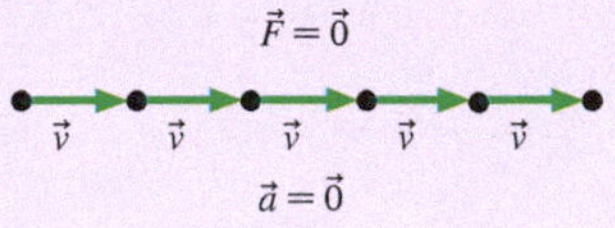


The first law tells us that an object that experiences no force will experience no acceleration.

### Newton's Second Law

An object with mass $m$ will undergo acceleration

$$\vec{a} = \frac{\vec{F}_{\text{net}}}{m}$$

where the net force $\vec{F}_{\text{net}} = \vec{F}_1 + \vec{F}_2 + \vec{F}_3 + \cdots$ is the vector sum of all the individual forces acting on the object.

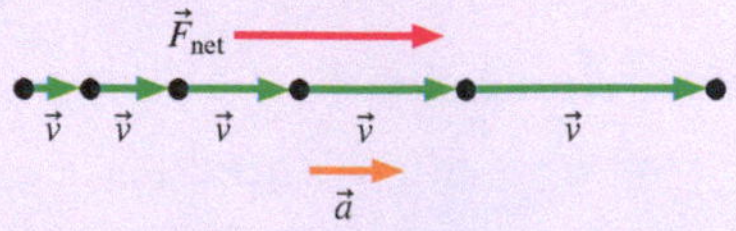


The second law tells us that a net force causes an object to accelerate. This is the relationship between force and motion. The acceleration points in the direction of $\vec{F}_{\text{net}}$.

### Newton's Third Law

Every force occurs as one member of an **action/reaction** pair of forces. The two members of an action/reaction pair:

- act on two *different* objects.
- point in opposite directions and are equal in magnitude:

$$\vec{F}_{\text{A on B}} = -\vec{F}_{\text{B on A}}$$

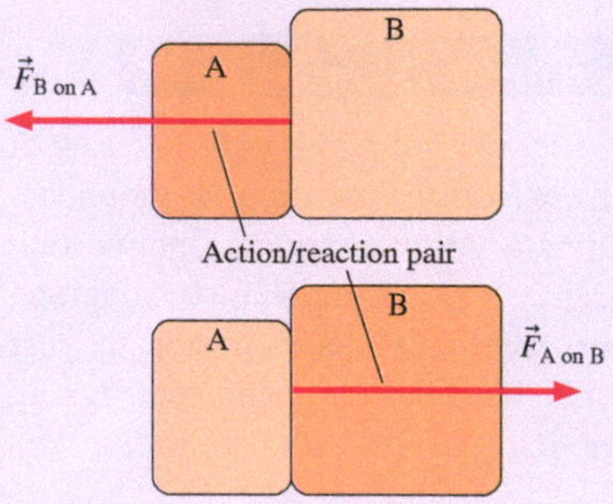


## IMPORTANT CONCEPTS

**Force** is a push or a pull on an object.

- Force is a vector, with a magnitude and a direction.
- A force requires an agent.
- A force is either a contact force or a long-range force.

The SI unit of force is the **newton** (N). A 1 N force will cause a 1 kg mass to accelerate at 1 m/s$^2$.

**Net force** is the vector sum of all the forces acting on an object.

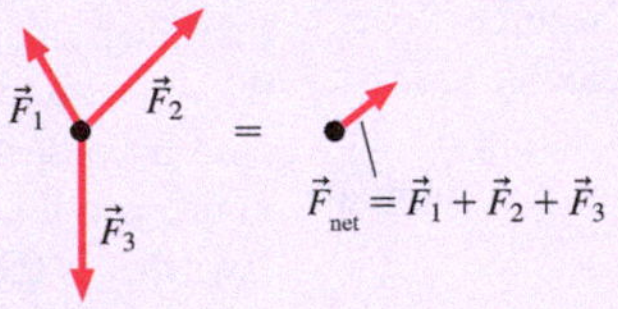


**Mass** is the property of an object that determines its resistance to acceleration.

If the same force is applied to objects A and B, then the ratio of their accelerations is related to the ratio of their masses as

$$\frac{a_A}{a_B} = \frac{m_B}{m_A}$$

The mass of objects can be determined in terms of their accelerations.

## APPLICATIONS

### Identifying Forces

Forces are identified by locating the points where other objects touch the object of interest. These are the points where contact forces are exerted. In addition, objects feel a long-range weight force.

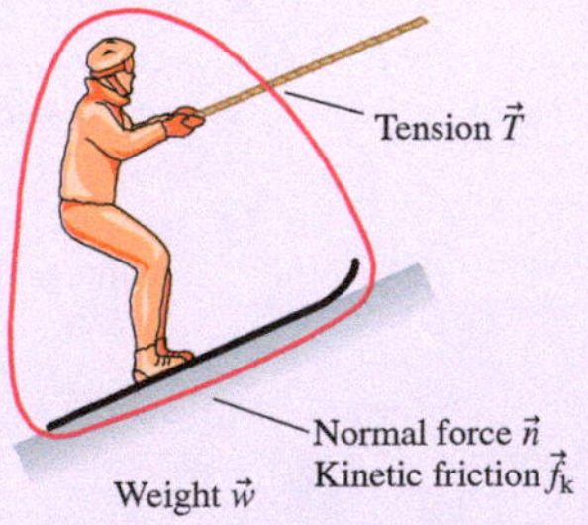


### Free-Body Diagrams

A free-body diagram represents the object as a particle at the origin of a coordinate system. Force vectors are drawn with their tails on the particle. The net force vector is drawn beside the diagram.

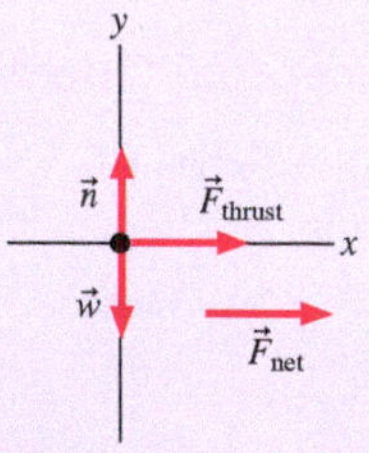

## Learning Objectives

After studying this chapter, you should be able to:

- Recognize and identify the forces acting on an object. *Conceptual Questions 4.14, 4.20; Problems 4.8, 4.9, 4.10, 4.11, 4.12*
- Combine multiple forces acting on an object. *Problems 4.4, 4.5, 4.6, 4.28*
- Draw a free-body diagram. *Problems 4.34, 4.35, 4.36, 4.37, 4.38*
- Understand the connection between force and motion. *Conceptual Questions 4.1, 4.3; Problems 4.2, 4.3, 4.20, 4.23, 4.24*
- Use Newton's second law. *Problems 4.14, 4.17, 4.21, 4.22, 4.25*
- Identify action/reaction pairs of forces on interacting objects. *Conceptual Questions 4.11, 4.17; Problems 4.43, 4.44, 4.45, 4.46*

## STOP TO THINK ANSWERS

**Chapter Preview Stop to Think: B.** The swan's velocity is to the left. Its speed is decreasing, so the acceleration is opposite the velocity, or to the right.

**Stop to Think 4.1: C.**

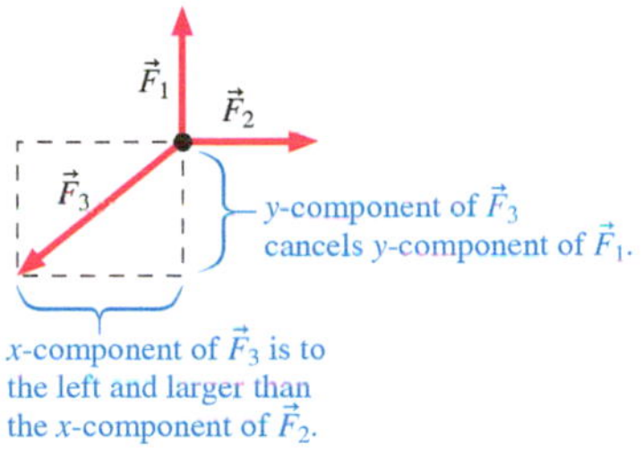


**Stop to Think 4.2: C.** The frog isn't moving, so a static friction force is keeping it at rest. If there was no friction, the weight force would cause the frog to slide down the slope. The static friction force opposes this, so it must be directed up the slope.

**Stop to Think 4.3: B.** The tension force is the force pulling the sled to the right. The friction force is opposing the motion, so it is directed to the left.

**Stop to Think 4.4: A, B, and D.** Friction and the normal force are the only contact forces. Nothing is touching the rock to provide a "force of the kick." We've agreed to ignore air resistance unless a problem specifically calls for it.

**Stop to Think 4.5: B.** Acceleration is proportional to force, so doubling the number of rubber bands doubles the acceleration of the original object from 2 $m/s^2$ to 4 $m/s^2$. But acceleration is also inversely proportional to mass. Doubling the mass cuts the acceleration in half, back to 2 $m/s^2$.

**Stop to Think 4.6: D.**

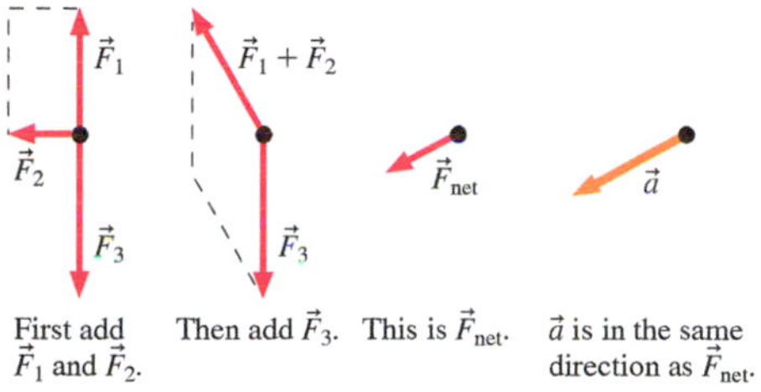


**Stop to Think 4.7: C.** The acceleration vector points downward as the elevator slows. $\vec{F}_{net}$ points in the same direction as $\vec{a}$, so $\vec{F}_{net}$ also points downward. This will be true if the tension is less than the weight: $T < w$.

**Stop to Think 4.8: C.** Newton's third law says that the force of A on B is *equal* and opposite to the force of B on A. This is always true. The mass of the objects isn't relevant, nor is the fact that the car and truck are accelerating.

**Video Tutor Solution** Chapter 4

# QUESTIONS

## Conceptual Questions

1. If an object is not moving, does that mean that there are no forces acting on it? Explain.
2. An object moves in a straight line at a constant speed. Is it true that there must be no forces of any kind acting on this object? Explain.
3. If you know all of the forces acting on a moving object, can you tell in which direction the object is moving? If the answer is Yes, explain how. If the answer is No, give an example.
4. Three arrows are shot horizontally. They have left the bow and are traveling parallel to the ground as shown in Figure Q4.4. Air resistance is negligible. Rank in order, from largest to smallest, the magnitudes of the *horizontal* forces $F_1$, $F_2$, and $F_3$ acting on the arrows. Some may be equal. State your reasoning.

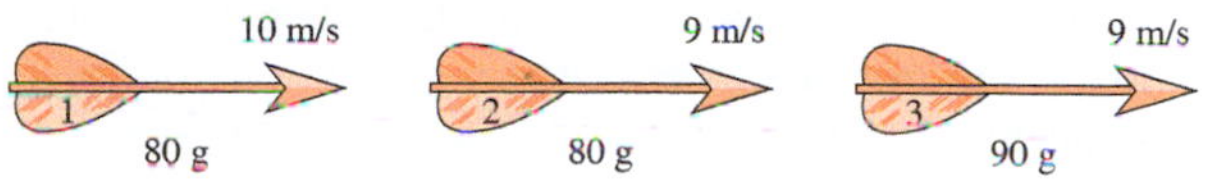


FIGURE Q4.4

Problem difficulty is labeled as | (straightforward) to ||||| (challenging). Problems labeled INT integrate significant material from earlier chapters; Problems labeled BIO are of biological or medical interest.

The eText icon indicates when there is a video tutor solution available for the chapter or for a specific problem. To launch these videos, log into your eText through Mastering™ Physics or log into the Study Area.

5. When your car accelerates away from a stop sign, you feel like you're being pushed back into your seat. Can you identify the force that is pushing you back? If not, why do you feel like you're being pushed back?
6. BIO Internal injuries in vehicular accidents may be due to what is called the "third collision." The first collision is the vehicle hitting the external object. The second collision is the person hitting something on the inside of the car, such as the dashboard or windshield. This may cause external lacerations. The third collision, possibly the most damaging to the body, is when organs, such as the heart or brain, hit the ribcage, skull, or other confines of the body, bruising the tissues on the leading edge and tearing the organ from its supporting structures on the trailing edge.
   a. Why is there a third collision? In other words, why are the organs still moving after the second collision?
   b. If the vehicle was traveling at 60 mph before the first collision, would the organs be traveling faster than, equal to, or slower than 60 mph just before the third collision?
7. Here's a great everyday use of the physics described in this chapter. If you are trying to get ketchup out of the bottle, the best way to do it is to turn the bottle upside down and give the bottle a sharp *upward* smack, forcing the bottle rapidly upward. Think about what subsequently happens to the ketchup, which is initially at rest, and use Newton's first law to explain why this technique is so successful.

8. a. Give an example of the motion of an object in which the frictional force on the object is directed opposite to the motion.
   b. Give an example of the motion of an object in which the frictional force on the object is in the same direction as the motion.
9. Suppose you are an astronaut in deep space, far from any source of gravity. You have two objects that look identical, but one has a large mass and the other a small mass. How can you tell the difference between the two?
10. Jonathan accelerates away from a stop sign. His eight-year-old daughter sits in the passenger seat. On whom does the back of the seat exert a greater force?
11. Suppose you are an astronaut on a spacewalk, far from any source of gravity. You find yourself floating alongside your spacecraft but 10 m away, with no propulsion system to get back to it. In your tool belt you have a hammer, a wrench, and a roll of duct tape. How can you get back to your spacecraft?
12. You're pushing horizontally on a large crate, but it won't budge. According to Newton's third law, the crate pushes back on you with the same magnitude of force that you exert on it. Suddenly the crate breaks free and you can now push it along the floor. Now is the force exerted on you by the crate greater than, equal to, or less than the force that you are exerting on the crate? Explain.
13. Josh and Taylor, standing face-to-face on frictionless ice, push off each other, causing each to slide backward. Josh is much bigger than Taylor. After the push, which of the two is moving faster?
14. A person sits on a sloped hillside. Is it ever possible to have the static friction force on this person point down the hill? Explain.
15. BIO Walking without slipping requires a static friction force between your feet (or footwear) and the floor. As described in this chapter, the force on your foot as you push off the floor is forward while the force exerted by your foot on the floor is backward. But what about your *other* foot, the one moved during a stride? What is the direction of the force on that foot as it comes into contact with the floor? Explain.
16. Figure 4.30b showed a situation in which the force of the road on the car's tire points forward. In other situations, the force points backward. Give an example of such a situation.
17. Two teams are engaged in a tug-of-war. According to Newton's third law, the red team pulls on the blue team with the same magnitude of force that the blue team pulls on the red team. How, then, can one team win?
18. A baseball pitcher throws a 100 mph fastball to batter Brandon Bopp. Bopp hits a 450 foot home run out of the park. When the ball contacts the bat, which force was greater: the force of the ball on the bat or the force of the bat on the ball?
19. The tire on this drag racer is severely twisted: The force of the road on the tire is quite large (most likely several times the weight of the car) and is directed forward as shown. Is the car speeding up or slowing down? Explain.

20. Suppose that, while in a squatting position, you stand on your hands, and then you pull up on your feet with a great deal of force. You are applying a large force to the bottoms of your feet, but no matter how strong you are, you will never be able to lift yourself off the ground. Use your understanding of force and motion to explain why this is not possible.

## Multiple-Choice Questions

21. | A block has acceleration $a$ when pulled by a string. If two identical blocks are glued together and pulled with twice the original force, their acceleration will be
    A. $(1/4)a$ B. $(1/2)a$ C. $a$ D. $2a$ E. $4a$
22. | A 5.0 kg block has an acceleration of 0.20 m/s$^2$ when a force is exerted on it. A second block has an acceleration of 0.10 m/s$^2$ when subject to the same force. What is the mass of the second block?
    A. 10 kg B. 5.0 kg C. 2.5 kg D. 7.5 kg
23. | Tennis balls experience a large drag force. A tennis ball is hit so that it goes straight up and then comes back down. The direction of the drag force is
    A. Always up. B. Up and then down.
    C. Always down. D. Down and then up.
24. || A group of students is making model cars that will be propelled by model rocket engines. These engines provide a nearly constant thrust force. The cars are light—most of the weight comes from the rocket engine—and friction and drag are very small. As the engine fires, it uses fuel, so it is much lighter at the end of the run than at the start. A student ignites the engine in a car, and the car accelerates. As the fuel burns and the car continues to speed up, the magnitude of the acceleration will
    A. Increase. B. Stay the same. C. Decrease.

25. | A person gives a box a shove so that it slides up a ramp, then reverses its motion and slides down. The direction of the force of friction is
   A. Always down the ramp.
   B. Up the ramp and then down the ramp.
   C. Always down the ramp.
   D. Down the ramp and then up the ramp.

26. || Craig is trying to push a heavy crate up a wooden ramp, but the crate won't budge. The direction of the static friction force acting on the crate is
   A. Up the ramp.
   B. Down the ramp.
   C. The friction force is zero.
   D. There is not enough information to tell.

27. | BIO INT As shown in the chapter, scallops use jet propulsion to move from one place to another. Their shells make them denser than water, so they normally rest on the ocean floor. If a scallop wishes to remain stationary, hovering a fixed distance above the ocean floor, it must eject water ______ so that the thrust force on the scallop is ______.
   A. upward, upward  B. upward, downward
   C. downward, upward  D. downward, downward

28. | A train engine pulls two identical cars behind it. Car 1 is attached directly to the engine, and car 2 is attached to car 1. The engine and cars are speeding up. The force of the engine on car 1 is ______ the force of car 1 on car 2.
   A. greater than  B. less than  C. equal to
   D. There is no force on either car.

29. | You're pushing a heavy filing cabinet across your office. You are tiring quickly and the cabinet is slowing down. The force that you exert on the cabinet is ______ the force that the cabinet exerts on you.
   A. greater than  B. less than  C. equal to

30. || Dave pushes his four-year-old son Thomas across the snow on a sled. As Dave pushes, Thomas speeds up. Which statement is true?
   A. The force of Dave on Thomas is larger than the force of Thomas on Dave.
   B. The force of Thomas on Dave is larger than the force of Dave on Thomas.
   C. Both forces have the same magnitude.
   D. It depends on how hard Dave pushes on Thomas.

31. | Figure Q4.31 shows block A sitting on top of block B. A constant force $\vec{F}$ is exerted on block B, causing block B to accelerate to the right. Block A rides on block B without slipping. Which statement is true?
   A. Block B exerts a friction force on block A, directed to the left.
   B. Block B exerts a friction force on block A, directed to the right.
   C. Block B does not exert a friction force on block A.

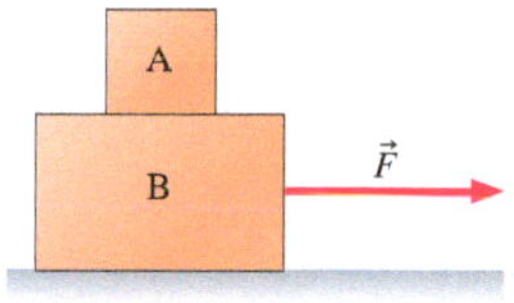


FIGURE Q4.31

# PROBLEMS

## Section 4.1 Motion and Forces

1. | BIO Whiplash injuries during an automobile accident are caused by the inertia of the head. If someone is wearing a seatbelt, her body will tend to move with the car seat. However, her head is free to move until the neck restrains it, causing damage to the neck. Brain damage can also occur.

   Figure P4.1 shows two sequences of head and neck motion for a passenger in an auto accident. One corresponds to a head-on collision, the other to a rear-end collision. Which is which? Explain.

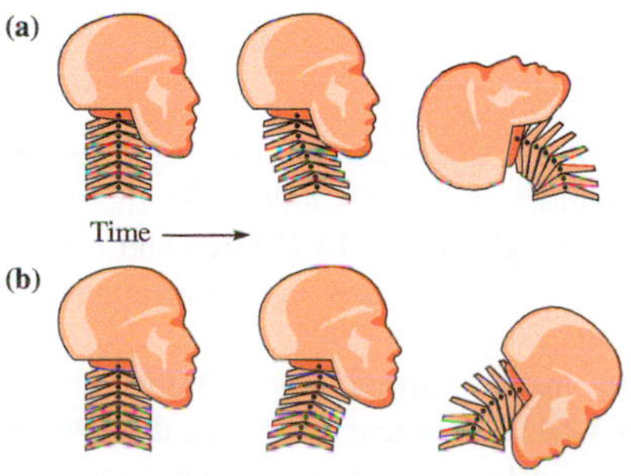


FIGURE P4.1

2. | BIO An automobile has a head-on collision. A passenger in the car experiences a compression injury to the brain. Is this injury most likely to be in the front or rear portion of the brain? Explain.

3. | In a head-on collision, an infant is much safer in a child safety seat when the seat is installed facing the rear of the car. Explain.

Problems 4 through 6 show two forces acting on an object at rest. Redraw the diagram, then add a third force that results in a net force of zero. Label the new force $\vec{F}_3$.

4. ||

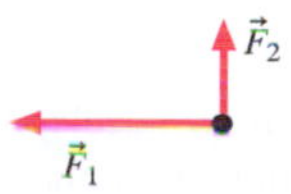


FIGURE P4.4

5. ||

FIGURE P4.5

6. ||

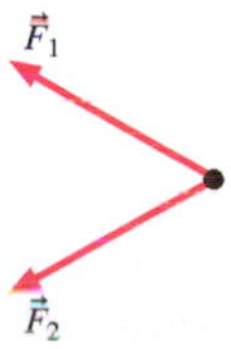


FIGURE P4.6

## Section 4.2 A Short Catalog of Forces

## Section 4.3 Identifying Forces

7. || A girl is swinging on a rope. Identify the forces acting on the girl at the very top of her swing.
8. || You're exploring a cave. Standing in a low cavern, you steady yourself by pushing straight up on the rock ceiling. Identify the forces acting on you.
9. || A baseball player is sliding into second base. Identify the forces on the baseball player.
10. || A jet plane is speeding down the runway during takeoff. Air resistance is not negligible. Identify the forces on the jet.
11. | A skier is sliding down a 15° slope. Friction is not negligible. Identify the forces on the skier.
12. || A falcon is hovering above the ground, then suddenly pulls in its wings and begins to fall toward the ground. Air resistance is not negligible. Identify the forces on the falcon.
BIO

## Section 4.4 What Do Forces Do?

13. |||| Figure P4.13 shows an acceleration-versus-force graph for three objects pulled by identical rubber bands. The mass of object 2 is 0.20 kg. What are the masses of objects 1 and 3? Explain your reasoning.

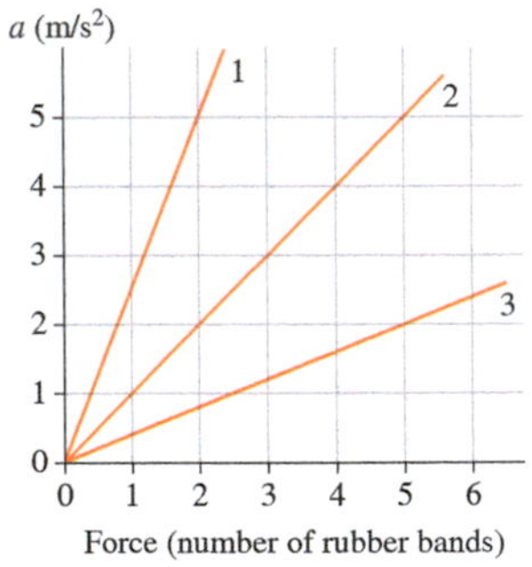


FIGURE P4.13

14. | A constant force applied to object A causes it to accelerate at 5 m/s$^2$. The same force applied to object B causes an acceleration of 3 m/s$^2$. Applied to object C, it causes an acceleration of 8 m/s$^2$.
    a. Which object has the largest mass?
    b. Which object has the smallest mass?
    c. What is the ratio of mass A to mass B $(m_A/m_B)$?
15. || A compact car has a maximum acceleration of 4.0 m/s$^2$ when it carries only the driver and has a total mass of 1200 kg. What is its maximum acceleration after picking up four passengers and their luggage, adding an additional 400 kg of mass?
16. || A student builds a rocket-propelled cart for a science project. Its acceleration is not quite high enough to win a prize, so he uses a larger rocket engine that provides 33% more thrust, although doing so increases the mass of the cart by 12%. By what percentage does the cart's acceleration increase?
17. | A constant force is applied to an object, causing the object to accelerate at 8.0 m/s$^2$. What will the acceleration be if
    a. The force is doubled?
    b. The object's mass is doubled?
    c. The force and the object's mass are both doubled?
    d. The force is doubled and the object's mass is halved?
18. ||| A man pulling an empty wagon causes it to accelerate at 1.4 m/s$^2$. What will the acceleration be if he pulls with the same force when the wagon contains a child whose mass is three times that of the wagon?
19. The Lamborghini Huracán has an initial acceleration of 0.75$g$. Its mass, with driver, is 1510 kg. If an 80 kg passenger rode along, what would the car's acceleration be?

## Section 4.5 Newton's Second Law

20. || Scallops eject water from their shells to provide a thrust force. The graph shows a smoothed graph of actual data for the initial motion of a 25 g scallop speeding up to escape a predator. What is the magnitude of the net force needed to achieve this motion? How does this force compare to the 0.25 N weight of the scallop?
BIO
INT

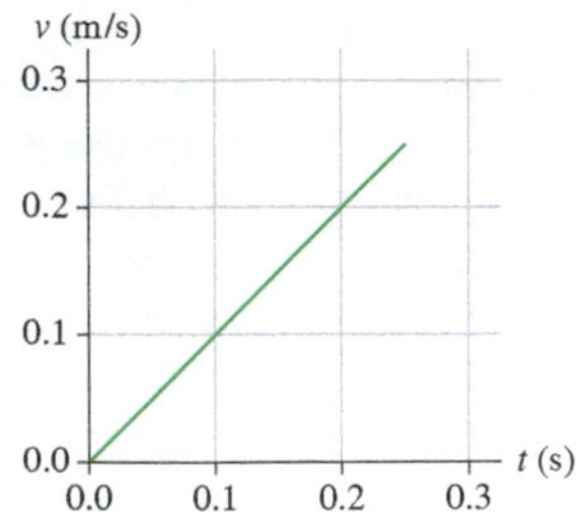


FIGURE P4.20

21. | Figure P4.21 shows an object's acceleration-versus-force graph. What is the object's mass?

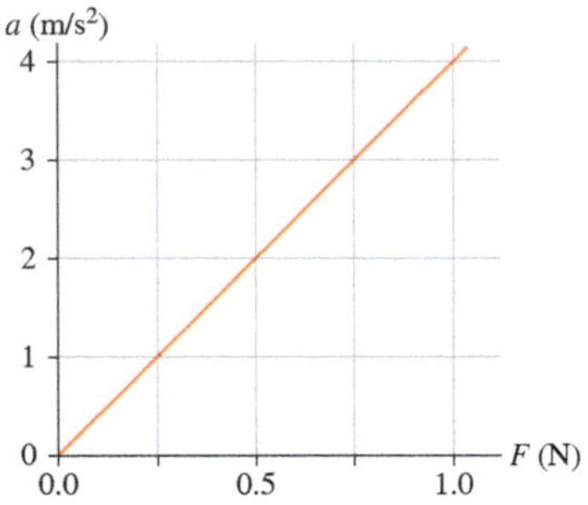


FIGURE P4.21

22. || In a crash test, a car is driven into a solid wall at a speed of 35 mph. Figure P4.22 shows the car's acceleration as it crashes into the wall. What is the maximum force experienced by the 1650 kg car?

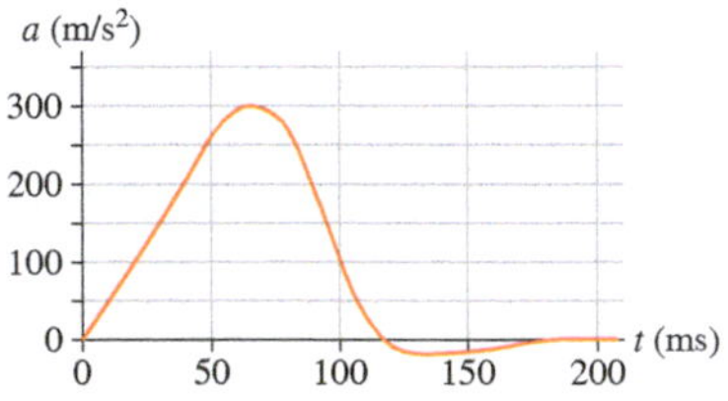


FIGURE P4.22

23. || Two children fight over a 200 g stuffed bear. The 25 kg boy pulls to the right with a 15 N force and the 20 kg girl pulls to the left with a 17 N force. Ignore all other forces on the bear (such as its weight).
    a. At this instant, can you say what the velocity of the bear is? If so, what are the magnitude and direction of the velocity?
    b. At this instant, can you say what the acceleration of the bear is? If so, what are the magnitude and direction of the acceleration?

24. || In t-ball, young players use a bat to hit a stationary ball off a INT stand. The 140 g ball has about the same mass as a baseball, but it is larger and softer. In one hit, the ball leaves the bat at 12 m/s after being in contact with the bat for 2.0 ms. Assume constant acceleration during the hit.
    a. What is the acceleration of the ball?
    b. What is the net force on the ball during the hit?
25. || A 55 kg ice skater is gliding along at 3.5 m/s. Five seconds INT later her speed has dropped to 2.9 m/s. What is the magnitude of the kinetic friction acting on her skates?
26. || The IKAROS spacecraft, launched in 2010, was designed to test the feasibility of *solar sails* for spacecraft propulsion. These large, ultralight sails are pushed on by the force of light from the sun, so the spacecraft doesn't need to carry any fuel. The force on IKAROS's sails was measured to be 1.12 mN. If this were the only force acting on the 290 kg spacecraft, by how much would its speed increase after 6 months of flight?
27. || The *head injury criterion* (HIC) is used to assess the likelihood BIO of head injuries arising from various types of collisions; an HIC greater than about 1000 s is likely to result in severe injuries or even death. The criterion can be written as HIC $= (a_{avg}/g)^{2.5}\Delta t$, where $a_{avg}$ is the average acceleration during the time $\Delta t$ that the head is being accelerated, and $g$ is the free-fall acceleration. Figure P4.27 shows a simplified graph of the net force on a crash dummy's 4.5 kg head as it hits the airbag during a automobile collision. What is the HIC in this collision?

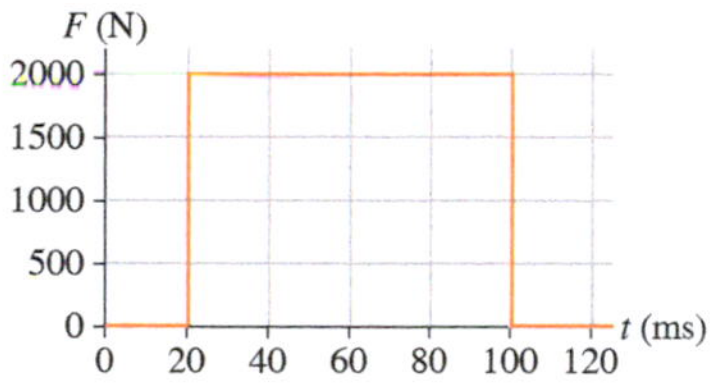


FIGURE P4.27

28. || The particle in Figure P4.28 is acted on by force $\vec{F}_1$ and a second force that is not shown. The particle's acceleration is also shown. Which of the forces $\vec{F}_2 - \vec{F}_4$ could be the second force? (There may be more than one correct answer.)

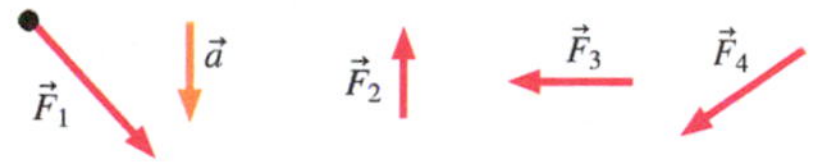


FIGURE P4.28

## Section 4.6 Free-Body Diagrams

Problems 29 through 31 show a free-body diagram. For each problem, (a) redraw the free-body diagram and (b) write a short description of a real object for which this is the correct free-body diagram. Use the situations described in Conceptual Examples 4.1, 4.2, and 4.3 as models of what a description should be like.

29. |

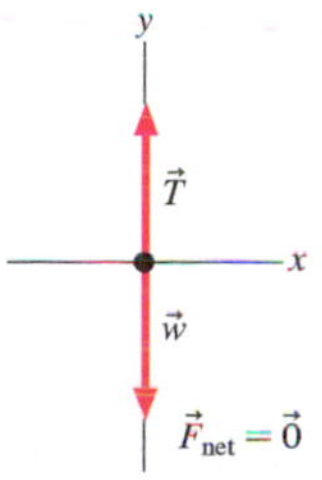


FIGURE P4.29

30. |

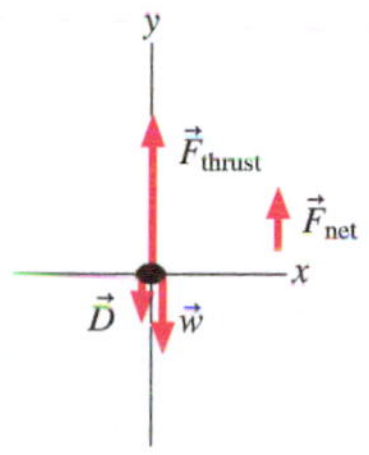


FIGURE P4.30

31. |

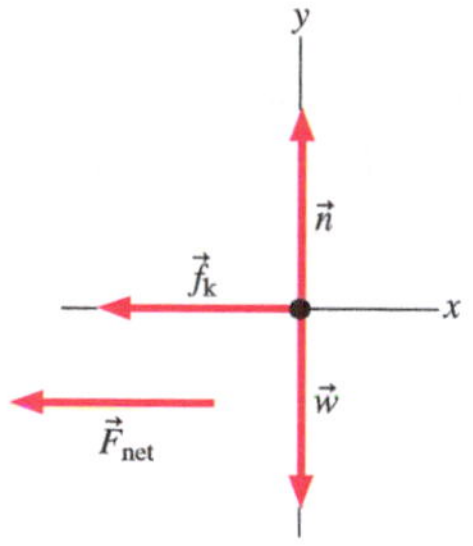


FIGURE P4.31

Problems 32 through 42 describe a situation. For each problem, identify all the forces acting on the object and draw a free-body diagram of the object.

32. You are leaning against a wall.
33. || Your car is accelerating from a stop.
34. || Your car is skidding to a stop from a high speed.
35. || Your physics textbook is sliding across the table.
36. || An ascending elevator, hanging from a cable, is coming to a stop.
37. | You are driving on the highway, and you come to a steep downhill section. As you roll down the hill, you take your foot off the gas pedal. You can ignore friction, but you can't ignore air resistance.
38. || You hold a picture motionless against a wall by pressing on it, as shown in Figure P4.38.

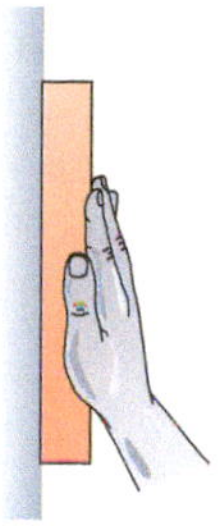

FIGURE P4.38

39. | A box is being dragged across the floor at a constant speed by a rope pulling horizontally on it. Friction is not negligible.
40. | A skydiver has his parachute open and is floating downward through the air at a constant speed.
41. || A cannonball has just been shot out of a cannon aimed 45° above the horizontal. Drag cannot be neglected.
42. || The light turns green and you accelerate rapidly forward so that you and your passenger feel "thrown back" into your seat. Consider your passenger to be the object.

## Section 4.7 Newton's Third Law

43. || Three ice skaters, numbered 1, 2, and 3, stand in a line, each with her hands on the shoulders of the skater in front. Skater 3, at the rear, pushes on skater 2. Identify all the action/reaction pairs of forces between the three skaters. Draw a free-body diagram for skater 2, in the middle. Assume the ice is frictionless.
44. | A girl stands on a sofa. Identify all the action/reaction pairs of forces between the girl and the sofa.
45. || A car is skidding to a stop on a level stretch of road. Identify all the action/reaction pairs of forces between the car and the road surface. Then draw a free-body diagram for the car.
46. ||| Squid use jet propulsion for rapid escapes. A squid pulls BIO water into its body and then rapidly ejects the water backward to propel itself forward. A 1.5 kg squid (not including water mass) can accelerate at 20 m/s$^2$ by ejecting 0.15 kg of water.
    a. What is the magnitude of the thrust force on the squid?
    b. What is the magnitude of the force on the water being ejected?
    c. What acceleration is experienced by the water?

Watch Video Solution Problems 4.53 and 4.65

## General Problems

47. | Redraw the motion diagram shown in Figure P4.47, then draw a vector beside it to show the direction of the net force acting on the object. Explain your reasoning.
INT

FIGURE P4.47

FIGURE P4.48

48. | Redraw the motion diagram shown in Figure P4.48, then draw a vector beside it to show the direction of the net force acting on the object. Explain your reasoning.
INT

49. | Redraw the motion diagram shown in Figure P4.49, then draw a vector beside it to show the direction of the net force acting on the object. Explain your reasoning.
INT

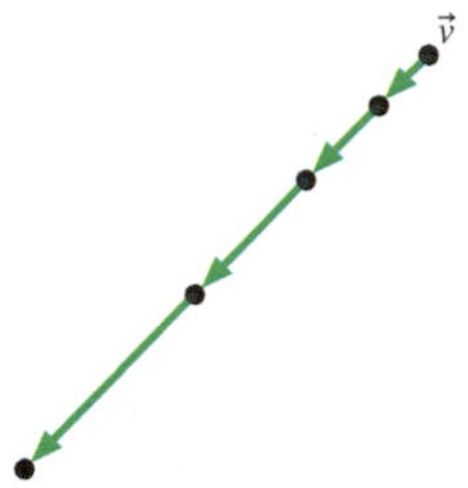

FIGURE P4.49

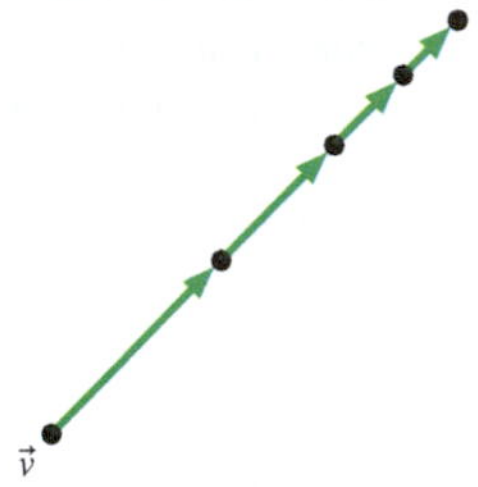

FIGURE P4.50

50. | Redraw the motion diagram shown in Figure P4.50, then draw a vector beside it to show the direction of the net force acting on the object. Explain your reasoning.
INT

51. ||||| A student draws the flawed free-body diagram shown in Figure P4.51 to represent the forces acting on a car traveling at constant speed on a level road. Identify the errors in the diagram, then draw a correct free-body diagram for this situation.

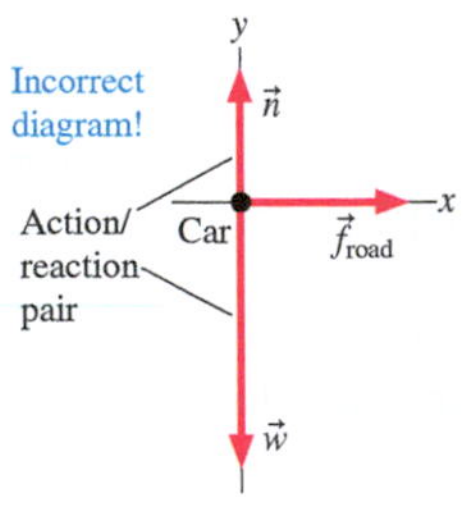


FIGURE P4.51

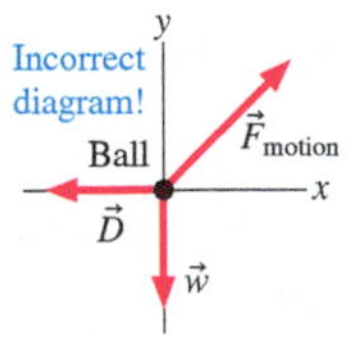


FIGURE P4.52

52. ||| A student draws the flawed free-body diagram shown in Figure P4.52 to represent the forces acting on a golf ball that is traveling upward and to the right a very short time after being hit off the tee. Air resistance is assumed to be relevant. Identify the errors in the diagram, then draw a correct free-body diagram for this situation.

Problems 53 through 65 describe a situation. For each problem, draw a motion diagram, a force identification diagram, and a free-body diagram.

53. || An elevator, suspended by a single cable, has just left the tenth floor and is speeding up as it descends toward the ground floor.
54. ||| A rocket is being launched straight up. Air resistance is not negligible.
55. ||| A jet plane is speeding down the runway during takeoff. Air resistance is not negligible.
56. || You've slammed on the brakes and your car is skidding to a stop while going down a 20° hill.
57. || A cave explorer is being lowered into a vertical shaft by a rope. He is slowing down as he nears the bottom.
58. || A basketball player is getting ready to jump, pushing off the ground and accelerating upward.
59. || A bale of hay sits on the bed of a trailer. The trailer is starting to accelerate forward, and the bale is slipping toward the back of the trailer.
60. ||| In the process of nailing up a heavy framed poster, a student pushes the poster straight in toward the wall; the poster is sliding downward at a constant speed.
61. ||| A spring-loaded gun shoots a plastic ball. The trigger has just been pulled and the ball is starting to move down the barrel. The barrel is horizontal.
62. || A person on a bridge throws a rock straight down toward the water. The rock has just been released.
63. ||| A gymnast has just landed on a trampoline. She's still moving downward as the trampoline stretches.
64. ||| A heavy box is in the back of a truck. The truck is accelerating to the right. Apply your analysis to the box.
65. || A bag of groceries is on the back seat of your car as you stop for a stop light. The bag does not slide. Apply your analysis to the bag.
66. || A car has a mass of 1500 kg. If the driver applies the brakes while on a gravel road, the maximum friction force that the tires can provide without skidding is about 7000 N. If the car is moving at 20 m/s, what is the shortest distance in which the car can stop safely?
67. ||| Researchers have measured the acceleration of racing greyhounds as a function of their speed; a simplified version of their results is shown in Figure P4.67. The acceleration at low speeds is constant and is limited by the fact that any greater acceleration would result in the dog pitching forward because of the force acting on its hind legs during its power stroke. At higher speeds, the dog's acceleration is limited by the maximum power its muscles can provide.
BIO
INT
    a. What is the agent of the force that causes the dog to accelerate?
    b. If the dog's mass is 32 kg, what is the average force acting on it during its initial acceleration phase?
    c. How far does the dog run in the first 4.0 s?

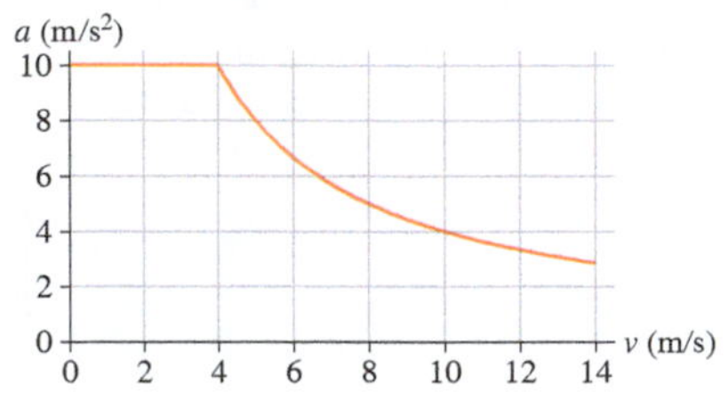


FIGURE P4.67

68. |||| In researching the forces that act on flying birds, scientists studied the motion of cockatiels' bodies as they fly in horizontal flight. As the birds beat their wings, their bodies move up and down. Figure P4.68 shows the vertical acceleration of a cockatiel's body during one wing beat, consisting of an upstroke followed by a downstroke.
BIO

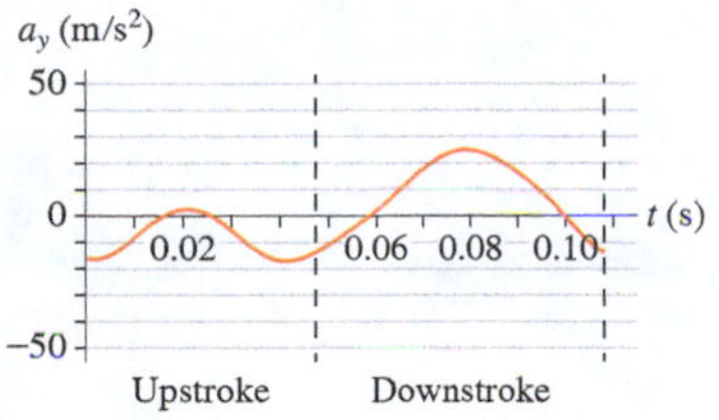


FIGURE P4.68

a. If the bird's mass is 87 g, what is the maximum net vertical force exerted on the bird?

b. The bird is held in level flight due to the force exerted on it by the air as the bird beats its wings. What is the maximum value of this force due to the air?

69. |||| The fastest pitched baseball was clocked at 47 m/s. If the pitcher exerted his force (assumed to be horizontal and constant) over a distance of 1.0 m, and a baseball has a mass of 145 g,
BIO

a. Draw a free-body diagram of the ball during the pitch.

b. What force did the pitcher exert on the ball during this record-setting pitch?

c. Estimate the force in part b as a fraction of the pitcher's weight.

70. ||| The trap-jaw ant, found throughout tropical South America, catches its prey by *very* rapidly closing its mandibles around its victim. Figure P4.70 shows the speed of one of its mandible jaws versus time in *micro*seconds.
BIO INT

a. What is the maximum acceleration of the ant's mandible?

b. The mass of a trap-jaw ant's mandible has been estimated to be about $1.3 \times 10^{-7}$ kg. Estimate the maximum force exerted by one mandible.

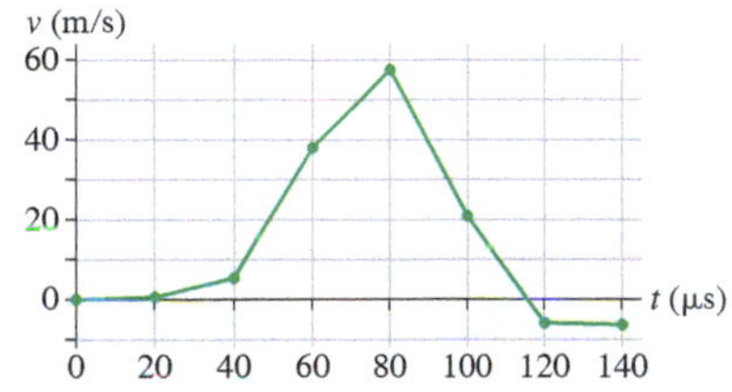


FIGURE P4.70

71. || The jumping ability of the African desert locust was measured by placing the insect on a *force plate,* a platform that can accurately measure the force that acts on it. When the locust jumped straight up, its acceleration was measured to follow the curve in Figure P4.71. What was the maximum force that this 0.50 g locust exerted on the force plate?
BIO

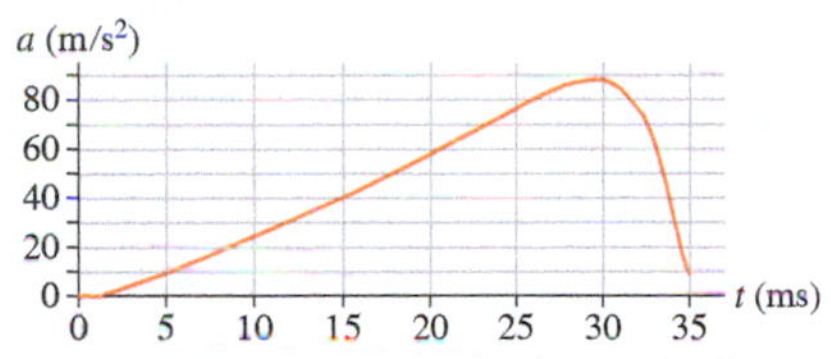


FIGURE P4.71

72. | The froghopper, champion leaper of the insect world, can jump straight up at 4.0 m/s. The jump itself lasts a mere 1.0 ms before the insect is clear of the ground.
BIO

a. Draw a free-body diagram of this mighty leaper while the jump is taking place.

b. While the jump is taking place, is the force that the ground exerts on the froghopper greater than, less than, or equal to the insect's weight? Explain.

73. || A beach ball is thrown straight up, and some time later it lands on the sand. Is the magnitude of the net force on the ball greater when it is going up or when it is on the way down? Or is it the same in both cases? Explain. Air resistance should not be neglected for a large, light object.

## MCAT-Style Passage Problems

### A Simple Solution for a Stuck Car

If your car is stuck in the mud and you don't have a winch to pull it out, you can use a piece of rope and a tree to do the trick. First, you tie one end of the rope to your car and the other to a tree, then pull as hard as you can on the middle of the rope, as shown in Figure P4.74a. This technique applies a force to the car much larger than the force that you can apply directly. To see why the car experiences such a large force, look at the forces acting on the center point of the rope, as shown in Figure P4.74b. The sum of the forces is zero, thus the tension is much greater than the force you apply. It is this tension force that acts on the car and, with luck, pulls it free.

74. | The sum of the three forces acting on the center point of the rope is assumed to be zero because

A. This point has a very small mass.
B. Tension forces in a rope always cancel.
C. This point is not accelerating.
D. The angle of deflection is very small.

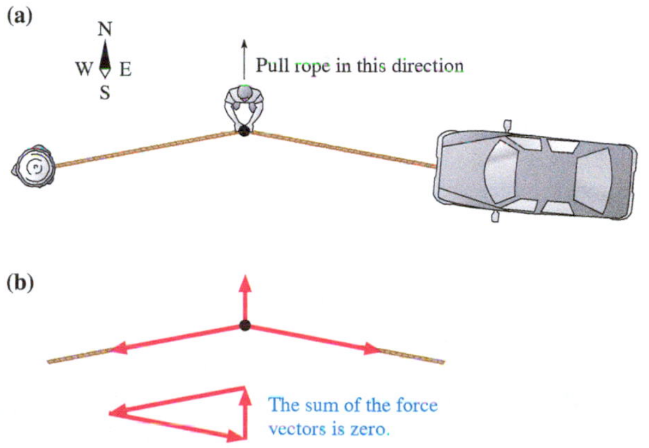


FIGURE P4.74

75. | When you are pulling on the rope as shown, what is the approximate direction of the tension force on the tree?

A. North B. South
C. East D. West

76. | Assume that you are pulling on the rope but the car is not moving. What is the approximate direction of the force of the mud on the car?

A. North B. South
C. East D. West

77. | Suppose your efforts work, and the car begins to move forward out of the mud. As it does so, the force of the car on the rope is

A. Zero.
B. Less than the force of the rope on the car.
C. Equal to the force of the rope on the car.
D. Greater than the force of the rope on the car.

# 5 Applying Newton's Laws

The sifaka is a type of lemur, found only in Madagascar, that is capable of spectacular leaps from tree to tree. Why does the sifaka spread out its arms and legs this way while making a leap?

## LOOKING AHEAD ➤

### Working with Forces

In this chapter you'll learn expressions for the different forces we've seen, and you'll learn how to use them to solve problems.

You'll learn how a balance between weight and drag forces leads to a maximum speed for a skydiver.

### Equilibrium Problems

The boy is pushing as hard as he can, but the sofa isn't going anywhere. It's in **equilibrium**—the sum of the forces on it is zero.

You'll learn to solve equilibrium problems by using the fact that there is no net force.

### Dynamics Problems

Newton's laws allow us to relate the forces acting on an object to its motion, and so to solve a wide range of **dynamics** problems.

This skier is picking up speed. You'll see how her acceleration is determined by the forces acting on her.

**GOAL** **To use Newton's laws to solve equilibrium and dynamics problems.**

## LOOKING BACK ◀

### Free-Body Diagrams

In Section 4.6 you learned to draw a free-body diagram showing the magnitudes and directions of the forces acting on an object.

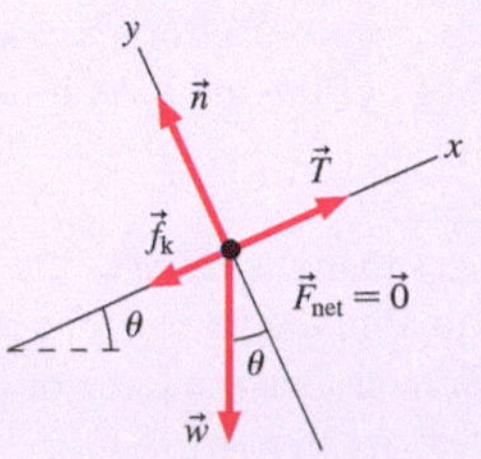


In this chapter, you'll use free-body diagrams as an essential problem-solving tool for single objects and interacting objects.

**STOP TO THINK**

An elevator is suspended from a cable. It is moving upward at a steady speed. Which is the correct free-body diagram for this situation?

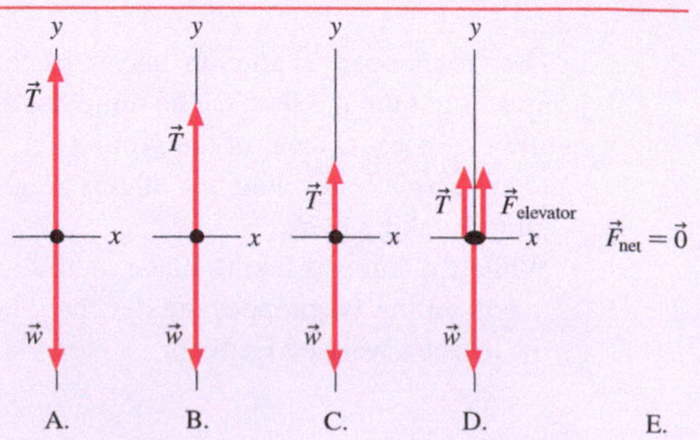

## 5.1 Equilibrium

In this chapter, we will use Newton's laws to solve force and motion problems. Before turning to a general treatment of such motion, let's examine the two simplest cases, in which the object is either at rest or moving in a straight line at a constant speed. Note that in both cases the object is not accelerating, so $\vec{a} = \vec{0}$.

We say that an object at rest is in **static equilibrium**; an object that is moving in a straight line at a constant speed is said to be in **dynamic equilibrium**. Because $\vec{a} = \vec{0}$ for both cases, Newton's second law $\vec{F}_{\text{net}} = m\vec{a}$ tells us that for an object to be in equilibrium, the net force $\vec{F}_{\text{net}}$ acting on it must be zero: $\vec{F}_{\text{net}} = \vec{0}$. Recall that $\vec{F}_{\text{net}}$ is the vector sum

$$\vec{F}_{\text{net}} = \vec{F}_1 + \vec{F}_2 + \vec{F}_3 + \cdots$$

where $\vec{F}_1$, $\vec{F}_2$, and so on are the individual forces, such as tension or friction, acting on the object. We found in ◀◀ SECTION 3.2 that vector sums can be evaluated in terms of the $x$- and $y$-components of the vectors; that is, the $x$-component of the net force is $(F_{\text{net}})_x = F_{1x} + F_{2x} + F_{3x} + \cdots$. If we restrict ourselves to problems where all the forces are in the $xy$-plane, then the equilibrium requirement $\vec{F}_{\text{net}} = \vec{0}$ is a shorthand way of writing two simultaneous equations:

$$(F_{\text{net}})_x = F_{1x} + F_{2x} + F_{3x} + \cdots = 0$$
$$(F_{\text{net}})_y = F_{1y} + F_{2y} + F_{3y} + \cdots = 0$$

Recall from your math classes that the Greek letter $\sum$ (sigma) stands for "the sum of." It will be convenient to abbreviate the sum of the $x$-components of all forces as

$$F_{1x} + F_{2x} + F_{3x} + \cdots = \sum F_x$$

With this notation, Newton's second law for an object in equilibrium, with $\vec{a} = \vec{0}$, can be written as the two equations

$$\sum F_x = ma_x = 0 \quad \text{and} \quad \sum F_y = ma_y = 0 \tag{5.1}$$

In equilibrium, the sums of the $x$- and $y$-components of the force are zero.

These equations are the basis for a general approach for solving equilibrium problems.

This human tower is in equilibrium because the net force on each man is zero.

### PROBLEM-SOLVING APPROACH 5.1 Equilibrium problems

***STRATEGIZE*** If an object is in equilibrium, the net force acting on it must be zero. We will use this fact to find the forces that keep it in equilibrium.

***PREPARE*** First check that the object is in equilibrium: Does $\vec{a} = \vec{0}$?

- An object at rest is in static equilibrium.
- An object moving at a constant velocity is in dynamic equilibrium.

Then identify all forces acting on the object and show them on a free-body diagram. Determine which forces you know and which you need to solve for.

***SOLVE*** An object in equilibrium must satisfy Newton's second law for the case where $\vec{a} = \vec{0}$. In component form, the requirement is

$$\sum F_x = ma_x = 0 \quad \text{and} \quad \sum F_y = ma_y = 0$$

You can find the force components that go into these sums directly from your free-body diagram. From these two equations, solve for the unknown forces in the problem.

***ASSESS*** Check that your result has the correct units, is reasonable, and answers the question.

The following examples show how to apply this Problem-Solving Approach to both static and dynamic equilibrium problems.

## Static Equilibrium

### EXAMPLE 5.1 Finding the forces on an orangutan

An orangutan weighing 500 N hangs from a vertical rope. What is the tension in the rope?

**STRATEGIZE** The orangutan is at rest, so it is in static equilibrium. The net force on it must then be zero. We will use this fact to find the tension.

**PREPARE** FIGURE 5.1 identifies the forces acting on the orangutan: the upward force of the tension in the rope and the downward, long-range force of gravity (the orangutan's weight $\vec{w}$).

FIGURE 5.1 The forces on an orangutan.

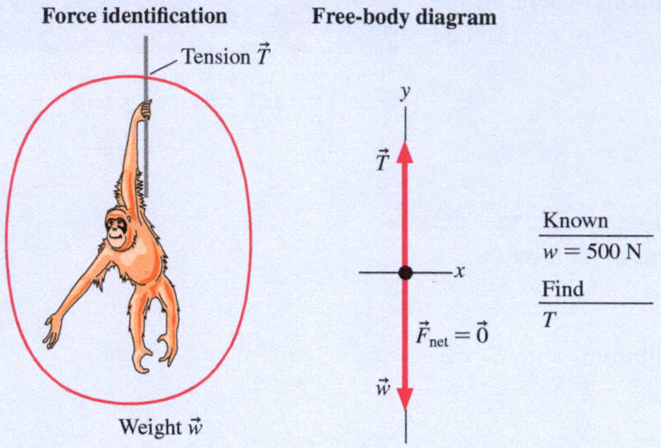


These forces are then shown on a free-body diagram, where it's noted that equilibrium requires $\vec{F}_{net} = \vec{0}$.

**SOLVE** Neither force has an $x$-component, so we need to examine only the $y$-components of the forces. In this case, the $y$-component of Newton's second law is

$$\sum F_y = T_y + w_y = ma_y = 0$$

You might have been tempted to write $T_y - w_y$ because the weight force points down. But remember that $T_y$ and $w_y$ are *components* of vectors and can thus be positive (for a vector such as $\vec{T}$ that points up) or negative (for a vector such as $\vec{w}$ that points down). The fact that $\vec{w}$ points down is taken into account when we *evaluate* the components—that is, when we write them in terms of the *magnitudes* $T$ and $w$ of the vectors $\vec{T}$ and $\vec{w}$.

Because the tension vector $\vec{T}$ points straight up, in the positive $y$-direction, its $y$-component is $T_y = T$. Because the weight vector $\vec{w}$ points straight down, in the negative $y$-direction, its $y$-component is $w_y = -w$. This is where the signs enter. With these components, Newton's second law becomes

$$T - w = 0$$

This equation is easily solved for the tension in the rope:

$$T = w = 500 \text{ N}$$

**ASSESS** It's not surprising that the tension in the rope equals the weight of the orangutan. That gives us confidence in our solution.

### EXAMPLE 5.2 Readying a wrecking ball

A wrecking ball weighing 2500 N hangs from a cable. Prior to swinging, it is pulled back to a 20° angle by a second, horizontal cable. What is the tension in the horizontal cable?

**STRATEGIZE** Because the ball is not yet moving, this is a static equilibrium problem; the net force acting on the ball must be zero.

**PREPARE** In FIGURE 5.2, we start by identifying all the forces acting on the ball: a tension force from each cable and the ball's weight. We've used different symbols $\vec{T}_1$ and $\vec{T}_2$ for the two different tension forces. We then construct a free-body diagram for these three forces, noting that $\vec{F}_{net} = m\vec{a} = \vec{0}$. We're looking for the magnitude $T_1$ of the tension force $\vec{T}_1$ in the horizontal cable.

FIGURE 5.2 Visual overview of a wrecking ball just before release.

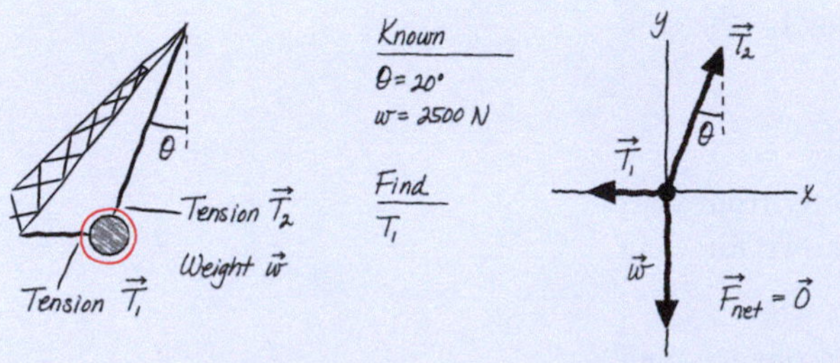


**SOLVE** The requirement of equilibrium is $\vec{F}_{net} = m\vec{a} = \vec{0}$. In component form, we have the two equations:

$$\sum F_x = T_{1x} + T_{2x} + w_x = ma_x = 0$$

$$\sum F_y = T_{1y} + T_{2y} + w_y = ma_y = 0$$

As always, we *add* the force components together. Now we're ready to write the components of each force vector in terms of the magnitudes and directions of those vectors. We learned how to do this in Section 3.3. With practice you'll learn to read the components directly off the free-body diagram, but to begin it's worthwhile to organize the components into a table.

| Force | Name of $x$-component | Value of $x$-component | Name of $y$-component | Value of $y$-component |
|---|---|---|---|---|
| $\vec{T}_1$ | $T_{1x}$ | $-T_1$ | $T_{1y}$ | 0 |
| $\vec{T}_2$ | $T_{2x}$ | $T_2 \sin\theta$ | $T_{2y}$ | $T_2 \cos\theta$ |
| $\vec{w}$ | $w_x$ | 0 | $w_y$ | $-w$ |

We see from the free-body diagram that $\vec{T}_1$ points along the negative $x$-axis, so $T_{1x} = -T_1$ and $T_{1y} = 0$. We need to be careful with our trigonometry as we find the components of $\vec{T}_2$. Remembering that the side adjacent to the angle is related to the cosine, we see that the vertical ($y$) component of $\vec{T}_2$ is $T_2 \cos\theta$. Similarly, the horizontal ($x$) component is $T_2 \sin\theta$. The weight vector points straight down, so its $y$-component is $-w$. Notice that negative signs enter as we evaluate the components of the vectors, *not* when we write Newton's second law. This is a critical aspect of solving force and motion problems. With these components, Newton's second law now becomes

$$-T_1 + T_2 \sin\theta + 0 = 0 \quad \text{and} \quad 0 + T_2 \cos\theta - w = 0$$

We can rewrite these equations as

$$T_1 = T_2 \sin\theta \quad \text{and} \quad T_2 \cos\theta = w$$

These are two simultaneous equations with two unknowns: $T_1$ and $T_2$. To eliminate $T_2$ from the two equations, we solve the second equation for $T_2$, giving $T_2 = w/\cos\theta$. Then we insert this expression for $T_2$ into the first equation to get

$$T_1 = \frac{w}{\cos\theta}\sin\theta = \frac{\sin\theta}{\cos\theta} w = w\tan\theta = (2500\ \text{N})\tan 20° = 910\ \text{N}$$

where we made use of the fact that $\tan\theta = \sin\theta/\cos\theta$.

**ASSESS** It seems reasonable that to pull the ball back to this modest angle, a force substantially less than the ball's weight is required.

## CONCEPTUAL EXAMPLE 5.3 Forces in static equilibrium

A rod is free to slide on a frictionless sheet of ice. One end of the rod is lifted by a string. If the rod is at rest, which diagram in FIGURE 5.3 shows the correct angle of the string?

FIGURE 5.3 Which is the correct angle of the string?

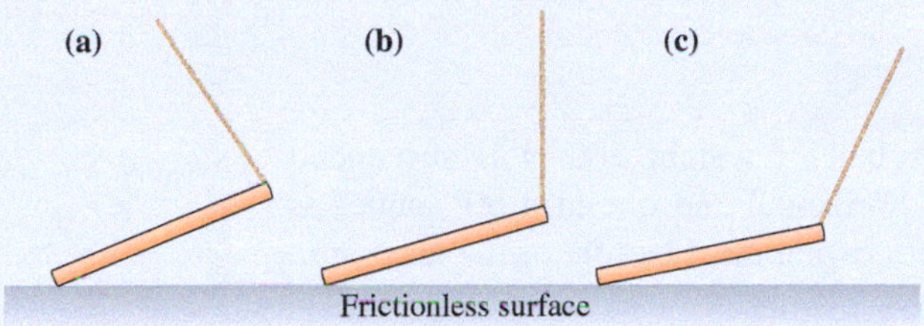


**REASON** Let's start by identifying the forces that act on the rod. In addition to the weight force, the string exerts a tension force and the ice exerts an upward normal force. What can we say about these forces? If the rod is to hang motionless, it must be in static equilibrium with $\sum F_x = ma_x = 0$ and $\sum F_y = ma_y = 0$. FIGURE 5.4 shows free-body diagrams for the three string orientations. Remember that tension always acts along the direction of the string and that the weight force always points straight down. The ice pushes up with a normal force perpendicular to the surface, but frictionless ice cannot exert any horizontal force. If the string is angled, we see that its horizontal component exerts a net force on the rod. Only in case b, where the tension and the string are vertical, can the net force be zero.

FIGURE 5.4 Free-body diagrams for three angles of the string.

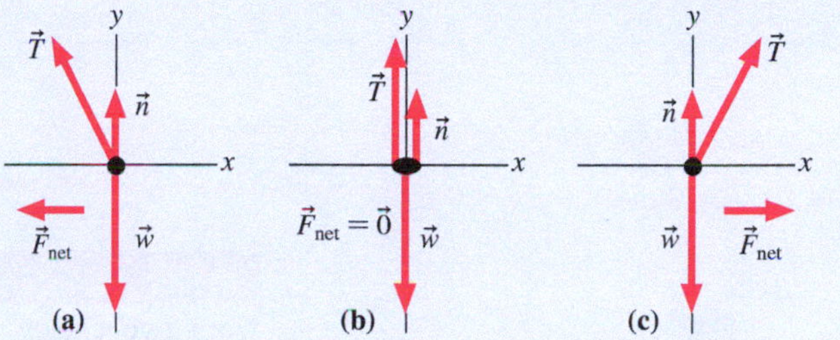


**ASSESS** If friction were present, the rod could in fact hang as in cases a and c. But without friction, the rods in these cases would slide until they came to rest as in case b.

# Dynamic Equilibrium

## EXAMPLE 5.4 Finding the tension in a rope while towing a car

A car with a mass of 1500 kg is being towed at a steady speed by a rope held at a 20° angle from the horizontal. A friction force of 320 N opposes the car's motion. What is the tension in the rope?

**STRATEGIZE** The car is moving in a straight line at a constant speed ($\vec{a} = \vec{0}$) so it is in dynamic equilibrium and must have $\vec{F}_{net} = m\vec{a} = \vec{0}$.

**PREPARE** FIGURE 5.5 shows the three contact forces acting on the car—the tension force $\vec{T}$, friction $\vec{f}$, and the normal force $\vec{n}$—and the long-range force of gravity $\vec{w}$. These four forces are shown on the free-body diagram.

FIGURE 5.5 Visual overview of a car being towed.

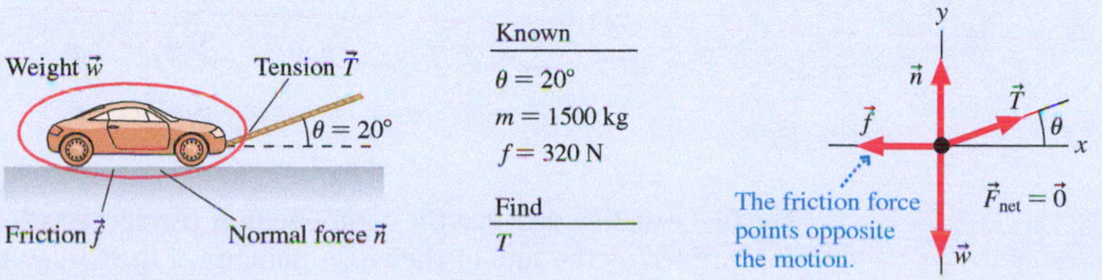


*Continued*

**SOLVE** This is still an equilibrium problem, even though the car is moving, so our problem-solving procedure is unchanged. With four forces, the requirement of equilibrium is

$$\sum F_x = n_x + T_x + f_x + w_x = ma_x = 0$$

$$\sum F_y = n_y + T_y + f_y + w_y = ma_y = 0$$

We can again determine the horizontal and vertical components of the forces by "reading" the free-body diagram. The results are shown in the table.

| Force | Name of *x*-component | Value of *x*-component | Name of *y*-component | Value of *y*-component |
|---|---|---|---|---|
| $\vec{n}$ | $n_x$ | 0 | $n_y$ | $n$ |
| $\vec{T}$ | $T_x$ | $T\cos\theta$ | $T_y$ | $T\sin\theta$ |
| $\vec{f}$ | $f_x$ | $-f$ | $f_y$ | 0 |
| $\vec{w}$ | $w_x$ | 0 | $w_y$ | $-w$ |

With these components, Newton's second law becomes

$$T\cos\theta - f = 0$$

$$n + T\sin\theta - w = 0$$

The first equation can be used to solve for the tension in the rope:

$$T = \frac{f}{\cos\theta} = \frac{320\text{ N}}{\cos 20°} = 340\text{ N}$$

to two significant figures. It turned out that we did not need the *y*-component equation in this problem. We would need it if we wanted to find the normal force $\vec{n}$.

**ASSESS** Had we pulled the car with a horizontal rope, the tension would need to exactly balance the friction force of 320 N. Because we are pulling at an angle, however, part of the tension in the rope pulls *up* on the car instead of in the forward direction. Thus we need a little more tension in the rope when it's at an angle, so our result seems reasonable.

**STOP TO THINK 5.1** A ball of weight 200 N is suspended from two cables, one horizontal and one at a 60° angle, as shown. Which of the following must be true of the tension in the angled cable?

A. $T > 200$ N  B. $T = 200$ N  C. $T < 200$ N

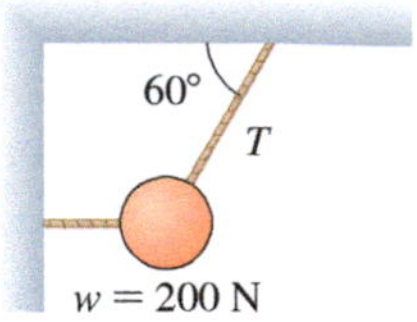


## 5.2 Dynamics and Newton's Second Law

Newton's second law is the essential link between force and motion. The essence of Newtonian mechanics can be expressed in two steps:

- The forces acting on an object determine its acceleration $\vec{a} = \vec{F}_{net}/m$.
- The object's motion can be found by using $\vec{a}$ in the equations of kinematics.

A video to support a section's topic is embedded in the eText.

**Video** Forces in Jumping

We want to develop an approach to solve a variety of problems in mechanics, but first we need to write the second law in terms of its components. To do so, let's first rewrite Newton's second law in the form

$$\vec{F}_{net} = \vec{F}_1 + \vec{F}_2 + \vec{F}_3 + \cdots = m\vec{a}$$

where $\vec{F}_1, \vec{F}_2, \vec{F}_3$, and so on are the forces acting on an object. To write the second law in component form merely requires that we use the *x*- and *y*-components of the acceleration. Thus Newton's second law, $\vec{F}_{net} = m\vec{a}$, is

$$\sum F_x = ma_x \quad \text{and} \quad \sum F_y = ma_y \tag{5.2}$$

Newton's second law in component form

The first equation says that **the component of the acceleration in the *x*-direction is determined by the sum of the *x*-components of the forces acting on the object.** A similar statement applies to the *y*-direction.

PROBLEM-SOLVING APPROACH 5.2

## Dynamics problems

Dynamics problems use Newton's second law as the connection between forces and kinematics.

**STRATEGIZE** There are two basic kinds of dynamics problems. If the forces on the object are known, you can use Newton's second law to find the acceleration and then, from kinematics, the object's position and velocity. In other cases, you can calculate the object's acceleration from kinematics; using the acceleration, you can apply Newton's second law to find the forces acting on the object. In both kinds of problems the approach to the solution is the same.

**PREPARE** Sketch a visual overview consisting of:

- A list of values that identifies known quantities and what the problem is trying to find.
- A force identification diagram to help you identify all the forces acting on the object.
- A free-body diagram that shows all the forces acting on the object.

If you'll need to use kinematics to find velocities or positions, you'll also need to sketch:

- A motion diagram to determine the direction of the acceleration.
- A pictorial representation that establishes a coordinate system, shows important points in the motion, and defines symbols.

It's OK to go back and forth between these steps as you visualize the situation.

**SOLVE** Write Newton's second law in component form as

$$\sum F_x = ma_x \qquad \text{and} \qquad \sum F_y = ma_y$$

You can find the components of the forces directly from your free-body diagram. Depending on the problem, either:

- Solve for the acceleration, then use kinematics to find velocities and positions.
- Use kinematics to determine the acceleration, then solve for unknown forces.

**ASSESS** Check that your result has the correct units, is reasonable, and answers the question.

Exercise 22

EXAMPLE 5.5

## Putting a golf ball

A golfer putts a 46 g ball with a speed of 3.0 m/s. Friction exerts a 0.020 N retarding force on the ball, slowing it down. Will her putt reach the hole, 10 m away?

**STRATEGIZE** We are given the friction force on the ball, which is an indication that we should first find the ball's acceleration using Newton's second law, then use kinematics to solve for the distance the ball rolls.

**PREPARE** FIGURE 5.6 is a visual overview of the problem. We've collected the known information, drawn a sketch, and identified what we want to find. The motion diagram shows that the ball is

FIGURE 5.6 Visual overview of a golf putt.

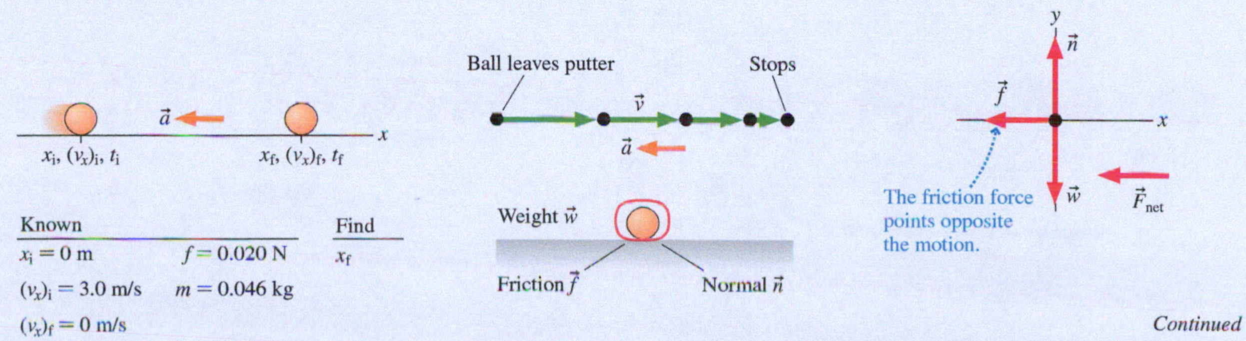


*Continued*

slowing down as it rolls to the right, so the acceleration vector points to the left. Next, we identify the forces acting on the ball and show them on a free-body diagram. Note that the net force points to the left, as it must because the acceleration points to the left.

**SOLVE** Newton's second law in component form is

$$\sum F_x = n_x + f_x + w_x = 0 - f + 0 = ma_x$$
$$\sum F_y = n_y + f_y + w_y = n + 0 - w = ma_y = 0$$

We've written the equations as sums, as we did with equilibrium problems, then "read" the values of the force components from the free-body diagram. The components are simple enough in this problem that we don't really need to show them in a table. It is particularly important to notice that we set $a_y = 0$ in the second equation. This is because the ball does not move in the $y$-direction, so it can't have any acceleration in the $y$-direction. This will be an important step in many problems.

The first equation is $-f = ma_x$, from which we find

$$a_x = -\frac{f}{m} = \frac{-(0.020\ \text{N})}{0.046\ \text{kg}} = -0.435\ \text{m/s}^2$$

To avoid rounding errors we keep an extra significant figure in this intermediate step in the calculation. The negative sign shows that the acceleration is directed to the left, as expected.

Now that we know the acceleration, we can use kinematics to find how far the ball will roll before stopping. We don't have any information about the time it takes for the ball to stop, so we'll use the kinematic equation $(v_x)_f^2 = (v_x)_i^2 + 2a_x(x_f - x_i)$. This gives

$$x_f = x_i + \frac{(v_x)_f^2 - (v_x)_i^2}{2a_x} = 0\ \text{m} + \frac{(0\ \text{m/s})^2 - (3.0\ \text{m/s})^2}{2(-0.435\ \text{m/s}^2)}$$
$$= 10.3\ \text{m}$$

If her aim is true, the ball will just make it into the hole.

**ASSESS** It seems reasonable that a ball putted on grass with an initial speed of 3 m/s—about jogging speed—would travel roughly 10 m.

## EXAMPLE 5.6 Towing a car with acceleration

A car with a mass of 1500 kg is being towed by a rope held at a 20° angle to the horizontal. A friction force of 320 N opposes the car's motion. What is the tension in the rope if the car goes from rest to 12 m/s in 10 s?

**STRATEGIZE** As you may have recognized, this problem is almost identical to Example 5.4. The difference is that the car is now accelerating, so it is no longer in equilibrium. We will first use kinematics to find the car's acceleration. Then we will use this acceleration in Newton's second law to find the tension in the rope.

**PREPARE** FIGURE 5.7 is a visual overview of the problem showing that the car's acceleration is directed to the right. The force identification diagram is the same as in Example 5.4, but the free-body diagram now indicates that there is a net force directed to the right, in the same direction as the acceleration.

**SOLVE** Newton's second law in component form is

$$\sum F_x = n_x + T_x + f_x + w_x = ma_x$$
$$\sum F_y = n_y + T_y + f_y + w_y = ma_y = 0$$

We've again used the fact that $a_y = 0$ for motion that is purely along the $x$-axis. The components of the forces were worked out in Example 5.4. With that information, Newton's second law in component form is

$$T\cos\theta - f = ma_x$$
$$n + T\sin\theta - w = 0$$

Because the car speeds up from rest to 12 m/s in 10 s, we can use kinematics to find the acceleration:

$$a_x = \frac{\Delta v_x}{\Delta t} = \frac{(v_x)_f - (v_x)_i}{t_f - t_i} = \frac{(12\ \text{m/s}) - (0\ \text{m/s})}{(10\ \text{s}) - (0\ \text{s})} = 1.2\ \text{m/s}^2$$

We can now use the first Newton's-law equation above to solve for the tension. We have

$$T = \frac{ma_x + f}{\cos\theta} = \frac{(1500\ \text{kg})(1.2\ \text{m/s}^2) + 320\ \text{N}}{\cos 20°} = 2300\ \text{N}$$

**ASSESS** The tension is substantially greater than the 340 N found in Example 5.4. It takes much more force to accelerate the car than to keep it rolling at a constant speed.

FIGURE 5.7 Visual overview of a car being towed.

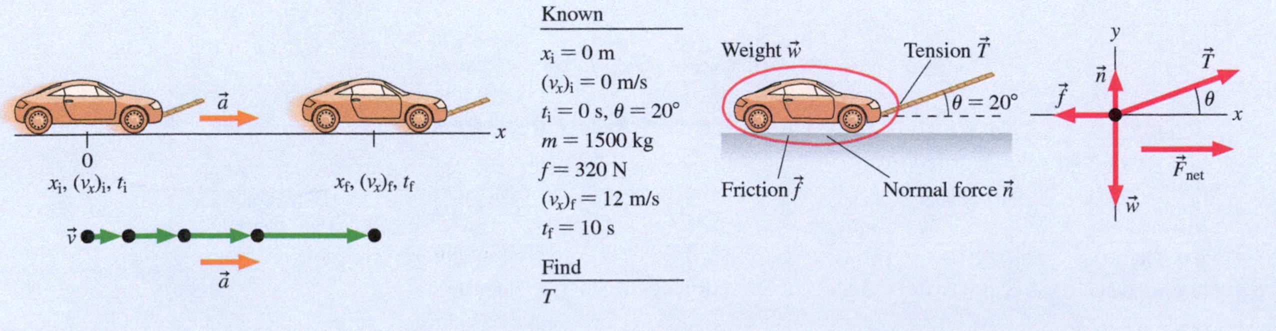

These first examples have shown all the details of our problem-solving approach. Our purpose has been to demonstrate how the approach is put into practice. Future examples will be briefer, but the basic *procedure* will remain the same.

**STOP TO THINK 5.2** A Martian lander is approaching the surface. It is slowing its descent by firing its rocket motor. Which is the correct free-body diagram for the lander?

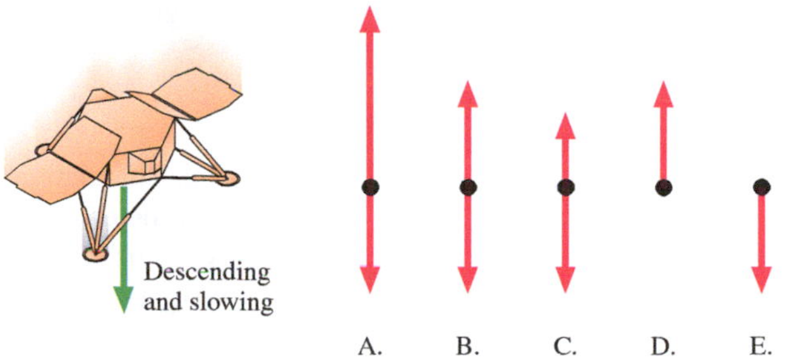


## 5.3 Mass and Weight

When the doctor asks what you weigh, what does she really mean? We do not make much distinction in our ordinary use of language between the terms "weight" and "mass," but in physics their distinction is of critical importance.

Mass, you'll recall from Chapter 4, is a quantity that describes an object's inertia, its tendency to resist being accelerated. Loosely speaking, it also describes the amount of matter in an object. Mass, measured in kilograms, is an intrinsic property of an object; it has the same value wherever the object may be and whatever forces might be acting on it.

Weight, however, is a *force*. Specifically, it is the gravitational force exerted on an object by a planet. Weight is a vector, not a scalar, and the vector's direction is always straight down. Weight is measured in newtons.

**Mass and weight are not the same thing,** but they are related. FIGURE 5.8 shows the free-body diagram of an object in free fall. The *only* force acting on this object is its weight $\vec{w}$, the downward pull of gravity. The object is in free fall, so, as we saw in ◀◀ SECTION 2.7, the acceleration is vertical, with $a_y = -g$, where $g$ is the free-fall acceleration, 9.80 m/s$^2$. Newton's second law for this object is thus

$$\sum F_y = -w = -mg$$

which tells us that

$$w = mg \tag{5.3}$$

The magnitude of the weight force, which we call simply "the weight," is directly proportional to the mass, with $g$ as the constant of proportionality.

FIGURE 5.8 The free-body diagram of an object in free fall.

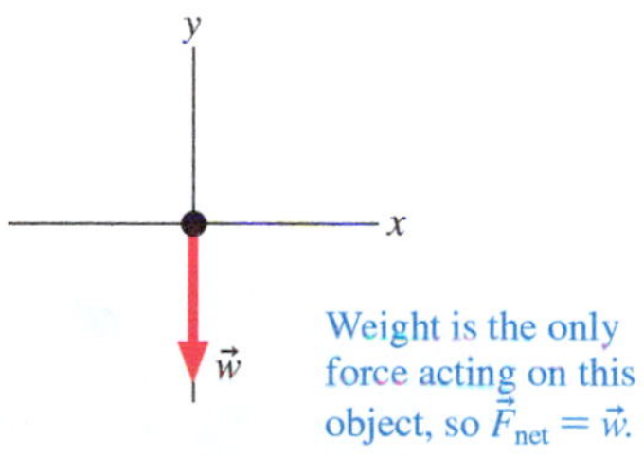


**NOTE** ▶ Although we derived the relationship between mass and weight for an object in free fall, the weight of an object is *independent* of its state of motion. Equation 5.3 holds for an object at rest on a table, sliding horizontally, or moving in any other way. ◀

Because an object's weight depends on $g$, and the value of $g$ varies from planet to planet, weight is not a fixed, constant property of an object. The value of $g$ at the surface of the moon is about one-sixth its earthly value, so an object on the moon would have only one-sixth its weight on earth. The object's weight on Jupiter would be greater than its weight on earth. Its mass, however, would be the same. The amount of matter has not changed, only the gravitational force exerted on that matter.

So, when the doctor asks what you weigh, she really wants to know your *mass*. That's the amount of matter in your body. You can't really "lose weight" by going to the moon, even though you would weigh less there!

On the moon, astronaut John Young jumped 2 feet straight up, despite his spacesuit that weighs 370 pounds on earth. On the moon, where $g = 1.6$ m/s$^2$, he and his suit together weighed only 90 pounds.

We need to make a clarification here. When you give your weight, you most likely give it in pounds, which is the unit of force in the English system. (We noted in Chapter 4 that the pound is defined as 1 lb = 4.45 N.) You might then "convert" this to kilograms. But a kilogram is a unit of mass, not a unit of force. An object that weighs 1 pound, meaning $w = mg = 4.45$ N, has a mass of

$$m = \frac{w}{g} = \frac{4.45 \text{ N}}{9.80 \text{ m/s}^2} = 0.454 \text{ kg}$$

This calculation is different from converting, for instance, feet to meters. Both feet and meters are units of length, and it's always true that 1 m = 3.28 ft. When you "convert" from pounds to kilograms, you are determining the mass that has a certain weight—two fundamentally different quantities—and this calculation depends on the value of $g$. But we are usually working on the earth, where we assume that $g = 9.80 \text{ m/s}^2$. In this case, a given mass always corresponds to the same weight, and we can use the relationships listed in TABLE 5.1.

**TABLE 5.1** Mass, weight, force

| |
|---|
| **Conversion between force units:** |
| 1 pound = 4.45 N |
| 1 N = 0.225 pound |
| **Correspondence between mass and weight, assuming $g = 9.80 \text{ m/s}^2$:** |
| 1 kg ↔ 2.20 lb |
| 1 lb ↔ 0.454 kg = 454 g |

**EXAMPLE 5.7 Typical masses and weights**

What are the weight, in N, and the mass, in kg, of a 90 pound gymnast, a 150 pound professor, and a 240 pound football player?

***STRATEGIZE*** Weight and mass are related by Equation 5.3.

***PREPARE*** We can use the conversions and correspondences in Table 5.1.

***SOLVE*** We use the correspondence between mass and weight just as we use the conversion factor between different forces:

$$w_{\text{gymnast}} = 90 \text{ lb} \times \frac{4.45 \text{ N}}{1 \text{ lb}} = 400 \text{ N} \qquad m_{\text{gymnast}} = 90 \text{ lb} \times \frac{0.454 \text{ kg}}{1 \text{ lb}} = 41 \text{ kg}$$

$$w_{\text{prof}} = 150 \text{ lb} \times \frac{4.45 \text{ N}}{1 \text{ lb}} = 670 \text{ N} \qquad m_{\text{prof}} = 150 \text{ lb} \times \frac{0.454 \text{ kg}}{1 \text{ lb}} = 68 \text{ kg}$$

$$w_{\text{player}} = 240 \text{ lb} \times \frac{4.45 \text{ N}}{1 \text{ lb}} = 1070 \text{ N} \qquad m_{\text{player}} = 240 \text{ lb} \times \frac{0.454 \text{ kg}}{1 \text{ lb}} = 110 \text{ kg}$$

***ASSESS*** We can use the information in this problem to assess the results of future problems. If you get an answer of 1000 N, you now know that this is approximately the weight of a football player, which can help with your assessment.

## Apparent Weight

The weight of an object is the force of gravity on that object. You may never have thought about it, but gravity is not a force that you can feel or sense directly. Your *sensation* of weight—how heavy you *feel*—is due to *contact forces* supporting you. As you read this, your sensation of weight is due to the normal force exerted on you by the chair in which you are sitting. The chair's surface touches you and activates nerve endings in your skin. You sense the magnitude of this force, and this is your sensation of weight. When you stand, you feel the contact force of the floor pushing against your feet. If you are hanging from a rope, you feel the friction force between the rope and your hands.

Let's define your **apparent weight** $w_{\text{app}}$ in terms of the force you feel:

$$w_{\text{app}} = \text{magnitude of supporting contact forces} \qquad (5.4)$$

Definition of apparent weight

**Physics students can't jump** If you ride in an elevator and try to jump up into the air just as the elevator starts to rise, you'll feel like you can hardly get off the ground because your apparent weight is greater than your actual weight. On a fast elevator with a large acceleration, it's like trying to jump with an extra 30 or 40 pounds!

If you are in equilibrium, your weight and apparent weight are generally the same. But if you undergo an acceleration, this is not necessarily the case. For instance, you feel "heavy" when an elevator you are riding in suddenly accelerates upward, and you feel lighter than normal as the upward-moving elevator brakes to a halt. Your true weight $w = mg$ has not changed during these events, but your *sensation* of your weight has.

Let's look at the details for this case. Imagine a man standing in an elevator as it accelerates upward. As FIGURE 5.9 shows, the only forces acting on the man are the upward normal force of the floor and the downward weight force. Because the man has an acceleration $\vec{a}$, according to Newton's second law there must be a net force acting on the man in the direction of $\vec{a}$.

FIGURE 5.9 A man in an accelerating elevator.

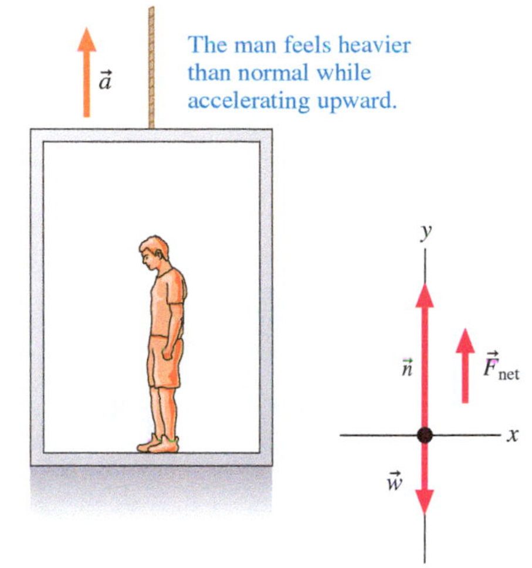


Looking at the free-body diagram in Figure 5.9, we see that the $y$-component of Newton's second law is

$$\sum F_y = n_y + w_y = n - w = ma_y \tag{5.5}$$

where $m$ is the man's mass. Solving Equation 5.5 for $n$ gives

$$n = w + ma_y \tag{5.6}$$

The normal force is the contact force supporting the man, so, given the definition of Equation 5.4, we can rewrite Equation 5.6 as

$$w_{app} = w + ma_y$$

For the case shown in Figure 5.9, where the elevator is accelerating upward, $a_y = +a$ so that $w_{app} = w + ma$. The apparent weight is greater than the man's weight $w$, so he feels heavier than normal. If, instead, the elevator has a downward acceleration, then $a_y = -a$ so that $w_{app} = w - ma$. The apparent weight is less than the man's weight, and the man feels lighter than normal.

The apparent weight isn't just a sensation, though. You can measure it with a scale. When you stand on a bathroom scale, the scale reading is the upward force of the scale on you. If you aren't accelerating (usually a pretty good assumption!), the upward force of the scale on you equals your weight, so the scale reading is your true weight. But if you are accelerating, this correspondence may not hold. In the example above, if the man were standing on a bathroom scale in the elevator, the contact force supporting him would be the upward force of the scale. This force is equal to the scale reading, so if the elevator is accelerating upward, the scale shows an increased weight: The scale reading is equal to $w_{app}$.

Apparent weight can be measured by a scale, so it's no surprise that apparent weight has real, physical implications. Astronauts are nearly crushed by their apparent weight during a rocket launch when $a$ is much greater than $g$. Much of the thrill of amusement park rides, such as roller coasters, comes from rapid changes in your apparent weight.

**EXAMPLE 5.8** **Finding a rider's apparent weight in an elevator**

Anjay's mass is 70 kg. He is standing on a scale in an elevator that is moving at 5.0 m/s. As the elevator slows to a stop, the scale reads 750 N. Before it stopped, was the elevator moving up or down? How long did the elevator take to come to rest?

**STRATEGIZE** The scale reads Anjay's apparent weight, which differs from his actual weight by $ma_y$. If his apparent weight is greater than his actual weight, then his acceleration is upward; if less, his acceleration is downward.

**PREPARE** The scale reading as the elevator comes to rest, 750 N, is Anjay's apparent weight. Anjay's actual weight is

$$w = mg = (70\text{ kg})(9.80\text{ m/s}^2) = 686\text{ N}$$

This is an intermediate step in the calculation, so we are keeping an extra significant figure.

**SOLVE** Because Anjay's apparent weight $w_{app}$ is greater than his actual weight $w$, his acceleration $a_y = (w_{app} - w)/m$ is *positive*. You learned in Chapter 2 that if an object is slowing, as is the case here, its velocity and acceleration vectors point in opposite directions, so the elevator's velocity must be *negative*—it is moving down.

We can use kinematics to find the stopping time. For this step we need the acceleration, which is

$$a_y = \frac{w_{app} - w}{m} = \frac{750\text{ N} - 686\text{ N}}{70\text{ kg}} = 0.91\text{ m/s}^2$$

Then we use the kinematic equation

$$(v_y)_f = (v_y)_i + a_y \Delta t$$

*Continued*

The elevator is initially moving downward, so $(v_y)_i = -5.0$ m/s, and it then comes to a halt, so $(v_y)_f = 0$. We know the acceleration, so the time interval is

$$\Delta t = \frac{(v_y)_f - (v_y)_i}{a_y} = \frac{0 - (-5.0\ \text{m/s})}{0.91\ \text{m/s}^2} = 5.5\ \text{s}$$

**ASSESS** Think back to your experiences riding elevators. If the elevator is moving downward and then comes to rest, you "feel heavy." This gives us confidence that our analysis of the motion is correct. And 5.0 m/s is a pretty fast elevator: At this speed, the elevator will be passing more than one floor per second. If you've been in a fast elevator in a tall building, you know that 5.5 s is reasonable for the time it takes for the elevator to slow to a stop.

## Weightlessness

Let's return to the elevator example. Suppose Anjay's apparent weight was 0 N—meaning that the scale read *zero*. Could this happen? Recall that Anjay's apparent weight is the contact force supporting him, so we are saying that the upward force from the scale is zero. If we use $W_{app} = 0$ in the equation for the apparent acceleration, we find

$$0 = w + ma_y = mg + ma_y$$

so that $a_y = -g = -9.8\ \text{m/s}^2$. This is a case we've seen before—it is free fall! This means that **a person (or any object) in free fall has zero apparent weight.**

Think about this carefully. Imagine a man inside an elevator that is in free fall (a frightening case, for sure!). This man is also in free fall, so he is falling at the exact same rate as the elevator. From the man's perspective, he is "floating" inside the elevator. A small push off the floor would send him floating toward the top of the elevator. He is what we call *weightless*.

"Weightless" does *not* mean "no weight." An object that is **weightless** has no *apparent* weight. The distinction is significant. The man's weight is still $mg$ because gravity is still pulling down on him, but he has no *sensation* of weight as he free falls. The term "weightless" is in a sense a poor one because it implies that objects have no weight. As we see, that is not the case.

You've seen videos of astronauts and various objects floating inside the International Space Station as it orbits the earth. If an astronaut tries to stand on a scale, it does not exert any force against her feet and reads zero. She is said to be weightless. But if the criterion to be weightless is to be in free fall, and if astronauts orbiting the earth are weightless, does this mean that they are in free fall? This is a very interesting question to which we shall return in Chapter 6.

**A weightless experience** As we learned in Chapter 3, objects undergoing projectile motion are in free fall. This specially adapted plane flies in the same parabolic trajectory as would a projectile with no air resistance. Objects inside, such as these passengers, thus move along a free-fall trajectory. They feel weightless, and then float with respect to the plane's interior, until the plane resumes normal flight, up to 30 seconds later.

**STOP TO THINK 5.3** You're bouncing up and down on a trampoline. After you have left the trampoline and are moving upward, your apparent weight is

A. More than your true weight.
B. Less than your true weight.
C. Equal to your true weight.
D. Zero.

# 5.4 Normal Forces

In Chapter 4 we saw that an object at rest on a table is subject to an upward force due to the table. This force is called the *normal force* because it is always directed normal, or perpendicular, to the surface of contact. As we saw, the normal force has its origin in the atomic "springs" that make up the surface. The harder the object bears down on the surface, the more these springs are compressed and the harder they push back. Thus the normal force *adjusts* itself so that the object stays on the surface without penetrating it. This fact is key in solving for the normal force.

**EXAMPLE 5.9** **Normal force on a pressed book**

A 1.2 kg book lies on a table. You press down on the book from above with a force of 15 N. What is the normal force acting on the book from the table below?

**STRATEGIZE** The book is in static equilibrium, so the net force acting on it is zero. We will use this fact to find the forces acting on the book.

**PREPARE** We need to identify the forces acting on the book and prepare a free-body diagram showing these forces. These steps are illustrated in FIGURE 5.10.

FIGURE 5.10 Finding the normal force on a book pressed from above.

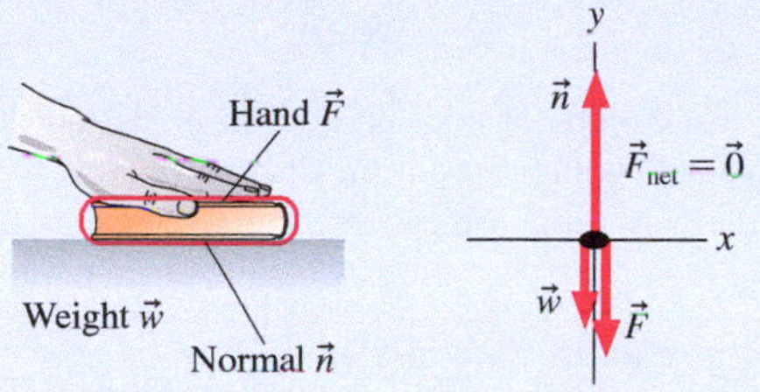


**SOLVE** Because the book is in static equilibrium, the net force on it must be zero. The only forces acting are in the $y$-direction, so Newton's second law is

$$\sum F_y = n_y + w_y + F_y = n - w - F = ma_y = 0$$

We learned in the last section that the weight force is $w = mg$. The weight of the book is thus

$$w = mg = (1.2\text{ kg})(9.8\text{ m/s}^2) = 12\text{ N}$$

With this information, we see that the normal force exerted by the table is

$$n = F + w = 15\text{ N} + 12\text{ N} = 27\text{ N}$$

**ASSESS** The magnitude of the normal force is *larger* than the weight of the book. From the table's perspective, the extra force from the hand pushes the book further into the atomic springs of the table. These springs then push back harder, giving a normal force that is greater than the weight of the book.

A common situation is an object on a ramp or incline. If friction is ignored, there are only two forces acting on the object: gravity and the normal force. However, we need to carefully work out the components of these two forces in order to solve dynamics problems. FIGURE 5.11a shows how. Be sure you avoid the two common errors shown in FIGURE 5.11b.

**KEY CONCEPT** FIGURE 5.11 The forces on an object on an incline.

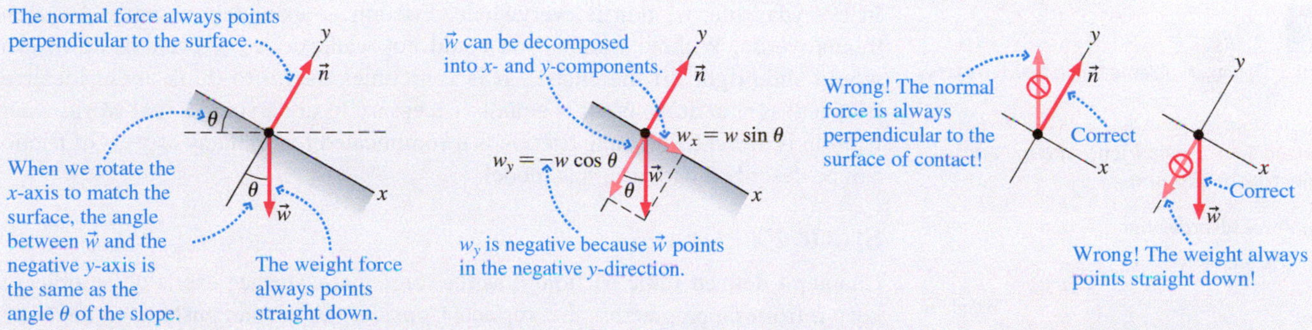


**STOP TO THINK 5.4** A mountain biker is climbing a steep 20° slope at a constant speed. The cyclist and bike have a combined weight of 800 N. Referring to Figure 5.11 for guidance, what can you say about the magnitude of the normal force of the ground on the bike?

A. $n > 800$ N
B. $n = 800$ N
C. $n < 800$ N

**EXAMPLE 5.10** **Acceleration of a downhill skier**

A skier slides down a steep 27° slope. On a slope this steep, friction is much smaller than the other forces acting on the skier and can be ignored. What is the skier's acceleration?

**STRATEGIZE** We will use Newton's second law in component form to find the acceleration. We should choose a coordinate system with the *x*-axis pointing down the slope. This greatly simplifies the analysis because then the skier does not move in the *y*-direction at all, making $a_y = 0$.

**PREPARE** FIGURE 5.12 is a visual overview. The free-body diagram is based on the information in Figure 5.11.

FIGURE 5.12 Visual overview of a downhill skier.

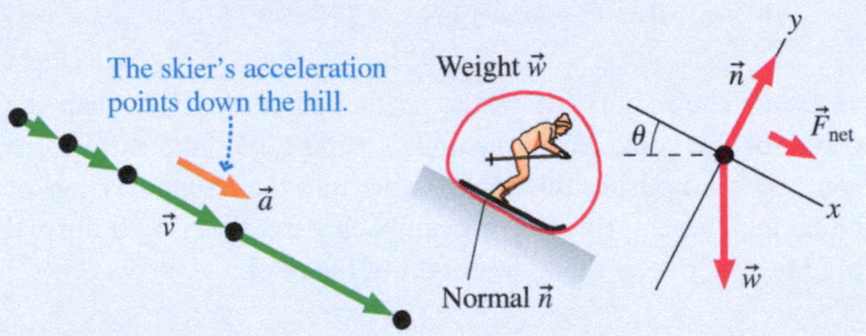


**SOLVE** We can now use Newton's second law in component form to find the skier's acceleration:

$$\sum F_x = w_x + n_x = ma_x$$
$$\sum F_y = w_y + n_y = ma_y$$

Because $\vec{n}$ points directly in the positive *y*-direction, $n_y = n$ and $n_x = 0$. Figure 5.11a showed the important fact that the angle between $\vec{w}$ and the negative *y*-axis is the *same* as the slope angle $\theta$. With this information, the components of $\vec{w}$ are $w_x = w\sin\theta = mg\sin\theta$ and $w_y = -w\cos\theta = -mg\cos\theta$, where we used the fact that $w = mg$. With these components in hand, Newton's second law becomes

$$\sum F_x = w_x + n_x = mg\sin\theta = ma_x$$
$$\sum F_y = w_y + n_y = -mg\cos\theta + n = ma_y = 0$$

In the second equation we used the fact that $a_y = 0$. The $m$ cancels in the first of these equations, leaving us with

$$a_x = g\sin\theta$$

This is the expression for acceleration on a frictionless surface that we presented, without proof, in Chapter 3. Now we've justified our earlier assertion. We can use this to calculate the skier's acceleration:

$$a_x = g\sin\theta = (9.8\ \text{m/s}^2)\sin 27° = 4.4\ \text{m/s}^2$$

**ASSESS** Our result shows that when $\theta = 0$, so that the slope is horizontal, the skier's acceleration is zero, as it should be. Further, when $\theta = 90°$ (a vertical slope), his acceleration is $g$, which makes sense because he's in free fall when $\theta = 90°$. Notice that the mass canceled out, so we didn't need to know the skier's mass. We first saw the formula for the acceleration in ◀◀ SECTION 3.4, but now we see the physical reasons behind it.

## 5.5 Friction

In everyday life, friction is everywhere. Friction is absolutely essential for many things we do. Without friction you could not walk, drive, or even sit down (you would slide right off the chair!). It is sometimes useful to think about idealized frictionless situations, but it is equally necessary to understand a real world where friction is present. Although friction is a complicated force, many aspects of friction can be described with a simple model.

**Video** Solving Problems Using Newton's Laws

FIGURE 5.13 Static friction keeps an object from slipping.

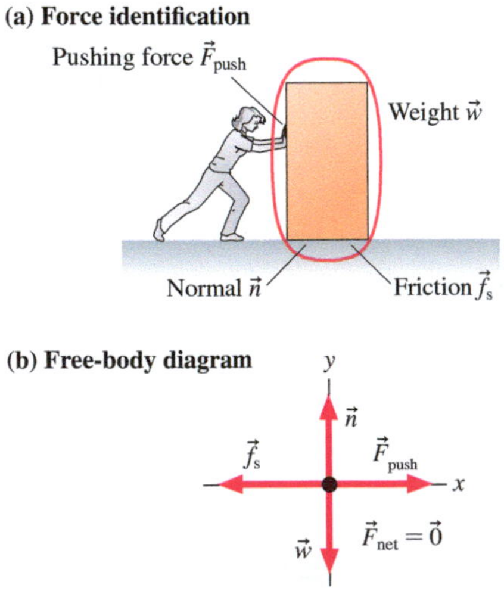


### Static Friction

Chapter 4 defined static friction $\vec{f}_s$ as the force that a surface exerts on an object to keep it from slipping across that surface. Consider the woman pushing on the box in FIGURE 5.13a. Because the box is not moving with respect to the floor, the woman's push to the right must be balanced by a static friction force $\vec{f}_s$ pointing to the left. This is the general rule for finding the *direction* of $\vec{f}_s$: Decide which way the object *would* move if there were no friction. The static friction force $\vec{f}_s$ then points in the opposite direction, to prevent motion relative to the surface.

Determining the *magnitude* of $\vec{f}_s$ is a bit trickier. Because the box is at rest, it's in static equilibrium. From the free-body diagram of FIGURE 5.13b, this means that the static friction force must exactly balance the pushing force, so that $f_s = F_{push}$. As shown in FIGURES 5.14a and 5.14b, the harder the woman pushes, the harder the friction force from the floor pushes back. If she reduces her pushing force, the friction force will automatically be reduced to match. Static friction acts in *response* to an applied force.

But there's clearly a limit to how big $\vec{f}_s$ can get. If the woman pushes hard enough, the box will slip and start to move across the floor. In other words, the static friction force has a *maximum* possible magnitude $f_{s\,max}$, as illustrated in FIGURE 5.14c. Experiments with friction show that $f_{s\,max}$ is proportional to the magnitude of the normal force between the surface and the object; that is,

$$f_{s\,max} = \mu_s n \tag{5.7}$$

where $\mu_s$ is called the **coefficient of static friction.** The coefficient is a number that depends on the materials from which the object and the surface are made. The higher the coefficient of static friction, the greater the "stickiness" between the object and the surface, and the harder it is to make the object slip. TABLE 5.2 lists some approximate values of coefficients of friction.

**NOTE** ► Equation 5.7 does not say $f_s = \mu_s n$. The value of $f_s$ depends on the force or forces that static friction has to balance to keep the object from moving. It can have any value from zero up to, but not exceeding, $\mu_s n$. ◄

So our rules for static friction are:

- The direction of static friction is such as to oppose motion.
- The magnitude $f_s$ of static friction adjusts itself so that the net force is zero and the object doesn't move.
- The magnitude of static friction cannot exceed the maximum value $f_{s\,max}$ given by Equation 5.7. If the friction force needed to keep the object stationary is greater than $f_{s\,max}$, the object slips and starts to move.

FIGURE 5.14 Static friction acts in response to an applied force.

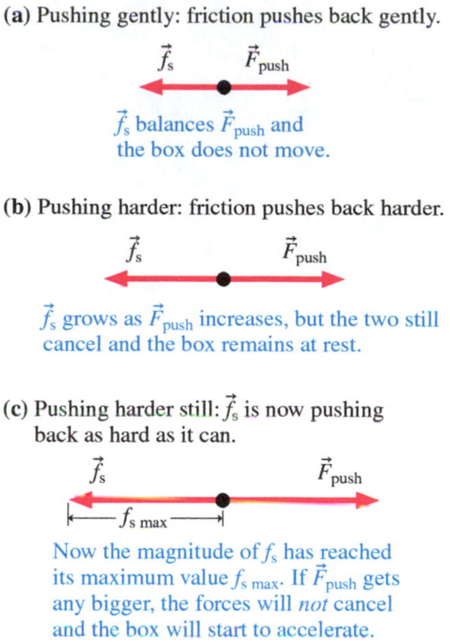
(a) Pushing gently: friction pushes back gently.
$\vec{f}_s$ $\vec{F}_{push}$
$\vec{f}_s$ balances $\vec{F}_{push}$ and the box does not move.
(b) Pushing harder: friction pushes back harder.
$\vec{f}_s$ $\vec{F}_{push}$
$\vec{f}_s$ grows as $\vec{F}_{push}$ increases, but the two still cancel and the box remains at rest.
(c) Pushing harder still: $\vec{f}_s$ is now pushing back as hard as it can.
$\vec{f}_s$ $\vec{F}_{push}$
$f_{s\,max}$
Now the magnitude of $f_s$ has reached its maximum value $f_{s\,max}$. If $\vec{F}_{push}$ gets any bigger, the forces will *not* cancel and the box will start to accelerate.

**Video** Figure 5.14

## Kinetic Friction

Once the box starts to slide, as in FIGURE 5.15, the static friction force is replaced by a kinetic (or sliding) friction force $\vec{f}_k$. Kinetic friction is in some ways simpler than static friction: The direction of $\vec{f}_k$ is always opposite to the direction in which an object slides across the surface, and experiments show that kinetic friction, unlike static friction, has a nearly *constant* magnitude, given by

$$f_k = \mu_k n \tag{5.8}$$

where $\mu_k$ is called the **coefficient of kinetic friction.** Equation 5.8 also shows that kinetic friction, like static friction, is proportional to the magnitude of the normal force $n$. Notice that **the magnitude of the kinetic friction force does not depend on how fast the object is sliding.**

TABLE 5.2 Coefficients of friction

| Materials | Static $\mu_s$ | Kinetic $\mu_k$ | Rolling $\mu_r$ |
|---|---|---|---|
| Rubber on concrete | 1.00 | 0.80 | 0.02 |
| Steel on steel (dry) | 0.80 | 0.60 | 0.002 |
| Steel on steel (lubricated) | 0.10 | 0.05 | |
| Wood on wood | 0.50 | 0.20 | |
| Wood on snow | 0.12 | 0.06 | |
| Ice on ice | 0.10 | 0.03 | |

FIGURE 5.15 The kinetic friction force is *opposite* to the direction of motion.

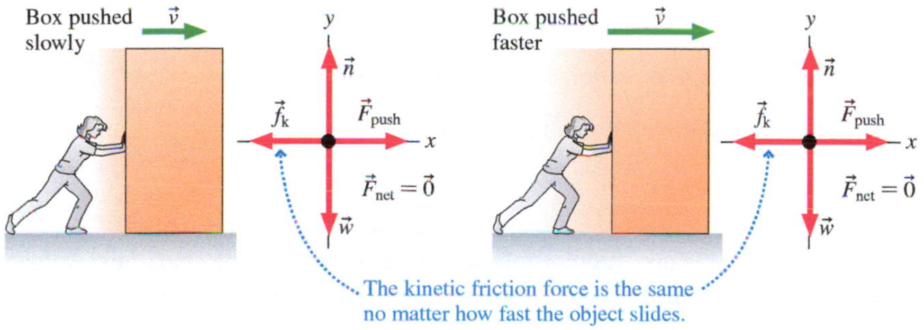


Table 5.2 includes approximate values of $\mu_k$. You can see that $\mu_k < \mu_s$, which explains why it is easier to keep a box moving than it is to start it moving.

FIGURE 5.16 The bottom of the wheel is stationary.

## Rolling Friction

When you are driving, if you slam on the brakes hard enough, your car tires slide against the road surface and leave skid marks. This is kinetic friction because the tire and the road are *sliding* against each other. A wheel *rolling* on a surface also experiences friction, but not kinetic friction: The portion of the wheel that contacts the surface is stationary with respect to the surface, not sliding. The photo in FIGURE 5.16 was taken with a stationary camera. Note that the part of the wheel touching the ground is not blurred, which indicates that this part of the wheel is not moving with respect to the ground.

The interaction between a rolling wheel and the road involves adhesion between and deformation of surfaces and can be quite complicated, but in many cases we can treat it like another type of friction force that opposes the motion, one defined by a **coefficient of rolling friction** $\mu_r$:

$$f_r = \mu_r n \tag{5.9}$$

Rolling friction acts very much like kinetic friction, but values of $\mu_r$ (see Table 5.2) are much lower than values of $\mu_k$. It's easier to roll something on wheels than to slide it!

**STOP TO THINK 5.5** Rank in order, from largest to smallest, the size of the friction forces $\vec{f}_A$ to $\vec{f}_E$ in the five different situations (one or more friction forces could be zero). The box and the floor are made of the same materials in all situations.

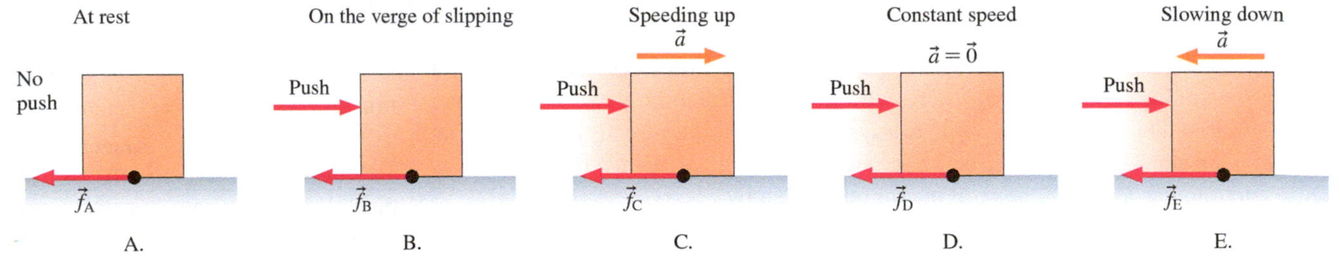

**To skid or not to skid** If you brake as hard as you can without skidding, the force that stops your car is the static friction force between your tires and the road. This force is bigger than the kinetic friction force, so if you skid, not only can you lose control of your car but you also end up taking a longer distance to stop. Antilock braking systems brake the car as hard as possible without skidding, stopping the car in the shortest possible distance while also retaining control.

## Working with Friction Forces

These ideas can be summarized in a *model* of friction:

$$\text{Static: } \vec{f}_s = (\text{magnitude} \le f_{s\,max} = \mu_s n, \text{ direction as necessary to prevent motion}) \tag{5.10}$$

$$\text{Kinetic: } \vec{f}_k = (\mu_k n, \text{ direction opposite the motion})$$

$$\text{Rolling: } \vec{f}_r = (\mu_r n, \text{ direction opposite the motion})$$

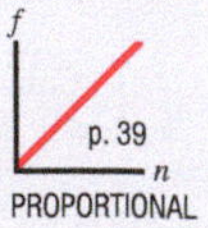


Here "motion" means "motion relative to the surface." The maximum value of static friction $f_{s\,max} = \mu_s n$ occurs when the object slips and begins to move.

NOTE ▸ Equations 5.10 are a "model" of friction, not a "law" of friction. These equations provide a reasonably accurate, but not perfect, description of how friction forces act. They are a simplification of reality that works reasonably well, which is what we mean by a "model." They are not a "law of nature" on a level with Newton's laws. ◂

**TACTICS BOX 5.1** **Working with friction forces**

❶ If the object is *not moving* relative to the surface it's in contact with, then the friction force is **static friction.** Draw a free-body diagram of the object. The *direction* of the friction force is such as to oppose sliding of the object relative to the surface. Then use Problem-Solving Approach 5.1 to solve for $f_s$. If $f_s$ is greater than $f_{s\,max} = \mu_s n$, then static friction cannot hold the object in place. The assumption that the object is at rest is not valid, and you need to redo the problem using kinetic friction.

❷ If the object is *sliding* relative to the surface, then **kinetic friction** is acting. From Newton's second law, find the normal force $n$. Equation 5.10 then gives the magnitude and direction of the friction force.

❸ If the object is *rolling* along the surface, then **rolling friction** is acting. From Newton's second law, find the normal force $n$. Equation 5.10 then gives the magnitude and direction of the friction force.

Exercises 20, 21

**Video** What the Physics? Tablecloth Trick

## EXAMPLE 5.11 Finding the force to slide a sofa

Carol wants to move her 32 kg sofa to a different room in the house. She places "sofa sliders," slippery disks with $\mu_k = 0.080$, under the feet of the sofa. She then pushes the sofa at a steady 0.40 m/s across the floor. How much force does she apply to the sofa?

***STRATEGIZE*** The sofa is sliding, so the friction force is kinetic friction. We will use the second step of Tactics Box 5.1.

***PREPARE*** Let's assume the sofa slides to the right. In this case, a kinetic friction force $\vec{f}_k$ opposes the motion by pointing to the left. In FIGURE 5.17 we identify the forces acting on the sofa and construct a free-body diagram.

FIGURE 5.17 Forces on a sofa being pushed across a floor.

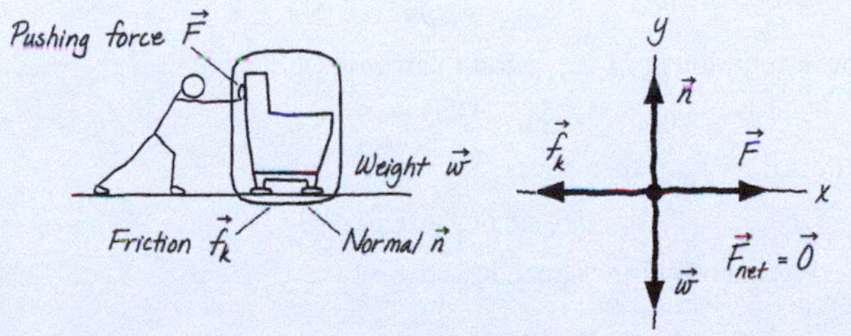


***SOLVE*** The sofa is moving at a constant speed, so it is in dynamic equilibrium with $\vec{F}_{net} = \vec{0}$. This means that the $x$- and $y$-components of the net force must be zero:

$$\sum F_x = n_x + w_x + F_x + (f_k)_x = 0 + 0 + F - f_k = 0$$

$$\sum F_y = n_y + w_y + F_y + (f_k)_y = n - w + 0 + 0 = 0$$

In the first equation, the $x$-component of $\vec{f}_k$ is equal to $-f_k$ because $\vec{f}_k$ is directed to the left. Similarly, in the second equation, $w_y = -w$ because the weight force points down.

From the first equation, we see that Carol's pushing force is $F = f_k$. To evaluate this, we need $f_k$. Here we can use our model for kinetic friction:

$$f_k = \mu_k n$$

Let's look at the vertical motion first. The second equation ultimately reduces to

$$n - w = 0$$

The weight force $w = mg$, so we can write

$$n = mg$$

This is a common result we'll see again. The force that Carol pushes with is equal to the friction force, and this depends on the normal force and the coefficient of kinetic friction, $\mu_k = 0.080$:

$$F = f_k = \mu_k n = \mu_k mg$$
$$= (0.080)(32\,\text{kg})(9.80\,\text{m/s}^2) = 25\,\text{N}$$

***ASSESS*** The speed with which Carol pushes the sofa does not enter into the answer. This makes sense because the kinetic friction force doesn't depend on speed. The final result of 25 N is a rather small force—only about $5\frac{1}{2}$ pounds—but we expect this because Carol has used slippery disks to move the sofa.

## CONCEPTUAL EXAMPLE 5.12 To push or pull a lawn roller?

A lawn roller is a heavy cylinder used to flatten a bumpy lawn, as shown in FIGURE 5.18. Is it easier to push or pull such a roller? Which is more effective for flattening the lawn: pushing or pulling? Assume that the pushing or pulling force is directed along the handle of the roller.

FIGURE 5.18 Pushing and pulling a lawn roller.

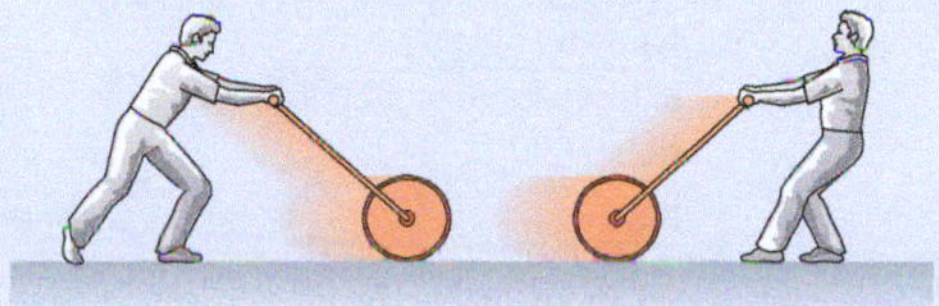

*Continued*

**REASON** FIGURE 5.19 shows free-body diagrams for the two cases. We assume that the roller is pushed at a constant speed so that it is in dynamic equilibrium with $\vec{F}_{\text{net}} = \vec{0}$. Because the roller does not move in the $y$-direction, the $y$-component of the net force must be zero. According to our model, the magnitude $f_r$ of rolling friction is proportional to the magnitude $n$ of the normal force. If we *push* on the roller, our pushing force $\vec{F}$ will have a downward $y$-component. To compensate for this, the normal force must increase and, because $f_r = \mu_r n$, the rolling friction will increase as well. This makes the roller harder to move. If we *pull* on the roller, the now upward $y$-component of $\vec{F}$ will lead to a *reduced* value of $n$ and hence of $f_r$. Thus the roller is easier to pull than to push.

However, the purpose of the roller is to flatten the soil. If the normal force $\vec{n}$ of the ground on the roller is greater, then by Newton's third law the force of the roller on the ground will be greater as well. So for smoothing your lawn, it's better to push.

FIGURE 5.19 Free-body diagrams for the lawn roller.

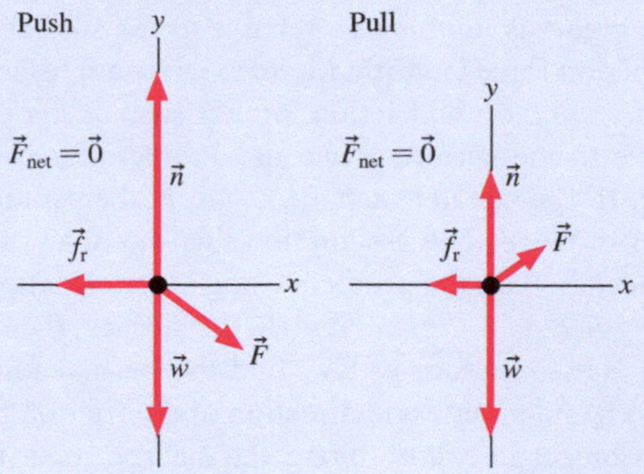


**ASSESS** You've probably experienced this effect while using an upright vacuum cleaner. The vacuum is harder to push on the forward stroke than when drawing it back.

## EXAMPLE 5.13 How to dump a file cabinet

A 50.0 kg steel file cabinet is in the back of a dump truck. The truck's bed, also made of steel, is slowly tilted. What is the magnitude of the static friction force on the cabinet when the bed is tilted 20°? At what angle will the file cabinet begin to slide?

**STRATEGIZE** Until it slides, the file cabinet is held in place by static friction, so we will use the first step of Tactics Box 5.1. To do so, we need the normal force, which we will find using the fact that the cabinet is in static equilibrium.

**PREPARE** From Tactics Box 5.1, we know that the file cabinet will slip when the static friction force reaches its maximum possible value $f_{\text{s max}}$. FIGURE 5.20 shows the visual overview when the truck bed is tilted at angle $\theta$. We can make the analysis easier if we tilt the coordinate system to match the bed of the truck. To prevent the file cabinet from slipping, the static friction force must point *up* the slope.

**SOLVE** Before it slips, the file cabinet is in static equilibrium. Newton's second law gives

$$\sum F_x = n_x + w_x + (f_s)_x = 0$$

$$\sum F_y = n_y + w_y + (f_s)_y = 0$$

From the free-body diagram we see that $f_s$ has only a negative $x$-component and that $n$ has only a positive $y$-component. We also have $w_x = w\sin\theta$ and $w_y = -w\cos\theta$. Thus the second law becomes

$$\sum F_x = w\sin\theta - f_s = mg\sin\theta - f_s = 0$$

$$\sum F_y = n - w\cos\theta = n - mg\cos\theta = 0$$

The $x$-component equation allows us to determine the magnitude of the static friction force when $\theta = 20°$:

$$f_s = mg\sin\theta = (50.0\text{ kg})(9.80\text{ m/s}^2)\sin 20° = 168\text{ N}$$

This value does not require that we know $\mu_s$. The coefficient of static friction enters only when we want to find the angle at which the file cabinet slips. Slipping occurs when the static friction force reaches its maximum value:

$$f_s = f_{\text{s max}} = \mu_s n$$

From the $y$-component of Newton's second law we see that $n = mg\cos\theta$. Consequently,

$$f_{\text{s max}} = \mu_s mg\cos\theta$$

The $x$-component of the second law gave

$$f_s = mg\sin\theta$$

Setting $f_s = f_{\text{s max}}$ then gives

$$mg\sin\theta = \mu_s mg\cos\theta$$

The $mg$ in both terms cancels, and we find

$$\frac{\sin\theta}{\cos\theta} = \tan\theta = \mu_s$$

$$\theta = \tan^{-1}\mu_s = \tan^{-1}(0.80) = 39°$$

**ASSESS** Steel doesn't slide all that well on unlubricated steel, so a fairly large angle is not surprising. The answer seems reasonable. It is worth noting that $n = mg\cos\theta$ in this example. A common error is to use simply $n = mg$. Be sure to evaluate the normal force within the context of each particular problem.

FIGURE 5.20 Visual overview of a file cabinet in a tilted dump truck.

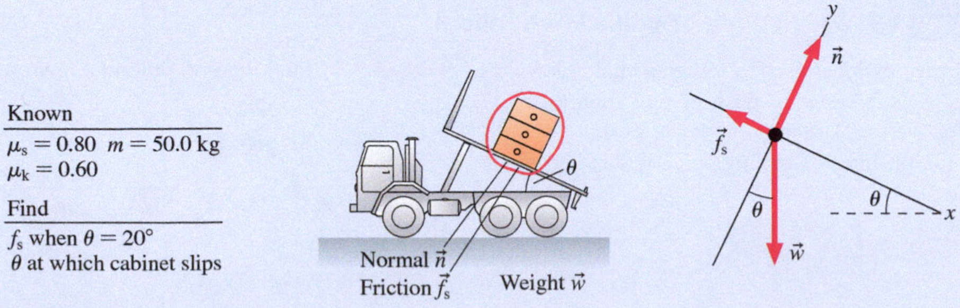

### Causes of Friction

It is worth taking a brief pause to look at the *causes* of friction. All surfaces, even those that are quite smooth to the touch, are very rough on a microscopic scale. When two objects are placed in contact, they do not make a smooth fit. Instead, as FIGURE 5.21 shows, the high points on one surface become jammed against the high points on the other surface, while the low points are not in contact at all. Only a very small fraction (typically $10^{-4}$) of the surface area is in actual contact. The amount of contact depends on how hard the surfaces are pushed together, which is why friction forces are proportional to $n$.

For an object to slip, you must push it hard enough to overcome the forces exerted at these contact points. Once the two surfaces are sliding against each other, their high points undergo constant collisions, deformations, and even brief bonding that lead to the resistive force of kinetic friction.

FIGURE 5.21 A microscopic view of friction.

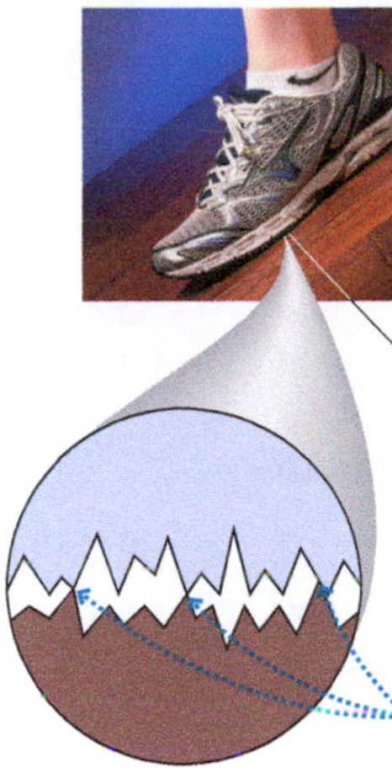


## 5.6 Drag

Drag is a force that opposes or retards the motion of an object as it moves through a *fluid*—a substance that flows, such as air or water. You experience drag forces due to the air—sometimes called *air resistance*—whenever you jog, bicycle, or drive your car. In biology, drag forces are extremely important for microorganisms moving in water. The drag force $\vec{D}$:

- Is opposite in direction to the velocity $\vec{v}$, as shown in FIGURE 5.22.
- Increases in magnitude as the object's speed increases.

FIGURE 5.22 The drag force is opposite the direction of motion.

In general, drag is more complex than friction, in part because the drag force has its origin in two rather different physical principles. Fortunately, in many cases of interest only one of these causes of drag is in effect; in such cases we can characterize drag through the use of fairly simple models. To understand which cause of drag applies in a given situation, we need to introduce the *Reynolds number*.

### Reynolds Number

We can illustrate the two basic causes of drag with examples. When a thrown baseball moves through the air, most of the drag—the force that resists its motion—occurs because the rapidly moving ball has to push away the air in front of it. This drag is thus due to *inertial forces*, forces that arise because the air's inertia makes it difficult to push through. In contrast, consider the motion of a small sphere slowly falling in a thick fluid such as honey. Here, the drag is mostly the result of *viscous forces* due to the properties of the honey itself: The honey "grabs" the surface of the sphere and tries to hold it back.

Which of these two causes of drag is most important depends on the particular situation. As we'll learn in the following discussion, for an object like the fast-moving baseball in air, where the drag is due to inertial forces, the drag force is proportional to $\rho v^2 L^2$, where $\rho$ (lowercase Greek rho) is the density of the fluid (here, air) through which the object moves, $v$ is its speed, and $L$ is a measure of its size (such as the diameter for a baseball). For an object moving through a thick fluid, where the drag is due to viscous forces, the drag force is proportional to $\eta L v$, where $\eta$ (lowercase Greek eta) is the **viscosity** of the fluid. The viscosity is larger for thick fluids such as honey than for more free-flowing ones such as water.

To characterize which kind of drag is relevant, we can formulate a number, called the **Reynolds number,** that is the *ratio* of the inertial force to the viscous force. Thus

$$\text{Reynolds number} = Re = \frac{\text{inertial forces}}{\text{viscous forces}} = \frac{\rho v^2 L^2}{\eta L v} = \frac{\rho v L}{\eta} \quad (5.11)$$

Written this way, the Reynolds number is *dimensionless;* that is, it has no units. When the Reynolds number is high, inertial drag is dominant; when the Reynolds number is low, viscous drag is more important.

**NOTE** ▶ The symbol for the Reynolds number is the two-letter combination $Re$. It is the only symbol in this text that uses two letters. ◀

In SI units, density is measured in $kg/m^3$ and viscosity in $Pa \cdot s$. Pa is the abbreviation for *pascal*, the SI unit of pressure, which is defined as $1\ Pa = 1\ N/m^2$. We'll have a lot more to say about pressure in later chapters. TABLE 5.3 lists the values of density and viscosity for some typical fluids. Viscosity is *very* dependent on temperature—think how quickly honey loses viscosity as you heat it—so the values shown are not appropriate at temperatures other than those listed.

**TABLE 5.3** Density and viscosity

| Fluid | $\rho$ (kg/m$^3$) | $\eta$(Pa · s) |
|---|---|---|
| Air (20°C at sea level) | 1.20 | $1.8 \times 10^{-5}$ |
| Ethyl alcohol (20°C) | 790 | $1.3 \times 10^{-3}$ |
| Olive oil (20°C) | 910 | $8.4 \times 10^{-2}$ |
| Water (20°C) | 1000 | $1.0 \times 10^{-3}$ |
| Water (40°C) | 1000 | $7.0 \times 10^{-4}$ |
| Honey (20°C) | 1400 | 10 |
| Honey (40°C) | 1400 | 1.7 |

The object's size $L$ is a characteristic or typical size of the object, which could be a height, width, or diameter. This seems a bit odd; for an object that is rectangular, the value of $Re$ depends on whether you select the width or the height for $L$. However, in practice, how you define $L$ makes very little difference. As you'll see, we want to know whether $Re$ is "high" or "low," values that differ by factors of 1000 or more. A difference of a factor of 2 or 3 will not affect our judgment about the size of $Re$. Said another way, our use of the Reynolds number needs only approximate values, not precise calculations.

As a simple example, consider a 75-mm-diameter baseball moving through 20°C air at a speed of 20 m/s. We can take the ball's characteristic size $L$ to be its diameter. The Reynolds number for this motion is

$$Re = \frac{\rho v L}{\eta} = \frac{(1.2\ \text{kg/m}^3)(20\ \text{m/s})(0.075\ \text{m})}{1.8 \times 10^{-5}\ \text{Pa} \cdot \text{s}} = 100{,}000$$

As you'll see, this is considered a very high Reynolds number.

**TABLE 5.4** Drag Coefficients

| Object | $C_D$ |
|---|---|
| Commercial airliner | 0.024 |
| Toyota Prius | 0.24 |
| Pitched baseball | 0.35 |
| Racing cyclist | 0.88 |
| Running person | 1.2 |

## Drag at High Reynolds Number

When the Reynolds number is greater than about 1000, it is considered high. A high Reynolds number means, as we've seen, that the drag arises not from viscosity but from inertial drag: The moving object has to push the fluid out of the way. As the example of the baseball suggests, the Reynolds number is high for most ordinary objects—balls, cars, planes—moving through air at ordinary speeds. $Re$ is also high for fish and larger objects moving through water.

For high Reynolds numbers, the drag force for motion through a fluid at speed $v$ is

$$\vec{D} = \left(\tfrac{1}{2} C_D \rho A v^2, \text{ direction opposite the motion}\right) \qquad (5.12)$$

Drag force for high Reynolds number, $Re > 1000$

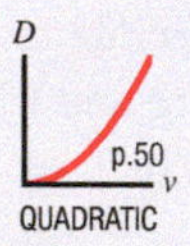


Here $\rho$ is again the density of the fluid, $A$ is the cross-section area of the object (in $m^2$) as defined below, and the dimensionless **drag coefficient** $C_D$ depends on the object's shape. More streamlined, aerodynamic shapes have lower values of $C_D$. TABLE 5.4 lists the drag coefficients for some common moving objects.

For a high Reynolds number, **the size of the drag force is proportional to the *square* of the object's speed.** If the speed doubles, drag increases by a factor of 4. This model of drag is often called **quadratic drag.**

The cross-section area $A$ is the two-dimensional projection that you see when an object is coming toward you. Many objects can be modeled as spheres or cylinders, so FIGURE 5.23 shows cross-section areas and drag coefficients for a sphere and for two different orientations of a cylinder.

**FIGURE 5.23** Cross-section areas and drag coefficients for a sphere and a cylinder.

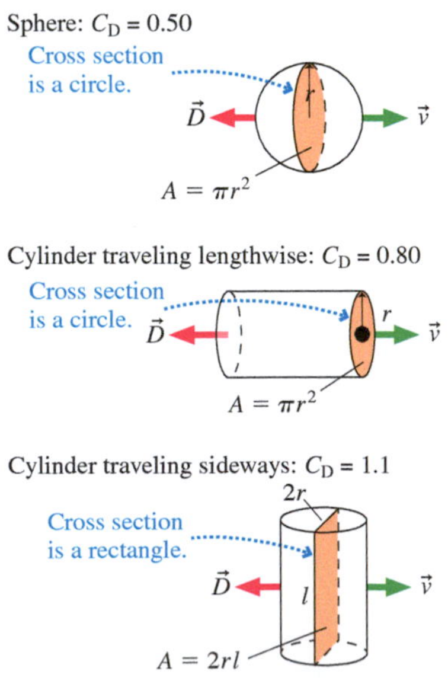

### EXAMPLE 5.14 The drag coefficient of a swimming penguin BIO

Biologists have estimated the drag coefficient of a swimming penguin by observing the rate at which a penguin's speed decreases in its glide phase, when it's not actively swimming and is slowing down. In one study, a gliding 4.8 kg Gentoo penguin has an acceleration of $-0.52\ \text{m/s}^2$ when its speed is 1.60 m/s. If its frontal area is $0.020\ \text{m}^2$, what is the penguin's drag coefficient?

**STRATEGIZE** When the penguin is gliding horizontally, the only force acting to slow it down is the drag force. Using Newton's second law, we will find the drag force from the penguin's acceleration and mass. Once the drag force is known, we will use Equation 5.12 to find the drag coefficient.

**PREPARE** In Equation 5.12, the density $\rho$ is that of the fluid through which the object moves, which is in this case water. From Table 5.3 the density of water is $1000\ \text{kg/m}^3$.

**SOLVE** Assume that the penguin is moving to the right along the $x$-axis. Then Newton's second law is

$$F_x = -D = ma_x$$

where the $x$-component of the drag is negative because it points to the left. We can solve for the magnitude of the drag force as

$$D = -ma_x = -(4.8\ \text{kg})(-0.52\ \text{m/s}^2) = 2.5\ \text{N}$$

Finally, from Equation 5.12 we can solve for the drag coefficient. We have

$$C_D = \frac{2D}{\rho A v^2} = \frac{2(2.5\ \text{N})}{(1000\ \text{kg/m}^3)(0.020\ \text{m}^2)(1.60\ \text{m/s})^2} = 0.098$$

**ASSESS** This drag coefficient is quite a bit better than that of the Toyota Prius in Table 5.4. This is reasonable given that penguins have a highly adapted, streamlined shape.

## Terminal Speed

Suppose an object starting from rest is pushed or pulled through a fluid by a constant force $\vec{F}$. Initially the speed is low and the drag force is small, as shown in FIGURE 5.24a, so the net force causes the object to speed up. But the drag force increases as the speed increases, and eventually the object reaches a speed at which the drag force has exactly the same magnitude as $\vec{F}$. Now the net force is zero, as shown in FIGURE 5.24b, so the object can no longer accelerate but will maintain this steady speed for as long as force $\vec{F}$ is applied. The steady, unchanging speed at which drag exactly counterbalances an applied force is called the object's **terminal speed.** In the case of an object fired at an initial speed *greater* than the terminal speed, the very large drag force slows the object until it is moving at the terminal speed.

FIGURE 5.24 An object reaches its terminal speed when the drag force exactly balances the applied force.

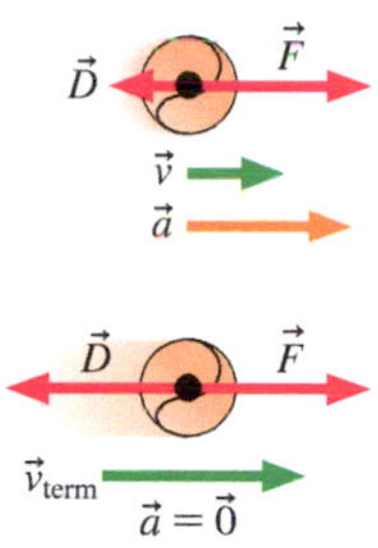


### EXAMPLE 5.15 The terminal speeds of a man and a mouse

A 75 kg skydiver and his 0.020 kg pet mouse falling after jumping from a plane are shown in FIGURE 5.25. Find the terminal speed of each.

FIGURE 5.25 A falling skydiver and mouse, and their cross-section areas.

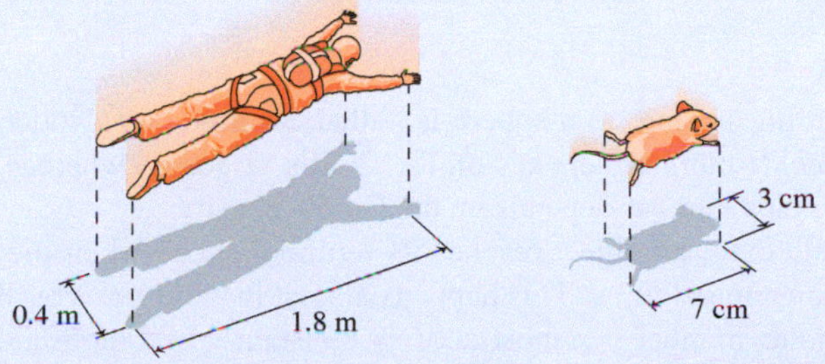


**STRATEGIZE** The force of gravity is a constant force pulling down on a falling object, while drag due to the air is an upward force. The terminal speed is reached when the force of gravity on each object is equal to the drag force. We will model both the man and the mouse as cylinders falling sideways.

**PREPARE** From Figure 5.23 we see that for a cylinder $C_D = 1.1$ and that a cylinder's cross-section area as seen from the side is $A = 2rl$. With the dimensions given, we can calculate $A_{man} = 0.72\ \text{m}^2$ and $A_{mouse} = 2.1 \times 10^{-3}\ \text{m}^2$. We assume that the two skydivers are at sufficiently low altitude that we can use the sea-level value of the density of air.

**SOLVE** The terminal speed is reached when $D = mg$, or

$$\frac{1}{2}C_D \rho A (v_{term})^2 = mg$$

from which we can find the terminal velocity as

$$v_{term} = \sqrt{\frac{2mg}{C_D \rho A}}$$

Thus our two skydivers have terminal speeds

$$v_{man} = \sqrt{\frac{2(75\ \text{kg})(9.8\ \text{m/s}^2)}{(1.1)(1.2\ \text{kg/m}^3)(0.72\ \text{m}^2)}} = 39\ \text{m/s}$$

$$v_{mouse} = \sqrt{\frac{2(0.020\ \text{kg})(9.8\ \text{m/s}^2)}{(1.1)(1.2\ \text{kg/m}^3)(2.1 \times 10^{-3}\ \text{m}^2)}} = 12\ \text{m/s}$$

*Continued*

**ASSESS** 39 m/s is about 85 mph. Reported terminal speeds for skydivers falling in the prone position are in the 100–120 mph range, so our simple model of the fall gives a result that is close but a bit too low. We've probably overestimated both $\rho$, because skydivers are at a high enough altitude that the air density is lower than at sea level, and $A$, because their legs are actually spread apart and allow air to flow between them. A more realistic, but also more complex, model would give a better prediction. The mouse, though, falls at a much more modest 12 m/s $\approx$ 25 mph. Small animals can usually survive a fall from *any* height. Many tree-dwelling animals, such as the sifaka in the photograph at the beginning of the chapter, can extend flaps of skin to increase their area and thus fall at an even slower speed.

NOTE ▸ Drag forces are fairly small for heavier objects, such as balls or bicycles, moving at not-too-fast speeds through air. To keeps things simple, **you can neglect drag in problems about objects moving through air unless the problem explicitly asks you to include it or asks about the terminal speed.** ◂

## Drag at Low Reynolds Number

A 10-$\mu$m-diameter dust particle settles to the ground at about 20 mm/s. It's not hard to calculate that the Reynolds number for the falling particle is $Re \approx 0.01$. This is an example of motion at low Reynolds number, which we define to be $Re < 1$. The fluid's viscosity, rather than its density, is the key factor for motion at low Reynolds number.

For low Reynolds number, the drag force for motion through a fluid at speed $v$ is

$$\vec{D} = (b\eta v, \text{ direction opposite the motion}) \tag{5.13}$$

where $b$ is a constant that depends on the size and shape of the object. For low Reynolds number, **the magnitude of the drag force is proportional to the object's speed.** If the speed doubles, drag also increases by a factor of 2. This model of drag is called **linear drag.**

NOTE ▸ Drag is more complex for intermediate values of Reynolds number, $1 < Re < 1000$. We don't consider those situations in this text. ◂

Linear drag is mostly applicable to very small objects, many of which can be reasonably modeled as spheres. That's fortunate because only for spheres does theory give us any guidance for the coefficient $b$. For a spherical object of radius $r$, it can be shown that $b = 6\pi r$. Thus

$$\vec{D}_{\text{sphere}} = (6\pi\eta r v, \text{ direction opposite the motion}) \tag{5.14}$$

Stokes' law for low Reynolds number, $Re < 1$

This expression for the linear drag on a sphere is called **Stokes' law.** Notice that linear drag at low Reynolds number depends on the fluid's viscosity, whereas quadratic drag at high Reynolds number depends on the fluid's density.

Just as with quadratic drag, an object reaches its terminal speed when the drag force exactly balances an applied force. This happens almost instantly, as you'll see, so motion at low Reynolds number is almost entirely constant-speed motion at the terminal speed. The force needed to propel a sphere at its terminal speed is

$$F = 6\pi\eta r v_{\text{term}} \tag{5.15}$$

We can find the terminal speed of a small sphere falling in air by using $F = mg$, as we did for high Reynolds number, but this approach doesn't quite work for objects falling in liquids. The reason is that the upward buoyant force of the liquid—a topic we'll study in Chapter 13—cannot be neglected.We'll return to this after we introduce buoyancy.

**EXAMPLE 5.16 Measuring the mass of a pollen grain** BIO

Pollen grains are very light. In one experiment to determine their mass, researchers dropped grains inside a clear glass cylinder and then watched their motion with a microscope. A 40-$\mu$m-diameter pollen grain was observed to fall at a rate of 5.3 cm/s. What is the mass in nanograms of this grain?

***STRATEGIZE*** We will model the grain as a sphere. At terminal speed, the force of gravity on the grain is equal to the drag force. Because this is a very small object moving quite slowly, its Reynolds number is very low, so the drag force is linear drag.

***PREPARE*** We'll assume the air in the cylinder was at a room temperature of 20°C and use the value for air viscosity in Table 5.3. In SI units, the terminal speed is $v_{term} = 0.053$ m/s.

***SOLVE*** We set the force of gravity equal to the linear drag force to get

$$mg = 6\pi\eta r v_{term}$$

We can solve for the mass as

$$\begin{aligned} m &= \frac{6\pi\eta r v_{term}}{g} \\ &= \frac{6\pi(1.8\times10^{-5}\ \text{Pa}\cdot\text{s})(20\times10^{-6}\ \text{m})(0.053\ \text{m/s})}{9.8\ \text{m/s}^2} \\ &= 3.7\times10^{-11}\ \text{kg} \end{aligned}$$

We can convert this result to nanograms (ng) as

$$m = 3.7\times10^{-11}\ \text{kg}\times\frac{1000\ \text{g}}{\text{kg}}\times\frac{1\ \text{ng}}{10^{-9}\ \text{g}} = 37\ \text{ng}$$

***ASSESS*** This extremely tiny mass is hard to assess. If we assume that the grain's density is similar to that of water (typical for living objects), then the mass of the 40-*um*-diameter sphere is about 30 ng, which is close to our result. Our answer is reasonable.

## Life at Low Reynolds Number BIO

From protozoa swimming in pond water to bacteria moving in intercellular fluid (essentially water), biology is filled with examples of motion at extremely low Reynolds number. For example, a 1-$\mu$m-diameter bacterium moving at a typical speed of 30 $\mu$m/s has $Re \approx 3\times10^{-5}$. Life at such low Reynolds number is very much at odds with our everyday experience.

The motion of your body and other everyday objects is at high Reynolds number, where inertia plays a central role. When you apply a force to an object, because of its inertia it takes time for it to gain speed. When you remove the force, drag forces don't cause the object to stop immediately; because of its inertia it takes some time for the object to coast to a stop. Newton's second law, which relates force and motion, is key to understanding motion at high Reynolds number.

Motion at extremely low Reynolds number couldn't be more different. After a force is applied, the object reaches its terminal speed $v_{term}$ *almost instantaneously* and then moves at constant speed. When the force is removed, the object comes to a dead halt *almost instantaneously*. Viscosity is so much more important than inertia that there's no coasting. The only way a tiny organism can move at all is to continuously apply a force—as do the cilia and flagella, the microscopic strands that propel microorganisms.

Suppose a tiny spherical organism of mass $m$ is moving at its terminal speed $v_{term}$ in response to an applied force $F$. If the applied force is removed, only the drag force remains. How long does it take to stop? We can estimate this time by using kinematics. If the drag force slows the organism with constant acceleration $a$, then we have

$$\Delta v = v_f - v_i = 0 - v_{term} = -a\Delta t = -\frac{F_{drag}}{m}\Delta t = \frac{-6\pi\eta r v_{term}}{\text{m}}\Delta t$$

from which we can find the time to stop as

$$\Delta t_{stop} = \frac{m}{6\pi\eta r} \tag{5.16}$$

In addition, during this time interval, the organism coasts a distance $\Delta x_{stop} = v_{term}\Delta t_{stop}$. (The acceleration is not actually constant because the drag force gets smaller as the organism slows down, but these stopping times and distances are still reasonable estimates.)

**EXAMPLE 5.17 A *Paramecium* coasts to a stop** BIO

*Paramecia,* common unicellular inhabitants of freshwater ponds, are about 100 $\mu$m in length and swim at about 200 $\mu$m/s. If a *Paramecium* suddenly stops swimming, how long does it take to come to a stop, and how far does it coast while stopping?

**STRATEGIZE** A *Paramecium*'s swimming speed is its terminal speed $v_{term}$ in response to the pushing force generated by beating its cilia. Although *Paramecia* are somewhat elongated, we'll simplify the situation by modeling the *Paramecium* as a 100-$\mu$m-diameter sphere. The density of a *Paramecium* is essentially that of water, so we can calculate its mass from the density of water and the volume of a sphere. We'll assume a moderate water temperature of 20°C.

**PREPARE** The mass is $m = \frac{4}{3}\pi r^3 \rho$, where $\rho$ is the density. We need to know the density of water and its viscosity at 20°C, both of which are given in Table 5.3.

**SOLVE** If we use the expression for the *Paramecium*'s mass in Equation 5.16, then the time it takes to come to a stop is

$$\Delta t_{stop} = \frac{\frac{4}{3}\pi r^3 \rho}{6\pi\eta r} = \frac{2r^2\rho}{9\eta} = \frac{2(50 \times 10^{-6}\text{ m})^2(1000\text{ kg/m}^3)}{9(1.0 \times 10^{-3}\text{ Pa}\cdot\text{s})}$$

$$= 5.6 \times 10^{-4}\text{ s} = 0.56\text{ ms}$$

During this time, the *Paramecium* travels distance

$$\Delta x_{stop} = v_{term}\Delta t_{stop} = (200 \times 10^{-6}\text{ m/s})(5.6 \times 10^{-4}\text{ s})$$

$$= 1.1 \times 10^{-7}\text{ m} = 0.11\ \mu\text{m}$$

**ASSESS** Stopping in less than 1 ms over a distance of 0.11 $\mu$m, which is only $\frac{1}{1000}$ the diameter of the organism, is, for all practical purposes, an instantaneous stop. This agrees with our assertion about the motion of microorganisms.

You can swim through water because your hands are able to push the water away from you, but swimming is much more challenging at very low Reynolds number. If you tried to swim through an extremely viscous fluid, like honey, your back-and-forth strokes would cause you to rock back and forth but there would be no net motion.

Microorganisms that need to move at very low Reynolds number had to evolve a different form of locomotion, their flagella and cilia. FIGURE 5.26 shows a simple model of a flagellum like the one a bacterium uses; it's basically a rotating corkscrew. The rotary motion—rather than the back-and-forth stroking motion of human swimming—enables the flagellum to apply a continuous push to the fluid. Thus the fluid, by Newton's third law, applies a continuous pushing force to the organism, and, as we've just seen, the organism almost instantaneously reaches its terminal speed and continues swimming at that constant speed.

Protozoa are propelled by cilia. The beating motion of cilia is more complex than the motion of a rotating flagellum, but the end result is the same: a constant push against the fluid. The essential point is that the physics of life at low Reynolds number dictates evolutionary constraints not faced by larger organisms.

FIGURE 5.26 A rotating flagellum exerts a continuous, propeller-like push on the fluid.

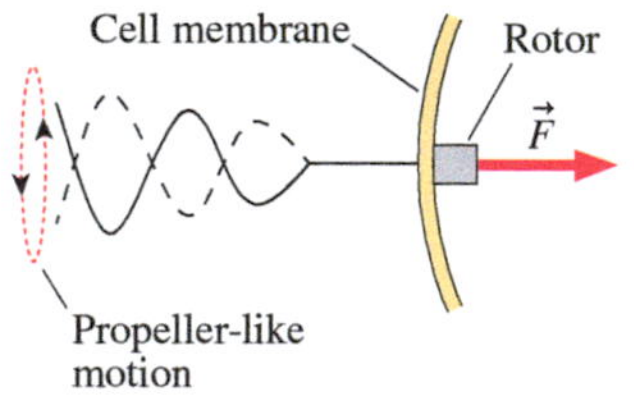


**STOP TO THINK 5.6** The terminal speed of a Styrofoam ball is 15 m/s. Suppose a Styrofoam ball is shot straight down with an initial speed of 30 m/s. Which velocity graph is correct?

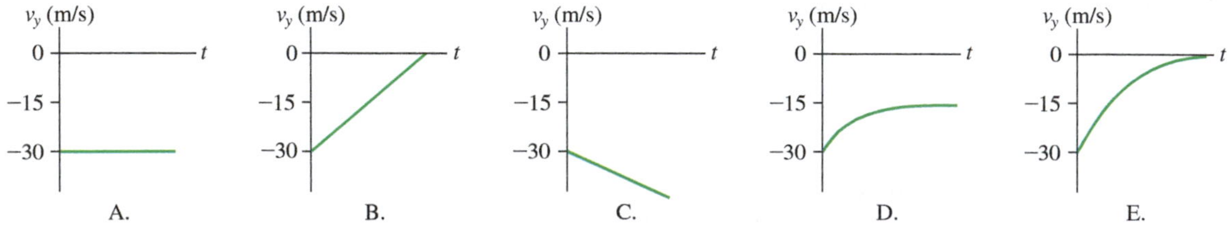


## Catalog of Forces Revisited

Before we continue, let's summarize the details of the different forces that we've seen so far, extending the catalog of forces that we presented in ◀◀SECTION 4.2.

**SYNTHESIS 5.1 A catalog of forces**

When solving mechanics problems, you'll often use the directions and details of the most common forces, outlined here.

**Weight**

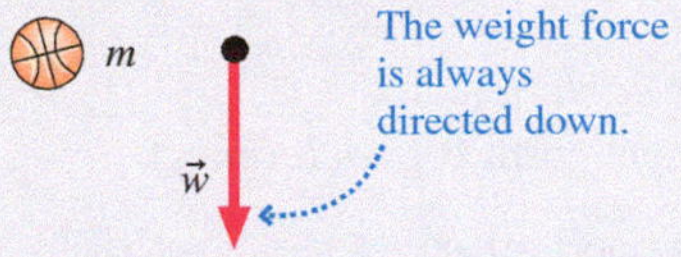

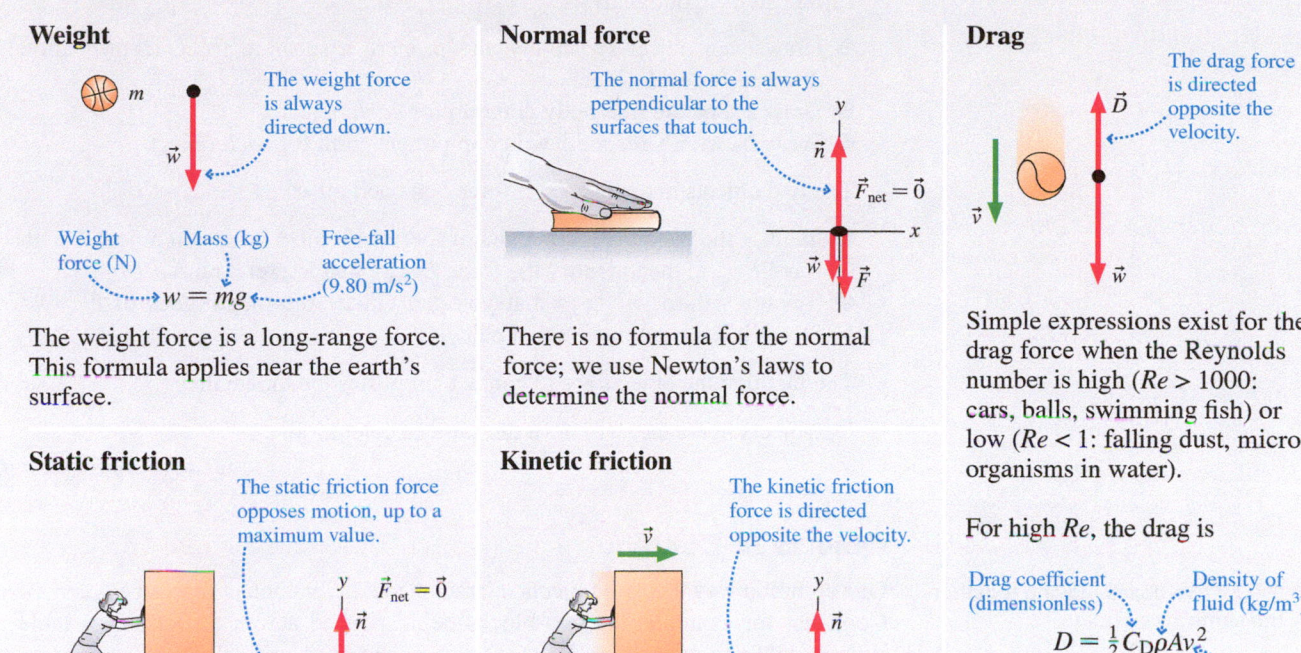

Weight force (N) · Mass (kg) · Free-fall acceleration (9.80 m/s²)

$$w = mg$$

The weight force is a long-range force. This formula applies near the earth's surface.

**Normal force**

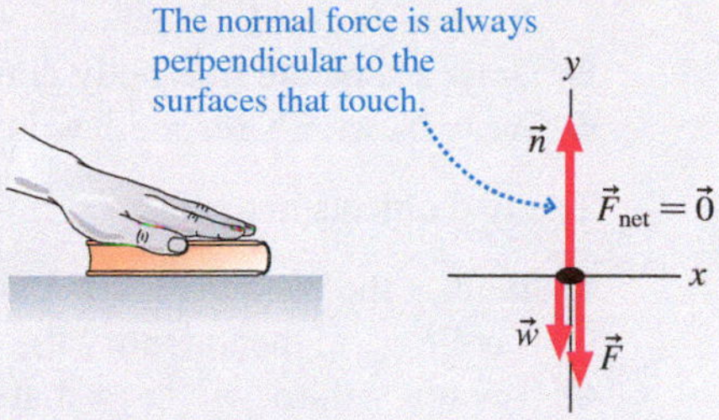


There is no formula for the normal force; we use Newton's laws to determine the normal force.

**Drag**

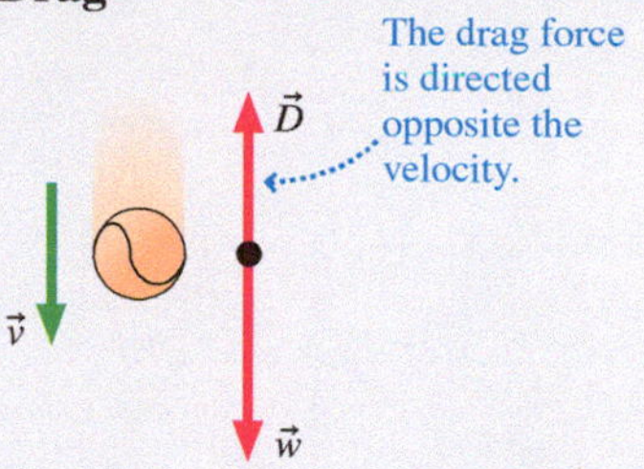


Simple expressions exist for the drag force when the Reynolds number is high ($Re > 1000$: cars, balls, swimming fish) or low ($Re < 1$: falling dust, microorganisms in water).

For high $Re$, the drag is

Drag coefficient (dimensionless) · Density of fluid (kg/m³) · Cross-section area (m²) · Speed (m/s)

$$D = \frac{1}{2} C_D \rho A v^2$$

For low $Re$, the drag on a spherical object is

$$D_{sphere} = 6\pi\eta r v$$

Viscosity (Pa·s) · Radius (m) · Speed (m/s)

**Static friction**

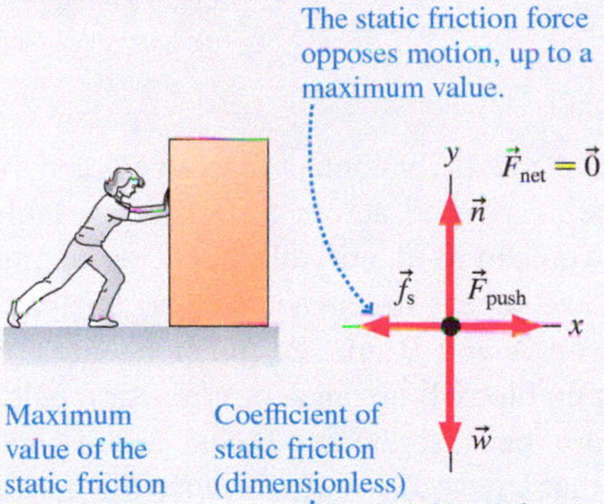


Maximum value of the static friction force (N) · Coefficient of static friction (dimensionless) · Normal force (N)

$$f_s \le f_{s\,max} = \mu_s n$$

**Kinetic friction**

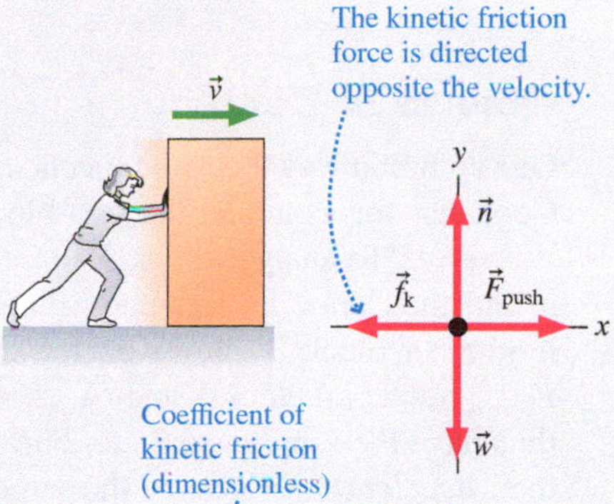


Coefficient of kinetic friction (dimensionless) · Normal force (N)

$$f_k = \mu_k n$$

## 5.7 Interacting Objects

Up to this point we have studied the dynamics of a single object subject to forces exerted on it by other objects. In Example 5.11, for instance, the sofa was acted upon by friction, normal, weight, and pushing forces that came from the floor, the earth, and the person pushing. As we've seen, such problems can be solved by an application of Newton's second law after all the forces have been identified.

But in Chapter 4 we found that real-world motion often involves two or more objects interacting with each other. We further found that forces always come in action/reaction *pairs* that are related by Newton's third law. To remind you, Newton's third law states:

- Every force occurs as one member of an action/reaction pair of forces. The two members of the pair always act on *different* objects.
- The two members of an action/reaction pair point in *opposite* directions and are *equal* in magnitude.

Our goal in this section is to learn how to apply the second *and* third laws to interacting objects.

**TACTICS BOX 5.2 Working with objects in contact**

When two objects are in contact and their motion is linked, we need to duplicate certain steps in our analysis:

1. Draw each object separately and prepare a separate force identification diagram for *each* object.
2. Draw a separate free-body diagram for *each* object.
3. Write Newton's second law in component form for *each* object.

The two objects in contact exert forces on each other:

4. Identify the action/reaction pairs of forces. If object A acts on object B with force $\vec{F}_{\text{A on B}}$, then identify the force $\vec{F}_{\text{B on A}}$ that B exerts on A.
5. Newton's third law says that you can equate the magnitudes of the two forces in each action/reaction pair.

The fact that the objects are in contact simplifies the kinematics:

6. Objects in contact will have the same acceleration.

Exercises 29, 30, 31

## Objects in Contact

FIGURE 5.27 Two boxes moving together have the same acceleration.

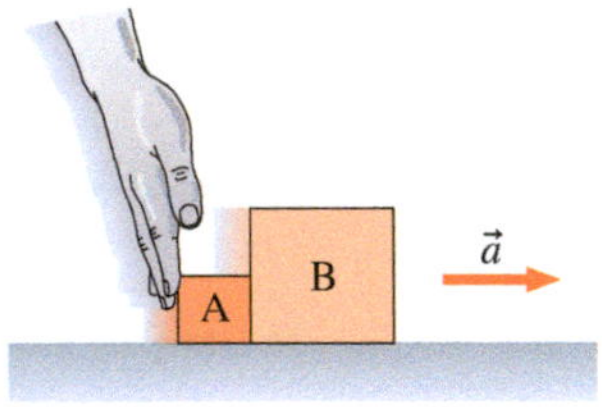


One common way that two objects interact is via direct contact forces between them. Consider, for example, the two blocks being pushed across a frictionless table in FIGURE 5.27. To analyze block A's motion, we need to identify all the forces acting on it and then draw its free-body diagram. We repeat the same steps to analyze the motion of block B. However, the forces on A and B are *not* independent: Forces $\vec{F}_{\text{B on A}}$ acting on block A and $\vec{F}_{\text{A on B}}$ acting on block B are an action/reaction pair and thus have the same magnitude. Furthermore, because the two blocks are in contact, their *accelerations* must be the same, so that $a_{\text{A}x} = a_{\text{B}x} = a_x$. Because the accelerations of both blocks are equal, we can omit the subscripts A and B and call both accelerations $a_x$.

These observations suggest that we can't solve for the motion of one block without considering the motion of the other block. Solving a motion problem thus means solving two problems in parallel.

**NOTE** ▶ Two steps are especially important when you draw the free-body diagrams. First, draw a *separate* diagram for each object. They need not have the same coordinate system. Second, show only the forces acting *on* that object. The force $\vec{F}_{\text{A on B}}$ goes on the free-body diagram of object B, but $\vec{F}_{\text{B on A}}$ goes on the diagram of object A. The two members of an action/reaction pair *always* appear on two different free-body diagrams—*never* on the same diagram. ◀

### EXAMPLE 5.18 Pushing two blocks

FIGURE 5.28 shows a 5.0 kg block A being pushed with a 3.0 N force. In front of this block is a 10 kg block B; the two blocks move together. What force does block A exert on block B?

***STRATEGIZE*** The two blocks are in contact and are moving together, so we will follow the steps in Tactics Box 5.2.

FIGURE 5.28 Two blocks are pushed by a hand.

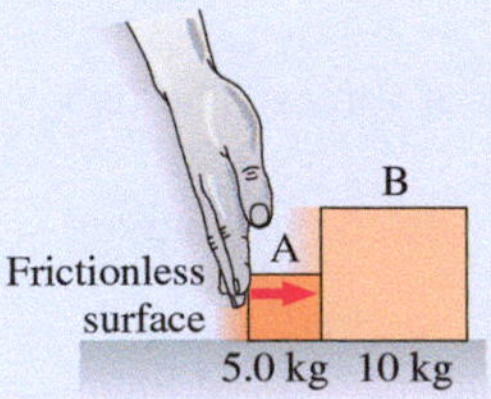


***PREPARE*** The visual overview of FIGURE 5.29 lists the known information and identifies $F_{\text{A on B}}$ as what we're trying to find. Then, following the steps of Tactics Box 5.2, we've drawn *separate* force identification diagrams and *separate* free-body diagrams for the two blocks. Both blocks have a weight force and a normal force, so we've used subscripts A and B to distinguish between them.

The force $\vec{F}_{\text{A on B}}$ is the contact force that block A exerts on B; it forms an action/reaction pair with the force $\vec{F}_{\text{B on A}}$ that block B exerts on A. Notice that force $\vec{F}_{\text{A on B}}$ is drawn acting on block B; it is the force *of* A *on* B. **Force vectors are always drawn on the**

**free-body diagram of the object that *experiences* the force,** not the object exerting the force. Because action/reaction pairs act in opposite directions, force $\vec{F}_{\text{B on A}}$ pushes backward on block A and appears on A's free-body diagram.

**SOLVE** We begin by writing Newton's second law in component form for each block. Because the motion is only in the $x$-direction, we need only the $x$-component of the second law. For block A,

$$\sum F_x = (F_{\text{H}})_x + (F_{\text{B on A}})_x = m_{\text{A}} a_{\text{A}x}$$

The force components can be "read" from the free-body diagram, where we see $\vec{F}_{\text{H}}$ pointing to the right and $\vec{F}_{\text{B on A}}$ pointing to the left. Thus

$$F_{\text{H}} - F_{\text{B on A}} = m_{\text{A}} a_{\text{A}x}$$

For B, we have

$$\sum F_x = (F_{\text{A on B}})_x = F_{\text{A on B}} = m_{\text{B}} a_{\text{B}x}$$

We have two additional pieces of information: First, Newton's third law tells us that $F_{\text{B on A}} = F_{\text{A on B}}$. Second, the boxes are in contact and must have the same acceleration $a_x$; that is, $a_{\text{A}x} = a_{\text{B}x} = a_x$. With this information, the two $x$-component equations become

$$F_{\text{H}} - F_{\text{A on B}} = m_{\text{A}} a_x$$

$$F_{\text{A on B}} = m_{\text{B}} a_x$$

Our goal is to find $F_{\text{A on B}}$, so we need to eliminate the unknown acceleration $a_x$. From the second equation, $a_x = F_{\text{A on B}}/m_{\text{B}}$. Substituting this into the first equation gives

$$F_{\text{H}} - F_{\text{A on B}} = \frac{m_{\text{A}}}{m_{\text{B}}} F_{\text{A on B}}$$

This can be solved for the force of block A on block B, giving

$$F_{\text{A on B}} = \frac{F_{\text{H}}}{1 + m_{\text{A}}/m_{\text{B}}} = \frac{3.0\ \text{N}}{1 + (5.0\ \text{kg})/(10\ \text{kg})} = \frac{3.0\ \text{N}}{1.5} = 2.0\ \text{N}$$

**ASSESS** Force $F_{\text{H}}$ accelerates both blocks, a total mass of 15 kg, but force $F_{\text{A on B}}$ accelerates only block B, with a mass of 10 kg. Thus it makes sense that $F_{\text{A on B}} < F_{\text{H}}$.

FIGURE 5.29 A visual overview of the two blocks.

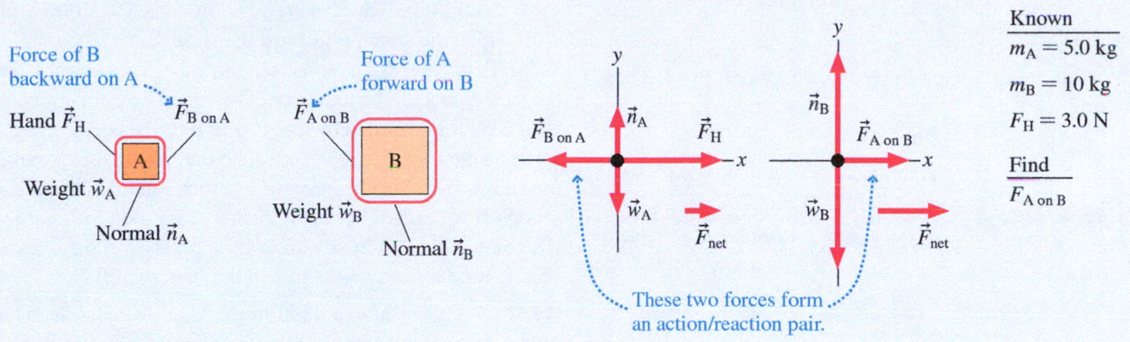


**STOP TO THINK 5.7** Boxes P and Q are sliding to the right across a frictionless table. The hand H is slowing them down. The mass of P is larger than the mass of Q. Rank in order, from largest to smallest, the *horizontal* forces on P, Q, and H.

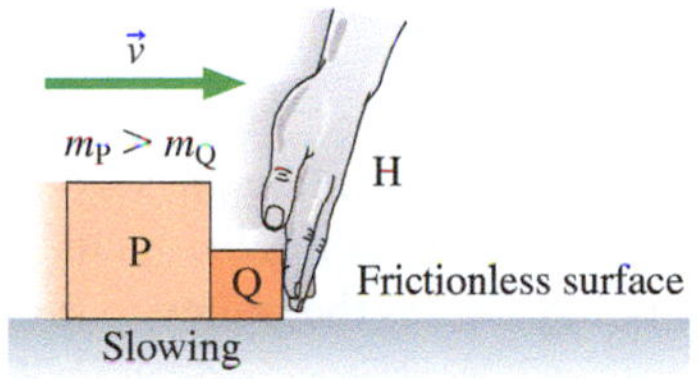


A. $F_{\text{Q on H}} = F_{\text{H on Q}} = F_{\text{P on Q}} = F_{\text{Q on P}}$
B. $F_{\text{Q on H}} = F_{\text{H on Q}} > F_{\text{P on Q}} = F_{\text{Q on P}}$
C. $F_{\text{Q on H}} = F_{\text{H on Q}} < F_{\text{P on Q}} = F_{\text{Q on P}}$
D. $F_{\text{H on Q}} = F_{\text{H on P}} > F_{\text{P on Q}}$

## 5.8 Ropes and Pulleys

Many objects are connected by strings, ropes, cables, and so on. We can learn several important facts about ropes and tension by considering the box being pulled by a rope in FIGURE 5.30. The rope in turn is being pulled by a hand that exerts a force $\vec{F}$ on the rope.

The box is pulled by the rope, so the box's free-body diagram shows a tension force $\vec{T}$. The *rope* is subject to two horizontal forces: the force $\vec{F}$ of the hand on the rope, and the force $\vec{F}_{\text{box on rope}}$ with which the box pulls back on the rope. In problems

FIGURE 5.30 A box being pulled by a rope.

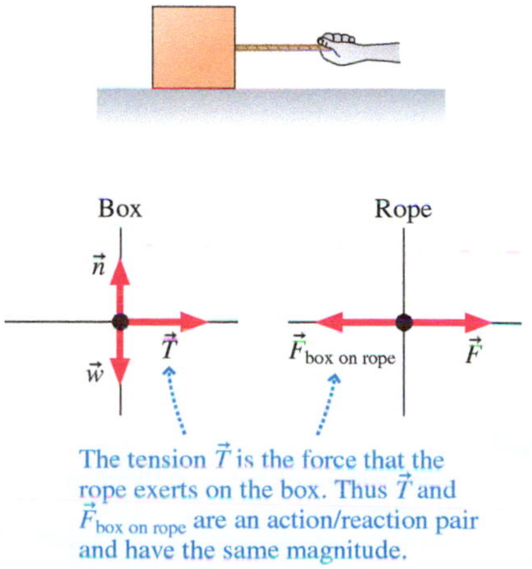

we'll consider, the mass of a string or rope is significantly less than the mass of the objects it pulls on, so we'll make the approximation—called the **massless string approximation**—that $m_{\text{rope}} = 0$. There is no weight force acting on the string, and so no supporting force is necessary. In this case, there are no forces that act along the vertical axis. $\vec{T}$ and $\vec{F}_{\text{box on rope}}$ form an action/reaction pair, so their magnitudes are equal: $F_{\text{box on rope}} = T$. Newton's second law *for the rope* is thus

$$\sum F_x = F - F_{\text{box on rope}} = F - T = m_{\text{rope}}a_x = 0 \qquad (5.17)$$

We've used the approximation $m_{\text{rope}} = 0$. In this case, we can say that $T = F$. Generally, **the tension in a massless string or rope equals the magnitude of the force pulling on the end of the string or rope.** As a result:

- A massless string or rope "transmits" a force undiminished from one end to the other: If you pull on one end of a rope with force $F$, the other end of the rope pulls on what it's attached to with a force of the same magnitude $F$.
- The tension in a massless string or rope is the same from one end to the other.

**CONCEPTUAL EXAMPLE 5.19 Pulling a rope**

FIGURE 5.31a shows a student pulling horizontally with a 100 N force on a rope that is attached to a wall. In FIGURE 5.31b, two students in a tug-of-war pull on opposite ends of a rope with 100 N each. Is the tension in the second rope larger than, smaller than, or the same as that in the first?

FIGURE 5.31 Pulling on a rope in two ways. Which produces a larger tension?

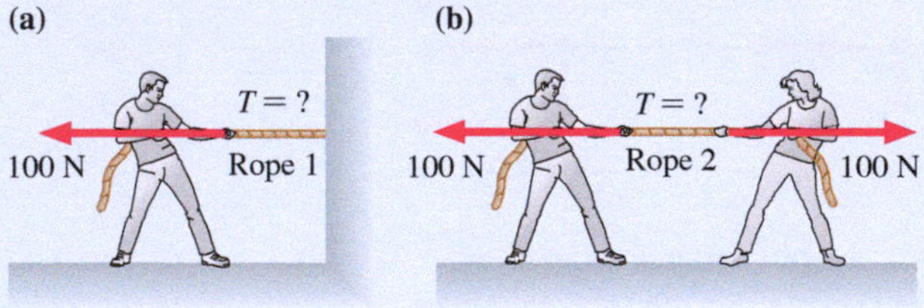


***REASON*** Surely pulling on a rope from both ends causes more tension than pulling on one end. Right? Before jumping to conclusions, let's analyze the situation carefully. We found previously that the force pulling on the end of a rope—here, the 100 N force exerted by the student—and the tension in the rope have the same magnitude. Thus, the tension in rope 1 is 100 N, the force with which the student pulls on the rope.

To find the tension in the second rope, consider the force that the *wall* exerts on the *first* rope. The first rope is in equilibrium, so the 100 N force exerted by the student must be balanced by a 100 N force on the rope from the wall. The first rope is being pulled from *both* ends by a 100 N force—the exact same situation as for the second rope, pulled by the students. A rope doesn't care whether it's being pulled on by a wall or by a person, so the tension in the second rope is the *same* as that in the first, or 100 N.

***ASSESS*** This example reinforces what we just learned about ropes: A rope pulls on the objects at each of its ends with a force equal in magnitude to the tension, and the external force applied to each end of the rope and the rope's tension have equal magnitude.

## Pulleys

Strings and ropes often pass over pulleys. FIGURE 5.32 shows block B dragging block A across a table as it falls. As the string moves, static friction between the string and the pulley causes the pulley to turn. If we assume that

- The string *and* the pulley are both massless, and
- There is no friction where the pulley turns on its axle,

then no net force is needed to accelerate the string or turn the pulley. In this case, **the tension in a massless string is unchanged by passing over a massless, frictionless pulley.** We'll assume such an *ideal pulley* for problems in this chapter.

FIGURE 5.32 An ideal pulley changes the direction in which a tension force acts, but not its magnitude.

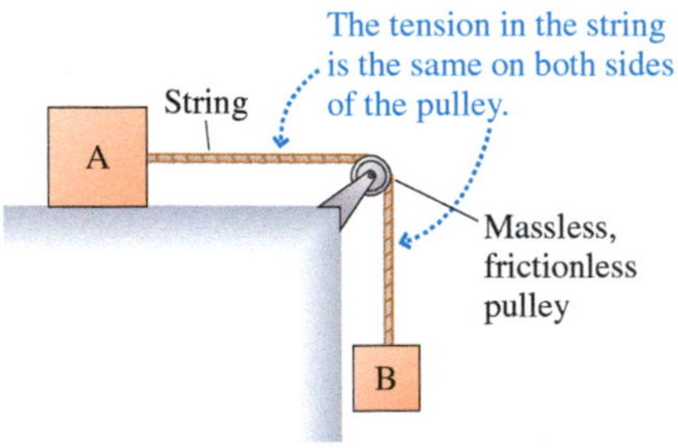


**TACTICS BOX 5.3 Working with ropes and pulleys**

For massless ropes or strings and massless, frictionless pulleys:

❶ If a force pulls on one end of a rope, the tension in the rope equals the magnitude of the pulling force.
❷ If two objects are connected by a rope, the tension is the same at both ends.
❸ If the rope passes over a pulley, the tension in the rope is unaffected.

Exercises 33, 34

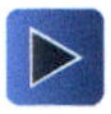

**Video** Tension in String between Hanging Weights

## EXAMPLE 5.20 Placing a leg in traction BIO

For serious fractures of the leg, the leg may need to have a stretching force applied to it to keep contracting leg muscles from forcing the broken bones together too hard. This is often done using *traction,* an arrangement of a rope, a weight, and pulleys as shown in FIGURE 5.33. The rope must make the same angle $\theta$ on both sides of the pulley so that the net force of the rope on the pulley is horizontally to the right, but $\theta$ can be adjusted to control the amount of traction. The doctor has specified 50 N of traction for this patient, with a 4.2 kg hanging mass. What is the proper angle $\theta$?

FIGURE 5.33 A leg in traction.

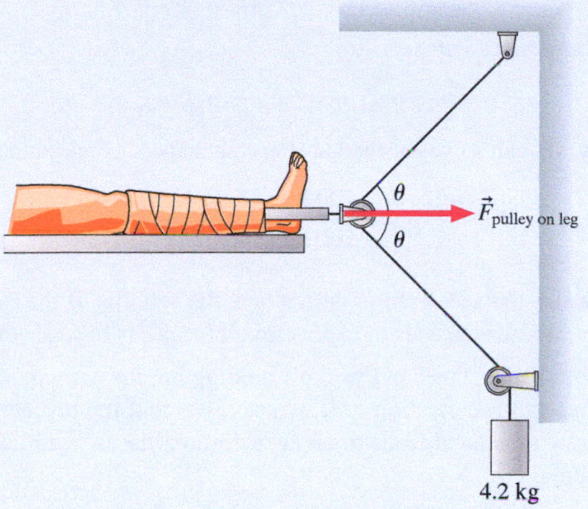


FIGURE 5.34 Free-body diagram for the pulley.

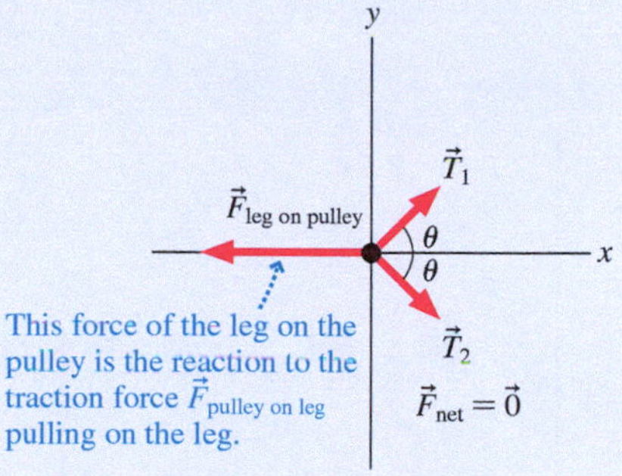


**STRATEGIZE** We will find the tension in the rope from the weight that hangs from it. The rope pulls on the upper pulley with *two tension forces, pointing in the two directions in which the rope leaves the pulley.*

**PREPARE** The pulley attached to the patient's leg is in static equilibrium, so the net force on it must be zero. FIGURE 5.34 shows a free-body diagram for the pulley, which we assume to be frictionless. Forces $\vec{T}_1$ and $\vec{T}_2$ are the tension forces of the rope as it pulls on the pulley. These forces are equal in magnitude for a frictionless pulley, and their combined pull is to the right. This force is balanced by the force $\vec{F}_{\text{leg on pulley}}$ of the patient's leg pulling to the left. The traction force $\vec{F}_{\text{pulley on leg}}$ forms an action/reaction pair with $\vec{F}_{\text{leg on pulley}}$, so 50 N of traction means that $\vec{F}_{\text{leg on pulley}}$ also has a magnitude of 50 N.

**SOLVE** Two important properties of ropes, given in Tactics Box 5.3, are that (1) the tension equals the magnitude of the force pulling on its end and (2) the tension is the same throughout the rope. Thus, if a hanging mass $m$ pulls on the rope with its weight $mg$, the tension along the entire rope is $T = mg$. For a 4.2 kg hanging mass, the tension is then $T = mg = 41.2$ N.

The pulley, in equilibrium, must satisfy Newton's second law for the case where $\vec{a} = \vec{0}$. Thus

$$\sum F_x = T_{1x} + T_{2x} + (F_{\text{leg on pulley}})_x = ma_x = 0$$

The tension forces both have the same magnitude $T$, and both are at angle $\theta$ from horizontal. The $x$-component of the leg force is negative because it's directed to the left. Then Newton's law becomes

$$2T\cos\theta - F_{\text{leg on pulley}} = 0$$

so that

$$\cos\theta = \frac{F_{\text{leg on pulley}}}{2T} = \frac{50\text{ N}}{82.4\text{ N}} = 0.607$$

$$\theta = \cos^{-1}(0.607) = 53°$$

**ASSESS** The traction force would approach $2mg = 82$ N if angle $\theta$ approached zero because the two tensions would pull in parallel. Conversely, the traction force would approach 0 N if $\theta$ approached 90°. Because the desired traction force is roughly halfway between 0 N and 82 N, an angle near 45° is reasonable.

## EXAMPLE 5.21 Lifting a stage set

A 200 kg set used in a play is stored in the loft above the stage. The rope holding the set passes up and over a pulley, then is tied backstage. The director tells a 100 kg stagehand to lower the set. When he unties the rope, the set falls and the unfortunate man is hoisted into the loft. What is the stagehand's acceleration?

**STRATEGIZE** The acceleration of each object—stagehand and set—is determined by the forces that act on it, so we will start by drawing a separate free-body diagram for each. From Tactics Box 5.3 we know that the tension force acting on each object has the same magnitude. In addition, their accelerations are related because they are connected by the rope.

**PREPARE** FIGURE 5.35 shows the visual overview. The objects of interest are the stagehand M and the set S, for which we've drawn separate free-body diagrams. Assume a massless rope and a massless, frictionless pulley. Tension forces $\vec{T}_S$ and $\vec{T}_M$ are due to a massless rope going over an ideal pulley, so their magnitudes are the same.

**SOLVE** From the two free-body diagrams, we can write Newton's second law in component form. For the man we have

$$\sum F_{My} = T_M - w_M = T_M - m_M g = m_M a_{My}$$

*Continued*

FIGURE 5.35 Visual overview for the stagehand and set.

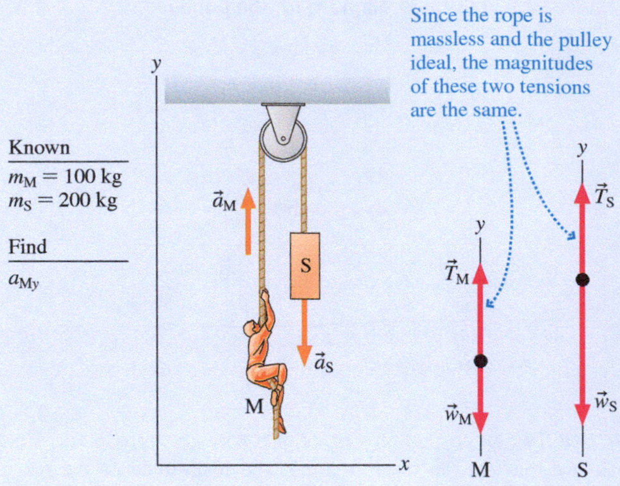


For the set we have

$$\sum F_{Sy} = T_S - w_S = T_S - m_S g = m_S a_{Sy}$$

Only the $y$-equations are needed. Because the stagehand and the set are connected by a rope, the upward distance traveled by one is the *same* as the downward distance traveled by the other. Thus the *magnitudes* of their accelerations must be the same, but, as Figure 5.35 shows, their *directions* are opposite. We can express this mathematically as $a_{Sy} = -a_{My}$. We also know that the two tension forces have equal magnitudes, which we'll call $T$. Inserting this information into the above equations gives

$$T - m_M g = m_M a_{My}$$

$$T - m_S g = -m_S a_{My}$$

These are simultaneous equations in the two unknowns $T$ and $a_{My}$. We can solve for $T$ in the first equation to get

$$T = m_M a_{My} + m_M g$$

Inserting this value of $T$ into the second equation then gives

$$m_M a_{My} + m_M g - m_S g = -m_S a_{My}$$

which we can rewrite as

$$(m_S - m_M)g = (m_S + m_M)a_{My}$$

Finally, we can solve for the hapless stagehand's acceleration:

$$a_{My} = \frac{m_S - m_M}{m_S + m_M} g = \left(\frac{100 \text{ kg}}{300 \text{ kg}}\right) \times 9.80 \text{ m/s}^2 = 3.3 \text{ m/s}^2$$

This is also the acceleration with which the set falls. If the rope's tension was needed, we could now find it from $T = m_M a_{My} + m_M g$.

**ASSESS** If the stagehand weren't holding on, the set would fall with free-fall acceleration $g$. It makes sense that the presence of the heavy stagehand leads to an acceleration that is significantly less than $g$.

## EXAMPLE 5.22 A not-so-clever bank robbery

Bank robbers have pushed a 1000 kg safe to a second-story floor-to-ceiling window. They plan to break the window, then lower the safe 3.0 m to their truck. Not being too clever, they stack up 500 kg of furniture, tie a rope between the safe and the furniture, and place the rope over a pulley. Then they push the safe out the window. What is the safe's speed when it hits the truck? The coefficient of kinetic friction between the furniture and the floor is 0.50.

**STRATEGIZE** We will again need to draw a separate free-body diagram for each object. The tension forces on the safe and the furniture are the same, and their accelerations are related because they are attached by the rope.

**PREPARE** The visual overview in FIGURE 5.36 establishes a coordinate system and defines the symbols that we need to calculate the safe's motion. The objects of interest are the safe S and the

FIGURE 5.36 Visual overview of the furniture and falling safe.

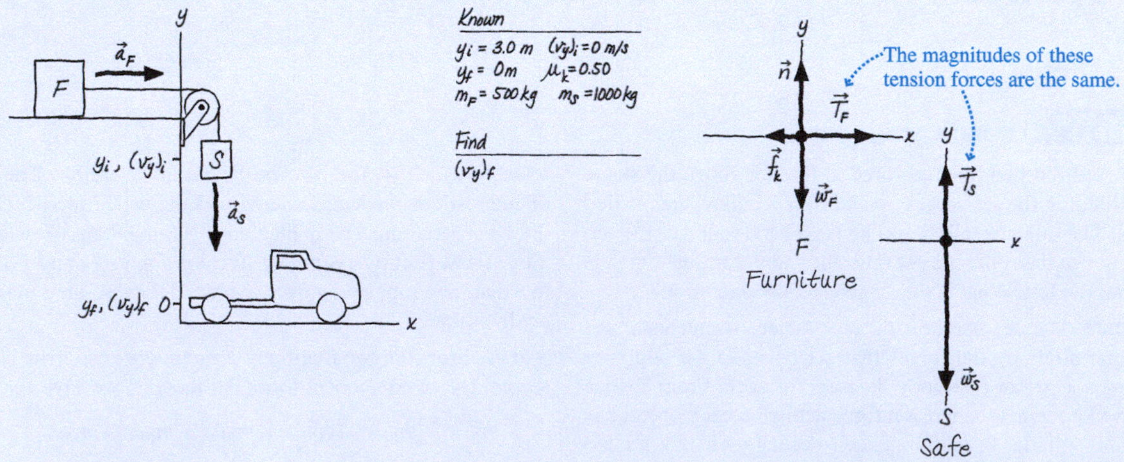

furniture F, which we model as particles. We assume a massless rope and a massless, frictionless pulley. The tension is then the same everywhere in the rope.

**SOLVE** We can write Newton's second law directly from the free-body diagrams. For the furniture,

$$\sum F_{Fx} = T_F - f_k = T - f_k = m_F a_{Fx}$$
$$\sum F_{Fy} = n - w_F = n - m_F g = 0$$

And for the safe,

$$\sum F_{Sy} = T_S - w_S = T - m_S g = m_S a_{Sy}$$

The safe and the furniture are tied together, so their accelerations have the same magnitude. But as the furniture slides to the right with positive acceleration $a_{Fx}$, the safe falls in the negative $y$-direction, so its acceleration $a_{Sy}$ is negative; we can express this mathematically as $a_{Fx} = -a_{Sy}$. We also have made use of the fact that $T_S = T_F = T$. We have one additional piece of information, the model of kinetic friction:

$$f_k = \mu_k n = \mu_k m_F g$$

where we used the $y$-equation of the furniture to deduce that $n = m_F g$. We substitute this result for $f_k$ into the $x$-equation of the furniture, then rewrite the furniture's $x$-equation and the safe's $y$-equation:

$$T - \mu_k m_F g = -m_F a_{Sy}$$
$$T - m_S g = m_S a_{Sy}$$

We have succeeded in reducing our knowledge to two simultaneous equations in the two unknowns $a_{Sy}$ and $T$. We subtract the second equation from the first to eliminate $T$:

$$(m_S - \mu_k m_F)g = -(m_S + m_F)a_{Sy}$$

Finally, we can solve for the safe's acceleration:

$$a_{Sy} = -\left(\frac{m_S - \mu_k m_F}{m_S + m_F}\right)g$$
$$= -\frac{1000\text{ kg} - 0.5(500\text{ kg})}{1000\text{ kg} + 500\text{ kg}} \times 9.80\text{ m/s}^2 = -4.9\text{ m/s}^2$$

Now we need to calculate the kinematics of the falling safe. Because the time of the fall is not known or needed, we can use

$$(v_y)_f{}^2 = (v_y)_i{}^2 + 2a_{Sy}\,\Delta y = 0 + 2a_{Sy}(y_f - y_i) = -2a_{Sy}y_i$$
$$(v_y)_f = \sqrt{-2a_{Sy}y_i} = \sqrt{-2(-4.9\text{ m/s}^2)(3.0\text{ m})} = 5.4\text{ m/s}$$

The value of $(v_y)_f$ is negative, but we only needed to find the speed, so we took the absolute value. It seems unlikely that the truck will survive the impact of the 1000 kg safe!

**ASSESS** 5.4 m/s, about 12 mph, seems reasonable for a falling safe that is dragging heavy furniture.

Newton's three laws form the cornerstone of the science of mechanics. These laws allowed scientists to understand many diverse phenomena, from the motion of a raindrop to the orbits of the planets. We will continue to develop Newtonian mechanics in the next few chapters because of its tremendous importance to the physics of everyday life. But it's worth keeping in the back of your mind that Newton's laws aren't the ultimate statement about motion. Later in this text we'll reexamine motion and mechanics from the perspective of Einstein's theory of relativity.

**STOP TO THINK 5.8** All three 50 kg blocks are at rest. Is the tension in rope 2 greater than, less than, or equal to the tension in rope 1?

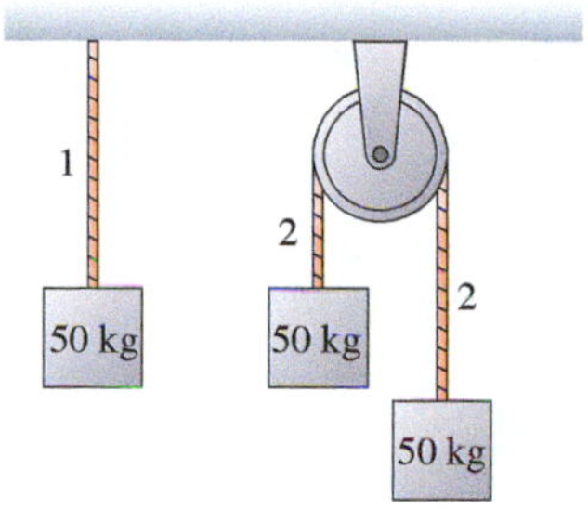

## INTEGRATED EXAMPLE 5.23 Stopping distances of a skidding car

A 1500 kg car is traveling at a speed of 30 m/s when the driver slams on the brakes and skids to a halt. Determine the stopping distance if the car is traveling up a 10° slope, down a 10° slope, or on a level road.

**STRATEGIZE** We will find the acceleration from the known forces and an application of Newton's second law. Then we will use kinematics to find the stopping distance.

**PREPARE** We represent the car as a particle and we use the model of kinetic friction. We want to solve the problem only once, not three separate times, so we'll leave the slope angle $\theta$ unspecified until the end.

FIGURE 5.37 shows the visual overview. We show the car sliding uphill, but these representations work equally well for a level or downhill slide if we let $\theta$ be zero or negative, respectively. We use a tilted coordinate system so that the motion is along the $x$-axis. The car *skids* to a halt, so we use the coefficient of *kinetic* friction for rubber on concrete from Table 5.2.

**SOLVE** Newton's second law and the model of kinetic friction are

$$\sum F_x = n_x + w_x + (f_k)_x$$
$$= 0 - mg\sin\theta - f_k = ma_x$$
$$\sum F_y = n_y + w_y + (f_k)_y$$
$$= n - mg\cos\theta + 0 = ma_y = 0$$

We've written these equations by "reading" the motion diagram and the free-body diagram. Notice that both components of the weight vector $\vec{w}$ are negative. $a_y = 0$ because the motion is entirely along the $x$-axis.

The second equation gives $n = mg\cos\theta$. Using this in the friction model, we find $f_k = \mu_k mg\cos\theta$. Inserting this result back into the first equation then gives

$$ma_x = -mg\sin\theta - \mu_k mg\cos\theta$$
$$= -mg(\sin\theta + \mu_k\cos\theta)$$
$$a_x = -g(\sin\theta + \mu_k\cos\theta)$$

This is a constant acceleration. Constant-acceleration kinematics gives

$$(v_x)_f^2 = 0 = (v_x)_i^2 + 2a_x(x_f - x_i) = (v_x)_i^2 + 2a_x x_f$$

which we can solve for the stopping distance $x_f$:

$$x_f = -\frac{(v_x)_i^2}{2a_x} = \frac{(v_x)_i^2}{2g(\sin\theta + \mu_k\cos\theta)}$$

Notice how the minus sign in the expression for $a_x$ canceled the minus sign in the expression for $x_f$. Evaluating our result at the three different angles gives the stopping distances:

$$x_f = \begin{cases} 48\text{ m} & \theta = 10° & \text{uphill} \\ 57\text{ m} & \theta = 0° & \text{level} \\ 75\text{ m} & \theta = -10° & \text{downhill} \end{cases}$$

The implications are clear about the danger of driving downhill too fast!

**ASSESS** 30 m/s ≈ 60 mph and 57 m ≈ 180 feet on a level surface. These are similar to the stopping distances you learned when you got your driver's license, so the results seem reasonable. Additional confirmation comes from noting that the expression for $a_x$ becomes $-g\sin\theta$ if $\mu_k = 0$. This is what you learned in Chapter 3 for the acceleration on a frictionless inclined plane.

FIGURE 5.37 Visual overview for a skidding car.

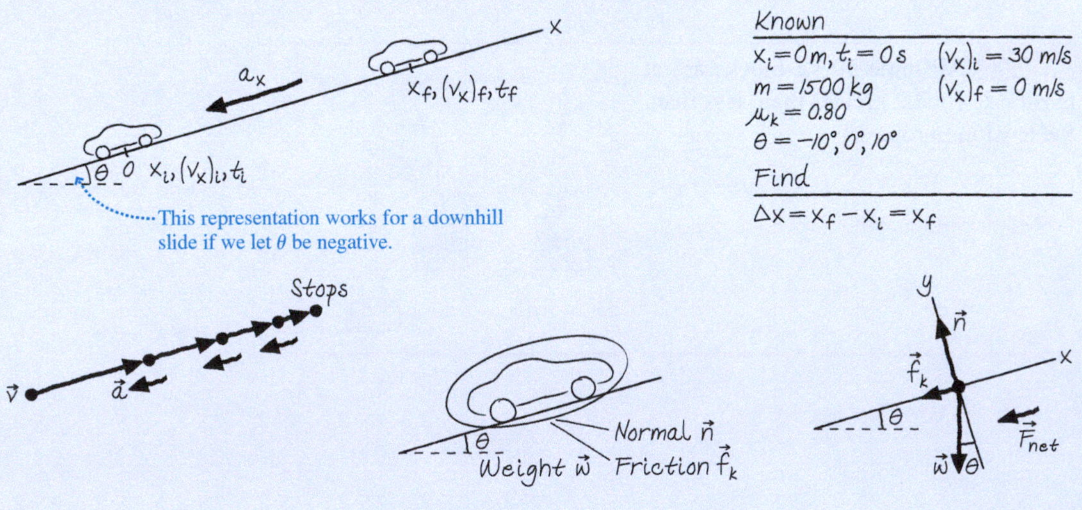

# SUMMARY

**GOAL** To use Newton's laws to solve equilibrium and dynamics problems.

## GENERAL STRATEGY

All examples in this chapter follow a four-part approach. You'll become a better problem solver if you adhere to it as you do the homework problems. The *Dynamics Worksheets* in the *Student Workbook* will help you structure your work in this way.

### Equilibrium Problems

Object at rest or moving at constant velocity.

**STRATEGIZE** The net force acting on an object in equilibrium is zero.

**PREPARE** Make simplifying assumptions.

- Check that the object is either at rest or moving with constant velocity ($\vec{a} = \vec{0}$).
- Identify forces and show them on a free-body diagram.

**SOLVE** Use Newton's second law in component form:

$$\sum F_x = ma_x = 0$$

$$\sum F_y = ma_y = 0$$

"Read" the components from the free-body diagram.

**ASSESS** Is your result reasonable?

### Dynamics Problems

Object accelerating.

**STRATEGIZE** Find the acceleration from the forces and Newton's second law, or use kinematics to find the acceleration and, from the second law, the forces.

**PREPARE** Make simplifying assumptions.

Make a **visual overview:**

- Sketch a pictorial representation.
- Identify known quantities and what the problem is trying to find.
- Identify all forces and show them on a free-body diagram.

**SOLVE** Use Newton's second law in component form:

$$\sum F_x = ma_x \quad \text{and} \quad \sum F_y = ma_y$$

"Read" the components of the vectors from the free-body diagram. If needed, use kinematics to find positions and velocities.

**ASSESS** Is your result reasonable?

## IMPORTANT CONCEPTS

Specific information about three important forces:

**Weight** $\vec{w} = (mg, \text{downward})$

**Friction** $\vec{f}_s = (0 \text{ to } \mu_s n, \text{direction as necessary to prevent motion})$

$\vec{f}_k = (\mu_k n, \text{direction opposite to the motion})$

$\vec{f}_r = (\mu_r n, \text{direction opposite to the motion})$

**Drag** Drag is directed opposite to the motion.

$D = \frac{1}{2}C_D\rho A v^2$, high $Re$

$D = 6\pi\eta r v$, sphere at low $Re$

Newton's laws are vector expressions. You must write them out by **components:**

$$(F_{net})_x = \sum F_x = ma_x$$

$$(F_{net})_y = \sum F_y = ma_y$$

For equilibrium problems, $a_x = 0$ and $a_y = 0$.

### Objects in contact

When two objects interact, you need to draw two separate free-body diagrams.

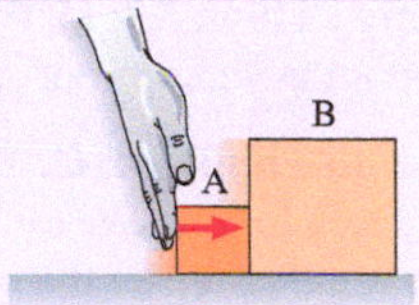


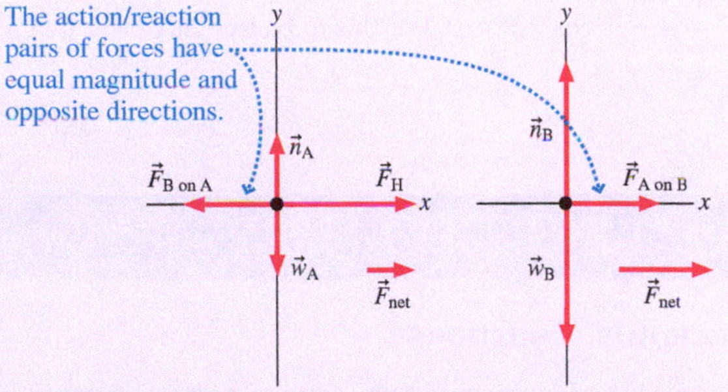


## APPLICATIONS

**Apparent weight** is the magnitude of the contact force supporting an object. It is what a scale reads, and it is your sensation of weight.

Apparent weight equals your true weight $w = mg$ only when the vertical acceleration is zero.

A falling object reaches **terminal speed** when the drag force exactly balances the weight force: $\vec{a} = \vec{0}$.

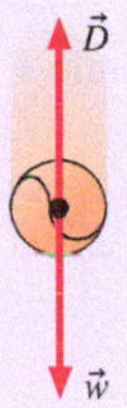


### Strings and pulleys

- A string or rope pulls what it's connected to with a force equal to its tension.
- The tension in a rope is equal to the force pulling on the rope.
- The tension in a massless rope is the same at all points in the rope.
- Tension does not change when a rope passes over a massless, frictionless pulley.

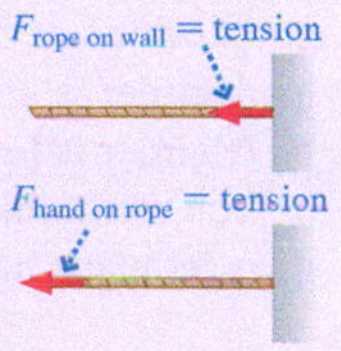

## Learning Objectives After studying this chapter, you should be able to:

- Solve problems about objects in equilibrium. *Conceptual Questions 5.1, 5.2; Problems 5.1, 5.2, 5.3, 5.6, 5.7*
- Use free-body diagrams, Newton's second law, and the problem-solving approach to solve dynamics problems. *Conceptual Question 5.6; Problems 5.10, 5.11, 5.12, 5.13, 5.14*
- Work with and distinguish between mass and weight. *Conceptual Questions 5.5, 5.13; Problems 5.16, 5.17, 5.18, 5.19, 5.20*
- Solve problems with sliding and rolling friction; understand how static friction can prevent motion. *Conceptual Questions 5.3, 5.4; Problems 5.24, 5.27, 5.28, 5.32, 5.33*
- Use the linear and quadratic models of drag to solve problems about motion through a fluid and to calculate terminal speeds. *Conceptual Questions 5.8, 5.15; Problems 5.35, 5.36, 5.37, 5.38, 5.40*
- Use Newton's third law to identify forces on and to solve problems about interacting objects. *Problems 5.41, 5.42, 5.43*
- Calculate the tension in ropes and solve problems involving ropes and pulleys. *Conceptual Questions 5.17, 5.20; Problems 5.44, 5.45, 5.46, 5.47, 5.48*

## STOP TO THINK ANSWERS

**Chapter Preview Stop to Think: B.** The elevator is moving at a steady speed, so $a = 0$ and therefore $F_{net} = 0$. There are two forces that act: the downward weight force and the upward tension force. Since $F_{net} = 0$, the magnitudes of these two forces must be equal.

**Stop to Think 5.1: A.** The ball is stationary, so the net force on it must be zero. Newton's second law for the vertical components of forces is $\Sigma F_y = T_y - w = 0$. The vertical component of the tension by itself is equal to $w$, so the tension must be greater than $w$.

**Stop to Think 5.2: A.** The lander is descending and slowing. The acceleration vector points upward, and so $\vec{F}_{net}$ points upward. This can be true only if the thrust has a larger magnitude than the weight.

**Stop to Think 5.3: D.** When you are in the air, there is no contact force supporting you, so your apparent weight is zero: You are weightless.

**Stop to Think 5.4: C.** The cyclist moves at a constant speed, so the net force is zero. We set the axes parallel to and perpendicular to the slope and compute the components of the weight force as in Figure 5.11. Newton's second law for the vertical components of forces is $\Sigma F_y = n - w \cos 20° = 0$. For this to be true, it's clear that $n < w$.

**Stop to Think 5.5: $f_B > f_C = f_D = f_E > f_A$.** Situations C, D, and E are all kinetic friction, which does not depend on either velocity or acceleration. Kinetic friction is less than the maximum static friction that is exerted in B. $f_A = 0$ because no friction is needed to keep the object at rest.

**Stop to Think 5.6: D.** The ball is shot *down* at 30 m/s, so $v_{0y} = -30$ m/s. This exceeds the terminal speed, so the upward drag force is *greater* than the downward weight force. Thus the ball *slows down* even though it is "falling." It will slow until $v_y = -15$ m/s, the terminal velocity, then maintain that velocity.

**Stop to Think 5.7: B.** $F_{\text{Q on H}} = F_{\text{H on Q}}$ and $F_{\text{P on Q}} = F_{\text{Q on P}}$ because these are action/reaction pairs. Box Q is slowing down and therefore must have a net force to the left. So from Newton's second law we also know that $F_{\text{H on Q}} > F_{\text{P on Q}}$.

**Stop to Think 5.8: Equal to.** Each block is hanging in equilibrium, with no net force, so the upward tension force is $mg$.

Video Tutor Solution Chapter 5

# QUESTIONS

## Conceptual Questions

1. An object is subject to two forces that do not point in opposite directions. Is it possible to choose their magnitudes so that the object is in equilibrium? Explain.
2. Are the objects described here in static equilibrium, dynamic equilibrium, or not in equilibrium at all?
   a. A girder is lifted at constant speed by a crane.
   b. A girder is lowered by a crane. It is slowing down.
   c. You're straining to hold a 200 lb barbell over your head.
   d. A jet plane has reached its cruising speed and altitude.
   e. A rock is falling into the Grand Canyon.
   f. A box in the back of a truck doesn't slide as the truck stops.
3. Boxes A and B in Figure Q5.3 both remain at rest. Is the friction force on A larger than, smaller than, or equal to the friction force on B? Explain.

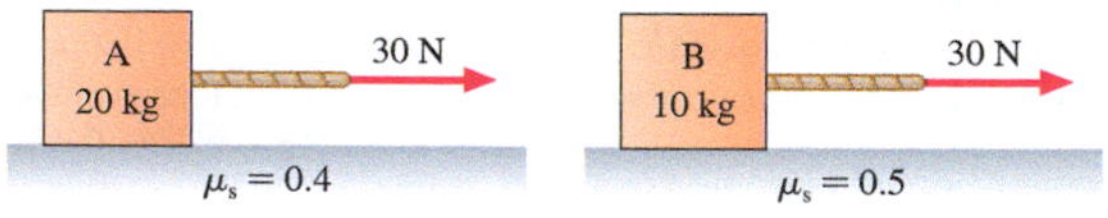


FIGURE Q5.3

Problem difficulty is labeled as | (straightforward) to ||||| (challenging). Problems labeled INT integrate significant material from earlier chapters; Problems labeled BIO are of biological or medical interest.

The eText icon indicates when there is a video tutor solution available for the chapter or for a specific problem. To launch these videos, log into your eText through Mastering™ Physics or log into the Study Area.

4. Boxes A and B in Figure Q5.3 are both sliding. Is the friction force on A larger than, smaller than, or equal to the friction force on B? Explain.
5. An astronaut takes his bathroom scale to the moon and then stands on it. Is the reading of the scale his true weight? Explain.
6. A light block of mass $m$ and a heavy block of mass $M$ are attached to the ends of a rope. A student holds the heavier block and lets the lighter block hang below it, as shown in Figure Q5.6. Then she lets go. Air resistance can be neglected.
   a. What is the tension in the rope while the blocks are falling, before either hits the ground?
   b. Would your answer be different if she had been holding the lighter block initially?

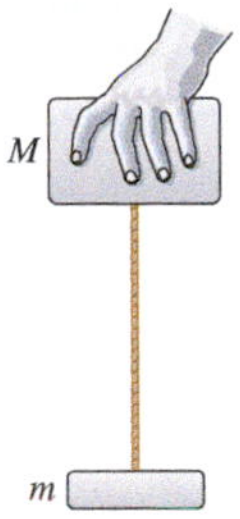

FIGURE Q5.6

7. a. Can the normal force on an object be directed horizontally? If not, why not? If so, provide an example.
   b. Can the normal force on an object be directed downward? If not, why not? If so, provide an example.
8. A ball is thrown straight up. Taking the drag force of air into account, does it take longer for the ball to travel to the top of its motion or for it to fall back down again?
9. You are going sledding with your friends, sliding down a snowy hill. Friction can't be ignored. Riding solo on your sled, you have a certain acceleration. Would the acceleration change if you let a friend ride with you, increasing the mass? Explain.
10. Suppose you are holding a box in front of you and away from your body by squeezing the sides, as shown in Figure Q5.10. Draw a free-body diagram showing all of the forces on the box. What is the force that is holding the box up, the force that is opposite the weight force?

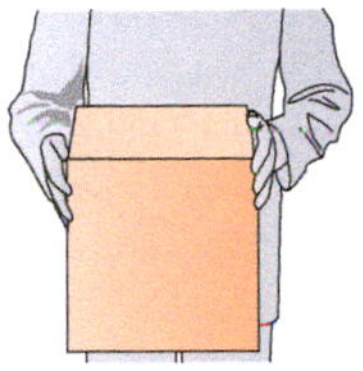
FIGURE Q5.10

11. You are walking up an icy slope. Suddenly your feet slip, and you start to slide backward. Will you slide at a constant speed, or will you accelerate?
12. Three objects move through the air as shown in Figure Q5.12. Rank in order, from largest to smallest, the three drag forces $D_1$, $D_2$, and $D_3$. Some may be equal. Give your answer in the form $A < B = C$ and state your reasoning.

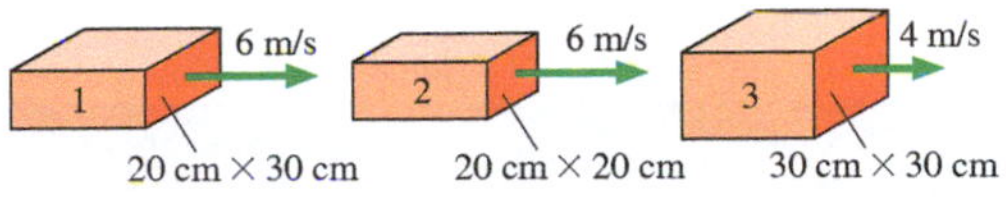

FIGURE Q5.12

13. A trampolinist flies straight up into the air, then back down onto the trampoline, where she comes instantaneously to rest when the trampoline is fully stretched. Rank in order, from largest to smallest, her apparent weight
   a. When she is instantaneously at rest.
   b. Just before she leaves the trampoline.
   c. When she is at the top of her flight.
14. Raindrops can fall at different speeds; some fall quite quickly, others quite slowly. Why might this be true?
15. An airplane moves through the air at a constant speed. The engines' thrust applies a force in the direction of motion, and this force is equal in magnitude and opposite in direction to the drag force. Reducing thrust will cause the plane to fly at a slower—but still constant—speed. Explain why this is so.
16. Is it possible for an object to travel in air faster than its terminal speed? If not, why not? If so, explain how this might happen.

For Questions 17 through 20, determine the tension in the rope at the point indicated with a dot.

- All objects are at rest.
- The strings and pulleys are massless, and the pulleys are frictionless.

17.
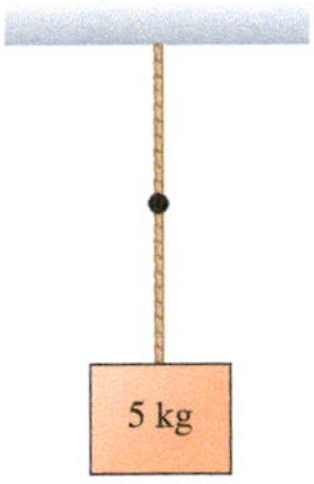

FIGURE Q5.17

18.
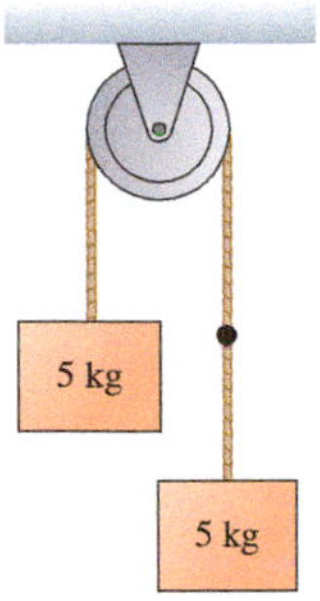

FIGURE Q5.18

19.
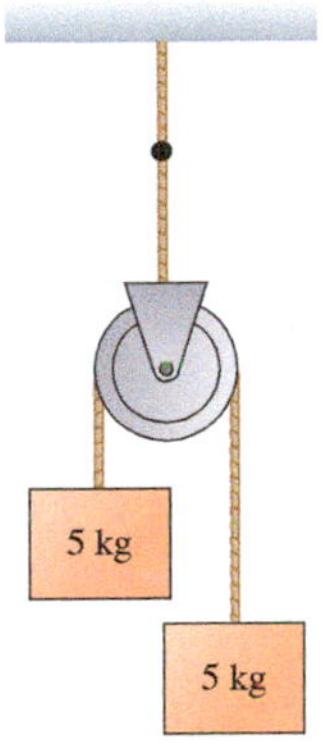

FIGURE Q5.19

20.
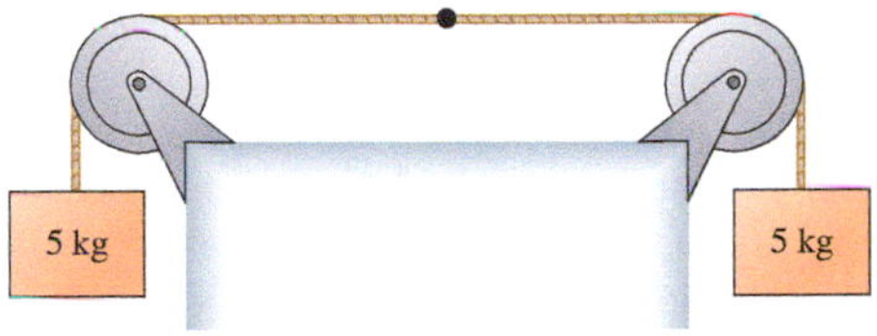

FIGURE Q5.20

21. In Figure Q5.21, block 2 is moving to the right. There is no friction between the floor and block 2, but there is a friction force between blocks 1 and 2. In which direction is the kinetic friction force on block 1? On block 2? Explain.

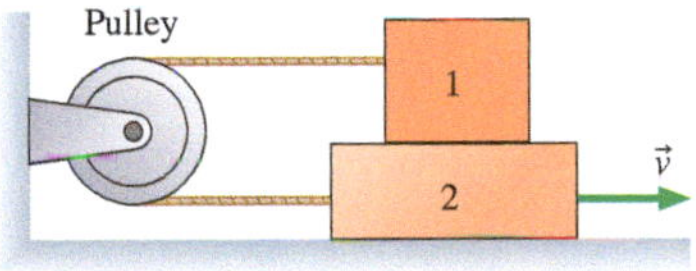

FIGURE Q5.21

## Multiple-Choice Questions

22. || The wood block in Figure Q5.22 is at rest on a wood ramp. In which direction is the static friction force on block 1?
   A. Up the slope.
   B. Down the slope.
   C. The friction force is zero.
   D. There's not enough information to tell.

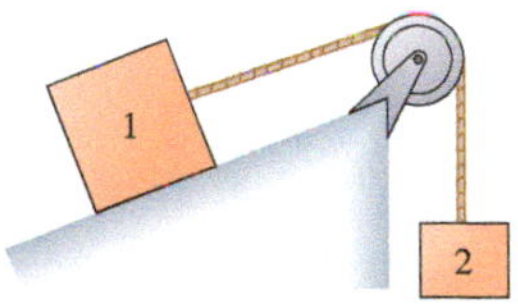

FIGURE Q5.22

23. || Wood block 1 in Figure Q5.22, which has a mass of 1.0 kg, is at rest on a wood ramp. The angle of the ramp is 20° above horizontal. What is the smallest mass of block 2 that will start block 1 sliding uphill?
   a. 0.35 kg   b. 0.83 kg   c. 12. kg
   d. 1.9 kg   e. 2.5 kg

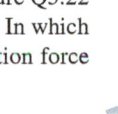

24. || A bicyclist rides down a hill with a 3.0° slope; he attains a terminal speed of 15.0 m/s. What terminal speed would he reach if he went down a 6.0° slope?
a. 17 m/s b. 21 m/s c. 26 m/s
d. 30 m/s e. 33 m/s

25. || A 2.0 kg ball is suspended by two light strings as shown in Figure Q5.25 . What is the tension $T$ in the angled string?
A. 9.5 N B. 15 N
C. 20 N D. 26 N
E. 30 N

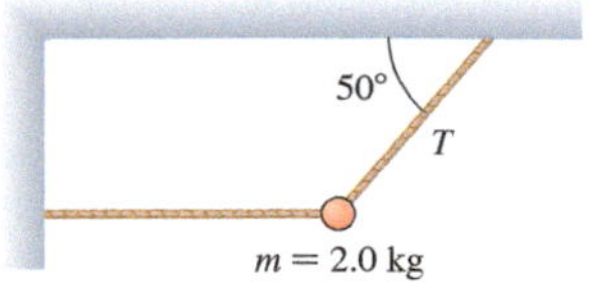

FIGURE Q5.25

26. | While standing in a low tunnel, you raise your arms and push against the ceiling with a force of 100 N. Your mass is 70 kg.
a. What force does the ceiling exert on you?
A. 10 N B. 100 N C. 690 N
D. 790 N E. 980 N
b. What force does the floor exert on you?
A. 10 N B. 100 N C. 690 N
D. 790 N E. 980 N

27. | A 5.0 kg dog sits on the floor of an elevator that is accelerating *downward* at 1.20 m/s$^2$.
a. What is the magnitude of the normal force of the elevator floor on the dog?
A. 34 N B. 43 N C. 49 N D. 55 N E. 74 N
b. What is the magnitude of the force of the dog on the elevator floor?
A. 4.2 N B. 49 N C. 55 N D. 43 N E. 74 N

28. | A 3.0 kg puck slides due east on a horizontal frictionless surface at a constant speed of 4.5 m/s. Then a force of magnitude 6.0 N, directed due north, is applied for 1.5 s. Afterward,
a. What is the northward component of the puck's velocity?
A. 0.50 m/s B. 2.0 m/s C. 3.0 m/s
D. 4.0 m/s E. 4.5 m/s
b. What is the speed of the puck?
A. 4.9 m/s B. 5.4 m/s C. 6.2 m/s
D. 7.5 m/s E. 11 m/s

29. | Eric has a mass of 60 kg. He is standing on a scale in an elevator that is accelerating downward at 1.7 m/s$^2$. What is the approximate reading on the scale?
A. 0 N B. 400 N C. 500 N D. 600 N

30. || The two blocks in Figure Q5.30 are at rest on frictionless surfaces. What must be the mass of the right block in order that the two blocks remain stationary?
A. 4.9 kg B. 6.1 kg C. 7.9 kg
D. 9.8 kg E. 12 kg

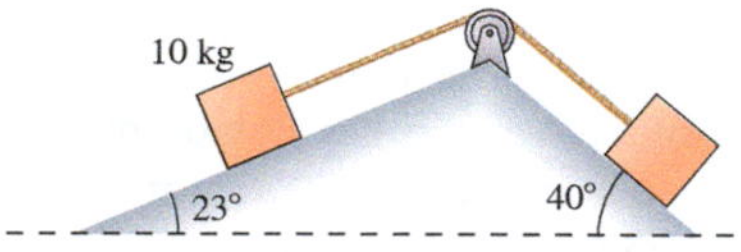

FIGURE Q5.30

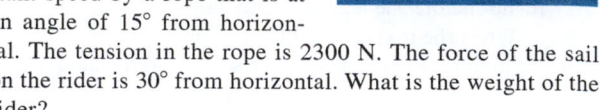

31. | A football player at practice pushes a 60 kg blocking sled across the field at a constant speed. The coefficient of kinetic friction between the grass and the sled is 0.30. How much force must he apply to the sled?
A. 18 N B. 60 N C. 180 N D. 600 N

32. | Two football players are pushing a 60 kg blocking sled across the field at a constant speed of 2.0 m/s. The coefficient of kinetic friction between the grass and the sled is 0.30. Once they stop pushing, how far will the sled slide before coming to rest?
A. 0.20 m B. 0.68 m C. 1.0 m D. 6.6 m

33. || Land Rover ads used to claim that their vehicles could climb a slope of 45°. For this to be possible, what must be the minimum coefficient of static friction between the vehicle's tires and the road?
A. 0.5 B. 0.7 C. 0.9 D. 1.0

34. || A truck is traveling at 30 m/s on a slippery road. The driver slams on the brakes and the truck starts to skid. If the coefficient of kinetic friction between the tires and the road is 0.20, how far will the truck skid before stopping?
A. 230 m B. 300 m C. 450 m D. 680 m

# PROBLEMS

## Section 5.1 Equilibrium

1. | The three ropes in Figure P5.1 are tied to a small, very light ring. Two of the ropes are anchored to walls at right angles, and the third rope pulls as shown. What are $T_1$ and $T_2$, the magnitudes of the tension forces in the first two ropes?

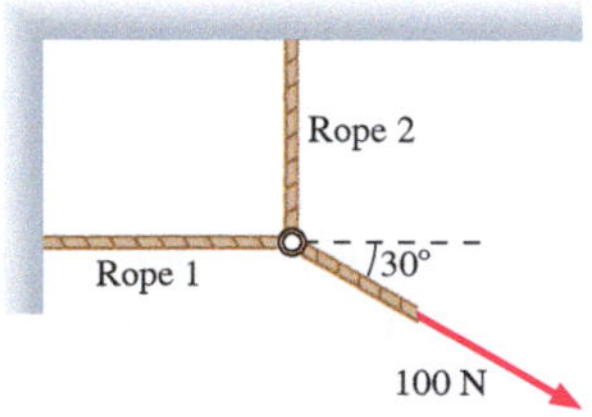

FIGURE P5.1

2. || In the sport of parasailing, a person is attached to a rope being pulled by a boat while hanging from a parachute-like sail. A rider is towed at a constant speed by a rope that is at an angle of 15° from horizontal. The tension in the rope is 2300 N. The force of the sail on the rider is 30° from horizontal. What is the weight of the rider?

3. || In the Skycoaster amusement park ride, riders are suspended from a tower by a long cable. A second cable then lifts them until they reach the starting position indicated in Figure P5.3. The lifting cable is then released, and the riders swing down the circular arc shown. If the four riders have a total mass of 270 kg, what are the tensions in the two cables just before release?

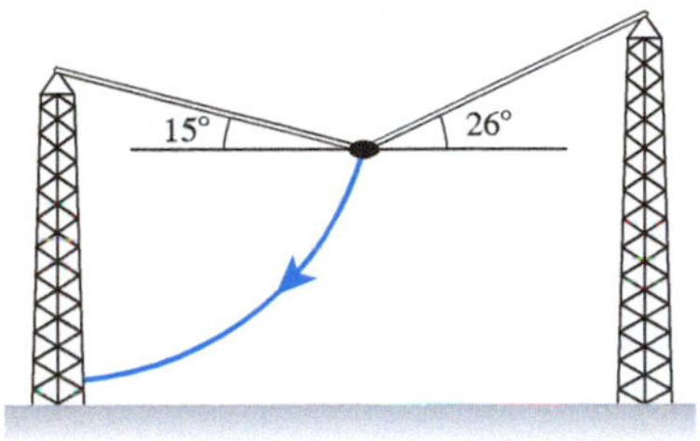


FIGURE P5.3

4. |||| A construction crew would like to support a 1000 kg steel beam with two angled ropes as shown in Figure P5.4. Their rope can support a maximum tension of 5600 N. Is this rope strong enough to do the job?

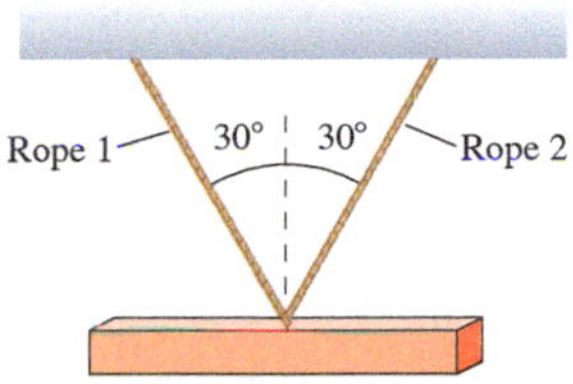


FIGURE P5.4

5. |||| When you bend your knee, the quadriceps muscle is stretched. This increases the tension in the quadriceps tendon attached to your kneecap (patella), which, in turn, increases the tension in the patella tendon that attaches your kneecap to your lower leg bone (tibia). Simultaneously, the end of your upper leg bone (femur) pushes outward on the patella. Figure P5.5 shows how these parts of a knee joint are arranged. What size force does the femur exert on the kneecap if the tendons are oriented as in the figure and the tension in each tendon is 60 N?

BIO

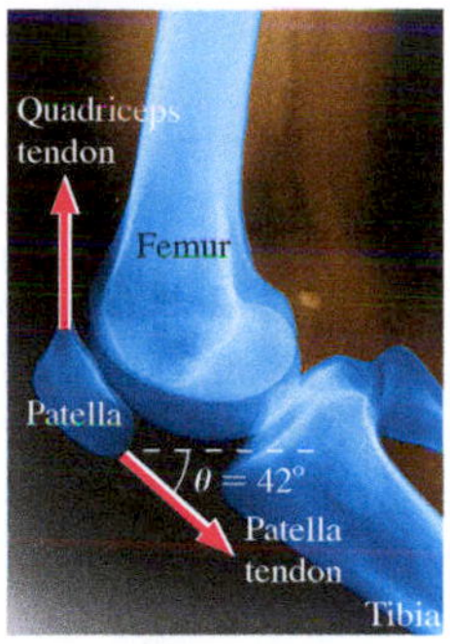


FIGURE P5.5

6. || An early submersible craft for deep-sea exploration was raised and lowered by a cable from a ship. When the craft was stationary, the tension in the cable was 6000 N. When the craft was lowered or raised at a steady rate, the motion through the water added an 1800 N drag force.
   a. What was the tension in the cable when the craft was being lowered to the seafloor?
   b. What was the tension in the cable when the craft was being raised from the seafloor?

7. || Bethany, who weighs 560 N, lies in a hammock suspended by ropes tied to two trees. One rope makes an angle of 45° with the ground; the other makes an angle of 30°. Find the tension in each of the ropes.

8. || A 65 kg student is walking on a slackline, a length of webbing stretched between two trees. The line stretches and so has a noticeable sag, as shown in Figure P5.8. At the point where his foot touches the line, the rope applies a tension force in each direction, as shown. What is the tension in the line?

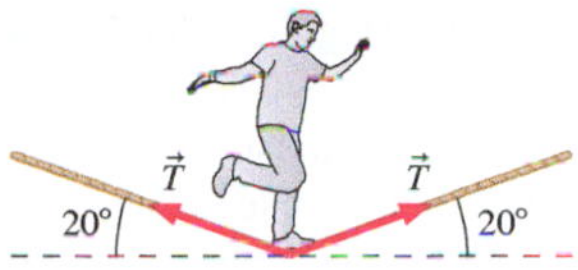


FIGURE P5.8

## Section 5.2 Dynamics and Newton's Second Law

9. ||| In the winter sport of curling, two teams alternate sliding 20 kg stones on an icy surface in an attempt to end up with the stone closest to the center of a target painted on the ice. During one turn, a player releases a stone that travels 27.9 m before coming to rest. The friction force acting on the stone is 2.0 N. What was the speed of the stone when the player released it?

INT

10. || The forces in Figure P5.10 are acting on a 2.0 kg object. What is $a_x$, the $x$-component of the object's acceleration?

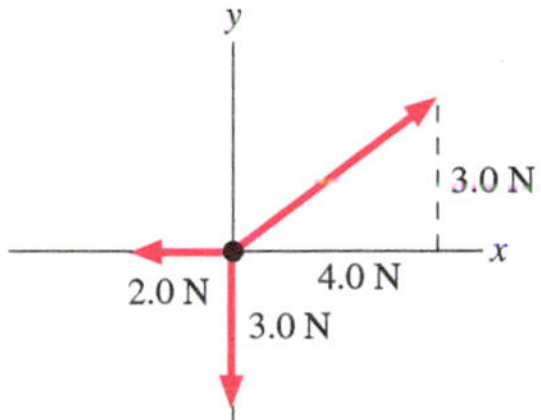


FIGURE P5.10

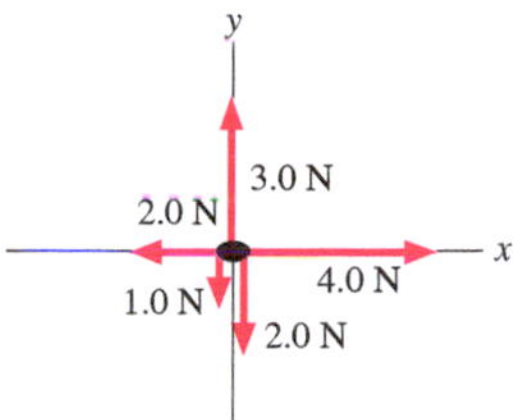


FIGURE P5.11

11. | The forces in Figure P5.11 are acting on a 2.0 kg object. Find the values of $a_x$ and $a_y$, the $x$- and $y$-components of the object's acceleration.

12. | A horizontal rope is tied to a 50 kg box on frictionless ice. What is the tension in the rope if
   a. The box is at rest?
   b. The box moves at a steady 5.0 m/s?
   c. The box has $v_x = 5.0$ m/s and $a_x = 5.0$ m/s$^2$?

13. ||| The acceleration of a baseball pitcher's hand as he delivers a pitch is extreme. For a professional player, this acceleration phase lasts only 50 ms, during which the ball's speed increases from 0 to about 90 mph, or 40 m/s. What is the force of the pitcher's hand on the 0.145 kg ball during this acceleration phase?

14. || In a head-on collision, a car stops in 0.10 s from a speed of 14 m/s. The driver has a mass of 70 kg, and is, fortunately, tightly strapped into his seat. What force is applied to the driver by his seat belt during that fraction of a second?

## Section 5.3 Mass and Weight

15. | An astronaut's weight on earth is 800 N. What is his weight on Mars, where $g = 3.76$ m/s$^2$?

16. | A woman has a mass of 55.0 kg.
   a. What is her weight on earth?
   b. What are her mass and her weight on the moon, where $g = 1.62$ m/s$^2$?

17. || The acceleration of the spacecraft in which the Apollo astronauts took off from the moon was 3.4 m/s$^2$. On the moon, $g = 1.6$ m/s$^2$. What was the apparent weight of a 75 kg astronaut during takeoff?

18. || a. How much force does an 80 kg astronaut exert on his chair while sitting at rest on the launch pad?
   b. How much force does the astronaut exert on his chair while accelerating straight up at 10 m/s$^2$?

Watch Video Solution Problems 5.19, 5.28, and 5.31

19. | It takes the elevator in a skyscraper 4.0 s to reach its cruising speed of 10 m/s. A 60 kg passenger gets aboard on the ground floor. What is the passenger's apparent weight
   a. Before the elevator starts moving?
   b. While the elevator is speeding up?
   c. After the elevator reaches its cruising speed?

20. || Riders on the Power Tower are launched skyward with an acceleration of $4g$, after which they experience a period of free fall. What is a 60 kg rider's apparent weight
   a. During the launch?
   b. During the period of free fall?

21. || Zach, whose mass is 80 kg, is in an elevator descending at 10 m/s. The elevator takes 3.0 s to brake to a stop at the first floor.
   a. What is Zach's apparent weight before the elevator starts braking?
   b. What is Zach's apparent weight while the elevator is braking?

22. ||| An astronaut lifts off in a rocket from the surface of the moon, where $g = 1.6\ \text{m/s}^2$. What vertical acceleration should his rocket have so that his apparent weight is equal to his true weight on earth?

23. ||| Figure P5.23 shows the
INT velocity graph of a 75 kg passenger in an elevator. What is the passenger's apparent weight at $t = 1.0$ s? At 5.0 s? At 9.0 s?

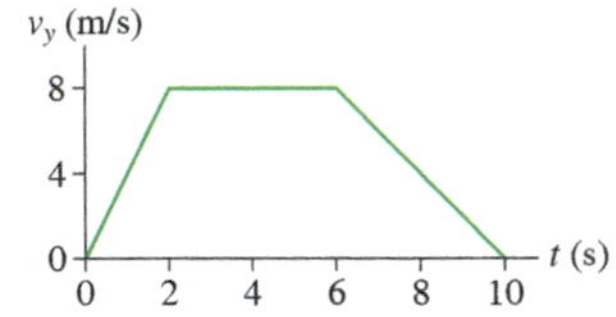


FIGURE P5.23

## Section 5.4 Normal Forces

24. | Mountain goats can easily scale slopes angled at 60° from
BIO horizontal. What are the normal force and the static friction force acting on a mountain goat that weighs 900 N and is standing on such a slope?

25. ||| A 23 kg child goes down a straight slide inclined 38° above horizontal. The child is acted on by his weight, the normal force from the slide, and kinetic friction.
   a. Draw a free-body diagram of the child.
   b. How large is the normal force of the slide on the child?

## Section 5.5 Friction

26. ||| A crate pushed along the floor with velocity $\vec{v}_i$ slides a distance $d$ after the pushing force is removed.
   a. If the mass of the crate is doubled but the initial velocity is not changed, what distance does the crate slide before stopping? Explain.
   b. If the initial velocity of the crate is doubled to $2\vec{v}_i$ but the mass is not changed, what distance does the crate slide before stopping? Explain.

27. || Two workers are sliding a 300 kg crate across the floor. One worker pushes forward on the crate with a force of 380 N while the other pulls in the same direction with a force of 350 N using a rope connected to the crate. Both forces are horizontal, and the crate slides with a constant speed. What is the crate's coefficient of kinetic friction on the floor?

28. ||| A 4000 kg truck is parked on a 7.0° slope. How big is the friction force on the truck?

29. ||| A 1000 kg car traveling at a speed of 40 m/s skids to a halt on wet concrete where $\mu_k = 0.60$. How long are the skid marks?

30. ||| A dump truck, whose bed is made of steel, holds an old steel watering trough. The bed of the truck is slowly raised until the trough begins to slide. What is the acceleration of the trough as it slides down the truck bed?

31. ||| A 10 kg crate is placed on a horizontal conveyor belt. The materials are such that $\mu_s = 0.50$ and $\mu_k = 0.30$.
   a. Draw a free-body diagram showing all the forces on the crate if the conveyer belt runs at constant speed.
   b. Draw a free-body diagram showing all the forces on the crate if the conveyer belt is speeding up.
   c. What is the maximum acceleration the belt can have without the crate slipping?
   d. If the acceleration of the belt exceeds the value determined in part c, what is the acceleration of the crate?

32. || It is friction that provides the force for a car to accelerate, so for high-performance cars the factor that limits acceleration isn't the *engine*; it's the *tires*. For typical rubber-on-concrete friction, what is the shortest time in which a car could accelerate from 0 to 60 mph?

33. || The rolling resistance for steel on steel is quite low; the coefficient of rolling friction is typically $\mu_r = 0.002$. Suppose a 180,000 kg locomotive is rolling at 10 m/s (just over 20 mph) on level rails. If the engineer disengages the engine, how much time will it take the locomotive to coast to a stop? How far will the locomotive move during this time?

34. || A vendor at the local art fair ties her tent to the concrete-filled coffee can shown in Figure P5.34. A stiff breeze comes up and the string becomes taut. What is the maximum value that the string tension can have before the can slips? The coefficient of static friction between the can and the ground is 0.60.

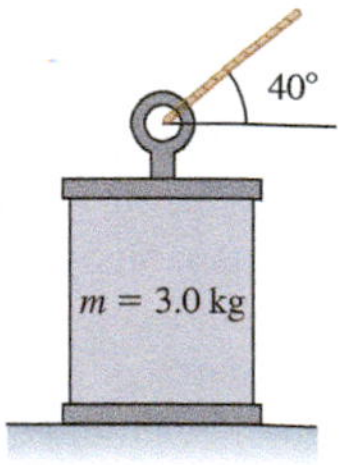


FIGURE P5.34

## Section 5.6 Drag

35. ||| Oceanographers use submerged sonar systems, towed by a cable from a ship, to map the ocean floor. In addition to their downward weight, there are buoyant forces and forces from the flowing water that allow them to travel in a horizontal path. One such submersible has a cross-section area of $1.3\ \text{m}^2$, a drag coefficient of 1.2, and, when towed at 5.1 m/s, the tow cable makes an angle of 30° with the horizontal. What is the tension in the cable?

36. || At its widest point, the diameter of a bottlenose dolphin is
BIO 0.50 m. Bottlenose dolphins are particularly sleek, having a drag coefficient of only about 0.090.
   a. What is the drag force acting on such a dolphin swimming at 7.5 m/s?
   b. Using the dolphin's diameter as its characteristic length, what is the Reynolds number as it swims at this speed in 20° C water?

37. ||| The most dangerous particles in polluted air are those with
BIO diameters less than 2.5 $\mu$m because they can penetrate deeply into the lungs. A 15-cm-tall closed container holds a sample of polluted air containing many spherical particles with a diameter of 2.5 $\mu$m and a mass of $1.4 \times 10^{-14}$ kg. How long does it take for all of the particles to settle to the bottom of the container?

38. | Running on a treadmill is slightly easier than running outside BIO because there is no drag force to work against. Suppose a 60 kg runner completes a 5.0 km race in 18 minutes. Use the cross-section area estimate of Example 5.15 to determine the drag force on the runner during the race. What is this force as a fraction of the runner's weight?

39. ||| What is the magnitude of the acceleration of a skydiver who is currently falling at one-half his eventual terminal speed?

40. | The air is less dense at higher elevations, so skydivers reach a high terminal speed. The highest recorded speed for a skydiver was achieved in a jump from a height of 39,000 m. At this elevation, the density of the air is only 4.3% of the surface density. Use the data from Example 5.15 to estimate the terminal speed of a skydiver at this elevation.

### Section 5.7 Interacting Objects

41. ||| A 1000 kg car pushes a 2000 kg truck that has a dead battery. When the driver steps on the accelerator, the drive wheels of the car push backward against the ground with a force of 4500 N.
   a. What is the magnitude of the force of the car on the truck?
   b. What is the magnitude of the force of the truck on the car?

42. || A 2200 kg truck has put its front bumper against the rear bumper of a 2400 kg SUV to give it a push. With the engine at full power and good tires on good pavement, the maximum forward force on the truck is 18,000 N.
   a. What is the maximum possible acceleration the truck can give the SUV?
   b. At this acceleration, what is the force of the SUV's bumper on the truck's bumper?

43. ||||| Blocks with masses of 1.0 kg, 2.0 kg, and 3.0 kg are lined up in a row on a frictionless table. All three are pushed forward by a 12 N force applied to the 1.0 kg block. How much force does the 2.0 kg block exert on (a) the 3.0 kg block and (b) the 1.0 kg block?

### Section 5.8 Ropes and Pulleys

44. ||| What is the tension in the rope of Figure P5.44?

45. |||| A house painter uses the chair-and-pulley arrangement of Figure P5.45 to lift himself up the side of a house. The painter's mass is 70 kg and the chair's mass is 10 kg. If the painter is at rest, what is the tension in the rope?

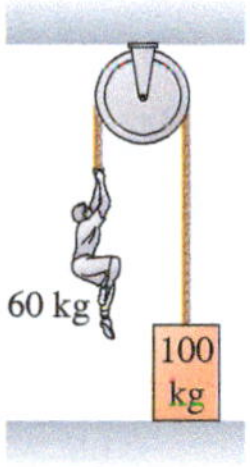

FIGURE P5.44

FIGURE P5.45

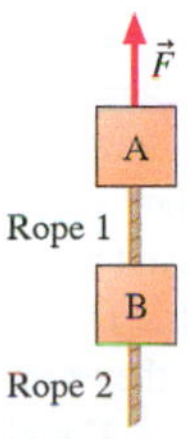

FIGURE P5.46

46. ||| Figure P5.46 shows two 1.00 kg blocks connected by a rope. A second rope hangs beneath the lower block. Both ropes have a mass of 250 g. The entire assembly is accelerated upward at 3.00 m/s$^2$ by force $\vec{F}$.
   a. What is $F$?
   b. What is the tension at the top end of rope 1?
   c. What is the tension at the bottom end of rope 1?
   d. What is the tension at the top end of rope 2?

47. || Each of 100 identical blocks sitting on a frictionless surface is connected to the next block by a massless string. The first block is pulled with a force of 100 N.
   a. What is the tension in the string connecting block 100 to block 99?
   b. What is the tension in the string connecting block 50 to block 51?

48. || A Jeep is stuck in the mud. The driver has a winch that can pull on its cable with a force of 40,000 N. The driver loops the cable through a pulley attached to a tree, then attaches the end of the cable to his Jeep, as shown in Figure P5.48. What is the maximum force that can be exerted on the Jeep by this cable arrangement?

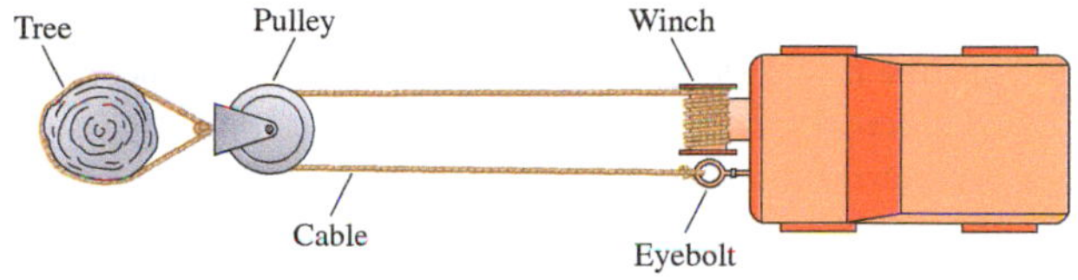

FIGURE P5.48

49. ||| A 500 kg piano is being lowered into position by a crane while two people steady it with ropes pulling to the sides. Bob's rope pulls to the left, 15° below horizontal, with 500 N of tension. Ellen's rope pulls toward the right, 25° below horizontal.
   a. What tension must Ellen maintain in her rope to keep the piano descending vertically at constant speed?
   b. What is the tension in the vertical main cable supporting the piano?

## General Problems

50. || Dana has a sports medal suspended by a long ribbon from her rearview mirror. As she accelerates onto the highway, she notices that the medal is hanging at an angle of 10° from the vertical.
   a. Does the medal lean toward or away from the windshield? Explain.
   b. What is her acceleration?

51. || Figure P5.51 shows the INT velocity graph of a 2.0 kg object as it moves along the $x$-axis. What is the net force acting on this object at $t = 1$ s? At 4 s? At 7 s?

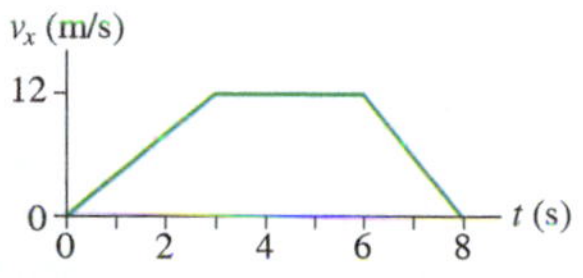

FIGURE P5.51

52. | Your forehead can withstand a force of about 6.0 kN before BIO fracturing, while your cheekbone can only withstand about 1.3 kN.
   a. If a 140 g baseball strikes your head at 30 m/s and stops in 0.0015 s, what is the magnitude of the ball's acceleration?
   b. What is the magnitude of the force that stops the baseball?
   c. What force does the baseball apply to your head? Explain.
   d. Are you in danger of a fracture if the ball hits you in the forehead? In the cheek?

53. | A 50 kg box hangs from a rope. What is the tension in the rope if
   a. The box is at rest?
   b. The box has $v_y = 5.0$ m/s and is speeding up at 5.0 m/s$^2$?

54. |||| BIO Scientists have studied how snakes grip and climb ropes. In one study, they found that an important characteristic of a rope is its "compliance"—that is, how easily the rope, while under tension, can be flexed. Figure P5.54 shows how scientists measured a rope's compliance by attaching it to two strings, each supporting an identical mass $m$. The strings contort the rope so that its middle section lies at angle $\theta$. For $\theta = 30°$ and $m = 100$ g, what are the tensions $T_1$ and $T_2$ in the upper and middle parts of the rope?

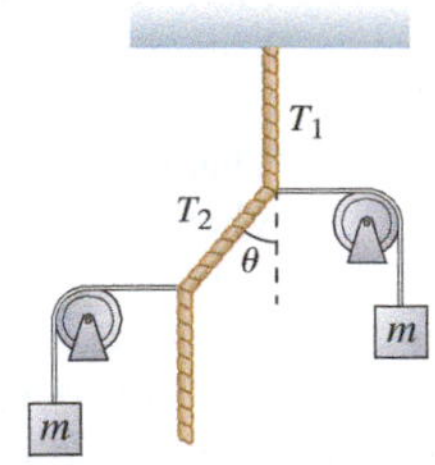

FIGURE P5.54

55. | A 50 kg box hangs from a rope. What is the tension in the rope if
   a. The box moves up at a steady 5.0 m/s?
   b. The box has $v_y = 5.0$ m/s and is slowing down at 5.0 m/s$^2$?

56. | INT A fisherman has caught a very large, 5.0 kg fish from a dock that is 2.0 m above the water. He is using lightweight fishing line that will break under a tension of 54 N or more. He is eager to get the fish to the dock in the shortest possible time. If the fish is at rest at the water's surface, what's the least amount of time in which the fisherman can raise the fish to the dock without losing it?

57. ||| Riders on the Tower of Doom, an amusement park ride, experience 2.0 s of free fall, after which they are slowed to a stop in 0.50 s. What is a 65 kg rider's apparent weight as the ride is coming to rest? By what factor does this exceed her actual weight?

58. | INT Just after launch, the space shuttle takes 8.0 s to reach a speed of 160 km/h. During this phase, what is the apparent weight of a 72 kg astronaut?

59. ||| BIO Seat belts and air bags save lives by reducing the forces exerted on the driver and passengers in an automobile collision. Cars are designed with a "crumple zone" in the front of the car. In the event of an impact, the passenger compartment decelerates over a distance of about 1 m as the front of the car crumples. An occupant restrained by seat belts and air bags decelerates with the car. By contrast, an unrestrained occupant keeps moving forward with no loss of speed (Newton's first law!) until hitting the dashboard or windshield, as we saw in Figure 4.2. These are unyielding surfaces, and the unfortunate occupant then decelerates over a distance of only about 5 mm.
   a. A 60 kg person is in a head-on collision. The car's speed at impact is 15 m/s. Estimate the net force on the person if he or she is wearing a seat belt and if the air bag deploys.
   b. Estimate the net force that ultimately stops the person if he or she is not restrained by a seat belt or air bag.
   c. How do these two forces compare to the person's weight?

60. ||| Corey, whose mass is 95 kg, stands on a bathroom scale in an elevator. The scale reads 830 N for the first 3.0 s after the elevator starts to move, then 930 N for the next 3.0 s. What is the elevator's velocity 6.0 s after starting?

61. ||| A 20,000 kg rocket has a rocket motor that generates $3.0 \times 10^5$ N of thrust.
   a. What is the rocket's initial upward acceleration?
   b. At an altitude of 5.0 km the rocket's acceleration has increased to 6.0 m/s$^2$. What mass of fuel has it burned?

62. ||| You've always wondered about the acceleration of the elevators in the 101-story-tall Empire State Building. One day, while visiting New York, you take your bathroom scale into the elevator and stand on it. The scale reads 150 lb as the door closes. The reading varies between 120 lb and 170 lb as the elevator travels 101 floors.
   a. What is the magnitude of the acceleration as the elevator starts upward?
   b. What is the magnitude of the acceleration as the elevator brakes to a stop?

63. |||| A 23 kg child goes down a straight slide inclined 38° above horizontal. The child is acted on by his weight, the normal force from the slide, kinetic friction, and a horizontal rope exerting a 30 N force as shown in Figure P5.63. How large is the normal force of the slide on the child?

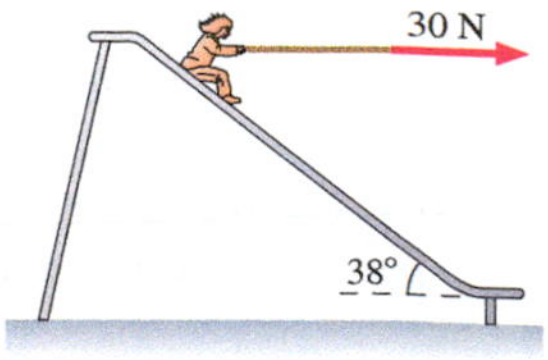

FIGURE P5.63

64. | BIO INT An impala is an African antelope capable of a remarkable vertical leap. In one recorded leap, a 45 kg impala went into a deep crouch, pushed straight up for 0.21 s, and reached a height of 2.5 m above the ground. To achieve this vertical leap, with what force did the impala push down on the ground? What is the ratio of this force to the antelope's weight?

65. | Josh starts his sled at the top of a 3.0-m-high hill that has a constant slope of 25°. After reaching the bottom, he slides across a horizontal patch of snow. Ignore friction on the hill, but assume that the coefficient of kinetic friction between his sled and the horizontal patch of snow is 0.050. How far from the base of the hill does he end up?

66. |||| BIO INT The standing vertical jump is a good test of an athlete's strength and fitness. The athlete goes into a deep crouch, then extends his legs rapidly; when his legs are fully extended, he leaves the ground and rises to his highest height. It is the force of the ground on the athlete during the extension phase that accelerates the athlete to the final speed with which he leaves the ground. A good jumper can exert a force on the ground equal to twice his weight. If his crouch is 60 cm deep, how far off the ground does he rise?

67. ||| A 70 kg bicyclist is coasting down a long hill with a 3.5° slope. He's moving quite rapidly, so air drag is important. His cross-section area is 0.32 m$^2$ and his drag coefficient is 0.88. What speed does he eventually reach, in mph?

68. ||| BIO Many birds can attain very high speeds when diving. Using radar, scientists measured the altitude of a barn swallow in a vertical dive; it dropped 208 m in 3.0 s. The mass of the swallow was estimated to be 0.018 kg, and its cross-section area as $5.6 \times 10^{-4}$ m$^2$. What was the drag coefficient for this swallow as it dove?

69. ||| INT The drag on a pitched baseball can be surprisingly large. Suppose a 145 g baseball with a diameter of 7.4 cm has an initial speed of 40.2 m/s (90 mph).
   a. What is the magnitude of the ball's acceleration due to the drag force?
   b. If the ball had this same acceleration during its entire 18.4 m trajectory, what would its final speed be?

70. | Small particulates can be removed from the emissions of a coal-fired power plant by *electrostatic precipitation*. The particles are given a small electric charge that draws them toward oppositely charged plates, where they stick. Consider a spherical particulate with a diameter of 1.0 $\mu$m. The electric force on this particle is $2.0 \times 10^{-13}$ N. What is the speed of such a particle? (The electric force is much greater than the particle's weight, which can be ignored.)

71. ||| You probably think of wet surfaces as being slippery. Surprisingly, the opposite is true for human skin, as you can demonstrate by sliding a dry versus a slightly damp fingertip along a smooth surface such as a desktop. Researchers have found that the static coefficient of friction between dry skin and steel is 0.27, while that between damp skin and steel can be as high as 1.4. Suppose a man holds a steel rod vertically in his hand, exerting a 400 N grip force on the rod. What is the heaviest rod he can hold without slipping if
BIO
   a. His hands are dry?
   b. His hands are wet?

72. || Researchers often use *force plates* to measure the forces that people exert against the floor during movement. A force plate works like a bathroom scale, but it keeps a record of how the reading changes with time. Figure P5.72 shows the data from a force plate as a woman jumps straight up and then lands.
BIO INT
   a. What was the vertical component of her acceleration during push-off?
   b. What was the vertical component of her acceleration while in the air?
   c. What was the vertical component of her acceleration during the landing?
   d. What was her speed as her feet left the force plate?
   e. How high did she jump?

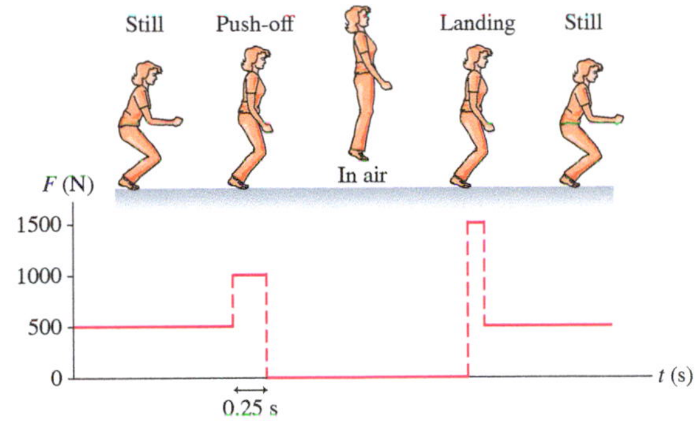

FIGURE P5.72

73. |||| A person with compromised pinch strength in his fingers can only exert a normal force of 6.0 N to either side of a pinch-held object, such as the book shown in Figure P5.73. What is the greatest mass book he can hold onto vertically before it slips out of his fingers? The coefficient of static friction of the surface between the fingers and the book cover is 0.80.

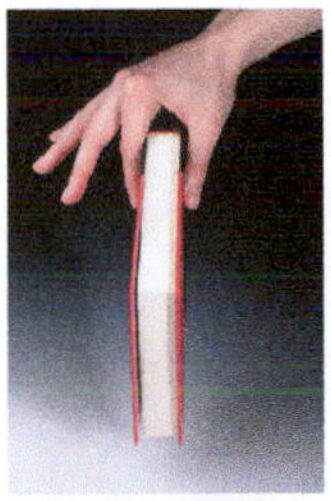
FIGURE P5.73

74. || It's possible for a determined group of people to pull an aircraft. Drag is negligible at low speeds, and the only force impeding motion is the rolling friction of the rubber tires on the concrete runway. In 2000, a team of 60 British police officers set a world record by pulling a Boeing 747, with a mass of 200,000 kg, a distance of 100 m in 53 s. The plane started at rest. Estimate the force with which each officer pulled on the plane, assuming constant pulling force and constant acceleration.
INT

75. ||| A 1.0 kg wood block is pressed against a vertical wood wall by a 12 N force as shown in Figure P5.75. If the block is initially at rest, will it move upward, move downward, or stay at rest?

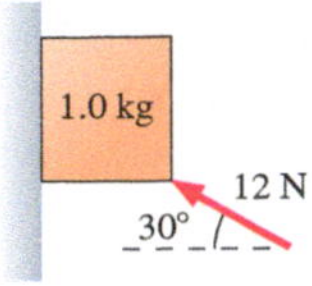

FIGURE P5.75

76. ||| A simple model shows how drawing a bow across a violin string causes the string to vibrate. As the bow moves across the string, static friction between the bow and the string pulls the string along with the bow. At some point, the tension pulling the string back exceeds the maximum static friction force and the string snaps back. This process repeats cyclically, causing the string's vibration. Assume the tension in a 0.33-m-long violin string is 50 N, and the coefficient of static friction between the bow and the string is $\mu_s = 0.80$. If the normal force of the bow on the string is 0.75 N, how far can the string be pulled before it slips if the string is bowed at its center?

77. ||| Two blocks are at rest on a frictionless incline, as shown in Figure P5.77. What are the tensions in the two strings?

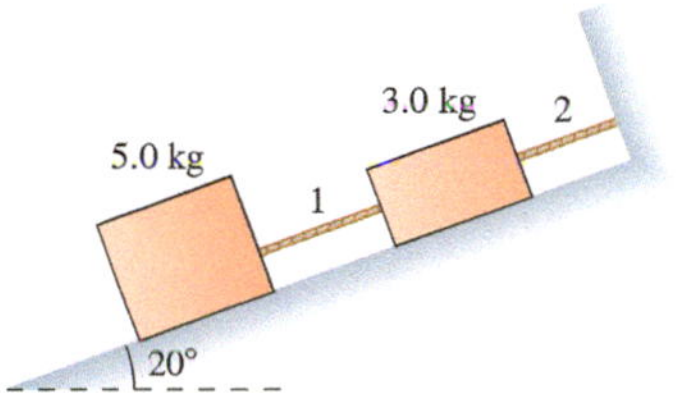

FIGURE P5.77

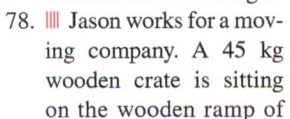

78. |||| Jason works for a moving company. A 45 kg wooden crate is sitting on the wooden ramp of his truck; the ramp is angled at 11°. What is the magnitude of the force, directed parallel to the ramp, that he needs to exert on the crate to get it to start moving
   a. Up the ramp?
   b. Down the ramp?

79. || Two identical 2.0 kg blocks are stacked as shown in Figure P5.79. The bottom block is free to slide on a frictionless surface. The coefficient of static friction between the blocks is 0.35. What is the maximum horizontal force that can be applied to the lower block without the upper block slipping?

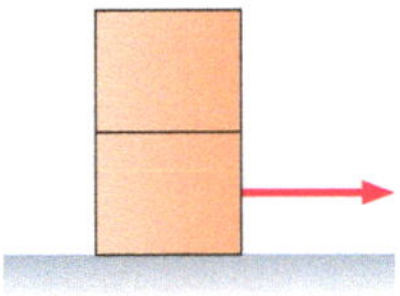
FIGURE P5.79

80. ||| Dana loads luggage into an airplane using a conveyor belt tilted at an angle of 20°. She places a few pieces of luggage on the belt before it starts to move, then she turns the belt on. It takes the belt 0.70 s to reach its top speed of 1.2 m/s. Does the luggage slip? Assume $\mu_s = 0.50$ between the luggage and the belt.

81. |||| In rock climbing, various rope and pulley systems have been devised to help haul up heavy loads, including injured climbers. A rescuer is hauling up an injured climber who weighs 660 N using the rope and pulley system shown in Figure P5.81. (The ropes in this figure are drawn at various angles for clarity, but you can assume they're all vertical.)
   a. What is the tension in rope 1, the rope that the rescuer pulls on?
   b. What is the tension in rope 2?

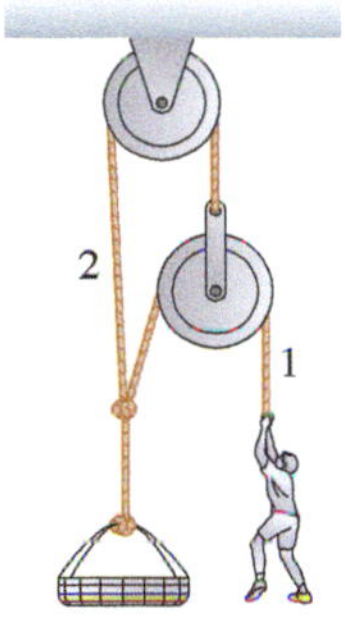

FIGURE P5.81

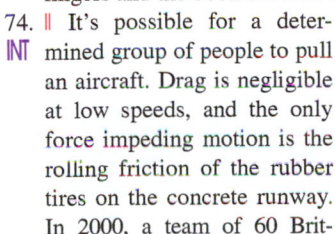

82. |||| Two blocks are connected by a string as in Figure P5.82. What is the upper block's acceleration if the coefficient of kinetic friction between the block and the table is 0.20?

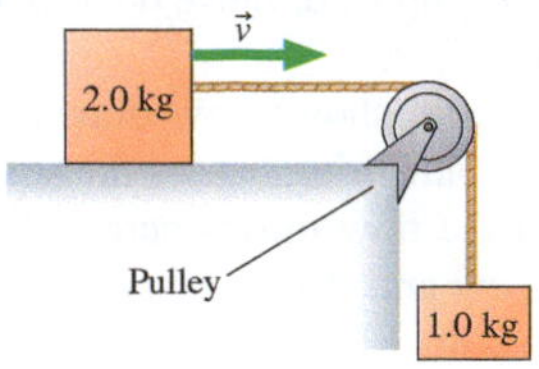


FIGURE P5.82

83. |||| The ramp in Figure P5.83 is frictionless. If the blocks are released from rest, which way does the 10 kg block slide, and what is the magnitude of its acceleration?

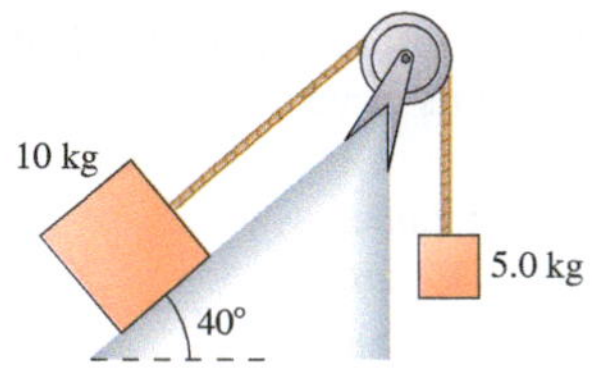


FIGURE P5.83

84. ||| The 100 kg block in Figure P5.84 takes 6.0 s to reach the floor after being released from rest. What is the mass of the block on the left?
INT

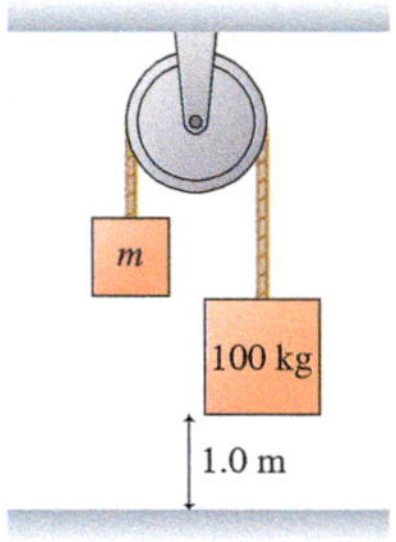


FIGURE P5.84

Watch Video Solution Problem 5.83

## MCAT-Style Passage Problems

### Sliding on the Ice

In the winter sport of curling, players give a 20 kg stone a push across a sheet of ice. The stone moves approximately 40 m before coming to rest. The final position of the stone, in principle, only depends on the initial speed at which it is launched and the force of friction between the ice and the stone, but team members can use brooms to sweep the ice in front of the stone to adjust its speed and trajectory a bit; they must do this without touching the stone. Judicious sweeping can lengthen the travel of the stone by 3 m.

85. | A curler pushes a stone to a speed of 3.0 m/s over a time of 2.0 s. Ignoring the force of friction, how much force must the curler apply to the stone to bring it up to speed?
A. 3.0 N B. 15 N C. 30 N D. 150 N

86. | The sweepers in a curling competition adjust the trajectory of the stone by
A. Decreasing the coefficient of friction between the stone and the ice.
B. Increasing the coefficient of friction between the stone and the ice.
C. Changing friction from kinetic to static.
D. Changing friction from static to kinetic.

87. | Suppose the stone is launched with a speed of 3 m/s and travels 40 m before coming to rest. What is the *approximate* magnitude of the friction force on the stone?
A. 0 N B. 2 N C. 20 N D. 200 N

88. | Suppose the stone's mass is increased to 40 kg, but it is launched at the same 3 m/s. Which one of the following is true?
A. The stone would now travel a longer distance before coming to rest.
B. The stone would now travel a shorter distance before coming to rest.
C. The coefficient of friction would now be greater
D. The force of friction would now be greater.

# 6 Circular Motion, Orbits, and Gravity

The horses are rounding a corner on a grassy course. Why do they lean into the turn this way?

## LOOKING AHEAD ▸

### Circular Motion

An object moving in a circle has an acceleration toward the center, so there must be a net force toward the center as well.

How much force does it take to swing the girl in a circle? You'll learn how to solve such problems.

### Apparent Forces

These riders feel like they're being pushed against the wall. But this force isn't a real force, it's an **apparent force.**

This apparent force makes the riders "feel heavy." You'll learn to calculate their apparent weight.

### Gravity and Orbits

The space station appears to float in space, but gravity is actually pulling down on it quite forcefully.

You'll learn **Newton's law of gravity**, and you'll see how the force of gravity keeps the station in orbit.

**GOAL** **To learn about motion in a circle, including orbital motion under the influence of a gravitational force.**

## LOOKING BACK ◂

### Centripetal Acceleration

In Section 3.7, you learned that an object moving in a circle at a constant speed experiences an acceleration directed toward the center of the circle.

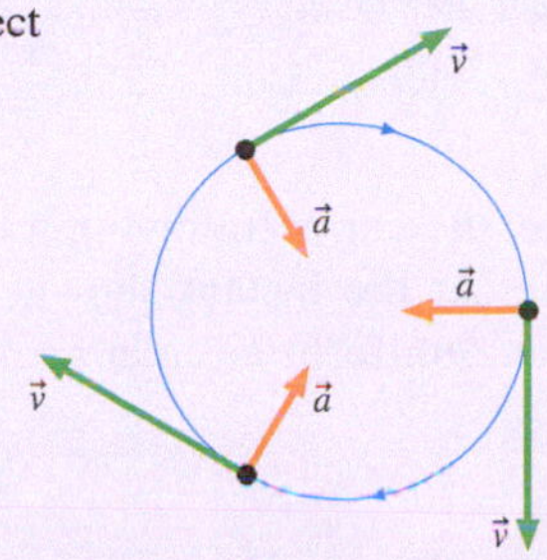


In this chapter, you'll learn how to extend Newton's second law, which relates acceleration to the forces that cause it, to this type of acceleration.

**STOP TO THINK**

A softball pitcher is throwing a pitch. At the instant shown, the ball is moving in a circular arc at a steady speed. At this instant, the acceleration is

A. Directed up.
B. Directed down.
C. Directed left.
D. Directed right.
E. Zero.

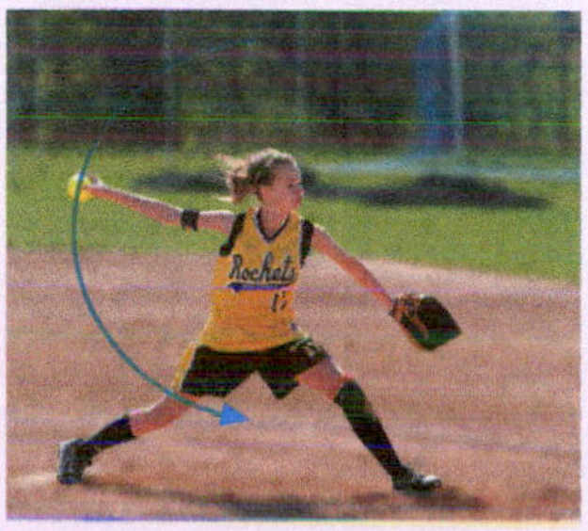

## 6.1 Uniform Circular Motion

Uniform circular motion at the fair.

The riders on this carnival ride are going in a circle at a constant speed, a type of motion that we've called uniform circular motion. We saw in ◀◀SECTION 3.7 that uniform circular motion results in an acceleration directed toward the center of the circle. According to Newton's second law, this means that there must be a net force directed toward the center of the circle. For these riders, it is the tension in the cables and the force of gravity that combine to provide this net force.

In this chapter, we'll look at objects moving in circles or circular arcs. We'll consider the details of the acceleration and the forces that provide this acceleration, combining our discussion and description of motion from Chapters 1–3 with our treatment of Newton's laws and dynamics from Chapters 4 and 5. For now, we consider objects that move at a *constant* speed; circular motion with *changing* speed will wait until Chapter 7.

### Velocity and Acceleration in Uniform Circular Motion

FIGURE 6.1 Velocity and acceleration for uniform circular motion.

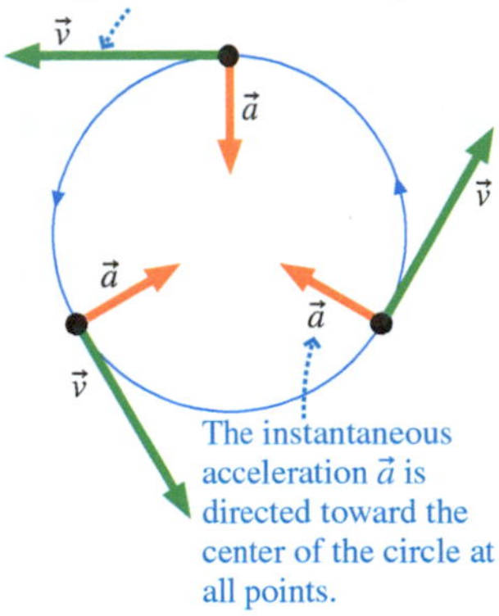


Although the *speed* of a particle in uniform circular motion is constant, its *velocity* is not constant because the *direction* of the motion is always changing. FIGURE 6.1 reminds you of the details: There is an acceleration at every point in the motion, with the acceleration vector $\vec{a}$ pointing toward the center of the circle. We called this the *centripetal acceleration,* and we showed that for uniform circular motion the acceleration is given by

$$a = \frac{v^2}{r} \tag{6.1}$$

Centripetal acceleration for uniform circular motion

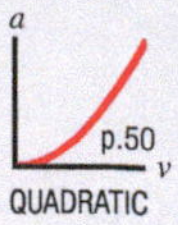


Thus an object's centripetal acceleration depends on both its speed and the distance from the center of its circular path.

**CONCEPTUAL EXAMPLE 6.1 A car rounding a corner**

A car is turning a tight corner at a constant speed. A top view of the motion is shown in FIGURE 6.2. The velocity vector for the car points to the east at the instant shown. What is the direction of the acceleration?

FIGURE 6.2 Top view of a car turning a corner.

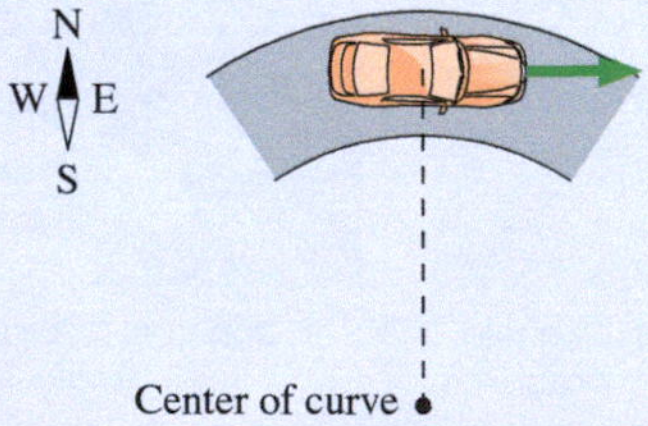


***REASON*** The curve that the car is following is a segment of a circle, so this is an example of uniform circular motion. For uniform circular motion, the acceleration is directed toward the center of the circle, which is to the south.

***ASSESS*** This acceleration is due to a change in direction, not a change in speed. And this matches your experience in a car: If you turn the wheel to the right—as the driver of this car is doing—your car then *changes* its motion toward the right, in the direction of the center of the circle.

NOTE ▶ In this example, the car is following a curve that is only a *segment* of a circle, not a *full* circle. At the instant shown, though, the motion follows a circular arc. You can have uniform circular motion without completing a full circle. ◀

## Period, Frequency, and Speed

Although an object needn't complete a full circle to be in uniform circular motion, in many of the cases we'll consider, objects will complete multiple full circles of motion, one after another. Since the motion is uniform, each time around the circle is just a repeat of the one before, so the motion is *periodic*.

The time interval it takes an object to go around a circle one time, completing one revolution (abbreviated rev), is called the **period** of the motion. Period is represented by the symbol $T$.

Rather than specify the time for one revolution, we can specify circular motion by its **frequency,** the number of revolutions per second, for which we use the symbol $f$. An object with a period of one-half second completes 2 revolutions each second. Similarly, an object can make 10 revolutions in 1 s if its period is one-tenth of a second. This shows that frequency is the inverse of the period:

$$f = \frac{1}{T} \tag{6.2}$$

Although frequency is often expressed as "revolutions per second," *revolutions* are not true units but merely the counting of events. Thus the SI unit of frequency is simply inverse seconds, or $s^{-1}$. Frequency may also be given in revolutions per minute (rpm) or another time interval, but these usually need to be converted to $s^{-1}$ before calculations are done.

FIGURE 6.3 shows an object moving at a constant speed in a circular path of radius $r$. We know the time for one revolution—one period $T$—and we know the distance traveled, so we can write an equation relating the speed to the period and radius:

FIGURE 6.3 Relating frequency and speed.

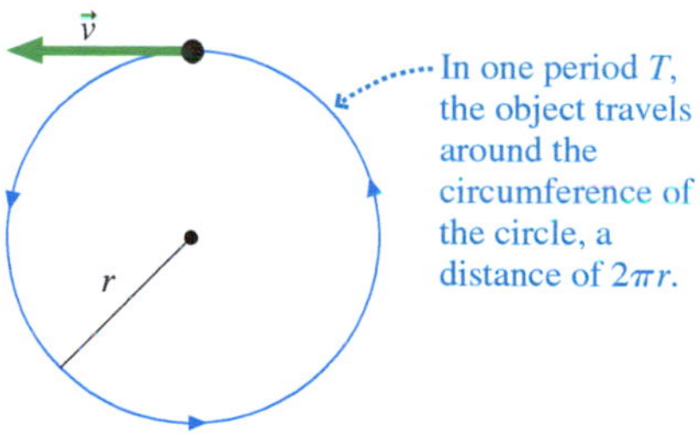


$$v = \frac{2\pi r}{T} \tag{6.3}$$

Given Equation 6.2 relating frequency and period, we can also write this equation as

$$v = 2\pi f r \tag{6.4}$$

We can combine this equation with Equation 6.1 for acceleration to get an expression for the centripetal acceleration in terms of the frequency or the period for the circular motion:

$$a = \frac{v^2}{r} = (2\pi f)^2 r = \left(\frac{2\pi}{T}\right)^2 r \tag{6.5}$$

### EXAMPLE 6.2 A spinning table saw blade

The circular blade of a table saw is 25 cm in diameter and spins at 3600 rpm. How much time is required for one revolution? How fast is one of the teeth at the edge of the blade moving? What is the tooth's acceleration?

**STRATEGIZE** Each tooth of the saw blade is undergoing uniform circular motion, so the period, frequency, speed, and acceleration of a tooth are all related by Equations 6.2–6.5.

**PREPARE** Before we get started, we need to do a couple of unit conversions. The diameter of the blade is 0.25 m, so its radius is 0.125 m. The frequency is given in rpm; we need to convert this to $s^{-1}$:

$$f = 3600\,\frac{\text{rev}}{\text{min}} \times \frac{1\ \text{min}}{60\ \text{s}} = 60\,\frac{\text{rev}}{\text{s}} = 60\ \text{s}^{-1}$$

**SOLVE** The time for one revolution is the period, which is given by Equation 6.2:

$$T = \frac{1}{f} = \frac{1}{60\ \text{s}^{-1}} = 0.017\ \text{s}$$

The speed of the tooth is given by Equation 6.4:

$$v = 2\pi f r = 2\pi(60\ \text{s}^{-1})(0.125\ \text{m}) = 47\ \text{m/s}$$

We can then use Equation 6.5 to find the acceleration:

$$a = (2\pi f)^2 r = (2\pi(60\ \text{s}^{-1}))^2(0.125\ \text{m}) = 1.8 \times 10^4\ \text{m/s}^2$$

**ASSESS** The speed of the tooth is extremely high: 47 m/s ≈ 100 mph! Still, this seems plausible for a high-speed saw. The acceleration is also remarkable, almost 2000$g$.

**EXAMPLE 6.3** **Finding the period of a carnival ride**

In the Quasar carnival ride, passengers travel in a horizontal 5.0-m-radius circle. For safe operation, the maximum sustained acceleration that riders may experience is 20 m/s$^2$, approximately twice the free-fall acceleration. What is the period of the ride when it is being operated at the maximum acceleration? How fast are the riders moving when the ride is operated at this period?

**STRATEGIZE** The passengers are in uniform circular motion, so their period, speed, and acceleration are related by Equations 6.4 and 6.5.

**PREPARE** We will assume that the cars on the ride are in uniform circular motion. The visual overview of FIGURE 6.4 shows a top view of the motion of the ride.

FIGURE 6.4 Visual overview for the Quasar carnival ride.

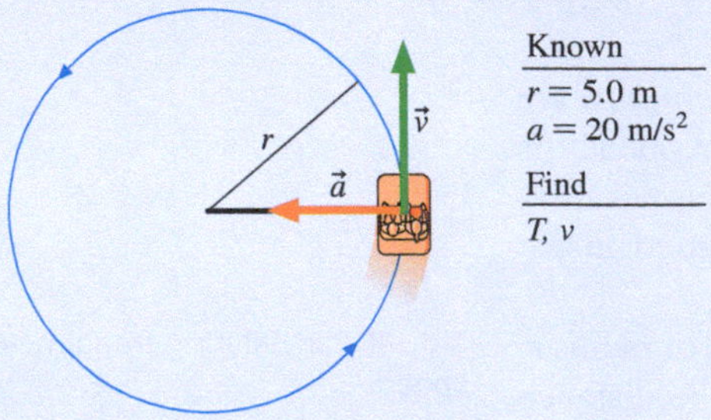


**SOLVE** Equation 6.5 shows that the acceleration increases with decreasing period: If the riders go around in a shorter time, the acceleration increases. Setting a maximum value for the acceleration means setting a minimum value for the period—as fast as the ride can safely go. We can rearrange Equation 6.5 to find the period in terms of the acceleration. Setting the acceleration equal to the maximum value gives the minimum period:

$$T = 2\pi\sqrt{\frac{r}{a}} = 2\pi\sqrt{\frac{5.0\text{ m}}{20\text{ m/s}^2}} = 3.1\text{ s}$$

We can then use Equation 6.4 to find the speed at which the riders move:

$$v = \frac{2\pi r}{T} = \frac{2\pi(5.0\text{ m})}{3.1\text{ s}} = 10\text{ m/s}$$

**ASSESS** One rotation in just over 3 seconds seems reasonable for a pretty zippy carnival ride. (The period for this particular ride is actually 3.7 s, so it runs a bit slower than the maximum safe speed.) But in this case we can do a quantitative check on our work. If we use our calculated velocity to find the acceleration using Equation 6.1, we find

$$a = \frac{v^2}{r} = \frac{(10\text{ m/s})^2}{5.0\text{ m}} = 20\text{ m/s}^2$$

This is the acceleration given in the problem statement, so we can have confidence in our work.

**Hurling the heavy hammer** Scottish games involve feats of strength. Here, a man is throwing a 30 lb hammer for distance. He starts by swinging the hammer rapidly in a circle. You can see from how he is leaning that he is providing a large force directed toward the center to produce the necessary centripetal acceleration. When the hammer is heading in the right direction, the man lets go. With no force directed toward the center, the hammer will stop going in a circle and fly in the chosen direction across the field.

**STOP TO THINK 6.1** Rank in order, from largest to smallest, the period of the motion of particles A to D.

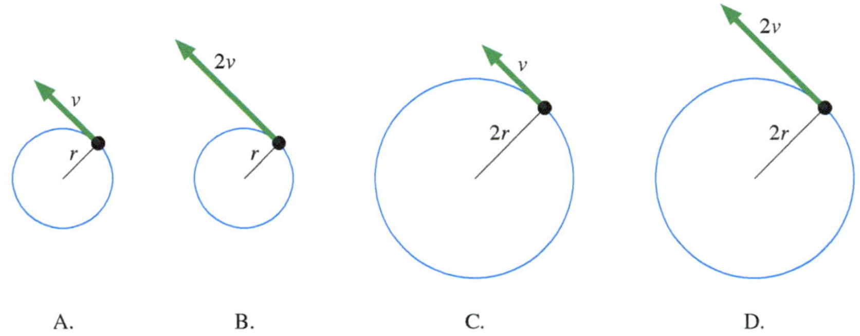


## 6.2 Dynamics of Uniform Circular Motion

Riders traveling around on a circular carnival ride are accelerating, as we have seen. Consequently, according to Newton's second law, the riders must have a net *force* acting on them.

We've already determined the acceleration of a particle in uniform circular motion—the centripetal acceleration of Equation 6.1. Newton's second law tells us what the net force must be to cause this acceleration:

$$\vec{F}_{\text{net}} = m\vec{a} = \left(\frac{mv^2}{r}\text{, toward center of circle}\right) \qquad (6.6)$$

Net force producing the centripetal acceleration of uniform circular motion

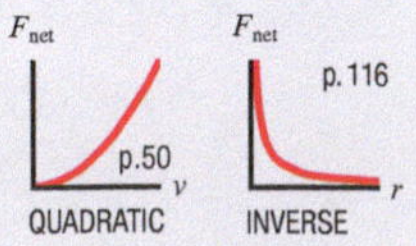


A video to support a section's topic is embedded in the eText.

**Video** Forces in Circular Motion

In other words, **a particle of mass $m$ moving at constant speed $v$ around a circle of radius $r$ must always have a net force of magnitude $mv^2/r$ pointing toward the center of the circle,** as in FIGURE 6.5. It is this net force that causes the centripetal acceleration of circular motion. Without such a net force, the particle would move off in a straight line tangent to the circle.

The force described by Equation 6.6 is not a *new* kind of force. The net force will be due to one or more of our familiar forces, such as tension, friction, or the normal force. Equation 6.6 simply tells us how the net force needs to act—how strongly and in which direction—to cause the particle to move with speed $v$ in a circle of radius $r$.

In each example of circular motion that we will consider in this chapter, a physical force or a combination of forces directed toward the center produces the necessary acceleration.

FIGURE 6.5 Net force for circular motion.

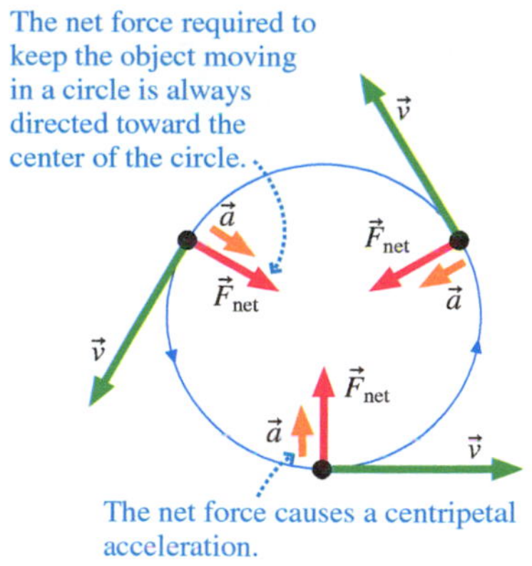


## CONCEPTUAL EXAMPLE 6.4 Forces on a car, part I

Engineers design curves on roads to be segments of circles. They also design dips and peaks in roads to be segments of circles with a radius that depends on expected speeds and other factors. A car is moving at a constant speed and goes into a dip in the road. At the very bottom of the dip, is the normal force of the road on the car greater than, less than, or equal to the car's weight?

FIGURE 6.6 Visual overview for the car in a dip.

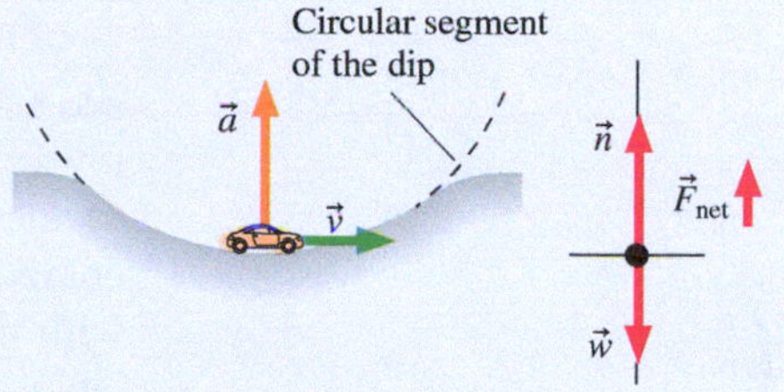


***REASON*** FIGURE 6.6 shows a visual overview of the situation. The car is accelerating, even though it is moving at a constant speed, because its direction is changing. When the car is at the bottom of the dip, the center of its circular path is directly above it and so its acceleration vector points straight up. The free-body diagram of Figure 6.6 identifies the only two forces acting on the car as the normal force, pointing upward, and its weight, pointing downward. Which is larger: $n$ or $w$?

Because $\vec{a}$ points upward, by Newton's second law there must be a net force on the car that also points upward. In order for this to be the case, the free-body diagram shows that the magnitude of the normal force must be *greater* than the weight.

***ASSESS*** You have probably experienced this situation. As you drive through a dip in the road, you feel "heavier" than normal. As discussed in Section 5.3, this is because your apparent weight—the normal force that supports you—is greater than your true weight.

## CONCEPTUAL EXAMPLE 6.5 Forces on a car, part II

A car is turning a corner at a constant speed, following a segment of a circle. What force provides the necessary centripetal acceleration?

FIGURE 6.7 Top views of a car turning a corner.

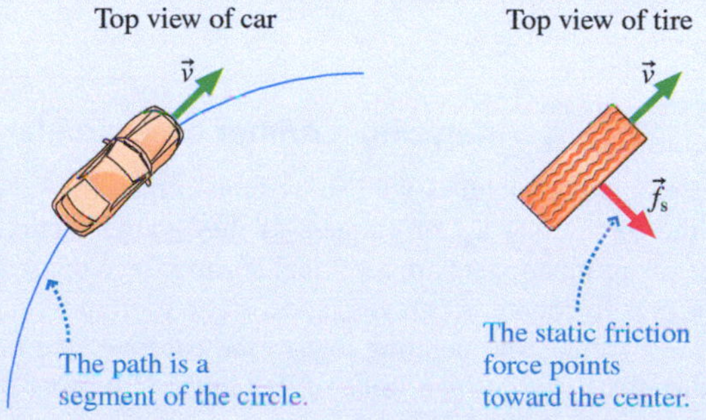


***REASON*** The car moves along a circular arc at a constant speed—uniform circular motion—for the quarter-circle necessary to complete the turn. We know that the acceleration is directed toward the center of the circle. What force or forces can we identify that provide this acceleration?

Imagine you are driving a car on a frictionless road, such as a very icy road. You would not be able to turn a corner. Turning the steering wheel would be of no use. The car would slide straight ahead, in accordance with both Newton's first law and the experience of anyone who has ever driven on ice! So it must be *friction* that causes the car to turn. The top view of the tire in FIGURE 6.7 shows the force on one of the car's tires as it turns a corner. It must be a *static* friction force, not kinetic, because the tires are not skidding: The points where the tires touch the road are not moving relative to the surface. If you skid, your car won't turn the corner—it will continue in a straight line!

***ASSESS*** This result agrees with your experience. You know that reduced friction makes it harder to turn corners, and you know as well that skidding on a curve is bad news. So it makes sense that static friction is the force at work.

Video Car Driving Over a Rise
Video Ball Leaves Circular Track

PROBLEM-SOLVING APPROACH 6.1 **Circular dynamics problems**

Circular motion involves an acceleration and thus a net force. We can therefore use techniques very similar to those we've already seen for other Newton's second-law problems.

***STRATEGIZE*** An object undergoing uniform circular motion has an acceleration directed toward the center of the circle. This acceleration, according to Newton's second law, is caused by a net force directed toward the center. The origin of this force is one or more of the forces we're familiar with, such as tension, weight, or friction.

***PREPARE*** Begin your visual overview with a pictorial representation in which you sketch the motion, define symbols, define axes, and identify what the problem is trying to find. There are two common situations:

- If the motion is in a horizontal plane, like a tabletop, draw the free-body diagram with the circle viewed edge-on, the $x$-axis pointing toward the center of the circle, and the $y$-axis perpendicular to the plane of the circle.
- If the motion is in a vertical plane, like a Ferris wheel, draw the free-body diagram with the circle viewed face-on, the $x$-axis pointing toward the center of the circle, and the $y$-axis tangent to the circle.

***SOLVE*** Newton's second law for uniform circular motion, $\vec{F}_{net} = (mv^2/r$, toward center of circle$)$, is a vector equation. Some forces act in the plane of the circle, some act perpendicular to the circle, and some may have components in both directions. In the coordinate system just described, with the $x$-axis pointing toward the center of the circle, Newton's second law is

$$\sum F_x = \frac{mv^2}{r} \qquad \text{and} \qquad \sum F_y = 0$$

That is, the net force toward the center of the circle has magnitude $mv^2/r$ while the net force perpendicular to the circle is zero. The components of the forces are found directly from the free-body diagram. Depending on the problem, either:

- Use the net force to determine the speed $v$, then use circular kinematics to find frequencies or other details of the motion.
- Use circular kinematics to determine the speed $v$, then solve for unknown forces.

***ASSESS*** Make sure your net force points toward the center of the circle. Check that your result has the correct units, is reasonable, and answers the question.

Exercise 9

EXAMPLE 6.6 **Analyzing a runner on a circular track** BIO

Kinesiologists—scientists who study human motion—can measure the forces that act on runners as they round a curve. The forces on a runner's feet are a vertical normal force that, on average, serves to counteract gravity, and a static friction force parallel to the ground and pointing toward the center of the runner's circular path. The average value of this inward-pointing force is measured to be 600 N for a 75 kg runner rounding a curve that has a 2.5 m radius. What is this runner's speed?

***STRATEGIZE*** We saw in Conceptual Example 6.5 that the force that causes the car's centripetal acceleration is the friction force between the road and the tires. The same situation holds for a runner moving in a circle: It is the static friction force between the track and the runner's shoes that causes the runner to move in a circle. Following the steps of Problem-Solving Approach 6.1, we will use Newton's second law to relate this force to the centripetal acceleration, from which we can find the runner's speed.

***PREPARE*** FIGURE 6.8 on the next page shows a visual overview of the problem. The main reason for the pictorial representation on the left is to illustrate the relevant geometry and to define the symbols that will be used. A circular dynamics problem usually does not have starting and ending points like a projectile problem, so subscripts such as $x_i$ or $y_f$ are usually not needed. Here we need to define the runner's speed $v$ and the radius $r$ of the circle.

FIGURE 6.8 A visual overview of the runner moving in a circular path.

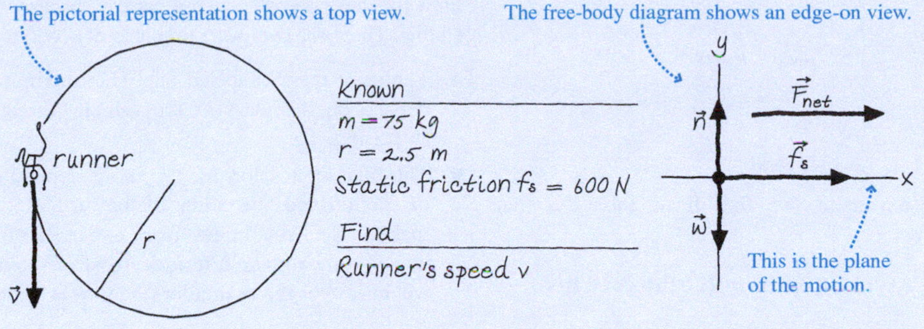


The free-body diagram shows the forces in the problem. Because the motion is in a horizontal plane, Problem-Solving Approach 6.1 tells us to draw the free-body diagram looking at the edge of the circle, with the $x$-axis pointing toward the center of the circle and the $y$-axis perpendicular to the plane of the circle. Three forces are acting on the runner: the weight force $\vec{w}$, the normal force of the ground $\vec{n}$, and the friction force $\vec{f}_s$.

**SOLVE** There is no net force in the $y$-direction, perpendicular to the circle, so $\vec{w}$ and $\vec{n}$ must be equal and opposite. There is a net force in the $x$-direction, toward the center of the circle, as there must be to cause the centripetal acceleration of circular motion. Only the friction force has an $x$-component, so Newton's second law is

$$\sum F_x = f_s = \frac{mv^2}{r}$$

We know the mass, the radius of the circle, and the friction force, so we can solve for $v$:

$$v = \sqrt{\frac{f_s r}{m}} = \sqrt{\frac{(600\text{ N})(2.5\text{ m})}{75\text{ kg}}} = 4.5\text{ m/s}$$

**ASSESS** The speed is about 10 mph, a reasonable value for a runner going around a sharp curve.

## EXAMPLE 6.7 Finding the maximum speed for a car to turn a corner

What is the maximum speed with which a 1500 kg car can make a turn around a curve of radius 20 m on a level (unbanked) road without sliding? (This radius turn is about what you might expect at a major intersection in a city.)

**STRATEGIZE** As Conceptual Example 6.5 showed, the force that causes the car's centripetal acceleration is the static friction between the car's tires and the road. If the car's speed is too high, the friction force needed to cause this centripetal force will exceed its maximum possible value and the car will slide.

**PREPARE** We start with the visual overview in FIGURE 6.9. The car moves along a circular arc at a constant speed—uniform circular motion—during the turn. The static friction force causes the centripetal acceleration, so this force must be directed toward the center of the circular arc. We've drawn the free-body diagram, shown from behind the car, with the friction force pointing in this direction. Because the motion is in a horizontal plane, we've again chosen an $x$-axis toward the center of the circle and a $y$-axis perpendicular to the plane of motion.

FIGURE 6.9 Visual overview of a car turning a corner.

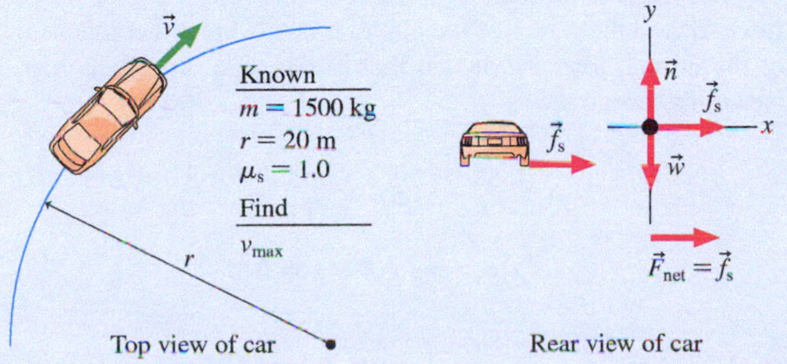


**SOLVE** The only force in the $x$-direction, toward the center of the circle, is static friction. Newton's second law along the $x$-axis is

$$\sum F_x = f_s = \frac{mv^2}{r}$$

This is the same equation as in the previous example, because the force toward the center of the circle in that example is also static friction.

Newton's second law in the $y$-direction is

$$\sum F_y = n - w = ma_y = 0$$

so that $n = w = mg$.

The car will slide when the static friction force reaches its maximum value. Recall from Equation 5.7 in Chapter 5 that this maximum force is given by

$$f_{s\,max} = \mu_s n = \mu_s mg$$

The maximum speed occurs when the static friction force reaches its maximum value, or when

$$f_{s\,max} = \frac{m{v_{max}}^2}{r}$$

*Continued*

Using the known value of $f_{s\,max}$, we find

$$\frac{mv_{max}{}^2}{r} = f_{s\,max} = \mu_s mg$$

Rearranging, we get

$$v_{max}{}^2 = \mu_s gr$$

For rubber tires on pavement, we find from Table 5.2 that $\mu_s = 1.0$. We then have

$$v_{max} = \sqrt{\mu_s gr} = \sqrt{(1.0)(9.8\ \text{m/s}^2)(20\ \text{m})} = 14\ \text{m/s}$$

**ASSESS** 14 m/s ≈ 30 mph, which seems like a reasonable upper limit for the speed at which a car can go around a curve without sliding. There are two other things to note about the solution:

- The car's mass canceled out. The maximum speed *does not* depend on the mass of the vehicle, though this may seem surprising.
- The final expression for $v_{max}$ *does* depend on the coefficient of friction and the radius of the turn. Both of these factors make sense. You know, from experience, that the speed at which you can take a turn decreases if $\mu_s$ is less (the road is wet or icy) or if $r$ is smaller (the turn is tighter).

FIGURE 6.10 Modifications to increase the maximum speed around a curve.

**(a)** Wings on a race car.

**(b)** A banked turn on a racetrack.

Because $v_{max}$ depends on $\mu_s$ and because $\mu_s$ depends on road conditions, the maximum safe speed through turns can vary dramatically. A car that easily handles a curve in dry weather can suddenly slide out of control when the pavement is wet. Icy conditions are even worse. If you lower the value of the coefficient of friction in Example 6.7 from 1.0 (dry pavement) to 0.1 (icy pavement), the maximum speed for the turn goes down to 4.4 m/s—about 10 mph!

Race cars turn corners at much higher speeds than normal passenger vehicles. One design modification of the *cars* to allow this is the addition of wings, as on the car in FIGURE 6.10a. The wings provide an additional force pushing the car *down* onto the pavement by deflecting air upward. This extra downward force increases the normal force, thus increasing the maximum static friction force and making faster turns possible.

There are also design modifications of the *track* that allow race cars to take corners at high speeds. If the track is banked by raising the outside edge of curved sections, the normal force can provide some of the force necessary to produce the centripetal acceleration, as we will see in the next example. The curves on racetracks may be quite sharply banked (FIGURE 6.10b). Curves on ordinary highways are often banked as well, though at more modest angles suiting the lower speeds.

## EXAMPLE 6.8 Finding a car's speed on a banked turn

A curve on a racetrack of radius 70 m is banked at a 15° angle. At what speed can a car take this curve without assistance from friction?

**STRATEGIZE** With no friction acting, it is the horizontal component of the normal force that causes the centripetal acceleration.

**PREPARE** After drawing the pictorial representation in FIGURE 6.11, we use the force identification diagram to find that, given that there is no friction acting, the only two forces are the normal force and the car's weight. We can then construct the free-body diagram, making sure that we draw the normal force perpendicular to the road's surface.

FIGURE 6.11 Visual overview for the car on a banked turn.

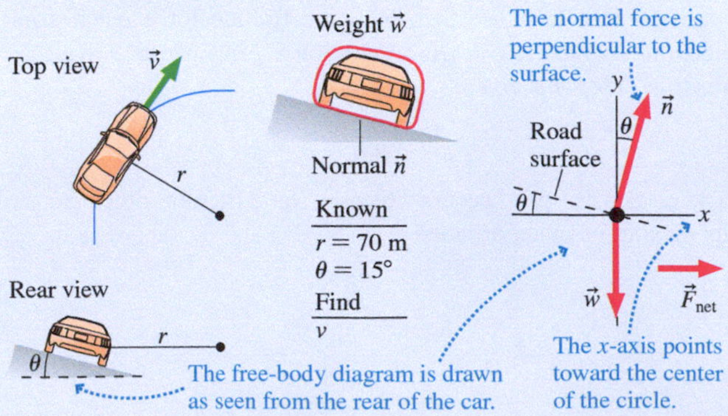


Even though the car is tilted, it is still moving in a *horizontal* circle. Thus, following Problem-Solving Approach 6.1, we choose the $x$-axis to be horizontal and pointing toward the center of the circle.

**SOLVE** Without friction, $n_x = n\sin\theta$ is the only component of force toward the center of the circle. It is this inward component of the normal force on the car that causes it to turn the corner. Newton's second law is

$$\sum F_x = n\sin\theta = \frac{mv^2}{r}$$

$$\sum F_y = n\cos\theta - w = 0$$

where $\theta$ is the angle at which the road is banked, and we've assumed that the car is traveling at the correct speed $v$. From the $y$-equation,

$$n = \frac{w}{\cos\theta} = \frac{mg}{\cos\theta}$$

Substituting this into the $x$-equation and solving for $v$ give

$$\left(\frac{mg}{\cos\theta}\right)\sin\theta = mg\tan\theta = \frac{mv^2}{r}$$

$$v = \sqrt{rg\tan\theta} = 14\text{ m/s}$$

**ASSESS** This is $\approx 30$ mph, a reasonable speed. Only at this exact speed can the turn be negotiated without reliance on friction forces.

The friction force provides the necessary centripetal acceleration not only for cars turning corners, but also for bicycles, horses, and, as we saw in Example 6.6, humans. The cyclists in FIGURE 6.12 are going through a tight turn; you can tell by how they lean. The road exerts both a vertical normal force and a horizontal friction force on their tires. A vector sum of these forces points at an angle. The cyclists lean to the side so that the sum of the road forces points along the line of their bikes and their bodies; this keeps them in balance. The horses in the photo that opened the chapter are leaning into the turn for a similar reason.

FIGURE 6.12 Road forces on a cyclist leaning into a turn.

## Maximum Walking Speed BIO

Humans and other two-legged animals have two basic gaits: walking and running. At slow speeds, you walk. When you need to go faster, you run. Why don't you just walk faster? There is an upper limit to the speed of walking, and this limit is set by the physics of circular motion.

Think about the motion of your body as you take a walking stride. You put one foot forward, then push off with your rear foot. Your body pivots over your front foot, and you bring your rear foot forward to take the next stride. As you can see in FIGURE 6.13a, the path that your body takes during this stride is the arc of a circle. **In a walking gait, your body is in circular motion as you pivot on your forward foot.**

A force toward the center of the circle is required for this circular motion, as shown in Figure 6.13. FIGURE 6.13b shows the forces acting on the woman's body during the midpoint of the stride: her weight, directed down, and the normal force of the ground, directed up. Newton's second law for the $x$-axis is

$$\sum F_x = w - n = \frac{mv^2}{r}$$

Because of her circular motion, the net force must point toward the center of the circle, or, in this case, down. In order for the net force to point down, the normal force must be *less* than her weight. Your body tries to "lift off" as it pivots over your foot, decreasing the normal force exerted on you by the ground. The normal force becomes smaller as you walk faster, but $n$ cannot be less than zero. Thus the maximum possible walking speed $v_{max}$ occurs when $n = 0$. Setting $n = 0$ in Newton's second law gives

$$w = mg = \frac{mv_{max}{}^2}{r}$$

Thus

$$v_{max} = \sqrt{gr} \qquad (6.7)$$

The maximum possible walking speed is limited by $r$, the length of the leg, and $g$, the free-fall acceleration. This formula is a good approximation of the maximum walking speed for humans and other animals. Giraffes, with their very long legs, can walk at high speeds. Animals such as mice with very short legs have such a low maximum walking speed that they rarely use this gait.

For humans, the length of the leg is approximately 0.7 m, giving $v_{max} \approx 2.6$ m/s $\approx$ 6 mph. You *can* walk this fast, though it becomes energetically unfavorable to walk at speeds above 4 mph. Most people make a transition to a running gait at about this speed.

FIGURE 6.13 Analysis of a walking stride.

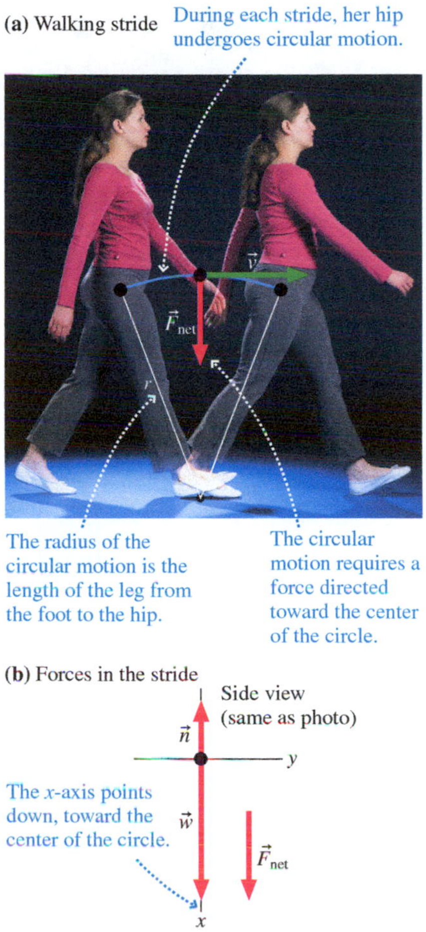

**STOP TO THINK 6.2** A block on a string spins in a horizontal circle on a frictionless table. Rank in order, from largest to smallest, the tensions $T_A$ to $T_E$ acting on the blocks A to E.

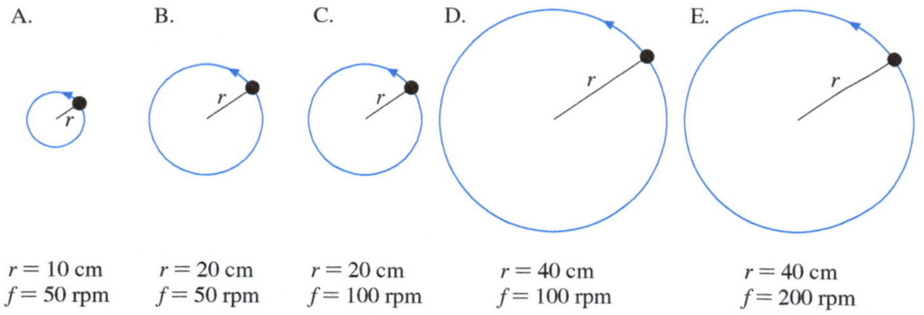


## 6.3 Apparent Forces in Circular Motion

FIGURE 6.14 Inside the Gravitron, a rotating circular room.

FIGURE 6.15 Bird's-eye view of a passenger in a car turning a corner.

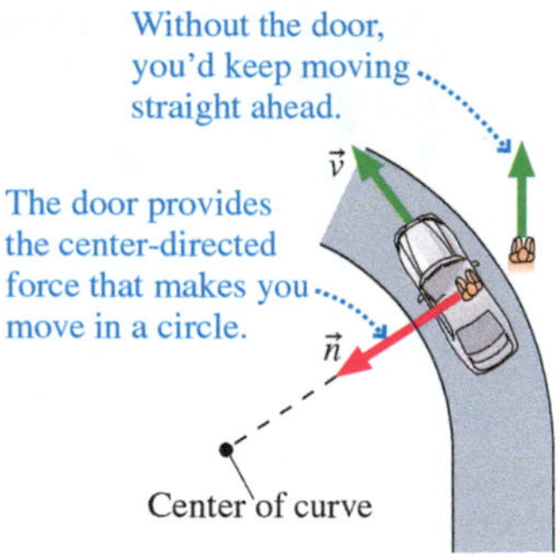


**Video** What the Physics? Shake It Off

FIGURE 6.14 shows a carnival ride that spins the riders around inside a large cylinder. The people are "stuck" to the inside wall of the cylinder! As you probably know from experience, the riders *feel* that they are being pushed outward, into the wall. But our analysis has found that an object in circular motion must have an *inward* force to create the centripetal acceleration. How can we explain this apparent difference?

### Centrifugal Force?

If you are a passenger in a car that turns a corner quickly, you may feel "thrown" by some mysterious force against the door. But is there really such a force? FIGURE 6.15 shows a bird's-eye view of you riding in a car as it makes a left turn. Just before the turn, you were moving in a straight line so, according to Newton's first law, as the car begins to turn, you want to continue moving along that same line. However, the door moves into your path and so runs into you! You feel the force of the door because it is this force, pushing *inward* toward the center of the curve, that is causing you to turn the corner. But you were not "thrown" into the door; the door ran into you.

A "force" that *seems* to push an object to the outside of a circle is called a *centrifugal force*. Despite having a name, there really is no such force. What you feel is your body trying to move ahead in a straight line (which would take you away from the center of the circle) as outside forces act to turn you in a circle. The only real forces, those that appear on free-body diagrams, are the ones pushing inward toward the center. **A centrifugal force will never appear on a free-body diagram and never be included in Newton's laws.**

With this in mind, let's revisit the rotating carnival ride. A person watching from above would see the riders in the cylinder moving in a circle with the walls providing the inward force that causes their centripetal acceleration. The riders *feel* as if they're being pushed outward because their natural tendency to move in a straight line is being resisted by the wall of the cylinder, which keeps getting in the way. But feelings aren't forces. The only actual force is the contact force of the cylinder wall pushing *inward*.

### Apparent Weight in Circular Motion

Imagine swinging a bucket of water over your head. If you swing the bucket fast enough, the water stays in. But you'll get a shower if you swing too slowly. Why does the water stay in the bucket? Or think about a roller coaster that does a loop-the-loop. How does the car stay on the track when it's upside down? You might have said that there was a centrifugal force holding the water in the bucket and the car on the track, but we have seen that there really isn't a centrifugal force. Analyzing these questions will tell us a lot about forces in general and circular motion in particular.

FIGURE 6.16a shows a roller coaster car going around a vertical loop-the-loop of radius $r$. If you've ever ridden a roller coaster, you know that your sensation of weight changes as you go over the crests and through the dips. To understand why, let's look at the forces on passengers going through the loop. To simplify our analysis, we will assume that the speed of the car stays constant as it moves through the loop.

FIGURE 6.16b shows a passenger's free-body diagram at the top and the bottom of the loop. Let's start by examining the forces on the passenger at the bottom of the loop. The only forces acting on her are her weight $\vec{w}$ and the normal force $\vec{n}$ of the seat pushing up on her. But recall from ◀◀ SECTION 5.3 that you don't feel the weight force. The force you *feel*, your apparent weight, is the magnitude of the contact force that supports you. Here the seat is supporting the passenger with the normal force $\vec{n}$, so her apparent weight is $w_{app} = n$. Based on our understanding of circular motion, we can say:

- She's moving in a circle, so there must be a net force directed toward the center of the circle—currently directly above her head—to provide the centripetal acceleration.
- The net force points *upward,* so it must be the case that $n > w$.
- Her apparent weight is $w_{app} = n$, so her apparent weight is greater than her true weight ($w_{app} > w$). Thus she "feels heavy" at the bottom of the circle.

This situation is the same as for the car driving through a dip in Conceptual Example 6.4. To analyze the situation quantitatively, we'll apply the steps of Problem-Solving Approach 6.1. As always, we choose the $x$-axis to point toward the center of the circle or, in this case, vertically upward. Then Newton's second law is

$$\sum F_x = n_x + w_x = n - w = \frac{mv^2}{r}$$

From this equation, the passenger's apparent weight is

$$w_{app} = n = w + \frac{mv^2}{r} \tag{6.8}$$

Her apparent weight at the bottom is *greater* than her true weight $w$, which agrees with your experience when you go through a dip or a valley.

Now let's look at the roller coaster car as it crosses the top of the loop. Things are a little trickier here. As Figure 6.16b shows, whereas the normal force of the seat pushes up when the passenger is at the bottom of the circle, it pushes *down* when she is at the top and the seat is above her. It's worth thinking carefully about this diagram to make sure you understand what it is showing.

The passenger is still moving in a circle, so there must be a net force *downward,* toward the center of the circle, to provide her centripetal acceleration. As always, we define the $x$-axis to be toward the center of the circle, so here the $x$-axis points vertically downward. Newton's second law gives

$$\sum F_x = n_x + w_x = n + w = \frac{mv^2}{r}$$

Note that $w_x$ is now *positive* because the $x$-axis is directed downward. We can solve for the passenger's apparent weight:

$$w_{app} = n = \frac{mv^2}{r} - w \tag{6.9}$$

If $v$ is sufficiently large, her apparent weight can exceed the true weight, just as it did at the bottom of the track.

But let's look at what happens if the car goes slower. Notice from Equation 6.9 that, as $v$ decreases, there comes a point when $mv^2/r = w$ and $n$ becomes zero. At that point, the seat is *not* pushing against the passenger at all! Instead, she is able to complete the circle because her weight force alone provides sufficient centripetal acceleration.

FIGURE 6.16 A roller coaster car going around a loop-the-loop.

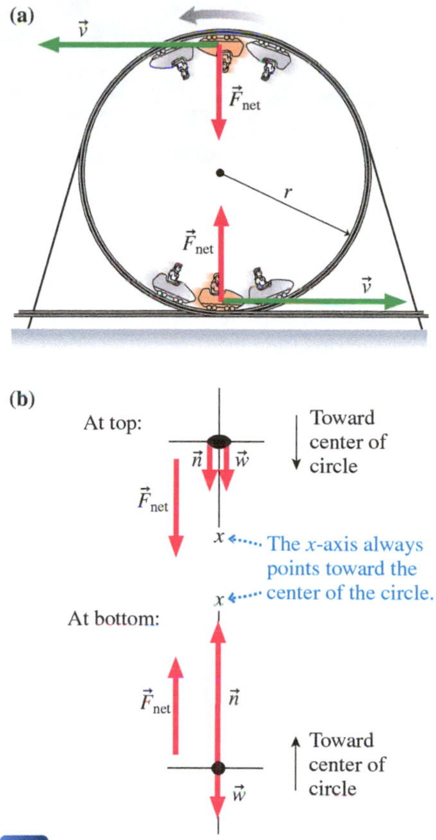


Video Figure 6.16

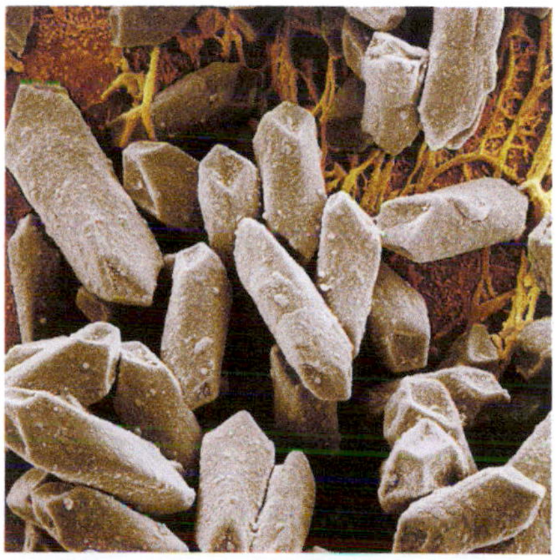

**When "down" is up** BIO You can tell, even with your eyes closed, what direction is down. Organs in your inner ear contain small crystals of calcium carbonate, called *otoliths.* These crystals are supported by a sensitive membrane. Your brain interprets "down" as the opposite of the direction of the normal force of the membrane on the otoliths. At the top of a loop in a roller coaster, this normal force is directed down, so your inner ear tells you that "down" is up. You are upside down, but it doesn't feel that way.

**A fast-spinning world** Saturn, a gas giant planet composed largely of fluid matter, is quite a bit larger than the earth. It also rotates much more quickly, completing one rotation in just under 11 hours. The rapid rotation decreases the apparent weight at the equator enough to distort the fluid surface. The planet has a noticeable oval shape, as the red circle shows. The diameter at the equator is 11% greater than the diameter at the poles.

The speed for which $n = 0$ is called the *critical speed* $v_c$. Because for $n$ to be zero we must have $mv_c^2/r = w$, the critical speed is

$$v_c = \sqrt{\frac{rw}{m}} = \sqrt{\frac{rmg}{m}} = \sqrt{gr} \tag{6.10}$$

What happens if the speed is slower than the critical speed? In this case, Equation 6.9 gives a *negative* value for $n$ if $v < v_c$. But that is physically impossible. The seat can push against the passenger ($n > 0$), but it can't *pull* on her, so the slowest possible speed is the speed for which $n = 0$ at the top. Thus, **the critical speed is the slowest speed at which the car can complete the circle.** If $v < v_c$, the passenger cannot turn the full loop but, instead, will fall from the car as a projectile! (This is why you're always strapped into a roller coaster.)

Water stays in a bucket swung over your head for the same reason. The bottom of the bucket pushes against the water to provide the inward force that causes circular motion. If you swing the bucket too slowly, the force of the bucket on the water drops to zero. At that point, the water leaves the bucket and becomes a projectile following a parabolic trajectory onto your head!

**EXAMPLE 6.9 How slow can you go?**

A motorcyclist in the Globe of Death, pictured here, rides in a 2.2-m-radius vertical loop. To keep control of the bike, the rider wants the normal force on his tires at the top of the loop to equal or exceed his and the bike's combined weight. What is the minimum speed at which the rider can take the loop?

***STRATEGIZE*** At the top of the loop both the normal force and the weight point downward. Together they provide the net force that determines the centripetal acceleration.

***PREPARE*** The visual overview for this problem is shown in FIGURE 6.17. In accordance with Problem-Solving Approach 6.1, we've chosen the $x$-axis to point toward the center of the circle.

FIGURE 6.17 Riding in a vertical loop around the Globe of Death.

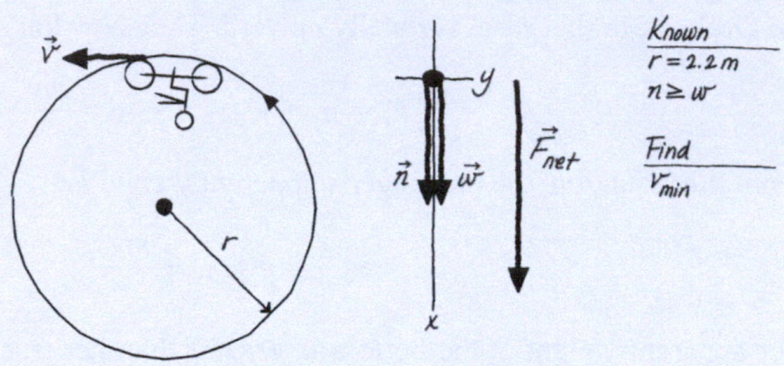


***SOLVE*** We will consider the forces at the top point of the loop. Because the $x$-axis points downward, Newton's second law is

$$\sum F_x = w + n = \frac{mv^2}{r}$$

The minimum acceptable speed occurs when $n = w$; thus

$$2w = 2mg = \frac{mv_{min}{}^2}{r}$$

Solving for the speed, we find

$$v_{min} = \sqrt{2gr} = \sqrt{2(9.8\text{ m/s}^2)(2.2\text{ m})} = 6.6\text{ m/s}$$

***ASSESS*** The minimum speed is $\approx 15$ mph, which isn't all that fast; the bikes can easily reach this speed. But normally several bikes are in the globe at one time. The big challenge is to keep all of the riders in the cage moving at this speed in synchrony. The period for the circular motion at this speed is $T = 2\pi r/v \approx 2$ s, leaving little room for error!

## Centrifuges BIO

The *centrifuge,* an important biological application of circular motion, is used to separate the components of a liquid that have different densities. Typically these are different types of cells, or the components of cells, suspended in water. You probably know that small particles suspended in water will eventually settle to the bottom. However, the downward motion due to gravity for extremely small objects such as cells is so slow that it could take days or even months for the cells to settle out. It's not practical to wait for biological samples to separate due to gravity alone.

The separation would go faster if the force of gravity could be increased. Although we can't change gravity, we can increase the apparent weight of objects in the sample by spinning them very fast, and that is what the centrifuge shown in FIGURE 6.18 does. The centrifuge produces centripetal accelerations that are thousands of times greater than free-fall acceleration. As the centrifuge effectively increases gravity to thousands of times its normal value, the cells or cell components settle out and separate by density in a matter of minutes or hours.

FIGURE 6.18 The operation of a centrifuge.

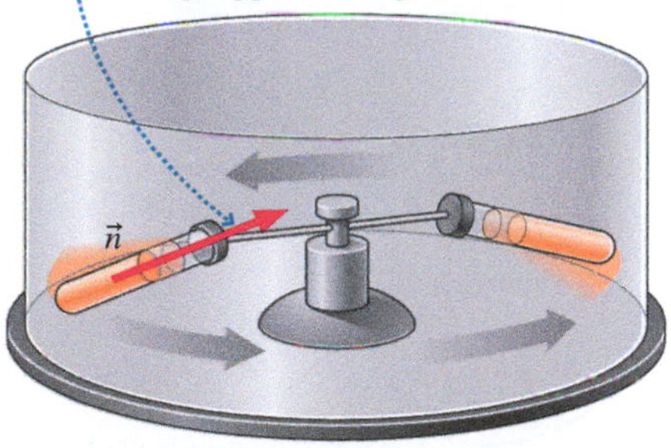


**EXAMPLE 6.10 Analyzing the ultracentrifuge**

An 18-cm-diameter ultracentrifuge produces an extraordinarily large centripetal acceleration of 250,000$g$, where $g$ is the free-fall acceleration due to gravity. What is its frequency in rpm? What is the apparent weight of a sample with a mass of 0.0030 kg?

***STRATEGIZE*** We will use Equation 6.5 for the relationship between the centripetal acceleration and the frequency and radius.

***PREPARE*** The acceleration in SI units is

$$a = 250{,}000(9.80\text{ m/s}^2) = 2.45 \times 10^6\text{ m/s}^2$$

The radius is half the diameter, or $r = 9.0\text{ cm} = 0.090\text{ m}$.

***SOLVE*** We can rearrange Equation 6.5 to find the frequency given the centripetal acceleration:

$$f = \frac{1}{2\pi}\sqrt{\frac{a}{r}} = \frac{1}{2\pi}\sqrt{\frac{2.45 \times 10^6\text{ m/s}^2}{0.090\text{ m}}} = 830\text{ rev/s}$$

Converting to rpm, we find

$$830\,\frac{\text{rev}}{\text{s}} \times \frac{60\text{ s}}{1\text{ min}} = 50{,}000\text{ rpm}$$

The acceleration is so high that every force is negligible except for the force that provides the centripetal acceleration. The net force is simply equal to the inward force, which is also the sample's apparent weight:

$$w_{\text{app}} = F_{\text{net}} = ma = (3.0 \times 10^{-3}\text{ kg})(2.45 \times 10^6\text{ m/s}^2) = 7.4 \times 10^3\text{ N}$$

The 3 gram sample has an effective weight of about 1700 pounds!

***ASSESS*** Because the acceleration is 250,000$g$, the apparent weight is 250,000 times the actual weight. This makes sense, as does the fact that we calculated a very high frequency, which is necessary to give the large acceleration.

**Human centrifuge** BIO If you spin your arm rapidly in a vertical circle, the motion will produce an effect like that in a centrifuge. The motion will assist outbound blood flow in your arteries and retard inbound blood flow in your veins. There will be a buildup of fluid in your hand that you will be able to see (and feel!) quite easily.

**STOP TO THINK 6.3** A car is rolling over the top of a hill at constant speed $v$. At this instant,

A. $n > w$.
B. $n < w$.
C. $n = w$.
D. We can't tell about $n$ without knowing $v$.

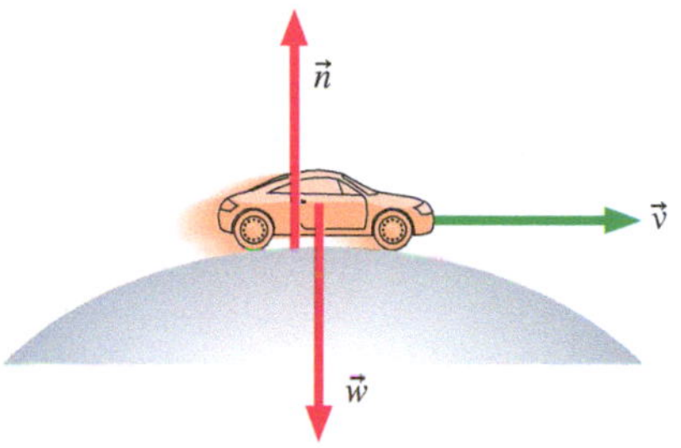


## 6.4 Circular Orbits and Weightlessness

The International Space Station orbits the earth in a circular path at a speed of over 17,000 miles per hour. What forces act on it? Why does it move in a circle? Before we start considering the physics of orbital motion, let's return, for a moment, to projectile motion. Projectile motion occurs when the only force on an object is gravity. Our analysis of projectiles made an implicit assumption that the earth is flat and that the free-fall acceleration, due to gravity, is everywhere straight down. This is an acceptable approximation for projectiles of limited range, such as baseballs or cannon balls, but there comes a point where we can no longer ignore the curvature of the earth.

## Orbital Motion

FIGURE 6.19 Projectiles being launched at increasing speeds from height $h$ on a smooth, airless planet.

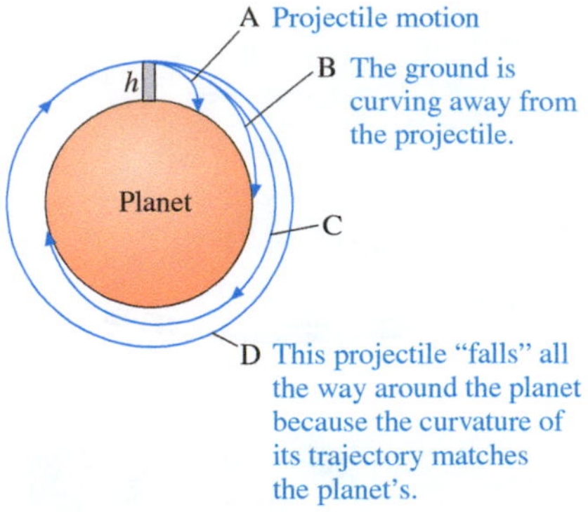


FIGURE 6.19 shows a perfectly smooth, spherical, airless planet with a vertical tower of height $h$. A projectile is launched from this tower with initial speed $v_i$ parallel to the ground. If $v_i$ is very small, as in trajectory A, the "flat-earth approximation" is valid and the problem is identical to Example 3.9 in which a dog ran off the end of a dock. The projectile simply falls to the ground along a parabolic trajectory.

As the initial speed $v_i$ is increased, it seems from the projectile's perspective that the ground is curving out from beneath it. It is still falling the entire time, always getting closer to the ground, but the distance that the projectile travels before finally reaching the ground—that is, its range—increases because the projectile must "catch up" with the ground that is curving away from it. Trajectories B and C are like this.

If the launch speed $v_i$ is sufficiently large, there comes a point at which the curve of the trajectory and the curve of the earth are parallel. In this case, the projectile "falls" but it never gets any closer to the ground! This is the situation for trajectory D. The projectile returns to the point from which it was launched, at the same speed at which it was launched, making a closed trajectory. Such a closed trajectory around a planet or star is called an **orbit.**

The most important point of this qualitative analysis is that, in the absence of air resistance, **an orbiting projectile is in free fall.** This is, admittedly, a strange idea, but one worth careful thought. An orbiting projectile is really no different from a thrown baseball or a dog jumping off a dock. The only force acting on it is gravity, but its tangential velocity is so great that the curvature of its trajectory matches the curvature of the earth. When this happens, the projectile "falls" under the influence of gravity but never gets any closer to the surface, which curves away beneath it.

When we first studied free fall in Chapter 2, we said that free-fall acceleration is always directed vertically downward. As we see in FIGURE 6.20, "downward" really means "toward the center of the earth." For a projectile in orbit, the direction of the force of gravity changes, always pointing toward the center of the earth.

FIGURE 6.20 The force of gravity is really directed toward the center of the earth.

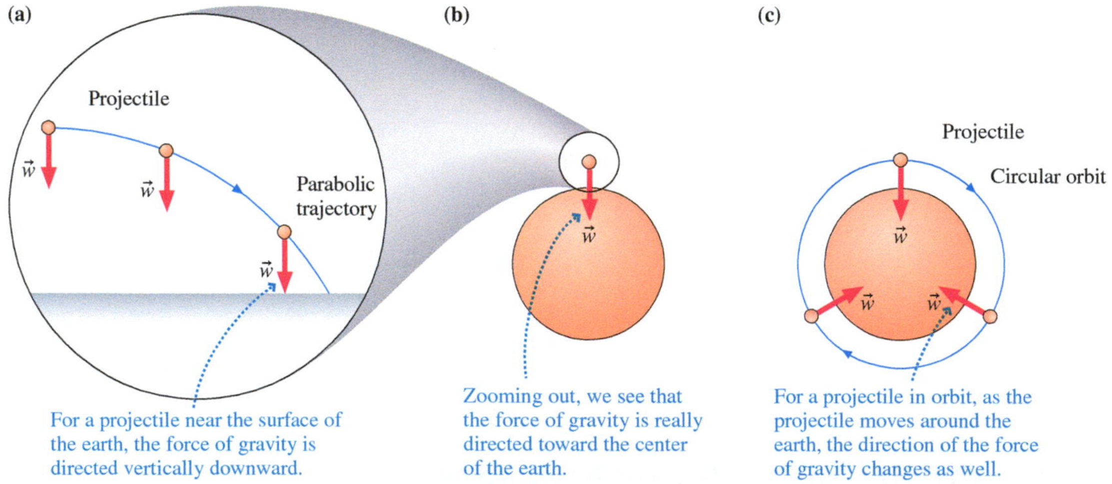


As you have learned, a force of constant magnitude that always points toward the center of a circle causes the centripetal acceleration of uniform circular motion. Because the only force acting on the orbiting projectile in Figure 6.20 is gravity, and we're assuming the projectile is very near the surface of the earth, we can write

$$a = \frac{F_{net}}{m} = \frac{w}{m} = \frac{mg}{m} = g \qquad (6.11)$$

An object moving in a circle of radius $r$ at speed $v_{orbit}$ will have this centripetal acceleration if

$$a = \frac{(v_{orbit})^2}{r} = g \quad (6.12)$$

That is, if an object moves parallel to the surface with the speed

$$v_{orbit} = \sqrt{gr} \quad (6.13)$$

then the free-fall acceleration provides exactly the centripetal acceleration needed for a circular orbit of radius $r$. An object with any other speed will not follow a circular orbit.

The earth's radius is $r = R_e = 6.37 \times 10^6$ m. The orbital speed of a projectile just skimming the surface of a smooth, airless earth is

$$v_{orbit} = \sqrt{gR_e} = \sqrt{(9.80\text{ m/s}^2)(6.37 \times 10^6\text{ m})} = 7900\text{ m/s} \approx 18{,}000\text{ mph}$$

We can use $v_{orbit}$ to calculate the period of the satellite's orbit:

$$T = \frac{2\pi r}{v_{orbit}} = 2\pi\sqrt{\frac{r}{g}} \quad (6.14)$$

For this earth-skimming orbit, $T = 5065$ s $= 84.4$ min.

Of course, actual satellites must orbit at some height above the surface to be above mountains and trees—and most of the atmosphere, so there is little drag. Most people envision the International Space Station orbiting far above the earth's surface, but the average height is just over 250 miles, giving a value of $r$ that is only 6% greater than the earth's radius—not too far from skimming the surface. At this slightly larger value of $r$, Equation 6.14 gives $T = 87$ min. In fact, the International Space Station orbits with a period of about 93 min, going around the earth more than 15 times each day.

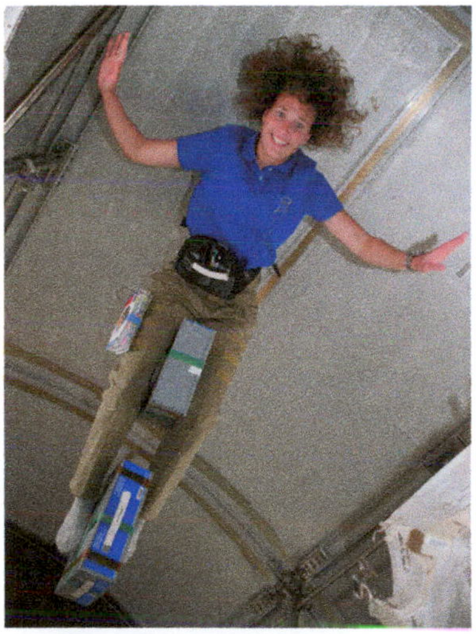

Zero apparent weight in space.

## Weightlessness in Orbit

When we discussed *weightlessness* in ◀◀ SECTION 5.3, we saw that it occurs during free fall. We asked whether astronauts and their spacecraft are in free fall. We can now give an affirmative answer: They are, indeed, in free fall. They are falling continuously around the earth, under the influence of only the gravitational force, but never getting any closer to the ground because the earth's surface curves beneath them. Weightlessness in space is no different from the weightlessness in a free-falling elevator. **Weightlessness does *not* occur from an absence of weight or an absence of gravity.** Instead, the astronaut, the spacecraft, and everything in it are "weightless" (i.e., their *apparent* weight is zero) because they are all falling together. We know that the free-fall acceleration doesn't depend on mass, so the astronaut and the station follow exactly the same orbit.

**Rotating space stations** BIO The weightlessness astronauts experience in orbit has serious physiological consequences. Astronauts who spend time in weightless environments lose bone and muscle mass and suffer other adverse effects. One solution is to introduce "artificial gravity." On a space station, the easiest way to do this would be to make the station rotate, producing an apparent weight. The designers of this space station model for the movie *2001: A Space Odyssey* made it rotate for just that reason.

## The Orbit of the Moon

The moon moves in an orbit around the earth that is approximately circular. The force that holds the moon in its orbit, that provides the necessary centripetal acceleration, is the gravitational attraction of the earth. The moon, like all satellites, is simply "falling" around the earth. But if we use the distance to the moon, $r = 3.84 \times 10^8$ m, in Equation 6.14 to predict the period of the moon's orbit, we get a period of approximately 11 hours. This is clearly wrong; you know that the period of the moon's orbit is about one month. What went wrong?

In using Equation 6.14, we assumed that the free-fall acceleration $g$ is the same at the distance of the moon as it is on or near the earth's surface. But if gravity is the force of the earth pulling on an object, it seems plausible that the size of that force, and thus the size of $g$, should diminish with increasing distance from the earth. And, indeed, the force of gravity does decrease with distance, in a manner we'll explore in the next section.

**STOP TO THINK 6.4** A satellite is in a low earth orbit. Which of the following changes would increase the orbital period?

A. Increasing the mass of the satellite
B. Increasing the height of the satellite about the surface
C. Increasing the value of $g$

## 6.5 Newton's Law of Gravity

Our current understanding of the force of gravity begins with Isaac Newton. The popular image of Newton coming to a key realization about gravity after an apple fell on his head is at least close to the truth: Newton himself said that the "notion of gravitation" came to him as he "sat in a contemplative mood" and "was occasioned by the fall of an apple."

The important notion that came to Newton is this: *Gravity is a universal force that affects all objects in the universe.* The force that causes the fall of an apple is the same force that keeps the moon in orbit. This is something widely accepted now, but at the time this was a revolutionary idea, and there were some important details for Newton to work out—in particular, the way that the force varies with distance.

### Gravity Obeys an Inverse-Square Law

Newton proposed that *every* object in the universe attracts *every other* object with a force that has the following properties:

1. The force is inversely proportional to the square of the distance between the objects.
2. The force is directly proportional to the product of the masses of the two objects.

FIGURE 6.21 The gravitational forces on masses $m_1$ and $m_2$.

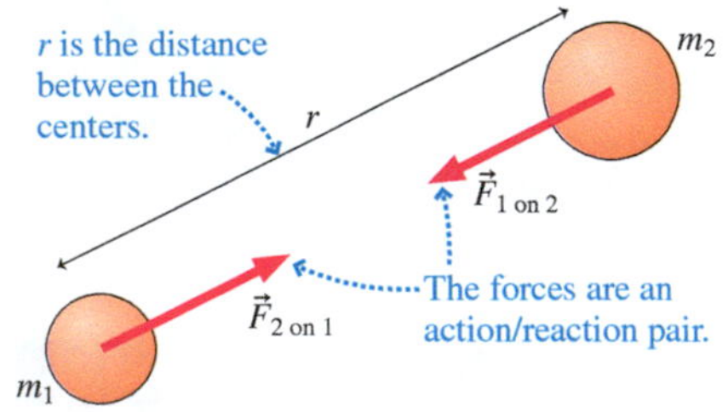


FIGURE 6.21 shows two spherical objects with masses $m_1$ and $m_2$ separated by distance $r$. Each object exerts an attractive force on the other, a force that we call the **gravitational force.** These two forces form an action/reaction pair, so $\vec{F}_{1\text{ on }2}$ is equal in magnitude and opposite in direction to $\vec{F}_{2\text{ on }1}$. The magnitude of the forces is given by Newton's law of gravity.

**Newton's law of gravity** If two objects with masses $m_1$ and $m_2$ are a distance $r$ apart, the objects exert attractive forces on each other of magnitude

$$F_{1\text{ on }2} = F_{2\text{ on }1} = \frac{Gm_1m_2}{r^2} \tag{6.15}$$

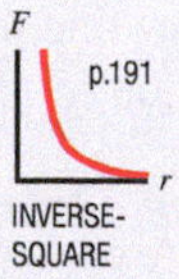


The forces are directed along the line joining the two objects.

The constant $G$ is called the **gravitational constant.** In SI units,

$$G = 6.67 \times 10^{-11}\ \text{N} \cdot \text{m}^2/\text{kg}^2$$

**NOTE** ► Strictly speaking, Newton's law of gravity applies to *particles* with masses $m_1$ and $m_2$. However, it can be shown that the law also applies to the force between two spherical objects if $r$ is the distance between their centers. ◄

As the distance $r$ between two objects increases, the gravitational force between them decreases. Because the distance appears squared in the denominator, Newton's law of gravity is what we call an **inverse-square** law. Doubling the distance between two masses causes the force between them to decrease by a factor of 4. This mathematical form is one we will see again, so it is worth our time to explore it in more detail.

## Inverse-square relationships

Two quantities have an **inverse-square relationship** if $y$ is inversely proportional to the *square* of $x$. We write the mathematical relationship as

$$y = \frac{A}{x^2}$$

$y$ is inversely proportional to $x^2$.

Here, $A$ is a constant. This relationship is sometimes written as $y \propto 1/x^2$.

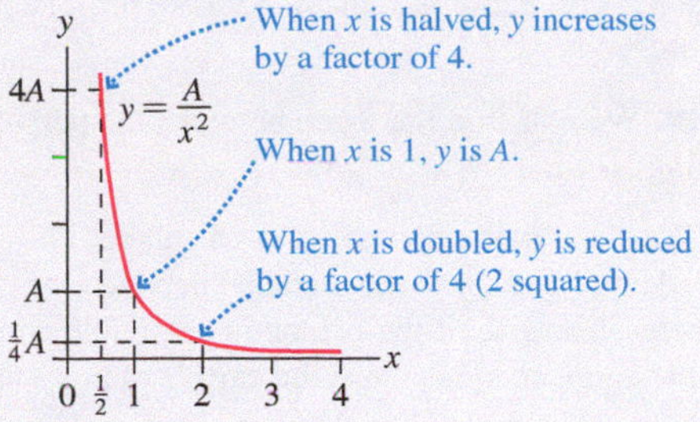


**SCALING** As the graph shows, inverse-square scaling means, for example:

- If you double $x$, you decrease $y$ by a factor of 4.
- If you halve $x$, you increase $y$ by a factor of 4.
- If you increase $x$ by a factor of 3, you decrease $y$ by a factor of 9.
- If you decrease $x$ by a factor of 3, you increase $y$ by a factor of 9.

Generally, **if *x increases* by a factor of $C$, *y decreases* by a factor of $C^2$.** If $x$ *decreases* by a factor of $C$, $y$ *increases* by a factor of $C^2$.

**RATIOS** For any two values of $x$—say, $x_1$ and $x_2$—we have

$$y_1 = \frac{A}{x_1^2} \quad \text{and} \quad y_2 = \frac{A}{x_2^2}$$

Dividing the $y_1$-equation by the $y_2$-equation, we find

$$\frac{y_1}{y_2} = \frac{A/x_1^2}{A/x_2^2} = \frac{A}{x_1^2}\frac{x_2^2}{A} = \frac{x_2^2}{x_1^2}$$

That is, the ratio of $y$-values is the inverse of the ratio of the squares of the corresponding values of $x$.

**LIMITS** As $x$ becomes large, $y$ becomes very small; as $x$ becomes small, $y$ becomes very large.

Exercises 19, 20

### CONCEPTUAL EXAMPLE 6.11 The gravitational force between two spheres

The gravitational force between two giant lead spheres is 0.010 N when the centers of the spheres are 20 m apart. What is the distance between their centers when the gravitational force between them is 0.160 N?

**REASON** We can solve this problem without knowing the masses of the two spheres. The key is to consider the ratios of forces and distances. Gravity is an inverse-square relationship; the force is related to the inverse square of the distance. The force *increases* by a factor of $(0.160\ \text{N})/(0.010\ \text{N}) = 16$, so the distance must *decrease* by a factor of $\sqrt{16} = 4$. The distance is thus $(20\ \text{m})/4 = 5.0\ \text{m}$.

**ASSESS** This type of ratio reasoning is a very good way to get a quick handle on the solution to a problem.

### EXAMPLE 6.12 Finding the gravitational force between two people

You are seated in your physics class next to another student 0.60 m away. Estimate the magnitude of the gravitational force between you. Assume that you each have a mass of 65 kg.

**STRATEGIZE** We will estimate the force using Newton's law of gravity.

**PREPARE** We model each of you as a sphere. This is not a particularly good model, but it will do for making an estimate. We then take 0.60 m as the distance between your centers.

**SOLVE** The gravitational force is given by Equation 6.15:

$$F_{\text{(you) on (other student)}} = \frac{Gm_{\text{you}}m_{\text{other student}}}{r^2}$$

$$= \frac{(6.67 \times 10^{-11}\ \text{N}\cdot\text{m}^2/\text{kg}^2)(65\ \text{kg})(65\ \text{kg})}{(0.60\ \text{m})^2}$$

$$= 7.8 \times 10^{-7}\ \text{N}$$

**ASSESS** The force is quite small, roughly the weight of one hair on your head. This seems reasonable; you don't normally sense this attractive force!

There is a gravitational force between all objects in the universe, but the gravitational force between two ordinary-sized objects is extremely small. Only when one (or both) of the masses is exceptionally large does the force of gravity become important. The downward force of the earth on you—your weight—is large because the earth has an enormous mass. And the attraction is mutual: By Newton's third law, you exert an upward force on the earth that is equal to your weight. However, the large mass of the earth makes the *effect* of this force on the earth negligible.

**EXAMPLE 6.13 Finding the gravitational force of the earth on a person**

What is the magnitude of the gravitational force of the earth on a 60 kg person? The earth has mass $5.98 \times 10^{24}$ kg and radius $6.37 \times 10^{6}$ m.

**STRATEGIZE** We will find the force between the person and the earth by using Newton's law of gravity.

**PREPARE** We again model the person as a sphere. The distance $r$ in Newton's law of gravity is the distance between the *centers* of the two spheres. The size of the person is negligible compared to the size of the earth, so we can use the earth's radius as $r$.

**SOLVE** The force of gravity on the person due to the earth can be computed using Equation 6.15:

$$F_{\text{earth on person}} = \frac{GM_e m}{R_e^2} = \frac{(6.67 \times 10^{-11}\ \text{N} \cdot \text{m}^2/\text{kg}^2)(5.98 \times 10^{24}\ \text{kg})(60\ \text{kg})}{(6.37 \times 10^{6}\ \text{m})^2} = 590\ \text{N}$$

**ASSESS** This force is exactly the same as we would calculate using the formula for the weight force, $w = mg$. This isn't surprising, though. Chapter 5 introduced the weight of an object as simply the "force of gravity" acting on it. Newton's law of gravity is a more fundamental law for calculating the force of gravity, but it's still the same force that we earlier called "weight."

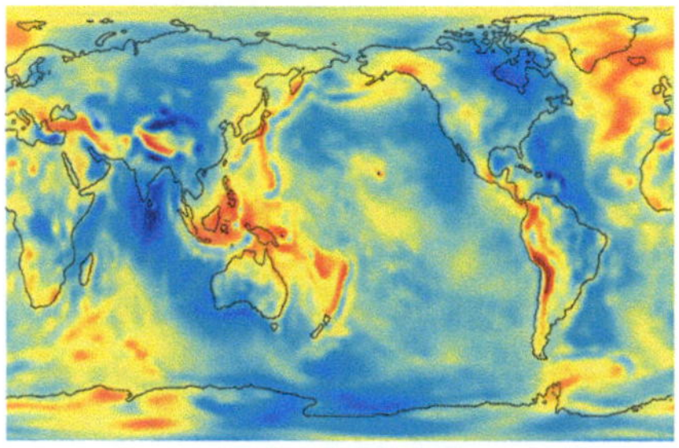

**Variable gravity** When we calculated the force of the earth's gravity, we assumed that the earth's shape and composition are uniform. In reality, unevenness in density and other factors create small variations in the earth's gravity, as shown in this image. Red means slightly stronger surface gravity; blue means slightly weaker. These variations are important for scientists who study the earth, but they are small enough that we can ignore them for the computations we'll do in this text.

**NOTE** ▶ We will use uppercase $R$ and $M$ to represent the large mass and radius of a star or planet, as we did in Example 6.13. ◀

## Gravity on Other Worlds

The force of gravitational attraction between the earth and you is responsible for your weight. If you traveled to another planet, your *mass* would be the same but your *weight* would vary, as we discussed in Chapter 5. Indeed, when astronauts ventured to the moon, television images showed them walking—and even jumping and skipping—with ease, even though they were wearing life-support systems with a mass greater than 80 kg, a visible reminder that the weight of objects is less on the moon. Let's consider why this is so.

**FIGURE 6.22** An astronaut weighing a mass on the moon.

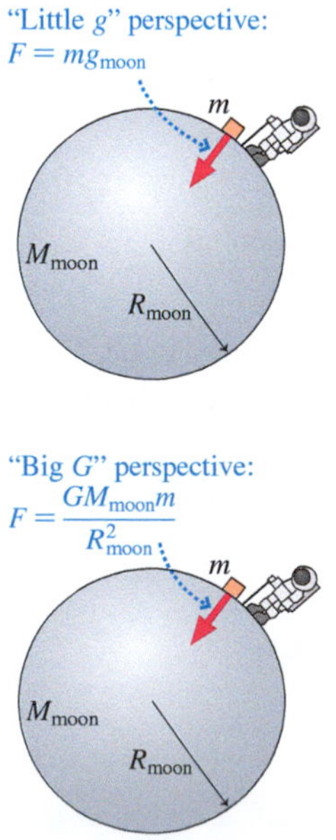


FIGURE 6.22 shows an astronaut on the moon weighing a rock of mass $m$. When we compute the weight of an object on the surface of the earth, we use the formula $w = mg$. We can do the same calculation for a mass on the moon, as long as we use the value of $g$ on the moon:

$$w = mg_{\text{moon}} \tag{6.16}$$

This is the "little $g$" perspective. Falling-body experiments on the moon give the value of $g_{\text{moon}}$ as 1.62 m/s$^2$.

But we can also take a "big $G$" perspective. The weight of the rock comes from the gravitational attraction of the moon, and we can compute this weight using Equation 6.15. The distance $r$ is the radius of the moon, which we'll call $R_{\text{moon}}$. Thus

$$F_{\text{moon on } m} = \frac{GM_{\text{moon}}m}{R^2_{\text{moon}}} \tag{6.17}$$

Because Equations 6.16 and 6.17 are two names and two expressions for the same force, we can equate the right-hand sides to find that

$$g_{\text{moon}} = \frac{GM_{\text{moon}}}{R^2_{\text{moon}}}$$

We have done this calculation for an object on the moon, but the result is completely general. At the surface of a planet (or a star), the free-fall acceleration $g$, a consequence of gravity, can be computed as

$$g_{\text{planet}} = \frac{GM_{\text{planet}}}{R^2_{\text{planet}}} \tag{6.18}$$

Free-fall acceleration on the surface of a planet

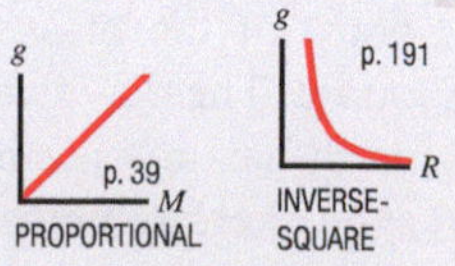

If we use values for the mass and the radius of the moon, we compute $g_{moon} = 1.62\ m/s^2$. This means that an object would weigh less on the moon than it would on the earth, where $g$ is $9.80\ m/s^2$. A 70 kg astronaut wearing an 80 kg spacesuit would weigh more than 330 lb on the earth but only 54 lb on the moon.

Equation 6.18 gives $g$ at the surface of a planet. More generally, imagine an object at distance $r > R$ from the center of a planet. Its free-fall acceleration at this distance is

$$g = \frac{GM}{r^2} \tag{6.19}$$

This more general result agrees with Equation 6.18 if $r = R$, but it allows us to determine the "local" free-fall acceleration at distances $r > R$. Equation 6.19 expresses Newton's idea that the size of $g$ should decrease as you get farther from the earth.

As you're flying in a jet airplane at a height of about 10 km, the free-fall acceleration is about 0.3% less than on the ground. At the height of the International Space Station, about 400 km, Equation 6.19 gives $g = 8.7\ m/s^2$, about 13% less than the free-fall acceleration on the earth's surface. If you use this slightly smaller value of $g$ in Equation 6.14 for the period of a satellite's orbit, you'll get the correct period of about 93 minutes. This value of $g$, only slightly less than the ground-level value, emphasizes the point that an object in orbit is not "weightless" due to the absence of gravity, but rather because it is in free fall.

**Walking on the moon** BIO We saw (Figure 6.13) that the maximum walking speed depends on leg length and the value of $g$. The low lunar gravity makes walking very easy but also makes it very slow: The maximum walking speed on the moon is about 1 m/s—a very gentle stroll! Walking at a reasonable pace was difficult for the Apollo astronauts, but the reduced weight made jumping quite easy. Videos from the surface of the moon often showed the astronauts getting from place to place by hopping or skipping—not for fun, but for speed and efficiency.

**EXAMPLE 6.14 Finding the speed required to orbit Deimos**

Mars has two moons, each much smaller than the earth's moon. The smaller of these two bodies, Deimos, isn't quite spherical, but we can model it as a sphere of radius 6.3 km. Its mass is $1.8 \times 10^{15}$ kg. At what speed would a projectile move in a very low orbit around Deimos?

**SOLVE** The free-fall acceleration at the surface of Deimos is small:

$$g_{Deimos} = \frac{GM_{Deimos}}{R^2_{Deimos}} = \frac{(6.67 \times 10^{-11}\ N \cdot m^2/kg^2)(1.8 \times 10^{15}\ kg)}{(6.3 \times 10^3\ m)^2} = 0.0030\ m/s^2$$

Given this, we can use Equation 6.13 to calculate the orbital speed:

$$v_{orbit} = \sqrt{gr} = \sqrt{(0.0030\ m/s^2)(6.3 \times 10^3\ m)} = 4.3\ m/s \approx 10\ mph$$

**ASSESS** This is quite slow. With a good jump, you could easily launch yourself into an orbit around Deimos!

**STOP TO THINK 6.5** Rank in order, from largest to smallest, the free-fall accelerations on the surfaces of the following planets.

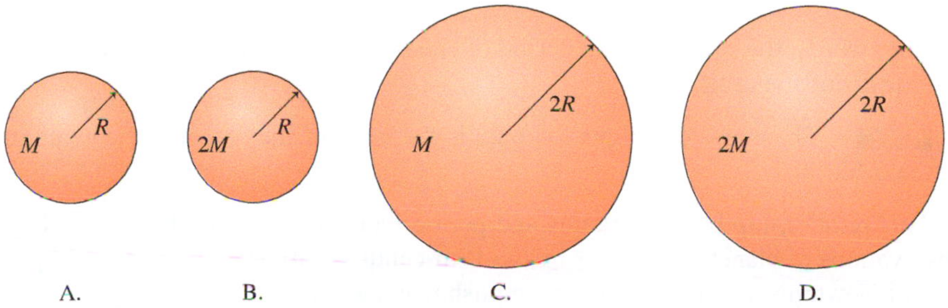


# 6.6 Gravity and Orbits

The planets of the solar system orbit the sun because the sun's gravitational pull, a force that points toward the center, causes the centripetal acceleration of circular motion. Mercury, the closest planet, experiences the largest acceleration, while Neptune, the most distant, has the smallest.

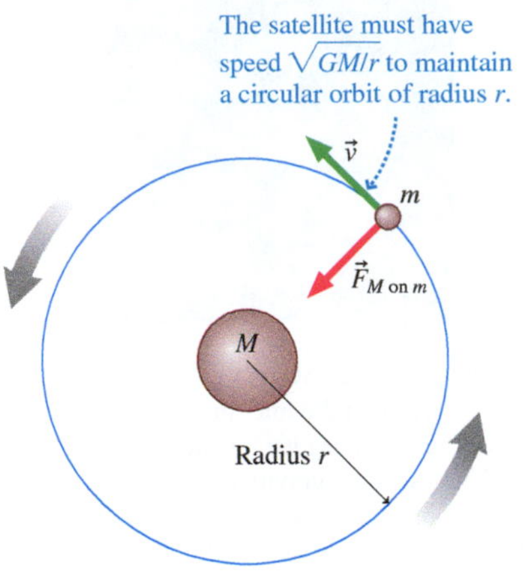


FIGURE 6.23 The orbital motion of a satellite is due to the force of gravity.

FIGURE 6.23 shows a large body of mass $M$ such as the sun, with a much smaller body such as the earth of mass $m$ orbiting it. The smaller body is called a **satellite,** even though it may be a planet orbiting the sun. Newton's second law tells us that $F_{M\text{ on }m} = ma$, where $F_{M\text{ on }m}$ is the gravitational force of the large body on the satellite and $a$ is the satellite's acceleration. $F_{M\text{ on }m}$ is given by Equation 6.15, and, because it's moving in a circular orbit, the satellite's acceleration is its centripetal acceleration, $mv^2/r$. Thus Newton's second law gives

$$F_{M\text{ on }m} = \frac{GMm}{r^2} = ma = \frac{mv^2}{r} \tag{6.20}$$

Solving for $v$, we find that the speed of a satellite in a circular orbit is

$$v = \sqrt{\frac{GM}{r}} \tag{6.21}$$

Speed of a satellite in a circular orbit of radius $r$ about a star or planet of mass $M$

A satellite must have this specific speed in order to maintain a circular orbit of radius $r$ about the larger mass $M$. If the velocity differs from this value, the orbit will become elliptical rather than circular. Notice that the orbital speed does not depend on the satellite's mass $m$. This is consistent with our previous discoveries that free-fall motion and projectile motion due to gravity are independent of the mass.

For a planet orbiting the sun, the period $T$ is the time to complete one full orbit. The relationship among speed, radius, and period is the same as for any circular motion, $v = 2\pi r/T$. Combining this with the value of $v$ for a circular orbit from Equation 6.21 gives

$$\sqrt{\frac{GM}{r}} = \frac{2\pi r}{T}$$

If we square both sides and rearrange, we find the period of a satellite:

$$T^2 = \left(\frac{4\pi^2}{GM}\right)r^3 \tag{6.22}$$

Relationship between the orbital period $T$ and radius $r$ for a satellite in a circular orbit around an object of mass $M$

In other words, **the square of the period of the orbit is proportional to the cube of the radius of the orbit.**

**NOTE** ▸ The mass $M$ in Equation 6.22 is the mass of the object at the center of the orbit. ◂

This relationship between radius and period had been deduced from naked-eye observations of planetary motions by the 17th-century astronomer Johannes Kepler. One of Newton's major scientific accomplishments was to use his law of gravity and his laws of motion to prove what Kepler had deduced from observations. Even today, Newton's law of gravity and equations such as Equation 6.22 are essential tools for the NASA engineers who launch probes to other planets in the solar system.

**NOTE** ▸ The table inside the back cover of this text contains astronomical information about the sun and the planets that will be useful for many of the end-of-chapter problems. Note that planets farther from the sun have longer periods, in agreement with Equation 6.22. ◂

**EXAMPLE 6.15 Locating a geostationary satellite**

Communication satellites appear to "hover" over one point on the earth's equator. A satellite that appears to remain stationary as the earth rotates is said to be in a *geostationary orbit.* What is the radius of the orbit of such a satellite?

***STRATEGIZE*** For the satellite to remain stationary with respect to the earth, the satellite's period must be 24 hours.

***PREPARE*** Converted to seconds, the 24-hour period of the satellite is $T = 8.64 \times 10^4$ s.

***SOLVE*** We solve for the radius of the orbit by rearranging Equation 6.22. The mass at the center of the orbit is the earth:

$$r = \left(\frac{GM_e T^2}{4\pi^2}\right)^{\frac{1}{3}}$$

$$= \left(\frac{(6.67 \times 10^{-11}\ \text{N} \cdot \text{m}^2/\text{kg}^2)(5.98 \times 10^{24}\ \text{kg})(8.64 \times 10^4\ \text{s})^2}{4\pi^2}\right)^{\frac{1}{3}}$$

$$= 4.23 \times 10^7\ \text{m}$$

***ASSESS*** This is a high orbit; the radius is about 7 times the radius of the earth. Recall that the radius of the International Space Station's orbit is only about 6% larger than that of the earth.

## Gravity on a Grand Scale

A spiral galaxy, similar to our Milky Way galaxy.

Although relatively weak, gravity is a long-range force. No matter how far apart two objects may be, there is a gravitational attraction between them. Consequently, gravity is the most ubiquitous force in the universe. It not only keeps your feet on the ground, but also is at work on a much larger scale. The Milky Way galaxy, the collection of stars of which our sun is a part, is held together by gravity. But why doesn't the attractive force of gravity simply pull all of the stars together?

The reason is that all of the stars in the galaxy are in orbit around the center of the galaxy. The gravitational attraction keeps the stars moving in orbits around the center of the galaxy rather than falling inward, much as the planets orbit the sun rather than falling into the sun. In the nearly 5 billion years that our solar system has existed, it has orbited the center of the galaxy approximately 20 times.

The galaxy as a whole doesn't rotate as a fixed object, though. All of the stars in the galaxy are different distances from the galaxy's center, and so orbit with different periods. Stars closer to the center complete their orbits in less time, as we would expect from Equation 6.22. As the stars orbit, their relative positions shift. Stars that are relatively near neighbors now could be on opposite sides of the galaxy at some later time.

The rotation of a *rigid body* like a wheel is much simpler. As a wheel rotates, all of the points keep the same relationship to each other. The rotational dynamics of such rigid bodies is a topic we will take up in the next chapter.

**STOP TO THINK 6.6** Each year, the radius of the moon's orbit increases by about 3.8 cm. How does this change affect the length of a month?

A. A month gets longer.
B. A month gets shorter.
C. The length of a month stays the same.

**INTEGRATED EXAMPLE 6.16 A hunter and his sling**

A Stone Age hunter stands on a cliff overlooking a flat plain. He places a 1.0 kg rock in a sling, ties the sling to a 1.0-m-long vine, then swings the rock in a horizontal circle around his head. The plane of the motion is 25 m above the plain below. The tension in the vine increases as the rock goes faster and faster. Suddenly, just as the tension reaches 200 N, the vine snaps. If the rock is moving toward the cliff at this instant, how far out on the plain (from the base of the cliff) will it land?

***STRATEGIZE*** This is a two-part problem. First, we will use circular-motion dynamics to calculate the speed of the rock as it moves in a circle. When the vine snaps, the rock will be moving horizontally with this same speed; we can then use projectile-motion kinematics to find how far it goes.

***PREPARE*** We start by modeling the rock as a particle in uniform circular motion. We can use Problem-Solving Approach 6.1 to analyze this part of the motion.

The force identification diagram of FIGURE 6.24a shows that the only contact force acting on the rock is the tension in the vine. Because the rock moves in a horizontal circle, you may be tempted to draw a free-body diagram like FIGURE 6.24b, where $\vec{T}$ is directed along the $x$-axis. You will quickly run into trouble, however, because in this diagram the net force has a downward $y$-component that would cause the rock to rapidly accelerate downward. But we know that it moves in a horizontal circle and that the net force must point toward the center of the circle. In this free-body diagram, the weight force $\vec{w}$ points straight down and is certainly correct, so the difficulty must be with $\vec{T}$.

*Continued*

FIGURE 6.24 Visual overview of a hunter swinging a rock.

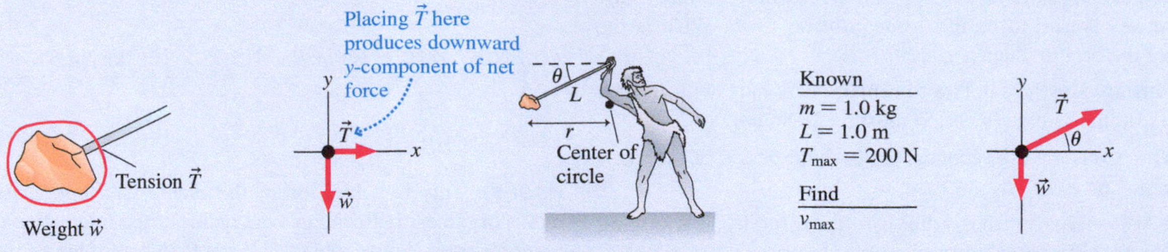


As an experiment, tie a small weight to a string, swing it over your head, and check the angle of the string. You will discover that the string is not horizontal but, instead, is angled downward. The sketch of FIGURE 6.24c labels this angle $\theta$. Notice that the rock moves in a *horizontal* circle, so the center of the circle is not at his hand. The $x$-axis points horizontally, to the center of the circle, but the tension force is directed along the vine. Thus the correct free-body diagram is the one in FIGURE 6.24d.

Once the vine breaks, the visual overview of the situation is shown in FIGURE 6.25. The important thing to note here is that the initial $x$-component of velocity is the speed the rock had an instant before the vine broke.

**SOLVE** From the free-body diagram of Figure 6.24d, Newton's second law for circular motion is

$$\sum F_x = T\cos\theta = \frac{mv^2}{r}$$
$$\sum F_y = T\sin\theta - mg = 0$$

where $\theta$ is the angle of the vine below the horizontal. We can use the $y$-equation to find the angle of the vine:

$$\sin\theta = \frac{mg}{T}$$

$$\theta = \sin^{-1}\left(\frac{mg}{T}\right) = \sin^{-1}\left(\frac{(1.0\text{ kg})(9.8\text{ m/s}^2)}{200\text{ N}}\right) = 2.81°$$

where we've evaluated the angle at the maximum tension of 200 N. The vine's angle of inclination is small but not zero.

Turning now to the $x$-equation, we find the rock's speed around the circle is

$$v = \sqrt{\frac{rT\cos\theta}{m}}$$

Be careful! The radius $r$ of the circle is not the length $L$ of the vine. You can see in Figure 6.24c that $r = L\cos\theta$. Thus

$$v = \sqrt{\frac{LT\cos^2\theta}{m}} = \sqrt{\frac{(1.0\text{ m})(200\text{ N})(\cos 2.81°)^2}{1.0\text{ kg}}} = 14.1\text{ m/s}$$

Because this is the horizontal speed of the rock just when the vine breaks, the initial velocity $(v_x)_i$ in the visual overview of the projectile motion, Figure 6.25, must be $(v_x)_i = 14.1$ m/s. Recall that a projectile has no horizontal acceleration, so the rock's final position is

$$x_f = x_i + (v_x)_i\,\Delta t = 0\text{ m} + (14.1\text{ m/s})\Delta t$$

where $\Delta t$ is the time the projectile is in the air. We're not given $\Delta t$, but we can find it from the vertical motion. For a projectile, the vertical motion is just free-fall motion, so we have

$$y_f = y_i + (v_y)_i\Delta t - \frac{1}{2}g(\Delta t)^2$$

The initial height is $y_i = 25$ m, the final height is $y_f = 0$ m, and the initial vertical velocity is $(v_y)_i = 0$ m/s. With these values, we have

$$0\text{ m} = 25\text{ m} + (0\text{ m/s})\Delta t - \frac{1}{2}(9.8\text{ m/s}^2)(\Delta t)^2$$

Solving this for $\Delta t$ gives

$$\Delta t = \sqrt{\frac{2(25\text{ m})}{9.8\text{ m/s}^2}} = 2.26\text{ s}$$

Now we can use this time to find

$$x_f = 0\text{ m} + (14.1\text{ m/s})(2.26\text{ s}) = 32\text{ m}$$

The rock lands 32 m from the base of the cliff.

**ASSESS** The circumference of the rock's circle is $2\pi r$, or about 6 m. At a speed of 14.1 m/s, the rock takes roughly half a second to go around once. This seems reasonable. The 32 m distance is about 100 ft, which seems easily attainable from a cliff over 75 feet high.

FIGURE 6.25 Visual overview of the rock in projectile motion.

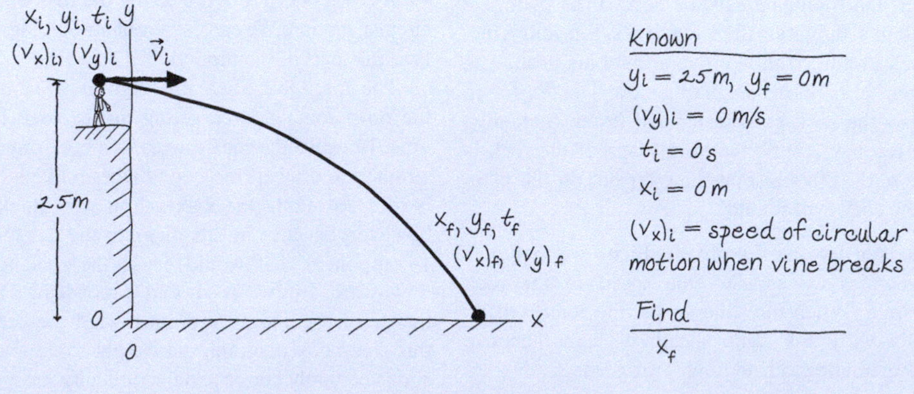

# SUMMARY

**GOAL** To learn about motion in a circle, including orbital motion under the influence of a gravitational force.

## GENERAL PRINCIPLES

### Uniform Circular Motion

An object moving in a circular path is in uniform circular motion if $v$ is constant.

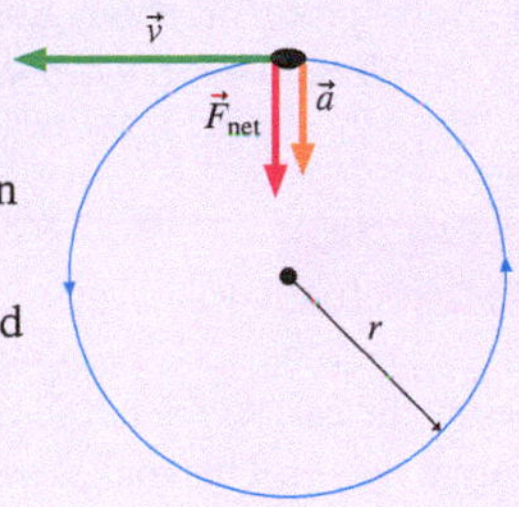


- The speed is constant, but the direction of motion is constantly changing.
- The **centripetal acceleration** is directed toward the center of the circle and has magnitude

$$a = \frac{v^2}{r}$$

- This acceleration requires a net force directed toward the center of the circle. Newton's second law for circular motion is

$$\vec{F}_{net} = m\vec{a} = \left(\frac{mv^2}{r}, \text{ toward center of circle}\right)$$

### Universal Gravitation

Two objects with masses $m_1$ and $m_2$ that are distance $r$ apart exert attractive gravitational forces on each other of magnitude

$$F_{1\,on\,2} = F_{2\,on\,1} = \frac{Gm_1m_2}{r^2}$$

where the gravitational constant is

$$G = 6.67 \times 10^{-11}\ \text{N} \cdot \text{m}^2/\text{kg}^2$$

This is **Newton's law of gravity.** Gravity is an inverse-square law.

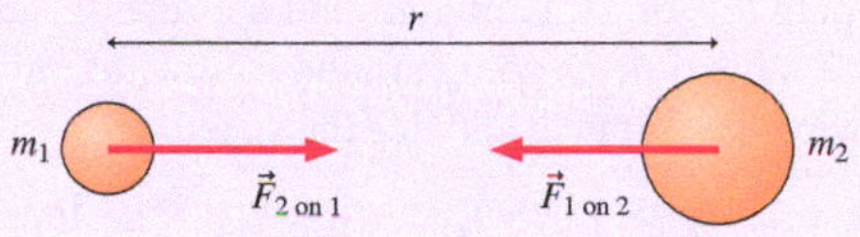

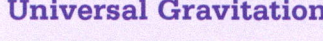

## IMPORTANT CONCEPTS

### Describing circular motion

For an object moving in a circle of radius $r$ at a constant speed $v$:

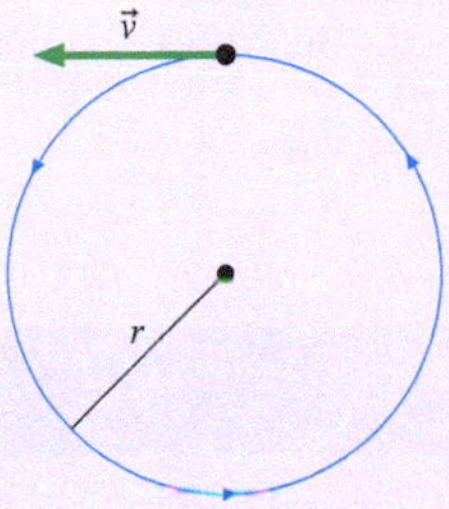


- The **period** $T$ is the time to go once around the circle:

$$T = \text{time for one revolution}$$

- The **frequency** $f$ is defined as the number of revolutions per second. It is defined in terms of the period:

$$f = \frac{1}{T}$$

- The frequency and period are related to the speed and the radius: $v = 2\pi fr = \dfrac{2\pi r}{T}$

### Planetary gravity

The gravitational attraction between a planet and a mass on the surface depends on the two masses and the distance to the center of the planet:

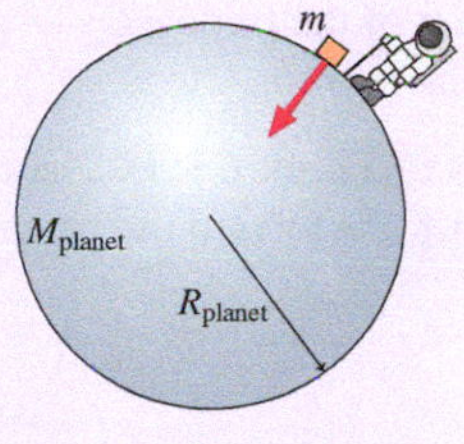


$$F_{planet\,on\,m} = \frac{GM_{planet}m}{R^2_{planet}}$$

We can use this to define a value of the free-fall acceleration at the surface of a planet:

$$g_{planet} = \frac{GM_{planet}}{R^2_{planet}}$$

## APPLICATIONS

### Apparent weight and weightlessness

Circular motion requires a net force pointing to the center. The apparent weight $w_{app} = n$ is usually not the same as the true weight $w$. $n$ must be $> 0$ for the object to be in contact with a surface.

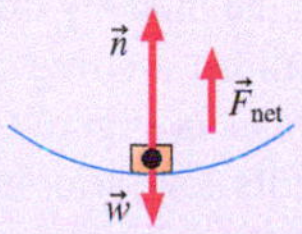


In orbital motion, the net force is provided by gravity alone. An astronaut and his spacecraft are both in free fall, so he feels weightless.

### Orbital motion

A **satellite** in a circular orbit of radius $r$ around an object of mass $M$ moves at a speed $v$ given by

$$v = \sqrt{\frac{GM}{r}}$$

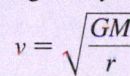

The period and radius are related as follows:

$$T^2 = \left(\frac{4\pi^2}{GM}\right)r^3$$

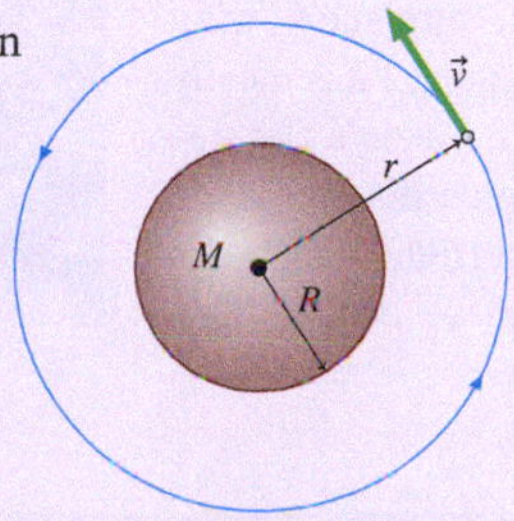


The speed of a satellite in a low orbit is

$$v = \sqrt{gr}$$

The orbital period is

$$T = 2\pi\sqrt{\frac{r}{g}}$$

## Learning Objectives After studying this chapter, you should be able to:

- Calculate period, frequency, and speed for objects in circular motion. *Conceptual Questions 6.1, 6.2; Problems 6.1, 6.2, 6.3, 6.9, 6.10*
- Use Newton's laws to solve dynamics problems for objects in uniform circular motion. *Conceptual Questions 6.8, 6.9; Problems 6.16, 6.17, 6.19, 6.20, 6.23*
- Understand the apparent weight of an object in circular motion. *Conceptual Questions 6.4, 6.12; Problems 6.22, 6.26, 6.28, 6.29*
- Analyze the circular orbits of planets and satellites. *Problems 6.7, 6.32, 6.33*
- Use Newton's law of gravity to calculate long-range gravitational forces. *Conceptual Questions 6.18, 6.19; Problems 6.36, 6.38, 6.39, 6.40, 6.41*
- Solve problems about gravity and orbits. *Conceptual Question 6.20; Problems 6.44, 6.47, 6.48, 6.49, 6.50*

## STOP TO THINK ANSWERS

**Chapter Preview Stop to Think: D.** The ball is in uniform circular motion. The acceleration is directed toward the center of the circle, which is to the right at the instant shown.

**Stop to Think 6.1: C > D = A > B.** Rearranging Equation 6.3 gives $T = \frac{2\pi r}{v}$. For the cases shown, speed is either $v$ or $2v$; the radius of the circular path is $r$ or $2r$. Going around a circle of radius $r$ at a speed $v$ takes the same time as going around a circle of radius $2r$ at a speed $2v$. It's twice the distance at twice the speed.

**Stop to Think 6.2: $T_E > T_D > T_C > T_B > T_A$.** The tension force provides the centripetal acceleration, and larger acceleration implies larger force. So the question reduces to one about acceleration: Rank the centripetal accelerations for these cases. Equation 6.5 shows that the acceleration is proportional to the radius of the circle and the square of the frequency, and so the acceleration increases steadily as we move from A to E.

**Stop to Think 6.3: B.** The car is moving in a circle, so there must be a net force toward the center of the circle. The center of the circle is below the car, so the net force must point downward. This can be true only if $w > n$. This makes sense; $n < w$, so the apparent weight is less than the true weight. The riders in the car "feel light"; if you've driven over a rise like this, you know that this is what you feel.

**Stop to Think 6.4: B.** The period of a satellite doesn't depend on the mass of the satellite. If you increase the height above the surface, you increase the radius of the orbit. This will result in an increased period. Increasing the value of $g$ would cause the period to decrease.

**Stop to Think 6.5: B > A > D > C.** The free-fall acceleration is proportional to the mass, but inversely proportional to the square of the radius.

**Stop to Think 6.6: A.** The length of a month is determined by the period of the moon's orbit. Equation 6.22 shows that as the moon gets farther away, the period of the orbit—and thus the length of a month—increases.

Video Tutor Solution Chapter 6

# QUESTIONS

## Conceptual Questions

1. A cyclist goes around a level, circular track at constant speed. Do you agree or disagree with the following statement? "Since the cyclist's speed is constant, her acceleration is zero." Explain.
2. In uniform circular motion, which of the following quantities are constant: speed, instantaneous velocity, centripetal acceleration, the magnitude of the net force?
3. The force on an airplane's wing from the air flowing past it is mostly perpendicular to the wing's flat surface; this is the force that holds the plane aloft in level flight. Given this fact, why do planes "bank" when making a turn?

4. Tarzan hangs from a vine without it breaking. But when he swings on the same vine starting from a high branch, the vine snaps at the bottom of his swing. Explain why this happens.
5. BIO Large birds like pheasants often walk short distances. Small birds like chickadees never walk. They either hop or fly. Why might this be?
6. When you drive fast on the highway with muddy tires, you can hear the mud flying off the tires into your wheel wells. Why does the mud fly off?
7. A ball on a string moves in a vertical circle as in Figure Q6.7. When the ball is at its lowest point, is the tension in the string greater than, less than, or equal to the ball's weight? Explain. (You may want to include a free-body diagram as part of your explanation.)

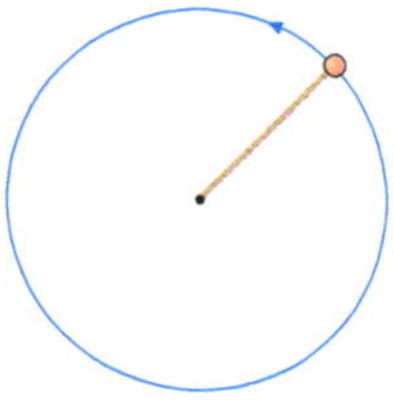

FIGURE Q6.7

Problem difficulty is labeled as | (straightforward) to ||||| (challenging). Problems labeled INT integrate significant material from earlier chapters; Problems labeled BIO are of biological or medical interest.

The eText icon indicates when there is a video tutor solution available for the chapter or for a specific problem. To launch these videos, log into your eText through Mastering™ Physics or log into the Study Area.

8. Give an everyday example of circular motion for which the centripetal acceleration is mostly or completely due to a force of the type specified: (a) Static friction. (b) Tension.
9. Give an everyday example of circular motion for which the centripetal acceleration is mostly or completely due to a force of the type specified: (a) Gravity. (b) Normal force.
10. On a roller coaster loop-the-loop the riders are upside-down at the top of the loop, while on a Ferris wheel the riders are upright at the top. Suppose a Ferris wheel and the loop of a roller coaster have the same diameter. If riders on each have the same magnitude of apparent weight at the top, which rider is moving faster? Explain.
11. A car coasts at a constant speed over a circular hill. Which of the free-body diagrams in Figure Q6.11 is correct? Explain.

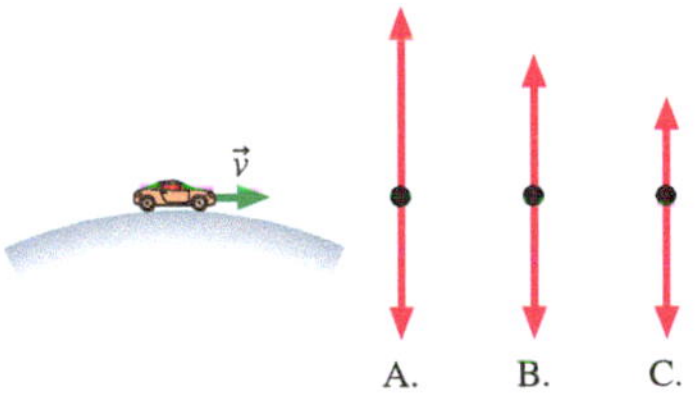


FIGURE Q6.11

12. In Figure Q6.11, at the instant shown, is the apparent weight of the car's driver greater than, less than, or equal to his true weight? Explain.
13. Riding in the back of a pickup truck can be very dangerous. If the truck turns suddenly, the riders can be thrown from the truck bed. Why are the riders ejected from the bed?
14. Playground swings move through an arc of a circle. When you are on a swing, and at the lowest point of your motion, is your apparent weight greater than, less than, or equal to your true weight? Explain.
15. Variation in your apparent weight is desirable when you ride a roller coaster; it makes the ride fun. However, too much variation over a short period of time can be painful. For this reason, the loops of real roller coasters are not simply circles like Figure 6.16a. A typical loop is shown in Figure Q6.15. The radius of the circle that matches the track at the top of the loop is much smaller than that of a matching circle at other places on the track. Explain why this shape gives a more comfortable ride than a circular loop.

FIGURE Q6.15

16. An object's apparent weight is slightly less when it is at the equator than when it is at the North Pole. Explain why this is so.
17. Why is it impossible for an astronaut inside an orbiting space station to go from one end to the other by walking normally?
18. If every object in the universe feels an attractive gravitational force due to every other object, why don't you feel a pull from someone seated next to you?
19. A mountain climber's weight is slightly less on the top of a tall mountain than at the base, though his mass is the same. Why?
20. The mass of Mars was not known until its two moons were discovered in 1877. Explain how observations of these moons—their distances from Mars and their orbital periods—could be used to find Mars's mass.

## Multiple-Choice Questions

21. | A ball on a string moves around a complete circle, once a second, on a frictionless, horizontal table. The tension in the string is measured to be 6.0 N. What would the tension be if the ball went around in only half a second?
A. 1.5 N B. 3.0 N C. 12 N D. 24 N
22. | As seen from above, a car rounds the curved path shown in Figure Q6.22 at a constant speed. Which vector best represents the net force acting on the car?

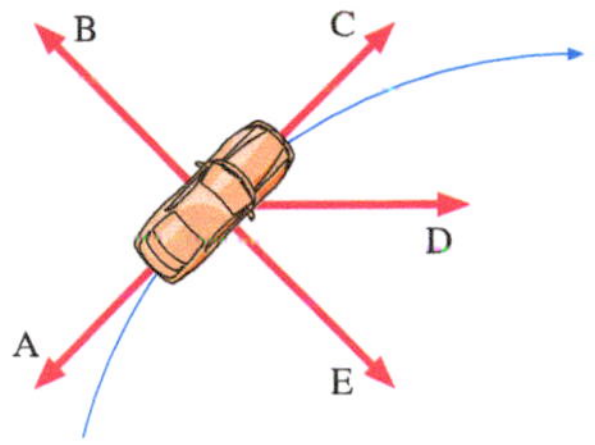


FIGURE Q6.22

23. | As we saw in the chapter, wings on race cars push them into the track. The increased normal force makes large friction forces possible. At one Formula One racetrack, cars turn around a half-circle with diameter 190 m at 68 m/s. For a 610 kg vehicle, the approximate minimum static friction force to complete this turn is
A. 6000 N B. 15,000 N C. 18,000 N
D. 24,000 N E. 30,000 N
24. || You swing a weight attached to a string in a vertical circle. At the top of the circle the string just barely goes slack for an instant. At this point the centripetal acceleration of the weight is
A. Greater than $g$.
B. Less than $g$.
C. Equal to $g$.
D. Can't answer without knowing the weight's speed.
25. || On a snowy day, when the coefficient of friction $\mu_s$ between a car's tires and the road is 0.50, the maximum speed that the car can go around a curve is 20 mph. What is the maximum speed at which the car can take the same curve on a sunny day when $\mu_s = 1.0$?
A. 20 mph B. 24 mph C. 28 mph
D. 33 mph E. 40 mph
26. | The cylindrical space station in Figure Q6.26, 200 m in diameter, rotates in order to provide artificial gravity of $g$ for the occupants. How much time does the station take to complete one rotation?
A. 3 s B. 20 s C. 28 s D. 32 s

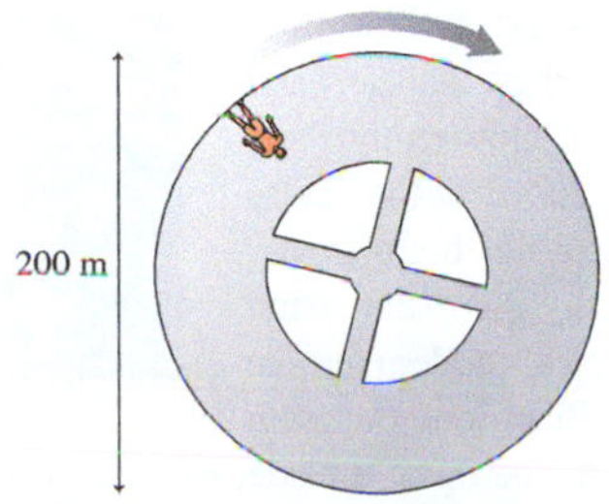


FIGURE Q6.26

27. ‖ The radius of Jupiter is 11 times that of earth, and the free-fall acceleration near its surface is 2.5 times that on earth. If we someday put a spacecraft in low Jupiter orbit, its orbital speed will be
   A. Greater than that for an earth satellite.
   B. The same as that for an earth satellite.
   C. Less than that for an earth satellite.
28. | A newly discovered planet has twice the mass and three times the radius of the earth. What is the free-fall acceleration at its surface, in terms of the free-fall acceleration $g$ at the surface of the earth?
   A. $\frac{2}{9}g$ B. $\frac{2}{3}g$ C. $\frac{3}{4}g$ D. $\frac{4}{3}g$
29. ‖ Suppose one night the radius of the earth doubled but its mass stayed the same. What would be an approximate new value for the free-fall acceleration at the surface of the earth?
   A. 2.5 $m/s^2$ B. 5.0 $m/s^2$ C. 10 $m/s^2$ D. 20 $m/s^2$
30. | Currently, the moon goes around the earth once every 27.3 days. If the moon could be brought into a new circular orbit with a smaller radius, its orbital period would be
   A. More than 27.3 days.
   B. 27.3 days.
   C. Less than 27.3 days.
31. ‖ Two planets orbit a star. You can ignore the gravitational interactions between the planets. Planet 1 has orbital radius $r_1$ and planet 2 has $r_2 = 4r_1$. Planet 1 orbits with period $T_1$. Planet 2 orbits with period
   A. $T_2 = \frac{1}{2}T_1$ B. $T_2 = 2T_1$ C. $T_2 = 4T_1$ D. $T_2 = 8T_1$

# PROBLEMS

## Section 6.1 Uniform Circular Motion

1. ‖ A 5.0-m-diameter merry-go-round is turning with a 4.0 s period. What is the speed of a child on the rim?
2. | The earth, with a radius of $6.4 \times 10^6$ m, rotates on its axis once a day. What is the speed of a person standing on the equator, due to the earth's rotation?
3. | An old-fashioned LP record rotates at $33\frac{1}{3}$ rpm.
   a. What is its frequency, in rev/s?
   b. What is its period, in seconds?
4. | A typical hard disk in a computer spins at 5400 rpm.
   a. What is the frequency, in rev/s?
   b. What is the period, in seconds?
5. | The hammer throw was one of the earliest Olympic events. In this event, a heavy ball attached to a chain is swung several times in a circular path until it is released. The winning athlete is the one who throws the ball the greatest distance. The last complete rotation of 2016 Olympic champion Anita Włodarczyk's final turn took only 0.43 s. The radius of the ball's path, including her extended arms, was 2.1 m.
   a. What was the frequency of this rotation?
   b. What was the speed of the ball?
   c. What was the ball's acceleration, in units of $g$?
6. ‖ The horse on a carousel is 4.0 m from the central axis.
   a. If the carousel rotates at 0.10 rev/s, how long does it take the horse to go around twice?
   b. How fast is a child on the horse going (in m/s)?
7. ‖| The radius of the earth's very nearly circular orbit around the sun is $1.50 \times 10^{11}$ m. Find the magnitude of the earth's (a) velocity and (b) centripetal acceleration as it travels around the sun. Assume a year of 365 days.
8. | Modern wind turbines are larger than they appear, and despite their apparently lazy motion, the speed of the blades tips can be quite high—many times higher than the wind speed. A typical modern turbine has blades 56 m long that spin at 13 rpm. At the tip of a blade, what are (a) the speed and (b) the centripetal acceleration?

9. | BIO The California sea lion is capable of making extremely fast, tight turns while swimming underwater. In one study, scientists observed a sea lion making a circular turn with a radius of 0.35 m while swimming at 4.2 m/s.
   a. What is the sea lion's centripetal acceleration, in units of $g$?
   b. What percentage is this acceleration of that of an F-15 fighter jet's maximum centripetal acceleration of $9g$?
10. | Baseball pitching machines are used to fire baseballs toward a batter for hitting practice. In one kind of machine, the ball is fed between two wheels that are rapidly rotating in opposite directions; as the ball is pulled in between the wheels, it rapidly accelerates up to the speed of the wheels' rims, at which point it is ejected. For a machine with 35-cm-diameter wheels, what rotational frequency (in rpm) do the wheels need to pitch a 90 mph fastball?

11. ‖ BIO Astronauts in the International Space Station must work out every day to counteract the effects of weightlessness. Researchers have investigated if riding a stationary bicycle while experiencing artificial gravity from a rotating platform gives any additional cardiovascular benefit. What frequency of rotation, in rpm, is required to give an acceleration of $1.4g$ to an astronaut's feet, if her feet are 1.1 m from the platform's rotational axis?
12. ‖ A typical running track is an oval with 74-m-diameter half circles at each end. A runner going once around the track covers a distance of 400 m. Suppose a runner, moving at a constant speed, goes once around the track in 1 min 40 s. What is her centripetal acceleration during the turn at each end of the track?

## Section 6.2 Dynamics of Uniform Circular Motion

13. ||||| Figure P6.13 is a bird's-eye view of particles on a string moving in horizontal circles on a tabletop. All are moving at the same speed. Rank in order, from largest to smallest, the tensions $T_1$ to $T_4$.

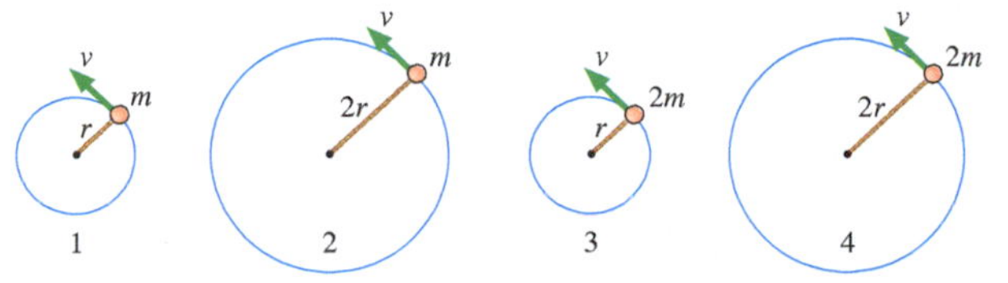


FIGURE P6.13

14. || In short-track speed skating, the track has straight sections and semicircles 16 m in diameter. Assume that a 65 kg skater goes around the turn at a constant 12 m/s.

   a. What is the horizontal force on the skater?
   b. What is the ratio of this force to the skater's weight?

15. ||| In addition to their remarkable top speeds of almost 60 mph,
BIO cheetahs have impressive cornering abilities. In one study, the maximum centripetal acceleration of a cheetah was measured to be 18 $m/s^2$. What minimum value of the coefficient of static friction between the ground and the cheetah's feet is necessary to provide this acceleration?

16. || A cyclist is rounding a 20-m-radius curve at 12 m/s. What is
INT the minimum possible coefficient of static friction between the bike tires and the ground?

17. || A 1500 kg car drives around a flat 200-m-diameter circular track at 25 m/s. What are the magnitude and direction of the net force on the car? What causes this force?

18. || A fast pitch softball player does a "windmill" pitch, illustrated
BIO in Figure P6.18, moving her hand through a circular arc to pitch a ball at 70 mph. The 0.19 kg ball is 50 cm from the pivot point at her shoulder. At the lowest point of the circle, the ball has reached its maximum speed.
   a. At the bottom of the circle, just before the ball leaves her hand, what is its centripetal acceleration?
   b. What are the magnitude and direction of the force her hand exerts on the ball at this point?

FIGURE P6.18

19. || One kind of baseball pitching machine works by rotating a light and stiff rigid rod about a horizontal axis until the ball is moving toward the target. Suppose a 144 g baseball is held 85 cm from the axis of rotation and released at the major league pitching speed of 85 mph.
   a. What is the ball's centripetal acceleration just before it is released?
   b. What is the magnitude of the net force that is acting on the ball just before it is released?

20. | A wind turbine has 12,000 kg blades that are 38 m long. The blades spin at 22 rpm. If we model a blade as a point mass at the midpoint of the blade, what is the inward force necessary to provide each blade's centripetal acceleration?

21. ||| You're driving your pickup truck around a curve that has a radius of 20 m. How fast can you drive around this curve before a steel toolbox slides on the steel bed of the truck?

22. || The spin cycle of a clothes washer extracts the water in clothing by greatly increasing the water's apparent weight so that it is efficiently squeezed through the clothes and out the holes in the drum. In a top loader's spin cycle, the 45-cm-diameter drum spins at 1200 rpm around a vertical axis. What is the apparent weight of a 1.0 g drop of water?

23. || Gibbons, small Asian apes, move by *brachiation,* swinging
BIO below a handhold to move forward to the next handhold. A 9.0 kg gibbon has an arm length (hand to shoulder) of 0.60 m. We can model its motion as that of a point mass swinging at the end of a 0.60-m-long, massless rod. At the lowest point of its swing, the gibbon is moving at 3.5 m/s. What upward force must a branch provide to support the swinging gibbon?

## Section 6.3 Apparent Forces in Circular Motion

24. ||| The passengers in a roller coaster car feel 50% heavier than their true weight as the car goes through a dip with a 30 m radius of curvature. What is the car's speed at the bottom of the dip?

25. ||| In the very Dutch sport of Fierljeppen, athletes run up to a long pole and then use it to vault across a canal. At the very top of his arc, a 55 kg vaulter is moving at 2.5 m/s and is 5.1 m from the bottom end of the pole. What vertical force does the pole exert on the vaulter?

26. || A roller coaster car is going over the top of a 15-m-radius circular rise. At the top of the hill, the passengers "feel light," with an apparent weight only 50% of their true weight. How fast is the coaster moving?

27. ||| As a roller coaster car crosses the top of a 40-m-diameter loop-the-loop, its apparent weight is the same as its true weight. What is the car's speed at the top?

28. || Unlike a roller coaster, the seats in a Ferris wheel swivel so that the rider is always seated upright. An 80-ft-diameter Ferris wheel rotates once every 24 s. What is the apparent weight of a 70 kg passenger at (a) the lowest point of the circle and (b) the highest point?

29. ||| You're driving your new sports car at 75 mph over the top of a hill that has a radius of curvature of 525 m. What fraction of your normal weight is your apparent weight as you crest the hill?

30. |||| In a car's suspension system, each wheel is connected to a INT vertical spring; these springs absorb shocks when the car travels on bumpy roads. In one car, each spring has a spring constant of $5.0 \times 10^4$ N/m. If this 1400 kg car is driven at 25 m/s through the bottom of a circular dip in the road that has a radius of 600 m, by how much do these springs compress compared to when the car is driven on a flat road?

31. ||| A typical laboratory centrifuge rotates at 4000 rpm. Test BIO INT tubes have to be placed into a centrifuge very carefully because of the very large accelerations.
   a. What is the acceleration at the end of a test tube that is 10 cm from the axis of rotation?
   b. For comparison, what is the magnitude of the acceleration a test tube would experience if stopped in a 1.0-ms-long encounter with a hard floor after falling from a height of 1.0 m?

### Section 6.4 Circular Orbits and Weightlessness

32. ||| A satellite orbiting the moon very near the surface has a period of 110 min. Use this information, together with the radius of the moon from the table on the inside of the back cover, to calculate the free-fall acceleration on the moon's surface.

33. | Many spacecraft have visited Mars over the years. Mars is smaller than Earth and has correspondingly weaker surface gravity. On Mars, the free-fall acceleration is only 3.8 m/s$^2$. What is the orbital period of a spacecraft in a low orbit near the surface of Mars?

### Section 6.5 Newton's Law of Gravity

34. |||| The centers of a 10 kg lead ball and a 100 g lead ball are separated by 10 cm.
   a. What gravitational force does each exert on the other?
   b. What is the ratio of this gravitational force to the weight of the 100 g ball?

35. || The gravitational force of a star on an orbiting planet 1 is $F_1$. Planet 2, which is twice as massive as planet 1 and orbits at twice the distance from the star, experiences gravitational force $F_2$. What is the ratio $F_2/F_1$? You can ignore the gravitational force between the two planets.

36. || The free-fall acceleration at the surface of planet 1 is 20 m/s$^2$. The radius and the mass of planet 2 are twice those of planet 1. What is the free-fall acceleration on planet 2?

37. |||| What is the ratio of the sun's gravitational force on you to the earth's gravitational force on you?

38. | Just before it landed on the moon, the Apollo 12 lunar lander INT had a mass of 7200 kg. What rocket thrust was necessary to have the lander touch down with zero acceleration?

39. || In recent years, astronomers have found planets orbiting nearby stars that are quite different from planets in our solar system. Kepler-12b, has a diameter that is 1.7 times that of Jupiter, but a mass that is only 0.43 that of Jupiter. What is the value of $g$ on this large, but low-density, world?

40. ||| The gravitational constant $G$ was first measured accurately by Henry Cavendish in 1798. He used an exquisitely sensitive balance to measure the force between two lead spheres whose centers were 0.23 m apart. One of the spheres had a mass of 158 kg, while the mass of the other sphere was 0.73 kg. What was the ratio of the gravitational force between these spheres to the weight of the lighter sphere?

41. || a. What is the gravitational force of the sun on the earth?
   b. What is the gravitational force of the moon on the earth?
   c. The moon's force is what percent of the sun's force?

42. || What is the value of $g$ on the surface of Saturn? Explain how such a low value is possible given Saturn's large mass—100 times that of Earth.

43. || What is the free-fall acceleration at the surface of (a) Mars and (b) Jupiter?

### Section 6.6 Gravity and Orbits

44. || In 2014, a space probe approached the rocky core of the comet Churyumov–Gerasimenko, which is only a few km in diameter. The probe then entered orbit around the comet at a distance of 30 km. The comet was found to have a mass of $1.0 \times 10^{13}$ kg. What was the orbital period of the probe around the comet, in earth days?

45. ||| Mars has two moons, Phobos and Deimos. Phobos orbits Mars at a distance of 9380 km from Mars's center, while Deimos orbits at 23,500 km from the center. What is the ratio of the orbital period of Deimos to that of Phobos?

46. ||| Mars's moon Phobos orbits the planet at a distance of 9380 km from its center, and it takes 7 hours and 39 minutes to complete one orbit. What is the ratio of Mars's mass to the mass of the earth?

47. || The *dwarf planet* Praamzius is estimated to have a diameter of about 300 km and orbits the sun at a distance of $6.4 \times 10^{12}$ m. What is its orbital period in years?

48. ||| Planet X orbits the star Omega with a "year" that is 200 earth days long. Planet Y circles Omega at four times the distance of Planet X. How long is a year on Planet Y?

49. ||| The International Space Station is in a 250-mile-high orbit. What are the station's orbital period, in minutes, and speed?

50. |||| An earth satellite moves in a circular orbit at a speed of 5500 m/s. What is its orbital period?

In recent years, scientists have discovered hundreds of planets orbiting other stars. Some of these planets are in orbits that are similar to that of earth, which orbits the sun ($M_{sun} = 1.99 \times 10^{30}$ kg) at a distance of $1.50 \times 10^{11}$ m, called 1 *astronomical unit* (1 au). Others have extreme orbits that are much different from anything in our solar system. Problems 51–53 relate to some of these planets that follow circular orbits around other stars.

51. ||| WASP-32b orbits with a period of only 2.7 days a star with a mass that is 1.1 times that of the sun. How many au from the star is this planet?

52. ||| HD 10180g orbits with a period of 600 days at a distance of 1.4 au from its star. What is the ratio of the star's mass to our sun's mass?

53. ||| Kepler-42c orbits at a very close 0.0058 au from a small star with a mass that is 0.13 that of the sun. How long is a "year" on this world?

## General Problems

54. || How fast must a plane fly along the earth's equator so that the sun stands still relative to the passengers? In which direction must the plane fly, east to west or west to east? Give your answer in both km/h and mph. The radius of the earth is 6400 km.

55. ||| The car in Figure P6.55 travels at a constant speed along the road shown. Draw vectors showing its acceleration at the three points A, B, and C, or write $\vec{a} = \vec{0}$. The lengths of your vectors should correspond to the magnitudes of the accelerations.

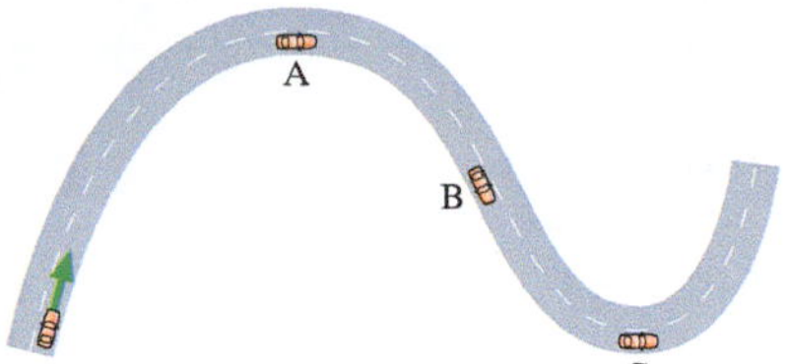


FIGURE P6.55

56. || A 75 kg man weighs himself at the north pole and at the equator. Which scale reading is higher? By how much? Assume the earth is a perfect sphere. Explain why the readings differ.

57. ||| A 1400 kg car drives at 27 m/s over a circular hill that has a radius of 430 m. At the point shown in Figure P6.57, what is the normal force on the car?

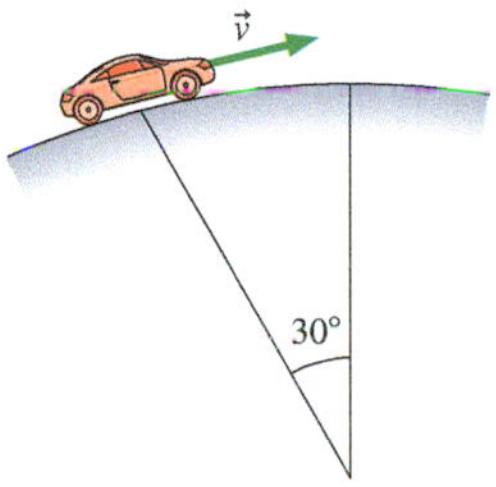


FIGURE P6.57

58. |||| A curve at a racetrack has a radius of 800 m and is banked at an angle of 7.0°. On a rainy day, the coefficient of friction between the cars' tires and the track is 0.50. What is the maximum speed at which a car could go around this curve without slipping?

59. |||| BIO It is well known that runners run more slowly around a curved track than a straight one. One hypothesis to explain this is that the *total* force from the track on a runner's feet—the magnitude of the *vector* sum of the normal force (that has average value $mg$ to counteract gravity) and the inward-directed friction force that causes the runner's centripetal acceleration—is greater when running around a curve than on a straight track. Runners compensate for this greater force by increasing the time their feet are in contact with the ground, which slows them down. For a sprinter running at 10 m/s around a curved track of radius 20 m, how much greater (as a percentage) is the average total force on their feet compared to when they are running in a straight line?

60. ||| BIO Biologists have studied the running ability of the northern quoll, a marsupial indigenous to Australia. In one set of experiments, they studied the maximum speed that quolls could run around a curved path without slipping. One quoll was running at 2.8 m/s around a curve with a radius of 1.2 m when it started to slip. What was the coefficient of static friction between the quoll's feet and the ground in this trial?

61. ||| BIO In a recent study of how mice negotiate turns, the mice ran around a circular 90° turn on a track with a radius of 0.15 m. The maximum speed measured for a mouse (mass = 18.5 g) running around this turn was 1.29 m/s. What is the minimum coefficient of friction between the track and the mouse's feet that would allow a turn at this speed?

62. ||| You are driving your car through a roundabout that has a radius of 9.0 m. Your physics textbook is lying on the seat next to you. What is the fastest speed at which you can go around the curve without the book sliding? The coefficient of static friction between the book and the seat is 0.30.

63. ||| In the Wall of Death carnival attraction, stunt motorcyclists ride around the inside of a large, 10-m-diameter wooden cylinder that has *vertical* walls. The coefficient of static friction between the riders' tires and the wall is 0.90. What is the minimum speed at which the motorcyclists can ride without slipping down the wall?

64. |||| In the swing carousel amusement park ride, riders sit in chairs that are attached by a chain to a large rotating drum. As the carousel turns, the riders move in a large circle with the chains tilted out from the vertical. In one such carousel, the riders move in a 16.5-m-radius circle and take 8.3 s to complete one revolution. What is the angle of the chains, as measured from the vertical?

65. || INT A 5.0 g coin is placed 15 cm from the center of a turntable. The coin has static and kinetic coefficients of friction with the turntable surface of $\mu_s = 0.80$ and $\mu_k = 0.50$. The turntable very slowly speeds up to 60 rpm. Does the coin slide off?

66. ||| In an old-fashioned amusement park ride, passengers stand inside a 3.0-m-tall, 5.0-m-diameter hollow steel cylinder with their backs against the wall. The cylinder begins to rotate about a vertical axis. Then the floor on which the passengers are standing suddenly drops away! If all goes well, the passengers will "stick" to the wall and not slide. Clothing has a static coefficient of friction against steel in the range 0.60 to 1.0 and a kinetic coefficient in the range 0.40 to 0.70. What is the minimum rotational frequency, in rpm, for which the ride is safe?

67. ||| INT The 0.20 kg puck on the frictionless, horizontal table in Figure P6.67 is connected by a string through a hole in the table to a hanging 1.20 kg block. With what speed must the puck rotate in a circle of radius 0.50 m if the block is to remain hanging at rest?

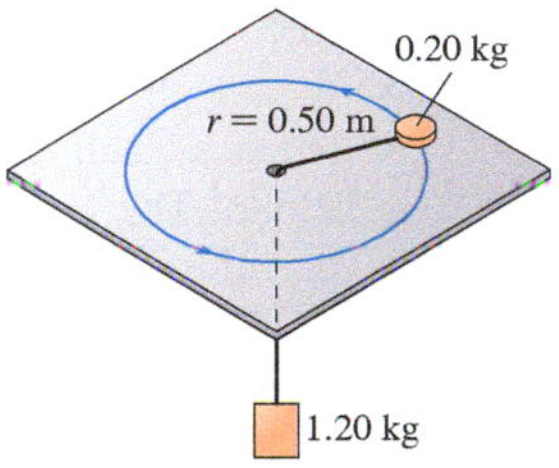


FIGURE P6.67

68. || While at the county fair, you decide to ride the Ferris wheel. Having eaten too many candy apples and elephant ears, you find the motion somewhat unpleasant. To take your mind off your stomach, you wonder about the motion of the ride. You estimate the radius of the big wheel to be 15 m, and you use your watch to find that each loop around takes 25 s.
   a. What are your speed and magnitude of your acceleration?
   b. What is the ratio of your apparent weight to your true weight at the top of the ride?
   c. What is the ratio of your apparent weight to your true weight at the bottom?

69. || A car drives over the top of a hill that has a radius of 50 m. What maximum speed can the car have without flying off the road at the top of the hill?

Watch Video Solution Problems 6.73 and 6.77

70. ||| The ultracentrifuge is an important tool for separating and BIO analyzing proteins in biological research. Because of the enor INT mous centripetal accelerations that can be achieved, the apparatus (see Figure 6.18) must be carefully balanced so that each sample is matched by another on the opposite side of the rotor shaft. Any difference in mass of the opposing samples will cause a net force in the horizontal plane on the shaft of the rotor; this force can actually be large enough to destroy the centrifuge. Suppose that a scientist makes a slight error in sample preparation, and one sample has a mass 10 mg greater than the opposing sample. If the samples are 10 cm from the axis of the rotor and the ultracentrifuge spins at 70,000 rpm, what is the magnitude of the net force on the rotor due to the unbalanced samples?

71. |||| A sensitive gravimeter at a mountain observatory finds that the free-fall acceleration is 0.0075 $m/s^2$ less than that at sea level. What is the observatory's altitude?

72. |||| In 2014, the Rosetta space probe reached the comet Churyumov–INT Gerasimenko. Although the comet's core is actually far from spherical, in this problem we'll model it as a sphere with a mass of $1.0 \times 10^{13}$ kg and a radius of 1.6 km. If a rock were dropped from a height of 1.0 m above the comet's surface, how long would it take to hit the surface?

73. ||| Planet Z is 10,000 km in diameter. The free-fall acceleration on Planet Z is 8.0 $m/s^2$.
   a. What is the mass of Planet Z?
   b. What is the free-fall acceleration 10,000 km above Planet Z's north pole?

74. ||| How long will it take a rock dropped from 2.0 m above the INT surface of Mars to reach the ground?

75. ||| A 20 kg sphere is at the origin and a 10 kg sphere is at INT $(x, y) = (20\text{ cm}, 0\text{ cm})$. At what point or points could you place a small mass such that the net gravitational force on it due to the spheres is zero?

76. ||| a. At what height above the earth is the free-fall acceleration 10% of its value at the surface?
   b. What is the speed of a satellite orbiting at that height?

77. | Mars has a small moon, Phobos, that orbits with a period of 7 h 39 min. The radius of Phobos' orbit is $9.4 \times 10^6$ m. Use only this information (and the value of $G$) to calculate the mass of Mars.

78. || You are the science officer on a visit to a distant solar system. Prior to landing on a planet you measure its diameter to be $1.80 \times 10^7$ m and its rotation period to be 22.3 h. You have previously determined that the planet orbits $2.20 \times 10^{11}$ m from its star with a period of 402 earth days. Once on the surface you find that the free-fall acceleration is 12.2 $m/s^2$. What are the masses of (a) the planet and (b) the star?

79. ||||| Europa, a satellite of Jupiter, is BIO believed to have a liquid ocean of water (with a possibility of life) beneath its icy surface. In planning a future mission to Europa, what is the fastest that an astronaut with legs of length 0.70 m could walk on the surface of Europa? Europa is 3100 km in diameter and has a mass of $4.8 \times 10^{22}$ kg.

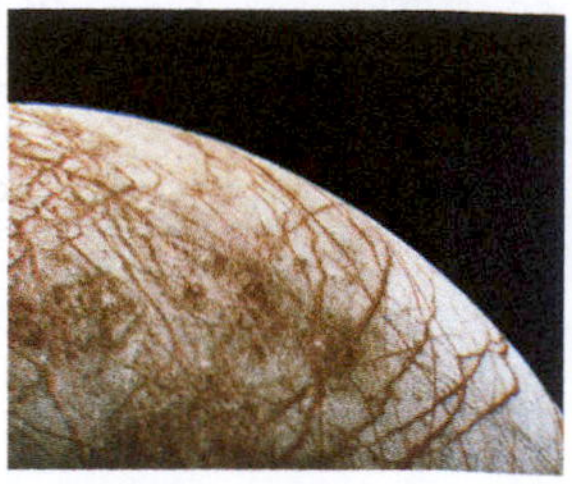

## MCAT-Style Passage Problems

### Orbiting the Moon

Suppose a spacecraft orbits the moon in a very low, circular orbit, just a few hundred meters above the lunar surface. The moon has a diameter of 3500 km, and the free-fall acceleration at the surface is 1.6 $m/s^2$.

80. | The direction of the net force on the craft is
   A. Away from the surface of the moon.
   B. In the direction of motion.
   C. Toward the center of the moon.
   D. Nonexistent, because the net force is zero.

81. | How fast is this spacecraft moving?
   A. 53 m/s
   B. 75 m/s
   C. 1700 m/s
   D. 2400 m/s

82. | How much time does it take for the spacecraft to complete one orbit?
   A. 38 min
   B. 76 min
   C. 110 min
   D. 220 min

83. | The material that comprises the side of the moon facing the earth is actually slightly more dense than the material on the far side. When the spacecraft is above a more dense area of the surface, the moon's gravitational force on the craft is a bit stronger. In order to stay in a circular orbit of constant height and speed, the spacecraft could fire its rockets while passing over the denser area. The rockets should be fired so as to generate a force on the craft
   A. Away from the surface of the moon.
   B. In the direction of motion.
   C. Toward the center of the moon.
   D. Opposite the direction of motion.

# 7 Rotational Motion

Design modifications make this cyclist and her racing bicycle more aerodynamic, which enables her to achieve higher speeds. But there are less visible modifications that permit greater acceleration as well—most important, the design of the tires and wheels. How do you design bicycle wheels to achieve greater acceleration?

## LOOKING AHEAD ➧

### Rotational Kinematics

The spinning roulette wheel isn't going anywhere, but it is moving. This is **rotational motion**.

You'll learn about angular velocity and other quantities we use to describe rotational motion.

### Torque

To start something moving, apply a force. To start something rotating, apply a **torque**, as this sailor is doing to the wheel.

You'll see that torque depends on how hard you push and also on *where* you push. A push far from the axle gives a large torque.

### Rotational Dynamics

The girl pushes on the outside edge of the merry-go-round, gradually increasing its rotation rate.

You'll learn a version of Newton's second law for rotational motion and use it to solve problems.

**GOAL** To understand the physics of rotating objects.

## LOOKING BACK ➧

### Circular Motion

In Chapter 6, you learned to describe *circular* motion in terms of period, frequency, velocity, and centripetal acceleration.

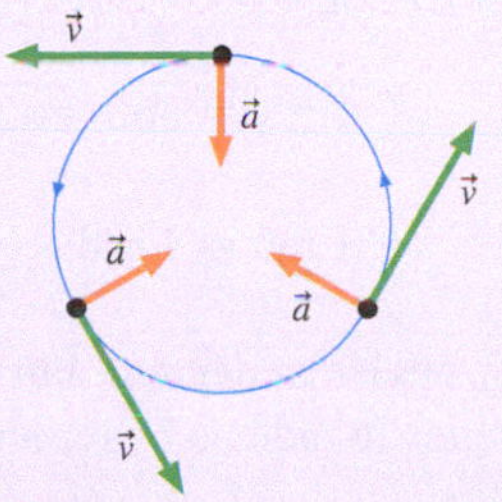


In this chapter, you'll learn to use angular velocity, angular acceleration, and other quantities that describe *rotational* motion.

**STOP TO THINK**

As an audio CD plays, the frequency at which the disk spins changes. At 210 rpm, the speed of a point on the outside edge of the disk is 1.3 m/s. At 420 rpm, the speed of a point on the outside edge is

A. 1.3 m/s  B. 2.6 m/s  
C. 3.9 m/s  D. 5.2 m/s

Moving blades on a wind turbine.

## 7.1 Describing Circular and Rotational Motion

The photo shows a spinning wind turbine. The blurring of the photo tells us that the blades are in motion. On a typical wind turbine, each blade takes 6 seconds to go around. But the outer parts of the blades appear more blurred than the parts nearer the hub; therefore these points are moving more rapidly. This means that each point on each blade completes one circle in 6 seconds, but different points move at different speeds. We'll need to extend our description of motion to handle this type of situation.

In this chapter, we'll consider **rotational motion,** the motion of objects that spin about an axis, like the blade assembly of the wind turbine. We'll introduce some new concepts, but we'll start by returning to a topic we first considered in ◀◀ SECTION 6.1—the motion of a particle in uniform circular motion, such as the motion of the tip of a wind turbine blade. We'll extend this treatment to describe the motion of the whole system. Doing this will require us to define some new quantities.

### Angular Position

FIGURE 7.1 A particle's angular position is described by angle $\theta$.

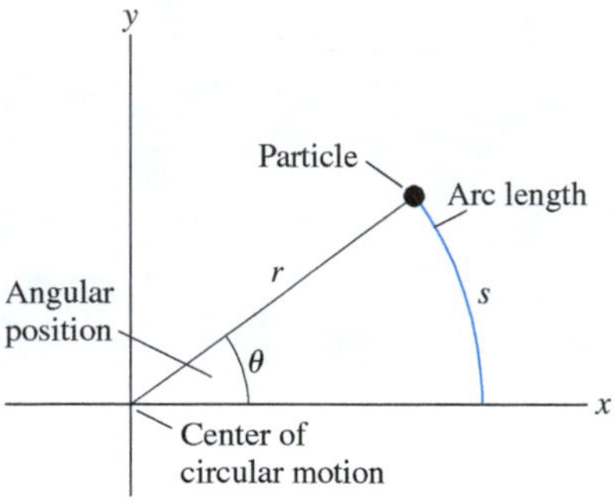


When we describe the motion of a particle as it moves in a circle, it is convenient to use the angle $\theta$ from the positive $x$-axis to describe the particle's location. This is shown in FIGURE 7.1. Because the particle travels in a circle with a fixed radius $r$, specifying $\theta$ completely locates the position of the particle. Thus we call angle $\theta$ the **angular position** of the particle.

We define $\theta$ to be positive when measured *counterclockwise* from the positive $x$-axis. An angle measured *clockwise* from the positive $x$-axis has a negative value. Rather than measure angles in degrees, mathematicians and scientists usually measure angle $\theta$ in the angular unit of *radians.* In Figure 7.1, we also show the **arc length** $s$, the distance that the particle has traveled along its circular path. We define the particle's angle $\theta$ in **radians** in terms of this arc length and the radius of the circle:

$$\theta\ (\text{radians}) = \frac{s}{r} \tag{7.1}$$

This is a sensible definition of an angle: The farther the particle has traveled around the circle (i.e., the greater $s$ is), the larger the angle $\theta$ in radians. The radian, abbreviated rad, is the SI unit of angle. An angle of 1 rad has an arc length $s$ exactly equal to the radius $r$. An important consequence of Equation 7.1 is that the arc length spanning the angle $\theta$ is

$$s = r\theta \tag{7.2}$$

**NOTE** ▶ Equation 7.2 is valid only if $\theta$ is measured in radians, not degrees. This very simple relationship between angle and arc length is one of the primary motivations for using radians. ◀

When a particle travels all the way around the circle—completing one *revolution,* abbreviated rev—the arc length it travels is the circle's circumference $2\pi r$. Thus the angle of a full circle is

$$\theta_{\text{full circle}} = \frac{s}{r} = \frac{2\pi r}{r} = 2\pi \text{ rad}$$

We can use this fact to write conversion factors among revolutions, radians, and degrees:

$$1 \text{ rev} = 360^\circ = 2\pi \text{ rad}$$

$$1 \text{ rad} = 1 \text{ rad} \times \frac{360^\circ}{2\pi \text{ rad}} = 57.3^\circ$$

We will often specify angles in degrees, but keep in mind that the SI unit is the radian. You can visualize angles in radians by remembering that 1 rad is just about 60°.

## Angular Displacement and Angular Velocity

For the *linear* motion you studied in Chapters 1 and 2, a particle with a larger velocity undergoes a greater displacement in each second than one with a smaller velocity, as FIGURE 7.2a shows. FIGURE 7.2b shows two particles undergoing uniform *circular* motion. The particle on the left is moving slowly around the circle; it has gone only one-quarter of the way around after 5 seconds. The particle on the right is moving much faster around the circle, covering half of the circle in the same 5 seconds. You can see that the particle to the right undergoes twice the **angular displacement** $\Delta\theta$ during each interval as the particle to the left. Its **angular velocity,** the angular displacement through which the particle moves each second, is twice as large. We use the symbol $\omega$ to represent angular velocity. This symbol is a lowercase Greek omega, not an ordinary $w$. The SI unit of angular velocity is rad/s.

In analogy with linear motion, where $v_x = \Delta x/\Delta t$, we thus define angular velocity as

$$\omega = \frac{\text{angular displacement}}{\text{time interval}} = \frac{\Delta\theta}{\Delta t} \tag{7.3}$$

Angular velocity of a particle in uniform circular motion

FIGURE 7.2 Comparing uniform linear and circular motion.

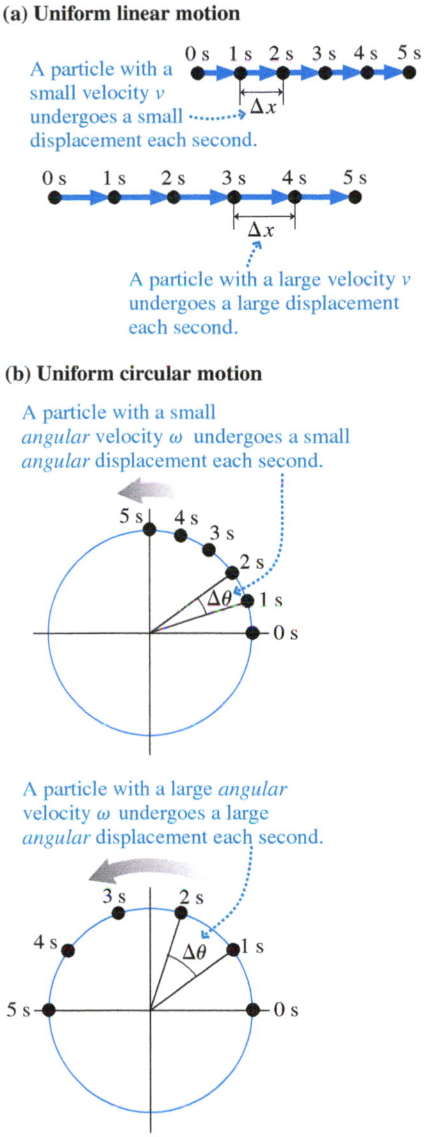


Figure 7.2a shows that the displacement $\Delta x$ of a particle in uniform linear motion changes by the same amount each second. Similarly, as Figure 7.2b shows, the *angular* displacement $\Delta\theta$ of a particle in uniform *circular* motion changes by the same amount each second. This means that **the angular velocity $\omega = \Delta\theta/\Delta t$ is constant for a particle moving with uniform circular motion.**

**EXAMPLE 7.1 Comparing angular velocities**

Find the angular velocities of the two particles in Figure 7.2b.

**STRATEGIZE** This is a straightforward calculation—the angular velocity is simply the change in the angular displacement divided by the time interval. The second particle moves through twice as large an angle as the first, so its angular velocity should be twice that of the first.

**PREPARE** For uniform circular motion, we can use any angular displacement $\Delta\theta$, as long as we use the corresponding time interval $\Delta t$. For each particle, we'll choose the angular displacement corresponding to the motion from $t = 0$ s to $t = 5$ s.

**SOLVE** The particle on the left travels one-quarter of a full circle during the 5 s time interval. We learned earlier that a full circle corresponds to an angle of $2\pi$ rad, so the angular displacement for this particle is $\Delta\theta = (2\pi \text{ rad})/4 = \pi/2$ rad. Thus its angular velocity is

$$\omega = \frac{\Delta\theta}{\Delta t} = \frac{\pi/2 \text{ rad}}{5 \text{ s}} = 0.314 \text{ rad/s}$$

The particle on the right travels halfway around the circle, or $\pi$ rad, in the 5 s interval. Its angular velocity is

$$\omega = \frac{\Delta\theta}{\Delta t} = \frac{\pi \text{ rad}}{5 \text{ s}} = 0.628 \text{ rad/s}$$

**ASSESS** The angular velocity of the second particle is double that of the first, as it should be. We should also check the scale of the answers. The angular velocity of the particle on the right is 0.628 rad/s, meaning that the particle travels through an angle of 0.628 rad each second. Because 1 rad $\approx 60°$, 0.628 rad is roughly 35°. In Figure 7.2b, the particle on the right appears to move through an angle of about this size during each 1 s time interval, so our answer is reasonable.

Angular velocity, like the velocity $v_x$ of one-dimensional motion, can be positive or negative. The signs for $\omega$ noted in FIGURE 7.3 are based on the convention that angles are positive when measured counterclockwise from the positive $x$-axis.

FIGURE 7.3 Positive and negative angular velocities.

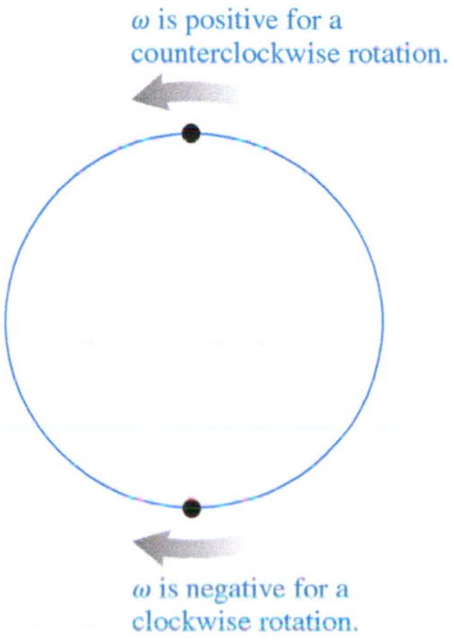

We've already noted how circular motion is analogous to linear motion, with angular variables replacing linear variables. Thus much of what you learned about linear kinematics and dynamics carries over to circular motion. For example, Equation 2.4 gave us a formula for computing a linear displacement during a time interval:

$$x_f - x_i = \Delta x = v_x \, \Delta t$$

You can see from Equation 7.3 that we can write a similar equation for an angular displacement:

$$\theta_f - \theta_i = \Delta\theta = \omega \, \Delta t \qquad (7.4)$$

Angular displacement for uniform circular motion

For linear motion, we use the term *speed* $v$ when we are not concerned with the direction of motion, *velocity* $v_x$ when we are. For circular motion, we define the **angular speed** to be the absolute value of the angular velocity, so that it's a positive quantity irrespective of the particle's direction of rotation. Although potentially confusing, it is customary to use the symbol $\omega$ for angular speed *and* for angular velocity. If the direction of rotation is not important, we will interpret $\omega$ to mean angular speed. In kinematic equations, such as Equation 7.4, $\omega$ is always the angular velocity, and you need to use a negative value for clockwise rotation.

**EXAMPLE 7.2** **Kinematics at the roulette wheel**

A small steel ball rolls counterclockwise around the inside of a stationary 30.0-cm-diameter roulette wheel, like the one shown in the chapter preview. The ball completes exactly 2 rev in 1.20 s.

a. What is the ball's angular velocity?
b. What is the ball's angular position at $t = 2.00$ s? Assume $\theta_i = 0$.

**STRATEGIZE** This problem is analogous to problems of motion along a line that we've seen. Instead of displacement and time, however, we are given angular displacement and time. The steps in the solution will exactly match the steps we would use for motion along a line.

**PREPARE** We can assume that the ball is in uniform circular motion, so its angular position is steadily changing. We are given the angular displacement for a certain time interval, so we can compute the angular velocity. Then, knowing the angular velocity, we can find the angular position at a future time.

**SOLVE**

a. The ball's angular velocity is $\omega = \Delta\theta/\Delta t$. We know that the ball completes 2 revolutions in 1.20 s and that each revolution corresponds to an angular displacement $\Delta\theta = 2\pi$ rad. Thus

$$\omega = \frac{2(2\pi \text{ rad})}{1.20 \text{ s}} = 10.47 \text{ rad/s}$$

We'll use this value in subsequent calculations, but our final result for the angular velocity should be reported to three significant figures, $\omega = 10.5$ rad/s. Because the rotation direction is counterclockwise, the angular velocity is positive.

b. The ball moves with constant angular velocity, so its angular position is given by Equation 7.4. Thus the ball's angular position at $t = 2.00$ s is

$$\theta_f = \theta_i + \omega \, \Delta t = 0 \text{ rad} + (10.47 \text{ rad/s})(2.00 \text{ s})$$
$$= 20.94 \text{ rad}$$

If we're interested in where the ball is at $t = 2.00$ s, we can write its angular position as an integer multiple of $2\pi$ (representing the number of complete revolutions the ball has made) plus a remainder:

$$\theta_f = 20.94 \text{ rad} = 3.333 \times 2\pi \text{ rad}$$
$$= 3 \times 2\pi \text{ rad} + 0.333 \times 2\pi \text{ rad}$$
$$= 3 \times 2\pi \text{ rad} + 2.09 \text{ rad}$$

In other words, at $t = 2.00$ s, the ball has completed 3 rev and is 2.09 rad = 120° into its fourth revolution. An observer would say that the ball's angular position is $\theta = 120°$.

**ASSESS** Since the ball completes 2 revolutions in 1.20 s, it seems reasonable that it completes 3.33 revolutions in 2.00 s.

The angular speed $\omega$ is closely related to the period $T$ and the frequency $f$ of the motion. If a particle in uniform circular motion moves around a circle once, the angular displacement is $\Delta\theta = 2\pi$ rad; the time interval is, by definition, $\Delta t = T$.

The angular speed, $\omega = \Delta\theta/\Delta t$, can thus be expressed in terms of the period $T$ or the frequency $f = 1/T$:

$$\omega = \frac{2\pi \text{ rad}}{T} = (2\pi \text{ rad})f \tag{7.5}$$

Angular speed in terms of period and frequency

**EXAMPLE 7.3** **Rotations in a car engine**

The crankshaft in your car engine is turning at 3000 rpm. What is the shaft's angular speed in rad/s?

**PREPARE** We'll need to convert rpm to rev/s and then use Equation 7.5.

**SOLVE** We convert rpm to rev/s by

$$\left(3000 \frac{\text{rev}}{\text{min}}\right)\left(\frac{1 \text{ min}}{60 \text{ s}}\right) = 50.0 \text{ rev/s}$$

Thus the crankshaft's angular speed is

$$\omega = (2\pi \text{ rad})f = (2\pi \text{ rad})(50.0 \text{ rev/s}) = 314 \text{ rad/s}$$

## Angular-Position and Angular-Velocity Graphs

For the one-dimensional motion you studied in Chapter 3, we found that position- and velocity-versus-time graphs were important and useful representations of motion. We can use the same kinds of graphs to represent angular motion. Let's begin by considering the motion of the roulette ball of Example 7.2. We found that it had angular velocity $\omega = 10.5$ rad/s, meaning that its angular *position* changed by $+10.5$ rad every second. This is exactly analogous to the one-dimensional motion problem of a car driving in a straight line with a velocity of 10.5 m/s, so that its position increases by 10.5 m each second. Using this analogy, we can construct the **angular position-versus-time graph** for the roulette ball shown in FIGURE 7.4.

The angular velocity is given by $\omega = \Delta\theta/\Delta t$. Graphically, this is the *slope* of the angular position-versus-time graph, just as the ordinary velocity is the slope of the position-versus-time graph. Thus we can create an **angular velocity-versus-time graph** by finding the slope of the corresponding angular position-versus-time graph.

FIGURE 7.4 Angular position for the ball on the roulette wheel.

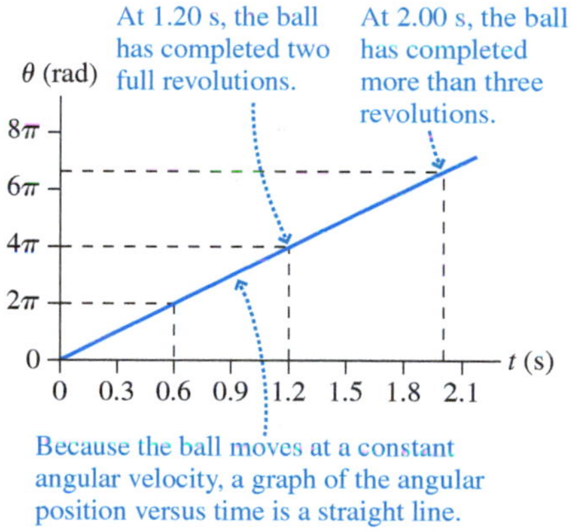


**EXAMPLE 7.4** **Graphing a bike ride**

Jake rides his bicycle home from campus. FIGURE 7.5 is the angular position-versus-time graph for a small rock stuck in the tread of his tire. First, draw the rock's angular velocity-versus-time graph, using rpm on the vertical axis. Then interpret the graphs with a story about Jake's ride.

**STRATEGIZE** There are three portions of the graph in Figure 7.5. For the two horizontal sections, there is no change in the angular position—the tire isn't moving. In between, there is a line with a positive slope. This indicates a steadily increasing angular position, so the tire is rotating at a constant angular velocity.

**PREPARE** The angular velocity $\omega$ is the slope of the angular position-versus-time graph. For the two horizontal segments of the graph, the slope is zero, so the angular speed is zero—as we've noted, the tire isn't moving. For the segment with the positive slope, computing the slope gives us the angular velocity.

FIGURE 7.5 Angular position-versus-time graph for Jake's bike ride.

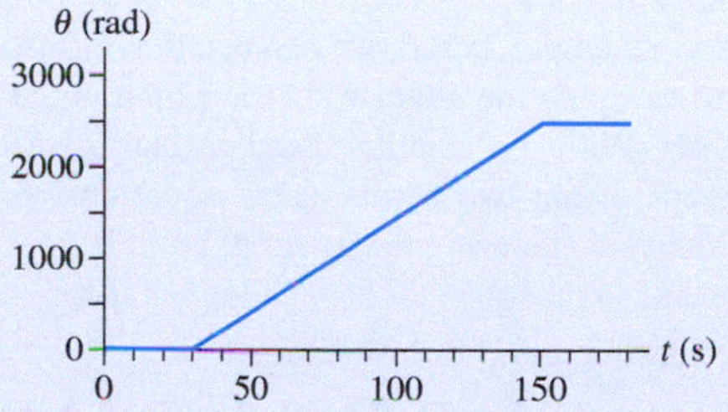


**SOLVE** Between $t = 30$ s and $t = 150$ s, an interval of 120 s, the rock's angular velocity (the slope of the angular position-versus-time graph) is

$$\omega = \text{slope} = \frac{2500 \text{ rad} - 0 \text{ rad}}{120 \text{ s}} = 20.8 \text{ rad/s}$$

*Continued*

Then we need to convert this to rpm:

$$\omega = \left(\frac{20.8 \text{ rad}}{1 \text{ s}}\right)\left(\frac{1 \text{ rev}}{2\pi \text{ rad}}\right)\left(\frac{60 \text{ s}}{1 \text{ min}}\right) = 200 \text{ rpm}$$

These values have been used to draw the angular velocity-versus-time graph of FIGURE 7.6. It looks like Jake waited 30 s for the light to change, then pedaled so that the bike wheel turned at a constant angular velocity of 200 rpm. 2.0 min later, he quickly braked to a stop for another 30-s-long red light.

**ASSESS** 200 rpm is about 3 revolutions per second. This seems reasonable for someone cycling at a pretty good clip.

FIGURE 7.6 Angular velocity-versus-time graph for Jake's bike ride.

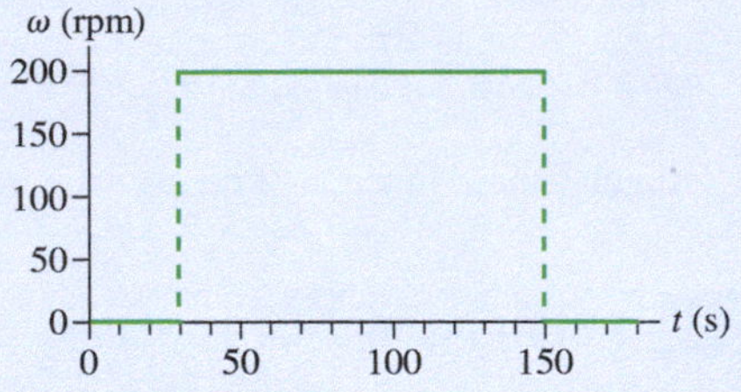


## Relating Speed and Angular Speed

Let's think back to the wind turbine example that opened the chapter. Different points on the blades move at different speeds. And we know that points farther from the axis—at larger values of $r$—move at higher speeds. In Chapter 6 we found that the speed of a particle moving with frequency $f$ around a circular path of radius $r$ is $v = 2\pi fr$. If we combine this result with Equation 7.5 for the angular speed, we find that speed $v$ and angular speed $\omega$ are related by

$$v = \omega r \tag{7.6}$$

Relationship between speed and angular speed

NOTE ▶ In Equation 7.6, **$\omega$ must be in units of rad/s.** If you are given a frequency in rev/s or rpm, you should convert it to an angular speed in rad/s. ◀

### EXAMPLE 7.5 Finding the speed at different points on a wind turbine blade

A large wind turbine (typical of the size and specifications of a turbine that you see in a modern wind farm) has three blades connected to a central hub. The blades are 50 m long and rotate at 10 rpm. What angular speed $\omega$ does this correspond to? What is the speed $v$ of the tip of a blade? What is the speed of a point 25 m from the hub?

**STRATEGIZE** This is a case of uniform rotational motion. The blades and the hub all move at the same angular speed, but points farther from the hub move at a higher speed. A point 50 m from the hub sweeps out a 50-m-radius circle 10 times each minute, so we expect that the speed will be quite high.

**PREPARE** FIGURE 7.7 shows the blades of the wind turbine, with blue circles illustrating the motion of two points, A and B, during one rotation. Each blade rotates once, so both points rotate through the same angle, 2π rad; we can use this information and the period to compute the angular speed, which is the same for the two points. With the angular speed in hand, we can compute the speeds of the two points using Equation 7.6. The figure clearly shows that, during the rotation, point B goes around a larger circle and so must be moving at a higher speed.

FIGURE 7.7 Rotating wind turbine blades.

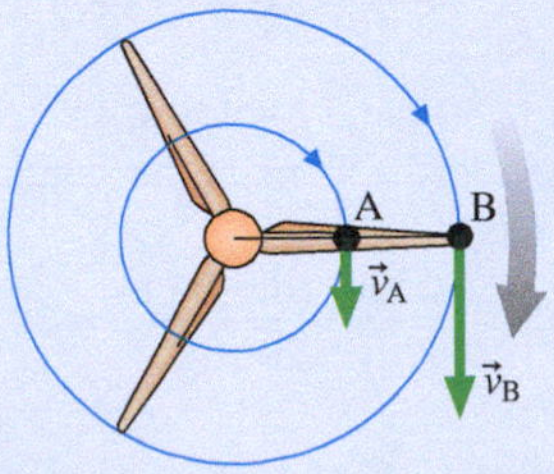


**SOLVE** The frequency is 10 rpm, so the blades rotate 10 times in 1 minute. The period is thus

$$T = \frac{1}{10} \text{ min} = 6.0 \text{ s}$$

Now we can find the angular speed:

$$\omega = \frac{2\pi \text{ rad}}{T} = \frac{2\pi}{6.0 \text{ s}} = 1.05 \text{ rad/s}$$

We can use the angular speed to find the speeds of points A and B:

$$v_A = \omega r = (1.05 \text{ rad/s})(25 \text{ m}) = 26 \text{ m/s}$$

$$v_B = \omega r = (1.05 \text{ rad/s})(50 \text{ m}) = 52 \text{ m/s}$$

**ASSESS** The computed angular speed corresponds to a rotation of about 1 rad in 1 second; 1 rad is about 60°, so this will take the blades around a 360° circle in about 6 seconds, which is just right. The tip of the blade travels in a circle of radius 50 m, which has a circumference of just over 300 m. The tip covers this distance in a mere 6.0 s, so we expect that $v_B$ will be just over 50 m/s, exactly what we found.

Now, let's do a reality check: 52 m/s is nearly 120 mph! Do the tips of the blades really move that fast? A quick investigation of manufacturer specifications reveals that they do. This seems surprising because 120 mph is faster than the wind blows except in extreme situations. But the action of the wind on the blades is similar to the action of the air on the wings of a moving plane: It's the lift force that pushes the blades around. An airplane can certainly slice through the air at 120 mph, so the final result makes sense, although it is initially surprising.

▶ **Why do clocks go clockwise?** In the northern hemisphere, the rotation of the earth causes the sun to follow a circular arc through the southern sky, rising in the east and setting in the west. The shadow cast by the sun thus sweeps in an arc from west to east, so the shadow on a sundial—the first practical timekeeping device—sweeps around the top of the dial from left to right. Early clockmakers used the same convention, which is how it came to be clockwise.

**STOP TO THINK 7.1** Which particle has angular position $5\pi/2$?

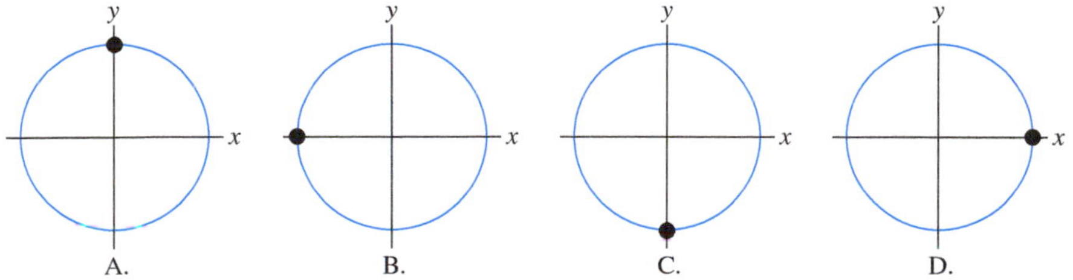


## 7.2 The Rotation of a Rigid Body

Think about the rotating wind turbine in Example 7.5. Until now, our study of physics has focused almost exclusively on the particle model, in which an entire object is represented as a single point in space. The particle model is adequate for understanding motion in a wide variety of situations, but we need to treat the turbine as an *extended object*—a system of particles for which the size and shape *do* make a difference and cannot be ignored.

A **rigid body** is an extended object whose size and shape do not change as it moves. For example, a bicycle wheel can be thought of as a rigid body. FIGURE 7.8 shows a rigid body as a collection of atoms held together by the rigid "massless rods" of molecular bonds.

Real molecular bonds are, of course, not perfectly rigid. That's why an object seemingly as rigid as a bicycle wheel can flex and bend. Thus Figure 7.8 is really a simplified *model* of an extended object, the **rigid-body model.** The rigid-body model is a very good approximation for many real objects of practical interest, such as wheels and axles.

FIGURE 7.9 illustrates the three basic types of motion of a rigid body: **translational motion, rotational motion,** and **combination motion.** We've already studied translational motion of a rigid body using the particle model. If a rigid body doesn't rotate, this model is often adequate for describing its motion. The rotational motion of a rigid body will be the main focus of this chapter. We'll also discuss an important case of combination motion—that of a *rolling* object—later in this chapter.

FIGURE 7.8 The rigid-body model of an extended object.

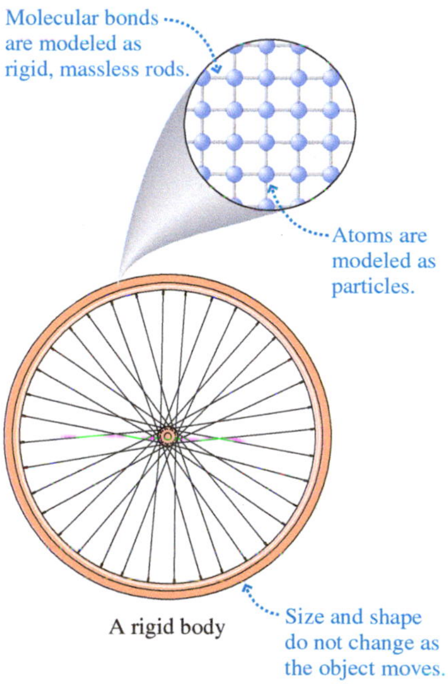


FIGURE 7.9 Three basic types of motion of a rigid body.

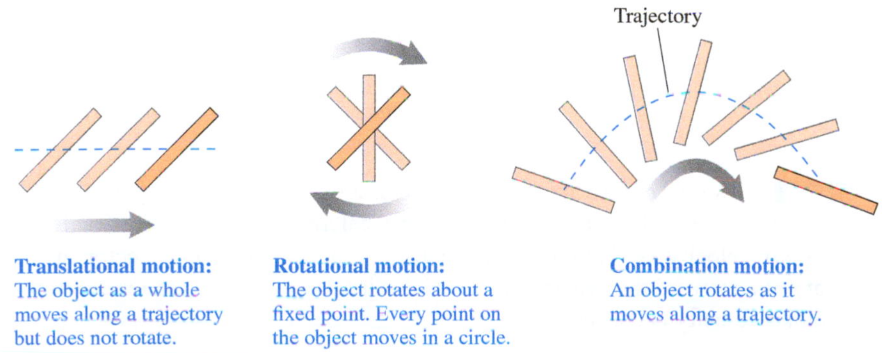

Trajectory
Translational motion: The object as a whole moves along a trajectory but does not rotate.
Rotational motion: The object rotates about a fixed point. Every point on the object moves in a circle.
Combination motion: An object rotates as it moves along a trajectory.

FIGURE 7.10 shows a wheel rotating on an axle. The figure shows the motion as the wheel rotates for a time interval $\Delta t$. Two points 1 and 2 on the wheel move different distances, so they move at different speeds. But these two points move through the same angle, so they move with the same angular velocity. In general, **every point on a rotating rigid body has the same angular velocity,** so we can refer to the angular velocity $\omega$ of the wheel.

FIGURE 7.10 All points on a wheel rotate with the same angular velocity.

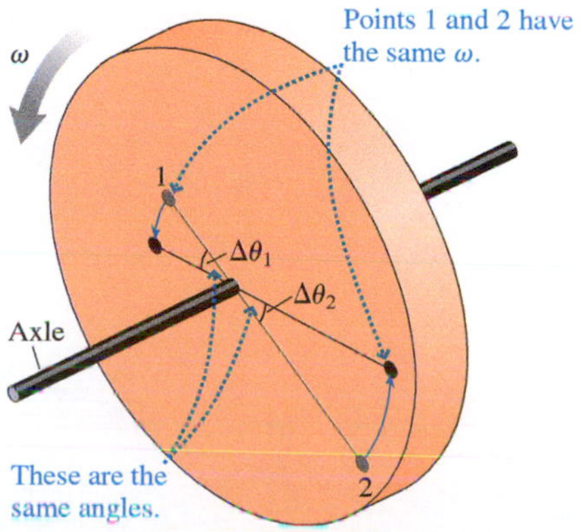

## Angular Acceleration

If you push on the edge of a bicycle wheel, it begins to rotate. If you continue to push, it rotates ever faster. Its angular velocity is *changing*. To understand the dynamics of rotating objects, we'll need to be able to describe this case of changing angular velocity—that is, the case of *nonuniform* circular motion.

FIGURE 7.11 A rotating wheel with a changing angular velocity.

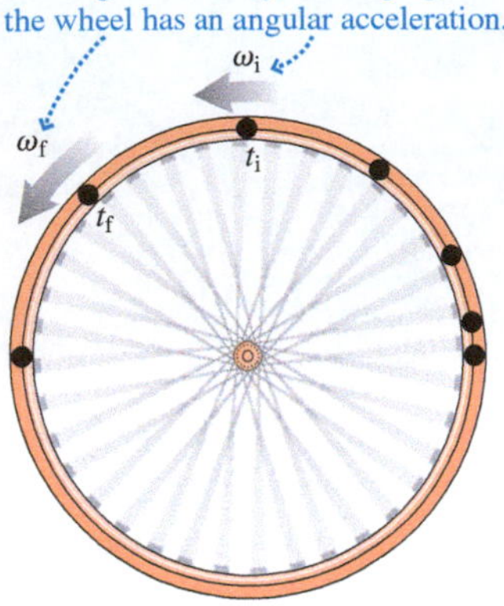


FIGURE 7.11 shows a bicycle wheel whose angular velocity is changing. The dot represents a particular point on the wheel at successive times. At time $t_i$ the angular velocity is $\omega_i$; at a later time $t_f = t_i + \Delta t$ the angular velocity has changed to $\omega_f$. The change in angular velocity during this time interval is

$$\Delta\omega = \omega_f - \omega_i$$

Recall that in Chapter 2 we defined the *linear* acceleration as

$$a_x = \frac{\Delta v_x}{\Delta t} = \frac{(v_x)_f - (v_x)_i}{\Delta t}$$

By analogy, we now define the **angular acceleration** as

$$\alpha = \frac{\text{change in angular velocity}}{\text{time interval}} = \frac{\Delta\omega}{\Delta t} \tag{7.7}$$

Angular acceleration for a particle in nonuniform circular motion

We use the symbol $\alpha$ (Greek alpha) for angular acceleration. Because the units of $\omega$ are rad/s, the units of angular acceleration are (rad/s)/s, or rad/s$^2$. From Equation 7.7, the sign of $\alpha$ is the same as the sign of $\Delta\omega$. FIGURE 7.12 shows how to determine the sign of $\alpha$. Be careful with the sign of $\alpha$; just as with linear acceleration, positive and negative values of $\alpha$ can't be interpreted as simply "speeding up" and "slowing down." Like $\omega$, the angular acceleration $\alpha$ is the same for every point on a rotating rigid body.

FIGURE 7.12 Determining the sign of the angular acceleration.

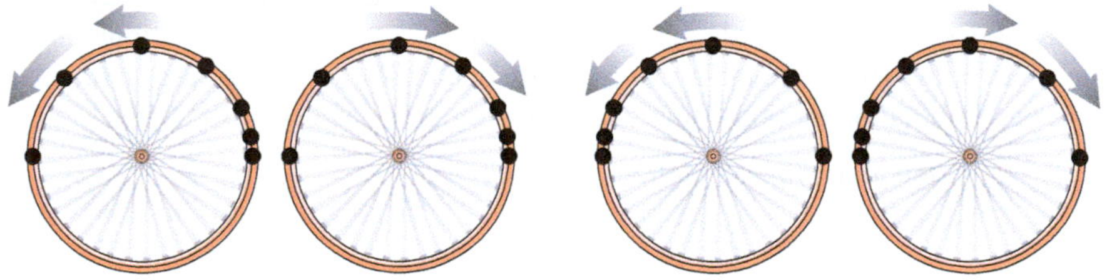


**NOTE** ▶ Don't confuse the angular acceleration with the centripetal acceleration introduced in Chapter 6. The angular acceleration indicates how rapidly the *angular* velocity is changing. The centripetal acceleration is a vector quantity that points toward the center of a particle's circular path; it is nonzero even if the angular velocity is constant. ◀

We noted that the solution to Example 7.2 for uniform circular motion was analogous to a solution to a problem of uniform motion along a line. Instead of position, we have angle; instead of velocity, we have angular velocity. In fact, the equations for circular motion that we use in this chapter are exactly analogous to the equations for motion along a line that you saw in Chapter 2, so we needn't derive any more equations; we can simply adapt ones you've seen before.

**SYNTHESIS 7.1 Linear and circular motion**

The variables and equations for linear motion have analogs for circular motion.

| | | Linear motion | Circular motion | |
|---|---|---|---|---|
| Variables | Position (m) | $x$ | $\theta$ | Angle (rad) |
| | Velocity (m/s) | $v_x = \dfrac{\Delta x}{\Delta t}$ | $\omega = \dfrac{\Delta \theta}{\Delta t}$ | Angular velocity (rad/s) |
| | Acceleration (m/s$^2$) | $a_x = \dfrac{\Delta v_x}{\Delta t}$ | $\alpha = \dfrac{\Delta \omega}{\Delta t}$ | Angular acceleration (rad/s$^2$) |
| Equations | Constant velocity | $\Delta x = v\Delta t$ | $\Delta \theta = \omega \Delta t$ | Constant angular velocity |
| | Constant acceleration | $\Delta v = a\Delta t$ | $\Delta \omega = \alpha \Delta t$ | Constant angular acceleration |
| | | $\Delta x = v\Delta t + \frac{1}{2}a(\Delta t)^2$ | $\Delta \theta = \omega_i \Delta t + \frac{1}{2}\alpha(\Delta t)^2$ | |

**EXAMPLE 7.6 Spinning up a computer disk**

The disk in a computer disk drive spins up from rest to a final angular speed of 5400 rpm in 2.00 s. What is the angular acceleration of the disk? At the end of 2.00 s, how many revolutions has the disk made?

**STRATEGIZE** This problem is clearly analogous to the linear motion problems we saw in Chapter 2. Something starts at rest and then moves with constant angular acceleration for a fixed interval of time. We will solve this problem by referring to Synthesis 7.1 and using equations that correspond to the equations we'd use to solve a similar linear motion problem.

**PREPARE** The initial angular velocity is $\omega_i = 0$ rad/s. The final angular velocity is $\omega_f = 5400$ rpm. In rad/s this is

$$\omega_f = \frac{5400 \text{ rev}}{\text{min}} \times \frac{1 \text{ min}}{60 \text{ s}} \times \frac{2\pi \text{ rad}}{1 \text{ rev}} = 565 \text{ rad/s}$$

**SOLVE** We start by finding the angular acceleration. We can do this using the definition of angular acceleration, the change in angular speed divided by the time interval:

$$\alpha = \frac{\Delta\omega}{\Delta t} = \frac{565 \text{ rad/s} - 0 \text{ rad/s}}{2.00 \text{ s}} = 282.5 \text{ rad/s}^2$$

We've kept an extra significant figure for later calculations, but we report our final result as $\alpha = 283$ rad/s$^2$.

Next, we use the equation for the angular displacement—analogous to the equation for linear displacement—during the period of constant angular acceleration to determine the angle through which the disk moves:

$$\begin{aligned}\Delta\theta &= \omega_i \Delta t + \tfrac{1}{2}\alpha(\Delta t)^2 \\ &= (0 \text{ rad/s})(2.00 \text{ s}) + \tfrac{1}{2}(282.5 \text{ rad/s}^2)(2.00 \text{ s})^2 \\ &= 565 \text{ rad}\end{aligned}$$

Each revolution corresponds to an angular displacement of $2\pi$, so we have

$$\begin{aligned}\text{number of revolutions} &= \frac{565 \text{ rad}}{2\pi \text{ rad/revolution}} \\ &= 90 \text{ revolutions}\end{aligned}$$

The disk completes 90 revolutions during the first 2 seconds.

**ASSESS** The disk spins up to 5400 rpm, which corresponds to 90 rev/s. If the disk spins at full speed for 2.0 s, it will undergo 180 revolutions. But it spins up from rest to this speed over a time of 2.0 s, so it makes sense that it undergoes half this number of revolutions as it gets up to speed.

## Graphs for Rotational Motion with Constant Angular Acceleration

In ◀◀ SECTION 2.5, we studied position, velocity, and acceleration graphs for motion with constant acceleration. We can extend that treatment to the current case of rotational motion. Because of the analogies between linear and angular quantities in Synthesis 7.1, the rules for graphing angular variables are identical to those for linear variables. In particular, **the angular velocity is the slope of the angular position-versus-time graph,** and **the angular acceleration is the slope of the angular velocity-versus-time graph.**

**EXAMPLE 7.7** **Graphing angular quantities**

FIGURE 7.13 shows the angular velocity-versus-time graph for the propeller of a ship.

a. Describe the motion of the propeller.
b. Draw the angular acceleration graph for the propeller.

FIGURE 7.13 The propeller's angular velocity.

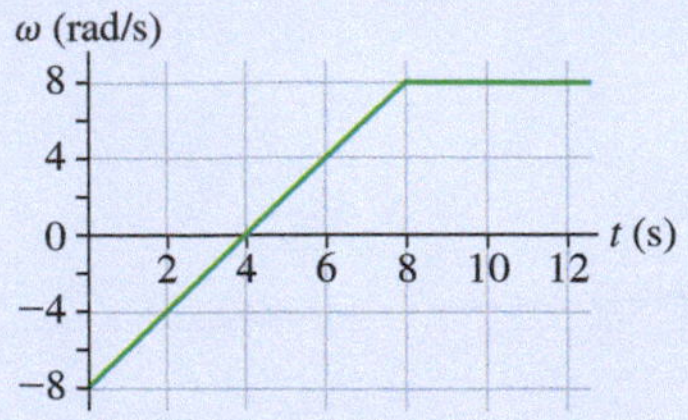


***STRATEGIZE*** We've solved analogous problems for motion in one dimension, and we will use the correspondence between the descriptions of linear and circular motion to guide our solution.

***PREPARE*** The graph shows the angular velocity as a function of time, so we can interpret the graph to describe the motion. We can then use the slope of the graph to determine the angular acceleration.

***SOLVE***

a. Initially the propeller has a negative angular velocity, so it is turning clockwise. It slows down until, at $t = 4$ s, it is instantaneously stopped. It then speeds up in the opposite direction until it is turning counterclockwise at a constant angular velocity.
b. The angular acceleration graph is the slope of the angular velocity graph. From $t = 0$ s to $t = 8$ s, the slope is

$$\frac{\Delta\omega}{\Delta t} = \frac{\omega_f - \omega_i}{\Delta t} = \frac{(8.0 \text{ rad/s}) - (-8.0 \text{ rad/s})}{8.0 \text{ s}} = 2.0 \text{ rad/s}^2$$

After $t = 8$ s, the slope is zero, so the angular acceleration is zero. This graph is plotted in FIGURE 7.14.

FIGURE 7.14 Angular acceleration graph for a propeller.

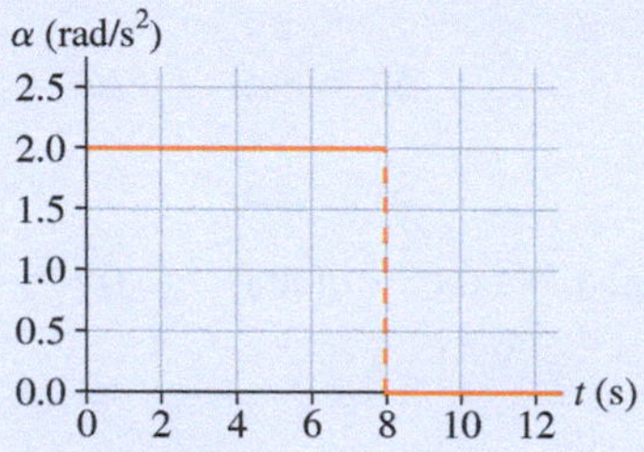


***ASSESS*** A comparison of these graphs with their linear analogs in Figure 2.25 suggests that we're on the right track.

FIGURE 7.15 Uniform and nonuniform circular motion.

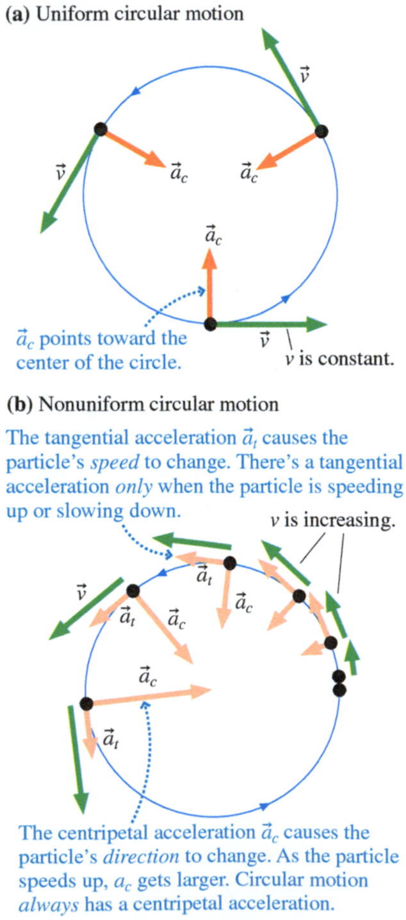


## Tangential Acceleration

As you learned in Chapter 6, and as FIGURE 7.15a reminds you, a particle undergoing uniform circular motion has an acceleration directed inward toward the center of the circle. This centripetal acceleration $\vec{a}_c$ is due to the change in the *direction* of the particle's velocity. Recall that the magnitude of the centripetal acceleration is $a_c = v^2/r = \omega^2 r$.

**NOTE** ▸ Centripetal acceleration will now be denoted $a_c$ to distinguish it from tangential acceleration $a_t$, discussed below. ◂

If a particle is undergoing angular acceleration, its angular speed is changing and, therefore, so is its speed. This means that the particle will have another component to its acceleration. FIGURE 7.15b shows a particle whose speed is increasing as it moves around its circular path. Because the *magnitude* of the velocity is increasing, this second component of the acceleration is directed *tangentially* to the circle, in the same direction as the velocity. This component of acceleration is called the **tangential acceleration.**

The tangential acceleration measures the rate at which the particle's speed around the circle increases. Thus its magnitude is

$$a_t = \frac{\Delta v}{\Delta t}$$

We can relate the tangential acceleration to the *angular* acceleration by using the relationship $v = \omega r$ between the speed of a particle moving in a circle of radius $r$ and its angular velocity $\omega$. We have

$$a_t = \frac{\Delta v}{\Delta t} = \frac{\Delta(\omega r)}{\Delta t} = \frac{\Delta\omega}{\Delta t} r$$

or, because $\alpha = \Delta\omega/\Delta t$ from Equation 7.7,

$$a_t = \alpha r \qquad (7.8)$$

Relationship between tangential and angular acceleration

We've seen that all points on a rotating rigid body have the same angular acceleration. From Equation 7.8, however, the centripetal and tangential accelerations of a point on a rotating object depend on the point's distance $r$ from the axis, so these accelerations are *not* the same for all points.

**STOP TO THINK 7.2** A ball on the end of a string swings in a horizontal circle once every second. State whether the magnitude of each of the following quantities is zero, constant (but not zero), or changing.

a. Velocity
b. Angular velocity
c. Centripetal acceleration
d. Angular acceleration
e. Tangential acceleration

## 7.3 Torque

We've seen that force is the cause of acceleration. But what about *angular* acceleration? What do Newton's laws tell us about rotational motion? To begin our study of rotational motion, we'll need to find a rotational equivalent of force.

Consider the common experience of pushing open a heavy door. FIGURE 7.16 is a top view of a door that is hinged on the left. Four forces are shown, all of equal strength. Which of these will be most effective at opening the door?

Force $\vec{F}_1$ will open the door, but force $\vec{F}_2$, which pushes straight toward the hinge, will not. Force $\vec{F}_3$ will open the door, but not as effectively as $\vec{F}_1$. What about $\vec{F}_4$? It is perpendicular to the door and it has the same magnitude as $\vec{F}_1$, but you know from experience that pushing close to the hinge is not as effective as pushing at the outer edge of the door.

The ability of a force to cause a rotation thus depends on three factors:

1. The magnitude $F$ of the force
2. The distance $r$ from the pivot—the axis about which the object can rotate—to the point at which the force is applied
3. The angle at which the force is applied

We can incorporate these three factors into a single quantity called the **torque $\tau$** (Greek tau). Loosely speaking, $\tau$ measures the effectiveness of a force at causing an object to rotate about a pivot. **Torque is the rotational equivalent of force.** In Figure 7.16, for instance, the torque $\tau_1$ due to $\vec{F}_1$ is greater than $\tau_4$ due to $\vec{F}_4$.

To make these ideas specific, FIGURE 7.17 shows a force $\vec{F}$ applied at one point of a wrench that's loosening a nut. Figure 7.17 defines the distance $r$ from the pivot to the point at which the force is applied; the **radial line,** the line starting at the pivot and extending through this point; and the angle $\phi$ (Greek phi) measured from the radial line to the direction of the force.

We saw in Figure 7.16 that force $\vec{F}_1$, which was directed perpendicular to the door, was effective in opening it, but force $\vec{F}_2$, directed toward the hinges, had no effect on its rotation. As shown in FIGURE 7.18, this suggests breaking the force $\vec{F}$ applied to the wrench into two component vectors: $\vec{F}_\perp$ directed perpendicular to the radial line, and $\vec{F}_\parallel$ directed parallel to it. Because $\vec{F}_\parallel$ points either directly toward or away from the pivot, it has no effect on the wrench's rotation, and thus contributes nothing to the torque. Only $\vec{F}_\perp$ tends to cause rotation of the wrench, so it is this component of the force that determines the torque.

**NOTE** ► The perpendicular component $\vec{F}_\perp$ is pronounced "F perpendicular" and the parallel component $\vec{F}_\parallel$ is "F parallel." ◄

We've seen that a force applied at a greater distance $r$ from the pivot has a greater effect on rotation, so we expect a larger value of $r$ to give a greater torque. We also

FIGURE 7.16 The four forces are the same strength, but they have different effects on the swinging door.

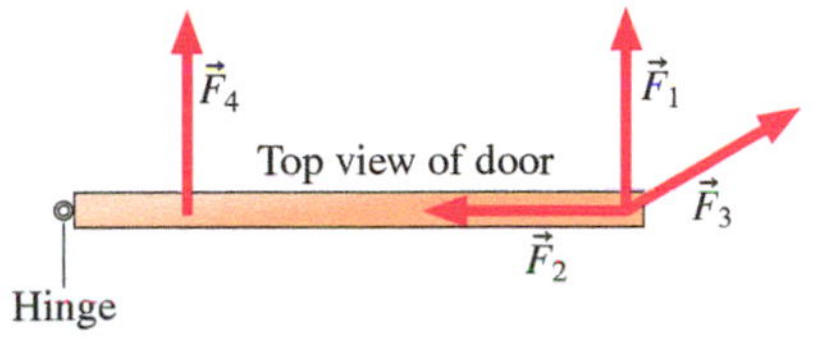


A video to support a section's topic is embedded in the eText.

**Video** Torque

FIGURE 7.17 Force $\vec{F}$ exerts a torque about the pivot point.

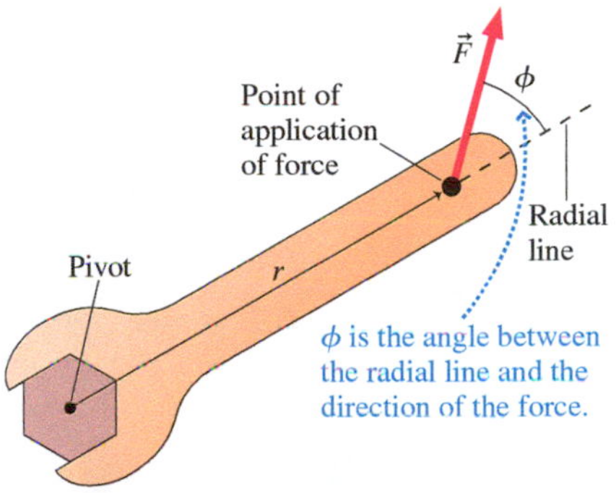


FIGURE 7.18 Torque is due to the component of the force perpendicular to the radial line.

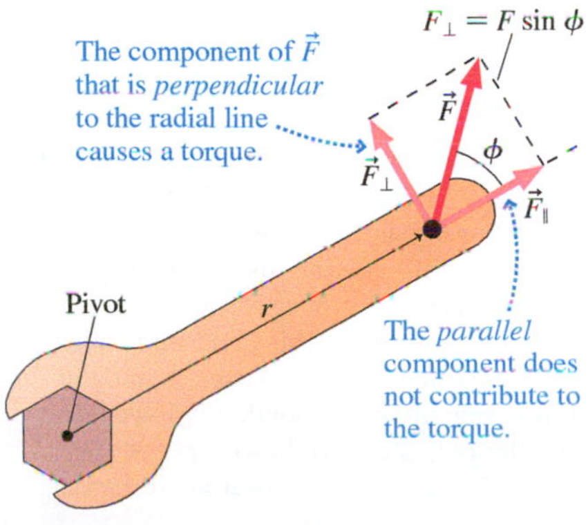

FIGURE 7.19 You can also calculate torque in terms of the moment arm between the pivot and the line of action.

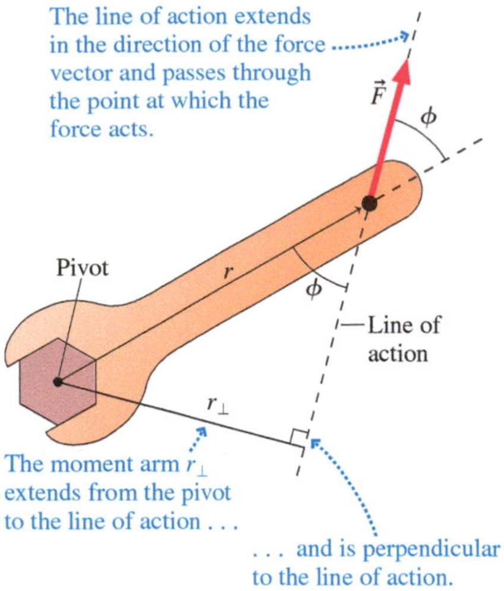

**Torque versus speed** To start and stop quickly, the basketball player needs to apply a large torque to her wheel. To make the torque as large as possible, the handrim—the outside wheel that she actually grabs—is almost as big as the wheel itself. The racer needs to move continuously at high speed, so his wheel spins much faster. To allow his hands to keep up, his handrim is much smaller than his chair's wheel, making its linear velocity correspondingly lower. The smaller radius means, however, that the torque he can apply is lower as well.

saw that only $\vec{F}_\perp$ contributes to the torque. Both these observations are contained in our first expression for torque:

$$\tau = rF_\perp \tag{7.9}$$

Torque due to a force with perpendicular component $F_\perp$ acting at a distance $r$ from the pivot

From this equation, we see that the SI unit of torque is the newton-meter, abbreviated N · m.

FIGURE 7.19 shows an alternative way to calculate torque. The line that is in the direction of the force, and passes through the point at which the force acts, is called the *line of action*. The perpendicular distance from this line to the pivot is called the **moment arm** (or *lever arm*) $r_\perp$. You can see from the figure that $r_\perp = r\sin\phi$. Further, Figure 7.18 showed that $F_\perp = F\sin\phi$. We can then write Equation 7.9 as $\tau = rF\sin\phi = F(r\sin\phi) = Fr_\perp$. Thus an equivalent expression for the torque is

$$\tau = r_\perp F \tag{7.10}$$

Torque due to a force $F$ with moment arm $r_\perp$

**CONCEPTUAL EXAMPLE 7.8** **Where do you put the pedals?**

It is hard to get going if you try to start your bike with the pedal at the highest point. Why is this?

***REASON*** Aided by the weight of the body, the greatest force can be applied to the pedal straight down. But with the pedal at the top, this force is exerted almost directly toward the pivot, causing only a small torque. We could say either that the perpendicular component of the force is small or that the moment arm is small.

***ASSESS*** If you've ever climbed a steep hill while standing on the pedals, you know that you get the greatest forward motion when one pedal is completely forward with the crank parallel to the ground. This gives the maximum possible torque because the force you apply is entirely perpendicular to the radial line, and the moment arm is as long as it can be.

We've seen that Equation 7.9 can be written as $\tau = rF_\perp = r(F\sin\phi)$, and Equation 7.10 as $\tau = r_\perp F = (r\sin\phi)F$. This shows that both methods of calculating torque lead to the same expression for torque—namely:

$$\tau = rF\sin\phi \tag{7.11}$$

where $\phi$ is the angle between the radial line and the direction of the force as illustrated in Figure 7.17.

**NOTE** ▶ Torque differs from force in a very important way. Torque is calculated or measured *about a particular point.* To say that a torque is 20 N · m is meaningless without specifying the point about which the torque is calculated. Torque can be calculated about any point, but its value depends on the point chosen because this choice determines $r$ and $\phi$. In practice, we usually calculate torques about a hinge, pivot, or axle. ◀

Equations 7.9–7.11 are three different ways of thinking about—and calculating—the torque due to a force. Equation 7.11 is the most general equation for torque, but

Equations 7.9 and 7.10 are generally more useful for practical problem solving. All three equations calculate the *same* torque, and all will give the same value.

### EXAMPLE 7.9 Torque to open a stuck door

Ryan is trying to open a stuck door. He pushes it at a point 0.75 m from the hinges with a 240 N force directed 20° away from being perpendicular to the door. There's a natural pivot point, the hinges. What torque does Ryan exert? How could he exert more torque?

**STRATEGIZE** This is a straightforward torque problem. But even for a problem as straightforward as this, we will start by drawing the visual overview in FIGURE 7.20. The figure shows the door, identifies the pivot point (the hinge, in this case), and shows the force, where it is applied and at what angle.

FIGURE 7.20 Ryan's force exerts a torque on the door.

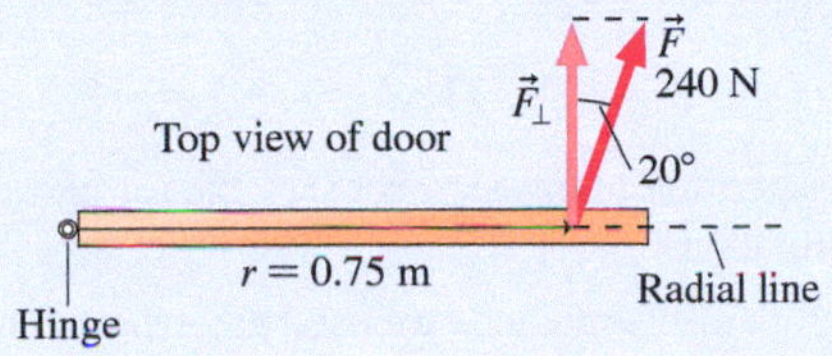


**PREPARE** In Figure 7.20 the radial line is shown drawn from the pivot—the hinge—through the point at which the force $\vec{F}$ is applied. We see that the component of $\vec{F}$ that is perpendicular to the radial line is $F_\perp = F\cos 20° = 226$ N. The distance from the hinge to the point at which the force is applied is $r = 0.75$ m.

**SOLVE** We can find the torque on the door from Equation 7.9:

$$\tau = rF_\perp = (0.75\text{ m})(226\text{ N}) = 170\text{ N}\cdot\text{m}$$

The torque depends on how hard Ryan pushes, where he pushes, and at what angle. If he wants to exert more torque, he could push at a point a bit farther out from the hinge, or he could push exactly perpendicular to the door. Or he could simply push harder!

**ASSESS** As you'll see by doing more problems, 170 N · m is a significant torque, but this makes sense if you are trying to free a stuck door.

Equations 7.9–7.11 give only the magnitude of the torque. But torque, like a force component, has a sign. **A torque that tends to rotate the object in a counterclockwise direction is positive, while a torque that tends to rotate the object in a clockwise direction is negative.** FIGURE 7.21 summarizes what we've said about the magnitude and the sign of a torque. Notice that a force pushing straight toward the pivot or pulling straight out from the pivot exerts *no* torque.

**Video** Torques and Moment Arms

**NOTE** ▶ When calculating a torque, you must supply the appropriate sign by observing the direction in which the torque acts. ◀

KEY CONCEPT FIGURE 7.21 Signs and strengths of the torque.

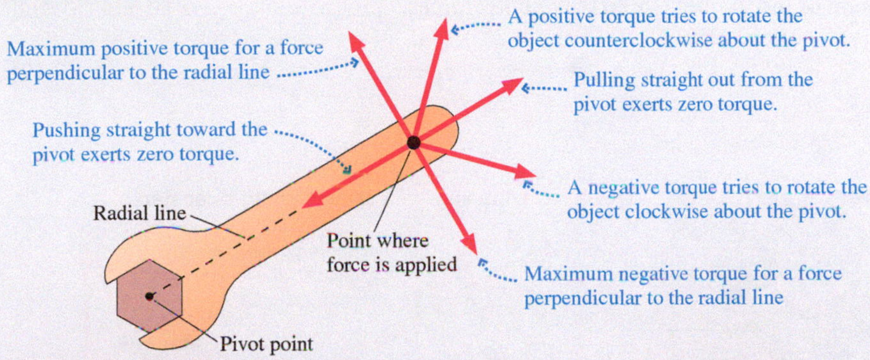

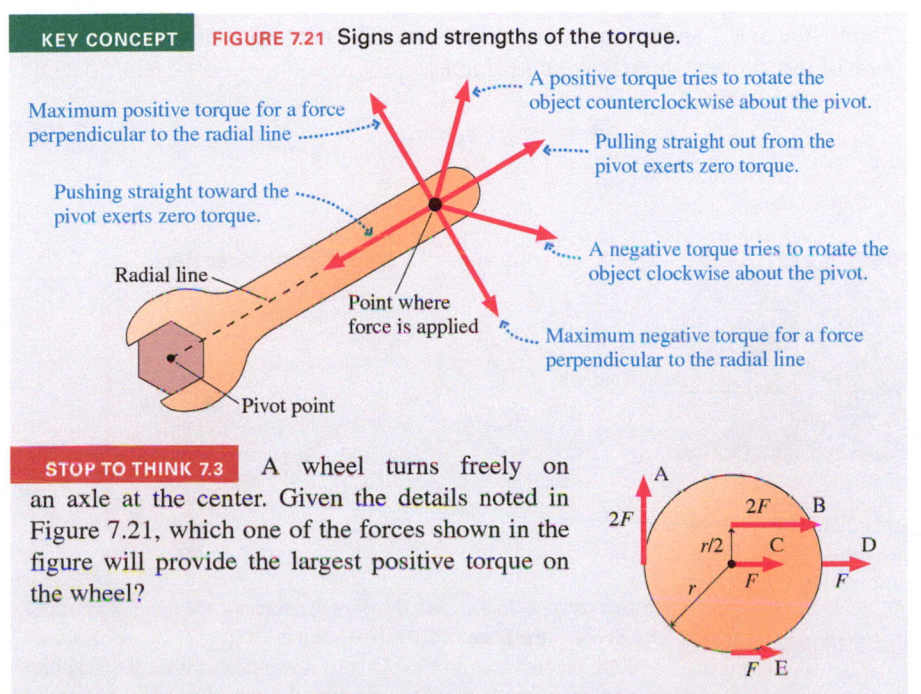

**STOP TO THINK 7.3** A wheel turns freely on an axle at the center. Given the details noted in Figure 7.21, which one of the forces shown in the figure will provide the largest positive torque on the wheel?

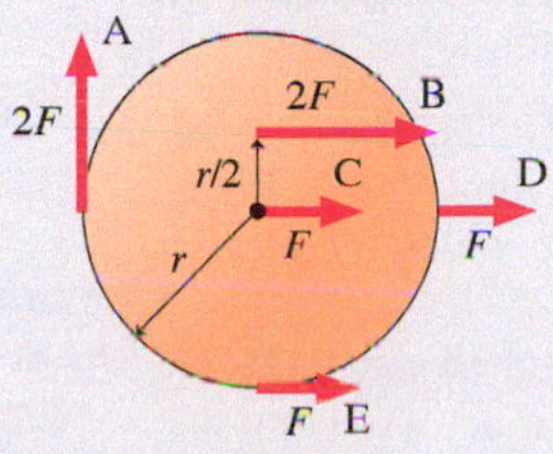

## EXAMPLE 7.10 Calculating the torque on a nut

Luis uses a 20-cm-long wrench to tighten a nut, turning it clockwise. The wrench handle is tilted 30° above the horizontal, and Luis pulls straight down on the end with a force of 100 N. How much torque does Luis exert on the nut?

**STRATEGIZE** We will start by drawing a visual overview that illustrates the geometry of the situation. There is a natural pivot point—the center of the nut that the wrench is turning.

**PREPARE** We make two diagrams in FIGURE 7.22 to illustrate the two different means of computing torque represented by Equations 7.9 and 7.10.

FIGURE 7.22 A wrench being used to turn a nut.

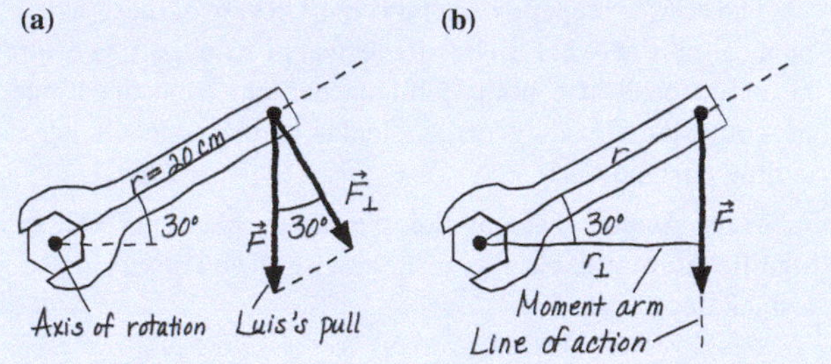


**SOLVE** According to Equation 7.9, the torque can be calculated as $\tau = rF_\perp$. From Figure 7.22a, we see that the perpendicular component of the force is

$$F_\perp = F\cos 30° = (100\text{ N})(\cos 30°) = 86.6\text{ N}$$

This gives a torque of

$$\tau = -rF_\perp = -(0.20\text{ m})(86.6\text{ N}) = -17\text{ N}\cdot\text{m}$$

We put in the minus sign because the torque is negative—it tries to rotate the nut in a *clockwise* direction.

Alternatively, we can use Equation 7.10 to find the torque. Figure 7.22b shows the moment arm $r_\perp$, the perpendicular distance from the pivot to the line of action. From the figure we see that

$$r_\perp = r\cos 30° = (0.20\text{ m})(\cos 30°) = 0.173\text{ m}$$

Then the torque is

$$\tau = -r_\perp F = -(0.173\text{ m})(100\text{ N}) = -17\text{ N}\cdot\text{m}$$

Again, we insert the minus sign because the torque acts to give a clockwise rotation.

**ASSESS** Both of the methods we used gave the same answer for the torque, as should be the case, and this gives us confidence in our results.

## Net Torque

FIGURE 7.23 The forces exert a net torque about the pivot point.

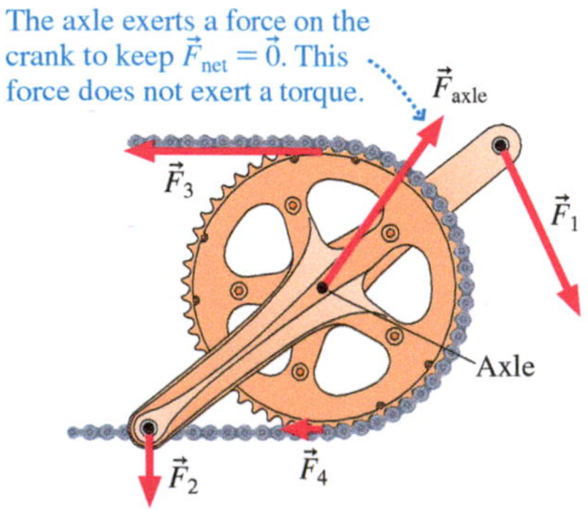


FIGURE 7.23 shows the forces acting on the crankset of a bicycle. Forces $\vec{F}_1$ and $\vec{F}_2$ are due to the rider pushing on the pedals, and $\vec{F}_3$ and $\vec{F}_4$ are tension forces from the chain. The crankset is free to rotate about a fixed axle, but the axle prevents it from having any translational motion with respect to the bike frame. It does so by exerting force $\vec{F}_{axle}$ on the object to balance the other forces and keep $\vec{F}_{net} = \vec{0}$.

Forces $\vec{F}_1$, $\vec{F}_2$, $\vec{F}_3$, and $\vec{F}_4$ exert torques $\tau_1$, $\tau_2$, $\tau_3$, and $\tau_4$ on the crank (measured about the axle), but $\vec{F}_{axle}$ does *not* exert a torque because it is applied at the pivot point—the axle—and so has zero moment arm. The *net* torque about the axle is the sum of the torques due to the *applied* forces:

$$\tau_{net} = \tau_1 + \tau_2 + \tau_3 + \tau_4 + \cdots = \sum \tau \qquad (7.12)$$

## EXAMPLE 7.11 Force in turning a capstan

A capstan is a device used on old sailing ships to raise the anchor. A sailor pushes the long lever, turning the capstan and winding up the anchor rope. If the capstan turns at a constant speed, the net torque on it, as we'll learn later in the chapter, is zero. The rope tension due to the weight of the anchor is 1500 N. The distance from the axis to the point on the lever where the sailor pushes is seven times the radius of the capstan around which the rope is wound.

With what force must the sailor push in order to turn the capstan at a constant speed, meaning that the net torque on the capstan is zero?

FIGURE 7.24 Top view of a sailor turning a capstan.

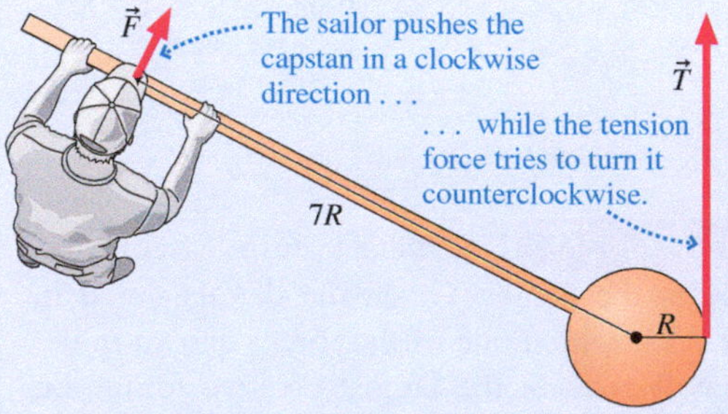


**STRATEGIZE** In this case, there are two forces—the force of the rope tension and the force of the sailor—that exert torques on the capstan. FIGURE 7.24 shows a top view looking down from above, showing the natural pivot (the center of the capstan), the geometry, and the forces.

**PREPARE** Figure 7.24 illustrates the situation. The rope pulls with a tension force $\vec{T}$ at distance $R$ from the axis of rotation. The sailor pushes with a force $\vec{F}$ at distance $7R$ from the axis. Both forces are perpendicular to their radial lines, so $\phi$ in Equation 7.11 is 90°.

**SOLVE** The torque due to the tension in the rope is

$$\tau_T = RT\sin 90° = RT$$

We don't know the capstan radius, so we'll just leave it as $R$ for now. This torque is positive because it tries to turn the capstan counterclockwise. The torque due to the sailor is

$$\tau_S = -(7R)F\sin 90° = -7RF$$

We put the minus sign in because this torque acts in the clockwise (negative) direction. The net torque is zero, so we have $\tau_T + \tau_S = 0$, or

$$RT - 7RF = 0$$

Note that the radius $R$ cancels, leaving

$$F = \frac{T}{7} = \frac{1500\text{ N}}{7} = 210\text{ N}$$

**ASSESS** 210 N is about 50 lb, a reasonable number to expect for the pushing force of a sailor. The force the sailor must exert is one-seventh the force the rope exerts: The long lever helps him lift the heavy anchor. In the HMS *Warrior*, built in 1860, it took 200 men turning the capstan to lift the huge anchor that weighed close to 55,000 N!

Note that forces $\vec{F}$ and $\vec{T}$ point in different directions. Their torques depend only on their directions with respect to their own radial lines, not on the directions of the forces with respect to each other. The force the sailor needs to apply remains unchanged as he circles the capstan.

**STOP TO THINK 7.4** Two forces act on the wheel shown. What third force, acting at point P, will make the net torque on the wheel zero?

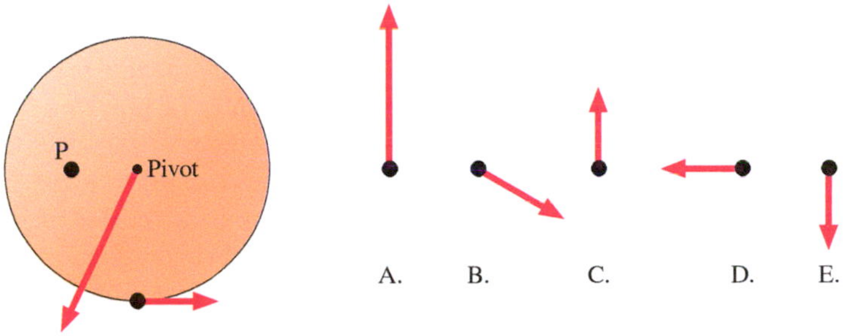


## 7.4 Gravitational Torque and the Center of Gravity

FIGURE 7.25 The center of gravity is the point where the weight appears to act.

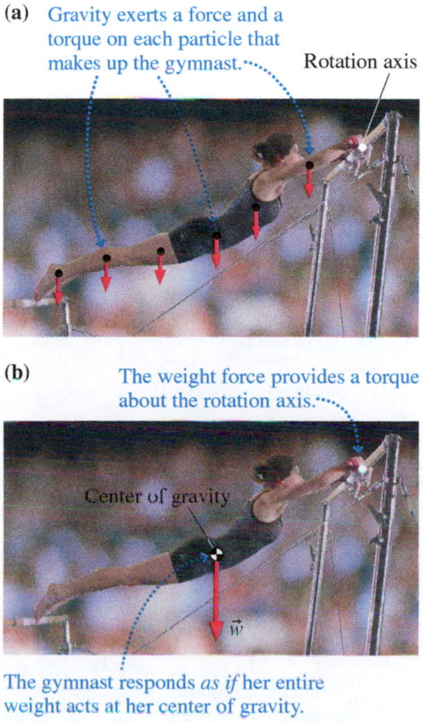


As the gymnast in FIGURE 7.25 pivots around the bar, a torque due to the force of gravity causes her to rotate toward a vertical position. A falling tree and a car hood slamming shut are other examples where gravity exerts a torque on an object. Stationary objects can also experience a torque due to gravity. A diving board experiences a gravitational torque about its fixed end. It doesn't rotate because of a counteracting torque provided by forces from the base at its fixed end.

We've learned how to calculate the torque due to a single force acting on an object. But gravity doesn't act at a single point on an object. It pulls downward on *every particle* that makes up the object, as shown for the gymnast in Figure 7.25a, and so each particle experiences a small torque due to the force of gravity that acts on it. The gravitational torque on the object as a whole is then the *net* torque exerted on all the particles. We won't prove it, but the gravitational torque can be calculated by assuming that the net force of gravity—that is, the object's weight $\vec{w}$—acts at a single special point on the object called its **center of gravity** (symbol ⊛). Then we can calculate the torque due to gravity by the methods we learned earlier for a single force ($\vec{w}$) acting at a single point (the center of gravity). Figure 7.25b shows how we can consider the gymnast's weight as acting at her center of gravity.

## EXAMPLE 7.12 The gravitational torque on a flagpole

A 3.2 kg flagpole extends from a wall at an angle of 25° from the horizontal. Its center of gravity is 1.6 m from the point where the pole is attached to the wall. What is the gravitational torque on the flagpole about the point of attachment?

**STRATEGIZE** In all the problems in this chapter, a visual overview is crucial to help you understand the details, so we will start with the diagram in FIGURE 7.26. We'll note dimensions, draw the weight vector at the proper point, and choose one of the approaches we've seen for calculating the torque.

**PREPARE** For the purpose of calculating gravitational torque, we can consider the entire weight of the pole as acting at the center of gravity. Because the moment arm $r_\perp$ is simple to visualize here, we'll use Equation 7.10 for the torque.

FIGURE 7.26 Visual overview of the flagpole.

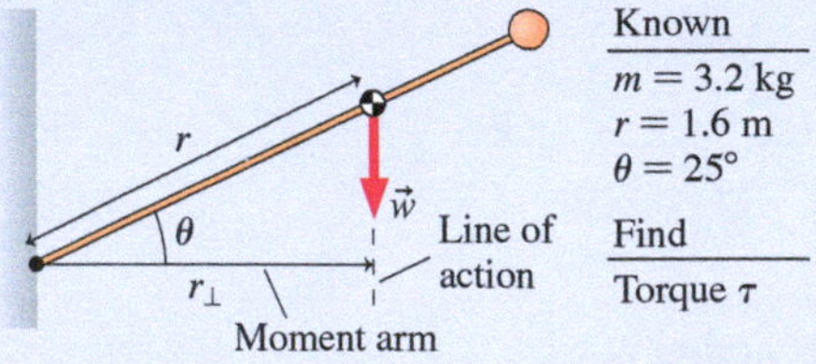


**SOLVE** From Figure 7.26, we see that the moment arm is $r_\perp = (1.6\text{ m})\cos 25° = 1.45\text{ m}$. Thus the gravitational torque on the flagpole, about the point where it attaches to the wall, is

$$\tau = -r_\perp w = -r_\perp mg = -(1.45\text{ m})(3.2\text{ kg})(9.8\text{ m/s}^2)$$
$$= -45\text{ N}\cdot\text{m}$$

We inserted the minus sign because the torque tries to rotate the pole in a clockwise direction. If the pole were attached to the wall by a hinge, the gravitational torque would cause the pole to fall. However, the actual rigid connection provides a counteracting (positive) torque to the pole that prevents this. The net torque is zero.

**ASSESS** A torque is a product of a force and a distance. The force is the weight of the pole, just over 30 N; the distance is the moment arm, just shy of 1.5 m. So 45 N · m is a reasonable result.

FIGURE 7.27 Suspending a ruler.

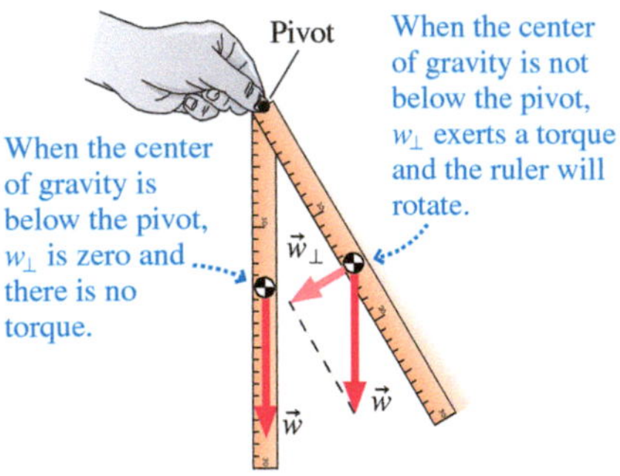


If you hold a ruler by its end so that it is free to pivot, you know that it will quickly rotate so that it hangs straight down. FIGURE 7.27 explains this result in terms of the center of gravity and gravitational torque. The center of gravity of the ruler lies at its center. If the center of gravity is directly below the pivot, there is no gravitational torque and the ruler will stay put. If you rotate the ruler to the side, the resulting gravitational torque will quickly pull it back until the center of gravity is again below the pivot. We've shown this for a ruler, but this is a general principle: **An object that is free to rotate about a pivot will come to rest with the center of gravity below the pivot point.**

If the center of gravity lies directly *above* the pivot, as in FIGURE 7.28, there is no torque due to the object's weight and it can remain balanced. However, if the object is even slightly displaced to either side, the gravitational torque will no longer be zero and the object will begin to rotate. This question of *balance*—the behavior of an object whose center of gravity lies above the pivot—will be explored in depth in Chapter 8.

FIGURE 7.28 Balancing a ruler.

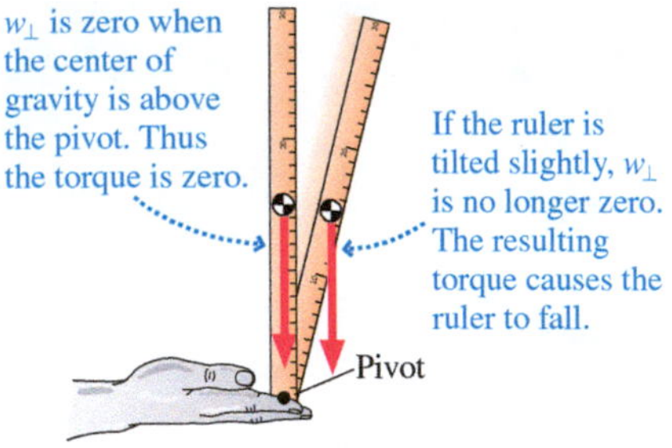


## Calculating the Position of the Center of Gravity

Because there's no gravitational torque when the center of gravity lies either directly above or directly below the pivot, it must be the case that **the torque due to gravity when the pivot is *at* the center of gravity is zero.** We can use this fact to find a general expression for the position of the center of gravity.

Consider the dumbbell shown in FIGURE 7.29. If we slide the triangular pivot back and forth until the dumbbell balances, the pivot must then be at the center of gravity (at position $x_{cg}$) and the torque due to gravity must therefore be zero. But we can calculate the gravitational torque directly by calculating and summing the torques about this point due to the two individual weights. Gravity acts on weight 1 with moment arm $r_1$, so the torque about the pivot at position $x_{cg}$ is

FIGURE 7.29 Finding the center of gravity of a dumbbell.

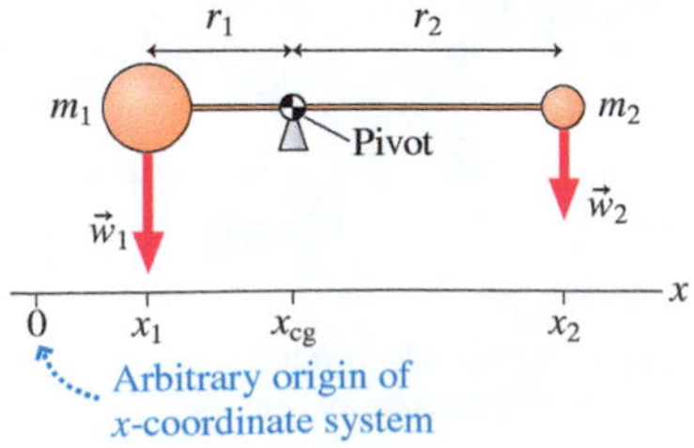


$$\tau_1 = r_1 w_1 = (x_{cg} - x_1)m_1 g$$

Similarly, the torque due to weight 2 is

$$\tau_2 = -r_2 w_2 = -(x_2 - x_{cg})m_2 g$$

This torque is negative because it tends to rotate the dumbbell in a clockwise direction. We've just argued that the net torque must be zero because the pivot is directly under the center of gravity, so

$$\tau_{net} = 0 = \tau_1 + \tau_2 = (x_{cg} - x_1)m_1 g - (x_2 - x_{cg})m_2 g$$

We can solve this equation for the position of the center of gravity $x_{cg}$:

$$x_{cg} = \frac{x_1 m_1 + x_2 m_2}{m_1 + m_2} \tag{7.13}$$

The following Tactics Box shows how Equation 7.13 can be generalized to find the center of gravity of *any* number of particles. If the particles don't all lie along the $x$-axis, then we'll also need to find the $y$-coordinate of the center of gravity.

**TACTICS BOX 7.1** **Finding the center of gravity**

1. Choose an origin for your coordinate system. You can choose any convenient point as the origin.
2. Determine the coordinates $(x_1, y_1), (x_2, y_2), (x_3, y_3), \ldots$ for the particles of masses $m_1, m_2, m_3, \ldots$, respectively.
3. The $x$-coordinate of the center of gravity is

$$x_{cg} = \frac{x_1 m_1 + x_2 m_2 + x_3 m_3 + \cdots}{m_1 + m_2 + m_3 + \cdots} \tag{7.14}$$

4. Similarly, the $y$-coordinate of the center of gravity is

$$y_{cg} = \frac{y_1 m_1 + y_2 m_2 + y_3 m_3 + \cdots}{m_1 + m_2 + m_3 + \cdots} \tag{7.15}$$

Exercises 12–14

**Video** Balancing a Meter Stick

Because the center of gravity depends on products such as $x_1 m_1$, objects with large masses count more heavily than objects with small masses. Consequently, **the center of gravity tends to lie closer to the heavier objects or particles** that make up the entire object.

**EXAMPLE 7.13** **Balancing a seesaw**

Emma and Noah are playing on a seesaw. Noah, mass 22 kg, is seated 1.6 m away from the pivot point. Emma, mass 25 kg, wants to sit at a point that will make the seesaw exactly balance on its pivot. How far away from the pivot must she sit? Assume that the mass of the seesaw itself is small enough to ignore.

**STRATEGIZE** We'll interpret the statement of Noah's position to mean that his center of gravity is 1.6 m from the pivot point; we'll mark the position of his center of gravity at that point. We are looking for the position of Emma's center of gravity, which should be placed so that the combined center of gravity for the children is right at the pivot.

**PREPARE** In FIGURE 7.30, we've defined coordinates so that the pivot point is at $x = 0$. Noah is 1.6 m from the pivot, so $x_N = -1.6$ m. We are looking for $x_E$, Emma's position, given that the combined center of gravity of the two children, with masses noted, is at the pivot.

**SOLVE** There are two masses, the masses of Noah and Emma, and the position of the center of gravity is at the pivot, so we can simplify Equation 7.14 for the position of the center of gravity:

$$x_{cg} = 0 = \frac{x_N m_N + x_E m_E}{m_N + m_E}$$

If the fraction on the right is equal to zero, the numerator must be zero, so

FIGURE 7.30 Children balance on a seesaw.

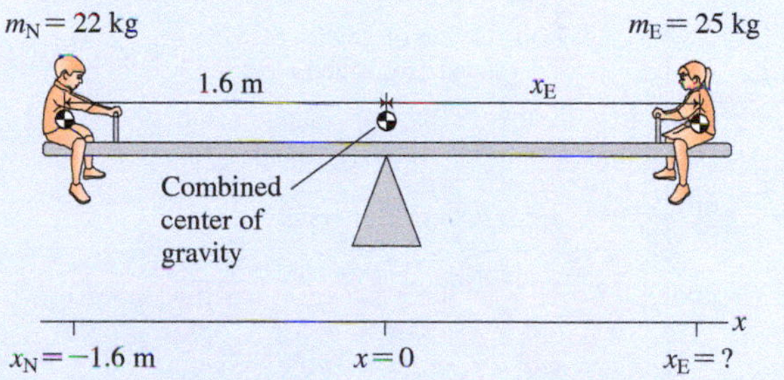


$$x_N m_N + x_E m_E = 0$$

$$x_E m_E = -x_N m_N$$

$$x_E = -x_N\left(\frac{m_N}{m_E}\right) = -(-1.6\text{ m})\frac{22\text{ kg}}{25\text{ kg}} = 1.4\text{ m}$$

Emma should sit 1.4 m from the pivot.

**ASSESS** If you've ever played on a seesaw, you know that to achieve balance the heavier person needs to sit closer to the pivot, so our answer makes sense. It also makes sense that the combined center of gravity is closer to the heavier child.

FIGURE 7.31 Body segment masses and centers of gravity for an 80 kg man.

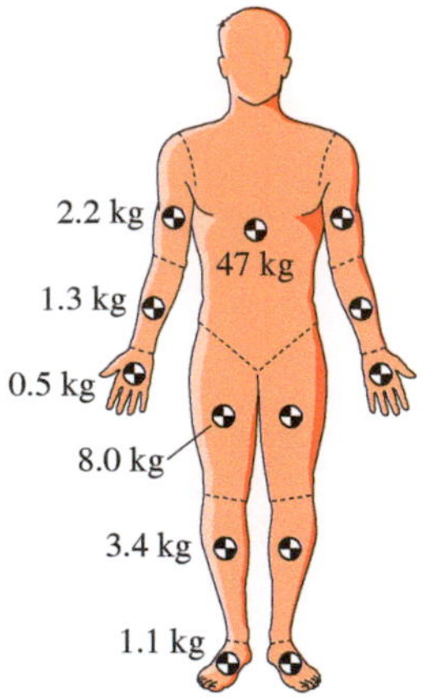


The center of gravity of an extended object can often be found by considering the object as made up of pieces, each with mass and center of gravity that are known or can be found. Then the coordinates of the entire object's center of gravity are given by Equations 7.14 and 7.15, with $(x_1, y_1)$, $(x_2, y_2)$, $(x_3, y_3)$, . . . the coordinates of the center of gravity of each piece and $m_1$, $m_2$, $m_3$, . . . their masses.

This method is used in biomechanics and kinesiology to calculate the center of gravity of the human body. FIGURE 7.31 shows how the body can be considered to be made up of segments, each of whose mass and center of gravity have been measured. The numbers shown are appropriate for a man with a total mass of 80 kg. For a given posture the positions of the segments and their centers of gravity can be found, and thus the whole-body center of gravity from Equations 7.14 and 7.15 (and a third equation for the $z$-coordinate). The next example explores a simplified version of this method.

## EXAMPLE 7.14 Finding the center of gravity of a gymnast

A gymnast performing on the rings holds himself in the pike position. FIGURE 7.32 shows how we can consider his body to be made up of two segments whose masses and center-of-gravity positions are shown. The upper segment includes his head, trunk, and arms, while the lower segment consists of his legs. Locate the overall center of gravity of the gymnast.

FIGURE 7.32 Centers of gravity of two segments of a gymnast.

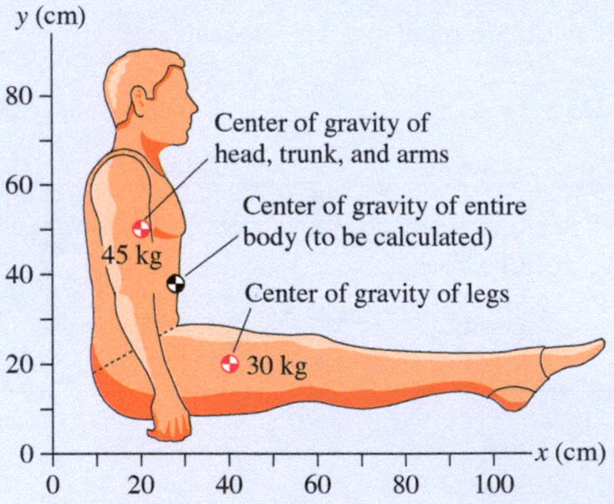


**STRATEGIZE** We will find the center of gravity of the gymnast by computing the combined center of gravity for the segments of his body. We will use Equation 7.14 for the $x$-coordinate and Equation 7.15 for the $y$-coordinate.

**PREPARE** From Figure 7.32 we can find the $x$- and $y$-coordinates of the segment centers of gravity:

$$x_{\text{trunk}} = 20 \text{ cm} \qquad y_{\text{trunk}} = 50 \text{ cm}$$

$$x_{\text{legs}} = 40 \text{ cm} \qquad y_{\text{legs}} = 20 \text{ cm}$$

**SOLVE** The $x$- and $y$-coordinates of the center of gravity are given by Equations 7.14 and 7.15:

$$x_{\text{cg}} = \frac{x_{\text{trunk}} m_{\text{trunk}} + x_{\text{legs}} m_{\text{legs}}}{m_{\text{trunk}} + m_{\text{legs}}}$$

$$= \frac{(20 \text{ cm})(45 \text{ kg}) + (40 \text{ cm})(30 \text{ kg})}{45 \text{ kg} + 30 \text{ kg}} = 28 \text{ cm}$$

and

$$y_{\text{cg}} = \frac{y_{\text{trunk}} m_{\text{trunk}} + y_{\text{legs}} m_{\text{legs}}}{m_{\text{trunk}} + m_{\text{legs}}}$$

$$= \frac{(50 \text{ cm})(45 \text{ kg}) + (20 \text{ cm})(30 \text{ kg})}{45 \text{ kg} + 30 \text{ kg}} = 38 \text{ cm}$$

**ASSESS** The center-of-gravity position of the entire body, shown in Figure 7.32, is closer to that of the heavier trunk segment than to that of the lighter legs. It also lies along a line connecting the two segment centers of gravity, just as it would for the center of gravity of two point particles. Note also that the gymnast's hands—the pivot point—are below his center of gravity, as we would expect for stability.

**STOP TO THINK 7.5** The balls are connected by very lightweight rods pivoted at the point indicated by a dot. The rod lengths are all equal except for A, which is twice as long. Rank in order, from least to greatest, the magnitudes of the net gravitational torques about the pivots for arrangements A to D.

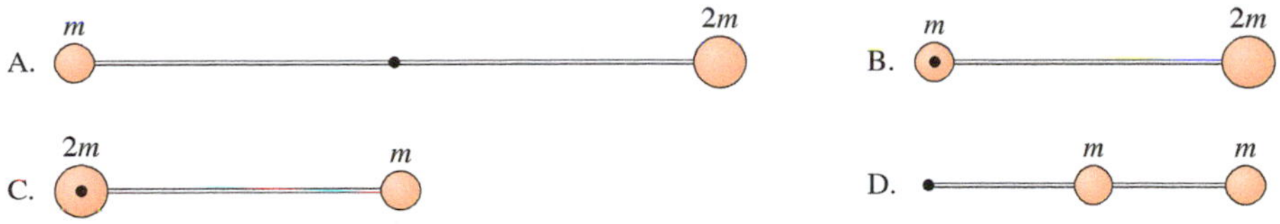


## 7.5 Rotational Dynamics and Moment of Inertia

In Section 7.3 we asked: What do Newton's laws tell us about rotational motion? We can now answer that question: **A torque causes an angular acceleration.** This is the rotational equivalent of our earlier assertion, for motion along a line, that a force causes an acceleration.

To see where this connection between torque and angular acceleration comes from, let's start by examining a *single particle* subject to a torque. FIGURE 7.33 shows a particle of mass $m$ attached to a lightweight, rigid rod of length $r$ that constrains the particle to move in a circle. The particle is subject to two forces. Because it's moving in a circle, there must be a force—here, the tension $\vec{T}$ from the rod—directed toward the center of the circle. As we learned in Chapter 6, this is the force responsible for changing the *direction* of the particle's velocity. The acceleration associated with this change in the particle's velocity is the centripetal acceleration $\vec{a}_c$.

FIGURE 7.33 A tangential force $\vec{F}$ exerts a torque on the particle and causes an angular acceleration.

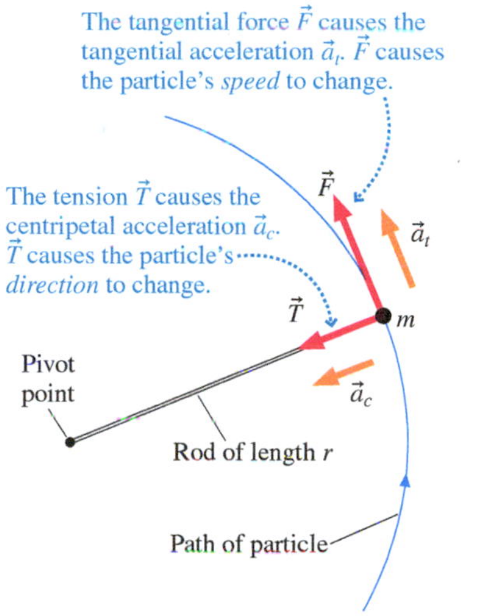


But the particle in Figure 7.33 is also subject to the force $\vec{F}$ that changes the *speed* of the particle. This force causes a tangential acceleration $\vec{a}_t$. Applying Newton's second law in the direction tangent to the circle gives

$$a_t = \frac{F}{m} \tag{7.16}$$

Now the tangential and angular accelerations are related by $a_t = \alpha r$, so we can rewrite Equation 7.16 as $\alpha r = F/m$, or

$$\alpha = \frac{F}{mr} \tag{7.17}$$

We can now connect this angular acceleration to the torque because force $\vec{F}$, which is perpendicular to the radial line, exerts torque

$$\tau = rF$$

With this relationship between $F$ and $\tau$, we can write Equation 7.17 as

$$\alpha = \frac{\tau}{mr^2} \tag{7.18}$$

Equation 7.18 gives a relationship between the torque on a single particle and its angular acceleration. Now all that remains is to expand this idea from a single particle to an extended object.

### Newton's Second Law for Rotation

FIGURE 7.34 shows a rigid body that undergoes rotation about a fixed and unmoving axis. According to the rigid-body model, we can think of the object as consisting of particles with masses $m_1, m_2, m_3, \ldots$ at fixed distances $r_1, r_2, r_3, \ldots$ from the axis.

FIGURE 7.34 The forces on a rigid body exert a torque about the rotation axis.

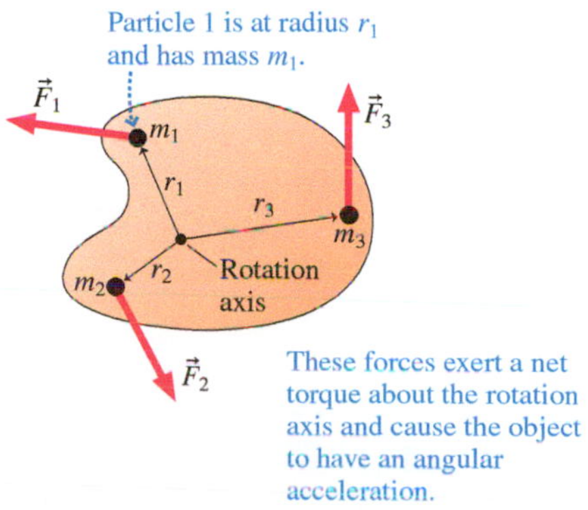

Suppose forces $\vec{F}_1$, $\vec{F}_2$, $\vec{F}_3, \ldots$ act on these particles. These forces exert torques around the rotation axis, so the object will undergo an angular acceleration $\alpha$. Because all the particles that make up the object rotate together, each particle has this *same* angular acceleration $\alpha$. Rearranging Equation 7.18 slightly, we can write the torques on the particles as

$$\tau_1 = m_1 r_1^2 \alpha \qquad \tau_2 = m_2 r_2^2 \alpha \qquad \tau_3 = m_3 r_3^2 \alpha$$

and so on for every particle in the object. If we add up all these torques, the *net* torque on the object is

$$\begin{aligned} \tau_{\text{net}} &= \tau_1 + \tau_2 + \tau_3 + \cdots = m_1 r_1^2 \alpha + m_2 r_2^2 \alpha + m_3 r_3^2 \alpha + \cdots \\ &= \alpha(m_1 r_1^2 + m_2 r_2^2 + m_3 r_3^2 + \cdots) = \alpha \sum m_i r_i^2 \end{aligned} \tag{7.19}$$

In factoring $\alpha$ out of the sum, we're making explicit use of the fact that every particle in a rotating rigid body has the *same* angular acceleration $\alpha$.

The quantity $\sum mr^2$ in Equation 7.19, which is the proportionality constant between angular acceleration and net torque, is called the object's **moment of inertia** $I$:

$$I = m_1 r_1^2 + m_2 r_2^2 + m_3 r_3^2 + \cdots = \sum m_i r_i^2 \tag{7.20}$$

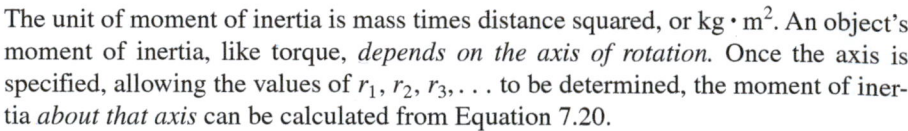

Moment of inertia of a collection of particles

The moment of inertia depends on the mass and also on where the mass is located. The granite ball in the photo has a mass of 8200 kg. It's also physically large, with much of the mass far from the center. For both of these reasons, the moment of inertia is extremely large. The sphere can freely spin on a thin layer of pressurized water. Even though the girl exerts a large torque, the extremely large moment of inertia means that the angular acceleration is very small.

The unit of moment of inertia is mass times distance squared, or kg · m². An object's moment of inertia, like torque, *depends on the axis of rotation.* Once the axis is specified, allowing the values of $r_1, r_2, r_3, \ldots$ to be determined, the moment of inertia *about that axis* can be calculated from Equation 7.20.

NOTE ▸ The word "moment" in "moment of inertia" and "moment arm" has nothing to do with time. It stems from the Latin *momentum,* meaning "motion." ◂

Substituting the moment of inertia into Equation 7.19 puts the final piece of the puzzle into place, giving us the fundamental equation for rigid-body dynamics:

**Newton's second law for rotation** An object that experiences a net torque $\tau_{\text{net}}$ about the axis of rotation undergoes an angular acceleration

$$\alpha = \frac{\tau_{\text{net}}}{I} \tag{7.21}$$

where $I$ is the moment of inertia of the object *about the rotation axis.*

In practice we often write $\tau_{\text{net}} = I\alpha$, but Equation 7.21 better conveys the idea that **a net torque is the cause of angular acceleration.** In the absence of a net torque ($\tau_{\text{net}} = 0$), the object has zero angular acceleration $\alpha$, so it either does not rotate ($\omega = 0$) or rotates with *constant* angular velocity ($\omega$ = constant).

## Interpreting the Moment of Inertia

Before rushing to calculate moments of inertia, let's get a better understanding of its meaning. First, notice that **moment of inertia is the rotational equivalent of mass.** It plays the same role in Equation 7.21 as does mass $m$ in the now-familiar $\vec{a} = \vec{F}_{\text{net}}/m$. Recall that objects with larger mass have a larger *inertia,* meaning that they're harder to accelerate. Similarly, an object with a larger moment of inertia is harder to get rotating: It takes a larger torque to spin up an object that has a larger moment of inertia than an object with a smaller moment of inertia. The fact that "moment of inertia" retains the word "inertia" reminds us of this.

**Video** Walking on a Tightrope

But why does the moment of inertia depend on the distances $r$ from the rotation axis? Think about trying to start a merry-go-round from rest, as shown in FIGURE 7.35. By pushing on the rim of the merry-go-round, you exert a torque on it, and its angular velocity begins to increase. If your friends sit at the rim of the merry-go-round, as in Figure 7.35a, their distances $r$ from the axle are large. Once you've spun up the merry-go-round to a reasonable angular velocity $\omega$, their speeds, given by $v = \omega r$, will be large as well. If your friends sit near the axle, however, as in Figure 7.35b, then $r$ is small and their speeds $v = \omega r$ will be small. In the first case, you've gotten your friends up to a higher speed, and it's no surprise that you will need to push harder to make that happen.

We can express this result using the concept of moment of inertia. In the first case, with large values of $r$, Equation 7.20 says that the moment of inertia will be large. In the second case, with small values of $r$, Equation 7.20 gives a small value of moment of inertia. And the larger moment of inertia means you need to apply a larger torque to produce a certain angular acceleration: You'll need to push harder.

Thus an object's moment of inertia depends not only on the object's mass but also on *how the mass is distributed* around the rotation axis. This is well known to bicycle racers. Every time a cyclist accelerates, she has to "spin up" the wheels and tires. The larger the moment of inertia, the more effort it takes and the smaller her acceleration. For this reason, racers use the lightest possible tires, and they put those tires on wheels that have been designed to keep the mass as close as possible to the center without sacrificing the necessary strength and rigidity.

Synthesis 7.1 connected quantities and equations for linear and circular motion. We can now do the same thing for linear and rotational dynamics.

FIGURE 7.35 Moment of inertia depends on both the mass and how the mass is distributed.

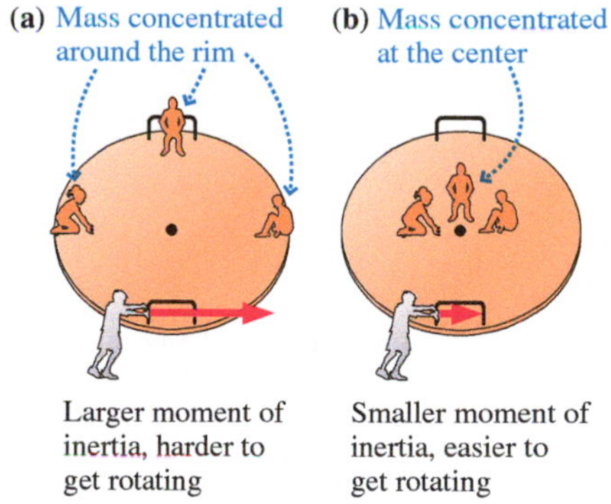


Video Figure 7.35

**SYNTHESIS 7.2 Linear and rotational dynamics**

The variables for linear dynamics have analogs for rotational dynamics. Newton's second law for rotational dynamics is expressed in terms of these variables.

| | Linear dynamics | Rotational dynamics |
|---|---|---|
| **Variables** | Net force (N) → $\vec{F}_{net}$<br>Mass (kg) → $m$<br>Acceleration (m/s$^2$) → $\vec{a}$ | $\tau_{net}$ ← Net torque (N · m)<br>$I$ ← Moment of inertia (kg · m$^2$)<br>$\alpha$ ← Angular acceleration (rad/s$^2$) |
| **Newton's second law** | $\vec{a} = \dfrac{\vec{F}_{net}}{m}$<br>Acceleration is caused by forces.<br>The larger the mass, the smaller the acceleration. | $\alpha = \dfrac{\tau_{net}}{I}$<br>Angular acceleration is caused by torques.<br>The larger the moment of inertia, the smaller the angular acceleration. |

▶ **Novel golf clubs** In recent years, manufacturers have introduced golf putters with heads that have high moments of inertia. When the putter hits the ball, the ball—by Newton's third law—exerts a force on the putter and thus exerts a torque that causes the head of the putter to rotate around the shaft. A large moment of inertia of the head will keep the resulting angular acceleration small, thus reducing unwanted rotation and allowing a truer putt.

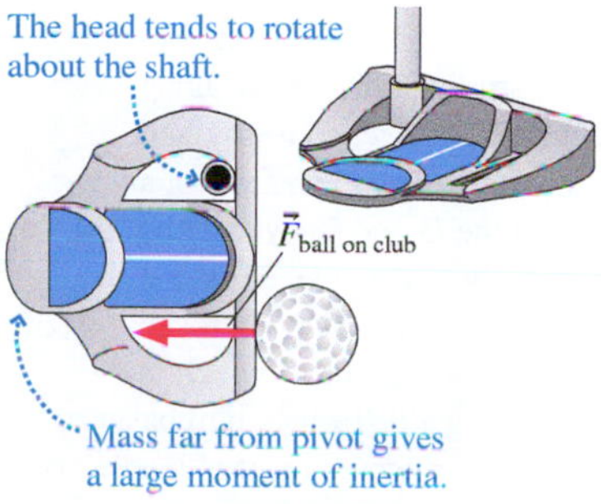


## The Moments of Inertia of Common Shapes

Newton's second law for rotational motion is easy to write, but we can't make use of it without knowing an object's moment of inertia. Unlike mass, we can't measure moment of inertia by putting an object on a scale. And although we can guess that

**TABLE 7.1** Moments of inertia of objects with uniform density and total mass $M$

| Object and axis | Picture | $I$ | Object and axis | Picture | $I$ |
|---|---|---|---|---|---|
| Thin rod (of any cross section), about center | 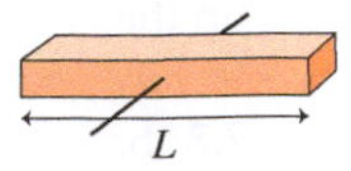 | $\frac{1}{12}ML^2$ | Cylinder or disk, about center | 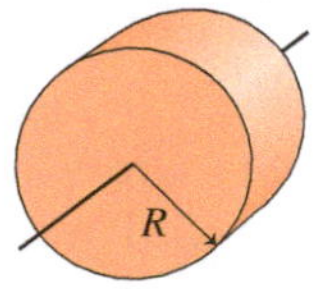 | $\frac{1}{2}MR^2$ |
| Thin rod (of any cross section), about end | 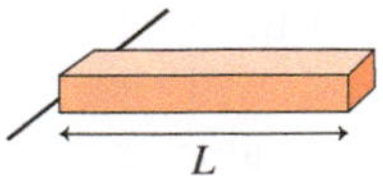 | $\frac{1}{3}ML^2$ | Cylindrical hoop, about center | 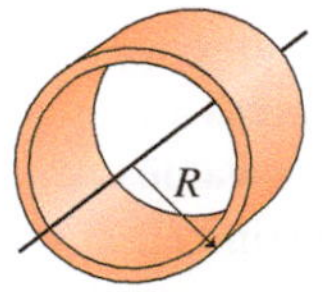 | $MR^2$ |
| Plane or slab, about center | 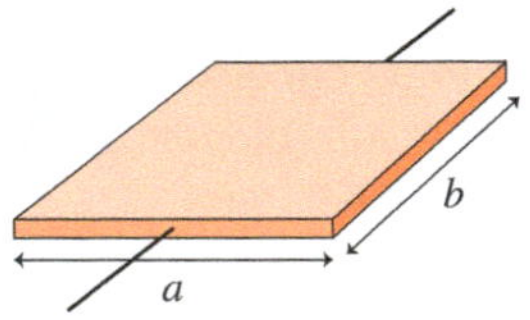 | $\frac{1}{12}Ma^2$ | Solid sphere, about diameter | 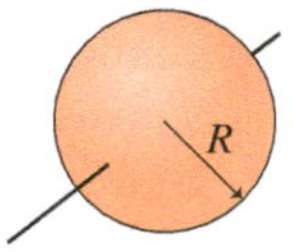 | $\frac{2}{5}MR^2$ |
| Plane or slab, about edge | 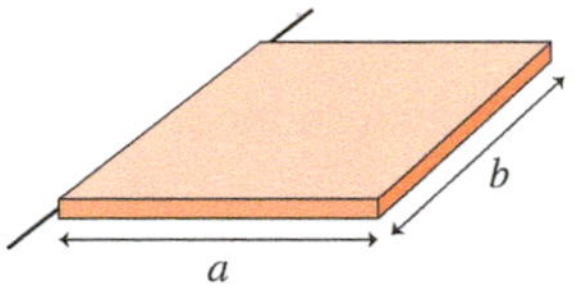 | $\frac{1}{3}Ma^2$ | Spherical shell, about diameter | 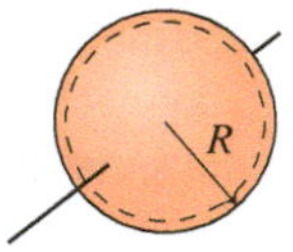 | $\frac{2}{3}MR^2$ |

the center of gravity of a symmetrical object is at the physical center of the object, we can *not* guess the moment of inertia of even a simple object. A short list of common moments of inertia is given in **TABLE 7.1**. We use a capital $M$ for the total mass of an extended object.

We can make some general observations about the moments of inertia in Table 7.1. For instance, the cylindrical hoop is composed of particles that are all the same distance $R$ from the axis. Thus each particle of mass $m$ makes the *same* contribution $mR^2$ to the hoop's moment of inertia. Adding up all these contributions gives

$$I = m_1R^2 + m_2R^2 + m_3R^2 + \cdots = (m_1 + m_2 + m_3 + \cdots)R^2 = MR^2$$

as given in the table. The solid cylinder of the same mass and radius has a *lower* moment of inertia than the hoop because much of the cylinder's mass is nearer its center. In the same way we can see why a slab rotated about its center has a lower moment of inertia than the same slab rotated about its edge: In the latter case, some of the mass is twice as far from the axis as the farthest mass in the former case. Those particles contribute *four times* as much to the moment of inertia, leading to an overall larger moment of inertia for the slab rotated about its edge.

## EXAMPLE 7.15 The point of the pole

The photo shows a high-wire walker striding confidently along a steel cable stretched nearly half a kilometer above the bottom of a gorge in the Grand Canyon. If he tips slightly to the side, a gravitational torque will start to rotate his body even farther—which is clearly not good. To help him balance, he carries a long, heavy pole. This increases the moment of inertia, which results in a smaller angular acceleration, giving him more time to recover his balance.

Model the man's body as a 1.8-m-long, 88 kg, uniform rod. The pole is 9.1 m long, with a mass of 20 kg. What is the moment of inertia of his body? Of the pole?

***STRATEGIZE*** We will model the man's body as a rod pivoted about the end. We will treat the pole as a rod as well, with a pivot at the center, where it is supported. We'll use the equations in Table 7.1 to find the moments of inertia of these shapes. The pole is intended to provide stability, so we expect the pole's moment of inertia to be significant compared to that of the man's body.

**PREPARE** We model both the man's body and the pole as rods. The difference is the location of the pivot. The man's body pivots about the end of the rod, his feet, so the moment of inertia is $\frac{1}{3}ML^2$, where $M$ is his mass and $L$ is his height. The pole pivots about the center, so the moment of inertia is $\frac{1}{12}ML^2$, where $M$ is the mass of the pole and $L$ its length.

**SOLVE** The moment of inertia of the man's body is

$$I_{\text{man}} = \tfrac{1}{3}ML^2 = \tfrac{1}{3}(88\text{ kg})(1.8\text{ m})^2 = 95\text{ kg}\cdot\text{m}^2$$

The moment of inertia of the pole is

$$I_{\text{pole}} = \tfrac{1}{12}ML^2 = \tfrac{1}{12}(20\text{ kg})(9.1\text{ m})^2 = 140\text{ kg}\cdot\text{m}^2$$

**ASSESS** Although the pole's mass is less than one-quarter of the man's mass, its moment of inertia is larger. The formula for moment of inertia includes the square of a length, so the pole's greater length more than makes up for its smaller mass. Holding the pole significantly increases the man's moment of inertia; the moment of inertia of the man holding the pole is 2.5 times that of the man alone. This results in a significant decrease in angular acceleration, providing much greater stability. The pole is flexible, and this provides an additional benefit. The downward sag of the pole lowers the position of the overall center of gravity, thus decreasing gravitational torque; this improves stability as well.

**STOP TO THINK 7.6** Four very lightweight disks of equal radii each have three identical heavy marbles glued to them as shown. Rank in order, from largest to smallest, the moments of inertia of the disks about the indicated axis.

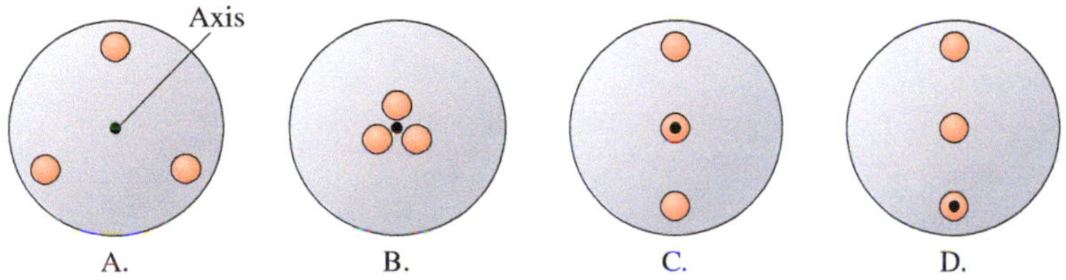


## 7.6 Using Newton's Second Law for Rotation

In this section we'll look at several examples of rotational dynamics for rigid bodies that rotate about a *fixed axis.* The restriction to a fixed axis avoids complications that arise for an object undergoing a combination of rotational and translational motion.

**PROBLEM-SOLVING APPROACH 7.1** **Rotational dynamics problems**

We can use a problem-solving approach for rotational dynamics that is very similar to the approach for linear dynamics in Chapter 5.

**STRATEGIZE** Model the object as a simple shape. What is rotating, and what point does it rotate around? Next, identify the forces that cause the rotation, and draw a visual overview to clarify the situation.

**PREPARE** Develop the visual overview by defining coordinates and symbols and listing known information. Specifically:

- Identify the axis about which the object rotates.
- Identify the forces and determine their distances from the axis.
- Calculate the torques caused by the forces, and find the signs of the torques.

**SOLVE** The mathematical representation is based on Newton's second law for rotational motion:

$$\tau_{\text{net}} = I\alpha \qquad \text{or} \qquad \alpha = \frac{\tau_{\text{net}}}{I}$$

- Find the moment of inertia either by direct calculation using Equation 7.20 or from Table 7.1 for common shapes of objects.
- Use rotational kinematics to find angular positions and velocities.

**ASSESS** Check that your result has the correct units, is reasonable, and answers the question.

Exercise 27

## EXAMPLE 7.16 Angular acceleration of a falling pole

In the caber toss, a contest of strength and skill that is part of Scottish games, contestants toss a heavy uniform pole, landing it on its end. A 5.9-m-tall pole with a mass of 79 kg has just landed on its end. It is tipped by 25° from the vertical and is starting to rotate about the end that touches the ground. Determine the angular acceleration. If we assume that the angular acceleration stays at this same value (an approximation, but it will give us a good sense of scale for this problem), how long will it take the pole to rotate by 5°?

**STRATEGIZE** We will follow the steps in Problem-Solving Approach 7.1. We'll start by modeling the pole as a uniform thin rod that rotates about one end under the influence of the gravitational torque. Our visual overview in FIGURE 7.36 notes the geometry, the forces that act, and where the forces act.

**PREPARE** Two forces are acting on the pole: the pole's weight $\vec{w}$, which acts at the center of gravity, and the force of the ground on the pole (not shown). This second force exerts no torque because it acts at the axis of rotation. The torque on the pole is thus due only to gravity. From the figure we see that this torque tends to rotate the pole in a counterclockwise direction, so the torque is positive. We can use this force to find the angular acceleration, which we can use to find the time it takes for the pole to rotate by 5°, which is 0.087 rad.

FIGURE 7.36 A falling pole undergoes an angular acceleration due to a gravitational torque.

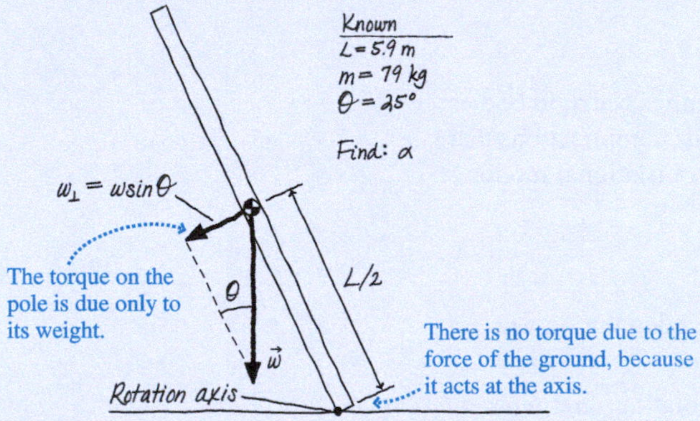


**SOLVE** The pole's center of gravity is at its center, a distance $L/2$ from the axis. You can see from the figure that the perpendicular component of $\vec{w}$ is $w_\perp = w\sin\theta$. Thus the torque due to gravity is

$$\tau_{\text{net}} = \left(\frac{L}{2}\right)w_\perp = \left(\frac{L}{2}\right)w\sin\theta = \frac{mgL}{2}\sin\theta$$

From Table 7.1, the moment of inertia of a thin rod rotated about its end is $I = \frac{1}{3}mL^2$. Thus, from Newton's second law for rotational motion, the angular acceleration is

$$\alpha = \frac{\tau_{\text{net}}}{I} = \frac{\frac{1}{2}mgL\sin\theta}{\frac{1}{3}mL^2} = \frac{3g\sin\theta}{2L}$$

$$= \frac{3(9.8\text{ m/s}^2)\sin 25^\circ}{2(5.9\text{ m})} = 1.1\text{ rad/s}^2$$

Once the pole starts to rotate, it will speed up. If we assume that the angular acceleration is constant, we can use the final equation in Synthesis 7.1 to calculate the time it takes for the pole to rotate by 0.087 rad. This is $\Delta\theta$; the pole starts at rest, $\omega_i = 0$, and we find

$$0.087\text{ rad} = (0)\Delta t + \tfrac{1}{2}\alpha(\Delta t)^2$$

$$\Delta t = 0.40\text{ s}$$

**ASSESS** The final result for the angular acceleration did not depend on the mass, as we might expect given the analogy with free-fall problems. And the final value for the angular acceleration is quite modest; the motion is fairly gentle. For this angular acceleration, it takes just under half a second to rotate by 5°. This is reasonable: You can see that the angular acceleration is inversely proportional to the length of the pole, and it's a long pole. The modest value of angular acceleration is fortunate—the caber is pretty heavy, and folks need some time to get out of the way when it topples!

## CONCEPTUAL EXAMPLE 7.17 Balancing a meter stick

You can easily balance a meter stick or a baseball bat on your palm. (Try it!) But it's nearly impossible to balance a pencil this way. Why?

**REASON** Suppose you've managed to balance a vertical stick on your palm, but then it starts to fall. You'll need to quickly adjust your hand to bring the stick back into balance. As Example 7.16 showed, the angular acceleration $\alpha$ of a thin rod is *inversely proportional* to $L$. Thus a long object like a meter stick topples much more slowly than a short one like a pencil. Your reaction time is fast enough to correct for a slowly tipping meter stick but not for a rapidly tipping pencil.

**ASSESS** If we double the length of a rod, its mass doubles and its center of gravity is twice as high, so the gravitational torque $\tau$ on it is four times as much. But because a rod's moment of inertia is $I = \frac{1}{3}ML^2$, the longer rod's moment of inertia will be *eight* times greater, so the angular acceleration will be only half as large. This gives you more time to move to put the support back under the center of gravity.

**EXAMPLE 7.18 Spinning up a flywheel**

When the electricity goes out on your campus, a backup source powers the servers that handle communications and the Internet. Rather than use batteries, many installations use a flywheel, a heavy rotating disk that spins, very rapidly, with nearly zero friction. An electric motor spins up the flywheel, which continues to spin with very little energy input. When the system needs to provide power, the flywheel's motion is used to turn an electric generator. A typical system has a 540 kg cylinder with a radius of 0.30 m. A small electric motor provides a constant 2.8 N · m torque to spin up the cylinder. If the cylinder starts at rest, how long does it take to reach its final spin rate of 13,000 rpm?

**STRATEGIZE** This is an angular dynamics problem. We will find the moment of inertia of the cylinder and then use the given torque to find the angular acceleration. We'll use this to find the time required to reach the final spin rate. Given the size and mass of the cylinder (half the mass of a small car), the relatively modest torque (one-sixth of the wrench torque in Example 7.10), and the very rapid spin (more than 200 rotations each second), we expect the motor to take quite a while to spin up the flywheel.

**PREPARE** We are given the spin rate in rpm; to do calculations, we need the angular speed in rad/s. The frequency in rev/s is

$$f = \left(13{,}000\ \frac{\text{rev}}{\text{min}}\right)\left(\frac{1\ \text{min}}{60\ \text{s}}\right) = 217\ \text{rev/s}$$

so the angular speed in rad/s is

$$\omega = (2\pi\ \text{rad})(217\ \text{rev/s}) = 1360\ \text{rad/s}$$

We'll keep an extra significant figure for the intermediate stages in the calculation.

**SOLVE** The flywheel is a cylinder rotated around the center. The moment of inertia is

$$I = \tfrac{1}{2}MR^2 = \tfrac{1}{2}(540\ \text{kg})(0.30\ \text{m})^2 = 24.3\ \text{kg}\cdot\text{m}^2$$

The angular acceleration is

$$\alpha = \frac{\tau}{I} = \frac{2.8\ \text{N}\cdot\text{m}}{24.3\ \text{kg}\cdot\text{m}^2} = 0.115\ \text{rad/s}^2$$

We can rearrange the definition of angular acceleration, Equation 7.7, to find the time to increase the angular speed to 1360 rad/s:

$$\Delta t = \frac{\Delta\omega}{\alpha} = \frac{1360\ \text{rad/s}}{0.115\ \text{rad/s}^2} = 11{,}800\ \text{s} = 3.3\ \text{hours}$$

**ASSESS** Given the modest torque, large flywheel, and rapid spin, we expected a long time, so our answer makes sense. The numbers given in the problem are real numbers for an actual system. The rated time for the initial spin-up is quite close to the calculated value.

## Constraints Due to Ropes and Pulleys

FIGURE 7.37 shows a rope passing over a pulley and connected to an object in linear motion. If the pulley turns *without the rope slipping on it,* then the rope's speed $v_{\text{rope}}$ must exactly match the speed of the rim of the pulley, which is $v_{\text{rim}} = \omega R$. If the pulley has an angular acceleration, the rope's acceleration $a_{\text{rope}}$ must match the *tangential* acceleration of the rim of the pulley, $a_t = \alpha R$.

The object attached to the other end of the rope has the same speed and acceleration as the rope. Consequently, the object must obey the constraints

$$v_{\text{obj}} = \omega R$$
$$a_{\text{obj}} = \alpha R \qquad (7.22)$$

Motion constraints for an object connected to a pulley of radius $R$ by a nonslipping rope

FIGURE 7.37 The rope's motion must match the motion of the rim of the pulley.

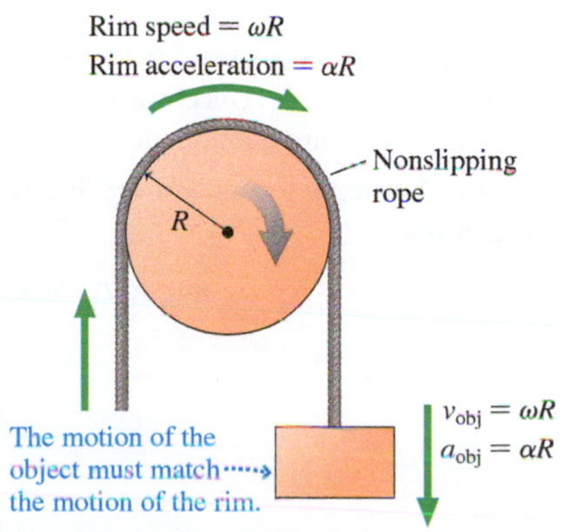

These constraints are similar to the acceleration constraints introduced in Chapter 5 for two objects connected by a string or rope.

**NOTE** ▶ The constraints are given as magnitudes. Specific problems will require you to specify signs that depend on the direction of motion and on the choice of coordinate system. ◀

## EXAMPLE 7.19 Time for a bucket to fall

Josh has just raised a 2.5 kg bucket of water using a well's winch when he accidentally lets go of the handle. The winch consists of a rope wrapped around a 3.0 kg, 4.0-cm-diameter cylinder, which rotates on an axle through the center. The bucket is released from rest 4.0 m above the water level of the well. How long does it take to reach the water?

**STRATEGIZE** There's a downward force on the bucket—the weight force. There is an upward force on the bucket, too—the rope tension. The tension force also pulls on the cylinder, so as the bucket falls, the cylinder speeds up. We can assume that the rope is massless and does not slip. No slipping means that the downward acceleration of the bucket matches the tangential acceleration of the rim of the cylinder, so we'll treat the fall of the bucket and the spin of the cylinder as two connected problems. FIGURE 7.38a gives a visual overview of the system, FIGURE 7.38b shows the free-body diagram for the bucket, and FIGURE 7.38c shows the tension force acting on the cylinder.

FIGURE 7.38 Visual overview of a falling bucket.

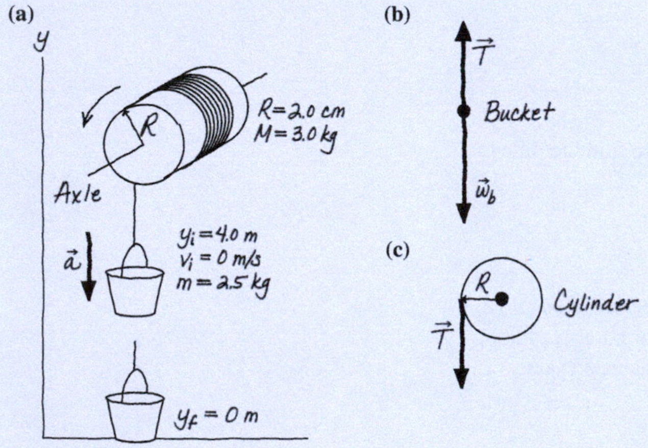


**PREPARE** The rope tension exerts an upward force on the bucket; the weight force acts downward. The rope tension exerts a downward force on the outer edge of the cylinder. The rope is massless, so these two tension forces have equal magnitudes, which we'll call $T$. Once we've analyzed the motion of the bucket, we can turn our attention to the cylinder.

**SOLVE** Newton's second law applied to the linear motion of the bucket is

$$ma_y = T - mg$$

where, as usual, the $y$-axis points upward. What about the cylinder? There are forces that act at the axle, but these forces don't exert a torque because they act at the rotation axis. The only torque comes from the rope tension. The moment arm for the tension is $r_\perp = R$, and the torque is positive because the rope turns the cylinder counterclockwise. Thus $\tau_{rope} = TR$, and Newton's second law for the rotational motion is

$$\alpha = \frac{\tau_{net}}{I} = \frac{TR}{\frac{1}{2}MR^2} = \frac{2T}{MR}$$

The moment of inertia of a cylinder rotating about a center axis was taken from Table 7.1.

The last piece of information we need is the constraint due to the fact that the rope doesn't slip. Equation 7.22 relates only the magnitudes of the linear and angular accelerations, but in this problem $\alpha$ is positive (counterclockwise acceleration), while $a_y$ is negative (downward acceleration). Hence

$$a_y = -\alpha R$$

Using $\alpha$ from the cylinder's equation in the constraint, we find

$$a_y = -\alpha R = -\frac{2T}{MR}R = -\frac{2T}{M}$$

Thus the tension is $T = -\frac{1}{2}Ma_y$. If we use this value of the tension in the bucket's equation, we can solve for the acceleration:

$$ma_y = -\tfrac{1}{2}Ma_y - mg$$
$$a_y = -\frac{g}{(1 + M/2m)} = -6.13 \text{ m/s}^2$$

The time to fall through $\Delta y = y_f - y_i = -4.0$ m is found from kinematics:

$$\Delta y = \tfrac{1}{2}a_y(\Delta t)^2$$
$$\Delta t = \sqrt{\frac{2\Delta y}{a_y}} = \sqrt{\frac{2(-4.0 \text{ m})}{-6.13 \text{ m/s}^2}} = 1.1 \text{ s}$$

**ASSESS** The expression for the acceleration gives $a_y = -g$ if $M = 0$. This makes sense because the bucket would be in free fall if there were no cylinder. When the cylinder has mass, the downward force of gravity on the bucket has to accelerate the bucket and spin the cylinder. Consequently, the acceleration is reduced and the bucket takes longer to fall.

## 7.7 Rolling Motion

Rolling is a *combination motion* in which an object rotates about an axis that is moving along a straight-line trajectory. For example, FIGURE 7.39 is a time-exposure photo of a rolling wheel with one lightbulb on the axis and a second lightbulb at the edge. The axis light moves straight ahead, but the edge light follows a curve called a *cycloid.*

FIGURE 7.39 The trajectories of the center of a wheel and of a point on the rim are seen in a time-exposure photograph.

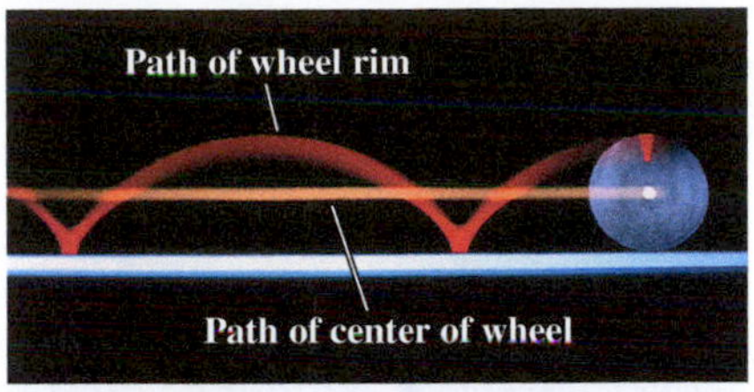


FIGURE 7.40 An object rolling through one revolution.

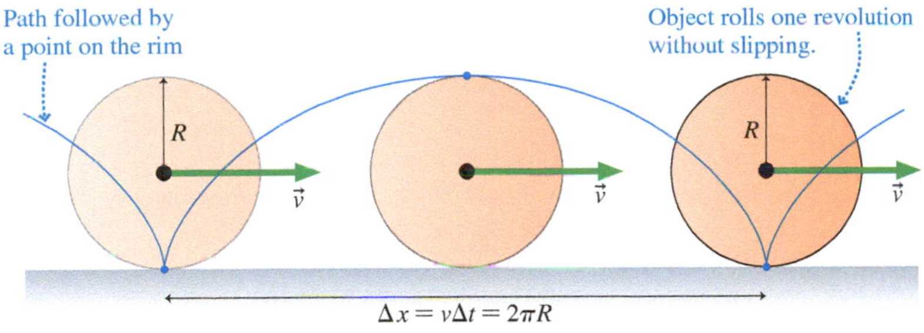


To understand rolling motion, consider FIGURE 7.40, which shows a round object—a wheel or a sphere—that rolls forward, *without slipping,* exactly one revolution. The point initially at the bottom follows the blue curve to the top and then back to the bottom. The overall position of the object is measured by the position $x$ of the object's center. Because the object doesn't slip, in one revolution the center moves forward exactly one circumference, so that $\Delta x = 2\pi R$. The time for the object to turn one revolution is its period $T$, so we can compute the speed of the object's center as

$$v = \frac{\Delta x}{T} = \frac{2\pi R}{T} \tag{7.23}$$

But $2\pi/T$ is the angular velocity $\omega$, as you learned in Chapter 6, which leads to

$$v = \omega R \tag{7.24}$$

Equation 7.24 is the **rolling constraint,** the basic link between translation and rotation for objects that roll without slipping.

We can find the velocity for any point on a rolling object by adding the velocity of that point when the object is in pure translation, without rolling, to the velocity of the point when the object is in pure rotation, without translating. FIGURE 7.41 shows how the velocity vectors at the top, center, and bottom of a rotating wheel are found in this way.

FIGURE 7.41 Rolling is a combination of translation and rotation.

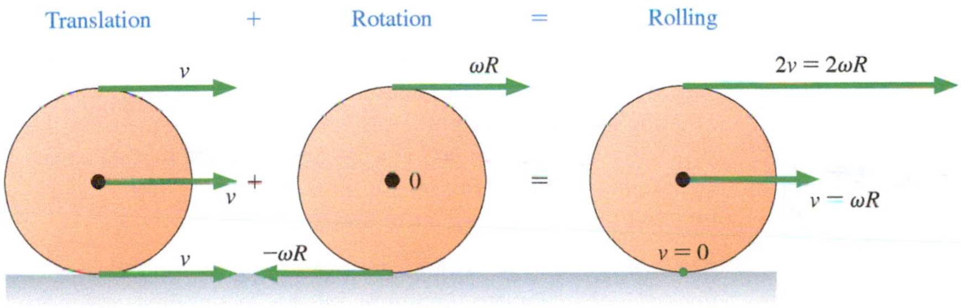


Thus the point at the top of the wheel has a forward speed of $v$ due to its translational motion plus a forward speed of $\omega R = v$ due to its rotational motion. The speed

**Ancient movers** The great stone *moai* of Easter Island were moved as far as 16 km from a quarry to their final positions. Archeologists believe that one possible method of moving these 14 ton statues was to place them on rollers. One disadvantage of this method is that the statues, placed on top of the rollers, move twice as fast as the rollers themselves. Thus rollers are continuously left behind and have to be carried back to the front and reinserted. Sadly, the indiscriminate cutting of trees for moving *moai* may have hastened the demise of this island civilization.

of a point at the top of a wheel is then $2v = 2\omega R$, or *twice* the speed of its center. On the other hand, the point at the bottom of the wheel, where it touches the ground, still has a forward speed of $v$ due to its translational motion. But its velocity due to rotation points *backward* with a magnitude of $\omega R = v$. Adding these, we find that the velocity of this lowest point is *zero*. In other words, **the point on the bottom of a rolling object is instantaneously at rest.**

Although this seems surprising, it is really what we mean by "rolling without slipping." If the bottom point had a velocity, it would be moving horizontally relative to the surface. In other words, it would be slipping or sliding across the surface. To roll without slipping, the bottom point, the point touching the surface, must be at rest.

## EXAMPLE 7.20 Rotating your tires

You are driving down the highway at 65 mph, which is 29 m/s. Your tires have a radius of 0.30 m.

a. How many times per second does each tire rotate?
b. What is the speed of a point at the top of a tire, relative to the ground?

***STRATEGIZE*** The questions concern your tires, which are in rolling motion. You've certainly noticed the tires on nearby cars as you drive down the highway; they are spinning at a pretty good clip, many times per second, so we will expect a large number for the tire's rotation frequency, the number of times per second the wheel goes around.

***PREPARE*** We can use Equation 7.24 to find the angular speed of the wheel given the speed of the car. If we know this, we can find the frequency of the rotation. Figure 7.41 shows us how to find the speed of a point at the top of a tire.

***SOLVE***

a. The car is moving at 29 m/s, so a rearrangement of Equation 7.24 gives the angular speed of the tire as

$$\omega = \frac{v}{R} = \frac{29\text{ m/s}}{0.30\text{ m}} = 97\text{ rad/s}$$

The frequency is

$$f = \frac{\omega}{2\pi\text{ rad}} = \frac{97\text{ rad/s}}{2\pi\text{ rad}} = 15\text{ rev/s}$$

b. The velocity of a point on the top of the tire is the sum of the velocity of the car and the velocity of the rotational motion of the wheel. Figure 7.41 shows us that this is twice the speed of the car, or 58 m/s.

***ASSESS*** The tire is spinning 15 times per second, a large number, as we expected. If you drive for an hour at highway speeds, your tires rotate more than 50,000 times!

**STOP TO THINK 7.7** A wheel rolls without slipping. Which is the correct velocity vector for point P on the wheel?

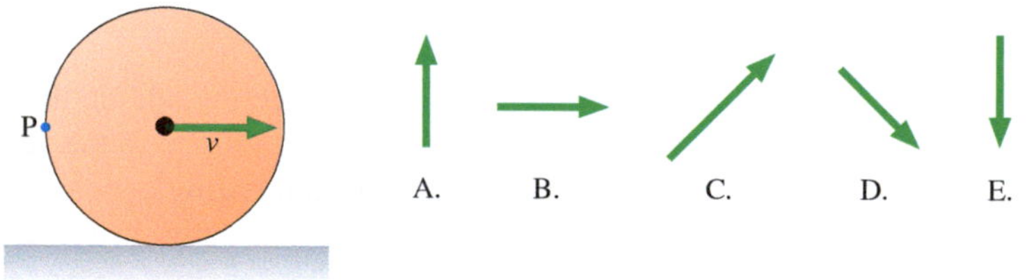


## INTEGRATED EXAMPLE 7.21 Spinning a gyroscope

A gyroscope is a top-like toy consisting of a heavy ring attached by light spokes to a central axle. The axle and ring are free to turn on bearings. To get the gyroscope spinning, a 30-cm-long string is wrapped around the 2.0-mm-diameter axle, then pulled with a constant force of 5.0 N. If the ring's diameter is 5.0 cm and its mass is 30 g, at what rate is it spinning, in rpm, once the string is completely unwound?

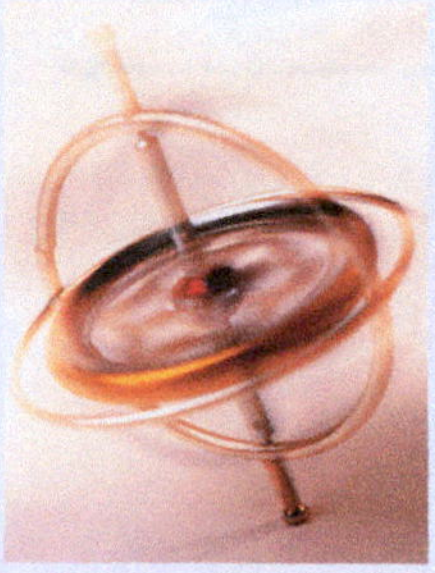

***STRATEGIZE*** This is a question about the ring's final angular velocity—how fast it is spinning when the string is unwound. The ring's angular acceleration is caused by the torque from the string. To find this angular velocity, we'll find the torque, calculate the moment of inertia of the ring, find the angular acceleration, and then use kinematics to find the final value.

***PREPARE*** FIGURE 7.42 is a visual overview of the problem that illustrates the geometry and provides the known information in the correct SI units.

FIGURE 7.42 Visual overview of a gyroscope being spun.

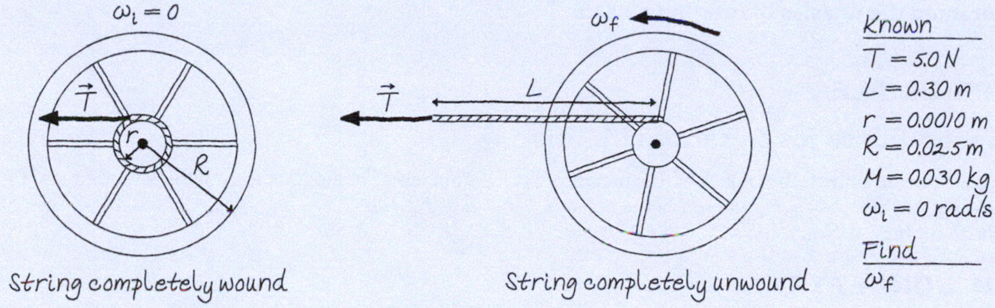


The string is light, so the tension in the string has the same magnitude as the force that pulls on the string, 5.0 N. Because the ring is heavy compared to the spokes and the axle, we model it as a cylindrical hoop, taking its moment of inertia from Table 7.1 to be $I = MR^2$. The radius $R$ is half the 5.0 cm ring diameter. The tension force acts at a distance $r$ from the axle, where $r$ is half the 2.0 mm axle diameter. The ring starts at rest, so the initial angular velocity is $\omega_i = 0$ rad/s. We are looking for the ring's final angular velocity, $\omega_f$.

***SOLVE*** The torque on the ring is due to the tension in the string. Because the string—and the line of action of the tension—is tangent to the axle, the moment arm of the tension force is the radius $r$ of the axle. Thus $\tau = r_\perp T = rT$. Now we can apply Newton's second law for rotational motion, Equation 7.21, to find the angular acceleration:

$$\alpha = \frac{\tau_{net}}{I} = \frac{rT}{MR^2} = \frac{(0.0010\text{ m})(5.0\text{ N})}{(0.030\text{ kg})(0.025\text{ m})^2} = 267\text{ rad/s}^2$$

We next use constant-angular-acceleration kinematics to find the final angular velocity. For the equation $\Delta\theta = \omega_i \Delta t + \frac{1}{2}\alpha \Delta t^2$ in Synthesis 7.1, we know $\alpha$ and $\omega_i$, and we should be able to find $\Delta\theta$ from the length of string unwound, but we don't know $\Delta t$. For the equation $\Delta\omega = \omega_f - \omega_i = \alpha \Delta t$, we know $\alpha$ and $\omega_i$, and $\omega_f$ is what we want to find, but again we don't know $\Delta t$. To find an equation that doesn't contain $\Delta t$, we first write

$$\Delta t = \frac{\omega_f - \omega_i}{\alpha}$$

from the second kinematic equation. Inserting this value for $\Delta t$ into the first equation gives

$$\Delta\theta = \omega_i \frac{\omega_f - \omega_i}{\alpha} + \frac{1}{2}\alpha\left(\frac{\omega_f - \omega_i}{\alpha}\right)^2$$

which can be simplified to

$$\omega_f^2 = \omega_i^2 + 2\alpha\,\Delta\theta$$

FIGURE 7.43 Relating the angle turned to the length of string unwound.

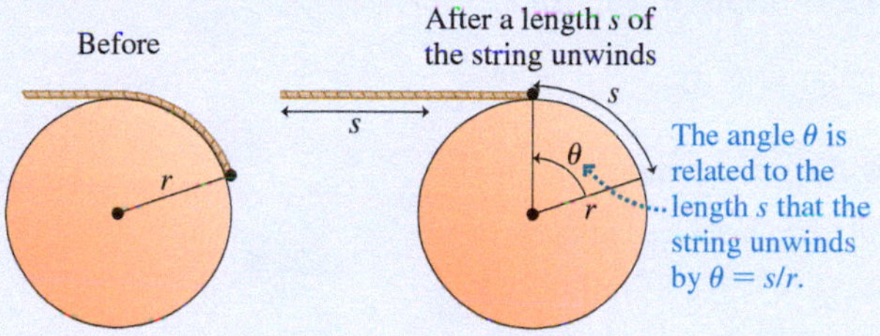


This equation, which is the rotational analog of the linear motion Equation 2.13, will allow us to find $\omega_f$ once $\Delta\theta$ is known.

FIGURE 7.43 shows how to find $\Delta\theta$. As a segment of string of length $s$ unwinds, the axle turns through an angle (based on the definition of radian measure) $\theta = s/r$. Thus as the whole string, of length $L$, unwinds, the axle (and the ring) turns through an angular displacement

$$\Delta\theta = \frac{L}{r} = \frac{0.30\text{ m}}{0.0010\text{ m}} = 300\text{ rad}$$

Now we can use our kinematic equation to find that

$$\omega_f^2 = \omega_i^2 + 2\alpha\,\Delta\theta = (0\text{ rad/s})^2 + 2(267\text{ rad/s}^2)(300\text{ rad})$$
$$= 160{,}000\ (\text{rad/s})^2$$

from which we find that $\omega_f = 400$ rad/s. Converting rad/s to rpm, we find that the gyroscope ring is spinning at

$$400\text{ rad/s} = \left(\frac{400\text{ rad}}{\text{s}}\right)\left(\frac{60\text{ s}}{1\text{ min}}\right)\left(\frac{1\text{ rev}}{2\pi\text{ rad}}\right) = 3800\text{ rpm}$$

***ASSESS*** This is fast, about the same as your car engine when it's on the highway, but if you've ever played with a gyroscope or a string-wound top, you know you can really get it spinning fast.

# SUMMARY

**GOAL** To understand the physics of rotating objects.

## GENERAL PRINCIPLES

### Newton's Second Law for Rotational Motion

If a net torque $\tau_{\text{net}}$ acts on an object, the object will experience an angular acceleration given by $\alpha = \tau_{\text{net}}/I$, where $I$ is the object's moment of inertia about the rotation axis.

This law is analogous to Newton's second law for linear motion, $\vec{a} = \vec{F}_{\text{net}}/m$.

## IMPORTANT CONCEPTS

### Describing circular motion

We define new variables for circular motion. By convention, counterclockwise is positive.

**Angular displacement:** $\Delta\theta = \theta_f - \theta_i$

**Angular velocity:** $\omega = \dfrac{\Delta\theta}{\Delta t}$

**Angular acceleration:** $\alpha = \dfrac{\Delta\omega}{\Delta t}$

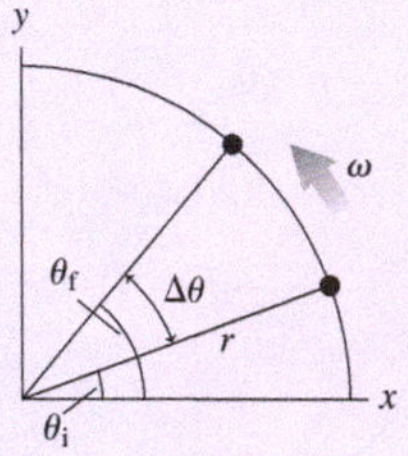


Angles are measured in radians:

$$1 \text{ rev} = 360° = 2\pi \text{ rad}$$

The angular velocity depends on the frequency and period:

$$\omega = \frac{2\pi}{T} = 2\pi f$$

### Torque

A force causes an object to undergo a linear acceleration; a torque causes an object to undergo an angular acceleration.

There are two interpretations of torque:

Interpretation 1: $\tau = rF_\perp$ Interpretation 2: $\tau = r_\perp F$

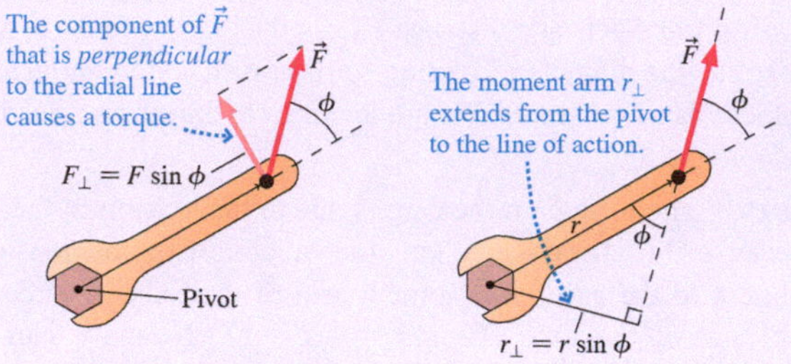


Both interpretations give the same expression for the magnitude of the torque: $\tau = rF \sin\phi$

### Relating linear and circular motion quantities

Linear and angular speeds are related by: $v = \omega r$

If the particle's speed is increasing, it will also have a tangential acceleration $\vec{a}_t$ directed tangent to the circle and an angular acceleration $\alpha$.

Angular and tangential accelerations are related by: $a_t = \alpha r$

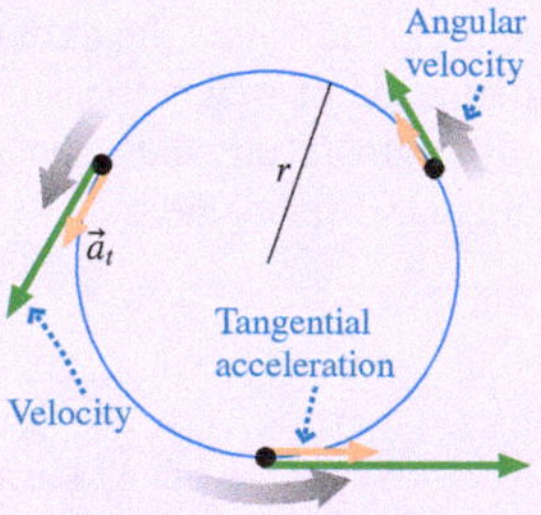


The **moment of inertia** is the rotational equivalent of mass. For an object made up of particles of masses $m_1, m_2, \ldots$ at distances $r_1, r_2, \ldots$ from the axis, the moment of inertia is

$$I = m_1 r_1^2 + m_2 r_2^2 + m_3 r_3^2 + \cdots$$

### Center of gravity

The **center of gravity** of an object is the point at which gravity can be considered to be acting.

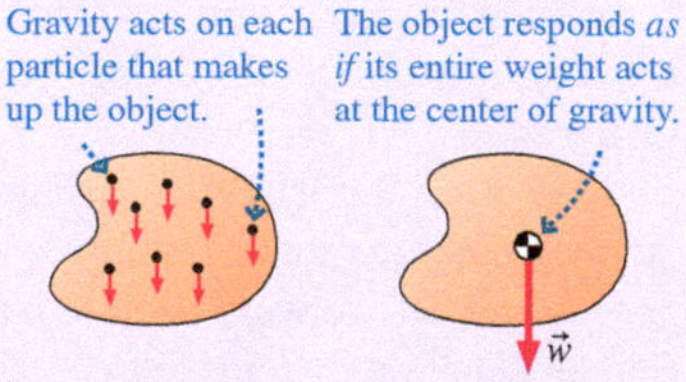

Gravity acts on each particle that makes up the object.
The object responds as if its entire weight acts at the center of gravity.
w

The **position of the center of gravity** depends on the distance $x_1, x_2, \ldots$ of each particle of mass $m_1, m_2, \ldots$ from the origin:

$$x_{\text{cg}} = \frac{x_1 m_1 + x_2 m_2 + x_3 m_3 + \cdots}{m_1 + m_2 + m_3 + \cdots}$$

## APPLICATIONS

### Moments of inertia of common shapes

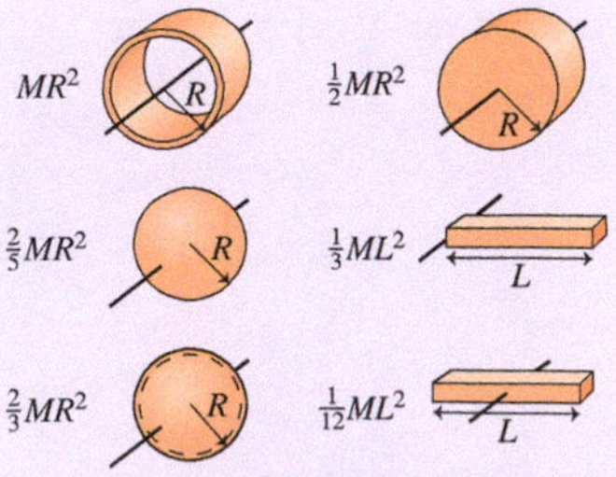


### Rotation about a fixed axis

When a net torque is applied to an object that rotates about a fixed axis, the object will undergo an **angular acceleration** given by

$$\alpha = \frac{\tau_{\text{net}}}{I}$$

If a rope unwinds from a pulley of radius $R$, the linear motion of an object tied to the rope is related to the angular motion of the pulley by

$$a_{\text{obj}} = \alpha R \qquad v_{\text{obj}} = \omega R$$

### Rolling motion

For an object that rolls without slipping,

$$v = \omega R$$

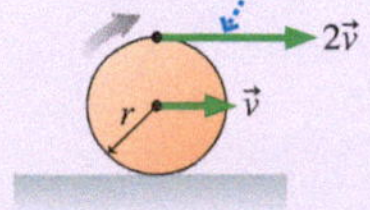

The velocity of a point at the top of the object is twice that of the center.
2v
r
v

## Learning Objectives After studying this chapter, you should be able to:

- Calculate angular velocity and interpret motion graphs for rotational motion. *Conceptual Question 7.2; Problems 7.3, 7.4, 7.5, 7.6, 7.9*
- Calculate angular acceleration and tangential acceleration. *Conceptual Question 7.10; Problems 7.13, 7.14, 7.15, 7.16, 7.17*
- Calculate the torque exerted on an extended object. *Conceptual Questions 7.4, 7.7; Problems 7.21, 7.22, 7.23, 7.24, 7.26*
- Determine an object's center of gravity and gravitational torque. *Conceptual Question 7.11; Problems 7.31, 7.33, 7.35, 7.37, 7.38*
- Calculate an object's moment of inertia. *Conceptual Questions 7.14, 7.18; Problems 7.39, 7.40, 7.42, 7.43, 7.44*
- Use Newton's second law for rotational motion to solve problems about rotational dynamics. *Conceptual Questions 7.12, 7.13; Problems 7.45, 7.47, 7.48, 7.49, 7.53*
- Analyze rolling motion. *Conceptual Question 7.17; Problems 7.57, 7.58, 7.59, 7.60*

## STOP TO THINK ANSWERS

**Chapter Preview Stop to Think: B.** The speed is proportional to the frequency. Doubling the frequency means doubling the speed.

**Stop to Think 7.1: A.** Because $5\pi/2$ rad $= 2\pi$ rad $+ \pi/2$ rad, the particle's position is one complete revolution ($2\pi$ rad) plus an extra $\pi/2$ rad. This extra $\pi/2$ rad puts the particle at position A.

**Stop to Think 7.2: a. constant (but not zero), b. constant (but not zero), c. constant (but not zero), d. zero, e. zero.** The angular velocity $\omega$ is constant. Thus the magnitudes of the velocity $v = \omega r$ and the centripetal acceleration $a_c = \omega^2 r$ are constant. This also means that the ball's angular acceleration $\alpha$ and tangential acceleration $a_t = \alpha r$ are both zero.

**Stop to Think 7.3: E.** Forces D and C act at or in line with the pivot and provide no torque. Of the others, only E tries to rotate the wheel counterclockwise, so it is the only choice that gives a positive nonzero torque.

**Stop to Think 7.4: A.** The force acting at the axis exerts no torque. Thus the third force needs to exert an equal but opposite torque to that exerted by the force acting at the rim. Force A, which has twice the magnitude but acts at half the distance from the axis, does so.

**Stop to Think 7.5: $\tau_B > \tau_D > \tau_A = \tau_C$.** The torques are $\tau_B = 2mgL$, $\tau_D = \frac{3}{2}mgL$, and $\tau_A = \tau_C = mgL$, where $L$ is the length of the rod in B.

**Stop to Think 7.6: $I_D > I_A > I_C > I_B$.** The moments of inertia are $I_B \approx 0$, $I_C = 2mr^2$, $I_A = 3mr^2$, and $I_D = mr^2 + m(2r)^2 = 5mr^2$.

**Stop to Think 7.7: C.** The velocity of P is the vector sum of $\vec{v}$ directed to the right and an upward velocity of the same magnitude due to the rotation of the wheel.

Video Tutor Solution Chapter 7

# QUESTIONS

## Conceptual Questions

1. The batter in a baseball game hits a home run. As he circles the bases, is his angular velocity positive or negative?
2. Viewed from somewhere in space above the north pole, would a point on the earth's equator have a positive or negative angular velocity due to the earth's rotation?
3. Figure Q7.3 shows four pulleys, each with a heavy and a light block strung over it. The blocks' velocities are shown. What are the signs (+ or −) of the angular velocity and angular acceleration of the pulley in each case?

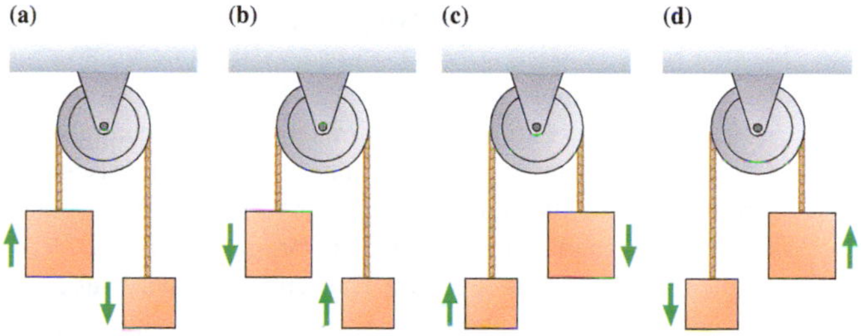

FIGURE Q7.3

Problem difficulty is labeled as | (straightforward) to ||||| (challenging). Problems labeled INT integrate significant material from earlier chapters; Problems labeled BIO are of biological or medical interest.

The eText icon indicates when there is a video tutor solution available for the chapter or for a specific problem. To launch these videos, log into your eText through Mastering™ Physics or log into the Study Area.

4. If you are using a wrench to loosen a very stubborn nut, you can make the job easier by using a "cheater pipe." This is a piece of pipe that slides over the handle of the wrench, as shown in Figure Q7.4, making it effectively much longer. Explain why this would help you loosen the nut.

FIGURE Q7.4

5. The torque needed to open a factory-sealed jar is about the same as the torque required to turn the wheel on a passenger car. You know that the force necessary to turn your car's steering wheel is less than the force needed to open a jar. How can the torques be the same?
6. The chain on a bicycle turns the rear wheel via a gear cluster, as shown in Figure Q7.6. The gears vary in size; the smallest gear has a radius of 2.3 cm, the largest a radius of 6.0 cm. If we assume a constant tension in the chain, which gear—the smallest or the largest—is the better choice for a rapid start?

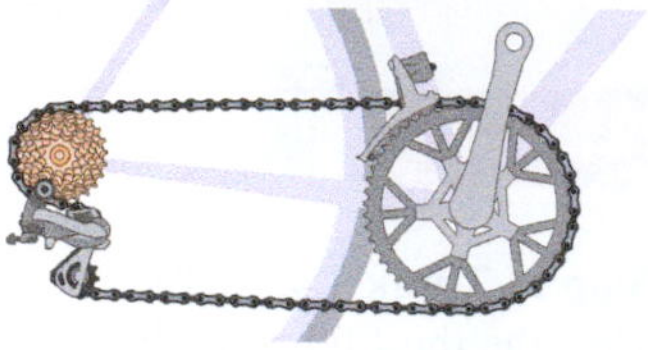

FIGURE Q7.6

7. Five forces are applied to a door, as seen from above in Figure Q7.7. For each force, is the torque about the hinge positive, negative, or zero?

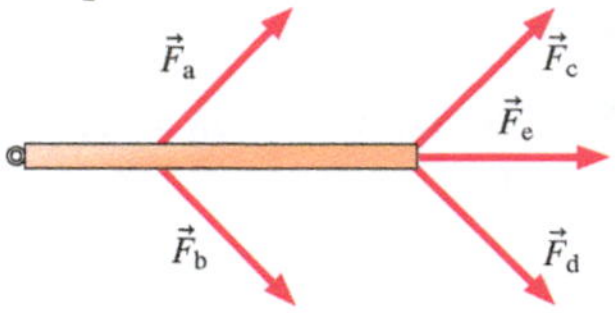


FIGURE Q7.7

8. A screwdriver with a very thick handle requires less force to operate than one with a very skinny handle. Explain why this is so.
9. If you have ever driven a truck, you likely found that it had a steering wheel with a larger diameter than that of a passenger car. Why is this?
10. A student gives a steady push to a ball at the end of a massless, rigid rod for 1 s, causing the ball to rotate clockwise in a horizontal circle as shown in Figure Q7.10. The rod's pivot is frictionless. Sketch a graph of the ball's angular velocity as a function of time for the first 3 s of the ball's motion. You won't be able to include numbers on the vertical axis, but your graph should have the correct sign and the correct shape.

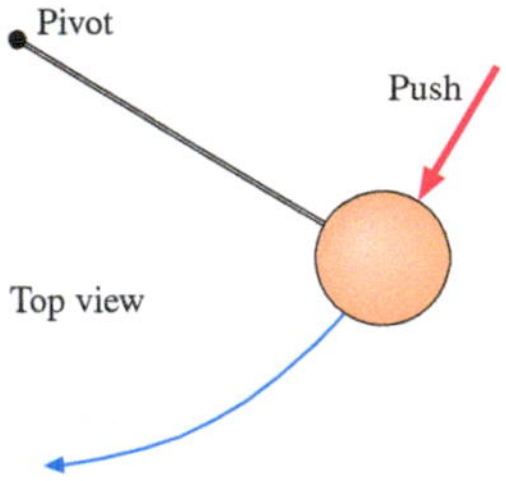


FIGURE Q7.10

11. You can use a simple technique to find the center of gravity of an irregular shape. Figure Q7.11 shows a cardboard cutout of the outline of the continental United States. The map is suspended from pivot 1 and allowed to hang freely; then a blue vertical line is drawn. The map is then suspended from pivot 2, hangs freely, and the red vertical line is drawn. The center of gravity lies at the intersection of the two lines. Explain how this technique works.

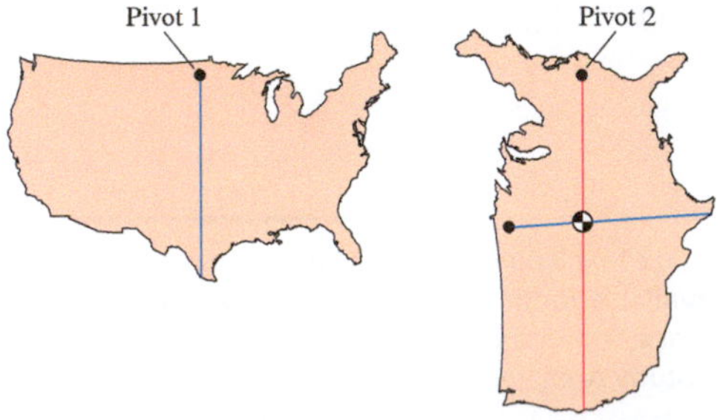


FIGURE Q7.11

12. If you grasp a hammer by its lightweight handle and wave it back and forth, and then grasp it by its much heavier head and wave it back and forth, as in the figure, you'll find that you can wave the hammer much more rapidly in the second case, when you grasp it by the head. Explain why this is so.

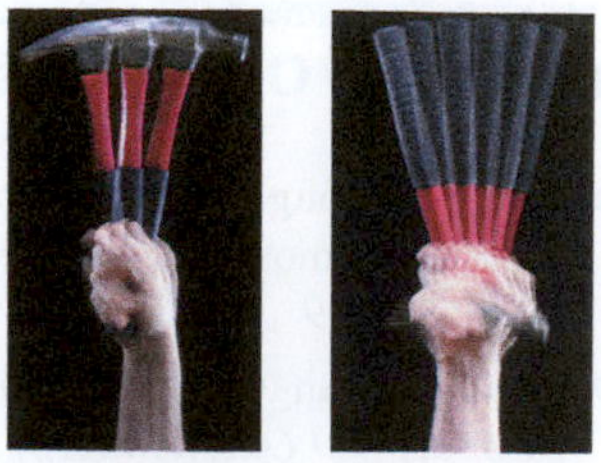

13. You have two identical-looking metal spheres of the same size and the same mass. One is solid; the other is hollow. If you place them on a ramp, the gravitational torques that make them roll down the slope are the same. But one sphere has a greater angular acceleration, so it reaches the bottom of the ramp first. Which one gets to the bottom of the ramp first, and why?
14. Getting on a so-called "tall bike" is harder than mounting a regular bike. But once a rider is in place, it is easier to balance on the tall bike, and the bike can be safely ridden at very slow speeds. Outline the physical principles that explain why you would expect balance to be easier on a taller bike.

15. Water parks often include a log rolling area, in which participants try to stay upright while balancing on top of floating logs that are free to rotate in the water. If you've tried this, you know that the larger the diameter of the log, the easier it is to balance on top. Explain why this is so.

16. The moment of inertia of a uniform rod about an axis through its center is $ML^2/12$. The moment of inertia about an axis at one end is $ML^2/3$. Explain *why* the moment of inertia is larger about the end than about the center.
17. The wheel in Figure Q7.17 is rolling to the right without slipping. Rank in order, from fastest to slowest, the *speeds* of the points labeled 1 through 5. Explain your reasoning.
18. With care, it's possible to walk on top of a barrel as it rolls. It is much easier to do this if the barrel is full than if it is empty. Explain why this is so.

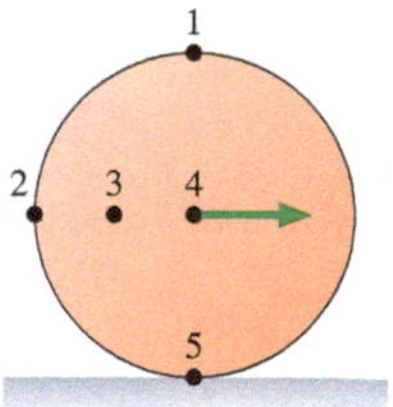


FIGURE Q7.17

## Multiple-Choice Questions

19. | A nut needs to be tightened with a wrench. Which force shown in Figure Q7.19 will apply the greatest torque to the nut?

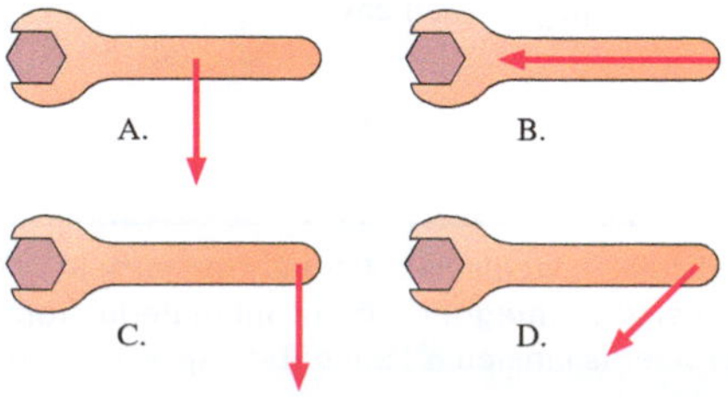


FIGURE Q7.19

20. | Specifications require the oil filter for your car to be tightened with a torque of 30 N · m. You are using a 15-cm-long oil filter wrench, and you apply a force at the very end of the wrench in the direction that produces maximum torque. How much force should you apply?
A. 2000 N B. 400 N C. 200 N D. 30 N

21. | A machine part is made up of two pieces, with centers of gravity shown in Figure Q7.21. Which point could be the center of gravity of the entire part?

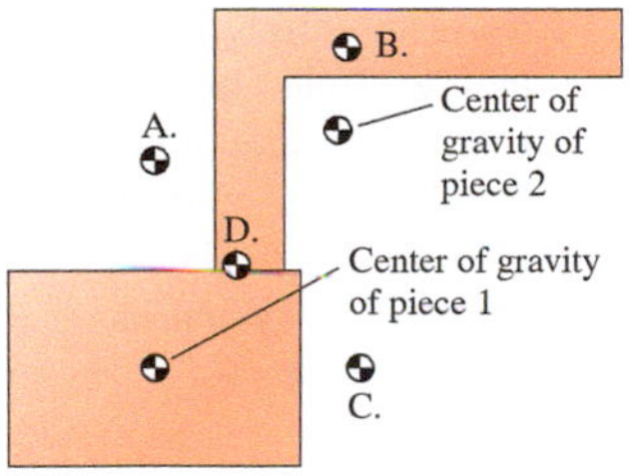

FIGURE Q7.21

22. | The two ends of the barbell shown in Figure Q7.22 are made of the same material. Which of the points shown is at the barbell's center of gravity?

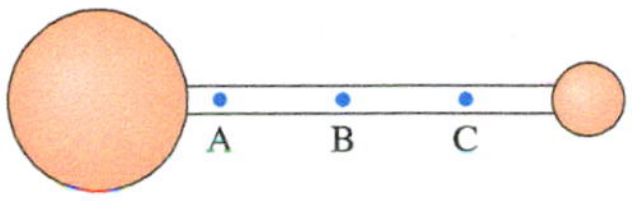

FIGURE Q7.22

23. | A man is holding the yoga position shown in Figure Q7.23. The center of gravity of the combination of his head, arms, and torso is shown as a red symbol on his torso; the center of mass of his legs and feet is a red symbol on his legs. Which of positions A through D corresponds to the center of mass of his entire body?
BIO

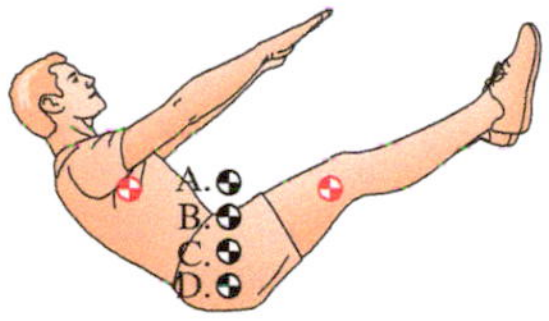

FIGURE Q7.23

24. || A typical compact disk has a mass of 15 g and a diameter of 120 mm. What is its moment of inertia about an axis through its center, perpendicular to the disk?
A. $2.7 \times 10^{-5}\ \text{kg} \cdot \text{m}^2$ B. $5.4 \times 10^{-5}\ \text{kg} \cdot \text{m}^2$
C. $1.1 \times 10^{-4}\ \text{kg} \cdot \text{m}^2$ D. $2.2 \times 10^{-4}\ \text{kg} \cdot \text{m}^2$

25. | Two horizontal rods are each held up by vertical strings tied to their ends. Rod 1 has length $L$ and mass $M$; rod 2 has length $2L$ and mass $2M$. Each rod then has one of its supporting strings cut, causing the rod to begin pivoting about the end that is still tied up. Which rod has a larger initial angular acceleration?
A. Rod 1 B. Rod 2
C. The initial angular acceleration is the same for both.

26. | A baseball bat has a thick, heavy barrel and a thin, light handle. If you want to hold a baseball bat on your palm so that it balances vertically, you should
A. Put the end of the handle in your palm, with the barrel up.
B. Put the end of the barrel in your palm, with the handle up.
C. The bat will be equally easy to balance in either configuration.

Questions 27 through 29 concern a classic figure-skating jump called the axel. A skater starts the jump moving forward as shown in Figure Q7.27, leaps into the air, and turns one-and-a-half revolutions before landing. The typical skater is in the air for about 0.5 s, and the skater's hands are located about 0.8 m from the rotation axis.

FIGURE Q7.27

27. || What is the approximate angular speed of the skater during the leap?
A. 2 rad/s B. 6 rad/s C. 9 rad/s D. 20 rad/s

28. | The skater's arms are fully extended during the jump. What is the approximate centripetal acceleration of the skater's hand?
A. $10\ \text{m/s}^2$ B. $30\ \text{m/s}^2$ C. $300\ \text{m/s}^2$ D. $450\ \text{m/s}^2$

29. | What is the approximate speed of the skater's hand?
A. 1 m/s B. 3 m/s C. 9 m/s D. 15 m/s

# PROBLEMS

## Section 7.1 Describing Circular and Rotational Motion

1. || What is the angular position in radians of the minute hand of a clock at (a) 5:00, (b) 7:15, and (c) 3:35?
2. | A child on a merry-go-round takes 3.0 s to go around once. What is his angular displacement during a 1.0 s time interval?
3. ||| What is the angular speed of the tip of the minute hand on a clock, in rad/s?
4. || An old-fashioned vinyl record rotates on a turntable at 45 rpm. What are (a) the angular speed in rad/s and (b) the period of the motion?
5. ||| The earth's radius is about 4000 miles. Kampala, the capital of Uganda, and Singapore are both nearly on the equator. The distance between them is 5000 miles.
   a. Through what angle do you turn, relative to the earth, if you fly from Kampala to Singapore? Give your answer in both radians and degrees.
   b. The flight from Kampala to Singapore takes 9 hours. What is the plane's angular speed relative to the earth?
6. | Two children are playing tetherball, in which a ball at the end of a cord spins around a pole. After a really good hit, the ball makes three complete revolutions in 2.0 s. What is the angular speed of the ball?
7. ||||| A turntable rotates counterclockwise at 78 rpm. A speck of dust on the turntable is at $\theta = 0.45$ rad at $t = 0$ s. What is the angle of the speck at $t = 8.0$ s? Your answer should be between 0 and $2\pi$ rad.
8. || A Ferris wheel on a California pier is 27 m high and rotates once every 32 seconds. When the wheel starts turning, you are at the very top.
   a. What is your angular position 75 seconds after the wheel starts turning, measured counterclockwise from the top? Express your answer as an angle between 0° and 360°.
   b. What is your speed $v$?

9. | Figure P7.9 shows the angular position of a potter's wheel.

   a. What is the angular displacement of the wheel between $t = 5$ s and $t = 15$ s?

   b. What is the angular velocity of the wheel at $t = 15$ s?

   c. What is the maximum speed of a point on the outside of the wheel, 15 cm from the axle?

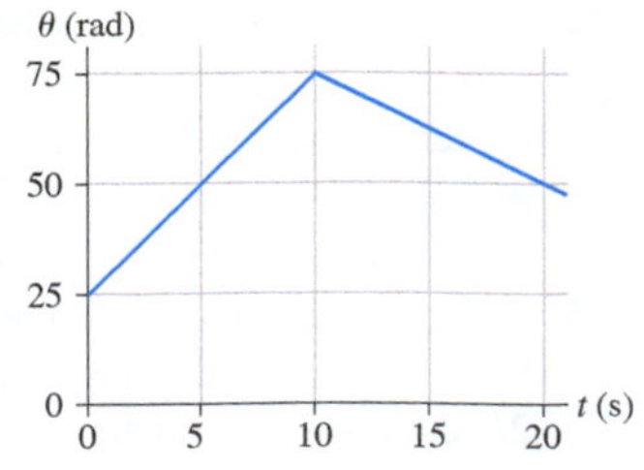

FIGURE P7.9

10. | Figure P7.10 shows a graph of the motion of the blades of a blender that is starts empty and is initially turned off. Then the power is turned on at the highest speed setting, some ice cubes are added, and the blades slow down. The blades are 3.0 cm long.

    a. When the blender is empty and running at high speed, how much time is required for a blade to turn by 90°?

    b. How many revolutions does a blade complete between 15 and 20 seconds?

    c. How fast is the tip of a blade moving when the blender is running empty?

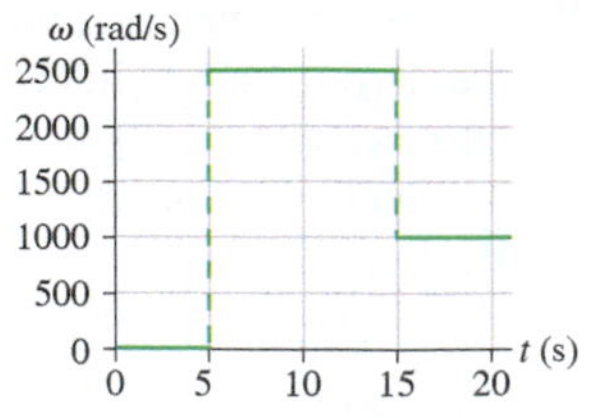

FIGURE P7.10

### Section 7.2 The Rotation of a Rigid Body

11. | The 1.00-cm-long second hand on a watch rotates smoothly.

    a. What is its angular velocity?

    b. What is the speed of the tip of the hand?

12. | The earth's radius is $6.37 \times 10^6$ m; it rotates once every 24 hours.

    a. What is the earth's angular speed?

    b. Viewed from a point above the north pole, is the angular velocity positive or negative?

    c. What is the speed of a point on the equator?

    d. What is the speed of a point on the earth's surface halfway between the equator and the pole? (Hint: What is the radius of the circle in which the point moves?)

13. ||| To throw a discus, the thrower holds it with a fully outstretched arm. Starting from rest, he begins to turn with a constant angular acceleration, releasing the discus after making one complete revolution. The diameter of the circle in which the discus moves is about 1.8 m. If the thrower takes 1.0 s to complete one revolution, starting from rest, what will be the speed of the discus at release?

14. || BIO The bigclaw snapping shrimp shown in Figure P7.14 is aptly named—it has one big claw that snaps shut with remarkable speed. The part of the claw that moves rotates through a 90° angle in 1.3 ms. If we assume that the claw is 1.5 cm long and that it undergoes a constant angular acceleration:

    a. What is the angular acceleration in $\text{rad/s}^2$?

    b. What is the final angular speed of the claw?

    c. What is the tangential acceleration of the tip of the claw?

    d. How fast is the tip of the claw moving at the end of its motion?

FIGURE P7.14

15. ||| A computer hard disk starts from rest, then speeds up with an angular acceleration of 190 $\text{rad/s}^2$ until it reaches its final angular speed of 7200 rpm. How many revolutions has the disk made 10.0 s after it starts up?

16. ||| BIO In a softball windmill pitch, the pitcher rotates her arm through just over half a circle, bringing the ball from a point above her shoulder and slightly forward to a release point below her shoulder and slightly forward. Figure P7.16 shows smoothed data for the angular velocity of the upper arm of a college softball pitcher doing a windmill pitch; at time $t = 0$ her arm is vertical and already in motion. For the first 0.15 s there is a steady increase in speed, leading to a final push with a greater acceleration during the final 0.05 s before the release. During the first 0.15 s of the pitch:

    a. What is the angular acceleration?

    b. If the ball is 0.60 m from her shoulder, what is the tangential acceleration of the ball? This is the key quantity here—it's a measure of how much the ball is speeding up. Express your answer in $\text{m/s}^2$ and in units of $g$.

    c. Through what angle does her arm rotate?

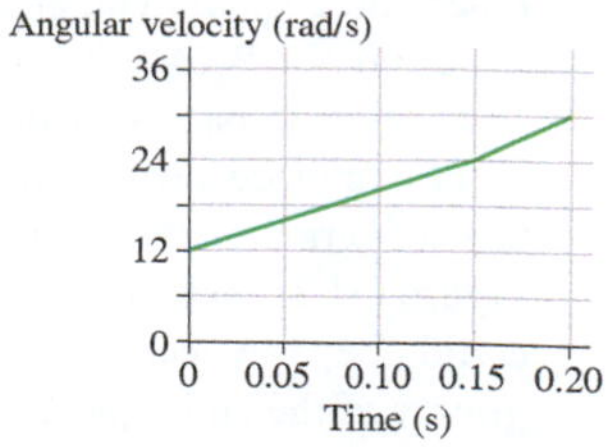

FIGURE P7.16

17. ||| The crankshaft in a race car goes from rest to 3000 rpm in 2.0 s.

    a. What is the crankshaft's angular acceleration?

    b. How many revolutions does it make while reaching 3000 rpm?

### Section 7.3 Torque

18. | Reconsider the situation in Example 7.10. If Luis pulls straight down on the end of a wrench that is in the same orientation but is 35 cm long, rather than 20 cm, what force must he apply to exert the same torque?

19. ||| Balls are attached to light rods and can move in horizontal circles as shown in Figure P7.19. Rank in order, from smallest to largest, the torques $\tau_1$ to $\tau_4$ about the centers of the circles. Explain.

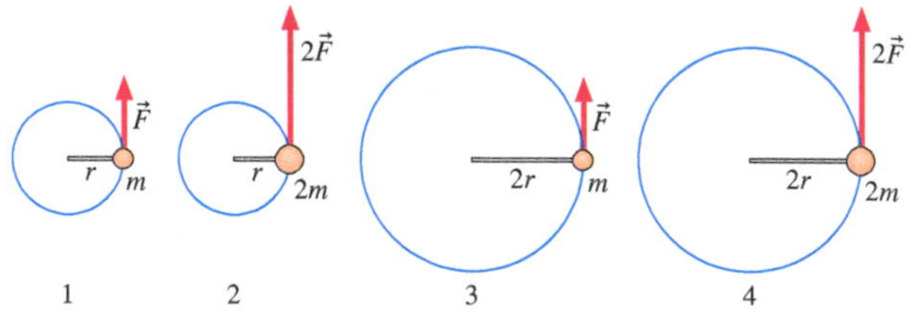

FIGURE P7.19

20. ||| BIO INT We can model a pine tree in the forest as having a compact canopy at the top of a relatively bare trunk. Wind blowing on the top of the tree exerts a horizontal force, and thus a torque that can topple the tree if there is no opposing torque. Suppose a tree's canopy presents an area of 9.0 $\text{m}^2$ to the wind centered at a height of 7.0 m above the ground. (These are reasonable values for forest trees.) If the wind blows at 6.5 m/s:

    a. What is the magnitude of the drag force of the wind on the canopy? Assume a drag coefficient of 0.50.

    b. What torque does this force exert on the tree, measured about the point where the trunk meets the ground?

21. | What is the net torque about the axle on the pulley in Figure P7.21?

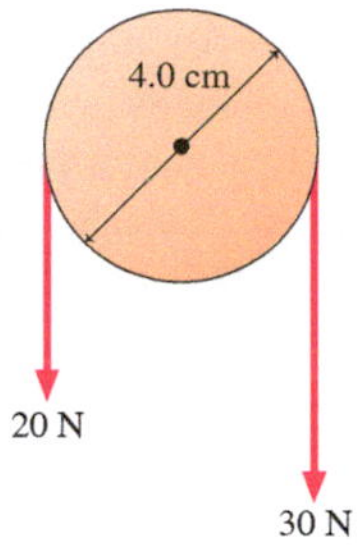

FIGURE P7.21

22. || The tune-up specifications of a car call for the spark plugs to be tightened to a torque of 38 N · m. You plan to tighten the plugs by pulling on the end of a 25-cm-long wrench. Because of the cramped space under the hood, you'll need to pull at an angle of 120° with respect to the wrench shaft. With what force must you pull?

23. || In Figure P7.23, force $\vec{F}_2$ acts half as far from the pivot as $\vec{F}_1$. What magnitude of $\vec{F}_2$ causes the net torque on the rod to be zero?

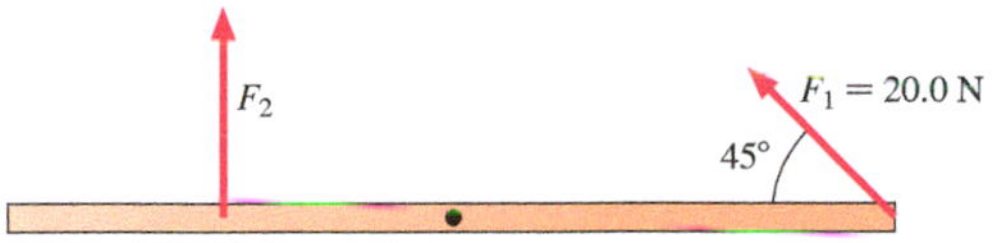

FIGURE P7.23

24. ||| A professor's office door is 0.91 m wide, 2.0 m high, and 4.0 cm thick; has a mass of 25 kg; and pivots on frictionless hinges. A "door closer" is attached to the door and the top of the door frame. When the door is open and at rest, the door closer exerts a torque of 5.2 N · m. What is the least force that you need to apply to the door to hold it open?

25. || What is the net torque on the bar shown in Figure P7.25, about the axis indicated by the dot?

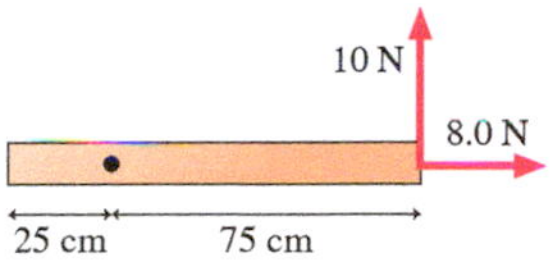

FIGURE P7.25

26. || A typical jar that has been tightened to a reasonable degree requires 2.0 N · m to open. If you grab a 7.0-cm-diameter jar lid with one hand so that your thumb and fingers exert equal magnitude forces on opposite sides of the lid, as in Figure P7.26, what is the magnitude of each of the forces?

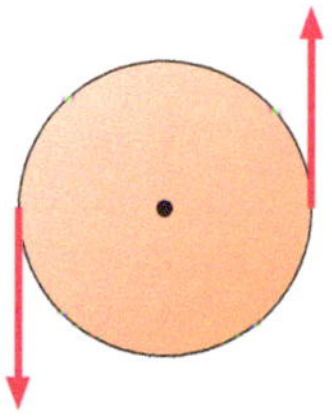
FIGURE P7.26

27. || What is the net torque on the bar shown in Figure P7.27, about the axis indicated by the dot?

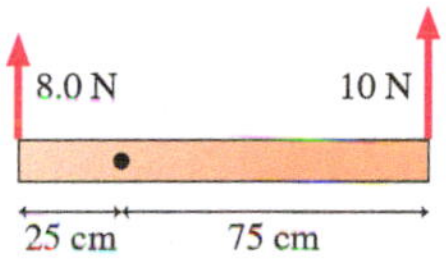

FIGURE P7.27

28. || A driver holds his hands on opposite sides of the 35-cm-diameter steering wheel in a modern sports car. A torque of 4.5 N · m is required to turn the wheel. If the driver applies an equal force on each side of the wheel, what is the minimum force each hand must supply?

29. || What is the net torque on the bar shown in Figure P7.29, about the axis indicated by the dot?

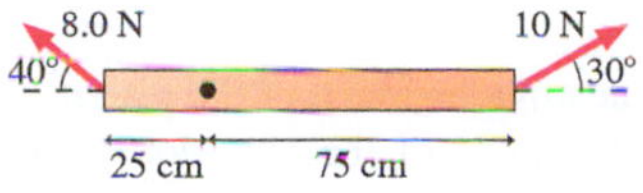

FIGURE P7.29

## Section 7.4 Gravitational Torque and the Center of Gravity

30. || When you stand quietly, you pivot back and forth a very small amount about your ankles. It's easier to maintain stability if you tip slightly forward, so that your center of gravity is slightly in front of your ankles. The 68 kg man shown in Figure P7.30 is 1.8 m tall; his center of gravity is 1.0 m above the ground and, during quiet standing, is 5.5 cm in front of his ankles. What is the magnitude of the gravitational torque about his ankles?
BIO
INT

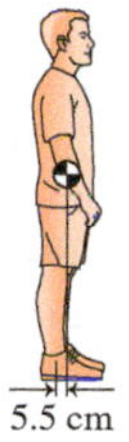

FIGURE P7.30

31. || The 2.0 kg, uniform, horizontal rod in Figure P7.31 is seen from the side. What is the gravitational torque about the point shown?

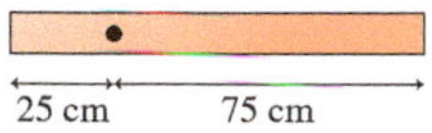

FIGURE P7.31

32. |||| A 4.00-m-long, 500 kg steel beam extends horizontally from the point where it has been bolted to the framework of a new building under construction. A 70.0 kg construction worker stands at the far end of the beam. What is the magnitude of the gravitational torque about the point where the beam is bolted into place?

33. ||| An athlete at the gym holds a 3.0 kg steel ball in his hand. His arm is 70 cm long and has a mass of 4.0 kg. What is the magnitude of the gravitational torque about his shoulder if he holds his arm
BIO
a. Straight out to his side, parallel to the floor?
b. Straight, but 45° below horizontal?

34. ||| The 2.0-m-long, 15 kg beam in Figure P7.34 is hinged at its left end. It is "falling" (rotating clockwise, under the influence of gravity), and the figure shows its position at three different times. What is the gravitational torque on the beam about an axis through the hinged end when the beam is at the
a. Upper position?
b. Middle position?
c. Lower position?

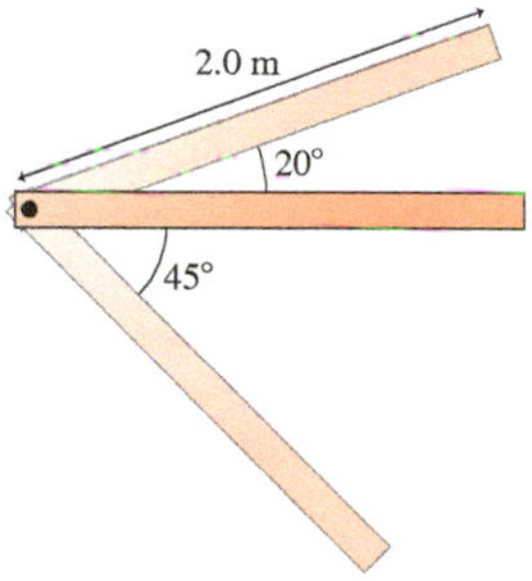

FIGURE P7.34

35. || Three identical coins lie on three corners of a square 10.0 cm on a side, as shown in Figure P7.35. Determine the $x$- and $y$-coordinates of the center of gravity of the group of three coins.

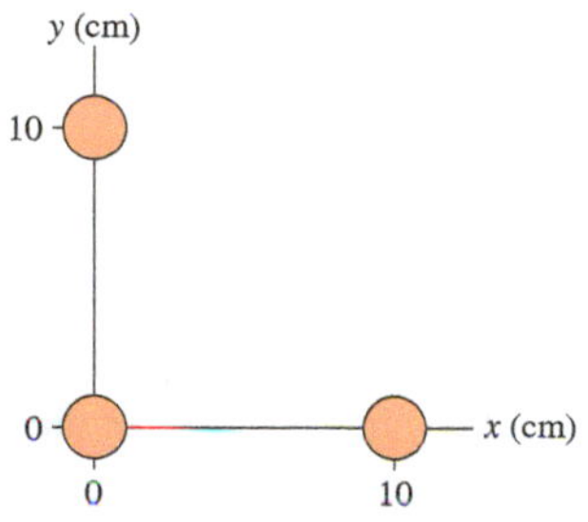

FIGURE P7.35

36. | Hold your arm outstretched so that it is horizontal. Estimate the mass of your arm and the position of its center of gravity. What is the gravitational torque on your arm in this position, computed around the shoulder joint?
BIO

37. || U.S. nickels have a mass of 5.00 g and are 1.95 mm thick. If you stack 3 nickels on a table, how far above the table is their center of gravity?

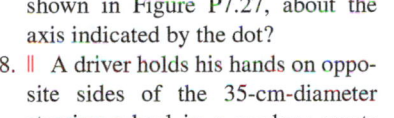

Watch Video Solution Problems 7.39 and 7.53

38. ||| Figure P7.38 shows two thin beams joined at right angles. The vertical beam is 15.0 kg and 1.00 m long and the horizontal beam is 25.0 kg and 2.00 m long.

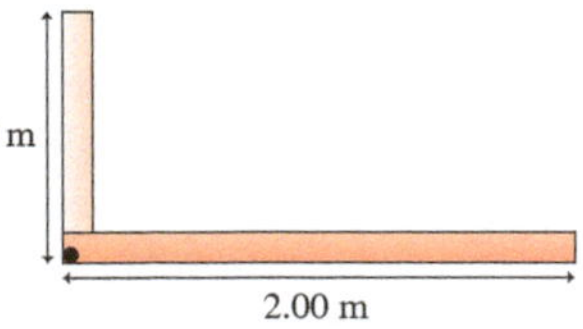


FIGURE P7.38

a. Find the center of gravity of the two joined beams. Express your answer in the form $(x, y)$, taking the origin at the corner where the beams join.
b. Calculate the gravitational torque on the joined beams about an axis through the corner. The beams are seen from the side.

### Section 7.5 Rotational Dynamics and Moment of Inertia

39. ||| A regulation table tennis ball is a thin spherical shell 40 mm in diameter with a mass of 2.7 g. What is its moment of inertia about an axis that passes through its center?

40. || Three pairs of balls are connected by very light rods as shown in Figure P7.40. Rank in order, from smallest to largest, the moments of inertia $I_1$, $I_2$, and $I_3$ about axes through the centers of the rods.

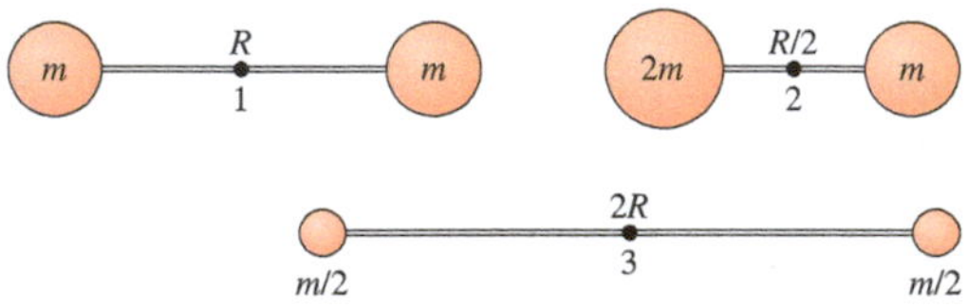


FIGURE P7.40

41. |||| A solid cylinder with a radius of 4.0 cm has the same mass as a solid sphere of radius $R$. If the cylinder and sphere have the same moment of inertia about their centers, what is the sphere's radius?

42. || We can model a small merry-go-round as a uniform circular disk with mass 88 kg and diameter 1.8 m. How many 22 kg children need to ride the merry-go-round, standing right at the outer edge, to double the moment of inertia of the system?

43. || A bicycle rim has a diameter of 0.65 m and a moment of inertia, measured about its center, of $0.19\ \text{kg}\cdot\text{m}^2$. What is the mass of the rim?

44. ||| A bowling ball is far from uniform. Lightweight bowling balls are made of a relatively low-density core surrounded by a thin shell with much higher density. A 7.0 lb (3.2 kg) bowling ball has a diameter of 0.216 m; 0.196 m of this is a 1.6 kg core, surrounded by a 1.6 kg shell. This composition gives the ball a higher moment of inertia than it would have if it were made of a uniform material. Given the importance of the angular motion of the ball as it moves down the alley, this has real consequences for the game.
a. Model a real bowling ball as a 0.196-m-diameter core with mass 1.6 kg plus a thin 1.6 kg shell with diameter 0.206 m (the average of the inner and outer diameters). What is the total moment of inertia?
b. How does your answer in part a compare to the moment of inertia of a uniform 3.2 kg ball with diameter 0.216 m?

### Section 7.6 Using Newton's Second Law for Rotation

45. | A small grinding wheel has a moment of inertia of $4.0 \times 10^{-5}\ \text{kg}\cdot\text{m}^2$. What net torque must be applied to the wheel for its angular acceleration to be $150\ \text{rad/s}^2$?

46. |||| a. What is the moment of inertia of the door in Problem 24?
b. If you let go of the open door, what is its angular acceleration immediately afterward?

47. | An object's moment of inertia is $2.0\ \text{kg}\cdot\text{m}^2$. Its angular velocity is increasing at the rate of 4.0 rad/s per second. What is the net torque on the object?

48. ||| The lightweight wheel on a road bike has a moment of inertia of $0.097\ \text{kg}\cdot\text{m}^2$. A mechanic, checking the alignment of the wheel, gives it a quick spin; it completes 5 rotations in 2.0 s. To bring the wheel to rest, the mechanic gently applies the disk brakes, which squeeze pads against a metal disk connected to the wheel. The pads touch the disk 7.1 cm from the axle, and the wheel slows down and stops in 1.5 s. What is the magnitude of the friction force on the disk?

49. |||| A 200 g, 20-cm-diameter plastic disk is spun on an axle through its center by an electric motor. What torque must the motor supply to take the disk from 0 to 1800 rpm in 4.0 s?

50. || The engine in a small airplane is specified to have a torque of $500\ \text{N}\cdot\text{m}$. This engine drives a 2.0-m-long, 40 kg single-blade propeller. On startup, how long does it take the propeller to reach 2000 rpm?

51. ||| A frictionless pulley, which can be modeled as a 0.80 kg solid cylinder with a 0.30 m radius, has a rope going over it, as shown in Figure P7.51. The tension in the rope is 10 N on one side and 12 N on the other. What is the angular acceleration of the pulley?

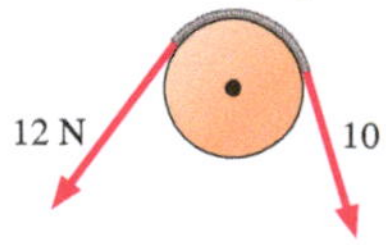


FIGURE P7.51

52. || A gymnast on the uneven parallel bars is at rest, tipped at a 45° angle from the vertical. The distance from her hands to her feet is 1.8 m. If we model her body as having a uniform cross section and assume that her center of gravity is midway between her hands and her feet, what is her initial angular acceleration?

53. ||| If you lift the front wheel of a poorly maintained bicycle off the ground and then start it spinning at 0.72 rev/s, friction in the bearings causes the wheel to stop in just 12 s. If the moment of inertia of the wheel about its axle is $0.30\ \text{kg}\cdot\text{m}^2$, what is the magnitude of the frictional torque?

54. || On page 224 there is a photograph of a girl pushing on a large stone sphere. The sphere has a mass of 8200 kg and a radius of 90 cm. Suppose that she pushes on the sphere tangent to its surface with a steady force of 50 N and that the pressured water provides a frictionless support. How long will it take her to rotate the sphere one time, starting from rest?

55. ||||| A toy top with a spool of diameter 5.0 cm has a moment of inertia of $3.0 \times 10^{-5}\ \text{kg}\cdot\text{m}^2$ about its rotation axis. To get the top spinning, its string is pulled with a tension of 0.30 N. How long does it take for the top to complete the first five revolutions? The string is long enough that it is wrapped around the top more than five turns.

56. ||||| A 1.5 kg block and a 2.5 kg block are attached to opposite ends of a light rope. The rope hangs over a solid, frictionless pulley that is 30 cm in diameter and has a mass of 0.75 kg. When the blocks are released, what is the acceleration of the lighter block?

### Section 7.7 Rolling Motion

57. ||| A bicycle with 0.80-m-diameter tires is coasting on a level road at 5.6 m/s. A small blue dot has been painted on the tread of the rear tire.
a. What is the angular speed of the tires?
b. What is the speed of the blue dot when it is 0.80 m above the road?
c. What is the speed of the blue dot when it is 0.40 m above the road?

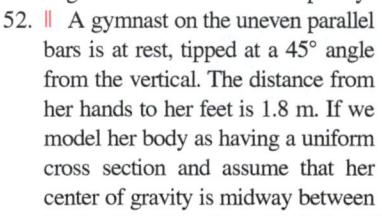

58. || A typical road bike wheel has a diameter of 70 cm including the tire. In a time trial, when a cyclist is racing along at 12 m/s:
   a. How fast is a point at the top of the tire moving?
   b. How fast, in rpm, are the wheels spinning?

59. || A 2.0-m-long slab of concrete is supported by rollers, as shown in Figure P7.59. If the slab is pushed to the right, it will move off the supporting rollers, one by one, on the left side. How far can the slab be moved before its center of gravity is to the right of the contact point with the rightmost roller—at which point the slab begins to tip?

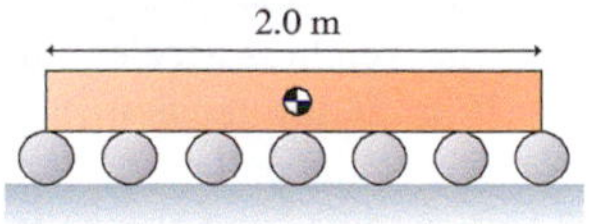


FIGURE P7.59

60. || A man in a barrel walking competition is moving along smoothly, with his barrel moving forward at 1.0 m/s.
   a. Think about how the man moves his legs. Is he walking forward or backward?
   b. From the point of view of the top of the barrel—the man's walking surface—how fast is he walking?

## General Problems

61. | Figure P7.61 shows the angular position-versus-time graph for a particle moving in a circle.
INT
   a. Write a description of the particle's motion.
   b. Draw the angular velocity-versus-time graph.

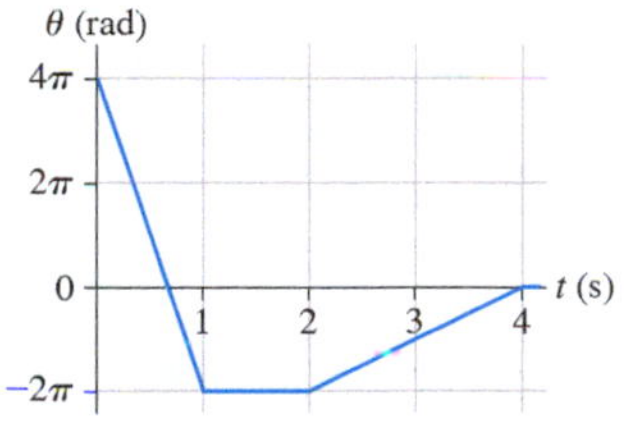


FIGURE P7.61

62. | The graph in Figure P7.62 shows the angular velocity of the crankshaft in a car. Draw a graph of the angular acceleration versus time. Include appropriate numerical scales on both axes.
INT

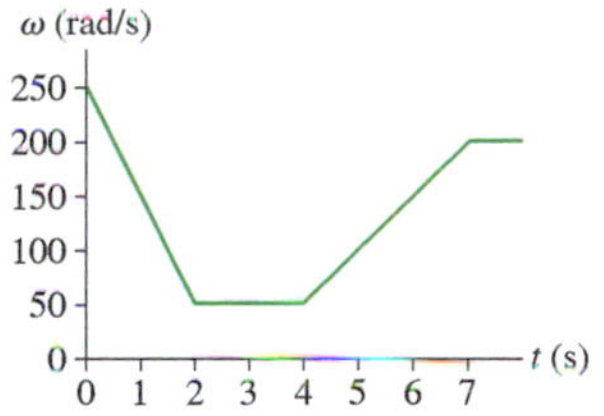


FIGURE P7.62

63. || A car with 58-cm-diameter tires accelerates uniformly from rest to 20 m/s in 10 s. How many times does each tire rotate?
INT

64. ||| The cable lifting an elevator is wrapped around a 1.0-m-diameter cylinder that is turned by the elevator's motor. The elevator is moving upward at a speed of 1.6 m/s. It then slows to a stop as the cylinder makes one complete turn at constant angular acceleration. How long does it take for the elevator to stop?

65. |||| The 20-cm-diameter disk in Figure P7.65 can rotate on an axle through its center. What is the net torque about the axle?

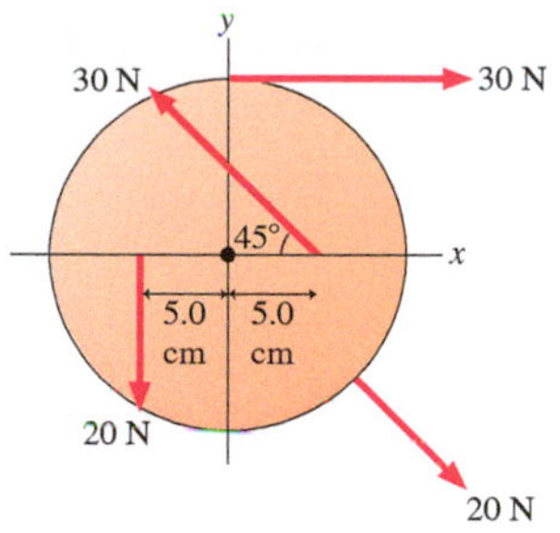


FIGURE P7.65

66. ||| A combination lock has a 1.0-cm-diameter knob that is part of the dial you turn to unlock the lock. To turn that knob, you grip it between your thumb and forefinger with a force of 0.60 N as you twist your wrist. Suppose the coefficient of static friction between the knob and your fingers is only 0.12 because some oil accidentally got onto the knob. What is the most torque you can exert on the knob without having it slip between your fingers?
INT

67. |||| A 70 kg man's arm, including the hand, can be modeled as a 75-cm-long uniform cylinder with a mass of 3.5 kg. In raising both his arms, from hanging down to straight up, by how much does he raise his center of gravity?
BIO

68. |||| The three masses shown in Figure P7.68 are connected by massless, rigid rods.
   a. Find the coordinates of the center of gravity.
   b. Find the moment of inertia about an axis that passes through mass A and is perpendicular to the page.
   c. Find the moment of inertia about an axis that passes through masses B and C.

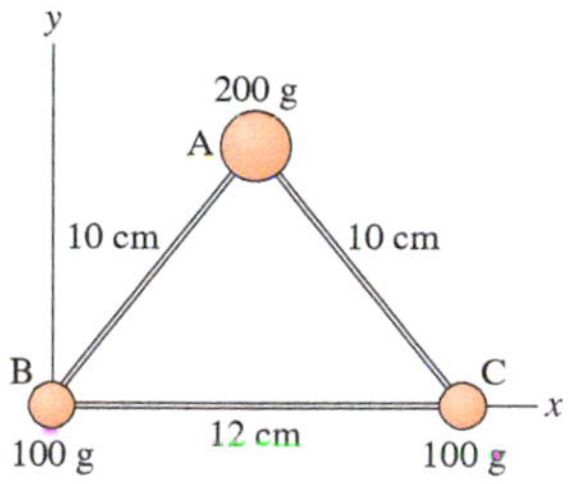


FIGURE P7.68

69. || A reasonable estimate of the moment of inertia of an ice skater spinning with her arms at her sides can be made by modeling most of her body as a uniform cylinder. Suppose the skater has a mass of 64 kg. One-eighth of that mass is in her arms, which are 60 cm long and 20 cm from the vertical axis about which she rotates. The rest of her mass is approximately in the form of a 20-cm-radius cylinder.
BIO
   a. Estimate the skater's moment of inertia to two significant figures.
   b. If she were to hold her arms outward, rather than at her sides, would her moment of inertia increase, decrease, or remain unchanged? Explain.

70. ||| A 17 g audio compact disk has a diameter of 12 cm. The disk spins under a laser that reads encoded data. The first track to be read is 2.3 cm from the axis; as the disk plays, the laser scans tracks farther and farther from the center. The part of the disk directly under the read head moves at a constant 1.2 m/s.
   a. When a disk is inserted, it takes 2.4 s to spin up from rest. What is the torque of the motor?
   b. As the disk plays, how does the angular speed change?

Watch Video Solution Problem 7.75

71. |||| The ropes in Figure P7.71 are each wrapped around a cylinder, and the two cylinders are fastened together. The smaller cylinder has a diameter of 10 cm and a mass of 5.0 kg; the larger cylinder has a diameter of 20 cm and a mass of 20 kg. What is the angular acceleration of the cylinders? Assume that the cylinders turn on a frictionless axle.

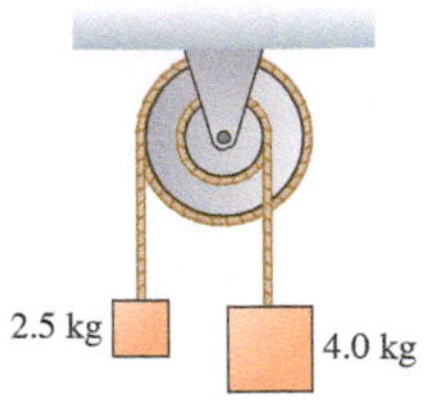


FIGURE P7.71

72. ||| A spin bike is an indoor bike that is designed to duplicate the feeling of regular road cycling. A typical spin bike has a very heavy flywheel. A friction pad or other brake provides damping that mirrors the drag and other resistive forces that a road cyclist experiences; the inertia of the wheel simulates the way a regular bike keeps moving even after you stop pedaling. A spin bike has a flywheel in two parts—a 12.5 kg disk with radius 0.23 m, and a 7.0 kg ring with mass concentrated at the outer edge of the disk. A friction pad exerts a force of 9.7 N on the outside of the disk. A cyclist is pedaling, spinning the disk at a typical 180 rpm. If she stops pedaling, how long will it take for the flywheel to come to a stop?

73. |||| Flywheels are large, massive wheels used to store energy. They can be spun up slowly, then the wheel's energy can be released quickly to accomplish a task that demands high power. An industrial flywheel has a 1.5 m diameter and a mass of 250 kg. A motor spins up the flywheel with a constant torque of 50 N · m. How long does it take the flywheel to reach top angular speed of 1200 rpm?

74. ||| A trap-jaw ant has mandibles that can snap shut with some force, as you might expect from its name. The formidable snap is good for more than capturing prey. When an ant snaps its jaws against the ground, the resulting force can launch the ant into the air. Here are typical data: An ant rotates its mandible, of length 1.30 mm and mass 130 $\mu$g (which we can model as a uniform rod rotated about its end), at a high angular speed. As the tip strikes the ground, it undergoes an angular acceleration of $3.5 \times 10^8$ rad/s$^2$. If we assume that the tip of the mandible hits perpendicular to the ground, what is the force on the tip? How does this compare to the weight of a 12 mg ant?
BIO
INT

75. |||| A 1.0 kg ball and a 2.0 kg ball are connected by a 1.0-m-long rigid, massless rod. The rod and balls are rotating clockwise about their center of gravity at 20 rpm. What torque will bring the balls to a halt in 5.0 s?

76. |||| A 1.5 kg block is connected by a rope across a 50-cm-diameter, 2.0 kg pulley, as shown in Figure P7.76. There is no friction in the axle, but there is friction between the rope and the pulley; the rope doesn't slip. The weight is accelerating upward at 1.2 m/s$^2$. What is the tension in the rope on the right side of the pulley?
INT

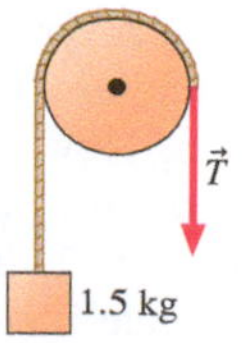


FIGURE P7.76

77. |||| A tradesman sharpens a knife by pushing it with a constant force against the rim of a grindstone. The 30-cm-diameter stone is spinning at 200 rpm and has a mass of 28 kg. The coefficient of kinetic friction between the knife and the stone is 0.20. If the stone slows steadily to 180 rpm in 10 s of grinding, what is the force with which the man presses the knife against the stone?
INT

## MCAT-Style Passage Problems

### The Bunchberry BIO

The bunchberry flower has the fastest-moving parts ever seen in a plant. Initially, the stamens are held by the petals in a bent position, storing energy like a coiled spring. As the petals release, the tips of the stamens fly up and quickly release a burst of pollen. Figure P7.78 shows the details of the motion. The tips of the stamens act like a catapult, flipping through a 60° angle; the times on the earlier photos show that this happens in just 0.30 ms. We can model a stamen tip as a 1.0-mm-long, 10 $\mu$g rigid rod with a 10 $\mu$g anther sac at one end and a pivot point at the opposite end. Though an oversimplification, we will model the motion by assuming the angular acceleration is constant throughout the motion.

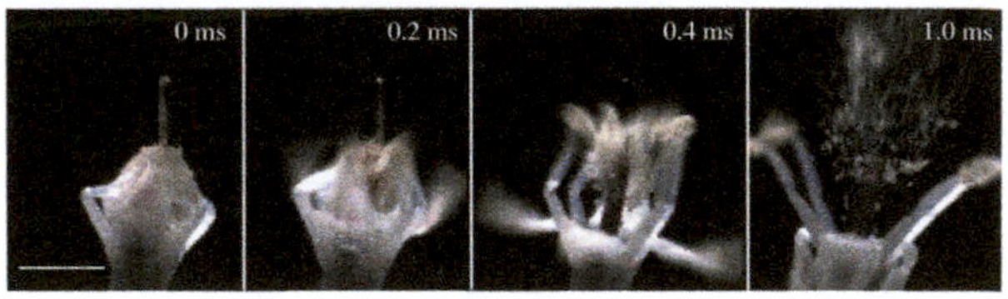


78. | What is the angular acceleration of the anther sac during the motion?
A. $3.5 \times 10^3$ rad/s$^2$ B. $7.0 \times 10^3$ rad/s$^2$
C. $1.2 \times 10^7$ rad/s$^2$ D. $2.3 \times 10^7$ rad/s$^2$

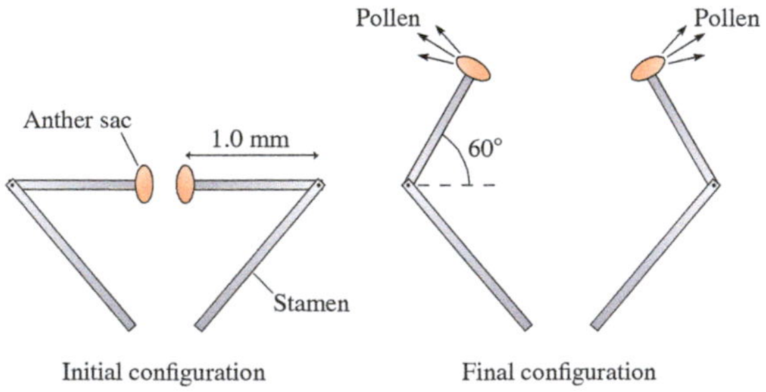


FIGURE P7.78

79. | What is the speed of the anther sac as it releases its pollen?
A. 3.5 m/s B. 7.0 m/s
C. 10 m/s D. 14 m/s

80. || How large is the "straightening torque"? (You can omit gravitational forces from your calculation; the gravitational torque is much less than this.)
A. $2.3 \times 10^{-7}$ N · m B. $3.1 \times 10^{-7}$ N · m
C. $2.3 \times 10^{-5}$ N · m D. $3.1 \times 10^{-5}$ N · m

### The Illusion of Flight

The grand jeté is a classic ballet maneuver in which a dancer executes a horizontal leap while moving her arms and legs up and then down. At the center of the leap, the arms and legs are gracefully extended, as we see in Figure P7.81a. The goal of the leap is to create the illusion of flight. As the dancer moves through the air, he or she is in free fall. In Chapter 3, we saw that this leads to projectile motion. But what part of the dancer follows the usual parabolic path? It won't come as a surprise to learn that it's the center of gravity. But when you watch a dancer leap through the air, you don't watch her center of gravity, you watch her head. If the translational motion of her head is horizontal—not parabolic—this creates the illusion that she is flying through the air, held up by unseen forces.

Figure P7.81b illustrates how the dancer creates this illusion. While in the air, she changes the position of her center of gravity relative to her body by moving her arms and legs up, then down. Her center of gravity moves in a parabolic path, but her head moves in a straight line. It's not flight, but it will appear that way, at least for a moment.

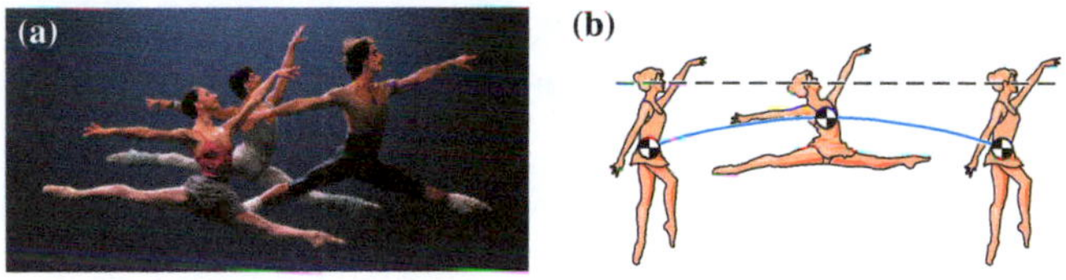


FIGURE P7.81

81. | To perform this maneuver, the dancer relies on the fact that the position of her center of gravity
 A. Is near the center of the torso.
 B. Is determined by the positions of her arms and legs.
 C. Moves in a horizontal path.
 D. Is outside of her body.

82. | Suppose you wish to make a vertical leap with the goal of getting your head as high as possible above the ground. At the top of your leap, your arms should be
 A. Held at your sides.
 B. Raised above your head.
 C. Outstretched, away from your body.

83. | When the dancer is in the air, is there a gravitational torque on her? Take the dancer's rotation axis to be through her center of gravity.
 A. Yes, there is a gravitational torque.
 B. No, there is not a gravitational torque.
 C. It depends on the positions of her arms and legs.

84. | In addition to changing her center of gravity, a dancer may change her moment of inertia. Consider her moment of inertia about a vertical axis through the center of her body. When she raises her arms and legs, this
 A. Increases her moment of inertia.
 B. Decreases her moment of inertia.
 C. Does not change her moment of inertia.

# 8 Equilibrium and Elasticity

How does a dancer balance so gracefully *en pointe*? And how does her foot withstand the great stresses concentrated on her toes? In this chapter we'll find answers to both these questions.

## LOOKING AHEAD ➠

### Static Equilibrium

As this cyclist balances on his back tire, the net force *and* the net torque on him must be zero.

You'll learn to analyze objects that are in **static equilibrium**.

### Springs

When a rider takes a seat, the spring is compressed and exerts a **restoring force,** pushing upward.

You'll learn to solve problems involving stretched and compressed springs.

### Properties of Materials

All materials have some "give"; if you pull on them, they stretch, and at some point they break.

Is this spider silk as strong as steel? You'll learn to think about what the question means, and how to answer it.

**GOAL** **To learn about the static equilibrium of extended objects, and the basic properties of springs and elastic materials.**

## LOOKING BACK ◀◀

### Torque

In Chapter 7, you learned to calculate the torque on an object due to an applied force.

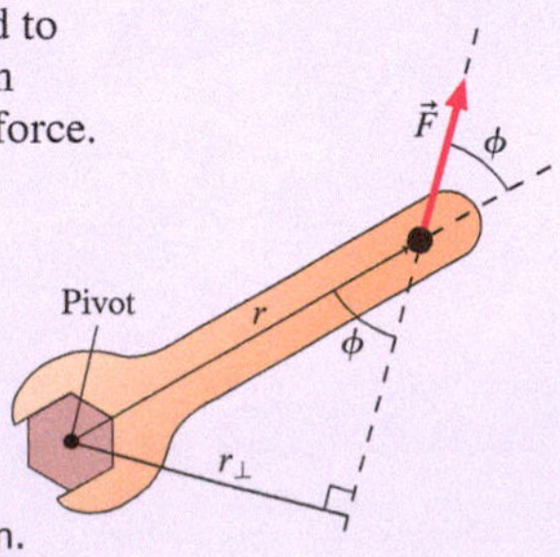


In this chapter, you'll extend your analysis to consider objects with many forces—and many torques—that act on them.

**STOP TO THINK**

An old-fashioned tire swing exerts a force on the branch and a torque about the point where the branch meets the trunk. If you hang the swing closer to the trunk, this will ______ the force and ______ the torque.

A. increase, increase
B. not change, increase
C. not change, not change
D. not change, decrease
E. decrease, not change
F. decrease, decrease

# 8.1 Torque and Static Equilibrium

We have now spent several chapters studying motion and its causes. In many disciplines, it is just as important to understand the conditions under which objects do *not* move. Buildings and dams must be designed such that they remain practically motionless, even when huge forces act on them. And joints in the body must sustain large forces when the body is supporting heavy loads, as in holding or carrying heavy objects.

Recall from ◀◀ SECTION 5.1 that an object at rest is in *static equilibrium.* As long as an object can be modeled as a *particle,* the condition necessary for static equilibrium is that the net force $\vec{F}_{net}$ on the particle is zero, as in FIGURE 8.1a, where the two forces applied to the particle balance and the particle can remain at rest.

But in Chapter 7 we moved beyond the particle model to study extended objects that can rotate. Consider, for example, the block in FIGURE 8.1b. In this case the two forces act along the same line, the net force is zero, and the block is in equilibrium. But what about the block in FIGURE 8.1c? The net force is still zero, but this time the block begins to rotate because the two forces exert a net *torque.* For an extended object, $\vec{F}_{net} = \vec{0}$ is not by itself enough to ensure static equilibrium. There is a second condition for static equilibrium of an extended object: The net torque $\tau_{net}$ on the object must also be zero.

If we write the net force in component form, the conditions for static equilibrium of an extended object are

FIGURE 8.1 A block with no net force acting on it may still be out of equilibrium.

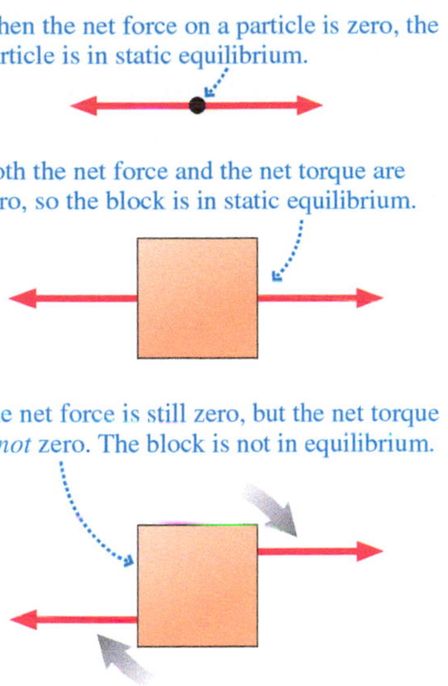

(a) When the net force on a particle is zero, the particle is in static equilibrium.
(b) Both the net force and the net torque are zero, so the block is in static equilibrium.
(c) The net force is still zero, but the net torque is *not* zero. The block is not in equilibrium.


$$\left.\begin{array}{l}\sum F_x = 0 \\ \sum F_y = 0\end{array}\right\} \text{ No net force} \qquad \sum \tau = 0 \;\} \text{ No net torque} \tag{8.1}$$

Conditions for static equilibrium of an extended object

Let's look at an example to see how we can use these conditions to analyze a physical situation.

**EXAMPLE 8.1 Lifting a load**

A participant in a "strongman" competition uses an old-fashioned device to move a large load with only muscle power. A 1 ton (910 kg) bucket of rocks is suspended from a sturdy, lightweight beam 0.85 m from a pivot. The man lifts the beam at its end, 3.6 m from the pivot, and holds it steady. How much force must the man apply? What is the force on the beam from the pivot?

***STRATEGIZE*** This is a static equilibrium problem—nothing is moving. The first step in our solution is to find an object to focus on. The way we've drawn the diagram in FIGURE 8.2, with the highlight color on the beam, hints at our choice. The object of interest in this case is the beam, not the man or the bucket. The beam is the element that links all of the pieces together; it touches the man, the bucket, and the pivot. The beam has different forces acting on it at different points, and it's not going anywhere. For the beam, the net force and the net torque must both be zero. Our next step is to draw the visual overview in FIGURE 8.3, in which we clearly identify both the forces that act on the beam and where they act. There is a downward force on the beam at the point where the bucket hangs, an upward force where the man supports the beam, and an upward force at the pivot. We will measure the distances from the pivot, which is the natural point about which to compute torques.

FIGURE 8.2 Using a beam to lift a load.

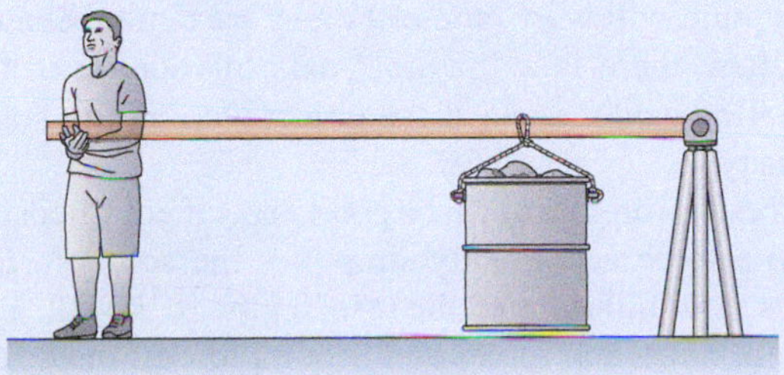

FIGURE 8.3 The forces acting on the beam.

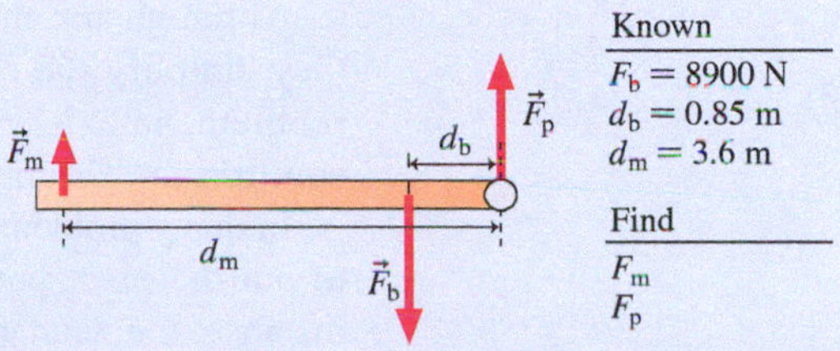


*Continued*

We've made some assumptions in drawing the diagram, assumptions that we will use to check our work at the end. The bucket is closer to the pivot than to the man, so we've drawn the force of the pivot as greater than the force of the man. And we expect that the total force supporting the beam—the upward force of the man plus the upward force of the pivot—will be equal to the weight of the bucket. We've drawn the arrows to match.

**PREPARE** We can add some details to the visual overview of Figure 8.3. The force of the bucket on the beam is equal to the weight of the bucket:

$$F_b = w_b = m_b g = (910 \text{ kg})(9.8 \text{ m/s}^2) = 8900 \text{ N}$$

This force acts a distance $d_b = 0.85$ m from the pivot. The man exerts a force 3.6 m from the pivot. We don't know the force of the man on the beam or the force of the pivot on the beam; these are the numbers we are trying to find.

**SOLVE** The first criterion for equilibrium is that there is no net force. There are no forces acting horizontally, so we consider the condition $\sum F_y = 0$, which gives

$$\sum F_y = F_m + F_p - F_b = 0$$

This clearly isn't enough information to solve the problem; there are two unknown forces in this equation. But we've got another condition, $\sum \tau = 0$. All the forces act perpendicular to the beam, so we can compute the torques as $\tau = rF_\perp$, where $F_\perp$ is the magnitude of each force. There is no torque from the force of the pivot because this force acts right at the pivot. The force of the bucket tries to rotate the beam counterclockwise, while the force of the man tries to rotate the beam clockwise. So our torque equation is

$$\sum \tau = F_b d_b - F_m d_m = 0$$

We know the value of every quantity in this equation except $F_m$, so we can calculate the force from the man:

$$F_m = F_b \frac{d_b}{d_m} = (8900 \text{ N})\frac{0.85 \text{ N}}{3.6 \text{ m}} = 2100 \text{ N}$$

We can now revisit the equation for forces to compute the force of the pivot:

$$F_p = F_b - F_m = 8900 \text{ N} - 2100 \text{ N} = 6800 \text{ N}$$

**ASSESS** The force the man must apply is 2100 N, the weight of about 200 kg. That's a lot of force, but it is a strongman competition, so this makes sense. This force is less than the force of the pivot, as we expected.

**STOP TO THINK 8.1** Which of these objects is in static equilibrium?

## Choosing the Pivot Point

In Chapter 7, we saw that the value of the torque depends on the choice of the pivot point. In Example 8.1, there was a natural choice for the pivot point, but we could have chosen another point. Would that choice have affected the result of our calculation?

Consider the hammer shown in FIGURE 8.4, supported on a pegboard by two pegs A and B. Because the hammer is in static equilibrium, the net torque around the pivot at peg A must be zero: The clockwise torque due to the weight $\vec{w}$ is exactly balanced by the counterclockwise torque due to the force $\vec{n}_B$ of peg B. (Recall that the torque due to $\vec{n}_A$ is zero because here $\vec{n}_A$ acts at the pivot A.) But if instead we take B as the pivot, the net torque is still zero. The counterclockwise torque due to $\vec{w}$ (with a large force but small moment arm) balances the clockwise torque due to $\vec{n}_A$ (with a small force but large moment arm). Indeed, **for an object in static equilibrium, the net torque about *every* point must be zero.** This means you can pick *any* point you wish as a pivot point for calculating the torque.

FIGURE 8.4 A hammer resting on two pegs.

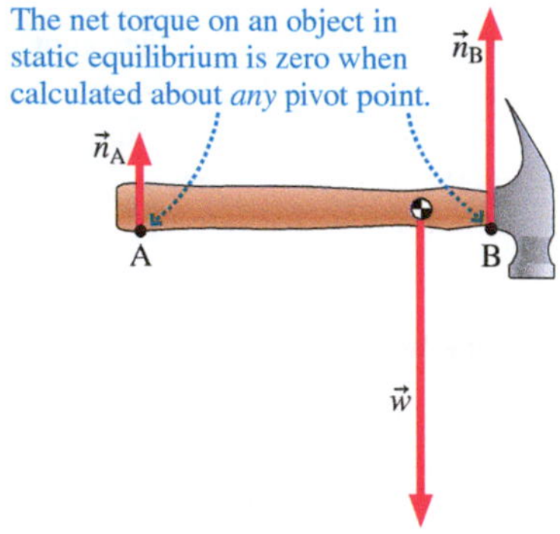


Although any choice of a pivot point will work, some choices are better because they simplify the calculations. Often, there is a "natural" axis of rotation in the problem, an axis about which rotation *would* occur if the object were not in static equilibrium. Example 8.1 is of this type.

In many problems, there are forces that are unknown or poorly specified. Choosing to put the pivot point where such a force acts greatly simplifies the solution. For instance, the woman in FIGURE 8.5 is in equilibrium as she rests on the rock wall. The force of the wall on her feet is a mix of normal forces and friction forces; the direction isn't well specified. The directions of the other two forces are well known: The tension force points along the rope, and the weight force points down. So a good choice of pivot point is where the woman's foot contacts the wall because this choice eliminates the torque due to the force of the wall on her foot, which isn't as well known as the other forces.

FIGURE 8.5 Choosing the pivot for a woman rappelling down a rock wall.

The torque due to $\vec{F}$ about this point is zero. This makes this point a good choice as the pivot.

## PROBLEM-SOLVING APPROACH 8.1 Static equilibrium problems

For situations involving static equilibrium, we can use the fact that there is no net force and no net torque as a basis for solving problems.

***STRATEGIZE*** Begin by making decisions:

- Decide on an object of interest. This should be an extended object with different forces acting on it at different points.
- Decide on a pivot point about which to compute torques.

Next, model the object as a simple shape, and begin to draw a visual overview of the system.

***PREPARE*** Add detail to your visual overview:

- Determine what forces act, and where they act. Add them to your diagram, and list values (if known) in your table.
- Determine the distances from the pivot to the points where forces act.

Think about the forces that act and the torques that they create. Think about whether the torques will be positive or negative. Remember that the torques due to any forces acting *at* the pivot are zero.

***SOLVE*** The mathematical steps are based on the conditions:

$$\vec{F}_{\text{net}} = \vec{0} \qquad \text{and} \qquad \tau_{\text{net}} = 0$$

- Write equations for $\sum F_x = 0$, $\sum F_y = 0$, and $\sum \tau = 0$.
- Solve the resulting equations.

***ASSESS*** Check that your result is reasonable and answers the question.

## EXAMPLE 8.2 Forces on a board resting on sawhorses

A board weighing 100 N sits across two sawhorses, as shown in FIGURE 8.6. What are the magnitudes of the normal forces of the sawhorses acting on the board?

FIGURE 8.6 A board sitting on two sawhorses.

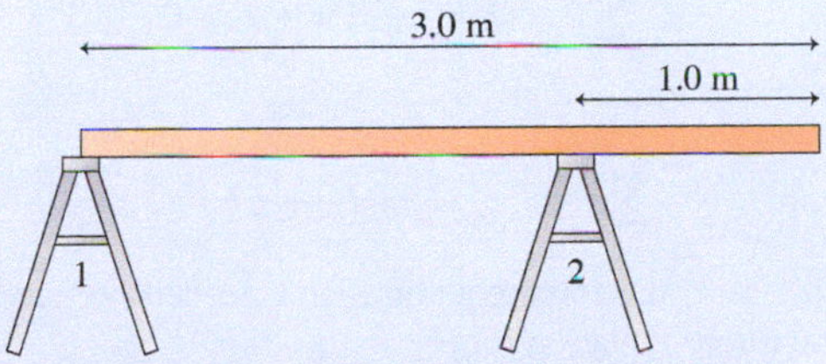


FIGURE 8.7 Visual overview of a board on two sawhorses.

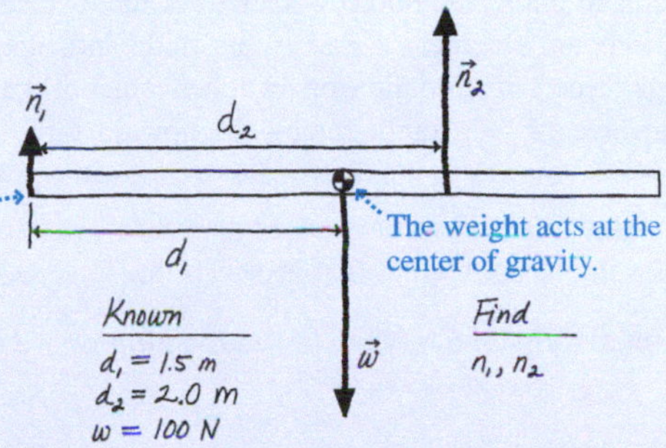


***STRATEGIZE*** We'll choose the object of interest to be the board. There are two supporting forces acting on it at separated points, plus the weight of the board itself. As discussed previously, a good choice for the pivot is a point at which an unknown force acts because that force contributes nothing to the torque. We know the weight of the board, but we don't know the forces from either of the sawhorses, so a good choice for the pivot is at one of the sawhorses. We'll choose the left end of the board, where sawhorse 1 supports the board.

The board and the forces acting on it are shown in FIGURE 8.7. We will simplify things by having the forces from the sawhorses act at single points, the center point of each sawhorse. The center of gravity of the board is closer to sawhorse 2, so we will expect the force from this sawhorse to be greater.

***PREPARE*** We add detail to the visual overview in Figure 8.7. $\vec{n}_1$ and $\vec{n}_2$ are the normal forces of the sawhorses supporting the

*Continued*

board, and $\vec{w}$ is the weight of the board, which acts at the center of gravity. The moment arm for $\vec{w}$ is $d_1 = 1.5$ m, half the board's length. Because $\vec{w}$ tends to rotate the board clockwise, its torque is negative. The moment arm for $\vec{n}_2$ is the distance of the second sawhorse from the pivot point, which is $d_2 = 2.0$ m. This force tends to rotate the board counterclockwise, so it exerts a positive torque.

**SOLVE** The board is in static equilibrium, so the net force $\vec{F}_{net}$ and the net torque $\tau_{net}$ must both be zero. The forces have only $y$-components, so the force equation is

$$\sum F_y = n_1 - w + n_2 = 0$$

As we've seen, the gravitational torque is negative and the torque from the upward force from the sawhorse is positive, so the torque equation is

$$\tau_{net} = -wd_1 + n_2 d_2 = 0$$

We now have two equations with the two unknowns $n_1$ and $n_2$. We can solve for $n_2$ in the torque equation and then substitute that result into the force equation. From the torque equation,

$$n_2 = \frac{d_1 w}{d_2} = \frac{(1.5\text{ m})(100\text{ N})}{2.0\text{ m}} = 75\text{ N}$$

The force equation is then $n_1 - 100\text{ N} + 75\text{ N} = 0$, which we can solve for $n_1$:

$$n_1 = w - n_2 = 100\text{ N} - 75\text{ N} = 25\text{ N}$$

**ASSESS** $n_2 > n_1$, so sawhorse 2 supports more weight, as we expected.

A video to support a section's topic is embedded in the eText.

**Video** Walking the Plank

The center of gravity of a uniform board is easy to see—it's right at the midpoint—but how about the center of gravity of your body? In Chapter 7, we saw a model for the human body that gave average positions of the center of gravity of different body segments for an 80 kg man, but these were average positions. Individuals vary quite a bit around these averages, as does the exact location of the overall center of gravity.

The location of your center of gravity depends on how you are posed, but consider the simple case of standing up straight with your arms at your side. Your center of gravity is approximately at the midpoint of your body, but to determine the position of your center of gravity with any accuracy, you need to measure it. The next example illustrates the use of a reaction board and a scale to make this determination. This is a standard measurement in biomechanics.

### EXAMPLE 8.3 Finding the center of gravity of the human body BIO

A woman weighing 600 N lies on a 2.5-m-long, 60 N reaction board with her feet over a fixed support. The scale on the right reads 250 N. What is the distance $d$ from the woman's feet to her center of gravity?

**STRATEGIZE** The problem is about the woman and the position of her center of gravity, but it makes sense to take the woman and the board, considered together, as the object of interest. As we show in the visual overview in FIGURE 8.8, there is an upward force from the scale, a downward weight force acting on the board (at the board's center of gravity, its midpoint), and a downward weight force acting on the woman at her center of gravity. We know these forces, but a fourth force—the upward force from the support—is unknown, so it makes sense to choose the pivot point here.

**PREPARE** We add details to Figure 8.8. The woman's weight is known, but not the position at which this force acts, the position of her center of gravity; this is what we are looking for. The known weight force of the board acts at its midpoint, and the scale exerts a known upward normal force at the end of the board. The two weight forces exert negative torques, and the normal force of the scale exerts a positive torque. There is no torque due to the force from the support because we have taken the pivot point here.

FIGURE 8.8 Visual overview of the reaction board and woman.

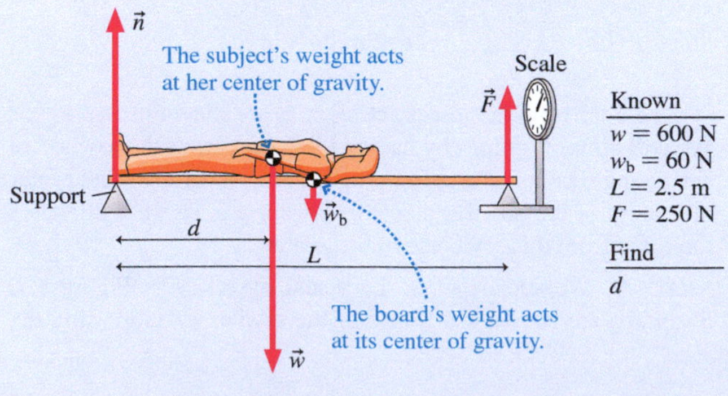


**SOLVE** Because the board and woman are in static equilibrium, the net force and net torque on them must be zero. The force equation reads

$$\sum F_y = n - w_b - w + F = 0$$

and the torque equation gives

$$\sum \tau = -\frac{L}{2}w_b - dw + LF = 0$$

In this case, the force equation isn't needed because we can solve the torque equation for $d$:

$$d = \frac{LF - \frac{1}{2}Lw_b}{w} = \frac{(2.5\text{ m})(250\text{ N}) - \frac{1}{2}(2.5\text{ m})(60\text{ N})}{600\text{ N}}$$
$$= 0.92\text{ m}$$

**ASSESS** If the woman is 5′ 6″ (1.68 m) tall, her center of gravity is $(0.92\text{ m})/(1.68\text{ m}) = 55\%$ of her height, or a little more than halfway up her body. This seems reasonable and this is, in fact, a typical value for women. The center of gravity for men tends to be a bit higher in the body.

## EXAMPLE 8.4 Will the ladder slip?

A 3.0-m-long ladder leans against a wall at an angle of 60° with respect to the floor. What is the minimum value of $\mu_s$, the coefficient of static friction with the ground, that will prevent the ladder from slipping? Assume that friction between the ladder and the wall is negligible.

**STRATEGIZE** The object of interest in this problem is straightforward to determine—it's the ladder. You know that the ladder would slide if there was no friction force; this force must be large enough to keep the ladder in static equilibrium. There is also a natural choice of pivot point in this problem. As FIGURE 8.9 shows, at the point where the ladder touches the ground, there are two forces—a normal force and a friction force. Placing the pivot point here means that there will be no torque for these forces, so the torque condition for equilibrium will be easily solved. Once we've completed the torque condition, we'll turn to the force conditions—one for the $x$-axis, one for the $y$-axis. At this point, friction will reenter the picture, as it must; the problem asks about the coefficient of friction. We are not given the weight of the ladder, so at some point in our process, the weight should cancel; this will be a good check on our work.

FIGURE 8.9 Visual overview of a ladder in static equilibrium.

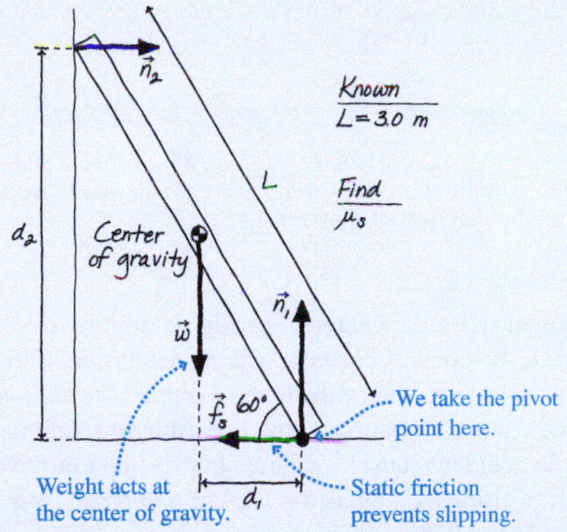


**PREPARE** Let's continue developing the details in the visual overview in Figure 8.9. The four vectors show where the forces act. It's easy to visualize the moment arm for the wall's normal force, $d_2$, and the moment arm for the weight force, $d_1$, so we compute torques using moment arms. Given the angles and distances noted in the figure, the two moment arms are

$$d_1 = \frac{1}{2}L\cos 60^\circ \quad d_2 = L\sin 60^\circ$$

We don't know the values of any of the forces, but we can write the relationship between the friction force and the normal force of the ground: $f_s \le \mu_s n_1$.

**SOLVE** There is no net torque about the pivot point we've chosen, and only two forces that exert torques. The weight force exerts a positive torque, $\tau_w = d_1 w$. The normal force exerts a negative torque, $\tau_n = -d_2 n_2$. Our torque condition is thus

$$\sum \tau = d_1 w - d_2 n_2 = 0$$

We can substitute the expressions we found for the distances and solve for $n_2$:

$$\frac{1}{2}(L\cos 60^\circ)w = (L\sin 60^\circ)n_2$$

$$n_2 = \frac{\frac{1}{2}(L\cos 60^\circ)w}{L\sin 60^\circ} = \frac{w}{2\tan 60^\circ}$$

There is no net torque, but there is also no net force. This gives us two additional conditions:

$$\sum F_x = n_2 - f_s = 0$$
$$\sum F_y = n_1 - w = 0$$

The second force condition simplifies to

$$n_1 = w$$

Our condition for the friction force thus becomes

$$f_s \le \mu_s w$$

The first force condition simplifies to

$$f_s = n_2$$

We then substitute the expression for $n_2$ that we found earlier, giving

$$f_s = \frac{w}{2\tan 60^\circ}$$

Finally, we combine our expressions for the friction force:

$$f_s = \frac{w}{2\tan 60^\circ} \le \mu_s w$$

The weight force cancels, and we simplify to

$$\mu_s \ge \frac{1}{2\tan 60^\circ} = 0.29$$

The minimum value of the static coefficient of friction is thus 0.29.

**ASSESS** The weight force did cancel out as we worked, which gives us confidence in our process. Our final result makes sense as well. You have certainly seen ladders leaning against walls at angles like this with no slipping, so we expect a final value for the coefficient of friction that's relatively modest and that most surfaces will satisfy. 0.29 is less than values we've seen for concrete and other surfaces, so our result makes sense.

**STOP TO THINK 8.2** A beam with a pivot on its left end is suspended from a rope. In which direction is the force of the pivot on the beam?

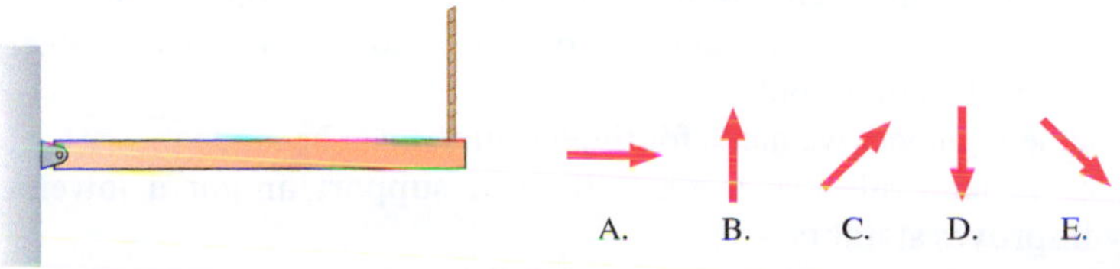

## 8.2 Stability and Balance

**Video** Center of Gravity and Stability

If you tilt a box up on one edge by a small amount and let go, it falls back down. If you tilt it too much, it falls over. And if you tilt it "just right," you can get the box to balance on its edge. What determines these three possible outcomes?

FIGURE 8.10 illustrates the idea with a car, but the results are general and apply in many situations. An extended object has a *base of support* on which it rests when in static equilibrium. If you tilt the object, one edge of the base of support becomes a pivot point. As long as the object's center of gravity remains over the base of support, torque due to gravity will rotate the object back toward its stable equilibrium position; we say that the object is **stable**. This is the situation in Figure 8.10b.

FIGURE 8.10 A car—or any object—will fall over when tilted too far.

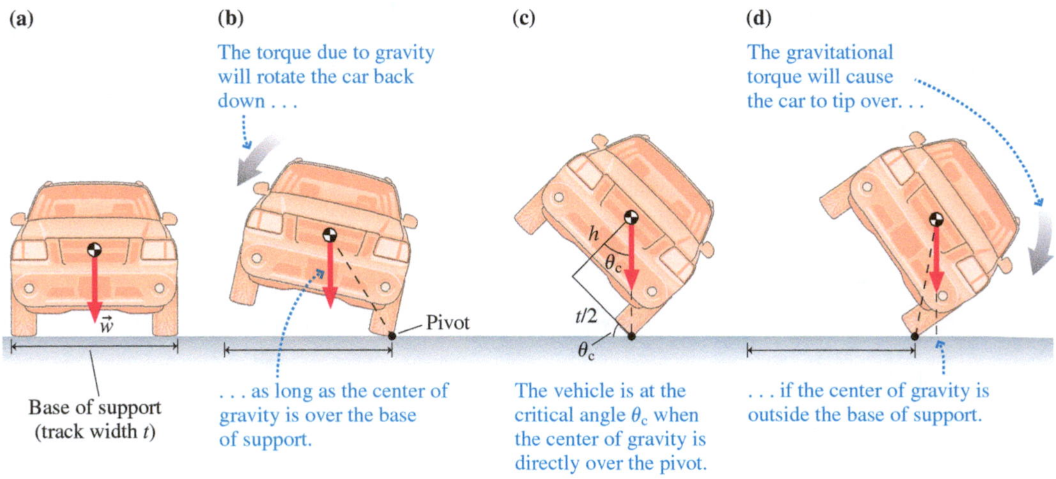


**Video** Figure 8.10

A *critical angle* $\theta_c$ is reached when the center of gravity is directly over the pivot point, as in Figure 8.10c. This is the point of balance, with no net torque. If the car continues to tip, the center of gravity moves outside the base of support, as in Figure 8.10d. Now, the gravitational torque causes a rotation in the opposite direction and the car rolls over; it is **unstable**. If an accident causes a vehicle to pivot up onto two wheels, it will roll back to an upright position as long as $\theta < \theta_c$, but it will roll over if $\theta > \theta_c$.

For vehicles, the distance between the tires—the base of support—is called the track width $t$. The height of the center of gravity is $h$. You can see from Figure 8.10c that the ratio of the two, the height-to-width ratio, determines when the critical angle is reached. An analysis of the triangle in the third image in Figure 8.10 shows how the critical angle depends on the track width $t$ and the height of the center of gravity $h$:

$$\theta_c = \tan^{-1}\left(\frac{\frac{1}{2}t}{h}\right)$$

Decreasing the height of the center of gravity or increasing the width of the base of support will increase the critical angle, leading to a more stable situation.

FIGURE 8.11 The passenger car has a lower center of gravity relative to its track width than does an SUV.

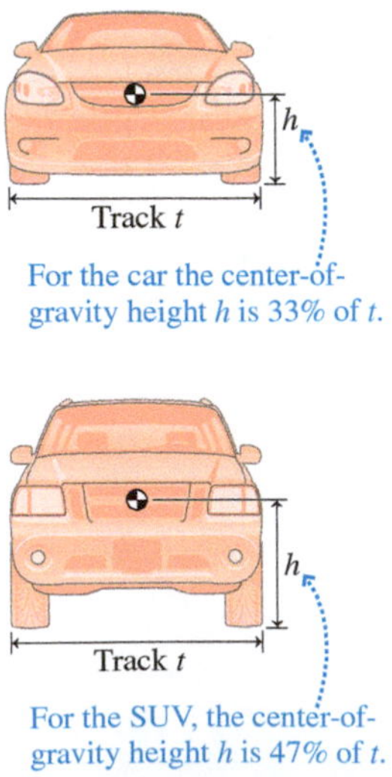


FIGURE 8.11 compares a passenger car and a sport utility vehicle (SUV). For the passenger car, with $h/t = 0.33$, the critical angle is $\theta_c = 57°$. But for the SUV, with $h/t = 0.47$, the critical angle is $\theta_c = 47°$. Loading an SUV with cargo further raises the center of gravity, especially if the roof rack is used, reducing $\theta_c$ even more. Various automobile safety groups have determined that a vehicle with $\theta_c > 50°$ is unlikely to roll over in an accident. A rollover becomes increasingly likely as $\theta_c$ is reduced below this threshold.

The same argument we made for tilted vehicles can be made for any object, leading to the general rule that **a wider base of support and/or a lower center of gravity improves stability.**

**CONCEPTUAL EXAMPLE 8.5** **Understanding ladder safety**

A man is standing on a ladder to paint a wall. He (unwisely) leans out from the side of the ladder, as in FIGURE 8.12. If he leans too far, the ladder will tip to the side. Explain why and when the ladder will tip.

***REASON*** For stability, the combined center of gravity of the man and the ladder must be over the ladder's base. When the man leans to the side, his center of gravity shifts. This changes the position of the combined center of gravity quite a bit; the man has more mass than the ladder, so the combined center of gravity is closer to his center of gravity than that of the ladder. FIGURE 8.13 shows that the situation is at a literal tipping point—if the man moves any farther to the left, the center of gravity will be to the left of the ladder's supports, and the ladder will tip.

***ASSESS*** This result makes sense to anyone who has used a ladder for such work; there is a very narrow lean angle for safe use. The design of the ladder, with a wider base, enhances safety, but only up to a point.

FIGURE 8.12 Unsafe lean on a ladder.

FIGURE 8.13 Analyzing the unsafe lean.

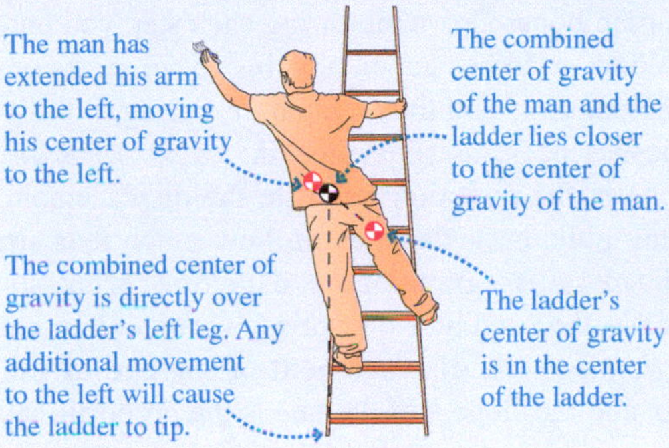

The man has extended his arm to the left, moving his center of gravity to the left.
The combined center of gravity of the man and the ladder lies closer to the center of gravity of the man.
The combined center of gravity is directly over the ladder's left leg. Any additional movement to the left will cause the ladder to tip.
The ladder's center of gravity is in the center of the ladder.


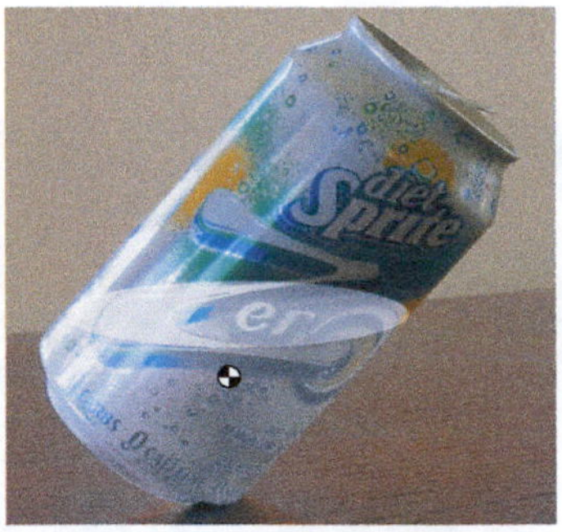

**Balancing a soda can** Try to balance a soda can—full or empty—on the narrow bevel at the bottom. It can't be done because, either full or empty, the center of gravity is near the center of the can. If the can is tilted enough to sit on the bevel, the center of gravity lies far outside this small base of support. But if you put about 2 ounces (60 ml) of water in an empty can, the center of gravity will be right over the bevel and the can will balance.

## Stability and Balance of the Human Body BIO

The human body is remarkable for its ability to constantly adjust its stance to remain stable on just two points of support. In walking, running, or even the simple act of rising from a chair, the position of the body's center of gravity is constantly changing. To maintain stability, we unconsciously adjust the positions of our arms and legs to keep our center of gravity over our base of support.

FIGURE 8.14a shows the body in its normal standing position. Notice that the center of gravity is well centered over the base of support (the feet), ensuring stability. If the person were now to stand on tiptoes *without* otherwise adjusting the body position, her center of gravity would fall behind the base of support, which is now the balls of the feet, and she would fall backward. To prevent this, as shown in FIGURE 8.14b, the body naturally leans forward, regaining stability by moving the center of gravity over the balls of the feet. Try this: Stand facing a wall with your toes touching the base of the wall. Now try standing on your toes. Your body can't move forward to keep your center of gravity over your toes, so you can't do it!

The chapter opened with a photo of a dancer who is performing a very delicate balancing act, adjusting the positions of her torso, head, arms, and legs so that her center of gravity is above a very small point of support. Now that you understand the basics of balance, you can see what a remarkable feat this is!

FIGURE 8.14 Standing on tiptoes.

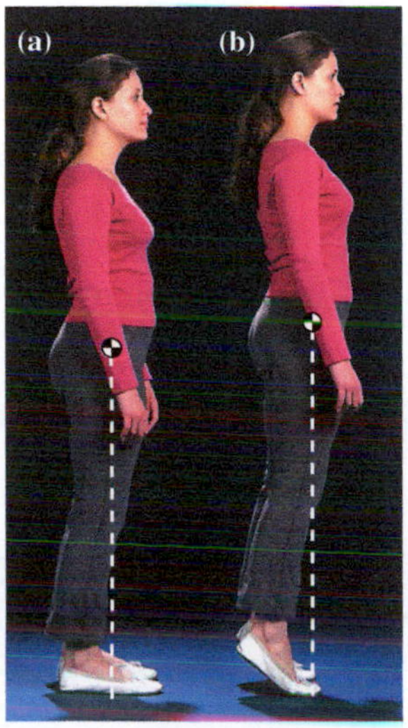


**STOP TO THINK 8.3** Rank in order, from least stable to most stable, the three objects shown in the figure. The positions of their centers of gravity are marked. (For the centers of gravity to be positioned like this, the objects must have a nonuniform composition.)

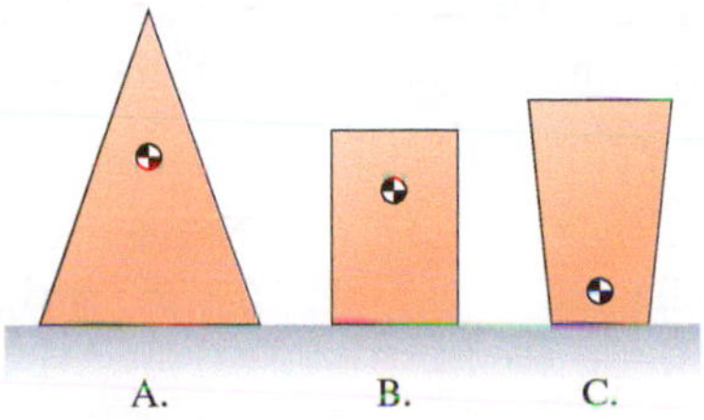

## 8.3 Springs and Hooke's Law

**Elasticity in action** A golf ball compresses quite a bit when struck. The restoring force that pushes the ball back into its original shape helps launch the ball off the face of the club, making for a longer drive.

We have assumed that objects in equilibrium maintain their shape as forces and torques are applied to them. In reality this is an oversimplification. Every solid object stretches, compresses, or deforms when a force acts on it. This change is easy to see when you press on a green twig on a tree, but even the largest branch on the tree will bend slightly under your weight.

If you stretch a rubber band, there is a force that tries to pull the rubber band back to its equilibrium, or unstretched, length. A force that restores a system to an equilibrium position is called a **restoring force.** Systems that exhibit such restoring forces are called **elastic.** The most basic examples of **elasticity** are things like springs and rubber bands. We introduced the spring force in ◀◀ SECTION 4.2. As we saw, if you stretch a spring, a tension-like force pulls back. Similarly, a compressed spring tries to re-expand to its equilibrium length. Elasticity and restoring forces are properties of much stiffer systems as well. The steel beams of a bridge bend slightly as you drive your car over it, but they are restored to equilibrium after your car passes by. Your leg bones flex a bit during each step you take.

When no forces act on a spring to compress or extend it, it will rest at its equilibrium length. If you then stretch the spring, it pulls back. If you compress the spring, it pushes back, as illustrated in FIGURE 8.15a. In general, the spring force always points in the direction opposite the displacement from equilibrium. How hard the spring pulls back depends on how much it is stretched, as shown in FIGURE 8.15b. FIGURE 8.15c is a graph of real data for a spring, showing the magnitude of the spring force as the stretch of the spring is varied. You can see that **the spring force is *proportional* to the displacement of the end of the spring.** This is a *linear relationship,* and the slope $k$ of the line is the proportionality constant:

$$F_{sp} = k\,\Delta x \qquad (8.2)$$

Compressing or stretching the spring twice as far results in a restoring force that is twice as large.

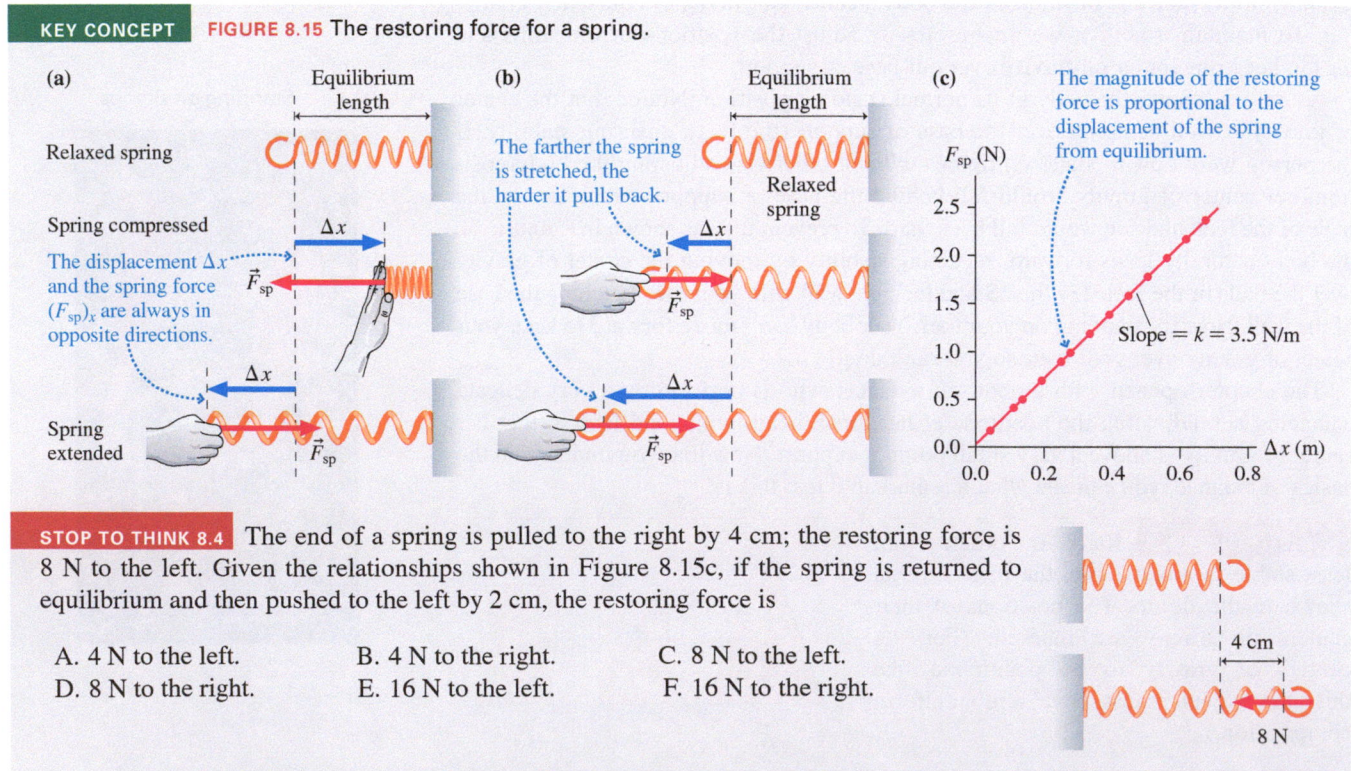

KEY CONCEPT **FIGURE 8.15** The restoring force for a spring.

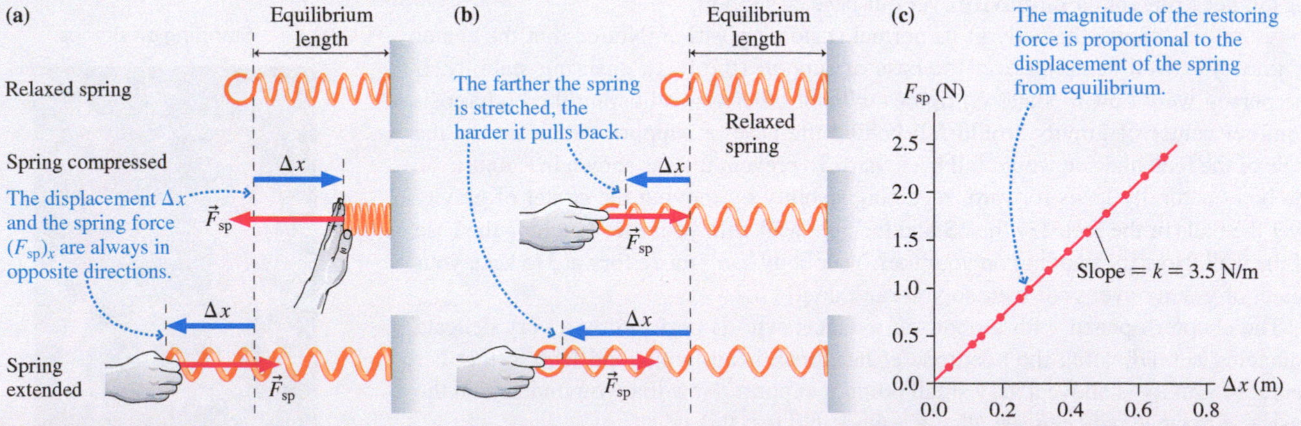


**STOP TO THINK 8.4** The end of a spring is pulled to the right by 4 cm; the restoring force is 8 N to the left. Given the relationships shown in Figure 8.15c, if the spring is returned to equilibrium and then pushed to the left by 2 cm, the restoring force is

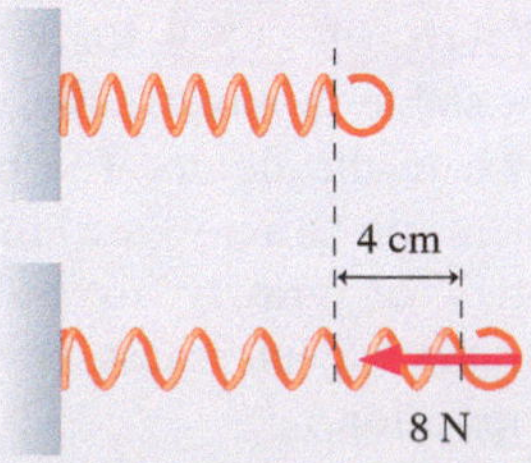


A. 4 N to the left.
B. 4 N to the right.
C. 8 N to the left.
D. 8 N to the right.
E. 16 N to the left.
F. 16 N to the right.

The value of the constant $k$ in Equation 8.2 depends on the spring. We call $k$ the **spring constant**; it has units of N/m. The spring constant $k$ is a property that characterizes a spring, just as the mass $m$ characterizes a particle. If $k$ is large, it takes a large pull to cause a significant stretch, and we call the spring a "stiff" spring. If $k$ is small, we can stretch the spring with very little force, and we call it a "soft" spring. Every spring has its own unique value of $k$. The spring constant for the data in Figure 8.15c can be determined from the slope of the straight line to be $k = 3.5$ N/m.

As Figure 8.15a shows, if the spring is compressed, $\Delta x$ is positive and, because $\vec{F}_{sp}$ points to the left, its component $(F_{sp})_x$ is negative. If the spring is stretched, however, $\Delta x$ is negative and, because $\vec{F}_{sp}$ points to the right, its component $(F_{sp})_x$ is positive. If we rewrite Equation 8.2 in terms of the *component* of the spring force, we get the most general form of the relationship between the restoring force and the displacement of the end of a spring, which is known as **Hooke's law**:

$$(F_{sp})_x = -k\Delta x \qquad (8.3)$$

*x*-component of the restoring force of the spring (N) → $(F_{sp})_x$

Displacement of the end of the spring (m) → $\Delta x$

Spring constant (N/m) → $k$

The negative sign says the restoring force and the displacement are in *opposite* directions.

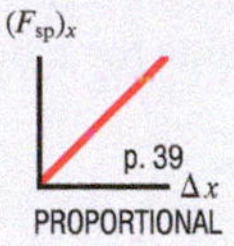


For motion in the vertical ($y$) direction, Hooke's law is $(F_{sp})_y = -k\,\Delta y$.

Hooke's law is not a true "law of nature" in the sense that Newton's laws are. It is actually just a model of a restoring force. It works extremely well for some springs, such as the one in Figure 8.15c, but less well for others. Hooke's law will fail for any spring that is compressed or stretched too far, as we'll see in the next section.

NOTE ▶ Just as we used massless strings, we will adopt the idealization of a *massless spring*. Though not a perfect description, it is a good approximation if the mass attached to a spring is much greater than the mass of the spring itself. ◀

## EXAMPLE 8.6 A spring in your step

Several companies manufacture running shoes that have thick heel pads containing steel springs to launch a runner back into the air after a heel strike. When a 70 kg runner rocks back on his heels so that all of his weight is supported by the heel springs, the springs each compress by 1.2 mm. When he is running hard and hits the ground on one heel, the force compressing the spring is 5.0 times his weight.

a. What is the spring constant $k$ of a heel spring?
b. By how much do the springs compress during a heel strike?

**STRATEGIZE** When the springs in the heels are compressed, they produce a restoring force that pushes upward on the runner's foot. When the runner rocks back on his heels, we can assume that half of the runner's weight is on each heel. We know the force that is compressing the spring and how much it compresses, so we will use those values to find the spring constant. With the spring constant in hand, we will find the compression in part b.

**PREPARE** The visual overview in FIGURE 8.16 shows the situation at one heel. The runner is in static equilibrium, so

$$\sum F_y = F_{sp} - \frac{1}{2}w = 0$$

$$F_{sp} = \frac{1}{2}mg$$

FIGURE 8.16 Visual overview of the heel spring.

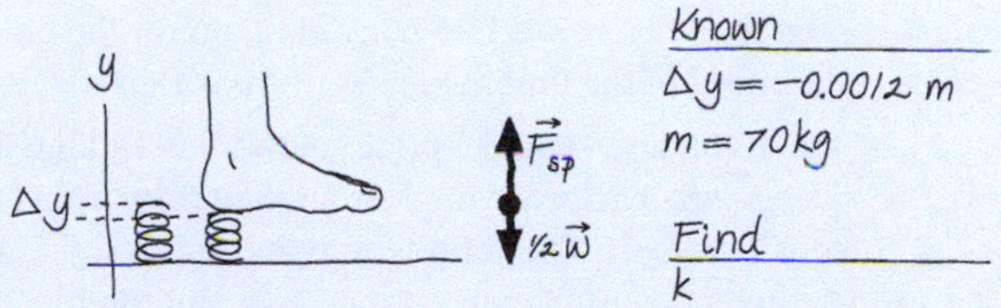


The spring's restoring force is determined by the compression, according to Equation 8.2:

$$F_{sp} = k\Delta y = k(0.0012\text{ m})$$

**SOLVE**

a. We know the restoring force of the spring, and we know how this is related to the compression. Combining these, we get

$$\frac{1}{2}(70\text{ kg})g = k(0.0012\text{ m})$$

$$k = 2.9 \times 10^5\text{ N/m}$$

*Continued*

b. In the second case, the restoring force is 5.0 times the weight. We use this force to find the new compression of the spring:

$$(5.0)mg = (29 \times 10^5 \text{ N/m})\Delta y$$
$$\Delta y = 0.012 \text{ m} = 1.2 \text{ cm}$$

During the greater force of a heel strike, the spring compresses by 1.2 cm.

**ASSESS** The spring is quite stiff, approximately 300,000 N/m. This is reasonable; the spring compresses only a little under the large force of the runner's weight. The force of the heel strike is 5 times the runner's weight, 10 times the original force, half the runner's weight. This gives a compression that is 10 times the original compression, 1.2 cm instead of 1.2 mm, as we'd expect.

### EXAMPLE 8.7 When does the block slip?

FIGURE 8.17 shows a spring attached to a 2.0 kg block. The other end of the spring is pulled by a motorized toy train that moves forward at 5.0 cm/s. The spring constant is 50 N/m, and the coefficient of static friction between the block and the surface is 0.60. The spring is at its equilibrium length at $t = 0$ s when the train starts to move. When does the block slip?

FIGURE 8.17 A toy train stretches the spring until the block slips.

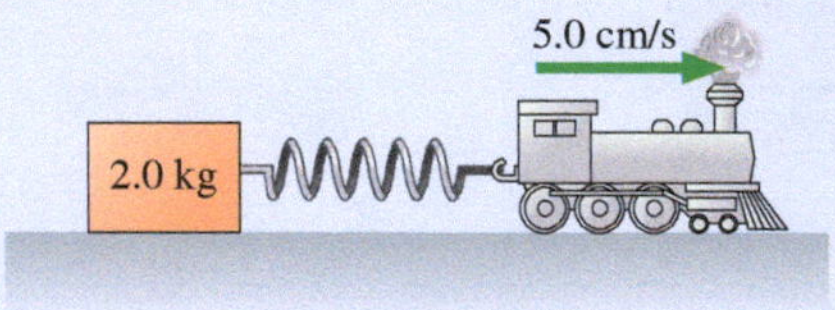


**STRATEGIZE** Recall that the tension in a massless string pulls equally at both ends of the string. The same is true for the spring force: It pulls (or pushes) equally at both ends. Imagine holding a rubber band with your left hand and stretching it with your right hand. Your left hand feels the pulling force, even though it is the right end of the rubber band that moves. This is the key to solving the problem: As the train moves, the spring stretches and the spring force increases. This spring force pulls to the right on the block. The static friction force can increase to keep the block in place, but there is a limit to the value of this force. At some point, the spring force will exceed the maximum value of the friction force, and then the block will slide.

**PREPARE** We model the block as a particle and the spring as a massless spring. FIGURE 8.18 is a free-body diagram for the block. We convert the speed of the train into m/s: $v = 0.050$ m/s.

**SOLVE** As the right end of the spring moves, stretching the spring, the spring pulls backward on the train *and* forward on the block with equal strength. The train is moving to the right, and so the spring force pulls to the left on the train. But the block is at the other end of the spring; the spring force pulls to the right

FIGURE 8.18 Free-body diagram for the block.

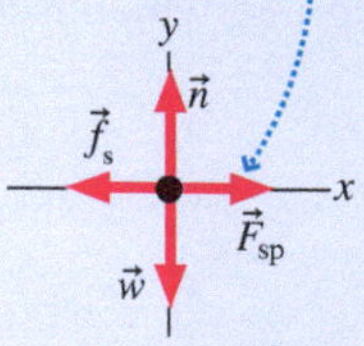


on the block, as shown in Figure 8.18. As the spring stretches, the static friction force on the block increases in magnitude to keep the block at rest. The block is in static equilibrium, so

$$\sum F_x = (F_{sp})_x + (f_s)_x = F_{sp} - f_s = 0$$

where $F_{sp}$ is the magnitude of the spring force. This magnitude is $F_{sp} = k\,\Delta x$, where $\Delta x = vt$ is the distance the train has moved. Thus

$$f_s = F_{sp} = k\,\Delta x$$

The block slips when the static friction force reaches its maximum value $f_{s\,max} = \mu_s n = \mu_s mg$. This occurs when the train has moved a distance

$$\Delta x = \frac{f_{s\,max}}{k} = \frac{\mu_s mg}{k} = \frac{(0.60)(2.0\text{ kg})(9.8\text{ m/s}^2)}{50\text{ N/m}} = 0.235\text{ m}$$

The time at which the block slips is

$$t = \frac{\Delta x}{v} = \frac{0.235\text{ m}}{0.050\text{ m/s}} = 4.7\text{ s}$$

**ASSESS** The result of about 5 s seems reasonable for a slowly moving toy train to stretch the spring enough for the block to slip.

Example 8.7 is a bit artificial, but it is also instructive. The train moves, stretching the spring, until the restoring force is larger than the friction force. At this point, the block slides forward. Once the block is sliding, the friction is kinetic, with a lower coefficient of friction, so the sliding continues for some time. When the block finally comes to rest, the friction is back to static, and the process begins again. This *stick-slip* process is important in applications ranging from the motion of arthritic joints to the gradual buildup of stress in the earth's crust before an earthquake.

**STOP TO THINK 8.5** A 1.0 kg weight is suspended from a spring, stretching it by 5.0 cm. How much does the spring stretch if the 1.0 kg weight is replaced by a 3.0 kg weight?

A. 5.0 cm B. 10.0 cm C. 15.0 cm D. 20.0 cm

## 8.4 Stretching and Compressing Materials

In Chapter 4 we noted that we can model most solid materials as being made of particle-like atoms connected by spring-like bonds. We can model a steel rod this way, as illustrated in FIGURE 8.19a. The spring-like bonds between the atoms in steel are quite stiff, but they can be stretched or compressed, meaning that even a steel rod is elastic. If you pull on the end of a steel rod, as in Figure 8.19a, you will slightly stretch the bonds between the particles that make it up, and the rod itself will stretch. The stretched bonds pull back on your hand with a restoring force that causes the rod to return to its original length when released. In this sense, the entire rod acts like a very stiff spring. As is the case for a spring, a restoring force is also produced by compressing the rod.

In FIGURE 8.19b, real data for a 1.0-m-long, 1.0-cm-diameter steel rod show that, just as for a spring, the restoring force is proportional to the change in length. However, the *scale* of the stretch of the rod and the restoring force is much different from that for a spring. It would take a force of 16,000 N to stretch the rod by only 1 mm, corresponding to a spring constant of $1.6 \times 10^7$ N/m! Steel is elastic, but under normal forces, it experiences only very small changes in dimension. Materials of this sort are called **rigid.**

The behavior of other materials, such as the rubber in a rubber band, can be different. A rubber band can be stretched quite far with a very small force, and then snaps back to its original shape when released. Materials that show large deformations with small forces are called **pliant.**

A rod's spring constant depends on several factors, as shown in FIGURE 8.20. First, we expect that a thick rod, with a large cross-section area $A$, will be more difficult to stretch than a thinner rod. Second, a rod with a long length $L$ will be easier to stretch by a given amount than a short rod (think of trying to stretch a rope by 1 cm—this would be easy to do for a 10-m-long rope, but it would be pretty hard for a piece of rope only 10 cm long). Finally, the stiffness of the rod will depend on the material that it's made of. Experiments bear out these observations, and it is found that the spring constant of the rod can be written as

$$k = \frac{YA}{L} \tag{8.4}$$

where the constant $Y$ is called **Young's modulus.** Young's modulus is a property of the *material* from which the rod is made—it does not depend on shape or size.

From Equation 8.2, the magnitude of the restoring force for a spring is related to the change in its length as $F_{sp} = k\,\Delta x$. Writing the change in the length of a rod as $\Delta L$, as shown in Figure 8.20, we can use Equation 8.4 to write the restoring force $F$ of a rod as

$$F = \frac{YA}{L}\Delta L \tag{8.5}$$

Equation 8.5 applies both to elongation (stretching) and to compression.

It's useful to rearrange Equation 8.5 in terms of two new ratios, the *stress* and the *strain*:

$$\frac{F}{A} = Y\left(\frac{\Delta L}{L}\right) \tag{8.6}$$

The ratio of force to cross-section area is called **stress.** The ratio of the change in length to the original length is called **strain.**

The strain is a measure of how much the rod stretches. The dimensionless ratio $\Delta L/L$ is the fractional change in the rod's length. If the rod's length changes by 1%, the strain is 0.01. The unit of stress is $N/m^2$. If the stress is due to stretching, we call it a **tensile stress.** Because strain is dimensionless, Young's modulus $Y$ has the same units as stress. TABLE 8.1 lists values of Young's modulus for several rigid materials. Large values of $Y$ characterize materials that are stiff. "Softer" materials have smaller values of $Y$.

If the rod is in static equilibrium, then the magnitude of the restoring force is equal to the force that is stretching the rod, and we can reframe this discussion with

FIGURE 8.19 Stretching a steel rod.

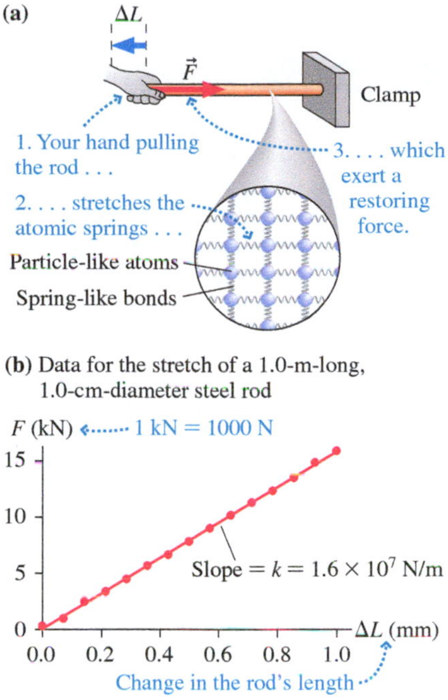


FIGURE 8.20 A rod stretched by length $\Delta L$.

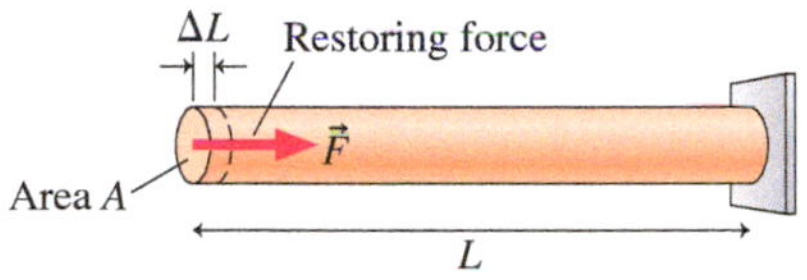


TABLE 8.1 Young's modulus for rigid materials

| Material | Young's modulus ($10^{10}$ N/m$^2$) |
|---|---|
| Cast iron | 20 |
| Steel | 20 |
| Silicon | 13 |
| Copper | 11 |
| Aluminum | 7 |
| Glass | 7 |
| Concrete | 3 |
| Wood (Douglas Fir) | 1 |

the strain—the deformation of the rod—resulting from an applied stress. The mathematics is the same, but this way of thinking is more intuitive in some cases. If we apply a force to a material, the deformation depends on the force we apply as well as the area over which the force is applied.

### EXAMPLE 8.8 Finding the stretch of a cable

A *Foucault pendulum* in a physics department (used to prove that the earth rotates) consists of a 120 kg steel ball that swings at the end of a 6.0-m-long steel cable. The cable has a diameter of 2.5 mm. When the ball was first hung from the cable, by how much did the cable stretch?

**STRATEGIZE** When the ball is suspended from the cable, it stretches—there is a strain. This results in a restoring force—there is a stress. The stress and the strain are related by Equation 8.6.

**PREPARE** The amount by which the cable stretches depends on the elasticity of the steel cable. Young's modulus for steel is given in Table 8.1 as $Y = 20 \times 10^{10}\ \text{N/m}^2$. Equation 8.6 uses the cross-section area of the cable, which is

$$A = \pi r^2 = \pi(0.00125\ \text{m})^2 = 4.91 \times 10^{-6}\ \text{m}^2$$

**SOLVE** Equation 8.6 relates the stretch of the cable $\Delta L$ to the restoring force $F$ and to the properties of the cable. Rearranging terms, we find that the cable stretches by

$$\Delta L = \frac{LF}{AY}$$

The restoring force of the cable is equal to the ball's weight:

$$F = w = mg = (120\ \text{kg})(9.8\ \text{m/s}^2) = 1180\ \text{N}$$

The change in length is thus

$$\Delta L = \frac{(6.0\ \text{m})(1180\ \text{N})}{(4.91 \times 10^{-6}\ \text{m}^2)(20 \times 10^{10}\ \text{N/m}^2)}$$

$$= 0.0072\ \text{m} = 7.2\ \text{mm}$$

**ASSESS** If you've ever strung a guitar with steel strings, you know that the strings stretch several millimeters with the force you can apply by turning the tuning pegs. So a stretch of 7 mm under a 120 kg load seems reasonable.

FIGURE 8.21 Stretch data for a steel rod.

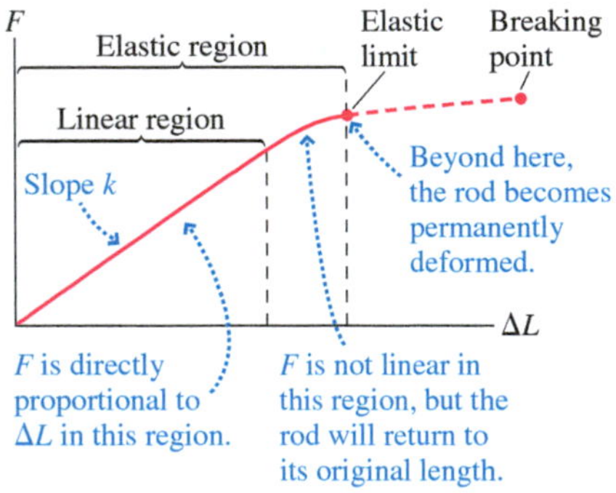


## Beyond the Elastic Limit

In the previous discussion, we found that if we stretch a rod by a small amount $\Delta L$, it will pull back with a restoring force $F$, according to Equation 8.5. But if we continue to stretch the rod, this simple linear relationship between $\Delta L$ and $F$ will eventually break down. FIGURE 8.21 is a graph of the rod's restoring force from the start of the stretch until the rod finally breaks.

As you can see, the graph has a *linear region,* the region where $F$ and $\Delta L$ are proportional to each other, obeying Hooke's law: $F = k\,\Delta L$. **As long as the stretch stays within the linear region, a solid rod acts like a spring and obeys Hooke's law.**

How far can you stretch the rod before damaging it? As long as the stretch is less than the **elastic limit,** the rod will return to its initial length $L$ when the force is removed. The elastic limit is the end of the **elastic region.** Stretching the rod beyond the elastic limit will permanently deform it, and the rod won't return to its original length. Finally, at a certain point the rod will reach a breaking point, where it will snap in two. The force that causes the rod to break depends on the area: A thicker rod can sustain a larger force. For a rod or cable of a particular material, we can determine an *ultimate stress,* also known as the **tensile strength,** the largest stress that the material can sustain before breaking:

$$\text{Tensile strength} = \frac{F_{max}}{A} \quad (8.7)$$

(Tensile strength: Largest stress that can be sustained (N/m²); $F_{max}$: Largest force that can be sustained (N); $A$: Cross-section area (m²))

TABLE 8.2 Tensile strengths of rigid materials

| Material | Tensile strength (N/m²) |
|---|---|
| Polypropylene | $20 \times 10^6$ |
| Glass | $60 \times 10^6$ |
| Cast iron | $150 \times 10^6$ |
| Aluminum | $400 \times 10^6$ |
| Steel | $1000 \times 10^6$ |

TABLE 8.2 lists values of tensile strength for rigid materials. When we speak of the *strength* of a material, we are referring to its tensile strength.

Objects can withstand a certain stress before deforming permanently. It's not the *force* that matters, it's the *stress*—the force divided by the area. Will a heavy weight dent a wood floor? Not if the weight is spread out over a large enough area, to minimize the stress.

We can rewrite Equation 8.7 to solve for the maximum force that can be sustained:

$$F_{max} = (\text{tensile strength})A \quad (8.8)$$

If a cable can't provide enough force to support a certain weight, you can either get a cable made of stronger stuff (higher tensile strength) or get a thicker cable (larger area).

**EXAMPLE 8.9 Finding the weight that a cable can support**

After a late night of studying physics, several 80 kg students decide it would be fun to swing on the Foucault pendulum of Example 8.8. What's the maximum number of students that the pendulum cable could support?

**STRATEGIZE** The weight that can be supported depends on the area and the tensile strength of the cable, both of which we know or can compute.

**PREPARE** The tensile strength for steel is given in Table 8.2 as $1000 \times 10^6\ \text{N/m}^2$, or $1.0 \times 10^9\ \text{N/m}^2$. The area of the cable is

$$A = \pi r^2 = \pi(0.00125\ \text{m})^2 = 4.91 \times 10^{-6}\ \text{m}^2$$

**SOLVE** We can use Equation 8.8 to find the maximum force the cable can support:

$$F_{\text{max}} = (1.0 \times 10^9\ \text{N/m}^2)(4.91 \times 10^{-6}\ \text{m}^2) = 4.9 \times 10^3\ \text{N}$$

This force is the weight of the heaviest mass the cable can support: $w = m_{\text{max}}g$. The maximum mass that can be supported is

$$m_{\text{max}} = \frac{F_{\text{max}}}{g} = 500\ \text{kg}$$

The ball has a mass of 120 kg, leaving 380 kg for the students. Four students have a mass of 320 kg, which is less than this value. But five students, totaling 400 kg, would cause the cable to break.

**ASSESS** Steel has a very large tensile strength, so it's reasonable that this very narrow wire can still support $4900\ \text{N} \approx 1100\ \text{lb}$.

## Elasticity of Biological Materials BIO

Suppose we take equal lengths of spider silk and steel wire, stretch each, and measure the restoring force of each until it breaks. The graph of stress versus strain might appear as in FIGURE 8.22.

The spider silk is certainly less stiff: For a given stress, the silk will stretch about 100 times farther than steel. Interestingly, though, spider silk and steel eventually fail at approximately the same stress. In this sense, spider silk is "as strong as steel." Many pliant biological materials share this combination of low stiffness and large tensile strength. These materials can undergo significant deformations without failing. Tendons, the walls of arteries, and the web of a spider are all quite strong but nonetheless capable of significant stretch.

Most bones in your body are made of two different kinds of bony material: dense and rigid compact bone on the outside, and porous, flexible spongy bone on the inside. FIGURE 8.23 shows a cross section of a typical bone. Compact bone and spongy bone have very different values of Young's modulus. Young's modulus for compact bone approaches that of concrete, so it is very rigid, with little ability to stretch or compress. In contrast, spongy bone has a much lower Young's modulus. Consequently, the elastic properties of bones can be well modeled as those of a hollow cylinder.

The structure of bones in birds actually approximates a hollow cylinder quite well. FIGURE 8.24 shows that a typical bone is a thin-walled tube of compact bone with a tenuous structure of spongy bone inside. Most of a cylinder's rigidity comes from the material near its surface. A hollow cylinder retains most of the rigidity of a solid one, but it is much lighter. Bird bones carry this idea to its extreme.

TABLE 8.3 gives values of Young's modulus for biological materials. Note the large difference between pliant and rigid materials. TABLE 8.4 shows the tensile strengths for biological materials. The values in Table 8.4 are for static forces—forces applied for a long time in a testing machine. Bone can withstand significantly greater stresses if the forces are applied for only a very short period of time.

**TABLE 8.3** Young's modulus for biological materials

| Material | Young's modulus ($10^{10}\ \text{N/m}^2$) |
|---|---|
| Tooth enamel | 6 |
| Compact bone | 1.6 |
| Spongy bone | 0.02–0.3 |
| Spider silk | 0.2 |
| Tendon | 0.15 |
| Cartilage | 0.0001 |
| Blood vessel (aorta) | 0.00005 |

**TABLE 8.4** Tensile strengths of biological materials

| Material | Tensile strength ($\text{N/m}^2$) |
|---|---|
| Spongy bone | $5 \times 10^6$ |
| Compact bone | $100 \times 10^6$ |
| Tendon | $100 \times 10^6$ |
| Spider silk | $1000 \times 10^6$ |

**FIGURE 8.22** Stress-versus-strain graphs for steel and spider silk.

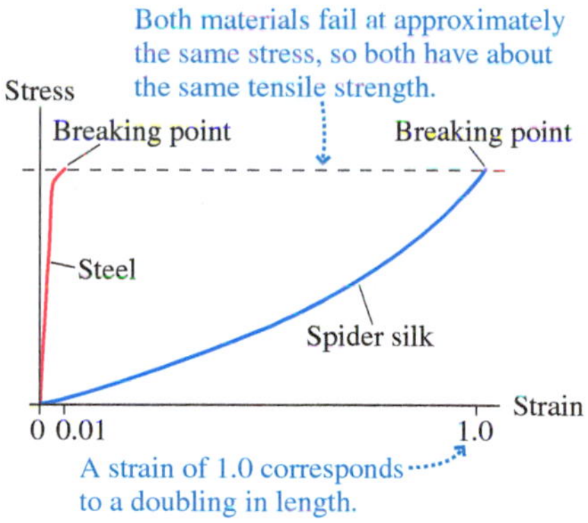


**FIGURE 8.23** Cross section of a long bone.

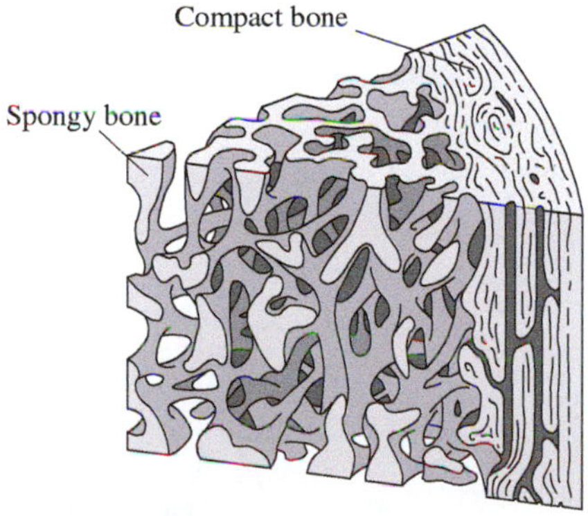


**FIGURE 8.24** Section of a bone from a bird.

Video What the Physics? Elastic Ligament

**EXAMPLE 8.10 Finding the compression of a bone** BIO

The femur, the long bone in the thigh, can be modeled as a tube of compact bone for most of its length. The cross-section area of the compact bone in the femur of a 70 kg person is $4.8 \times 10^{-4}\ \text{m}^2$, a typical value.

a. If this person supports his entire weight on one leg, what fraction of the tensile strength of the bone does this stress represent?
b. By what fraction of its length does the femur shorten?

**PREPARE** The stress on the femur is $F/A$. Here $F$, the force compressing the femur, is the person's weight, so $F = mg$. The fractional change $\Delta L/L$ in the femur is the strain, which we can find using Equation 8.6, taking the value of Young's modulus for compact bone from Table 8.3.

**SOLVE**

a. The person's weight is $mg = (70\ \text{kg})(9.8\ \text{m/s}^2) = 690\ \text{N}$. The resulting stress on the femur is

$$\frac{F}{A} = \frac{690\ \text{N}}{4.8 \times 10^{-4}\ \text{m}^2} = 1.4 \times 10^6\ \text{N/m}^2$$

A stress of $1.4 \times 10^6\ \text{N/m}^2$ is 1.4% of the tensile strength of compact bone given in Table 8.4.

b. We can compute the strain as

$$\frac{\Delta L}{L} = \left(\frac{1}{Y}\right)\frac{F}{A} = \left(\frac{1}{1.6 \times 10^{10}\ \text{N/m}^2}\right)(1.4 \times 10^6\ \text{N/m}^2)$$
$$= 8.8 \times 10^{-5} \approx 0.0001$$

The femur compression is $\Delta L \approx 0.0001L$, or $\approx 0.01\%$ of its length. (The femur is far from a uniform structure, so we've expressed the answer as an approximate result to one significant figure.)

**ASSESS** It makes sense that, under ordinary standing conditions, the stress on the femur is only a percent or so of the maximum value it can sustain.

FIGURE 8.25 Rear legs of model skeletons of a cat (left) and an armadillo (right).

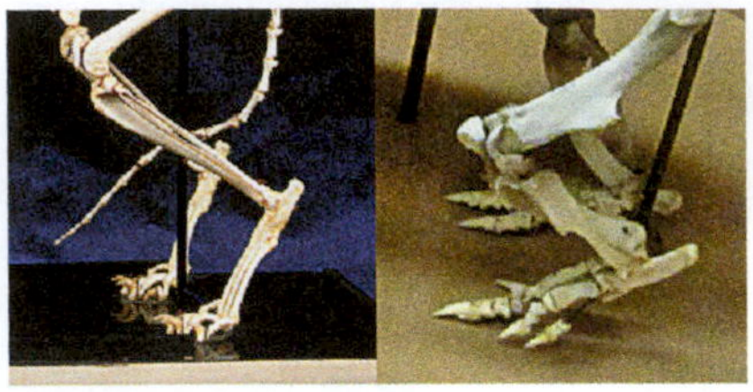

The dancer in the chapter-opening photo stands *en pointe,* balanced delicately on the tip of her shoe with her entire weight supported on a very small area. The stress on the bones in her toes is very large, but it is still much less than the tensile strength of bone.

**STOP TO THINK 8.6** A 10 kg mass is hung from a 1-m-long cable, causing the cable to stretch by 2 mm. Suppose a 10 kg mass is hung from a 2 m length of the same cable. By how much does the cable stretch?

A. 0.5 mm B. 1 mm C. 2 mm D. 3 mm E. 4 mm

## 8.5 Forces and Torques in the Body BIO

FIGURE 8.26 The operation of a nutcracker.

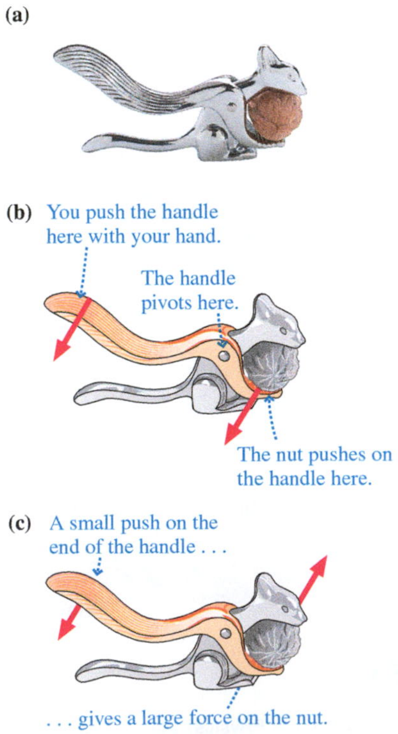


FIGURE 8.25 shows two mounted skeletons, one of a cat and one of an armadillo. Which one is a jumper and which is a digger? It's easy to spot differences between the two skeletons. One obvious difference is the thicker bones of the armadillo, with clear protrusions for muscle attachment. This is a creature that is built for strength. The cat, however, is built for speed, with lighter bones and longer limbs.

It's not just the *thickness* of the bones that speaks to the strength-versus-speed issue—there are also details of the *shapes* of the bones. Look at the ankle joints on the two animals. In particular, note the long extension of the bones of the foot on the armadillo skeleton. This is the attachment for the tendon that pushes the toes downward. In the armadillo, this attachment is quite far back from the joint, which gives a large torque for a given muscle force. This leads to greater force at the toes, a good thing for an animal that gets its food by digging. But there's a cost: Having the attachment point so close to the joint permits large forces but limits the possible speed. In the skeleton of the cat, an agile jumper, we see that the attachment point is much closer to the joint. This permits a greater range of motion as well as more rapid motion. The armadillo is built for strength; the cat is built for speed.

### Mechanical Advantage

FIGURE 8.26a shows a nutcracker. Why do you need a tool to crack a nut? In a word: force. It takes a good deal of force to crack a nut, more than you can easily provide with your muscles alone.

How does using a nutcracker give more force? This is a static equilibrium problem that we can solve. Let's look not at the nut, but at the handle, which pivots about the rivet. The handle is in static equilibrium, so we can use the techniques of this chapter to understand the nutcracker. As FIGURE 8.26b shows, when you push down on the handle, the handle experiences a force from the nut as well. If we take the pivot as the point about which we compute torques, there is a counterclockwise torque from the force of your hand, and a clockwise torque from the force of the nut. The magnitudes of these two torques must be the same, so the force from the nut (which is applied much closer to the pivot) must be much larger than the force of your hand. This is the point of the nutcracker, as FIGURE 8.26c shows. The nut pushes on the handle with a large force, so the handle exerts an equally large force on the nut. A small force on the handle leads to a large force on the nut. For the particular situation illustrated in Figure 8.26, the force on the nut is 3 times the force from your hand. In this way, the tool has amplified the force that you are able to bring to bear on the nut.

There are many hand tools that have a similar action, for which modest forces far from the pivot lead to large forces near the pivot. Can and bottle openers, pliers, shovels, pry bars—all of these rely on this principle, which we might call *leverage*, or **mechanical advantage.** If the nutcracker enables you apply 3 times the force you can apply with your hand, we say that it has a mechanical advantage of 3.

## Strength Versus Range of Motion and Speed

Now let's look at the operation of your ankle joint. As we'll see, the geometry is the same as that of the nutcracker, with one crucial difference—the muscle force is applied close to the pivot, not far away.

### EXAMPLE 8.11 Forces in the ankle joint BIO

For the purposes of this problem, we'll assume a simple model of the operation of your foot and ankle. We'll assume that your foot pivots as a single structure about a single pivot in the ankle. This simple model will help us explore the operation of the ankle joint and draw some interesting conclusions.

Let's take your foot as the object of interest. When you stand on tiptoe, your foot pivots about your ankle. As shown in FIGURE 8.27, the forces on one foot are an upward force on your toes from the floor, a downward force on your ankle from the lower leg bone, and an upward force on the heel of your foot from your Achilles tendon. Suppose a 61 kg woman stands on one foot, on tiptoe, with the sole of her foot making a 25° angle with the floor; the distances are as shown in Figure 8.27. What is the magnitude of the tension force in the tendon? By what fraction does this force exceed the woman's weight? What is the magnitude of the force in the ankle joint?

FIGURE 8.27 Forces on the foot when standing on tiptoe.

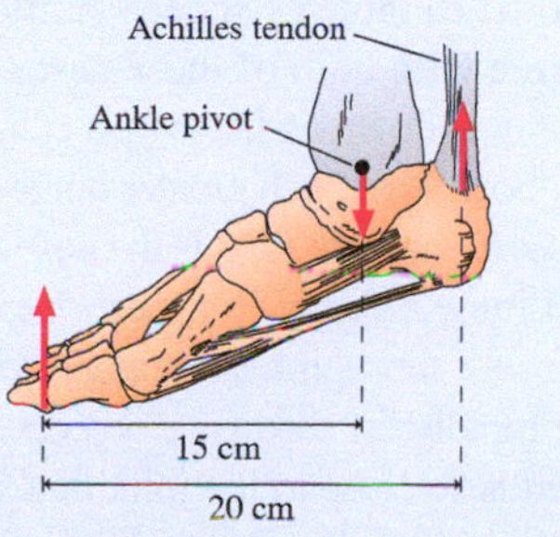


**STRATEGIZE** As we've noted, the object of interest is the foot—this is the object that pivots, that has forces applied at different points. We'll look at the static equilibrium of the foot given the forces that act on it. We'll take as our pivot point the ankle pivot. We are given the perpendicular distance to the pivot—the moment arm—for each force, so we'll compute torques using $\tau = r_{\perp}F$.

**PREPARE** There are three forces that act on the foot, as shown in the simplified view of the situation in FIGURE 8.28. The force of the floor has only a $y$-component; the same is true of the tendon force. There must be no net force along either the $x$- or $y$-axes, so we've drawn the force at the ankle as vertical as well.

The magnitude of the upward force from the floor is equal to the woman's weight, which we round to two significant figures:

$$F_{\text{floor}} = w = (61\ \text{kg})(9.8\ \text{m/s}^2) = 600\ \text{N}$$

FIGURE 8.28 Simplified view of the forces on the foot.

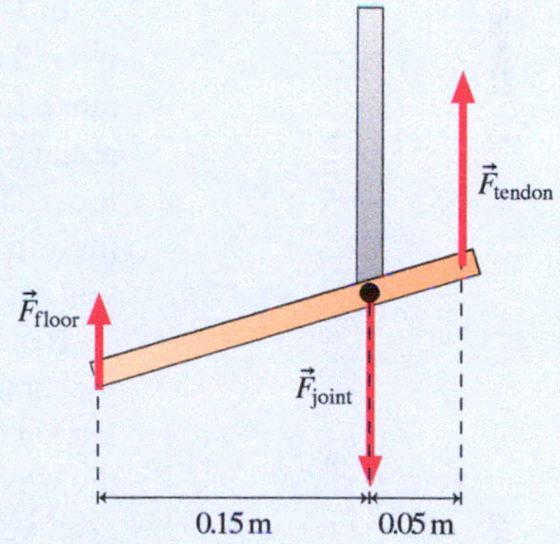


This force acts at a perpendicular distance of 15 cm, or 0.15 m, from the pivot and is directed upward. The force from the tendon is simply the tension in the tendon, which is what we are solving for:

$$F_{\text{tendon}} = T$$

This force acts at a perpendicular distance of 5 cm, or 0.05 m, from the pivot; this force is directed upward as well. The final force is the force at the pivot, the force in the joint, which we call $F_{\text{joint}}$. This is the force with which the bones of the upper leg press on the bones of the foot. This force acts at the pivot and so contributes no torque; it must be directed downward to balance the two upward forces.

**SOLVE** The foot is in static equilibrium, so it experiences no net force and no net torque. The force at the pivot contributes no torque, so the torque condition reduces to

$$\sum \tau = 0 = (r_{\perp})_{\text{tendon}} F_{\text{tendon}} - (r_{\perp})_{\text{floor}} F_{\text{floor}}$$
$$= (0.05\ \text{m})T - (0.15\ \text{m})(600\ \text{N})$$

*Continued*

We can solve for the tension in the tendon:

$$T = 1800\ \text{N}$$

This is 3 times the woman's weight. Now, let's look at the force condition $\sum F_y = 0$ for static equilibrium:

$$\sum F_y = 0 = F_{\text{floor}} - F_{\text{joint}} + F_{\text{tendon}} = 600\ \text{N} - F_{\text{joint}} + 1800\ \text{N}$$

We can solve this equation for the force in the joint:

$$F_{\text{joint}} = 2400\ \text{N}$$

This force is 4 times her weight.

**ASSESS** In terms of basic physics, our results make sense. The force of the tendon acts closer to the pivot than the force of the floor, so we expect this force to be larger than her weight.

In the last step of the example, we noted that the result made sense in terms of basic physics. But it does raise certain questions. For the woman to lift her body on her toes, her tendon—and the muscles that pull on it—must pull with a force that is triple her body weight. This is the opposite of the situation we saw with the nutcracker, where a small applied force resulted in a large force on the nut. Lifting your weight by standing on your toes requires your muscles to provide a force that is 3 times your weight, which seems a bit surprising. If the attachment point of the tendon was farther from the joint, it would take less muscle force for her to raise her body. Isn't that desirable?

There are animals—like the armadillo—that do have the tendon connected farther from the ankle pivot. For such an animal, a given muscle force results in a larger force at the toes. This is very useful for the armadillo, who digs as it searches for food. But there's a cost. A muscle can contract by no more than 1/3 of its resting length. The farther from the pivot the muscle or attached tendon pulls, the smaller the range of motion in the limb. Muscles also have a maximum contraction speed. The farther from the pivot the muscle or attached tendon pulls, the slower the resulting motion of the limb.

In Example 8.11, the distance to the tendon attachment was 1/3 of the distance from the ankle to the toes. For the armadillo, the tendon attaches at about 1/2 the distance from the ankle to the toes. This makes sense for a digger that needs large forces at the toes. For a cat, the tendon attaches at about 1/5 the distance from the ankle to the toes. This means a cat isn't likely to be a great digger, but it has a bigger range of motion and is apt to be capable of speedy motion of its paws. If you are a cat owner, you are likely to agree with both of these assessments.

Bend your foot up and down as far as you can, and note the range of motion. Now do the same with your elbow. The much greater range of motion in the elbow joint requires the tendon to attach closer to the joint than is the case for your ankle. This reduces the force at the hand for a given muscle force. But there are considerations other than strength: You need your elbow to move through a large angle for normal use, and our ancestors needed their hands to move rapidly for throwing objects or throwing a punch. In the case of the elbow, the attachment is so close to the joint that very large muscle and tendon forces are required, in some cases near the limits of what muscle and tendon can support.

### EXAMPLE 8.12 Analyzing forces and torques in the elbow BIO

In weightlifting, where the applied forces are large, holding the body in static equilibrium requires muscle forces that are quite large indeed. In the strict curl event, a standing athlete lifts a barbell by moving only his forearms, which pivot at the elbow. Record lifts are in the range of 110 kg. FIGURE 8.29 shows the arm bones and the main lifting muscle when the forearm is horizontal. The distance from the tendon to the elbow joint is 4.0 cm, and from the barbell to the elbow 35 cm. For a typical tendon cross-section area of $1.3 \times 10^{-4}\ \text{m}^2$, what percentage of the maximum tension the tendon can support is required to perform a 110 kg strict curl?

**STRATEGIZE** We'll solve the problem for one arm, which we assume is lifting half the weight. The forearm is in motion, but the speed and changes in speed are quite modest given the forces at work, so we will treat this as a static equilibrium problem. The object of interest is the forearm. We know the forces and where they act, and the elbow is a natural pivot. All of the forces are perpendicular to the arm, so there is no need to take components.

Solving the static equilibrium problem will allow us to find the tension in the tendon. We will then compare this with the maximum possible tension, computed with the known area of the tendon and the tensile strength of tendon given in Table 8.4.

**PREPARE** FIGURE 8.30 shows a simplified model of the situation. There are three forces at work. The magnitude of the force of the tendon is just the tension in the tendon, which we are trying to find:

$$F_{\text{tendon}} = T$$

This force acts a perpendicular distance of 0.040 m from the pivot. The force of the barbell is the weight of half the mass:

$$F_{\text{barbell}} = \left(\frac{110\ \text{kg}}{2}\right) g = 540\ \text{N}$$

FIGURE 8.29 The arm lifting a barbell.

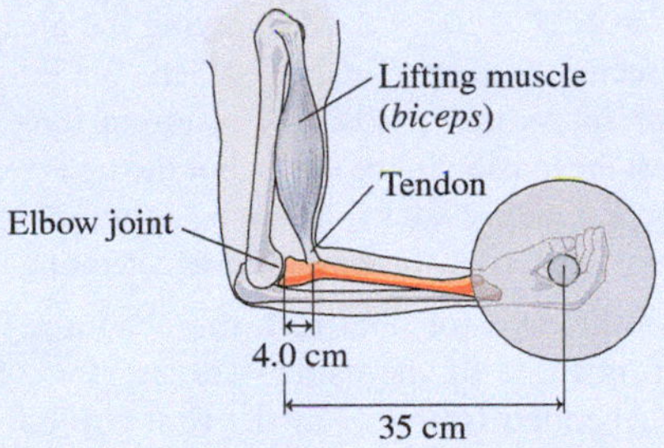


FIGURE 8.30 A simplified model of the arm and weight.

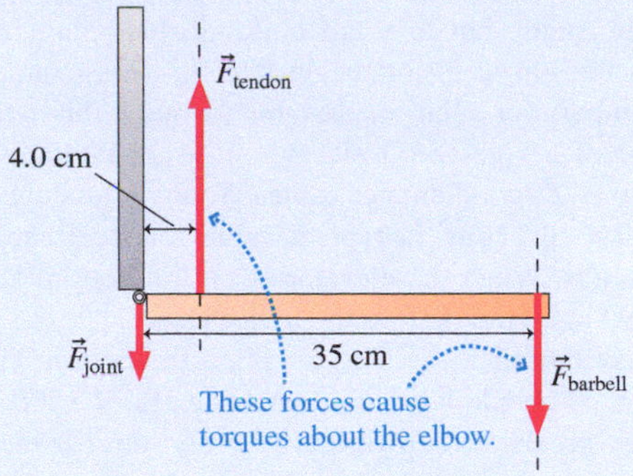


This force acts 35 cm from the pivot. The force in the joint acts at the pivot, so it does not contribute a torque.

***SOLVE*** The tension in the tendon tries to rotate the arm counterclockwise, so it produces a positive torque. The torque due to the barbell, which tries to rotate the arm in a clockwise direction, is negative. For static equilibrium, the magnitudes of the two torques must be equal. Given the forces and distances we identified, we can write

$$T(4.0 \text{ cm}) = (540 \text{ N})(35 \text{ cm})$$

We can solve this equation for the tension in the tendon. The distances appear in a ratio, so the units cancel; there is no need for unit conversion:

$$T = (540 \text{ N})\frac{35 \text{ cm}}{4.0 \text{ cm}} = 4700 \text{ N}$$

The tendon tension comes from the muscles, which must provide a force nearly 9 times the weight lifted! For this lift, the biceps in the arms are pulling with a combined force of about 1 ton, which makes this impressive lift seem even more amazing.

The tendon sustains a very large tension force. The maximum possible tendon tension is fixed by the cross-section area of the tendon and the tensile strength given in Table 8.4:

$$T_{max} = (100 \times 10^6 \text{ N/m}^2)(1.3 \times 10^{-4} \text{ m}^2) = 13{,}000 \text{ N}$$

The required tension for the lift is 36% of the maximum possible tension.

***ASSESS*** The large value for the tendon tension makes sense, given the problem statement, as does the fact that the tension is a significant fraction of what the tendon can support. The lift is possible, but it's nearing the limit of what the tissues of the body can do.

Let's look at one more example, the motion of the jaw. A typical person can generate a bite force of 1200 N at the second molars, a force that is probably greater than the person's weight. The masseter muscle that provides most of the force to close your jaw isn't a particularly large muscle, but its attachment is quite favorable for providing large forces, as FIGURE 8.31a shows. The force vector shows the approximate line of force of the masseter muscle. The line of the force is about 5 cm from the pivot, compared to about 7 cm from the molars. This means that the force at the molars is nearly equal to the full force of the muscle. This is a dramatic difference from the previous examples. There's a trade-off, though—you may be capable of great bite strength, but your jaw has a very limited range of motion and you have limited bite speed. FIGURE 8.31b shows a dog jaw along with the pivot and the approximate line of force of the masseter. The prominent canine teeth at the front of the jaw are much farther from the pivot than the muscle, so the canine teeth are well adapted for rapid, slashing bites. Cats have much shorter jaws than dogs. What does this imply about their bite speed and force?

**Making light work of moving** BIO
The tendon force is so large in Example 8.12 because the weight is supported much farther from the elbow than the point where the tendon attaches. If the weight is supported closer to the elbow, the downward torque of the weight is much less, reducing the necessary tendon and muscle force. In the picture, two people are using lifting straps to carry a heavy appliance. The straps hang very close to the elbow, so the required muscle force to support the weight is much less than it would otherwise be.

FIGURE 8.31 Jaw motion in a human and a dog.

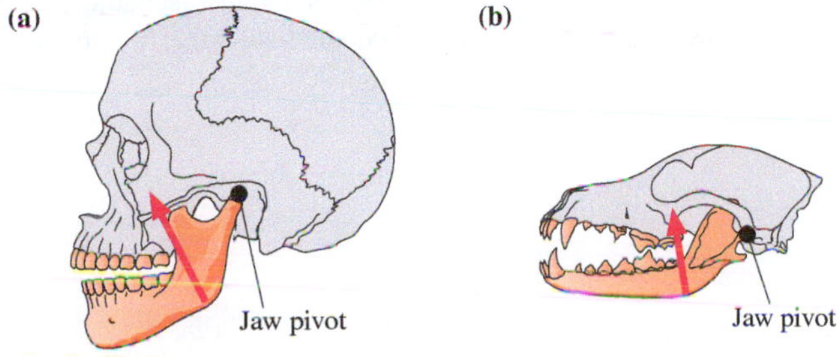

## INTEGRATED EXAMPLE 8.13 Elevator cable stretch

The steel cables that hold elevators stretch only a very small fraction of their length, but in a tall building this small fractional change can add up to a noticeable stretch. This example uses realistic numbers for such an elevator to make this point. The 2300 kg car of a high-speed elevator in a tall building is supported by six 1.27-cm-diameter cables. Young's modulus for the cables is $10 \times 10^{10}$ N/m$^2$, a typical value for steel cables with multiple strands. When the elevator is on the bottom floor, the cables rise 90 m up the shaft to the motor above.

On a busy morning, the elevator is on the bottom floor and fills up with 20 people who have a total mass of 1500 kg. The elevator then accelerates upward at 2.3 m/s$^2$ until it reaches its cruising speed. How much do the cables stretch due to the weight of the car alone? How much additional stretch occurs when the passengers are in the car? And, what is the total stretch of the cables while the elevator is accelerating? In all cases, you can ignore the mass of the cables.

**STRATEGIZE** The restoring force of the cable is the force that supports the elevator. We will determine the force that supports the elevator, and then use this force to determine the stretch of the cable.

**PREPARE** We can compute the stretch of the cables by rewriting Equation 8.6, the equation relating stress and strain, to get

$$\Delta L = \frac{LF}{YA}$$

Here $L$ is the length of the cables, $F$ is the restoring force exerted by the cables, and $Y$ is Young's modulus, which we're given. Although there are six cables, we can imagine them combined into one cable with a cross-section area $A$ six times that of each individual cable. Each cable has radius 0.00635 m and cross-section area $\pi r^2 = 1.27 \times 10^{-4}$ m$^2$. Multiplying the area by 6 gives a total cross-section area $A = 7.62 \times 10^{-4}$ m$^2$.

FIGURE 8.32 shows the details. The restoring force exerted by the cable is just the tension in the cable. For the first two questions, the cable supports the elevator in static equilibrium; for the third, there is a net force because the elevator is accelerating upward.

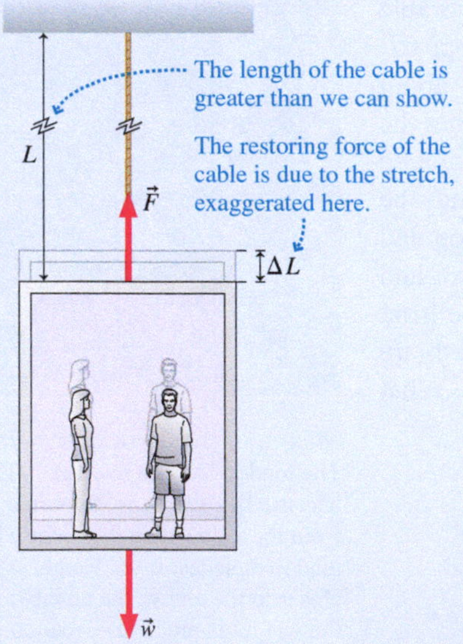


FIGURE 8.32 Details of the cable stretch and the forces acting on the elevator.

**SOLVE** When the elevator is at rest, the net force is zero and so the restoring force of the cable—the tension—is equal to the weight suspended from it. For the first two questions, the forces are

$$F_1 = m_{\text{car}}g = (2300\text{ kg})(9.8\text{ m/s}^2) = 22{,}500\text{ N}$$

$$F_2 = m_{\text{car+passengers}}g = (2300\text{kg} + 1500\text{kg})(9.8\text{m/s}^2) = 37{,}200\text{N}$$

These forces stretch the cable by

$$\Delta L_1 = \frac{(90\text{ m})(22{,}500\text{ N})}{(10 \times 10^{10}\text{ N/m}^2)(7.62 \times 10^{-4}\text{ m}^2)} = 0.027\text{ m} = 2.7\text{ cm}$$

$$\Delta L_2 = \frac{(90\text{ m})(37{,}200\text{ N})}{(10 \times 10^{10}\text{ N/m}^2)(7.62 \times 10^{-4}\text{ m}^2)} = 0.044\text{ m} = 4.4\text{ cm}$$

The additional stretch when the passengers board is

$$4.4\text{ cm} - 2.7\text{ cm} = 1.7\text{ cm}$$

When the elevator is accelerating upward, the tension in the cables must increase. Newton's second law for the vertical motion is

$$\Sigma F_y = F - w = ma_y$$

The restoring force, or tension, is thus

$$F = w + ma_y = mg + ma_y = (3800\text{ kg})(9.8\text{ m/s}^2 + 2.3\text{ m/s}^2)$$
$$= 46{,}000\text{ N}$$

Thus right at the start of the motion, when the full 90 m of cable is still deployed, we find that the cables are stretched by

$$\Delta L = \frac{(90\text{ m})(46{,}000\text{ N})}{(10 \times 10^{10}\text{ N/m}^2)(7.62 \times 10^{-4}\text{ m}^2)} = 0.054\text{ m} = 5.4\text{ cm}$$

**ASSESS** When the passengers enter the car, the cable stretches by 1.7 cm, or about two-thirds of an inch. This is large enough to notice (as we might expect given the problem statement) but not large enough to cause concern to the passengers. The total stretch for a fully loaded elevator accelerating upward is 5.4 cm, just greater than 2 inches. This is not unreasonable for cables that are nearly as long as a football field—the fractional change in length is still quite small.

# SUMMARY

**GOAL** To learn about the static equilibrium of extended objects, and the basic properties of springs and elastic materials.

## GENERAL PRINCIPLES

### Static Equilibrium

An object in **static equilibrium** must have no net force on it and no net torque. Mathematically, we express this as

$$\sum F_x = 0$$
$$\sum F_y = 0$$
$$\sum \tau = 0$$

Since the net torque is zero about *any* point, the pivot point for calculating the torque can be chosen at any convenient location.

### Springs and Hooke's Law

When a spring is stretched or compressed, it exerts a force proportional to the change $\Delta x$ in its length but in the opposite direction. This is known as **Hooke's law:**

$$(F_{sp})_x = -k\,\Delta x$$

The constant of proportionality $k$ is called the **spring constant.** It is larger for a "stiff" spring.

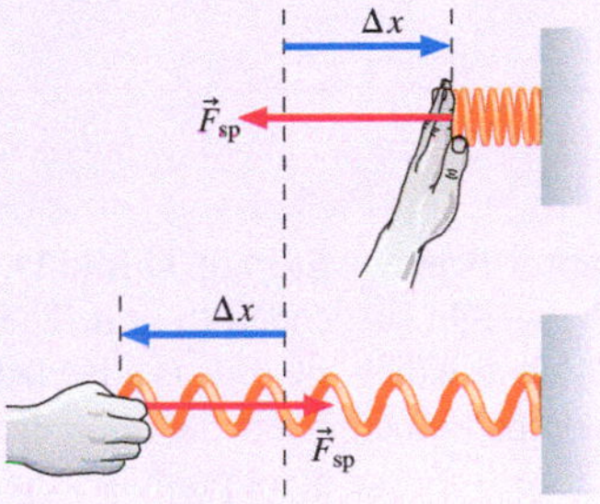


## IMPORTANT CONCEPTS

### Stability

An object is **stable** if its center of gravity is over its base of support; otherwise, it is **unstable.**

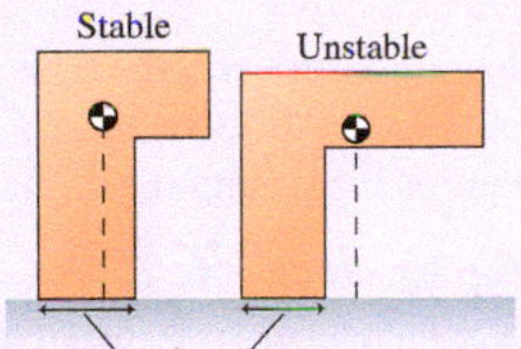


If an object is tipped, it will reach the limit of its stability when its center of gravity is over the edge of the base. This defines the **critical angle** $\theta_c$.

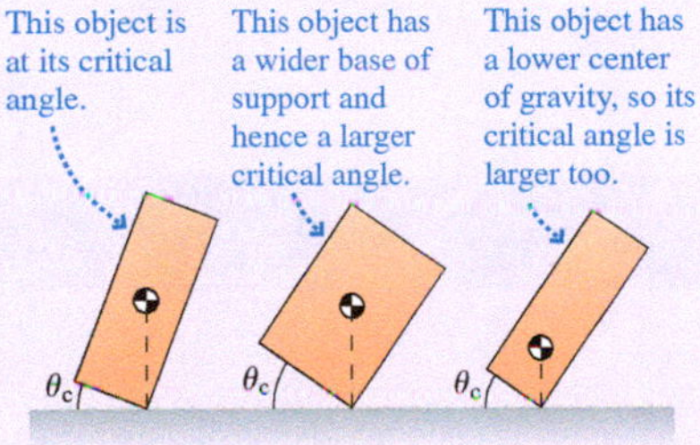


Greater stability is possible with a lower center of gravity or a broader base of support.

### Elastic materials and Young's modulus

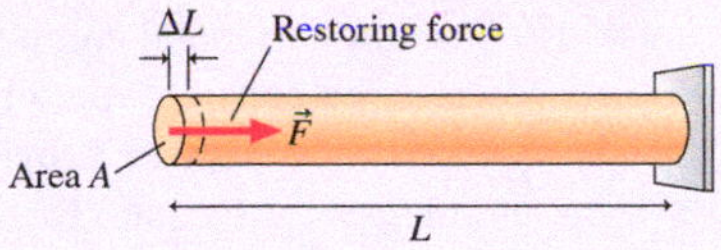


A solid rod illustrates how materials respond when stretched or compressed.

$$\left(\frac{F}{A}\right) = Y\left(\frac{\Delta L}{L}\right)$$

**Stress** is the restoring force of the rod divided by its cross-section area. **Strain** is the fractional change in the rod's length. Young's modulus

This equation can also be written as

$$F = \left(\frac{YA}{L}\right)\Delta L$$

This is the "spring constant" $k$ for the rod.

showing that a rod obeys Hooke's law and acts like a very stiff spring.

## APPLICATIONS

### Forces and torques in the body

Muscles and tendons apply the forces and torques needed to maintain static equilibrium. These forces may be quite large.

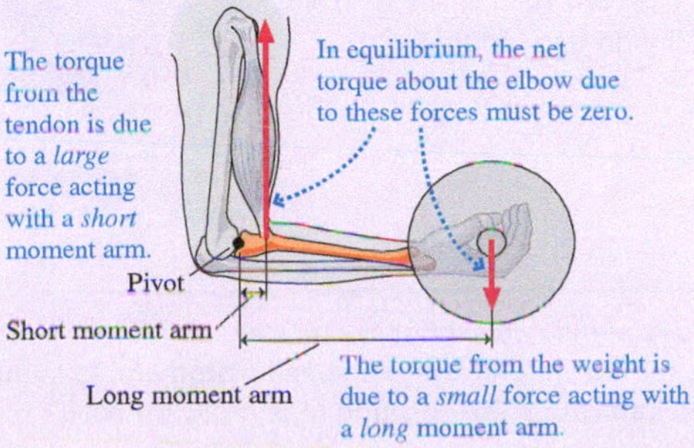

The torque from the tendon is due to a *large* force acting with a *short* moment arm.
In equilibrium, the net torque about the elbow due to these forces must be zero.
Pivot
Short moment arm
Long moment arm
The torque from the weight is due to a *small* force acting with a *long* moment arm.

### The elastic limit and beyond

If a rod or other object is not stretched too far, when released it will return to its original shape.

If stretched too far, an object will permanently deform and finally break. The stress at which an object breaks is its **tensile stress.**

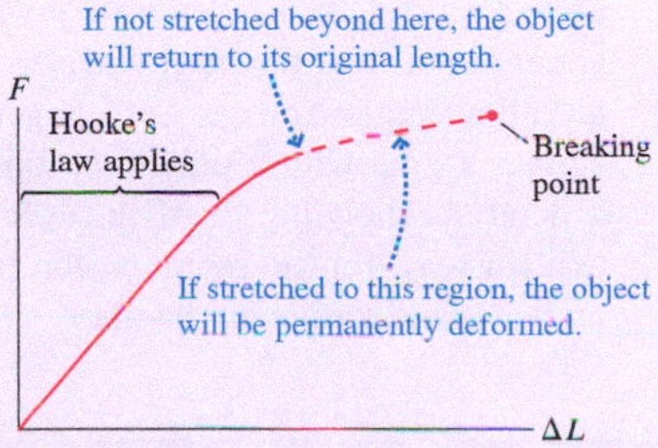

If not stretched beyond here, the object will return to its original length.
F
Hooke's law applies
Breaking point
If stretched to this region, the object will be permanently deformed.
ΔL

## Learning Objectives After studying this chapter, you should be able to:

- Use force and torque to solve static equilibrium problems. *Conceptual Questions 8.2, 8.15; Problems 8.3, 8.4, 8.7, 8.10, 8.47*
- Determine an object's stability. *Conceptual Questions 8.6, 8.7; Problems 8.15, 8.16, 8.17, 8.18, 8.19*
- Use Hooke's law to calculate the force exerted by a spring. *Conceptual Questions 8.10, 8.11; Problems 8.23, 8.24, 8.25, 8.27, 8.28*
- Use stress, strain, and Young's modulus to calculate the elastic properties of materials. *Conceptual Questions 8.12, 8.17; Problems 8.30, 8.31, 8.32, 8.33, 8.36, 8.41*

## STOP TO THINK ANSWERS

**Chapter Preview Stop to Think: D.** The tension in the rope is equal to the weight suspended from it. The force does not change with the position at which the rope is attached. But the torque about the point where the branch meets the trunk depends on both this force and the distance from the pivot point. As the distance decreases, so does the torque.

**Stop to Think 8.1: D.** Only object D has both zero net force and zero net torque.

**Stop to Think 8.2: B.** The tension in the rope and the weight have no horizontal component. To make the net force zero, the force due to the pivot must also have no horizontal component, so we know it points either up or down. Now consider the torque about the point where the rope is attached. The tension provides no torque. The weight exerts a counterclockwise torque. To make the net torque zero, the pivot force must exert a *clockwise* torque, which it can do only if it points *up*.

**Stop to Think 8.3: B, A, C.** The critical angle $\theta_c$, shown in the figure, measures how far the object can be tipped before falling. B has the smallest critical angle, followed by A, then C.

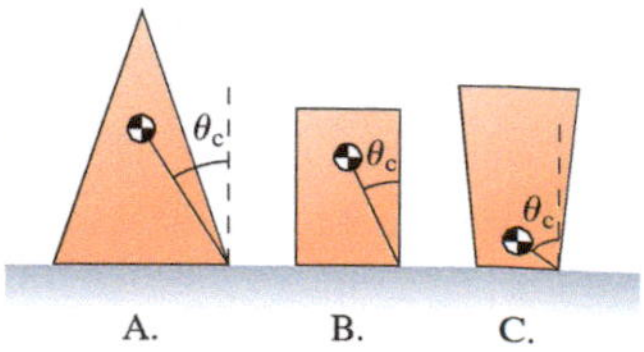


**Stop to Think 8.4: B.** The spring is now compressed, so the restoring force will be to the right. The end of the spring is displaced by half of what it was in the earlier case, so the restoring force will be half what it was as well.

**Stop to Think 8.5: C.** The restoring force of the spring is proportional to the stretch. Increasing the restoring force by a factor of 3 requires increasing the stretch by a factor of 3.

**Stop to Think 8.6: E.** The cables have the same diameter, and the force is the same, so the stress is the same in both cases. This means that the strain, $\Delta L/L$, is the same. The 2 m cable will experience twice the change in length of the 1 m cable.

Video Tutor Solution Chapter 8

# QUESTIONS

## Conceptual Questions

1. Sketch a force acting at point P in Figure Q8.1 that would make the rod be in static equilibrium. Is there only one such force?
2. Could a ladder on a level floor lean against a wall in static equilibrium if there were no friction forces? Explain.
3. If you are using a rope to raise a tall mast, attaching the rope to the middle of the mast as in Figure Q8.3a gives a very small torque about the base of the mast when the mast is at a shallow angle. You can get a larger torque by adding a pole with a pulley on top, as in Figure Q8.3b. Draw a diagram showing all of the forces acting on the mast and explain why, for the same tension in the rope, adding this pole increases the torque on the mast.

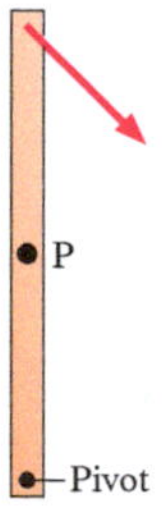


FIGURE Q8.1

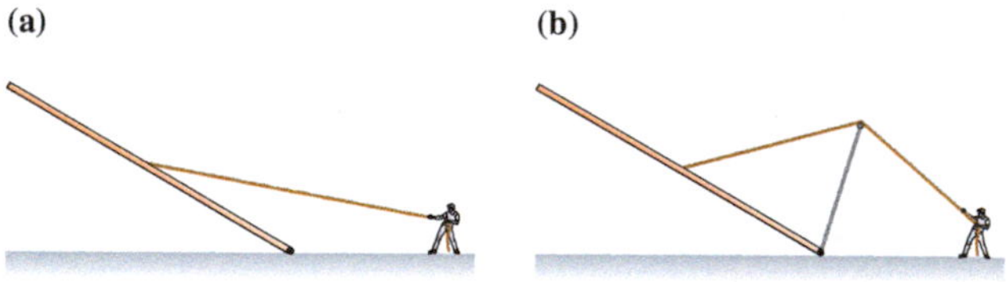


FIGURE Q8.3

4. As divers stand on tiptoes on the edge of a diving platform, in preparation for a high dive, as shown in Figure Q8.4, they usually extend their arms in front of them. Why do they do this?

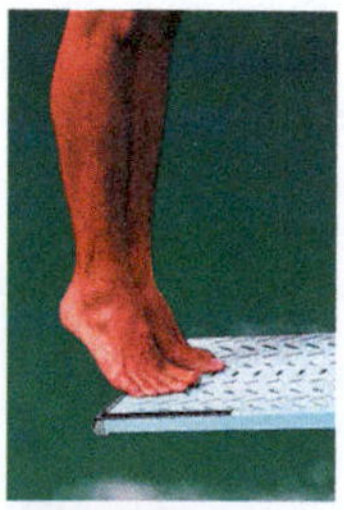

FIGURE Q8.4

Problem difficulty is labeled as | (straightforward) to ||||| (challenging). Problems labeled INT integrate significant material from earlier chapters; Problems labeled BIO are of biological or medical interest.

The eText icon indicates when there is a video tutor solution available for the chapter or for a specific problem. To launch these videos, log into your eText through Mastering™ Physics or log into the Study Area.

5. Where are the centers of gravity of the two people doing the classic yoga poses shown in Figure Q8.5?

FIGURE Q8.5

6. You must lean quite far forward as you rise from a chair (try it!). Explain why.
7. If you stand with your back and heels against a wall, and bend forward at the waist, you will topple forward. Explain why.
8. A spring exerts a 10 N force after being stretched by 1 cm from its equilibrium length. By how much will the spring force *increase* if the spring is stretched from 4 cm away from equilibrium to 5 cm from equilibrium?
9. A spring is attached to the floor and pulled straight up by a string. The string's tension is measured. The graph in Figure Q8.9 shows the tension in the spring as a function of the spring's length $L$. What is the spring constant?

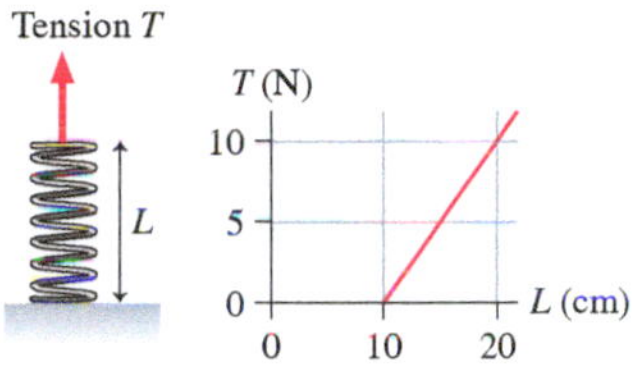


FIGURE Q8.9

10. A typical mattress has a network of springs that provide support. If you sit on a mattress, the springs compress. A heavier person compresses the springs more than a lighter person. Use the properties of springs and spring forces to explain why.
11. Take a spring and cut it in half to make two springs. How does the spring constant of the smaller springs relate to that of the original spring?
12. A wire is stretched to its breaking point by a 5000 N force. A longer wire made of the same material has the same diameter. Is the force that will stretch it right to its breaking point larger than, smaller than, or equal to 5000 N? Explain.
13. If you put a heavy load in a wheelbarrow, the force you apply to the handles is much less than the weight of the load and the wheelbarrow. To lift with the least force, you put the load toward the front of the wheelbarrow, near to the front wheel. Explain how this reduces the force to support the handles.

14. BIO When you carry shopping bags, rather than grasp the handles with your hand as in Figure Q8.14a, you might choose to put them over your arm and slide the handle toward your elbow as in Figure Q8.14b. Explain why this leads to less muscle effort to carry the bags and less force in your elbow joint.

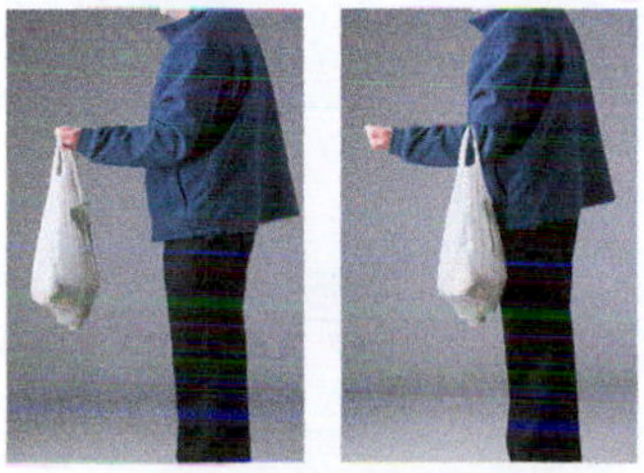

FIGURE Q8.14

15. BIO You can apply a much larger bite force with your molars, in the back of your mouth, than with your incisors, at the front of your mouth. Explain why this is so.
16. BIO An isometric exercise is one in which the joint angle does not change during the application of muscle force. For instance, you can join your hands together, pushing down with one hand and pushing up with the other. In each arm, the muscles that provide the force are connected to the bones of the arm by tendons. If you increase the forces but keep the angles the same, the muscles will still contract. Explain how this is possible.
17. Steel nails are rigid and unbending. Steel wool is soft and squishy. How would you account for this difference?

## Multiple-Choice Questions

18. || Two children hold opposite ends of a lightweight, 1.8-m-long horizontal pole with a water bucket hanging from it. The older child supports twice as much weight as the younger child. How far is the bucket from the older child?
A. 0.3 m B. 0.6 m
C. 0.9 m D. 1.2 m
19. || The uniform rod in Figure Q8.19 has a weight of 14.0 N. What is the magnitude of the normal force on the surface?
A. 7 N B. 14 N
C. 20 N D. 28 N

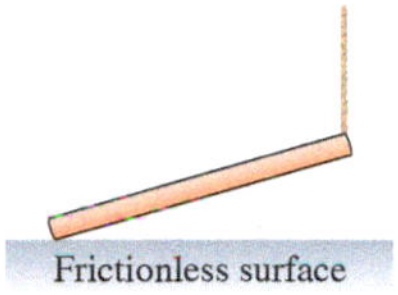


FIGURE Q8.19

20. | A student lies on a very light, rigid board with a scale under each end. Her feet are directly over one scale, and her body is positioned as shown in Figure Q8.20. The two scales read the values shown in the figure. What is the student's weight?

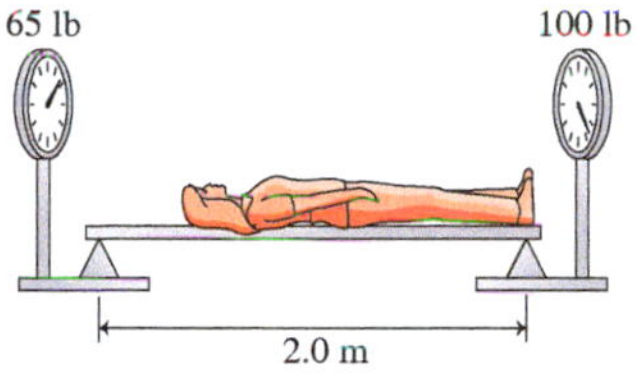


FIGURE Q8.20

A. 65 lb B. 75 lb C. 100 lb D. 165 lb
21. | For the student in Figure Q8.20, approximately how far from her feet is her center of gravity?
A. 0.6 m B. 0.8 m C. 1.0 m D. 1.2 m

Questions 22 through 24 use the information in the following paragraph and figure.

When you stand on a trampoline, the surface depresses below equilibrium, and the surface pushes up on you, as the data for a real trampoline in Figure Q8.22 show. The linear variation of the force as a function of distance means that we can model the restoring force as that of a spring.

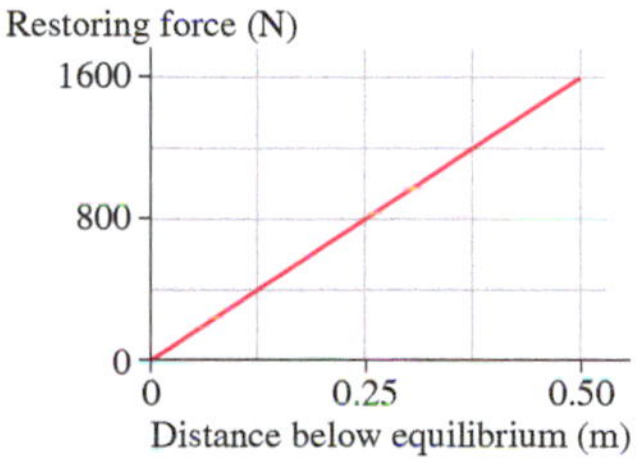


FIGURE Q8.22

22. || What is the spring constant of the trampoline?
A. 800 N/m B. 1600 N/m C. 2400 N/m D. 3200 N/m
23. || A 65 kg gymnast stands on the trampoline. How much does the center of the trampoline compress?
A. 5.0 cm B. 10 cm C. 15 cm D. 20 cm
24. || A 65 kg gymnast jumps on the trampoline. At the lowest point of her motion, the trampoline is depressed 0.50 m below equilibrium. What is her approximate upward acceleration at this instant?
A. 10 m/s$^2$ B. 15 m/s$^2$ C. 20 m/s$^2$ D. 25 m/s$^2$

Watch Video Solution Problem 8.9

25. ‖ Some climbing ropes are designed to noticeably stretch under a load to lessen the forces in a fall. A weight is attached to 10 m length of climbing rope, which then stretches by 20 cm. Now, this single rope is replaced by a doubled rope—two pieces of rope next to each other. How much does the doubled rope stretch?
    A. 5.0 cm
    B. 10 cm
    C. 20 cm
    D. 40 cm

26. ‖ Some climbing ropes are designed to noticeably stretch under a load to lessen the forces in a fall. A 10 m length of climbing rope is supporting a climber; this causes the rope to stretch by 60 cm. If the climber is supported by a 20 m length of the same rope, by how much does the rope stretch?
    A. 30 cm B. 60 cm C. 90 cm D. 120 cm

27. ‖ You have a heavy piece of equipment hanging from a 1.0-mm-diameter wire. Your supervisor asks that the length of the wire be doubled without changing how far the wire stretches. What diameter must the new wire have?
    A. 1.0 mm B. 1.4 mm C. 2.0 mm D. 4.0 mm

# PROBLEMS

## Section 8.1 Torque and Static Equilibrium

1. ‖ A 64 kg student stands on a very light, rigid board that rests on a bathroom scale at each end, as shown in Figure P8.1. What is the reading on each of the scales?

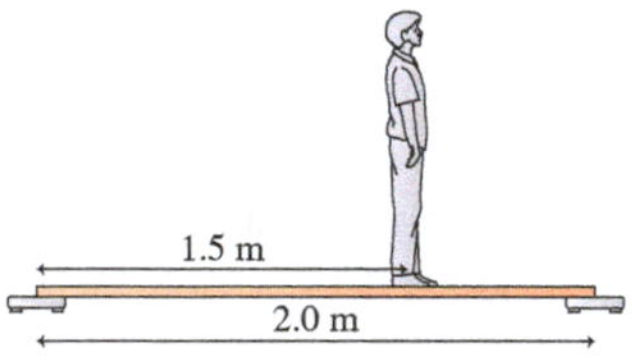

FIGURE P8.1

2. ||| Suppose the student in Figure P8.1 is 54 kg, and the board being stood on has a 10 kg mass. What is the reading on each of the scales?

3. ‖ How close to the right edge of the 56 kg picnic table shown in Figure P8.3 can a 70 kg man stand without the table tipping over? **Hint:** When the table is just about to tip, what is the force of the ground on the table's left leg?

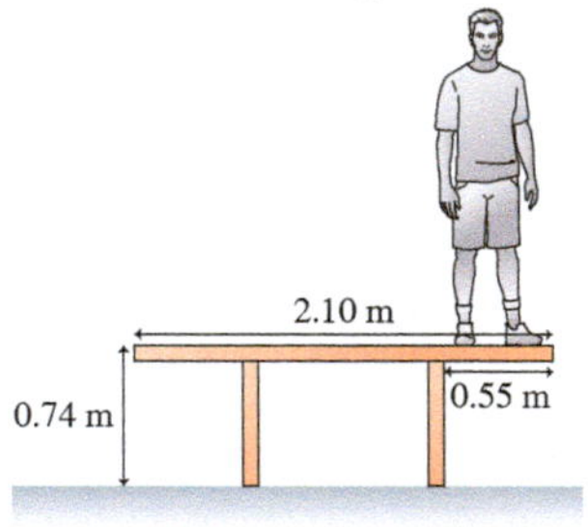

FIGURE P8.3

4. ‖ In Figure P8.4, a 70 kg man walks out on a 10 kg beam that rests on, but is not attached to, two supports. When the beam just starts to tip, what is the force exerted on the beam by the right support?

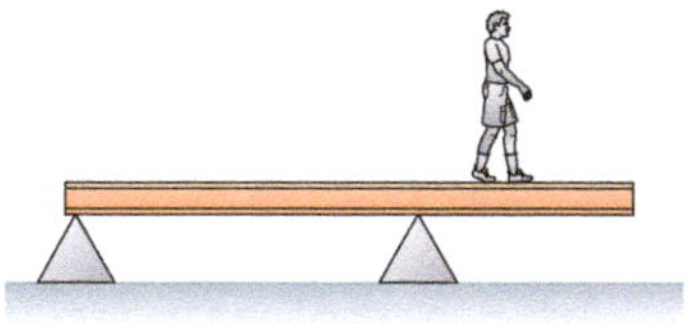
FIGURE P8.4

5. ||| You're carrying a 3.6-m-long, 25 kg pole to a construction site when you decide to stop for a rest. You place one end of the pole on a fence post and hold the other end of the pole 35 cm from its tip. How much force must you exert to keep the pole motionless in a horizontal position?

6. | A typical horse weighs 5000 N. The distance between the front and rear hooves and the distance from the rear hooves to the center of mass for a typical horse are shown in Figure P8.6. What fraction of the horse's weight is borne by the front hooves?
BIO

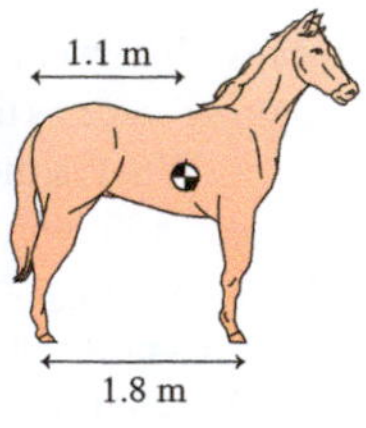

FIGURE P8.6

7. | A vendor hangs an 8.0 kg sign in front of his shop with a cable held away from the building by a lightweight pole. The pole is free to pivot about the end where it touches the wall, as shown in Figure P8.7. What is the tension in the cable?

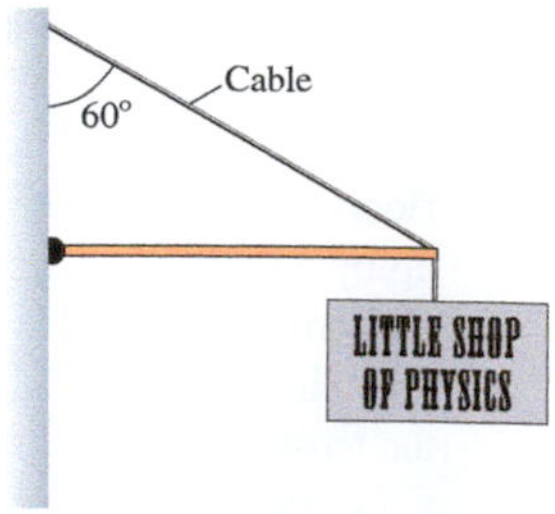

FIGURE P8.7

8. ‖ A hippo's body is 4.0 m long with front and rear feet located as in Figure P8.8. The hippo carries 60% of its weight on its front feet. How far from its tail is the hippo's center of gravity?
BIO

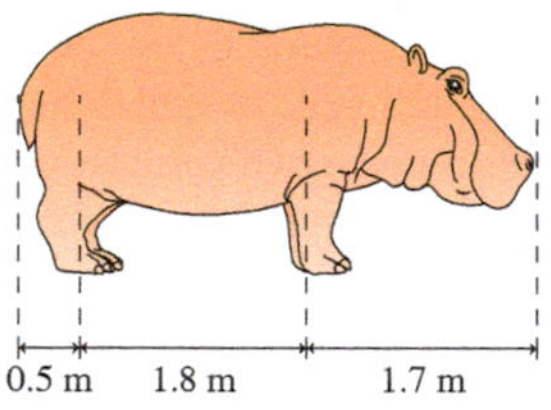

FIGURE P8.8

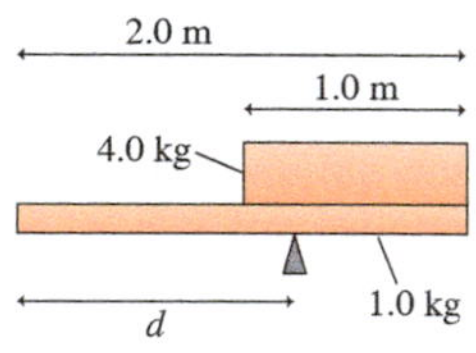

FIGURE P8.9

9. ||| The two objects in Figure P8.9 are balanced on the pivot. What is distance $d$?

10. ‖ A bicycle mechanic is checking a road bike's chain. He applies a 45 N force to a pedal at the angle shown in Figure P8.10 while keeping the wheel from rotating. The pedal is 17 cm from the center of the crank; the gear has a diameter of 16 cm. What is the tension in the chain?

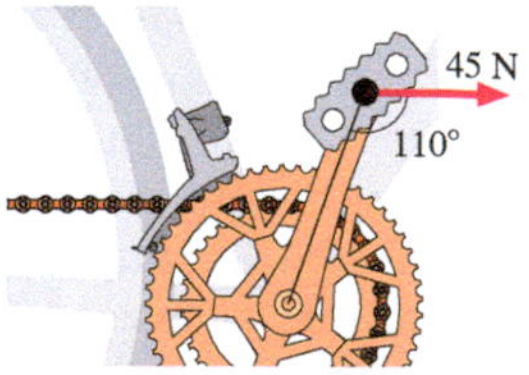

FIGURE P8.10

11. ||| A 60 kg diver stands at the end of a 30 kg springboard, as shown in Figure P8.11. The board is attached to a hinge at the left end but simply rests on the right support. What is the magnitude of the vertical force exerted by the hinge on the board?

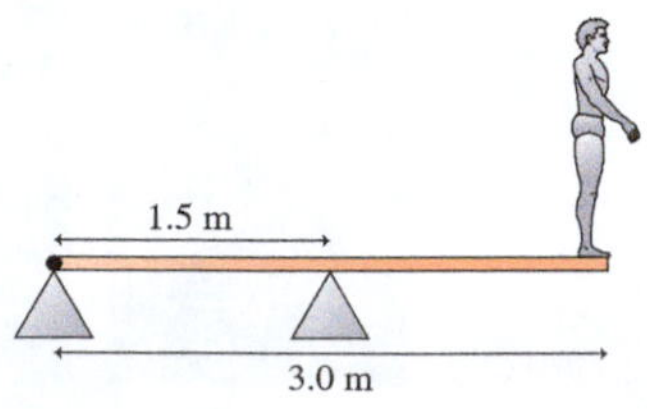

FIGURE P8.11

12. ||| A bike chain can support a tension of no more than 9800 N. The pedal connects to a crank 17 cm from the axle, and the gear pulling the chain has a 9.1 cm radius. When riding at a constant speed, with the crank and pedal horizontal, as in Figure P8.12, what is the maximum force that can be applied to the pedal before the chain breaks?

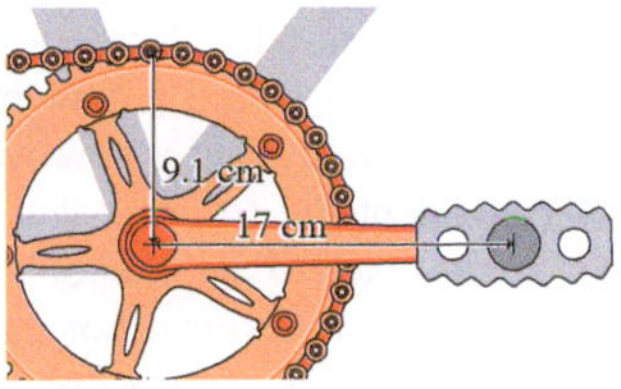

FIGURE P8.12

13. || A uniform beam of length 1.0 m and mass 10 kg is attached to a wall by a cable, as shown in Figure P8.13. The beam is free to pivot at the point where it attaches to the wall. What is the tension in the cable?

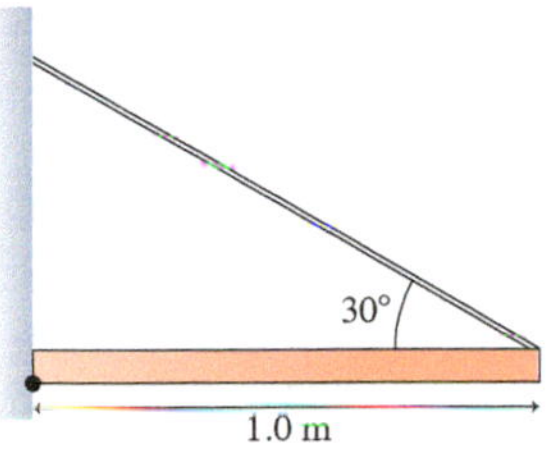

FIGURE P8.13

14. ||| The towers holding small wind turbines are often raised and lowered for easy servicing of the turbine. Figure P8.14 shows a 1000 kg wind turbine mounted on the end of a 24-m-long, 700 kg tower that connects to a support column at a pivot. A piston connected 3.0 m from the pivot applies the force needed to raise or lower the tower. At the instant shown, the wind turbine is being raised at a very slow, constant speed. What magnitude force is the piston applying?

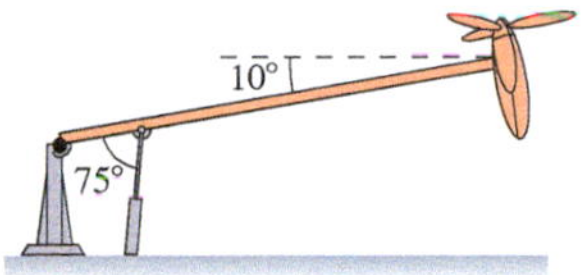

FIGURE P8.14

## Section 8.2 Stability and Balance

15. | A standard four-drawer filing cabinet is 52 inches high and 15 inches wide. If it is evenly loaded, the center of gravity is at the center of the cabinet. A worker is tilting a filing cabinet to the side to clean under it. To what angle can he tilt the cabinet before it tips over?

16. | A double-decker London bus might be in danger of rolling over in a highway accident, but at the low speeds of its urban environment, it's plenty stable. The track width is 2.05 m. With no passengers, the height of the center of gravity is 1.45 m, rising to 1.73 m when the bus is loaded to capacity. What are the critical angles for both the unloaded and loaded bus?

17. | The stability of a vehicle is often rated by the *static stability factor*, which is one-half the track width divided by the height of the center of gravity above the road. A typical SUV has a static stability factor of 1.2. What is the critical angle?

18. | A magazine rack has a center of gravity 16 cm above the floor, as shown in Figure P8.18. Through what maximum angle, in degrees, can the rack be tilted without falling over?

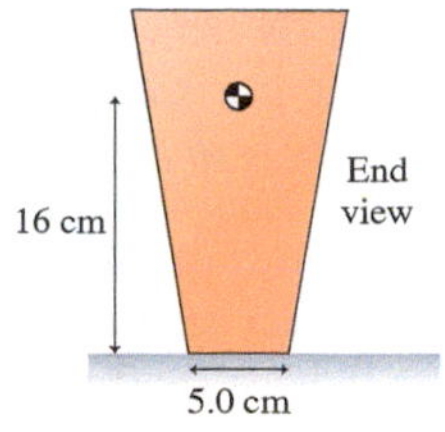

FIGURE P8.18

19. || A car manufacturer claims that you can drive its new vehicle across a hill with a 47° slope before the vehicle starts to tip. If the vehicle is 2.0 m wide, how high is its center of gravity?

20. || A thin 2.00 kg box rests on a 6.00 kg board that hangs over the end of a table, as shown in Figure P8.20. How far can the center of the box be from the end of the table before the board begins to tilt?

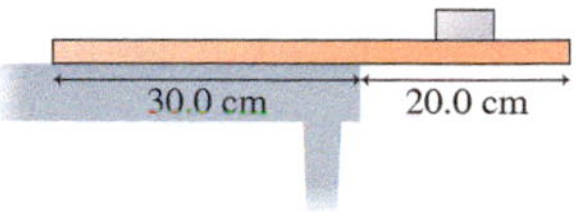

FIGURE P8.20

## Section 8.3 Springs and Hooke's Law

21. | An orthodontic spring, connected between the upper and
BIO lower jaws, is adjusted to provide no force with the mouth open. When the patient closes her mouth, however, the spring compresses by 6.0 mm. What force is exerted if the spring constant is 160 N/m?

22. || The four wheels of a car are connected to the car's body by spring assemblies that let the wheels move up and down over bumps and dips in the road. When a 68 kg (about 150 lb) person sits on the left front fender of a small car, this corner of the car dips by about 1.2 cm (about ½ in). If we treat the spring assembly as a single spring, what is the approximate spring constant?

23. | Experiments using "optical tweezers" measure the elasticity
BIO of individual DNA molecules. For small enough changes in length, the elasticity has the same form as that of a spring. A DNA molecule is anchored at one end, then a force of 1.5 nN $(1.5 \times 10^{-9}\text{ N})$ pulls on the other end, causing the molecule to stretch by 5.0 nm $(5.0 \times 10^{-9}\text{ m})$. What is the spring constant of that DNA molecule?

24. || A passenger railroad car has a total of 8 wheels. Springs on each wheel compress—slightly—when the car is loaded. Ratings for the car give the stiffness per wheel (the spring constant, treating the entire spring assembly as a single spring) as $2.8 \times 10^7$ N/m. When 30 passengers, each with average mass 80 kg, board the car, how much does the car move down on its spring suspension? Assume that each wheel supports 1/8 the weight of the car.

25. || One end of a 10-cm-long spring is attached to the ceiling. When a 2.0 kg mass is hung from the other end, the spring stretches to a length of 15 cm.
    a. What is the spring constant?
    b. How long is the spring when a 3.0 kg mass is suspended from it?

26. | A scale used to weigh fish consists of a spring hung from a support. The spring's equilibrium length is 10.0 cm. When a 4.0 kg fish is suspended from the end of the spring, it stretches to a length of 12.4 cm.
    a. What is the spring constant $k$ for this spring?
    b. If an 8.0 kg fish is suspended from the spring, what will be the length of the spring?

27. || A spring has an unstretched length of 10 cm. It exerts a restoring force $F$ when stretched to a length of 11 cm.
    a. For what total stretched length of the spring is its restoring force $3F$?
    b. At what compressed length is the restoring force $2F$?

28. || A spring stretches 5.0 cm when a 0.20 kg block is hung from it. If a 0.70 kg block replaces the 0.20 kg block, how far does the spring stretch?

29. | You need to make a spring scale to measure the mass of objects hung from it. You want each 1.0 cm length along the scale to correspond to a mass difference of 0.10 kg. What should be the value of the spring constant?

## Section 8.4 Stretching and Compressing Materials

30. || Dynamic climbing ropes are designed to be quite pliant, allowing a falling climber to slow down over a long distance. The graph in Figure P8.30 shows force-versus-strain data for an 11-mm-diameter climbing rope. What is the Young's modulus for this rope?

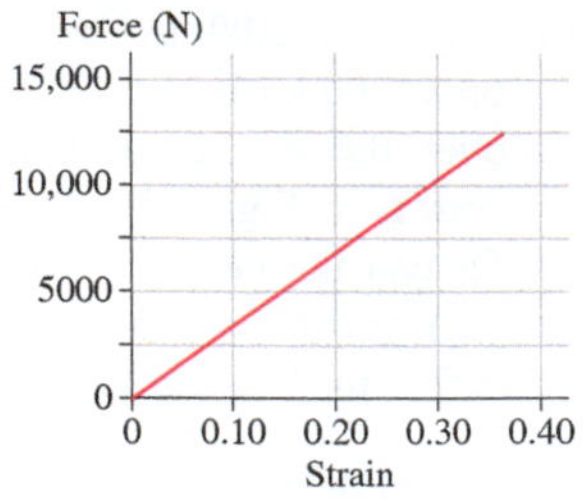


FIGURE P8.30

31. || A force stretches a wire by 1.0 mm.
   a. A second wire of the same material has the same cross section and twice the length. How far will it be stretched by the same force?
   b. A third wire of the same material has the same length and twice the diameter as the first. How far will it be stretched by the same force?

32. || Static climbing ropes are designed to be relatively stiff so that they stretch less than dynamic ropes. To meet a certain specification, an 11-mm-diameter rope must experience a maximum elongation of 5.0% when supporting a 150 kg load. What is the minimum Young's modulus?

33. |||| What hanging mass will stretch a 2.0-m-long, 0.50-mm-diameter steel wire by 1.0 mm?

34. ||| An 80-cm-long, 1.0-mm-diameter steel guitar string must be tightened to a tension of 2.0 kN by turning the tuning screws. By how much is the string stretched?

35. ||| A mineshaft has an ore elevator hung from a single braided cable of diameter 2.5 cm. Young's modulus of the cable is $10 \times 10^{10}$ N/m$^2$. When the cable is fully extended, the end of the cable is 800 m below the support. How much does the fully extended cable stretch when 1000 kg of ore is loaded?

36. || The normal force of the ground on the foot can reach three
BIO times a runner's body weight when the foot strikes the pavement. By what amount does the 52-cm-long femur of an 80 kg runner compress at this moment? The cross-section area of the bone of the femur can be taken as $5.2 \times 10^{-4}$ m$^2$.

37. |||| A three-legged wooden bar stool made out of solid Douglas fir has legs that are 2.0 cm in diameter. When a 75 kg man sits on the stool, by what percent does the length of the legs decrease? Assume, for simplicity, that the stool's legs are vertical and that each bears the same load.

38. || To penetrate armor, a projectile's point concentrates force
BIO in a small area, creating a stress large enough that the armor
INT fails. A species of jellyfish launches a pointed needle that can penetrate the hard shell of a crustacean. The rapid deceleration on impact creates a 32 $\mu$N force on the tip, which has a very small 15 nm radius. What is the resulting stress? How does this compare to the ultimate stress of steel?



39. ||| A 3.0-m-tall, 50-cm-diameter concrete column supports a 200,000 kg load. By how much is the column compressed?

40. || You've just put a new wood floor in your house. An object
BIO will dent the flooring if the stress—the force divided by the
INT area—exerted by the object is great enough. Who is more likely to dent your floor: a 50 kg woman in high-heeled shoes (assume a circular heel pad 0.50 cm in diameter) with all of her weight on one heel or a 5000 kg African elephant (assume a circular contact area of 40 cm in diameter for one foot) standing on all four feet?

41. || A glass optical fiber in a communications system has a diameter of 9.0 $\mu$m.
   a. What maximum tension could this fiber support without breaking?
   b. Assume that the fiber stretches in a linear fashion until the instant it breaks. By how much will a 10-m-long fiber have stretched when it is at the breaking point?

42. ||| If you tethered a space station to the earth by a long cable,
INT you could get to space in an elevator that rides up the cable—much simpler and cheaper than riding to space on a rocket. There's one big problem, however: There is no way to create a cable that is long enough. The cable would need to reach 36,000 km upward, to the height where a satellite orbits at the same speed as the earth rotates; a cable this long made of ordinary materials couldn't even support its own weight. Consider a steel cable suspended from a point high above the earth. The stress in the cable is highest at the top; it must support the weight of cable below it. What is the greatest length the cable could have without failing? For the purposes of this problem, you can ignore the variation in gravity near the surface of the earth. **Hint:** The mass of the cable is the volume of the cable multiplied by the density. The density of steel is 7900 kg/m$^3$.

43. || The Achilles tendon connects the muscles in your calf to the
BIO back of your foot. When you are sprinting, your Achilles tendon alternately stretches, as you bring your weight down onto your forward foot, and contracts to push you off the ground. A 70 kg runner has an Achilles tendon that is 15 cm long with a typical $1.1 \times 10^{-4}$ m$^2$ area.
   a. By how much will the runner's Achilles tendon stretch if the maximum force on it is 8.0 times his weight, a typical value while running?
   b. What fraction of the tendon's length does this correspond to?

## Section 8.5 Forces and Torques in the Body

44. || A woman is pushing a load in a wheelbarrow, as in Figure P8.44. The combined mass of the wheelbarrow and the load is 110 kg, with a center of gravity 0.25 m behind the axle. The woman supports the wheelbarrow at the handles, 1.1 m behind the axle.
   a. What is the force required to support the wheelbarrow?
   b. What fraction of the weight of the wheelbarrow and the load does this force represent?

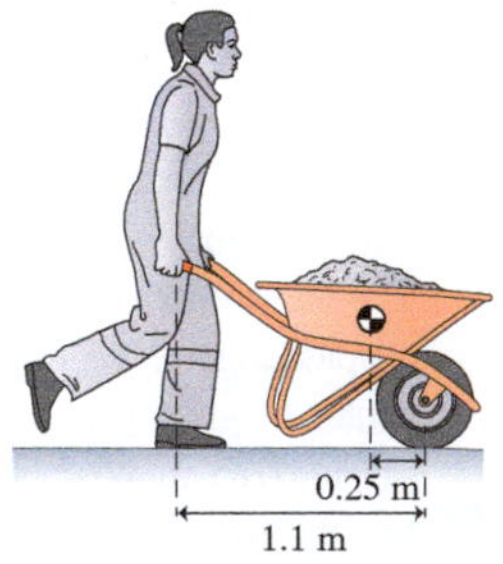


FIGURE P8.44

45. ‖ Figure P8.45 shows the operation of a garlic press. The lower part of the press is held steady, and the upper handle is pushed down, thereby crushing a garlic clove through a screen. Approximate distances are shown in the figure. If the user exerts a 12 N force on the upper handle, estimate the force on the clove.

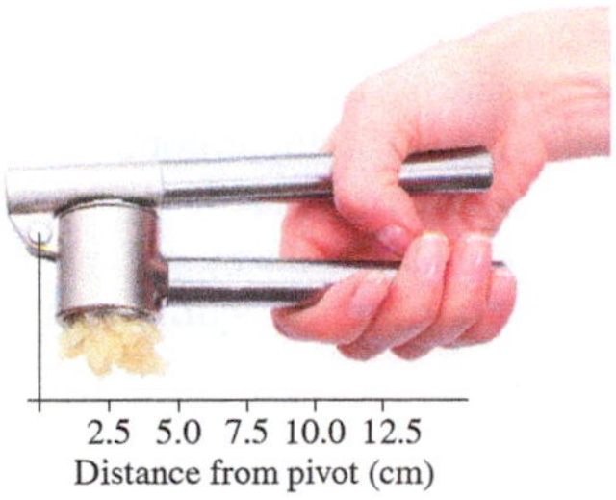

FIGURE P8.45

46. ||| Consider a rower in a scull as in Figure P8.46. The oars aren't accelerating, and they are rotating at a constant speed, so the net force and net torque on the oars are zero. An oar is 2.8 m long, and the rower pulls with a 250 N force on the handle, which is 0.92 m from the pivot.

FIGURE P8.46

a. Assume that the oar touches the water at its very end. What is the drag force from the water on the oar? Assume that the oar is perpendicular to the boat, and that the force of the rower and the drag force are both perpendicular to the oar.
b. Given that both oars are the same, what is the total force propelling the boat forward?

47. ‖ BIO Hold your upper arm vertical and your lower arm horizontal with your hand palm-down on a table, as shown in Figure P8.47. If you now push down on the table, you'll feel that your triceps muscle has contracted and is trying to pivot your lower arm about the elbow joint. If a person with the arm dimensions shown pushes down hard with a 90 N force (about 20 lb), what force must the triceps muscle provide? You can ignore the mass of the arm and hand in your calculation.

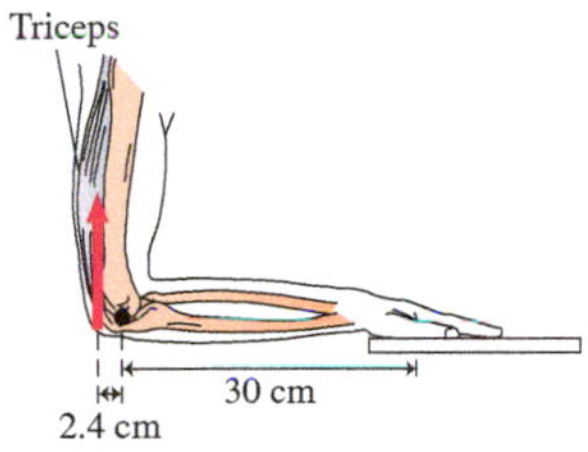

FIGURE P8.47

48. ‖ BIO If you stand on one foot while holding your other leg up behind you, your muscles apply a force to hold your leg in this raised position. We can model this situation as in Figure P8.48. The leg pivots at the knee joint, and the force that holds the leg up is provided by a tendon attached to the lower leg as shown. Assume that the lower leg and the foot have a combined mass of 4.0 kg, and that their combined center of gravity is at the center of the lower leg.
a. How much force must the tendon exert to keep the leg in this position?
b. As you hold your leg in this position, the upper leg exerts a force on the lower leg at the knee joint. What are the magnitude and direction of this force?

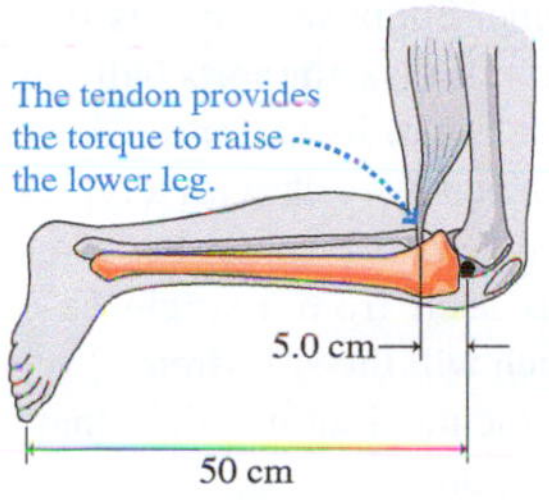

The tendon provides
the torque to raise
the lower leg.
5.0 cm
50 cm

FIGURE P8.48

49. ‖ BIO If you hold your arm outstretched with palm upward, as in Figure P8.49, the force to keep your arm from falling comes from your deltoid muscle. The arm of a typical person has mass 4.0 kg and the distances and angles shown in the figure.
a. What force must the deltoid muscle provide to keep the arm in this position?
b. By what factor does this force exceed the weight of the arm?

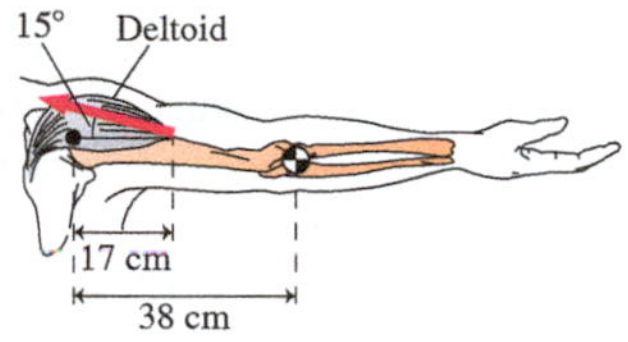

FIGURE P8.49

50. ||| BIO Dogs—like many animals—stand and walk on their toes. A photo of the rear foot of a dog is shown in Figure P8.50a; Figure P8.50b shows the bones of the leg and foot along with relevant distances. The colored element corresponds to your foot, and the connection with the leg corresponds to your ankle. The Achilles tendon pulls on the end of the foot, along a line 4.0 cm from the ankle. What is the tension in the tendon if a 20 kg dog is supporting ¼ of its weight on one rear foot?

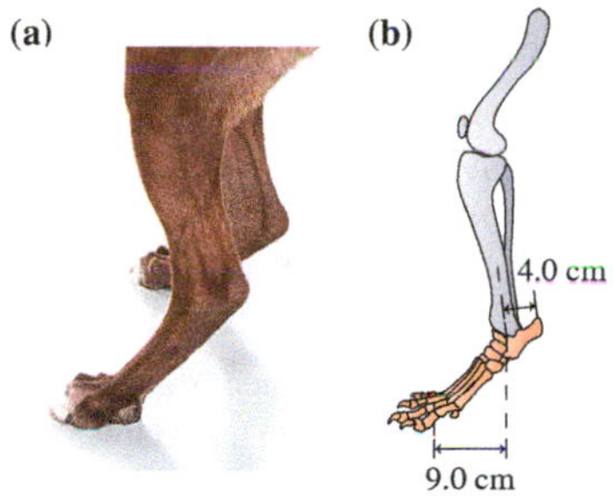

FIGURE P8.50

## General Problems

51. |||| A 3.0-m-long rigid beam with a mass of 100 kg is supported at each end, as shown in Figure P8.51. An 80 kg student stands 2.0 m from support 1. How much upward force does each support exert on the beam?

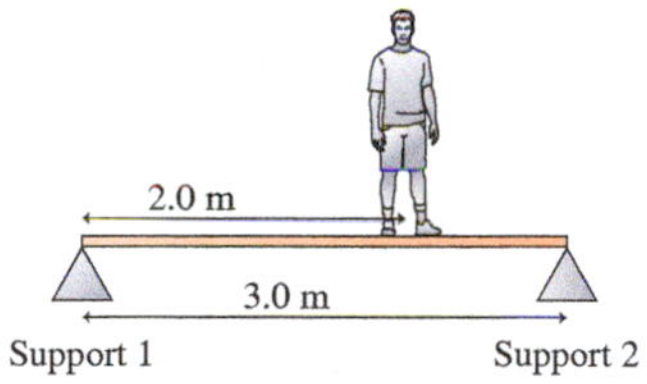

FIGURE P8.51

52. ||| An 80 kg construction worker sits down 2.0 m from the end of a 1450 kg steel beam to eat his lunch, as shown in Figure P8.52. The cable supporting the beam is rated at 15,000 N. Should the worker be worried?

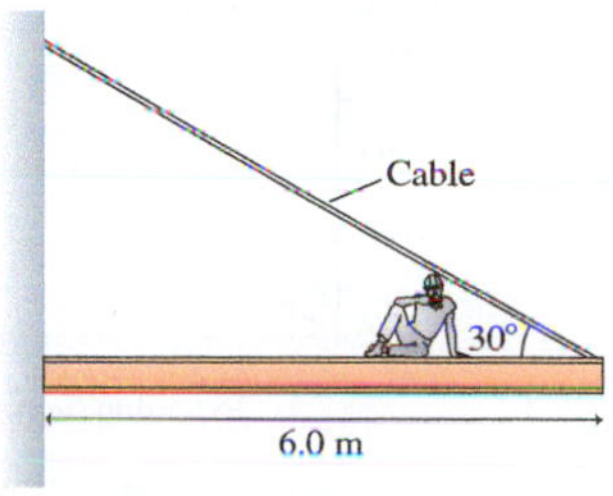

FIGURE P8.52

53. ||| A man is attempting to raise a 7.5-m-long, 28 kg flagpole that has a hinge at the base by pulling on a rope attached to the top of the pole, as shown in Figure P8.53. With what force does the man have to pull on the rope to hold the pole motionless in this position?

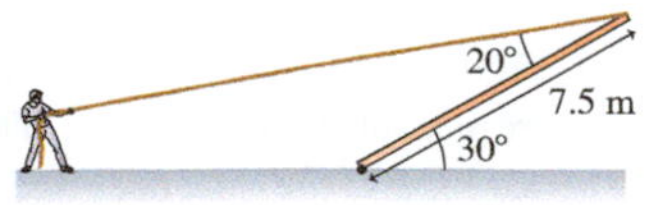

FIGURE P8.53

54. ||| An 85 kg man stands in a very strong wind moving at 14 m/s
BIO at torso height. As you know, he will need to lean in to the
INT wind, and we can model the situation to see why. Assume that the man has a mass of 85 kg, with a center of gravity 1.0 m above the ground. The action of the wind on his torso, which we approximate as a cylinder 50 cm wide and 90 cm long centered 1.2 m above the ground, produces a force that tries to tip him over backward. To keep from falling over, he must lean forward.
   a. What is the magnitude of the torque provided by the wind force? Take the pivot point at his feet. Assume that he is standing vertically.
   b. At what angle to the vertical must the man lean to provide a gravitational torque that is equal to this torque due to the wind force?

55. |||| A 40 kg, 5.0-m-long beam is supported by, but not attached to, the two posts in Figure P8.55. A 20 kg boy starts walking along the beam. How close can he get to the right end of the beam without it tipping?

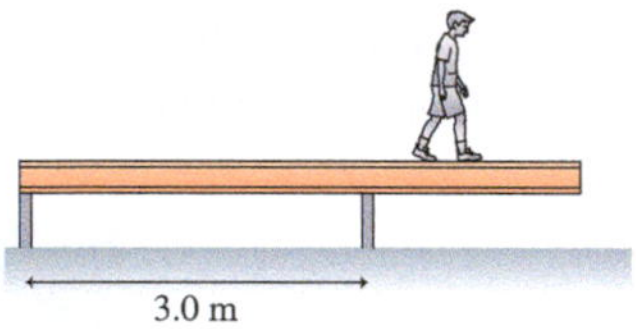

FIGURE P8.55

56. ||| Two identical, side-by-side springs with spring constant 240 N/m support a 2.00 kg hanging box. Each spring supports the same weight. By how much is each spring stretched?

57. | A 5.0 kg mass hanging from a spring scale is slowly lowered onto a vertical spring, as shown in Figure P8.57. The scale reads in newtons.
   a. What does the spring scale read just before the mass touches the lower spring?
   b. The scale reads 20 N when the lower spring has been compressed by 2.0 cm. What is the value of the spring constant for the lower spring?
   c. At what compression distance will the scale read zero?

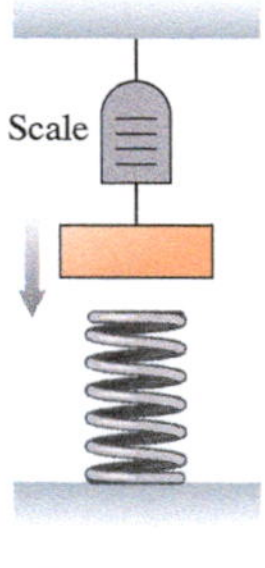

FIGURE P8.57

58. | DNA molecules are typically folded
BIO tightly. Stretching a strand of DNA means straightening it, and the molecules resist this straightening. Investigators can attach beads to the ends of a strand of DNA and, using "optical tweezers," measure the force required to produce a certain extension. Data for the stretch of a 3500 base pair strand of DNA approximately follow the line in the graph in Figure P8.58. What is the spring constant for this strand of DNA?

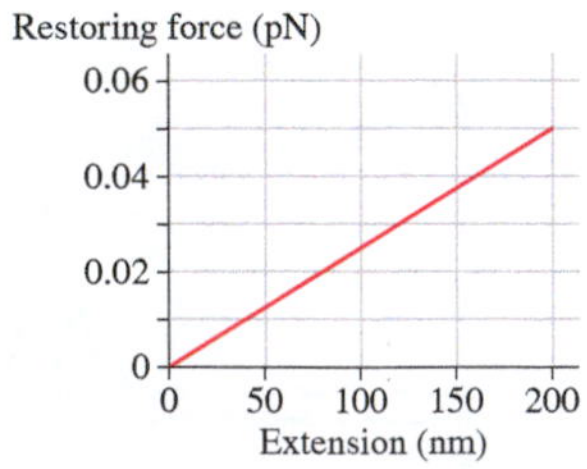

FIGURE P8.58

Watch Video Solution Problems 8.57 and 8.61

59. || Figure P8.59 shows two springs attached to a block that can slide on a frictionless surface. In the block's equilibrium position, the left spring is compressed by 2.0 cm.
   a. By how much is the right spring compressed?
   b. What is the net force on the block if it is moved 15 cm to the right of its equilibrium position?

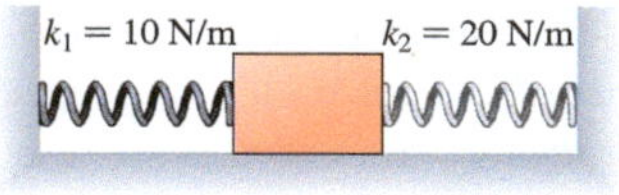

FIGURE P8.59

60. ||| A 25 kg child bounces on a pogo stick. The pogo stick has
INT a spring with spring constant $2.0 \times 10^4$ N/m. When the child makes a nice big bounce, she finds that at the bottom of the bounce she is accelerating *upward* at 9.8 m/s$^2$. How much is the spring compressed?

61. ||| Figure P8.61 shows a lightweight plank supported at its right end by a 7.0-mm-diameter rope with a tensile strength of $6.0 \times 10^7$ N/m$^2$.
   a. What is the maximum force that the rope can support?
   b. What is the greatest distance, measured from the pivot, that the center of gravity of an 800 kg piece of heavy machinery can be placed without snapping the rope?

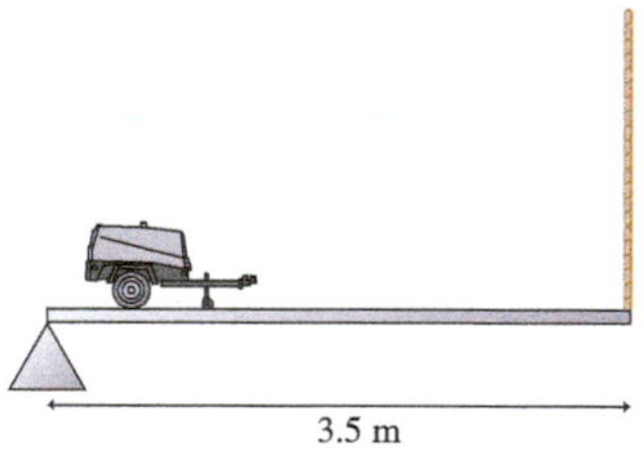

FIGURE P8.61

62. || In the hammer throw, an
INT athlete spins a heavy mass in a circle at the end of a cable before releasing it for distance. For male athletes, the "hammer" is a mass of 7.3 kg at the end of a 1.2 m cable, which is typically a 3.0-mm-diameter steel cable. A world-class thrower can get the hammer up to a speed of 29 m/s. If an athlete swings the mass in a horizontal circle centered on the handle he uses to hold the cable
   a. What is the tension in the cable?
   b. How much does the cable stretch?

63. ||| There is a disk of cartilage between each pair of vertebrae
BIO in your spine. Suppose a disk is 0.50 cm thick and 4.0 cm in diameter. If this disk supports half the weight of a 65 kg person, by what fraction of its thickness does the disk compress?

64. || Orb spiders make silk with a typical diameter of 0.15 mm.
BIO a. A typical large orb spider has a mass of 0.50 g. If this spider suspends itself from a single 12-cm-long strand of silk, by how much will the silk stretch?
   b. What is the maximum weight that a single thread of this silk could support?

65. ||| Larger animals have sturdier bones than smaller animals.
BIO A mouse's skeleton is only a few percent of its body weight, compared to 16% for an elephant. To see why this must be so, recall, from Example 8.10, that the stress on the femur for a man standing on one leg is 1.4% of the bone's tensile strength. Suppose we scale this man up by a factor of 10 in all dimensions, keeping the same body proportions. Use the data for Example 8.10 to compute the following.
   a. Both the inside and outside diameter of the femur, the region of compact bone, will increase by a factor of 10. What will be the new cross-section area?
   b. The man's body will increase by a factor of 10 in each dimension. What will be his new mass?
   c. If the scaled-up man now stands on one leg, what fraction of the tensile strength is the stress on the femur?

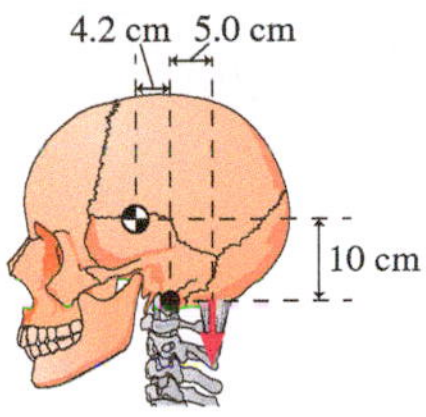

FIGURE P8.66

66. ||| The main muscles that hold your
BIO head upright attach to your spine in back of the point where your head pivots on your neck. Figure P8.66 shows typical numbers for the distance from the pivot to the muscle attachment point and the distance from the pivot to the center of gravity of the head. The muscles pull down to keep your head upright. If the muscle relaxes—if, for instance, you doze in one of your classes besides Physics—your head tips forward. In the questions that follow, assume that your head has a mass of 4.8 kg, and that you maintain the relative angle between your head and your spine.
   a. With the head held level, as in Figure P8.66, what muscle force is needed to keep a 4.8 kg head upright?
   b. If you tip your body forward so that your spine is level with the ground, what muscle force is needed to keep your head in the same orientation relative to the spine?
   c. If you tip your body backward, you will reach a point where no muscle force is needed to keep your head upright. For the distances given in Figure P8.66, at what angle does this balance occur?

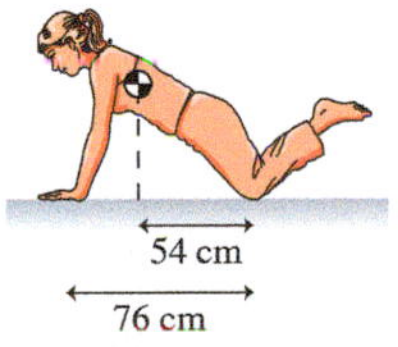

FIGURE P8.67

67. ||| A woman weighing 580 N does a
BIO pushup from her knees, as shown in Figure P8.67. What are the normal forces of the floor on (a) each of her hands and (b) each of her knees?

68. |||| When you bend over, a series of large muscles, the erector
BIO spinae, pull on your spine to hold you up. Figure P8.68 shows a simplified model of the spine as a rod of length $L$ that pivots at its lower end. In this model, the center of gravity of the 320 N weight of the upper torso is at the center of the spine. The 160 N weight of the head and arms acts at the top of the spine. The erector spinae muscles are modeled as a single muscle that acts at an 12° angle to the spine. Suppose the person in Figure P8.68 bends over to an angle of 30° from the horizontal.
   a. What is the tension in the erector muscle?
      **Hint:** Align your $x$-axis with the axis of the spine.
   b. A force from the pelvic girdle acts on the base of the spine. What is the component of this force in the direction of the spine? (This large force is the cause of many back injuries).

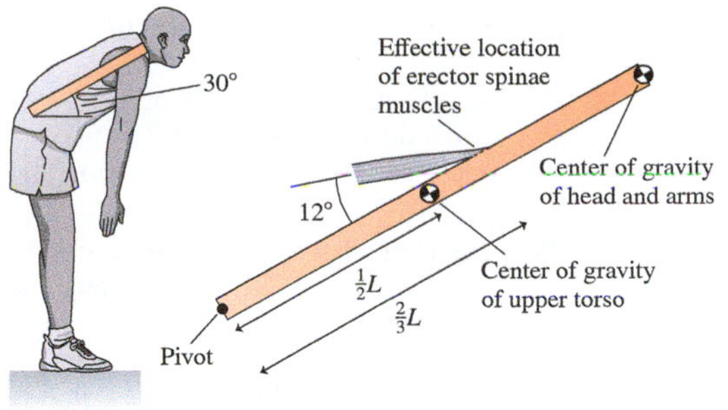

FIGURE P8.68

## MCAT-Style Passage Problems

### Elasticity of a Ligament BIO

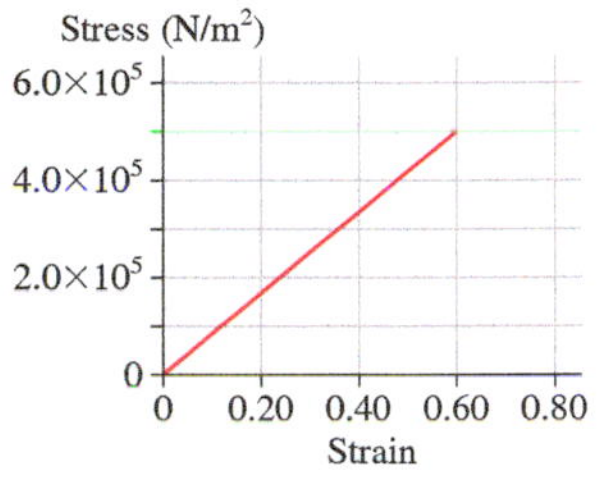

FIGURE P8.69

The nuchal ligament in a horse supports the weight of the horse's head. This ligament is much more elastic than a typical ligament, stretching from 15% to 45% longer than its resting length as a horse's head moves up and down while it runs. This stretch of the ligament stores energy, making locomotion more efficient. Measurements on a segment of ligament show a linear stress-versus-strain relationship until the stress approaches 0.80. Smoothed data for the stretch are shown in Figure P8.69.

69. || What is the approximate Young's modulus for the ligament?
   A. $4 \times 10^5$ N/m$^2$
   B. $8 \times 10^5$ N/m$^2$
   C. $3 \times 10^6$ N/m$^2$
   D. $6 \times 10^6$ N/m$^2$

70. || The segment of ligament tested has a resting length of 40 mm. How long is the ligament at a strain of 0.60?
   A. 46 mm
   B. 52 mm
   C. 58 mm
   D. 64 mm

71. || Suppose the ligament has a circular cross section. For a certain ligament, an investigator measures the restoring force at a strain of 0.40. If the ligament is replaced with one that has twice the diameter, by what factor does the restoring force increase?
   A. 1.4
   B. 2
   C. 4
   D. 8

72. || The volume of the ligament stays the same as it stretches, so the cross-section area decreases as the length increases. Given this, how would a force $F$ versus change in length $\Delta L$ curve appear?

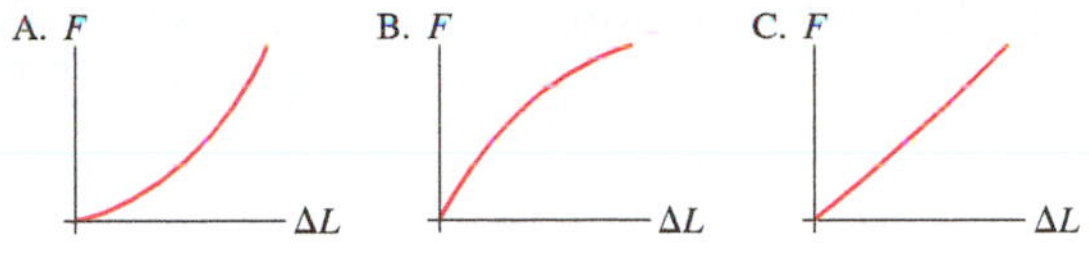

# PART I SUMMARY Force and Motion

## KNOWLEDGE STRUCTURE I Force and Motion

**BASIC GOALS**
How can we describe motion?
How does an object respond to a force?
How do systems interact?
What is the nature of the force of gravity?
How can we analyze the motion and deformation of extended objects?

**CROSS-CUTTING CONCEPTS**
Newton's laws are the connection between forces and motion. Using these laws, we can understand how particles move on a straight line, or along curved paths like that of a rider on a Ferris wheel or a baseball in flight. Newton's laws are also the key to understanding the motion of rotating objects.

**GENERAL PRINCIPLES**

**Newton's first law** An object with no forces acting on it will remain at rest or move in a straight line at a constant speed.

**Newton's second law** $\vec{F}_{\text{net}} = m\vec{a}$

**Newton's third law** $\vec{F}_{\text{A on B}} = -\vec{F}_{\text{B on A}}$

**Newton's law of gravity** $F_{1\text{ on }2} = F_{2\text{ on }1} = \dfrac{Gm_1m_2}{r^2}$

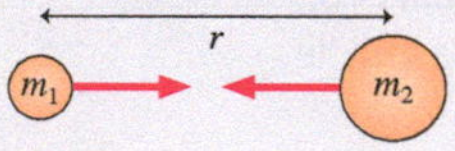


**Kinematics** is the mathematical description of motion.

**Velocity** is the rate of change of position: $v_x = \dfrac{\Delta x}{\Delta t}$

**Acceleration** is the rate of change of velocity: $a_x = \dfrac{\Delta v_x}{\Delta t}$

The **kinematic equations** for motion with constant acceleration are

$$(v_x)_f = (v_x)_i + a_x\Delta t$$
$$x_f = x_i + (v_x)_i\Delta t + \frac{1}{2}a_x(\Delta t)^2$$
$$(v_x)_f^2 = (v_x)_i^2 + 2a_x\Delta x$$

**Projectile motion** consists of vertical motion with $a_y = -g$ and horizontal motion with constant velocity $(v_x)_i$.

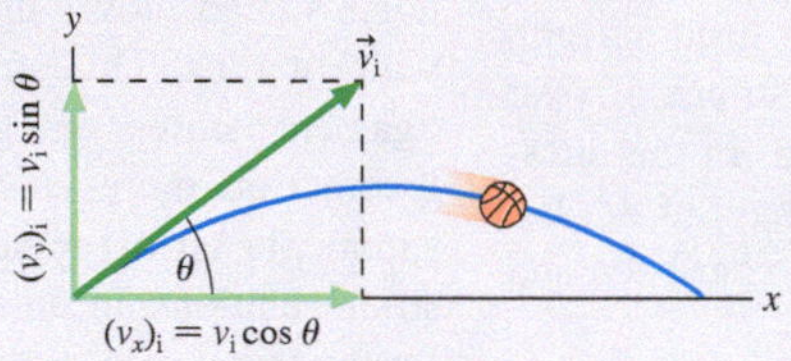


$$x_f = x_i + (v_x)_i\Delta t \qquad y_f = y_i + (v_y)_i\Delta t - \frac{1}{2}g(\Delta t)^2$$
$$(v_x)_f = (v_x)_i = \text{constant} \qquad (v_y)_f = (v_y)_i - g\Delta t$$

In **uniform circular motion** the acceleration points toward the center of the circle.

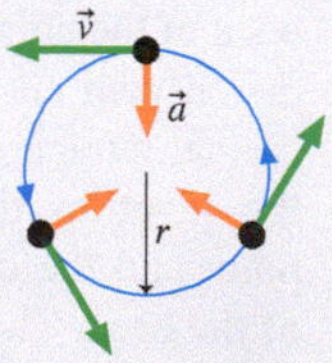


The frequency $f$, period $T$, angular velocity $\omega$, speed $v$, and acceleration $a$ are related by

$$f = \frac{1}{T} \qquad \omega = 2\pi f$$
$$v = \omega r = \frac{2\pi r}{T} \qquad a = \frac{v^2}{r} = \omega^2 r$$

**Dynamics** is the study of **forces** and how they cause motion.

**Common forces:**

**Weight** $\vec{w} = (mg, \text{downward})$

**Friction** $\vec{f}_k = (\mu_k n, \text{opposite motion})$

$\vec{f}_s = (0 \text{ to } \mu_s n, \text{direction to prevent motion})$

**Drag** $\vec{D} = (\frac{1}{2}C_D\rho A v^2, \text{opposite motion})$

**Spring** $\vec{F}_{sp} = (-k\Delta x, \text{opposite displacement})$

A **free-body diagram** shows all the forces acting on an object.

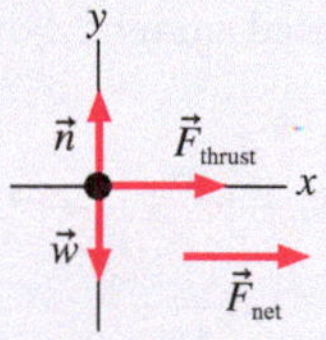


Dynamics problems are solved by writing Newton's second law in component form:

$$\sum F_x = F_{\text{thrust}} = ma_x$$
$$\sum F_y = n\text{-}mg = ma_y$$

**Circular-motion dynamics:** To provide the centripetal acceleration, there must be a net force directed toward the center of the circle.

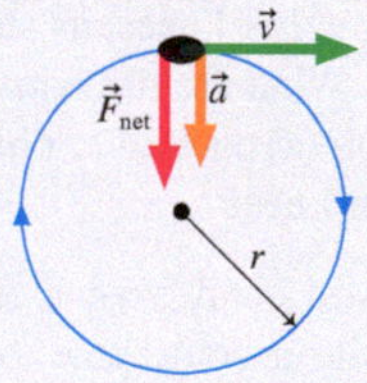


$$\vec{F}_{\text{net}} = \left(\frac{mv^2}{r}, \text{toward center of circle}\right)$$

### Rigid bodies

An object's **moment of inertia** is the rotational equivalent of mass:

$$I = \sum m_i r_i^2$$

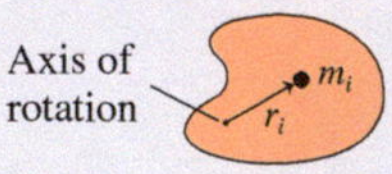


Its **center of gravity** is the point at which gravity acts:

$$x_{cg} = \frac{x_1m_1 + x_2m_2 + \cdots}{m_1 + m_2 + \cdots}$$

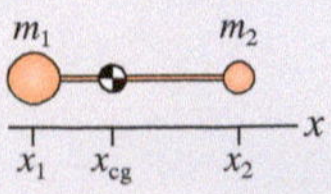

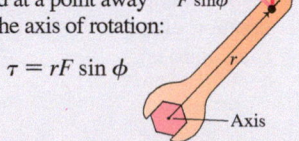

and similarly for $y_{cg}$.

### Torque and rotational dynamics

**Torque** is the rotational analog of force. A torque acts when a force is applied at a point away from the axis of rotation:

$$\tau = rF\sin\phi$$

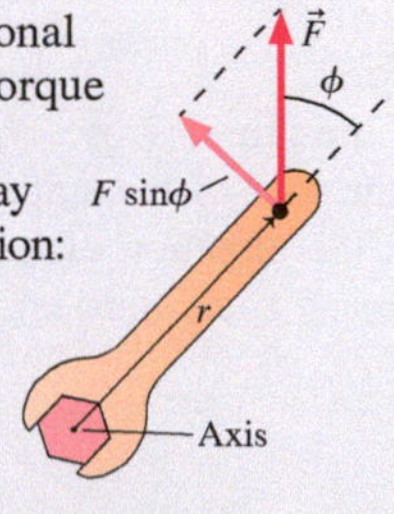


### Rotation about a fixed axis

When a net torque is applied to an object with moment of inertia $I$ that can rotate about a fixed axis, the object undergoes an **angular acceleration** given by

$$\alpha = \frac{\tau_{\text{net}}}{I}$$

ONE STEP BEYOND

## Dark Matter and the Structure of the Universe

The idea that the earth exerts a gravitational force on us is something we now accept without questioning. But when Isaac Newton developed this idea to show that the gravitational force also holds the moon in its orbit, it was a remarkable, ground-breaking insight. It changed the way that we look at the universe we live in.

Newton's laws of motion and gravity are tools that allow us to continue Newton's quest to better understand our place in the cosmos. But it sometimes seems that the more we learn, the more we realize how little we actually know and understand.

Here's an example. Advances in astronomy over the past 100 years have given us great insight into the structure of the universe. But everything our telescopes can see appears to be only a small fraction of what is out there. Approximately 80% of the mass in the universe is *dark matter*—matter that gives off no light or other radiation that we can detect. Everything that we have ever seen through a telescope is merely the tip of the cosmic iceberg.

What is this dark matter? Black holes? Neutrinos? Some form of exotic particle? We simply aren't sure. It could be any of these, or all of them—or something entirely different that no one has yet dreamed of. You might wonder how we know that such matter exists if no one has seen it. Even though we can't directly observe dark matter, we see its effects. And you now know enough physics to understand why.

Whatever dark matter is, it has mass, and so it has gravity. This picture of the Andromeda galaxy shows a typical spiral galaxy structure: a dense collection of stars in the center surrounded by a disk of stars and other matter. This is the shape of our own Milky Way galaxy.

The spiral Andromeda galaxy.

This structure is reminiscent of the structure of the solar system: a dense mass (the sun) in the center surrounded by a disk of other matter (the planets, asteroids, and comets). The sun's gravity keeps the planets in their orbits, but the planets would fall into the sun if they were not in constant motion around it. The same is true of a spiral galaxy; everything in the galaxy orbits its center. Our solar system orbits the center of our galaxy with a period of about 200 million years.

The orbital speed of an object depends on the mass that pulls on it. If you analyze our sun's motion about the center of the Milky Way, or the motion of stars in the Andromeda galaxy about its center, you find that the orbits are much faster than they should be, based on how many stars we see. There must be some other mass present.

There's another problem with the orbital motion of stars around the center of their galaxies. We know that the orbital speeds of planets decrease with distance from the sun; Neptune orbits at a much slower speed than the earth. We might expect something similar for galaxies: Stars farther from the center should orbit at reduced speeds. But they don't. As we measure outward from the center of the galaxy, the orbital speed stays about the same—even as we get to the edge of the visible disk. There must be some other mass—the invisible dark matter—exerting a gravitational force on the stars. This dark matter, which far outweighs the matter we can see, seems to form a halo around the centers of galaxies, providing the gravitational force necessary to produce the observed rotation. Other observations of the motions of galaxies with respect to each other verify this basic idea.

On a cosmic scale, the picture is even stranger. The universe is currently expanding. The mutual gravitational attraction of all matter—regular and dark—in the universe should slow this expansion. But recent observations of the speeds of distant galaxies imply that the expansion of the universe is accelerating, so there must be yet another component to the universe, something that "pushes out." The best explanation at present is that the acceleration is caused by *dark energy*. The nature of dark matter isn't known, but the nature of dark energy is even more mysterious. If current theories hold, it's the most abundant stuff in the universe. And we don't know what it is.

This sort of mystery is what drives scientific investigation. It's what drove Newton to wonder about the connection between the fall of an apple and the motion of the moon, and what drove investigators to develop all of the techniques and theories you will learn about in the coming chapters.

*The following questions are related to the passage "Dark Matter and the Structure of the Universe" on the previous page.*

1. As noted in the passage, our solar system orbits the center of the Milky Way galaxy in about 200 million years. If there were no dark matter in our galaxy, this period would be
   A. Longer.
   B. The same.
   C. Shorter.
2. Saturn is approximately 10 times as far away from the sun as the earth. This means that its orbital acceleration is ________ that of the earth.
   A. 1/10
   B. 1/100
   C. 1/1000
   D. 1/10,000
3. Saturn is approximately 10 times as far away from the sun as the earth. If dark matter changed the orbital properties of the planets so that Saturn had the same orbital speed as the earth, Saturn's orbital acceleration would be ________ that of the earth.
   A. 1/10
   B. 1/100
   C. 1/1000
   D. 1/10,000
4. Which of the following might you expect to be an additional consequence of the fact that galaxies contain more mass than expected?
   A. The gravitational force between galaxies is greater than expected.
   B. Galaxies appear less bright than expected.
   C. Galaxies are farther away than expected.
   D. There are more galaxies than expected.

*The following passages and associated questions are based on the material of Part I.*

### Animal Athletes BIO

Different animals have very different capacities for running. A horse can maintain a top speed of 20 m/s for a long distance but has a maximum acceleration of only 6.0 m/s$^2$, half what a good human sprinter can achieve with a block to push against. Greyhounds, dogs especially bred for feats of running, have a top speed of 17 m/s, but their acceleration is much greater than that of the horse. Greyhounds are particularly adept at turning corners at a run.

FIGURE I.1

5. If a horse starts from rest and accelerates at the maximum value until reaching its top speed, how much time elapses, to the nearest second?
   A. 1 s B. 2 s
   C. 3 s D. 4 s
6. If a horse starts from rest and accelerates at the maximum value until reaching its top speed, how far does it run, to the nearest 10 m?
   A. 40 m B. 30 m
   C. 20 m D. 10 m
7. A greyhound on a racetrack turns a corner at a constant speed of 15 m/s with an acceleration of 7.1 m/s$^2$. What is the radius of the turn?
   A. 40 m B. 30 m
   C. 20 m D. 10 m
8. A human sprinter of mass 70 kg starts a run at the maximum possible acceleration, pushing backward against a block set in the track. What is the force of his foot on the block?
   A. 1500 N B. 840 N
   C. 690 N D. 420 N
9. In the photograph of the greyhounds in Figure I.1, what is the direction of the net force on each dog?
   A. Up
   B. Down
   C. Left, toward the outside of the turn
   D. Right, toward the inside of the turn

### Sticky Liquids BIO

At small scales, viscous drag becomes very important. To a paramecium (Figure I.2), a single-celled animal that can propel itself through water with fine hairs on its body, swimming through water feels like swimming through honey would to you. We can model a paramecium as a sphere of diameter 50 $\mu$m, with a mass of $6.5 \times 10^{-11}$ kg.

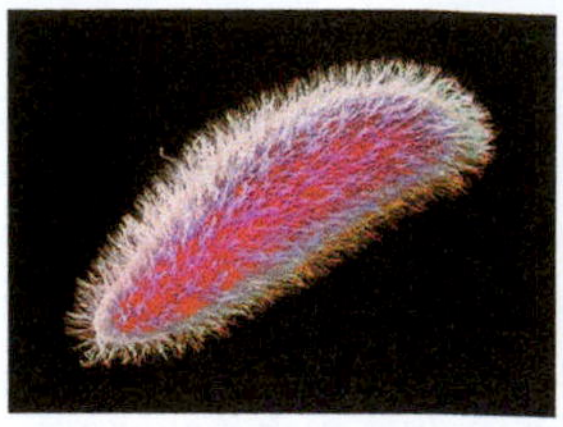

FIGURE I.2

10. A paramecium swimming at a constant speed of 0.25 mm/s ceases propelling itself and slows to a stop. At the instant it stops swimming, what is the magnitude of its acceleration?
    A. 0.2$g$ B. 0.5$g$
    C. 2$g$ D. 5$g$
11. If the acceleration of the paramecium in Problem 10 were to stay constant as it came to rest, approximately how far would it travel before stopping?
    A. 0.02 $\mu$m B. 0.2 $\mu$m
    C. 2 $\mu$m D. 20 $\mu$m
12. If the paramecium doubles its swimming speed, how does this change the drag force?
    A. The drag force decreases by a factor of 2.
    B. The drag force is unaffected.
    C. The drag force increases by a factor of 2.
    D. The drag force increases by a factor of 4.
13. You can test the viscosity of a liquid by dropping a steel sphere into it and measuring the speed at which it sinks. For viscous fluids, the sphere will rapidly reach a terminal speed. At this terminal speed, the net force on the sphere is
    A. Directed downward.
    B. Zero.
    C. Directed upward.

## Pulling Out of a Dive BIO

Falcons are excellent fliers that can reach very high speeds by diving nearly straight down. To pull out of such a dive, a falcon extends its wings and flies through a circular arc that redirects its motion. The forces on the falcon that control its motion are its weight and an upward lift force—like an airplane—due to the air flowing over its wings. At the bottom of the arc, as in Figure I.3, a falcon can easily achieve an acceleration of 15 m/s$^2$.

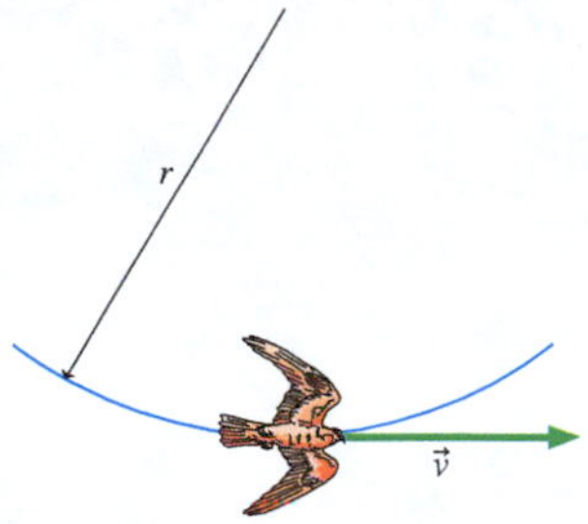

FIGURE I.3

14. At the bottom of the arc, as in Figure I.3, what is the direction of the net force on the falcon?
    A. To the left, opposite the motion
    B. To the right, in the direction of the motion
    C. Up
    D. Down
    E. The net force is zero.
15. Suppose the falcon weighs 8.0 N and is turning with an acceleration of 15 m/s$^2$ at the lowest point of the arc. What is the magnitude of the upward lift force at this instant?
    A. 8.0 N B. 12 N
    C. 16 N D. 20 N
16. A falcon starts from rest, does a free-fall dive from a height of 30 m, and then pulls out by flying in a circular arc of radius 50 m. Which segment of the motion has a higher acceleration?
    A. The free-fall dive
    B. The circular arc
    C. The two accelerations are equal.

## Bending Beams

If you bend a rod down, it compresses the lower side of the rod and stretches the top, resulting in a restoring force. Figure I.4 shows a beam of length $L$, width $w$, and thickness $t$ fixed at one end and free to move at the other. Deflecting the end of the beam causes a restoring force $F$ at the end of the beam. The magnitude of the restoring force $F$ depends on the dimensions of the beam, the Young's modulus $Y$ for the material, and the deflection $d$. For small values of the deflection, the restoring force is

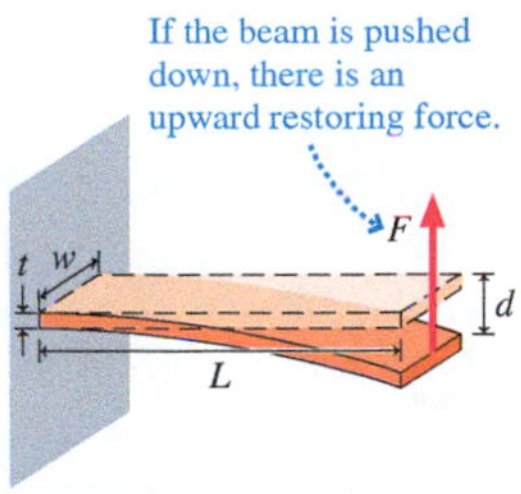

If the beam is pushed down, there is an upward restoring force.
F
t
w
d
L

FIGURE I.4

$$F = \left[\frac{Ywt^3}{4L^3}\right]d$$

This is similar to the formula for the restoring force of a spring, with the quantity in brackets playing the role of the spring constant $k$.

When a 70 kg man stands on the end of a springboard (a type of diving board), the board deflects by 4.0 cm.

17. If a 35 kg child stands at the end of the board, the deflection is
    A. 1.0 cm B. 2.0 cm
    C. 3.0 cm D. 4.0 cm
18. A 70 kg man jumps up and lands on the end of the board, deflecting it by 12 cm. At this instant, what is the approximate magnitude of the upward force the board exerts on his feet?
    A. 700 N B. 1400 N
    C. 2100 N D. 2800 N
19. If the board is replaced by one that is half the length but otherwise identical, how much will it deflect when a 70 kg man stands on the end?
    A. 0.50 cm B. 1.0 cm
    C. 2.0 cm D. 4.0 cm

## Additional Integrated Problems

20. You go to the playground and slide down the slide, a 3.0-m-long ramp at an angle of 40° with respect to horizontal. The pants that you've worn aren't very slippery; the coefficient of kinetic friction between your pants and the slide is $\mu_k = 0.45$. A friend gives you a very slight push to get you started. How long does it take you to reach the bottom of the slide?
21. If you stand on a scale at the equator, the scale will read slightly less than your true weight due to your circular motion with the rotation of the earth.
    a. Draw a free-body diagram to show why this is so.
    b. By how much is the scale reading reduced for a person with a true weight of 800 N?
22. Dolphins and other sea creatures can leap to great heights by swimming straight up and exiting the water at a high speed. A 210 kg dolphin leaps straight up to a height of 7.0 m. When the dolphin reenters the water, drag from the water brings it to a stop in 1.5 m. Assuming that the force of the water on the dolphin stays constant as it slows down,
    a. How much time does it take for the dolphin to come to rest?
    b. What is the force of the water on the dolphin as it is coming to rest?

PART

# II Conservation Laws

A peregrine falcon pulls in its wings and begins a nearly vertical dive. Speeding up as it falls, the falcon can reach a speed of 80 m/s or higher, moving faster than any other animal. When the falcon strikes a slower-moving bird at this high speed, the force of the resulting impact may exceed 100 times the falcon's weight, making for a very effective strike. In the next section, we'll learn how analyze the falcon's fall and the forces in an impact using the principles of conservation of energy and momentum.

OVERVIEW

# Why Some Things Stay the Same

Part I of this textbook was about *change*. Simple observations show us that most things in the world around us are changing. Even so, there are some things that *don't* change even as everything else is changing around them. Our emphasis in Part II will be on things that stay the same.

Consider, for example, a strong, sealed box in which you have replaced all the air with a mixture of hydrogen and oxygen. The mass of the box plus the gases inside is 600.0 g. Now, suppose you use a spark to ignite the hydrogen and oxygen. As you know, this is an explosive reaction, with the hydrogen and oxygen combining to create water—and quite a bang. But the strong box contains the explosion and all of its products.

What is the mass of the box after the reaction? The gas inside the box is different now, but a careful measurement would reveal that the mass hasn't changed—it's still 600.0 g! We say that the mass is *conserved*. Of course, this is true only if the box has stayed sealed. For conservation of mass to apply, the system must be *closed*.

## Conservation Laws

A closed system of interacting objects has another remarkable property. Each system is characterized by a certain number, and no matter how complex the interactions, the value of this number never changes. This number is called the *energy* of the system, and the fact that it never changes is called the *law of conservation of energy*. It is, perhaps, the single most important physical law ever discovered.

The law of conservation of energy is much more general than Newton's laws. Energy can be converted to many different forms, and, in all cases, the total energy stays the same:

- Gasoline, diesel, and jet engines convert the energy of a fuel into the mechanical energy of moving pistons, wheels, and gears.
- A solar cell converts the electromagnetic energy of light into electric energy.
- An organism converts the chemical energy of food into a variety of other forms of energy, including kinetic energy, sound energy, and thermal energy.

Energy will be *the* most important concept throughout the remainder of this text and much of Part II will focus on understanding what energy is and how it is used.

But energy is not the only conserved quantity. We will begin Part II with the study of two other quantities that are conserved in a closed system: *momentum* and *angular momentum*. Their conservation will help us understand a wide range of physical processes, from the forces when two rams butt heads to the graceful spins of ice skaters.

Conservation laws will give us a new and different *perspective* on motion. Some situations are most easily analyzed from the perspective of Newton's laws, but others make much more sense when analyzed from a conservation-law perspective. An important goal of Part II is to learn which perspective is best for a given problem.

# 9 Momentum

Male rams butt heads at high speeds in a ritual to assert their dominance. How can the force of this collision be minimized so as to avoid damage to their brains?

## LOOKING AHEAD ➤

### Impulse

This golf club delivers an **impulse** to the ball as the club strikes it.

You'll learn that a longer-lasting, stronger force delivers a greater impulse to an object.

### Momentum and Impulse

The impulse delivered by the player's head *changes* the ball's **momentum.**

You'll learn how to calculate this momentum change using the **impulse-momentum theorem.**

### Conservation of Momentum

As the octopus sends out a jet of water, the momentum of the octopus plus the water stays the same. The momentum is **conserved.**

You'll learn a powerful new *before-and-after* problem-solving approach using this **law of conservation of momentum.**

**GOAL** **To learn about impulse, momentum, and a new problem-solving approach based on conservation laws.**

## LOOKING BACK ◄

### Newton's Third Law

In Section 4.7 you learned about Newton's third law. In this chapter, you'll apply this law in order to understand the conservation of momentum.

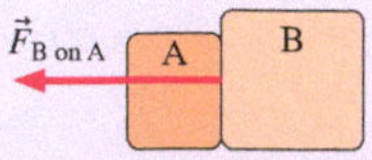


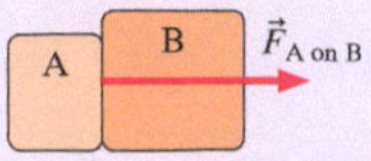


Newton's third law states that the force that object B exerts on A has *equal magnitude* but is *directed opposite to* the force that A exerts on B.

**STOP TO THINK**

A hammer hits a nail. The force of the nail on the hammer is

A. Greater than the force of the hammer on the nail.
B. Less than the force of the hammer on the nail.
C. Equal to the force of the hammer on the nail.
D. Zero.

## 9.1 Impulse

A **collision** is a short-duration interaction between two objects. The collision between a tennis ball and a racket, or your foot and a soccer ball, may seem instantaneous to your eye, but that is a limitation of your perception. The sequence of high-speed photos of a soccer kick shown in FIGURE 9.1 reveals that the ball is compressed as the foot begins its contact. It takes time to compress the ball, and more time for the ball to re-expand as it leaves the foot.

FIGURE 9.1 A sequence of high-speed photos of a soccer ball being kicked.

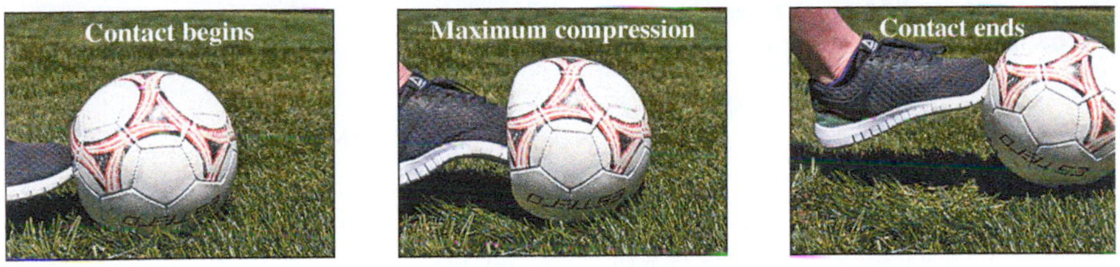


The duration of a collision depends on the materials from which the objects are made. The harder the objects, the shorter the contact time. A collision between two pool balls lasts less than 1 ms; the collision between the foot and the soccer ball in Figure 9.1 lasted more than 20 ms.

Let's look at the soccer ball kick in Figure 9.1 again. As the foot and the ball first come into contact, as shown in the left frame, the ball is just beginning to compress. By the middle frame of the figure, the ball has sped up and become greatly compressed. Finally, as shown in the right frame, the ball, now moving very fast, is again only slightly compressed. The ball is elastic, as you know; when you compress it, it exerts a restoring force equal to the force of the foot on the ball. Greater compression implies a greater force, just as for a spring.

FIGURE 9.2 is a graph of force versus time for a typical soccer ball kick. The force is zero until the foot first contacts the ball, rises quickly to a maximum value, and then falls back to zero as the ball leaves the foot. Thus there is a well-defined duration $\Delta t$ of the force. A large force like this exerted during a short interval of time is called an **impulsive force.** The forces of a hammer on a nail and of a bat on a baseball are other examples of impulsive forces.

A harder kick (i.e., a taller force curve) or a kick of longer duration (a wider force curve) causes the ball to leave the kicker's foot with a higher speed; that is, the *effect* of the kick is larger. A taller or wider force-versus-time curve has a larger *area* between the curve and the axis (i.e., the area "under" the force curve is larger), so we can say that **the effect of an impulsive force is proportional to the area under the force-versus-time curve.** This area, shown in FIGURE 9.3a, is the **impulse** $J$ of the force.

Impulsive forces can be complex, and the shape of the force-versus-time graph often changes in a complicated way. Consequently, it is often useful to think of the collision in terms of an *average* force $F_{avg}$. As FIGURE 9.3b shows, $F_{avg}$ is defined to be the constant force that has the same duration $\Delta t$ and the same area under the force curve as the real force. You can see from the figure that the area under the force curve can be written simply as $F_{avg}\,\Delta t$. Thus

$$\text{impulse } J = \text{area under the force curve} = F_{avg}\,\Delta t \qquad (9.1)$$

Impulse due to a force acting for a duration $\Delta t$

From Equation 9.1 we see that impulse has units of N · s, but N · s are equivalent to kg · m/s. We'll see shortly why the latter are the preferred units for impulse.

FIGURE 9.2 The force on a soccer ball changes rapidly.

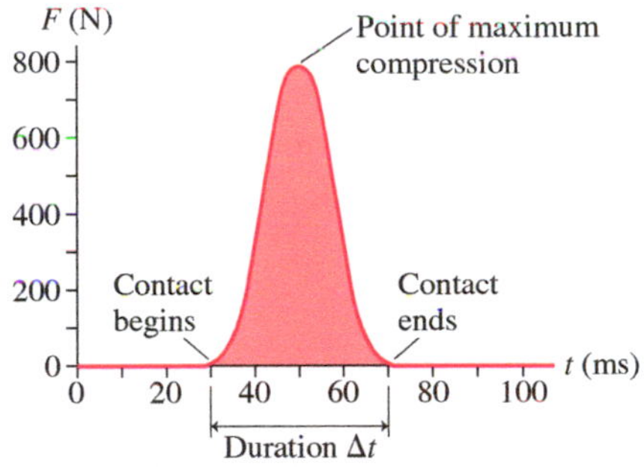


FIGURE 9.3 Looking at the impulse graphically.

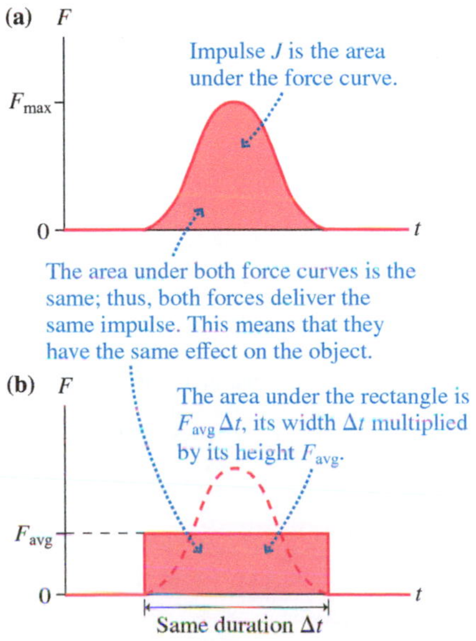
(a) F
Impulse J is the area under the force curve.
F max
0
t
The area under both force curves is the same; thus, both forces deliver the same impulse. This means that they have the same effect on the object.
(b) F
The area under the rectangle is F avg Δt, its width Δt multiplied by its height F avg.
F avg
0
t
Same duration Δt

So far, we've been assuming the force is directed along a coordinate axis, such as the $x$-axis. In this case impulse is a *signed* quantity—it can be positive or negative. A positive impulse results from an average force directed in the positive $x$-direction (that is, $F_{avg}$ is positive), while a negative impulse is due to a force directed in the negative $x$-direction ($F_{avg}$ is negative). More generally, the impulse is a *vector* quantity pointing in the direction of the average force vector:

$$\vec{J} = \vec{F}_{avg}\,\Delta t \tag{9.2}$$

**EXAMPLE 9.1 Finding the impulse on a bouncing ball**

FIGURE 9.4 shows smoothed force-versus-time data for the bounce of a rubber ball.

a. What is the impulse on the ball?
b. What is the average force on the ball?

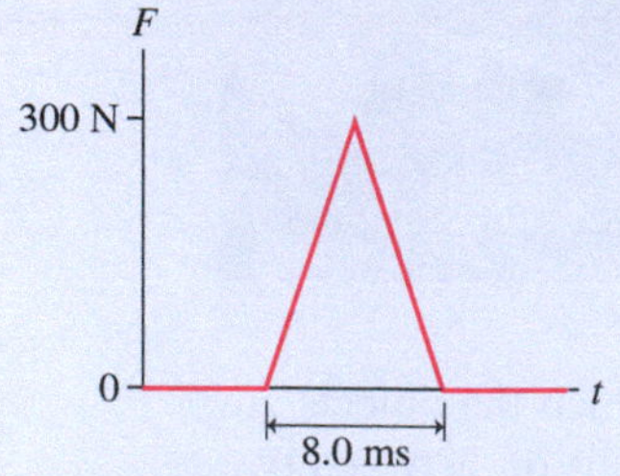


FIGURE 9.4 The force of the floor on a bouncing ball.

***STRATEGIZE*** The impulse is the area under the force curve. After we compute the area, and thus the impulse, we will calculate the average force using Equation 9.1.

***PREPARE*** Here the shape of the graph is triangular, so we'll need to use the fact that the area of a triangle is $\frac{1}{2} \times$ height $\times$ base.

***SOLVE***

a. The impulse is

$$J = \frac{1}{2}(300\ \text{N})(0.0080\ \text{s}) = 1.2\ \text{N}\cdot\text{s} = 1.2\ \text{kg}\cdot\text{m/s}$$

b. We can rearrange Equation 9.1 to find the average force that would give this same impulse:

$$F_{avg} = \frac{J}{\Delta t} = \frac{1.2\ \text{N}\cdot\text{s}}{0.0080\ \text{s}} = 150\ \text{N}$$

***ASSESS*** In this example, the average value of the force is half the maximum value. This is not surprising for a triangular force because the area of a triangle is half the base times the height.

**STOP TO THINK 9.1** The two graphs show smoothed force-versus-time data for two collisions. Which force delivers the greater impulse?

A. Force A
B. Force B
C. Both forces deliver the same impulse.

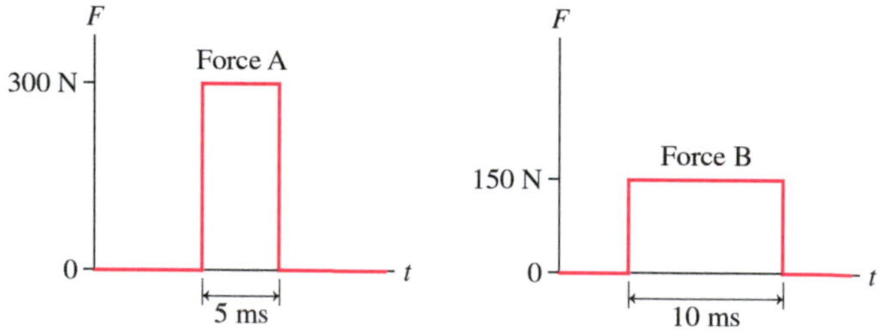


## 9.2 Momentum and the Impulse-Momentum Theorem

We've noted that the effect of an impulsive force depends on the impulse delivered to the object. The effect also depends on the object's mass. Our experience tells us that giving a kick to a heavy object will change its velocity much less than giving the same kick to a light object. We want now to find a quantitative relationship for impulse, mass, and velocity change.

FIGURE 9.5 The stick exerts an impulse on the puck, changing its speed.

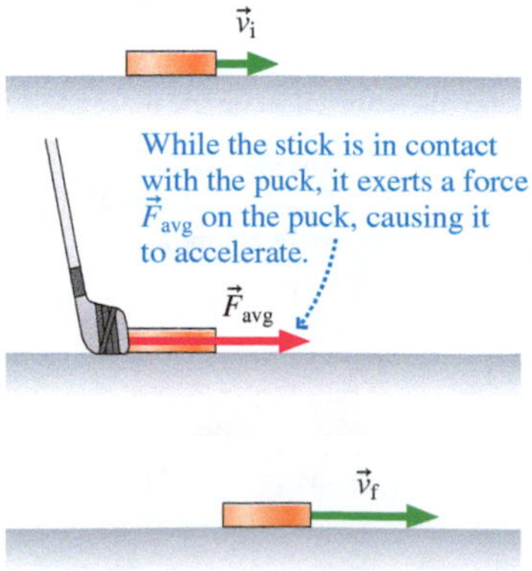


### Momentum

Consider the puck of mass $m$ in FIGURE 9.5, sliding with an initial velocity $\vec{v}_i$. It is struck by a hockey stick that delivers an impulse $\vec{J} = \vec{F}_{avg}\,\Delta t$ to the puck. After the impulse, the puck leaves the stick with a final velocity $\vec{v}_f$. How is this final velocity related to the initial velocity?

From Newton's second law, the average acceleration of the puck during the time the stick is in contact with it is

$$\vec{a}_{avg} = \frac{\vec{F}_{avg}}{m} \tag{9.3}$$

The average acceleration is related to the change in the velocity by

$$\vec{a}_{avg} = \frac{\Delta\vec{v}}{\Delta t} = \frac{\vec{v}_f - \vec{v}_i}{\Delta t} \quad (9.4)$$

Combining Equations 9.3 and 9.4, we have

$$\frac{\vec{F}_{avg}}{m} = \vec{a}_{avg} = \frac{\vec{v}_f - \vec{v}_i}{\Delta t}$$

or, rearranging,

$$\vec{F}_{avg}\,\Delta t = m\vec{v}_f - m\vec{v}_i \quad (9.5)$$

We recognize the left side of this equation as the impulse $\vec{J}$. The right side is the *change* in the quantity $m\vec{v}$. This quantity, the product of the object's mass and velocity, is called the **momentum** of the object. The symbol for momentum is $\vec{p}$:

$$\vec{p} = m\vec{v} \quad (9.6)$$

Momentum of an object of mass $m$ and velocity $\vec{v}$

From Equation 9.6, the units of momentum are those of mass times velocity, or kg · m/s. We noted previously that kg · m/s are the preferred units of impulse. Now we see that the reason for that preference is to match the units of momentum.

FIGURE 9.6 shows that the momentum $\vec{p}$ is a *vector* quantity that points in the same direction as the velocity vector $\vec{v}$. Like any vector, $\vec{p}$ can be decomposed into $x$- and $y$-components. Equation 9.6, which is a vector equation, is a shorthand way to write the two equations

$$p_x = mv_x$$
$$p_y = mv_y \quad (9.7)$$

**NOTE** ▸ One of the most common errors in momentum problems is failure to use the correct signs. The momentum component $p_x$ has the same sign as $v_x$. Just like velocity, momentum is positive for an object moving to the right (on the $x$-axis) or up (on the $y$-axis), but *negative* for an object moving to the left or down. ◂

The *magnitude* of an object's momentum is simply the product of the object's mass and speed, or $p = mv$. A heavy, fast-moving object will have a great deal of momentum, while a light, slow-moving object will have very little. Two objects with very different masses can have similar values of momentum if their speeds are very different as well. TABLE 9.1 gives some typical values of the momentum of various moving objects.

## The Impulse-Momentum Theorem

We can now write Equation 9.5 in terms of impulse and momentum:

$$\vec{J} = \vec{p}_f - \vec{p}_i = \Delta\vec{p} \quad (9.8)$$

Impulse-momentum theorem

where $\vec{p}_i = m\vec{v}_i$ is the object's initial momentum, $\vec{p}_f = m\vec{v}_f$ is its final momentum after the impulse, and $\Delta\vec{p} = \vec{p}_f - \vec{p}_i$ is the *change* in its momentum. This expression is known as the **impulse-momentum theorem**. It states that **an impulse delivered to an object causes the object's momentum to change.** That is, the *effect* of an impulsive force is to change the object's momentum from $\vec{p}_i$ to

$$\vec{p}_f = \vec{p}_i + \vec{J} \quad (9.9)$$

A video to support a section's topic is embedded in the eText.

**Video** Force and Momentum Change

FIGURE 9.6 An object's momentum vector $\vec{p}$ can be decomposed into $x$- and $y$-components.

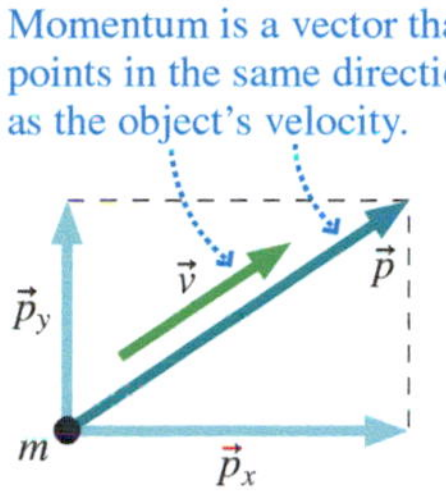


TABLE 9.1 Some typical values of momentum (approximate)

| Object | Mass (kg) | Speed (m/s) | Momentum (kg · m/s) |
|---|---|---|---|
| Falling raindrop | $2 \times 10^{-5}$ | 5 | $10^{-4}$ |
| Bullet | 0.004 | 500 | 2 |
| Pitched baseball | 0.15 | 40 | 6 |
| Running person | 70 | 3 | 200 |
| Car on highway | 1000 | 30 | $3 \times 10^4$ |

**Legging it** BIO A frog making a jump wants to gain as much momentum as possible before leaving the ground. This means that he wants the greatest impulse $J = F_{avg}\,\Delta t$ delivered to him by the ground. There is a maximum force that muscles can exert, limiting $F_{avg}$. But the time interval $\Delta t$ over which the force is exerted can be greatly increased by having long legs. Many animals that are good jumpers have particularly long legs.

Equation 9.8 can also be written in terms of its $x$- and $y$-components as

$$\begin{aligned} J_x &= \Delta p_x = (p_x)_f - (p_x)_i = m(v_x)_f - m(v_x)_i \\ J_y &= \Delta p_y = (p_y)_f - (p_y)_i = m(v_y)_f - m(v_y)_i \end{aligned} \tag{9.10}$$

The impulse-momentum theorem is illustrated by two examples in FIGURE 9.7. In Figure 9.7a, the putter strikes the ball, exerting a force on it and delivering an impulse $\vec{J} = \vec{F}_{avg}\,\Delta t$. Notice that the direction of the impulse is the same as that of the force. Because $\vec{p}_i = \vec{0}$ in this situation, we can use the impulse-momentum theorem to find that the ball leaves the putter with momentum $\vec{p}_f = \vec{p}_i + \vec{J} = \vec{J}$.

**NOTE** ► You can think of the putter as changing the ball's momentum by transferring momentum to it as an impulse. Thus we say the putter *delivers* an impulse to the ball, and the ball *receives* an impulse from the putter. ◄

The soccer player in Figure 9.7b presents a more complicated case. Here, the initial momentum of the ball is directed downward to the left. The impulse delivered to it by the player's head, upward to the right, is strong enough to reverse the ball's motion and send it off in a new direction. The graphical addition of vectors in Figure 9.7b again shows that $\vec{p}_f = \vec{p}_i + \vec{J}$.

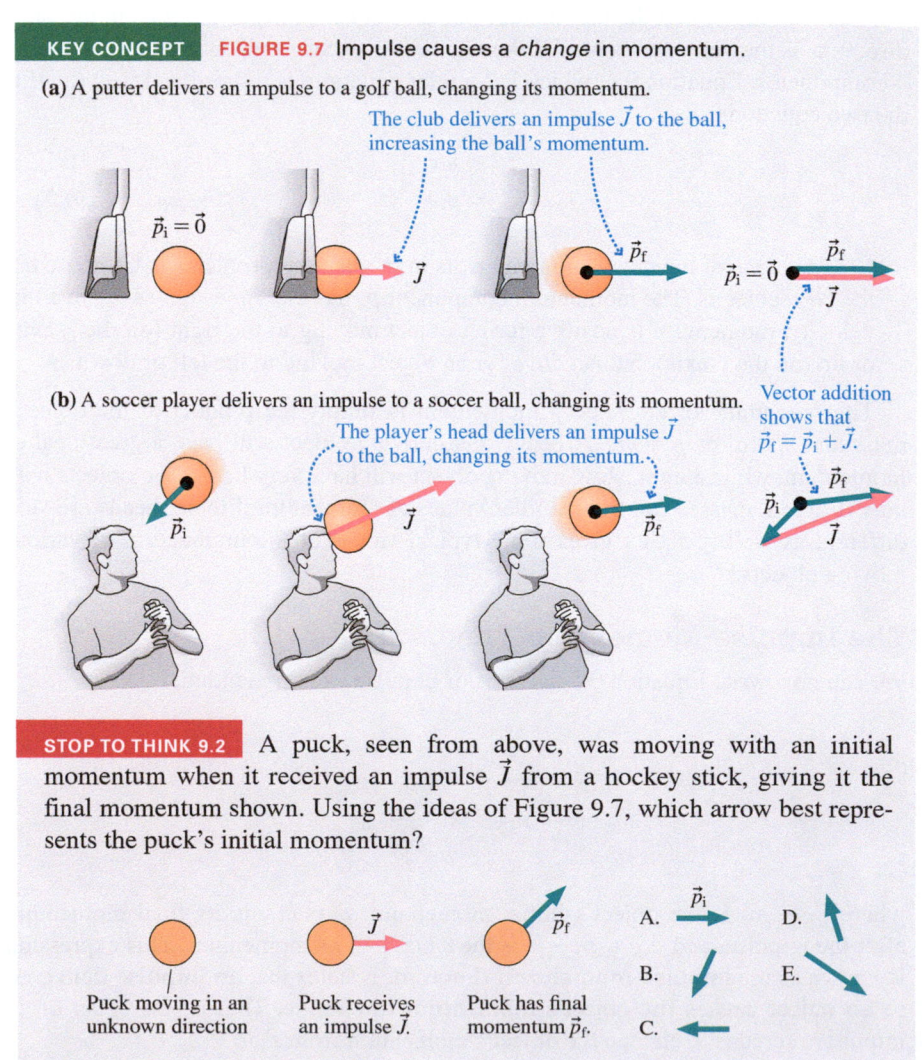

KEY CONCEPT **FIGURE 9.7** Impulse causes a *change* in momentum.

(a) A putter delivers an impulse to a golf ball, changing its momentum.

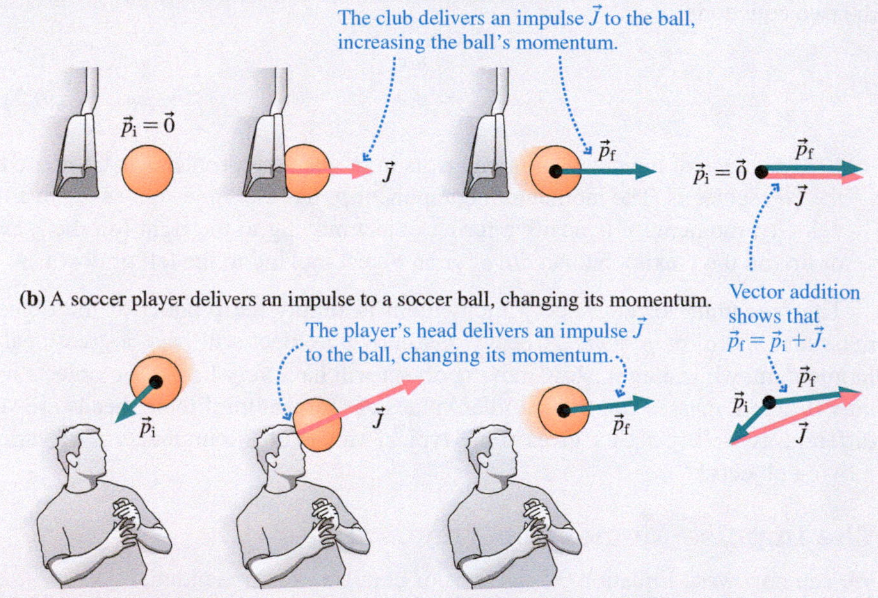


**STOP TO THINK 9.2** A puck, seen from above, was moving with an initial momentum when it received an impulse $\vec{J}$ from a hockey stick, giving it the final momentum shown. Using the ideas of Figure 9.7, which arrow best represents the puck's initial momentum?

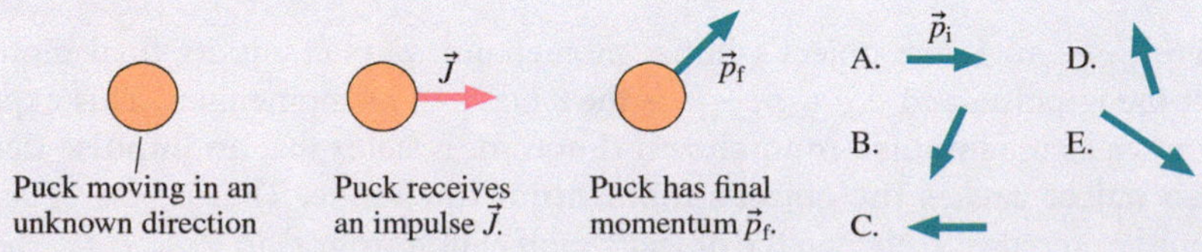

## EXAMPLE 9.2 Calculating the change in momentum

A ball of mass $m = 0.25$ kg rolling to the right at 1.3 m/s strikes a wall and rebounds to the left at 1.1 m/s. What is the change in the ball's momentum? What is the impulse delivered to it by the wall?

**STRATEGIZE** The impulse-momentum theorem deals with the change in momentum due to an impulse. The details of how the impulse is delivered don't matter. In problems like this one, we don't need to worry about the details of the collision, only the change from the initial state to the final state. To solve such problems, we'll introduce a new type of visual overview, one in which we consider the situation before the interaction and then after the interaction. This before-and-after visual overview is shown in FIGURE 9.8. In this case, we'll also include a diagram showing the force during the collision with the wall. We won't include such diagrams in all cases, but it's useful to visualize the force that causes the momentum change. The force is directed to the left, as we know it must be.

FIGURE 9.8 Visual overview for a ball bouncing off a wall.

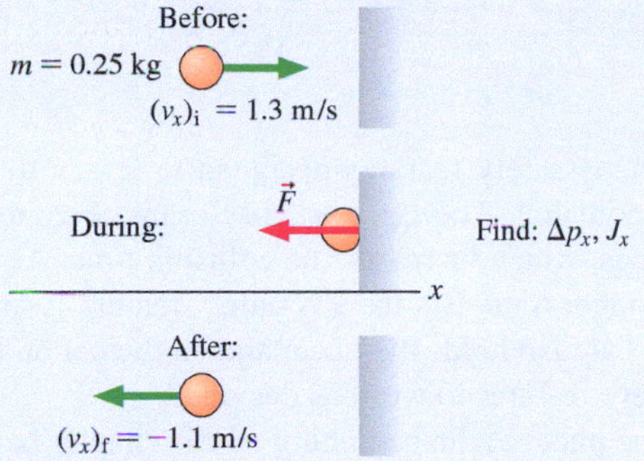


**PREPARE** The ball is moving along the $x$-axis, so we write the momentum in component form, as in Equation 9.7. The change in momentum is then the difference between the final and initial values of the momentum. By the impulse-momentum theorem, the impulse is equal to this change in momentum.

**SOLVE** The $x$-component of the initial momentum is

$$(p_x)_i = m(v_x)_i = (0.25 \text{ kg})(1.3 \text{ m/s}) = 0.325 \text{ kg}\cdot\text{m/s}$$

The $y$-component of the momentum is zero both before and after the bounce. After the ball rebounds, the $x$-component is

$$(p_x)_f = m(v_x)_f = (0.25 \text{ kg})(-1.1 \text{ m/s}) = -0.275 \text{ kg}\cdot\text{m/s}$$

It is particularly important to notice that the $x$-component of the momentum, like that of the velocity, is negative. This indicates that the ball is moving to the *left*. The change in momentum is

$$\Delta p_x = (p_x)_f - (p_x)_i = (-0.275 \text{ kg}\cdot\text{m/s}) - (0.325 \text{ kg}\cdot\text{m/s})$$
$$= -0.60 \text{ kg}\cdot\text{m/s}$$

By the impulse-momentum theorem, the impulse delivered to the ball by the wall is equal to this change, so

$$J_x = \Delta p_x = -0.60 \text{ kg}\cdot\text{m/s}$$

**ASSESS** The impulse is negative, indicating that the force that causes the impulse is pointing to the left, which we noted must be the case.

## SYNTHESIS 9.1 Momentum and impulse

A moving object has momentum. A force acting on an object delivers an *impulse* that changes the object's momentum.

The **momentum** of an object is the product of its mass and its velocity.

$$\vec{p} = m\vec{v}$$

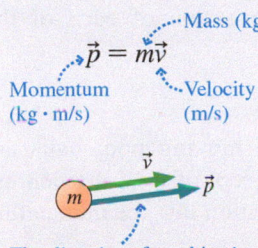

Momentum (kg · m/s); Mass (kg); Velocity (m/s)

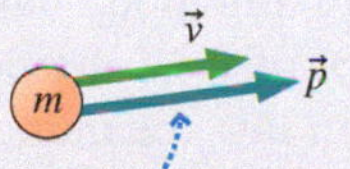


The *direction* of an object's momentum is the same as its velocity.

The force acting on an object . . . delivers an **impulse** $J$ that is equal to the area under the force curve.

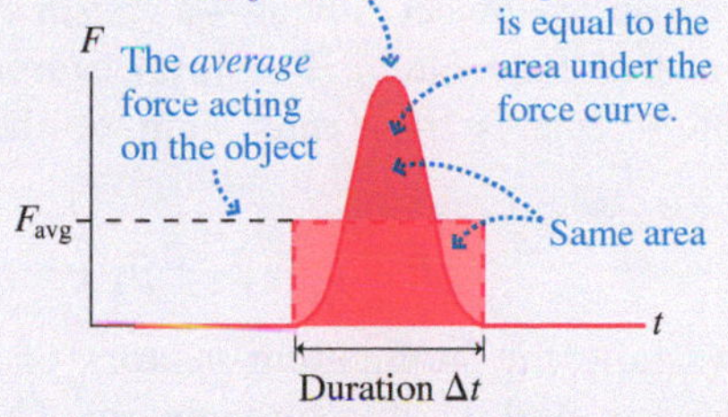


$$\vec{J} = \vec{F}_{avg}\,\Delta t$$

Impulse (kg · m/s); Average force (N); Duration (s)

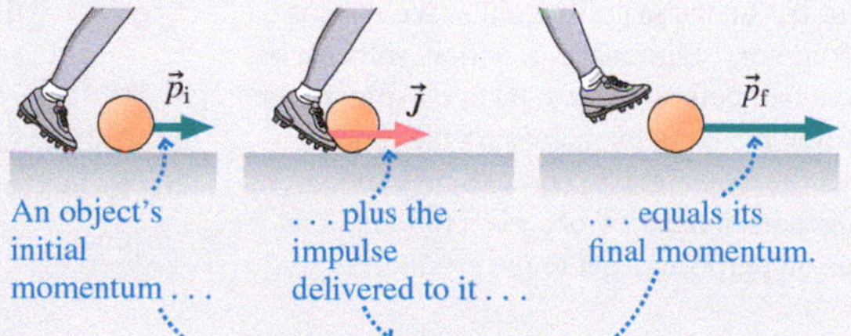


An object's initial momentum . . . plus the impulse delivered to it . . . equals its final momentum.

$$\vec{p}_i + \vec{J} = \vec{p}_f$$

This relationship can also be written in terms of the *change* in the object's momentum:

$$\vec{J} = \vec{p}_f - \vec{p}_i = \Delta\vec{p}$$

The impulse delivered . . . equals the change in the momentum.

The rams in the chapter-opening photo butt heads—and their heads go from moving to standing still. A change in momentum occurs, which means an impulse. An impulse means a force, and a large force on the head can cause injury. How do the rams keep from getting hurt? A similar situation is at work in a car crash—the car goes from moving to standing still, as do the driver and the passengers in the car. There is a change in momentum; this is unavoidable. What strategies reduce the chances of injury to the occupants of the car?

In these examples, the object has momentum $\vec{p}_i$ just before impact and zero momentum after (i.e., $\vec{p}_f = \vec{0}$). The impulse-momentum theorem tells us that

$$\vec{J} = \vec{F}_{avg}\,\Delta t = \Delta\vec{p} = \vec{p}_f - \vec{p}_i = -\vec{p}_i$$

or

$$\vec{F}_{avg} = -\frac{\vec{p}_i}{\Delta t} \qquad (9.11)$$

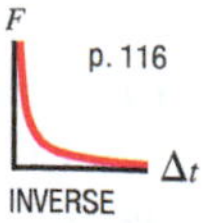


That is, the average force needed to stop an object is *inversely proportional* to the duration $\Delta t$ of the collision. **If the duration of the collision can be increased, the force of the impact will be decreased.**

**CONCEPTUAL EXAMPLE 9.3 A stretch in time**

The seatbelts in your car are made of webbing that is designed to have a significant stretch when supporting large forces. Why is this useful?

***REASON*** Suppose a car crashes into a fixed barrier and comes to rest. The driver will come to rest as well. Increasing the time over which the driver comes to rest will decrease the force on the driver. If the seatbelt can stretch, the driver can still be moving after the car has stopped—the driver can come to rest over a longer time. Increasing the time reduces the force, and thus the chance for injury.

***ASSESS*** This makes sense. If the seatbelt could not stretch at all, a driver or passenger would stop much more suddenly, which would increase the force and the likelihood of injury.

**A spiny cushion** BIO The spines of a hedgehog obviously help protect it from predators. But they serve another function as well. If a hedgehog falls from a tree—a not uncommon occurrence—it simply rolls itself into a ball before it lands. Its thick spines then cushion the blow by increasing the time it takes for the animal to come to rest. Indeed, hedgehogs have been observed to fall out of trees on purpose to get to the ground!

Modern automobiles have many safety features designed to lessen the force on occupants of the car during a collision. The car itself has crumple zones, regions designed to fail under large forces, which increases the collision time. Air bags also bring occupants to rest over a longer time. But the key safety feature is the seatbelt, which stretches to reduce force but also holds the occupants in the car and positions them correctly for the other safety features to work as designed.

The butting rams shown in the photo at the beginning of this chapter have adaptations that allow them to collide at high speeds without injury to their brains. The cranium has a double wall to prevent skull injuries, and there is a thick spongy mass that increases the time it takes for the brain to come to rest upon impact, reducing the magnitude of the force on the brain.

## Total Momentum

If we have more than one object moving—a *system* of objects—then the system as a whole has an overall momentum. The **total momentum** $\vec{P}$ (note the capital $P$) of a system of objects is the vector sum of the momentum of each of the objects:

$$\vec{P} = \vec{p}_1 + \vec{p}_2 + \vec{p}_3 + \cdots$$

FIGURE 9.9 shows how the momentum vectors of three moving pool balls are graphically added to find the total momentum. The concept of total momentum will be of key importance when we discuss the conservation law for momentum in Section 9.4.

FIGURE 9.9 The total momentum of three pool balls.

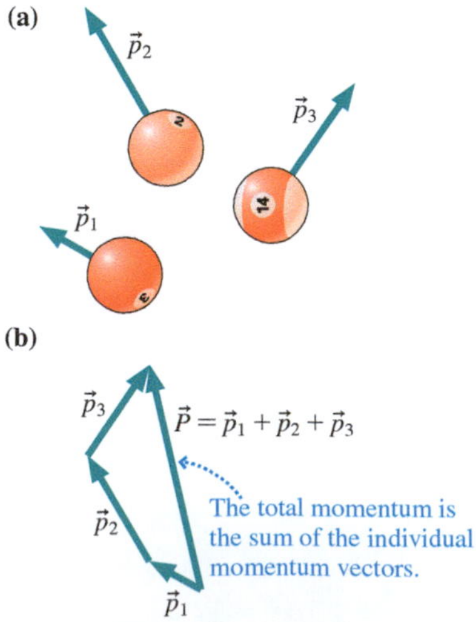


**STOP TO THINK 9.3** The cart's change of momentum is

A. −30 kg·m/s
B. −20 kg·m/s
C. −10 kg·m/s
D. 10 kg·m/s
E. 20 kg·m/s
F. 30 kg·m/s

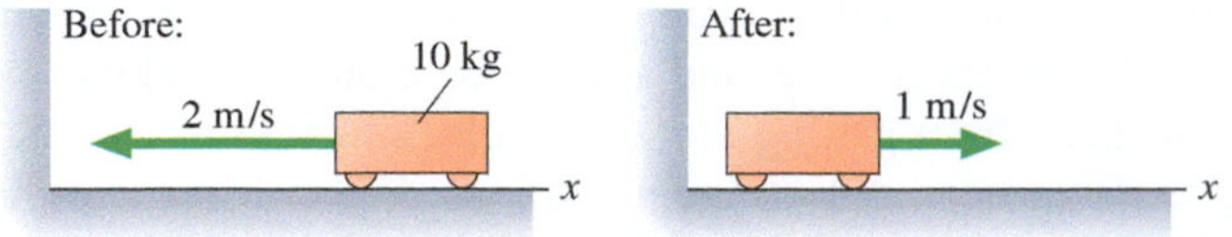

## 9.3 Solving Impulse and Momentum Problems

The visual overviews and free-body diagrams that you learned to draw in Chapters 1–8 were oriented toward the use of Newton's laws and a subsequent kinematic analysis. Now that we are working with momentum, a conserved quantity, we want to consider the situation before and after an impact or other event. We'll draw a before-and-after visual overview, an idea we introduced in Example 9.2.

**TACTICS BOX 9.1 Drawing a before-and-after visual overview**

1. **Sketch the situation.** Use two drawings, labeled "Before" and "After," to show the objects *immediately before* they interact and again *immediately after* they interact.
2. **Establish a coordinate system.** Select your axes to match the motion.
3. **Define symbols.** Define symbols for the masses and for the velocities before and after the interaction. Position and time are not needed.
4. **List known information.** List the values of quantities known from the problem statement or that can be found quickly with simple geometry or unit conversions. Before-and-after pictures are usually simpler than the pictures you used for dynamics problems, so listing known information on the sketch is often adequate.
5. **Identify the desired unknowns.** What quantity or quantities will allow you to answer the question? These should have been defined as symbols in step 3.

Exercises 9–11

**EXAMPLE 9.4 Force in hitting a baseball**

A 150 g baseball is thrown with a speed of 20 m/s. It is hit straight back toward the pitcher at a speed of 40 m/s. The impulsive force of the bat on the ball has the shape shown in FIGURE 9.10. What is the *maximum* force $F_{max}$ that the bat exerts on the ball? What is the *average* force that the bat exerts on the ball?

FIGURE 9.10 The interaction force between the baseball and the bat.

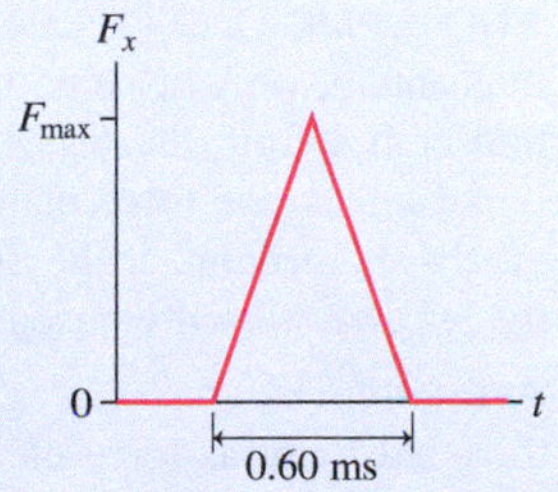


***STRATEGIZE*** When a baseball collides with the bat, the impulse from the bat changes the momentum of the ball. We'll compute the change in momentum and use this value to figure out the details of the impulse.

***PREPARE*** To help determine the change in momentum, we start with the before-and-after visual overview in FIGURE 9.11. We develop the visual overview following the steps of Tactics Box 9.1. $F_x$, the force of the bat on the ball in Figure 9.10, is positive, so the force is directed to the right. A force to the right means that the ball came from the left and rebounds to the right, as we've drawn.

***SOLVE*** In the last several chapters we've started the mathematical solution with Newton's second law. Now we want to use the impulse-momentum theorem:

$$\Delta p_x = J_x = \text{area under the force curve}$$

FIGURE 9.11 A before-and-after visual overview.

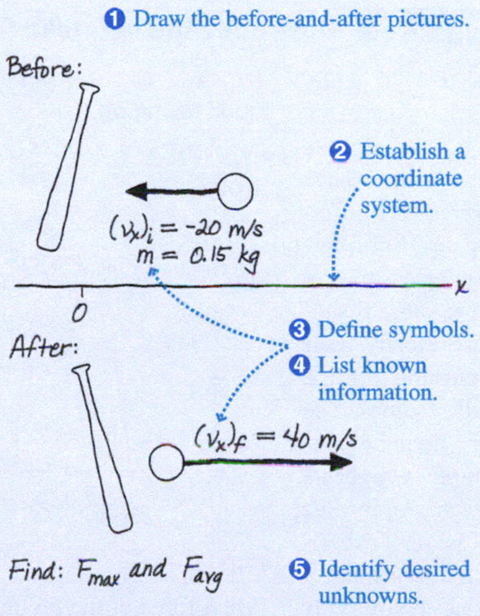


We know the velocities before and after the collision, so we can find the change in the ball's momentum:

$$\Delta p_x = m(v_x)_f - m(v_x)_i = (0.15 \text{ kg})(40 \text{ m/s} - (-20 \text{ m/s}))$$
$$= 9.0 \text{ kg} \cdot \text{m/s}$$

*Continued*

The force curve is a triangle with height $F_{max}$ and width 0.60 ms. As in Example 9.1, the area under the curve is

$$J_x = \text{area} = \frac{1}{2} \times F_{max} \times (6.0 \times 10^{-4}\text{ s})$$
$$= (F_{max})(3.0 \times 10^{-4})$$

According to the impulse-momentum theorem, $\Delta p_x = J_x$, so we have

$$9.0\text{ kg} \cdot \text{m/s} = (F_{max})(3.0 \times 10^{-4}\text{ s})$$

Thus the *maximum* force is

$$F_{max} = \frac{9.0\text{ kg} \cdot \text{m/s}}{3.0 \times 10^{-4}\text{ s}} = 30{,}000\text{ N}$$

Using Equation 9.1, we find that the *average* force, which depends on the collision duration $\Delta t = 6.0 \times 10^{-4}$ s, has a smaller value:

$$F_{avg} = \frac{J_x}{\Delta t} = \frac{\Delta p_x}{\Delta t} = \frac{9.0\text{ kg} \cdot \text{m/s}}{6.0 \times 10^{-4}\text{ s}} = 15{,}000\text{ N}$$

***ASSESS*** $F_{max}$ is a large force, but we expect a large force for such a short collision time and a large change in momentum.

## The Impulse Approximation

When two objects interact during a collision or other brief interaction, such as that between the bat and ball of Example 9.4, the forces *between* them are generally quite large. Other forces may also act on the interacting objects, but usually these forces are *much* smaller than the interaction forces. In Example 9.4, for example, the 1.5 N weight of the ball is vastly less than the 30,000 N force of the bat on the ball. We can reasonably ignore these small forces *during* the brief time of the impulsive force. Doing so is called the **impulse approximation.**

When we use the impulse approximation, $(p_x)_i$ and $(p_x)_f$—and $(v_x)_i$ and $(v_x)_f$—are the values of the momentum (and velocity) *immediately* before and *immediately* after the collision. For example, the velocities in Example 9.4 are those of the ball just before and after it collides with the bat. The values will subsequently change as other forces, such as weight and drag, exert their influence.

### EXAMPLE 9.5 How does the bird take flight?

Scientists who study birds that make a rapid takeoff from a standing start have discovered that different species use different strategies to get airborne. Some species use mostly their legs, pushing off the ground; some use mostly their wings, with rapid wingbeats giving a large lift force. In one experiment, a 190 g quail took off essentially straight up from a force plate that measured the normal force of the ground on the quail. The net upward force on the quail is plotted in FIGURE 9.12. The quail left the ground at 3.5 m/s.

FIGURE 9.12 Force versus time for a quail taking off.

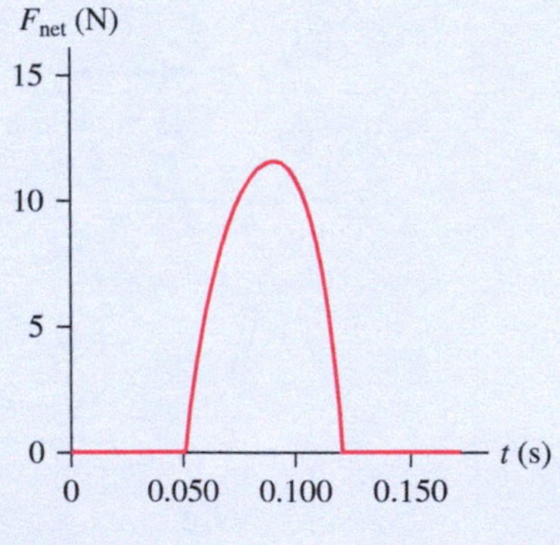


a. Approximate the impulse by estimating the area under the force-versus-time curve.
b. What is the change in momentum of the quail as it leaves the ground?
c. Given the values you found, as the quail executes a rapid takeoff, does it use mostly its legs or mostly its wings?

***STRATEGIZE*** Once we determine the impulse and the change in momentum, we will judge what fraction of the takeoff momentum comes from the legs. If the quail uses only its legs, the impulse from the force plate will be equal to the change in the quail's momentum. If the change in momentum is greater than the impulse, this difference is due to the lift force from the wings.

***PREPARE*** The graph in Figure 9.12 is the net upward force on the quail from the force plate. The force plate pushes up on the quail because the quail pushes down with its legs. The area under the curve is a measure of the impulse delivered to the quail as a consequence of its pushing off with its legs. In FIGURE 9.13a, we use a triangle to approximate the area under the curve to let us compute the impulse. FIGURE 9.13b is a before-and-after visual overview of the quail taking off.

***SOLVE*** a. The area of the triangle in Figure 9.13a approximates the area under the force-versus-time curve, which is the impulse $J_y$:

$$J_y = \text{area under curve} = \frac{1}{2} \times \text{base} \times \text{height}$$
$$= \frac{1}{2}(0.125\text{ s} - 0.045\text{ s})(13\text{ N}) = 0.52\text{ kg} \cdot \text{m/s}$$

b. The quail starts at rest, so the change in momentum is the final momentum:

$$\Delta p_y = (p_y)_f = (0.19\text{ kg})(3.5\text{ m/s}) = 0.67\text{ kg} \cdot \text{m/s}$$

FIGURE 9.13 Estimating the impulse and determining the change in momentum of the quail.

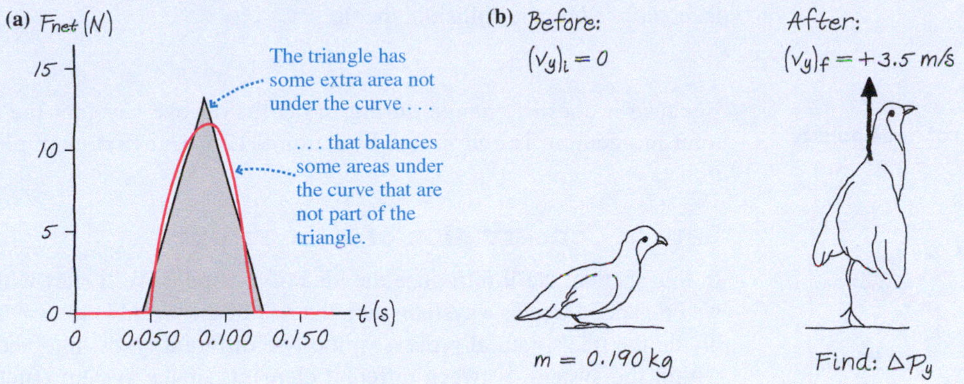


c. The impulse is nearly 80% of the change in momentum, so most of the change in momentum is provided by the impulse from the legs, with only a small amount coming from the lift force provided by the wings.

**ASSESS** Quail tend to make short flights, they spend most of their time on the ground, and they are good runners, so it's reasonable to expect them to rely mostly on their legs to get airborne. Our results seem reasonable.

**STOP TO THINK 9.4** A 10 g rubber ball and a 10 g clay ball are each thrown at a wall with equal speeds. The rubber ball bounces; the clay ball sticks. Which ball receives the greater impulse from the wall?

A. The clay ball receives a greater impulse because it sticks.
B. The rubber ball receives a greater impulse because it bounces.
C. They receive equal impulses because they have equal values of momentum.
D. Neither receives an impulse because the wall doesn't move.

## 9.4 Conservation of Momentum

Let's consider two objects, such as the rams shown in the opening photo of this chapter, that interact during the brief moment of a collision. During a collision, two objects exert forces on each other that vary in a complex way. We usually don't even know the magnitudes of these forces. Using Newton's second law to predict the outcome of such a collision would thus be a daunting challenge. However, by using Newton's *third* law in the language of impulse and momentum, we'll find that it's possible to describe the *outcome* of a collision—the final speeds and directions of the colliding objects—in a simple way.

FIGURE 9.14 shows two balls initially headed toward each other. The balls collide, then bounce apart. The forces during the collision, when the balls are interacting, form an action/reaction pair $\vec{F}_{1\text{ on }2}$ and $\vec{F}_{2\text{ on }1}$. For now, we'll continue to assume that the motion is one dimensional along the $x$-axis.

During the collision, the impulse $J_{2x}$ delivered to ball 2 by ball 1 is the average value of $\vec{F}_{1\text{ on }2}$ multiplied by the collision time $\Delta t$. Likewise, the impulse $J_{1x}$ delivered to ball 1 by ball 2 is the average value of $\vec{F}_{2\text{ on }1}$ multiplied by $\Delta t$. Because $\vec{F}_{1\text{ on }2}$ and $\vec{F}_{2\text{ on }1}$ form an action/reaction pair, they have equal magnitudes but opposite directions. As a result, the two impulses $J_{1x}$ and $J_{2x}$ are also equal in magnitude but opposite in sign, so that $J_{1x} = -J_{2x}$.

According to the impulse-momentum theorem, the change in the momentum of ball 1 is $\Delta p_{1x} = J_{1x}$ and the change in the momentum of ball 2 is $\Delta p_{2x} = J_{2x}$. Because $J_{1x} = -J_{2x}$, the change in the momentum of ball 1 is equal in magnitude but opposite in sign to the change in momentum of ball 2. If ball 1's momentum increases by a

FIGURE 9.14 A collision between two balls.

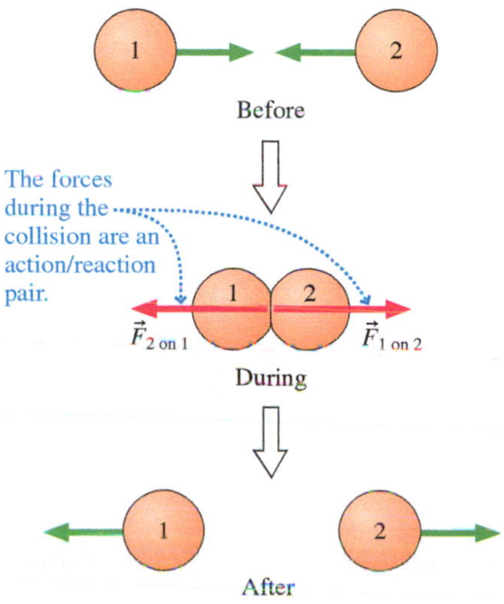

certain amount during the collision, ball 2's momentum will *decrease* by exactly the same amount. This implies that the total momentum $P_x = p_{1x} + p_{2x}$ of the two balls is *unchanged* by the collision; that is,

$$(P_x)_f = (P_x)_i \tag{9.12}$$

Because it doesn't change during the collision, we say that the *x*-component of total momentum is *conserved*. Equation 9.12 is our first example of a *conservation law*.

FIGURE 9.15 A system of three objects.

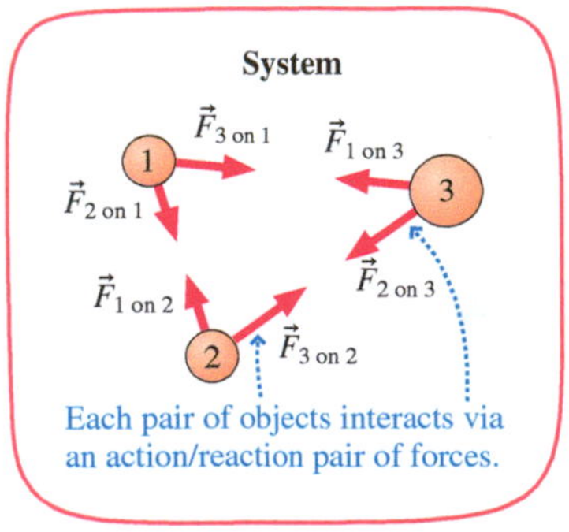


## Law of Conservation of Momentum

In this section, we'll introduce the idea of a *system*. We'll start with an example: If we consider a tree as a system, then the rest of the world is the *environment*. We can divide the tree's natural processes into two different types: interactions that happen within the system, between different elements of the system (such as transport of water from the roots to the leaves), and interactions between the system and the environment (such as light from the sun falling on the leaves).

This approach of dividing the world into system and environment will be very important to us. As we'll see in future chapters, how we define the system can dramatically change the way that we approach the solution to a problem. And the boundaries aren't always clear, as we can see by considering the example of the tree. Bacteria live on the roots of the tree, facilitating the uptake of nutrients. Are these bacteria part of the system or part of the environment? We could define the system either way depending on the questions we are asking.

**Those are the breaks** At the start of a game of pool, a player hits the cue ball into the triangle of stationary numbered balls. The forces between the balls are much larger than friction or the other forces at work, so we can use the impulse approximation. If we define the system as the collection of all the balls on the table, then the only forces that matter are internal forces. This means that the total momentum of the system is the same before and after the break. In the photo you can see the balance between motion to the left and to the right; there is no momentum along the horizontal axis, either before or after the break.

Let's continue our discussion with a simple system, three objects that interact with each other—you might imagine three balls on a pool table. We've drawn the system in FIGURE 9.15, including a red line around the three objects to denote the boundary of the system. Arrows show the forces between the three objects. Forces that act only between objects within the system are called **internal forces.** Forces that come from agents outside the system are called **external forces.** When one of the balls smacks into a second ball, this is an internal force; it's a force between two objects of the system. When someone uses a cue to strike one of the balls, this is an external force; it's from an agent outside the system.

Each pair of objects in the system in Figure 9.15 interacts via forces that are an action/reaction pair. Exactly as for the two-object collision, the change in momentum of object 2 due to the force from object 3 is equal in magnitude, but opposite in direction, to the change in object 3's momentum due to object 2. The net change in the momentum of these two objects due to their interaction forces is thus zero. The same argument holds for every pair, with the result that, no matter how complicated the forces between the objects, **there is no change in the total momentum $\vec{P}$ of the system.** The total momentum of the system remains constant: It is *conserved*. We can generalize this principle as follows: **The total momentum of a system subject to only internal forces is conserved.**

FIGURE 9.16 A system of objects subject to external forces.

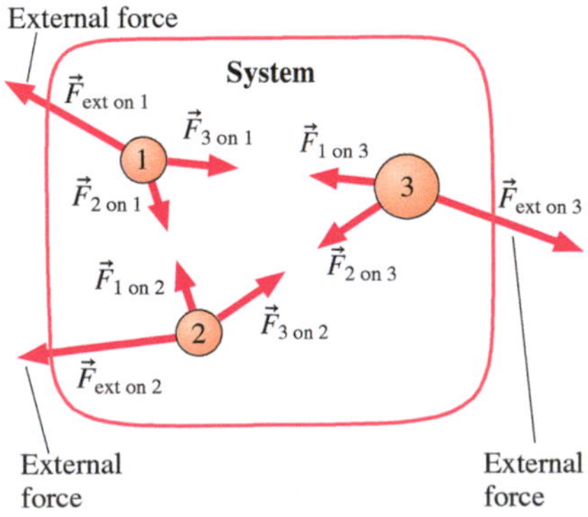


In FIGURE 9.16 we show the same three-object system of Figure 9.15, but now with *external* forces acting on the three objects. These external forces *can* change the momentum of the system. During a time interval $\Delta t$, for instance, the external force $\vec{F}_{\text{ext on }1}$ acting on object 1 changes its momentum, according to the impulse-momentum theorem, by $\Delta\vec{p}_1 = (\vec{F}_{\text{ext on }1})\Delta t$. The momentum of each of the other two objects changes similarly. Thus the change in the total momentum is

$$\begin{aligned}\Delta\vec{P} &= \Delta\vec{p}_1 + \Delta\vec{p}_2 + \Delta\vec{p}_3 \\ &= (\vec{F}_{\text{ext on }1}\Delta t) + (\vec{F}_{\text{ext on }2}\Delta t) + (\vec{F}_{\text{ext on }3}\Delta t) \\ &= (\vec{F}_{\text{ext on }1} + \vec{F}_{\text{ext on }2} + \vec{F}_{\text{ext on }3})\,\Delta t \\ &= \vec{F}_{\text{net}}\,\Delta t\end{aligned} \tag{9.13}$$

where $\vec{F}_{\text{net}}$ is the net force due to *external forces*.

Equation 9.13 has a very important implication in the case where the net force on a system is zero: **If $\vec{F}_{net} = \vec{0}$, the *total* momentum $\vec{P}$ of the system does not change.** The total momentum remains constant, *regardless* of whatever interactions are going on *inside* the system. With no net external force acting that can change its momentum, we call a system with $\vec{F}_{net} = \vec{0}$ an **isolated system.**

The importance of these results is sufficient to elevate them to a law of nature, alongside Newton's laws.

**Law of conservation of momentum** The total momentum $\vec{P}$ of an isolated system is a constant. Interactions within the system do not change the system's total momentum.

**NOTE** ▶ It is worth emphasizing the critical role of Newton's third law in the derivation of Equation 9.13. The law of conservation of momentum is a direct consequence of the fact that interactions within an isolated system are action/reaction pairs. ◀

Mathematically, the law of conservation of momentum for an isolated system is

$$\vec{P}_f = \vec{P}_i \tag{9.14}$$

Law of conservation of momentum for an isolated system

**The total momentum after an interaction is equal to the total momentum before the interaction.** Because Equation 9.14 is a vector equation, the equality is true for each of the components of the momentum vector; that is,

$$\underbrace{(p_{1x})_f + (p_{2x})_f + (p_{3x})_f + \cdots}_{\text{Final momentum}} = \underbrace{(p_{1x})_i + (p_{2x})_i + (p_{3x})_i + \cdots}_{\text{Initial momentum}} \quad (x\text{-component})$$

$$(p_{1y})_f + (p_{2y})_f + (p_{3y})_f + \cdots = (p_{1y})_i + (p_{2y})_i + (p_{3y})_i + \cdots \quad (y\text{-component}) \tag{9.15}$$

(Terms labeled Object 1, Object 2, Object 3.)

## EXAMPLE 9.6 Speed of ice skaters pushing off

Two ice skaters, Sandra and David, stand facing each other on frictionless ice. Sandra has a mass of 45 kg, David a mass of 80 kg. They then push off from each other. After the push, Sandra moves off at a speed of 2.2 m/s. What is David's speed?

***STRATEGIZE*** We will consider the system to be the two skaters. For each skater, the upward normal force of the ice balances the downward weight force to make $F_{net} = \vec{0}$. We will thus consider the two skaters to be an isolated system. The momentum of the system is therefore conserved—the total momentum of the skaters before and after the push will be the same. Even though we know nothing about the details of the forces between the two skaters—how hard they push or for how long—we will solve this problem by looking at the "before" and "after" cases and considering conservation of momentum.

***PREPARE*** We start with FIGURE 9.17, which shows a before-and-after visual overview for the two skaters. The total momentum before they push off is $\vec{P}_i = \vec{0}$ because both skaters are at rest. Consequently, the total momentum will still be $\vec{0}$ *after* they push off.

FIGURE 9.17 Before-and-after visual overview for two skaters pushing off from each other.

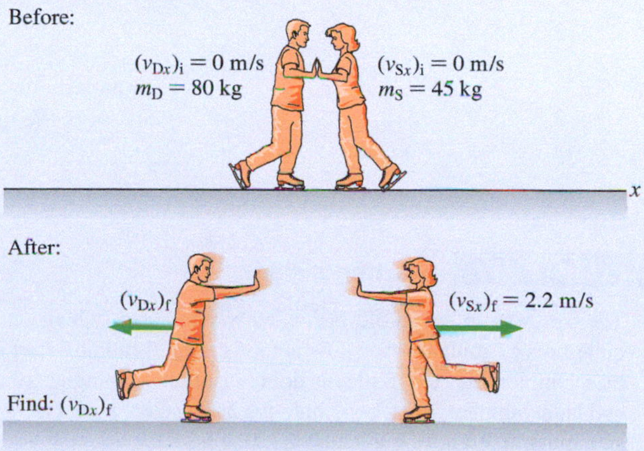


*Continued*

**SOLVE** Since the motion is only in the $x$-direction, we need to consider only $x$-components of momentum. We write Sandra's initial momentum as $(p_{Sx})_i = m_S(v_{Sx})_i$, where $m_S$ is her mass and $(v_{Sx})_i$ her initial velocity. Similarly, we write David's initial momentum as $(p_{Dx})_i = m_D(v_{Dx})_i$. Both of the skaters have zero initial momentum because both are initially at rest.

We can now apply the mathematical statement of momentum conservation, Equation 9.15. Writing the final momentum of Sandra as $m_S(v_{Sx})_f$ and that of David as $m_D(v_{Dx})_f$, we have

$$\underbrace{m_S(v_{Sx})_f + m_D(v_{Dx})_f}_{\text{The skaters' final momentum . . .}} = \underbrace{m_S(v_{Sx})_i + m_D(v_{Dx})_i}_{\text{. . . equals their initial momentum . . .}} = \underbrace{0}_{\text{. . . which was zero.}}$$

Solving for $(v_{Dx})_f$, we find

$$(v_{Dx})_f = -\frac{m_S}{m_D}(v_{Sx})_f = -\frac{45\text{ kg}}{80\text{ kg}} \times 2.2\text{ m/s} = -1.2\text{ m/s}$$

David moves backward with a *speed* of 1.2 m/s.

Notice that we didn't need to know any details about the force between David and Sandra in order to find David's final speed. Conservation of momentum *mandates* this result.

**ASSESS** It seems reasonable that Sandra, whose mass is less than David's, has the greater final speed.

## PROBLEM-SOLVING APPROACH 9.1 Conservation of momentum problems

We can use the law of conservation of momentum to relate the values of velocity and momentum of objects *after* an interaction to their values *before* the interaction.

**STRATEGIZE** Clearly define the *system*.

- If possible, choose a system that is isolated ($\vec{F}_{net} = \vec{0}$) or within which the interactions are sufficiently short and intense that you can ignore external forces for the duration of the interaction (the impulse approximation). Momentum is then conserved.
- If it's not possible to choose an isolated system, try to divide the problem into parts such that momentum is conserved during one segment of the motion. Other segments of the motion can be analyzed using Newton's laws or, as you'll learn in Chapter 10, conservation of energy.

**PREPARE** Following Tactics Box 9.1, draw a before-and-after visual overview. Define symbols that will be used in the problem, list known values, and identify what you're trying to find.

**SOLVE** The mathematical representation is based on the law of conservation of momentum, Equations 9.15. Because we generally want to solve for the velocities of objects, we usually use Equations 9.15 in the equivalent form

$$m_1(v_{1x})_f + m_2(v_{2x})_f + \cdots = m_1(v_{1x})_i + m_2(v_{2x})_i + \cdots$$
$$m_1(v_{1y})_f + m_2(v_{2y})_f + \cdots = m_1(v_{1y})_i + m_2(v_{2y})_i + \cdots$$

**ASSESS** Check that your result has the correct units, is reasonable, and answers the question.

Exercise 17

## EXAMPLE 9.7 Starting a sled

The best way to get an old-fashioned wooden sled (which can be quite heavy) going on the snow is to get a good running start and then jump on the sled. Bailey is doing just that, running at 4.0 m/s and launching her 26 kg body onto the 5.9 kg sled. How fast is the sled moving just after she lands on it?

**STRATEGIZE** When Bailey lands on the sled and grabs onto it, many forces are at work. But if we choose the system as Bailey + sled, the forces involved in this collision—friction forces between Bailey and the sled, the force of her holding the handles—are *internal* forces. The normal force balances the weight of both Bailey and the sled, so there is no net vertical force on the system. There is a small friction force from the runners of the sled on the snow, but if we consider the motion just after Bailey lands on the sled, we can ignore this force; it is much smaller than the internal forces. We will

FIGURE 9.18 Before-and-after visual overview of Bailey and the sled.

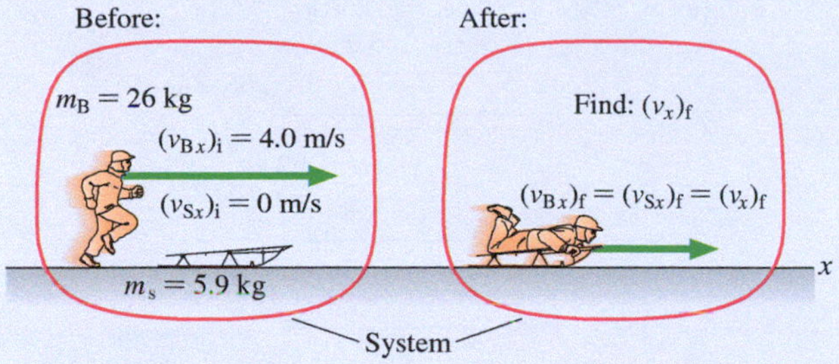


consider Bailey + sled as an isolated system, and the total momentum of the system will be conserved.

**PREPARE** The visual overview in FIGURE 9.18 has two parts, before and after, with the system clearly labeled. Before Bailey lands on the sled, she is moving with velocity $(v_{Bx})_i$ but the sled is stationary. After she lands on the sled, she and the sled move at a common final velocity, $(v_x)_f$.

**SOLVE** To determine the final velocity, we use conservation of momentum. The total momentum of the system before and after is the same: $(P_x)_f = (P_x)_i$. The initial momentum is the sum of the momentum of each of the individual objects. Since the sled is initially at rest, we can simplify:

$$(P_x)_i = m_B(v_{Bx})_i + m_S(v_{Sx})_i = m_B(v_{Bx})_i$$

The final momentum is the combined mass of Bailey and the sled moving at the common speed:

$$(P_x)_f = (m_B + m_S)(v_x)_f$$

The final momentum is equal to the initial momentum, $(P_x)_f = (P_x)_i$, so we can write

$$(m_B + m_S)(v_x)_f = m_B(v_{Bx})_i$$

Using known values, we can solve this equation for the final speed:

$$(v_x)_f = \left(\frac{m_B}{m_B + m_S}\right)(v_{Bx})_i = \left(\frac{26\text{ kg}}{26\text{ kg} + 5.9\text{ kg}}\right)(4.0\text{ m/s}) = 3.3\text{ m/s}$$

The speed of the sled is 3.3 m/s after Bailey jumps on—plenty fast for a good start to an epic run down the hill!

**ASSESS** The final speed is just a bit less than Bailey's initial speed, which makes sense. Bailey's mass is much greater than that of the sled.

Notice how straightforward this example was! No forces, no kinematic equations, no simultaneous equations. Why didn't we think of this before? Although conservation laws are indeed powerful, they can answer only certain questions. If we had wanted to know whether Bailey slid across the sled, or the magnitude of the force on her hands, or the acceleration of the sled, or how long the collision took, we would not have been able to find answers on the basis of the conservation law. There is a price to pay for finding a simple connection between before and after, and that price is the loss of information about the details of the interaction. If we are satisfied with knowing only about before and after, then conservation laws are a straightforward way to proceed.

## It Depends on the System

The first step in Problem-Solving Approach 9.1 is to clearly define the *system*. This is worth emphasizing because many problem-solving errors arise from trying to apply momentum conservation to an inappropriate system. **The goal is to choose a system whose momentum will be conserved.** Even then, it is the *total* momentum of the system that is conserved, not the momentum of any individual object within the system.

Video Conservation of Linear Momentum

In Example 9.7, the question asked for the speed of the sled after Bailey lands on it. Suppose we had chosen to consider the sled alone as the system. The sled experiences a force from Bailey. This is an external force, so we can't consider the sled as an isolated system and conservation of momentum isn't a useful tool. (Indeed, we know that the sled can't be an isolated system; it speeds up, so its momentum changes.) That's why, in our solution, we added Bailey to the mix, taking the system as Bailey plus the sled. We can ignore the external forces, so momentum is conserved. The sled's momentum increases and Bailey's momentum decreases, but the increase in the sled's momentum is exactly equal to the decrease in Bailey's momentum. We were able to quickly calculate the final speed of Bailey and the sled.

## Explosions

An **explosion,** where the particles of the system move apart after a brief, intense interaction, is the opposite of a collision. The explosive forces, which could be from an expanding spring or from expanding hot gases, are *internal* forces. If the system is isolated, its total momentum during the explosion will be conserved.

Video What the Physics? Rocket Chair

**EXAMPLE 9.8 Recoil speed of a toy rifle**

A 30 g ball is fired from a 1.2 kg spring-loaded toy rifle with a speed of 15 m/s. What is the recoil speed of the rifle?

**STRATEGIZE** As the ball moves down the barrel, there are complicated forces exerted on the ball and on the rifle. However, if we take the system to be ball + rifle, these are *internal* forces that do not change the total momentum.

The *external* forces of the rifle's and ball's weights are balanced by the external force exerted by the person holding the rifle, so $\vec{F}_{net} = \vec{0}$. This is an isolated system and the law of conservation of momentum applies.

**PREPARE** FIGURE 9.19 shows a visual overview before and after the ball is fired. We assume the ball is fired in the $+x$-direction.

**SOLVE** The $x$-component of the total momentum is $P_x = p_{Bx} + p_{Rx}$. Everything is at rest before the trigger is pulled, so the initial momentum is zero. After the trigger is pulled, the internal force of the spring pushes the ball down the barrel *and* pushes the rifle backward. Conservation of momentum gives

$$(P_x)_f = m_B(v_{Bx})_f + m_R(v_{Rx})_f = (P_x)_i = 0$$

Solving for the rifle's velocity, we find

$$(v_{Rx})_f = -\frac{m_B}{m_R}(v_{Bx})_f = -\frac{0.030\text{ kg}}{1.2\text{ kg}} \times 15\text{ m/s} = -0.38\text{ m/s}$$

FIGURE 9.19 Before-and-after visual overview for a toy rifle.

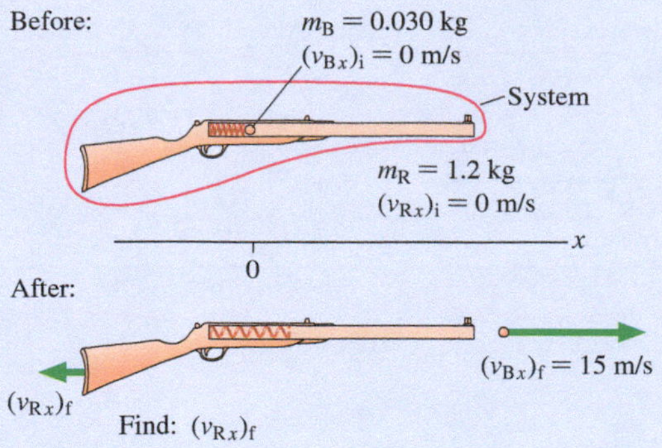


The minus sign indicates that the rifle's recoil is to the left. The recoil *speed* is 0.38 m/s.

**ASSESS** The bullet is much lighter than the rifle, so it makes sense that the recoil speed of the rifle is much less than the speed of the bullet. Real rifles fire their bullets at much higher velocities, and their recoil is correspondingly higher. Shooters need to brace themselves against the "kick" of the rifle back against their shoulder.

FIGURE 9.20 Rocket propulsion is an example of conservation of momentum.

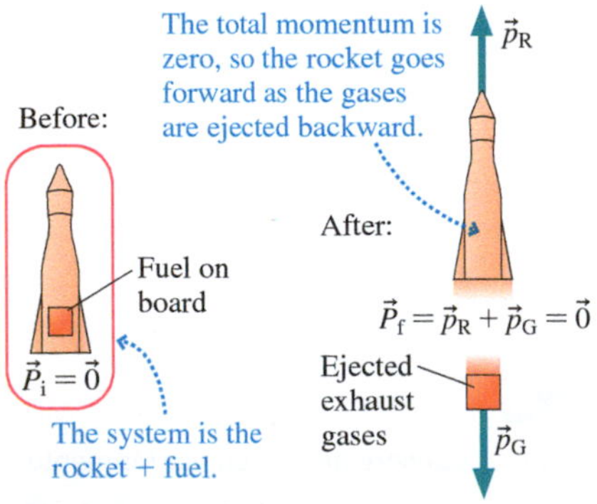


This would be a very difficult problem to solve using Newton's laws. But Example 9.8 is straightforward when approached from the before-and-after perspective of a conservation law. The selection of ball + rifle as the "system" was the critical step. For momentum conservation to be useful, we had to select a system in which the complicated forces due to the spring and to friction were all internal forces. The rifle by itself is *not* an isolated system, so its momentum is *not* conserved.

Much the same reasoning explains how a rocket or jet aircraft accelerates. FIGURE 9.20 shows a rocket with a parcel of fuel on board. Burning converts the fuel to hot gases that are expelled from the rocket motor. If we choose rocket + gases to be the system, then the burning and expulsion are internal forces. In deep space there are no other forces, so the total momentum of the rocket + gases system must be conserved. The rocket gains forward velocity and momentum as the exhaust gases are shot out the back, but the *total* momentum of the system remains zero.

Many people find it hard to understand how a rocket can accelerate in the vacuum of space because there is nothing to "push against." Thinking in terms of momentum, you can see that the rocket does not push against anything *external,* but only against the gases that it pushes out the back. In return, in accordance with Newton's third law, the gases push forward on the rocket.

**Squid propulsion** BIO Squids use a form of jet propulsion to make quick movements to escape enemies or catch prey. The squid draws in water through a pair of valves in its outer sheath, or mantle, and then quickly expels the water through a funnel, propelling the squid backward.

**STOP TO THINK 9.5** An explosion in a rigid pipe shoots three balls out of its ends. A 6 g ball comes out the right end. A 4 g ball comes out the left end with twice the speed of the 6 g ball. From which end, left or right, does the third ball emerge?

## 9.5 Inelastic Collisions

A rubber ball dropped on the floor bounces—it's *elastic*—but a ball of clay sticks to the floor without bouncing; we call such a collision *inelastic*. A golf club hitting a golf ball causes the ball to rebound away from the club (elastic), but a bullet striking a block of wood becomes embedded in the block (inelastic).

A collision in which the two objects stick together and move with a common final velocity is called a **perfectly inelastic collision.** The clay hitting the floor and the bullet embedding itself in the wood are examples of perfectly inelastic collisions. Other examples include railroad cars coupling together upon impact and darts hitting a dart board. FIGURE 9.21 emphasizes the fact that the two objects have a common final velocity after they collide. (We have drawn the combined object moving to the right, but it could have ended up moving to the left, or not moving at all, depending on the objects' masses and initial velocities.)

In other collisions, the two objects bounce apart. We've looked at some examples of these kinds of collisions, but a full analysis requires some ideas about energy. We will learn more about collisions and energy in Chapter 10.

FIGURE 9.21 A perfectly inelastic collision.

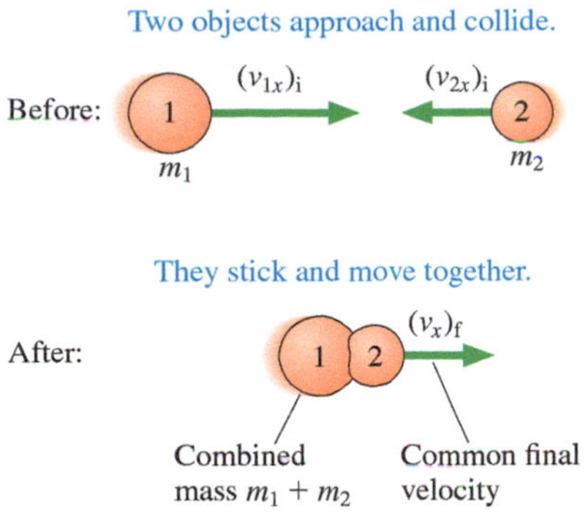


## EXAMPLE 9.9 A perfectly inelastic collision of railroad cars

In assembling a train from several railroad cars, two of the cars, with masses $2.0 \times 10^4$ kg and $4.0 \times 10^4$ kg, are rolled toward each other. When they meet, they couple and stick together. The lighter car has an initial speed of 1.5 m/s; the collision causes it to reverse direction at 0.25 m/s. What was the initial speed of the heavier car?

**STRATEGIZE** We will define the system to be the two railroad cars. The large forces between the cars when they collide are internal forces. Friction and other external forces are much smaller. We will use the impulse approximation and consider the system of the two cars as isolated, so total momentum is conserved in the collision. The cars stick together, so this is a perfectly inelastic collision.

**PREPARE** FIGURE 9.22 shows a visual overview. We've chosen to let the $2.0 \times 10^4$ kg car (car 1) start out moving to the right, so $(v_{1x})_i$ is a positive 1.5 m/s. The cars move to the left after the collision, so their common final velocity is $(v_x)_f = -0.25$ m/s. You can see that velocity $(v_{2x})_i$ must be negative in order to "turn around" both cars.

FIGURE 9.22 Before-and-after visual overview for two train cars colliding on a track.

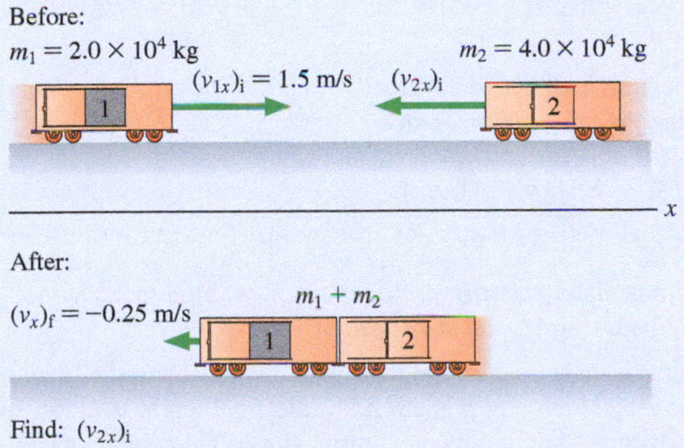


**SOLVE** The law of conservation of momentum, $(P_x)_f = (P_x)_i$, is

$$(m_1 + m_2)(v_x)_f = m_1(v_{1x})_i + m_2(v_{2x})_i$$

where we make use of the fact that the combined mass $m_1 + m_2$ moves together after the collision. We can solve for the initial velocity of the $4.0 \times 10^4$ kg car:

$$(v_{2x})_i = \frac{(m_1 + m_2)(v_x)_f - m_1(v_{1x})_i}{m_2}$$

$$= \frac{(6.0 \times 10^4\ \text{kg})(-0.25\ \text{m/s}) - (2.0 \times 10^4\ \text{kg})(1.5\ \text{m/s})}{4.0 \times 10^4\ \text{kg}}$$

$$= -1.1\ \text{m/s}$$

The negative sign, which we anticipated, indicates that the heavier car started out moving to the left. The initial *speed* of the car, which we were asked to find, is 1.1 m/s.

**ASSESS** Our final result makes sense. The final speed of the two cars is only one-sixth the initial speed of the lighter car. The small final momentum means that the initial momentum must also be small, so the magnitude of the initial momentum of the heavier car should be about the same as that of the lighter car. The heavier car has twice the mass, so its speed should be about half that of the lighter car, just as we found.

**STOP TO THINK 9.6** The two objects shown collide and stick together. After the collision, the combined objects

A. Move to the right as shown.
B. Move to the left.
C. Are at rest.

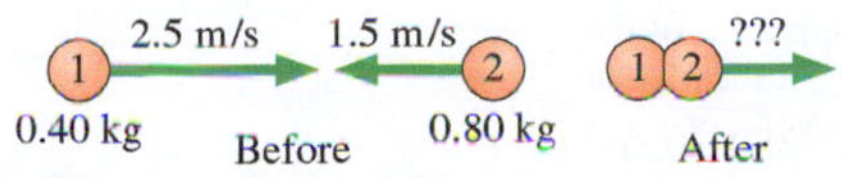

## 9.6 Momentum and Collisions in Two Dimensions

Our examples thus far have been confined to motion along a one-dimensional axis; now we'll consider motion in a plane. The total momentum $\vec{P}$ is the *vector* sum of the momentum $\vec{p} = m\vec{v}$ of each of the individual objects. Consequently, as Equations 9.15 showed, momentum is conserved only if each component of $\vec{P}$ is conserved:

$$(p_{1x})_f + (p_{2x})_f + (p_{3x})_f + \cdots = (p_{1x})_i + (p_{2x})_i + (p_{3x})_i + \cdots$$
$$(p_{1y})_f + (p_{2y})_f + (p_{3y})_f + \cdots = (p_{1y})_i + (p_{2y})_i + (p_{3y})_i + \cdots$$

The steps of Problem-Solving Approach 9.1 still apply.

### EXAMPLE 9.10 Analyzing a peregrine falcon strike BIO

Peregrine falcons often grab their prey in flight. A falcon, flying at 18 m/s, swoops down at a 45° angle from behind a pigeon flying horizontally at 9.0 m/s. The falcon has a mass of 0.80 kg and the pigeon a mass of 0.36 kg. What are the speed and direction of the falcon (now holding the pigeon) immediately after impact?

**STRATEGIZE** We'll take the system to be falcon + pigeon. The external forces at work are negligible compared to the internal forces of the impact, so the total momentum of the falcon + pigeon system is conserved. This is a perfectly inelastic collision because after the collision the birds move at a common velocity.

**PREPARE** For a two-dimensional collision, conservation of momentum means that the $x$-component of the total momentum before the collision must equal the $x$-component of the total momentum after the collision, and similarly for the $y$-components. FIGURE 9.23 is a before-and-after visual overview.

FIGURE 9.23 Before-and-after visual overview for a falcon catching a pigeon.

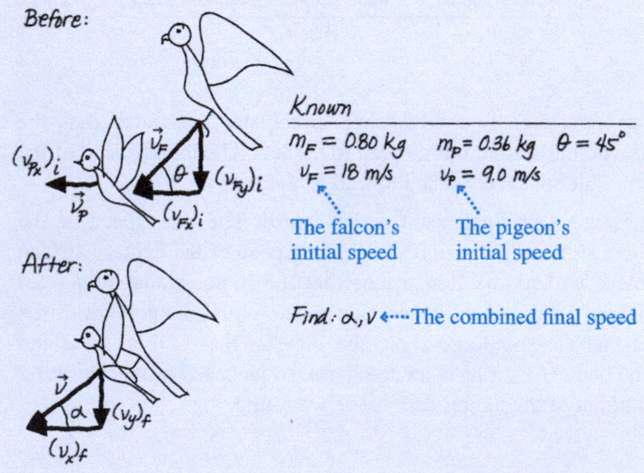


**SOLVE** We start by finding the $x$- and $y$-components of the momentum before the collision. For the $x$-component we have

$$(P_x)_i = m_F(v_{Fx})_i + m_P(v_{Px})_i = m_F(-v_F \cos\theta) + m_P(-v_P)$$

(The $x$-component of the initial momentum . . . equals the $x$-component of the initial momentum of the falcon . . . plus the $x$-component of the initial momentum of the pigeon. Both velocity components are negative, since they point to the left.)

$$= (0.80\text{ kg})(-18\text{ m/s})(\cos 45°) + (0.36\text{ kg})(-9.0\text{ m/s})$$
$$= -13.4\text{ kg}\cdot\text{m/s}$$

Similarly, for the $y$-component of the initial momentum we have

$$(P_y)_i = m_F(v_{Fy})_i + m_P(v_{Py})_i = m_F(-v_F \sin\theta) + 0$$
$$= (0.80\text{ kg})(-18.0\text{ m/s})(\sin 45°) = -10.2\text{ kg}\cdot\text{m/s}$$

After the collision, the two birds move with a common velocity $\vec{v}$ that is directed at an angle $\alpha$ from the horizontal. The $x$-component of the final momentum is then

$$(P_x)_f = (m_F + m_P)(v_x)_f$$

Momentum conservation requires $(P_x)_f = (P_x)_i$, so

$$(v_x)_f = \frac{(P_x)_i}{m_F + m_P} = \frac{-13.4\text{ kg}\cdot\text{m/s}}{(0.80\text{ kg}) + (0.36\text{ kg})} = -11.6\text{ m/s}$$

Similarly, $(P_y)_f = (P_y)_i$ gives

$$(v_y)_f = \frac{(P_y)_i}{m_F + m_P} = \frac{-10.2\text{ kg}\cdot\text{m/s}}{(0.80\text{ kg}) + (0.36\text{ kg})} = -8.79\text{ m/s}$$

From the figure we see that $\tan\alpha = (v_y)_f/(v_x)_f$, so that

$$\alpha = \tan^{-1}\left(\frac{(v_y)_f}{(v_x)_f}\right) = \tan^{-1}\left(\frac{-8.79\text{ m/s}}{-11.6\text{ m/s}}\right) = 37°$$

The magnitude of the final velocity (i.e., the speed) can be found from the Pythagorean theorem as

$$v = \sqrt{(v_x)_f^2 + (v_y)_f^2}$$
$$= \sqrt{(-11.6\text{ m/s})^2 + (-8.79\text{ m/s})^2} = 15\text{ m/s}$$

Thus immediately after impact the falcon, with its meal, is moving 37° below horizontal at a speed of 15 m/s.

**ASSESS** It makes sense that the falcon slows down after catching the slower-moving pigeon. Also, the final angle is closer to the horizontal than the falcon's initial angle. This seems reasonable because the pigeon was initially flying horizontally, making the total momentum vector more horizontal than the direction of the falcon's initial momentum.

It's instructive to examine the collision in this example with a picture of the momentum vectors. The vectors $\vec{p}_F$ and $\vec{p}_P$ before the collision, and their sum $\vec{P} = \vec{p}_F + \vec{p}_P$, are shown in FIGURE 9.24. You can see that the total momentum vector makes a 37° angle with the negative $x$-axis. The individual values of momentum change in the collision, *but the total momentum does not.*

FIGURE 9.24 The momentum vectors of the falcon strike.

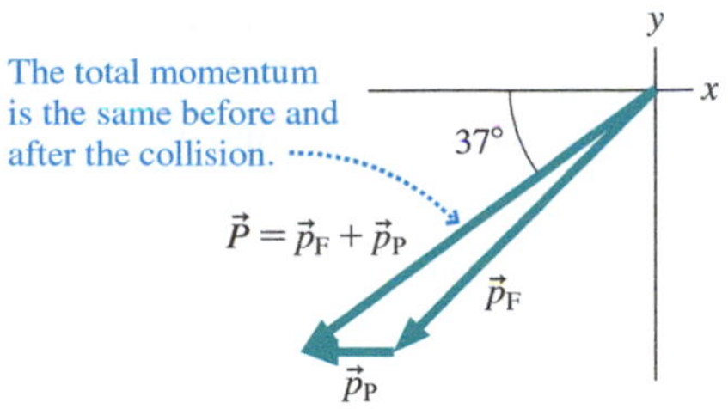


## 9.7 Angular Momentum

For a single object, we can think of the law of conservation of momentum as an alternative way of stating Newton's first law. Rather than saying that an object will continue to move in a straight line at constant velocity unless acted on by a net force, we can say that the momentum of an isolated object is conserved. Both express the idea that an object moving in a straight line tends to "keep going" unless something acts on it to change its motion.

Next, we want to come up with a rotational analog for the conservation of momentum, just as we found rotational analogs for motion (angular velocity, angular acceleration) and force (torque). Momentum isn't conserved for an object that is moving in a circle. Momentum is a vector, and the momentum of an object in circular motion changes as the direction of motion changes. Nonetheless, a ball that is moving in a circle at the end of a string tends to "keep going" in a circular path. And a spinning bicycle wheel would keep turning if it were not for friction. The quantity that expresses this idea for circular motion is called *angular momentum.*

Video Angular Momentum

Let's look at the example of pushing a merry-go-round, as in FIGURE 9.25. If you push tangentially to the rim, you are applying a torque to the merry-go-round. As you learned in ◀◀ SECTION 7.5, the merry-go-round's angular speed will continue to increase for as long as you apply this torque. The harder (greater torque) or longer (greater time) you push, the greater the increase in the angular velocity.

FIGURE 9.25 By applying a torque to the merry-go-round, the girl is increasing its angular momentum.

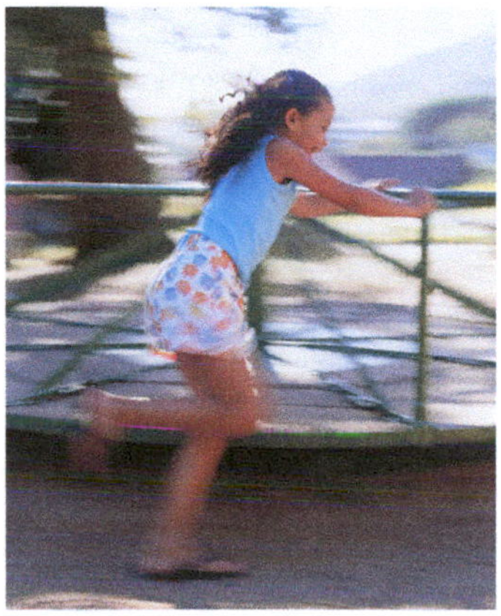

Suppose you apply a constant torque $\tau_{net}$ to the merry-go-round for a time $\Delta t$. By how much will the merry-go-round's angular speed increase? In Section 7.5 we found that the angular acceleration $\alpha$ is given by the rotational equivalent of Newton's second law, or

$$\alpha = \frac{\tau_{net}}{I} \tag{9.16}$$

where $I$ is the merry-go-round's moment of inertia.

The angular acceleration is the rate of change of the angular velocity, so

$$\alpha = \frac{\Delta\omega}{\Delta t} \tag{9.17}$$

Setting Equations 9.16 and 9.17 equal to each other gives

$$\frac{\Delta\omega}{\Delta t} = \frac{\tau_{net}}{I}$$

or, rearranging,

$$\tau_{net}\,\Delta t = I\,\Delta\omega \tag{9.18}$$

If you recall the impulse-momentum theorem for *linear* motion, which is

$$\vec{F}_{net}\,\Delta t = m\,\Delta\vec{v} = \Delta\vec{p} \tag{9.19}$$

you can see that Equation 9.18 is an analogous statement about rotational motion. Because the quantity $I\omega$ is evidently the rotational equivalent of $m\vec{v}$, the linear momentum $\vec{p}$, it seems reasonable to define the **angular momentum** $L$ to be

$$L = I\omega \tag{9.20}$$

Angular momentum of an object with moment of inertia $I$ rotating at angular velocity $\omega$

The SI units of angular momentum are those of moment of inertia times angular velocity, or kg · m$^2$/s.

**TABLE 9.2** Rotational and linear dynamics

| Rotational dynamics | Linear dynamics |
|---|---|
| Torque $\tau_{net}$ | Force $\vec{F}_{net}$ |
| Moment of inertia $I$ | Mass $m$ |
| Angular velocity $\omega$ | Velocity $\vec{v}$ |
| Angular momentum $L = I\omega$ | Linear momentum $\vec{p} = m\vec{v}$ |

Just as an object in linear motion can have a large momentum by having either a large mass or a high speed, a rotating object can have a large angular momentum by having a large moment of inertia or a large angular velocity. The merry-go-round in Figure 9.25 has a larger angular momentum if it's spinning fast than if it's spinning slowly. Also, the merry-go-round (large $I$) has a much larger angular momentum than a toy top (small $I$) spinning with the same angular velocity.

**TABLE 9.2** summarizes the analogies between linear and rotational quantities that you learned in Chapter 7 and adds the analogy between linear momentum and angular momentum.

## Conservation of Angular Momentum

Having now defined angular momentum, we can write Equation 9.18 as

$$\tau_{net}\,\Delta t = \Delta L \tag{9.21}$$

in exact analogy with its linear dynamics equivalent, Equation 9.19. This equation implies that the change in the angular momentum of an object is proportional to the net torque applied to the object. If the net external torque on an object is *zero,* then the change in the angular momentum is zero as well. That is, a rotating object will continue to rotate with *constant* angular momentum—to "keep going"—unless acted upon by an external torque. We can state this conclusion as the *law of conservation of angular momentum*:

**Law of conservation of angular momentum** The angular momentum of a rotating object subject to no net external torque ($\tau_{net} = 0$) is a constant. The final angular momentum $L_f$ is equal to the initial angular momentum $L_i$.

For a system that consists of more than one rotating object we define the **total angular momentum** as the sum of the angular momentum of each of the objects in the system. If no net external torque acts on the system, the mathematical statement of the law of conservation of angular momentum is then

Final (f) and initial (i) moment of inertia and angular velocity of object 1

$$\overbrace{(I_1)_f(\omega_1)_f} + \underbrace{(I_2)_f(\omega_2)_f} + \cdots = \overbrace{(I_1)_i(\omega_1)_i} + \underbrace{(I_2)_i(\omega_2)_i} + \cdots \tag{9.22}$$

Final (f) and initial (i) moment of inertia and angular velocity of object 2

We will look at problems involving multiple objects in which the angular momentum before and after an interaction is the same. The solution is similar to what we've done for conservation of momentum in one and two dimensions.

## Varying Moment of Inertia

There is one aspect of angular momentum that is not analogous to linear momentum. As we've noted, for an isolated object, the *linear* momentum $\vec{p} = m\vec{v}$ is constant. The object's mass can't change, so the velocity can't change either. The object keeps moving at a constant speed. For an isolated object, the *angular* momentum $L = I\omega$ is constant as well. The object's mass can't change, but it can experience a change in its moment of inertia because the *distribution* of mass can change. This means that **an isolated, rotating object can experience a change in angular velocity.**

The classic example is a spinning figure skater. A skater in a spin on the ice experiences very little friction. The normal force balances her weight force, so she is, to a good approximation, an isolated system. If she starts the spin with her arms and legs far out from the axis of her spin, she'll start with a large moment of inertia. As she pulls her arms and legs in, her moment of inertia decreases, so her angular velocity increases. FIGURE 9.26 illustrates this idea. By dramatically reducing their moment of inertia, world-class skaters can increase the angular speed of a spin by a factor of 5 or more.

FIGURE 9.26 A spinning figure skater.

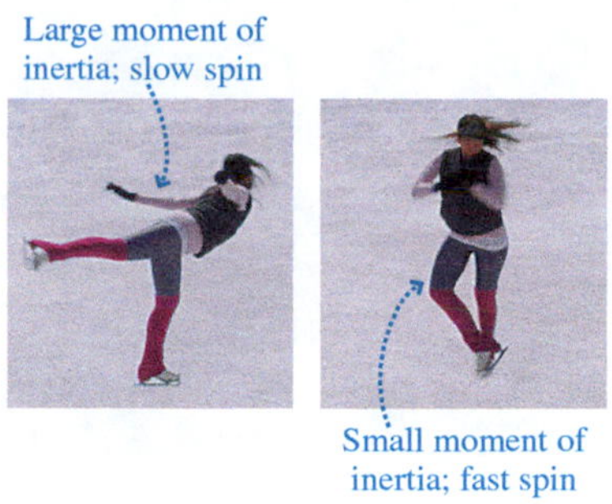


Video Figure 9.26

## EXAMPLE 9.11 Period of a merry-go-round

Joey, whose mass is 36 kg, stands at the center of a 200 kg merry-go-round that is rotating once every 2.5 s. While it is rotating, Joey walks out to the edge of the merry-go-round, 2.0 m from its center. What is the rotational period of the merry-go-round when Joey gets to the edge?

**STRATEGIZE** Take the system to be Joey + merry-go-round and assume frictionless bearings. There is no external torque on this system, so the angular momentum of the system will be conserved.

**PREPARE** As shown in the visual overview of FIGURE 9.27, we model the merry-go-round as a uniform disk of radius $R = 2.0$ m. From Table 7.1, the moment of inertia of a disk is $I_{\text{disk}} = \frac{1}{2}MR^2$. If we model Joey as an object of mass $m$, his moment of inertia is zero when he is at the center, but it increases to $mR^2$ when he reaches the edge.

FIGURE 9.27 Visual overview of the merry-go-round.

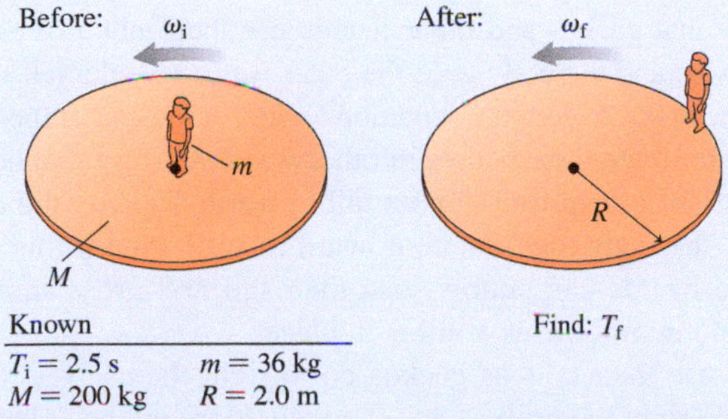


**SOLVE** The mathematical statement of the law of conservation of angular momentum is Equation 9.22. The initial angular momentum is

$$L_i = (I_{\text{Joey}})_i(\omega_{\text{Joey}})_i + (I_{\text{disk}})_i(\omega_{\text{disk}})_i = 0 \cdot \omega_i + \frac{1}{2}MR^2\omega_i = \frac{1}{2}MR^2\omega_i$$

Here we have used the fact that both Joey and the disk have the same initial angular velocity, which we have called $\omega_i$. Similarly, the final angular momentum is

$$L_f = (I_{\text{Joey}})_f\omega_f + (I_{\text{disk}})_f\omega_f = mR^2\omega_f + \frac{1}{2}MR^2\omega_f = \left(mR^2 + \frac{1}{2}MR^2\right)\omega_f$$

where $\omega_f$ is the common final angular velocity of both Joey and the disk.

The law of conservation of angular momentum states that $L_f = L_i$, so that

$$\left(mR^2 + \frac{1}{2}MR^2\right)\omega_f = \frac{1}{2}MR^2\omega_i$$

Canceling the $R^2$ terms from both sides and solving for $\omega_f$ give

$$\omega_f = \left(\frac{M}{M + 2m}\right)\omega_i$$

The initial angular velocity is related to the initial period of rotation $T_i$ by

$$\omega_i = \frac{2\pi}{T_i} = \frac{2\pi}{2.5\text{ s}} = 2.51\text{ rad/s}$$

Thus the final angular velocity is

$$\omega_f = \left(\frac{200\text{ kg}}{200\text{ kg} + 2(36\text{ kg})}\right)(2.51\text{ rad/s}) = 1.85\text{ rad/s}$$

When Joey reaches the edge, the period of the merry-go-round has increased to

$$T_f = \frac{2\pi}{\omega_f} = \frac{2\pi}{1.85\text{ rad/s}} = 3.4\text{ s}$$

**ASSESS** The merry-go-round rotates *more slowly* after Joey moves out to the edge. This makes sense because if the system's moment of inertia increases, as it does when Joey moves out, the angular velocity must decrease to keep the angular momentum constant.

**The eye of a hurricane** As air masses from the slowly rotating outer zones are drawn toward the low-pressure center, their moment of inertia decreases. Because the angular momentum of these air masses is conserved, their speed must *increase* as they approach the center, leading to the high wind speeds near the center of the storm.

Solving Example 9.11 using Newton's laws would be quite difficult. We would have to deal with internal forces, such as Joey's feet against the merry-go-round, and other complications. Problems like this show the true power of conservation laws to solve problems. As long as we are worried only about the before and after states, we can use conservation laws to make a connection between the two without being concerned about the details of the interactions.

## CONCEPTUAL EXAMPLE 9.12 Free falling, with a twist

Geckos have feet that stick to surfaces, which gives them a remarkable ability to climb, including the ability to scale smooth vertical surfaces. Sometimes, though, their feet fail them, and they fall. No matter their orientation when this happens, geckos can use their stout tails to flip themselves over and land on their feet. Suppose a gecko is falling, upside down, and is not rotating. How can it possibly right itself before landing?

**REASON** When the gecko begins to fall, it is not rotating—its angular momentum is zero. If we treat the gecko as the system, its angular momentum can't change. The gecko can't start rotating. Instead, let's consider the system as being made up of two parts, the body of the gecko and the stout tail. The total angular momentum of the body + tail system is zero. If the gecko spins its tail in one direction, the body will spin in the opposite direction. When the gecko's body reaches the correct orientation, the gecko simply stops rotating its tail. The gecko's body will then stop rotating as well, and the gecko will land on its feet. FIGURE 9.28 outlines how the gecko might do this.

FIGURE 9.28 Motion of a falling gecko.

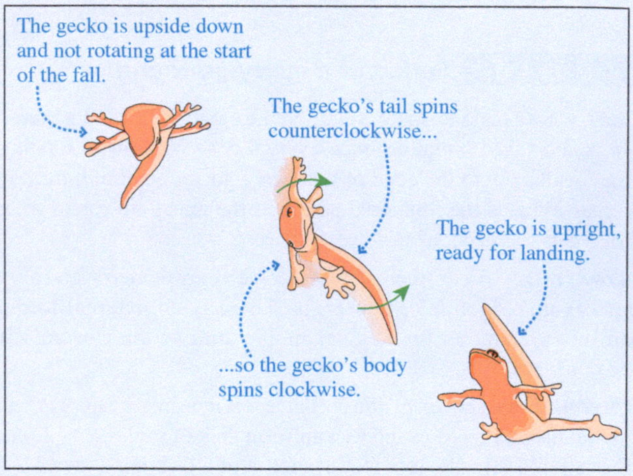

The gecko is upside down and not rotating at the start of the fall.
The gecko's tail spins counterclockwise...
The gecko is upright, ready for landing.
...so the gecko's body spins clockwise.

**ASSESS** This description makes sense. Lizards certainly use their tails for balance, and the flip described would be a natural extension of this ability. We were told that the gecko has a stout tail; having a large mass—and therefore a large moment of inertia—in the tail would be necessary for the gecko to right itself as described.

In fact, video studies show that geckos and other lizards use their tails just as we described. The tail spins one way, and the body goes the other way. Once the tail stops spinning, so does the body, now in the preferred orientation to land at the end of the fall.

Lizards also use their tails to rotate their bodies in other ways. A lizard that needs to raise its body in the middle of a leap can flex its tail upward. To keep the total angular momentum constant, the body must rotate upward as well. And if this puts the tail in harm's way, some lizards can simply lose their tail and grow another, keeping acrobatics to a minimum until the new tail is in place.

You might think that cats use their tails as geckos do to right themselves when they fall. In fact, falling cats do right themselves, but they rely more on their flexible bodies than their fluffy tails. Cats can bend in the middle and complete a complicated twisting motion that leads to a rotation of the body. Tailless cats can right themselves just fine, but the same is not true for geckos.

**Acrobatic bats** Bats have heavier wings than other fliers. This makes flight more energetically expensive for bats than for birds, but it enhances maneuverability. Twisting their heavy wings this way and that allows the bats to reorient their bodies in flight, much as the gecko spins its substantial tail to reposition itself. This is a real asset to an animal that must fly in tight spaces and complete complex maneuvers such as landing on a ceiling perch.

**STOP TO THINK 9.7** The left figure shows two boys of equal mass standing halfway to the edge on a turntable that is freely rotating at angular speed $\omega_i$. They then walk to the positions shown in the right figure. The final angular speed $\omega_f$ is

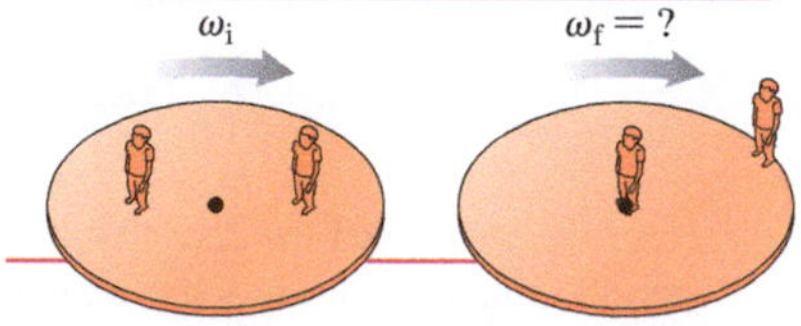


A. Greater than $\omega_i$. B. Less than $\omega_i$. C. Equal to $\omega_i$.

## INTEGRATED EXAMPLE 9.13 Aerial firefighting

A forest fire is easiest to attack when it's just getting started. In remote locations, this often means using airplanes to rapidly deliver large quantities of water and fire suppressant to the blaze.

The "Superscooper" is an amphibious aircraft that can pick up a 6000 kg load of water by skimming over the surface of a river or lake and scooping water directly into its storage tanks. As it approaches the water's surface at a speed of 35 m/s, an empty Superscooper has a mass of 13,000 kg.

a. It takes the plane 12 s to pick up a full load of water. If we ignore the force on the plane due to the thrust of its propellers, what is its speed immediately after picking up the water?
b. What is the impulse delivered to the plane by the water?
c. What is the average force of the water on the plane?
d. The plane then flies over the fire zone at 40 m/s. It releases water by opening doors in the belly of the plane, allowing the water to fall straight down with respect to the plane. What is the plane's speed after dropping the water if it takes 5.0 s to do so?

**STRATEGIZE** We can solve part a, and later part d, using conservation of momentum, following Problem-Solving Approach 9.1. We'll need to choose the system with care, so that $\vec{F}_{net} = \vec{0}$. The plane alone is not an appropriate system for using conservation of momentum: As the plane scoops up the water, the water exerts a large external drag force on the plane, so $\vec{F}_{net}$ is definitely not zero. Instead, we should choose the plane *and* the water it is going to scoop up as the system. Then there are no external forces in the $x$-direction, and the net force in the $y$-direction is zero, since neither plane nor water accelerates appreciably in this direction during the scooping process. The complicated forces between plane and water are now *internal* forces that do not change the total momentum of the plane + water system.

**PREPARE** With the system chosen, we follow the steps of Tactics Box 9.1 to prepare the before-and-after visual overview shown in FIGURE 9.29.

FIGURE 9.29 Visual overview of the plane and water.

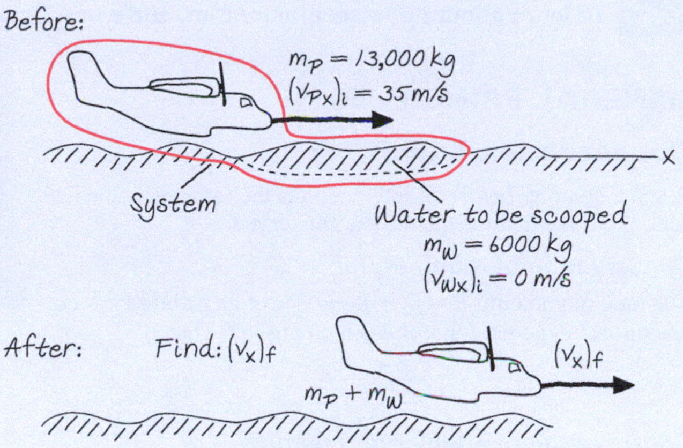


Parts b and c are impulse-and-momentum problems, so to solve them we'll use the impulse-momentum theorem, Equation 9.8. The impulse-momentum theorem considers the dynamics of a *single* object—here, the plane—subject to external forces—in this case, from the water.

**SOLVE** a. The $x$-component of the law of conservation of momentum is

$$(P_x)_f = (P_x)_i$$

or

$$(m_P + m_W)(v_x)_f = m_P(v_{Px})_i + m_W(v_{Wx})_i = m_P(v_{Px})_i + 0$$

Here we've used the facts that the initial velocity of the water is zero and that the final situation, as in an inelastic collision, has the combined mass of the plane and water moving with the same velocity $(v_x)_f$. Solving for $(v_x)_f$, we find

$$(v_x)_f = \frac{m_P(v_{Px})_i}{m_P + m_W} = \frac{(13{,}000 \text{ kg})(35 \text{ m/s})}{(13{,}000 \text{ kg}) + (6000 \text{ kg})} = 24 \text{ m/s}$$

b. The impulse-momentum theorem is $J_x = \Delta p_x$, where $\Delta p_x = m_P \Delta v_x$ is the change in the plane's momentum. Thus

$$J_x = m_P \Delta v_x = m_P[(v_x)_f - (v_{Px})_i]$$
$$= (13{,}000 \text{ kg})(24 \text{ m/s} - 35 \text{ m/s}) = -1.4 \times 10^5 \text{ kg} \cdot \text{m/s}$$

c. From Equation 9.1, the definition of impulse, we have

$$(F_{avg})_x = \frac{J_x}{\Delta t} = \frac{-1.4 \times 10^5 \text{ kg} \cdot \text{m/s}}{12 \text{ s}} = -12{,}000 \text{ N}$$

d. Because the water drops straight down *relative to the plane*, it has the same $x$-component of velocity immediately after being dropped as before being dropped. That is, simply opening the doors doesn't cause the water to speed up or slow down horizontally, so the water's horizontal momentum doesn't change upon being dropped. Because the total momentum of the plane + water system is conserved, the momentum of the plane doesn't change either. The plane's speed after the drop is still 40 m/s.

**ASSESS** The mass of the water is nearly half that of the plane, so the significant decrease in the plane's velocity as it scoops up the water is reasonable. The force of the water on the plane is large, but is still only about 10% of the plane's weight, $mg = 130{,}000$ N, so the answer seems to be reasonable.

# SUMMARY

**GOAL** To learn about impulse, momentum, and a new problem-solving approach based on conservation laws.

## GENERAL PRINCIPLES

### Conservation Laws

When a quantity *before* an interaction is the same *after* the interaction, we say that the quantity is **conserved.**

#### Conservation of momentum

The total momentum $\vec{P} = \vec{p}_1 + \vec{p}_2 + \cdots$ of an **isolated system**—one on which no net force acts—is a constant. Thus

$$\vec{P}_\text{f} = \vec{P}_\text{i}$$

#### Conservation of angular momentum

The angular momentum $L$ of a rotating object or system of objects subject to zero net external torque is a constant. Thus

$$L_\text{f} = L_\text{i}$$

This can be written in terms of the initial and final moments of inertia $I$ and angular velocities $\omega$ as

$$(I_1)_\text{f}(\omega_1)_\text{f} + (I_2)_\text{f}(\omega_2)_\text{f} + \cdots = (I_1)_\text{i}(\omega_1)_\text{i} + (I_2)_\text{i}(\omega_2)_\text{i} + \cdots$$

### Solving Momentum Conservation Problems

***STRATEGIZE*** Choose an isolated system or a system that is isolated during at least part of the problem.

***PREPARE*** Draw a visual overview of the system before and after the interaction.

***SOLVE*** Write the law of conservation of momentum in terms of vector components:

$$(p_{1x})_\text{f} + (p_{2x})_\text{f} + \cdots = (p_{1x})_\text{i} + (p_{2x})_\text{i} + \cdots$$
$$(p_{1y})_\text{f} + (p_{2y})_\text{f} + \cdots = (p_{1y})_\text{i} + (p_{2y})_\text{i} + \cdots$$

In terms of masses and velocities, this is

$$m_1(v_{1x})_\text{f} + m_2(v_{2x})_\text{f} + \cdots = m_1(v_{1x})_\text{i} + m_2(v_{2x})_\text{i} + \cdots$$
$$m_1(v_{1y})_\text{f} + m_2(v_{2y})_\text{f} + \cdots = m_1(v_{1y})_\text{i} + m_2(v_{2y})_\text{i} + \cdots$$

***ASSESS*** Is the result reasonable?

## IMPORTANT CONCEPTS

**Momentum** $\vec{p} = m\vec{v}$

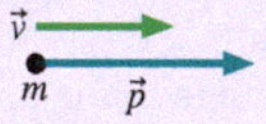


**Impulse** $J_x$ = area under force curve

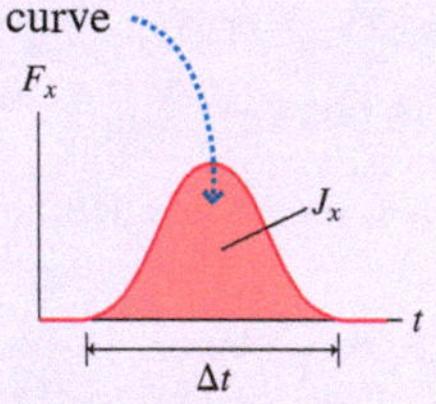


Impulse and momentum are related by the **impulse-momentum theorem**

$$\Delta p_x = J_x$$

This is an alternative statement of Newton's second law.

**Angular momentum** $L = I\omega$ is the rotational analog of linear momentum $\vec{p} = m\vec{v}$.

**System** A group of interacting objects

**Isolated system** A system on which the net external force is zero

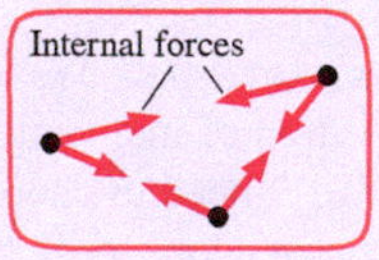


**Before-and-after visual overview**

- Define the system.
- Use two drawings to show the system *before* and *after* the interaction.
- List known information and identify what you are trying to find.

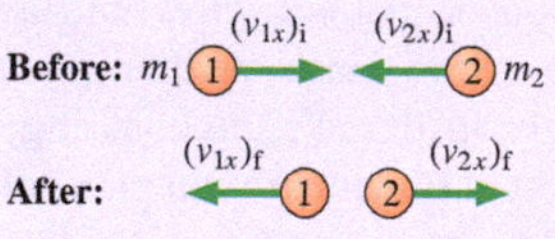


## APPLICATIONS

**Collisions** Two or more objects come together. In a perfectly inelastic collision, they stick together and move with a common final velocity.

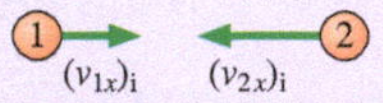


**Explosions** Two or more objects move away from each other.

**Two dimensions** Both the $x$- and $y$-components of the total momentum $\vec{P}$ must be conserved, giving two simultaneous equations.

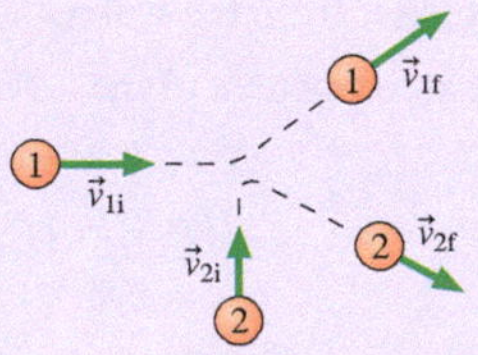

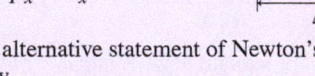

## Learning Objectives After studying this chapter, you should be able to:

- Calculate the momentum of and impulse on an object. *Conceptual Questions 9.1, 9.3; Problems 9.1, 9.2, 9.5, 9.7*
- Use the impulse-momentum theorem to solve impulse and momentum problems. *Conceptual Questions 9.4, 9.10; Problems 9.3, 9.4, 9.6, 9.9, 9.16*
- Use Problem-Solving Approach 9.1 to solve conservation of momentum problems. *Conceptual Questions 9.9, 9.12; Problems 9.10, 9.13, 9.15, 9.20, 9.23*
- Apply conservation of momentum to collisions and explosions. *Conceptual Questions 9.7, 9.17; Problems 9.19, 9.26, 9.28, 9.30, 9.33*
- Use momentum in two dimensions. *Problems 9.35, 9.36, 9.37, 9.38, 9.39*
- Understand and use angular momentum and its conservation. *Conceptual Questions 9.18, 9.19; Problems 9.40, 9.41, 9.42, 9.43, 9.45*

## STOP TO THINK ANSWERS

**Chapter Preview Stop to Think: C.** The force of the hammer on the nail and the force of the nail on the hammer are the two members of an action/reaction pair and thus, according to Newton's third law, must have the same magnitude.

**Stop to Think 9.1: C.** Impulse equals the area under the force-versus-time curve or, for these rectangular graphs, the force times the duration $\Delta t$. Force A has twice the magnitude of force B, but the collision lasts for half the time. Hence both curves have the same area and the same impulse.

**Stop to Think 9.2: D.** We know that $\vec{p}_f = \vec{p}_i + \vec{J}$. As shown in the figure at right, the only initial momentum vector that satisfies this relationship is vector D.

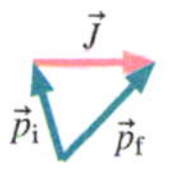


**Stop to Think 9.3: F.** The cart is initially moving in the negative $x$-direction, so $(p_x)_i = -20\ \text{kg}\cdot\text{m/s}$. After it bounces, $(p_x)_f = 10\ \text{kg}\cdot\text{m/s}$. Thus $\Delta p_x = (10\ \text{kg}\cdot\text{m/s}) - (-20\ \text{kg}\cdot\text{m/s}) = 30\ \text{kg}\cdot\text{m/s}$.

**Stop to Think 9.4: B.** The clay ball goes from $(v_x)_i = v$ to $(v_x)_f = 0$, so $J_{clay} = \Delta p_x = -mv$. The rubber ball rebounds, going from $(v_x)_i = v$ to $(v_x)_f = -v$ (same speed, opposite direction). Thus $J_{rubber} = \Delta p_x = -2mv$. The rubber ball has a greater momentum change, and this requires a greater impulse.

**Stop to Think 9.5: Right end.** The balls started at rest, so the total momentum of the system is zero. It's an isolated system, so the total momentum after the explosion is still zero. The 6 g ball has momentum $6v$. The 4 g ball, with velocity $-2v$, has momentum $-8v$. The combined momentum of these two balls is $-2v$. In order for $P$ to be zero, the third ball must have a *positive* momentum $(+2v)$ and thus a positive velocity.

**Stop to Think 9.6: B.** The momentum of object 1 is $(0.40\ \text{kg})(2.5\ \text{m/s}) = 1.0\ \text{kg}\cdot\text{m/s}$, while that of object 2 is $(0.80\ \text{kg})(-1.5\ \text{m/s}) = -1.2\ \text{kg}\cdot\text{m/s}$. The total momentum is then $1.0\ \text{kg}\cdot\text{m/s} - 1.2\ \text{kg}\cdot\text{m/s} = -0.2\ \text{kg}\cdot\text{m/s}$. Because it's negative, the total momentum, and hence the final velocity of the objects, is directed to the left.

**Stop to Think 9.7: B.** Angular momentum $L = I\omega$ is conserved. Both boys have mass $m$ and initially stand distance $R/2$ from the axis. Thus the initial moment of inertia is $I_i = I_{disk} + 2 \times m(R/2)^2 = I_{disk} + \frac{1}{2}mR^2$. The final moment of inertia is $I_f = I_{disk} + 0 + mR^2$, because the boy standing at the axis contributes nothing to the moment of inertia. Because $I_f > I_i$ we must have $\omega_f < \omega_i$.

**Video Tutor Solution** Chapter 9

# QUESTIONS

## Conceptual Questions

1. You are bowling with a 10-pound ball that you throw at 10 mph, and you've decided that you need to hit the pins with a ball that has more momentum. What are two possible changes you could make?
2. Rank in order, from largest to smallest, the magnitudes of the momentum $p_{1x}$ through $p_{5x}$ of the objects presented in Figure Q9.2. Explain.

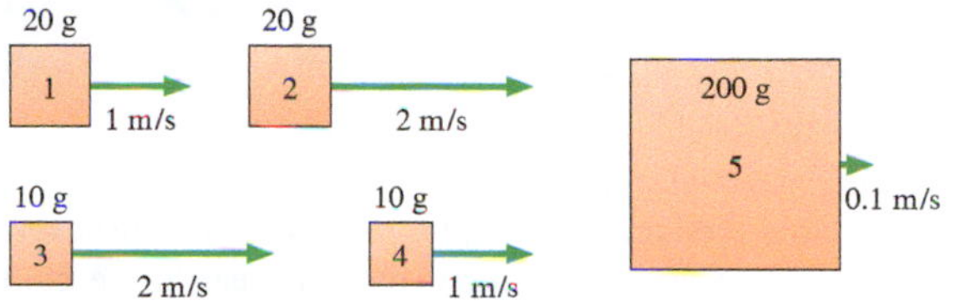


FIGURE Q9.2

Problem difficulty is labeled as | (straightforward) to ||||| (challenging). Problems labeled INT integrate significant material from earlier chapters; Problems labeled BIO are of biological or medical interest.

The eText icon indicates when there is a video tutor solution available for the chapter or for a specific problem. To launch these videos, log into your eText through Mastering™ Physics or log into the Study Area.

3. Two pucks, of mass $m$ and $4m$, lie on a frictionless table. Equal forces are used to push both pucks forward a distance of 1 m.
   a. Which puck takes longer to travel the distance? Explain.
   b. Which puck has the greater momentum upon completing the distance? Explain.
4. When you catch a water balloon, it's best to start with your hand in motion, moving with the balloon, and then gradually slow it to rest. Why is this approach desirable?
5. When you leap down from a high perch, you have a gentler landing with less force if you bend your knees as you land. Use the principles of impulse and momentum to explain why this approach reduces the force on you.
6. Students in a technology class are racing cars propelled by carbon dioxide cartridges. A burst of gas is released when the seal of the cartridge is broken, and this propels the car down the track. The cartridges are designed to provide a certain impulse. Use the concepts of impulse and momentum to explain why such a cartridge will propel a lighter car to a higher speed.
7. A stationary firecracker explodes into three pieces. One piece travels off to the east; a second travels to the north. Which of the vectors of Figure Q9.7 could be the velocity of the third piece? Explain.

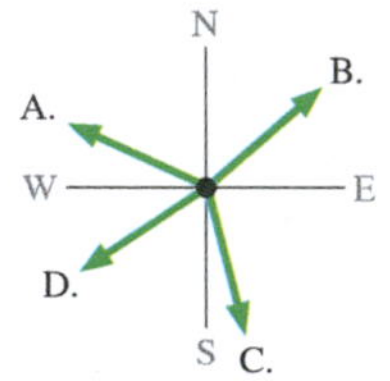


FIGURE Q9.7

8. Two students stand at rest, facing each other on frictionless skates. They then start tossing a heavy ball back and forth between them. Describe their subsequent motion.
9. A 2 kg cart rolling to the right at 3 m/s runs into a 3 kg cart rolling to the left. After the collision, both carts are stationary. What was the original speed of the 3 kg cart?
10. Automobiles are designed with "crumple zones" intended to collapse in a collision. Why would a manufacturer design part of a car so that it collapses in a collision?
11. You probably know that it feels better to catch a baseball if you are wearing a padded glove. Explain why this is so, using the ideas of momentum and impulse.
12. In the early days of rocketry, some people claimed that rockets couldn't fly in outer space as there was no air for the rockets to push against. Suppose you were an early rocket scientist and met someone who made this argument. How would you convince the person that rockets could travel in space?

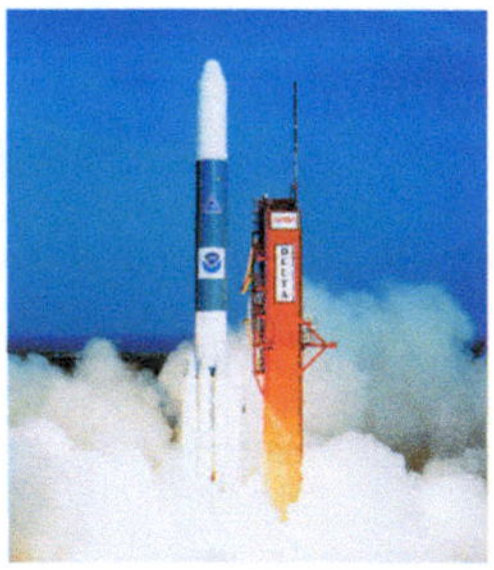

13. In the past, asteroids striking the earth have produced disastrous results. If we discovered an asteroid on a collision course with the earth, we could, in principle, deflect it and avoid an impact by focusing a laser on the surface. Intense surface heating from the laser could cause surface material to be ejected into space at high speed. How would this deflect the asteroid?
14. Two ice skaters, Megan and Jason, push off from each other on frictionless ice. Jason's mass is twice that of Megan.
   a. Which skater, if either, experiences the greater impulse during the push? Explain.
   b. Which skater, if either, has the greater speed after the push-off? Explain.
15. Suppose a rubber ball and a steel ball collide. Which, if either, receives the larger impulse? Explain.
16. While standing still on a basketball court, you throw the ball to a teammate. Why do you not move backward as a result? Is the law of conservation of momentum violated?
17. To win a prize at the county fair, you're trying to knock down a bowling pin by hitting it with a thrown object. Should you choose to throw a rubber ball, which will bounce off the pin, or a beanbag, which will strike the pin and not bounce? Assume the ball and beanbag have equal size and weight. Explain.
18. In the playground game of tetherball, a ball tied to a rope circles a pole, wrapping the rope around the pole is it goes. Opponents compete to wrap the ball around the pole in one direction or the other; the game ends when the rope is entirely wrapped around the pole and the ball touches the pole. If you get the ball going around in one direction, as the rope shortens and the ball goes in smaller and smaller circles, the ball goes around with a shorter and shorter period. Explain why this happens.

19. Monica stands at the edge of a circular platform that is slowly rotating on a frictionless axle. She then walks toward the opposite edge, passing through the platform's center. Describe the motion of the platform as Monica makes her trip.
20. If the earth warms significantly, the polar ice caps will melt. Water will move from the poles, near the earth's rotation axis, and will spread out around the globe. In principle, this will change the length of the day. Why? Will the length of the day increase or decrease?
21. If you jump from a diving board and find that your body is rotating forward, you can spin your arms to stop the rotation and make your body land feet first. Explain in which direction you would rotate your arms and how this would help.
22. A dancer leaping into the air and trying for the maximum possible vertical height will move her arms upward as she is pushing off with her legs, putting her arms in rapid upward motion before she leaves the ground. Explain how this approach can give greater height.

## Multiple-Choice Questions

23. || Model rocket engines are rated by their thrust force and by the impulse they provide. You can use this information to determine the time interval at which the engines fire. Two rocket engines provide the same impulse. The first engine provides 6 N of thrust for 2 s; the second provides 4 N of thrust. For how long does this second engine fire?
   A. 1 s B. 2 s C. 3 s D. 4 s
24. | Curling is a sport played with 20 kg stones that slide across an ice surface. Suppose a curling stone sliding at 1 m/s strikes another, stationary stone and comes to rest in 2 ms. Approximately how much force is there on the stone during the impact?
   A. 200 N B. 1000 N C. 2000 N D. 10,000 N
25. | Two balls are hung from cords. The first ball, of mass 1.0 kg, is pulled to the side and released, reaching a speed of 2.0 m/s at the bottom of its arc. Then, as shown in Figure Q9.25, it hits and sticks to another ball. The speed of the pair just after the collision is 1.2 m/s. What is the mass of the second ball?
   A. 0.67 kg B. 2.0 kg C. 1.7 kg D. 1.0 kg

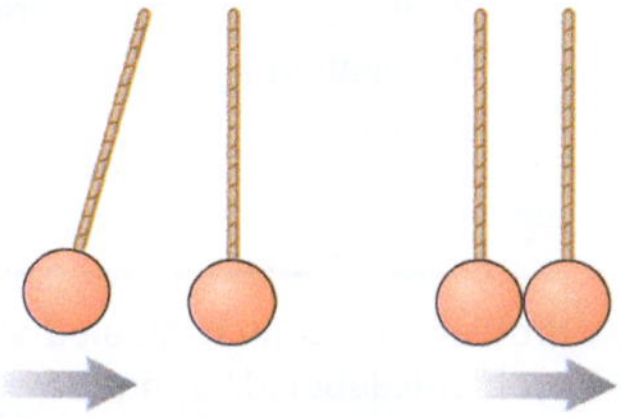

FIGURE Q9.25

26. | Figure Q9.26 shows two blocks sliding on a frictionless surface. Eventually the smaller block catches up with the larger one, collides with it, and sticks. What is the speed of the two blocks after the collision?
A. $v_i/2$ B. $4v_i/5$ C. $v_i$ D. $5v_i/4$ E. $2v_i$

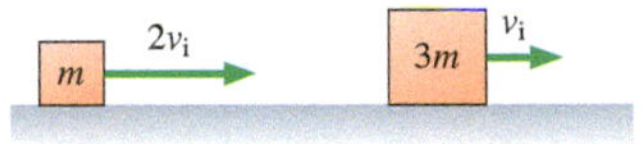


FIGURE Q9.26

27. | Two friends are sitting in a stationary canoe. At $t = 3.0$ s the person at the front tosses a sack to the person in the rear, who catches the sack 0.2 s later. Which plot in Figure Q9.27 shows the velocity of the boat as a function of time? Positive velocity is forward, negative velocity is backward. Neglect any drag force on the canoe from the water.

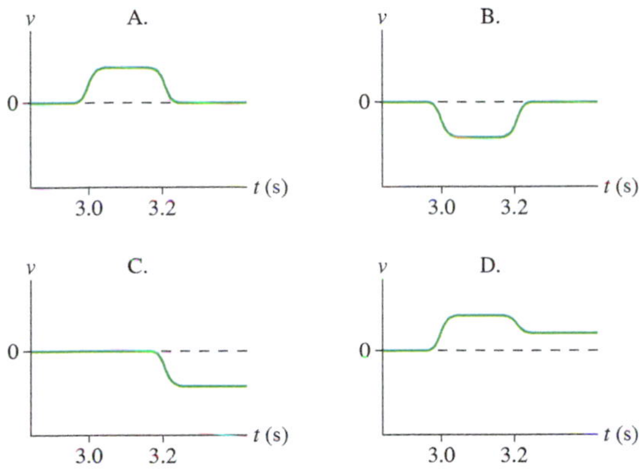


FIGURE Q9.27

28. || A 4.0-m-diameter playground merry-go-round, with a moment of inertia of 400 kg · m$^2$, is freely rotating with an angular velocity of 2.0 rad/s. Ryan, whose mass is 80 kg, runs on the ground around the outer edge of the merry-go-round in the opposite direction to its rotation. Still moving, he jumps directly onto the rim of the merry-go-round, bringing it (and himself) to a halt. How fast was Ryan running when he jumped on?
A. 2.0 m/s
B. 4.0 m/s
C. 5.0 m/s
D. 7.5 m/s
E. 10 m/s

29. || A disk rotates freely on a vertical axis with an angular velocity of 30 rpm. An identical disk rotates above it in the same direction about the same axis, but without touching the lower disk, at 20 rpm. The upper disk then drops onto the lower disk. After a short time, because of friction, they rotate together. The final angular velocity of the disks is
A. 50 rpm
B. 40 rpm
C. 25 rpm
D. 20 rpm
E. 10 rpm

# PROBLEMS

### Section 9.1 Impulse

### Section 9.2 Momentum and the Impulse-Momentum Theorem

1. | At what speed do a bicycle and its rider, with a combined mass of 100 kg, have the same momentum as a 1500 kg car traveling at 1.0 m/s?
2. | Two physics students are doing a side competition during a game of bowling, seeing who can toss a ball with the larger momentum. The first bowler throws a 4.5 kg ball at 7.2 m/s. A second bowler throws a 6.4 kg ball. What speed must she beat to win the competition?
3. | A 57 g tennis ball is served at 45 m/s. If the ball started from rest, what impulse was applied to the ball by the racket?
4. || A large raindrop—the type that lands with a definite splat—has a mass of 0.014 g and hits your roof at a speed of 8.1 m/s.
   a. What is the magnitude of the impulse delivered to your roof?
   b. If the raindrop comes to rest in 0.37 ms, what is the magnitude of the average force of the impact?
5. || A student throws a 120 g snowball at 7.5 m/s at the side of the schoolhouse, where it hits and sticks. What is the magnitude of the average force on the wall if the duration of the collision is 0.15 s?
6. || Model rocket engines are rated by the impulse that they deliver when they fire. A particular engine is rated to deliver an impulse of 3.5 kg · m/s. The engine powers a 120 g rocket, including the mass of the engine. What is the final speed of the rocket once the engine has fired? (Ignore the change in mass as the engine fires and ignore the weight force during the short duration firing of the engine.)
7. |||| In Figure P9.7, what value of $F_{max}$ gives an impulse of 6.0 N · s?

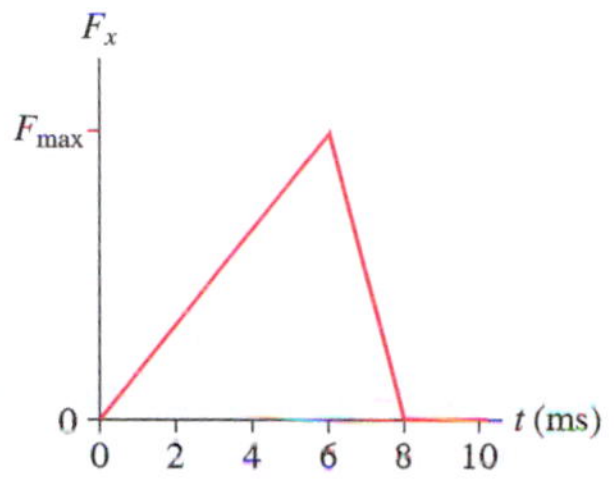


FIGURE P9.7

### Section 9.3 Solving Impulse and Momentum Problems

8. || A billiard ball of mass 0.28 kg hits a second, identical ball at a speed of 7.2 m/s and comes to rest as the second ball flies off. The collision takes 250 $\mu$s. What is the average force on the first ball? The second ball?

Watch Video Solution Problem 9.25

9. || Use the impulse-momentum theorem to find how long a stone falling straight down takes to increase its speed from 5.5 m/s to 10.4 m/s.

10. || A trap-jaw ant snaps its mandibles shut at very high speed, a
BIO good trait for catching small prey. But an ant can also slam its mandibles into the ground; the resulting force can launch the ant into the air for a quick escape. A 12 mg ant hits the ground with an average force of 47 mN for a time of 0.13 ms; these are all typical values. At what speed does it leave the ground?

11. |||| A 60 g tennis ball with an initial speed of 32 m/s hits a wall and rebounds with the same speed. Figure P9.11 shows the force of the wall on the ball during the collision. What is the value of $F_{max}$, the maximum value of the contact force during the collision?

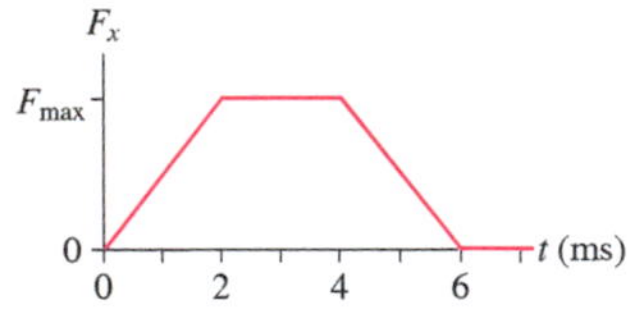


FIGURE P9.11

12. || Investigators studying the
BIO effect of hitting a soccer ball with the head are using a force plate to look at the forces in ball collisions; the force when the ball hits a player's head will be similar. A 0.43 kg ball is launched at a force plate at 16 m/s. Smoothed data for the force versus time for the collision are shown in Figure P9.12. At what speed does the ball rebound from the plate?

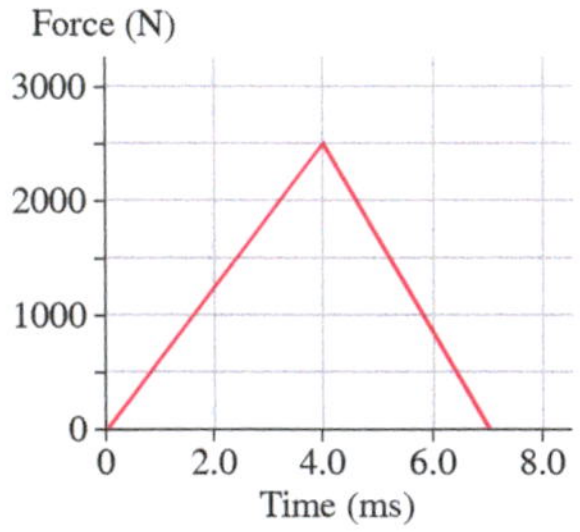


FIGURE P9.12

13. || A child is sliding on a sled at 1.5 m/s to the right. You stop the sled by pushing on it for 0.50 s in a direction opposite to its motion. If the mass of the child and sled is 35 kg, what average force do you need to apply to stop the sled? Use the concepts of impulse and momentum.

14. || Ferns spread spores instead of seeds, and some ferns eject
BIO the spores at surprisingly high speeds. One species accelerates 1.4 μg spores to a 4.5 m/s ejection speed in a time of 1.0 ms. What impulse is provided to the spores? What is the average force on a spore?

15. | As part of a safety investigation, two 1400 kg cars traveling at 20 m/s are crashed into different barriers. Find the average forces exerted on (a) the car that hits a line of water barrels and takes 1.5 s to stop, and (b) the car that hits a concrete barrier and takes 0.10 s to stop.

16. || Climbing ropes stretch when they catch a falling climber,
INT thus increasing the time it takes the climber to come to rest and reducing the force on the climber. In one standardized test of ropes, an 80 kg mass falls 4.8 m before being caught by a 2.5-m-long rope. If the net force on the mass must be kept below 11 kN, what is the minimum time for the mass to come to rest at the end of the fall?

17. || In a Little League baseball game, the 145 g ball reaches the batter with a speed of 15.0 m/s. The batter hits the ball, and it leaves his bat with a speed of 20.0 m/s in exactly the opposite direction.
a. What is the magnitude of the impulse delivered by the bat to the ball?
b. If the bat is in contact with the ball for 1.5 ms, what is the magnitude of the average force exerted by the bat on the ball?

18. || The flowers of the bunchberry plant open with astonishing force
BIO and speed, with pollen grains accelerating from rest to 7.5 m/s.
a. What impulse is delivered to a $1.0 \times 10^{-10}$ kg pollen grain?
b. What is the average force if this impulse is delivered in 0.30 ms?
c. How does this force compare to the weight of the pollen grain?

## Section 9.4 Conservation of Momentum

19. |||| A small, 100 g cart is moving at 1.20 m/s on a frictionless track when it collides with a larger, 1.00 kg cart at rest. After the collision, the small cart recoils at 0.850 m/s. What is the speed of the large cart after the collision?

20. || In principle, when you fire a rifle, the recoil should push you backward. How big a push will it give? Let's find out by doing a calculation in a very artificial situation. Suppose a man standing on frictionless ice fires a rifle horizontally. The mass of the man together with the rifle is 70 kg, and the mass of the bullet is 10 g. If the bullet leaves the muzzle at a speed of 500 m/s, what is the final speed of the man? Given your result, do you expect to be pushed backward when you fire a rifle?

21. ||| A 2.7 kg block of wood sits on a frictionless table. A 3.0 g bullet, fired horizontally at a speed of 500 m/s, goes completely through the block, emerging at a speed of 220 m/s. What is the speed of the block immediately after the bullet exits?

22. || Squid rely on jet propulsion when a rapid escape is neces-
BIO sary. A 1.5 kg squid at rest pulls 0.10 kg of water into its mantle, then ejects this water at a remarkable 45 m/s. Right after this ejection, how fast is the squid moving?

23. || A 10,000 kg railroad car is rolling at 2.00 m/s when a 4000 kg load of gravel is suddenly dropped in. What is the car's speed just after the gravel is loaded?

24. || Some species of jellyfish use jet propulsion to get around, a
BIO much gentler form of jet propulsion than squid use. A small jellyfish takes water into its bell; the total mass of jellyfish and water is 3.2 g. The jellyfish then rapidly ejects 1.3 g of water, achieving a speed of 0.070 m/s. What is the speed of the ejected water?

25. | A 55 kg hunter, standing on frictionless ice, shoots a 42 g bullet at a speed of 620 m/s. What is the recoil speed of the hunter?

26. || If you free the cork in a highly pressurized champagne bottle, the resulting launch of the cork will, in principle, cause the bottle to recoil. A filled champagne bottle has a mass of 1.8 kg. The cork has a mass of 7.5 g and is launched at 20 m/s. If the bottle could move freely, with what speed would it recoil? Is this something you are likely to notice?

## Section 9.5 Inelastic Collisions

27. || A 300 g bird flying along at 6.0 m/s sees a 10 g insect heading straight toward it with a speed of 30 m/s. The bird opens its mouth wide and enjoys a nice lunch. What is the bird's speed immediately after swallowing?

28. || Peregrine falcons frequently grab prey birds from the air, as
BIO in Example 9.10. Sometimes they strike at high enough speeds that the force of the impact disables prey birds. A 480 g peregrine falcon high in the sky spies a 240 g pigeon some distance below. The falcon slows to a near stop, then goes into a dive—called a stoop—and picks up speed as she falls. The falcon reaches a vertical speed of 45 m/s before striking the pigeon, which we can assume is stationary. The falcon strikes the pigeon and grabs it in her talons. The collision between the birds lasts 0.015 s.
a. What is the final speed of the falcon and pigeon?
b. What is the average force on the pigeon during the impact?

29. ‖ A 71 kg baseball player jumps straight up to catch a hard-hit ball. If the 140 g ball is moving horizontally at 28 m/s, and the catch is made when the ballplayer is at the highest point of his leap, what is his speed immediately after stopping the ball?
30. ‖ In a football game, a 90 kg receiver leaps straight up in the air to catch the 0.42 kg ball the quarterback threw to him at a vigorous 21 m/s, catching the ball at the highest point in his jump. Right after catching the ball, how fast is the receiver moving? Is he likely to notice this recoil?
31. ||| A kid at the junior high cafeteria wants to propel an empty milk carton along a lunch table by hitting it with a 3.0 g spit ball. If he wants the speed of the 20 g carton just after the spit ball hits it to be 0.30 m/s, at what speed should his spit ball hit the carton?
32. | The parking brake on a 2000 kg Cadillac has failed, and it is rolling slowly, at 1 mph, toward a group of small children. Seeing the situation, you realize you have just enough time to drive your 1000 kg Volkswagen head-on into the Cadillac and save the children. With what speed should you impact the Cadillac to bring it to a halt?
33. | A 2.0 kg block slides along a frictionless surface at 1.0 m/s. A second block, sliding at 4.0 m/s, collides with the first from behind and sticks to it. The final velocity of the combined blocks is 2.0 m/s. What was the mass of the second block?
34. ‖ Erica (36 kg) and Danny (47 kg) are bouncing on a trampoline. Just as Erica reaches the high point of her bounce, Danny is moving upward past her at 4.1 m/s. At that instant he grabs hold of her. What is their speed just after he grabs her?

## Section 9.6 Momentum and Collisions in Two Dimensions

35. |||| A 20 g ball of clay traveling east at 3.0 m/s collides with a 30 g ball of clay traveling north at 2.0 m/s. What are the speed and the direction of the resulting 50 g blob of clay?
36. ||| Casey is driving a 1600 kg car toward the east. She goes through an intersection at a speed of 16 m/s (approximately 35 mph), the speed limit on both roads of the intersection. Kerry is driving a car of mass 1200 kg into the intersection, going north, and doesn't see or doesn't heed a red light, and slams into Casey's car. The cars lock together and skid to a stop. Later, the two review the scene with the police. Skid marks from the instant after the collision reveal that the two cars were moving exactly northeast. Kerry claims to have been driving at the speed limit, but Casey says that Kerry seemed to be going over the speed limit before the collision. Who is correct? Use the concept of conservation of momentum to make your case.
37. ‖ Two objects collide and bounce apart. Figure P9.37 shows the initial momentum of each object and the final momentum of object 2. What is the final momentum of object 1? Show your answer by copying the figure and drawing the final momentum vector on the figure.

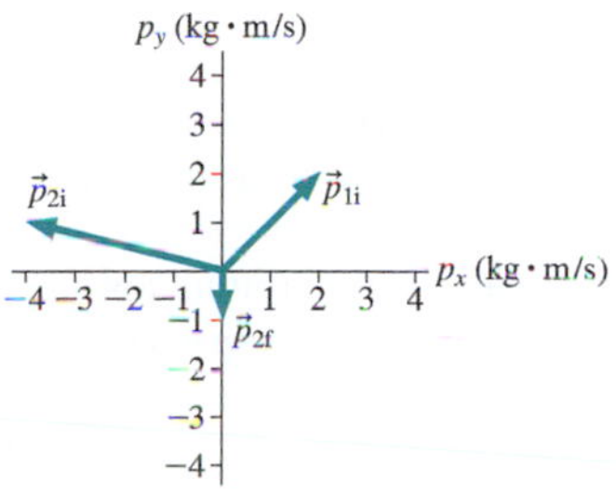


FIGURE P9.37

38. ‖ A 20 g ball of clay traveling east at 2.0 m/s collides with a 30 g ball of clay traveling 30° south of west at 1.0 m/s. What are the speed and direction of the resulting 50 g blob of clay?
39. ‖ A firecracker in a coconut blows the coconut into three pieces. Two pieces of equal mass fly off south and west, perpendicular to each other, at 20 m/s. The third piece has twice the mass of the other two. What are the speed and direction of the third piece?

## Section 9.7 Angular Momentum

40. ‖ What is the angular momentum about the axle of the 500 g rotating bar in Figure P9.40?

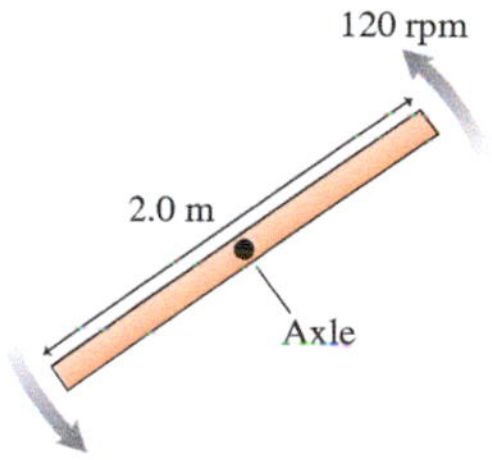


FIGURE P9.40

41. |||| What is the angular momentum about the axle of the 2.0 kg, 4.0-cm-diameter rotating disk in Figure P9.41?

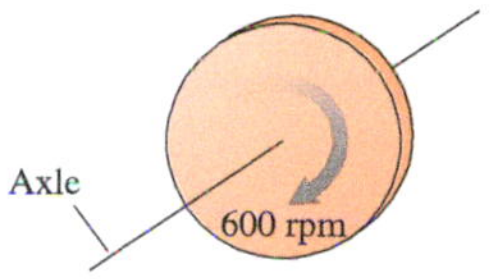


FIGURE P9.41

42. | Divers change their body position in midair while rotating BIO about their center of mass. In one dive, the diver leaves the board with her body nearly straight, then tucks into a somersault position. If the moment of inertia of the diver in a straight position is 14 kg · m$^2$ and in a tucked position is 4.0 kg · m$^2$, by what factor does her angular speed increase?
43. ‖ Ice skaters often end their performances with spin turns, BIO where they spin very fast about their center of mass with their arms folded in and legs together. Upon ending, their arms extend outward, proclaiming their finish. Not quite as noticeably, one leg goes out as well. Suppose that the moment of inertia of a skater with arms out and one leg extended is 3.2 kg · m$^2$ and for arms and legs in is 0.80 kg · m$^2$. If she starts out spinning at 5.0 rev/s, what is her angular speed (in rev/s) when her arms and one leg open outward?
44. ||| A diver leaves the platform with her body straight. BIO INT Her body is in a relatively slow rotation, with an angular speed of 4.0 rad/s. She then tucks into a pike position, with her body essentially folded in half. We can use a simple model to understand what happens next. First, model her 50 kg, 1.8 m body as uniform. Next, assume that when she goes into a pike position, she really does fold her body exactly in half. In terms of this model,

a. What is her initial moment of inertia?
b. What is her moment of inertia in the pike position?
c. What is her angular speed in the pike position?
d. How many rotations does she complete in the 1.3 s that she holds the pike position?

45. ||| INT What is the angular momentum of the moon around the earth? The moon's mass is $7.4 \times 10^{22}$ kg and it orbits $3.8 \times 10^{8}$ m from the earth.

## General Problems

46. ||| What is the impulse on a 3.0 kg object that experiences the force described by the graph in Figure P9.46?

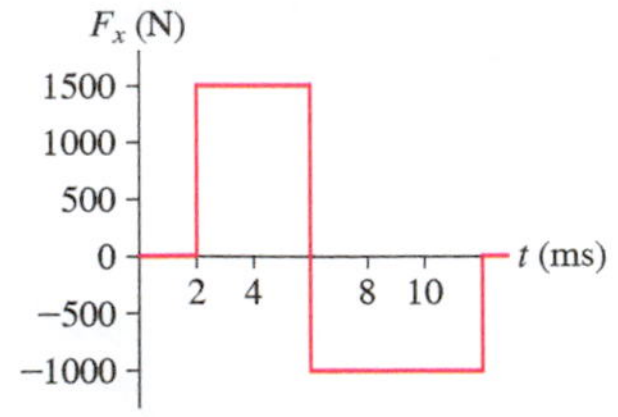


FIGURE P9.46

47. ||| A 600 g air-track glider collides with a spring at one end of the track. Figure P9.47 shows the glider's velocity and the force exerted on the glider by the spring. How long is the glider in contact with the spring?

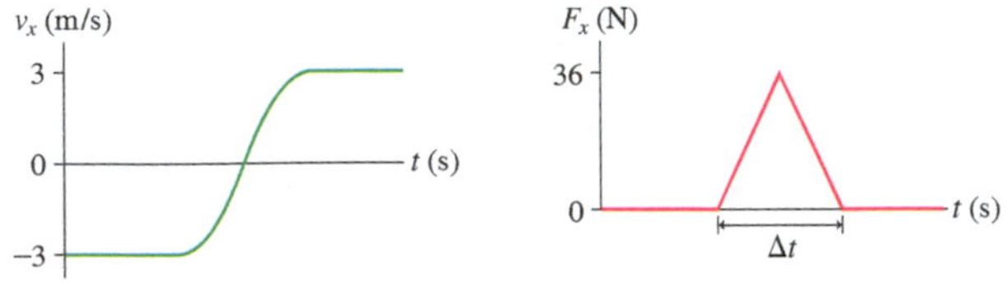


FIGURE P9.47

48. || Far in space, where gravity is negligible, a 425 kg rocket traveling at 75.0 m/s in the positive $x$-direction fires its engines. Figure P9.48 shows the thrust force as a function of time. The mass lost by the rocket during these 30.0 s is negligible.

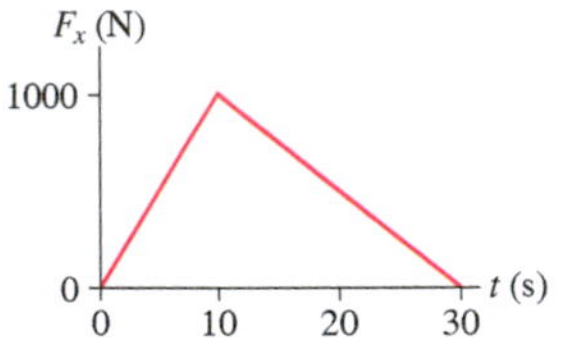


FIGURE P9.48

a. What impulse does the engine impart to the rocket?
b. At what time does the rocket reach its maximum speed? What is the maximum speed?

49. |||| INT A 200 g ball is dropped from a height of 2.0 m, bounces on a hard floor, and rebounds to a height of 1.5 m. Figure P9.49 shows the impulse received from the floor. What maximum force does the floor exert on the ball?

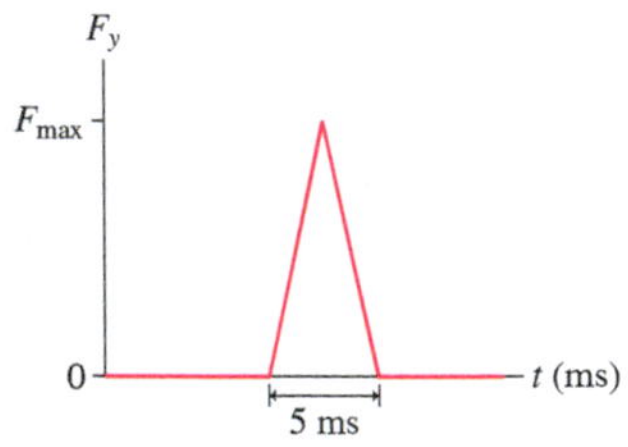


FIGURE P9.49

50. |||| INT A 200 g ball is dropped from a height of 2.0 m and bounces on a hard floor. The force on the ball from the floor is shown in Figure P9.50. How high does the ball rebound?

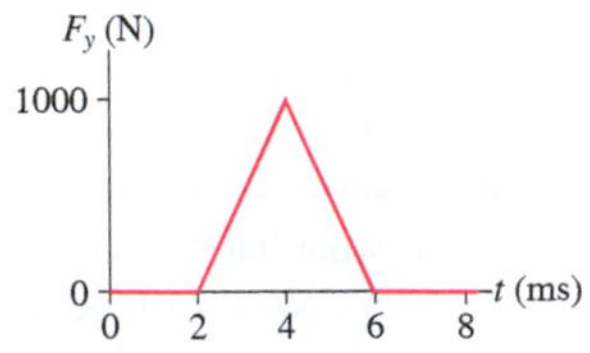


FIGURE P9.50

51. ||| INT Figure P9.51 is a graph of the force exerted by the floor on a woman making a vertical jump. At what speed does she leave the ground?
**Hint:** The force of the floor is not the only force acting on the woman.

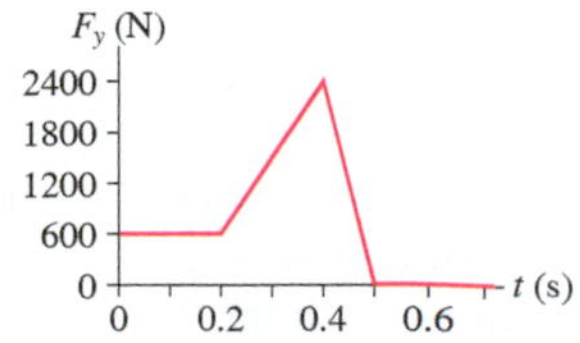


FIGURE P9.51

52. | BIO INT We've seen that squid can escape from predators by ejecting water. Some squid do this at the surface of the ocean, thus launching themselves into the air—a particularly effective escape strategy. Suppose a 36 kg squid (not including water) at rest at the surface of the water brings in and quickly ejects 3.0 kg of water to achieve a takeoff speed of 3.5 m/s; these are typical numbers.

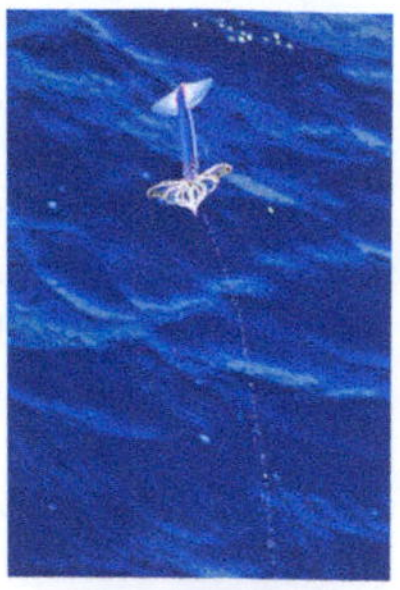

a. At what speed does the squid eject the water?
b. If we ignore lift and drag forces as the squid flies through the air, what is the maximum horizontal range that the squid can achieve before splashing down?

53. ||| A 140 g baseball is moving horizontally to the right at 35 m/s when it is hit by the bat. The ball flies off to the left at 55 m/s, at an angle of 25° above the horizontal. What are the magnitude and direction of the impulse that the bat delivers to the ball?

54. ||| A tennis player swings her 1000 g racket with a speed of 10 m/s. She hits a 60 g tennis ball that was approaching her at a speed of 20 m/s. The ball rebounds at 40 m/s.
a. How fast is her racket moving immediately after the impact? You can ignore the interaction of the racket with her hand for the brief duration of the collision.
b. If the tennis ball and racket are in contact for 10 ms, what is the average force that the racket exerts on the ball?

55. || A 20 g ball of clay is thrown horizontally at 30 m/s toward a 1.0 kg block sitting at rest on a frictionless surface. The clay hits and sticks to the block.
a. What is the speed of the block and clay right after the collision?
b. Use the block's initial and final speeds to calculate the impulse the clay exerts on the block.
c. Use the clay's initial and final speeds to calculate the impulse the block exerts on the clay.
d. Does $\vec{J}_{\text{block on clay}} = -\vec{J}_{\text{clay on block}}$?

56. || Dan is gliding on his skateboard at 4.0 m/s. He jumps backward off the skateboard, kicking the skateboard forward at 8.0 m/s. How fast is Dan going as his feet hit the ground? Dan's mass is 50 kg and the skateboard's mass is 5.0 kg.

57. ||| INT Ethan, whose mass is 80 kg, stands at one end of a very long, stationary wheeled cart that has a mass of 500 kg. He then starts sprinting toward the other end of the cart. He soon reaches his top speed of 8.0 m/s, measured relative to the cart. What is the cart's speed when Ethan reaches top speed?

58. || Three identical train cars, coupled together, are rolling east at 2.0 m/s. A fourth car traveling east at 4.0 m/s catches up with the three and couples to make a four-car train. A moment later, the train cars hit a fifth car that was at rest on the tracks, and it couples to make a five-car train. What is the speed of the five-car train?

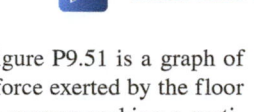

59. | A 110 kg linebacker running at 2.0 m/s and an 82 kg quarterback running at 3.0 m/s have a head-on collision in midair. The linebacker grabs and holds onto the quarterback. Who ends up moving forward after they hit?

60. || INT Most geologists believe that the dinosaurs became extinct 65 million years ago when a large comet or asteroid struck the earth, throwing up so much dust that the sun was blocked out for a period of many months. Suppose an asteroid with a diameter of 2.0 km and a mass of $1.0 \times 10^{13}$ kg hits the earth with an impact speed of $4.0 \times 10^4$ m/s.
   a. What is the earth's recoil speed after such a collision? (Use a reference frame in which the earth was initially at rest.)
   b. What percentage is this of the earth's speed around the sun? (Use the astronomical data inside the back cover.)

61. |||| INT Two ice skaters, with masses of 75 kg and 55 kg, stand facing each other on a 15-m-wide frozen river. The skaters push off against each other, glide backward straight toward the river's edges, and reach the edges at exactly the same time. How far did the 75 kg skater glide?

62. || INT Two ice skaters, with masses of 50 kg and 75 kg, are at the center of a 60-m-diameter circular rink. The skaters push off against each other and glide to opposite edges of the rink. If the heavier skater reaches the edge in 20 s, how long does the lighter skater take to reach the edge?

63. ||| One billiard ball is shot east at 2.00 m/s. A second, identical billiard ball is shot west at 1.00 m/s. The balls have a glancing collision, not a head-on collision, deflecting the second ball by 90° and sending it north at 1.41 m/s. What are the speed and direction of the first ball after the collision?

64. |||| INT A 10 g bullet is fired into a 10 kg wood block that is at rest on a wood table. The block, with the bullet embedded, slides 5.0 cm across the table. What was the speed of the bullet?

65. ||| BIO A typical raindrop is much more massive than a mosquito and much faster than a mosquito flies. How does a mosquito survive the impact? Recent research has found that the collision of a falling raindrop with a mosquito is a perfectly inelastic collision. That is, the mosquito is "swept up" by the raindrop and ends up traveling along with the raindrop. Once the relative speed between the mosquito and the raindrop is zero, the mosquito is able to detach itself from the drop and fly away.
   a. A hovering mosquito is hit by a raindrop that is 40 times as massive and falling at 8.2 m/s, a typical raindrop speed. How fast is the raindrop, with the attached mosquito, falling immediately afterward if the collision is perfectly inelastic?
   b. Because a raindrop is "soft" and deformable, the collision duration is a relatively long 8.0 ms. What is the mosquito's average acceleration, in $g$'s, during the collision? The peak acceleration is roughly twice the value you found, but the mosquito's rigid exoskeleton allows it to survive accelerations of this magnitude. In contrast, humans cannot survive an acceleration of more than about $10g$.

66. |||| INT A 15 g bullet is fired at 610 m/s into a 4.0 kg block that sits at the edge of a 75-cm-high table. The bullet embeds itself in the block and carries it off the table. How far from the point directly below the table's edge does the block land?

67. |||| INT A 1500 kg weather rocket accelerates upward at $10.0 \text{ m/s}^2$. It explodes 2.00 s after liftoff and breaks into two fragments, one twice as massive as the other. Photos reveal that the lighter fragment traveled straight up and reached a maximum height of 530 m. What were the speed and direction of the heavier fragment just after the explosion?

68. ||| INT A canoe is designed to have very little drag when it moves along its length. Riley, mass 52 kg, sits in a 21 kg canoe in the middle of a lake. She dives into the water off the front of the canoe, along the axis of the canoe. She dives forward at 1.6 m/s relative to the boat. Just after her leap,
   a. How fast is she moving relative to the water?
   b. How fast is the canoe moving relative to the water?

69. || Two 500 g blocks of wood are 2.0 m apart on a frictionless table. A 10 g bullet is fired at 400 m/s toward the blocks. It passes all the way through the first block, then embeds itself in the second block. The speed of the first block immediately afterward is 6.0 m/s. What is the speed of the second block after the bullet stops?

70. ||| A water pipe in a building delivers 1000 liters (with mass 1000 kg) of water per second. The water is moving through the pipe at 1.4 m/s. The pipe has a 90° bend, and the pipe will require a supporting structure, called a *thrust block*, at the bend, as in Figure P9.70. We can use the ideas of momentum and impulse to understand why. Each second, 1000 kg of water moving at $v_x = 1.4$ m/s changes direction to move at $v_y = 1.4$ m/s.

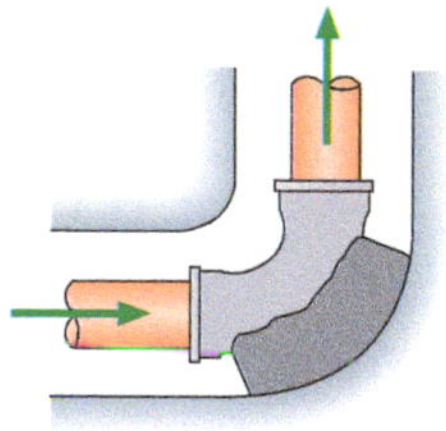

FIGURE P9.70

   a. What are the magnitude and direction of the change in momentum of the 1000 kg of water?
   b. What are the magnitude and direction of the necessary impulse?
   c. This impulse takes place over 1.0 s. What is the necessary force?

71. || A spaceship of mass $2.0 \times 10^6$ kg is cruising at a speed of $5.0 \times 10^6$ m/s when the antimatter reactor fails, blowing the ship into three pieces. One section, having a mass of $5.0 \times 10^5$ kg, is blown straight backward with a speed of $2.0 \times 10^6$ m/s. A second piece, with mass $8.0 \times 10^5$ kg, continues forward at $1.0 \times 10^6$ m/s. What are the direction and speed of the third piece?

72. ||| At the county fair, Chris throws a 0.15 kg baseball at a 2.0 kg wooden milk bottle, hoping to knock it off its stand and win a prize. The ball bounces straight back at 20% of its incoming speed, knocking the bottle straight forward. What is the bottle's speed, as a percentage of the ball's incoming speed?

73. ||| Figure P9.73 shows a collision between three balls of clay. The three hit simultaneously and stick together. What are the speed and direction of the resulting blob of clay?

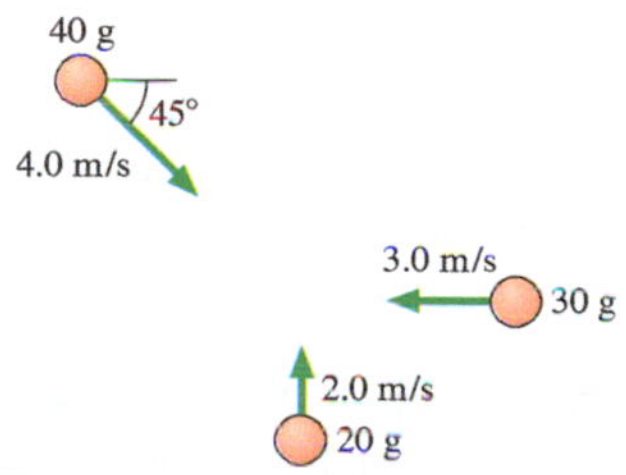


FIGURE P9.73

74. ||| The carbon isotope $^{14}$C is used for carbon dating of archeological artifacts. $^{14}$C (mass $2.34 \times 10^{-26}$ kg) decays by the process known as *beta decay* in which the nucleus emits an electron (the beta particle) and a subatomic particle called a neutrino. In one such decay, the electron and the neutrino are emitted at right angles to each other. The electron (mass $9.11 \times 10^{-31}$ kg) has a speed of $5.00 \times 10^7$ m/s and the neutrino has a momentum of $8.00 \times 10^{-24}$ kg·m/s. What is the recoil speed of the nucleus?

75. ||| INT Figure P9.75 shows a 100 g puck revolving at 100 rpm on a 20-cm-radius circle on a frictionless table. A string attached to the puck passes through a hole in the middle of the table. The end of the string below the table is then slowly pulled down until the puck is revolving in a 10-cm-radius circle. How many revolutions per minute does the puck make at this new radius?

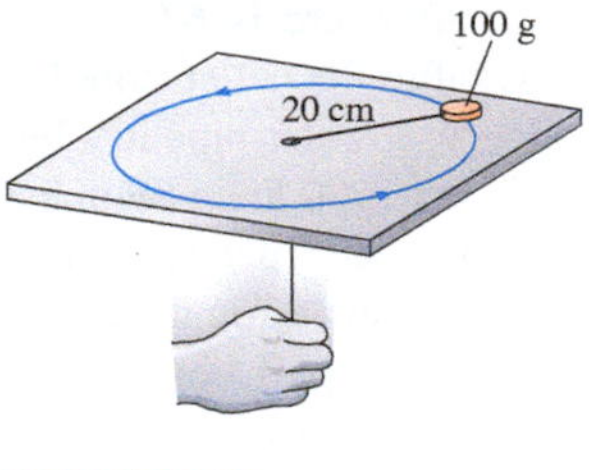


FIGURE P9.75

76. || A 2.0 kg, 20-cm-diameter turntable rotates at 100 rpm on frictionless bearings. Two 500 g blocks fall from above, hit the turntable simultaneously at opposite ends of a diagonal, and stick. What is the turntable's angular speed, in rpm, just after this event?

77. || Joey, from Example 9.11, stands at rest at the outer edge of the frictionless merry-go-round of Figure 9.27. The merry-go-round is also at rest. Joey then begins to run around the perimeter of the merry-go-round, finally reaching a constant speed, measured relative to the ground, of 5.0 m/s. What is the final angular speed of the merry-go-round?

78. ||| A 3.0-m-diameter merry-go-round with a mass of 250 kg is spinning at 20 rpm. John runs around the merry-go-round at 5.0 m/s, in the same direction that it is turning, and jumps onto the outer edge. John's mass is 30 kg. What is the merry-go-round's angular speed, in rpm, after John jumps on?

79. ||| Disk A, with a mass of 2.0 kg and a radius of 40 cm, rotates clockwise about a frictionless vertical axle at 30 rev/s. Disk B, also 2.0 kg but with a radius of 20 cm, rotates counterclockwise about that same axle, but at a greater height than disk A, at 30 rev/s. Disk B slides down the axle until it lands on top of disk A, after which they rotate together. After the collision, what is their common angular speed (in rev/s) and in which direction do they rotate?

## MCAT-Style Passage Problems

### Hitting a Golf Ball

Consider a golf club hitting a golf ball. To a good approximation, we can model this as a collision between the rapidly moving head of the golf club and the stationary golf ball, ignoring the shaft of the club and the golfer.

A golf ball has a mass of 46 g. Suppose a 200 g club head is moving at a speed of 40 m/s just before striking the golf ball. After the collision, the golf ball's speed is 60 m/s.

80. | What is the momentum of the club + ball system right before the collision?
A. 1.8 kg·m/s B. 8.0 kg·m/s
C. 3220 kg·m/s D. 8000 kg·m/s

81. | Immediately after the collision, the momentum of the club + ball system will be
A. Less than before the collision.
B. The same as before the collision.
C. Greater than before the collision.

82. | A manufacturer makes a golf ball that compresses more than a traditional golf ball when struck by a club. How will this affect the average force during the collision?
A. The force will decrease.
B. The force will not be affected.
C. The force will increase.

83. | By approximately how much does the club head slow down as a result of hitting the ball?
A. 4 m/s B. 6 m/s C. 14 m/s D. 26 m/s

### Analyzing a Dance Move BIO INT

Understanding the details of timing and forces in motion can improve the performance of athletes, including dancers. For Problems 84–87, consider the forces involved in a ballet jump called a sauté demi plié. Figure P9.84a shows the sequence of moves in the jump. The dancer starts upright, then quickly bends her knees, moving downward. After she reaches the bottom of this dip, she extends her legs, pushing herself upward. After this upward push, she leaves the ground, beginning a short period of time in the air. Figure P9.84b is a slightly idealized graph of the net force on a 42 kg dancer executing this move.

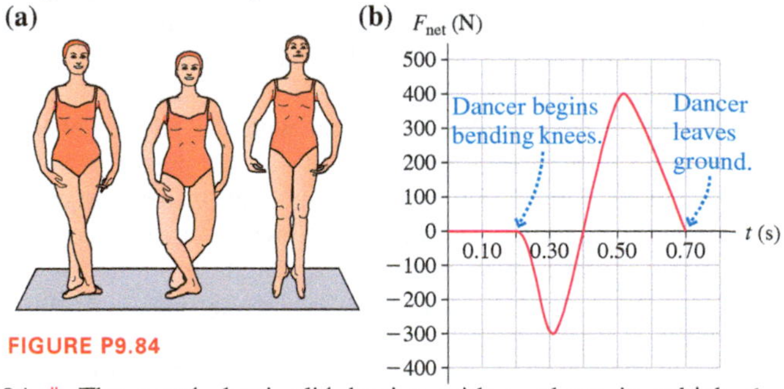


FIGURE P9.84

84. || The sauté demi plié begins with a phase in which the net force on the dancer is negative. During this phase of the jump,
A. The normal force of the floor on her is zero.
B. The normal force of the floor on her is less than her weight but greater than zero.
C. The normal force of the floor on her is equal to her weight.
D. The normal force of the floor on her is greater than her weight.

85. || At what time does the dancer reach the lowest point of her motion, when her speed is zero?
A. 0.20 s
B. 0.40 s
C. Between 0.40 s and 0.70 s
D. After 0.70 s

86. || What is the approximate net impulse on the dancer, from the moment she begins bending her knees to the instant she leaves the floor?
A. 15 kg·m/s$^2$ B. 30 kg·m/s$^2$
C. 60 kg·m/s$^2$ D. 90 kg·m/s$^2$

87. || To the nearest m/s, how fast is the dancer moving when she leaves the floor?
A. 1 m/s B. 2 m/s C. 3 m/s D. 4 m/s

# 10 Energy and Work

A greyhound can rapidly go from a standstill to a very speedy run—meaning a very rapid increase in kinetic energy, the energy of motion. How does the greyhound's ability to convert energy from one form to another compare to that of other animals?

## LOOKING AHEAD ▸

### Forms of Energy

This dolphin has lots of **kinetic energy** as it leaves the water. At its highest point its energy is mostly **potential energy.**

You'll learn about several of the most important forms of energy—kinetic, potential, and thermal.

### Work and Energy

The woman does **work** on the jack, applying a force to the handle and pushing it down. This is a *transfer* of energy into the system, increasing the potential energy as the car is lifted.

You'll learn how to calculate the work done by a force, and how this work is related to the *change* in a system's energy.

### Conservation of Energy

As they slide, their potential energy decreases and their kinetic energy increases, but their total energy is unchanged: It is **conserved.**

How fast will they be moving when they reach the bottom? You'll use a new before-and-after analysis to find out.

**GOAL** To introduce the concept of energy and to learn a new problem-solving strategy based on conservation of energy.

## LOOKING BACK ◂

### Motion with Constant Acceleration

In Chapter 2 you learned how to describe the motion of a particle that has a constant acceleration. In this chapter, you'll use the constant-acceleration equations to connect work and energy.

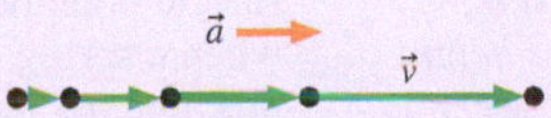

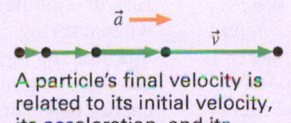

A particle's final velocity is related to its initial velocity, its acceleration, and its displacement by

$$(v_x)_f^2 = (v_x)_i^2 + 2a_x\,\Delta x$$

**STOP TO THINK**

A car pulls away from a stop sign with a constant acceleration. After traveling 10 m, its speed is 5 m/s. What will its speed be after traveling 40 m?

A. 10 m/s
B. 20 m/s
C. 30 m/s
D. 40 m/s

# 10.1 The Basic Energy Model

Energy. It's a word you hear all the time. We use chemical energy to heat our homes and bodies, electric energy to run our lights and computers, and solar energy to grow our crops and forests. We're told to use energy wisely and not to waste it. Athletes and weary students consume "energy bars" and "energy drinks."

But just what is energy? The concept of energy has grown and changed over time, and it is not easy to define in a general way just what energy is. Rather than starting with a formal definition, we'll let the concept of energy expand slowly over the course of several chapters. In this chapter we introduce several fundamental forms of energy, including kinetic energy, potential energy, and thermal energy. Our goal is to understand the characteristics of energy, how energy is used, and, especially important, how energy is transformed from one form into another. Understanding these transformations will allow us to understand and explore a wide variety of physical phenomena. Anything that happens involves a transformation of energy from one form to another, so the range of topics we'll consider in this chapter is extensive.

In solving problems, we'll use a key fact about energy: **Energy is neither created nor destroyed: If one form of energy in a system decreases, it must appear in an equal amount in another form.** Many scientists consider this law of conservation of energy to be the most important of all the laws of nature.

## Systems and Forms of Energy

FIGURE 10.1 A system and its energies.

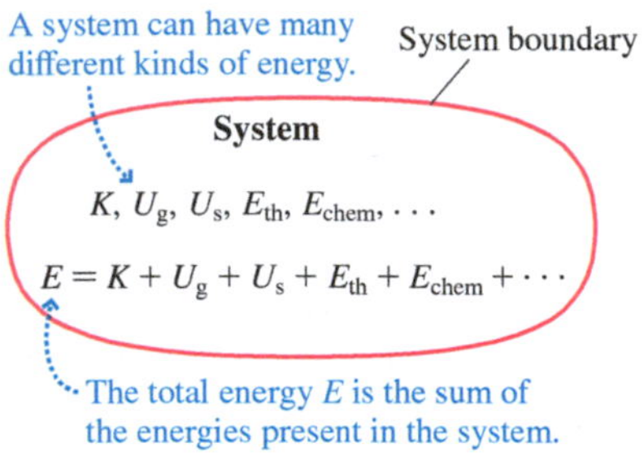


In Chapter 9 we introduced the idea of a *system* of interacting objects. A system can be as simple as a falling acorn or as complex as a city. But whether simple or complex, every system in nature has associated with it a quantity we call its **total energy** $E$. The total energy is the sum of the different kinds of energies present in the system. In the table below, we give a brief overview of some of the more important forms of energy; in the rest of the chapter, we'll look at several of these forms of energy in greater detail.

A system may have many of these kinds of energy at one time. For instance, a moving car has kinetic energy of motion, chemical energy stored in its gasoline, thermal energy in its hot engine, and many other forms of energy. FIGURE 10.1 illustrates the idea that the total energy of the system, $E$, is the *sum* of all the different energies present in the system:

$$E = K + U_g + U_s + E_{th} + E_{chem} + \cdots \qquad (10.1)$$

The energies shown in this sum are the forms of energy in which we'll be most interested in this and the next chapter. The ellipses ($\cdots$) stand for other forms of energy, such as nuclear or electric, that also might be present. We'll treat these and others in later chapters.

### Some important forms of energy

**Kinetic energy $K$**

Kinetic energy is the energy of *motion.* All moving objects have kinetic energy. The heavier an object and the faster it moves, the more kinetic energy it has. The wrecking ball in this picture is effective in part because of its large kinetic energy.

**Gravitational potential energy $U_g$**

Gravitational potential energy is *stored* energy associated with an object's *height above the ground.* As this coaster ascends, energy is stored as gravitational potential energy. As it descends, this stored energy is converted into kinetic energy.

**Elastic or spring potential energy $U_s$**

Elastic potential energy is energy stored when a spring or other elastic object, such as this archer's bow, is *stretched.* This energy can later be transformed into the kinetic energy of the arrow.

*Continued*

**Thermal energy $E_{th}$**

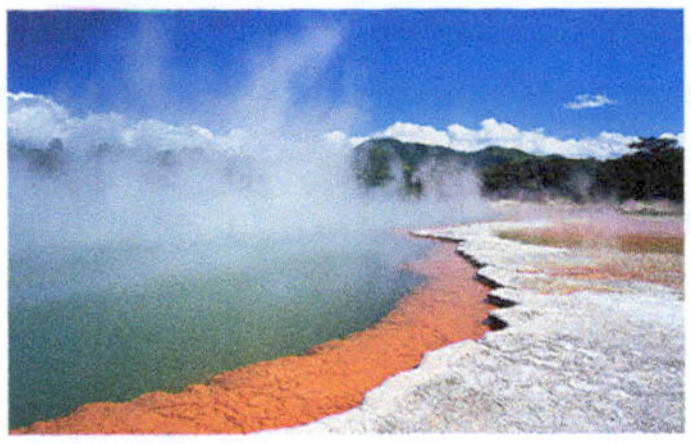

Hot objects have more *thermal energy* than cold ones because the molecules in a hot object jiggle around more than those in a cold object. Thermal energy is the sum of the microscopic kinetic and potential energies of all the molecules in an object.

**Chemical energy $E_{chem}$**

Electric forces cause atoms to bind together to make molecules. Energy can be stored in these bonds, energy that can later be released as the bonds are rearranged during chemical reactions. All animals eat, taking in chemical energy to provide energy to move muscles and fuel processes of the body.

**Nuclear energy $E_{nuclear}$**

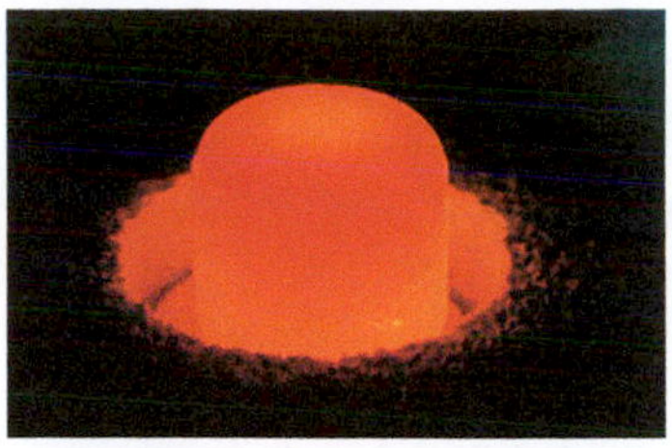

The forces that hold together the particles in the nucleus of the atom are much stronger than the electric forces that hold together molecules, so they store a great deal more energy. Certain nuclei break apart into smaller fragments, releasing some of this *nuclear energy*. The energy is transformed into the kinetic energy of the fragments and then into thermal energy.

## Energy Transformations

If the amounts of each form of energy never changed, the world would be a very dull place. What makes the world interesting is that **energy of one kind can be transformed into energy of another kind.** The following table illustrates a few common energy transformations. In this table, we use an arrow $\rightarrow$ as a shorthand way of representing an energy transformation.

### Some energy transformations

**A weightlifter lifts a barbell over her head**

The barbell has much more gravitational potential energy when high above her head than when on the floor. To lift the barbell, she transforms chemical energy in her body into gravitational potential energy of the barbell.

$$E_{chem} \rightarrow U_g$$

**A base runner slides into the base**

When running, he has lots of kinetic energy. After sliding, he has none. His kinetic energy is transformed mainly into thermal energy: The ground and his legs are slightly warmer.

$$K \rightarrow E_{th}$$

**A burning campfire**

The wood contains considerable chemical energy. When the carbon in the wood combines chemically with oxygen in the air, this chemical energy is transformed largely into thermal energy of the hot gases and embers.

$$E_{chem} \rightarrow E_{th}$$

**A springboard diver**

Here's a two-step energy transformation. At the instant shown, the board is flexed to its maximum extent, so that elastic potential energy is stored in the board. Soon this energy will begin to be transformed into kinetic energy; then, as the diver rises into the air and slows, this kinetic energy will be transformed into gravitational potential energy.

$$U_s \rightarrow K \rightarrow U_g$$

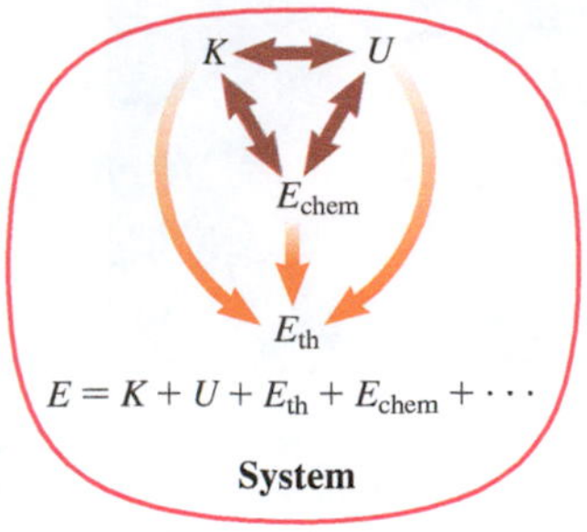


FIGURE 10.2 Energy transformations within the system.

FIGURE 10.2 reinforces the idea that **energy transformations are changes of energy *within* the system from one form to another.** (The $U$ in this figure is a generic potential energy; it could be gravitational potential energy $U_g$, spring potential energy $U_s$, or some other form of potential energy.) There are two types of arrows in the figure. The arrow between $K$ and $U$ is a two-way arrow; it's easy to transform energy back and forth between these forms. When the springboard diver goes up in the air, his kinetic energy is transformed into gravitational potential energy; when he comes back down, this process is reversed. But the arrow between $K$ and $E_{\text{th}}$ is a one-way arrow pointing toward $E_{\text{th}}$. When the runner slides into the base, his kinetic energy is transformed into thermal energy. This process doesn't spontaneously reverse, although this would certainly make baseball a more exciting game. In Chapter 11, we'll see that it is possible to transform thermal energy into other forms, but it's not easy, and there are real limitations.

A video to support a section's topic is embedded in the eText.

**Video** The Basic Energy Model

## Energy Transfers and Work

We've just seen that energy *transformations* occur between forms of energy *within* a system. But every physical system also interacts with the world around it—that is, with its *environment.* In the course of these interactions, the system can exchange energy with the environment. **An exchange of energy between system and environment is called an energy *transfer.*** There are two primary energy-transfer processes: **work,** the *mechanical* transfer of energy to or from a system by pushing or pulling on it, and **heat,** the *nonmechanical* transfer of energy from the environment to the system (or vice versa) because of a temperature difference between the two.

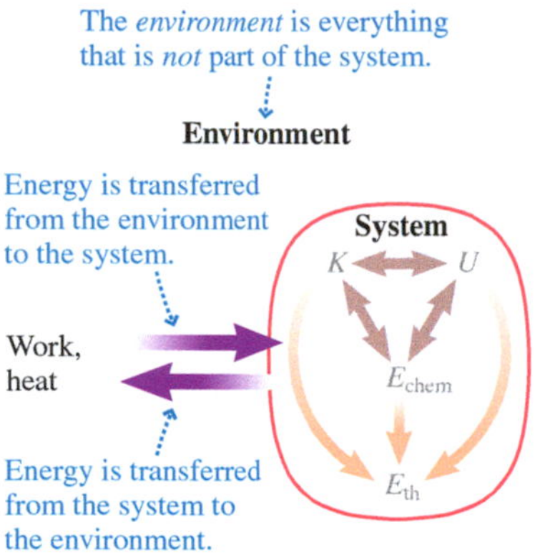


FIGURE 10.3 Work and heat are energy transfers into and out of the system.

FIGURE 10.3, which we call the **basic energy model,** shows how our energy model is modified to include energy transfers into and out of the system as well as energy transformations within the system. In this chapter we'll consider energy transfers by means of work; the concept of heat will be developed in Chapters 11 and 12.

"Work" is a common word in the English language, with many meanings. When you first think of work, you probably think of physical effort or the job you do to make a living. In physics, "work" is the process of *transferring* energy from the environment to a system, or from a system to the environment, by the application of mechanical forces—pushes and pulls—to the system. Once the energy has been transferred to the system, it can appear in many forms. Exactly what form it takes depends on the details of the system and how the forces are applied. The table below gives three examples of energy transfers due to work. We use $W$ as the symbol for work.

**Energy transfers: work**

**Putting a shot**

**The system:** The shot

**The environment:** The athlete

As the athlete pushes on the shot to get it moving, he is doing work on the system; that is, he is transferring energy from himself to the shot. The energy transferred to the system appears as kinetic energy.

**The transfer:** $W \rightarrow K$

**Striking a match**

**The system:** The match and matchbox

**The environment:** The hand

As the hand quickly pulls the match across the box, the hand does work on the system, increasing its thermal energy. The match head becomes hot enough to ignite.

**The transfer:** $W \rightarrow E_{\text{th}}$

**Firing a slingshot**

**The system:** The slingshot

**The environment:** The boy

As the boy pulls back on the elastic bands, he does work on the system, increasing its elastic potential energy.

**The transfer:** $W \rightarrow U_s$

Notice that in each example on the preceding page, the environment applies a force while the system undergoes a *displacement*. Energy is transferred as work only when the system *moves* while the force acts. A force applied to a stationary object, such as when you push against a wall, transfers no energy to the object and thus does no work.

**NOTE** ▸ In the table on the preceding page, energy is being transferred *from* the athlete *to* the shot by the force of his hand. We say he "does work" on the shot. We speak similarly for the other examples. The hand does work on the match and matchbox, and the boy does work on the slingshot. ◂

## The Law of Conservation of Energy

Work done on a system represents energy that is transferred into or out of the system. This transferred energy *changes* the system's energy by exactly the amount of work $W$ that was done. Writing the change in the system's energy as $\Delta E$, we can represent this idea mathematically as

$$\Delta E = W \tag{10.2}$$

Now the total energy $E$ of a system is, according to Equation 10.1, the sum of the different energies present in the system. Thus the change in $E$ is the sum of the *changes* in the different energies present. Then Equation 10.2 gives what is called the *work-energy equation:*

**The work-energy equation** The total energy of a system changes by the amount of work done on it:

$$\Delta E = \Delta K + \Delta U_g + \Delta U_s + \Delta E_{th} + \Delta E_{chem} + \cdots = W \tag{10.3}$$

**NOTE** ▸ Equation 10.3, the work-energy equation, is the mathematical representation of the basic energy model of Figure 10.3. Together, they are the heart of what the subject of energy is all about. ◂

Suppose we have an **isolated system,** one that is separated from its surrounding environment in such a way that no energy is transferred into or out of the system. This means that *no work is done on the system.* The energy within the system may be transformed from one form into another, but it is a deep and remarkable fact of nature that, during these transformations, the total energy of an isolated system—the *sum* of all the individual kinds of energy—remains *constant,* as shown in FIGURE 10.4. We say that **the total energy of an isolated system is *conserved.***

For an isolated system, we must set $W = 0$ in Equation 10.3, leading to the following statement of the *law of conservation of energy:*

**Law of conservation of energy** The total energy of an isolated system remains constant:

$$\Delta E = \Delta K + \Delta U_g + \Delta U_s + \Delta E_{th} + \Delta E_{chem} + \cdots = 0 \tag{10.4}$$

FIGURE 10.4 An isolated system.

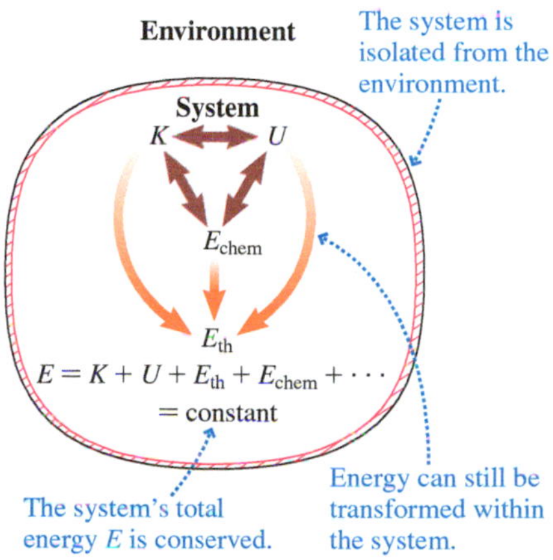


The law of conservation of energy is similar to the law of conservation of momentum. A system's momentum changes when an external force acts on it, but the total momentum of an *isolated* system doesn't change. Similarly, a system's energy changes when external forces do work on it, but the total energy of an *isolated* system doesn't change.

In solving momentum problems, we adopted a new before-and-after perspective: The momentum *after* an interaction was the same as the momentum *before* the interaction. We will introduce a similar before-and-after perspective for energy that will lead to an extremely powerful problem-solving strategy.

**STOP TO THINK 10.1** A roller coaster slows as it goes up a hill. The energy transformation is

A. $U_g \to K$  B. $U_g \to E_{th}$  C. $K \to U_g$  D. $K \to E_{th}$

## 10.2 Work

We've seen that work is the transfer of energy to or from a system by the application of forces exerted on the system by the environment. Thus work is done on a system by forces *outside* the system; we call such forces *external forces*. Only external forces can change the energy of a system. *Internal forces*—forces between objects *within* the system—cause energy transformations within the system but don't change the system's total energy. In order for energy to be transferred as work, the system must undergo a displacement—it must *move*—during the time that the force is applied.

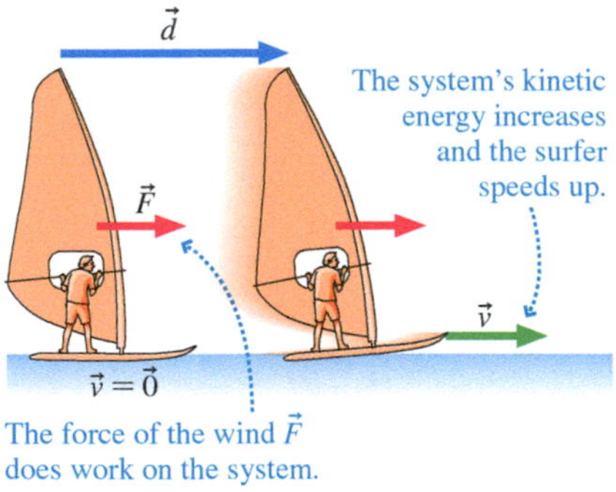


FIGURE 10.5 The force of the wind does work on the system, increasing its kinetic energy $K$.

Consider a system consisting of a windsurfer at rest, as shown on the left in FIGURE 10.5. Let's assume that there is no friction or drag force acting on the board. Initially the system has no kinetic energy. But if a force from outside the system, such as the force due to the wind, begins to act on the system, the surfer will begin to speed up, and his kinetic energy will increase. In terms of energy transfers, we would say that the energy of the system has increased because of the work done on the system by the force of the wind.

What determines how much work is done by the force of the wind? First, we note that the greater the distance over which the wind pushes the surfer, the faster the surfer goes, and the more his kinetic energy increases. This implies a greater transfer of energy. So, **the larger the displacement, the greater the work done.** Second, if the wind pushes with a stronger force, the surfer speeds up more rapidly, and the change in his kinetic energy is greater than with a weaker force. **The stronger the force, the greater the work done.**

This suggests that the amount of energy transferred to a system by a force $\vec{F}$—that is, the amount of work done by $\vec{F}$—depends on both the magnitude $F$ of the force *and* the displacement $d$ of the system. Many experiments of this kind have established that the amount of work done by $\vec{F}$ is *proportional* to both $F$ and $d$. For the simplest case described above, where the force $\vec{F}$ is constant and points in the direction of the object's displacement, the expression for the work done is found to be

$$W = Fd \tag{10.5}$$

Work done by a constant force $\vec{F}$ in the direction of a displacement $\vec{d}$

The unit of work, that of force multiplied by distance, is N · m. This unit is so important that it has been given its own name, the **joule** (rhymes with *cool*). We define:

$$1 \text{ joule} = 1 \text{ J} = 1 \text{ N} \cdot \text{m}$$

Because work is simply energy being transferred, **the joule is the unit of *all* forms of energy.** Note that work, unlike momentum, is a *scalar* quantity—it has a magnitude but not a direction.

## EXAMPLE 10.1 Working like a dog

A dog in a weight-pulling competition tugs a sled 4.9 m across a snowy track at a constant speed. The force needed to keep the sled moving is 350 N. How much work does the dog do? Where does this energy go?

**STRATEGIZE** Let's take the system to be the sled + snow. The friction force between the runners and the snow is thus an internal force, so it won't change the total energy of the system, just the form of the energy. But the rope extends outside the system; this is an external interaction, so the tension force of the rope does work on the sled as it moves. Since the dog pulls on the end of the rope, we can say, informally, that the dog does work on the system.

**PREPARE** We'll continue, as we did with momentum problems in Chapter 9, with a before-and-after visual overview, shown in FIGURE 10.6. The tension force is in the direction of the sled's motion, so we can use Equation 10.5 to calculate the work that the dog does on the sled.

**SOLVE** The work done is

$$W = Fd = (350\text{ N})(4.9\text{ m}) = 1700\text{ J}$$

The dog does work on the system, but the kinetic energy doesn't increase (the sled doesn't speed up) and the gravitational potential energy doesn't increase (the track is level). The energy the dog puts into the system goes to increasing the system's thermal energy as friction warms up the runners and the snow.

**ASSESS** 1700 J is a decent amount of energy, as we'll see, but pulling with a 350 N force (about 80 pounds) for a distance of 4.9 m (about 16 feet) sounds like a lot of work, so our result makes sense.

FIGURE 10.6 A dog pulling a loaded sled.

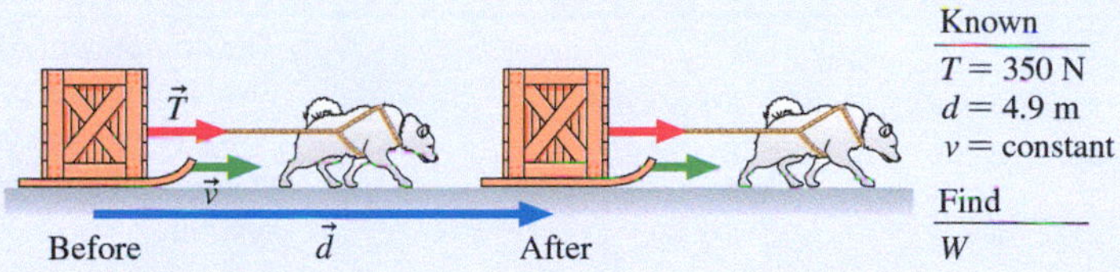


## Work by Forces at an Angle to the Displacement

A force does the greatest possible amount of work on an object when the force points in the same direction as the object's displacement. Less work is done when the force acts at an angle to the displacement. To see this, consider the kite buggy of FIGURE 10.7a, pulled along a horizontal path by the angled force of the kite string $\vec{F}$. As shown in FIGURE 10.7b, we can divide $\vec{F}$ into a component $F_\perp$ perpendicular to the motion, and a component $F_\parallel$ parallel to the motion. Only the parallel component acts to accelerate the rider and increase her kinetic energy, so only the parallel component does work on the rider. From Figure 10.7b, we see that if the angle between $\vec{F}$ and the displacement is $\theta$, then the parallel component is $F_\parallel = F\cos\theta$. So, when the force acts at an angle $\theta$ to the direction of the displacement, we have

$$W = F_\parallel d = Fd\cos\theta \qquad (10.6)$$

Work done by a constant force $\vec{F}$ at an angle $\theta$ to the displacement $\vec{d}$

FIGURE 10.7 The force on the kite buggy is at an angle to the displacement.

(a)

(b)

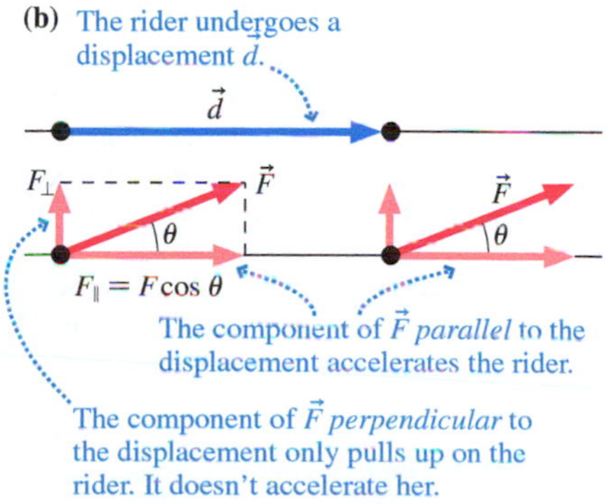


Notice that this more general definition of work agrees with Equation 10.5 if $\theta = 0°$.

Tactics Box 10.1 shows how to calculate the work done by a force at any angle to the direction of motion. The system illustrated is a block sliding on a frictionless, horizontal surface, so that only the kinetic energy is changing. However, the same relationships hold for any object undergoing a displacement.

The quantities $F$ and $d$ are always positive, so **the sign of $W$ is determined entirely by the angle $\theta$ between the force and the displacement.** Note that Equation 10.6, $W = Fd\cos\theta$, is valid for any angle $\theta$. In three special cases, $\theta = 0°$, $\theta = 90°$, and $\theta = 180°$, however, there are simple versions of Equation 10.6 that you can use. These are noted in Tactics Box 10.1.

## TACTICS BOX 10.1 Calculating the work done by a constant force

| Direction of force relative to displacement | Angles and work done | Sign of $W$ | Energy transfer |
|---|---|---|---|
| 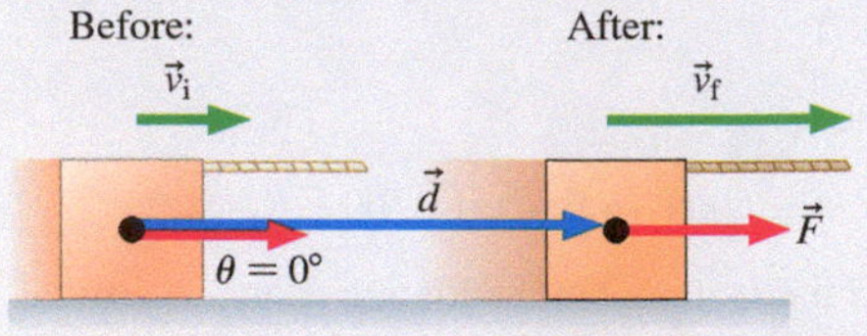 | $\theta = 0°$<br>$\cos\theta = 1$<br>$W = Fd$ | + | The force is in the direction of motion. The block has its greatest positive acceleration. $K$ increases the most:<br>**Maximum energy transfer *into* the system.** |
| 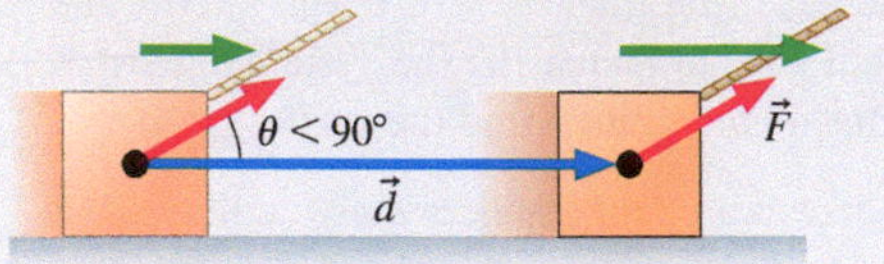 | $\theta < 90°$<br>$W = Fd\cos\theta$ | + | The component of force parallel to the displacement is less than $F$. The block has a smaller positive acceleration. $K$ increases less:<br>**Decreased energy transfer *into* the system.** |
| 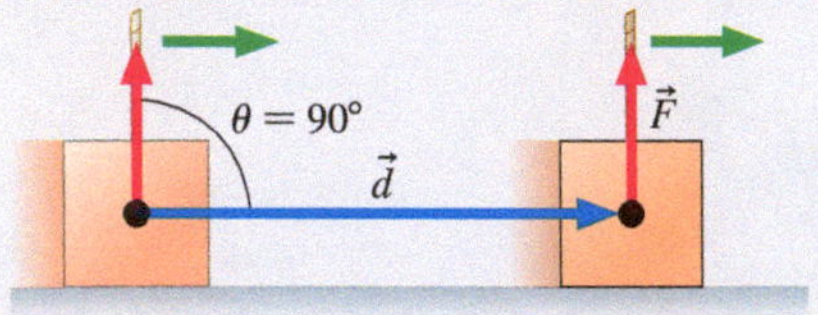 | $\theta = 90°$<br>$\cos\theta = 0$<br>$W = 0$ | 0 | There is no component of force in the direction of motion. The block moves at constant speed. No change in $K$:<br>**No energy transferred.** |
| 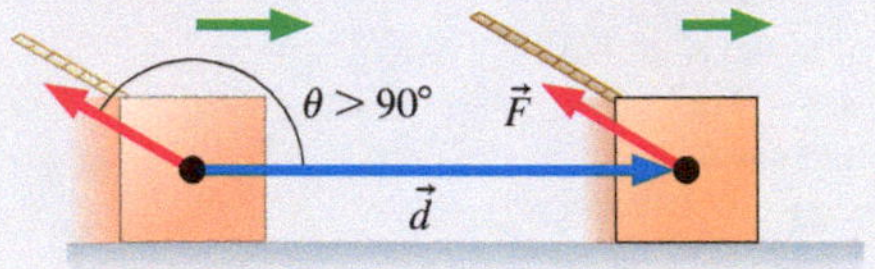 | $\theta > 90°$<br>$W = Fd\cos\theta$ | − | The component of force parallel to the displacement is opposite the motion. The block slows down, and $K$ decreases:<br>**Decreased energy transfer *out of* the system.** |
| 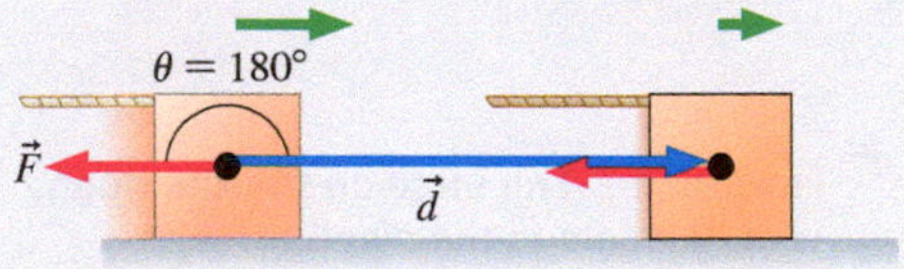 | $\theta = 180°$<br>$\cos\theta = -1$<br>$W = -Fd$ | − | The force is directly opposite the motion. The block has its greatest deceleration. $K$ decreases the most:<br>**Maximum energy transfer *out of* the system.** |

Exercises 5–6

### EXAMPLE 10.2 Work done in pulling a suitcase

It's 120 m from one gate to another in the airport. You use a strap inclined upward at a 45° angle to pull your suitcase through the airport. The tension in the strap is 20 N. How much work do you do?

**STRATEGIZE** Let's take the system to be the suitcase + floor. As with the dog sled, friction forces (in the wheels or between the wheels and the floor) are internal forces. Both the strap and you are forces outside the system. The tension force of the strap does work on the suitcase as it rolls. Since you are the one pulling the strap, this is, ultimately, energy provided by you.

**PREPARE** FIGURE 10.8 is a before-and-after visual overview showing the suitcase and the strap. The force is at an angle to the displacement, so we must use Equation 10.6 to calculate the work.

**SOLVE** The tension force does work

$$W = Td\cos\theta = (20\text{ N})(120\text{ m})\cos(45°) = 1700\text{ J}$$

FIGURE 10.8 A suitcase pulled by a strap.

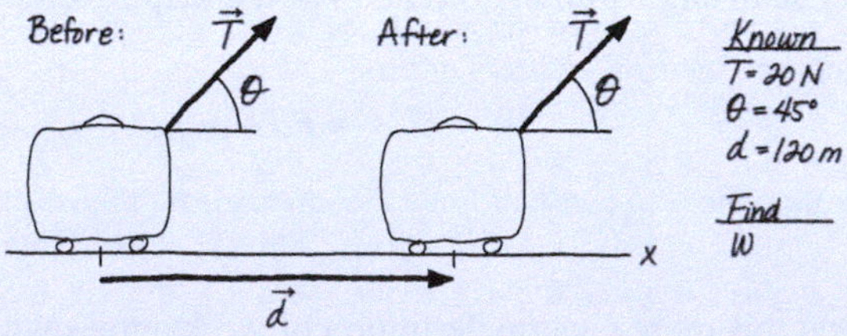


**ASSESS** This is the same amount of work that the dog did pulling the sled. The force is much less, but the distance is much greater, so this result makes sense.

**CONCEPTUAL EXAMPLE 10.3** **Work done by a parachute**

A drag racer is slowed by a parachute. What is the sign of the work done?

***REASON*** The drag force on the drag racer is shown in FIGURE 10.9, along with the dragster's displacement as it slows. The force points in the direction opposite the displacement, so the angle $\theta$ in Equation 10.6 is 180°. Then $\cos\theta = \cos(180°) = -1$. Because $F$ and $d$ in Equation 10.6 are magnitudes, and hence positive, the work $W = Fd\cos\theta = -Fd$ done by the drag force is *negative*.

FIGURE 10.9 The force acting on a drag racer.

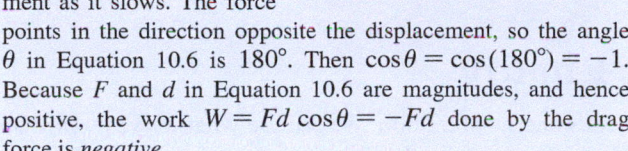

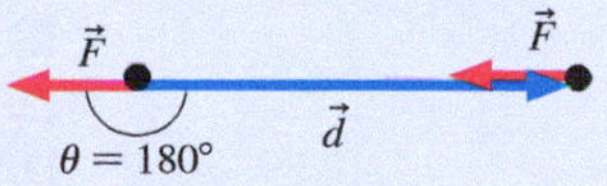


***ASSESS*** Applying Equation 10.3 to this situation, we have

$$\Delta K = W$$

because the only system energy that changes is the racer's kinetic energy $K$. Because the kinetic energy is decreasing, its change $\Delta K$ is negative. This agrees with the sign of $W$. This example illustrates the general principle that **negative work represents a transfer of energy out of the system.**

If several forces act on an object that undergoes a displacement, each does work on the object. The **total** (or **net**) **work** $W_{total}$ is the sum of the work done by each force. The total work represents the total energy transfer *to* the system from the environment (if $W_{total} > 0$) or *from* the system to the environment (if $W_{total} < 0$).

## Forces That Do No Work

The fact that a force acts on an object doesn't mean that the force will do work on the object. The table below shows three common cases where a force does no work.

**Forces that do no work**

**If the object undergoes no displacement while the force acts, no work is done.**

This can sometimes seem counterintuitive. The weightlifter struggles mightily to hold the barbell over his head. But during the time the barbell remains stationary, he does no work on it because its displacement is zero. Why then is it so hard for him to hold it there? Your muscles use energy to apply a force even if there is no displacement and thus no work. We'll talk about the energy that you use to perform a task in Chapter 11.

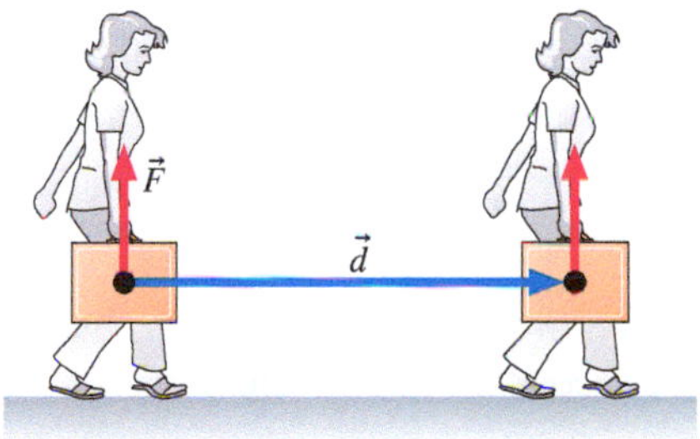


**A force perpendicular to the displacement does no work.**

The woman exerts only a vertical force on the briefcase she's carrying. This force has no component in the direction of the displacement, so the briefcase moves at a constant velocity and its kinetic energy remains constant. Since the energy of the briefcase doesn't change, it must be that no energy is being transferred to it as work.

(This is the case where $\theta = 90°$ in Tactics Box 10.1.)

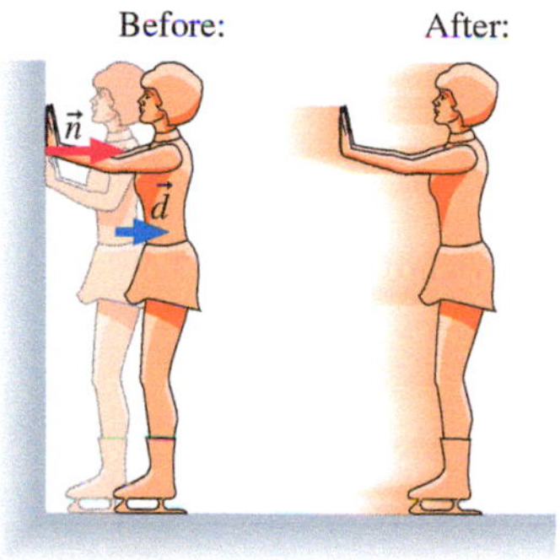


**If the part of the object on which the force acts undergoes no displacement, no work is done.**

Even though the wall pushes on the skater with a normal force $\vec{n}$ and she undergoes a displacement $\vec{d}$, the wall does no work on her, because the point of her body on which $\vec{n}$ acts—her hands—undergoes no displacement. This makes sense: How could energy be transferred as work from an inert, stationary object? The energy to get the skater moving comes, as you know, from her muscles. This is an internal transformation; chemical energy in her muscles is converted to kinetic energy of her motion.

**STOP TO THINK 10.2** Which force does the most work?

A. The 10 N force
B. The 8 N force
C. The 6 N force
D. They all do the same amount of work.

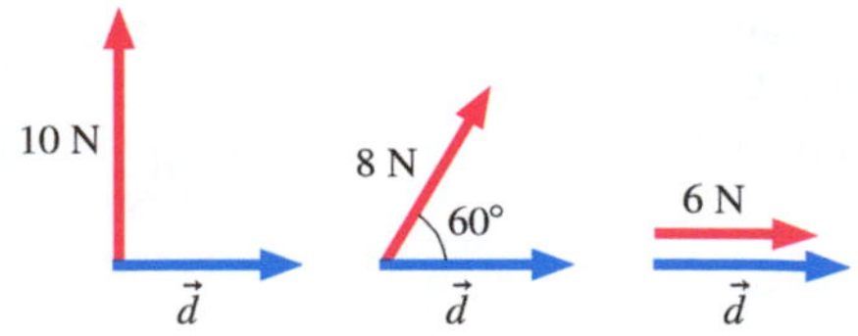


## 10.3 Kinetic Energy

Kinetic energy is an object's energy of motion. We can use what we've learned about work, and some simple kinematics, to find quantitative expressions for kinetic energy.

We'll start with the case of an object in motion along a line. Such an object has **translational kinetic energy.** In Chapter 7, we introduced the idea of rotational motion: Objects can be in motion even if they aren't going anywhere. An object, like the blade of a wind turbine, rotating about a fixed axis has **rotational kinetic energy,** the kinetic energy of the rotational motion.

### Translational Kinetic Energy

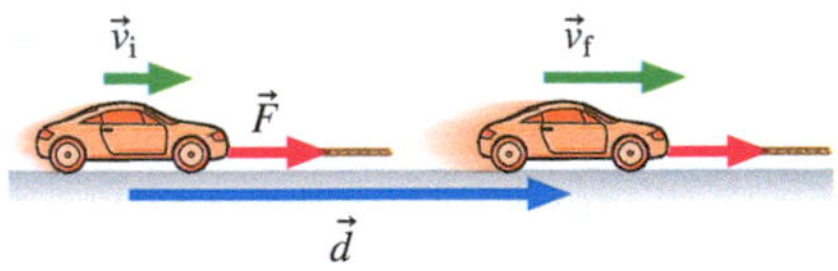


FIGURE 10.10 The work done by the tow rope increases the car's kinetic energy.

Consider a car being pulled by a tow rope, as in FIGURE 10.10. The rope pulls with a constant force $\vec{F}$ while the car undergoes a displacement $\vec{d}$, so the force does work $W = Fd$ on the car. If we ignore friction and drag, the work done by $\vec{F}$ is transferred entirely into the car's energy of motion—its kinetic energy. In this case, the change in the car's kinetic energy is given by the work-energy equation, Equation 10.3, as

$$W = \Delta K = K_f - K_i \tag{10.7}$$

Using kinematics, we can find another expression for the work done, in terms of the car's initial and final speeds. Recall from ◀SECTION 2.5 the kinematic equation

$$v_f^2 = v_i^2 + 2a\,\Delta x$$

Applied to the motion of our car, $\Delta x = d$ is the car's displacement and, from Newton's second law, the acceleration is $a = F/m$. Thus we can write

$$v_f^2 = v_i^2 + \frac{2Fd}{m} = v_i^2 + \frac{2W}{m}$$

where we have replaced $Fd$ with the work $W$. If we now solve for the work, we find

$$W = \frac{1}{2}m(v_f^2 - v_i^2) = \frac{1}{2}mv_f^2 - \frac{1}{2}mv_i^2$$

If we compare this result with Equation 10.7, we see that

$$K_f = \frac{1}{2}mv_f^2 \quad \text{and} \quad K_i = \frac{1}{2}mv_i^2$$

In general, then, an object of mass $m$ moving with speed $v$ has kinetic energy

$$K = \frac{1}{2}mv^2 \qquad (10.8)$$

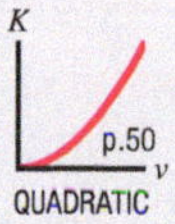


Kinetic energy of an object of mass $m$ moving with speed $v$

From Equation 10.8, the units of kinetic energy are those of mass times speed squared, or $\text{kg} \cdot (\text{m/s})^2$. But

$$1\ \text{kg} \cdot (\text{m/s})^2 = \underbrace{1\ \text{kg} \cdot (\text{m/s}^2)}_{1\ \text{N}} \cdot \text{m} = 1\ \text{N} \cdot \text{m} = 1\ \text{J}$$

We see that the units of kinetic energy are the same as those of work, as they must be. **TABLE 10.1** gives some approximate kinetic energies.

**TABLE 10.1** Some approximate kinetic energies

| Object | Kinetic energy |
|---|---|
| Ant walking | $1 \times 10^{-8}$ J |
| Coin dropped 1 m | $5 \times 10^{-3}$ J |
| Person walking | 70 J |
| Fastball, 100 mph | 150 J |
| Bullet | 5000 J |
| Car, 60 mph | $5 \times 10^{5}$ J |
| Supertanker, 20 mph | $2 \times 10^{10}$ J |

## EXAMPLE 10.4 Finding the work to set a boat in motion

At a history center, an old canal boat is pulled by two draft horses. It doesn't take much force to keep the boat moving; the drag force is quite small. But it takes some work to get the 55,000 kg boat up to speed! The horses can pull with a steady force and put a 1400 N tension in the rope that connects to the boat. The rope is straight and level. The boat starts from rest, and the horses pull steadily as they begin their walk down the tow-path. How much distance do the horses cover as they bring the boat up to its final speed of 0.70 m/s?

**STRATEGIZE** Let's take the system to be the boat. We could include the water, but since we can ignore the drag force (we're told that it's small), it's not important to do so. The rope is not part of the system, so the tension force does work on the boat. It's this work, which comes from energy provided by the horses, that increases the kinetic energy, and thus the speed, of the boat. We'll consider the initial state to be the boat at rest, the final state to be the boat in motion at its final speed.

**PREPARE** **FIGURE 10.11** is a before-and-after visual overview of the situation. The work that is done by the rope will change the energy of the system, so we can use Equation 10.3, the work-energy equation. Because the only thing that changes is the speed, the only form of energy that changes is the kinetic energy, so we can simplify the equation to

$$\Delta K = W$$

This makes sense—the work done changes the kinetic energy of the boat. The tension force is in the direction of the motion, so the work done is $W = Td$. The boat starts at rest, with kinetic energy equal to zero, so the change in kinetic energy is just the final kinetic energy: $\Delta K = \frac{1}{2}mv_f^2$.

**SOLVE** With the details noted, the work-energy equation reduces to

$$\frac{1}{2}mv_f^2 = Td$$

We are looking for the distance the horses pull the boat:

$$d = \frac{mv_f^2}{2T} = \frac{(55{,}000\ \text{kg})(0.70\ \text{m/s})^2}{2(1400\ \text{N})} = 9.6\ \text{m}$$

**ASSESS** This distance is about 30 feet. This seems a reasonable distance; the horses would be pulling for several strides as they get the boat up to speed.

**FIGURE 10.11** Getting the canal boat up to speed.

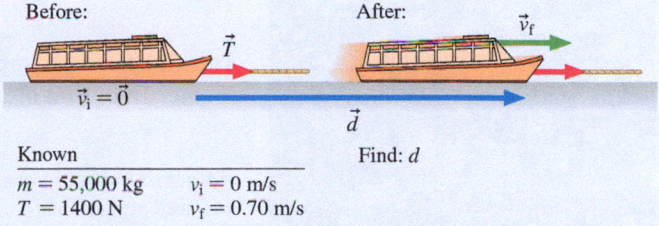

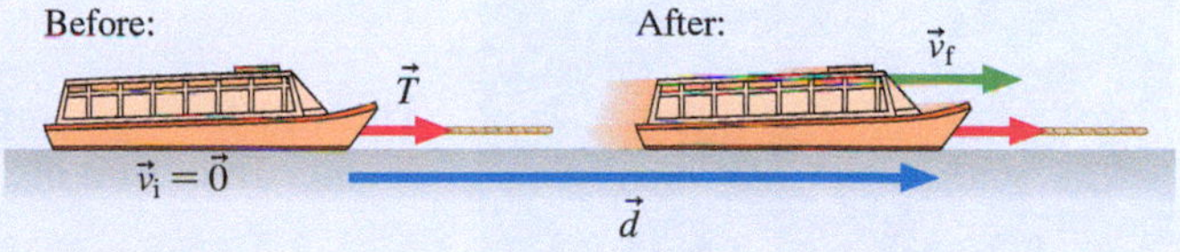


Known
$m = 55{,}000$ kg $\quad v_i = 0$ m/s
$T = 1400$ N $\quad v_f = 0.70$ m/s

Find: $d$

**STOP TO THINK 10.3** Rank in order, from greatest to least, the kinetic energies of the sliding pucks.

FIGURE 10.12 Rotational kinetic energy of a spinning wind turbine.

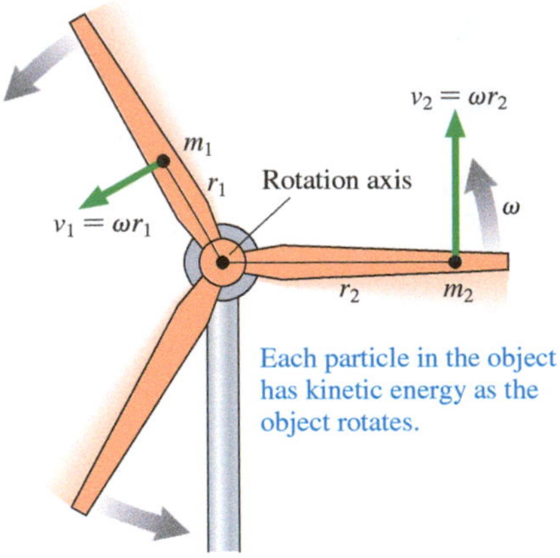


## Rotational Kinetic Energy

FIGURE 10.12 shows the rotating blades of a wind turbine. Although the blades have no overall translational motion, each particle in the blades is moving and hence has kinetic energy. Adding up the kinetic energy of all the particles that make up the blades, we find that the blades have rotational kinetic energy, the kinetic energy due to rotation.

In Figure 10.12, we focus on the motion of two particles in the wind turbine blades. The blade assembly rotates with angular velocity $\omega$. Recall from ◀◀SECTION 7.1 that a particle moving with angular velocity $\omega$ in a circle of radius $r$ has a speed $v = \omega r$. Thus particle 1, which rotates in a circle of radius $r_1$, moves with speed $v_1 = r_1\omega$ and so has kinetic energy $\frac{1}{2}m_1v_1^2 = \frac{1}{2}m_1r_1^2\omega^2$. Similarly, particle 2, which rotates in a circle with a larger radius $r_2$, has kinetic energy $\frac{1}{2}m_2r_2^2\omega^2$. The object's rotational kinetic energy is the sum of the kinetic energies of *all* the particles:

$$K_{\text{rot}} = \frac{1}{2}m_1r_1^2\omega^2 + \frac{1}{2}m_2r_2^2\omega^2 + \cdots = \frac{1}{2}\left(\sum mr^2\right)\omega^2$$

You will recognize the term in parentheses as our old friend, the moment of inertia $I$. Thus the rotational kinetic energy is

$$K_{\text{rot}} = \frac{1}{2}I\omega^2 \tag{10.9}$$

Rotational kinetic energy of an object with moment of inertia $I$ and angular velocity $\omega$

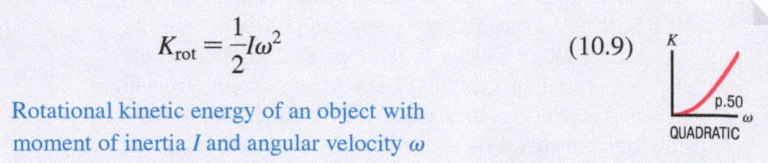

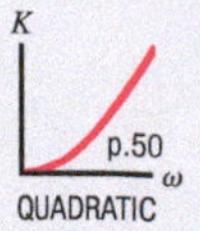


NOTE ▶ Rotational kinetic energy is *not* a new form of energy. It is the ordinary kinetic energy of motion, only now expressed in a form that is especially convenient for rotational motion. Comparison with the familiar $\frac{1}{2}mv^2$ shows again that the moment of inertia $I$ is the rotational equivalent of mass. ◀

A rolling object, such as a wheel, is undergoing both rotational *and* translational motions. Consequently, its total kinetic energy is the sum of its rotational and translational kinetic energies:

$$K = K_{\text{trans}} + K_{\text{rot}} = \frac{1}{2}mv^2 + \frac{1}{2}I\omega^2 \tag{10.10}$$

This illustrates an important fact: **The kinetic energy of a rolling object is always greater than that of a nonrotating object moving at the same speed.**

◀ **Rotational recharge** A promising new technology would replace spacecraft batteries that need periodic and costly replacement with a *flywheel*—a cylinder rotating at a very high angular speed. Energy from solar panels is used to speed up the flywheel, which stores energy as rotational kinetic energy that can then be converted back into electric energy as needed.

### EXAMPLE 10.5 Where should you trim the weight?

Any time a cyclist stops, it will take energy to get moving again. Using less energy to get going means more energy is available to go farther or go faster, so racing cyclists want their bikes to be as light as possible. It's particularly important to have lightweight wheels, as this example will show. Consider two bikes that have the same total mass but different mass wheels. Bike 1 has a 10.0 kg frame and two 1.00 kg wheels; bike 2 has a 9.00 kg frame and two 1.50 kg wheels. Both bikes thus have the same 12.0 kg total mass. What is the kinetic energy of each bike when they are moving at 12.0 m/s? Most of the weight of the tire and wheel is at the rim, so we can model each wheel as a hoop.

**STRATEGIZE** As the bike moves, the wheels rotate. The bike has translational kinetic energy, but the wheels have both translational and rotational kinetic energy. If the bike is moving at speed $v$, we know from Chapter 7 that the wheels rotate at $\omega = v/R$, where $R$ is the radius of a wheel.

**PREPARE** Each bike's frame has only translational kinetic energy $K_{frame} = \frac{1}{2}mv^2$, where $m$ is the mass of the frame. The kinetic energy of each rolling wheel is given by Equation 10.10. From Table 7.1, we find that $I$ for a hoop is $MR^2$, where $M$ is the mass of one wheel.

**SOLVE** From Equation 10.10 the kinetic energy of each rolling wheel is

$$K_{wheel} = \frac{1}{2}Mv^2 + \frac{1}{2}I\omega^2 = \frac{1}{2}Mv^2 + \frac{1}{2}\underbrace{(MR^2)}_{I}\underbrace{\left(\frac{v}{R}\right)^2}_{\omega^2} = Mv^2$$

Then the total kinetic energy of a bike is

$$K = K_{frame} + 2K_{wheel} = \frac{1}{2}mv^2 + 2Mv^2$$

The factor of 2 in the second term occurs because each bike has two wheels. Thus the kinetic energies of the two bikes are

$$K_1 = \frac{1}{2}(10.0\text{ kg})(12.0\text{ m/s})^2 + 2(1.00\text{ kg})(12.0\text{ m/s})^2$$
$$= 1010\text{ J}$$
$$K_2 = \frac{1}{2}(9.00\text{ kg})(12.0\text{ m/s})^2 + 2(1.50\text{ kg})(12.0\text{ m/s})^2$$
$$= 1080\text{ J}$$

The kinetic energy of bike 2 is about 7% higher than that of bike 1. Note that the radius of the wheels was not needed in this calculation.

**ASSESS** We were told that it's particularly important for cyclists to have lightweight wheels, so this result makes sense. Both of the bikes in the example have the same total mass, but the one with lighter wheels takes less energy to get moving. Shaving a little extra weight off your bike's wheels is more useful than taking that same weight off the bike's frame.

## 10.4 Potential Energy

When two or more objects in a system interact, it is sometimes possible to *store* energy in the system in a way that the energy can be easily recovered. For instance, the earth and a ball interact by the gravitational force between them. If the ball is lifted up into the air, energy is stored in the ball + earth system, energy that can later be recovered as kinetic energy when the ball is released and falls. Similarly, a spring is a system made up of countless atoms that interact via their atomic "springs." If we push a box against a spring, energy is stored that can be recovered when the spring later pushes the box across the table. This sort of stored energy is called **potential energy,** since it has the *potential* to be converted into other forms of energy, such as kinetic or thermal energy.

NOTE ▸ Potential energy is really a property of a *system*, but we often speak informally of the potential energy of an *object*. We might say, for instance, that raising a ball increases its potential energy. This is fine as long as we remember that this energy is really stored in the ball + earth system. ◂

The forces due to gravity and springs are special in that they allow for the storage of energy. Other interaction forces do not. When a dog pulls a sled, the sled interacts with the ground via the force of friction, and the work that the dog does on the sled is converted into thermal energy. The energy is *not* stored up for later recovery—it slowly diffuses into the environment and cannot be recovered.

FIGURE 10.13 Lifting a book increases the system's gravitational potential energy.

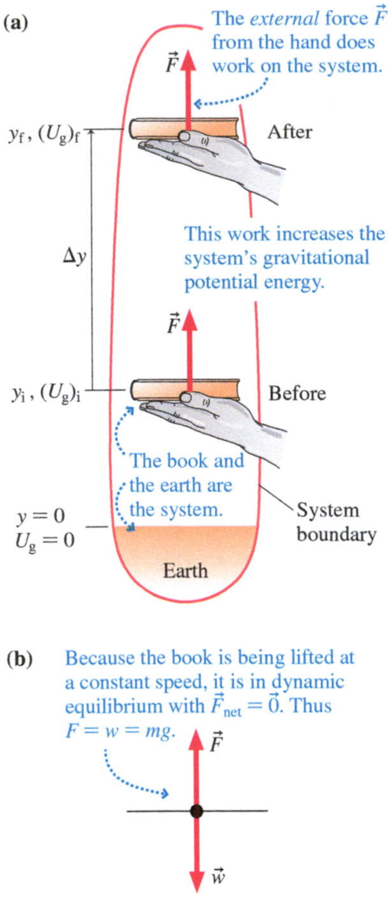


## Gravitational Potential Energy

To find an expression for **gravitational potential energy $U_g$**, let's consider the system of the book and the earth shown in FIGURE 10.13a. The book is lifted at a constant speed from its initial position at $y_i$ to a final height $y_f$. The lifting force of the hand is external to the system and so does work $W$ on the system, increasing its energy. The book is lifted at a constant speed, so its kinetic energy doesn't change. Because there's no friction, the book's thermal energy doesn't change either. Thus the work done goes entirely into increasing the gravitational potential energy of the system. According to Equation 10.3, the work-energy equation, this can be written as $\Delta U_g = W$. Because $\Delta U_g = (U_g)_f - (U_g)_i$, Equation 10.3 can be written

$$(U_g)_f = (U_g)_i + W \tag{10.11}$$

The work done is $W = Fd$, where $d = \Delta y = y_f - y_i$ is the vertical distance that the book is lifted. From the free-body diagram of FIGURE 10.13b, we see that $F = mg$. Thus $W = mg\,\Delta y$, and so

$$(U_g)_f = (U_g)_i + mg\Delta y \tag{10.12}$$

Because our final height was greater than our initial height, $\Delta y$ is positive and $(U_g)_f > (U_g)_i$. **The higher the object is lifted, the greater the gravitational potential energy in the object + earth system.**

We can express Equation 10.12 in terms of the change in potential energy, $\Delta U_g = (U_g)_f - (U_g)_i$:

$$\Delta U_g = mg\Delta y \tag{10.13}$$

If we lift a 1.5 kg book up by $\Delta y = 2.0$ m, we increase the system's gravitational potential energy by $\Delta U_g = (1.5\text{ kg})(9.8\text{ m/s}^2)(2.0\text{ m}) = 29.4$ J. This increase is *independent* of the book's starting height: The gravitational potential energy increases by 29.4 J whether we lift the book 2.0 m starting at sea level or starting at the top of the Washington Monument. This illustrates an important general fact about *every* form of potential energy: **Only *changes* in potential energy are significant.**

Because of this fact, we are free to choose a *reference level* where we define $U_g$ to be zero. Our expression for $U_g$ is particularly simple if we choose this reference level to be at $y = 0$. We then have

$$U_g = mgy \tag{10.14}$$

Gravitational potential energy of an object of mass $m$ at height $y$ (assuming $U_g = 0$ when the object is at $y = 0$)

**EXAMPLE 10.6 Racing up a skyscraper**

In the Empire State Building Run-Up, competitors race up the 1576 steps of the Empire State Building, climbing a total vertical distance of 320 m. How much gravitational potential energy does a 70 kg racer gain during this race?

Racers head up the staircase in the Empire State Building Run-Up.

***STRATEGIZE*** We'll take the system to be the racer + earth so that we can consider gravitational potential energy.

***PREPARE*** We are asked for the change in gravitational potential energy as the racer goes up the stairs, so we need only consider the change in height, which is given. We can use Equation 10.13 to compute the change in potential energy during the run.

***SOLVE*** As the racer goes up the stairs, her change in gravitational potential energy is

$$\Delta U_g = mg\Delta y = (70\text{ kg})(9.8\text{ m/s}^2)(320\text{ m}) = 2.2 \times 10^5\text{ J}$$

***ASSESS*** This is a lot of energy. According to Table 10.1, it's comparable to the energy of a speeding car. But the difference in height is pretty great, so this seems reasonable. In Chapter 11, we'll consider how much food energy you'd need to consume to fuel this climb.

An important conclusion from Equation 10.14 is that gravitational potential energy depends only on the height of the object above the reference level $y = 0$, not on the object's horizontal position. To understand why, consider carrying a briefcase while walking on level ground at a constant speed. As shown in the table on page 317, the vertical force of your hand on the briefcase is *perpendicular* to the displacement. No work is done on the briefcase, so its gravitational potential energy remains constant as long as its height above the ground doesn't change.

This idea can be applied to more complicated cases, such as the 82 kg hiker in FIGURE 10.14. His gravitational potential energy depends *only* on his height $y$ above the reference level. Along path A, it's the same value $U_g = mgy = 80$ kJ at any point where he is at height $y = 100$ m above the reference level. If he had instead taken path B, his gravitational potential energy at $y = 100$ m would be the same 80 kJ. It doesn't matter *how* he gets to the 100 m elevation; his potential energy at that height is always the same. **Gravitational potential energy depends only on the *height* of an object and not on the path the object took to get to that position.** This fact will allow us to use the law of conservation of energy to easily solve a variety of problems that would be very difficult to solve using Newton's laws alone.

FIGURE 10.14 The hiker's gravitational potential energy depends only on his height above the $y = 0$ m reference level.

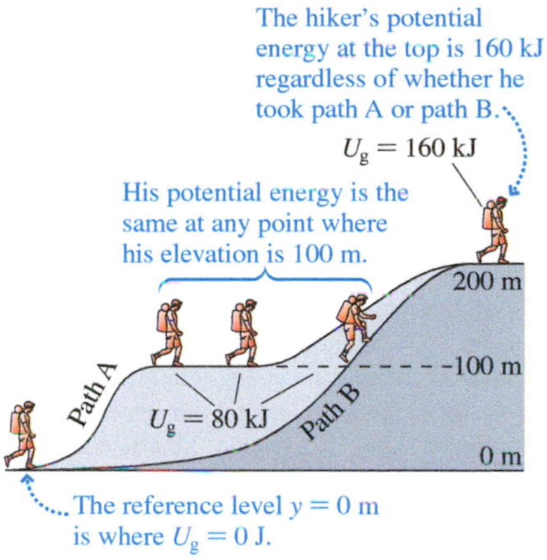


**STOP TO THINK 10.4** Rank in order, from largest to smallest, the gravitational potential energies of identical balls 1 through 4.

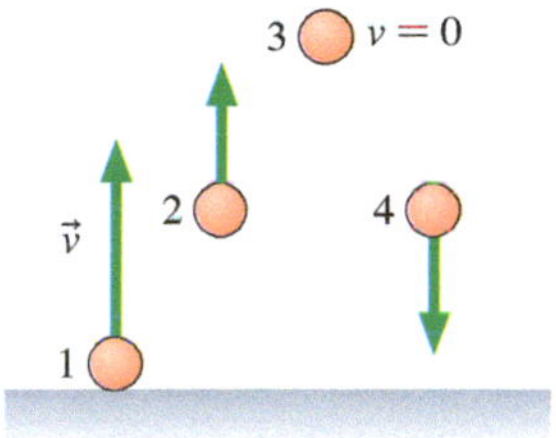


## Elastic Potential Energy

Energy can also be stored in a compressed or extended spring as **elastic** (or **spring**) **potential energy** $U_s$. We can find out how much energy is stored in a spring by using an external force to slowly compress the spring. This external force does work on the spring, transferring energy to the spring. Since only the elastic potential energy of the spring is changing, Equation 10.3 becomes

$$\Delta U_s = W \qquad (10.15)$$

That is, we can find out how much elastic potential energy is stored in the spring by calculating the amount of work needed to compress the spring.

FIGURE 10.15 shows a spring being compressed by a hand. In ◀◀ SECTION 8.3 we found that the force the spring exerts on the hand is $F_s = -k\,\Delta x$ (Hooke's law), where $\Delta x$ is the displacement of the end of the spring from its equilibrium position and $k$ is the spring constant. In Figure 10.15 we have set the origin of our coordinate system at the equilibrium position. The displacement from equilibrium $\Delta x$ is therefore equal to $x$, and the spring force is then $-kx$. By Newton's third law, the force that the hand exerts on the spring is thus $F = +kx$.

FIGURE 10.15 The force required to compress a spring is not constant.

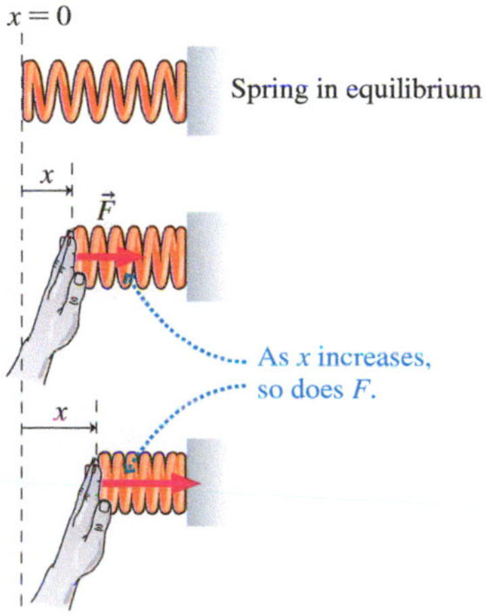


As the hand pushes the end of the spring from its equilibrium position to a final position $x$, the applied force increases from 0 to $kx$. This is not a constant force, so we can't use Equation 10.5, $W = Fd$, to find the work done. However, it seems reasonable to calculate the work by using the *average* force in Equation 10.5. Because the force varies from $F_i = 0$ to $F_f = kx$, the average force used to compress the spring is $F_{avg} = \frac{1}{2}kx$. Thus the work done by the hand is

$$W = F_{avg}d = F_{avg}x = \left(\frac{1}{2}kx\right)x = \frac{1}{2}kx^2$$

This work is stored as potential energy in the spring, so we can use Equation 10.15 to find that as the spring is compressed, the elastic potential energy increases by

$$\Delta U_s = \frac{1}{2}kx^2$$

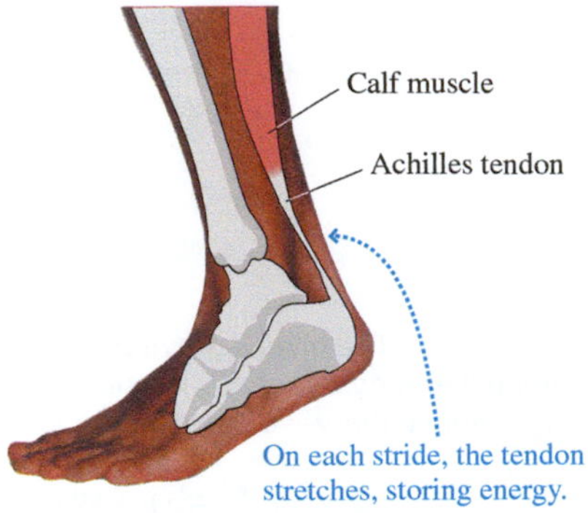


**Spring in your step** BIO When you run, your feet repeatedly stop and start; when your foot strikes the ground, it comes to rest, losing kinetic energy. About 35% of this decrease in kinetic energy is stored as elastic potential energy in the stretchable Achilles tendon of the lower leg. On each plant of the foot, the tendon is stretched, storing some energy. The tendon springs back as you push off the ground again, helping to propel you forward and returning the stored elastic potential energy back to kinetic energy, recovering this energy that would otherwise be lost and thus increasing your efficiency.

Just as in the case of gravitational potential energy, we have found an expression for the *change* in $U_s$, not $U_s$ itself. Again, we are free to set $U_s = 0$ at any convenient spring extension. An obvious choice is to set $U_s = 0$ at the point where the spring is in equilibrium, neither compressed nor stretched—that is, at $x = 0$. With this choice we have

$$U_s = \frac{1}{2}kx^2 \tag{10.16}$$

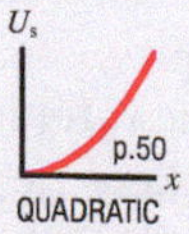


Elastic potential energy of a spring displaced a distance $x$ from equilibrium (assuming $U_s = 0$ when the end of the spring is at $x = 0$)

**NOTE** ▶ Because $U_s$ depends on the *square* of the displacement $x$, $U_s$ is the same whether $x$ is positive (the spring is compressed as in Figure 10.15) or negative (the spring is stretched). ◀

**EXAMPLE 10.7 Finding the energy stored in a stretched tendon** BIO

We noted that your Achilles tendon stretches when you run, and this stores some energy—energy that is returned to you when you push off the ground with your foot. **FIGURE 10.16** shows smoothed data for restoring force versus extension for the Achilles tendon and attached muscle in a female subject. When she runs, at one point in her stride the stretch reaches a maximum of 0.50 cm. What energy is stored for this stretch?

**FIGURE 10.16** Force data for the stretch of the Achilles tendon.

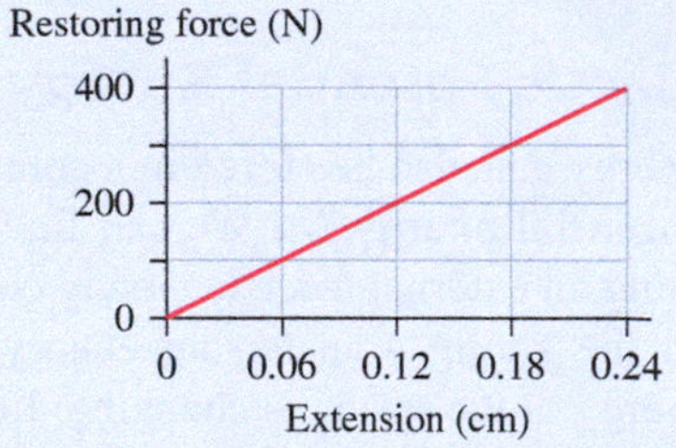


***STRATEGIZE*** The force increases linearly with extension, so we can model the tendon as a spring. If we find the spring constant, we can compute the stored energy using Equation 10.16.

***PREPARE*** The spring constant $k$ is the slope of the graph in Figure 10.16. At the top right, the line goes through a point that is easy to read off the axes. Using this point, we determine the slope to be

$$k = \text{slope} = \frac{400\text{ N}}{0.0024\text{ m}} = 1.67 \times 10^5\text{ N/m}$$

***SOLVE*** With this spring constant, the energy stored for a 0.50 cm (0.0050 m) stretch is

$$U_s = \frac{1}{2}kx^2 = \frac{1}{2}(1.67 \times 10^5\text{ N/m})(0.0050\text{ m})^2 = 2.1\text{ J}$$

***ASSESS*** Table 10.1 gives 70 J as the kinetic energy for a person walking; 2.1 J is a few percent of that value. This is reasonable: When you run, the only part of your body that stops and starts is your feet; it's a fraction of this energy that is recovered, and we expect this to be a small fraction of the kinetic energy of your body. However, saving even this small amount of energy is useful; over a long run of many steps, it will add up!

**STOP TO THINK 10.5** When a spring is stretched by 5 cm, its elastic potential energy is 1 J. What will its elastic potential energy be if it is *compressed* by 10 cm?

A. −4 J B. −2 J C. 2 J D. 4 J

## 10.5 Thermal Energy

We noted earlier that thermal energy is related to the microscopic motion of the atoms of an object. As FIGURE 10.17 shows, the atoms in a hot object jiggle around their average positions more than the atoms in a cold object. This has two consequences. First, each atom is on average moving faster in the hot object. This means that each atom has a higher *kinetic energy*. Second, each atom in the hot object tends to stray farther from its equilibrium position, leading to a greater stretching or compressing of the spring-like molecular bonds. This means that each atom has on average a higher *potential energy*. The potential energy stored in any one bond and the kinetic energy of any one atom are both exceedingly small, but there are incredibly many bonds and atoms. The sum of all these microscopic potential and kinetic energies is what we call **thermal energy $E_{th}$**. Increasing an object's thermal energy corresponds to increasing its temperature.

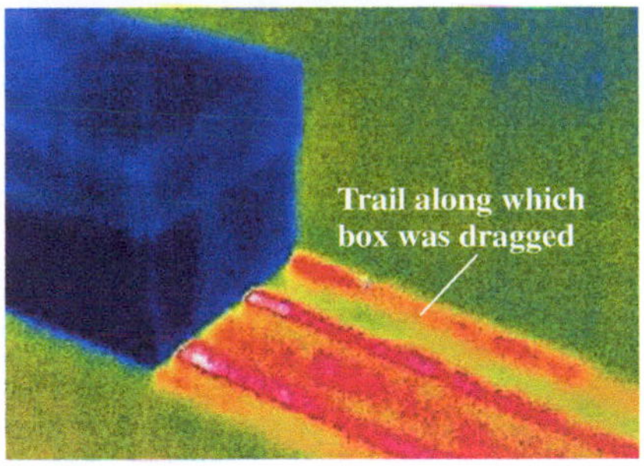


FIGURE 10.18 A thermograph of a box that's been dragged across the floor.

FIGURE 10.17 An atomic view of thermal energy.

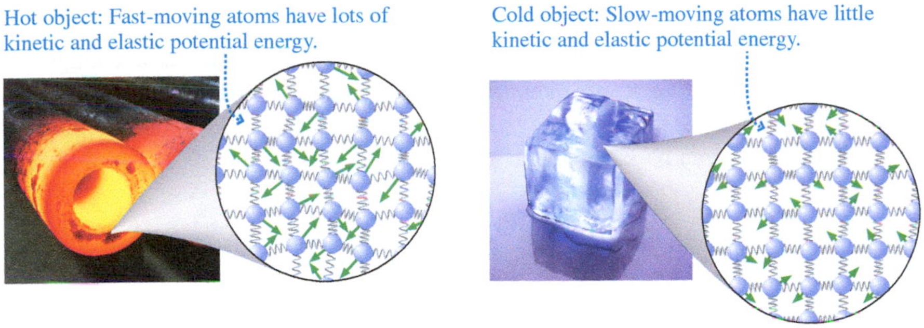

Hot object: Fast-moving atoms have lots of kinetic and elastic potential energy.
Cold object: Slow-moving atoms have little kinetic and elastic potential energy.

FIGURE 10.18 shows a thermograph of a heavy box and the floor across which it has just been dragged. In this image, cool areas appear in shades of blue and green, warm areas in shades of red. You can clearly see that the patch of floor that the box has been dragged across is much warmer than the box or the rest of the floor. Dragging the box across the floor caused the thermal energy of the system to increase.

This increase in thermal energy is a general feature of any system in which there is friction between sliding objects. An atomic-level explanation is illustrated in FIGURE 10.19. The interaction between the surfaces that leads to the force of friction also leads to increased thermal energy of the sliding surfaces.

We can find a quantitative expression for the change in thermal energy by considering the case of the box pulled by a rope at a constant speed. Let's consider the system to be the box + floor. As the box is pulled across the floor, the rope exerts a constant forward force $\vec{F}$ on the box, while the friction force $\vec{f}_k$ exerts a constant force on the box that is directed backward. Because the box moves at a constant speed, the magnitudes of these two forces are equal: $F = f_k$. As the box moves through a displacement $d = \Delta x$, the rope does work $W = F\Delta x = f_k \Delta x$ on the box. This work represents energy transferred into the system, so the system's energy must increase. The box's kinetic energy and gravitational potential energy don't change, so the increased energy must be in the form of thermal energy $E_{th}$. Because all of the work into the system shows up as increased thermal energy, we can say that

$$\Delta E_{th} = f_k \Delta x \tag{10.17}$$

The increased thermal energy is distributed between the two surfaces, the box and the floor, the two elements of the system. Although we arrived at Equation 10.17 by considering energy transferred into the system via work done by an external force, the equation is equally valid for the transformation of energy into thermal energy when, for instance, an object slides to a halt on a rough surface.

FIGURE 10.19 How friction causes an increase in thermal energy.

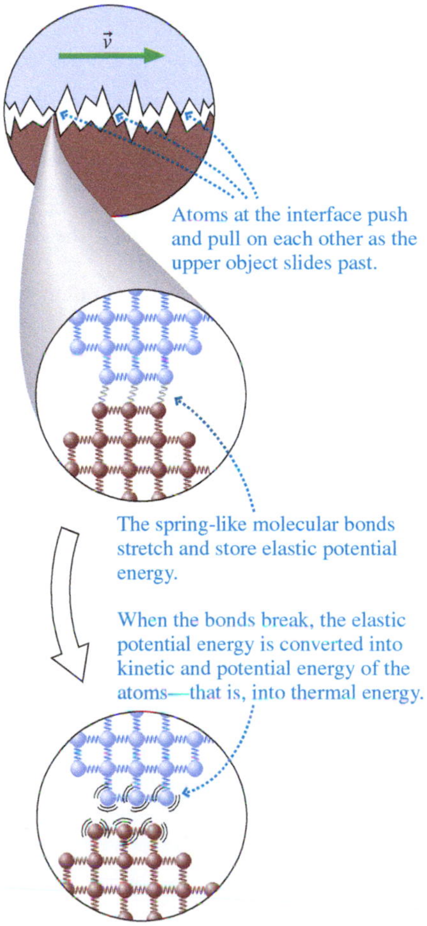


Video Figure 10.19

If an object moves through air, water, or another fluid at a constant speed, there is a constant drag force opposite the motion, so we can do an analysis similar to that for the friction force. If we consider the object + fluid to be the system, as the object moves at a constant speed, the drag force $D$ transforms energy into thermal energy:

$$\Delta E_{\text{th}} = D\Delta x \tag{10.18}$$

The change in thermal energy is mostly a change in the air or the water: Collisions between the object and particles of the fluid cause the particles to move more quickly, thus increasing thermal energy.

The work-energy equation, Equation 10.3, states that the change in the total energy of a system equals the energy transferred to or from the system as work. If we consider only those forms of energy that are typically transformed during the motion of ordinary objects—kinetic energy $K$, gravitational and elastic potential energies $U_g$ and $U_s$, and thermal energy $E_{\text{th}}$—then the work-energy equation can be written as

$$\Delta K + \Delta U + \Delta E_{\text{th}} = W \tag{10.19}$$

**NOTE** ▸ We've written the change in potential energy as a single potential energy change $\Delta U$. Depending on the situation, we can interpret this as a change in gravitational potential energy, a change in elastic potential energy, or a combination of the two. In later chapters, we'll add additional forms of potential energy, and this equation can be adapted accordingly. ◂

**EXAMPLE 10.8 How much energy does it take to swim a kilometer?**

How much energy is required for a 70 kg swimmer to complete a 1.0 km swim at a steady 1.4 m/s? We can assume typical data for a swimmer moving through the water: frontal area 0.080 m$^2$, drag coefficient 0.45, density of water 1000 kg/m$^3$.

***STRATEGIZE*** Moving through the water at a constant speed means continuously replacing the energy that drag transforms into thermal energy. We can compute the change in thermal energy during the swim to find energy that the swimmer must supply. The speed is constant, so we can use Equation 10.18 to find the change in thermal energy.

***PREPARE*** We'll need to compute the drag force before we can find the necessary energy. A swimmer moving through the water has a very large Reynold's number, so we can use Equation 5.12 to compute the drag force, as shown in the next column:

$$D = \frac{1}{2}C_{\text{D}}\rho A v^2$$
$$= \frac{1}{2}(0.45)(1000 \text{ kg/m}^3)(0.080 \text{ m}^2)(1.4 \text{ m/s})^2 = 35 \text{ N}$$

***SOLVE*** With the drag force in hand, we can find the energy converted to thermal energy:

$$\Delta E_{\text{th}} = D\Delta x = (35 \text{ N})(1000 \text{ m}) = 3.5 \times 10^4 \text{ J}$$

This is the energy that the swimmer must supply, which is all converted to thermal energy. The net effect of swimming laps is to warm up the water in the pool!

***ASSESS*** This is a lot of energy, about one-sixth of the energy needed to climb the Empire State Building, which seems reasonable.

We've done a few calculations of the energy required for certain tasks. But there's another factor to consider—efficiency. How much energy would your body actually *use* in completing these tasks? The swimmer needs to supply $3.5 \times 10^4$ J to move through the water. But how much metabolic energy will it cost the swimmer to provide this energy? Swimming is, for humans, a reasonably inefficient form of locomotion, so the energy used by the swimmer is quite a bit greater than the value we found. We'll return to this issue in Chapter 11.

**STOP TO THINK 10.6** A block with an initial kinetic energy of 4.0 J comes to rest after sliding 1.0 m. How far would the block slide if it had 8.0 J of initial kinetic energy?

A. 1.4 m B. 2.0 m C. 3.0 m D. 4.0 m

## 10.6 Conservation of Energy

Just as for momentum conservation, we will develop a before-and-after perspective for energy conservation. We'll limit our consideration for now to kinetic energy, potential energy (both gravitational and elastic), thermal energy, and work. We then note that $\Delta K = K_f - K_i$ and $\Delta U = U_f - U_i$. Now we can rewrite Equation 10.19 as a rule that we can use to solve problems:

$$K_f + U_f + \Delta E_{th} = K_i + U_i + W \qquad (10.20)$$

Before-and-after work-energy equation

**Video** Breaking Boards

Equation 10.20 states that a system's final energy, including any change in the system's thermal energy, equals its initial energy plus any energy added to the system as work. This equation is the basis for a powerful problem-solving approach.

**Video** Chin Basher?

**NOTE** ▶ We don't write $\Delta E_{th}$ as $(E_{th})_f - (E_{th})_i$ in Equation 10.20 because the initial and final values of the thermal energy are typically unknown; only their difference $\Delta E_{th}$ can be measured. ◀

In Section 10.1 we introduced the idea of an isolated system—one in which no work is done on the system and no energy is transferred into or out of the system. In that case, $W = 0$ in Equation 10.20, so the final energy, including any change in thermal energy, equals the initial energy:

$$K_f + U_f + \Delta E_{th} = K_i + U_i$$

The following table shows how to choose an isolated system for common situations.

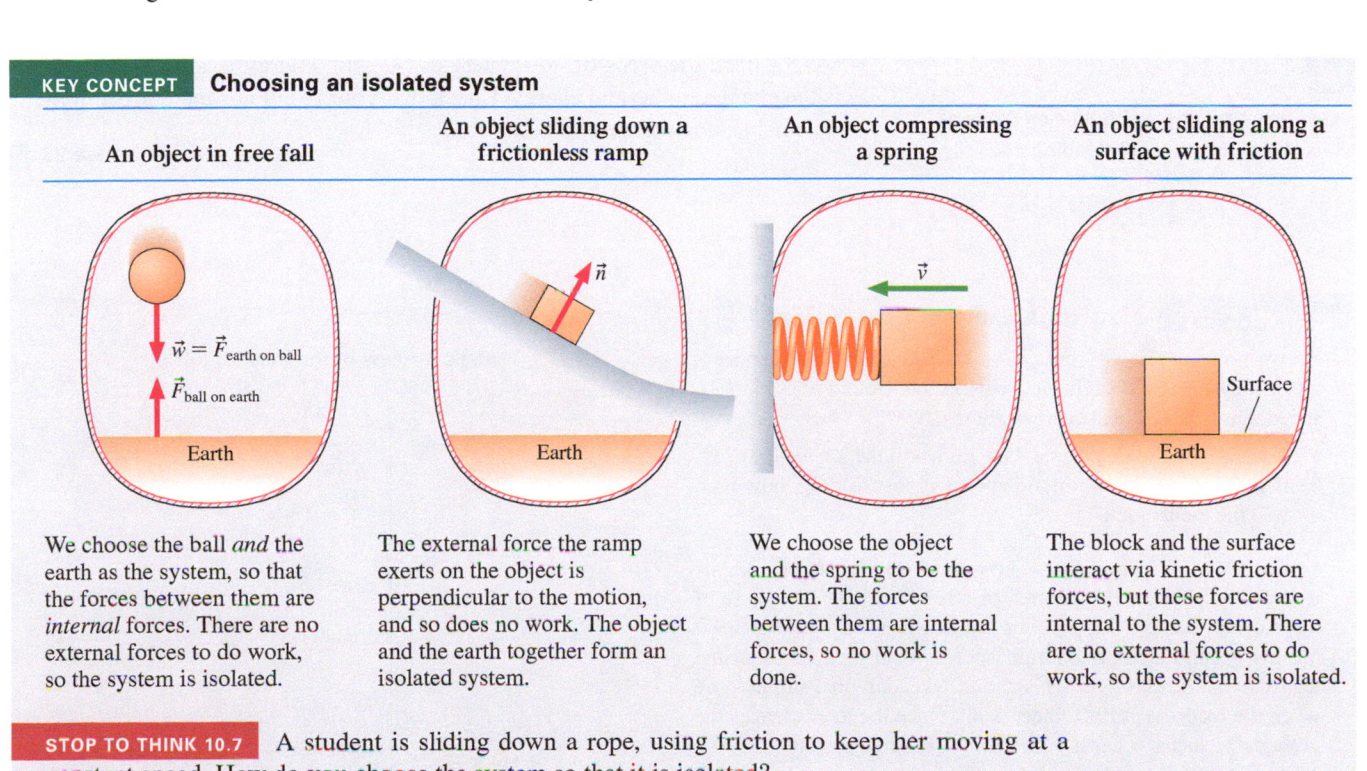

**KEY CONCEPT** **Choosing an isolated system**

| An object in free fall | An object sliding down a frictionless ramp | An object compressing a spring | An object sliding along a surface with friction |
|---|---|---|---|

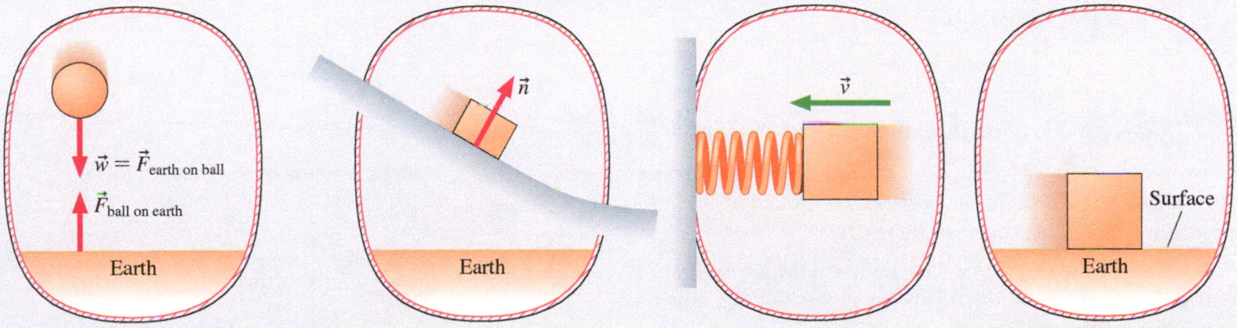


| | | | |
|---|---|---|---|
| We choose the ball *and* the earth as the system, so that the forces between them are *internal* forces. There are no external forces to do work, so the system is isolated. | The external force the ramp exerts on the object is perpendicular to the motion, and so does no work. The object and the earth together form an isolated system. | We choose the object and the spring to be the system. The forces between them are internal forces, so no work is done. | The block and the surface interact via kinetic friction forces, but these forces are internal to the system. There are no external forces to do work, so the system is isolated. |

**STOP TO THINK 10.7** A student is sliding down a rope, using friction to keep her moving at a constant speed. How do you choose the system so that it is isolated?

### Using the Law of Conservation of Energy

Now that we have mathematical expressions for different forms of energy and a general before-and-after equation expressing the law of conservation of energy, we have all the tools we need to formulate a problem-solving approach. We'll sketch out the details and then use it to solve a range of problems.

**Spring into action** BIO A locust can jump as far as 1 meter, an impressive distance for such a small animal. To make such a jump, its legs must extend much more rapidly than muscles can ordinarily contract. Thus, instead of using its muscles to make the jump directly, the locust uses them to more slowly stretch an internal "spring" near its knee joint. This stores elastic potential energy in the spring. When the muscles relax, the spring is suddenly released, and its energy is rapidly converted into kinetic energy of the insect.

**Video** What the Physics? Bouncing Balloons
**Video** What the Physics? Bending and Breaking Boards

## PROBLEM-SOLVING APPROACH 10.1 Conservation of energy problems

The work-energy equation and the law of conservation of energy relate a system's *final* energy to its *initial* energy. We can solve for initial and final heights, speeds, and displacements from these energies.

***STRATEGIZE*** The first step in a conservation of energy problem is to choose the system. This means thinking about the forces and the energies involved. We'll consider the situation before and after a process or an interaction, so we must also decide on the initial and final states.

***PREPARE*** As we did for momentum problems, we'll start with a before-and-after visual overview, as outlined in Tactics Box 9.1. Note the known quantities, and identify what you're trying to find.

***SOLVE*** Apply Equation 10.20:

$$K_f + U_f + \Delta E_{th} = K_i + U_i + W$$

Start with this general equation, then specialize to the case at hand:

- Use the appropriate form or forms of potential energy.
- If the system is isolated, no work is done. Set $W = 0$.
- If there is no friction, drag, or similar force, set $\Delta E_{th} = 0$.

Depending on the problem, you'll need to calculate the initial and/or final values of these energies. You can then solve for the unknown energies, and from these any unknown speeds (from $K$), heights and distances (from $U_g$ and $U_s$), or displacements, friction, or drag forces.

***ASSESS*** Check the signs of your energies. Kinetic energy is always positive, as is the change in thermal energy. Check that your result has the correct units, is reasonable, and answers the question.

Exercise 23

## EXAMPLE 10.9 How high can the locust jump? BIO

As we noted, a desert locust is an excellent jumper. Suppose a 2.0 g locust leaps straight up, leaving the ground at 3.1 m/s, a speed that a desert locust can easily reach.

a. If we ignore the drag force, how high will the locust jump?
b. If 20% of the initial kinetic energy is lost to drag, how high will the locust jump?

***STRATEGIZE*** We don't know how much the locust extends its legs as it pushes off the ground, or other details of this phase of the motion. For us, the problem starts when the locust is leaving the ground at its maximum speed. We'll then consider the locust's motion through the air, and take our final point to be when the locust is at its highest point. Once the locust leaves the ground, the locust + earth + air form an isolated system.

***PREPARE*** The before-and-after visual overview is shown in FIGURE 10.20. We'll describe the situation using Equation 10.20 with $W = 0$, because we've identified an isolated system. For the initial and final points we chose our working equation becomes

$$K_f + (U_g)_f + \Delta E_{th} = K_i + (U_g)_i$$

FIGURE 10.20 Visual overview of the locust's jump.

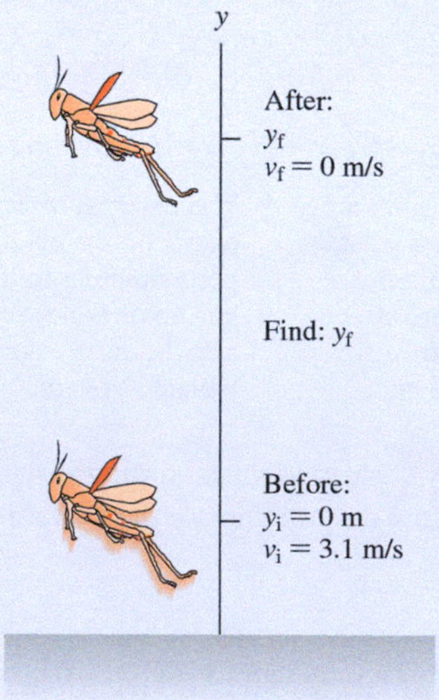

***SOLVE*** a. For this part, we ignore drag, so $\Delta E_{\text{th}} = 0$. We then substitute expressions for the various forms of energy to find

$$\frac{1}{2}mv_f^2 + mgy_f = \frac{1}{2}mv_i^2 + mgy_i$$

The mass appears in every term, so we can cancel out this factor. We'll take our starting position to be $y_i = 0$ and note that the final velocity $v_f = 0$ to simplify further:

$$gy_f = \frac{1}{2}v_i^2$$

$$y_f = \frac{v_i^2}{2g} = \frac{(3.1\ \text{m/s})^2}{2(9.8\ \text{m/s}^2)} = 0.49\ \text{m}$$

b. For this part of the problem, we assume that 20% of the initial kinetic energy is lost to drag. The "lost" energy is transformed into thermal energy, so $\Delta E_{\text{th}} = (0.20)K_i$. Our working equation then becomes

$$K_f + (U_g)_f + (0.20)K_i = K_i + (U_g)_i$$

$$K_f + (U_g)_f = (0.80)K_i + (U_g)_i$$

Simplifying as we did before, we find

$$gy_f = (0.80)\frac{1}{2}v_i^2$$

$$y_f = (0.80)\frac{v_i^2}{2g} = (0.80)\frac{(3.1\ \text{m/s})^2}{2(9.8\ \text{m/s}^2)} = 0.39\ \text{m}$$

***ASSESS*** We noted that the desert locust can jump a horizontal distance of 1 m, so a vertical leap of half a meter seems reasonable. Notice that the locust's mass didn't enter into our calculation; this isn't a surprise, given that the motion is free fall. This also gives us confidence in our solution.

## EXAMPLE 10.10 To push or not to push?

The Summit Plummet is an extreme water slide—one of the steepest and fastest in the world. Riders drop 36 m from the start until they hit a run-out at the bottom. If you give yourself a good push at the start, so that you begin your plunge moving at 2.0 m/s, how fast are you moving when you get to the bottom? How fast would you be moving if you skipped the push? The slide is steep and slippery, so assume that you can ignore friction and drag forces.

***STRATEGIZE*** We'll take the system to be the rider + earth. The initial state has the rider moving at 2.0 m/s at a height of 36 m above the bottom of the slide; the final state is at the bottom.

***PREPARE*** The visual overview in FIGURE 10.21 shows the initial and final states and the slide in between. The exact shape of the slide doesn't matter; we care only about the difference in height. We'll describe the situation using Equation 10.20 with $W = 0$, because this is an isolated system—once you start the ride, no one is giving you a push! There is no elastic potential energy, only gravitational, and we can ignore friction, so our working equation becomes

$$K_f + (U_g)_f = K_i + (U_g)_i$$

FIGURE 10.21 Visual overview of the trip down the water slide.

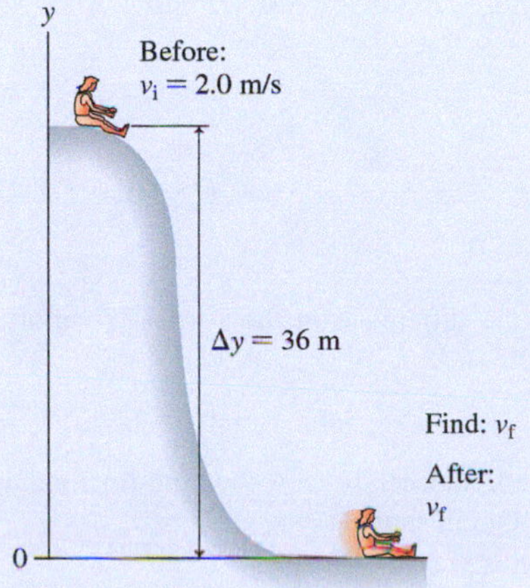


***SOLVE*** If we write our working equation in terms of the change in potential energy, we can express this in terms of the change in height using Equation 10.13:

$$K_f = K_i + ((U_g)_i - (U_g)_f) = K_i + \Delta U_g = K_i + mg\Delta y$$

We rewrite the kinetic energy in terms of speed and then solve for the final speed:

$$\frac{1}{2}mv_f^2 = \frac{1}{2}mv_i^2 + mg\Delta y$$

$$v_f = \sqrt{v_i^2 + 2g\Delta y} = \sqrt{(2.0\ \text{m/s})^2 + 2(9.8\ \text{m/s}^2)(36\ \text{m})} = 27\ \text{m/s}$$

This is pretty speedy! Now, suppose you skip the initial push. How much does this change the final result? The calculation is the same but with $v_i = 0$:

$$v_f = \sqrt{2g\Delta y} = \sqrt{2(9.8\ \text{m/s}^2)(36\ \text{m})} = 27\ \text{m/s}$$

We get exactly the same result to 2 significant figures: greater precision is not warranted given the approximations we've made. Push or not—the final result is about the same! Most of your energy at the end of the ride comes from the change in potential energy, not your initial push.

***ASSESS*** We weren't given the mass of the rider, but the mass canceled along the way, which gives us confidence in the process. The final result, 27 m/s (about 60 mph), is pretty fast. But this is an extreme slide, and a website for this slide claims that you can expect to reach 60 mph, so the result of the calculation is reasonable, even if actually riding the slide isn't.

## EXAMPLE 10.11 Speed of a spring-launched ball

A spring-loaded toy gun is used to launch a 10 g plastic ball. The spring, which has a spring constant of 10 N/m, is compressed by 10 cm as the ball is pushed into the barrel. When the trigger is pulled, the spring is released and shoots the ball back out horizontally. What is the ball's speed as it leaves the barrel? Assume that friction is negligible.

**STRATEGIZE** Let's take the system to be the ball + spring. The initial state has the compressed spring touching the stationary ball; the final state is the expanded spring and the ball in motion.

**PREPARE** The visual overview is shown in FIGURE 10.22. We have chosen the origin of the coordinate system to be the equilibrium position of the free end of the spring, making $x_i = -10$ cm and $x_f = 0$ cm. Work is done on the spring during the compression, but during the time the spring is expanding, the ball + spring is an isolated system, so $W = 0$. We are ignoring friction, so $\Delta E_{th} = 0$. Because the launch is horizontal, we can ignore changes in gravitational potential energy. With these assumptions, the work-energy equation becomes

FIGURE 10.22 Before-and-after visual overview of a ball being shot out of a spring-loaded toy gun.

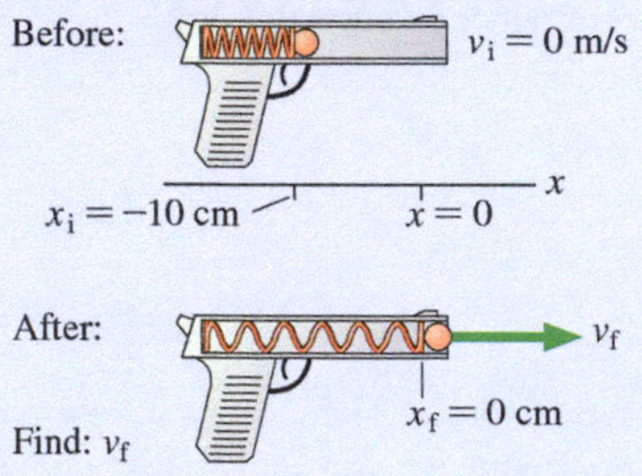


$$K_f + (U_s)_f = K_i + (U_s)_i$$

**SOLVE** We can use expressions for kinetic energy and elastic potential energy to rewrite this equation as

$$\tfrac{1}{2}mv_f^2 + \tfrac{1}{2}kx_f^2 = \tfrac{1}{2}mv_i^2 + \tfrac{1}{2}kx_i^2$$

We know that $x_f = 0$ m and $v_i = 0$ m/s, so this simplifies to

$$\tfrac{1}{2}mv_f^2 = \tfrac{1}{2}kx_i^2$$

It is now straightforward to solve for the ball's speed:

$$v_f = \sqrt{\frac{kx_i^2}{m}} = \sqrt{\frac{(10\ \text{N/m})(-0.10\ \text{m})^2}{0.010\ \text{kg}}} = 3.2\ \text{m/s}$$

**ASSESS** The ball moves pretty slowly, which we expect for a toy gun. Our result seems reasonable.

This is not a problem that we could have easily solved with Newton's laws. The acceleration is not constant, and we have not learned how to handle the kinematics of variable acceleration. But with conservation of energy, this was a straightforward problem.

## EXAMPLE 10.12 The thermal energy of a trip down a slide

Quinn is at rest at the top of a playground slide. The main part of the slide is 5.0 m long, and it is tipped at a 30° angle. Quinn starts sliding, moving down the tipped section, and then the slide levels out so that he leaves the slide at 2.0 m/s, moving horizontally. If Quinn's mass is 24 kg, how much thermal energy is deposited in his trousers and in the slide?

**STRATEGIZE** We'll choose Quinn + earth + slide to be the system. The initial point will be when Quinn is motionless at the top of the slide; the final point will be when Quinn has descended the slide and is moving off the end. As Quinn goes down the slide, potential energy decreases and kinetic energy increases. But there is clearly friction, so that his speed is slower than it would be otherwise. Some of the energy is transformed into thermal energy of his trousers and the slide, internal to the system. It's this change in thermal energy that we'll solve for.

**PREPARE** The visual overview in FIGURE 10.23 shows the initial and final points of the motion. We've chosen an isolated system, so $W = 0$. The only form of potential energy is gravitational potential energy, so the work-energy equation reduces to

$$K_f + (U_g)_f + \Delta E_{th} = K_i + (U_g)_i$$

FIGURE 10.23 Visual overview for motion down the slide.

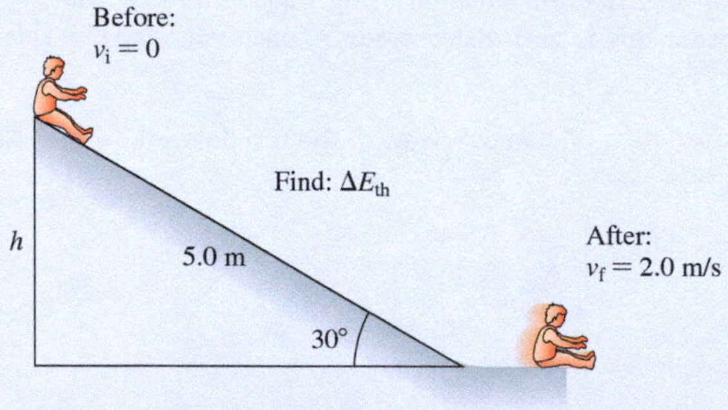


We can find the difference in the vertical position $h$ from the geometry of the slide:

$$h = (5.0\ \text{m})\sin 30° = 2.5\ \text{m}$$

We'll take the initial height as $y_i = h$, the final height as $y_f = 0$, so $(U_g)_f = 0$. Quinn starts at rest, so $K_i = 0$.

***SOLVE*** With all of these parts in hand we can simplify the work-energy equation further and then solve for $\Delta E_{th}$:

$$\Delta E_{th} = (U_g)_i - K_f = mgy - \frac{1}{2}mv_f^2 = 588\ J - 48\ J = 540\ J$$

We've found values for the change in potential energy and the final kinetic energy so that you can see the relative magnitudes. The values in the problem are typical values for a slide—about an 8 foot drop, kids launched off the end at a slow jogging pace. During the slide, the most important energy transformation that takes place is the increase in thermal energy. The slide is mostly about warming things up rather than getting kids up to speed!

***ASSESS*** If you remember going down the slide as a child, you no doubt remember the appreciable warming during the motion, so our result makes sense.

## SYNTHESIS 10.1 Energy and its conservation

The energies present in an isolated system can transform from one kind into another, but the total energy is *conserved*. The unit of all types of energy is the **joule** (J).

**Kinetic energy** is the energy of motion.

$$K = \frac{1}{2}mv^2$$

Mass (kg); Velocity (m/s)

**Gravitational potential energy** is stored energy associated with an object's height above the ground.

$$U_g = mgy$$

Free-fall acceleration; Mass (kg); Height (m) above a reference level $y = 0$

**Elastic potential energy** is stored energy associated with a stretched or compressed spring.

$$U_s = \frac{1}{2}kx^2$$

Spring constant (N/m); Displacement of end of spring from equilibrium (m)

**Work** is the transfer of energy into or out of a system by an external force:

$$W = F_{\parallel}d$$

Work into (+) or out of (−) a system; Force parallel to motion; Displacement

The before-and-after work-energy equation captures the **law of conservation of energy**:

$$K_f + U_f + \Delta E_{th} = K_i + U_i + W$$

Final kinetic and potential energy plus change in thermal energy; Initial kinetic and potential energy plus energy transferred by work

**STOP TO THINK 10.8** At the water park, Katie slides down each of the frictionless slides shown. At the top, she is given a push so that she has the same initial speed each time. At the bottom of which slide is she moving the fastest?

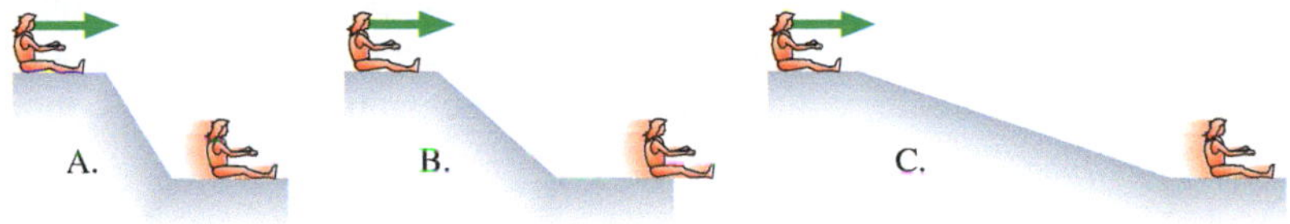


A. Slide A
B. Slide B
C. Slide C
D. Her speed is the same at the bottom of all three slides.

**Video** Loop-the-Loop

## 10.7 Energy Diagrams

Energy is a central concept in physics, but it's also crucial for understanding chemistry, biology, and other sciences. In this section and the following one, we'll develop different means of describing energy that connect to these other subjects that you are likely studying or have studied. It's not that energy is different in chemistry, but the way it is treated, the language used to describe it, may be.

In this section, we'll consider isolated systems for which there is no friction or drag. In this case, there is no work and there is no change in thermal energy, so the work-energy equation becomes

$$K_f + U_f = K_i + U_i$$

In other words, the sum of the kinetic and potential energy is constant. We'll define the total energy $E$ as the sum of these two quantities. For the systems we consider in this section,

$$E = K + U = \text{constant}$$

Kinetic energy depends on an object's speed, but potential energy depends on its *position*. A tossed ball's gravitational potential energy depends on its height $y$, while the elastic potential energy of a compressed spring depends on the displacement $x$. Other potential energies also depend in some way on position. A graph showing a system's potential energy and total energy as a function of position is called an **energy diagram.** We'll spend some time learning about energy diagrams so that we can use them to think about bonds and chemical reactions in the next section.

FIGURE 10.24 The energy diagram of a ball in free fall.

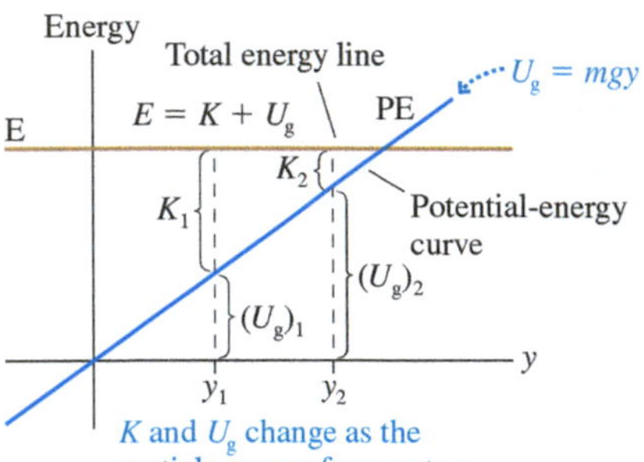


FIGURE 10.24 is the energy diagram of a ball in free fall. This is a bit different from most graphs we've seen. It doesn't include time; the horizontal axis is the vertical position $y$, and the vertical axis represents energy. The lines on the graph show different energies as a function of the vertical position. The gravitational potential energy increases with the vertical position; the mathematical relationship is $U_g = mgy$, and a graph of $mgy$ versus $y$ is a straight line through the origin with slope $mg$. The resulting blue *potential-energy curve* is labeled PE. The tan line labeled E is the system's total energy. This line is always horizontal because the sum of kinetic and potential energy is the same at every point.

Suppose we consider a ball that is at a vertical position $y_1$ and is moving upward. When the ball is at height $y_1$, the distance from the axis up to the potential-energy curve is the potential energy $(U_g)_1$ at that position. Because $K_1 = E - (U_g)_1$, the kinetic energy is represented graphically as the distance between the potential-energy curve and the total energy line. Now, the ball continues to rise. Some time later it is at height $y_2$. The energy diagram shows that the potential energy $(U_g)_2$ has increased while the ball's kinetic energy $K_2$ has decreased, as we know must be the case. Kinetic energy has been transformed into potential energy, but their sum has not changed.

NOTE ▶ In graphs like this, the potential-energy curve PE is determined by the physical properties of the system—for example, the mass or the spring constant. But the total energy line E is under your control. If you change the *initial conditions*, such as throwing the ball upward with a different speed or compressing a spring by a different amount, the total energy line will appear at a different position. We can thus use an energy diagram to see how changing the initial conditions affects the subsequent motion. ◀

FIGURE 10.25 The energy diagram of a mass on a horizontal spring.

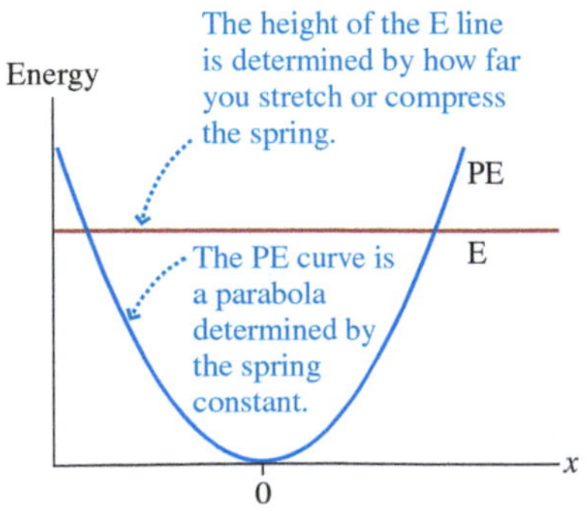


FIGURE 10.25 is the energy diagram of a mass on a horizontal spring. In this case, the blue potential-energy curve $U_s = \frac{1}{2}kx^2$ is a parabola centered at $x = 0$, the equilibrium position of the end of the spring. The blue PE curve is determined by the spring constant; we can't change it. But we can set the tan E line to any height we wish by stretching or compressing the spring to different lengths. The figure shows one possible E line.

Suppose you pull the mass out to position $x_R$ and release it from rest. FIGURE 10.26 shows a five-frame "movie" of the subsequent motion. Initially, in frame a, the energy is entirely potential—the energy of a stretched spring—so the E line has been drawn to cross the PE line at $x_a = x_R$. This is the graphical statement that initially $E = U_s$ and $K = 0$.

The restoring force pulls the mass toward the origin. In frame b, where the mass has reached $x_b$, the potential energy has decreased while the kinetic energy—the

FIGURE 10.26 A five-frame movie of a mass oscillating on a spring.

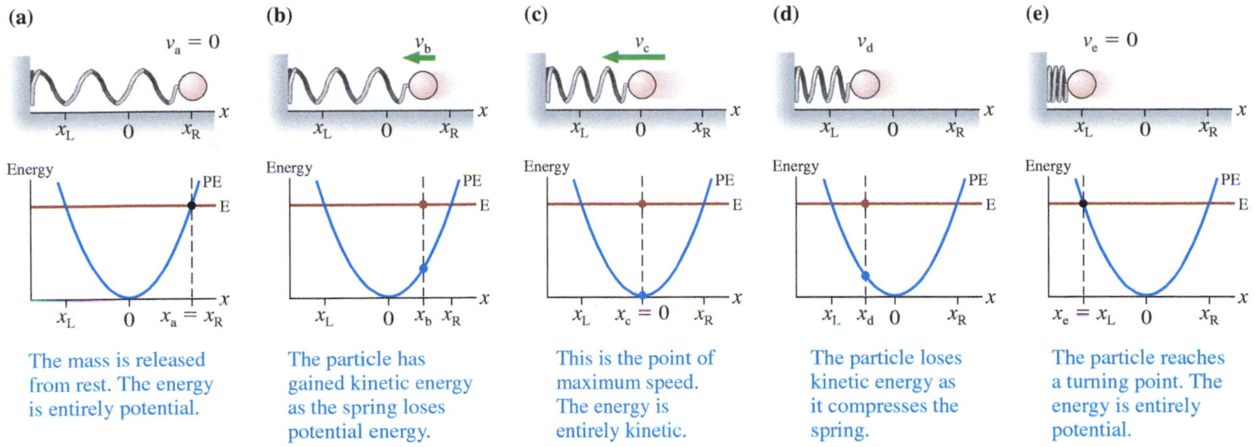


distance *above* the PE curve—has increased. Notice that the total energy—the brown dot—hasn't changed. The mass continues to speed up until it reaches maximum speed at $x_c = 0$, where the PE curve is at a minimum and the distance above the PE curve is maximum. At position $x_d$, the mass has started to slow down as it begins to transform kinetic energy back into elastic potential energy.

The mass continues moving to the left until, in frame e, it reaches position $x_L$, where the total energy line crosses the potential-energy curve. This point, where $K = 0$ and the energy is entirely potential, is a *turning point* where the mass reverses direction. A mass would need negative kinetic energy to be to the left of $x_L$, and that's not physically possible. You should be able to see, from the energy diagram, that the mass will *oscillate* back and forth between positions $x_L$ and $x_R$, having maximum kinetic energy (and thus maximum speed) each time it passes through $x = 0$.

Now, let's consider a different initial condition. Suppose you pull the mass out to a greater initial distance. You've increased the potential energy in the system, and thus the total energy. The tan E line is now at a greater height, and it will intersect the PE graph at two points that are farther from the equilibrium point. With this new initial condition, the mass will oscillate back and forth between two points at a greater distance from equilibrium.

## Interpreting Energy Diagrams

The lessons we learn from Figure 10.26 are true for any energy diagram:

- At any position, the distance from the axis to the PE curve is the object's potential energy. The distance from the PE curve to the E line is its kinetic energy.
- The object cannot be at a position where the PE curve is above the E line.
- A position where the E line crosses the PE curve is a turning point where the object reverses direction.
- If the E line crosses the PE curve at two positions, the object will oscillate between those two positions. Its speed will be maximum at the position where the PE curve is a minimum.

**CONCEPTUAL EXAMPLE 10.13 Interpreting an energy diagram**

FIGURE 10.27 is a more general energy diagram. We don't know how this potential energy was created, but we can still use the energy diagram to understand how a particle with this potential energy will move. Suppose a particle begins at rest at the position shown in the figure and is then released. Describe its subsequent motion.

***REASON*** We've added details to the graph and sketched out details of the motion in FIGURE 10.28. The particle is at rest at the starting point, so $K = 0$ and the total energy is equal to the potential energy. We can draw the E line through this point. The PE curve tells us the particle's potential energy at each position.

*Continued*

FIGURE 10.27 A more general energy diagram.

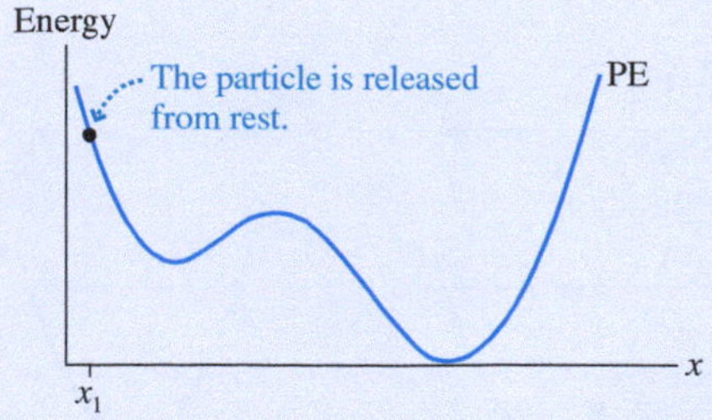


The distance between the PE curve and the E line is the particle's kinetic energy. The particle cannot move to the left because that would require the PE curve to go above the E line, so it begins moving to the right. The particle speeds up from $x_1$ to $x_2$ because $U$ decreases and thus $K$ must increase. It then slows down (but doesn't stop) from $x_2$ to $x_3$ as it goes over the "potential-energy hill." It speeds up after $x_3$ until it reaches maximum speed at $x_4$, where the PE curve is a minimum. The particle then steadily slows from $x_4$ to $x_5$ as kinetic energy is transformed into an increasing potential energy. Position $x_5$ is a turning point, a position where the E line crosses the PE curve. The particle is instantaneously at rest and then reverses direction. Because the E line crosses the PE curve at both $x_1$ and $x_5$ the particle will oscillate back and forth between these two points, speeding up and slowing down as described.

FIGURE 10.28 The motion of the particle in the potential of Figure 10.27.

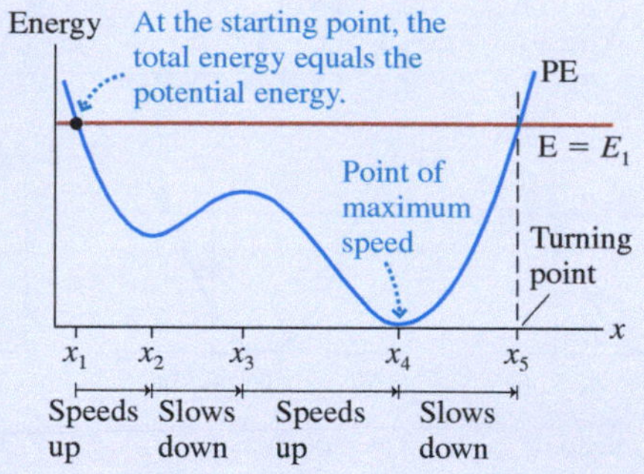


**ASSESS** Our results make sense. The particle is moving fastest where the PE line is lowest, as it must, and it turns around where the E and PE lines cross, meaning $K = 0$ and the particle is at rest.

## Equilibrium Positions

Positions $x_2$, $x_3$, and $x_4$ in Figure 10.28, where the potential energy has a local minimum or maximum, are special positions. Consider the particle at position $x_3$ with energy $E_3$ in FIGURE 10.29a. Its energy is entirely potential energy and its kinetic energy is zero. It must be *at rest*—in equilibrium.

But suppose this particle is slightly disturbed—a tiny push to the right or left—giving it a very small amount of kinetic energy. The particle will begin to move away from $x_3$, moving away faster and faster because the potential energy decreases—and thus kinetic energy increases—on both sides of $x_3$. The situation is analogous to trying to balance a marble on the top of a hill; we can do so if the positioning is absolutely perfect, but any small displacement or disturbance will cause the marble to roll down the hill. An equilibrium position for which any small disturbance drives the particle away from equilibrium is called a point of **unstable equilibrium.** **Any local maximum in the PE curve is a point of unstable equilibrium.**

In contrast, consider a particle at position $x_2$ with energy $E_2$ in FIGURE 10.29b. Its kinetic energy is zero and, as we just discussed, the particle must be at rest. Position $x_2$ is also an equilibrium position, this time for a particle with energy $E_2$. What happens if this particle is slightly disturbed, raising the E line by a very small amount? Now the E line will intersect the PE curve just slightly to either side of $x_2$. These

FIGURE 10.29 Positions of (a) unstable and (b) stable equilibrium.

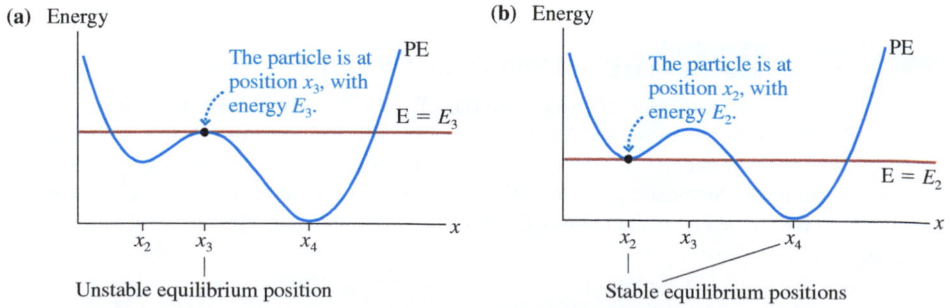

intersections are turning points, so the particle will undergo a very small oscillation centered on $x_2$, rather like a marble in the bottom of a bowl. An equilibrium for which a small disturbance causes only a small oscillation around the equilibrium position is called a point of **stable equilibrium.** You should recognize that **any local minimum in the PE curve is a point of stable equilibrium.** Position $x_4$ is also a point of stable equilibrium—in this case for a particle with $E_4 = 0$.

In the next section, we'll see how ideas about stable and unstable equilibrium help us understand molecular bonds and chemical reactions.

**STOP TO THINK 10.9** The figures below show blue PE curves and tan E lines for four identical particles. Which particle has the highest maximum speed?

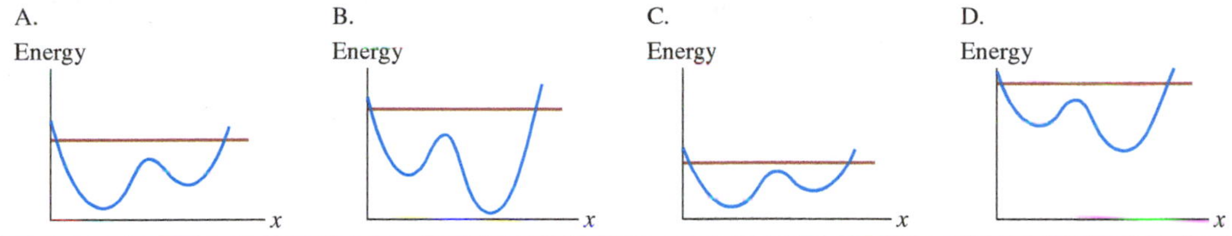


## 10.8 Molecular Bonds and Chemical Energy

With few exceptions, the materials of everyday life are made of atoms bound together into larger molecules. The *molecular bond* that holds two atoms together is an electric interaction between the atoms' negative electrons and positive nuclei. The electric force, like the gravitational force, is a force that can store energy. Fortunately, we don't need to know any details about electric potential energy—a topic we'll take up in Chapter 21—to deduce the energy diagram of a molecular bond.

We've noted that molecular bonds are somewhat analogous to springs: The normal force when an object rests on a table arises from the compression of spring-like bonds, and thermal energy is due, in part, to spring-like vibrations of atoms around an equilibrium position. This suggests that the energy diagram of two atoms connected by a molecular bond should look similar to the Figure 10.25 energy diagram of a mass on a spring.

FIGURE 10.30 shows the experimentally determined energy diagram of the diatomic molecule HCl (hydrogen chloride). Distance $x$ is the *atomic separation*, the distance between the hydrogen and chlorine atoms. Note the very small distances: 1 nm $= 10^{-9}$ m. The left side of the PE curve looks very much like the PE curve of a spring; the right side starts out similar to the PE curve of a spring and then levels off. We can interpret and understand this potential-energy diagram by using what we learned in Section 10.7.

FIGURE 10.30 The energy diagram of the diatomic molecule HCl.

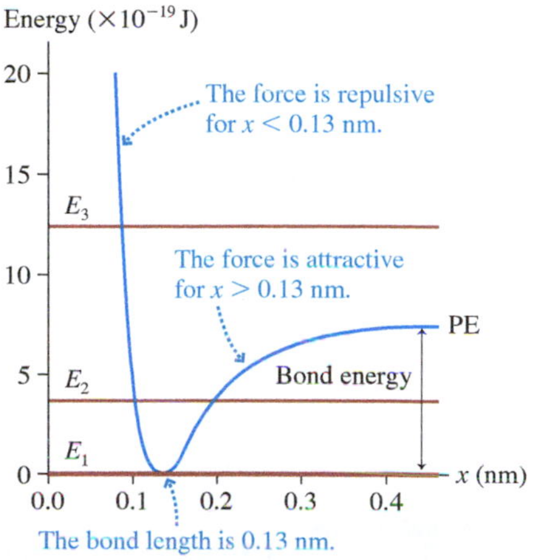


There is a clear minimum of the potential energy curve, a position of stable equilibrium. We've set the potential energy equal to zero at this position. If the total energy is zero as well, as is the case for the total energy line $E_1$, the atoms will rest at this separation. They will have no kinetic energy—no molecular vibration—and will form a molecule with an atomic separation of 0.13 nm. This is the **bond length** of HCl.

If we try to push the atoms closer together, the potential energy rises very rapidly. Physically, this is an electric repulsion between the negative electrons orbiting each atom, but it's analogous to the increasingly strong repulsive force we get when we compress a spring. Thus the PE curve to the left of the equilibrium position looks very much like the PE curve of a spring.

There are also attractive forces between two atoms. These can be the attractive force between two oppositely charged ions, as is the case for HCl; the attractive forces of covalent bonds when electrons are shared; or even weak *polarization forces* that are related to the static electricity force by which a comb that has been brushed through your hair attracts small pieces of paper. For any of these, the attractive force resists if we try to pull the atoms apart—analogous to stretching a spring—and thus potential energy increases to the right. The equilibrium position, with minimum potential energy, is the separation at which the repulsive force between electrons and the attractive force are exactly balanced.

The repulsive force gets stronger as we push the atoms closer together, but the attractive force gets *weaker* as we pull them farther apart. If we pull too hard, the bond breaks and the atoms come apart. Consequently, the PE curve becomes *less steep* as $x$ increases, eventually leveling off when the atoms are so far apart that they cease interacting with each other. This difference between the attractive and repulsive forces explains the asymmetric PE curve in Figure 10.30.

It turns out, for quantum physics reasons, that a molecule cannot have $E = 0$ and thus cannot simply rest at the equilibrium position. By requiring the molecule to have some energy, as for the total energy line $E_2$ in Figure 10.30, we see that the atoms will oscillate back and forth between two turning points where the total energy line crosses the PE curve. This is a *molecular vibration*, and atoms held together by a molecular bond are constantly vibrating. For an HCl molecule with energy as $E_2 = 3.5 \times 10^{-19}$ J, illustrated, the distance between the atoms oscillates between roughly 0.10 nm and 0.18 nm.

As we've seen, an object's thermal energy is the sum of the energies of all the moving and vibrating atoms and molecules. Increasing a system's thermal energy increases the energy of each molecule. If we imagine the line $E_2$ in Figure 10.30 being raised, we can see that increased thermal energy, and thus increased temperature, corresponds to molecules vibrating more vigorously, with larger amplitude and more kinetic energy.

Suppose the molecule's energy is increased to $E_3 = 12.5 \times 10^{-19}$ J. This could happen, for example, if the molecule absorbs some light. We can see from the energy diagram that the molecules will keep moving apart. By raising the molecule's energy to $E_3$, we've broken the molecular bond. The breaking of molecular bonds by the absorption of light is called **photodissociation.** Light-mediated reactions, from sun tanning to photosynthesis to vision, are very similar to photodissociation but involve conformational changes in macromolecules—which require energy—rather than the actual breaking of bonds.

The **bond energy** is the minimum energy required to break a bond when the molecule's energy corresponds to a "room temperature" of 25°C. Bond energy is shown on an energy diagram as the vertical distance from the total energy line to the potential-energy "plateau" on the right side of the diagram. HCl molecules at room temperature have energy $E \approx 0.04 \times 10^{-19}$ J, barely distinguishable from zero energy. We can see in Figure 10.30 that the bond energy of HCl is approximately $7.5 \times 10^{-19}$ J.

**NOTE** ▸ Chemists and biologists usually quote molecular energies in kJ/mol or kcal/mol. Physicists prefer to work directly with the energy per molecule or energy per bond in J. To find energies in kJ/mol, simply multiply the bond energy by Avogadro's number. For example, the $7.5 \times 10^{-19}$ J bond energy of HCl becomes 450 kJ/mol. ◂

**EXAMPLE 10.14 Does the photon have enough energy?**

An energy diagram for molecular oxygen, $O_2$, is shown in FIGURE 10.31. A germicidal lamp for sterilizing equipment uses short-wavelength ultraviolet radiation at 185 nm. At this wavelength, each photon, or quantum, of ultraviolet light has $10.7 \times 10^{-19}$ J of energy. If a molecule of $O_2$ at room temperature absorbs one photon of light from the lamp, does this provide enough energy to split the molecule? If so, what will be the kinetic energy of the atoms after they have separated?

FIGURE 10.31 The energy diagram for molecular oxygen, $O_2$.

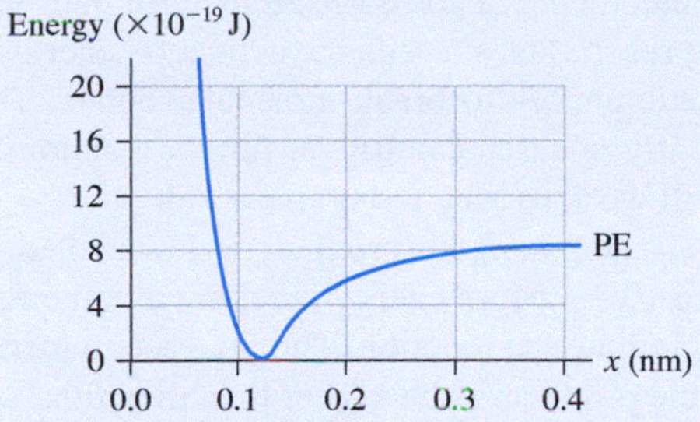


FIGURE 10.32 Comparing the photon energy to the potential energy for $O_2$.

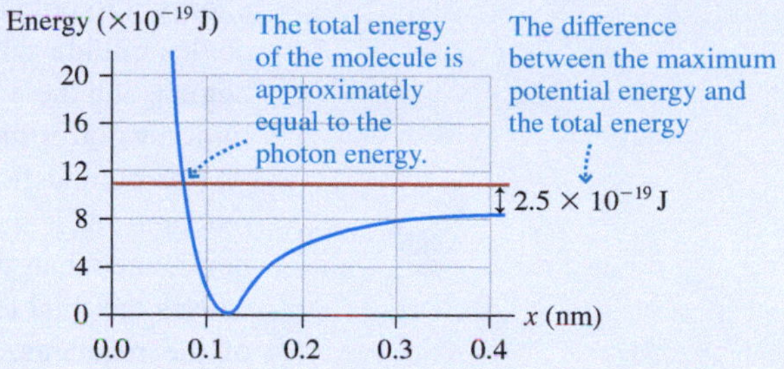


***STRATEGIZE*** We can use the molecular energy diagram in Figure 10.31 to determine what happens at different energies.

***PREPARE*** We'll assume that the room-temperature energy is similar to that for HCl, very close to zero. After the molecule absorbs the photon, the energy of the molecule will be nearly equal to the photon energy.

***SOLVE*** FIGURE 10.32 is the $O_2$ energy diagram with a total energy line added that corresponds to the photon energy. The total energy is much greater than the maximum of the potential energy on the right side of the graph. After absorbing the photon, the two oxygen molecules will separate. The residual kinetic energy will be equal to the difference between the total energy and the maximum of the potential energy, which we estimate to be $2.5 \times 10^{-19}$ J.

***ASSESS*** Germicidal lamps at this wavelength are known to produce $O_3$, ozone, a very reactive form of oxygen—so it's clear that the photons have enough energy to break apart the normally stable $O_2$ molecules. Therefore, our answer makes sense.

## Chemical Reactions

Energy ideas are key to understanding what happens during a chemical reaction. The basic idea of any chemical reaction—involving simple diatomic molecules or large biological macromolecules—is that some molecular bonds are broken and new molecular bonds are formed. For example, a simple reaction that we can symbolize as $AB + CD \rightarrow AC + BD$ requires the bonds of molecules AB and CD to be broken and then new bonds to form between atoms A and C and between atoms B and D. And, as we've just seen, it takes energy to break molecular bonds.

FIGURE 10.33 is the energy diagram of a chemical reaction. It's very much like the energy diagrams we've been using, but with one important difference: The position coordinate of the horizontal axis has been replaced with an abstract **reaction coordinate.** The reaction coordinate is not a physical quantity that could be measured; instead, it shows in a general sense the progress of bond breaking and bond formation as a reaction moves from reactants, on the left, to products, on the right.

FIGURE 10.33 A reaction energy diagram.

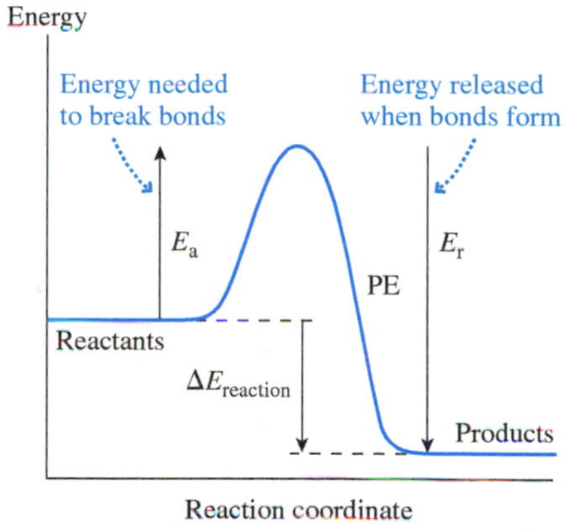


All reaction energy diagrams have a large hump, or *energy barrier*, in the middle. This represents the energy required to break the bonds of the reactant molecules. For the reaction to take place, the reactants must increase their potential energy by the amount $E_a$, called the **activation energy.** Graphically, the activation energy is the height of the energy barrier above the initial potential energy of the reactants.

How does this happen? The reactant molecules have thermal energy, which means that the individual molecules are moving around and vibrating. When molecules collide, this energy can be transformed into the increased potential energy of stretched bonds. If the thermal energy is too low, the increased potential energy is less than the activation energy, meaning that the bonds don't break and the reaction doesn't occur. Wood and oxygen don't react at room temperature, even though the reaction—combustion—is energetically favorable, because the reactants don't have enough thermal energy to allow bond breaking during collisions. In essence, the total energy line is lower than the energy barrier, so there's a turning point in the reaction coordinate.

To burn wood, you must substantially increase the thermal energy of at least a portion of the fuel. You can do this with the high-temperature flame from a match. Some of the reactant molecules can then, via collisions, transform their large kinetic energy into potential energy that reaches or exceeds the activation energy—the molecules collide with enough kinetic energy to break molecular bonds. The reaction begins, and the energy subsequently released can trigger further reactions. Once you light a wood splint, the flame will work its way along the wood.

For combustion, the potential energy of the products is lower than that of the reactants, as is the case in Figure 10.33. More energy is released in the formation of new bonds than was required to initiate the reaction. The increase in thermal energy causes the final temperature of the products to be higher than the initial temperature of the reactants. Such reactions are called **exothermic reactions.** In contrast, an **endothermic reaction** releases less energy than was required to initiate it. Such reactions require a continuous input of energy to keep going.

NOTE ▸ Chemists and biologists often describe reactions in terms of what is called *free energy*. Free energy is a more appropriate description when some or all of the energy released in a reaction is used to do work rather than increasing the thermal energy. This is often the case in biology, where the energy released by a reaction does work by moving molecules around or changing the configuration of macromolecules instead of simply heating up the cell. We'll leave the definition and use of free energy to your chemistry and biology classes, simply noting that the analysis of a reaction in terms of free energy is exactly the same as the analysis presented here. ◂

## Reaction Rates and Catalysts

The reaction energy diagram tells us nothing about the *rate of reaction*—how fast a reaction proceeds. A more detailed theory, which you will study in chemistry, finds that the rate of reaction increases *exponentially* as the activation energy decreases. An exponential change means that even a small decrease in activation energy can produce a large increase in the rate of reaction.

The role of a catalyst is to provide an alternate reaction pathway with a lower activation energy, thus dramatically speeding up the rate of reaction. FIGURE 10.34 shows an exothermic reaction, one that is energetically favorable but where the energy barrier is so high that this reaction will not happen at room temperature because the reaction rate is essentially zero. In Figure 10.34, we see that a catalyst offers an alternate pathway whose activation energy is easily exceeded by room-temperature molecules. A catalyst can dramatically increase reaction rates.

FIGURE 10.34 A reaction energy diagram for a chemical reaction with and without a catalyst.

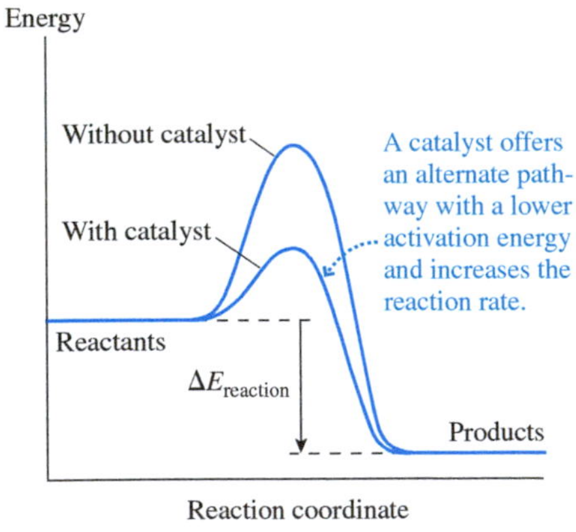


Most of biochemistry is mediated by catalysts in the form of *enzymes*. Processes such as respiration, photosynthesis, and protein synthesis involve energetically favorable exothermic (also called *exergonic*) reactions, but the activation energy is so high that the reactants, on their own, would react barely, if at all, at normal temperatures. Enzymes catalyze these reactions, allowing them to proceed at a rate sufficient for cellular functions.

## Chemical Energy

The law of conservation of energy includes the term $\Delta E_{\text{chem}}$, the change in chemical energy. Physics usually focuses on systems in which chemical energy is not important, but energy conservation also has to apply to chemistry and biology. Chemical energy is simply a name for the total electric potential energy stored in all the molecular bonds of a system. If there are no reactions, the chemical energy doesn't change and we can ignore it, because only energy *changes* enter into the law of energy conservation.

If there are chemical reactions, then the breaking and creation of molecular bonds change the system's chemical energy. In Figure 10.33 the energy of the products is lower than the energy of the reactants. To make the reaction go, potential energy must be increased by the activation energy $E_a$—the energy of breaking bonds. Once the reaction is over the energy barrier, the formation of new bonds releases energy $E_r$; that is, potential energy is transformed into thermal energy. The difference, $\Delta E_{\text{reaction}} = E_r - E_a$, is the net energy released in *one* reaction due to the change in bonds.

Any realistic system has a vast number of chemical reactions taking place. The change in chemical energy is simply the total energy released by all these reactions. If N reactions take place, then

$$\Delta E_{\text{chem}} = N\Delta E_{\text{reaction}} \tag{10.21}$$

In biological systems, the production of chemical energy via the catalyzed reactions of respiration powers the cellular machinery and maintains body temperature.This change in energy will be an important part of the story in Chapter 11.

## 10.9 Energy in Collisions

In Chapter 9 we studied collisions between two objects. We found that if no external forces are acting on the objects, the total *momentum* of the objects will be conserved. Now we wish to study what happens to *energy* in collisions.

Let's first re-examine a perfectly inelastic collision. We studied just such a collision in Example 9.8. Recall that in such a collision the two objects stick together and then move with a common final velocity. What happens to the energy?

### EXAMPLE 10.15 How much energy is transformed in a collision between railroad cars?

**FIGURE 10.35** shows two train cars that move toward each other, collide, and couple together. In Example 9.8, we used conservation of momentum to find the final velocity shown in Figure 10.35 from the given initial velocities. How much thermal energy is created in this collision?

**FIGURE 10.35** Before-and-after visual overview of a collision between two train cars.

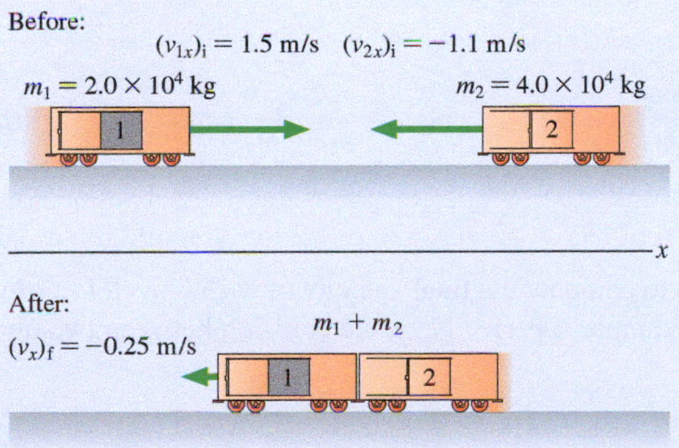


**STRATEGIZE** We'll choose our system to be the two cars. The initial state is the instant before the collision; the final state is the instant just after.

**PREPARE** This is an isolated system, so $W = 0$. Because the track is horizontal, there is no change in potential energy. Thus the work-energy equation reduces to

$$K_f + \Delta E_{th} = K_i$$

Energy is conserved, but *kinetic* energy is not; it will be lower after the collision than before.

**SOLVE** The initial kinetic energy is

$$K_i = \frac{1}{2}m_1(v_{1x})_i^2 + \frac{1}{2}m_2(v_{2x})_i^2$$

$$= \frac{1}{2}(2.0 \times 10^4 \text{ kg})(1.5 \text{ m/s})^2 + \frac{1}{2}(4.0 \times 10^4 \text{ kg})(-1.1 \text{ m/s})^2$$

$$= 4.7 \times 10^4 \text{ J}$$

Because the cars stick together and move as a single object with mass $m_1 + m_2$, the final kinetic energy is

$$K_f = \frac{1}{2}(m_1 + m_2)(v_x)_f^2$$

$$= \frac{1}{2}(6.0 \times 10^4 \text{ kg})(-0.25 \text{ m/s})^2 = 1900 \text{ J}$$

From the conservation of energy equation above, we find that the thermal energy increases by

$$\Delta E_{th} = K_i - K_f = 4.7 \times 10^4 \text{ J} - 1900 \text{ J} = 4.5 \times 10^4 \text{ J}$$

This amount of the initial kinetic energy is transformed into thermal energy during the impact of the collision.

**ASSESS** The cars are moving much more slowly after the collision than before, so we expect that most of the kinetic energy is transformed into thermal energy, just as we observed.

## Elastic Collisions

FIGURE 10.36 A tennis ball collides with a racket. Notice that the ball is compressed and the strings are stretched.

FIGURE 10.36 shows a collision of a tennis ball with a racket. The ball is compressed and the racket strings stretch as the two collide, then the ball expands and the strings rebound as the two are pushed apart. In the language of energy, the kinetic energy of the objects is transformed into the elastic potential energy of the ball and strings, then back into kinetic energy as the two objects spring apart. If *all* of the kinetic energy is stored as elastic potential energy, and *all* of the elastic potential energy is transformed back into the post-collision kinetic energy of the objects, then mechanical energy is conserved. A collision in which mechanical energy is conserved is called a **perfectly elastic collision.**

Needless to say, most real collisions fall somewhere between perfectly elastic and perfectly inelastic. A rubber ball bouncing on the floor might "lose" 20% of its kinetic energy on each bounce and return to only 80% of the height of the preceding bounce. But collisions between two very hard objects, such as two pool balls or two steel balls, come close to being perfectly elastic. And collisions between microscopic particles, such as atoms or electrons, can be perfectly elastic.

FIGURE 10.37 A perfectly elastic collision.

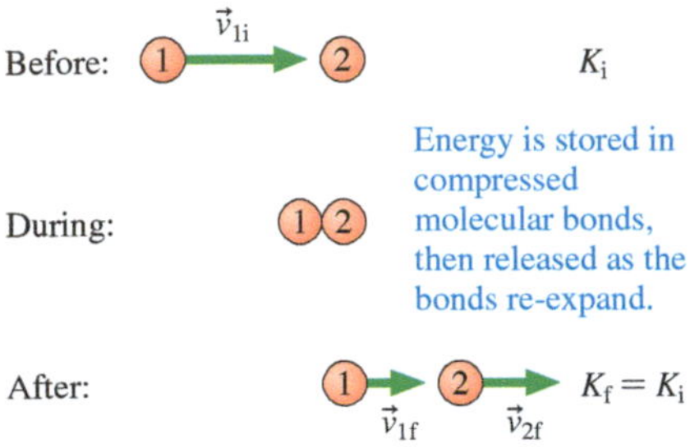


FIGURE 10.37 shows a head-on, perfectly elastic collision of a ball of mass $m_1$, having initial velocity $(v_{1x})_i$, with a ball of mass $m_2$ that is initially at rest. The balls' velocities after the collision are $(v_{1x})_f$ and $(v_{2x})_f$. These are velocities, not speeds, and have signs. Ball 1, in particular, might bounce backward and have a negative value for $(v_{1x})_f$.

The collision must obey two conservation laws: conservation of momentum (obeyed in any collision) and conservation of mechanical energy (because the collision is perfectly elastic). Although the energy is transformed into potential energy during the collision, the mechanical energy before and after the collision is purely kinetic energy. Thus,

$$\text{momentum conservation:} \quad m_1(v_{1x})_i = m_1(v_{1x})_f + m_2(v_{2x})_f$$

$$\text{energy conservation:} \quad \frac{1}{2}m_1(v_{1x})_i^2 = \frac{1}{2}m_1(v_{1x})_f^2 + \frac{1}{2}m_2(v_{2x})_f^2$$

Momentum conservation alone is not sufficient to analyze the collision because there are two unknowns: the two final velocities. That is why we did not consider perfectly elastic collisions in Chapter 9. Energy conservation gives us another condition. The complete solution of these two equations involves straightforward but rather lengthy algebra. We'll just give the solution here:

$$(v_{1x})_f = \frac{m_1 - m_2}{m_1 + m_2}(v_{1x})_i \qquad (v_{2x})_f = \frac{2m_1}{m_1 + m_2}(v_{1x})_i \tag{10.22}$$

Perfectly elastic collision with object 2 initially at rest

Equations 10.22 allow us to compute the final velocity of each object. Let's look at a common and important example: a perfectly elastic collision between two objects of equal mass.

**EXAMPLE 10.16 Finding the aftermath of a collision between air hockey pucks**

On an air hockey table, a moving puck, traveling to the right at 2.3 m/s, makes a head-on collision with an identical puck at rest. What is the final velocity of each puck?

***PREPARE*** The before-and-after visual overview is shown in FIGURE 10.38. We've shown the final velocities in the picture, but we don't really know yet which way the pucks will move. Because one puck was initially at rest, we can use Equations 10.22 to find the final velocities of the pucks. The pucks are identical, so we have $m_1 = m_2 = m$.

***SOLVE*** We use Equations 10.22 with $m_1 = m_2 = m$ to get

$$(v_{1x})_f = \frac{m - m}{m + m}(v_{1x})_i = 0 \text{ m/s}$$

$$(v_{2x})_f = \frac{2m}{m + m}(v_{1x})_i = (v_{1x})_i = 2.3 \text{ m/s}$$

**FIGURE 10.38** A moving puck collides with a stationary puck.

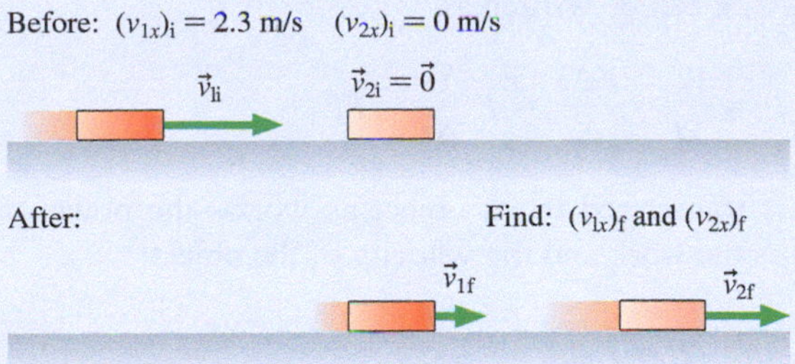


The incoming puck stops dead, and the initially stationary puck goes off with the same velocity that the incoming one had.

***ASSESS*** You can see that momentum and energy are conserved: The incoming puck's momentum and energy are completely transferred to the outgoing puck. If you've ever played pool, you've probably seen this sort of collision when you hit a ball head-on with the cue ball. The cue ball stops and the other ball picks up the cue ball's velocity.

**STOP TO THINK 10.10** A small ball with mass $M$ is at rest. It is then struck by a ball with twice the mass, moving at speed $v_0$. The situation after the collision is shown in the figure. Is this possible?

A. Yes
B. No, because momentum is not conserved
C. No, because energy is not conserved
D. No, because neither momentum nor energy is conserved

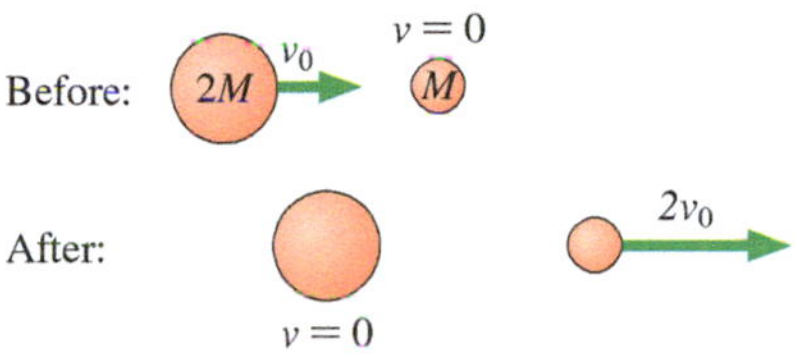


## 10.10 Power

We've now studied how energy can be transformed from one kind into another and how it can be transferred between the environment and the system as work. In many situations we would like to know *how quickly* the energy is transformed or transferred. Is a transfer of energy very rapid, or does it take place over a long time? In passing a truck, your car needs to transform a certain amount of the chemical energy in its fuel into kinetic energy. It makes a *big* difference whether your engine can do this in 20 s or 60 s!

The question How quickly? implies that we are talking about a *rate*. For example, the velocity of an object—how fast it is going—is the *rate of change* of position. So, when we raise the issue of how fast the energy is transformed, we are talking about the *rate of transformation* of energy. Suppose in a time interval $\Delta t$ an amount of energy $\Delta E$ is transformed from one form into another. The rate at which this energy is transformed is called the **power** $P$ and is defined as

$$P = \frac{\Delta E}{\Delta t} \tag{10.23}$$

Power when an amount of energy $\Delta E$ is transformed in a time interval $\Delta t$

The unit of power is the **watt,** which is defined as 1 watt = 1 W = 1 J/s.

Power also measures the rate at which energy is transferred into or out of a system as work $W$. If work $W$ is done in time interval $\Delta t$, the rate of energy *transfer* is

$$P = \frac{W}{\Delta t} \tag{10.24}$$

Power when an amount of work $W$ is done in a time interval $\Delta t$

If a person, animal, vehicle, or device is transforming or transferring energy at a rate of 3 J/s, we say that it has an **output power** of 3 W.

Both these cars take about the same energy to reach 60 mph, but the race car gets there in a much shorter time, so its *power* is much greater.

The English unit of power is the *horsepower.* The conversion factor to watts is

1 horsepower = 1 hp = 746 W

Many common appliances, such as motors, are rated in hp.

We can express Equation 10.24 in a different form. If in the time interval $\Delta t$ an object undergoes a displacement $\Delta x$, the work done by a force acting on the object is $W = F\Delta x$. Then Equation 10.24 can be written as

$$P = \frac{W}{\Delta t} = \frac{F\Delta x}{\Delta t} = F\frac{\Delta x}{\Delta t} = Fv$$

The rate at which energy is transferred to an object as work—the power—is the product of the force that does the work and the velocity of the object:

$$P = Fv \qquad (10.25)$$

Rate of energy transfer due to a force $F$ acting on an object moving at velocity $v$

**EXAMPLE 10.17 Finding the output power for a weightlifter** BIO

A 100 kg weightlifter performs a lift called a clean and jerk, raising a 190 kg bar from the ground to a height of 1.9 m in a time of 1.8 s. What is his output power?

**STRATEGIZE** We'll take the system to be the weightlifter + bar + earth. We'll assume that the bar is stationary before and after the lift, so the relevant energy change is the increase in gravitational potential energy of the bar. This is an isolated system, so the change is an internal transformation. We'll use Equation 10.23 to compute the power.

**PREPARE** The change in potential energy depends on the change in height:

$$\Delta U_g = mg\Delta y = (190\text{ kg})(9.8\text{ m/s}^2)(1.9\text{ m}) = 3540\text{ J}$$

**SOLVE** The power of the transformation is

$$P = \frac{\Delta E}{\Delta t} = \frac{3540\text{ J}}{1.8\text{ s}} = 2000\text{ W}$$

**ASSESS** This is a lot of power—about 2.7 horsepower! But we'd expect such a large output power for a large weightlifter.

The 100 kg weightlifter has a very large output power, which isn't surprising. How about someone smaller?

**EXAMPLE 10.18 Finding the output power for a sprinter** BIO

A 50 kg sprinter accelerates from 0 to 11 m/s in 3.0 s. What is the output power for this rapid start?

**STRATEGIZE** We can take the system to be the runner + earth. Let's assume that the track is level, so there is no change in potential energy, only a change in kinetic energy. We can safely ignore drag and other forces, so this is isolated system and the change is an internal transformation.

**PREPARE** The initial kinetic energy is zero, so the change in kinetic energy is equal to the final kinetic energy:

$$\Delta K = K_f = \frac{1}{2}mv_f^2 = \frac{1}{2}(50\text{ kg})(11\text{ m/s})^2 = 3000\text{ J}$$

**SOLVE** The power of the transformation is

$$P = \frac{\Delta E}{\Delta t} = \frac{3000\text{ J}}{3.0\text{s}} = 1000\text{ W}$$

**ASSESS** This is a lot of power—about 1.3 horsepower—but less than for the weightlifter, which makes sense. In this case, as for the weightlifter, the power came from the athlete's muscles.

A 100 kg weightlifter can produce more output power than a 50 kg sprinter, which makes sense. It's worthwhile to consider what we'll call the **specific power**—the output power divided by the mass of the person (or animal, device, or machine) doing the transformation or the work:

$$\text{specific power} = \frac{\text{power of a transformation or a transfer}}{\text{mass of agent causing the transformation or transfer}}$$

Let's compute the specific power for the weightlifter and the sprinter:

$$\text{sprinter:} \qquad \text{specific power} = \frac{1000\ \text{W}}{50\ \text{kg}} = 20\ \text{W/kg}$$

$$\text{weightlifter:} \qquad \text{specific power} = \frac{2000\ \text{W}}{100\ \text{kg}} = 20\ \text{W/kg}$$

The numbers for the sprinter and the weightlifter are at the extreme end of what humans are capable of; these are numbers typical of world-class athletes. It's interesting to note that the specific power for both cases is about the same. Humans in peak condition who are skilled at athletic pursuits are capable of short bursts of about 20 W/kg. Larger athletes can produce more output power, but the power per kilogram is about the same. As we'll see in the next chapter, humans can't sustain this level of power output; this number applies for only short bursts that use the large muscles of the body. Sustained activities such as cycling or swimming correspond to specific powers of perhaps 5 W/kg for elite athletes.

Smaller animals are generally capable of higher specific powers. A bushbaby, a 200 g primate that gets around by executing rapid leaps in the trees it calls home, is able to push off with its legs with sufficient force to accelerate to 6.7 m/s in 0.16 s, corresponding to a specific power of 140 W/kg. This is the upper end of what can be accomplished with muscle power alone. The leap of the 2.0 g desert locust that we considered earlier has an even higher specific power, but this power comes from springs in the legs; muscles alone could not get the locust to such a high speed in such a short time. Many of the impressive jumpers of the insect world, from fleas to springtails, use energy storage systems corresponding to springs to power their leaps.

Of course, the notion of specific power can be applied to other systems as well. We can do a similar calculation for a passenger car, either starting from rest or climbing a hill, to find a specific power in the range of 90 W/kg. This is an interesting measure of a vehicle.

**STOP TO THINK 10.11** Four students run up the stairs in the times shown. Rank in order, from largest to smallest, their power outputs $P_A$ through $P_D$.

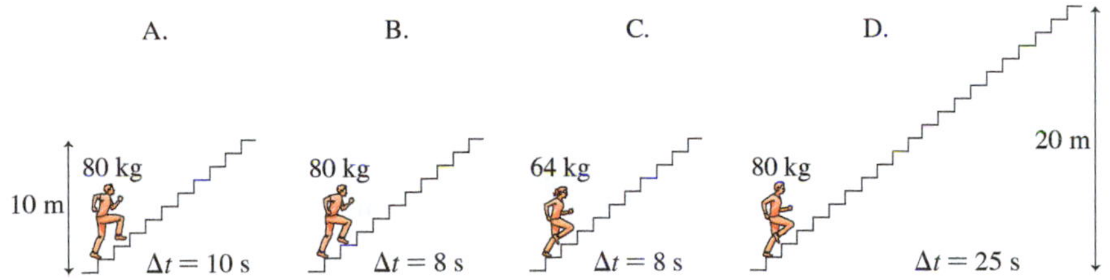

## INTEGRATED EXAMPLE 10.19 Stopping a runaway truck

A truck's brakes can overheat and fail while descending mountain highways, leading to an extremely dangerous runaway truck. Some highways have *runaway-truck ramps* to safely bring out-of-control trucks to a stop. These uphill ramps are covered with a deep bed of gravel. The uphill slope and the large coefficient of rolling friction as the tires sink into the gravel bring the truck to a safe halt.

A 22,000 kg truck heading down a 3.5° slope at 20 m/s ($\approx$45 mph) suddenly has its brakes fail. Fortunately, there's a runaway-truck ramp 600 m ahead. The ramp slopes upward at an angle of 10°, and the coefficient of rolling friction between the truck's tires and the loose gravel is $\mu_r = 0.40$. Ignore air resistance and rolling friction as the truck rolls down the highway.

a. Use conservation of energy to find how far along the ramp the truck travels before stopping.
b. By how much does the thermal energy of the truck and ramp increase as the truck stops?

**STRATEGIZE** We'll follow the steps of Problem-Solving Approach 10.1. We start by defining the system as the truck + ramp + earth. Gravitational potential energy will be part of the solution, and the change in thermal energy will be an internal transformation.

**PREPARE** FIGURE 10.39 shows a before-and-after visual overview. Because we're going to need to determine friction forces to calculate the increase in thermal energy, we've also drawn a free-body diagram for the truck as it moves up the ramp. One slight complication is that the $y$-axis of free-body diagrams is drawn perpendicular to the slope, whereas the calculation of gravitational potential energy needs a vertical $y$-axis to measure height.

FIGURE 10.39 Visual overview of the runaway truck.

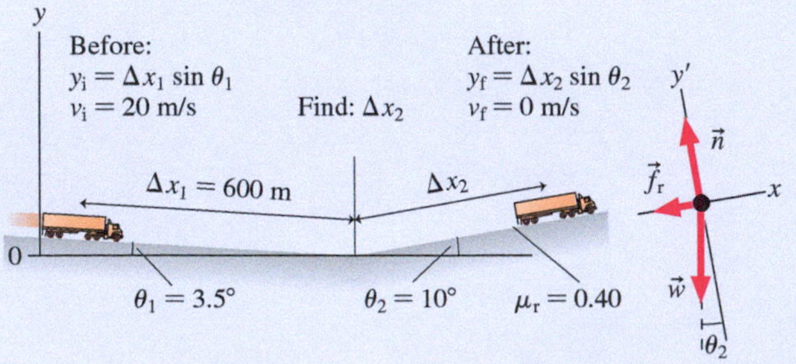


We've dealt with this by labeling the free-body diagram axis the $y'$-axis.

**SOLVE** a. The work-energy equation for the motion of the truck, from the moment its brakes fail to when it finally stops, is

$$K_f + (U_g)_f + \Delta E_{th} = K_i + (U_g)_i$$

Because friction is present only along the ramp, thermal energy will increase only as the truck moves up the ramp. This thermal energy is then given by $\Delta E_{th} = f_r\,\Delta x_2$, because $\Delta x_2$ is the length of the ramp. The conservation of energy equation then is

$$\frac{1}{2}mv_f^2 + mgy_f + f_r\,\Delta x_2 = \frac{1}{2}mv_i^2 + mgy_i$$

From Figure 10.39 we have $y_i = \Delta x_1 \sin\theta_1$, $y_f = \Delta x_2 \sin\theta_2$, and $v_f = 0$, so the equation becomes

$$mg\,\Delta x_2 \sin\theta_2 + f_r\,\Delta x_2 = \frac{1}{2}mv_i^2 + mg\,\Delta x_1 \sin\theta_1$$

To find $f_r = \mu_r n$ we need to find the normal force $n$. The free-body diagram shows that

$$\sum F_{y'} = n - mg\cos\theta_2 = a_{y'} = 0$$

from which $f_r = \mu_r n = \mu_r mg\cos\theta_2$. With this result for $f_r$, our conservation of energy equation is

$$mg\,\Delta x_2 \sin\theta_2 + \mu_r mg\cos\theta_2\,\Delta x_2 = \frac{1}{2}mv_i^2 + mg\,\Delta x_1 \sin\theta_1$$

which, after we divide both sides by $mg$, simplifies to

$$\Delta x_2 \sin\theta_2 + \mu_r \cos\theta_2 \Delta x_2 = \frac{v_i^2}{2g} + \Delta x_1 \sin\theta_1$$

Solving this for $\Delta x_2$ gives

$$\Delta x_2 = \frac{\dfrac{v_i^2}{2g} + \Delta x_1 \sin\theta_1}{\sin\theta_2 + \mu_r \cos\theta_2}$$

$$= \frac{\dfrac{(20\text{ m/s})^2}{2(9.8\text{ m/s}^2)} + (600\text{ m})(\sin 3.5°)}{\sin 10° + 0.40(\cos 10°)} = 100\text{ m}$$

b. We know that $\Delta E_{th} = f_r\,\Delta x_2 = (\mu_r mg\cos\theta_2)\Delta x_2$, so that

$$\Delta E_{th} = (0.40)(22{,}000\text{ kg})(9.8\text{ m/s}^2)(\cos 10°)(100\text{ m})$$
$$= 8.5 \times 10^6\text{ J}$$

**ASSESS** It seems reasonable that a truck that speeds up as it rolls 600 m downhill takes only 100 m to stop on a steeper, high-friction ramp. At the top of the hill the truck's kinetic energy is $K_i = \frac{1}{2}mv_i^2 = \frac{1}{2}(22{,}000\text{ kg})(20\text{ m/s})^2 = 4.4 \times 10^6$ J, which is of the same order of magnitude as $\Delta E_{th}$. Our answer is reasonable.

# SUMMARY

**GOAL** To introduce the concept of energy and to learn a new problem-solving strategy based on conservation of energy.

## GENERAL PRINCIPLES

### Basic Energy Model

Within a system, energy can be **transformed** between various forms.

Energy can be **transferred** into or out of a system in two basic ways:

- **Work:** The transfer of energy by mechanical forces
- **Heat:** The nonmechanical transfer of energy from a hotter to a colder object

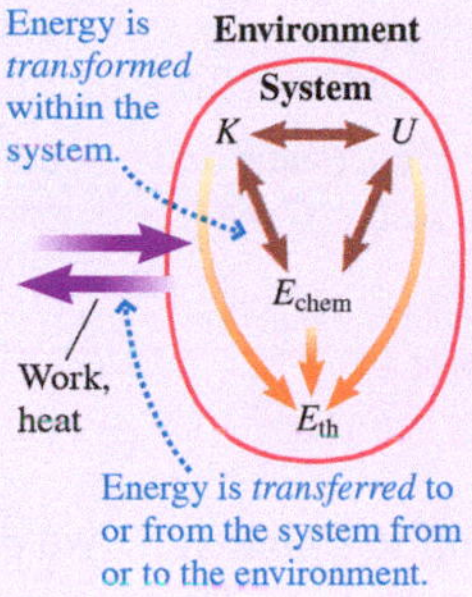


### Conservation of Energy

When work $W$ is done on a system, the system's total energy changes by the amount of work done. In mathematical form, this is the **work-energy equation:**

$$\Delta E = \Delta K + \Delta U_g + \Delta U_s + \Delta E_{th} + \Delta E_{chem} + \cdots = W$$

A system is isolated when no energy is transferred into or out of the system. This means the work is zero, giving the **law of conservation of energy:**

$$\Delta K + \Delta U_g + \Delta U_s + \Delta E_{th} + \Delta E_{chem} + \cdots = 0$$

### Solving Energy Transfer and Energy Conservation Problems

***STRATEGIZE*** Choose the system. Determine the initial and final states.

***PREPARE*** Draw a before-and-after visual overview.

***SOLVE*** Use the before-and-after version of the work-energy equation:

$$K_f + U_f + \Delta E_{th} = K_i + U_i + W$$

Start with this general equation, then specialize to the case at hand:

- Use the appropriate form or forms of potential energy.
- If the system is isolated, set $W = 0$.
- If there is no friction or drag, set $\Delta E_{th} = 0$.

***ASSESS*** See if the numbers make sense—and if the numbers add up. Energy is conserved, and kinetic energy and the change in thermal energy are always positive.

## IMPORTANT CONCEPTS

**Kinetic energy** is an energy of motion:

$$K = \underbrace{\tfrac{1}{2}mv^2}_{\text{Translational}} + \underbrace{\tfrac{1}{2}I\omega^2}_{\text{Rotational}}$$

**Potential energy** is energy stored in a system of interacting objects.

- **Gravitational potential energy:** $U_g = mgy$
- **Elastic potential energy:** $U_s = \dfrac{1}{2}kx^2$

**Thermal energy** is the sum of the microscopic kinetic and potential energies of all the molecules in an object. The hotter an object, the more thermal energy it has. When kinetic (sliding) friction is present, the increase in the thermal energy is $\Delta E_{th} = f_k \Delta x$. When the drag force is present, the increase in the thermal energy is $\Delta E_{th} = D\Delta x$.

**Work** is the process by which energy is transferred to or from a system by the application of mechanical forces.

If a particle moves through a displacement $\vec{d}$ while acted upon by a constant force $\vec{F}$, the force does work

$$W = F_{\parallel}d = Fd\cos\theta$$

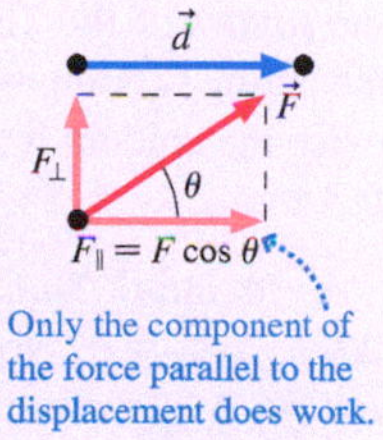


## APPLICATIONS

**Energy diagrams** are a useful way to analyze physical systems.

This curve shows the potential energy of a two-atom molecule as a function of the atomic separation.

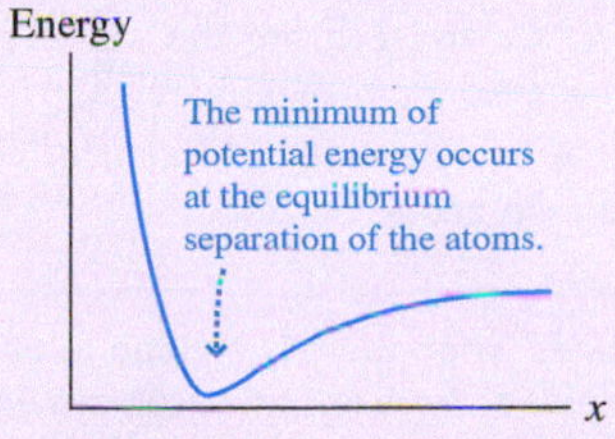


**Perfectly elastic collisions** Both mechanical energy and momentum are conserved.

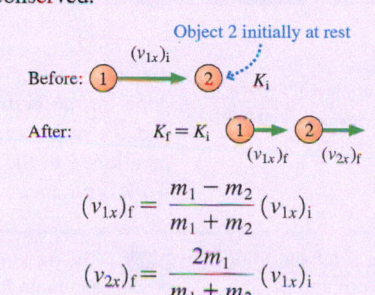

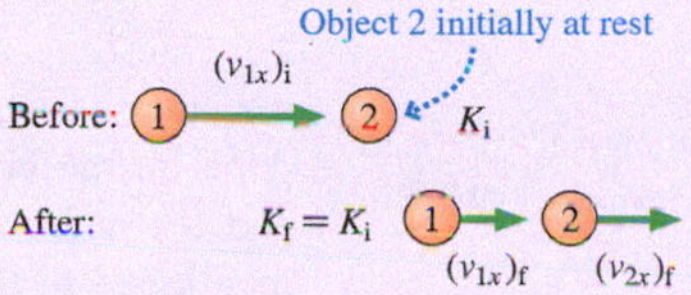


$$(v_{1x})_f = \frac{m_1 - m_2}{m_1 + m_2}(v_{1x})_i$$

$$(v_{2x})_f = \frac{2m_1}{m_1 + m_2}(v_{1x})_i$$

**Power** is the rate at which energy is transformed . . .

$$P = \frac{\Delta E}{\Delta t}$$

$\Delta E$: Amount of energy transformed; $\Delta t$: Time required to transform it

. . . or at which work is done.

$$P = \frac{W}{\Delta t}$$

$W$: Amount of work done; $\Delta t$: Time required to do work

## Learning Objectives After studying this chapter, you should be able to:

- Calculate the work done on an object. *Problems 10.1, 10.2, 10.3, 10.4, 10.5, 10.6, 10.7*
- Calculate an object's kinetic and potential energy. *Conceptual Questions 10.17, 10.18; Problems 10.14, 10.17, 10.20, 10.23, 10.25*
- Understand and calculate the change in thermal energy. *Conceptual Question 10.12; Problems 10.29, 10.30, 10.31, 10.32, 10.34*
- Use the problem-solving approach to solve conservation of energy problems. *Conceptual Questions 10.20, 10.22; Problems 10.36, 10.37, 10.39, 10.41, 10.48*
- Draw and interpret energy diagrams. *Conceptual Question 10.24; Problems 10.51, 10.52*
- Interpret and use molecular bond energies. *Problems 10.53, 10.54*
- Apply energy and momentum conservation to elastic collisions. *Problems 10.55, 10.56, 10.58*
- Understand and calculate power. *Problems 10.60, 10.61, 10.62, 10.65, 10.68*

## STOP TO THINK ANSWERS

**Chapter Preview Stop to Think: A.** Because the car starts from rest, $v_i = 0$ and the kinematic equation is $(v_x)_f^2 = 2a_x\Delta x$, so that $(v_x)_f = \sqrt{2a_x\Delta x}$. Thus the speed is proportional to the square root of the displacement. If, as in this question, the displacement increases by a factor of 4, the speed only doubles. So the speed will increase from 5 m/s to 10 m/s.

**Stop to Think 10.1: C.** The coaster slows. Its kinetic energy is decreasing because kinetic energy is transformed into gravitational potential energy as the coaster climbs the hill.

**Stop to Think 10.2: C.** $W = Fd\cos\theta$. The 10 N force at 90° does no work at all. $\cos 60° = \frac{1}{2}$, so the 8 N force does less work than the 6 N force.

**Stop to Think 10.3: B > D > A = C.** $K = \frac{1}{2}mv^2$. Using the given masses and velocities, we find $K_A = 2.0$ J, $K_B = 4.5$ J, $K_C = 2.0$ J, $K_D = 4.0$ J.

**Stop to Think 10.4: $(U_g)_3 > (U_g)_2 = (U_g)_4 > (U_g)_1$.** Gravitational potential energy depends only on height, not speed.

**Stop to Think 10.5: D.** The potential energy of a spring depends on the *square* of the displacement $x$, so the energy is positive whether the spring is compressed or extended. If the spring is compressed by twice the amount it had been stretched, the energy will increase by a factor of $2^2 = 4$. So the energy will be $4 \times 1\text{ J} = 4$ J.

**Stop to Think 10.6: B.** We can use conservation of energy to write $\Delta K + \Delta E_{th} = 0$. Now if the initial kinetic energy doubles, so does $\Delta K$, so $\Delta E_{th}$ must double as well. But $\Delta E_{th} = f_k \Delta x$, so if $\Delta E_{th}$ doubles, then $\Delta x$ doubles to 2.0 m.

**Stop to Think 10.7:** We define the system as the student, the rope, and the earth. The friction force and the weight force are internal to the system, so it is an isolated system.

**Stop to Think 10.8: D.** In all three cases, Katie has the same initial kinetic energy and potential energy. Thus her energy must be the same at the bottom of the slide in all three cases. Because she has only kinetic energy at the bottom, her speed there must be the same in all three cases as well.

**Stop to Think 10.9: B.** The kinetic energy is the difference between the total energy and the potential energy. This is highest at the bottom of the right well in B.

**Stop to Think 10.10: C.** The initial momentum is $(2M)v_0 + 0$, and the final momentum is $0 + M(2v_0)$. These are equal, so momentum is conserved. The initial kinetic energy is $\frac{1}{2}(2M){v_0}^2 = M{v_0}^2$, and the final kinetic energy is $\frac{1}{2}M(2v_0)^2 = 2M{v_0}^2$. The final kinetic energy is *greater* than the initial kinetic energy, so this collision is not possible. (If the final kinetic energy had been less than the initial kinetic energy, the collision could be possible because the difference in energy could be converted into thermal energy.)

**Stop to Think 10.11: $P_B > P_A = P_C > P_D$.** The power here is the rate at which each runner's internal chemical energy is converted into gravitational potential energy. The change in gravitational potential energy is $mg\Delta y$, so the power is $mg\Delta y/\Delta t$. For runner A, the ratio $m\Delta y/\Delta t$ equals $(80\text{ kg})(10\text{ m})/(10\text{ s}) = 80\text{ kg}\cdot\text{m/s}$. For C, the ratio is also 80 kg·m/s. For B, it's 100 kg·m/s, while for D the ratio is 64 kg·m/s.

**Video Tutor Solution** Chapter 10

# QUESTIONS

## Conceptual Questions

1. The brake shoes of your car are made of a material that can tolerate very high temperatures without being damaged. Why is this so?

For Questions 2 through 9, give a specific example of a system with the energy transformation shown. In these questions, $W$ is the work done on the system, and $K$, $U$, and $Eth$ are the kinetic, potential, and thermal energies of the system, respectively. Any energy not mentioned in the transformation is assumed to remain constant; if work is not mentioned, it is assumed to be zero.

Problem difficulty is labeled as | (straightforward) to ||||| (challenging). Problems labeled INT integrate significant material from earlier chapters; Problems labeled BIO are of biological or medical interest.

The eText icon indicates when there is a video tutor solution available for the chapter or for a specific problem. To launch these videos, log into your eText through Mastering™ Physics or log into the Study Area.

2. $W \rightarrow K$
3. $W \rightarrow U$
4. $K \rightarrow U$
5. $K \rightarrow W$
6. $U \rightarrow K$
7. $W \rightarrow \Delta E_{th}$
8. $U \rightarrow \Delta E_{th}$
9. $K \rightarrow \Delta E_{th}$
10. A ball of putty is dropped from a height of 2 m onto a hard floor, where it sticks. What object or objects need to be included within the system if the system is to be isolated during this process?
11. A diver leaps from a high platform, speeds up as she falls, and then slows to a stop in the water. How do you define the system so that the energy changes are all transformations internal to an isolated system?
12. When your hands are cold, you can rub them together to warm them. Explain the energy transformations that make this possible.
13. Puck B has twice the mass of puck A. Starting from rest, both pucks are pulled the same distance across frictionless ice by strings with the same tension.
    a. Compare the final kinetic energies of pucks A and B.
    b. Compare the final speeds of pucks A and B.
14. To change a tire, you need to use a jack to raise one corner of your car. While doing so, you happen to notice that pushing the jack handle down 20 cm raises the car only 0.2 cm. Use energy concepts to explain why the handle must be moved so far to raise the car by such a small amount.
15. You drop two balls from a tower, one of mass $m$ and the other of mass $2m$. Just before they hit the ground, which ball, if either, has the larger kinetic energy? Explain.
16. If you fall and skid to a stop on a carpeted floor, you can get a rug burn. Much of this discomfort comes from abrasion, but there can also be a real burn where the skin was too hot. How does this happen?
17. A roller coaster car rolls down a frictionless track, reaching speed $v$ at the bottom.
    a. If you want the car to go twice as fast at the bottom, by what factor must you increase the height of the track?
    b. Does your answer to part a depend on whether the track is straight or not? Explain.
18. A spring gun shoots out a plastic ball at speed $v$. The spring is then compressed twice the distance it was on the first shot.
    a. By what factor is the spring's potential energy increased?
    b. By what factor is the ball's speed increased? Explain.
19. A baseball pitcher can throw a baseball (mass 0.14 kg) much faster than a football quarterback can throw a football (mass 0.42 kg). Use energy concepts to explain why you would expect this to be true.
20. Sandy and Chris stand on the edge of a cliff and throw identical mass rocks at the same speed. Sandy throws her rock horizontally while Chris throws his upward at an angle of 45° to the horizontal. Are the rocks moving at the same speed when they hit the ground, or is one moving faster than the other? If one is moving faster, which one? Explain.
21. A solid cylinder and a hollow cylinder have the same mass, same radius, and turn on frictionless, horizontal axles. (The hollow cylinder has lightweight spokes connecting it to the axle.) A rope is wrapped around each cylinder and tied to a block. The blocks have the same mass and are held the same height above the ground as shown in Figure Q10.21. Both blocks are released simultaneously. The ropes do not slip. Which block hits the ground first? Or is it a tie? Explain.

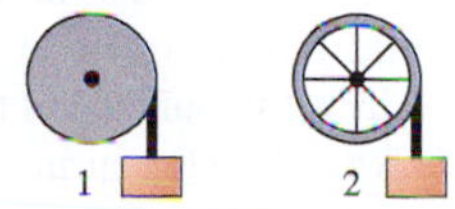

FIGURE Q10.21

22. A bowler tosses a ball without spin. The ball slides down the alley. At some point, friction with the alley makes the ball start to roll; eventually, it rolls without sliding. When the ball reaches this point, it is moving at a lower speed than the original toss. Use energy concepts to give two reasons for this change.
23. BIO Ferns that eject spores generally do so in pairs, with two spores flying off in opposite directions. The structure from which the spores are launched is quite lightweight. If it takes a certain amount of energy to eject each spore, explain how launching the spores in pairs provides for the greatest initial launch speed for each spore.
24. Figure Q10.24 shows a potential-energy diagram for a particle. The particle is at rest at point A and is then given a slight nudge to the right. Describe the subsequent motion.

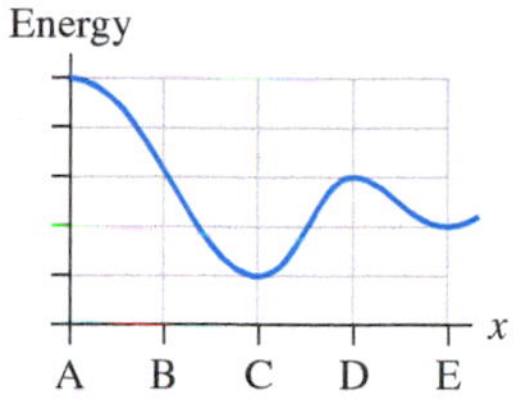

FIGURE Q10.24

## Multiple-Choice Questions

25. || A roller coaster starts from rest at its highest point and then descends on its (frictionless) track. Its speed is 30 m/s when it reaches ground level. What was its speed when its height was half that of its starting point?
    A. 11 m/s B. 15 m/s C. 21 m/s D. 25 m/s
26. | A woman uses a pulley and a rope to raise a 20 kg weight to a height of 2 m. If it takes 4 s to do this, about how much power is she supplying?
    A. 100 W B. 200 W C. 300 W D. 400 W
27. | A hockey puck sliding along frictionless ice with speed $v$ to the right collides with a horizontal spring and compresses it by 2.0 cm before coming to a momentary stop. What will be the spring's maximum compression if the same puck hits it at a speed of $2v$?
    A. 2.0 cm B. 2.8 cm C. 4.0 cm
    D. 5.6 cm E. 8.0 cm
28. || A block slides down a smooth ramp, starting from rest at a height $h$. When it reaches the bottom it's moving at speed $v$. It then continues to slide up a second smooth ramp. At what height is its speed equal to $v/2$?
    A. $h/4$ B. $h/2$ C. $3h/4$ D. $2h$
29. | A wrecking ball is suspended from a 5.0-m-long cable that makes a 30° angle with the vertical. The ball is released and swings down. What is the ball's speed at the lowest point?
    A. 7.7 m/s B. 4.4 m/s C. 3.6 m/s D. 3.1 m/s
30. || BIO A dog can provide sufficient power to pull a sled with a 60 N force at a steady 2.0 m/s. Suppose the dog is hitched to a different sled that requires 120 N to move at a constant speed. How fast can the dog pull this second sled?
    A. 0.50 m/s B. 1.0 m/s C. 1.5 m/s D. 2.0 m/s
31. ||| Most of the energy you expend in cycling is dissipated by the drag force. If you double your speed, you increase the drag force by a factor of 4. This increases the power to cycle at this greater speed by what factor?
    A. 2 B. 4 C. 8 D. 16

Watch Video Solution Problem 10.5

# PROBLEMS

## Section 10.2 Work

1. || A 2.0 kg book is lying on a 0.75-m-high table. You pick it up and place it on a bookshelf 2.3 m above the floor. During this process,
   a. How much work does gravity do on the book?
   b. How much work does your hand do on the book?
2. || The two ropes seen in Figure P10.2 are used to lower a 255 kg piano exactly 5 m from a second-story window to the ground. How much work is done by each of the three forces?

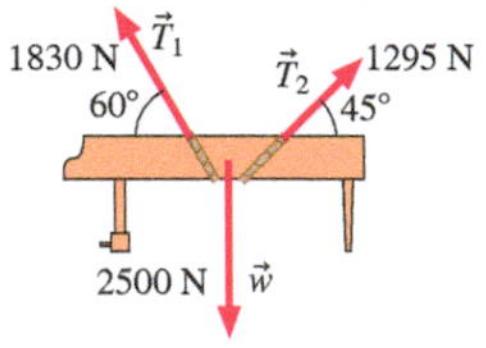


FIGURE P10.2

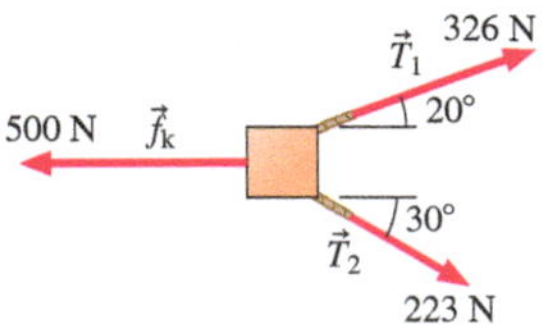


FIGURE P10.3

3. | The two ropes shown in the bird's-eye view of Figure P10.3 are used to drag a crate exactly 3 m across the floor. How much work is done by each of the ropes on the crate?
4. | You are pulling a child in a wagon. The rope handle is inclined upward at a 60° angle. The tension in the handle is 20 N. How much work do you do if you pull the wagon 100 m at a constant speed?
5. | A boy flies a kite with the string at a 30° angle to the horizontal. The tension in the string is 4.5 N. How much work does the string do on the boy if the boy
   a. Stands still?
   b. Walks a horizontal distance of 11 m away from the kite?
   c. Walks a horizontal distance of 11 m toward the kite?
6. || A typical muscle fiber is 2.0 cm long and has a cross-section BIO area of $3.1 \times 10^{-9}$ m$^2$. When the muscle fiber is stimulated, it pulls with a force of 1.2 mN. What is the work done by the muscle fiber as it contracts to a length of 1.6 cm?
7. || A crate slides down a ramp that makes a 20° angle with the ground. To keep the crate moving at a steady speed, Paige pushes back on it with a 68 N horizontal force. How much work does Paige do on the crate as it slides 3.5 m down the ramp?

## Section 10.3 Kinetic Energy

8. || A wind turbine works by slowing the air that passes its blades and converting much of the extracted kinetic energy to electric energy. A large wind turbine has 45-m-radius blades. In typical conditions, 92,000 kg of air moves past the blades every second. If the air is moving at 12 m/s before it passes the blades and the wind turbine extracts 40% of this kinetic energy, how much energy is extracted every second?
9. || At what speed does a 1000 kg compact car have the same kinetic energy as a 20,000 kg truck going 25 km/h?
10. | A 60 kg runner in a sprint moves at 11 m/s. A 60 kg cheetah in a sprint moves at 33 m/s. By what factor does the kinetic energy of the cheetah exceed that of the human runner?
11. | A car is traveling at 10 m/s.
    a. How fast would the car need to go to double its kinetic energy?
    b. By what factor does the car's kinetic energy increase if its speed is doubled to 20 m/s?
12. || The opposite of a wind turbine is an electric fan: The electric energy that powers the fan is converted to the kinetic energy of moving air. A fan is putting 1.0 J of kinetic energy into the air every second. Then the fan speed is increased by a factor of 2. Air moves through the fan faster, so the fan moves twice as much air at twice the speed. How much kinetic energy goes into the air every second?
13. | How fast would an 80 kg man need to run in order to have the same kinetic energy as an 8.0 g bullet fired at 400 m/s?
14. ||| A fielder tosses a 0.15 kg baseball at 32 m/s at a 30° angle to the horizontal. What is the ball's kinetic energy at the start of its motion? What is the kinetic energy at the highest point of its arc?
15. ||| Sam's job at the amusement park is to slow down and bring to a stop the boats in the log ride. If a boat and its riders have a mass of 1200 kg and the boat drifts in at 1.2 m/s, how much work does Sam do to stop it?
16. || A school has installed a modestly-sized wind turbine. The three blades are 4.6 m long; each blade has a mass of 45 kg. You can assume that the blades are uniform along their lengths. When the blades spin at 240 rpm, what is the kinetic energy of the blade assembly?
17. ||| The turntable in a microwave oven has a moment of inertia of 0.040 kg · m$^2$ and rotates continuously, making a complete revolution every 4.0 s. What is its kinetic energy?
18. || A typical meteor that hits the earth's upper atmosphere has a mass of only 2.5 g, about the same as a penny, but it is moving at an impressive 40 km/s. As the meteor slows, the resulting thermal energy makes a glowing streak across the sky, a shooting star. The small mass packs a surprising punch. At what speed would a 900 kg compact car need to move to have the same kinetic energy?

19. |||| An energy storage system based on a flywheel (a rotating disk) can store a maximum of 4.0 MJ when the flywheel is rotating at 20,000 revolutions per minute. What is the moment of inertia of the flywheel?

## Section 10.4 Potential Energy

20. || The lowest point in Death Valley is 85.0 m below sea level. The summit of nearby Mt. Whitney has an elevation of 4420 m. What is the change in gravitational potential energy of an energetic 65.0 kg hiker who makes it from the floor of Death Valley to the top of Mt. Whitney?
21. | The world's fastest humans can reach speeds of about 11 m/s. In order to increase his gravitational potential energy by an amount equal to his kinetic energy at full speed, how high would such a sprinter need to climb?
22. | A 72 kg bike racer climbs a 1200-m-long section of road that has a slope of 4.3°. By how much does his gravitational potential energy change during this climb?
23. || A 1000 kg wrecking ball hangs from a 15-m-long cable. The ball is pulled back until the cable makes an angle of 25° with the vertical. By how much has the gravitational potential energy of the ball changed?

24. || How far must you stretch a spring with $k = 1000$ N/m to store 200 J of energy?

25. || How much energy can be stored in a spring with a spring constant of 500 N/m if its maximum possible stretch is 20 cm?

26. || The spring in a retractable ballpoint pen is 1.8 cm long, with a 300 N/m spring constant. When the pen is retracted, the spring is compressed by 1.0 mm. When you click the button to extend the pen, you compress the spring by an additional 6.0 mm. How much energy is required to extend the pen?

27. |||| The elastic energy stored in your tendons can contribute up
BIO to 35% of your energy needs when running. Sports scientists have studied the change in length of the knee extensor tendon in sprinters and nonathletes. They find (on average) that the sprinters' tendons stretch 41 mm, while nonathletes' stretch only 33 mm. The spring constant for the tendon is the same for both groups, 33 N/mm. What is the difference in maximum stored energy between the sprinters and the nonathletes?

28. || Scallops use muscles
BIO to close their shells.
INT Opening the shell is another story—muscles can only pull, they can't push. Instead of muscles, the shell is opened by a spring, a pad of a very elastic biological material called abductin. When the shell closes, the pad compresses; a restoring force then pushes the shell back open. The energy to open the shell comes from the elastic energy that was stored when the shell was closed. Figure P10.28 shows smoothed data for the restoring force of an abductin pad versus the compression. When the shell closes, the pad compresses by 0.15 mm. How much elastic potential energy is stored?

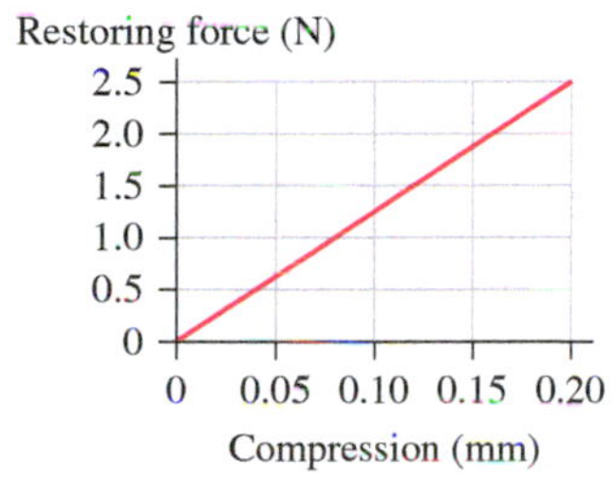

FIGURE P10.28

## Section 10.5 Thermal Energy

29. || Mark pushes his broken car 150 m down the block to his friend's house. He has to exert a 110 N horizontal force to push the car at a constant speed. How much thermal energy is created in the tires and road during this short trip?

30. || When you skid to a stop on your bike, you can significantly
INT heat the small patch of tire that rubs against the road surface. Suppose a person skids to a stop by hitting the brake on his back tire, which supports half the 80 kg combined mass of the bike and rider, leaving a skid mark that is 40 cm long. Assume a coefficient of kinetic friction of 0.80. How much thermal energy is deposited in the tire and the road surface?

31. ||| A 900 N crate slides 12 m down a ramp that makes an angle of 35° with the horizontal. If the crate slides at a constant speed, how much thermal energy is created?

32. || If you slide down a rope, it's possible to create enough ther-
INT mal energy to burn your hands or your legs where they grip the rope. Suppose a 40 kg child slides down a rope at a playground, descending 2.0 m at a constant speed. How much thermal energy is created as she slides down the rope?

33. ||| A 25 kg child slides down a playground slide at a *constant speed.* The slide has a height of 3.0 m and is 7.0 m long. Using the law of conservation of energy, find the magnitude of the kinetic friction force acting on the child.

34. || Some runners train with parachutes that trail behind them
BIO to provide a large drag force. These parachutes are designed
INT to have a large drag coefficient. One model expands to a square 1.8 m on a side, with a drag coefficient of 1.4. A runner completes a 200 m run at 5.0 m/s with this chute trailing behind. How much thermal energy is added to the air by the drag force?

## Section 10.6 Conservation of Energy

35. || A boy reaches out of a window and tosses a ball straight up with a speed of 10 m/s. The ball is 20 m above the ground as he releases it. Use conservation of energy to find
   a. The ball's maximum height above the ground.
   b. The ball's speed as it passes the window on its way down.
   c. The speed of impact on the ground.

36. | The famous cliff divers of Acapulco leap from a perch 35 m
INT above the ocean. How fast are they moving when they reach the water surface? What happens to their kinetic energy as they slow to a stop in the water?

37. ||| What minimum speed does a 100 g puck need to make it to the top of a frictionless ramp that is 3.0 m long and inclined at 20°?

38. | You can, in an emergency, start a manual transmission car by putting it in neutral, letting the car roll down a hill to pick up speed, then putting it in gear and quickly letting out the clutch. If the car needs to be moving at 3.5 m/s for this to work, how high a hill do you need? (You can ignore friction and drag.)

39. ||| A 1500 kg car is approaching the hill shown in Figure P10.39 at 10 m/s when it suddenly runs out of gas.
   a. Can the car make it to the top of the hill by coasting?
   b. If your answer to part a is yes, what is the car's speed after coasting down the other side?

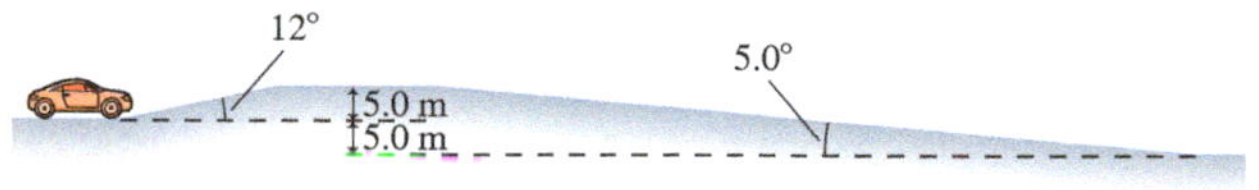

FIGURE P10.39

40. | A 480 g peregrine falcon reaches a
BIO speed of 75 m/s in a vertical dive called a stoop. If we assume that the falcon speeds up under the influence of gravity only, what is the minimum height of the dive needed to achieve this speed?

41. || A fireman of mass 80 kg slides down a pole. When he reaches the bottom, 4.2 m below his starting point, his speed is 2.2 m/s. By how much has thermal energy increased during his slide?

42. || A 20 kg child slides down a 3.0-m-high playground slide. She starts from rest, and her speed at the bottom is 2.0 m/s.
   a. What energy transfers and transformations occur during the slide?
   b. What is the total change in the thermal energy of the slide and the seat of her pants?

43. || A hockey puck is given an initial speed of 5.0 m/s. If the coefficient of kinetic friction between the puck and the ice is 0.05, how far does the puck slide before coming to rest? Solve this problem using conservation of energy.

44. ||| Monica pulls her daughter Jessie in a bike trailer. The trailer and Jessie together have a mass of 25 kg. Monica starts up a 100-m-long slope that's 4.0 m high. On the slope, Monica's bike pulls on the trailer with a constant force of 8.0 N. They start out at the bottom of the slope with a speed of 5.3 m/s. What is their speed at the top of the slope?

45. || In the winter activity of tubing, riders slide down snow-covered slopes while sitting on large inflated rubber tubes. To get to the top of the slope, a rider and his tube, with a total mass of 80 kg, are pulled at a constant speed by a tow rope that maintains a constant tension of 340 N. How much thermal energy is created in the slope and the tube during the ascent of a 30-m-high, 120-m-long slope?

46. || BIO INT Mosses don't spread by dispersing seeds; they disperse tiny spores. The spores are so small that they will stay aloft and move with the wind, but getting them to be windborne requires the moss to shoot the spores upward. Some species do this by using a spore-containing capsule that dries out and shrinks. The pressure of the air trapped inside the capsule increases. At a certain point, the capsule pops, and a stream of spores is ejected upward at 3.6 m/s, reaching an ultimate height of 20 cm. What fraction of the initial kinetic energy is converted to the final potential energy? What happens to the "lost" energy?

47. || A cyclist is coasting at 12 m/s when she starts down a 450-m-long slope that is 30 m high. The cyclist and her bicycle have a combined mass of 70 kg. A steady 12 N drag force due to air resistance acts on her as she coasts all the way to the bottom. What is her speed at the bottom of the slope?

48. || INT When you stand on a trampoline, the surface depresses below equilibrium, and the surface pushes up on you, as the data for a real trampoline in Figure P10.48 show. The linear variation of the force as a function of distance means that we can model the restoring force as that of a spring. A 72 kg gymnast jumps on the trampoline. At the lowest point of his motion, he is 0.80 m below equilibrium. If we assume that all of the energy stored in the trampoline goes into his motion, how high above this lowest point will he rise?

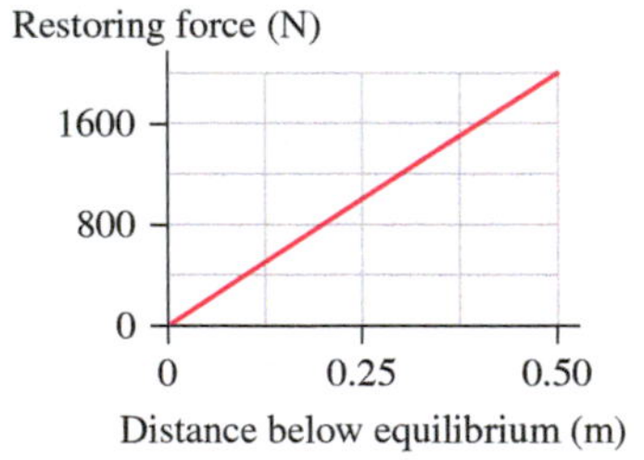

FIGURE P10.48

49. || The 5.0-m-long rope in Figure P10.49 hangs vertically from a tree right at the edge of a ravine. A woman wants to use the rope to swing to the other side of the ravine. She runs as fast as she can, grabs the rope, and swings out over the ravine.

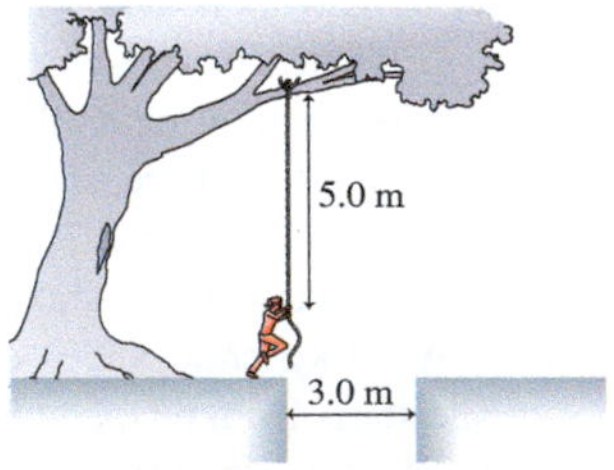

FIGURE P10.49

a. As she swings, what energy conversion is taking place?
b. When she's directly over the far edge of the ravine, how much higher is she than when she started?
c. Given your answers to parts a and b, how fast must she be running when she grabs the rope in order to swing all the way across the ravine?

50. ||| INT The Special Olympics raises money through "plane pull" events in which teams of 25 people compete to see who can pull a 74,000 kg airplane 3.7 m across the tarmac. The inertia of the plane is an issue—but so is the 14,000 N rolling friction force that works against the teams. If a team pulls with a constant force and moves the plane 3.7 m in 6.1 s (an excellent time), what fraction of the team's work goes to kinetic energy and what fraction goes to thermal energy?

### Section 10.7 Energy Diagrams

51. || Figure P10.51 is the potential-energy diagram for a 20 g particle that is released from rest at $x = 1.0$ m.
a. Will the particle move to the right or to the left? How can you tell?
b. What is the particle's maximum speed? At what position does it have this speed?
c. Where are the turning points of the motion?

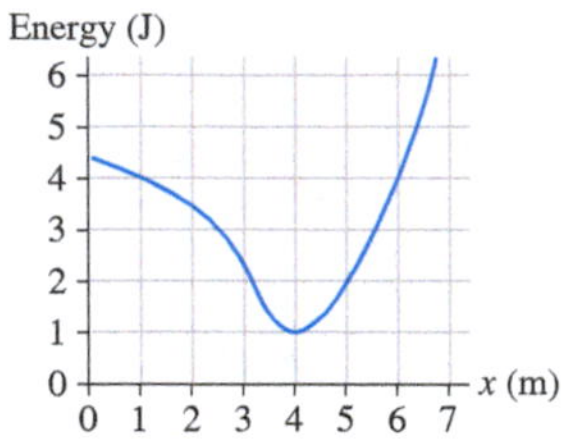

FIGURE P10.51

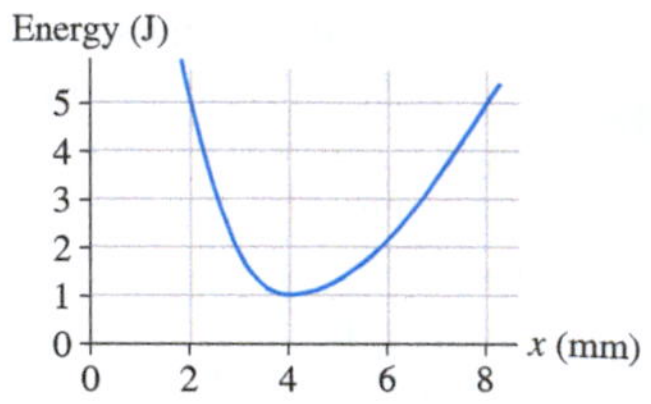

FIGURE P10.52

52. || For the potential-energy diagram in Figure P10.52, what is the maximum speed of a 2.0 g particle that oscillates between $x = 2.0$ mm and $x = 8.0$ mm?

### Section 10.8 Molecular Bonds and Chemical Energy

At normal temperatures and pressures, hydrogen gas is composed of $H_2$ molecules. An energy diagram for a hydrogen molecule appears in Figure P10.53. Use this information to answer Problems 10.53 and 10.54.

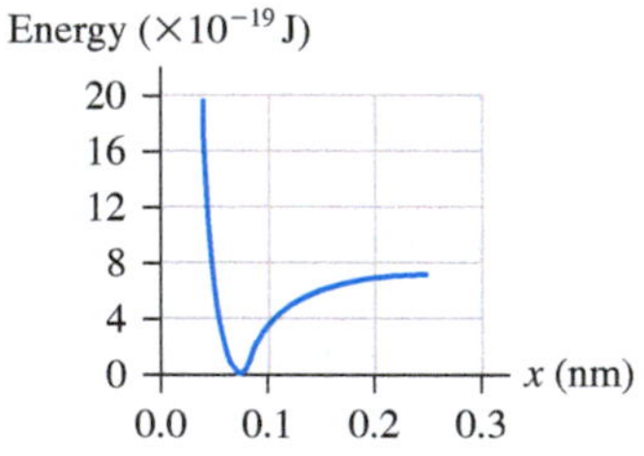

FIGURE P10.53

53. | How far apart are the individual atoms in a molecule of $H_2$?

54. || What energy photon is needed to dissociate a molecule of $H_2$?

### Section 10.9 Energy in Collisions

55. || A 50 g marble moving at 2.0 m/s strikes a 20 g marble at rest. What is the speed of each marble immediately after the collision? Assume the collision is perfectly elastic and the marbles collide head-on.

56. || Ball 1, with a mass of 100 g and traveling at 10 m/s, collides head-on with ball 2, which has a mass of 300 g and is initially at rest. What are the final velocities of each ball if the collision is (a) perfectly elastic? (b) perfectly inelastic?

57. | An air-track glider undergoes a perfectly inelastic collision with an identical glider that is initially at rest. What fraction of the first glider's initial kinetic energy is transformed into thermal energy in this collision?

58. | Two balls undergo a perfectly elastic head-on collision, with one ball initially at rest. If the incoming ball has a speed of 200 m/s, what are the final speed and direction of each ball if
   a. The incoming ball is *much* more massive than the stationary ball?
   b. The stationary ball is *much* more massive than the incoming ball?

### Section 10.10 Power

59. || a. How much work must you do to push a 10 kg block of steel across a steel table at a steady speed of 1.0 m/s for 3.0 s? The coefficient of kinetic friction for steel on steel is 0.60.
   b. What is your power output while doing so?

60. || A shooting star is actually the track of a meteor, typically a small chunk of debris from a comet that has entered the earth's atmosphere. As the drag force slows the meteor down, its kinetic energy is converted to thermal energy, leaving a glowing trail across the sky. A typical meteor has a surprisingly small mass, but what it lacks in size it makes up for in speed. Assume that a meteor has a mass of 1.5 g and is moving at an impressive 50 km/s, both typical values. What power is generated if the meteor slows down over a typical 2.1 s? Can you see how this tiny object can make a glowing trail that can be seen hundreds of kilometers away?

61. | a. How much work does an elevator motor do to lift a 1000 kg elevator a height of 100 m at a constant speed?
   b. How much power must the motor supply to do this in 50 s at constant speed?

62. || A 500 kg horse can provide a steady output power of 750 W (that is, 1 horsepower) when pulling a load. How about a 38 kg sled dog? Data show that a 38 kg dog can pull a sled that requires a pulling force of 60 N at a steady 2.2 m/s. What are the specific power values for the dog and the horse? What is the minimum number of dogs needed to provide the same power as one horse?
BIO

63. ||| A 1000 kg sports car accelerates from 0 to 30 m/s in 10 s. What is the average power of the engine?

64. || A world-class sprinter running a 100 m dash was clocked at 5.4 m/s 1.0 s after starting running and at 9.8 m/s 1.5 s later. In which of these time intervals, 0 to 1.0 s or 1.0 s to 2.5 s, was his output power greater?
BIO

65. || An elite Tour de France cyclist can maintain an output power of 450 W during a sustained climb. At this output power, how long would it take an 85 kg cyclist (including the mass of his bike) to climb the famed 1100-m-high Alpe d'Huez mountain stage?
BIO

66. || A 70 kg human sprinter can accelerate from rest to 10 m/s in 3.0 s. During the same time interval, a 30 kg greyhound can accelerate from rest to 20 m/s. What is the specific power for each of these athletes?
BIO

67. || A 710 kg car drives at a constant speed of 23 m/s. It is subject to a drag force of 500 N. What power is required from the car's engine to drive the car
   a. On level ground?
   b. Up a hill with a slope of 2.0°?

68. || A 95 kg quarterback accelerates a 0.42 kg ball from rest to 24 m/s in 0.083 s. What is the specific power for this toss?
BIO

69. ||| An elevator weighing 2500 N ascends at a constant speed of 8.0 m/s. How much power must the motor supply to do this?

70. || Humans can produce an output power as great as 20 W/kg during extreme exercise. Sloths are not so energetic. At its maximum speed, a 4.0 kg sloth can climb a height of 6.0 m in 2.0 min. What's the specific power for this climb?
BIO

## General Problems

71. || A 550 kg elevator accelerates upward at 1.2 m/s$^2$ for the first 15 m of its motion. How much work is done during this part of its motion by the cable that lifts the elevator?
INT

72. || The energy yield of a nuclear weapon is often defined in terms of the equivalent mass of a conventional explosive. 1 ton of a conventional explosive releases 4.2 GJ. A typical nuclear warhead releases 250,000 times more, so the yield is expressed as 250 kilotons. That is a staggering explosion, but the asteroid impact that wiped out the dinosaurs was significantly greater. Assume that the asteroid was a sphere 10 km in diameter, with a density of 2500 kg/m$^3$ and moving at 30 km/s. What energy was released at impact, in joules and in kilotons?

73. || A 2.3 kg box, starting from rest, is pushed up a ramp by a 10 N force parallel to the ramp. The ramp is 2.0 m long and tilted at 17°. The speed of the box at the top of the ramp is 0.80 m/s. Consider the system to be the box + ramp + earth.
   a. How much work $W$ does the force do on the system?
   b. What is the change $\Delta K$ in the kinetic energy of the system?
   c. What is the change $\Delta U_g$ in the gravitational potential energy of the system?
   d. What is the change $\Delta E_{th}$ in the thermal energy of the system?

74. ||| A 55 kg skateboarder wants to just make it to the upper edge of a "half-pipe" with a radius of 3.0 m, as shown in Figure P10.74. What speed does he need at the bottom if he will coast all the way up? The skateboarder isn't a simple particle: Assume that his mass in a deep crouch is concentrated 0.75 m from the half-pipe. If he remains in that position all the way up, what initial speed does he need to reach the upper edge?

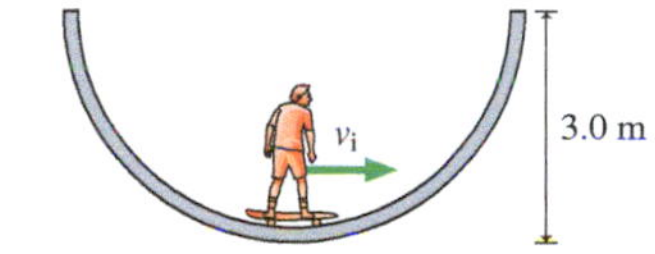

FIGURE P10.74

75. ||| Fleas have remarkable jumping ability. A 0.50 mg flea, jumping straight up, would reach a height of 40 cm if there were no air resistance. In reality, air resistance limits the height to 20 cm.
BIO
   a. What is the flea's kinetic energy as it leaves the ground?
   b. At its highest point, what fraction of the initial kinetic energy has been converted to potential energy?

76. ||| You are driving your 1500 kg car at 20 m/s down a hill with a 5.0° slope when a deer suddenly jumps out onto the roadway. You slam on your brakes, skidding to a stop. How far do you skid before stopping if the kinetic friction force between your tires and the road is 1.2 × 10$^4$ N? Solve this problem using conservation of energy.

Watch Video Solution Problem 10.79

77. || A 20 kg child is on a swing that hangs from 3.0-m-long chains, as shown in Figure P10.77. What is her speed $v_i$ at the bottom of the arc if she swings out to a 45° angle before reversing direction?

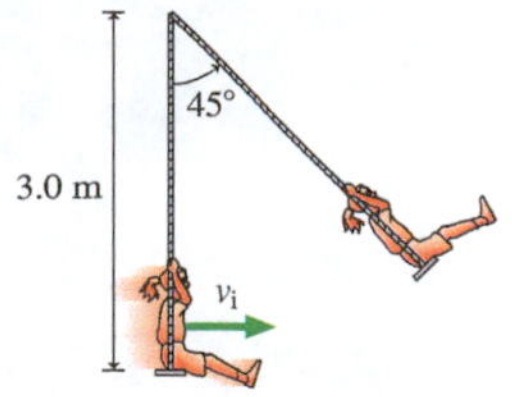


FIGURE P10.77

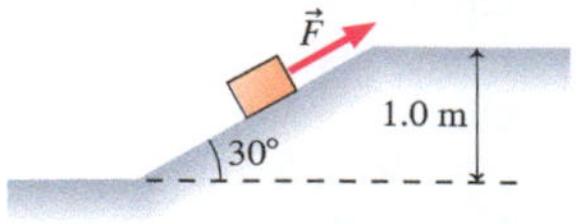


FIGURE P10.78

78. || Suppose you lift a 20 kg box by a height of 1.0 m.
   a. How much work do you do in lifting the box?
   Instead of lifting the box straight up, suppose you push it up a 1.0-m-high ramp that makes a 30° degree angle with the horizontal, as shown in Figure P10.78. Being clever, you choose a ramp with no friction.
   b. How much force $F$ is required to push the box straight up the slope at a constant speed?
   c. How long is the ramp?
   d. Use your force and distance results to calculate the work you do in pushing the box up the ramp. How does this compare to your answer to part a?

79. | The sledder shown in Figure P10.79 starts from the top of a frictionless hill and slides down into the valley. What initial speed $v_i$ does the sledder need to just make it over the next hill?

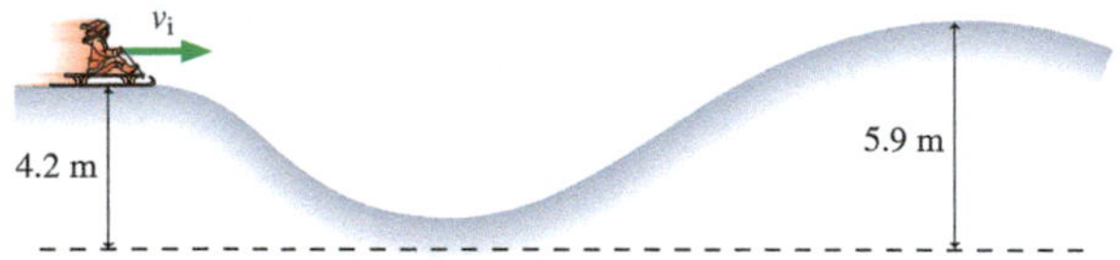


FIGURE P10.79

80. ||||| In a physics lab experiment, a spring clamped to the table
INT shoots a 20 g ball horizontally. When the spring is compressed 20 cm, the ball travels horizontally 5.0 m and lands on the floor 1.5 m below the point at which it left the spring. What is the spring constant?

81. ||||| The maximum energy a bone can absorb without breaking is
BIO surprisingly small. For a healthy human of mass 60 kg, experimental data show that the leg bones of both legs can absorb about 200 J.
   a. From what maximum height could a person jump and land rigidly upright on both feet without breaking his legs? Assume that all the energy is absorbed in the leg bones in a rigid landing.
   b. People jump from much greater heights than this; explain how this is possible.
   **Hint:** Think about how people land when they jump from greater heights.

82. || In an amusement park water slide, people slide down an
INT essentially frictionless tube. The top of the slide is 3.0 m above the bottom where they exit the slide, moving horizontally, 1.2 m above a swimming pool. What horizontal distance do they travel from the exit point before hitting the water? Does the mass of the person make any difference?

83. ||| You have been asked to design a "ballistic spring system" to measure the speed of bullets. A bullet of mass $m$ is fired into a block of mass $M$. The block, with the embedded bullet, then slides across a frictionless table and collides with a horizontal spring whose spring constant is $k$. The opposite end of the spring is anchored to a wall. The spring's maximum compression $d$ is measured.
   a. Find an expression for the bullet's initial speed $v_B$ in terms of $m$, $M$, $k$, and $d$.
   **Hint:** This is a two-part problem. The bullet's collision with the block is an inelastic collision. What quantity is conserved in an inelastic collision? Subsequently the block hits a spring on a frictionless surface. What quantity is conserved in this collision?
   b. What was the speed of a 5.0 g bullet if the block's mass is 2.0 kg and if the spring, with $k = 50$ N/m, was compressed by 10 cm?
   c. What fraction of the bullet's initial kinetic energy is "lost"? Where did it go?

84. || Boxes A and B in Figure P10.84 have masses of 12.0 kg and 4.0 kg, respectively. The two boxes are released from rest. Use conservation of energy to find the boxes' speed when box B has fallen a distance of 0.50 m. Assume a frictionless upper surface.

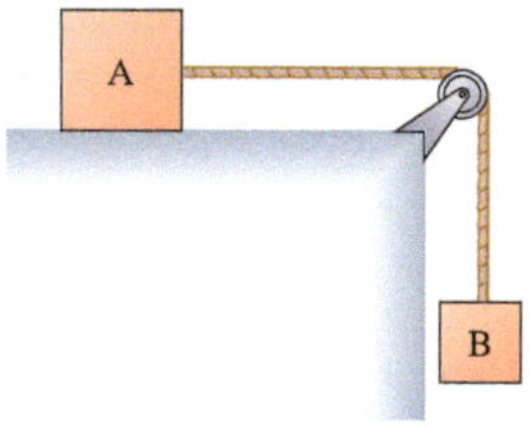


FIGURE P10.84

85. || Two coupled boxcars are rolling along at 2.5 m/s when they
INT collide with and couple to a third, stationary boxcar.
   a. What is the final speed of the three coupled boxcars?
   b. What fraction of the cars' initial kinetic energy is transformed into thermal energy?

86. ||| A 50 g ball of clay traveling at 6.5 m/s hits and sticks to a
INT 1.0 kg block sitting at rest on a frictionless surface.
   a. What is the speed of the block after the collision?
   b. Show that the mechanical energy is *not* conserved in this collision. What percentage of the ball's initial kinetic energy is "lost"? Where did this kinetic energy go?

87. || A package of mass $m$ is
INT released from rest at a warehouse loading dock and slides down a 3.0-m-high frictionless chute to a waiting truck. Unfortunately, the truck driver went on a break without having removed the previous package, of mass $2m$, from the bottom of the chute as shown in Figure P10.87.

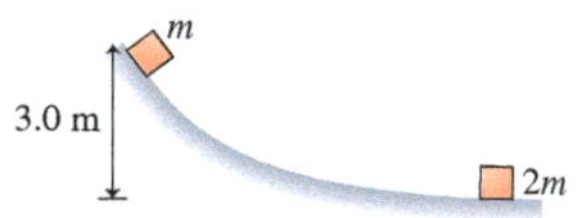


FIGURE P10.87

   a. Suppose the packages stick together. What is their common speed after the collision?
   b. Suppose the collision between the packages is perfectly elastic. To what height does the package of mass $m$ rebound?

88. ||| Swordfish are capable of stunning output power for short
BIO bursts. A 650 kg swordfish has a cross-section area of 0.92 $m^2$
INT and a drag coefficient of 0.0091—exceptionally low due to a number of adaptations. Such a fish can sustain a speed of 30 m/s for a few seconds. Assume seawater has a density of 1026 $kg/m^3$. What is the specific power for motion at this high speed?

89. || The mass of an elevator and its occupants is 1200 kg. The electric motor that lifts the elevator can provide a maximum power of 15 kW. What is the maximum constant speed at which this motor can lift the elevator?

## MCAT-Style Passage Problems

### Tennis Ball Testing

A tennis ball bouncing on a hard surface compresses and then rebounds. The details of the rebound are specified in tennis regulations. Tennis balls, to be acceptable for tournament play, must have a mass of 57.5 g. When dropped from a height of 2.5 m onto a concrete surface, a ball must rebound to a height of 1.4 m. During impact, the ball compresses by approximately 6 mm.

90. | How fast is the ball moving when it hits the concrete surface? (Ignore air resistance.)
A. 5 m/s B. 7 m/s
C. 25 m/s D. 50 m/s

91. | If the ball accelerates uniformly when it hits the floor, what is its approximate acceleration as it comes to rest before rebounding?
A. 1000 $m/s^2$ B. 2000 $m/s^2$
C. 3000 $m/s^2$ D. 4000 $m/s^2$

92. | The ball's kinetic energy just after the bounce is less than just before the bounce. In what form does this lost energy end up?
A. Elastic potential energy
B. Gravitational potential energy
C. Thermal energy
D. Rotational kinetic energy

93. | By approximately what percent does the kinetic energy decrease?
A. 35% B. 45% C. 55% D. 65%

94. | When a tennis ball bounces from a racket, the ball loses approximately 30% of its kinetic energy to thermal energy. A ball that hits a racket at a speed of 10 m/s will rebound with approximately what speed?
A. 8.5 m/s B. 7.0 m/s C. 4.5 m/s D. 3.0 m/s

### Work and Power in Cycling

When you ride a bicycle at constant speed, almost all of the energy you expend goes into the work you do against the drag force of the air. In this problem, assume that *all* of the energy expended goes into working against drag. As we saw in Section 5.6, the drag force on an object is approximately proportional to the square of its speed with respect to the air. For this problem, assume that $F \propto v^2$ exactly and that the air is motionless with respect to the ground unless noted otherwise. Suppose a cyclist and her bicycle have a combined mass of 60 kg and she is cycling along at a speed of 5 m/s.

95. | If the drag force on the cyclist is 10 N, how much energy does she use in cycling 1 km?
A. 6 kJ B. 10 kJ C. 50 kJ D. 100 kJ

96. | Under these conditions, how much power does she expend as she cycles?
A. 10 W B. 50 W C. 100 W D. 200 W

97. | If she doubles her speed to 10 m/s, how much energy does she use in cycling 1 km?
A. 20 kJ B. 40 kJ C. 200 kJ D. 400 kJ

98. | How much power does she expend when cycling at that speed?
A. 100 W B. 200 W C. 400 W D. 1000 W

99. | Upon reducing her speed back down to 5 m/s, she hits a headwind of 5 m/s. How much power is she expending now?
A. 100 W B. 200 W C. 500 W D. 1000 W

# 11 Using Energy

The odd hopping gait of a kangaroo has a very practical purpose: It lets the kangaroo cover great distances with minimal energy input. How is this efficiency possible?

## LOOKING AHEAD

### Energy Use in the Body

The energy to run, to move, and to stay warm comes from the chemical energy in the food you eat.

You'll learn to calculate how much energy your body requires to complete a range of tasks.

### Temperature and Heat

This tea kettle gets hotter because of the transfer of energy from the burner as **heat.**

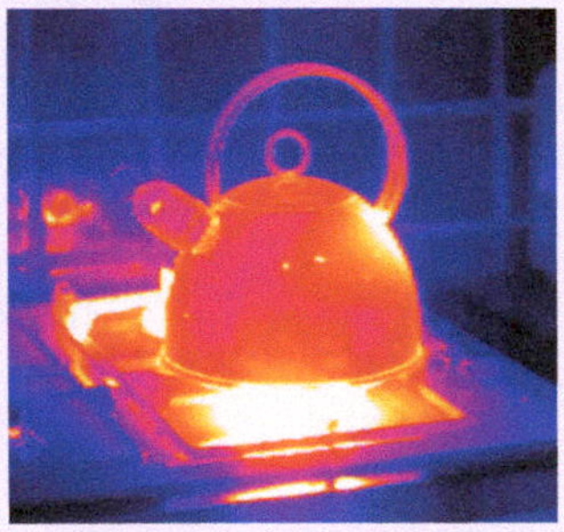

You'll learn that *heat* is energy transferred due to a temperature difference between objects.

### Heat Engines

A **heat engine** is a device, such as this Icelandic geothermal power plant, that transforms thermal energy into useful work.

You'll learn how to calculate the maximum efficiency for converting thermal energy into work.

**GOAL** To learn about practical energy transformations and transfers, and the limits on how efficiently energy can be used.

## LOOKING BACK

### The Basic Energy Model

The basic energy model you learned about in Chapter 10 emphasized work and mechanical energy. In this chapter, we'll focus on thermal energy, chemical energy, and energy transfers in the form of heat.

Work and heat are energy transfers that change the system's total energy. If the system is isolated, the total energy is *conserved.*

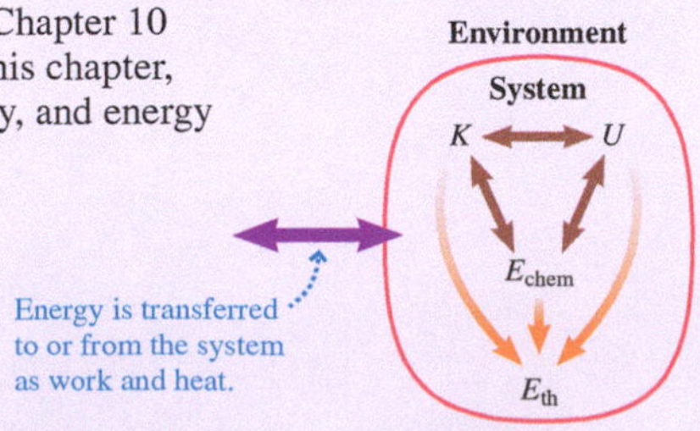


**STOP TO THINK**

Christina throws a javelin into the air. As she propels it forward from rest, she does 270 J of work on it. At its highest point, its gravitational potential energy has increased by 70 J. What is the javelin's kinetic energy at this point?

A. 270 J  B. 340 J  C. 200 J
D. −200 J  E. −340 J

# 11.1 Transforming Energy

As we saw in Chapter 10, energy can't be created or destroyed; it can only be converted from one form to another. When we say we are *using energy,* we mean that we are transforming it, such as transforming the chemical energy of food into the kinetic energy of your body. The following table revisits the idea of energy transformations, considering some realistic situations that have interesting theoretical and practical limitations.

**Energy transformations**

Light energy hitting a solar cell on top of this walkway light is converted to electric energy and then stored as chemical energy in a battery.

At night, the battery's chemical energy is converted to electric energy that is then converted to light energy in a light- emitting diode.

Light energy is absorbed by photosynthetic pigments in soybean plants, which use this energy to create concentrated chemical energy.

The soybeans are harvested and their oil is used to make candles. When the candle burns, the stored chemical energy is transformed into light energy and thermal energy.

A wind turbine converts the translational kinetic energy of moving air into electric energy.

Miles away, this energy can be used by a fan, which transforms the electric energy back into the kinetic energy of moving air.

In each of the processes in the table, energy is transformed from one form into another and another, ending up in the same form as it began. But some energy appears to have been "lost" along the way. The garden light certainly shines with a glow that is much less bright than the sun that shined on it. When sunlight falls on the soybean plants, only a small fraction of the light energy ends up as chemical energy in the plants. No energy is lost, of course; we know that total energy is conserved. But at each stage in the processes in the table, much—or most—of the energy is converted into forms that are less useful to us.

The table also illustrates another key point: We are broadening our scope to include forms of energy beyond those we considered in Chapter 10, including radiant energy (the light that hits the solar cell and the plants) and electric energy (energy that is transferred from the wind turbine to the fan).

Recall the work-energy equation from Chapter 10:

$$\Delta E = \Delta K + \Delta U + \Delta E_{\text{th}} + \Delta E_{\text{chem}} + \cdots = W \qquad (11.1)$$

This equation includes work, an energy transfer. Work can be positive or negative; it's positive when energy is transferred into a system, negative when energy is transferred out of a system. In Chapter 10 we defined work as a mechanical transfer of energy; in this chapter we'll broaden our definition of work to include electric energy into and out of motors and generators.

Let's look at a practical example of a chain of energy transformations, focusing on the energy that is "lost" and the energy that remains. Electric power companies sometimes use excess electric energy to pump water uphill to a reservoir. They reclaim some of this energy when demand increases by using the water to generate electricity as it flows back down. FIGURE 11.1 shows the energy transformations in a simple case. We assume 100 J of electric energy at the start and define the system as the water and the network of pipes, pumps, and generators that make up the plant. The electric energy that is used to run the pump is a work input; the electric energy that is extracted at the end is a work output. Pumping the water uphill requires 100 J of work, but this

FIGURE 11.1 Energy transformations for a pumped storage system.

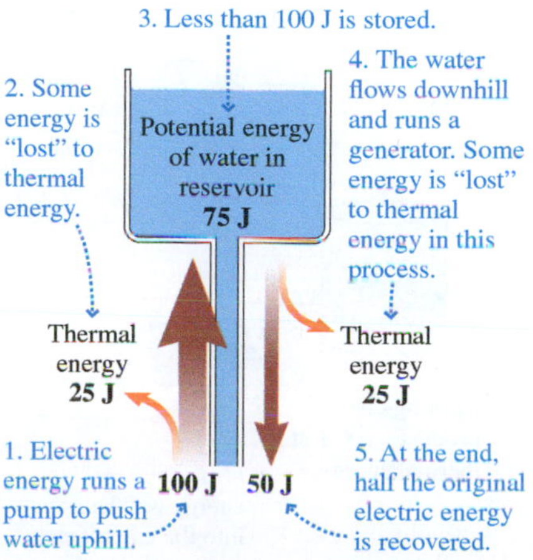

3. Less than 100 J is stored.
2. Some energy is "lost" to thermal energy.
Potential energy of water in reservoir
75 J
4. The water flows downhill and runs a generator. Some energy is "lost" to thermal energy in this process.
Thermal energy
25 J
Thermal energy
25 J
1. Electric energy runs a pump to push water uphill.
100 J
50 J
5. At the end, half the original electric energy is recovered.

increases the potential energy of the water by only 75 J, with 25 J transformed into thermal energy by friction in the pump and other causes. When the water flows back downhill, a generator converts some—but not all—of the potential energy to electric energy. Ultimately, a 100 J input results in a 50 J output, with a 50 J increase in thermal energy of the system. This transformation into thermal energy is an irreversible change. There is no practical way to recover this thermal energy and convert it back to electric energy. **The energy isn't lost, but it is lost to our use.**

In Chapter 10, we considered cases in which the losses were small or could be ignored. In this chapter, we'll focus on more realistic situations in which this is not the case, and we'll explore and explain the reasons why.

## Efficiency

To *get* 50 J of energy out of your pumped storage plant, you would actually need to *pay* to put in 100 J of energy, as we saw in Figure 11.1. Because only 50% of the energy input is returned as useful energy output, we can say that this plant has an efficiency of 50%. Generally, we can define efficiency as

$$e = \frac{\text{what you get}}{\text{what you had to pay}} \tag{11.2}$$

General definition of efficiency

These cooling towers release thermal energy from a coal-fired power plant.

**The larger the energy losses in a system, the lower its efficiency.**

Reductions in efficiency can arise from two different sources:

- **Process limitations.** In some cases, energy losses may be due to practical details of an energy transformation process. You could, in principle, design a process that entailed smaller losses.
- **Fundamental limitations.** In other cases, energy losses are due to physical laws that cannot be circumvented. The best process that could theoretically be designed will have less than 100% efficiency. As we will see, these limitations result from the difficulty of transforming thermal energy into other forms of energy.

As we'll see, if you are of average weight, walking up a typical flight of stairs increases your body's potential energy by about 1800 J. But a measurement of the energy used by your body to climb the stairs would show that your body uses about 7200 J of chemical energy to complete this task. The 1800 J increase in potential energy is "what you get"; the 7200 J your body uses is "what you had to pay." We can use Equation 11.2 to compute the efficiency for this action:

$$e = \frac{1800\text{ J}}{7200\text{ J}} = 0.25 = 25\%$$

This relatively low efficiency is due to *process limitations*. The efficiency is less than 100% because of the biochemistry of how your food is metabolized and the biomechanics of how you move.

The electric energy you use daily must be generated from other sources. In a coal-fired power plant, chemical energy is converted to thermal energy by burning, and much of the resulting thermal energy is converted to electric energy. In fact, as we'll see, less than half of the energy is transformed into electric energy; *most* of the thermal energy from the burning coal is simply exhausted into the environment.

A typical power-plant cycle is shown in FIGURE 11.2. "What you get" is the energy output, the 40 J of electric energy. "What you had to pay" is the energy input, the 100 J of chemical energy, giving an efficiency of

$$e = \frac{40\text{ J}}{100\text{ J}} = 0.40 = 40\%$$

Only 40% of the energy in the coal is used to generate electricity; 60% goes to warming the area around the plant. Why would you design a plant to waste more than half of the energy input?

FIGURE 11.2 Energy transformations in a coal-fired power plant.

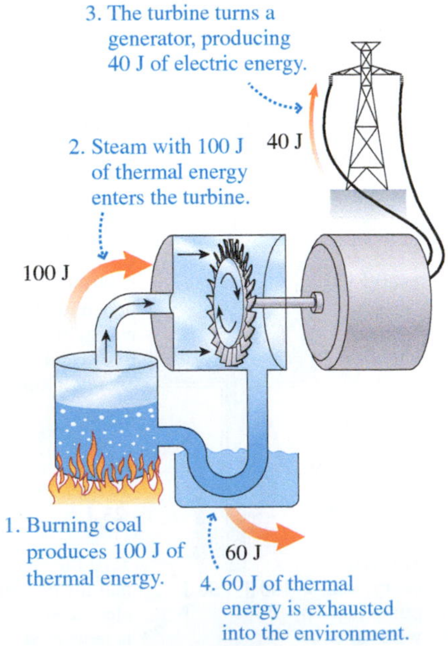

Unlike your stair-climbing efficiency, the rather modest power-plant efficiency turns out to be largely due to a *fundamental limitation:* Thermal energy cannot be transformed into other forms of energy with 100% efficiency. In later sections, we will explore the fundamental properties of thermal energy that make this so.

## PROBLEM-SOLVING APPROACH 11.1 Energy efficiency problems

The efficiency of a process is the ratio of the useful energy output (what you get) to the original energy input (what you had to pay).

**STRATEGIZE** To start, you need to decide what to identify as the input and the output:

1. Choose what energy is "what you get." This could be the useful energy output of an engine or process or the work that is done in completing a process. For example, when you climb a flight of stairs, "what you get" is your change in potential energy.
2. Decide what energy is "what you had to pay." This will generally be the total energy input needed for an engine, task, or process. For example, when you climb a flight of stairs, "what you had to pay" is the chemical energy used by your body.

**PREPARE** You may need to do additional calculations:

- Compute values for "what you get" and "what you had to pay."
- Be certain that all energy values are in the same units.

**SOLVE** Compute the efficiency using $e = \frac{\text{what you get}}{\text{what you had to pay}}$.

**ASSESS** Check your answer to see if it is reasonable, given what you know about typical efficiencies for the process under consideration.

## EXAMPLE 11.1 Determining the efficiency of lightbulbs

A 75 W incandescent bulb and a 15 W compact fluorescent bulb, as shown in FIGURE 11.3, each produce 3.0 W of visible-light energy. What are the efficiencies of these two types of bulbs for converting electric energy into light?

FIGURE 11.3 Incandescent and compact fluorescent bulbs.

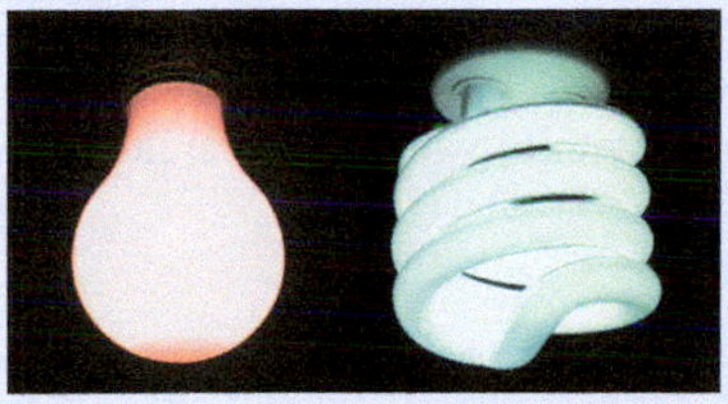

**STRATEGIZE** In this case, what you have to pay is the energy to light the bulb, the electric energy input. (If you purchase your electricity from the local utility, this is, literally, what you need to pay.) The energy input is the rating of the bulb: 15 W for the compact fluorescent bulb and 75 W for the incandescent bulb. The purpose of a lightbulb is to produce light, so what you get is light output, the 3.0 W of visible-light energy.

**PREPARE** The problem statement doesn't give values for energy; we are given values for power. But 15 W is 15 J/s, so we can consider the energy in or out in 1 second. In 1 second, the visible-light output for each bulb is 3.0 J. In 1 second, the energy input for the compact fluorescent bulb is 15 J; for the incandescent bulb, it is 75 J.

**SOLVE** The efficiencies of the two bulbs are computed using the energy in 1 second:

$$e(\text{compact fluorescent bulb}) = \frac{3.0\text{ J}}{15\text{ J}} = 0.20 = 20\%$$

$$e(\text{incandescent bulb}) = \frac{3.0\text{ J}}{75\text{ J}} = 0.040 = 4\%$$

**ASSESS** Both bulbs produce the same visible-light output, but the compact fluorescent bulb does so with a significantly lower energy input, so it is more efficient, as we know to be the case.

## EXAMPLE 11.2 How much water flows through the dam?

A large hydroelectric power plant uses water from a reservoir to turn a generator. The water leaves the generator below the dam, 60 m below the level of the reservoir. The plant uses an 85% efficient generator to supply 1.0 GW of electricity. What mass of water flows through the plant each second?

**STRATEGIZE** What you get is clear: electric energy. We are given the plant's power output, so we'll consider the electric-energy output in 1 second. The plant's power output is 1.0 GW, so in 1 second, the plant puts out $1.0 \times 10^9$ J of electric energy. The change in potential energy of the water is what you had to pay. The ratio of these two values is the plant's efficiency, which is given.

**PREPARE** If water of mass $m$ goes through the plant, the change in potential energy is

$$\Delta U_g = mg\Delta y = mg(-60\text{ m})$$

The water loses potential energy, so this is a negative number. The energy used by the generator is $mg(60\text{ m})$.

**SOLVE** The efficiency of the plant is 0.85; this is the ratio of what you get to what you had to pay, so we can write

$$\frac{1.0 \times 10^9\text{ J}}{mg(60\text{ m})} = 0.85$$

We can solve this equation for the mass of water:

$$m = \frac{1.0 \times 10^9\text{ J}}{g(60\text{ m})(0.85)} = 2.0 \times 10^6\text{ kg}$$

The density of water is 1000 kg/m$^3$, so this corresponds to 2000 m$^3$ of water.

**ASSESS** In 1 second, enough water flows through the plant to fill two Olympic-size swimming pools. But 1.0 GW is a lot of power, enough for a city of a million people, so this result makes sense. These are actual numbers for a real plant, and this much water really passes through the dam's giant generators.

**STOP TO THINK 11.1** Crane 1 uses 10 kJ of energy to lift a 50 kg box to the roof of a building. Crane 2 uses 20 kJ to lift a 100 kg box the same distance. Which crane is more efficient?

A. Crane 1
B. Crane 2
C. Both cranes have the same efficiency.

## 11.2 Energy in the Body BIO

FIGURE 11.4 Energy of the body, considered as the system.

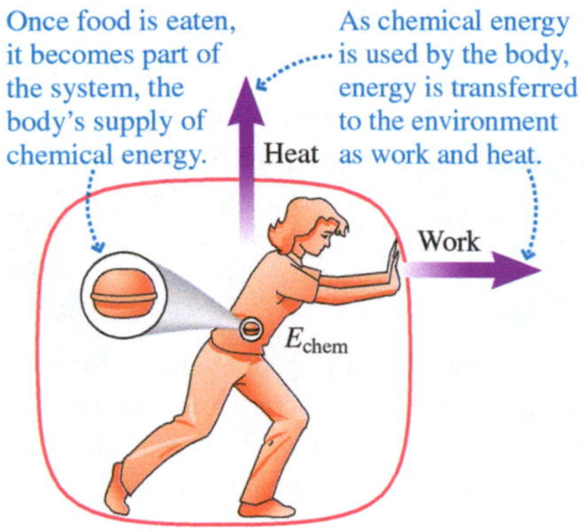


In this section, we will look at energy in the body, which will give us the opportunity to explore a number of different energy transformations and transfers in a practical context. FIGURE 11.4 shows the body considered as the system for energy analysis. The chemical energy in food provides the necessary energy input for your body to function. It is this energy that is used for energy transfers with the environment.

### Getting Energy from Food: Energy Inputs

When you walk up a flight of stairs, where does the energy come from to increase your body's potential energy? At some point, the energy came from the food you ate, but what were the intermediate steps? The chemical energy in food is made available to the cells in the body by a two-step process. First, the digestive system breaks down food into simpler molecules such as glucose, a simple sugar. These molecules are delivered via the bloodstream to cells in the body, where they are metabolized by combining with oxygen, as in Equation 11.3:

Glucose from the digestion of food combines with oxygen that is breathed in to produce . . . . . . carbon dioxide, which is exhaled; water, which can be used by the body; and energy.

$$\underset{\text{Glucose}}{C_6H_{12}O_6} + \underset{\text{Oxygen}}{6O_2} \longrightarrow \underset{\text{Carbon dioxide}}{6CO_2} + \underset{\text{Water}}{6H_2O} + \text{energy} \tag{11.3}$$

This metabolism releases energy, much of which is stored in a molecule called adenosine triphosphate, or ATP. Cells in the body use this ATP to do all the work of life: Muscle cells use it to contract, nerve cells use it to produce electrical signals.

**EXAMPLE 11.3 Computing basic biochemical efficiency** BIO

Power for cellular processes comes from the conversion of ATP to ADP, adenosine diphosphate. The cell then recycles the ADP molecules, using chemical energy from glucose to convert the ADP back to ATP. Converting one molecule of ATP to ADP releases $5.1 \times 10^{-20}$ J. Metabolizing one molecule of glucose releases $4.7 \times 10^{-18}$ J. If everything goes smoothly in the biochemical pathways that use this energy to produce ATP, metabolizing one glucose molecule produces 38 molecules of ATP. What is the overall efficiency of this energy conversion process?

***STRATEGIZE*** What you get is the energy of 38 molecules of ATP; what you have to pay is the energy from the metabolism of one molecule of glucose.

***PREPARE*** The metabolism of one molecule of glucose releases $4.7 \times 10^{-18}$ J; each molecule of ATP can release $5.1 \times 10^{-20}$ J.

***SOLVE*** The efficiency is

$$e = \frac{\text{what you get}}{\text{what you had to pay}} = \frac{(38)(5.1 \times 10^{-20}\text{ J})}{4.7 \times 10^{-18}\text{ J}} = 0.41$$

***ASSESS*** We know that the efficiency of this process must be less than 1.0; we also know that it must be greater than 0.25, which is the efficiency of the body in using energy to climb stairs. Our result falls between these two extremes, which gives us confidence in our result.

Metabolizing 1 g of glucose (or any other carbohydrate) from food or body stores releases approximately 17 kJ of energy; 1 g of fat from food or body stores provides 37 kJ. TABLE 11.1 compares the energy contents of carbohydrates and other foods with other common sources of chemical energy.

It is possible to measure the chemical energy content of food by burning it. Burning food may seem quite different from metabolizing it, but if glucose is burned, the chemical formula for the reaction is again Equation 11.3—the two reactions are the same. Burning food transforms its chemical energy into thermal energy, which we can measure. Thermal energy is often measured in units of calories (cal) rather than joules; 1.00 calorie is equivalent to 4.19 joules. The energy content of food is usually given in Calories (Cal, with a capital "C"); 1 Calorie (also called a "food calorie") is equal to 1000 calories. We usually express energy in joules, so it's worthwhile to sum up the relationships among these different units:

$$1.00 \text{ Cal} = 1000 \text{ cal} = 4190 \text{ J} = 4.19 \text{ kJ}$$
$$1.00 \text{ kJ} = 1000 \text{ J} = 239 \text{ cal} = 0.239 \text{ Cal}$$

TABLE 11.1 Energy in fuels

| Fuel | Energy in 1 g of fuel (in kJ) |
|---|---|
| Hydrogen | 121 |
| Gasoline | 44 |
| Fat (in food) | 37 |
| Coal | 27 |
| Carbohydrates (in food) | 17 |
| Wood chips | 15 |

**EXAMPLE 11.4 How much energy does a soda provide?** BIO

A 12 oz can of soda contains 40 g (or a bit less than 1/4 cup) of sugar, a simple carbohydrate. What is the chemical energy in joules? How many Calories is this?

***SOLVE*** From Table 11.1, 1 g of sugar contains 17 kJ of energy, so 40 g contains

$$40 \text{ g} \times \frac{17 \times 10^3 \text{ J}}{1 \text{ g}} = 68 \times 10^4 \text{ J} = 680 \text{ kJ}$$

Converting to Calories, we get

$$680 \text{ kJ} = (680 \text{ kJ})\frac{1.00 \text{ Cal}}{4.19 \text{ kJ}} = 160 \text{ Cal}$$

***ASSESS*** 160 Calories is a typical value for the energy content of a 12 oz can of soda (check the nutrition label on one to see), so this result seems reasonable.

**Counting calories** BIO Most dry foods burn quite well, as this photo of corn chips illustrates. You could set food on fire to measure its energy content, but this isn't really necessary. The chemical energies of the basic components of food (carbohydrates, proteins, fats) have been carefully measured—by burning—in a device called a *calorimeter.* Foods are analyzed to determine their composition, and their chemical energy can then be calculated.

The first item on the nutrition label on packaged foods is Calories—a measure of the chemical energy in the food. (In Europe, where SI units are standard, you will find the energy contents listed in kJ.) The energy contents of some common foods are given in TABLE 11.2.

TABLE 11.2 Energy content of foods

| Food | Energy content in Cal | Energy content in kJ | Food | Energy content in Cal | Energy content in kJ |
|---|---|---|---|---|---|
| Carrot (large) | 30 | 125 | Slice of pizza | 300 | 1260 |
| Fried egg | 100 | 420 | Frozen burrito | 350 | 1470 |
| Apple (large) | 125 | 525 | Apple pie slice | 400 | 1680 |
| Beer (can) | 150 | 630 | Fast-food meal: | | |
| BBQ chicken wing | 180 | 750 | burger, fries, | | |
| Latte (whole milk) | 260 | 1090 | drink (large) | 1350 | 5660 |

## Using Energy in the Body: Energy Outputs

You know that your body uses energy when you exercise. But even at rest, your body uses energy to build and repair tissue, digest food, and keep warm. TABLE 11.3 lists the amounts of power used by different tissues in the resting body. All told, your body uses 100 W, or 100 J per second, when at rest, and all of this energy is

TABLE 11.3 Energy use at rest

| Organ | Resting power (W) of 68 kg individual |
|---|---|
| Liver | 26 |
| Brain | 19 |
| Kidneys | 11 |
| Heart | 7 |
| Skeletal muscle | 18 |
| Remainder of body | 19 |
| **Total** | **100** |

FIGURE 11.5 The mask measures the oxygen used by the athlete.

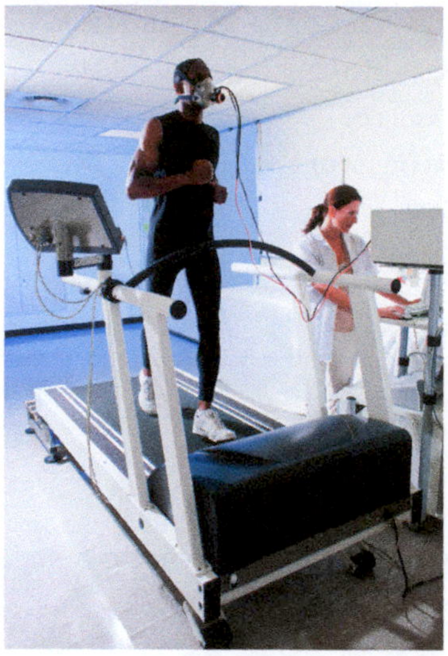

TABLE 11.4 Metabolic power use during activities

| Activity | Metabolic power (W) of 68 kg individual |
|---|---|
| Typing | 125 |
| Ballroom dancing | 250 |
| Walking at 5 km/h | 380 |
| Cycling at 15 km/h | 480 |
| Swimming at a fast crawl | 800 |
| Running at 15 km/h | 1150 |

FIGURE 11.6 Climbing a set of stairs.

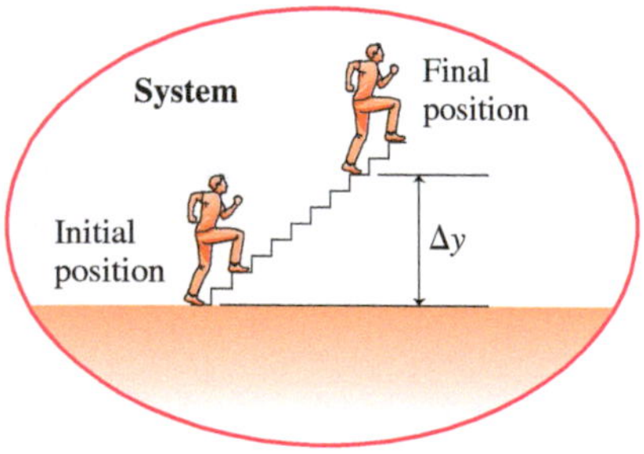


ultimately converted to thermal energy. Your body stays at a constant temperature, so this energy is dissipated as heat into the environment. This process has practical consequences. If you take physics in a class of 100 people in a lecture hall, the class, as a whole, dissipates 10,000 W into the room—the equivalent of seven electric space heaters running at full blast. In the winter, this extra heat is welcome; in the summer, the air conditioning system must work harder to keep the room cool.

Your cells use ATP to power cellular processes, so the cells of your body must continually replenish their ATP stores with metabolic reactions such as the one shown in Equation 11.3. This requires oxygen, so it is possible to measure the energy used by the body, at rest or engaged in exercise, by measuring the oxygen used. Using a respiratory apparatus, as shown in FIGURE 11.5, physiologists can precisely measure the body's energy use by measuring how much oxygen the body is taking up. The device determines the body's *total metabolic energy use*—all of the energy used by the body while performing an activity. This total includes all of the body's basic processes such as breathing plus whatever additional energy is needed to perform the activity. This corresponds to measuring "what you had to pay."

The metabolic energy used in an activity depends on an individual's size, level of fitness, and other variables. But we can make reasonable estimates for the power used in various activities for a typical individual. Some values are given in TABLE 11.4.

## Efficiency of the Human Body

Suppose you climb a set of stairs at a constant speed, as in FIGURE 11.6. What is your body's efficiency for this process? To find out, let's apply the work-energy equation. We note first that no work is done on you: There is no external input of energy, as there would be if you took the elevator. Furthermore, if you climb at a constant speed, there's no change in your kinetic energy. However, your gravitational potential energy clearly increases as you climb, as does the overall thermal energy of you and perhaps the surrounding air. This latter fact is something you know well: If you climb several sets of stairs, you certainly warm up in the process! And, finally, your body must use chemical energy to power your muscles for the climb.

For this case of climbing stairs, then, the work-energy equation reduces to

$$\Delta E_{\text{chem}} + \Delta U_{\text{g}} + \Delta E_{\text{th}} = 0 \tag{11.4}$$

Thermal energy and gravitational potential energy are increasing, so $\Delta E_{\text{th}}$ and $\Delta U_{\text{g}}$ are positive; chemical energy is being used, so $\Delta E_{\text{chem}}$ is a negative number. We can get a better feeling about what is happening by rewriting Equation 11.4 as

$$|\Delta E_{\text{chem}}| = \Delta U_{\text{g}} + \Delta E_{\text{th}} \tag{11.5}$$

The *magnitude* of the change in the chemical energy is equal to the sum of the changes in the gravitational potential and thermal energies. Chemical energy from your body is converted into potential energy and thermal energy; in the final position, you are at a greater height and your body is slightly warmer.

Earlier in the chapter, we noted that the efficiency for stair climbing is about 25%. Let's see where that number comes from.

1. *What you get.* What you get is the change in potential energy: You have raised your body to the top of the stairs. If you climb a flight of stairs of vertical height $\Delta y$, the increase in potential energy is $\Delta U_{\text{g}} = mg\Delta y$. Assuming a mass of 68 kg and a change in height of 2.7 m (about 9 ft, a reasonable value for a flight of stairs), we compute

$$\Delta U_{\text{g}} = (68 \text{ kg})(9.8 \text{ m/s}^2)(2.7 \text{ m}) = 1800 \text{ J}$$

2. *What you had to pay.* The cost is the metabolic energy your body used in completing the task. As we've seen, physiologists can measure directly how much energy $|\Delta E_{\text{chem}}|$ your body uses to perform a task. A typical value for climbing a flight of stairs is

$$|\Delta E_{\text{chem}}| = 7200 \text{ J}$$

Given the definition of efficiency in Equation 11.2, we can compute an efficiency for climbing the stairs:

$$e = \frac{\Delta U_{\text{g}}}{|\Delta E_{\text{chem}}|} = \frac{1800 \text{ J}}{7200 \text{ J}} = 0.25 = 25\%$$

For the types of activities we will consider in this chapter, such as running, walking, and cycling, the body's efficiency is typically in the range of 20–30%. **We will generally use a value of 25% for the body's efficiency for our calculations.** Efficiency varies from individual to individual and from activity to activity, but this rough approximation will be sufficient for our purposes.

The metabolic power values given in Table 11.4 represent the energy *used by the body* while these activities are being performed. Given that we assume an efficiency of 25%, the body's actual *useful power output* is then quite a bit less than this. The table's value for cycling at 15 km/h (a bit less than 10 mph) is 480 W. If we assume that the efficiency for cycling is 25%, the actual power going to forward propulsion will be only 120 W. An elite racing cyclist whizzing along at 35 km/h is using about 300 W for forward propulsion, so his metabolic power is about 1200 W.

The energy you use per second while running is proportional to your speed; running twice as fast takes approximately twice as much power. But running twice as fast takes you twice as far in the same time, so the energy you use to run a certain *distance* doesn't depend on how fast you run! Running a marathon takes approximately the same amount of energy whether you complete it in 2 hours, 3 hours, or 4 hours; it is only the power that varies.

NOTE ▶ It is important to remember the distinction between the metabolic energy used to perform a task and the work done in a physics sense; these values can be quite different. Your muscles use power when applying a force, even when there is no motion. Holding a weight above your head involves no external work, but it clearly takes metabolic power to keep the weight in place! ◀

This distinction—between the work done (what you get) and the energy used by the body (what you had to pay)—is important to keep in mind when you do calculations on energy used by the body. We'll consider two different types of problems:

- **Type 1: "What you get" can be quantified.** For some tasks, such as climbing stairs, we can easily compute the energy change that is the outcome of the task; that is, we can compute what you get. We assume an efficiency of 25%, so the energy used by the body ("what you had to pay") is 4 times this amount. There is no need to use table data; you can do the calculation directly.
- **Type 2: "What you get" can't be easily quantified.** For other tasks, it is not so easy to find the energy that corresponds to "what you get." If you ride your bike at a constant speed on a level course, you don't change your potential energy or your kinetic energy. In energy terms, there is no good way to compute "what you get." In such cases, we consider data from metabolic studies, such as the data in Table 11.4. This is the actual power used by the body to complete a task—in other words, "what you had to pay." We assume an efficiency of 25%, so the useful power output ("what you get") is ¼ of this value.

**High heating costs?** BIO The daily energy use of mammals is much higher than that of reptiles, largely because mammals use energy to maintain a constant body temperature. A 40 kg timber wolf uses approximately 19,000 kJ during the course of a day. A Komodo dragon, a reptilian predator of the same size, uses only 2100 kJ.

## CONCEPTUAL EXAMPLE 11.5 Determining energy transformations in weightlifting BIO

A weightlifter lifts a 50 kg bar from the floor to a position over his head and back to the floor again 10 times in succession. At the end of this exercise, what energy transformations have taken place?

**REASON** We will take the system to be the weightlifter plus the bar. The environment does no work on the system, and we will assume that no heat is transferred from the system to the environment. The bar has returned to its starting position and is not moving, so there has been no change in potential or kinetic energy. The equation for energy conservation is thus $\Delta E_{\text{chem}} + \Delta E_{\text{th}} = 0$. This equation tells us that $\Delta E_{\text{chem}}$ must be negative. This makes sense because the muscles *use* chemical energy—depleting the body's store—each time the bar is raised or lowered. Ultimately, all of this energy is transformed into thermal energy.

**ASSESS** You know that you warm up when you are working out, so this result makes sense. Most exercises in the gym—lifting weights, running on a treadmill—involve only the transformation of chemical energy into thermal energy.

## EXAMPLE 11.6 Determining the energy usage for a cyclist BIO

A cyclist pedals for 20 min at a speed of 15 km/h. How much metabolic energy is required? How much energy is used for forward propulsion?

**STRATEGIZE** This is a Type 2 problem. There is no easy way to compute what you get—there is no change in potential or kinetic energy. As the cyclist moves, all of the chemical energy the body uses ends up as thermal energy in the cyclist's body or in the air. We'll use the value in Table 11.4 for the energy used. This is the energy the body uses, so it is the input—what you had to pay. If we assume an efficiency of 25%, the energy output is ¼ of this.

**PREPARE** Table 11.4 gives a value of 480 W for the metabolic power used in cycling at 15 km/h. 480 W is the power used by the body; the power going into forward propulsion is much less than this. The cyclist uses energy at this rate for 20 min, or 1200 s.

**SOLVE** We know the power and the time, so we can compute the energy needed by the body as follows:

$$\Delta E = P\Delta t = (480\ \text{J/s})(1200\ \text{s}) = 580\ \text{kJ}$$

If we assume an efficiency of 25%, then only 25%, or 140 kJ, of this energy is used for forward propulsion. The remainder goes into thermal energy.

**ASSESS** This result is reasonable. 15 km/h isn't that speedy; most recreational cyclists could keep up this pace. So we don't expect the energy expenditure to be all that great—a glance at Table 11.2 shows that 580 kJ is slightly more than the energy available in a large apple.

Your body is really quite efficient. The typical 25% efficiency that we've used might not sound that great, but it's better than the efficiency of a typical automobile or that of other devices that use chemical energy as a power source. This efficiency is something to be aware of, though, if you have overindulged and plan to "work off" the calories. Based on the previous examples, the energy in a typical fast-food meal would power the cyclist for more than 3 hours!

## EXAMPLE 11.7 How high can you climb? BIO

How many flights of stairs could you climb on the energy contained in a 12 oz can of soda? During the climb, how much energy is transformed into thermal energy? Assume that your mass is 68 kg and that a flight of stairs has a vertical height of 2.7 m.

**STRATEGIZE** This is a Type 1 problem. It's easy to quantify what you get—a change in potential energy. This is the energy output. What you had to pay is the energy input, the energy in the soda. We will determine the energy in the soda. Then, if we assume a typical efficiency, 25% of this energy will go to the increase in potential energy, which we will use to determine the height. The rest will go to thermal energy.

**PREPARE** In Example 11.4, we determined that the energy content of a can of soda is 680 kJ. This is the chemical energy input. We considered the work-energy equation for this case earlier in the section; we can repeat Equation 11.5, noting the relative sizes of the different energies:

$$|\Delta E_{\text{chem}}| = \Delta U_{\text{g}} + \Delta E_{\text{th}}$$

This is "what you get," the change in potential energy. It is 25% of the total. (pointing to $\Delta U_{\text{g}}$)

The balance, 75% of the total, is transformed into thermal energy. (pointing to $\Delta E_{\text{th}}$)

**SOLVE** At 25% efficiency, the amount of chemical energy transformed into increased potential energy is

$$\Delta U_{\text{g}} = (0.25)(680 \times 10^3\ \text{J}) = 1.7 \times 10^5\ \text{J}$$

Because $\Delta U_{\text{g}} = mg\Delta y$, the height gained is

$$\Delta y = \frac{\Delta U_{\text{g}}}{mg} = \frac{1.7 \times 10^5\ \text{J}}{(68\,\text{kg})(9.8\ \text{m/s}^2)} = 255\ \text{m}$$

With each flight of stairs having a height of 2.7 m, the number of flights climbed is

$$\frac{255\ \text{m}}{2.7\ \text{m}} \approx 94\ \text{flights}$$

The balance of the energy is transformed into thermal energy:

$$\Delta E_{\text{th}} = |\Delta E_{\text{chem}}| - \Delta U_{\text{g}} = 6.8 \times 10^5\ \text{J} - 1.7 \times 10^5\ \text{J}$$
$$= 5.1 \times 10^5\ \text{J} = 510\ \text{kJ}$$

**ASSESS** 94 flights is almost enough to get to the top of the Empire State Building! But this makes sense; there is a lot of energy in a can of soda. And if you climb several sets of stairs, you know that your body warms up, so the large increase in thermal energy makes sense as well.

## Energy Storage

If the energy that the body gets from food is not used, it is stored. A small amount of energy needed for immediate use is stored as ATP. A larger amount of energy is stored as chemical energy of simple carbohydrates in muscle tissue and the liver. A healthy adult might store 400 g of these carbohydrates, which is a little more carbohydrate than is typically consumed in one day.

If the energy input from food continuously exceeds the energy outputs of the body, this energy will be stored in the form of fat under the skin and around the organs. From an energy point of view, gaining weight is simply explained!

**EXAMPLE 11.8 How far can you run? BIO**

The body stores about 400 g of carbohydrates. Approximately how far could a 68 kg runner travel on this stored energy?

***STRATEGIZE*** This is a Type 2 problem; we can't easily quantify what you get, so we'll use the value from Table 11.4 to determine the power. We will use the data in Table 11.1 to determine the energy in the carbohydrate stored in the body. From the energy available and the rate at which the body uses this energy, we can determine how long the energy will last, and then we will calculate the distance the runner can travel.

***PREPARE*** Table 11.1 gives a value of 17 kJ per g of carbohydrate. The 400 g of carbohydrates in the body contain an energy of

$$E_{\text{chem}} = (400\text{ g})(17 \times 10^3\text{ J/g}) = 6.8 \times 10^6\text{ J}$$

***SOLVE*** Table 11.4 gives the power used in running at 15 km/h as 1150 W. The time that the stored chemical energy will last at this rate is

$$\Delta t = \frac{\Delta E_{\text{chem}}}{P} = \frac{6.8 \times 10^6\text{ J}}{1150\text{ W}} = 5.91 \times 10^3\text{ s} = 1.64\text{ h}$$

And the distance that can be covered during this time at 15 km/h is

$$\Delta x = v\Delta t = (15\text{ km/h})(1.64\text{ h}) = 25\text{ km}$$

to two significant figures.

***ASSESS*** A marathon is longer than this—just over 42 km. Even with "carbo loading" before the event (eating high-carbohydrate meals), many marathon runners "hit the wall" before the end of the race as they reach the point where they have exhausted their store of carbohydrates. Given that this is a problem for long-distance runners, our answer makes sense. But if runners deplete their carbohydrate reserves, how can they finish the race? The body has other energy stores (in fats, for instance), but the rate that they can be drawn on is much lower.

## Energy and Locomotion

When you walk at a constant speed on level ground, your kinetic energy is constant. Your potential energy is also constant. So why does your body need energy to walk? Where does this energy go?

We use energy to walk because of mechanical inefficiencies in our gait. FIGURE 11.7 shows how the speed of your foot typically changes during each stride. The kinetic energy of your leg and foot increases, only to go to zero at the end of the stride. The kinetic energy is mostly transformed into thermal energy in your muscles and in your shoes. This thermal energy is lost; it can't be used for making more strides.

This inefficiency is a process limitation. It's possible to do better. Footwear can be designed to minimize the loss of kinetic energy to thermal energy. A spring in the sole of the shoe can store potential energy, which can be returned to kinetic energy during the next stride. Such a spring will make the collision with the ground more elastic. We saw in Chapter 10 that the tendons in the ankle store a certain amount of energy during a stride; very stout tendons in the legs of kangaroos store energy even more efficiently. Their peculiar hopping gait is quite efficient at high speeds.

**FIGURE 11.7** Human locomotion analysis.

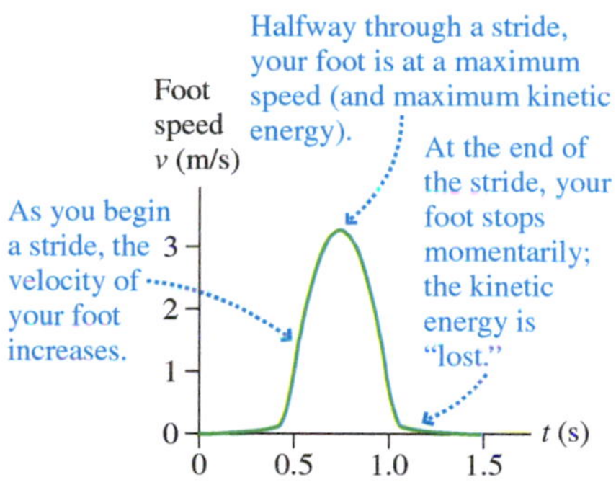


▸ **Where do you wear the weights? BIO** If you wear a backpack with a mass equal to 1% of your body mass, your energy expenditure for walking will increase by 1%. But if you wear ankle weights with a combined mass of 1% of your body mass, the increase in energy expenditure is 6%, because you must repeatedly accelerate this extra mass. If you want to "burn more fat," wear the weights on your ankles, not on your back! If you are a runner who wants to shave seconds off your time in the mile, you might try lighter shoes.

**STOP TO THINK 11.2** A runner is moving at a constant speed on level ground. Chemical energy in the runner's body is being transformed into other forms of energy. Most of the chemical energy is transformed into

A. Kinetic energy. B. Potential energy. C. Thermal energy.

## 11.3 Temperature, Thermal Energy, and Heat

What do you mean when you say something is "hot"? Do you mean that it has a high temperature? Or do you mean that it has a lot of thermal energy? Or maybe that the object in some way contains "heat"? Are all of these definitions the same? Let's give some thought to the meanings of temperature, thermal energy, and heat, and the relationship between them.

### Relating Thermal Energy and Temperature

FIGURE 11.8 The brake disks on a race car can reach very high temperatures during rapid braking.

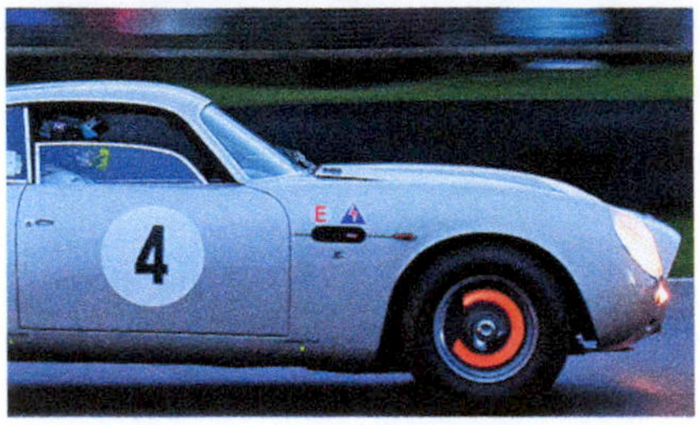

As we saw in ◀◀ SECTION 10.5, the thermal energy of an object is the sum of the kinetic and elastic potential energies of the object's atoms. When we add thermal energy, the atoms move more quickly and the bonds between atoms stretch. This change in thermal energy also leads to a higher temperature.

When a race car driver hits the brakes, the kinetic energy of the moving vehicle is transformed into thermal energy of the brake disks. This increases the temperature of the brake disks, as the photo in FIGURE 11.8 shows; the disks are at a high enough temperature to glow like the burner on an electric stove. The thermal energy of the disks must increase as the car comes to rest, but it's useful to limit the resulting change in temperature. In practice, race car designers can do this by making one simple change: using more massive brake disks. How does increasing the disks' mass keep them at a lower temperature? At this point, we'll need to carefully consider the meanings of thermal energy and temperature. We'll do this by considering a gas, but the results are general and apply to all phases of matter.

FIGURE 11.9 Atoms of a gas at low temperature and high temperature.

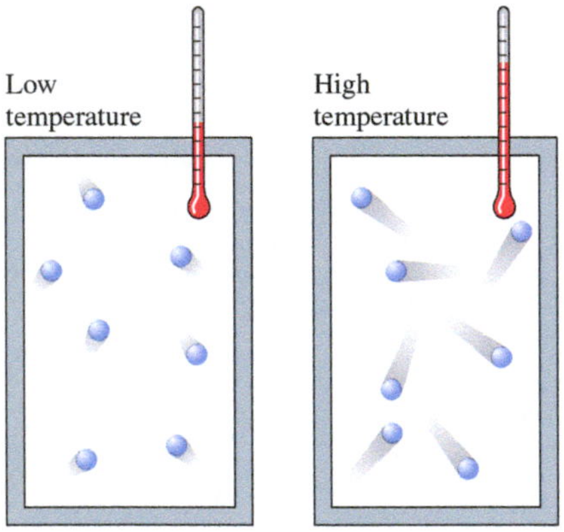


FIGURE 11.9 shows two samples of gas, one at a low temperature and one at a high temperature. You know what this means: The atoms in the high-temperature gas move at a higher speed than the atoms in the low-temperature gas. You can also see that the gas sample on the right has a higher thermal energy than the gas sample on the left. The two samples have the same number of atoms, but the atoms in the sample on the right have higher kinetic energy.

Now let's do a thought experiment: Add more and more atoms to the sample on the left, with each new atom having the same speed as the atoms already there. This clearly increases the thermal energy of the system—it increases the total energy of the atoms—but it won't increase the temperature.

The temperature of a system does not depend on the size of the system. If you mix together two glasses of water, each at a temperature of 20°C, you will have a larger volume of water at the same temperature of 20°C. The combined volume has more atoms, and therefore more total thermal energy, but each atom is moving just about as it was before, and so the average kinetic energy per atom is unchanged. It is this average kinetic energy of the atoms that is related to the temperature. **The temperature of an object is a measure of the *average* energy of the atoms that make up the object.**

Thermal energy is the *total* energy of the atoms; temperature is the *average* energy. This is an important distinction.

**CONCEPTUAL EXAMPLE 11.9 How does increased mass lead to decreased temperature?**

A race car driver notices that during periods of aggressive driving, the brake disks on her car reach temperatures that are too high. She switches to new disks that are made of the same material but have twice the mass, and the temperature of the disks now stays in a reasonable range. How did increasing the mass of the disks solve the problem?

***REASON*** During braking, kinetic energy of the car is transferred to thermal energy in the brake disks. If the mass of the disks is increased, this energy is shared over a larger number of atoms, so the average energy of each atom increases by less. The addition of thermal energy therefore causes a smaller temperature rise.

***ASSESS*** This explanation makes sense. You know, from experience, that it takes more energy to raise the temperature of larger objects, so a given energy input will cause a smaller temperature change in a larger object.

## Temperature Scales

A common glass-tube thermometer works by allowing a small volume of mercury or alcohol to expand or contract when placed in contact with a "hot" or "cold" object. Other thermometers work in different ways, but all—at a microscopic level—depend on the speed of the object's atoms as they collide with the atoms in the thermometer. All thermometers are sampling the average kinetic energy of the atoms in the object.

The temperature scale used in scientific work (and in almost every country in the world) is the *Celsius scale*. As you likely know, the Celsius scale is defined so that the freezing point of water is 0°C and the boiling point is 100°C. The units of the Celsius temperature scale are "degrees Celsius," which we abbreviate as °C. The Fahrenheit scale, still widely used in the United States, is related to the Celsius scale by

$$T(°\text{C}) = \frac{5}{9}(T(°\text{F}) - 32°) \qquad T(°\text{F}) = \frac{9}{5}T(°\text{C}) + 32° \tag{11.6}$$

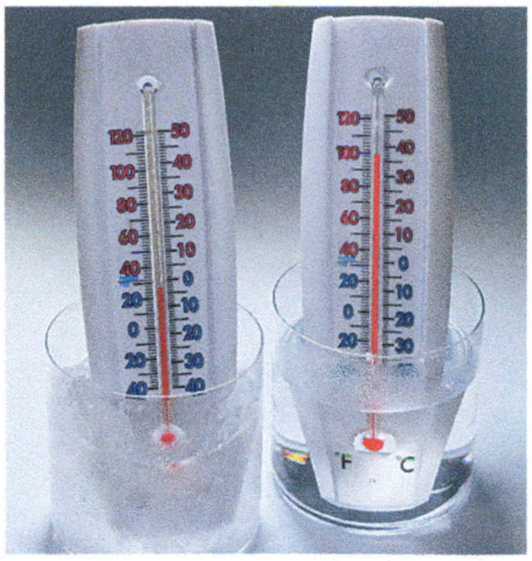

Thermal expansion of the liquid in the thermometer pushes it higher when immersed in hot water than in ice water.

Both the Celsius and the Fahrenheit scales have a zero point that is arbitrary—simply an agreed-upon convention—and both allow negative temperatures. But we can define a temperature scale for which the temperature is always positive. Consider the gas samples in Figure 11.9. If we slow the atoms down, this reduces the average kinetic energy of the atoms and so reduces the temperature. But there's a limit. If the atoms slow to a stop, the kinetic energy is zero; no lower temperature is possible. This is the temperature we call **absolute zero.** If we take this temperature as the zero of our temperature scale, temperatures below zero are not possible.

This is how zero is defined on the temperature scale called the *Kelvin scale:* **Zero degrees is the point at which the kinetic energy of atoms is zero.** All temperatures on the Kelvin scale are positive, so it is often called an *absolute temperature scale.* The units are "kelvin" (not degrees kelvin!), abbreviated K.

The spacing between divisions on the Kelvin scale is the same as that on the Celsius scale; the only difference is the position of the zero point. Absolute zero—the temperature at which atoms would cease moving—is −273°C. The conversion between Celsius and Kelvin temperatures is therefore quite straightforward:

$$T(\text{K}) = T(°\text{C}) + 273 \qquad T(°\text{C}) = T(\text{K}) - 273 \tag{11.7}$$

One Celsius degree and one kelvin are the same. This means that a temperature *difference* is the same on both scales:

$$\Delta T(\text{K}) = \Delta T(°\text{C})$$

On the Kelvin scale, the freezing point of water at 0°C is $T = 0 + 273 = 273$ K. A 30°C warm summer day is $T = 303$ K on the Kelvin scale. FIGURE 11.10 gives a side-by-side comparison of these scales.

**NOTE** ▸From now on, we will use the symbol $T$ for temperature in kelvin. We will denote other scales by showing the units in parentheses. In the equations in this chapter and the rest of the text, ***T* should be interpreted as a temperature in kelvin.** ◂

**FIGURE 11.10** Celsius and Kelvin temperature scales.

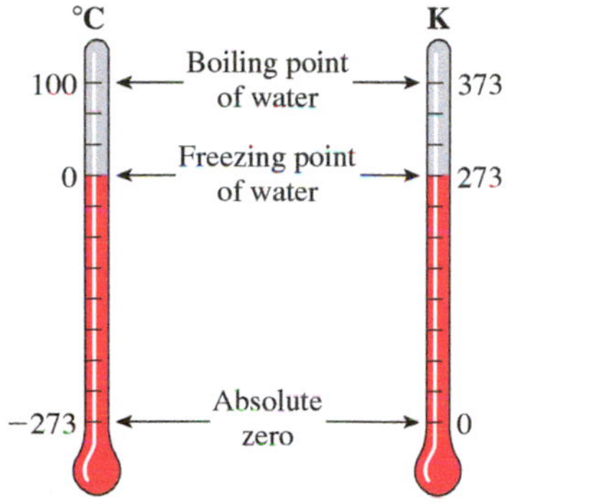


Temperature *differences* are the same on the Celsius and Kelvin scales. The temperature difference between the freezing point and boiling point of water is 100°C or 100 K.

### EXAMPLE 11.10 Finding the temperature on different scales

The coldest temperature ever measured on earth was −129°F, in Antarctica. What is this in °C and K?

***SOLVE*** We first use Equation 11.6 to convert the temperature to the Celsius scale:

$$T(°\text{C}) = \frac{5}{9}(-129° - 32°) = -89°\text{C}$$

We can then use Equation 11.7 to convert this to kelvin:

$$T = -89 + 273 = 184 \text{ K}$$

***ASSESS*** This is cold, but still far warmer than the coldest temperatures achieved in the laboratory, which are very close to 0 K.

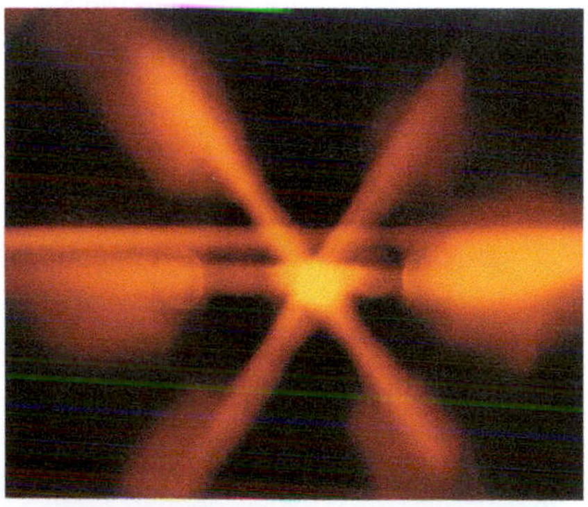

**Optical molasses** It isn't possible to reach absolute zero, where the atoms would be still, but it is possible to get quite close by slowing the atoms down directly. These crossed laser beams produce what is known as "optical molasses." As we will see in Chapter 28, light is made of photons, which carry energy and momentum. Interactions of the atoms of a diffuse gas with the photons cause the atoms to slow down. In this manner atoms can be slowed to speeds that correspond to a temperature as cold as $5 \times 10^{-10}$ K!

**STOP TO THINK 11.3** Two samples of a gas, sample 1 and sample 2, have the same thermal energy. Sample 1 has twice as many atoms as sample 2. What can we say about the temperatures of the two samples?

A. $T_1 > T_2$ B. $T_1 = T_2$ C. $T_1 < T_2$

## What Is Heat?

FIGURE 11.11 Raising the temperature of a pan of water by heating it.

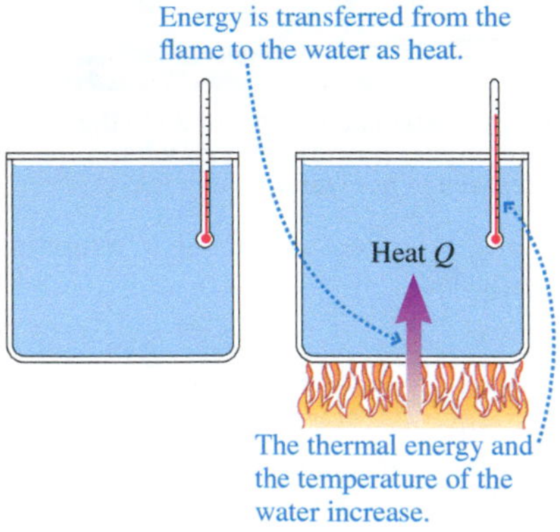


In Chapter 10, we saw that a system could exchange energy with the environment through two different means: work and heat. Work was treated in detail in Chapter 10; now it is time to look at the transfer of energy by heat.

Heat is a more elusive concept than work. We use the word "heat" very loosely in the English language, often as synonymous with "hot." We might say, on a very hot day, "This heat is oppressive." If your apartment is cold, you may say, "Turn up the heat." It's time to develop more precise language to discuss these concepts.

Suppose you put a pan of cold water on the stove. If you light the burner so that there is a hot flame under the pan, as in FIGURE 11.11, the temperature of the water increases. We know that this increased temperature means that the water has more thermal energy, so this process must have *transferred* energy to the system. This energy transferred from the hotter flame to the cooler water is what we call heat. In general, **heat is energy transferred between two objects because of a temperature difference between them.** Heat always flows from the hotter object to the cooler one, never in the opposite direction. If there is no temperature difference, no energy is transferred as heat. We use $Q$ as the symbol for heat.

## An Atomic Model of Heat

FIGURE 11.12 Collisions at a barrier transfer energy from faster molecules to slower molecules.

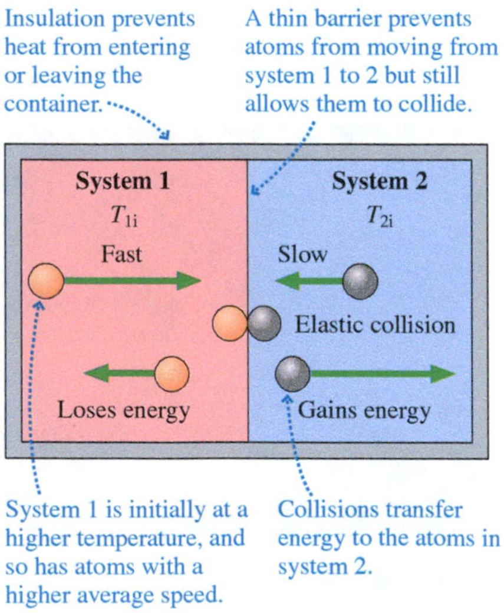


Let's consider an atomic model to explain why thermal energy is transferred from higher temperatures to lower. FIGURE 11.12 shows a rigid, insulated container that is divided into two sections by a very thin barrier. Each side is filled with a gas of the same kind of atoms. The left side, which we'll call system 1, is at an initial temperature $T_{1i}$. System 2 on the right is at an initial temperature $T_{2i}$. We imagine the barrier to be so thin that atoms can collide at the boundary as if the barrier were not there, yet it prevents atoms from moving from one side to the other.

Suppose that system 1 is initially at a higher temperature: $T_{1i} > T_{2i}$. This means that the atoms in system 1 have a higher average kinetic energy. Figure 11.12 shows a fast atom and a slow atom approaching the barrier from opposite sides. They undergo a perfectly elastic collision at the barrier. Although no net energy is lost in a perfectly elastic collision, the faster atom loses energy while the slower one gains energy. There is an energy *transfer* from the faster atom's side to the slower atom's side.

Because the atoms in system 1 are, on average, more energetic than the atoms in system 2, the collisions transfer energy from system 1 to system 2. The net energy transfer, from all collisions, is from the warmer system 1 to the cooler system 2. This transfer of energy is heat; **thermal energy is transferred from the faster atoms on the warmer side to the slower atoms on the cooler side.**

This transfer will continue until a stable situation is reached. This is a situation we call **thermal equilibrium.** How do the systems "know" when they've reached thermal equilibrium? Energy transfer continues until the atoms on both sides of the barrier have the *same average kinetic energy.* Once the average kinetic energies are the same, individual collisions will still transfer energy from one side to the other. But since both sides have atoms with the same average kinetic energies, the amount of energy transferred from 1 to 2 will equal the amount transferred from 2 to 1. Once the average kinetic energies are the same, there is no net energy transfer.

As we've seen, the average kinetic energy of the atoms in a system is a measure of the system's temperature. If two systems exchange energy until their atoms have the same average kinetic energy, we can say that

$$T_{1f} = T_{2f} = T_f$$

That is, heat is transferred until the two systems reach a common final temperature; this is the state of thermal equilibrium. We considered a rather artificial system in this case, but the result is quite general: **Two systems placed in thermal contact will transfer thermal energy from hot to cold until their final temperatures are the same.** This process is illustrated in FIGURE 11.13.

Heat is a transfer of energy. The sign that we use for transfers is defined in FIGURE 11.14. In the process of Figure 11.13, $Q_1$ is negative because system 1 loses energy; $Q_2$ is positive because system 2 gains energy. No energy escapes from the container, so all of the energy that was lost by system 1 was gained by system 2. We can write that as

$$Q_2 = -Q_1$$

The heat energy lost by one system is gained by the other.

FIGURE 11.13 Two systems in thermal contact exchange thermal energy.

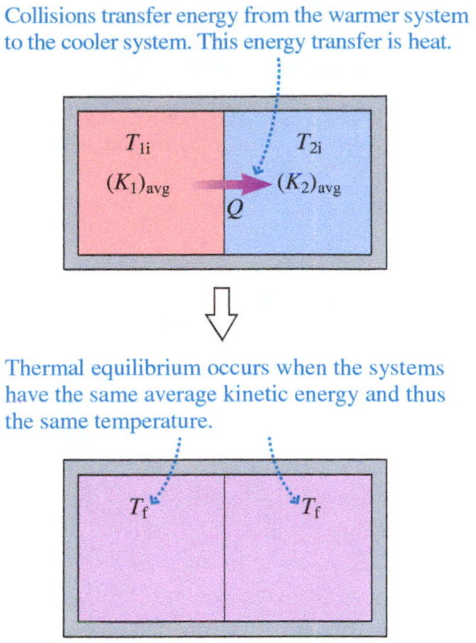


## 11.4 The First Law of Thermodynamics

We need to broaden our work-energy equation, Equation 11.1, to include energy transfers in the form of heat. The sign conventions for heat, as shown in Figure 11.14, have the same form as those for work—a positive value means a transfer into the system, a negative value means a transfer out of the system. Thus, when we include heat $Q$ in the work-energy equation, it appears on the right side of the equation along with work $W$:

$$\Delta K + \Delta U + \Delta E_{th} + \Delta E_{chem} + \cdots = W + Q \qquad (11.8)$$

This equation is our most general statement to date about the conservation of energy; it includes all of the energy transfers and transformations we have discussed.

In Chapter 10, we focused on systems where the potential and kinetic energies could change, such as a sled moving down a hill. Earlier in this chapter, we looked at the body, where the chemical energy changes. Now lets consider systems in which only the thermal energy changes, systems that aren't moving, that aren't changing chemically, but whose temperatures can change. Such systems are the province of **thermodynamics,** the study of thermal energy and heat and their relationships to other forms of energy and energy transfer. The question of how to keep your house cool in the summer is a question of thermodynamics. If energy is transferred into your house, the thermal energy increases and the temperature rises. To reduce the temperature, you must transfer energy out of the house. This is the purpose of an air conditioner.

If we consider cases in which only thermal energy changes, Equation 11.8 can be simplified. This simpler version is a statement of conservation of energy for systems in which only the thermal energy changes; we call it the **first law of thermodynamics:**

FIGURE 11.14 The sign of $Q$.

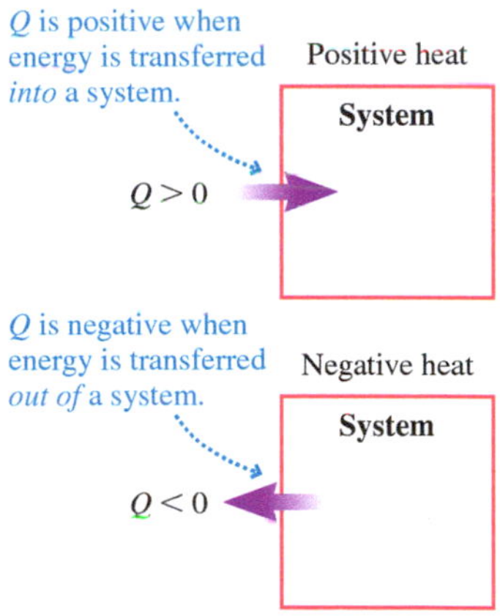


A video to support a section's topic is embedded in the eText.
**Video** Work and Thermal Energy in Gases

> **First law of thermodynamics** For systems in which only the thermal energy changes, the change in thermal energy is equal to the energy transferred into or out of the system as work $W$, heat $Q$, or both:
>
> $$\Delta E_{th} = W + Q \qquad (11.9)$$

Work and heat are the two ways by which energy can be transferred between a system and its environment, thereby changing the system's energy. And, in thermodynamic systems, the only energy that changes is the thermal energy. Whether this energy increases or decreases depends on the signs of $W$ and $Q$. The possible energy transfers between a system and the environment are illustrated in FIGURE 11.15.

FIGURE 11.15 Energy transfers in a thermodynamic system.

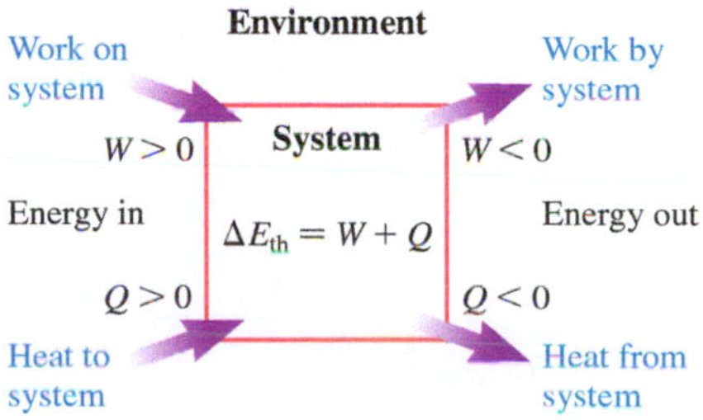

**CONCEPTUAL EXAMPLE 11.11** **What happens when you compress a gas?**

Suppose a gas is in an insulated container, so that no heat energy can escape. If a piston is used to compress the gas, what happens to the temperature of the gas?

**REASON** The piston applies a force to the gas, and there is a displacement. This means that work is done on the gas by the piston $(W > 0)$. No thermal energy can be exchanged with the environment, meaning $Q = 0$. Since energy is transferred into the system, the thermal energy of the gas must increase. This means that the temperature must increase as well.

**ASSESS** This result makes sense in terms of everyday observations you may have made. When you use a bike pump to inflate a tire, the pump and the tire get warm. This temperature increase is largely due to the warming of the air by the work you do compressing the gas.

**EXAMPLE 11.12** **Why does food heat up in a blender?**

If you mix food in a blender, the electric motor does work on the system, which consists of the food inside the container. This work can noticeably warm up the food. Suppose the blender motor runs at a power of 250 W for 40 s. During this time, 2000 J of heat flow from the now-warmer food to its cooler surroundings. By how much does the thermal energy of the food increase?

**STRATEGIZE** This is a case where only the thermal energy of the system changes, so we will use the first law of thermodynamics, Equation 11.9. The motor does work on the system, and some energy leaves the system as heat. The difference between the two is the change in the thermal energy of the system.

**PREPARE** We can find the work done by the motor from the power it uses and the time it runs. Equation 10.24 relates work to power and time. The work done on the system is

$$W = P\Delta t = (250\ \text{W})(40\ \text{s}) = 10{,}000\ \text{J}$$

Because heat *leaves* the system, its sign is negative, so $Q = -2000\ \text{J}$.

**SOLVE** With these values in hand, the first law of thermodynamics gives

$$\Delta E_{\text{th}} = W + Q = 10{,}000\ \text{J} - 2000\ \text{J} = 8000\ \text{J}$$

**ASSESS** It seems reasonable that the work done by the powerful motor rapidly increases the thermal energy, while thermal energy only slowly leaks out as heat. The increased thermal energy of the food implies an increased temperature. If you run a blender long enough, the food can actually start to steam, as the photo shows.

## Energy-Transfer Diagrams

**FIGURE 11.16** Energy-transfer diagrams.

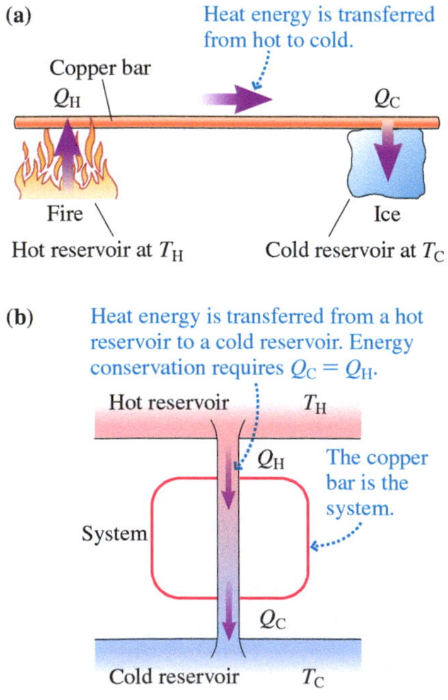


Suppose you drop a hot rock into the ocean. Heat is transferred from the rock to the ocean until the rock and ocean are the same temperature. Although the ocean warms up ever so slightly, $\Delta T_{\text{ocean}}$ is so small as to be completely insignificant.

An **energy reservoir** is an object or a part of the environment so large that, like the ocean, its temperature does not noticeably change when heat is transferred between the system and the reservoir. A reservoir at a higher temperature than the system is called a *hot reservoir.* A vigorously burning flame is a hot reservoir for small objects placed in the flame. A reservoir at a lower temperature than the system is called a *cold reservoir.* The ocean is a cold reservoir for the hot rock. We will use $T_H$ and $T_C$ to designate the temperatures of the hot and cold reservoirs.

Heat energy is transferred between a system and a reservoir if they have different temperatures. We will define

$$Q_H = \text{amount of heat transferred to or from a hot reservoir}$$

$$Q_C = \text{amount of heat transferred to or from a cold reservoir}$$

**By definition, $Q_H$ and $Q_C$ are *positive* quantities.**

FIGURE 11.16a shows a heavy copper bar placed between a hot reservoir (at temperature $T_H$) and a cold reservoir (at temperature $T_C$). Heat $Q_H$ is transferred from the hot reservoir into the copper, and heat $Q_C$ is transferred from the copper into the cold reservoir. FIGURE 11.16b is an **energy-transfer diagram** for this process. The hot reservoir is generally drawn at the top, the cold reservoir at the bottom, and the system—the copper bar in this case—between them. The reservoirs and the system are connected by "pipes" that show the energy transfers. Figure 11.16b shows heat $Q_H$ being transferred into the system and $Q_C$ being transferred out.

FIGURE 11.17 illustrates an important fact about heat transfers that we have discussed: Spontaneous transfers go in one direction only, from hot to cold. This is an important result that has significant practical implications.

FIGURE 11.17 An impossible energy transfer.

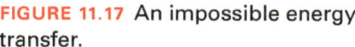

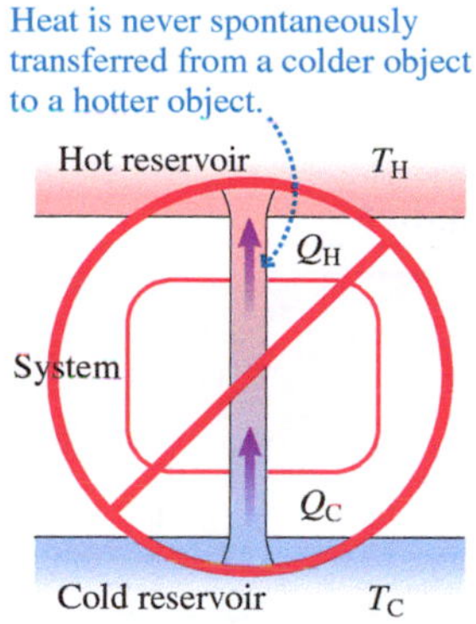


**CONCEPTUAL EXAMPLE 11.13 Keeping your cool** BIO

Why—in physics terms—is it more taxing on the body to exercise in very hot weather?

***REASON*** Your body continuously converts chemical energy to thermal energy, as we have seen. In order to maintain a constant body temperature, your body must continuously transfer heat to the environment. This is a simple matter in cool weather when heat is spontaneously transferred to the environment, but when the air temperature is higher than your body temperature, your body cannot cool itself this way and must use other mechanisms to transfer this energy, such as perspiring. These mechanisms require additional energy expenditure.

***ASSESS*** Strenuous exercise in hot weather can easily lead to a rise in body temperature if the body cannot exhaust heat quickly enough.

**STOP TO THINK 11.4** You have driven your car for a while and now turn off the engine. Your car's radiator is at a higher temperature than the air around it. Considering the radiator as the system, as the radiator cools down we can say that

A. $Q > 0$ B. $Q = 0$ C. $Q < 0$

## 11.5 Heat Engines

In the early stages of the industrial revolution, most of the energy needed to run mills and factories came from water power. Water in a high reservoir will naturally flow downhill. A waterwheel can be used to harness this natural flow of water to produce some useful energy because some of the potential energy lost by the water as it flows downhill can be converted into other forms.

It is possible to do something similar with heat. Thermal energy is naturally transferred from a hot reservoir to a cold reservoir; it is possible to take some of this energy as it is transferred and convert it to other forms. This is the job of a device known as a **heat engine.**

The energy-transfer diagram of FIGURE 11.18a illustrates the basic physics of a heat engine. It takes in energy as heat from the hot reservoir, turns some of it into useful work, and exhausts the balance as waste heat into the cold reservoir. Any heat engine has exactly the same schematic.

FIGURE 11.18 The operation of a heat engine.

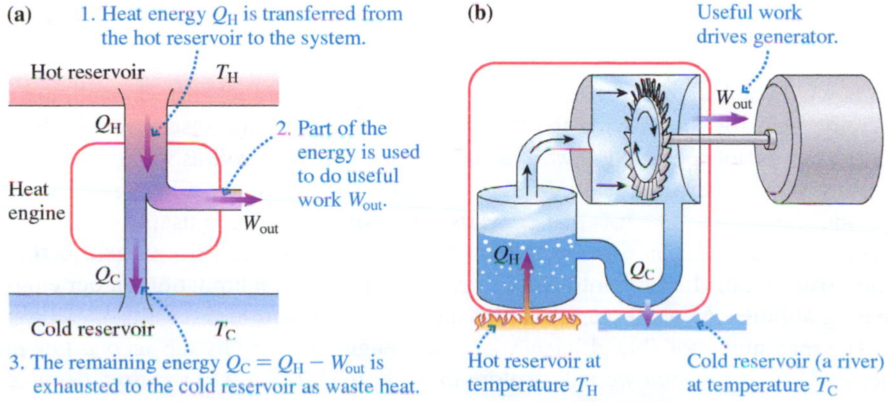


**Video** Figure 11.18

FIGURE 11.18b shows the directions of the energy flows and the locations of the reservoirs for a real heat engine, the power plant discussed earlier in this chapter. The cold reservoir for a power plant can be a river or lake, as shown, or simply the atmosphere, to which heat $Q_C$ is transferred by cooling towers.

Most of the energy that you use daily comes from the conversion of chemical energy into thermal energy and the subsequent conversion of that energy into other forms. The following table shows some common examples of heat engines.

**Heat engines**

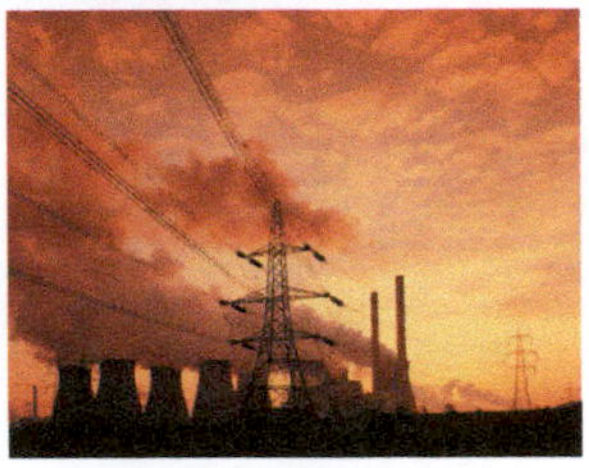

Much of the electricity that you use was generated by heat engines. Natural gas or other fossil fuels are burned to produce high-temperature, high-pressure steam. The steam does work by spinning a turbine attached to a generator, which produces electricity. Some of the energy of the steam is extracted in this way, but more than half simply flows "downhill" and is deposited in a cold reservoir, often a lake or a river.

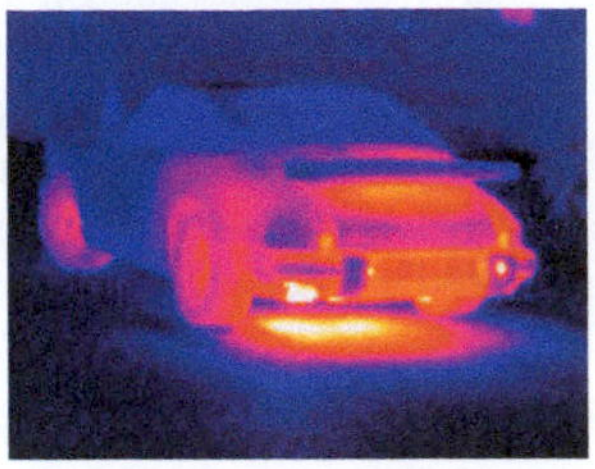

Your car gets the energy it needs to run from the chemical energy in gasoline. The gasoline is burned; the resulting hot gases are the hot reservoir. Some of the thermal energy is converted into the kinetic energy of the moving vehicle, but more than 90% is lost as heat to the surrounding air via the radiator and the exhaust, as shown in this thermogram.

There are many small, simple heat engines that are part of things you use daily. This fan, which can be put on top of a wood stove, uses the thermal energy of the stove to provide power to drive air around the room. Where are the hot and cold reservoirs in this device?

We assume, for a heat engine, that the engine's thermal energy doesn't change. This means that there is no net energy transfer into or out of the heat engine. Because energy is conserved, we can say that the useful work extracted is equal to the difference between the heat energy transferred from the hot reservoir and the heat exhausted into the cold reservoir:

$$W_{out} = Q_H - Q_C$$

The energy input to the engine is $Q_H$ and the energy output is $W_{out}$.

NOTE ▸ We earlier defined $Q_H$ and $Q_C$ to be positive quantities. We also define the energy output of a heat engine $W_{out}$ to be a positive quantity. For heat engines, the directions of the transfers will always be clear, and we will take all of these basic quantities to be positive. ◂

We can use the definition of efficiency, from earlier in the chapter, to compute the heat engine's efficiency:

$$e = \frac{\text{what you get}}{\text{what you had to pay}} = \frac{W_{out}}{Q_H} = \frac{Q_H - Q_C}{Q_H} \qquad (11.10)$$

FIGURE 11.19 A perfect (and impossible!) heat engine.

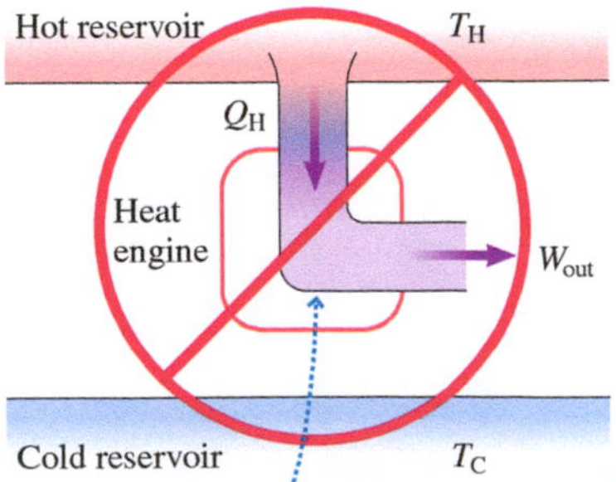


$Q_H$ is what you had to pay because this is the energy of the fuel burned—and, quite literally, paid for—to provide the high temperature of the hot reservoir. The heat energy that is not converted to work ends up in the cold reservoir as waste heat.

Why should we waste energy this way? Why don't we make a heat engine like the one shown in FIGURE 11.19 that converts 100% of the heat into useful work? The surprising answer is that we can't. **No heat engine can operate without exhausting some fraction of the heat into a cold reservoir.** This isn't a limitation on our engineering abilities. As we'll see, it's a fundamental law of nature.

The maximum possible efficiency of a heat engine is fixed by the *second law of thermodynamics,* which we will explore in Section 11.7. We will not do a detailed derivation, but simply note that the second law gives the theoretical maximum efficiency of an ideal heat engine as

$$e_{max} = 1 - \frac{T_C}{T_H} \quad (11.11)$$

Maximum possible efficiency of a heat engine; $T_C$: Temperature of cold reservoir; $T_H$: Temperature of hot reservoir. Both $T_C$ and $T_H$ must be in *kelvin*.

The maximum possible efficiency of any heat engine is fixed by the ratio of the temperatures of the hot and cold reservoirs. We can increase the efficiency of a heat engine by increasing the temperature of the hot reservoir or by decreasing the temperature of the cold reservoir.

**NOTE** ▶ The actual efficiency of real heat engines is usually noticeably less than the theoretical maximum. Equation 11.10 is a statement of energy conservation; it applies to *any* heat engine. Equation 11.11 applies only to an *ideal* heat engine; it's the upper limit of what is possible. ◀

The work done by a heat engine is always less than the heat input: $W < Q_H$. Consequently, there *must* be heat $Q_C$ exhausted to the cold reservoir. We can increase the efficiency and reduce the quantity of heat that is "wasted" this way, but we can't eliminate it. This is a *fundamental* limit on efficiency due to a law of nature. No matter how well you design your power plant, its efficiency can't exceed the limit of Equation 11.11.

## EXAMPLE 11.14 Determining the theoretical and actual efficiencies for a heat engine

Iceland has both high geothermal activity, with high temperatures near the surface, and abundant cold surface water. Iceland has many power plants that take advantage of the proximity of these natural hot and cold reservoirs. One plant uses an underground source at 122°C as the hot reservoir and a nearby lake at 5°C as the cold reservoir. The plant draws 16 MW from the hot reservoir to produce 1.8 MW of electricity. How does the actual efficiency of the plant compare to the theoretical maximum efficiency?

**STRATEGIZE** Computing the actual efficiency will be straightforward; it's just the ratio of what you get (the electricity output) to what you had to pay (the heat input). We will determine the theoretical efficiency from the temperatures of the hot and cold reservoirs, which are given.

**PREPARE** The plant's operation is specified in terms of power, but Equation 11.10 uses energy, so we can consider the energy transferred or transformed in 1 second. In 1 second, the plant draws 16 MJ from the hot reservoir and produces 1.8 MJ of electric energy.

To compute the maximum possible efficiency, we need the absolute temperatures of the hot and cold reservoirs:

$$T_H = 122°C = 395 \text{ K} \qquad T_H = 5°C = 278 \text{ K}$$

**SOLVE** The actual efficiency of the plant is

$$e = \frac{W_{out}}{Q_H} = \frac{1.8 \text{ MJ}}{16 \text{ MJ}} = 0.11$$

The maximum possible efficiency for the plant is given by Equation 11.11:

$$e_{max} = 1 - \frac{T_C}{T_H} = 1 - \frac{278 \text{ K}}{395 \text{ K}} = 0.30$$

The theoretical maximum efficiency is nearly triple the actual efficiency.

**ASSESS** The actual efficiency is less than the maximum efficiency, as it must be, and this gives us confidence in our result. It's perhaps not surprising that this particular plant's efficiency is so much less than the theoretical maximum, because there is no cost for the energy input. There might not be a huge incentive to increase efficiency if you don't literally have to pay for "what you had to pay."

There are fundamental limits that specify that some of the energy in a heat engine must be deposited in a cold reservoir, but there is no law of nature that says this energy must be wasted. It's common practice in Europe for power plants to use the heat that they would otherwise exhaust into the environment to heat nearby homes. Another option is shown in the photo of the Icelandic geothermal power plant in the chapter preview. The lagoon shown is the cold reservoir of the power plant in the background; the excess heat from the plant's operation warms the water, making it a very popular bathing spot. The cold reservoir of this particular heat engine is the country's largest tourist attraction!

**STOP TO THINK 11.5** Which of the following changes (there may be more than one) would increase the maximum theoretical efficiency of a heat engine?

A. Increase $T_H$  B. Increase $T_C$  C. Decrease $T_H$  D. Decrease $T_C$

## 11.6 Heat Pumps

The inside of your refrigerator is colder than the air in your kitchen, so heat will always flow from the room to the inside of the refrigerator, warming it. This happens every time you open the door and as heat "leaks" through the refrigerator's walls. To keep your refrigerator cool, you need some way to move this heat back out to the warmer room. Transferring heat energy from a cold reservoir to a hot reservoir—the opposite of the natural direction—is the job of a **heat pump.**

If you have a container of hot water next to a container of cold water, heat will flow from the hot water to the cold. But a heat pump can cause the heat to move the other way. The water dispenser in FIGURE 11.20 has containers for hot and cold water. A heat pump between them moves heat from the cold water to the hot water. The energy used to chill the cold water results in increased thermal energy in the hot water. Paying for cold water gets you hot water for free.

FIGURE 11.20 A heat pump provides hot and cold water.

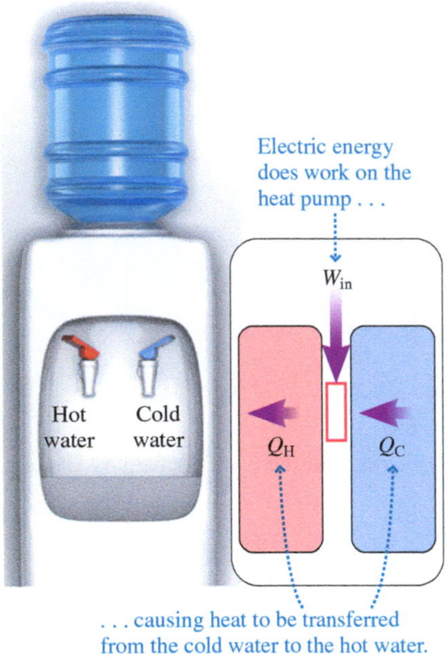


Heat pumps are most often used for cooling, by transferring heat to another location. The heat pump in a refrigerator transfers heat from the cold air inside the refrigerator to the warmer air in the room. An air conditioner is a heat pump that performs a similar task, transferring heat from the cool air inside a house (or a car) to the warmer air outside. But a heat pump can also be used for heating; on a cold winter day, an electric heat pump can extract heat from the cold outside air to warm the inside of a house. A reversible heat pump can do both—move heat out of the house in the summer and into the house in the winter.

In all of these cases, we are moving energy against the natural direction it would flow. This requires an energy input—work must be done—as shown in the energy-transfer diagram of a heat pump in FIGURE 11.21. Energy must be conserved, so the heat deposited in the hot side must equal the sum of the heat removed from the cold side and the work input:

FIGURE 11.21 The operation of a heat pump.

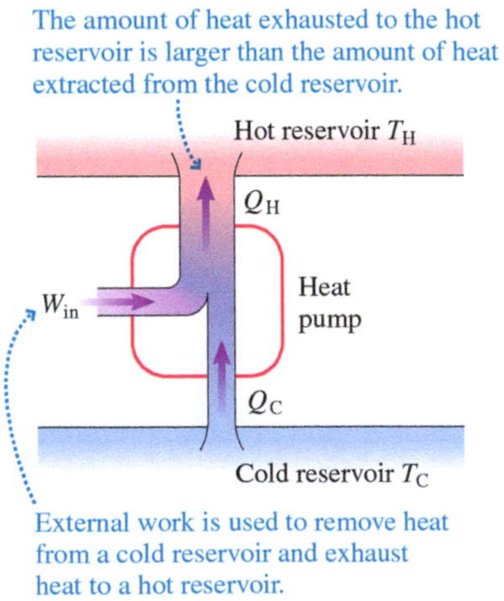


$$Q_H = Q_C + W_{in}$$

For heat pumps, rather than compute efficiency, we compute an analogous quantity called the **coefficient of performance (COP).** A refrigerator uses a heat pump for cooling, removing heat from a cold reservoir to keep it cold. As Figure 11.21 shows, we must do work to make this happen. If we use a heat pump for cooling, we define the coefficient of performance as

$$\text{COP} = \frac{\text{what you get}}{\text{what you had to pay}} = \frac{\text{energy removed from cold reservoir}}{\text{work required to perform the transfer}} = \frac{Q_C}{W_{in}} \tag{11.12}$$

The second law of thermodynamics limits the efficiency of a heat pump just as it limits the efficiency of a heat engine. The maximum possible coefficient of performance is related to the temperatures of the hot and cold reservoirs:

$$\text{COP}_{max} = \frac{T_C}{T_H - T_C} \tag{11.13}$$

Theoretical maximum coefficient of performance of a heat pump used for cooling

We can also use a heat pump for heating, moving heat from a cold reservoir to a hot reservoir to keep it warm. In that case, we define the coefficient of performance as

$$\mathrm{COP} = \frac{\text{what you get}}{\text{what you had to pay}} = \frac{\text{energy added to hot reservoir}}{\text{work required to perform the transfer}} = \frac{Q_H}{W_{in}} \quad (11.14)$$

In this case, the maximum possible coefficient of performance is

$$\mathrm{COP}_{max} = \frac{T_H}{T_H - T_C} \quad (11.15)$$

Theoretical maximum coefficient of performance of a heat pump used for heating

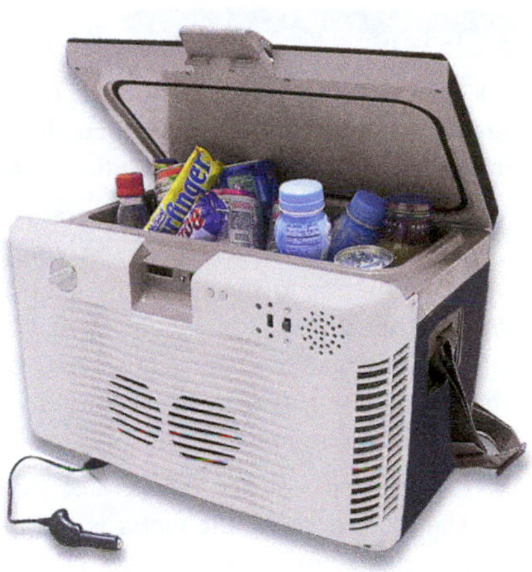

**Hot or cold lunch?** Small coolers like this use *Peltier devices* that run off the 12 V electrical system of your car. They can transfer heat from the interior of the cooler to the outside, keeping food or drinks cool. But Peltier devices, like some other heat pumps, are reversible; switching the direction of current reverses the direction of heat transfer, causing heat to be transferred into the interior and keeping your lunch warm!

In both cases, **a larger coefficient of performance means a more efficient heat pump.** Unlike the efficiency of a heat engine, which must be less than 1, the COP of a heat pump can be—and usually is—greater than 1.

**NOTE** ▸ As for heat engines, the actual efficiency of real heat pumps is usually noticeably less than the theoretical maximum. Equations 11.12 and 11.14 are statements of energy conservation; they apply to *any* heat pump. Equations 11.13 and 11.15 apply only to *ideal* heat pumps; they give the upper limits of what is possible. ◂

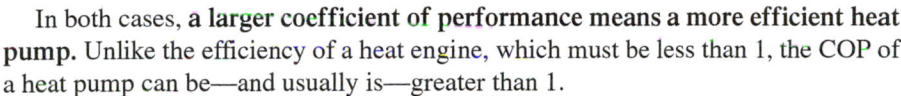

**EXAMPLE 11.15** **How much energy does it take to heat the house?**

A heat pump moves heat from the chilly 8°C air outside a house to the toastier 21°C air inside. The actual coefficient of performance of the system is 3.2, a typical value.

a. If the system adds 11 kJ to the house each second, how much electric energy is required?

b. How much energy would it take to add 11 kJ each second if the heat pump ran at the maximum theoretical efficiency?

***STRATEGIZE*** This heat pump is used for heating, so we will use Equation 11.14 to compute the energy required for the system. We will use Equation 11.15 to find the maximum possible COP.

***PREPARE*** To compute the maximum possible COP, we need to know the absolute temperatures of the cold and hot reservoirs:

$$T_H = 21°\mathrm{C} = 294\ \mathrm{K} \qquad T_H = 8°\mathrm{C} = 281\ \mathrm{K}$$

***SOLVE*** We rearrange Equation 11.14 to find $W_{in}$, the electric energy required, for the given COP of 3.2:

$$W_{in} = \frac{Q_H}{\mathrm{COP}} = \frac{11\ \mathrm{kJ}}{3.2} = 3.4\ \mathrm{kJ}$$

The maximum possible COP for the heat pump is given by Equation 11.15:

$$\mathrm{COP}_{max} = \frac{T_H}{T_H - T_C} = \frac{294\ \mathrm{K}}{294\ \mathrm{K} - 281\ \mathrm{K}} = 23$$

With this greater efficiency, it would take much less energy to add 11 kJ to the house:

$$W_{in} = \frac{Q_H}{\mathrm{COP}_{max}} = \frac{11\ \mathrm{kJ}}{23} = 0.48\ \mathrm{kJ}$$

***ASSESS*** The 3.4 kJ input of electric energy is much less than the 11 kJ of heat added to the house, so we have confidence in our results—that's the way a heat pump is supposed to work. The actual COP of 3.2 is much less than the theoretical maximum, so it's no surprise that a heat pump running at its theoretical maximum efficiency can achieve the same result with much less energy input.

A heat pump like the one in the example is a great idea if you live in an area where electric heat is common. An electric space heater uses 3.4 kJ of electricity to add 3.4 kJ of heat to your house. The heat pump uses this 3.4 kJ of electric energy to add 11 kJ of heat to your house—a clear improvement. But a heat pump working at the maximum possible efficiency could add 78 kJ to your house for this 3.4 kJ input. There are practical limitations of design, but the big difference between the theoretical and actual efficiencies makes it clear that we could be doing better.

**CONCEPTUAL EXAMPLE 11.16** **Keeping your cool?**

It's a hot day, and your apartment is rather warm. Your roommate suggests cooling off the apartment by keeping the door of the refrigerator open. Will this help the situation?

***REASON*** The energy-transfer diagram for this process is shown in FIGURE 11.22. The heat pump in your refrigerator moves heat from its interior to the air in the apartment as usual. But, with the door open, heat flows spontaneously from the apartment back into the refrigerator. Note, however, that more heat enters the apartment than leaves it—the open refrigerator is actually warming up the apartment!

***ASSESS*** Even a closed refrigerator slowly warms the room it's in. The work done by the motor is always transformed into extra heat exhausted to the hot reservoir.

FIGURE 11.22 Energy-transfer diagram for an open refrigerator.

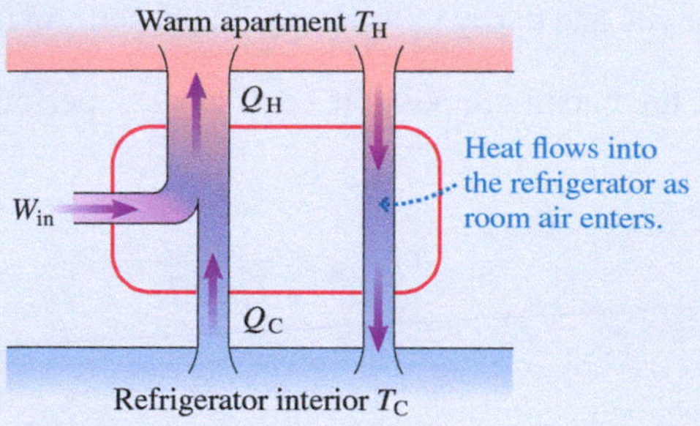


**STOP TO THINK 11.6** Which of the following changes would allow your refrigerator to use less energy to run? (There may be more than one correct answer.)

A. Increasing the temperature inside the refrigerator
B. Increasing the temperature of the kitchen
C. Decreasing the temperature inside the refrigerator
D. Decreasing the temperature of the kitchen

## 11.7 Entropy and the Second Law of Thermodynamics

Throughout the chapter, we have noticed certain trends and certain limitations in energy transformations and transfers. Heat is transferred spontaneously from hot to cold, not from cold to hot. The spontaneous transfer of heat from hot to cold is an example of an **irreversible** process, a process that can happen in only one direction. Why are some processes irreversible? The spontaneous transfer of heat from cold to hot would not violate any law of physics that we have seen to this point, but it is never observed. There must be another law of physics that prevents it.

### Reversible and Irreversible Processes

At a microscopic level, collisions between molecules are completely reversible. In FIGURE 11.23 we see two possible movies of a collision between two gas molecules, one forward and the other backward. You can't tell by looking which is really going forward and which is being played backward. Nothing in either collision looks wrong, and no measurements you might make on either would reveal any violations of Newton's laws. Interactions at the molecular level are **reversible** processes.

FIGURE 11.23 Molecular collisions are reversible.

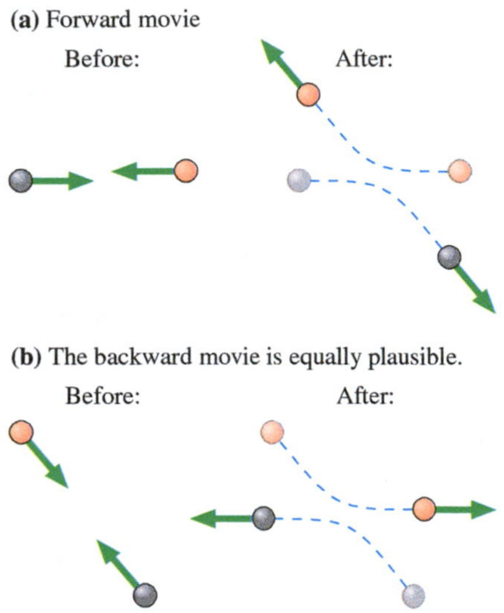


At a macroscopic level, it's a different story. FIGURE 11.24 shows two possible movies of the collision of a car with a barrier. One movie is being run forward, the other backward. The backward movie of Figure 11.24b is obviously wrong. But what has been violated in the backward movie? To have the car return to its original shape and spring away from the wall would not violate any laws of physics we have so far discussed.

FIGURE 11.24 Macroscopic collisions are not reversible.

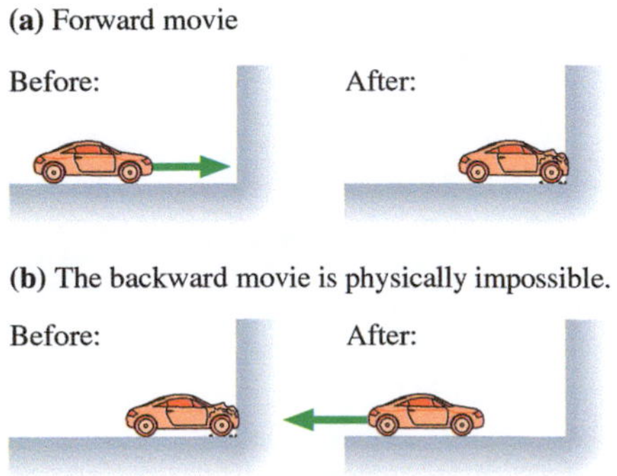


If microscopic motions are all reversible, how can macroscopic phenomena such as the car crash end up being irreversible? If reversible collisions can cause heat to be transferred from hot to cold, why do they never cause heat to be transferred from cold to hot?

## Thermal Energy Spreads

A fundamental difference between microscopic and macroscopic objects is that macroscopic objects have thermal energy, the kinetic and potential energies of all the molecules that make up the object. A macroscopic object can store energy internally, but a point-like particle cannot. Thermal energy didn't appear in our study of the mechanics of particles; it became an issue only when we considered macroscopic objects.

A macroscopic object contains a vast number of atoms and molecules. The thermal energy is the total energy of all these particles, but knowing the thermal energy tells us nothing about how the energy is distributed. FIGURE 11.25a shows a container of gas in which one molecule is moving while all the others are at rest. The thermal energy of the gas is simply the kinetic energy of the one moving molecule. This is one possible way that the energy could be distributed, but this is not a stable, equilibrium situation. The moving molecule will quickly collide with other molecules, giving them kinetic energy while the original fast molecule loses energy. Eventually, we expect that every atom in the gas will be moving, something like the situation shown in FIGURE 11.25b. The thermal energy is the same in both cases; it's just distributed differently.

FIGURE 11.25 Two different distributions of the thermal energy of a gas.

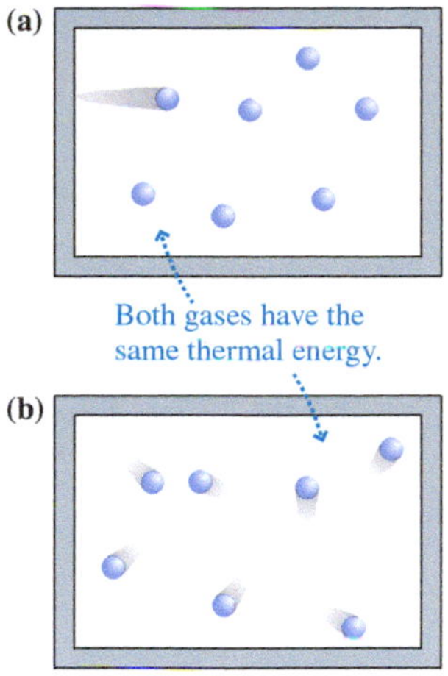


Let's think about what happened here. Initially, the system's thermal energy was concentrated in a single molecule. As time went on, collisions and interactions caused the system's energy to spread out, or disperse, until it was shared among all the constituents. And this spreading will not spontaneously reverse. Ordinary molecular collisions cause a concentrated energy to spread out, but we never see a system spontaneously concentrating dispersed energy into a single constituent. **The spreading of energy is an irreversible process.**

All thermal interactions involve the spreading of energy. Suppose a hot object and a cold object are brought into thermal contact, as in FIGURE 11.26. The thermal energy is initially concentrated in the hot object; it is not shared equally. As we've seen, collisions at the boundary transfer energy from the more energetic molecules on the warmer side to the less energetic molecules on the cooler side until the thermal energy is shared equally among the constituents of the combined system. After some time, thermal equilibrium is reached, which means that no additional sharing or spreading is possible.

FIGURE 11.26 A thermal interaction causes energy to spread until it is shared equally.

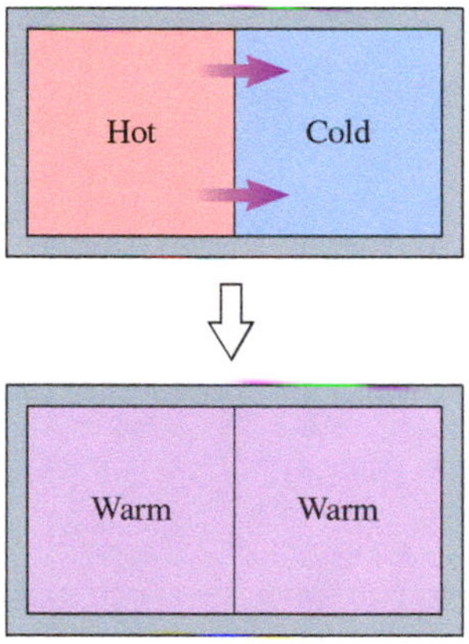


The *spreading* of thermal energy happens spontaneously—it's simply what happens when macroscopic systems interact. The *concentration* of thermal energy does not happen spontaneously. You could imagine that energy continues to be transferred from left to right in Figure 11.26, making the right side hotter than the left by concentrating the energy in the right side, but this never happens in practice.

## Entropy

Scientists use the term **entropy** to quantify the spread or dispersal of thermal energy. The formal definition is fairly technical, but a qualitative discussion will capture the key elements. A system in which energy is concentrated—not spread out—has low entropy. The initial situations in Figures 11.25 and 11.26 are states of low entropy. In each case, the system spontaneously evolved to a state in which the energy is more spread out. These are states of higher entropy; the entropy increases as thermal energy becomes more spread out. Because equilibrium is a state in which thermal energy is maximally dispersed, or shared among all the constituents, **the entropy of a system is maximum when the system is in equilibrium.**

This observation about how thermal energy spreads and how macroscopic systems evolve irreversibly toward equilibrium is a new law of physics, the **second law of thermodynamics,** that is usually stated in terms of entropy:

> **Second law of thermodynamics** The entropy of an isolated system never decreases. The entropy either increases until the system reaches equilibrium or, if the system began in equilibrium, stays the same. Entropy is maximum when the system is in equilibrium.

NOTE ▶ The qualifier "isolated" is crucial. We could reduce a system's entropy by reaching in from the outside, perhaps using tiny tweezers to give most of the thermal energy to a few fast molecules while freezing the others in place. Similarly, we can transfer heat from cold to hot by using a refrigerator. The second law is about what a system can or cannot do spontaneously, on its own, without outside intervention. ◀

Thermodynamics is a science of energy, and we need two laws of energy to understand how macroscopic systems behave. We can restate the two laws of energy as follows:

- **First law of thermodynamics:** Energy is conserved. It can be transferred or transformed, but the total amount does not change.
- **Second law of thermodynamics:** Energy spreads out. An isolated system evolves until the thermal energy is maximally dispersed among all the system's constituents.

It is the second law that creates the "arrow of time" that we observe in irreversible processes. Systems can evolve over time only in ways that continue to spread out thermal energy, thus increasing entropy, not in ways that concentrate thermal energy, thus decreasing entropy.

Understanding the thermal interaction of Figure 11.26 requires not only the first law—energy is transferred from the hotter to the colder object without change—but also the second law. Energy would also be conserved if it were transferred from the colder object to the hotter one, thus increasing their temperature difference, but that cannot happen because it would lower the overall entropy by reducing the energy spread. The evolution toward thermal equilibrium, with equal temperatures, is the route of increasing entropy—increasing energy spread—and that is the only evolution allowed by the second law.

## Reasoning with Entropy

If we add heat to a system, we increase its thermal energy. By increasing the thermal energy, we also increase the ability of the system to disperse thermal energy, so we increase the entropy. Similarly, if we take heat out of a system, we reduce the ability of the system to disperse thermal energy, so we reduce the entropy.

FIGURE 11.27 The free expansion of a gas increases its entropy.

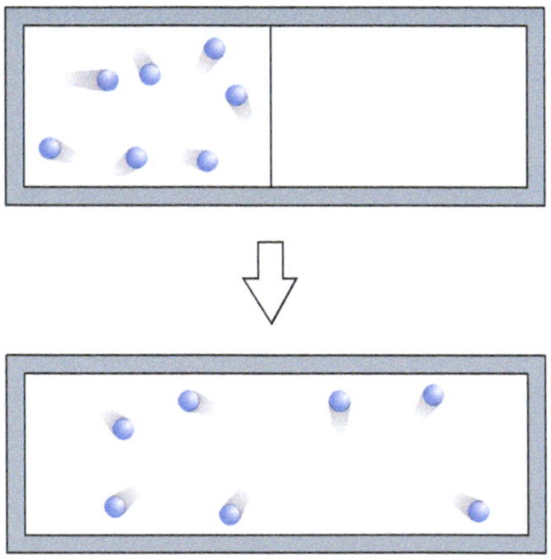

The thermal energy hasn't changed, but the entropy has.

The entropy can also change if there is no change in thermal energy. FIGURE 11.27 shows a container of a gas with the molecules kept on one side by a partition. The container is an isolated system; no heat is transferred to or from the environment. If the partition is removed, the gas expands to fill the cylinder—what we call a *free expansion*. The volume has changed and the pressure will change. But if we think about energy, we know that this expansion does not change the gas temperature. No work is done because the expanding gas isn't pushing against anything, and no heat is transferred from the environment, so the first law of thermodynamics tells us there's no change in thermal energy and hence no change in temperature. But the system's entropy has increased! The thermal energy may not have changed in value, but it's more spatially dispersed because each molecule has more room to move around. And a larger dispersal of energy corresponds to greater entropy. The fact that entropy increases makes the free expansion an irreversible process; the reverse process of all the gas molecules moving into the left half of the cylinder never occurs spontaneously.

Something else that changes is the ability of the gas to do work. Instead of simply removing the partition, we could have allowed the compressed gas to do work by pushing on a piston. This would have transferred some of the thermal energy to the environment by doing mechanical work. The thermal energy is unchanged by the free expansion, but the gas now has less ability to do work. In general, our ability to extract useful work from thermal energy depends not only on the quantity of thermal energy but also on the system's entropy. Increased entropy, with the energy more dispersed, means less ability to do work.

This is why biochemical reactions are analyzed using free energy, which, loosely speaking, is the amount of work that can be done. Any increase in entropy lessens the work that can be done and thus lessens the free energy even if the total energy is unchanged.

**CONCEPTUAL EXAMPLE 11.17** **What is the entropy change when ice melts?**

A phase change, such as melting or freezing, occurs at a fixed temperature. What happens to the entropy of the water if a block of frozen water at 0°C becomes liquid water at 0°C?

***REASON*** The entropy can differ, even at the same temperature, if one of the phases of matter—solid or liquid in this case—is able to spread out the energy more. Ice is a solid, so water molecules are frozen into a lattice. Molecules jiggle around—that's what thermal energy is—but their energy is confined to small volumes around the lattice positions. In contrast, the molecules in liquid water can move freely to all points in space. This allows the thermal energy to be more spatially dispersed, just as it was for the free expansion of a gas. Thus the entropy increases as ice melts.

***ASSESS*** This result makes sense. You know that you need to add heat to melt ice, so you expect this change to increase the entropy.

Of all the forms of energy we've seen, only thermal energy is associated with entropy. As long as thermal energy isn't involved, you can freely and reversibly convert between different forms of energy, as we've seen. When you toss a ball into the air, the kinetic energy is converted to potential energy as the ball rises, then back to kinetic energy as the ball falls. But when some form of energy is transformed into thermal energy, entropy has increased, and the second law of thermodynamics tells us that this change is irreversible. When you drop a ball on the floor, it bounces several times and then stops moving. The kinetic energy of the ball in motion has become thermal energy of the slightly warmer ball at rest. This process won't reverse—the ball won't suddenly cool and jump into the air! This process would conserve energy, but it would reduce entropy, and so would violate this new law of physics.

**CONCEPTUAL EXAMPLE 11.18** **Why are hybrid vehicles more efficient?**

Hybrid vehicles are powered by a gasoline engine paired with an electric motor and batteries. Hybrids get much better mileage in the stop and go of city driving than conventional vehicles do. When you brake to a stop in a conventional car, friction converts the kinetic energy of the car's motion into thermal energy in the brakes. In a typical hybrid car, depressing the brake pedal will connect a generator that converts much of the car's kinetic energy into chemical energy in a battery. Explain how this makes a hybrid vehicle more efficient.

***REASON*** When energy is transformed into thermal energy, the increase in entropy makes this change irreversible. When you brake a conventional car, the kinetic energy is transformed into the thermal energy of hot brakes and is lost to your use. In a hybrid vehicle, the kinetic energy is converted into chemical energy in a battery. This change is reversible; when the car starts again, the energy can be transformed back into kinetic energy.

***ASSESS*** This makes sense. Whenever energy is converted to thermal energy, it is in some sense "lost," which reduces efficiency. The hybrid vehicle avoids this transformation, and so is more efficient.

**STOP TO THINK 11.7** Which of the following processes does not involve a change in entropy?

A. An electric heater raises the temperature of a cup of water by 20°C.
B. A ball rolls up a ramp, decreasing in speed as it rolls higher.
C. A basketball is dropped from 2 m and bounces until it comes to rest.
D. The sun shines on a black surface and warms it.

## 11.8 Systems, Energy, and Entropy

We commented earlier that no heat engine can be 100% efficient, transforming all of the input heat into useful work. Doing so wouldn't violate energy conservation, but it would violate the second law of thermodynamics. Now we are in a position to explain why.

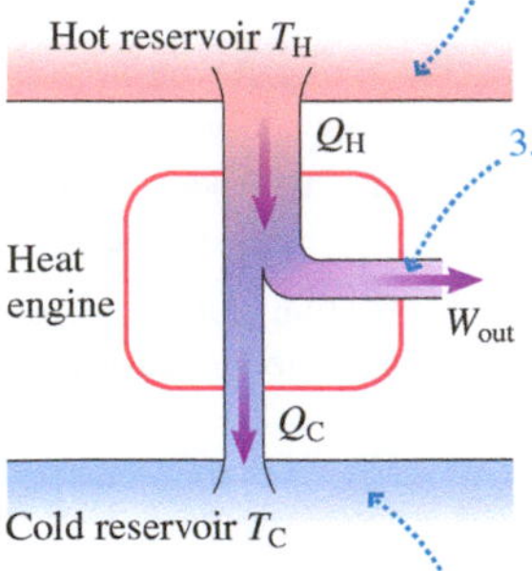


FIGURE 11.28 Entropy considerations for a heat engine.

FIGURE 11.28 is another look at the energy diagram of a heat engine. The engine runs in a closed cycle and returns to the same state time after time, like the pistons in your car engine going through a closed cycle but returning to the same position. The entropy of the engine increases and decreases at various points in the cycle, but *over one full cycle* there's no change in the engine's entropy.

That's not true for the hot and cold reservoirs. The hot reservoir is steadily losing energy, the energy that it transfers to the engine. Less energy means less ability to disperse energy, so the entropy of the hot reservoir is steadily decreasing. Conversely, the entropy of the cold reservoir is steadily increasing as it receives the exhaust energy from the engine, increasing the cold reservoir's ability to disperse energy. The work output of the engine doesn't involve an entropy change because it doesn't involve thermal energy.

A 100% efficient heat engine would transform all the input heat $Q_H$ into work and have no exhaust $(Q_C = 0)$. With no exhaust, and thus no increase of entropy in the cold reservoir, the net operation of the heat engine would involve a decreasing entropy of the hot reservoir with no corresponding entropy increase. The second law forbids an overall decrease in entropy, so this is not possible.

The most efficient possible heat engine is one that has zero change in entropy. This is the efficiency specified in Equation 11.11 for an ideal heat engine, the maximum efficiency allowed by the second law of thermodynamics.

## The Conservation of Energy and Energy Conservation

We have all heard that it is important to "conserve energy." We turn off lights when we leave rooms, drive our cars less, turn down our thermostats. But this brings up an interesting question: If we have a law of conservation of energy, which states that energy can't be created or destroyed, what do we really mean by "conserving energy"?

We started this chapter looking at energy transformations. We saw that whenever energy is transformed, some of it is "lost." And now we know what this means: The energy isn't really lost, but it is converted into thermal energy. This change is irreversible; thermal energy can't be efficiently converted back into other forms of energy.

We aren't, as a society or as a planet, running out of energy. We can't! What we can run out of is high-quality sources of energy. Oil is a good example. A gallon of gasoline contains a great deal of chemical energy. It is easily transported, and it is easily burned to generate heat, electricity, or motion. When you burn gasoline in your car, you don't use up its energy—you simply convert its chemical energy into thermal energy. You thus decrease the amount of high-quality chemical energy in the world and increase the supply of thermal energy. The amount of energy is still the same; it's just in a less useful form.

Perhaps the best way to "conserve energy" is to concentrate on efficiency, to reduce "what you had to pay." More efficient lightbulbs, more efficient cars—all of these use less energy to produce the same final result.

**Sealed, but not isolated** BIO This glass container is a completely sealed system containing living organisms, shrimp and algae. But the organisms will live and grow for many years. The reason this is possible is that the glass sphere, though sealed, is not an *isolated* system. Energy can be transferred in and out as light and heat. If the container were placed in a darkened room, the organisms would quickly perish.

## Entropy and Life BIO

The second law of thermodynamics is sometimes interpreted as saying that "things run down"—that gradually other forms of energy are converted to thermal energy and that entropy increases to the point where no useful work can be done. But living organisms seem to violate this rule:

- Plants grow from simple seeds to complex entities.
- Single-celled fertilized eggs grow into complex adult organisms.
- Over the last billion years or so, life has evolved from simple unicellular organisms to the very complex forms we see today.

Rather than running down, many living systems seem to be ramping up and becoming more complex. How can this be?

There is an important qualification to the second law of thermodynamics: It applies to isolated systems, systems that do not exchange energy with their environment. The situation is different if energy is transferred into or out of the system.

Consider your body. Every day, you take in chemical energy in the food you eat. As you use this energy, most of it ends up as thermal energy that you exhaust as heat into the environment, thereby increasing the entropy of the environment. An energy diagram of this situation is shown in FIGURE 11.29. The entropy of your body stays approximately constant, but the entropy of the environment is steadily increasing. It's the energy flow *through* your body that allows you to maintain your body in an organized, low-entropy state. There's no violation of the second law because the total entropy—that of you plus that of the environment—is increasing.

To develop, grow, or evolve, organisms take in high-quality, low-entropy energy in the form of food or sunlight, then exhaust thermal energy to the environment. This continuous exchange of energy with the environment, which increases environmental entropy, makes your life—and all life—possible without violating physical laws.

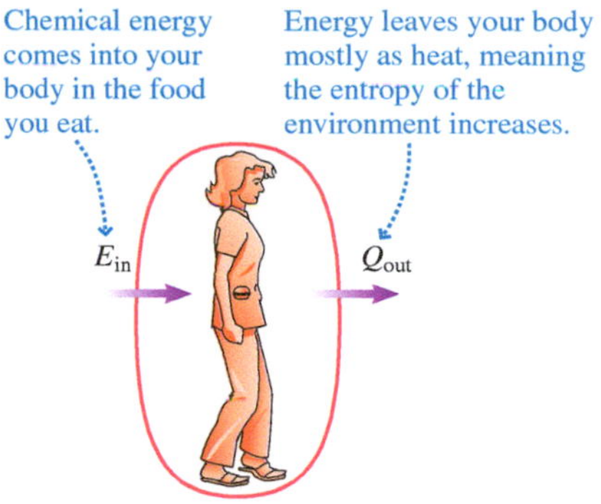


FIGURE 11.29 A thermodynamic view of the body.

## Entropy, Free Energy, and Spontaneity

We'll conclude with a more technical section for those who have learned in chemistry to use free energy to analyze chemical reactions. We'll assume some background knowledge from your other courses in this section; we'll assume that you are familiar with the concepts of free energy and enthalpy.

A reaction can occur spontaneously only if the free energy decreases in going from reactants to products. Technically, the free energy used in reaction analysis is the *Gibbs free energy*, hence the symbol $G$. The requirement for a spontaneous reaction is $\Delta G < 0$. The meaning of this requirement is this: It's the second law of thermodynamics telling us that spontaneous processes require an increase in entropy.

But the entropy of what? In this case, the entropy of the universe! We can't apply the second law to the reactants and products only because they are not an isolated system; they transfer heat to the environment in an exothermic reaction or absorb heat from the environment in an endothermic reaction. So let's expand our system to be the reaction plus the surrounding environment that is affected by the reaction. We can call this the universe, although in practice nothing beyond the immediate surroundings is likely to change.

By definition, the universe is an isolated system that does not interact with anything outside it. The second law *does* apply to the universe, so any spontaneous process that takes place in the universe must obey

$$\Delta S_{univ} > 0$$

where $S$ is the symbol for entropy. For our purposes, the universe consists of the chemicals involved in the reaction and the environment surrounding the reaction. If we make the reasonable assumption that entropies add (an assumption that can be proven), then $S_{univ} = S_{chem} + S_{environ}$. In that case, the second law requires

$$\Delta S_{chem} + \Delta S_{environ} > 0 \qquad (11.16)$$

A reaction may temporarily change the temperature of part of the system, but the end state of the reaction is when the reaction products and the environment have re-equilibrated to a common final temperature. Let's assume that the environment is so large that its temperature is not affected by the reactions, so overall the reaction takes place at constant temperature $T$, the temperature of the environment. Although the environment's temperature doesn't change, its entropy changes because heat is transferred to (or from) the environment by the reaction. If the environment receives heat $Q_{environ}$ at temperature $T$, its entropy changes by $\Delta S_{environ} = Q_{environ}/T$, a result you learned in chemistry.

Because the universe is isolated, any heat gained by the environment is heat lost by the chemicals (e.g., heat released in the reaction), so $Q_{environ} = -Q_{chem}$. Thus Equation 11.16 can be written as

$$\Delta S_{chem} - \frac{Q_{chem}}{T} > 0$$

Now we have the second-law requirement for spontaneity written in terms of only the chemicals involved in the reaction and the temperature.

Chemical reactions are also assumed to occur at constant pressure. At constant pressure, the heat of the reaction is given by the change in enthalpy $H$ between the reactants and the products: $Q_{\text{chem}} = \Delta H_{\text{chem}}$. Thus spontaneity requires

$$\Delta S_{\text{chem}} - \frac{\Delta H_{\text{chem}}}{T} > 0$$

We can now drop the subscript, knowing that both the entropy and the enthalpy are those of the chemicals. If we first multiply through by $T$ and then by $-1$ (which changes $>$ to $<$), we have

$$\Delta H - T\Delta S = \Delta(H - TS) < 0$$

But $H - TS$ is the definition of the free energy: $G = H - TS$. Thus the second-law requirement for a spontaneous process—namely, that the entropy must increase and cannot decrease—is

$$\Delta G < 0$$

The free-energy requirement for a spontaneous reaction is about entropy and the second law. A reaction can occur spontaneously if there is an overall increase in entropy.

## INTEGRATED EXAMPLE 11.19 Determining the efficiency of an automobile

In the absence of external forces, Newton's first law tells us that a car would continue at a constant speed once it was moving. So why can't you get your car up to speed and then take your foot off the gas? Because there *are* external forces. At highway speeds, nearly all of the force opposing your car's motion is the drag force of the air. (Rolling friction is much smaller than drag at highway speeds, so we'll ignore it.) FIGURE 11.30 shows that as your car moves through the air it pushes the air aside. Doing so takes energy, and your car's engine must keep running to replace this lost energy.

Sports cars, with their aerodynamic shape, are often those that lose the least energy to drag. A typical sports car has a 350 hp engine, a drag coefficient of 0.30, and a low profile with the area of the front of the car being a modest 1.8 m$^2$. Such a car gets about 25 miles per gallon of gasoline at a highway speed of 30 m/s (just over 65 mph).

Suppose this car is driven 25 miles on a level stretch of road at 30 m/s. It will consume 1 gallon of gasoline, which contains $1.4 \times 10^8$ J of chemical energy. What is the car's efficiency for this trip?

FIGURE 11.30 A wind-tunnel test shows airflow around a car.

**STRATEGIZE** We'll consider the car as the system. As the car moves through the air, the drag force is taking energy away. It's this energy that must be replenished with chemical energy from the fuel.

**PREPARE** We can use what we learned about the drag force in Chapter 5 to compute the amount of energy needed to move the car forward through the air. We can use this value in Equation 11.2 as "what you get"; it is the minimum amount of energy that *could* be used to move the car forward at this speed. We can then use the mileage data to calculate "what you had to pay"; this is the energy actually used by the car's engine. Once we have these two pieces of information, we can calculate efficiency.

**SOLVE** Heat and work both play a role in this process, so we need to use our most general equation about energy and energy conservation, Equation 11.8:

$$\Delta K + \Delta U + \Delta E_{\text{th}} + \Delta E_{\text{chem}} + \cdots = W + Q$$

The car is moving at a constant speed, so its kinetic energy is not changing. The road is assumed to be level, so there is no change in potential energy. Once the car is warmed up, its temperature will be constant, so its thermal energy isn't changing. Equation 11.8 reduces to

$$\Delta E_{\text{chem}} = W + Q$$

The car's engine is using chemical energy as it burns fuel. The engine transforms some of this energy into work—the propulsion force pushing the car forward against the opposing force of air drag. But the engine also transforms much of the chemical energy into "waste heat" that is transferred to the environment through the radiator and the exhaust. $\Delta E_{\text{chem}}$ is negative because the amount of energy stored in the gas tank is decreasing as gasoline is burned. $W$ and $Q$ are also negative, according to the sign convention of Figure 11.15, because these energies are being transferred *out* of the system to the environment.

In Chapter 5 we saw that the drag force on an object moving at a speed $v$ is given by

$$\vec{D} = \left(\frac{1}{2}C_D \rho A v^2, \text{ direction opposite the motion}\right)$$

where $C_D$ is the drag coefficient, $\rho$ is the density of air (approximately 1.2 kg/m$^3$), and $A$ is the area of the front of the car. The drag force on the car moving at 30 m/s is

$$D = \frac{1}{2}(0.30)(1.2\ \text{kg/m}^3)(1.8\ \text{m}^2)(30\ \text{m/s})^2 = 292\ \text{N}$$

The drag force opposes the motion. To move the car forward at constant speed, and thus with no *net* force, requires a propulsion force $F = 292$ N in the forward direction. In Chapter 10 we found that the power—the rate of energy expenditure—to move an object at speed $v$ with force $F$ is $P = Fv$. Thus the power the car must supply—at the wheels—to keep the car moving down the highway is

$$P = Fv = (292\ \text{N})(30\ \text{m/s}) = 8760\ \text{W}$$

Let's compare this to the engine power by converting to horsepower:

$$P = 8760\ \text{W}\left(\frac{1\ \text{hp}}{746\ \text{W}}\right) = 12\ \text{hp}$$

Only a small fraction of the engine's 350 horsepower is needed to keep the car moving at highway speeds.

The distance traveled is

$$\Delta x = 25\ \text{mi} \times \frac{1.6\ \text{km}}{1\ \text{mi}} \times \frac{1000\ \text{m}}{1\ \text{km}} = 40{,}000\ \text{m}$$

and the time required to travel it is $\Delta t = (40{,}000\ \text{m})/(30\ \text{m/s}) = 1333$ s. Thus the minimum energy needed to travel 40 km—the energy needed simply to push the air aside—is

$$E_{\text{min}} = P\Delta t = (8760\ \text{W})(1333\ \text{s}) = 1.17 \times 10^7\ \text{J}$$

In traveling this distance, the car uses 1 gallon of gas, with $1.4 \times 10^8$ J of chemical energy. This is, quite literally, what you had to pay to drive this distance. Thus the car's efficiency is

$$e = \frac{\text{what you get}}{\text{what you had to pay}} = \frac{1.17 \times 10^7\ \text{J}}{1.4 \times 10^8\ \text{J}} = 0.084 = 8.4\%$$

**ASSESS** The efficiency of the car is quite low, even compared to other engines that we've seen. Nonetheless, our calculation agrees reasonably well with actual measurements. Gasoline-powered vehicles simply are inefficient, which is one factor favoring more efficient alternative vehicles. Smaller mass and better aerodynamic design improve the efficiency of vehicles, but a large part of the inefficiency of a gasoline-powered vehicle is inherent in the thermodynamics of the engine itself and in the complex drive train needed to transfer the engine's power to the wheels.

# SUMMARY

**GOAL** To learn about practical energy transformations and transfers, and the limits on how efficiently energy can be used.

## GENERAL PRINCIPLES

### Energy and Efficiency

When energy is transformed from one form into another, some may be "lost," usually to thermal energy, due to practical or theoretical constraints. This limits the efficiency of processes. We define **efficiency** as

Efficiency: Used for heat engines → $e = \dfrac{\text{what you get}}{\text{what you had to pay}} = \text{COP}$ ← COP: Used for heat pumps

### Entropy and Irreversibility

Thermal energy tends to spread out. This spreading is irreversible; once the energy has spread, it doesn't go back. This means that heat is transferred from a hot object to a cold object until the two are at equilibrium; once a system is at equilibrium, it will stay there.

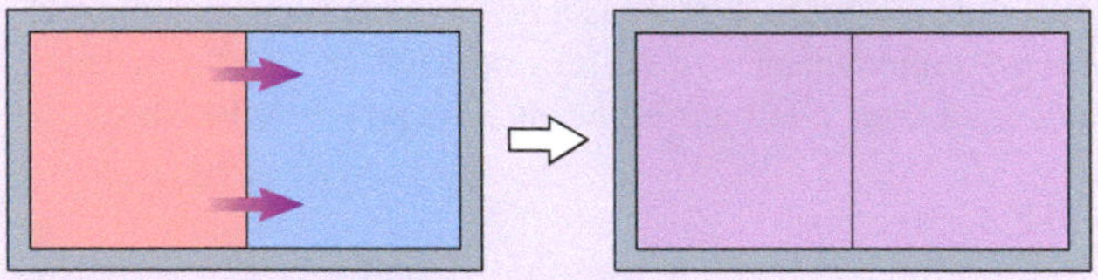

### The Laws of Thermodynamics

The **first law of thermodynamics** is a statement of conservation of energy for systems in which only thermal energy changes:

$$\Delta E_{th} = W + Q$$

The **second law of thermodynamics** specifies the way that isolated systems can evolve:

**The entropy of an isolated system never decreases.**

This law has practical consequences:

- Heat energy spontaneously flows only from hot to cold.
- A transformation of energy into thermal energy is irreversible.
- No heat engine can be 100% efficient.

**Heat** is energy transferred between two objects because they are at different temperatures. Energy will be transferred until thermal equilibrium is reached.

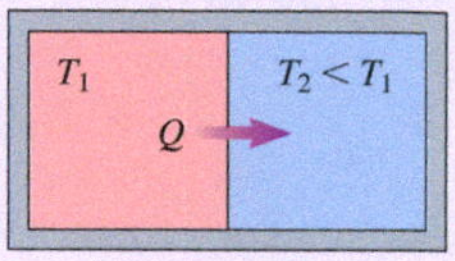


## IMPORTANT CONCEPTS

### Thermal energy

- For a gas, the thermal energy is the **total kinetic energy** of motion of the atoms.
- Thermal energy is random kinetic energy and so is associated with entropy.

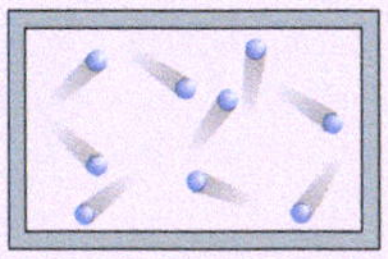

### Temperature

- For a gas, temperature is related to the **average kinetic energy** of the motion of the atoms.
- Two systems are in **thermal equilibrium** if they are at the same temperature. No heat energy is transferred.

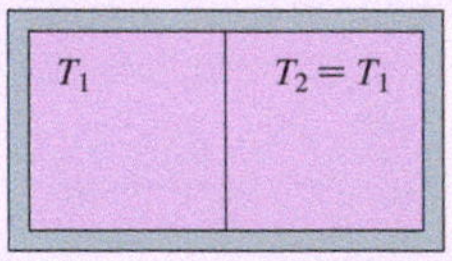


A **heat engine** converts thermal energy from a hot reservoir into useful work. Some heat is exhausted into a cold reservoir, limiting efficiency.

$$e_{max} = 1 - \frac{T_C}{T_H}$$

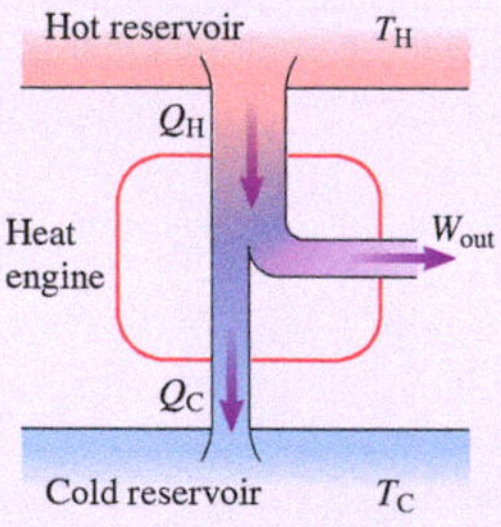


A **heat pump** uses an energy input to transfer heat from a cold side to a hot side. The **coefficient of performance** is analogous to efficiency. For cooling, its limit is

$$\text{COP}_{max} = \frac{T_C}{T_H - T_C}$$

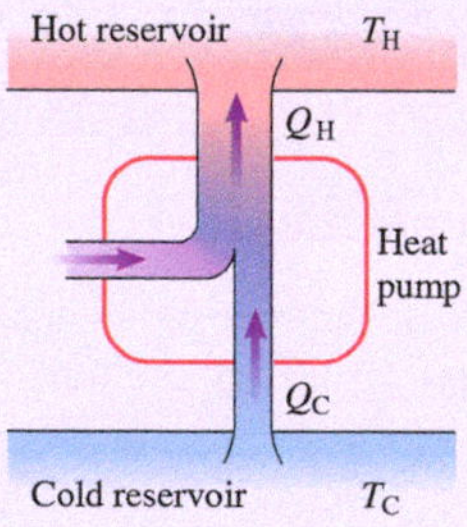


## APPLICATIONS

### Efficiencies

**Energy in the body** Cells in the body metabolize chemical energy in food. Efficiency for most actions is about 25%.

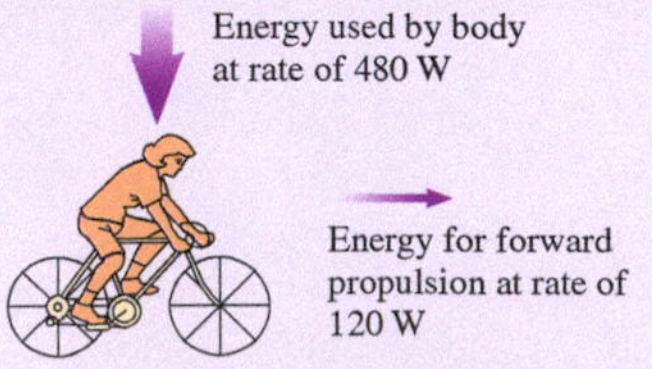


**Power plants** A typical power plant converts about 1/3 of the energy input into useful work. The rest is exhausted as waste heat.

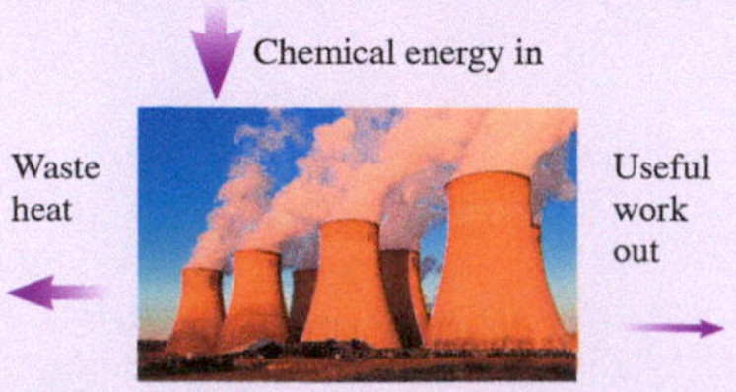


### Temperature scales

Zero on the **Kelvin temperature scale** is the temperature at which the kinetic energy of atoms is zero. This is **absolute zero**. The conversion from °C to K is

$$T(\text{K}) = T(°\text{C}) + 273$$

▶ All temperatures in equations must be in kelvin. ◀

## Learning Objectives After studying this chapter, you should be able to:

- Identify transformations of energy and calculate their efficiencies. *Conceptual Questions 11.1, 11.7; Problems 11.2 , 11.3, 11.4, 11.5, 11.6*
- Understand how energy is used and stored in the body. *Conceptual Questions 11.2, 11.4; Problems 11.7, 11.10 , 11.11, 11.13, 11.16*
- Distinguish between work, heat, temperature, and thermal energy. *Conceptual Questions 11.12, 11.19; Problems 11.28, 11.31, 11.32, 11.34, 11.35*
- Apply the first law of thermodynamics to heat engines, heat pumps, and refrigerators. *Conceptual Questions 11.24, 11.25; Problems 11.36, 11.42, 11.47, 11.48, 11.51*
- Understand the concept of entropy and the limitations that the second law of thermodynamics places on energy transformations. *Conceptual Questions 11.28, 11.30; Problems 11.52, 11.53*

## STOP TO THINK ANSWERS

**Chapter Preview Stop to Think: C.** The work Christina does on the javelin is transferred into kinetic energy, so the javelin has 270 J of kinetic energy as it leaves her hand. As the javelin rises, some of its kinetic energy is transformed into gravitational potential energy, but the total energy remains constant at 270 J. Thus at its highest point, the javelin's kinetic energy is 270 J − 70 J = 200 J.

**Stop to Think 11.1: C.** In each case, what you get is the potential-energy change of the box. Crane 2 lifts a box with twice the mass the same distance as crane 1, so you get twice as much energy with crane 2. How about what you have to pay? Crane 2 uses 20 kJ, crane 1 only 10 kJ. Comparing crane 1 and crane 2, we find crane 2 has twice the energy out for twice the energy in, so the efficiencies are the same.

**Stop to Think 11.2: C.** As the body uses chemical energy from food, approximately 75% is transformed into thermal energy. Also, kinetic energy of motion of the legs and feet is transformed into thermal energy with each stride. Most of the chemical energy is transformed into thermal energy.

**Stop to Think 11.3: C.** Samples 1 and 2 have the same thermal energy, which is the total kinetic energy of all the atoms. Sample 1 has twice as many atoms, so the average energy per atom, and thus the temperature, must be less.

**Stop to Think 11.4: C.** The radiator is at a higher temperature than the surrounding air. Thermal energy is transferred out of the system to the environment, so $Q < 0$.

**Stop to Think 11.5: A, D.** The efficiency is fixed by the ratio of $T_C$ to $T_H$. Decreasing this ratio increases efficiency; the heat engine will be more efficient with a hotter hot reservoir or a colder cold reservoir.

**Stop to Think 11.6: A, D.** The closer the temperatures of the hot and cold reservoirs, the more efficient the heat pump can be. (It is also true that having the two temperatures be closer will cause less thermal energy to "leak" out.) Any change that makes the two temperatures closer will allow the refrigerator to use less energy to run.

**Stop to Think 11.7: B.** In this case, kinetic energy is transformed into potential energy; there is no entropy change. In the other cases, energy is transformed into thermal energy, meaning entropy increases.

Video Tutor Solution Chapter 11

# QUESTIONS

## Conceptual Questions

1. Rub your hands together vigorously. What happens? Discuss the energy transfers and transformations that take place.
2. BIO Describe the energy transfers and transformations that occur from the time you sit down to breakfast until you've completed a fast bicycle ride.
3. BIO According to Table 11.4, cycling at 15 km/h requires less metabolic energy than running at 15 km/h. Suggest reasons why this is the case.
4. BIO Walking up a set of stairs and running up the stairs require approximately the same metabolic energy. Explain why you'd expect this to be the case.
5. BIO A pronghorn, the fastest North American animal, is capable of running at 18 m/s (40 mph) for 10 minutes, after which it must slow down. The time limit isn't because the pronghorn runs out of energy; it's because the pronghorn's temperature rises, and it must stop to cool down. Why does the pronghorn's temperature rise during the run?

FIGURE Q11.5

Problem difficulty is labeled as | (straightforward) to ||||| (challenging). Problems labeled INT integrate significant material from earlier chapters; Problems labeled BIO are of biological or medical interest.

The eText icon indicates when there is a video tutor solution available for the chapter or for a specific problem. To launch these videos, log into your eText through Mastering™ Physics or log into the Study Area.

6. BIO When the hoof of a galloping horse hits the ground, the digital flexor tendon in its lower leg may stretch by 5% in length, a significant stretch for this 45 cm tendon. The tendon is elastic; most—but not all—of the energy stored in the stretch is returned, which allows for efficient locomotion. In an exercising horse, the tendon may reach a temperature of 46°C, much higher than the horse's 38°C body temperature. How can you explain this temperature rise?
7. For most automobiles, the number of miles per gallon decreases as highway speed increases. Fuel economy drops as speeds increase from 55 to 65 mph, then decreases further as speeds increase to 75 mph. Explain why this is the case.
8. BIO If you want to jump as high as possible, it's best to move downward quickly to a deep crouch, stretching tendons and muscles, before pushing off and leaving the ground rather than simply pushing off from a stationary crouch. Explain how this provides additional energy for the jump.
9. When a spacecraft returns to earth, its surfaces reach a very high temperature during its high-speed reentry into the atmosphere. Is this temperature rise due to the transfer of energy as heat? Explain.
10. When you grind wheat to make flour, the flour comes out very warm—you can actually scorch the flour if you grind too quickly. Explain how this temperature rise comes about.
11. One end of a short aluminum rod is in a campfire and the other end is in a block of ice, as shown in Figure Q11.11. The temperature at every point in the rod has reached a steady value. If 100 J of energy are transferred from the fire to the rod, how much energy goes from the rod into the ice?

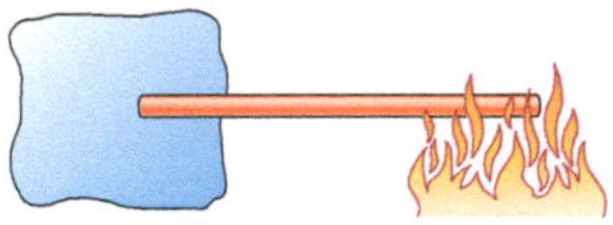

FIGURE Q11.11

12. Two blocks of copper, one of mass 1 kg and the second of mass 3 kg, are at the same temperature. Which block has more thermal energy? If the blocks are placed in thermal contact, will the thermal energy of the blocks change? If so, how?

For Questions 13 through 18, give a specific example of a process that has the energy changes and transfers described. (For example, if the question states "$\Delta E_{th} > 0$, $W = 0$," you are to describe a process that has an increase in thermal energy and no transfer of energy by work. You could write "Heating a pan of water on the stove.")

13. $\Delta E_{th} < 0$, $W = 0$
14. $\Delta E_{th} > 0$, $Q = 0$
15. $\Delta E_{th} < 0$, $Q = 0$
16. $\Delta E_{th} > 0$, $W \neq 0$, $Q \neq 0$
17. $\Delta E_{th} < 0$, $W \neq 0$, $Q \neq 0$
18. $\Delta E_{th} = 0$, $W \neq 0$, $Q \neq 0$
19. If you leave a fan running, it actually warms up the air in the room, a fact that surprises many people. Explain why a fan would be expected to warm the air that passes through it.
20. A fire piston—an impressive physics demonstration—ignites a fire without matches. The operation is shown in Figure Q11.20. A wad of cotton is placed at the bottom of a sealed syringe with a tight-fitting plunger. When the plunger is rapidly depressed, the air temperature in the syringe rises enough to ignite the cotton. Explain why the air temperature rises, and why the plunger must be pushed in very quickly.

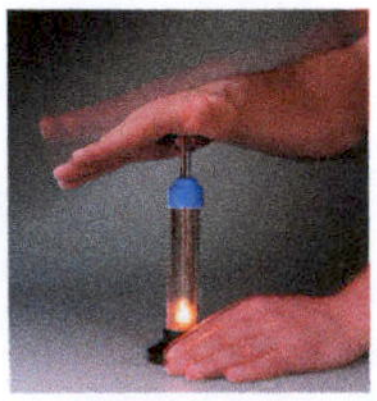

FIGURE Q11.20

21. In a gasoline engine, fuel vapors are ignited by a spark. In a diesel engine, a fuel–air mixture is drawn in, then rapidly compressed to as little as 1/20 the original volume, in the process increasing the temperature enough to ignite the fuel–air mixture. Explain why the temperature rises during the compression.
22. If you hold a rubber band loosely between two fingers and then stretch it, you can tell by touching it to the sensitive skin of your forehead that stretching the rubber band has increased its temperature. If you then let the rubber band rest against your forehead, it soon returns to its original temperature. What are the signs of $W$ and $Q$ for the entire process?
23. Your car's engine is a heat engine; it converts the thermal energy from burning fuel into energy to move your car and power its systems. On a cold winter day, you needn't feel guilty about cranking up the heat in your car; running the heater doesn't cost any additional energy beyond the small amount needed to run the fan. Explain why this is so.
24. The geothermal power plant at Brady Hot Springs in Nevada uses the outside air as the cold reservoir. The plant has a higher electricity output in the morning than it does in the evening. Why would you expect this to be the case?
25. You can save money on electricity if you put your refrigerator in the basement, which is usually cooler than the rest of your house. Explain.
26. The ground temperature a few meters below the surface is fairly constant throughout the year and is near the average value of the air temperature. In areas in which the air temperature drops very low in the winter, the exterior unit of a heat pump designed for heating is sometimes buried underground in order to use the earth as a thermal reservoir. Why is it worthwhile to bury the heat exchanger, even if the underground unit costs more to purchase and install than one above ground?
27. Electric vehicles increase speed by using an electric motor that draws energy from a battery. When the vehicle slows, the motor runs as a generator, recharging the battery. Explain why this means that an electric vehicle can be more efficient than a gasoline-fueled vehicle.
28. When the sun's light hits the earth, the temperature rises. Is there an entropy change to accompany this transformation? Explain.
29. A company markets an electric heater that is described as 100% efficient at converting electric energy to thermal energy. Does this violate the second law of thermodynamics?
30. Living creatures generally have a higher temperature than their environment. Explain why this makes sense.
31. A reaction happens in a perfectly sealed and insulated container. Does the total energy of the system decrease? The free energy?

## Multiple-Choice Questions

32. || On clear day, sunlight delivers approximately 1000 J each second to a 1 $m^2$ surface; smaller areas receive proportionally less—half the area receives half the energy. A 12% efficient solar cell, a square 15 cm on a side, is in bright sunlight. How much electric power does it produce?
A. 0.9 W B. 1.8 W C. 2.7 W D. 3.6 W
33. || For locations that have no electric service, companies are designing bicycle-powered generators that are nearly 100% efficient that can be used to charge cell phones or power lights. If you are the person chosen to pedal and your friends need a total of 150 W of electric power, how much metabolic power will you use as you pedal?
A. 150 W B. 300 W C. 450 W D. 600 W

34. ||| A person is walking on level ground at constant speed. What energy transformation is taking place?
   A. Chemical energy is being transformed to thermal energy.
   B. Chemical energy is being transformed to kinetic energy.
   C. Chemical energy is being transformed to kinetic energy and thermal energy.
   D. Chemical energy and thermal energy are being transformed to kinetic energy.
35. | A person walks 1 km, turns around, and runs back to where he started. Compare the energy used and the power during the two segments.
   A. The energy used and the power are the same for both.
   B. The energy used while walking is greater, the power while running is greater.
   C. The energy used while running is greater, the power while running is greater.
   D. The energy used is the same for both segments, the power while running is greater.
36. || 200 J of heat is added to two gases, each in a sealed container. Gas 1 is in a rigid container that does not change volume. Gas 2 expands as it is heated, pushing out a piston that lifts a small weight. Which gas has the greater increase in its thermal energy?
   A. Gas 1 B. Gas 2
   C. Both gases have the same increase.
37. || An electric power plant uses energy from burning coal to generate steam at 450°C. The plant is cooled by 20°C water from a nearby river. If burning coal provides 100 MJ of heat, what is the theoretical minimum amount of heat that must be transferred to the river during the conversion of heat to electric energy?
   A. 100 MJ B. 90 MJ C. 60 MJ D. 40 MJ
38. || A refrigerator's freezer compartment is set at –10°C; the kitchen is 24°C. What is the theoretical minimum amount of electric energy necessary to pump 1.0 J of energy out of the freezer compartment?
   A. 0.89 J B. 0.87 J C. 0.13 J D. 0.11 J

# PROBLEMS

## Section 11.1 Transforming Energy

1. || A 10% efficient engine accelerates a 1500 kg car from rest to 15 m/s. How much energy is transferred to the engine by burning gasoline?
2. || Diesel engines give more miles per gallon than gasoline engines, but some of this is due to the higher energy content of diesel fuel. At highway speeds, it takes 0.20 MJ to move an aerodynamic car 1.0 km. At highway speeds, with a gasoline engine, a car gets 16 km per liter of fuel; with a diesel engine under the hood, the car gets 19 km per liter of fuel. One liter of diesel contains 36 MJ; 1 liter of gasoline contains only 32 MJ. What is the efficiency of each of these two engines?
3. | A typical photovoltaic cell delivers $4.0 \times 10^{-3}$ W of electric energy when illuminated with $1.2 \times 10^{-1}$ W of light energy. What is the efficiency of the cell?
4. ||| When the Glen Canyon hydroelectric power plant in Arizona is running at capacity, 690 $m^3$ of water flows through the dam each second. The water is released 220 m below the top of the reservoir. If the generators that the dam employs are 90% efficient, what is the maximum possible electric power output?
5. || An individual white LED (light-emitting diode) has an efficiency of 20% and uses 1.0 W of electric power. How many LEDs must be combined into one light source to give a total of 1.6 W of visible-light output (comparable to the light output of a 40 W incandescent bulb)? What total power is necessary to run this LED light source?
6. || A typical wind turbine extracts 40% of the kinetic energy of the wind that blows through the area swept by the blades. For a large turbine, 110,000 kg of air moves past the blades at 15 m/s every second. If the wind turbine extracts 40% of this kinetic energy, and if 80% of this energy is converted to electric energy, what is the power output of the generator?

## Section 11.2 Energy in the Body

7. ||| In an average human, basic life processes require energy to be supplied at a steady rate of 100 W. What daily energy intake, in Calories, is required to maintain these basic processes? BIO
8. | We noted that, under ideal conditions, metabolizing one molecule of glucose can form 38 molecules of ATP. A more realistic value is 30 molecules of ATP. For this lower number, what is the efficiency of the process of converting the chemical energy of glucose to that of ATP? BIO
9. | An "energy bar" contains 22 g of carbohydrates. How much energy is this in joules? In calories? In Calories? BIO
10. | Jessie and Jaime complete a 5.0 km race. Each has a mass of 68 kg. Jessie runs the race at 15 km/h; Jaime walks it at 5 km/h. How much metabolic energy does each use to complete the course? BIO
11. | A sleeping 68 kg man has a metabolic power of 71 W. How many Calories does he burn during an 8.0 hour sleep? BIO
12. | Tessa and Jody, each of mass 68 kg, go out for some exercise together. Tessa runs at 15 km/h; Jody cycles alongside at the same speed. After 20 minutes, how much metabolic energy has each used? BIO
13. ||| An "energy bar" contains 22 g of carbohydrates. If the energy bar was his only fuel, how far could a 68 kg person walk at 5.0 km/h? BIO
14. || Each time he does one pushup, Jose, who has a mass of 75 kg, raises his center of mass by 25 cm. He completes an impressive 150 pushups in 5 minutes, exercising at a steady rate. BIO
   a. If we assume that lowering his body has no energetic cost, what is his metabolic power during this workout?
   b. In fact, it costs Jose a certain amount of energy to lower his body—about half of what it costs to raise it. If you include this in your calculation, what is his metabolic power?
15. ||||| For how long would a 68 kg athlete have to swim at a fast crawl to use all the energy available in a typical fast-food meal of burger, fries, and a drink? BIO
16. || The basis of muscle action is the power stroke of the myosin protein pulling on an actin filament. It takes the energy of one molecule of ATP, $5.1 \times 10^{-20}$ J, to produce a displacement of 10 nm against a force of 1.0 pN. What is the efficiency? BIO INT
17. ||||| The label on a candy bar says 400 Calories. Assuming a typical efficiency for energy use by the body, if a 60 kg person were to use the energy in this candy bar to climb stairs, how high could she go? BIO

Watch Video Solution Problems 11.19 and 11.23

18. ||| Some jellyfish use jet propulsion to get around; the efficiency of this process can be quite low, as a calculation will show. A 1.9 g jellyfish is at rest. It takes water into its bell, then rapidly ejects 1.3 g of water, achieving a speed of 0.070 m/s. If we define "what you get" as the kinetic energy of the jellyfish, and "what you had to pay" as the total kinetic energy of the jellyfish and the water, what is the efficiency? If we then add in a 25% metabolic efficiency, what's the overall efficiency of this form of locomotion? BIO INT

19. ||| A weightlifter curls a 30 kg bar, raising it each time a distance of 0.60 m. How many times must he repeat this exercise to burn off the energy in one slice of pizza? BIO

20. || In an extreme marathon, participants run a total of 100 km; world-class athletes maintain a pace of 15 km/h. How many 230 Calorie energy bars would be required to fuel such a run for a 68 kg athlete? BIO

21. ||| A weightlifter works out at the gym each day. Part of her routine is to lie on her back and lift a 40 kg barbell straight up from chest height to full arm extension, a distance of 0.50 m. BIO
    a. How much work does the weightlifter do to lift the barbell one time?
    b. If the weightlifter does 20 repetitions a day, what total energy does she expend on lifting, assuming a typical efficiency for energy use by the body.
    c. How many 400 Calorie donuts can she eat a day to supply that energy?

22. || A 10 kg migratory swan cruises at 20 m/s. A calculation that takes into account the necessary forces shows that this motion requires 200 W of mechanical power. If we assume an efficiency similar to humans, a reasonable assumption, then the metabolic power of the swan is significantly higher than this. The swan doesn't stop to eat during a long day of flying; it gets the energy it needs from fat stores. Assuming an efficiency similar to humans, after 12 hours of flight, BIO INT
    a. How far has the swan traveled?
    b. How much metabolic energy has it used?
    c. What fraction of its body mass does it lose?

23. ||| Suppose your body was able to use the chemical energy in gasoline. How far could you pedal a bicycle at 15 km/h on the energy in 1 gal of gas? (1 gal of gas has a mass of 3.2 kg.) BIO

Questions 24 through 26 use the following information.

Flapping flight is very energy intensive. A wind tunnel test on an 89 g starling showed that the bird used 12 W of metabolic power to fly at 11 m/s.

24. || A starling flies 1.0 km at 11 m/s. How many grams of carbohydrates does the bird need to consume to fuel this flight? BIO

25. || A migrating starling flies steadily at 11 m/s for 1.0 h, using energy from its fat stores. How many grams of fat does it burn? BIO

26. || What is the specific metabolic power for starling flight? How does this compare to the specific metabolic power for a 68 kg human running at 15 km/h? BIO INT

## Section 11.3 Temperature, Thermal Energy, and Heat

27. | Helium has the lowest boiling point of any substance, at 4.2 K. What is this temperature in °C and °F?

28. | The planet Mercury's surface temperature varies from 700 K during the day to 90 K at night. What are these values in °C and °F?

29. | A piece of metal at 100°C has its Celsius temperature doubled. By what factor does its kelvin temperature increase?

30. | A normal human body temperature can range from 33.2°C to 38.2°C. This range includes variations with gender, time of day, and where the measurement is taken. What is this range in °F? BIO

## Section 11.4 The First Law of Thermodynamics

31. || 500 J of work are done on a system in a process that decreases the system's thermal energy by 200 J. How much energy is transferred to or from the system as heat?

32. || Your roommate leaves a 120 W fan running in your apartment. Over the course of an hour, how much thermal energy does the fan add to the air?

33. | 600 J of heat energy are transferred to a system that does 400 J of work. By how much does the system's thermal energy change?

34. || Compressing air to fill a scuba tank warms it up. A dive shop compensates by putting the tank in a tub of water, keeping the tank and the gas inside it at a constant temperature as it is filled. If the system is the tank, what are the signs of $W$ and $Q$ for this process? What can you say about the relative magnitudes of $W$ and $Q$?

35. | 10 J of heat are removed from a gas sample while it is being compressed by a piston that does 20 J of work. What is the change in the thermal energy of the gas? Does the temperature of the gas increase or decrease?

## Section 11.5 Heat Engines

36. | A heat engine extracts 55 kJ from the hot reservoir and exhausts 40 kJ into the cold reservoir. What are (a) the work done and (b) the efficiency?

37. || A heat engine does 20 J of work while exhausting 30 J of waste heat. What is the engine's efficiency?

38. | A power plant running at 35% efficiency generates 300 MW of electric power. At what rate (in MW) is heat energy exhausted to the river that cools the plant?

39. ||| A heat engine operating between energy reservoirs at 20°C and 600°C has 30% of the maximum possible efficiency. How much energy does this engine extract from the hot reservoir to do 1000 J of work?

40. | A newly proposed device for generating electricity from the sun is a heat engine in which the hot reservoir is created by focusing sunlight on a small spot on one side of the engine. The cold reservoir is ambient air at 20°C. The designer claims that the efficiency will be 60%. What minimum hot-reservoir temperature, in °C, would be required to produce this efficiency?

41. | Converting sunlight to electricity with solar cells has an efficiency of ≈15%. It's possible to achieve a higher efficiency (though currently at higher cost) by using concentrated sunlight as the hot reservoir of a heat engine. Each dish in Figure P11.41 concentrates sunlight on one side of a heat engine, producing a hot-reservoir temperature of 650°C. The cold reservoir, ambient air, is approximately 30°C. The actual working efficiency of this device is ≈30%. What is the theoretical maximum efficiency?

FIGURE P11.41

42. A California geothermal power plant generates 4.3 MW by extracting 24 MW of heat from an underground source at 108°C and then discharging waste heat into the 17°C air.
    a. What is the actual efficiency of the plant?
    b. What is the theoretical efficiency?
43. Each second, a nuclear power plant generates 2000 MJ of thermal energy from nuclear reactions in the reactor's core. This energy is used to boil water and produce high-pressure steam at 300°C. The steam spins a turbine, which produces 700 MJ of electric power, then the steam is condensed and the water is cooled to 30°C before starting the cycle again.
    a. What is the maximum possible efficiency of the plant?
    b. What is the plant's actual efficiency?
44. A 32% efficient electric power plant produces 900 MJ of electric energy per second and discharges waste heat into 20°C ocean water. Suppose the waste heat could be used to heat homes during the winter instead of being discharged into the ocean. A typical American house requires an average 20 kW for heating. How many homes could be heated with the waste heat of this one power plant?

## Section 11.6 Heat Pumps

45. A refrigerator takes in 20 J of work and exhausts 50 J of heat. What is the refrigerator's coefficient of performance?
46. Air conditioners are rated by their coefficient of performance at 80°F inside temperature and 95°F outside temperature. An efficient but realistic air conditioner has a coefficient of performance of 3.2. What is the maximum possible coefficient of performance?
47. 50 J of work are done on a refrigerator with a coefficient of performance of 4.0. How much heat is (a) extracted from the cold reservoir and (b) exhausted to the hot reservoir?
48. INT A NATO base in northern Norway is warmed with a heat pump that uses 7.0°C ocean water as the cold reservoir. Heat extracted from the ocean water warms fluid to 80°C; this warmed fluid is used to heat the building. When the system is working at full capacity, 2000 kW of heat are delivered to the building at the cost of 600 kW of electric energy.
    a. What is the actual coefficient of performance of the system?
    b. What is the theoretical maximum coefficient of performance of the system?
49. Find the maximum possible coefficient of performance for a heat pump used to heat a house in a northerly climate in winter. The inside is kept at 20°C while the outside is −20°C.
50. The inside of your refrigerator is approximately 0°C. Heat from the inside of your refrigerator is deposited into the air in your kitchen, which has a temperature of approximately 20°C. At these operating temperatures, what is the maximum possible coefficient of performance of your refrigerator?
51. 250 students sit in an auditorium listening to a physics lecture. Because they are thinking hard, each is using 125 W of metabolic power, slightly more than they would use at rest. An air conditioner with a COP of 5.0 is being used to keep the room at a constant temperature. What minimum electric power must be used to operate the air conditioner?

## Section 11.7 Entropy and the Second Law of Thermodynamics

## Section 11.8 Systems, Energy, and Entropy

52. Which, if any, of the heat engines in Figure P11.52 below violate (a) the first law of thermodynamics or (b) the second law of thermodynamics? Explain.

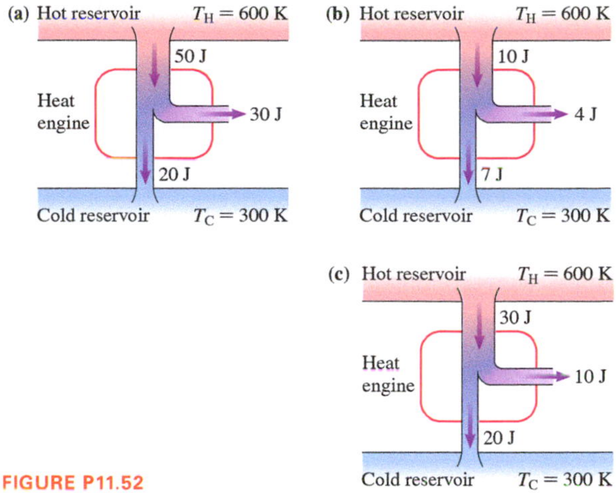


FIGURE P11.52

53. Which, if any, of the refrigerators in Figure P11.53 below violate (a) the first law of thermodynamics or (b) the second law of thermodynamics? Explain.

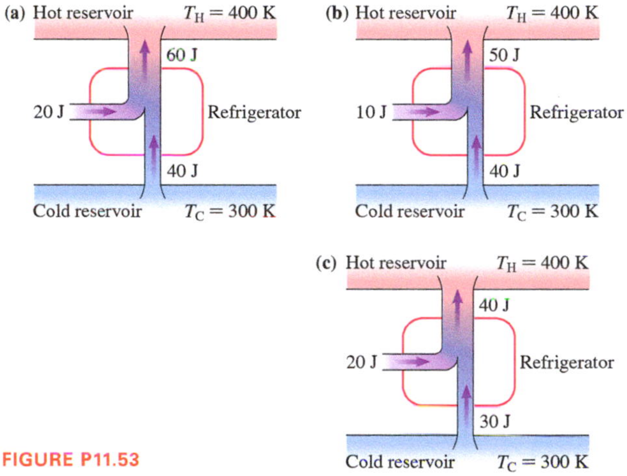


FIGURE P11.53

54. BIO Entropy can be used as a measure of the overall health of an ecosystem—higher entropy correlates with better health. A thermal image of a landscape, where color indicates temperature, can be used for a quick assessment. What would be the thermal signature of a healthy ecosystem?

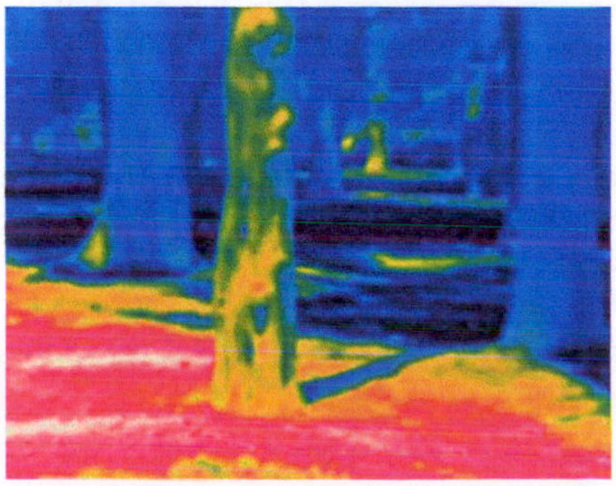

Watch Video Solution Problem 11.66

## General Problems

55. ||||| How many slices of pizza must you eat to walk for 1.0 h at a speed of 5.0 km/h? (Assume your mass is 68 kg.) BIO

56. | The winning time for the 2005 annual race up 86 floors of the Empire State Building was 10 min and 49 s. The winner's mass was 60 kg. BIO
 a. If each floor was 3.7 m high, what was the winner's change in gravitational potential energy?
 b. If the efficiency in climbing stairs is 25%, what total energy did the winner expend during the race?
 c. How many food Calories did the winner "burn" in the race?
 d. Of those Calories, how many were converted to thermal energy?
 e. What was the winner's metabolic power in watts during the race up the stairs?

57. || The record time for a Tour de France cyclist to ascend the famed 1100-m-high Alpe d'Huez was 37.5 min, set by Marco Pantani in 1997. Pantani and his bike had a mass of 65 kg.
 a. How many Calories did he expend during this climb?
 b. What was his average metabolic power during the climb?

58. ||| Championship swimmers take about 22 s and about 30 arm strokes to move through the water in a 50 m freestyle race. BIO INT
 a. From Table 11.4, a swimmer's metabolic power is 800 W. If the efficiency for swimming is 25%, how much energy is expended moving through the water in a 50 m race?
 b. If half the energy is used in arm motion and half in leg motion, what is the energy expenditure per arm stroke?
 c. Model the swimmer's hand as a paddle. During one arm stroke, the paddle moves halfway around a 90-cm-radius circle. If all the swimmer's forward propulsion during an arm stroke comes from the hand pushing on the water and none from the arm (somewhat of an oversimplification), what is the average force of the hand on the water?

59. ||||| A 68 kg hiker walks at 5.0 km/h up a 7% slope. What is the necessary metabolic power? **Hint:** You can model her power needs as the sum of the power to walk on level ground plus the power needed to raise her body by the appropriate amount. BIO

60. || To make your workouts more productive, you can get an electrical generator that you drive with the rear wheel of your bicycle when it is mounted in a stand. BIO
 a. Your laptop charger uses 75 W. What is your body's metabolic power use while running the generator to power your laptop charger, given the typical efficiency for such tasks? Assume 100% efficiency for the generator.
 b. Your laptop takes 1 hour to recharge. If you run the generator for 1 hour, how much energy does your body use? Express your result in joules and in Calories.

61. || The resistance of an exercise bike is often provided by a generator; that is, the energy that you expend is used to generate electric energy, which is then dissipated. Rather than dissipate the energy, it could be used for practical purposes. BIO
 a. A typical person can maintain a steady energy expenditure of 400 W on a bicycle. Assuming a typical efficiency for the body, and a generator that is 80% efficient, what useful electric power could you produce with a bicycle-powered generator?
 b. How many people would need to ride bicycle generators simultaneously to power a 400 W TV in the gym?

62. || Smaller mammals use proportionately more energy than larger mammals; that is, it takes more energy per gram to power a mouse than a human. A typical mouse has a mass of 20 g and, at rest, needs to consume 3.0 Cal each day for basic body processes. BIO
 a. If a 68 kg human used the same energy per kg of body mass as a mouse, how much energy would be needed each day?
 b. What resting power does this correspond to? How much greater is this than the resting power noted in the chapter?

63. || Larger animals use proportionately less energy than smaller animals; that is, it takes less energy per kg to power an elephant than to power a human. A 5000 kg African elephant requires about 70,000 Cal for basic needs for one day. BIO
 a. If a 68 kg human required the same energy per kg of body mass as an elephant, how much energy would be required each day?
 b. What resting power does this correspond to? How much less is this than the resting power noted in the chapter?

64. ||| An engine does 10 J of work and exhausts 15 J of waste heat.
 a. What is the engine's efficiency?
 b. If the cold-reservoir temperature is 20°C, what is the minimum possible temperature in °C of the hot reservoir?

65. ||| An engine operating at maximum theoretical efficiency whose cold-reservoir temperature is 7°C is 40% efficient. By how much should the temperature of the hot reservoir be increased to raise the efficiency to 60%?

66. ||| An engineer claims to have measured the characteristics of a heat engine that takes in 100 J of thermal energy and produces 50 J of useful work. Is this engine possible? If so, what is the smallest possible ratio of the temperatures (in kelvin) of the hot and cold reservoirs?

67. |||| A typical coal-fired power plant burns 300 metric tons of coal *every hour* to generate $2.7 \times 10^6$ MJ of electric energy. 1 metric ton = 1000 kg; 1 metric ton of coal has a volume of 1.5 $m^3$. The heat of combustion of coal is 28 MJ/kg. Assume that *all* heat is transferred from the fuel to the boiler and that *all* the work done in spinning the turbine is transformed into electric energy.
 a. Suppose the coal is piled up in a 10 m × 10 m room. How tall must the pile be to operate the plant for one day?
 b. What is the power plant's efficiency?

68. || Driving on asphalt roads entails very little rolling resistance, so most of the energy of the engine goes to overcoming air resistance. But driving slowly in dry sand is another story. If a 1500 kg car is driven in sand at 5.0 m/s, the coefficient of rolling friction is 0.06. In this case, nearly all of the energy that the car uses to move goes to overcoming rolling friction, so you can ignore air drag in this problem. INT
 a. What propulsion force is needed to keep the car moving forward at a constant speed?
 b. What power is required for propulsion at 5.0 m/s?
 c. If the car gets 15 mpg when driving on sand, what is the car's efficiency? One gallon of gasoline contains $1.4 \times 10^8$ J of chemical energy.

69. ||| Air conditioners sold in the United States are given a seasonal energy-efficiency ratio (SEER) rating that consumers can use to compare different models. A SEER rating is the ratio of heat pumped to energy input, similar to a COP but using English units, so a higher SEER rating means a more efficient model. You can determine the COP of an air conditioner by dividing the SEER rating by 3.4. For typical inside and outside temperatures when you'd be using air conditioning, estimate the theoretical maximum SEER rating of an air conditioner. (New air conditioners must have a SEER rating that exceeds 13, quite a bit less than the theoretical maximum, but there are practical issues that reduce efficiency.)

70. || The surface waters of tropical oceans are at a temperature of 27°C while water at a depth of 1200 m is at 3°C. It has been suggested these warm and cold waters could be the energy reservoirs for a heat engine, allowing us to do work or generate electricity from the thermal energy of the ocean. What is the maximum efficiency possible of such a heat engine?

71. || The light energy that falls on a square meter of ground over the course of a typical sunny day is about 20 MJ. The average rate of electric energy consumption in one house is 1.0 kW.
   a. On average, how much energy does one house use during each 24 h day?
   b. If light energy to electric energy conversion using solar cells is 15% efficient, how many square miles of land must be covered with solar cells to supply the electric energy for 250,000 houses? Assume there is no cloud cover.

## MCAT-Style Passage Problems

### Kangaroo Locomotion BIO

Kangaroos have very stout tendons in their legs that can be used to store energy. When a kangaroo lands on its feet, the tendons stretch, transforming kinetic energy of motion to elastic potential energy. Much of this energy can be transformed back into kinetic energy as the kangaroo takes another hop. The kangaroo's peculiar hopping gait is not very efficient at low speeds but is quite efficient at high speeds.

Figure P11.72 shows the energy cost of human and kangaroo locomotion. The graph shows oxygen uptake (in mL/s) per kg of body mass, allowing a direct comparison between the two species.

For humans, the energy used per second (i.e., power) is proportional to the speed. That is, the human curve nearly passes through the origin, so running twice as fast takes approximately twice as much power. For a hopping kangaroo, the graph of energy use has only a very small slope. In other words, the energy used per second changes very little with speed. Going faster requires very little additional power. Treadmill tests on kangaroos and observations in the wild have shown that they do not become winded at any speed at which they are able to hop. No matter how fast they hop, the necessary power is approximately the same.

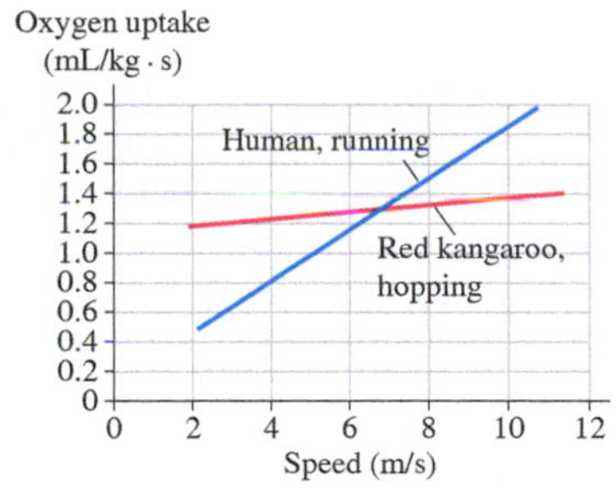


**FIGURE P11.72** Oxygen uptake (a measure of energy use per second) for a running human and a hopping kangaroo.

72. | A person runs 1 km. How does his speed affect the total energy needed to cover this distance?
   A. A faster speed requires less total energy.
   B. A faster speed requires more total energy.
   C. The total energy is about the same for a fast speed and a slow speed.

73. | A kangaroo hops 1 km. How does its speed affect the total energy needed to cover this distance?
   A. A faster speed requires less total energy.
   B. A faster speed requires more total energy.
   C. The total energy is about the same for a fast speed and a slow speed.

74. | At a speed of 4 m/s,
   A. A running human is more efficient than an equal-mass hopping kangaroo.
   B. A running human is less efficient than an equal-mass hopping kangaroo.
   C. A running human and an equal-mass hopping kangaroo have about the same efficiency.

75. | At approximately what speed would a human use half the power of an equal-mass kangaroo moving at the same speed?
   A. 3 m/s  B. 4 m/s  C. 5 m/s  D. 6 m/s

76. | At what speed does the hopping motion of the kangaroo become more efficient than the running gait of a human?
   A. 3 m/s  B. 5 m/s  C. 7 m/s  D. 9 m/s

### Animal Power

Measurements in the 1800s showed that a draft horse could provide a maximum steady output power over the course of a working day; that's how the "horsepower," about 750 W, came to be defined. Another animal that's used for pulling loads for which there are good data is the sled dog. A 38 kg dog can provide power to pull a sled with a force of 60 N at a steady 2.2 m/s. Dogs and horses have approximately the same 25% metabolic efficiency as humans.

77. || A horse is working steadily, providing 750 W of output
BIO power. How much food energy does the horse need to work for
INT 1 hour, to the nearest MJ?
   A. 3 MJ  B. 5 MJ  C. 9 MJ  D. 11 MJ

78. || How many sled dogs would be needed to provide 1 horse-
BIO power?
INT A. 6  B. 12  C. 14  D. 20

79. || A dog pulls a sled that requires a 120 N pulling force. How
BIO fast can the dog pull the sled, to the nearest m/s?
INT A. 1 m/s  B. 2 m/s  C. 3 m/s  D. 4 m/s

## KNOWLEDGE STRUCTURE II Conservation Laws

**BASIC GOALS** How is the system "after" an interaction related to the system "before"?
What quantities are conserved, and under what conditions?
Why are some energy changes more efficient than others?

**CROSS-CUTTING CONCEPTS** We divide the world into two parts: the **system** and the **environment**—everything else. The system is a collection of elements that work together, that share energy and momentum. A system is isolated if it does not exchange energy or momentum with the environment.

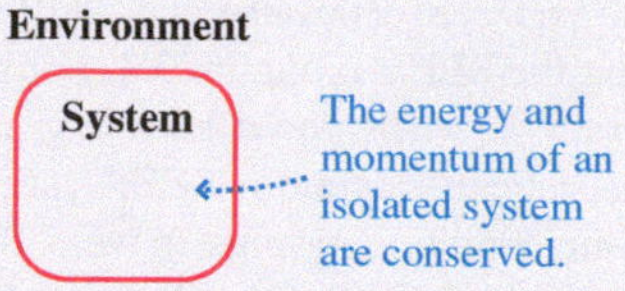


**GENERAL PRINCIPLES** **Law of conservation of momentum** For an isolated system, $\vec{P}_f = \vec{P}_i$.
**Law of conservation of energy** For an isolated system, there is no change in the system's energy:

$$\Delta K + \Delta U_g + \Delta U_s + \Delta E_{th} + \Delta E_{chem} + \cdots = 0$$

Energy can be exchanged with the environment as work or heat:

$$\Delta K + \Delta U_g + \Delta U_s + \Delta E_{th} + \Delta E_{chem} + \cdots = W + Q$$

**Laws of thermodynamics** First law: If only thermal energy changes, $\Delta E_{th} = W + Q$.
Second law: The entropy of an isolated system always increases.

**BASIC PROBLEM-SOLVING APPROACH** Draw a visual overview for the system "before" and "after"; then use the conservation of momentum or energy equation to relate the two. If necessary, calculate impulse, work, or both.

### Momentum and impulse

In a collision, the total momentum

$$\vec{P} = \vec{p}_1 + \vec{p}_2 = m_1\vec{v}_1 + m_2\vec{v}_2$$

is the same before and after.

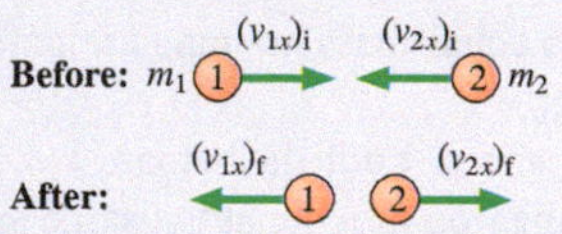


A force can change the momentum of an object. The change is the **impulse:** $\Delta p_x = J_x$.

$J_x =$ area under force curve

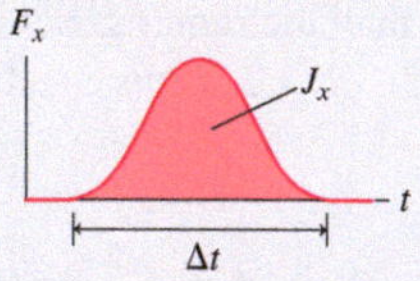


### Basic model of energy

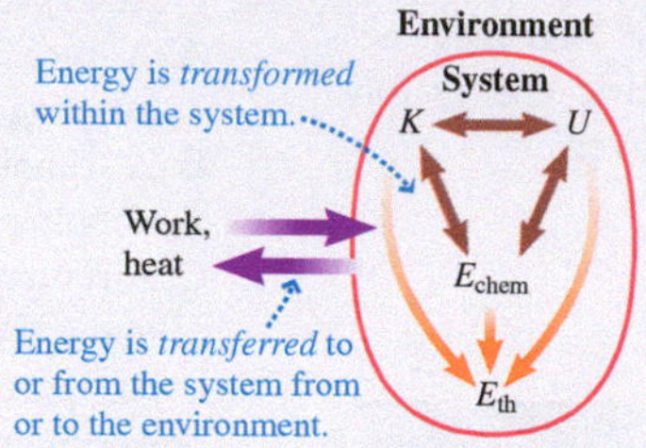


Work $W = F_\parallel d$ is done by the component of a force parallel to a displacement.

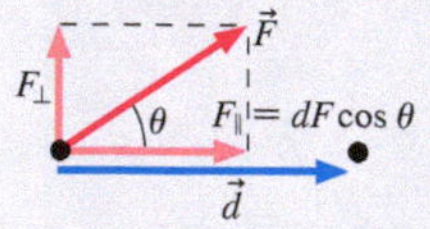


### Limitations on energy transfers and transformations

Thermal energy is random kinetic energy. Changing other forms of energy to thermal energy is **irreversible.**

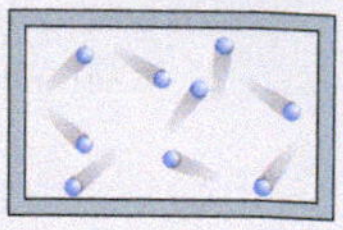

When energy is transformed from one form into another, some may be "lost" as thermal energy. This limits efficiency:

$$\text{efficiency: } e = \frac{\text{what you get}}{\text{what you had to pay}}$$

A heat engine can convert thermal energy to useful work. The efficiency must be less than 100%.

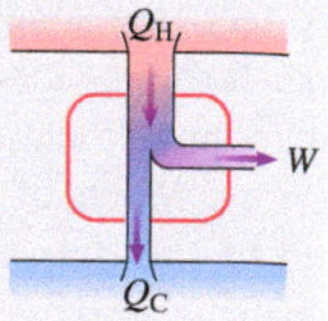


### Momentum

Momentum is the product of the mass and the velocity of an object. Momentum is a vector quantity.

$$\vec{p} = m\vec{v}$$

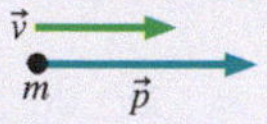


The units of momentum are kg · m/s.

### Energy

Energy comes in a variety of different forms. Some can be easily quantified:

| | |
|---|---|
| **Kinetic energy** | $K = \frac{1}{2}mv^2$ |
| **Gravitational potential energy** | $U_g = mgy$ |
| **Elastic potential energy** | $U_s = \frac{1}{2}kx^2$ |

Other important forms are:

| | |
|---|---|
| **Thermal energy** | $E_{th}$ |
| **Chemical energy** | $E_{chem}$ |

Energy is measured in joules, with 1 J = 1 N · m.

### Relating Force and Momentum

The change in momentum, the impulse, is related to the average force in a collision:

$$\Delta p_x = J_x = F_{avg}\Delta t$$

### Work-Energy Equation

For a system before and after a transformation or a transfer by work:

$$K_f + U_f + \Delta E_{th} = K_i + U_i + W$$

### Power

Power is the rate at which energy is transformed or work is done:

$$P = \frac{\Delta E}{\Delta t} \qquad P = \frac{W}{\Delta t}$$

Power is measured in watts, with 1 W = 1 J/s.

## Order Out of Chaos

The second law of thermodynamics specifies that "the future" is the direction of entropy increase. But, as we have seen, this doesn't mean that systems must invariably run down or decay. You don't need to look far to find examples of systems that spontaneously evolve to a state of greater order.

A snowflake is a perfect example. As water freezes, the random motion of water molecules is transformed into the orderly arrangement of a crystal. The entropy of the snowflake is less than that of the water vapor from which it formed. Has the second law of thermodynamics been turned on its head?

The entropy of the water molecules in the snowflake certainly decreases, but the water doesn't freeze as an isolated system. For it to freeze, heat energy must be transferred from the water to the surrounding air. The entropy of the air increases by *more* than the entropy of the water decreases. Thus the *total* entropy of the water + air system increases when a snowflake is formed, just as the second law predicts. If the system isn't isolated, its entropy can decrease without violating the second law as long as the entropy increases somewhere else.

Systems that become *more* ordered as time passes, and in which the entropy decreases, are called *self-organizing systems.* These systems can't be isolated. It is common in self-organizing systems to find a substantial flow of energy *through* the system. Your body takes in chemical energy from food, makes use of that energy, and then gives waste heat back to the environment. It is this energy flow that allows systems to develop a high degree of order and a very low entropy. The entropy of the environment undergoes a significant *increase* so as to let selected subsystems decrease their entropy and become more organized.

Self-organizing systems don't violate the second law of thermodynamics, but this fact doesn't really explain their existence. If you toss a coin, no law of physics says that you can't get heads 100 times in a row—but you don't expect this to happen. Can we show that self-organization isn't just possible, but likely?

Let's look at a simple example. Suppose you heat a shallow dish of oil at the bottom, while holding the temperature of the top constant. When the temperature difference between the top and the bottom of the dish is small, heat is transferred from the bottom to the top by conduction. But convection begins when the temperature difference becomes large enough. The pattern of convection needn't be random, though; it can develop in a stable, highly ordered pattern, as we see in the figure. Convection is a much more efficient means of transferring energy than conduction, so the spread of energy is *increased* as a result of the development of these ordered *convection cells.*

The development of the convection cells is an example of self-organization. The roughly $10^{23}$ molecules in the fluid had been moving randomly but now have begun behaving in a very orderly fashion. But there is more to the story. The convection cells transfer energy from the hot lower side of the dish to the cold upper side. This hot-to-cold energy transfer increases the entropy of the surrounding environment, as we have seen. In becoming more organized, the system has become more effective at transferring heat, resulting in a greater rate of entropy increase! Order has arisen out of disorder in the system, but the net result is a more rapid increase of the entropy of the universe.

Convection cells are thus a thermodynamically favorable form of order. We should expect this, because convection cells aren't confined to the laboratory. We see them in the sun, where they transfer energy from lower levels to the surface, and in the atmosphere of the earth, where they give rise to some of our most dramatic weather.

Self-organizing systems are a very active field of research in physical and biological sciences. The 1977 Nobel Prize in chemistry was awarded to the Belgian scientist Ilya Prigogine for his studies of *nonequilibrium thermodynamics,* the basic science underlying self-organizing systems. Prigogine and others have shown how energy flow through a system can, when the conditions are right, "bring order out of chaos." And this spontaneous ordering is not just possible—it can be probable. The existence and evolution of self-organizing systems, from thunderstorms to life on earth, might just be nature's preferred way of increasing entropy in the universe.

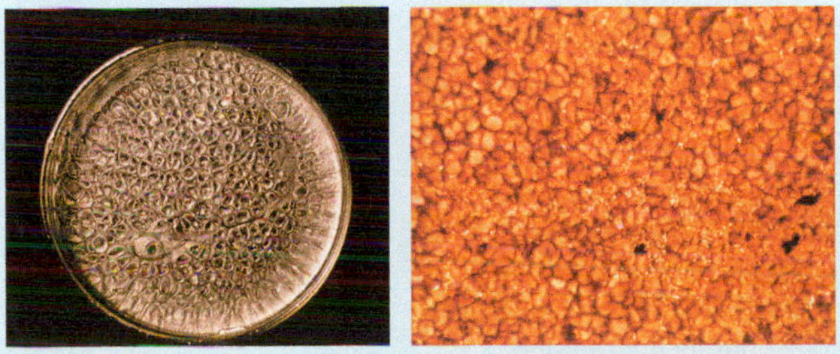

Convection cells in a shallow dish of oil heated from below (left) and in the sun (right). In both, warmer fluid is rising (lighter color) and cooler fluid is sinking (darker color).

*The following questions are related to the passage "Order Out of Chaos" on the previous page.*

1. When water freezes to make a snowflake crystal, the entropy of the water
   A. Decreases.
   B. Increases.
   C. Does not change.
2. When thermal energy is transferred from a hot object to a cold object, the overall entropy
   A. Decreases.
   B. Increases.
   C. Does not change.
3. Do convection cells represent a reversible process?
   A. Yes, because they are orderly.
   B. No, because they transfer thermal energy from hot to cold.
   C. It depends on the type of convection cell.
4. In an isolated system far from thermal equilibrium, as time passes,
   A. The total energy stays the same; the total entropy stays the same.
   B. The total energy decreases; the total entropy increases.
   C. The total energy stays the same; the total entropy increases.
   D. The total energy decreases; the total entropy stays the same.

*The following passages and associated questions are based on the material of Part II.*

### Big Air

A new generation of pogo sticks lets a rider bounce more than 2 meters off the ground by using elastic bands to store energy. When the pogo's plunger hits the ground, the elastic bands stretch as the pogo and rider come to rest. At the low point of the bounce, the stretched bands start to contract, pushing out the plunger and launching the rider into the air. For a total mass of 80 kg (rider plus pogo), a stretch of 0.40 m launches a rider 2.0 m above the starting point.

5. If you were to jump to the ground from a height of 2 meters, you'd likely injure yourself. But a pogo rider can do this repeatedly, bounce after bounce. How does the pogo stick make this possible?
   A. The elastic bands absorb the energy of the bounce, keeping it from hurting the rider.
   B. The elastic bands warm up as the rider bounces, absorbing dangerous thermal energy.
   C. The elastic bands simply convert the rider's kinetic energy to potential energy.
   D. The elastic bands let the rider come to rest over a longer time, meaning less force.
6. Assuming that the elastic bands stretch and store energy like a spring, how high would the 80 kg pogo and rider go for a stretch of 0.20 m?
   A. 2.0 m  B. 1.5 m  C. 1.0 m  D. 0.50 m
7. Suppose a much smaller rider (total mass of rider plus pogo of 40 kg) mechanically stretched the elastic bands of the pogo by 0.40 m, then got on the pogo and released the bands. How high would this unwise rider go?
   A. 8.0 m  B. 6.0 m  C. 4.0 m  D. 3.0 m
8. A pogo and rider of 80 kg total mass at the high point of a 2.0 m jump will drop 1.6 m before the pogo plunger touches the ground, slowing to a stop over an additional 0.40 m as the elastic bands stretch. What approximate average force does the pogo stick exert on the ground during the landing?
   A. 4000 N  B. 3200 N  C. 1600 N  D. 800 N
9. Riders can use fewer elastic bands, reducing the effective spring constant of the pogo. The maximum stretch of the bands is still 0.40 m. Reducing the number of bands will
   A. Reduce the force on the rider and give a lower jump height.
   B. Not change the force on the rider but give a lower jump height.
   C. Reduce the force on the rider but give the same jump height.
   D. Make no difference to the force on the rider or the jump height.

### Testing Tennis Balls

Tennis balls are tested by being dropped from a height of 2.5 m onto a concrete floor. The 57 g ball hits the ground, compresses, then rebounds. A ball will be accepted for play if it rebounds to a height of about 1.4 m; it will be rejected if the bounce height is much more or much less than this.

10. Consider the sequence of energy transformations in the bounce. When the dropped ball is motionless on the floor, compressed, and ready to rebound, most of the energy is in the form of
    A. Kinetic energy.
    B. Gravitational potential energy.
    C. Thermal energy.
    D. Elastic potential energy.
11. If a ball is "soft," it will spend more time in contact with the floor and won't rebound as high as it is supposed to. The force on the floor of the "soft" ball is ________ the force on the floor of a "normal" ball.
    A. Greater than
    B. The same as
    C. Less than
12. Suppose a ball is dropped from 2.5 m and rebounds to 1.4 m.
    a. How fast is the ball moving just before it hits the floor?
    b. What is the ball's speed just after leaving the floor?
    c. What happens to the "lost" energy?
    d. If the time of the collision with the floor is 6.0 ms, what is the average force on the ball during the impact?

### Squid Propulsion BIO

Squid usually move by using their fins, but they can utilize a form of jet propulsion, ejecting water at high speed to rocket them backward, as shown in Figure II.1. A 4.0 kg squid can slowly draw in and then quickly eject 0.30 kg of water. The water is ejected in 0.10 s at a speed of 10 m/s. This gives the squid a quick burst of speed to evade predators or catch prey.

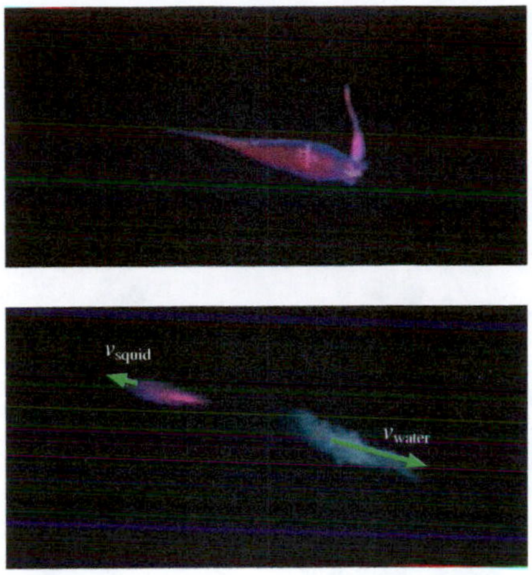


FIGURE II.1

13. What is the speed of the squid immediately after the water is ejected?
    A. 10 m/s  B. 7.5 m/s  C. 1.3 m/s  D. 0.75 m/s
14. What is the squid's approximate acceleration, in $g$'s?
    A. $10g$  B. $7.5g$  C. $1.0g$  D. $0.75g$
15. What is the average force on the water during the jet?
    A. 100 N  B. 30 N  C. 10 N  D. 3.0 N
16. This form of locomotion is speedy, but is it efficient? The energy that the squid expends goes two places: the kinetic energy of the squid and the kinetic energy of the water. Think about how to define "what you get" and "what you had to pay"; then calculate an efficiency for this particular form of locomotion. (You can ignore biomechanical efficiency for this problem.)

### Teeing Off

A golf club has a lightweight flexible shaft with a heavy block of wood or metal (called the head of the club) at the end. A golfer making a long shot off the tee uses a driver, a club whose 300 g head is much more massive than the 46 g ball it will hit. The golfer swings the driver so that the club head is moving at 40 m/s just before it collides with the ball. The collision is so rapid that it can be treated as the collision of a moving 300 g mass (the club head) with a stationary 46 g mass (the ball); the shaft of the club and the golfer can be ignored. The collision takes 5.0 ms, and the ball leaves the tee with a speed of 63 m/s.

17. What is the change in momentum of the ball during the collision?
    A. 1.4 kg · m/s  B. 1.8 kg · m/s
    C. 2.9 kg · m/s  D. 5.1 kg · m/s
18. What is the speed of the club head immediately after the collision?
    A. 30 m/s  B. 25 m/s  C. 19 m/s  D. 11 m/s
19. Is this a perfectly elastic collision?
    A. Yes
    B. No
    C. There is insufficient information to make this determination.
20. If we define the kinetic energy of the club head before the collision as "what you had to pay" and the kinetic energy of the ball immediately after as "what you get," what is the efficiency of this energy transfer?
    A. 0.54  B. 0.46  C. 0.38  D. 0.27

### Additional Integrated Problems

21. Football players measure their acceleration by seeing how fast they can sprint 40 yards (37 m). A zippy player can, from a standing start, run 40 yards in 4.1 s, reaching a top speed of about 11 m/s. For an 80 kg player, what is the average power output for this sprint?
    A. 300 W  B. 600 W  C. 900 W  D. 1200 W
22. The unit of horsepower was defined by considering the power output of a typical horse. Working-horse guidelines in the 1900s called for them to pull with a force equal to 10% of their body weight at a speed of 3.0 mph. For a typical working horse of 1200 lb, what power does this represent in W and in hp?
23. A 100 kg football player is moving at 6.0 m/s to the east; a 130 kg player is moving at 5.0 m/s to the west. They meet, each jumping into the air and grabbing the other player. While they are still in the air, which way is the pair moving, and how fast?
24. A swift blow with the hand can break a pine board. As the hand hits the board, the kinetic energy of the hand is transformed into elastic potential energy of the bending board; if the board bends far enough, it breaks. Applying a force to the center of a particular pine board deflects the center of the board by a distance that increases in proportion to the force. Ultimately the board breaks at an applied force of 800 N and a deflection of 1.2 cm.
    a. To break the board with a blow from the hand, how fast must the hand be moving? Use 0.50 kg for the mass of the hand.
    b. If the hand is moving this fast and comes to rest in a distance of 1.2 cm, what is the average force on the hand?
25. A child's sled has rails that slide with little friction across the snow. Logan has an old wooden sled with heavy iron rails that has a mass of 10 kg—quite a bit for a 30 kg child! Logan runs at 4.0 m/s and leaps onto the stationary sled and holds on tight as it slides forward. The impact time with the sled is 0.25 s.
    a. Immediately after Logan jumps on the sled, how fast is it moving?
    b. What was the force on the sled during the impact?
    c. How much energy was "lost" in the impact? Where did this energy go?

PART

# III Properties of Matter

Male emperor penguins overwinter on the Antarctic ice, going long periods without eating in bitterly cold surroundings. Minimizing heat loss to the environment is crucial because replacing lost energy draws down precious reserves. In this section, we'll consider properties of matter, including thermal energy and its transfer as heat. We'll learn to understand and appreciate some of the adaptations these remarkable birds use to keep warm during the winter.

OVERVIEW

## Beyond the Particle Model

The first 11 chapters of this text have made extensive use of the *particle model* in which we represent objects as point masses. The particle model is especially useful for describing how discrete objects move through space and how they interact with each other. Whether a ball is made of metal or wood is irrelevant to calculating its trajectory.

But there are many situations where the distinction between metal and wood is crucial. If you toss a metal ball and a wood ball into a pond, one sinks and the other floats. If you stir a pan on the stove with a metal spoon, it can quickly get too hot to hold unless it has a wooden handle.

Wood and metal have different physical properties. So do air and water. Our goal in Part III is to describe and understand the similarities and differences of different materials. To do so, we must go beyond the particle model and dig deeper into the nature of matter.

### Macroscopic Physics

In Part III, we will be concerned with systems that are solids, liquids, or gases. Properties such as pressure, temperature, specific heat, and viscosity are characteristics of the system as a whole, not of the individual particles. Solids, liquids, and gases are often called *macroscopic* systems—the prefix "macro" (the opposite of "micro") meaning "large." We'll make sense of the behavior of these macroscopic properties by considering a microscopic view in which we think of these systems as collections of particle-like atoms. This "micro-to-macro" development will be a key piece of the following chapters.

In the coming chapters, we'll consider a wide range of practical questions, such as:

- How do the temperature and pressure of a system change when you heat it? Why do some materials respond quickly, others slowly?
- What are the mechanisms by which a system exchanges heat energy with its environment? Why does blowing on a cup of hot coffee cause it to cool off?
- Why are there three phases of matter—solids, liquids, gases? What happens during a phase change?
- Why do some objects float while others, with the same mass, sink? What keeps a massive steel ship afloat?
- What are the laws of motion of a flowing liquid? How do they differ from the laws governing the motion of a particle?

Both Newton's laws and the law of conservation of energy will remain important tools—they are, after all, the basic laws of physics—but we'll have to learn how they apply to macroscopic systems.

It should come as no surprise that an understanding of macroscopic systems and their properties is essential for understanding the world around us. Biological systems, from cells to ecosystems, are macroscopic systems exchanging energy with their environment. On a larger scale, energy transport on earth gives us weather, and the exchange of energy between the earth and space determines our climate.

# 12 Thermal Properties of Matter

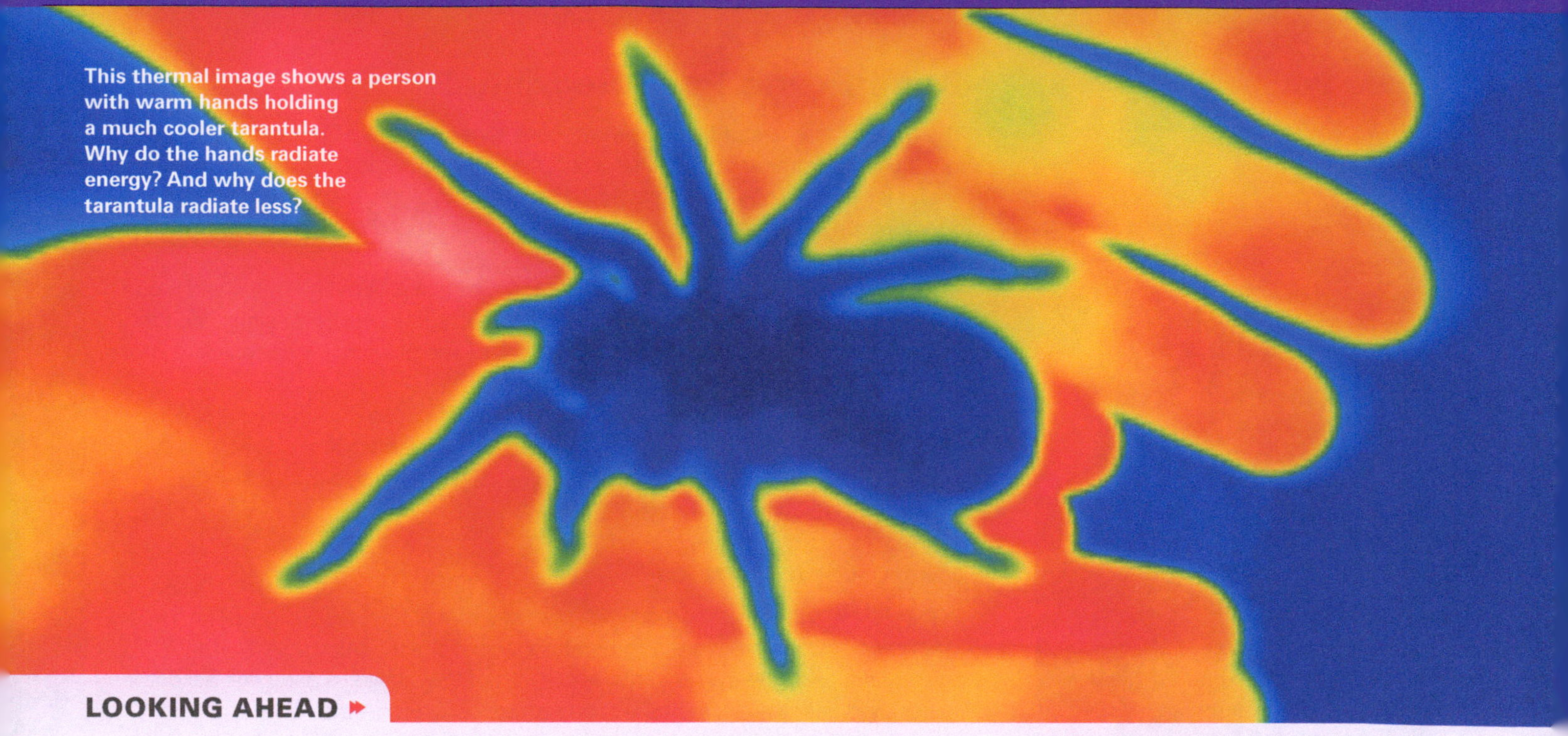

This thermal image shows a person with warm hands holding a much cooler tarantula. Why do the hands radiate energy? And why does the tarantula radiate less?

## LOOKING AHEAD ▸▸

### The Ideal Gas

The high pressure in a car tire is due to the countless collisions between the air molecules inside and the tire's walls.

You'll learn how gas properties are related to the microscopic motion of the gas molecules.

### Heat and Temperature

Adding ice cools your drink as heat is transferred from the warm drink to the cold ice; even more heat is used to melt the ice.

You'll learn how to compute the temperature changes that occur when heat is transferred or during a **phase change** such as melting.

### Heat Transfer

Its large ears keep an elephant cool. Blood flowing through large vessels loses heat to the environment, returning to the body at a lower temperature.

You'll learn about the heat-transfer mechanisms of **conduction, convection,** and **radiation.**

**GOAL** To use the atomic model of matter to explain many properties of matter associated with heat and temperature.

## LOOKING BACK ◂◂

### Heat

In Section 11.4 you learned about heat and the first law of thermodynamics. In this chapter we will explore some of the consequences of transferring heat to or from a system, and doing work on the system.

You learned that a system's energy can be changed by doing work on it *or* by transferring heat to it.

**STOP TO THINK**

A blender does 5000 J of work on the food in its bowl. During the time the blender runs, 2000 J of heat is transferred from the warm food to the cooler environment. What is the change in the thermal energy of the food?

A. +2000 J
B. +3000 J
C. +7000 J
D. −2000 J
E. −3000 J

# 12.1 The Atomic Model of Matter

We began exploring the concepts of thermal energy, temperature, and heat in Chapter 11, but many unanswered questions remain. How do the properties of matter depend on temperature? When you add heat to a system, by how much does its temperature change? And how is heat transferred to or from a system?

These are questions about the *macroscopic* state of systems, but we'll start our exploration by looking at a *microscopic* view, the atomic model that we've used to explain friction, elastic forces, and the nature of thermal energy. In this chapter, we'll use the atomic model to understand and explain the thermal properties of matter.

As you know, each element and most compounds can exist as a gas, liquid, or solid. You've certainly learned about these three **phases** of matter earlier in your education. An atomic view of the three phases is shown in FIGURE 12.1.

FIGURE 12.1 Atomic models of the three phases of matter: gas, liquid, and solid.

In a gas, the particles are in random motion and interact only through elastic collisions.
In a liquid, the particles have weak bonds that keep them close together. The particles can slide around each other, so the liquid can flow.
In a rigid solid, the particles are connected by relatively stiff spring-like bonds.

Our atomic model makes a simplification that is worth noting. The basic particles in Figure 12.1 are drawn as simple spheres; no mention is made of the nature of the particles. In a real gas, the basic particles might be helium atoms or nitrogen molecules. In a solid, the basic particles might be the gold atoms that make up a bar of gold or the water molecules that make up ice. But, because many of the properties of gases, liquids, and solids do not depend on the exact nature of the particles that make them up, it's often a reasonable assumption to ignore these details and just consider the gas, liquid, or solid as being made up of simple spherical *particles*.

## Atomic Mass and Atomic Mass Number

Before we see how the atomic model explains the thermal properties of matter, we need to remind you of some "atomic accounting." Recall that atoms of different elements have different masses. The mass of an atom is determined primarily by its most massive constituents: the protons and neutrons in its nucleus. The *sum* of the number of protons and the number of neutrons is the **atomic mass number** $A$:

$$A = \text{number of protons} + \text{number of neutrons}$$

$A$, which by definition is an integer, is written as a leading superscript on the atomic symbol. For example, the primary isotope of carbon, with six protons (which is what makes it carbon) and six neutrons, has $A = 12$ and is written $^{12}C$. The radioactive isotope $^{14}C$, used for carbon dating of archeological finds, contains six protons and eight neutrons.

The **atomic mass** scale is established by defining the mass of $^{12}C$ to be exactly 12 u, where u is the symbol for the *atomic mass unit*. That is, $m(^{12}C) = 12$ u. In kg, the atomic mass unit is

$$1 \text{ u} = 1.66 \times 10^{-27} \text{ kg}$$

Atomic masses are all very nearly equal to the integer atomic mass number $A$. For example, the mass of $^{1}H$, with $A = 1$, is $m = 1.0078$ u. For our present purposes, it will be sufficient to use the integer atomic mass numbers as the values of the atomic mass. That is, we'll use $m(^{1}H) = 1$ u, $m(^{4}He) = 4$ u, and $m(^{16}O) = 16$ u. For molecules, the **molecular mass** is the sum of the atomic masses of the atoms that form the molecule. Thus the molecular mass of the molecule $O_2$, the constituent of oxygen gas, is $m(O_2) = 2m(^{16}O) = 32$ u.

**NOTE** ▸ An element's atomic mass number is *not* the same as its atomic number. The *atomic number*, which gives the element's position in the periodic table, is the number of protons. ◂

TABLE 12.1 lists the atomic mass numbers of some of the elements that we'll use for examples and homework problems. A complete periodic table, including atomic masses, is found in Appendix B.

TABLE 12.1 Some atomic mass numbers

| Element | Symbol | $A$ |
|---|---|---|
| Hydrogen | $^{1}H$ | 1 |
| Helium | $^{4}He$ | 4 |
| Carbon | $^{12}C$ | 12 |
| Nitrogen | $^{14}N$ | 14 |
| Oxygen | $^{16}O$ | 16 |
| Neon | $^{20}Ne$ | 20 |
| Aluminum | $^{27}Al$ | 27 |
| Argon | $^{40}Ar$ | 40 |
| Lead | $^{207}Pb$ | 207 |

## The Definition of the Mole

One way to specify the amount of substance in a system is to give its mass. Another way, one connected to the number of atoms, is to measure the amount of substance in *moles*. **1 mole of substance, abbreviated 1 mol, is $6.02 \times 10^{23}$ basic particles.**

TABLE 12.2 Monatomic and diatomic gases

| Monatomic | Diatomic |
|---|---|
| Helium (He) | Hydrogen ($H_2$) |
| Neon (Ne) | Nitrogen ($N_2$) |
| Argon (Ar) | Oxygen ($O_2$) |

The basic particle depends on the substance. Helium is a **monatomic gas,** meaning that the basic particle is the helium atom. Thus $6.02 \times 10^{23}$ helium atoms are 1 mol of helium. But oxygen gas is a **diatomic gas** because the basic particle is the two-atom diatomic molecule $O_2$. 1 mol of oxygen gas contains $6.02 \times 10^{23}$ *molecules* of $O_2$ and thus $2 \times 6.02 \times 10^{23}$ oxygen atoms. TABLE 12.2 lists the monatomic and diatomic gases that we will use for examples and problems.

The number of basic particles per mole of substance is called **Avogadro's number** $N_A$. The value of Avogadro's number is thus

$$N_A = 6.02 \times 10^{23} \text{ mol}^{-1}$$

The number $n$ of moles in a substance containing $N$ basic particles is

$$n = \frac{N}{N_A} \tag{12.1}$$

Moles of a substance in terms of the number of basic particles

One mole of helium, sulfur, copper, and mercury.

The **molar mass** of a substance, $M_{mol}$, is the mass *in grams* of 1 mol of substance. To a good approximation, the numerical value of the molar mass equals the numerical value of the atomic or molecular mass. That is, the molar mass of He, with $m = 4$ u, is $M_{mol}(\text{He}) = 4$ g/mol, and the molar mass of diatomic $O_2$ is $M_{mol}(O_2) = 32$ g/mol.

You can use the molar mass to determine the number of moles. In one of the few instances where the proper units are *grams* rather than kilograms, the number of moles contained in a system of mass $M$ consisting of atoms or molecules with molar mass $M_{mol}$ is

$$n = \frac{M \text{ (in grams)}}{M_{mol}} \tag{12.2}$$

Moles of a substance in terms of its mass

**EXAMPLE 12.1 Determining quantities of oxygen**

A system contains 100 g of oxygen. How many moles does it contain? How many molecules?

***SOLVE*** The diatomic oxygen molecule $O_2$ has molar mass $M_{mol} = 32$ g/mol. From Equation 12.2,

$$n = \frac{100 \text{ g}}{32 \text{ g/mol}} = 3.1 \text{ mol}$$

Each mole contains $N_A$ molecules, so the total number is $N = nN_A = 1.9 \times 10^{24}$ molecules.

## Volume

The volume $V$ of a macroscopic system is the amount of space the system occupies. The SI unit of volume is $m^3$. Nonetheless, both $cm^3$ and, to some extent, liters (L) are widely used metric units of volume. In most cases, you *must* convert these to $m^3$

before doing calculations. Some important conversions for volumes are listed here. Although it is true that 1 m = 100 cm, it is not true that 1 $m^3$ = 100 $cm^3$. FIGURE 12.2 shows that the volume conversion factor is 1 $m^3 = 10^6$ $cm^3$.

$$1\ \text{m}^3 = 1000\ \text{liters} = 10^6\ \text{cm}^3 \qquad 1\ \text{liter} = 1000\ \text{cm}^3 \qquad 1\ \text{cm}^3 = 1\ \text{mL}$$

FIGURE 12.2 There are $10^6$ $cm^3$ in 1 $m^3$.

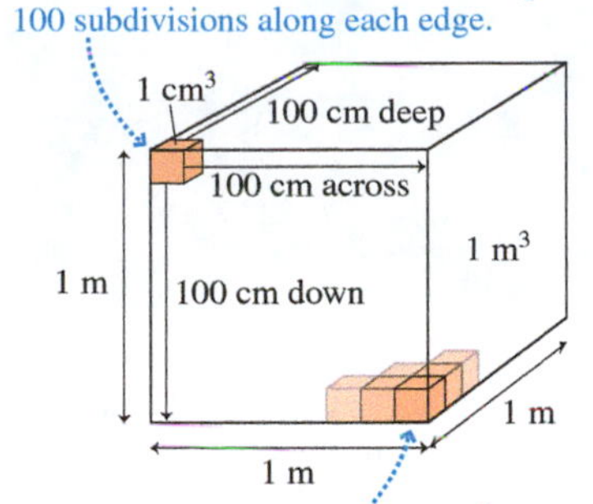


**STOP TO THINK 12.1** Which system contains more atoms: 5 mol of helium ($A = 4$) or 1 mol of neon ($A = 20$)?

A. Helium B. Neon C. They have the same number of atoms.

## 12.2 The Atomic Model of an Ideal Gas

Solids and liquids are nearly incompressible because the atomic particles are in close contact with each other. Gases, in contrast, are highly compressible because the atomic particles are far apart. When we work with gases, we'll assume a set of simplifications that we call the **ideal-gas model,** summed up in FIGURE 12.3. It turns out that most real gases at the temperatures and pressures we work with can be explained quite well using this model; the simplifications are not far from reality. We discussed thermal energy in the context of a gas in ◀◀ SECTION 11.3. The thermal energy of a gas is simply the total kinetic energy of the atoms of the gas; the temperature is a measure of the average kinetic energy of the atoms. Adding heat to a gas makes its atoms move faster, increasing the thermal energy and the temperature.

FIGURE 12.3 The ideal-gas model.

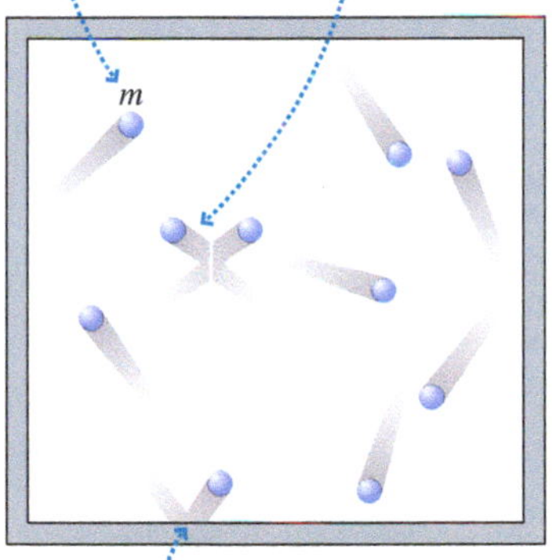

The gas is made up of a large number N of particles of mass m, each moving randomly.
The particles are quite far from each other and interact only rarely when they collide.
m
The collisions of the particles with each other (and with walls of the container) are elastic; no energy is lost in these collisions.

For an ideal gas, there is a straightforward relationship between the temperature and the average kinetic energy per atom $K_{avg}$:

$$T = \frac{2}{3}\frac{K_{avg}}{k_B} \tag{12.3}$$

where $k_B$ is a constant known as **Boltzmann's constant.** Its value is

$$k_B = 1.38 \times 10^{-23}\ \text{J/K}$$

We can rearrange Equation 12.3 to give the average kinetic energy in terms of the temperature:

$$K_{avg} = \frac{3}{2}k_B T \tag{12.4}$$

The thermal energy of an ideal gas consisting of $N$ atoms is the sum of the kinetic energies of the individual atoms:

$$E_{th} = NK_{avg} = \frac{3}{2}Nk_B T \tag{12.5}$$

Thermal energy of an ideal gas of $N$ atoms

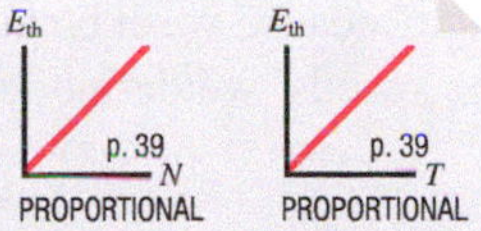


For an ideal gas, **thermal energy is directly proportional to temperature.** Consequently, a change in the thermal energy of an ideal gas is proportional to a change in temperature:

$$\Delta E_{th} = \frac{3}{2}Nk_B \Delta T \tag{12.6}$$

**Is it cold in space?** The International Space Station orbits in the upper thermosphere, about 400 km above the surface of the earth. There is still a trace of atmosphere left at this altitude, and it has quite a high temperature—over 1000°C. Although the average speed of the air molecules here is high, there are so few air molecules present that the thermal energy is extremely low.

**EXAMPLE 12.2 Finding the energy needed to warm up a room**

A large bedroom contains about $1 \times 10^{27}$ molecules of air. If we model the air as an ideal gas, how much energy is required to raise the temperature of the air in the room by 5°C.

**STRATEGIZE** We can use Equation 12.6 to determine the change in thermal energy corresponding to the noted change in temperature.

**PREPARE** The actual temperature of the gas doesn't matter—only the change. The temperature increase is given as 5°C, implying a change in the absolute temperature by the same amount: $\Delta T = 5\text{ K}$.

**SOLVE** We can use Equation 12.6 to calculate the amount by which the room's thermal energy must be increased:

$$\Delta E_{\text{th}} = \frac{3}{2} N k_{\text{B}} \Delta T = \frac{3}{2}(1 \times 10^{27})(1.38 \times 10^{-23}\text{ J/K})(5\text{ K}) = 1 \times 10^{5}\text{ J} = 100\text{ kJ}$$

This is the energy we would have to supply—probably in the form of heat from a furnace—to raise the temperature.

**ASSESS** 100 kJ isn't that much energy. Table 11.2 showed it to be less than the food energy in a carrot! This seems reasonable because you know that your furnace can quickly warm up the air in a room. Heating up the walls and furnishings is another story.

## Molecular Speeds and Temperature

FIGURE 12.4 The distribution of molecular speeds in nitrogen gas at 20°C.

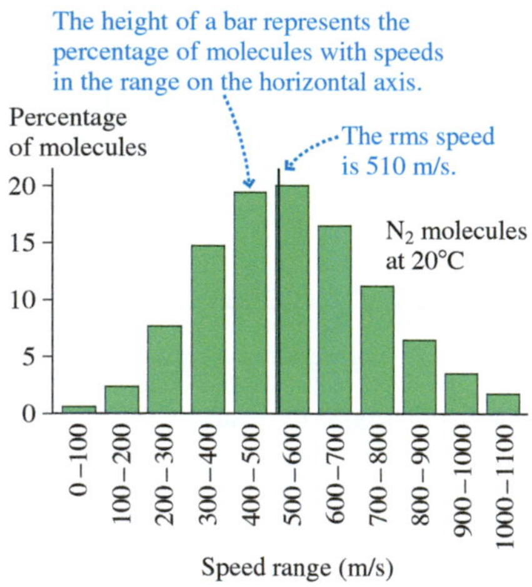


The atomic model of an ideal gas is based on random motion, so it's no surprise that the individual atoms in a gas are moving at different speeds. FIGURE 12.4 shows data from an experiment to measure the molecular speeds in nitrogen gas at 20°C. The results are presented as a histogram, a bar chart in which the height of the bar indicates what percentage of the molecules have a speed in the range of speeds shown below the bar. For example, 16% of the molecules have speeds in the range from 600 m/s to 700 m/s. The most probable speed, as judged from the tallest bar, is $\sim 500\text{ m/s}$. This is quite fast: $\sim 1200$ mph!

Because temperature is proportional to the average kinetic energy of the atoms, it will be useful to calculate the average kinetic energy for this distribution. An individual atom of mass $m$ and velocity $v$ has kinetic energy $K = \frac{1}{2}mv^2$. Recall that the average of a series of measurements is found by adding all the values and then dividing the total by the number of data points. Thus, we can find the average kinetic energy by adding up all the kinetic energies of all the atoms and then dividing the total by the number of atoms:

$$K_{\text{avg}} = \frac{\sum \frac{1}{2}mv^2}{N} = \frac{1}{2}m\frac{\sum v^2}{N} = \frac{1}{2}m(v^2)_{\text{avg}} \tag{12.7}$$

The quantity $\sum v^2/N$ is the sum of the values of $v^2$ for all the atoms divided by the number of atoms. By definition, this is the average of the *squares* of all the individual speeds, which we've written $(v^2)_{\text{avg}}$.

The square root of this average is about how fast a typical atom in the gas is moving. Because we'll be taking the square root of the average, or mean, of the square of the speeds, we define the **root-mean-square speed** as

$$v_{\text{rms}} = \sqrt{(v^2)_{\text{avg}}} = \text{speed of a typical atom} \tag{12.8}$$

The root-mean-square speed is often referred to as the *rms speed*. The rms speed isn't the average speed of atoms in the gas; it's the speed of an atom with the average kinetic energy. But the average speed and the rms speed are very nearly equal, so we'll interpret an rms speed as telling us the speed of a typical atom in the gas.

Rewriting Equation 12.7 in terms of $v_{\text{rms}}$ gives the average kinetic energy per atom:

$$K_{\text{avg}} = \frac{1}{2}mv_{\text{rms}}^2 \tag{12.9}$$

We can relate the temperature to the speeds of the atoms by substituting Equation 12.9 into Equation 12.4:

$$T = \frac{1}{3}\frac{mv_{\text{rms}}^2}{k_B} \tag{12.10}$$

Solving Equation 12.10 for the rms speed of the atoms, we find that

$$v_{\text{rms}} = \sqrt{\frac{3k_B T}{m}} \tag{12.11}$$

rms speed of an atom of mass $m$ in an ideal gas at temperature $T$

**NOTE** ▸ You must use absolute temperature, in kelvin, to compute rms speeds. ◂

We have been considering ideal gases made of atoms, but our results are equally valid for real gases made of either atoms (such as helium, He) or molecules (such as oxygen, $O_2$). At a given temperature, Equation 12.11 shows that the speed of atoms or molecules in a gas varies with the atomic or molecular mass. A gas with lighter atoms will have faster atoms, on average, than a gas with heavier atoms. The equation also shows that higher temperatures correspond to faster atomic or molecular speeds. The rms speed is proportional to the *square root* of the temperature. This is a new mathematical form that we will see again, so we will take a look at its properties.

**Martian airsicles** The atmosphere of Mars is mostly carbon dioxide. At night, the temperature may drop so low that the molecules in the atmosphere will slow down enough to stick together—the atmosphere actually freezes. The frost on the surface in this image from the Viking 2 lander is composed partially of frozen carbon dioxide.

## Square-root relationships

Two quantities are said to have a **square-root relationship** if $y$ is proportional to the square root of $x$. We write the mathematical relationship as

$$y = A\sqrt{x}$$

$y$ is proportional to the square root of $x$

The graph of a square-root relationship is a parabola that has been rotated by 90°.

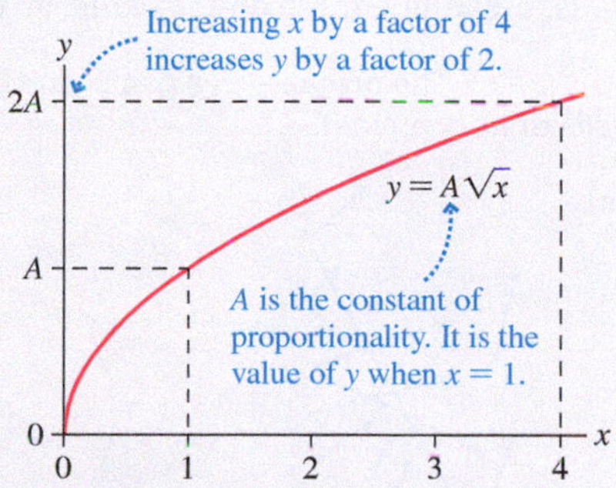


**SCALING** If $x$ has the initial value $x_1$, then $y$ has the initial value $y_1$. Changing $x$ from $x_1$ to $x_2$ changes $y$ from $y_1$ to $y_2$. The ratio of $y_2$ to $y_1$ is

$$\frac{y_2}{y_1} = \frac{A\sqrt{x_2}}{A\sqrt{x_1}} = \sqrt{\frac{x_2}{x_1}}$$

which is the square root of the ratio of $x_2$ to $x_1$.

- If you increase $x$ by a factor of 4, you increase $y$ by a factor of $\sqrt{4} = 2$.
- If you decrease $x$ by a factor of 9, you decrease $y$ by a factor of $\sqrt{9} = 3$.

These examples illustrate a general rule:

**Changing $x$ by a factor of $c$ changes $y$ by a factor of $\sqrt{c}$.**

Exercises 7, 8

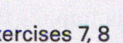

### EXAMPLE 12.3 Finding the speeds of air molecules

Most of the earth's atmosphere is the gas nitrogen, which consists of molecules, $N_2$. At the coldest temperature ever recorded on earth, –95°C, what is the root-mean-square speed of the nitrogen molecules? Does the temperature at the earth's surface ever get high enough that a typical molecule is moving at twice this speed? (The highest temperature ever recorded on earth was 54°C.)

**STRATEGIZE** The first part of the problem is a straightforward calculation. The second part is best done by considering ratios. The rms speed varies as the square root of the temperature, so we will determine the factor by which we'd need to increase the temperature to double the rms speed.

*Continued*

**PREPARE** We can use the periodic table to determine that the mass of a nitrogen atom is 14 u. A molecule consists of two atoms, so its mass is 28 u. Thus the molecular mass in SI units (i.e., kg) is

$$m = 28\text{ u} \times \frac{1.66 \times 10^{-27}\text{ kg}}{1\text{ u}} = 4.6 \times 10^{-26}\text{ kg}$$

The problem statement gives two temperatures we'll call $T_1$ and $T_2$; we need to express these in kelvin. The lowest temperature ever observed on earth is $T_1 = -95 + 273 = 178$ K; the highest temperature is $T_2 = 54 + 273 = 327$ K.

**SOLVE** We use Equation 12.11 to find $v_{rms}$ for the nitrogen molecules at $T_1$:

$$v_{rms} = \sqrt{\frac{3k_B T_1}{m}} = \sqrt{\frac{3(1.38 \times 10^{-23}\text{ J/K})(178\text{ K})}{4.6 \times 10^{-26}\text{ kg}}} = 400\text{ m/s}$$

Because the rms speed is proportional to the square root of the temperature, doubling the rms speed would require increasing the temperature by a factor of 4. The ratio of the highest temperature ever recorded to the lowest temperature ever recorded is less than this:

$$\frac{T_2}{T_1} = \frac{327\text{ K}}{178\text{ K}} = 1.8$$

The temperature at the earth's surface is never high enough that a typical nitrogen molecule would move at twice the computed speed.

**ASSESS** We can use the square-root relationship to assess our computed result for the molecular speed. Figure 12.4 shows an rms speed of 510 m/s for nitrogen molecules at 20°C, or 293 K. Temperature $T_1$ is about 0.6 of this, so we'd expect to compute a speed that is lower by the square root of 0.6 about 0.8, which is what we found.

## Pressure

Everyone has some sense of the concept of *pressure*. If you get a hole in your bicycle tire, the higher-pressure air inside comes squirting out. It's hard to get the lid off a vacuum-sealed jar because of the low pressure inside. But just what is pressure?

Let's take an atomic-scale view of pressure, defining it in terms of the motion of particles of a gas. Suppose we have a sample of gas in a container with rigid walls. As particles in the gas move around, they sometimes collide with and bounce off the walls, creating a force on the walls, as illustrated in FIGURE 12.5a.

FIGURE 12.5 The pressure in a gas is due to the force of the particles colliding with the walls of its container.

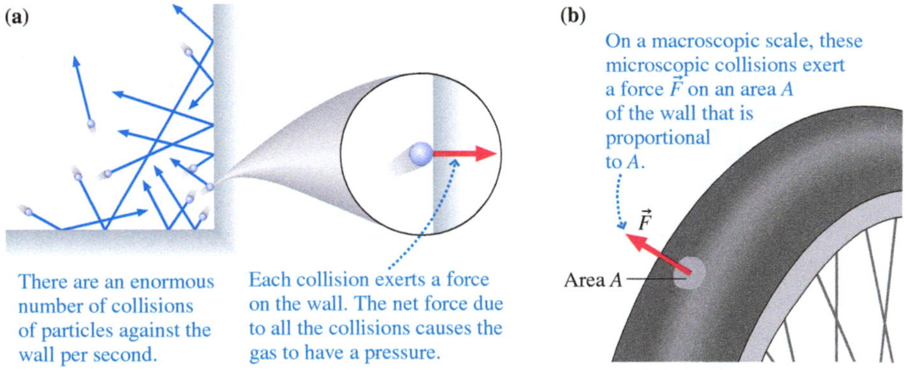

(a)
There are an enormous number of collisions of particles against the wall per second.
Each collision exerts a force on the wall. The net force due to all the collisions causes the gas to have a pressure.
(b)
On a macroscopic scale, these microscopic collisions exert a force $\vec{F}$ on an area $A$ of the wall that is proportional to $A$.
$\vec{F}$
Area $A$


These countless microscopic collisions are what lead to the pressure in the gas. On a small patch of the container wall with surface area $A$, these collisions result in a continuous macroscopic force of magnitude $F$ directed perpendicular to the wall, as shown for the bicycle tire in FIGURE 12.5b. If the size of the patch is doubled, then twice as many particles will hit it every second, leading to a doubling of the force. This implies that the force is proportional to the area of the patch, and so the *ratio* $F/A$ is constant. We define this ratio to be the **pressure** $p$ of the gas:

$$p = \frac{F}{A} \tag{12.12}$$

Definition of pressure in a gas

You can see from Equation 12.12 that a gas exerts a force of magnitude

$$F = pA \tag{12.13}$$

on a surface of area $A$.

**NOTE** ▶ We sometimes talk informally about "the force exerted by the pressure," but it's really the gas that exerts the force, not the pressure. ◀

From its definition, you can see that pressure has units of N/m$^2$. The SI unit of pressure is the **pascal,** defined as

$$1 \text{ pascal} = 1 \text{ Pa} = 1 \frac{\text{N}}{\text{m}^2}$$

A pascal is a very small pressure, so we usually see pressures in kilopascals, where 1 kPa = 1000 Pa.

The pressure with the most significance for our daily lives is the pressure of the atmosphere, caused by the microscopic collisions of the air molecules that surround us. The pressure of the atmosphere varies with altitude and the weather, but the global average pressure at sea level, called the *standard atmosphere,* is

$$1 \text{ standard atmosphere} = 1 \text{ atm} = 101{,}300 \text{ Pa} = 101.3 \text{ kPa}$$

Just as we measured acceleration in units of $g$, we will often measure pressure in units of atm.

In the United States, pressure is often expressed in pounds per square inch, or psi. When you measure pressure in the tires on your car or bike, you probably use a gauge that reads in psi. The conversion factor is

$$1 \text{ atm} = 14.7 \text{ psi}$$

The total force on the surface of your body due to the pressure of the atmosphere is over 40,000 pounds. Why doesn't this enormous force pushing in simply crush you? The key is that there is also a force pushing out. FIGURE 12.6a shows an empty plastic soda bottle. The bottle isn't a very sturdy structure, but the inward force due to the pressure of the atmosphere outside the bottle doesn't crush it because there is an equal pressure due to air inside the bottle that pushes out. The forces pushing on both sides are quite large, but they exactly balance and so there is no net force.

A net pressure force is exerted only where there's a pressure *difference* between the two sides of a surface. FIGURE 12.6b shows a surface of area $A$ with a pressure difference $\Delta p = p_2 - p_1$ between the two sides. The net pressure force is

$$F_{\text{net}} = F_2 - F_1 = p_2 A - p_1 A = A(p_2 - p_1) = A\,\Delta p$$

This is the force that holds the lid on a vacuum-sealed jar, where the pressure inside is less than the pressure outside.

Decreasing the number of molecules in a container decreases the pressure because there are fewer collisions with the walls. The pressure in a completely empty container would be $p = 0$ Pa. This is called a *perfect vacuum.* A perfect vacuum cannot be achieved because it's impossible to remove every molecule from a region of space. In practice, a **vacuum** is an enclosed space in which $p \ll 1$ atm. Using $p = 0$ Pa is then a good approximation.

The basic principle behind pressure measurements is expressed by Equation 12.12. A pressure gauge measures the force exerted by the gas on a known area. The actual or **absolute pressure** $p$ is directly proportional to this force. Because the effects of pressure depend on pressure differences, most gauges measure not the absolute pressure but what is called the **gauge pressure** $p_g$, the *difference* between the absolute pressure and atmospheric pressure. Most gauges measure this difference directly because they measure a force on a surface. FIGURE 12.7 shows how this works for a tire gauge. The gauge measures the difference between the pressure inside the tire and the pressure outside the tire, which is just the pressure of the atmosphere, $p_{\text{atmos}}$.

The calculations we'll do later in the chapter involve absolute pressure, but tire pressures and other common pressures are reported as gauge pressures, so we'll often need to convert between the two. The gauge pressure is, as in the case of the

A video to support a section's topic is embedded in the eText.

**Video** Force and Pressure

FIGURE 12.6 The net force depends on the pressure difference.

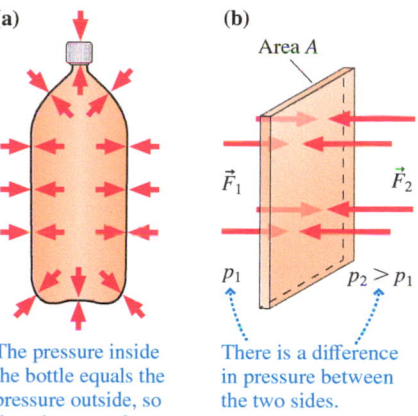

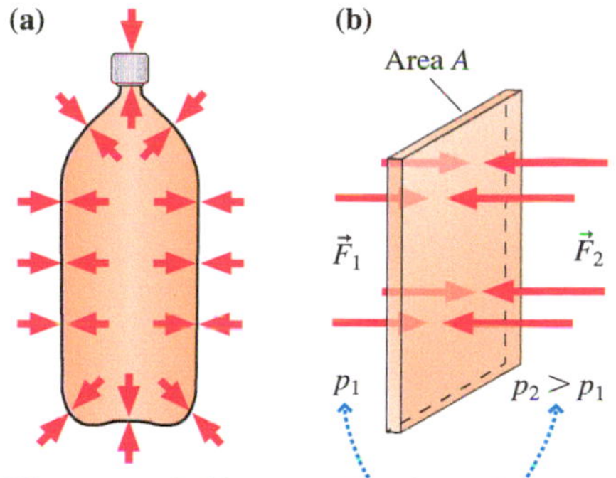

**Too little pressure too fast** BIO The rockfish is a popular game fish that can be caught at ocean depths up to 1000 ft. The great pressure at these depths is balanced by an equally large pressure inside the fish's gas-filled *swim bladder*. If the fish is hooked and rapidly raised to the lower pressure of the surface, the pressure inside the swim bladder is suddenly much greater than the pressure outside, causing the swim bladder to expand dramatically. When reeled in from a great depth, these fish seldom survive.

FIGURE 12.7 A tire gauge measures the difference between the tire's pressure and atmospheric pressure.

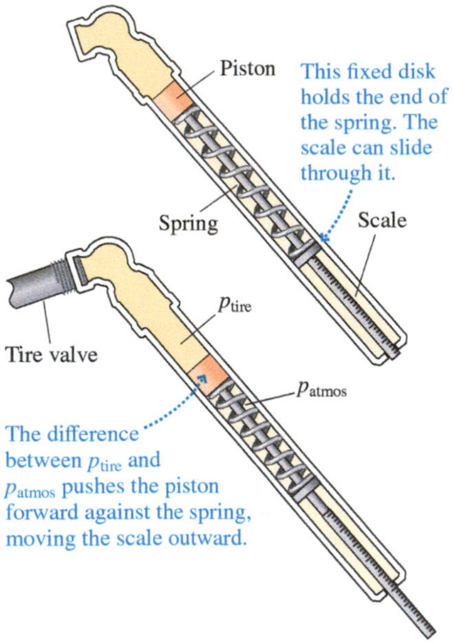
Piston
This fixed disk holds the end of the spring. The scale can slide through it.
Spring
Scale
$p_{tire}$
Tire valve
$p_{atmos}$
The difference between $p_{tire}$ and $p_{atmos}$ pushes the piston forward against the spring, moving the scale outward.

tire gauge, the difference between an absolute pressure and atmospheric pressure, so we can write

$$p_g = p - p_{atmos} \qquad p = p_g + p_{atmos} \tag{12.14}$$

If you are at sea level, you can assume that $p_{atmos} = 1.0$ atm; at other elevations, you'll need to use the local pressure.

**EXAMPLE 12.4 Finding the force due to a pressure difference**

Patients suffering from decompression sickness may be treated in a hyperbaric oxygen chamber filled with oxygen at greater than atmospheric pressure. A cylindrical chamber with flat end plates of diameter 0.75 m is filled with oxygen to a gauge pressure of 27 kPa. What is the resulting force on the end plate of the cylinder?

**STRATEGIZE** The net force $F_{net} = A\Delta p$ on the end plate depends on $\Delta p$, the difference between the pressure inside and the pressure outside the chamber. This is the gauge pressure, 27 kPa.

**PREPARE** The end plates are circular, with area $A = \pi r^2 = \pi(0.375 \text{ m})^2 = 0.442 \text{ m}^2$.

**SOLVE** The pressure difference results in a net force

$$F_{net} = A\,\Delta p = (0.442 \text{ m}^2)(27{,}000 \text{ Pa}) = 12 \text{ kN}$$

**ASSESS** The area of the end plate is large, so we expect a large force, which is exactly what we get—the force is equivalent to the weight of about 15 people. If you've seen chambers like this, you know that the end plate is fastened in place with stout bolts. It is remarkable to think that this large force results from the collisions of tiny air molecules with the plate!

## The Ideal-Gas Law

We can use the fact that the pressure in a gas is due to the collisions of particles with the walls to make some qualitative predictions. FIGURE 12.8 presents a few such predictions.

FIGURE 12.8 Relating gas pressure to other variables.

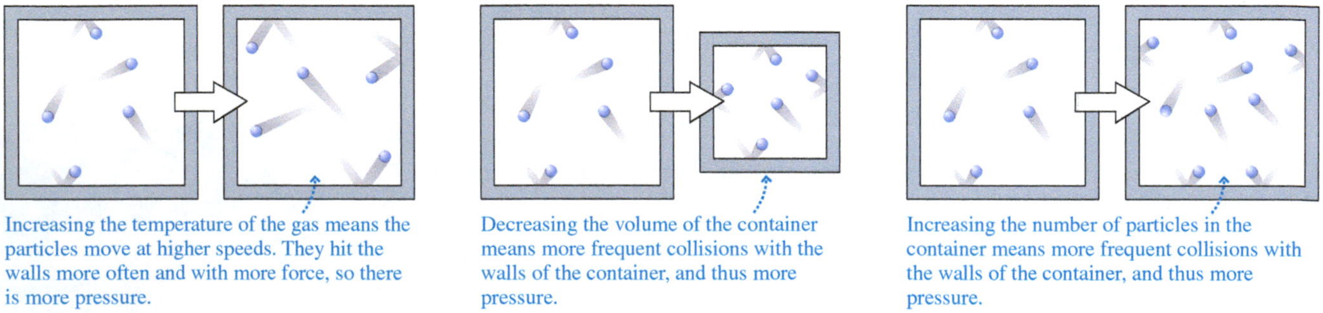
Increasing the temperature of the gas means the particles move at higher speeds. They hit the walls more often and with more force, so there is more pressure.
Decreasing the volume of the container means more frequent collisions with the walls of the container, and thus more pressure.
Increasing the number of particles in the container means more frequent collisions with the walls of the container, and thus more pressure.

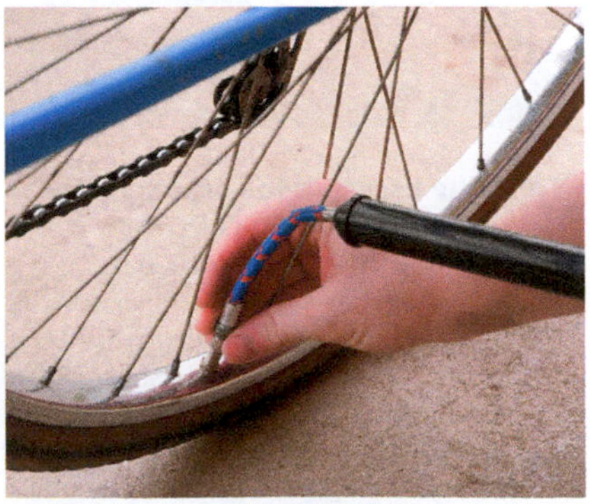

Pumping up a bicycle tire adds more particles to the fixed volume of the tire, increasing the pressure.

Based on the reasoning in Figure 12.8, we expect the following proportionalities:

- Pressure should be proportional to the temperature of the gas: $p \propto T$.
- Pressure should be inversely proportional to the volume of the container: $p \propto 1/V$.
- Pressure should be proportional to the number of gas particles: $p \propto N$.

In fact, careful experiments back up each of these predictions, leading to a single equation that expresses these proportionalities:

$$p = C\frac{NT}{V}$$

The proportionality constant $C$ turns out to be none other than Boltzmann's constant $k_B$, which allows us to write

$$pV = Nk_{\mathrm{B}}T \tag{12.15}$$

Ideal-gas law, version 1

Equation 12.15 is known as the **ideal-gas law.**

Equation 12.15 is written in terms of the number $N$ of particles in the gas, whereas the ideal-gas law is stated in chemistry in terms of the number $n$ of moles. But the change is straightforward to make. The number of particles is $N = nN_A$, so we can rewrite Equation 12.15 as

$$pV = nN_{\mathrm{A}}k_{\mathrm{B}}T = nRT \tag{12.16}$$

Ideal-gas law, version 2

In this version of the equation, the proportionality constant—known as the *gas constant*—is

$$R = N_{\mathrm{A}}k_{\mathrm{B}} = 8.31 \text{ J/mol} \cdot \text{K}$$

The units may seem unusual, but the product of Pa and $m^3$, the units of $pV$, is equivalent to J.

Let's review the meanings and the units of the various quantities in the ideal-gas law:

$$pV = nRT$$

- $p$: Absolute pressure (Pa)
- $V$: Volume of the sample or container of gas ($m^3$)
- $n$: Number of moles in the sample or container of gas
- $R$: Gas constant, 8.31 J/mol · K
- $T$: Temperature in kelvin (K)

**EXAMPLE 12.5 Finding the volume of a mole of a gas**

What volume is occupied by 1 mole of an ideal gas at a pressure of 1.00 atm and a temperature of 0°C?

***STRATEGIZE*** We are given the pressure, the temperature, and the number of moles. We will use version 2 of the ideal-gas law to find the volume.

***PREPARE*** The first step in ideal-gas law calculations is to convert all quantities to SI units:

$$p = 1.00 \text{ atm} = 101.3 \times 10^3 \text{ Pa}$$
$$T = 0 + 273 = 273 \text{ K}$$

***SOLVE*** We rearrange version 2 of the ideal-gas law equation to compute

$$V = \frac{nRT}{p} = \frac{(1.00 \text{ mol})(8.31 \text{ J/mol} \cdot \text{K})(273 \text{ K})}{101.3 \times 10^3 \text{ Pa}} = 0.0224 \text{ m}^3$$

We recall from earlier in the chapter that $1.00 \text{ m}^3 = 1000 \text{ L}$, so we can write

$$V = 22.4 \text{ L}$$

***ASSESS*** At this temperature and pressure, we find that the volume of 1 mole of a gas is 22.4 L, a result you might recall from chemistry. When we do calculations using gases, it will be useful to keep this volume in mind to see if our answers make physical sense.

**STOP TO THINK 12.2** A sample of ideal gas is in a sealed container. The temperature of the gas and the volume of the container are both increased. What other properties of the gas necessarily change? (More than one answer may be correct.)

A. The rms speed of the gas atoms
B. The thermal energy of the gas
C. The pressure of the gas
D. The number of molecules of gas

## 12.3 Ideal-Gas Processes

Suppose you measure the pressure in the tires on your car on a cold morning. How much will the tire pressure increase after the tires warm up in the sun and the temperature of the air in the tires has increased? We will solve this problem later in the chapter, but, for now, note these properties of this process:

- The quantity of gas is fixed. No air is added to or removed from the tire.
- There is a well-defined initial state. The initial values of pressure, volume, and temperature will be designated $p_i$, $V_i$, and $T_i$.
- There is a well-defined final state in which the pressure, volume, and temperature have values $p_f$, $V_f$, and $T_f$.

Processes with these properties are called **ideal-gas processes.**

For gases in sealed containers, the number of moles (and the number of molecules) does not change. In that case, the ideal-gas law can be written as

$$\frac{pV}{T} = nR = \text{constant}$$

The values of the variables in the initial and final states are then related by

$$\frac{p_f V_f}{T_f} = \frac{p_i V_i}{T_i} \tag{12.17}$$

Initial and final states for an ideal gas in a sealed container

This before-and-after relationship between the two states, reminiscent of a conservation law, will be valuable for solving many problems.

NOTE ▶ Because pressure and volume appear on both sides of the equation, pressure and volume can be in any units; we don't necessarily need to convert to SI units. However, temperature *must* be in K. Unit-conversion factors for pressure and volume are multiplicative factors, and the same factor on both sides of the equation cancels. But the conversion from K to °C is an *additive* factor, and additive factors in the denominator don't cancel. ◀

FIGURE 12.9 The state of the gas and ideal-gas processes can be shown on a $pV$ diagram.

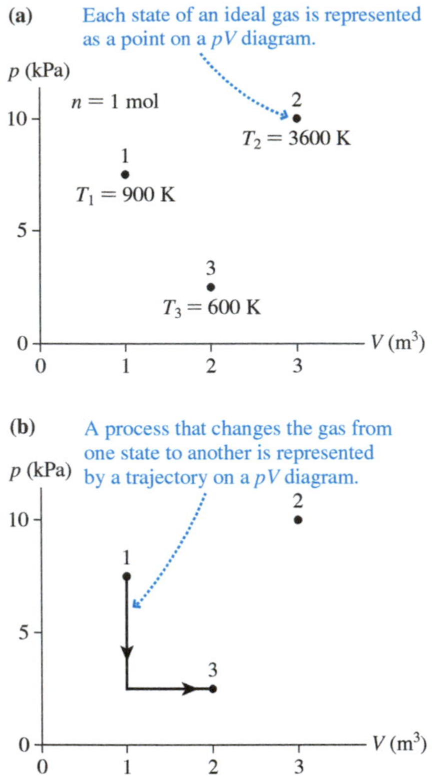


### *pV* Diagrams

It's useful to represent ideal-gas processes on a graph called a ***pV* diagram.** The important idea behind a $pV$ diagram is that each point on the graph represents a single, unique state of the gas. This may seem surprising because a point on a graph specifies only the values of pressure and volume. But knowing $p$ and $V$, and assuming that $n$ is known for a sealed container, we can find the temperature from the ideal-gas law. Thus each point on a $pV$ diagram actually represents a triplet of values $(p, V, T)$ specifying the state of the gas.

For example, FIGURE 12.9a is a $pV$ diagram showing three states of a system consisting of 1 mol of gas. The values of $p$ and $V$ can be read from the axes, then the temperature at that point calculated from the ideal-gas law. An ideal-gas process—a process that changes the state of the gas by, for example, heating it or compressing it—can be represented as a "trajectory" in the $pV$ diagram. FIGURE 12.9b shows one possible process by which the gas of Figure 12.9a is changed from state 1 to state 3.

FIGURE 12.10 A constant-volume process.

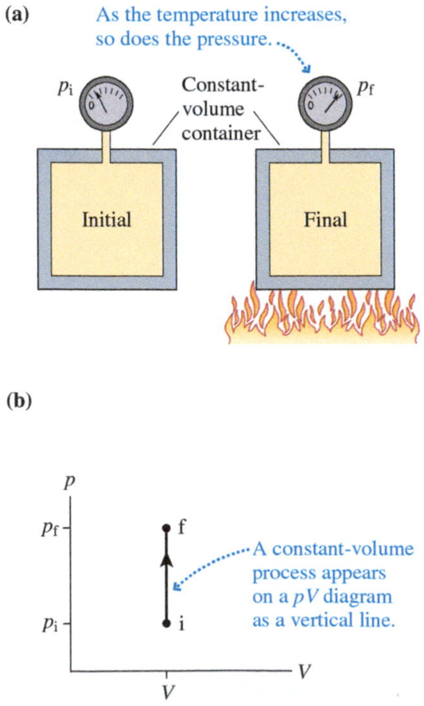


### Constant-Volume Processes

Suppose you have a gas in the closed, rigid container shown in FIGURE 12.10a. Warming the gas will raise its pressure without changing its volume. This is an example of a **constant-volume process.** $V_f = V_i$ for a constant-volume process.

Because the value of $V$ doesn't change, this process is shown as the vertical line $i \rightarrow f$ on the $pV$ diagram of FIGURE 12.10b. **A constant-volume process appears on a pV diagram as a vertical line.**

**EXAMPLE 12.6 Computing tire pressure on a hot day**

The pressure in a car tire measures 30.0 psi on a cool morning when the air temperature is 0°C. After the day warms up and bright sun shines on the black tire, the temperature of the air inside the tire reaches 30°C. What is the tire pressure at this temperature?

**STRATEGIZE** A tire is (to a good approximation) a sealed container with constant volume, so this is a constant-volume process.

**PREPARE** We weren't told anything about elevation, so we assume that the atmospheric pressure is 1.0 atm. When we solve this problem, we don't need to convert units, but we do need to be sure that we are using the correct form of the pressure. The measured tire pressure is a gauge pressure, but the ideal-gas law requires an absolute pressure. We can use Equation 12.14 to convert to absolute pressure. The initial pressure is

$$p_i = (p_g)_i + 1.00\text{ atm} = 30.0\text{ psi} + 14.7\text{ psi} = 44.7\text{ psi}$$

Temperatures must be in kelvin, so we convert:

$$T_i = 0°\text{C} + 273 = 273\text{ K}$$
$$T_f = 30°\text{C} + 273 = 303\text{ K}$$

**SOLVE** We can use Equation 12.17 to solve for the final pressure. In this equation, we divide both sides by $V_f$, and then cancel the ratio of the two volumes, which is equal to 1 for this constant-volume process:

$$p_f = p_i \frac{V_i}{V_f}\frac{T_f}{T_i} = p_i \frac{T_f}{T_i}$$

The units for $p_f$ will be the same as those for $p_i$, so we can keep the initial pressure in psi. The pressure at the higher temperature is

$$p_f = 44.7\text{ psi} \times \frac{303\text{ K}}{273\text{ K}} = 49.6\text{ psi}$$

This is an absolute pressure, but the problem asks for the measured pressure in the tire—a gauge pressure. Converting to gauge pressure gives

$$(p_g)_f = p_f - 1.00\text{ atm} = 49.6\text{ psi} - 14.7\text{ psi} = 34.9\text{ psi}$$

**ASSESS** The temperature has changed by 30 K, which is a bit more than 10% of the initial temperature, so we expect a large change in pressure. Our result seems reasonable, and it has practical implications: If you check the pressure in your tires when they are at a particular temperature, don't expect the pressure to be the same when conditions change!

## Constant-Pressure Processes

Many gas processes take place at a constant, unchanging pressure. A constant-pressure process is also called an **isobaric process.** For a constant-pressure process $p_f = p_i$.

One way to produce a constant-pressure process is shown in FIGURE 12.11a, where a gas is sealed in a cylinder by a lightweight, tight-fitting cap—a *piston*—that is free to slide up and down. In fact, the piston *will* slide up or down, compressing or expanding the gas inside, until it reaches the position at which $p_{gas} = p_{ext}$. That's the equilibrium position for the piston, the position at which the upward force $F_{gas} = p_{gas}A$, where $A$ is the area of the face of the piston, exactly balances the downward force $F_{ext} = p_{ext}A$ due to the external pressure $p_{ext}$. The gas pressure in this situation is thus equal to the external pressure. As long as the external pressure doesn't change, neither can the gas pressure inside the cylinder.

FIGURE 12.11 A constant-pressure process.

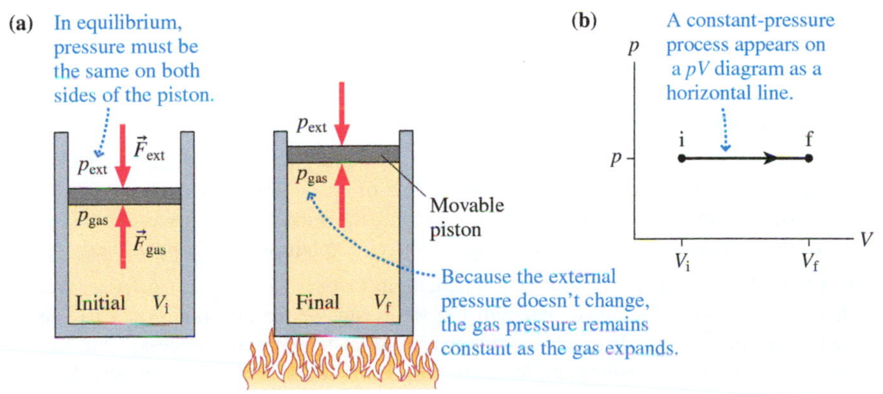


Suppose we heat the gas in the cylinder. The gas pressure doesn't change because the pressure is controlled by the unchanging external pressure, not by the temperature. But as the temperature rises, the faster-moving atoms cause the gas to expand, pushing the piston outward. Because the pressure is always the same, this is a constant-pressure process, with a trajectory as shown in FIGURE 12.11b. **A constant-pressure process appears on a *pV* diagram as a horizontal line.**

**EXAMPLE 12.7** **Finding the final volume of a compressed gas**

A gas in a cylinder with a movable piston occupies 50.0 $\text{cm}^3$ at 50°C. The gas is cooled at constant pressure until the temperature is 10°C. What is the final volume?

**STRATEGIZE** The pressure of the gas doesn't change, so this is a constant-pressure process, with $p_i/p_f = 1$.

**PREPARE** The temperatures must be in kelvin, so we convert:

$$T_i = 50°\text{C} + 273 = 323\text{ K}$$

$$T_f = 10°\text{C} + 273 = 283\text{ K}$$

**SOLVE** We can use Equation 12.17 to solve for $V_f$:

$$V_f = V_i \frac{p_i}{p_f}\frac{T_f}{T_i} = 50.0\text{ cm}^3 \times 1 \times \frac{283\text{ K}}{323\text{ K}} = 43.8\text{ cm}^3$$

**ASSESS** Our answer makes sense. The gas cools, so we expect a smaller volume. The cooling is—in absolute temperature terms—relatively modest, so we expect a modest change in volume.

In this example and in Example 12.6, we have not converted pressure and volume units because these multiplicative factors cancel. But we did convert temperature to kelvin because this *additive* factor does *not* cancel.

## Constant-Temperature Processes

FIGURE 12.12 A constant-temperature process.

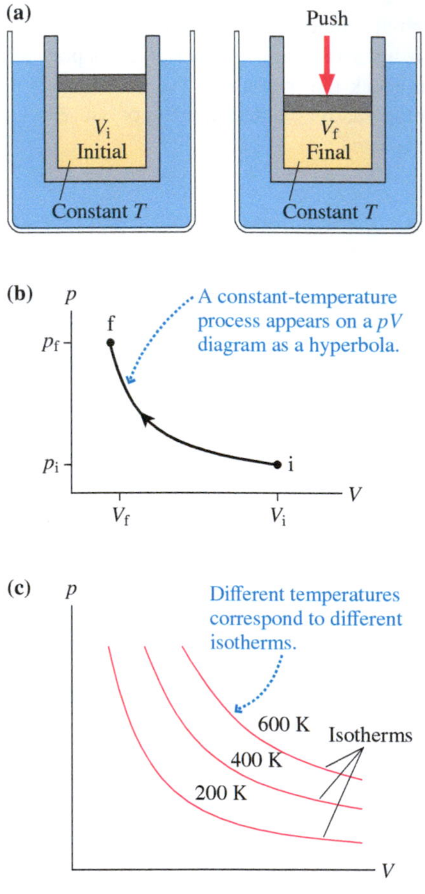


A constant-temperature process is also called an **isothermal process.** For a constant-temperature process $T_f = T_i$. One possible constant-temperature process is illustrated in FIGURE 12.12a. A piston is being pushed down to compress a gas, but the gas cylinder is submerged in a large container of liquid that is held at a constant temperature. If the piston is pushed *slowly,* then heat-energy transfer through the walls of the cylinder will keep the gas at the same temperature as the surrounding liquid. This is an *isothermal compression.* The reverse process, with the piston slowly pulled out, is an *isothermal expansion.*

Representing a constant-temperature process on the $pV$ diagram is a little more complicated than the two preceding processes because both $p$ and $V$ change. As long as $T$ remains fixed, we have the relationship

$$p = \frac{nRT}{V} = \frac{\text{constant}}{V} \tag{12.18}$$

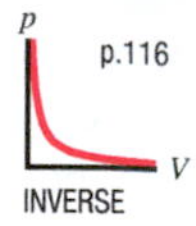


**Because there is an inverse relationship between $p$ and $V$, the graph of a constant-temperature process is a *hyperbola.***

Figure 12.12a shows an isothermal compression, which is represented graphically as the hyperbola i → f in FIGURE 12.12b. An isothermal expansion would move in the opposite direction along the hyperbola. The graph of an isothermal process is known as an **isotherm.**

The location of the hyperbola depends on the value of $T$. If we use a higher constant temperature for the process in Figure 12.12a, the isotherm will move farther from the origin of the $pV$ diagram. FIGURE 12.12c shows three isotherms for this process at three different temperatures. A gas undergoing a constant-temperature process will move along the isotherm for the appropriate temperature.

**EXAMPLE 12.8** **Why does the bag expand?**

Residents of Colorado, where it's easy to drive rapidly to different elevations, are familiar with this scenario: You buy a sealed bag of chips at a low elevation and then drive your car up into the mountains. As you climb, the atmospheric pressure drops, the air inside the bag expands, and the bag gets "puffy." Explain what is happening.

**REASON** The chip bag is sealed—no gas gets in or out—so we can treat this as an ideal-gas process. Folks in the car will use the heater or the air conditioner to keep the temperature in the car steady, so this is a constant-temperature process. Chip bags aren't very sturdy, so the pressure inside the bag is the same as the pressure outside the bag. As the pressure outside the bag drops, so does the pressure inside the bag. The volume of air in the bag is inversely proportional to the pressure, so as the pressure drops, the volume of the air increases and the bag expands.

**ASSESS** The result makes sense. A change in pressure also occurs on an airplane, so you may have seen something similar during a flight, or you may have had bottles with caps that pop off and contents pushed out as the gas inside expands.

**EXAMPLE 12.9 Finding the volume of air in the lungs** BIO

A snorkeler takes a deep breath at the surface, filling his lungs with 4.0 L of air. He then descends to a depth of 5.0 m, where the pressure is 0.50 atm higher than at the surface. At this depth, what is the volume of air in the snorkeler's lungs?

***STRATEGIZE*** Most of the snorkeler's air is nitrogen, which his body doesn't use. Oxygen is used by the body, but for every oxygen molecule taken from the lungs, a molecule of carbon dioxide is returned. The ideal-gas law doesn't care what the molecules are, just the total number. The total number of gas molecules stays approximately constant, and we can treat this as an ideal-gas process. The air stays at body temperature, so this is a constant-temperature process with $T_f = T_i$, so $T_f/T_i = 1$. As the snorkeler descends, the pressure inside his lungs rises to match the pressure of the surrounding water because the body can't sustain large pressure differences between inside and out.

***PREPARE*** We need absolute pressures for our calculation. At the surface, the pressure of the air inside the snorkeler's lungs is 1.0 atm—it's the pressure at sea level. The additional pressure at the noted depth is 0.50 atm, so the absolute pressure at this depth is 1.5 atm.

***SOLVE*** The ideal-gas law for a sealed container (the lungs) gives

$$V_f = V_i \frac{p_i}{p_f}\frac{T_f}{T_i} = 4.0\text{ L} \times \frac{1.0\text{ atm}}{1.5\text{ atm}} \times 1 = 2.7\text{ L}$$

Notice that we didn't need to convert pressure to SI units. As long as the units are the same in the numerator and the denominator, they cancel.

***ASSESS*** The air has a smaller volume at the higher pressure, as we would expect. The air inside your lungs does compress—significantly!—when you dive below the surface.

## Thermodynamics of Ideal-Gas Processes

Chapter 11 introduced the first law of thermodynamics, and we saw that heat and work are just two different ways to add energy to a system. We've been considering the changes when we heat gases, but now we want to consider the other form of energy transfer—work.

When gases expand, they can do work by pushing against a piston. This is how the engine under the hood of your car works: When the spark plug fires in a cylinder in the engine, it ignites the gaseous fuel-air mixture inside. The hot gas expands, pushing the piston out and, through various mechanical linkages, turning the wheels of your car. Energy is transferred out of the gas as work; we say that the gas does work on the piston. Similarly, the gas in Figure 12.11a does work as it pushes on and moves the piston.

You learned in ◀◀ SECTION 10.2 that the work done by a constant force $F$ in pushing an object a distance $d$ is $W = Fd$. Let's apply this idea to a gas. FIGURE 12.13a shows a gas cylinder sealed at one end by a movable piston. Force $\vec{F}_{gas}$ is due to the gas pressure and has magnitude $F_{gas} = pA$. Force $\vec{F}_{ext}$, perhaps a force applied by a piston rod, is equal in magnitude and opposite in direction to $\vec{F}_{gas}$. The gas pressure would blow the piston out if the external force weren't there!

Suppose the gas expands at constant pressure, pushing the piston outward from $x_i$ to $x_f$, a distance $d = x_f - x_i$, as shown in FIGURE 12.13b. As it does, the force due to the gas pressure does work

$$W_{gas} = F_{gas}d = (pA)(x_f - x_i) = p(x_fA - x_iA)$$

But $x_iA$ is the cylinder's initial volume $V_i$ (recall that the volume of a cylinder is the length times the area of the base) and $x_fA$ is the final volume $V_f$. Thus the work done is

$$W_{gas} = p(V_f - V_i) = p\Delta V \tag{12.19}$$

Work done by a gas in a constant-pressure process

where $\Delta V$ is the *change* in volume.

Equation 12.19 has a particularly simple interpretation on a $pV$ diagram. As FIGURE 12.14a on the next page shows, $p\Delta V$ is the "area under the $pV$ graph" between $V_i$ and $V_f$. Although we've shown this result for only a constant-pressure process, it turns out to be true for all ideal-gas processes. That is, as FIGURE 12.14b shows,

$$W_{gas} = \text{area under the } pV \text{ graph between } V_i \text{ and } V_f$$

FIGURE 12.13 The expanding gas does work on the piston.

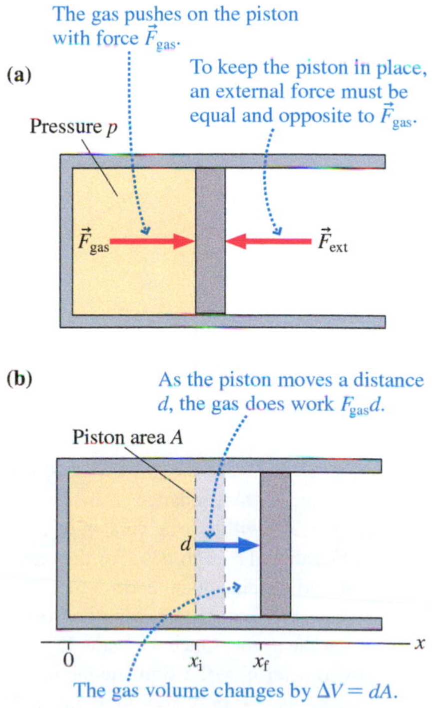

FIGURE 12.14 Calculating the work done in an ideal-gas process.

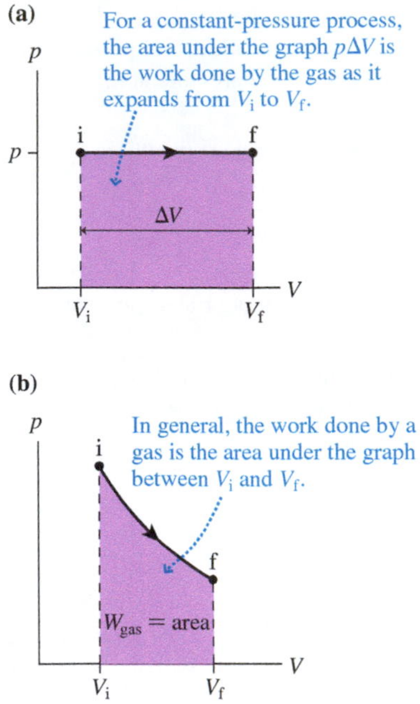


There are a few things to clarify:

- In order for the gas to do work, its volume must change. No work is done in a constant-volume process.
- The simple relationship of Equation 12.19 applies only to constant-pressure processes. For any other ideal-gas process, you must use the geometry of the $pV$ diagram to calculate the area under the graph.
- To calculate work, pressure must be in Pa and volume in $m^3$. The product of Pa (which is $N/m^2$) and $m^3$ is $N \cdot m$. But $1\ N \cdot m$ is 1 J—the unit of work and energy.
- $W_{gas}$ is positive if the gas expands ($\Delta V > 0$). The gas does work by pushing against the piston. In this case, the work done is energy transferred out of the system, and the energy of the gas decreases. $W_{gas}$ is negative if the piston compresses the gas ($\Delta V < 0$) because the force $\vec{F}_{gas}$ is opposite the displacement of the piston. Energy is transferred into the system as work, and the energy of the gas increases. We often say "work is done *on* the gas," but this just means that $W_{gas}$ is negative.

In the first law of thermodynamics, $\Delta E_{th} = Q + W$, $W$ is the work done by the environment—that is, by force $\vec{F}_{ext}$ acting on the system. But $\vec{F}_{ext}$ and $\vec{F}_{gas}$ are equal and opposite forces, as we noted previously, so the work done by the environment is the negative of the work done by the gas: $W = -W_{gas}$. Consequently, the first law of thermodynamics can be written as

$$\Delta E_{th} = Q - W_{gas} \tag{12.20}$$

The thermal energy of an ideal gas depends only on its temperature as $E_{th} = \frac{3}{2}Nk_B T$. Comparing Equations 12.15 and 12.16 shows that $Nk_B$ is equal to $nR$, so we can write the change in thermal energy of an ideal gas as

$$\Delta E_{th} = \frac{3}{2}Nk_B\,\Delta T = \frac{3}{2}nR\,\Delta T \tag{12.21}$$

**EXAMPLE 12.10 Finding the heat for an ideal-gas process**

A cylinder with a movable piston contains 0.016 mol of helium. A researcher expands the gas via the process illustrated in FIGURE 12.15. To achieve this, does she need to heat the gas? If so, how much heat energy must be added or removed?

FIGURE 12.15 $pV$ diagram for Example 12.10.

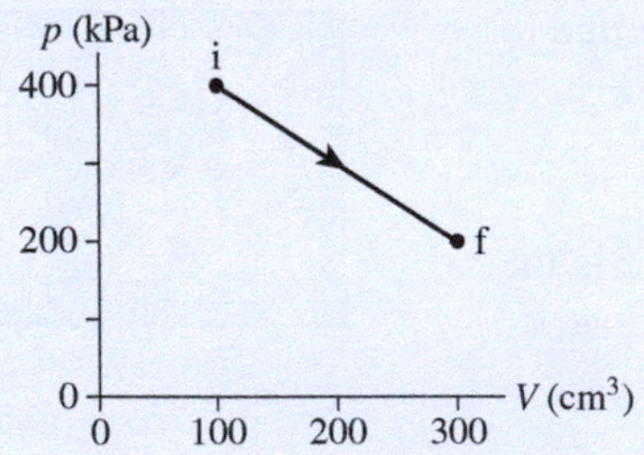


**STRATEGIZE** As the gas expands, it does work on the piston. Its temperature may change as well, implying a change in thermal energy. We will use the version of the first law of thermodynamics in Equation 12.20 to describe the energy changes in the gas. We'll first find the change in thermal energy by computing the temperature change, then calculate how much work is done by looking at the area under the graph. Once we know $W_{gas}$ and $\Delta E_{th}$, we will use the first law to determine the sign and the magnitude of the heat—telling us whether heat energy goes in or out, and how much.

**PREPARE** The graph tells us the pressure and the volume, so we can use the ideal-gas law to compute the temperature at the initial and final points. To do this we need the volumes in SI units. Reading the initial and final volumes on the graph and converting, we find

$$V_i = 100\ cm^3 \times \frac{1\ m^3}{10^6\ cm^3} = 1.0 \times 10^{-4}\ m^3$$

$$V_f = 300\ cm^3 \times \frac{1\ m^3}{10^6\ cm^3} = 3.0 \times 10^{-4}\ m^3$$

**SOLVE** The initial and final temperatures are found using the ideal-gas law:

$$T_i = \frac{p_i V_i}{nR} = \frac{(4.0 \times 10^5\ Pa)(1.0 \times 10^{-4}\ m^3)}{(0.016\ mol)(8.31\ J/mol \cdot K)} = 300\ K$$

$$T_f = \frac{p_f V_f}{nR} = \frac{(2.0 \times 10^5\ Pa)(3.0 \times 10^{-4}\ m^3)}{(0.016\ mol)(8.31\ J/mol \cdot K)} = 450\ K$$

The temperature increases, and so must the thermal energy. We can use Equation 12.21 to compute this change:

$$\Delta E_{th} = \frac{3}{2}(0.016\ mol)(8.31\ J/mol \cdot K)(450\ K - 300\ K) = 30\ J$$

The other piece of the puzzle is to compute the work done. We do this by finding the area under the graph for the process. FIGURE 12.16 shows that we can do this calculation by viewing the area as a triangle on top of a rectangle. Notice that the areas are in joules because they are the product of Pa and $m^3$.

The total work is

$$W_{gas} = \text{area of triangle} + \text{area of rectangle}$$
$$= 20\text{ J} + 40\text{ J} = 60\text{ J}$$

Now we can use the first law as written in Equation 12.20 to find the heat:

$$Q = \Delta E_{th} + W_{gas} = 30\text{ J} + 60\text{ J} = 90\text{ J}$$

This is a positive number, so—using the conventions introduced in Chapter 11—we see that 90 J of heat energy must be added to the gas.

FIGURE 12.16 The work done by the expanding gas is the total area under the graph.

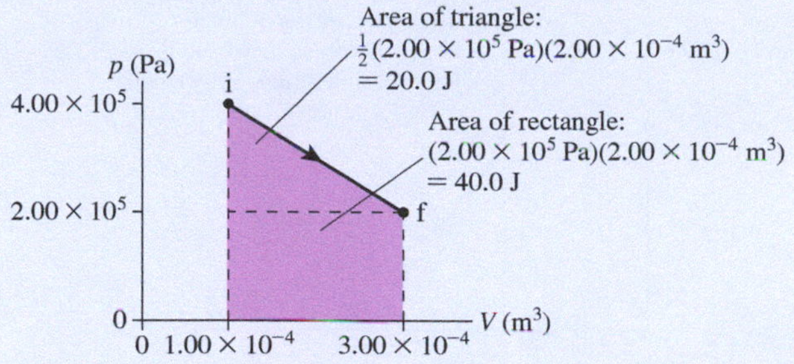


***ASSESS*** The gas does work—a loss of energy—but its temperature increases, so it makes sense that heat energy must be added.

## Adiabatic Processes

You may have noticed that when you pump up a bicycle tire with a hand pump, the pump gets warm. We noted the reason for this in Chapter 11: When you press down on the handle of the pump, a piston in the pump's chamber compresses the gas, doing work on it. According to the first law of thermodynamics, doing work on the gas increases its thermal energy. So the gas temperature goes up, and heat is then transferred through the walls of the pump to your hand.

Now suppose you compress a gas in an insulated container, so that no heat is exchanged with the environment, or you compress a gas so quickly that there is no time for heat to be transferred. In either case, $Q = 0$. If a gas process has $Q = 0$, for either a compression or an expansion, we call this an **adiabatic process.**

An expanding gas does work, so $W_{gas} > 0$. If the expansion is adiabatic, meaning $Q = 0$, then the first law of thermodynamics as written in Equation 12.20 tells us that $\Delta E_{th} < 0$. Temperature is proportional to thermal energy, so the temperature will decrease as well. **An adiabatic expansion lowers the temperature of a gas.** If the gas is compressed, work is done on the gas ($W_{gas} < 0$). If the compression is adiabatic, the first law of thermodynamics implies that $\Delta E_{th} > 0$ and thus that the temperature increases. **An adiabatic compression raises the temperature of a gas.** Adiabatic processes allow you to use work, rather than heat, to change the temperature of a gas.

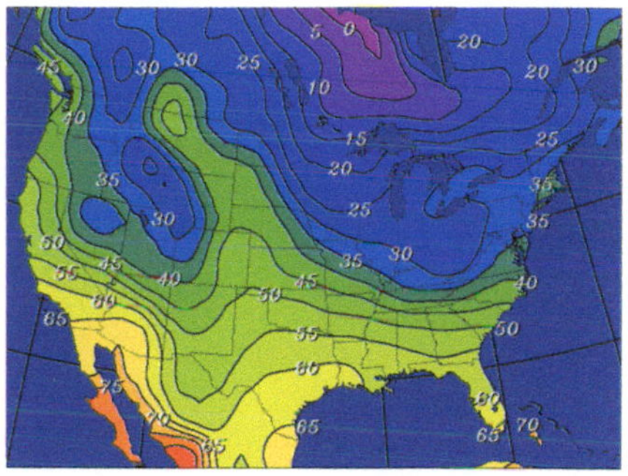

**Warm mountain winds** This image shows surface temperatures (in °F) in North America on a winter day. Notice the bright green area of unseasonably warm temperatures extending north and west from the center of the continent. On this day, a strong westerly wind, known as a Chinook wind, was blowing down off the Rocky Mountains, rapidly moving from high elevations (and low pressures) to low elevations (and higher pressures). The air was rapidly compressed as it descended. The compression was so rapid that no heat was exchanged with the environment, so this was an adiabatic process that significantly increased the air temperature.

**CONCEPTUAL EXAMPLE 12.11 What is the shape of the curve?**

FIGURE 12.17 shows the $pV$ diagram of a gas undergoing an isothermal compression from point 1 to point 2. Sketch how the $pV$ diagram would look if the gas were compressed from point 1 to the same final pressure by a rapid adiabatic compression.

***REASON*** An adiabatic compression increases the temperature of the gas as the work done on the gas is transformed into thermal energy. Consequently, as seen in FIGURE 12.18 on the next page, the curve of the adiabatic compression cuts across the isotherms to end on a higher-temperature isotherm when the gas pressure reaches $p_2$.

FIGURE 12.17 $pV$ diagram for an isothermal compression.

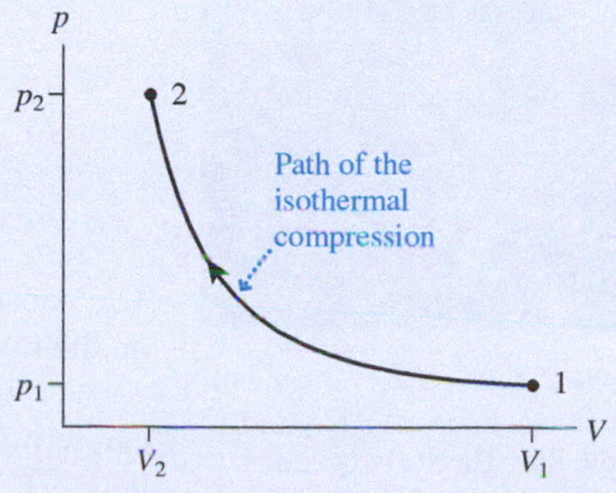


Continued

FIGURE 12.18 $pV$ diagram for an adiabatic compression.

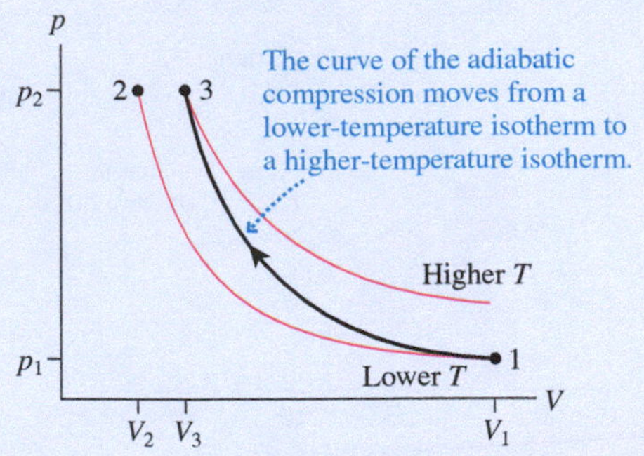


**ASSESS** In an isothermal compression, heat energy is transferred out of the gas so that the gas temperature stays the same. This heat transfer doesn't happen in an adiabatic compression, so we'd expect the gas to have a higher final temperature. In general, the temperature at the final point of an adiabatic compression is higher than at the starting point. Similarly, an adiabatic expansion ends on a lower-temperature isotherm.

**STOP TO THINK 12.3** What is the ratio $T_f/T_i$ for this process?

A. 1/4
B. 1/2
C. 1 (no change)
D. 2
E. 4
F. There is not enough information to decide.

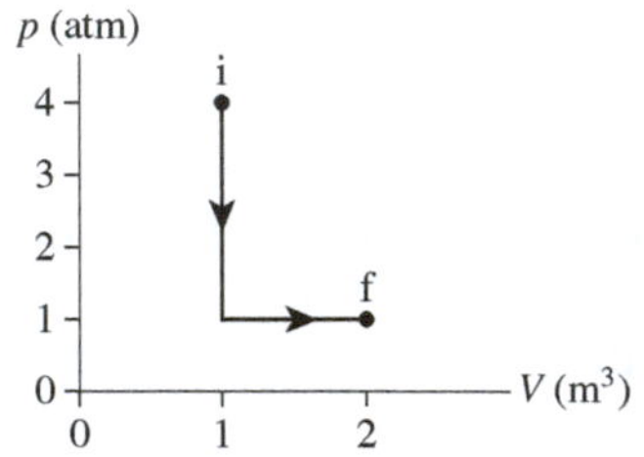


## 12.4 Thermal Expansion

The bonds between atoms in solids and liquids mean that solids and liquids are much less compressible than gases, as we've noted. But raising the temperature of a solid or a liquid does produce a small, measurable change in volume. This **thermal expansion** underlies many practical phenomena.

FIGURE 12.19 Increasing the temperature of a solid causes it to expand.

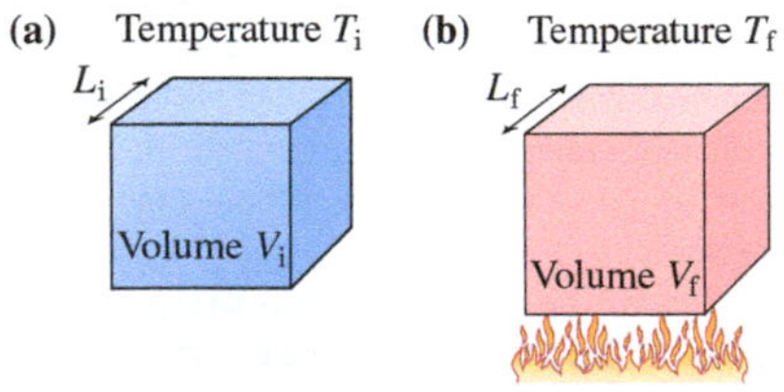


FIGURE 12.19a shows a cube of material, initially at temperature $T_i$. The edge of the cube has length $L_i$, and the cube's volume is $V_i$. We then heat the cube, increasing its temperature to $T_f$, as shown in FIGURE 12.19b. The cube's edge length increases to $L_f$, and its volume increases to $V_f$.

For most substances, the change in volume $\Delta V = V_f - V_i$ is linearly related to the change in temperature $\Delta T = T_f - T_i$ by

$$\Delta V = \beta V_i \Delta T \tag{12.22}$$

Volume thermal expansion

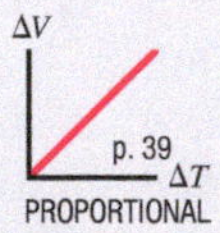


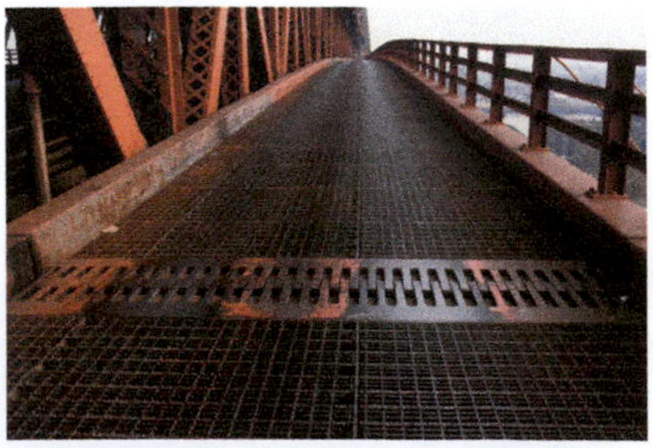

**Expanding spans** A long steel bridge will slightly increase in length on a hot day and decrease on a cold day. Thermal expansion joints let the bridge's length change without causing the roadway to buckle.

The constant $\beta$ is known as the **coefficient of volume expansion.** Its value depends on the material the object is made of. $\Delta T$ is measured in K, so the units of $\beta$ are $K^{-1}$.

As a solid's volume increases, each of its linear dimensions increases as well. We can write a similar expression for this linear thermal expansion. If an object of initial length $L_i$ undergoes a temperature change $\Delta T$, its length changes to $L_f$. The change in length, $\Delta L = L_f - L_i$, is given by

$$\Delta L = \alpha L_i \Delta T \quad (12.23)$$

Linear thermal expansion

The constant $\alpha$ is the **coefficient of linear expansion.** Note that Equations 12.22 and 12.23 apply equally well to thermal *contractions,* in which case both $\Delta T$ and $\Delta V$ (or $\Delta L$) are negative.

Values of $\alpha$ and $\beta$ for common materials are listed in TABLE 12.3. The volume expansion of a liquid can be measured, but, because a liquid can change shape, we don't assign a coefficient of linear expansion to a liquid. The values of $\alpha$ and $\beta$ are very small, so the changes in length and volume, $\Delta L$ and $\Delta V$, are always a small fraction of their original values.

TABLE 12.3 Coefficients of linear and volume thermal expansion at 20°C

| Substance | Linear $\alpha$ ($K^{-1}$) | Volume $\beta$ ($K^{-1}$) |
|---|---|---|
| Glass | $9 \times 10^{-6}$ | $27 \times 10^{-6}$ |
| Iron or steel | $12 \times 10^{-6}$ | $36 \times 10^{-6}$ |
| Concrete | $12 \times 10^{-6}$ | $36 \times 10^{-6}$ |
| Aluminum | $23 \times 10^{-6}$ | $69 \times 10^{-6}$ |
| Water | | $210 \times 10^{-6}$ |
| Ethyl alcohol | | $1100 \times 10^{-6}$ |
| Air (and other gases) | | $3400 \times 10^{-6}$ |

NOTE ► The expressions for thermal expansion are only approximate expressions that apply over a limited range of temperatures. They are what we call **empirical formulas;** they are a good fit to measured data, but they do not represent any underlying fundamental law. ◄

## EXAMPLE 12.12 How much closer to space?

The height of the Space Needle, a steel observation tower in Seattle, is 180 meters on a 0°C winter day. How much taller is it on a hot summer day when the temperature is 30°C?

***STRATEGIZE*** The steel expands because of an increase in temperature. We are interested in the change in height, so we'll consider the linear expansion.

***PREPARE*** The expansion depends on the difference in temperature, which is

$$\Delta T = T_f - T_i = 30°C - 0°C = 30°C = 30\text{ K}$$

***SOLVE*** The coefficient of linear expansion is given in Table 12.3; we can use this value in Equation 12.23 to compute the increase in height:

$$\Delta L = \alpha L_i \Delta T = (12 \times 10^{-6}\text{ K}^{-1})(180\text{ m})(30\text{ K}) = 0.065\text{ m}$$

***ASSESS*** You don't notice buildings getting taller on hot days, so we expect the final answer to be small. The change is a small fraction of the height of the tower, as we expect. Compared to 180 m, an expansion of 6.5 cm is not something you would easily notice—but it isn't negligible. The thermal expansion of structural elements in towers and bridges must be accounted for in the design to avoid damaging stresses. When designers failed to properly account for thermal stresses in the marble panels cladding the Amoco Building in Chicago, all 43,000 panels had to be replaced, at great cost.

## CONCEPTUAL EXAMPLE 12.13 What happens to the hole?

A metal plate has a circular hole in it. As the plate is heated, does the hole get larger or smaller?

***REASON*** As the plate expands, you might think the hole would shrink (with the metal expanding into the hole). But suppose we took a metal plate and simply drew a circle where a hole could be cut. On heating, the plate and the marked area both expand, as we see in FIGURE 12.20. We could cut a hole on the marked line before or after heating; the size of the hole would be larger in the latter case. Therefore, the size of the hole must expand as the plate expands.

***ASSESS*** When an object undergoes thermal expansion (or contraction), all dimensions increase (or decrease) by the *same percentage*.

FIGURE 12.20 The thermal expansion of a hole.

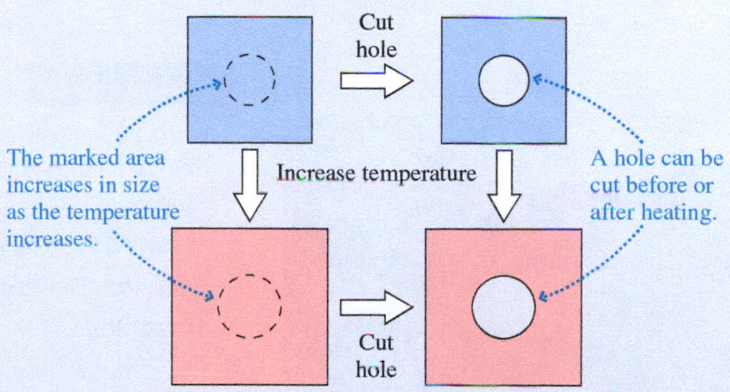


STOP TO THINK 12.4 An aluminum ring is tight around a solid iron rod. If we wish to loosen the ring to remove it from the rod, we should

A. Increase the temperature of the ring and rod.
B. Decrease the temperature of the ring and rod.

## 12.5 Specific Heat and Heat of Transformation

If you hold a glass of cold water in your hand, the heat from your hand will raise the temperature of the water. The heat from your hand will also melt an ice cube; melting is an example of what we will call a **phase change,** a change of state from one phase to another. In this section, we'll consider these two types of changes.

### Specific Heat

Adding 4190 J of heat energy to 1 kg of water raises its temperature by 1 K. If you are fortunate enough to have 1 kg of gold, you need only 129 J of heat to raise its temperature by 1 K. The amount of heat that raises the temperature of 1 kg of a substance by 1 K is called the **specific heat** of that substance. The symbol for specific heat is $c$. Water has specific heat $c_{\text{water}} = 4190$ J/kg · K, and the specific heat of gold is $c_{\text{gold}} = 129$ J/kg · K. Specific heat depends only on the material from which an object is made. TABLE 12.4 lists the specific heats of some common liquids and solids.

**TABLE 12.4** Specific heats of solids and liquids

| Substance | $c$ (J/kg · K) |
|---|---|
| **Solids** | |
| Lead | 128 |
| Gold | 129 |
| Copper | 385 |
| Iron | 449 |
| Aluminum | 900 |
| Water ice | 2090 |
| Mammalian body | 3400 |
| **Liquids** | |
| Mercury | 140 |
| Ethyl alcohol | 2400 |
| Water | 4190 |

If heat $c$ is required to raise the temperature of 1 kg of a substance by 1 K, then heat $Mc$ is needed to raise the temperature of mass $M$ by 1 K and $Mc\,\Delta T$ is needed to raise the temperature of mass $M$ by $\Delta T$. In general, the heat needed to bring about a temperature change $\Delta T$ is

$$Q = Mc\,\Delta T \qquad (12.24)$$

Heat needed to produce a temperature change $\Delta T$ for mass $M$ with specific heat $c$

$Q$ can be either positive (temperature goes up) or negative (temperature goes down).

It takes more heat energy to change the temperature of a substance with a large specific heat than to change the temperature of a substance with a small specific heat. Water, with a very large specific heat, is slow to warm up and slow to cool down. This large "thermal inertia" of water is essential for the biological processes of life.

**Video** Heating Water and Aluminum

**Temperate lakes** At night, the large specific heat of water prevents the temperature of a body of water from dropping nearly as much as that of the surrounding air. Early in the morning, water vapor evaporating from a warm lake quickly condenses in the colder air above, forming mist. During the day, the opposite happens: The air becomes much warmer than the water.

**EXAMPLE 12.14 How much energy is needed to run a fever?**

A 70 kg student catches the flu, and his body temperature increases from 37.0°C (98.6°F) to 39.0°C (102.2°F). How much energy is required to raise his body's temperature?

***STRATEGIZE*** The increase in temperature requires the addition of energy.

***PREPARE*** The change in temperature $\Delta T$ is 2.0°C, or 2.0 K.

***SOLVE*** Raising the temperature of the body uses energy supplied internally from the chemical reactions of the body's metabolism, which transfer heat to the body. The specific heat of the body is given in Table 12.4 as 3400 J/kg · K. We can use Equation 12.24 to find the necessary heat energy:

$$Q = Mc\,\Delta T = (70\text{ kg})(3400\text{ J/kg}\cdot\text{K})(2.0\text{ K}) = 4.8 \times 10^5\text{ J}$$

***ASSESS*** The body is mostly water, with a large specific heat, and the mass of the body is large, so we'd expect a large amount of energy to be necessary. Looking back to Chapter 11, we see that this is approximately the energy in a large apple, or the amount of energy required to walk 1 mile.

## Phase Changes

Suppose you remove a few ice cubes from the freezer and place them in a sealed container with a thermometer. Then, as in FIGURE 12.21a, you put a steady flame under the container. We'll assume that the heating is done slowly so that the inside of the container always has a single, uniform temperature.

FIGURE 12.21b shows a graph of the temperature as a function of time. In stage 1, the ice steadily warms without melting until it reaches 0°C. During stage 2, the temperature remains fixed at 0°C for an extended period of time during which the ice melts. As the ice is melting, both the ice temperature and the liquid water temperature remain at 0°C. Even though the system is being heated, the temperature doesn't begin to rise again until all the ice has melted.

KEY CONCEPT **FIGURE 12.21** The temperature as a function of time as water is transformed from solid to liquid to gas.

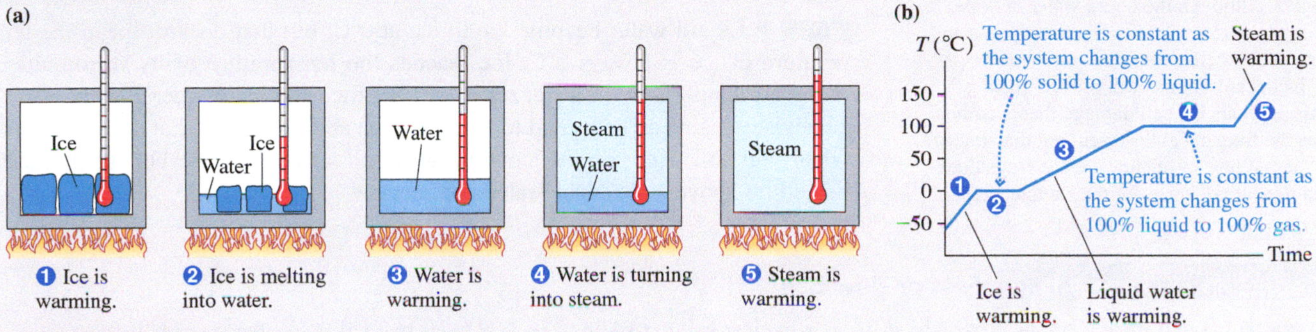


**STOP TO THINK 12.5** In Figure 12.21, by comparing the slope of the graph during the time the liquid water is warming to the slope as steam is warming, we can say that

A. The specific heat of water is larger than that of steam.
B. The specific heat of water is smaller than that of steam.
C. The specific heat of water is equal to that of steam.
D. The slope of the graph is not related to the specific heat.

At the end of stage 2, the ice has completely melted into liquid water. During stage 3, the flame slowly warms this liquid water, raising its temperature until it reaches 100°C. During stage 4, the liquid water remains at 100°C as it turns into water vapor—steam—which also remains at 100°C. Even though the system is being heated, the temperature doesn't begin to rise again until all the water has been converted to steam. Finally, stage 5 begins, in which the system is pure steam whose temperature rises continuously from 100°C.

**NOTE** ▶ In everyday language, the three phases of water are called *ice, water,* and *steam.* The term "water" implies the liquid phase. Scientifically, these are the solid, liquid, and gas phases of the compound called *water.* When we are working with different phases of water, we'll use the term "water" in the scientific sense of a collection of $H_2O$ molecules. We'll say either *liquid* or *liquid water* to denote the liquid phase. ◀

**CONCEPTUAL EXAMPLE 12.15 Strategy for cooling a drink**

If you have a warm soda that you wish to cool, is it more effective to add 25 g of liquid water at 0°C or 25 g of water ice at 0°C?

**REASON** If you add liquid water at 0°C, heat will be transferred from the soda to the water, raising the temperature of the water and lowering that of the soda. If you add water ice at 0°C, heat will first be transferred from the soda to the ice to melt it, transforming the 0°C ice to 0°C liquid water, then will be transferred to the liquid water to raise its temperature. Thus more thermal energy will be removed from the soda, giving it a lower final temperature, if ice is used rather than liquid water.

**ASSESS** This makes sense because you know that this is what you do in practice. To cool a drink, you drop in an ice cube.

**Frozen frogs** BIO It seems impossible, but common wood frogs survive the winter with much of their bodies frozen. When you dissolve substances in water, the freezing point lowers. Although the liquid water *between* cells in the frogs' bodies freezes, the water *inside* their cells remains liquid because of high concentrations of dissolved glucose. This prevents the cell damage that accompanies the freezing and subsequent thawing of tissues. When spring arrives, the frogs thaw and appear no worse for their winter freeze.

The temperature at which—if the thermal energy is increased—a solid becomes a liquid is called the **melting point;** if the thermal energy is instead decreased, a liquid becomes a solid at the **freezing point.** Melting and freezing are *phase changes.* A system at the melting point is in **phase equilibrium,** meaning that any amount of solid can coexist with any amount of liquid. Raise the temperature ever so slightly and the entire system soon becomes liquid. Lower it slightly and it all becomes solid.

You can see another region of phase equilibrium of water in Figure 12.21b at 100°C. This is a phase equilibrium between the liquid phase and the gas phase, and any amount of liquid can coexist with any amount of gas at this temperature. As heat is added to the system, the temperature stays the same. The added energy is used to break bonds between the liquid molecules, allowing them to move into the gas phase. The temperature at which a gas becomes a liquid is called the **condensation point;** the temperature at which a liquid becomes a gas is the **boiling point.**

NOTE ► Liquid water becomes solid ice at 0°C, but that doesn't mean the temperature of ice is always 0°C. Ice reaches the temperature of its surroundings. If the air temperature in a freezer is −20°C, then the ice temperature is −20°C. Likewise, steam can be heated to temperatures above 100°C. That doesn't happen when you boil water on the stove because the steam escapes, but steam can be heated far above 100°C in a sealed container. ◄

**CONCEPTUAL EXAMPLE 12.16 Fast or slow boil?**

You are cooking pasta on the stove; the water is at a slow boil. Will the pasta cook more quickly if you turn up the burner on the stove so that the water is at a fast boil?

***REASON*** Water boils at 100°C; no matter how vigorously the water is boiling, the temperature is the same. It is the temperature of the water that determines how fast the cooking takes place. Adding heat at a faster rate will make the water boil away more rapidly but will not change the temperature—and will not alter the cooking time.

***ASSESS*** This result may seem counterintuitive, but you can try the experiment next time you cook pasta!

## Heat of Transformation

Lava—molten rock—undergoes a phase change from liquid to solid when it contacts liquid water; the transfer of heat to the water causes the water to undergo a phase change from liquid to gas.

In Figure 12.21b, the phase changes appeared as horizontal line segments on the graph. During these segments, heat is being transferred to the system but the temperature isn't changing. The thermal energy continues to increase during a phase change, but, as noted, the additional energy goes into breaking molecular bonds rather than speeding up the molecules. **A phase change is characterized by a change in thermal energy without a change in temperature.**

The amount of heat energy that causes 1 kg of a substance to undergo a phase change is called the **heat of transformation** of that substance. For example, laboratory experiments show that 333,000 J of heat are needed to melt 1 kg of ice at 0°C. The symbol for heat of transformation is $L$. The heat required for the entire system of mass $M$ to undergo a phase change is

$$Q = ML \tag{12.25}$$

"Heat of transformation" is a generic term that refers to any phase change. Two particular heats of transformation are the **heat of fusion** $L_f$, the heat of transformation between a solid and a liquid, and the **heat of vaporization** $L_v$, the heat of transformation between a liquid and a gas. The heat needed for these phase changes is

$$Q = \begin{cases} \pm ML_f & \text{Heat needed to melt/freeze mass } M \\ \pm ML_v & \text{Heat needed to boil/condense mass } M \end{cases} \tag{12.26}$$

The ± indicates that heat must be added to the system during melting or boiling but removed from the system during freezing or condensing. **You must explicitly include the minus sign when it is needed.** Carefully consider the nature of the phase change and whether heat is added or removed. When an ice cube melts in your hand, heat goes from your hand to the ice cube. In your freezer, where the temperature is below the freezing point of water, liquid water freezes as heat goes from the liquid water to the surrounding air. If you put a pan of water on the stove, you must add heat to make the water boil. When you breathe on a mirror, water vapor from your lungs condenses on the glass, and heat goes from the water vapor into the cooler glass.

TABLE 12.5 lists some heats of transformation. Notice that the heat of vaporization is always much higher than the heat of fusion.

**TABLE 12.5** Melting and boiling temperatures and heats of transformation at standard atmospheric pressure

| | Solid to Liquid | | Liquid to Gas | |
|---|---|---|---|---|
| Substance | Melting Point $T_m$ (°C) | Heat of Fusion $L_f$ (J/kg) | Boiling Point $T_b$ (°C) | Heat of Vaporization $L_v$ (J/kg) |
| Nitrogen ($N_2$) | −210 | $0.26 \times 10^5$ | −196 | $1.99 \times 10^5$ |
| Ethyl alcohol | −114 | $1.09 \times 10^5$ | 78 | $8.79 \times 10^5$ |
| Mercury | −39 | $0.11 \times 10^5$ | 357 | $2.96 \times 10^5$ |
| Water | 0 | $3.33 \times 10^5$ | 100 | $22.6 \times 10^5$ |
| Lead | 328 | $0.25 \times 10^5$ | 1750 | $8.58 \times 10^5$ |

**EXAMPLE 12.17 How much energy is needed to melt a popsicle?**

A girl eats a 45 g frozen popsicle that was taken out of a −10°C freezer. How much energy does her body use to bring the popsicle up to body temperature?

***STRATEGIZE*** We will assume that the popsicle is pure water. There are three parts to the problem, corresponding to stages 1–3 in Figure 12.21: The popsicle must be warmed to 0°C, the popsicle must melt, and then the resulting water must be warmed to body temperature. We will figure out how much heat is needed for each stage; the sum is the total energy required.

***PREPARE*** Normal body temperature is 37°C. The specific heats of ice and liquid water are given in Table 12.4; the heat of fusion of water is given in Table 12.5.

***SOLVE*** The heat needed to warm the frozen water by $\Delta T = 10°\text{C} = 10\text{ K}$ to the melting point is

$$Q_1 = Mc_{\text{ice}}\,\Delta T = (0.045\text{ kg})(2090\text{ J/kg}\cdot\text{K})(10\text{ K}) = 940\text{ J}$$

Note that we use the specific heat of water ice, not liquid water, in this equation. Melting 45 g of ice requires heat

$$Q_2 = ML_f = (0.045\text{ kg})(3.33 \times 10^5\text{ J/kg}) = 15{,}000\text{ J}$$

The liquid water must now be warmed to body temperature; this requires heat

$$Q_3 = Mc_{\text{water}}\,\Delta T = (0.045\text{ kg})(4190\text{ J/kg}\cdot\text{K})(37\text{ K})$$
$$= 7000\text{ J}$$

The total energy is the sum of these three values: $Q_{\text{total}} = 23{,}000$ J.

***ASSESS*** More energy is needed to melt the ice than to warm the water, as we would expect. A commercial popsicle has 40 Calories, which is about 170 kJ. Roughly 15% of the chemical energy in this frozen treat is used to bring it up to body temperature!

## Evaporation

Water boils at 100°C. But individual molecules of water can move from the liquid phase to the gas phase at lower temperatures. This process is known as **evaporation.** Water evaporates as sweat from your skin at a temperature well below 100°C. Different particles in a liquid move at different speeds, as they do in a gas. At any temperature, some molecules will be moving fast enough to go into the gas phase. And they will do so, carrying away thermal energy as they go. The molecules that leave the liquid are the ones that have the highest kinetic energy, so evaporation reduces the average kinetic energy (and thus the temperature) of the liquid left behind.

**Keeping your cool** BIO Humans (and cattle and horses) have sweat glands, so we can perspire to moisten our skin, allowing evaporation to cool our bodies. Animals that do not perspire can also use evaporation to keep cool. Dogs, goats, rabbits, and even birds pant, evaporating water from their respiratory passages. Elephants spray water on their skin; other animals may lick their fur.

**The opposite of sweating** You are sitting outside, enjoying a cold beverage and you notice that your glass "sweats"—drops of water collect on the outside. Your glass isn't really sweating, of course. The water doesn't come from inside the glass but from outside, as water vapor from the air condenses onto the sides of the glass. Heat leaves the water vapor when it condenses—and so heat is added to your glass. Sweating may keep you cool due to evaporation, but a "sweating" glass means condensation, and this means a warming drink.

**Video** What the Physics? Fighting Fire with Evaporation

The evaporation of water, both from sweat and in moisture you exhale, is one of the body's methods of exhausting the excess heat of metabolism to the environment, allowing you to maintain a steady body temperature. The heat to evaporate a mass $M$ of water is $Q = ML_v$, and this amount of heat is removed from your body. However, the heat of vaporization $L_v$ is a little larger than the value for boiling. At a skin temperature of 30°C, the heat of vaporization of water is

$$L_v(\text{at } 30°) = 24 \times 10^5 \text{ J/kg}$$

This is 6% higher than the value at 100°C in Table 12.5. You should use this value for $L_v$ when you consider the heat transfer due to evaporation of perspiration from the skin and similar situations.

**EXAMPLE 12.18 Computing heat loss by perspiration** BIO

The human body can produce approximately 30 g of perspiration per minute. At what rate is it possible to exhaust heat by the evaporation of perspiration?

***STRATEGIZE*** We are looking for a rate of exhausting heat, so we are looking for a power in watts, an energy divided by a time. We'll assume that perspiration is pure water, which is a good approximation. The energy is the heat required to evaporate 30 g of water from the skin; the time is 1 minute.

***PREPARE*** The water is evaporated from the skin, which we assume is at 30°C. The value for the heat of transformation is therefore $L_v = 24 \times 10^5$ J/kg.

***SOLVE*** The evaporation of 30 g of perspiration at normal body temperature requires heat energy

$$Q = ML_v = (0.030 \text{ kg})(24 \times 10^5 \text{ J/kg}) = 7.2 \times 10^4 \text{ J}$$

This is the heat lost per minute; the rate of heat loss is

$$\frac{Q}{\Delta t} = \frac{7.2 \times 10^4 \text{ J}}{60 \text{ s}} = 1200 \text{ W}$$

***ASSESS*** Given the metabolic power required for different activities, as listed in Chapter 11, this is sufficient to keep the body cool even when exercising in hot weather—as long as the person drinks enough water to keep up this rate of perspiration.

**STOP TO THINK 12.6** 1 kg of barely molten lead, at 328°C, is poured into a large beaker holding liquid water right at the boiling point, 100°C. What is the mass of the water that will be boiled away as the lead solidifies?

A. 0 kg B. <1 kg C. 1 kg D. >1 kg

## 12.6 Calorimetry

FIGURE 12.22 Cooling hot coffee with ice.

If you've put an ice cube into a hot drink to cool it quickly, you engaged, in a trial-and-error way, in a practical aspect of heat transfer known as **calorimetry,** the quantitative measurement of the heat transferred between systems or evolved in reactions. You know that heat energy will be transferred from the hot drink into the cold ice cube, reducing the temperature of the drink, as shown in FIGURE 12.22.

Let's make this qualitative picture more precise. FIGURE 12.23 shows two systems that can exchange heat with each other but that are isolated from everything else. Suppose they start at different temperatures $T_1$ and $T_2$. As you know, heat energy will be transferred from the hotter to the colder system until they reach a common final temperature $T_f$. In the cooling coffee example of Figure 12.22, the coffee is system 1, the ice is system 2, and the insulating barrier is the mug.

The insulation prevents any heat energy from being transferred to or from the environment, so energy conservation tells us that any energy leaving the hotter system must enter the colder system. The concept is straightforward, but to state the idea mathematically we need to be careful with signs.

Let $Q_1$ be the energy transferred to system 1 as heat. $Q_1$ is positive if energy *enters* system 1, negative if energy *leaves* system 1. Similarly, $Q_2$ is the energy transferred to system 2. The fact that the systems are merely exchanging energy can be written as $|Q_1| = |Q_2|$. But $Q_1$ and $Q_2$ have opposite signs, so $Q_1 = -Q_2$. No energy is exchanged with the environment, so it makes more sense to write this relationship as

$$Q_{net} = Q_1 + Q_2 = 0 \tag{12.27}$$

**NOTE** ▶ The signs are very important in calorimetry problems. $\Delta T$ is always $T_f - T_i$, so $\Delta T$ and $Q$ are negative for any system whose temperature decreases. The proper sign of $Q$ for any phase change must be supplied *by you,* depending on the direction of the phase change. ◀

**FIGURE 12.23** Heat is being transferred from system 1 to system 2, so $Q_2 > 0$ and $Q_1 < 0$.

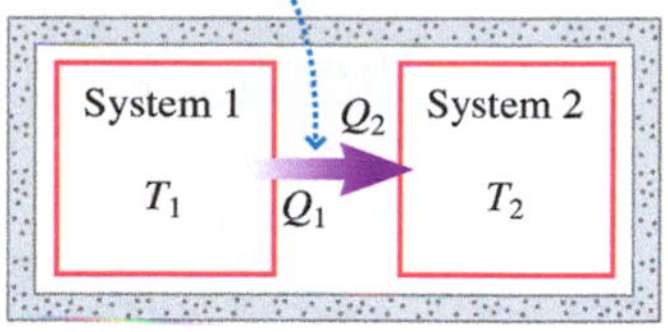


## PROBLEM-SOLVING APPROACH 12.1 Calorimetry problems

When two systems are brought into thermal contact, we use calorimetry to find the heat transferred between them and their final equilibrium temperature.

***STRATEGIZE*** Identify the individual interacting systems. Assume that they are isolated from the environment.

***PREPARE*** List the known information and identify what you need to find. Convert all quantities to SI units.

***SOLVE*** The statement of energy conservation is

$$Q_{net} = Q_1 + Q_2 + \cdots = 0$$

- For systems that undergo a temperature change, $Q_{\Delta T} = Mc(T_f - T_i)$. Be sure to have the temperatures $T_i$ and $T_f$ in the correct order.
- For systems that undergo a phase change, $Q_{phase} = \pm ML$. Supply the correct sign by observing whether energy enters or leaves the system during the transition.
- Some systems may undergo a temperature change *and* a phase change. Treat the changes separately. The heat energy is $Q = Q_{\Delta T} + Q_{phase}$.

***ASSESS*** The final temperature should be between the initial temperatures. A $T_f$ that is higher or lower than all initial temperatures is an indication that something is wrong, usually a sign error.

## EXAMPLE 12.19 Using calorimetry to identify a metal

200 g of an unknown metal is heated to 90.0°C, then dropped into 50.0 g of water at 20.0°C in an insulated container. The water temperature rises within a few seconds to 27.7°C, then changes no further. Identify the metal.

***STRATEGIZE*** The interacting systems are the metal and the water. We'll assume that the two systems quickly come to a final temperature, so there is no time for heat to be exchanged with the environment.

***PREPARE*** We label the temperatures as follows: The initial temperature of the metal is $T_m$; the initial temperature of the water is $T_w$. The common final temperature is $T_f$. For water, $c_w = 4190$ J/kg·K is known from Table 12.4. Only the specific heat $c_m$ of the metal is unknown.

*Continued*

**SOLVE** Energy conservation requires that $Q_w + Q_m = 0$. Using $Q = Mc(T_f - T_i)$ for each, we have

$$Q_w + Q_m = M_w c_w (T_f - T_w) + M_m c_m (T_f - T_m) = 0$$

This is solved for the unknown specific heat:

$$c_m = \frac{-M_w c_w (T_f - T_w)}{M_m (T_f - T_m)}$$
$$= \frac{-(0.0500\text{ kg})(4190\text{ J/kg}\cdot\text{K})(27.7°\text{C} - 20.0°\text{C})}{(0.200\text{ kg})(27.7°\text{C} - 90.0°\text{C})}$$
$$= 129\text{ J/kg}\cdot\text{K}$$

Referring to Table 12.4, we find we have either 200 g of gold or, if we made an ever-so-slight experimental error, 200 g of lead!

**ASSESS** The temperature of the unknown metal changed much more than the temperature of the water. This means that the specific heat of the metal must be much less than that of water, which is exactly what we found.

## EXAMPLE 12.20 Calorimetry with a phase change

Your 500 mL diet soda, with a mass of 500 g, is at 20°C, room temperature, so you cool it by adding 100 g of ice from the −20°C freezer. Does all the ice melt? If so, what is the final temperature? If not, what fraction of the ice melts? Assume that you have a well-insulated cup.

**STRATEGIZE** The interacting systems are the soda and the ice; heat will be transferred from the soda to the ice. You have a well-insulated cup, so we'll assume that the systems are isolated. We will also assume that the diet soda has the same specific heat as water; this is a good approximation.

There are two possible outcomes. If all the ice melts as the soda cools, then $T_f > 0°\text{C}$. It's also possible that the soda will cool to 0°C before all the ice has melted, leaving the ice and liquid in equilibrium at 0°C. We will need to distinguish between these outcomes before we can complete the solution.

**PREPARE** The soda has mass $M_{soda} = 0.50$ kg and an initial temperature of 20°C. The ice has mass $M_{ice} = 0.10$ kg and an initial temperature of −20°C. Since we are looking for a temperature difference, we can keep the temperatures in Celsius. We are looking for the final temperature of the combined system.

**SOLVE** Let's first calculate the heat needed to melt all the ice and leave it as liquid water at 0°C. To do so, we must warm the ice by 20 K to 0°C, then change it to water. The heat input for this two-stage process is

$$Q_{melt} = \underbrace{M_{ice}c_{ice}(20\text{ K})}_{\text{This is the energy to raise the temperature of the ice from }-20°\text{C to }0°\text{C. }\Delta T = 20\text{ K.}} + \underbrace{M_{ice}L_f}_{\text{This is the energy to melt the ice once it reaches }0°\text{C.}} = 37{,}500\text{ J}$$

where $L_f$ is the heat of fusion of water.

$Q_{melt}$ is a *positive* quantity because we must *add* heat to melt the ice. Next, let's calculate how much heat energy will leave the 500 g soda if it cools all the way to 0°C:

$$Q_{cool} = M_{soda}c_{water}(-20\text{ K}) = -41{,}900\text{ J}$$

where $\Delta T = -20$ K because the temperature decreases. Because $|Q_{cool}| > Q_{melt}$, the soda has sufficient energy to melt all the ice. Hence the final state will be all liquid at $T_f > 0$. (Had we found $|Q_{cool}| < Q_{melt}$, then the final state would have been an ice-liquid mixture at 0°C.)

Energy conservation requires $Q_{ice} + Q_{soda} = 0$. The heat $Q_{ice}$ consists of three terms: warming the ice to 0°C, melting the ice to water at 0°C, then warming the 0°C water to $T_f$. The mass will still be $M_{ice}$ in the last of these steps because it is the "ice system," but we need to use the specific heat of *liquid water.* Thus

$$Q_{ice} + Q_{soda} = [M_{ice}c_{ice}(20\text{ K}) + M_{ice}L_f + M_{ice}c_{water}(T_f - 0°\text{C})] + M_{soda}c_{water}(T_f - 20°\text{C}) = 0$$

We've already done part of the calculation, allowing us to write

$$37{,}500\text{ J} + M_{ice}c_{water}(T_f - 0°\text{C}) + M_{soda}c_{water}(T_f - 20°\text{C}) = 0$$

Solving for $T_f$ gives

$$T_f = \frac{20M_{soda}c_{water} - 37{,}500\text{ J}}{M_{ice}c_{water} + M_{soda}c_{water}} = 1.8°\text{C}$$

**ASSESS** A good deal of ice has been put in the soda, so it ends up being cooled nearly to the freezing point, as we might expect.

**STOP TO THINK 12.7** 1 kg of lead at 100°C is dropped into a container holding 1 kg of water at 0°C. Once the lead and water reach thermal equilibrium, the final temperature is

A. $<50$°C B. 50°C C. $>50$°C

## 12.7 Specific Heats of Gases

Just as for solids and liquids, heating a gas changes its temperature. But by how much? FIGURE 12.24 shows two isotherms on the $pV$ diagram for a gas. Processes A and B, which start on the $T_i$ isotherm and end on the $T_f$ isotherm, have the *same* temperature change $\Delta T = T_f - T_i$, so we might expect them both to require the same amount of heat. But it turns out that process A, which takes place at constant volume, requires *less* heat than does process B, which occurs at constant pressure. The reason is that work is done in process B but not in process A.

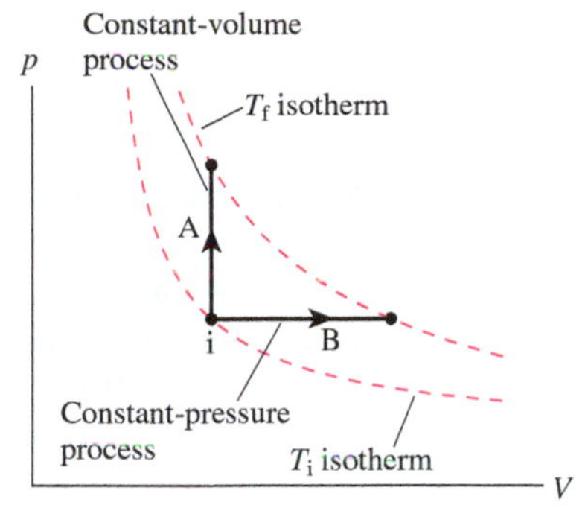


FIGURE 12.24 Processes A and B have the same $\Delta T$ and the same $\Delta E_{th}$, but they require different amounts of heat.

It is useful to define two different versions of the specific heat of gases: one for constant-volume processes and one for constant-pressure processes. We will define these as *molar* specific heats because we usually do gas calculations using moles instead of mass. The quantity of heat needed to change the temperature of $n$ moles of gas by $\Delta T$ is, for a constant-volume process,

$$Q = nC_V \Delta T \quad (12.28)$$

and for a constant-pressure process,

$$Q = nC_P \Delta T \quad (12.29)$$

$C_V$ is the **molar specific heat at constant volume,** and $C_P$ is the **molar specific heat at constant pressure.** TABLE 12.6 gives the values of $C_V$ and $C_P$ for some common gases. The units are J/mol · K. The values for air are essentially equal to those for $N_2$.

TABLE 12.6 Molar specific heats of gases (J/mol · K) at 20°C

| Gas | $C_P$ | $C_V$ |
|---|---|---|
| **Monatomic Gases** | | |
| He | 20.8 | 12.5 |
| Ne | 20.8 | 12.5 |
| Ar | 20.8 | 12.5 |
| **Diatomic Gases** | | |
| $H_2$ | 28.7 | 20.4 |
| $N_2$ | 29.1 | 20.8 |
| $O_2$ | 29.2 | 20.9 |

It's interesting that all the monatomic gases in Table 12.6 have the same values for $C_P$ and for $C_V$. Why should this be? Monatomic gases are really close to ideal, so let's go back to our atomic model of the ideal gas. We know that the thermal energy of an ideal gas of $N$ atoms is $E_{th} = \frac{3}{2}Nk_BT = \frac{3}{2}nRT$. If the temperature of an ideal gas changes by $\Delta T$, its thermal energy changes by

$$\Delta E_{th} = \frac{3}{2}nR\,\Delta T \quad (12.30)$$

If we keep the volume of the gas constant, so that no work is done, this energy change can come only from heat, so

$$Q = \frac{3}{2}nR\,\Delta T \quad (12.31)$$

Comparing Equation 12.31 with the definition of molar specific heat in Equation 12.28, we see that the molar specific heat at constant volume must be

$$C_V\,(\text{monatomic gas}) = \frac{3}{2}R = 12.5 \text{ J/mol}\cdot\text{K} \quad (12.32)$$

This *predicted* value from the ideal-gas model is exactly the *measured* value of $C_V$ for the monatomic gases in Table 12.6, a good check that the model we have been using is correct.

If you heat a gas in a sealed container so that there is no change in volume, then no work is done. But if you heat a sample of gas in a cylinder with a piston to keep it at constant pressure, the gas must expand, and it will do work as it expands. The expression for $\Delta E_{th}$ in Equation 12.30 is still valid, but now, according to the first law of thermodynamics, $Q = \Delta E_{th} + W_{gas}$. The work done by the gas in a constant-pressure process is $W_{gas} = p\,\Delta V$, so the heat required is

$$Q = \Delta E_{th} + W_{gas} = \frac{3}{2}nR\,\Delta T + p\,\Delta V \quad (12.33)$$

The ideal-gas law, $pV = nRT$, implies that if $p$ is constant and only $V$ and $T$ change, then $p\Delta V = nR\,\Delta T$. Using this result in Equation 12.33, we find that the heat needed to change the temperature by $\Delta T$ in a constant-pressure process is

$$Q = \frac{3}{2}nR\,\Delta T + nR\,\Delta T = \frac{5}{2}nR\,\Delta T$$

A comparison with the definition of molar specific heat shows that

$$C_P = \frac{5}{2}R = 20.8 \text{ J/mol}\cdot\text{K} \tag{12.34}$$

This is larger than $C_V$, as expected, and in perfect agreement for all the monatomic gases in Table 12.6.

**EXAMPLE 12.21 Finding the work done by an expanding gas**

A typical weather balloon is made of a thin latex envelope that takes very little force to stretch, so the pressure inside the balloon is approximately equal to atmospheric pressure. Suppose a weather balloon filled with 180 mol of helium is waiting for launch on a cold morning at a high-altitude station. The balloon warms in the sun, which raises the temperature of the gas from 0°C to 30°C. As the balloon expands, how much work is done by the expanding gas?

**STRATEGIZE** The work done is equal to $p\Delta V$, but we don't know the pressure (it's not sea level and we don't know the altitude) and we don't know the volume of the balloon. Instead, we'll use the first law of thermodynamics. We can compute how much heat energy is transferred to the balloon as it warms because this is a temperature change at constant pressure, and we can compute how much the thermal energy of the gas increases because we know $\Delta T$.

**PREPARE** We can rewrite Equation 12.20 as

$$W_{\text{gas}} = Q - \Delta E_{\text{th}}$$

The change in temperature of the gas is 30°C, so $\Delta T = 30$ K.

**SOLVE** The heat required to increase the temperature of the gas is given by Equation 12.29:

$$Q = nC_P\Delta T = (180\,\text{mol})(20.8\text{ J/mol}\cdot\text{K})(30\,\text{K}) = 112\text{ kJ}$$

The change in thermal energy depends on the change in temperature according to Equation 12.21:

$$\Delta E_{\text{th}} = \frac{3}{2}nR\Delta T = \frac{3}{2}(180\text{ mol})(8.31\text{ J/mol}\cdot\text{K})(30\text{ K})$$
$$= 67.3\text{ kJ}$$

The work done by the expanding balloon is just the difference between these two values:

$$W_{\text{gas}} = Q - \Delta E_{\text{th}} = 112\text{ kJ} - 67.3\text{ kJ} = 45\text{ kJ}$$

**ASSESS** The numbers are large—it's a lot of heat and a large change in thermal energy—but it's a big balloon with a lot of gas, so this seems reasonable.

FIGURE 12.25 Thermal energy of a diatomic gas.

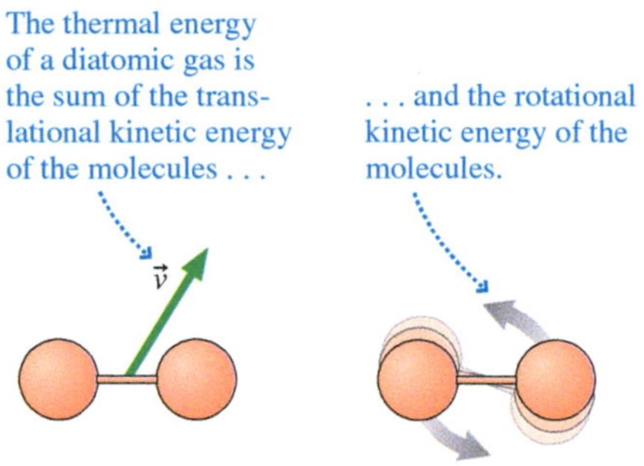


Now let's turn to the diatomic gases in Table 12.6. The molar specific heats are higher than for monatomic gases, and our atomic model explains why. The thermal energy of a monatomic gas consists exclusively of the translational kinetic energy of the atoms; heating a monatomic gas simply means that the atoms move faster. The thermal energy of a diatomic gas is more than just the translational energy, as shown in FIGURE 12.25. Heating a diatomic gas makes the molecules move faster, but it also causes them to rotate more rapidly. Energy goes into the translational kinetic energy of the molecules (thus increasing the temperature), but some goes into rotational kinetic energy. Because some of the added heat goes into rotation and not into translation, the specific heat of a diatomic gas is higher than that of a monatomic gas, as we see in Table 12.6.

## 12.8 Heat Transfer

You feel warmer when the sun is shining on you, colder when you are sitting on a cold concrete bench or when a stiff wind is blowing. This is due to the transfer of heat. Although we've talked about heat and heat transfers a lot in these last two chapters, we haven't said much about *how* heat is transferred from a hotter object to a colder object. There are four basic mechanisms, described in the following table, by which objects exchange heat with other objects or their surroundings. Evaporation was treated in an earlier section; in this section we will consider the other mechanisms.

**Heat-transfer mechanisms**

When two objects are in direct physical contact, such as the soldering iron and the circuit board, heat is transferred by *conduction.* **Energy is transferred by direct contact.**

This special photograph shows air currents near a flame. Hot gases from the combustion rise, carrying thermal energy in a process known as *convection.* **Energy is transferred by the bulk motion of molecules with high thermal energy.**

The lamp shines on the lambs huddled below, warming them. The energy is transferred by infrared *radiation,* a form of electromagnetic waves. **Energy is transferred by electromagnetic waves.**

When you blow on a cup of cocoa, this increases the rate of *evaporation,* rapidly cooling it. **Energy is transferred by the removal of molecules with high thermal energy.**

## Conduction

If you hold a metal spoon in a cup of hot coffee, the handle of the spoon soon gets warm. Thermal energy is transferred along the spoon from the coffee to your hand. The difference in temperature between the two ends drives this heat transfer by a process known as **conduction.** Conduction is the transfer of thermal energy directly through a physical material.

FIGURE 12.26 shows a copper rod placed between a hot reservoir (a fire) and a cold reservoir (a block of ice). We can use our atomic model to see how thermal energy is transferred along the rod by the interaction between atoms in the rod; fast-moving atoms at the hot end transfer energy to slower-moving atoms at the cold end.

Suppose we set up a series of experiments to measure the heat $Q$ transferred through various rods. We would find the following trends in our data:

- $Q$ increases if the temperature difference $\Delta T$ between the hot end and the cold end is increased.
- $Q$ increases if the cross-section area $A$ of the rod is increased.
- $Q$ decreases if the length $L$ of the rod is increased.
- Some materials (such as metals) transfer heat quite readily. Other materials (such as wood) transfer very little heat.

**FIGURE 12.26** Conduction of heat in a solid rod.

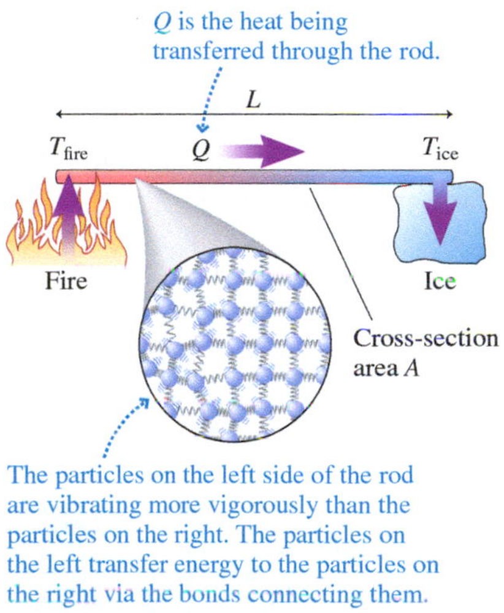


The final observation is one that is familiar to you: If you are stirring a pot of hot soup on the stove, you generally use a wood or plastic spoon rather than a metal one.

These experimental observations about heat conduction can be summarized in a single formula. If heat $Q$ is transferred in a time interval $\Delta t$, the *rate* of heat transfer (joules per second, or watts) is $Q/\Delta t$. For a material of cross-section area $A$ and length $L$, spanning a temperature difference $\Delta T$, the rate of heat transfer is

$$\frac{Q}{\Delta t} = \left(\frac{kA}{L}\right)\Delta T \qquad (12.35)$$

Rate of conduction of heat across a temperature difference

The quantity $k$, which characterizes whether the material is a good or a poor conductor of heat, is called the **thermal conductivity** of the material. Because the heat transfer rate J/s is a *power,* measured in watts, the units of $k$ are W/m · K. Values of $k$

**Cold feet, warm heart** BIO A penguin standing on the cold Antarctic ice loses very little heat through its feet. Its thick skin with limited thermal conductivity helps, but more important, the penguin's feet are very cold, which minimizes heat loss by conduction. The penguin can have cold feet so close to the warmth of the body because of an adaptation called *countercurrent heat exchange*. Arteries carrying warm blood to the feet run next to veins carrying cold blood back from the feet. Heat goes from the arteries to the veins, cooling the blood going to the feet and warming the blood going to the body—so the feet stay cool while the body stays warm.

for some common materials are listed in TABLE 12.7; a larger number for $k$ means a material is a better conductor of heat.

TABLE 12.7 Thermal conductivity values (measured at 20°C)

| Material | $k$ (W/m · K) | Material | $k$ (W/m · K) |
|---|---|---|---|
| Diamond | 1000 | Skin | 0.50 |
| Silver | 420 | Muscle | 0.46 |
| Copper | 400 | Fat | 0.21 |
| Iron | 72 | Wood | 0.2 |
| Stainless steel | 14 | Carpet | 0.04 |
| Ice | 1.7 | Fur, feathers | 0.02–0.06 |
| Concrete | 0.8 | Air (27°C, 100 kPa) | 0.026 |
| Plate glass | 0.75 | | |

The weak bonds between molecules make most biological materials poor conductors of heat. Fat is a worse conductor than muscle, so sea mammals have thick layers of fat for insulation. The bodies of land mammals are insulated with fur, and those of birds with feathers. Both fur and feathers trap a good deal of air, so their conductivity is similar to that of air, as Table 12.7 shows.

### EXAMPLE 12.22 Warming the bench

At the start of the section we noted that you "feel cold" when you sit on a cold concrete bench. "Feeling cold" really means that your body is losing a significant amount of heat. How significant? Suppose you are sitting on a 10°C concrete bench. You are wearing thin clothing that provides negligible insulation. In this case, most of the insulation that protects your body's core (temperature 37°C) from the cold of the bench is provided by a 1.0-cm-thick layer of fat on the part of your body that touches the bench. (The thickness varies from person to person, but this is a reasonable average value.) A good estimate of the area of contact with the bench is 0.10 m$^2$. Given these details, what is the rate of heat loss by conduction?

***STRATEGIZE*** Heat is lost to the bench by conduction through the fat layer, so we will compute the rate of heat loss by using Equation 12.35.

***PREPARE*** The thickness of the conducting layer is 0.010 m, the area is 0.10 m$^2$, and the thermal conductivity of fat is given in Table 12.7. The temperature difference is the difference between your body's core temperature (37°C) and the temperature of the bench (10°C), a difference of 27°C, or 27 K.

***SOLVE*** We have all of the data we need to use Equation 12.35 to compute the rate of heat loss:

$$\frac{Q}{\Delta t}=\left(\frac{(0.21\ \text{W/m}\cdot\text{K})(0.10\ \text{m}^2)}{0.010\ \text{m}}\right)(27\ \text{K})=57\ \text{W}$$

***ASSESS*** 57 W is more than half your body's resting power, which we learned in Chapter 11 is approximately 100 W. That's a significant loss, so your body will feel cold, a result that seems reasonable if you've ever sat on a cold bench for any length of time.

**A feather coat** BIO A penguin's short, dense feathers serve a different role than the flight feathers of other birds: They trap air to provide thermal insulation. The fluffy feathers of the juvenile penguin makes this trapping of air very clear, but the smooth covering of the adults works in a similar manner.

## Convection

In conduction, faster-moving atoms transfer thermal energy to adjacent atoms. But in fluids such as water or air, there is a more efficient means to move energy: by transferring the faster-moving atoms themselves. When you place a pan of cold water on a burner on the stove, it's heated on the bottom. This heated water expands and becomes less dense than the water above it, so it rises to the surface while cooler, denser water sinks to take its place. This transfer of thermal energy by the motion of a fluid is known as **convection.**

Convection is usually the main mechanism for heat transfer in fluid systems. On a small scale, convection mixes the pan of water that you heat on the stove; on a large scale, convection is responsible for making the wind blow and ocean currents circulate. Air is a very poor thermal conductor, but it is very effective at transferring energy by convection. To use air for thermal insulation, it is necessary to trap the air in small pockets to limit convection. And that's exactly what feathers, fur, double-paned windows, and fiberglass insulation do.

## Radiation

You *feel* the warmth from the glowing red coals in a fireplace. On a cool day, you prefer to sit in the sun rather than the shade so that the sunlight keeps you warm. In both cases heat energy is being transferred to your body in the form of **radiation.**

Radiation consists of electromagnetic waves—a topic we will explore further in later chapters—that transfer energy from the object that emits the radiation to the object that absorbs it. All warm objects emit radiation in this way. Objects near room temperature emit radiation in the invisible infrared part of the electromagnetic spectrum. Thermal images, like the one of the teapot in FIGURE 12.27, are made with cameras having special infrared-sensitive detectors that allow them to "see" radiation that our eyes don't respond to. Hotter objects, such as glowing embers, still emit most of their radiant energy in the infrared, but some is emitted as visible red light.

FIGURE 12.27 A thermal image of a teapot.

NOTE ► The word "radiation" comes from "radiate," which means "to beam." You have likely heard the word used to refer to x rays and radioactive materials. This is not the sense we use in this chapter. Here, we use "radiation" to mean electromagnetic waves that "beam" from an object. ◄

Radiation is a significant part of the energy balance that keeps your body at the proper temperature. You can change your body temperature by absorbing more radiation (sitting next to a fire) or by emitting more radiation (taking off your hat and scarf to expose more skin).

Video Figure 12.27

The energy radiated by an object shows a strong dependence on temperature. In the photo that opened the chapter, the person's warm hands emit quite a bit more energy than the cooler spider. We can quantify the dependence of radiated energy on temperature. If heat energy $Q$ is radiated in a time interval $\Delta t$ by an object with surface area $A$ and absolute temperature $T$, the *rate* of heat transfer $Q/\Delta t$ (joules per second) is found to be

$$\frac{Q}{\Delta t} = e\sigma AT^4 \tag{12.36}$$

Rate of heat transfer by radiation at temperature $T$ (Stefan's law)

Video Energy Transfer by Radiation

Quantities in this equation are defined as follows:

- $e$ is the **emissivity** of a surface, a measure of the effectiveness of radiation. The value of $e$ ranges from 0 to 1. Human skin of any color is a very effective radiator at body temperature, with $e = 0.97$.
- $T$ is the absolute temperature in kelvin.
- $A$ is the surface area in $\text{m}^2$.
- $\sigma$ is a constant with the value $\sigma = 5.67 \times 10^{-8}\ \text{W/m}^2 \cdot \text{K}^4$.

Notice the very strong fourth-power dependence on temperature. Doubling the absolute temperature of an object increases the radiant heat transfer by a factor of 16!

An adult human with bare skin in a room at a comfortable temperature has a skin temperature of approximately 33°C, or 306 K. A typical value for the skin's surface area is 1.8 $\text{m}^2$. With these values and the emissivity of skin noted previously, we can calculate the rate of heat transfer via radiation from the skin:

$$\frac{Q}{\Delta t} = e\sigma AT^4 = (0.97)\left(5.67 \times 10^{-8}\,\frac{\text{W}}{\text{m}^2 \cdot \text{K}^4}\right)(1.8\ \text{m}^2)(306\ \text{K})^4 = 870\ \text{W}$$

As we learned in Chapter 11, the body at rest generates approximately 100 W of thermal energy. If the body radiated energy at 870 W, it would quickly cool. At this rate of emission, the body temperature would drop by 1°C every 7 minutes! Clearly, there must be some mechanism to balance this emitted radiation, and there is: the radiation *absorbed* by the body.

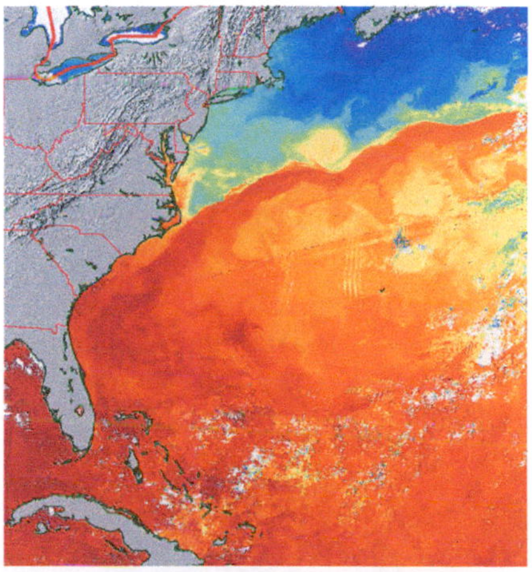

**Global heat transfer** This satellite image shows radiation emitted by the waters of the ocean off the east coast of the United States. You can clearly see the warm waters of the Gulf Stream. The satellite can see this radiation from the earth because it readily passes through the atmosphere into space. This radiation is also the only way the earth can cool. This has consequences for the energy balance of the earth, as we will discuss in One Step Beyond at the end of Part VI.

When you sit in the sun, your skin warms due to the radiation you absorb. Even if you are not in the sun, you are absorbing the radiation emitted by the objects surrounding you. Suppose an object at temperature $T$ is surrounded by an environment at temperature $T_0$. The *net* rate at which the object radiates heat energy—that is, radiation emitted minus radiation absorbed—is

$$\frac{Q_{net}}{\Delta t} = e\sigma A(T^4 - T_0^4) \tag{12.37}$$

This makes sense. An object should have no net energy transfer by radiation if it's in thermal equilibrium $(T = T_0)$ with its surroundings. Note that the emissivity $e$ appears for absorption as well; objects that are good emitters are also good absorbers.

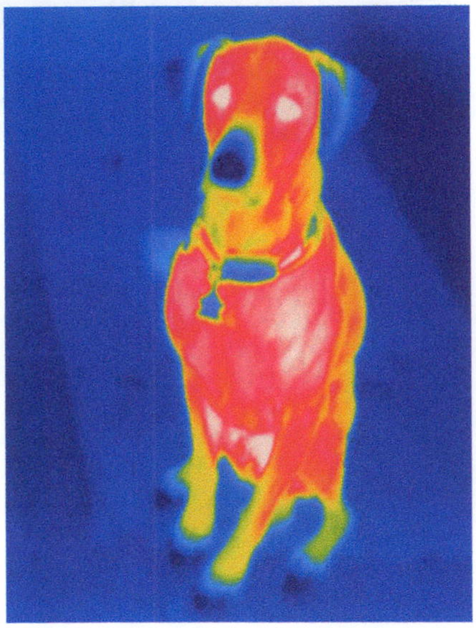

BIO The infrared radiation emitted by a dog is captured in this image. His cool nose and paws radiate much less energy than the rest of his body.

**EXAMPLE 12.23 Determining energy loss by radiation for the body** BIO

A person with a skin temperature of 33°C is in a room at 24°C. What is the net rate of heat transfer by radiation?

**STRATEGIZE** We know the temperature of the skin and the environment, so we will compute the net rate of transfer using Equation 12.37.

**PREPARE** Body temperature is $T = 33 + 273 = 306$ K; the temperature of the room is $T_0 = 24 + 273 = 297$ K.

**SOLVE** The net radiation rate is

$$\frac{Q_{net}}{\Delta t} = e\sigma A(T^4 - T_0^4)$$

$$= (0.97)\left(5.67 \times 10^{-8}\frac{\text{W}}{\text{m}^2 \cdot \text{K}^4}\right)(1.8\ \text{m}^2)[(306\ \text{K})^4 - (297\ \text{K})^4] = 98\ \text{W}$$

**ASSESS** This is a reasonable value, roughly matching your resting metabolic power. When you are dressed (little convection) and sitting on wood or plastic (little conduction), radiation is your body's primary mechanism for dissipating the excess thermal energy of metabolism.

**STOP TO THINK 12.8** Suppose you are an astronaut in the vacuum of space, hard at work in your sealed spacesuit. The only way that you can transfer heat to the environment is by

A. Conduction. B. Convection. C. Radiation. D. Evaporation.

## 12.9 Diffusion

In 1827, the botanist Robert Brown was observing pollen grains through a microscope. He noticed that pollen grains suspended in water were not stationary, as expected, but moved about continuously in a jittery, erratic fashion rather like that shown in FIGURE 12.28. This apparently random motion, today called **Brownian motion,** is caused by the ceaseless bombardment of the pollen grains by individual water molecules.

FIGURE 12.28 The random motion of a pollen grain undergoing Brownian motion.

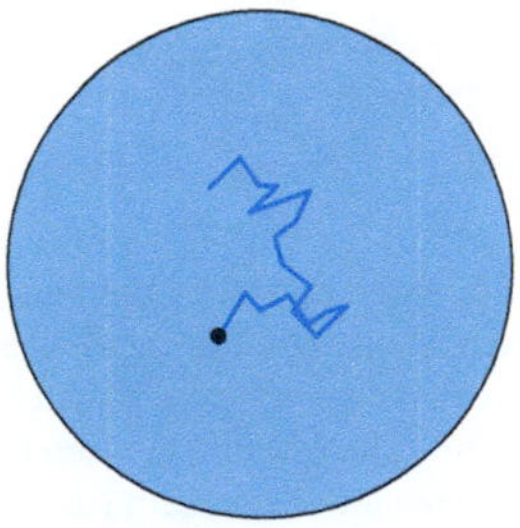

Brownian motion is an example of a **random walk.** Suppose you stood at the origin of an $xy$-coordinate system, turned through a randomly chosen angle, and then took a step with a randomly chosen length. Wherever you end up, you repeat the process again. And again. You would be engaging in a random walk, and your path would look something like the path of the particle in Figure 12.28.

We can't see or track individual molecules, but their motion—the motion of thermal energy—is also a random walk. At atmospheric pressure, a molecule in air can move only about 80 nm before it collides with another molecule. Each collision sends the molecule off in a new direction, and the trajectory of an individual molecule also looks similar to Figure 12.28—a sequence of straight lines of varying lengths connected by random turns.

Imagine using a pipette to put a very tiny droplet of dye molecules in a beaker of water. The droplet will gradually disperse into the water. This net movement of molecules from a region of higher concentration (the initial droplet) to a region of lower concentration (the water) is called **diffusion.** Microscopically, diffusion is simply the random walks of the dye molecules as they collide both with each other and with the more numerous water molecules.

FIGURE 12.29 shows 50 molecules initially grouped so close together that we can't distinguish them. Then each begins a random walk, and their positions are shown at $t = 1$ s and $t = 9$ s. We can see that the molecules are diffusing outward. A random step is just as likely to go right as to go left, or to go up as to go down, so the *average position* of all the molecules is $(0, 0)$. However, the *average distance* from the origin is not zero. And it increases with time.

We earlier defined the root-mean-square speed $v_{rms}$. Similarly, we can define a **root-mean-square distance** $r_{rms}$ as

$$r_{rms} = \sqrt{(r^2)_{avg}} = \text{distance of a typical molecule} \qquad (12.38)$$

The rms distance is a rough measure of how far the molecules have diffused. Figure 12.29 shows the rms distance at $t = 9$ s. A mathematical analysis of a random walk in three dimensions finds

$$r_{rms} = \sqrt{6Dt} \qquad (12.39)$$

where $D$ is called the **diffusion constant.** Its value, with units of $m^2/s$, depends on the type of molecules that are diffusing, the type of molecules that they are diffusing through, and the temperature and pressure. TABLE 12.8 lists some typical values. We notice two trends in the values. First, larger molecules have smaller diffusion constants and thus diffuse more slowly. Second, the constants for diffusion in air are roughly $10^4$ times larger than are those for diffusion in water. This shouldn't be surprising: Water is much denser than air, so the distances between molecular collisions—the lengths of the steps of the random walk—are much shorter.

If a group of molecules is released close together, how long will it take them to diffuse a distance $L$? There's no definitive answer to this question because the diffusing molecules don't all arrive at once. However, we've seen that $r_{rms}$ is about how far the molecules have diffused, so we can estimate the time needed to diffuse a distance $L$ by setting $r_{rms} = L$ in Equation 12.39 and solving for time:

$$t = \frac{L^2}{6D} \qquad (12.40)$$

Estimate of time to diffuse a distance $L$ with diffusion constant $D$

The time depends on the square of the distance, so an increase in the distance results in a much larger increase in time, as Example 12.24 illustrates.

FIGURE 12.29 Random walks cause a group of molecules to diffuse outward.

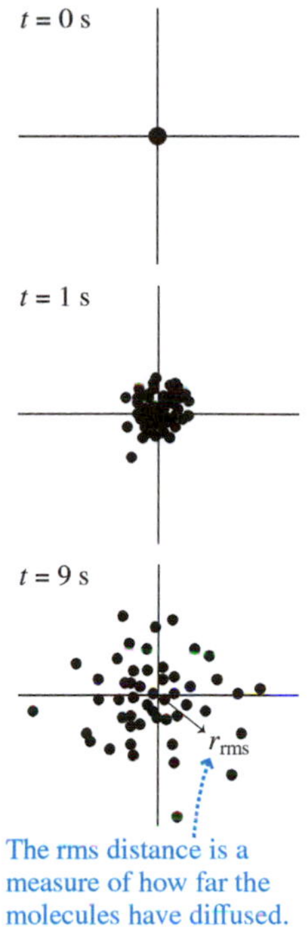


TABLE 12.8 Diffusion constants

| Substance | $D(m^2/s)$ |
|---|---|
| **Diffusion in air, 1 atm, 25°C** | |
| $O_2$ | $1.8 \times 10^{-5}$ |
| $CO_2$ | $1.9 \times 10^{-5}$ |
| $H_2O$ | $2.5 \times 10^{-5}$ |
| **Diffusion in water, 25°C** | |
| Acetylcholine | $4.0 \times 10^{-10}$ |
| Glucose | $7.1 \times 10^{-10}$ |
| Urea | $1.3 \times 10^{-9}$ |
| $CO_2$ | $2.0 \times 10^{-9}$ |
| $O_2$ | $2.3 \times 10^{-9}$ |

## EXAMPLE 12.24 Diffusion of oxygen through water BIO

A cell doesn't need a circulatory system, but your body does. Let's do a quick calculation to see why. A typical cell has a diameter of $10\,\mu m$. The smallest mammals in the world, shrews, are about 10 mm across. Compute the diffusion time for oxygen molecules through water at 25°C for these two distances.

***STRATEGIZE*** We will use Equation 12.40 to find the diffusion times for the two distances.

*Continued*

***PREPARE*** Table 12.8 gives the diffusion constant of oxygen through water at 25°C as $D = 2.3 \times 10^{-9}\,\mathrm{m^2/s}$.

***SOLVE*** For the two distances we're considering, we find

$$\text{For } L = 10\,\mu\text{m},\ t = \frac{(10 \times 10^{-6}\,\mathrm{m^2})^2}{6(2.3 \times 10^{-9}\,\mathrm{m^2/s})} = 0.0072\,\mathrm{s}$$

$$\text{For } L = 10\,\text{mm},\ t = \frac{(10 \times 10^{-3}\,\mathrm{m^2})^2}{6(2.3 \times 10^{-9}\,\mathrm{m^2/s})} = 7200\,\mathrm{s}$$

The time for oxygen to diffuse across the full diameter of a cell is about 7 ms; this is fast enough that there is no need for active transport. For a size typical of the smallest mammals, the diffusion time is much longer—about 2 hours! Clearly, mammals need a circulatory system to deliver oxygen to the cells of the body.

***ASSESS*** Diffusion works for a cell, but not for a shrew. This result makes sense. We know that all but the smallest animals have circulatory systems. The diffusion time increases as the *square* of the distance, so it takes 1,000,000 times longer to diffuse 1000 times as far!

Your circulatory system delivers oxygen to different parts of your body, but once the oxygen reaches the cells, diffusion can take over. The cellular transport of oxygen, nutrients, and larger macromolecules is often by passive diffusion:

- Air in the alveoli of the lungs, the small gas-filled sacs through which gas exchange with the blood takes place, is separated from blood in the pulmonary capillaries by a membrane that is about 1 $\mu$m thick. Oxygen in the lungs rapidly diffuses through this membrane into the capillaries, while excess carbon dioxide in the capillaries diffuses into the lungs and is exhaled.
- Oxygen, glucose, and other molecules diffuse across the cell wall from capillaries into cells. Waste products, such as carbon dioxide or urea, diffuse from the cell into the capillaries for removal.
- A nerve impulse moves across a synapse when a neurotransmitter, such as acetylcholine, is released from the pre-synaptic membrane and diffuses across the synaptic cleft to the post-synaptic membrane. The transmission takes less than a microsecond because the synaptic cleft is so narrow.

The body needs active transport systems, such as the circulatory system, to move molecules through distances of more than a few cell diameters because diffusion is not an effective mechanism at these scales.

**FIGURE 12.30** Diffusion of molecules across a concentration difference.

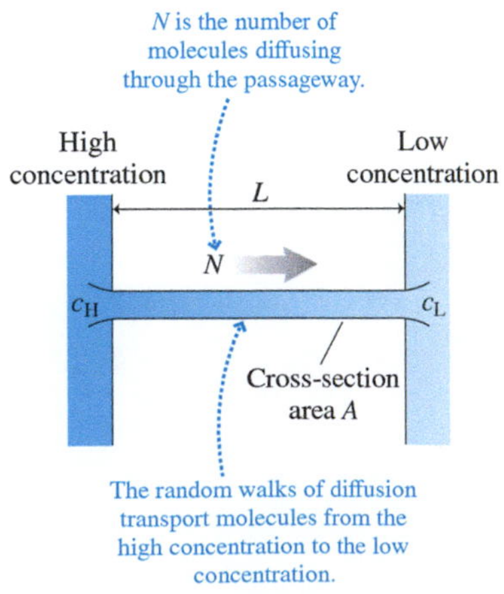


## Fick's Law

Heat conduction *transfers energy* between reservoirs with different temperatures. Similarly, diffusion *transfers matter* between reservoirs with different concentrations.

FIGURE 12.30 shows two reservoirs, a distance $L$ apart, holding different concentrations of some substance: a high concentration $c_H$ and a low concentration $c_L$. When $N$ molecules occupy volume $V$, their **concentration** is

$$c = \frac{N}{V} \tag{12.41}$$

The concentration is the number of molecules per cubic meter, with units $m^{-3}$.

**NOTE** ▸ Chemists and biologists measure concentration in other units, such as molecules per cubic centimeter, moles per liter, or parts per million. The concepts in this section still apply, but the diffusion constant may be in different units. ◂

Suppose the reservoirs are connected by a passageway with cross-section area $A$ through which molecules diffuse. Diffusion is a random walk, so at any instant there are molecules moving in both directions. However, more molecules move from the high concentration toward the low concentration than in the other direction, so diffusion causes a net transfer of molecules when there is a *concentration difference* $\Delta c = c_H - c_L$. If $N$ molecules are transferred in a time interval $\Delta t$, the *rate* of transfer (molecules per second) is $N/\Delta t$.

According to **Fick's law,** which can be derived from an analysis of random walks, the rate of molecule transfer is

$$\frac{N}{\Delta t} = \left(\frac{DA}{L}\right)\Delta c \tag{12.42}$$

Rate of transfer of molecules across a concentration difference

where $D$ is the diffusion constant. Fick's law is exactly analogous to the heat-conduction equation, Equation 12.35, with the diffusion constant $D$ playing the same role as the thermal conductivity constant $k$.

**Wrinkly respiration** BIO The hellbender salamander, native to the eastern United States, can be over half a meter long, with a mass of 2 kilograms. Remarkably, this large amphibian doesn't have lungs or gills. Instead, its extracts oxygen from water by diffusion through its skin. The wrinkles on the sides of the salamander aren't just for looks; they increase the surface area of its skin so that diffusion can deliver sufficient oxygen.

**EXAMPLE 12.25 Carbon dioxide diffusion in photosynthesis** BIO

Carbon dioxide makes up about 0.040% of the atmosphere. Carbon dioxide enters a green leaf, where photosynthesis occurs, by diffusing through small pores called *stomata*, pores with a depth of about 10 $\mu$m. When the stomata are open, we can assume that their total area is 2.0% of the leaf area. Photosynthesis consumes $CO_2$, and the concentration within the leaf is about half the atmospheric value. Estimate, in both molecules/s and $\mu$mol/s, the rate of $CO_2$ uptake by a leaf with an area of $7.0 \times 10^{-4}\,\text{m}^2$.

**STRATEGIZE** The atmosphere and the leaf are reservoirs with different concentrations of $CO_2$ connected by pores through which matter transfer can occur via diffusion. Fick's law describes the rate of $CO_2$ uptake.

**PREPARE** From Table 12.8, the diffusion constant of carbon dioxide through air at 25°C is $D = 1.9 \times 10^{-5}\,\text{m}^2/\text{s}$. The stomata depth is the length across which diffusion occurs, so $L = 10 \times 10^{-6}\,\text{m}$, and the total area of the stomata is 2.0% of the leaf area, or $1.4 \times 10^{-5}\,\text{m}^2$.

Now, let's consider the concentration. We can use the ideal-gas law to determine the number of molecules in a cubic meter of air at 1 atm ($1.013 \times 10^6$ Pa) pressure and a temperature of 25°:

$$N = \frac{pV}{k_B T} = \frac{(1.013 \times 10^5\,\text{Pa})(1.0\,\text{m}^3)}{(1.38 \times 10^{-23}\,\text{J/K})(298\,\text{K})} = 2.5 \times 10^{25}$$

$CO_2$ represents 0.040% of the molecules, so the concentration of $CO_2$ in the high concentration reservoir is

$$c_H = (0.00040)\frac{N}{V} = (0.00040)\frac{2.5 \times 10^{25}}{1.0\,\text{m}^3} = 1.0 \times 10^{22}\,\text{m}^{-3}$$

We're told that $c_L \approx \frac{1}{2}c_H$, so the concentration difference is $\Delta c \approx \frac{1}{2}c_H$.

**SOLVE** We use Fick's law to calculate the rate at which molecules diffuse from the air into the leaf. The area is the total area of the stomata; the length across which diffusion occurs is the length of the stomata:

$$\frac{N}{\Delta t} = \left(\frac{DA}{L}\right)\Delta c$$

$$= \left(\frac{(1.9 \times 10^{-5}\,\text{m}^2/\text{s})(1.4 \times 10^{-5}\,\text{m}^2)}{10 \times 10^{-6}\,\text{m}}\right)(0.50 \times 10^{22}\,\text{m}^{-3})$$

$$= 1.3 \times 10^{17}\,\text{molecules/s}$$

We can convert to moles by dividing by Avogadro's number:

$$\frac{N}{\Delta t} = \frac{1.3 \times 10^{17}\,\text{molecules/s}}{6.02 \times 10^{23}\,\text{molecules/mol}}$$

$$= 2.2 \times 10^{-7}\,\text{mol/s} = 0.22\,\mu\text{mol/s}$$

**ASSESS** $CO_2$ uptake by plants is more complicated than we've assumed, so it's no surprise that our final number is a bit lower than typical values, but it is of the correct order of magnitude.

## INTEGRATED EXAMPLE 12.26 Breathing in cold air BIO

On a cold day, breathing costs your body energy; as the cold air comes into contact with the warm tissues of your lungs, the air warms due to heat transferred from your body. The thermal image of the person inflating a balloon in FIGURE 12.31 shows that exhaled air is quite a bit warmer than the surroundings.

FIGURE 12.31 A thermal image of a person blowing up a balloon on a cold day.

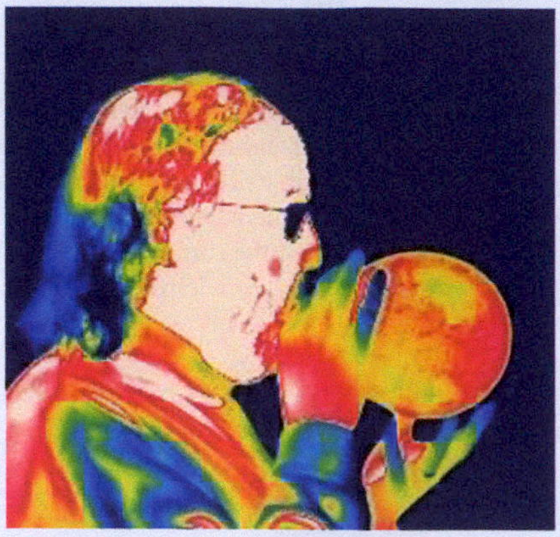

FIGURE 12.32 shows this process for a frosty $-10°C$ (14°F) day. The inhaled air warms to nearly the temperature of the interior of your body, 37°C. When you exhale, some heat is retained by the body, but most is lost; the exhaled air is still about 30°C.

FIGURE 12.32 Breathing warms the air.

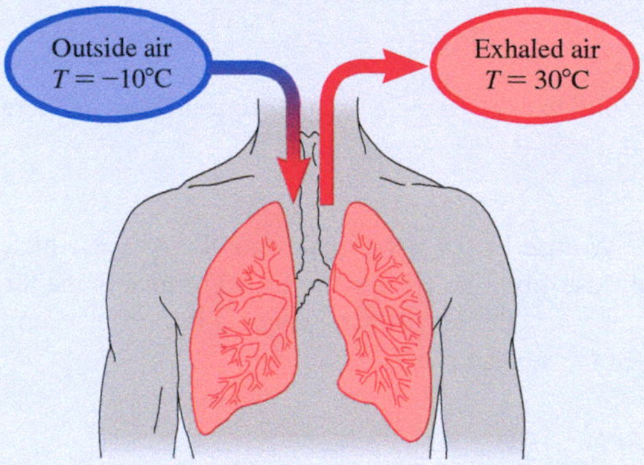


Your lungs hold several liters of air, but only a small part of the air is exchanged during each breath. A typical person takes 12 breaths each minute, with each breath drawing in 0.50 L of outside air. If the air warms up from $-10°C$ to 30°C, what is the volume of the air exhaled with each breath? What fraction of your body's resting power goes to warming the air? (Note that gases are exchanged as you breathe—oxygen to carbon dioxide—but to a good approximation the number of atoms, and thus the number of moles, stays the same.) Consider only the energy required to warm the air, not the energy lost to evaporation from the tissues of the lungs.

**STRATEGIZE** There are two parts to the problem. First, we'll figure out the increase in volume of the air as it is warmed. Next, we'll figure out the amount of energy needed to warm the air.

**PREPARE** When your body warms the air, its temperature increases, so its volume will increase as well. The initial and final states represented in Figure 12.32 are at atmospheric pressure, so we will treat this change as a constant-pressure process. We need the absolute temperatures of the initial and final states:

$$T_i = -10°C + 273 = 263\ K$$
$$T_f = 30°C + 273 = 303\ K$$

Because the initial and final pressures are the same, we can find the energy needed to warm the air by computing the heat needed to raise the temperature of a gas at constant pressure. The change in temperature is $+40°C$, so we can use $\Delta T = 40$ K. Air is a mix of nitrogen and oxygen with a small amount of other gases. $C_P$ for nitrogen and oxygen is the same to two significant figures, so we assume the gas has $C_P = 29$ J/mol · K.

**SOLVE** The change in volume of the air in your lungs is a constant-pressure process. 0.50 L of air is breathed in; this is $V_i$. When the temperature increases, so does the volume. The gas isn't in a sealed container, but we are considering the same "parcel" of gas before and after, so we can use the ideal-gas law to find the volume after the temperature increase:

$$V_f = V_i \frac{p_i}{p_f} \frac{T_f}{T_i} = (0.50\ L) \times 1 \times \frac{303\ K}{263\ K} = 0.58\ L$$

The volume increases by a little over 10%, from 0.50 L to 0.58 L.

Now we can move on to the second part: determining the energy required. We need the number of moles of gas, which we can compute from the ideal-gas law:

$$n = \frac{pV}{RT} = \frac{(101.3 \times 10^3\ Pa)(0.50 \times 10^{-3}\ m^3)}{(8.31\ J/mol \cdot K)(263\ K)} = 0.023\ mol$$

In doing this calculation, we used $1\ m^3 = 1000$ L to convert the 0.50 L volume to $m^3$. Now we can compute the heat needed to warm one breath, using Equation 12.29:

$$Q(\text{one breath}) = nC_P\,\Delta T = (0.023\ mol)(29\ J/mol \cdot K)(40\ K) = 27\ J$$

If you take 12 breaths a minute, a single breath takes 1/12 of a minute, or 5.0 s. Thus the heat power your body supplies to warm the incoming air is

$$P = \frac{Q}{\Delta t} = \frac{27\ J}{5.0\ s} = 5.4\ W$$

At rest, your body typically uses 100 W, so this is just over 5% of your body's resting power. You breathe in a good deal of air each minute and warm it by quite a bit, but the specific heat of air is small enough that the energy required is reasonably modest.

**ASSESS** The air expands slightly and a small amount of energy goes into heating it. If you've been outside on a cold day, you know that neither change is dramatic, so this final result seems reasonable. The energy loss is noticeable but reasonably modest; other forms of energy loss are more important when you are outside on a cold day.

# SUMMARY

**GOAL** To use the atomic model of matter to explain many properties of matter associated with heat and temperature.

## GENERAL PRINCIPLES

### Atomic Model

We model matter as being made of simple basic particles. The relationship of these particles to each other defines the phase.

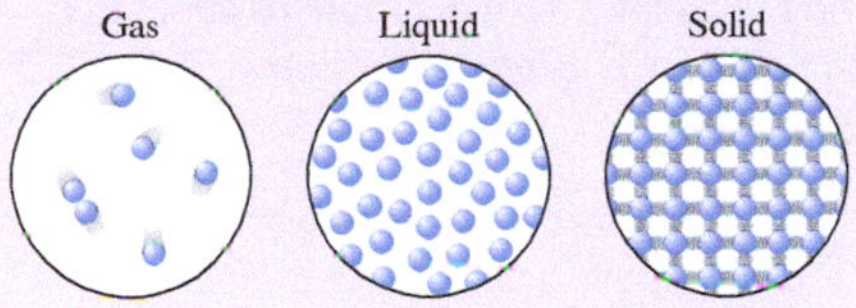


The atomic model explains thermal expansion, specific heat, and heat transfer.

### Atomic Model of a Gas

Macroscopic properties of gases can be explained in terms of the atomic model of the gas. The speed of the particles is related to the temperature:

$$v_{rms} = \sqrt{\frac{3k_BT}{m}}$$

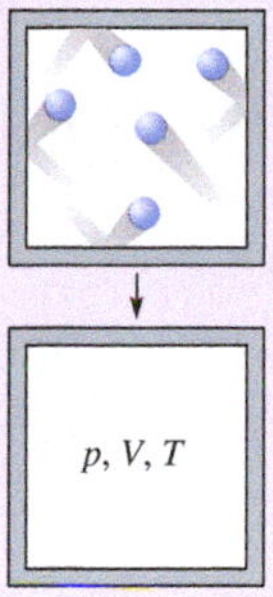


The collisions of particles with each other and with the walls of the container determine the pressure.

### Ideal-Gas Law

The ideal-gas law relates the pressure, volume, and temperature in a sample of gas. We can express the law in terms of the number of atoms or the number of moles in the sample:

$$pV = Nk_BT$$
$$pV = nRT$$

For a gas process in a sealed container,

$$\frac{p_iV_i}{T_i} = \frac{p_fV_f}{T_f}$$

## IMPORTANT CONCEPTS

### Effects of heat transfer

A system that is heated can either change temperature or change phase.

The **specific heat** $c$ of a material is the heat required to raise 1 kg by 1 K.

$$Q = Mc\,\Delta T$$

The **heat of transformation** is the energy necessary to change the phase of 1 kg of a substance. Heat is added to change a solid to a liquid or a liquid to a gas; heat is removed to reverse these changes.

$$Q = \begin{cases} \pm ML_f \text{ (melt/freeze)} \\ \pm ML_v \text{ (boil/condense)} \end{cases}$$

The **molar specific heat** of a gas depends on the process.

For a constant-volume process: $Q = nC_V\,\Delta T$

For a constant-pressure process: $Q = nC_P\,\Delta T$

### Mechanisms of heat transfer

An object can transfer heat to other objects or to its environment:

**Conduction** is the transfer of heat by direct physical contact.

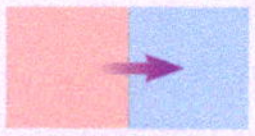

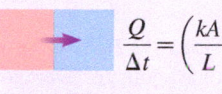

$$\frac{Q}{\Delta t} = \left(\frac{kA}{L}\right)\Delta T$$

**Convection** is the transfer of heat by the motion of a fluid.

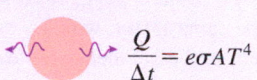

**Radiation** is the transfer of heat by electromagnetic waves.

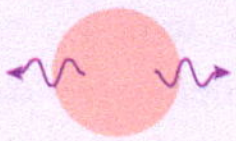

$$\frac{Q}{\Delta t} = e\sigma AT^4$$

A ***pV* (pressure-volume) diagram** is a useful means of looking at a process involving a gas.

A **constant-volume** process follows a vertical line between initial and final states.

A **constant-pressure** process follows a horizontal line between initial and final states.

A **constant-temperature** process follows a curved path between initial and final states.

The work done by a gas is the area under the graph.

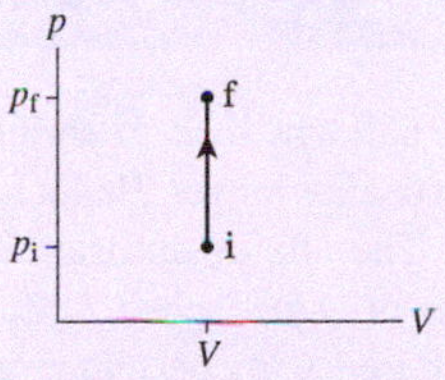


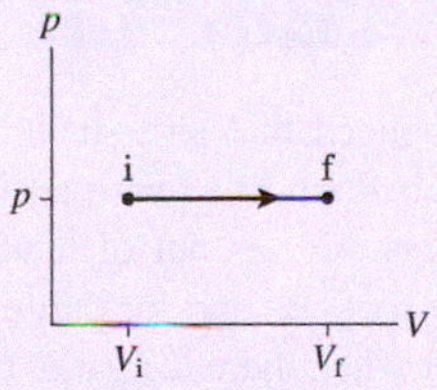


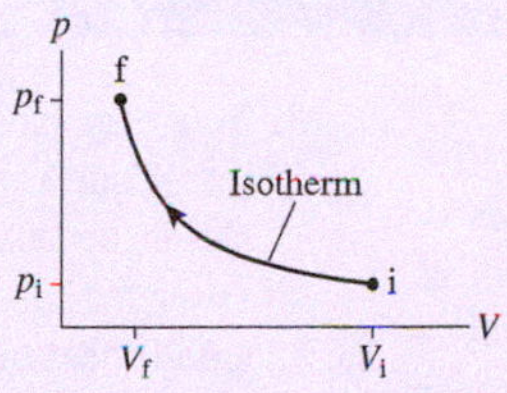


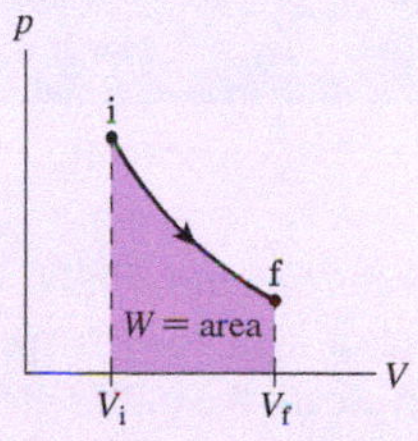


### Diffusion

The random motion of molecules causes diffusion from an area of high concentration to an area of low concentration.

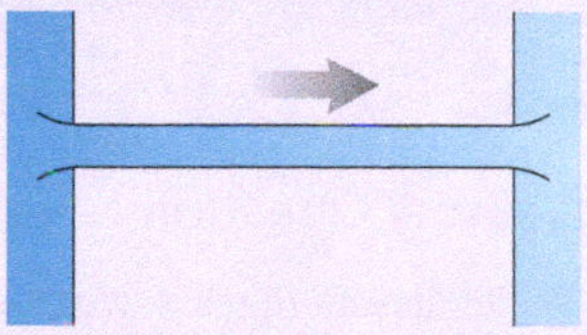

The time for diffusion varies with the square of the distance over which the diffusion occurs:

$$t = \frac{L^2}{6D}$$

## APPLICATIONS

**Thermal expansion** Objects experience an increase in volume and an increase in length when their temperature changes:

$$\Delta V = \beta V_i\,\Delta T \qquad \Delta L = \alpha L_i\,\Delta T$$

**Calorimetry** When two systems interact thermally, they come to a common final temperature determined by

$$Q_{net} = Q_1 + Q_2 = 0$$

The number of **moles** is

$$n = \frac{M\text{ (in grams)}}{M_{mol}}$$

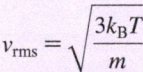

## Learning Objectives After studying this chapter, you should be able to:

- Work with moles and atomic masses. *Conceptual Question 12.1; Problems 12.1, 12.2, 12.3, 12.4, 12.5*
- Understand the atomic model of gas temperature and pressure. *Conceptual Questions 12.2, 12.5; Problems 12.8, 12.14, 12.17, 12.19, 12.22*
- Use the ideal-gas law, *pV* diagrams, and the first law of thermodynamics to solve problems about ideal-gas processes. *Conceptual Questions 12.9, 12.20; Problems 12.29, 12.31, 12.32, 12.36, 12.38*
- Calculate the thermal expansion of solids and liquids. *Conceptual Questions 12.11, 12.12; Problems 12.42, 12.43, 12.44, 12.45*
- Solve calorimetry and phase-change problems for solids, liquids, and gases. *Conceptual Questions 12.14, 12.16; Problems 12.49, 12.60, 12.61, 12.65, 12.68*
- Calculate the rate of heat transfer via conduction and radiation. *Conceptual Questions 12.23, 12.24; Problems 12.73, 12.74, 12.75, 12.76, 12.77*
- Use random walks and Fick's law to solve problems about the diffusion of gases and liquids. *Conceptual Questions 12.28, 12.29; Problems 12.80, 12.81, 12.82*

## STOP TO THINK ANSWERS

**Chapter Preview Stop to Think: B.** We invoke the first law of thermodynamics, $\Delta E_{th} = W + Q$. The work done by the blender is positive because it adds energy to the food. But the heat transfer is negative because energy is leaving the food as heat. Thus $\Delta E_{th} = 5000\text{ J} - 2000\text{ J} = 3000\text{ J}$.

**Stop to Think 12.1: A.** Both helium and neon are monatomic gases, where the basic particles are atoms. 5 mol of helium contain 5 times as many atoms as 1 mol of neon, though both samples have the same mass.

**Stop to Think 12.2: A, B.** An increase in temperature means that the atoms have a larger average kinetic energy and will thus have a larger rms speed. Because the thermal energy of the gas is simply the total kinetic energy of the atoms, this must increase as well. The pressure *could* change, but it's not required; $T$ and $V$ could increase by the same factor, which would keep $p$ constant. The container is sealed, so the number of molecules does not change.

**Stop to Think 12.3: B.** The product $pV/T$ is constant. During the process, $pV$ decreases by a factor of 2, so $T$ must decrease by a factor of 2 as well.

**Stop to Think 12.4: A.** The thermal expansion coefficients of aluminum are greater than those of iron. Heating the rod and the ring will expand the outer diameter of the rod and the inner diameter of the ring, but the ring's expansion will be greater.

**Stop to Think 12.5: A.** Because heat is being added at a constant rate, the slope tells us the ratio of the temperature change $\Delta T$ to the heat added $Q$; that is, the slope is proportional to $\Delta T/Q$. But by the definition of specific heat, this ratio is equal to $1/Mc$. Thus the segment with the *lesser* slope (water) has the *larger* specific heat $c$.

**Stop to Think 12.6: B.** To solidify the lead, heat must be removed; this heat boils the water. The heat of vaporization of water is 10 times the heat of fusion of lead, so much less than 1 kg of water vaporizes as 1 kg of lead solidifies.

**Stop to Think 12.7: A.** The lead cools and the water warms as heat is transferred from the lead to the water. The specific heat of water is much larger than that of lead, so the temperature change of the water is much less than that of the lead.

**Stop to Think 12.8: C.** With a sealed suit and no matter around you, there is no way to transfer heat to the environment except by radiation.

Video Tutor Solution Chapter 12

# QUESTIONS

## Conceptual Questions

1. Which has more mass, a mole of Ne gas or a mole of $N_2$ gas?
2. If you launch a projectile upward with a high enough speed, its kinetic energy is sufficient to allow it to escape the earth's gravity—it will go up and not come back down. Given enough time, hydrogen and helium gas atoms in the earth's atmosphere will escape, so these elements are not present in our atmosphere. Explain why hydrogen and helium atoms have the necessary speed to escape but why other elements, such as oxygen and nitrogen, do not.
3. You may have noticed that latex helium balloons tend to shrink rather quickly; a balloon filled with air lasts a lot longer. Balloons shrink because gas diffuses out of them. The rate of diffusion is faster for smaller particles and for particles of higher speed. Diffusion is also faster when there is a large difference in concentration between two sides of a membrane. Given these facts, explain why an air-filled balloon lasts longer than a helium balloon.
4. If you double the typical speed of the molecules in a gas, by what factor does the pressure change? Give a simple explanation why the pressure changes by this factor.

Problem difficulty is labeled as | (straightforward) to ||||| (challenging). Problems labeled INT integrate significant material from earlier chapters; Problems labeled BIO are of biological or medical interest.

The eText icon indicates when there is a video tutor solution available for the chapter or for a specific problem. To launch these videos, log into your eText through Mastering™ Physics or log into the Study Area.

5. Two gases have the same number of molecules per cubic meter ($N/V$) and the same rms speed. The molecules of gas 2 are more massive than the molecules of gas 1.
   a. Do the two gases have the same pressure? If not, which is larger?
   b. Do the two gases have the same temperature? If not, which is larger?
6. If the temperature $T$ of an ideal gas doubles, by what factor does the average kinetic energy of the atoms change?
7. A bottle of helium gas and a bottle of argon gas contain equal numbers of atoms at the same temperature. Which bottle, if either, has the greater total thermal energy?
8. A gas cylinder contains 1.0 mol of helium at a temperature of 20°C. A second identical cylinder contains 1.0 mol of neon at 20°C. The helium atoms are moving with a higher average speed, but the gas pressure in the two containers is the same. Explain how this is possible.
9. A gas is in a sealed container. By what factor does the gas pressure change if
   a. The volume is doubled and the temperature is tripled?
   b. The volume is halved and the temperature is tripled?
10. A gas is in a sealed container. By what factor does the gas temperature change if
   a. The volume is doubled and the pressure is tripled?
   b. The volume is halved and the pressure is tripled?
11. You need to precisely measure the dimensions of a large wood panel for a construction project. Your metal tape measure was left outside for hours in the sun on a hot summer day, and now the tape is so hot it's painful to pick up. How will your measurements differ from those taken by your coworker, whose tape stayed in the shade? Explain.
12. A common trick for opening a stubborn lid on a jar is to run very hot water over the lid for a short time. Explain how this helps to loosen the lid.

13. You fill up a steel container with cool water to the very top. Now you place the container in the sun. As the container warms up, the water expands, but so does the container. Does the water overflow? Explain.
14. Materials A and B have equal densities, but A has a larger specific heat than B. You have 100 g cubes of each material. Cube A, initially at 0°C, is placed in good thermal contact with cube B, initially at 200°C. The cubes are inside a well-insulated container where they don't interact with their surroundings. Is their final temperature greater than, less than, or equal to 100°C? Explain.
15. Two containers hold equal masses of nitrogen gas at equal temperatures. You supply 10 J of heat to container A while not allowing its volume to change, and you supply 10 J of heat to container B while not allowing its pressure to change. Afterward, is temperature $T_A$ greater than, less than, or equal to $T_B$? Explain.
16. You need to raise the temperature of a gas by 10°C. To use the smallest amount of heat energy, should you heat the gas at constant pressure or at constant volume? Explain.
17. The interior of the earth is at a higher temperature than it would be if it were cooling steadily; there must be internal sources of heat. Current models of the earth's interior suggest that one source of heat is the crystallization of liquid iron. Explain how this process provides heat to the earth's interior.
18. If you are exposed to water vapor at 100°C, you are likely to experience a worse burn than if you are exposed to liquid water at 100°C. Why is water vapor more damaging than liquid water at the same temperature?
19. When night falls, the temperature of the earth's surface starts to drop. On a cool night, dew starts to form on the grass as water vapor condenses. Once dew starts to form, the rate of temperature decrease slows. Explain why this change occurs.
20. A sample of ideal gas is in a cylinder with a movable piston. 600 J of heat is added to the gas in an isothermal process. As the gas expands, pushing against the piston, how much work does it do?
21. A student is heating chocolate in a pan on the stove. He uses a cooking thermometer to measure the temperature of the chocolate and sees it varies as shown in Figure Q12.21. Describe what is happening to the chocolate in each of the three portions of the graph.

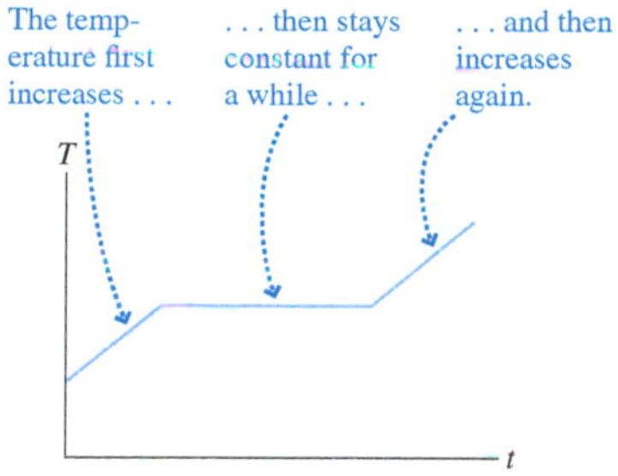


FIGURE Q12.21

22. If you bake a cake at high elevation, where atmospheric pressure is lower than at sea level, you will need to adjust the recipe. You will need to cook the cake for a longer time, and you will need to add less baking powder. (Baking powder is a leavening agent. As it heats, it releases gas bubbles that cause the cake to rise.) Explain why those adjustments are necessary.
23. Snakes and other reptiles that live on land warm up in the sun. Their bodies can reach temperatures well above the ambient air temperature. But fish don't do this. Most fish are at nearly the same temperature as the water in which they swim. How do you explain the difference?
24. Male emperor penguins spend the long Antarctic night on the ice. They keep from freezing with exquisite thermal insulation. Their insulation is so effective that the surface temperature of their feathers can actually be *lower* than that of the surrounding air. Explain how this is possible.
25. A student is asked to sketch a $pV$ diagram for a gas that goes through a cycle consisting of (a) an isobaric expansion, (b) a constant-volume reduction in temperature, and (c) an isothermal process that returns the gas to its initial state. The student draws the diagram shown in Figure Q12.25. What, if anything, is wrong with the student's diagram?

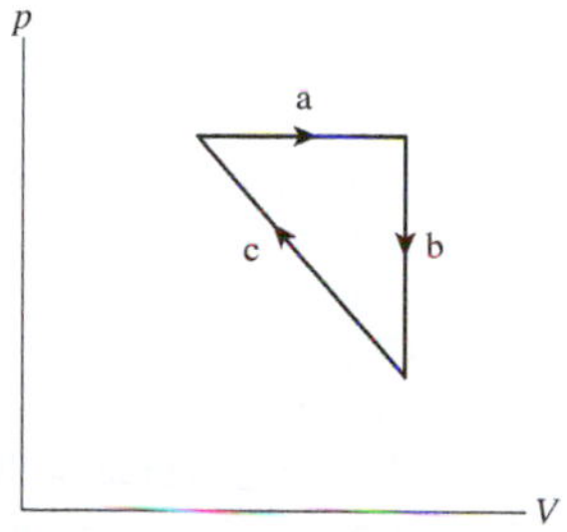


FIGURE Q12.25

26. You can buy an expensive frying pan that is made of copper but has a handle made of stainless steel. Why are different metals used for the part of the pan that touches the stove and for the part that touches your hand?

27. At night, the surface of the earth cools, mostly by radiation. The surface radiates energy upward; the much cooler sky above radiates much less. Patches of ground under trees can stay warmer, which is why, if you live somewhere that has cold, clear nights, you may have noticed some mornings when there is frost on open patches of ground but not under trees. Why does being under a tree keep the ground warmer?
28. The diffusion constants in Table 12.8 are given at 25°C. In fact, diffusion constants vary significantly with temperature. Explain why you would expect this to be true. Would you expect the values to increase or decrease as the temperature rises?
29. If a person's lungs are compromised, they can use supplemental oxygen to increase the concentration of oxygen in their lungs. Explain how this helps increase oxygen uptake.

## Multiple-Choice Questions

30. | A tire is inflated to a gauge pressure of 35 psi. The absolute pressure in the tire is
    A. Less than 35 psi.
    B. Equal to 35 psi.
    C. Greater than 35 psi.
31. | The number of atoms in a container is increased by a factor of 2 while the temperature is held constant. The pressure
    A. Decreases by a factor of 4.
    B. Decreases by a factor of 2.
    C. Stays the same.
    D. Increases by a factor of 2.
    E. Increases by a factor of 4.
32. ||| A gas is compressed by an isothermal process that decreases its volume by a factor of 2. In this process, the pressure
    A. Does not change.
    B. Increases by a factor of less than 2.
    C. Increases by a factor of 2.
    D. Increases by a factor of more than 2.
33. | The thermal energy of a container of helium gas is halved. What happens to the temperature, in kelvin?
    A. It decreases to one-fourth its initial value.
    B. It decreases to one-half its initial value.
    C. It stays the same.
    D. It increases to twice its initial value.
34. ||| A gas is compressed by an adiabatic process that decreases its volume by a factor of 2. In this process, the pressure
    A. Does not change.
    B. Increases by a factor of less than 2.
    C. Increases by a factor of 2.
    D. Increases by a factor of more than 2.
35. | A cup of water is heated with a heating coil that delivers 100 W of heat. In one minute, the temperature of the water rises by 20°C. What is the mass of the water?
    A. 72 g B. 140 g
    C. 720 g D. 1.4 kg
36. | Three identical beakers each hold 1000 g of water at 20°C. 100 g of liquid water at 0°C is added to the first beaker, 100 g of ice at 0°C is added to the second beaker, and the third beaker gets 100 g of aluminum at 0°C. The contents of which container end up at the lowest final temperature?
    A. The first beaker.
    B. The second beaker.
    C. The third beaker.
    D. All end up at the same temperature.
37. || Steam at 100°C causes worse burns than liquid water at BIO 100°C. This is because
    A. The steam is hotter than the water.
    B. Heat is transferred to the skin as steam condenses.
    C. Steam has a higher specific heat than water.
    D. Evaporation of liquid water on the skin causes cooling.
38. || Oxygen diffuses across the membranes that separate the air BIO in the alveoli in the lungs from the blood in the capillaries. If you double the thickness of the separating membrane, by what factor will the diffusion time increase?
    A. $\sqrt{2}$ B. 2
    C. $2\sqrt{2}$ D. 4

# PROBLEMS

### Section 12.1 The Atomic Model of Matter

1. | Which contains the most moles: 10 g of hydrogen gas, 100 g of carbon, or 500 g of lead?
2. |||| How many grams of water ($H_2O$) have the same number of oxygen atoms as 1.0 mol of oxygen gas?
3. |||| How many atoms of hydrogen are in 100 g of hydrogen peroxide ($H_2O_2$)?
4. || How many cubic millimeters ($mm^3$) are in 1 L?
5. || A box is 200 cm wide, 40 cm deep, and 3.0 cm high. What is its volume in $m^3$?

### Section 12.2 The Atomic Model of an Ideal Gas

6. | What is the typical speed of a helium atom in a child's balloon at a room temperature of 20°C?
7. | An ideal gas is at 20°C. If we double the average kinetic energy of the gas atoms, what is the new temperature in °C?
8. || A typical helium balloon contains 1.1 g of helium gas. At 20°C room temperature, what is the total kinetic energy of the helium in the balloon?
9. || An ideal gas is at 20°C. The gas is cooled, reducing the thermal energy by 10%. What is the new temperature in °C?
10. || An ideal gas at 0°C consists of $1.0 \times 10^{23}$ atoms. 10 J of thermal energy are added to the gas. What is the new temperature in °C?
11. || An ideal gas at 20°C consists of $2.2 \times 10^{22}$ atoms. 4.3 J of thermal energy are removed from the gas. What is the new temperature in °C?
12. |||| Total lung capacity of a typical adult is approximately 5.0 L. BIO Approximately 20% of the air is oxygen. At sea level and at an average body temperature of 37°C, how many moles of oxygen do the lungs contain at the end of an inhalation?

13. |||| Many cultures around the world still use a simple weapon
BIO called a blowgun, a tube with a dart that fits tightly inside. A sharp breath into the end of the tube launches the dart. When exhaling forcefully, a healthy person can supply air at a gauge pressure of 6.0 kPa. What force does this pressure exert on a dart in a 1.5-cm-diameter tube?

14. ||| When you stifle a sneeze, you can damage delicate tissues
BIO because the pressure of the air that is not allowed to escape may rise by up to 45 kPa. If this extra pressure acts on the inside of your 8.4-mm-diameter eardrum, what is the outward force?

15. | What is the absolute pressure of the air in your car's tires, in psi, when your pressure gauge indicates they are inflated to 35.0 psi? Assume you are at sea level.

16. || For a normal car riding on tires with relatively flexible sidewalls, the weight of the car is held up, in large measure, by the pressure of the air in the tires. If you look at one of your car's tires, you'll note that the tire is flattened slightly to make a rectangle where it touches the ground. The area of the resulting "contact patch" depends on the pressure in the tires. To a good approximation, the upward normal force of the ground (which we can assume is equal to ¼ of the car's weight) on this patch of the tire is equal to the downward pressure force on the patch.
    a. Suppose you inflate your 2000 kg car's tires to the recommended pressure, as measured by a gauge. The resulting contact patch is 18 cm wide and 12 cm long. What does the gauge read?
    b. If you let a bit of air out of your tire, what happens to the area of the contact patch?

17. || Mars has an atmosphere composed almost entirely of carbon dioxide, with an average temperature of −63°C. What is the rms speed of a molecule in Mars's atmosphere?

18. || We've discussed elevators that are raised with cables, but many
INT elevators use pistons. A pump provides fluid at very high pressure to the piston; the resulting force on the piston raises the elevator and passengers. For a typical piston elevator, the mass of the elevator and passengers is 2000 kg and the radius of the piston is a surprisingly small 4.2 cm. The elevator begins to move upward, accelerating at 0.80 m/s$^2$. What is the pressure in the piston?

19. | The lowest pressure ever obtained in a laboratory setting is $4.0 \times 10^{-11}$ Pa. At this pressure, how many molecules of air would there be in a 20°C experimental chamber with a volume of 0.090 m$^3$?

20. ||| The pressure inside a champagne bottle can be quite high
INT and can launch a cork explosively. Suppose you open a bottle at sea level. The absolute pressure inside a champagne bottle is 6 times atmospheric pressure; the cork has a mass of 7.5 g and a diameter of 18 mm. Assume that once the cork starts to move, the only force that matters is the pressure force. What is the acceleration of the cork?

21. || Helium has the lowest condensation point of any substance; the gas liquefies at 4.2 K. 1.0 L of liquid helium has a mass of 125 g. What is the volume of this amount of helium in gaseous form at STP (1 atm and 0°C)?

22. || Some jellyfish can eject stingers with remarkable force and
BIO speed. One species does this by building up a 15 MPa gauge
INT pressure that pushes against the base of a 2.0-$\mu$m-diameter stylet, forcing it outward.
    a. What force does the excess pressure exert on the stylet?
    b. The mass set into motion is a tiny $1.0 \times 10^{-12}$ kg. What is the resulting acceleration?

23. || A college student is working on her physics homework in her dorm room. Her room contains a total of $6.0 \times 10^{26}$ gas molecules. As she works, her body is converting chemical energy into thermal energy at a rate of 125 W. If her dorm room were an isolated system (dorm rooms can certainly feel like that) and if all of this thermal energy were transferred to the air in the room, by how much would the temperature increase in 10 min?

24. || Party stores sell small tanks containing 30 g of helium gas. If you use such a tank to fill 0.010 m$^3$ foil balloons (which don't stretch, and so have an internal pressure that is very close to atmospheric pressure), how many balloons can you expect to fill?

25. || On average, each person in the industrialized world is responsible for the emission of 10,000 kg of carbon dioxide ($CO_2$) every year. This includes the $CO_2$ that you generate directly, by burning fossil fuels to operate your car or your furnace, as well as the $CO_2$ generated on your behalf by electric generating stations and manufacturing plants. $CO_2$ is a greenhouse gas that contributes to climate change. What is the volume, in m$^3$, of 10,000 kg of $CO_2$ at 20°C and 1.0 atm pressure? To put this in perspective, the volume of a typical hot-air balloon is 3000 m$^3$.

### Section 12.3 Ideal-Gas Processes

26. ||| Suppose you start your day in Denver, on a cool 10°C spring day. The local atmospheric pressure is 85 kPa. You fill your car's tires until the gauge shows 210 kPa (about 30 psi). You then drive up to Fairplay, Colorado, where the atmospheric pressure is lower—70 kPa—and the temperature drops to 0°C. When you finish lunch, you check your tire pressure. What does the gauge read?

27. || A cylinder contains 3.0 L of oxygen at 300 K and 2.4 atm. The gas is heated, causing a piston in the cylinder to move outward. The heating causes the temperature to rise to 600 K and the volume of the cylinder to increase to 9.0 L. What is the final gas pressure?

28. || On a chilly 10°C day, you quickly take a deep breath—all
BIO your lungs can hold, 4.0 L. The air warms to your body temperature of 37°C. If the air starts at a pressure of 1.0 atm, and you hold the volume of your lungs constant (a good approximation) and the number of molecules in your lungs stays constant as well (also a good approximation), what is the increase in pressure inside your lungs?

29. ||| 0.10 mol of argon gas is admitted to an evacuated 50 cm$^3$ container at 20°C. The gas then undergoes heating at constant volume to a temperature of 300°C.
    a. What is the final pressure of the gas?
    b. Show the process on a $pV$ diagram. Include a proper scale on both axes.

30. || A football is inflated in the locker room before the game. The air warms as it is pumped, so it enters the ball at a temperature of 27°C. The ball is inflated to a gauge pressure of 13 psi. The ball is used for play at 10°C. Once the ball cools, what is the pressure in the ball? Assume that atmospheric pressure is 14.7 psi.

31. | 0.10 mol of argon gas is admitted to an evacuated 50 cm$^3$ container at 20°C. The gas then undergoes an isobaric heating to a temperature of 300°C.
    a. What is the final volume of the gas?
    b. Show the process on a $pV$ diagram. Include a proper scale on both axes.

Watch Video Solution Problem 12.38

32. 0.10 mol of argon gas is admitted to an evacuated 50 $cm^3$ container at 20°C. The gas then undergoes an isothermal expansion to a volume of 200 $cm^3$.
    a. What is the final pressure of the gas?
    b. Show the process on a $pV$ diagram. Include a proper scale on both axes.
33. 0.0040 mol of gas undergoes the process shown in Figure P12.33.
    a. What type of process is this?
    b. What are the initial and final temperatures?

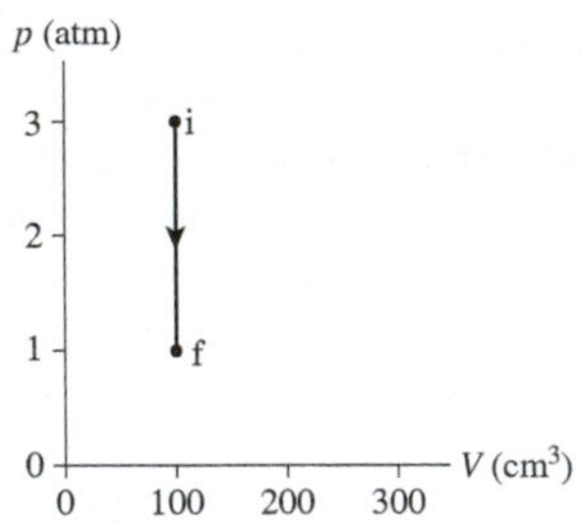


FIGURE P12.33

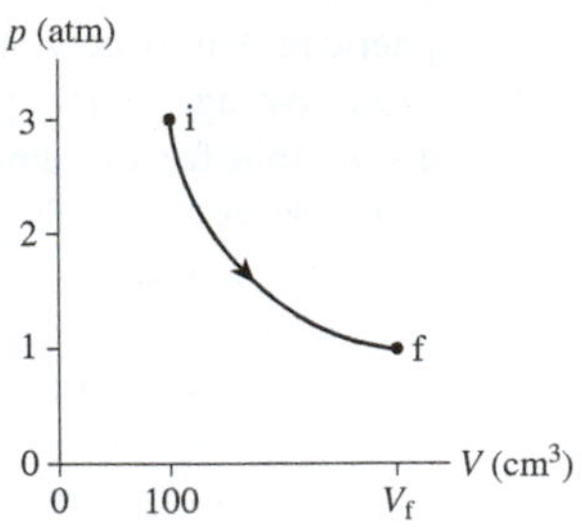


FIGURE P12.34

34. 0.0040 mol of gas follows the hyperbolic trajectory shown in Figure P12.34.
    a. What type of process is this?
    b. What are the initial and final temperatures?
    c. What is the final volume $V_f$?
35. A gas with an initial temperature of 900°C undergoes the process shown in Figure P12.35.
    a. What type of process is this?
    b. What is the final temperature?
    c. How many moles of gas are there?

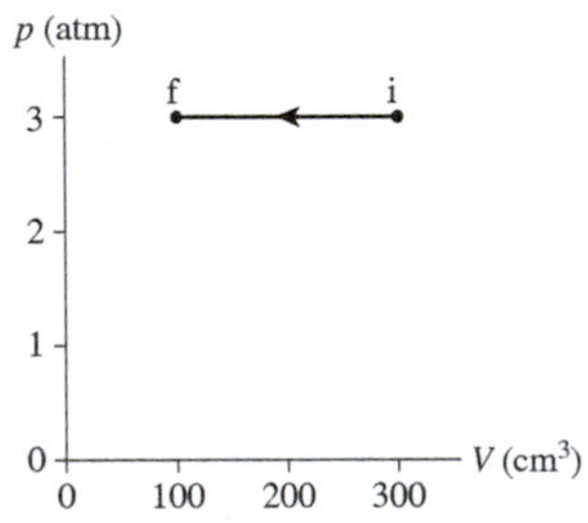


FIGURE P12.35

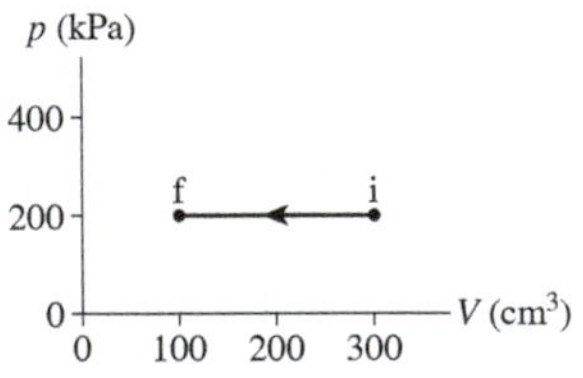


FIGURE P12.36

36. How much work is done on the gas in the process shown in Figure P12.36?
37. A 1.0 $cm^3$ air bubble is released from the sandy bottom of a warm, shallow sea, where the gauge pressure is 1.5 atm. The bubble rises slowly enough that the air inside remains at the same constant temperature as the water.
    a. What is the volume of the bubble as it reaches the surface?
    b. As the bubble rises, is heat energy transferred from the water to the bubble or from the bubble to the water? Explain.
38. A weather balloon rises through the atmosphere, its volume expanding from 4.0 $m^3$ to 12 $m^3$ as the temperature drops from 20°C to –10°C. If the initial gas pressure inside the balloon is 1.0 atm, what is the final pressure?
39. On a cool morning, when the temperature is 15°C, you measure the pressure in your car tires to be 30 psi. After driving 20 mi on the freeway, the temperature of your tires is 45°C. What pressure will your tire gauge now show?
40. BIO Many fish maintain buoyancy with a gas-filled swim bladder. The pressure inside the swim bladder is the same as the outside water pressure, so when a fish descends to a greater depth, the gas compresses. Adding gas to restore the original volume requires energy. A fish at a depth where the absolute pressure is 3.0 atm has a swim bladder with the desired volume of $5.0 \times 10^{-4}$ $m^3$. The fish now descends to a depth where the absolute pressure is 5.0 atm.
    a. The gas in the swim bladder is always the same temperature as the fish's body. What is the volume of the swim bladder at the greater depth?
    b. The fish remains at the greater depth, slowly adding gas to the swim bladder to return it to its desired volume. How much work is required?
41. Suppose you inflate your car tires to 35 psi on a 20°C day. Later, the temperature drops to 0°C. What is the pressure in your tires now?

## Section 12.4 Thermal Expansion

42. When you apply the brakes on your car, the kinetic energy of your vehicle is transformed into thermal energy in your brake disks. During a mountain descent, a 28.00-cm-diameter iron brake disk heats up from 30°C to 180°C. What is the diameter of the disk after it heats up?
43. The length of a steel beam increases by 0.73 mm when its temperature is raised from 22°C to 35°C. What is the length of the beam at 22°C?
44. In the chapter, you saw that bridges often have expansion joints to account for the changes in size due to temperature changes. Suppose a bridge is supported by steel girders; the central span has a girder that is 30 m long. If the girder warms from 0°C to 30°C, by how much does its length increase?
45. Older railroad tracks in the U.S. are made of 12-m-long pieces of steel. When the tracks are laid, gaps are left between the sections to prevent buckling when the steel thermally expands. If a track is laid at 16°C, how large should the gaps be if the track is not to buckle when the temperature is as high as 50°C?
46. At 20°C, the hole in an aluminum ring is 2.500 cm in diameter. You need to slip this ring over a steel shaft that has a room-temperature diameter of 2.506 cm. To what common temperature should the ring and the shaft be heated so that the ring will just fit onto the shaft?
47. The temperature of an aluminum disk is increased by 120°C. By what percentage does its volume increase?

## Section 12.5 Specific Heat and Heat of Transformation

48. BIO INT When people exercise, they report feeling exhaustion when their body temperature rises to 39.7°C. A 68 kg man is running at 15 km/h on a hot, muggy day—so hot and muggy that he can't lose any heat to the environment. His body temperature starts at 37.0°C. How long can he run before he feels exhausted?
49. How much energy must be removed from a 200 g block of ice to cool it from 0°C to −30°C?
50. Suppose that the 800 W of radiation in a microwave oven is absorbed by 250 g of water in a very lightweight cup. How long will it take for the water to warm up from 20°C to 80°C?
51. a. 100 J of heat energy are transferred to 20 g of mercury initially at 20°C. By how much does the temperature increase?
    b. How much heat is needed to raise the temperature of 20 g of water by the same amount?

52. || Dogs keep themselves cool by panting, rapidly breathing air in and out. Panting results in evaporation from moist tissues of the airway and lungs, which cools the animal. Measurements show that, on a 35°C day with a relative humidity of 50%, a 12 kg dog loses 1.0 g of water per minute if it is panting vigorously. What rate of heat loss, in watts, does this achieve?
BIO

53. || The maximum amount of water an adult in temperate climates can perspire in one hour is typically 1.8 L. However, after several weeks in a tropical climate the body can adapt, increasing the maximum perspiration rate to 3.5 L/h. At what rate, in watts, is energy being removed when perspiring that rapidly? Assume all of the perspired water evaporates. At body temperature, the heat of vaporization of water is $L_v = 24 \times 10^5$ J/kg.
BIO

54. || A pronghorn antelope can run at a remarkable 18 m/s for up to 10 minutes, almost triple the speed that an elite human runner can maintain. For a 32 kg pronghorn, this requires an astonishing 3.4 kW of metabolic power, which leads to a significant increase in body temperature. If the pronghorn had no way to exhaust heat to the environment, by how much would its body temperature increase during this run? (In fact, it will lose some heat, so the rise won't be this dramatic, but it will be quite noticeable, requiring adaptations that keep the pronghorn's brain cooler than its body in such circumstances.)
BIO INT

55. || Alligators and other reptiles don't use enough metabolic energy to keep their body temperatures constant. They cool off at night and must warm up in the sun in the morning. Suppose a 300 kg alligator with an early-morning body temperature of 25°C is absorbing radiation from the sun at a rate of 1200 W. How long will the alligator need to warm up to a more favorable 30°C? (Assume that the specific heat of the reptilian body is the same as that of the mammalian body.)
BIO

56. ||| The Bodø NATO base in northern Norway uses a heat pump to extract heat from ocean water. 7.0°C ocean water is continuously drawn into the system, heat is extracted, and the water is returned to the ocean at 3.0°C. When the system is working at full capacity, 2000 kW of heat is delivered to the building at the cost of 600 kW of electric energy.
INT
   a. How much energy is removed from the water (the cold reservoir) each second?
   b. How much water moves through the system each second?

57. || When air is inhaled, it quickly becomes saturated with water vapor as it passes through the moist upper airways. When a person breathes dry air, about 25 mg of water are exhaled with each breath. At 12 breaths/min, what is the rate of energy loss due to evaporation? Express your answer in both watts and Calories per day. At body temperature, the heat of vaporization of water is $L_v = 24 \times 10^5$ J/kg.

58. || Over the course of a day, 4.0 kg of water evaporates from the leaves of a corn plant.
BIO
   a. How much energy is required to evaporate the water? (Assume that the temperature of the leaves is 30°C.)
   b. If the plant is active for 12 hours, how much power does this correspond to? You can think of this as the necessary power to drive transport in the plant.

59. ||| It is important for the body to have mechanisms to effectively cool itself; if not, moderate exercise could easily increase body temperatures to dangerous levels. Suppose a 70 kg man runs on a treadmill for 30 min, using a metabolic power of 1000 W. Assume that all of this power goes to thermal energy in the body. If he couldn't perspire or otherwise cool his body, by how much would his body temperature rise during this exercise?

60. ||| What minimum heat is needed to bring 100 g of water at 20°C to the boiling point and completely boil it away?

## Section 12.6 Calorimetry

61. || Brewed coffee is often too hot to drink right away. You can cool it with an ice cube, but this dilutes it. Or you can buy a device that will cool your coffee without dilution—a 200 g aluminum cylinder that you take from your freezer and place in a mug of hot coffee. If the cylinder is cooled to –20°C, a typical freezer temperature, and then dropped into a large cup of coffee (essentially water, with a mass of 500 g) at 85°C, what is the final temperature of the coffee?

62. || If you pour whisky over ice, the ice will cool the drink, but it will also dilute it. A solution is to use whisky stones. Suppose Ernest pours 55 g of whisky at 22°C room temperature, and then adds three whisky stones to cool it. Each stone is a 32 g soapstone cube that is stored in the freezer at –15°C. The specific heat of soapstone is 980 J/kg · K; the specific heat of whisky is 3400 J/kg · K. What is the final temperature of the whisky?
INT

63. ||| 30 g of copper pellets are removed from a 300°C oven and immediately dropped into 100 mL of water at 20°C in an insulated cup. What will the new water temperature be?

64. ||| It's possible to boil water by adding hot rocks to it, a technique that has been used in many societies over time. If you heat a rock in the fire, you can easily get it to a temperature of 500°C. If you use granite or other similar stones, the specific heat is about 800 J/kg · K. If 5.0 kg of water at 10°C is in a leak-proof vessel, what minimum number of 1.0 kg stones must be added to bring the water to a boil?

65. ||| A copper block is removed from a 300°C oven and dropped into 1.00 kg of water at 20.0°C. The water quickly reaches 25.5°C and then remains at that temperature. What is the mass of the copper block?

66. || If you ask for an iced coffee, the barista will add ice cubes to hot coffee to cool it down. To cool 350 mL of coffee at 90°C to 0°C, what is the minimum mass of 0°C ice required?

67. || If a person has a dangerously high fever, submerging her in ice water is a bad idea, but an ice pack can help to quickly bring her body temperature down. How many grams of ice at 0°C will be melted in bringing down a 60 kg patient's fever from 40°C to 39°C?
BIO

## Section 12.7 Specific Heats of Gases

68. | A container holds 1.0 g of argon at a pressure of 8.0 atm.
   a. How much heat is required to increase the temperature by 100°C at constant volume?
   b. How much will the temperature increase if this amount of heat energy is transferred to the gas at constant pressure?

69. || A container holds 1.0 g of oxygen at a pressure of 8.0 atm.
   a. How much heat is required to increase the temperature by 100°C at constant pressure?
   b. How much will the temperature increase if this amount of heat energy is transferred to the gas at constant volume?

Watch Video Solution Problem 12.73

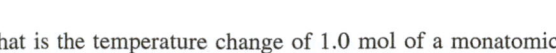

70. | What is the temperature change of 1.0 mol of a monatomic gas if its thermal energy is increased by 1.0 J?

71. | Heating 2.5 mol of neon in a rigid container causes the temperature to increase by 15°C. By how much would the temperature increase if the same amount of heat was added at a constant pressure?

72. || You draw in a deep breath on a chilly day, inhaling 3.0 L of
BIO 0°C air. If the pressure in your lungs is a constant 1.0 atm, how much heat must your body supply to warm the air to your 37°C internal body temperature?

## Section 12.8 Heat Transfer

73. ||| A 1.8-cm-thick wood floor covers a 4.0 m × 5.5 m room. The subfloor on which the flooring sits is at a temperature of 16.2°C, while the air in the room is at 19.6°C. What is the rate of heat conduction through the floor?

74. |||| A stainless-steel-bottomed kettle, its bottom 24 cm in diameter and 1.6 mm thick, sits on a burner. The kettle holds boiling water, and energy flows into the water from the kettle bottom at 800 W. What is the temperature of the bottom surface of the kettle?

75. ||| Seals may cool themselves by
BIO using *thermal windows,* patches on their bodies with much higher than average surface temperature. Suppose a seal has a 0.030 $m^2$ thermal window at a temperature of 30°C. If the seal's surroundings are a frosty −10°C, what is the net rate of energy loss by radiation? Assume an emissivity equal to that of a human.

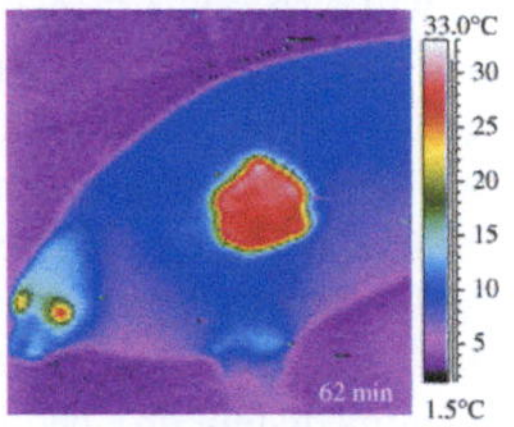


76. || Electronics and inhabitants of the International Space Station generate a significant amount of thermal energy that the station must get rid of. The only way that the station can exhaust thermal energy is by radiation, which it does using thin, 1.8-m-by-3.6-m panels that have a working temperature of about 6°C. How much power is radiated from each panel? Assume that the panels are in the shade so that the absorbed radiation will be negligible. Assume that the emissivity of the panels is 1.0. **Hint:** Don't forget that the panels have two sides!

77. || The glowing filament in a lamp is radiating energy at a rate of 60 W. At the filament's temperature of 1500°C, the emissivity is 0.23. What is the surface area of the filament?

78. ||| A 30 kg male emperor penguin under a clear sky in the
BIO Antarctic winter loses very little heat to the environment by convection; its feathers provide very good insulation. It does lose some heat through its feet to the ice, and some heat due to evaporation as it breathes; the combined power is about 12 W. The outside of the penguin's body is a chilly –22°C, but its surroundings are an even chillier –38°C. The penguin's surface area is 0.56 $m^2$, and its emissivity is 0.97.
   a. What is the net rate of energy loss by radiation?
   b. If the penguin has a 45 W basal metabolic rate, will it feel warm or cold under these circumstances?

79. ||| If you lie on the ground at night with no cover, you get cold
BIO rather quickly. Much of this is due to energy loss by radiation. At night in a dry climate, the temperature of the sky can drop to −40°C. If you are lying on the ground with thin clothing that provides little insulation, the surface temperature of your skin and clothes will be about 30°C. Estimate the net rate at which your body loses energy by radiation to the night sky under these conditions. **Hint:** What area should you use?

## Section 12.9 Diffusion

80. || A nerve impulse is propagated across a synapse when the
BIO neurotransmitter acetylcholine is released from the pre-synaptic membrane and diffuses across the 20-nm-wide synaptic cleft to the post-synaptic membrane.
   a. How long does it take a nerve signal to cross a synapse? You can assume that the synaptic fluid is essentially water.
   b. What is the speed of transmission, in m/s?

81. || The small capillaries in the lungs are
BIO in close contact with the alveoli. A red blood cell takes up oxygen during the 0.5 s that it squeezes through a capillary at the surface of an alveolus. What is the diffusion time for oxygen across the 1-$\mu$m-thick membrane separating air from blood? Assume that the diffusion coefficient for oxygen in tissue is $2 \times 10^{-11}$ $m^2/s$. Give your answer to 1 significant figure.

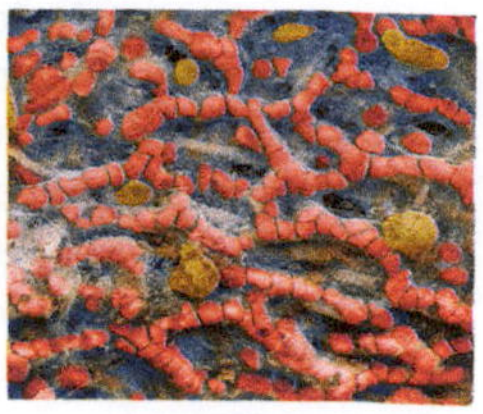

82. ||| The partial pressure of oxygen in the lungs is about 150 mm
BIO of Hg. (The partial pressure is the pressure of the oxygen alone, if all other gases were removed.) This corresponds to a concentration of $5.3 \times 10^{24}$ molecules per $m^3$. In the oxygen-depleted blood entering the pulmonary capillaries, the concentration is $1.4 \times 10^{24}$ molecules per $m^3$. The blood is separated from air in the alveoli of the lungs by a 1-$\mu$m-thick membrane. What is the rate of transfer of oxygen to the blood through the $5 \times 10^{-9}$ $m^2$ surface area of one alveolus? Give your answer in both molecules/s and $\mu$mol/s. Assume The diffusion coefficient for oxygen in tissue is $2 \times 10^{-11}$ $m^2/s$. Give your answer to 1 significant figure.

## General Problems

83. || A compressed-air cylinder is known to fail if the pressure exceeds 110 atm. A cylinder that was filled to 25 atm at 20°C is stored in a warehouse. Unfortunately, the warehouse catches fire and the temperature reaches 950°C. Does the cylinder explode?

84. ||| 80 J of work are done on the gas in the process shown in Figure P12.84. What is $V_f$ in $cm^3$?

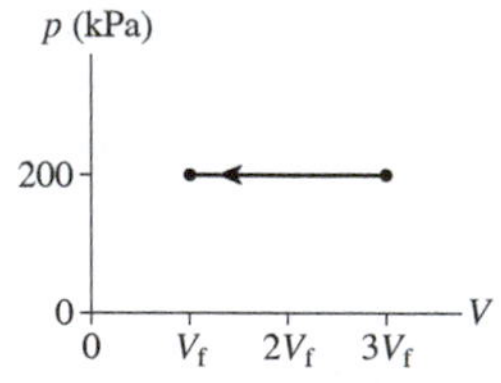


FIGURE P12.84

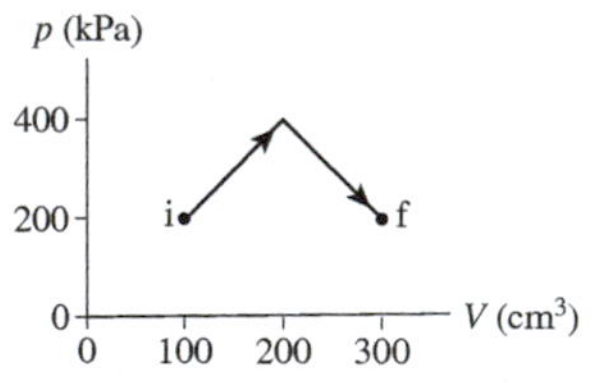


FIGURE P12.85

85. ||| How much work is done by the gas in the process shown in Figure P12.85?

86. | 0.10 mol of gas undergoes the process 1 → 2 shown in Figure P12.86.
   a. What are temperatures $T_1$ and $T_2$?
   b. What type of process is this?
   c. The gas undergoes constant-volume heating from point 2 until the pressure is restored to the value it had at point 1. What is the final temperature of the gas?

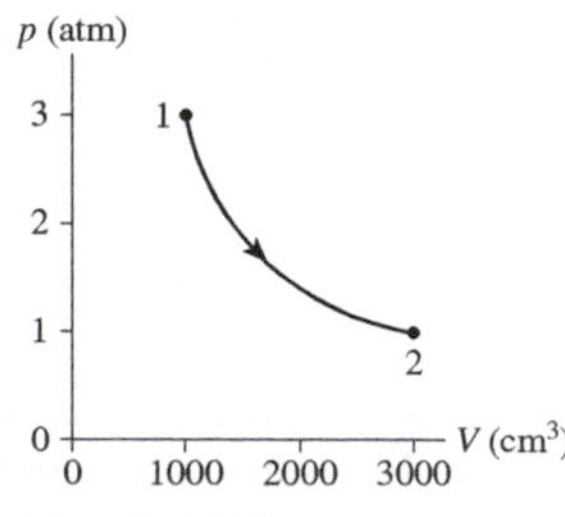


FIGURE P12.86

87. || Your heart does work to pump blood through your body. BIO INT Each minute, 5.0 L of blood travels through your circulatory system. The pressure drops from 16 kPa as it exits the heart to approximately zero as it returns to the heart. We can calculate the work required to move a fluid volume $V$ through a pressure difference with a formula very similar to the one you saw for work in gases:
   a. What is the work done to pump the blood?
   b. What is the output power of the heart?

88. ||| A 5.0-m-diameter garden pond holds $5.9 \times 10^3$ kg of water. 400 W/m$^2$ of solar energy strikes the pond's surface, meaning each square meter of the pond's surface is absorbing 400 W. How many hours will it take for the pond to warm from 15°C to 25°C?

89. || James Joule (after whom the unit of energy is named) claimed that the water at the bottom of Niagara Falls should be warmer than the water at the top, 51 m above the bottom. He reasoned that the falling water would transform its gravitational potential energy at the top into thermal energy at the bottom, where turbulence brings the water almost to a halt. If this transformation is the only process occurring, how much warmer will the water at the bottom be?

90. |||| Susan, whose mass is 68 kg, climbs 59 m to the top of the BIO INT Cape Hatteras lighthouse.
   a. During the climb, by how much does her potential energy increase?
   b. For a typical efficiency of 25%, what metabolic energy does she require to complete the climb?
   c. When exercising, the body must perspire and use other mechanisms to cool itself to avoid potentially dangerous increases on body temperature. If we assume that Susan doesn't perspire or otherwise cool herself and that all of the "lost" energy goes into increasing her body temperature, by how much would her body temperature increase during this climb?

91. ||| A typical nuclear reactor generates 1000 MW of electric energy. In doing so, it produces "waste heat" at a rate of 2000 MW, and this heat must be removed from the reactor. Many reactors are sited next to large bodies of water so that they can use the water for cooling. Consider a reactor where the intake water is at 18°C. State regulations limit the temperature of the output water to 30°C so as not to harm aquatic organisms. How many kilograms of cooling water have to be pumped through the reactor each minute?

92. ||| An often-overlooked consequence of power generation is INT "thermal pollution" from excess heat deposited in the environment. The heating can be considerable. Here are some typical numbers: A nuclear power plant generates 1.0 GW of electric power with an operating efficiency of 35%. The reactor has a single-pass cooling system; water from a nearby river is brought in, warmed, and returned to the river. When operating at full power, the plant takes in 90 m$^3$ of water per second, comparable to the flow of a small river. If the river water comes in at 10°C, at what temperature does it emerge?

93. |||| A 68 kg woman cycles at a constant 15 km/h. All of the met- BIO INT abolic energy that does not go to forward propulsion is converted to thermal energy in her body. If the only way her body has to keep cool is by evaporation, how many kilograms of water must she lose to perspiration each hour to keep her body temperature constant?

94. ||| A 1200 kg car traveling at 60 mph quickly brakes to a halt. INT The kinetic energy of the car is converted to thermal energy of the disk brakes. The brake disks (one per wheel) are iron disks with a mass of 4.0 kg. Estimate the temperature rise in each disk as the car stops.

95. ||| A 5000 kg African elephant has a resting metabolic rate of BIO INT 2500 W. On a hot day, the elephant's environment is likely to be nearly the same temperature as the animal itself, so cooling by radiation is not effective. The only plausible way to keep cool is by evaporation, and elephants spray water on their body to accomplish this.

If this is the only possible means of cooling, how many kilograms of water per hour must be evaporated from an elephant's skin to keep it at a constant temperature?

96. ||| When you put on the brakes on your bicycle, friction heats INT the steel rims of your wheels. Could this heating be a problem? Suppose a 65 kg cyclist with a 15 kg bike is descending Trail Ridge Road in Rocky Mountain National Park, going down a 7.0% grade and thus losing 7.0 m in height for every 100 m of travel along the road. If the cyclist keeps a constant speed of 6.0 m/s, and we assume that all of the "lost" energy ends up as thermal energy in the two steel rims, each of mass 0.80 kg, by how much does the temperature of each rim rise in 1.0 minute?

97. || An experiment measures the temperature of a 200 g substance while steadily supplying heat to it. Figure P12.97 shows the results of the experiment. What are (a) the specific heat of the liquid phase and (b) the heat of vaporization?

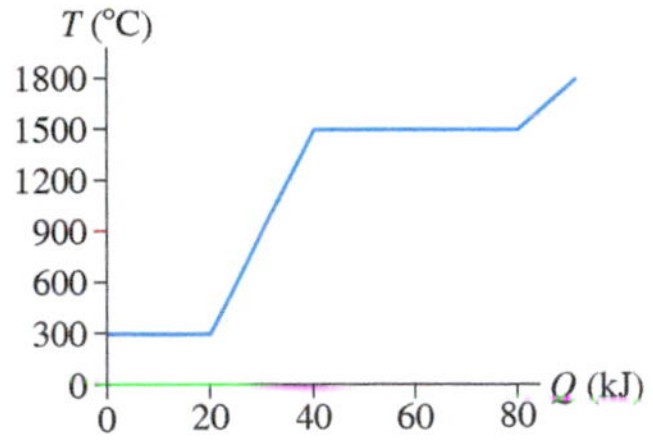


FIGURE P12.97

98. |||| Your 300 mL cup of coffee is too hot to drink when served at 90°C. What is the mass of an ice cube, taken from a −20°C freezer, that will cool your coffee to a pleasant 60°C?

99. || A gas is compressed from 600 cm$^3$ to 200 cm$^3$ at a constant pressure of 400 kPa. At the same time, 100 J of heat energy is transferred out of the gas. What is the change in thermal energy of the gas during this process?

100. |||| An expandable cube, initially 20 cm on each side, contains 3.0 g of helium at 20°C. 1000 J of heat energy are transferred to this gas. What are (a) the final pressure if the process is at constant volume and (b) the final volume if the process is at constant pressure?

101. ||| 0.10 mol of a monatomic gas follows the process shown in Figure P12.101.
   a. How much heat energy is transferred to or from the gas during process 1 → 2?
   b. How much heat energy is transferred to or from the gas during process 2 → 3?
   c. What is the total change in thermal energy of the gas?

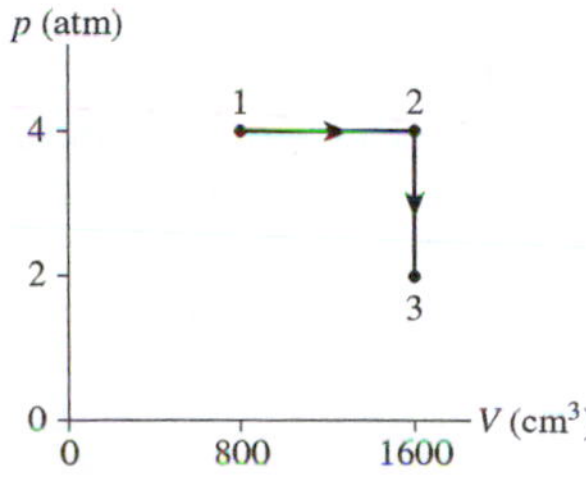


FIGURE P12.101

102. || A monatomic gas follows the process $1 \rightarrow 2 \rightarrow 3$ shown in Figure P12.102. How much heat is needed for (a) process $1 \rightarrow 2$ and (b) process $2 \rightarrow 3$?

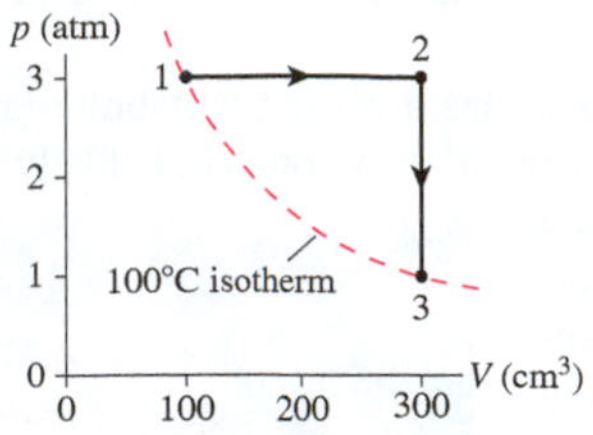


FIGURE P12.102

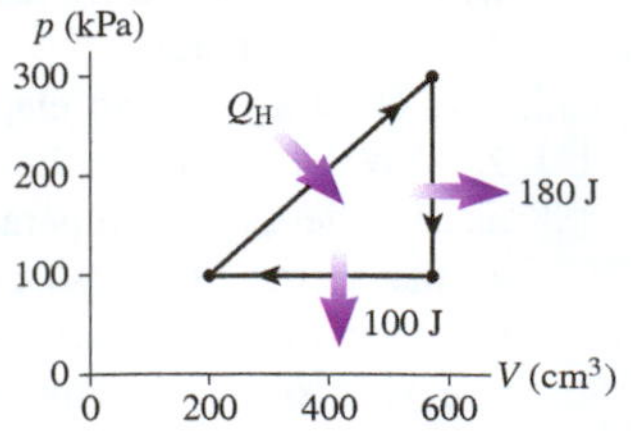


FIGURE P12.103

103. ||| What are (a) the heat $Q_H$ extracted from the hot reservoir and
INT (b) the efficiency for a heat engine described by the $pV$ diagram of Figure P12.103?

104. | Homes are often insulated with fiberglass insulation in their walls and ceiling. The thermal conductivity of fiberglass is 0.040 W/m·K. Suppose that the total surface area of the walls and roof of a windowless house is 370 $m^2$ and that the thickness of the insulation is 10 cm. At what rate does heat leave the house on a day when the outside temperature is 30°C colder than the inside temperature?

105. || The surface area of an adult human is about 1.8 $m^2$. Suppose
BIO a person with a skin temperature of 34°C is standing with bare skin in a room where the air is 25°C but the walls are 17°C.

a. There is a "dead-air" layer next to your skin that acts as insulation. If the dead-air layer is 5.0 mm thick, what is the person's rate of heat loss by conduction?
b. What is the person's net radiation loss to the walls? The emissivity of skin is 0.97.
c. Does conduction or radiation contribute more to the total rate of energy loss?
d. If the person is metabolizing food at a rate of 155 W, does the person feel comfortable, chilly, or too warm?

## MCAT-Style Passage Problems

### Thermal Properties of the Oceans

Seasonal temperature changes in the ocean only affect the top layer of water, to a depth of 500 m or so. This "mixed" layer is thermally isolated from the cold, deep water below. The average temperature of this top layer of the world's oceans, which has area $3.6 \times 10^8$ km², is approximately 17°C.

In addition to seasonal temperature changes, the oceans have experienced an overall warming trend over the last century that is expected to continue as the earth's climate changes. A warmer ocean means a larger volume of water; the oceans will rise. Suppose the average temperature of the top layer of the world's oceans were to increase from a temperature $T_i$ to a temperature $T_f$. The area of the oceans will not change, as this is fixed by the size of the ocean basin, so any thermal expansion of the water will cause the water level to rise, as shown in Figure P12.106. The original volume is the product of the original depth and the surface area, $V_i = Ad_i$. The change in volume is given by $\Delta V = A\,\Delta d$.

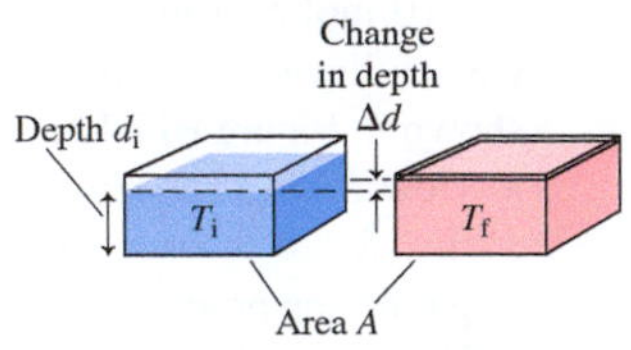


FIGURE P12.106

106. | If the top 500 m of ocean water increased in temperature from 17°C to 18°C, what would be the resulting rise in ocean height?

A. 0.11 m  B. 0.22 m
C. 0.44 m  D. 0.88 m

107. | Approximately how much energy would be required to raise the temperature of the top layer of the oceans by 1°C? (1 $m^3$ of water has a mass of 1000 kg.)

A. $1 \times 10^{24}$ J  B. $1 \times 10^{21}$ J
C. $1 \times 10^{18}$ J  D. $1 \times 10^{15}$ J

108. | An increase in temperature of the water may cause other changes. An increase in surface water temperature is likely to ________ the rate of evaporation.

A. Increase  B. Not affect  C. Decrease

109. | The ocean is mostly heated from the top, by light from the sun. The warmer surface water doesn't mix much with the colder deep ocean water. This lack of mixing can be ascribed to a lack of

A. Conduction.  B. Convection.
C. Radiation.  D. Evaporation.

### The Work of Breathing

We've seen that the area *under* a $pV$ graph is the work done in an ideal-gas process. If the process follows a closed curve, the work is the area *inside* the curve. The graph in Figure P12.110 shows the gauge pressure in the lungs versus the volume of gas in the lungs for a person who is taking rapid, deep breaths. During one complete breath, the pressure-versus-volume data trace out the curve shown, in the direction of the arrows. The energy expended in one complete breath is represented by the shaded area inside the curve. If you graph pressure-versus-volume data for normal breathing, the upper and lower lines are much closer together and the volume range is much smaller.

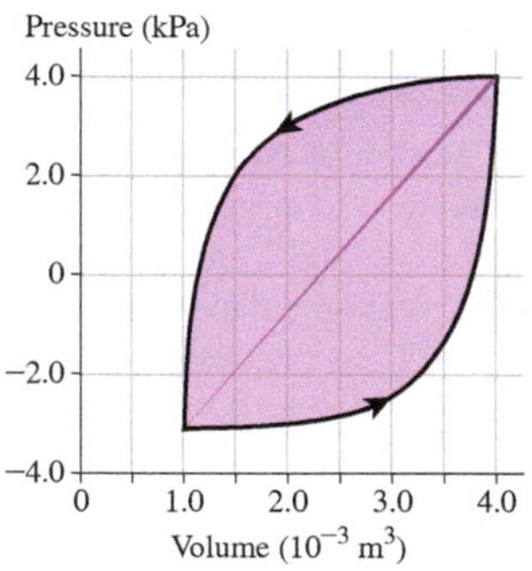


FIGURE P12.110

110. || In the graph in Figure P12.110, the ________ line corresponds to inhalation; the ________ line corresponds to
INT
BIO exhalation.

A. upper, lower  B. lower, upper

111. || For the graph in the figure, approximately how much energy
INT is required for one complete breath?
BIO

A. 5 J  B. 15 J
C. 25 J  D. 35 J

112. || For one cycle of normal breathing, we expect the energy
INT expended to be ________ for the cycle illustrated in the figure.
BIO

A. more than  B. the same as  C. less than

# 13 Fluids

The albatross must build up speed before it can take off. Why does this graceful flier require such an awkward start to get airborne?

## LOOKING AHEAD ➤

### Pressure in Liquids

A liquid's pressure increases with depth. The high pressure at the base of this water tower pushes water throughout the city.

You'll learn about **hydrostatics**—how liquids behave when they're in equilibrium.

### Buoyancy

These students are competing in a concrete canoe contest. How can such heavy, dense objects stay afloat?

You'll learn how to find the **buoyant force** on an object in a fluid using **Archimedes' principle**.

### The Circulatory System

This woman is measuring her blood pressure. Why does she wear the cuff where she does?

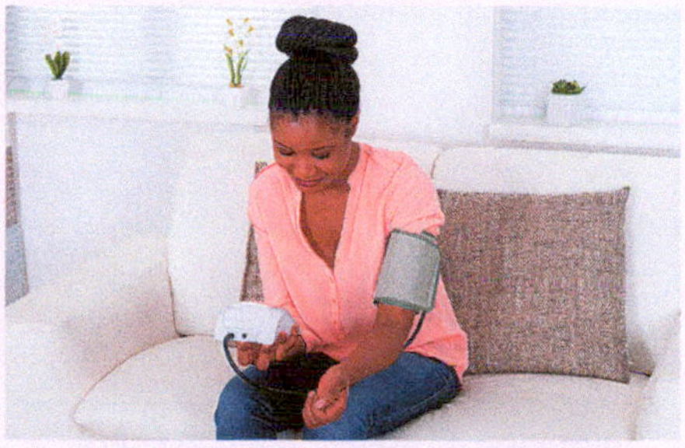

You'll explore how the principles of **fluid dynamics** apply to blood flow in arteries, capillaries, and veins.

**GOAL** To understand the static and dynamic properties of fluids.

## LOOKING BACK ◀

### Equilibrium

In Section 5.1, you learned that for an object to be at rest—in *static equilibrium*—the net force on it must be zero. We'll use the principle of equilibrium in this chapter to understand how an object floats.

This mountain goat is in equilibrium: Its weight is balanced by the normal force of the rock.

**STOP TO THINK**

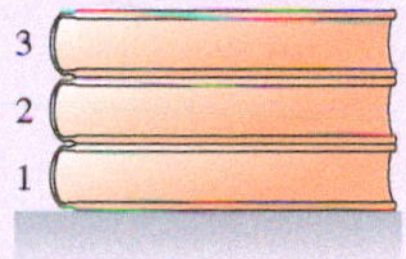


Three identical books are stacked vertically. The normal force of book 1 on book 2 is

A. Equal to the weight of one book.
B. Less than the weight of one book.
C. Greater than the weight of one book.

FIGURE 13.1 Simple atomic-level models of gases and liquids.

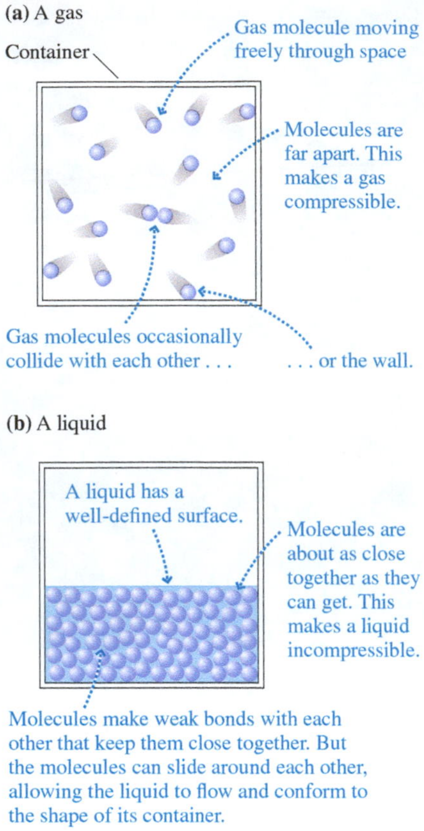

(a) A gas
Container
Gas molecule moving freely through space
Molecules are far apart. This makes a gas compressible.
Gas molecules occasionally collide with each other . . .
. . . or the wall.
(b) A liquid
A liquid has a well-defined surface.
Molecules are about as close together as they can get. This makes a liquid incompressible.
Molecules make weak bonds with each other that keep them close together. But the molecules can slide around each other, allowing the liquid to flow and conform to the shape of its container.

TABLE 13.1 Densities of fluids at 1 atm pressure

| Substance | $\rho$ ($kg/m^3$) |
|---|---|
| Hydrogen gas (20°C) | 0.083 |
| Helium gas (20°C) | 0.166 |
| Air (20°C) | 1.20 |
| Air (0°C) | 1.28 |
| Gasoline | 680 |
| Ethyl alcohol | 790 |
| Oil (typical) | 900 |
| Water | 1000 |
| Seawater | 1030 |
| Blood (whole) | 1060 |
| Glycerin | 1260 |
| Mercury | 13,600 |

# 13.1 Fluids and Density

A **fluid** is simply a substance that flows. Because they flow, fluids take the shape of their containers rather than retaining a shape of their own. Gases and liquids are both fluids, and their similarities are often more important than their differences.

As you learned in ◀◀ SECTION 12.2, a gas, as shown in FIGURE 13.1a, is a system in which each molecule moves freely through space until, on occasion, it collides with another molecule or with the wall of the container. Gases are *compressible;* that is, the volume of a gas is easily increased or decreased, a consequence of the "empty space" between the molecules in a gas.

Liquids are more complicated than either gases or solids. Liquids, like solids, are essentially *incompressible.* This property tells us that the molecules in a liquid, as in a solid, are about as close together as they can get. At the same time, a liquid flows and deforms to fit the shape of its container. The fluid nature of a liquid tells us that the molecules are free to move around. Together, these observations suggest the model of a liquid shown in FIGURE 13.1b.

## Density

Suppose we have several blocks of copper, each of different size. Each block has a different mass $m$ and a different volume $V$. Nonetheless, all the blocks are copper, so there should be some quantity that has the *same* value for all the blocks, telling us, "This is copper, not some other material." The most important such parameter is the *ratio* of mass to volume, which we call the **mass density** $\rho$ (lowercase Greek rho):

$$\rho = \frac{m}{V} \tag{13.1}$$

Mass density of an object of mass $m$ and volume $V$

Therefore, an object of mass density $\rho$ and volume $V$ has mass

$$m = \rho V \tag{13.2}$$

The SI units of mass density are $kg/m^3$. Nonetheless, units of $g/cm^3$ are widely used. You need to convert these to SI units before doing most calculations:

$$1\ g/cm^3 = 1000\ kg/m^3$$

The mass density is usually called simply "the density" if there is no danger of confusion. Density is independent of the object's size. That is, mass and volume are parameters that characterize a *specific piece* of some substance—say, copper—whereas density characterizes the substance itself. All pieces of copper have the same density, which differs from the density of almost any other substance.

TABLE 13.1 provides a short list of the densities of various fluids. Notice the enormous difference between the densities of gases and liquids. Gases have lower densities because the molecules in gases are farther apart than in liquids. Also, the density of a liquid varies only slightly with temperature because its molecules are always nearly in contact. The density of a gas, such as air, has a larger variation with temperature because it's easy to change the already large distance between the molecules.

What does it *mean* to say that the density of gasoline is 680 $kg/m^3$? Recall in Chapter 1 we discussed the meaning of the word "per." We found that it meant "for each," so that 2 miles per hour means you travel 2 miles *for each* hour that passes. In the same way, saying that the density of gasoline is 680 kg per cubic meter means that there are 680 kg of gasoline *for each* 1 cubic meter of the liquid. If we have 2 $m^3$ of gasoline, each will have a mass of 680 kg, so the total mass will be $2 \times 680\ kg = 1360\ kg$.

**EXAMPLE 13.1 Weighing the air in a living room**

What is the mass of air in a living room with dimensions $4.0\text{ m} \times 6.0\text{ m} \times 2.5\text{ m}$?

**STRATEGIZE** We will compute the volume of air in the room, and with this figure, use the density to compute the mass.

**PREPARE** Table 13.1 gives air density at a temperature of 20°C, which is about room temperature.

**SOLVE** The room's volume is

$$V = (4.0\text{ m}) \times (6.0\text{ m}) \times (2.5\text{ m}) = 60\text{ m}^3$$

The mass of the air is

$$m = \rho V = (1.20\text{ kg/m}^3)(60\text{ m}^3) = 72\text{ kg}$$

**ASSESS** This is perhaps more mass—about that of an adult person—than you might have expected from a substance that hardly seems to be there. For comparison, a swimming pool this size would contain 60,000 kg of water.

**STOP TO THINK 13.1** A piece of glass is broken into two pieces of different size. Rank in order, from largest to smallest, the densities of pieces 1, 2, and 3.

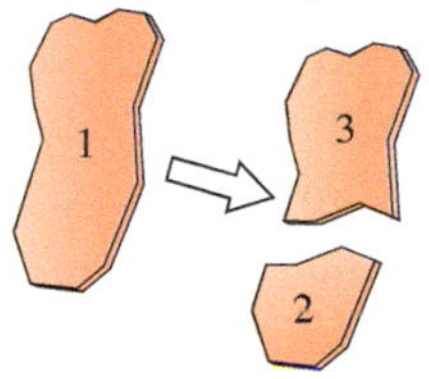

1
3
2


## 13.2 Pressure

In ◀◀ SECTION 12.2, you learned how a gas exerts a force on the walls of its container. Liquids also exert forces on the walls of their containers, as shown in FIGURE 13.2, where a force $\vec{F}$ due to the liquid pushes against a small area $A$ of the wall. Just as for a gas, we define the pressure at this point in the fluid to be the ratio of the force to the area on which the force is exerted:

$$p = \frac{F}{A} \tag{13.3}$$

This is the same as Equation 12.12 of Chapter 12. Recall also from Chapter 12 that the SI unit of pressure, the pascal, is defined as

$$1\text{ pascal} = 1\text{ Pa} = 1\,\frac{\text{N}}{\text{m}^2}$$

The force due to a fluid's pressure pushes not only on the walls of its container, but on *all* parts of the fluid itself. If we punch holes in a container of water, the water spurts out from the holes, as in FIGURE 13.3. It is the force due to the pressure of the water behind each hole that pushes the water forward through the holes.

Let's imagine measuring the pressure at different points within a fluid using the simple pressure-measuring device shown in FIGURE 13.4a. Because we know the spring constant $k$ and the area $A$, we can determine the pressure by measuring the compression of the spring. FIGURE 13.4b shows the results of some simple experiments.

FIGURE 13.2 The fluid presses against area $A$ with force $\vec{F}$.

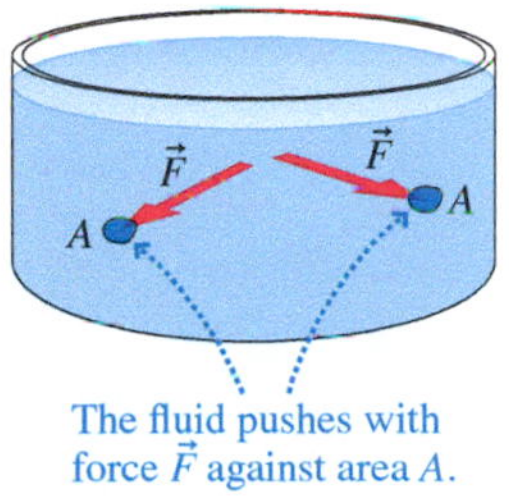

F
F
A
A
The fluid pushes with force F against area A.


FIGURE 13.3 Pressure pushes the water *sideways*, out of the holes.

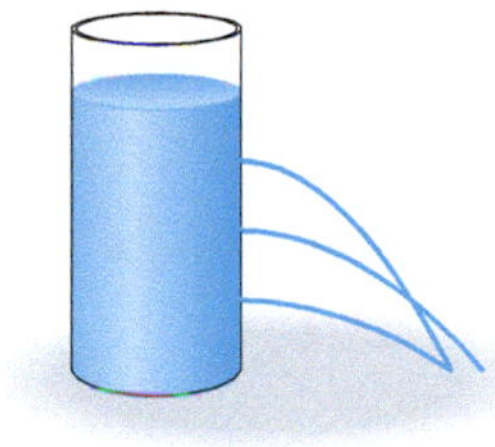

FIGURE 13.4 Learning about pressure.

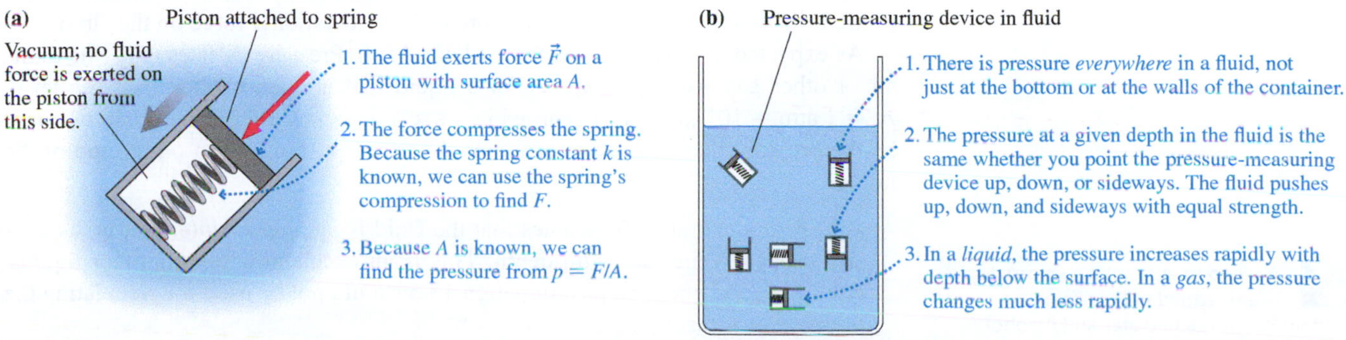

(a)
Piston attached to spring
Vacuum; no fluid force is exerted on the piston from this side.
1. The fluid exerts force F on a piston with surface area A.
2. The force compresses the spring. Because the spring constant k is known, we can use the spring's compression to find F.
3. Because A is known, we can find the pressure from p = F/A.
(b)
Pressure-measuring device in fluid
1. There is pressure everywhere in a fluid, not just at the bottom or at the walls of the container.
2. The pressure at a given depth in the fluid is the same whether you point the pressure-measuring device up, down, or sideways. The fluid pushes up, down, and sideways with equal strength.
3. In a liquid, the pressure increases rapidly with depth below the surface. In a gas, the pressure changes much less rapidly.

The first statement in Figure 13.4b emphasizes again that pressure exists at *all* points within a fluid, not just at the walls of the container. You may recall that tension exists at *all* points in a string, not only at its ends where it is tied to an object. We understood tension as the different parts of the string *pulling* against each other. Pressure is an analogous idea, except that the different parts of a fluid are *pushing* against each other.

The spring-and-piston gauge shown in Figure 13.4 measures the actual or *absolute* pressure $p$, an idea introduced in Chapter 12. But you also learned in ◀◀ SECTION 12.2 that most practical pressure gauges, such as tire gauges, measure the *gauge pressure* $p_g$, which is the amount the actual pressure is above atmospheric pressure; that is, the gauge pressure is $p_g = p - p_{atmos}$.

## Pressure in Liquids

If we pour a liquid into a container, the force of gravity pulls the liquid down, causing it to fill the bottom of the container. It is this force of gravity—that is, the weight of the liquid—that is responsible for the pressure in a liquid. Pressure increases with depth in a liquid because the liquid below is being squeezed by all the liquid above, as well as the pressure of the air above the liquid.

**FIGURE 13.5** Measuring the pressure at depth $d$ in a liquid.

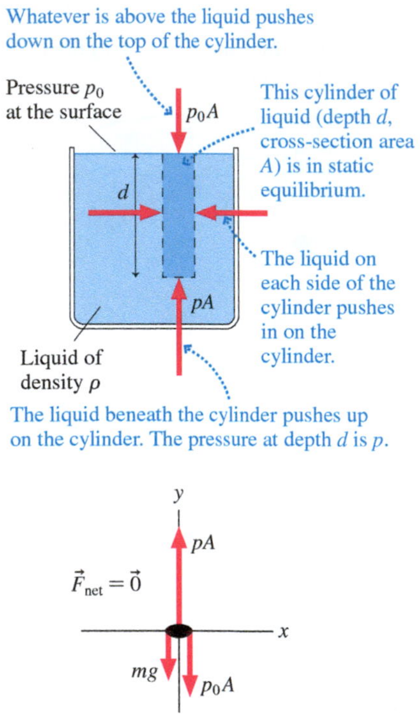


Free-body diagram of the column of liquid. The horizontal forces cancel and are not shown.

We'd like to determine the pressure at a depth $d$ below the surface of the liquid. We will assume that the liquid is at rest; flowing liquids will be considered later in this chapter. The darker shaded cylinder of liquid in FIGURE 13.5 extends from the surface to depth $d$. This cylinder, like the rest of the liquid, is in static equilibrium with $\vec{F}_{net} = \vec{0}$. Several forces act on this cylinder: its weight $mg$, a downward force $p_0A$ due to the pressure $p_0$ at the surface of the liquid, an upward force $pA$ due to the liquid beneath the cylinder pushing up on the bottom of the cylinder, and inward-directed forces due to the liquid pushing in on the sides of the cylinder. The forces due to the liquid pushing on the cylinder are a consequence of our earlier observation that different parts of a fluid push against each other. Pressure $p$, the pressure at the bottom of the cylinder, is what we're trying to find.

The horizontal forces cancel each other. The upward force balances the two downward forces, so

$$pA = p_0A + mg \tag{13.4}$$

The liquid is a cylinder of cross-section area $A$ and height $d$. Its volume is $V = Ad$ and its mass is $m = \rho V = \rho Ad$. Substituting this expression for the mass of the liquid into Equation 13.4, we find that the area $A$ cancels from all terms. The pressure at depth $d$ in a liquid is then

$$p = p_0 + \rho gd \tag{13.5}$$

Pressure of a liquid with density $\rho$ at depth $d$

Because of our assumption that the liquid is at rest, the pressure given by Equation 13.5 is called the **hydrostatic pressure.** The fact that $g$ appears in Equation 13.5 reminds us that the origin of this pressure is the gravitational force on the fluid.

As expected, $p = p_0$ at the surface, where $d = 0$. Pressure $p_0$ is usually due to the air or other gas above the liquid. For a liquid that is open to the air at sea level, $p_0 = 1$ atm $= 101$ kPa, as we learned in ◀◀ SECTION 12.2. In other situations, $p_0$ might be the pressure due to a piston or a closed surface pushing down on the top of the liquid.

**NOTE** ▶ Equation 13.5 assumes that the fluid is *incompressible;* that is, its density $\rho$ doesn't increase with depth. This is an excellent assumption for liquids, but not a good one for a gas. Equation 13.5 should not be used for calculating the pressure of a gas. ◀

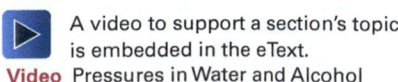

A video to support a section's topic is embedded in the eText.
**Video** Pressures in Water and Alcohol

### EXAMPLE 13.2 The pressure on a submarine

A submarine cruises at a depth of 300 m. What is the pressure at this depth? Give the answer in both pascals and atmospheres.

**STRATEGIZE** Pressure varies with depth below the surface. We will use Equation 13.5 to find the pressure at the depth of the submarine.

**PREPARE** $p_0$ is the pressure at the surface, which is 101 kPa. The density of seawater is given in Table 13.1 as 1030 kg/m$^3$.

**SOLVE** The pressure at depth $d = 300$ m is

$$\begin{aligned} p &= p_0 + \rho g d \\ &= (1.01 \times 10^5\ \text{Pa}) + (1030\ \text{kg/m}^3)(9.80\ \text{m/s}^2)(300\ \text{m}) \\ &= 3.13 \times 10^6\ \text{Pa} \end{aligned}$$

Converting the answer to atmospheres gives

$$p = (3.13 \times 10^6\ \text{Pa}) \times \frac{1\ \text{atm}}{1.013 \times 10^5\ \text{Pa}} = 30.9\ \text{atm}$$

**ASSESS** At this great depth, we expect a large pressure, so our result makes sense.

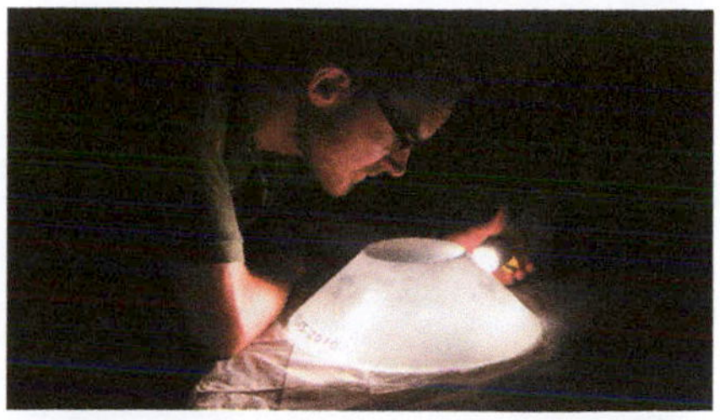

**Thick windows** The deeper you dive in the ocean, the greater the pressure. At the 4500 m depths the research submarine *Alvin* reaches, the pressure is greater than 450 atm! The pilots certainly want to look outside, but putting a window in a craft at this depth is no mean feat. In the photo, a technician inspects the window that will fill a viewport. The tapered shape will have its larger face toward the sea. The water pressure will push the window firmly into its conical seat, making a tight seal.

**FIGURE 13.6** Liquid in connected tubes.

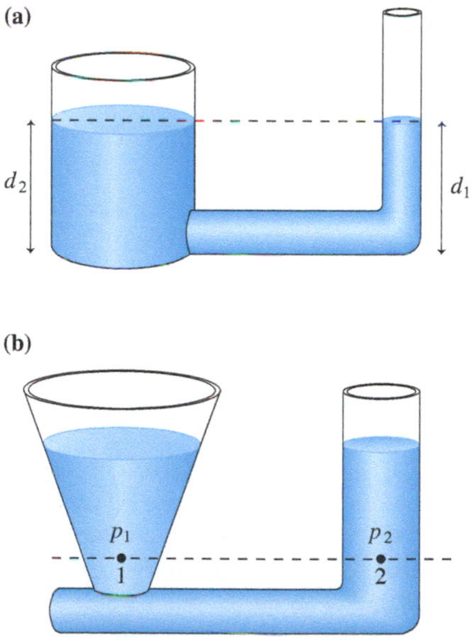


**FIGURE 13.6a** shows two connected tubes. It's certainly true that the larger volume of liquid in the wide tube weighs more than the liquid in the narrow tube. You might think that this extra weight would push the liquid in the narrow tube higher than in the wide tube. But it doesn't. If $d_1$ were larger than $d_2$, then, according to the hydrostatic pressure equation, the pressure at the bottom of the narrow tube would be higher than the pressure at the bottom of the wide tube. This *pressure difference* would cause the liquid to *flow* from right to left until the heights were equal. Thus a first conclusion: **A connected liquid in hydrostatic equilibrium rises to the same height in all open regions of the container.**

**FIGURE 13.6b** shows two connected tubes of different shape. The conical tube holds more liquid above the dashed line, so you might think that $p_1 > p_2$. But it isn't. Both points are at the same depth; thus $p_1 = p_2$. If $p_1$ were greater than $p_2$, the pressure at the bottom of the left tube would be greater than the pressure at the bottom of the right tube. This would cause the liquid to flow until the pressures were equal. Thus a second conclusion: **In hydrostatic equilibrium, the pressure is the same at all points on a horizontal line through a connected liquid of a single kind.**

**NOTE** ▸ Both of these conclusions are restricted to liquids in hydrostatic equilibrium. The situation is entirely different for flowing fluids, as we'll see later on. ◂

### EXAMPLE 13.3 Pressure in a closed tube

Water fills the tube shown in **FIGURE 13.7**. What is the pressure at the top of the closed tube?

**FIGURE 13.7** A bent tube closed at one end.

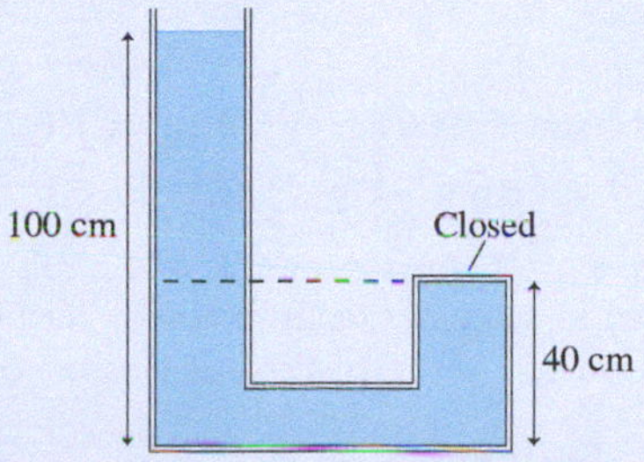


**STRATEGIZE** This is a liquid in hydrostatic equilibrium. The closed tube is not an open region of the container, so the water cannot rise to an equal height. Nevertheless, the pressure is still the same at all points on a horizontal line. In particular, the pressure at the top of the closed tube equals the pressure in the open tube at the height of the dashed line.

**PREPARE** The top of the container is open to the atmosphere. We assume that $p_0 = 1$ atm $= 101$ kPa.

**SOLVE** A point 40 cm above the bottom of the open tube is at a depth of 60 cm. The pressure at this depth is

$$\begin{aligned} p &= p_0 + \rho g d \\ &= (1.01 \times 10^5\ \text{Pa}) + (1000\ \text{kg/m}^3)(9.80\ \text{m/s}^2)(0.60\ \text{m}) \\ &= 1.07 \times 10^5\ \text{Pa} = 1.06\ \text{atm} \end{aligned}$$

**ASSESS** The water column that creates this pressure is not very tall, so it makes sense that the pressure is only a little higher than atmospheric pressure.

We can draw one more conclusion from the hydrostatic pressure equation $p = p_0 + \rho g d$. If we change the pressure at the surface to $p_1 = p_0 + \Delta p$, so that $\Delta p$ is the *change* in pressure, then the pressure at a point at a depth $d$ becomes

$$p' = p_1 + \rho g d = (p_0 + \Delta p) + \rho g d = (p_0 + \rho g d) + \Delta p = p + \Delta p$$

That is, the pressure at depth $d$ changes by the same amount as it did at the surface. This idea is called *Pascal's principle:*

**Video** Water Level in Pascal's Vases

**Pascal's principle** If the pressure at one point in an incompressible fluid is changed, the pressure at every other point in the fluid changes by the same amount.

We now have enough information to formulate a set of rules for working with hydrostatic problems.

**TACTICS BOX 13.1 Hydrostatics**

1. **Draw a picture.** Show open surfaces, pistons, boundaries, and other features that affect pressure. Include height and area measurements and fluid densities. Identify the points at which you need to find the pressure.
2. **Determine the pressure $p_0$ at surfaces.**
   - **Surface open to the air:** $p_0 = p_{atmos}$, usually 1 atm.
   - **Surface in contact with a gas:** $p_0 = p_{gas}$.
   - **Closed surface:** $p_0 = F/A$, where $F$ is the force that the surface, such as a piston, exerts on the fluid.
3. **Use horizontal lines.** The pressure in a connected fluid (of one kind) is the same at any point along a horizontal line.
4. **Allow for gauge pressure.** Pressure gauges read $p_g = p - p_{atmos}$.
5. **Use the hydrostatic pressure equation:** $p = p_0 + \rho g d$.

Exercises 5–7

**EXAMPLE 13.4 Pressure in a tube with two liquids**

A U-shaped tube is closed at one end; the other end is open to the atmosphere. Water fills the side of the tube that includes the closed end, while oil, floating on the water, fills the side of the tube open to the atmosphere. The two liquids do not mix. The height of the oil above the point where the two liquids touch is 75 cm, while the height of the closed end of the tube above this point is 25 cm. What is the gauge pressure at the closed end?

***STRATEGIZE*** Following the steps in Tactics Box 13.1, we will start by drawing the picture shown in FIGURE 13.8. We assume that the pressure at the open surface of the oil is $p_0 = 1$ atm. Pressures $p_1$ and $p_2$ are the same because they are on a horizontal line that connects two points in the *same* fluid. (The pressure at point A is *not* equal to $p_3$, even though point A and the closed end are on the same horizontal line, because the two points are in *different* fluids.)

We will apply the hydrostatic pressure equation twice: once to find the pressure $p_1$ by its known depth below the open end at pressure $p_0$, and again to find the pressure $p_3$ at the closed end once we know $p_2$ a distance $d$ below it.

FIGURE 13.8 A tube containing two different liquids.

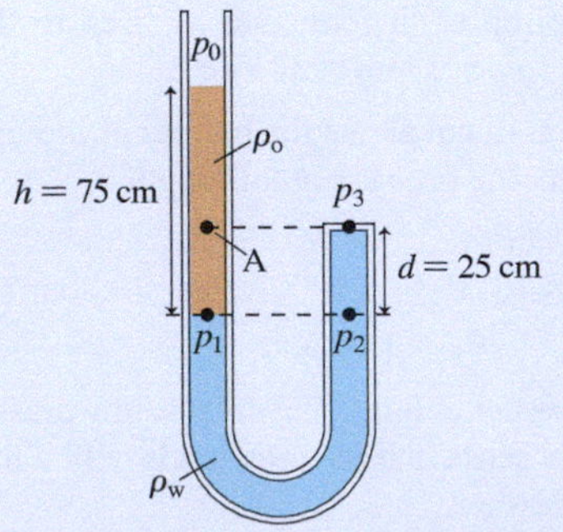


***PREPARE*** We need the densities of water and oil, which are found in Table 13.1 to be $\rho_w = 1000$ kg/m$^3$ and $\rho_o = 900$ kg/m$^3$.

***SOLVE*** The pressure at point 1, 75 cm below the open end, is

$$\begin{aligned} p_1 &= p_0 + \rho_o g h \\ &= 1 \text{ atm} + (900 \text{ kg/m}^3)(9.8 \text{ m/s}^2)(0.75 \text{ m}) \\ &= 1 \text{ atm} + 6620 \text{ Pa} \end{aligned}$$

(We will keep $p_0 = 1$ atm separate in this result because we'll eventually need to subtract exactly 1 atm to calculate the gauge pressure.) We can also use the hydrostatic pressure equation to find

$$\begin{aligned} p_2 &= p_3 + \rho_w g d \\ &= p_3 + (1000 \text{ kg/m}^3)(9.8 \text{ m/s}^2)(0.25 \text{ m}) \\ &= p_3 + 2450 \text{ Pa} \end{aligned}$$

But we know that $p_2 = p_1$, so

$$p_3 = p_2 - 2450 \text{ Pa} = p_1 - 2450 \text{ Pa}$$
$$= 1 \text{ atm} + 6620 \text{ Pa} - 2450 \text{ Pa}$$
$$= 1 \text{ atm} + 4200 \text{ Pa}$$

The gauge pressure at point 3, the closed end of the tube, is $p_3 - 1$ atm or 4200 Pa.

**ASSESS** The oil's open surface is 50 cm higher than the water's closed surface. Their densities are not too different, so we expect a pressure difference of roughly $\rho g(0.50 \text{ m}) = 5000$ Pa. This is not too far from our answer, giving us confidence that it's correct.

## Atmospheric Pressure

We live at the bottom of a "sea" of air that extends up many kilometers. As FIGURE 13.9 shows, there is no well-defined top to the atmosphere; it just gets less and less dense with increasing height until reaching zero in the vacuum of space. Nonetheless, 99% of the air in the atmosphere is below about 30 km.

If we recall that a gas like air is compressible, we can see why the atmosphere becomes less dense with increasing altitude. In a liquid, pressure increases with depth because of the weight of the liquid above. The same holds true for the air in the atmosphere, but because the air is compressible, the weight of the air above compresses the air below, increasing its density. At high altitudes there is very little air above to push down, so the density is less.

We learned in Section 12.2 that the global average sea-level pressure, the *standard atmosphere,* is 1 atm = 101,300 Pa. The standard atmosphere, usually referred to simply as "atmospheres," is a commonly used unit of pressure. But it is not an SI unit, so you must convert atmospheres to pascals before doing most calculations with pressure.

**NOTE** ▶ Unless you happen to live right at sea level, the atmospheric pressure around you is not exactly 1 atm. A pressure gauge must be used to determine the actual atmospheric pressure. For simplicity, this text will assume that the pressure of the air is $p_{atmos} = 1$ atm unless stated otherwise. ◀

Atmospheric pressure varies not only with altitude, but also with changes in the weather. Local winds and weather are largely determined by the presence and movement of air masses of differing pressure. You may have seen weather maps like the one shown in FIGURE 13.10 on the evening news. The letters H and L denote regions of high and low atmospheric pressure.

FIGURE 13.9 Atmospheric pressure and density.

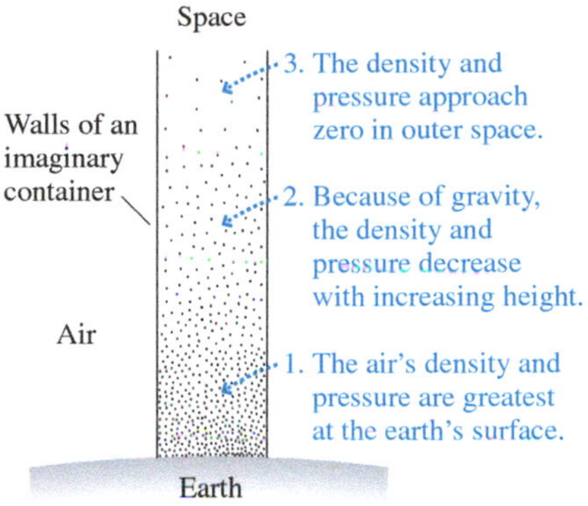


FIGURE 13.10 High- and low-pressure zones on a weather map.

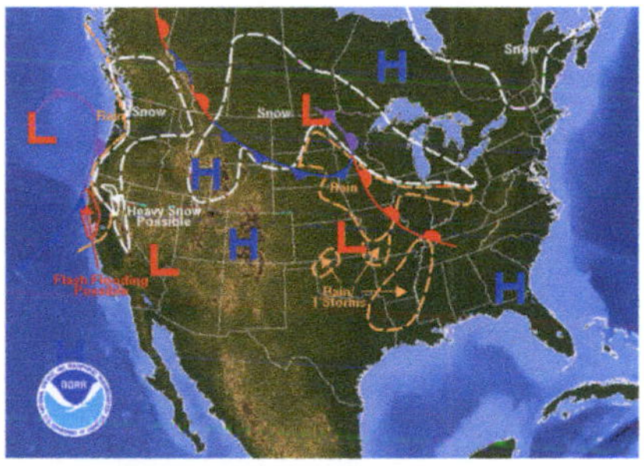


## Measuring Atmospheric Pressure

FIGURE 13.11a shows a glass tube, sealed at the bottom, that has been completely filled with a liquid. If we temporarily seal the top end, we can invert the tube and place it in a beaker of the same liquid. When the temporary seal is removed, some, but not all, of the liquid runs out, leaving a liquid column in the tube that is a height $h$ above the surface of the liquid in the beaker. This device, shown in FIGURE 13.11b, is a **barometer.** What does it measure? And why doesn't *all* the liquid in the tube run out?

We can analyze the barometer using the steps of Tactics Box 13.1. Point 1 in Figure 13.11b is open to the atmosphere, so $p_1 = p_{atmos}$. The pressure at point 2 is the pressure due to the weight of the liquid in the tube plus the pressure of the gas above the liquid. But in this case there is no gas above the liquid! Because the tube had been completely full of liquid when it was inverted, the space left behind when the liquid ran out is essentially a vacuum, with $p_0 = 0$. Thus pressure $p_2$ is simply $\rho gh$.

Because points 1 and 2 are on a horizontal line, and the liquid is in hydrostatic equilibrium, the pressures at these two points must be equal. Equating these two pressures gives

$$p_{atmos} = \rho gh \quad (13.6)$$

Thus we can measure the atmosphere's pressure by measuring the height of the liquid column in a barometer.

FIGURE 13.11 A barometer.

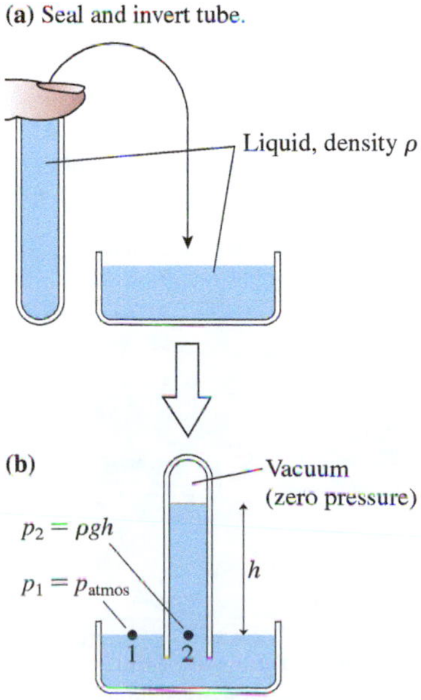

Equation 13.6 shows that the liquid height is $h = p_{atmos}/\rho g$. If a barometer were made using water, with $\rho = 1000\ kg/m^3$, the liquid column would be

$$h = \frac{101{,}300\ \text{Pa}}{(1000\ \text{kg/m}^3)(9.8\ \text{m/s}^2)} \approx 10\ \text{m}$$

which is impractical. Instead, mercury, with its high density of 13,600 $kg/m^3$, is usually used. The average air pressure at sea level causes a column of mercury in a mercury barometer to stand 760 mm above the surface.

Because of the importance of mercury-filled barometers in measuring pressure, the height of a column of mercury in millimeters is a common unit of pressure. From our discussion of barometers, 760 millimeters of mercury (abbreviated "mm Hg") corresponds to a pressure of 1 atm.

### Pressure Units

In practice, pressure is measured in a number of different units. This plethora of units and abbreviations has arisen historically as scientists and engineers working on different subjects (liquids, high-pressure gases, low-pressure gases, weather, etc.) developed what seemed to them the most convenient units. These units continue in use through tradition, so it is necessary to convert back and forth among them. TABLE 13.2 gives the basic conversions.

TABLE 13.2 Pressure units

| Unit | Abbreviation | Conversion to Pa | Uses |
|---|---|---|---|
| pascal | Pa | | SI unit: 1 Pa = 1 $N/m^2$ used in most calculations |
| atmosphere | atm | 1 atm = 101.3 kPa | general |
| millimeters of mercury | mm Hg | 1 mm Hg = 133 Pa | gases and barometric pressure |
| inches of mercury | in | 1 in = 3.39 kPa | barometric pressure in U.S. weather forecasting |
| pounds per square inch | psi | 1 psi = 6.89 kPa | U.S. engineering and industry |

**STOP TO THINK 13.2** A U-shaped tube is open to the atmosphere on both ends. Water is poured into the tube, followed by oil, which floats on the water because it is less dense than water. Which figure shows the correct equilibrium configuration?

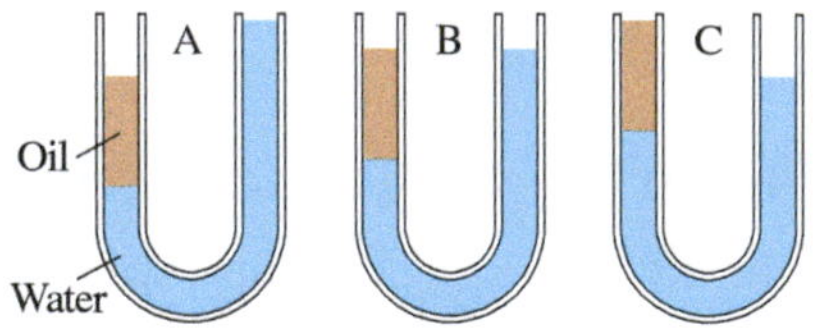


## 13.3 Buoyancy

Wood floats on the surface of a lake. A penny with a mass of a few grams sinks, but a massive steel barge floats. How can we understand these diverse phenomena?

An air mattress floats effortlessly on the surface of a swimming pool. But if you've ever tried to push an air mattress underwater, you know it is nearly impossible. As you push down, the water pushes up. This upward force of a fluid is called the **buoyant force.**

The basic reason for the buoyant force is straightforward. FIGURE 13.12 shows a cylinder submerged in a liquid. The pressure in the liquid increases with depth, so the pressure at the bottom of the cylinder is greater than at the top. Both cylinder ends have equal area, so force $\vec{F}_{up}$ is greater than force $\vec{F}_{down}$. (Remember that pressure forces push in *all* directions.) Consequently, the pressure in the liquid exerts a *net upward force* on the cylinder of magnitude $F_{net} = F_{up} - F_{down}$. This is the buoyant force.

FIGURE 13.12 The buoyant force arises because the fluid pressure at the bottom of the cylinder is greater than that at the top.

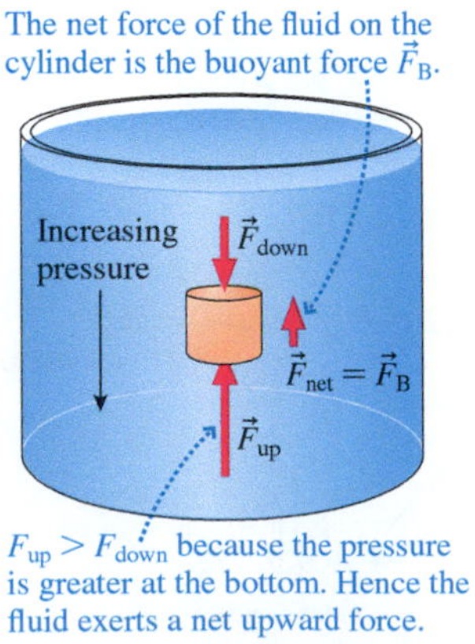

The submerged cylinder illustrates the idea in a simple way, but the result is not limited to cylinders or to liquids. Suppose we isolate a parcel of fluid of arbitrary shape and volume by drawing an imaginary boundary around it, as shown in **FIGURE 13.13a.** This parcel is in static equilibrium. Consequently, the parcel's weight force pulling it down must be balanced by an upward force. The upward force, which is exerted on this parcel of fluid by the surrounding fluid, is the buoyant force $\vec{F}_B$. The buoyant force matches the weight of the fluid: $F_B = w$.

Now imagine that we remove this parcel of fluid and instantaneously replace it with an object having exactly the same shape and size, as shown in **FIGURE 13.13b.** Because the buoyant force is exerted by the *surrounding* fluid, and the surrounding fluid hasn't changed, the buoyant force on this new object is *exactly the same* as the buoyant force on the parcel of fluid that we removed.

**KEY CONCEPT** **FIGURE 13.13** The buoyant force on an object is the same as the buoyant force on an equal volume of fluid.

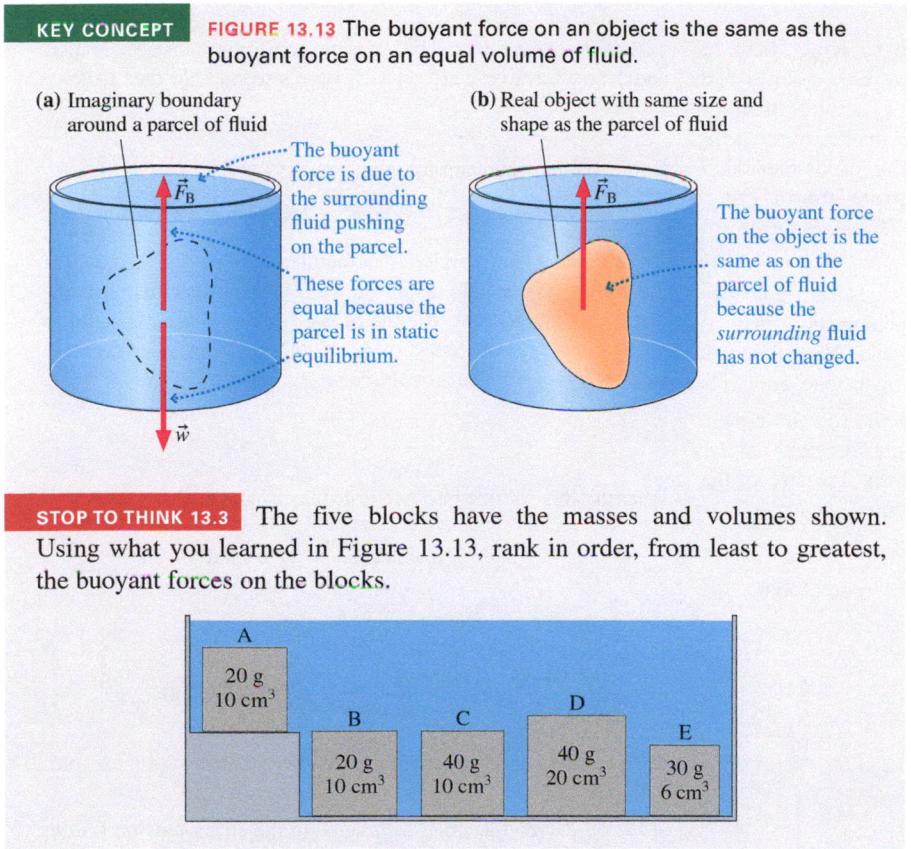

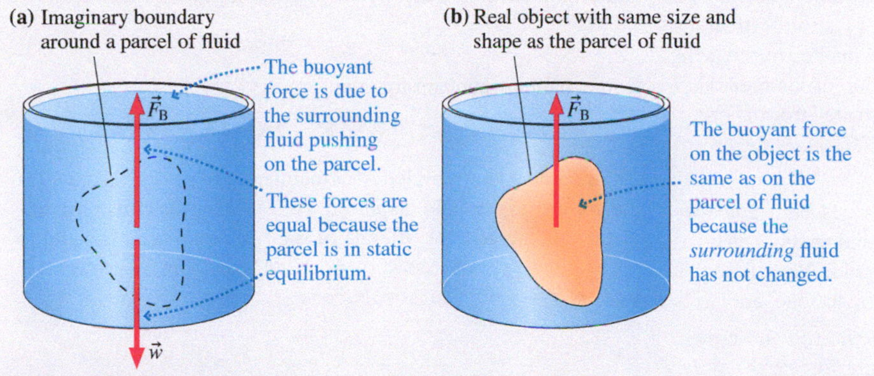


**STOP TO THINK 13.3** The five blocks have the masses and volumes shown. Using what you learned in Figure 13.13, rank in order, from least to greatest, the buoyant forces on the blocks.

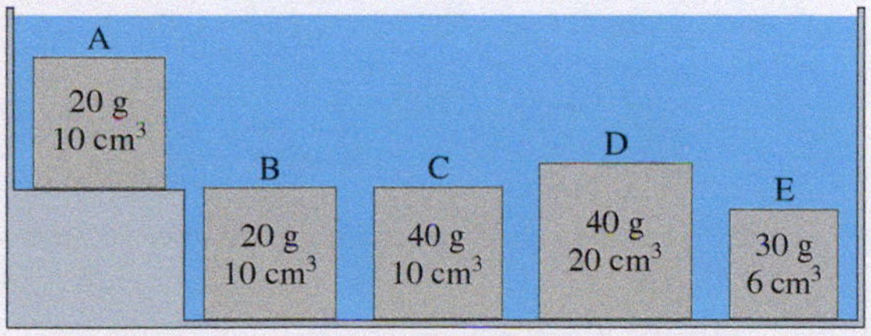


When an object (or a portion of an object) is immersed in a fluid, it *displaces* fluid that would otherwise fill that region of space. The displaced fluid's volume is exactly the volume of the portion of the object that is immersed in the fluid. Figure 13.13 leads us to conclude that the magnitude of the upward buoyant force matches the weight of this displaced fluid.

This idea was first recognized by the ancient Greek mathematician and scientist Archimedes, and today we know it as *Archimedes' principle:*

**Archimedes' principle** A fluid exerts an upward buoyant force $\vec{F}_B$ on an object immersed in or floating on the fluid. The magnitude of the buoyant force equals the weight of the fluid displaced by the object.

Suppose the fluid has density $\rho_f$ and the object displaces a volume $V_f$ of fluid. The mass of the displaced fluid is then $m_f = \rho_f V_f$ and so its weight is $w_f = \rho_f V_f g$. Thus Archimedes' principle in equation form is

$$F_B = \rho_f V_f g \tag{13.7}$$

Buoyant force on an object displacing volume $V_f$ of fluid of density $\rho_f$

Video Weighing Weights in Water

**NOTE** ▶ It is important to distinguish the density and volume of the displaced fluid from the density and volume of the object. To do so, we use subscript f for the fluid and o for the object. ◀

If the object is completely submerged, the volume of fluid displaced is the same as the volume of the object, so $V_f = V_o$.

### EXAMPLE 13.5 Is the crown gold?

Legend has it that Archimedes was asked by King Hiero of Syracuse to determine whether a crown was pure gold or had been adulterated with a less valuable metal by an unscrupulous goldsmith. It was this problem that led Archimedes to the principle that bears his name. In a modern version of his method, a crown weighing 8.30 N is suspended underwater from a string. The tension in the string is measured to be 7.81 N. Is the crown pure gold?

**STRATEGIZE** To discover whether the crown is pure gold, we will proceed as Archimedes did—by determining its density $\rho_o$ and comparing that to the known density of gold. We will do this by analyzing the forces on the crown, including the buoyant force.

**PREPARE** FIGURE 13.14 shows the forces acting on the crown. In addition to the familiar tension and weight forces, the water exerts an upward buoyant force on the crown. The size of the buoyant force is given by Archimedes' principle.

FIGURE 13.14 The forces acting on the submerged crown.

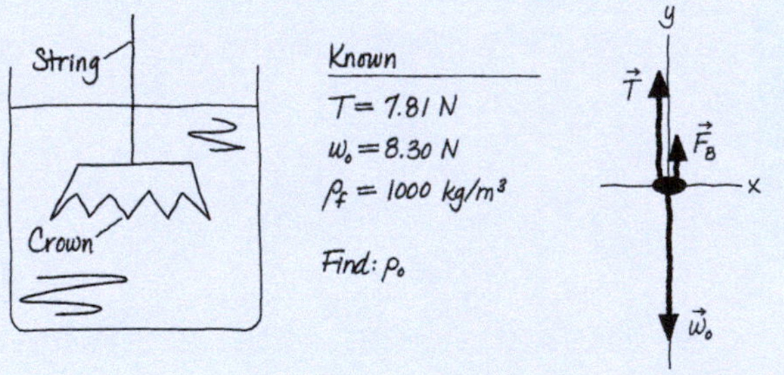


**SOLVE** Because the crown is in static equilibrium, its acceleration and the net force on it are zero. Newton's second law then reads

$$\sum F_y = F_B + T - w_o = 0$$

from which the buoyant force is

$$F_B = w_o - T = 8.30\ \text{N} - 7.81\ \text{N} = 0.49\ \text{N}$$

According to Archimedes' principle, $F_B = \rho_f V_f g$, where $V_f$ is the volume of the fluid displaced. Here, where the crown is completely submerged, the volume of the fluid displaced is equal to the volume $V_o$ of the crown. Now the crown's weight is $w_o = m_o g = \rho_o V_o g$, so its volume is

$$V_o = \frac{w_o}{\rho_o g}$$

Inserting this volume into Archimedes' principle gives

$$F_B = \rho_f V_o g = \rho_f \left(\frac{w_o}{\rho_o g}\right) g = \frac{\rho_f}{\rho_o} w_o$$

or, solving for $\rho_o$,

$$\rho_o = \frac{\rho_f w_o}{F_B} = \frac{(1000\ \text{kg/m}^3)(8.30\ \text{N})}{0.49\ \text{N}} = 17{,}000\ \text{kg/m}^3$$

The crown's density is considerably lower than that of pure gold, which is 19,300 kg/m$^3$. The crown is not pure gold.

**ASSESS** Given the small difference in the string tension for the two cases, the buoyant force is very small. Therefore, we expect a large density and our answer makes sense.

## Float or Sink?

If we *hold* an object underwater and then release it, it rises to the surface, sinks, or remains "hanging" in the water. How can we predict which it will do? Whether it heads for the surface or the bottom depends on whether the upward buoyant force $F_B$ on the object is larger or smaller than the downward weight force $w_o$.

The magnitude of the buoyant force is $\rho_f V_f g$. The weight of a uniform object, such as a block of steel, is simply $\rho_o V_o g$. But a compound object, such as a scuba diver, may have pieces of varying density. If we define the **average density** to be $\rho_{avg} = m_o/V_o$, the weight of a compound object can be written as $w_o = \rho_{avg} V_o g$.

Video What the Physics? Buoyant Balloons

Comparing $\rho_f V_f g$ to $\rho_{avg} V_o g$, and noting that $V_f = V_o$ for an object that is fully submerged, we see that an object floats or sinks depending on whether the fluid density $\rho_f$ is larger or smaller than the object's average density $\rho_{avg}$. If the densities are equal, the object is in static equilibrium and hangs motionless. This is called **neutral buoyancy.** These conditions are summarized in Tactics Box 13.2.

**TACTICS BOX 13.2 Finding whether an object floats or sinks**

❶ Object sinks

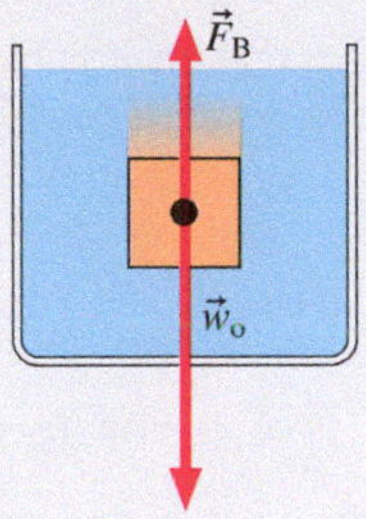


An object sinks if it weighs more than the fluid it displaces—that is, if its average density is greater than the density of the fluid:

$$\rho_{avg} > \rho_f$$

❷ Object floats

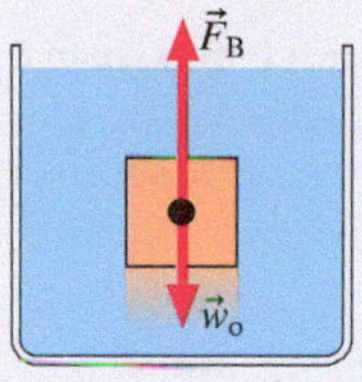


An object rises to the surface if it weighs less than the fluid it displaces—that is, if its average density is less than the density of the fluid:

$$\rho_{avg} < \rho_f$$

❸ Object has neutral buoyancy

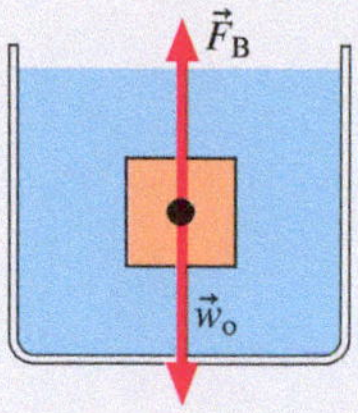


An object hangs motionless if it weighs exactly the same as the fluid it displaces—that is, if its average density equals the density of the fluid:

$$\rho_{avg} = \rho_f$$

Exercises 10–12

Steel is denser than water, so a chunk of steel sinks. Oil is less dense than water, so oil floats on water. Scuba divers use weighted belts to adjust their average density to match the density of the water, so they maintain their depth in the water without exerting effort.

If we release a block of wood underwater, the net upward force causes the block to shoot to the surface. Then what? To understand floating, let's begin with a *uniform* object such as the block shown in FIGURE 13.15. This object contains nothing tricky, like indentations or voids. Because it's floating, it must be the case that $\rho_o < \rho_f$.

Now that the object is floating, it's in static equilibrium. Thus, the upward buoyant force, given by Archimedes' principle, exactly balances the downward weight of the object; that is,

$$F_B = \rho_f V_f g = w_o = \rho_o V_o g \tag{13.8}$$

For a floating object, the volume of the displaced fluid is *not* the same as the volume of the object. In fact, we can see from Equation 13.8 that the volume of fluid displaced by a floating object of uniform density is

$$V_f = \frac{\rho_o}{\rho_f} V_o \tag{13.9}$$

which is *less* than $V_o$ because $\rho_o < \rho_f$.

NOTE ▶ Equation 13.9 applies only to *uniform* objects. It does not apply to boats, hollow spheres, or other objects of nonuniform composition. ◀

Video Buoyancy and Density Part 1

FIGURE 13.15 A floating object is in static equilibrium.

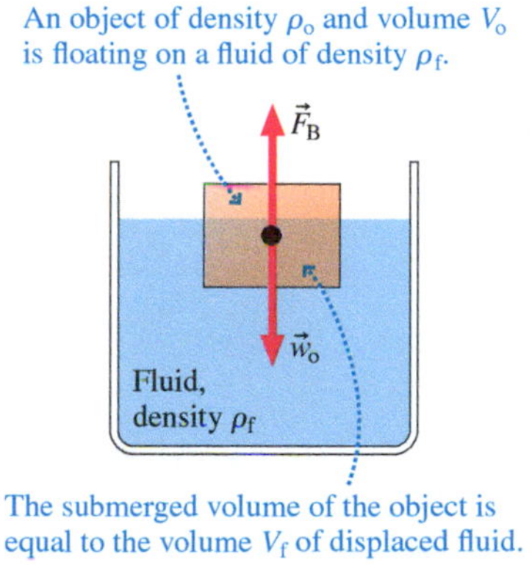


Video Figure 13.15

▶ **Hidden depths** Most icebergs break off glaciers and are fresh-water ice with a density of 917 kg/m$^3$. The density of seawater is 1030 kg/m$^3$. Thus

$$V_f = \frac{917\ \text{kg/m}^3}{1030\ \text{kg/m}^3} V_o = 0.89 V_o$$

$V_f$, the volume of the displaced water, is also the volume of the iceberg that is underwater. The saying "90% of an iceberg is underwater" is correct!

**CONCEPTUAL EXAMPLE 13.6** **Which has the greater buoyant force?**

A block of iron sinks to the bottom of a vessel of water while a block of wood of the *same size* floats. On which is the buoyant force greater?

**REASON** The buoyant force is equal to the volume of water displaced. The iron block is completely submerged, so it displaces a volume of water equal to its own volume. The wood block floats, so it displaces only the fraction of its volume that is under water, which is *less* than its own volume. The buoyant force on the iron block is therefore greater than on the wood block.

**ASSESS** This result may seem counterintuitive, but remember that the iron block sinks because of its high density, while the wood block floats because of its low density. A smaller buoyant force is sufficient to keep it floating.

**EXAMPLE 13.7** **Measuring the density of an unknown liquid**

You need to determine the density of an unknown liquid. You notice that a wooden block floats in this liquid with 4.6 cm of the side of the block submerged. When the block is placed in water, it also floats but with 5.8 cm submerged. What is the density of the unknown liquid?

**STRATEGIZE** We will write expressions for the volume submerged—the volume of fluid displaced—for the two cases. The block floats in both liquids, so we write Equation 13.9 for the two cases. We will then combine the two equations to solve for the density of the fluid.

**PREPARE** It's useful to draw a picture to visualize the situation. We don't know the area of the block, so we simply label it *A*. FIGURE 13.16 shows the block and submerged lengths $h_u$ in the unknown liquid and $h_w$ in water.

FIGURE 13.16 A wooden block floating in two liquids.

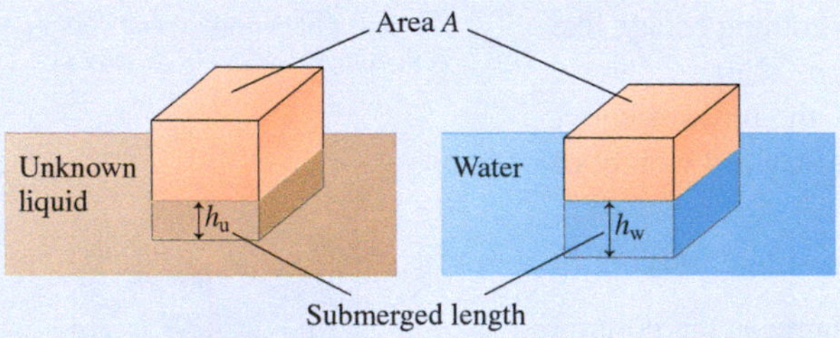


**SOLVE** The block displaces volume $V_u = Ah_u$ of the unknown liquid. Thus

$$V_u = Ah_u = \frac{\rho_o}{\rho_u}V_o$$

Similarly, the block displaces volume $V_w = Ah_w$ of the water, leading to

$$V_w = Ah_w = \frac{\rho_o}{\rho_w}V_o$$

Because there are two fluids, we've used subscripts w for water and u for the unknown in place of the fluid subscript f. The product $\rho_o V_o$ appears in both equations. In the first $\rho_o V_o = \rho_u A h_u$, and in the second $\rho_o V_o = \rho_w A h_w$. Equating the right-hand sides gives

$$\rho_u A h_u = \rho_w A h_w$$

The area *A* cancels, and the density of the unknown liquid is

$$\rho_u = \frac{h_w}{h_u}\rho_w = \frac{5.8\text{ cm}}{4.6\text{ cm}}1000\text{ kg/m}^3 = 1300\text{ kg/m}^3$$

Comparison with Table 13.1 shows that the unknown liquid is likely to be glycerin.

**ASSESS** The object floats slightly higher in the unknown liquid than in water, so we expect a density slightly greater than that of water, as we found.

FIGURE 13.17 How a boat floats.

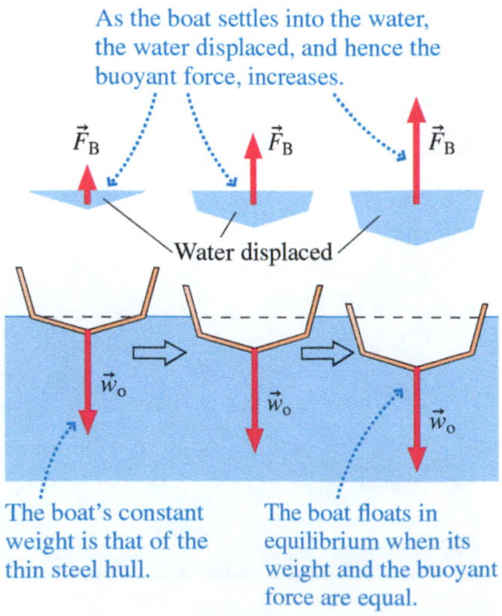


## Boats and Balloons

A chunk of steel sinks, so how does a steel-hulled boat float? As we've seen, an object floats if the upward buoyant force—the weight of the displaced water—balances the weight of the object. A boat is really a large hollow shell whose weight is determined by the volume of steel in the hull. As FIGURE 13.17 shows, the volume of water displaced by a shell is *much* larger than the volume of the hull itself. As a boat settles into the water, it sinks until the weight of the displaced water exactly matches the boat's weight. It is then in static equilibrium, so it floats at that level.

The concept of buoyancy and flotation applies to all fluids, not just liquids. An object immersed in a gas such as air feels a buoyant force as well. Because the density of air is so low, this buoyant force is generally negligible. Nonetheless, even though the buoyant force due to air is small, an object will float in air if it weighs less than the air that it displaces. For a balloon to float, it must be filled with a gas that has a *lower* density than that of air. The following example illustrates how this works.

**EXAMPLE 13.8** **How fast will the balloon rise?**

A 2.0 g latex balloon is filled with helium. When it is completely inflated, its shape is approximately spherical with a diameter of 24 cm. If the balloon is released, what is its initial upward acceleration?

**STRATEGIZE** The balloon plus the helium inside it has a certain mass; this leads to a downward weight force. But the balloon displaces a certain amount of air, and this leads to an upward buoyant force. We're told that the balloon accelerates upward, so we know that the upward buoyant force exceeds the downward weight force, as in FIGURE 13.18. This net force will produce an upward acceleration.

FIGURE 13.18 The forces acting on a helium balloon.

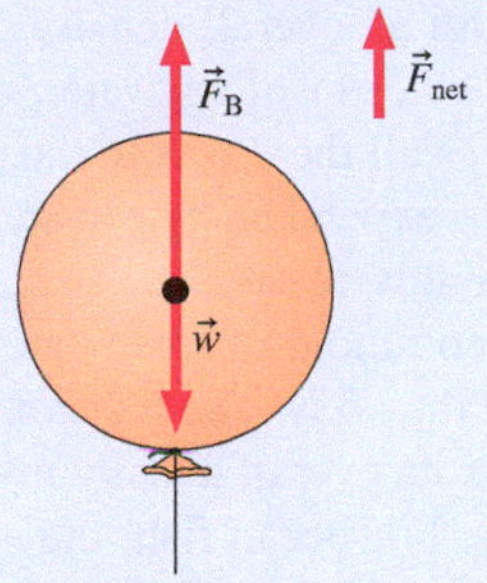


**PREPARE** We assume that the atmospheric pressure is 1 atm; we can also assume that the pressure inside the balloon is the same. Table 13.1 lists the densities for helium and for air at 1 atm pressure: $\rho_{\rm He} = 0.166\ {\rm kg/m^3}$ and $\rho_{\rm air} = 1.20\ {\rm kg/m^3}$. The volume of the balloon in $\rm m^3$ is

$$V = \frac{4}{3}\pi r^3 = \frac{4}{3}\pi(0.12\ {\rm m})^3 = 0.00724\ {\rm m^3}$$

The thickness of the latex is negligible, so this is the volume of helium in the balloon as well as $V_{\rm f}$, the volume of fluid displaced.

**SOLVE** The total mass is the sum of the mass of the latex balloon (0.0020 kg) and the mass of the helium:

$$m = 0.0020\ {\rm kg} + \rho_{\rm He}V = 0.0020\ {\rm kg} + (0.166\ {\rm kg/m^3})(0.00724\ {\rm m^3}) = 0.0032\ {\rm kg}$$

The weight force is

$$w = (0.0032\ {\rm kg})(9.8\ {\rm m/s^2}) = 0.031\ {\rm N}$$

The upward buoyant force is

$$F_{\rm B} = \rho_{\rm f}Vg = (1.20\ {\rm kg/m^3})(0.00724\ {\rm m^3})(9.8\ {\rm m/s^2}) = 0.085\ {\rm N}$$

The net force is the difference of these two forces:

$$F_{\rm net} = F_{\rm B} - w = 0.085\ {\rm N} - 0.031\ {\rm N} = 0.054\ {\rm N}$$

The initial acceleration of the balloon is

$$a = \frac{F_{\rm net}}{m} = \frac{0.054\ {\rm N}}{0.0032\ {\rm kg}} = 17\ {\rm m/s^2}$$

**ASSESS** The buoyant force is much greater than the weight force, so we expect a very large initial acceleration. If you've ever been holding a helium balloon and loosened your grip for just a second, you've no doubt seen this in action!

**Hot air rising** A hot-air balloon is filled with a low-density gas: hot air! You learned in Chapter 12 that gases expand upon heating, which lowers their density. The air at the top of a hot-air balloon is surprisingly toasty—about 100°C. The density of the heated air is about 80% that of room-temperature air. The weight of the air displaced significantly exceeds the weight of the air inside the balloon, allowing the balloon to lift the weight of the passengers and the basket.

**Video** Buoyancy and Density Part 2

TABLE 13.3 Densities of body components

| Body Component | Density ($\rm kg/m^3$) |
|---|---|
| Fat | 0.90 |
| Water | 1.00 |
| Blood | 1.05 |
| Muscle | 1.06 |
| Bone | 1.28 |

**Masters of density** BIO A manatee may look chubby, but these mammals live in fresh water and must dive for their food. Too much fat would make them positively buoyant, which would cost them energy to forage. A typical body fat percentage for a manatee is about 7%, comparable to elite human athletes, and this reduces the energy cost for diving. Manatees have other adaptations to reduce the energy cost of moving in the water. They carefully adjust the amount of residual air in their lungs to achieve nearly neutral buoyancy. They can also move air from one lung to the other to roll, and from the top of the lungs to the bottom to tip forward for a dive.

## Buoyancy and Bodies BIO

Different components of your body have different densities, as shown in TABLE 13.3. Of the different body components, only fat has a density lower than that of water. An average college student's body is about 20% fat, with the balance a mix of water, muscle, blood, and bone. With this typical fat percentage, the overall density of the body is about 1050 $\rm kg/m^3$. Submerged in water, a person with this density would sink. But if the person takes a deep breath, she reduces her overall density to approximately 990 $\rm kg/m^3$ and will float. This is a common experience for most folks: After taking a deep breath, you float; if you exhale as completely as possible, you sink.

In Example 13.5, we saw that the density of an object could be determined by weighing it both underwater and in air. The same thing can be done with a person. Since fat has a lower density than muscle or bone, a lower overall body density implies a greater proportion of body fat. To determine a person's density, he is first weighed in air and then lowered completely into water, asked to exhale as completely as possible, and weighed again. It's then a reasonably straightforward calculation to determine the percentage of body fat. Empirical formulas take into account the air space that remains in the body and expected percentages of the other elements to give an estimate of the amount of fat in a person's body.

If an animal lives in water, then buoyancy is a fact of life that it must deal with. The body of a typical fish is similar to yours. It is made of muscle, skin, bone, and fat, with an overall density of about 1050 kg/m$^3$. This is greater than the density of fresh water or of seawater, and fish have no lungs filled with air to compensate. If this was all there was to the story, fish would have a tendency to sink, so they would need to expend energy to stay at the same depth.

For this reason, all fish have structures that provide additional buoyancy. The most common, in freshwater and saltwater fish, is a *swim bladder,* a gas-filled sac inside the body. The fish controls the volume of gas in the swim bladder, and thereby adjusts its overall density until it is a reasonably close match to the water in which it swims. For ocean fish, the swim bladder might take up 1–3% of the volume of the body; for freshwater fish, 5–7% is more typical. A fish must continually adjust the amount of gas in the swim bladder as it moves to greater and lesser depths; changes in pressure change the volume of the gas.

**EXAMPLE 13.9** **How big a bladder?**

With its swim bladder deflated, a 9.5 kg yellowfin tuna has a density of 1050 kg/m$^3$. It is swimming near the surface of the ocean. What volume of gas must the fish have in its swim bladder to achieve neutral buoyancy?

***STRATEGIZE*** For neutral buoyancy, the density of the fish must equal the density of seawater.

***PREPARE*** The density of seawater is given in Table 13.1 as 1030 kg/m$^3$. We can use the fish's density and its mass to find its volume $V_f$:

$$V_f = \frac{9.5 \text{ kg}}{1050 \text{ kg/m}^3} = 0.00905 \text{ m}^3$$

***SOLVE*** The density of the fish including the swim bladder is the ratio of the mass of the fish (we can ignore the mass of the air in the swim bladder) to the volume of the fish plus the volume of the swim bladder $V_b$. We can set this equal to the density of seawater:

$$\frac{9.5 \text{ kg}}{0.00905 \text{ m}^3 + V_b} = 1030 \text{ kg/m}^3$$

Solving for $V_b$, we find

$$V_b = 0.00017 \text{ m}^3$$

***ASSESS*** This is about 2% of the fish's volume, which is typical for an ocean fish, so our result seems reasonable. A fish would typically maintain a slight negative buoyancy, so real data for tuna show a volume slightly less than this.

Sharks don't have swim bladders; neither do marine mammals. Sharks and many marine mammals have areas of low-density fats that reduce their average density. Sperm whales have a great deal of low-density oil in an organ in their heads. There is evidence that these deep-diving whales adjust the temperature of the oil to change its density. They cool the oil to make them negatively buoyant at the start of a dive; they then warm the oil during the dive so they achieve positive buoyancy when it is time to resurface.

**STOP TO THINK 13.4** An ice cube is floating in a glass of water that is filled to the brim. When the ice cube melts, the water level will

A. Fall. B. Stay the same. C. Rise, causing the water to spill.

## 13.4 Fluids in Motion

We've focused thus far on fluid statics, but it's time to turn our attention to fluids in motion—water coming out of your tap, the wind blowing over the land. Fluid flow is a complex subject. Many aspects of fluid flow, especially turbulence and the formation of eddies, are still not well understood and are areas of current research. We can avoid most of the complexity by making certain assumptions. In our discussion, we'll assume:

- **Fluids are incompressible.** This is a very good assumption for liquids, but it also holds reasonably well for a moving gas, such as air. For instance, even when a 100 mph wind slams into a wall, its density changes by only about 1%.
- **The flow is steady.** That is, the fluid velocity at each point in the fluid is constant; it does not fluctuate or change with time. Flow under these conditions is called *laminar flow*, and it is distinguished from *turbulent flow*.

The rising smoke in FIGURE 13.19 begins as laminar flow, recognizable by its smooth contours, but at some point it undergoes a transition to turbulent flow. Our model of fluids can be applied to the laminar flow, but not to the turbulent flow. This limitation isn't as limiting as it might seem. The motion of blood in your circulatory system, for instance, is always laminar if things are working as they should—turbulence spells trouble. Turbulence in blood vessels makes a very particular sound in a stethoscope, which allows a physician listening to your heart and lungs to detect potentially serious problems in a routine physical exam.

FIGURE 13.19 Rising smoke changes from laminar flow to turbulent flow.

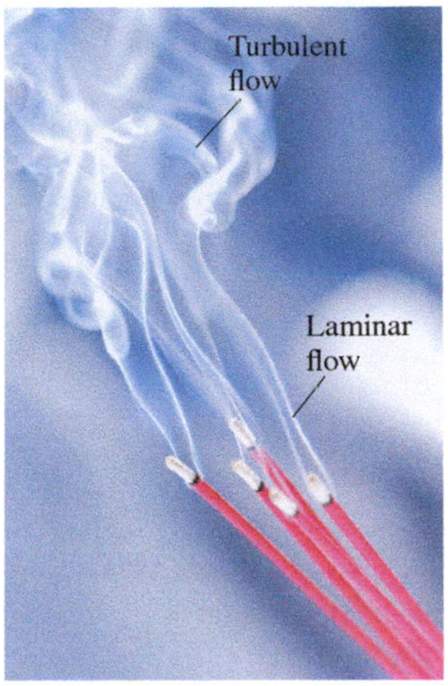


## The Role of Viscosity

Water flows much more easily than honey, which has a much higher *viscosity*, a quality you learned about in ◀◀ SECTION 5.6. For some flows—such as the flow of water in a river—we can ignore viscosity. Viscosity is resistance to flow, and assuming that we can ignore viscosity is analogous to assuming that we can ignore friction, or that we can ignore drag. For other flows—such as the motion of blood in the small vessels—viscosity is very important.

The different situations are distinguished by the value of the *Reynolds number*, which you also learned about in ◀◀ SECTION 5.6. The Reynolds number is the ratio of inertial forces to viscous forces for an object moving through a fluid, but it can be applied to the motion of fluids as well. For computing the Reynolds number for a fluid flow, the length scale is the depth of the flow in a channel or the diameter of a pipe carrying a flow. If the Reynolds number is large, the viscous forces are small compared to the other forces at work and can be ignored. For water flowing at 1 m/s in a river that is 2 m deep, the Reynolds number is

$$Re = \frac{\rho v L}{\eta} = \frac{(1000\ \text{kg/m}^3)(1\ \text{m/s})(2\ \text{m})}{1 \times 10^{-3}\ \text{Pa}\cdot\text{s}} = 2{,}000{,}000$$

This is clearly large enough that we needn't worry about viscosity. However, for blood (viscosity $3.5 \times 10^{-3}\ \text{Pa}\cdot\text{s}$) moving in an arteriole with a diameter of 0.00020 m at a speed of 0.0030 m/s, the Reynolds number is

$$Re = \frac{\rho v L}{\eta} = \frac{(1050\ \text{kg/m}^3)(0.0030\ \text{m/s})(0.00020\ \text{m})}{3.5 \times 10^{-3}\ \text{Pa}\cdot\text{s}} = 0.18$$

In this case, the motion is dominated by viscosity.

The Reynolds number also determines the transition between laminar and turbulent flow. We'll see an example when we treat fluid flow in the circulatory system later in the chapter.

We'll start our treatment of fluid flow by considering properties that apply regardless of viscosity, regardless of the value of the Reynolds number for the flow.

## The Equation of Continuity

Consider a fluid flowing through a tube—oil through a pipe or blood through an artery. If the tube's diameter changes, as happens in FIGURE 13.20, what happens to the speed of the fluid?

When you squeeze a toothpaste tube, the volume of toothpaste that emerges matches the amount by which you reduce the volume of the tube. An *incompressible* fluid flowing through a rigid tube or pipe acts the same way. Fluid is neither created

FIGURE 13.20 Flow speed changes through a tapered tube.

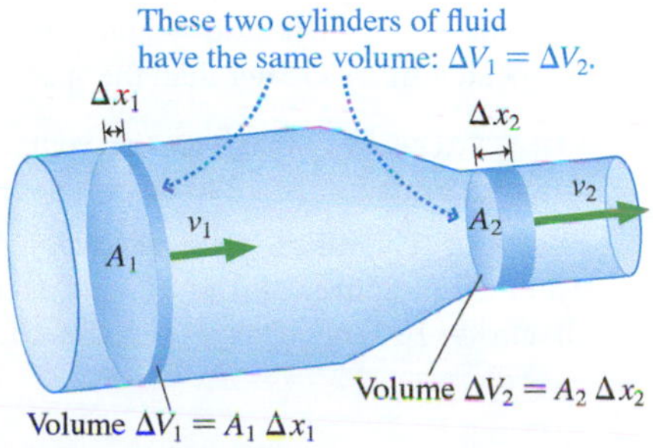

nor destroyed within the tube, and there's no place to store any extra fluid introduced into the tube. If volume $V$ enters the tube during some interval of time $\Delta t$, then an equal volume of fluid must leave the tube.

To see the implications of this idea, suppose all the molecules of the fluid in Figure 13.20 are moving to the right with speed $v_1$ at a point where the cross-section area is $A_1$. Farther along the tube, where the cross-section area is $A_2$, their speed is $v_2$. During an interval of time $\Delta t$, the molecules in the wider section move forward a distance $\Delta x_1 = v_1 \Delta t$ and those in the narrower section move $\Delta x_2 = v_2 \Delta t$. Because the fluid is incompressible, the volumes $\Delta V_1$ and $\Delta V_2$ must be equal; that is,

$$\Delta V_1 = A_1 \Delta x_1 = A_1 v_1 \Delta t = \Delta V_2 = A_2 \Delta x_2 = A_2 v_2 \Delta t \qquad (13.10)$$

Dividing both sides of the equation by $\Delta t$ gives the **equation of continuity:**

**Video** Continuity

$$v_1 A_1 = v_2 A_2 \qquad (13.11)$$

The equation of continuity relating the speed $v$ of an incompressible fluid to the cross-section area $A$ of the tube in which it flows

Equations 13.10 and 13.11 say that **the volume of an incompressible fluid entering one part of a tube or pipe must be matched by an equal volume leaving downstream.**

An important consequence of the equation of continuity is that **flow is faster in narrower parts of a tube, slower in wider parts.** You're familiar with this conclusion from many everyday observations. The garden hose shown in FIGURE 13.21a squirts water farther after you put a nozzle on it. This is because the narrower opening of the nozzle gives the water a higher exit speed. Water flowing from the faucet shown in FIGURE 13.21b picks up speed as it falls. As a result, the flow tube "necks down" to a smaller diameter.

FIGURE 13.21 The speed of the water is inversely proportional to the diameter of the stream.

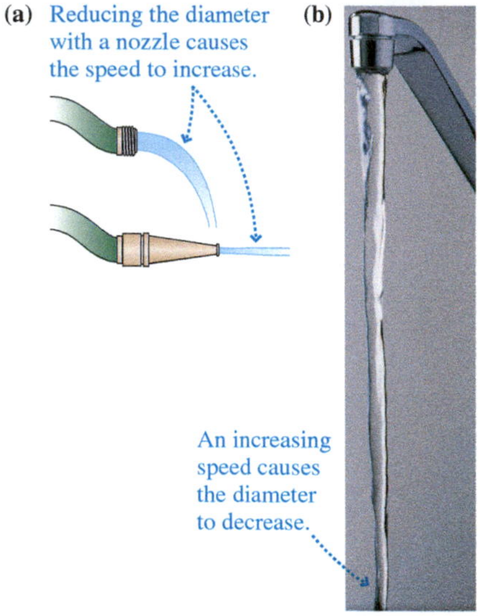


The *rate* at which fluid flows through the tube—volume per second—is $\Delta V/\Delta t$. This is called the **volume flow rate** $Q$. We can see from Equation 13.10 that

$$Q = \frac{\Delta V}{\Delta t} = vA \qquad (13.12)$$

Volume flow rate for liquid moving at speed $v$ through a tube of cross-section area $A$

The SI units of $Q$ are $m^3/s$, although in practice $Q$ may be measured in $cm^3/s$, liters per minute, or, in the United States, gallons per minute and cubic feet per minute. Another way to express the meaning of the equation of continuity is to say that **the volume flow rate is constant at all points in a tube.**

**EXAMPLE 13.10 Speed of water through a hose**

A garden hose has an inside diameter of 16 mm. The hose can fill a 10 L bucket in 20 s.

a. What is the speed of the water out of the end of the hose?
b. What diameter nozzle would cause the water to exit with a speed 4 times greater than the speed inside the hose?

**STRATEGIZE** We are given the volume flow rate; we will use this with Equation 13.12 to determine the speed of the flow.

**PREPARE** The volume flow rate is $Q = \Delta V/\Delta t = (10\text{ L})/(20\text{ s}) = 0.50\text{ L/s}$. To convert this to SI units, recall that $1\text{ L} = 1000\text{ mL} = 10^3\text{ cm}^3 = 10^{-3}\text{ m}^3$. Thus $Q = 5.0 \times 10^{-4}\text{ m}^3/\text{s}$.

**SOLVE** a. The speed of the water is

$$v = \frac{Q}{A} = \frac{Q}{\pi r^2} = \frac{5.0 \times 10^{-4}\text{ m}^3/\text{s}}{\pi(0.0080\text{ m})^2} = 2.5\text{ m/s}$$

b. The quantity $Q = vA$ remains constant as the water flows through the hose and then the nozzle. To increase $v$ by a factor of 4, $A$ must be reduced by a factor of 4. The cross-section area depends on the square of the diameter, so the area is reduced by a factor of 4 if the diameter is reduced by a factor of 2. Thus the necessary nozzle diameter is 8 mm.

**ASSESS** This seems like a reasonable speed for water through a garden hose.

### Representing Fluid Flow: Streamlines and Fluid Elements

Representing the flow of fluid is more complicated than representing the motion of a point particle because fluid flow is the collective motion of a vast number of particles. FIGURE 13.22 gives us an idea of one possible fluid-flow representation. Here smoke is being used to help engineers visualize the airflow around a car in a wind tunnel. The smoothness of the flow tells us this is laminar flow. But notice also how the individual smoke trails retain their identity. They don't cross or get mixed together. Each smoke trail represents a *streamline* in the fluid.

Imagine that we could inject a tiny colored drop of water into a stream of water undergoing laminar flow. Because the flow is steady and the water is incompressible, this colored drop would maintain its identity as it flowed along. The path or trajectory followed by this "particle of fluid" is called a **streamline.** Smoke particles mixed with the air allow us to see the streamlines in the wind-tunnel photograph of Figure 13.22. FIGURE 13.23 illustrates three important properties of streamlines.

FIGURE 13.22 Streamlines in the laminar airflow around a car in a wind tunnel.

FIGURE 13.23 Particles in a fluid move along streamlines.

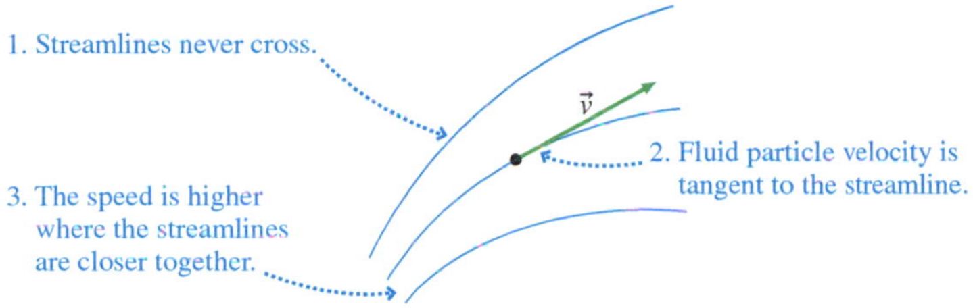


As we study the motion of a fluid, it is often also useful to consider a small *volume* of fluid, a volume containing many particles of fluid. Such a volume is called a **fluid element.** FIGURE 13.24 shows two important properties of a fluid element. Unlike a particle, a fluid element has an actual shape and volume. Although the shape of a fluid element may change as it moves, the equation of continuity requires that its volume remain constant.

FIGURE 13.24 Motion of a fluid element.

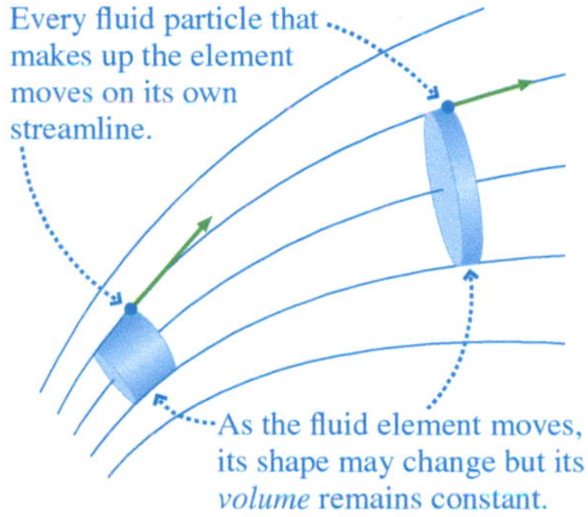

Every fluid particle that makes up the element moves on its own streamline.
As the fluid element moves, its shape may change but its *volume* remains constant.

**STOP TO THINK 13.5**

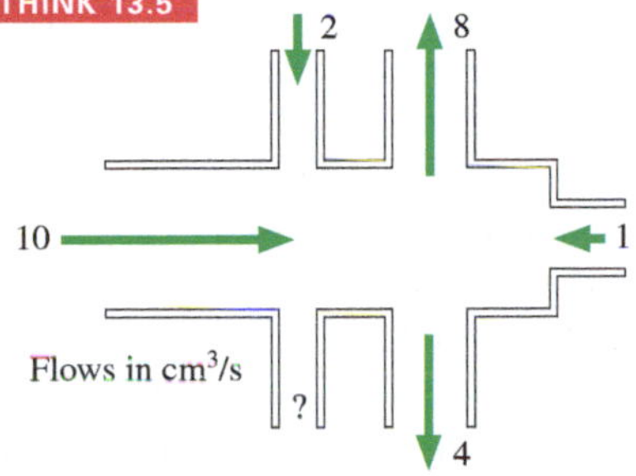


The figure shows volume flow rates (in cm$^3$/s) for all but one tube. What is the volume flow rate through the unmarked tube? Is the flow direction in or out?

## 13.5 Fluid Dynamics

The equation of continuity describes a moving fluid but doesn't tell us anything about *why* the fluid is in motion. To understand the dynamics, consider the fluid moving from left to right through the tube shown in FIGURE 13.25. The fluid moves at a steady speed $v_1$ in the wider part of the tube. In accordance with the equation of continuity, its speed is a higher, but steady, $v_2$ in the narrower part of the tube. If we follow a fluid element through the tube, we see that it undergoes an *acceleration* from $v_1$ to $v_2$ in the tapered section of the tube. According to Newton's second law, there must be a net force acting on this fluid element to accelerate it.

FIGURE 13.25 A motion diagram of a fluid element moving through a narrowing tube.

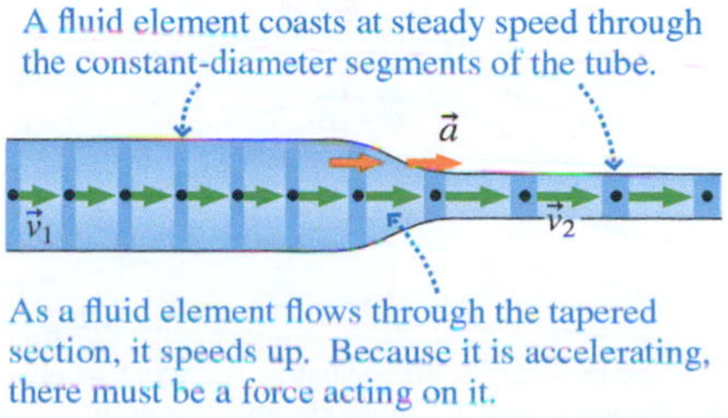

FIGURE 13.26 The net force on a fluid element due to pressure points from high to low pressure.

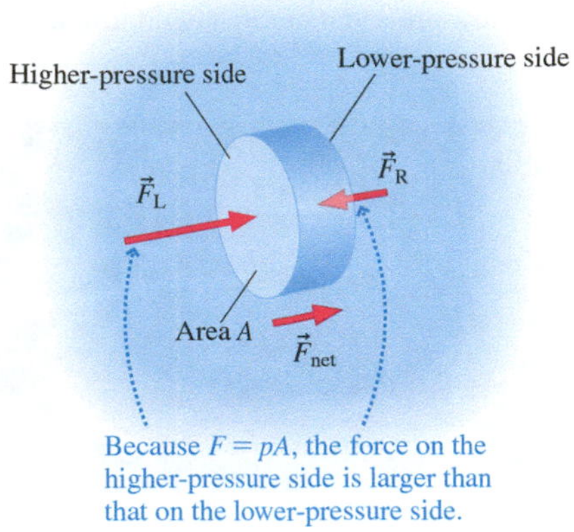


FIGURE 13.27 A Venturi tube measures flow speeds of a fluid.

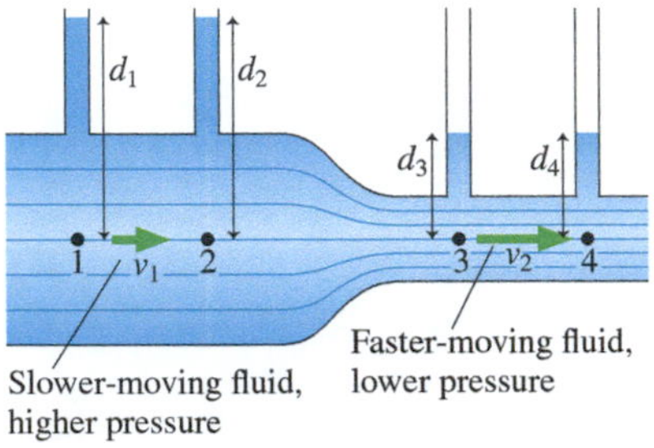


FIGURE 13.28 Wind speeds up as it crests a hill.

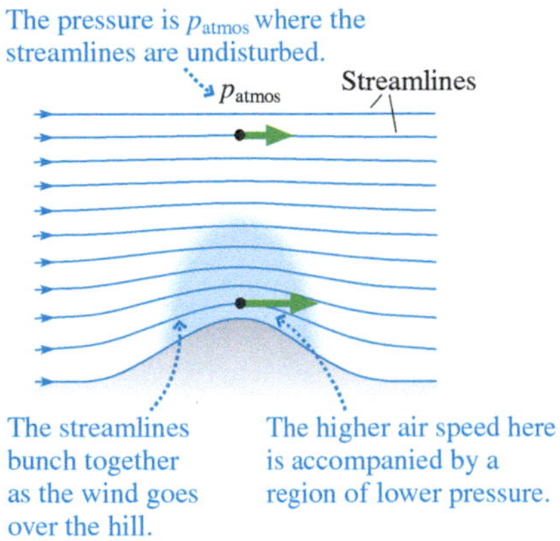


FIGURE 13.29 Airflow over a wing generates lift by creating unequal pressures above and below.

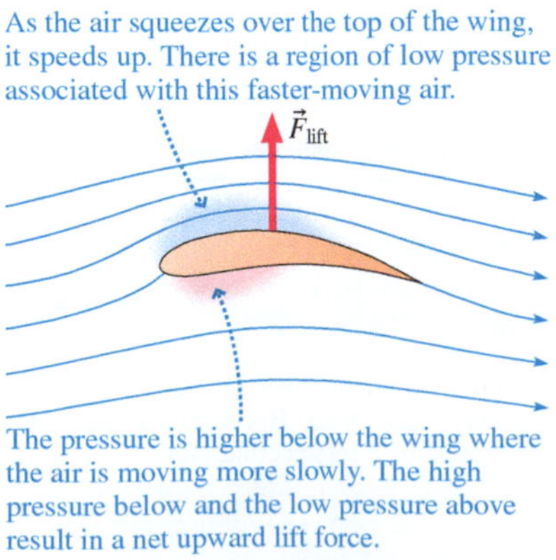

As the air squeezes over the top of the wing, it speeds up. There is a region of low pressure associated with this faster-moving air.
$\vec{F}_{lift}$
The pressure is higher below the wing where the air is moving more slowly. The high pressure below and the low pressure above result in a net upward lift force.


What is the origin of this force? There are no external forces, and the horizontal motion rules out gravity. Instead, the fluid element is pushed from both ends by the *surrounding fluid*—that is, by *pressure forces.* The fluid element with cross-section area $A$ in FIGURE 13.26 has a higher pressure on its left side than on its right. Thus the force $F_L = p_L A$ of the fluid pushing on its left side—a force pushing to the right—is greater than the force $F_R = p_R A$ of the fluid on its right side. The net force, which points from the higher-pressure side of the element to the lower-pressure side, is

$$F_{net} = F_L - F_R = (p_L - p_R)A = A\,\Delta p$$

In other words, there's a net force on the fluid element, causing it to change speed, if and only if there's a pressure *difference* $\Delta p$ between the two faces.

Thus, in order to accelerate the fluid elements in Figure 13.25 through the neck, the pressure $p_1$ in the wider section of the tube must be higher than the pressure $p_2$ in the narrower section. When the pressure is changing from one point in a fluid to another, we say that there is a **pressure gradient** in that region. We can also say that pressure forces are caused by pressure gradients, so if we can ignore viscosity, **a fluid accelerates wherever there is a pressure gradient.**

As a result, **the pressure is higher at a point along a streamline where the fluid is moving slower, lower where the fluid is moving faster.** This property of fluids was discovered in the 18th century by the Swiss scientist Daniel Bernoulli and is called the **Bernoulli effect.**

**NOTE** ▶ It is important to realize that it is the change in pressure from high to low that *causes* the fluid to speed up. A high fluid speed doesn't *cause* a low pressure any more than a fast-moving particle causes the force that accelerated it. ◀

This relationship between pressure and fluid speed can be used to measure the speed of a fluid with a device called a *Venturi tube.* FIGURE 13.27 shows a simple Venturi tube suitable for a flowing liquid. The high pressure at point 1, where the fluid is moving slowly, causes the fluid to rise in the vertical pipe to a total height $d_1$. Because there's no vertical motion of the fluid, we can use the hydrostatic pressure equation to find that the pressure at point 1 is $p_1 = p_0 + \rho g d_1$. At point 3, on the same streamline, the fluid is moving faster and the pressure is lower; thus the fluid in the vertical pipe rises to a lower height $d_3$. The pressure *difference* is $\Delta p = \rho g(d_1 - d_3)$, so the pressure difference across the neck of the pipe can be found by measuring the difference in heights of the fluid. We'll see later in this section how to relate this pressure difference to the increase of speed.

## Applications of the Bernoulli Effect

Consider the flow of air over a hill, as shown in FIGURE 13.28. Far to the left, away from the hill, the wind blows at a constant speed, so its streamlines are equally spaced. But as the air moves over the hill, the hill forces the streamlines to bunch together so that the air speeds up. According to the Bernoulli effect, there exists a zone of *low pressure* at the crest of the hill, where the air is moving the fastest.

Using these ideas, we can understand *lift*, the upward force on the wing of a moving airplane that makes flight possible. FIGURE 13.29 shows an airplane wing, seen in cross section, for an airplane flying to the left. The figure is drawn in the reference frame of the airplane, so that the wing appears stationary and the air flows past it to the right. The shape of the wing is such that, just as for the hill in Figure 13.28, the streamlines of the air must squeeze together as they pass over the wing. This increased speed is, by the Bernoulli effect, accompanied by a region of low pressure above the wing. The air speed below the wing is actually slowed, so that there's a region of high pressure below. This high pressure pushes up on the wing, while the low-pressure air above it presses down, but less strongly. The result is a net upward force—the lift. The same principles are at work on the wings of soaring birds. The lift force results from the motion of air over the wings, so the albatross in the chapter-opening photo must build up some speed before it generates enough lift to fly.

## Bernoulli's Equation

We've seen that a pressure gradient causes a fluid to accelerate horizontally. Not surprisingly, gravity can also cause a fluid to speed up or slow down if the fluid changes elevation. These are the key ideas of fluid dynamics. Now we would like to make these ideas quantitative by finding a numerical relationship for pressure, height, and the speed of a fluid. We can do so by applying a version of the work-energy equation you learned in ◀◀ SECTION 10.6,

$$\Delta K + \Delta U = W$$

where $U$ is the gravitational potential energy and $W$ is the work done by other forces—in this case, pressure forces. Recall that we're still ignoring viscosity, so there's no dissipation of energy to thermal energy.

FIGURE 13.30a shows fluid flowing through a tube. The tube narrows from cross-section area $A_1$ to area $A_2$ as it bends uphill. Let's concentrate on the large volume of the fluid that is shaded in Figure 13.30a. **This moving fluid element will be our system for the purpose of applying conservation of energy.**

To use conservation of energy, we need to draw a before-and-after overview. The "before" situation is shown in FIGURE 13.30b. A short time $\Delta t$ later, the system has moved along the tube a bit, as shown in the "after" drawing of FIGURE 13.30c. Because the tube is not of uniform diameter, the two ends of the fluid system do not move the same distance during $\Delta t$: The lower end moves distance $\Delta x_1$, while the upper end moves $\Delta x_2$. Thus the system moves *out of* a cylindrical volume $\Delta V_1 = A_1\,\Delta x_1$ at the lower end and *into* a volume $\Delta V_2 = A_2\,\Delta x_2$ at the upper end. The equation of continuity tells us that these two volumes must be the same, so $A_1\,\Delta x_1 = A_2\,\Delta x_2 = \Delta V$.

From the "before" situation to the "after" situation, the system *loses* the kinetic and potential energy it originally had in the volume $\Delta V_1$ but *gains* the kinetic and potential energy in the volume $\Delta V_2$ that it later occupies. (The energy it has in the region between these two small volumes is unchanged.) Let's find the kinetic energy in each of these small volumes. The mass of fluid in each cylinder is $m = \rho\,\Delta V$, where $\rho$ is the density of the fluid. The kinetic energies of the two small volumes 1 and 2 are then

$$K_1 = \frac{1}{2}\underbrace{\rho\,\Delta V}_{m}v_1^2 \quad \text{and} \quad K_2 = \frac{1}{2}\underbrace{\rho\,\Delta V}_{m}v_2^2$$

Thus the net *change* in kinetic energy is

$$\Delta K = K_2 - K_1 = \frac{1}{2}\rho\,\Delta V v_2^2 - \frac{1}{2}\rho\,\Delta V v_1^2$$

Similarly, the net change in the gravitational potential energy of our fluid system is

$$\Delta U = U_2 - U_1 = \rho\,\Delta V g y_2 - \rho\,\Delta V g y_1$$

The final piece of our conservation of energy treatment is the work done on the system by the rest of the fluid. As the fluid system moves, positive work is done on it by the force $\vec{F}_1$ due to the pressure $p_1$ of the fluid to the left of the system, while negative work is done on the system by the force $\vec{F}_2$ due to the pressure $p_2$ of the fluid to the right of the system. In ◀◀ SECTION 10.2 you learned that the work done by a force $F$ acting on a system as it moves through a displacement $\Delta x$ is $F\,\Delta x$. Thus the positive work is

$$W_1 = F_1\,\Delta x_1 = (p_1 A_1)\,\Delta x_1 = p_1(A_1\,\Delta x_1) = p_1\,\Delta V$$

Similarly the negative work is

$$W_2 = -F_2\,\Delta x_2 = -(p_2 A_2)\,\Delta x_2 = -p_2(A_2\,\Delta x_2) = -p_2\,\Delta V$$

Thus the *net* work done on the system is

$$W = W_1 + W_2 = p_1\,\Delta V - p_2\,\Delta V = (p_1 - p_2)\,\Delta V$$

We can now use these expressions for $\Delta K$, $\Delta U$, and $W$ to write the energy equation as

$$\underbrace{\frac{1}{2}\rho\,\Delta V v_2^2 - \frac{1}{2}\rho\,\Delta V v_1^2}_{\Delta K} + \underbrace{\rho\,\Delta V g y_2 - \rho\,\Delta V g y_1}_{\Delta U} = \underbrace{(p_1 - p_2)\,\Delta V}_{W}$$

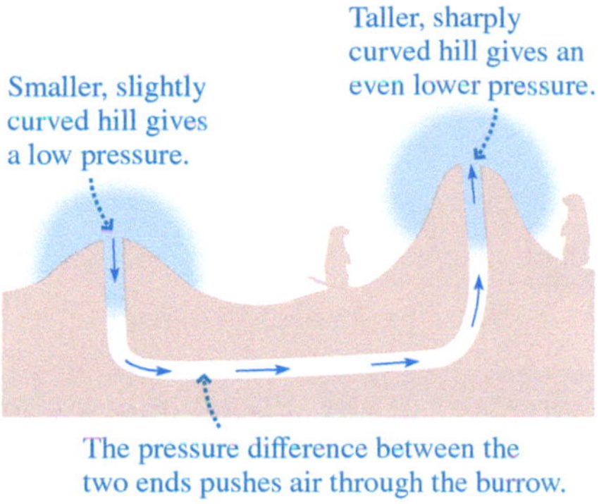

Smaller, slightly curved hill gives a low pressure.
Taller, sharply curved hill gives an even lower pressure.
The pressure difference between the two ends pushes air through the burrow.

**Nature's air conditioning** BIO Prairie dogs ventilate their underground burrows with the same aerodynamic forces and pressures that give airplanes lift. The two entrances to their burrows are surrounded by mounds, one higher than the other. When the wind blows across these mounds, the pressure is reduced at the top. The taller mound, with its greater curvature, has the lower pressure of the two entrances. Air then is pushed through the burrow toward this lower-pressure side.

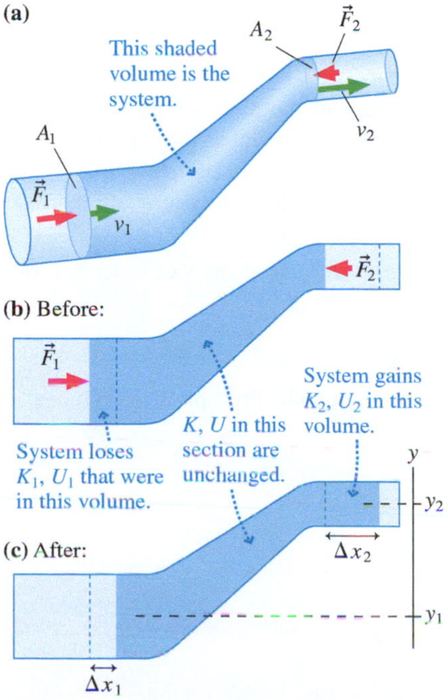


FIGURE 13.30 A fluid flowing through a tube.

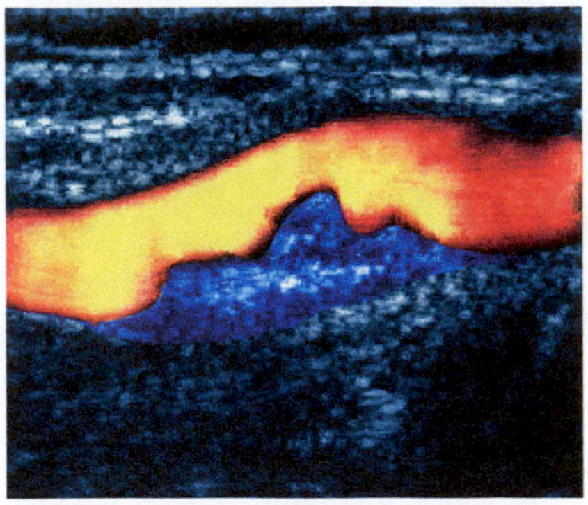

◀ **Living under pressure** BIO When plaque builds up in major arteries, dangerous drops in blood pressure can result. *Doppler ultrasound* uses sound waves to detect the velocity of flowing blood. The image shows the blood flow through a carotid artery with significant plaque buildup; yellow indicates a higher blood velocity than red. Once the velocities are known at two points along the flow, Bernoulli's equation can be used to deduce the corresponding pressure drop.

The $\Delta V$'s cancel, and we can rearrange the remaining terms to get **Bernoulli's equation,** which relates fluid quantities at two points along a streamline:

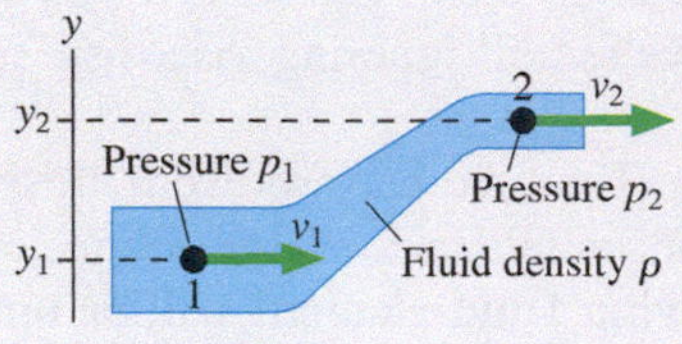

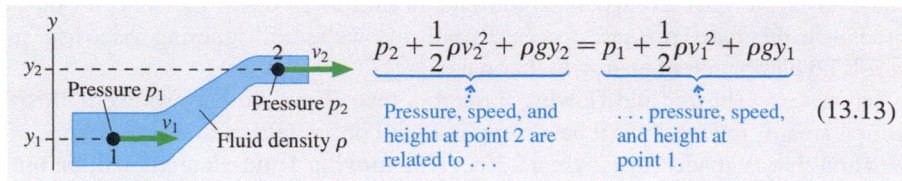

$$p_2 + \frac{1}{2}\rho v_2^2 + \rho g y_2 = p_1 + \frac{1}{2}\rho v_1^2 + \rho g y_1 \qquad (13.13)$$

Pressure, speed, and height at point 2 are related to . . . / . . . pressure, speed, and height at point 1.

Video What the Physics? Air Bender

Equation 13.13, a quantitative statement of the ideas we developed earlier in this section, is really nothing more than a statement about work and energy. Using Bernoulli's equation is very much like using the law of conservation of energy. Rather than identifying a "before" and "after," we want to identify two points on a streamline.

## EXAMPLE 13.11 Pressure in an irrigation system

Water flows through the pipes shown in FIGURE 13.31. The water's speed through the lower pipe is 5.0 m/s, and a pressure gauge reads 75 kPa. What is the reading of the pressure gauge on the upper pipe?

**FIGURE 13.31** The water pipes of an irrigation system.

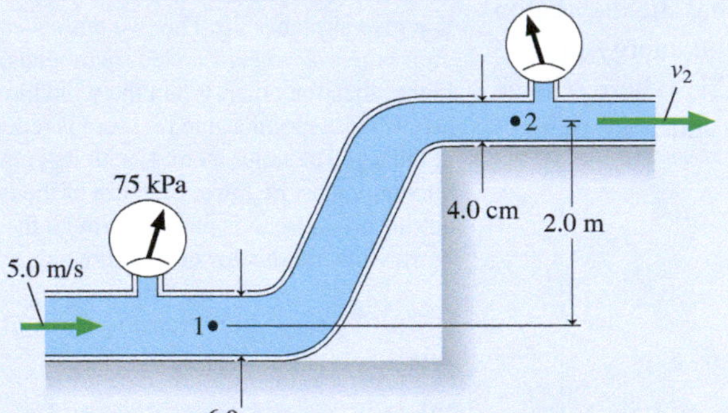


***STRATEGIZE*** We will consider a streamline that goes along the middle of the pipe, connecting point 1 in the lower pipe with point 2 in the upper pipe. Bernoulli's equation will allow us to find differences in pressure along the flow. We can also use the equation of continuity to relate the speed of the flow to the area at different points.

***PREPARE*** The density of water is 1000 kg/m$^3$.

***SOLVE*** Bernoulli's equation, Equation 13.13, relates the pressures, fluid speeds, and heights at points 1 and 2. It can be used to solve for the pressure $p_2$ at point 2:

$$p_2 = p_1 + \frac{1}{2}\rho v_1^2 - \frac{1}{2}\rho v_2^2 + \rho g y_1 - \rho g y_2$$
$$= p_1 + \frac{1}{2}\rho(v_1^2 - v_2^2) + \rho g(y_1 - y_2)$$

All quantities on the right are known except $v_2$, and that is where the equation of continuity will be useful. The cross-section areas and water speeds at points 1 and 2 are related by

$$v_1 A_1 = v_2 A_2$$

from which we find

$$v_2 = \frac{A_1}{A_2}v_1 = \frac{r_1^2}{r_2^2}v_1 = \frac{(0.030\text{ m})^2}{(0.020\text{ m})^2}(5.0\text{ m/s}) = 11.25\text{ m/s}$$

The pressure at point 1 is $p_1 = 75\text{ kPa} + 1\text{ atm} = 176{,}300\text{ Pa}$. We can now use the above expression for $p_2$ to calculate $p_2 = 105{,}900\text{ Pa}$. This is the absolute pressure; the pressure gauge on the upper pipe is

$$p_2 = 105{,}900\text{ Pa} - 1\text{ atm} = 4.6\text{ kPa}$$

***ASSESS*** We find a lower pressure at point 2; this makes sense. Reducing the pipe size decreases the pressure because it makes $v_2 > v_1$. Gaining elevation also reduces the pressure.

Video Bernoulli's Principle: Venturi Tubes

**STOP TO THINK 13.6** Rank in order, from highest to lowest, the liquid heights $h_1$ to $h_4$ in tubes 1 to 4. The airflow is from left to right.

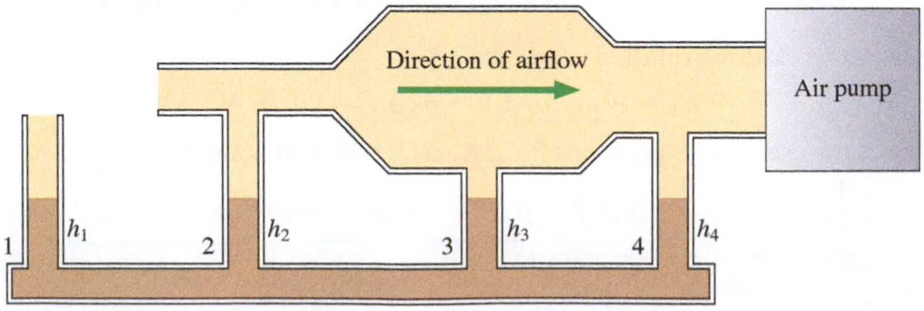

## 13.6 Viscosity and Poiseuille's Equation

In this section, we'll begin to include the effects of viscosity. Viscosity is important to understanding many real-world situations, from blood flow to the flight of birds to throwing a curveball.

If there is no viscosity, a fluid will "coast" at constant speed through a constant-diameter tube with no change in pressure. That's why the fluid heights are the same in the first and second pressure-measuring columns on the Venturi tube of Figure 13.27. But as FIGURE 13.32 shows, a viscous fluid requires a *pressure difference* between the ends of a tube to keep the fluid moving at constant speed. As the fluid moves, viscosity works like friction, converting some of the energy of the flow into thermal energy. The pressure difference does work to keep the energy of the flow constant. A greater viscosity means that more energy is lost and a greater pressure difference is required to maintain the flow. Think about how much harder you have to suck on a straw to drink a thick milkshake than to drink water or to pull air through the straw.

FIGURE 13.32 A viscous fluid needs a pressure difference to keep it moving.

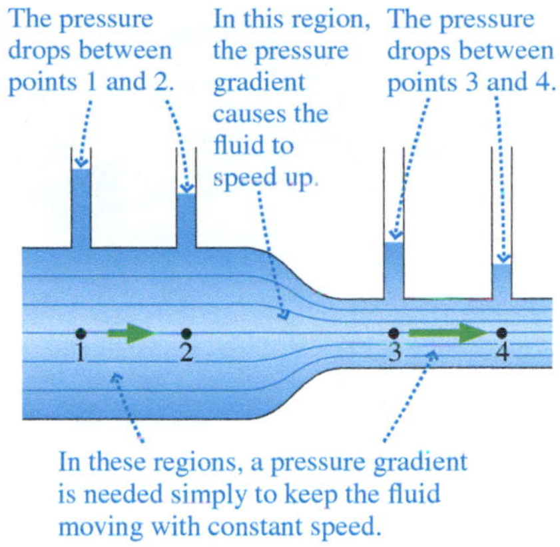


FIGURE 13.33 shows a viscous fluid flowing with a constant average speed $v_{avg}$ through a tube of length $L$ and cross-section area $A$. Experiments show that the pressure difference needed to keep the fluid moving is proportional to $v_{avg}$ and to $L$ and inversely proportional to $A$. We can write

$$\Delta p = 8\pi\eta \frac{Lv_{avg}}{A} \tag{13.14}$$

where $\eta$ is the viscosity.

FIGURE 13.33 The pressure difference needed to keep the fluid flowing is proportional to the fluid's viscosity.

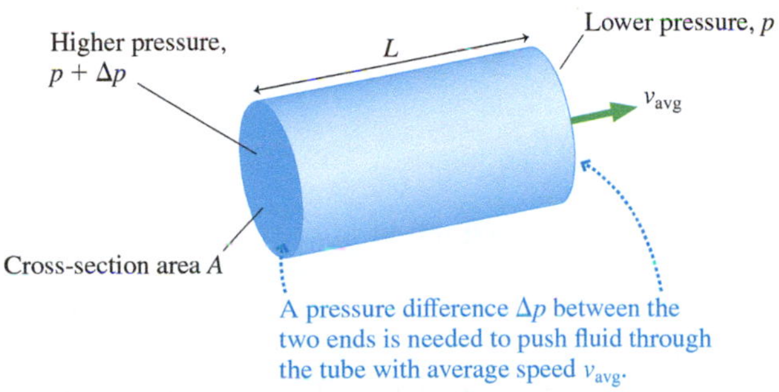


Equation 13.14 makes sense. A more viscous fluid needs a larger pressure difference to push it through the tube; if there is no viscosity, $\eta = 0$, and the fluid will keep flowing without any pressure difference. We can also see from Equation 13.14 that the units of viscosity are Pa · s. TABLE 13.4 gives values of $\eta$ for some common fluids, a wider selection than we saw in Chapter 5. Note that the viscosity of many liquids decreases *very* rapidly with temperature. Cold oil hardly flows at all, but hot oil pours almost like water.

TABLE 13.4 Viscosities of fluids

| Fluid | $\eta$ (Pa · s) | Fluid | $\eta$ (Pa · s) |
|---|---|---|---|
| Air 20°C | $1.8 \times 10^{-5}$ | Motor oil: | |
| Water: | | −30°C | $3 \times 10^{5}$ |
| 20°C | $1.0 \times 10^{-3}$ | 40°C | 0.07 |
| 40°C | $0.7 \times 10^{-3}$ | 100°C | 0.01 |
| 60°C | $0.5 \times 10^{-3}$ | Molasses: | |
| | | 20°C | 50 |
| Whole blood 37°C | $3.5 \times 10^{-3}$ | 50°C | 2 |

## Poiseuille's Equation

FIGURE 13.34 Viscosity alters the velocities of the fluid particles.

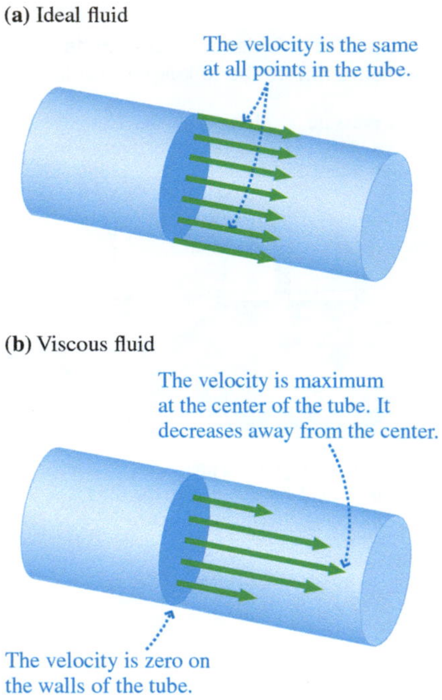


Viscosity has a profound effect on how a fluid moves through a tube. FIGURE 13.34a shows that in an ideal fluid, all fluid particles move with the same speed $v$, the speed that appears in the equation of continuity. For a viscous fluid, FIGURE 13.34b shows that the fluid moves fastest in the center of the tube. The speed decreases as we move away from the center of the tube until it reaches zero on the walls of the tube. That is, the layer of fluid in contact with the tube doesn't move at all. Whether it is water through pipes or blood through arteries, the fact that the fluid at the outer edges "lingers" and barely moves allows deposits to build on the inside walls of a tube.

Although we can't characterize the flow of a viscous liquid by a single flow speed $v$, we can still define the *average* flow speed. Suppose a fluid with viscosity $\eta$ flows through a circular pipe with radius $R$ and cross-section area $A = \pi R^2$. From Equation 13.14, a pressure difference $\Delta p$ between the ends of the pipe causes the fluid to flow with average speed

$$v_{\text{avg}} = \frac{R^2}{8\eta L}\Delta p \qquad (13.15)$$

The average flow speed is directly proportional to the pressure difference; for the fluid to flow twice as fast, we would need to double the pressure difference between the ends of the pipe.

Equation 13.12 defined the volume flow rate $Q = \Delta V/\Delta t$ and found that $Q = vA$ for an ideal fluid. For viscous flow, where $v$ isn't constant throughout the fluid, we simply need to replace $v$ with the average speed $v_{\text{avg}}$ found in Equation 13.15. Using $A = \pi R^2$ for a circular tube, we see that a pressure difference $\Delta p$ causes a volume flow rate

$$Q = v_{\text{avg}}A = \frac{\pi R^4 \Delta p}{8\eta L} \qquad (13.16)$$

Poiseuille's equation for viscous flow through a tube of radius $R$ and length $L$

This result is called **Poiseuille's equation** after the 19th-century French scientist Jean Poiseuille who first performed this calculation.

One surprising result of Poiseuille's equation is the very strong dependence of the flow on the tube's radius; the volume flow rate is proportional to the *fourth* power of $R$. If we double the radius of a tube, the flow rate will increase by a factor of $2^4 = 16$.

### EXAMPLE 13.12 What causes the pressure drop?

When water goes through the small holes in a showerhead, the total area for the flow decreases, so the speed of the flow increases. The gauge pressure drops from the value in the water line behind the showerhead to zero after it exits; a greater difference in pressure permits a greater increase in speed and a more vigorous shower.

The pressure at the showerhead is less than where the supply line connects to the city's water supply; a difference in pressure is necessary to move the water through the pipe. A greater flow requires a greater drop in pressure. If someone else starts to use water when you are taking a shower, there can be a small but noticeable pressure drop. You'll notice this by the reduction in flow of your shower.

Let's look at some typical numbers. Suppose you are taking a shower after school. Your home's water travels 40 m from the city's main line to your house, through a pipe of radius 0.0072 m. The water is a chilly 10°C, so the viscosity is $1.3 \times 10^{-3}$ Pa · s, slightly greater than the value given for 20°C in Table 13.4. The gauge pressure at the showerhead is 200 kPa, enough to provide a vigorous shower. You are the only person in the house using water, and the flow rate of your shower is $2.0 \times 10^{-4}\ \text{m}^3/\text{s}$. But then your roommate, in another bathroom, flushes a toilet. As the toilet's tank refills, the supply line is supplying a total flow of $6.0 \times 10^{-4}\ \text{m}^3/\text{s}$. What is the new pressure at the shower head? Assume that there is no appreciable change in height throughout the system.

**STRATEGIZE** We will assume that the pressure at the start of the supply line, where it meets the city water supply, is constant. We will use the flow rate and the characteristics of the pipe to find the pressure difference necessary to drive the flow. This will give us the pressure at the start of the supply line. The higher flow will require a greater pressure difference, leading to a lower pressure at the showerhead.

***PREPARE*** All of the quantities in the problem are given in the proper SI units, so no unit conversions are necessary.

***SOLVE*** We rearrange Equation 13.16 to give the pressure difference in terms of the flow rate:

$$\Delta p = \frac{8\eta LQ}{\pi R^4} = \frac{8(1.3 \times 10^{-3}\ \text{Pa}\cdot\text{s})(40\ \text{m})(2.0 \times 10^{-4}\ \text{m}^3/\text{s})}{\pi(0.0072\ \text{m})^4}$$
$$= 9.9\ \text{kPa} \approx 10\ \text{kPa}$$

This is the pressure difference across the supply line, so the city water supply has a pressure of 210 kPa. At the higher flow rate, the difference in pressure is

$$\Delta p = \frac{8\eta LQ}{\pi R^4} = \frac{8(1.3 \times 10^{-3}\ \text{Pa}\cdot\text{s})(40\ \text{m})(6.0 \times 10^{-4}\ \text{m}^3/\text{s})}{\pi(0.0072\ \text{m})^4}$$
$$= 30\ \text{kPa}$$

With the toilet and the shower both using water, the pressure drops by 30 kPa, so the pressure at the showerhead is

$$210\ \text{kPa} - 30\ \text{kPa} = 180\ \text{kPa}$$

***ASSESS*** This is a small but noticeable change in pressure from the original situation, exactly as expected.

**STOP TO THINK 13.7** A viscous fluid flows through the pipe shown. The three marked segments are of equal length. Across which segment is the pressure difference the greatest?

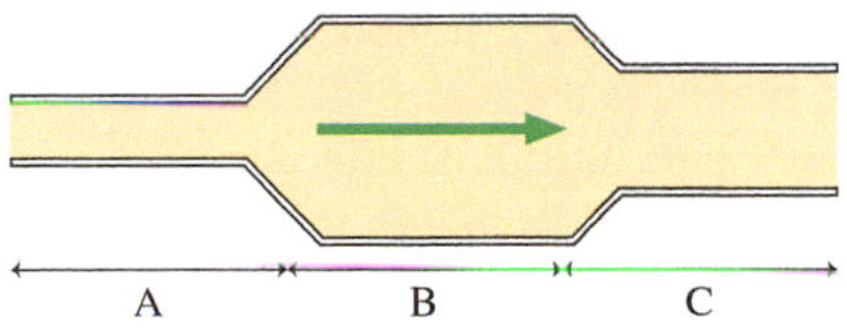


## 13.7 The Circulatory System BIO

Once every second, more or less, your heart beats, and the contraction of the chambers of your heart sends blood flowing through your body's arteries, capillaries, and veins. In this section, we'll use the principles that we've learned in earlier sections to explain and explore the motion of blood through your circulatory system.

We'll focus on the motion of blood through your body, from your heart's left ventricle to the right atrium; we won't examine the pulmonary circulation through your lungs. As usual, we'll make a simplified model of the system. We'll assume, as we've done before, that the fluid is incompressible (a very good approximation) and that the flow is laminar. The relationships seen earlier in the chapter will therefore apply. In the larger vessels, we'll be able to assume that the effect of viscosity is negligible; in the smaller vessels, it isn't.

**FIGURE 13.35** Blood pressure during one cycle of a heartbeat.

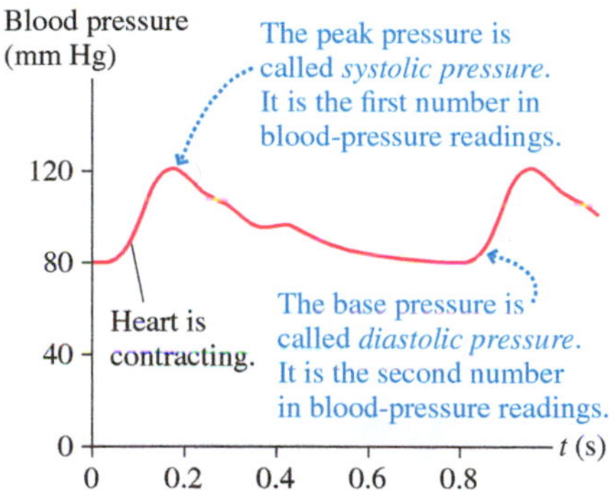


### Blood Pressure

When your heart muscle contracts, the pressure of the blood increases, and blood leaves the left ventricle and moves into the large artery called the aorta. **FIGURE 13.35** is a pressure graph showing how the blood pressure at the heart changes during one cycle of the heartbeat. Blood pressure is a gauge pressure, the pressure in excess of the pressure of the atmosphere, and it is traditionally measured in mm Hg. The graph shows a fairly typical blood pressure for a healthy young adult, with a *systolic* (peak) pressure of 120 mm Hg and a *diastolic* (base) pressure of 80 mm Hg—that is, a blood pressure of "120 over 80."

**FIGURE 13.36** Measuring blood pressure.

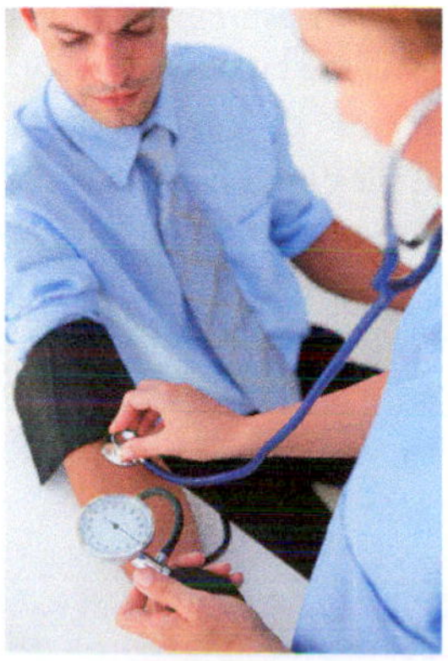

If your blood pressure is too high, it can cause many negative health effects. If your blood pressure is too low, it may indicate other health problems. For this reason, a visit to the doctor usually starts with a measurement of blood pressure. As shown in **FIGURE 13.36**, blood pressure is measured with a cuff that goes around the upper arm. A nurse pumps air into the cuff while using a stethoscope to listen to the flow of blood in a large artery and reading a gauge that shows the pressure in the cuff. Your arteries are flexible; if the pressure of the air in the cuff is greater than the pressure of the blood in the artery, the greater pressure outside the artery squeezes it shut, stopping blood flow. The nurse pumps air into the cuff until the pressure is well above the systolic pressure so that the cuff squeezes the artery shut and cuts off the blood flow. He or she then slowly lets air leak out of the cuff, listening for telltale sounds of blood flow in your artery. When the cuff pressure drops below the systolic pressure, the pressure pulse during each beat of your heart forces the artery to open briefly and a squirt of blood goes through. The nurse records the pressure when he or she hears the blood start to flow. This is your systolic pressure. This pulsing of the blood through your artery lasts until the cuff pressure reaches the diastolic pressure, when the artery remains open continuously and the blood flows smoothly. This transition is easily heard in the stethoscope, and the nurse records your diastolic pressure.

If we need only an average value, we'll take 100 mm Hg, the average of typical systolic and diastolic values, as a typical blood pressure. Our equations require SI units, so we'll remind you of the conversion of mm Hg to Pa:

$$1 \text{ mm Hg} = 133 \text{ Pa}$$

The clinically interesting measure for blood pressure is the pressure at the heart, but we've noted that blood pressure is measured in an artery in the arm. Are these two pressures the same? We know that there can be a difference in pressure across a tube or pipe due to the effects of viscosity; there can also be a difference in pressure at different points in a fluid due to differences in elevation.

### EXAMPLE 13.13 How much does blood pressure change? BIO

A woman is lying down with her arms horizontal. Along her upper arm, her brachial artery stretches 20 cm with a diameter of 4.0 mm. Blood moves through the artery at an average speed of 9.0 cm/s. What is the pressure drop along the 20 cm length?

**STRATEGIZE** Any pressure difference along the artery isn't due to a height difference because the artery is horizontal. Therefore, any pressure difference must be due to the viscosity of the blood, so we will compute the pressure difference using Equation 13.14.

**PREPARE** The viscosity of blood is given in Table 13.4 as $3.5 \times 10^{-3}$ Pa · s. We are looking for the pressure drop along 20 cm of artery, so the pressure drop along a tube of length $L = 0.20$ m. The radius is $R = 0.0020$ m and the fluid moves at an average speed $v_{avg} = 0.090$ m/s.

**SOLVE** Using the relevant values in Equation 13.14 gives the pressure drop as

$$\Delta p = \frac{8\pi(3.5 \times 10^{-3}\ \text{Pa}\cdot\text{s})(0.20\ \text{m})(0.090\ \text{m/s})}{\pi(0.0020\ \text{m})^2}$$
$$= 130\ \text{Pa} = 0.95\ \text{mm Hg}$$

**ASSESS** We know that, clinically, the blood pressure in the arm is taken to be the same as the blood pressure at the heart, so we'd expect a very small difference in pressure along large arteries such as the brachial artery. And, indeed, the number that we found is a small fraction of the average blood pressure—less than 1%—and certainly small compared to the rapid fluctuations in blood pressure that occur naturally.

There is only a minimal change in pressure due to viscous losses along your arteries. But there can certainly be differences in pressure due to differences in elevation. The blood in your arteries is a connected fluid, so the pressures are lower at greater heights. In the blood pressure measurement shown in Figure 13.36, the cuff is placed on the arm at the level of the heart so that there is no difference in pressure. How much error is introduced if this is not the case?

### EXAMPLE 13.14 Computing the change in blood pressure with height BIO

A patient is lying in bed. Suppose she raises her arm for a blood pressure measurement, with the cuff placed 15 cm (about 6 inches) above her heart. How much of an error does this introduce in the measurement? Ignore any pressure drop due to the viscosity of the blood.

**STRATEGIZE** Blood flows from the heart through the aorta and other arteries until it reaches the artery compressed by the cuff. We will use Equation 13.5 to determine the pressure at different depths in this connected fluid.

**PREPARE** In Equation 13.5, we can take $p_0$ as the pressure at the cuff; the pressure $p$ is the pressure at the heart, $d = 0.15$ m below the cuff.

**SOLVE** The pressure at the cuff is lower than the pressure at the heart by

$$p - p_0 = \rho g d = (1050\ \text{kg/m}^3)(9.8\ \text{m/s}^2)(0.15\ \text{m})$$
$$= 1540\ \text{Pa} \approx 12\ \text{mm Hg}$$

**ASSESS** This is a significant difference—more than 10 times the difference in pressure due to viscous losses. To avoid introducing such an error, the blood pressure cuff must be placed at the level of the heart.

Blood pressure decreases with increasing height above the heart. When you are upright, the blood pressure in your brain is noticeably lower than the pressure at your heart. For animals with long necks, this difference in pressure is correspondingly greater.

▸ **Extreme blood pressure** BIO A giraffe's head can be more than 2 meters above its heart. If the blood pressure in the brain is greater than zero (as it must be), the pressure at the heart must be quite high. A typical blood pressure for a resting giraffe is 240/160 mm Hg—a number that would raise serious alarms in a human!

## The Arteries and Capillaries

In the human body, blood pumped from the heart to the body starts its journey in a single large artery, the aorta. The flow then branches into smaller blood vessels, the large arteries that feed the head, the trunk, and the limbs. These branch into still smaller arteries, which then branch into a network of much smaller arterioles, which branch further into the capillaries. FIGURE 13.37 shows a schematic outline of the circulation, with average values for the diameters of the individual vessels, the total cross-section area of all of each type of vessel considered together, and the pressure in these vessels, assuming that the person is lying down so that there is no pressure change due to differences in elevation.

FIGURE 13.37 Schematic overview of blood flow in the circulatory system.

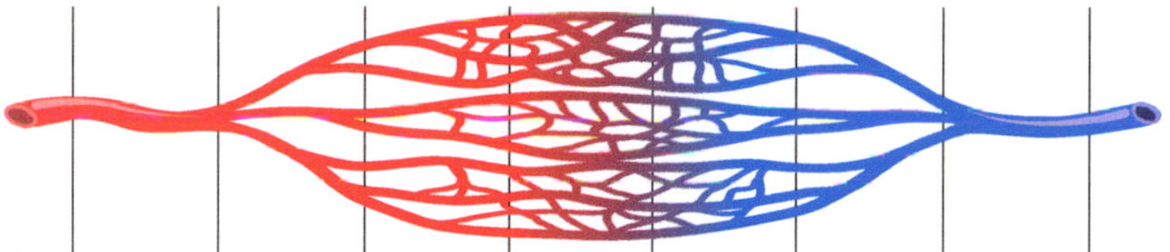

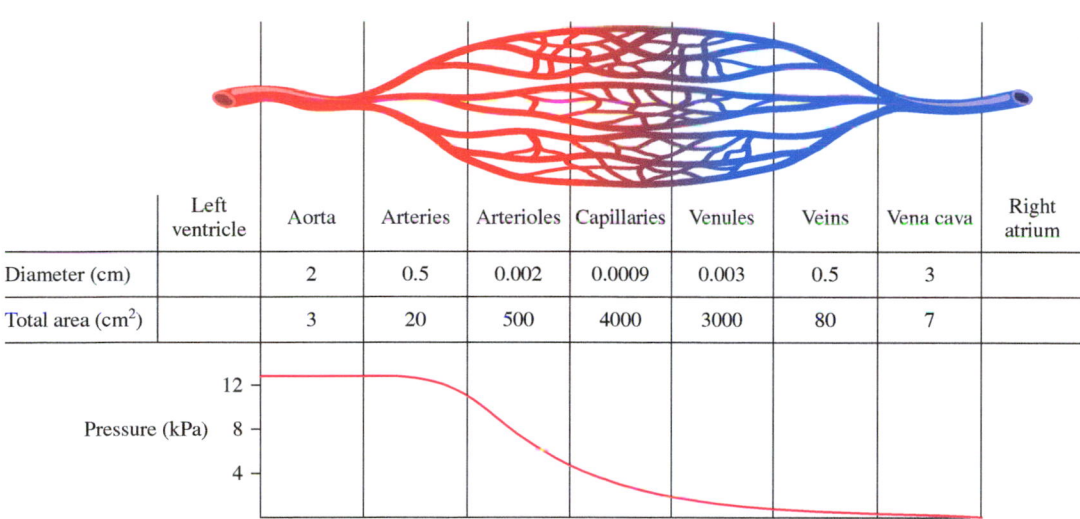

| | Left ventricle | Aorta | Arteries | Arterioles | Capillaries | Venules | Veins | Vena cava | Right atrium |
|---|---|---|---|---|---|---|---|---|---|
| Diameter (cm) | | 2 | 0.5 | 0.002 | 0.0009 | 0.003 | 0.5 | 3 | |
| Total area ($cm^2$) | | 3 | 20 | 500 | 4000 | 3000 | 80 | 7 | |

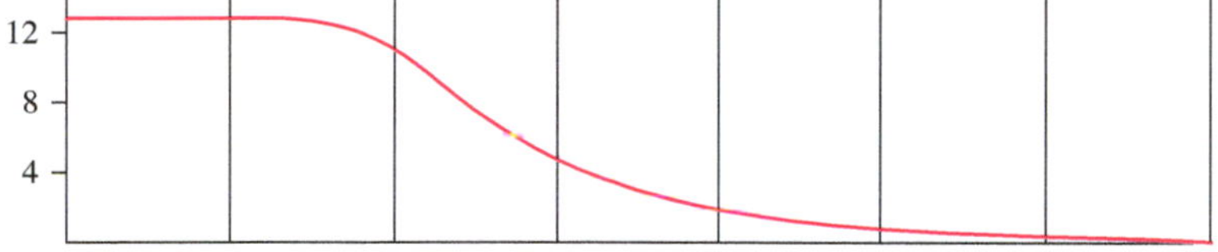


The numbers in the table are averages from different sources, representing variations among individuals of different sizes, ages, and levels of physical conditioning. The numbers are given to one significant figure; we really can't claim more precision. Nonetheless, a few broad trends emerge:

- As we've seen, there is only a very small change in pressure across the larger arteries. The pressure begins to drop only when the blood enters the smaller arteries and, even more so, the arterioles, where viscosity starts to have a significant effect. Most of the pressure drop occurs in these smaller vessels. Changes in the size of the small arteries and arterioles have significant effects on blood flow and on blood pressure.
- As blood moves from the aorta to the arteries and then to the arterioles, the diameter of the individual vessels decreases but the total area of all of the vessels increases. There is a high degree of branching; going from the aorta to the arterioles, the area of an individual vessel decreases by a factor of 1,000,000, but the area of all of the vessels considered together increases by a factor of 200. This implies that the single aorta eventually branches to 200,000,000 arterioles!

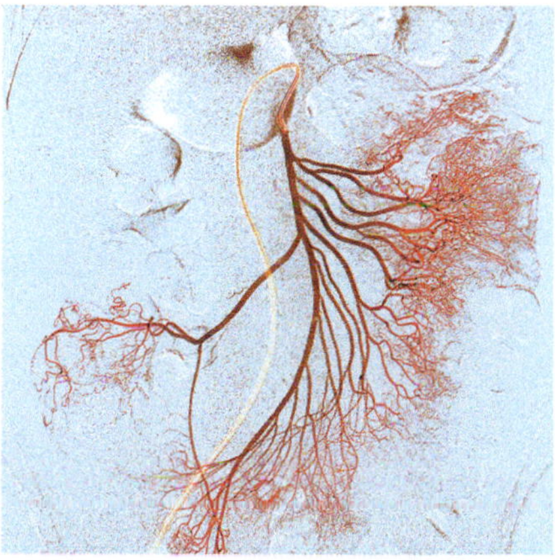

This preserved section of blood vessels shows the tremendous increase in number and in total area as blood vessels branch from large arteries to arterioles. One large artery gives rise to thousands of smaller vessels.

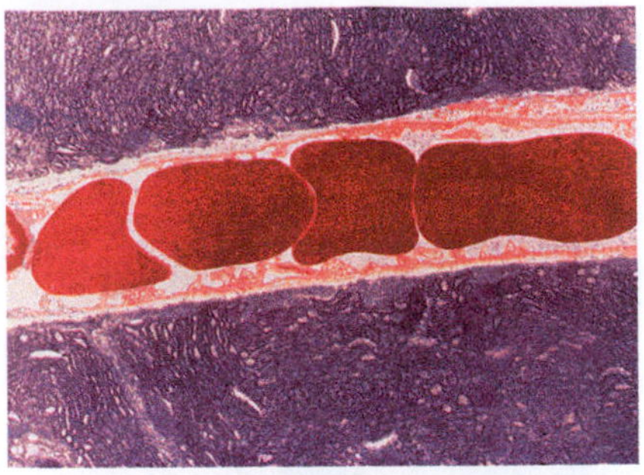

**Single file, please** BIO The capillaries are about as small as they can be because red blood cells can barely fit through. The red blood cells move through one after another as shown in the photo; they even deform to fit the contours of the capillary. This puts the cells close to the walls of the capillary, which enhances diffusion. This ordering actually reduces the viscosity of the blood and so reduces the pressure needed to move blood through these vessels.

- As the total cross-section area for the flow increases, the continuity equation indicates that the flow speed must decrease. As the flow branches from a large vessel to a number of smaller vessels, the area of the branches exceeds that of the initial vessel, so the speed decreases.

If we assume that 5 liters of blood move through the circulatory system in 1 minute, the volume flow rate is

$$Q = \frac{5 \text{ liters}}{1 \text{ minute}} = \frac{0.005 \text{ m}^3}{60 \text{ s}} = 8.3 \times 10^{-5} \text{ m}^3/\text{s}$$

All of this blood goes through the aorta, for which Figure 13.37 shows a diameter of 2 cm. The speed of the flow can be computed using Equation 13.12:

$$v = \frac{Q}{A} = \frac{8.3 \times 10^{-5} \text{ m}^3/\text{s}}{\pi(0.010 \text{ m})^2} = 0.26 \text{ m/s}$$

This is a reasonable number for the average flow speed. As the vessels branch and the total area increases, the velocity decreases, as we've discussed.

This decrease in velocity is important for the primary purpose of blood—delivering oxygen to and removing carbon dioxide from cells in the body. These gases diffuse across the thin walls of the tiniest vessels, the capillaries. The low speed of blood flow in these tiny, thin vessels allows more time for this diffusion to take place. Once gas exchange has occurred, the blood is collected by a venule to be returned to the heart.

**EXAMPLE 13.15 Determining the details of the capillaries** BIO

Figure 13.37 gives typical values for the diameter of a capillary as well as the total cross-section area of all the capillaries together. A typical capillary has a length of 1.0 mm. If the heart is pumping blood at a typical 5 liters per minute, how much time does it take for a red blood cell to move through a capillary?

**STRATEGIZE** Given the volume flow rate and the cross-section area, we can find the flow speed in the capillaries, which we will use to find the time a blood cell takes to traverse a capillary.

**PREPARE** The total cross-section area of all of the capillaries is given in Figure 13.37 as $4000 \text{ cm}^2 = 0.40 \text{ m}^2$. We showed previously that a flow rate of 5 liters per minute corresponds to $8.3 \times 10^{-5} \text{ m}^3/\text{s}$.

**SOLVE** We can find the speed of blood through the capillaries using Equation 13.12:

$$v = \frac{Q}{A} = \frac{8.3 \times 10^{-5} \text{ m}^3/\text{s}}{0.40 \text{ m}^2} = 2.1 \times 10^{-4} \text{ m/s}$$

At this speed, it takes a blood cell a time

$$\Delta t = \frac{\Delta x}{v} = \frac{0.0010 \text{ m}}{2.1 \times 10^{-4} \text{ m/s}} = 4.8 \text{ s} \approx 5 \text{ s}$$

to traverse a capillary.

**ASSESS** The relatively long time for the blood to traverse the capillary makes sense; there must be sufficient time for diffusion to occur.

We've assumed that the flow in the circulatory system is laminar. For flow in tubes, the transition between laminar and turbulent flow occurs at a Reynolds number of about 2000. For tubes, the appropriate length parameter is the diameter of the tube, so for the flow in the aorta just described, the Reynolds number is

$$Re = \frac{\rho v L}{\eta} = \frac{(1050 \text{ kg/m}^3)(0.26 \text{ m/s})(0.020 \text{ m})}{3.5 \times 10^{-3} \text{ Pa} \cdot \text{s}} = 1600$$

This is just below the threshold for turbulence to develop. If there is narrowing in the large arteries, speeds increase, and the combination of these two factors can put *Re* above the critical value, leading to turbulent flow. A physician can detect turbulent flow from its characteristic sound. This is one thing a physician is screening for when listening to your heart; the telltale sound of turbulent flow is a sign of trouble.

**EXAMPLE 13.16 Using the difference in blood speeds to compute a blood pressure drop** BIO

If the aortic valve, through which blood exits the left ventricle, becomes narrowed, it can lead to significant health problems. One of the diagnostic criteria is a drop in pressure as the blood traverses the valve. If this pressure drop exceeds 40 mm Hg, a diagnosis of severe aortic stenosis (a stenosis is a narrowing of a blood vessel) is warranted. Clinicians can use Doppler echocardiography, a noninvasive test using ultrasound, to measure the speeds of the blood upstream and downstream of the valve. The speeds can then be used to compute a corresponding pressure drop. In one patient, as the left ventricle is ejecting blood, measurements show speeds of 1.1 m/s upstream of the valve and 3.5 m/s just past the valve. Does the patient meet the criterion for severe aortic stenosis?

**STRATEGIZE** In the aorta and other large blood vessels, we can ignore losses due to viscosity, so we will treat this problem using Bernoulli's equation, Equation 13.13.

**PREPARE** We can ignore any change in vertical position. We are looking for the difference in pressure before $(p_1)$ and after $(p_2)$ the valve, so Equation 13.13 reduces to

$$\Delta p = p_1 - p_2 = \frac{1}{2}\rho v_2^2 - \frac{1}{2}\rho v_1^2$$

**SOLVE** The drop in pressure is

$$\begin{aligned}\Delta p = p_1 - p_2 &= \frac{1}{2}\rho(v_2^2 - v_1^2)\\ &= \frac{1}{2}(1050\ \text{kg/m}^3)\Big((3.5\ \text{m/s})^2 - (1.1\ \text{m/s})^2\Big)\\ &= 5.8\ \text{kPa} = 44\ \text{mm Hg}\end{aligned}$$

This is above the threshold, so a positive diagnosis is warranted.

**ASSESS** The clinical data show a significant increase in speed and a corresponding drop in pressure as the blood exits the valve, so it's no surprise that this person has a serious condition. In a normal heart, there is some increase in speed as the blood exits the aortic valve. This causes a reduction in pressure, but normally the pressure increases again as the flow slows down on entering the aorta. If the increase in speed is too great, as it is for this patient, the flow may cross the boundary between laminar and turbulent flow, with negative consequences.

## The Veins

The pressure difference along the arteries keeps the blood moving. But this doesn't work for the veins. Figure 13.37 shows that, when the blood reaches the veins, the pressure is quite low. How then does the blood make its way back to the heart? It can't be a pressure difference along the veins that keeps the blood moving; the pressures are simply too low. The veins must have other structures that keep the blood moving.

To further prove this point, let's consider the veins in your lower legs and feet. When you are lying down, the pressure in these veins is nearly zero. When you are upright, this pressure rises. If you are walking or exercising, the venous pressure rises to about 30 mm Hg. What would the pressure be if the veins were simple tubes?

**EXAMPLE 13.17 What is the pressure in the veins?** BIO

A woman's feet are 1.3 m below her heart. If we assume her veins are simple tubes, what is the blood pressure in the veins of her feet?

**STRATEGIZE** When the blood returns to the right atrium, the pressure is approximately zero. If the veins from the feet to the heart are simple tubes, then the blood in the veins is a connected fluid, so the pressure will increase with distance below the heart.

**PREPARE** We can use Equation 13.5 to find the pressure at the feet. We are looking at the pressure at $d = 1.3$ m, the distance below the heart.

**SOLVE** Equation 13.5 gives the venous pressure in the feet as

$$\begin{aligned}p &= p_0 + \rho g d = 0 + (1050\ \text{kg/m}^3)(9.8\ \text{m/s}^2)(1.3\ \text{m})\\ &= 13.4\ \text{kPa} = 100\ \text{mm Hg}\end{aligned}$$

**ASSESS** The necessary pressure is far above what is observed, so the veins can't be simple tubes; there must be more to the story.

The veins of the legs, and other veins in the body, have one-way valves. The contraction of muscles surrounding the veins squeezes them, thereby moving the blood. The valves ensure that this motion is toward the heart, and they prevent the blood from moving back down the veins when the muscles relax. If you sit or stand for a long time, this system can't do its job, and blood can collect in your legs. (Your veins are quite expandable; they change size to accommodate changes in blood volume and blood flow.) At some point, you'll want to get up and move around, using your leg muscles, which pushes the blood back toward the heart.

◂ **Walking improves the circulation** BIO A horse has no muscles in its lower legs or hoofs that aid in blood return to the heart. Instead, the normal force of the ground on each hoof compresses a network of veins in the hoof, pushing blood upward, where valves in the veins prevent return flow. When a horse walks, the changing forces on each hoof make this system work as a pump, efficiently returning blood to the heart.

## INTEGRATED EXAMPLE 13.18 An intravenous transfusion BIO

At the hospital, a patient often receives fluids via an intravenous (IV) infusion. A bag of the fluid is held at a fixed height above the patient's body. The fluid then travels down a large-diameter, flexible tube to a catheter—a short tube with a small diameter—inserted into the patient's vein.

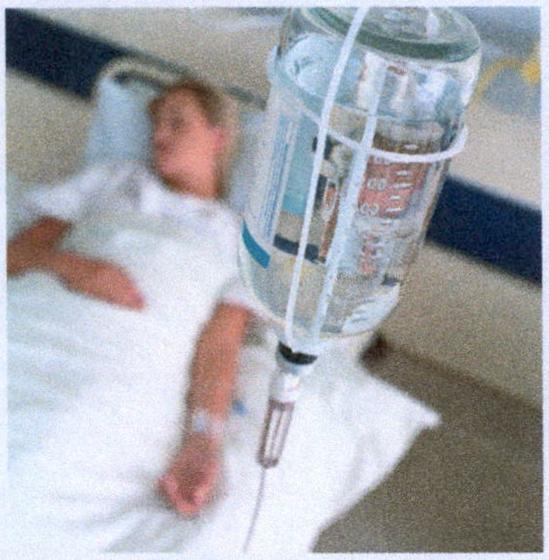

1.0 L of saline solution, with a density of 1020 kg/m$^3$ and a viscosity of $1.1 \times 10^{-3}$ Pa·s, is to be infused into a patient in 8.0 h. The catheter is 30 mm long and has an inner diameter of 0.30 mm. The pressure in the patient's vein is 10 mm Hg. How high above the patient should the bag be positioned to get the desired flow rate?

**STRATEGIZE** We're concerned with the flow of a viscous fluid. According to Poiseuille's equation, the flow rate depends inversely on the fourth power of a tube's radius. The tube from the elevated bag to the catheter has a large diameter, while the diameter of the catheter is small. Thus, we expect the flow rate to be determined entirely by the flow through the narrow catheter; the wide tube has a negligible effect on the rate. We will determine the pressure difference across the catheter that is necessary to provide the needed flow. Then we will use this to determine the necessary pressure in the bag, and thus the height at which it must be placed.

**PREPARE** FIGURE 13.38 shows a sketch of the situation, defines variables, and lists the known information.

To use Poiseuille's equation for the catheter, we need the pressure difference $\Delta p$ between the ends of the catheter. We know the pressure on the end of the catheter in the patient's vein is $p_v = 10$ mm Hg or, converting to SI units using Table 13.2,

$$p_v = 10 \text{ mm Hg} \times \frac{101 \times 10^3 \text{ Pa}}{760 \text{ mm Hg}} = 1330 \text{ Pa}$$

This is a gauge pressure, the pressure in excess of 1 atm. The pressure on the fluid side of the catheter is due to the hydrostatic pressure of the saline solution filling the bag and the flexible tube leading to the catheter. This pressure is given by the hydrostatic pressure equation $p = p_0 + \rho g d$, where $d$ is the "depth" of the catheter below the bag. Thus we'll use $\Delta p$ to find $d$.

**SOLVE** The desired volume flow rate is

$$Q = \frac{\Delta V}{\Delta t} = \frac{1.0 \text{ L}}{8.0 \text{ h}} = 0.125 \text{ L/h}$$

FIGURE 13.38 Visual overview of an IV transfusion.

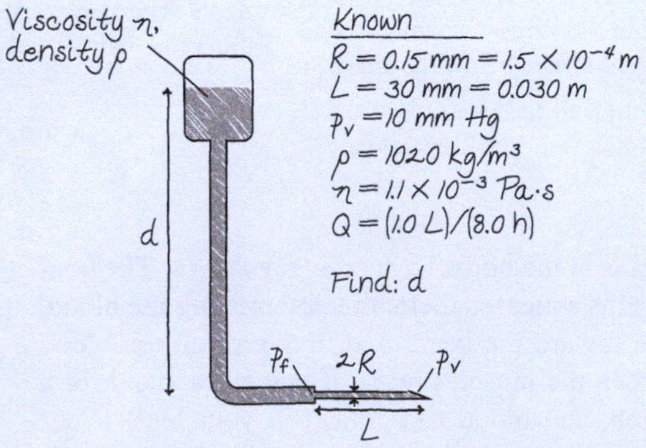

Converting to SI units using $1.0 \text{ L} = 1.0 \times 10^{-3} \text{ m}^3$, we have

$$Q = 0.125 \frac{\text{L}}{\text{h}} \times \frac{1.0 \times 10^{-3} \text{ m}^3}{\text{L}} \times \frac{1 \text{ h}}{3600 \text{ s}} = 3.47 \times 10^{-8} \text{ m}^3/\text{s}$$

Poiseuille's equation for viscous fluid flow is

$$Q = \frac{\pi R^4 \Delta p}{8\eta L}$$

Thus the pressure difference needed between the ends of the tube to produce the desired flow rate $Q$ is

$$\Delta p = \frac{8\eta L Q}{\pi R^4}$$

$$= \frac{8(1.1 \times 10^{-3} \text{ Pa} \cdot \text{s})(0.030 \text{ m})(3.47 \times 10^{-8} \text{ m}^3/\text{s})}{\pi(1.5 \times 10^{-4} \text{ m})^4}$$

$$= 5760 \text{ Pa}$$

Now $\Delta p$ is the difference between the fluid pressure $p_f$ at one end of the catheter and the vein pressure $p_v$ at the other end: $\Delta p = p_f - p_v$. We know $p_v$, so

$$p_f = p_v + \Delta p = 1330 \text{ Pa} + 5760 \text{ Pa} = 7090 \text{ Pa}$$

This pressure, like the vein pressure, is a gauge pressure. The true hydrostatic pressure at the catheter is $p = 1 \text{ atm} + 7090 \text{ Pa}$. But the hydrostatic pressure in the fluid is

$$p = p_0 + \rho g d = 1 \text{ atm} + \rho g d$$

We see that $\rho g d = 7090$ Pa. Solving for $d$, we find the required elevation of the bag above the patient's arm:

$$d = \frac{p_f}{\rho g} = \frac{7090 \text{ Pa}}{(1020 \text{ kg/m}^3)(9.8 \text{ m/s}^2)} = 0.71 \text{ m}$$

**ASSESS** This height of almost a meter seems reasonable for the height of an IV bag. In practice, the bag can be raised or lowered to adjust the fluid flow rate.

# SUMMARY

**GOAL** To understand the static and dynamic properties of fluids.

## GENERAL PRINCIPLES

### Fluid Statics

#### Gases

- Freely moving particles
- Compressible
- Pressure mainly due to particle collisions with walls

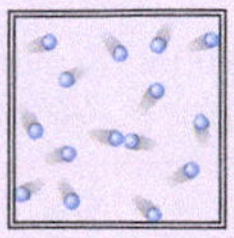

#### Liquids

- Loosely bound particles
- Incompressible
- Pressure due to the weight of the liquid

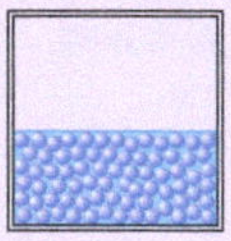

- Hydrostatic pressure at depth $d$ is $p = p_0 + \rho g d$
- The pressure is the same at all points on a horizontal line through a liquid (of one kind) in hydrostatic equilibrium

### Fluid Dynamics

We assume that fluids are incompressible, flow is laminar

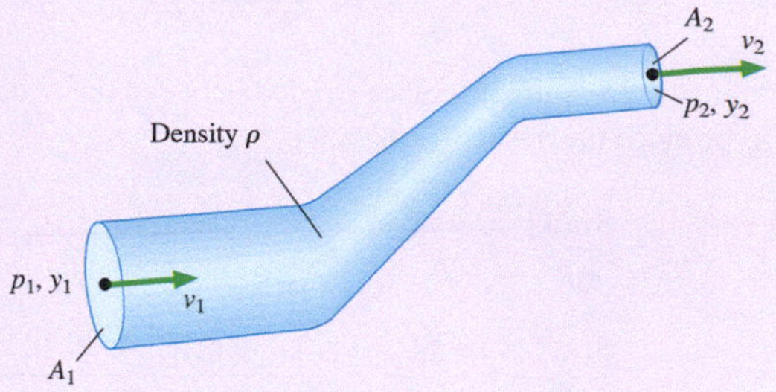


**Equation of continuity**

**Volume flow rate** $Q = \dfrac{\Delta V}{\Delta t} = v_1 A_1 = v_2 A_2$

**Bernoulli's equation** is a statement of energy conservation:

$$p_1 + \frac{1}{2}\rho v_1{}^2 + \rho g y_1 = p_2 + \frac{1}{2}\rho v_2{}^2 + \rho g y_2$$

**Poiseuille's equation** governs viscous flow through a tube:

$$Q = v_{avg}A = \frac{\pi R^4 \Delta p}{8\eta L}$$

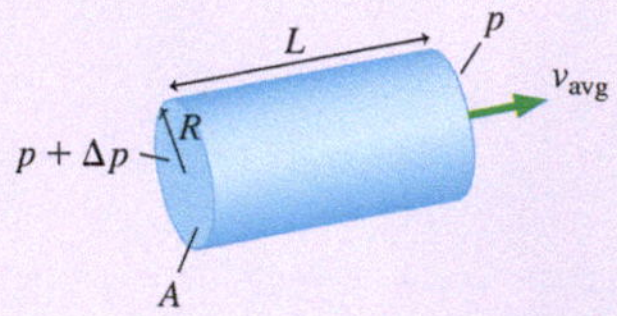


## IMPORTANT CONCEPTS

**Density** $\rho = m/V$, where $m$ is mass and $V$ is volume.

**Pressure** $p = F/A$, where $F$ is force magnitude and $A$ is the area on which the force acts.

- Pressure exists at all points in a fluid.
- Pressure pushes equally in all directions.
- Gauge pressure $p_g = p - 1$ atm.

**Viscosity** $\eta$ is the property of a fluid that makes it resist flowing.

**Representing fluid flow**
**Streamlines** are the paths of individual fluid particles.

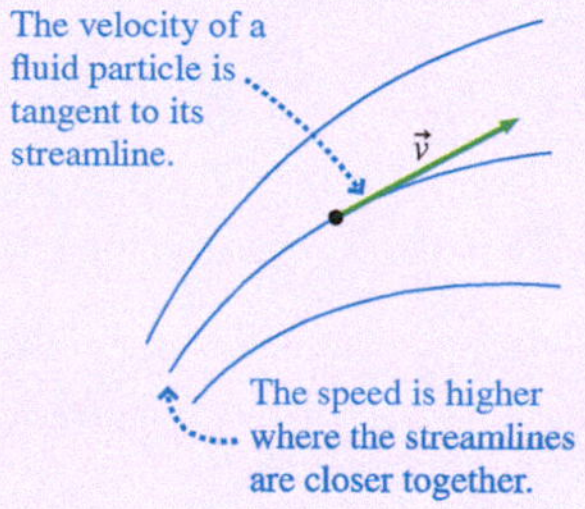


**Fluid elements** contain a fixed volume of fluid. Their shape may change as they move.

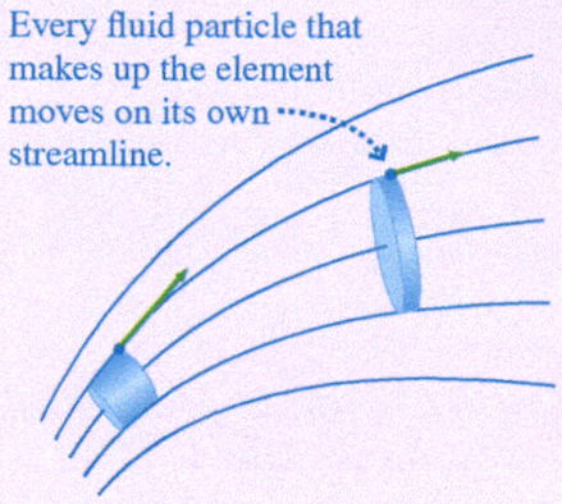

Every fluid particle that makes up the element moves on its own streamline.

## APPLICATIONS

**Buoyancy** is the upward force of a fluid on an object immersed in the fluid.

**Archimedes' principle:** The magnitude of the buoyant force equals the weight of the fluid displaced by the object.

| | | |
|---|---|---|
| Sink: | $\rho_{avg} > \rho_f$ | $F_B < w_o$ |
| Float: | $\rho_{avg} < \rho_f$ | $F_B > w_o$ |
| Neutrally buoyant: | $\rho_{avg} = \rho_f$ | $F_B = w_o$ |

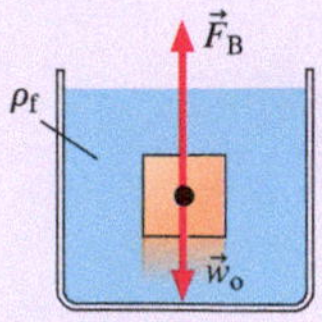


**Barometers** measure atmospheric pressure. Atmospheric pressure is related to the height of the liquid column by $p_{atmos} = \rho g h$.

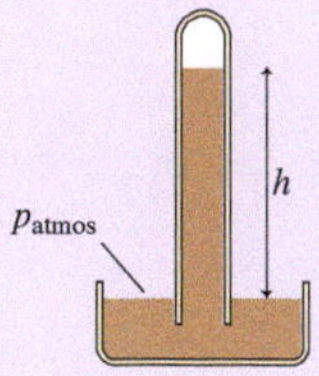


Our understanding of fluids can be applied to the motion of blood in the **circulatory system.**

- The flow is laminar under normal circumstances.
- Blood in the large arteries forms a connected fluid, so height differences matter, but viscosity can be ignored.
- In the small arteries and arterioles, viscosity is a major factor.

## Learning Objectives After studying this chapter, you should be able to:

- Work with the density of gases and liquids. *Conceptual Questions 13.2, 13.3; Problems 13.2, 13.3, 13.4, 13.5, 13.6*
- Calculate and use the pressure in a liquid. *Conceptual Questions 13.8, 13.9; Problems 13.8, 13.9, 13.10, 13.17, 13.18*
- Use Archimedes' principle to understand buoyancy. *Conceptual Questions 13.19, 13.20; Problems 13.23, 13.24, 13.25, 13.26, 13.27*
- Use the equation of continuity to solve problems about fluid flow. *Conceptual Questions 13.28, 13.29; Problems 13.32, 13.33, 13.34, 13.35*
- Use Bernoulli's and Poiseuille's equations to solve problems about fluid dynamics. *Conceptual Questions 13.30, 13.31; Problems 13.35, 13.39, 13.41, 13.42, 13.45*
- Understand the motion of blood in the circulatory system. *Problems 13.47, 13.48, 13.49, 13.50, 13.51*

## STOP TO THINK ANSWERS

**Chapter Preview Stop to Think: C.** Consider books 2 and 3 to be a single object in equilibrium. Then the downward weight force of these two books must be balanced by the upward normal force of book 1 on book 2, so that the normal force is equal to the weight of *two* books.

**Stop to Think 13.1: $\rho_1 = \rho_2 = \rho_3$.** Density depends only on what the object is made of, not how big the pieces are.

**Stop to Think 13.2: C.** Because points 1 and 2 are at the same height and are connected by the same fluid (water), they must be at the same pressure. This means that these two points can support the same weight of fluid above them. The oil is less dense than water, so this means a taller column of oil can be supported on the left side than of water on the right side.

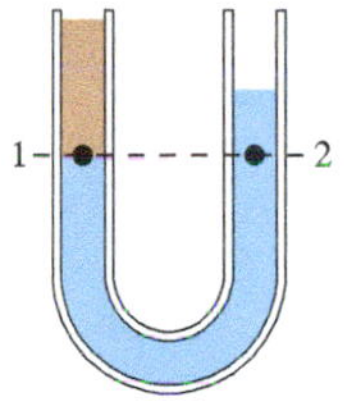


**Stop to Think 13.3: $(F_B)_D > (F_B)_A = (F_B)_B = (F_B)_C > (F_B)_E$.** The buoyant force on an object is equal to the weight of the fluid it displaces. Block D has the greatest volume and thus displaces the greatest amount of fluid. Blocks A, B, and C have equal volumes, so they displace equal amounts of fluids, but less than D displaces. And block E has the smallest volume and displaces the least amount of fluid. The buoyant force on an object does not depend on the object's mass, or its position in the fluid.

**Stop to Think 13.4: B.** The weight of the displaced water equals the weight of the ice cube. When the ice cube melts and turns into water, that amount of water will exactly fill the volume that the ice cube is now displacing.

**Stop to Think 13.5: 1 $cm^3/s$ out.** The fluid is incompressible, so the sum of what flows in must match the sum of what flows out. 13 $cm^3/s$ is known to be flowing in while 12 $cm^3/s$ flows out. An additional 1 $cm^3/s$ must flow out to achieve balance.

**Stop to Think 13.6: $h_2 > h_4 > h_3 > h_1$.** The liquid level is higher where the pressure is lower. The pressure is lower where the flow speed is higher. The flow speed is highest in the narrowest tube, zero in the open air.

**Stop to Think 13.7: A.** All three segments have the same volume flow rate $Q$. According to Poiseuille's equation, the segment with the smallest radius $R$ has the greatest pressure difference $\Delta p$.

Video Tutor Solution Chapter 13

# QUESTIONS

## Conceptual Questions

1. Which has the greater density, 1 g of mercury or 1000 g of water?
2. A $1 \times 10^{-3}$ $m^3$ chunk of material has a mass of 3 kg.
   a. What is the material's density?
   b. Would a $2 \times 10^{-3}$ $m^3$ chunk of the same material have the same mass? Explain.
   c. Would a $2 \times 10^{-3}$ $m^3$ chunk of the same material have the same density? Explain.
3. Object 1 has an irregular shape. Its density is 4000 $kg/m^3$.
   a. Object 2 has the same shape and dimensions as object 1, but it is twice as massive. What is the density of object 2?
   b. Object 3 has the same mass and the same *shape* as object 1, but its size in all three dimensions is twice that of object 1. What is the density of object 3?
4. BIO When you get a blood transfusion the bag of blood is held above your body, but when you donate blood the collection bag is held below. Why is this?
5. BIO To explore the bottom of a 10-m-deep lake, your friend Tom proposes to get a long garden hose, put one end on land and the other in his mouth for breathing underwater, and descend into the depths. Susan, who overhears the conversation, reacts with horror and warns Tom that he will not be able to inhale when he is at the lake bottom. Why is Susan so worried?
6. If you let go of a helium balloon, it quickly rises. As it rises, the balloon gets larger and larger until it pops. Why does the balloon expand as it rises?

Problem difficulty is labeled as | (straightforward) to ||||| (challenging). Problems labeled INT integrate significant material from earlier chapters; Problems labeled BIO are of biological or medical interest.

The eText icon indicates when there is a video tutor solution available for the chapter or for a specific problem. To launch these videos, log into your eText through Mastering™ Physics or log into the Study Area.

7. Scuba divers are warned that if they must make a rapid ascent, they should exhale on the way up. If a diver rapidly ascends to the surface with lungs full of air, his lungs could be damaged. Explain why this is so.
8. Rank in order, from largest to smallest, the pressures at A, B, and C in Figure Q13.8. Explain.
9. Refer to Figure Q13.8. Rank in order, from largest to smallest, the pressures at D, E, and F. Explain.

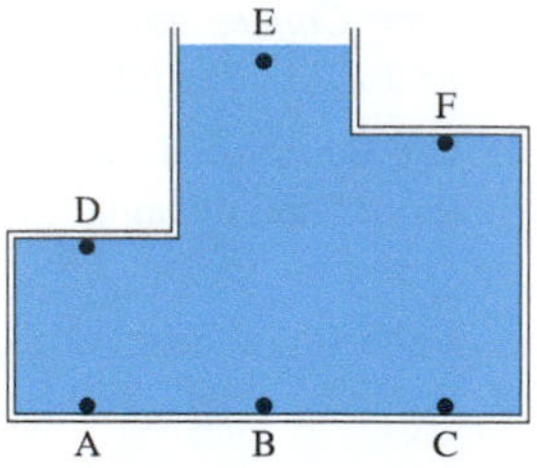

FIGURE Q13.8

10. A steel cylinder at sea level contains air at a high pressure. Attached to the tank are two gauges, one that reads absolute pressure and one that reads gauge pressure. The tank is then brought to the top of a mountain. For each of the two gauges, does the pressure reading decrease, increase, or remain the same? Does the difference between the two readings decrease, increase, or remain the same? Explain.
11. In Figure Q13.11, A and B are rectangular tanks full of water. They have equal heights and equal side lengths (the dimension into the page), but different widths.

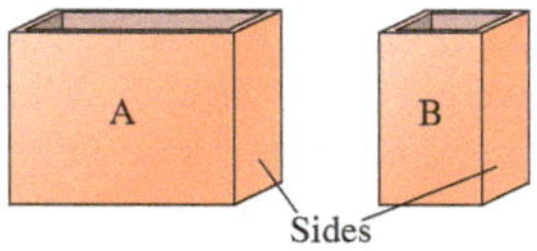

FIGURE Q13.11

    a. Compare the forces the water exerts on the bottoms of the tanks. Is $F_A$ larger, smaller, or equal to $F_B$? Explain.
    b. Compare the forces the water exerts on the sides of the tanks. Is $F_A$ larger, smaller, or equal to $F_B$? Explain.
12. Imagine a square column of the atmosphere, 1 m on a side, that extends all the way to the top of the atmosphere. How much does this column of air weigh in newtons?
13. Water expands when heated. Suppose a beaker of water is heated from 10°C to 90°C. Does the pressure at the bottom of the beaker increase, decrease, or stay the same? Explain.
14. In Figure Q13.14, is $p_A$ larger, smaller, or equal to $p_B$? Explain.

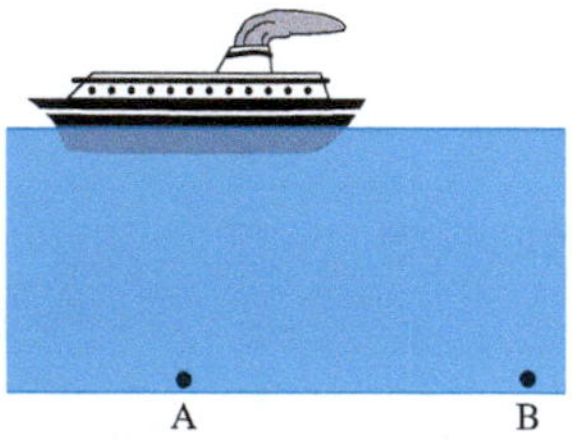

FIGURE Q13.14

15. A beaker of water rests on a scale. A metal ball is then lowered into the beaker using a string tied to the ball. The ball doesn't touch the sides or bottom of the beaker, and no water spills from the beaker. Does the scale reading decrease, increase, or stay the same? Explain.
16. Rank in order, from largest to smallest, the densities of objects A, B, and C in Figure Q13.16. Explain.

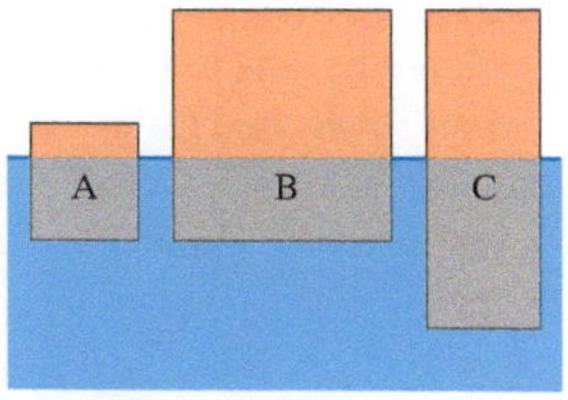

FIGURE Q13.16

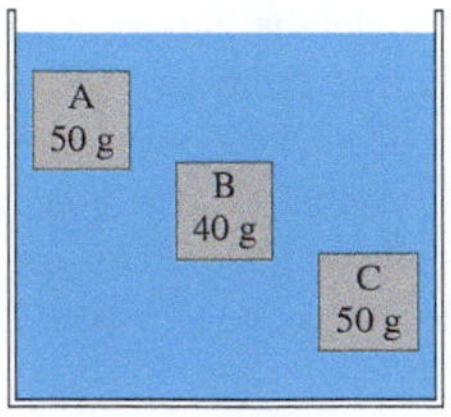

FIGURE Q13.17

17. Objects A, B, and C in Figure Q13.17 have the same volume. Rank in order, from largest to smallest, the sizes of the buoyant forces $F_A$, $F_B$, and $F_C$ on A, B, and C. Explain.
18. Refer to Figure Q13.17. Now A, B, and C have the same density, but still have the masses given in the figure. Rank in order, from largest to smallest, the sizes of the buoyant forces on A, B, and C. Explain.
19. A heavy lead block and a light aluminum block of equal size sit at rest at the bottom of a pool of water. Is the buoyant force on the lead block greater than, less than, or equal to the buoyant force on the aluminum block? Explain.
20. When you place an egg in water, it sinks. If you start adding salt to the water, after some time the egg floats. Explain.
21. The water of the Dead Sea is extremely salty, which gives it a very high density of 1240 kg/m$^3$. Explain why a person floats much higher in the Dead Sea than in ordinary water.

22. Fish can adjust their buoyancy with an organ called the *swim bladder*. The swim bladder is a flexible gas-filled sac; the fish can increase or decrease the amount of gas in the swim bladder so that it stays neutrally buoyant—neither sinking nor floating. Suppose the fish is neutrally buoyant at some depth and then goes deeper. What needs to happen to the volume of air in the swim bladder? Will the fish need to add or remove gas from the swim bladder to maintain its neutral buoyancy?
BIO
23. Freshwater fish tend to have larger swim bladders than saltwater fish. Explain why you would expect this to be true.
BIO
24. Elephant seals do deep dives to forage for food. The energy used by the seals in diving depends on their body composition; for example, fatter seals use less energy to swim to the surface at the end of a dive, allowing them to take longer, more productive dives. Explain why you'd expect this to be the case.
BIO

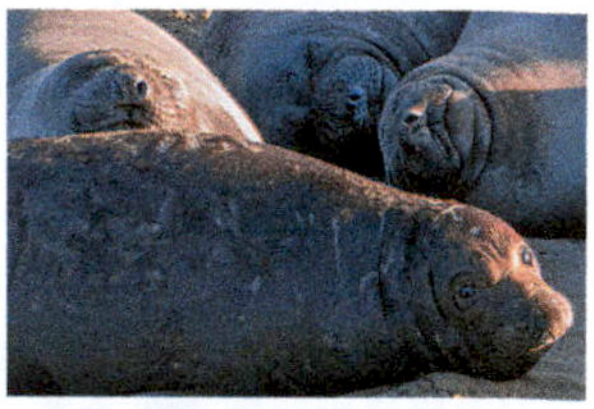

25. A higher level of hemoglobin in the blood increases the blood's density. This is the basis for a simple test that can be used to see if a prospective blood donor has a high enough hemoglobin level to donate safely. A drop of blood is placed in a copper sulfate solution, and the time for the drop to sink to the bottom is measured. If this time is too long, the hemoglobin level is too low. Explain how this test works.
BIO
26. Early airships—craft that relied on large volumes of low-density gas to rise from the ground—were filled with hydrogen, but safety concerns necessitated a switch to helium. When an airship made the switch, how did this affect its cargo capacity? Explain.
27. The air in a hot-air balloon can be heated only so much; the temperature of the gas inside the balloon can only be so high before damaging the nylon envelope. This means that hot-air balloons can lift a larger load on a cool morning than on a warm afternoon. Explain.

28. A liquid with negligible viscosity flows through the pipe shown in Figure Q13.28. This is an overhead view.
    a. Rank in order, from largest to smallest, the flow speeds $v_1$ to $v_4$ at points 1 to 4. Explain.
    b. Rank in order, from largest to smallest, the pressures $p_1$ to $p_4$ at points 1 to 4. Explain.

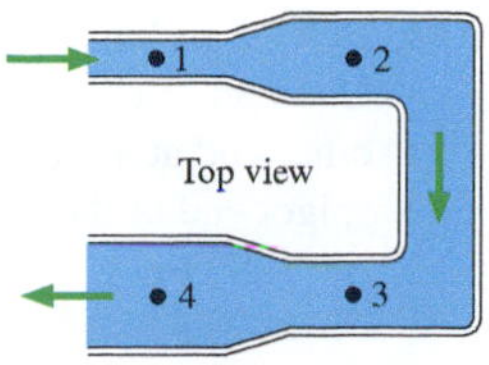

FIGURE Q13.28

29. You are looking for a deep spot to take a swim in a local river. The width of the river is about the same all along its length, but there are places where the water flows quickly and places where it flows slowly. Which is a better choice for a swim?
30. Is it possible for a fluid in a tube to flow in the direction from low pressure to high pressure? If so, give an example. If not, why not?
31. Wind blows over the house shown in Figure Q13.31. A window on the ground floor is open. Is there an airflow through the house? If so, does the air flow in the window and out the chimney, or in the chimney and out the window? Explain.

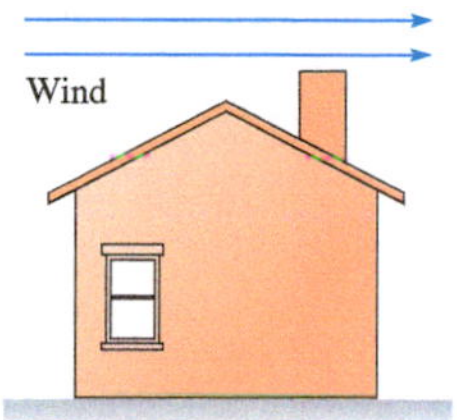

FIGURE Q13.31

## Multiple-Choice Questions

32. | In Figure Q13.32, water is slowly poured into the container until the water level has risen into tubes 1, 2, and 3. The water doesn't overflow from any of the tubes. How do the water depths in the three columns compare?
    A. $d_1 > d_2 > d_3$ B. $d_1 < d_2 < d_3$
    C. $d_1 = d_2 = d_3$ D. $d_1 = d_2 > d_3$
    E. $d_1 < d_2 = d_3$

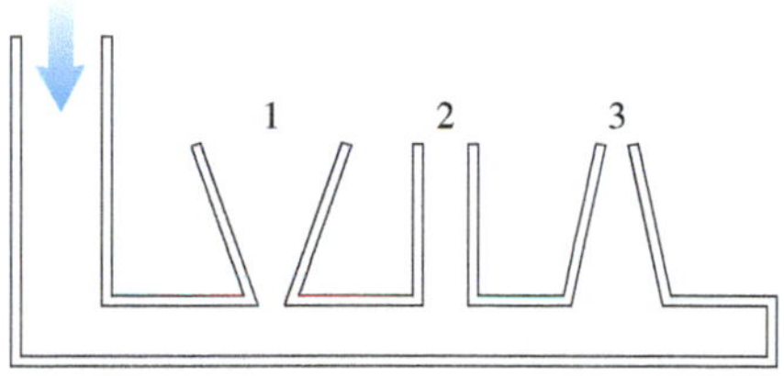

FIGURE Q13.32

33. | Figure Q13.33 shows a 100 g block of copper ($\rho = 8900\ kg/m^3$) and a 100 g block of aluminum ($\rho = 2700\ kg/m^3$) connected by a massless string that runs over two massless, frictionless pulleys. The two blocks exactly balance, since they have the same mass. Now suppose that the whole system is submerged in water. What will happen?
    A. The copper block will fall, the aluminum block will rise.
    B. The aluminum block will fall, the copper block will rise.
    C. Nothing will change.
    D. Both blocks will rise.

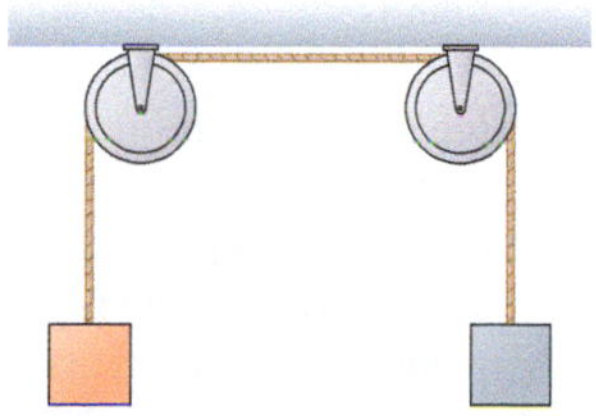

FIGURE Q13.33

34. | Masses A and B rest on very light pistons that enclose a fluid, as shown in Figure Q13.34. There is no friction between the pistons and the cylinders they fit inside. Which of the following is true?
    A. Mass A is greater. B. Mass B is greater.
    C. Mass A and mass B are the same.

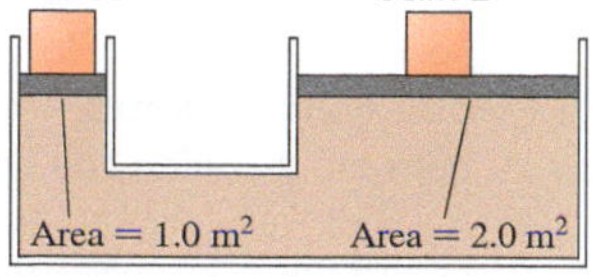

FIGURE Q13.34

35. | BIO If you dive underwater, you notice an uncomfortable pressure on your eardrums due to the increased pressure. The human eardrum has an area of about 70 $mm^2$($7 \times 10^{-5}\ m^2$), and it can sustain a force of about 7 N without rupturing. If your body had no means of balancing the extra pressure (which, in reality, it does), what would be the maximum depth you could dive without rupturing your eardrum?
    A. 0.3 m B. 1 m C. 3 m D. 10 m
36. || An 8.0 lb bowling ball has a diameter of 8.5 inches. When lowered into water, this ball will
    A. Float. B. Sink. C. Have neutral buoyancy.
37. | A large beaker of water is filled to its rim with water. A block of wood is then carefully lowered into the beaker until the block is floating. In this process, some water is pushed over the edge and collects in a tray. The weight of the water in the tray is
    A. Greater than the weight of the block.
    B. Less than the weight of the block.
    C. Equal to the weight of the block.
38. ||| A 55 g soapstone cube—a whisky stone—is used to chill a glass of whisky. Soapstone has a density of 3000 $kg/m^3$, whisky a density of 940 $kg/m^3$. What is the approximate normal force of the bottom of the glass on a single stone?
    A. 0.1 N B. 0.2 N C. 0.3 N D. 0.4 N
39. || BIO At the surface of a freshwater spring, a manatee with a small amount of air in its lungs is neutrally buoyant. The manatee now dives to a depth of 10 m. At this greater depth, the buoyant force is
    A. Less than the weight force.
    B. The same as the weight force.
    C. Greater than the weight force.
40. || BIO The density of a typical hippo's body is 1030 $kg/m^3$, so it will sink in fresh water. What is the buoyant force on a submerged 1500 kg hippo?
    A. 14,000 N B. 14,300 N
    C. 14,600 N D. 14,900 N
41. | An object floats in water, with 75% of its volume submerged. What is its approximate density?
    A. 250 $kg/m^3$ B. 750 $kg/m^3$
    C. 1000 $kg/m^3$ D. 1250 $kg/m^3$
42. | A syringe is being used to squirt water as shown in Figure Q13.42. The water is ejected from the nozzle at 10 m/s. At what speed is the plunger of the syringe being depressed?
    A. 0.01 m/s
    B. 0.1 m/s
    C. 1 m/s
    D. 10 m/s

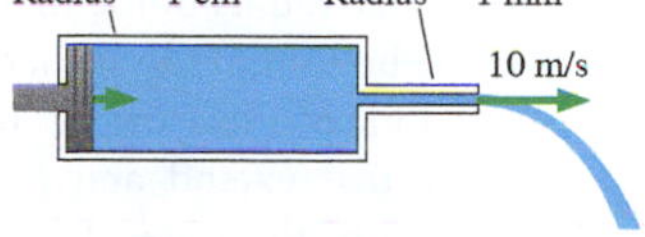

FIGURE Q13.42

43. || Water flows through a 4.0-cm-diameter horizontal pipe at a speed of 1.3 m/s. The pipe then narrows down to a diameter of 2.0 cm. Ignoring viscosity, what is the pressure difference between the wide and narrow sections of the pipe?
A. 850 Pa B. 3400 Pa C. 9300 Pa
D. 12,700 Pa E. 13,500 Pa

44. || A 15-m-long garden hose has an inner diameter of 2.5 cm. One end is connected to a spigot; 20°C water flows from the other end at a rate of 1.2 L/s. What is the gauge pressure at the spigot end of the hose?
A. 1900 Pa B. 2700 Pa C. 4200 Pa
D. 5800 Pa E. 7300 Pa

# PROBLEMS

## Section 13.1 Fluids and Density

1. || A 100 mL beaker holds 120 g of liquid. What is the liquid's density in SI units?
2. | A 68 kg college student's body has a typical density of 1050 $kg/m^2$. What is his volume? BIO
3. | A standard gold bar stored at Fort Knox, Kentucky, is 7.00 inches long, 3.63 inches wide, and 1.75 inches tall. Gold has a density of 19,300 $kg/m^3$. What is the mass of such a gold bar?
4. || A 5.0 kg freshwater fish at the surface of a lake is neutrally buoyant. If the density of its body with its swim bladder deflated is 1060 $kg/m^3$, what volume of gas must be in the swim bladder for the fish to be neutrally buoyant? BIO
5. || Air enclosed in a sphere has density $\rho = 1.4\ kg/m^3$. What will the density be if the radius of the sphere is halved, compressing the air within?
6. || Fat cells in humans are composed almost entirely of pure triglycerides with an average density of about 900 $kg/m^3$. If 20% of the mass of a 70 kg student's body is fat (a typical value), what is the total volume of the fat in his body? BIO

## Section 13.2 Pressure

7. || The deepest point in the ocean is 11 km below sea level, deeper than Mt. Everest is tall. What is the pressure in atmospheres at this depth?
8. || James Cameron piloted a submersible craft to the bottom of the Challenger Deep, the deepest point on the ocean's floor, 11,000 m below the surface. What was the total inward force on the 1.1-m-diameter pilot sphere in which Cameron sat?
9. || A tall cylinder contains 25 cm of water. Oil is carefully poured into the cylinder, where it floats on top of the water, until the total liquid depth is 40 cm. What is the gauge pressure at the bottom of the cylinder?
10. ||| A 35-cm-tall, 5.0-cm-diameter cylindrical beaker is filled to its brim with water. What is the downward force of the water on the bottom of the beaker?
11. || The gauge pressure at the bottom of a cylinder of liquid is $p_g = 0.40$ atm. The liquid is poured into another cylinder with twice the radius of the first cylinder. What is the gauge pressure at the bottom of the second cylinder?
12. || Snorkelers breathe through tubes that extend above the surface of the water. In principle, a snorkeler could go deeper with a longer tube, but the extra pressure at the greater depth would make breathing too difficult. If a snorkeler can develop a pressure in her lungs that is 10 kPa below the pressure outside her body, what is the longest snorkel she could use in the ocean?

13. |||| A research submarine has a 20-cm-diameter window 8.0 cm thick. The manufacturer says the window can withstand forces up to $1.0 \times 10^6$ N. What is the submarine's maximum safe depth in seawater? The pressure inside the submarine is maintained at 1.0 atm.
14. |||| The highest that George can suck water up a very long straw is 2.0 m. (This is a typical value.) What is the lowest pressure that he can maintain in his mouth? BIO
15. ||| What is the gas pressure inside the box shown in Figure P13.15?

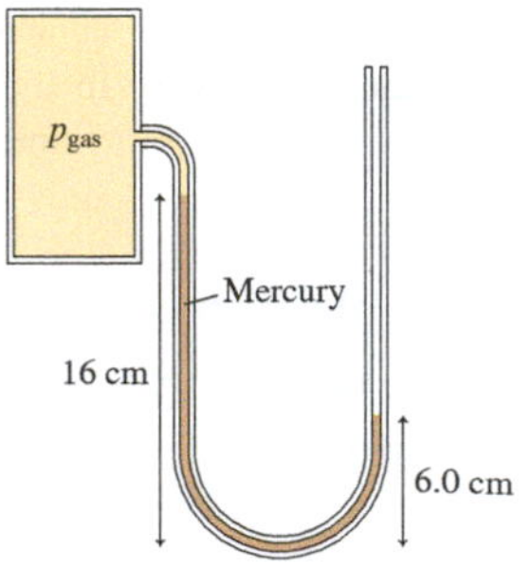

FIGURE P13.15

16. || A city uses a water tower to store water for times of high demand. When demand is light, water is pumped into the tower. When demand is heavy, water can flow from the tower without overwhelming the pumps. To provide water at a typical 350 kPa gauge pressure, how tall must the tower be?
17. || The container shown in Figure P13.17 is filled with oil. It is open to the atmosphere on the left.
    a. What is the pressure at point A?
    b. What is the pressure difference between points A and B? Between points A and C?

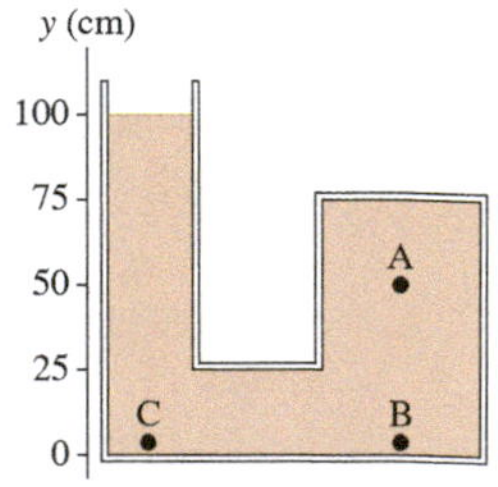

FIGURE P13.17

18. ||| Glycerin is poured into an open U-shaped tube until the height in both sides is 20 cm. Ethyl alcohol is then poured into one arm until the height of the alcohol column is 20 cm. The two liquids do not mix. What is the difference in height between the top surface of the glycerin and the top surface of the alcohol?
19. ||| A U-shaped tube, open to the air on both ends, contains mercury. Water is poured into the left arm until the water column is 10.0 cm deep. How far upward from its initial position does the mercury in the right arm rise?
20. | What is the height of a water barometer at atmospheric pressure?

## Section 13.3 Buoyancy

21. ||| A cargo barge is loaded in a saltwater harbor for a trip up a freshwater river. If the rectangular barge is 3.0 m by 20.0 m and sits 0.80 m deep in the harbor, how deep will it sit in the river?

22. || Hippos spend much of their lives in water, but amazingly, they don't swim. They have, like manatees, very little body fat. The density of a hippo's body is approximately 1030 $kg/m^3$, so it sinks to the bottom of the freshwater lakes and rivers it frequents—and then it simply walks on the bottom. A 1500 kg hippo is completely submerged, standing on the bottom of a lake. What is the approximate value of the upward normal force on the hippo?

23. || A 10 cm × 10 cm × 10 cm wood block with a density of 700 $kg/m^3$ floats in water.
    a. What is the distance from the top of the block to the water if the water is fresh?
    b. If it's seawater?
24. || The classic Goodyear blimp is essentially a helium balloon—a big one, containing 5700 $m^3$ of helium. If the envelope and gondola have a total mass of 4300 kg, what is the maximum cargo load when the blimp flies at a sea-level location? Assume an air temperature of 20°C.
25. || What is the tension in the string in Figure P13.25?

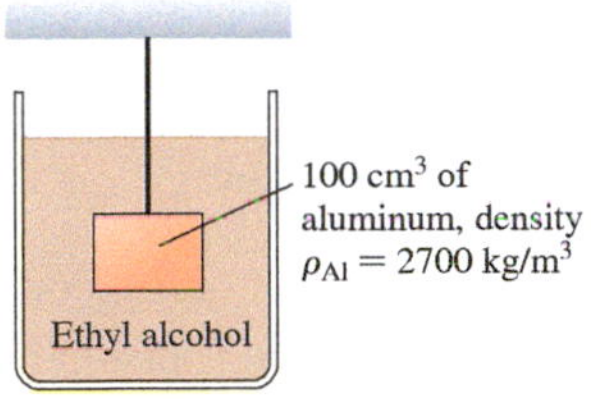

FIGURE P13.25

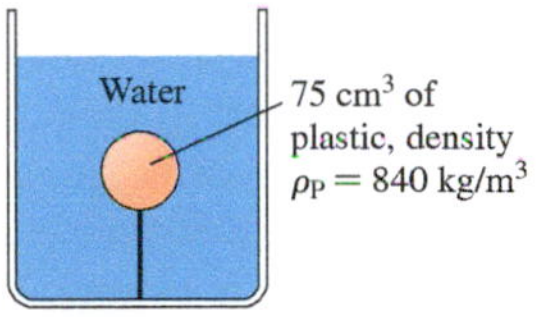

FIGURE P13.26

26. || What is the tension in the string in Figure P13.26?
27. ||| To determine an athlete's body fat, she is weighed first in air
BIO and then again while she's completely underwater, as discussed on page 466. It is found that she weighs 690 N when weighed in air and 42 N when weighed underwater. What is her average density?
28. || Sharks are generally negatively buoyant; the upward buoyant
BIO force is less than the weight force. This is one reason sharks tend to swim continuously; water moving past their fins causes a lift force that keeps sharks from sinking. A 92 kg bull shark has a density of 1040 $kg/m^3$.
    a. What lift force must the shark's fins provide if the shark is swimming in seawater?
    b. Bull sharks often swim into freshwater rivers. What lift force is required in a river?
29. ||| Styrofoam has a density of 32 $kg/m^3$. What is the maximum mass that can hang without sinking from a 50-cm-diameter Styrofoam sphere in water? Assume the volume of the mass is negligible compared to that of the sphere.
30. ||| A 2.5 g latex balloon is filled with 2.4 g of helium. When
INT filled, the balloon is a 30-cm-diameter sphere. When released, the balloon accelerates upward until it reaches a terminal speed. What is this speed? Assume an air density of 1.2 $kg/m^3$.
31. |||| Calculate the buoyant force due to the surrounding air on a man weighing 800 N. Assume his average density is the same as that of water.

## Section 13.4 Fluids in Motion

32. || Your low-flow showerhead is delivering water at $1.1 \times 10^{-4}$ $m^3/s$, about 1.8 gallons per minute. If this is the only water being used in your house, how fast is the water moving through your house's water supply line, which has a diameter of 0.021 m (about ¾ of an inch)?
33. || Water flowing through a 2.0-cm-diameter pipe can fill a 300 L bathtub in 5.0 min. What is the speed of the water in the pipe?
34. |||| A pump is used to empty a 6000 L wading pool. The water exits the 2.5-cm-diameter hose at a speed of 2.1 m/s. How long will it take to empty the pool?
35. || A 1.0-cm-diameter pipe widens to 2.0 cm, then narrows to 0.50 cm. Liquid flows through the first segment at a speed of 4.0 m/s.
    a. What are the speeds in the second and third segments?
    b. What is the volume flow rate through the pipe?

## Section 13.5 Fluid Dynamics

36. || During typical urination, a man releases about 400 mL of
BIO urine in about 30 seconds through the urethra, which we can model as a tube 4 mm in diameter and 20 cm long. Assume that urine has the same density as water, and that viscosity can be ignored for this flow.
    a. What is the flow speed in the urethra?
    b. If we assume that the fluid is released at the same height as the bladder and that the fluid is at rest in the bladder (a reasonable approximation), what bladder pressure would be necessary to produce this flow? (In fact, there are additional factors that require additional pressure; the actual pressure is higher than this.)
37. || What does the top pressure gauge in Figure P13.37 read?

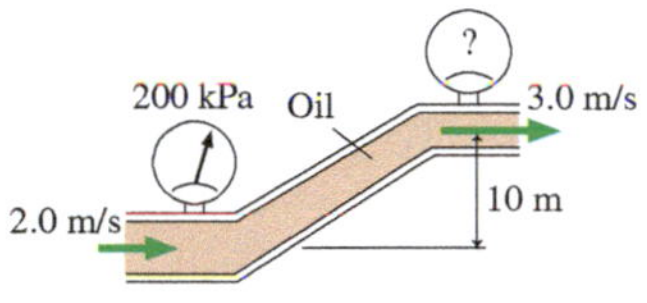

FIGURE P13.37

38. ||| When you urinate, you
BIO increase pressure in your bladder to produce the flow. For an elephant, gravity does the work. An elephant urinates at a remarkable rate of 0.0060 $m^3$ (a bit over a gallon and a half) per second. Assume that the urine exits 1.0 m below the bladder and passes through the urethra, which we can model as a tube of diameter 8.0 cm and length 1.2 m. Assume that urine has the same density as water, and that viscosity can be ignored for this flow.
    a. What is the speed of the flow?
    b. If we assume that the liquid is at rest in the bladder (a reasonable assumption) and that the pressure where the urine exits is equal to atmospheric pressure, what does Bernoulli's equation give for the pressure in the bladder? (In fact, the pressure is higher than this; other factors are at work. But you can see that no increase in bladder pressure is needed!)
39. ||| A rectangular trough, 2.0 m long, 0.60 m wide, and 0.45 m deep, is completely full of water. One end of the trough has a small drain plug right at the bottom edge. When you pull the plug, at what speed does water emerge from the hole?
40. |||| The 3.0-cm-diameter water line in Figure P13.40 splits into two 1.0-cm-diameter pipes. All pipes are circular and at the same elevation. At point A, the water speed is 2.0 m/s and the gauge pressure is 50 kPa. What is the gauge pressure at point B?

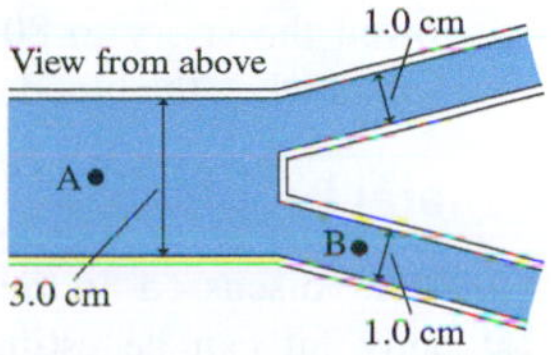

FIGURE P13.40

Watch Video Solution Problems 13.56 and 13.63

## Section 13.6 Viscosity and Poiseuille's Equation

41. |||| What pressure difference is required between the ends of a 2.0-m-long, 1.0-mm-diameter horizontal tube for 40°C water to flow through it at an average speed of 4.0 m/s?

42. || A milkshake has a viscosity of 0.50 Pa · s. To drink this shake through a straw of diameter 0.56 cm and length 22 cm, you need to reduce the pressure at the top of the straw to less than atmospheric pressure. If you want to drain a 480 mL shake in 2.0 minutes, what pressure difference is needed? You can ignore the height difference between the top and the bottom of the straw.

43. |||| Water flows at 0.25 L/s through a 10-m-long garden hose 2.5 cm in diameter that is lying flat on the ground. The temperature of the water is 20°C. What is the gauge pressure of the water where it enters the hose?

44. || How easy is it to breathe through a straw? When you breathe BIO deeply, you pull in 4.0 L of air in about 3.0 s. This requires a pressure difference of about 4.0 kPa between the air in your lungs and the outside air. What additional pressure difference is required to pull 20°C air through a straw that is 22 cm long and 0.56 cm in diameter?

45. || Figure P13.45 shows a water-filled syringe with a 4.0-cm-long needle. What is the gauge pressure of the water at the point P, where the needle meets the wider chamber of the syringe?

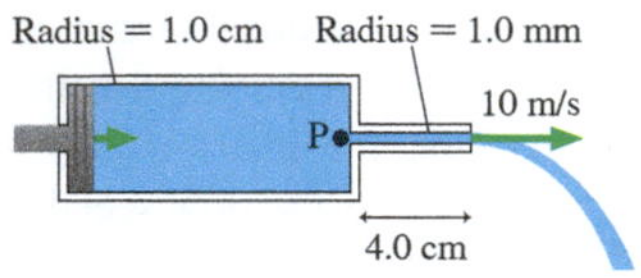

FIGURE P13.45

## Section 13.7 The Circulatory System

46. || When you hold your hands at your sides, you may have noticed BIO that the veins sometimes bulge—the height difference between your heart and your hands produces increased pressure in the veins. The same thing happens in the arteries. Estimate the distance that your hands are below your heart. If the average arterial pressure at your heart is a typical 100 mm Hg, what is the average arterial pressure in your hands when they are held at your side?

47. || The top of your head is about 30 cm above your heart. What BIO is the blood pressure difference between your heart and the top of your head?

48. || To keep blood from pooling in their lower legs on plane trips, BIO some people wear compression socks. These socks are sold by the pressure they apply; a typical rating is 20 mm Hg. Over what vertical distance can this pressure move the blood?

49. || Using the data in Figure 13.37 for the venules, and assuming BIO a typical blood flow rate of 5.0 L per minute,
    a. What is the flow speed in the venules?
    b. What is the pressure difference across a 1.0 cm length of venule?

50. || Using the data in Figure 13.37 for the arterioles, and assuming BIO a typical blood flow rate of 5.0 L per minute,
    a. What is the flow speed in the arterioles?
    b. What is the pressure difference across a 1.0 cm length of arteriole?

51. ||| Blood in a carotid artery carrying blood to the head is moving BIO at 0.15 m/s when it reaches a section where plaque has narrowed the artery to 80% of its diameter. What pressure drop occurs when the blood reaches this narrow section?

## General Problems

52. |||| As discussed in Section 13.3, a person's percentage of BIO body fat can be estimated by weighing the person both in air and underwater, from which the average density $\rho_{avg}$ can be calculated. A widely used equation that relates average body density to fat percentage is the *Siri equation,* % body fat $= (495/\rho_{avg}) - 450$, where $\rho_{avg}$ is measured in $g/cm^3$. One person's weight in air is 947 N while his weight underwater is 44 N. What is his percentage of body fat?

53. ||| The density of aluminum is $2700\ kg/m^3$. How many atoms INT are in a $2.0\ cm \times 2.0\ cm \times 2.0\ cm$ cube of aluminum?

54. |||| An oil layer floats on 85 cm of water in a tank. The absolute pressure at the bottom of the tank is 112.0 kPa. How thick is the oil?

55. || a. In Figure P13.55, how much force does the fluid exert on the end of the cylinder at A?
    b. How much force does the fluid exert on the end of the cylinder at B?

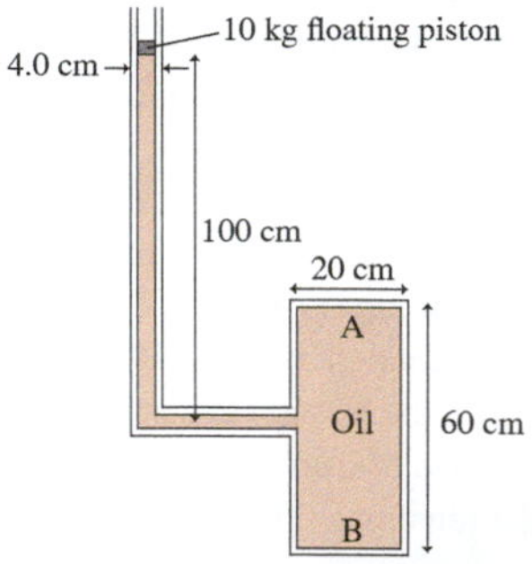

FIGURE P13.55

56. ||| A diver 50 m deep in 10°C fresh water exhales a 1.0-cm-INT diameter bubble. What is the bubble's diameter just as it reaches the surface of the lake, where the water temperature is 20°C? **Hint:** Assume that the air bubble is always in thermal equilibrium with the surrounding water.

57. || A sphere completely submerged in water is tethered to the bottom with a string. The tension in the string is one-third the weight of the sphere. What is the density of the sphere?

58. || You need to determine the density of a ceramic statue. If you suspend it from a spring scale, the scale reads 28.4 N. If you then lower the statue into a tub of water so that it is completely submerged, the scale reads 17.0 N. What is the density?

59. || A 5.0 kg rock whose density is $4800\ kg/m^3$ is suspended by a string such that half of the rock's volume is under water. What is the tension in the string?

60. || A flat slab of styrofoam, with a density of $32\ kg/m^3$, floats on a lake. What minimum volume must the slab have so that a 40 kg boy can sit on the slab without it sinking?

61. |||| A 2.0 mL syringe has an inner diameter of 6.0 mm, a needle BIO inner diameter of 0.25 mm, and a plunger pad diameter (where you place your finger) of 1.2 cm. A nurse uses the syringe to inject medicine into a patient whose blood pressure is 140/100. Assume the liquid is an ideal fluid.
    a. What is the minimum force the nurse needs to apply to the syringe?
    b. The nurse empties the syringe in 2.0 s. What is the flow speed of the medicine through the needle?

62. ||| A child's water pistol shoots water through a 1.0-mm-INT diameter hole. If the pistol is fired horizontally 70 cm above the ground, a squirt hits the ground 1.2 m away. What is the volume flow rate during the squirt? Ignore air resistance.

63. || The leaves of a tree lose water to the atmosphere via the process BIO of *transpiration.* A particular tree loses water at the rate of $3 \times 10^{-8}\ m^3/s$; this water is replenished by the upward flow of sap through vessels in the trunk. This tree's trunk contains about 2000 vessels, each 100 $\mu$m in diameter. What is the speed of the sap flowing in the vessels?

64. ‖ A hurricane wind blows across a 6.00 m × 15.0 m flat roof at a speed of 130 km/h.
    a. Is the air pressure above the roof higher or lower than the pressure inside the house? Explain.
    b. What is the pressure difference?
    c. How much force is exerted on the roof? If the roof cannot withstand this much force, will it "blow in" or "blow out"?

65. ||||| Water flows from the pipe shown in Figure P13.65 with a speed of 4.0 m/s.
    a. What is the water pressure as it exits into the air?
    b. What is the height $h$ of the standing column of water?

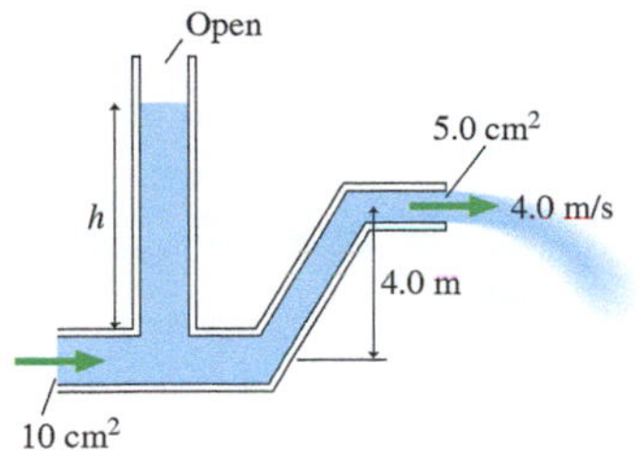


FIGURE P13.65

66. |||| Air at 20°C flows through the tube shown in Figure P13.66. Assume that air is an ideal fluid.
    a. What are the air speeds $v_1$ and $v_2$ at points 1 and 2?
    b. What is the volume flow rate?

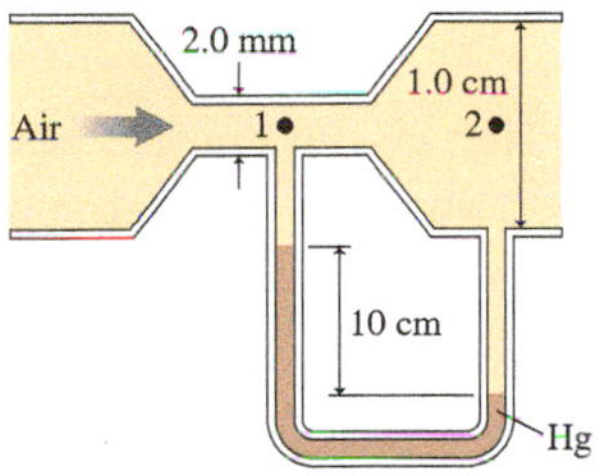


FIGURE P13.66

67. ||| Air at 20°C flows through the tube shown in Figure P13.67 at a rate of 1200 cm³/s. Assume that air is an ideal fluid. What is the height $h$ of mercury in the right side of the U-tube?
INT

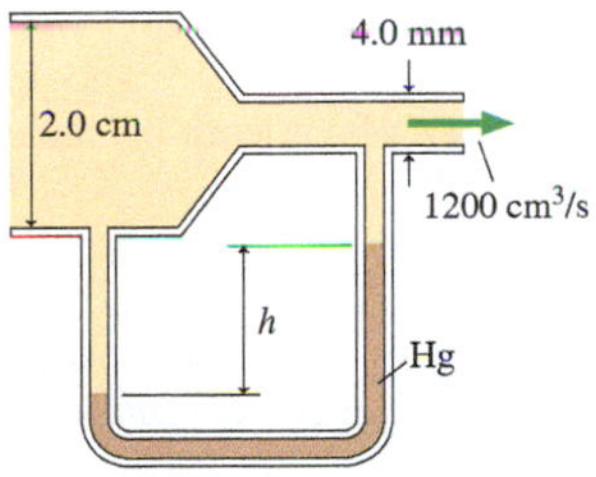


FIGURE P13.67

68. |||| There are two carotid arteries that feed blood to the brain, one on each side of the neck and head. One patient's carotid arteries are each 11.2 cm long and have an inside diameter of 5.2 mm. Near the middle of the left artery, however, is a 2.0-cm-long *stenosis*, a section of the artery with a smaller diameter of 3.4 mm. For the same blood flow rate, what is the ratio of the pressure drop along the patient's left carotid artery to the drop along his right artery?
BIO

69. ||||| Figure P13.69 shows a section of a long tube that narrows near its open end to a diameter of 1.0 mm. Water at 20°C flows out of the open end at 0.020 L/s. What is the gauge pressure at point P, where the diameter is 4.0 mm?

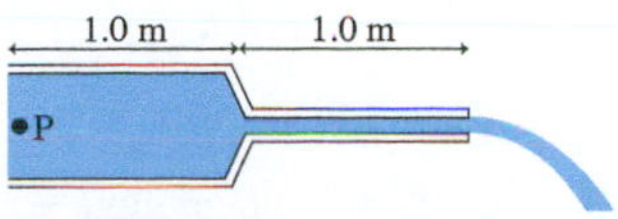


FIGURE P13.69

70. ‖ Smoking tobacco is bad for your circulatory health. In an attempt to maintain the blood's capacity to deliver oxygen, the body increases its red blood cell production, and this increases the viscosity of the blood. In addition, nicotine from tobacco causes arteries to constrict.
BIO

    For a nonsmoker, normal blood flow requires a pressure difference of 8.0 mm Hg between the two ends of an artery. Assume that smoking increases viscosity by 10% and reduces the arterial diameter to 90% of its previous value. For a smoker, what pressure difference is needed to maintain the same blood flow?

71. ||| A stiff, 10-cm-long tube with an inner diameter of 3.0 mm is attached to a small hole in the side of a tall beaker. The tube sticks out horizontally. The beaker is filled with 20°C water to a level 45 cm above the hole, and it is continually topped off to maintain that level. What is the volume flow rate through the tube?

## MCAT-Style Passage Problems

### Blood Pressure and Blood Flow BIO

As blood goes from the left ventricle through the arteries and veins of the human body, both its speed and pressure change. The arteries and arterioles can either constrict, reducing the area, or dilate, increasing the area, in response to certain conditions. Both of these changes can affect blood flow and blood pressure. An artery can also develop a permanent narrow area (stenosis) or a permanent wide area (aneurysm). Both of these changes can have significant health consequences.

72. | Suppose that in response to some stimulus a small blood vessel narrows to 90% of its original diameter. If there is no change in the pressure across the vessel, what is the ratio of the new volume flow rate to the original flow rate?
    A. 0.66 B. 0.73 C. 0.81 D. 0.90

73. | Sustained exercise can increase the blood flow rate of the heart by a factor of 5 with only a modest increase in blood pressure. This is a large change in flow. Although several factors come into play, which of the following physiological changes would most plausibly account for such a large increase in flow with a small change in pressure?
    A. A decrease in the viscosity of the blood
    B. Dilation of the smaller blood vessels to larger diameters
    C. Dilation of the aorta to larger diameter
    D. An increase in the oxygen carried by the blood

74. ‖ A patient has developed a narrowing of the coronary arteries that provide blood to the heart. If the rate of blood flow through the arteries is to stay the same,
BIO
    A. The length of the arteries must decrease as well.
    B. The speed of the flow must not change.
    C. The pressure difference across the arteries must stay the same, requiring a decrease in blood pressure.
    D. The pressure difference across the arteries must increase, requiring an increase in blood pressure.

75. ‖ The nicotine in tobacco causes the arteries to constrict. A rapid constriction of the arteries likely leads to
BIO
    A. A rapid decrease in blood pressure.
    B. No change in blood pressure.
    C. A rapid increase in blood pressure.

76. ‖ A patient has developed an aneurysm in the aorta, a short section where the diameter is twice the normal diameter. In the aneurysm, the speed of the blood is ______ than in the section before the aneurysm, and the pressure is ________ than in the section before the aneurysm.
BIO
    A. Greater, greater B. Greater, less
    C. Less, greater D. Less, less

## KNOWLEDGE STRUCTURE III Properties of Matter

**BASIC GOALS** How can we describe the macroscopic flows of energy and matter in heat transfer and fluid flow?

### CROSS-CUTTING CONCEPTS

**Flows of energy and matter**

A difference in temperature leads to transfer of heat: **conduction.**

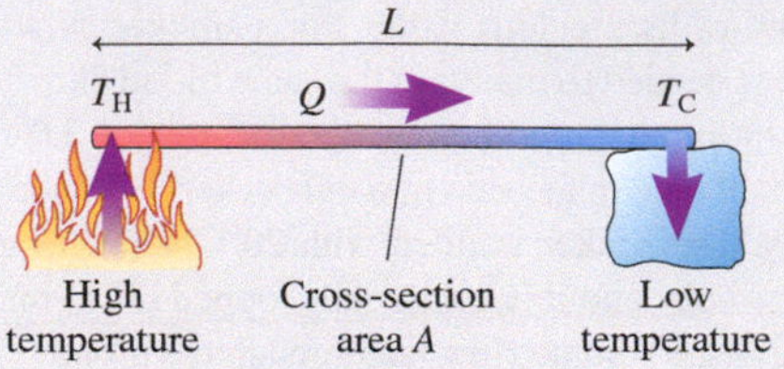


$$\frac{Q}{\Delta t} = \left(\frac{kA}{L}\right)\Delta T$$

A difference in concentration leads to a transfer of matter: **diffusion.**

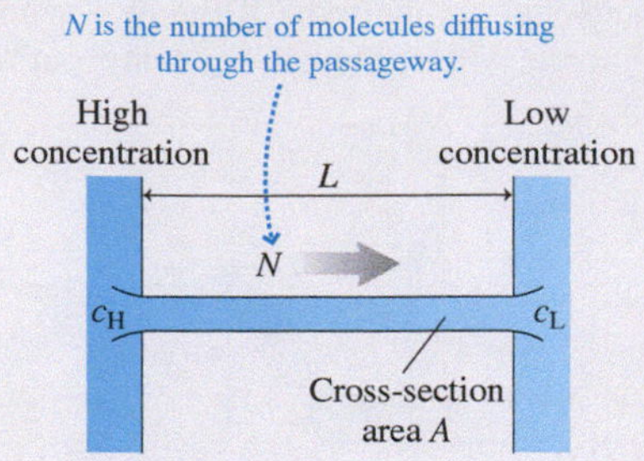


$$\frac{N}{\Delta t} = \left(\frac{DA}{L}\right)\Delta c$$

### GENERAL PRINCIPLES Phases of matter

**Gas:** Particles are freely moving. Thermal energy is kinetic energy of motion of particles. Pressure is due to collisions of particles with walls of container.

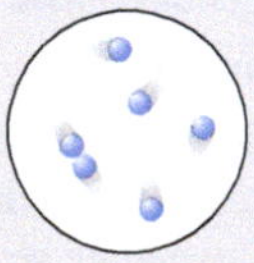

**Liquid:** Particles are loosely bound, but flow is possible. Heat is transferred by convection and conduction. Pressure is due to gravity.

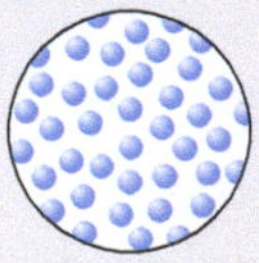

**Solid:** Particles are joined by spring-like bonds. Increasing temperature causes expansion. Heat is transferred by conduction only.

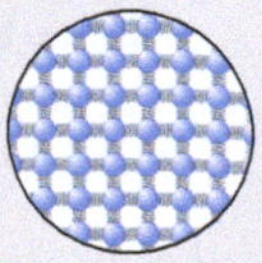

Heat must be added to change a solid to a liquid and a liquid to a gas. Heat must be removed to reverse these changes.

$$Q = \begin{cases} \pm ML_f & \text{Heat needed to melt/freeze mass } M \\ \pm ML_v & \text{Heat needed to boil/condense mass } M \end{cases}$$

### Fluids and pressure

A **fluid** is a substance that can flow, a liquid or a gas. A fluid exerts **pressure** on the walls of its container.

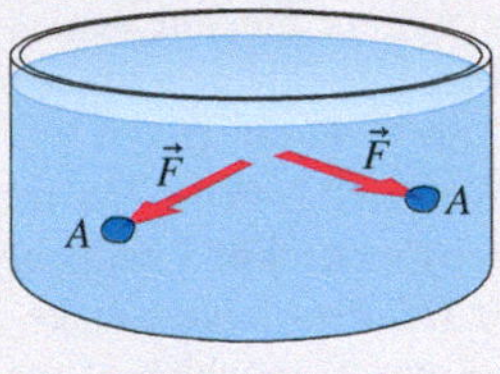


The pressure is the ratio of the force to the area over which this force acts.

$$p = \frac{F}{A}$$

The gauge pressure in a container is the difference in pressure between the inside and the outside. For containers of gases, this is usually atmospheric pressure.

$$p_g = p - p_{atmos}$$

The pressure in the gas laws is the absolute pressure, the actual pressure in a container.

### Ideal-gas processes

We can graph processes on a $pV$ diagram. Work done by the gas is the area under the graph.

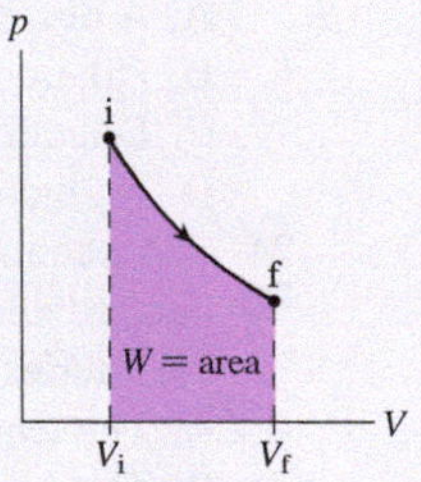


For a gas process in a sealed container, the ideal-gas law relates values of pressure, volume, and temperature:

$$\frac{p_i V_i}{T_i} = \frac{p_f V_f}{T_f}$$

- Constant-volume process $V = \text{constant}$
- Isothermal process $T = \text{constant}$
- Isobaric process $p = \text{constant}$
- Adiabatic process $Q = 0$

### Heat and heat transfer

The **specific heat** $c$ of a material is the heat required to raise the temperature of 1 kg by 1 K:

$$Q = Mc\,\Delta T$$

When two or more systems interact thermally, they come to a common final temperature determined by

$$Q_{net} = Q_1 + Q_2 + Q_3 + \cdots = 0$$

**Conduction** transfers heat by direct physical contact.
**Convection** transfers heat by the motion of a fluid.
**Radiation** transfers heat by electromagnetic waves.

### Fluid statics

**Archimedes' principle:** The magnitude of the buoyant force equals the weight of the fluid displaced by the object.

An object sinks if it is more dense than the fluid in which it is submerged; if it is less dense, it floats.

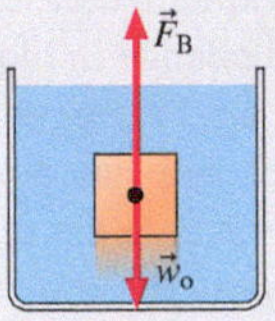


Pressure increases with depth in a liquid. The pressure at depth $d$ is

$$p = p_0 + \rho g d$$

### Fluid flow

Ideal fluid flow is laminar, incompressible, and nonviscous. The **equation of continuity** is

$$v_1 A_1 = v_2 A_2$$

**Bernoulli's equation** is

$$p_1 + \tfrac{1}{2}\rho v_1^{\,2} + \rho g y_1 = p_2 + \tfrac{1}{2}\rho v_2^{\,2} + \rho g y_2$$

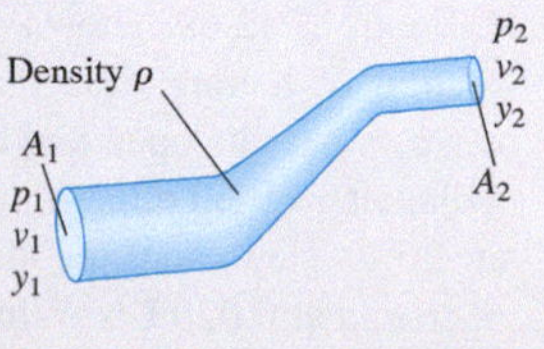

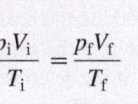

## Size and Life

Physicists look for simple models and general principles that underlie and explain diverse physical phenomena. In the first 13 chapters of this textbook, you've seen that just a handful of general principles and laws can be used to solve a wide range of problems. Can this approach have any relevance to a subject like biology? It may seem surprising, but there *are* general "laws of biology" that apply, with quantitative accuracy, to organisms as diverse as elephants and mice.

Let's look at an example. An elephant uses more metabolic power than a mouse. This is not surprising, as an elephant is much bigger. But recasting the data shows an interesting trend. When we looked at the energy required to raise the temperature of different substances, we considered *specific heat*. The "specific" meant that we considered the heat required for 1 kilogram. For animals, rather than metabolic rate, we can look at the *specific metabolic rate,* the metabolic power used *per kilogram* of tissue. If we factor out the mass difference between a mouse and an elephant, are their specific metabolic powers the same?

In fact, the specific metabolic rate varies quite a bit among mammals, as the graph of specific metabolic rate versus mass shows. But there is an interesting trend: All of the data points lie on a single smooth curve. In other words, there really is a biological *law* we can use to predict a mammal's metabolic rate knowing only its mass $M$. In particular, the specific metabolic rate is proportional to $M^{-0.25}$. Because a 4000 kg elephant is 160,000 times more massive than a 25 g mouse, the mouse's specific metabolic power is $(160{,}000)^{0.25} = 20$ times that of the elephant. A law that shows how a property scales with the size of a system is called a *scaling law*.

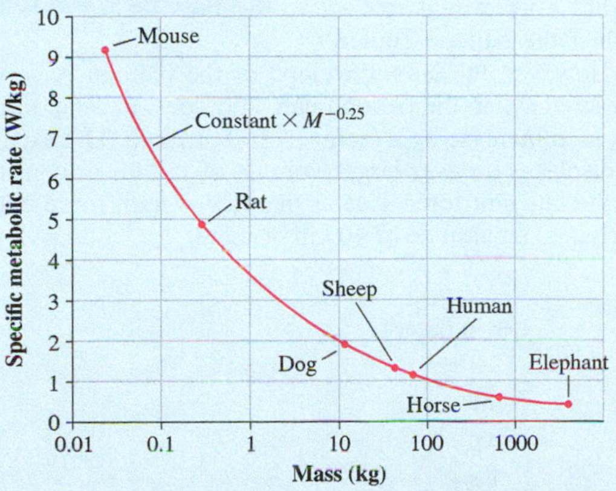


Specific metabolic rate as a function of body mass follows a simple scaling law.

A similar scaling law holds for birds, reptiles, and even bacteria. Why should a single simple relationship hold true for organisms that range in size from a single cell to a 100 ton blue whale? Interestingly, no one knows for sure. It is a matter of current research to find out just what this and other scaling laws tell us about the nature of life.

Perhaps the metabolic-power scaling law is a result of heat transfer. In Chapter 12, we noted that all metabolic energy used by an animal ends up as heat, which must be transferred to the environment. A 4000 kg elephant has 160,000 times the mass of a 25 g mouse, but it has only about 3000 times the surface area. The heat transferred to the environment depends on the surface area; the more surface area, the greater the rate of heat transfer. An elephant with a mouse-sized metabolism simply wouldn't be able to dissipate heat fast enough—it would quickly overheat and die.

If heat dissipation were the only factor limiting metabolism, we can show that the specific metabolic rate should scale as $M^{-0.33}$, quite different from the $M^{-0.25}$ scaling observed. Clearly, another factor is at work. Exactly what underlies the $M^{-0.25}$ scaling is still a matter of debate, but some recent analysis suggests the scaling is due to limitations not of heat transfer but of fluid flow. Cells in mice, elephants, and all mammals receive nutrients and oxygen for metabolism from the bloodstream. Because the minimum size of a capillary is about the same for all mammals, the structure of the circulatory system must vary from animal to animal. The human aorta has a diameter of about 1 inch; in a mouse, the diameter is approximately 1/20th of this. Thus a mouse has fewer levels of branching to smaller and smaller blood vessels as we move from the aorta to the capillaries. The smaller blood vessels in mice mean that viscosity is more of a factor throughout the circulatory system. The circulatory system of a mouse is quite different from that of an elephant.

A model of specific metabolic rate based on blood-flow limitations predicts a $M^{-0.25}$ law, exactly as observed. The model also makes other testable predictions. For example, the model predicts that the smallest possible mammal should have a body mass of about 1 gram—exactly the size of the smallest shrew. Even smaller animals have different types of circulatory systems; in the smallest animals, nutrient transport is by diffusion alone. But the model can be extended to predict that the specific metabolic rate for these animals will follow a scaling law similar to that for mammals, exactly as observed. It is too soon to know if this model will ultimately prove to be correct, but it's indisputable that there are large-scale regularities in biology that follow mathematical relationships based on the laws of physics.

*The following questions are related to the passage "Size and Life" on the previous page.* BIO

1. A typical timber wolf has a mass of 40 kg, a typical jackrabbit a mass of 2.5 kg. Given the scaling law presented in the passage, we'd expect the specific metabolic rate of the jackrabbit to be higher by a factor of
   A. 2 B. 4 C. 8 D. 16
2. A typical timber wolf has a mass of 40 kg, a typical jackrabbit a mass of 2.5 kg. Given the scaling law presented in the passage, we'd expect the wolf to use _______ times more energy than a jackrabbit in the course of a day.
   A. 2 B. 4 C. 8 D. 16
3. Given the data of the graph, approximately how much energy, in Calories, would a 200 g rat use during the course of a day?
   A. 10 B. 20 C. 100 D. 200
4. All other things being equal, species that inhabit cold climates tend to be larger than related species that inhabit hot climates. For instance, the Alaskan hare is the largest North American hare, with a typical mass of 5.0 kg, double that of a jackrabbit. A likely explanation is that
   A. Larger animals have more blood flow, allowing for better thermoregulation.
   B. Larger animals need less food to survive than smaller animals.
   C. Larger animals have larger blood volumes than smaller animals.
   D. Larger animals lose heat less quickly than smaller animals.
5. The passage proposes that there are quantitative "laws" of biology that have their basis in physical principles, using the scaling of specific metabolic rate with body mass as an example. Which of the following regularities among animals might also be an example of such a "law"?
   A. As a group, birds have better color vision than mammals.
   B. Reptiles have a much lower specific metabolic rate than mammals.
   C. Predators tend to have very good binocular vision; prey animals tend to be able to see over a very wide angle.
   D. Jump height varies very little among animals. Nearly all animals, ranging in size from a flea to a horse, have a maximum vertical leap that is quite similar.

*The following passages and associated questions are based on the material of Part III.*

### Keeping Your Cool BIO

A 68 kg cyclist is pedaling down the road at 15 km/h, using a total metabolic power of 480 W. A certain fraction of this energy is used to move the bicycle forward, but the balance ends up as thermal energy in his body, which he must get rid of to keep cool. On a very warm day, conduction, convection, and radiation transfer little energy, and so he does this by perspiring, with the evaporation of water taking away the excess thermal energy.

6. If the cyclist reaches his 15 km/h cruising speed by rolling down a hill, what is the approximate height of the hill?
   A. 22 m B. 11 m C. 2 m D. 1 m
7. As he cycles at a constant speed on level ground, at what rate is chemical energy being converted to thermal energy in his body, assuming a typical efficiency of 25% for the conversion of chemical energy to the mechanical energy of motion?
   A. 480 W B. 360 W C. 240 W D. 120 W
8. To keep from overheating, the cyclist must get rid of the excess thermal energy generated in his body. If he cycles at this rate for 2 hours, how many liters of water must he perspire, to the nearest 0.1 liter?
   A. 0.4 L B. 0.9 L C. 1.1 L D. 1.4 L
9. Being able to exhaust this thermal energy is very important. If he isn't able to get rid of any of the excess heat, by how much will the temperature of his body increase in 10 minutes of riding, to the nearest 0.1°C?
   A. 0.3°C B. 0.6°C C. 0.9°C D. 1.2°C

### Weather Balloons

The data used to generate weather forecasts are gathered by hundreds of weather balloons launched from sites throughout the world. A typical balloon is made of latex and filled with hydrogen. A packet of sensing instruments (called a *radiosonde*) transmits information back to earth as the balloon rises into the atmosphere. At the beginning of its flight, the average density of the weather balloon package (total mass of the balloon plus cargo divided by their volume) is less than the density of the surrounding air, so the balloon rises. As it does, the density of the surrounding air decreases, as shown in Figure III.1. The balloon will rise to the point at which the buoyant force of the air exactly balances its weight. This would not be very high if the balloon couldn't expand. However, the latex envelope of the balloon is very thin and very stretchy, so the balloon can, and does, expand, allowing the volume to increase by a factor of 100 or more. The expanding balloon displaces an ever-larger volume of the lower-density air, keeping the buoyant force greater than the weight force until the balloon rises to an altitude of 40 km or more.

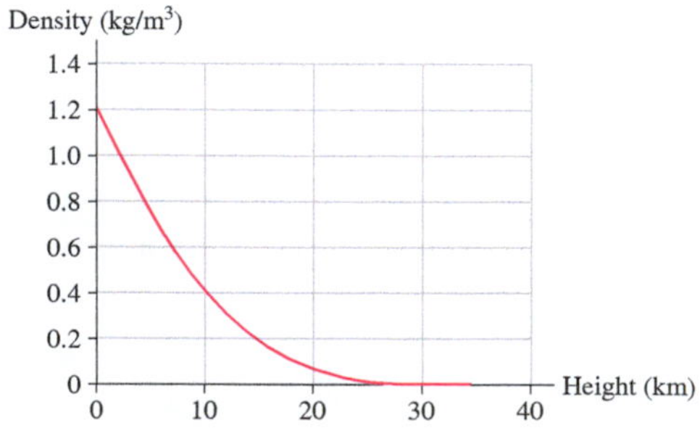


FIGURE III.1

10. A balloon launched from sea level has a volume of approximately 4 $m^3$. What is the approximate buoyant force on the balloon?
A. 50 N B. 40 N
C. 20 N D. 10 N
11. A balloon launched from sea level with a volume of 4 $m^3$ will have a volume of about 12 $m^3$ on reaching an altitude of 10 km. What is the approximate buoyant force now?
A. 50 N B. 40 N
C. 20 N D. 10 N
12. The balloon expands as it rises, keeping the pressures inside and outside the balloon approximately equal. If the balloon rises slowly, heat transfers will keep the temperature inside the same as the outside air temperature. A balloon with a volume of 4.0 $m^3$ is launched at sea level, where the atmospheric pressure is 100 kPa and the temperature is 15°C. It then rises slowly to a height of 5500 m, where the pressure is 50 kPa and the temperature is –20°C. What is the volume of the balloon at this altitude?
A. 5.0 $m^3$ B. 6.0 $m^3$
C. 7.0 $m^3$ D. 8.0 $m^3$
13. If the balloon rises quickly, so that no heat transfer is possible, the temperature inside the balloon will drop as the gas expands. If a 4.0 $m^3$ balloon is launched at a pressure of 100 kPa and rapidly rises to a point where the pressure is 50 kPa, the volume of the balloon will be
A. Greater than 8.0 $m^3$.
B. 8.0 $m^3$.
C. Less than 8.0 $m^3$.
14. At the end of the flight, the radiosonde is dropped and falls to earth by parachute. Suppose the parachute achieves its terminal speed at a height of 30 km. As it descends into the atmosphere, how does the terminal speed change?
A. It increases.
B. It stays the same.
C. It decreases.

### Passenger Balloons

Long-distance balloon flights are usually made using a hot-air-balloon/helium-balloon hybrid. The balloon has a sealed, flexible chamber of helium gas that expands or contracts to keep the helium pressure approximately equal to the air pressure outside. The helium chamber sits on top of an open (that is, air can enter or leave), constant-volume chamber of propane-heated air. Assume that the hot air and the helium are kept at a constant temperature by burning propane.

15. A balloon is launched at sea level, where the air pressure is 100 kPa. The helium has a volume of 1000 $m^3$ at this altitude. What is the volume of the helium when the balloon has risen to a height where the atmospheric pressure is 33 kPa?
A. 330 $m^3$ B. 500 $m^3$ C. 1000 $m^3$ D. 3000 $m^3$
16. A balloon is launched at sea level, where the air pressure is 100 kPa. The density in the hot-air chamber is 1.0 $kg/m^3$. What is the density of the air when the balloon has risen to a height where the atmospheric pressure is 33 kPa?
A. 3.0 $kg/m^3$
B. 1.0 $kg/m^3$
C. 0.66 $kg/m^3$
D. 0.33 $kg/m^3$
17. A balloon is at a height of 5.0 km and is descending at a constant rate. The buoyancy force is directed ____________; the drag force is directed ____________.
A. Up, up
B. Up, down
C. Down, up
D. Down, down

### Additional Integrated Problems

18. BIO When you exhale, all of the air in your lungs must exit through the trachea. If you exhale through your nose, this air subsequently leaves through your nostrils. The area of your nostrils is less than that of your trachea. How does the speed of the air in the trachea compare to that in the nostrils?
19. BIO Sneezing requires an increase in pressure of the air in the lungs; a typical sneeze might result in an extra pressure of 7.0 kPa. Estimate how much force this exerts on the diaphragm, the large muscle at the bottom of the ribcage.
20. A 20 kg block of aluminum sits on the bottom of a tank of water. How much force does the block exert on the bottom of the tank?
21. BIO We've seen that fish can control their buoyancy through the use of a swim bladder, a gas-filled organ inside the body. You can assume that the gas pressure inside the swim bladder is roughly equal to the external water pressure. A fish swimming at a particular depth adjusts the volume of its swim bladder to give it neutral buoyancy. If the fish swims upward or downward, the changing water pressure causes the bladder to expand or contract. Consequently, the fish must adjust the quantity of gas to restore the original volume and thus reestablish neutral buoyancy. Consider a large, 7.0 kg striped bass with a volume of 7.0 L. When neutrally buoyant, 7.0% of the fish's volume is taken up by the swim bladder. Assume a body temperature of 15°C.
a. How many moles of air are in the swim bladder when the fish is at a depth of 80 ft?
b. What will the volume of the swim bladder be if the fish ascends to a 50 ft depth without changing the quantity of gas?
c. To return the swim bladder to its original size, how many moles of gas must be removed?

PART IV

# Oscillations and Waves

Wolves are social animals, and they howl to communicate over distances of several miles with other members of their pack. How are such sounds made? How do they travel through the air? And how are other wolves able to hear these sounds from such a great distance?

OVERVIEW

# Motion That Repeats Again and Again

Up to this point in the text, we have generally considered processes that have a clear starting and ending point, such as a car accelerating from rest to a final speed, or a solid being heated from an initial to a final temperature. In Part IV, we begin to consider processes that are *periodic*—they repeat. A child on a swing, a boat bobbing on the water, and even the repetitive bass beat of a rock song are *oscillations* that happen over and over without a starting or ending point. The *period*, the time for one cycle of the motion, will be a key parameter for us to consider as we look at oscillatory motion.

Our first goal will be to develop the language and tools needed to describe oscillations, ranging from the swinging of the bob of a pendulum clock to the bouncing of a car on its springs. Once we understand oscillations, we will extend our analysis to consider oscillations that travel—*waves*.

## The Wave Model

We've had great success modeling the motion of complex objects as the motion of one or more particles. We were even able to explain the macroscopic properties of matter, such as pressure and temperature, in terms of the motion of the atomic particles that comprise all matter.

Now it's time to explore another way of looking at nature, the *wave model*. Familiar examples of waves include

- Ripples on a pond.
- The sound of thunder.
- The swaying ground of an earthquake.
- A vibrating guitar string.
- The colors of a rainbow.

Despite the great diversity of types and sources of waves, there is a single, elegant physical theory that can describe them all. Our exploration of wave phenomena will call upon water waves, sound waves, and light waves for examples, but our goal will be to emphasize the unity and coherence of the ideas that are common to *all* types of waves. As was the case with the particle model, we will use the wave model to explain a wide range of phenomena.

## When Waves Collide

The collision of two particles is a dramatic event. Energy and momentum are transferred as the two particles head off in different directions. Something much gentler happens when two waves come together—the two waves pass through each other unchanged. Where they overlap, we get a *superposition* of the two waves. We will finish our discussion of waves by analyzing the standing waves that result from the superposition of two waves traveling in opposite directions. The physics of standing waves will allow us to understand how your vocal tract can produce such a wide range of sounds, and how your ears are able to analyze them.

# 14 Oscillations

A gibbon swings below a branch, moving forward by pivoting below one handhold and then the next. How does an understanding of oscillatory motion allow us to analyze the swinging motion of a gibbon—and the walking motion of other animals?

## LOOKING AHEAD

### Motion That Repeats

When the girl moves down, the springy ropes pull up. This restoring force produces an **oscillation:** one bounce after another.

You'll see many examples of systems with restoring forces that lead to oscillatory motion.

### Simple Harmonic Motion

The sand records the motion of the oscillating pendulum. The **sinusoidal** shape tells us that this is **simple harmonic motion.**

All oscillations show a similar form. You'll learn to describe and analyze oscillating systems.

### Resonance

When you make a system oscillate at its **natural frequency,** you can get a large amplitude. We call this **resonance.**

You'll learn how resonance of a membrane in the inner ear lets you determine the **pitch** of a musical note.

**GOAL** To understand systems that oscillate with simple harmonic motion.

## LOOKING BACK

### Springs and Restoring Forces

In Chapter 8, you learned that a stretched spring exerts a restoring force proportional to the stretch:

$$F_{sp} = -k\Delta x$$

In this chapter, you'll see how this linear restoring force leads to an oscillation, with a frequency determined by the spring constant $k$.

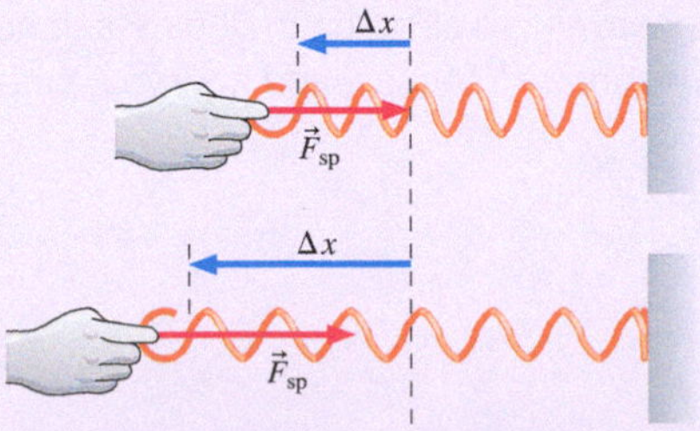


**STOP TO THINK**

A hanging spring has length 10 cm. A 100 g mass is hung from the spring, stretching it to 12 cm. What will be the length of the spring if this mass is replaced by a 200 g mass?

A. 14 cm B. 16 cm
C. 20 cm D. 24 cm

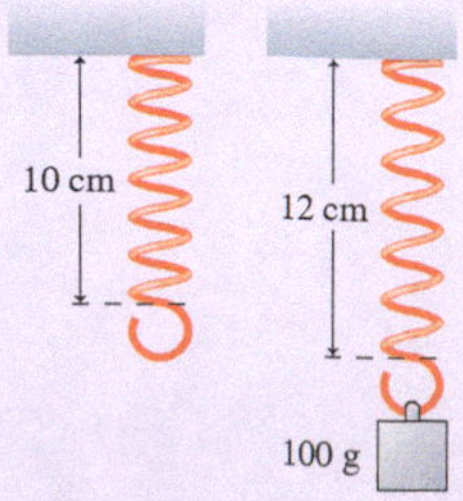

## 14.1 Equilibrium and Oscillation

Consider a marble that is free to roll inside a spherical bowl, as shown in FIGURE 14.1. The marble has an **equilibrium position** at the bottom of the bowl where it will rest with no net force on it. If you push the marble away from equilibrium, the marble's weight leads to a net force directed back toward the equilibrium position. We call this a **restoring force** because it acts to restore equilibrium. The magnitude of this restoring force increases if the marble is moved farther away from the equilibrium position.

If you pull the marble to the side and release it, it doesn't just roll back to the bottom of the bowl and stay put. It keeps on moving, rolling up and down each side of the bowl, repeatedly moving through its equilibrium position, as we see in FIGURE 14.2. We call such repetitive motion an **oscillation.** This oscillation is a result of an interplay between the restoring force and the marble's inertia, something we will see in all of the oscillations we consider.

We'll start our description by noting the most important fact about oscillatory motion: It repeats. Any oscillation is characterized by a *period,* the time for the motion to repeat. We met the concepts of period and frequency when we studied circular motion in Chapter 6. As a starting point then, let's review these ideas and see how they apply to oscillatory motion.

FIGURE 14.1 Equilibrium and restoring forces for a ball in a bowl.

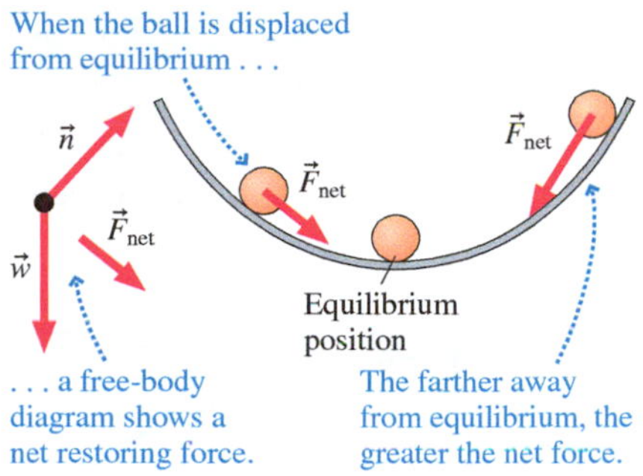


FIGURE 14.2 The oscillating motion of a ball rolling in a bowl.

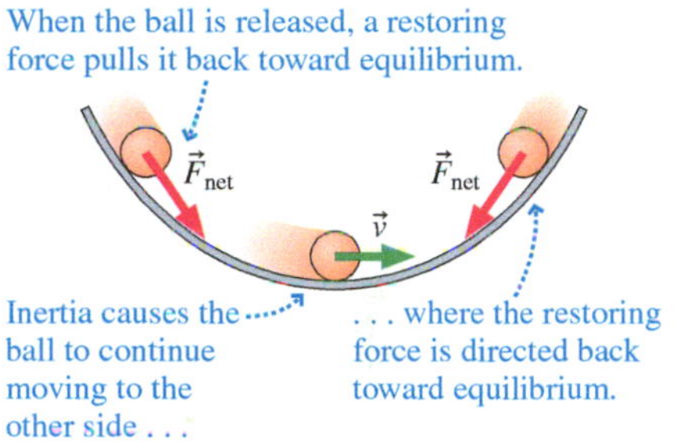


### Frequency and Period

An electrocardiogram (ECG), such as the one shown in FIGURE 14.3, is a record of the electrical signals of the heart as it beats. We will explore the ECG in some detail in Chapter 21. Although the shape of a typical ECG is rather complex, notice that it has a *repeating pattern.* For this or any oscillation, the time to complete one full cycle is called the **period** of the oscillation. Period is given the symbol $T$.

An equivalent piece of information is the number of cycles, or oscillations, completed per second. If the period is $\frac{1}{10}$ s, then the oscillator can complete 10 cycles in 1 second. Conversely, an oscillation period of 10 s allows only $\frac{1}{10}$ of a cycle to be completed per second. In general, $T$ seconds per cycle implies that $1/T$ cycles are completed each second. The number of cycles per second is called the **frequency** $f$ of the oscillation. The relationship between frequency and period is therefore

$$f = \frac{1}{T} \quad \text{or} \quad T = \frac{1}{f} \tag{14.1}$$

The units of frequency are **hertz,** abbreviated Hz. By definition,

$$1 \text{ Hz} = 1 \text{ cycle per second} = 1 \text{ s}^{-1}$$

We will frequently deal with very rapid oscillations and make use of the units shown in TABLE 14.1.

NOTE ▶ Uppercase and lowercase letters *are* important. 1 MHz is 1 megahertz = $10^6$ Hz, but 1 mHz is 1 millihertz = $10^{-3}$ Hz! ◀

FIGURE 14.3 BIO An electrocardiogram has a well-defined period.

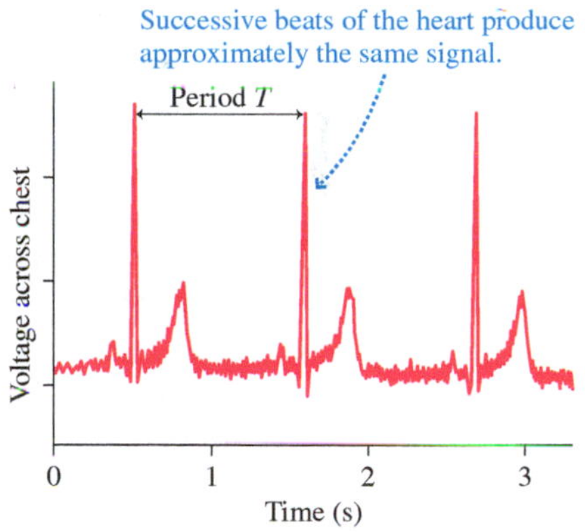


**EXAMPLE 14.1 Finding the frequency and period of a radio station**

An FM radio station broadcasts an oscillating radio wave at a frequency of 100 MHz. What is the period of the oscillation?

***SOLVE*** The frequency $f$ of oscillations in the radio transmitter is 100 MHz = $1.0 \times 10^8$ Hz. The period is the inverse of the frequency; hence,

$$T = \frac{1}{f} = \frac{1}{1.0 \times 10^8 \text{ Hz}} = 1.0 \times 10^{-8} \text{ s} = 10 \text{ ns}$$

TABLE 14.1 Common units of frequency

| Frequency | Period |
|---|---|
| $10^3$ Hz = 1 kilohertz = 1 kHz | 1 ms |
| $10^6$ Hz = 1 megahertz = 1 MHz | 1 $\mu$s |
| $10^9$ Hz = 1 gigahertz = 1 GHz | 1 ns |

## Simple Harmonic Motion

Let's make a graph of the motion of the marble in a bowl, with positions to the right of equilibrium positive and positions to the left of equilibrium negative. FIGURE 14.4 shows a series of "snapshots" of the motion and the corresponding points on the graph. This graph has the form of a *cosine function.* A graph or a function that has the form of a sine or cosine function is called **sinusoidal.** A sinusoidal oscillation is called **simple harmonic motion,** often abbreviated SHM.

FIGURE 14.4 Constructing a position-versus-time graph for a marble rolling in a bowl.

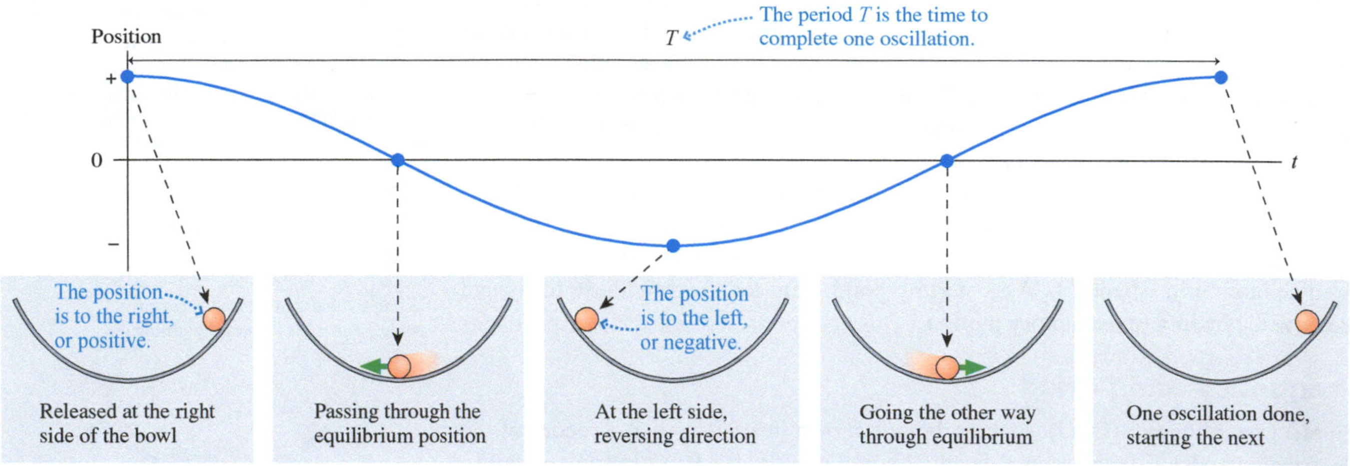


A video to support a section's topic is embedded in the eText.

**Video** Oscillations

A marble rolling in the bottom of a bowl undergoes simple harmonic motion, as does a car bouncing on its springs. SHM is very common, but most cases of SHM can be modeled as one of two simple systems: an object with mass (or simply a "mass") oscillating on a spring or a pendulum swinging back and forth. Examples are shown in the following table.

**Examples of simple harmonic motion**

| Oscillating system | | Related real-world example BIO | |
|---|---|---|---|
| **Mass on a spring** 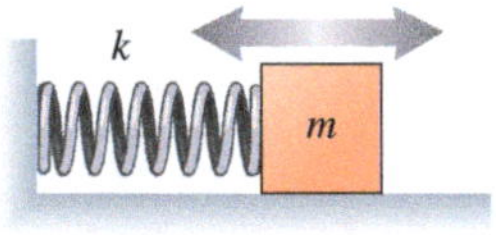  | The mass oscillates back and forth due to the restoring force of the spring. The period depends on the mass and the stiffness of the spring. | **Vibrations in the ear** 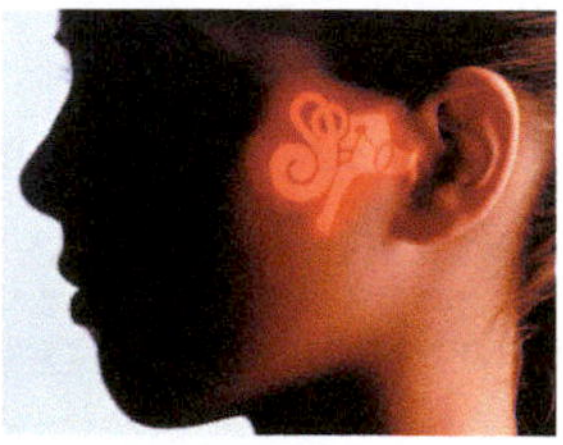 | Sound waves entering the ear cause the oscillation of a membrane in the cochlea, part of the inner ear. The vibration can be modeled as a mass on a spring. The period of oscillation of a segment of the membrane depends on mass (the thickness of the membrane) and stiffness (the rigidity of the membrane). |
| **Pendulum** 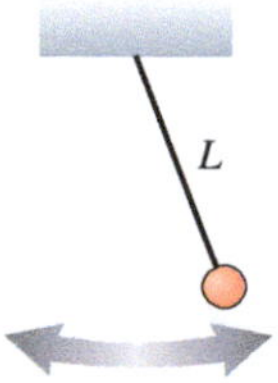  | The mass oscillates back and forth due to the restoring gravitational force. The period depends on the length of the pendulum and the free-fall acceleration $g$. | **Motion of legs while walking**  | The motion of a walking animal's legs can be modeled as pendulum motion. The rate at which the legs swing depends on the length of the legs and the free-fall acceleration $g$. |

**STOP TO THINK 14.1** Two oscillating systems have periods $T_1$ and $T_2$, with $T_1 < T_2$. How are the frequencies of the two systems related?

A. $f_1 < f_2$ B. $f_1 = f_2$ C. $f_1 > f_2$

## 14.2 Linear Restoring Forces and SHM

**FIGURE 14.5** shows a glider that rides with very little friction on an air track. There is a spring connecting the glider to the end of the track. When the spring is neither stretched nor compressed, the net force on the glider is zero. The glider just sits there—this is the equilibrium position.

If the glider is now displaced from this equilibrium position by $\Delta x$, the spring exerts a force back toward equilibrium—a restoring force. In ◀◀SECTION 8.3, we found that the spring force is given by Hooke's law: $(F_{sp})_x = -k\,\Delta x$, where $k$ is the spring constant. (Recall that a "stiffer" spring has a larger value of $k$.) If we set the origin of our coordinate system at the equilibrium position, the displacement from equilibrium $\Delta x$ is equal to $x$; thus the spring force can be written as $(F_{sp})_x = -kx$.

The net force on the glider is simply the spring force, so we can write

$$(F_{net})_x = -kx \qquad (14.2)$$

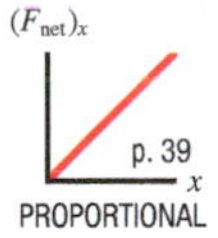


The negative sign tells us that this is a restoring force because the force is in the direction opposite the displacement. If we pull the glider to the right ($x$ is positive), the force is to the left (negative)—back toward equilibrium.

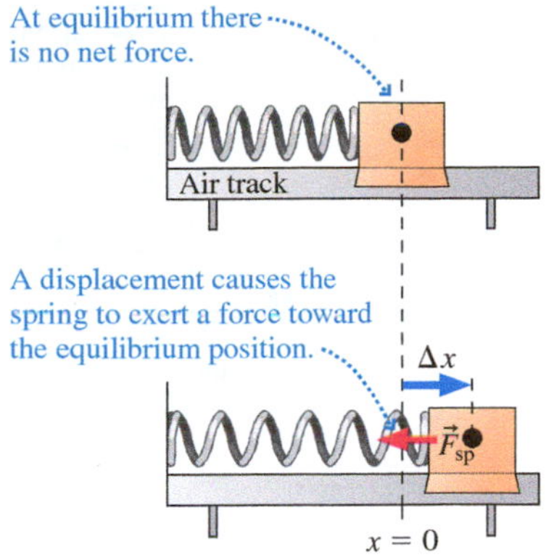


**FIGURE 14.5** The restoring force on an air-track glider attached to a spring.

This is a **linear restoring force;** that is, **the net force is toward the equilibrium position and is proportional to the distance from equilibrium.**

### Horizontal Motion of a Mass on a Spring

If we pull the air-track glider of Figure 14.5 a short distance to the right and release it, it will oscillate back and forth. **FIGURE 14.6** shows actual data from an experiment in which the position of a glider was measured 20 times every second. This is a position-versus-time graph that has been rotated 90° from its usual orientation in order for the $x$-axis to match the motion of the glider.

The object's maximum displacement from equilibrium is called the **amplitude** $A$ of the motion. The object's position oscillates between $x = -A$ and $x = +A$.

**NOTE** ▶ The amplitude is the distance from equilibrium to the maximum, *not* the distance from the minimum to the maximum. ◀

The graph of the position is sinusoidal, so this is simple harmonic motion. **Oscillation about an equilibrium position with a linear restoring force is always simple harmonic motion.**

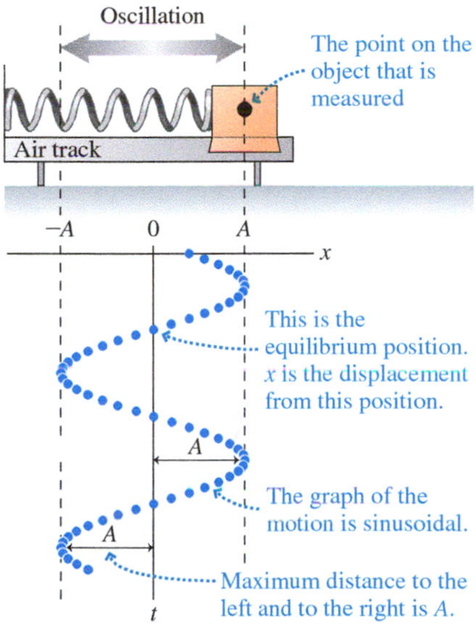


**FIGURE 14.6** An experiment showing the horizontal oscillation of an air-track glider.

### Vertical Motion of a Mass on a Spring

**FIGURE 14.7** shows a block of mass $m$ hanging from a spring with spring constant $k$. Will this mass-spring system have the same simple harmonic motion as the horizontal system we just saw, or will gravity add an additional complication? An important fact to notice is that the equilibrium position of the block is *not* where the spring is at its unstretched length. At the equilibrium position of the block, where it hangs motionless, the spring has stretched by $\Delta L$.

Finding $\Delta L$ is a static-equilibrium problem in which the upward spring force balances the downward weight force of the block. The $y$-component of the spring force is given by Hooke's law:

$$(F_{sp})_y = k\,\Delta L \qquad (14.3)$$

Newton's first law for the block in equilibrium is

$$(F_{net})_y = (F_{sp})_y + w_y = k\,\Delta L - mg = 0 \qquad (14.4)$$

from which we can find

$$\Delta L = \frac{mg}{k} \qquad (14.5)$$

This is the distance the spring stretches when the block is attached to it.

**FIGURE 14.7** The equilibrium position of a mass on a vertical spring.

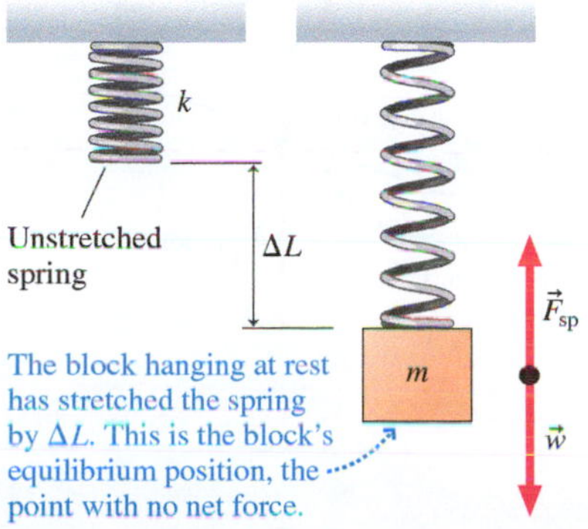

**FIGURE 14.8** Displacing the block from its equilibrium position produces a restoring force.

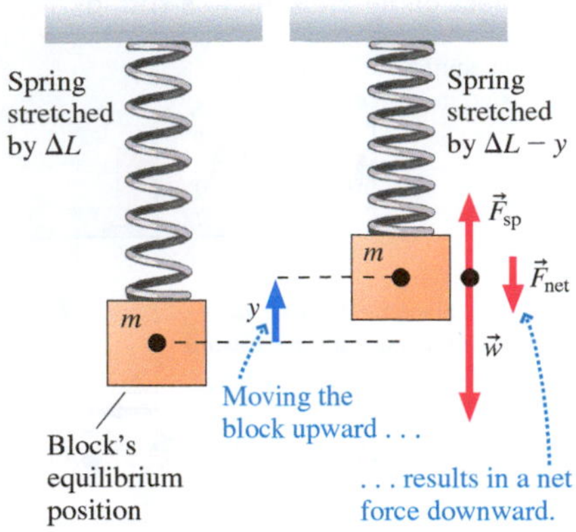


**Video** Basic Oscillation Problems

Suppose we now displace the block from this equilibrium position, as shown in **FIGURE 14.8.** We've placed the origin of the $y$-axis at the block's equilibrium position in order to be consistent with our analyses of oscillations throughout this chapter. If the block moves upward, as in the figure, the spring gets shorter compared to its equilibrium length, but the spring is still *stretched* compared to its unstretched length in Figure 14.7. When the block is at position $y$, the spring is stretched by an amount $\Delta L - y$ and hence exerts an *upward* spring force $F_{sp} = k(\Delta L - y)$. The net force on the block at this point is

$$(F_{net})_y = (F_{sp})_y + w_y = k(\Delta L - y) - mg = (k\Delta L - mg) - ky \tag{14.6}$$

But $k\,\Delta L - mg = 0$, from Equation 14.4, so the net force on the block is

$$(F_{net})_y = -ky \tag{14.7}$$

Equation 14.7 for a mass hung from a spring has the same form as Equation 14.2 for the horizontal spring, where we found $(F_{net})_x = -kx$. That is, the restoring force for vertical oscillations is identical to the restoring force for horizontal oscillations. **The role of gravity is to determine where the equilibrium position is, but it doesn't affect the restoring force for displacement from the equilibrium position.** Because it has a linear restoring force, **a mass on a vertical spring oscillates with simple harmonic motion.**

## The Pendulum

The chapter opened with a picture of a gibbon, whose body swings back and forth below a tree branch. The motion of the gibbon is essentially that of a **pendulum,** a mass (or *bob*) suspended from a pivot point by a light string or rod. A pendulum oscillates about its equilibrium position, but is this simple harmonic motion? To answer this question, we need to examine the restoring force on the pendulum. If the restoring force is linear, the motion will be simple harmonic.

**FIGURE 14.9a** shows a mass $m$ attached to a string of length $L$ and free to swing back and forth. The pendulum's position can be described by either the arc of length $s$ or the angle $\theta$, both of which are zero when the pendulum hangs straight down. Because angles are measured counterclockwise, $s$ and $\theta$ are positive when the pendulum is to the right of center, negative when it is to the left.

**FIGURE 14.9** Describing the motion of and force on a pendulum.

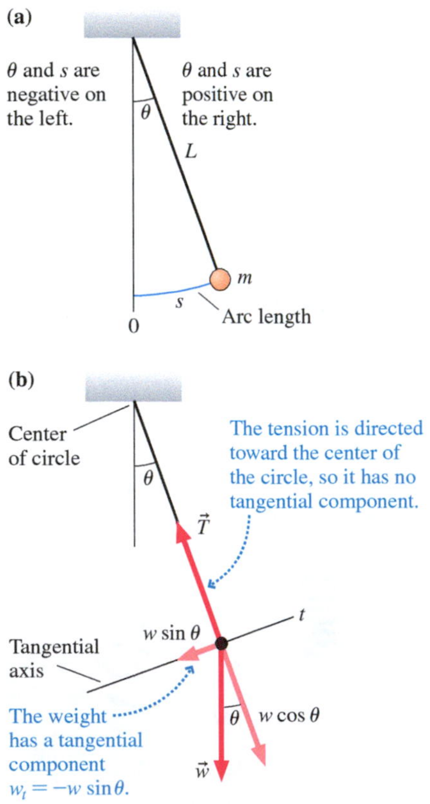


As shown on the free-body diagram of **FIGURE 14.9b,** two forces are acting on the mass: the string tension $\vec{T}$ and the weight $\vec{w}$. The motion is along a circular arc. We choose a coordinate system centered on the mass with one axis along the radius of the circle (i.e., along the string) and the other tangent to the circle. We resolve the tension and weight forces into two components, a *tangential* component parallel to the motion (and denoted with subscript $t$) and a component directed along the string toward the center of the circular arc.

The mass must move along a circular arc, as noted. As Figure 14.9b shows, the net force in this direction is the tangential component of the weight:

$$(F_{net})_t = \sum F_t = w_t = -mg\sin\theta \tag{14.8}$$

This is the restoring force pulling the mass back toward the equilibrium position.

This equation becomes much simpler if we restrict the pendulum's oscillations to *small angles* of about 10° (0.17 rad) or less. For such small angles, if the angle $\theta$ is in radians, it turns out that

$$\sin\theta \approx \theta$$

This result, which applies only when the angle is in radians, is called the **small-angle approximation.**

**NOTE** ▶ For an angle of 10°, $\sin\theta$ differs from $\theta$ by only 0.5%. But even for a 45° angle the error is only 10%. Unless stated otherwise, you can use the small-angle approximation for all problems in this text. ◀

Recall from ◀◀ **SECTION 7.1** that the angle is related to the arc length by $\theta = s/L$. Using this and the small-angle approximation, we can write the restoring force as

$$(F_{net})_t = -mg\sin\theta \approx -mg\theta = -mg\frac{s}{L} = -\left(\frac{mg}{L}\right)s \tag{14.9}$$

The net force is directed toward equilibrium, and it is linearly proportional to the displacement $s$ from equilibrium. **The force on a pendulum is a linear restoring force for small angles, so the pendulum will undergo simple harmonic motion.**

**STOP TO THINK 14.2** A ball is hung from a rope, making a pendulum. When it is pulled 5° to the side, the restoring force is 1.0 N. What will be the magnitude of the restoring force if the ball is pulled 10° to the side?

A. 0.5 N B. 1.0 N C. 1.5 N D. 2.0 N

## 14.3 Describing Simple Harmonic Motion

Now that we know what *causes* simple harmonic motion, we can develop graphical and mathematical descriptions. If we do this for one system, we can adapt the description to any system. The details will vary, but the form of the motion will stay the same. The following table shows one period of the oscillation of a mass on a vertical spring.

**KEY CONCEPT Details of oscillatory motion**

A mass is suspended from a vertical spring with an equilibrium position at $y = 0$. The mass is lifted by a distance $A$ and released.

1. The mass starts at its maximum positive displacement, $y = A$. The velocity is zero, but the acceleration is negative because there is a net downward force.
2. The mass is now moving downward, so the velocity is negative. As the mass nears equilibrium, the restoring force—and thus the magnitude of the acceleration—decreases.
3. At this time the mass is moving downward with its maximum speed. It's at the equilibrium position, so the net force—and thus the acceleration—is zero.
4. The velocity is still negative but its magnitude is decreasing, so the acceleration is positive.
5. The mass has reached the lowest point of its motion, a **turning point.** The spring is at its maximum extension, so there is a net upward force and the acceleration is positive.
6. The mass has begun moving upward; the velocity and acceleration are positive.
7. The mass is passing through the equilibrium position again, in the opposite direction, so it has a positive velocity. There is no net force, so the acceleration is zero.
8. The mass continues moving upward. The velocity is positive but its magnitude is decreasing, so the acceleration is negative.
9. The mass is now back at its starting position. This is another turning point. The mass is at rest but will soon begin moving downward, and the cycle will repeat.

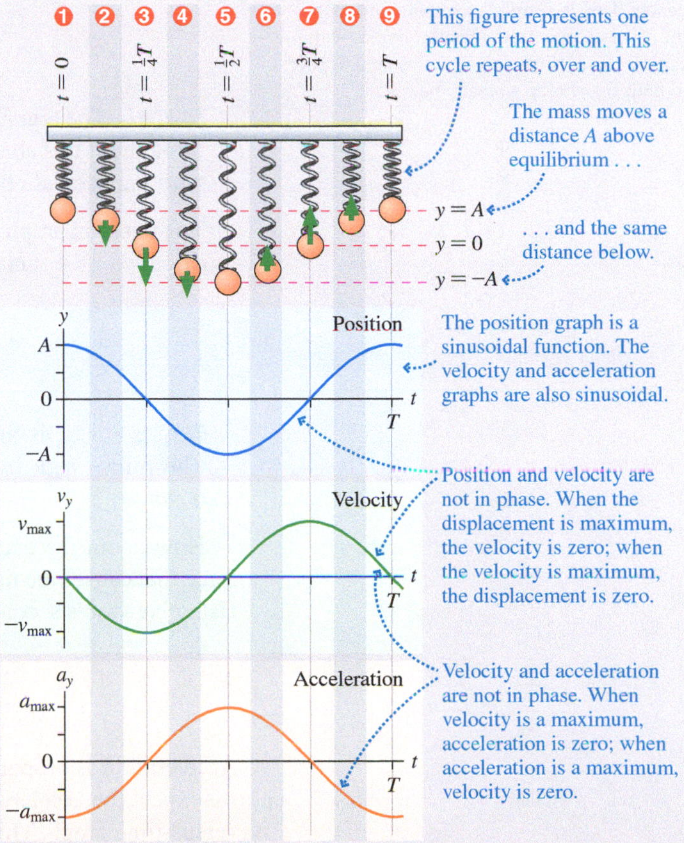


**STOP TO THINK 14.3** The graphs in this table apply to pendulum motion as well as the motion of a mass on a spring. A pendulum in a clock has a period of 2.0 seconds. You pull the pendulum to the right—a positive displacement—and let it go; we call this time $t = 0$ s. At what time will the pendulum (a) first be at its maximum negative displacement, (b) first have its maximum speed, (c) first have its maximum positive velocity, and (d) first have its maximum positive acceleration?

A. 0.5 s B. 1.0 s C. 1.5 s D. 2.0 s

**Keeping the beat** The metal rod in a metronome swings back and forth, making a loud click each time it passes through the center. This is simple harmonic motion, so the motion repeats, cycle after cycle, with the same period. Musicians use this steady click to help them keep a steady beat.

The graphs in the previous table are for an oscillation in which the object just happened to be at $y = A$ at $t = 0$. You can certainly imagine a different set of *initial conditions,* with the object at $y = -A$ or somewhere in the middle of an oscillation. This would change the starting point of the graphs but not their overall appearance. The oscillation repeats over and over, so the exact choice of starting point isn't crucial; the results we have outlined are general. Keep in mind that all the features of the motion of a mass on a spring also apply to other examples of simple harmonic motion.

The position-versus-time graph in the table on the preceding page is a cosine curve. We can write the object's position as

$$x(t) = A\cos\left(\frac{2\pi t}{T}\right) \tag{14.10}$$

where the notation $x(t)$ indicates that the position $x$ is a *function* of time $t$. Because $\cos(2\pi \text{ rad}) = \cos(0 \text{ rad})$, we see that the position at time $t = T$ is the same as the position at $t = 0$. In other words, this is a cosine function with period $T$. We can write Equation 14.10 in an alternative form. Because the oscillation frequency is $f = 1/T$, we can write

$$x(t) = A\cos(2\pi f t) \tag{14.11}$$

**NOTE** ► The argument $2\pi f t$ of the cosine function is in *radians.* That will be true throughout this chapter. Don't forget to set your calculator to radian mode before working oscillation problems. ◄

The position graph is a cosine function; the velocity graph is an upside-down sine function with the same period $T$. The velocity $v_x$ can thus be written as

$$v_x(t) = -v_{\max}\sin\left(\frac{2\pi t}{T}\right) = -v_{\max}\sin(2\pi f t) \tag{14.12}$$

**NOTE** ► $v_{\max}$ is the maximum *speed* and thus is inherently a *positive* number. The minus sign in Equation 14.12 is needed to turn the sine function upside down. ◄

How about the acceleration? As we saw earlier in Equation 14.2, the restoring force that causes the mass to oscillate with simple harmonic motion is $(F_{\text{net}})_x = -kx$. Using Newton's second law, we see that this force causes an acceleration

$$a_x = \frac{(F_{\text{net}})_x}{m} = -\frac{k}{m}x \tag{14.13}$$

Acceleration is proportional to the position $x$, but with a minus sign. Consequently we expect the acceleration-versus-time graph to be an inverted form of the position-versus-time graph. This is, indeed, what we find; the acceleration-versus-time graph presented in the earlier table is clearly an upside-down cosine function with the same period $T$. We can write the acceleration as

$$a_x(t) = -a_{\max}\cos\left(\frac{2\pi t}{T}\right) = -a_{\max}\cos(2\pi f t) \tag{14.14}$$

In this and coming chapters, we will use sine and cosine functions extensively, so we will summarize some of the key aspects of mathematical relationships using these functions.

## Sinusoidal relationships

A quantity that oscillates in time and can be written

$$x = A\sin\left(\frac{2\pi t}{T}\right)$$

or

$$x = A\cos\left(\frac{2\pi t}{T}\right)$$

is called a **sinusoidal function** with **period** $T$. The argument of the functions, $2\pi t/T$, is in radians.

The graphs of both functions have the same shape, but they have different initial values at $t = 0$ s.

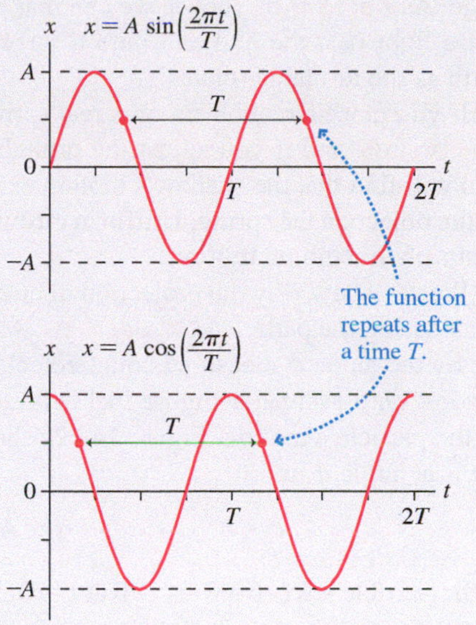


***LIMITS*** If $x$ is a sinusoidal function, then $x$ is:

- *Bounded*—it can take only values between $A$ and $-A$.
- *Periodic*—it repeats the same sequence of values over and over again. Whatever value $x$ has at time $t$, it has the same value at $t + T$.

***SPECIAL VALUES*** The function $x$ has special values at certain times:

| | $t=0$ | $t=\frac{1}{4}T$ | $t=\frac{1}{2}T$ | $t=\frac{3}{4}T$ | $t=T$ |
|---|---|---|---|---|---|
| $x = A\sin(2\pi t/T)$ | 0 | $A$ | 0 | $-A$ | 0 |
| $x = A\cos(2\pi t/T)$ | $A$ | 0 | $-A$ | 0 | $A$ |

Exercise 6

**EXAMPLE 14.2** **Understanding the motion of a glider on a spring**

An air-track glider oscillates horizontally on a spring at a frequency of 0.50 Hz. Suppose the glider is pulled to the right of its equilibrium position by 12 cm and then released. Where will the glider be 1.0 s after its release? What is its velocity at this point?

***STRATEGIZE*** The motion of the glider will be sinusoidal. It is released from rest at $x = 12$ cm, so its amplitude is 12 cm.

***PREPARE*** The frequency is 0.50 Hz, so the period is $T = 1/f = 2.0$ s. The glider is released at maximum extension from the equilibrium position, meaning that we can take this point to be $t = 0$.

***SOLVE*** 1.0 s is exactly half the period. As the graph of the motion in FIGURE 14.10 shows, half a cycle brings the glider to its left turning point, 12 cm to the left of the equilibrium position. The velocity at this point is zero.

FIGURE 14.10 Position-versus-time graph for the glider.

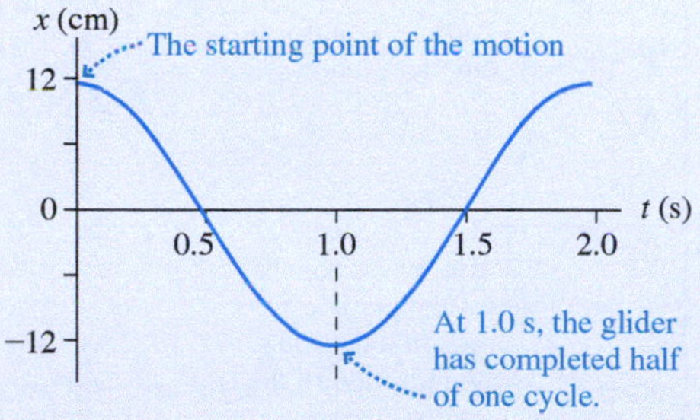


***ASSESS*** Drawing a graph was an important step that helped us make sense of the motion.

## Connecting SHM to Uniform Circular Motion

Both circular motion and simple harmonic motion are motions that repeat. Many of the concepts we have used for describing simple harmonic motion were introduced in our study of circular motion. We will use this connection to extend our knowledge of the kinematics of circular motion to simple harmonic motion.

FIGURE 14.11 Projecting the circular motion of a rotating ball.

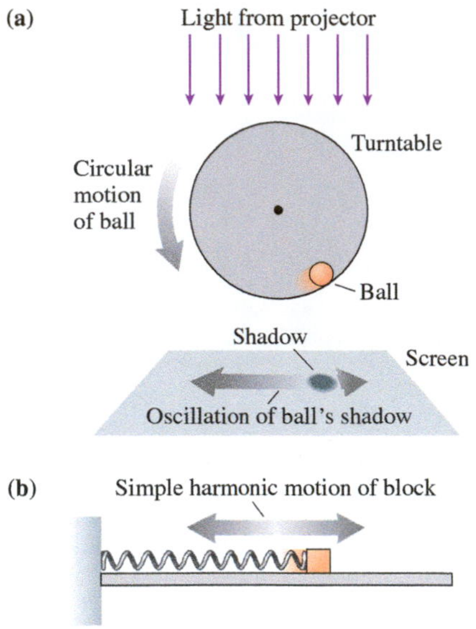


We can demonstrate the relationship between circular and simple harmonic motion with a simple experiment. Suppose a turntable has a small ball glued to the edge. As FIGURE 14.11a shows, we can make a "shadow movie" of the ball by projecting a light past the ball and onto a screen. The ball's shadow oscillates back and forth as the turntable rotates.

If you place a real object on a real spring directly below the shadow, as shown in FIGURE 14.11b, and if you adjust the turntable to have the same period as the spring, you will find that the shadow's motion exactly matches the simple harmonic motion of the object on the spring. **Uniform circular motion projected onto one dimension is simple harmonic motion.**

We can show why the projection of circular motion is simple harmonic motion by considering the particle in FIGURE 14.12a. As in ◀◀ SECTION 7.1, we can locate the particle by the angle $\phi$ measured counterclockwise from the $x$-axis. Projecting the ball's shadow onto a screen in Figure 14.11a is equivalent to observing just the $x$-component of the particle's motion. Figure 14.12a shows that the $x$-component, when the particle is at angle $\phi$, is

$$x = A\cos\phi$$

If the particle starts from $\phi_0 = 0$ at $t = 0$, its angle at a later time $t$ is

$$\phi = \omega t$$

where $\omega$ is the particle's *angular velocity,* as defined in Section 7.1. Recall that the angular velocity is related to the frequency $f$ by

$$\omega = 2\pi f$$

The particle's $x$-component can therefore be expressed as

$$x(t) = A\cos(2\pi f t) \tag{14.15}$$

This is identical to Equation 14.11 for the position of a mass on a spring! **The $x$-component of a particle in uniform circular motion is simple harmonic motion.**

FIGURE 14.12 Projections of position, velocity, and acceleration.

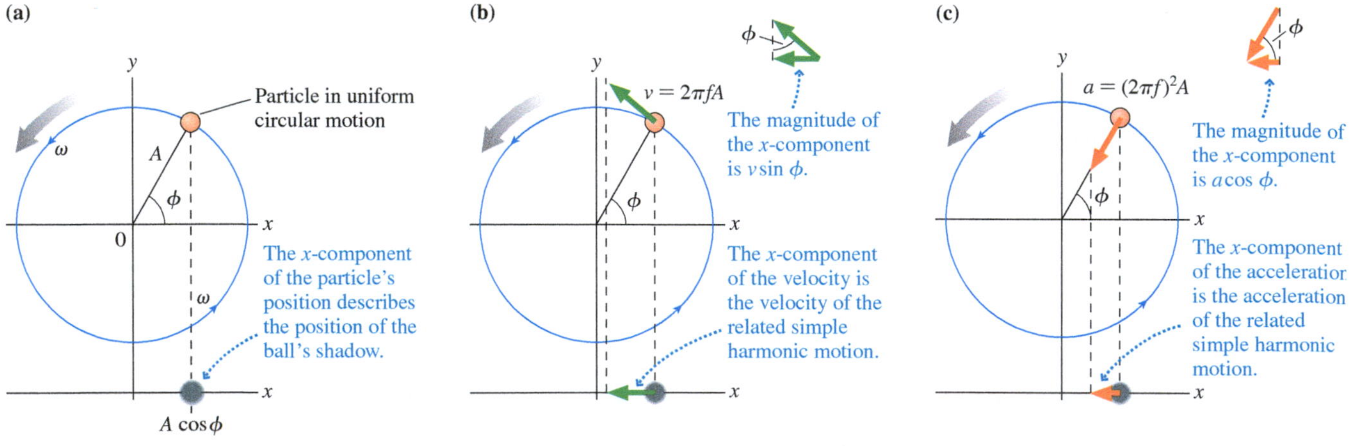


We can use this correspondence to deduce more details. FIGURE 14.12b shows the velocity vector tangent to the circle. The magnitude of the velocity vector is the particle's speed. Recall from ◀◀ SECTION 6.1 that the speed of a particle in circular motion with radius $A$ and frequency $f$ is $v = 2\pi fA$. Thus the $x$-component of the velocity vector, which is pointing in the negative $x$-direction, is

$$v_x = -v\sin\phi = -(2\pi f)A\sin(2\pi f t)$$

According to the correspondence between circular motion and simple harmonic motion, this is the velocity of an object in simple harmonic motion. This is, indeed, exactly Equation 14.12, which we deduced from the graph of the motion, if we define the maximum speed to be

$$v_{max} = 2\pi f A \tag{14.16}$$

FIGURE 14.12c shows the centripetal acceleration for the circular motion. The magnitude of the centripetal acceleration is $a = v^2/A = (2\pi f)^2 A$. The $x$-component of the acceleration vector, which is the acceleration for simple harmonic motion, is

$$a_x = -a\cos\phi = -(2\pi f)^2 A\cos(2\pi f t)$$

The maximum acceleration is thus

$$a_{max} = (2\pi f)^2 A \tag{14.17}$$

Once you establish that motion is simple harmonic, you can describe it with the equations developed above, which we summarize in Synthesis 14.1. **If you know the amplitude and the frequency, the motion is completely specified.**

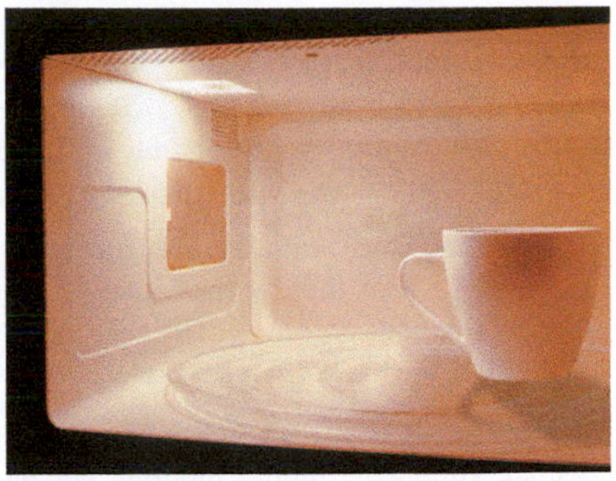

**SHM in your microwave** The next time you are warming a cup of water in a microwave oven, try this: As the turntable rotates, moving the cup in a circle, stand in front of the oven with your eyes level with the cup and watch it, paying attention to the side-to-side motion. You'll see something like the turntable demonstration. The cup's apparent motion is the horizontal component of the turntable's circular motion—simple harmonic motion!

## SYNTHESIS 14.1 Describing simple harmonic motion

The position, velocity, and acceleration of objects undergoing simple harmonic motion are sinusoidal functions whose amplitudes and phases are related in a particular way.

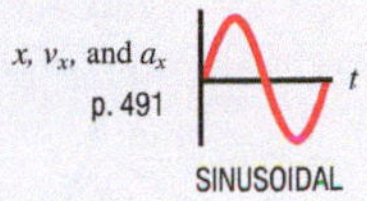


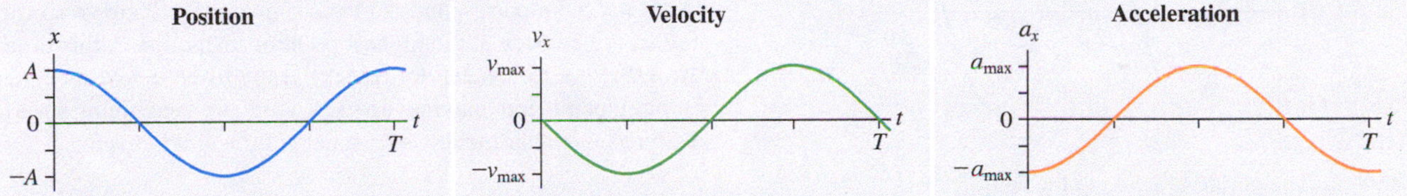


At time $t$, the displacement, velocity, and acceleration are given by:

$x(t) = A\cos(2\pi f t)$ $\quad$ $v_x(t) = -v_{max}\sin(2\pi f t)$ $\quad$ $a_x(t) = -a_{max}\cos(2\pi f t)$

The maximum values of the displacement, velocity, and acceleration are determined by the amplitude $A$ and the frequency $f$:

$x_{max} = A$ $\quad$ $v_{max} = 2\pi f A$ $\quad$ $a_{max} = (2\pi f)^2 A$

## EXAMPLE 14.3 Measuring the sway of a tall building in the wind

The John Hancock Center in Chicago is 100 stories high. Strong winds can cause the building to sway, as is the case with all tall buildings. On particularly windy days, the top of the building oscillates with an amplitude of 40 cm ($\approx$16 in) and a period of 7.7 s. What are the maximum speed and acceleration of the top of the building?

**STRATEGIZE** We will use the relationships for amplitude, velocity, and acceleration from Synthesis 14.1.

**PREPARE** We assume that the oscillation of the building is simple harmonic motion with amplitude $A = 0.40$ m. The frequency can be computed from the period:

$$f = \frac{1}{T} = \frac{1}{7.7\text{ s}} = 0.13\text{ Hz}$$

**SOLVE** We can use the equations for maximum velocity and acceleration in Synthesis 14.1 to compute:

$$v_{max} = 2\pi f A = 2\pi(0.13\text{ Hz})(0.40\text{ m}) = 0.33\text{ m/s}$$

$$a_{max} = (2\pi f)^2 A = [2\pi(0.13\text{ Hz})]^2(0.40\text{ m}) = 0.27\text{ m/s}^2$$

In terms of the free-fall acceleration, the maximum acceleration is $a_{max} = 0.027g$.

**ASSESS** The acceleration is quite small, as you would expect; if it were large, building occupants would certainly complain! Even if they don't notice the motion directly, office workers on high floors of tall buildings may experience a bit of nausea when the oscillations are large because the acceleration affects the equilibrium organs in the inner ear.

**EXAMPLE 14.4 Analyzing the motion of a hanging toy**

A classic children's toy consists of a wooden animal suspended from a spring. If you lift the toy up by 10 cm and let it go, it will gently bob up and down, completing 4 oscillations in 10 seconds.

a. What is the oscillation frequency?
b. When does the toy first reach its maximum speed, and what is this speed?
c. What are the position and velocity 4.0 s after you release the toy?

**STRATEGIZE** The toy is a mass on a vertical spring, so it undergoes simple harmonic motion. We will need to use the relationships from Synthesis 14.1, using the variable $y$ instead of $x$ because the motion here is vertical.

**PREPARE** The toy begins its motion at its maximum positive displacement of 10 cm, or 0.10 m; this is the amplitude of the motion. Because the toy is released at maximum positive displacement, we can set the moment of release as $t = 0$ s. The toy completes 4 oscillations in 10 seconds, so the period is 1/4 of 10 seconds, or $T = 2.5$ s. FIGURE 14.13 shows the first two cycles of the oscillation.

FIGURE 14.13 A position graph for the spring toy.

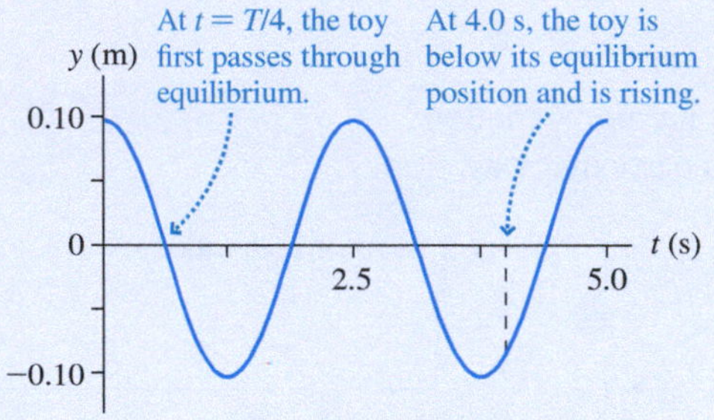


**SOLVE** a. The period is 2.5 s, so the frequency is

$$f = \frac{1}{T} = \frac{1}{2.5\ \text{s}} = 0.40\ \text{oscillation/s} = 0.40\ \text{Hz}$$

b. Figure 14.13 shows that the toy first passes through the equilibrium position at $t = T/4 = (2.5\ \text{s})/4 = 0.62\ \text{s}$. This is a point of maximum speed, which we calculate using the maximum speed equation from Synthesis 14.1:

$$v_{\text{max}} = 2\pi f A = 2\pi(0.40\ \text{Hz})(0.10\ \text{m}) = 0.25\ \text{m/s}$$

c. 4.0 s after release is $t = 4.0$ s. Figure 14.13 shows that the toy is below the equilibrium position and rising at this time, so we expect that the position is negative and the velocity is positive. We can use expressions from Synthesis 14.1 to find the position and velocity at this time:

$$\begin{aligned} y(t = 4.0\ \text{s}) &= A\cos(2\pi f t) \\ &= (0.10\ \text{m})\cos[2\pi(0.40\ \text{Hz})(4.0\ \text{s})] \\ &= -0.081\ \text{m} \end{aligned}$$

$$\begin{aligned} v_y(t = 4.0\ \text{s}) &= -v_{\text{max}}\sin(2\pi f t) \\ &= -(0.25\ \text{m/s})\sin[2\pi(0.40\ \text{Hz})(4.0\ \text{s})] \\ &= 0.15\ \text{m/s} \end{aligned}$$

*Remember that your calculator must be in radian mode to do calculations like these.* The toy is below its equilibrium position $(y < 0)$ and moving upward $(v_y > 0)$, as we expected from the graph in Figure 14.13.

**ASSESS** 5.0 s corresponds to two complete oscillations, so the toy would be back at the highest point of its motion at this time. At 4.0 s, it's reasonable to expect the toy to be below its equilibrium point and moving upward. And the maximum speed, 0.25 m/s, is rather modest—reasonable for a children's toy.

**STOP TO THINK 14.4** The figures show four identical oscillators at different points in their motion. Which is moving fastest at the time shown?

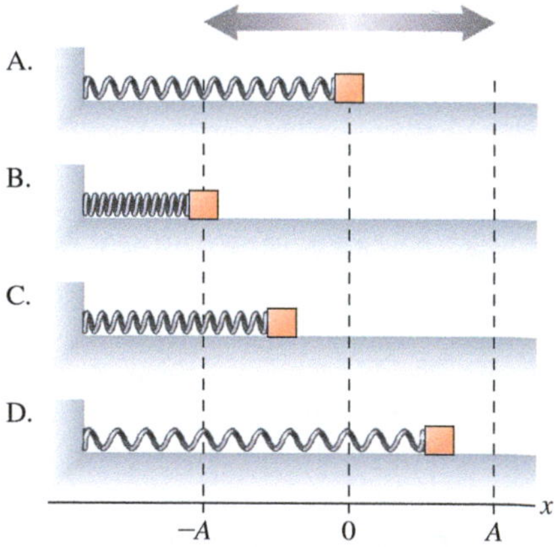

Elastic cords lead to the up-and-down motion of a bungee jump.

## 14.4 Energy in Simple Harmonic Motion

A bungee jumper falls, increasing in speed, until the elastic cords attached to his ankles start to stretch. The kinetic energy of his motion is transformed into the elastic potential energy of the cords. Once the cords reach their maximum stretch, his velocity reverses. He rises as the elastic potential energy of the cords is transformed back into kinetic energy. If he keeps bouncing (simple harmonic motion), this transformation happens again and again.

This interplay between kinetic and potential energy is very important for understanding simple harmonic motion. Let's explore these energy transformations by looking at the five diagrams in FIGURE 14.14a, which show the position and velocity of a mass on a spring at successive points in time.

The object begins at rest, with the spring at a maximum extension; here, the kinetic energy is zero and the potential energy is a maximum. As the spring contracts, the object speeds up until it reaches the center point of its oscillation, the equilibrium point. At this point, the potential energy is zero and the kinetic energy is a maximum. As the object continues to move, it slows down as it compresses the spring. Eventually, it reaches the turning point, where its instantaneous velocity is zero. At this point, the kinetic energy is zero and the potential energy is again a maximum.

Now suppose that the object has mass $m$, the spring has spring constant $k$, and that the motion takes place on a frictionless surface. You learned in ◀◀ SECTION 10.4 that the elastic potential energy of a spring stretched by a distance $x$ from its equilibrium position is

$$U = \frac{1}{2}kx^2 \tag{14.18}$$

Because the surface is frictionless there is no energy loss to thermal energy, so conservation of energy for this system can be written

$$E = K + U = \frac{1}{2}mv^2 + \frac{1}{2}kx^2 = \text{constant} \tag{14.19}$$

FIGURE 14.14b shows a graph of the potential energy, kinetic energy, and total energy for the object as it moves. You can see that, as the object goes through its motion, energy is transformed from potential to kinetic and then back to potential. At maximum displacement, with $x = \pm A$ and $v_x = 0$, the energy is purely potential, so the potential energy has its maximum value:

$$E(\text{at } x = \pm A) = U_{\text{max}} = \frac{1}{2}kA^2 \tag{14.20}$$

At $x = 0$, where $v_x = \pm v_{\text{max}}$, the energy is purely kinetic, so the kinetic energy has its maximum value:

$$E(\text{at } x = 0) = K_{\text{max}} = \frac{1}{2}m(v_{\text{max}})^2 \tag{14.21}$$

FIGURE 14.14 Energy transformations for a mass on a spring.

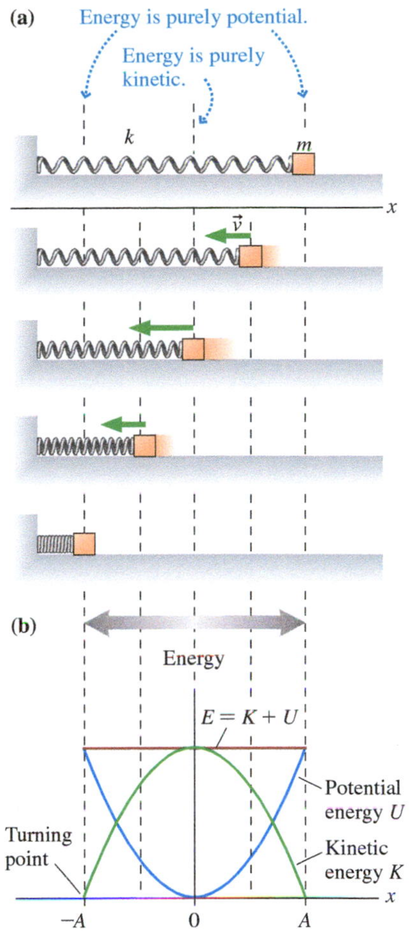


## CONCEPTUAL EXAMPLE 14.5 Energy changes for a playground swing

You are at the playground, undergoing simple harmonic motion on a swing. Describe the changes in energy that occur during one cycle of the motion, starting from when you are at the farthest forward point, motionless, and just about to swing backward.

***REASON*** The motion of the swing is that of a pendulum, with you playing the role of the pendulum bob. The energy at different points of the motion is illustrated in FIGURE 14.15. When you are motionless and at the farthest forward point, you are raised up; your energy is entirely gravitational potential energy. As you swing back, your potential energy decreases and your kinetic energy increases, reaching a maximum when the swing is at the lowest point. The swing continues to move backward; you rise up, transforming kinetic energy into potential energy. The process then reverses as you move forward.

***ASSESS*** This description matches the experience of anyone who has been on a swing. You know that you are momentarily motionless at the highest points, and that the motion is fastest at the lowest point.

FIGURE 14.15 Energy at different points of the motion of a swing.

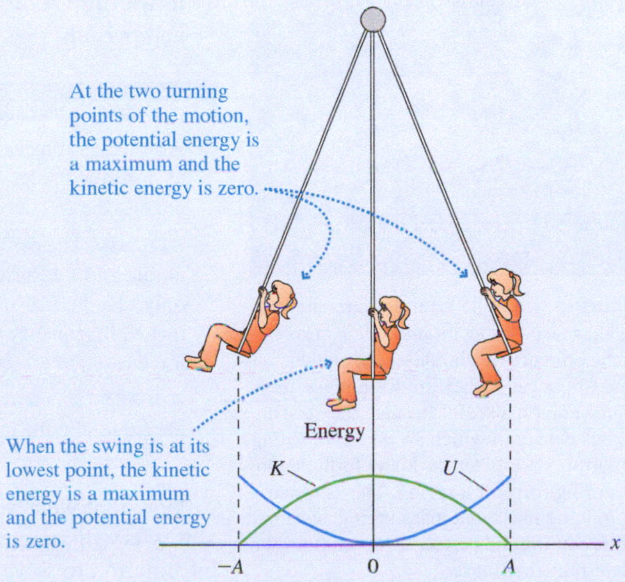

## Finding the Frequency for Simple Harmonic Motion

We can use what we know about energy in simple harmonic motion to deduce other details of the motion. Let's return to the mass on a spring of Figure 14.14. The graph in Figure 14.14b shows energy being transformed back and forth between kinetic and potential energy. At the turning points, the energy is purely potential; at the equilibrium point, the energy is purely kinetic. Because the total energy doesn't change, the maximum kinetic energy given in Equation 14.21 must be equal to the maximum potential energy given in Equation 14.20:

$$\frac{1}{2}m(v_{\text{max}})^2 = \frac{1}{2}kA^2 \qquad (14.22)$$

By solving Equation 14.22 for the maximum speed, we can see that it is related to the amplitude by

$$v_{\text{max}} = \sqrt{\frac{k}{m}}A \qquad (14.23)$$

Earlier we found that

$$v_{\text{max}} = 2\pi fA \qquad (14.24)$$

Comparing Equations 14.23 and 14.24, we see that the frequency, and thus the period, of an oscillating mass on a spring is determined by the spring constant $k$ and the object's mass $m$:

$$f = \frac{1}{2\pi}\sqrt{\frac{k}{m}} \quad \text{and} \quad T = 2\pi\sqrt{\frac{m}{k}} \qquad (14.25)$$

Frequency and period of SHM
for mass $m$ on a spring with spring constant $k$

Video Figure 14.16

FIGURE 14.16 Frequency dependence on mass and spring stiffness.

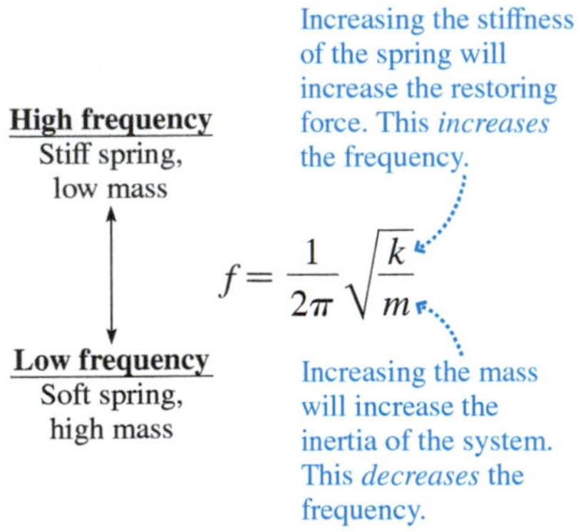

High frequency
Stiff spring, low mass
Low frequency
Soft spring, high mass
Increasing the stiffness of the spring will increase the restoring force. This increases the frequency.
Increasing the mass will increase the inertia of the system. This decreases the frequency.
f = 1/2π √(k/m)


We can make two observations about these equations:

- **The frequency and period of simple harmonic motion are determined by the physical properties of the oscillator.** The frequency and period of a mass on a spring are determined by (1) the mass and (2) the stiffness of the spring, as shown in FIGURE 14.16. This dependence of frequency and period on a force term and an inertia term will also apply to other oscillators.
- **The frequency and period of simple harmonic motion do not depend on the amplitude $A$.** A small oscillation and a large oscillation have the same frequency and period.

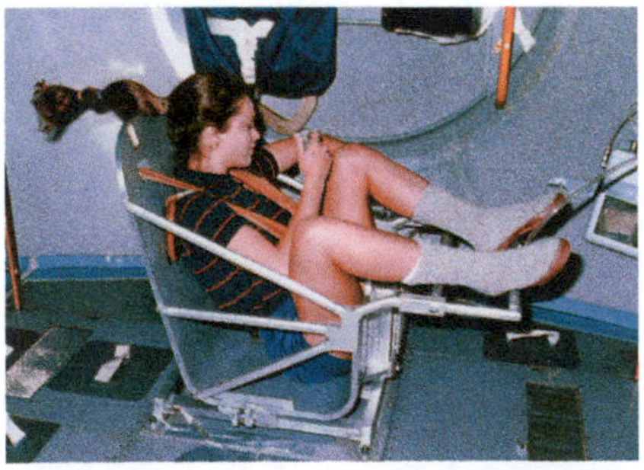

**Measuring mass in space** Astronauts on extended space flights monitor their mass to track the effects of weightlessness on their bodies. But because they are weightless, they can't just hop on a scale! Instead, they use an ingenious device in which an astronaut sitting on a platform oscillates back and forth due to the restoring force of a spring. The astronaut is the moving mass in a mass-spring system, so by measuring the period of her motion, she can determine her mass.

### CONCEPTUAL EXAMPLE 14.6 Changing mass, changing period

An astronaut measures her mass each day using the Body Mass Measurement Device, as described at left. During an 8-day flight, her mass steadily decreases. How does this change the frequency of her oscillatory motion on the device?

**REASON** The period and frequency of a mass-spring system depend on the mass of the object and the spring constant. The spring constant of the device won't change, so the only change that matters is the change in the astronaut's mass. Equation 14.25 shows that the frequency is proportional to $\sqrt{k/m}$, so a decrease in her mass will cause an increase in the frequency. The oscillation will be a bit more rapid.

**ASSESS** This makes sense. The force of the spring—which causes the oscillation—is the same, but the mass to be accelerated is less. We expect a higher frequency.

Now that we have a complete description of simple harmonic motion in one system, we will summarize the details in Tactics Box 14.1, showing how to use this information to solve oscillation problems.

## TACTICS BOX 14.1 Identifying and analyzing simple harmonic motion

1. If the net force acting on a particle is a linear restoring force, the motion is simple harmonic motion around the particle's equilibrium position.
2. The position, velocity, and acceleration as a function of time are given in Synthesis 14.1. The equations are given in terms of $x$, but they can be written in terms of $y$, $\theta$, or some other variable if the situation calls for it.
3. The amplitude $A$ is the maximum value of the displacement from equilibrium. The maximum speed and the maximum magnitude of the acceleration are given in Synthesis 14.1.
4. The frequency $f$ (and hence the period $T = 1/f$) depends on the physical properties of the particular oscillator, but $f$ does *not* depend on $A$.

   For a mass on a spring, the frequency is given by $f = \dfrac{1}{2\pi}\sqrt{\dfrac{k}{m}}$.
5. The sum of potential energy and kinetic energy is constant. As the oscillation proceeds, energy is transformed from kinetic into potential energy and then back again.

Exercises 10–11

## EXAMPLE 14.7 Finding the frequency of a mass hanging from a spring

A spring has an unstretched length of 10.0 cm. A 25 g mass is hung from the spring, stretching it to a length of 15.0 cm. If the mass is pulled down and released so that it oscillates, what will be the frequency of the oscillation?

**STRATEGIZE** The spring provides a linear restoring force, so the motion will be simple harmonic, as noted in Tactics Box 14.1. The oscillation frequency depends on the spring constant, which we will determine from the stretch of the spring.

**PREPARE** FIGURE 14.17 gives a visual overview of the situation.

**SOLVE** When the mass hangs at rest, after stretching the spring to 15 cm, the net force on it must be zero. Thus the magnitude of the upward spring force equals the downward weight, giving $k\Delta L = mg$. The spring constant is thus

$$k = \frac{mg}{\Delta L} = \frac{(0.025\text{ kg})(9.8\text{ m/s}^2)}{0.050\text{ m}} = 4.9\text{ N/m}$$

Now that we know the spring constant, we can compute the oscillation frequency:

$$f = \frac{1}{2\pi}\sqrt{\frac{k}{m}} = \frac{1}{2\pi}\sqrt{\frac{4.9\text{ N/m}}{0.025\text{ kg}}} = 2.2\text{ Hz}$$

FIGURE 14.17 Visual overview of a mass suspended from a spring.

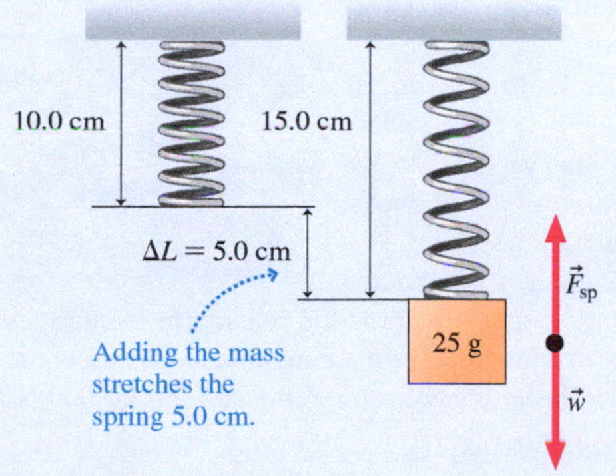


**ASSESS** 2.2 Hz is 2.2 oscillations per second. This seems like a reasonable frequency for a mass on a spring. A frequency in the kHz range (thousands of oscillations per second) would have been suspect!

## EXAMPLE 14.8 Weighing DNA molecules BIO

It has recently become possible to "weigh" individual DNA molecules by measuring the influence of their mass on a nanoscale oscillator. FIGURE 14.18 shows a thin rectangular cantilever etched out of silicon. The cantilever has a mass of $3.7 \times 10^{-16}$ kg. If pulled down and released, the end of the cantilever vibrates with simple harmonic motion, moving up and down like a diving board after a jump. When the end of the cantilever is bathed with DNA molecules whose ends have been modified to bind to a surface, one or more molecules may attach to the end of the cantilever. The addition of their mass causes a very slight—but measurable—decrease in the oscillation frequency.

FIGURE 14.18 A nanoscale cantilever.

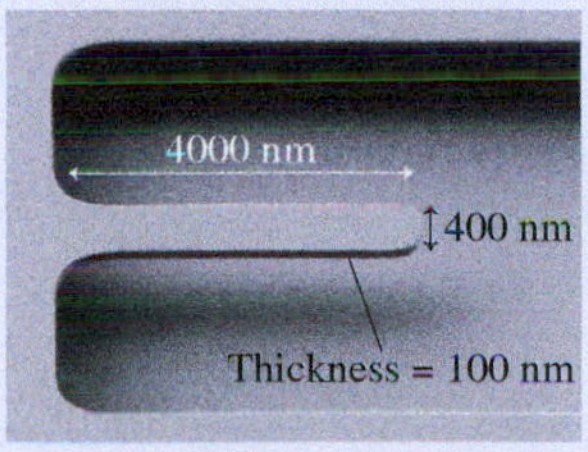


A vibrating cantilever of mass $M$ can be modeled as a simple block of mass $\frac{1}{3}M$ attached to a spring. (The factor of $\frac{1}{3}$ arises from the moment of inertia of a bar pivoted at one end: $I = \frac{1}{3}ML^2$.) Neither the mass nor the spring constant can be determined very accurately—perhaps to only two significant figures—but the

*Continued*

oscillation frequency can be measured with very high precision simply by counting the oscillations. In one experiment, the cantilever was initially vibrating at exactly 12 MHz. Attachment of a DNA molecule caused the frequency to decrease by 50 Hz. What was the mass of the DNA molecule?

**STRATEGIZE** According to Equation 14.25, the oscillation frequency of a mass on a spring depends on the mass and the spring constant. When the DNA attaches, increasing the mass, the spring constant remains constant. We will use this fact to determine the change in mass.

**PREPARE** We model the cantilever as a block of mass $m = \frac{1}{3}M = 1.2 \times 10^{-16}$ kg oscillating on a spring with spring constant $k$. When the mass increases to $m + m_{\text{DNA}}$, the oscillation frequency decreases from $f_0 = 12{,}000{,}000$ Hz to $f_1 = 11{,}999{,}950$ Hz.

**SOLVE** We can solve Equation 14.25 for the spring constant to get $k = m(2\pi f)^2$. The spring constant $k_0$ before the mass is added is equal to the spring constant $k_1$ after it's added, so we have

$$k_0 = m(2\pi f_0)^2 = k_1 = (m + m_{\text{DNA}})(2\pi f_1)^2$$

The $2\pi$ terms cancel, and we can rearrange this equation to give

$$\frac{m + m_{\text{DNA}}}{m} = 1 + \frac{m_{\text{DNA}}}{m} = \left(\frac{f_0}{f_1}\right)^2 = \left(\frac{12{,}000{,}000\text{ Hz}}{11{,}999{,}950\text{ Hz}}\right)^2 = 1.0000083$$

Subtracting 1 from both sides gives

$$\frac{m_{\text{DNA}}}{m} = 0.0000083$$

and thus

$$m_{\text{DNA}} = 0.0000083m = (0.0000083)(1.2 \times 10^{-16}\text{ kg}) = 1.0 \times 10^{-21}\text{ kg} = 1.0 \times 10^{-18}\text{ g}$$

**ASSESS** This is a reasonable mass for a DNA molecule. It's a remarkable technical achievement to be able to measure a mass this small. With further improvements in sensitivity, scientists will be able to determine the number of base pairs in a strand of DNA simply by weighing it!

## EXAMPLE 14.9 A gymnast jumps on a springboard

In the vault event, the gymnast begins her routine by jumping on a *springboard*, a platform with springs underneath. In one vault, a 45 kg gymnast jumps 0.60 m above the springboard's surface. She then lands on the springboard, which compresses it. What is the maximum compression of the springboard? How much time does it take for the board to fully compress? The spring constant of the board is 50,000 N/m.

**STRATEGIZE** We will use conservation of energy to solve this problem. At the moment the gymnast is at her highest point above the board, the system's energy is entirely her gravitational potential energy. Later, when the board is fully compressed, the energy is her (lower) gravitational potential energy plus the potential energy of the spring (we ignore the small mass of the board). By conservation of energy, these two energies are equal. We can calculate the compression time by treating the mass-spring system as an oscillator undergoing simple harmonic motion.

**PREPARE** Once the gymnast has landed on the platform, we can model the system as a mass (the gymnast) on a spring (the platform). FIGURE 14.19 gives a visual overview of the problem that we can use for conservation of energy. We can measure both the gymnast's position $y$ and the platform's compression $x$ from the platform's uncompressed position.

**SOLVE** The gymnast is instantaneously at rest at both the beginning and end of the motion, so she has no kinetic energy at those points. We also note that her final position is the same as the maximum compression of the platform—that is, $x_f$. Thus we can write conservation of energy as

$$\underbrace{mgx_f}_{(U_g)_f} + \underbrace{\tfrac{1}{2}kx_f^2}_{(U_{sp})_f} = \underbrace{mgy_i}_{(U_g)_i}$$

FIGURE 14.19 A gymnast jumps onto a springboard.

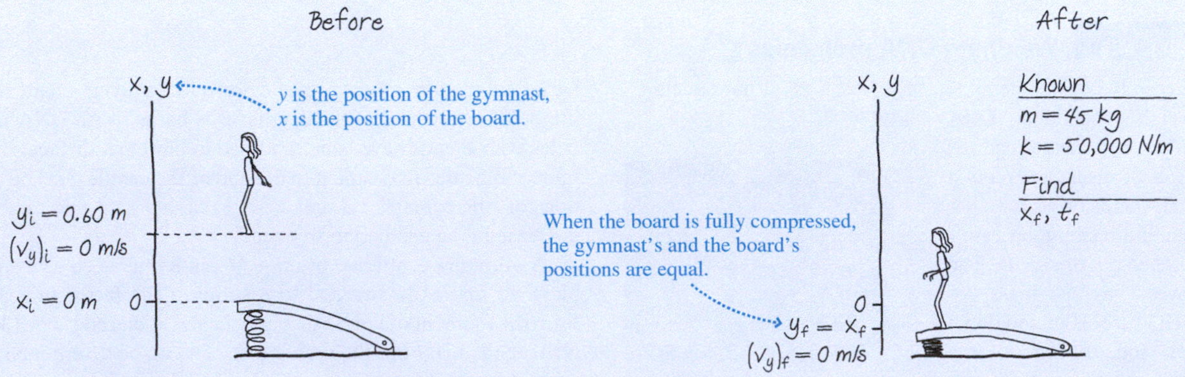

This is a quadratic equation in the variable $x_f$. Using the quadratic equation, the solution is

$$x_f = \frac{-mg \pm \sqrt{(mg)^2 + 4\left(\frac{1}{2}k\right)(mgy_i)}}{k}$$

Inserting the values $m = 45$ kg, $g = 9.8$ m/s$^2$, $k = 50{,}000$ N/m, and $y_i = 0.60$ m, we find that $x_i$ equals either $+9.4$ cm (from the " + " sign in the quadratic equation) or $-11.2$ cm (from the " − " sign). Because the board is *compressed*, so that $x_f < 0$, we know that the correct answer is $x_f = -11.2$ cm. (The positive solution corresponds to the artificial case where the gymnast is instantaneously at rest after moving back up on the springboard; this could happen only if her feet became attached to the board so that she couldn't continue upward on her own!)

To find the time for the compression to occur, we note that the compression is $\frac{1}{4}$ of a cycle of simple harmonic motion, so the time to completely compress the board occurs at $t_f = \frac{1}{4}T$. The period is computed using Equation 14.25, so we write

$$t_f = \frac{1}{4}T = \frac{1}{4}2\pi\sqrt{\frac{m}{k}} = \frac{\pi}{2}\sqrt{\frac{45\text{ kg}}{50{,}000\text{ N/m}}} = 0.047\text{ s}$$

This is the time required for the springboard to compress.

**ASSESS** The compression distance of around 11 cm seems reasonable. The time to compress the board is short, but the springs in the board are very stiff so this seems plausible. It is interesting that the time to compress the springboard—that is, the time it takes for the gymnast to stop—depends on only her mass and the spring constant; it does not depend on how fast she was moving when she landed on the board, or how far the board was compressed. This is a feature of simple harmonic motion for a mass on a spring: The period depends on only the mass and the spring constant, not on other details of the motion.

We can think of the gymnast's landing on the springboard as a *collision* between her and a spring-like, or *elastic*, object. Many real-world collisions involve a moving object that is stopped by an elastic object. In other cases (such as a bouncing rubber ball) the object itself is elastic. The time of the collision is reasonably constant, independent of the speed of the collision, for the reasons noted in Example 14.9.

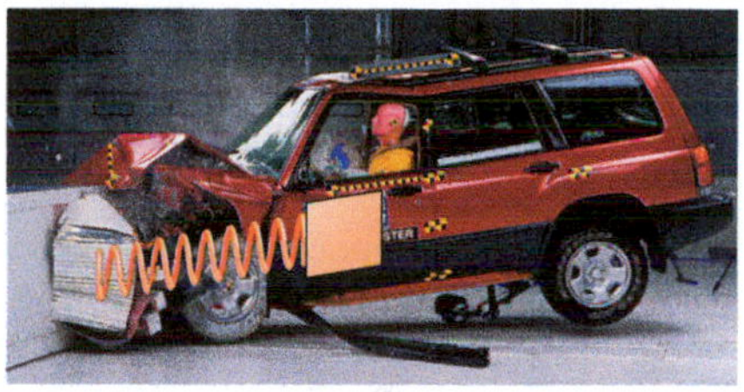

**Automobile collision times** When a car hits an obstacle, it takes approximately 0.1 s to come to rest, regardless of the initial speed. The crumpling of the front of a car is complex, but for many cars the force is approximately proportional to the displacement during the compression, so we can model the body of the car as a mass and the front of the car as a spring. In this model, the time for the car to come to rest does not depend on the initial speed.

**STOP TO THINK 14.5** Four mass-spring systems have masses and spring constants shown here. Rank in order, from highest to lowest, the frequencies of the oscillations.

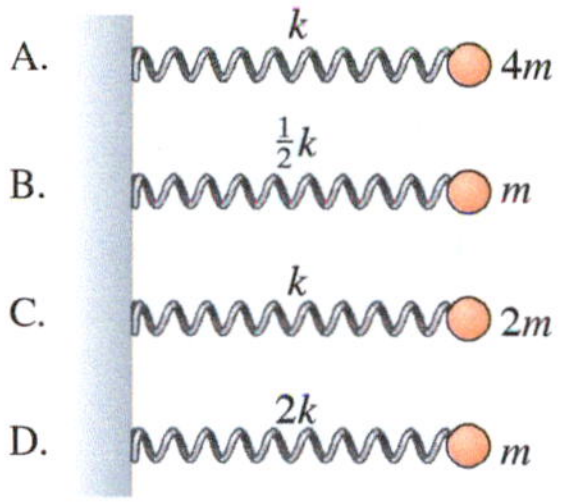


## 14.5 Pendulum Motion

As we've already seen, a simple pendulum—a mass at the end of a string or a rod that is free to pivot—is another system that exhibits simple harmonic motion. Everything we have learned about the mass on a spring can be applied to the pendulum as well.

In the first part of the chapter, we looked at the restoring force in the pendulum. For a pendulum of length $L$ displaced by an arc length $s$, as in FIGURE 14.20, the tangential restoring force is

$$(F_{\text{net}})_t = -\frac{mg}{L}s \tag{14.26}$$

**NOTE** ▸ Recall that this equation holds accurately only for small angles. ◂

This linear restoring force has exactly the same form as the net force in a mass-spring system, but with the constants $mg/L$ in place of the constant $k$. Given this, we can quickly deduce the essential features of pendulum motion by replacing $k$, wherever it occurs in the oscillating spring equations, with $mg/L$.

**FIGURE 14.20** A simple pendulum.

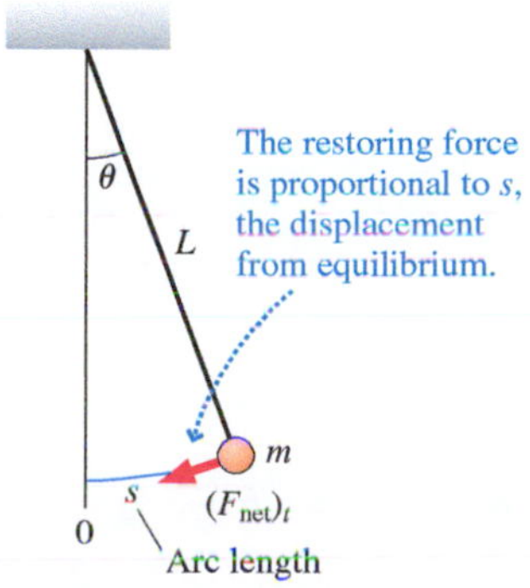

- The oscillation of a pendulum is simple harmonic motion, so the equations of motion can be written for the arc length or the angle:

$$s(t) = A\cos(2\pi ft) \qquad \text{or} \qquad \theta(t) = \theta_{max}\cos(2\pi ft)$$

- The frequency can be obtained from the equation for the frequency of the mass on a spring by substituting $mg/L$ in place of $k$:

$$f = \frac{1}{2\pi}\sqrt{\frac{g}{L}} \quad \text{and} \quad T = 2\pi\sqrt{\frac{L}{g}} \tag{14.27}$$

Frequency and period of a pendulum of length $L$ with free-fall acceleration $g$

- As for a mass on a spring, the frequency does not depend on the amplitude. Note also that **the frequency, and hence the period, is independent of the mass.** It depends only on the length of the pendulum.
- Equation 14.16 showed that the maximum speed for a particle in simple harmonic motion is $v_{max} = 2\pi fA$. Using the frequency $f$ from Equation 14.27, we find that the maximum speed of a pendulum bob is

$$v_{max} = \sqrt{\frac{g}{L}}A = \sqrt{gL}\,\theta_{max} \tag{14.28}$$

The dependence of the pendulum frequency on length and gravity is summed up in FIGURE 14.21. Adjusting the period of a pendulum is a matter of adjusting the length, which is a mechanical operation that can be performed with great precision. Clocks based on the period of a pendulum were the most accurate timepieces available until well into the 20th century.

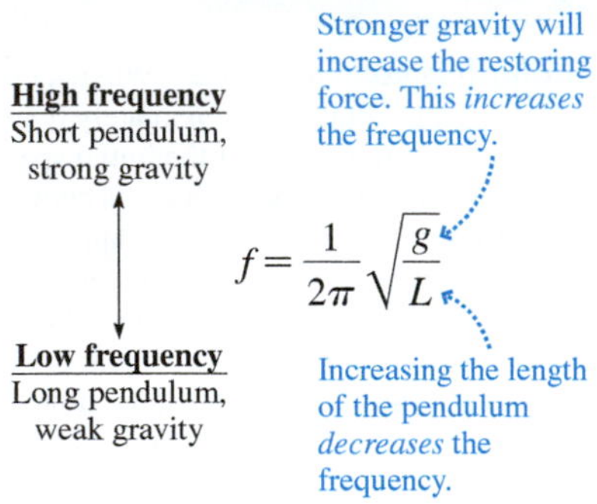


FIGURE 14.21 Frequency dependence on length and gravity for a pendulum.

**EXAMPLE 14.10 Designing a pendulum for a clock**

A grandfather clock is designed so that one swing of the pendulum in either direction takes 1.00 s. What is the length of the pendulum?

**STRATEGIZE** The period of a pendulum depends only on its length, according to Equation 14.27.

**PREPARE** One period of the pendulum is two swings, so the period is $T = 2.00$ s.

**SOLVE** The period is independent of the mass and depends only on the length. From Equation 14.27,

$$T = \frac{1}{f} = 2\pi\sqrt{\frac{L}{g}}$$

Solving for $L$, we find

$$L = g\left(\frac{T}{2\pi}\right)^2 = (9.80\ \text{m/s}^2)\left(\frac{2.00\ \text{s}}{2\pi}\right)^2 = 0.993\ \text{m}$$

**ASSESS** A pendulum clock with a "tick" or "tock" each second requires a long pendulum of about 1 m—which is why these clocks were originally known as "tall case clocks."

## Physical Pendulums and Locomotion

In Chapter 6, we computed maximum walking speed using the ideas of circular motion. We can also model the motion of your legs during walking as pendulum motion. When you walk, you push off with your rear leg and then let it swing forward for the next stride. Your leg swings forward under the influence of gravity—like a pendulum.

Try this: Stand on one leg, and gently swing your free leg back and forth. There is a certain frequency at which it will naturally swing. This is your leg's pendulum frequency. Now, try swinging your leg at twice this frequency. You can do it, but it is very difficult. The muscles that move your leg back and forth aren't very strong because under normal circumstances they don't need to apply much force.

A pendulum, like your leg, whose mass is distributed along its length is known as a **physical pendulum.** FIGURE 14.22 shows a simple pendulum and a physical pendulum of the same length. The position of the center of gravity of the physical pendulum is at a distance $d$ from the pivot.

Finding the frequency is a rotational motion problem similar to those we considered in Chapter 7. We would expect the frequency to depend on the moment of inertia and the distance to the center of gravity as follows:

- The moment of inertia $I$ is a measure of an object's resistance to rotation. Increasing the moment of inertia while keeping other variables equal should cause the frequency to decrease. In an expression for the frequency of the physical pendulum, we would expect $I$ to appear in the denominator.
- When the pendulum is pushed to the side, a gravitational torque pulls it back. The greater the distance $d$ of the center of gravity from the pivot point, the greater the torque. Increasing this distance while keeping the other variables constant should cause the frequency to increase. In an expression for the frequency of the physical pendulum, we would expect $d$ to appear in the numerator.

A careful analysis of the motion of the physical pendulum produces a result for the frequency that matches these expectations:

$$f = \frac{1}{2\pi}\sqrt{\frac{mgd}{I}} \tag{14.29}$$

Frequency of a physical pendulum of mass $m$, moment of inertia $I$, with center of gravity distance $d$ from the pivot

**FIGURE 14.22** A simple pendulum and a physical pendulum of equal length.

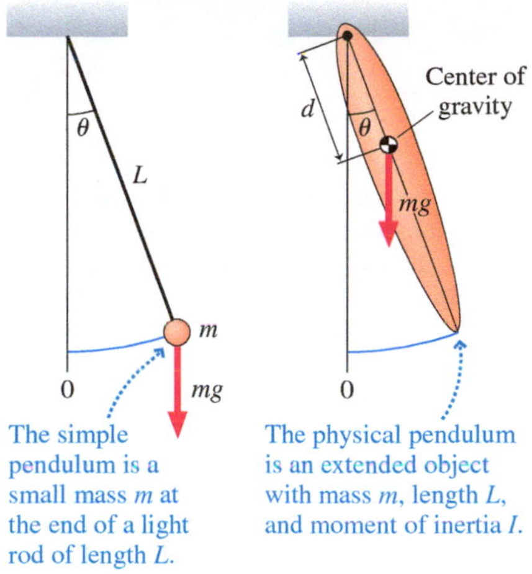


**EXAMPLE 14.11 Finding the frequency of a swinging leg** BIO

A student in a biomechanics lab measures the length of his leg, from hip to heel, to be 0.90 m. What is the frequency of the pendulum motion of the student's leg? What is the period?

**STRATEGIZE** Equation 14.29 for the period of a physical pendulum contains the center-of-gravity distance $d$ and the moment of inertia $I$. We'll need to make some simplifying assumptions to estimate these lengths.

**PREPARE** We can model a human leg reasonably well as a rod of uniform cross section, pivoted at one end (the hip). Recall from Chapter 7 that the moment of inertia of a rod pivoted about its end is $\frac{1}{3}mL^2$. The center of gravity of a uniform leg is at the midpoint, so $d = L/2$.

**SOLVE** The frequency of a physical pendulum is given by Equation 14.29. Before we put in numbers, we will use symbolic relationships and simplify:

$$f = \frac{1}{2\pi}\sqrt{\frac{mgd}{I}} = \frac{1}{2\pi}\sqrt{\frac{mg(L/2)}{\frac{1}{3}mL^2}} = \frac{1}{2\pi}\sqrt{\frac{3}{2}\frac{g}{L}}$$

The expression for the frequency is similar to that for the simple pendulum, but with an additional numerical factor of 3/2 inside the square root. The numerical value of the frequency is

$$f = \frac{1}{2\pi}\sqrt{\left(\frac{3}{2}\right)\left(\frac{9.8\ \text{m/s}^2}{0.90\ \text{m}}\right)} = 0.64\ \text{Hz}$$

The period is

$$T = \frac{1}{f} = 1.6\ \text{s}$$

**ASSESS** Notice that we didn't need to know the mass of the leg to find the period. The period of a physical pendulum does not depend on the mass, just as it doesn't for the simple pendulum. The period depends only on the *distribution* of mass. When you walk, swinging your free leg forward to take another stride corresponds to half a period of this pendulum motion. For a period of 1.6 s, this is 0.80 s. For a normal walking pace, one stride in just under one second sounds about right.

As you walk, your legs swing as physical pendulums as you bring them forward. The frequency is fixed by the length of your legs and their distribution of mass; it doesn't depend on amplitude. Consequently, you don't increase your walking speed by taking more rapid steps—changing the frequency is quite difficult. You simply

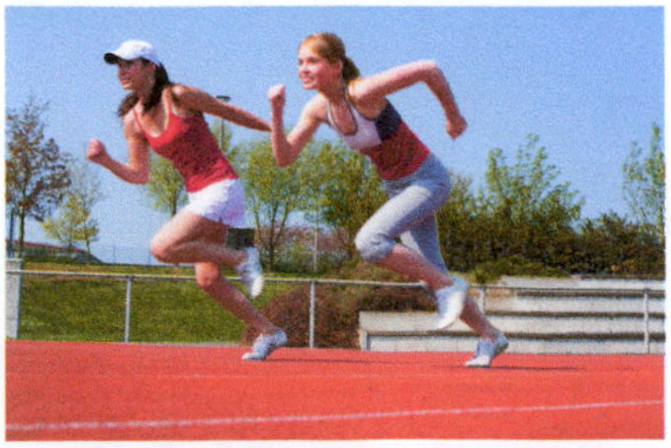

**How do you hold your arms?** BIO You maintain your balance when walking or running by moving your arms back and forth opposite the motion of your legs. You hold your arms so that the natural period of their pendulum motion matches that of your legs. At a normal walking pace, your arms are extended and naturally swing at the same period as your legs. When you run, your gait is more rapid. To decrease the period of the pendulum motion of your arms to match, you bend them at the elbows, shortening their effective length and increasing the natural frequency of oscillation.

take longer strides, changing the amplitude but not the frequency. At some point a limit is reached, and you use another gait—you start to run.

As we saw in the photo that opens the chapter, gibbons move through the forest canopy with a hand-over-hand swinging motion called *brachiation*. The body swings under a pivot point where a hand grips a tree branch, a clear example of pendulum motion. A brachiating ape will increase its speed by taking bigger swings. But, just as for walking, at some point a limit is reached, and gibbons use a different gait, launching themselves from branch to branch through the air.

**STOP TO THINK 14.6** A pendulum clock is made with a metal rod. It keeps perfect time at a temperature of 20°C. At a higher temperature, the metal rod lengthens. How will this change the clock's timekeeping?

A. The clock will run fast; the dial will be ahead of the actual time.
B. The clock will keep perfect time.
C. The clock will run slow; the dial will be behind the actual time.

## 14.6 Damped Oscillations

A real pendulum clock must have some energy input; otherwise, the oscillation of the pendulum would slowly decrease in amplitude due to air resistance. If you strike a bell, the oscillation will soon die away as energy is lost to sound waves in the air and dissipative forces within the metal of the bell.

Unless energy is continually added to an oscillator, its amplitude will decrease—sometimes very slowly, but other times quite quickly—as its mechanical energy is transformed into the thermal energy of the oscillator and its environment. An oscillation that runs down and stops is called a **damped oscillation.**

For a pendulum, the main energy loss is due to air resistance, which we called the *drag force* in Chapter 4. When we learned about the drag force, we noted that it depends on velocity: The faster the motion, the bigger the drag force. For this reason, the decrease in amplitude of an oscillating pendulum will be fastest at the start of the motion. For a pendulum or other oscillator with modest damping, we end up with a graph of motion like that in FIGURE 14.23a. The maximum displacement, $x_{max}$, decreases with time. As the oscillation decays, the *rate* of the decay decreases; the difference between successive peaks is less.

If we plot a smooth curve that connects the peaks of successive oscillations (we call such a curve an *envelope*), we get the dashed line shown in FIGURE 14.23b. It's possible, using calculus, to show that $x_{max}$ decreases with time as

$$x_{max}(t) = Ae^{-t/\tau} \tag{14.30}$$

where $e \approx 2.718$ is the base of the natural logarithm and $A$ is the *initial* amplitude. This steady decrease of $x_{max}$ with time is called an **exponential decay.**

The constant $\tau$ (lowercase Greek tau) in Equation 14.30 is called the **time constant.** After one time constant has elapsed—that is, at $t = \tau$—the maximum displacement $x_{max}$ has decreased to

$$x_{max}(\text{at } t = \tau) = Ae^{-1} = \frac{A}{e} \approx 0.37A$$

In other words, the oscillation has decreased after one time constant to about 37% of its initial value. The time constant $\tau$ measures the "characteristic time" during which damping causes the amplitude of the oscillation to decay away. An oscillation that decays quickly has a small time constant, whereas a "lightly damped" oscillator, which decays very slowly, has a large time constant.

FIGURE 14.23 The motion of a damped oscillator.

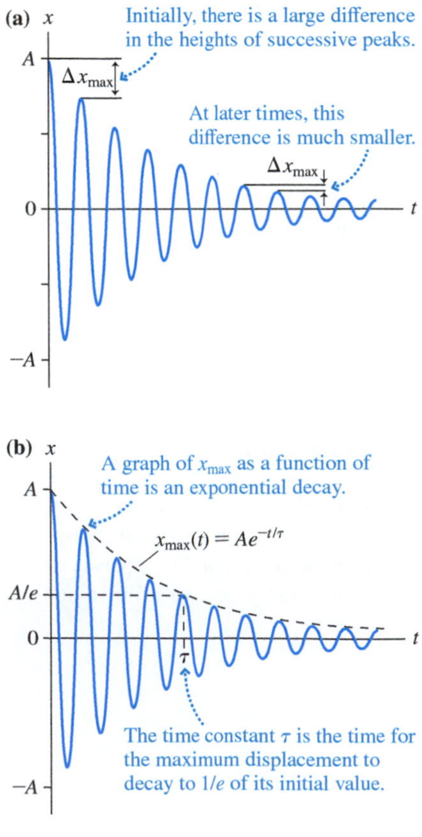

(a) x
Initially, there is a large difference in the heights of successive peaks.
A
Δx_max
At later times, this difference is much smaller.
Δx_max
0
t
−A
(b) x
A graph of x_max as a function of time is an exponential decay.
A
x_max(t) = Ae^(−t/τ)
A/e
0
τ
t
The time constant τ is the time for the maximum displacement to decay to 1/e of its initial value.
−A

Because we will see exponential decay again, we will look at it in more detail.

## Exponential decay

**Exponential decay** occurs when a quantity $y$ is proportional to the number $e$ taken to the power $-t/\tau$. The quantity $\tau$ is known as the **time constant.** We write this mathematically as

$$y = Ae^{-t/\tau}$$

$y$ is proportional to $e^{-t/\tau}$

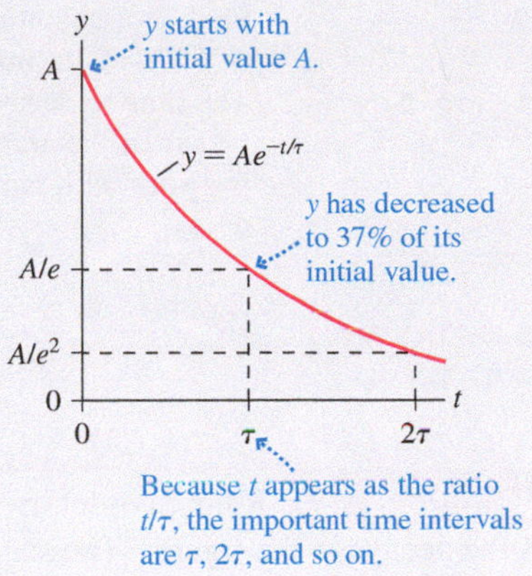


***SCALING*** Whenever $t$ increases by one time constant, $y$ decreases by a factor of $1/e$. For instance:

- At time $t = 0$, $y = A$.
- Increasing time to $t = \tau$ reduces $y$ to $A/e$.
- A further increase to $t = 2\tau$ reduces $y$ by another factor of $1/e$ to $A/e^2$.

Generally, we can say:

$$\text{At } t = n\tau, y \text{ has the value } A/e^n.$$

***LIMITS*** As $t$ becomes large, $y$ becomes very small and approaches zero.

Exercises 14–17

The damped oscillation shown in Figure 14.23 continues for a long time. The amplitude isn't zero after one time constant, or two, or three.... Mathematically, the oscillation never ceases, though the amplitude will eventually be so small as to be undetectable. For practical purposes, we can speak of the time constant $\tau$ as the *lifetime* of an oscillation—a measure of about how long it takes to decay. The best way to measure the relative size of the time constant is to compare it to the period. If $\tau \gg T$, the oscillation persists for many, many periods and the amplitude decrease from cycle to cycle is quite small. The oscillation of a bell after it is struck has $\tau \gg T$; the sound continues for a long time. Other oscillatory systems have very short time constants, as noted in the description of a car's suspension.

**Damping smoothes the ride** A car's wheels are attached to the car's body with springs so that the wheels can move up and down as the car moves over an uneven road. The car-spring system oscillates with a typical period of just under 1 second. You don't want the car to continue bouncing after hitting a bump, so a shock absorber provides damping. The time constant for the damping is about the same length as the period, so that the oscillation damps quickly.

### EXAMPLE 14.12 Finding a clock's decay time

The pendulum in a grandfather clock has a period of 1.00 s. If the clock's driving spring is allowed to run down, damping due to friction will cause the pendulum to slow to a stop. If the time constant for this decay is 300 s, how long will it take for the pendulum's swing to be reduced to half its initial amplitude?

***STRATEGIZE*** The time constant of 300 s is much greater than the 1.00 s period, so this is an example of modest damping as described by Equation 14.30. We are asked "how long," which implies that we want to find a time $t$. Because $t$ is in the exponent in Equation 14.30, we'll need to use logarithms to find it.

***PREPARE*** We need to know the time constant of 300 s, but the 1.00 s period of the pendulum is not needed.

***SOLVE*** Equation 14.30 gives an expression for the decay of the maximum displacement of a damped harmonic oscillator:

$$x_{max}(t) = Ae^{-t/\tau}$$

As noted in Tactics Box 14.1, we can write this equation equally well in terms of the pendulum's angle in the form

$$\theta_{max}(t) = \theta_i e^{-t/\tau}$$

where $\theta_i$ is the initial angle of swing. We wish to find the time $t$ at which the amplitude has decayed to half its initial value. At this time,

$$\theta_{max}(t) = \theta_i e^{-t/\tau} = \frac{1}{2}\theta_i$$

*Continued*

The $\theta_i$ cancels, giving $e^{-t/\tau} = \frac{1}{2}$.

To solve this for $t$, we take the natural logarithm of both sides and use the logarithm property $\ln(e^a) = a$:

$$\ln(e^{-t/\tau}) = -\frac{t}{\tau} = \ln\left(\frac{1}{2}\right) = -\ln 2$$

In the last step we used the property $\ln(1/b) = -\ln b$. Now we can solve for $t$:

$$t = \tau \ln 2$$

The time constant was specified as $\tau = 300$ s, so

$$t = (300 \text{ s})(0.693) = 208 \text{ s}$$

It will take 208 s, or about 3.5 min, for the oscillations to decay by half after the spring has run down.

**ASSESS** The time is less than the time constant, which makes sense. The time constant is the time for the amplitude to decay to 37% of its initial value; we are looking for the time to decay to 50% of its initial value, which should be a shorter time. The time to decay to $\frac{1}{2}$ of the initial amplitude, $t = \tau \ln 2$, could be called the *half-life*. We will see this expression again when we work with radioactivity, another example of an exponential decay.

**STOP TO THINK 14.7** Rank in order, from largest to smallest, the time constants $\tau_A$ to $\tau_D$ of the decays in the figures. The scales on all the graphs are the same.

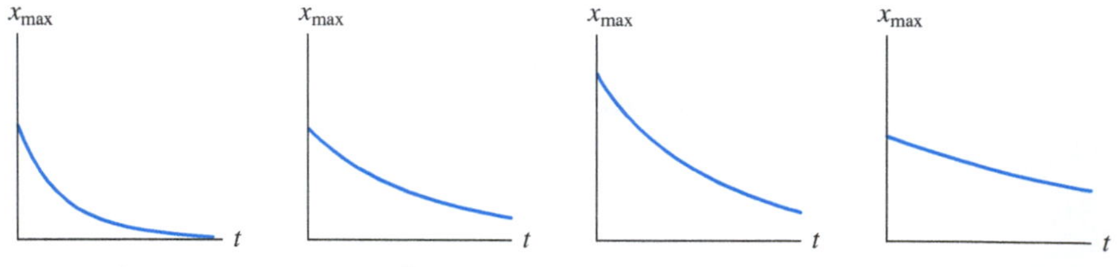


## 14.7 Driven Oscillations and Resonance

If you jiggle a cup of water, the water sloshes back and forth. This is an example of an oscillator (the water in the cup) subjected to a periodic external force (from your hand). This motion is called a **driven oscillation.**

We can give many examples of driven oscillations. The electromagnetic coil on the back of a loudspeaker cone provides a periodic magnetic force to drive the cone back and forth, causing it to send out sound waves. Earthquakes cause the surface of the earth to move back and forth; this motion causes buildings to oscillate, possibly producing damage or collapse.

Consider an oscillating system that, when left to itself, oscillates at a frequency $f_0$. We will call this the **natural frequency** of the oscillator. $f_0$ is simply the frequency of the system if it is displaced from equilibrium and released.

◀ **Serious sloshing** Water in Canada's Bay of Fundy oscillates back and forth with a period of 12 hours—nearly equal to the period of the moon's tidal force, which gives two daily high tides 12.5 hours apart. This *resonance*, a close match between the bay's natural frequency and the moon's driving frequency, produces a huge tidal amplitude. Low tide can be 16 m below high tide, leaving boats high and dry.

Suppose that this system is now subjected to a *periodic* external force of frequency $f_{ext}$. This frequency, which is called the **driving frequency,** is completely independent of the oscillator's natural frequency $f_0$. Somebody or something in the environment selects the frequency $f_{ext}$ of the external force, causing the force to push on the system $f_{ext}$ times every second. The external force causes the oscillation of the system, so it will oscillate at $f_{ext}$, the driving frequency, not at its natural frequency $f_0$.

Let's return to the example of the cup of water. If you nudge the cup, you will notice that the water sloshes back and forth at a particular frequency; this is the natural frequency $f_0$. Now shake the cup at some frequency; this is the driving frequency $f_{ext}$. As you shake the cup, the oscillation amplitude of the water depends very sensitively on the frequency $f_{ext}$ of your hand. If the driving frequency is near the natural frequency of the system, the oscillation amplitude may become so large that water splashes out of the cup.

Any driven oscillator will show a similar dependence of amplitude on the driving frequency. Suppose a mass on a spring has a natural frequency $f_0 = 2$ Hz. We can use an external force to push and pull on the mass at frequency $f_{ext}$, measure the amplitude of the resulting oscillation, and then repeat this over and over for many different driving frequencies. A graph of amplitude versus driving frequency, such as the one in FIGURE 14.24, is called the oscillator's **response curve.**

In Figure 14.24, for frequencies substantially lower or higher than the oscillator's natural frequency, the system oscillates, but its amplitude is very small. The system simply does not respond well to a driving frequency that differs much from $f_0$. As the driving frequency gets closer and closer to the natural frequency, the amplitude of the oscillation rises dramatically. After all, $f_0$ is the frequency at which the system "wants" to oscillate, so it is quite happy to respond to a driving frequency near $f_0$. Hence the amplitude reaches a maximum when the driving frequency matches the system's natural frequency: $f_{ext} = f_0$. This large-amplitude response to a driving force whose frequency matches the natural frequency of the system is a phenomenon called **resonance.** Within the context of driven oscillations, the natural frequency $f_0$ is often called the **resonance frequency.**

The amplitude can become exceedingly large when the frequencies match, especially if there is very little damping. FIGURE 14.25 shows the response curve of the oscillator of Figure 14.24 with different amounts of damping. Three different graphs are plotted, each with a different time constant for damping. The three graphs have damping that ranges from $\tau = 50T$ (very little damping) to $\tau = 2T$ (significant damping).

▶ **Simple harmonic music** A wine glass has a natural frequency of oscillation and a very small amount of damping. A tap on the rim of the glass causes it to "ring" like a bell. If you moisten your finger and slide it gently around the rim of the glass, it will stick and slip in quick succession. With some practice you can match the stick-slip to the frequency of oscillation of the glass. The resulting resonance can lead to a very loud sound. Adding water changes the frequency, so you can tune a set of glasses to make an unusual musical instrument.

FIGURE 14.24 The response curve shows the amplitude of a driven oscillator at frequencies near its natural frequency $f_0 = 2$ Hz.

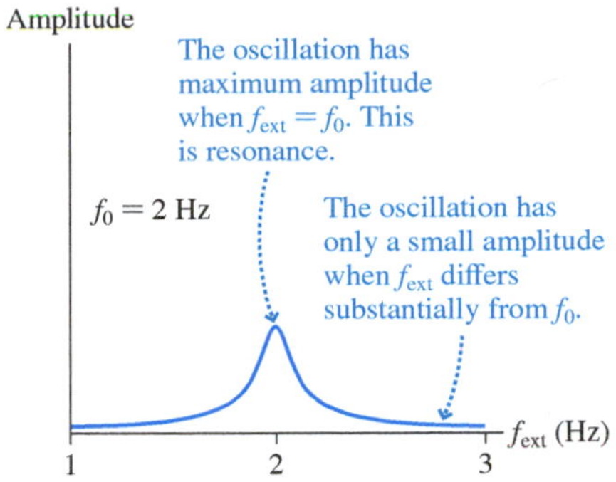


FIGURE 14.25 The response curve becomes taller and narrower as the damping is reduced.

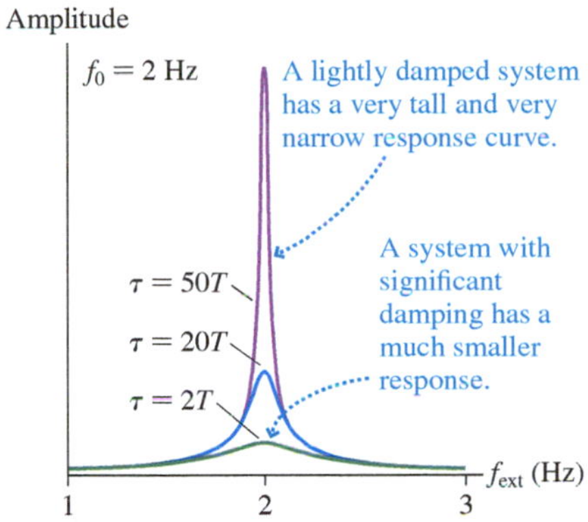

**Video** Pendulum Resonance

**CONCEPTUAL EXAMPLE 14.13** **Fixing an unwanted resonance in a railroad car**

Railroad cars have a natural frequency at which they rock side to side. This can lead to problems on certain stretches of track that have bumps where the rails join. If the joints alternate sides, with a bump on the left rail and then on the right, a train car moving down the track is bumped one way and then the other. In some cases, bumps have caused rocking with amplitude large enough to derail the train. A train moving down the track at a certain speed is experiencing a large amplitude of oscillation due to alternating joints in the track. How can the driver correct this potentially dangerous situation?

***REASON*** The large amplitude of oscillation is produced by a resonance, a match between the frequency at which the train car rocks back and forth and the frequency at which the car hits the bumps. To eliminate this resonance, the driver must either reduce the speed of the train—decreasing the driving frequency—or increase the speed of the train—thus increasing the driving frequency.

***ASSESS*** It's perhaps surprising that increasing the speed of the train could produce a smoother ride. But increasing the frequency at which the train hits the bumps will eliminate the match with the natural rocking frequency just as surely as decreasing the speed.

## Resonance and Hearing BIO

Resonance in a system means that certain frequencies produce a large response and others do not. The phenomenon of resonance is responsible for the frequency discrimination of the ear.

As we will see in the next chapter, sound is a vibration in air. FIGURE 14.26 provides an overview of the structures by which sound waves that enter the ear produce vibrations in the cochlea, the coiled, fluid-filled, sound-sensing organ of the inner ear.

FIGURE 14.26 The structures of the ear.

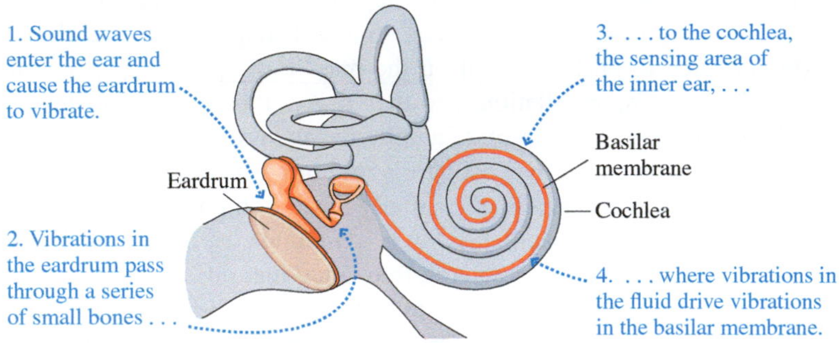

1. Sound waves enter the ear and cause the eardrum to vibrate.
2. Vibrations in the eardrum pass through a series of small bones . . .
Eardrum
3. . . . to the cochlea, the sensing area of the inner ear, . . .
Basilar membrane
Cochlea
4. . . . where vibrations in the fluid drive vibrations in the basilar membrane.


FIGURE 14.27 Resonance plays a role in determining the frequencies of sounds we hear.

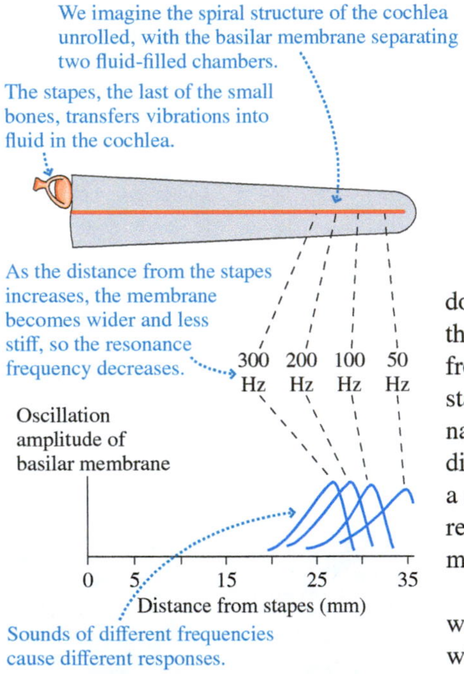


FIGURE 14.27 shows a very simplified model of the cochlea. As a sound wave travels down the cochlea, it causes a large-amplitude vibration of the basilar membrane at the point where the membrane's natural oscillation frequency matches the sound frequency—a resonance. Lower-frequency sound causes a response farther from the stapes. Sensitive hair cells on the membrane sense the vibration and send nerve signals to your brain. The fact that different frequencies produce maximal response at different positions allows your brain to very accurately determine frequency because a small shift in frequency causes a detectable change in the position of the maximal response. People with no musical training can listen to two notes and easily determine which is at a higher pitch.

We now know a bit about how your ear responds to the vibration of a sound wave—but how does this vibration get from a source to your ear? This is a topic we will consider in the next chapter, when we look at waves, oscillations that travel.

### INTEGRATED EXAMPLE 14.14 Analyzing the motion of a rope swing

A rope hangs down from a high tree branch at the edge of a river. Josh, who has mass 65 kg, trots at 2.5 m/s to the edge of the river, grabs the rope 7.2 m below where it is tied to the branch, and swings out over the river.

a. What is the minimum time that Josh must hang on to make it back to shore?
b. What is the maximum tension in the rope?

**STRATEGIZE** This is pendulum motion, with Josh as the mass at the end of a 7.2-m-long rope. The time for him to swing out and return to shore is one-half the pendulum's period, which we will find from Equation 14.27.

Once Josh starts swinging, he moves in a circular arc, so he will have a centripetal acceleration. This acceleration is caused by the tension and the weight force acting on him. We will use the ideas of circular motion dynamics to relate the tension to his speed.

**PREPARE** Josh grabs the rope, swings out, and comes back to shore, as shown in FIGURE 14.28. As he swings out, his speed decreases until he reaches the greatest distance from shore. When he returns to shore, he has the same speed as when he left. As we noted, the time for this motion is half the oscillation period of the pendulum.

FIGURE 14.28 Pendulum motion of a person swinging on a rope.

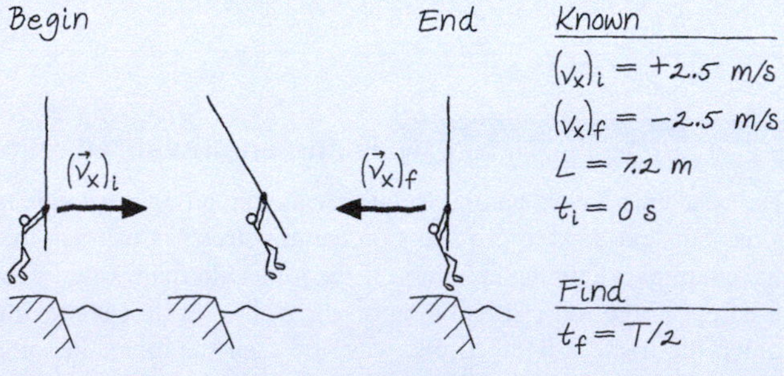

Once Josh starts swinging in a circular arc, the force required to produce his centripetal acceleration is the difference between the tension in the rope and the radial component of Josh's weight. The tension will be largest when the centripetal acceleration is largest, which is at the lowest point of the swing, where his speed is the greatest. In FIGURE 14.29 we have prepared a visual overview, including a free-body diagram, showing Josh as he returns to this lowest point. Notice that we use the symbol $T_s$ for the tension in the rope so as not to confuse it with the symbol $T$ for period.

FIGURE 14.29 A circular motion view gives insight into forces.

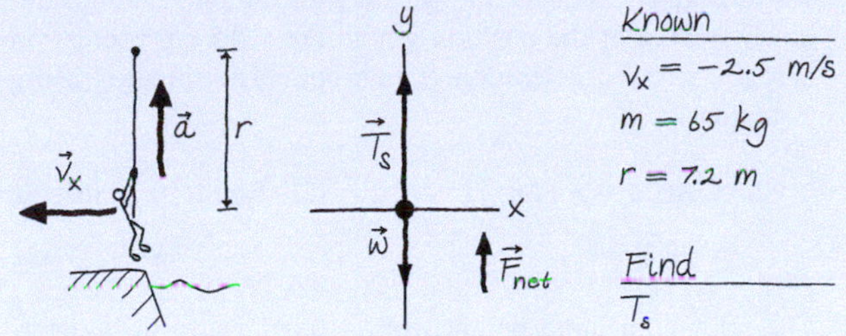


***SOLVE*** a. The period for the pendulum motion is

$$T = 2\pi\sqrt{\frac{L}{g}} = 2\pi\sqrt{\frac{7.2\text{ m}}{9.8\text{ m/s}^2}} = 5.4\text{ s}$$

As shown in Figure 14.28, Josh must hold on until $t_f = T/2 = 2.7$ s.

b. At the lowest point of the motion, where Josh's speed is the highest, the centripetal acceleration is directed upward along the $y$-axis and is (from Equation 3.20)

$$a_y = \frac{v^2}{r} = \frac{(2.5\text{ m/s})^2}{7.2\text{ m}} = 0.87\text{ m/s}^2$$

As outlined in Figure 14.29, Newton's second law for motion along the $y$-axis is

$$\Sigma F_y = (T_s)_y + w_y = T_s - mg = ma_y$$

We know the acceleration, so we can solve for the magnitude of the tension:

$$T_s = mg + ma_y = (65\text{ kg})(9.8\text{ m/s}^2 + 0.87\text{ m/s}^2) = 690\text{ N}$$

***ASSESS*** Both of our answers seem reasonable. A time of 2.4 s to swing out and back sounds about right. And the speed is small and the rope is long, so the magnitude of the centripetal acceleration is modest. We thus expect the tension in the rope to be only slightly larger than Josh's weight, just as we found.

## INTEGRATED EXAMPLE 14.15 Springboard diving

Flexible diving boards designed for large deflections are called springboards. If a diver jumps up and lands on the end of the board, the resulting deflection of the diving board produces a linear restoring force that launches him into the air. But if the diver simply bobs up and down on the end of the board, we can effectively model his motion as that of a mass oscillating on a spring.

A light and flexible springboard deflects by 15 cm when a 65 kg diver stands on its end. He then jumps and lands on the end of the board, depressing it by a total of 25 cm, after which he moves up and down with the oscillations of the end of the board.

a. What is the frequency of the oscillation?
b. What is the maximum speed of his up-and-down motion?

Suppose the diver then drives the motion of the board with his legs, gradually increasing the amplitude of the oscillation. At some point the oscillation becomes large enough that his feet leave the board.

c. What is the amplitude of the oscillation when the diver just becomes airborne at one point of the cycle? What is the acceleration at this point?
d. A diver leaving a springboard can achieve a much greater height than a diver jumping from a fixed platform. Use energy concepts to explain how the spring of the board allows a greater vertical jump.

***STRATEGIZE*** We will model the diver on the board as a mass on a spring; we will consider the spring itself to be massless. As we've seen, the oscillation frequency is determined by the spring constant and the mass. The mass of the diver is given; we will determine the spring constant from the deflection of the springboard when the diver stands on the end.

As the diver oscillates up and down, the board will never rise above its undeflected position. To see this, imagine pushing down on a vertical spring, then slowly letting your hand rise. The spring will follow your hand until the spring reaches its undeflected position; at this point it cannot follow your hand further unless you actively pull on it.

***PREPARE*** FIGURE 14.30 is a sketch of the oscillation that will help us visualize the motion. The equilibrium position, with the diver

FIGURE 14.30 Position-versus-time graph for the springboard.

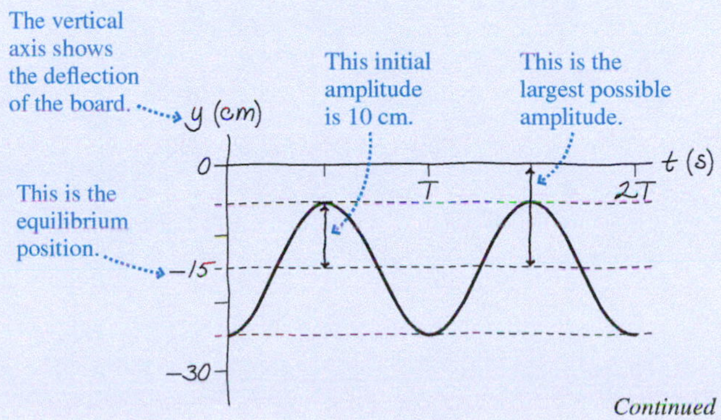


Continued

standing motionless on the end of the board, corresponds to a deflection of 15 cm. When the diver jumps on the board, the total deflection is 25 cm, which means a deflection of an additional 10 cm beyond the equilibrium position, so 10 cm is the amplitude of the subsequent oscillation. When the board rises, it won't bend beyond its undeflected position, so the maximum possible amplitude is 15 cm.

**NOTE** ▸ We've started the graph at the lowest point of the motion. Because we'll use only the equations for the maximum values of the speed and the acceleration, not the full equations that describe the motion, the exact starting point isn't critical. ◂

***SOLVE***

a. **FIGURE 14.31** shows the forces on the diver as he stands motionless at the end of the board. The net force on him is zero, $\vec{F}_{net} = \vec{F}_{sp} + \vec{w} = \vec{0}$, so the two forces have equal magnitudes and we can write

$$F_{sp} = w$$

A linear restoring force means that the board obeys Hooke's law for a spring: $F_{sp} = k\,\Delta y$, where $k$ is the spring constant. Thus the equilibrium equation is

$$k\,\Delta y = mg$$

Solving for the spring constant, we find

$$k = \frac{mg}{\Delta y} = \frac{(65\text{ kg})(9.8\text{ m/s}^2)}{0.15\text{ m}} = 4.25 \times 10^3\text{ N/m}$$

**FIGURE 14.31** Forces on a springboard diver at rest at the end of the board.

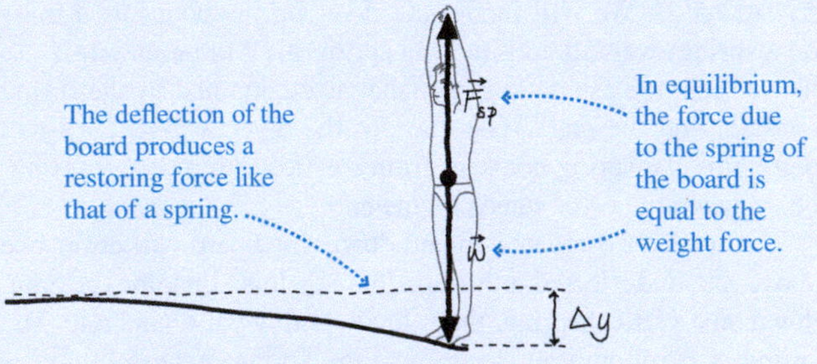


The frequency of the oscillation depends on the diver's mass and the spring constant of the board:

$$f = \frac{1}{2\pi}\sqrt{\frac{k}{m}} = \frac{1}{2\pi}\sqrt{\frac{4.25 \times 10^3\text{ N/m}}{65\text{ kg}}} = 1.29\text{ Hz}$$

b. The maximum speed of the oscillation is given by Equation 14.16:

$$v_{max} = 2\pi f A = 2\pi(1.29\text{ Hz})(0.10\text{ m}) = 0.81\text{ m/s}$$

c. We can see from Figure 14.30 that an amplitude of 15 cm returns the board to its undeflected position. At this point, the board exerts no upward force—no supporting normal force—on the diver, so the diver loses contact with the board. (His apparent weight becomes zero.) The acceleration at this point in the motion has the maximum possible magnitude but is negative because the acceleration graph is an upside-down version of the position graph. For a 15 cm oscillation amplitude, the acceleration at this point is computed using Equation 14.17:

$$a = -a_{max} = -(2\pi f)^2 A = [2\pi(1.29\text{ Hz})]^2(0.15\text{ m})$$
$$= -9.8\text{ m/s}^2$$

d. The maximum jump height from a fixed platform is determined by the maximum speed at which a jumper leaves the ground. During a jump, chemical energy in the muscles is transformed into kinetic energy. This kinetic energy is transformed into potential energy as the jumper rises, back into kinetic energy as he falls, then into thermal energy as he hits the ground. The springboard recaptures and stores this kinetic energy rather than letting it degrade to thermal energy; a diver can jump up once and land on the board, storing the energy of his initial jump as elastic potential energy of the bending board. Then, as the board rebounds, turning the stored energy back into kinetic energy, the diver can push off from this moving platform, transforming even more chemical energy and thus further increasing his kinetic energy. This allows the diver to get "two jumps worth" of chemical energy in a single jump.

***ASSESS*** We learned in earlier chapters that an object loses contact with a surface—like a car coming off the track in a loop-the-loop—when its apparent weight becomes zero: $w_{app} = 0$. And in Chapter 5 we found that the apparent weight of an object in vertical motion becomes zero when $a = -g$, that is, when it enters free fall. This correspondence is a good check on our work. Because part c has the answer we expect, we have confidence in our earlier steps.

# SUMMARY

**GOAL** To understand systems that oscillate with simple harmonic motion.

## GENERAL PRINCIPLES

### Frequency and Period

SHM occurs when a **linear restoring force** acts to return a system to an equilibrium position. The frequency and period depend on the details of the oscillator, but **the frequency and period do not depend on the amplitude.**

**Mass on spring**

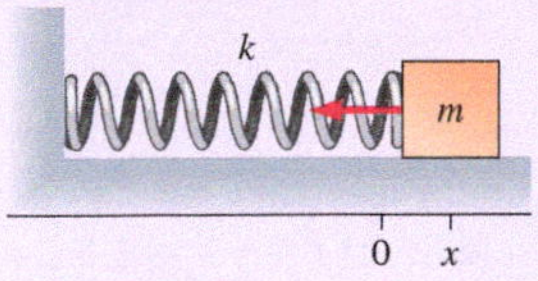


$(F_{net})_x = -kx$

The frequency and period of a mass on a spring depend on the mass and the spring constant: They are the same for horizontal and vertical systems.

$$f = \frac{1}{2\pi}\sqrt{\frac{k}{m}} \qquad T = 2\pi\sqrt{\frac{m}{k}}$$

**Pendulum**

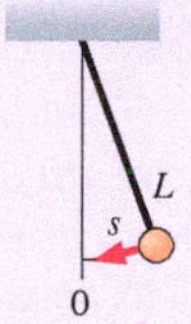

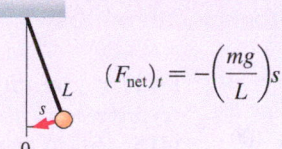

$(F_{net})_t = -\left(\frac{mg}{L}\right)s$

The frequency and period of a pendulum depend on the length and the free-fall acceleration. They do not depend on the mass.

$$f = \frac{1}{2\pi}\sqrt{\frac{g}{L}} \qquad T = 2\pi\sqrt{\frac{L}{g}}$$

### Energy

If there is no friction or dissipation, kinetic and potential energies are alternately transformed into each other in SHM, with the sum of the two conserved.

$$E = \frac{1}{2}mv_x^2 + \frac{1}{2}kx^2 = \frac{1}{2}m{v_{max}}^2 = \frac{1}{2}kA^2$$

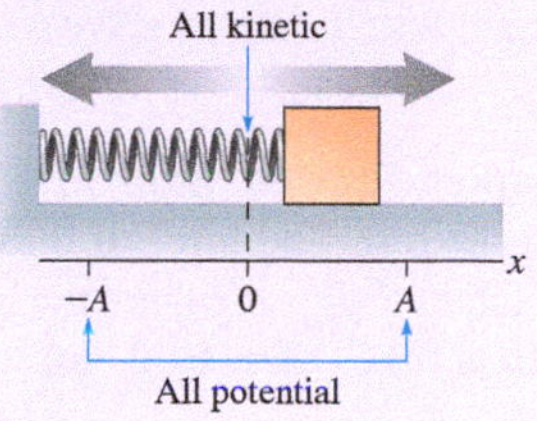


## IMPORTANT CONCEPTS

### Oscillation

An **oscillation** is a repetitive motion about an equilibrium position. The **amplitude** $A$ is the maximum displacement from equilibrium. The period $T$ is the time for one cycle. We may also characterize an oscillation by its frequency $f$.

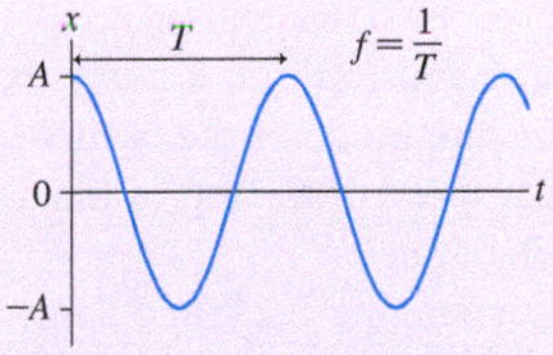


### Simple Harmonic Motion (SHM)

SHM is an oscillation that is described by a sinusoidal function. All systems that undergo SHM can be described by the same functional forms.

**Position-versus-time** is a cosine function.

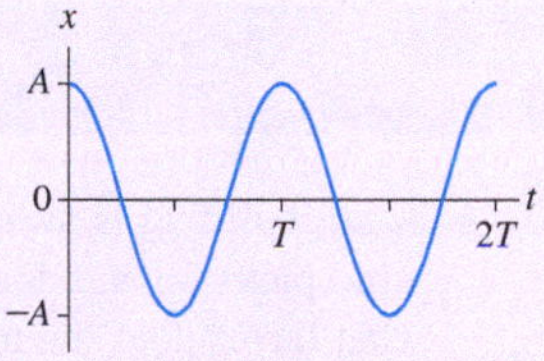


$x(t) = A\cos(2\pi ft)$

**Velocity-versus-time** is an inverted sine function.

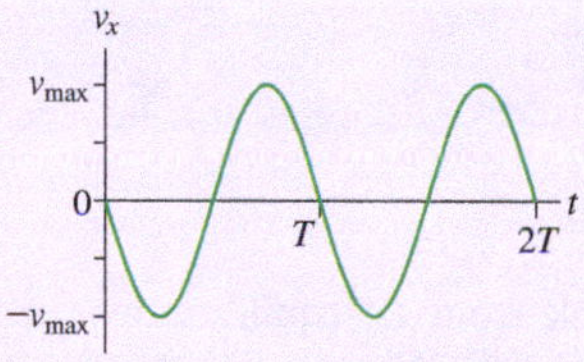


$v_x(t) = -[(2\pi f)A]\sin(2\pi ft)$

**Acceleration-versus-time** is an inverted cosine function.

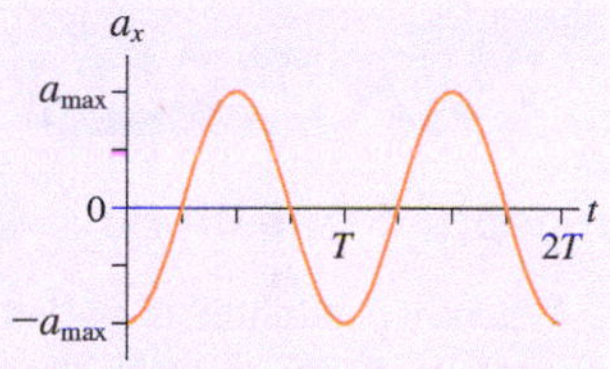


$a_x(t) = -[(2\pi f)^2 A]\cos(2\pi ft)$

## APPLICATIONS

### Damping

Simple harmonic motion with damping (due to drag) decreases in amplitude over time. The **time constant** $\tau$ determines how quickly the amplitude decays.

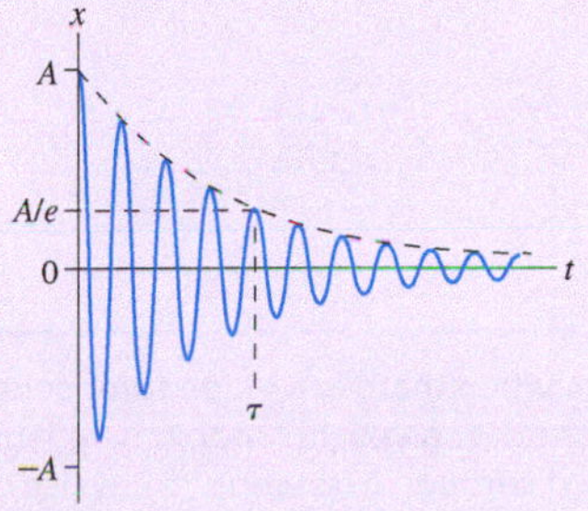


### Resonance

A system that oscillates has a **natural frequency** of oscillation $f_0$. **Resonance** occurs if the system is driven with a frequency $f_{ext}$ that matches this natural frequency. This may produce a large amplitude of oscillation.

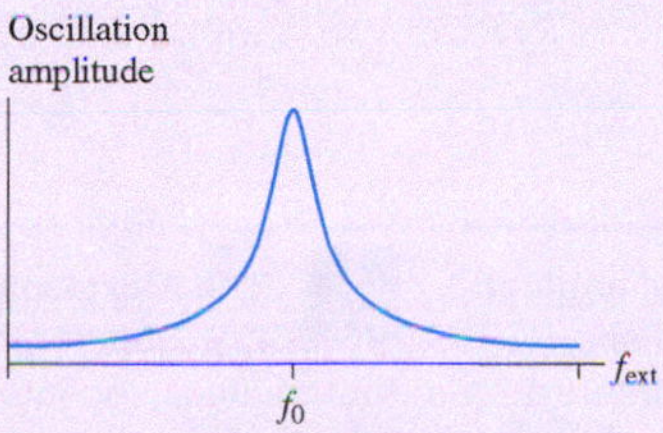


### Physical pendulum

A **physical pendulum** is a pendulum with mass distributed along its length. The frequency depends on the position of the center of gravity and the moment of inertia.

The motion of legs during walking can be described using a physical pendulum model.

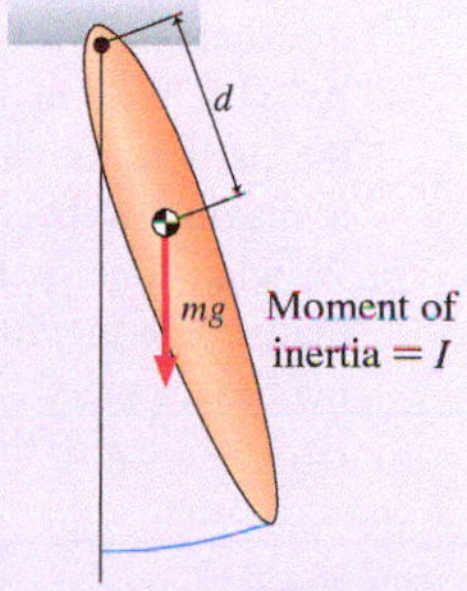


$$f = \frac{1}{2\pi}\sqrt{\frac{mgd}{I}}$$

## Learning Objectives After studying this chapter, you should be able to:

- Work with the period, frequency, and amplitude of oscillatory motion. *Conceptual Questions 14.1, 14.2; Problems 14.1, 14.2, 14.3, 14.4, 14.5*
- Use the graphical and mathematical representations of simple harmonic motion. *Conceptual Questions 14.8, 14.11; Problems 14.6, 14.7, 14.8, 14.12, 14.18*
- Apply energy conservation to simple harmonic motion. *Conceptual Questions 14.6, 14.7; Problems 14.19, 14.21, 14.23, 14.24, 14.26*
- Solve problems about simple and physical pendulums. *Conceptual Questions 14.14, 14.16; Problems 14.27, 14.28, 14.30, 14.33, 14.38*
- Use the concepts of damping and resonance. *Conceptual Questions 14.20, 14.21; Problems 14.39, 14.40, 14.41, 14.44, 14.47*

## STOP TO THINK ANSWERS

**Chapter Preview Stop to Think: A.** If the weight doubles, so will the restoring force of the spring. This also means doubling the stretch of the spring—not the length, but the *change* in length. 100 g stretches the spring by 2 cm, so 200 g will stretch it by 4 cm.

**Stop to Think 14.1: C.** The frequency is inversely proportional to the period, so a shorter period implies a higher frequency.

**Stop to Think 14.2: D.** The restoring force is proportional to the displacement. If the displacement from equilibrium is doubled, the force is doubled as well.

**Stop to Think 14.3: B, A, C, B.** The motion of the pendulum exactly matches the graphs in the table. The pendulum starts at rest with a maximum positive displacement and then undergoes simple harmonic motion. By looking at the graphs, we can see that the pendulum (a) first reaches its maximum negative displacement at $t = \frac{1}{2}T = 1.0$ s, (b) first reaches its maximum speed at $t = \frac{1}{4}T = 0.5$ s, (c) first reaches its maximum positive velocity at $t = \frac{3}{4}T = 1.5$ s, and (d) first has its maximum positive acceleration at $t = \frac{1}{2}T = 1.0$ s.

**Stop to Think 14.4: A.** The maximum speed occurs when the mass passes through its equilibrium position.

**Stop to Think 14.5: $f_D > f_C = f_B > f_A$.** The frequency is determined by the ratio of $k$ to $m$.

**Stop to Think 14.6: C.** The increase in length will cause the frequency to decrease and thus the period will increase. The time between ticks will increase, so the clock will run slow.

**Stop to Think 14.7: $\tau_D > \tau_B = \tau_C > \tau_A$.** The time constant is the time to decay to 37% of the initial height. The time constant is independent of the initial height.

Video Tutor Solution Chapter 14

# QUESTIONS

## Conceptual Questions

1. When a pendulum is pulled back from its equilibrium position by 10°, the restoring force is 1.0 N. When it is pulled back to 30°, the force increases to 2.8 N. If the pendulum is then released from 30°, will its motion be periodic? Will its motion be simple harmonic? Explain.
2. A person's heart rate is given in beats per minute. Is this a period or a frequency?
3. Figure Q14.3 shows the position-versus-time graph of a particle in SHM.
   a. At what time or times is the particle moving right at maximum speed?
   b. At what time or times is the particle moving left at maximum speed?
   c. At what time or times is the particle instantaneously at rest?

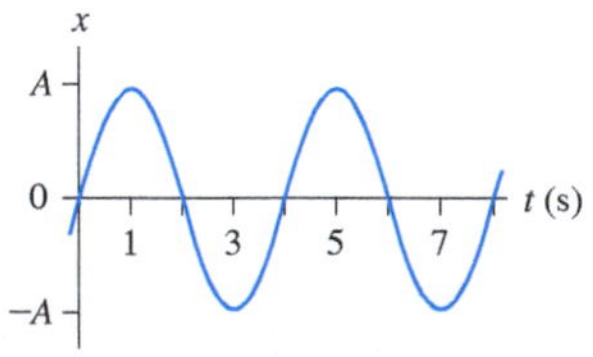

FIGURE Q14.3

4. A tall building is swaying back and forth on a gusty day. The wind picks up and doubles the amplitude of the oscillation. By what factor does the maximum speed of the top of the building increase? The maximum acceleration?
5. A child is on a swing, gently swinging back and forth with a maximum angle of 5°. A friend gives her a small push so that she now swings with a maximum angle of 10°. By what factor does this increase her maximum speed?
6. A block oscillating on a spring has an amplitude of 20 cm. What will be the amplitude if the maximum kinetic energy is doubled?
7. A block oscillating on a spring has a maximum kinetic energy of 2.0 J. What will be the maximum kinetic energy if the amplitude is doubled? Explain.

Problem difficulty is labeled as | (straightforward) to ||||| (challenging). Problems labeled INT integrate significant material from earlier chapters; Problems labeled BIO are of biological or medical interest.

The eText icon indicates when there is a video tutor solution available for the chapter or for a specific problem. To launch these videos, log into your eText through Mastering™ Physics or log into the Study Area.

8. A mass hanging from a spring undergoes vertical simple harmonic motion.
   a. Where in the motion is the magnitude of the net force equal to zero?
   b. Where in the motion is the velocity equal to zero?
   c. Where in the motion does the acceleration have its greatest magnitude?
   d. Where in the motion is the spring force equal to zero?
9. For the graph in Figure Q14.9, determine the frequency $f$ and the oscillation amplitude $A$.

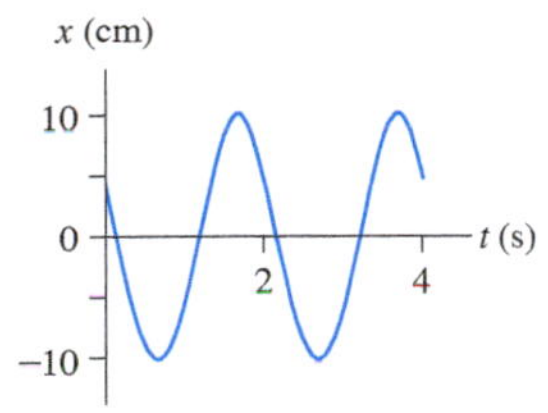


FIGURE Q14.9

10. For the graph in Figure Q14.10, determine the frequency $f$ and the oscillation amplitude $A$.

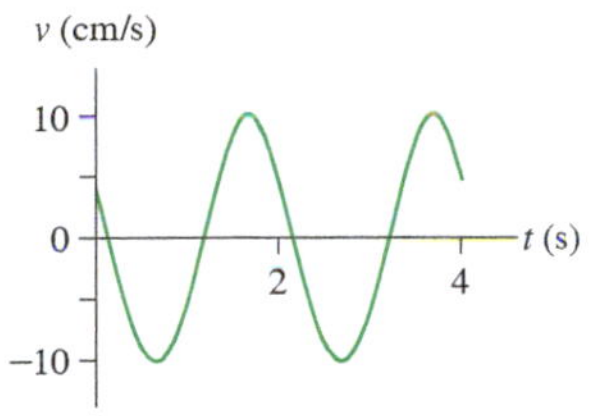


FIGURE Q14.10

11. A block oscillating on a spring has period $T = 2.0$ s.
    a. What is the period if the block's mass is doubled?
    b. What is the period if the value of the spring constant is quadrupled?
    c. What is the period if the oscillation amplitude is doubled while $m$ and $k$ are unchanged?

    **Note:** You do not know values for either $m$ or $k$. Do *not* assume any particular values for them. The required analysis involves thinking about ratios.
12. If a mass on a spring moving horizontally were taken to the moon, how would its frequency change? What about a mass on a spring moving vertically? A pendulum?
13. BIO Flies flap their wings at frequencies much too high for pure muscle action. A hypothesis for how they achieve these high frequencies is that the flapping of their wings is the driven oscillation of a mass-spring system. One way to test this is to trim a fly's wings. If the oscillation of the wings can be modeled as a mass-spring system, how would this change the frequency of the wingbeats?
14. Denver is at a higher elevation than Miami; the free-fall acceleration is slightly less at this higher elevation. If a pendulum clock keeps perfect time in Miami, will it run fast or slow in Denver? Explain.
15. If you want to play a tune on wine glasses, you'll need to adjust the oscillation frequencies by adding water to the glasses. This changes the mass that oscillates (more water means more mass) but not the restoring force, which is determined by the stiffness of the glass itself. If you need to raise the frequency of a particular glass, should you add water or remove water?
16. A young girl and her mother are swinging on a swing set. Who, if either, has the longer period of oscillation? Explain.

17. Two identical pendulums are pulled back from their equilibrium positions. One is pulled back by 5° and the other by 10°. They are released simultaneously. Which pendulum gets to the bottom of its swing first?
18. BIO Gibbons move through the trees by swinging from successive handholds, as we have seen. To increase their speed, gibbons may bring their legs close to their bodies. How does this help them move more quickly?
19. BIO Sprinters push off from the ball of their foot, then bend their knee to bring their foot up close to the body as they swing their leg forward for the next stride. Why is this an effective strategy for running fast?

20. BIO Humans have a range of hearing of approximately 20 Hz to 20 kHz. Mice have auditory systems similar to humans, but all of the physical elements are smaller. Given this, would you expect mice to have a higher or lower frequency range than humans? Explain.
21. In 1831, soldiers marched across the Broughton Bridge in England in "lock step"—that is, marching in time together. As they marched, the bridge began to bounce in time with their footsteps; the amplitude of this bouncing became larger and larger until the bridge suddenly collapsed. Explain why the lock step of the soldiers caused this collapse and why afterward soldiers were ordered to "break step," or march at their own individual rates, as they crossed bridges.
22. BIO We've seen that stout tendons in the legs of hopping kangaroos store energy. When a kangaroo lands, much of the kinetic energy of motion is converted to elastic energy as the tendons stretch, returning to kinetic energy when the kangaroo again leaves the ground. If a hopping kangaroo increases its speed, it spends more time in the air with each bounce, but the contact time with the ground stays approximately the same. Explain why you would expect this to be the case.

## Multiple-Choice Questions

23. | A spring has an unstretched length of 20 cm. A 100 g mass hanging from the spring stretches it to an equilibrium length of 30 cm.
    a. Suppose the mass is pulled down to where the spring's length is 40 cm. When it is released, it begins to oscillate. What is the amplitude of the oscillation?
       A. 5.0 cm B. 10 cm C. 20 cm D. 40 cm
    b. For the data given above, what is the frequency of the oscillation?
       A. 0.10 Hz B. 0.62 Hz C. 1.6 Hz D. 10 Hz
    c. Suppose this experiment were done on the moon, where the free-fall acceleration is approximately 1/6 of that on the earth. How would this change the frequency of the oscillation?
       A. The frequency would decrease.
       B. The frequency would increase.
       C. The frequency would stay the same.

24. | | Figure Q14.24 represents the motion of a mass on a spring.
a. What is the period of this oscillation?
A. 12 s B. 24 s C. 36 s D. 48 s E. 50 s

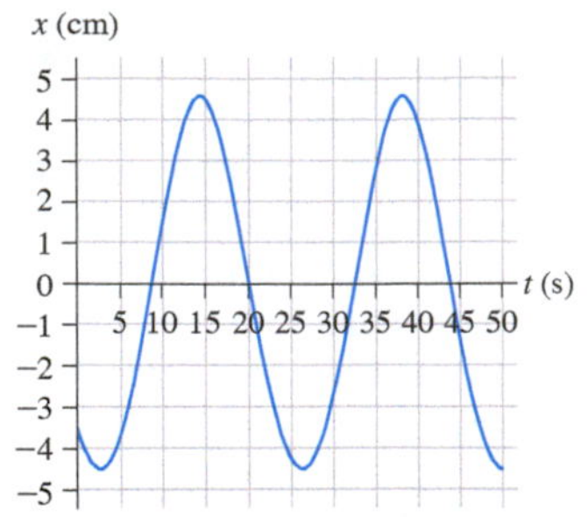


FIGURE Q14.24

b. What is the amplitude of the oscillation?
A. 1.0 cm
B. 2.5 cm
C. 4.5 cm
D. 5.0 cm
E. 9.0 cm

c. What is the position of the mass at time $t = 30$ s?
A. −4.5 cm
B. −2.5 cm
C. 0.0 cm
D. 4.5 cm
E. 30 cm

d. When is the first time the velocity of the mass is zero?
A. 0 s
B. 2 s
C. 8 s
D. 10 s
E. 13 s

e. At which of these times does the kinetic energy have its maximum value?
A. 0 s
B. 8 s
C. 13 s
D. 26 s
E. 30 s

25. | A ball of mass $m$ oscillates on a spring with spring constant $k = 200$ N/m. The ball's position is $x = (0.350 \text{ m}) \cos(15.0t)$, with $t$ measured in seconds.
a. What is the amplitude of the ball's motion?
A. 0.175 m B. 0.350 m C. 0.700 m
D. 7.50 m E. 15.0 m
b. What is the frequency of the ball's motion?
A. 0.35 Hz B. 2.39 Hz C. 5.44 Hz
D. 6.28 Hz E. 15.0 Hz
c. What is the value of the mass $m$?
A. 0.45 kg B. 0.89 kg C. 1.54 kg
D. 3.76 kg E. 6.33 kg
d. What is the total mechanical energy of the oscillator?
A. 1.65 J B. 3.28 J C. 6.73 J
D. 10.1 J E. 12.2 J
e. What is the ball's maximum speed?
A. 0.35 m/s B. 1.76 m/s C. 2.60 m/s
D. 3.88 m/s E. 5.25 m/s

26. | A 0.20 kg mass on a horizontal spring is pulled back 2.0 cm and released. If, instead, a 0.40 kg mass were used in this same experiment, the total energy of the system would
A. Double. B. Remain the same. C. Be half as large.

27. | A 0.20 kg mass on a horizontal spring is pulled back a certain distance and released. The maximum speed of the mass is measured to be 0.28 m/s. If, instead, a 0.40 kg mass were used in this same experiment, the maximum speed would be
A. 0.14 m/s B. 0.20 m/s C. 0.28 m/s
D. 0.40 m/s E. 0.56 m/s

28. | A heavy brass ball is used to make a pendulum with a period of 5.5 s. How long is the cable that connects the pendulum ball to the ceiling?
A. 4.7 m B. 6.2 m
C. 7.5 m D. 8.7 m

29. | Very loud sounds can damage hearing by injuring the
BIO vibration-sensing hair cells on the basilar membrane. Suppose a person has injured hair cells on a segment of the basilar membrane close to the stapes. What type of sound is most likely to have produced this particular pattern of damage?
A. Loud music with a mix of different frequencies
B. A very loud, high-frequency sound
C. A very loud, low-frequency sound

# PROBLEMS

### Section 14.1 Equilibrium and Oscillation
### Section 14.2 Linear Restoring Forces and SHM

1. | When a guitar string plays the note "A," the string vibrates at 440 Hz. What is the period of the vibration?

2. | In the aftermath of an intense earthquake, the earth as a whole "rings" with a period of 54 minutes. What is the frequency (in Hz) of this oscillation?

3. | In taking your pulse, you count 75 heartbeats in 1 min. What
BIO are the period (in s) and frequency (in Hz) of your heart's oscillations?

4. | A spring scale hung from the ceiling stretches by 6.4 cm when a 1.0 kg mass is hung from it. The 1.0 kg mass is removed and replaced with a 1.5 kg mass. What is the stretch of the spring?

5. | A heavy steel ball is hung from a cord to make a pendulum. The ball is pulled to the side so that the cord makes a 5° angle with the vertical. Holding the ball in place takes a force of 20 N. If the ball is pulled farther to the side so that the cord makes a 10° angle, what force is required to hold the ball?

### Section 14.3 Describing Simple Harmonic Motion

6. | | An air-track glider attached to a spring oscillates between the 10 cm mark and the 60 cm mark on the track. The glider completes 10 oscillations in 33 s. What are the (a) period, (b) frequency, (c) amplitude, and (d) maximum speed of the glider?

7. ||| An air-track glider is attached to a spring. The glider is pulled to the right and released from rest at $t = 0$ s. It then oscillates with a period of 2.0 s and a maximum speed of 40 cm/s.
   a. What is the amplitude of the oscillation?
   b. What is the glider's position at $t = 0.25$ s?
8. | What are the (a) amplitude and (b) frequency of the oscillation shown in Figure P14.8?

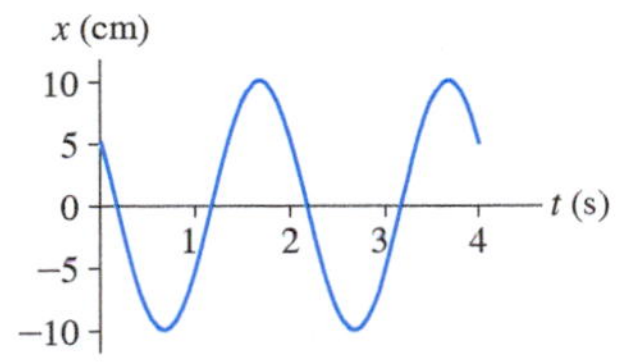

FIGURE P14.8

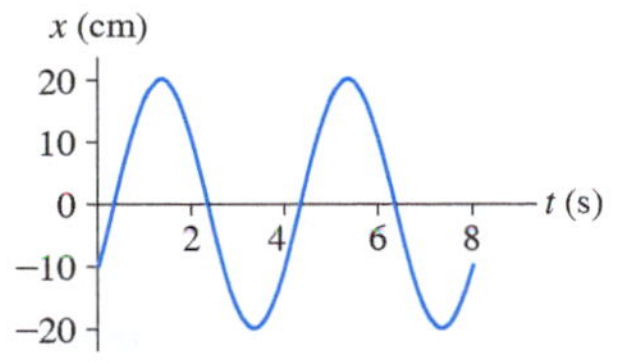

FIGURE P14.9

9. | What are the (a) amplitude and (b) frequency of the oscillation shown in Figure P14.9?
10. ||| A block with a mass of 0.28 kg is attached to a horizontal spring. The block is pulled back from its equilibrium position until the spring exerts a force of 1.0 N on the block. When the block is released, it oscillates with a frequency of 1.2 Hz. How far was the block pulled back before being released?
11. ||| The acceleration of an oscillator undergoing simple harmonic motion is described by the equation $a_x(t) = -(18.0 \text{ m/s}^2)\cos(30t)$, where the time $t$ is measured in seconds. What is the amplitude of this oscillator?
12. || BIO The end of a nanoscale cantilever used for weighing DNA molecules (see Example 14.8) oscillates at 8.8 MHz with an amplitude of 5.0 nm. What are (a) the maximum speed and (b) the maximum acceleration of the end of the cantilever?
13. || Some passengers on an ocean cruise may suffer from motion sickness as the ship rocks back and forth on the waves. At one position on the ship, passengers experience a vertical motion of amplitude 1 m with a period of 15 s.
   a. To one significant figure, what is the maximum acceleration of the passengers during this motion?
   b. What fraction is this of $g$?
14. ||| The New England Merchants Bank Building in Boston is 152 m high. On windy days it sways with a frequency of 0.17 Hz, and the acceleration of the top of the building can reach 2.0% of the free-fall acceleration, enough to cause discomfort for occupants. What is the total distance, side to side, that the top of the building moves during such an oscillation?
15. || BIO The motion of a nightingale's wingtips can be modeled as simple harmonic motion. In one study, the tips of a bird's wings were found to move up and down with an amplitude of 8.8 cm and a period of 0.82 s. What are the wingtips' (a) maximum speed and (b) maximum acceleration?
16. | In a loudspeaker, an electromagnetic coil rapidly drives a paper cone back and forth, sending out sound waves. If the cone of a loudspeaker moves sinusoidally at 1.2 kHz with an amplitude of 3.5 $\mu$m, what are the cone's maximum speed and acceleration?
17. | BIO We can model the motion of a bumblebee's wing as simple harmonic motion. A bee beats its wings 250 times per second, and the wing tip moves at a maximum speed of 2.5 m/s. What is the amplitude of the wing tip's motion?
18. || BIO Hummingbirds may seem fragile, but their wings are capable of sustaining very large forces and accelerations. Figure P14.18 shows data for the vertical position of the wing tip of a rufous hummingbird. What is the maximum acceleration of the wing tips, in m/s$^2$ and in units of $g$?

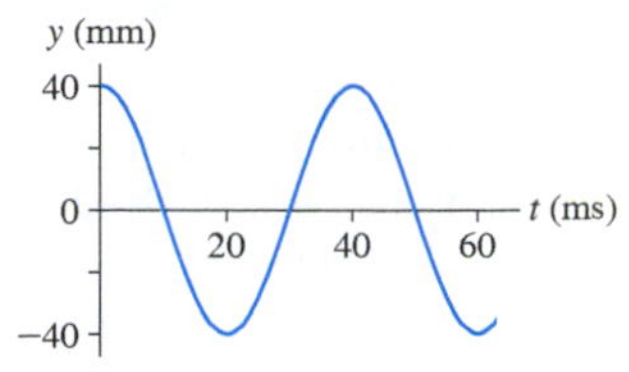

FIGURE P14.18

## Section 14.4 Energy in Simple Harmonic Motion

19. ||| a. When the displacement of a mass on a spring is $\frac{1}{2}A$, what fraction of the mechanical energy is kinetic energy and what fraction is potential energy?
   b. At what displacement, as a fraction of $A$, is the energy half kinetic and half potential?
20. ||| A 1.0 kg block is attached to a spring with spring constant 16 N/m. While the block is sitting at rest, a student hits it with a hammer and almost instantaneously gives it a speed of 40 cm/s. What are
   a. The amplitude of the subsequent oscillations?
   b. The block's speed at the point where $x = \frac{1}{2}A$?
21. | A block attached to a spring with unknown spring constant oscillates with a period of 2.00 s. What is the period if
   a. The mass is doubled?
   b. The mass is halved?
   c. The amplitude is doubled?
   d. The spring constant is doubled?
   Parts a to d are independent questions, each referring to the initial situation.
22. || A 200 g air-track glider is attached to a spring. The glider is pushed 10.0 cm against the spring, then released. A student with a stopwatch finds that 10 oscillations take 12.0 s. What is the spring constant?
23. ||| The position of a 50 g oscillating mass is given by $x(t) = (2.0 \text{ cm})\cos(10t)$, where $t$ is in seconds. Determine:
   a. The amplitude.
   b. The period.
   c. The spring constant.
   d. The maximum speed.
   e. The total energy.
   f. The velocity at $t = 0.40$ s.
24. ||| A 50-cm-long spring is suspended from the ceiling. A 250 g mass is connected to the end and held at rest with the spring unstretched. The mass is released and falls, stretching the spring by 20 cm before coming to rest at its lowest point. It then continues to oscillate vertically.
   a. What is the spring constant?
   b. What is the amplitude of the oscillation?
   c. What is the frequency of the oscillation?
25. || A 200 g mass attached to a horizontal spring oscillates at a frequency of 2.0 Hz. At one instant, the mass is at $x = 5.0$ cm and has $v_x = -30$ cm/s. Determine:
   a. The period.
   b. The amplitude.
   c. The maximum speed.
   d. The total energy.
26. || A 507 g mass oscillates with an amplitude of 10.0 cm on a spring whose spring constant is 20.0 N/m. Determine:
   a. The period.
   b. The maximum speed.
   c. The total energy.

## Section 14.5 Pendulum Motion

27. | A mass on a string of unknown length oscillates as a pendulum with a period of 4.00 s. What is the period if
    a. The mass is doubled?
    b. The string length is doubled?
    c. The string length is halved?
    d. The amplitude is halved?

    Parts a to d are independent questions, each referring to the initial situation.

28. | The mass in a pendulum clock completes one full back-and-forth oscillation in 1.00 s. What is the length of the rod?

29. ||| A 200 g ball is tied to a string. It is pulled to an angle of 8.00° and released to swing as a pendulum. A student with a stopwatch finds that 10 oscillations take 12.0 s. How long is the string?

30. || The free-fall acceleration on the moon is $1.62\ \text{m/s}^2$. What is the length of a pendulum whose period on the moon matches the period of a 2.00-m-long pendulum on the earth?

31. | Astronauts on the first trip to Mars take along a pendulum that has a period on earth of 1.50 s. The period on Mars turns out to be 2.45 s. Use this data to calculate the Martian free-fall acceleration.

32. ||| A building is being knocked down with a wrecking ball, which is a big metal sphere that swings on a 10-m-long cable. You are (unwisely!) standing directly beneath the point from which the wrecking ball is hung when you notice that the ball has just been released and is swinging directly toward you. How much time do you have to move out of the way?

33. ||| BIO Interestingly, there have been several studies using cadavers to determine the moment of inertia of human body parts by letting them swing as a pendulum about a joint. In one study, the center of gravity of a 5.0 kg lower leg was found to be 18 cm from the knee. When pivoted at the knee and allowed to swing, the oscillation frequency was 1.6 Hz. What was the moment of inertia of the lower leg?

34. ||| A child on a swing set swings back and forth with a period of 3.3 s and an amplitude of 25°. What is the maximum speed of the child as she swings?

35. ||| You and your friends find a rope that hangs down 15 m from a high tree branch right at the edge of a river. You find that you can run, grab the rope, swing out over the river, and drop into the water. You run at 2.0 m/s and grab the rope, launching yourself out over the river. How long must you hang on if you want to drop into the water at the greatest possible distance from the edge?

36. ||| BIO In a study designed to better understand the dynamics of walking, a subject stood with one leg at rest, and swung the other leg at various frequencies. The metabolic power expended, in watts per kilogram of body weight, is shown in Figure P14.36. Estimate the length of this subject's legs.

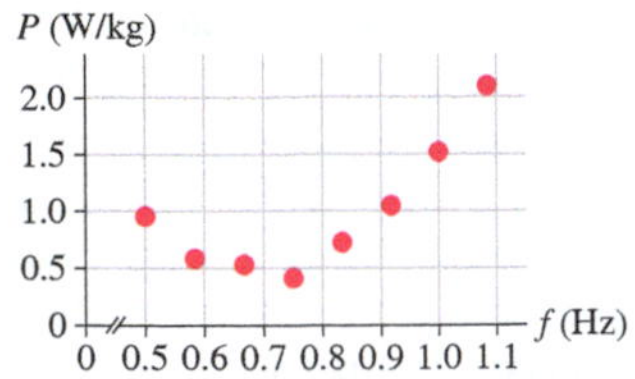


FIGURE P14.36

37. |||| BIO An elephant's legs have a reasonably uniform cross section from top to bottom, and they are quite long, pivoting high on the animal's body. When an elephant moves at a walk, it uses very little energy to bring its legs forward, simply allowing them to swing like pendulums. For fluid walking motion, this time should be half the time for a complete stride; as soon as the right leg finishes swinging forward, the elephant plants the right foot and begins swinging the left leg forward.

    a. An elephant has legs that stretch 2.3 m from its shoulders to the ground. How much time is required for one leg to swing forward after completing a stride?
    b. What would you predict for this elephant's stride frequency? That is, how many steps per minute will the elephant take?

38. ||| In a science museum, you may have seen a Foucault pendulum, which is used to demonstrate the rotation of the earth. In one museum's pendulum, the 110 kg bob swings from a 15.8-m-long cable with an amplitude of 5.0°.
    a. What is the period of this pendulum?
    b. What is the bob's maximum speed?
    c. What is the pendulum's maximum kinetic energy?
    d. When the bob is at its maximum displacement, how much higher is it than when it is at its equilibrium position?

## Section 14.6 Damped Oscillations

39. | The amplitude of an oscillator decreases to 36.8% of its initial value in 10.0 s. What is the value of the time constant?

40. || A physics department has a Foucault pendulum, a long-period pendulum suspended from the ceiling. The pendulum has an electric circuit that keeps it oscillating with a constant amplitude. When the circuit is turned off, the oscillation amplitude decreases by 50% in 22 minutes. What is the pendulum's time constant? How much additional time elapses before the amplitude decreases to 25% of its initial value?

41. || Calculate and draw an accurate displacement graph from $t = 0$ s to $t = 10$ s of a damped oscillator having a frequency of 1.0 Hz and a time constant of 4.0 s.

42. || A small earthquake starts a lamppost vibrating back and forth. The amplitude of the vibration of the top of the lamppost is 6.5 cm at the moment the quake stops, and 8.0 s later it is 1.8 cm.
    a. What is the time constant for the damping of the oscillation?
    b. What was the amplitude of the oscillation 4.0 s after the quake stopped?

43. ||| BIO The common field cricket makes its characteristic loud chirping sound using a specialized vibrating structure in its wings. The motion of this structure—and the sound intensity that it produces—can be modeled as a damped oscillation. The sound intensity of such a cricket is shown in Figure P14.43.
    a. What is the frequency of the oscillations?
    b. What is the time constant for the decay of these oscillations?

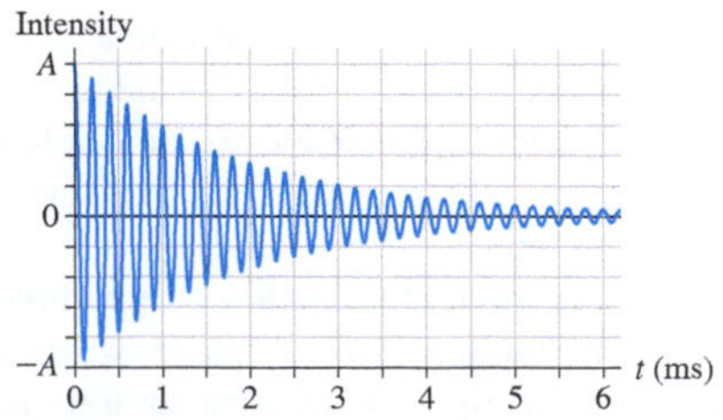


FIGURE P14.43

44. A damped pendulum has a period of 0.66 s and a time constant of 4.1 s. How many oscillations will this pendulum make before its amplitude has decreased to 20% of its initial amplitude?

45. When you drive your car over a bump, the springs connecting the wheels to the car compress. Your shock absorbers then damp the subsequent oscillation, keeping your car from bouncing up and down on the springs. Figure P14.45 shows real data for a car driven over a bump. We can model this as a damped oscillation, although this model is far from perfect. Estimate the frequency and time constant in this model.

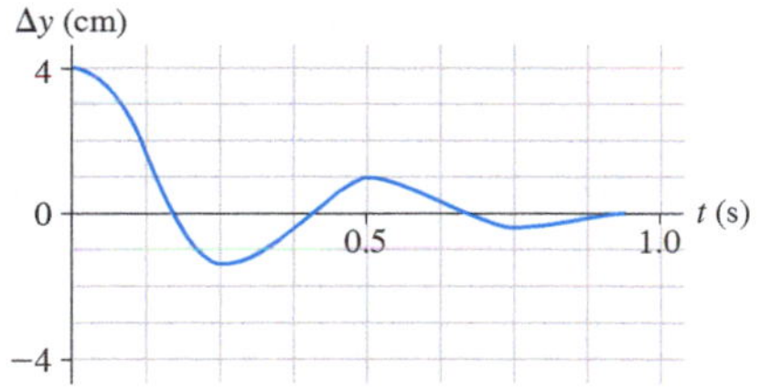

FIGURE P14.45

### Section 14.7 Driven Oscillations and Resonance

46. Taipei 101 (a 101-story building in Taiwan) is sited in an area that is prone to earthquakes and typhoons, both of which can lead to dangerous oscillations of the building. To reduce the maximum amplitude, the building has a *tuned mass damper*, a 660,000 kg mass suspended from 42-m-long cables that oscillates at the same natural frequency as the building. When the building sways, the pendulum swings, reaching an amplitude of 75 cm in strong winds or tremors. Damping the motion of the mass reduces the maximum amplitude of oscillation of the building.

a. What is the period of oscillation of the building?
b. During strong winds, how fast is the pendulum moving when it passes through the equilibrium position?

47. A 25 kg child sits on a 2.0-m-long rope swing. You are going to give the child a small, brief push at regular intervals. If you want to increase the amplitude of her motion as quickly as possible, how much time should you wait between pushes?

48. Your car rides on springs, so it will have a natural frequency of oscillation. Figure P14.48 shows data for the amplitude of motion of a car driven at different frequencies. The car is driven at 20 mph over a washboard road with bumps spaced 10 feet apart; the resulting ride is quite bouncy. Should the driver speed up or slow down for a smoother ride?

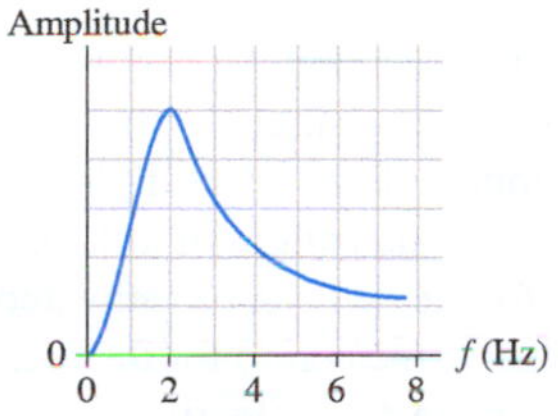

FIGURE P14.48

49. BIO Vision is blurred if the head is vibrated at 29 Hz because the vibrations are resonant with the natural frequency of the eyeball held by the musculature in its socket. If the mass of the eyeball is 7.5 g, a typical value, what is the effective spring constant of the musculature attached to the eyeball?

## General Problems

50. A spring has an unstretched length of 12 cm. When an 80 g ball is hung from it, the length increases by 4.0 cm. Then the ball is pulled down another 4.0 cm and released.
a. What is the spring constant of the spring?
b. What is the period of the oscillation?
c. Draw a position-versus-time graph showing the motion of the ball for three cycles of the oscillation. Let the equilibrium position of the ball be $y = 0$. Be sure to include appropriate units on the axes so that the period and the amplitude of the motion can be determined from your graph.

51. A 0.40 kg ball is suspended from a spring with spring constant 12 N/m. If the ball is pulled down 0.20 m from the equilibrium position and released, what is its maximum speed while it oscillates?

52. When Kayla stands on her trampoline, it sags by 0.21 m. Now she starts bouncing. How much time elapses between the instant when she first lands on the trampoline's surface and when she passes the same point on the way up?

53. A spring with spring constant 15.0 N/m hangs from the ceiling. A ball is suspended from the spring and allowed to come to rest. It is then pulled down 6.00 cm and released. If the ball makes 30 oscillations in 20.0 s, what are its (a) mass and (b) maximum speed?

54. A block attached to a horizontal spring is pulled back a certain distance from equilibrium, then released from rest at $t = 0$ s. If the frequency of the block is 0.72 Hz, what is the earliest time after the block is released that its kinetic energy is exactly one-half of its potential energy?

55. INT A circular cylinder has a diameter of 2.0 cm and a mass of 10 g. It floats in water with its long axis perpendicular to the water's surface. It is pushed down into the water by a small distance and released; it then bobs up and down. What is the oscillation frequency?

56. A spring is hung from the ceiling. When a coffee mug is attached to its end, the spring stretches 2.0 cm before reaching its new equilibrium length. The mug is then pulled down slightly and released. What is the frequency of oscillation?

57. On your first trip to Planet X you happen to take along a 200 g mass, a 40.0-cm-long spring, a meter stick, and a stopwatch. You're curious about the free-fall acceleration on Planet X, where ordinary tasks seem easier than on earth, but you can't find this information in your Visitor's Guide. One night you suspend the spring from the ceiling in your room and hang the mass from it. You find that the mass stretches the spring by 31.2 cm. You then pull the mass down 10.0 cm and release it. With the stopwatch you find that 10 oscillations take 14.5 s. Can you now satisfy your curiosity?

58. An object oscillating on a spring has the velocity graph shown in Figure P14.58. Draw a velocity graph if the following changes are made.
a. The amplitude is doubled and the frequency is halved.
b. The amplitude and spring constant are kept the same, but the mass is quadrupled.

Parts a and b are independent questions, each starting from the graph shown.

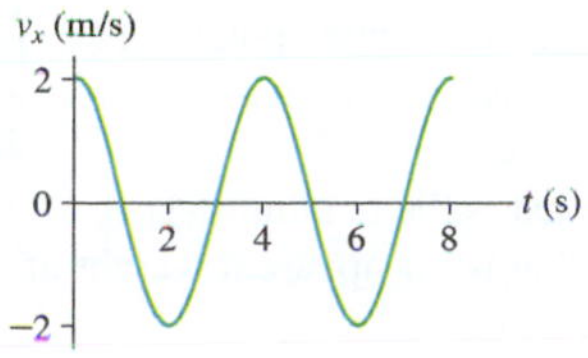

FIGURE P14.58

59. || The two graphs in Figure P14.59 are for two different vertical mass-spring systems.
   a. What is the frequency of system A? What is the first time at which the mass has maximum speed while traveling in the upward direction?
   b. What is the period of system B? What is the first time at which the mechanical energy is all potential?
   c. If both systems have the same mass, what is the ratio $k_A/k_B$ of their spring constants?

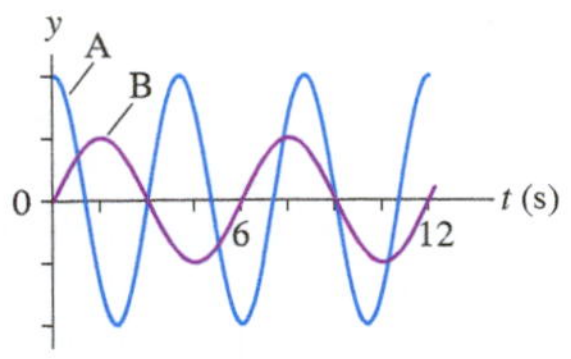

FIGURE P14.59

60. ||| As we've seen, astronauts measure their mass by measuring the period of oscillation when sitting in a chair connected to a spring. The Body Mass Measurement Device on Skylab, a 1970s space station, had a spring constant of 606 N/m. The empty chair oscillated with a period of 0.901 s. What is the mass of an astronaut who oscillates with a period of 2.09 s when sitting in the chair?
BIO

61. ||||| A 100 g ball attached to a spring with spring constant 2.50 N/m oscillates horizontally on a frictionless table. Its velocity is 20.0 cm/s when $x = -5.00$ cm.
   a. What is the amplitude of oscillation?
   b. What is the speed of the ball when $x = 3.00$ cm?

62. || The ultrasonic transducer used in a medical ultrasound imaging device is a very thin disk ($m = 0.10$ g) driven back and forth in SHM at 1.0 MHz by an electromagnetic coil.
BIO
   a. The maximum restoring force that can be applied to the disk without breaking it is 40,000 N. What is the maximum oscillation amplitude that won't rupture the disk?
   b. What is the disk's maximum speed at this amplitude?

63. || A compact car has a mass of 1200 kg. When empty, the car bounces up and down on its springs 2.0 times per second. What is the car's oscillation frequency when it is carrying four 70 kg passengers?

64. ||| A car with a total mass of 1400 kg (including passengers) is driving down a washboard road with bumps spaced 5.0 m apart. The ride is roughest—that is, the car bounces up and down with the maximum amplitude—when the car is traveling at 6.0 m/s. What is the spring constant of the car's springs?

65. || A 500 g air-track glider attached to a spring with spring constant 10 N/m is sitting at rest on a frictionless air track. A 250 g glider is pushed toward it from the far end of the track at a speed of 120 cm/s. It collides with and sticks to the 500 g glider. What are the amplitude and period of the subsequent oscillations?
INT

66. |||| Tarzan, who has a mass of 80 kg, holds onto the end of a vine that is at a 12° angle from the vertical. He steps off his branch and, just at the bottom of his swing, he grabs onto his chimp friend Cheetah, whose mass is 40 kg. What is the maximum angle the rope reaches as Tarzan swings to the other side?
INT

67. ||| In the Pirate Boat ride at the amusement park, riders swing back and forth in a pendulum-like "boat." The distance from the boat to the pivot point is 13 m, and the maximum angle the boat reaches is 40°.
INT

   a. What is the maximum speed the boat attains?
   b. What is the apparent weight of a 55 kg rider at the bottom of the arc?

68. || The typical American man has a leg length of 0.85 m and walks at a speed of 1.4 m/s. A giraffe's legs are 1.8 m long. At what speed do you expect a giraffe to walk? **Hint:** An animal's speed is proportional to the length of its legs times the frequency of its strides.
BIO

69. |||| Suppose that during each step, the leg of the student in Example 14.11 swings through a total distance of 2.0 m. At the end of the swing, this foot rests on the ground for 0.2 s before the other leg begins its swing.
BIO
   a. At what speed does this student walk? (Think carefully about how far forward the student moves at each step.)
   b. The swinging leg reaches its maximum speed at the bottom of its arc. How many times faster is this maximum leg speed (measured with respect to the ground) than the average walking speed?

70. |||| A physical pendulum consists of a uniform rod that can swing freely from one end, with a small, heavy bob attached to the other end. If the length of the rod is 1.35 m, and the mass of the bob and the rod are both 1.0 kg, what is the period of this pendulum?
INT

71. |||| The pendulum on a grandfather clock has a period of 2.00 s. If the clock is not wound, the pendulum's amplitude begins to decay at a rate of 0.53% each pendulum period.
   a. What is the time constant of this pendulum?
   b. What percentage of the pendulum's energy is lost each period?

72. |||| Orangutans can move by brachiation, swinging like a pendulum beneath successive handholds. If an orangutan has arms that are 0.90 m long and repeatedly swings to a 20° angle, taking one swing immediately after another, estimate how fast it is moving in m/s.
BIO
INT

73. |||| An infant's toy has a 120 g wooden animal hanging from a spring. If pulled down gently, the animal oscillates up and down with a period of 0.50 s. His older sister pulls the spring a bit more than intended. She pulls the animal 30 cm below its equilibrium position, then lets go. The animal flies upward and detaches from the spring right at the animal's equilibrium position. If the animal does not hit anything on the way up, how far above its equilibrium position will it go?
INT

74. |||| A jellyfish can propel itself with jets of water pushed out of its bell, a flexible structure on top of its body. The elastic bell and the water it contains function as a mass-spring system, greatly increasing efficiency. Normally, the jellyfish emits one jet right after the other, but we can get some insight into the jet system by looking at a single jet thrust. Figure P14.74 shows a graph of the motion of one point in the wall of the bell for such a single jet; this is the pattern of a damped oscillation. The spring constant for the bell can be estimated to be 1.2 N/m.
BIO

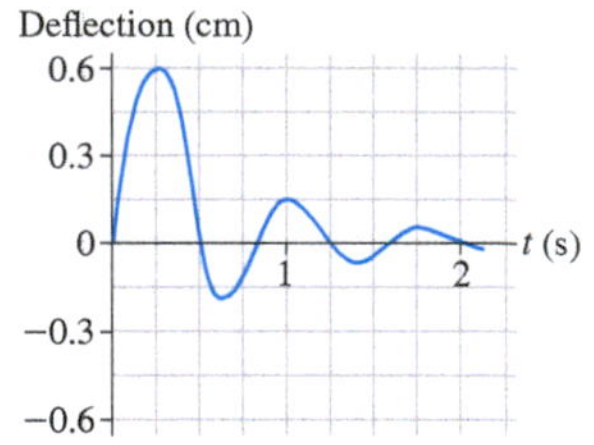

FIGURE P14.74

   a. What is the period for the oscillation?
   b. Estimate the effective mass participating in the oscillation. This is the mass of the bell itself plus the mass of the water.
   c. Consider the peaks of positive displacement in the graph. By what factor does the amplitude decrease over one period? Given this, what is the time constant for the damping?

75. |||| A 200 g oscillator in a vacuum chamber has a frequency of 2.0 Hz. When air is admitted, the oscillation decreases to 60% of its initial amplitude in 50 s. How many oscillations will have been completed when the amplitude is 30% of its initial value?

76. ||| While seated on a tall bench, extend your lower leg a small amount and then let it swing freely about your knee joint, with no muscular engagement. It will oscillate as a damped pendulum. Figure P14.76 is a graph of the lower leg angle versus time in such an experiment. Estimate (a) the period and (b) the time constant for this oscillation.
BIO

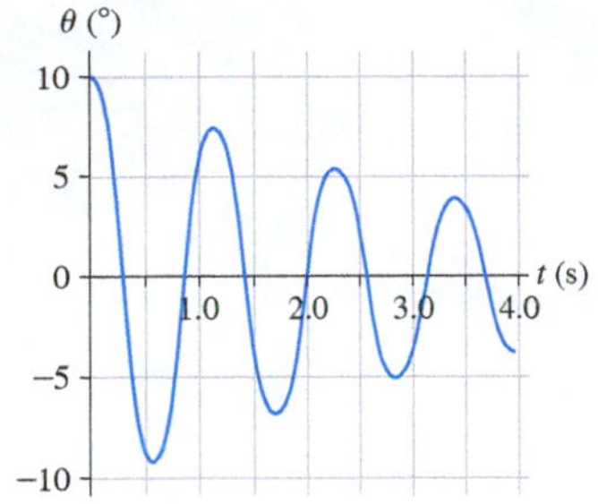


FIGURE P14.76

## MCAT-Style Passage Problems

### The Spring in Your Step BIO

In Chapter 10, we saw that a runner's Achilles tendon will stretch like a spring and then rebound, storing and returning energy during a step. We can model this as the simple harmonic motion of a mass-spring system. When the foot rolls forward, the tendon spring begins to stretch as the weight moves to the ball of the foot, transforming kinetic energy into elastic potential energy. This is the first phase of an oscillation. The spring then rebounds, converting potential energy to kinetic energy as the foot lifts off the ground. The oscillation is fast: Sprinters running a short race keep each foot in contact with the ground for about 0.10 second, and some of that time corresponds to the heel strike and subsequent rolling forward of the foot.

77. | We can make a static measurement to deduce the spring constant to use in the model. If a 61 kg woman stands on a low wall with her full weight on the ball of one foot and the heel free to move, the stretch of the Achilles tendon will cause her center of gravity to lower by about 2.5 mm. What is the spring constant?
A. $1.2 \times 10^4$ N/m
B. $2.4 \times 10^4$ N/m
C. $1.2 \times 10^5$ N/m
D. $2.4 \times 10^5$ N/m

78. | If, during a stride, the stretch causes her center of mass to lower by 10 mm, what is the stored energy?
A. 3.0 J
B. 6.0 J
C. 9.0 J
D. 12 J

79. | If we imagine a full cycle of the oscillation, with the woman bouncing up and down and the tendon providing the restoring force, what will her oscillation period be?
A. 0.10 s
B. 0.15 s
C. 0.20 s
D. 0.25 s

80. || Given what you have calculated for the period of the full oscillation in this model, what is the landing-to-liftoff time for the stretch and rebound of the sprinter's foot?
A. 0.050 s
B. 0.10 s
C. 0.15 s
D. 0.20 s

### Web Spiders and Oscillations BIO

All spiders have special organs that make them exquisitely sensitive to vibrations. Web spiders detect vibrations of their web to determine what has landed in their web, and where.

In fact, spiders carefully adjust the tension of strands to "tune" their web. Suppose an insect lands and is trapped in a web. The silk of the web serves as the spring in a spring-mass system while the body of the insect is the mass. The frequency of oscillation depends on the restoring force of the web and the mass of the insect. Spiders respond more quickly to larger—and therefore more valuable—prey, which they can distinguish by the web's oscillation frequency.

Suppose a 12 mg fly lands in the center of a horizontal spider's web, causing the web to sag by 3.0 mm.

81. | Assuming that the web acts like a spring, what is the spring constant of the web?
A. 0.039 N/m
B. 0.39 N/m
C. 3.9 N/m
D. 39 N/m

82. | Modeling the motion of the fly on the web as a mass on a spring, at what frequency will the web vibrate when the fly hits it?
A. 0.91 Hz
B. 2.9 Hz
C. 9.1 Hz
D. 29 Hz

83. | If the web were vertical rather than horizontal, how would the frequency of oscillation be affected?
A. The frequency would be higher.
B. The frequency would be lower.
C. The frequency would be the same.

84. | Spiders are more sensitive to oscillations at higher frequencies. For example, a low-frequency oscillation at 1 Hz can be detected for amplitudes down to 0.1 mm, but a high-frequency oscillation at 1 kHz can be detected for amplitudes as small as 0.1 $\mu$m. For these low- and high-frequency oscillations, we can say that
A. The maximum acceleration of the low-frequency oscillation is greater.
B. The maximum acceleration of the high-frequency oscillation is greater.
C. The maximum accelerations of the two oscillations are approximately equal.

# 15 Traveling Waves and Sound

This bat's ears are much more prominent than its eyes. It appears that hearing is a much more important sense than sight for bats. How does a bat use sound waves to locate prey?

## LOOKING AHEAD

### Traveling Waves

Shaking one end of the spring up and down causes a disturbance—a **wave**—to travel along the spring.

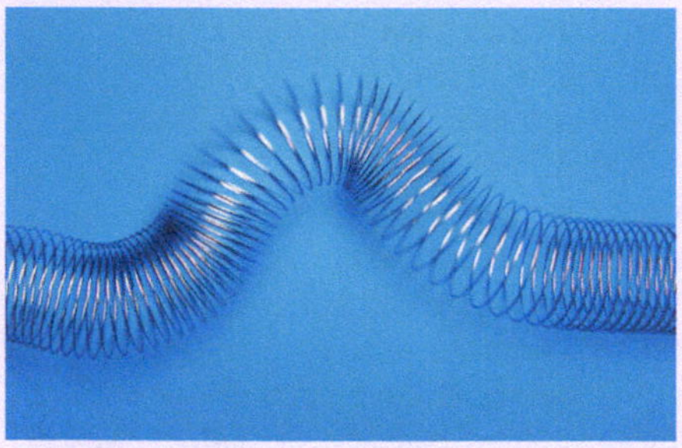

You'll learn the **wave model** that describes phenomena ranging from light waves to earthquake waves.

### Describing Waves

A wave moves out from the vibrating rod. The shape of the wave is *sinusoidal,* a form we've seen for oscillations.

The terms and equations used to describe waves are closely related to those for oscillations, as you'll see.

### Energy and Intensity

All waves carry energy. The energy of a laser beam can be used to burn away unwanted parts of the cornea in eye surgery.

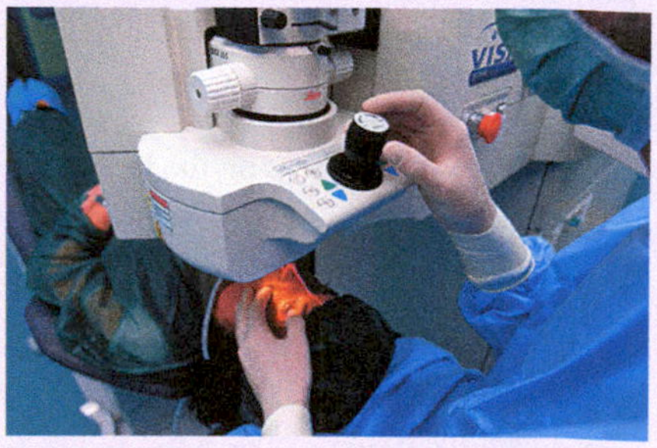

You'll learn to calculate **intensity,** a measure of how spread out or concentrated a wave's energy is.

**GOAL** To learn the basic properties of traveling waves.

## LOOKING BACK

### Simple Harmonic Motion

In Chapter 14, you learned to use the terminology of simple harmonic motion to describe oscillations.

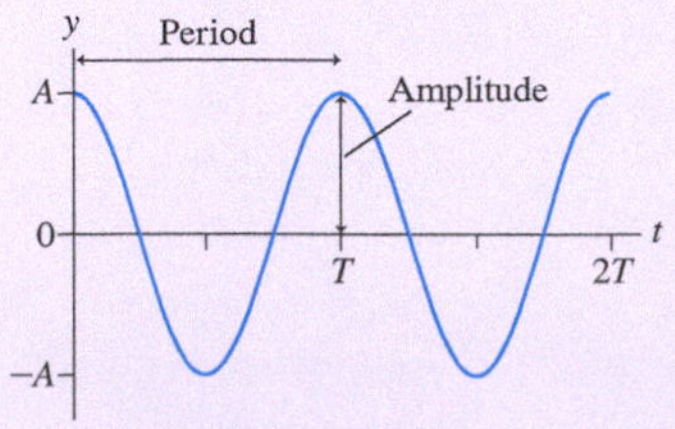


The terms you learned—such as period, frequency, and amplitude—apply to descriptions of wave motion as well.

**STOP TO THINK**

A wooden toy hangs from a spring. When you pull it down and release it, it reaches the highest point of its motion after 1.0 s. What is the frequency of the oscillation?

A. 2.0 Hz
B. 1.5 Hz
C. 1.0 Hz
D. 0.5 Hz

# 15.1 The Wave Model

The *particle model* that we have been using since Chapter 1 allowed us to simplify the treatment of motion of complex objects by considering them to be particles. Balls, cars, and rockets obviously differ from one another, but the general features of their motions are well described by treating them as particles. As we saw in Chapter 3, a ball or a rock or a car flying through the air will undergo the same motion. The particle model helps us understand this underlying simplicity.

In this chapter we will introduce the basic properties of waves with a **wave model** that emphasizes those aspects of wave behavior common to all waves. Although sound waves, water waves, and radio waves are clearly different, the wave model will allow us to understand many of the important features they share.

The wave model is built around the idea of a **traveling wave,** which is an organized disturbance that travels with a well-defined wave speed. This definition seems straightforward, but we must understand several new terms to gain a complete understanding of the concept of a traveling wave.

## Mechanical Waves

**Mechanical waves** are waves that involve the motion of a substance through which they move, the **medium.** For example, the medium of a water wave is the water, the medium of a sound wave is the air, and the medium of a wave on a stretched string is the string.

As a wave passes through a medium, the atoms that make up the medium are displaced from equilibrium, much like pulling a spring away from its equilibrium position. This is a **disturbance** of the medium. The water ripples of FIGURE 15.1 are a disturbance of the water's surface.

FIGURE 15.1 Ripples on a pond are a traveling wave.

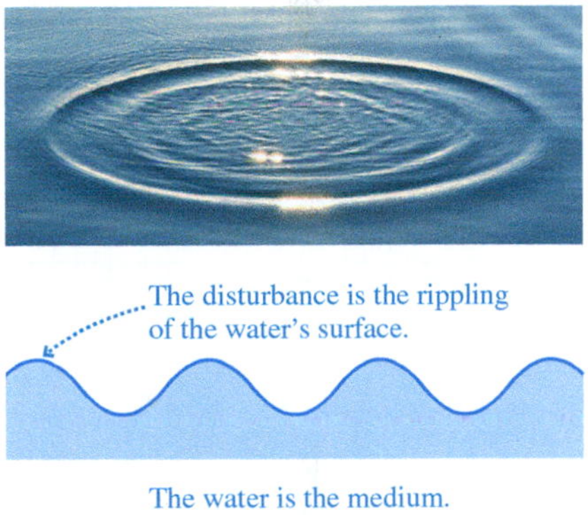


A wave disturbance is created by a *source.* The source of a wave might be a rock thrown into water, your hand plucking a stretched string, or an oscillating loudspeaker cone pushing on the air. Once created, the disturbance travels outward through the medium at the **wave speed** $v$. This is the speed with which a ripple moves across the water or a pulse travels down a string.

The disturbance propagates through the medium, and a wave does transfer *energy,* but **the medium as a whole does not travel!** The ripples on the pond (the disturbance) move outward from the splash of the rock, but there is no outward flow of water. Likewise, the particles of a string oscillate up and down but do not move in the direction of a pulse traveling along the string. **A wave transfers energy, but it does not transfer any material or substance outward from the source.**

## Electromagnetic and Matter Waves

Mechanical waves require a medium, but there are waves that do not. Two important types of such waves are electromagnetic waves and matter waves.

**Electromagnetic waves** are waves of an *electromagnetic field.* Electromagnetic waves are very diverse, including visible light, radio waves, microwaves, and x rays. Electromagnetic waves require no material medium and can travel through a vacuum; light can travel through empty space, though sound cannot. At this point, we have not defined what an "electromagnetic field" is, so we won't worry about the precise nature of what is "waving" in electromagnetic waves. The wave model can describe many of the important aspects of these waves without a detailed description of their exact nature. We'll look more closely at electromagnetic waves in Chapter 25, once we have a full understanding of electric and magnetic fields.

One of the most significant discoveries of the 20th century was that material particles, such as electrons and atoms, have wave-like characteristics. We will learn how a full description of matter at an atomic scale requires an understanding of such **matter waves** in Chapter 28.

You may have been at a sporting event in which spectators do "The Wave." The wave moves around the stadium, but the spectators (the medium, in this case) stay right where they are. This is a clear example of the principle that a wave does not transfer any material.

### Transverse and Longitudinal Waves

Most waves fall into two general classes: *transverse* and *longitudinal.* For mechanical waves, these terms describe the relationship between the motion of the particles that carry the wave and the motion of the wave itself, as the following table shows.

**Two types of wave motion**

**A transverse wave**

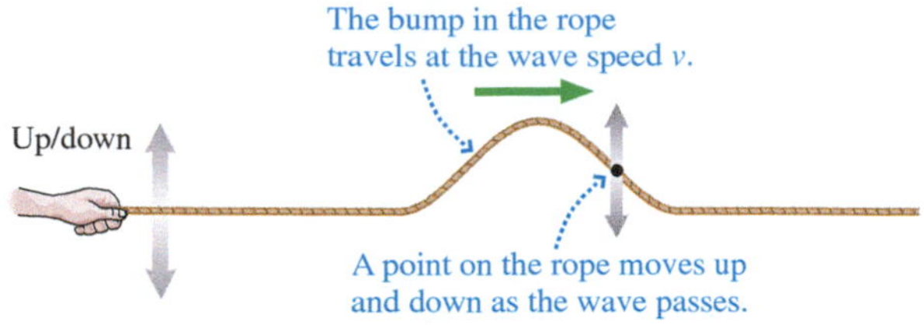


For mechanical waves, a **transverse wave** is a wave in which the particles in the medium move *perpendicular* to the direction in which the wave travels. Shaking the end of a stretched string up and down creates a wave that travels along the string in a horizontal direction while the particles that make up the string oscillate vertically.

**A longitudinal wave**

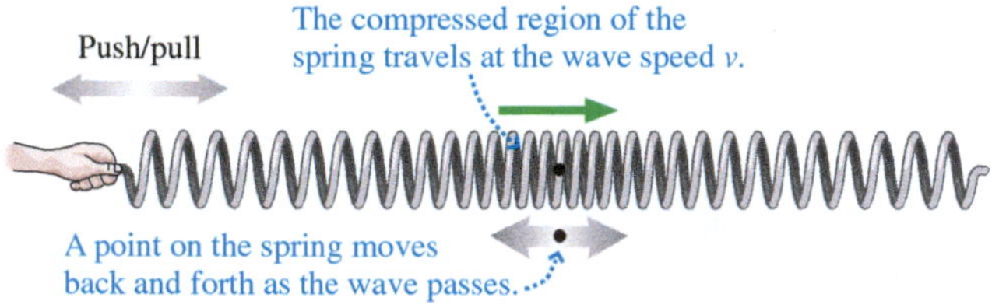


In a **longitudinal wave,** the particles in the medium move *parallel* to the direction in which the wave travels. Quickly moving the end of a spring back and forth sends a wave—in the form of a compressed region—down the spring. The particles that make up the spring oscillate horizontally as the wave passes.

**FIGURE 15.2** Different types of earthquake waves.

The passage of a longitudinal P wave expands and compresses the ground. The motion is parallel to the direction of travel of the wave.

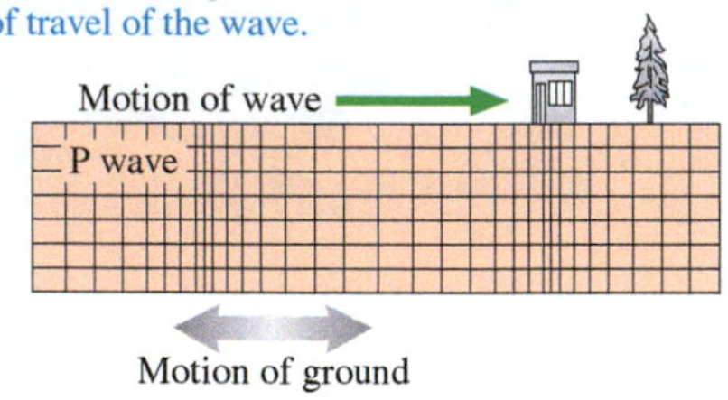


The passage of a transverse S wave moves the ground up and down. The motion is perpendicular to the direction of travel of the wave.

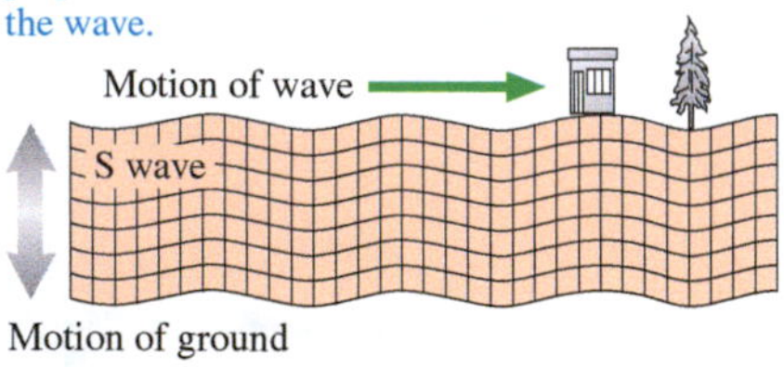


A video to support a section's topic is embedded in the eText.

**Video** Traveling Waves

The rapid motion of the earth's crust during an earthquake can produce a disturbance that travels through the earth. The two most important types of earthquake waves are S waves (which are transverse) and P waves (which are longitudinal), as shown in **FIGURE 15.2.** The longitudinal P waves are faster, but the transverse S waves are more destructive. Residents of a city a few hundred kilometers from an earthquake will feel the resulting P waves as much as a minute before the S waves, giving them a crucial early warning.

**STOP TO THINK 15.1** Spectators at a sporting event do "The Wave," as shown in the photo on the preceding page. Is this a transverse or longitudinal wave?

## 15.2 Traveling Waves

When you drop a pebble in a pond, waves travel outward. But how does this happen? How does a mechanical wave travel through a medium? In answering this question, we must be careful to distinguish the motion of the wave from the motion of the particles that make up the medium. The wave itself is not a particle, so we cannot apply Newton's laws to the wave. However, we can use Newton's laws to examine how the medium responds to a disturbance.

### Waves on a String

**FIGURE 15.3** shows a transverse *wave pulse* traveling to the right along a stretched string. Imagine watching a little dot on the string as a wave pulse passes by. As the pulse approaches from the left, the string near the dot begins to curve. Once the string

**FIGURE 15.3** The motion of a string as a wave passes.

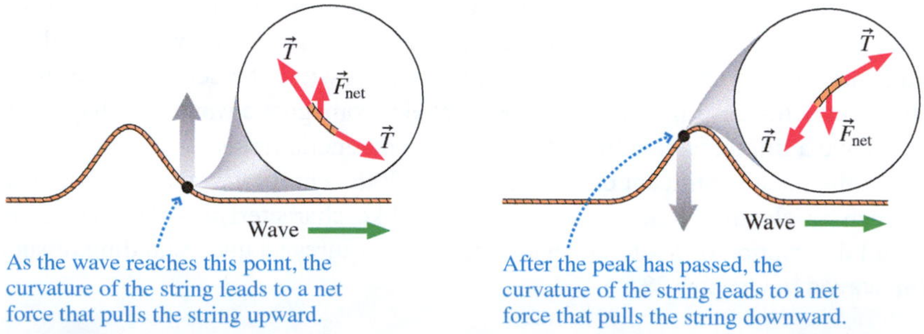

curves, the tension forces pulling on a small segment of string no longer cancel each other. As the wave passes, the curvature of the string leads to a net force that first pulls each little piece of the string up and then, after the pulse passes, back down. Each point on the string moves perpendicular to the motion of the wave, so **a wave on a string is a transverse wave.**

No new physical principles are required to understand how this wave moves. The motion of a pulse along a string is a direct consequence of the tension acting on the segments of the string. An external force created the pulse, but **once started, the pulse continues to move because of the internal dynamics of the medium.**

## Sound Waves

Next, let's see how a sound wave in air is created using a loudspeaker. When the loudspeaker cone in FIGURE 15.4a moves forward, it compresses the air in front of it, as shown in FIGURE 15.4b. The *compression* is the disturbance that travels forward through the air. This is much like the sharp push on the end of the spring on the preceding page, so **a sound wave is a longitudinal wave.** We usually think of sound waves as traveling in air, but sound can travel through any gas, through liquids, and even through solids.

Just as the motion of a wave on a string is determined by the physics of a string under tension, the motion of a sound wave in air is determined by the physics of gases that we explored in Chapter 12. Once created, the wave in Figure 15.4b will propagate forward; its motion is entirely determined by the properties of the air.

FIGURE 15.4 A sound wave produced by a loudspeaker.

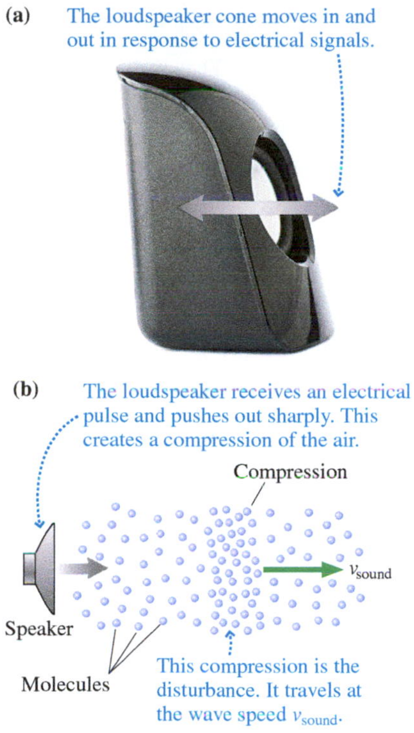


Video Figure 15.4

## Wave Speed Is a Property of the Medium

The above discussions of waves on a string and sound waves in air show that the motion of these waves depends on the properties of the medium. In fact, we'll see that **the wave speed does not depend on the shape or size of the pulse, how the pulse was generated, or how far it has traveled**—only the medium that carries the wave. Let's examine this idea in more detail for both waves on a string and sound waves.

What properties of a string determine the speed of waves traveling along the string? The only likely candidates are the string's mass, length, and tension. Because a pulse doesn't travel faster on a longer and thus more massive string, neither the total mass $m$ nor the total length $L$ is important. Instead, the speed depends on the mass-to-length *ratio,* which is called the **linear density** $\mu$ of the string:

$$\mu = \frac{m}{L} \tag{15.1}$$

Linear density characterizes the *type* of string we are using. A fat string has a larger value of $\mu$ than a skinny string made of the same material. Similarly, a steel wire has a larger value of $\mu$ than a plastic string of the same diameter.

How does the speed of a wave on a string vary with the tension and the linear density? Using what we know about forces and motion, we can make some predictions:

- A string with a greater tension responds more rapidly, so the wave will move at a higher speed. **Wave speed increases with increasing tension.**
- A string with a greater linear density has more inertia. It will respond less rapidly, so the wave will move at a lower speed. **Wave speed decreases with increasing linear density.**

A full analysis of the motion of the string leads to an expression for the speed of a wave that shows both of these trends:

$$v_{string} = \sqrt{\frac{T_s}{\mu}} \tag{15.2}$$

Wave speed on a stretched string with tension $T_s$ and linear density $\mu$

The subscript s on the symbol $T_s$ for the string's tension will distinguish it from the symbol $T$ for the *period* of oscillation.

**Sensing water waves** BIO The African clawed frog has a simple hunting strategy: It sits and "listens" for prey animals. It detects water waves, not sound waves, using an array of sensors called the *lateral line organ* on each side of its body. An incoming wave reaches different parts of the organ at different times, and this allows the frog to determine where the waves come from.

Every point on a wave pulse travels with the speed given by Equation 15.2. We can increase the wave speed either by *increasing* the string's tension (making it tighter) or by *decreasing* the string's linear density (making it skinnier). We'll examine the implications for stringed musical instruments in Chapter 16.

**EXAMPLE 15.1** **When does the spider sense its lunch?** BIO

All spiders are very sensitive to vibrations. An orb spider will sit at the center of its large, circular web and monitor radial threads for vibrations created when an insect lands. Assume that these threads are made of silk with a linear density of $1.0 \times 10^{-5}$ kg/m under a tension of 0.40 N, both typical numbers. If an insect lands in the web 30 cm from the spider, how long will it take for the spider to find out?

**STRATEGIZE** When the insect hits the web, a wave pulse will be transmitted along the silk fibers. The speed of the wave depends on the properties of the silk.

**PREPARE** The speed of a wave on a string is given by Equation 15.2.

**SOLVE** First, we determine the speed of the wave:

$$v = \sqrt{\frac{T_s}{\mu}} = \sqrt{\frac{0.40\ \text{N}}{1.0 \times 10^{-5}\ \text{kg/m}}} = 200\ \text{m/s}$$

The time for the wave to travel a distance $d = 30$ cm to reach the spider is

$$\Delta t = \frac{d}{v} = \frac{0.30\ \text{m}}{200\ \text{m/s}} = 1.5\ \text{ms}$$

**ASSESS** Spider webs are made of very light strings under significant tension, so the wave speed is quite high and we expect a short travel time—important for the spider to quickly respond to prey caught in the web. Our answer makes sense.

What properties of a gas determine the speed of a sound wave traveling through the gas? It seems plausible that the speed of a sound pulse is related to the speed with which the molecules of the gas move—faster molecules should mean a faster sound wave. In ◀◀ SECTION 12.2, we found that the typical speed of an atom of mass $m$, the root-mean-square speed, is

$$v_{\text{rms}} = \sqrt{\frac{3k_B T}{m}}$$

where $k_B$ is Boltzmann's constant and $T$ the absolute temperature in kelvin. A thorough analysis finds that the sound speed is slightly less than this rms speed, but has the same dependence on the temperature and the molecular mass:

$$v_{\text{sound}} = \sqrt{\frac{\gamma k_B T}{m}} = \sqrt{\frac{\gamma R T}{M}} \quad (15.3)$$

Sound speed in a gas at temperature $T$

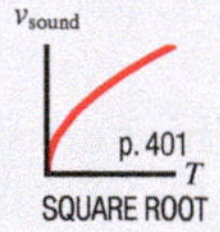


In Equation 15.3, $R$ is the gas constant, $M$ is the molar mass (kg per mol), and $\gamma$ is a constant that depends on the gas: $\gamma = 1.67$ for monatomic gases such as helium, $\gamma = 1.40$ for diatomic gases such as nitrogen and oxygen, and $\gamma \approx 1.3$ for a triatomic gas such as carbon dioxide or water vapor.

Certain trends in Equation 15.3 are worth mentioning:

- The speed of sound in air (and other gases) increases with temperature. **For calculations in this chapter, you can use the speed of sound in air at 20°C, 343 m/s, unless otherwise specified.**
- At a given temperature, the speed of sound increases as the molecular mass of the gas decreases. Thus the speed of sound in room-temperature helium is faster than that in room-temperature air.
- The speed of sound doesn't depend on the pressure or the density of the gas.

TABLE 15.1 lists the speeds of sound in various materials. The bonds between atoms in liquids and solids result in higher sound speeds in these phases of matter. Generally, sound waves travel faster in liquids than in gases, and faster in solids than in liquids.

**TABLE 15.1** The speed of sound

| Medium | Speed (m/s) |
|---|---|
| Air (0°C) | 331 |
| Air (20°C) | 343 |
| Helium (0°C) | 970 |
| Ethyl alcohol | 1170 |
| Water | 1480 |
| Human tissue (ultrasound) | 1540 |
| Lead | 1200 |
| Aluminum | 5100 |
| Granite | 6000 |
| Diamond | 12,000 |

The speed of sound in a solid depends on its density and stiffness. Light, stiff solids (such as diamond) transmit sound at very high speeds.

### EXAMPLE 15.2 The speed of sound on Mars

On a typical Martian morning, the very thin atmosphere (which is almost entirely carbon dioxide) is a frosty $-100°C$. What is the speed of sound on such a morning? At what approximate temperature would the speed be double this value?

***STRATEGIZE*** Equation 15.3 gives the speed of sound in terms of the molar mass and the absolute temperature. We will assume that the atmosphere is composed of pure $CO_2$.

***PREPARE*** The molar mass of $CO_2$ is the sum of the molar masses of the constituents: 12 g/mol for C and 16 g/mol for each O, giving $M = 12 + 16 + 16 = 44\text{ g/mol} = 0.044\text{ kg/mol}$. The absolute temperature is $T = 173$ K. Carbon dioxide is a triatomic gas, so $\gamma = 1.3$.

***SOLVE*** The speed of sound with the noted conditions is

$$v_{\text{sound}} = \sqrt{\frac{\gamma RT}{M}} = \sqrt{\frac{(1.3)(8.31\text{ J/mol}\cdot\text{K})(173\text{ K})}{0.044\text{ kg/mol}}} = 210\text{ m/s}$$

Rather than do a separate calculation to determine the temperature required for the higher speed, we can make an argument using proportionality. The speed is proportional to the square root of the temperature, so doubling the speed requires an increase in the temperature by a factor of 4 to about 690 K, or about 420°C.

***ASSESS*** The speed of sound doesn't depend on pressure, so even though the atmosphere is "thin" we needn't adjust our calculation. The speed of sound is lower for heavier molecules and colder temperatures, so we expect that the speed of sound on Mars, with its cold, carbon-dioxide atmosphere, will be lower than that on earth, just as we found.

Electromagnetic waves, such as light, travel at a much higher speed than do mechanical waves. As we'll discuss in Chapter 25, all electromagnetic waves travel at the same speed in a vacuum. We call this speed the **speed of light,** which we represent with the symbol $c$. The value of the speed of light in a vacuum is

$$v_{\text{light}} = c = 3.00 \times 10^8\text{ m/s} \tag{15.4}$$

This is almost one million times the speed of sound in air! At this speed, light could circle the earth 7.5 times in one second.

**NOTE** ▶ The speed of electromagnetic waves is lower when they travel through a material, but this value for the speed of light in a vacuum is also good for electromagnetic waves traveling through air. ◀

### EXAMPLE 15.3 How far away was the lightning?

During a thunderstorm, you see a flash from a lightning strike. 8.0 seconds later, you hear the crack of the thunder. How far away did the lightning strike?

***STRATEGIZE*** Two different kinds of waves are involved, with very different wave speeds. The flash of the lightning generates light waves; these waves travel from the point of the strike to your position almost instantaneously. The strike also generates the sound waves that you hear as thunder; these waves travel much more slowly.

***PREPARE*** The time for light to travel 1 mile (1610 m) is

$$\Delta t = \frac{d}{v} = \frac{1610\text{ m}}{3.00 \times 10^8\text{ s}} = 5.37 \times 10^{-6}\text{ s} \approx 5\ \mu\text{s}$$

We are given the time to an accuracy of only 0.1 s, so it's clear that we can ignore the travel time for the light flash! The delay between the flash and the thunder is simply the time it takes for the sound wave to travel.

***SOLVE*** We assume that the speed of sound has its room-temperature (20°C) value of 343 m/s. During the time between seeing the flash and hearing the thunder, the sound travels a distance

$$d = v\Delta t = (343\text{ m/s})(8.0\text{ s}) = 2.7 \times 10^3\text{ m} = 2.7\text{ km}$$

***ASSESS*** This seems reasonable. As you know from casual observations of lightning storms, an 8-second delay between the flash of the lightning and the crack of the thunder means a strike that is close but not too close. A few km seems reasonable.

**Video** What the Physics? Speed of Sound

**Distance to a lightning strike** Sound travels approximately 1 km in 3 s, or 1 mi in 5 s. When you see a lightning flash, start counting seconds. When you hear the thunder, stop counting. Divide the result by 3, and you will have the approximate distance to the lightning strike in kilometers; divide by 5 and you will have the approximate distance in miles.

**STOP TO THINK 15.2** Suppose you shake the end of a stretched string to produce a wave. Which of the following actions will increase the speed of the wave down the string? There may be more than one correct answer; if so, give all that are correct.

A. Move your hand up and down more quickly as you generate the wave.
B. Move your hand up and down a greater distance as you generate the wave.
C. Use a heavier string of the same length, under the same tension.
D. Use a lighter string of the same length, under the same tension.
E. Stretch the string tighter to increase the tension.
F. Loosen the string to decrease the tension.

## 15.3 Graphical and Mathematical Descriptions of Waves

Describing waves and their motion takes a bit more thought than describing the motion of particles. An object described by the particle model, such as a car or a runner, is at a particular position $x$ at a single instant of time $t$. This means we can describe a particle using a function of time, such as $x(t)$ or $v(t)$. Unlike a particle, a wave is not localized at a single point; it exists—and is changing with time—throughout the entire medium through which it travels. Thus to describe waves mathematically, we will need to use functions of both time *and* space, such as $y(x, t)$.

Before starting in on a mathematical description of waves, however, we want to get a good *visual* understanding of wave motion. To do so, we'll study two kinds of wave graphs: *snapshot graphs,* which depict the shape of the wave at one instant of time, and *history graphs,* which show the motion over time of a single point of the wave.

**NOTE** ▸ The analysis that follows is for a wave on a string, which is easy to visualize, but the results apply to any traveling wave. ◂

FIGURE 15.5 A snapshot graph of a wave pulse on a string.

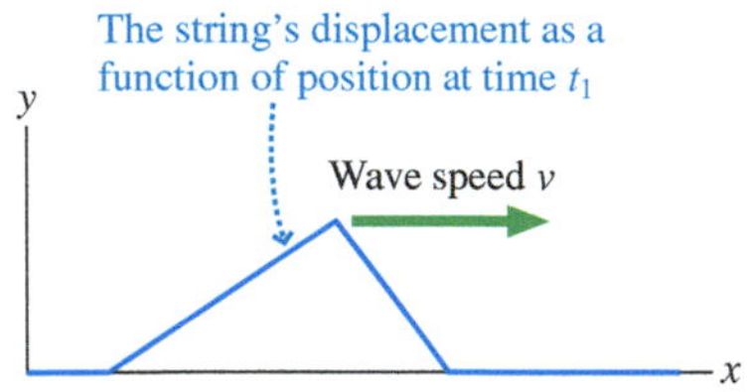


FIGURE 15.6 A sequence of snapshot graphs shows the wave in motion.

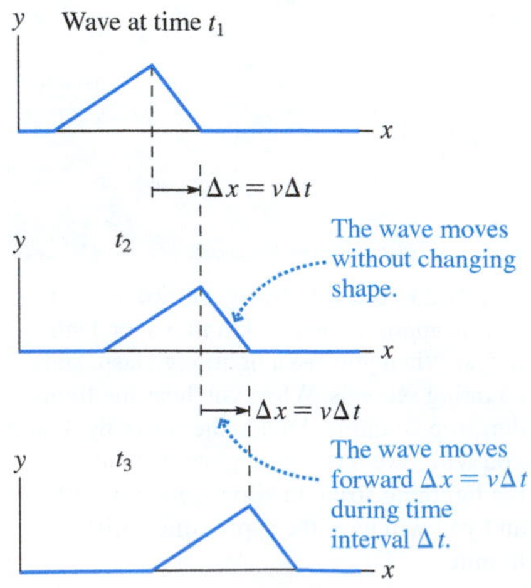


### Snapshot and History Graphs

Consider the wave pulse shown moving along a stretched string in FIGURE 15.5. (We will consider somewhat artificial triangular and square wave pulses in this section to clearly show the edges of the pulse.) The graph shows the string's displacement $y$ at a particular instant of time $t_1$ as a function of position $x$ along the string. This is a "snapshot" of the wave, much like what you might make by using a camera to take a picture of the wave at the time $t_1$. A graph that shows the wave's displacement as a function of position at a single instant of time is called a **snapshot graph.**

As the wave moves, we can take more snapshots. FIGURE 15.6 shows a sequence of snapshot graphs as the wave of Figure 15.5 continues to move. These are like successive frames from a video, reminiscent of the sequences of pictures we saw in Chapter 1. The wave pulse moves forward a distance $\Delta x = v\Delta t$ during each time interval $\Delta t$; that is, the wave moves at a constant speed.

A snapshot graph shows the motion of the *wave*, but that's only half the story. Now we want to consider the motion of the *medium.* FIGURE 15.7a shows four snapshot graphs of a wave as it travels on a string. In each of the graphs we've placed a dot, located at position $x_1$, at one point on the string. As the wave travels horizontally, the dot moves vertically because a wave on a string is a transverse wave. We use the vertical positions of the dots from the snapshot graphs to construct the graph in FIGURE 15.7b, which shows the motion of this one point on the string. We call this a **history graph** because it shows the history—the time evolution—of a particular point in the medium. The snapshot graphs are pictures of the waves at particular instants in time, but the history graph isn't: It's a record of the motion of *one point* in the medium over time.

Notice that the snapshot graphs show the steeper edge of the wave on the *right* but the history graph has the steeper edge on the *left*. As the wave moves toward the dot, the steep leading edge of the wave causes the dot to rise quickly. Later, the shallower trailing edge causes the dot to descend more slowly. On the history graph, as on any displacement-versus-time graph, earlier times are to the left and later times are to the right. The rapid rise when the wave hits the dot is at an early time and so appears on the left side of the Figure 15.7b history graph, whereas the slow descent of the dot occurs later and so appears to the right side of the graph. This reversal will not be present in all cases, though; you'll need to consider each situation individually.

**KEY CONCEPT** FIGURE 15.7 Constructing a history graph.

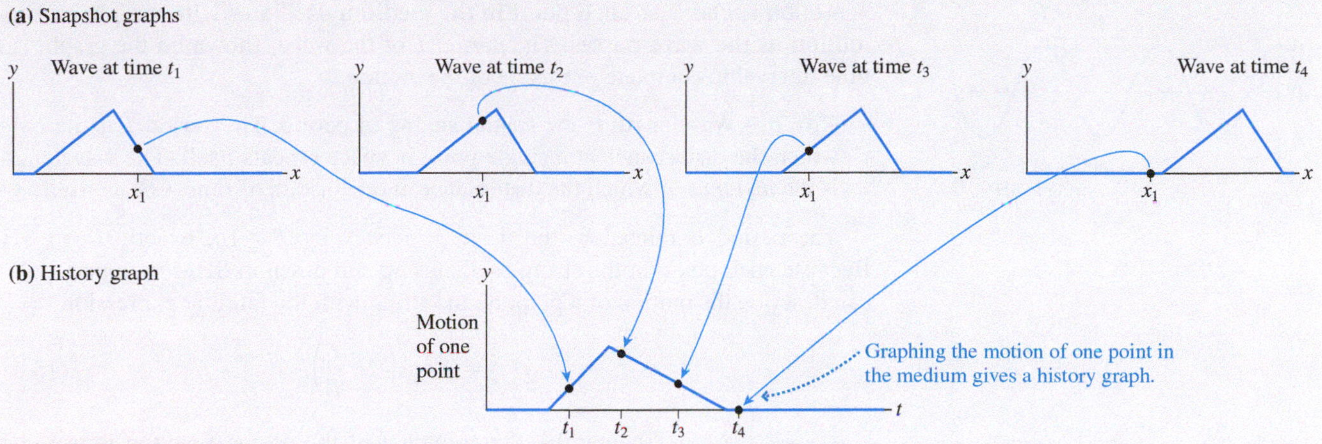

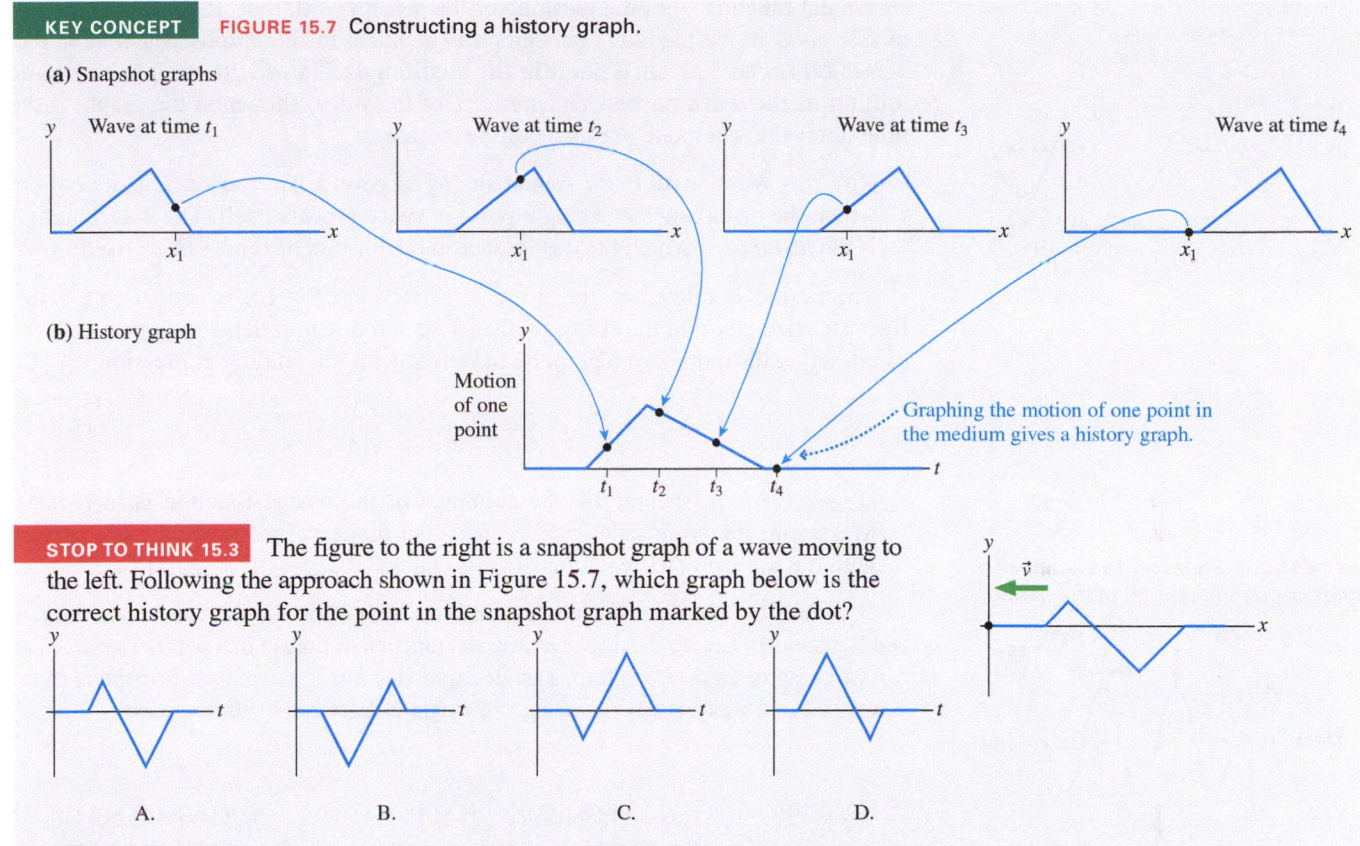

**STOP TO THINK 15.3** The figure to the right is a snapshot graph of a wave moving to the left. Following the approach shown in Figure 15.7, which graph below is the correct history graph for the point in the snapshot graph marked by the dot?

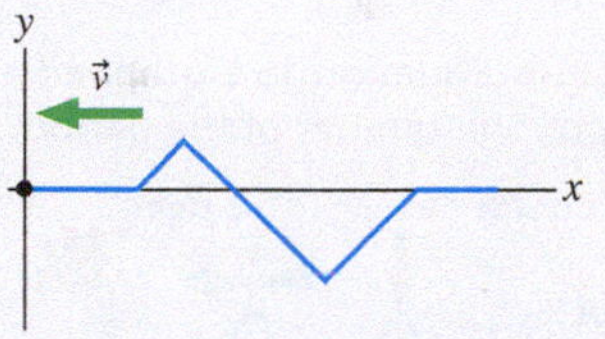


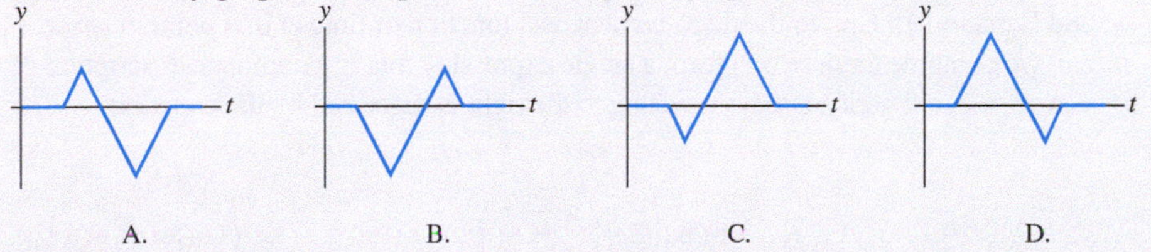


## The Mathematical Description of Sinusoidal Waves

Waves can come in many different shapes, but for the mathematical description of wave motion we will focus on a particular shape, the **sinusoidal wave.** This is the type of wave produced by a source that oscillates with simple harmonic motion. A loudspeaker cone that oscillates in SHM radiates a sinusoidal sound wave.

The pair of snapshot graphs in FIGURE 15.8 shows two successive views of a string carrying a sinusoidal wave, revealing the motion of the wave as it moves to the right. We define the **amplitude** $A$ of the wave to be the maximum value of the displacement. The crests of the wave—the high points—have displacement $y_{crest} = A$, and the troughs—the low points—have displacement $y_{trough} = -A$. Because the wave is produced by a source undergoing SHM, which is periodic, the wave is periodic as well. As you move from left to right along the wave frozen in time in the top snapshot graph of Figure 15.8, the disturbance repeats itself over and over. The distance spanned by one cycle of the motion is called the **wavelength** of the wave. Wavelength is symbolized by $\lambda$ (lowercase Greek lambda) and, because it is a length, it is measured in units of meters. The wavelength is shown in Figure 15.8 as the distance between two crests, but it could equally well be the distance between two troughs. As time passes, the wave moves to the right; comparing the two snapshot graphs in Figure 15.8 makes this motion apparent.

FIGURE 15.8 Snapshot graphs show the motion of a sinusoidal wave.

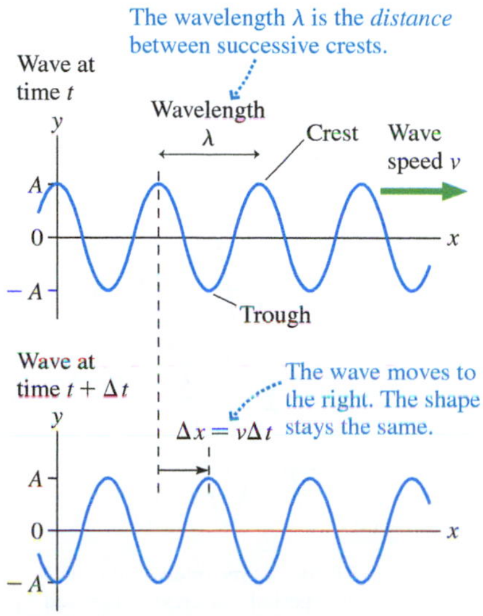

The snapshot graphs of Figure 15.8 show that the wave, at one instant in time, is a sinusoidal function of the distance $x$ along the wave, with wavelength $\lambda$. At the time represented by the top graph of Figure 15.8, the displacement is given by

$$y(x) = A\cos\left(2\pi\frac{x}{\lambda}\right) \tag{15.5}$$

FIGURE 15.9 A history graph shows the motion of a point on a string carrying a sinusoidal wave.

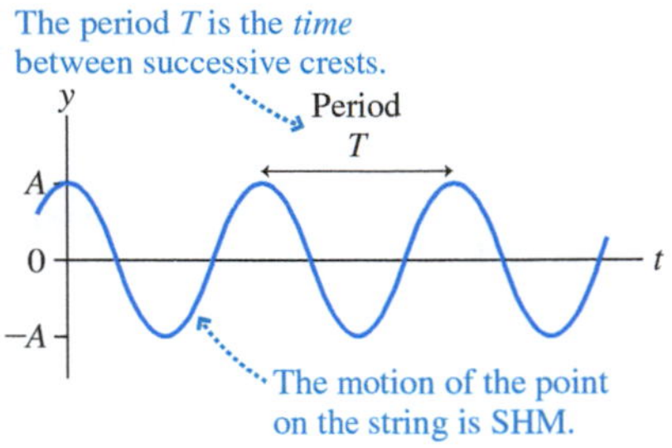


Next, let's look at the motion of a point in the medium as this wave passes. FIGURE 15.9 shows a history graph for a point on a string as the sinusoidal wave of Figure 15.8 passes by. This graph has exactly the same shape as the snapshot graphs of Figure 15.8—it's a sinusoidal function—but the meaning of the graph is different: It shows the motion of *one point* in the medium. This graph is identical to the graphs you first saw in ◀ SECTION 14.3 because **each point in the medium oscillates with simple harmonic motion as the wave passes.** The *period* $T$ of the wave, shown on the graph, is the time interval to complete one cycle of the motion.

NOTE ▶ Wavelength is the spatial analog of period. The period $T$ is the *time* in which the disturbance at a single point in space repeats itself. The wavelength $\lambda$ is the *distance* in which the disturbance at one instant of time repeats itself. ◀

The period is related to the wave *frequency* by $T = 1/f$, exactly as in SHM. Because each point on the string oscillates up and down in SHM with period $T$, we can describe the motion of a point on the string with the familiar expression

$$y(t) = A\cos\left(2\pi\frac{t}{T}\right) \tag{15.6}$$

NOTE ▶ As in Chapter 14, the argument of the cosine function is in radians. Make sure that your calculator is in radian mode before starting a calculation with the trigonometric equations in this chapter. ◀

FIGURE 15.10 Equation 15.7 graphed at intervals of one-quarter of the period.

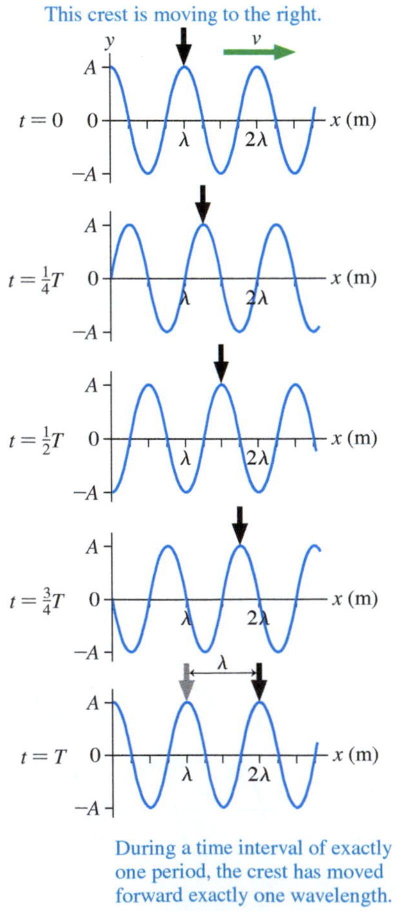


Equation 15.5 gives the displacement as a function of position at one instant in time, and Equation 15.6 gives the displacement as a function of time at one point in space. How can we combine these two to form a single expression that is a complete description of the wave? As we'll verify, a wave traveling to the right is described by the equation

$$y(x, t) = A\cos\left(2\pi\left(\frac{x}{\lambda} - \frac{t}{T}\right)\right) \tag{15.7}$$

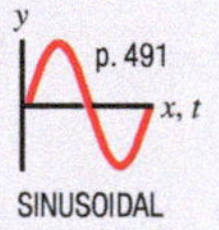


Displacement of a traveling wave moving to the right with amplitude $A$, wavelength $\lambda$, and period $T$

The notation $y(x, t)$ indicates that the displacement $y$ is a function of the *two* variables $x$ and $t$. We must specify both where ($x$) and when ($t$) before we can calculate the displacement of the wave.

We can understand why this expression describes a wave traveling to the right by looking at FIGURE 15.10. In the figure, we have graphed Equation 15.7 at five instants of time, each separated by one-quarter of the period $T$, to make five snapshot graphs. The crest marked with the arrow represents one point on the wave. As the time $t$ increases, so does the position $x$ of this point—the wave moves to the right. One full period has elapsed between the first graph and the last. During this time, the wave has moved by one wavelength. And, as the wave has passed, each point in the medium has undergone one complete oscillation.

For a wave traveling to the left, we have a slightly different form:

$$y(x, t) = A\cos\left(2\pi\left(\frac{x}{\lambda} + \frac{t}{T}\right)\right) \tag{15.8}$$

Displacement of a traveling wave moving to the left

**NOTE** ► A wave moving to the right (the $+x$-direction) has a $-$ in the expression, while a wave moving to the left (the $-x$-direction) has a $+$. Remember that the sign is *opposite* the direction of travel. ◄

**EXAMPLE 15.4 Determining the rise and fall of a moving boat**

A boat is moving to the right at 5.0 m/s with respect to the water. An ocean wave is moving to the left, opposite to the motion of the boat. The waves have 2.0 m between the top of the crests and the bottom of the troughs. The period of the waves is 8.3 s, and their wavelength is 110 m. At one instant, the boat sits on a crest of the wave. 20 s later, what is the vertical displacement of the boat?

**STRATEGIZE** Because the wave is moving to the left, the wave height as a function of position and time is given by Equation 15.8. From the boat's speed, we will find its position $x$ at time $t = 20$ s, and then use Equation 15.8 to find its height.

**PREPARE** We begin with a visual overview, as in FIGURE 15.11. Let $t = 0$ be the instant the boat is on the crest, and draw a snapshot graph of the traveling wave at that time. The boat begins at a crest of the wave, so we see that the boat can start the problem at $x = 0$.

FIGURE 15.11 Visual overview for the boat.

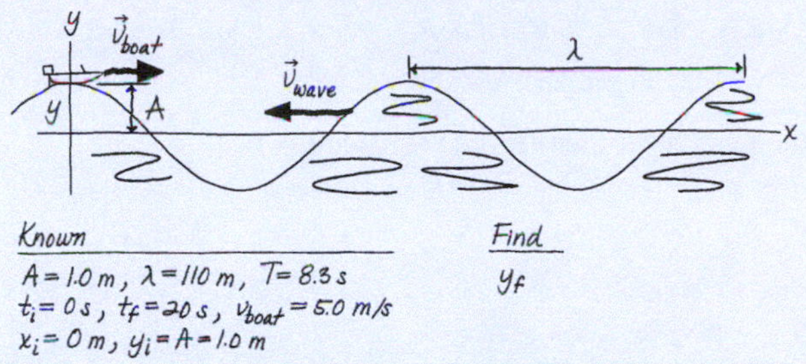


The distance between the high and low points of the wave is 2.0 m; the amplitude is half this, so $A = 1.0$ m. The wavelength and period are given in the problem.

**SOLVE** The boat is moving to the right at 5.0 m/s. At $t_f = 20$ s the boat is at position

$$x_f = (5.0\ \text{m/s})(20\ \text{s}) = 100\ \text{m}$$

We need to find the wave's displacement at this position and time. Substituting known values for amplitude, wavelength, and period into Equation 15.8, for a wave traveling to the left, we obtain the following equation for the wave:

$$y(x, t) = (1.0\ \text{m})\cos\left(2\pi\left(\frac{x}{110\ \text{m}} + \frac{t}{8.3\ \text{s}}\right)\right)$$

At $t_f = 20$ s and $x_f = 100$ m, the boat's displacement on the wave is

$$y_f = y(\text{at } 100\ \text{m}, 20\ \text{s}) = (1.0\ \text{m})\cos\left(2\pi\left(\frac{100\ \text{m}}{110\ \text{m}} + \frac{20\ \text{s}}{8.3\ \text{s}}\right)\right)$$
$$= -0.42\ \text{m}$$

Don't forget that your calculator must be in radian mode when you make your final computation!

**ASSESS** The final displacement is negative—which means that the boat is in a trough of a wave, not underwater.

## The Fundamental Relationship for Sinusoidal Waves

In Figure 15.10, one critical observation is that the wave crest marked by the arrow has moved one full wavelength between the first graph and the last. That is, **during a time interval of exactly one period $T$, each crest of a sinusoidal wave travels forward a distance of exactly one wavelength $\lambda$.** Because speed is distance divided by time, the wave speed must be

Video Traveling Waves Problems

$$v = \frac{\text{distance}}{\text{time}} = \frac{\lambda}{T} \tag{15.9}$$

Because $f = 1/T$, it is customary to write Equation 15.9 in the form

$$v = \lambda f \tag{15.10}$$

Relationship between velocity, wavelength, and frequency for sinusoidal waves

Although Equation 15.10 has no special name, it is *the* fundamental relationship for sinusoidal waves. When using it, keep in mind the *physical* meaning that a wave moves forward a distance of one wavelength during a time interval of one period.

At this point, it's worthwhile to bring together the details about wave motion we've learned in order to review and to make connections.

## SYNTHESIS 15.1 Wave motion

Wave speed is determined by the medium.

For a wave on a stretched string:

$$v_{\text{string}} = \sqrt{\frac{T_s}{\mu}}$$

$T_s$: String tension (N)
$\mu$: String mass divided by length (kg/m)

For a sound wave in a gas:

$$v_{\text{sound}} = \sqrt{\frac{\gamma k_B T}{m}}$$

$T$: Temperature (K)
$m$: Molecular mass (kg)

For example, at room temperature in air:

$$v_{\text{sound}}(20\ °\text{C}) = 343\ \text{m/s}$$

In a vacuum or air, the speed of light is

$$v_{\text{light}} = 3.00 \times 10^8\ \text{m/s}$$

A sinusoidal wave moves one wavelength in one period, giving the fundamental relationship

$$v = \lambda f = \frac{\lambda}{T}$$

$v$: Speed of the wave (m/s)
$f$: Frequency of the wave (Hz)
$\lambda$: Wavelength (m)
$T$: Period of the wave (s)

As a sinusoidal wave travels, each point in the medium moves with simple harmonic motion.

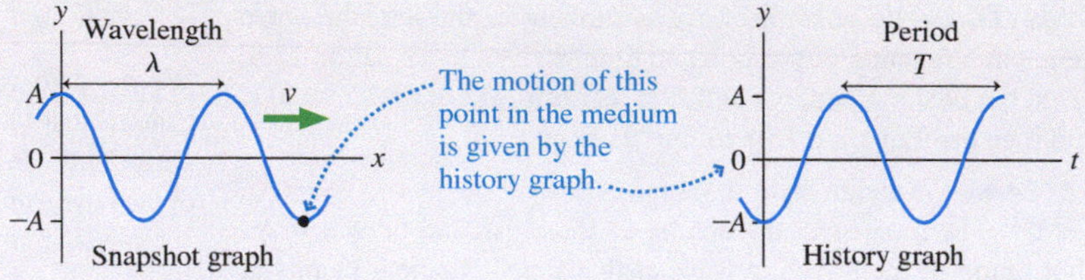


The displacement of a traveling sinusoidal wave is given by the equations

$$y(x, t) = A \cos\left(2\pi\left(\frac{x}{\lambda} - \frac{t}{T}\right)\right)$$

Wave moving to the right

$$y(x, t) = A \cos\left(2\pi\left(\frac{x}{\lambda} + \frac{t}{T}\right)\right)$$

Wave moving to the left

These expressions are appropriate for a wave that is at its maximum ($y = A$) at the origin ($x = 0$) at $t = 0$.

## EXAMPLE 15.5 Writing the equation for a wave

A sinusoidal wave with an amplitude of 1.5 cm and a frequency of 100 Hz travels at 200 m/s in the positive $x$-direction. Write the equation for the wave's displacement as it travels.

**STRATEGIZE** The wave is moving to the right, so its equation will have the form of Equation 15.7. We will need the amplitude $A$, the wavelength $\lambda$, and the period $T$.

**PREPARE** The problem statement gives the characteristics of the wave: $A = 1.5\ \text{cm} = 0.015\ \text{m}$, $v = 200\ \text{m/s}$, and $f = 100\ \text{Hz}$.

**SOLVE** The amplitude is given in the problem statement. To find the wavelength, we can use the fundamental relationship for sinusoidal waves:

$$\lambda = \frac{v}{f} = \frac{200\ \text{m/s}}{100\ \text{Hz}} = 2.0\ \text{m}$$

The period can be calculated from the frequency:

$$T = \frac{1}{f} = \frac{1}{100\ \text{Hz}} = 0.010\ \text{s}$$

With these values in hand, we can write the equation for the wave's displacement:

$$y(x, t) = (0.015\ \text{m}) \cos\left(2\pi\left(\frac{x}{2.0\ \text{m}} - \frac{t}{0.010\ \text{s}}\right)\right)$$

**ASSESS** If the speed of a wave is known, you can use the fundamental relationship for sinusoidal waves to find the wavelength if you are given the frequency, or the frequency if you are given the wavelength. You'll often need to do this in the early stages of problems that you solve.

**STOP TO THINK 15.4** Three waves travel to the right with the same speed. Which wave has the highest frequency? All three graphs have the same horizontal scale.

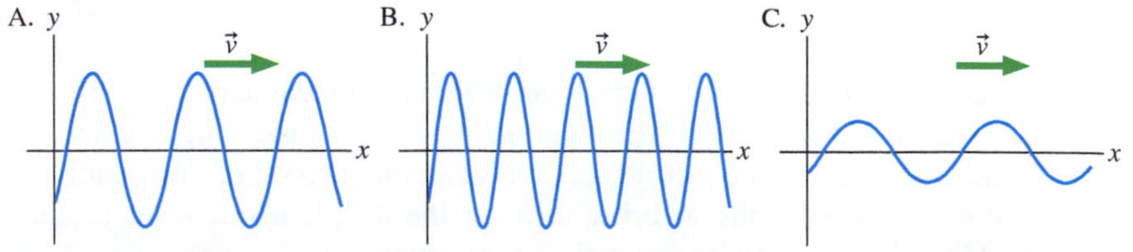

## 15.4 Sound and Light Waves

Think about how you are experiencing the world right now. Chances are, your senses of sight and sound are hard at work, detecting and interpreting light and sound waves from the world around you.

### Sound Waves

We saw in Figure 15.4 how a loudspeaker creates a sound wave. If the loudspeaker cone moves with simple harmonic motion, it will create a sinusoidal sound wave, as illustrated in FIGURE 15.12. Each time the cone moves forward, it pushes the air molecules closer together, creating a region of higher pressure. A half cycle later, as the cone moves backward, the air has room to expand and the pressure decreases. These regions of higher and lower pressure are called **compressions** and **rarefactions,** respectively.

As Figure 15.12 suggests, it is often most convenient and informative to think of a sound wave as a pressure wave. As the graph of the pressure shows, the pressure oscillates sinusoidally around the atmospheric pressure $p_{atmos}$. This is a snapshot graph of the wave at one instant of time, and the distance between two adjacent crests (two points of maximum compression) is the wavelength $\lambda$.

When the wave reaches your ear, the oscillating pressure causes your eardrums to vibrate. This vibration is transferred through your inner ear to the cochlea, where it is sensed, as we learned in Chapter 14. Humans with normal hearing are able to detect sinusoidal sound waves with frequencies between about 20 Hz and 20,000 Hz, or 20 kHz. Low frequencies are perceived as a "low pitch" bass note, while high frequencies are heard as a "high pitch" treble note.

FIGURE 15.12 A sound wave is a pressure wave.

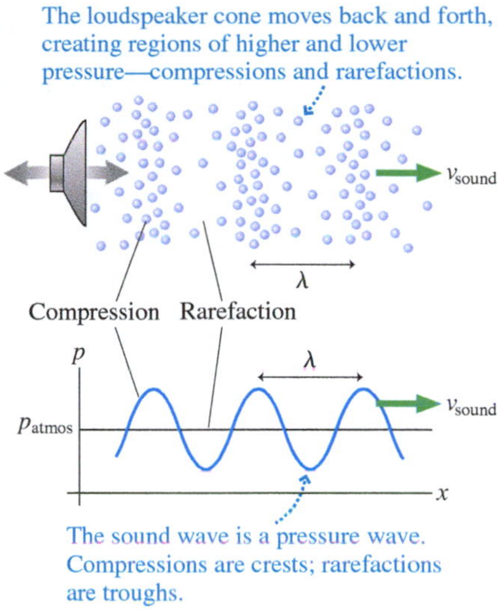


**EXAMPLE 15.6 Range of wavelengths of sound**

What are the wavelengths of sound waves at the limits of human hearing and at the midrange frequency of 500 Hz? Notes sung by human voices are near 500 Hz, as are notes played by striking keys near the center of a piano keyboard.

**STRATEGIZE** We will solve for the wavelengths using the fundamental relationship $v = \lambda f$.

**PREPARE** We will do our calculation at room temperature, 20°C, so we will use $v = 343$ m/s for the speed of sound.

**SOLVE** For the three wavelengths, the fundamental relationship gives

$$f = 20\text{ Hz}: \quad \lambda = \frac{v}{f} = \frac{343\text{ m/s}}{20\text{ Hz}} = 17\text{ m}$$

$$f = 500\text{ Hz}: \quad \lambda = \frac{v}{f} = \frac{343\text{ m/s}}{500\text{ Hz}} = 0.69\text{ m}$$

$$f = 20\text{ kHz}: \quad \lambda = \frac{v}{f} = \frac{343\text{ m/s}}{20 \times 10^3\text{ Hz}} = 0.017\text{ m} = 1.7\text{ cm}$$

**ASSESS** The wavelength of a 20 kHz note is only 1.7 cm. At the other extreme, a 20 Hz note has a huge wavelength of 17 m! A wave moves forward one wavelength during a time interval of one period, and a wave traveling at 343 m/s can move 17 m during the $\frac{1}{20}$ s period of a 20 Hz note.

It is well known that dogs are sensitive to high frequencies that humans cannot hear. Other animals also have quite different ranges of hearing than humans; some examples are listed in TABLE 15.2.

TABLE 15.2 Range of hearing for selected animals

| Animal | Range of hearing (Hz) |
|---|---|
| Elephant | <5–12,000 |
| Owl | 200–12,000 |
| Human | 20–20,000 |
| Dog | 30–45,000 |
| Mouse | 1000–90,000 |
| Bat | 2000–100,000 |
| Porpoise | 75–150,000 |

Elephants communicate over great distances with vocalizations at frequencies far too low for us to hear. There are animals that use frequencies well above the range of our hearing as well. High-frequency sounds are useful for *echolocation*—emitting a pulse of sound and listening for its reflection. Bats, which generally feed at night, rely much more on their hearing than their sight. They find and catch insects by echolocation, emitting loud chirps whose reflections are detected by their large, sensitive ears. The frequencies that they use are well above the range of our hearing; we call such sound **ultrasound.** Why do bats and other animals use high frequencies for this purpose?

In Chapter 19, we will look at the *resolution* of optical instruments. The finest detail that your eye—or any optical instrument—can detect is limited by the wavelength of light. Shorter wavelengths allow for the imaging of smaller details.

The same limitations apply to the acoustic image of the world made by bats. In order to sense fine details of their surroundings, bats must use sound of very short wavelength (and thus high frequency). Other animals that use echolocation, such as porpoises, also produce and sense high-frequency sounds.

Sound travels very well through tissues in the body, and the reflections of sound waves from different tissues can be used to create an image of the body's interior. The fine details necessary for a clinical diagnosis require the short wavelengths of ultrasound. You have certainly seen ultrasound images taken during pregnancy, where the use of x rays is clearly undesirable.

**EXAMPLE 15.7** **Ultrasonic frequencies in medicine** BIO

To make a sufficiently detailed ultrasound image of a fetus in its mother's uterus, a physician has decided that a wavelength of 0.50 mm is needed. What frequency is required?

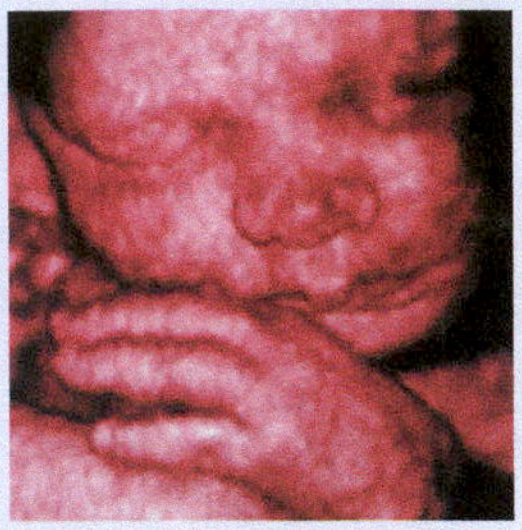

Computer processing of an ultrasound image shows fine detail.

**STRATEGIZE** We need to remember that the speed of sound in tissue is different from the speed in air.

**PREPARE** The speed of ultrasound in the body is given in Table 15.1 as 1540 m/s.

**SOLVE** We can use the fundamental relationship for sinusoidal waves to calculate

$$f = \frac{v}{\lambda} = \frac{1540\ \text{m/s}}{0.50 \times 10^{-3}\ \text{m}} = 3.1 \times 10^6\ \text{Hz} = 3.1\ \text{MHz}$$

**ASSESS** This is a reasonable result. Clinical ultrasound uses frequencies in the range of 1–20 MHz. Lower frequencies have greater penetration; higher frequencies (and thus shorter wavelengths) show finer detail.

## Light and Other Electromagnetic Waves

A light wave is an electromagnetic wave, an oscillation of the electromagnetic field. Other electromagnetic waves, such as radio waves, microwaves, and ultraviolet light, have the same physical characteristics as light waves even though we cannot sense them with our eyes. As we saw earlier, all electromagnetic waves travel through a vacuum (or air) with the same speed: $c = 3.00 \times 10^8\ \text{m/s}$.

The wavelengths of light are extremely short. Visible light is an electromagnetic wave with a wavelength (in air) of roughly 400 nm ($= 400 \times 10^{-9}$ m) to 700 nm ($=700 \times 10^{-9}$ m). Each wavelength is perceived as a different color. Longer wavelengths, in the 600–700 nm range, are seen as orange or red light; shorter wavelengths, in the 400–500 nm range, are seen as blue or violet light.

FIGURE 15.13 shows that the visible spectrum is a small slice out of the much broader **electromagnetic spectrum.** We will have much more to say about light and the rest of the electromagnetic spectrum in future chapters.

FIGURE 15.13 The electromagnetic spectrum from $10^6$ Hz to $10^{18}$ Hz.

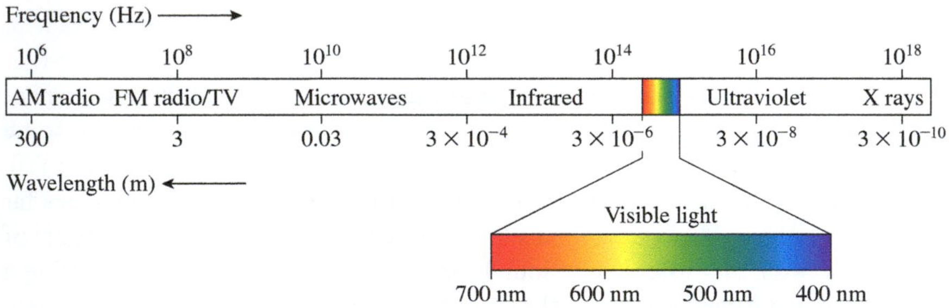


**EXAMPLE 15.8** **Finding the frequency of microwaves**

The wavelength of microwaves in a microwave oven is 12 cm. What is the frequency of the waves?

**STRATEGIZE** The frequency is related to the wavelength by the fundamental relationship $f = v/\lambda$.

**PREPARE** Microwaves are electromagnetic waves, so their speed is the speed of light, $c = 3.00 \times 10^8\ \text{m/s}$.

**SOLVE** Using the fundamental relationship, we find

$$f = \frac{v}{\lambda} = \frac{3.00 \times 10^8\ \text{m/s}}{0.12\ \text{m}} = 2.5 \times 10^9\ \text{Hz} = 2.5\ \text{GHz}$$

**ASSESS** This is a high frequency, but the speed of the waves is also very high, so our result seems reasonable.

We've just used the same equation to describe ultrasound and microwaves, which brings up a remarkable point: The wave model we've been developing applies equally well to all types of waves. Features such as wavelength, frequency, and speed are characteristics of waves in general. At this point we don't yet know any details about electromagnetic fields, yet we can use the wave model to say some significant things about the properties of electromagnetic waves.

The 2.5 GHz frequency of microwaves in an oven that we found in Example 15.8 is similar to the frequencies of other devices. Many cordless phones work at a frequency of 2.4 GHz, and cell phones at just under 2.0 GHz. Although these frequencies are close to those of the waves that heat your food in a microwave oven, the *intensity* is much less—which brings us to the topic of the next section.

**STOP TO THINK 15.5** Comparing two different types of electromagnetic waves, infrared and ultraviolet, we can say that

A. Infrared has a longer wavelength and higher frequency than ultraviolet.
B. Infrared has a shorter wavelength and higher frequency than ultraviolet.
C. Infrared has a longer wavelength and lower frequency than ultraviolet.
D. Infrared has a shorter wavelength and lower frequency than ultraviolet.

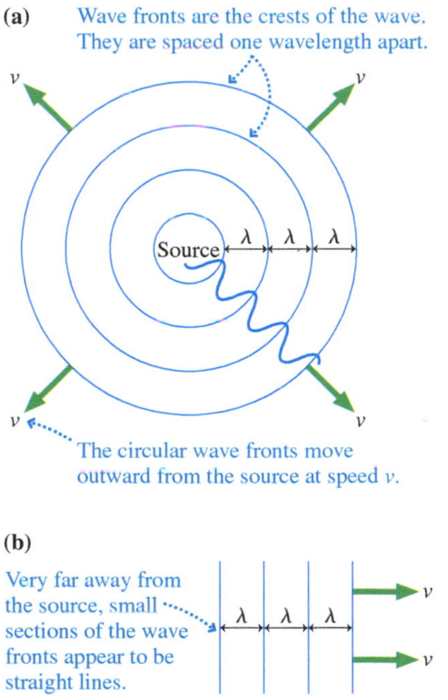


**FIGURE 15.14** The wave fronts of a circular or spherical wave.

## 15.5 Energy and Intensity

A traveling wave transfers energy from one point to another. The sound wave from a loudspeaker sets your eardrum into motion. Light waves from the sun warm the earth and, if focused with a lens, can start a fire. The *power* of a wave is the rate, in joules per second, at which the wave transfers energy. As you learned in ◀◀ SECTION 10.10, power is measured in watts. A person singing or shouting as loud as possible is emitting energy in the form of sound waves at a rate of about 1 W, or 1 J/s. In this section, we will learn how to characterize the power of waves. A first step in doing so is to understand how waves change as they spread out.

### Circular, Spherical, and Plane Waves

If you take a photograph of ripples spreading on a pond and mark the location of the *crests* on the photo, your picture looks like FIGURE 15.14a. The lines that locate the crests are **wave fronts,** and they are spaced precisely one wavelength apart. A wave like this is called a **circular wave,** a two-dimensional wave that spreads across a surface.

Although the wave fronts are circles, you would hardly notice the curvature if you observed a small section of the wave front very far away from the source. The wave fronts would appear to be parallel lines, still spaced one wavelength apart and traveling at speed $v$ as in FIGURE 15.14b.

Many waves, such as sound waves or light waves, move in three dimensions. Loudspeakers and lightbulbs emit **spherical waves.** The crests of the wave form a series of spherical shells separated by the wavelength $\lambda$. In essence, the waves are three-dimensional ripples. It is useful to draw wave-front diagrams such as Figure 15.14a, but now the circles are slices through the spherical shells locating the wave crests.

If you observe a spherical wave far from its source, the small piece of the wave front that you can see is a little patch on the surface of a very large sphere. If the radius of the sphere is large, you will not notice the curvature and this patch of the wave front appears to be a plane. FIGURE 15.15 illustrates the idea of a **plane wave.**

**FIGURE 15.15** A plane wave.

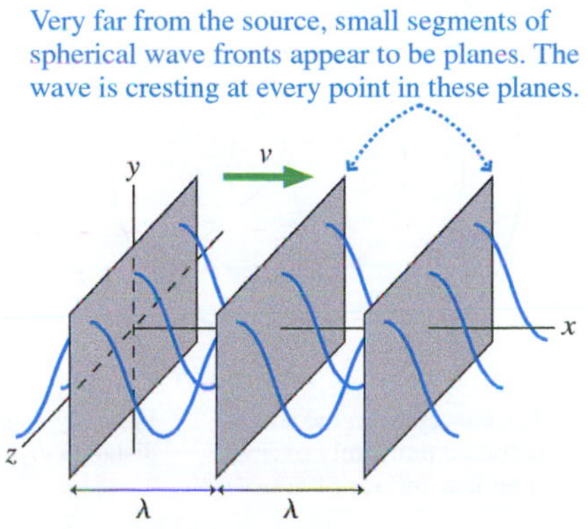

Very far from the source, small segments of spherical wave fronts appear to be planes. The wave is cresting at every point in these planes.
y
v
x
z
λ
λ

## Power, Energy, and Intensity

Imagine doing two experiments with a lightbulb that emits 2 W of visible light. In the first, you use this single lightbulb to illuminate a large lecture hall. In the second experiment, you use the same lightbulb to illuminate a small closet. The power emitted by the bulb is the same in both cases, but, as you know, the walls of the closet appear much brighter than those of the lecture hall because the same power is concentrated in a much smaller area. We say that the light striking the closet walls is more *intense* than the light striking the walls of the lecture hall. Similarly, a loudspeaker that beams its sound forward into a small area produces a louder sound in that area than a speaker of equal power that radiates the sound in all directions. Quantities such as brightness and loudness depend not only on the rate of energy transfer, or power, but also on the *area* that receives that power.

FIGURE 15.16 Plane waves of power $P$ impinge on area $a$.

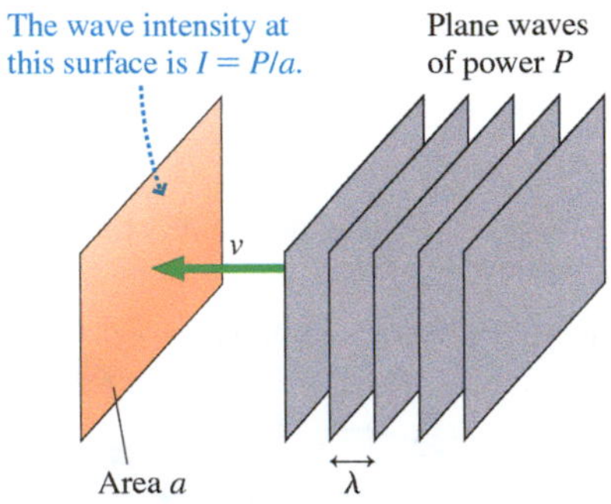


FIGURE 15.16 shows a wave impinging on a surface of area $a$. The surface is perpendicular to the direction in which the wave is traveling. This might be a real physical surface, such as your eardrum or a solar cell, but it could equally well be a mathematical surface in space that the wave passes right through. If the wave has power $P$, we define the **intensity** $I$ of the wave as

$$I = \frac{P}{a} \tag{15.11}$$

The SI units of intensity are $\text{W/m}^2$. Because intensity is a power-to-area ratio, a wave concentrated onto a small area has a higher intensity than a wave of equal power that is spread out over a large area.

NOTE ▶ In this chapter we will use $a$ for area to avoid confusion with amplitude, for which we use the symbol $A$. ◀

**EXAMPLE 15.9 The intensity of a laser beam**

A bright, tightly focused laser pointer emits 1.0 mW of light power into a beam that is 1.0 mm in diameter. What is the intensity of the laser beam?

***STRATEGIZE*** The laser's intensity is given by Equation 15.11.

***SOLVE*** The light waves of the laser beam pass through a circle of diameter 1.0 mm. The intensity of the laser beam is

$$I = \frac{P}{a} = \frac{P}{\pi r^2} = \frac{0.0010\text{ W}}{\pi(0.00050\text{ m})^2} = 1300\text{ W/m}^2$$

***ASSESS*** This intensity is roughly equal to the intensity of sunlight at noon on a summer day. Such a high intensity for a low-power source may seem surprising, but the area is very small, so the energy is packed into a tiny spot. You know that the light from a laser pointer won't burn you but you don't want the beam to shine into your eye, so an intensity similar to that of sunlight seems reasonable.

FIGURE 15.17 A source emitting uniform spherical waves.

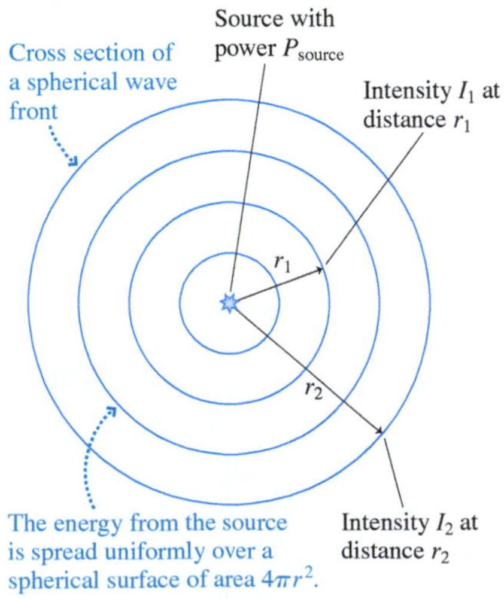


Sound from a loudspeaker and light from a lightbulb become less intense as you get farther from the source. This is because spherical waves spread out to fill larger and larger volumes of space. If a source of spherical waves radiates uniformly in all directions, then, as FIGURE 15.17 shows, the power at distance $r$ is spread uniformly over the surface of a sphere of radius $r$. The surface area of a sphere is $a = 4\pi r^2$, so the intensity of a uniform spherical wave is

$$I = \frac{P_{source}}{4\pi r^2} \tag{15.12}$$

Intensity at distance $r$ of a spherical wave from a source of power $P_{source}$

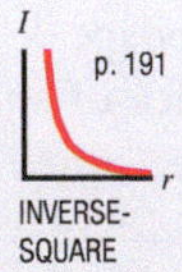


p. 191

The inverse-square dependence of $r$ is really just a statement of energy conservation. The source emits energy at the rate of $P$ joules per second. The energy is spread over a larger and larger area as the wave moves outward. Consequently, the *energy per area* must decrease in proportion to the surface area of a sphere.

If the intensity at distance $r_1$ is $I_1 = P_{\text{source}}/4\pi r_1{}^2$ and the intensity at $r_2$ is $I_2 = P_{\text{source}}/4\pi r_2^2$, then you can see that the intensity *ratio* is

$$\frac{I_1}{I_2} = \frac{r_2^2}{r_1^2} \tag{15.13}$$

You can use Equation 15.13 to compare the intensities at two distances from a source without needing to know the power of the source.

**NOTE** ▶ Wave intensities are strongly affected by reflections and absorption. Equations 15.12 and 15.13 apply to situations such as the light from a star and the sound from a firework exploding high in the air. Indoor sound does *not* obey a simple inverse-square law because of the many reflecting surfaces. ◀

**The better to hear you with** BIO The great grey owl has its ears on the front of its face, hidden behind its facial feathers. Its round face works like a radar dish, collecting the energy of sound waves over a large area and "funneling" it into the ears. This allows owls to sense very quiet sounds.

### EXAMPLE 15.10 Intensity of sunlight on Mars

The intensity of sunlight on the earth's surface is approximately 1000 W/m$^2$ at noon on a summer day. Mars orbits at a distance from the sun approximately 1.5 times that of earth.

a. Assuming similar absorption of energy by the Martian atmosphere, what would you predict for the intensity of sunlight at noon during the Martian summer?
b. The Sojourner rover, an early Mars rover, had a rectangular array of solar cells approximately 0.60 m long and 0.37 m wide. What is the maximum solar energy this array could capture?
c. If we assume a solar-to-electric conversion efficiency of 18%, typical of high-quality solar cells, what is the maximum useful power the solar cells could produce?

**STRATEGIZE** We aren't given the power emitted by the sun or the distances from the sun to earth and the sun to Mars, but we can find the intensity at the surface of Mars by using ratios of distances and intensities. Once we know the intensity, we will use the area of the solar array to determine the energy that can be captured.

**PREPARE** The ratio $(r_{\text{earth}}/r_{\text{Mars}})^2$ is equal to $(1/1.5)^2$.

**SOLVE** a. According to Equation 15.13,

$$I_{\text{Mars}} = I_{\text{earth}}\frac{r_{\text{earth}}^2}{r_{\text{Mars}}^2} = (1000\ \text{W/m}^2)\left(\frac{1}{1.5}\right)^2 = 440\ \text{W/m}^2$$

b. The area of the solar array is

$$a = (0.60\ \text{m})(0.37\ \text{m}) = 0.22\ \text{m}^2$$

If the solar cells are turned to face the sun, the full area can capture energy. We can rewrite Equation 15.11 to give the power capture by the solar array in terms of area and intensity:

$$P = Ia = (440\ \text{W/m}^2)(0.22\ \text{m}^2) = 100\ \text{W}$$

That is, the cells capture solar energy at the rate of 100 joules per second.

c. Not all of this energy is converted to electricity. Multiplying by the 18% efficiency, we get

$$P = (100\ \text{W})(0.18) = 18\ \text{W}$$

**ASSESS** The intensity of sunlight is quite a bit less on Mars than on earth, as we would expect given the greater distance of Mars from the sun. And the efficiency of solar cells is modest, so the final result—though small—seems reasonable. Indeed, the peak power from the rover's solar array was just less than this.

**STOP TO THINK 15.6** A plane wave, a circular wave, and a spherical wave all have the same intensity. Each of the waves travels the same distance. Afterward, which wave has the highest intensity?

A. The plane wave B. The circular wave C. The spherical wave

## 15.6 Loudness of Sound

Ten guitars playing in unison sound only about twice as loud as one guitar. Generally, **increasing the sound intensity by a factor of 10 results in an increase in perceived loudness by a factor of approximately 2.** The difference in intensity between the quietest sound you can detect and the loudest you can safely hear is a factor of 1,000,000,000,000! A normal conversation has 10,000 times the sound intensity of a whisper, but it sounds only about 16 times as loud.

The loudness of sound is measured by a quantity called the **sound intensity level.** Because of the wide range of intensities we can hear, and the fact that the

**The loudest animal in the world** BIO
The blue whale is the largest animal in the world, up to 30 m (about 100 ft) long, weighing 150,000 kg or more. It is also the loudest. At close range in the water, the 10–30 second calls of the blue whale would be intense enough to damage tissues in your body. Their loud, low-frequency calls can be heard by other whales hundreds of miles away.

difference in perceived loudness is much less than the actual difference in intensity, the sound intensity level is measured on a *logarithmic scale.* In this section we will explain what this means. The units of sound intensity level (i.e., of loudness) are *decibels,* a word you have likely heard.

## The Decibel Scale

There is a lower limit to the intensity of sound that a human can hear. The exact value varies among individuals and with frequency, but an average value for the lowest-intensity sound that can be heard in an extremely quiet room is

$$I_0 = 1.0 \times 10^{-12}\ \text{W/m}^2$$

This intensity is called the *threshold of hearing.*

It's logical to place the zero of our loudness scale at the threshold of hearing. All other sounds can then be referenced to this intensity. To create a loudness scale, we define the *sound intensity level,* expressed in **decibels** (dB), as

$$\beta = (10\ \text{dB})\log_{10}\left(\frac{I}{I_0}\right) \tag{15.14}$$

Sound intensity level in decibels for a sound of intensity $I$

$\beta$ is the lowercase Greek letter beta. The decibel is named after Alexander Graham Bell, inventor of the telephone. Sound intensity level is dimensionless, since it's formed from the ratio of two intensities, so decibels are actually just a *name* to remind us that we're dealing with an intensity *level* rather than a true intensity.

Equation 15.14 takes the base-10 logarithm of the intensity ratio $I/I_0$. As a reminder, logarithms work like this:

If you express a number as a power of 10 . . . . . . the logarithm is the exponent.

$$\log_{10}(1000) = \log_{10}(10^3) = 3$$

Right at the threshold of hearing, where $I = I_0$, the sound intensity level is

$$\beta = (10\ \text{dB})\log_{10}\left(\frac{I_0}{I_0}\right) = (10\ \text{dB})\log_{10}(1) = (10\ \text{dB})\log_{10}(10^0) = 0\ \text{dB}$$

The threshold of hearing corresponds to 0 dB, as we wanted.

We can find the intensity from the sound intensity level by taking the inverse of the $\log_{10}$ function. Recall, from the definition of the base-10 logarithm, that $10^{\log(x)} = x$. Applying this to Equation 15.14, we find

$$I = (I_0)10^{(\beta/10\ \text{dB})} \tag{15.15}$$

TABLE 15.3 **Sound intensity levels and intensities of common environmental sounds**

| Sound | $\beta$ (dB) | $I$ (W/m$^2$) |
|---|---|---|
| Threshold of hearing | 0 | $1.0 \times 10^{-12}$ |
| Person breathing, at 3 m | 10 | $1.0 \times 10^{-11}$ |
| A whisper, at 1 m | 20 | $1.0 \times 10^{-10}$ |
| Classroom during test, no talking | 30 | $1.0 \times 10^{-9}$ |
| Residential street, no traffic | 40 | $1.0 \times 10^{-8}$ |
| Quiet restaurant | 50 | $1.0 \times 10^{-7}$ |
| Normal conversation, at 1 m | 60 | $1.0 \times 10^{-6}$ |
| Busy traffic | 70 | $1.0 \times 10^{-5}$ |
| Vacuum cleaner, for user | 80 | $1.0 \times 10^{-4}$ |
| Niagara Falls, at viewpoint | 90 | $1.0 \times 10^{-3}$ |
| Pneumatic hammer, at 2 m | 100 | 0.010 |
| Home stereo at max volume | 110 | 0.10 |
| Rock concert | 120 | 1.0 |
| Threshold of pain | 130 | 10 |

TABLE 15.3 lists the sound intensity levels and intensities for a number of typical sounds. Notice that the sound intensity level increases by 10 dB each time the actual intensity increases by a factor of 10. For example, the sound intensity level increases from 70 dB to 80 dB when the sound intensity increases from $10^{-5}$ W/m$^2$ to $10^{-4}$ W/m$^2$. A 20 dB increase in the sound intensity level means a factor of 100 increase in intensity; 30 dB a factor of 1000. We found earlier that sound is perceived as "twice as loud" when the intensity increases by a factor of 10. In terms of decibels, we can say that the apparent loudness of a sound doubles with each 10 dB increase in the sound intensity level.

The range of sounds in Table 15.3 is very wide; the top of the scale, 130 dB, represents 10 trillion times the intensity of the quietest sound you can hear. Vibrations of this intensity will injure the delicate sensory apparatus of the ear and cause pain. Exposure to less intense sounds also is not without risk. A fairly short exposure to 120 dB can cause damage to the hair cells in the ear, but lengthy exposure to sound intensity levels of over 85 dB can produce damage as well.

Before we begin to look at examples that use sound intensity level, let's summarize what we've learned about the energy carried by waves.

## SYNTHESIS 15.2 Wave power and intensity

We have different measures of the energy carried by waves.

The **intensity** of a wave is the power per square meter of area:

Intensity (W/m²) · Power (W) · Area (m²)

$$I = \frac{P}{a}$$

When a wave hits a surface, the power absorbed in area $a$ is

$$P = Ia$$

Spherical waves spread over the surface of a sphere of increasing size.

The **power** of the source is the rate at which it emits energy (J/s, or W).

Distance from the source (m)

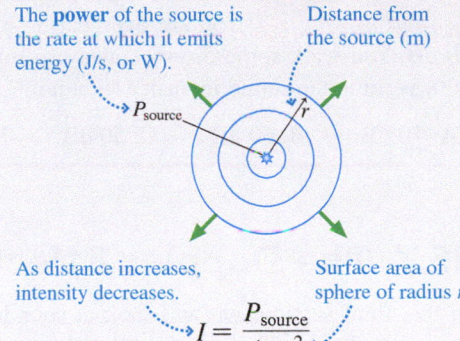

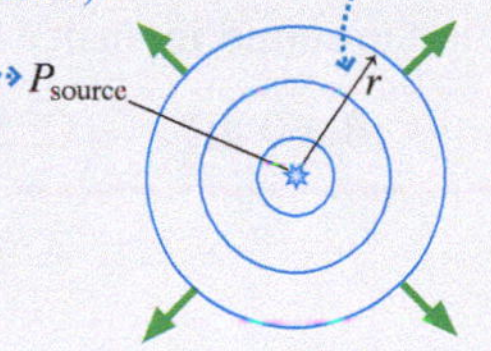


As distance increases, intensity decreases.

Surface area of sphere of radius $r$

$$I = \frac{P_{source}}{4\pi r^2}$$

We compute **sound intensity level** from relative intensity:

Sound intensity level (dB) · Intensity (W/m²)

$$\beta = (10\text{ dB})\log_{10}\left(\frac{I}{I_0}\right)$$

The smallest intensity humans can sense

$$I_0 = 1.0 \times 10^{-12}\text{ W/m}^2$$

We can also compute intensity from sound intensity level:

$$I = (I_0)10^{(\beta/10\text{ dB})}$$

### EXAMPLE 15.11 Finding the loudness of a shout

A person shouting at the top of his lungs emits about 1.0 W of energy as sound waves. What is the sound intensity level 1.0 m from such a person?

**STRATEGIZE** We assume that the shouting person emits a spherical sound wave. Synthesis 15.2 gives the details for the decrease in intensity with distance and the calculation of the resulting sound intensity level.

**SOLVE** At a distance of 1.0 m, the sound intensity is

$$I = \frac{P}{4\pi r^2} = \frac{1.0\text{ W}}{4\pi(1.0\text{ m})^2} = 0.080\text{ W/m}^2$$

Thus the sound intensity level is

$$\beta = (10\text{ dB})\log_{10}\left(\frac{0.080\text{ W/m}^2}{1.0 \times 10^{-12}\text{ W/m}^2}\right) = 110\text{ dB}$$

**ASSESS** This is quite loud (compare with values in Table 15.3), as you might expect.

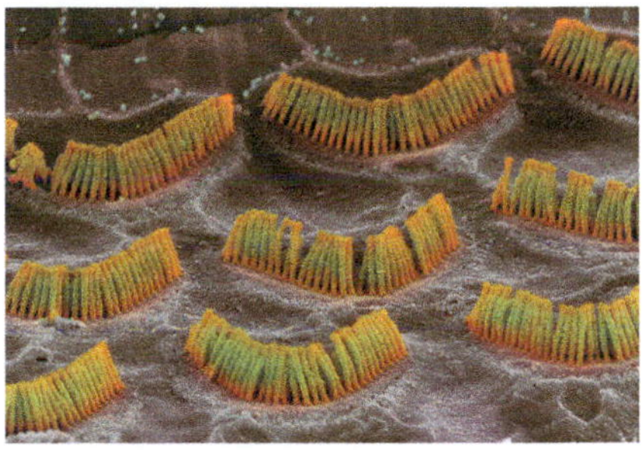

**Hearing hairs** BIO This electron microscope picture shows the hair cells in the cochlea of the ear that are responsible for sensing sound. Even very tiny vibrations transmitted into the fluid of the cochlea deflect the hairs, triggering a response in the cells. Motion of the basilar membrane by as little as 0.5 nm, about 5 atomic diameters, can produce an electrical response in the hair cells. With this remarkable sensitivity, it is no wonder that loud sounds can damage these structures.

**NOTE** ▶ In calculating the sound intensity level, be sure to use the $\log_{10}$ button on your calculator, not the natural logarithm button. On many calculators, $\log_{10}$ is labeled LOG and the natural logarithm is labeled LN. ◀

### EXAMPLE 15.12 How far away can you hear a conversation?

The sound intensity level 1.0 m from a person talking in a normal conversational voice is 60 dB. Suppose you are outside, 100 m from the person speaking. If it is a very quiet day with minimal background noise, are you able to hear him or her?

**STRATEGIZE** We know how sound intensity changes with distance, but not sound intensity level, so we need to break this problem into three steps. First, we will convert the sound intensity level at 1.0 m into intensity, using Equation 15.15. Second, we will compute the intensity at a distance of 100 m, using Equation 15.13. Finally, we will convert this result back to a sound intensity level, using Equation 15.14, so that we can judge the loudness.

**PREPARE** We denote the intensity at 1 m by $I(1\text{ m})$ and that at 100 m by $I(100\text{ m})$.

**SOLVE** The intensity is

$$I(1\text{ m}) = (1.0 \times 10^{-12}\text{ W/m}^2)10^{(60\text{ dB}/10\text{ dB})} = 1.0 \times 10^{-6}\text{ W/m}^2$$

*Continued*

The intensity at 100 m can be found using Equation 15.13:

$$\frac{I(100\text{ m})}{I(1\text{ m})} = \frac{(1\text{ m})^2}{(100\text{ m})^2}$$

$$I(100\text{ m}) = I(1\text{ m})(1.0 \times 10^{-4}) = 1.0 \times 10^{-10}\text{ W/m}^2$$

This intensity corresponds to a sound intensity level

$$\beta = (10\text{ dB})\log_{10}\left(\frac{1.0 \times 10^{-10}\text{ W/m}^2}{1.0 \times 10^{-12}\text{ W/m}^2}\right) = 20\text{ dB}$$

This is above the threshold for hearing—about the level of a whisper—so it can be heard on a very quiet day.

**ASSESS** The sound is well within what the ear can detect. However, normal background noise is rarely less than 40 dB, which would make the conversation difficult to decipher. Our result thus seems reasonable based on experience. The sound is at a level that can theoretically be detected, but it is much quieter than the ambient level of sound, so it's not likely to be noticed.

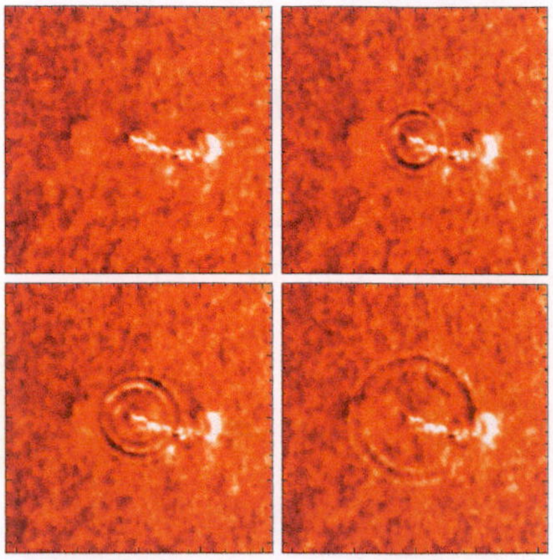

**Catching a wave on the sun** The Doppler effect shifts the frequency of light emitted from the sun's surface. If the surface is rising, the light is shifted to higher frequencies; these positions are shown darker. If the surface is falling, the light is shifted to lower frequencies and is shown lighter. This series of images of the sun's surface shows a wave produced by the disruption of a solar flare.

**STOP TO THINK 15.7** You are overhearing a very heated conversation that registers 80 dB. You walk some distance away so that the intensity decreases by a factor of 100. What is the sound intensity level now?

A. 70 dB B. 60 dB C. 50 dB D. 40 dB E. 30 dB F. 20 dB

## 15.7 The Doppler Effect and Shock Waves

In this final section, we will look at sounds from moving objects and for moving observers. You've likely noticed that the pitch of an ambulance's siren gets higher as it approaches you and drops as it goes past you. A higher pitch suddenly becomes a lower pitch. This change in frequency, which is due to the motion of the ambulance, is called the *Doppler effect.* A more dramatic effect happens when an object moves faster than the speed of sound. The crack of a whip is a *shock wave* produced when the tip moves at a *supersonic* speed. These examples—and much of this section—concern sound waves, but the phenomena we will explore apply generally to all waves.

### Sound Waves from a Moving Source

FIGURE 15.18a shows a source of sound waves moving away from Pablo and toward Nancy at a steady speed $v_s$. The subscript s indicates that this is the speed of the source, not the speed of the waves. The source is emitting sound waves of frequency $f_s$ as it travels. Part a of the figure shows the positions of the source at times $t = 0$, $T$, $2T$, and $3T$, where $T = 1/f_s$ is the period of the waves.

FIGURE 15.18 A motion diagram showing the wave fronts emitted by a source as it moves to the right at speed $v_s$.

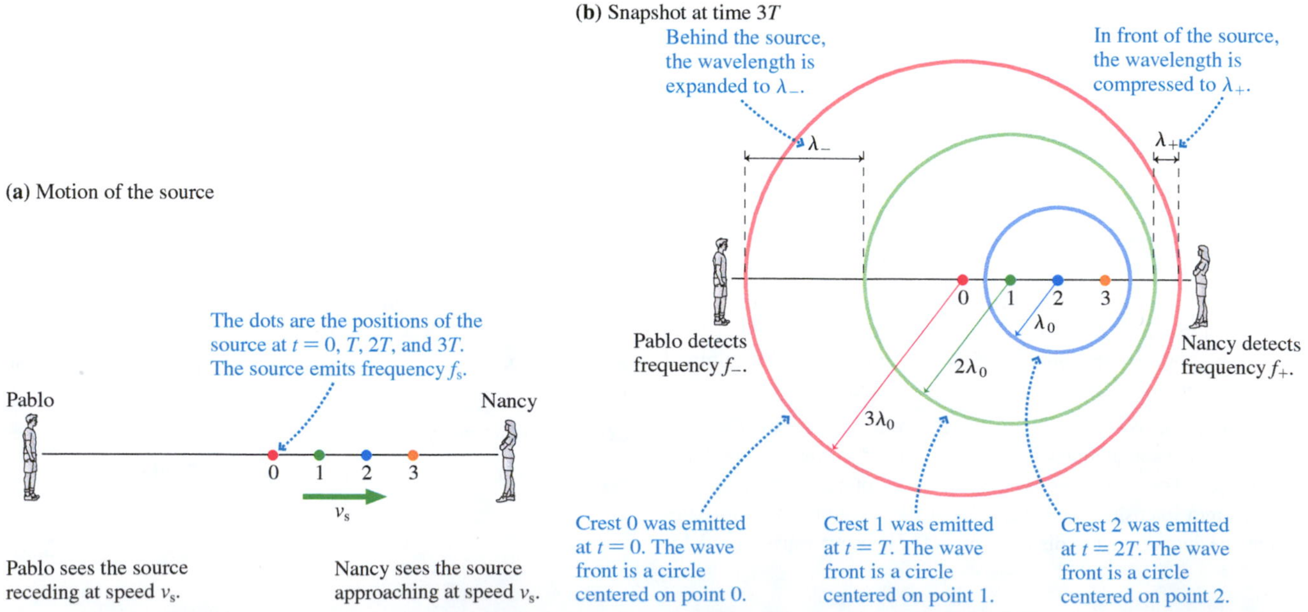

After a wave crest leaves the source, its motion is governed by the properties of the medium. The motion of the source cannot affect a wave that has already been emitted. Thus each circular wave front in FIGURE 15.18b is centered on the point from which it was emitted. You can see that the wave crests are bunched up in the direction in which the source is moving and are stretched out behind it. The distance between one crest and the next is one wavelength, so the wavelength $\lambda_+$ that Nancy measures is *less* than the wavelength $\lambda_s = v/f_s$ that would be emitted if the source were at rest. Similarly, $\lambda_-$ behind the source is larger than $\lambda_s$.

These crests move through the medium at the wave speed $v$. Consequently, the frequency $f_+ = v/\lambda_+$ detected by the observer whom the source is approaching is *higher* than the frequency $f_s$ emitted by the source. Similarly, $f_- = v/\lambda_-$ detected behind the source is *lower* than frequency $f_s$. This change of frequency when a source moves relative to an observer is called the **Doppler effect.** The frequency heard by a stationary observer depends on whether the observer sees the source approaching or receding:

**Approaching source**

$$f_+ = \frac{f_s}{1 - v_s/v}$$

Frequency of the source (Hz) → $f_s$; Speed of the *waves* (m/s) → $v$; Speed of the *source* of the waves (m/s) → $v_s$; The observed frequency is *increased.*

**Receding source**

$$f_- = \frac{f_s}{1 + v_s/v} \qquad (15.16)$$

The observed frequency is *decreased.*

As expected, $f_+ > f_s$ (the frequency is higher) for an approaching source because the denominator is less than 1, and $f_- < f_s$ (the frequency is lower) for a receding source.

**EXAMPLE 15.13 How fast are the police driving?**

A police siren has a frequency of 550 Hz as the police car approaches you and 450 Hz after it has passed you and is moving away. How fast are the police traveling?

***STRATEGIZE*** The siren's frequency is altered by the Doppler effect. The frequency is $f_+$ as the car approaches and $f_-$ as it moves away. We will write two equations for these frequencies and solve for the speed of the police car, $v_s$.

***SOLVE*** Because our goal is to find $v_s$, we rewrite Equations 15.16 as

$$f_s = \left(1 + \frac{v_s}{v}\right) f_- \quad \text{and} \quad f_s = \left(1 - \frac{v_s}{v}\right) f_+$$

Subtracting the second equation from the first, we get

$$0 = f_- - f_+ + \frac{v_s}{v}(f_- + f_+)$$

Now we can solve for the speed $v_s$:

$$v_s = \frac{f_+ - f_-}{f_+ + f_-} v = \frac{100\ \text{Hz}}{1000\ \text{Hz}} 343\ \text{m/s} = 34\ \text{m/s}$$

***ASSESS*** This is pretty fast (about 75 mph) but reasonable for a police car speeding with the siren on.

## A Stationary Source and a Moving Observer

Suppose the police car in Example 15.13 is at rest while you drive toward it at 34 m/s. You might think that this is equivalent to having the police car move toward you at 34 m/s, but there is an important difference. Mechanical waves move through a medium, and the Doppler effect depends not just on how the source and the observer move with respect to each other but also on how they move with respect to the medium. The frequency heard by a moving observer depends on whether the observer is approaching or receding from the source:

**Approaching observer**

$$f_+ = \left(1 + \frac{v_o}{v}\right) f_s$$

Speed of the *observer* (m/s) → $v_o$; Frequency of the source (Hz) → $f_s$; Speed of the *waves* (m/s) → $v$; The observed frequency is *increased.*

**Receding observer**

$$f_- = \left(1 - \frac{v_o}{v}\right) f_s \qquad (15.17)$$

The observed frequency is *decreased.*

The greater the distance to a galaxy, the faster it moves away from us. This photo shows galaxies at distances of up to 12 billion light years. The great distances imply large red shifts, which make the light from the most distant galaxies appear distinctly red.

## The Doppler Effect for Light Waves

If a source of light waves is receding from you, the wavelength $\lambda_-$ that you detect is longer than the wavelength $\lambda_s$ emitted by the source. Because the wavelength is shifted toward the red end of the visible spectrum, the longer wavelengths of light, this effect is called the **red shift.** Similarly, the light you detect from a source moving toward you is **blue shifted** to shorter wavelengths. For objects moving at normal speeds, this is a small effect; you won't see a Doppler shift of the flashing light of the police car in the previous example!

Distant galaxies *all* have a distinct red shift—all distant galaxies are moving away from us. How can we make sense of this observation? The most straightforward explanation is that the galaxies of the universe are *all* moving apart from each other. Extrapolating backward in time brings us to a point when all the matter of the universe—and even space itself, according to the theory of relativity—began rushing out of a primordial fireball. Many observations and measurements since have given support to the idea that the universe began in a *Big Bang* about 14 billion years ago.

## Frequency Shift on Reflection from a Moving Object

A wave striking a barrier or obstacle can *reflect* and travel back toward the source of the wave, a process we will examine more closely in the next chapter. For sound, the reflected wave is called an *echo.* A bat uses echolocation to determine the distance to a flying insect by measuring the time between the emission of an ultrasonic chirp and the detection of the echo from the insect. But if the bat is to catch the insect, it's just as important to know where the insect is going—its velocity. The bat can figure this out by noting the *frequency shift* in the reflected wave, another application of the Doppler effect.

Suppose a sound wave of frequency $f_s$ travels toward a moving object. The object would see the wave's frequency Doppler shifted to a higher frequency $f_+$, as given by Equation 15.17. The wave reflected back toward the sound source by the moving object is also Doppler shifted to a higher frequency because, to the source, the reflected wave is coming from a moving object. Thus the echo from a moving object is "double Doppler shifted." If the object's speed $v_o$ is much lower than the wave speed $v$ ($v_o \ll v$), the frequency shift of waves reflected from a moving object is

$$\Delta f = \pm 2 f_s \frac{v_o}{v} \tag{15.18}$$

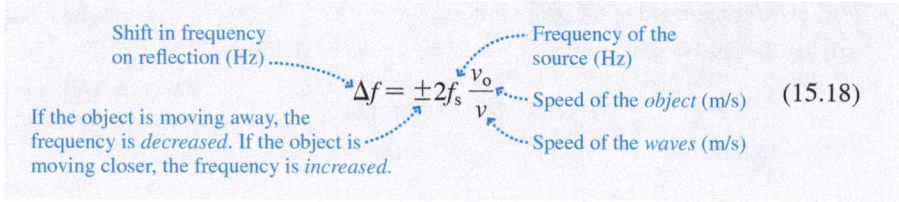

Shift in frequency on reflection (Hz): $\Delta f$

If the object is moving away, the frequency is *decreased.* If the object is moving closer, the frequency is *increased.*: $\pm$

Frequency of the source (Hz): $f_s$

Speed of the *object* (m/s): $v_o$

Speed of the *waves* (m/s): $v$

Notice that there is no shift (i.e., the reflected wave has the same frequency as the emitted wave) when the reflecting object is at rest.

Frequency shift on reflection is observed for all types of waves. Radar units emit pulses of radio waves and observe the reflected waves. The time between the emission of a pulse and its return gives an object's position. The change in frequency of the returned pulse gives the object's speed. This is the principle behind the radar guns used by traffic police, as well as the Doppler radar images you have seen in weather reports.

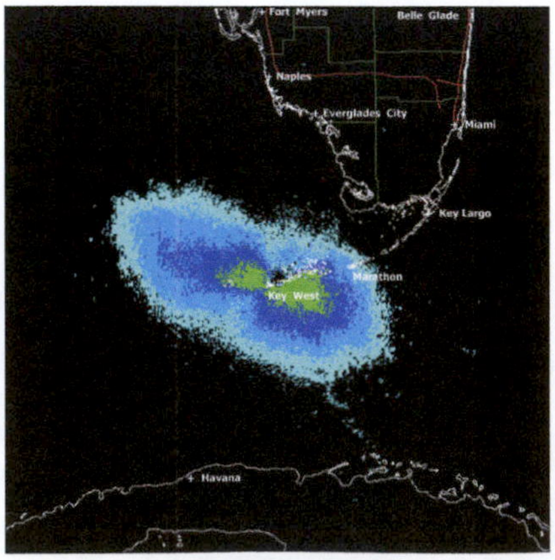


◂ **Keeping track of wildlife with radar** Doppler radar is tuned to measure only those reflected radio waves that have a frequency shift, eliminating reflections from stationary objects and showing only objects in motion. Televised Doppler radar images of storms are made using radio waves reflected from moving water droplets. But this Doppler radar image of an area off the tip of Florida was made on a clear night with no rain. The blue and green patch isn't a moving storm, it's a moving flock of birds. Migratory birds frequently move at high altitudes at night, so Doppler radar is an excellent tool for analyzing their movements.

**EXAMPLE 15.14 The Doppler blood flowmeter** BIO

If an ultrasound source is pressed against the skin, the sound waves reflect off tissues in the body. Some of the sound waves reflect from blood cells moving through arteries toward or away from the source, producing a frequency shift in the reflected wave. A biomedical engineer is designing a *Doppler blood flowmeter* to measure blood flow in an artery where a typical flow speed is known to be 0.60 m/s. What ultrasound frequency should she use to produce a frequency shift of 1500 Hz when this flow is detected?

**STRATEGIZE** The ultrasound is reflecting from a moving object—the blood—so Equation 15.18 will apply.

**PREPARE** In Equation 15.18, $v_0$ is the speed of the blood, or 0.60 m/s, and $v$ is the speed of sound in human tissue, or 1540 m/s.

**SOLVE** We can rewrite Equation 15.18 to calculate the frequency of the emitter:

$$f_s = \left(\frac{\Delta f}{2}\right)\left(\frac{v}{v_{blood}}\right)$$

The values on the right side are all known. Thus the required ultrasound frequency is

$$f_s = \left(\frac{1500\text{ Hz}}{2}\right)\left(\frac{1540\text{ m/s}}{0.60\text{ m/s}}\right) = 1.9\text{ MHz}$$

**ASSESS** Doppler units to measure blood flow in deep tissues actually work at about 2.0 MHz, so our answer is reasonable.

If we add a frequency-shift measurement to an ultrasound imaging unit, like the one in Example 15.7, we have a device—called *Doppler ultrasound*—that can show not only structure but also motion. *Doppler ultrasound* is a very valuable tool in cardiology because an image can reveal the motion of the heart muscle and the blood in addition to the structure of the heart itself.

## Shock Waves

When sound waves are emitted by a moving source, the frequency is shifted, as we have seen. Now, let's look at what happens when the source speed $v_s$ exceeds the wave speed $v$ and the source "outruns" the waves it produces.

Earlier in this section, in Figure 15.18, we looked at a motion diagram of waves emitted by a moving source. FIGURE 15.19 is the same diagram, but in this case the source is moving faster than the waves. This motion causes the waves to overlap. (Compare Figure 15.19 to Figure 15.18.) The amplitudes of the overlapping waves add up to produce a very large amplitude wave—a **shock wave.**

Anything that moves faster than the speed of sound in air will create a shock wave. FIGURE 15.20a is a specialized photograph of a **supersonic** (faster than the speed of sound) jet airplane. The photograph is enhanced to show the shock waves that emanate from the bow, wings, and tail of the aircraft. This shock wave travels along with the jet. If the jet flew over you, the passing of the shock wave would produce a **sonic boom,** a loud noise that sounds like a clap of thunder. The crack of a whip is a sonic boom as well, though on a much smaller scale.

FIGURE 15.19 Waves emitted by a source traveling faster than the speed of the waves in a medium produce a shock wave.

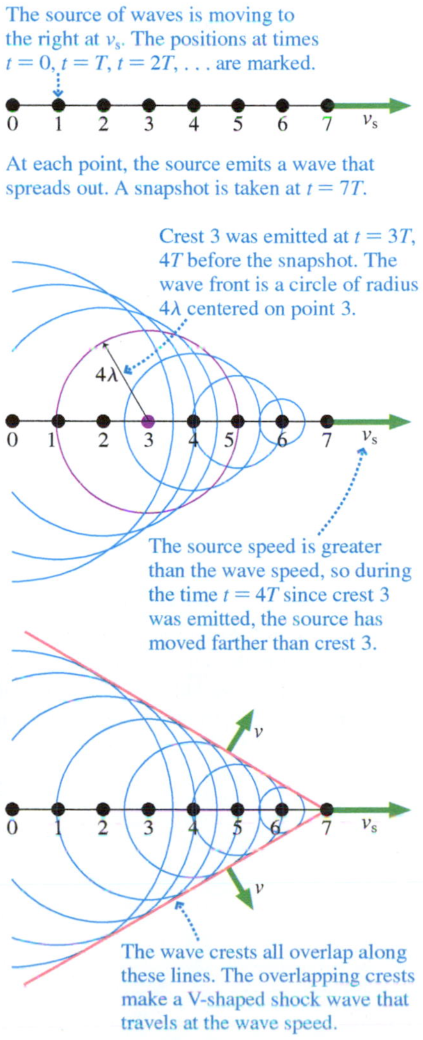


FIGURE 15.20 Extreme and everyday examples of shock waves.

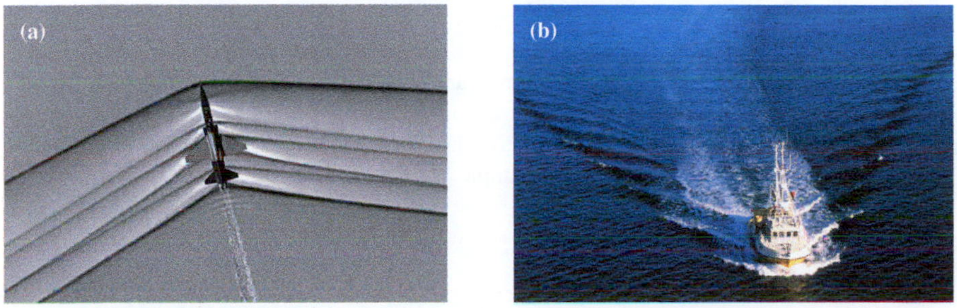


Shock waves are not necessarily an extreme phenomenon. A boat (or even a duck!) can easily travel faster than the speed of the water waves it creates; the resulting wake, with its characteristic "V" shape as shown in FIGURE 15.20b, is really a shock wave.

The source of waves is moving to the right at $v_s$. The positions at times $t = 0, t = T, t = 2T, \dots$ are marked.

At each point, the source emits a wave that spreads out. A snapshot is taken at $t = 7T$.

Crest 3 was emitted at $t = 3T$, $4T$ before the snapshot. The wave front is a circle of radius $4\lambda$ centered on point 3.

The source speed is greater than the wave speed, so during the time $t = 4T$ since crest 3 was emitted, the source has moved farther than crest 3.

The wave crests all overlap along these lines. The overlapping crests make a V-shaped shock wave that travels at the wave speed.

**STOP TO THINK 15.8** Amy and Zack are both listening to a source of sound waves that is moving to the right. Compare the frequencies each hears.

A. $f_{Amy} > f_{Zack}$
B. $f_{Amy} = f_{Zack}$
C. $f_{Amy} < f_{Zack}$

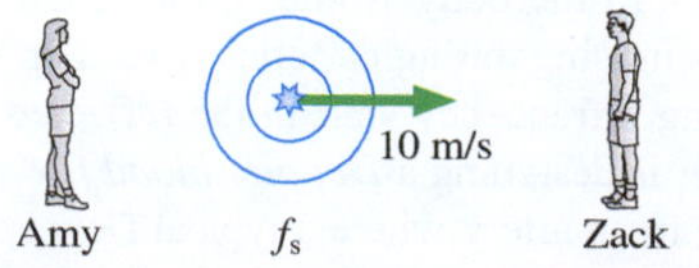


## INTEGRATED EXAMPLE 15.15 Shaking the ground

Earthquakes are dramatic slips of the earth's crust. You may feel the waves generated by an earthquake even if you're some distance from the *epicenter,* the point where the slippage occurs. Earthquake waves are usually complicated, but some of the long-period waves shake the ground with motion that is approximately simple harmonic motion. FIGURE 15.21 shows the vertical positions of the ground recorded at a distant monitoring station following an earthquake that hit Japan in 2003. This particular wave traveled with a speed of 3500 m/s.

a. Was the wave transverse or longitudinal?
b. What were the wave's frequency and wavelength?
c. What were the maximum speed and the maximum acceleration of the ground during this earthquake wave?
d. Intense earthquake waves produce accelerations greater than the free-fall acceleration of gravity. How does this wave compare?

**FIGURE 15.21** The vertical motion of the ground during the passage of an earthquake wave.

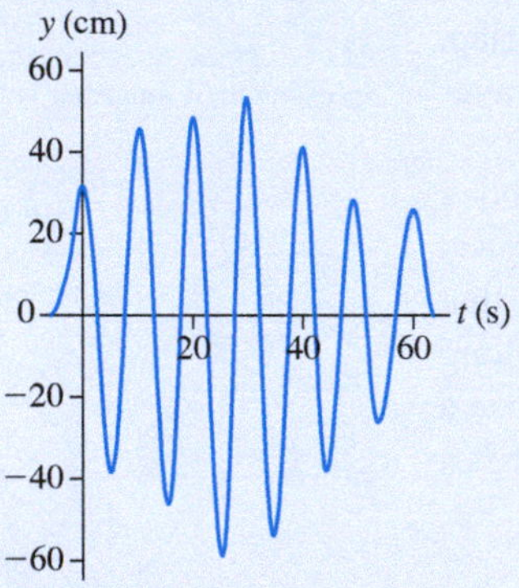


***STRATEGIZE*** Figure 15.21 is a history graph showing the motion in the *y*-direction (i.e., the *vertical* motion) of one point of the medium—the ground. The graph is approximately sinusoidal, so we will model the motion of the ground as simple harmonic motion and the traveling wave as a sinusoidal wave. Consequently, we will use the fundamental relationships for sinusoidal waves to relate the wavelength, frequency, and speed.

***PREPARE*** If we look at the middle portion of the wave shown in Figure 15.21, we can estimate the amplitude to be 0.50 m (it varies, but this is a reasonable average over a few cycles). We can also see from the graph that 6 cycles of the oscillation took 60 s, so the period was $T = 10$ s.

***SOLVE***

a. As the graph shows, the motion of the ground was vertical, perpendicular to the horizontal motion of the wave traveling along the ground, so this was a transverse wave.

b. The period was 10 s, so the frequency was $f = 1/T = 1/10\text{ s} = 0.10$ Hz. The speed of the wave was 3500 m/s, giving a wavelength of

$$\lambda = \frac{v}{f} = \frac{3500\text{ m/s}}{0.10\text{ Hz}} = 35{,}000\text{ m} = 35\text{ km}$$

c. The motion of the ground was simple harmonic motion with frequency 0.10 Hz and amplitude 0.50 m. We can compute the maximum speed and acceleration using relationships from Chapter 14:

$$v_{max} = 2\pi f A = (2\pi)(0.10\text{ Hz})(0.50\text{ m}) = 0.31\text{ m/s}$$
$$a_{max} = (2\pi f)^2 A = [2\pi(0.10\text{ Hz})]^2(0.50\text{ m}) = 0.20\text{ m/s}^2$$

d. This was a reasonably gentle earthquake wave, with an acceleration of the ground much less than the free-fall acceleration of gravity:

$$a_{max}\text{ (in units of } g) = \frac{0.20\text{ m/s}^2}{9.8\text{ m/s}^2} = 0.020g$$

***ASSESS*** The wavelength is quite long, as we might expect for such a fast wave with a long period. Given the relatively small amplitude and long period, it's no surprise that the maximum speed and acceleration are relatively modest. You'd certainly feel the passage of this wave, but it wouldn't knock buildings down. This value is less than the acceleration for the sway of the top of the building in Example 14.3 in Chapter 14!

# SUMMARY

**GOAL** To learn the basic properties of traveling waves.

## GENERAL PRINCIPLES

### The Wave Model

This model is based on the idea of a **traveling wave,** which is an organized disturbance traveling at a well-defined **wave speed** $v$.

- In **transverse waves** the particles of the medium move *perpendicular* to the direction in which the wave travels.

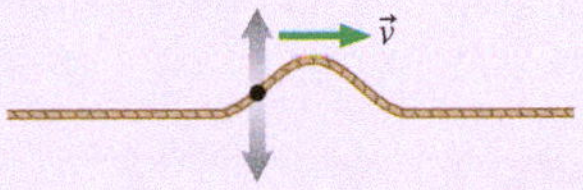


- In **longitudinal waves** the particles of the medium move *parallel* to the direction in which the wave travels.

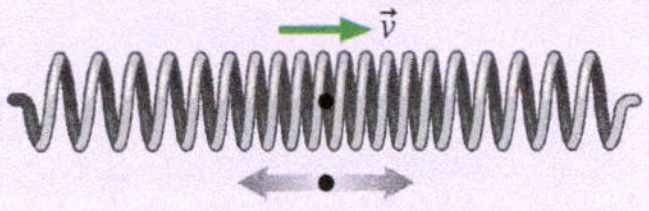


A wave transfers energy, but there is no material or substance transferred.

**Mechanical waves** require a material **medium**. The speed of the wave is a property of the medium, not the wave. The speed does not depend on the size or shape of the wave.

- For a **wave on a string,** the string is the medium.

$$v_{string} = \sqrt{\frac{T_s}{\mu}}$$

- A **sound wave** is a wave of compressions and rarefactions of a medium such as air.

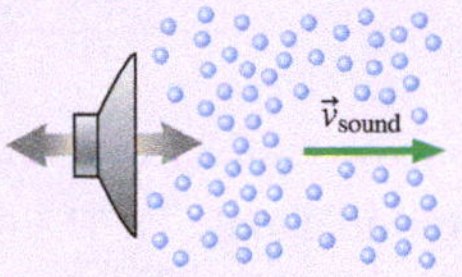


In a gas:

$$v_{sound} = \sqrt{\frac{\gamma RT}{M}}$$

**Electromagnetic waves** are waves of the electromagnetic field. They do not require a medium. All electromagnetic waves travel at the same speed in a vacuum, $c = 3.00 \times 10^8$ m/s.

## IMPORTANT CONCEPTS

### Graphical representation of waves

A **snapshot graph** is a picture of a wave at one instant in time. For a periodic wave, the **wavelength** $\lambda$ is the distance between crests.

Fixed $t$:

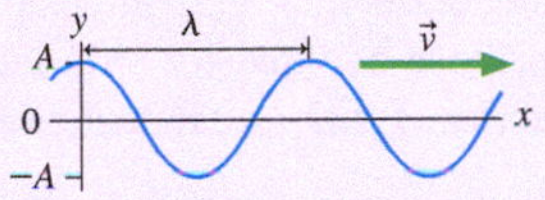


A **history graph** is a graph of the displacement of one point in a medium versus time. For a periodic wave, the **period** $T$ is the time between crests.

Fixed $x$:

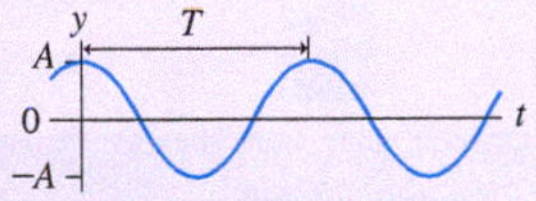


### Mathematical representation of waves

**Sinusoidal waves** are produced by a source moving with simple harmonic motion. The equation for a sinusoidal wave is a function of position and time:

$$y(x, t) = A\cos\left(2\pi\left(\frac{x}{\lambda} \pm \frac{t}{T}\right)\right)$$

+: wave travels to left

−: wave travels to right

For sinusoidal waves:

$$T = \frac{1}{f} \qquad v = \lambda f$$

The **intensity** of a wave is the ratio of the power to the area:

$$I = \frac{P}{a}$$

For a **spherical wave** the power decreases with the surface area of the spherical **wave fronts:**

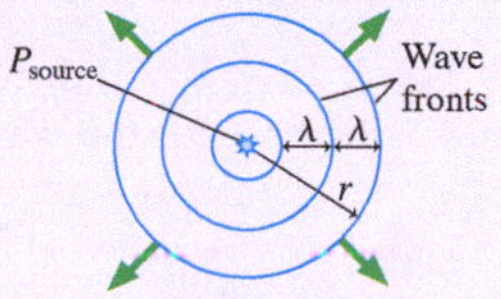


$$I = \frac{P_{source}}{4\pi r^2}$$

## APPLICATIONS

The loudness of a sound is given by the **sound intensity level.** This is a logarithmic function of intensity and is in units of **decibels** (dB).

The sound intensity level, in dB, is given by

$$\beta = (10\text{ dB})\log_{10}\left(\frac{I}{I_0}\right)$$

where $I_0 = 1.0 \times 10^{-12}$ W/m$^2$ is the *threshold of hearing,* the quietest sound that we can hear. A sound at the threshold of hearing corresponds to 0 dB.

The **Doppler effect** is a shift in frequency when there is relative motion of a wave source (frequency $f_s$, wave speed $v$) and an observer.

**Moving source, stationary observer:**

Receding source:

$$f_- = \frac{f_s}{1 + v_s/v}$$

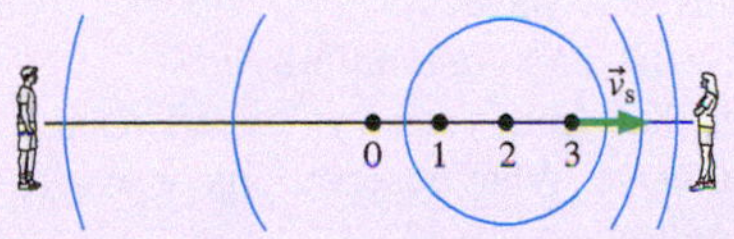


Approaching source:

$$f_+ = \frac{f_s}{1 - v_s/v}$$

**Moving observer, stationary source:**

Approaching the source: $f_+ = \left(1 + \frac{v_o}{v}\right)f_s$

Moving away from the source: $f_- = \left(1 - \frac{v_o}{v}\right)f_s$

**Reflection from a moving object:**

For $v_o \ll v$, $\Delta f = \pm 2f_s\frac{v_o}{v}$

When an object moves faster than the wave speed in a medium, a **shock wave** is formed.

## Learning Objectives After studying this chapter, you should be able to:

- Use the wave model to understand the basic properties of waves. *Conceptual Questions 15.1, 15.2; Problem 15.8*
- Calculate the speeds of waves on strings and the speeds of sound waves. *Conceptual Questions 15.4, 15.5; Problems 15.1, 15.2, 15.3, 15.5, 15.7*
- Use the graphical and mathematical representations of waves. *Conceptual Questions 15.10, 15.11; Problems 15.9, 15.14, 15.16, 15.17, 15.18*
- Solve problems about sound and light waves. *Conceptual Questions 15.12, 15.13; Problems 15.24, 15.26, 15.27, 15.28, 15.30*
- Apply the ideas of energy and power to waves. *Conceptual Question 15.14; Problems 15.31, 15.32, 15.35, 15.36, 15.37*
- Calculate the loudness of sound. *Conceptual Question 15.16; Problems 15.40, 15.41, 15.42, 15.43, 15.44*
- Understand and use the Doppler effect. *Conceptual Questions 15.17, 15.18; Problems 15.51, 15.52, 15.53, 15.54, 15.55*

## STOP TO THINK ANSWERS

**Chapter Preview Stop to Think: D.** The toy goes from the lowest point to its highest point in 1.0 s. This is one-half of a full oscillation, so the period is 2.0 s, giving a frequency of 0.5 Hz.

**Stop to Think 15.1: Transverse.** The wave moves horizontally through the crowd, but individual spectators move up and down, transverse to the motion of the wave.

**Stop to Think 15.2: D, E.** Shaking your hand faster or farther will change the shape of the wave, but this will not change the wave speed; the speed is a property of the medium. Changing the linear density of the string or its tension will change the wave speed. To increase the speed, you must decrease the linear density or increase the tension.

**Stop to Think 15.3: A.** As the wave moves toward the origin, this point moves up at a constant speed for a short distance, reverses and moves down for a greater distance until the displacement is negative, and then moves back to where it started. Graph A shows this motion.

**Stop to Think 15.4: B.** All three waves have the same speed, so the frequency is highest for the wave that has the shortest wavelength. (Imagine the three waves moving to the right. The one with the crests closest together has the crests passing by most rapidly.)

**Stop to Think 15.5: C.** Figure 15.13 shows that infrared is to the left of the visible light region of the spectrum and ultraviolet is to the right. The upper scale on the diagram shows that infrared thus has a lower frequency than ultraviolet. All electromagnetic waves travel at the same speed, so infrared must have a longer wavelength.

**Stop to Think 15.6: A.** The plane wave does not spread out, so its intensity will be constant. The other two waves spread out, so their intensity will decrease.

**Stop to Think 15.7: B.** If the intensity decreases by a factor of 10, this corresponds to a decrease in sound intensity level of 10 dB. $100 = 10 \times 10$, so a decrease of intensity by a factor of 100 corresponds to two 10 dB decreases, for a total decrease of 20 dB.

**Stop to Think 15.8: C.** The source is moving toward Zack, so he observes a higher frequency. The source is moving away from Amy, so she observes a lower frequency.

Video Tutor Solution Chapter 15

# QUESTIONS

## Conceptual Questions

1. a. In your own words, define what a *transverse wave* is.
   b. Give an example of a wave that, from your own experience, you know is a transverse wave. What observations or evidence tells you this is a transverse wave?
2. a. In your own words, define what a *longitudinal wave* is.
   b. Give an example of a wave that, from your own experience, you know is a longitudinal wave. What observations or evidence tells you this is a longitudinal wave?
3. The wave pulses shown in Figure Q15.3 travel along the same string. Rank in order, from largest to smallest, their wave speeds $v_1$, $v_2$, and $v_3$. Explain.

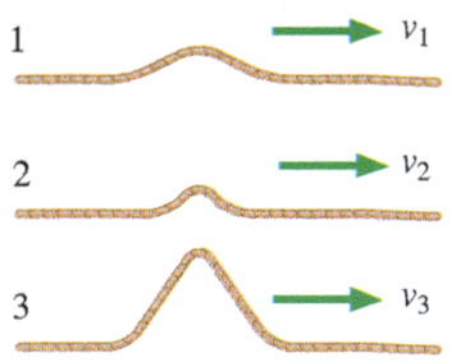

FIGURE Q15.3

4. A loudspeaker emits a sound wave at a particular frequency. If the intensity of the wave is doubled, how does the speed of the wave change? What if the frequency is doubled?

Problem difficulty is labeled as | (straightforward) to ||||| (challenging). Problems labeled INT integrate significant material from earlier chapters; Problems labeled BIO are of biological or medical interest.

The eText icon indicates when there is a video tutor solution available for the chapter or for a specific problem. To launch these videos, log into your eText through Mastering$^{TM}$ Physics or log into the Study Area.

5. A wave pulse travels along a string at a speed of 200 cm/s. What will be the speed if:
   a. The string's tension is doubled?
   b. The string's mass is quadrupled (but its length is unchanged)?
   c. The string's length is quadrupled (but its mass is unchanged)?
   d. The string's mass and length are both quadrupled?

   Note that parts a–d are independent and refer to changes made to the original string.
6. BIO Harbor seals, like many animals, determine the direction from which a sound is coming by sensing the difference in arrival times at their two ears. A small difference in arrival times means that the object is in front of the seal; a larger difference means it is to the left or right. There is a minimum time difference that a seal can sense, and this leads to a limitation on a seal's direction sense. Seals can distinguish between two sounds that come from directions 3° apart in air, but this increases to 9° in water. Explain why you would expect a seal's directional discrimination to be worse in water than in air.
7. Home professionals like realtors use ultrasonic measuring tools to quickly measure the size of rooms. These instruments work by sending out a pulse of ultrasonic sound, then measuring the time it takes for this pulse to reflect off a wall and return to the device. If you used such a device on a particularly hot day, would you measure the length of a given room to be longer or shorter than on a cool day?
8. When in air, a waterproof speaker emits sound waves that have a frequency of 1000 Hz. When the speaker is lowered into water, does the frequency of the sound increase, decrease, or remain the same? Does the wavelength of the sound increase, decrease, or remain the same?
9. Figure Q15.9 shows a history graph of the motion of one point on a string as a wave traveling to the left passes by. Sketch a snapshot graph for this wave.

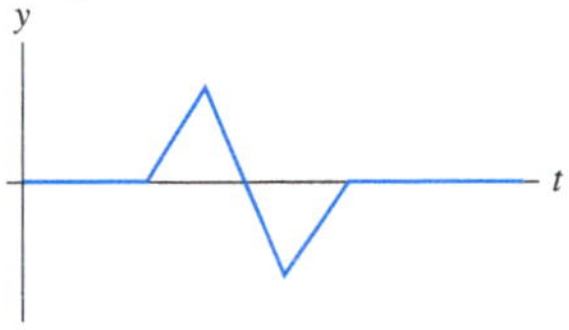


FIGURE Q15.9

10. Figure Q15.10 shows a history graph *and* a snapshot graph for a wave pulse on a string. They describe the same wave from two perspectives.
    a. In which direction is the wave traveling? Explain.
    b. What is the speed of this wave?

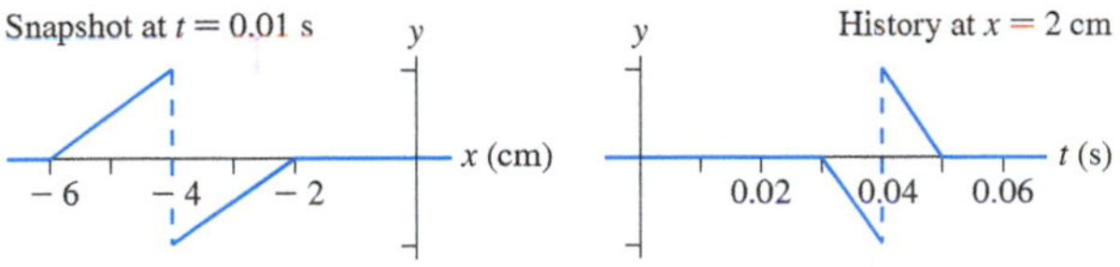


FIGURE Q15.10

11. Rank in order, from largest to smallest, the wavelengths $\lambda_1$ to $\lambda_3$ for sound waves having frequencies $f_1 = 100$ Hz, $f_2 = 1000$ Hz, and $f_3 = 10{,}000$ Hz. Explain.
12. BIO Bottlenose dolphins use echolocation pulses with a frequency of about 100 kHz, higher than the frequencies used by most bats. Why might you expect these water-dwelling creatures to use higher echolocation frequencies than bats?

13. BIO Some bat species have auditory systems that work best over a narrow range of frequencies. To account for this, the bats adjust the sound frequencies they emit so that the returning, Doppler-shifted sound pulse is in the correct frequency range. As a bat increases its forward speed, should it increase or decrease the frequency of the emitted pulses to compensate?
14. A laser beam has intensity $I_0$.
    a. What is the intensity, in terms of $I_0$, if a lens focuses the laser beam to 1/10 its initial diameter?
    b. What is the intensity, in terms of $I_0$, if a lens defocuses the laser beam to 10 times its initial diameter?
15. A wave pulse travels along a horizontal string. As the pulse passes a point on the string, the point moves vertically up and then back down again. How does the vertical speed of the point compare to the speed of the wave?
16. The volume control on a stereo is designed so that three clicks of the dial increase the output by 10 dB. How many clicks are required to increase the power output of the loudspeakers by a factor of 100?
17. A bullet can travel at a speed of over 1000 m/s. When a bullet is fired from a rifle, the actual firing makes a distinctive sound, but people at a distance may hear a second, different sound that is even louder. Explain the source of this sound.
18. You are standing at $x = 0$ m, listening to seven identical sound sources described by Figure Q15.18. At $t = 0$ s, all seven are at $x = 343$ m and moving as shown below. The sound from all seven will reach your ear at $t = 1$ s. Rank in order, from highest to lowest, the seven frequencies $f_1$ to $f_7$ that you hear at $t = 1$ s. Explain.

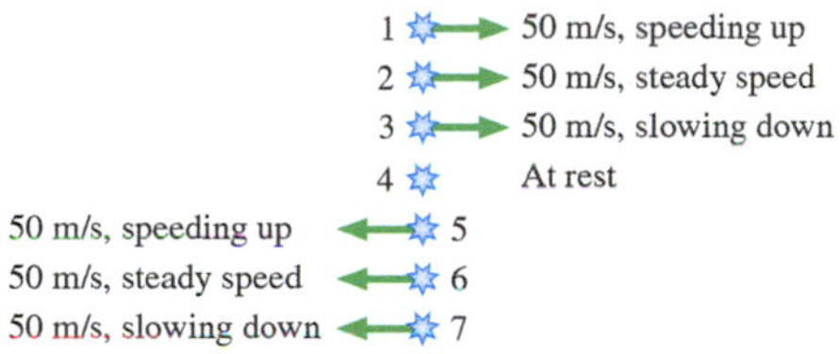


FIGURE Q15.18

19. Police radar guns work by measuring the change in frequency of a reflected pulse of electromagnetic waves. A gun reads correctly only if a car is moving directly toward or away from the person making the measurement. Explain why this is so.

## Multiple-Choice Questions

20. || An AM radio listener is located 5.0 km from the radio station. If he is listening to the radio at the frequency of 660 kHz, how many wavelengths fit in the distance from the station to his house?

    A. 11 B. 22 C. 33 D. 132
21. BIO | The probe used in a medical ultrasound examination emits sound waves in air that have a wavelength of 0.12 mm. What is the wavelength of the sound waves in the patient?

    A. 0.027 mm B. 0.12 mm C. 0.26 mm D. 0.54 mm
22. BIO | Ultrasound can be used to deliver energy to tissues for therapy. It can penetrate tissue to a depth approximately 200 times its wavelength. What is the approximate depth of penetration of ultrasound at a frequency of 5.0 MHz?

    A. 0.29 mm B. 1.4 cm C. 6.2 cm D. 17 cm

23. | A sinusoidal wave traveling on a string has a period of 0.20 s, a wavelength of 32 cm, and an amplitude of 3 cm. The speed of this wave is

A. 0.60 cm/s. B. 6.4 cm/s. C. 15 cm/s. D. 160 cm/s.

24. || Two strings of different linear density are joined together and pulled taut. A sinusoidal wave on these strings is traveling to the right, as shown in Figure Q15.24. When the wave goes across the boundary from string 1 to string 2, the frequency is unchanged. What happens to the velocity?

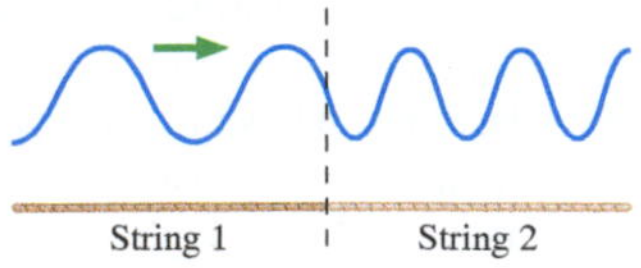

FIGURE Q15.24

A. The velocity increases.
B. The velocity stays the same.
C. The velocity decreases.

25. || You stand at $x = 0$ m, listening to a sound that is emitted at frequency $f_s$. Figure Q15.25 shows the frequency you hear during a four-second interval. Which of the following describes the motion of the sound source?

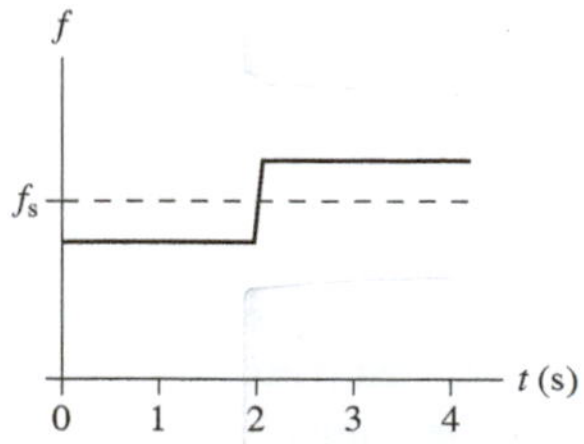

FIGURE Q15.25

A. It moves from left to right and passes you at $t = 2$ s.
B. It moves from right to left and passes you at $t = 2$ s.
C. It moves toward you but doesn't reach you. It then reverses direction at $t = 2$ s.
D. It moves away from you until $t = 2$ s. It then reverses direction and moves toward you but doesn't reach you.

# PROBLEMS

## Section 15.1 The Wave Model

## Section 15.2 Traveling Waves

1. || The wave speed on a string under tension is 200 m/s. What is the speed if the tension is doubled?
2. || The wave speed on a string is 150 m/s when the tension is 75.0 N. What tension will give a speed of 180 m/s?
3. || The back wall of an auditorium is 26.0 m from the stage. If you are seated in the middle row, how much time elapses between a sound from the stage reaching your ear directly and the same sound reaching your ear after reflecting from the back wall?
4. |||| Bats sense objects in the dark by echolocation, in which they emit very short pulses of sound and then listen for their echoes off the objects. A bat is flying directly toward a wall 50 m away when it emits a pulse. 0.28 s later it receives the pulse. What is the bat's speed? BIO
5. ||| A scientist measures the speed of sound in a monatomic gas to be 449 m/s at 20°C. What element does this gas consist of?
6. | A medical ultrasound imaging system sends out a steady stream of very short pulses. To simplify analysis, the reflection of one pulse should be received before the next is transmitted. If the system is being used to create an image of tissue 12 cm below the skin, what is the minimum time between pulses? How many pulses per second does this correspond to? BIO
7. || An earthquake 45 km from a city produces P and S waves that travel outward at 5000 and 3000 m/s, respectively. Once city residents feel the shaking of the P wave, how much time do they have before the S wave arrives?

## Section 15.3 Graphical and Mathematical Descriptions of Waves

8. || A stationary boat in the ocean is experiencing waves from a storm. The waves move at 56 km/h and have a wavelength of 160 m, both typical values. The boat is at the crest of a wave. How much time elapses until the boat is first at the trough of a wave?
9. || Figure P15.9 is a snapshot graph of a wave at $t = 0$ s. Draw the history graph for this wave at $x = 6$ m, for $t = 0$ s to 6 s.

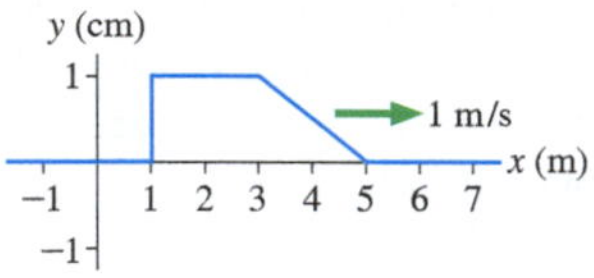

FIGURE P15.9

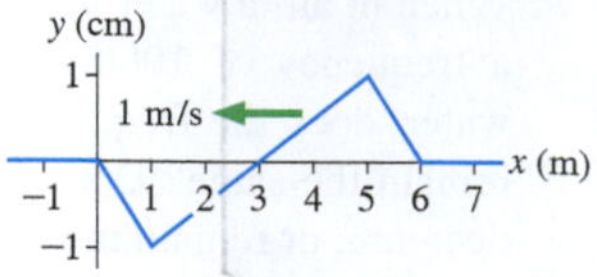

FIGURE P15.10

10. || Figure P15.10 is a snapshot graph of a wave at $t = 2$ s. Draw the history graph for this wave at $x = 0$ m, for $t = 0$ s to 8 s.
11. | Figure P15.10 is a snapshot graph of a wave at $t = 2$ s. What is the maximum speed reached by a particle on the string?
12. ||| Figure P15.12 is a history graph at $x = 0$ m of a wave moving to the right at 1 m/s. Draw a snapshot graph of this wave at $t = 1$ s.

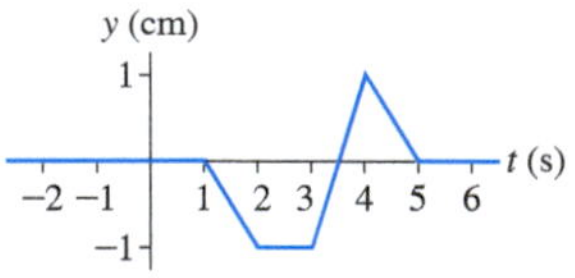

FIGURE P15.12

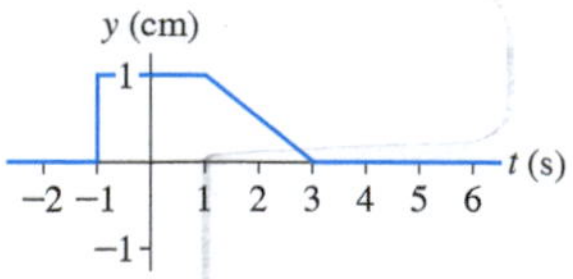

FIGURE P15.13

13. ||| Figure P15.13 is a history graph at $x = 2$ m of a wave moving to the left at 1 m/s. Draw the snapshot graph of this wave at $t = 0$ s.
14. ||| Figure P15.14 shows history graphs of two different points on a string as a wave pulse moves along the string. The blue curve is the history graph for the point at $x = 1.0$ cm, and the green curve is for the point at $x = -3.0$ cm. What is the velocity (including the correct sign for its direction) of this wave?

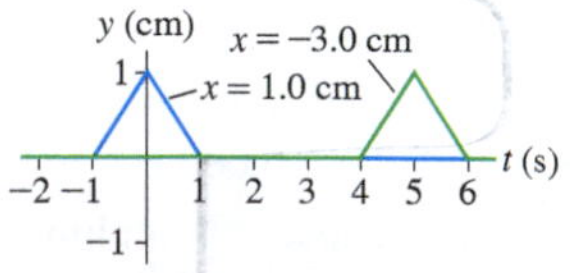

FIGURE P15.14

15. | A sinusoidal wave has period 0.20 s and wavelength 2.0 m. What is the wave speed?
16. | A sinusoidal wave travels with speed 200 m/s. Its wavelength is 4.0 m. What is its frequency?
17. || The motion detector used in a physics lab sends and receives 40 kHz ultrasonic pulses. A pulse goes out, reflects off the object being measured, and returns to the detector. The lab temperature is 20°C.
 a. What is the wavelength of the waves emitted by the motion detector?
 b. How long does it take for a pulse that reflects off an object 2.5 m away to make a round trip?
18. | The displacement of a wave traveling in the positive $x$-direction is $y(x, t) = (3.5\text{ cm}) \times \cos(2.7x - 92t)$, where $x$ is in m and $t$ is in s. What are the (a) frequency, (b) wavelength, and (c) speed of this wave?
19. ||| A traveling wave has displacement given by $y(x, t) = (2.0\text{ cm}) \times \cos(2\pi x - 4\pi t)$, where $x$ is measured in cm and $t$ in s.
 a. Draw a snapshot graph of this wave at $t = 0$ s.
 b. On the same set of axes, use a dotted line to show the snapshot graph of the wave at $t = 1/8$ s.
 c. What is the speed of the wave?
20. | Figure P15.20 is a snapshot graph of a wave at $t = 0$ s. What are the amplitude, wavelength, and frequency of this wave?

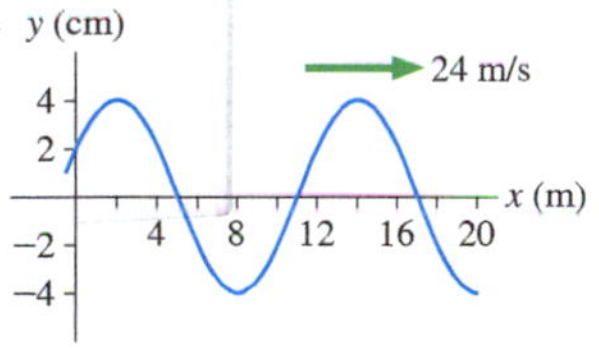

FIGURE P15.20

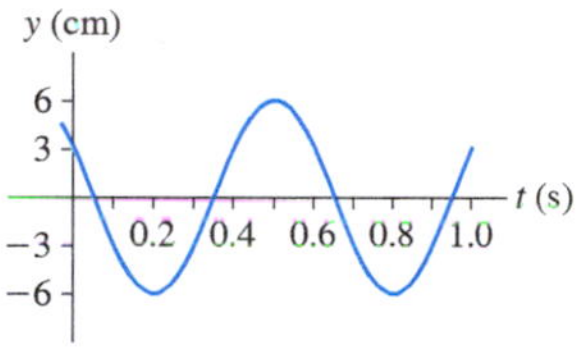

FIGURE P15.21

21. | Figure P15.21 is a history graph at $x = 0$ m of a wave moving to the right at 2 m/s. What are the amplitude, frequency, and wavelength of this wave?
22. || Figure P15.22 shows snapshot (left) and history (right) graphs for a wave traveling on a string. The snapshot graph shows the wave at $t = 0$ s; the history graph shows the displacement of the point on the string at $x = 1.5$ cm, indicated by the dot in the snapshot graph. What is the speed of the wave, and in which direction is it traveling?

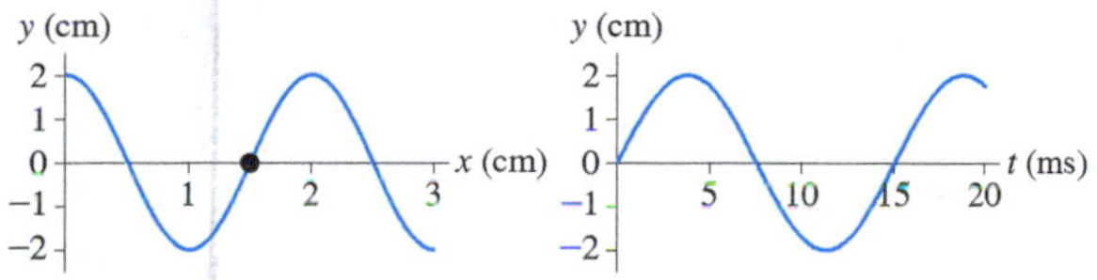

FIGURE P15.22

23. |||| A sinusoidal wave moving to the left has a wavelength of 5.0 cm and a frequency of 50 Hz. At $t = 0$ s, the wave has a crest at $x = 0$ cm. What is the earliest time after $t = 0$ s at which there is a crest at the position $x = 3.0$ cm?

### Section 15.4 Sound and Light Waves

24. | People with very good pitch discrimination can very quickly BIO determine what note they are listening to. The note on the musical scale called $C_6$ (two octaves above middle C) has a frequency of 1050 Hz. Some trained musicians can identify this note after hearing only 12 cycles of the wave. How much time does this correspond to?
25. | A dolphin emits ultrasound at 100 kHz and uses the timing BIO of reflections to determine the position of objects in the water. What is the wavelength of this ultrasound?
26. | Elephants can communicate over distances as far as 6 km by BIO using very low-frequency sound waves. What is the wavelength of a 10 Hz sound wave emitted by an elephant?
27. || a. What is the frequency of blue light that has a wavelength of 450 nm?
 b. What is the frequency of red light that has a wavelength of 650 nm?
28. | Research vessels at sea can create images of their surroundings by sending out sound waves and measuring the time until they detect echoes. This image of a shipwreck on the ocean bottom was made from the surface with 600 kHz ultrasound.

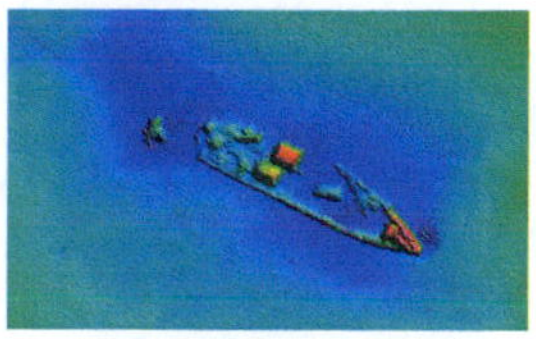

 a. What was the wavelength?
 b. How deep is the shipwreck if echoes were detected 0.42 s after the sound waves were emitted?
29. | A bullet shot from a rifle travels at 1000 m/s. What is the elapsed time between when the bullet strikes a target 500 m away, and when the sound of the gunshot reaches the target?
30. || a. An FM radio station broadcasts at a frequency of 101.3 MHz. What is the wavelength?
 b. What is the frequency of a sound source that produces the same wavelength in 20°C air?

### Section 15.5 Energy and Intensity

31. || Sound is detected when a sound wave causes the eardrum BIO INT to vibrate (see Figure 14.26). Typically, the diameter of the eardrum is about 8.4 mm in humans. When someone speaks to you in a normal tone of voice, the sound intensity at your ear is approximately $1.0 \times 10^{-6}\text{ W/m}^2$. How much energy is delivered to your eardrum each second?
32. ||| At a rock concert, the sound intensity 1.0 m in front of the BIO INT bank of loudspeakers is $0.10\text{ W/m}^2$. A fan is 30 m from the loudspeakers. Her eardrums have a diameter of 8.4 mm. How much sound energy is transferred to each eardrum in 1.0 second?
33. ||| The intensity of electromagnetic waves from the sun is $1.4\text{ kW/m}^2$ just above the earth's atmosphere. Eighty percent of this reaches the surface at noon on a clear summer day. Suppose you model your back as a 30 cm × 50 cm rectangle. How many joules of solar energy fall on your back as you work on your tan for 1.0 h?
34. |||| A sun-like star is barely visible to naked-eye observers on earth when it is a distance of 7.0 light years, or $6.6 \times 10^{16}$ m, away. The sun emits a power of $3.8 \times 10^{26}$ W. Using this information, at what distance would a candle that emits a power of 0.20 W just be visible?
35. ||| A large solar panel on a spacecraft in earth orbit produces 1.0 kW of power when the panel is turned toward the sun. What power would the solar cell produce if the spacecraft were in orbit around Saturn, 9.5 times as far from the sun?
36. ||| Solar cells convert the energy of incoming light to electric energy; a good quality cell operates at an efficiency of 15%. Each person in the United States uses energy (for lighting, heating, transportation, etc.) at an average rate of 11 kW. Although sunlight varies with season and time of day, solar energy falls on the United States at an average intensity of $200\text{ W/m}^2$. Assuming you live in an average location, what total solar-cell area would you need to provide all of your energy needs with energy from the sun?

37. II LASIK eye surgery uses pulses of laser light to shave off BIO tissue from the cornea, reshaping it. A typical LASIK laser emits a 1.0-mm-diameter laser beam with a wavelength of 193 nm. Each laser pulse lasts 15 ns and contains 1.0 mJ of light energy.
   a. What is the power of one laser pulse?
   b. During the very brief time of the pulse, what is the intensity of the light wave?

38. II Using a dish-shaped mirror, a *solar cooker* concentrates the sun's energy onto a pot for cooking. A cooker with a 1.5-m-diameter dish focuses the sun's energy onto a pot with a diameter of 25 cm. Given that the intensity of sunlight is about 1000 W/m$^2$,

   a. How much solar power does the dish capture?
   b. What is the intensity at the base of the pot?

39. I The world's most powerful laser is the LFEX laser in Japan. It can produce a 2 petawatt ($2 \times 10^{15}$ W) laser pulse that last for 1 ps. The laser is focused onto a small spot that is 30 $\mu$m in diameter. What is the light intensity within this spot?

### Section 15.6 Loudness of Sound

40. I What is the sound intensity level of a sound with an intensity of $3.0 \times 10^{-6}$ W/m$^2$?

41. III What is the sound intensity of a whisper at a distance of 2.0 m, in W/m$^2$? What is the corresponding sound intensity level in dB?

42. III The record for the world's loudest burp is 109.9 dB, mea-BIO sured at a distance of 2.5 m from the burper. Assuming that this sound was emitted as a spherical wave, what was the power emitted by the burper during his record burp?

43. II The sound intensity from a jack hammer breaking concrete is 2.0 W/m$^2$ at a distance of 2.0 m from the point of impact. This is sufficiently loud to cause permanent hearing damage if the operator doesn't wear ear protection. What are (a) the sound intensity and (b) the sound intensity level for a person watching from 50 m away?

44. III A concert loudspeaker suspended high off the ground emits 35 W of sound power. A small microphone with a 1.0 cm$^2$ area is 50 m from the speaker. What are (a) the sound intensity and (b) the sound intensity level at the position of the microphone?

45. III The African cicada is the world's loudest insect, producing a BIO sound intensity level of 107 dB at a distance of 0.50 m. What is the intensity of its sound (in W/m$^2$) as heard by someone standing 3.0 m away?

46. III From a distance of 4.0 m, a bystander listens to a jackhammer breaking concrete. How far would he need to move from the jackhammer so that its perceived loudness decreases by a factor of 8?

47. II A rock band playing an outdoor concert produces sound at 120 dB 5.0 m away from their single working loudspeaker. What is the sound intensity level 35 m from the speaker?

48. III Your ears are sensitive to differences in pitch, but they are not BIO very sensitive to differences in intensity. You are not capable of detecting a difference in sound intensity level of less than 1 dB. By what factor does the sound intensity increase if the sound intensity level increases from 60 dB to 61 dB?

49. II 30 seconds of exposure to 115 dB sound can damage your BIO hearing, but a much quieter 94 dB may begin to cause damage after 1 hour of continuous exposure. You are going to an outdoor concert, and you'll be standing near a speaker that emits 50 W of acoustic power as a spherical wave. What minimum distance should you be from the speaker to keep the sound intensity level below 94 dB?

50. III When you speak, your voice sounds 10 dB louder to someone BIO standing directly in front of you than to someone at the same distance but directly behind you. What is the ratio of the intensity of your voice for someone in front of you to the intensity for someone behind you?

### Section 15.7 The Doppler Effect and Shock Waves

51. I An opera singer in a convertible sings a note at 600 Hz while cruising down the highway at 90 km/h. What is the frequency heard by
   a. A person standing beside the road in front of the car?
   b. A person standing beside the road behind the car?

52. II An osprey's call is a distinct whistle at 2200 Hz. An osprey BIO calls while diving at you, to drive you away from her nest. You hear the call at 2300 Hz. How fast is the osprey approaching?

53. II A whistle you use to call your hunting dog has a frequency of 21 kHz, but your dog is ignoring it. You suspect the whistle may not be working, but you can't hear sounds above 20 kHz. To test it, you ask a friend to blow the whistle, then you hop on your bicycle. In which direction should you ride (toward or away from your friend) and at what minimum speed to know if the whistle is working?

54. I An echocardiogram uses 4.4 MHz ultrasound to measure blood BIO flow in the aorta. The blood is moving away from the probe at 1.4 m/s. What is the frequency shift of the reflected ultrasound?

55. I A friend of yours is loudly singing a single note at 400 Hz while driving toward you at 25.0 m/s on a day when the speed of sound is 340 m/s.
   a. What frequency do you hear?
   b. What frequency does your friend hear if you suddenly start singing at 400 Hz?

56. III The frequency of light emitted from hydrogen present in the Andromeda galaxy has been found to be 0.10% higher than that from hydrogen measured on earth. Is this galaxy approaching or receding from the earth, and at what speed?

57. III A Doppler blood flow unit emits ultrasound at 5.0 MHz. BIO What is the frequency shift of the ultrasound reflected from blood moving in an artery at a speed of 0.20 m/s?

58. III A train whistle is heard at 300 Hz as the train approaches town. The train cuts its speed in half as it nears the station, and the sound of the whistle is then 290 Hz. What is the speed of the train before and after slowing down?

## General Problems

59. III At the 18 km cruising altitude of Concorde, a passenger aircraft that flew at twice the speed of sound, the temperature was $-57°$C. What was the Concorde's cruising speed?

60. III A 2.0-m-long string is under 20 N of tension. A pulse travels the length of the string in 50 ms. What is the mass of the string?

61. III A female orb spider has a mass of 0.50 g. She is suspended BIO INT from a tree branch by a 1.1 m length of 0.0020-mm-diameter silk. Spider silk has a density of 1300 kg/m$^3$. If you tap the branch and send a vibration down the thread, how long does it take to reach the spider?

62. || BIO A spider spins a web with silk threads of density 1300 $kg/m^3$ and diameter 3.0 $\mu$m. A typical tension in the radial threads of such a web is 7.0 mN. Suppose a fly hits this web. Which will reach the spider first: the very slight sound of the impact or the disturbance traveling along the radial thread of the web?

63. ||| INT In 2003, an earthquake in Japan generated 1.1 Hz waves that traveled outward at 7.0 km/s. 200 km to the west, seismic instruments recorded a maximum acceleration of $0.25g$ along the east-west axis.
   a. How much time elapsed between the earthquake and the first detection of the waves?
   b. Was this a transverse or a longitudinal wave?
   c. What was the wavelength?
   d. What was the maximum horizontal displacement of the ground as the wave passed?

64. ||| BIO A coyote can locate a sound source with good accuracy by comparing the arrival times of a sound wave at its two ears. Suppose a coyote is listening to a bird whistling at 1000 Hz. The bird is 3.0 m away, directly in front of the coyote's right ear. The coyote's ears are 15 cm apart.
   a. What is the difference in the arrival times of the sound at the left ear and the right ear?
   b. What is the ratio of this times difference to the period of the sound wave?

   **Hint:** You are looking for the difference between two numbers that are nearly the same. What does this near equality imply about the necessary precision during intermediate stages of the calculation?

65. ||| A wave travels along a steel string with a speed of 22 m/s. At some point along the string, its diameter doubles. What is the speed of the wave in this thicker part?

66. || INT The string in Figure P15.66 has a linear density of $7.2 \times 10^{-5}$ kg/m. What is the speed of a wave on this string?

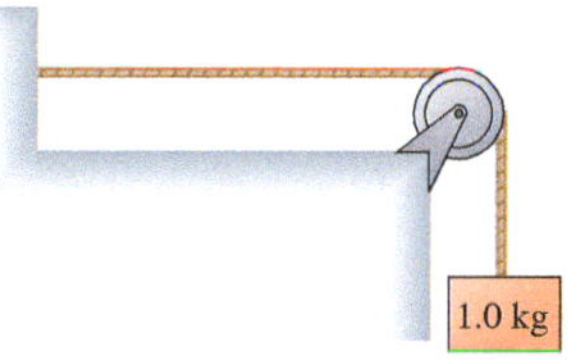


FIGURE P15.66

67. || BIO Low-frequency vertical oscillations are one possible cause of motion sickness, with 0.30 Hz having the strongest effect. Your boat is bobbing in place at just the right frequency to cause you the maximum discomfort. The water wave that is bobbing the boat has crests that are 30 m apart.
   a. What is the speed of the waves?
   b. What will be the boat's vertical oscillation frequency if you drive the boat at 5.0 m/s in the direction of the oncoming waves?

68. ||| INT Figure P15.68 shows two snapshot graphs taken 10 ms apart, with the blue curve being the first snapshot. What are the (a) wavelength, (b) speed, (c) frequency, and (d) amplitude of this wave?

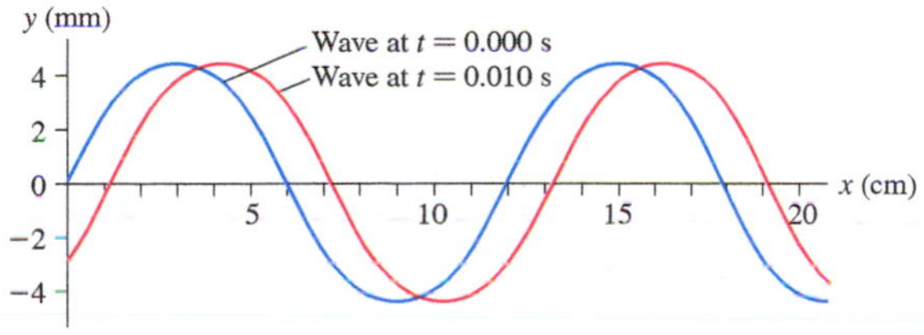


FIGURE P15.68

69. ||| The pressure in a sound wave in steel is given by $p(x, t) = p_{atm} + p_0 \cos(2.4x - (1.4 \times 10^4)t)$, where $p_{atm}$ is atmospheric pressure, $p_0$ is the amplitude of the wave, $x$ is in m, and $t$ in s. What are the speed and frequency of this wave?

70. ||| A wave on a string is described by $y(x, t) = (3.0\text{ cm}) \times \cos[2\pi(x/(2.4\text{ m}) + t/(0.20\text{ s}))]$, where $x$ is in m and $t$ in s.
   a. In what direction is this wave traveling?
   b. What are the wave speed, frequency, and wavelength?
   c. At $t = 0.50$ s, what is the displacement of the string at $x = 0.20$ m?

71. | Write the $y$-equation for a wave traveling in the negative $x$-direction with wavelength 50 cm, speed 4.0 m/s, and amplitude 5.0 cm.

72. ||| INT A point on a string undergoes simple harmonic motion as a sinusoidal wave passes. When a sinusoidal wave with speed 24 m/s, wavelength 30 cm, and amplitude of 1.0 cm passes, what is the maximum speed of a point on the string?

73. ||| BIO The threshold of hearing—the lowest-intensity sound a person can hear—depends on, among many factors, the frequency of the sound. During a hearing exam, the hearing specialist creates an *audiogram*, a graph of the patient's hearing threshold (in dB) versus sound frequency; thresholds above 20 dB indicate hearing loss. For the audiogram shown in Figure P15.73, what is the ratio of the sound intensity (in $W/m^2$) of the faintest sound the patient can hear at 1 kHz to that of the faintest sound she can hear at 6 kHz?

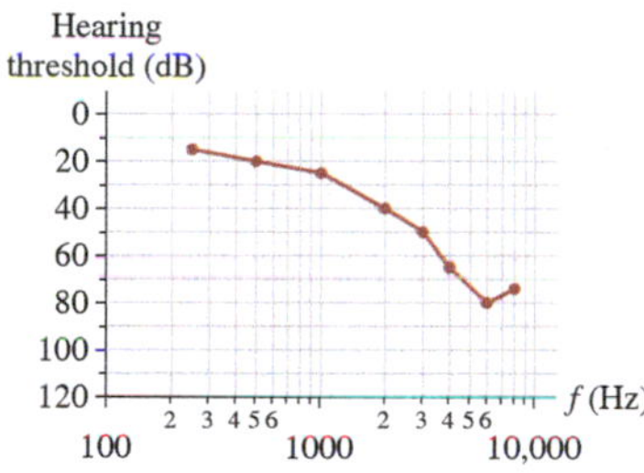


FIGURE P15.73

74. | The total power consumption by all humans on earth is approximately $10^{13}$ W. Let's compare this to the power of incoming solar radiation. The intensity of radiation from the sun at the top of the atmosphere is 1380 $W/m^2$. The earth's radius is $6.37 \times 10^6$ m.
   a. What is the total solar power received by the earth?
   b. By what factor does this exceed the total human power consumption?

75. | INT A dark blue cylindrical bottle is 22 cm high and has a diameter of 7.0 cm. It is filled with water. The bottle absorbs 60% of the light that shines on it as it lies on its side in the noonday sun, with intensity 1000 $W/m^2$. By how much will the temperature of the water increase in 5.0 min if there's negligible heat loss to the surrounding air?

76. || BIO Assume that the opening of the ear canal has a diameter of 7.0 mm. For this problem, you can ignore any focusing of energy into the opening by the pinna, the external folds of the ear.
   a. How much sound power is "captured" by one ear at 0 dB, the threshold of hearing?
   b. How much energy does one ear "capture" during 1 hour of listening to a lecture delivered at a conversational volume of 60 dB?

77. | The sound intensity 50 m from a wailing tornado siren is 0.10 $W/m^2$. What is the sound intensity level 300 m from the siren?

Watch Video Solution Problems 15.79 and 15.81

78. || One of the loudest sound generators ever created is the INT Danley Sound Labs Matterhorn. When run at full power, the device uses an input power of 40,000 W. The output power registers 94 dB at 250 m. What is the efficiency of this device—that is, the ratio of output power to input power?

79. ||| A harvest mouse can detect sounds below the threshold of BIO human hearing, as quiet as $-10$ dB. Suppose you are sitting in a field on a very quiet day while a harvest mouse sits nearby. A very gentle breeze causes a leaf 1.5 m from your head to rustle, generating a faint sound right at the limit of your ability to hear it. The sound of the rustling leaf is also right at the threshold of hearing of the harvest mouse. How far is the harvest mouse from the leaf?

80. |||| A speaker at an open-air concert emits 600 W of sound power, radiated equally in all directions.
    a. What is the intensity of the sound 5.0 m from the speaker?
    b. What sound intensity level would you experience there if you did not have any protection for your ears?
    c. Earplugs you can buy in the drugstore have a noise reduction rating of 23 decibels. If you are wearing those earplugs but your friend Phil is not, how far from the speaker should Phil stand to experience the same loudness as you?

81. | A physics professor demonstrates the Doppler effect by tying INT a 600 Hz sound generator to a 1.0-m-long rope and whirling it around her head in a horizontal circle at 100 rpm. What are the highest and lowest frequencies heard by a student in the classroom? Assume the room temperature is 20°C.

82. ||||| When the heart pumps blood into the aorta, the *pressure* BIO INT *gradient*—the difference between the blood pressure inside the heart and the blood pressure in the artery—is an important diagnostic measurement. A direct measurement of the pressure gradient is difficult, but an indirect determination can be made by inferring the pressure difference from a measurement of velocity. Blood is essentially at rest in the heart; when it leaves and enters the aorta, it speeds up significantly and—according to Bernoulli's equation—the pressure must decrease. A doctor using 2.5 MHz ultrasound measures a 6000 Hz frequency shift as the ultrasound reflects from blood ejected from the heart.
    a. What is the speed of the blood in the aorta?
    b. What is the difference in blood pressure between the inside of the heart and the aorta? Assume that the patient is lying down and that there is no difference in height as the blood moves from the heart into the aorta.

## MCAT-Style Passage Problems

### Echolocation BIO

As discussed in the chapter, many species of bats find flying insects by emitting pulses of ultrasound and listening for the reflections. This technique is called *echolocation*. Bats possess several adaptations that allow them to echolocate very effectively.

83. | Although we can't hear them, the ultrasonic pulses are very loud. In order not to be deafened by the sound they emit, bats can temporarily turn off their hearing. Muscles in the ear cause the bones in their middle ear to separate slightly, so that they don't transmit vibrations to the inner ear. After an ultrasound pulse ends, a bat can hear an echo from an object a minimum of 1 m away. Approximately how much time after a pulse is emitted is the bat ready to hear its echo?
    A. 0.5 ms B. 1 ms C. 3 ms D. 6 ms

84. | Bats are sensitive to very small changes in frequency of the reflected waves. What information does this allow them to determine about their prey?
    A. Size B. Speed C. Distance D. Species

85. | Some bats have specially shaped noses that focus ultrasound echolocation pulses in the forward direction. Why is this useful?
    A. Increasing intensity reduces the time delay for a reflected pulse.
    B. The energy of the pulse is concentrated in a smaller area, so the intensity is larger; reflected pulses will have a larger intensity as well.
    C. Increasing intensity allows the bat to use a lower frequency and still have the same spatial resolution.

86. | Some bats utilize a sound pulse with a rapidly decreasing frequency. A decreasing-frequency pulse has
    A. Decreasing wavelength.
    B. Decreasing speed.
    C. Increasing wavelength.
    D. Increasing speed.

# 16 Superposition and Standing Waves

The didgeridoo is a musical instrument played by aboriginal tribes in Australia. It consists of a hollow tube that is open at both ends. How can the player get a wide range of notes out of such a simple device?

## LOOKING AHEAD ▸▸

### Superposition

Where the two water waves meet, the motion of the water is the sum, or **superposition,** of the waves.

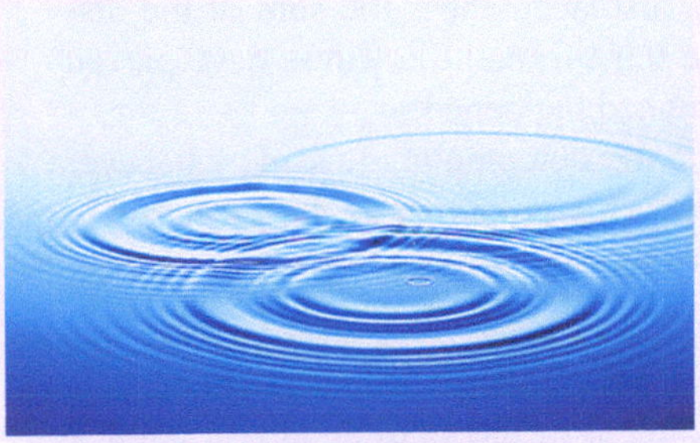

You'll learn how this **interference** can be constructive or destructive, leading to larger or smaller amplitudes.

### Standing Waves

The superposition of waves on a string can lead to a wave that oscillates in place—a **standing wave.**

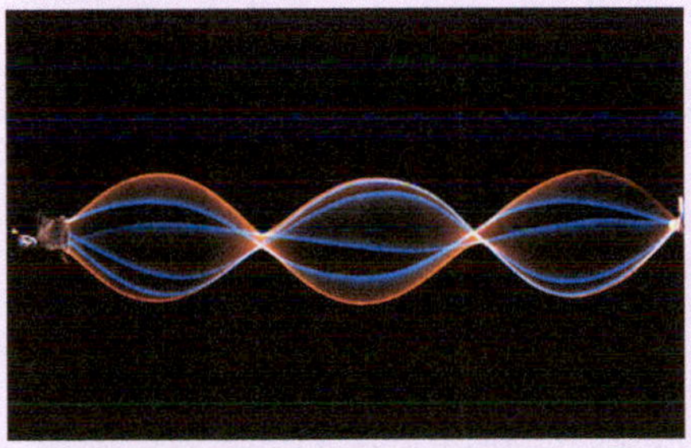

You'll learn the patterns of standing waves on strings and standing sound waves in tubes.

### Speech and Hearing

Changing the shape of your mouth alters the pattern of standing sound waves in your vocal tract.

You'll learn how your vocal tract produces, and your ear interprets, different mixes of waves.

**GOAL** To use the idea of superposition to understand the phenomena of interference and standing waves.

## LOOKING BACK ◂◂

### Traveling Waves

In Chapter 15 you learned the properties of traveling waves and relationships among the variables that describe them.

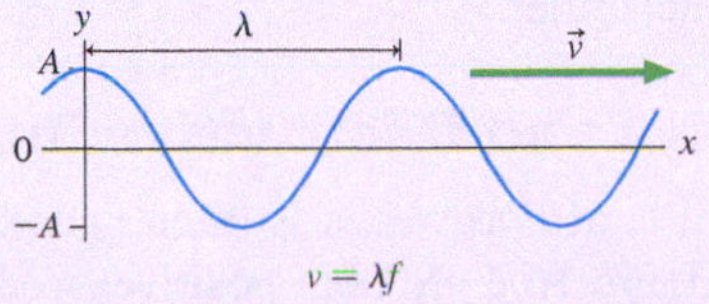


In this chapter, you'll extend the analysis to understand the interference of waves and the properties of standing waves.

**STOP TO THINK**

A 170 Hz sound wave in air has a wavelength of 2.0 m. The frequency is now doubled to 340 Hz. What is the new wavelength?

A. 4.0 m
B. 3.0 m
C. 2.0 m
D. 1.0 m

## 16.1 The Principle of Superposition

FIGURE 16.1a shows two baseball players, Alan and Bill, at batting practice. Unfortunately, someone has turned the pitching machines so that pitching machine A throws baseballs toward Bill while machine B throws toward Alan. If two baseballs are launched at the same time and with the same speed, they collide at the crossing point and bounce away. Two baseballs cannot occupy the same point of space at the same time.

FIGURE 16.1 Two baseballs cannot pass through each other. Two waves can.

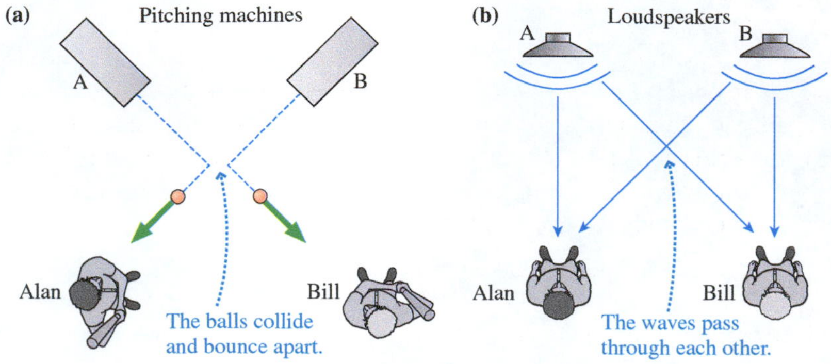


FIGURE 16.2 Two waves on a stretched string pass through each other.

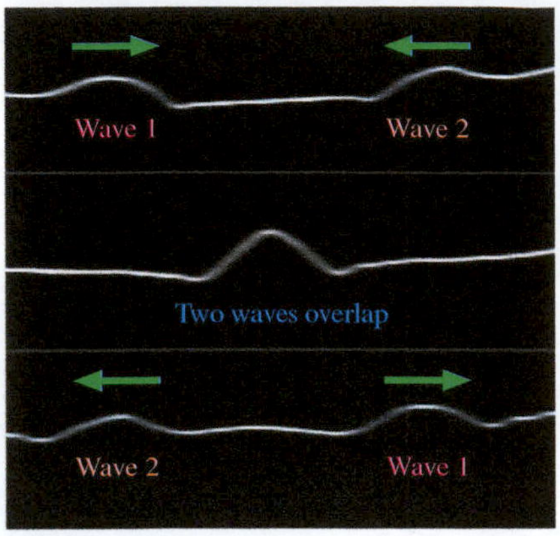


But unlike baseballs, sound waves *can* pass directly through each other. In FIGURE 16.1b, Alan and Bill are listening to the stereo system in the locker room after practice. Both hear the sound from each speaker perfectly, without any distortion or missing sound. Evidently the sound wave that travels from speaker A toward Bill passes right through the wave that travels from speaker B toward Alan, with no effect on either wave. This is a basic property of waves.

What happens to the medium at a point where two waves are present simultaneously? What is the displacement of the medium at this point? FIGURE 16.2 shows a sequence of photos of two waves traveling along a stretched string. In the first photo, the waves are approaching each other. In the second, the waves overlap; you can see that the displacement of the string is larger than it was for either of the individual waves. A careful measurement reveals that the displacement is the sum of the displacements of the two individual waves. In the third frame, the waves have passed through each other and continue on as if nothing had happened.

This result is not limited to stretched strings; the outcome is the same whenever two waves of any type pass through each other. This is known as the *principle of superposition:*

**Principle of superposition** When two or more waves are *simultaneously* present at a single point in space, the displacement of the medium at that point is the sum of the displacements due to each individual wave.

FIGURE 16.3 The superposition of two waves on a string as they pass through each other.

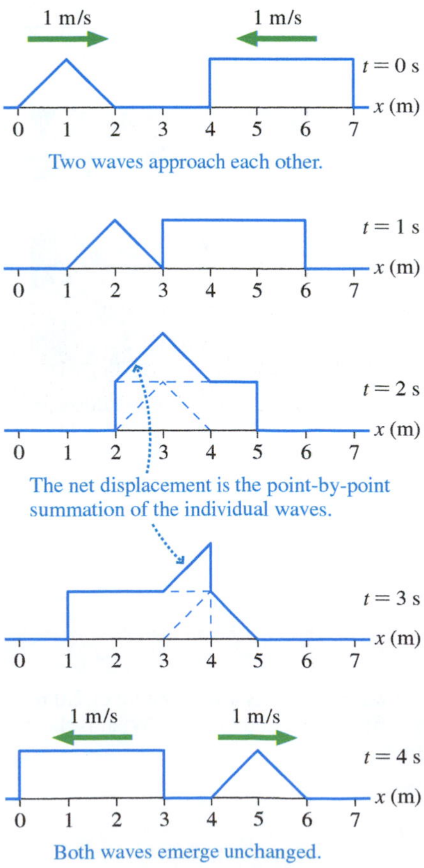


Let's illustrate this principle with an idealized example. FIGURE 16.3 shows five snapshot graphs taken 1 s apart of two waves traveling at the same speed (1 m/s) in opposite directions along a string. The displacement of each wave is shown as a dashed line. The solid line is the sum *at each point* of the two displacements at that point. This is the displacement that you would actually observe as the two waves pass through each other.

### Constructive and Destructive Interference

The superposition of two waves is often called **interference.** The displacements of the waves in Figure 16.3 are both positive, so the total displacement of the medium where they overlap is larger than it would be due to either of the waves separately. We call this **constructive interference.**

FIGURE 16.4 shows another series of snapshot graphs of two counterpropagating waves—waves traveling in opposite directions—but this time one has a negative displacement. The principle of superposition still applies, but now the displacements are opposite to each other. The displacement of the medium where the waves overlap is *less* than it would be due to either of the waves separately. We call this **destructive interference.**

In the series of graphs in Figure 16.4 the displacement of the medium at $x = 3.5$ m is always zero. The positive displacement of the wave traveling to the right and the negative displacement of the wave traveling to the left always exactly cancel at this spot. The complete cancellation at one point of two waves traveling in opposite directions is something we will see again. When the displacements of the waves cancel, where does the energy of the wave go? We know that the waves continue on unchanged after their interaction, so no energy is dissipated. Consider the graph at $t = 1.5$ s. There is no net displacement at any point of the medium at this instant, but the *string is rapidly moving vertically.* The energy of the waves hasn't vanished—it is in the form of the kinetic energy of the medium.

FIGURE 16.4 Two waves with opposite displacements produce destructive interference.

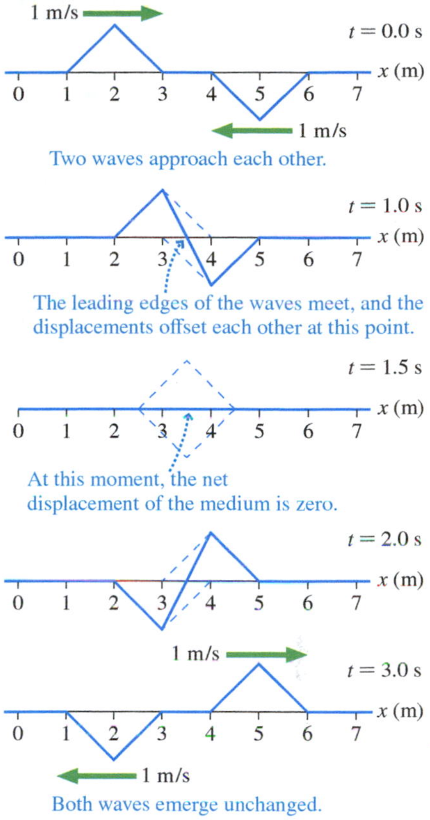


**STOP TO THINK 16.1** Two pulses on a string approach each other at speeds of 1 m/s. What is the shape of the string at $t = 6$ s?

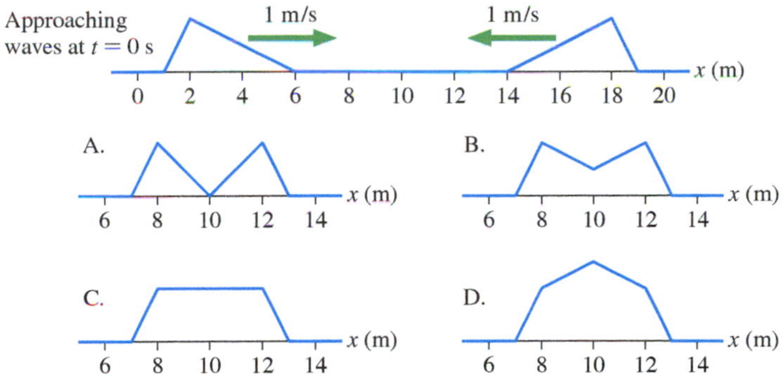


A video to support a section's topic is embedded in the eText.
**Video** What the Physics? Destructive Interference

## 16.2 Standing Waves

When you pluck a guitar string or a rubber band stretched between your fingers, you create waves. But how is this possible? There isn't really anywhere for the waves to go, because the string or the rubber band is held between two fixed ends. FIGURE 16.5 shows a strobe photograph of waves on a stretched elastic cord. This is a wave, though it may not look like one because it doesn't "travel" either right or left. Waves that are "trapped" between two boundaries, like those in the photo or on a guitar string, are what we call *standing waves.* **Individual points on the string oscillate up and down, but the wave itself does not travel.** It is called a **standing wave** because the crests and troughs "stand in place" as it oscillates. As we'll see, a standing wave isn't a totally new kind of wave; it is simply the superposition of two traveling waves moving in opposite directions.

FIGURE 16.5 The motion of a standing wave on a string.

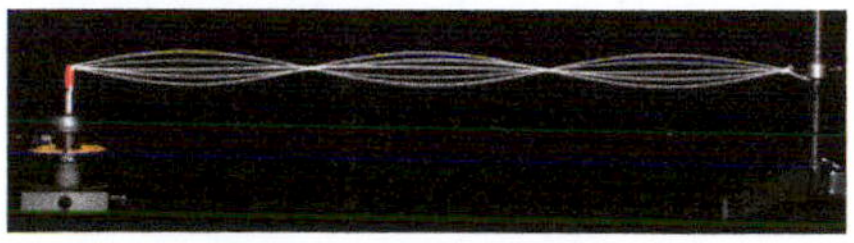

### Superposition Creates a Standing Wave

Suppose we have a string on which two sinusoidal waves of equal wavelength and amplitude travel in opposite directions, as in FIGURE 16.6a (next page). When the waves meet, the displacement of the string will be a superposition of these two waves. FIGURE 16.6b shows nine snapshot graphs, at intervals of $\frac{1}{8}T$, of the two waves as they move through each other. The red and orange dots identify particular crests of each of the waves to help you see that the red wave is traveling to the right and the orange wave to the left. At *each point,* the net displacement of the medium is found by adding the red displacement and the orange displacement. The resulting blue wave is the superposition of the two traveling waves.

FIGURE 16.6 Two sinusoidal waves traveling in opposite directions.

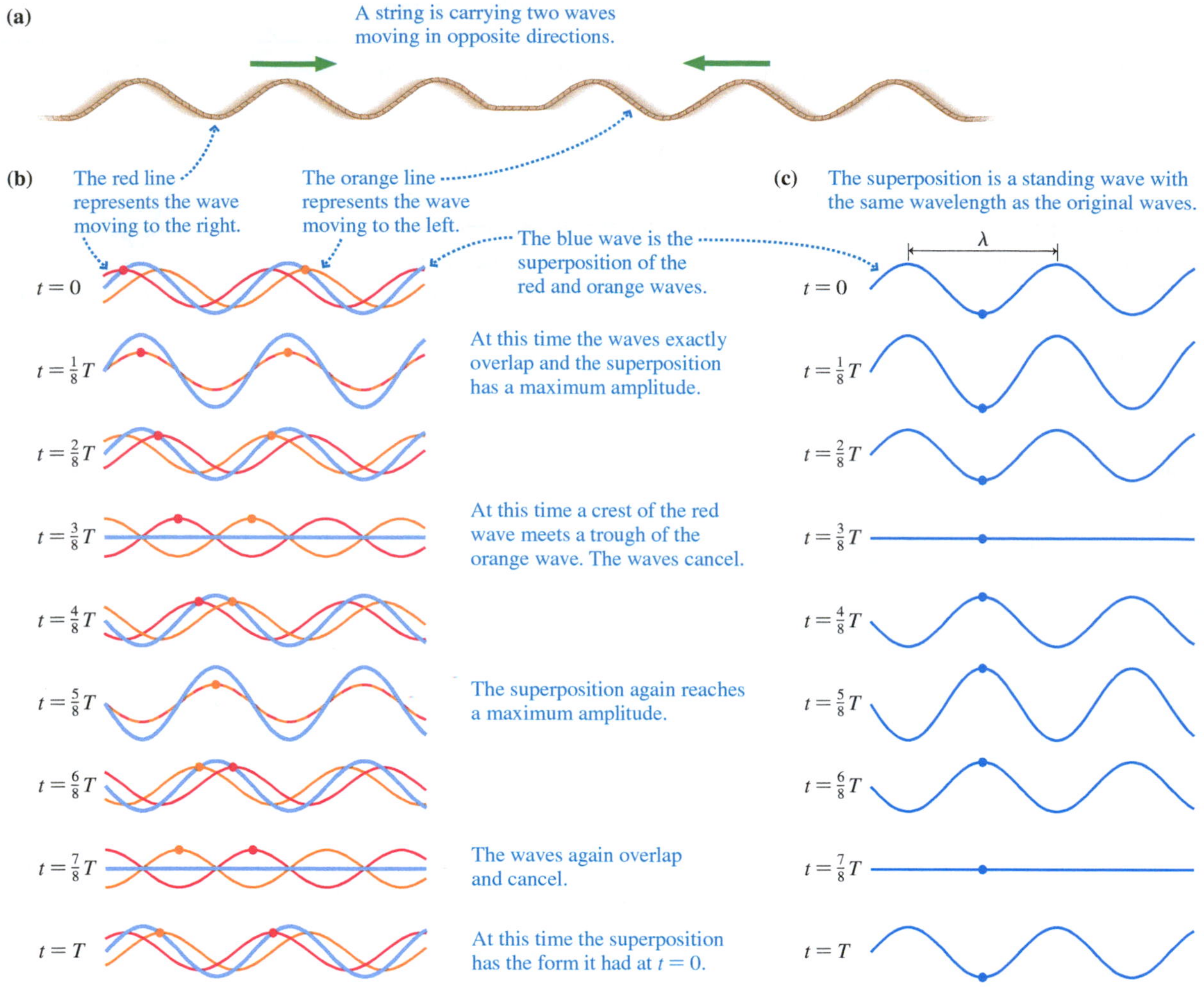


FIGURE 16.6c shows just the superposition of the two waves. This is the wave that you would actually observe in the medium. The blue dot shows that the wave in Figure 16.6c is moving neither right nor left. The superposition of the two counter-propagating traveling waves is a standing wave. Notice that the wavelength of the standing wave, the distance between two crests or two troughs, is the same as the wavelengths of the two traveling waves that combine to produce it.

FIGURE 16.7 Superimposing multiple snapshot graphs of a standing wave clearly shows the nodes and antinodes.

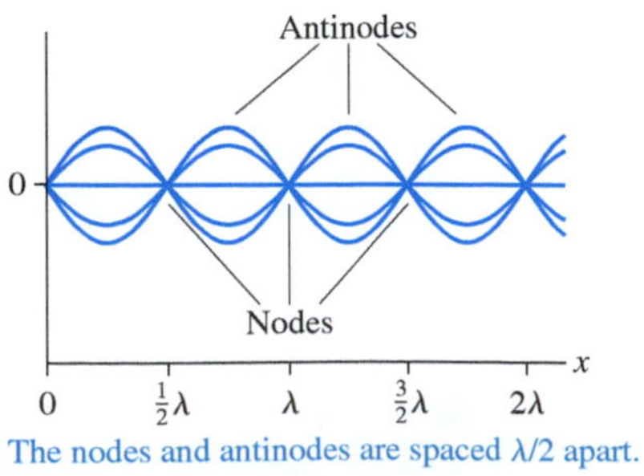


## Nodes and Antinodes

In FIGURE 16.7 we have superimposed the nine snapshot graphs of Figure 16.6c into a single graphical representation of this standing wave. The graphs at different times overlap, much like the photos of the string at different times in the strobe photograph of Figure 16.5. The motion of individual points on the standing wave is now clearly seen. A striking feature of a standing-wave pattern is points that *never move!* These points, which are spaced $\lambda/2$ apart, are called **nodes.** Halfway between the nodes are points where the particles in the medium oscillate with maximum displacement. These points of maximum amplitude are called **antinodes,** and you can see that they are also spaced $\lambda/2$ apart. This means that **the wavelength of a standing wave is** twice **the distance between successive nodes or successive antinodes.**

It seems surprising and counterintuitive that some particles in the medium have no motion at all. This happens for the same reason we saw in Figure 16.4: The two waves exactly offset each other at that point. Look carefully at the two traveling waves in Figure 16.6b. You will see that the nodes occur at points where at *every instant* of time the displacements of the two traveling waves have equal magnitudes but *opposite signs.* Thus the superposition of the displacements at these points is always zero—they are points of destructive interference. The antinodes have large displacements. They correspond to points where the two displacements have equal magnitudes and the *same sign* at all times. Constructive interference at these points gives a displacement twice that of each individual wave.

The intensity of a wave is largest at points where it oscillates with maximum amplitude. FIGURE 16.8 shows that the points of maximum intensity along the standing wave occur at the antinodes; the intensity is zero at the nodes. For a standing sound wave, the loudness varies from zero (no sound) at the nodes to a maximum at the antinodes and then back to zero. The key idea is that **the intensity is maximum at points of constructive interference and zero at points of destructive interference.**

FIGURE 16.8 Intensity of a standing wave.

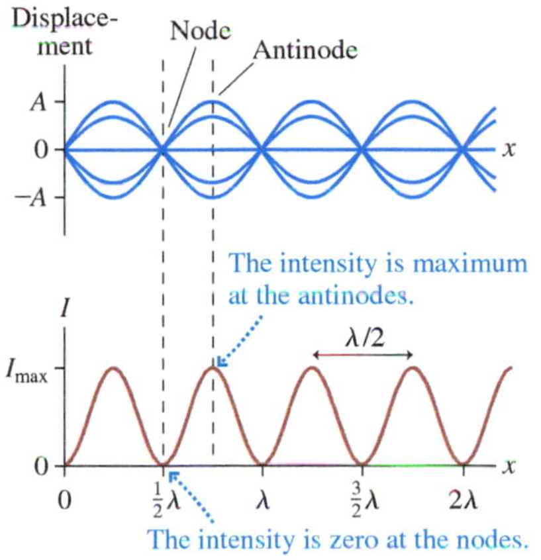


**EXAMPLE 16.1 Setting up a standing wave**

Two children hold an elastic cord at each end. Each child shakes her end of the cord 2.0 times per second, sending waves at 3.0 m/s toward the middle, where the two waves combine to create a standing wave. What is the distance between adjacent nodes?

***SOLVE*** The distance between adjacent nodes is $\lambda/2$. The wavelength, frequency, and speed are related as $v = \lambda f$, as we saw in Chapter 15, so the wavelength is

$$\lambda = \frac{v}{f} = \frac{3.0\ \text{m/s}}{2.0\ \text{Hz}} = 1.5\ \text{m}$$

The distance between adjacent nodes is $\lambda/2$ and thus is 0.75 m.

**STOP TO THINK 16.2** A standing wave is set up on a string. A series of snapshots of the wave are superimposed to produce the diagram at right. What is the wavelength?

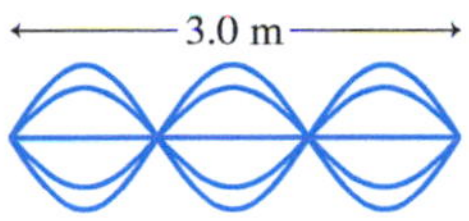


A. 6.0 m B. 4.0 m C. 3.0 m D. 2.0 m E. 1.0 m

## 16.3 Standing Waves on a String

The oscillation of a guitar string is a standing wave. A standing wave is naturally produced on a string when both ends are fixed (i.e., tied down), as in the case of a guitar string or the string in the photo of Figure 16.5. We also know that a standing wave is produced when there are two counterpropagating traveling waves. But you don't shake both ends of a guitar string to produce the standing wave! How do we actually get two traveling waves on a string with both ends fixed? Before we can answer this question, we need a brief explanation of what happens when a traveling wave encounters a boundary or a discontinuity.

### Reflections

We know that light reflects from mirrors; it can also reflect from the surface of a pond or from a pane of glass. As we saw in Chapter 15, sound waves reflect as well; that's how an echo is produced. To understand reflections, we'll look at waves on a string, but the results can be applied to other waves as well.

Suppose we have a string that is attached to a wall or other fixed support, as in FIGURE 16.9. The wall is what we will call a *boundary*—it's the end of the medium. When the pulse reaches this boundary, it reflects, moving away from the wall. *All* the wave's energy is reflected; hence **the amplitude of a wave reflected from a boundary is unchanged.** Figure 16.9 shows that the amplitude doesn't change when the pulse reflects, but the pulse is inverted.

FIGURE 16.9 A wave reflects when it encounters a boundary.

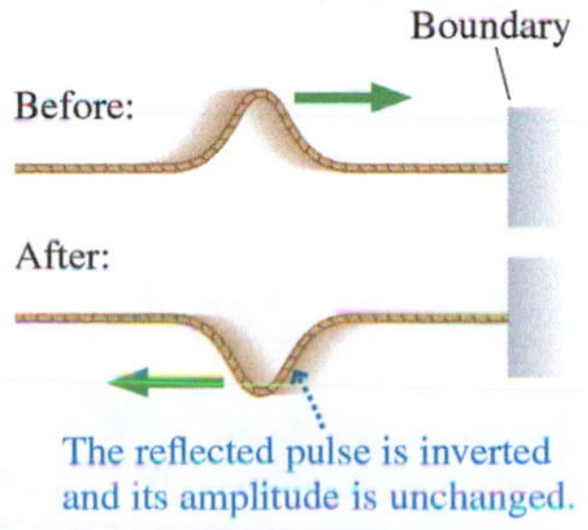

**Through the glass darkly** A piece of window glass is a discontinuity to a light wave, so it both transmits and reflects light. The left photo shows the windows in a brightly lit room at night. The small percentage of the interior light that reflects from windows is more intense than the transmitted light from outside, so the reflection dominates and the windows show a mirror-like reflection of the room. Turning the room lights out reveals the transmitted light from outside.

Waves also reflect from what we will call a *discontinuity,* a point where there is a change in the properties of the medium. FIGURE 16.10a shows a discontinuity where a string with a larger linear density on the left connects with a string with a smaller linear density on the right. The tension is the same in both strings, so the wave speed is slower on the left, faster on the right. Whenever a wave encounters a discontinuity, some of the wave's energy is *transmitted* forward and some is *reflected* backward. Because energy must be conserved, both the transmitted and the reflected pulses have a smaller amplitude than the initial pulse in this case.

FIGURE 16.10 The reflection of a wave at a discontinuity.

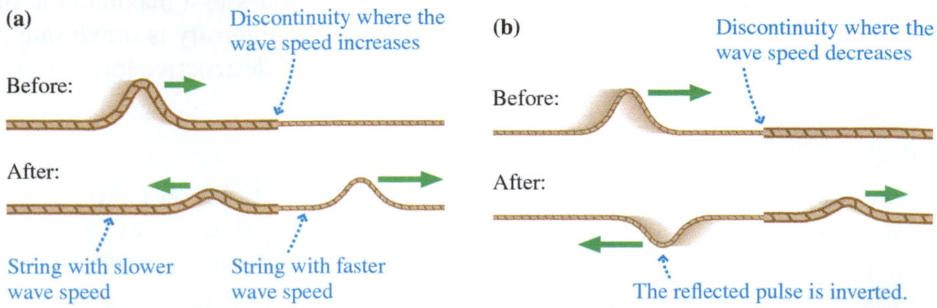


In FIGURE 16.10b, an incident wave encounters a discontinuity at which the wave speed decreases. Once again, some of the wave's energy is transmitted and some is reflected.

In Figure 16.10a, the reflected pulse is right-side up. The string on the right is light and provides little resistance, so the junction moves up and down as the pulse passes. This motion of the string is like the original "snap" of the string that started the pulse, so the reflected pulse has the same orientation as the original pulse. In Figure 16.10b, the string on the right is more massive, so it looks more like the fixed boundary in Figure 16.9 and, as in that situation, the reflected pulse is inverted.

## Creating a Standing Wave

FIGURE 16.11 Reflections at the two boundaries cause a standing wave on the string.

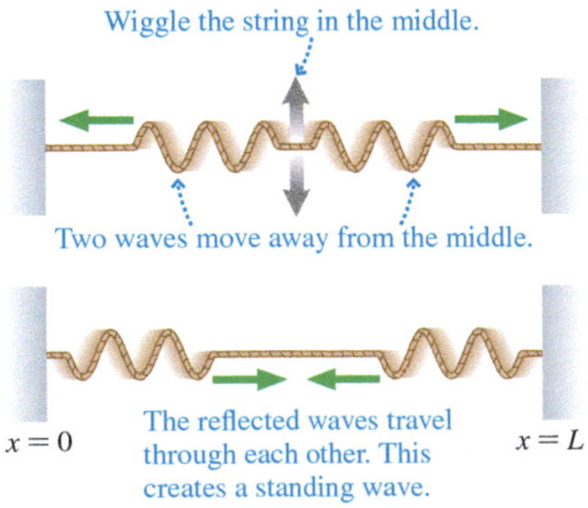


Now that we understand reflections, let's look at how to create a standing wave. FIGURE 16.11 shows a string of length $L$ that is tied at $x = 0$ and $x = L$. This string has *two* boundaries where reflections can occur. If you wiggle the string in the middle, sinusoidal waves travel outward in both directions and soon reach the boundaries, where they reflect. The reflections at the ends of the string cause two waves of *equal amplitude and wavelength* to travel in opposite directions along the string. As we've just seen, these are the conditions that cause a standing wave!

What kind of standing waves might develop on the string? There are two conditions that must be met:

- Because the string is fixed at the ends, the displacements at $x = 0$ and $x = L$ must be zero at all times. Stated another way, we require nodes at both ends of the string.
- We know that standing waves have a spacing of $\lambda/2$ between nodes. This means that the nodes must be equally spaced.

FIGURE 16.12 shows the first three possible waves that meet these conditions. These are called the standing-wave **modes** of the string. To help quantify the possible waves, we can assign a **mode number** $m$ to each. The first wave in Figure 16.12, with a node at each end, has mode number $m = 1$. The next wave is $m = 2$, and so on.

NOTE ▸ Figure 16.12 shows only the first three modes, for $m = 1$, $m = 2$, and $m = 3$. But there are many more modes, for all possible values of $m$. ◂

The distance between adjacent nodes is $\lambda/2$, so the different modes have different wavelengths. For the first mode in Figure 16.12, the distance between nodes is the length of the string, so we can write

$$\lambda_1 = 2L$$

The subscript identifies the mode number; in this case $m = 1$. For $m = 2$, the distance between nodes is $L/2$; this means that $\lambda_2 = L$. Generally, for any mode $m$ the wavelength is given by the equation

$$\lambda_m = \frac{2L}{m} \qquad m = 1, 2, 3, 4, \cdots \tag{16.1}$$

Wavelengths of standing-wave modes of a string of length $L$

These are the only possible wavelengths for standing waves on the string. **A standing wave can exist on the string *only* if its wavelength is one of the values given by Equation 16.1.**

NOTE ▶ Other wavelengths, which would be perfectly acceptable wavelengths for a traveling wave, cannot exist as a *standing* wave of length $L$ because they do not meet the constraint of having a node at each end of the string. ◀

If standing waves are possible for only certain wavelengths, then only specific oscillation frequencies are allowed. Because $\lambda f = v$ for a sinusoidal wave, the oscillation frequency corresponding to wavelength $\lambda_m$ is

$$f_m = \frac{v}{\lambda_m} = \frac{v}{2L/m} = m\left(\frac{v}{2L}\right) \qquad m = 1, 2, 3, 4, \ldots \tag{16.2}$$

Frequencies of standing-wave modes of a string of length $L$

You can tell by looking at a particular standing wave what mode number it corresponds to. FIGURE 16.13 shows the first three standing-wave modes for a wave on a string fixed at both ends, with mode numbers, wavelengths, and frequencies labeled. **The mode number $m$ is equal to the number of antinodes of the standing wave,** so you can determine the mode by counting the number of antinodes (*not* the number of nodes). Once you know the mode number, you can use the formulas for wavelength and frequency to determine these values.

FIGURE 16.12 The first three possible standing waves on a string of length $L$.

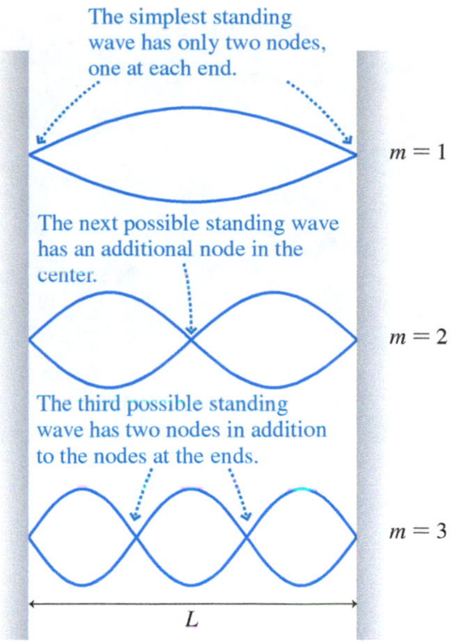


Video Figure 16.12

KEY CONCEPT FIGURE 16.13 Possible standing-wave modes of a string fixed at both ends.

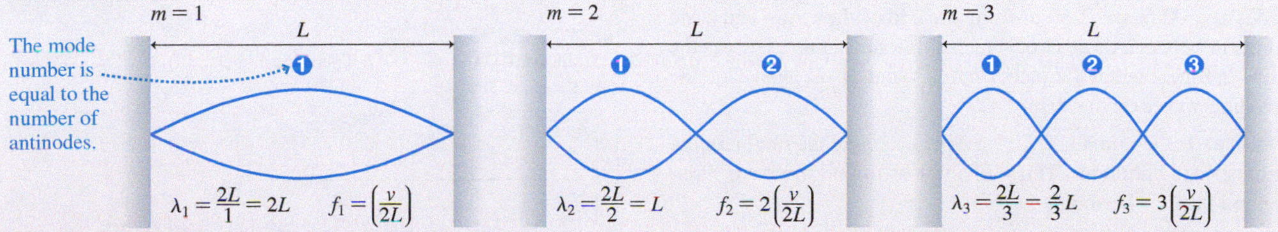


STOP TO THINK 16.3 A 2.0-m-long string carries a standing wave as in the figure at right. Extend the pattern and the formulas shown in Figure 16.13 to determine the mode number and the wavelength of this particular standing-wave mode.

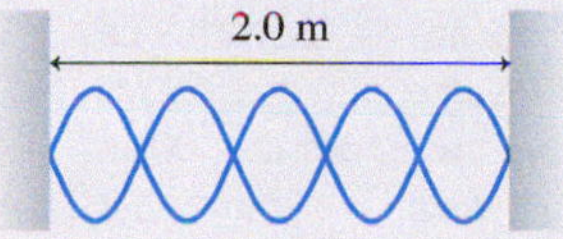


A. $m = 6, \lambda = 0.67$ m
B. $m = 6, \lambda = 0.80$ m
C. $m = 5, \lambda = 0.80$ m
D. $m = 5, \lambda = 1.0$ m
E. $m = 4, \lambda = 0.80$ m
F. $m = 4, \lambda = 1.0$ m

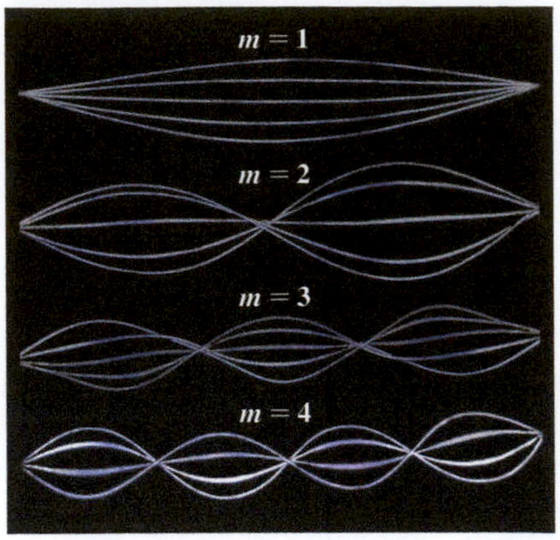


FIGURE 16.14 Resonant modes of a stretched string.

In ◀◀ SECTION 14.7, we looked at the concept of *resonance*. A mass on a spring has a certain frequency at which it "wants" to oscillate. If the system is driven at its resonance frequency, it will develop a large amplitude of oscillation. A stretched string will support standing waves, meaning it has a series of frequencies at which it "wants" to oscillate: the frequencies of the different standing-wave modes. We can call these **resonant modes,** or more simply, **resonances.** A small oscillation of a stretched string at a frequency near one of its resonant modes will cause it to develop a standing wave with a large amplitude. FIGURE 16.14 shows photographs of the first four standing-wave modes on a string, corresponding to four different driving frequencies.

NOTE ▶ When we draw standing-wave modes, as in Figure 16.13, we usually show only the *envelope* of the wave, the greatest extent of the motion of the string. The string's motion is actually continuous and goes through all intermediate positions as well, as we see from the time-exposure photographs of standing waves in Figure 16.14. ◀

## The Fundamental Frequency and Higher Harmonics

An inspection of the frequencies of the first three standing-wave modes of a stretched string in Figure 16.13 reveals an interesting pattern. The first mode has frequency

$$f_1 = \frac{v}{2L} \qquad (16.3)$$

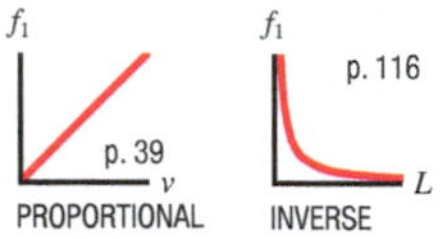


We call this the **fundamental frequency** of the string. All of the other modes have frequencies that are multiples of this fundamental frequency. We can rewrite Equation 16.2 in terms of the fundamental frequency as

$$f_m = mf_1 \qquad m = 1, 2, 3, 4, \ldots \qquad (16.4)$$

**The allowed standing-wave frequencies are all integer multiples of the fundamental frequency.** This sequence of possible frequencies is called a set of **harmonics.** The fundamental frequency $f_1$ is also known as the *first harmonic,* the $m = 2$ wave at frequency $f_2$ is called the *second harmonic,* the $m = 3$ wave is called the *third harmonic,* and so on. The frequencies above the fundamental frequency, the harmonics with $m = 2, 3, 4, \ldots$, are referred to as the **higher harmonics.**

EXAMPLE 16.2 **Identifying harmonics on a string**

A 2.50-m-long string vibrates as a 100 Hz standing wave with nodes at 1.00 m and 1.50 m from one end of the string and at no points in between these two. Which harmonic is this? What is the string's fundamental frequency? And what is the speed of the traveling waves on the string?

***STRATEGIZE*** From a sketch, we will determine the total number of nodes and antinodes. This will allow us to find the mode number and thus the harmonic.

***PREPARE*** We begin with the visual overview in FIGURE 16.15, in which we sketch this particular standing wave and note the known and unknown quantities. We set up an $x$-axis with one end of the string at $x = 0$ m and the other end at $x = 2.50$ m. The ends of the string are nodes, and there are nodes at 1.00 m and 1.50 m as well, with no nodes in between. We know that standing-wave nodes are equally spaced, so there must be other nodes on the string, as shown in Figure 16.15a. Figure 16.15b is a sketch of the standing-wave mode with this node structure.

FIGURE 16.15 A visual overview of the string.

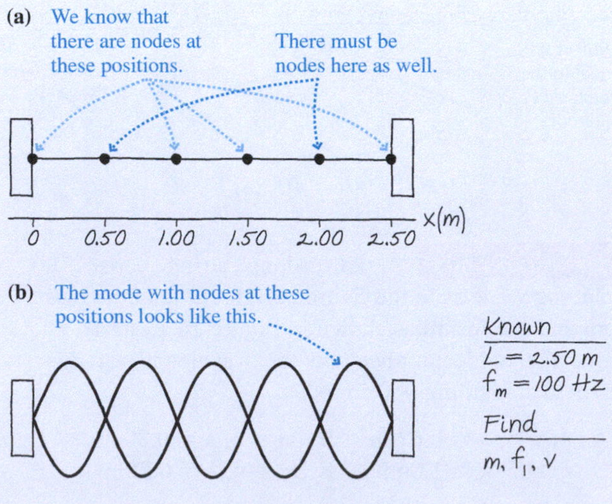

**SOLVE** We count the number of antinodes of the standing wave to deduce the mode number; this is mode $m = 5$. This is the fifth harmonic. The frequencies of the harmonics are given by $f_m = mf_1$, so the fundamental frequency is

$$f_1 = \frac{f_5}{5} = \frac{100\text{ Hz}}{5} = 20\text{ Hz}$$

The wavelength of the fundamental mode is $\lambda_1 = 2L = 2(2.50\text{ m}) = 5.00\text{ m}$, so we can find the wave speed using the fundamental relationship for sinusoidal waves:

$$v = \lambda_1 f_1 = (5.00\text{ m})(20\text{ Hz}) = 100\text{ m/s}$$

**ASSESS** We can calculate the speed of the wave using any possible mode, which gives us a way to check our work. The distance between successive nodes is $\lambda/2$. Figure 16.15 shows that the nodes are spaced by 0.50 m, so the wavelength of the $m = 5$ mode is 1.00 m. The frequency of this mode is 100 Hz, so we calculate

$$v = \lambda_5 f_5 = (1.00\text{ m})(100\text{ Hz}) = 100\text{ m/s}$$

This is the same speed that we calculated earlier, which gives us confidence in our results.

## Stringed Musical Instruments

Stringed musical instruments, such as the guitar, the piano, and the violin, all have strings that are fixed at both ends and tightened to create tension. A disturbance is generated on the string by plucking, striking, or bowing. The disturbance creates standing waves on the string.

In Chapter 15, we saw that the speed of a wave on a stretched string depends on $T_s$, the tension in the string, and $\mu$, the linear density, as $v = \sqrt{T_s/\mu}$. Combining this with Equation 16.3, we find that the fundamental frequency is

$$f_1 = \frac{v}{2L} = \frac{1}{2L}\sqrt{\frac{T_s}{\mu}} \tag{16.5}$$

When you pluck or bow a string, you initially excite a wide range of frequencies. However, resonance sees to it that the only frequencies that persist are those of the possible standing waves. The string will support a wave of the fundamental frequency $f_1$ plus waves of all the higher harmonics $f_2, f_3, f_4$, and so on. Your brain interprets the sound as a musical note of frequency $f_1$; the higher harmonics determine the *tone quality,* a concept we will explore later in the chapter.

For instruments like the guitar or the violin, the strings are all the same length and under approximately the same tension. The strings have different frequencies because they differ in linear density. The lower-pitched strings are "fat" while the higher-pitched strings are "skinny." This difference changes the frequency by changing the wave speed.

**Standing waves on a bridge** This photo shows the Tacoma Narrows suspension bridge on the day in 1940 when it experienced a catastrophic oscillation that led to its collapse. Aerodynamic forces caused the amplitude of a particular resonant mode of the bridge to increase dramatically until the bridge failed. In this photo, the red line shows the original line of the deck of the bridge. You can clearly see the large amplitude of the oscillation and the node at the center of the span.

**CONCEPTUAL EXAMPLE 16.3 Tuning and playing a guitar**

Although a guitar's strings are all the same length, a player can produce many different notes by pressing the string against frets, metal bars on the neck of the guitar, as shown in FIGURE 16.16. The fret becomes the new end of the string, making the effective length shorter.

FIGURE 16.16 Guitar frets.

a. A guitar player plucks a string to play a note. He then presses down on a fret to make the string shorter. Does the new note have a higher or lower frequency?

b. The frequency of one string is too low. (Musically, we say the note is "flat.") How must the tension be adjusted to bring the string to the right frequency?

**REASON**

a. The fundamental frequency, the note we hear, is $f_1 = (1/2L)\sqrt{T_s/\mu}$. Because $f_1$ is inversely proportional to $L$, decreasing the string length increases the frequency.

b. Because $f$ is proportional to the square root of $T_s$, the player must increase the tension to increase the fundamental frequency.

**ASSESS** If you watch a guitarist, you can see that he plays higher notes by moving his fingers to shorten the strings.

**EXAMPLE 16.4 Setting the tension in a guitar string**

The fifth string on a guitar plays the musical note A, at a frequency of 110 Hz. On a typical guitar, this string is stretched between two fixed points 0.640 m apart, and this length of string has a mass of 2.86 g. What is the tension in the string?

**STRATEGIZE** Equation 16.5 relates the frequency to the string tension.

**PREPARE** Strings sound at their fundamental frequency, so 110 Hz is $f_1$.

*Continued*

***SOLVE*** The linear density of the string is

$$\mu = \frac{m}{L} = \frac{2.86 \times 10^{-3}\ \text{kg}}{0.640\ \text{m}} = 4.47 \times 10^{-3}\ \text{kg/m}$$

We can rearrange Equation 16.5 for the fundamental frequency to solve for the tension in terms of the other variables:

$$T_s = (2Lf_1)^2\mu = [2(0.640\ \text{m})(110\ \text{Hz})]^2(4.47 \times 10^{-3}\ \text{kg/m})$$
$$= 88.6\ \text{N}$$

***ASSESS*** If you have ever strummed a guitar, you know that the tension is quite large, so this result seems reasonable. If each of the guitar's six strings has approximately the same tension, the total force on the neck of the guitar is a bit more than 500 N.

## Standing Electromagnetic Waves

**FIGURE 16.17** A laser contains a standing light wave between two parallel mirrors.

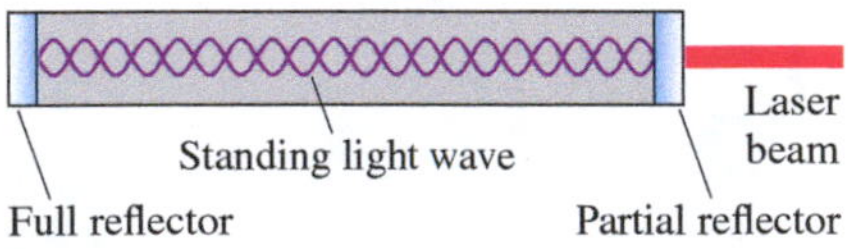


The standing-wave descriptions we've found for a vibrating string are valid for any transverse wave, including an electromagnetic wave. Standing light waves can be established between two parallel mirrors that reflect the light back and forth. The mirrors are boundaries, analogous to the boundaries at the ends of a string. This is how a laser works. The two facing mirrors in FIGURE 16.17 form a *laser cavity.*

Because the mirrors act exactly like the points to which a string is tied, the light wave must have a node at the surface of each mirror. (To allow some of the light to escape the laser cavity and form the *laser beam,* one of the mirrors lets some of the light through. This doesn't affect the node.)

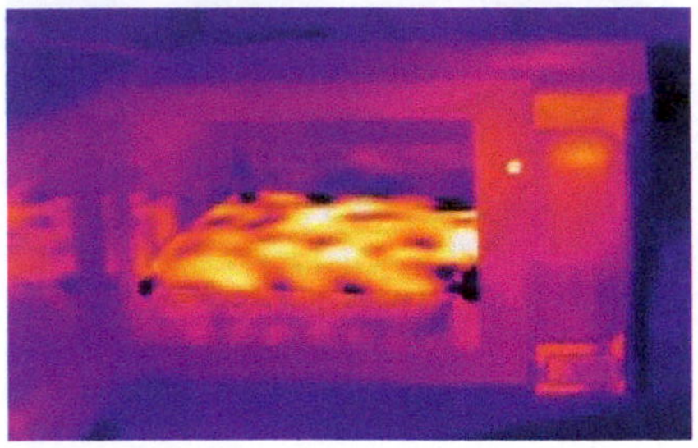

**Microwave modes** A microwave oven uses microwaves with a wavelength of about 12 cm to heat food. The inside walls of a microwave oven are reflective to microwaves, so we have the correct conditions to set up a standing wave. A standing wave has high intensity at the antinodes and low intensity at the nodes, so your oven has hot spots and cold spots, as we see in this thermal image showing the interior of an oven. A turntable in a microwave oven keeps the food moving so that no part of your dinner remains at a node or an antinode.

**EXAMPLE 16.5 Finding the mode number for a laser**

A helium-neon laser emits light of wavelength $\lambda = 633$ nm. A typical cavity for such a laser is 15.0 cm long. What is the mode number of the standing wave in this cavity?

***STRATEGIZE*** Just like a standing wave on a string, the standing light wave has nodes at each end, so Equation 16.1 applies.

***SOLVE*** The standing light wave in a laser cavity has a mode number $m$ that is roughly

$$m = \frac{2L}{\lambda} = \frac{2 \times 0.150\ \text{m}}{633 \times 10^{-9}\ \text{m}} = 474{,}000$$

***ASSESS*** The wavelength of light is very short, so we'd expect there to be many nodes within the length of the cavity. A high mode number seems reasonable.

**STOP TO THINK 16.4** The illustration to the right shows a standing wave on a string. Which of the modes shown below (on the same string) has twice the frequency of the original wave?

Original standing wave

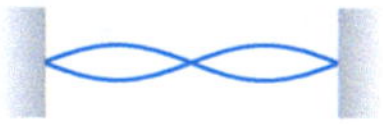

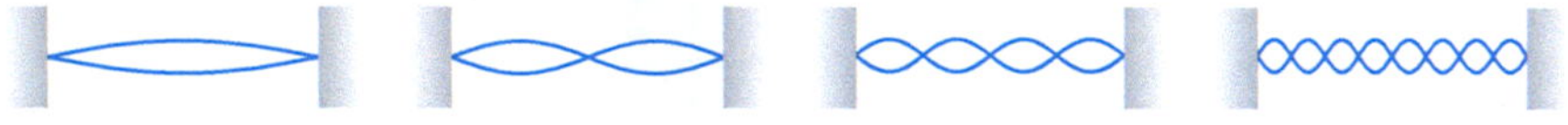


# 16.4 Standing Sound Waves

**Video** Standing Sound Waves

Wind instruments like flutes, trumpets, and didgeridoos work very differently from stringed instruments. The player blows air into one end of a tube, producing standing waves of *sound* that make the notes we hear.

You saw in ◀◀ SECTION 15.4 that a sound wave is a longitudinal pressure wave. The air molecules oscillate back and forth parallel to the direction in which the wave is traveling, creating *compressions* (regions of higher pressure) and *rarefactions* (regions of lower pressure). Consider a sound wave confined to a long, narrow column of air, such as the air in a tube. A wave traveling down the tube eventually reaches the end, where it encounters the atmospheric pressure of the surrounding environment. This is a discontinuity, much like the small rope meeting the big rope in Figure 16.10b. Part of the wave's energy is transmitted out into the environment, allowing you to hear the sound, and part is reflected back into the tube. Reflections at both ends of the tube create waves traveling both directions inside the tube, and their superposition, like that of the reflecting waves on a string, is a standing wave.

We start by looking at a sound wave in a tube open at both ends. What kind of standing waves can exist in such a tube? Because the ends of the tube are open to the atmosphere, the pressure at the ends is fixed at atmospheric pressure and cannot vary. This is analogous to a stretched string that is fixed at the end. As a result, **the open end of a column of air must be a node of the pressure wave.**

FIGURE 16.18a shows a column of air open at both ends. We call this an *open-open tube*. Just as the antinodes of a standing wave on a string are points where the string oscillates with maximum displacement, the antinodes of a standing sound wave are where the pressure has the largest variation, creating, alternately, the maximum compression and the maximum rarefaction. In Figure 16.18a, the air molecules squeeze together on the left side of the tube. Then, in FIGURE 16.18b, half a cycle later, the air molecules squeeze together on the right side. The varying density creates a variation in pressure across the tube, as the graphs show. FIGURE 16.18c combines the information of Figures 16.18a and 16.18b into a graph of the pressure of the standing sound wave in the tube.

As the standing wave oscillates, the air molecules "slosh" back and forth along the tube with the wave frequency, alternately pushing together (maximum pressure at the antinode) and pulling apart (minimum pressure at the antinode). This makes sense, because sound is a longitudinal wave in which the air molecules oscillate parallel to the tube.

FIGURE 16.18 The $m = 2$ standing sound wave inside an open-open column of air.

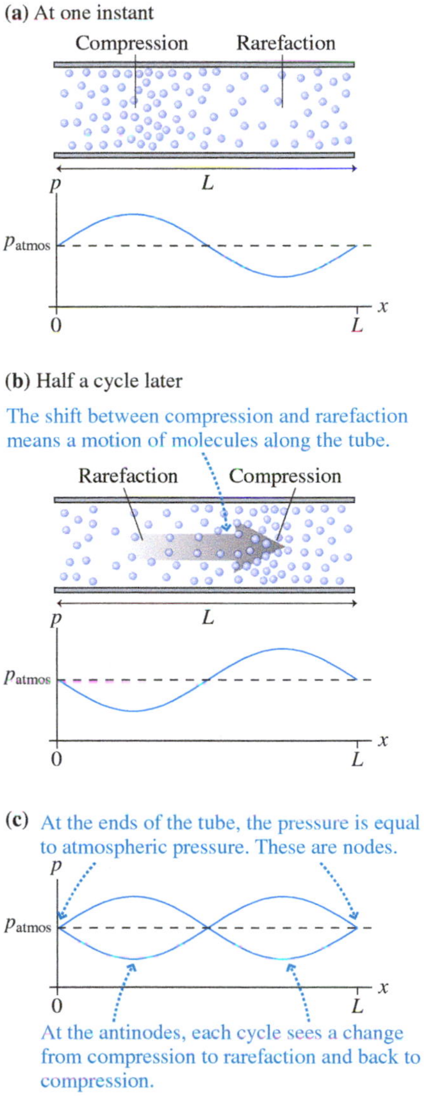


**NOTE** ▶ The variation in pressure from atmospheric pressure in a real standing sound wave is much smaller than Figure 16.18 implies. When we display graphs of the pressure in sound waves, we won't generally graph the pressure $p$. Instead, we will graph $\Delta p$, the variation from atmospheric pressure. ◀

Many musical instruments, such as a flute, can be modeled as open-open tubes. The flutist blows across one end to create a standing wave inside the tube, and a note of this frequency is emitted from both ends of the flute. The possible standing waves in tubes, like standing waves on strings, are resonances of the system. A gentle puff of air across the mouthpiece of a flute can cause large standing waves at these resonant frequencies.

Other instruments work differently from the flute. A trumpet or a clarinet is a column of air open at the bell end but *closed* by the player's lips at the mouthpiece. To be complete in our treatment of sound waves in tubes, we need to consider tubes that are closed at one or both ends. At a closed end, the air molecules can alternately rush toward the wall, creating a compression, and then rush away from the wall, leaving a rarefaction. Thus **a closed end of an air column is an antinode of pressure.**

FIGURE 16.19 on the next page shows graphs of the first three standing-wave modes of a tube open at both ends (an open-open tube), a tube closed at both ends (a *closed-closed tube*), and a tube open at one end but closed at the other (an *open-closed tube*), all with the same length $L$. These are graphs of the pressure wave, with a node at open ends and an antinode at closed ends. The standing wave in the closed-closed tube looks like the wave in the open-open tube except that the positions of the nodes and antinodes are interchanged. In both cases there are $m$ half-wavelength segments between the ends; thus the wavelengths and frequencies of an open-open tube and a closed-closed tube are the same as those of a string tied at both ends:

$$\lambda_m = \frac{2L}{m} \qquad f_m = m\left(\frac{v}{2L}\right) = mf_1 \qquad m = 1, 2, 3, 4, \ldots \tag{16.6}$$

Wavelengths and frequencies of standing sound wave modes in an open-open or closed-closed tube

**FIGURE 16.19** The first three standing sound wave modes in columns of air with different ends. These graphs show the pressure variation in the tube.

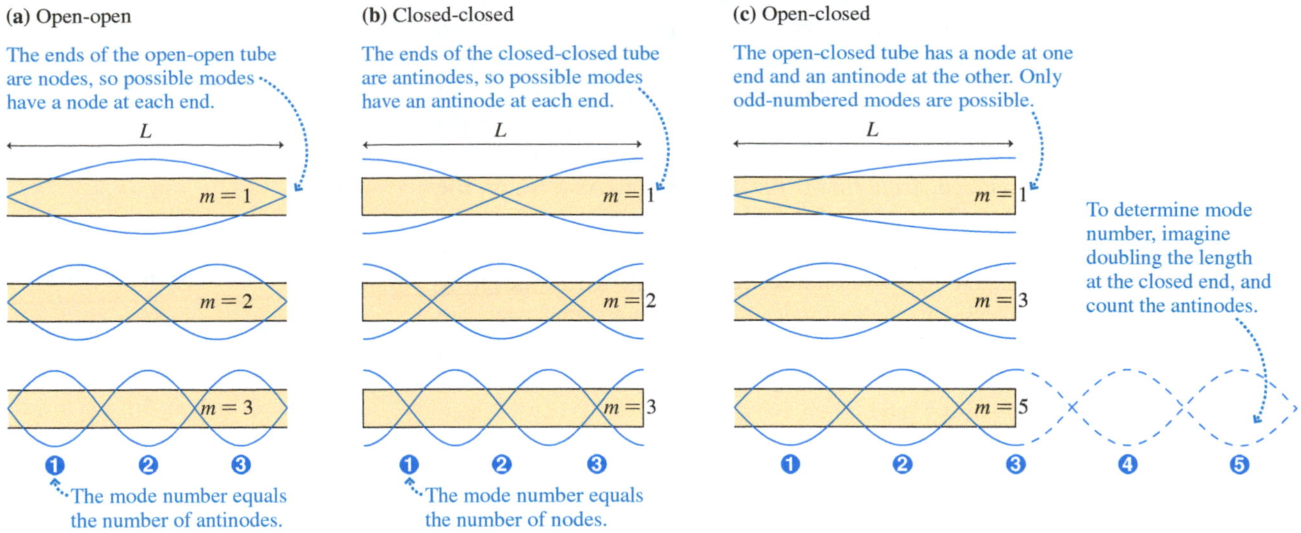


The open-closed tube is different, as we can see from Figure 16.19c. The $m = 1$ mode has a node at one end and an antinode at the other, and so has only one-quarter of a wavelength in a tube of length $L$. The $m = 1$ wavelength is $\lambda_1 = 4L$, twice the $m = 1$ wavelength of an open-open or a closed-closed tube. Consequently, **the fundamental frequency of an open-closed tube is half that of an open-open or a closed-closed tube of the same length.**

The wavelength of the next mode of the open-closed tube is $4L/3$. Because this is $1/3$ of $\lambda_1$, we assign $m = 3$ to this mode. The wavelength of the subsequent mode is $4L/5$, so this is $m = 5$. In other words, an open-closed tube allows only odd-numbered modes. Consequently, the possible wavelengths and frequencies are

$$\lambda_m = \frac{4L}{m} \qquad f_m = m\left(\frac{v}{4L}\right) = mf_1 \qquad m = 1, 3, 5, 7, \ldots \tag{16.7}$$

Wavelengths and frequencies of standing sound wave modes in an open-closed tube

**NOTE** ► Because sound is a pressure wave, the graphs of Figure 16.19 are *not* "pictures" of the wave as they are for a string wave. The graphs show the pressure variation versus position $x$. The tube itself is shown merely to indicate the location of the open and closed ends, but the diameter of the tube is *not* related to the amplitude of the wave. ◄

At this point, it's worthwhile to compare and contrast the details we've seen for different types of standing waves. We can also formulate a general strategy for solving standing-wave problems.

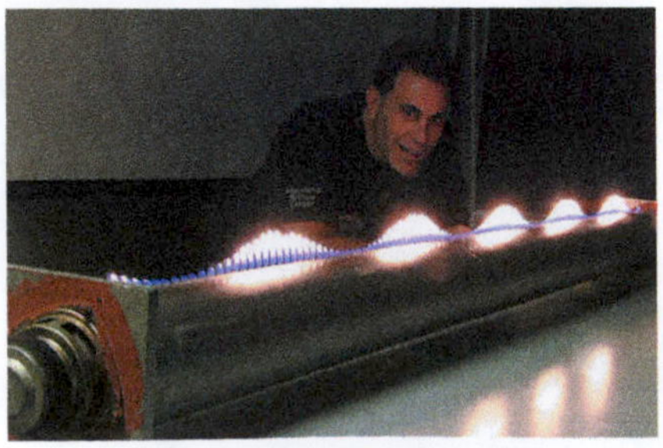

◄ **Fiery interference** In this apparatus, a speaker at one end of the metal tube emits a sinusoidal wave. The wave reflects from the other end, which is closed, to make a counter-propagating wave and set up a standing sound wave in the tube. The tube is filled with propane gas that exits through small holes on top. The burning propane allows us to easily discern the nodes and the antinodes of the standing sound wave.

## SYNTHESIS 16.1 Standing-wave modes

There are only two sets of frequency/wavelength relationships for standing-wave modes.

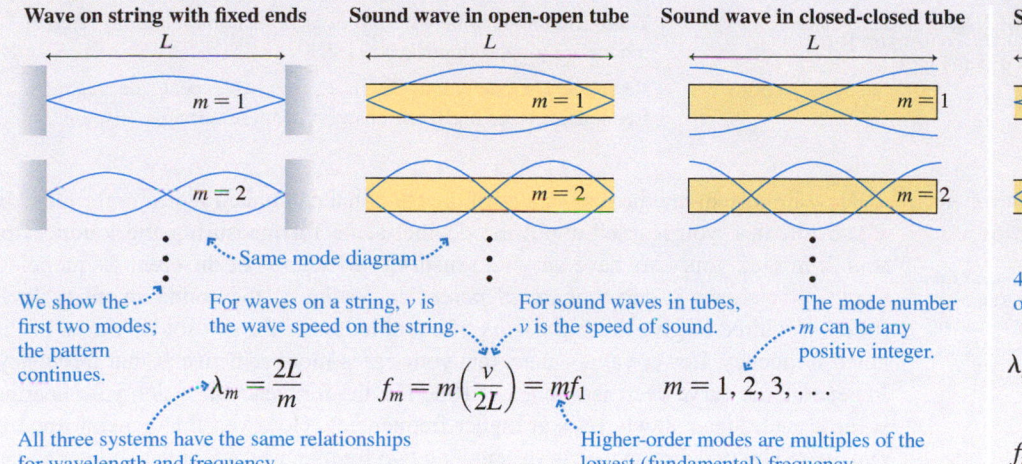

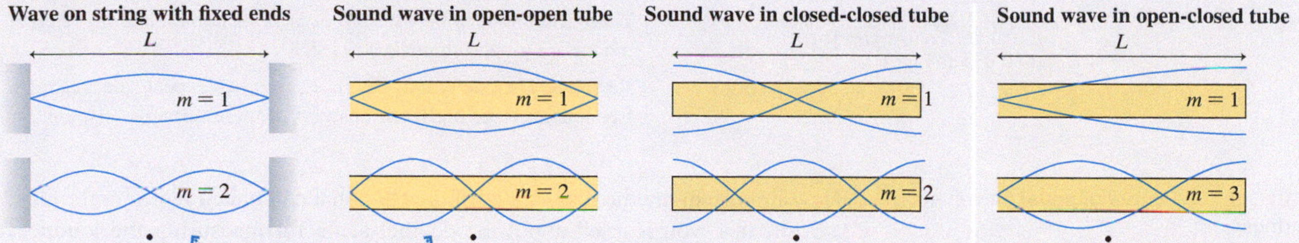


Same mode diagram

We show the first two modes; the pattern continues.

For waves on a string, $v$ is the wave speed on the string.

For sound waves in tubes, $v$ is the speed of sound.

The mode number $m$ can be any positive integer.

$$\lambda_m = \frac{2L}{m} \qquad f_m = m\left(\frac{v}{2L}\right) = mf_1 \qquad m = 1, 2, 3, \ldots$$

All three systems have the same relationships for wavelength and frequency.

Higher-order modes are multiples of the lowest (fundamental) frequency.

$4L$ instead of $2L$

Open-closed tubes have only odd-numbered modes.

$$\lambda_m = \frac{4L}{m} \qquad m = 1, 3, 5, \ldots$$

$$f_m = m\left(\frac{v}{4L}\right) = mf_1$$

## PROBLEM-SOLVING APPROACH 16.1 Standing waves

We can use the same general approach for any standing wave.

***STRATEGIZE*** Standing waves occur whenever a wave is confined between two boundaries. In this case, only certain modes, with particular wavelengths and frequencies, can exist.

***PREPARE***

- For sound waves, determine what sort of pipe or tube you have: open-open, closed-closed, or open-closed.
- For string or light waves, the ends will be fixed points.
- For other types of standing waves, such as electromagnetic or water waves, the mode diagram will be similar to one of the cases outlined in Synthesis 16.1. You can then work by analogy with waves on strings or sound waves in tubes.
- Determine known values: length of the tube or string, frequency, wavelength, positions of nodes or antinodes.
- It may be useful to sketch a visual overview, including a picture of the relevant mode.

***SOLVE*** Once you have determined the mode diagram, you can use the appropriate set of equations in Synthesis 16.1 to find the wavelength and frequency. These equations work for any type of wave.

***ASSESS*** Does your final answer seem reasonable? Is there another way to check on your results? For example, the frequency times the wavelength for any mode should equal the wave speed—you can check to see that it does.

## EXAMPLE 16.6 Resonances of the ear canal BIO

The eardrum, which transmits vibrations to the sensory organs of your ear, lies at the end of the ear canal. As FIGURE 16.20 shows, the ear canal in adults is about 2.5 cm in length. What frequency standing waves can occur within the ear canal that are within the range of human hearing? The speed of sound in the warm air of the ear canal is 350 m/s.

***STRATEGIZE*** The ear canal acts as a tube that supports standing sound waves.

***PREPARE*** We proceed according to the steps in Problem-Solving Approach 16.1. We can treat the ear canal as an open-closed tube: open to the atmosphere at the external end, closed by the eardrum at the other end. The possible standing-wave modes appear as in Figure 16.19c. The length of the tube is 2.5 cm.

FIGURE 16.20 The anatomy of the ear.

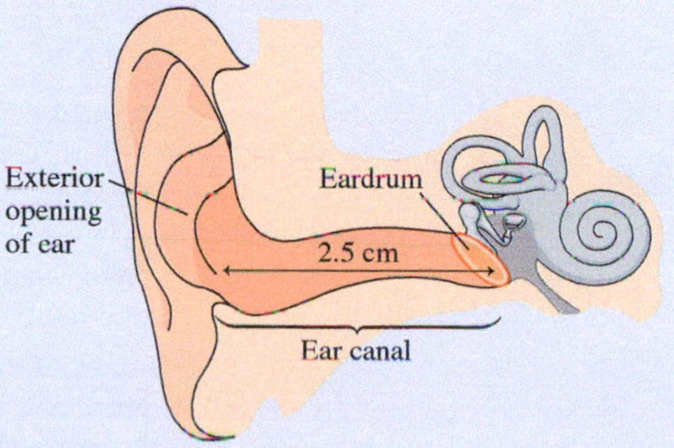


*Continued*

**SOLVE** Equation 16.7 gives the allowed frequencies in an open-closed tube. We are looking for the frequencies in the range 20 Hz–20,000 Hz. The fundamental frequency is

$$f_1 = \frac{v}{4L} = \frac{350\ \text{m/s}}{4(0.025\ \text{m})} = 3500\ \text{Hz}$$

The higher harmonics are odd multiples of this frequency:

$$f_3 = 3(3500\ \text{Hz}) = 10{,}500\ \text{Hz}$$

$$f_5 = 5(3500\ \text{Hz}) = 17{,}500\ \text{Hz}$$

These three modes lie within the range of human hearing; higher modes are greater than 20,000 Hz.

**ASSESS** The ear canal is short, so we expect the resonant frequencies to be high; our answers seem reasonable.

FIGURE 16.21 A curve of equal perceived loudness.

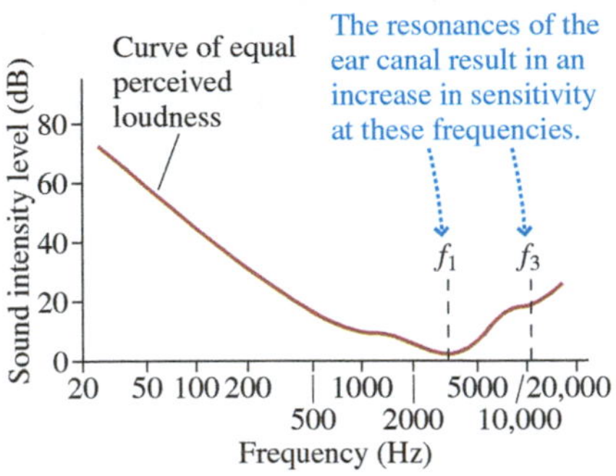


How important are the resonances of the ear canal calculated in Example 16.6? In ◀ SECTION 15.6, you learned about the decibel scale for measuring the loudness of sound. In fact, your ears have varying sensitivity to sounds of different frequencies. FIGURE 16.21 shows a curve of *equal perceived loudness,* the sound intensity level (in dB) required to give the *impression* of equal loudness for sinusoidal waves of the noted frequency. Lower values mean that your ear is more sensitive at that frequency. In general, the curve decreases to about 1000 Hz, the frequency at which your hearing is most acute, then slowly rises at higher frequencies. However, this general trend is punctuated by two dips in the curve, showing two frequencies at which a quieter sound produces the same perceived loudness. As you can see, these two dips correspond to the resonances $f_1$ and $f_3$ of the ear canal. Incoming sounds at these frequencies produce a larger oscillation, resulting in an increased sensitivity to these frequencies.

## Wind Instruments

With a wind instrument, blowing into the mouthpiece creates a standing sound wave inside a tube of air. The player changes the notes by using her fingers to cover holes or open valves, changing the effective length of the tube. The first open hole becomes a node because the tube is open to the atmosphere at that point. The fact that the holes are on the side, rather than literally at the end, makes very little difference. The length of the tube determines the standing-wave resonances, and thus the musical note that the instrument produces.

Many wind instruments have a "buzzer" at one end of the tube, such as a vibrating reed on a saxophone or clarinet, or the musician's vibrating lips on a trumpet or trombone. Buzzers like these generate a continuous range of frequencies rather than single notes, which is why they sound like a "squawk" if you play on just the mouthpiece without the rest of the instrument. When the buzzer is connected to the body of the instrument, most of those frequencies cause little response. But the frequencies from the buzzer that match the resonant frequencies of the instrument cause the buildup of large amplitudes at these frequencies—standing-wave resonances. The combination of these frequencies makes the musical note that we hear.

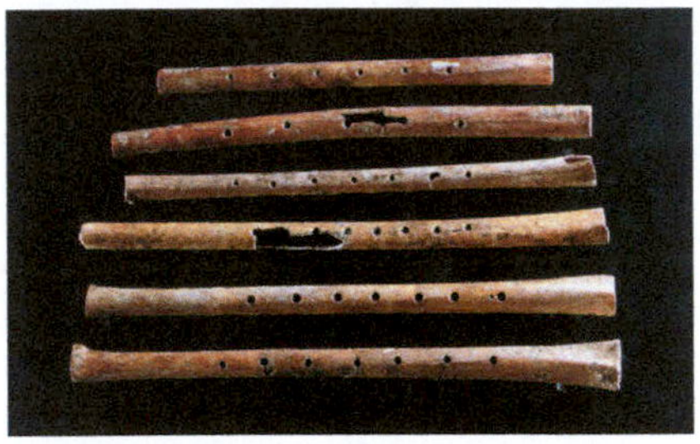

**Truly classical music** The oldest known musical instruments are bone flutes from burial sites in central China. The flutes in the photo are up to 9000 years old and are made from naturally hollow bones from crowned cranes. The positions of the holes determine the frequencies that the flutes can produce. Soon after the first flutes were created, the design was standardized so that different flutes would play the same notes—including the notes in the modern Chinese musical scale.

**CONCEPTUAL EXAMPLE 16.7 Comparing the flute and the clarinet**

A flute and a clarinet have about the same length, but the lowest note that can be played on the clarinet is much lower than the lowest note that can be played on the flute. Why is this?

**REASON** A flute is an open-open tube; the frequency of the fundamental mode is $f_1 = v/2L$. A clarinet is open at one end, but the player's lips and the reed close it at the other end. The clarinet is thus an open-closed tube with a fundamental frequency $f_1 = v/4L$. This is about half the fundamental frequency of the flute, so the lowest note on the clarinet has a much lower pitch. In musical terms, it's about an octave lower than the flute.

**ASSESS** A quick glance at Synthesis 16.1 shows that the wavelength of the lowest mode of the open-closed tube is longer than that of the open-open tube, so we expect a lower frequency for the clarinet.

A clarinet has a lower pitch than a flute because its lowest mode has a longer wavelength and thus a lower frequency. But the higher harmonics are different as well. An open-open tube like a flute has all of the harmonics; an open-closed tube like a clarinet has only those with odd mode numbers. This gives the two instruments a very different tone quality—it's easy to distinguish the sound of a flute from that of a clarinet. We'll explore this connection between the harmonics an instrument produces and its tone quality in the next section.

**EXAMPLE 16.8** **The importance of warming up**

Wind instruments have an adjustable joint to change the tube length. Players know that they may need to adjust this joint to stay in tune—that is, to stay at the correct frequency. To see why, suppose a "cold" flute plays the note A at 440 Hz when the air temperature is 20°C.

a. How long is the tube? At 20°C, the speed of sound in air is 343 m/s.
b. As the player blows air through the flute, the air inside the instrument warms up. Once the air temperature inside the flute has risen to 32°C, increasing the speed of sound to 350 m/s, what is the frequency?
c. At the higher temperature, how must the length of the tube be changed to bring the frequency back to 440 Hz?

**STRATEGIZE** A flute is an open-open tube with fundamental frequency $f_1 = v/2L$. We know the speed of sound $v$ and the frequency $f_1$, so we can find the tube length $L$. A higher speed of sound will then lead to a higher frequency. Finally, using the higher speed of sound, we can find the new length of the tube that will give the required 440 Hz frequency.

**SOLVE**

a. At 20°C, the length corresponding to 440 Hz is

$$L = \frac{v}{2f_1} = \frac{343\ \text{m/s}}{2(440\ \text{Hz})} = 0.390\ \text{m}$$

b. As the speed of sound increases, the frequency changes to

$$f_1(\text{at }32°\text{C}) = \frac{350\ \text{m/s}}{2(0.390\ \text{m})} = 449\ \text{Hz}$$

c. To bring the flute back into tune, the length must be increased to give a frequency of 440 Hz with a speed of 350 m/s. The new length is

$$L = \frac{v}{2f_1} = \frac{350\ \text{m/s}}{2(440\ \text{Hz})} = 0.398\ \text{m}$$

Thus the flute must be increased in length by 8 mm.

**ASSESS** A small change in the absolute temperature produces a correspondingly small change in the speed of sound. We expect that this will require a small change in length, so our answer makes sense.

**STOP TO THINK 16.5** A tube that is open at both ends supports a standing wave with harmonics at 300 Hz and 400 Hz, with no harmonics between. What is the fundamental frequency of this tube?

A. 50 Hz B. 100 Hz C. 150 Hz D. 200 Hz E. 300 Hz

## 16.5 Speech and Hearing BIO

When you hear a note played on a guitar, it sounds very different from the same note played on a trumpet. And you have perhaps been to a lecture in which the speaker talked at essentially the same pitch the entire time—but you could still understand what was being said. Clearly, there is more to your brain's perception of sound than pitch alone. How do you tell the difference between a guitar and a trumpet? How do you distinguish between an "oo" vowel sound and an "ee" vowel sound at the same pitch?

## The Frequency Spectrum

To this point, we have pictured sound waves as sinusoidal waves, with a well-defined frequency. In fact, most of the sounds that you hear are not pure sinusoidal waves. Most sounds are a mix, or superposition, of different frequencies. For example, we have seen how certain standing-wave modes are possible on a stretched string. When you pluck a string on a guitar, you generally don't excite just one standing-wave mode—you simultaneously excite many different modes.

If you play the note "middle C" on a guitar, the fundamental frequency is 262 Hz. There will be a standing wave at this frequency, but there will also be standing waves at the frequencies 524 Hz, 786 Hz, 1048 Hz, . . . , all the higher harmonics predicted by Equation 16.4.

FIGURE 16.22 The frequency spectrum and a graph of the sound wave of a guitar playing a note with fundamental frequency 262 Hz.

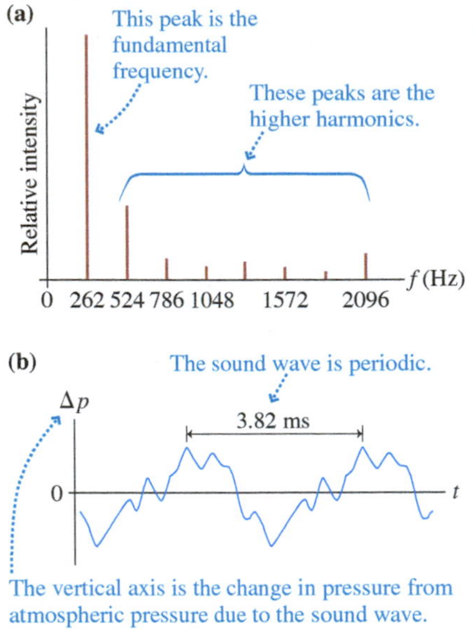


FIGURE 16.22a is a bar chart showing all the frequencies present in the sound of the vibrating guitar string. The height of each bar shows the relative intensity of that harmonic. A bar chart showing the relative intensities of the different frequencies is called the **frequency spectrum** of the sound.

When your brain interprets the mix of frequencies from the guitar in Figure 16.22a, it identifies the fundamental frequency as the *pitch.* 262 Hz corresponds to middle C, so you will identify the pitch as middle C, even though the sound consists of many different frequencies. Your brain uses the higher harmonics to determine the **tone quality,** which is also called the *timbre.* The tone quality—and therefore the higher harmonics—is what makes a middle C played on a guitar sound quite different from a middle C played on a trumpet. The frequency spectrum of a trumpet would show a very different pattern of the relative intensities of the higher harmonics.

The sound wave produced by a guitar playing middle C is shown in FIGURE 16.22b. The sound wave is periodic, with a period of 3.82 ms that corresponds to the 262 Hz fundamental frequency. **The higher harmonics don't change the period of the sound wave; they change only its shape.** The sound wave of a trumpet playing middle C would also have a 3.82 ms period, but its shape would be entirely different.

The didgeridoo, the musical instrument shown at the start of the chapter, consists of a hollow stem or branch. It is an open-closed tube like a clarinet or trumpet. The player blows air through his lips pressed at the end, but he also adds sounds from his vocal cords. This changes the mix of standing-wave modes that are produced, permitting a wide range of sounds.

FIGURE 16.23 The frequency spectrum from the vocal cords, and after passing through the vocal tract.

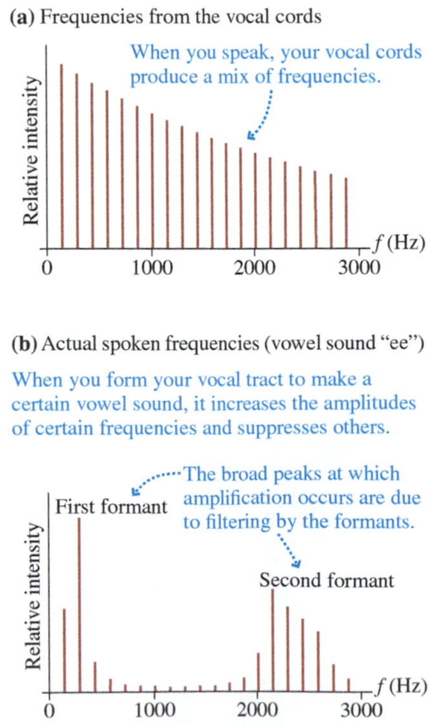


## Vowels and Formants

Try this: Keep your voice at the same pitch and say the "ee" sound, as in "beet," then the "oo" sound, as in "boot." Pay attention to how you reshape your mouth as you move back and forth between the two sounds. The two vowel sounds are at the same pitch, but they sound quite different. The difference in sound arises from the difference in the higher harmonics, just as for musical instruments. As you speak, you adjust the properties of your vocal tract to produce different mixes of harmonics that make the "ee," "oo," "ah," and other vowel sounds.

Speech begins with the vibration of your vocal cords, stretched bands of tissue in your throat. The vibration is similar to that of a wave on a stretched string. In ordinary speech, the average fundamental frequency for adult males and females is about 150 Hz and 250 Hz, respectively, but you can change the vibration frequency by changing the tension of your vocal cords. That's how you make your voice higher or lower as you sing.

Your vocal cords produce a mix of different frequencies as they vibrate—the fundamental frequency and a rich mixture of higher harmonics. If you put a microphone in your throat and measured the sound waves right at your vocal cords, the frequency spectrum would appear as in FIGURE 16.23a.

There is more to the story though. Before reaching the opening of your mouth, sound from your vocal cords must pass through your vocal tract—a series of hollow cavities including your throat, mouth, and nose. The vocal tract acts like a series of tubes, and, as in any tube, certain frequencies will set up standing-wave resonances.

The rather broad standing-wave resonances of the vocal tract are called **formants.** Harmonics of your vocal cords at or near the formant frequencies will be amplified; harmonics far from a formant will be suppressed. FIGURE 16.23b shows the formants of an adult male making an "ee" sound. The filtering of the vocal cord harmonics by the formants is clear.

You can change the shape and frequencies of the formants, and thus the sound you make, by changing the shape and length of your vocal tract. You do this by changing your mouth opening and the shape and position of your tongue. The first two formants for an "ee" sound are at roughly 270 Hz and 2300 Hz, but for an "oo" sound they are 300 Hz and 870 Hz. The much lower second formant of the "oo" emphasizes midrange frequencies, making a "calming" sound, while the more strident sound of "ee" comes from enhancing the higher frequencies.

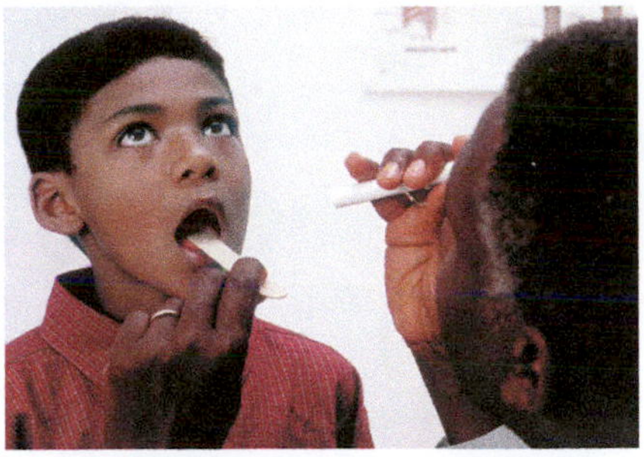

**Saying "ah"** BIO Why, during a throat exam, does a doctor ask you to say "ah"? This particular vowel sound is formed by opening the mouth and the back of the throat wide—giving a clear view of the tissues of the throat.

**CONCEPTUAL EXAMPLE 16.9 High-frequency hearing loss** BIO

As you age, your hearing sensitivity will decrease. For most people, the loss of sensitivity is greater for higher frequencies. The loss of sensitivity at high frequencies may make it difficult to understand what others are saying. Why is this?

***REASON*** It is the high-frequency components of speech that allow us to distinguish different vowel sounds. A decrease in sensitivity to these higher frequencies makes it more difficult to make such distinctions.

***ASSESS*** This result makes sense. In Figure 16.23b, the lowest frequency is less than 200 Hz, but the second formant is over 2000 Hz. There's a big difference between what we hear as the pitch of someone's voice and the frequencies we use to interpret speech.

**STOP TO THINK 16.6** If you speak at a certain pitch, then hold your nose and continue speaking at the same pitch, your voice sounds very different. This is because

A. The fundamental frequency of your vocal cords has changed.
B. The frequencies of the harmonics of your vocal cords have changed.
C. The pattern of resonance frequencies of your vocal tract has changed, thus changing the formants.

FIGURE 16.24 Interference of waves from two sources.

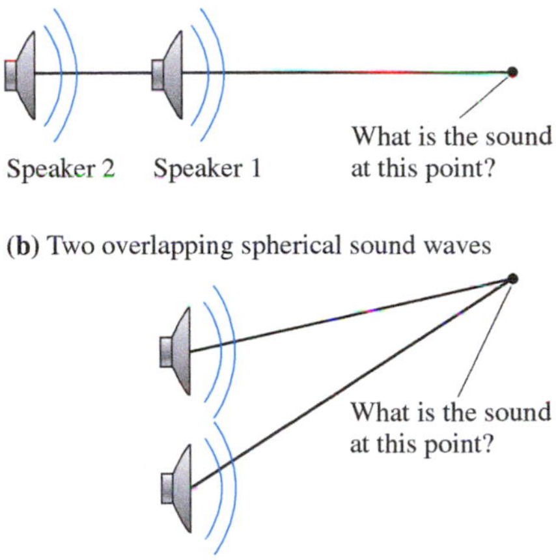


## 16.6 The Interference of Waves from Two Sources

Perhaps you have seen headphones that offer "active noise reduction." When you turn on the headphones, they produce sound that somehow *cancels* noise from the external environment. How does adding sound to a system make it quieter?

We began the chapter by noting that waves, unlike particles, can pass through each other. Where they do, the principle of superposition tells us that the displacement of the medium is the sum of the displacements due to each wave acting alone. Consider the two loudspeakers in FIGURE 16.24, both emitting sound waves with the same frequency. In Figure 16.24a, sound from loudspeaker 2 passes loudspeaker 1, then two overlapped sound waves travel to the right along the $x$-axis. What sound is heard at the point indicated with the dot? And what about at the dot in Figure 16.24b, where the speakers are side by side? These are two cases we will consider in this section. Although we'll use sound waves for our discussion, the results apply to all waves.

### Interference Along a Line

FIGURE 16.25 shows traveling waves from two loudspeakers spaced exactly one wavelength apart. The graphs are slightly displaced vertically from each other so that you can see what each wave is doing, but the *physical situation* is one in which the waves

FIGURE 16.25 Constructive interference of two waves traveling along the $x$-axis.

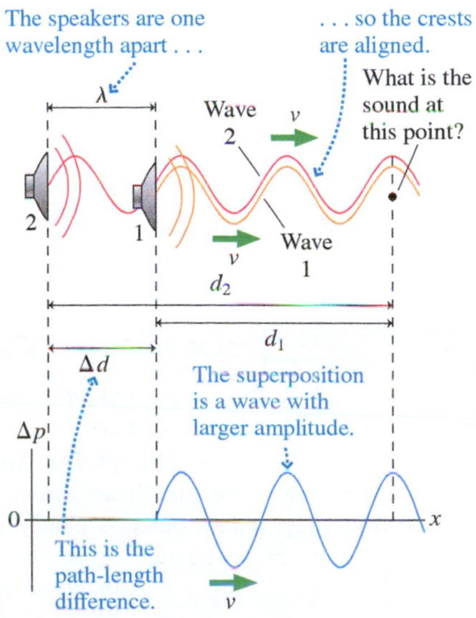

are traveling *on top of* each other. We assume that the two speakers emit sound waves of identical frequency $f$, wavelength $\lambda$, and amplitude $A$.

According to the principle of superposition, at every point along the line the net sound pressure wave will be the sum of the pressures from the individual waves. Because the two speakers are separated by one wavelength, the two waves are aligned crest-to-crest and trough-to-trough. Waves aligned this way are said to be **in phase;** waves that are in phase march along "in step" with each other. The superposition is a traveling wave with wavelength $\lambda$ and twice the amplitude of the individual waves. This is constructive interference.

NOTE ▶ Textbook illustrations like Figure 16.25 can be misleading because they're frozen in time. The net sound wave in Figure 16.25 is a *traveling* wave, moving to the right with the speed of sound $v$. It differs from the two individual waves only by having twice the amplitude. This is not a standing wave with nodes and antinodes that remain in one place. ◀

If $d_1$ and $d_2$ are the distances from loudspeakers 1 and 2 to a point at which we want to know the combined sound wave, their difference $\Delta d = d_2 - d_1$ is called the **path-length difference.** It is the *extra* distance traveled by wave 2 on the way to the point where the two waves are combined. In Figure 16.25, we see that constructive interference results from a path-length difference $\Delta d = \lambda$. But increasing $\Delta d$ by an additional $\lambda$ would produce exactly the same result, so we will also have constructive interference for $\Delta d = 2\lambda$, $\Delta d = 3\lambda$, and so on. In other words, **two waves will be in phase and will produce constructive interference any time their path-length difference is a whole number of wavelengths.**

FIGURE 16.26 Destructive interference of two waves traveling along the $x$-axis.

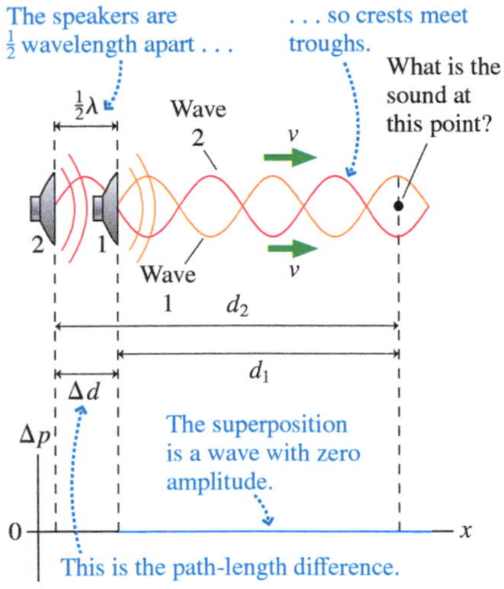


Video Out-of-Phase Speakers

In FIGURE 16.26, the two speakers are separated by half a wavelength. Now the crests of one wave align with the troughs of the other, and the waves march along "out of step" with each other. We say that the two waves are **out of phase.** When two waves are out of phase, they are equal and opposite at every point. Consequently, the sum of the two waves is zero *at every point.* This is destructive interference.

The destructive interference of Figure 16.26 results from a path-length difference $\Delta d = \frac{1}{2}\lambda$. Again, increasing $\Delta d$ by an additional $\lambda$ would produce a picture that looks exactly the same, so we will also have destructive interference for $\Delta d = 1\frac{1}{2}\lambda$, $\Delta d = 2\frac{1}{2}\lambda$, and so on. That is, **two waves will be out of phase and will produce destructive interference any time their path-length difference is a whole number of wavelengths plus half a wavelength.**

In summation, for two identical sources of waves, constructive interference occurs when the path-length difference is

$$\Delta d = m\lambda \qquad m = 0, 1, 2, 3, \ldots \tag{16.8}$$

and destructive interference occurs when the path-length difference is

$$\Delta d = \left(m + \frac{1}{2}\right)\lambda \qquad m = 0, 1, 2, 3, \ldots \tag{16.9}$$

NOTE ▶ The path-length difference needed for constructive or destructive interference depends on the wavelength and hence the frequency. If one particular frequency interferes destructively, another may not. ◀

## EXAMPLE 16.10 Interference of sound from two speakers

Susan stands directly in front of two speakers that are in line with each other. The farther speaker is 6.0 m from her; the closer speaker is 5.0 m away. The speakers are connected to the same 680 Hz sound source, and Susan hears the sound loud and clear. The frequency of the source is slowly increased until, at some point, Susan can no longer hear it. What is the frequency when this cancellation occurs? Assume that the speed of sound in air is 340 m/s.

**STRATEGIZE** Susan hears the sound "loud and clear" at the initial frequency of 680 Hz, which indicates constructive interference with $\Delta d = m\lambda_{\text{initial}}$. From the information given, we can find $\Delta d$ and $\lambda_{\text{initial}}$ and thus $m$. Destructive interference will occur when the frequency is increased (and the wavelength decreased) until $\Delta d = (m + \frac{1}{2})\lambda_{final}$; this will determine the final wavelength $\lambda_{final}$ and hence the final frequency.

***PREPARE*** We start with a visual overview of the situation, as shown in FIGURE 16.27. (The speakers are actually in line with each other, but we've displaced them vertically for clarity.) The sound waves from the two speakers overlap at Susan's position. The path-length difference—the extra distance traveled by the wave from speaker 1—is just the difference in the distances from the speakers to Susan's position. In this case,

$$\Delta d = d_2 - d_1 = 6.0 \text{ m} - 5.0 \text{ m} = 1.0 \text{ m}$$

Initially, with a 680 Hz tone and a 340 m/s sound speed, the wavelength is

$$\lambda_{initial} = \frac{v}{f} = \frac{340 \text{ m/s}}{680 \text{ Hz}} = 0.50 \text{ m}$$

FIGURE 16.27 Visual overview of the loudspeakers.

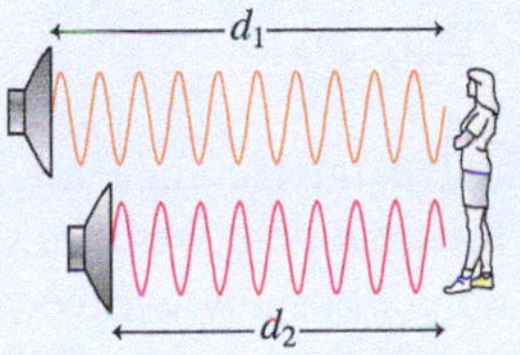


Known
$d_1 = 6.0$ m
$d_2 = 5.0$ m
$v = 340$ m/s

Find
$f$

***SOLVE*** We can find the value of $m$ from

$$m = \frac{\Delta d}{\lambda_{initial}} = \frac{1.0 \text{ m}}{0.50 \text{ m}} = 2$$

We see that $m$ is a whole number, as expected for constructive interference. As the frequency increases and the wavelength decreases, at some point the criterion for destructive interference, $\Delta d = (m + \frac{1}{2})\lambda_{final} = 2.5\lambda_{final}$, will be met. We thus have

$$\lambda_{final} = \frac{\Delta d}{2.5} = \frac{1.0 \text{ m}}{2.5} = 0.40 \text{ m}$$

This corresponds to a frequency of

$$f = \frac{340 \text{ m/s}}{0.40 \text{ m}} = 850 \text{ Hz}$$

***ASSESS*** 850 Hz is an increase of 170 Hz from the original 680 Hz, an increase of one-fourth of the original frequency. This makes sense: Originally, 2 cycles of the wave "fit" in the 1.0 m path-length difference; now, 2.5 cycles "fit," an increase of one-fourth of the original.

Another interesting and important case of interference, illustrated in FIGURE 16.28, occurs when one loudspeaker emits a sound wave that is *the exact inverse* of the wave from the other speaker. If the speakers are side by side, so that $\Delta d = 0$, the superposition of these two waves will result in destructive interference; they will completely cancel. This destructive interference does not require the waves to have any particular frequency or any particular shape.

Headphones with *active noise reduction* use this technique. A microphone on the outside of the headphones measures ambient sound. A circuit inside the headphones produces an inverted version of the microphone signal and sends it to the headphone speakers. The ambient sound and the inverted version of the sound from the speakers arrive at the ears together and interfere destructively, reducing the sound intensity. In this case, *adding* sound results in a *lower* overall intensity inside the headphones!

FIGURE 16.28 Opposite waves cancel.

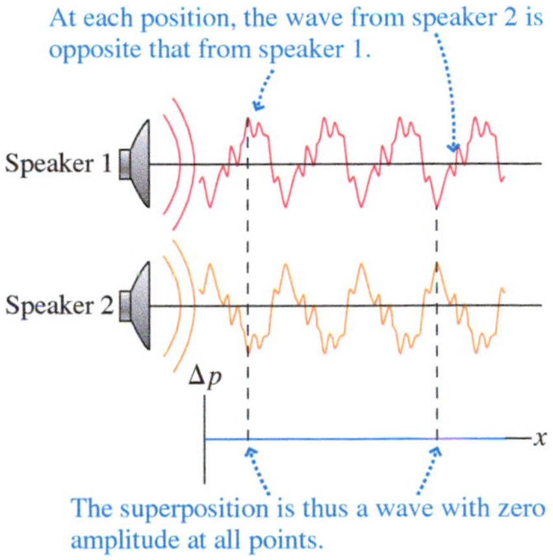


## Interference of Spherical Waves

Interference along a line illustrates the idea of interference, but it's not very realistic. In practice, sound waves from a loudspeaker or light waves from a lightbulb spread out as spherical waves. FIGURE 16.29 shows a wave-front diagram for a spherical wave. Recall that the wave fronts represent the *crests* of the wave and are spaced by the wavelength $\lambda$. Halfway between two wave fronts is a trough of the wave. What happens when two spherical waves overlap? For example, imagine two loudspeakers emitting identical sound waves radiating in all directions. FIGURE 16.30 on the next page shows the wave fronts of the two waves. This is a static picture, of course, so you have to imagine the wave fronts spreading out as new circular rings are born at the speakers. The waves overlap as they travel, and, as was the case in one dimension, this causes interference.

Consider a particular point like that marked by the red dot in Figure 16.30. The two waves each have a crest at this point, so there is constructive interference here. But at other points, such as that marked by the black dot, a crest overlaps a trough, so this is a point of destructive interference.

FIGURE 16.29 A spherical wave.

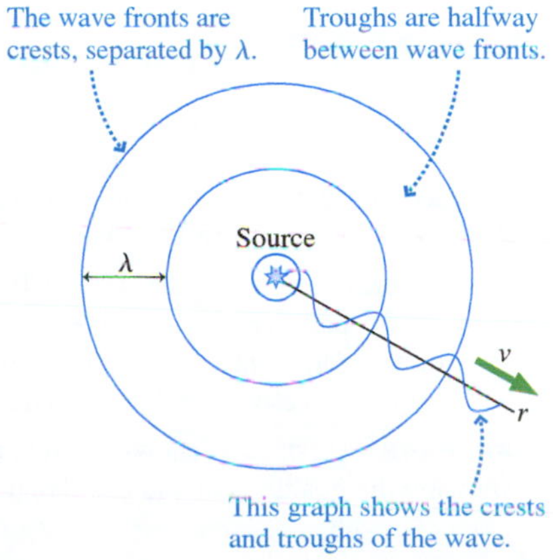

FIGURE 16.30 The overlapping wave patterns of two sources.

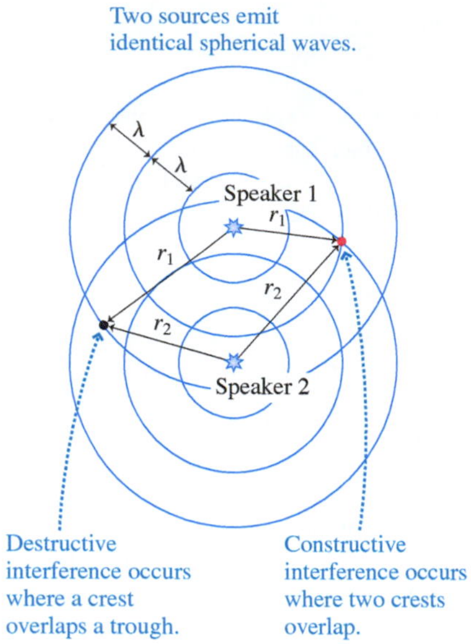


Notice—simply by counting the wave fronts—that the red dot is three wavelengths from speaker 2 ($r_2 = 3\lambda$) but only two wavelengths from speaker 1 ($r_1 = 2\lambda$). The path-length difference of the two waves arriving at the red dot is $\Delta r = r_2 - r_1 = \lambda$. That is, the wave from speaker 2 has to travel one full wavelength more than the wave from speaker 1, so the waves are in phase (crest aligned with crest) and interfere constructively. You should convince yourself that $\Delta r$ is a *whole number of wavelengths* at every point where two wave fronts intersect.

Similarly, the path-length difference at the black dot, where the interference is destructive, is $\Delta r = \frac{1}{2}\lambda$. As with interference along a line, destructive interference results when the path-length difference is a whole number of wavelengths plus half a wavelength.

Thus the general rule for determining whether there is constructive or destructive interference at any point is the same for spherical waves as for waves traveling along a line. For identical sources, constructive interference occurs when the path-length difference is

$$\Delta r = m\lambda \qquad m = 0, 1, 2, 3, \ldots \tag{16.10}$$

Destructive interference occurs when the path-length difference is

$$\Delta r = \left(m + \frac{1}{2}\right)\lambda \qquad m = 0, 1, 2, 3, \ldots \tag{16.11}$$

The conditions for constructive and destructive interference are the same for spherical waves as for waves along a line. And the treatment we have seen for sound waves can be applied to any wave, as we have noted. For any two wave sources, the following Tactics Box sums up how to determine whether the interference at a point is constructive or destructive.

### TACTICS BOX 16.1 Identifying constructive and destructive interference

1. Identify the path length from each source to the point of interest. Compute the path-length difference $\Delta r = |r_2 - r_1|$.
2. Find the wavelength, if it is not specified.
3. If the path-length difference is a whole number of wavelengths ($\lambda, 2\lambda, 3\lambda, \ldots$), crests are aligned with crests and there is constructive interference.
4. If the path-length difference is a whole number of wavelengths plus a half wavelength ($1\frac{1}{2}\lambda, 2\frac{1}{2}\lambda, 3\frac{1}{2}\lambda, \ldots$), crests are aligned with troughs and there is destructive interference.

Exercises 9, 10

**NOTE** ▸ Keep in mind that interference is determined by $\Delta r$, the path-length *difference,* not by $r_1$ or $r_2$. ◂

### EXAMPLE 16.11 Is the sound loud or quiet?

Two speakers are 3.0 m apart and play identical tones of frequency 170 Hz. Sam stands directly in front of one speaker at a distance of 4.0 m. Is this a loud spot or a quiet spot? Assume that the speed of sound in air is 340 m/s.

***STRATEGIZE*** Following the steps of Tactics Box 16.1, we need to compute the path-length difference, $r_2 - r_1$, and determine whether it is of the form $m\lambda$ (constructive interference, loud sound) or $(m + \frac{1}{2})\lambda$ (destructive interference, quiet sound).

***PREPARE*** FIGURE 16.31 shows a visual overview of the situation, showing the positions of and path lengths from each speaker.

FIGURE 16.31 Visual overview of two speakers.

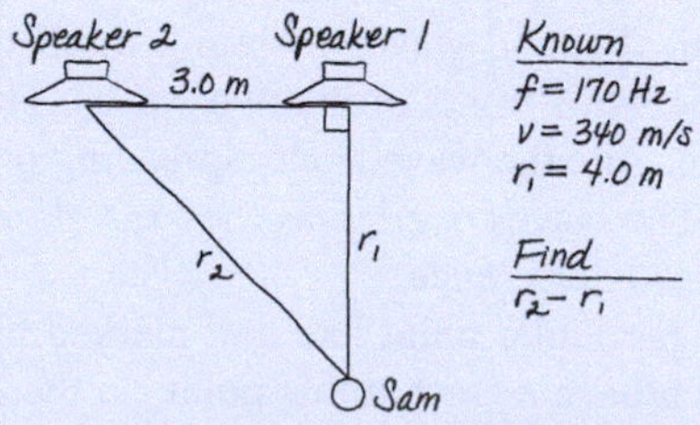

**SOLVE** We start by computing the path-length difference. $r_1$, $r_2$, and the distance between the speakers form a right triangle, so we can use the Pythagorean theorem to find

$$r_2 = \sqrt{(4.0\text{ m})^2 + (3.0\text{ m})^2} = 5.0\text{ m}$$

Thus the path-length difference is

$$\Delta r = r_2 - r_1 = 1.0\text{ m}$$

Next, we compute the wavelength:

$$\lambda = \frac{v}{f} = \frac{340\text{ m/s}}{170\text{ Hz}} = 2.0\text{ m}$$

The path-length difference is $\frac{1}{2}\lambda$, so this is a point of destructive interference. Sam is at a quiet spot.

You are regularly exposed to sound from two separated sources: stereo speakers. When you walk across a room in which a stereo is playing, why don't you hear a pattern of loud and soft sounds? First, we don't listen to single frequencies. Music is a complex sound wave with many frequencies, but only one frequency at a time satisfies the condition for constructive or destructive interference. Most of the sound frequencies are not affected. Second, reflections of sound waves from walls and furniture make the situation much more complex than the idealized two-source picture in Figure 16.30. Sound wave interference can be heard, but it takes careful selection of a pure tone and a room with no hard, reflecting surfaces. Interference that's rather tricky to demonstrate with sound waves is easy to produce with light waves, as we'll see in the next chapter.

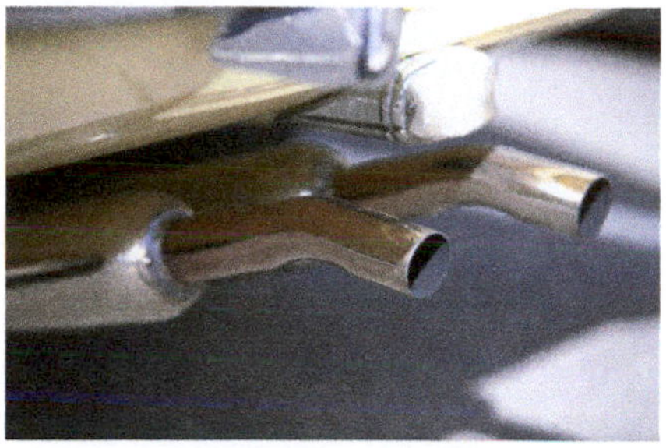

**Taming and tuning exhaust noise**
The wide section of pipe at the end of this exhaust system is a *resonator*. It's a tube that surrounds the exhaust pipe. One end is open to the exhaust system and the other end (the end seen here) is closed. Sound waves in the exhaust enter, reflect from the closed end, and reenter the exhaust pipe. If the length of the tube is just right, the reflected wave is out of phase with the sound wave in the pipe, producing destructive interference. The resonator is tuned to eliminate the loudest frequencies from the engine, but other frequencies will produce constructive interference, enhancing them and giving the exhaust a certain "note."

**STOP TO THINK 16.7** These speakers emit identical sound waves with a wavelength of 1.0 m. At the point indicated, is the interference constructive, destructive, or something in between?

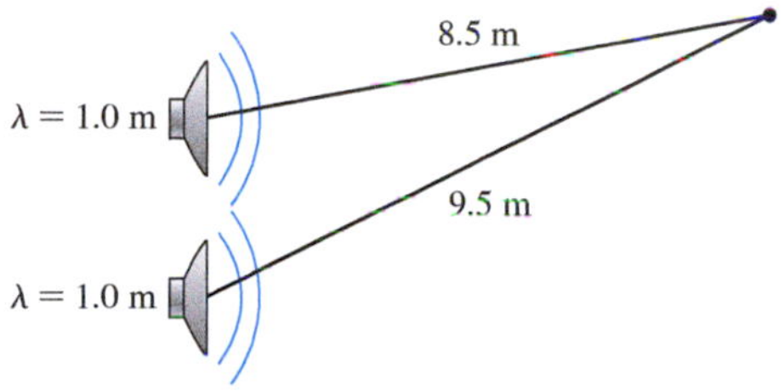


## 16.7 Beats

Suppose two sinusoidal waves are traveling toward your ear, as shown in FIGURE 16.32. The two waves have the same amplitude but slightly different frequencies: The orange wave has a slightly higher frequency (and thus a slightly shorter wavelength) than the red wave. This slight difference causes the waves to combine in a manner that alternates between constructive and destructive interference. Their superposition, drawn in blue below the two waves, is a wave whose amplitude shows a

FIGURE 16.32 Creating beats by the superposition of two sound waves with slightly different frequencies.

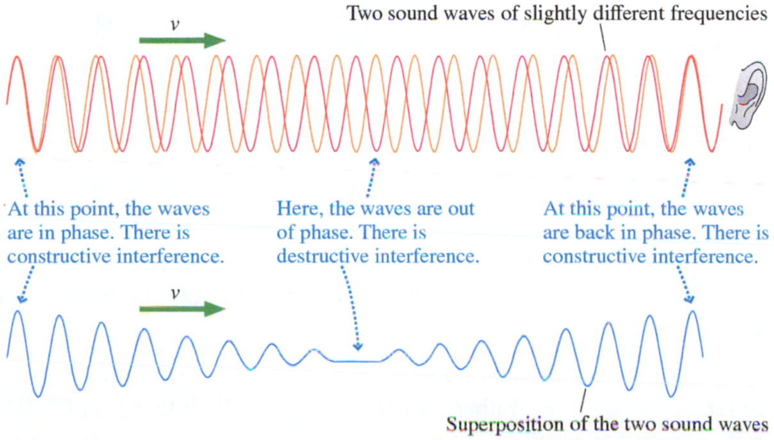

periodic variation. As the waves reach your ear, you will hear a single tone whose intensity is *modulated.* That is, the sound goes up and down in volume, loud, soft, loud, soft, . . . , making a distinctive sound pattern called **beats.**

Suppose the two waves have frequencies $f_1$ and $f_2$ that differ only slightly, so that $f_1 \approx f_2$. A complete mathematical analysis would show that the air oscillates against your eardrum at frequency

$$f_{osc} = \frac{1}{2}(f_1 + f_2)$$

This is the *average* of $f_1$ and $f_2$, and it differs little from either since the two frequencies are nearly equal. Further, the intensity of the sound is modulated at a frequency called the *beat frequency:*

$$f_{beat} = |f_1 - f_2| \qquad (16.12)$$

The beat frequency is the *difference* between the two individual frequencies.

FIGURE 16.33 is a history graph of the wave at the position of your ear. You can see both the sound wave oscillation at frequency $f_{osc}$ and the much slower intensity oscillation at frequency $f_{beat}$. Frequency $f_{osc}$ determines the pitch you hear, while $f_{beat}$ determines the frequency of the loud-soft-loud modulations of the sound intensity.

Musicians can use beats to tune their instruments. If one flute is properly tuned at 440 Hz but another plays at 438 Hz, the flutists will hear two loud-soft-loud beats per second. The second flutist is "flat" and needs to shorten her flute slightly to bring the frequency up to 440 Hz.

Many measurement devices use beats to determine an unknown frequency by comparing it to a known frequency. For example, Chapter 15 described a Doppler blood flowmeter that used the Doppler shift of ultrasound reflected from moving blood to determine its speed. The meter determines this very small frequency shift by combining the emitted wave and the reflected wave and measuring the resulting beat frequency. The beat frequency is equal to the shift in frequency on reflection, exactly what is needed to determine the blood speed.

FIGURE 16.33 The modulated sound of beats.

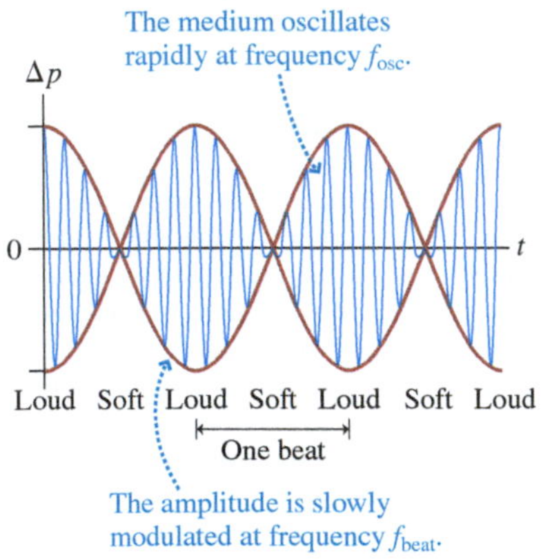


**EXAMPLE 16.12 Detecting bats using beats** BIO

The little brown bat is a common bat species in North America. It emits echolocation pulses at a frequency of 40 kHz, well above the range of human hearing. To allow observers to "hear" these bats, the bat detector shown in FIGURE 16.34 combines the bat's sound wave at frequency $f_1$ with a wave of frequency $f_2$ from a tunable oscillator. The resulting beat frequency is isolated with a filter, then amplified and sent to a loudspeaker. To what frequency should the tunable oscillator be set to produce an audible beat frequency of 3 kHz?

**STRATEGIZE** The oscillator's frequency needs to be close to that of the bat's pulses, so that the resulting beat frequency is low enough for humans to hear.

**SOLVE** The beat frequency is $f_{beat} = |f_1 - f_2|$, so the oscillator frequency and the bat frequency need to *differ* by 3 kHz. An oscillator frequency of either 37 kHz or 43 kHz will work nicely.

**ASSESS** These two frequencies are close to 40 kHz, a requirement to have a clear beat frequency.

FIGURE 16.34 The operation of a bat detector.

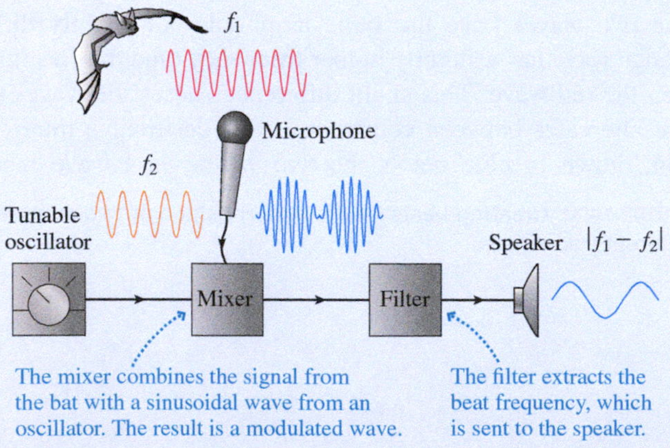


**STOP TO THINK 16.8** You hear three beats per second when two sound tones are generated. The frequency of one tone is known to be 610 Hz. The frequency of the other is

A. 604 Hz
B. 607 Hz
C. 613 Hz
D. 616 Hz
E. Either A or D
F. Either B or C

## INTEGRATED EXAMPLE 16.13 The size of a dog determines the sound of its growl BIO

The sounds of the human vocal system result from the interplay of two different oscillations: the oscillation of the vocal cords and the standing-wave resonances of the vocal tract. A dog's vocalizations are based on similar principles, but the canine vocal tract is simpler than that of a human. When a dog growls or howls, the shape of the vocal tract is essentially a tube closed at the larynx and open at the lips.

All dogs growl at a low pitch because the fundamental frequency of the vocal cords is quite low. But the growls of small dogs and big dogs differ because they have very different formants. The frequency of the formants is determined by the length of the vocal tract, and the vocal-tract length is a pretty good measure of the size of a dog. A larger dog has a longer vocal tract and a correspondingly lower-frequency formant, which sends an important auditory message to other dogs.

The masses of two different dogs and the frequencies of their formants are given in TABLE 16.1.

a. What are the approximate vocal-tract lengths of these two dogs? Assume that the speed of sound is 350 m/s at a dog's body temperature.
b. Growls aren't especially loud; at a distance of 1.0 m, a dog's growl is about 60 dB—the same as normal conversation. What is the acoustic power emitted by a 60 dB growl?
c. The lower-pitched growl of the Doberman will certainly sound more menacing, but which dog's 60 dB growl sounds louder to a human? **Hint:** Assume that most of the acoustic energy is at frequencies near the first formant; then look at the curve of equal perceived loudness in Figure 16.21.

TABLE 16.1 Mass and acoustic data for two growling dogs

| Breed | Mass (kg) | First formant (Hz) | Second formant (Hz) |
|---|---|---|---|
| West Highland Terrier (Westie) | 8.0 | 650 | 1950 |
| Doberman | 38 | 350 | 1050 |

***STRATEGIZE*** We will model the dog's vocal tract as an open-closed tube.

***PREPARE*** FIGURE 16.35 shows an idealized model of the vocal tract. The formants correspond to the standing-wave resonances of this system. The first two modes are shown in the figure. The first formant corresponds to $m = 1$; the second formant corresponds to $m = 3$ because an open-closed tube has only odd harmonics. Knowing the frequencies of the formants allows us to determine the length of the tube that produces them.

FIGURE 16.35 A model of the canine vocal tract.

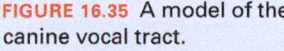

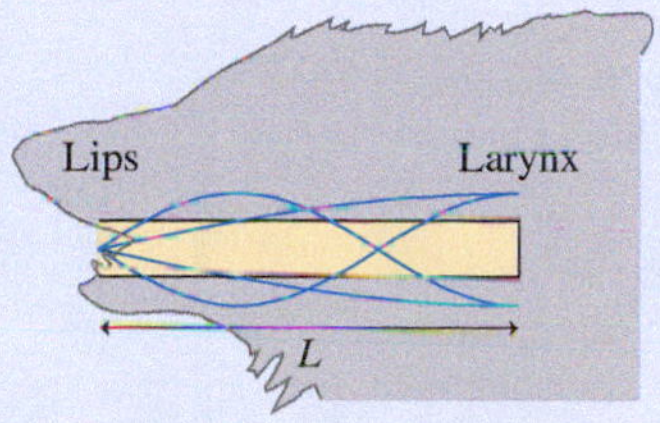


For the question about loudness, we can determine the sound intensity from the sound intensity level. Knowing the distance, we can use this value to determine the emitted acoustic power.

***SOLVE***

a. We can use the frequency of the first formant to find the length of the vocal tract. The standing-wave frequencies of an open-closed tube are given by

$$f_m = m\frac{v}{4L} \qquad m = 1, 3, 5, \ldots$$

Rearranging this to solve for $L$, with $m = 1$, we find

$$L(\text{Westie}) = m\frac{v}{4f_m} = (1)\frac{350 \text{ m/s}}{4(650 \text{ Hz})} = 0.13 \text{ m}$$

$$L(\text{Doberman}) = m\frac{v}{4f_m} = (1)\frac{350 \text{ m/s}}{4(350 \text{ Hz})} = 0.25 \text{ m}$$

The length is greater for the Doberman, as we would predict, given the relative masses of the dogs.

b. Equation 15.15 lets us compute the sound intensity for a given sound intensity level:

$$I = (1.0 \times 10^{-12} \text{ W/m}^2)10^{(\beta/10 \text{ dB})}$$

$$= (1.0 \times 10^{-12} \text{ W/m}^2)10^{(60 \text{ dB}/10 \text{ dB})} = 1.0 \times 10^{-6} \text{ W/m}^2$$

This is the intensity at a distance of 1.0 m. The sound spreads out in all directions, so we can use Equation 15.12, $I = P_{\text{source}}/4\pi r^2$, to compute the power of the source:

$$P_{\text{source}} = I \cdot 4\pi r^2 = (1.0 \times 10^{-6} \text{ W/m}^2) \cdot 4\pi(1.0 \text{ m})^2 = 13\ \mu\text{W}$$

A growl may sound menacing, but that's not because of its power!

c. Figure 16.21 is a curve of equal perceived loudness. The curve steadily decreases until reaching $\approx 3500$ Hz, so the acoustic power needed to produce the same sensation of loudness decreases with frequency up to this point. That is, below 3500 Hz your ear is more sensitive to higher frequencies than to lower frequencies. Much of the acoustic power of a dog growl is concentrated at the frequencies of the first two formants. The Westie's growl has its energy at higher frequencies than that of the Doberman, so the Westie's growl will sound louder—though the lower formants of the Doberman's growl will make it sound more formidable.

***ASSESS*** The lengths of the vocal tract that we calculated—13 cm (5 in) for a small terrier and 25 cm (10 in) for a Doberman—seem reasonable given the size of the dogs. We can also check our work by looking at the second formant, corresponding to $m = 3$; using this harmonic in the second expression in Equation 16.7, we find

$$L(\text{Westie}) = m\frac{v}{4f_m} = (3)\frac{350 \text{ m/s}}{4(1950 \text{ Hz})} = 0.13 \text{ m}$$

$$L(\text{Doberman}) = m\frac{v}{4f_m} = (3)\frac{350 \text{ m/s}}{4(1050 \text{ Hz})} = 0.25 \text{ m}$$

This exact match to our earlier calculations gives us confidence in our model and in our results.

**Video** Harmonics and Voices

# SUMMARY

**GOAL** To use the idea of superposition to understand the phenomena of interference and standing waves.

## GENERAL PRINCIPLES

### Principle of Superposition

The displacement of a medium when more than one wave is present is the sum of the displacements due to each individual wave.

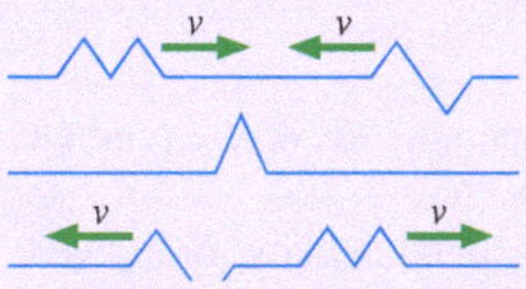


### Interference

The superposition of two or more waves into a single wave is called interference.

**Constructive interference** occurs when crests are aligned with crests and troughs with troughs; the waves are *in phase*. It occurs when the path-length difference $\Delta d$ is a whole number of wavelengths.

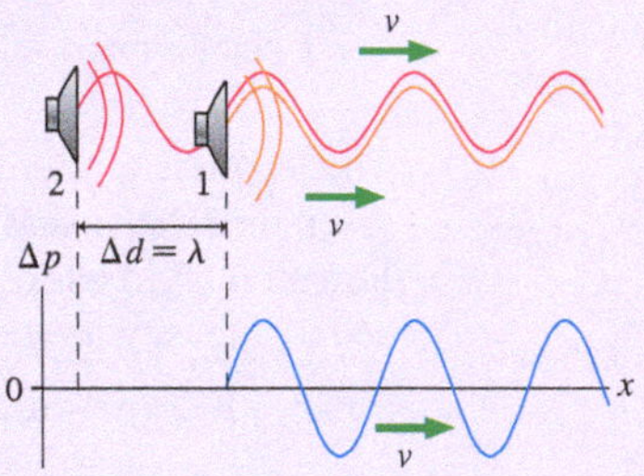


**Destructive interference** occurs when crests are aligned with troughs; the waves are *out of phase*. It occurs when the path-length difference $\Delta d$ is a whole number of wavelengths plus half a wavelength.

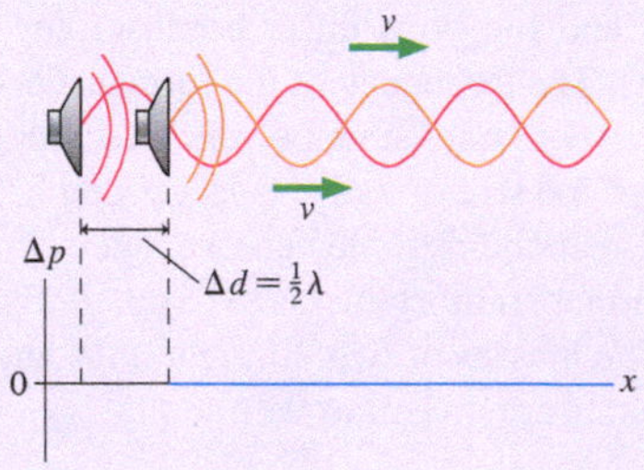


## IMPORTANT CONCEPTS

### Standing Waves

Two identical traveling waves moving in opposite directions create a standing wave.

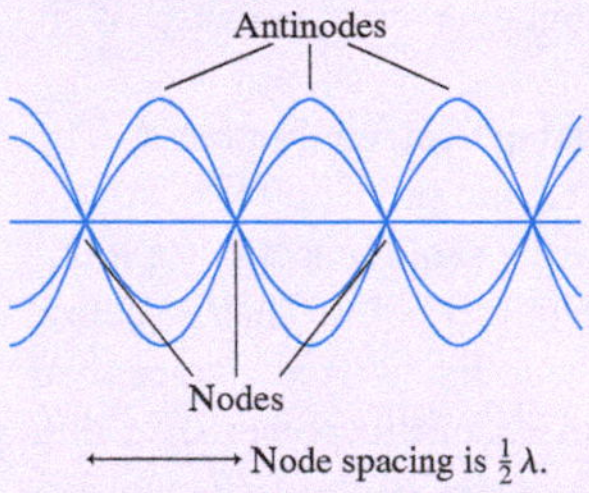


The boundary conditions determine which standing-wave frequencies and wavelengths are allowed. The allowed standing waves are **modes** of the system.

A **standing wave on a string** has a node at each end. Possible modes:

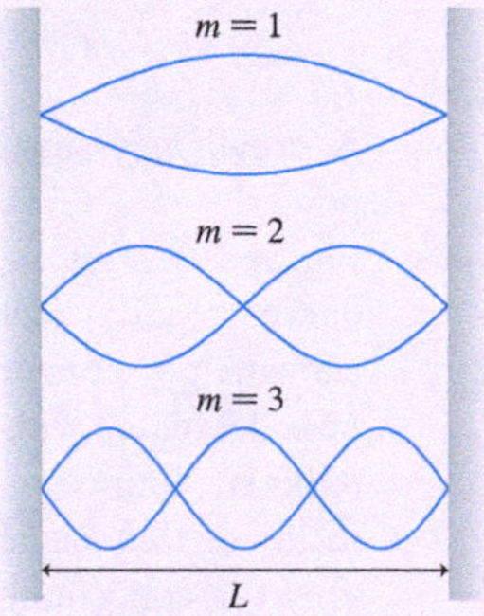


$$\lambda_m = \frac{2L}{m} \qquad f_m = m\left(\frac{v}{2L}\right) = mf_1$$

$$m = 1, 2, 3, \ldots$$

A **standing sound wave in a tube** can have different boundary conditions: open-open, closed-closed, or open-closed.

**Open-open**

$$f_m = m\left(\frac{v}{2L}\right)$$

$$m = 1, 2, 3, \ldots$$

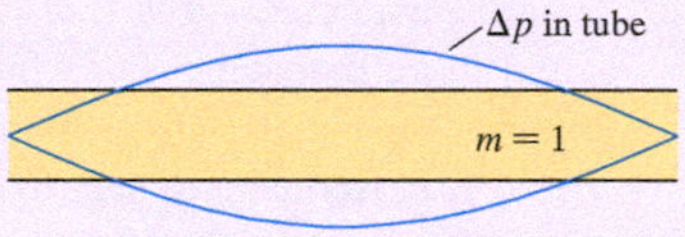


**Closed-closed**

$$f_m = m\left(\frac{v}{2L}\right)$$

$$m = 1, 2, 3, \ldots$$

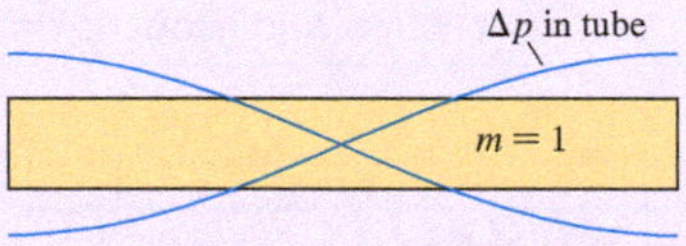


**Open-closed**

$$f_m = m\left(\frac{v}{4L}\right)$$

$$m = 1, 3, 5, \ldots$$

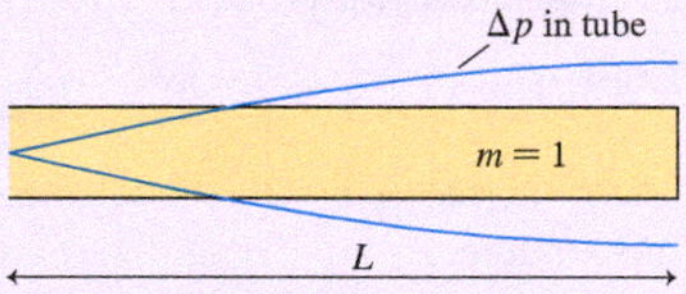

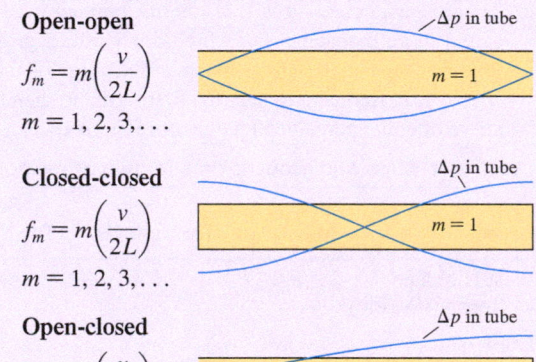

## APPLICATIONS

**Beats** (loud-soft-loud-soft modulations of intensity) are produced when two waves of slightly different frequencies are superimposed.

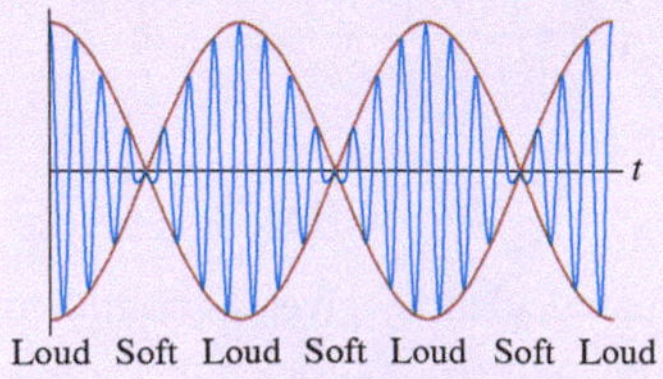


$$f_{beat} = |f_1 - f_2|$$

Standing waves are multiples of a **fundamental frequency,** the frequency of the lowest mode. The higher modes are the higher **harmonics.**

For sound, the fundamental frequency determines the perceived **pitch;** the higher harmonics determine the **tone quality.**

Our vocal cords create a range of harmonics. The mix of higher harmonics is changed by our vocal tract to create different vowel sounds.

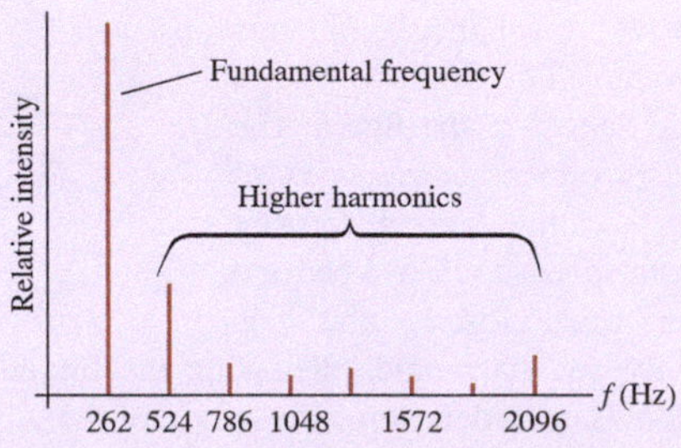

## Learning Objectives After studying this chapter, you should be able to:

- Understand and apply the principle of superposition. *Conceptual Question 16.2; Problems 16.1, 16.2, 16.3, 16.4, 16.5*
- Solve problems about standing waves on a string. *Conceptual Questions 16.10, 16.16; Problems 16.8, 16.9, 16.12, 16.15, 16.16*
- Solve problems about standing sound waves. *Conceptual Questions 16.9, 16.13; Problems 16.20, 16.22, 16.23, 16.26, 16.32*
- Apply wave concepts to speech and hearing. *Conceptual Questions 16.14, 16.17; Problems 16.33, 16.34, 16.35, 16.36*
- Calculate the constructive and destructive interference of waves from two sources. *Conceptual Questions 16.3, 16.11; Problems 16.37, 16.40, 16.42, 16.43, 16.44*
- Calculate the beat frequency of two simultaneous waves. *Problems 16.45, 16.46, 16.47, 16.48, 16.49*

## STOP TO THINK ANSWERS

**Chapter Preview Stop to Think: D.** The frequency, wavelength, and speed are related by $v = \lambda f$. The wave speed is the speed of sound, which stays the same, so doubling the frequency means halving the wavelength to 1.0 m.

**Stop to Think 16.1: C.** The figure shows the two waves at $t = 6$ s and their superposition. The superposition is the *point-by-point* addition of the displacements of the two individual waves.

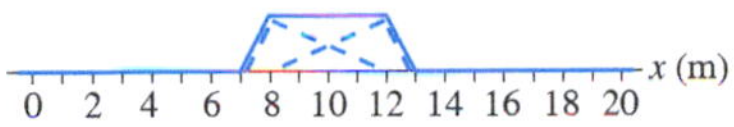


**Stop to Think 16.2: D.** There is a node at each end of the string and two nodes in the middle, so the distance between nodes is 1.0 m. The distance between successive nodes is $\lambda/2$, so the wavelength is 2.0 m.

**Stop to Think 16.3: C.** There are five antinodes, so the mode number is $m = 5$. The equation for the wavelength then gives $\lambda = 2L/m = 2(2.0\text{ m})/5 = 0.80$ m.

**Stop to Think 16.4: C.** Standing-wave frequencies are $f_m = mf_1$. The original wave has frequency $f_2 = 2f_1$ because it has two antinodes. The wave with frequency $2f_2 = 4f_1$ is the $m = 4$ mode with four antinodes.

**Stop to Think 16.5: B.** 300 Hz and 400 Hz are not $f_1$ and $f_2$ because 400 Hz $\neq 2 \times 300$ Hz. Instead, both are multiples of the fundamental frequency. Because the difference between them is 100 Hz, we see that $f_3 = 3 \times 100$ Hz and $f_4 = 4 \times 100$ Hz. Thus $f_1 = 100$ Hz.

**Stop to Think 16.6: C.** Holding your nose does not affect the way your vocal cords vibrate, so it won't change the fundamental frequency or the harmonics of the vocal-cord vibrations. But it will change the frequencies of the formants of the vocal system: The nasal cavities will be open-closed tubes instead of open-open tubes. The frequencies that are amplified are altered, and your voice sounds quite different.

**Stop to Think 16.7: Constructive interference.** The path-length difference is $\Delta r = 1.0\text{ m} = \lambda$. Interference is constructive when the path-length difference is a whole number of wavelengths.

**Stop to Think 16.8: F.** The beat frequency is the difference between the two frequencies.

**Video Tutor Solution** Chapter 16

# QUESTIONS

## Conceptual Questions

1. Light can pass easily through water and through air, but light will reflect from the surface of a lake. What does this tell you about the speed of light in air and in water?

2. A small wave pulse and a large wave pulse approach each other on a string; the large pulse is moving to the right. Some time after the pulses have met and passed each other, which of the following statements is correct? (More than one answer may be correct.) Explain.
   A. The large pulse continues unchanged, moving to the right.
   B. The large pulse continues moving to the right but is smaller in amplitude.
   C. The small pulse is reflected and moves off to the right with its original amplitude.
   D. The small pulse is reflected and moves off to the right with a smaller amplitude.
   E. The two pulses combine into a single pulse moving to the right.

Problem difficulty is labeled as | (straightforward) to ||||| (challenging). Problems labeled INT integrate significant material from earlier chapters; Problems labeled BIO are of biological or medical interest.

The eText icon indicates when there is a video tutor solution available for the chapter or for a specific problem. To launch these videos, log into your eText through Mastering™ Physics or log into the Study Area.

3. You are listening to music from a loudspeaker. Then a second speaker is turned on. Is it possible that the music you now hear is quieter than it was with only the first speaker playing? Explain.
4. A guitarist finds that the pitch of one of her strings is slightly flat—the frequency is a bit too low. Should she increase or decrease the tension of the string? Explain.
5. BIO Certain illnesses inflame your vocal cords, causing them to swell. How does this affect the pitch of your voice? Explain.
6. If you're swimming underwater and knock two rocks together, you will hear a very loud noise. But if your friend above the water knocks two rocks together, you'll barely hear the sound. Explain.
7. Figure Q16.7 shows a standing sound wave in a tube of air that is open at both ends.

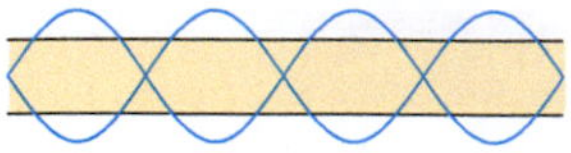

FIGURE Q16.7

   a. Which mode (value of $m$) standing wave is this?
   b. Is the air vibrating horizontally or vertically?
8. A typical flute is about 66 cm long. A piccolo is a very similar instrument, though it is smaller, with a length of about 32 cm. How does the pitch of a piccolo compare to that of a flute?
9. Some pipes on a pipe organ are open at both ends, others are closed at one end. For pipes that play low-frequency notes, there is an advantage to using pipes that are closed at one end. What is the advantage?
10. A washtub bass is a simple instrument consisting of a string stretched between a pole and a metal washtub. The musician can play different notes by pulling back more or less on the pole. Explain how this tuning method works.

11. You are standing directly between two loudspeakers playing identical 350 Hz tones. Describe what you will hear if you start walking along a line perpendicular to the line between the speakers.
12. If you pour liquid in a tall, narrow glass, you may hear sound with a steadily rising pitch. What is the source of the sound, and why does the pitch rise as the glass fills?
13. BIO When you speak after breathing helium, in which the speed of sound is much greater than in air, your voice sounds quite different. The frequencies emitted by your vocal cords do not change since they are determined by the mass and tension of your vocal cords. So what *does* change when your vocal tract is filled with helium rather than air?
14. Figure Q16.14 shows frequency spectra of the same note played on a flute (modeled as an open-open tube) and on a clarinet (a closed-open tube). Which figure corresponds to the flute, and which to the clarinet? Explain.

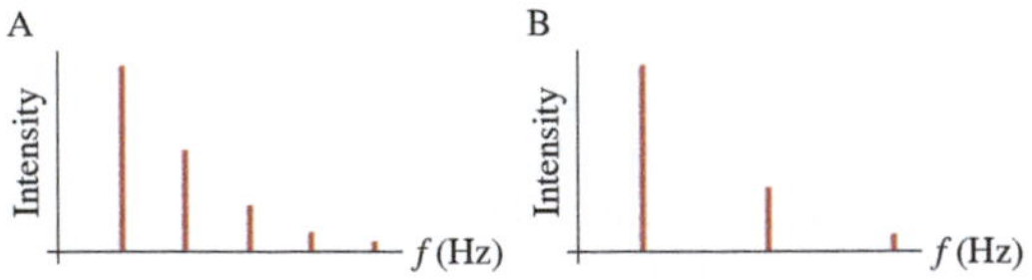


FIGURE Q16.14

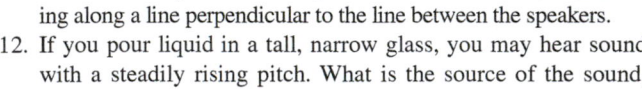

15. A synthesizer is a keyboard instrument that can be made to sound like a flute, a trumpet, a piano—or any other musical instrument. Pressing a key sets the fundamental frequency of the note that is produced. Other settings that the player can adjust change the sound quality of the synthesizer without altering the fundamental frequency. How is this change accomplished? What is being adjusted?
16. Figure Q16.16 shows a standing wave on a string. Point A on the string is shown at time $t = 0$. In terms of the wave's period $T$, what are the earliest times after $t = 0$ that points B, C, and D reach the positions shown?

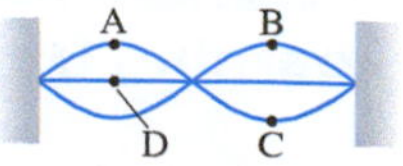


FIGURE Q16.16

17. BIO A small boy and a grown woman both speak at approximately the same pitch. Nonetheless, it's easy to tell which is which from listening to the sounds of their voices. How are you able to make this determination?

## Multiple-Choice Questions

Questions 18 through 20 refer to the snapshot graph of Figure Q16.18.

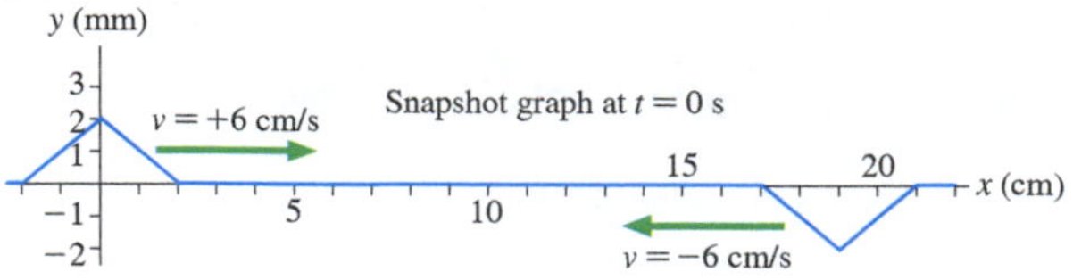


FIGURE Q16.18

18. | At $t = 1$ s, what is the displacement $y$ of the string at $x = 7$ cm?
   A. −1.0 mm B. 0 mm C. 0.5 mm
   D. 1.0 mm E. 2.0 mm
19. | At $x = 3$ cm, what is the earliest time that $y$ will equal 2 mm?
   A. 0.5 s B. 0.7 s C. 1.0 s
   D. 1.5 s E. 2.5 s
20. | At $t = 1.5$ s, what is the value of $y$ at $x = 10$ cm?
   A. −2.0 mm B. −1.0 mm C. −0.5 mm
   D. 0 mm E. 1.0 mm
21. | In a tube, standing-wave modes are found at 200 Hz and 400 Hz. The tube could *not* be
   A. Open-closed. B. Open-open.
   C. Closed-closed. D. It could be any of these.
22. | A student in her physics lab measures the standing-wave modes of a tube. The lowest frequency that makes a resonance is 20 Hz. As the frequency is increased, the next resonance is at 60 Hz. What will be the next resonance after this?
   A. 80 Hz B. 100 Hz C. 120 Hz D. 180 Hz
23. | An organ pipe is tuned to exactly 384 Hz when the temperature in the room is 20°C. Later, when the air has warmed up to 25°C, the frequency is
   A. Greater than 384 Hz. B. 384 Hz.
   C. Less than 384 Hz.
24. | BIO Resonances of the ear canal lead to increased sensitivity of hearing, as we've seen. Dogs have a much longer ear canal—5.2 cm—than humans. What are the two lowest frequencies at which dogs have an increase in sensitivity? The speed of sound in the warm air of the ear is 350 m/s.
   A. 1700 Hz, 3400 Hz B. 1700 Hz, 5100 Hz
   C. 3400 Hz, 6800 Hz D. 3400 Hz, 10,200 Hz
25. | The frequency of the lowest standing-wave mode on a 1.0-m-long string is 20 Hz. What is the wave speed on the string?
   A. 10 m/s B. 20 m/s C. 30 m/s D. 40 m/s
26. | Suppose you pluck a string on a guitar and it produces the note A at a frequency of 440 Hz. Now you press your finger down on the string against one of the frets, making this point the new end of the string. The newly shortened string has 4/5 the length of the full string. When you pluck the string, its frequency will be
   A. 350 Hz B. 440 Hz C. 490 Hz D. 550 Hz

# PROBLEMS

## Section 16.1 The Principle of Superposition

1. | Figure P16.1 is a snapshot graph at $t = 0$ s of two waves on a taut string approaching each other at 1 m/s. Draw six snapshot graphs, stacked vertically, showing the string at 1 s intervals from $t = 1$ s to $t = 6$ s.

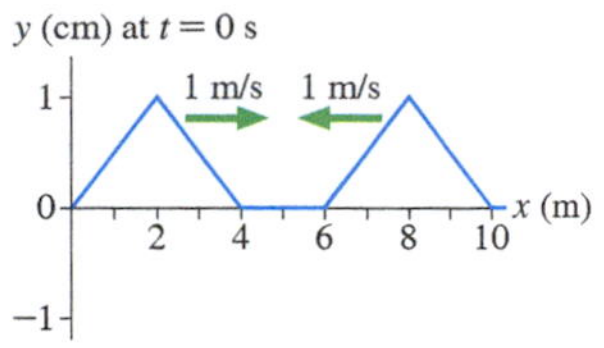


FIGURE P16.1

2. ||| Figure P16.2 is a snapshot graph at $t = 0$ s of two waves approaching each other at 1 m/s. Draw two snapshot graphs, stacked vertically, showing the string at $t = 2$ s and 3 s.

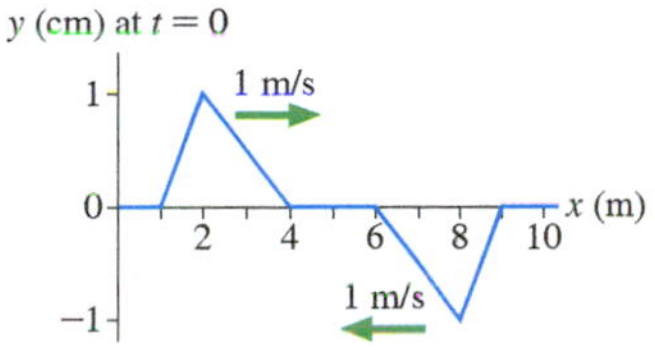


FIGURE P16.2

3. ||| Figure P16.2 is a snapshot graph at $t = 0$ s of two waves approaching each other at 1 m/s. Draw a history graph of the point of the string at $x = 4$ cm.
4. || Figure P16.4a is a snapshot graph at $t = 0$ s of two waves on a string approaching each other at 1 m/s. At what time was the snapshot graph in Figure P16.4b taken?

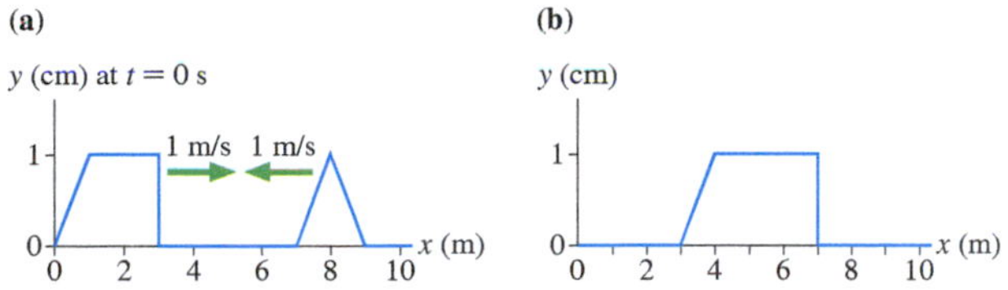


FIGURE P16.4

5. | Figure P16.5 is a snapshot graph at $t = 0$ s of two waves on a string approaching each other at 1 m/s. List the values of the displacement of the string at $x = 5.0$ cm at 1 s intervals from $t = 0$ s to $t = 6$ s.

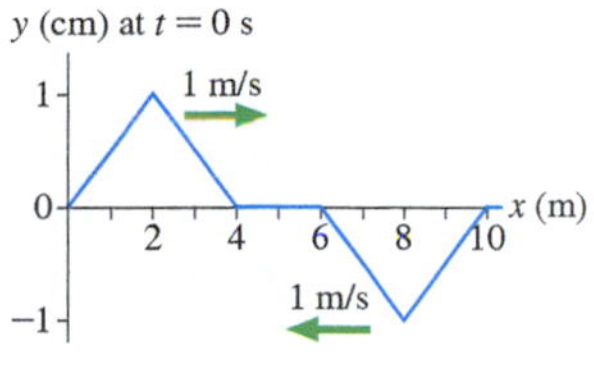


FIGURE P16.5

## Section 16.2 Standing Waves

## Section 16.3 Standing Waves on a String

6. |||| Figure P16.6 is a snapshot graph at $t = 0$ s of a pulse on a string moving to the right at 1 m/s. The string is fixed at $x = 3$ m. Draw a snapshot graph of the string at time $t = 1.25$ s.

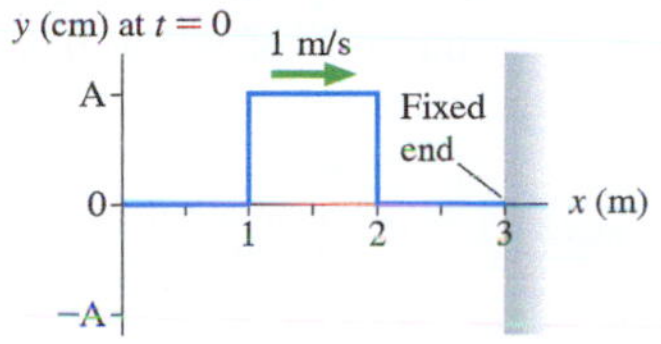


FIGURE P16.6

7. ||| At $t = 0$ s, a small "upward" (positive $y$) pulse centered at $x = 6.0$ m is moving to the right on a string with fixed ends at $x = 0.0$ m and $x = 10.0$ m. The wave speed on the string is 4.0 m/s. At what time will the string next have the same appearance that it did at $t = 0$ s?
8. ||| You are holding one end of an elastic cord that is fastened to a wall 3.0 m away. You begin shaking the end of the cord at 3.5 Hz, creating a continuous sinusoidal wave of wavelength 1.0 m. How much time will pass until a standing wave fills the entire length of the string?
9. ||| A 2.0-m-long string is fixed at both ends and tightened until the wave speed is 40 m/s. What is the frequency of the standing wave shown in Figure P16.9?

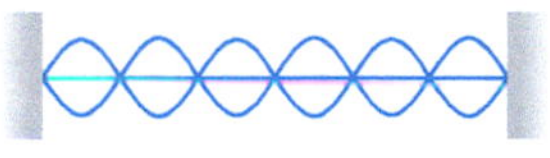

FIGURE P16.9

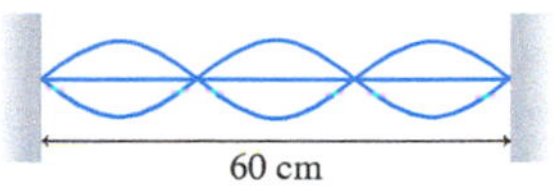


FIGURE P16.10

10. || Figure P16.10 shows a standing wave oscillating at 100 Hz on a string. What is the wave speed?
11. || A bass guitar string is 89 cm long with a fundamental frequency of 30 Hz. What is the wave speed on this string?
12. || The four strings of a bass guitar are 0.865 m long and are tuned to the notes G (98 Hz), D (73.4 Hz), A (55 Hz), and E (41.2 Hz). In one bass guitar, the G and D strings have a linear mass density of 5.8 g/m, and the A and E strings have a linear mass density of 26.8 g/m. What is the total force exerted by the strings on the neck?
13. || a. What are the three longest wavelengths for standing waves on a 240-cm-long string that is fixed at both ends?
    b. If the frequency of the second-longest wavelength is 50.0 Hz, what is the frequency of the third-longest wavelength?
14. ||| The G string on a guitar is 59 cm long and has a fundamental frequency of 196 Hz. A guitarist can play different notes by pushing the string against various frets, which changes the string's length. The first fret from the neck gives A♭ (207.65 Hz); the second fret gives A (220 Hz). How far apart are the first and second frets?
15. || A guitar string with a linear density of 2.0 g/m is stretched between supports that are 60 cm apart. The string is observed to form a standing wave with three antinodes when driven at a frequency of 420 Hz. What are (a) the frequency of the fifth harmonic of this string and (b) the tension in the string?
16. || Some guitarists like the feel of a set of strings that all have the same tension. For such a guitar, the G string (196 Hz) has a mass density of 0.31 g/m. What is the mass density of the A string (110 Hz)?
17. || The lowest note on a grand piano has a frequency of 27.5 Hz. The entire string is 2.00 m long and has a mass of 400 g. The vibrating section of the string is 1.90 m long. What tension is needed to tune this string properly?
18. | An experimenter finds that standing waves on a string fixed at both ends occur at 24 Hz and 32 Hz, but at no frequencies in between.
    a. What is the fundamental frequency?
    b. Draw the standing-wave pattern for the string at 32 Hz.

19. II Ocean waves of wavelength 26 m are moving directly toward a concrete barrier wall at 4.4 m/s. The waves reflect from the wall, and the incoming and reflected waves overlap to make a lovely standing wave with an antinode at the wall. (Such waves are a common occurrence in certain places.) A kayaker is bobbing up and down with the water at the first antinode out from the wall. How far from the wall is she? What is the period of her up-and-down motion?

### Section 16.4 Standing Sound Waves

20. I The lowest frequency in the audible range is 20 Hz. What are the lengths of (a) the shortest open-open tube and (b) the shortest open-closed tube needed to produce this frequency?

21. IIII BIO To study the physical basis of underwater hearing in frogs, scientists used a vertical tube filled with water to a depth of 1.4 m. A microphone at the bottom of the tube was used to create standing sound waves in the water column. Frogs were lowered to different depths where the standing waves created large or small pressure variations. Because the microphone creates the sound, the bottom of the tube is a pressure antinode; the water's surface, fixed at atmospheric pressure, is a node.
    a. What is the fundamental frequency of this water-filled tube?
    b. A frog sits on a platform located 0.28 m from the bottom. What is the lowest frequency that would result in a sound node at this point?

22. I The vuvuzela is a simple horn, typically 0.65 m long, that fans use to make noise at sporting events. What is the frequency of the fundamental note produced by a vuvuzela?

23. I The world's longest organ pipe, in the Boardwalk Hall Auditorium in Atlantic City, is 64 feet long. What is the fundamental frequency of this open-open pipe?

24. I An organ pipe is made to play a low note at 27.5 Hz, the same as the lowest note on a piano. Assuming a sound speed of 343 m/s, what length open-open pipe is needed? What length open-closed pipe would suffice?

25. II INT The speed of sound in room temperature (20°C) air is 343 m/s; in room temperature helium, it is 1010 m/s. The fundamental frequency of an open-closed tube is 315 Hz when the tube is filled with air. What is the fundamental frequency if the air is replaced with helium?

26. I BIO *Parasaurolophus* was a dinosaur whose distinguishing feature was a hollow crest on the head. The 1.5-m-long hollow tube in the crest had connections to the nose and throat, leading some investigators to hypothesize that the tube was a resonant chamber for vocalization. If you model the tube as an open-closed system, what are the first three resonant frequencies? Assume a speed of sound of 350 m/s.

27. II A drainage pipe running under a freeway is 30.0 m long. Both ends of the pipe are open, and wind blowing across one end causes the air inside to vibrate.
    a. If the speed of sound on a particular day is 340 m/s, what will be the fundamental frequency of air vibration in this pipe?
    b. What is the frequency of the lowest harmonic that would be audible to the human ear?
    c. What will happen to the frequency in the later afternoon as the air begins to cool?

28. I The pan flute is a musical instrument consisting of a number of closed-end tubes of different lengths. When the musician blows over the open ends, each tube plays a different note. The longest pipe is 0.33 m long. What is the frequency of the note it plays?

29. I BIO Although the vocal tract is quite complicated, we can make a simple model of it as an open-closed tube extending from the opening of the mouth to the diaphragm, the large muscle separating the abdomen and the chest cavity. What is the length of this tube if its fundamental frequency equals a typical speech frequency of 200 Hz? Assume a sound speed of 350 m/s. Does this result for the tube length seem reasonable, based on observations on your own body?

30. I You know that you sound better when you sing in the shower. This has to do with the amplification of frequencies that correspond to the standing-wave resonances of the shower enclosure. A shower enclosure is created by adding glass doors and tile walls to a standard bathtub, so the enclosure has the dimensions of a standard tub, 0.75 m wide and 1.5 m long. Standing sound waves can be set up along either axis of the enclosure. What are the lowest two frequencies that correspond to resonances on each axis of the shower? These frequencies will be especially amplified. Assume a sound speed of 343 m/s.

31. IIII BIO A child has an ear canal that is 1.3 cm long. At what sound frequencies in the audible range will the child have increased hearing sensitivity?

32. II When a sound wave travels directly toward a hard wall, the incoming and reflected waves can combine to produce a standing wave. There is an antinode right at the wall, just as at the end of a closed tube, so the sound near the wall is loud. You are standing beside a brick wall listening to a 50 Hz tone from a distant loudspeaker. How far from the wall must you move to find the first quiet spot? Assume a sound speed of 340 m/s.

### Section 16.5 Speech and Hearing

33. II BIO The first formant of your vocal system can be modeled as the resonance of an open-closed tube, the closed end being your vocal cords and the open end your lips. Estimate the frequency of the first formant from the graph of Figure 16.23, and then estimate the length of the tube of which this is a resonance. Does your result seem reasonable?

34. II When you voice the vowel sound in "hat," you narrow the opening where your throat opens into the cavity of your mouth so that your vocal tract appears as two connected tubes. The first is in your throat, closed at the vocal cords and open at the back of the mouth. The second is the mouth itself, open at the lips and closed at the back of the mouth—a different condition than for the throat because of the relatively larger size of the cavity. The corresponding formant frequencies are 800 Hz (for the throat) and 1500 Hz (for the mouth). What are the lengths of these two cavities? Assume a sound speed of 350 m/s. BIO

35. III The first and second formants when you make an "ee" vowel sound are approximately 270 Hz and 2300 Hz. The speed of sound in your vocal tract is approximately 350 m/s. If you breathe a mix of oxygen and helium (as deep-sea divers often do), the speed increases to 750 m/s. With this mix of gases, what are the frequencies of the two formants? BIO INT

36. III Figure P16.36 shows the two lowest resonances recorded in the vocal tract of the eastern towhee, a small songbird. BIO
a. Is this bird's vocal tract better modeled as an open-open tube or an open-closed tube?
b. Estimate the length of the towhee's vocal tract.

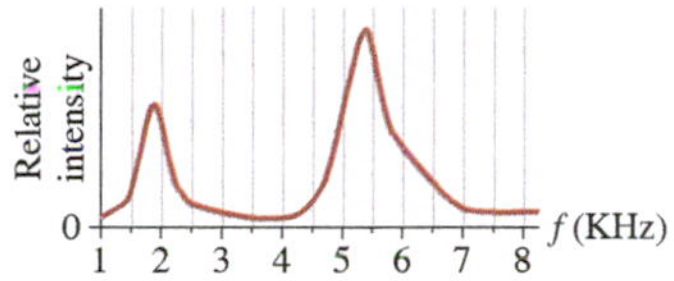

FIGURE P16.36

## Section 16.6 The Interference of Waves from Two Sources

37. III Two loudspeakers emit identical sound waves along the $x$-axis. The sound at a point on the axis has maximum intensity when the speakers are 20 cm apart. The sound intensity decreases as the distance between the speakers is increased, reaching zero at a separation of 30 cm.
a. What is the wavelength of the sound?
b. If the distance between the speakers continues to increase, at what separation will the sound intensity again be a maximum?

38. III Two loudspeakers in a 20°C room emit 686 Hz sound waves along the $x$-axis. What is the smallest distance between the speakers for which the interference of the sound waves for a point on the axis is destructive?

39. IIII Two loudspeakers, 1.0 m apart, emit sound waves with the same frequency along the positive $x$-axis. Victor, standing on the axis to the right of the speakers, hears no sound. As the frequency is slowly tripled, Victor hears the sound go through the sequence loud-soft-loud-soft-loud before becoming quiet again. What was the original sound frequency?

40. II In noisy factory environments, it's possible to use a loudspeaker to cancel persistent low-frequency machine noise at the position of one worker. The details of practical systems are complex, but we can present a simple example that gives you the idea. Suppose a machine 5.0 m away from a worker emits a persistent 80 Hz hum. To cancel the sound at the worker's location with a speaker that exactly duplicates the machine's hum, how far from the worker should the speaker be placed? Assume a sound speed of 340 m/s.

41. IIII Two identical loudspeakers separated by distance $d$ emit 170 Hz sound waves along the $x$-axis. As you walk along the axis, away from the speakers, you don't hear anything even though both speakers are on. What are three possible values for $d$? Assume a sound speed of 340 m/s.

42. I Figure P16.42 shows the circular wave fronts emitted by two sources. Make a table with rows labeled P, Q, and R and columns labeled $r_1$, $r_2$, $\Delta r$, and C/D. Fill in the table for points P, Q, and R, giving the distances as multiples of $\lambda$ and indicating, with a C or a D, whether the interference at that point is constructive or destructive.

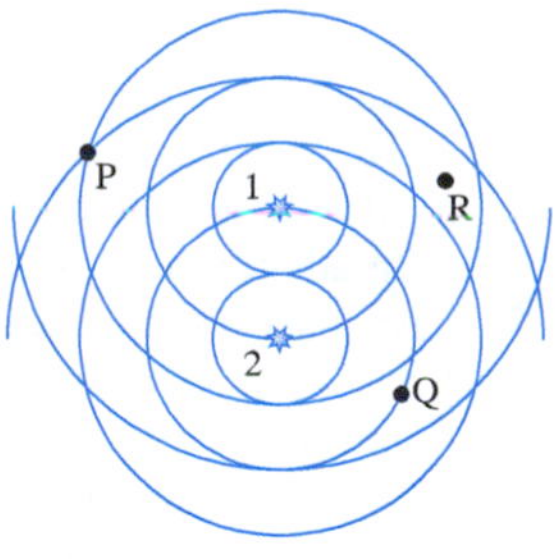

FIGURE P16.42

43. III Two identical loudspeakers 2.0 m apart are emitting 1800 Hz sound waves into a room where the speed of sound is 340 m/s. Is the point 4.0 m directly in front of one of the speakers, perpendicular to the line joining the speakers, a point of maximum constructive interference, perfect destructive interference, or something in between?

44. II Two identical loudspeakers 2.0 m apart are emitting sound waves into a room where the speed of sound is 340 m/s. Abby is standing 5.0 m in front of one of the speakers, perpendicular to the line joining the speakers, and hears a maximum in the intensity of the sound. What is the lowest possible frequency of sound for which this is possible?

## Section 16.7 Beats

45. III Musicians can use beats to tune their instruments. One flute is properly tuned and plays the musical note A at exactly 440 Hz. A second player sounds the same note and hears that her instrument is slightly "flat" (that is, at too low a frequency). Playing at the same time as the first flute, she hears two loud-soft-loud beats per second. What is the frequency of her instrument?

46. III A student waiting at a stoplight notices that her turn signal, which has a period of 0.85 s, makes one blink exactly in sync with the turn signal of the car in front of her. The blinker of the car ahead then starts to get ahead, but 17 s later the two are exactly in sync again. What is the period of the blinker of the other car?

47. I Two strings are adjusted to vibrate at exactly 200 Hz. Then the tension in one string is increased slightly. Afterward, three beats per second are heard when the strings vibrate at the same time. What is the new frequency of the string that was tightened?

48. III Figure P16.48 shows the superposition of two sound waves. What are their frequencies?

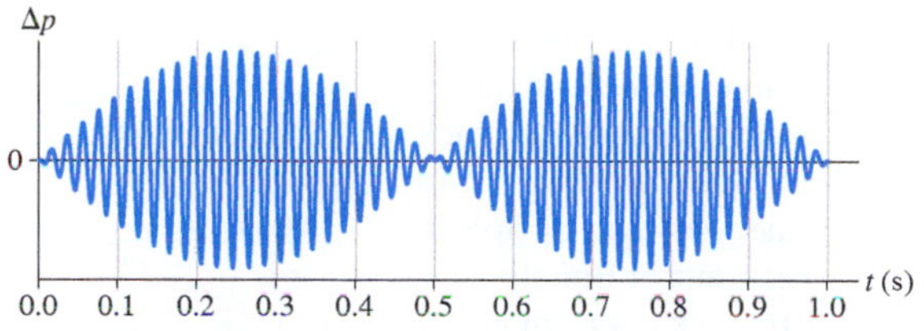

FIGURE P16.48

49. II A flute player hears four beats per second when she compares her note to a 523 Hz tuning fork (the note C). She can match the frequency of the tuning fork by pulling out the "tuning joint" to lengthen her flute slightly. What was her initial frequency?

## General Problems

50. | The fundamental frequency of a standing wave on a 1.0-m-long string is 440 Hz. What would be the wave speed of a pulse moving along this string?

51. ||| BIO INT In addition to producing images, ultrasound can be used to heat tissues of the body for therapeutic purposes. An emitter is placed against the surface of the skin; the amplitude of the ultrasound wave at this point is quite large. When a sound wave hits the boundary between soft tissue and bone, most of the energy is reflected. The boundary acts like the closed end of a tube, which can lead to standing waves. Suppose 0.70 MHz ultrasound is directed through a layer of tissue with a bone 0.55 cm below the surface. Will standing waves be created? Explain.

52. ||| INT An 80-cm-long steel string with a linear density of 1.0 g/m is under 200 N tension. It is plucked and vibrates at its fundamental frequency. What is the wavelength of the sound wave that reaches your ear in a 20°C room?

53. |||| BIO INT Tendons are, essentially, elastic cords stretched between two fixed ends; as such, they can support standing waves. These resonances can be undesirable. The Achilles tendon connects the heel with a muscle in the calf. A woman has a 20-cm-long tendon with a cross-section area of 110 $\text{mm}^2$. The density of tendon tissue is 1100 $\text{kg/m}^3$. For a reasonable tension of 500 N, what will be the resonant frequencies of her Achilles tendon?

54. |||| INT Two loudspeakers directly face each other 30 m apart, with the left speaker positioned at $x = 0$ m. The pressure of the sound wave emitted by the left speaker is described by the equation $\Delta p_L = p_0 \cos(1.90x - 630t)$, while that from the right speaker is given by $\Delta p_R = p_0 \cos(1.90x + 630t)$, where $x$ is measured in m and $t$ in s. What is the point nearest to the left speaker at which there is a node in the sound wave?

55. |||| BIO INT Spiders may "tune" strands of their webs to give enhanced response at frequencies corresponding to the frequencies at which desirable prey might struggle. Orb web silk has a typical diameter of 0.0020 mm, and spider silk has a density of 1300 $\text{kg/m}^3$. To give a resonance at 100 Hz, to what tension must a spider adjust a 12-cm-long strand of silk?

56. || INT A particularly beautiful note reaching your ear from a rare Stradivarius violin has a wavelength of 39.1 cm. The room is slightly warm, so the speed of sound is 344 m/s. If the string's linear density is 0.600 g/m and the tension is 150 N, how long is the vibrating section of the violin string?

57. || INT A 12 kg hanging sculpture is suspended by a 90-cm-long, 5.0 g steel wire. When the wind blows hard, the wire hums at its fundamental frequency. What is the frequency of the hum?

58. || Lake Erie is prone to remarkable *seiches*—standing waves that slosh water back and forth in the lake basin from the west end at Toledo to the east end at Buffalo. Figure P16.58 shows smoothed data for the displacement from normal water levels along the lake at the high point of one particular seiche. 3 hours later the water was at normal levels throughout the basin; 6 hours later the water was high in Toledo and low in Buffalo.

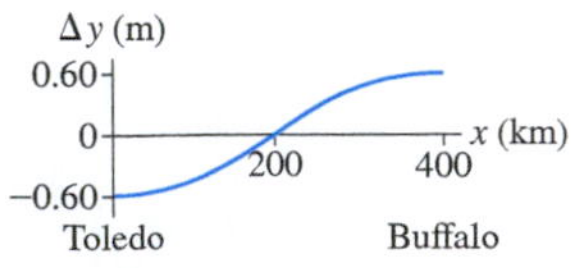

FIGURE P16.58

a. What is the wavelength of this standing wave?
b. What is the frequency?
c. What is the wave speed?

59. ||||| INT A guitar player can change the frequency of a string by "bending" it—pushing it along a fret that is perpendicular to its length. This stretches the string, increasing its tension and its frequency. The G string on a guitar is 64 cm long and has a tension of 74 N. The guitarist pushes this string down against a fret located at the center of the string, which gives it a frequency of 392 Hz. He then bends the string, pushing with a force of 4.0 N so that it moves 8.0 mm along the fret. What is the new frequency?

60. ||| INT A carbon-dioxide laser emits infrared light with a wavelength of 10.6 $\mu$m.
a. What is the length of a tube that will oscillate in the $m = 100{,}000$ mode?
b. What is the frequency?
c. Imagine a pulse of light bouncing back and forth between the ends of the tube. How many round trips will the pulse make in each second?

61. || A 40-cm-long tube has a 40-cm-long insert that can be pulled in and out, as shown in Figure P16.61. A vibrating tuning fork is held next to the tube. As the insert is slowly pulled out, the sound from the tuning fork creates standing waves in the tube when the total length $L$ is 42.5 cm, 56.7 cm, and 70.9 cm. What is the frequency of the tuning fork? The air temperature is 20°C.

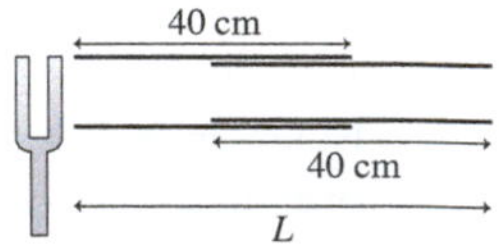

FIGURE P16.61

62. || The width of a particular microwave oven is exactly right to support a standing-wave mode. Measurements of the temperature across the oven show that there are cold spots at each edge of the oven and at three spots in between. The wavelength of the microwaves is 12 cm. How wide is the oven?

63. ||| Two loudspeakers located along the $x$-axis as shown in Figure P16.63 produce sounds of equal frequency. Speaker 1 is at the origin, while the location of speaker 2 can be varied by a remote control wielded by the listener. He notices maxima in the sound intensity when speaker 2 is located at $x = 0.75$ m and 1.00 m, but at no points in between. What is the frequency of the sound? Assume the speed of sound is 340 m/s.

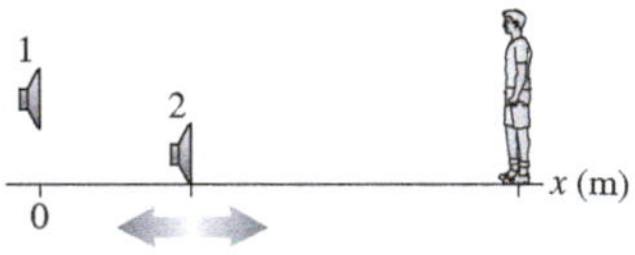

FIGURE P16.63

64. | Two loudspeakers 42.0 m apart and facing each other emit identical 115 Hz sinusoidal sound waves in a room where the sound speed is 345 m/s. Susan is walking along a line between the speakers. As she walks, she finds herself moving through loud and quiet spots. If Susan stands 19.5 m from one speaker, is she standing at a quiet spot or a loud spot?

65. ||| You are standing 2.50 m directly in front of one of the two loudspeakers shown in Figure P16.65. They are 3.00 m apart and both are playing a 686 Hz tone in phase. As you begin to walk directly away from the speaker, at what distances from the speaker do you hear a *minimum* sound intensity? The room temperature is 20°C.

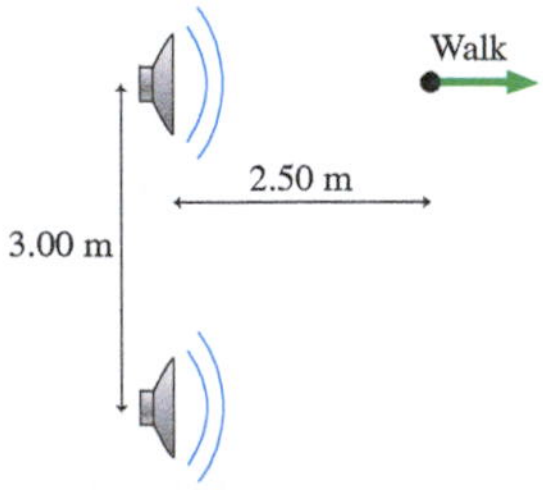

FIGURE P16.65

66. ||| Two loudspeakers, 4.0 m apart and facing each other, play identical sounds of the same frequency. You stand halfway between them, where there is a maximum of sound intensity. Moving from this point toward one of the speakers, you encounter a minimum of sound intensity when you have moved 0.25 m.
    a. What is the frequency of the sound?
    b. If the frequency is then increased while you remain 0.25 m from the center, what is the first frequency for which that location will be a maximum of sound intensity?
67. || Piano tuners tune pianos by listening to the beats between the harmonics of two different strings. When properly tuned, the note A should have the frequency 440 Hz and the note E should be at 659 Hz. The tuner can determine this by listening to the beats between the third harmonic of the A and the second harmonic of the E. A tuner first tunes the A string very precisely by matching it to a 440 Hz tuning fork. She then strikes the A and E strings simultaneously and listens for beats between the harmonics. What beat frequency indicates that the E string is properly tuned?
68. ||| A flutist assembles her flute in a room where the speed of sound is 342 m/s. When she plays the note A, it is in perfect tune with a 440 Hz tuning fork. After a few minutes, the air inside her flute has warmed to where the speed of sound is 346 m/s.
    a. How many beats per second will she hear if she now plays the note A as the tuning fork is sounded?
    b. How far does she need to extend the "tuning joint" of her flute to be in tune with the tuning fork?
69. || Police radars determine speed by measuring the shift of radio
INT waves reflected by a moving vehicle. They do so by determining the beat frequency between the reflected wave and the 10.5 GHz emitted wave. Some units can be calibrated by using a tuning fork; holding a vibrating fork in front of the unit causes the display to register a speed corresponding to the vibration frequency. A tuning fork is labeled "55 mph." What is the frequency of the tuning fork?
70. || A Doppler blood flowmeter emits ultrasound at a frequency
BIO of 5.0 MHz. What is the beat frequency between the emitted
INT waves and the waves reflected from blood cells moving away from the emitter at 0.15 m/s?
71. | An ultrasound unit is being used to measure a patient's heart-
BIO beat by combining the emitted 2.0 MHz signal with the sound
INT waves reflected from the moving tissue of one point on the heart. The beat frequency between the two signals has a maximum value of 520 Hz. What is the maximum speed of the heart tissue?

## MCAT-Style Passage Problems

### Harmonics and Harmony

You know that certain musical notes sound good together—harmonious—whereas others do not. This harmony is related to the various harmonics of the notes.

The musical notes C (262 Hz) and G (392 Hz) make a pleasant sound when played together; we call this consonance. As Figure P16.72 shows, the harmonics of the two notes are either far from each other or very close to each other (within a few Hz). This is the key to consonance: harmonics that are spaced either far apart or very close. The close harmonics have a beat frequency of a few Hz that is perceived as pleasant. If the harmonics of two notes are close but not too close, the rather high beat frequency between the two is quite unpleasant. This is what we hear as dissonance. Exactly how much a difference is maximally dissonant is a matter of opinion, but harmonic separations of 30 or 40 Hz seem to be quite unpleasant for most people.

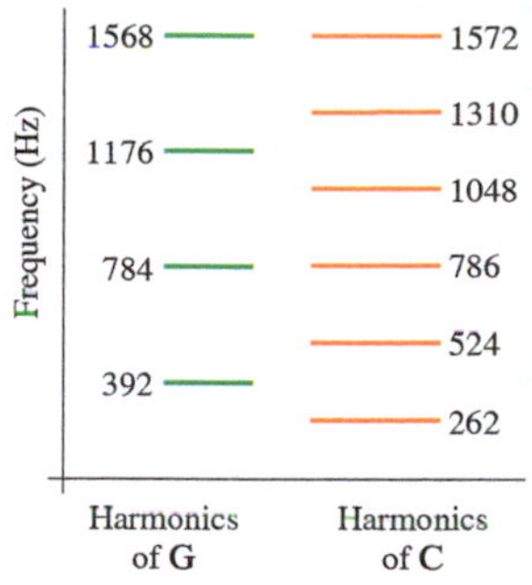


FIGURE P16.72

72. | What is the beat frequency between the second harmonic of G and the third harmonic of C?
    A. 1 Hz B. 2 Hz C. 4 Hz D. 6 Hz
73. | Would a G-flat (frequency 370 Hz) and a C played together be consonant or dissonant?
    A. Consonant B. Dissonant
74. | An organ pipe open at both ends is tuned so that its fundamental frequency is a G. How long is the pipe?
    A. 43 cm B. 87 cm C. 130 cm D. 173 cm
75. | If the C were played on an organ pipe that was open at one end and closed at the other, which of the harmonic frequencies in Figure P16.72 would be present?
    A. All of the harmonics in the figure would be present.
    B. 262, 786, and 1310 Hz
    C. 524, 1048, and 1572 Hz
    D. 262, 524, and 1048 Hz

# PART IV SUMMARY Oscillations and Waves

### KNOWLEDGE STRUCTURE IV

**BASIC GOALS**

How can we describe oscillatory motion?
How does a wave travel through a medium?
What are the distinguishing features of waves?
What happens when two waves meet?

**CROSS-CUTTING CONCEPTS**

The mathematics describing oscillations are *universal,* describing a mass on a spring, a gibbon swinging from a branch, and the speed of a walking person. Similarly, the same theory of waves works for stringed musical instruments, ground motion during an earthquake, and the warming energy of the sun.

**GENERAL PRINCIPLES**

Simple harmonic motion occurs when a linear restoring force acts to return a system to equilibrium. The period of simple harmonic motion depends on physical properties of the system but not the amplitude. All sinusoidal waves, whether water waves, sound waves, or light waves, have the same functional form. When two waves meet, they pass through each other, combining where they overlap by superposition.

## Simple harmonic motion

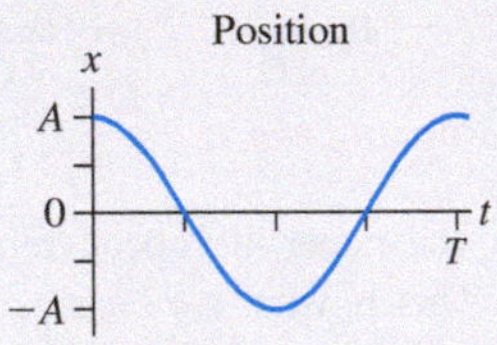


$x(t) = A\cos(2\pi ft)$

$x_{max} = A$

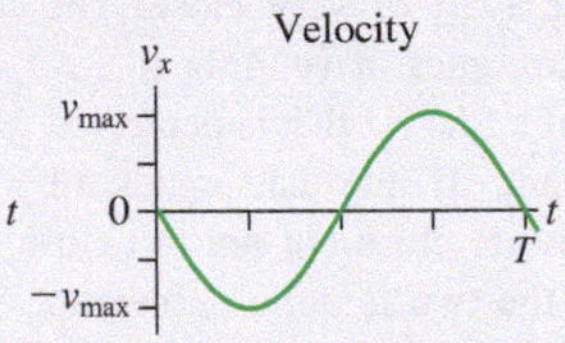


$v_x(t) = -v_{max}\sin(2\pi ft)$

$v_{max} = 2\pi fA$

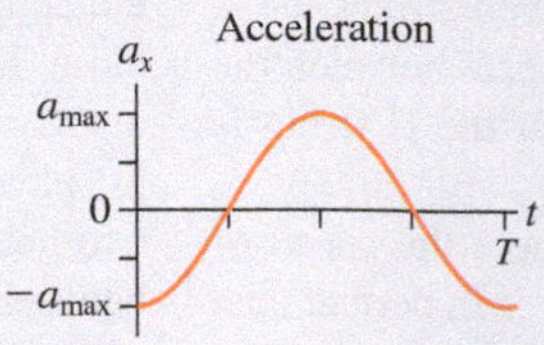


$a_x(t) = -a_{max}\sin(2\pi ft)$

$a_{max} = (2\pi f)^2 A$

The frequency of a mass on a spring depends on the spring constant and the mass:

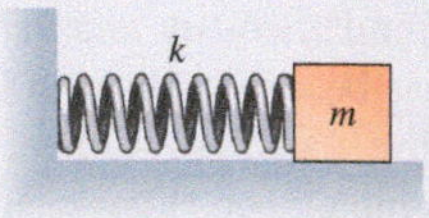


$$f = \frac{1}{2\pi}\sqrt{\frac{k}{m}}$$

The frequency of a pendulum depends on its length and the free-fall acceleration:

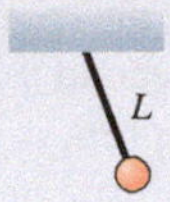


$$f = \frac{1}{2\pi}\sqrt{\frac{g}{L}}$$

## Traveling waves

A **mechanical wave** travels through a medium at a well-defined wave speed $v$.

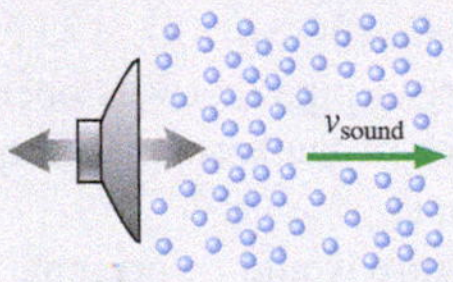

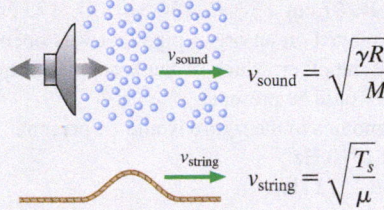

$$v_{sound} = \sqrt{\frac{\gamma RT}{M}}$$

$$v_{string} = \sqrt{\frac{T_s}{\mu}}$$

**Light waves** travel through a vacuum at speed $c = 3.00 \times 10^8$ m/s.

A **snapshot graph** shows the wave at a particular instant in time:

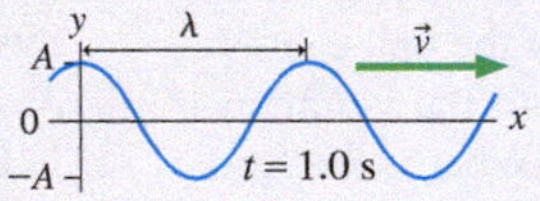


A **history graph** shows the displacement of the wave versus time at one point in the medium:

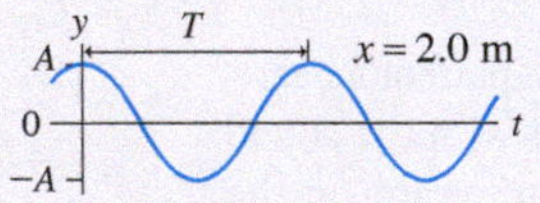


The speed $v$, wavelength $\lambda$, period $T$, and frequency $f$ of a wave are related:

$$T = \frac{1}{f} \qquad v = \lambda f$$

A sinusoidal traveling wave is described by the equation

$$y = A\cos\left(2\pi\left(\frac{x}{\lambda} \pm \frac{t}{T}\right)\right)$$

+: wave travels to the left
−: wave travels to the right

## Superposition and interference

The superposition of two waves is the sum of the displacements of the two individual waves.

**Constructive interference** occurs when crests are aligned with crests and troughs with troughs. Constructive interference occurs when the path-length difference is a whole number of wavelengths.

**Destructive interference** occurs when crests are aligned with troughs. Destructive interference occurs when the path-length difference is a whole number of wavelengths plus half a wavelength.

## Standing waves

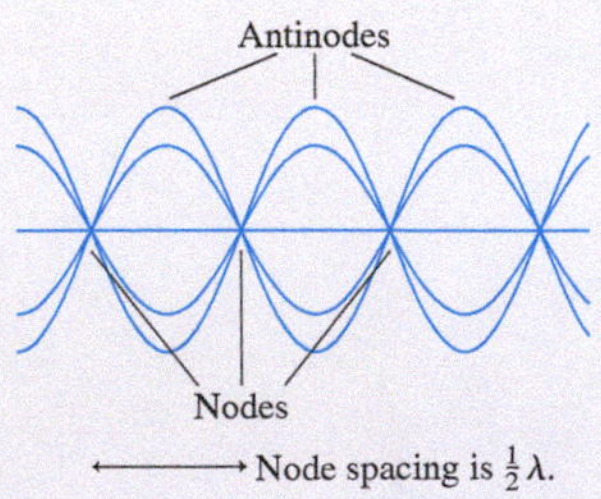


Standing waves are due to the superposition of two traveling waves moving in opposite directions.

The boundary conditions determine which standing-wave frequencies and wavelengths are allowed. The allowed standing waves are modes of the system.

For a string of length $L$, the modes have wavelength and frequency

$$\lambda_m = \frac{2L}{m} \qquad f_m = m\left(\frac{v_{string}}{2L}\right) = mf_1 \qquad m = 1, 2, 3, 4, \ldots$$

## Waves in the Earth and the Ocean

In December 2004, a large earthquake off the coast of Indonesia produced a devastating water wave, called a *tsunami,* that caused tremendous destruction thousands of miles away from the earthquake's epicenter. The tsunami was a dramatic illustration of the energy carried by waves.

It was also a call to action. Many of the communities hardest hit by the tsunami were struck hours after the waves were generated, long after seismic waves from the earthquake that passed through the earth had been detected at distant recording stations, long after the possibility of a tsunami was first discussed. With better detection and more accurate models of how a tsunami is formed and how a tsunami propagates, the affected communities could have received advance warning. The study of physics may seem an abstract undertaking with few practical applications, but on this day a better scientific understanding of these waves could have averted tragedy.

Let's use our knowledge of waves to explore the properties of a tsunami. In Chapter 15, we saw that a vigorous shake of one end of a rope causes a pulse to travel along it, carrying energy as it goes. The earthquake that produced the Indian Ocean tsunami of 2004 caused a sudden upward displacement of the seafloor that produced a corresponding rise in the surface of the ocean. This was the *disturbance* that produced the tsunami, very much like a quick shake on the end of a rope. The resulting wave propagated through the ocean, as we see in the figure.

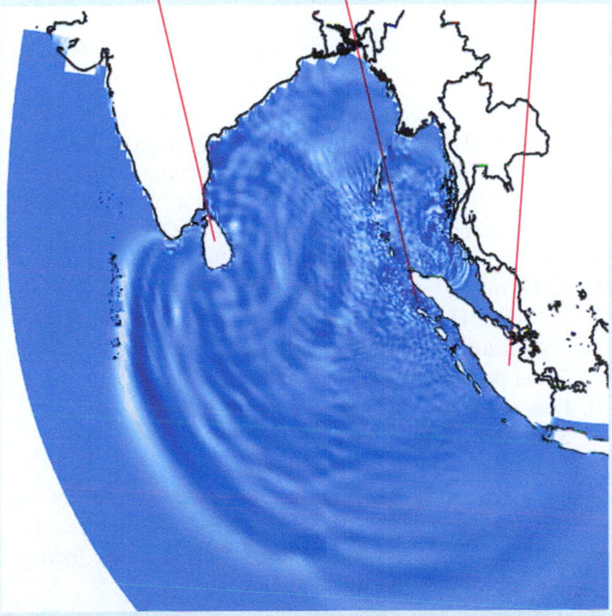


One frame from a computer simulation of the Indian Ocean tsunami three hours after the earthquake that produced it. The disturbance propagating outward from the earthquake is clearly seen, as are wave reflections from the island of Sri Lanka.

This simulation of the tsunami looks much like the ripples that spread when you drop a pebble into a pond. But there is a big difference—the scale. The fact that you can see the individual waves on this diagram that spans 5000 km is quite revealing. To show up so clearly, the individual wave pulses must be very wide—up to hundreds of kilometers from front to back.

A tsunami is actually a "shallow water wave," even in the deep ocean, because the depth of the ocean is much less than the width of the wave. Consequently, a tsunami travels differently than normal ocean waves. In Chapter 15 we learned that wave speeds are fixed by the properties of the medium. That is true for normal ocean waves, but the great width of the wave causes a tsunami to "feel the bottom." Its wave speed is determined by the depth of the ocean: The greater the depth, the greater the speed. In the deep ocean, a tsunami travels at hundreds of kilometers per hour, much faster than a typical ocean wave. Near shore, as the ocean depth decreases, so does the speed of the wave.

The height of the tsunami in the open ocean was about half a meter. Why should such a small wave—one that ships didn't even notice as it passed—be so fearsome? Again, it's the *width* of the wave that matters. Because a tsunami is the wave motion of a considerable mass of water, great energy is involved. As the front of a tsunami wave nears shore, its speed decreases, and the back of the wave moves faster than the front. Consequently, the width decreases. The water begins to pile up, and the wave dramatically increases in height.

The Indian Ocean tsunami had a height of up to 15 m when it reached shore, with a width of up to several kilometers. This tremendous mass of water was still moving at high speed, giving it a great deal of energy. A tsunami reaching the shore isn't like a typical wave that breaks and crashes. It is a kilometers-wide wall of water that moves onto the shore and just keeps on coming. In many places, the water reached 2 km inland.

The impact of the Indian Ocean tsunami was devastating, but it was the first tsunami for which scientists were able to use satellites and ocean sensors to make planet-wide measurements. An analysis of the data has helped us better understand the physics of these ocean waves. We won't be able to stop future tsunamis, but with a better knowledge of how they are formed and how they travel, we will be better able to warn people to get out of their way.

# PART IV PROBLEMS

*The following questions are related to the passage "Waves in the Earth and the Ocean" on the previous page.*

1. Rank from fastest to slowest the following waves according to their speed of propagation:
   A. An earthquake wave B. A tsunami
   C. A sound wave in air D. A light wave
2. The increase in height as a tsunami approaches shore is due to
   A. The increase in frequency as the wave approaches shore.
   B. The increase in speed as the wave approaches shore.
   C. The decrease in speed as the wave approaches shore.
   D. The constructive interference with the wave reflected from shore.
3. In the middle of the Indian Ocean, the tsunami referred to in the passage was a train of pulses approximating a sinusoidal wave with speed 200 m/s and wavelength 150 km. What was the approximate period of these pulses?
   A. 1 min B. 3 min
   C. 5 min D. 15 min
4. If a train of pulses moves into shallower water as it approaches a shore,
   A. The wavelength increases.
   B. The wavelength stays the same.
   C. The wavelength decreases.
5. The tsunami described in the passage produced a very erratic pattern of damage, with some areas seeing very large waves and nearby areas seeing only small waves. Which of the following is a possible explanation?
   A. Certain areas saw the wave from the primary source, others only the reflected waves.
   B. The superposition of waves from the primary source and reflected waves produced regions of constructive and destructive interference.
   C. A tsunami is a standing wave, and certain locations were at nodal positions, others at antinodal positions.

*The following passages and associated questions are based on the material of Part IV.*

### Deep-Water Waves

Water waves are called *deep-water waves* when the depth of the water is much greater than the wavelength of the wave. The speed of deep-water waves depends on the wavelength as follows:

$$v = \sqrt{\frac{g\lambda}{2\pi}}$$

Suppose you are on a ship at rest in the ocean, observing the crests of a passing sinusoidal wave. You estimate that the crests are 75 m apart.

6. Approximately how much time elapses between one crest reaching your ship and the next?
   A. 3 s
   B. 5 s
   C. 7 s
   D. 12 s
7. The captain starts the engines and sails directly opposite the motion of the waves at 4.5 m/s. Now how much time elapses between one crest reaching your ship and the next?
   A. 3 s
   B. 5 s
   C. 7 s
   D. 12 s
8. In the deep ocean, a longer-wavelength wave travels faster than a shorter-wavelength wave. Thus, a higher-frequency wave travels ______ a lower-frequency wave.
   A. Faster than
   B. At the same speed as
   C. Slower than

### Attenuation of Ultrasound BIO

Ultrasound is absorbed in the body; this complicates the use of ultrasound to image tissues. The intensity of a beam of ultrasound decreases by a factor of 2 after traveling a distance of 40 wavelengths. Each additional travel of 40 wavelengths results in a decrease by another factor of 2.

9. A beam of 1.0 MHz ultrasound begins with an intensity of 1000 W/m$^2$. After traveling 12 cm through tissue with no significant reflection, the intensity is about
   A. 750 W/m$^2$ B. 500 W/m$^2$
   C. 250 W/m$^2$ D. 125 W/m$^2$
10. A physician is making an image with ultrasound of initial intensity 1000 W/m$^2$. When the frequency is set to 1.0 MHz, the intensity drops to 500 W/m$^2$ at a certain depth in the patient's body. What will be the intensity at this depth if the physician changes the frequency to 2.0 MHz?
    A. 750 W/m$^2$ B. 500 W/m$^2$
    C. 250 W/m$^2$ D. 125 W/m$^2$
11. A physician is using ultrasound to make an image of a patient's heart. Increasing the frequency will provide
    A. Better penetration and better resolution.
    B. Less penetration but better resolution.
    C. More penetration but worse resolution.
    D. Less penetration and worse resolution.
12. A physician is using Doppler ultrasound to measure the motion of a patient's heart. The device measures the beat frequency between the emitted and the reflected waves. Increasing the frequency of the ultrasound will
    A. Increase the beat frequency.
    B. Not affect the beat frequency.
    C. Decrease the beat frequency.

## Measuring the Speed of Sound

A student investigator is measuring the speed of sound by looking at the time for a brief, sinusoidal pulse from a loudspeaker to travel down a tube, reflect from the closed end, and reach a microphone. The apparatus is shown in Figure IV.1a; typical data recorded by the microphone are graphed in Figure IV.1b. The first pulse is the sound directly from the loudspeaker; the second pulse is the reflection from the closed end. A portion of the returning wave reflects from the open end of the tube and makes another round trip before being detected by the microphone; this is the third pulse seen in the data.

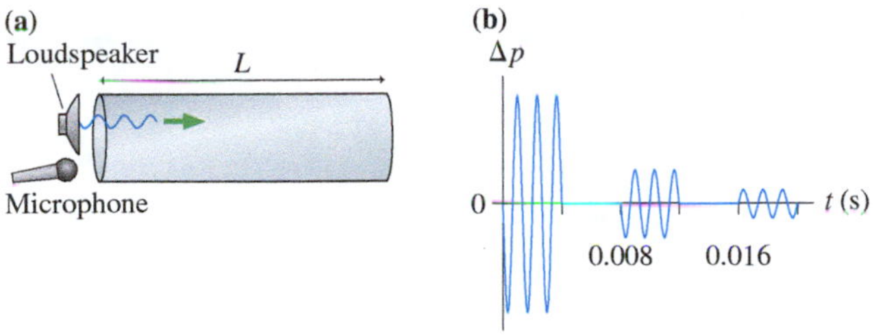


FIGURE IV.1

13. What was the approximate frequency of the sound wave used in this experiment?
    A. 250 Hz
    B. 500 Hz
    C. 750 Hz
    D. 1000 Hz
14. What can you say about the reflection of sound waves at the ends of a tube?
    A. Sound waves are inverted when reflected from both open and closed tube ends.
    B. Sound waves are inverted when reflected from a closed end, not inverted when reflected from an open end.
    C. Sound waves are inverted when reflected from an open end, not inverted when reflected from a closed end.
    D. Sound waves are not inverted when reflected from either open or closed tube ends.
15. What was the approximate length of the tube?
    A. 0.35 m
    B. 0.70 m
    C. 1.4 m
    D. 2.8 m
16. An alternative technique to determine sound speed is to measure the frequency of a standing wave in the tube. What is the wavelength of the lowest resonance of this tube?
    A. $L/2$
    B. $L$
    C. $2L$
    D. $4L$

## In the Swing

A rope swing is hung from a tree right at the edge of a small creek. The rope is 5.0 m long; the creek is 3.0 m wide.

17. You sit on the swing, and your friend gives you a gentle push so that you swing out over the creek. How long will it be until you swing back to where you started?
    A. 4.5 s B. 3.4 s
    C. 2.2 s D. 1.1 s
18. Now you switch places with your friend, who has twice your mass. You give your friend a gentle push so that he swings out over the creek. How long will it be until he swings back to where he started?
    A. 4.5 s B. 3.4 s
    C. 2.2 s D. 1.1 s
19. Your friend now pushes you over and over, so that you swing higher and higher. At some point you are swinging all the way across the creek—at the top point of your arc you are right above the opposite side. How fast are you moving when you get back to the lowest point of your arc?
    A. 6.3 m/s B. 5.4 m/s
    C. 4.4 m/s D. 3.1 m/s

## Additional Integrated Problems

20. BIO The jumping gait of the kangaroo is efficient because energy is stored in the stretch of stout tendons in the legs; the kangaroo literally bounces with each stride. We can model the bouncing of a kangaroo as the bouncing of a mass on a spring. A 70 kg kangaroo hits the ground, the tendons stretch to a maximum length, and the rebound causes the kangaroo to leave the ground approximately 0.10 s after its feet first touch.
    a. Modeling this as the motion of a mass on a spring, what is the period of the motion?
    b. Given the kangaroo mass and the period you've calculated, what is the spring constant?
    c. If the kangaroo speeds up, it must bounce higher and farther with each stride, and so must store more energy in each bounce. How does this affect the time and the amplitude of each bounce?
21. BIO A brand of earplugs reduces the sound intensity level by 27 dB. By what factor do these earplugs reduce the acoustic intensity?
22. BIO Sperm whales, just like bats, use echolocation to find prey. A sperm whale's vocal system creates a single sharp click, but the emitted sound consists of several equally spaced clicks of decreasing intensity. Researchers use the time interval between the clicks to estimate the size of the whale that created them. Explain how this might be done.

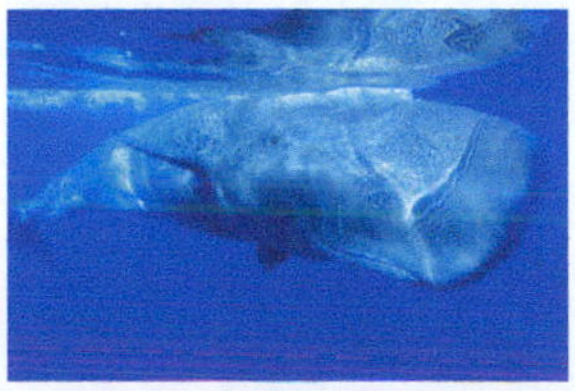

    **Hint:** The head of a sperm whale is complex, with air pockets at either end.

PART

# V Optics

Alligators, like other animals that are active during the night, have a layer behind the retina of the eye that reflects light. This reflection causes light to pass through the retina a second time, which gives the alligator increased vision sensitivity in low-light conditions. If you take a flash photo of a gator at night, this layer reflects light back toward you, as this image clearly shows. In the coming section, we'll explore the visual systems of humans and other animals and demonstrate how the right combination of transparent tissues at the back of the eye can lead to a strong reflection—the "eyeshine" that you've seen in dogs, cats, and other animals.

# OVERVIEW

## Light Is a Wave

Isaac Newton is best known for his studies of mechanics and the three laws that bear his name, but he also did important early work on optics. He was the first person to carefully study how a prism breaks white light into colors. Newton was a strong proponent of the "corpuscle" theory of light, arguing that light consists of a stream of particles.

That's one possible model of light. But, as you will see, the beautiful colors of a peacock's feathers and the shimmery rainbow of a soap bubble both require us to model light as a wave, not a particle. The wave theory we developed in Part IV will be put to good use in Part V as we begin our investigation of light and optics with an analysis of the *wave model* of light.

### The Ray Model

Newton was correct in his observation that light seems to travel in straight lines, something we wouldn't expect a wave to do. Consequently, our investigations of how light works will be greatly aided by another model of light, the *ray model,* in which light travels in straight lines, bounces from mirrors, and is bent by lenses.

The ray model will be an excellent tool for analyzing many of the practical applications of optics. When you look in a mirror, you see an *image* of yourself that appears to be behind the mirror. We will use the ray model of light to determine just how it is that mirrors and lenses form images. At the same time, we will need to reconcile the wave and ray models, learning how they are related to each other and when it is appropriate to use each.

### Working with Light

The nature of light is quite subtle and elusive. In Parts VI and VII, we will turn to the question of just what light is. As we will see, light has both wave-like *and* particle-like aspects. For now, however, we will set this question aside and work with the wave and ray models to develop a practical understanding of light. This will lead us, in Chapter 19, to an analysis of some common optical instruments. We will explore how a camera captures images and how telescopes and microscopes work.

Ultimately, the fact that you are reading this text is due to the optics of the first optical instrument you ever used, your eye! We will investigate the optics of the eye, learn how the cornea and lens bend light to create an image on your retina, and see how glasses or contact lenses can be used to correct the image should it be out of focus.

# 17 Wave Optics

The vivid colors of this humming-bird's feathers have a sheen unlike that of ordinary pigments, and they change strikingly depending on the angle at which they're viewed. How does light interact with the feathers to produce this bright display?

## LOOKING AHEAD ▸▸

### The Wave Model of Light

The varying colors reflected from this DVD can be understood using the **wave model** of light, which is the focus of this chapter.

You'll learn how the wave model explains the interaction of light with small-scale structures.

### Interference

The bright colors of peacock feathers result from the **interference** of light reflecting from multiple layers in the feathers.

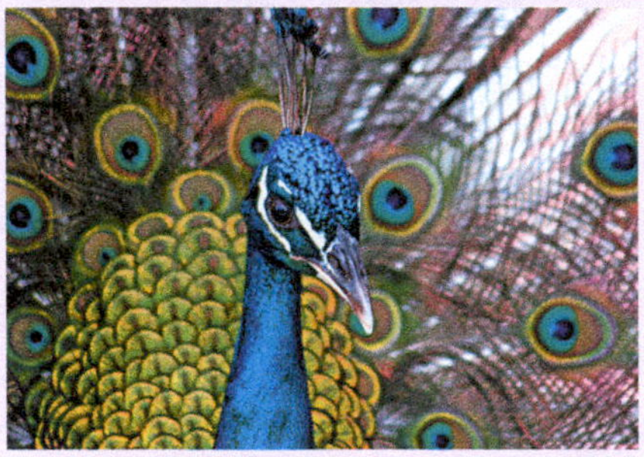

You'll learn how multiple light waves can interfere constructively or destructively.

### Diffraction

The light waves passing this sewing needle are slightly bent, or *diffracted,* causing the bands seen in the image.

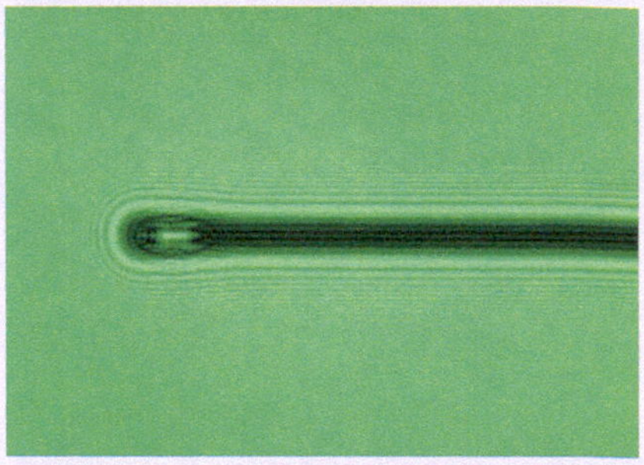

You'll learn how **diffraction** occurs for light passing through a narrow slit or a small, circular aperture.

**GOAL** To understand and apply the wave model of light.

## LOOKING BACK ◂◂

### Interference of Two Waves

In Section 16.6, you studied the interference of sound waves. We'll use these same ideas when we study the interference of light waves in this chapter.

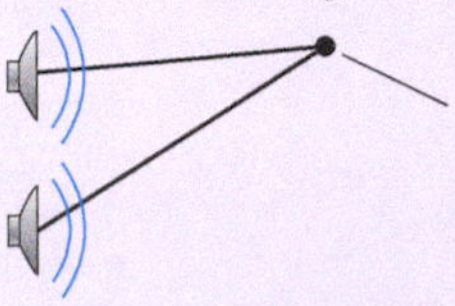

You learned that waves interfere constructively at a point if their path-length difference is an integer number of wavelengths.

**STOP TO THINK**

Sound waves spread out from two speakers; the circles represent crests of the spreading waves. The interference is

A. Constructive at both points 1 and 2.
B. Destructive at both points 1 and 2.
C. Constructive at 1, destructive at 2.
D. Destructive at 1, constructive at 2.

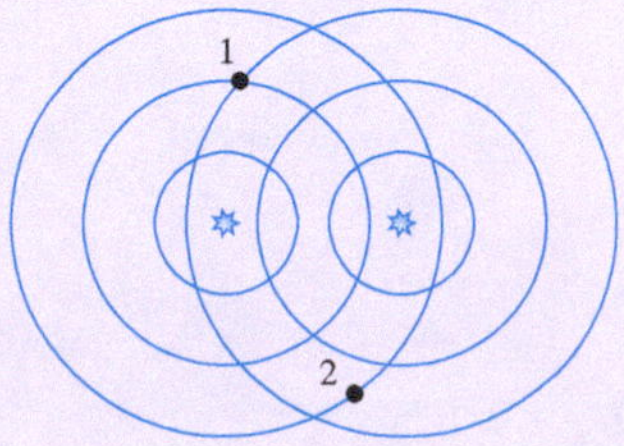

# 17.1 What Is Light?

Under some circumstances, light acts like particles traveling in straight lines. But change the circumstances, and light shows the same kinds of wave-like behavior as sound waves or water waves. Change the circumstances yet again, and light exhibits behavior that has characteristics of both waves and particles.

Rather than an all-encompassing "theory of light," it will be better to develop three **models of light,** detailed in the following table. Each model successfully explains the behavior of light within a certain range of physical situations. Each of these models will be developed in the coming chapters.

**Three models of light**

| The Wave Model | The Ray Model | The Photon Model |
|---|---|---|
| Under many circumstances, light exhibits the same behavior as sound or water waves. The bright colors of a hummingbird and the blue of the sky are best understood in terms of the wave model of light. Some aspects of the wave model of light were introduced in Chapters 15 and 16, and the wave model is the primary focus of this chapter. The study of light as a wave is called **wave optics.** | The properties of prisms, mirrors, lenses, and optical instruments such as telescopes and microscopes are best understood in terms of *light rays,* straight-line paths that light follows. The ray model of light, the basis of **ray optics,** is the subject of the next chapter. | Modern technology is increasingly reliant on quantum physics. In the quantum world, light consists of *photons* that have both wave-like and particle-like properties. Photons are the *quanta* of light. Much of the quantum theory of light is beyond the scope of this text, but we will take a peek at the important ideas in Chapters 25 and 28. |

## The Propagation of Light Waves

FIGURE 17.1 shows sunlight passing through an opening and making a sharp-edged shadow as it falls upon the floor. This behavior is what you would expect if light consisted of rays that travel in straight paths.

Is this behavior, where light seems to travel in straight lines, consistent with the motion of a wave? Let's use ordinary water waves to illustrate some basic properties of wave motion common to all waves, including light. FIGURE 17.2 shows a water wave, as seen looking straight down on the water surface. The wave enters from the top of the picture, then passes though a window-like opening in a barrier. After passing through the opening, the wave *spreads out* to fill the space behind the opening. This spreading of a wave is the phenomenon called **diffraction.** Diffraction is a sure sign that whatever is passing through the opening is a wave.

The straight-line travel of light appears to be incompatible with this spreading of a wave. Notice, however, that the width of the opening in Figure 17.2 is only slightly larger than the wavelength of the water wave. As FIGURE 17.3 shows, something quite different occurs when we make the opening much wider. Rather than spreading out, now the wave continues to move straight forward, with a well-defined boundary between where the wave is moving and its "shadow," where there is no wave. This is just the behavior observed for light in Figure 17.1.

FIGURE 17.1 Light passing through an opening makes a sharp-edged shadow.

FIGURE 17.2 A water wave passing through a narrow opening in a barrier.

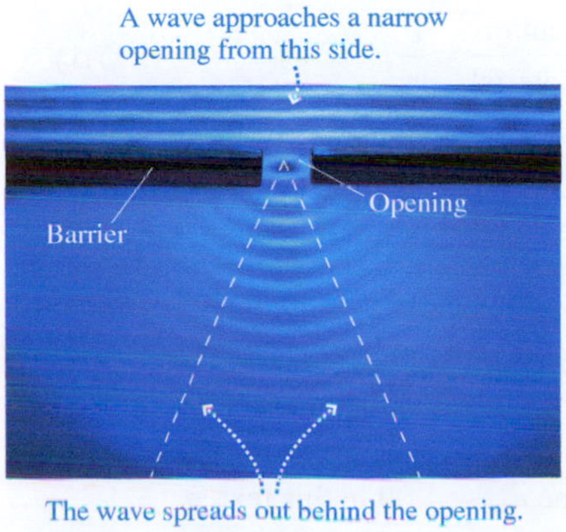

A wave approaches a narrow opening from this side.
Opening
Barrier
The wave spreads out behind the opening.

FIGURE 17.3 A water wave passing through a wide opening.

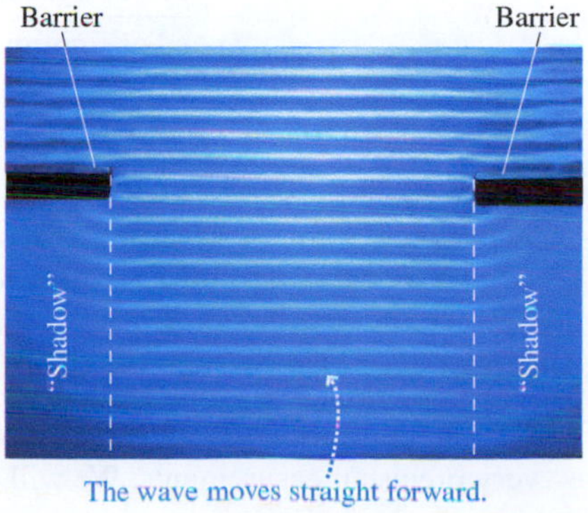

Whether a wave spreads out or travels straight ahead, with sharp shadows on both sides, evidently depends on the size of the objects with which the wave interacts. The spreading of diffraction becomes noticeable only when an opening or object is "narrow," comparable in size to the wavelength of the wave. And the wavelength of light is *extremely* short. When light waves interact with everyday-sized objects, such as the opening in Figure 17.1, the situation is like that of the water wave in Figure 17.3. The wave travels straight ahead, and we'll be able to use the ray model of light. Only when the size of an object or an opening approaches the wavelength of light does diffraction become important.

## Light Is an Electromagnetic Wave

If light is a wave, what is it that is waving? As we briefly noted in Chapter 15, light-consists of very rapidly oscillating electric and magnetic fields: It is an *electromagnetic wave.* We will examine the nature of electromagnetic waves in more detail in Part VI after we introduce the ideas of electric and magnetic fields. For now we can say that light waves are a "self-sustaining oscillation of the electromagnetic field." Being self-sustaining means that electromagnetic waves require *no material medium* in order to travel. Fortunately, we can learn about the wave properties of light without having to understand electromagnetic fields.

Recall from ◀◀ SECTION 15.4 that all electromagnetic waves, including light waves, travel in a vacuum at the same speed, called the *speed of light.* Its value is

$$v_{\text{light}} = c = 3.00 \times 10^8 \text{ m/s}$$

where the symbol $c$ is used to designate the speed of light.

Recall also that the wavelengths of light are extremely small, ranging from about 400 nm for violet light to 700 nm for red light. Electromagnetic waves with wavelengths outside this range are not visible to the human eye. A prism is able to spread the different wavelengths apart, from which we learn that "white light" is all the colors, or wavelengths, combined. The spread of colors seen with a prism, or in a rainbow, is called the *visible spectrum.*

If the wavelengths of light are incredibly small, the oscillation frequencies are unbelievably high. The frequency for a 600 nm wavelength of light is

$$f = \frac{v}{\lambda} = \frac{3.00 \times 10^8 \text{ m/s}}{600 \times 10^{-9} \text{ m}} = 5.00 \times 10^{14} \text{ Hz}$$

This is a trillion $(10^{12})$ higher than sound frequencies.

## The Index of Refraction

Light waves travel with speed $c$ in a vacuum, but they slow down as they pass through transparent materials such as water or glass or even, to a very slight extent, air. The slowdown is a consequence of interactions between the electromagnetic field of the wave and the electrons in the material. The speed of light in a material is characterized by the material's **index of refraction** $n$, defined as

$$n = \frac{\text{speed of light in a vacuum}}{\text{speed of light in the material}} = \frac{c}{v} \tag{17.1}$$

The index of refraction determined by the speed of light in a material

where $v$ is the speed of light in the material. The index of refraction of a material is always greater than 1 because $v$ is always less than $c$. A vacuum has $n = 1$ exactly. TABLE 17.1 lists the indices of refraction for several materials. Liquids and solids have higher indices of refraction than gases, simply because they have a much higher density of atoms for the light to interact with.

TABLE 17.1 Typical indices of refraction

| Material | Index of refraction |
|---|---|
| Vacuum | 1 exactly |
| Air | 1.0003 |
| Water | 1.33 |
| Glass | 1.50 |
| Diamond | 2.42 |

NOTE ▶ An accurate value for the index of refraction of air is relevant only in very precise measurements. We will assume $n_{\text{air}} = 1.00$ in this text. ◀

If the speed of a light wave changes as it enters into a transparent material, such as glass, what happens to the light's frequency and wavelength? Because $v = \lambda f$, either $\lambda$ or $f$ or both have to change when $v$ changes.

As an analogy, think of a sound wave in the air as it impinges on the surface of a pool of water. As the air oscillates back and forth, it periodically pushes on the surface of the water. These pushes generate the compressions of the sound wave that continues on into the water. Because each push of the air causes one compression of the water, the frequency of the sound wave in the water must be *exactly the same* as the frequency of the sound wave in the air. In other words, **the frequency of a wave does not change as the wave moves from one medium to another.**

FIGURE 17.4 shows a light wave passing through a transparent material with index of refraction $n$. As the wave travels through a vacuum it has wavelength $\lambda_{vac}$ and frequency $f_{vac}$ such that $\lambda_{vac} f_{vac} = c$. In the material, $\lambda_{mat} f_{mat} = v = c/n$. The frequency does not change as the wave enters ($f_{mat} = f_{vac}$), so the wavelength must change. The wavelength in the material is

$$\lambda_{mat} = \frac{v}{f_{mat}} = \frac{c}{nf_{mat}} = \frac{c}{nf_{vac}} = \frac{\lambda_{vac}}{n} \tag{17.2}$$

**The wavelength in the transparent material is shorter than the wavelength in a vacuum.** This makes sense. Suppose a marching band is marching at one step per second at a speed of 1 m/s. Suddenly they slow their speed to $\frac{1}{2}$ m/s but maintain their march at one step per second. The only way to go slower while marching at the same pace is to take *smaller steps*. When a light wave enters a material, the only way it can go slower while oscillating at the same frequency is to have a *shorter wavelength*.

FIGURE 17.4 Light passing through a transparent material with index of refraction $n$.

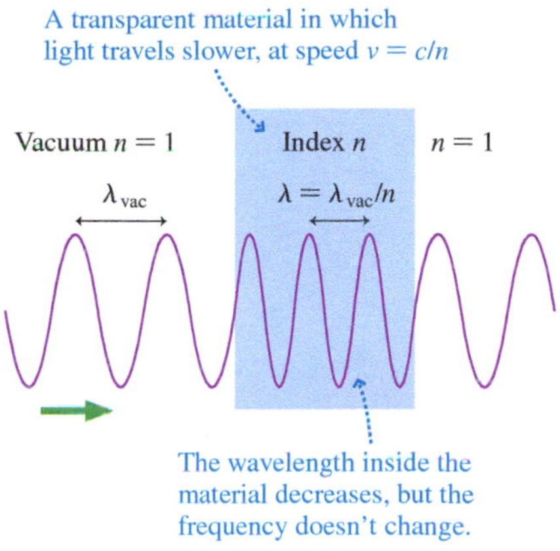


## EXAMPLE 17.1 Analyzing light traveling through glass

Orange light with a wavelength of 600 nm is incident on a 1.00-mm-thick glass microscope slide.

a. What is the light speed in the glass?
b. How many wavelengths of the light are inside the slide?

**STRATEGIZE** When the light enters the glass, its speed decreases; so does its wavelength. We will compute both of these changes using the index of refraction of glass. Once we know the wavelength, we will determine how many wavelengths fit inside the thickness of the slide.

**PREPARE** Table 17.1 gives the index of refraction of glass as $n_{glass} = 1.50$.

**SOLVE**

a. We rearrange Equation 17.1 to find the speed of light in the glass:

$$v_{glass} = \frac{c}{n_{glass}} = \frac{3.00 \times 10^8 \text{ m/s}}{1.50} = 2.00 \times 10^8 \text{ m/s}$$

b. Because $n_{air} = 1.00$, the wavelength of the light is the same in air and vacuum: $\lambda_{vac} = \lambda_{air} = 600$ nm. Thus the wavelength inside the glass is

$$\lambda_{glass} = \frac{\lambda_{vac}}{n_{glass}} = \frac{600 \text{ nm}}{1.50} = 400 \text{ nm} = 4.00 \times 10^{-7} \text{ m}$$

$N$ wavelengths span a distance $d = N\lambda$, so the number of wavelengths in $d = 1.00$ mm is

$$N = \frac{d}{\lambda} = \frac{1.00 \times 10^{-3} \text{ m}}{4.00 \times 10^{-7} \text{ m}} = 2500$$

**ASSESS** The wavelength of light is quite small, so we expect a large number of wavelengths to fit inside the slide.

**STOP TO THINK 17.1** A light wave travels through three transparent materials of equal thickness. Rank in order, from the highest to lowest, the indices of refraction $n_1$, $n_2$, and $n_3$.

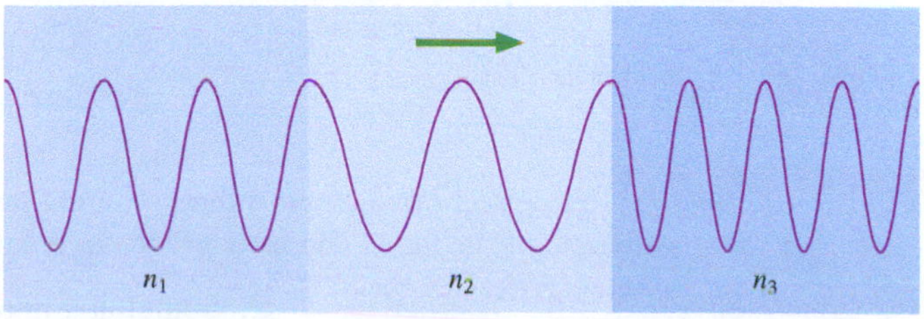

## 17.2 The Interference of Light

In ◀◀ SECTION 16.6 you learned that waves of equal wavelength emitted from two sources—such as the sound waves emitted from two loudspeakers—can overlap and *interfere,* leading to a large amplitude where the waves interfere constructively and a small or even zero amplitude where they interfere destructively. Interference is inherently a wave phenomenon, so if light acts like a wave we should be able to observe the interference of light waves.

To do so we need two sources of light waves with exactly the same wavelength. This is difficult to arrange with conventional light sources. Instead, consider the situation shown in FIGURE 17.5. Laser light passes through a narrow opening—a *slit*—that is only 0.1 mm wide, comparable to the width of a human hair. This situation is similar to Figure 17.2, where water waves spread out after passing through a narrow opening. When light waves pass through a narrow slit, they too spread out behind the slit, just as the water wave did behind the opening in the barrier. This diffraction of the light waves means that a sufficiently narrow slit acts as a source of light waves that spread out behind it.

FIGURE 17.5 Light, just like a water wave, spreads out behind a slit in a screen if the slit is sufficiently small.

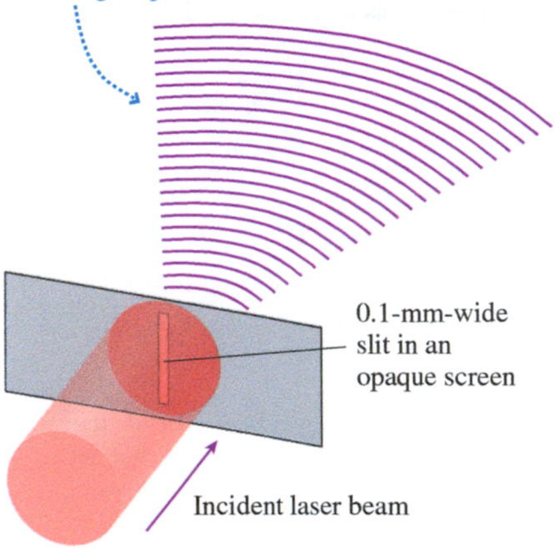


### Young's Double-Slit Experiment

To observe interference, we need *two* light sources whose waves can overlap and interfere. FIGURE 17.6a shows an experiment in which a laser beam is aimed at an opaque screen containing two long, narrow slits that are very close together. This pair of slits is called a **double slit,** and in a typical experiment they are $\approx 0.1$ mm wide and spaced $\approx 0.5$ mm apart. We will assume that the laser beam illuminates both slits equally, and any light passing through the slits strikes a viewing screen. Such a double-slit experiment was first performed by Thomas Young in 1801, using sunlight instead of a laser beam.

What should we expect to see on the viewing screen? FIGURE 17.6b is a view from above the experiment, looking down on the top ends of the slits and the top edge of the viewing screen. Because the slits are very narrow, **light spreads out behind each slit** as it did in Figure 17.5, and these two spreading waves overlap in the region between the slits and the screen.

KEY CONCEPT FIGURE 17.6 A double-slit interference experiment.

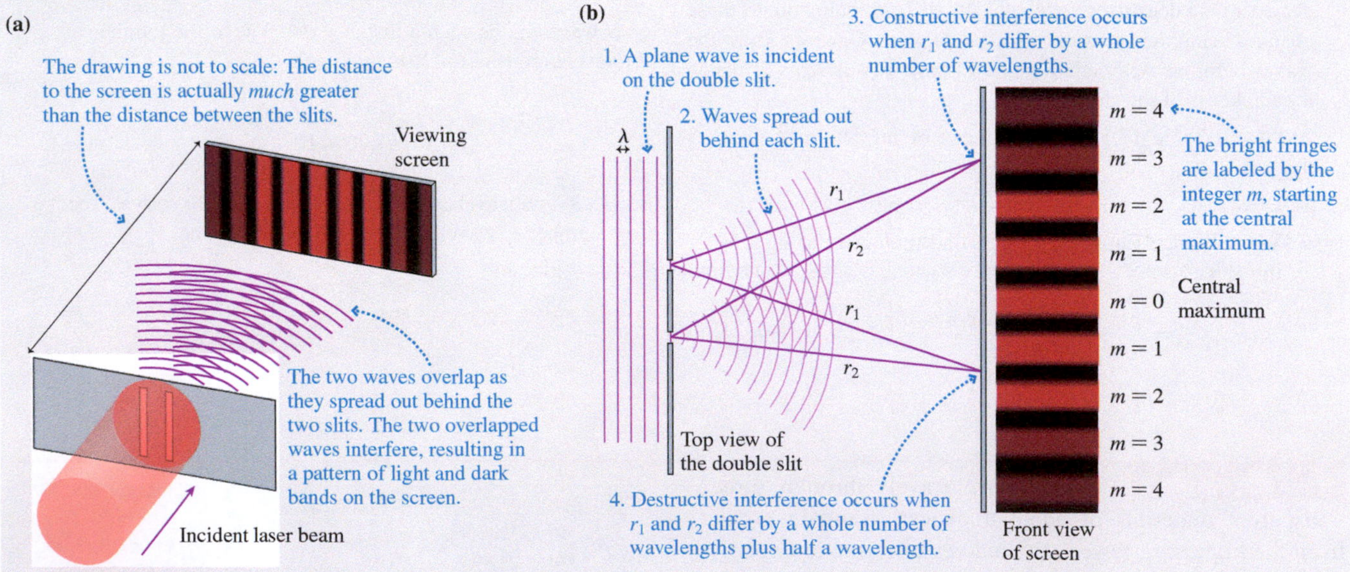


STOP TO THINK 17.2 In Figure 17.6b, suppose that for some point P on the screen $r_1 = 5{,}002{,}248.5\lambda$ and $r_2 = 5{,}002{,}251.5\lambda$, where $\lambda$ is the wavelength of the light. The interference at point P is

A. Constructive. B. Destructive. C. Something in between.

In Chapter 16 we found that *constructive* interference occurs at a given point when the distances $r_1$ and $r_2$ from speakers 1 and 2 to that position differ by a *whole number* of wavelengths. For light waves, we thus expect constructive interference to occur, and the intensity of light on the screen to be high, when the distances $r_1$ and $r_2$ from the two slits to a point on the screen differ by a whole number of wavelengths. And, just as we found for sound waves, destructive interference for light waves occurs at positions on the screen for which $r_1$ and $r_2$ differ by a whole number of wavelengths plus half a wavelength. At these positions the screen will be dark.

The photograph in Figure 17.6b shows how the screen looks. As we move along the screen, the difference $\Delta r = r_2 - r_1$ alternates between being a whole number of wavelengths and a whole number of wavelengths plus half a wavelength, leading to a series of alternating bright and dark bands of light called **interference fringes.** The bright fringes are numbered by an integer $m = 1, 2, 3, \ldots$, going outward from the center. The brightest fringe, at the midpoint of the viewing screen, has $m = 0$ and is called the **central maximum.**

## Analyzing Double-Slit Interference

Figure 17.6b showed qualitatively that interference is produced behind a double slit by the overlap of the light waves spreading out behind each opening. Now let's analyze the experiment more carefully. FIGURE 17.7a shows the geometry of a double-slit experiment in which the spacing between the two slits is $d$ and the distance to the viewing screen is $L$. **We will assume that $L$ is *very* much larger than $d$.**

Our goal is to determine if the interference at a particular point on the screen is constructive, destructive, or something in between. As we've just noted, constructive interference between two waves from identical sources occurs at points for which the path-length difference $\Delta r = r_2 - r_1$ is an integer number of wavelengths, which we can write as

$$\Delta r = m\lambda \qquad m = 0, 1, 2, 3, \ldots \qquad (17.3)$$

Thus the interference at a particular point is constructive, producing a bright fringe, if $\Delta r = m\lambda$ at that point.

We need to find the specific positions on the screen where $\Delta r = m\lambda$. Point P in Figure 17.7a is a distance $r_1$ from one slit and $r_2$ from the other. We can specify point P either by its distance $y$ from the center of the viewing screen or by the angle $\theta$ shown in Figure 17.7a; angle $\theta$ and distance $y$ are related by

$$y = L \tan \theta \qquad (17.4)$$

Because the screen is very far away compared to the spacing between the slits, the two paths to point P are virtually parallel, and the small triangle that is shaded green in the enlargement of FIGURE 17.7b is a right triangle whose angle is also $\theta$. The path-length difference between the two waves is the short side of this triangle, so

$$\Delta r = d \sin \theta \qquad (17.5)$$

Bright fringes due to constructive interference then occur at angles $\theta_m$ such that

$$\Delta r = d \sin \theta_m = m\lambda \qquad m = 0, 1, 2, 3, \ldots \qquad (17.6)$$

We have added the subscript $m$ to denote that $\theta_m$ is the angle of the $m$th bright fringe, starting with $m = 0$ at the center.

The center of the viewing screen at $y = 0$ is equally distant from both slits, so $\Delta r = 0$. This point of constructive interference, with $m = 0$, is the bright fringe identified as the central maximum in Figure 17.6b. The path-length difference increases as you move away from the center of the screen, and the $m = 1$ fringes occur at the positions where $\Delta r = 1\lambda$. That is, one wave has traveled exactly one wavelength farther than the other. In general, **the $m$th bright fringe occurs where one wave has traveled m wavelengths farther than the other and thus $\Delta r = m\lambda$.**

FIGURE 17.7 Geometry of the double-slit experiment.

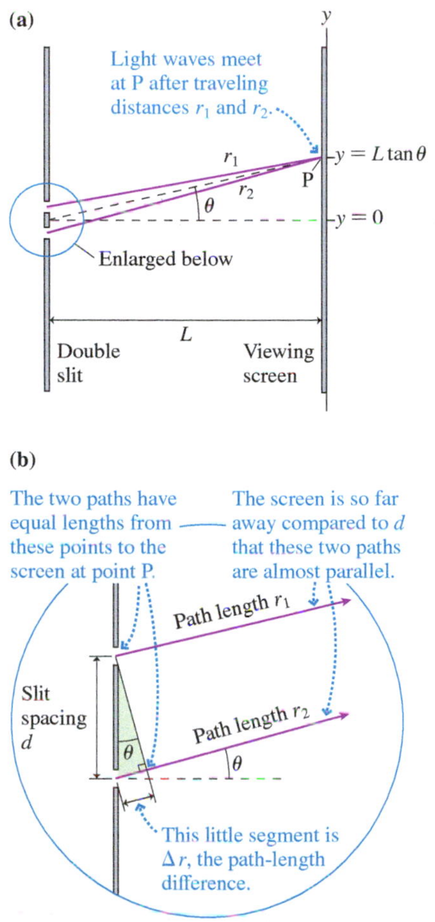

This condition is satisfied for the angles given by Equation 17.6. Our condition for the angles at which bright fringes are observed is thus

$$\sin\theta_m = m\left(\frac{\lambda}{d}\right) \qquad m = 0, 1, 2, 3, \ldots \tag{17.7}$$

Angles of bright fringes for double-slit interference with slit spacing $d$

We can also specify the distance of the bright fringes from the center of the pattern. The angles in a double-slit experiment are always very small; in practice, $\theta_m < 1°$. For such small angles, we learned in ◀◀ SECTION 14.2 the small-angle approximation, $\sin\theta \cong \theta$, where $\theta$ must be in radians. We can therefore say that $\theta_m = m(\lambda/d)$. Using this angle in Equation 17.4, we find that the $m$th bright fringe occurs at position

$$y_m = \frac{m\lambda L}{d} \qquad m = 0, 1, 2, 3, \ldots \tag{17.8}$$

Positions of bright fringes for double-slit interference at screen distance $L$

The interference pattern is symmetrical, so there is an $m$th bright fringe at the same distance on both sides of the center. You can see this in Figure 17.6b.

**EXAMPLE 17.2 How far do the waves travel?**

Light from a helium-neon laser ($\lambda = 633$ nm) illuminates two slits spaced 0.40 mm apart. A viewing screen is 2.0 m behind the slits. A bright fringe is observed at a point 9.5 mm from the center of the screen. What is the fringe number $m$, and how much farther does the wave from one slit travel to this point than the wave from the other slit?

***STRATEGIZE*** A bright fringe is observed when one wave has traveled an integer number of wavelengths farther than the other. Thus we know that $\Delta r$ must be $m\lambda$, where $m$ is an integer.

***PREPARE*** We can find $m$ from Equation 17.8.

***SOLVE*** Solving Equation 17.8 for $m$ gives

$$m = \frac{y_m d}{\lambda L} = \frac{(9.5 \times 10^{-3}\ \text{m})(0.40 \times 10^{-3}\ \text{m})}{(633 \times 10^{-9}\ \text{m})(2.0\ \text{m})} = 3$$

Then the extra distance traveled by one wave compared to the other is

$$\Delta r = m\lambda = 3(633 \times 10^{-9}\ \text{m}) = 1.9 \times 10^{-6}\ \text{m}$$

***ASSESS*** We expect that the path-length differences in two-slit interference are generally very small, just a few wavelengths of light, so the result is reasonable.

Equation 17.8 predicts that **the interference pattern is a series of equally spaced bright lines** on the screen, exactly as shown in Figure 17.6b. How do we know the fringes are equally spaced? The **fringe spacing** between fringe $m$ and fringe $m + 1$ is

$$\Delta y = y_{m+1} - y_m = \frac{(m+1)\lambda L}{d} - \frac{m\lambda L}{d}$$

which simplifies to

$$\Delta y = \frac{\lambda L}{d} \tag{17.9}$$

Spacing between any two adjacent bright fringes

Because $\Delta y$ is independent of $m$, *any* two adjacent bright fringes have the same spacing.

The dark fringes are bands of destructive interference. You learned in Chapter 16 that destructive interference occurs at positions where the path-length difference of the waves is a whole number of wavelengths plus half a wavelength:

$$\Delta r = \left(m + \frac{1}{2}\right)\lambda \qquad m = 0, 1, 2, 3, \ldots \tag{17.10}$$

We can use Equation 17.6 for $\Delta r$ and the small-angle approximation to find that the dark fringes are located at positions

$$y'_m = \left(m + \frac{1}{2}\right)\frac{\lambda L}{d} \qquad m = 0, 1, 2, 3, \ldots \tag{17.11}$$

Positions of dark fringes for double-slit interference

We have used $y'_m$, with a prime, to distinguish the location of the $m$th minimum from the $m$th maximum at $y_m$. You can see from Equation 17.11 that **the dark fringes are located exactly halfway between the bright fringes.** FIGURE 17.8 summarizes the symbols we use to describe two-slit interference.

FIGURE 17.9 is a graph of the double-slit intensity versus $y$. Notice the unusual orientation of the graph, with the intensity increasing toward the left so that the $y$-axis can match the experimental layout. You can see that the intensity oscillates between dark fringes, where the intensity is zero, and equally spaced bright fringes of maximum intensity. The maxima occur at positions where $y_m = m\lambda L/d$.

FIGURE 17.8 Symbols used to describe two-slit interference.

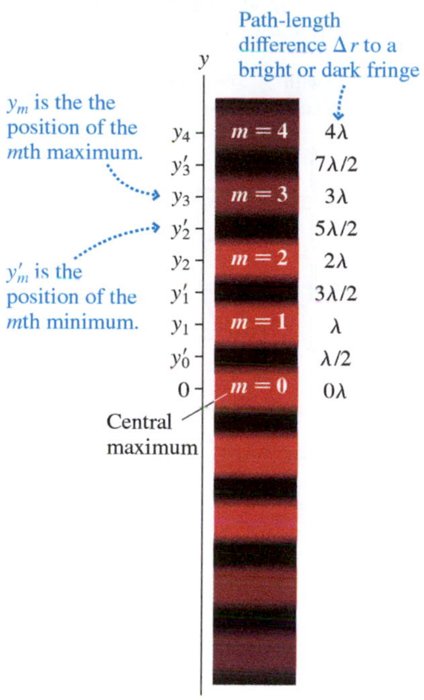


FIGURE 17.9 Intensity of the interference fringes in the double-slit experiment.

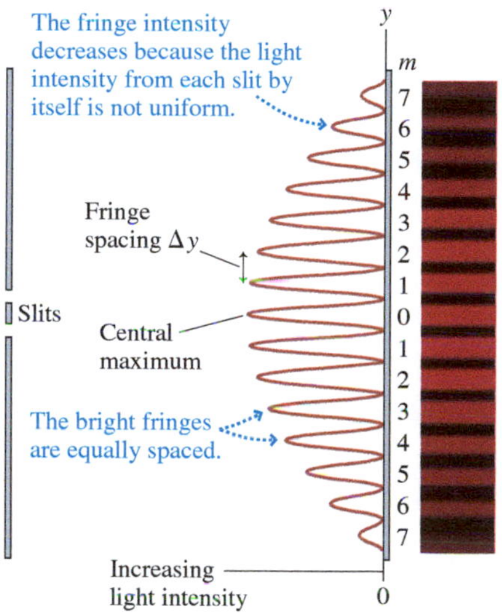


## EXAMPLE 17.3 Measuring the wavelength of light

A double-slit interference pattern is observed on a screen 1.0 m behind two slits spaced 0.30 mm apart. From the center of one particular fringe to the center of the ninth bright fringe from this one is 1.6 cm. What is the wavelength of the light?

***STRATEGIZE*** It is not always obvious which fringe is the central maximum. Slight imperfections in the slits can make the interference fringe pattern less than ideal. However, we do not need to identify the $m = 0$ fringe because we can make use of the fact, expressed in Equation 17.9, that the fringe spacing $\Delta y$ is uniform. The interference pattern looks like the photograph of Figure 17.6b.

***PREPARE*** The fringe spacing is

$$\Delta y = \frac{1.6 \text{ cm}}{9} = 1.78 \times 10^{-3} \text{ m}$$

***SOLVE*** Using this fringe spacing in Equation 17.9, we find that the wavelength is

$$\lambda = \frac{d}{L}\Delta y = \frac{3.0 \times 10^{-4} \text{ m}}{1.0 \text{ m}}(1.78 \times 10^{-3} \text{ m})$$
$$= 5.3 \times 10^{-7} \text{ m} = 530 \text{ nm}$$

It is customary to express the wavelengths of visible light in nanometers. Be sure to do this as you solve problems.

***ASSESS*** We've noted that visible light spans the wavelength range 400–700 nm, so finding a wavelength in this range is reasonable.

**STOP TO THINK 17.3** Light of wavelength $\lambda_1$ illuminates a double slit, and interference fringes are observed on a screen behind the slits. When the wavelength is changed to $\lambda_2$, the fringes get closer together. Is $\lambda_2$ larger or smaller than $\lambda_1$?

## 17.3 The Diffraction Grating

FIGURE 17.10 Top view of a diffraction grating with $N = 10$ slits.

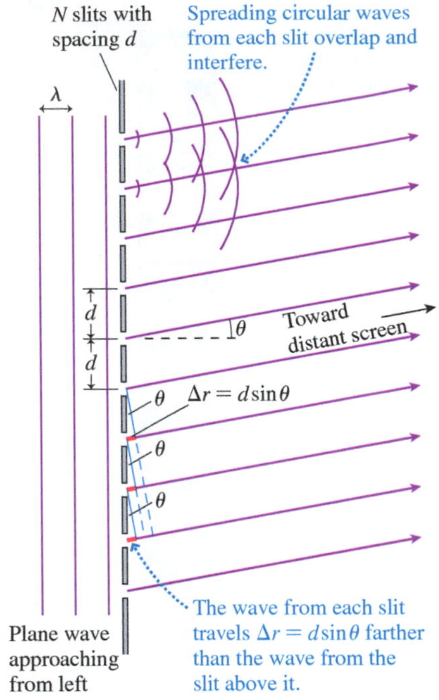


Suppose we were to replace the double slit with an opaque screen that has $N$ closely spaced slits. When illuminated from one side, each of these slits becomes the source of a light wave that diffracts, or spreads out, behind the slit. Such a multi-slit device is called a **diffraction grating.** The light intensity pattern on a screen behind a diffraction grating is due to the interference of $N$ overlapped waves.

FIGURE 17.10 shows a diffraction grating in which $N$ slits are equally spaced a distance $d$ apart. This is a top view of the grating, as we look down on the experiment, and the slits extend above and below the page. Only 10 slits are shown here, but a practical grating will have hundreds or even thousands of slits. Suppose a plane wave of wavelength $\lambda$ approaches from the left. The crest of a plane wave arrives *simultaneously* at each of the slits, causing the wave emerging from each slit to be *in phase* with the wave emerging from every other slit—that is, all the emerging waves crest and trough simultaneously. Each of these emerging waves spreads out, just like the light wave in Figure 17.5, and after a short distance they all overlap with each other and interfere.

We want to know how the interference pattern will appear on a screen behind the grating. The light wave at the screen is the superposition of $N$ waves, from $N$ slits, as they spread and overlap. As we did with the double slit, we'll assume that the distance $L$ to the screen is very large in comparison with the slit spacing $d$; hence the path followed by the light from one slit to a point on the screen is *very nearly* parallel to the path followed by the light from neighboring slits. You can see in Figure 17.10 that the wave from one slit travels distance $\Delta r = d \sin\theta$ farther than the wave from the slit above it and $\Delta r = d \sin\theta$ less than the wave below it. This is the same reasoning we used in Figure 17.7 to analyze the double-slit experiment.

FIGURE 17.11 Interference for a grating with five slits.

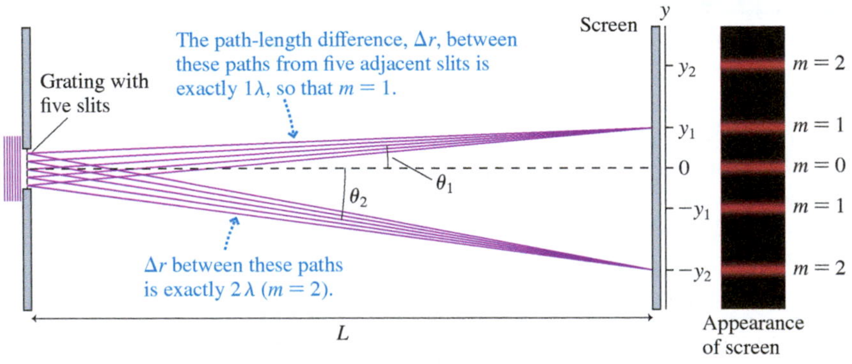


Figure 17.10 was a magnified view of the slits. FIGURE 17.11 steps back to where we can see the viewing screen, for a grating with five slits. If the angle $\theta$ is such that $\Delta r = d \sin\theta = m\lambda$, where $m$ is an integer, then the light wave arriving at the screen from one slit will travel *exactly* $m$ wavelengths more or less than light from the two slits next to it, so these waves will be *exactly in phase* with each other. But each of those waves is in phase with waves from the slits next to them, and so on until we reach the end of the grating. In other words, **$N$ light waves, from $N$ different slits, will *all* be in phase with each other when they arrive at a point on the screen at angle $\theta_m$ such that**

$$\sin\theta_m = m\left(\frac{\lambda}{d}\right) \qquad m = 0, 1, 2, 3, \ldots \tag{17.12}$$

Angles of bright fringes due to a diffraction grating with slits distance $d$ apart

This is the same equation as for double-slit interference, which makes sense: The condition for constructive interference is met for pairs of slits across the grating. In the case of a diffraction grating, though, the angles for the bright fringes are not necessarily small, so we can't use the small-angle approximation.

The pattern on the screen will have bright constructive-interference fringes at the values of $\theta_m$ given by Equation 17.12. When this happens, we say that the light is "diffracted at angle $\theta_m$." In some cases, it's easier to measure distances instead of angles. The geometry for this pattern is the same as it was for the double-slit case, so we can again use Equation 17.4 to specify the positions of the bright fringes. We can't use the small-angle approximation, so the position $y_m$ of the $m$th maximum is given as

$$y_m = L \tan \theta_m \tag{17.13}$$

Positions of bright fringes due to a diffraction grating distance $L$ from screen

The integer $m$ is called the **order** of the diffraction. Because $d$ is usually very small, it is customary to characterize a grating by the number of *lines per millimeter*. Here "line" is synonymous with "slit," so the number of lines per millimeter is simply the inverse of the slit spacing $d$ in millimeters. Equation 17.12 requires a value of $d$ in meters, so you'll need to make this conversion as well.

In Equation 17.12, larger values of the wavelength $\lambda$ imply larger diffraction angles $\theta_m$. When white light, a mix of all colors, shines through a diffraction grating, different colors diffract at different angles—the colors of light are spread into a spectrum, a rainbow.

**EXAMPLE 17.4** **Exploring diffraction from a feather** BIO

If you gently spread the barbs on a feather and look through it at a light source, you'll see a clear pattern of rainbows due to diffraction by the feather's barbules, which are small, evenly spaced structures visible in the micrograph in FIGURE 17.12. If the barbules are spaced at 50 per mm, a typical value, what is the first-order diffraction angle for blue light of 450 nm and red light of 650 nm?

FIGURE 17.12 The diffraction of the light seen through a feather is due to diffraction from the barbules.

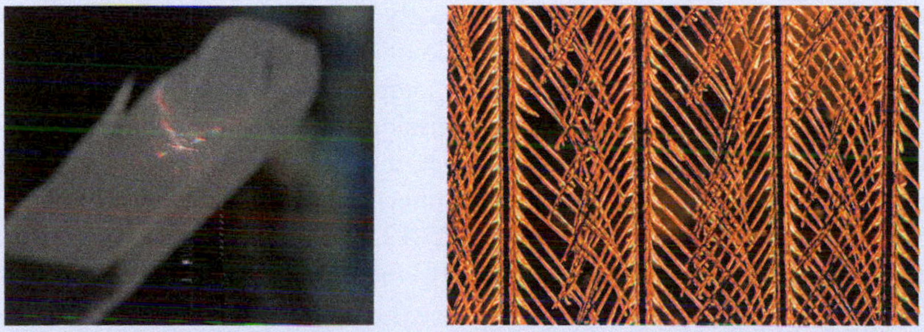

***STRATEGIZE*** The barbules are evenly spaced, with open spaces between them, so they function as a diffraction grating. We will find the diffraction angles using Equation 17.12.

***PREPARE*** We are told that there are 50 barbules per mm. In meters, the grating spacing $d$ is

$$d = \frac{1}{50}\text{ mm} = 0.020\text{ mm} = 2.0 \times 10^{-5}\text{ m}$$

*Continued*

**SOLVE** For the two colors noted, the first-order diffraction angle is

$$\sin\theta_1(\text{blue}) = \frac{\lambda}{d} = \frac{450 \times 10^{-9}\text{ m}}{2.0 \times 10^{-5}\text{ m}} = 0.0225 \qquad \sin\theta_1(\text{red}) = \frac{\lambda}{d} = \frac{650 \times 10^{-9}\text{ m}}{2.0 \times 10^{-5}\text{ m}} = 0.0325$$

$$\theta_1(\text{blue}) = 1.3° \qquad \theta_1(\text{red}) = 1.9°$$

**ASSESS** In the photo of the diffraction pattern from the feather barbules, the rainbow pattern has a small spread, so a small angle seems reasonable.

There is an important difference between the intensity pattern of double-slit interference shown in Figure 17.8 and the intensity pattern of a multiple-slit diffraction grating shown in Figure 17.11: The bright fringes of a diffraction grating are *much* narrower. In general, as the number of slits $N$ increases, the bright fringes get narrower and brighter. This trend is shown in FIGURE 17.13 for gratings with two slits (double-slit interference), 10 slits, and 50 slits. For a practical diffraction grating, which may have thousands of closely spaced slits, the interference pattern consists of a small number of *very* bright and *very* narrow fringes while most of the screen remains dark.

FIGURE 17.13 The intensity on the screen due to three diffraction gratings. Notice that the intensity axes have different scales.

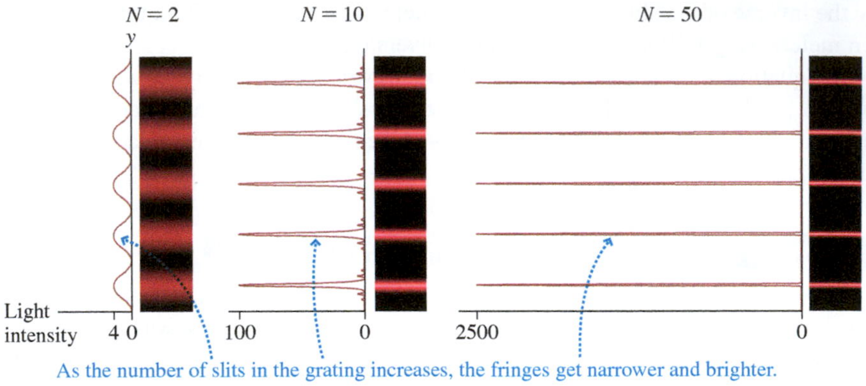


FIGURE 17.14 A diffraction grating separates two different wavelengths of light.

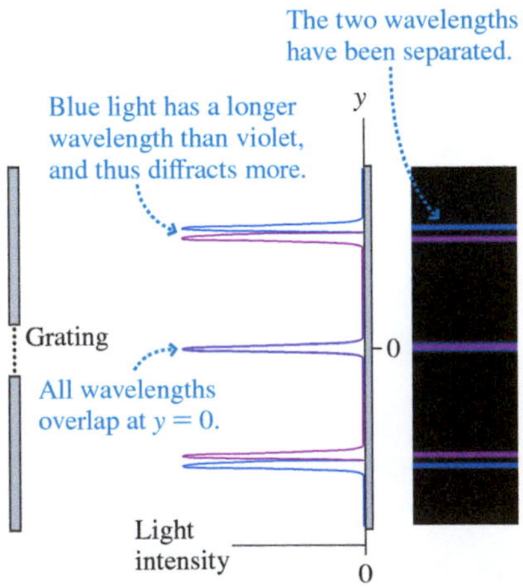


A video to support a section's topic is embedded in the eText.

Video Figure 17.14

## Spectroscopy

As we'll see in Chapter 29, each atomic element in the periodic table, if appropriately excited by light, electricity, or collisions with other atoms, emits light at only certain well-defined wavelengths. By accurately measuring these wavelengths, we can deduce the various elements in a sample of unknown composition. Molecules also emit light that is characteristic of their composition. The science of measuring the wavelengths of atomic and molecular emissions is called **spectroscopy.**

Because their bright fringes are so distinct, diffraction gratings are an ideal tool for spectroscopy. Suppose the light incident on a grating consists of two slightly different wavelengths. According to Equation 17.12, each wavelength will diffract at a slightly different angle and, if $N$ is sufficiently large, we'll see two distinct fringes on the screen. FIGURE 17.14 illustrates this idea.

**EXAMPLE 17.5 Measuring wavelengths emitted by sodium atoms**

Light from a sodium lamp passes through a diffraction grating that has 1000 slits per millimeter. The interference pattern is viewed on a screen 1.000 m behind the grating. Two bright yellow fringes are visible 72.88 cm and 73.00 cm from the central maximum. What are the wavelengths of these two fringes?

**STRATEGIZE** This situation is similar to that in Figure 17.14. Light passes through the diffraction grating, leading to two sets of fringes that correspond to the two wavelengths. The two fringes are very close together, so we expect the wavelengths to be only slightly different.

**PREPARE** No other yellow fringes are mentioned, so we assume these two fringes correspond to the first-order diffraction ($m = 1$).

**SOLVE** The distance $y_m$ of a bright fringe from the central maximum is related to the diffraction angle by $y_m = L \tan \theta_m$. Thus the diffraction angles of these two fringes are

$$\theta_1 = \tan^{-1}\left(\frac{y_1}{L}\right) = \begin{cases} 36.085° & \text{fringe at 72.88 cm} \\ 36.129° & \text{fringe at 73.00 cm} \end{cases}$$

These angles must satisfy the interference condition $\sin(\theta_1) = \lambda/d$, so the wavelengths are

$$\lambda = d \sin \theta_1$$

What is $d$? If a 1 mm length of the grating has 1000 slits, then the spacing from one slit to the next must be 1/1000 mm, or $d = 1.00 \times 10^{-6}$ m. Thus the wavelengths creating the two bright fringes are

$$\lambda = d \sin \theta_1 = \begin{cases} 589.0 \text{ nm} & \text{fringe at 72.88 cm} \\ 589.6 \text{ nm} & \text{fringe at 73.00 cm} \end{cases}$$

**ASSESS** In Chapter 15 you learned that yellow light has a wavelength of about 600 nm, so our answer is reasonable. The two wavelengths are nearly equal, as expected, which gives us further confidence in our results.

Instruments that measure and analyze spectra, called *spectrophotometers,* are widely used in chemistry, biology, and medicine. Because each molecule has a distinct spectrum—a "fingerprint"—spectroscopy is used to identify specific biomolecules in tissue, drugs in urine, and chlorophyll in seawater.

## Reflection Gratings

We have analyzed what is called a *transmission grating,* with many parallel slits. We also want to consider *reflection gratings.* The simplest reflection grating, shown in FIGURE 17.15, is a mirror with hundreds or thousands of narrow, parallel grooves cut into the surface. The grooves divide the surface into many parallel reflective stripes, each of which, when illuminated, becomes the source of a spreading wave. Thus an incident light wave is divided into $N$ overlapped waves. The interference pattern is exactly the same as the interference pattern of light transmitted through $N$ parallel slits, and so **Equation 17.13 applies to reflection gratings as well as to transmission gratings.**

Video Reflection Grating

The rainbow of colors seen on the surface of a CD or DVD is an everyday display of this phenomenon. The surface of a DVD is smooth plastic with a mirror-like reflective coating. As shown in FIGURE 17.16, billions of microscopic holes, each about 320 nm in diameter, are "burned" into the surface with a laser. The presence or absence of a hole at a particular location on the disk is interpreted as the 0 or 1 of

FIGURE 17.16 A DVD's colors are caused by diffraction.

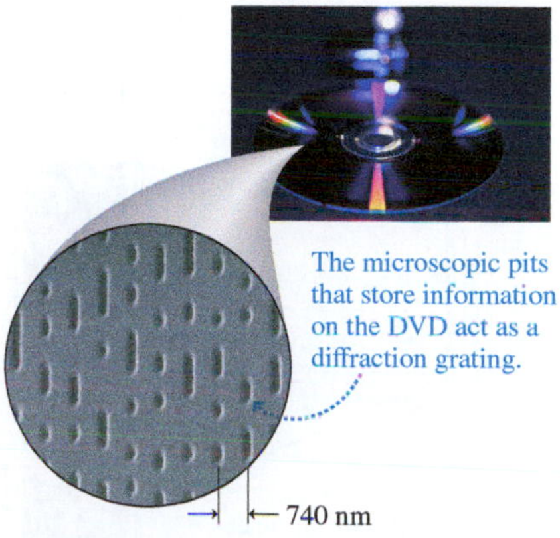

The microscopic pits that store information on the DVD act as a diffraction grating.
740 nm

FIGURE 17.15 A reflection grating.

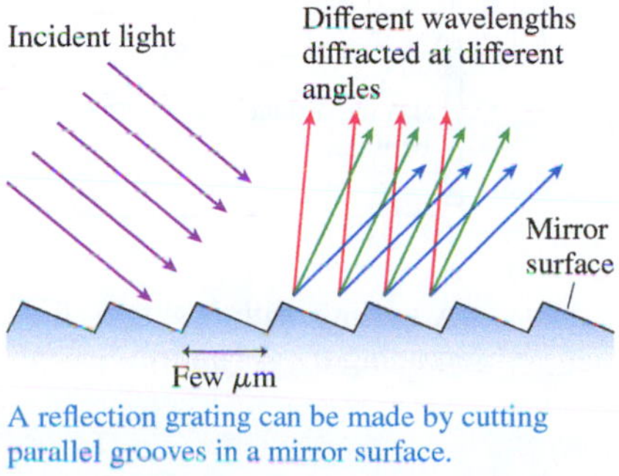

digitally encoded information. But from an optical perspective, the array of holes in a shiny surface is a two-dimensional version of the reflection grating shown in Figure 17.15.

Nature has many examples of reflection gratings as well, as the next example shows.

## EXAMPLE 17.6 Beetle colors BIO

The hard outer coating of many beetles has microscopic structures that produce amazing colors. In some ground beetles, parallel ridges in the cuticle function as a reflection grating, the source of the rainbow colors in the reflection from the beetle in FIGURE 17.17. An analysis of the grating spacing can provide valuable taxonomic information.

A researcher illuminated a specimen of ground beetle with a beam of yellow light with $\lambda = 570$ nm. The resulting diffraction produced clear maxima at 13° and 25°. What is the spacing of the ridges that produces this diffraction effect?

**FIGURE 17.17** The prominent rainbow stripe is due to diffraction from a reflection grating in the beetle's cuticle.

**STRATEGIZE** The ridges are working as a reflection grating. The spacing between the ridges is the grating spacing.

**PREPARE** We can assume that the angles correspond to the first two orders, so $\theta_1 = 13°$ and $\theta_2 = 25°$. We can do a separate calculation of the grating spacing $d$ for each of the angles.

**SOLVE** We rearrange Equation 17.12 to solve for the spacing $d$ for the two angles given:

$$m = 1: d = \frac{\lambda}{\sin\theta_1} = \frac{570 \times 10^{-9}\ \text{m}}{\sin(13°)} = 2.5\ \mu\text{m}$$

$$m = 2: d = 2\frac{\lambda}{\sin\theta_2} = 2\frac{570 \times 10^{-9}\ \text{m}}{\sin(25°)} = 2.7\ \mu\text{m}$$

The two orders of diffraction are produced by the same grating, so they should give the same spacing. The details given in the problem are real numbers from a research paper; the fact that the results do not agree exactly means that there was some uncertainty in the measurement. Our best estimate of the grating spacing is the average of the two results, 2.6 $\mu$m.

**ASSESS** The grating spacing is comparable to other gratings we've seen, and our two results are a reasonable match to each other, so we can have confidence in our results.

## SYNTHESIS 17.1 Double-slit interference and diffraction gratings

The physical principles underlying double-slit interference and diffraction gratings are the same: A bright fringe occurs on a screen when the path-length difference between waves is an integer number $m = 0, 1, 2, 3\ldots$ times the wavelength $\lambda$.

### Double-slit interference

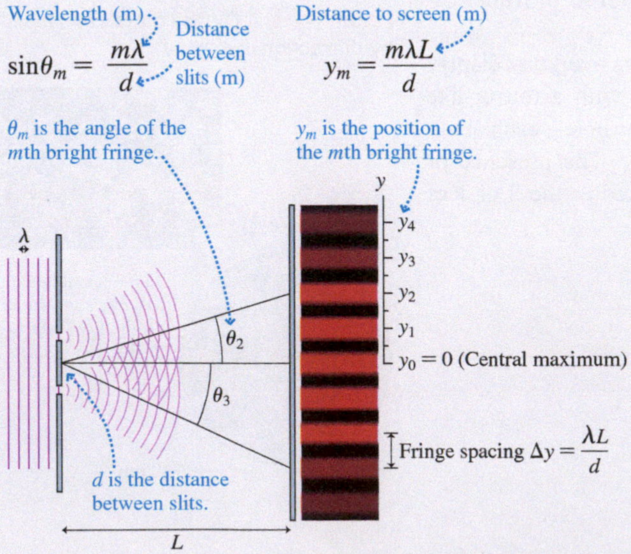


### Diffraction gratings

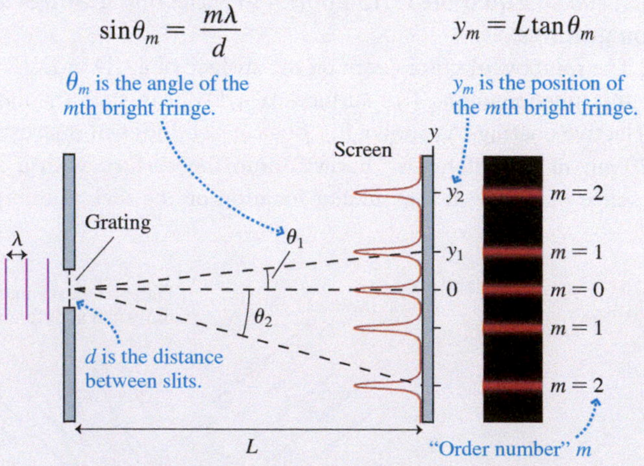


A grating with $N$ slits or lines per mm has slit spacing $d = (1\ \text{mm})/N$.

**STOP TO THINK 17.4** White light passes through a diffraction grating and forms rainbow patterns on a screen behind the grating. For each rainbow,

A. The red side is on the right, the violet side on the left.
B. The red side is on the left, the violet side on the right.
C. The red side is closest to the center of the screen, the violet side is farthest from the center.
D. The red side is farthest from the center of the screen, the violet side is closest to the center.

FIGURE 17.18 Two reflections are visible in the window, one from each surface.

## 17.4 Thin-Film Interference

In ◀◀ SECTION 16.6 you learned about the interference of sound waves in one dimension. Depending on whether they are in phase or out of phase, two sound waves of the same frequency, traveling in the same direction, can undergo constructive or destructive interference. Light waves can also interfere in this way. Equal-frequency light waves are produced when *partial reflection* at a boundary splits a light wave into a reflected wave and a transmitted wave. The interference of light waves reflected from the two boundaries of a thin film, such as the thin film of water that makes a soap bubble, is called **thin-film interference.** Thin-film coatings, less than 1 $\mu$m thick, are used for the antireflection coatings on the lenses in cameras, microscopes, and other optical equipment.

### Interference of Reflected Light Waves

As you know, and as we discussed in Chapter 16, a light wave encountering a piece of glass is partially transmitted and partially reflected. In fact, a light wave is partially reflected from *any* boundary between two transparent media with different indices of refraction. Thus light is partially reflected not only from the front surface of a sheet of glass, but from the back surface as well, as it exits from the glass into the air. This leads to the *two* reflections seen in FIGURE 17.18.

Another important aspect of wave reflections was shown for strings in Figure 16.10b of the last chapter. If a wave moves from a string with a higher wave speed to a string with a lower wave speed, the reflected wave is *inverted* with respect to the incoming wave. It is not inverted if the wave moves from a string with a lower wave speed to a string with a higher wave speed.

The same thing happens for light waves. When a light wave moves from a medium with a higher light speed (lower index of refraction) to a medium with a lower light speed (higher index of refraction), the reflected wave is inverted. This inversion of the wave, called a *phase change,* is equivalent to adding an extra half-wavelength $\lambda/2$ to the distance the wave travels. You can see this in FIGURE 17.19, where a reflected wave with a phase change is compared to a reflection without a phase change. In summary, we can say that **a light wave undergoes a phase change if it reflects from a boundary at which the index of refraction increases.** There's no phase change at a boundary where the index of refraction decreases.

FIGURE 17.19 Reflected waves with and without a phase change.

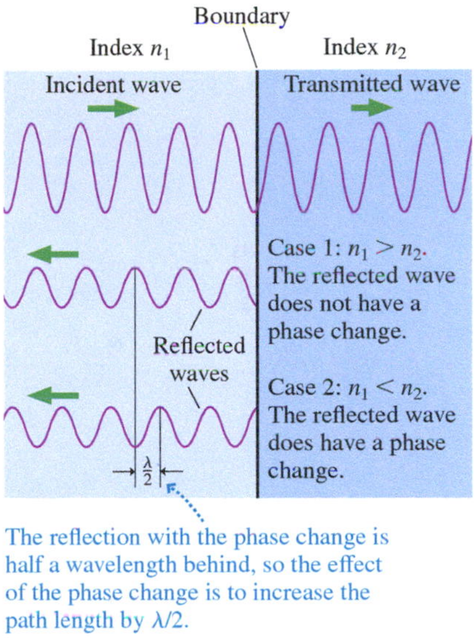


Consider a thin, transparent film with thickness $t$ and index of refraction $n$ coated onto a piece of glass. FIGURE 17.20 shows a light wave of wavelength $\lambda$ approaching the film. Most of the light is transmitted into the film, but, as we've seen, a bit is reflected off the first (air-film) boundary. Further, a bit of the wave that continues into the film is reflected off the second (film-glass) boundary. The two reflected waves, which have exactly the same frequency, travel back out into the air where they overlap and interfere. As we learned in Chapter 16, the two reflected waves will interfere constructively to cause a *strong reflection* if they are *in phase* (i.e., if their crests overlap). If the two reflected waves are *out of phase,* with the crests of one

FIGURE 17.20 In thin-film interference, two reflections, one from the film and one from the glass, overlap and interfere.

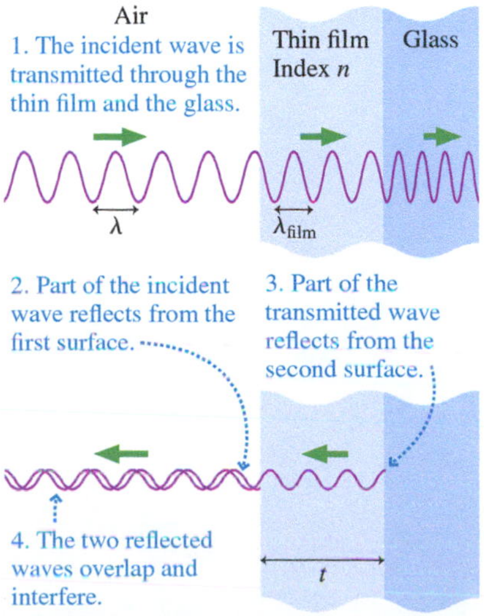

wave overlapping the troughs of the other, they will interfere destructively to cause a *weak reflection* or, if their amplitudes are equal, *no reflection* at all.

We found the interference of two sound waves to be constructive if their path-length difference is $\Delta d = m\lambda$ and destructive if $\Delta d = \left(m + \frac{1}{2}\right)\lambda$, where $m$ is an integer. The same idea holds true for reflected light waves, for which the path-length difference is the extra distance traveled by the wave that reflects from the second surface. Because this wave travels twice through a film of thickness $t$, the path-length difference is $\Delta d = 2t$.

We noted previously that the phase change when a light wave reflects from a boundary with a higher index of refraction is equivalent to adding an extra half-wavelength to the distance traveled. This leads to two situations:

1. If *neither* or *both* waves have a phase change due to reflection, the net addition to the path-length difference is zero. The *effective path-length difference* is $\Delta d_{\text{eff}} = 2t$.
2. If only *one* wave has a phase change due to reflection, the effective path-length difference is increased by one half-wavelength to $\Delta d_{\text{eff}} = 2t + \frac{1}{2}\lambda$.

The interference of the two reflected waves is then constructive if $\Delta d_{\text{eff}} = m\lambda_{\text{film}}$ and destructive if $\Delta d_{\text{eff}} = \left(m + \frac{1}{2}\right)\lambda_{\text{film}}$. Why $\lambda_{\text{film}}$? Because the extra distance is traveled inside the film, so we need to compare $2t$ to the wavelength in the film. Further, the film's index of refraction is $n$, so the wavelength in the film is $\lambda_{\text{film}} = \lambda/n$, where $\lambda$ is the wavelength of the light in vacuum or air.

With this information, we can write the conditions for constructive and destructive interference of the light waves reflected by a thin film:

$$2t = m\frac{\lambda}{n} \qquad m = 0, 1, 2, \ldots \tag{17.14}$$

Condition for constructive interference with either 0 or 2 reflective phase changes
Condition for destructive interference with only 1 reflective phase change

$$2t = \left(m + \frac{1}{2}\right)\frac{\lambda}{n} \qquad m = 0, 1, 2, \ldots \tag{17.15}$$

Condition for destructive interference with either 0 or 2 reflective phase changes
Condition for constructive interference with only 1 reflective phase change

**NOTE** ▸ Equations 17.14 and 17.15 give the film thicknesses that yield constructive or destructive interference. At other thicknesses, the waves will interfere neither fully constructively nor fully destructively, and the reflected intensity will fall somewhere between these two extremes. ◂

**NOTE** ▸ Equations 17.14 and 17.15 have an index variable $m$ that can take on a range of values, including $m = 0$. In Equation 17.14, setting $m = 0$ leads to zero thickness of the film. You can't have a film with zero thickness, but you can have a film whose thickness is nearly zero—a thickness much less than the wavelength of light. But this is a special case; when we ask you to calculate the thinnest film that will produce a given result, you needn't consider the case of nearly zero thickness unless you are advised to do so. ◂

These conditions are the basis of a procedure to analyze thin-film interference.

**TACTICS BOX 17.1 Analyzing thin-film interference**

Follow the light wave as it passes through the film. The wave reflecting from the second boundary travels an extra distance $2t$.

1. Note the indices of refraction of the three media: the medium before the film, the film itself, and the medium beyond the film. The first and third may be the same. There's a reflective phase change at any boundary where the index of refraction increases.
2. If *neither* or *both* reflected waves undergo a phase change, the phase changes cancel and the effective path-length difference is $\Delta d_{\text{eff}} = 2t$. Use Equation 17.14 for constructive interference and 17.15 for destructive interference.
3. If *only one* wave undergoes a phase change, the effective path-length difference is $\Delta d_{\text{eff}} = 2t + \frac{1}{2}\lambda$. Use Equation 17.14 for destructive interference and 17.15 for constructive interference.

Exercises 12, 13

**EXAMPLE 17.7 Designing an antireflection coating**

To keep unwanted light from reflecting from the surface of eyeglasses or other lenses, a thin film of a material with an index of refraction $n = 1.38$ is coated onto the plastic lens ($n = 1.55$). What is the thinnest film that will minimize reflection for $\lambda = 550$ nm, the middle of the visible-light spectrum?

The glasses on the top have an antireflection coating on them. Those on the bottom do not.

**STRATEGIZE** The film coats the plastic lens, so there is a reflection from the top of the film, where it meets the air, and the bottom of the film, where it meets the plastic. We want to find the thickness for which there is destructive interference between these two reflections, minimizing the overall reflection.

**PREPARE** We follow the steps of Tactics Box 17.1. As the light traverses the film, it first reflects at the front surface of the coating. Here, the index of refraction increases from that of air ($n = 1.00$) to that of the film ($n = 1.38$), so there is a reflective phase change. The light then reflects from the rear surface of the coating. The index of refraction again increases from that of the film ($n = 1.38$) to that of the plastic ($n = 1.55$). With two phase changes, Tactics Box 17.1 tells us that we should use Equation 17.15 for destructive interference.

**SOLVE** We can solve Equation 17.15 for the thickness $t$ that causes destructive interference:

$$t = \frac{\lambda}{2n}\left(m + \frac{1}{2}\right)$$

The thinnest film is the one for which $m = 0$, which gives

$$t = \frac{550 \text{ nm}}{2(1.38)} \times \frac{1}{2} = 100 \text{ nm}$$

**ASSESS** Interference effects occur when path-length differences are on the order of a wavelength, so our answer of 100 nm seems reasonable.

A thin film can *reduce* reflection, but it can also *enhance* reflection. This is important in biological systems. It's possible to make something like a mirror, with very strong reflections at certain wavelengths, with transparent materials. You have certainly seen the reflected light from the eyes of a cat or dog at night. This "eyeshine" is the reflection of light from a layer at the back of the eye called the *tapetum lucidum* (Latin for "bright carpet"). The tapetum is a common structure in the eyes of animals that must see in low light. Light that passes through the retina is reflected by the tapetum back through the cells of the retina, giving these cells a second chance to detect the light.

## EXAMPLE 17.8 What light is reflected? BIO

Sharks and related fish have a very well-developed tapetum. FIGURE 17.21 shows a camera flash reflected from a shark's eye back toward the camera. This reflected light is much brighter than the diffuse reflection from the body of the shark.

FIGURE 17.21 Eyeshine in a flash photo of a shark.

FIGURE 17.22 shows the structure responsible for the enhanced reflection. It consists of two layers of nearly transparent cells (whose index of refraction is essentially that of water) with a stack of guanine crystals sandwiched between them. Light is reflected from the interface at both sides of the stack of crystals, and the reflection is especially strong if there is constructive interference between these two reflections. If the layer of crystals is 80 nm thick, for what visible-light wavelengths is there constructive interference?

FIGURE 17.22 Structure of the tapetum in a shark.

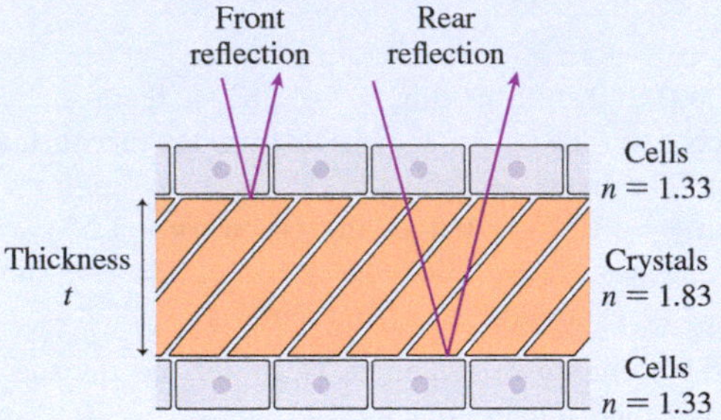


**STRATEGIZE** We are looking for constructive interference between two reflections, so we'll use the steps of Tactics Box 17.1. The film in this case is the layer of crystals. We are looking for a visible-light wavelength, so we'll look for a wavelength or wavelengths between 400 nm and 700 nm.

**PREPARE** For the front reflection in Figure 17.22, the index increases at the boundary, so there is a reflective phase change. For the rear reflection, the index decreases, so there is no reflective phase change. We are thus looking for constructive interference with 1 reflective phase change, so we can use Equation 17.15.

**SOLVE** We rearrange Equation 17.15 to solve for the wavelength, using $n = 1.83$, the index of refraction of the crystals:

$$\lambda = \frac{2tn}{(m + 1/2)} = \frac{2(80 \times 10^{-9}\text{ m})(1.83)}{(m + 1/2)}$$

The first possible solution is for $m = 0$:

$$\lambda(m = 0) = 590\text{ nm}$$

For $m = 1$ and higher values, we get wavelengths beyond the visible-light spectrum, so this is our final result. The reflection will be enhanced for wavelengths around 590 nm, which corresponds to yellow light.

**ASSESS** In Figure 17.21, the reflection looks distinctly yellow. Our calculation matches the real world, which gives us confidence in our results.

**Video** Biological Mirrors

FIGURE 17.23 Light and dark fringes caused by thin-film interference due to the air layer between two microscope slides.

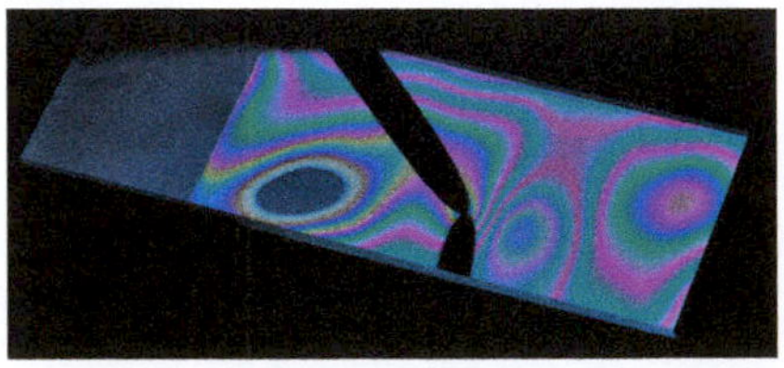

Other animals that benefit from increased vision in low-light conditions also have a tapetum. The details of the structures that lead to the reflection vary, so the colors vary as well. In dogs, the reflection tends to be bluish; in alligators, as you saw in the photo at the start of Part V, it is reddish.

To this point, we've looked at the reflection from thin films. But a thin layer of air sandwiched between two glass surfaces also exhibits thin-film interference due to the waves that reflect off both interior air-glass boundaries. FIGURE 17.23 shows two microscope slides being pressed together. The light and dark "fringes" occur because the slides are not exactly flat and they touch each other only at a few points. Everywhere else there is a thin layer of air between them. At some points, the air layer's thickness is such as to give constructive interference (light fringes), while at other places its thickness gives destructive interference (dark fringes).

Another feature of thin-film interference is also illustrated by Figure 17.23. The fringes aren't just light and dark; they are also colored. The slides are illuminated with white light, which, as we've seen, spans a range of wavelengths. For certain thicknesses of the air film, some wavelengths will experience destructive interference, and other wavelengths will experience constructive interference. This gives the reflection different colors at different locations, because the thickness of the air layer varies from place to place. The materials responsible for the reflection—glass and air—are both transparent, but these colorless materials produce a reflection that shows colors that are very bright indeed.

## Structural Color from Thin Films

You've certainly noticed the bright colors in soap bubbles. When illuminated by white light from the sun, the reflections from the film that defines the bubble, the interfaces between the soapy water and the air, result in constructive and destructive interference that suppresses some wavelengths and enhances others. The watery film is colorless, but interference can produce bright colors, with different thicknesses leading to different colors. This is an everyday example of **structural color**, color that depends not on pigment, but on an object's structure.

The colors of a soap bubble.

### CONCEPTUAL EXAMPLE 17.9 What color is the film?

FIGURE 17.24 shows a soap film in a metal ring that is held vertically. Explain why the colors shift as you move down the film, and why the film at the top reflects no light at all.

FIGURE 17.24 A soap film in a metal ring.

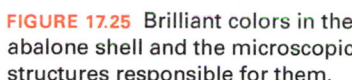

**REASON** Because of the pull of gravity, the film is thicker near the bottom and thinner at the top. Now, consider a greenish band on the film. At that thickness, the wavelengths of green light experience constructive interference, while those of red and blue light experience destructive interference. If we move to a lower position on the film, the thickness increases, and so the path-length difference increases and the longer wavelengths of yellow or red light experience constructive interference—the color changes.

The very top of the film is extremely thin, thinner than the wavelength of light. When the film is very thin, there is almost no path-length difference between the waves reflected off the front and the back of the film. However, the wave reflected off the back undergoes a reflective phase change and is out of phase with the wave reflected off the front. The two waves thus always interfere destructively, no matter what their wavelength. Thus all reflection is suppressed and the film appears dark.

**ASSESS** The absence of reflected light at the top of the film may seem surprising, but our explanation makes sense given what we know about thin films and reflection, which gives us confidence in our reasoning.

The brilliant interference colors produced by structures are generally more intense than those produced by pigments. Some of the most brightly colored animals and plants have little or no pigment. The brilliant colors of peacock feathers shown in the photo at the start of the chapter result from the interaction of light with structures that are themselves quite dull in color. The mother of pearl, or *nacre,* layer inside an abalone shell is made largely of transparent materials, but the interaction of light with complex structures produces the bright, striking colors shown in FIGURE 17.25a. A micrograph of a cross section of the thin nacre layer in FIGURE 17.25b shows stacks of approximately 500-nm-thick plates that produce strong reflections for certain colors and suppress reflections for others. Slight variations in thickness and spacing produce dramatic color variations across the surface.

FIGURE 17.25 Brilliant colors in the abalone shell and the microscopic structures responsible for them.

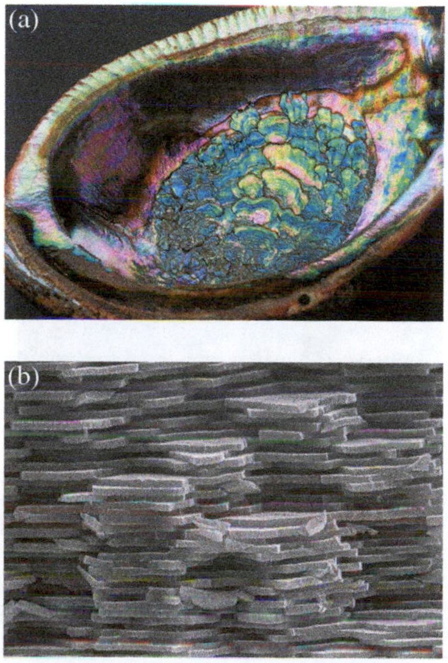


### EXAMPLE 17.10 Why so blue? BIO

The rainforests of South Asia are home to the peacock fern, whose striking blue leaves, shown in FIGURE 17.26, stand out sharply in the green of the forest floor. This plant isn't a true fern, but it does share a property with peacock feathers—the blue isn't from pigment, but is an example of structural color. This color develops only in ferns that grow in low light under a forest canopy, where light is in short supply and the light that reaches there is tipped toward the red end of the spectrum. The bluish leaves have membranes near their outer surface with thickness 70 nm and index of refraction $n = 1.45$. The layers above and below the membranes have a lower index of refraction. What is the longest wavelength of visible light for which there is constructive interference?

FIGURE 17.26 The blue leaves of the peacock fern.

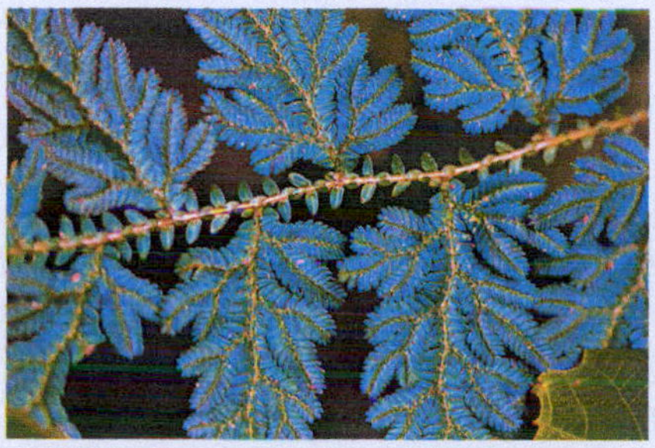

*Continued*

**STRATEGIZE** We are looking for constructive interference between the reflections from the two sides of the membrane. We will follow the steps of Tactics Box 17.1.

**PREPARE** The layers above and below the membrane have a lower index of refraction than the membrane, so there is a phase change for the first reflection but not for the second. We are looking for constructive interference with one reflective phase change, so we can use Equation 17.15.

**SOLVE** We rearrange Equation 17.15 to find the wavelength:

$$\lambda = \frac{2tn}{\left(m + \frac{1}{2}\right)}$$

We are looking for the longest wavelength of visible light. Larger values of $m$ lead to shorter wavelengths, so we start by assuming that $m = 0$. This gives

$$\lambda = 4tn = 4(70 \times 10^{-9}\ \text{m})(1.45) = 410\ \text{nm}$$

This wavelength is at the edge of the visible-light spectrum, so this is our final result. Any other value of $m$ will give wavelengths beyond the visible-light spectrum.

**ASSESS** The layer is quite thin, so we expect the wavelength to be short. And a short-wavelength reflection gives a blue color, so our result matches reality.

The membrane layer gives the leaves of the peacock fern a lovely color, but this isn't the reason for the layer. Reflection is enhanced for short-wavelength blue light; for other wavelengths, it is suppressed. For blue light, the membrane causes increased reflection, which reduces absorption. Since there isn't much blue light under the canopy, this isn't a big loss. For longer wavelengths, where most of the light energy is, the membrane suppresses reflection, which enhances absorption. The net effect is that more light energy is absorbed and is available for photosynthesis. Losing a little blue gains the leaf a lot of red, so the peacock fern shows its true colors in the shaded canopy. In sunnier locations, the intensities of red and blue light are more similar, so this adaptation isn't useful and the leaves are the usual green color.

FIGURE 17.27 Changing color with angle.

The glass tile in FIGURE 17.27 has a transparent thin-film layer sandwiched between glass layers. There is no pigment in the tile; it is made of transparent materials. The color comes from thin-film interference. Notice that different locations on the tile appear different colors. What is the source of this variation? For this photo, a diffuse white light source was above the tile. At the top edge of the tile, the reflected angle is large, so the light has traveled through the thin film at an oblique angle, leading to a large value of $\Delta d_{\text{eff}}$. This leads to destructive interference for the long wavelengths of red light. Eliminating the red light leaves blue and green, so the reflection appears blue-green at the top of the tile. At the bottom edge of the tile, the reflected angle is smaller, and so is $\Delta d_{\text{eff}}$. This leads to destructive interference for the shorter wavelengths of green light. Eliminating the green light leaves blue and red, so the reflection appears magenta.

This change in color with angle is a hallmark of structural color, as is the metallic, shiny nature of the reflection. Such colors are often called **iridescent,** which means "displaying a play of lustrous colors like those of the rainbow." The feathers of the hummingbird shown in the chapter opener result from thin-film interference in air layers in structures in the feathers. The interference conditions change sharply with angle, leading the colors to shift rapidly on these fast-moving birds. The feathers appear bright green and blue in the photo; the next second, they might appear brown, or purple, or—in some species—red. The colors have almost certainly evolved to make these birds more attractive to potential mates. The bright flashes of color attract your attention, and they apparently attract the attention of the opposite sex as well.

Iridescence is a clear indication that the source of a color is structural. But there's another tipoff: If an animal has any blue on it, it's almost certainly because of structural color. Blue pigments are very rare in nature. Some animals get their blue colors from interference or diffraction, but others get their blue coloring from *scattering,* the random redirection of incoming light. If the structures that cause the scattering are small compared to the wavelengths of light, the short wavelengths scatter much more than the long wavelengths. Air molecules in the atmosphere scatter the light from the sun; this leads to the blue of the sky because the short wavelengths of blue are scattered much more than the long wavelengths of red. Scattering from small particles leads to the blue of the sky and also the blue colors of blue eyes, the blue spruce, and the bluebird.

**Blue like the sky** The only pigment in the Eastern Bluebird's feathers is brown, but tiny air pockets and other structures in the feathers scatter incoming light. Short-wavelength light is scattered strongly; other colors are absorbed. The net result is a brilliant blue that is, truly, the same blue as the sky in which the bird flies.

**STOP TO THINK 17.5** Reflections from a thin layer of air between two glass plates cause constructive interference for a particular wavelength of light $\lambda$. By how much must the thickness of this layer be increased for the interference to be destructive?

A. $\lambda/8$ B. $\lambda/4$ C. $\lambda/2$ D. $\lambda$

## 17.5 Single-Slit Diffraction

We opened this chapter with a photograph of a water wave passing through a hole in a barrier, then spreading out on the other side. The preview to this chapter showed how light, after passing a narrow needle, also spreads out on the other side. We're now ready to look at the details of this type of diffraction.

FIGURE 17.28 again shows the experimental arrangement for observing the diffraction of light through a narrow slit of width $a$. Diffraction through a tall, narrow slit of width $a$ is known as **single-slit diffraction.** A viewing screen is placed a distance $L$ behind the slit, and we will assume that $L \gg a$. The light pattern on the viewing screen consists of a *central maximum* flanked by a series of weaker **secondary maxima** and dark fringes. Notice that the central maximum is significantly broader than the secondary maxima. It is also significantly brighter than the secondary maxima, although that is hard to tell here because this photograph has been overexposed to make the secondary maxima show up better.

FIGURE 17.28 A single-slit diffraction experiment.

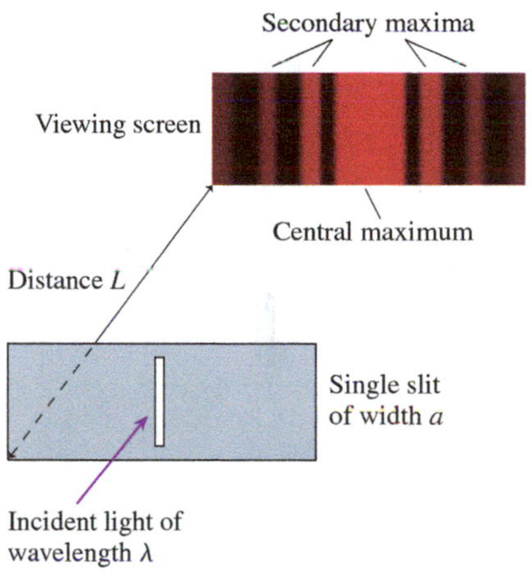


### Huygens' Principle

Our analysis of the superposition of waves from distinct sources, such as two loudspeakers or the two slits in a double-slit experiment, has tacitly assumed that the sources are *point sources,* with no measurable extent. To understand single-slit diffraction, we need to think about the propagation of an *extended* wave front. This problem was first considered by the Dutch scientist Christiaan Huygens, a contemporary of Newton.

In ◀◀ SECTION 15.5 you learned how wave *fronts*—the "crests" of a wave—evolve with time in plane and spherical waves. Huygens developed a geometrical model to visualize how *any* wave, such as a wave passing through a narrow slit, evolves. **Huygens' principle** has two parts:

1. Each point on a wave front is the source of a spherical *wavelet* that spreads out at the wave speed.
2. At a later time, the shape of the wave front is the curve that is tangent to all the wavelets.

FIGURE 17.29 on the next page illustrates Huygens' principle for a plane wave and a spherical wave. As you can see, the curve tangent to the wavelets of a plane wave is a plane that has propagated to the right. The curve tangent to the wavelets of a spherical wave is a larger sphere.

FIGURE 17.29 Huygens' principle applied to the propagation of plane waves and spherical waves.

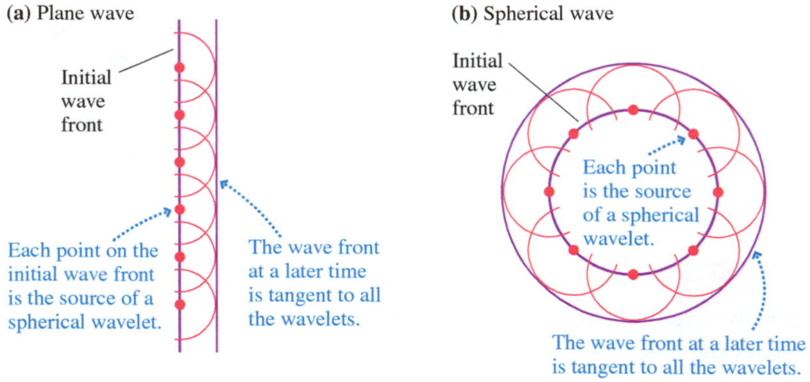


## Analyzing Single-Slit Diffraction

FIGURE 17.30a shows a wave front passing through a narrow slit of width $a$. According to Huygens' principle, each point on the wave front can be thought of as the source of a spherical wavelet. These wavelets overlap and interfere, producing the diffraction pattern seen on the viewing screen.

FIGURE 17.30 Each point on the wave front is a source of spherical wavelets. The superposition of these wavelets produces the diffraction pattern on the screen.

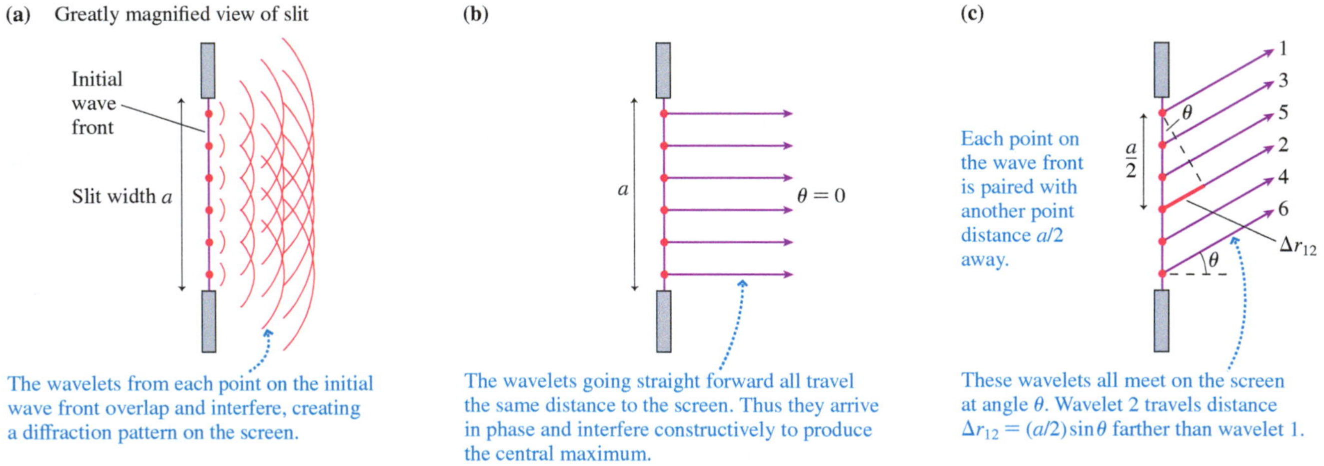

Water waves can be seen diffracting behind the "slit" between the breakwaters. The wave pattern can be understood using Huygens' principle.

FIGURE 17.30b shows the paths of several wavelets as they travel straight ahead to the central point on the screen. (The screen is *very* far to the right in this magnified view of the slit.) The paths to the screen are very nearly parallel to each other; thus all the wavelets travel the same distance and arrive at the screen *in phase* with each other. The *constructive interference* between these wavelets produces the central maximum of the diffraction pattern at $\theta = 0$.

The situation is different at points away from the center of the screen. Wavelets 1 and 2 in FIGURE 17.30c start from points that are distance $a/2$ apart. If the angle is such that $\Delta r_{12}$, the extra distance traveled by wavelet 2, happens to be $\lambda/2$, then wavelets 1 and 2 arrive out of phase and interfere destructively. But if $\Delta r_{12}$ is $\lambda/2$, then the difference $\Delta r_{34}$ between paths 3 and 4 and the difference $\Delta r_{56}$ between paths 5 and 6 are also $\lambda/2$. Those pairs of wavelets also interfere destructively. The superposition of all the wavelets produces perfect destructive interference.

Figure 17.30c happens to show six wavelets, but our conclusion is valid for any number of wavelets. The key idea is that **every point on the wave front can be paired with another point that is distance *a*/2 away.** If the path-length difference is $\lambda/2$, the wavelets that originate at these two points will arrive at the screen out of phase and interfere destructively. When we sum the displacements of all $N$ wavelets, they will—pair by pair—add to zero. The viewing screen at this position will be dark. This is the main idea of the analysis, one worth thinking about carefully.

You can see from Figure 17.30c that $\Delta r_{12} = (a/2)\sin\theta$. This path-length difference will be $\lambda/2$, the condition for destructive interference, if

$$\Delta r_{12} = \frac{a}{2}\sin\theta_1 = \frac{\lambda}{2} \qquad (17.16)$$

or, equivalently, $\sin\theta_1 = \lambda/a$.

We can extend this idea to find other angles of perfect destructive interference. Suppose each wavelet is paired with another wavelet from a point $a/4$ away. If $\Delta r$ between these wavelets is $\lambda/2$, then all $N$ wavelets will again cancel in pairs to give complete destructive interference. The angle $\theta_2$ at which this occurs is found by replacing $a/2$ in Equation 17.16 with $a/4$, leading to the condition $a\sin\theta_2 = 2\lambda$. This process can be continued, and we find that the general condition for complete destructive interference is

$$a\sin\theta_p = p\lambda \qquad p = 1, 2, 3, \ldots \qquad (17.17)$$

The condition for *dark fringes* in single-slit diffraction with slit width *a*

For any situation of single-slit diffraction you are likely to encounter, $\theta_p \ll 1$ rad (about 60°), so we can use the small-angle approximation $\sin\theta_p \cong \theta_p$ (with $\theta_p$ in radians) to simplify Equation 17.17:

$$\theta_p = p\frac{\lambda}{a} \qquad p = 1, 2, 3, \ldots \qquad (17.18)$$

Equations 17.17 and 17.18 let us determine the angles to the dark minima in the diffraction pattern of a single slit. Notice that $p = 0$ is explicitly *excluded.* $p = 0$ corresponds to the straight-ahead position at $\theta = 0$, but you saw in Figures 17.5 and 17.30b that $\theta = 0$ is the central *maximum,* not a minimum.

**NOTE** ▶ Equation 17.17 is *mathematically* the same as the condition for the $m$th *maximum* of the double-slit interference pattern. But the physical meaning here is quite different. Equation 17.17 locates the *minima* (dark fringes) of the single-slit diffraction pattern. ◀

In between the dark fringes, the intensity varies smoothly. A graph of the intensity pattern is shown in FIGURE 17.31. You can see the bright central maximum at $\theta = 0$, the weaker secondary maxima, and the dark points of destructive interference at the angles given by Equation 17.17. Compare this graph to the photograph of Figure 17.28 and make sure you see the agreement between the two.

**FIGURE 17.31** A graph of the intensity of a single-slit diffraction pattern.

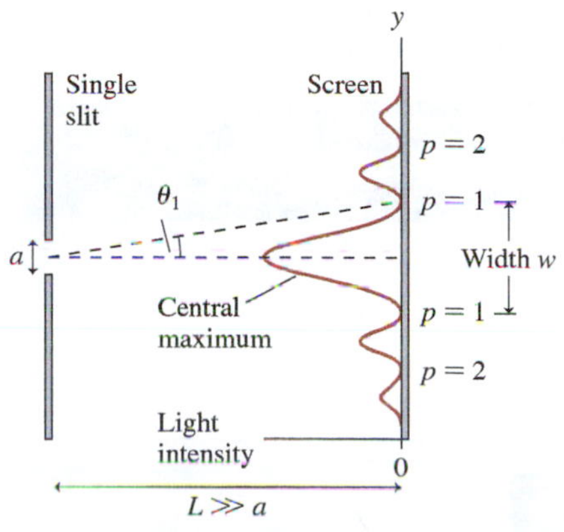


## The Width of a Single-Slit Diffraction Pattern

We'll often find it useful to measure positions on the screen. The position of the $p$th dark fringe, at angle $\theta_p$, is $y_p = L\tan\theta_p$, where $L$ is the distance from the slit to the viewing screen. Using Equation 17.18 for $\theta_p$ and the small-angle approximation

$\tan\theta_p \approx \theta_p$, we find that the dark fringes in the single-slit diffraction pattern are located at

$$y_p = \frac{p\lambda L}{a} \qquad p = 1, 2, 3, \ldots \tag{17.19}$$

Positions of dark fringes for single-slit diffraction with screen distance $L$

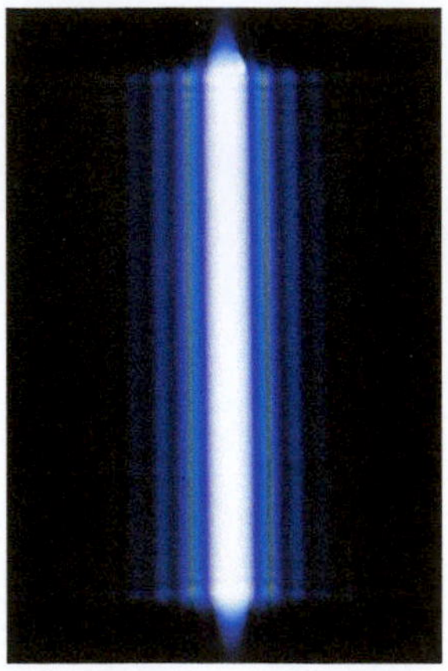

The central maximum of this single-slit diffraction pattern appears white because it is overexposed. The width of the central maximum is clear.

Again, $p = 0$ is explicitly excluded because the midpoint on the viewing screen is the central maximum, not a dark fringe.

A single-slit diffraction pattern is dominated by the central maximum, which is much brighter than the secondary maxima. The width $w$ of the central maximum, shown in Figure 17.31, is defined as the distance between the two $p = 1$ minima on either side of the central maximum. Because the pattern is symmetrical, the width is simply $w = 2y_1$. This is

$$w = \frac{2\lambda L}{a} \tag{17.20}$$

Width of the central maximum for single-slit diffraction

An important implication of Equation 17.20 is that a narrower slit (smaller $a$) causes a *wider* diffraction pattern. **The smaller the opening a wave squeezes through, the *more* it spreads out on the other side.**

### EXAMPLE 17.11 Finding the width of a slit

Light from a helium-neon laser ($\lambda = 633$ nm) passes through a narrow slit and is seen on a screen 2.0 m behind the slit. The first minimum in the diffraction pattern is 1.2 cm from the middle of the central maximum. How wide is the slit?

**STRATEGIZE** The light passes through a single slit, so the pattern is a single-slit diffraction pattern.

**PREPARE** The first minimum in a single-slit diffraction pattern corresponds to $p = 1$. The position of this minimum is given as $y_1 = 1.2$ cm. We can then use Equation 17.19 to find the slit width $a$.

**SOLVE** Equation 17.19 gives

$$a = \frac{p\lambda L}{y_p} = \frac{(1)(633 \times 10^{-9}\ \text{m})(2.0\ \text{m})}{0.012\ \text{m}}$$
$$= 1.1 \times 10^{-4}\ \text{m} = 0.11\ \text{mm}$$

**ASSESS** This value is typical of the slit widths used to observe single-slit diffraction.

FIGURE 17.32 The diffraction pattern from a laser shining on a wire.

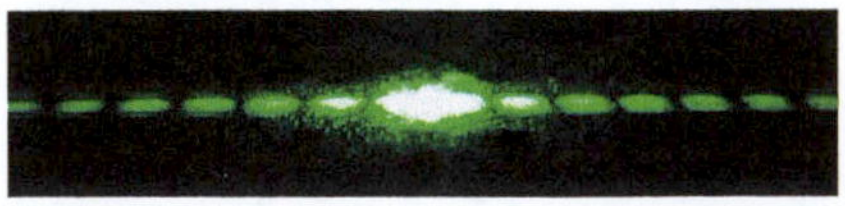

If a laser beam shines on a thin wire that blocks the beam itself, the pattern that appears on a screen behind the wire shows up as in FIGURE 17.32. You'll note that this image appears the same as the single-slit diffraction pattern. One of the surprising facts about diffraction, known as *Babinet's principle,* is that the diffraction pattern from an opaque body is the same as that from a hole or slit of the same size and shape. The diffraction pattern from a 10 $\mu$m wire is the same as that from a 10 $\mu$m slit, and all of the equations that work for slits will work just as well for wires. This can be very practical for measuring the size of small objects; it may be easier to measure distances on the diffraction pattern than to measure physical dimensions on the scale where diffraction is important. Here's something worth noting about the diffraction pattern from a wire or other opaque body: The center of the diffraction pattern is bright—it's the central maximum. This is a clear indication of the wave nature of light. For a small object and the right conditions, the center of the object's shadow is a bright spot or line!

Video What the Physics? Edge Diffraction

**EXAMPLE 17.12 Using diffraction to measure the width of a hair**

A student is doing a lab experiment in which she uses diffraction to measure the width of a shaft of her hair. She shines a 530 nm green laser pointer on a single hair, which produces a diffraction pattern on a screen 1.2 m away. The width of the central maximum of the pattern is 14 mm.

a. What is the thickness of the hair?
b. If she chose a wider shaft of hair, how would this change the width of the central maximum?

**STRATEGIZE** The diffraction pattern produced by a hair of width $a$ will be the same as that from a slit of width $a$.

**PREPARE** We are given the width of the central maximum, so we can use Equation 17.20 to analyze the situation, with $a$ as the width of the hair instead of the width of a slit.

**SOLVE**

a. We rearrange Equation 17.20 to solve for $a$, the width of the hair:

$$a = \frac{2\lambda L}{w} = \frac{2(530 \times 10^{-9}\ \text{m})(1.2\ \text{m})}{14 \times 10^{-3}\ \text{m}} = 91\ \mu\text{m}$$

b. For a single slit, we know that a wider slit gives a narrower pattern; the same is true for the hair.

**ASSESS** This is a fairly typical value for the thickness of a hair, so our result is reasonable.

**STOP TO THINK 17.6** The figure shows two single-slit diffraction patterns. The distance between the slit and the viewing screen is the same in both cases. Which of the following could be true?

A. The slits are the same for both; $\lambda_1 > \lambda_2$
B. The slits are the same for both; $\lambda_2 > \lambda_1$
C. The wavelengths are the same for both; $a_1 > a_2$
D. The wavelengths are the same for both; $a_2 > a_1$

$\lambda_1$

$\lambda_2$

## 17.6 Circular-Aperture Diffraction

Diffraction occurs if a wave passes through an opening of any shape. A common situation of practical importance is diffraction of a wave by a **circular aperture.**

Consider some examples. A loudspeaker cone generates sound by the rapid oscillation of a diaphragm, but the sound wave must pass through the circular aperture defined by the outer edge of the speaker cone before it travels into the room beyond. This is diffraction by a circular aperture. Telescopes and microscopes are the reverse. Light waves from outside need to enter the instrument. To do so, they must pass through a circular lens. In fact, the performance limit of optical instruments is determined by the diffraction of the circular openings through which the waves must pass. This is an issue we'll look at more closely in Chapter 19.

FIGURE 17.33 shows a circular aperture of diameter $D$. Light waves passing through this aperture spread out to generate a *circular* diffraction pattern. You should compare this to Figure 17.28 for a single slit to note the similarities and differences. The diffraction pattern still has a *central maximum,* now circular, and it is surrounded by a series of secondary bright fringes. Most of the intensity is contained within the central maximum.

FIGURE 17.33 The diffraction of light by a circular opening.

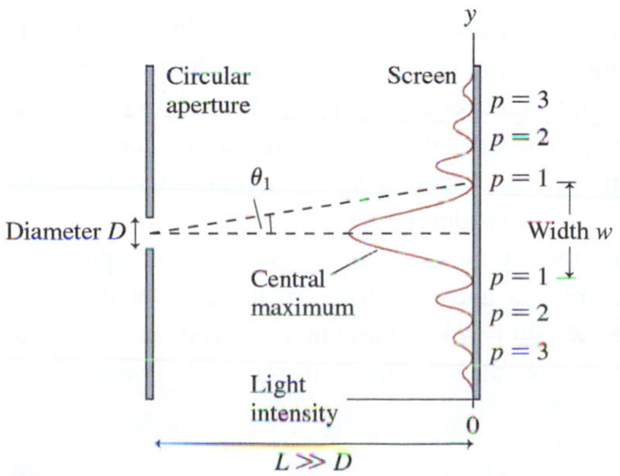


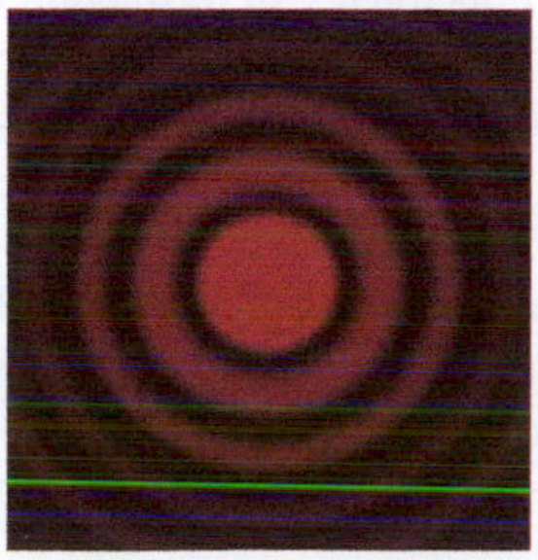

Angle $\theta_1$ locates the first minimum in the intensity, where there is perfect destructive interference. A mathematical analysis of circular diffraction finds that

$$\theta_1 = \frac{1.22\lambda}{D} \tag{17.21}$$

where $D$ is the *diameter* of the circular opening. Equation 17.21 has assumed the small-angle approximation, which is valid for all practical cases we'll consider.

The width of the central maximum on a screen a distance $L$ from the aperture is twice the distance from the center of the pattern to the first minimum; this is

$$w = 2y_1 = 2L \tan\theta_1 \tag{17.22}$$

Assuming the small-angle approximation, we can replace $\tan\theta_1$ with the value of $\theta_1$ from Equation 17.21 to find

$$w = \frac{2.44\lambda L}{D} \tag{17.23}$$

Width of central maximum for diffraction from a circular aperture of diameter $D$

The diameter of the diffraction pattern increases with distance $L$, showing that light spreads out behind a circular aperture, but it decreases if the size $D$ of the aperture is increased.

**EXAMPLE 17.13** **Finding the right viewing distance**

Light from a helium-neon laser ($\lambda = 633$ nm) passes through a 0.50-mm-diameter hole. How far away should a viewing screen be placed to observe a diffraction pattern whose central maximum is 3.0 mm in diameter?

**STRATEGIZE** The size of the central maximum is determined by the wavelength and the aperture size, which are given, and the distance to the screen, which we will determine.

**PREPARE** We can rearrange Equation 17.23 to solve for $L$, the distance to the screen.

**SOLVE** The screen distance is

$$L = \frac{wD}{2.44\lambda} = \frac{(3.0\times10^{-3}\text{ m})(5.0\times10^{-4}\text{ m})}{2.44(633\times10^{-9}\text{ m})} = 0.97\text{ m}$$

**ASSESS** The hole is relatively large compared to the wavelength of light, so the spreading will be relatively small. To see a central maximum of the noted size, we expect a large screen distance.

FIGURE 17.34 The diffraction pattern from a small particle.

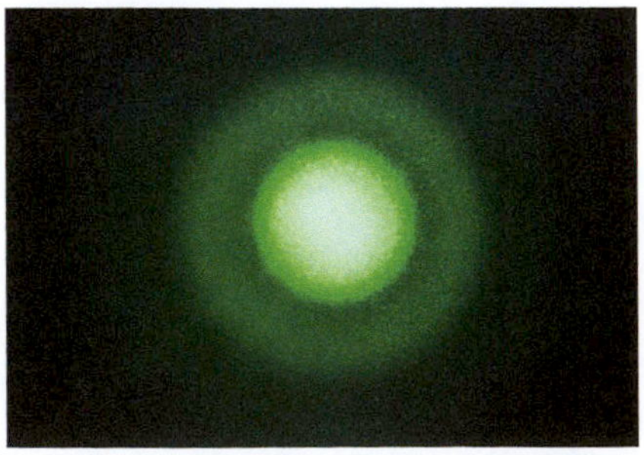

In the previous section, we saw that the diffraction pattern from a wire was the same as for a slit of the same size. The principle applies here as well; the diffraction pattern from a circular obstruction is the same as for a circular aperture. FIGURE 17.34 shows an image of the diffraction pattern for laser light around a small, spherical particle. The appearance matches the pattern of Figure 17.33. The pattern is dominated by the bright central maximum. The width of the central maximum is determined by the size of the particle; Equation 17.23 applies, with $D$ the diameter of the obstruction. This permits the use of diffraction to measure the size of circular or spherical objects.

Notice that the center of the diffraction pattern resulting from light passing by a small particle in Figure 17.34 is a bright spot, just as we saw a bright fringe at the center of the diffraction pattern from a wire. Explaining this observation requires us to use the wave nature of light. In the next two chapters, we'll consider the interaction of light with objects much larger than the wavelength. There, the ray model of light will be more appropriate for describing how light reflects from mirrors and refracts from lenses. But the wave model will reappear in Chapter 19 when we study telescopes and microscopes. We'll find that the *resolution* of these instruments has a fundamental limit set by the wave nature of light.

**INTEGRATED EXAMPLE 17.14** **Laser range finding**

Scientists use *laser range finding* to measure the distance to the moon with great accuracy. A very brief (100 ps) laser pulse, with a wavelength of 532 nm, is fired at the moon, where it reflects off an array of 100 4.0-cm-diameter mirrors placed there by Apollo 14 astronauts in 1971. The reflected laser light returns to earth, where it is collected by a telescope and detected. The average earth-moon distance is 384,000 km.

The laser beam spreads out on its way to the moon because of diffraction, reaching the mirrors with an intensity of 300 W/m$^2$. The reflected beam spreads out even more on its way back because of diffraction due to the circular aperture of the mirrors.

a. What is the round-trip time for the laser pulse to travel to the moon and back?
b. If we want to measure the distance to the moon to an accuracy of 1.0 cm, how accurately must the arrival time of the returning pulse be measured?
c. Because of the spread of the beam due to diffraction, the light arriving at earth from one of the mirrors will be spread over a circular spot. Estimate the diameter of this spot.
d. What is the intensity of the laser beam when it arrives back at the earth?

***STRATEGIZE*** The spreading out of the beam due to diffraction will increase the area of the beam, thus reducing its intensity. Once we've determined how much the beam spreads out, we will determine the area over which the beam is spread, and thus the intensity.

***PREPARE*** When the light reflects from one of the circular mirrors, diffraction causes it to spread out as shown in FIGURE 17.35. The width $w$ of the central maximum for circular-aperture diffraction is given by Equation 17.23.

FIGURE 17.35 The geometry of the returning laser beam.

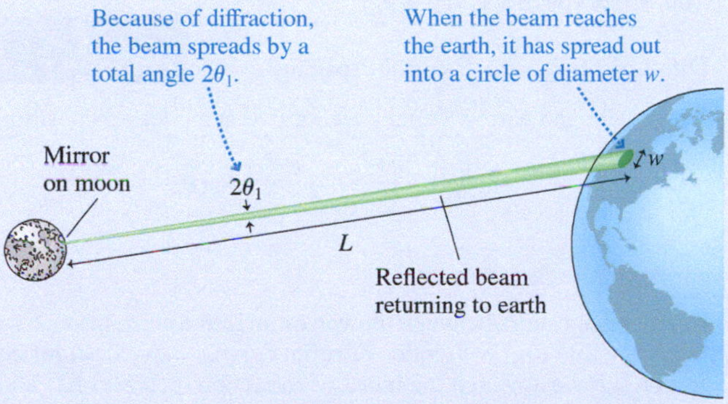


Because we know the intensity of the laser beam as it strikes the mirrors, and we can easily find a mirror's area, we can use Equation 15.11 to calculate the power of the beam as it leaves a mirror. The intensity of the beam when it arrives back on earth will be much lower and can also be found from Equation 15.11.

***SOLVE***

a. The round-trip distance is $2L$. Thus the round-trip travel time for the pulse, traveling at speed $c$, is

$$\Delta t = \frac{2L}{c} = \frac{2(3.84 \times 10^8 \text{ m})}{3.00 \times 10^8 \text{ m/s}} = 2.56 \text{ s}$$

b. If we wish to measure the moon's distance from the earth to an accuracy of $\pm 1.0$ cm, then, because the laser beam travels both to and from the moon, we need to know the round-trip distance to an accuracy of $\pm 2.0$ cm. The time it takes light to travel $\Delta x = 2.0$ cm is

$$\Delta t = \frac{\Delta x}{c} = \frac{0.020 \text{ m}}{3.00 \times 10^8 \text{ m/s}} = 6.6 \times 10^{-11 \text{ s}} = 66 \text{ ps}$$

Thus the arrival time of the pulses must be timed to an accuracy of about 70 ps.

c. The light arriving at the moon reflects from circular mirrors of diameter $D$. Diffraction by these circular apertures causes the returning light to spread out with angular width $2\theta_1$, where $\theta_1$ is the angle of the first minimum on either side of the central maximum. Equation 17.23 found that the width of the central maximum—the diameter of the circular spot of light when it reaches the earth—is

$$w = \frac{2.44\lambda L}{D} = \frac{2.44(532 \times 10^{-9} \text{ m})(3.84 \times 10^8 \text{ m})}{0.040 \text{ m}}$$
$$= 12{,}000 \text{ m}$$

d. The light is reflected from circular mirrors of radius $r = D/2$ and area $a = \pi r^2 = \pi D^2/4$. The power reflected from one mirror is, according to Equation 15.11,

$$P = Ia = I\pi\frac{D^2}{4} = (300 \text{ W/m}^2)\pi\frac{(0.040 \text{ m})^2}{4} = 0.38 \text{ W}$$

When the pulse returns to earth, it is now spread over the large area $\pi w^2/4$. Thus the intensity at the earth's surface from one mirror reflection is

$$I_1 = \frac{P}{\pi\frac{w^2}{4}} = \frac{0.38 \text{ W}}{\pi\frac{(12{,}000 \text{ m})^2}{4}} = 3.3 \times 10^{-9} \text{ W/m}^2$$

There are 100 reflecting mirrors in the array, so the total intensity reaching earth is $I = 100I_1 = 3.3 \times 10^{-7}$ W/m$^2$. A large telescope is needed to detect this very small intensity.

***ASSESS*** We expect the return pulse to have a very small intensity, so our result makes sense. Despite the difficult challenge of detecting this very weak signal, the accuracy of these measurements is astounding: The latest experiments can measure the instantaneous distance to the moon to $\pm 1$ mm, a precision of 3 parts in a trillion!

# SUMMARY

**GOAL** To understand and apply the wave model of light.

## GENERAL PRINCIPLES

### The Wave Model

The wave model considers light to be a wave propagating through space. Interference and diffraction are important. The wave model is appropriate when light interacts with objects whose size is comparable to the wavelength of light, or roughly less than about 0.1 mm.

**Huygens' principle** says that each point on a wave front is the source of a spherical wavelet. The wave front at a later time is tangent to all the wavelets.

## IMPORTANT CONCEPTS

The **index of refraction** of a material determines the speed of light in that material: $v = c/n$. The index of refraction of a material is always greater than 1, so that $v$ is always less than $c$.

The wavelength $\lambda$ in a material with index of refraction $n$ is *shorter* than the wavelength $\lambda_{vac}$ in a vacuum: $\lambda = \lambda_{vac}/n$.

The *frequency* of light does not change as it moves from one material to another.

**Diffraction** is the spreading of a wave after it passes through an opening.

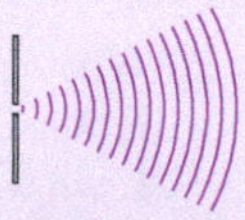

Constructive and destructive **interference** are due to the overlap of two or more waves as they spread behind openings.

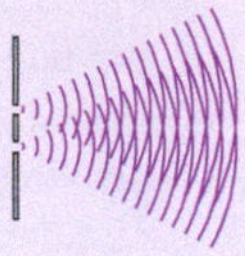

## APPLICATIONS

### Diffraction from a single slit

A single slit of width $a$ has a bright **central maximum** of width

$$w = \frac{2\lambda L}{a}$$

that is flanked by weaker **secondary maxima.**

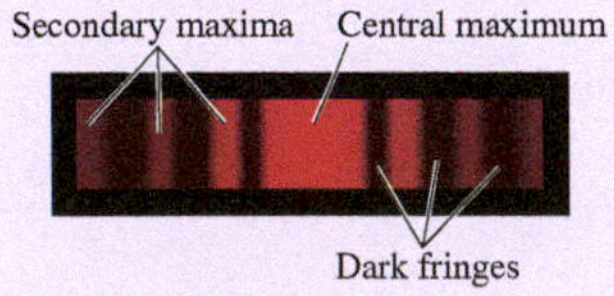


Dark fringes are located at angles such that

$$a \sin\theta_p = p\lambda \qquad p = 1, 2, 3, \ldots$$

If $\lambda/a \ll 1$, then from the small-angle approximation,

$$\theta_p = \frac{p\lambda}{a} \qquad y_p = \frac{p\lambda L}{a}$$

### Circular aperture of diameter $D$

A bright central maximum of diameter

$$w = \frac{2.44\lambda L}{D}$$

is surrounded by circular secondary maxima. The first dark fringe is located at

$$\theta_1 = \frac{1.22\lambda}{D} \qquad y_1 = \frac{1.22\lambda L}{D}$$

For an aperture of any shape, a smaller opening causes a greater spreading of the wave behind the opening.

### Interference from multiple slits

Waves overlap as they spread out behind slits. Bright fringes are seen on the viewing screen at positions where the path-length difference $\Delta r$ between successive slits is equal to $m\lambda$, where $m$ is an integer.

**Double slit** with separation $d$

Equally spaced bright fringes are located at

$$\sin\theta_m = \frac{m\lambda}{d} \qquad y_m = \frac{m\lambda L}{d} \qquad m = 0, 1, 2, \ldots$$

The **fringe spacing** is $\Delta y = \dfrac{\lambda L}{d}$

**Diffraction grating** with slit spacing $d$

Very bright and narrow fringes are located at angles and positions

$$\sin\theta_m = \frac{m\lambda}{d} \qquad y_m = L\tan\theta_m$$

### Thin-film interference

Interference occurs between the waves reflected from the two surfaces of a thin film with index of refraction $n$. A wave that reflects from a surface at which the index of refraction increases has a phase change.

| Interference | 0 or 2 phase changes | 1 phase change |
|---|---|---|
| Constructive | $2t = m\dfrac{\lambda}{n}$ | $2t = \left(m + \dfrac{1}{2}\right)\dfrac{\lambda}{n}$ |
| Destructive | $2t = \left(m + \dfrac{1}{2}\right)\dfrac{\lambda}{n}$ | $2t = m\dfrac{\lambda}{n}$ |

## Learning Objectives After studying this chapter, you should be able to:

- Recognize and use the three models of light. *Conceptual Questions 17.10, 17.21*
- Understand how light waves propagate in a vacuum and in media. *Conceptual Questions 17.2, 17.5; Problems 17.2, 17.3, 17.4, 17.5, 17.6*
- Calculate the interference-fringe pattern of a double slit. *Conceptual Questions 17.7, 17.8; Problems 17.8, 17.9, 17.10, 17.12, 17.13*
- Work with diffraction gratings, and apply diffraction gratings to problems in spectroscopy. *Conceptual Questions 17.9, 17.12; Problems 17.15, 17.16, 17.17, 17.18, 17.20*
- Understand how interference occurs in thin, transparent films. *Conceptual Questions 17.15, 17.16; Problems 17.24, 17.26, 17.27, 17.28, 17.31*
- Calculate the diffraction patterns of single slits and circular apertures. *Conceptual Questions 17.11, 17.22; Problems 17.33, 17.36, 17.37, 17.43, 17.44*

## STOP TO THINK ANSWERS

**Chapter Preview Stop to Think: C.** At point 1, two crests overlap, leading to a large wave amplitude; this is constructive interference. At point 2, the crest of one wave overlaps a trough of the other wave, so the amplitude is zero; this is destructive interference.

**Stop to Think 17.1: $n_3 > n_1 > n_2$.** $\lambda = \lambda_{vac}/n$, so a shorter wavelength corresponds to a higher index of refraction.

**Stop to Think 17.2: A.** The type of interference observed depends on the path-length *difference* $\Delta r = r_2 - r_1$, not on the values of $r_1$ and $r_2$ alone. Here, $\Delta r = 3\lambda$, so the path-length difference is an integer number of wavelengths, giving constructive interference.

**Stop to Think 17.3: Smaller.** The fringe spacing $\Delta y$ is directly proportional to the wavelength $\lambda$.

**Stop to Think 17.4: D.** Longer wavelengths have larger diffraction angles. Red light has a longer wavelength than violet light, so red light is diffracted farther from the center.

**Stop to Think 17.5: B.** An extra path difference of $\lambda/2$ must be added to change from constructive to destructive interference. In thin-film interference, one wave passes *twice* through the film. To increase the path length by $\lambda/2$, the thickness needs to be increased by only one-half this, or $\lambda/4$.

**Stop to Think 17.6: B or C.** The width of the central maximum, which is proportional to $\lambda/a$, has increased. This could occur either because the wavelength has increased or because the slit width has decreased.

Video Tutor Solution Chapter 17

# QUESTIONS

## Conceptual Questions

1. The frequency of a light wave in air is $5.3 \times 10^{14}$ Hz. Is the frequency of this wave higher, lower, or the same after the light enters a piece of glass?
2. Rank in order the following according to their speeds, from slowest to fastest: (i) 425-nm-wavelength light through a pane of glass, (ii) 500-nm-wavelength light through air, (iii) 540-nm-wavelength light through water, (iv) 670-nm-wavelength light through a diamond, and (v) 670-nm-wavelength light through a vacuum.
3. DVDs and Blu-ray disks store information in patterns that are read by laser light. The shorter the wavelength of the light, the closer the data tracks can be placed on the disk. A Blu-ray player uses violet light; a DVD player uses red light. Which disk can store more information?
4. The wavelength of a light wave is 700 nm in air; this light appears red. If this wave enters a pool of water, its wavelength becomes $\lambda_{air}/n = 530$ nm. If you were swimming underwater, the light would still appear red. Given this, what property of a wave determines its color?
5. Increasing the density of a material tends to increase the index of refraction. Does light travel faster in seawater or in fresh water?
6. A double-slit interference experiment shows fringes on a screen. The entire experiment is then immersed in water. Do the fringes on the screen get closer together, farther apart, remain the same, or disappear entirely? Explain.
7. In a double-slit interference experiment, interference fringes are observed on a distant screen. The width of both slits is then doubled without changing the distance between their centers.
   a. What happens to the spacing of the fringes? Explain.
   b. What happens to the intensity of the bright fringes? Explain.

Problem difficulty is labeled as | (straightforward) to ||||| (challenging). Problems labeled INT integrate significant material from earlier chapters; Problems labeled BIO are of biological or medical interest.

The eText icon indicates when there is a video tutor solution available for the chapter or for a specific problem. To launch these videos, log into your eText through Mastering™ Physics or log into the Study Area.

8. Figure Q17.8 shows the viewing screen in a double-slit experiment with monochromatic light. Fringe C is the central maximum.
   a. What will happen to the fringe spacing if the wavelength of the light is decreased?
   b. What will happen to the fringe spacing if the spacing between the slits is decreased?
   c. What will happen to the fringe spacing if the distance to the screen is decreased?
   d. Suppose the wavelength of the light is 500 nm. How much farther is it from the dot on the screen in the center of fringe E to the left slit than it is from the dot to the right slit?

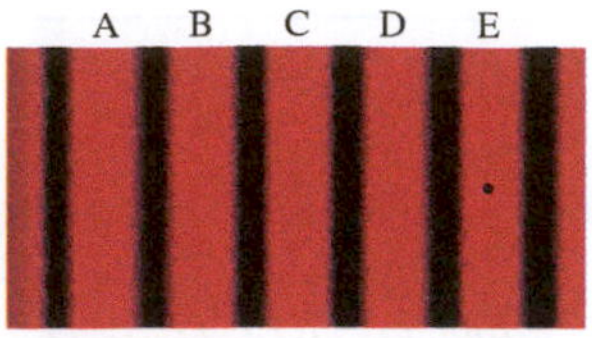

FIGURE Q17.8

9. Figure Q17.8 is the interference pattern seen on a viewing screen behind 2 slits. Suppose the 2 slits were replaced by 20 slits having the same spacing $d$ between adjacent slits.
   a. Would the number of fringes on the screen increase, decrease, or stay the same?
   b. Would the fringe spacing increase, decrease, or stay the same?
   c. Would the width of each fringe increase, decrease, or stay the same?
   d. Would the brightness of each fringe increase, decrease, or stay the same?
10. Figure Q17.10 shows the light intensity on a viewing screen behind a single slit of width $a$. The light's wavelength is $\lambda$. Is $\lambda < a$, $\lambda = a$, $\lambda > a$, or is it not possible to tell? Explain.

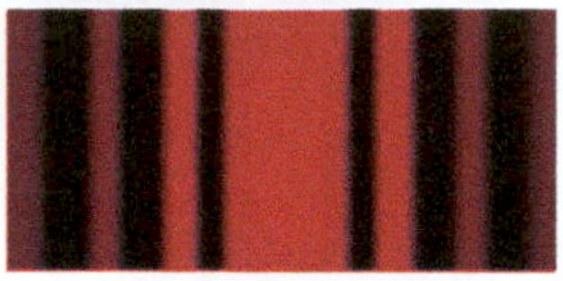
FIGURE Q17.10

11. Figure Q17.11 shows the light intensity on a viewing screen behind a circular aperture. What happens to the width of the central maximum if
   a. The wavelength is increased?
   b. The diameter of the aperture is increased?
   c. How will the screen appear if the aperture diameter is less than the light wavelength?

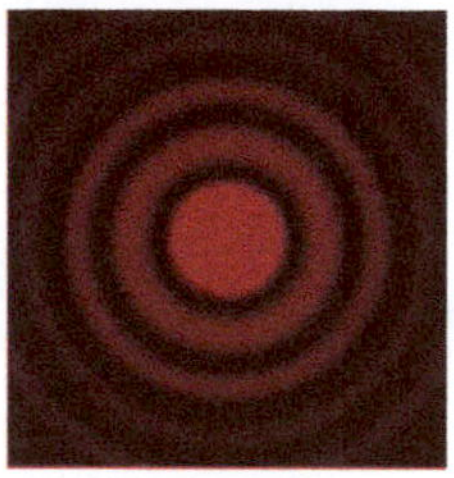
FIGURE Q17.11

12. If you look at the light spectrum reflected from the surface of a DVD compared to the spectrum reflected from the surface of a CD, you'll see that the rainbows from the DVD are more spread out, with greater angular separation. What does this tell you about the relative track spacing on these two types of disks?
13. If you look through a piece of very fine fabric at a tiny white light source, you will see a rainbow pattern. Explain the source of the pattern.
14. Figure Q17.14 shows a light wave incident on and passing through a thin soap film. Reflections from the front and back surfaces of the film create smaller waves (not shown in the figure) that travel to the left of the film, where they interfere. Is the interference constructive, destructive, or something in between? Explain.

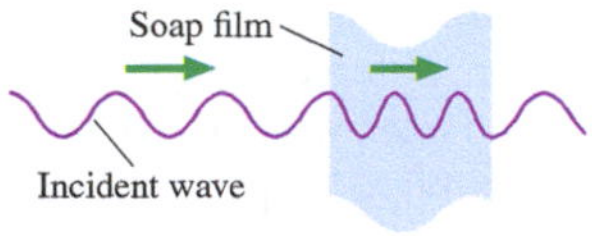

FIGURE Q17.14

15. Antireflection coatings for glass usually have an index of refraction that is less than that of glass. Explain how this permits a thinner coating.
16. Solar cells generally have an antireflection coating. Explain how this increases their efficiency.
17. An oil film on top of water has one patch that is much thinner than the wavelength of visible light. The index of refraction of the oil is less than that of water. Will the reflection from that extremely thin part of the film be bright or dark? Explain.
18. BIO The distinctly blue shade of a blue spruce results from the scattering of light by small waxy particles that coat the leaves. Explain how this scattering can protect the leaves from damage by short-wavelength ultraviolet light while still permitting the passage of longer wavelengths of light for photosynthesis.

19. Should the antireflection coating of a microscope objective lens designed for use with ultraviolet light be thinner, thicker, or the same thickness as the coating on a lens designed for visible light?
20. Example 17.7 showed that a thin film whose thickness is one-quarter of the wavelength of light in the film serves as an antireflection coating when coated on glass. In Example 17.7, $n_{\text{film}} < n_{\text{glass}}$. If a quarter-wave thickness film with $n_{\text{film}} > n_{\text{glass}}$ were used instead, would the film still serve as an antireflection coating? Explain.
21. You are standing against the wall near a corner of a large building. A friend is standing against the wall that is around the corner from you. You can't see your friend. How is it that you can hear her when she talks to you?
22. An investigator is using a laser to illuminate a distant target. He decides that he needs a smaller beam, so he puts a pinhole directly in front of the laser. He finds that this actually spreads the beam out, making matters worse. Explain what is happening.

## Multiple-Choice Questions

23. | Light of wavelength 500 nm in air enters a glass block with index of refraction $n = 1.5$. When the light enters the block, which of the following properties of the light will not change?
   A. The speed of the light
   B. The frequency of the light
   C. The wavelength of the light
24. | The frequency of a light wave in air is $4.6 \times 10^{14}$ Hz. What is the wavelength of this wave after it enters a pool of water?
   A. 300 nm B. 490 nm C. 650 nm D. 870 nm
25. | Light passes through a diffraction grating with a slit spacing of 0.001 mm. A viewing screen is 100 cm behind the grating. If the light is blue, with a wavelength of 450 nm, at about what distance from the center of the interference pattern will the first-order maximum appear?
   A. 5 cm B. 25 cm C. 50 cm D. 100 cm
26. || Blue light of wavelength 450 nm passes through a diffraction grating with a slit spacing of 0.001 mm and makes an interference pattern on the wall. How many bright fringes will be seen?
   A. 1 B. 3 C. 5 D. 7

27. | Yellow light of wavelength 590 nm passes through a diffraction grating and makes an interference pattern on a screen 80 cm away. The first bright fringes are 1.9 cm from the central maximum. How many lines per mm does this grating have?
A. 20 B. 40 C. 80 D. 200

28. | Light passes through a 10-$\mu$m-wide slit and is viewed on a screen 1 m behind the slit. If the width of the slit is narrowed, the band of light on the screen will
A. Become narrower.
B. Become wider.
C. Stay about the same.

29. || Reflected light from a thin film of oil gives constructive interference for light with a wavelength inside the film of $\lambda_{\text{film}}$. By how much would the film thickness need to be increased to give destructive interference?
A. $2\lambda_{\text{film}}$ B. $\lambda_{\text{film}}$ C. $\lambda_{\text{film}}/2$ D. $\lambda_{\text{film}}/4$

30. || A laser shines through a pinhole, and it makes a diffraction pattern on a screen behind. The area of the circular hole is reduced by a factor of 2. What can you say about the new intensity?
A. It is less than 1/2 the original intensity.
B. It is 1/2 the original intensity.
C. Is is the same as the original intensity.
D. It is greater than the original intensity.

31. | You want to estimate the diameter of a very small circular pinhole that you've made in a piece of aluminum foil. To do so, you shine a red laser pointer ($\lambda = 632$ nm) at the hole and observe the diffraction pattern on a screen 3.5 m behind the foil. You measure the width of the central maximum to be 15 mm. What is the diameter of the hole?
A. 0.18 mm B. 0.29 mm C. 0.36 mm D. 1.1 mm

# PROBLEMS

## Section 17.1 What Is Light?

1. ||| How long does it take light to travel through a 3.0-mm-thick piece of window glass?

2. | a. How long (in ns) does it take light to travel 1.0 m in a vacuum?
b. What distance does light travel in water, glass, and diamond during the time that it travels 1.0 m in a vacuum?

3. |||| A 5.0-cm-thick layer of oil ($n = 1.46$) is sandwiched between a 1.0-cm-thick sheet of glass and a 2.0-cm-thick sheet of polystyrene plastic ($n = 1.59$). How long (in ns) does it take light incident perpendicular to the glass to pass through this 8.0-cm-thick sandwich?

4. || A light wave has a 670 nm wavelength in air. Its wavelength in a transparent solid is 420 nm.
a. What is the speed of light in this solid?
b. What is the light's frequency in the solid?

5. || A helium-neon laser beam has a wavelength in air of 633 nm. It takes 1.38 ns for the light to travel through 30.0 cm of an unknown liquid. What is the wavelength of the laser beam in the liquid?

6. | Data are carried by pulses of light through long, transparent fibers—fiber-optic cables. Long lengths of fiber carry signals 100 km between amplifier stations. How long does it take a signal to travel this distance if the index of refraction of the fiber is 1.45?

## Section 17.2 The Interference of Light

7. | Two narrow slits 50 $\mu$m apart are illuminated with light of wavelength 500 nm. The light shines on a screen 1.2 m distant. What is the angle of the $m = 2$ bright fringe? How far is this fringe from the center of the pattern?

8. ||| Light from a sodium lamp ($\lambda = 589$ nm) illuminates two narrow slits. The fringe spacing on a screen 150 cm behind the slits is 4.0 mm. What is the spacing (in mm) between the two slits?

9. || Two narrow slits are illuminated by light of wavelength $\lambda$. The slits are spaced 20 wavelengths apart. What is the angle, in radians, between the central maximum and the $m = 1$ bright fringe?

10. || Figure P17.10 shows the fringes observed in a double-slit interference experiment when the two slits are illuminated by white light. The central maximum is white because all of the colors overlap. This is not true for the other fringes. The $m = 1$ fringe clearly shows bands of color, with red appearing farther from the center of the pattern, and blue closer. If the slits that create this pattern are 20 $\mu$m apart and are located 0.85 m from the screen, what are the $m = 1$ distances from the central maximum for red (700 nm) and violet (400 nm) light?

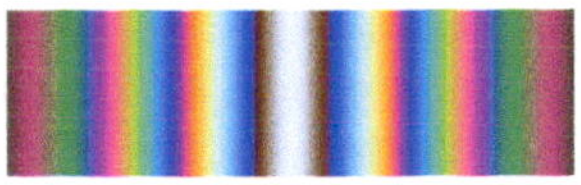

FIGURE P17.10

11. || A double-slit experiment is performed with light of wavelength 600 nm. The bright interference fringes are spaced 1.8 mm apart on the viewing screen. What will the fringe spacing be if the light is changed to a wavelength of 400 nm?

12. || Light illuminating a pair of slits contains two wavelengths, 500 nm and an unknown wavelength. The 10th bright fringe of the unknown wavelength overlaps the 9th bright fringe of the 500 nm light. What is the unknown wavelength?

13. || Two narrow slits are 0.12 mm apart. Light of wavelength 550 nm illuminates the slits, causing an interference pattern on a screen 1.0 m away. Light from each slit travels to the $m = 1$ maximum on the right side of the central maximum. How much farther did the light from the left slit travel than the light from the right slit?

## Section 17.3 The Diffraction Grating

14. | A laser beam of wavelength 670 nm shines through a diffraction grating that has 750 lines/mm. Sketch the pattern that appears on a screen 1.0 m behind the grating, noting distances on your drawing and explaining where these numbers come from.

15. ||| A diffraction grating with 750 slits/mm is illuminated by light that gives a first-order diffraction angle of 34.0°. What is the wavelength of the light?

16. || A commercial diffraction grating has 500 lines per mm. When a student shines a 530 nm laser through this grating, how many bright spots could be seen on a screen behind the grating?

17. |||| A 1.0-cm-wide diffraction grating has 1000 slits. It is illuminated by light of wavelength 550 nm. What are the angles of the first two diffraction orders?

Watch Video Solution Problems 17.27 and 17.31

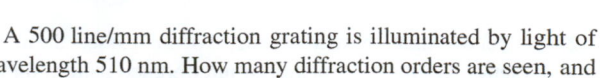

18. ||| A physics instructor wants to project a spectrum of visible-light colors from 400 nm to 700 nm as part of a classroom demonstration. She shines a beam of white light through a diffraction grating that has 500 lines per mm, projecting a pattern on a screen 2.4 m behind the grating.
    a. How wide is the spectrum that corresponds to $m = 1$?
    b. How much distance separates the end of the $m = 1$ spectrum and the start of the $m = 2$ spectrum?
19. || The human eye can readily detect wavelengths from about 400 nm to 700 nm. If white light illuminates a diffraction grating having 750 lines/mm, over what range of angles does the visible $m = 1$ spectrum extend?
20. || BIO Glass catfish, tropical fish popular with hobbyists, have no pigment, and matching of the index of refraction of their tissues leaves them largely transparent—you can see through them. When a beam of white light illuminates these fish, diffraction of light from evenly spaced striations in muscle fibers produces a rainbow pattern. As the muscles contract, the striations get closer together, and this affects the diffraction pattern. This phenomenon has been used to study muscle contraction in living fish as they swim. Investigators set up a chamber with moving water where the fish swam steadily to stay stationary. The investigators then passed a beam of laser light of wavelength 632 nm through the fish's muscle tissue as it swam. Muscle action produced periodic changes in the distances between striations, which ranged from 1.87 to 1.94 $\mu$m. Investigators measured the change of position of a bright spot corresponding to $m = 1$ on a screen. If the screen was 30 cm behind the fish, what was the distance spanned by the diffraction spot as it moved back and forth? The screen was in the tank with the fish, so that the entire path of the laser was in water and tissue with an index of refraction close to that of water. The properties of the diffraction pattern were thus determined by the wavelength in water.

21. ||| A diffraction grating with 600 lines/mm is illuminated with light of wavelength 500 nm. A very wide viewing screen is 2.0 m behind the grating.
    a. What is the distance between the two $m = 1$ fringes?
    b. How many bright fringes can be seen on the screen?
22. ||| BIO Figure P17.22 shows a microscopic view of muscle tissue. In the figure, structures called *sarcomeres* are bordered by ridges. The regular pattern of the ridges means that the muscle can operate as a reflection grating, and a measurement of the resulting diffraction pattern can give a measurement of the length of the sarcomeres, which is the distance between the ridges. This has been used to provide measurements of sarcomere length during exercise, as the sarcomeres lengthen and shorten. In one case, an investigator shined a 632 nm laser on exposed muscle tissue in a patient's forearm as he moved his wrist back and forth through a 100° angle, contracting and then stretching the muscle. The resulting diffraction pattern was projected onto a screen 2.4 cm from the muscle. The investigator measured the distance between the two $m = 2$ fringes on either side of the central maximum. This length varied from 1.9 cm to 2.7 cm. What were the minimum and maximum values of the sarcomere length?

FIGURE P17.22

23. || A 500 line/mm diffraction grating is illuminated by light of wavelength 510 nm. How many diffraction orders are seen, and what is the angle of each?

## Section 17.4 Thin-Film Interference

24. || What is the thinnest film of $MgF_2$ ($n = 1.38$) on glass that produces a strong reflection for orange light with a wavelength of 600 nm?
25. |||| A very thin oil film ($n = 1.25$) floats on water ($n = 1.33$). What is the thinnest film that produces a strong reflection for green light with a wavelength of 500 nm?
26. || BIO Antireflection coatings can be used on the *inner* surfaces of eyeglasses to reduce the reflection of stray light into the eye, thus reducing eyestrain.
    a. A 90-nm-thick coating is applied to the lens. What must be the coating's index of refraction to be most effective at 480 nm? Assume that the coating's index of refraction is less than that of the lens.
    b. If the index of refraction of the coating is 1.38, what thickness should the coating be so as to be most effective at 480 nm? The thinnest possible coating is best.
27. || Solar cells are given antireflection coatings to maximize their efficiency. Consider a silicon solar cell ($n = 3.50$) coated with a layer of silicon dioxide ($n = 1.45$). What is the minimum coating thickness that will minimize the reflection at the wavelength of 700 nm, where solar cells are most efficient?
28. || BIO The blue-ringed octopus reveals the bright blue rings that give it its name as a warning display. The rings have a stack of reflectin (a protein used for structural color in many cephalopods) plates with index of refraction $n = 1.59$ separated by cells with index $n = 1.37$. The plates have thickness 62 nm. What is the longest wavelength, in air, of light that will give constructive interference from opposite sides of the reflecting plates?

29. ||| A thin film of $MgF_2$ ($n = 1.38$) coats a piece of glass. Constructive interference is observed for the reflection of light with wavelengths of 500 nm and 625 nm. What is the thinnest film for which this can occur?
30. || BIO Pigments don't survive fossilization; even though we have fossil skin from dinosaurs, we don't know what color they were. But fossilization does preserve structure. Specimens from a rare cache of 50-million-year-old beetle fossils still show the microscopic layers that produced structural colors in the living creatures, and we can deduce the colors from an understanding of thin-film interference. One fossil showed 80 nm plates of fossilized chitin (modern samples have index of refraction $n = 1.56$) embedded in fossilized tissue (for which we can assume $n = 1.33$). What is the longest wavelength for which there is constructive interference for reflections from opposite sides of the chitin layers? If this wavelength is enhanced by reflection, what color was the beetle?
31. | A soap bubble is essentially a thin film of water surrounded by air. The colors you see in soap bubbles are produced by interference. What visible wavelengths of light are strongly reflected from a 390-nm-thick soap bubble? What color would such a soap bubble appear to be?

32. ||| Mourning doves have a small patch of iridescent feathers. The color is produced by a 330-nm-thick layer of keratin ($n = 1.56$) with air on both sides that is found around the edge of the feather barbules. For what wavelength or wavelengths would this structure produce constructive interference?
BIO

### Section 17.5 Single-Slit Diffraction

33. || A helium-neon laser ($\lambda = 633$ nm) illuminates a single slit and is observed on a screen 1.50 m behind the slit. The distance between the first and second minima in the diffraction pattern is 4.75 mm. What is the width (in mm) of the slit?

34. ||| For a demonstration, a professor uses a razor blade to cut a thin slit in a piece of aluminum foil. When she shines a laser pointer ($\lambda = 680$ nm) through the slit onto a screen 5.5 m away, a diffraction pattern appears. The bright band in the center of the pattern is 8.0 cm wide. What is the width of the slit?

35. || A 0.50-mm-wide slit is illuminated by light of wavelength 500 nm. What is the width of the central maximum on a screen 2.0 m behind the slit?

36. || Early investigators (including Thomas Young) measured the thickness of wool fibers using diffraction. One early instrument used a collimated beam of 560 nm light to produce a diffraction pattern on a screen placed 30 cm from a single wool fiber. If the fiber's diameter was 16 $\mu$m, what was the width of the central maximum on the screen?
BIO

37. ||| The second minimum in the diffraction pattern of a 0.10-mm-wide slit occurs at 0.70°. What is the wavelength of the light?

38. || Quality control systems have been developed to remotely measure the diameter of wires using diffraction. A wire with a stated diameter of 170 $\mu$m blocks the beam of a 633 nm laser, producing a diffraction pattern on a screen 50.0 cm distant. The width of the central maximum is measured to be 3.77 mm. The wire should have a diameter within 1% of the stated value. Does this wire pass the test?

### Section 17.6 Circular-Aperture Diffraction

39. ||| A 0.50-mm-diameter hole is illuminated by light of wavelength 500 nm. What is the width of the central maximum on a screen 2.0 m behind the slit?

40. || Light from a helium-neon laser ($\lambda = 633$ nm) passes through a circular aperture and is observed on a screen 4.0 m behind the aperture. The width of the central maximum is 2.5 cm. What is the diameter (in mm) of the hole?

41. ||| You want to photograph a circular diffraction pattern whose central maximum has a diameter of 1.0 cm. You have a helium-neon laser ($\lambda = 633$ nm) and a 0.12-mm-diameter pinhole. How far behind the pinhole should you place the viewing screen?

42. || Investigators measure the size of fog droplets using the diffraction of light. A camera records the diffraction pattern on a screen as the droplets pass in front of a laser, and a measurement of the size of the central maximum gives the droplet size. In one test, a 690 nm laser creates a pattern on a screen 30 cm from the droplets. If the central maximum of the pattern is 0.28 cm in diameter, how large is the droplet?

43. || Infrared light of wavelength 2.5 $\mu$m illuminates a 0.20-mm-diameter hole. What is the angle of the first dark fringe in radians? In degrees?

44. || Diffraction can be used to provide a quick test of the size of red blood cells. Blood is smeared onto a slide, and a laser shines through the slide. The size of the cells is very consistent, so the multiple diffraction patterns overlap and produce an overall pattern that is similar to what a single cell would produce. Ideally, the diameter of a red blood cell should be between 7.5 and 8.0 $\mu$m. If a 633 nm laser shines through a slide and produces a pattern on a screen 24.0 cm distant, what range of sizes of the central maximum should be expected? Values outside this range might indicate a health concern and warrant further study.
BIO

## General Problems

45. ||| An advanced computer sends information to its various parts via infrared light pulses traveling through silicon fibers ($n = 3.50$). To acquire data from memory, the central processing unit sends a light-pulse request to the memory unit. The memory unit processes the request, then sends a data pulse back to the central processing unit. The memory unit takes 0.50 ns to process a request. If the information has to be obtained from memory in 2.00 ns, what is the maximum distance the memory unit can be from the central processing unit?

46. |||| Figure P17.46 shows the light intensity on a screen behind a double slit. The slit spacing is 0.20 mm and the wavelength of the light is 600 nm. What is the distance from the slits to the screen?

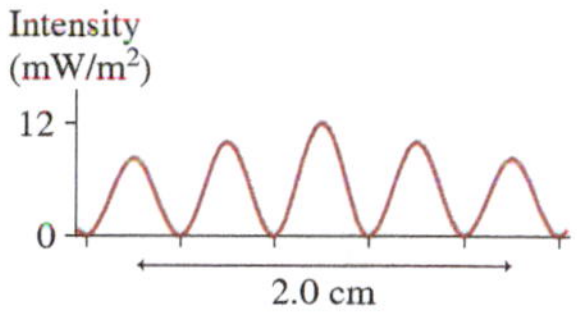


FIGURE P17.46

47. ||||| Figure P17.46 shows the light intensity on a screen behind a double slit. The slit spacing is 0.20 mm and the screen is 2.0 m behind the slits. What is the wavelength of the light?

48. || Figure P17.48 shows the light intensity on a screen 2.5 m behind a double slit. The wavelength of the light is 532 nm. What is the spacing between the slits?

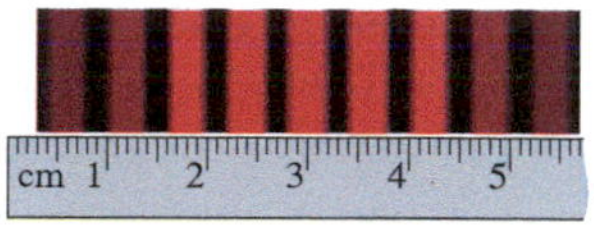


FIGURE P17.48

49. ||| The two most prominent wavelengths in the light emitted by a hydrogen discharge lamp are 656 nm (red) and 486 nm (blue). Light from a hydrogen lamp illuminates a diffraction grating with 500 lines/mm, and the light is observed on a screen 1.50 m behind the grating. What is the distance between the first-order red and blue fringes?

50. |||| White light (400–700 nm) is incident on a 600 line/mm diffraction grating. What is the width of the first-order rainbow on a screen 2.0 m behind the grating?

51. |||| A miniature spectrometer used for chemical analysis has a diffraction grating with 800 slits/mm set 25.0 mm in front of the detector "screen." The detector can barely distinguish two bright lines that are 30 $\mu$m apart in the first-order spectrum. What is the *resolution* of the spectrometer at a wavelength of 600 nm? That is, if two distinct wavelengths can barely be distinguished, one of them being 600.0 nm, what is the wavelength difference $\Delta\lambda$ between the two?

52. ||| Figure P17.52 shows the interference pattern on a screen 1.0 m behind an 800 line/mm diffraction grating. What is the wavelength of the light?

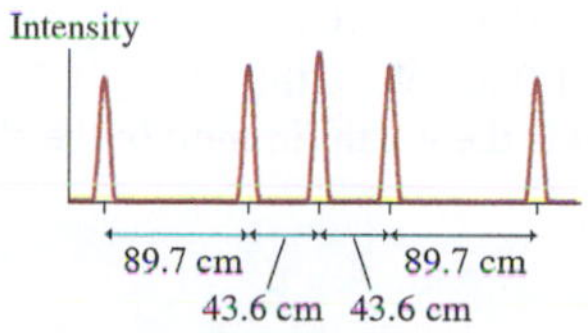


FIGURE P17.52

Watch Video Solution Problem 17.60

53. ||| Figure P17.52 shows the interference pattern on a screen 1.0 m behind a diffraction grating. The wavelength of the light is 600 nm. How many lines per millimeter does the grating have?

54. |||| The shiny surface of a CD is imprinted with millions of tiny pits, arranged in a pattern of thousands of essentially concentric circles that act like a reflection grating when light shines on them. You decide to determine the distance between those circles by aiming a laser pointer (with $\lambda = 680$ nm) perpendicular to the disk and measuring the diffraction pattern reflected onto a screen 1.5 m from the disk. The central bright spot you expected to see is blocked by the laser pointer itself. You do find two other bright spots separated by 1.4 m, one on either side of the missing central spot. The rest of the pattern is apparently diffracted at angles too great to show on your screen. What is the distance between the circles on the CD's surface?

55. ||| The wings of some beetles have closely spaced parallel lines
BIO of melanin, causing the wing to act as a reflection grating. Suppose sunlight shines straight onto a beetle wing. If the melanin lines on the wing are spaced 2.0 $\mu$m apart, what is the first-order diffraction angle for green light ($\lambda = 550$ nm)?

56. ||| Light emitted by element X passes through a diffraction grating that has 1200 slits/mm. The interference pattern is observed on a screen 75.0 cm behind the grating. First-order maxima are observed at distances of 56.2 cm, 65.9 cm, and 93.5 cm from the central maximum. What are the wavelengths of light emitted by element X?

57. |||| Light of a single wavelength is incident on a diffraction grating with 500 slits/mm. Several bright fringes are observed on a screen behind the grating, including one at 45.7° and one next to it at 72.6°. What is the wavelength of the light?

58. || A sheet of glass is coated with a 500-nm-thick layer of oil ($n = 1.42$).
   a. For what *visible* wavelengths of light do the reflected waves interfere constructively?
   b. For what *visible* wavelengths of light do the reflected waves interfere destructively?
   c. What is the color of reflected light? What is the color of transmitted light?

59. |||| A laboratory dish, 20 cm in diameter, is half filled with
INT water. One at a time, 0.50 $\mu$L drops of oil from a micropipette are dropped onto the surface of the water, where they spread out into a uniform thin film. After the first drop is added, the intensity of 600 nm light reflected from the surface is very low. As more drops are added, the reflected intensity increases, then decreases again to a minimum after a total of 13 drops have been added. What is the index of refraction of the oil?

60. ||| You need to use your cell phone, which broadcasts an 830 MHz signal, but you're in an alley between two massive, radio-wave-absorbing buildings that have only a 15 m space between them. What is the angular width, in degrees, of the electromagnetic wave after it emerges from between the buildings?

61. ||| Figure P17.61 shows the light intensity on a screen behind a single slit. The wavelength of the light is 500 nm and the screen is 1.0 m behind the slit. What is the width (in mm) of the slit?

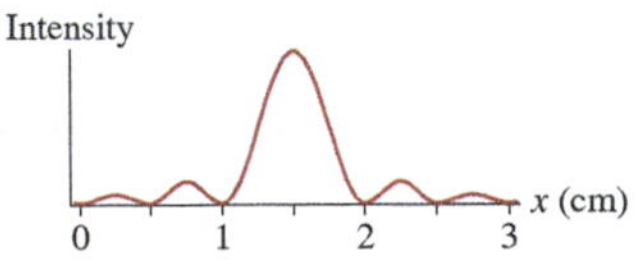

FIGURE P17.61

62. ||| Figure P17.61 shows the light intensity on a screen behind a single slit. The wavelength of the light is 600 nm and the slit width is 0.15 mm. What is the distance from the slit to the screen?

63. || Figure P17.63 shows the light intensity on a screen 2.5 m behind an aperture. The aperture is illuminated with light of wavelength 600 nm.
   a. Is the aperture a single slit or a double slit? Explain.
   b. If the aperture is a single slit, what is its width? If it is a double slit, what is the spacing between the slits?

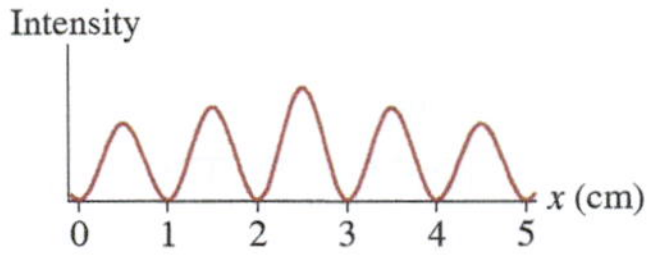

FIGURE P17.63

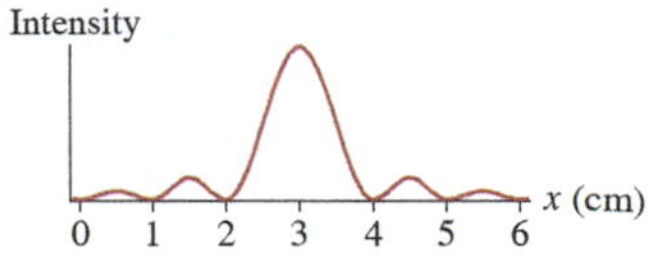

FIGURE P17.64

64. || Figure P17.64 shows the light intensity on a screen 2.5 m behind an aperture. The aperture is illuminated with light of wavelength 600 nm.
   a. Is the aperture a single slit or a double slit? Explain.
   b. If the aperture is a single slit, what is its width? If it is a double slit, what is the spacing between the slits?

65. ||| One day, after pulling down your window shade, you notice that sunlight is passing through a pinhole in the shade and making a small patch of light on the far wall. Having recently studied optics in your physics class, you're not too surprised to see that the patch of light seems to be a circular diffraction pattern. It appears that the central maximum is about 3 cm across, and you estimate that the distance from the window shade to the wall is about 3 m. Knowing that the average wavelength of sunlight is about 500 nm, estimate the diameter of the pinhole.

66. |||| A radar for tracking aircraft broadcasts a 12 GHz micro-
INT wave beam from a 2.0-m-diameter circular radar antenna. From a wave perspective, the antenna is a circular aperture through which the microwaves diffract.
   a. What is the diameter of the radar beam at a distance of 30 km?
   b. If the antenna emits 100 kW of power, what is the average microwave intensity at 30 km?

67. ||||| A helium-neon laser ($\lambda = 633$ nm), shown in Figure P17.67, is built with a glass tube of inside diameter 1.0 mm. One mirror is partially transmitting to allow the laser beam out. An electrical discharge in the tube causes it to glow like a neon light. From an optical perspective, the laser beam is a light wave that diffracts out through a 1.0-mm-diameter circular opening.
   a. Explain why a laser beam can't be *perfectly* parallel, with no spreading.
   b. The angle $\theta_1$ to the first minimum is called the *divergence angle* of a laser beam. What is the divergence angle of this laser beam?
   c. What is the diameter (in mm) of the laser beam after it travels 3.0 m?
   d. What is the diameter of the laser beam after it travels 1.0 km?

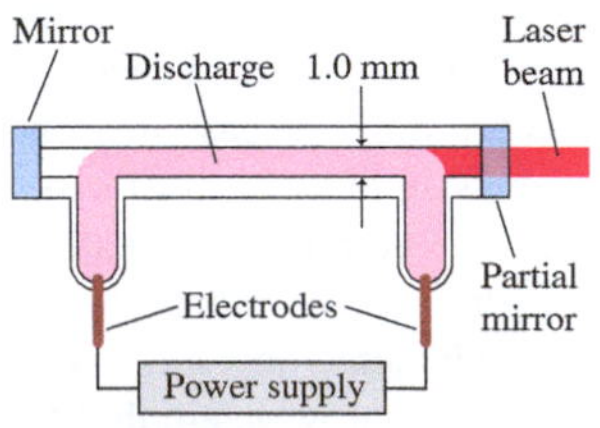

FIGURE P17.67

68. || In the laser range-finding experiments of Example 17.14, the laser beam fired toward the moon spreads out as it travels because it diffracts through a circular exit as it leaves the laser. In order for the reflected light to be bright enough to detect, the laser spot on the moon must be no more than 1 km in diameter. Staying within this diameter is accomplished by using a special large-diameter laser. If $\lambda = 532$ nm, what is the minimum diameter of the circular opening from which the laser beam emerges? The earth-moon distance is 384,000 km.

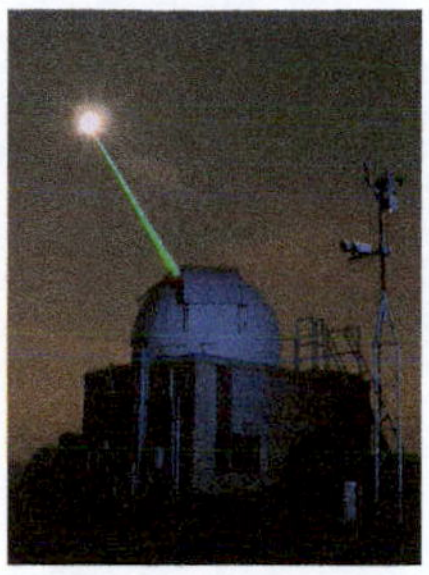

## MCAT-Style Passage Problems

### The Blue Morpho Butterfly BIO

The brilliant blue color of a blue morpho butterfly is, like the colors of peacock feathers, due to interference. Figure P17.69a shows an easy way to demonstrate this: If a drop of the clear solvent acetone is placed on the wing of a blue morpho butterfly, the color changes from a brilliant blue to an equally brilliant green—returning to blue once the acetone evaporates. There would be no change if the color were due to pigment.

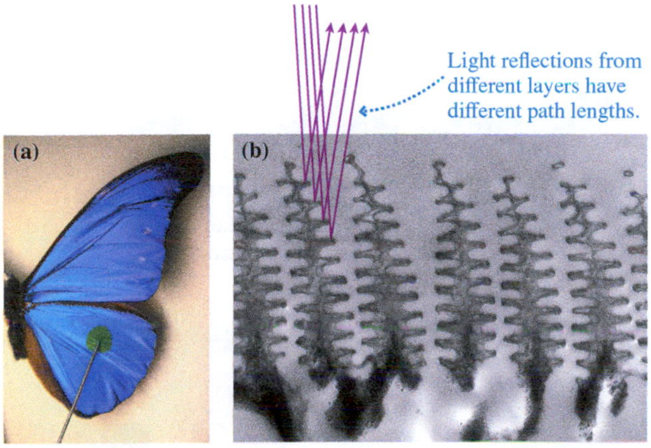


FIGURE P17.69

A cross section of a scale from the wing of a blue morpho butterfly reveals the source of the butterfly's color. As Figure P17.69b shows, the scales are covered with structures that look like small Christmas trees. Light striking the wings reflects from different layers of these structures, and the differing path lengths cause the reflected light to interfere constructively or destructively, depending on the wavelength. For light at normal incidence, blue light experiences constructive interference while other colors undergo destructive interference and cancel. Acetone fills the spaces in the scales with a fluid of index of refraction $n = 1.38$; this changes the conditions for constructive interference and results in a change in color.

69. | The coloring of the blue morpho butterfly is protective. As the butterfly flaps its wings, the angle at which light strikes the wings changes. This causes the butterfly's color to change and makes it difficult for a predator to follow. This color change is because
   A. A diffraction pattern appears only at certain angles.
   B. The index of refraction of the wing tissues changes as the wing flexes.
   C. The motion of the wings causes a Doppler shift in the reflected light.
   D. As the angle changes, the differences in paths among light reflected from different surfaces change, resulting in constructive interference for a different color.

70. | The change in color when acetone is placed on the wing is due to the difference between the indices of refraction of acetone and air. Consider light of some particular color. In acetone,
   A. The frequency of the light is less than in air.
   B. The frequency of the light is greater than in air.
   C. The wavelength of the light is less than in air.
   D. The wavelength of the light is greater than in air.

71. | The scales on the butterfly wings are actually made of a transparent material with index of refraction 1.56. Light reflects from the surface of the scales because
   A. The scales' index of refraction is different from that of air.
   B. The scales' index of refraction is similar to that of glass.
   C. The scales' density is different from that of air.
   D. Different colors of light have different wavelengths.

# 18 Ray Optics

These thin beams of light at this laser show are well described using ray optics. How do these light beams behave when they reflect from shiny surfaces or pass through transparent materials?

## LOOKING AHEAD

### Reflection

Light rays can bounce, or **reflect,** off a surface. Rays from the goose's head reflect from the water, forming an upside-down image.

You'll learn how the **law of reflection** can be used to understand image formation by mirrors.

### Refraction

The two images of the frog are due to **refraction,** the bending of light rays as they travel from one material into another.

You'll learn **Snell's law** for refraction and how images can be formed by refraction.

### Lenses and Mirrors

Rays refracting at the surfaces of this lens form a magnified **image** of the person behind it.

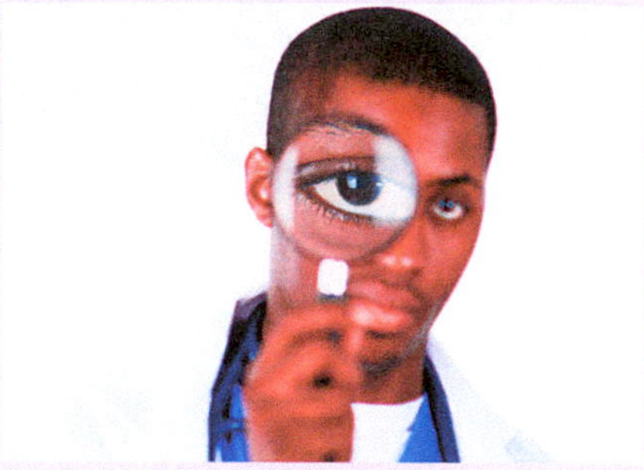

You'll learn how to locate and characterize the images formed by lenses and mirrors.

**GOAL** To understand and apply the ray model of light.

## LOOKING BACK

### The Ray Model of Light

In Chapter 17, you learned that light spreads out as it passes through a narrow slit, but travels straight forward through wide openings.

Waves travel in a straight path —a *ray*—behind a wide slit.

In this chapter, you'll study the behavior of light in the *ray model,* applicable when light interacts with objects of everyday size such as mirrors or lenses.

**STOP TO THINK**

The dark screen has a 2-mm-diameter hole. The bulb is the only source of light. What do you see on the viewing screen?

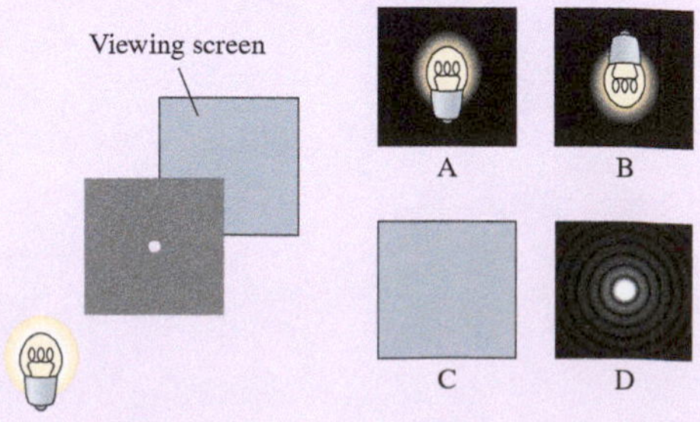

# 18.1 The Ray Model of Light

A flashlight makes a beam of light through the night's darkness, sunbeams stream into a darkened room through a small hole in the shade, and laser beams are even more well defined. Our everyday experience that light travels in straight lines is the basis of the ray model of light.

The ray model is an oversimplification of reality, but nonetheless is very useful within its range of validity. As we saw in Chapter 17, diffraction and other wave aspects of light are important only for apertures and objects comparable in size to the wavelength of light. Because the wavelength is so small, typically 0.5 $\mu$m, the wave nature of light is not apparent when light interacts with ordinary-sized objects. The ray model of light, which ignores diffraction, is valid as long as any apertures through which the light passes (lenses, mirrors, holes, and the like) are larger than about 1 mm.

To begin, let us define a **light ray** as a line in the direction along which light energy is flowing. A light ray is an abstract idea, not a physical entity or a "thing." Any narrow beam of light, such as the laser beam in FIGURE 18.1, is actually a bundle of many parallel light rays. You can think of a single light ray as the limiting case of a laser beam whose diameter approaches zero. Laser beams are good approximations of light rays, certainly adequate for demonstrating ray behavior, but any real laser beam is a bundle of many parallel rays.

FIGURE 18.1 A laser beam is a bundle of parallel light rays.

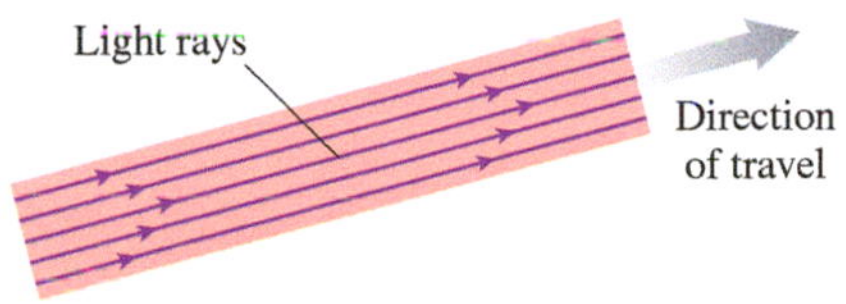


The following table outlines five basic ideas and assumptions of the ray model of light.

### The ray model of light

**Light rays are straight lines.**

Light travels through a vacuum or a transparent material in straight lines called light rays. The speed of light in a material is $v = c/n$, where $n$ is the index of refraction of the material.

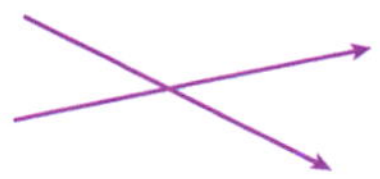

**Light rays can cross.**

Light rays do not interact with each other. Two rays can cross without either being affected in any way.

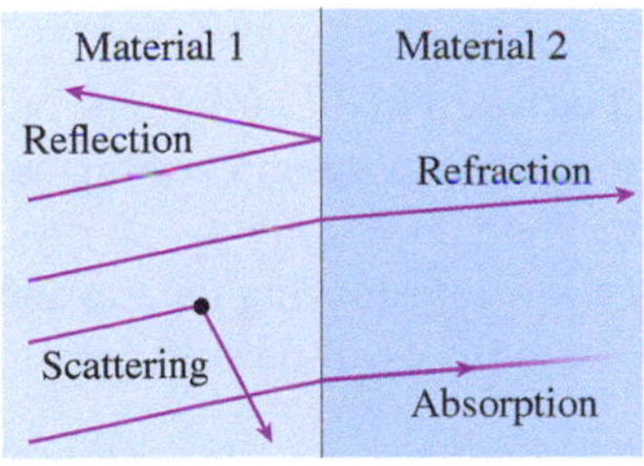


**A light ray goes on forever unless it interacts with matter.**

A light ray continues forever unless it has an interaction with matter that causes the ray to change direction or to be absorbed. Light interacts with matter in four different ways:

- At an interface between two materials, light can be *reflected, refracted,* or both.
- Within a material, light can be either *scattered* or *absorbed.*

These interactions are discussed later in the chapter.

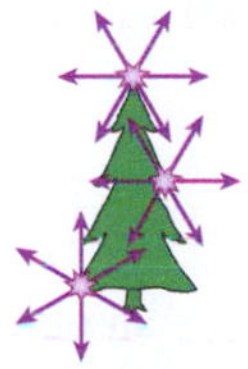

**An object is a source of light rays.**

An **object** is a source of light rays. Rays originate from *every* point on the object, and each point sends rays in *all* directions. Objects may be self-luminous—they create light rays—or they may be reflective objects that reflect only rays that originate elsewhere.

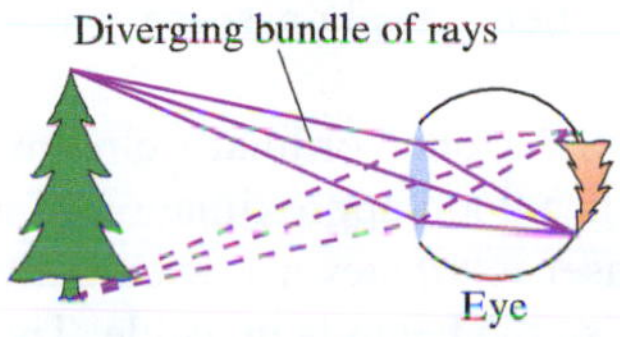


**The eye sees by focusing a bundle of rays.**

The eye sees an object when *diverging* bundles of rays from each point on the object enter the pupil and are focused to an image on the retina. Imaging is discussed later in the chapter, and the eye will be treated in much greater detail in Chapter 19.

## Sources of Light Rays

In the ray model, there are two kinds of objects. **Self-luminous objects** (or *sources*) directly create light rays. Self-luminous objects include lightbulbs and the sun. Other objects, such as a piece of paper or a tree, are **reflective objects** that reflect rays originating in self-luminous objects. The table below shows four important kinds of self-luminous sources.

**Self-luminous objects**

**A ray source**

Since a light ray is an idealization, there are no true ray sources. Still, the thin beam of a laser is often a good approximation of a single ray.

**A point source**

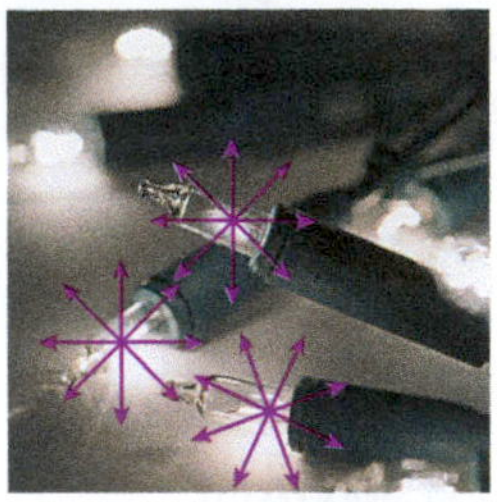

A point source is also an idealized source of light. It is infinitely small and emits light rays in every direction. The tiny filaments of these bulbs approximate point sources.

**An extended source**

This is the most common light source. The *entire surface* of an extended source is luminous, so that **every point of an extended source acts as a point source.** Lightbulbs, flames, and the sun are extended sources.

**A parallel-ray source**

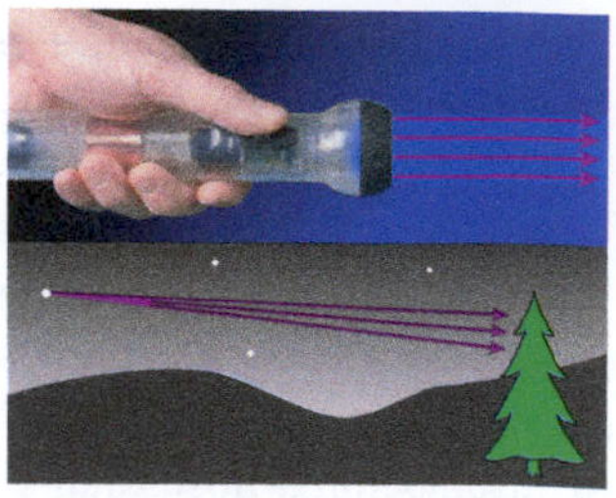

Certain sources, such as flashlights and movie projectors, produce a bundle of parallel rays. Rays from a very distant object, such as a star, are very nearly parallel.

Reflective objects, such as a newspaper, a face, or a mirror, can also be considered as sources of light rays. However, the origin of these rays is not in the object itself. Instead, light rays from a self-luminous object strike a reflective object and "bounce" off of it. These rays can then illuminate other objects, or enter our eyes and form images of the reflective object, just as rays from self-luminous objects do.

## Ray Diagrams

FIGURE 18.2 A ray diagram simplifies the situation by showing only a few rays.

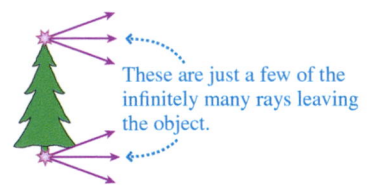

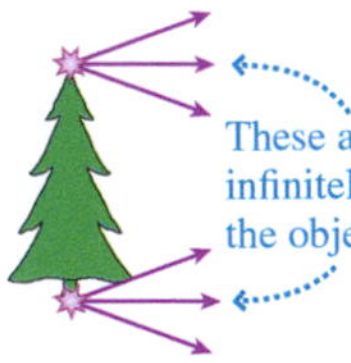

These are just a few of the infinitely many rays leaving the object.


Rays originate from *every* point on an object and travel outward in *all* directions, but a diagram trying to show all these rays would be hopelessly messy and confusing. To simplify the picture, we usually use a **ray diagram** that shows only a few rays. For example, FIGURE 18.2 is a ray diagram showing only a few rays leaving the top and bottom points of the object and traveling to the right. These rays will be sufficient to show us how the object is imaged by lenses or mirrors.

**NOTE** ▶ Ray diagrams are the basis for a *visual overview* that we'll use throughout this chapter. The rays shown on the diagram are a subset of the infinitely many rays leaving the object. ◀

FIGURE 18.3 A laser beam traveling through air is invisible.

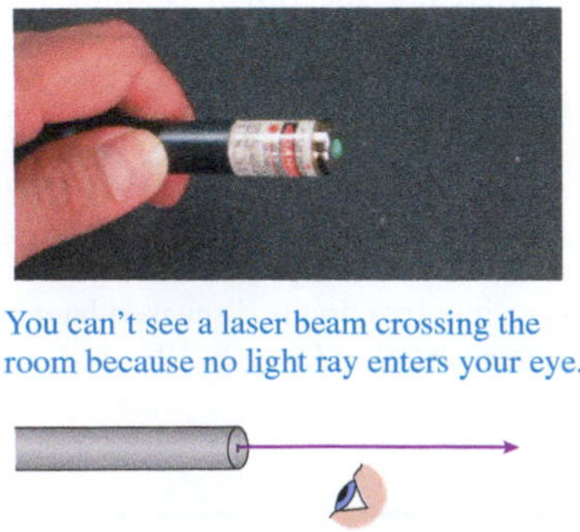

You can't see a laser beam crossing the room because no light ray enters your eye.


## Seeing Objects

How do we *see* an object? The eye works by focusing an image of an object on the retina, a process we'll examine in Chapter 19. **In order for our eye to see an object, rays from that object must enter the eye.** This idea helps explain some subtle points about seeing.

Can you see a laser beam traveling across the room? Under ordinary circumstances, the answer is no. As we've seen, a laser beam is a good approximation of a single ray. This ray travels in a straight line from the laser to whatever it eventually strikes. As FIGURE 18.3 shows, no light ray enters the eye, so the beam is invisible. The same argument holds for a parallel-ray source.

Everyone can see a point source or an extended source, though, as shown in FIGURE 18.4. Because a point source emits rays in *every* direction, some of these rays will enter the eye no matter where it is located. Thus a point source is visible to everyone looking at it. And, since every point on the surface of an extended source is itself a point source, all parts of an extended source (not blocked by something else) can be viewed by all observers as well.

We can also use our simple model of seeing to explain how we see nonluminous objects. Most ordinary objects—paper, skin, grass—reflect incident light in every direction, a process called **diffuse reflection.** FIGURE 18.5 illustrates the idea. Single rays are broken into many weaker rays that leave in all directions, a process called **scattering.**

Scattered light is what allows you to read a book by lamplight. As shown in Figure 18.5, every point on the surface of the page is struck by a ray (or rays) from the lamp. Then, because of diffuse reflection, these rays scatter in every direction; some of the scattered rays reach your eye, allowing you to see the page.

It is possible to make a laser beam visible by scattering it from very small particles suspended in air. These particles can be smoke, dust, or water droplets such as fog. FIGURE 18.6 shows that as the beam strikes such a particle, it scatters rays in every direction. Some of these rays enter the eye, making each particle in the path of the beam visible and outlining the beam's path across the room.

FIGURE 18.4 Point and extended sources can be seen by all observers.

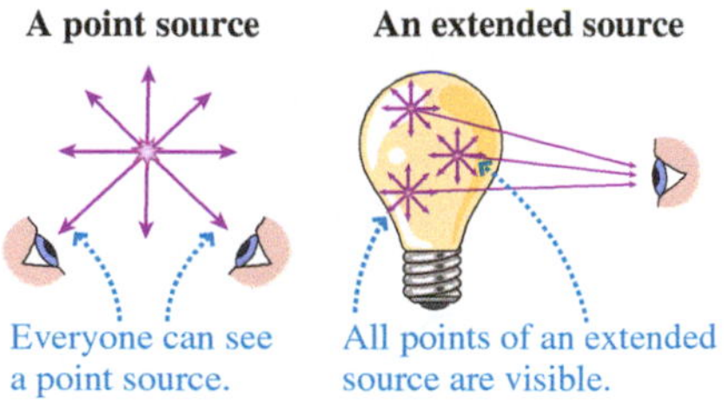


FIGURE 18.5 Reading a book by scattered light.

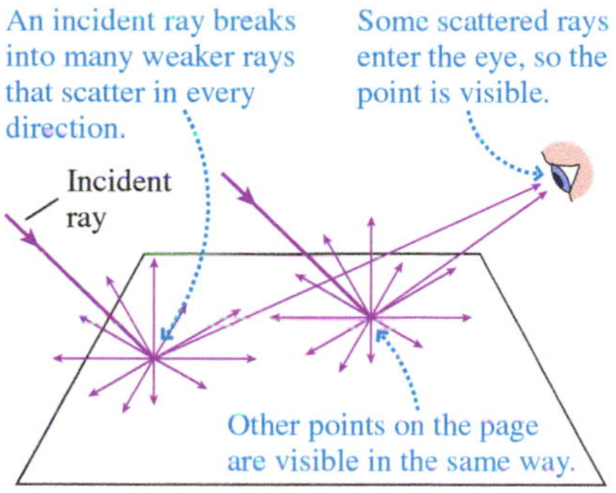


FIGURE 18.6 A laser beam is visible if it travels through smoke or dust.

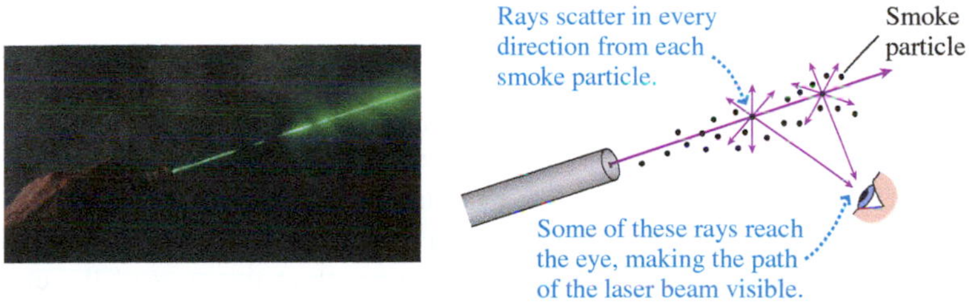


A video to support a section's topic is embedded in the eText.

Video Scattering

In Figure 18.6, you can see the laser beam because it scatters from smoke particles. Smoke particles and fog droplets are large compared to the wavelength of light, so all wavelengths of light are equally likely to scatter. This is why a cloud of smoke or a patch of fog appears white. But, as we saw in Chapter 17, if light scatters from objects that are much smaller than the wavelength of light, the short wavelengths of light are much more likely to scatter than the longer wavelengths in the spectrum.

Suppose the sun is near the horizon, so the rays of light from the sun are nearly horizontal, as in FIGURE 18.7. If you look up at the sky, you don't see the light from the sun unless the light is redirected toward your eye by scattering from air molecules. This scattering from air molecules is quite weak, but across 100 km of atmosphere, the amount of light scattered adds up, and a good deal of light is sent your way. As Figure 18.7 shows, the light rays that reach your eyes are predominantly from the short-wavelength end of the spectrum, so the sky appears blue. Scattering from small structures is responsible for the color of blue eyes; there is no blue pigment. If your eyes are blue, the blue of your eyes is literally the blue of the sky.

FIGURE 18.7 Scattering of light is responsible for the blue of the sky.

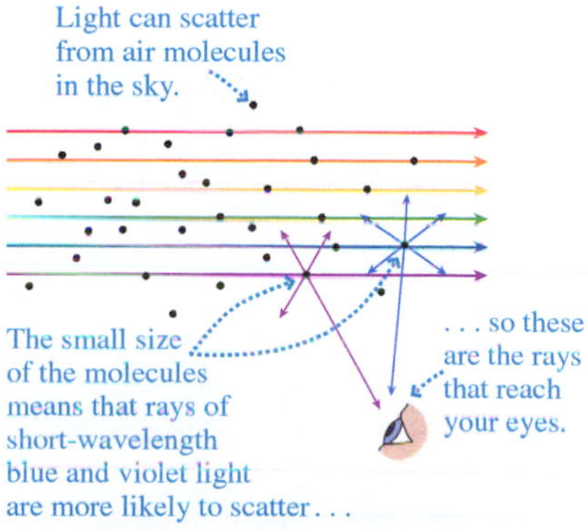


## Shadows

Suppose an opaque object (such as a cardboard disk) is placed between a source of light and a screen. The object intercepts some of the rays, leaving a dark area behind it.

FIGURE 18.8 Shadows produced by point and extended sources of light.

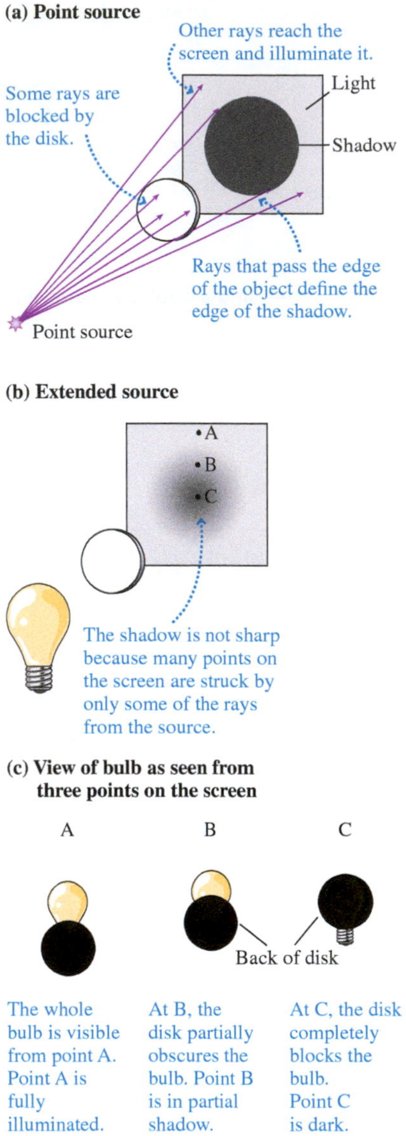


Other rays travel on to the screen and illuminate it. The simplest shadows are those cast by a point source of light, as in FIGURE 18.8a. With a point source, the shadow is completely dark, and the edges of the shadow are sharp.

Shadows cast by extended sources are more complicated. An extended source is a collection of a large number of point sources, each of which casts its own shadow. However, as shown in FIGURE 18.8b, the patterns of shadow and light from each point overlap and thus the shadow region is no longer sharp. FIGURE 18.8c shows the view of the *bulb* as seen from three points on the screen. Depending on the size of the source, there is often a true shadow that no light reaches, surrounded by a fuzzy region of increasing brightness.

**STOP TO THINK 18.1** The only source of light in a room is the lightbulb shown. An opaque disk is placed in front of the bulb. A screen is then placed successively at positions A, B, and C. At which screen position(s) does the disk cast a shadow that is completely dark at the center?

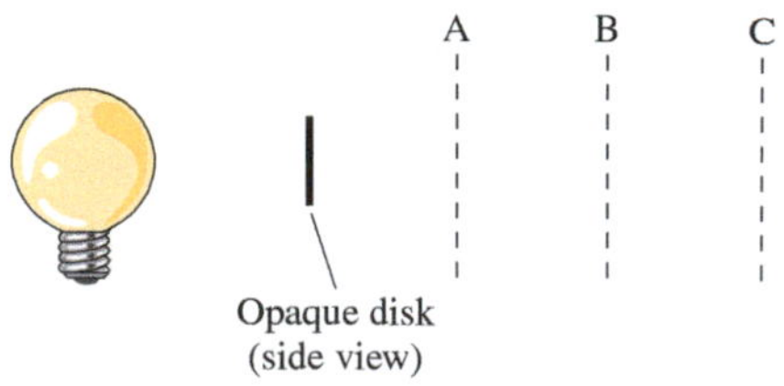


## 18.2 Reflection

As you know, light is reflected from smooth metal surfaces, like the silvered backs of mirrors. You see your reflection in the bathroom mirror first thing every morning as well as reflections in your car's rearview mirror as you drive to school. As you learned in Section 16.3, light also reflects from a discontinuity such as a boundary between two different materials as we saw in the photo at the start of the chapter, where the image of the goose was reflected from the still surface of a pond. Reflection from such smooth boundaries and smooth, shiny surfaces is called **specular reflection.**

FIGURE 18.9 Specular reflection of light.

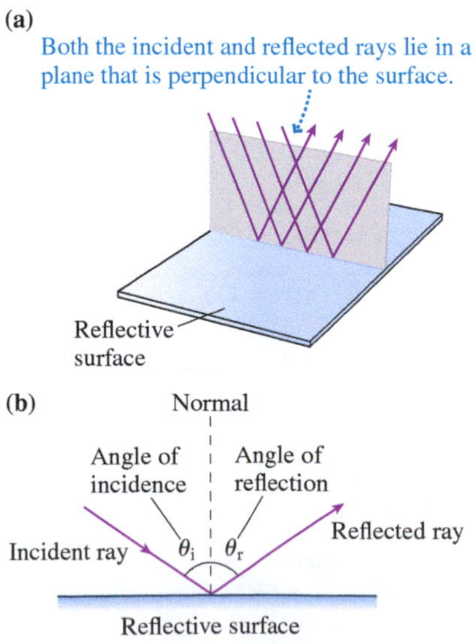


FIGURE 18.9a shows a bundle of parallel light rays reflecting from a mirror-like surface. You can see that the incident and reflected rays are both in a plane that is normal, or perpendicular, to the reflective surface. A three-dimensional perspective accurately shows the relationship between the light rays and the surface, but figures such as this are hard to draw by hand. Instead, it is customary to represent reflection with the simpler visual overview of FIGURE 18.9b. In this figure:

- The incident and reflected rays are in the plane of the page. The reflective surface extends into and out of the page.
- A *single* light ray represents the entire bundle of parallel rays. This is oversimplified, but it keeps the figure and the analysis clear.

The angle $\theta_i$ between the incident ray and a line perpendicular to the surface—the *normal* to the surface—is called the **angle of incidence.** Similarly, the **angle of reflection** $\theta_r$ is the angle between the reflected ray and the normal to the surface. The **law of reflection,** easily demonstrated with simple experiments, states:

1. The incident ray and the reflected ray are both in the same plane, which is perpendicular to the surface, and
2. The angle of reflection equals the angle of incidence: $\theta_r = \theta_i$.

**NOTE** ▸ Optics calculations *always* use the angle measured from the normal, not the angle between the ray and the surface. ◂

**EXAMPLE 18.1 Light reflecting from a mirror**

A full-length mirror on a closet door is 2.0 m tall. The bottom touches the floor. A bare lightbulb hangs 1.0 m from the closet door, 0.5 m above the top of the mirror. How long is the streak of reflected light across the floor?

**STRATEGIZE** We will treat the lightbulb as a point source, then trace rays of light from the bulb and use the law of reflection to see where they strike the floor. The range of the rays that strike the floor will be the length of the streak.

**PREPARE** FIGURE 18.10 is a visual overview of the light rays. We need to consider only the two rays that strike the edges of the mirror. All other reflected rays will fall between these two.

FIGURE 18.10 Visual overview of light rays reflecting from a mirror.

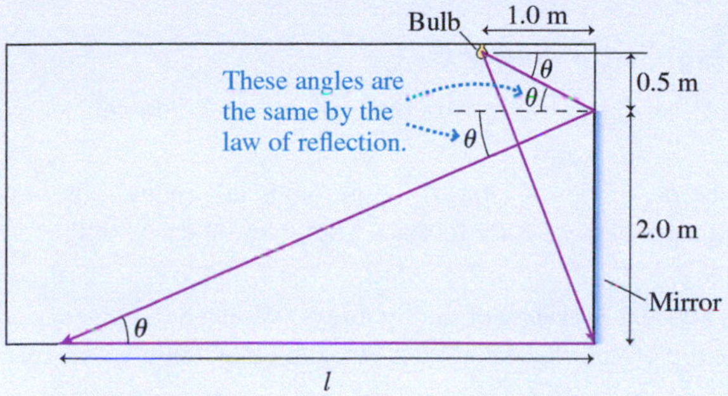


**SOLVE** The ray that strikes the bottom of the mirror reflects from it and hits the floor just where the mirror meets the floor. For the top ray, Figure 18.10 has used the law of reflection to set the angle of reflection equal to the angle of incidence; we call both $\theta$. By simple geometry, the other angles shown are also equal to $\theta$. From the small triangle at the upper right,

$$\theta = \tan^{-1}\left(\frac{0.5\ \text{m}}{1.0\ \text{m}}\right) = 26.6°$$

But we also have $\tan\theta = (2.0\ \text{m})/l$, or

$$l = \frac{2.0\ \text{m}}{\tan\theta} = \frac{2.0\ \text{m}}{\tan 26.6°} = 4.0\ \text{m}$$

Since the lower ray struck right at the mirror's base, the total length of the reflected streak is 4.0 m.

**ASSESS** The visual overview was drawn more or less to scale, and the length of the streak on the floor seems about twice as long as the mirror, so we can have confidence in the results of our calculation.

## Diffuse Reflection

If you magnify the surface of a diffuse reflector, you'll find that on the microscopic scale it is quite rough. The law of reflection $\theta_r = \theta_i$ is still obeyed at each point, but the irregularities of the surface cause the reflected rays to leave in many random directions. This situation is shown in FIGURE 18.11. Diffuse reflection is much more common than the mirror-like specular reflection.

FIGURE 18.11 Diffuse reflection from an irregular surface.

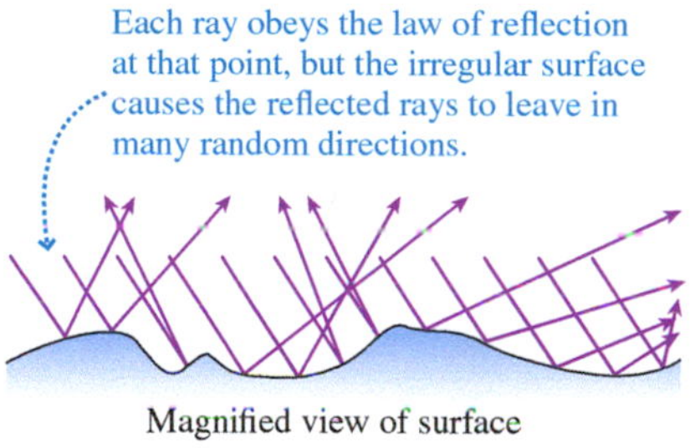


## The Plane Mirror

When you look at yourself in a mirror, how does this happen—how does the reflection of light lead to your seeing an image of yourself? FIGURE 18.12a shows rays from point source P reflecting from a flat mirror, called a **plane mirror**. Consider the two particular rays shown in FIGURE 18.12b. The horizontal ray strikes the mirror at a zero angle of incidence, so it reflects straight back with a zero angle of reflection. The second, angled ray reflects with $\theta_r = \theta_i$. As indicated by the dashed lines, both reflected rays appear to have come from a point P′. But because our argument applies to any incoming ray, *all* reflected rays appear to be coming from point P′, as FIGURE 18.12c shows. We call P′, the point from which the reflected rays diverge, the **virtual image** of P. The image is "virtual" in the sense that no rays actually leave P′, which is in darkness behind the mirror. But as far as your eye is concerned, the light rays act exactly *as if* the light really originated at P′. So while you may say "I see P in the mirror," what you are actually seeing is the virtual image of P.

FIGURE 18.12 The light rays reflecting from a plane mirror.

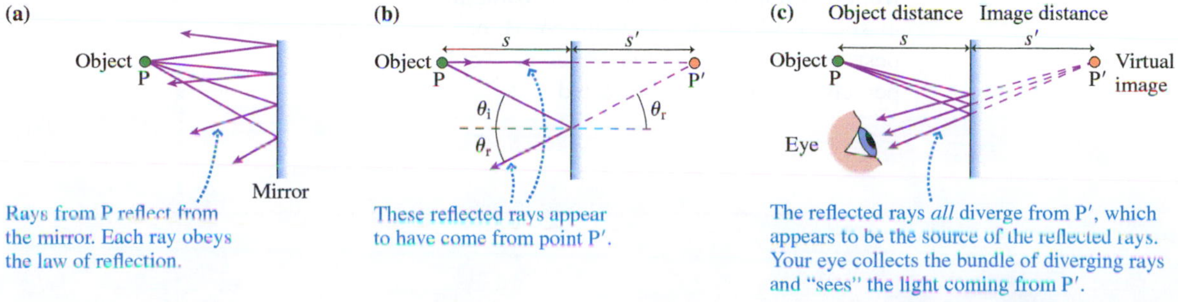

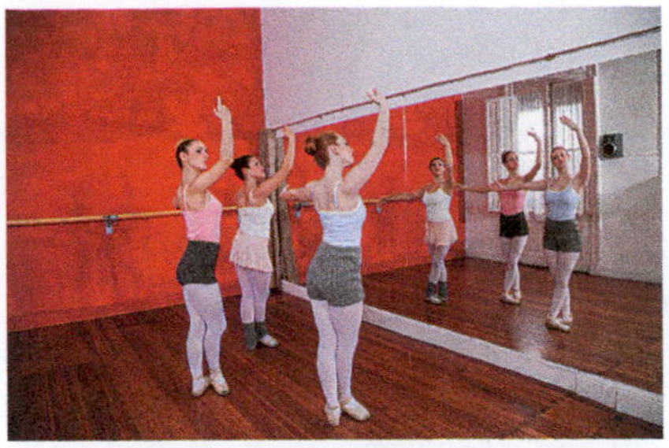

The dancers stand in front of the mirror. You can see their images in the mirror. Each image is the same distance behind the mirror as the dancer is in front of it.

**FIGURE 18.13** Viewing an extended object.

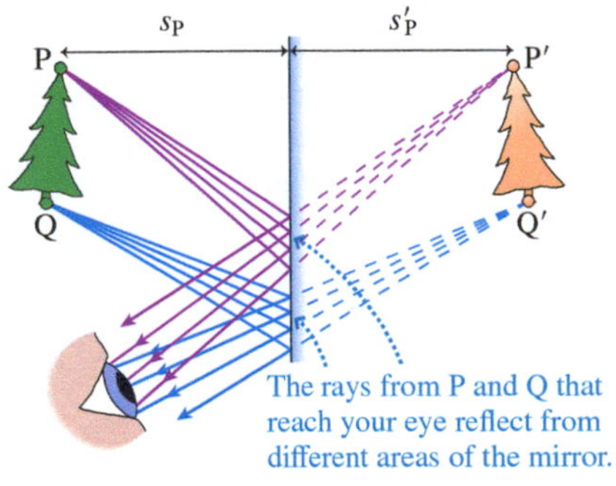


**Video** What the Physics? Ghost Truck

In Figure 18.12b, simple geometry dictates that P′ is the same distance behind the mirror as P is in front of the mirror. That is, the **image distance** $s'$ is equal to the **object distance** $s$:

$$s' = s \qquad \text{(plane mirror)} \qquad (18.1)$$

For an extended object, such as the one in FIGURE 18.13, each point on the object has a corresponding image point an equal distance on the opposite side of the mirror. The eye captures and focuses diverging bundles of rays from each point of the image in order to see the full image in the mirror. Two facts are worth noting:

- Rays from each point on the object spread out in all directions and strike *every point* on the mirror. Only a very few of these rays enter your eye, but the other rays are real and might be seen by other observers.
- Rays from points P and Q enter your eye after reflecting from *different* areas of the mirror. This is why you can't always see the full image of an object in a very small mirror.

**EXAMPLE 18.2 How high is the mirror?**

If your height is $h$, what is the shortest mirror on the wall in which you can see your full image?

**STRATEGIZE** We'll use the ray model of light. If the mirror is tall enough, there will be a path for rays from all parts of your body to reach your eyes, and you will be able to see your full image.

**PREPARE** FIGURE 18.14 is a visual overview of the light rays. We need to consider only the two rays that leave the top of your head and your feet and reflect into your eye.

**FIGURE 18.14** Visual overview of light rays from your head and feet reflecting into your eye.

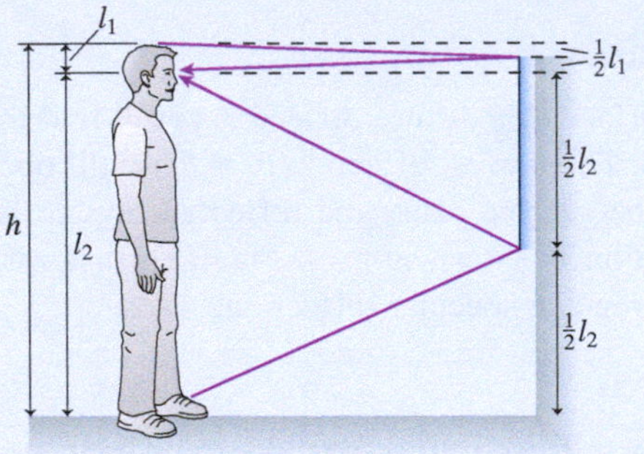


**SOLVE** Let the distance from your eyes to the top of your head be $l_1$ and the distance to your feet be $l_2$. Your height is $h = l_1 + l_2$. A light ray from the top of your head that reflects from the mirror at $\theta_r = \theta_i$ and enters your eye must, by congruent triangles, strike the mirror a distance $\frac{1}{2}l_1$ above your eyes. Similarly, a ray from your foot to your eye strikes the mirror a distance $\frac{1}{2}l_2$ below your eyes. The distance between these two points on the mirror is $\frac{1}{2}l_1 + \frac{1}{2}l_2 = \frac{1}{2}h$. A ray from anywhere else on your body will reach your eye if it strikes the mirror between these two points. Thus the shortest mirror in which you can see your full reflection is $\frac{1}{2}h$.

**ASSESS** At no point in our derivation did we use the distance from you to the mirror. This makes sense. Getting closer to the mirror doesn't allow you to see more of your image—it just gets you closer to the image. If you doubt this result, try it!

**STOP TO THINK 18.2** An object is placed in front of a mirror. The observer is positioned as shown. Which of the points, A, B, or C, best indicates where the observer would perceive the image to be located?

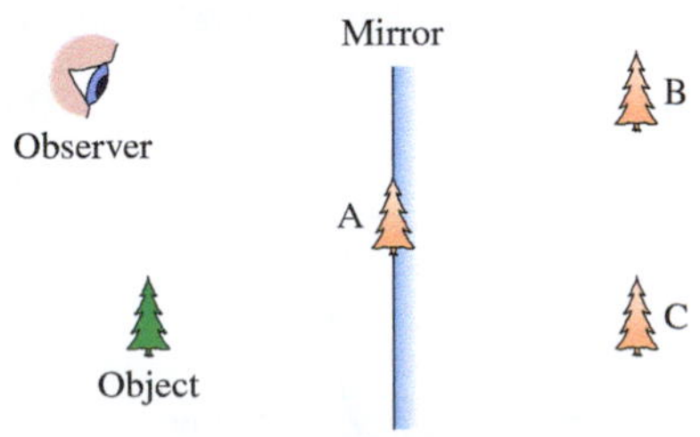

## 18.3 Refraction

FIGURE 18.15 shows light passing through a glass prism. The light goes from the air into the glass, then back into the air. Two things happen when a light ray crosses the boundary between the air and the glass:

- Part of the light *reflects* from the boundary, obeying the law of reflection. This is how you see reflections from pools of water or storefront windows, even though water and glass are transparent.
- Part of the light continues into the glass. It is *transmitted* rather than reflected, but the transmitted ray changes direction as it crosses the boundary. The transmission of light from one medium to another, but with a change in direction, is called **refraction.**

NOTE ▸ The transparent material through which light travels is called the medium (plural *media*). ◂

In Figure 18.15, the ray changes direction as the light enters the glass, and also as it leaves. This is the central concept we'll deal with when we consider refraction, the change in direction that occurs when light crosses from one medium into another.

Reflection from the boundary between transparent media is usually weak. Most of the light is refracted and passes through the boundary; usually only a small fraction is reflected. Our goal in this section is to understand refraction, so we will usually ignore the weak reflection and focus on the transmitted light.

FIGURE 18.16a shows the refraction of light rays from a parallel beam of light, such as a laser beam, and rays from a point source. These pictures remind us that an infinite number of rays are incident on the boundary, but our analysis will be simplified if we focus on a single light ray. FIGURE 18.16b is a ray diagram showing the refraction of a single ray at a boundary between medium 1 and medium 2. Let the angle between the ray and the normal be $\theta_1$ in medium 1 and $\theta_2$ in medium 2. Just as for reflection, the angle between the incident ray and the normal is the *angle of incidence*. The angle on the transmitted side, *measured from the normal,* is called the **angle of refraction.** Notice that $\theta_1$ is the angle of incidence in Figure 18.16b but is the angle of refraction in FIGURE 18.16c, where the ray is traveling in the opposite direction.

In 1621, Dutch scientist Willebrord Snell proposed a mathematical statement of the "law of refraction" or, as we know it today, Snell's law. If a ray refracts between medium 1 and medium 2, which have indices of refraction $n_1$ and $n_2$, the ray angles $\theta_1$ and $\theta_2$ in the two media are related by

$$n_1 \sin\theta_1 = n_2 \sin\theta_2 \qquad (18.2)$$

Snell's law for refraction between two media

Notice that Snell's law does not mention which is the incident angle and which the refracted angle.

TABLE 18.1 on the next page lists the indices of refraction for several media. The $n$ in the law of refraction, Equation 18.2, is the same index of refraction $n$ we studied in ◂◂ SECTION 17.1. There we found that the index of refraction determines the speed of a light wave in a medium according to $v = c/n$. Huygens' principle can be used to show that Snell's law is a *consequence* of the change in the speed of light as it moves across a boundary between media.

### Examples of Refraction

Look back at Figure 18.16. As the ray in Figure 18.16b moves from medium 1 to medium 2, where $n_2 > n_1$, it bends closer to the normal. In Figure 18.16c, where the

FIGURE 18.15 A light beam refracts twice in passing through a glass prism.

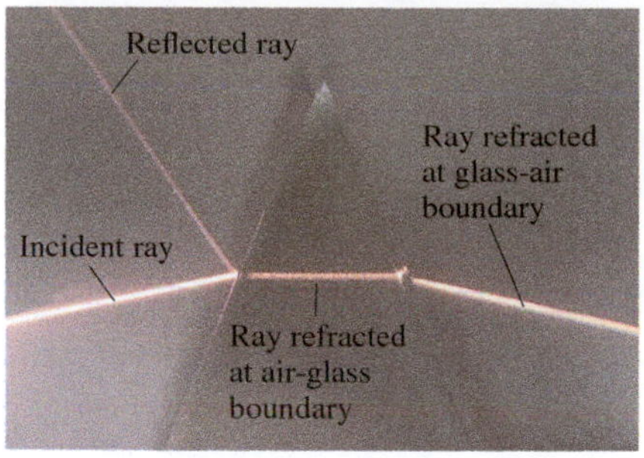


Video Index of Refraction

FIGURE 18.16 Refraction of light rays.

(a) Refraction of parallel and point-source rays

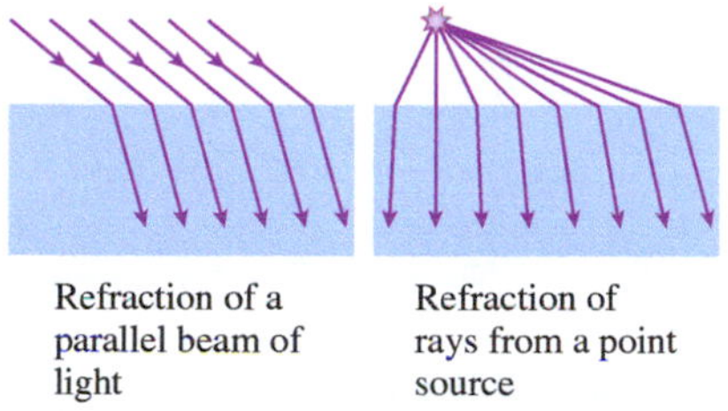


(b) Refraction from lower-index medium to higher-index medium

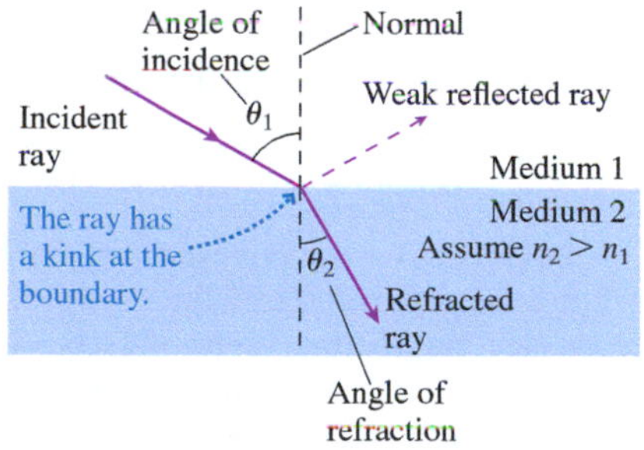


(c) Refraction from higher-index medium to lower-index medium

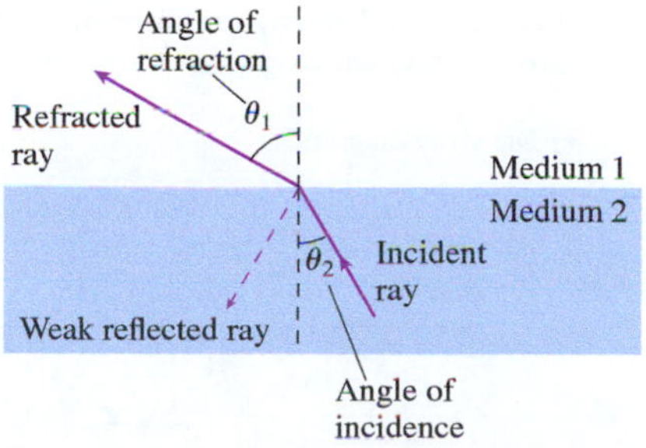

**TABLE 18.1** Indices of refraction

| Medium | $n$ |
|---|---|
| Vacuum | 1 exactly |
| Air (actual) | 1.0003 |
| Air (accepted)* | 1.00 |
| Water | 1.33 |
| Oil | 1.46 |
| Glass (typical) | 1.50 |
| Polystyrene plastic | 1.59 |
| Cubic zirconia | 2.18 |
| Diamond | 2.42 |

*Use this value in problems.

ray moves from medium 2 to medium 1, it bends away from the normal. This is a general conclusion that follows from Snell's law:

- When a ray is transmitted into a material with a higher index of refraction, it bends to make a smaller angle with the normal.
- When a ray is transmitted into a material with a lower index of refraction, it bends to make a larger angle with the normal.

This rule becomes a central idea in a procedure for analyzing refraction problems.

**TACTICS BOX 18.1 Analyzing refraction**

❶ **Draw a ray diagram.** Represent the light beam with one ray.

❷ **Draw a line normal (perpendicular) to the boundary.** Do this at each point where the ray intersects a boundary.

❸ **Show the ray bending in the correct direction.** The angle is larger on the side with the smaller index of refraction. This is the qualitative application of Snell's law.

❹ **Label angles of incidence and refraction.** Measure all angles from the normal.

❺ **Use Snell's law.** Calculate the unknown angle or unknown index of refraction.

Exercises 10–13

◀ **Erasing boundaries** Light reflects and refracts at the boundary between two transparent media, but only if there is a difference in their indices of refraction. You can make nearly invisible repairs to glass objects by using glue with an index of refraction that matches the index of the glass being repaired. Then light hits the boundary between the glue and the glass and continues without reflecting or refracting, so there is no way to tell that the boundary is there.

**EXAMPLE 18.3 Deflecting a laser beam**

A laser beam is aimed at a 1.0-cm-thick sheet of glass at an angle 30° above the glass.

a. What is the laser beam's direction of travel in the glass?
b. What is its direction in the air on the other side of the glass?

**STRATEGIZE** We will solve this problem following the steps in Tactics Box 18.1.

**PREPARE** We can represent the laser beam with a single ray and use the ray model of light. FIGURE 18.17 is a visual overview in which the first four steps of Tactics Box 18.1 have been identified. Notice that the angle of incidence must be measured from the normal, so $\theta_1 = 60°$, not the 30° value given in the problem. The index of refraction of glass is taken from Table 18.1. The index of refraction of air is assumed to be $n_{\text{air}} = 1$.

**FIGURE 18.17** The ray diagram of a laser beam passing through a sheet of glass.

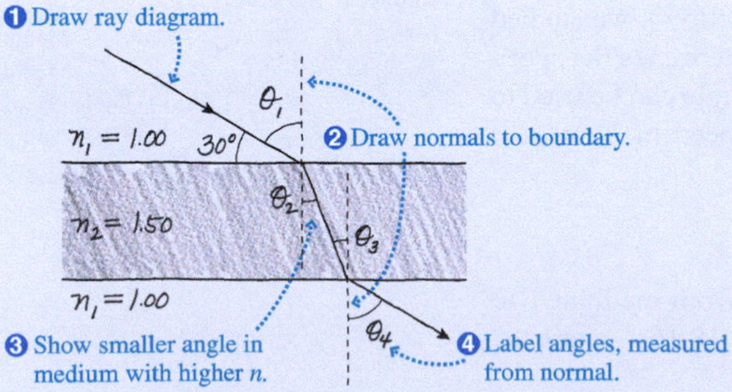


**SOLVE**

a. Snell's law, the final step in the Tactics Box, is $n_1 \sin\theta_1 = n_2 \sin\theta_2$. Using $\theta_1 = 60°$, we find that the direction of travel in the glass is

$$\theta_2 = \sin^{-1}\left(\frac{n_1 \sin\theta_1}{n_2}\right) = \sin^{-1}\left(\frac{\sin 60°}{1.5}\right)$$
$$= \sin^{-1}(0.577) = 35.3°$$

or 35° to two significant figures.

b. Snell's law at the second boundary is $n_2 \sin\theta_3 = n_1 \sin\theta_4$. You can see from Figure 18.17 that the interior angles are equal: $\theta_3 = \theta_2 = 35.3°$. Thus the ray emerges back into the air traveling at angle

$$\theta_4 = \sin^{-1}\left(\frac{n_2 \sin\theta_3}{n_1}\right) = \sin^{-1}(1.5 \sin 35.3°)$$
$$= \sin^{-1}(0.867) = 60°$$

This is the same as $\theta_1$, the original angle of incidence.

**ASSESS** As expected, the laser beam bends toward the normal as it moves into the higher-index glass, and away from the normal as it moves back into air. The beam exits the glass still traveling in the same direction as it entered, but its path is *displaced.* This is a general result for light traveling through a medium with parallel sides. As the glass becomes thinner, the displacement becomes less.

**EXAMPLE 18.4** **Measuring the index of refraction**

FIGURE 18.18 shows a laser beam deflected by a 30°-60°-90° prism. What is the prism's index of refraction?

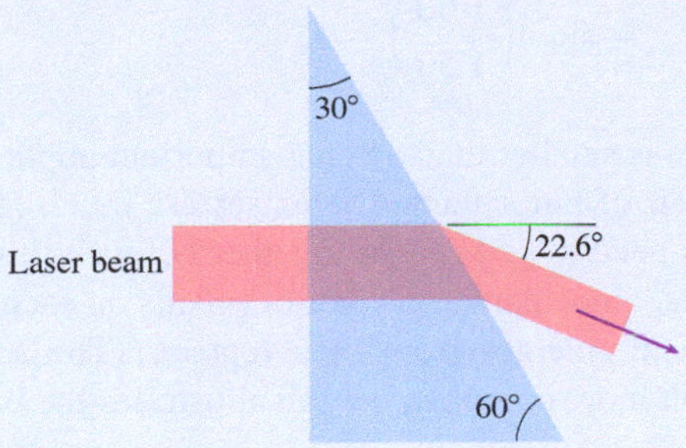


FIGURE 18.18 A prism deflects a laser beam.

**STRATEGIZE** We will represent the laser beam with a single ray and use the ray model of light.

**PREPARE** FIGURE 18.19 uses the steps of Tactics Box 18.1 to draw a ray diagram. The ray is incident perpendicular to the front face of the prism ($\theta_i = 0°$); thus it is transmitted through the first boundary without deflection. At the second boundary it is especially important to *draw the normal to the surface* at the point of incidence and to *measure angles from the normal.*

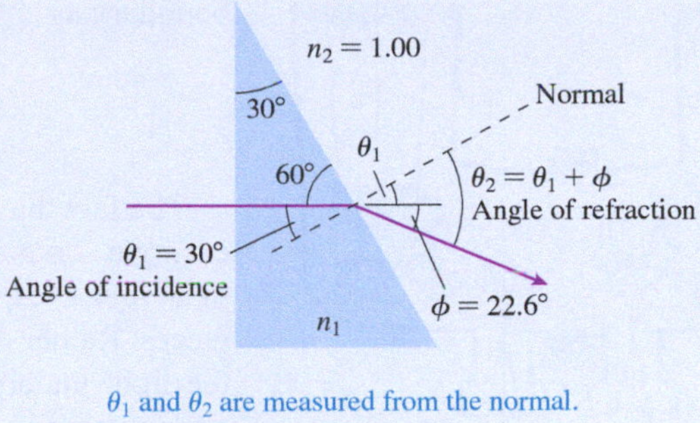


FIGURE 18.19 Visual overview of a laser beam passing through the prism.

**SOLVE** From the geometry of the triangle we find that the laser's angle of incidence on the hypotenuse of the prism is $\theta_1 = 30°$, the same as the apex angle of the prism. The ray exits the prism at angle $\theta_2$ such that the deflection is $\phi = \theta_2 - \theta_1 = 22.6°$. Thus $\theta_2 = 52.6°$. Knowing both angles and $n_2 = 1.00$ for air, we use Snell's law to find $n_1$:

$$n_1 = \frac{n_2 \sin \theta_2}{\sin \theta_1} = \frac{1.00 \sin 52.6°}{\sin 30°} = 1.59$$

**ASSESS** Referring to the indices of refraction in Table 18.1, we see that the prism could be made of polystyrene plastic.

## Total Internal Reflection

What would have happened in Example 18.4 if the prism angle had been 45° rather than 30°? The light rays would approach the rear surface of the prism at an angle of incidence $\theta_1 = 45°$. When we try to calculate the angle of refraction at which the ray emerges into the air, we find

$$\sin \theta_2 = \frac{n_1}{n_2} \sin \theta_1 = \frac{1.59}{1.00} \sin 45° = 1.12$$

$$\theta_2 = \sin^{-1}(1.12) = ???$$

Angle $\theta_2$ cannot be computed because the sine of an angle can't be greater than 1. The ray is unable to refract through the boundary. Instead, 100% of the light *reflects* from the boundary back into the prism. This process is called **total internal reflection,** often abbreviated TIR. That it really happens is illustrated in FIGURE 18.20. Here, three light beams strike the surface of the water at increasing angles of incidence. The two beams with the smallest angles of incidence refract out of the water, but the beam with the largest angle of incidence undergoes total internal reflection at the water's surface.

FIGURE 18.21 shows several rays leaving a point source in a medium with index of refraction $n_1$. The medium on the other side of the boundary has $n_2 < n_1$. As we've seen, crossing a boundary into a material with a lower index of refraction causes the ray to bend away from the normal. Two things happen as angle $\theta_1$ increases. First, the refraction angle $\theta_2$ approaches 90°. Second, the fraction of the light energy that is transmitted decreases while the fraction reflected increases.

A **critical angle** $\theta_c$ is reached when $\theta_2 = 90°$. Snell's law becomes $n_1 \sin \theta_c = n_2 \sin 90°$, or

$$\theta_c = \sin^{-1}\left(\frac{n_2}{n_1}\right) \qquad (18.3)$$

Critical angle of incidence for total internal reflection

FIGURE 18.20 One of the three beams of light undergoes total internal reflection.

FIGURE 18.21 Refraction and reflection of rays as the angle of incidence increases.

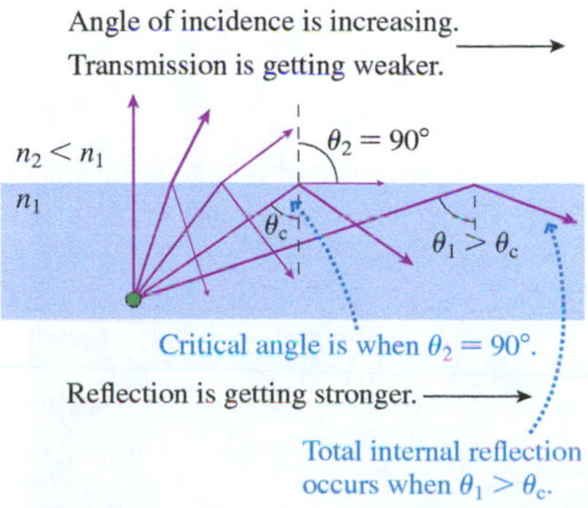


**Video** Figure 18.21

FIGURE 18.22 Binoculars and other optical instruments make use of total internal reflection.

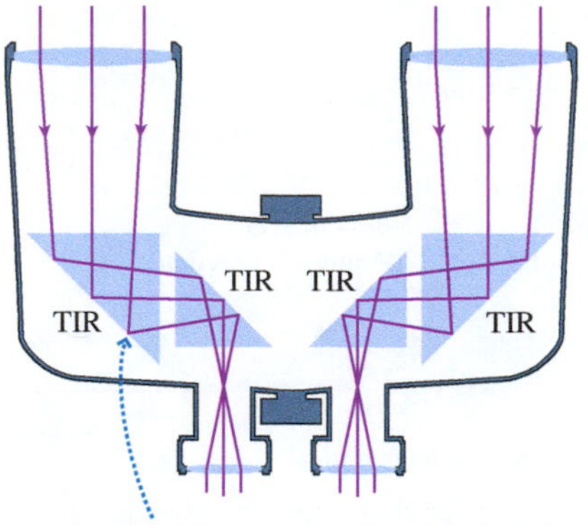


The refracted light vanishes at the critical angle and the reflection becomes 100% for any angle $\theta_1 > \theta_c$. The critical angle is well defined because of our assumption that $n_2 < n_1$. **There is no critical angle and no total internal reflection if $n_2 > n_1$.**

We can compute the critical angle in a typical piece of glass at the glass-air boundary as

$$\theta_{c\ glass} = \sin^{-1}\left(\frac{1.00}{1.50}\right) = 42°$$

The fact that the critical angle is smaller than 45° has important applications. For example, FIGURE 18.22 shows a pair of binoculars. The lenses are much farther apart than your eyes, so the light rays need to be brought together before exiting the eyepieces. Rather than using mirrors, binoculars use a pair of prisms on each side. Thus the light undergoes two TIRs and emerges from the eyepiece. (The actual prism arrangement in binoculars is a bit more complex, but this illustrates the basic idea.)

**EXAMPLE 18.5 Where can you see the starfish?**

A woman lies on a paddleboard floating on the still surface of a tropical lagoon, scanning the water below. A starfish rests on the sandy bottom of the lagoon, 3.0 m below the surface. What is the diameter of the circle on the surface of the water inside which the woman can see the starfish?

**STRATEGIZE** If we can see something, light rays from the object must be reaching our eyes. So we will treat the starfish as a source of light rays and consider the path of the rays to see if they make it from the water into the air. We'll draw rays coming off at all angles; only some of the rays refract into the air where they can be seen from above.

**PREPARE** We can model the starfish as a point source. We then sketch the path of the rays from the starfish as in FIGURE 18.23. Rays that strike the surface at an angle greater than the critical angle undergo total internal reflection back down into the water. The diameter of the circle of light is the distance $D$ between the two points at which rays strike the surface at the critical angle.

**SOLVE** From trigonometry, the circle diameter is $D = 2h \tan(\theta_c)$, where $h$ is the depth of the water. The critical angle for the water-air boundary is $\theta_c = \sin^{-1}(1.00/1.33) = 48.8°$. The diameter of the circle is thus

$$D = 2(3.0\text{ m}) \tan(48.8°) = 6.9\text{ m}$$

**ASSESS** This result seems reasonable if you've ever had the experience of looking for objects below the surface of the water. You can see objects directly below you, but you can't see objects off to the side.

FIGURE 18.23 Visual overview of the rays of light from the starfish.

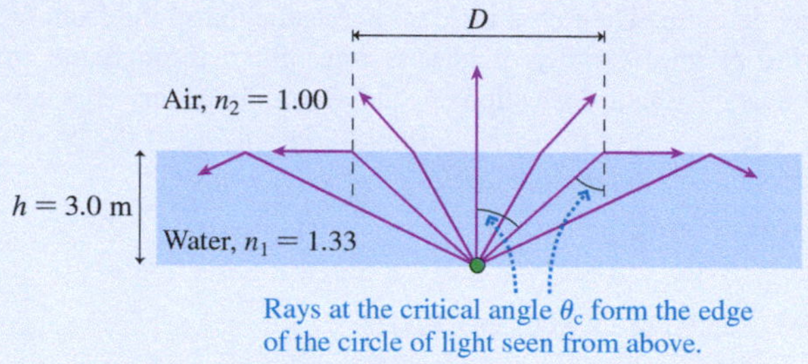


Now, let's think about the view from below the surface. If the woman instead rests on the bottom of the lagoon and looks up, what does she see?

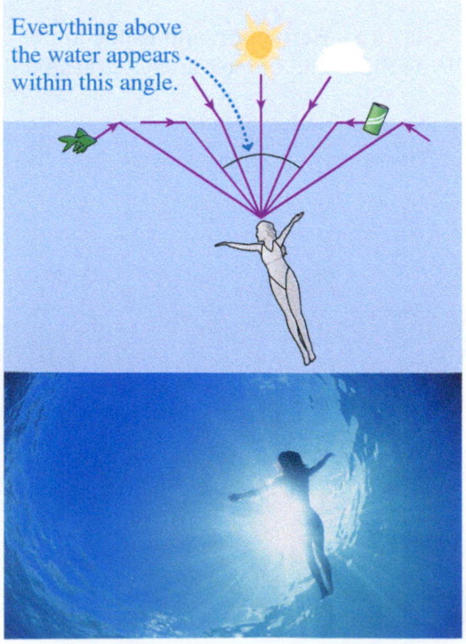

Everything above the water appears within this angle.

◀ **Snell's window** We can understand what a diver sees when she's underwater by reversing the direction of all the rays in Figure 18.23 in Example 18.5. The drawing shows that she can see the sun overhead and clouds at larger angles. She can even see objects sitting at the waterline—but they appear at the edge of a circle as she looks up. Anything outside of this circle is a reflection of something in the water. The photo shows what she sees: a bright circle from the sky above—*Snell's window*—surrounded by the dark reflection of the water below.

## Fiber Optics

The most important modern application of total internal reflection is the transmission of light through optical fibers. FIGURE 18.24a on the next page shows a laser beam shining into the end of a long, narrow-diameter glass fiber. The light rays pass easily from the air into the glass, but they then strike the inside wall of the fiber at an angle of incidence $\theta_1$ approaching 90°. This is much larger than the critical angle, so the laser beam undergoes TIR and remains inside the glass. The laser beam continues to "bounce" its way down the fiber as if the light were inside a pipe. Indeed, optical fibers are sometimes called "light pipes." The rays have an angle of incidence *smaller* than the critical angle ($\theta_1 \approx 0$) when they finally reach the flat end of the fiber; thus they refract out without difficulty and can be detected.

While a simple glass fiber can transmit light, dependence on a glass-air boundary is not sufficiently reliable for commercial use. Any small scratch on the side of the fiber alters the rays' angle of incidence and allows leakage of light. **FIGURE 18.24b** shows the construction of a practical optical fiber. A small-diameter glass *core* is surrounded by a layer of glass *cladding*. The glasses used for the core and the cladding have $n_{core} > n_{cladding}$. Thus, light undergoes TIR at the core-cladding boundary and remains confined within the core. This boundary is not exposed to the environment and hence retains its integrity even under adverse conditions.

Optical fibers have important applications in medical diagnosis and treatment. Thousands of small fibers can be fused together to make an *endoscope,* a flexible bundle capable of transmitting high-resolution images along its length. During *arthroscopic surgery,* surgeons perform operations on injured joints using an endoscope inserted through a small incision. The endoscope allows the surgeon to observe the procedure, which is performed with instruments inserted through another incision. The recovery time for such surgery is usually much shorter than for conventional operations that require a full incision to expose the interior of the joint.

**FIGURE 18.24** Light rays are confined within an optical fiber by TIR.

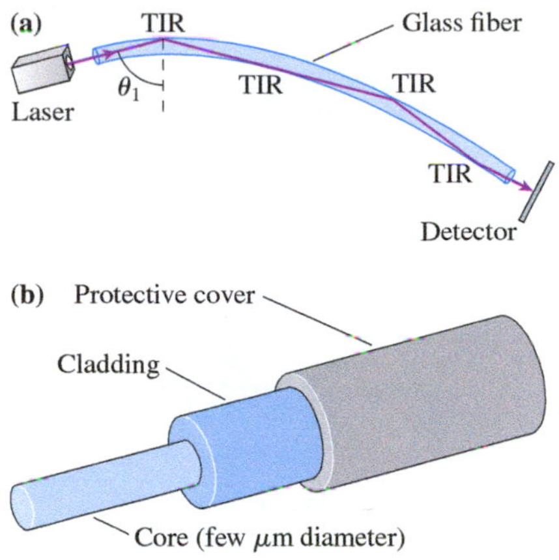


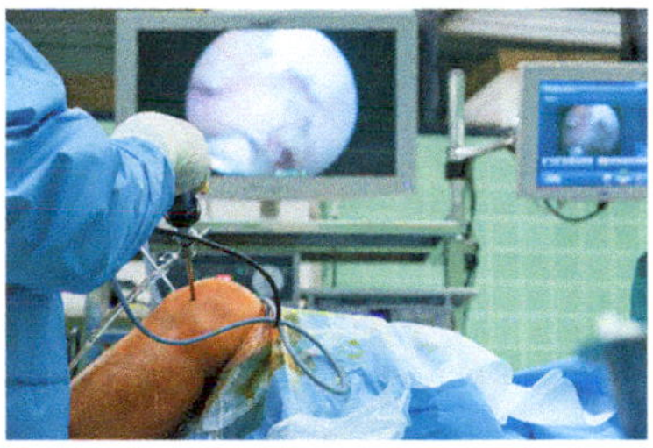

Arthroscopic surgery using an endoscope.

BIO

**STOP TO THINK 18.3** A light ray travels from medium 1 to medium 3 as shown. For these media,

A. $n_3 > n_1$
B. $n_3 = n_1$
C. $n_3 < n_1$
D. We can't compare $n_1$ to $n_3$ without knowing $n_2$.

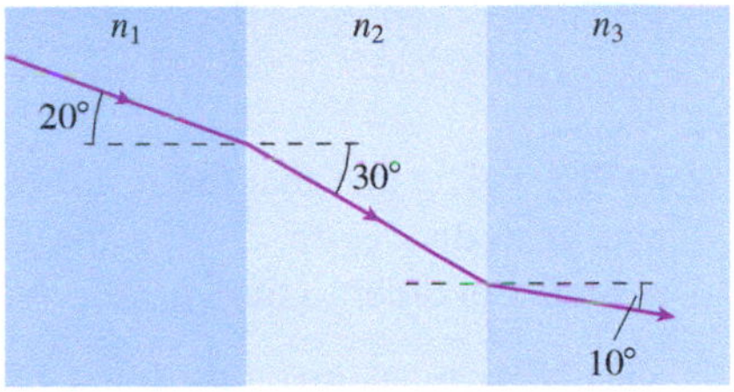


## 18.4 Image Formation by Refraction

**FIGURE 18.25a** shows a photograph of a ruler as seen through the front of an aquarium tank. The part of the ruler below the waterline appears *closer* than the part that is above water. **FIGURE 18.25b** shows why this is so. Rays that leave point P on the ruler refract away from the normal at the water-air boundary. (The thin glass wall of the aquarium has little effect on the refraction of the rays and can be ignored.) To your eye, outside the aquarium, these rays appear to diverge not from the object at point P, but instead from point P′ that is *closer* to the boundary. The same argument holds for every point on the ruler, so that **the ruler appears closer than it really is because of refraction of light at the boundary.**

We found that the rays reflected from a mirror diverge from a point that is not the object point. We called that point a *virtual image.* Similarly, if rays from an object point P refract at a boundary between two media such that the rays then diverge from a point P′ and *appear* to come from P′, we call P′ a virtual image of point P. The virtual image of the ruler is what you see.

Let's examine this image formation a bit more carefully. **FIGURE 18.26** on the next page shows a boundary between two transparent media that have indices of refraction $n_1$ and $n_2$. Point P, a source of light rays, is the object. Point P′, from which the rays *appear* to diverge, is the virtual image of P. The figure assumes $n_1 > n_2$, but this assumption isn't necessary. Distance *s,* measured from the boundary, is the object distance. Our goal is to determine the image distance $s'$.

The line through the object and perpendicular to the boundary is called the **optical axis.** Consider a ray that leaves the object at angle $\theta_1$ with respect to the optical axis. $\theta_1$ is also the angle of incidence at the boundary, where the ray refracts into the second medium at angle $\theta_2$. By tracing the refracted ray backward, we can see that $\theta_2$ is also the angle between the refracted ray and the optical axis at point P′.

**FIGURE 18.25** Refraction causes an object in an aquarium to appear closer than it really is.

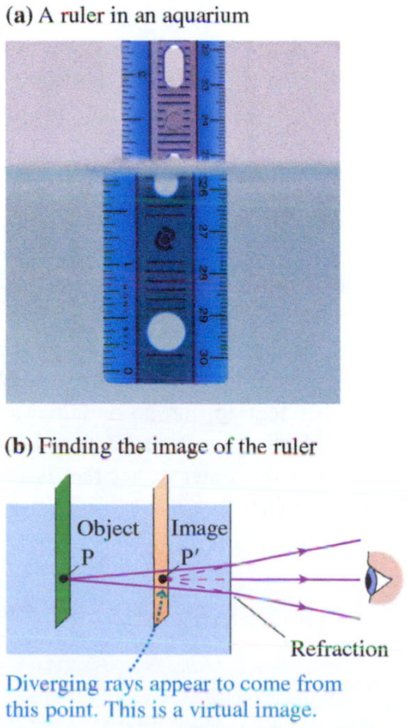

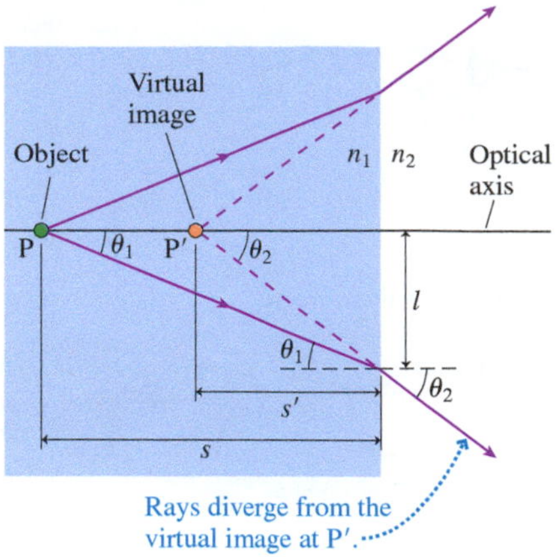


FIGURE 18.26 Finding the virtual image P′ of an object at P.

The distance $l$ is common to both the incident and the refracted rays, and we can see that $l = s \tan\theta_1 = s' \tan\theta_2$. Thus

$$s' = \frac{\tan\theta_1}{\tan\theta_2}s \tag{18.4}$$

Snell's law relates the sines of angles $\theta_1$ and $\theta_2$; that is,

$$\frac{\sin\theta_1}{\sin\theta_2} = \frac{n_2}{n_1} \tag{18.5}$$

In practice, the angle between any of these rays and the optical axis is very small because the pupil of your eye is very much smaller than the distance between the object and your eye. (The angles in the figure have been greatly exaggerated.) The small-angle approximation $\sin\theta \approx \tan\theta \approx \theta$, where $\theta$ is in radians, is therefore applicable. Consequently,

$$\frac{\tan\theta_1}{\tan\theta_2} \approx \frac{\sin\theta_1}{\sin\theta_2} = \frac{n_2}{n_1} \tag{18.6}$$

Using this result in Equation 18.4, we find that the image distance is

$$s' = \frac{n_2}{n_1}s \tag{18.7}$$

The index of refraction of the medium that the object is in ($n_1$). The index of refraction of the medium that the observer is in ($n_2$).

**NOTE** ► The fact that the result for $s'$ is independent of $\theta_1$ implies that *all* rays appear to diverge from the same point P′. This property of the diverging rays is essential in order to have a well-defined image. ◄

This section has given us a first look at image formation via refraction. We will extend this idea to image formation with lenses in the next section.

**EXAMPLE 18.6 Where are your feet?**

Rowan is walking in a shallow, clear bay, in still water just over her knees. When she looks down at her feet in the sand, she notes that they appear closer to her than usual. Normally, when she tips her head forward to look at her feet, her feet are 1.60 m from her eyes. How far away do her feet appear when they are under 0.60 m of water?

**STRATEGIZE** There is refraction at the surface of the water that will lead to a difference between the image distance, the perceived distance to her feet, and the object distance, the actual distance to her feet.

**PREPARE** Rowan's feet are the object; she observes with her eyes. In Equation 18.7, $n_1$ is the index of refraction of the medium the object is in—the water—so $n_1 = 1.33$. $n_2$ is the index of the medium the observer is in—the air—so $n_2 = 1$. The quantities $s$ and $s'$ in Equation 18.7 are the distances of the object and the image from the boundary, the surface of the water; these distances are shown in FIGURE 18.27. We are asked to find how far away her feet appear, so our final answer must include the distance $d$ from the water surface to her eyes. The distance from the surface of the water to her feet is $s = 0.60$ m; the distance from her eyes to her feet is $d + s = 1.60$ m, so $d = 1.00$ m.

**SOLVE** We use Equation 18.7 to find the image distance $s'$:

$$s' = \frac{n_2}{n_1}s = \frac{1}{1.33}(0.60\text{ m}) = 0.45\text{ m}$$

FIGURE 18.27 Visual overview of a woman looking at her feet under the water.

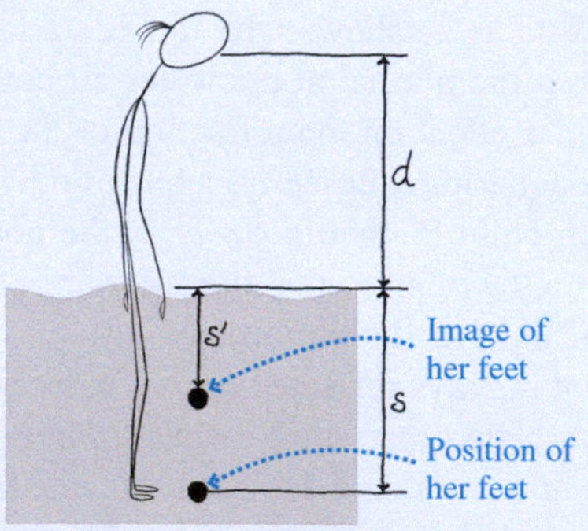


The total distance to the image of her feet is thus

$$d + s' = 1.00\text{ m} + 0.45\text{ m} = 1.45\text{ m}$$

This is noticeably closer than the actual distance.

**ASSESS** If you have ever looked down at your feet while standing in the water, you have observed this effect. Your feet really do seem too close!

# 18.5 Thin Lenses: Ray Tracing

A **lens** is a transparent object that uses refraction of light rays at *curved* surfaces to form an image. In this section we want to establish a pictorial method of understanding image formation. This method is called *ray tracing.*

FIGURE 18.28 Converging and diverging lenses.

(a) Converging lenses, which are thicker in the center than at the edges, refract parallel rays toward the optical axis.

(b) Diverging lenses, which are thinner in the center than at the edges, refract parallel rays away from the optical axis.

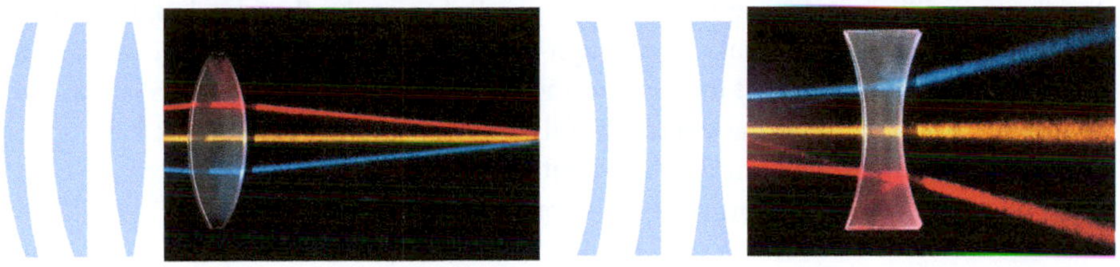

Video Real and Virtual Images

FIGURE 18.28 shows parallel light rays entering two different lenses. The lens in Figure 18.28a, called a **converging lens,** causes the rays to refract *toward* the optical axis. FIGURE 18.29 shows how this works. An incoming ray refracts toward the optical axis at both the first, air-to-glass boundary *and* the second, glass-to-air boundary. FIGURE 18.30a shows that *all* incoming rays initially parallel to the optical axis converge at the *same* point, the **focal point** of the lens. The distance of the focal point from the lens is called the **focal length** *f* of the lens.

If the parallel rays approached the lens from the right side instead, they would focus to a point on the left side of the lens, indicating that there is a focal point on *each* side of the lens. Both focal points are the same distance *f* from the lens. The focal point on the side from which the light is incident is the *near focal point;* the focal point on the other side is the *far focal point.*

The lens in Figure 18.28b, called a **diverging lens,** causes the rays to refract *away* from the axis. A diverging lens also has two focal points, although these are not as obvious as for a converging lens. FIGURE 18.30b clarifies the situation. A backward projection of the diverging rays shows that they all *appear* to have started from the same point. This is the near focal point of a diverging lens, and its distance from the lens is the focal length of the lens. For both types of lenses, **the focal length is the distance from the lens to the point at which rays parallel to the optical axis converge or from which they appear to diverge.**

The focal points and focal length are properties of the particular lens itself. The curvature of its surfaces and its index of refraction determine its focal length. Further, for a converging lens, only rays initially parallel to the optical axis converge at the focal point. Rays not initially parallel to the axis may also converge at a different point, but this point is not the focal point. For a diverging lens, only rays initially parallel to the optical axis appear to diverge from the focal point.

NOTE ▶ For now, we take the focal length *f* to be a positive quantity. Later, when we introduce the *thin-lens equation,* we'll find that the focal length is sometimes negative. ◀

FIGURE 18.29 Both surfaces of a converging lens bend an incident ray toward the optical axis.

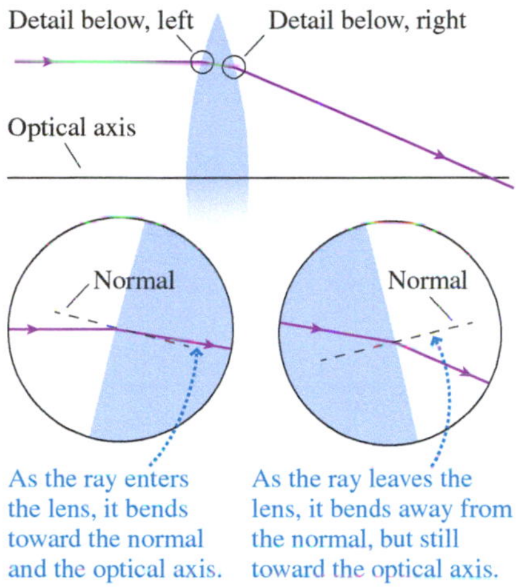

Detail below, left
Detail below, right
Optical axis
Normal
Normal
As the ray enters the lens, it bends toward the normal and the optical axis.
As the ray leaves the lens, it bends away from the normal, but still toward the optical axis.

FIGURE 18.30 The focal point and focal length of converging and diverging lenses.

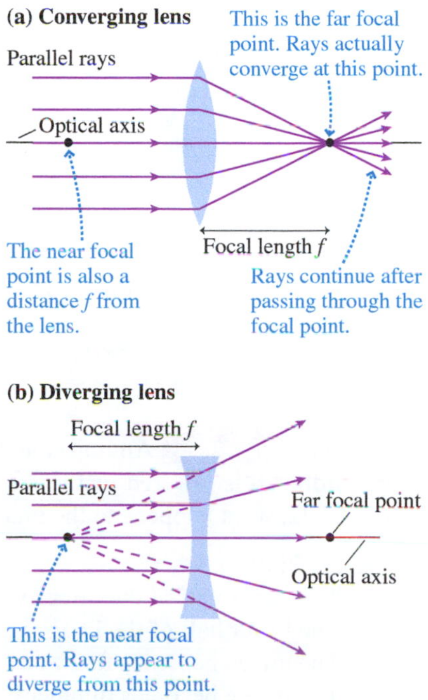


## Converging Lenses

These basic observations about lenses are enough for us to understand image formation by a **thin lens,** an idealized lens whose thickness is zero and that lies entirely in a plane called the **lens plane.** Within this *thin-lens approximation,* **all refraction occurs as the rays cross the lens plane, and all distances are measured from the lens plane.** Fortunately, the thin-lens approximation is quite good for most practical applications of lenses.

FIGURE 18.31 Three important sets of rays passing through a thin, converging lens.

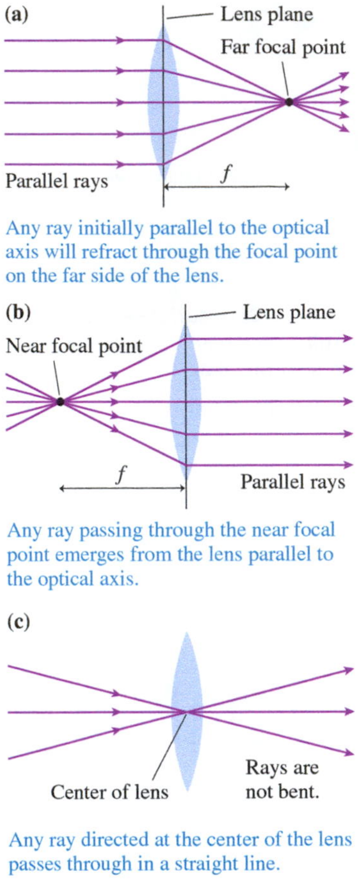


**NOTE** ► We'll *draw* lenses as if they have a thickness, because it will serve as a visual reminder of the lens type, but our analysis will not depend on the details of the drawing. ◄

FIGURE 18.31 shows three important sets of light rays passing through a thin, converging lens. Part (a) is familiar from Figure 18.30a. If the direction of each of the rays in Figure 18.31a is reversed, Snell's law tells us that each ray will exactly retrace its path and emerge from the lens parallel to the optical axis. This leads to Figure 18.31b, which is the "mirror image" of part (a).

Figure 18.31c shows three rays passing through the *center* of the lens. At the center, the two sides of a lens are very nearly parallel to each other. Earlier, in Example 18.3, we found that a ray passing through a piece of glass with parallel sides is *displaced* but *not bent* and that the displacement is small if the thickness is small. Consequently, a ray through the center of a thin lens, which we model as having zero thickness, is neither bent nor displaced but travels in a straight line.

These three situations form the basis for ray tracing.

## Real Images

FIGURE 18.32 shows a lens and an object whose distance $s$ from the lens is larger than the focal length. The figure illustrates the principles of ray tracing that we'll use to determine the location of an image formed by the lens. For now, we'll focus on one point on the object, labeled P. Rays leave this point in all directions, but we've shown only a few, and we'll further restrict our attention to three *special rays*, labeled in the figure, that have particular properties. The first special ray is parallel to the optical axis before the lens, so it goes through the far focal point. The second special ray goes through the center of the lens and so is not deflected. The third special ray goes through the near focal point, so it exits the lens parallel to the optical axis. These three rays converge at point P′ on the far side of the lens.

We've singled out three special rays, and drawn a few more, but *all* of the rays from point P on the object that pass through the lens are refracted by the lens so as to

**KEY CONCEPT** FIGURE 18.32 Rays from an object point P are refracted by the lens and converge to a real image at point P′.

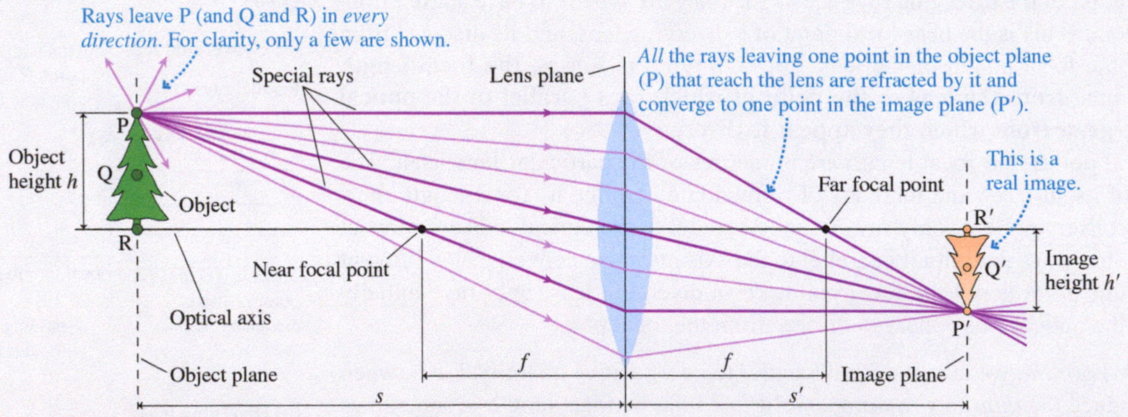


**STOP TO THINK 18.4** An object and lens are positioned to form a well-focused image on a viewing screen. Then a piece of cardboard is lowered just in front of the lens to cover the top half of the lens. Using what you've learned from Figure 18.32, what happens to the image on the screen?

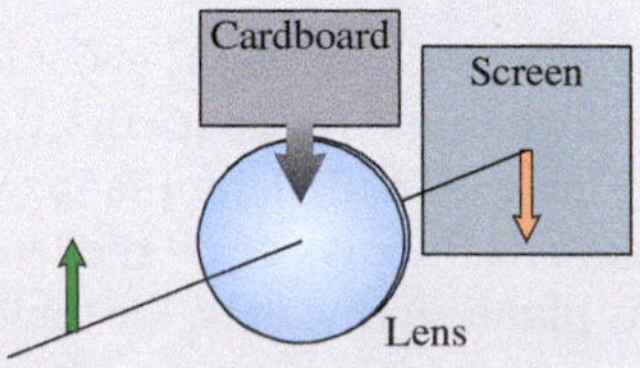


A. Nothing
B. The upper half of the image vanishes.
C. The lower half of the image vanishes.
D. The image becomes fuzzy and out of focus.
E. The image becomes dimmer but remains in focus.

converge at point P′ on the opposite side of the lens, at a distance $s'$ from the lens. When rays diverge from point P on the object and interact with a lens such that the refracted rays *converge* at point P′, then we call P′ a **real image** of point P. Contrast this with our prior definition of a *virtual image* as a point from which rays appear to *diverge,* but through which no rays actually pass.

In Figure 18.32, all the rays from point Q on the object converge at point Q′, and all the rays from point R converge at point R′. Ultimately, each point on the object gives rise to a corresponding point on the image. All object points that are in the same plane, the **object plane,** converge to image points in the **image plane.** Points Q and R in the object plane of Figure 18.32 have image points Q′ and R′ in the same plane as point P′. Once we locate *one* point in the image plane, such as point P′, we know that the full image lies in the same plane.

There are two additional observations to make about Figure 18.32. First, as also seen in FIGURE 18.33, the image is upside down with respect to the object. This is called an **inverted image,** and it is a standard characteristic of real-image formation. Second, rays from point P fill the entire lens surface, so that all portions of the lens contribute to the image. A larger lens will "collect" more rays and thus make a brighter image.

FIGURE 18.34 is a close-up view of the rays and images very near the image plane. The rays don't stop at P′ unless we place a screen in the image plane. When we do so, we see a sharp, well-focused image on the screen. If a screen is placed in front of or behind the image plane, an image is produced on the screen, but it's blurry and out of focus.

**NOTE** ▶ Our ability to see a real image on a screen sets real images apart from virtual images. But keep in mind that we need not *see* a real image in order to *have* an image. A real image exists at a point in space where the rays converge even if there's no viewing screen in the image plane. ◀

Figure 18.32 highlights the three "special rays" that are based on the three situations of Figure 18.31. Notice that these three rays alone are sufficient to locate the image point P′. That is, we don't need to draw all the rays shown in Figure 18.32. If you use graph paper and draw your lines carefully, as outlined in the following Tactics Box, you can locate the image with good accuracy.

**TACTICS BOX 18.2 Ray tracing for a converging lens**

1. **Draw an optical axis.** Use graph paper or a ruler! Establish an appropriate scale.
2. **Center the lens on the axis.** Draw the lens plane perpendicular to the axis through the center of the lens. Mark and label the focal points at distance $f$ on either side.
3. **Represent the object with an upright arrow at distance $s$.** It's usually best to place the base of the arrow on the axis and to draw the arrow smaller than the radius of the lens.
4. **Draw the three "special rays" from the tip of the arrow.** Use a straightedge or a ruler. The rays refract at the lens plane, *not* at the surfaces of the lens.
   a. A ray initially parallel to the axis refracts through the far focal point.
   b. A ray that enters the lens along a line through the near focal point emerges parallel to the axis.
   c. A ray through the center of the lens does not bend.
5. **Extend the rays until they converge.** The rays converge at the image point. Draw the rest of the image in the image plane. If the base of the object is on the axis, then the base of the image will also be on the axis.
6. **Measure the image distance $s'$.** Also, if needed, measure the image height relative to the object height. The magnification can be found from Equation 18.9.

Exercise 19

Video Partially Covering a Lens

FIGURE 18.33 The lamp's image is upside down.

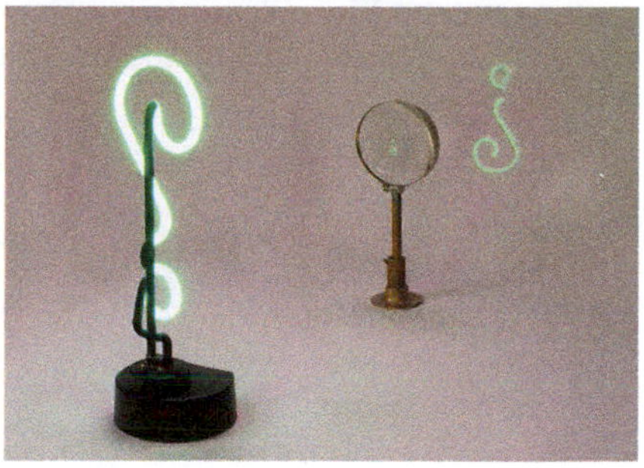

FIGURE 18.34 A close-up look at the rays and images near the image plane.

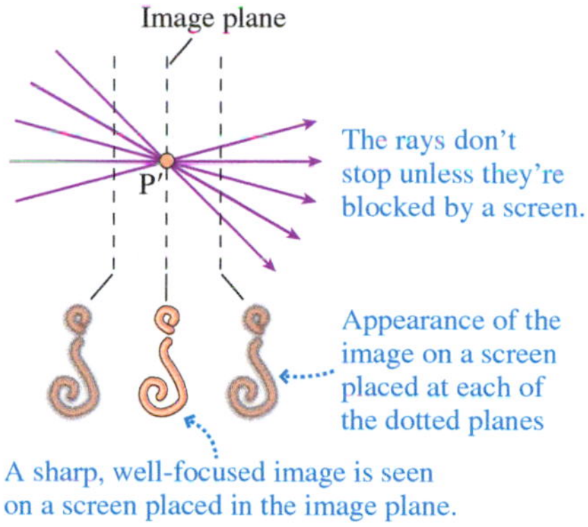

**EXAMPLE 18.7 Finding the image of a flower**

A 4.0-cm-diameter flower is 200 cm from the 50-cm-focal-length lens of a camera. How far should the plane of the camera's light detector be placed behind the lens to record a well-focused image? What is the diameter of the image on the detector?

**STRATEGIZE** We will use ray tracing, following the steps of Tactics Box 18.2, to find the location of the image.

**PREPARE** We start with an accurate drawing of the situation, scaled so that 25 cm corresponds to 1 cm on the diagram.

**SOLVE** FIGURE 18.35 shows the ray-tracing diagram and the steps of Tactics Box 18.2. The image has been drawn in the plane where the three special rays converge. You can see *from the drawing* that the image distance is $s' \approx 65$ cm. This is where the detector needs to be placed to record a focused image. The heights of the object and image are labeled $h$ and $h'$. These are, respectively, the diameter of the flower and the diameter of the flower's image. The ray through the center of the lens is a straight line; thus the object and image both subtend the same angle $\theta$. From similar triangles,

$$\frac{h'}{s'} = \frac{h}{s}$$

Solving for $h'$ gives

$$h' = h\frac{s'}{s} = (4.0 \text{ cm})\frac{65 \text{ cm}}{200 \text{ cm}} = 1.3 \text{ cm}$$

The flower's image has a diameter of 1.3 cm.

**ASSESS** The ray-tracing process we have outlined has a built-in check. The first two rays converge at a point; the third ray should converge at the same point. If it doesn't, there is an error in our work. In this case, all three rays converge at the same point, so we can have confidence in our result.

FIGURE 18.35 Ray-tracing diagram for the image of a flower.

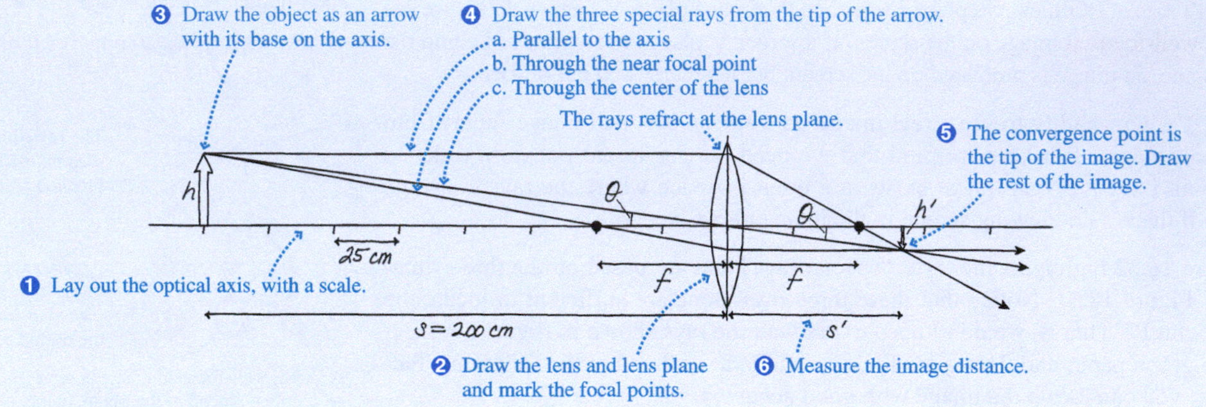


## Magnification

The image can be either larger or smaller than the object, depending on the location and focal length of the lens. Because the image height scales with that of the object, we're usually interested in the *ratio* $h'/h$ of the image height to the object height. This ratio is greater than 1 when the image is taller than the object, and less than 1 when the image is shorter than the object.

But there's more to a description of the image than just its size. We also want to know its *orientation* relative to the object; that is, is the image upright or inverted? Image size and orientation information are usually combined in a single number, the **magnification** $m$. The *absolute value* of the magnification is the ratio of image height to object height:

$$|m| = \frac{h'}{h} \tag{18.8}$$

The *sign* of the magnification gives the orientation. A negative value of $m$ indicates that the image is inverted relative to the object; a positive value of $m$ indicates that the image is upright.

The magnification is related to the image and object distances. You saw in Example 18.7 that the heights and positions of the image and the object are related, with $h'/h = s'/s$. We can replace the ratio of heights in Equation 18.8 with the ratio of image and object distances. We also add a negative sign so that the sign of the magnification matches the orientation. In Example 18.7 for instance, the image was inverted, so $m$ must be negative in this case. With these changes, our expression for magnification becomes:

$$m = -\frac{s'}{s} \tag{18.9}$$

Magnification of a lens or mirror

As an example, the magnification in Example 18.7 is

$$m = -\frac{s'}{s} = -\frac{65\text{ cm}}{200\text{ cm}} = -0.33$$

which indicates that the image is 33% the size of the object and, because of the minus sign, is inverted. This matches the diagram nicely, as it should.

**NOTE** ▶ Equation 18.9 applies to real or virtual images produced by both lenses and mirrors. Although $s$ and $s'$ are both positive in Example 18.7, leading to a negative magnification and an inverted image, we'll see that this is *not* the case for virtual images. ◀

**STOP TO THINK 18.5** A lens produces a sharply focused, inverted image on a screen. What will you see on the screen if the lens is removed?

A. The image will be inverted and blurry.
B. The image will be upright and sharp.
C. The image will be upright and blurry.
D. The image will be much dimmer but otherwise unchanged.
E. There will be no image at all.

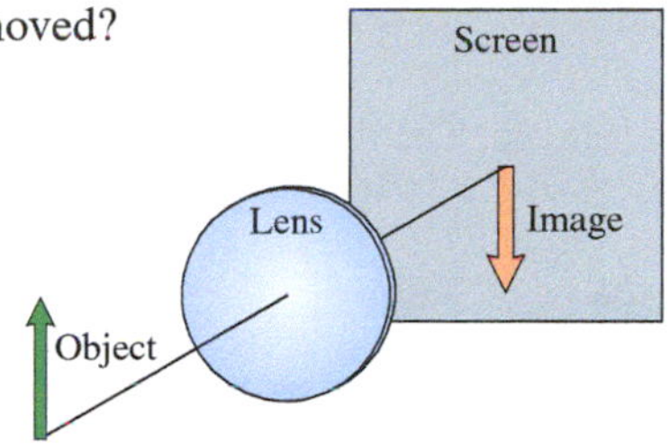


FIGURE 18.36 Rays from an object at distance $s < f$ are refracted by the lens and form a virtual image at point P′.

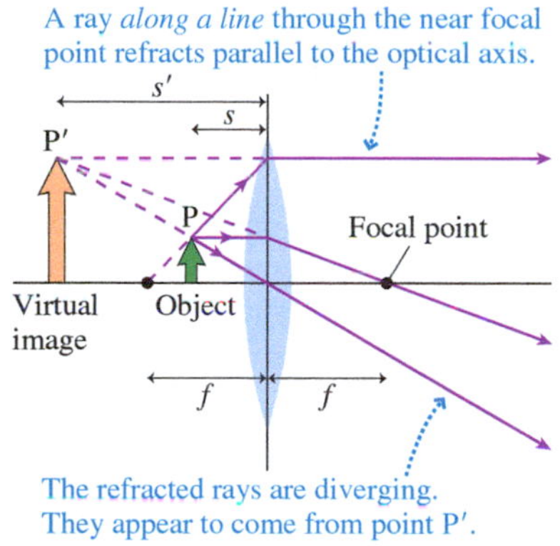


## Virtual Images

The preceding section considered a converging lens with the object at distance $s > f$; the object was outside the focal point. What if the object is inside the focal point, at distance $s < f$? FIGURE 18.36 shows ray tracing to locate the image for this situation.

The special rays initially parallel to the axis and through the center of the lens present no difficulties. However, a ray through the near focal point would travel toward the left and would never reach the lens! Referring back to Figure 18.31b, we can see that the rays emerging parallel to the axis entered the lens *along a line* passing through the near focal point. It's the angle of incidence on the lens that is important, not whether the light ray actually passes through the focal point. This was the basis for the wording of step 4b in Tactics Box 18.2 and is the third special ray shown in Figure 18.36.

The object is closer to the lens than the focal point, so the rays that reach the lens are strongly diverging. Refraction by the converging lens will reduce the degree of divergence, but this isn't sufficient to make the rays converge. The three rays appear to *diverge* from point P′. This is the situation we found for rays reflecting from a mirror and for the rays refracting out of an aquarium. Point P′ is a *virtual image* of the object point P. Furthermore, it is an **upright image,** having the same orientation as the object.

The refracted rays, which are all to the right of the lens, *appear* to come from P′, but none of the rays were ever at that point. No image would appear on a screen placed in the image plane at P′. So what good is a virtual image?

Your eye collects and focuses bundles of diverging rays. Thus, as FIGURE 18.37a shows, you can "see" a virtual image by looking through the lens. This is exactly what you do with a magnifying glass, producing a scene like the one in FIGURE 18.37b.

FIGURE 18.37 A converging lens is a magnifying glass when the object distance is $< f$.

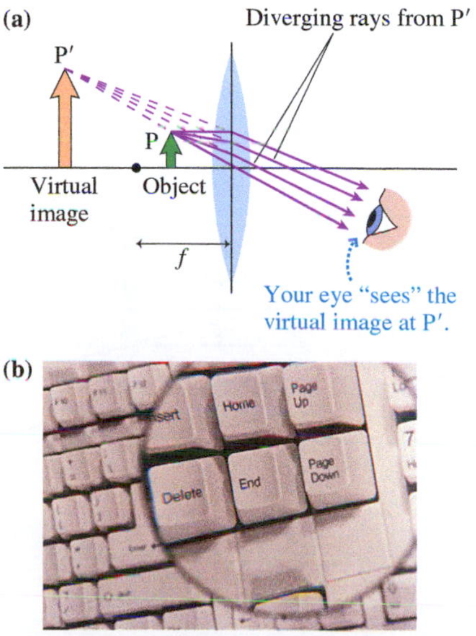

**NOTE** ► Rays from an object passing through a converging lens can converge to form a real image if the object distance is greater than the focal length of the lens: $s > f$. If $s < f$, the rays leaving a converging lens do not converge, and the result is a virtual image. ◄

Because a virtual image is upright, the magnification $m = -s'/s$ is positive. This means that the ratio $s'/s$ must be *negative*. We can ensure this if **we define the image distance $s'$ to be negative for a virtual image,** which indicates that the image is on the *same* side of the lens as the object. This is our first example of a **sign convention** for the various distances that appear in understanding image formation from lenses and mirrors. We'll have more to say about sign conventions when we study the thin-lens equation in a later section.

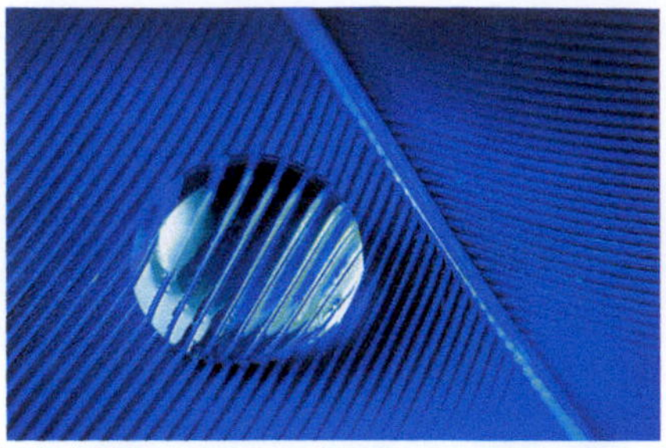

◄ **Water lens** Small droplets of water bead up on a feather. These droplets have the shape of a converging lens, which make a virtual image of the feather underneath, providing a clearer view of the feather's structure. This is the same effect at work in the image on the cover of the text.

**EXAMPLE 18.8 Finding the image for a converging lens**

To see a flower better, you hold a 6.0-cm-focal-length magnifying glass 4.0 cm from the flower. What is the magnification?

***PREPARE*** The flower is in the object plane. We use ray tracing to locate the image. Once the image distance is known, we can use Equation 18.9 to find the magnification.

***SOLVE*** FIGURE 18.38 shows the ray-tracing diagram. The three special rays diverge from the lens, but we can use a straightedge to extend the rays backward to the point from which they diverge. This point, the image point, is seen to be 12 cm to the left of the lens. Because this is a virtual image, the image distance is $s' = -12$ cm. From Equation 18.9 the magnification is

$$m = -\frac{s'}{s} = -\frac{-12\text{ cm}}{4.0\text{ cm}} = 3.0$$

FIGURE 18.38 Ray-tracing diagram for a magnifying glass.

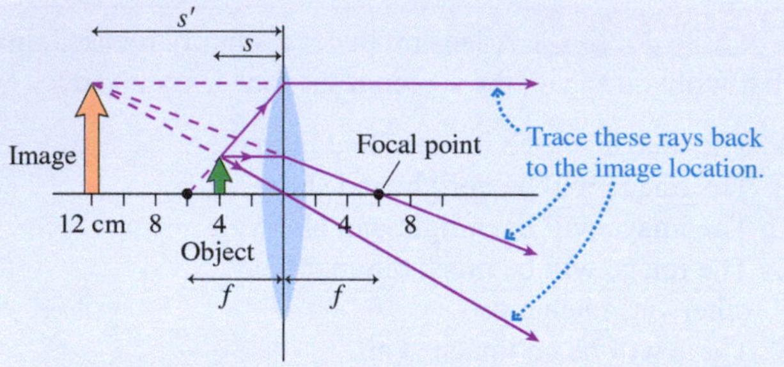


***ASSESS*** The three special rays converge to a single point, and the image orientation and size in our ray-tracing diagram match the computed magnification—the image is upright and much larger than the object—so we have confidence in our results.

## Diverging Lenses

As Figure 18.28b showed, a *diverging lens* is one that is thinner at its center than at its edges. FIGURE 18.39 shows three important sets of rays passing through a diverging lens. These are based on Figures 18.28b and 18.30b, where you saw that rays initially parallel to the axis diverge after passing through a diverging lens.

FIGURE 18.39 Three important sets of rays passing through a thin, diverging lens.

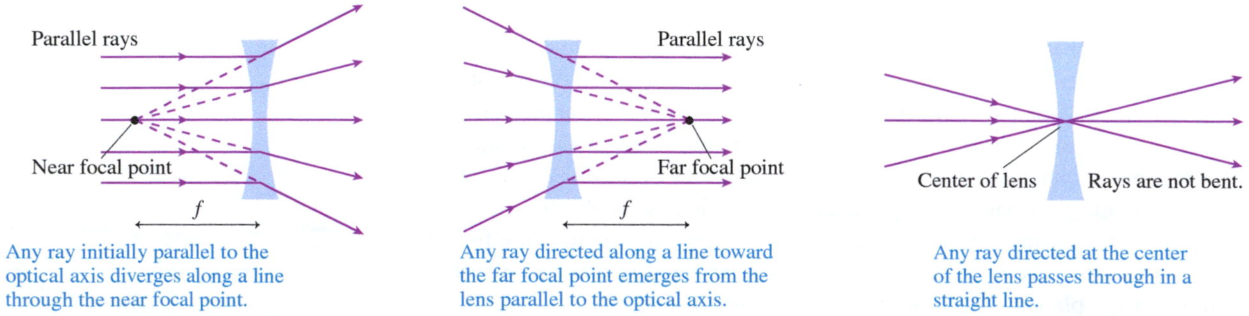


Ray tracing follows the steps of Tactics Box 18.2 for a converging lens *except* that two of the three special rays in step 4 are different.

**TACTICS BOX 18.3 Ray tracing for a diverging lens**

❶–❸ **Follow steps 1 through 3 of Tactics Box 18.2.**

❹ **Draw the three "special rays" from the tip of the arrow.** Use a straightedge or a ruler. The rays refract at the lens plane.

a. A ray parallel to the axis diverges along a line through the near focal point.
b. A ray along a line toward the far focal point emerges parallel to the axis.
c. A ray through the center of the lens does not bend.

❺ **Trace the diverging rays backward.** The point from which they are diverging is the image point, which is always a virtual image.

❻ **Measure the image distance** $s'$, which, because the image is virtual, we will take as a negative number. Also, if needed, measure the image height relative to the object height. The magnification can be found from Equation 18.9.

**EXAMPLE 18.9 Finding the image location for a diverging lens**

A diverging lens with a focal length of 50 cm is placed 100 cm from a flower. Where is the image? What is its magnification?

***PREPARE*** The flower is in the object plane. We use ray tracing to locate the image. Then we use Equation 18.9 to find the magnification.

***SOLVE*** FIGURE 18.40 shows the ray-tracing diagram. The three special rays (labeled a, b, and c to match Tactics Box 18.3) do not converge. However, they can be traced backward to an intersection ≈33 cm to the left of the lens. Because the rays appear to diverge from the image, this is a virtual image and $s' < 0$. The magnification is

$$m = -\frac{s'}{s} = -\frac{-33\text{ cm}}{100\text{ cm}} = 0.33$$

The image, which can be seen by looking *through* the lens, is one-third the size of the object and upright.

FIGURE 18.40 Ray-tracing diagram for a diverging lens.

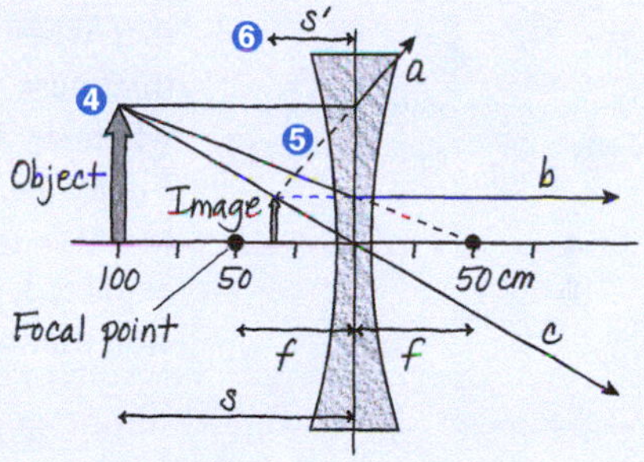


***ASSESS*** The three special rays appear to diverge from a single point, and the image orientation and size in our ray-tracing diagram match the computed magnification —the image is upright and smaller than the object—so we have confidence in our results.

Diverging lenses *always* make virtual images and, for this reason, are rarely used to make images this way. But diverging lenses can be an important element of optical systems, and they are very important for vision correction. If you are nearsighted, the lenses in your glasses are diverging lenses, as we'll see in the next chapter.

## 18.6 Image Formation with Spherical Mirrors

Curved mirrors can also be used to form images. Such mirrors are commonly used in telescopes, security and rearview mirrors, and searchlights. Their images can be analyzed with ray diagrams similar to those used with lenses. We'll consider only the important case of **spherical mirrors,** whose surfaces are sections of a sphere.

FIGURE 18.41 shows parallel light rays approaching two spherical mirrors. The mirror shown in Figure 18.41a, where the edges curve toward the light source, is often called a *concave mirror.* As the figure shows, parallel rays reflect off the shiny front surface of the mirror and pass through a single point on the optical axis—the rays are brought together at a focal point, as was the case for a converging lens. For this reason, we refer to a mirror with this shape as a **converging mirror.** The optical behavior of this mirror is exactly analogous to that of a converging lens. The mirror shown in Figure 18.41b, where the edges curve away from the light source, is often called a *convex mirror.* Parallel rays that reflect off its surface spread apart, as was

FIGURE 18.41 The focal point and focal length of converging and diverging mirrors.

(a)

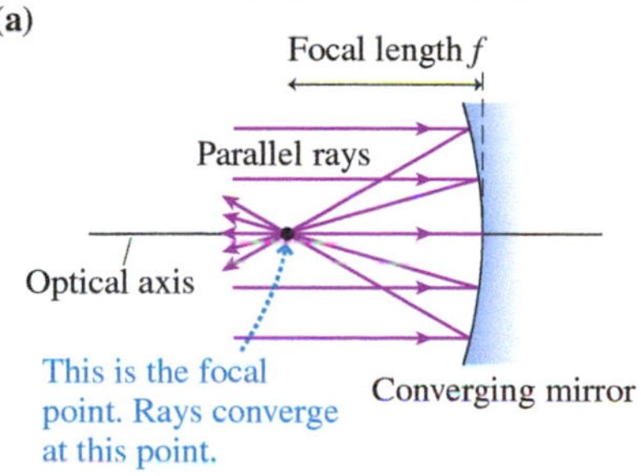


(b)

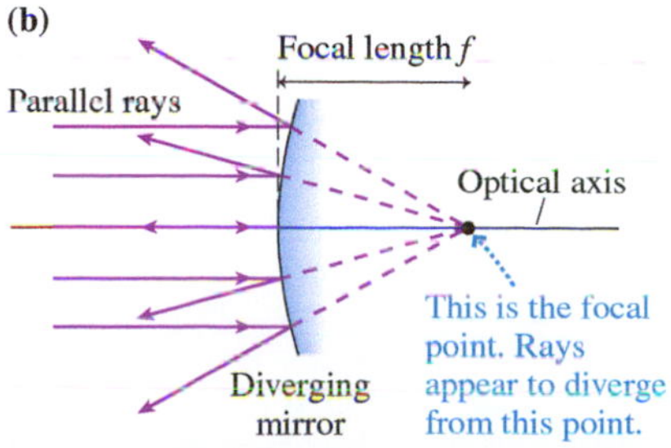

FIGURE 18.42 Three special rays for a converging mirror.

(a)

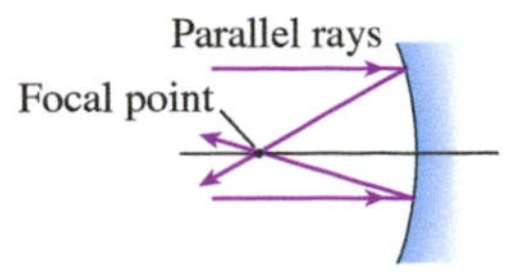


Any ray initially parallel to the optical axis will reflect through the focal point.

(b)

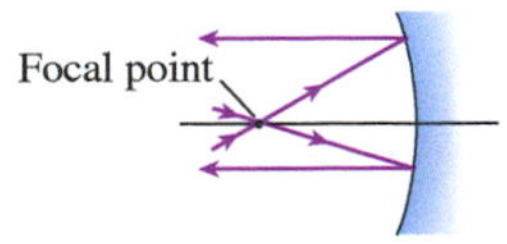


Any ray passing through the focal point will, after reflection, emerge parallel to the optical axis.

(c)

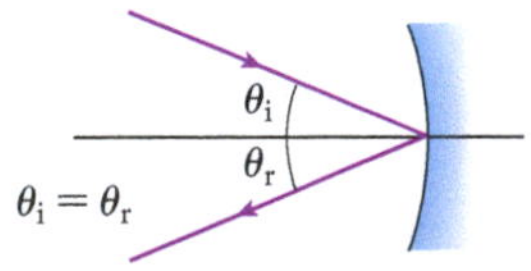


Any ray directed at the center of the mirror will reflect at an equal angle on the opposite side of the optical axis.

the case for a diverging lens, so we refer to a mirror with this shape as a **diverging mirror.** The rays of light appear to diverge from a focal point behind the mirror, similar to the situation with a diverging lens. This mirror will produce virtual images similar to those made by a diverging lens.

## Converging Mirrors

To understand image formation by a converging mirror, consider the three special rays shown in FIGURE 18.42. These rays are closely related to those used for ray tracing with converging lenses. Figure 18.42a shows two incoming rays parallel to the optical axis. As Figure 18.41a showed, these rays reflect off the mirror and pass through the focal point.

Figure 18.42b is the same as Figure 18.42a, but with the directions of the rays reversed. Here we see that rays passing through the focal point emerge parallel to the axis. Finally, Figure 18.42c shows what happens to a ray that is directed toward the center of the mirror. Right at its center, the surface of the mirror is perpendicular to the optical axis. The law of reflection then tells us that the incoming ray will reflect at the same angle, but on the opposite side of the optical axis.

Let's begin by considering the case where the object's distance $s$ from the mirror is greater than the focal length ($s > f$), as shown in FIGURE 18.43. The three special rays just discussed are enough to locate the position and size of the image. Recall that when ray tracing a thin lens, although we drew the lens as having an actual thickness, the rays refracted at an imaginary plane centered on the lens. Similarly, when ray tracing mirrors, the incoming rays reflect off the **mirror plane** as shown in Figure 18.43, not off the curved surface of the mirror. We see that the image is *real* because rays converge at the image point P′. Further, the image is *inverted.* As expected, the image formation is analogous to that for a converging lens. The expression that we derived previously for magnification applies here as well.

FIGURE 18.43 A real image formed by a converging mirror.

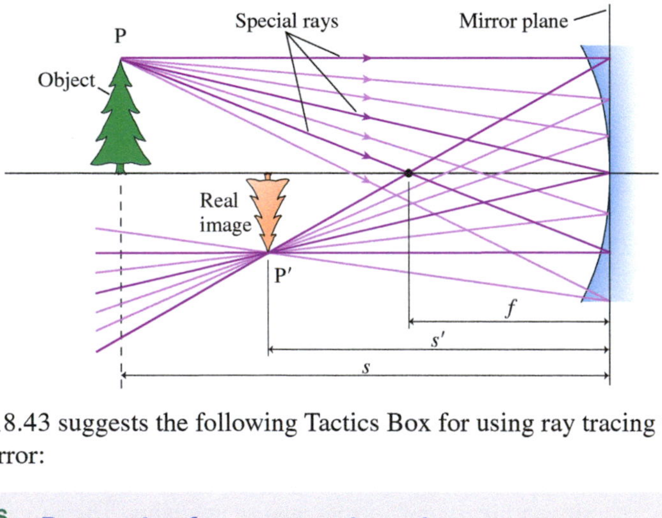


Figure 18.43 suggests the following Tactics Box for using ray tracing with a converging mirror:

**TACTICS BOX 18.4 Ray tracing for a converging mirror**

❶ **Draw an optical axis.** Use graph paper or a ruler! Establish an appropriate scale.

❷ **Center the mirror on the axis.** Mark and label the focal point at distance $f$ from the mirror's surface. Draw the mirror plane through the mirror's center, perpendicular to the axis.

❸ **Represent the object with an upright arrow at distance $s$.** It's usually best to place the base of the arrow on the axis and to draw the arrow smaller than the radius of the mirror.

❹ **Draw the three "special rays" from the tip of the arrow.** Use a straightedge or a ruler. The rays should reflect off the mirror plane.

a. A ray parallel to the axis reflects through the focal point.
b. An incoming ray that passes through the focal point reflects back parallel to the axis.
c. A ray that strikes the center of the mirror reflects at an equal angle on the opposite side of the optical axis.

❺ **Extend the rays until they converge.** The rays converge at the image point. Draw the rest of the image in the image plane. If the base of the object is on the axis, then the base of the image will also be on the axis.

❻ **Measure the image distance $s'$.** Also, if needed, measure the image height relative to the object height. The magnification can be found from Equation 18.9.

Exercises 21a, 22

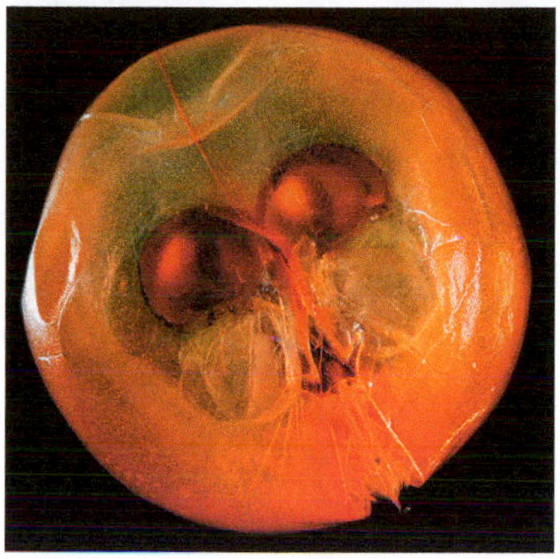

▶ **Look into my eyes** BIO The eyes of most animals use lenses to focus an image. *Gigantocypris,* a deep-sea crustacean, is unusual in that it uses two converging mirrors to focus light onto its retina. Because it lives at depths where no sunlight penetrates, *Gigantocypris* is thought to use its mirror eyes to hunt bioluminescent animals.

## EXAMPLE 18.10 Finding the image location for a converging mirror

A 3.0-cm-high object is located 60 cm from a converging mirror. The mirror's focal length is 40 cm. Use ray tracing to find the position, height, and magnification of the image.

**STRATEGIZE** The image is farther from the mirror than the focal length, so we will expect the image to be real and inverted as for a converging lens and as in Figure 18.43.

**PREPARE** FIGURE 18.44 shows the ray-tracing diagram and the steps of Tactics Box 18.4.

**SOLVE** After preparing a careful drawing, we can use a ruler to find that the image position is $s' \approx 120$ cm. The magnification is thus

$$m = -\frac{s'}{s} \approx -\frac{120\text{ cm}}{60\text{ cm}} = -2.0$$

The negative sign indicates that the image is inverted. The image height is thus twice the object height, or $h' \approx 6$ cm.

**ASSESS** The three special rays converge at a single point, and the image is real and inverted, as we expected. This gives us confidence in our results.

FIGURE 18.44 Ray-tracing diagram for a converging mirror.

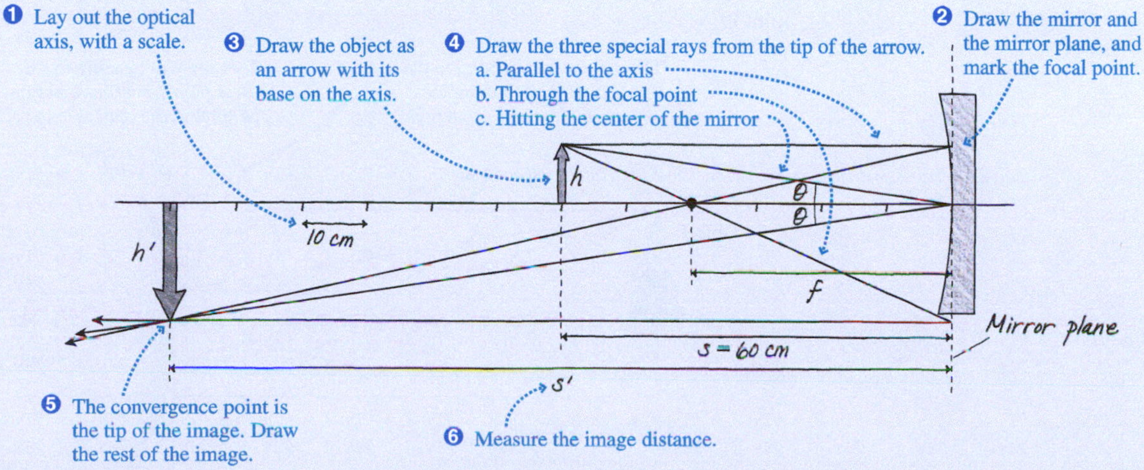

FIGURE 18.45 Reflection from a silvered ball.

If the object is inside the focal point ($s < f$), ray tracing can be used to show that the image is a virtual image. This situation is analogous to the formation of a virtual image by a lens when the object is inside the focal point.

## Diverging Mirrors

If you look at your reflection in a silvered ball, such as a tree ornament, you will notice that your image appears right-side up but quite small, as in FIGURE 18.45. This makes sense. Given its shape, the ball works like a diverging mirror, so we'd expect a small, upright image like the image produced by a diverging lens. We can use ray tracing to verify this.

Once more, there are three special rays we can use to find the location of the image. These rays are shown in FIGURE 18.46; they are similar to the special rays we've already studied in other situations.

FIGURE 18.46 Three special rays for a diverging mirror.

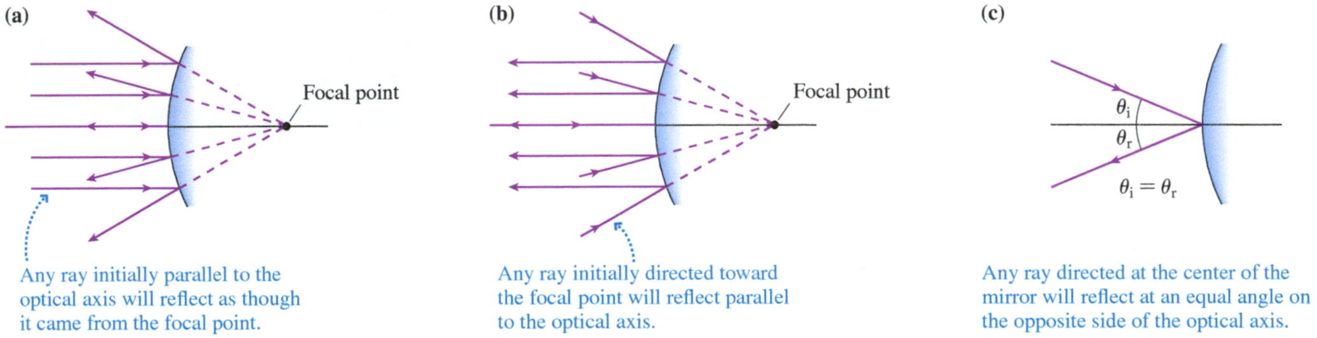


We can use these three special rays to find the image of an object, as shown in FIGURE 18.47. We see that the image is virtual—no actual rays converge at the image point P′. Instead, diverging rays *appear* to have come from this point. The image is also upright and much smaller than the object, as is the case for a diverging lens.

FIGURE 18.47 Rays from point P reflect from the mirror and appear to have come from point P′.

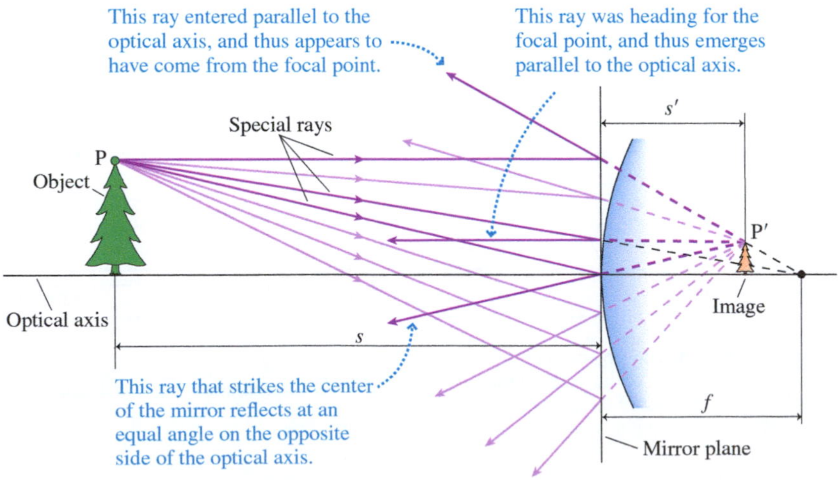

These observations form the basis of the following Tactics Box.

**TACTICS BOX 18.5 Ray tracing for a diverging mirror**

❶–❸ **Follow steps 1 through 3 of Tactics Box 18.4.**

❹ **Draw the three "special rays" from the tip of the arrow.** Use a straightedge or a ruler. The rays should reflect off the mirror plane.

a. A ray parallel to the axis reflects as though it came from the focal point.
b. A ray initially directed toward the focal point reflects parallel to the axis.
c. A ray that strikes the center of the mirror reflects at an equal angle on the opposite side of the optical axis.

❺ **Extend the emerging rays *behind the mirror* until they converge.** The point of convergence is the image point. Draw the rest of the image in the image plane. If the base of the object is on the axis, then the base of the image will also be on the axis.

❻ **Measure the image distance $s'$.** Also, if needed, measure the image height relative to the object height. The magnification can be found from Equation 18.9.

Exercises 21b, 23

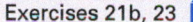

The small image in a diverging mirror allows more of the store to be seen.

**Video** Diverging Mirror

Diverging mirrors are used for a variety of safety and monitoring applications, such as passenger-side rearview mirrors and the round mirrors used in stores to keep an eye on the customers. The idea behind such mirrors can be understood from Figure 18.47. When an object is reflected in a diverging mirror, the image appears smaller. Because the image is, in a sense, a miniature version of the object, you can *see much more of it* within the edges of the mirror than you could with an equal-sized flat mirror. This wide-angle view is clearly useful for checking traffic behind you or for checking up on your store.

**CONCEPTUAL EXAMPLE 18.11 Driver and passenger mirrors**

The rearview mirror on the driver's side of a car is a plane (flat) mirror, while the mirror on the passenger's side is a diverging mirror. Why is this?

***REASON*** The driver needs a wide field of view from both mirrors. He sits close to the driver-side mirror, so it appears large and can reflect a fairly wide view of what's behind. The passenger-side mirror is quite far from the driver, so it appears relatively small. If it were flat, it would offer only a narrow view. The diverging mirror gives a smaller image, which allows the driver to see more of what's behind the car in this smaller area.

***ASSESS*** The next time you are riding on the passenger side of a car, compare the view in the two mirrors; this new perspective will let you see the benefit of making the more distant mirror a diverging mirror. There is one problem with the image in the diverging mirror, though. The image is smaller than the object, so the driver may perceive it to be more distant than it actually is. This fact is the reason for the warning on the passenger-side mirror: "Objects may be closer than they appear."

**STOP TO THINK 18.6** A converging mirror of focal length $f$ forms an image of the moon. Where is the image located?

A. At the mirror's surface
B. Almost exactly a distance $f$ behind the mirror
C. Almost exactly a distance $f$ in front of the mirror
D. At a distance behind the mirror equal to the distance of the moon in front of the mirror

## 18.7 The Thin-Lens Equation

Ray tracing is an important tool for quickly grasping the overall positions and sizes of an object and its image. For more precise work, however, we would like a mathematical expression that relates the three fundamental quantities of an optical system: the focal length $f$ of the lens or mirror, the object distance $s$, and the image distance $s'$. We can find such an expression by considering the converging lens in FIGURE 18.48. Two of the special rays are shown: one initially parallel to the optical axis that then passes through the far focal point, and the other traveling undeviated through the center of the lens.

FIGURE 18.48 Deriving the thin-lens equation.

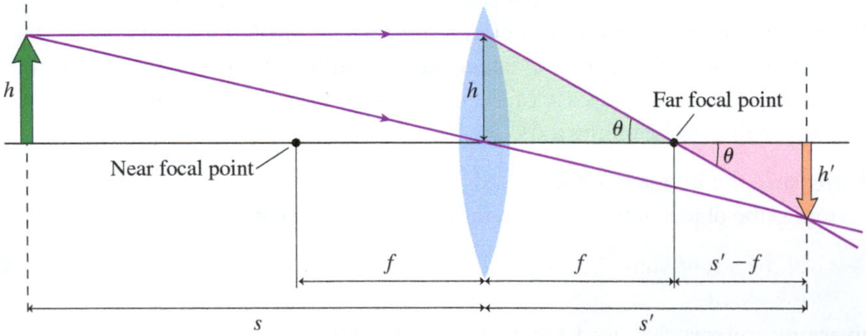


Consider the two right triangles highlighted in green and pink. Because they both have one 90° angle and a second angle $\theta$ that is the same for both, the two triangles are *similar.* This means that they have the same shape, although their sizes may be different. For similar triangles, the ratios of any two similar sides are the same. Thus we have

$$\frac{h'}{h} = \frac{s' - f}{f} \tag{18.10}$$

Further, we found in Example 18.7 that

$$\frac{h'}{h} = \frac{s'}{s} \tag{18.11}$$

Combining Equations 18.10 and 18.11 gives

$$\frac{s'}{s} = \frac{s' - f}{f}$$

We then divide both sides by $s'$:

$$\frac{1}{s} = \frac{s' - f}{s'f} = \frac{1}{f} - \frac{1}{s'}$$

which we can write as

$$\frac{1}{s} + \frac{1}{s'} = \frac{1}{f} \tag{18.12}$$

Thin-lens equation (also works for mirrors) relating object and image distances to focal length

This equation, the **thin-lens equation,** relates the three important quantities $f$, $s$, and $s'$. In particular, if we know the focal length of a lens and the object's distance from the lens, we can use the thin-lens equation to find the position of the image.

**NOTE** ▸ Although we derived the thin-lens equation for a converging lens that produced a real image, it works equally well for *any* image—real or virtual—produced by both converging and diverging lenses. And, in spite of its name, the thin-lens equation also describes the images formed by mirrors. ◂

It's worth checking that the thin-lens equation describes what we already know about lenses. In Figure 18.31a, we saw that rays initially parallel to the optical axis are focused at the focal point of a converging lens. Initially parallel rays come from an object extremely far away, with $s \rightarrow \infty$. Because $1/\infty = 0$, the thin-lens equation tells us that the image distance is $s' = f$, as we expected. Or suppose an object is located right at the focal point, with $s = f$. Then, according to Equation 18.12, $1/s' = 1/f - 1/s = 0$. This implies that the image distance is infinitely far away ($s' = \infty$), so the rays leave the lens parallel to the optical axis. Indeed, this is what Figure 18.31b showed. Now, it's true that no real object or image can be at infinity. But if either the object or image is more than several focal lengths from the lens ($s \gg f$ or $s' \gg f$), then it's an excellent approximation to consider the distance to be infinite, the rays to be parallel to the axis, and the reciprocal ($1/s$ or $1/s'$) in the thin-lens equation to be zero.

## Sign Conventions for Lenses and Mirrors

We've already noted that the image distance $s'$ is positive for real images and negative for virtual images. In the thin-lens equation, the sign of the focal length can also be either positive or negative, depending on the type of lens or mirror. Synthesis 18.1 shows the sign conventions that we will use in this text and brings together what we've learned about lenses, mirrors, and the images they create.

### SYNTHESIS 18.1 Lenses, mirrors, and their sign conventions

There are three distinct image situations that can occur for lenses and for mirrors. We outline these here, and give the sign of each quantity. Note that the object distance $s$ is always positive.

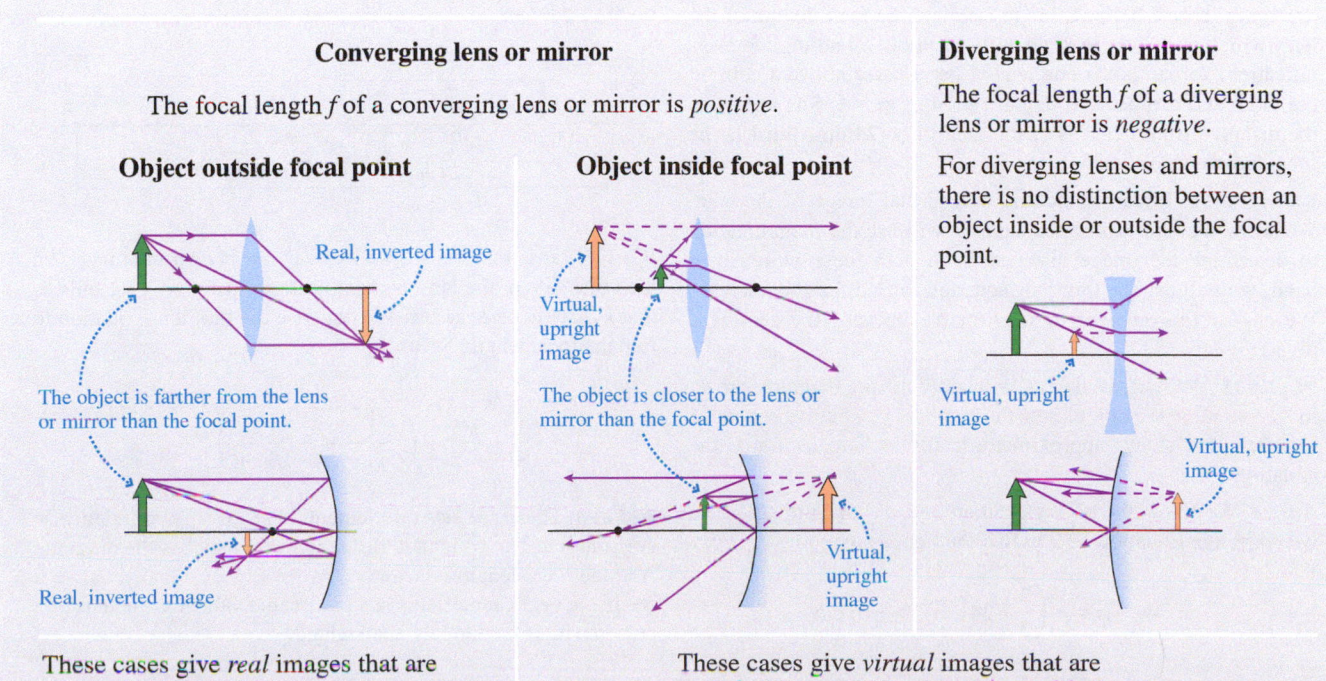

**Converging lens or mirror**

The focal length $f$ of a converging lens or mirror is *positive*.

**Object outside focal point**

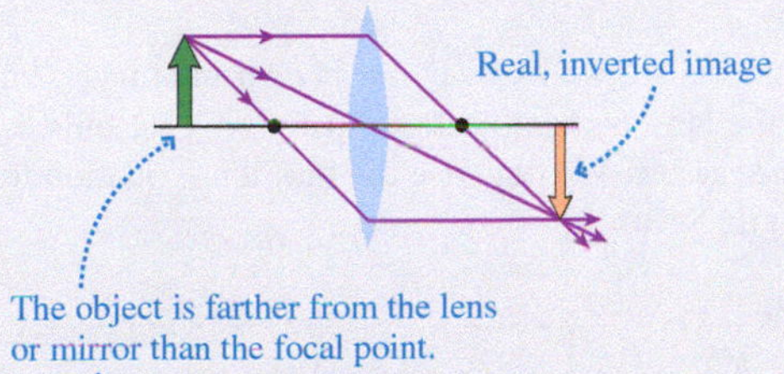


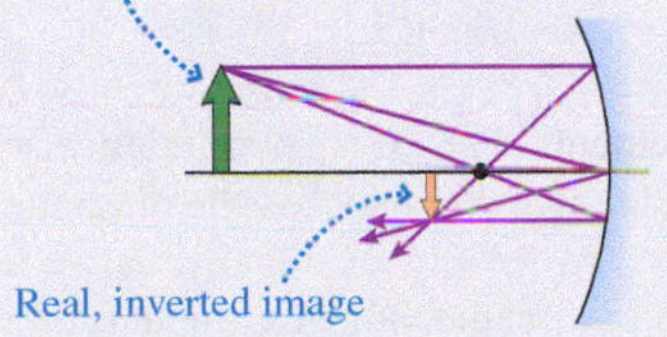


**Object inside focal point**

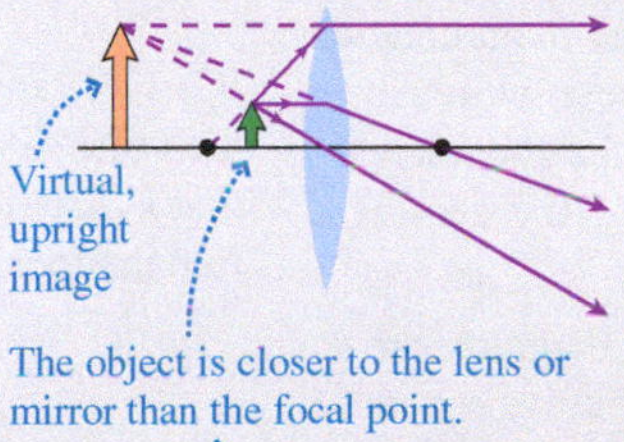


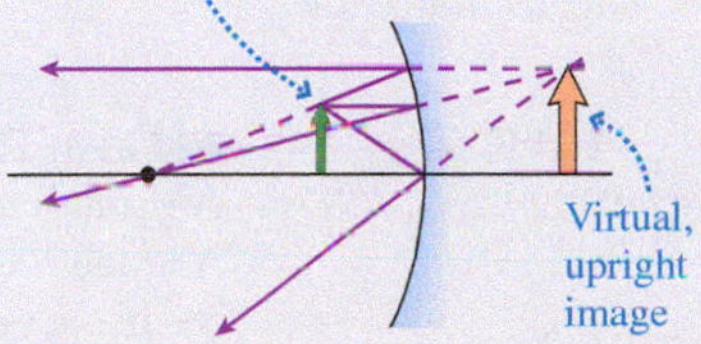


**Diverging lens or mirror**

The focal length $f$ of a diverging lens or mirror is *negative*.

For diverging lenses and mirrors, there is no distinction between an object inside or outside the focal point.

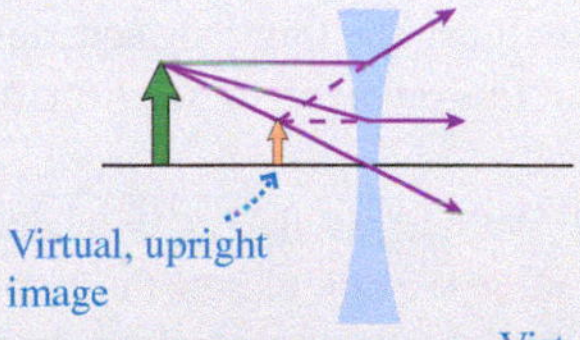


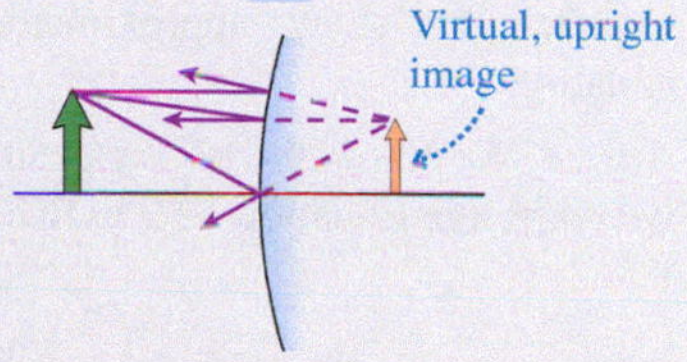


These cases give *real* images that are *inverted* (the magnification $m$ is *negative*). The image distance $s'$ is *positive*.

These cases give *virtual* images that are *upright* (the magnification $m$ is *positive*). The image distance $s'$ is *negative*.

NOTE ▸ When the thin-lens equation is used, the focal length must be taken as positive or negative according to Synthesis 18.1. ◂

## EXAMPLE 18.12 Analyzing a magnifying lens

A stamp collector uses a magnifying lens that sits 2.0 cm above the stamp. The magnification is 4. What is the focal length of the lens?

**STRATEGIZE** Since the image is larger than the object, the lens must be a converging lens. The collector looks through the lens to see the image, so we know that the image is virtual. This is a situation we've seen before. We know that the object is closer to the lens than the focal point and that the image is upright.

**PREPARE** Even though we have an equation to solve for the various distances, it's still a good idea to start with a ray-tracing diagram to help us make sense of the situation. The diagram is shown in FIGURE 18.49.

**SOLVE** A virtual image is upright, so $m = +4$. The magnification is related to the object and image distances as $m = -s'/s$; thus

$$s' = -4s = -4(2.0\text{ cm}) = -8.0\text{ cm}$$

We can use $s$ and $s'$ in the thin-lens equation to find the focal length:

$$\frac{1}{f} = \frac{1}{s} + \frac{1}{s'} = \frac{1}{2.0\text{ cm}} + \frac{1}{-8.0\text{ cm}} = 0.375\text{ cm}^{-1}$$

FIGURE 18.49 Ray-tracing diagram of a magnifying lens.

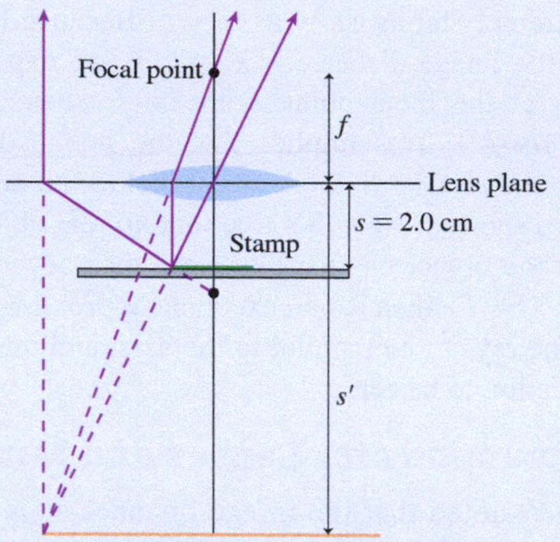


Thus

$$f = \frac{1}{0.375\text{ cm}^{-1}} = 2.7\text{ cm}$$

**ASSESS** $f > 2$ cm, as expected because the object has to be inside the focal point.

## EXAMPLE 18.13 Finding the focal length of a diverging lens

If you are nearsighted, the lenses that correct your vision are diverging lenses. It's possible to estimate the focal length of the lens, which correlates with your prescription, by noting the magnification. Jordan holds one lens of her glasses above a coin on the table and estimates a magnification of $m = \frac{2}{3}$. She measures the distance from her lens to the coin to be 24 cm. What is the focal length of the lens?

**STRATEGIZE** Jordan is looking at a virtual image of the coin. We know the object distance, and we will use the magnification to determine the image distance. With both these numbers in hand, we will use the thin-lens equation to find the focal length. We expect this number to be negative because the lens is a diverging lens.

**PREPARE** We start, as always, with a ray-tracing diagram, shown in FIGURE 18.50. We don't know the numbers yet, but we can make a sketch that shows approximate details to help us frame the situation.

**SOLVE** We know the object position and the magnification, so we rearrange Equation 18.9 to find the image position:

$$s' = -ms = -\left(\frac{2}{3}\right)(24\text{ cm}) = -16\text{ cm}$$

FIGURE 18.50 Ray-tracing diagram for the image in a diverging lens.

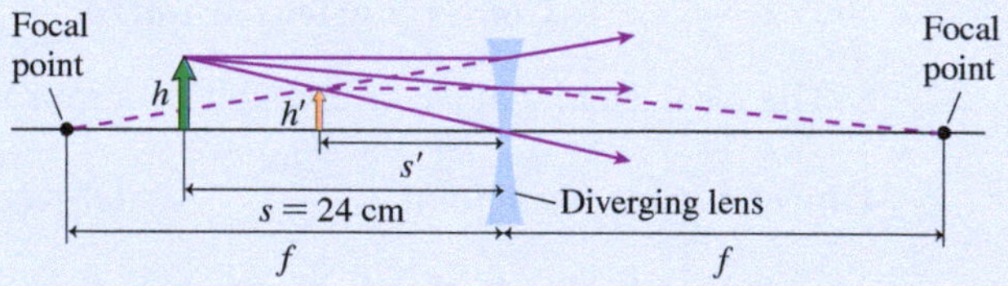


It makes sense that $s' < 0$; we know that it is a virtual image, on the same side of the lens as the object, so Synthesis 18.1 tells us that $s'$ should be negative. We now use the thin-lens equation to find the focal length:

$$f = \frac{1}{\left(\frac{1}{s} + \frac{1}{s'}\right)} = \frac{1}{\left(\frac{1}{24\text{ cm}} + \frac{1}{-16\text{ cm}}\right)} = -48\text{ cm}$$

**ASSESS** There are several elements that give us confidence in our solution. The computed distance for $s'$ matches the diagram. And the focal length is negative, as we expect. Finally, as we'll see in the next chapter, this is a reasonable value for the lens of a person who is moderately nearsighted.

## EXAMPLE 18.14 Finding the mirror image of a candle

A converging mirror has a focal length of 2.4 m. Where should a candle be placed so that its image is inverted and twice as large as the object?

**STRATEGIZE** The basic principles of image formation are the same for lenses and mirrors, so we'll solve this problem as we would for a lens. A converging mirror works similarly to a converging lens.

**PREPARE** The image is inverted, so we know that it is a real image, and it will be on the same side of the mirror as the object. The magnification will be negative, so we can set $m = -2$. The image is twice as large as the object, so we know that the image distance will be twice the object distance. This is enough information to sketch the ray-tracing diagram shown in FIGURE 18.51. We can use our diagram to estimate a value for $s$; it seems to be about half again as large as $f$. We can use this to assess our final result.

FIGURE 18.51 Ray-tracing diagram for the image in a converging mirror.

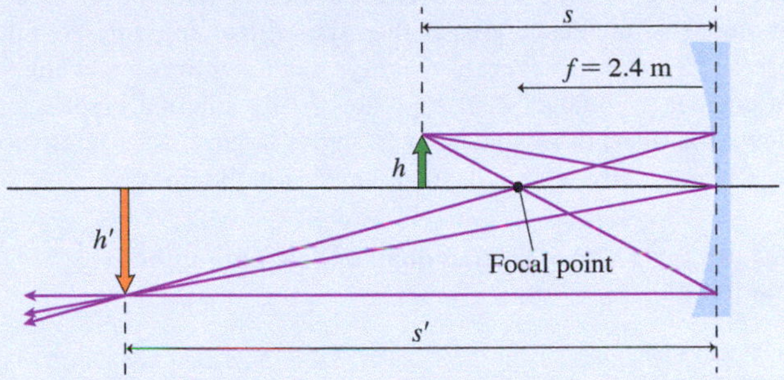


**SOLVE** We use Equation 18.9 to relate $s$ and $s'$:

$$s' = -ms$$

We insert this expression for $s'$ into the thin-lens equation:

$$\frac{1}{f} = \frac{1}{s} + \frac{1}{s'} = \frac{1}{s} + \frac{1}{-ms} = \frac{1}{s} + \frac{1}{-(-2)s} = \frac{1}{s} + \frac{1}{2s} = \frac{3}{2s}$$

from which we get

$$s = \frac{3}{2}f = \frac{3}{2}(2.4\text{ m}) = 3.6\text{ m}$$

The candle should be placed 3.6 m in front of the mirror.

**ASSESS** The final result for the object distance is $s = 1.5f$, just what we expected from our ray-tracing diagram. Even though making the diagram wasn't strictly necessary to solve the problem, the diagram helped us visualize the situation and check our work.

**STOP TO THINK 18.7** A candle is placed in front of a converging lens. A well-focused image of the flame is seen on a screen on the opposite side of the lens. If the candle is moved farther away from the lens, how must the screen be adjusted to keep showing a well-focused image?

A. The screen must be moved closer to the lens.
B. The screen must be moved farther away from the lens.
C. The screen does not need to be moved.

## INTEGRATED EXAMPLE 18.15 Optical fiber imaging BIO

An *endoscope* is a narrow bundle of optical fibers that can be inserted through a bodily opening or a small incision to view the interior of the body. As FIGURE 18.52 shows, an *objective* lens focuses a real image onto the entrance face of the fiber bundle. Individual fibers—using total internal reflection—transport the light to the exit face, where it emerges. The doctor observes a magnified image of the exit face by viewing it through an *eyepiece* lens.

A lens and light sources on the business end of an endoscope. The fiber bundle is in the black sheath.

FIGURE 18.52 An endoscope.

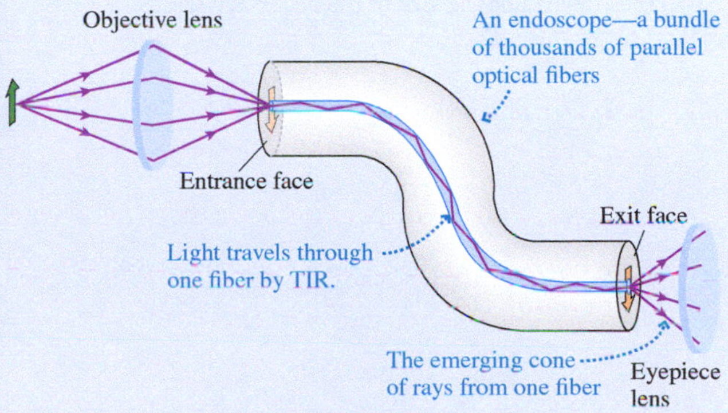


*Continued*

A single optical fiber, as discussed in Section 18.3 and shown in FIGURE 18.53, consists of a higher-index glass core surrounded by a lower-index cladding layer. To remain in the fiber, light rays propagating through the core must strike the cladding boundary at angles of incidence greater than the critical angle $\theta_c$. As Figure 18.53 shows, this means that rays that enter the core at angles of incidence smaller than $\theta_{max}$ are totally internally reflected down the fiber; those that enter at angles larger than $\theta_{max}$ are not totally internally reflected, and escape from the fiber.

FIGURE 18.53 Cross section of an optical fiber (greatly magnified).

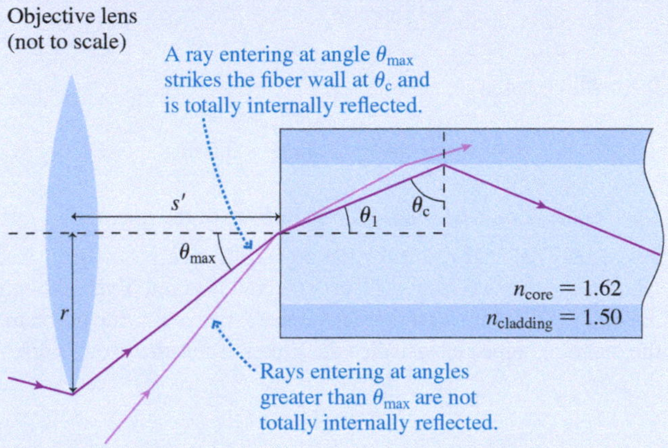


The objective lens of an endoscope must be carefully matched to the fiber. Ideally, the lens diameter is such that rays from the edge of the lens enter the fiber at $\theta_{max}$, as shown in Figure 18.53. A lens larger than this is not useful because rays from its outer regions will enter the fiber at an angle larger than $\theta_{max}$ and will thus escape from the fiber. A lens smaller than this will gather less light.

a. What is $\theta_{max}$ for the fiber shown in Figure 18.53?
b. A typical objective lens is 3.0 mm in diameter and can focus on an object 3.0 mm in front of it. What focal length should the lens have so that rays from its edge just enter the fiber at angle $\theta_{max}$?
c. What is the magnification of this lens?

***STRATEGIZE*** We will solve this problem by working backward from the situation inside the fiber. Once we know the critical angle, we will determine the entrance angle, and, knowing this, we will then determine the size of the necessary lens.

***PREPARE*** From Figure 18.53 we can find the critical angle $\theta_c$ and then use geometry to find $\theta_1$. Snell's law can then be used to find $\theta_{max}$. Once $\theta_{max}$ is known, we can find the image distance $s'$, and, because we know that the object distance is $s = 3$ mm, we can use the thin-lens equation to solve for $f$.

***SOLVE***

a. The critical angle for total internal reflection is given by Equation 18.3:

$$\theta_c = \sin^{-1}\left(\frac{n_2}{n_1}\right) = \sin^{-1}\left(\frac{n_{cladding}}{n_{core}}\right) = \sin^{-1}\left(\frac{1.50}{1.62}\right) = 67.8^\circ$$

Because of the right triangle, this ray's angle of refraction from the fiber's face is $\theta_1 = 90^\circ - 67.8^\circ = 22.2^\circ$. Then, by Snell's law, this ray enters from the air at angle $\theta_{max}$ such that

$$n_{air}\sin\theta_{max} = 1.00\sin\theta_{max} = n_{core}\sin\theta_1 = 1.62(\sin 22.2^\circ) = 0.612$$

Thus

$$\theta_{max} = \sin^{-1}(0.612) = 37.7^\circ$$

b. As Figure 18.53 shows, the distance of the lens from the fiber's face—the image distance $s'$—is related to the lens radius $r = 1.5$ mm by

$$\frac{r}{s'} = \tan\theta_{max}$$

Thus the image distance is

$$s' = \frac{r}{\tan\theta_{max}} = \frac{1.5\text{ mm}}{\tan 37.7^\circ} = 1.94\text{ mm}$$

Then the thin-lens equation gives

$$\frac{1}{f} = \frac{1}{s} + \frac{1}{s'} = \frac{1}{3.0\text{ mm}} + \frac{1}{1.94\text{ mm}} = 0.849\text{ mm}^{-1}$$

so that

$$f = \frac{1}{0.849\text{ mm}^{-1}} = 1.18\text{ mm}$$

c. The magnification is

$$m = -\frac{s'}{s} = -\frac{1.94\text{ mm}}{3.0\text{ mm}} = -0.65$$

***ASSESS*** The object distance of 3.0 mm is greater than the 1.2 mm focal length we calculated, as must be the case when a converging lens produces a real image.

# SUMMARY

**GOAL** To understand and apply the ray model of light.

## GENERAL PRINCIPLES

### Reflection

Law of reflection: $\theta_r = \theta_i$

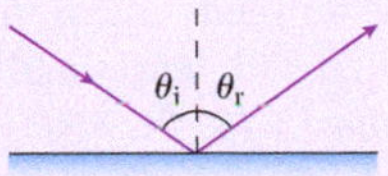


Reflection can be **specular** (mirror-like) or **diffuse** (from rough surfaces).

Plane mirrors: A virtual image is formed at P′ with $s' = s$, where $s$ is the **object distance** and $s'$ is the **image distance.**

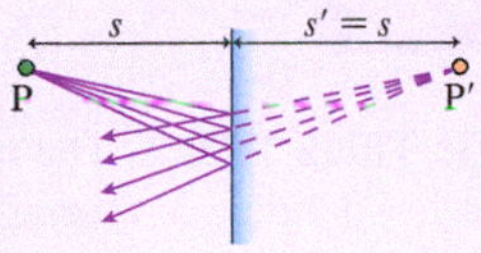


### Refraction

**Snell's law** of refraction:

$$n_1 \sin\theta_1 = n_2 \sin\theta_2$$

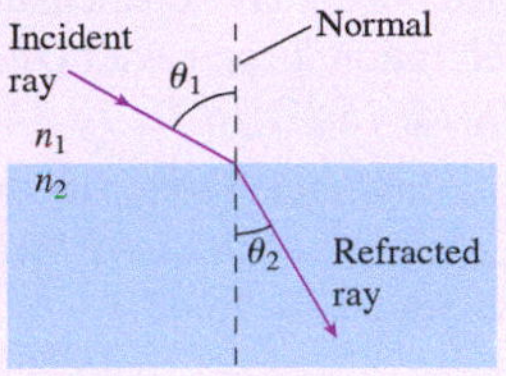


**Index of refraction** is $n = c/v$. The ray is closer to the normal on the side with the larger index of refraction.

If $n_2 < n_1$, **total internal reflection** (TIR) occurs when the angle of incidence $\theta_1$ is greater than $\theta_c = \sin^{-1}(n_2/n_1)$.

## IMPORTANT CONCEPTS

### The ray model of light

Light travels along straight lines, called **light rays,** at speed $v = c/n$.

A light ray continues forever unless an interaction with matter causes it to reflect, refract, scatter, or be absorbed.

Light rays come from self-luminous or reflective **objects.** Each point on the object sends rays in all directions.

**Ray diagrams** use only a few select rays to represent all the rays emitted by an object.

In order for the eye to see an object (or image), rays from the object or image must enter the eye.

### Image formation

If rays diverge from P and, after interacting with a lens or mirror, *appear* to diverge from P′ without actually passing through P′, then P′ is a **virtual image** of P.

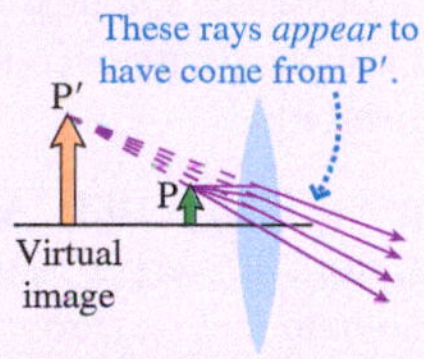


If rays diverge from P and interact with a lens or mirror so that the refracted rays *converge* at P′, then P′ is a **real image** of P. Rays actually pass through a real image.

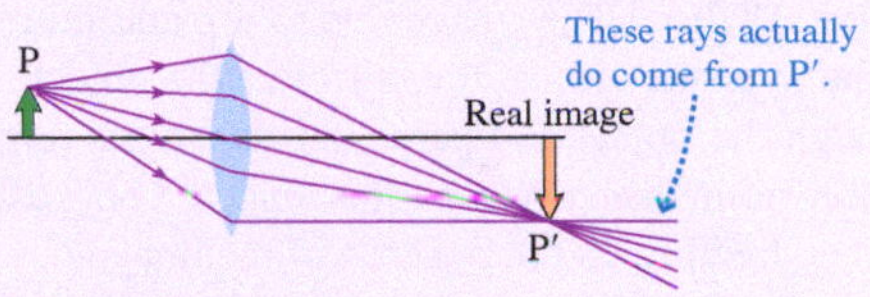


## APPLICATIONS

### Ray tracing for lenses

Three special rays in three basic situations:

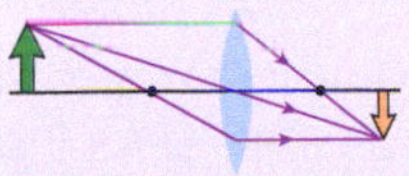

Converging lens
Real image

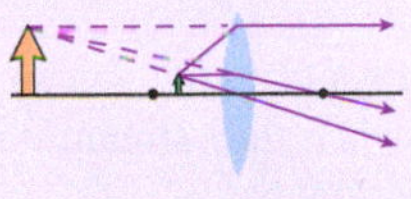

Converging lens
Virtual image

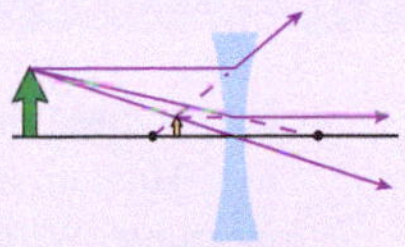

Diverging lens
Virtual image

### Ray tracing for mirrors

Three special rays in three basic situations:

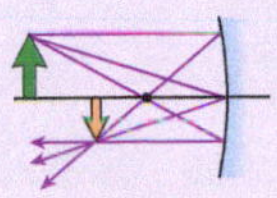

Converging mirror
Real image

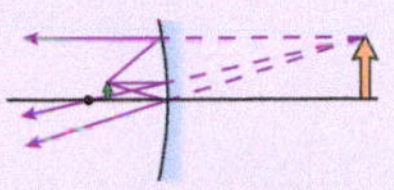

Converging mirror
Virtual image

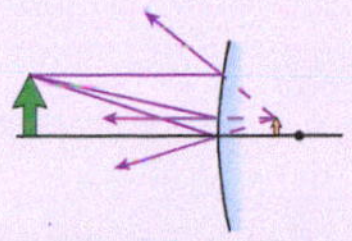

Diverging mirror
Virtual image

### The thin-lens equation

For a lens or curved mirror, the object distance $s$, the image distance $s'$, and the focal length $f$ are related by the thin-lens equation:

$$\frac{1}{s} + \frac{1}{s'} = \frac{1}{f}$$

The **magnification** of a lens or mirror is $m = -s'/s$.

**Sign conventions** for the thin-lens equation:

| Quantity | Positive when | Negative when |
|---|---|---|
| $s$ | Always | Not treated here |
| $s'$ | *Real* image; on opposite side of a lens from object, or in front of a mirror | *Virtual* image; on same side of a lens as object, or behind a mirror |
| $f$ | Converging lens or mirror | Diverging lens or mirror |
| $m$ | Image is upright. | Image is inverted. |

## Learning Objectives After studying this chapter, you should be able to:

- Apply the ray model of light and draw ray diagrams. *Conceptual Questions 18.1, 18.2; Problems 18.1, 18.2, 18.3*
- Analyze specular reflection with the law of reflection. *Conceptual Questions 18.4, 18.8; Problems 18.4, 18.5, 18.6, 18.8, 18.11*
- Use Snell's law to calculate refraction and total internal reflection. *Conceptual Questions 18.10, 18.11; Problems 18.13, 18.15, 18.17, 18.22, 18.24*
- Use ray tracing to locate the images of lenses and mirrors. *Conceptual Questions 18.13, 18.16; Problems 18.32, 18.36, 18.38, 18.40, 18.43*
- Distinguish between real and virtual images and understand the sign convention for images. *Conceptual Questions 18.12, 18.26; Problems 18.30, 18.31, 18.33, 18.37, 18.42*
- Calculate the magnification of an optical system. *Problems 18.34, 18.40*
- Use the thin-lens equation to solve problems about image formation with lenses and spherical mirrors. *Conceptual Question 18.19; Problems 18.44, 18.45, 18.48, 18.49, 18.60*

## STOP TO THINK ANSWERS

**Chapter Preview Stop to Think: B.** The diameter of the hole is much greater than the wavelength of light, so the ray model is applicable and light will travel in rays from the bulb to the screen. Light rays from the top of the bulb will go through the hole and hit near the bottom of the screen, while rays from the bottom of the bulb will go through the hole and hit near the top of the screen. When all the rays are taken into account, the image of the bulb on the screen will be inverted.

**Stop to Think 18.1: A and B.** Rays from the very top and bottom of the bulb that go through the top and bottom of the disk define the regions with no light (shadow) and those with light. The shadow is complete at only positions A and B.

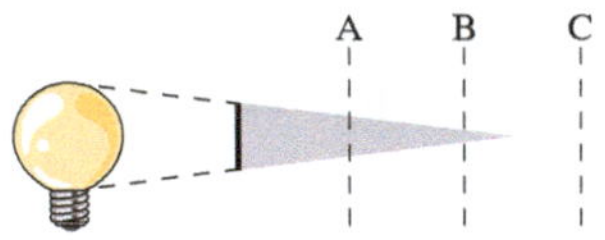


**Stop to Think 18.2: C.** The image due to a plane mirror is always located on the opposite side of the mirror as the object, along a line passing through the object and perpendicular to the mirror, and the same distance from the mirror as the object. The position of the observer has no bearing on the position of the image.

**Stop to Think 18.3: A.** The ray travels closer to the normal in both media 1 and 3 than in medium 2, so $n_1$ and $n_3$ are both greater than $n_2$. The angle is smaller in medium 3 than in medium 1, so $n_3 > n_1$.

**Stop to Think 18.4: E.** From Figure 18.32, the image at P′ is created by many rays from P that go through *all* parts of the lens. If the upper half of the lens is obscured, all the rays that go through the lower half still focus at P′. The image remains sharp but, because of the fewer rays reaching it, becomes dimmer.

**Stop to Think 18.5: E.** The rays from the object are diverging. Without a lens, the rays cannot converge to form any kind of image on the screen.

**Stop to Think 18.6: C.** For a converging mirror, the focal length $f$ is the distance from the mirror at which incoming parallel rays meet. The moon is so distant that rays from any point on the moon are very nearly parallel. Thus the image of the moon would be very nearly at a distance $f$ in front of the mirror.

**Stop to Think 18.7: A.** The thin-lens equation is $1/s + 1/s' = 1/f$. The focal length of the lens is fixed. Because $1/s$ gets smaller as $s$ is increased, $1/s'$ must get larger to compensate. Thus $s'$ must decrease.

Video Tutor Solution Chapter 18

# QUESTIONS

## Conceptual Questions

1. During a solar eclipse, the sun—a small but extended source—casts a shadow of the moon on the earth. Explain why the moon's shadow has a dark center surrounded by a region of increasing brightness.

2. Can you see the rays from the sun on a clear day? Why or why not? How about when they stream through a forest on a foggy morning? Why or why not?
3. On the earth, you can see the ground in someone's shadow; on the moon, you can't—the shadow is deep black. Explain the difference.

Problem difficulty is labeled as | (straightforward) to ||||| (challenging). Problems labeled INT integrate significant material from earlier chapters; Problems labeled BIO are of biological or medical interest.

The eText icon indicates when there is a video tutor solution available for the chapter or for a specific problem. To launch these videos, log into your eText through Mastering™ Physics or log into the Study Area.

4. On a day with fresh snow, you might notice that shadows—where the direct light from the sun is blocked—have a distinctly bluish cast. Explain why you might expect this.

5. If you take a walk on a summer night along a dark, unpaved road in the woods, with a flashlight pointing at the ground several yards ahead to guide your steps, any water-filled potholes are noticeable because they appear much darker than the surrounding dry road. Explain why.
6. You are looking at the image of a pencil in a mirror, as shown in Figure Q18.6.
   a. What happens to the image you see if the top half of the mirror, down to the midpoint, is covered with a piece of cardboard? Explain.
   b. What happens to the image you see if the bottom half of the mirror is covered with a piece of cardboard?

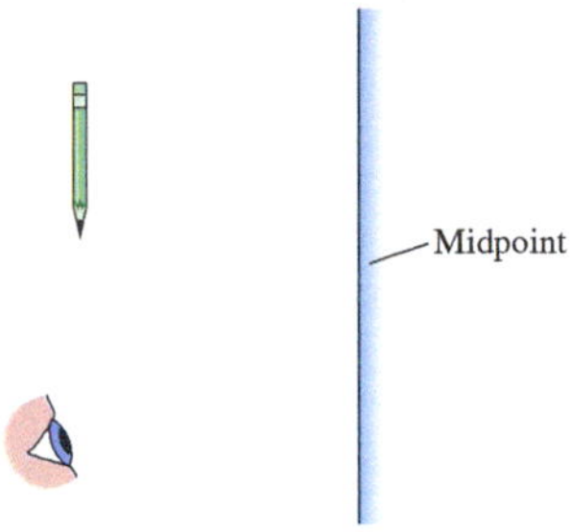


FIGURE Q18.6

7. In *The Toilet of Venus* by Velázquez (see Figure Q18.7), we can see the face of Venus in the mirror. Can she see her own face in the mirror, when the mirror is held as shown in the picture? If yes, explain why; if not, what does she see instead?

FIGURE Q18.7

Velazquez, Diego Rodriguez (1599–1660). The Toilet of Venus ('The Rokeby Venus'), 1647–51. Oil on canvas, 122.5 × 177 cm. Presented by The Art Fund, 1906 (NG2057). National Gallery. © National Gallery, London / Art Resource, NY.

8. The light spots that form a circle around the outside of the fountain in Figure Q18.8 come from a light source in the center of the fountain. Explain how the light makes it from the center of the fountain to the circles of light on the ground where the water hits.

FIGURE Q18.8

9. In Manet's *A Bar at the Folies-Bergère* (see Figure Q18.9), the reflection of the barmaid is visible in the mirror behind her. Is this the reflection you would expect if the mirror's surface is parallel to the bar? Where is the man seen facing her in the mirror actually standing?

FIGURE Q18.9

Edouard Manet (1832–1883), "A Bar at the Folies-Bergere," 1881/82. Oil on Canvas. 37 13/16″ × 51″ (90 × 130 cm). Courtauld Institute Galleries, London. AKG-Images.

10. If you look at a fish through the corner of a rectangular aquarium, you sometimes see two images, one on each side of the corner, as shown in Figure Q18.10. Sketch a top view of the situation and trace the path of some of the light rays that reach your eye from the fish to show how this can happen.

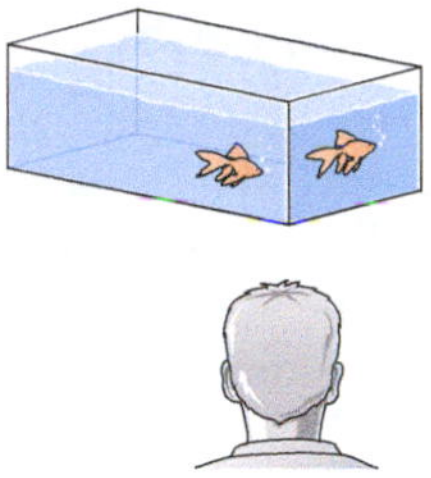

FIGURE Q18.10

11. You are looking straight into the front of an aquarium. You see a fish off to your right. Is the fish actually in the direction you're looking, farther to the right, or farther to the left? Explain.
12. If you hold a spoon in front of your face so that you see your image in the bowl of the spoon, your image is upright when you hold the spoon close to your face, but inverted when you hold the spoon far away. Explain why this change occurs.
13. a. Consider *one* point on an object near a lens. What is the minimum number of rays needed to locate its image point?
    b. For each point on the object, how many rays from this point actually strike the lens and refract to the image point?
14. A converging lens and a converging mirror have the same focal length in air. Which one has a longer focal length if they are used underwater?
15. Figure Q18.15 shows a photograph of droplets of water suspended from the stem of a plant in a garden, with a flower behind. There appear to be small flowers in each of the droplets; explain what is happening here.

FIGURE Q18.15

16. Figure Q18.16 shows droplets of water beaded up on a leaf. The light is coming from the left. The droplets work like a lens, focusing sunlight to bright spots behind the droplets. Use the ray model of light to explain why each bright spot is surrounded by a dark ring.

FIGURE Q18.16

17. A lens can be used to start a fire by focusing an image of the sun onto a piece of flammable material. All other things being equal, would a lens with a short focal length or a long focal length be better as a fire starter? Explain.
18. A piece of transparent plastic is molded into the shape of a converging lens, but it is hollow inside and has only a thin plastic wall. If immersed in water, would this air-filled lens act as a converging lens, act as a diverging lens, or not act as a lens at all?
19. From where you stand one night, you see the moon directly over a nearby streetlamp. You use a converging lens to get the moon's image sharply focused on a card; when you do so, the image of the lamp is a little out of focus. To bring the lamp's image into sharp focus, do you need to move the card closer to the lens or farther from the lens?

## Multiple-Choice Questions

Questions 20 through 22 are concerned with the situation sketched in Figure Q18.20, in which a beam of light in the air encounters a transparent block with index of refraction $n = 1.53$. Some of the light is reflected and some is refracted.

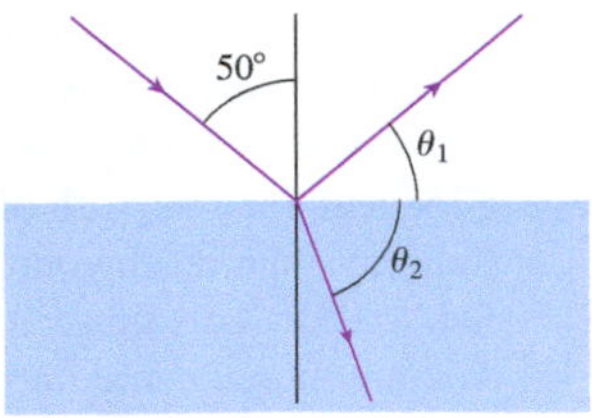

FIGURE Q18.20

20. | What is $\theta_1$?
 A. 40° B. 45°
 C. 50° D. 90°
21. | What is $\theta_2$?
 A. 20° B. 30°
 C. 50° D. 60°
22. | Is there an angle of incidence between 0° and 90° such that all of the light will be reflected?
 A. Yes, at an angle greater than 50°
 B. Yes, at an angle less than 50°
 C. No
23. | A 2.0-m-tall man is 5.0 m from the converging lens of a camera. His image appears on a detector that is 50 mm behind the lens. How tall is his image on the detector?
 A. 10 mm B. 20 mm
 C. 25 mm D. 50 mm
24. || You are 2.4 m from a plane mirror, and you would like to take a picture of yourself in the mirror. You need to manually adjust the focus of the camera by dialing in the distance to what you are photographing. What distance do you dial in?
 A. 1.2 m B. 2.4 m
 C. 3.6 m D. 4.8 m
25. | As shown in Figure Q18.25, an object is placed in front of a diverging mirror. At what position is the image located?

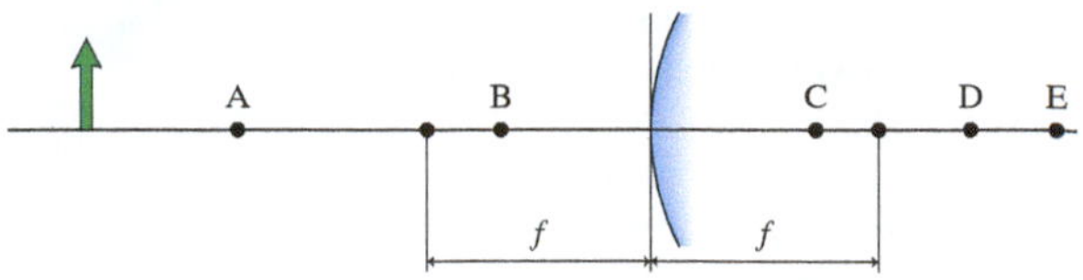

FIGURE Q18.25

26. | A virtual image of an object can be formed by
 A. A converging lens B. A diverging lens
 C. A plane mirror D. A converging mirror
 E. Any of the above
27. | An object is 40 cm from a converging lens with a focal length of 30 cm. A real image is formed on the other side of the lens, 120 cm from the lens. What is the magnification?
 A. −2.0 B. −3.0 C. −4.0 D. −1.33 E. −0.33
28. | The lens in Figure Q18.28 is used to produce a real image of a candle flame. What is the focal length of the lens?
 A. 9.0 cm
 B. 12 cm
 C. 24 cm
 D. 36 cm
 E. 48 cm

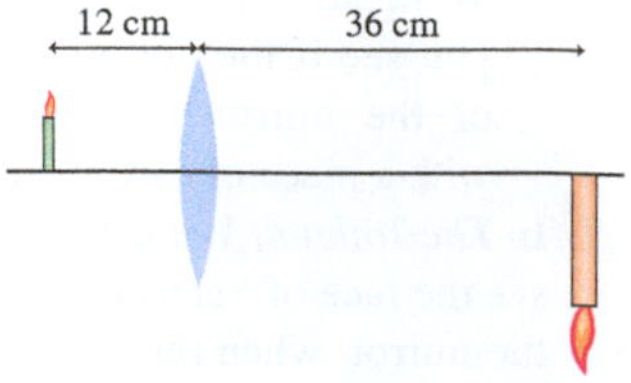

FIGURE Q18.28

29. | You look at yourself in a diverging mirror. Your image is
 A. Upright. B. Inverted.
 C. It's impossible to tell without knowing how far you are from the mirror and its focal length.

# PROBLEMS

### Section 18.1 The Ray Model of Light

1. || A 5.0-ft-tall girl stands on level ground. The sun is 25° above the horizon. How long is her shadow?
2. | In Figure P18.2, the compact flame of a candle casts a 15-cm-high shadow of a 8.2-cm-tall tree cutout. The candle is 3.2 cm from the cutout; how far is the candle from the wall?

FIGURE P18.2

3. |||| A point source of light illuminates an aperture 2.00 m away. A 12.0-cm-wide bright patch of light appears on a screen 1.00 m behind the aperture. How wide is the aperture?

### Section 18.2 Reflection

4. | The mirror in Figure P18.4 deflects a horizontal laser beam by 60°. What is the angle $\phi$?

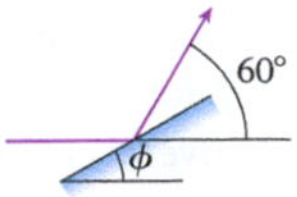

FIGURE P18.4

5. |||| It is 165 cm from your eyes to your toes. You're standing 200 cm in front of a tall mirror. How far is it from your eyes to the image of your toes?
6. || Figure P18.6 shows an object O in front of a plane mirror. Draw rays from the object that reflect from the mirror to determine from which locations A–D the object's image is visible.

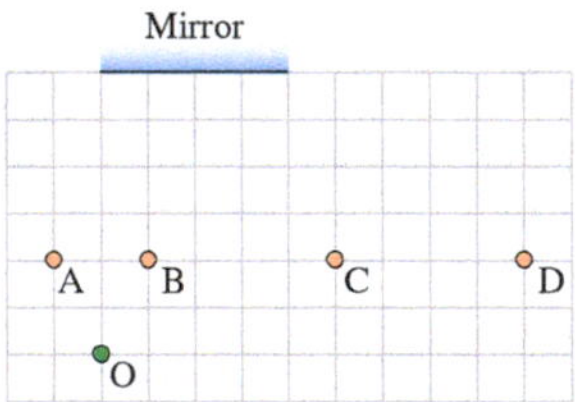

FIGURE P18.6

7. || A ray of light enters the region between two mirrors, as shown in Figure P18.7. How many times does the light reflect before exiting the space between the two mirrors?

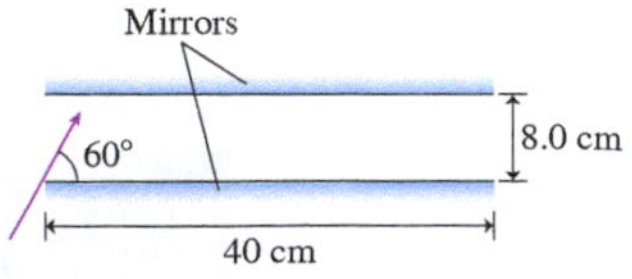

FIGURE P18.7

8. || You are standing 1.5 m from a mirror, and you want to use a classic camera to take a photo of yourself. This camera requires you to select the distance of whatever you focus on. What distance do you choose?
9. ||| Starting 3.5 m from a department store mirror, Suzanne walks toward the mirror at 1.5 m/s for 2.0 s. How far is Suzanne from her image in the mirror after 2.0 s?
INT

Watch Video Solution Problem 18.12

10. || The lightbulb in Figure P18.10 is 50 cm from a mirror. It emits 1.5 W of visible light. A small barrier blocks the direct rays of light from the bulb from reaching a sensor 70 cm to the right, but not the reflected rays. What is the light intensity at the sensor?
INT

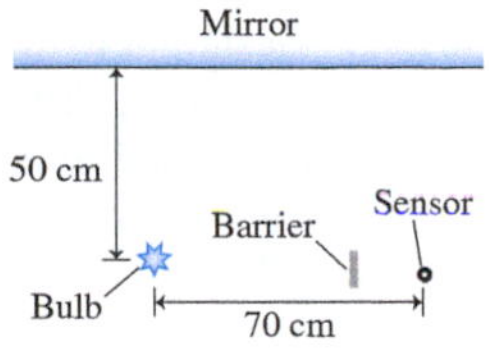

FIGURE P18.10

11. ||| A ray of light impinges on a mirror as shown in Figure P18.11. A second mirror is fastened at 90° to the first. After striking both mirrors, at what angle relative to the incoming ray does the outgoing ray emerge?

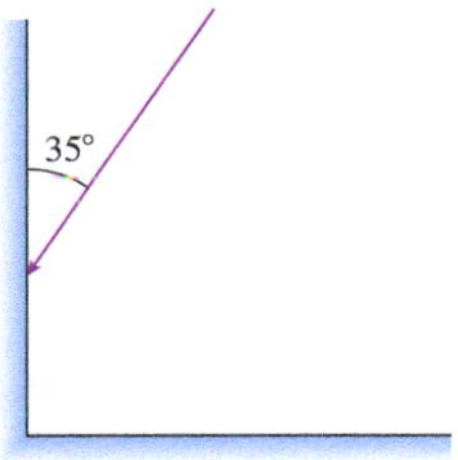

FIGURE P18.11

## Section 18.3 Refraction

12. ||| An underwater diver sees the sun 50° above horizontal. How high is the sun above the horizon to a fisherman in a boat above the diver?
13. || A laser beam in air is incident on a liquid at an angle of 37° with respect to the normal. The laser beam's angle in the liquid is 26°. What is the liquid's index of refraction?
14. ||| The sun is 60° above the horizon. Rays from the sun strike the still surface of a pond and cast a shadow of a stick that is stuck in the sandy bottom of the pond. If the stick is 10 cm tall, how long is the shadow?
15. ||| A 1.0-cm-thick layer of water stands on a horizontal slab of glass. A light ray in the air is incident on the water 60° from the normal. After entering the glass, what is the ray's angle from the normal?
16. || Figure P18.16 shows a ray of light entering an equilateral prism, with all sides and angles equal to each other. The ray traverses the prism parallel to the bottom and emerges at the same angle at which it entered. If $\theta = 40°$, what is the index of refraction of the prism?

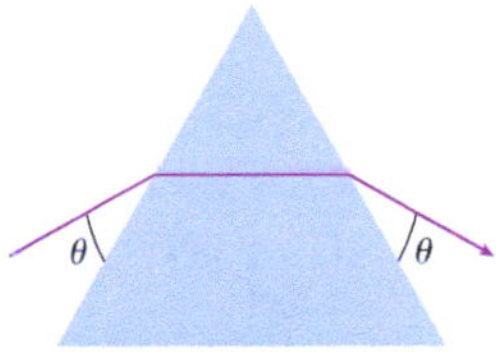

FIGURE P18.16

17. ||| A 4.0-m-wide swimming pool is filled to the top. The bottom of the pool becomes completely shaded in the afternoon when the sun is 20° above the horizon. How deep is the pool?
18. || You are on a snorkeling trip. Deep below the water, you look up at the surface of the water. Right at sunset, at what angle from the vertical do you see the sun?
19. || A ray of light traveling through air encounters a 1.2-cm-thick sheet of glass at a 35° angle of incidence. How far does the light ray travel in the glass before emerging on the far side?
20. || A thin glass rod is submerged in oil. What is the critical angle for light traveling inside the rod?
21. || A light ray travels inside a horizontal plate of glass, striking its upper surface at an angle of incidence of 60°. This ray is totally internally reflected at the glass-air boundary. A liquid is then poured on top of the glass. What is the largest index of refraction that the liquid could have such that the ray is still totally internally reflected?
22. ||| A typical diamond is cut as shown in Figure P18.22. A ray of light has entered the flat window at the top of the diamond perpendicular to the surface, as shown in the figure. Analyze the path of the ray for the next two interactions with the surfaces of the diamond.

FIGURE P18.22

23. || A light ray travels inside a block of sodium fluoride that has index of refraction $n = 1.33$ as shown in Figure P18.23. The ray strikes the vertical wall at the critical angle, totally reflects, and then emerges into the air above the block. What is the angle $\theta_2$ at which the ray emerges?

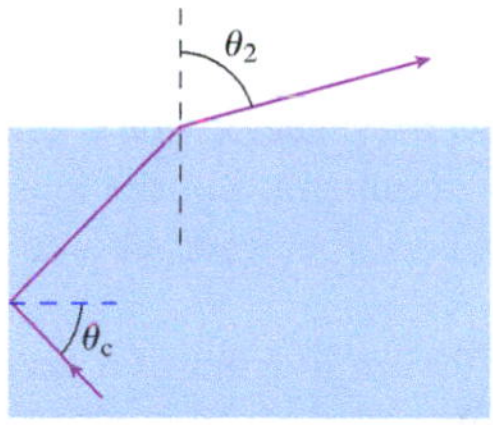

FIGURE P18.23

24. | Canola oil is less dense than water, so it floats on water, but its index of refraction is 1.47, higher than that of water. When you are adding oil and water to a bottle to make salad dressing, you notice a silvery reflection of light from the boundary between the oil and water. What is the critical angle for light going from the oil into the water?

## Section 18.4 Image Formation by Refraction

25. ||| A biologist keeps a specimen of his favorite beetle embedded in a cube of polystyrene plastic. The hapless bug appears to be 2.0 cm within the plastic. What is the beetle's actual distance beneath the surface?
26. || The composition of the ancient atmosphere can be determined by analyzing bubbles of air trapped in amber, which is fossilized tree resin. (This is one way we know that the air in the time of the dinosaurs was richer in oxygen than our current atmosphere.) An air bubble appears to be 7.2 mm below the flat surface of a piece of amber, which has index of refraction 1.54. How long a needle is required to reach the bubble?

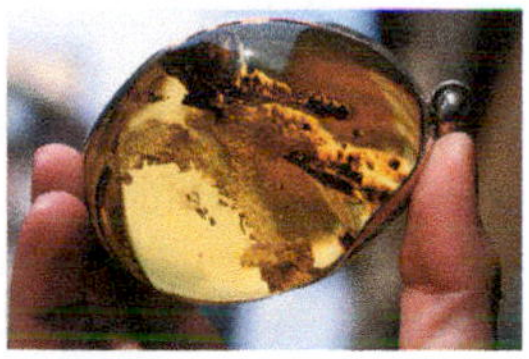

27. || A fish in a flat-sided aquarium sees a can of fish food on the counter. To the fish's eye, the can looks to be 30 cm outside the aquarium. What is the actual distance between the can and the aquarium? (You can ignore the thin glass wall of the aquarium.)
28. || A 1.8-m-tall diver is standing completely submerged on the bottom of a swimming pool, in 3.0 m of water. You are sitting on the end of the diving board, almost directly over her. How tall does the diver appear to be?

Watch Video Solution Problems 18.29, 18.36, and 18.45

29. | A swim mask has a pocket of air between your eyes and the flat glass front.

a. If you look at a fish while swimming underwater with a swim mask on, does the fish appear closer or farther than it really is? Draw a ray diagram to explain.
b. Does the fish see your face closer or farther than it really is? Draw a ray diagram to explain.

### Section 18.5 Thin Lenses: Ray Tracing

30. | An object is 30 cm in front of a converging lens with a focal length of 10 cm. Use ray tracing to determine the location of the image. Is the image upright or inverted? Is it real or virtual?
31. | An object is 6.0 cm in front of a converging lens with a focal length of 10 cm. Use ray tracing to determine the location of the image. Is the image upright or inverted? Is it real or virtual?
32. || In order to start a fire, a camper turns a lens toward the sun to focus its rays on a piece of wood. The lens has a 10 cm focal length. Draw a ray diagram of the lens and the incoming light rays to show where the wood should be placed for the best effect.
33. || An object is 20 cm in front of a diverging lens with a focal length of $-10$ cm. Use ray tracing to determine the location of the image. Is the image upright or inverted? Is it real or virtual?
34. || You are using a converging lens to look at a splinter in your finger. The lens has a 9.0 cm focal length, and you place the splinter 6.0 cm from the lens. How far from the lens is the image? What is the magnification?
35. | An object is 15 cm in front of a diverging lens with a focal length of $-10$ cm. Use ray tracing to determine the location of the image. Is the image upright or inverted? Is it real or virtual?
36. | A 2.0-cm-tall object is located 8.0 cm in front of a converging lens with a focal length of 10 cm. Use ray tracing to determine the location and height of the image. Is the image upright or inverted? Is it real or virtual?

### Section 18.6 Image Formation with Spherical Mirrors

37. | A converging cosmetic mirror has a focal length of 40 cm. A 5-cm-long mascara brush is held upright 20 cm from the mirror. Use ray tracing to determine the location and height of its image. Is the image upright or inverted? Is it real or virtual?
38. || A photographer took this image of himself in a converging mirror. Give some thought to what you are seeing here: One of the hands is his left hand; the other is the image of his left hand. Given the relative sizes and positions of his hand and the image of his hand, how far from the mirror did he place his hand to take this photo? Express your answer in terms of $f$, the mirror's focal length.

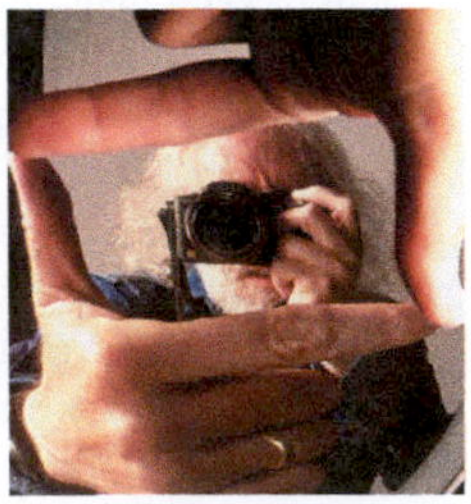

FIGURE P18.38

39. | A lightbulb is 60 cm from a converging mirror with a focal length of 20 cm. Use ray tracing to determine the location of its image. Is the image upright or inverted? Is it real or virtual?
40. || A flashlight uses a small lightbulb placed in front of a converging mirror. The light from the bulb should reflect from the mirror and emerge as a tight beam of light—a series of parallel rays. Where should the bulb be placed relative to the mirror?
41. || A dentist uses a curved mirror to view the back side of teeth on the upper jaw. Suppose she wants an erect image with a magnification of 2.0 when the mirror is 1.2 cm from a tooth. (Treat this problem as though the object and image lie along a straight line.) Use ray tracing to decide whether a converging or diverging mirror is needed, and to estimate its focal length.
BIO
42. || A diverging mirror, like the passenger-side rearview mirror on a car, has a focal length of $-2.0$ *m*. An object is 4.0 m from the mirror. Use ray tracing to determine the location of its image. Is the image upright or inverted? Is it real or virtual?
43. ||| An object is 12 cm in front of a diverging mirror. The mirror creates an image that is 75% as tall as the object. Use ray tracing to find the distance of the focal point from the mirror.

### Section 18.7 The Thin-Lens Equation

For Problems 44 through 53, calculate the image position and height.

44. | A 2.0-cm-tall object is 40 cm in front of a converging lens that has a 20 cm focal length.
45. || A 1.0-cm-tall object is 10 cm in front of a converging lens that has a 30 cm focal length.
46. || A 2.0-cm-tall object is 15 cm in front of a converging lens that has a 20 cm focal length.
47. | A 1.0-cm-tall object is 75 cm in front of a converging lens that has a 30 cm focal length.
48. | A 2.0-cm-tall object is 15 cm in front of a diverging lens that has a $-20$ cm focal length.
49. | A 1.0-cm-tall object is 60 cm in front of a diverging lens that has a $-30$ cm focal length.
50. | A 3.0-cm-tall object is 15 cm in front of a diverging mirror that has a $-25$ cm focal length.
51. || A 3.0-cm-tall object is 45 cm in front of a diverging mirror that has a $-25$ cm focal length.
52. | A 3.0-cm-tall object is 15 cm in front of a converging mirror that has a 25 cm focal length.
53. | A 3.0-cm-tall object is 45 cm in front of a converging mirror that has a 25 cm focal length.
54. || A toy insect viewer for kids consists of a plastic container with a lens in the lid. The lid is 12 cm from the bottom of the container. The lens produces a magnified image of an insect on the bottom of the container. If the magnification is 4.0, what is the focal length of the lens?
55. | At what distance from a converging mirror with a 35 cm focal length should an object be placed so that its image is the same distance from the mirror as the object?
56. ||| The sun is 150,000,000 km from earth; its diameter is 1,400,000 km. A student uses a 4.0-cm-diameter lens with $f = 10$ cm to cast an image of the sun on a piece of paper.
INT
a. Where should the paper be placed to get a sharp image?
b. What is the diameter of the image on the paper?
c. The intensity of the incoming sunlight is 1050 $W/m^2$. What is the power of the light captured by the lens?
d. What is the intensity of sunlight in the projected image? Assume that all of the light captured by the lens is focused into the image.

57. || The illumination lights in an operating room use a converging mirror to focus an image of a bright lamp onto the surgical site. One such light has a mirror with a focal length of 15 cm. If the patient is 1.0 m from the mirror, where should the lamp be placed relative to the mirror?

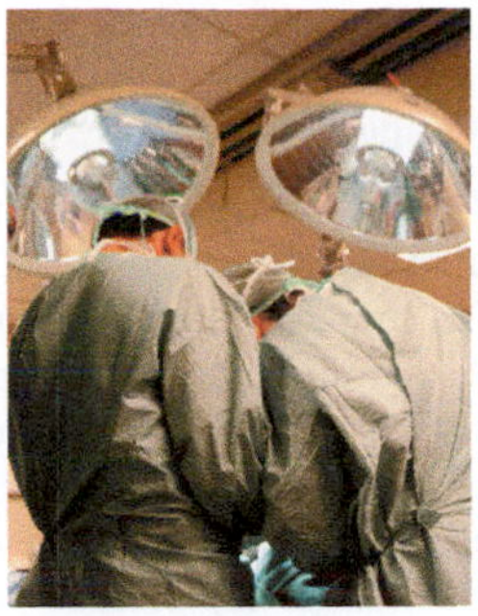

58. ||| The sun is 150,000,000 km from earth; its diameter is 1,400,000 km. For a science project on solar power, a student uses a 24-cm-diameter converging mirror with a focal length of 45 cm to focus sunlight onto an object. This casts an image of the sun on the object. For the most intense heat, the image of the sun should be in focus.
INT
   a. Where should the object be placed?
   b. What is the diameter of the image?
   c. The intensity of the incoming sunlight is 1050 W/m$^2$. What is the total power of the light captured by the mirror?
   d. What is the intensity of sunlight in the projected image? Assume that all of the light captured by the mirror is focused into the image.

59. || The moon is $3.5 \times 10^6$ m in diameter and $3.8 \times 10^8$ m from the earth's surface. The 1.2-m-focal-length converging mirror of a telescope focuses an image of the moon onto a detector. What is the diameter of the moon's image?

60. ||| Consider a typical diverging passenger-side mirror with a focal length of −80 cm. A 1.5-m-tall cyclist on a bicycle is 25 m from the mirror. You are 1.0 m from the mirror, and suppose, for simplicity, that the mirror, you, and the cyclist all lie along a line.
   a. How far are you from the image of the cyclist?
   b. What is the image height?

## General Problems

61. || You slowly back away from a plane mirror at a speed of 0.10 m/s. With what speed does your image appear to be moving away from you?
INT

62. || At what angle $\phi$ should the laser beam in Figure P18.62 be aimed at the mirrored ceiling in order to hit the midpoint of the far wall?

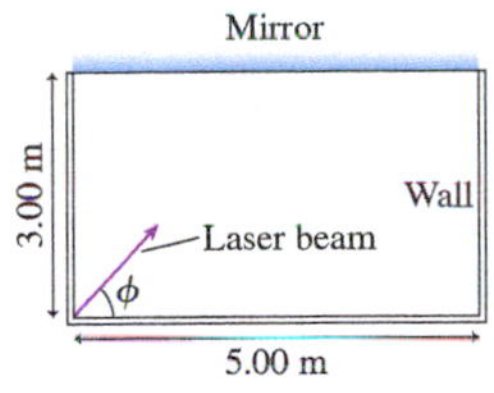


FIGURE P18.62

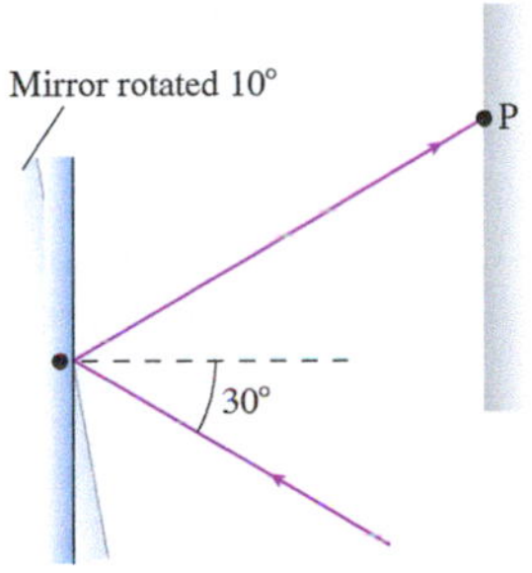


FIGURE P18.63

63. |||| A laser beam is incident on a mirror at an angle of 30°, as shown in Figure P18.63. It reflects off the mirror and strikes a wall 2.0 m away at point P. By what distance does the laser spot on the wall move if the mirror is rotated by 10°?

64. ||| The place you get your hair cut has two nearly parallel mirrors 5.0 m apart. As you sit in the chair, your head is 2.0 m from the nearer mirror. Looking toward this mirror, you first see your face and then, farther away, the back of your head. (The mirrors need to be slightly nonparallel for you to be able to see the back of your head, but you can treat them as parallel in this problem.) How far away does the back of your head appear to be? Neglect the thickness of your head.

65. || What is the angle of incidence in air of a light ray whose angle of refraction in glass is half the angle of incidence?

66. ||| Figure P18.66 shows a light ray incident on a glass cylinder. What is the angle $\alpha$ of the ray after it has entered the cylinder?

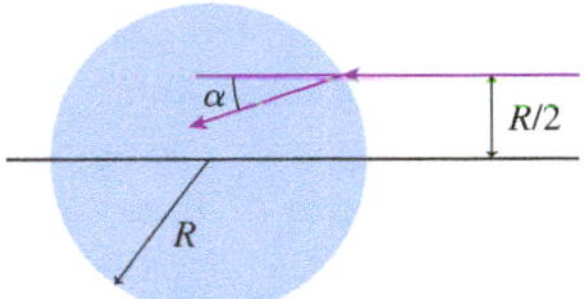


FIGURE P18.66

67. |||| It's nighttime, and you've dropped your goggles into a swimming pool that is 3.0 m deep. If you hold a laser pointer 1.0 m directly above the edge of the pool, you can illuminate the goggles if the laser beam enters the water 2.0 m from the edge. How far are the goggles from the edge of the pool?

68. ||| Figure P18.68 shows a meter stick lying on the bottom of a 100-cm-long tank with its zero mark against the left edge. You look into the tank at a 30° angle, with your line of sight just grazing the upper left edge of the tank. What mark do you see on the meter stick if the tank is (a) empty, (b) half full of water, and (c) completely full of water?

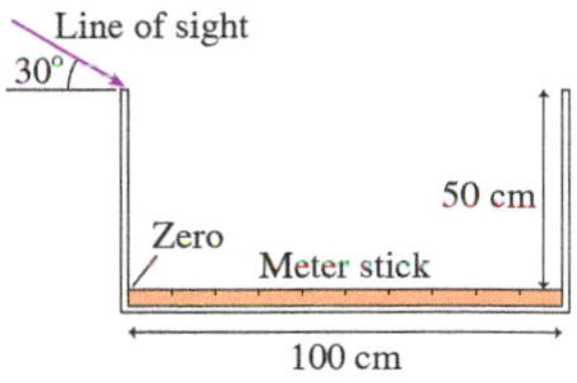


FIGURE P18.68

69. ||| What is the exit angle $\theta$ from the glass prism in Figure P18.69?

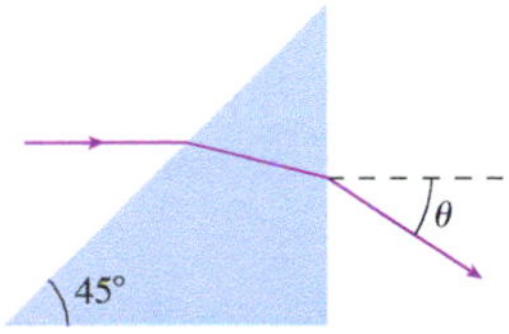


FIGURE P18.69

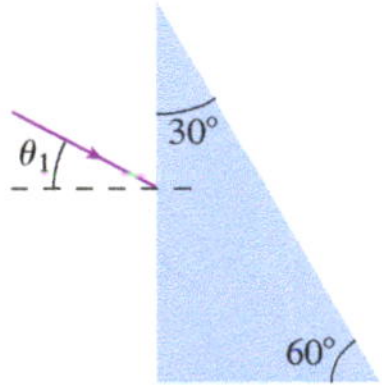


FIGURE P18.70

70. |||| What is the smallest angle $\theta_1$ for which a laser beam will undergo total internal reflection on the hypotenuse of the glass prism in Figure P18.70?

71. ||| A 1.0-cm-thick layer of water stands on a horizontal slab of glass. Light from within the glass is incident on the glass-water boundary. What is the maximum angle of incidence for which a light ray can emerge into the air above the water?

72. |||| The glass core of an optical fiber has index of refraction 1.60. The index of refraction of the cladding is 1.48. What is the maximum angle between a light ray and the wall of the core if the ray is to remain inside the core?

73. ||| A 150-cm-tall diver is standing completely submerged on the bottom of a swimming pool full of water. From your point of view out of the water, on the edge of the pool, how tall does the diver appear to be?

74. || To a fish, the 4.00-mm-thick aquarium walls appear only 3.50 mm thick. What is the index of refraction of the walls?

75. || A microscope is focused on an amoeba. When a 0.15-mm-thick cover glass ($n = 1.50$) is placed over the amoeba, by how far must the microscope objective be moved to bring the organism back into focus? Must it be raised or lowered?
BIO

76. || You need to use a 24-cm-focal-length lens to produce an inverted image twice the height of an object. At what distance from the object should the lens be placed?

77. |||| A nearsighted person might correct his vision by wearing diverging lenses with focal length $f = -50$ cm. When wearing his glasses, he looks not at actual objects but at the virtual images of those objects formed by his glasses. Suppose he looks at a 12-cm-long pencil held vertically 2.0 m from his glasses. Use ray tracing to determine the location and height of the image.
BIO

78. || A 1.5-cm-tall object is 90 cm in front of a diverging lens that has a 45 cm focal length. Use ray tracing to find the position and height of the image. To do this accurately, use a ruler or paper with a grid. Determine the image distance and image height by making measurements on your diagram.

79. ||| A 2.0-cm-tall candle flame is 2.0 m from a wall. You happen to have a lens with a focal length of 32 cm. How many places can you put the lens to form a well-focused image of the candle flame on the wall? For each location, what are the height and orientation of the image?

80. || A 2.0-cm-diameter spider is 2.0 m from a wall. Determine the focal length and position (measured from the wall) of a lens that will make a half-size image of the spider on the wall.

81. ||| Figure P18.81 shows a meter stick held lengthwise along the optical axis of a converging mirror. How long is the image of the meter stick?

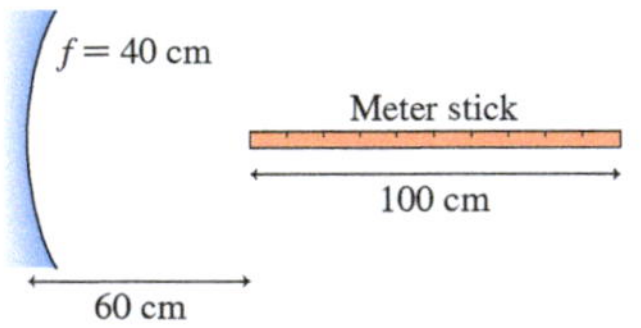


FIGURE P18.81

82. || A slide projector needs to create a 98-cm-high image of a 2.0-cm-tall slide. The screen is 300 cm from the slide.
   a. What focal length does the lens need? Assume that it is a thin lens.
   b. How far should you place the lens from the slide?

## MCAT-Style Passage Problems

### Mirages

There is an interesting optical effect you have likely noticed while driving along a flat stretch of road on a sunny day. A small, distant dip in the road appears to be filled with water. You may even see the reflection of an oncoming car. But, as you get closer, you find no puddle of water after all; the shimmering surface vanishes, and you see nothing but empty road. It was only a *mirage,* the name for this phenomenon.

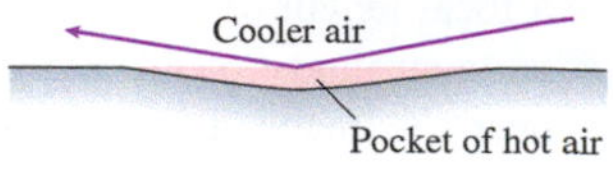


FIGURE P18.83

The mirage is due to the different index of refraction of hot and cool air. The actual bending of the light rays that produces the mirage is subtle, but we can make a simple model as follows. When air is heated, its density decreases and so does its index of refraction. Consequently, a pocket of hot air in a dip in a road has a lower index of refraction than the cooler air above it. Incident light rays with large angles of incidence (that is, nearly parallel to the road, as shown in Figure P18.83) experience total internal reflection. The mirage that you see is due to this reflection. As you get nearer, the angle goes below the critical angle and there is no more total internal reflection; the "water" disappears!

83. | The pocket of hot air appears to be a pool of water because
   A. Light reflects at the boundary between the hot and cool air.
   B. Its density is close to that of water.
   C. Light refracts at the boundary between the hot and cool air.
   D. The hot air emits blue light that is the same color as the daytime sky.

84. | Which of these changes would allow you to get closer to the mirage before it vanishes?
   A. Making the pocket of hot air nearer in temperature to the air above it
   B. Looking for the mirage on a windy day, which mixes the air layers
   C. Increasing the difference in temperature between the pocket of hot air and the air above it
   D. Looking at it from a greater height above the ground

85. | If you could clearly see the image of an object that was reflected by a mirage, the image would appear
   A. Magnified.
   B. With up and down reversed.
   C. Farther away than the object.
   D. With right and left reversed.

# 19 Optical Instruments

This anableps is called the "four-eyed fish." How must the upper half of its eye differ from the lower half so that it has clear vision both above and below the waterline at the same time?

## LOOKING AHEAD ➤

### The Human Eye

Our most important optical instruments are our own eyes, which use a lens to focus light onto the light-sensitive retina.

You'll learn how near- and farsightedness can be corrected with eyeglasses or contact lenses.

### Optical Instruments

A converging lens is the simplest magnifier. We'll also study microscopes and telescopes.

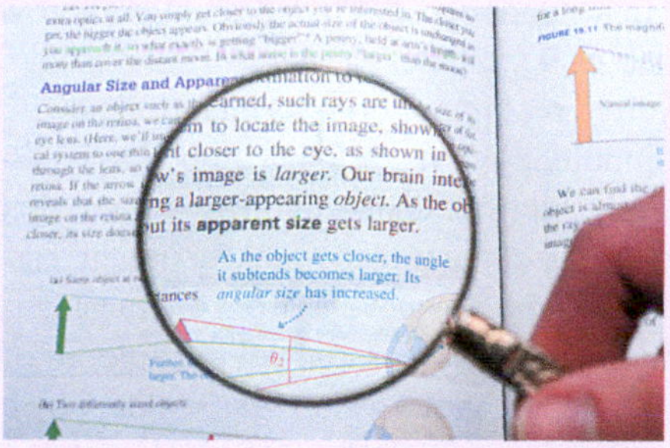

You'll learn how optical instruments can be designed to magnify objects up to a thousand times.

### Optical Resolution

This magnified image of chromosomes is slightly blurry because of a limit to the microscope's **resolution** due to diffraction.

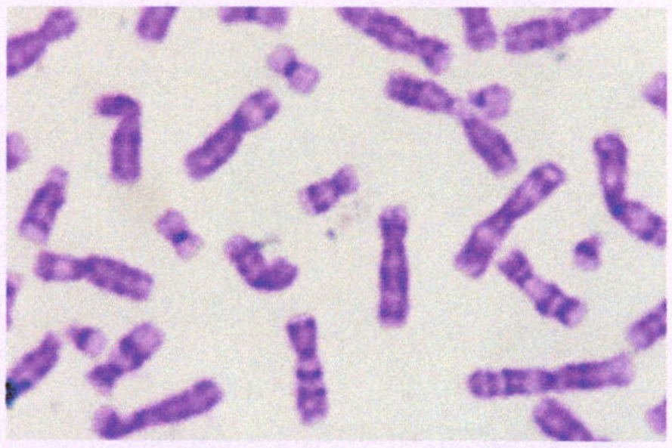

You'll learn that a microscope cannot resolve features much smaller than the wavelength of light.

**GOAL** To understand how common optical instruments work.

## LOOKING BACK ◀

### Image Formation by Lenses

In Section 18.5, you learned how a single lens can form a real image of an object. In this chapter, we'll study how *combinations* of two lenses can form highly magnified images, such as those in microscopes and telescopes.

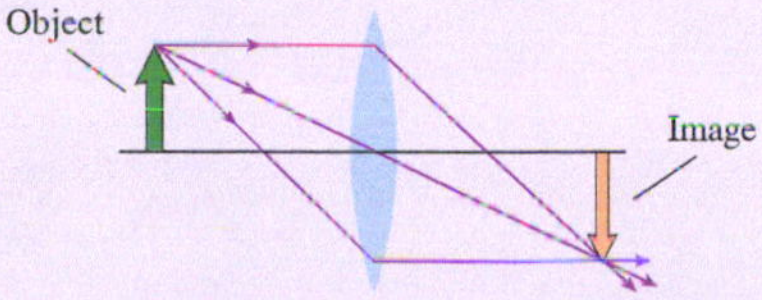


A real image is one through which light rays actually pass. It's on the opposite side of the lens from the object.

**STOP TO THINK**

A converging lens creates a real, inverted image. For this to occur, the object must be

A. Closer to the lens than the focal point.
B. Farther from the lens than the focal point.
C. At the focal point.

**A pinhole eye** BIO The chambered nautilus is the only animal with a true pinhole "camera" as an eye. Light rays that pass through the small opening form a crude image on the back surface of the eye, where the rays strike light-sensitive cells. The image isn't great, but the overall visual acuity of the nautilus is similar to that of similar creatures.

## 19.1 The Camera

This chapter will investigate a number of optical instruments in which a combination of lenses and mirrors performs a useful function. We'll start with an instrument familiar to everyone: the camera. A **camera** is a device that projects a real image onto a plane surface, where the image can be recorded onto film or, in today's digital cameras, an electronic detector. Although modern cameras are marvels of optical engineering, decent images can be produced using only a light-proof box with a small hole punched in it. Such a **pinhole camera** is shown in FIGURE 19.1a. FIGURE 19.1b uses the ray model of light passing through a small hole to illustrate how the pinhole camera works. Each point on an object emits light rays in all directions, but, ideally, only one of these rays passes through the hole and reaches the film. Each point on the object thus illuminates just one point on the film, forming the image. As the figure illustrates, the geometry of the rays causes the image to be upside down.

FIGURE 19.1 A pinhole camera.

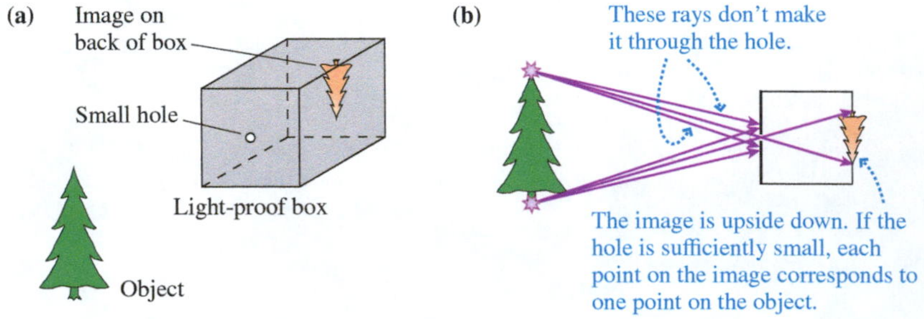


Actually, in practice, each *point* on the object illuminates a small but finite *patch* on the film. This is because the finite size of the hole allows several rays from each point on the object to pass through at slightly different angles. As a result, the image is slightly blurred. Maximum sharpness is achieved by making the hole smaller and smaller, which makes the image dimmer and dimmer. (Diffraction also becomes an issue if the hole gets too small.) A real pinhole camera has to accept a small amount of blurring as the trade-off for having an image bright enough to be practical.

FIGURE 19.2 A camera.

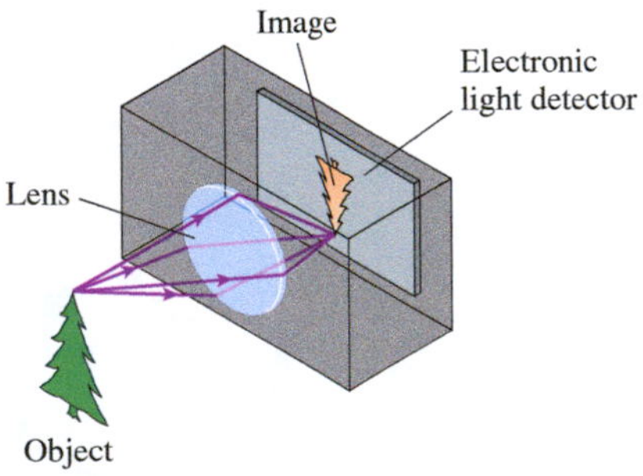


A video to support a section's topic is embedded in the eText.
**Video** Figure 19.2

In a standard camera, a dramatic improvement is made possible by using a *lens* in place of a pinhole. FIGURE 19.2 shows how a camera's converging lens projects an inverted real image onto its electronic detector, just as a pinhole camera does. Unlike a pinhole, however, a lens can be large, letting in plenty of light while still giving a sharply focused image. Not shown are the *shutter,* an opaque barrier that is briefly moved out of the way in order for light to pass through the lens, and the *diaphragm,* a set of leaves that can move in from the outside of the lens to effectively reduce its diameter. The diaphragm helps control the amount of light that reaches the camera's detector.

FIGURE 19.3 shows light rays from an object passing through the lens and converging at the image plane B. Here, a single point P on the object focuses to a single point P′ in the image plane. If the electronic detector is located in this plane, a sharp image will form on it. If, however, the detector had been located a bit in *front* of the image

FIGURE 19.3 Focusing a camera.

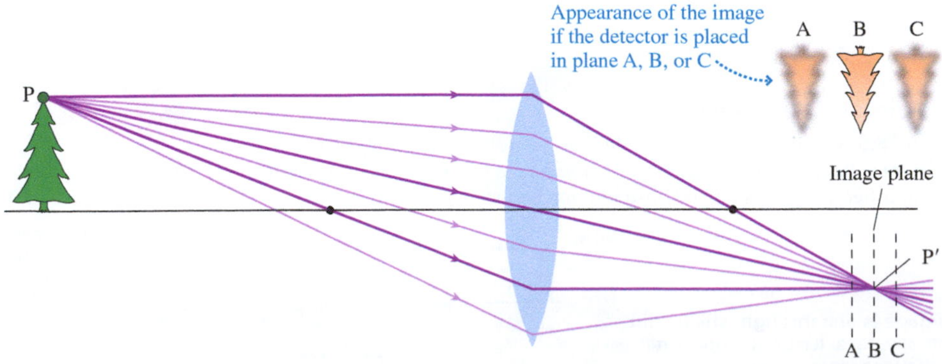

plane, at position A, rays from point P would not yet have completely converged and would form a small blurry *circle* on the detector instead of a sharp point. Thus the image would appear blurred, as shown. Similarly, if the detector is placed at C, *behind* the image plane, the rays will be diverging from their perfect focus and again form a blurry image.

Thus to get a sharp image, the detector must be accurately located in the image plane. Figure 19.3 shows that one way to do this is to move the detector until it coincides with the image plane. More commonly, however, a camera is **focused** by moving the *lens* either toward or away from the detector plane until the image is sharp. In either case, the lens-detector distance is varied.

**Video** Telephoto Lens

### EXAMPLE 19.1 Focusing a camera

A digital camera whose lens has a focal length of 8.0 mm is used to take a picture of an object 30 cm away. If the object is 2.4 cm high, how large will the image appear on the detector once the camera is focused on the object?

**STRATEGIZE** As shown in Figure 19.3, the image will be in focus when the detector is in the image plane, so we need to find the image distance. Once we know the image and object distances, we will determine the magnification.

**PREPARE** We can use the thin-lens equation, Equation 18.12, to find the image distance $s'$ given the lens's focal length, $f = 8.0$ mm, and the distance from the lens to the object, $s = 30$ cm. With the image distance in hand, we can determine the magnification and then use this to determine the image height.

**SOLVE** We rearrange the thin-lens equation to solve for the image distance $s'$:

$$\frac{1}{s'} = \frac{1}{f} - \frac{1}{s} = \frac{1}{0.0080\text{ m}} - \frac{1}{0.30\text{ m}} = 122\text{ m}^{-1}$$

$$s' = \frac{1}{122\text{ m}^{-1}} = 8.2\text{ mm}$$

This is the lens-detector distance. The magnification is given by Equation 18.9:

$$m = -\frac{s'}{s} = -\frac{0.0082\text{ m}}{0.30\text{ m}} = -0.027$$

The negative sign tells us that the image is inverted, as we know it must be. The absolute magnitude of the magnification, 0.027, is the ratio of the image and object heights, so

$$h' = h(0.027) = (2.4\text{ cm})(0.027) = 0.065\text{ m} = 0.65\text{ mm}$$

**ASSESS** The object distance is much greater than the focal length of the lens, so we expect that the image distance will be nearly equal to the focal length, as we found. The image distance is much smaller than the object distance, so the image will be much smaller than the object, as we found. The detector must have very fine structure to record detail in an image of this small size.

In addition to adjusting the lens position, there are other adjustments you can make to take a good photo. The detector needs a certain amount of light, but not too much. You can adjust the amount of light by adjusting the shutter speed, changing the amount of time that light is allowed to enter. Most cameras also allow you to adjust the *aperture*—the size of the opening through which the light comes—by adjusting the diaphragm in front of the lens. If the aperture is wide, it lets in more light, which is good if the light is dim. If the aperture is narrow, it lets in less light, a good choice for taking photos on a bright day. Narrowing the aperture also makes the focus sharper. If the aperture is very small, you have essentially made a pinhole camera—an image would form even if the lens wasn't there! Objects form sharp images at a range of distances, so the image focus is more forgiving of an inexact placement of the lens. A similar effect is at work in your eye: If your vision is compromised, you might squint—narrowing the opening to your eye—to sharpen the focus.

If a digital picture is magnified enough, you can see the individual pixels that make it up.

**FIGURE 19.4** A CCD chip used in a digital camera.

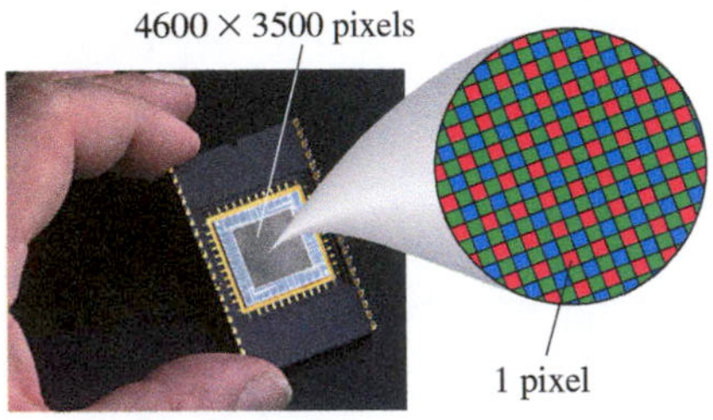


The detector in a digital camera is usually a *charge-coupled device,* or **CCD.** A CCD consists of a rectangular array of many millions of small detectors called *pixels.* When light hits one of these pixels, it generates an electric charge proportional to the light intensity. Thus an image is recorded on the CCD in terms of little packets of charge. After the CCD has been exposed, the charges are read out, the signal levels are digitized, and the picture is stored in the digital memory of the camera.

**FIGURE 19.4** shows a CCD "chip" and, schematically, the magnified appearance of the pixels on its surface. To record color information, different pixels are covered by red, green, or blue filters; a pixel covered by a green filter, for instance, records only the intensity of the green light hitting it. Later, the camera's microprocessor interpolates nearby colors to give each pixel an overall true color. The structure of the retina of the eye is remarkably similar, as we'll see in Chapter 25.

**STOP TO THINK 19.1** The screen in a pinhole camera is moved farther away from the pinhole. The image on the screen will

A. Become larger.
B. Become smaller.
C. Remain the same size.

## 19.2 The Human Eye BIO

**FIGURE 19.5** The human eye.

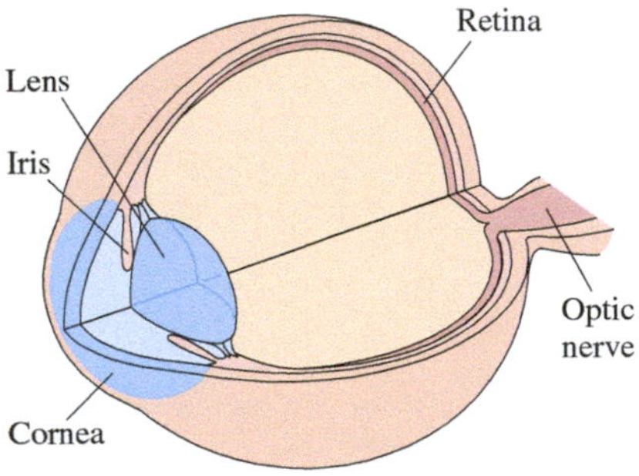


The human eye functions much like a camera. Like the camera, it has three main functional groups: an optical system to focus the incoming light, a diaphragm to adjust the amount of light entering the eye, and a light-sensitive surface to detect the resulting image. The parts of the eye that make up these three groups are shown in **FIGURE 19.5.** The *cornea* and the *lens* are together responsible for refracting incoming light rays and producing an image. The adjustable *iris* determines how much light enters the eye, in much the same way as does the diaphragm of a camera. And the *retina* is the light-sensitive surface on which the image is formed. The retina is the biological equivalent of the CCD in a digital camera.

### Refractive Power

The lenses used for vision correction, as well as other types of lenses, are often characterized by their **refractive power.** The refractive power of a lens is the inverse of its focal length:

$$P = \frac{1}{f} \tag{19.1}$$

Refractive power of a lens with focal length $f$

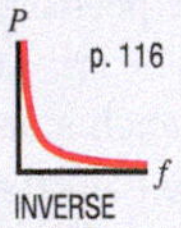


The SI unit of refractive power is the *diopter,* abbreviated D, which is defined as $1\text{ D} = 1\text{ m}^{-1}$. To compute power in diopters, we must express the focal length in meters. To match the sign convention for focal length from Chapter 18, a converging lens has a positive power, a diverging lens a negative power.

**EXAMPLE 19.2 Relating power and focal length**

A typical pair of off-the-shelf reading glasses has +2.0 D lenses. What is the focal length of these lenses?

***SOLVE*** We rearrange Equation 19.1 to find the focal length:

$$f = \frac{1}{+2.0\text{ D}} = \frac{1}{+2.0\text{ m}^{-1}} = +0.50\text{ m}$$

The glasses have converging lenses with a focal length of 50 cm.

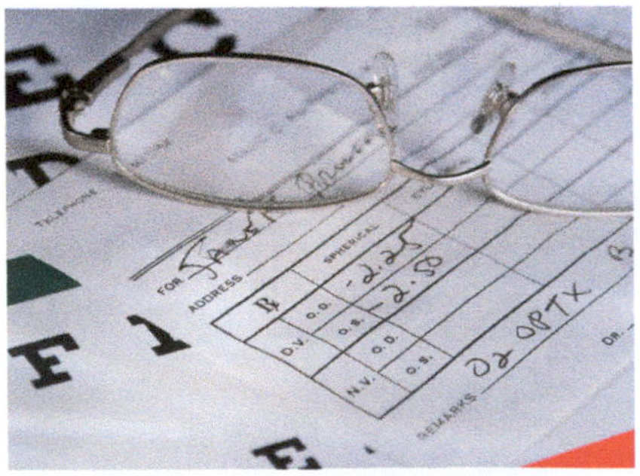

**I can see clearly now** When writing eyeglass prescriptions, optometrists don't write the D because the lens maker already knows that prescriptions are in diopters. The optometrist's prescription for this patient is −2.25 D for the right eye (top) and −2.50 D for the left eye (bottom). The negative numbers mean that the lenses are diverging lenses, which correct for nearsightedness, as we'll see.

Refractive power is a measure of how much the lens bends rays of light. A high-power lens bends light rays through a large angle, leading to a short focal length. A low-power lens bends the rays through a small angle, leading to a long focal length.

FIGURE 19.6a shows the focusing of parallel rays of light by two thin lenses that have refractive powers $P_1$ and $P_2$. If we place the two lenses right next to each other, as in FIGURE 19.6b, the refractive power of the combination is the sum of the two powers:

$$P_{\text{total}} = P_1 + P_2 = \frac{1}{f_1} + \frac{1}{f_2} \tag{19.2}$$

Total refractive power for the combination of two thin lenses

This useful relationship—that refractive powers add for lenses in contact—is the reason that refractive power is used to characterize corrective lenses.

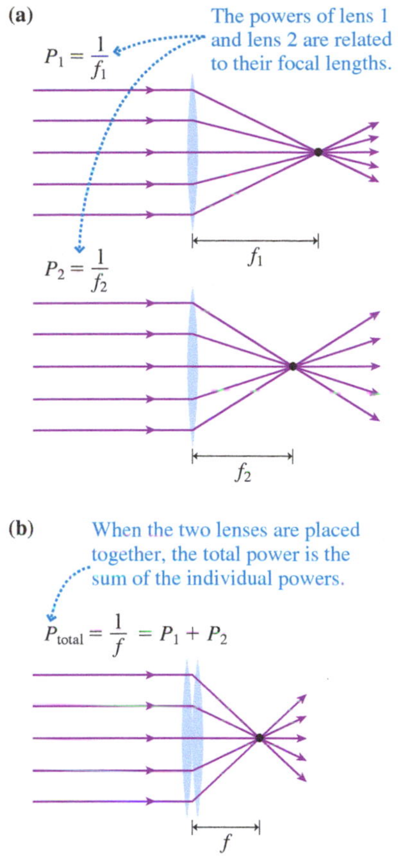

FIGURE 19.6 Combining two lenses.

**EXAMPLE 19.3 Finding the focal length of a combination of lenses**

A magnifier intended for very close work comes with two lenses, one with power +10 D and the other with power +20 D. The two lenses can be stacked to combine their powers for greater magnification. What is the focal length of the combination of the two lenses?

***SOLVE*** The power of the combination of the two lenses is the sum of the two powers:

$$P_{\text{total}} = P_1 + P_2 = +30\text{ D}$$

This gives a focal length of

$$f = \frac{1}{P_{\text{total}}} = 0.033\text{ m} = 3.3\text{ cm}$$

***ASSESS*** If you've ever used a magnifier like this, you know that you must hold it quite close to the object you are viewing—the focal length is short—so our result makes sense.

## The Optical System of the Eye

Like a camera, the eye works by focusing incoming rays onto a light-sensitive surface, the retina. The primary structures that determine the optics of the eye are shown in FIGURE 19.7. Rays of light enter the eye; they are refracted as they enter the cornea and then as they pass through the lens. Most of the eye's refraction occurs at the surface of the cornea because of the strong curvature of the cornea and the large difference between the indices of refraction on either side of the surface. Light refracts less as it passes through the lens because the lens's index of refraction doesn't differ much from that of the fluid in which it is embedded. The lens's role is to adjust the refractive power of the optical system, the combination of elements that focus light on the retina. The shape of the cornea is fixed, but the shape of the lens is changed through the action of the ciliary muscle.

FIGURE 19.7 Focusing of light in the eye.

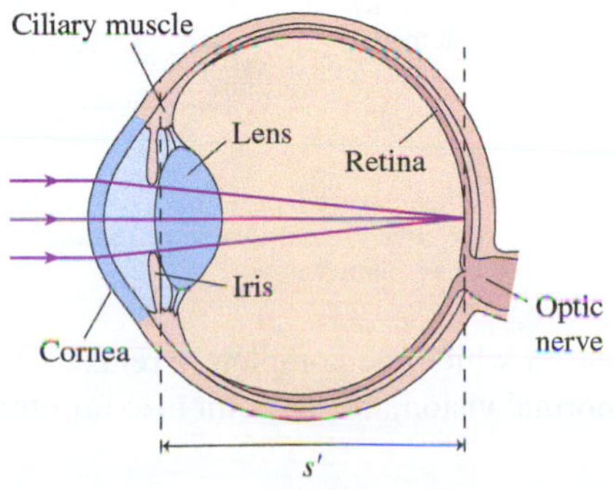

The optical system of the eye is complex, but for our purposes it's reasonable to model the eye's focusing properties as being due to a single thin lens whose shape can change. That will allow us to use the thin-lens equation of Chapter 18. When the eye is relaxed—that is, when the ciliary muscle is relaxed—the lens is shaped to bring parallel rays of light to a focus right at the retina. Parallel rays of light come from an object at a great distance. If object distance $s$ is very large, $1/s$ is very small, so the thin-lens equation becomes

$$\frac{1}{f} = \frac{1}{s} + \frac{1}{s'} \approx \frac{1}{s'} \tag{19.3}$$

The distance from the lens to the retina is approximately 1.7 cm, so we can say that $s' \approx 0.017$ m. If we use this number in Equation 19.3, we find

$$\frac{1}{f} \approx \frac{1}{s'} = 60 \text{ m}^{-1}$$

When the eye is relaxed, the power of the optical system is thus

$$P(\text{relaxed eye}) = \frac{1}{f} \approx 60 \text{ D}$$

Of this 60 D of refractive power, approximately 40 D comes from the cornea and 20 D comes from the lens. If the lens is surgically removed, which it often is for people who have the clouding of the lens characteristic of cataracts, the cornea alone still provides a marginal level of vision.

NOTE ▸ There is quite a bit of variation among individuals. These numbers don't apply to every eye, but they are representative values that we can use to assess our work. ◂

## Focusing and Accommodation

If you focus on a distant object, your eyes are relaxed; the optical system of your eye has a total refractive power of about 60 D, and a crisp image forms on your retina. But, under normal circumstances, you are continually adjusting your gaze to view objects at different distances. A camera adjusts its focus by changing the distance from the lens to the detector, but this doesn't happen in your eye. In your eye, $s'$, the distance from the lens to the retina, stays the same. To focus at different distances, the ciliary muscle in the eye changes the shape of the lens, adjusting its refractive power, as shown in FIGURE 19.8. This process of changing the lens shape as the eye focuses at different distances is called **accommodation.**

**Video** The Eye

FIGURE 19.8 Accommodation by the eye.

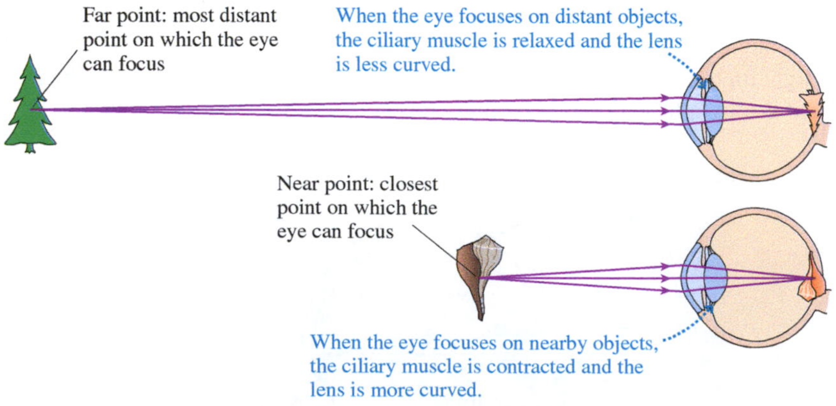


The most distant point on which the completely relaxed eye can focus is called the eye's **far point.** For normal vision, the far point is at infinity. The closest point on which the eye can focus, with the ciliary muscle fully contracted, is called the **near point.** Objects closer than the near point cannot be brought into sharp focus.

When we solve problems that involve vision, it makes sense to work in terms of the refractive power instead of the focal length. Since $P = 1/f$, we'll use the following version of the thin-lens equation:

$$P = \frac{1}{s} + \frac{1}{s'} \tag{19.4}$$

In Equation 19.4, $s$ is the distance to the object and $s'$ is the distance to the retina, which is fixed. Focusing on objects at different distances requires a change in the refractive power of the visual system. With the eye fully relaxed, the refractive power is at a minimum; we'll call this $P_{min}$. With the maximum accommodation, with the eye focused on an object at the near point, the optical system of the eye has the largest possible refractive power, $P_{max}$.

**EXAMPLE 19.4 What is the eye's range of accommodation?** BIO

A typical 20-year-old student has a far point of infinity and a near point of about 10 cm. (Try it and see. Pick one eye, and see how close to your eye you can hold this book while keeping it in sharp focus.) What change in power is needed to focus at the 10 cm near point?

***STRATEGIZE*** We'll solve this image problem using the version of the thin-lens equation in Equation 19.4. We can use this equation to determine the necessary power for sharp focus at different distances.

***PREPARE*** The far point of the eye is infinity. When the ciliary muscle is relaxed and the eye is focused on an object at infinity, the thin-lens equation gives the refractive power of the relaxed eye. This is the minimum power:

$$P_{min} = \frac{1}{s} + \frac{1}{s'} = \frac{1}{s'}$$

***SOLVE*** To focus at 10 cm, the ciliary muscle contracts and the lens assumes a more curved shape. This is the near point, so the maximum power is

$$P_{max} = \frac{1}{s} + \frac{1}{s'} = \frac{1}{0.10\text{ m}} + \frac{1}{s'} = 10\text{ D} + \frac{1}{s'}$$

The change in power to move from the far point to the near point is the difference between the minimum and maximum powers:

$$\Delta P = P_{max} - P_{min} = \left(10\text{ D} + \frac{1}{s'}\right) - \left(\frac{1}{s'}\right) = 10\text{ D}$$

***ASSESS*** The final power is greater than the initial power, which makes sense; the rays of light from the close object are diverging, and a greater degree of refraction is needed to bring them to a focus. This range of accommodation is typical for someone of college age.

When you are 20, your lens is flexible enough, and your ciliary muscle strong enough, to give you an additional 10 D of refractive power. As you age, you'll find that your range of accommodation decreases. A good average value for the general population is 4 D, corresponding to a near point of 25 cm. Once you reach age 60, this value drops to about 1 D of refractive power. 1 D of accommodation gives a near point of 1 meter; objects held closer than this—like a menu or the screen on a phone—can't be seen clearly. To see objects at a closer distance, you'll need a visual aid. You'll need reading glasses.

**Video** What the Physics? Fixing Focus

We saw, in Equation 19.2, that the refractive power of two lenses placed close to each other is simply the sum of the refractive powers of the individual lenses. Vision correction involves placing a lens in front of the eye, either the lens in a pair of eyeglasses or a contact lens. The lens is close enough to the eye that the total refractive power will be the sum of the refractive power of the eye and the refractive power of the lens. We'll solve problems of vision correction by looking at the refractive power of the optical system of the eye and then deciding what additional refractive power is needed to achieve a desired result.

## Presbyopia

The loss of range of accommodation, and the resulting change in the near point, that comes with age is known as **presbyopia**—which means, more or less, "elder eyes."

### EXAMPLE 19.5 What reading glasses are needed? BIO

A 50-year-old man has good distance vision, but his near point is 50 cm. To bring this to a more comfortable 25 cm, what power reading glasses should he buy? With the reading glasses on, what is his new far point?

**STRATEGIZE** We will use the thin-lens equation in Equation 19.4 to determine the range of accommodation that the man has and the power he needs for a closer near point. We will then use the equation to determine how the additional power affects his far point.

**PREPARE** We can prepare by finding the man's range of accommodation. His distance vision is fine; we can assume a far point of infinity. The refractive power of his relaxed eye is thus

$$P_{\text{min}} = \frac{1}{s'}$$

His near point is 50 cm, which corresponds to a maximum refractive power of

$$P_{\text{max}} = \frac{1}{0.50\text{ m}} + \frac{1}{s'} = 2.0\text{ D} + \frac{1}{s'}$$

**SOLVE** The man is capable of adjusting his eye's lens to give only an additional 2.0 D of refractive power, enough to focus at 50 cm. To focus at a more reasonable 25 cm, he needs a total power of

$$P = \frac{1}{0.25\text{ m}} + \frac{1}{s'} = 4.0\text{ D} + \frac{1}{s'} = P_{\text{max}} + 2.0\,D$$

This is the power that he needs; $P_{\text{max}}$ is what he has, the current maximum power of his visual system. The necessary correction is the difference between what he needs and what he has, which is +2.0 D. He should buy reading reading glasses with this prescription. The power of this lens is added to the power of his optical system to bring his near point closer. With +2.0 D reading glasses he'll be able to clearly focus on objects 25 cm away.

Wearing the reading glasses comes at a cost, though; with the glasses on he'll be able to see close objects, but he won't be able to see clearly at a distance. The glasses add 2.0 D to the power of his optical system at all times. With his eye relaxed, the addition of this power brings the power of his optical system to

$$P = P_{\text{min}} + 2.0\text{ D} = \frac{1}{s'} + 2.0\text{ D}$$

The larger power decreases the value of $s$ that corresponds to his far point:

$$P = \frac{1}{s'} + 2.0\text{ D} = \frac{1}{s} + \frac{1}{s'}$$

$$s = 0.50\text{ m} = 50\text{ cm}$$

**ASSESS** The final result for the required prescription is a typical value for reading glasses. And, if you see folks who wear reading glasses for close work, you'll often see them getting around the problem of compromised distance vision by looking over the top of their glasses.

## Myopia and Hyperopia

Looking through reading glasses gives good close vision; looking over the top gives good distance vision.

At some point, you *will* develop presbyopia—this particular change in your optical system is an inevitable consequence of aging. You will need reading glasses for close work. But many people need vision correction much earlier. You'd like your near point to be at 25 cm or so; you'd like your far point to be at infinity. But not everyone is so fortunate.

A person who is *farsighted* can see distant objects (but even then must use some accommodation rather than a relaxed eye), but his near point is farther than 25 cm so he cannot focus on nearby objects. This condition, called **hyperopia,** is due to a mismatch between the focusing power of the optical system and the length of the eyeball—the power of the lens and cornea is too small for the length of the eyeball.

As your eye focuses on close objects, the power of the optical system is increased. For a hyperopic individual, the point of maximum accommodation is reached at a distance that is far from the eye, as illustrated in FIGURE 19.9a. Moving the object to a more reasonable 25 cm distance leads to an image that is out of focus, as in FIGURE 19.9b—the rays of light are not brought together enough to converge on the retina. The solution, as shown in FIGURE 19.9c, is to add an additional converging lens, with eyeglasses or a contact lens, that increases the power of the optical system and brings the near point closer. A lens with higher power will bring the near point closer, but there's a limit; at some point, distant vision will be compromised, as we saw in Example 19.5. So an optometrist will prescribe a lens that gives good distance vision with the eye completely relaxed. This will bring the near point closer while eliminating the need for accommodation when viewing at a distance.

A person who is *nearsighted* can clearly see nearby objects when her eye is relaxed (and extremely close objects by using accommodation), but no amount of

KEY CONCEPT

**FIGURE 19.9** Hyperopia.

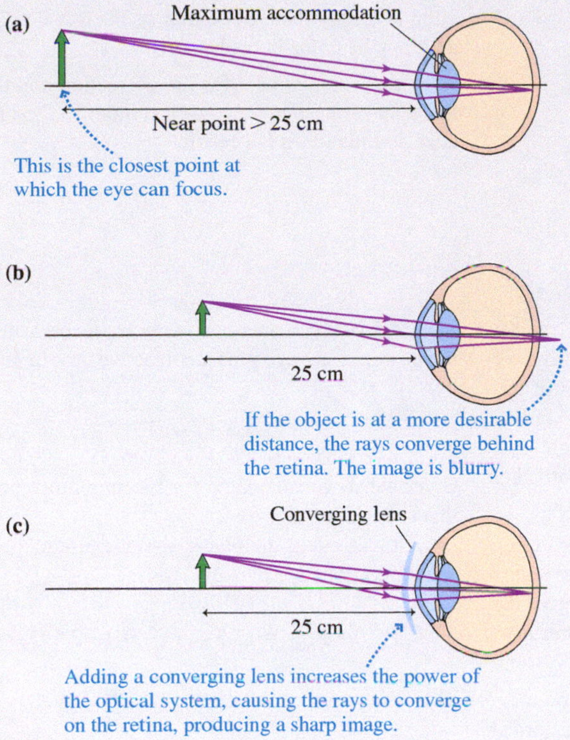


**FIGURE 19.10** Myopia.

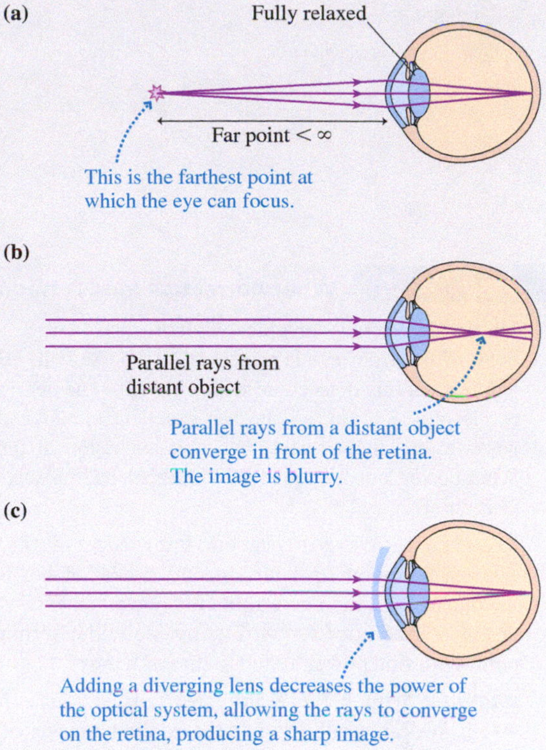


**STOP TO THINK 19.2** Certain eye conditions increase the curvature of the cornea, which increases the power of this element of the optical system. If a person with good vision experiences an increase in the curvature of the cornea, this renders the person __________ and requires vision correction with a ________ lens.

A. Myopic, converging
B. Myopic, diverging
C. Hyperopic, converging
D. Hyperopic, diverging

relaxation allows her to see distant objects. This condition, called **myopia,** is also due to a mismatch between the focusing power of the optical system and the length of the eyeball—the power of the optical system is too great for the length of the eyeball.

**Video** Vision Differences

For a myopic individual the fully relaxed eye is focused at a far point that is not all that far away, as FIGURE 19.10a shows. When the myopic person looks at a distant object, the parallel rays of light are brought together too strongly and so converge in front of the retina, as illustrated in FIGURE 19.10b, instead of on the retina as desired. To correct myopia, an optometrist will reduce the total power of the optical system by adding a diverging lens, as shown in FIGURE 19.10c. This will bring objects at a distance into focus, although this does move the near point farther from the eye.

## EXAMPLE 19.6 What is the uncorrected near point? BIO

Sanjay's vision is hyperopic. The prescription for his right eye is +3.0 D, which gives him a near point of 25 cm. What is the near point for this eye when he is not wearing his glasses?

***STRATEGIZE*** We will use our knowledge of the correction and the near point to determine the power of his uncorrected vision. Then we will use the power to determine his uncorrected near point.

***PREPARE*** With the optical system of his eye at its maximum power $P_{max}$ and with the corrective lens in place, Sanjay's near point is 25 cm. The combined power is $P_{max}$ plus the power of the lens. We can use this to find $P_{max}$:

$$P_{max} + 3.0\text{ D} = \frac{1}{0.25\text{ m}} + \frac{1}{s'} = 4.0\text{ D} + \frac{1}{s'}$$

$$P_{max} = 1.0\text{ D} + \frac{1}{s'}$$

*Continued*

**SOLVE** With his eye at maximum accommodation, $s$ is the near point. This distance is

$$P_{max} = 1.0\text{ D} + \frac{1}{s'} = \frac{1}{s} + \frac{1}{s'}$$

$$s = 1.0\text{ m}$$

You can see why Sanjay needs glasses—his uncorrected near point is longer than the length of his arms, so he can't focus on anything he holds in his hands!

**ASSESS** The numbers for the correction and the near-point distance make sense in terms of Example 19.5 on presbyopia, so we have confidence in our result.

### EXAMPLE 19.7 What corrective lens is needed? BIO

Martina's vision is myopic; without vision correction, the far point of her right eye is a modest 40 cm. As with other myopic individuals, this does come with a benefit—the near point of this eye is only 8.0 cm, so she has excellent close vision. Martina wears a lens that corrects her distance vision in the right eye. What power lens is required? With this lens in place, what is her near point?

**STRATEGIZE** We will use our knowledge of the correction and the far point to determine the power of her uncorrected vision. We will then determine the power necessary to give the desired far point of $s = \infty$. Finally, we will determine her near point with this power added to that of her eye's optical system.

**PREPARE** With her eye fully relaxed, Martina's far point is 40 cm. We can use this to find the minimum power of her optical system:

$$P_{min} = \frac{1}{0.40\text{ m}} + \frac{1}{s'} = 2.5\text{ D} + \frac{1}{s'}$$

The maximum power comes at her uncorrected near point:

$$P_{max} = \frac{1}{0.080\text{ m}} + \frac{1}{s'} = 12.5\text{ D} + \frac{1}{s'}$$

**SOLVE** Martina wants her far point to be at infinity; she wants the power of her eye plus a corrective lens to be

$$P_{min} + P_{lens} = \frac{1}{\infty} + \frac{1}{s'} = \frac{1}{s'}$$

Given what we calculated for the minimum power $P_{min}$, we see that she needs $P_{lens} = -2.5$ D.

With this lens in place, the maximum power of her fully accommodated eye plus the lens is

$$P_{max} + P_{lens} = 12.5\text{ D} + \frac{1}{s'} + (-2.5\text{ D}) = 10\text{ D} + \frac{1}{s'}$$

This gives a near point $s$ of

$$P_{max} + P_{lens} = 10\text{ D} + \frac{1}{s'} = \frac{1}{s} + \frac{1}{s'}$$

$$s = 0.10\text{ m} = 10\text{ cm}$$

Wearing the lens moves her near point a bit farther away, but this is a worthwhile trade-off given how much it extends her far point!

**ASSESS** We found a negative power for her prescription, and the magnitude of the correction is fairly typical, so we can have confidence in our results.

## Other Types of Vision Correction and Vision Enhancement

People who wear glasses or contacts are generally nearsighted, which means that the power of their optical system is too large for the length of the eyeball. An eyeglass lens or a contact lens placed directly on the eye, can correct for this. But, in recent years, LASIK and other forms of surgical vision correction have become common. In LASIK, a laser removes some of the thickness of the cornea near the center, flattening it and reducing the refractive power.

If you are old enough to develop presbyopia, but you also require correction for myopia, there's an easy fix: For close work, take off your glasses—the too-large power of your optical system brings close objects into focus. A more elegant solution is to wear *bifocals,* glasses that have lenses of different powers at the top and bottom. A large negative correction at the top gives good distance vision; a smaller negative correction at the bottom is used to focus on close objects. Modern eyewear usually includes progressive lenses in which the power varies vertically across the lens, giving a range of possible distances for comfortable focus.

If you find yourself working for long periods of time at a computer, you might find it useful to get "computer glasses." The lenses put your far point right at the screen, so your eyes can be fully relaxed as you work.

Nearly everyone has some degree of **astigmatism,** generally because the shape of the cornea isn't uniform—it's not a perfectly spherical lens as we've assumed. This means that vertical and horizontal lines focus at different distances. Astigmatism is easily corrected with glasses or contacts. If your prescription has two values for each eye, one labeled "sphere" and one labeled "cylinder," the refractive power of the lens is the former; the latter is a measure of the correction needed for astigmatism.

Your eyes work very well in air, but your vision is quite blurry underwater. When you swim underwater, the difference in the refractive indices of the cornea and the water is too small to allow significant refraction—you lose most of the 40 D of refractive power provided by the cornea. If you want to see underwater, you can wear goggles so that the surface of the cornea is in contact with air, not water.

To account for the reduced refractive power of the cornea underwater, many fish have sharply curved corneas that provide significant refractive power despite the small difference in the indices of refraction. The anableps fish shown at the beginning of the chapter lives at the water's surface and needs to see both above and below the water at the same time. To focus simultaneously on objects on both sides of the waterline, it has evolved a very asymmetrical cornea that is more strongly curved below the waterline.

**Moving between worlds** BIO Penguins need to see very well both above and below the water. Their eyes differ from yours in that almost all the refractive power is provided by the lens; the cornea is nearly flat and provides very little of the power of the optical system. Penguins experience a much smaller change in vision when they move between air and water, and their very flexible lens can easily accommodate to correct for the difference.

**STOP TO THINK 19.3** As people age, the length of their eyeballs may decrease slightly. If there is no change in the lens or the cornea, a person who always had good vision then has

A. Presbyopia
B. Hyperopia
C. Myopia

## 19.3 The Magnifier

When you move closer to an object, the image is larger on your retina. There is a limit to how close you can get, though; for distances closer than your near point, you can't bring the image into focus.

There's a simple solution: You can use a magnifier—a magnifying glass—to get closer. As we saw in ◀◀ SECTION 18.5, a magnifier is a simple converging lens that has a positive power. If you hold the magnifier right next to your eye, the magnifier's power adds to that of your eye. This brings your near point closer, just as it does for a person who has presbyopia or hyperopia, and this allows you to bring the object closer.

**EXAMPLE 19.8 How much closer?** BIO

Kerry's eye has a near point of 25 cm—not close enough to see the fine details in an old photo. She places a +16 D lens right next to her eye. How close can she bring the photo now?

***STRATEGIZE*** We know Kerry's near point and we will use it to find the maximum power of the optical system of her eye. To this, we'll add +16 D and then find the new near point.

***PREPARE*** Kerry's near point is 25 cm, so the maximum power of her optical system is

$$P_{max} = \frac{1}{0.25\text{ m}} + \frac{1}{s'} = 4.0\text{ D} + \frac{1}{s'}$$

***SOLVE*** To this power, Kerry adds the +16 D of the lens. The maximum power of her optical system with the lens in place is $20\text{ D} + 1/s'$. This increased power lets her focus at a much smaller distance:

$$20\text{ D} + \frac{1}{s'} = \frac{1}{s} + \frac{1}{s'}$$

$$s = \frac{1}{20\text{ m}^{-1}} = 5.0\text{ cm}$$

This distance is smaller than her near point by a factor of 5. Bringing the object closer by a factor of 5 increases the image size by a factor of 5, so she'll have a much better view.

***ASSESS*** 5 cm is about 2 inches, about right for the near point when using a good magnifier, so our result makes sense.

Example 19.8 shows one way to use a magnifier, but you don't need to put the lens right next to your eye. You can place the lens close to the object and look from a distance. This still produces a magnified image, as illustrated by the photo in the chapter preview. Explaining this observation will require us to develop some new concepts that will be of use when we look at other optical instruments.

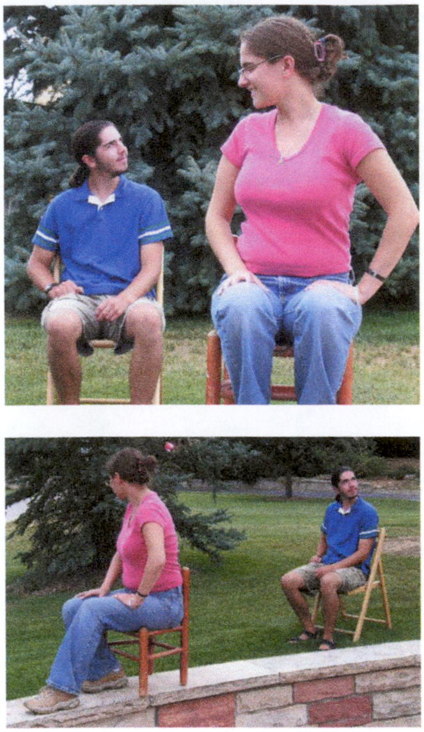

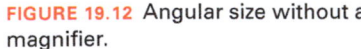

**Movie magic?** Even if the more distant of two equally sized objects *appears* smaller, we don't usually believe it actually *is* smaller because there are abundant visual clues that tell our brain that it's farther away. If those clues are removed, however, the brain readily accepts the illusion that the farther object is smaller. The technique of *forced perspective* is a special effect used in movies to give this illusion. Here, Camelia, who is actually closer to the camera, looks like a giant compared to Kevin. The lower photo shows how the trick was done.

## Angular Size and Apparent Size

When you get closer to an object, it appears bigger. Obviously the actual size of the object is unchanged as you approach it, so what exactly is getting "bigger"? If you hold an aspirin tablet at arm's length, it will just cover the disk of the full moon, despite the fact that the moon is considerably larger. In what sense is the aspirin tablet the same size as the moon? Consider an object such as the green arrow in FIGURE 19.11a. To find the size of its image on the retina, we can trace the two rays shown that go through the center of the lens. As we've learned, such rays are undeviated as they pass through the lens, so we can use them to locate the image, shown in green, on the retina. If the arrow is then brought closer to the eye, as shown in red, ray tracing reveals that the arrow's image is *larger.* Our brain interprets a larger image on the retina as representing a larger-appearing *object.* As the object is moved closer, its size doesn't change, but its **apparent size** gets larger.

FIGURE 19.11 How the apparent size of an object is determined.

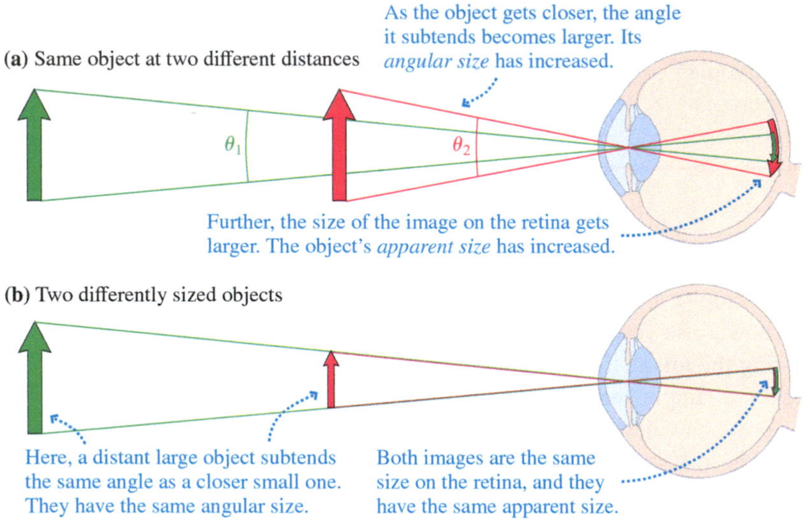

(a) Same object at two different distances
As the object gets closer, the angle it subtends becomes larger. Its *angular size* has increased.
$\theta_1$
$\theta_2$
Further, the size of the image on the retina gets larger. The object's *apparent size* has increased.
(b) Two differently sized objects
Here, a distant large object subtends the same angle as a closer small one. They have the same angular size.
Both images are the same size on the retina, and they have the same apparent size.


Angles $\theta_1$ and $\theta_2$ in Figure 19.11a are the angles *subtended* by the green and red arrows. The angle subtended by an object is called its **angular size.** As you can see from the figure, **objects that subtend a larger angle appear larger to the eye.** It's also possible, as FIGURE 19.11b shows, for objects with different actual sizes to have the same angular size and thus the same apparent size, like the aspirin tablet and the moon.

## Using a Magnifier

FIGURE 19.12 Angular size without a magnifier.

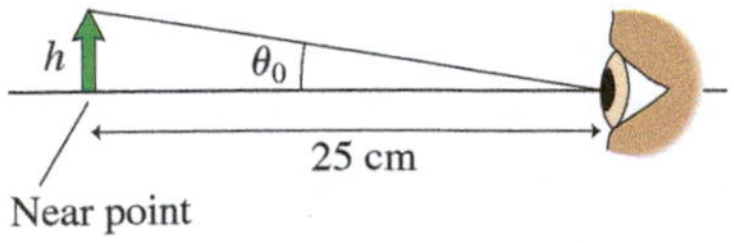


The eye cannot focus on an object closer than its near point, which we will take as 25 cm, a good average value for the general population. Thus an in-focus object has its maximum angular size $\theta_0$ when the object is at the near point, as shown in FIGURE 19.12. The geometry of this figure shows that $\tan\theta_0 = h/0.25\text{ m}$. If this angle is fairly small, which it usually is, we can use the small-angle approximation $\tan\theta_0 \approx \theta_0$ to write

$$\theta_0 \approx \frac{h}{0.25\text{ m}}$$

How does a magnifier lead to an angular size larger than this? In the way that a magnifier is usually used, the lens is held such that the object is at or just inside the lens's focal point. As shown graphically in FIGURE 19.13, this produces a virtual image that is quite far from the lens. Your eye, looking through the lens, "sees" the virtual image. This is a convenient image location, because your eye's muscles are fully relaxed when looking at a distant image. Thus you can use the magnifier in this way for a long time without eye strain.

FIGURE 19.13 The magnifier.

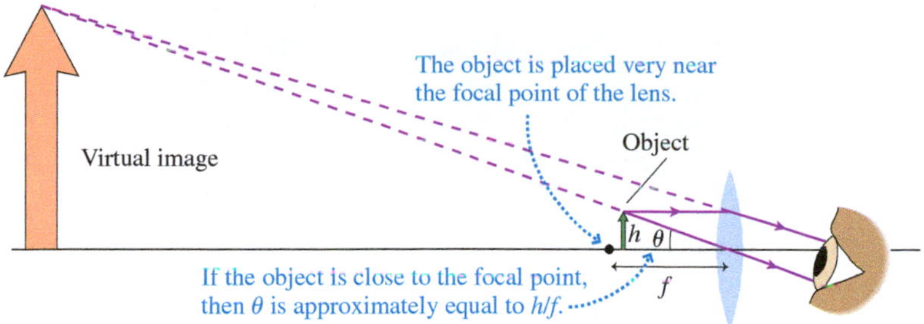


We can find the angular size of the image using Figure 19.13. Suppose that the object is almost exactly at the focal point, a distance $f$ from the lens. Then, tracing the ray that goes through the lens's center, we can see that the angular size $\theta$ of the image is such that $\tan\theta = h/f$ or, again using the small-angle approximation,

$$\theta \approx \frac{h}{f} \tag{19.5}$$

Thus the angular size $\theta$ when using the magnifier is larger than that without the magnifier by a factor of

$$M = \frac{\theta}{\theta_0} = \frac{h/f}{h/0.25\text{ m}} = \frac{0.25\text{ m}}{f} \tag{19.6}$$

$M$ is called the **angular magnification** of the magnifier. With a lens of short focal length it is possible to get magnifications as high as about 20.

## EXAMPLE 19.9 Finding the focal length and power of a magnifier

When you buy reading glasses, the converging lenses are rated by their refractive power. When you buy a magnifier, the converging lens is rated by its magnification. If you buy a magnifying lens that is said to provide "5× magnification," what is the focal length of this lens? What is its refractive power?

**STRATEGIZE** 5× magnification means that $M = 5$. We will use this value to determine the lens's focal length.

**PREPARE** The magnification is related to the focal length by Equation 19.6.

**SOLVE** We rearrange Equation 19.6 to find the focal length:

$$f = \frac{0.25\text{ m}}{M} = \frac{0.25\text{ m}}{5} = 0.050\text{ m}$$

The focal length is 5.0 cm. Then the refractive power is

$$P = \frac{1}{f} = \frac{1}{0.050\text{ m}} = 20\text{ D}$$

**ASSESS** 20 D is a reasonable value for a strong magnifier.

## CONCEPTUAL EXAMPLE 19.10 The angular size of a magnified image

An object is placed right at the focal point of a magnifier. How does the apparent size of the image depend on where the *eye* is placed relative to the lens?

**REASON** When the object is precisely at the focal point of the lens, Equation 19.5 holds exactly. The angular size is equal to *h*/*f* *independent* of the position of the eye. Thus the object's apparent size is independent of the eye's position as well. FIGURE 19.14 on the next page shows a calculator at the focal point of a magnifier. The apparent size of the COS button is the same whether the camera taking the picture is close to or far from the lens.

*Continued*

FIGURE 19.14 Viewing a magnifier with the object at its focus.

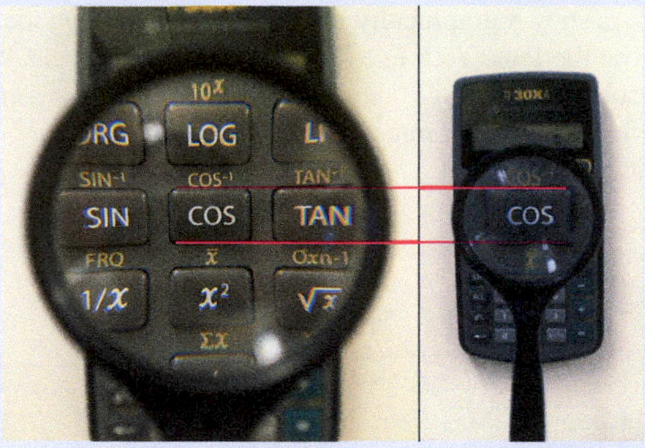


***ASSESS*** When the object is at the magnifier's focus, we've seen that the image is at infinity. The situation is similar to observing any "infinitely" distant object, such as the moon. If you walk closer to or farther from the moon, its apparent size doesn't change at all. The same holds for a virtual *image* at infinity: Its apparent size is independent of the point from which you observe it.

**STOP TO THINK 19.4** A student tries to use a *diverging* lens as a magnifier. She observes a coin placed at the focal point of the lens. She sees

A. An upright image, smaller than the object.
B. An upright image, larger than the object.
C. An inverted image, smaller than the object.
D. An inverted image, larger than the object.
E. A blurry image.

## 19.4 The Microscope

To get higher magnifications than are possible using a simple magnifier, a *combination* of lenses must be used. This is how microscopes and telescopes are constructed. A simple rule governs how two lenses work in combination: **The image from the first lens acts as the object for the second lens.** The following example illustrates this rule.

### EXAMPLE 19.11 Finding the image for two lenses in combination

A 5.0-cm-focal-length converging lens is 16.0 cm in front of a −10.0-cm-focal-length diverging lens. A 4.0-cm-tall object is placed 11.0 cm in front of the converging lens. What are the position and size of the final image?

***STRATEGIZE*** We'll start with a ray-tracing diagram to help us understand the situation and tell us what to expect for an answer. The first lens is a converging lens, with the object at a distance greater than the focal length, so this lens will produce a real image. This real image will serve as the object for the second lens.

***PREPARE*** In FIGURE 19.15a we use the three special rays of the converging lens to locate the image it forms. The image produced by this lens is a real image that falls between the two lenses. We then use this image as the object for the second lens in FIGURE 19.15b. The final image is inverted, virtual, about half the size of the object, and roughly 4 cm to the left of the diverging lens.

FIGURE 19.15 Two lenses in combination.

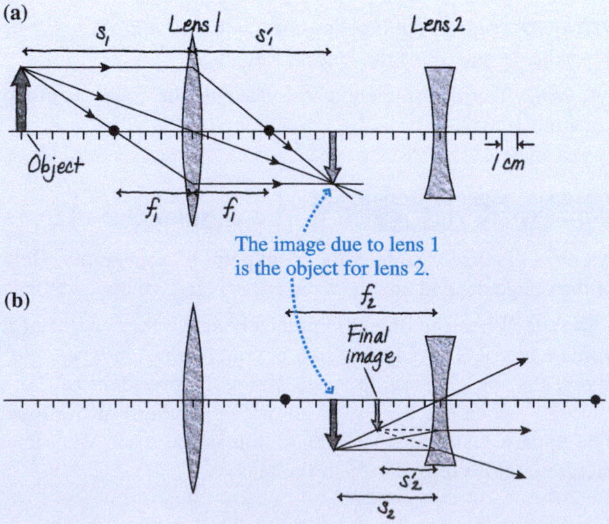

Mathematically, we can use the thin-lens equation to find the image location and size due to the first lens, then use the first lens as the object for the second lens in a second use of the lens equation.

**SOLVE** We first solve for the image due to lens 1. We have

$$\frac{1}{s_1'} = \frac{1}{f_1} - \frac{1}{s_1} = \frac{1}{5.0\text{ cm}} - \frac{1}{11.0\text{ cm}}$$

from which $s_1' = 9.17$ cm. Because this is a positive image distance, the image is real and located to the right of the first lens. The magnification of the first lens is $m_1 = -s_1'/s_1 = -(9.17\text{ cm})/(11.0\text{ cm}) = -0.833$.

This image is the object for the second lens. Because it is 9.17 cm to the right of the first lens, and the lenses are 16.0 cm apart, it is 16.0 cm − 9.17 cm = 6.83 cm in front of the second lens. Thus $s_2 = 6.83$ cm.

Applying the thin-lens equation again, we have

$$\frac{1}{s_2'} = \frac{1}{f_2} - \frac{1}{s_2} = \frac{1}{-10.0\text{ cm}} - \frac{1}{6.83\text{ cm}}$$

from which we find $s_2' = -4.06$ cm, or −4.1 cm to two significant figures. Because this image distance is negative, the image is virtual and located to the left of lens 2, as shown in Figure 19.15b. The magnification of the second lens is $m_2 = -s_2'/s_2 = -(-4.06\text{ cm})/(6.83\text{ cm}) = 0.594$.

Thus the final image size is

$$h_2' = m_2 h_2 = m_2 h_1' = m_2(m_1 h_1) = m_1 m_2 h_1$$
$$= (-0.833)(0.594)(4.0\text{ cm}) = -2.0\text{ cm}$$

Here we used the fact that the object height $h_2$ of the second lens is equal to the image height $h_1'$ of the first lens.

**ASSESS** Our calculated values for the final image height and position match our diagram, so we can have confidence in our solution.

A microscope, whose major parts are shown in FIGURE 19.16, attains a magnification of up to 1000 by using two lenses in combination. A specimen to be observed is placed on the *stage* of the microscope, directly beneath the **objective lens** (or simply the **objective**), a converging lens with a relatively short focal length. The objective creates a magnified real image that is further enlarged by the **eyepiece,** a lens used as an ordinary magnifier. In most modern microscopes, a prism bends the paths of the rays from the object so that the eyepiece is at a comfortable angle. The objective and eyepiece lenses are generally compound lenses; there may also be additional lenses along the optical path. In what follows, we'll consider a simplified version of a microscope with simple lenses and no prism.

FIGURE 19.16 A microscope.

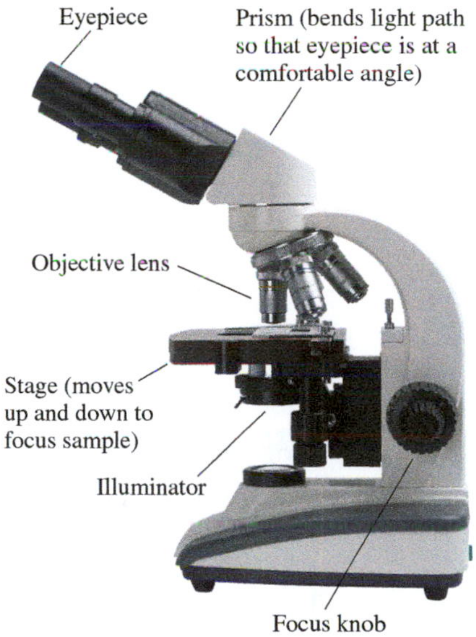


Let's examine the magnification process in more detail. In FIGURE 19.17 we have drawn a microscope tilted horizontally. The object distance is just slightly greater than the focal length $f_o$ of the objective lens, so the objective forms a highly magnified real image of the object at a distance $s' = L$. In working microscopes this distance has been standardized; a typical value is $L = 160$ mm. We will assume this value in all calculations unless we state otherwise. Most microscopes, such as the one shown in Figure 19.16, are focused by moving the sample stage up and down, using the focusing knob, until the object distance is correct for placing the image at $L$.

FIGURE 19.17 A horizontal view of the optics in a microscope.

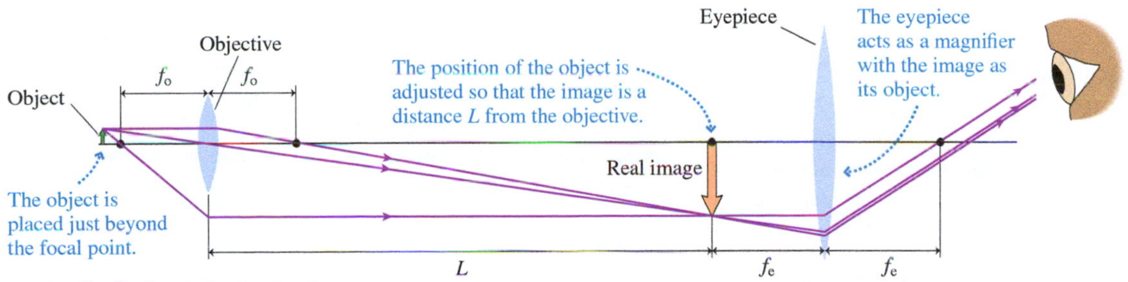


From Equation 18.9, the magnification of the objective lens is

$$m_o = -\frac{s'}{s} \approx -\frac{L}{f_o} \qquad (19.7)$$

Here we used the fact that the image distance $s'$ is equal to $L$ and the object distance $s$ is very close to the focal length $f_o$ of the objective. The minus sign tells us that the image is inverted with respect to the object.

The object is quite close to the focal point of the objective lens, so the image made by the objective is much larger than the object. This enlarged image is viewed by the eyepiece, which acts as a magnifier, providing a further degree of angular magnification. The total magnification is the product of the two magnifications, that of the objective and that of the eyepiece. The angular magnification of the eyepiece is given by Equation 19.6: $M_e = (0.25\text{ m})/f_e$. Together, the objective and eyepiece produce total angular magnification

$$M = m_o M_e = -\frac{L}{f_o}\frac{0.25\text{ m}}{f_e} \tag{19.8}$$

The minus sign shows that the image seen in a microscope is inverted.

### EXAMPLE 19.12 Finding the focal length of a microscope objective

A microscope objective is labeled "20×." What is its focal length?

**PREPARE** "20×" means that the objective has a magnification $m_o$ of −20. We can use Equation 19.7 with $L$ as 160 mm.

**SOLVE** From Equation 19.7 we have

$$f_o = -\frac{L}{m_o} = -\frac{160\text{ mm}}{-20} = 8.0\text{ mm}$$

**ASSESS** This focal length is significantly shorter than the length $L$, in agreement with Figure 19.17.

The magnification of a microscope objective is called its "power." Thus we would say that the 20× objective of Example 19.12 has a power of 20. (The power or magnification of a microscope objective should not be confused with the refractive power of a lens.) Many microscopes have a set of objectives that can be pivoted into place to change the overall magnification. A complete set of objectives might include 5×, 10×, 20×, 40×, and 100×. Eyepieces are also specified by their magnification, or "power," and give additional magnification in the range of 10× to 20×.

### EXAMPLE 19.13 Viewing blood cells BIO

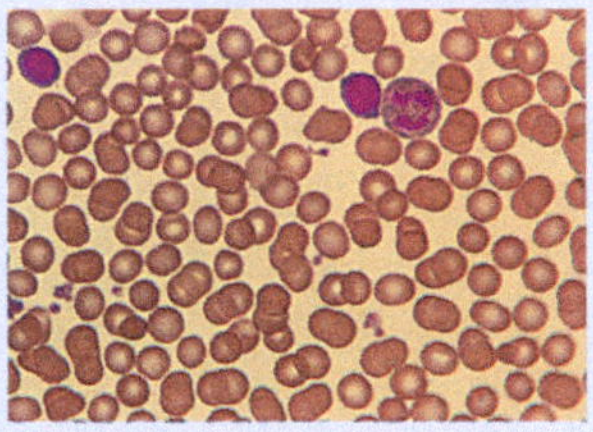

A pathologist inspects a sample of 7-μm-diameter human blood cells under a microscope. She selects a 40× objective and a 10× eyepiece. To give a sense of the magnification the microscope provides, compute the size of an object, viewed at a 25 cm distance, that would have the same apparent size as a blood cell seen through the microscope.

**STRATEGIZE** The definition of angular magnification given by Equation 19.8 compares the magnified angular size of the object being viewed to that of an object seen at a near-point distance of 25 cm. If we multiply the size of the object by the magnification, this gives us exactly what we want to know—the size of an object held at 25 cm that would have the same angular size as the image.

**PREPARE** The microscope's angular magnification is $M = -(40) \times (10) = -400$.

**SOLVE** The magnified cells will have the same apparent size as an object $400 \times 7\ \mu\text{m} \approx 3$ mm in diameter seen from a distance of 25 cm.

**ASSESS** 3 mm is about the size of a capital O in this text, so a blood cell seen through the microscope will have about the same apparent size as an O seen from a comfortable reading distance. This is approximately the size of the red blood cells in the micrograph shown here, so it gives a reasonable idea of how the blood cells will appear in the microscope.

**STOP TO THINK 19.5** A biologist changes the objective lens of a microscope, replacing a 20× objective with a 10× objective. To keep the magnification the same, the focal length of the eyepiece must

A. Be doubled.
B. Be halved.
C. Remain the same.
D. The magnification cannot stay the same if the objective power is changed.

# 19.5 The Telescope

The microscope magnifies small objects that can be placed near its objective lens. A *telescope* is used to magnify distant objects. The two-lens arrangement shown in FIGURE 19.18 is similar to that of the microscope, but the objective lens has a long focal length instead of the very short focal length of a microscope objective. Because the object is very far away ($s \approx \infty$), the converging objective lens forms a real image of the distant object at the lens's focal point. A second lens, the eyepiece, is then used as a simple magnifier to enlarge this real image for final viewing by the eye.

FIGURE 19.18 The telescope.

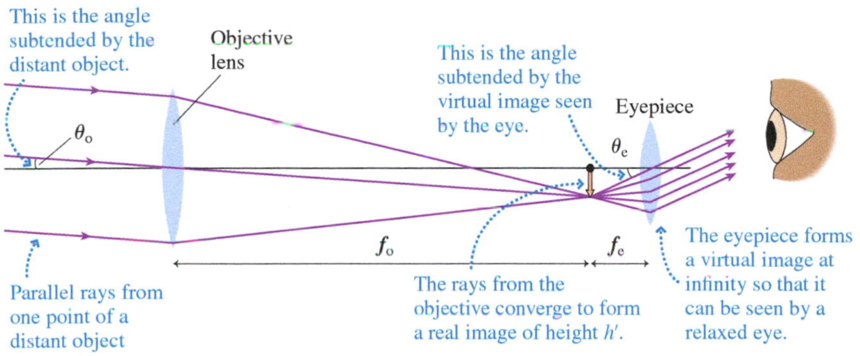


We can use Figure 19.18 to find the magnification of a telescope. The original object subtends an angle $\theta_o$. Because the object is distant, its image is formed in the focal plane of the objective lens, a distance $f_o$ from the objective. From the geometry of Figure 19.18, the height of this image (negative because it's inverted) is

$$h' \approx -f_o\theta_o$$

where we have used the small-angle approximation $\tan\theta_o \approx \theta_o$. This image is now the object for the eyepiece lens, which functions as a magnifier. The height of the "object" it views is $h'$, so, from Equation 19.5, the angular size $\theta_e$ of the virtual image formed by the eyepiece is

$$\theta_e = \frac{h'}{f_e} = \frac{-f_o\theta_o}{f_e}$$

where $f_e$ is the focal length of the eyepiece. The telescope's angular magnification is the ratio of the angular size seen when looking through the telescope to that seen without the telescope, so we have

$$M = \frac{\theta_e}{\theta_o} = -\frac{f_o}{f_e} \tag{19.9}$$

The minus sign indicates that we see an upside-down image when we look through a simple telescope. This is not a problem when we look at astronomical objects, but it could be disconcerting to a bird watcher. More sophisticated telescope designs produce an upright image.

To get a high magnification with a telescope, the focal length of the objective should be large and that of the eyepiece small. Contrast this with the magnification of a microscope, Equation 19.8, which is high when the focal lengths of both objective and eyepiece are small.

**Location, location, location** If you want a clear view of the heavens, having a large magnification isn't necessarily helpful. When you magnify your view of an astronomical object, you also magnify the effects of atmospheric turbulence—visible to the naked eye as the twinkling of stars—that obscures the details of objects being observed. Higher magnification makes a larger image, but also a blurrier image. Modern telescopes actively correct for this problem, but the best approach is simply to avoid it. The Hubble Space Telescope orbits the earth high above the atmosphere. With its 2.4-m-diameter mirror, it has produced some of the most spectacular images of astronomical objects, such as this gas cloud surrounding the star V838 Monocerotis.

**EXAMPLE 19.14** **How large is the image?**

A typical hobby telescope has an objective lens with a 700 mm focal length and an eyepiece with a 25 mm focal length. The moon has an angular size of about 0.5° when viewed without optical aid. What angle does the moon subtend in the telescope?

***STRATEGIZE*** We will use Equation 19.9 to determine the magnification. Then we will multiply the moon's angular size by the magnification to find its apparent size in the telescope.

***PREPARE*** The magnification of the telescope with the noted eyepiece is

$$M = -\frac{700 \text{ mm}}{25 \text{ mm}} = -28$$

***SOLVE*** With this magnification, the angular size of the moon is $(28) \times (0.5°) = 14°$.

***ASSESS*** This is plenty of magnification for a good view. If you are viewing this text at the original size at a typical reading distance, this is twice the angle the photo of the moon subtends from your point of view.

FIGURE 19.19 A reflecting telescope.

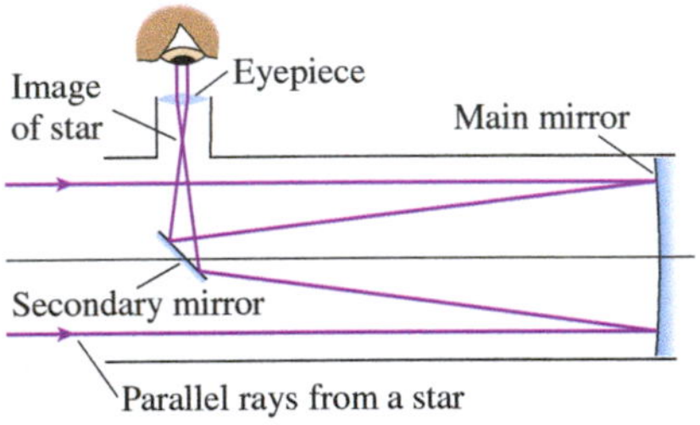


The telescope shown in Figure 19.18 uses a lens as its objective and hence is known as a *refracting telescope*. It is also possible to make a *reflecting telescope* with a converging mirror instead of a lens, as shown in FIGURE 19.19. One problem with this arrangement is that the image is formed in front of the mirror where it's hard to magnify with an eyepiece without getting one's head in the way. Newton, who built the first such telescope, used a small angled plane mirror, called a *secondary mirror,* to deflect the image to an eyepiece on the side of the telescope.

Many celestial objects are reasonably large but quite dim. The Andromeda galaxy, some 2 million light years from earth, appears 4 times wider than the full moon in the sky, but its low brightness makes it challenging to find in the night sky. Seeing this galaxy in detail means collecting a lot of light, which requires a large area for a telescope's lens or mirror. To photograph even more distant objects, even larger areas are required, so modern astronomical telescopes are quite large. The Subaru Telescope in Hawaii is the world's largest single-mirror telescope; its mirror has a diameter of 8.3 m (27 ft)! There is simply no way to make or support a lens this large, so all modern astronomical telescopes are reflecting telescopes. No one spends time actually looking through these telescopes. Imaging devices record the light, and these devices generally ride in front of the mirror, recording the image directly, with no secondary mirror.

## 19.6 Color and Dispersion

*I procured me a triangular glass prism to try therewith the celebrated phenomena of colors.*
Isaac Newton

One of the most obvious visual aspects of light is color. Color is a *perception,* not a physical quantity. Color is associated with the wavelength of light. The fact that we see light with a wavelength of 650 nm as "red" tells us how our visual system responds to electromagnetic waves of this wavelength.

### Color

People have known since antiquity that we can break white light into colors by using irregularly shaped crystals or pieces of glass. A common idea was that the glass or crystal somehow altered the properties of the light by *adding* color to the light. Newton suggested a different explanation. He first passed a sunbeam through a prism, producing the familiar rainbow of light. We say that the prism *disperses* the light. Newton's novel idea, shown in FIGURE 19.20a, was to use a second prism, inverted with respect to the first, to "reassemble" the colors. He found that the light that emerged from the second prism was a beam of pure white light.

But the emerging light beam is white only if *all* the rays are allowed to move between the two prisms. Blocking some of the rays with small obstacles, as in FIGURE 19.20b, causes the emerging light beam to have color. This suggests that color is associated with the light itself, not with anything that the prism is "doing" to the light. Newton tested this idea by inserting a small aperture between the prisms to pass only the rays of a particular color, such as green. If the prism alters the properties of light, then the second prism should change the green light to other colors. Instead, the light emerging from the second prism is unchanged from the green light entering the prism.

These and similar experiments show that:

- What we perceive as white light is a mixture of all colors. White light can be dispersed into its various colors and, equally important, mixing all the colors produces white light.
- The index of refraction of a transparent material differs slightly for different colors of light. Glass has a slightly higher index of refraction for violet light than for green light or red light. Consequently, different colors of light refract at slightly different angles. A prism does not alter the light or add anything to the light; it simply causes the different colors that are inherent in white light to follow slightly different trajectories.

FIGURE 19.20 Newton used prisms to study color.

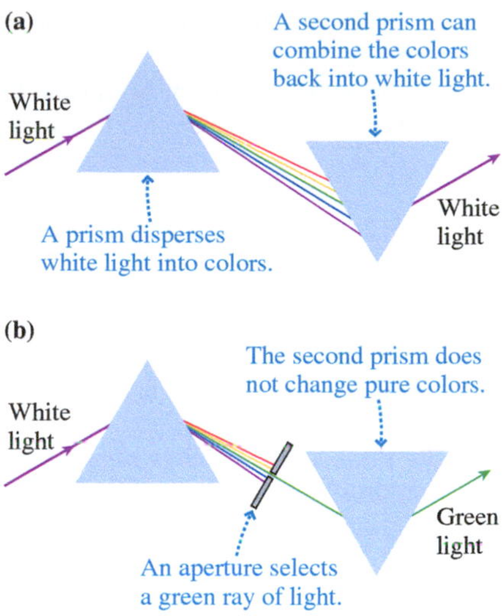


## Dispersion

In 1801 Thomas Young performed his two-slit interference experiment, which first showed that different colors are associated with light of different wavelengths. The longest wavelengths are perceived as red light and the shortest wavelengths are perceived as violet light. TABLE 19.1 is a brief summary of the *visible spectrum* of light.

The slight variation of index of refraction with wavelength is known as **dispersion.** FIGURE 19.21 shows the *dispersion curves* of two common glasses. Notice that $n$ **is *higher* when the wavelength is *shorter*;** thus violet light refracts more than red light.

TABLE 19.1 A brief summary of the visible spectrum of light

| Color | Approximate wavelength |
|---|---|
| Deepest red | 700 nm |
| Red | 650 nm |
| Yellow | 600 nm |
| Green | 550 nm |
| Blue | 450 nm |
| Deepest violet | 400 nm |

FIGURE 19.21 Dispersion curves show how the index of refraction varies with wavelength.

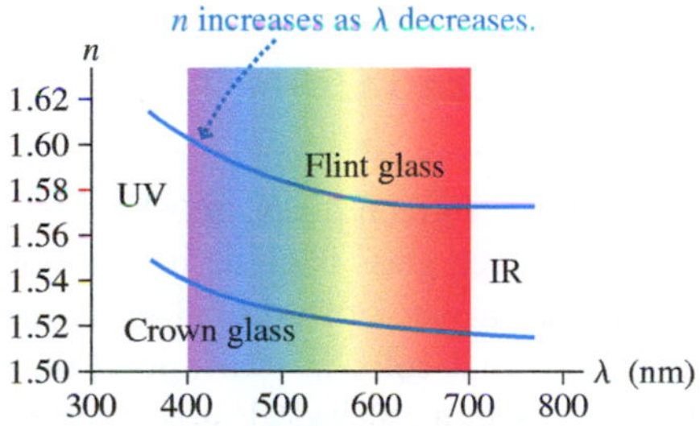


**EXAMPLE 19.15 Dispersing light with a prism**

Example 18.4 in Chapter 18 found that a ray incident on a 30° prism is deflected by 22.6° if the prism's index of refraction is 1.59. Suppose that this is the index of refraction of deep violet light, and that deep red light has an index of refraction of 1.54.

a. What is the deflection angle for deep red light?
b. If a beam of white light is dispersed by this prism, how wide is the rainbow spectrum on a screen 2.0 m away?

***STRATEGIZE*** First, look back at Figure 18.19 in Example 18.4 to see the geometry. We determined the deflection angle for deep violet light, and we will follow the same approach to find the angle for deep red light. Once we know the two angles, we will find the angular spread.

***PREPARE*** In both cases, the ray is incident on the hypotenuse of the prism at $\theta_1 = 30°$.

***SOLVE***

a. If $n_1 = 1.54$ for deep red light, the refraction angle is

$$\theta_2 = \sin^{-1}\left(\frac{n_1 \sin\theta_1}{n_2}\right) = \sin^{-1}\left(\frac{1.54 \sin 30°}{1.00}\right) = 50.35°$$

*Continued*

Example 18.4 showed that the deflection angle is $\phi = \theta_2 - \theta_1$, so deep red light is deflected by $\phi_{red} = 20.35°$. This angle is slightly smaller than the deflection angle for violet light, $\phi_{violet} = 22.60°$.

b. The entire spectrum is spread between $\phi_{red} = 20.35°$ and $\phi_{violet} = 22.60°$. The angular spread is

$$\delta = \phi_{violet} - \phi_{red} = 2.25° = 0.0393 \text{ rad}$$

c. At distance *r*, the spectrum spans an arc length

$$s = r\delta = (2.0 \text{ m})(0.0393 \text{ rad}) = 0.0785 \text{ m} = 7.9 \text{ cm}$$

**ASSESS** Notice that we needed three significant figures for $\phi_{red}$ and $\phi_{violet}$ in order to determine $\delta$, the *difference* between the two angles, to two significant figures. The angle is so small that there's no appreciable difference between arc length and a straight line. The spectrum is 7.9 cm wide at a distance of 2.0 m.

## Rainbows

You have no doubt seen a rainbow after an evening rainstorm. FIGURE 19.22a shows that the basic cause of the rainbow is a combination of refraction, reflection, and dispersion.

FIGURE 19.22 Light seen in a rainbow has undergone refraction + reflection + refraction in a raindrop.

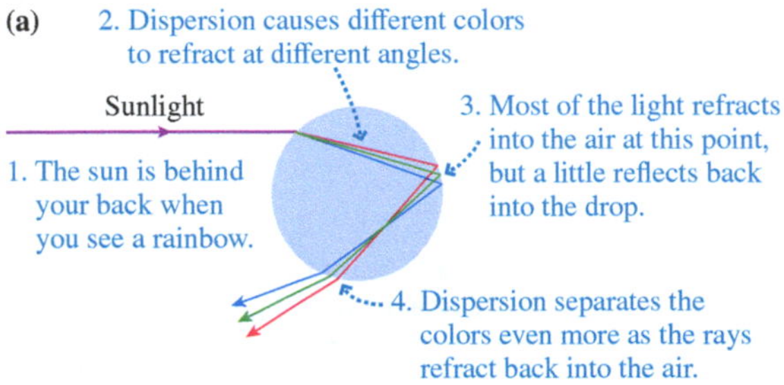

(a)
2. Dispersion causes different colors to refract at different angles.

1. The sun is behind your back when you see a rainbow.

3. Most of the light refracts into the air at this point, but a little reflects back into the drop.

4. Dispersion separates the colors even more as the rays refract back into the air.


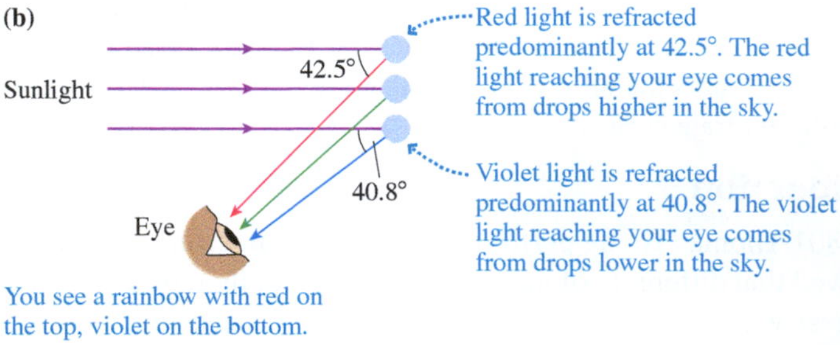

(b)
Sunlight
42.5°
40.8°
Eye
Red light is refracted predominantly at 42.5°. The red light reaching your eye comes from drops higher in the sky.
Violet light is refracted predominantly at 40.8°. The violet light reaching your eye comes from drops lower in the sky.
You see a rainbow with red on the top, violet on the bottom.


The rays leaving the drop in Figure 19.22a are spreading apart, so they can't all reach your eye. As FIGURE 19.22b shows, a ray of red light reaching your eye comes from a drop *higher* in the sky than a ray of violet light. A rainbow is a mosaic; each droplet that forms the rainbow sends a particular wavelength of light toward you. You need to look higher in the sky to see the droplets that send rays of red light toward your eye, lower in the sky to see the droplets that send violet light toward your eye.

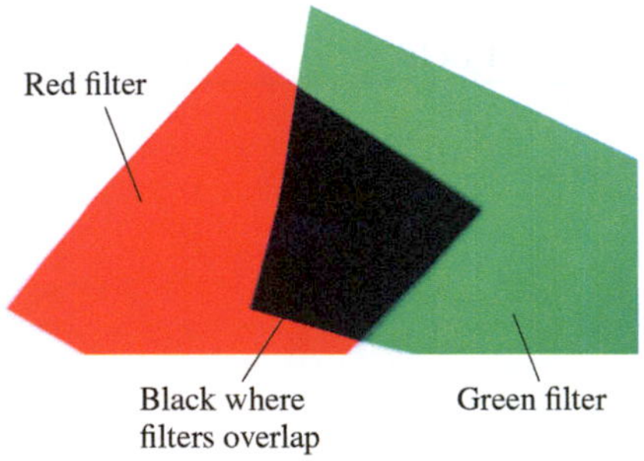

Black where filters overlap


No light at all passes through both a green and a red filter.

## Colored Filters and Colored Objects

White light that passes through a piece of green glass emerges as green light. A possible explanation would be that the green glass *adds* "greenness" to the white light, but Newton found otherwise. Green glass is green because it *removes* any light that is "not green." More precisely, a piece of colored glass *absorbs* all wavelengths except those of one color, and that color is transmitted through the glass without hindrance. We can think of a piece of colored glass or plastic as a *filter* that removes all wavelengths except a chosen few.

### CONCEPTUAL EXAMPLE 19.16 Filtering light

White light passes through a green filter and is observed on a screen. Describe how the screen will look if a second green filter is placed between the first filter and the screen. Describe how the screen will look if a red filter is placed between the green filter and the screen.

**REASON** The first filter removes all light except for wavelengths near 550 nm that we perceive as green light. A second green filter doesn't have anything to do. The non-green wavelengths have already been removed, and the green light emerging from the first filter will pass through the second filter without difficulty. The screen will continue to be green and its intensity will not change. A red filter, by contrast, absorbs all wavelengths except those near 650 nm. The red filter will absorb the green light, and *no* light will reach the screen. The screen will be dark.

Opaque objects appear colored by virtue of *pigments* that absorb light of some wavelengths but *reflect* light of other wavelengths. For example, red paint contains pigments that reflect light of wavelengths near 650 nm while absorbing all other wavelengths.

As an example, FIGURE 19.23 shows the absorption curve of *chlorophyll.* Chlorophyll is essential for photosynthesis in green plants. The chemical reactions of photosynthesis use blue/violet light and red light, so chlorophyll absorbs these wavelengths. When white light falls on leaves that contain chlorophyll, the red and blue are absorbed; what's left is green, which gives the leaves of chlorophyll-containing plants their characteristic color. When you look at the green leaves on a tree, you're seeing the light that was reflected because it *wasn't* needed for photosynthesis.

FIGURE 19.23 The absorption curve of chlorophyll.

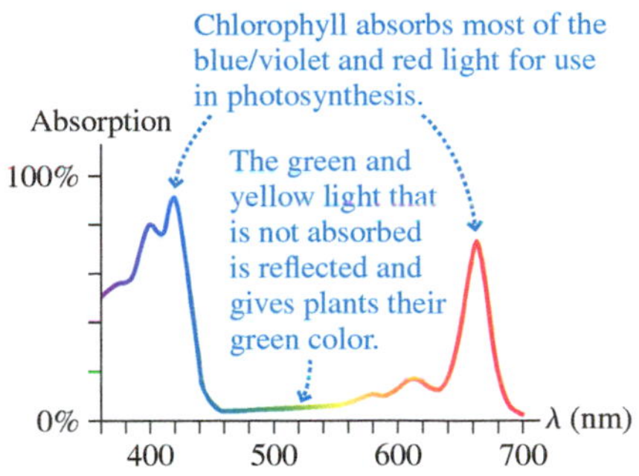


**STOP TO THINK 19.6** A swatch of pure red fabric is viewed through a green filter. The fabric appears

A. Red. B. Green. C. Yellow. D. Black.

## 19.7 Resolution of Optical Instruments

Suppose you want to study the *E. coli* bacterium. It's quite small, about 2 $\mu$m long and 0.5 $\mu$m wide. You might imagine that you could pair a 150× objective (the highest magnification available) with a 25× eyepiece to get a total magnification of 3750. At that magnification, the *E. coli* would appear about 8 mm across—about the size of Lincoln's head on a penny—with much fine detail revealed. But if you tried this, you'd be disappointed. Although you would see the general shape of a bacterium, you wouldn't be able to make out any details. All real optical instruments are limited in the details they can observe. Some limits are practical: Lenses are never perfect; they suffer from **aberrations.** But even a perfect lens would have a fundamental limit to the smallest details that could be seen. As we'll see, this limit is set by the diffraction of light, and so is intimately related to the wave nature of light itself. Together, lens aberrations and diffraction set a limit on an optical system's **resolution**—its ability to make out the fine details of an object.

### Aberrations

Consider the simple lens shown in FIGURE 19.24 imaging an object located at infinity, so that the incoming rays are parallel. An ideal lens would focus all the rays to a single point. However, for a real lens with spherical surfaces, the rays that pass near the lens's center come to a focus a bit farther from the lens than those that pass near its edge. There is no single focal point; even at the "best" focus the image is a bit blurred. This inability of a real lens to focus perfectly is called **spherical aberration.**

FIGURE 19.24 Spherical aberration.

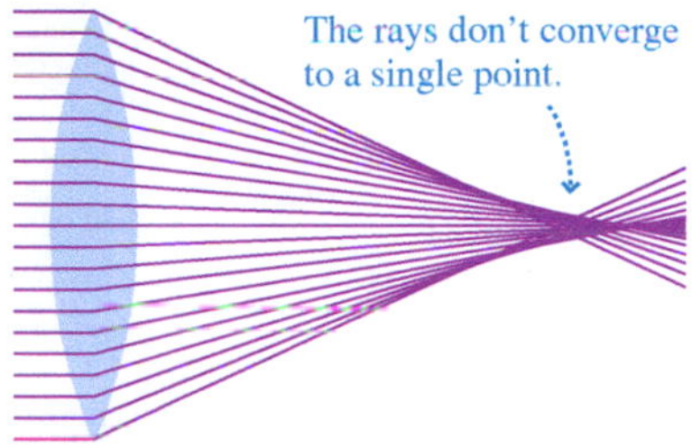


A careful examination of Figure 19.24 shows that the outer rays are most responsible for the poor focus. Consequently, the effects of spherical aberration can be minimized by using a diaphragm to pass only rays near the optical axis. This "stopping down" of a lens improves its imaging characteristics at the expense of its light-gathering capabilities. The cornea and the lens of your eye each have some spherical aberration. In bright light, your pupil contracts, which reduces the effect; the aberration is more apparent in dim light, when your pupil is wide open. Laser surgery to correct myopia may introduce some spherical aberration, which is why some people who have had laser vision correction may experience difficulties with their vision at night.

As we learned in the previous section, glass has *dispersion;* that is, the index of refraction of glass varies slightly with wavelength. The higher a lens's index of refraction, the more it bends incoming light rays. Because the index of refraction for violet light is higher than that for red light, a lens's focal length is slightly shorter for violet light than for red light. Consequently, different colors of light come to a focus at slightly different distances from the lens. If red light is sharply focused on a viewing screen, then blue and violet wavelengths are not well focused. This imaging error, illustrated in FIGURE 19.25, is called **chromatic aberration.**

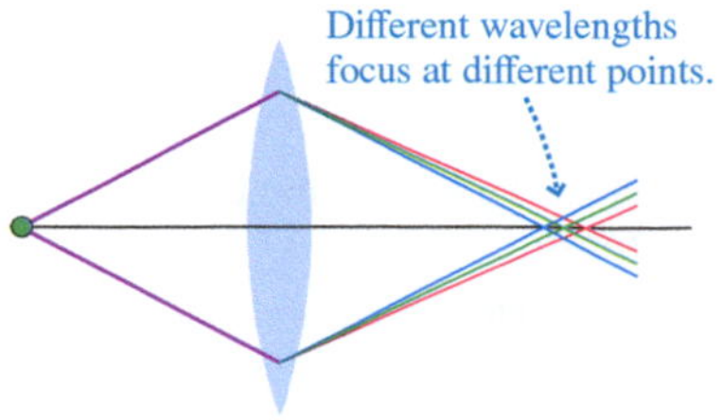


FIGURE 19.25 Chromatic aberration.

## Correcting Aberrations

Single lenses always have aberrations of some kind. For high-quality optics, such as those used in microscopes or telescopes, the aberrations are minimized by using a careful *combination* of lenses. An important example is the **achromatic doublet** (achromatic = "without color"), two lenses used in combination to greatly reduce chromatic aberration. FIGURE 19.26 shows how this works. A converging lens is paired with a weaker diverging lens; the combination has an overall positive refractive power and so is converging. However, the glasses are chosen so that the weaker diverging lens has a greater dispersion than the stronger converging lens. In this way the colors of white light, separated at first by the converging lens, are brought back together by the diverging lens. Achromatic doublets also minimize spherical aberration. Real microscope objectives are even more complex, but are based on the same principle as the achromatic doublet.

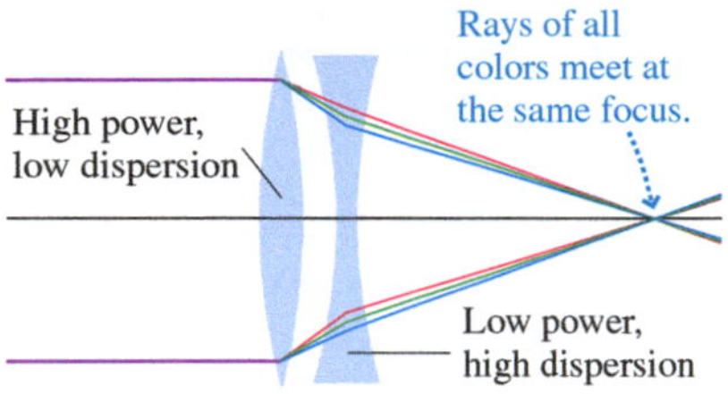


FIGURE 19.26 An achromatic lens.

## Resolution and the Wave Nature of Light

Modern lenses can be well corrected for aberrations, so they might be expected to focus perfectly. According to the ray model of light, a perfect lens should focus parallel rays to a single point in the focal plane. However, we've already hinted that there's a more fundamental limit to the performance of an optical instrument, a limit set by the wave nature of light.

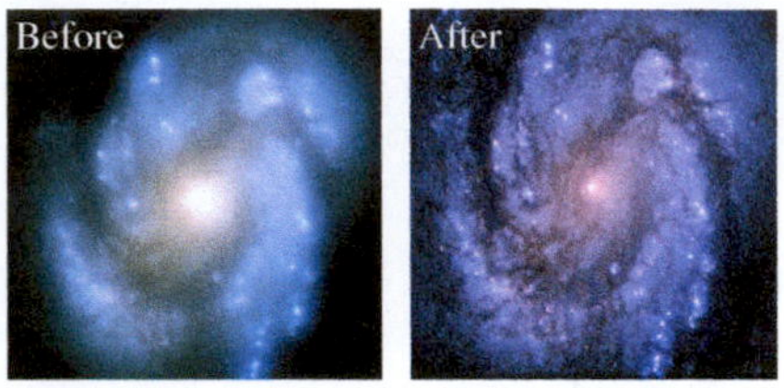


◂ **Eyeglasses in space** After launch, it was discovered that the mirror of the Hubble Space Telescope had been ground to the wrong shape, giving it severe spherical aberration. In a later service mission, corrective optics—in essence, very high-tech glasses—were put in place to correct for this spherical aberration. The photos to the right show an image of a galaxy before and after the corrective optics were added.

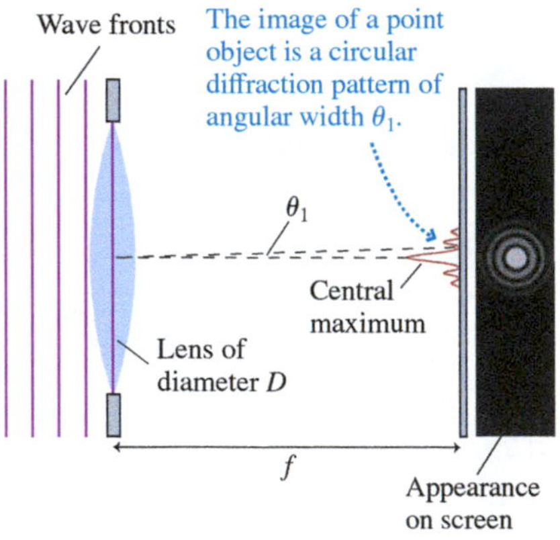


FIGURE 19.27 The image of a distant point source is a circular diffraction pattern.

FIGURE 19.27 shows a plane wave from a distant point source such as a star being focused by a lens of diameter $D$. Only those waves that pass *through* the lens can be focused, so the lens acts like a circular aperture of diameter $D$ in an opaque barrier. In other words, the lens both focuses *and diffracts* light waves.

You learned in ◂◂ SECTION 17.6 that a circular aperture produces a diffraction pattern with a bright central maximum surrounded by dimmer circular fringes. Consequently, as Figure 19.27 shows, light from a distant point source focuses not to a perfect point but, instead, to a small circular diffraction pattern. Equation 17.21 in Chapter 17 gave the angle $\theta_1$ of the outer edge of the central maximum as

$$\theta_1 = \frac{1.22\lambda}{D} \qquad (19.10)$$

Because the wavelength of light $\lambda$ is so much shorter than the lens diameter $D$ of an ordinary lens, the angular size of the central maximum is very small—but it is not zero.

The pupil in your eye is a circular opening, so when you look at a distant point light source, the image on your retina—even ignoring focusing errors or aberrations—is a fuzzy circle of light, not a sharp point. In Chapter 17, we learned that the width of the central maximum is $w \simeq 2.44\lambda L/D$. The wavelength in this equation is the wavelength inside the eye, where the light goes through the pupil, so the full equation is

$$w \simeq \frac{2.44\lambda L}{nD} \qquad (19.11)$$

$n$ is the index of refraction of the fluid surrounding the lens, which is about the same as that of water. $L$ is the distance from the pupil to the retina. We can estimate the size of the central maximum by assuming a typical eyeball length, $L = 1.7$ cm.

**EXAMPLE 19.17 How large is the image?** BIO

If your pupil is open to a typical diameter of 3 mm, what is the size of the image on your retina when you look at the tiny dot of a 530 nm green laser pointer on a distant screen?

**STRATEGIZE** We will assume that the tiny dot of the laser pointer is a point source. We are looking for the size of the central maximum of the diffraction when the light from this source passes through the pupil.

**PREPARE** The opening that produces the diffraction is the pupil, so $D = 0.0030$ m. We assume an index of refraction $n = 1.33$.

**SOLVE** We use Equation 19.11 to find the width of the central maximum:

$$w \simeq \frac{(2.44)(530 \times 10^{-9}\ \text{m})(0.017\ \text{m})}{(1.33)(0.0030\ \text{m})} = 5.5 \times 10^{-6}\ \text{m} \simeq 6\ \mu\text{m}$$

**ASSESS** As we'll see, this size makes sense given the structure of your retina.

Your vision is sharpest in the central part of the retina, an area called the *fovea.* This area contains a high density of light-sensing *photoreceptors,* in this case the color-sensing cone cells. The center-to-center spacing of these cells is approximately 3 $\mu$m. If the image of the laser pointer spot in Example 19.17 is centered on one cone cell, then the edge of the central maximum—the first dark fringe—is at the center of the next cone cell, as shown in FIGURE 19.28. This makes sense; we'd expect the size of the central maximum to closely match the spacing of the cone cells. Packing the cone cells any tighter wouldn't increase visual acuity, and packing them less tightly wouldn't take full advantage of the possible resolution of the eye. Increasing the pupil size doesn't improve resolution, because this increases the importance of aberrations. For pupil sizes larger than a few mm, resolution of the eye isn't limited by diffraction—it is limited by aberrations.

FIGURE 19.28 An image of cone cells in the fovea with a scale and the central maximum of the diffraction pattern overlaid.

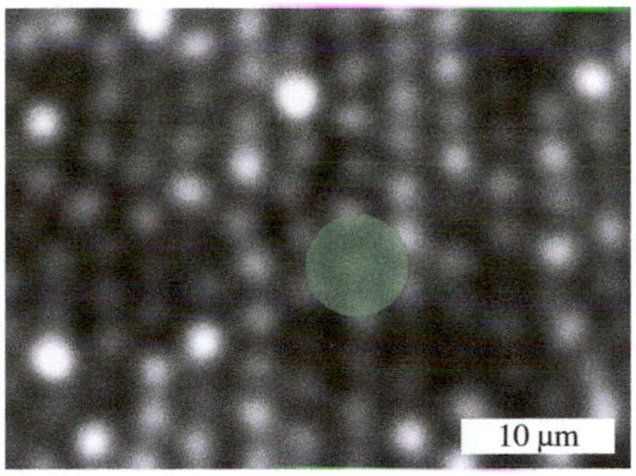


The fact that light is focused to a small spot, not a perfect point, has important consequences for how well a telescope can resolve two stars separated by only a small angle in the sky. FIGURE 19.29a shows how two nearby stars would appear in a telescope. Instead of perfect points, they appear as two diffraction images. Nonetheless, because they're clearly two separate stars, we say they are *resolved.* FIGURE 19.29b shows two stars that are closer together. Here the two diffraction patterns overlap, and it is becoming difficult to see them as two independent stars: They are barely resolved. The two very nearby stars in FIGURE 19.29c are so close together that we can't resolve them at all.

FIGURE 19.29 The resolution of a telescope.

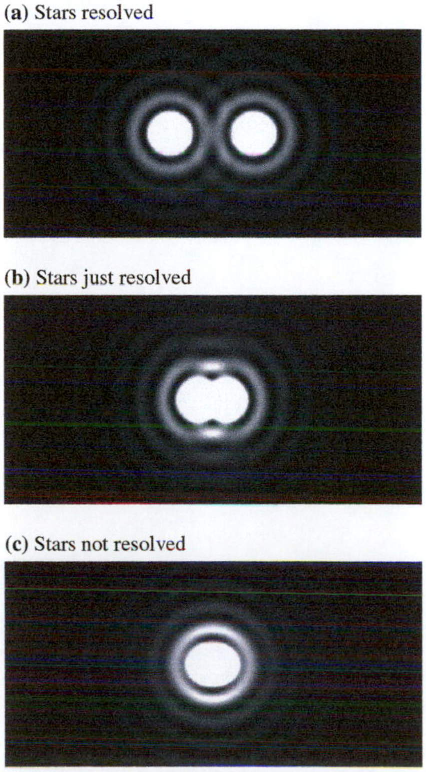


How close can the two diffraction patterns be before we can no longer resolve them? One of the major scientists of the 19th century, Lord Rayleigh, studied this problem and suggested a reasonable rule that today is called **Rayleigh's criterion.** In Figure 19.29b, where the two stars are just resolved, *the central maximum of the diffraction pattern of one star lies on top of the first dark fringe of the diffraction pattern of the other star.* Because the angle between the central maximum and the first dark fringe is $\theta_1$, the centers of the two stars are separated by angle $\theta_1 = 1.22\lambda/D$. Thus Rayleigh's criterion is:

> Two objects are resolvable if they are separated by an angle $\theta$ that is greater than $\theta_1 = 1.22\lambda/D$. If their angular separation is less than $\theta_1$, then they are not resolvable. If their separation is equal to $\theta_1$, then they are just barely resolvable.

For telescopes, the angle $\theta_1 = 1.22\lambda/D$ is called the *angular resolution* of the telescope. The angular resolution depends only on the lens diameter and the wavelength; the magnification is not a factor. Two overlapped, unresolved images will remain overlapped and unresolved no matter what the magnification. For visible light, where $\lambda$ is pretty much fixed, the only parameter over which the astronomer has any control is the diameter of the lens or mirror of the telescope.

## The Resolution of a Microscope

A microscope differs from a telescope in that it magnifies objects that are very close to the lens, not far away. Nonetheless, the wave nature of light still sets a limit on the ultimate resolution of a microscope. FIGURE 19.30 on the next page shows the objective

FIGURE 19.30 The resolution of a microscope.

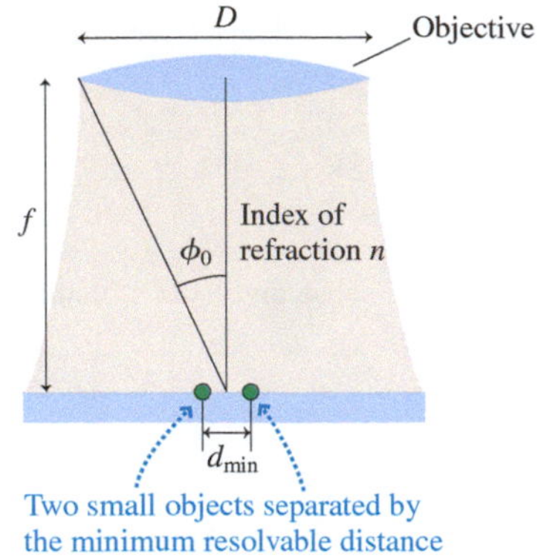


lens of a microscope that is observing two small objects. An analysis based on Rayleigh's criterion finds that the smallest resolvable separation between the two objects is

$$d_{min} = \frac{0.61\lambda}{n \sin \phi_0} \tag{19.12}$$

Here, $\phi_0$, defined in Figure 19.30, is the angular size of the objective lens and $n$ is the index of refraction of the medium between the objective lens and the specimen being observed. Usually this medium is air, so that $n = 1$, but biologists often use an *oil-immersion microscope* in which this space is filled with oil that has $n \approx 1.5$. From Equation 19.12, we can see that this higher value of $n$ reduces $d_{min}$, allowing objects that are closer together to be resolved.

We define the **numerical aperture** (NA) of the objective lens when immersed in a fluid of index $n$ as

$$\text{NA} = n \sin \phi_0$$

The minimum resolvable distance of a microscope, also called its **resolving power** (RP), is given by

$$\text{RP} = d_{min} = \frac{0.61\lambda_0}{\text{NA}} \tag{19.13}$$

The resolving power . . . . . . is the smallest resolvable separation between two objects.
Wavelength of light in vacuum
Numerical aperture of objective lens

The lower the resolving power, the *better* the objective is at seeing small details.

In principle, it would appear from Equation 19.13 that the resolving power of a microscope could be made as low as desired simply by increasing the numerical aperture. But there are severe practical limits on how high the numerical aperture can be made. The highest possible numerical aperture for a high-magnification 100× objective used in air is about 0.95. For an oil-immersion objective, the numerical aperture might be as high as 1.3. With such an objective, the resolving power would be

$$\text{RP} \approx 0.5\lambda_0$$

This illustrates the fundamental fact that **the minimum resolving power of a microscope, and thus the size of the smallest detail observable, is about half the wavelength of light.** This is a *fundamental limit* set by the wave nature of light. For $\lambda \approx 400$ nm, at the short-wavelength edge of the visible spectrum, the maximum possible resolving power is RP $\approx 200$ nm.

FIGURE 19.31 shows an actual micrograph of the bacillus *E. coli.* The size of a wavelength in the middle of the visible-light spectrum, 500 nm, is shown. This is about the same as the width of a bacterium. The smallest resolved features are about half of this. A higher magnification would not reveal any more detail because this micrograph is at the resolution limit set by the diffraction of light.

FIGURE 19.31 Optical and electron micrographs of *E. coli.* BIO

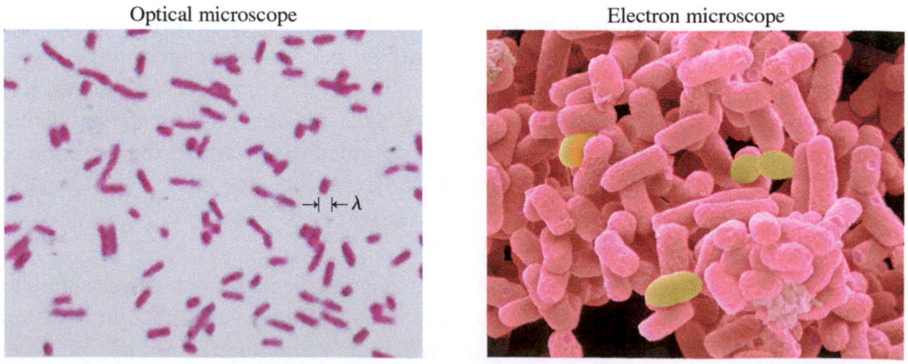

In contrast, the **electron microscope** micrograph of *E. coli* shows a wealth of detail that is not seen in the optical picture. In Chapter 28 we'll find out why the resolving power of an electron microscope is so much lower than that of an optical microscope.

**▸ The anatomy of a microscope objective**
The two most important specifications of a microscope objective, its magnification and its numerical aperture, are prominently displayed on its barrel. Many biological studies are conducted through a cover glass. This cover glass can introduce spherical aberration, blurring the image. By turning the correction collar, you can adjust this objective to correct for the exact thickness of the cover glass used.

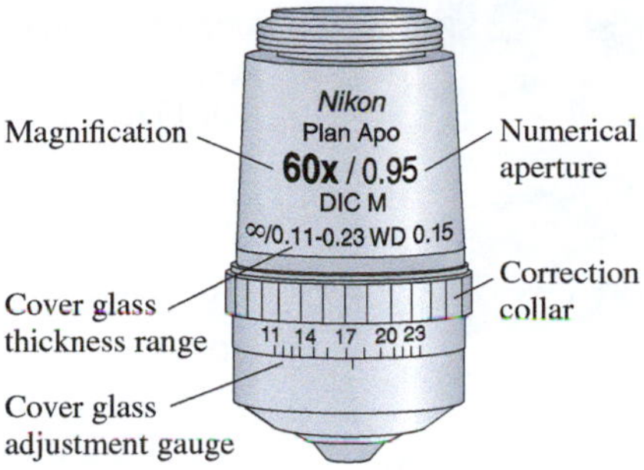


**EXAMPLE 19.18 Finding the resolving power of a microscope**

A microscope objective lens has a diameter of 6.8 mm and a focal length of 4.0 mm. For a sample viewed in air, what is the resolving power of this objective in red light? In blue light?

***STRATEGIZE*** We will use Equation 19.13 to find the resolving power. We'll need the numerical aperture of the objective, given as $\mathrm{NA} = n \sin \phi_0$.

***PREPARE*** From the geometry of Figure 19.30,

$$\tan \phi_0 = \frac{D/2}{f} = \frac{3.4\text{ mm}}{4.0\text{ mm}} = 0.85$$

from which $\phi_0 = \tan^{-1} 0.85 = 40.4°$ and $\sin \phi_0 = \sin 40.4° = 0.65$. Hence the numerical aperture is (since $n = 1$ in air)

$$\mathrm{NA} = n \sin \phi_0 = 1 \times 0.65 = 0.65$$

***SOLVE*** Once we have a value for NA, we calculate the resolving power using Equation 19.13:

$$\mathrm{RP} = \frac{0.61\lambda_0}{0.65} = 0.94\lambda_0$$

Wavelengths of different colors of light were listed in Table 19.1. For red light, with $\lambda_0 = 650$ nm, RP = 610 nm, while blue light, with $\lambda_0 = 450$ nm, has RP = 420 nm.

***ASSESS*** The resolving power is on the same order as the wavelength, as we expect, and the resolving power is lower for the shorter wavelength, so our results seem reasonable.

**STOP TO THINK 19.7** Four lenses are used as microscope objectives, all for light that has the same wavelength $\lambda$. Rank in order, from highest to lowest, the resolving powers $\mathrm{RP}_1$ to $\mathrm{RP}_4$ of the lenses.

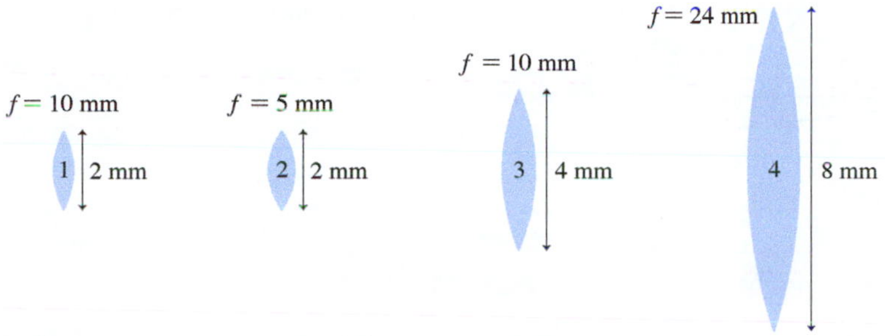

## INTEGRATED EXAMPLE 19.19 Determining the visual acuity of a kestrel BIO

Like most birds of prey, the American kestrel has excellent eyesight. The smallest angular separation between two objects that an eye can resolve is called its *visual acuity*. A smaller visual acuity means better eyesight because objects that are closer together can be resolved.

The kestrel's eye is shorter than yours, with a distance of 0.90 cm from lens to retina. The diameter of its pupil is about the same as yours, 3.0 mm. Its retina has photoreceptors that are more closely spaced, about 2 $\mu$m apart. For the purposes of this problem, we'll model the optical system very simply, with a single lens projecting an image onto the retina, and we'll assume that the eye is filled with fluid that has an index of refraction the same as that of water.

a. Laboratory measurements indicate that the kestrel can just resolve two small objects that have an angular separation of only 0.013°. How does this result compare with the visual acuity predicted by Rayleigh's criterion? Take the wavelength of light in air to be 550 nm.
b. What is the distance on the retina between the images of two small objects that can just be resolved? How does this distance compare to the 2 $\mu$m distance between two photoreceptors?
c. The kestrel's excellent visual acuity allows it to detect small motions. If a single object moves laterally to produce two subsequent images that have the noted angular separation, the kestrel can, in principle, detect it. Suppose the kestrel is hunting on the ground directly below its 18-m-high perch in a tree. How far would a mouse need to move for the kestrel to notice this motion?

**STRATEGIZE** We are interested in the kestrel's ability to resolve two closely spaced objects. To help visualize the situation, we'll start with the ray-tracing diagram in FIGURE 19.32. $\theta$ is the smallest angle that the kestrel can resolve, $s$ and $s'$ are the object and image distances, $d$ is the distance between the two objects, and $d'$ is the distance between the two images.

FIGURE 19.32 The kestrel's eye observing two closely spaced objects.

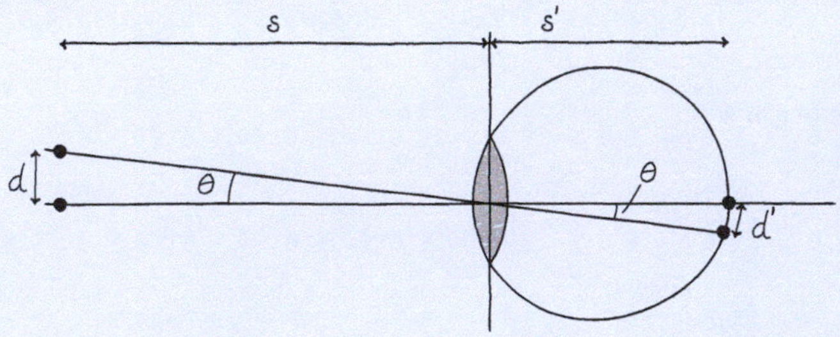


**PREPARE** Inside the eye, the wavelength of light is shorter than its wavelength in air because of the index of refraction of the liquid within the eye. Thus for part (a) we will use $\lambda = (550 \text{ nm})/1.33 = 410 \text{ nm}$ in Rayleigh's criterion. For part (b), Figure 19.32 shows that, within the small-angle approximation, the distance $d'$ on the retina between the images of two small objects subtending an angle $\theta$ is simply $d' = \theta s'$, where $\theta$ is in radians. Similarly, for part (c), the distance $d$ between the two small objects subtending an angle $\theta$ is simply $d = \theta s$.

**SOLVE**

a. For a perfect lens, with no aberrations, the smallest resolvable angular separation between two objects is given by Rayleigh's criterion as

$$\theta = \frac{1.22\lambda}{D} = \frac{(1.22)(410 \times 10^{-9} \text{ m})}{3.0 \times 10^{-3} \text{ m}} = 1.7 \times 10^{-4} \text{ rad}$$

Recalling that there are 360° in $2\pi$ rad, this angle is

$$\theta = (1.7 \times 10^{-4} \text{ rad}) \times \frac{360°}{2\pi \text{ rad}} = 0.0097° \simeq 0.01°$$

This is only slightly less than the experimentally observed resolution. Kestrels are able to resolve images at separations near the Rayleigh limit, a testament to the excellence of their vision.

b. The measured angle of 0.013° corresponds to $2.3 \times 10^{-4}$ rad. Thus the distance between the images of the two small objects is

$$d' = \theta s' = (2.3 \times 10^{-4} \text{ rad})(9.0 \text{ mm})$$
$$= 2.1 \times 10^{-3} \text{ mm} = 2.1 \ \mu\text{m}$$

This is just about the same as the photoreceptor distance. This is about as good as it gets; the light from the two objects falls on adjacent photoreceptors. Better resolution wouldn't be possible without packing photoreceptors more tightly in the fovea.

c. The distance $d$ is computed similarly to $d'$, with an object distance $s = 18$ m:

$$d = \theta s = (2.3 \times 10^{-4} \text{ rad})(18 \text{ m})$$
$$= 4.1 \text{ mm}$$

This distance is very small—about $\frac{1}{6}$ inch—and the kestrel can, in principle, spot this motion from a perch 60 feet above the ground. You might not notice motion at the limits of your resolution, but a raptor's optical system is well adapted to detect changes in the visual field corresponding to such small movements.

**ASSESS** We know that raptors have excellent vision, so the fact that our calculations show that their vision works at almost the theoretical limits gives us confidence in our results.

# SUMMARY

**GOAL** To understand how common optical instruments work.

## IMPORTANT CONCEPTS

The **refractive power** $P$ of a lens is the inverse of its focal length: $P = 1/f$. Refractive power is measured in diopters:

$$1\text{ D} = 1\text{ m}^{-1}$$

### Lenses in combination

When two lenses are placed next to each other, the refractive power of the pair is the sum of the individual refractive powers:

$$P = P_1 + P_2$$

When two lenses work together to form an image, the image from the first lens serves as the object for the second lens.

### Resolution of optical instruments

The **resolution** of a telescope or microscope is limited by imperfections, or **aberrations,** in the optical elements, and by the more fundamental limits imposed by diffraction.

For a *microscope,* the minimum resolvable distance between two objects is

$$d_{min} = \frac{0.61\lambda}{\text{NA}}$$

For a *telescope,* the minimum resolvable angular separation between two objects is

$$\theta_1 = \frac{1.22\lambda}{D}$$

### Color and dispersion

The eye perceives light of different wavelengths as having different colors.

**Dispersion** is the dependence of the index of refraction $n$ of a transparent medium on the wavelength of light: Long wavelengths have the lowest $n$, short wavelengths the highest $n$.

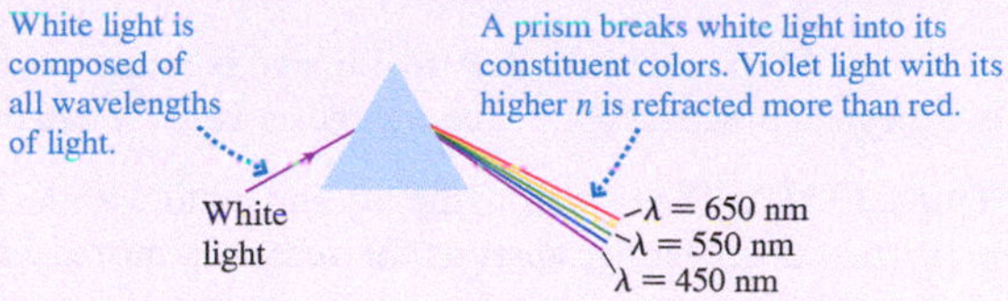

White light is composed of all wavelengths of light.
A prism breaks white light into its constituent colors. Violet light with its higher n is refracted more than red.
White light
λ = 650 nm
λ = 550 nm
λ = 450 nm

### Angular and apparent size

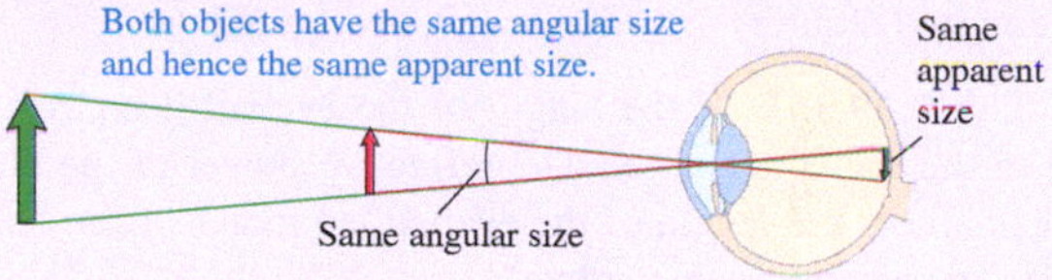


## APPLICATIONS

### The camera and the eye

Both the camera and the eye work by focusing an image on a light-sensitive surface.

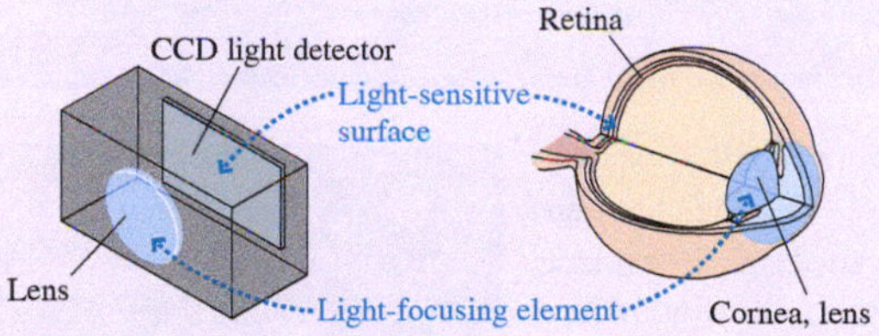


The camera focuses by changing the lens-detector distance, while the eye focuses by changing the focal length of its lens.

### The telescope

**The telescope** magnifies distant objects. The objective lens creates a real image of the distant object. This real image is then magnified by the eyepiece lens, which acts as a simple magnifier. The angular magnification is $M = -f_o/f_e$.

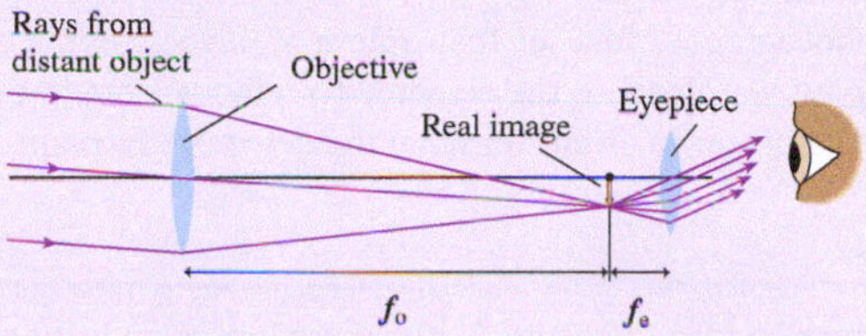


### The magnifier

Without a lens, an object cannot be viewed closer than the eye's near point of ≈0.25 m. Its angular size $\theta_0$ is $h/0.25$ m.

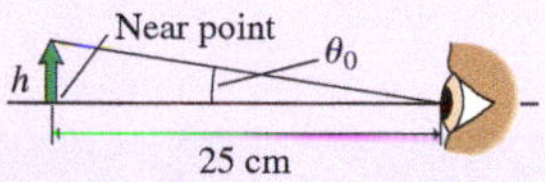


If the object is now placed at the focal point of a converging lens, its angular size is increased to $\theta = h/f$.

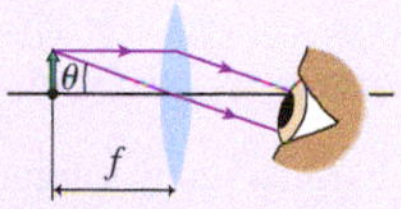


The angular magnification is $M = \theta/\theta_0 = 0.25\text{ m}/f$.

### The microscope

**The microscope** magnifies a small, nearby object. The objective lens creates a real image of the object. This real image is then further magnified by the eyepiece lens, which acts as a simple magnifier. The angular magnification is

$$M = -\frac{L \times 0.25\text{ m}}{f_o f_e}$$

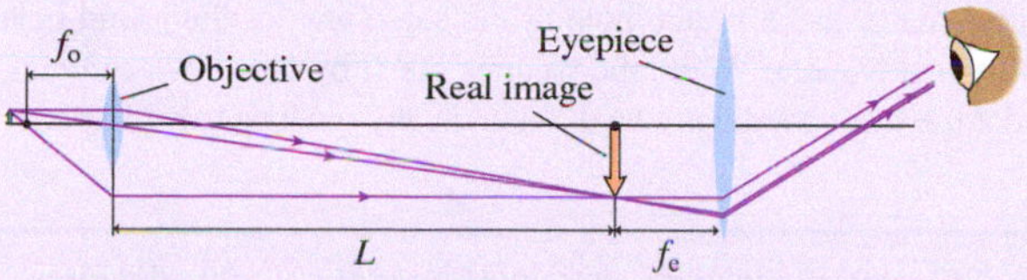

## Learning Objectives After studying this chapter, you should be able to:

- Understand how a camera works. *Conceptual Questions 19.4, 19.5; Problems 19.1, 19.2, 19.4, 19.5, 19.6*
- Analyze the optics of the eye, and specify the appropriate lenses to correct vision defects. *Conceptual Questions 19.12, 19.17; Problems 19.17, 19.18, 19.19, 19.23, 19.25*
- Work with magnifiers and angular magnification. *Conceptual Question 19.6; Problems 19.27, 19.28, 19.29, 19.30, 19.31*
- Analyze simple microscopes and telescopes. *Conceptual Questions 19.14, 19.16; Problems 19.33, 19.35, 19.41, 19.42, 19.43*
- Understand the origin and significance of color. *Conceptual Question 19.15; Problems 19.46, 19.47, 19.48*
- Use Rayleigh's criterion to calculate the resolution limits of an optical system. *Conceptual Question 19.15; Problems 19.49, 19.50, 19.51, 19.52, 19.53*

## STOP TO THINK ANSWERS

**Chapter Preview Stop to Think: B.** A real image is formed only if the object is beyond the focal point. See Synthesis 18.1 for details.

**Stop to Think 19.1: A.** The rays from the top and bottom of the tree in Figure 19.1b will get farther apart as the screen is moved back, resulting in a larger image.

**Stop to Think 19.2: B.** If the power of the optical system is increased, images will form in front of the retina. The person is myopic and will require correction with a diverging lens.

**Stop to Think 19.3: B.** If the length of the eyeball decreases, the distance to the retina is shorter. The refractive power of the lens is then too small for the length of the eyeball, so images form behind the retina. The person has hyperopia.

**Stop to Think 19.4: A.** Ray tracing shows why:

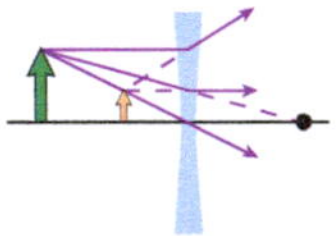

**Stop to Think 19.5: B.** The total magnification is the product of the objective magnification $m_o$ and the eyepiece angular magnification $M_e$. If $m_o$ is halved, from 20× to 10×, $M_e$ must be doubled. Because $M_e$ is inversely proportional to the eyepiece focal length, the focal length of the eyepiece must be halved.

**Stop to Think 19.6: D.** The fabric is "pure red," so it reflects only red light. A green filter lets through only green light, so it blocks all of the reflected light. No light from the fabric can pass through the filter, so it appears black.

**Stop to Think 19.7: $RP_1 > RP_4 > RP_2 = RP_3$.** The resolving power is $RP = 0.61\lambda/\sin\phi_0$ for objectives used in air ($n = 1$), so the resolving power is higher (worse resolution) when the angle $\phi_0$ is smaller. From Figure 19.30 you can see that $\phi_0$ is smaller when the ratio $D/f$ is smaller. These ratios are $(D/f)_1 = 1/5$, $(D/f)_2 = 2/5$, $(D/f)_3 = 2/5$, and $(D/f)_4 = 1/3$.

Video Tutor Solution Chapter 19

# QUESTIONS

## Conceptual Questions

1. On a sunny summer day, with the sun overhead, you can stand under a tree and look on the ground at the pattern of light that has passed through gaps between the leaves. You may see illuminated circles of varying brightness. Why are there circles, when the gaps between the leaves have irregular shapes?

2. Suppose you have two pinhole cameras. The first has a small round hole in the front of the camera. The second is identical in every regard, except that it has a square hole of the same area as the round hole in the first camera. Would the pictures taken by these two cameras, under the same conditions, be different in any obvious way? Explain.
3. Figure Q19.3 shows the viewing screen in an English camera obscura, a windowless room with a lens in one wall and a screen on the opposite wall, 3 meters from the lens. What can you say about the focal length of the lens?

FIGURE Q19.3

4. A telephoto lens—designed to produce large images of distant objects—sticks out quite far from the front of the camera. Explain why this is so.
5. A photographer focuses his camera on his subject. The subject then moves closer to the camera. To refocus, should the lens be moved closer to or farther from the detector? Explain.

Problem difficulty is labeled as | (straightforward) to ||||| (challenging). Problems labeled INT integrate significant material from earlier chapters; Problems labeled BIO are of biological or medical interest.

The eText icon indicates when there is a video tutor solution available for the chapter or for a specific problem. To launch these videos, log into your eText through Mastering™ Physics or log into the Study Area.

6. The object for a magnifier is usually placed very close to the focal point of the lens, creating a virtual image that can be viewed with the relaxed eye. But the object could be placed so that the image is at the eye's near-point distance. In that case, the image can be viewed only by using the full accommodation of the eye. With this arrangement, is the angular magnification greater than, equal to, or less than the magnification when the image is far away? Explain.
7. A nature photographer taking a close-up shot of an insect replaces the standard lens on his camera with a lens that has a shorter focal length and is positioned farther from the CCD detector. Explain why he does this.
8. The CCD detector in a certain camera has a width of 8 mm. The photographer realizes that with the lens she is currently using, she can't fit the entire landscape she is trying to photograph into her picture. Should she switch to a lens with a longer or shorter focal length? Explain.
9. BIO All humans have what is known as a *blind spot,* where the optic nerve exits the eye and no light-sensitive cells exist. To locate your blind spot, look at the figure of the cross. Close your left eye and place your index finger on the cross. Slowly move your finger to the left while following it with your right eye. At a certain point the cross will disappear. Is your right eye's blind spot on the right or left side of your retina? Explain.

10. BIO Suppose you wanted special glasses designed to wear underwater, without a face mask. Should the glasses use a converging or diverging lens in order for you to be able to focus under water? Explain.
11. BIO Young children usually have a fair degree of hyperopia. As they grow, their eyeballs lengthen and the hyperopia usually clears up. Given what you know about accommodation, explain why young children, though hyperopic, don't need glasses.
12. BIO When you swim underwater, your vision is compromised, as we've seen. The same is true of water-dwelling animals that venture up into the air.
    a. When you open your eyes underwater, are you nearsighted or farsighted?
    b. When a dolphin pokes its head out of the water, is it nearsighted or farsighted?
13. BIO Children of the Moken tribe spend much of their days foraging for food underwater. Their eyes are adapted to this; their lenses are capable of more accommodation and, when they dive, their pupils contract. Explain how each of these changes improves their underwater vision.
14. You have lenses with the following focal lengths: $f = 25$ mm, 50 mm, 100 mm, and 200 mm. Which lens or pair of lenses would you use, and in what arrangement, to get the highest-power magnifier, microscope, and telescope? Explain.
15. A friend lends you the eyepiece of his microscope to use on your own microscope. He claims that since his eyepiece has the same diameter as yours but twice the focal length, the resolving power of your microscope will be doubled. Is his claim valid? Explain.
16. A student makes a microscope using an objective lens and an eyepiece. If she moves the lenses closer together, does the microscope's magnification increase or decrease? Explain.
17. BIO Is the wearer of the glasses in Figure Q19.17 nearsighted or farsighted? How can you tell?
18. A red card is illuminated by red light. What color does it appear to be? What if it's illuminated by blue light?

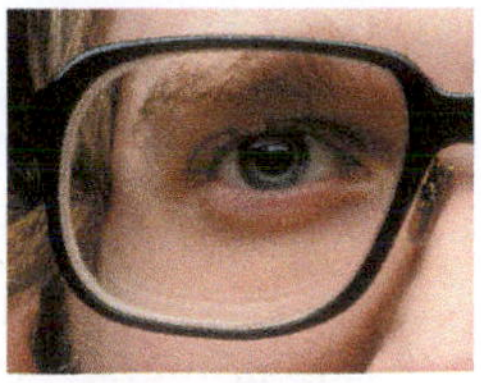

FIGURE Q19.17

## Multiple-Choice Questions

19. | A microscope has a tube length of 20 cm. What combination of objective and eyepiece focal lengths will give an overall magnification of 100?
    A. 1.5 cm, 3 cm B. 2 cm, 2 cm
    C. 1 cm, 5 cm D. 3 cm, 8 cm
20. || The distance between the objective and eyepiece of a telescope is 55 cm. The focal length of the eyepiece is 5.0 cm. What is the angular magnification of this telescope?
    A. −10 B. −11 C. −50 D. −275
21. | BIO A nearsighted person has a near point of 20 cm and a far point of 40 cm. When he is wearing glasses to correct his distant vision, what is his near point?
    A. 10 cm B. 20 cm C. 40 cm D. 1.0 m
22. | BIO A nearsighted person has a near point of 20 cm and a far point of 40 cm. What refractive power lens is necessary to correct this person's vision to allow her to see distant objects?
    A. −5.0 D B. −2.5 D C. +2.5 D D. +5.0 D
23. | BIO A 60-year-old man has a near point of 100 cm. What refractive power reading glasses would he need to focus on a newspaper held at a comfortable distance of 40 cm?
    A. −2.5 D B. −1.5 D C. +1.5 D D. +2.5 D
24. || BIO A cataract is a clouding or opacity that develops in the eye's lens. In extreme cases, the lens of the eye may need to be removed. This would have the effect of leaving a person
    A. Nearsighted.
    B. Farsighted.
    C. Neither nearsighted nor farsighted.
25. || BIO Everyone's vision is compromised underwater, but some can see a bit better than others. Which of the following individuals would have the clearest underwater vision?
    A. A person with vision that requires no correction.
    B. A person with myopia.
    C. A person with hyperopia.
    D. A person with presbyopia.
26. | In a darkened room, red light shines on a red cup, a white card, and a blue toy. The cup, card, and toy will appear, respectively,
    A. Red, red, blue.
    B. Red, white, blue.
    C. Red, red, black.
    D. Red, black, blue.
27. || An amateur astronomer looks at the moon through a telescope with a 15-cm-diameter objective. What is the minimum separation between two objects on the moon that she can resolve with this telescope? Assume her eye is most sensitive to light with a wavelength of 550 nm.
    A. 120 m B. 1.7 km C. 26 km D. 520 km

Watch Video Solution Problem 19.7

# PROBLEMS

### Section 19.1 The Camera

1. | Suppose you point a pinhole camera at a 15-m-tall tree that is 75 m away. If the detector is 22 cm behind the pinhole, what will be the size of the tree's image on the detector?
2. | A student has built a 20-cm-long pinhole camera for a science fair project. She wants to photograph the Washington Monument, which is 167 m (550 ft) tall, and to have the image on the detector be 5.0 cm high. How far should she stand from the Washington Monument?
3. || A pinhole camera is made from an 80-cm-long box with a small hole in one end. If the hole is 5.0 m from a 1.8-m-tall person, how tall will the image of the person on the detector be?
4. || You are taking a picture of a giraffe that is standing far away from you. The image is just too small, so you swap the 50-mm-focal-length lens in your camera for a 600 mm telephoto lens. By what factor does this increase the size of the image?
5. || A photographer uses his camera, whose lens has a 50 mm focal length, to focus on an object 2.0 m away. He then wants to take a picture of an object that is 40 cm away. How far, and in which direction, must the lens move to focus on this second object?
6. || Turning the barrel of a 50-mm-focal-length lens on a manual-focus camera moves the lens closer to or farther from the sensor to focus on objects at different distances. The lens has a stated range of focus from 0.45 m to infinity. How far does the lens move between these two extremes?
7. || An older camera has a lens with a focal length of 50 mm and uses 36-mm-wide film. Using this camera, a photographer takes a picture of the Golden Gate Bridge that completely spans the width of the film. Now he wants to take a picture of the bridge using his digital camera with its 12-mm-wide CCD detector. What focal length should this camera's lens have for the image of the bridge to cover the entire detector?
8. || In old Polaroid cameras, the image was projected on film that developed inside the camera. The developed image was then ejected from the camera as a printed photograph. The standard film size for one popular camera was 79 mm square. The film was 116 mm behind the lens. If you wanted a picture of your 1.6-m-tall friend to fill half the frame, how far away from you did she need to stand?
9. ||| In Figure P19.9 the camera lens has a 50 mm focal length. How high is the man's well-focused image on the CCD detector?

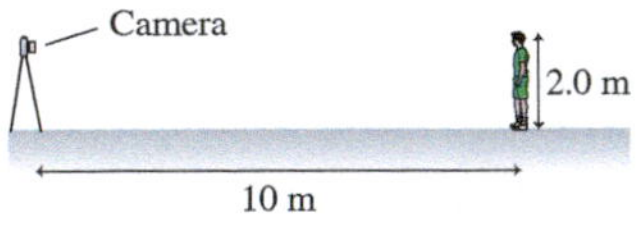


FIGURE P19.9

### Section 19.2 The Human Eye

10. | A lens with $f = +15$ cm is paired with a lens with $f = -20$ cm. What is the focal length of the combination?
11. | Two converging lenses with focal lengths of 20 cm and 24 cm are combined to make a single lens. What is the focal length of the combination?
12. || Oscar has an old-fashioned lantern projector that uses a candle and a lens to project a slide onto the wall. The lens is 1.3 cm from the slide and projects an image onto a screen 80 cm away. Oscar adds a lens to the original lens to allow him to project an image onto a screen 30 cm away.
    a. What is the power of the original lens?
    b. What is the desired power of the combined lenses?
    c. What power lens must Oscar add?
13. || A +2.0 D lens is being used to make an image of a distant object on a screen. The image is 2.4 cm tall. A second +2.0 D lens is added to the first, and the lens combination is moved to refocus the image. How tall is the new image?
14. | BIO If the retina is 1.7 cm from the lens in the eye, how large is the image on the retina of a person of height 1.5 m standing 8.0 m away?
15. || BIO Measure your near point by bringing this page up to the closest distance at which the image is still crisp—not the closest at which you can still read the letters, but the closest at which they remain sharp. (If you wear glasses or contacts, keep them on.) Measure the distance to determine your near point. What is your range of accommodation?
16. || BIO At a distance of 6 meters, a person with average vision is able to clearly read letters 1.0 cm high. Approximately how large do the letters appear on the retina? (Assume that the retina is 1.7 cm from the lens.)
17. || BIO A farsighted person has a near point of 50 cm. What strength lens, in diopters, is needed to bring his near point to 25 cm?
18. ||| BIO Rachel has good distant vision but has a touch of presbyopia. Her near point is 0.60 m. When she wears +2.0 D reading glasses, what is her near point? Her far point?
19. | BIO A nearsighted woman has a far point of 300 cm. What kind of lens, converging or diverging, should be prescribed for her to see distant objects more clearly? What refractive power should the lens have?
20. ||| BIO A nearsighted person has a near point of 12 cm and a far point of 40 cm. What power corrective lens is needed for her to have clear distant vision? With this corrective lens in place, what is her new near point?
21. || BIO Martin has severe myopia, with a far point of only 17 cm. He wants to get glasses that he'll wear while using his computer, whose screen is 65 cm away. What refractive power is needed for Martin to view the screen with his eyes relaxed?
22. || BIO Mary's glasses have +4.0 D converging lenses. This gives her a near point of 20 cm. What is the location of her near point when she is not wearing her glasses?
23. ||| BIO Rank the following people from the most nearsighted to the most farsighted, indicating any ties:
    A. Bernie has a prescription of +2.0 D.
    B. Carol needs diverging lenses with a focal length of −0.35 m.
    C. Maria Elena wears converging lenses with a focal length of 0.50 m.
    D. Janet has a prescription of +2.5 D.
    E. Warren's prescription is −3.2 D.
24. || BIO With −5.0 D corrective lenses, Juliana's distant vision is quite sharp. She has a pair of −3.5 D computer glasses that puts her computer screen right at her far point. How far away is her computer?

25. || The rod and cone cells in the central part of the retina—the fovea—are packed closer together, giving a more detailed view. This area of increased rod and cone density has a diameter of about 1.5 mm. When you read a book, you want the image of the text you are reading to fall on the fovea. If you hold a book 30 cm from your eyes, how wide is the spot on the page whose image just fills the fovea? (Assume that the retina is 1.7 cm from the lens.)
BIO

## Section 19.3 The Magnifier

26. || A jeweler is wearing a 20 D magnifying lens directly in front of his eye. If his near point is a typical 25 cm, how close can he hold a gem that he is inspecting?
BIO
27. ||| Oliver has had a stamp collection since he was a boy. In those days, holding a stamp 10 cm from his eye gave him a clear image. Now, his near point has receded to 90 cm, so he holds a magnifying lens directly in front of his eye to let him bring stamps closer. To the nearest diopter, what power lens enables him to focus on a stamp 10 cm away?
BIO
28. ||| Anna holds a 12 D magnifier directly in front of her eye to get a close look at a 19-mm-diameter penny. What is the closest possible distance that she can hold the coin to have it appear in focus? At this distance, how large does the image of the coin appear on her retina? (Assume a typical 25 cm near point and a distance of 1.7 cm between the lens and the retina.)
BIO
29. || The diameter of a penny is 19 mm. As we've seen, the moon subtends an angle of approximately 0.5° in the sky. How far from your eye must a penny be held so that it has the same apparent size as the moon?
30. || When you solder electronic components, you want to keep your face away from the work but still be able to see it clearly, so you use a soldering station with a built-in lens. Clips hold the work you are soldering, and a flexible arm holds the lens above the work. You view the work from some distance, but the area under the lens is magnified enough that you can see fine details. A typical magnifier is rated at 3×. How far from the work should the lens be placed?

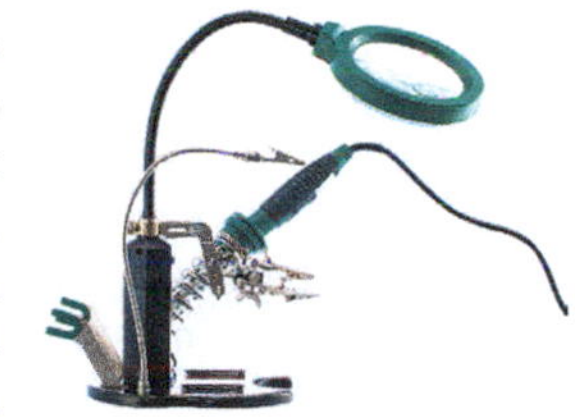

31. | A magnifier has a magnification of 5×. How far from the lens should an object be placed so that its (virtual) image is at the near-point distance of 25 cm?
32. || People with compromised vision can buy full-page magnifiers that enlarge printed materials to a size that they can more easily read. A typical magnifier is rated at 2× magnification and is held in a stand above the page. How far from the page is the magnifier?

## Section 19.4 The Microscope

33. || A student microscope has an objective lens that is labeled with a magnification of 10×. Assume a length $L = 120$ mm.
   a. What is the focal length of the objective lens?
   b. What focal length eyepiece lens is needed to give an overall magnification of 150×?
34. ||| You are using a microscope with a 10× eyepiece. What focal length of the objective lens will give a total magnification of 200×? Assume a length $L = 160$ mm.
35. |||| A forensic scientist is using a microscope with a 15x objective and a 5× eyepiece to examine a hair from a crime scene. How far from the objective is the hair? Assume a length $L = 160$ mm.
36. ||| A microscope has an 8.0-mm-focal-length objective. For the microscope to be in focus, how far should the objective lens be from the specimen? Assume a length $L = 160$ mm.
37. || The objective lens of a microscope has a focal length of 5.0 mm. What eyepiece focal length will give the microscope an overall angular magnification of 350? Assume a length $L = 160$ mm.
38. ||| A student microscope has a 10× objective and a 15× eyepiece. A student is using the microscope to view a thin slice of an apple. The cells in the apple have a diameter of 0.10 mm. Assume a length $L = 160$ mm.
   a. How far from the objective lens should the apple slice be placed?
   b. How does the angular size of a cell when viewed through the microscope compare to the angular size of a 19-mm-diameter penny at a distance of 25 cm?

## Section 19.5 The Telescope

39. || For the combination of two identical lenses shown in Figure P19.39, find the position, size, and orientation of the final image of the 2.0-cm-tall object.

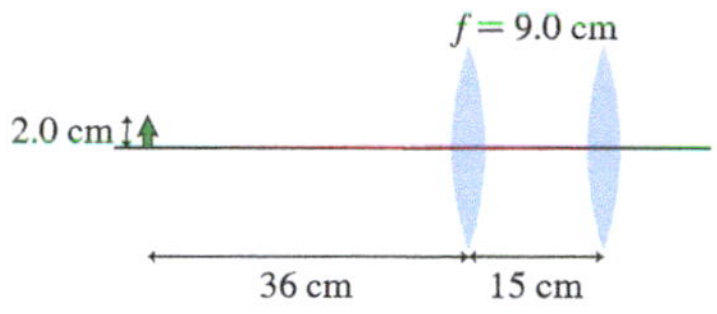


FIGURE P19.39

40. || For the combination of two lenses shown in Figure P19.40, find the position, size, and orientation of the final image of the 1.0-cm-tall object.

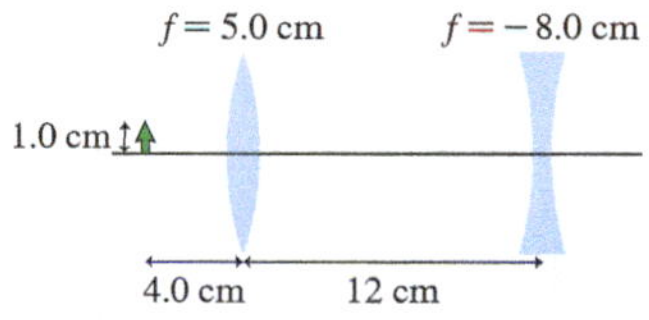


FIGURE P19.40

41. || A researcher is trying to shoot a tranquilizer dart at a 2.0-m-tall rhino that is 150 m away. Its angular size as seen through the rifle telescope is 9.1°. What is the magnification of the telescope?
42. || The objective lens of the refracting telescope at the Lick Observatory in California has a focal length of 57 ft.
   a. What is the refractive power of this lens?
   b. What focal length (mm) eyepiece would give a magnification of 1000× for this telescope?
43. ||| You use your 8× binoculars to focus on a yellow-rumped warbler (length 14 cm) in a tree 18 m away from you. What angle (in degrees) does the image of the warbler subtend on your retina?
44. || Use the astronomical data provided in this text to compute the angular size of Jupiter as seen from earth when the distance from earth to Jupiter is the same as the distance from the sun to Jupiter. Your telescope's objective lens has a focal length of 700 mm, and you use a 10 mm eyepiece to view Jupiter. What is the angular size of the image?

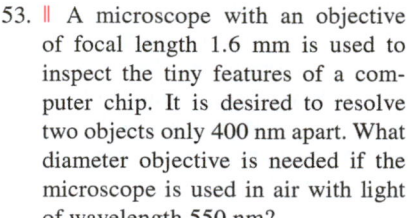

45. || Your telescope has a 1000-mm-focal-length objective. You use your telescope to take a photo of the full moon, which has an angular size of 0.52° that night. You position a CCD detector so that it captures the image produced by the objective lens. What is the diameter of that image?

### Section 19.6 Color and Dispersion

46. |||| A narrow beam of light with wavelengths from 450 nm to 700 nm is incident perpendicular to one face of a prism made of crown glass, for which the index of refraction ranges from $n = 1.533$ to $n = 1.517$ for those wavelengths. The light strikes the opposite side of the prism at an angle of 40°. What is the angular spread of the beam as it leaves the prism?

47. ||| A ray of white light strikes the surface of a 4.0-cm-thick slab of flint glass as shown in Figure P19.47. As the ray enters the glass, it is dispersed into its constituent colors. Estimate how far apart the rays of deepest red and deepest violet light are as they exit the bottom surface. Which exiting ray is closer to point P?

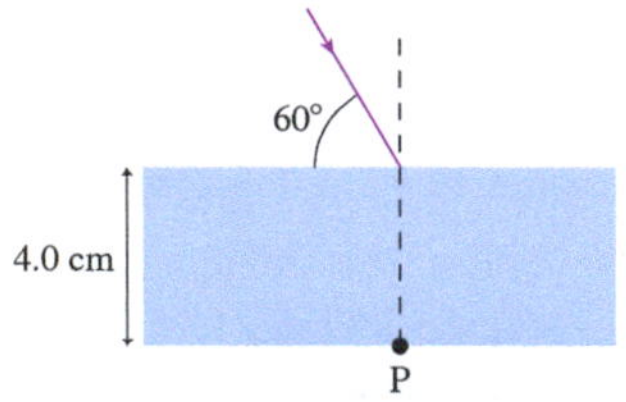

FIGURE P19.47

48. |||| A ray of red light, for which $n = 1.54$, and a ray of violet light, for which $n = 1.59$, travel through a piece of glass. They meet right at the boundary between the glass and the air, and emerge into the air as one ray with an angle of refraction of 22.5°. What is the angle between the two rays in the glass?

### Section 19.7 Resolution of Optical Instruments

49. | The near point for your myopic uncle is 10 cm. Your own vision is normal; that is, your near point is 25 cm. Suppose you and your uncle hold dimes (which are 1.7 cm in diameter) at your respective near points.
   a. For you, what is the dime's angular size, in radians?
   b. For your uncle, what is the dime's angular size, in radians?
   c. Do these calculations suggest any benefit to nearsightedness?

50. ||| Two lightbulbs are 1.0 m apart. From what distance can these lightbulbs be marginally resolved by a small telescope with a 4.0-cm-diameter objective lens? Assume that the lens is limited only by diffraction and $\lambda = 600$ nm.

51. | A 1.0-cm-diameter microscope objective has a focal length of 2.8 mm. It is used with light of wavelength of 550 nm.
   a. What is the objective's resolving power if used in air?
   b. What is the resolving power of the objective if it is used in an oil-immersion microscope with $n_{oil} = 1.45$?

52. || INT Makers of cameras often use pixel count as an indicator of quality, and add sensors with more and more pixels. But the number of pixels is only one determinant of image quality. If the width of the central maximum of a diffraction pattern is larger than twice the spacing between the pixels, then packing the pixels more densely won't change the image quality. It's hard to get a camera with good optical quality in a smartphone; the lens must be small, so diffraction will be an issue. One smartphone manufacturer includes a camera with 20.7 MP, which sounds very good. But let's look at the numbers: The distance between the pixels is 1.2 $\mu$m; the focal length of the lens is 4.8 mm, with a 2.2-mm-diameter aperture. For 600 nm light, what is the width of the central maximum? Could a sensor with more widely spaced pixels provide similar image quality?

53. || A microscope with an objective of focal length 1.6 mm is used to inspect the tiny features of a computer chip. It is desired to resolve two objects only 400 nm apart. What diameter objective is needed if the microscope is used in air with light of wavelength 550 nm?

54. ||| Suppose you are on a planet orbiting a star 10 light years away. (One light year is the distance light travels in one year.) At this distance, what is the minimum diameter for an objective mirror that will just barely resolve the earth and the sun? Assume a wavelength of 600 nm. (Astronomers have discovered many worlds around nearby stars, but this has been done by indirect means. It's a tall order to obtain a picture of a planet around a star. The problem isn't resolution; it's the great difference in brightness between the two objects.)

## General Problems

55. |||| Jason uses a lens with a focal length of 10.0 cm as a magnifier by holding it right up to his eye. He is observing an object that is 8.0 cm from the lens. What is the angular magnification of the lens used this way if Jason's near-point distance is 25 cm?

56. | A magnifier is labeled "5×." What would its magnification be if used by a person with a near-point distance of 50 cm?

57. || A 20× microscope objective is marked NA = 0.40. What is the diameter of the objective lens? Assume a length $L = 160$ mm.

58. |||| INT Two converging lenses with focal lengths of 40 cm and 20 cm are 10 cm apart. A 2.0-cm-tall object is 15 cm in front of the 40-cm-focal-length lens.
   a. Use ray tracing to find the position and height of the image. To do this accurately use a ruler or paper with a grid. Determine the image distance and image height by making measurements on your diagram.
   b. Calculate the image height and position relative to the second lens. Compare with your ray-tracing answers in part a.

59. ||| INT A converging lens with a focal length of 40 cm and a diverging lens with a focal length of −40 cm are 160 cm apart. A 2.0-cm-tall object is 60 cm in front of the converging lens.
   a. Use ray tracing to find the position and height of the image. To do this, accurately use a ruler or paper with a grid. Determine the image distance and image height by making measurements on your diagram.
   b. Calculate the image height and image position relative to the second lens. Compare with your answers in part a.

60. || A lens with a focal length of 25 cm is placed 40 cm in front of a lens with a focal length of 5.0 cm. How far from the second lens is the final image of an object infinitely far from the first lens? Is this image in front of or behind the second lens?

61. || A microscope with a 5× objective lens images a 1.0-mm-diameter specimen. What is the diameter of the real image of this specimen formed by the objective lens?

62. || Your task in physics lab is to make a microscope from two lenses. One lens has a focal length of 10 cm, the other a focal length of 3.0 cm. You plan to use the more powerful lens as the objective, and you want its image to be 16 cm from the lens, as in a standard biological microscope.
   a. How far should the objective lens be from the object to produce a real image 16 cm from the objective?
   b. What will be the magnification of your microscope?

63. || The objective lens and the eyepiece lens of a telescope are 1.0 m apart. The telescope has an angular magnification of 50. Find the focal lengths of the eyepiece and the objective.

64. ||| Your telescope has an objective lens with a focal length of 1.0 m. You point the telescope at the moon, then realize that the eyepiece is missing. You can still see the real image of the moon formed by the objective lens if you place your eye a little past the image so as to view the rays diverging from the image plane, just as rays would diverge from an object at that location. What is the angular magnification of the moon if you view its real image from 25 cm away, your near-point distance?

65. ||| Martha is viewing a distant mountain with a telescope that has a 120-cm-focal-length objective lens and an eyepiece with a 2.0 cm focal length. She sees a bird that's 60 m distant and wants to observe it. To do so, she has to refocus the telescope. By how far and in which direction (toward or away from the objective) must she move the eyepiece in order to focus on the bird?

66. || Susan is quite nearsighted; without her glasses, her far point is 35 cm and her near point is 15 cm. Her glasses allow her to view distant objects with her eye relaxed. With her glasses on, what is the closest object on which she can focus?

67. ||| A spy satellite uses a telescope with a 2.0-m-diameter mirror. It orbits the earth at a height of 220 km. What minimum spacing must there be between two objects on the earth's surface if they are to be resolved as distinct objects by this telescope? Assume the telescope's resolution is limited only by diffraction and that it records 500 nm light.

68. ||||| Two stars have an angular separation of $3.3 \times 10^{-6}$ rad. What diameter telescope objective is necessary to just resolve these two stars, using light with a wavelength of 650 nm?

69. ||| Frank is nearsighted and his glasses require a prescription of $-1.5$ D. One day he can't find his glasses, but he does find an older pair with a prescription of $-1.0$ D. What is the most distant object that Frank can focus on while wearing this older pair of glasses?

70. ||| All eyeglass lenses have chromatic aberration. This is usu-
BIO ally expressed as *dioptric spread.* For a typical $+6.0$ D lens, the focusing power for red light is $+5.90$, the power for green light is $+6.00$ D, and the power for blue light is $+6.10$. This gives a dioptric spread of 0.20 diopter. Suppose a myopic person is observing a distant object through this lens, and the green light of the image is in perfect focus. Assume a distance $s' = 1.7000$ cm. How far in front of or behind the retina is the point of sharp focus for red light? For blue light? (In practice, this amount of dioptric spread is fine; eyeglass wearers don't notice anything less than a spread of 0.30 diopter.)

71. ||||| What is the angular resolution of the Hubble Space Tele- scope's 2.4-m-diameter mirror when viewing light with a wavelength of 550 nm? The resolution of a reflecting telescope is calculated exactly the same as for a refracting telescope.

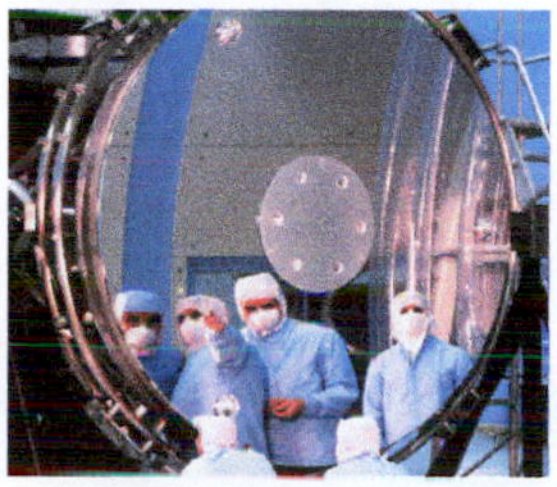

72. | The Hubble Space Telescope has a mirror diameter of 2.4 m. Suppose the telescope was used to photograph the surface of the moon from a distance of $3.8 \times 10^{8}$ m. What is the distance between two objects that the telescope can barely resolve? Assume the wavelength of light is 600 nm.

73. ||| Once dark adapted, the pupil of your eye is approximately
BIO 7 mm in diameter. The headlights of an oncoming car are 120 cm apart. If the resolution of your eye is limited only by diffraction, at what distance are the two headlights marginally resolved? Assume the light's wavelength in air is 600 nm and the index of refraction inside the eye is 1.33. (Your eye is not really good enough to resolve headlights at this distance, due to aberrations in the lens.)

74. |||| The normal human eye has maximum visual acuity
BIO with a pupil size of about 3 mm. For larger pupils, acuity decreases due to increasing aberrations; for smaller pupils, acuity decreases due to increasing effects of diffraction. If your pupil diameter is 2.0 mm, as it would be in fairly bright light, what is the smallest diameter circle that you can barely see as a circle, rather than just a dot, if the circle is at your near point, 25 cm from your eye? Assume the light's wavelength in air is 600 nm and the index of refraction inside the eye is 1.33.

75. ||| Microtubules are structures in cells that maintain cell
BIO shape and facilitate the movement of molecules within the cell. They are long, hollow cylinders with a diameter of about 25 nm. It is possible to incorporate fluorescent molecules into microtubules; when illuminated by ultraviolet light, the fluorescent molecules emit visible light that can be imaged by the optical system of a microscope. If the emitted light has a wavelength of 500 nm and the NA of the microscope objective is 1.4, can a biologist looking through the microscope tell whether she is looking at a single microtubule or at two microtubules lying side by side?

## MCAT-Style Passage Problems

### Surgical Vision Correction BIO

The optics of your visual system have a total refractive power of about $+60$ D—about $+20$ D from the lens in your eye and $+40$ D from the curved shape of your cornea. Surgical procedures to correct vision generally do not work on the lens; they work to reshape the cornea. In the most common procedure, a laser is used to remove tissue from the center of the cornea, reducing its curvature. This change in shape can correct certain kinds of vision problems.

76. | Flattening the cornea would be a good solution for someone who was
A. Nearsighted. B. Farsighted.
C. Either nearsighted or farsighted.

77. | Suppose a woman has a far point of 50 cm. How much should the refractive power of her cornea be changed to correct her vision?
A. $-2.0$ D
B. $-1.0$ D
C. $+1.0$ D
D. $+2.0$ D

78. | The length of your eye decreases slightly as you age, making the lens a bit closer to the retina. Suppose a man had his vision surgically corrected at age 30. At age 70, once his eyes had decreased slightly in length, he would be
A. Nearsighted. B. Farsighted.
C. Neither nearsighted nor farsighted.

## KNOWLEDGE STRUCTURE V

**BASIC GOALS** What are the consequences of the wave nature of light?
In the ray model, how do light rays refract and reflect to form images?

### CROSS-CUTTING CONCEPTS: THE NATURE OF LIGHT

Light is an electromagnetic wave. In vacuum, light travels at

$$c = 3.00 \times 10^8 \text{ m/s}$$

In a medium, the speed is determined by the index of refraction:

$$v = \frac{c}{n}$$

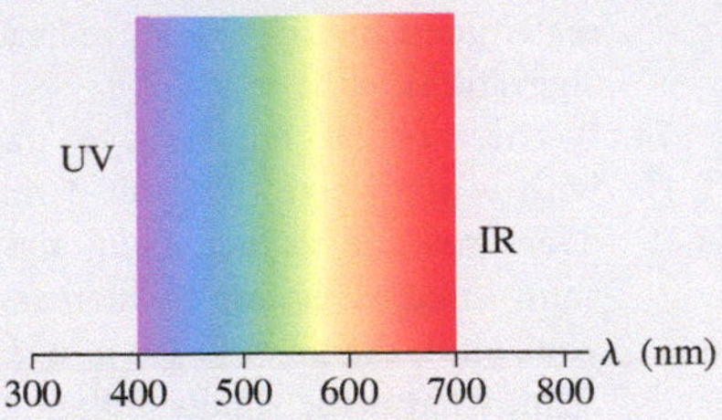


The color of light is determined by the wavelength; red light has a longer wavelength than blue.

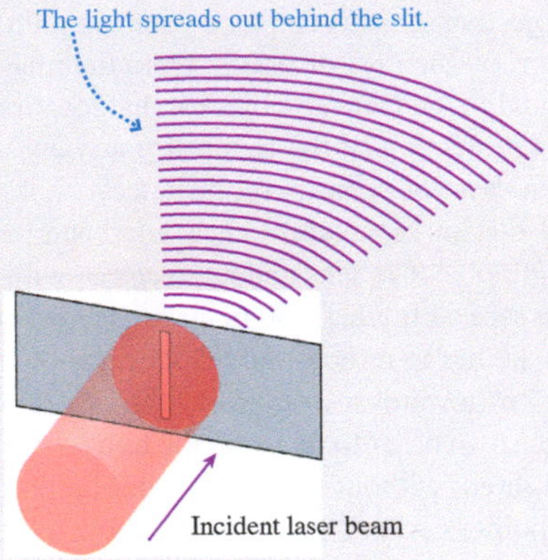

The light spreads out behind the slit.
Incident laser beam

When the scale of objects is small, the wave nature of light is important.

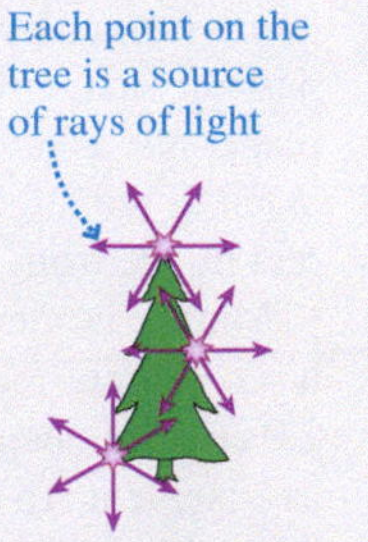


When the scale of objects is large, the wave nature of light can be ignored.

### GENERAL PRINCIPLES: MODELS OF LIGHT

Light is understood using two models, the wave model, in which light exhibits wave properties such as interference and diffraction, and the ray model, in which light travels in straight lines until it reflects or refracts.

### Wave model

- Light spreads out when it passes through a narrow opening. This is **diffraction.**

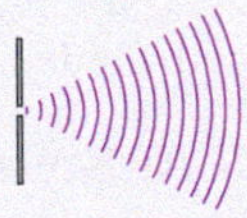

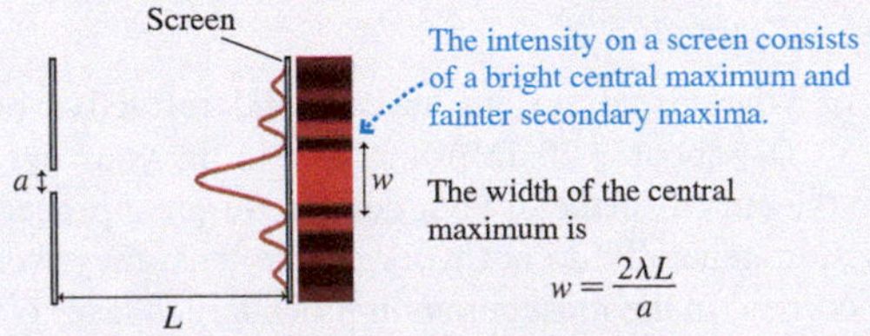


- Light waves from multiple slits in a screen **interfere** where they overlap. The light intensity is large where the interfering waves are in phase, and small where they are out of phase.

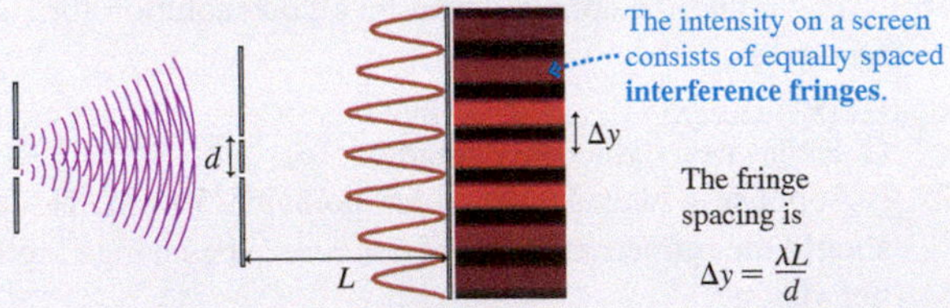


- Light waves reflected from the two surfaces of a thin transparent film also interfere.

### The resolution of optical instruments

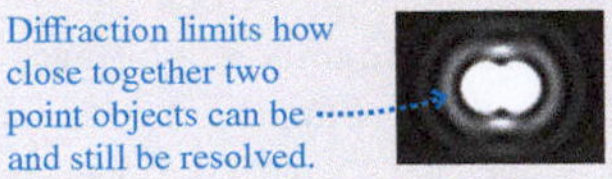

Diffraction limits how close together two point objects can be and still be resolved.

For a *telescope,* the minimum resolvable angular separation between two objects is

$$\theta_1 = \frac{1.22\lambda}{D}$$

For a *microscope,* the minimum resolvable distance between two objects is

$$d_{min} = \frac{0.61\lambda}{\text{NA}}$$

where the *numerical aperture* (NA) is a characteristic of the microscope objective.

### Ray model

- Light travels out from its source in straight lines, called **rays.**
- Rays reflect off a surface between two media, obeying the **law of reflection,** $\theta_i = \theta_r$.

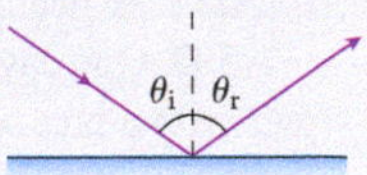


- Light rays change direction as they cross the surface between two media. The angles of incidence and refraction are related by **Snell's law:**

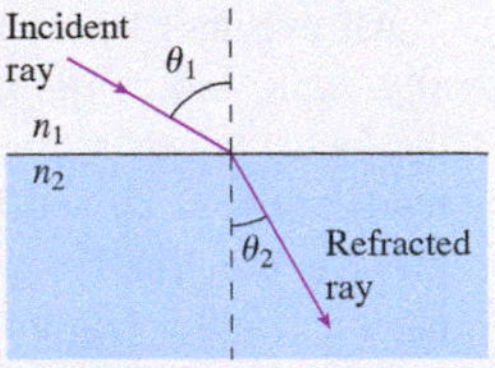


$$n_1 \sin\theta_1 = n_2 \sin\theta_2$$

where $n$ is the **index of refraction.** The speed of light in a transparent material is $v = c/n$.

### Image formation by lenses and mirrors

A lens or mirror has a characteristic **focal length** $f$. Rays parallel to the optical axis come to focus a distance $f$ from the lens or mirror.

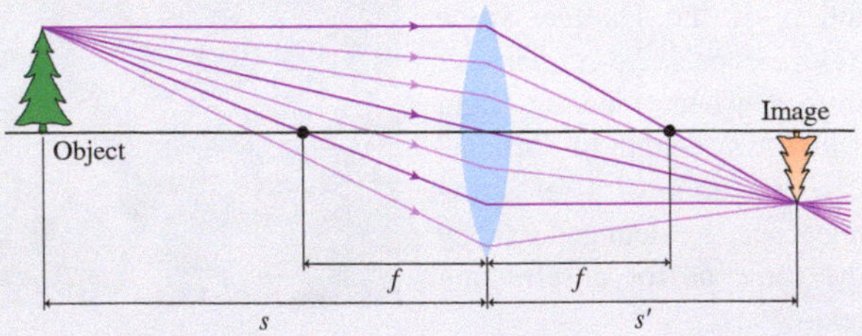


The **object distance** $s$, the **image distance** $s'$, and the focal length are related by the *thin-lens equation,* which also works for mirrors:

$$\frac{1}{s} + \frac{1}{s'} = \frac{1}{f}$$

## Scanning Confocal Microscopy

Although modern microscopes are marvels of optical engineering, their basic design is not too different from the 1665 compound microscope of Robert Hooke. Recently, advances in optics, lasers, and computer technology have made practical a new kind of optical microscope, the *scanning confocal microscope.* This microscope is capable of taking images of breathtaking clarity.

The figure shows the microscope's basic principle of operation. The left part of the figure shows how the translucent specimen is illuminated by light from a laser. The laser beam is converted to a diverging bundle of rays by suitable optics, reflected off a mirror, then directed through a microscope objective lens to a focus within the sample. The microscope objective focuses the laser beam to a very small ($\approx 0.5\ \mu m$) spot. Note that light from the laser passes through other regions of the specimen but, because the rays are not focused in those regions, they are not as intensely illuminated as is the point at the focus. This is the first important aspect of the design: Very intensely illuminate one very small volume of the sample while leaving other regions only weakly illuminated.

As shown in the right half of the figure, light is reflected from all illuminated points in the sample and passes back through the objective lens. The mirror that had reflected the laser light downward is actually a *partially transparent mirror* that reflects 50% of the light and transmits 50%. Thus half of the light reflected upward from the sample passes through the mirror and is focused on a screen containing a small hole. Because of the hole, only light rays that emanate from the brightly illuminated volume in the sample can completely pass through the hole and reach the light detector behind it. Rays from other points in the sample either miss the hole completely or are out of focus when they reach the screen, so that only a small fraction of them pass through the hole. This second key design aspect limits the detected light to only those rays that are emitted from the point in the sample at which the laser light was originally focused.

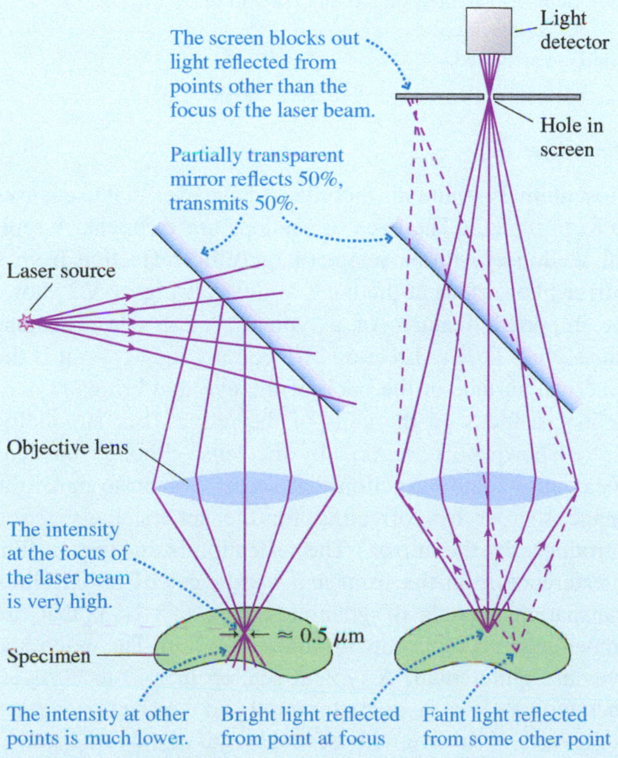


A confocal microscope.

So we see that (a) the point in the sample that is at the focus of the objective is much more intensely illuminated than any other point, so it reflects more rays than any other point, and (b) the hole serves to further limit the detected rays to only those that emanate from the focus. Taken together, these design aspects ensure the detected light comes from a very small, very well-defined volume in the sample.

The microscope as shown would only be useful for examining one small point in the sample. To make an actual *image,* the objective is *scanned* across the sample while the intensity is recorded by a computer. This procedure builds up an image of the sample one *scan line* at a time. The final result is a picture of the sample in the very narrow plane in which the laser beam is focused. Different planes within the sample can be imaged by moving the objective up or down before scanning. With this method, three dimensional images of a specimen can be produced.

The improvement in contrast and resolution over conventional microscopy can be striking. The images show a section of a mouse kidney taken using conventional and confocal microscopy. Because light reflected from all parts of the specimen reaches the camera in a conventional microscope, that image appears blurred and has low contrast. The confocal microscope image represents a single plane or slice of the sample, and many details become apparent that are invisible in the conventional image.

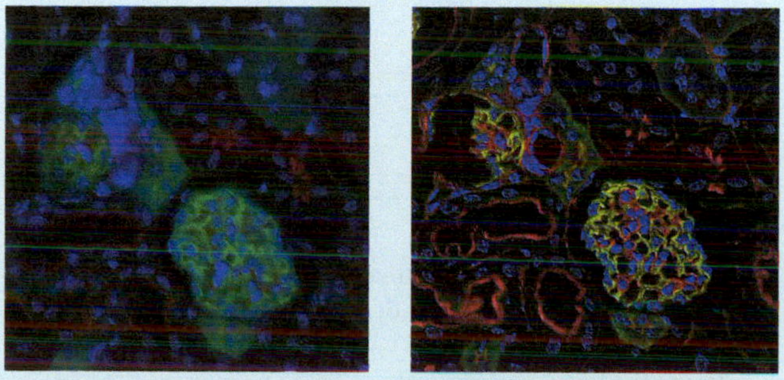

A section of fluorescently stained mouse kidney imaged using standard optical microscopy (left) and scanning confocal microscopy (right).

*The following questions are related to the passage "Scanning Confocal Microscopy" on the previous page.*

1. A laser beam consists of parallel rays of light. To convert this light to the diverging rays required for a scanning confocal microscope requires
   A. A converging lens.
   B. A diverging lens.
   C. Either a converging or a diverging lens.
2. If, because of a poor-quality objective, the light from the laser illuminating the sample in a scanning confocal microscope is focused to a larger spot,
   A. The image would be dimmer because the light illuminating the point imaged would be dimmer.
   B. The image would be blurry because light from more than one point would reach the detector.
   C. The image would be dimmer and blurry—both of the above problems would exist.
3. The resolution of a scanning confocal microscope is limited by diffraction, just as for a regular microscope. In principle, switching to a laser with a shorter wavelength would provide
   A. Greater resolution.
   B. Lesser resolution.
   C. The same resolution.
4. In the optical system shown in the passage, the distance from the source of the diverging light rays to the sample is ________ the distance from the sample to the screen.
   A. greater than
   B. the same as
   C. less than

*The following passages and associated questions are based on the material of Part V.*

## Horse Sense BIO

The ciliary muscles in a horse's eye can make only small changes to the shape of the lens, so a horse can't change the shape of the lens to focus on objects at different distances as humans do. Instead, a horse relies on the fact that its eyes aren't spherical. As Figure V.1 shows, different points at the back of the eye are at somewhat different distances from the front of the eye. We say that the eye has a "ramped retina"; images that form on the top of the retina are farther from the cornea and lens than those that form at lower positions. The horse uses this ramped retina to focus on objects at different distances, tipping its head so that light from an object forms an image at a vertical location on the retina that is at the correct distance for sharp focus.

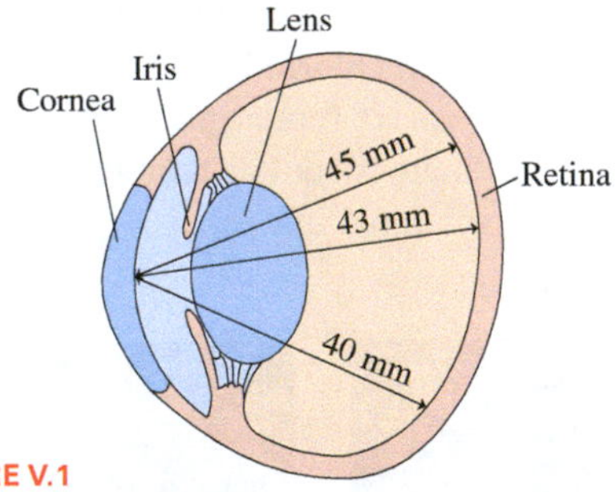


FIGURE V.1

5. In a horse's eye, the image of a close object will be in focus
   A. At the top of the retina.
   B. At the bottom of the retina.
6. In a horse's eye, the image of a distant object will be in focus
   A. At the top of the retina.
   B. At the bottom of the retina.
7. A horse is looking straight ahead at a person who is standing quite close. The image of the person spans much of the vertical extent of the retina. What can we say about the image on the retina?
   A. The person's head is in focus; the feet are out of focus.
   B. The person's feet are in focus; the head is out of focus.
   C. The person's head and feet are both in focus.
   D. The person's head and feet are both out of focus.
8. Certain medical conditions can change the shape of a horse's eyeball; these changes can affect vision. If the lens and cornea are not changed but all of the distances in Figure V.1 are increased slightly, then the horse will be
   A. Nearsighted.
   B. Farsighted.
   C. Unable to focus clearly at any distance.

## Mirror Eyes BIO

Most animals—humans included—have eyes that use lenses to form images. The eyes of scallops are different. A typical scallop eye forms images largely by reflection from a mirror-like surface at the back of the eye. Figure V.2 shows the important features of a typical scallop eye. The lens causes very little redirection of incoming light rays; it is the spherical surface in the back of the eye that brings rays of light to a focus on the cells of the retina. (For simplicity, we've shown no refraction by the lens, although the lens does cause some refraction that seems to help to make the image sharper by correcting for the spherical aberration introduced by the mirror.) The reflection is due to thin-film interference from the front and back faces of 80-nm-thick transparent crystals of guanine, index $n = 1.83$, that are embedded in cytoplasm with index $n = 1.34$. The individual eyes are quite small. A typical scallop has 40 to 60 eyes, each with a 450-$\mu$m–diameter pupil and a reflecting surface at the back of the eye with a focal length of only 200 $\mu$m.

The unusual imaging system of the scallop eye makes it very sensitive to light. The ratio of the focal length to the aperture of an optical system is known as the *f-number*. A smaller f-number implies greater light sensitivity. A telescope optimized for light gathering might have an f-number of 4; the scallop's f-number of less than 0.5 means that its small eyes work well in very dim light.

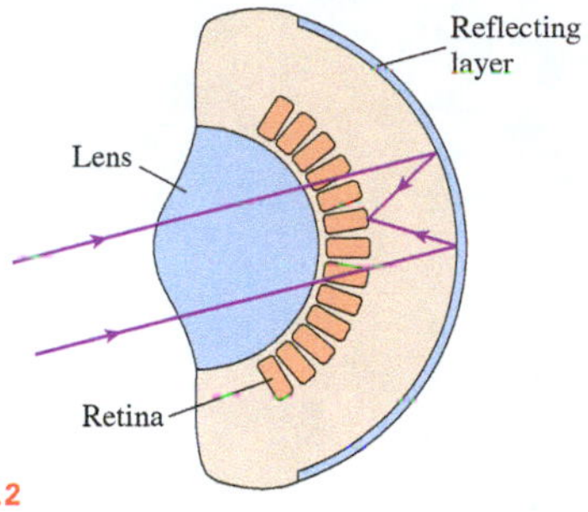


FIGURE V.2

9. There is very little refraction when the light enters the lens of the scallop eye. This means that
   A. The index of refraction of the lens is much greater than that of water.
   B. The index of refraction of the lens is approximately equal to that of water.
   C. The index of refraction of the lens is much less than that of water.
10. Consider the reflection of light from the front and back faces of the guanine crystals. What is the number of reflective phase changes?
    A. 0
    B. 1
    C. 2
11. What is the longest wavelength for which there is a strong reflection from the guanine crystals, to the nearest 100 nm?
    A. 400 nm
    B. 500 nm
    C. 600 nm
    D. 700 nm
12. The image formed on the scallop's retina is
    A. Real and inverted.
    B. Real and upright.
    C. Virtual and inverted.
    D. Virtual and upright.
13. For the typical pupil diameter of 3.0 mm, what is the approximate f-number for a human eye?
    A. 2
    B. 4
    C. 6
    D. 8

### Pupil Size BIO

The lens of your eye changes to focus on objects at different distances; your eye's pupil changes to adapt to different light levels. This causes changes in your vision as well. Suppose you are in a dark movie theater, with your pupils fully dilated to 8.0 mm in diameter. Now you emerge into the bright daylight, and your pupils quickly contract to 2.0 mm in diameter.

14. When your pupil contracts, by what factor does this change decrease the total power of the light entering your eye?
    A. 2
    B. 4
    C. 8
    D. 16
15. When your pupil contracts, how does this affect the spherical aberration of your eye's optical system?
    A. The spherical aberration is increased.
    B. The spherical aberration does not change.
    C. The spherical aberration decreases.
16. When your pupil contracts, how does this affect the angular resolution of your eye?
    A. The resolution increases.
    B. The resolution does not change.
    C. The resolution decreases.

### Additional Integrated Problems

17. BIO The pupil of your eye is smaller in bright light than in dim light. Explain how this makes images seen in bright light appear sharper than images seen in dim light.
18. BIO People with good vision can make out an 8.8-mm-tall letter on an eye chart at a distance of 6.1 m. Approximately how large is the image of the letter on the retina? Assume that the distance from the lens to the retina is 24 mm.
19. A photographer uses a lens with $f = 50$ mm to form an image of a distant object on the CCD detector in a digital camera. The image is 1.2 mm high, and the intensity of light on the detector is 2.5 W/m$^2$. She then switches to a lens with $f = 300$ mm that is the same diameter as the first lens. What are the height of the image and the intensity now?
20. Sound and other waves undergo diffraction just as light does. Suppose a loudspeaker in a 20°C room is emitting a steady tone of 1200 Hz. A 1.0-m-wide doorway in front of the speaker diffracts the sound wave. A person on the other side walks parallel to the wall in which the door is set, staying 12 m from the wall. When he is directly in front of the doorway, he can hear the sound clearly and loudly. As he continues walking, the sound intensity decreases. How far must he walk from the point where he was directly in front of the door until he reaches the first quiet spot?

PART VI

# Electricity and Magnetism

To make a quick escape, a squid ejects water from its mantle, causing it to rocket backward. The signal from the brain that causes the mantle to contract is an electrical signal. This signal is transmitted along a special nerve fiber that is capable of extremely fast signal transmission. In this section, we'll learn the basic principles of electricity that explain signal transmission in the nervous system of squid and in humans.

# OVERVIEW

## Charges, Currents, and Fields

The early Greeks discovered that a piece of amber that has been rubbed briskly can attract feathers or small pieces of straw. They also found that certain stones from the region they called *Magnesia* can pick up pieces of iron. These first experiences with the forces of electricity and magnetism began a chain of investigations that has led to today's high-speed computers, lasers, fiber-optic communications, and magnetic resonance imaging, as well as mundane modern-day miracles such as the lightbulb.

The development of a successful electromagnetic theory, which occupied the leading physicists of Europe for most of the 19th century, led to sweeping revolutions in both science and technology. The complete formulation of the theory of the electromagnetic field has been called by no less than Einstein "the most important event in physics since Newton's time."

The basic phenomena of electricity and magnetism are not as familiar to most people as are those of mechanics. We will deal with this lack of experience by placing a large emphasis on these basic phenomena. We will begin where the Greeks did, by looking at the forces between objects that have been briskly rubbed, exploring the concept of *electric charge.* We will first make systematic observations of how charges behave, and this will lead us to consider the forces between charges and how charges behave in different materials. *Electric current,* whether it is used for lighting a lightbulb or changing the state of a computer memory element, is simply a controlled motion of charges through conducting materials. One of our goals will be to understand how charges move through electric circuits.

When we turn to magnetic behavior, we will again start where the Greeks did, noting how magnets stick to some metals. Magnets also affect compass needles. And, as we will see, an electric current can affect a compass needle in exactly the same way as a magnet. This observation shows the close connection between electricity and magnetism, which leads us to the phenomenon of *electromagnetic waves.*

Our theory of electricity and magnetism will introduce the entirely new concept of a *field.* Electricity and magnetism are about the long-range interactions of charges, both static charges and moving charges, and the field concept will help us understand how these interactions take place.

### Microscopic Models

The field theory provides a macroscopic perspective on the phenomena of electricity and magnetism, but we can also take a microscopic view. At the microscopic level, we want to know what charges are, how they are related to atoms and molecules, and how they move through various kinds of materials. Electromagnetic waves are composed of electric and magnetic fields. The interaction of electromagnetic waves with matter can be analyzed in terms of the interactions of these fields with the charges in matter. When you heat food in a microwave oven, you are using the interactions of electric and magnetic fields with charges in a very fundamental way.

# 20 Electric Fields and Forces

When bees visit flowers, their foraging might seem random, but there are patterns. A bee can tell, from a distance, that a flower was visited recently by another bee, so it can skip this flower and go to another. How is the bee able to make this determination?

## LOOKING AHEAD

### Charges and Coulomb's Law

A comb rubbed through your hair attracts a thin stream of water. The **charge model** of electricity explains this force.

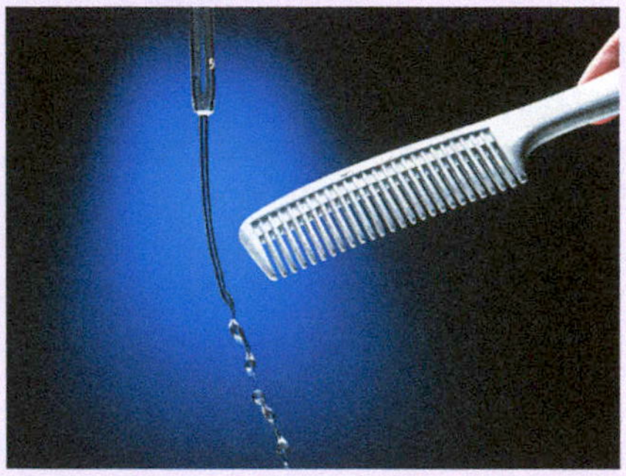

You'll learn to use **Coulomb's law** to calculate the force between two charged particles.

### The Electric Field

Charges create an **electric field** around them. In thunderclouds, the field can be strong enough to ionize air, causing lightning.

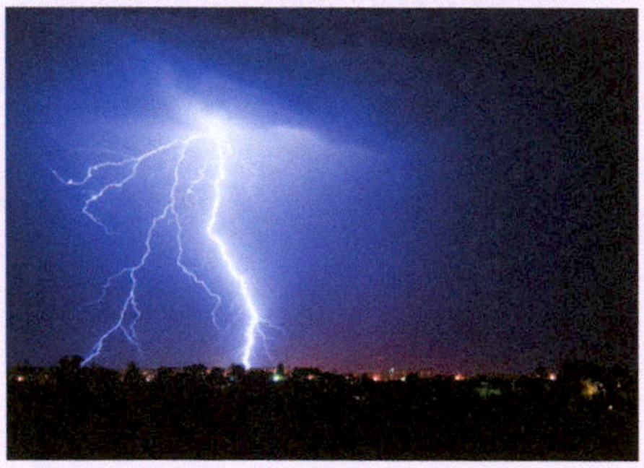

You'll learn how to calculate the electric field for several important arrangements of charges.

### Forces in Electric Fields

Charged DNA fragments in this electrophoresis gel are separated for analysis as they migrate at different rates in an applied electric field.

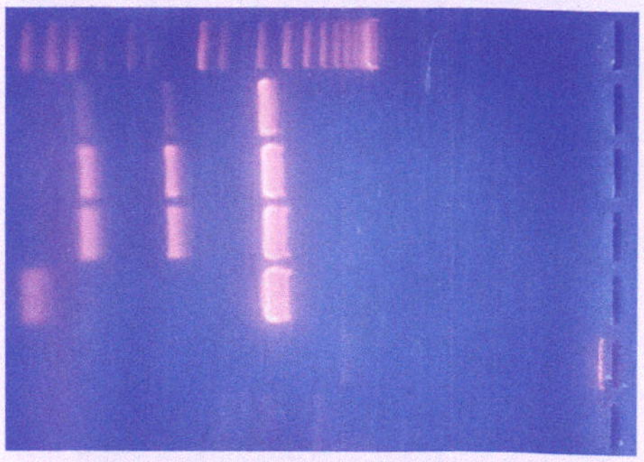

You'll learn how electric fields exert forces and torques on charged particles.

**GOAL** To develop a basic understanding of electric phenomena in terms of charges, forces, and fields.

## LOOKING BACK

### Vectors and Components

In Section 3.2 you studied how a vector could be resolved into its component vectors.

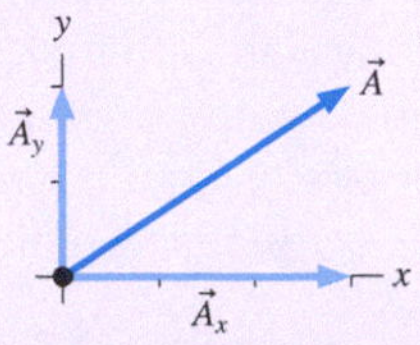


You learned that a vector $\vec{A}$ can be represented as the sum of its component vectors $\vec{A}_x$ and $\vec{A}_y$.

Electric forces and electric fields are vectors, so you will need to use vector components to solve electric force and field problems.

**STOP TO THINK**

The tension in the rope is 100 N. Given that $\sin 30° = 0.50$ and $\cos 30° = 0.87$, the $x$- and $y$-components of the tension are

A. $-87$ N, 50 N
B. 87 N, 50 N
C. $-50$ N, 87 N
D. 50 N, $-87$ N
E. 87 N, $-50$ N

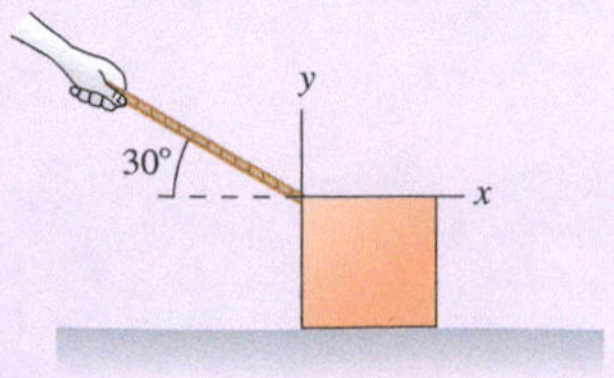

## 20.1 Charges and Forces

You can receive a mildly unpleasant shock and produce a little spark if you touch a metal doorknob after scuffing your shoes across a carpet. A plastic comb that you've run through your hair will pick up bits of paper and other small objects. In both of these cases, two objects are *rubbed* together. Why should rubbing an object cause forces and sparks? What kind of forces are these? These are the questions with which we begin our study of electricity.

Our first goal is to develop a model for understanding electric phenomena in terms of *charges* and *forces*. We will later use our contemporary knowledge of atoms to understand electricity on a microscopic level, but the basic concepts of electricity make *no* reference to atoms or electrons. The theory of electricity was well established long before the electron was discovered.

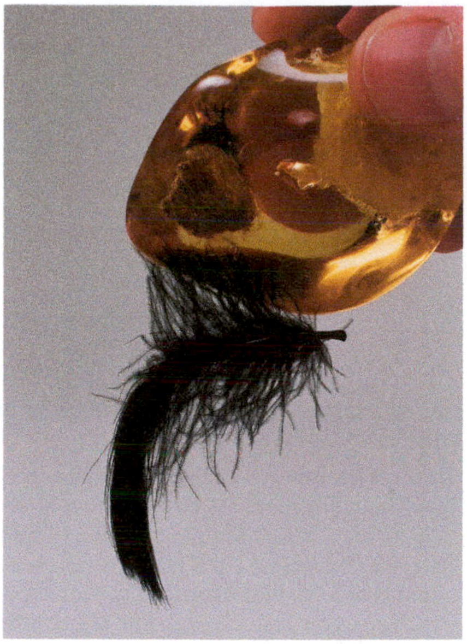

The ancient Greeks first noted the electrical nature of matter by observing amber, a form of fossilized tree resin. When rubbed with fur, amber buttons would attract bits of feather, hair, or straw. The Greek word for amber, *elektron*, is the source of our words "electric," "electricity," and—of course—"electron."

### Experimenting with Charges

Let's imagine working in a laboratory where we can make observations of electric phenomena. This is a modest lab, much like one you would have found in the year 1800. The major tools are:

- A number of plastic and glass rods, each several inches long. These can be held in your hand or suspended by threads from a support.
- Pieces of wool and silk.
- Small metal spheres, an inch or two in diameter, on wood stands.

We will manipulate and use these tools with the goal of developing a theory to explain the phenomena we see. The experiments and observations described below are very much like those of early investigators of electric phenomena.

**Discovering electricity I**

**Experiment 1**

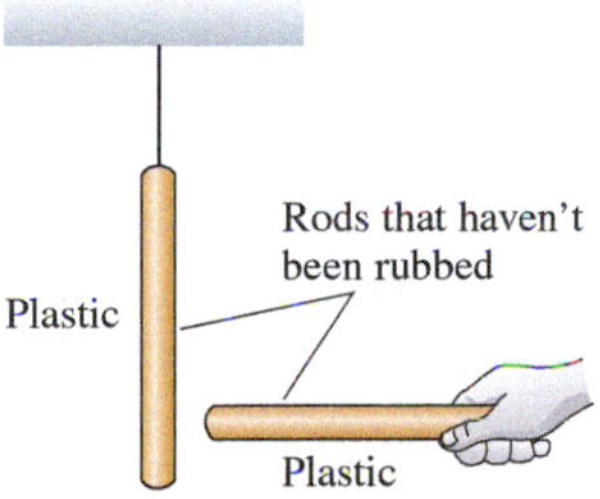


Take a plastic rod that has been undisturbed for a long period of time and hang it by a thread. Pick up another undisturbed plastic rod and bring it close to the hanging rod. Nothing happens to either rod.

**Interpretation:** There are no special electrical properties to these undisturbed rods. We say that they are **neutral.**

**Experiment 2**

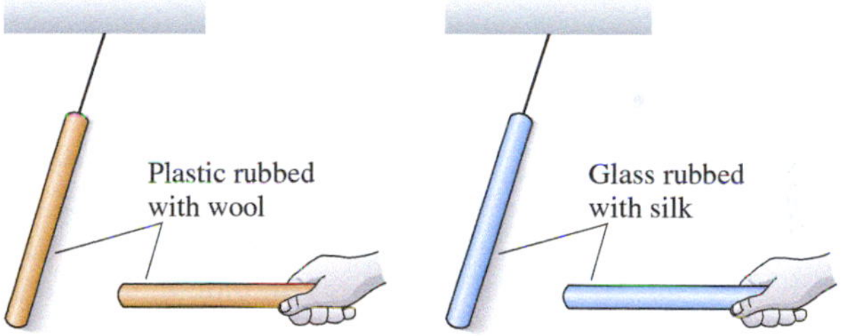


Vigorously rub both the hanging plastic rod and the handheld plastic rod with wool. Now the hanging rod *moves away* from the handheld rod when you bring the two close together. Rubbing two glass rods with silk produces the same result: The two rods repel each other.

**Interpretation:** Rubbing a rod somehow changes its properties so that forces now act between two such rods. We call this process of rubbing **charging** and say that the rubbed rod is *charged,* or that it has *acquired a* ***charge.***

Experiment 2 shows that there is a *long-range repulsive force* (i.e., a force requiring no contact) between two identical objects that have been charged in the *same* way, such as two plastic rods both rubbed with wool or two glass rods rubbed with silk. The force between charged objects is called the **electric force.** We have seen a long-range force before, gravity, but the gravitational force is always attractive. This is the first time we've observed a repulsive long-range force. However, the electric force is not always repulsive, as the next experiment shows.

### Discovering electricity II

**Experiment 3**

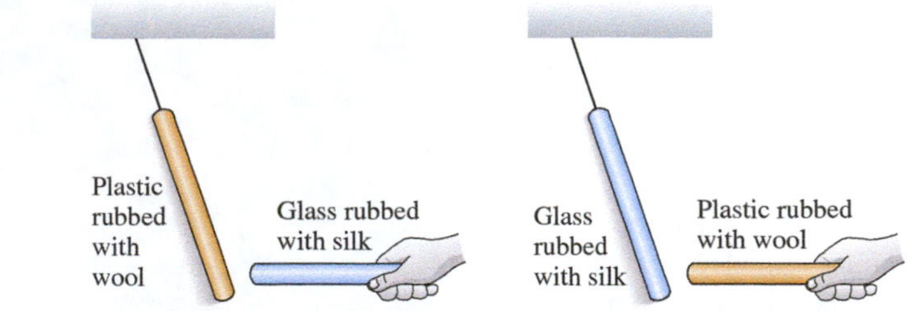


Bring a glass rod that has been rubbed with silk close to a hanging plastic rod that has been rubbed with wool. These two rods *attract* each other.

**Interpretation:** We can explain this experiment as well as Experiment 2 by assuming that there are two *different* kinds of charge that a material can acquire. We *define* the kind of charge acquired by a glass rod as *positive* charge, and that acquired by a plastic rod as *negative* charge. Then these two experiments can be summarized as **like charges** (positive/positive or negative/negative) exert repulsive forces on each other, while **opposite charges** (positive/negative) exert attractive forces on each other.

**Experiment 4**

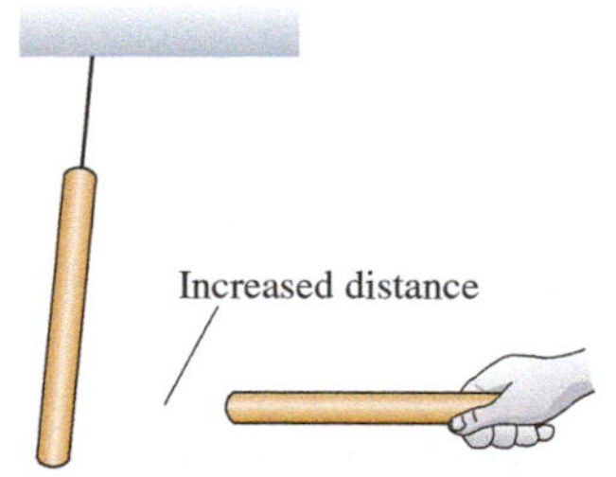


- If the two rods are held farther from each other, the force between them decreases.
- The strength of the force is greater for rods that have been rubbed more vigorously.

**Interpretation:** Like the gravitational force, the electric force decreases with the distance between the charged objects. And, the greater the charge on the two objects, the greater the force between them.

Although we showed experimental results for only plastic rods rubbed with wool and glass rods rubbed with silk, further experiments show that there are *only* two kinds of charge, positive and negative. For instance, when you rub a balloon on your hair, the balloon becomes negatively charged, while nylon rubbed with a polyester cloth becomes positively charged.

## Visualizing Charge

FIGURE 20.1 Visualizing charge.

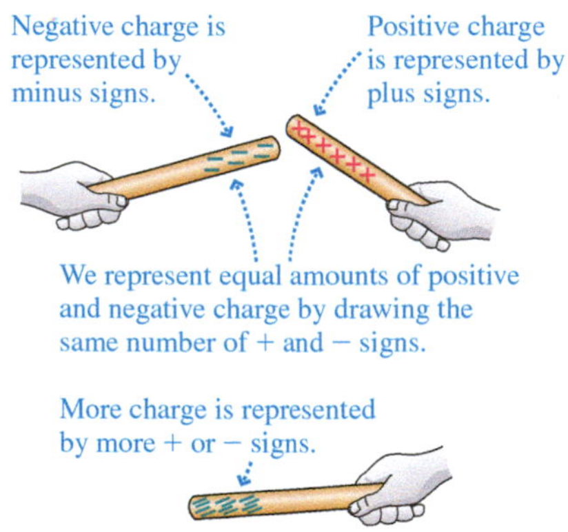

Negative charge is represented by minus signs.
Positive charge is represented by plus signs.
We represent equal amounts of positive and negative charge by drawing the same number of + and − signs.
More charge is represented by more + or − signs.

Diagrams are going to be an important tool for understanding and explaining charges and the forces between charged objects. FIGURE 20.1 shows how to draw a *charge diagram,* which gives a schematic picture of the distribution of charge on an object. It's important to realize that the + and − signs drawn in Figure 20.1 do not represent "individual" charges. At this point, we are thinking of charge only as something that can be acquired by an object by rubbing, so in charge diagrams the + and − signs represent where the charge is only in a general way. In Section 20.2 we'll look at an atomic view of charging and learn about the single microscopic charges of protons and electrons.

We can gain an important insight into the nature of charge by investigating what happens when we bring together a plastic rod and the wool used to charge it, as the following experiment shows.

### Discovering electricity III

**Experiment 5**

Start with a neutral, uncharged hanging plastic rod and a piece of wool. Rub the plastic rod with the wool, then hold the wool close to the rod. The rod is *attracted* to the wool.

**Interpretation:** From Experiment 3 we know that the plastic rod has a negative charge. Because the wool attracts the rod, the wool must have a *positive* charge.

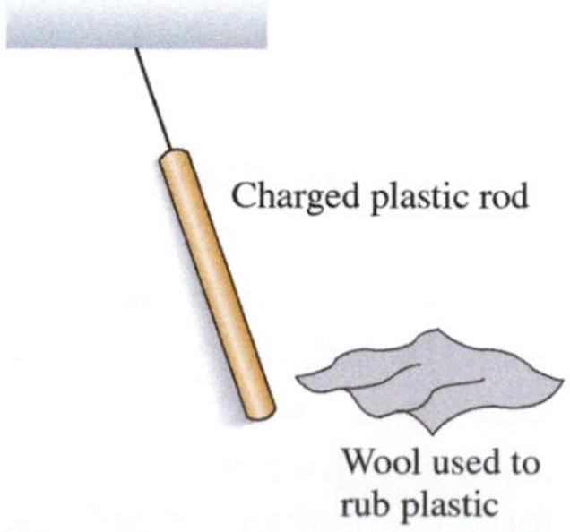

Experiment 5 shows that when a plastic rod is rubbed by wool, not only does the plastic rod acquire a negative charge, but the wool used to rub it acquires a positive charge. This observation can be explained if we postulate that a neutral object is not one that has no charge at all; rather, **a neutral object contains** *equal amounts* **of positive and negative charge.** Just as in ordinary addition, where $2 + (-2) = 0$, equal amounts of opposite charge "cancel," leaving no overall or *net* charge.

In this model, an object becomes positively charged if the amount of positive charge on it exceeds the amount of negative charge; the mathematical analogy of this is $2 + (-1) = +1$. Similarly, an object is negatively charged when the amount of negative charge on it is greater than the amount of positive charge, analogous to $2 + (-3) = -1$.

As FIGURE 20.2 shows, the rubbing process works by *transferring* charge from one object to the other. When the two are rubbed together, negative charge is transferred from the wool to the rod. This clearly leaves the rod with an excess of negative charge. But the wool, having lost some of its negative charge to the rod, now has an *excess* of positive charge, leaving it positively charged. (We'll see in Section 20.2 why it is usually negative charge that moves.)

A video to support a section's topic is embedded in the eText.
**Video** Charges and Forces: Demonstrations

FIGURE 20.2 How a plastic rod and wool acquire charge during the rubbing process.

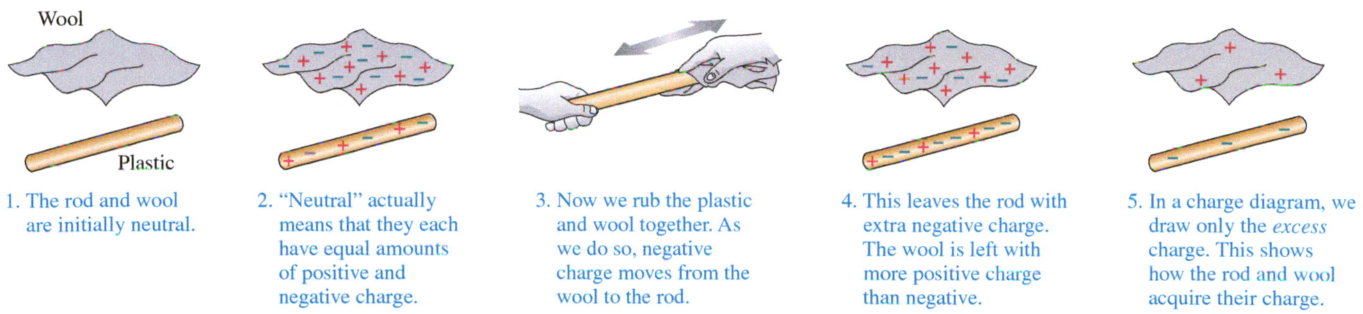

Wool
Plastic
1. The rod and wool are initially neutral.
2. "Neutral" actually means that they each have equal amounts of positive and negative charge.
3. Now we rub the plastic and wool together. As we do so, negative charge moves from the wool to the rod.
4. This leaves the rod with extra negative charge. The wool is left with more positive charge than negative.
5. In a charge diagram, we draw only the *excess* charge. This shows how the rod and wool acquire their charge.


There is another crucial fact about charge implicit in Figure 20.2: Nowhere in the rubbing process was charge either created or destroyed. Charge was merely transferred from one place to another. It turns out that this fact is a fundamental law of nature, the **law of conservation of charge.** If a certain amount of positive charge appears somewhere, an equal amount of negative charge must appear elsewhere so that the net charge doesn't change.

The results of our experiments, and our interpretation of them in terms of positive and negative charge, can be summarized in the following **charge model.**

**Charge model, part I** The basic postulates of our model are:

- Frictional forces, such as rubbing, add something called charge to an object or remove it from the object. The process itself is called *charging*. More vigorous rubbing produces a larger quantity of charge.
- There are two kinds of charge, positive and negative.
- Two objects with *like charge* (positive/positive or negative/negative) exert repulsive forces on each other. Two objects with *opposite charge* (positive/negative) exert attractive forces on each other. We call these *electric forces.*
- The force between two charged objects is a long-range force. The magnitude of the force increases as the quantity of charge increases and decreases as the distance between the charges increases.
- *Neutral* objects have an *equal mixture* of positive and negative charge.
- The rubbing process charges the objects by *transferring* charge from one to the other. The objects acquire equal but opposite charges.
- Charge is conserved: It cannot be created or destroyed.

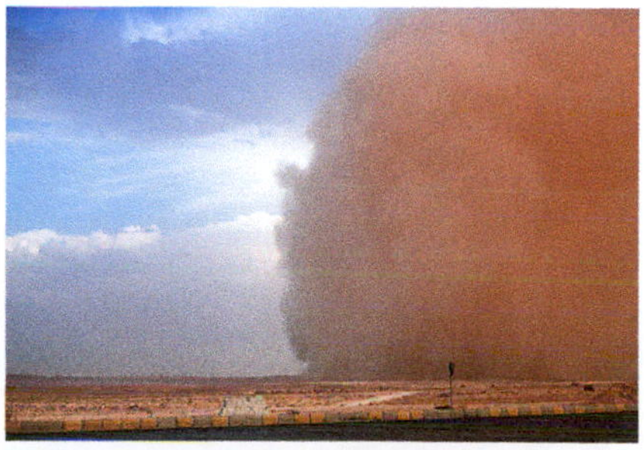

**Dust in the wind** When sand and dust grains collide as they blow across a sandy surface, the contact separates charges. The grains develop a charge; so does the surface. In some cases, a repulsive force between the charged surface and the charged particles can cause a dramatic increase in the number of particles leaving the surface. When you see images of an intense sandstorm, know that it might be more than the wind that's lifting up the grains—it could also be the electric forces between them.

## Insulators and Conductors

Experiments 2, 3, and 5 involved a transfer of charge from one object to another. Let's do some more experiments with charge to look at how charge *moves* on different materials.

**Discovering electricity IV**

**Experiment 6**

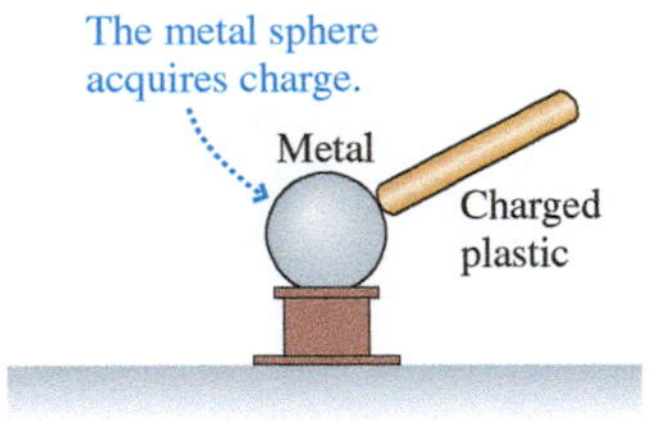


Charge a plastic rod by rubbing it with wool. Touch a neutral metal sphere with the rubbed area of the rod. The metal sphere then repels a charged, hanging plastic rod. The metal sphere appears to have acquired a charge of the same sign as the plastic rod.

**Experiment 7**

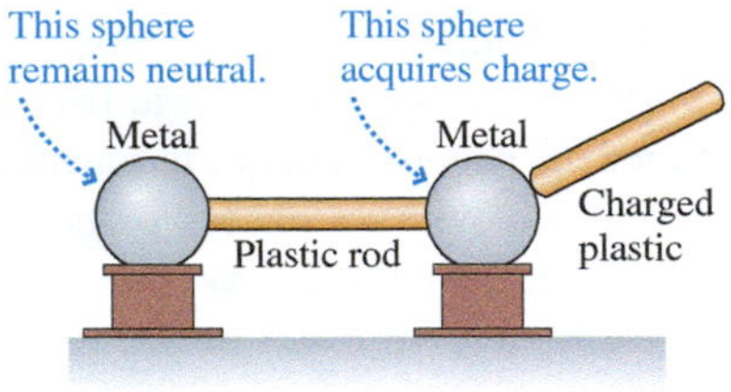


Place two metal spheres close together with a plastic rod connecting them. Charge a second plastic rod, by rubbing, and touch it to one of the metal spheres. Afterward, the metal sphere that was touched repels a charged, hanging plastic rod. The other metal sphere does not.

**Experiment 8**

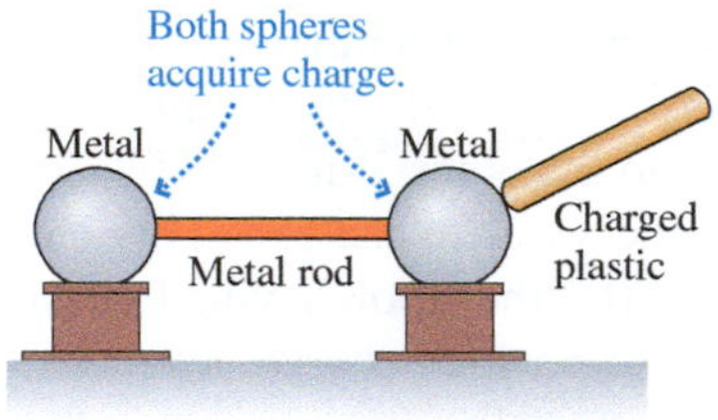


Repeat Experiment 7 with a metal rod connecting the two metal spheres. Touch one metal sphere with a charged plastic rod. Afterward, *both* metal spheres repel a charged, hanging plastic rod.

Our final set of experiments has shown that charge can be transferred from one object to another only when the objects *touch.* Contact is required. Removing charge from an object, which you can do by touching it, is called **discharging.**

In Experiments 7 and 8, charge is transferred from the charged rod to the metal sphere as the two are touched together. In Experiment 7, the other sphere remains neutral, indicating that no charge moved along the plastic rod connecting the two spheres. In Experiment 8, by contrast, the other sphere is found to be charged; evidently charge has moved along the metal rod that connects the spheres, transferring some charge from the first sphere to the second. We define **conductors** as those materials through or along which charge easily moves, and **insulators** as those materials on or in which charges remain immobile. Glass and plastic are insulators; metal is a conductor.

This new information allows us to add more postulates to our charge model:

### Charge model, part II

- There are two types of materials. Conductors are materials through or along which charge easily moves. Insulators are materials on or in which charges remain fixed in place.
- Charge can be transferred from one object to another by contact.

**NOTE** ► Both insulators and conductors can be charged. They differ in the ability of charge to *move.* ◄

◄ A dry day, a plastic slide, and a child with clothes of the right fabric lead to a lovely demonstration of electric charges and forces. The rubbing of the child's clothes on the slide has made her build up charge. The body is a good conductor, so the charges spread across her body and her hair. The resulting repulsion produces a dramatic result!

**CONCEPTUAL EXAMPLE 20.1** **Transferring charge**

In Experiment 8, touching a metal sphere with a charged plastic rod caused a second metal sphere, connected by a metal rod to the first, to become charged with the same type of charge as the rod. Use the postulates of the charge model to construct a charge diagram for the process.

**REASON** We need the following ideas from the charge model:

- **Charge is transferred upon contact.** The plastic rod was charged by rubbing with wool, giving it a negative charge. The charge doesn't move around on the rod, an insulator, but some of the charge is transferred to the metal upon contact.
- **Metal is a conductor.** Once in the metal, which is a conductor, the charges are free to move around.
- **Like charges repel.** Because like charges repel, these negative charges quickly move as far apart as they possibly can. Some move through the connecting metal rod to the second sphere. Consequently, the second sphere acquires a net negative charge. The repulsive forces drive the negative charges as far apart as they can possibly get, causing them to end up on the *surfaces* of the conductors.

The charge diagram in FIGURE 20.3 illustrates these three steps.

FIGURE 20.3 A charge diagram for Experiment 8.

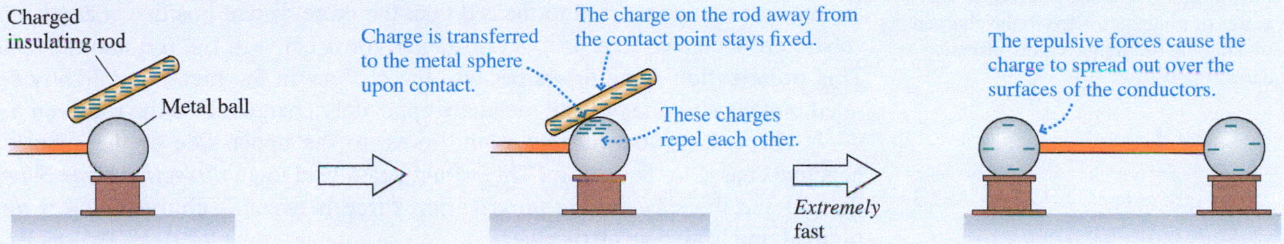


In Conceptual Example 20.1, once the charge is placed on the conductor it rapidly distributes itself over the conductor's surface. This movement of charge is *extremely* fast. Other than this very brief interval during which the charges are adjusting, the charges on an isolated conductor are in static equilibrium with the charges at rest. This condition is called **electrostatic equilibrium.**

**CONCEPTUAL EXAMPLE 20.2** **Drawing a charge diagram for an electroscope**

Many electricity demonstrations are carried out with the help of an *electroscope* like the one shown in FIGURE 20.4. Touching the sphere at the top of an electroscope with a charged plastic rod causes the leaves to fly apart and remain hanging at an angle. Use charge diagrams to explain why.

**REASON** We will use the charge model and our understanding of insulators and conductors to make a series of charge diagrams in FIGURE 20.5 that show the charging of the electroscope.

**ASSESS** The charges move around, but, because charge is conserved, the total number of negative charges doesn't change from picture to picture.

FIGURE 20.4 A charged electroscope.

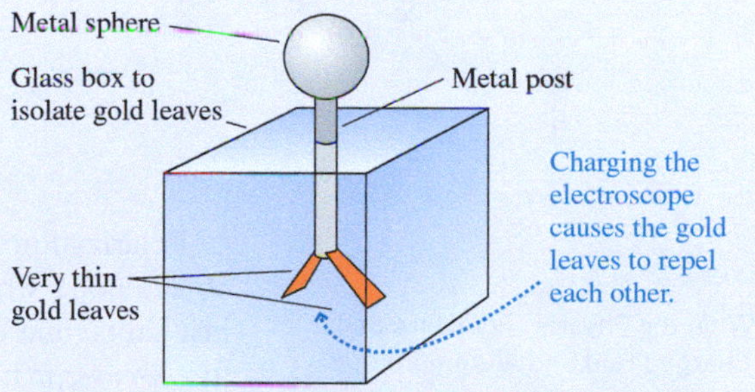


FIGURE 20.5 Charging an electroscope.

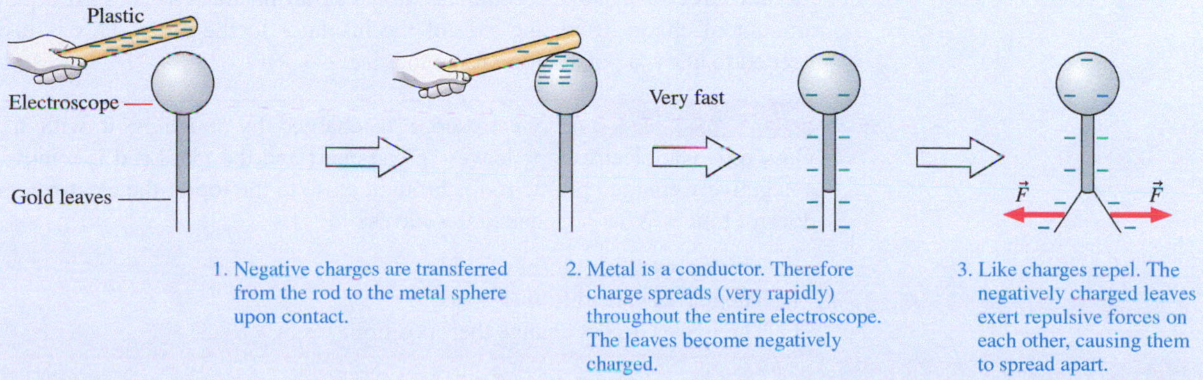

Plastic
Electroscope
Gold leaves
Very fast
$\vec{F}$
$\vec{F}$
1. Negative charges are transferred from the rod to the metal sphere upon contact.
2. Metal is a conductor. Therefore charge spreads (very rapidly) throughout the entire electroscope. The leaves become negatively charged.
3. Like charges repel. The negatively charged leaves exert repulsive forces on each other, causing them to spread apart.

**Picking up pollen** BIO Rubbing a rod with a cloth gives the rod an electric charge. In a similar fashion, the rapid motion of a bee's wings through the air gives the bee a small positive electric charge. As small pieces of paper are attracted to a charged rod, so are tiny grains of pollen attracted to the charged bee, making the bee a much more effective pollinator.

## Polarization

At the beginning of this chapter we showed a picture of a small feather being picked up by a piece of amber that had been rubbed with fur. The amber was charged by rubbing, but the feather had not been rubbed—it was *neutral.* How can our charge model explain the attraction of a neutral object toward a charged one?

Although a feather is an insulator, it's easiest to understand this phenomenon by first considering how a neutral *conductor* is attracted to a charged object. FIGURE 20.6 shows how this works. Because the charged rod doesn't touch the sphere, no charge is added to or removed from the sphere. Instead, the rod attracts some of the sphere's negative charge to the side of the sphere near the rod. This leaves a deficit of negative charge on the opposite side of the sphere, so that side is now positively charged. This slight *separation* of the positive and negative charge in a neutral object when a charged object is brought near is called **charge polarization.**

Figure 20.6 also shows that, because the negative charges at the top of the sphere are more strongly attracted to the rod than the more distant positive charges on the sphere are repelled, there is a net *attractive* force between the rod and the sphere. This **polarization force** arises because the charges in the metal are slightly separated, *not* because the rod and metal are oppositely charged. Had the rod been *negatively* charged, positive charge would move to the upper side of the sphere and negative charge to the bottom. This would again lead to an *attractive* force between the rod and the sphere. **The polarization force between a charged object and a neutral one is always attractive.**

FIGURE 20.6 Why a neutral metal object is attracted to a charged object.

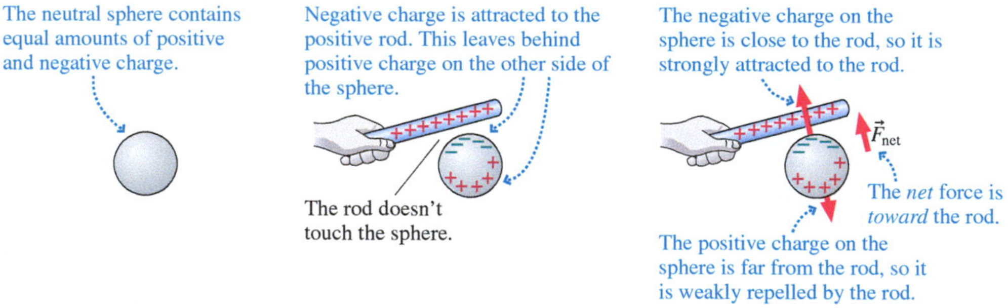


**Video** What the Physics? Franklin's Bells
**Video** Charged Rod and Aluminum Can

Polarization explains why forces arise between a charged object and a metal object along which charge can freely move. But the feathers attracted to amber are insulators, and charge can't move through an insulator. Nevertheless, the attractive force between a charged object and an insulator is also a polarization force. As we'll learn in the next section, the charge in each *atom* that makes up an insulator can be slightly polarized. Although the charge separation in one atom is exceedingly small, the net effect over all the countless atoms in an insulator is to shift a perceptible amount of charge from one side of the insulator to the other. This is just what's needed to allow a polarization force to arise.

**STOP TO THINK 20.1** An electroscope is charged by touching it with a positive glass rod. The electroscope leaves spread apart and the glass rod is removed. Then a negatively charged plastic rod is brought close to the top of the electroscope, but it doesn't touch. What happens to the leaves?

A. The leaves move closer together.
B. The leaves spread farther apart.
C. The leaves do not change their position.

# 20.2 Charges, Atoms, and Molecules

We have been speaking about giving objects positive or negative charge without explaining what is happening at an atomic level. You already know that the basic constituents of atoms—the nucleus and the electrons surrounding it—are charged. In this section we will connect our observations of the previous section with our understanding of the atomic nature of matter.

Our current model of the atom is that it is made up of a very small and dense positively charged *nucleus* that contains positively charged *protons* as well as neutral particles called *neutrons*. The nucleus is surrounded by much-less-massive negatively charged *electrons* that form an **electron cloud,** as illustrated in FIGURE 20.7. The atom is held together by the attractive electric force between the positive nucleus and the negative electrons.

Experiments show that **charge, like mass, is an inherent property of electrons and protons.** It's no more possible to have an electron without charge than it is to have an electron without mass.

FIGURE 20.7 Our modern view of the atom.

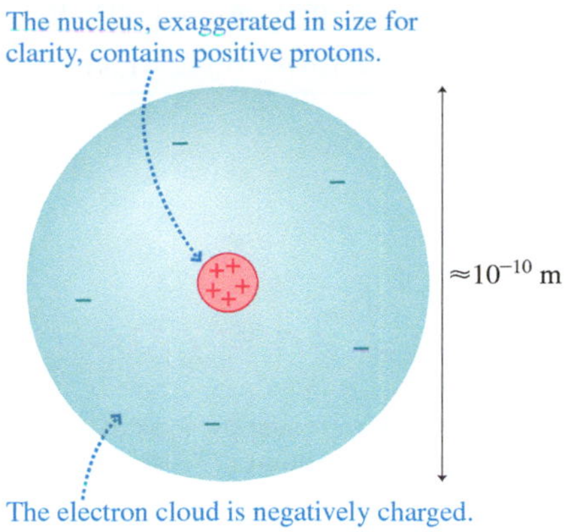


## An Atomic View of Charging

Electrons and protons are the basic charges in ordinary matter. **There are no other sources of charge.** Consequently, the observations we made in Section 20.1 need to be explained in terms of electrons and protons.

Experimentally, electrons and protons are found to have charges of opposite sign but *exactly* equal magnitude. Thus, because charge is due to electrons and protons, **an object is charged if it has an unequal number of electrons and protons.** An object with a negative charge has more electrons than protons; an object with a positive charge has more protons than electrons. Most macroscopic objects have an *equal number* of protons and electrons. Such an object has no *net* charge; we say it is *electrically neutral.*

In practice, objects acquire a positive charge not by gaining protons but by losing electrons. Protons are *extremely* tightly bound within the nucleus and cannot be added to or removed from atoms. Electrons, however, are bound much more loosely than the protons and can be removed with little effort.

The process of removing an electron from the electron cloud of an atom is called **ionization.** An atom that is missing an electron is called a *positive ion.* Some atoms can accommodate an *extra* electron and thus become a *negative ion.* FIGURE 20.8 shows positive and negative ions.

FIGURE 20.8 Positive and negative ions.

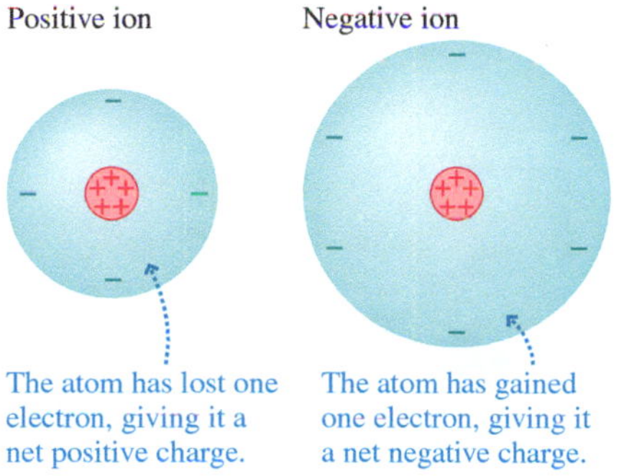


The charging processes we observed in Section 20.1 involved rubbing and friction. The forces of friction often cause molecular bonds at the surface to break as two materials slide past each other. Molecules are electrically neutral, but FIGURE 20.9 shows that *molecular ions* can be created when one of the bonds in a large molecule is broken. If the positive molecular ions remain on one material and the negative ions on the other, one of the objects being rubbed ends up with a net positive charge and the other with a net negative charge. This is how a plastic rod is charged by rubbing with wool or a comb is charged by passing through your hair.

FIGURE 20.9 Charging by friction may result from molecular ions produced as bonds are broken.

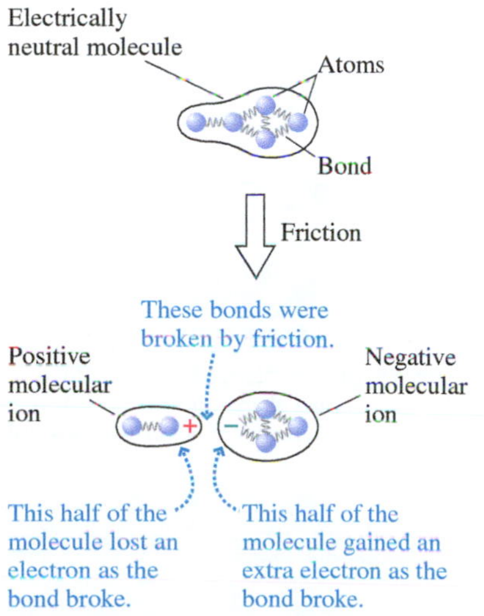


## Charge Conservation

Charge is represented by the symbol $q$ (or sometimes $Q$). The SI unit of charge is the **coulomb** (C), named for French scientist Charles Coulomb, one of many scientists investigating electricity in the late 18th century.

Protons and electrons have the same amount of charge, but of opposite signs. We use the symbol $e$ for the **fundamental charge,** the magnitude of the charge of an electron or a proton. The fundamental charge $e$ has been measured to have the value

$$e = 1.60 \times 10^{-19}\ \text{C}$$

TABLE 20.1 Protons and electrons

| Particle | Mass (kg) | Charge (C) |
|---|---|---|
| Proton | $1.67 \times 10^{-27}$ | $+e = 1.60 \times 10^{-19}$ |
| Electron | $9.11 \times 10^{-31}$ | $-e = -1.60 \times 10^{-19}$ |

TABLE 20.1 lists the masses and charges of protons and electrons.

NOTE ▶ The amount of charge produced by rubbing plastic or glass rods is typically in the range 1 nC ($10^{-9}$ C) to 100 nC ($10^{-7}$ C). This corresponds to an excess or deficit of $10^{10}$ to $10^{12}$ electrons. But, because of the enormous number of atoms in a macroscopic object, this represents an excess or deficit of only perhaps 1 electron in $10^{13}$. ◀

That charge is associated with electrons and protons explains why charge is conserved. Because electrons and protons are neither created nor destroyed in ordinary processes, their associated charge is conserved as well.

## Insulators and Conductors

FIGURE 20.10 A microscopic look at an insulator and a conductor.

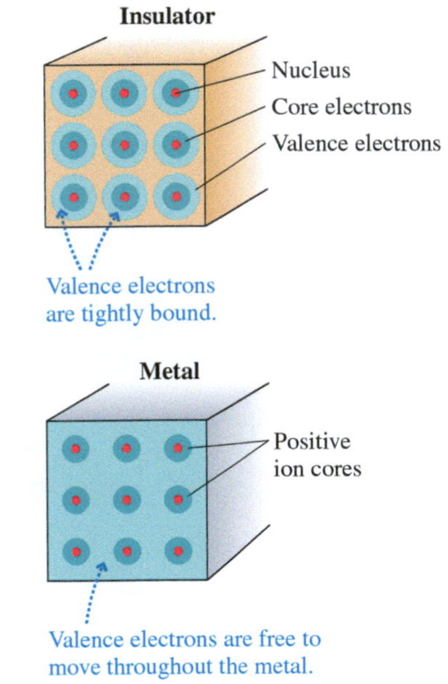


FIGURE 20.10 looks inside an insulator and a metallic conductor. The electrons in the insulator are all tightly bound to the positive nuclei and so are not free to move around. Charging an insulator by friction leaves patches of molecular ions on the surface, but these patches are immobile.

In metals, the outer atomic electrons (called the *valence electrons* in chemistry) are only weakly bound to the nuclei. As the atoms come together to form a solid, these outer electrons become detached from their parent nuclei and are free to wander about through the entire solid. The solid *as a whole* remains electrically neutral, because we have not added or removed any electrons, but the electrons are now rather like a negatively charged gas or liquid—what physicists like to call a **sea of electrons**—permeating an array of positively charged **ion cores.** However, although the electrons are highly mobile *within* the metal, they are still weakly bound to the ion cores and will not leave the metal.

## Electric Dipoles

FIGURE 20.11 An induced electric dipole.

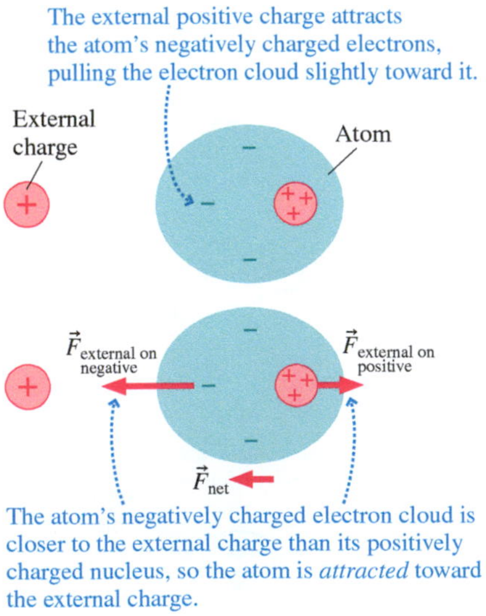


In the last section we noted that an insulator, such as a feather, becomes polarized when brought near a charged object. We can use an atomic description of matter to see why.

Consider what happens if we bring a positive charge near a neutral atom. As FIGURE 20.11 shows, the charge polarizes the atom by attracting the electron cloud while repelling the nucleus. The polarization of just one atom is a very small effect, but there are an enormous number of atoms in an insulator. Added together, their net polarization—and the resulting polarization force—can be quite significant. This is how the rubbed amber picks up a feather, exerting an upward polarization force on it larger than the downward force of gravity.

Two equal but opposite charges with a separation between them are called an **electric dipole.** In this case, where the polarization is caused by the external charge, the atom has become an *induced electric dipole.* Because the negative end of the dipole is slightly closer to the positive charge, the attractive force on the negative end slightly exceeds the repulsive force on the positive end, and there is a net force toward the external charge. If the charge on a piece of amber causes the atoms in a feather to become induced electric dipoles, the net attractive force can be enough to lift the feather.

## Hydrogen Bonding

Some molecules have an asymmetry in their charge distribution that makes them *permanent electric dipoles.* An important example is the water molecule. Bonding between the hydrogen and oxygen atoms results in an unequal sharing of charge that, as shown in FIGURE 20.12, leaves the hydrogen atoms with a small positive charge and the oxygen atom with a small negative charge.

When two water molecules are close, the attractive electric force between the positive hydrogen atom of one molecule and the negative oxygen atom of the second molecule can form a weak bond, called a **hydrogen bond,** as illustrated in FIGURE 20.13. These weak bonds result in a certain "stickiness" between water molecules that is

**FIGURE 20.12** A water molecule is a permanent electric dipole.

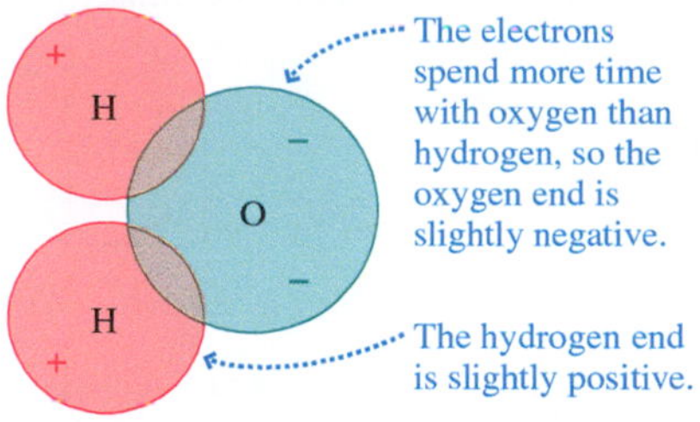

H
O
H
+
+
–
–
The electrons spend more time with oxygen than hydrogen, so the oxygen end is slightly negative.
The hydrogen end is slightly positive.

**FIGURE 20.13** Hydrogen bonds between water molecules.

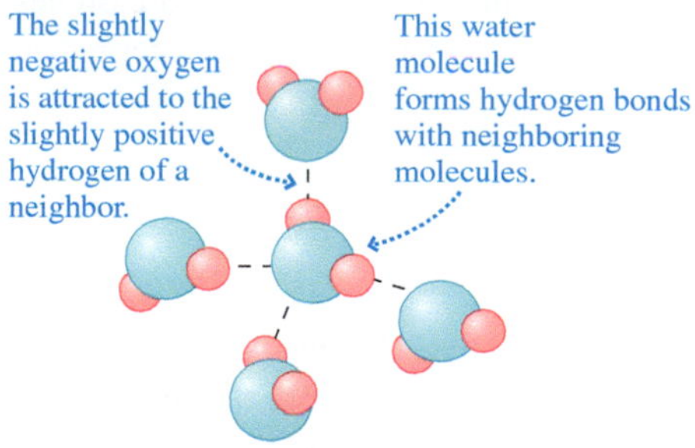

The slightly negative oxygen is attracted to the slightly positive hydrogen of a neighbor.
This water molecule forms hydrogen bonds with neighboring molecules.

responsible for many of water's special properties, including its expansion on freezing, the wide range of temperatures over which it is liquid, and its high heat of vaporization.

Hydrogen bonds are extremely important in biological systems. As you know, the DNA molecule has the structure of a double helix. Information in DNA is coded in the *nucleotides,* the four molecules guanine, thymine, adenine, and cytosine. The nucleotides on one strand of the DNA helix form hydrogen bonds with the nucleotides on the opposite strand.

The nucleotides bond only in certain pairs: Cytosine always forms a bond with guanine, adenine with thymine. This preferential bonding is crucial to DNA replication. When the two strands of DNA are taken apart, each separate strand of the DNA forms a template on which another complementary strand can form, creating two identical copies of the original DNA molecule.

The preferential bonding of nucleotide base pairs in DNA is explained by hydrogen bonding. In each of the nucleotides, the hydrogen atoms have a small positive charge, oxygen and nitrogen a small negative charge. The positive hydrogen atoms on one nucleotide attract the negative oxygen or nitrogen atoms on another. As the detail in **FIGURE 20.14** shows, the geometry of the nucleotides allows cytosine to form a hydrogen bond only with guanine, adenine only with thymine.

**FIGURE 20.14** BIO Hydrogen bonds in DNA base pairs.

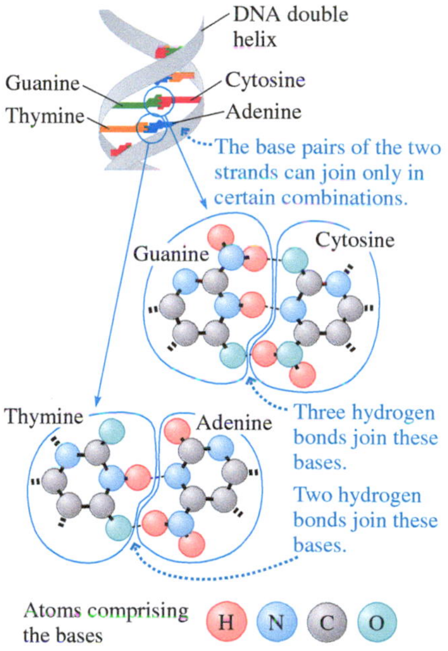


**STOP TO THINK 20.2** Rank in order, from most positive to most negative, the charges $q_A$ to $q_E$ of these five systems.

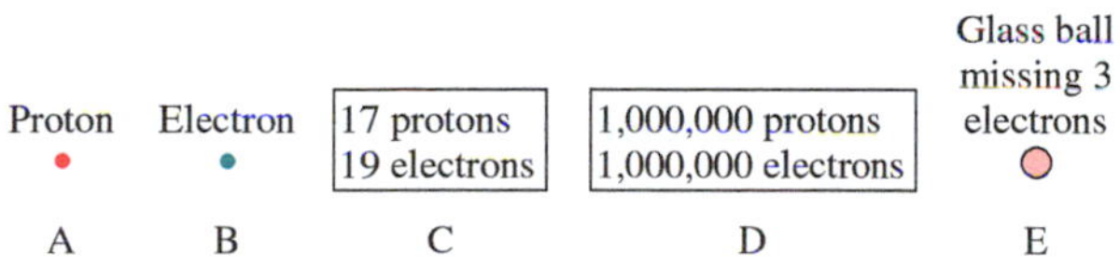


## 20.3 Coulomb's Law

The last two sections established a *model* of charges and electric forces. This model is very good at explaining electric phenomena and providing a general understanding of electricity. Now we need to become quantitative. Experiment 4 in Section 20.1 found that the electric force increases for objects with more charge and decreases as charged objects are moved farther apart. The force law that describes this behavior is known as *Coulomb's law.*

In the mathematical formulation of Coulomb's law, we will use the magnitude of the charge only, not the sign. We show this by using the absolute value notation we used earlier in the text. $|q|$ therefore represents the magnitude of the charge. It is always a positive number, whether the charge is positive or negative.

**Video** Charges and Forces: Warm-Ups

### Coulomb's law

**Magnitude:** If two charged particles that have charges $q_1$ and $q_2$ are a distance $r$ apart, the particles exert forces on each other of magnitude

$$F_{1\text{ on }2} = F_{2\text{ on }1} = \frac{K|q_1||q_2|}{r^2} \tag{20.1}$$

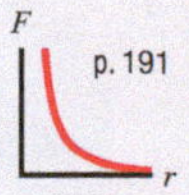

p. 191
INVERSE-SQUARE

where the charges are in coulombs (C), and $K = 8.99 \times 10^9\ \text{N} \cdot \text{m}^2/\text{C}^2$ is called the **electrostatic constant.** These forces are an action/reaction pair, equal in magnitude and opposite in direction. It is customary to round $K$ to $9.0 \times 10^9\ \text{N} \cdot \text{m}^2/\text{C}^2$ for all but extremely precise calculations, and we will do so.

**Direction:** The forces are directed along the line joining the two particles. The forces are *repulsive* for two like charges and *attractive* for two opposite charges.

FIGURE 20.15 Attractive and repulsive forces between charged particles.

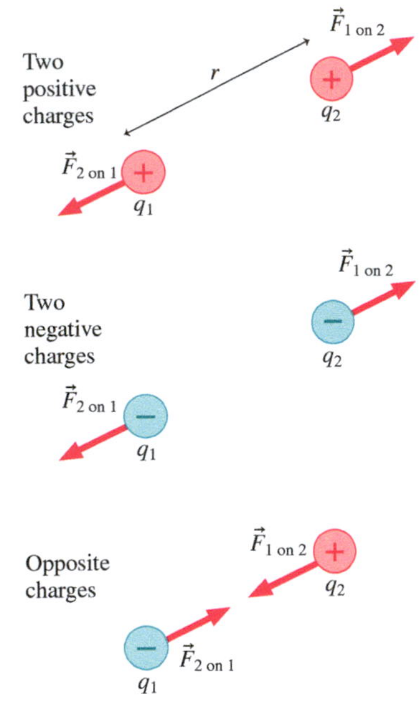


We sometimes speak of the "force between charge $q_1$ and charge $q_2$," but keep in mind that we are really dealing with charged *objects* that also have a mass, a size, and other properties. Charge is not some disembodied entity that exists apart from matter. Coulomb's law describes the force between charged *particles.*

**NOTE** ▸ Coulomb's law applies only to *point charges.* A point charge is an idealized material object with charge and mass but with no size or extension. For practical purposes, two charged objects can be modeled as point charges if they are much smaller than the separation between them. ◂

Coulomb's law looks much like Newton's law of gravity, which you studied in ◂◂ SECTION 6.5, but there is a key difference: The charge $q$ can be either positive or negative, so the forces can be attractive or repulsive. Consequently, the absolute value signs in Equation 20.1 are especially important. The first part of Coulomb's law gives only the *magnitude* of the force, which is always positive. The direction must be determined from the second part of the law. FIGURE 20.15 shows the forces between different combinations of positive and negative charges.

## Using Coulomb's Law

Coulomb's law is a force law, and forces are vectors. **Electric forces, like other forces, can be superimposed.** If multiple charges 1, 2, 3, ... are present, the *net* electric force on charge $j$ due to all other charges is therefore the sum of all the individual forces due to each charge; that is,

$$\vec{F}_{\text{net}} = \vec{F}_{1\text{ on }j} + \vec{F}_{2\text{ on }j} + \vec{F}_{3\text{ on }j} + \cdots \tag{20.2}$$

where each of the forces $\vec{F}_{i\text{ on }j}$ is given by Equation 20.1.

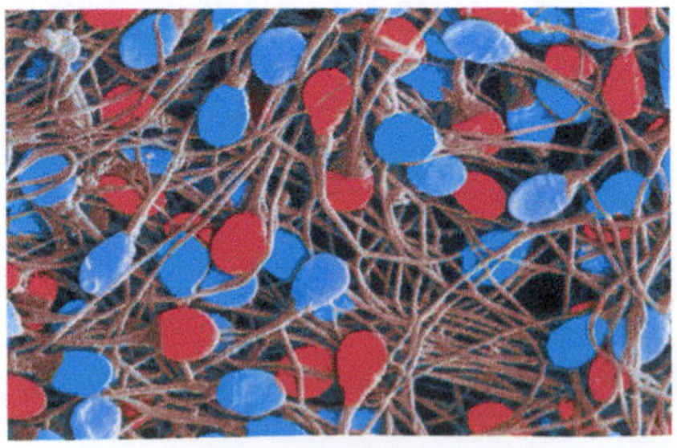

◂ **Separating the girls from the boys** BIO Sperm cells can be sorted according to whether they contain an X or a Y chromosome. The cells are put into solution, and the solution is forced through a nozzle, which breaks the solution into droplets. Suppose a droplet contains a sperm cell. An optical test measures which type of chromosome, X or Y, the cell has. The droplet is then given a positive charge if it has an X sperm cell, negative if it contains a Y. The droplets fall between two oppositely charged plates where they are pushed left or right depending on their charge—separating the X from the Y.

## PROBLEM-SOLVING APPROACH 20.1 Electric forces and Coulomb's law

We can use Coulomb's law to find the electric force on a charged particle due to one or more other charged particles.

**STRATEGIZE** Identify point charges or objects that can be modeled as point charges. Think about the forces involved, their likely directions and magnitudes. Think about what you are trying to find and the likely details of the result.

**PREPARE** Create a visual overview in which you establish a coordinate system, show the positions of the charges, show the force vectors on the charges, and define distances and angles.

**SOLVE** The magnitude of the force between point charges is given by Coulomb's law:

$$F_{1\text{ on }2} = F_{2\text{ on }1} = \frac{K|q_1||q_2|}{r^2}$$

Use your visual overview as a guide to the use of this law:

- Show the directions of the forces—repulsive for like charges, attractive for opposite charges—on the visual overview.
- Draw the lengths of your force vectors according to Coulomb's law: A particle with a greater charge, or one that is closer, will lead to a longer force vector.
- When possible, do graphical vector addition on the visual overview. While not exact, it tells you the type of answer you should expect.
- Write each force vector in terms of its $x$- and $y$-components, then add the components to find the net force. Use the visual overview to determine which components are positive and which are negative.

**ASSESS** Check that your result has the correct units, is reasonable, and answers the question.

Exercise 15

## EXAMPLE 20.3 Adding electric forces in one dimension

Two +10 nC charged particles are 2.0 cm apart on the $x$-axis. What is the net force on a +1.0 nC charge midway between them? What is the net force if the charged particle on the right is replaced by a −10 nC charge?

**STRATEGIZE** We will proceed using the steps of Problem-Solving Approach 20.1. We will model the charged particles as point charges.

**PREPARE** The visual overview of FIGURE 20.16 establishes a coordinate system and shows the forces $\vec{F}_{1\text{ on }3}$ and $\vec{F}_{2\text{ on }3}$. Figure 20.16a shows a +10 nC charge on the right; Figure 20.16b shows a −10 nC charge.

FIGURE 20.16 A visual overview of the forces for the two cases.

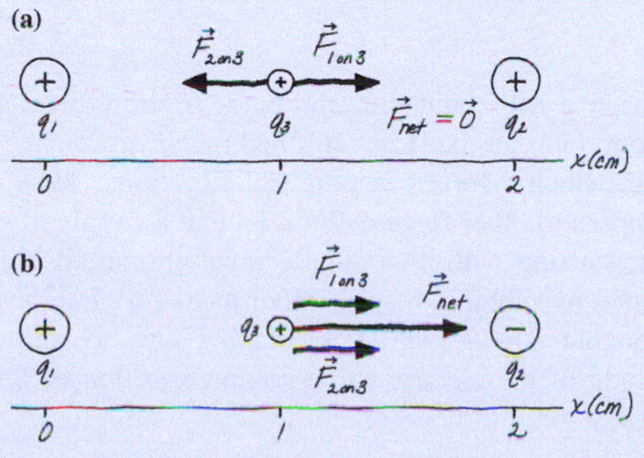


**SOLVE** Electric forces are vectors, and the net force on $q_3$ is the *vector* sum $\vec{F}_{\text{net}} = \vec{F}_{1\text{ on }3} + \vec{F}_{2\text{ on }3}$. Charges $q_1$ and $q_2$ each exert a repulsive force on $q_3$, but these forces are equal in magnitude and opposite in direction. Consequently, $\vec{F}_{\text{net}} = \vec{0}$. The situation changes if $q_2$ is negative, as in Figure 20.16b. In this case, the two forces are equal in magnitude but in the *same* direction, so $\vec{F}_{\text{net}} = 2\vec{F}_{1\text{ on }3}$. The magnitude of the force is given by Coulomb's law. The force due to $q_1$ is

$$F_{1\text{ on }3} = \frac{K|q_1||q_3|}{r_{13}^2}$$

$$= \frac{(9.0\times10^9\ \text{N}\cdot\text{m}^2/\text{C}^2)(10\times10^{-9}\ \text{C})(1.0\times10^{-9}\ \text{C})}{(0.010\ \text{m})^2}$$

$$= 9.0\times10^{-4}\ \text{N}$$

There is an equal force due to $q_2$, so the net force on the 1.0 nC charge is $\vec{F}_{\text{net}} = (1.8\times10^{-3}\ \text{N, to the right})$.

**ASSESS** This example illustrates the important idea that electric forces are *vectors*. An important part of assessing our answer is to see if it is "reasonable." In the second case, the net force on the charge is approximately 1 mN. Generally, charges of a few nC separated by a few cm experience forces in the range from a fraction of a mN to several mN. With this guideline, the answer appears to be reasonable.

## EXAMPLE 20.4 Adding electric forces in two dimensions

Three charged particles with $q_1 = -50$ nC, $q_2 = +50$ nC, and $q_3 = +30$ nC are placed as shown in FIGURE 20.17. What is the net force on charge $q_3$ due to the other two charges?

FIGURE 20.17 The arrangement of the charges.

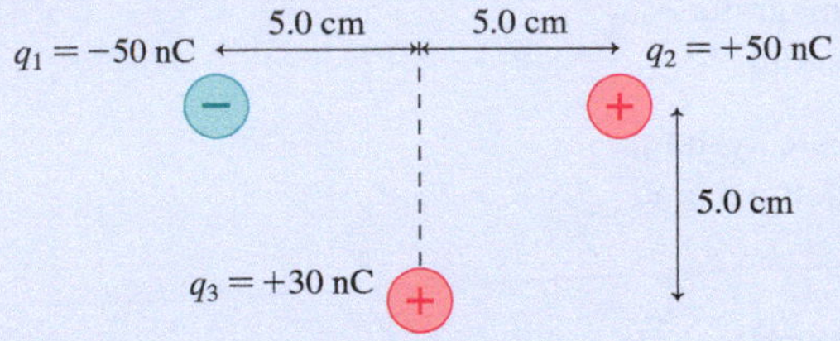


**STRATEGIZE** We will solve for the net force using the steps of Problem-Solving Approach 20.1.

**PREPARE** We start with the visual overview shown in FIGURE 20.18a. We have defined a coordinate system, with charge $q_3$ at the origin. We have drawn the forces on charge $q_3$, with directions determined by the signs of the charges. We can see from the geometry that the forces $\vec{F}_{1\text{ on }3}$ and $\vec{F}_{2\text{ on }3}$ are at the angles noted in the figure. The vector addition in FIGURE 20.18b shows the anticipated direction of the net force; this will be a good check on our final result. The distance between charges $q_1$ and $q_3$ is the same as that between charges $q_2$ and $q_3$; this distance is $r = \sqrt{(5.0\text{ cm})^2 + (5.0\text{ cm})^2} = 7.07$ cm. Because the magnitudes of $q_1$ and $q_2$ are equal and they are equidistant from $q_3$, we expect the magnitudes of the forces they exert to be equal, as we have drawn in Figure 20.18.

FIGURE 20.18 A visual overview of the charges and forces.

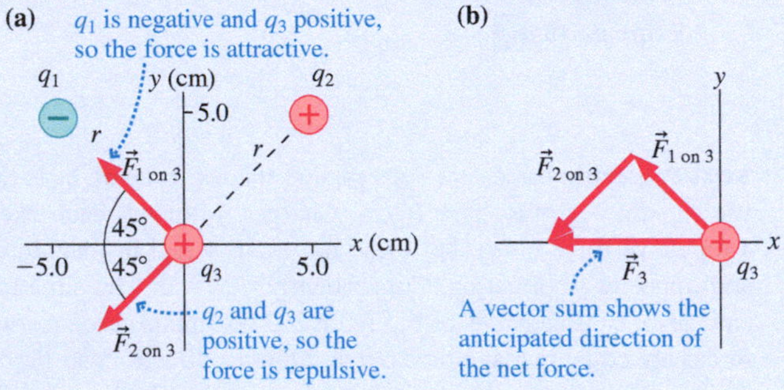


**SOLVE** We are interested in the net force on charge $q_3$. Let's start by using Coulomb's law to compute the magnitudes of the two forces on charge $q_3$:

$$F_{1\text{ on }3} = \frac{K|q_1||q_3|}{r^2}$$

$$= \frac{(9.0 \times 10^9\ \text{N}\cdot\text{m}^2/\text{C}^2)(50 \times 10^{-9}\ \text{C})(30 \times 10^{-9}\ \text{C})}{(0.0707\ \text{m})^2}$$

$$= 2.7 \times 10^{-3}\ \text{N}$$

The magnitudes of the charges and the distance are the same for $F_{2\text{ on }3}$, so

$$F_{2\text{ on }3} = \frac{K|q_2||q_3|}{r^2} = 2.7 \times 10^{-3}\ \text{N}$$

The components of these forces are illustrated in FIGURE 20.19a.

FIGURE 20.19 The net force on $q_3$ is to the left.

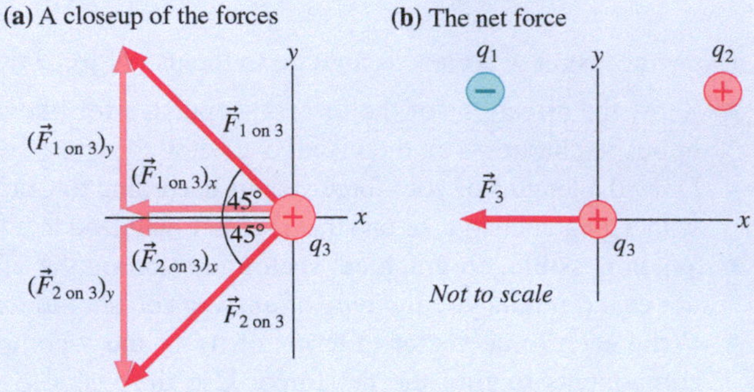


Computing values for the components, we find

$$(F_{1\text{ on }3})_x = -(2.7 \times 10^{-3}\ \text{N}) \cos 45° = -1.9 \times 10^{-3}\ \text{N}$$
$$(F_{1\text{ on }3})_y = (2.7 \times 10^{-3}\ \text{N}) \sin 45° = 1.9 \times 10^{-3}\ \text{N}$$
$$(F_{2\text{ on }3})_x = -(2.7 \times 10^{-3}\ \text{N}) \cos 45° = -1.9 \times 10^{-3}\ \text{N}$$
$$(F_{2\text{ on }3})_y = -(2.7 \times 10^{-3}\ \text{N}) \sin 45° = -1.9 \times 10^{-3}\ \text{N}$$

Next, we add components of the net force:

$$F_{3x} = (F_{1\text{ on }3})_x + (F_{2\text{ on }3})_x = -1.9 \times 10^{-3}\ \text{N} - 1.9 \times 10^{-3}\ \text{N}$$
$$= -3.8 \times 10^{-3}\ \text{N}$$
$$F_{3y} = (F_{1\text{ on }3})_y + (F_{2\text{ on }3})_y$$
$$= +1.9 \times 10^{-3}\ \text{N} - 1.9 \times 10^{-3}\ \text{N} = 0$$

Thus the net force, as shown in FIGURE 20.19b, is

$$\vec{F}_3 = (3.8 \times 10^{-3}\ \text{N},\ -x\text{-direction})$$

**ASSESS** The net force is directed to the left, as we anticipated. The magnitude of the net force, a few mN, seems reasonable as well.

You've had a lot of experience with the weight force, the friction force, and other forces that we've talked about, but you likely haven't had much experience with the electric force. How large are electric forces in practical situations? How does the magnitude of this force compare to other forces? We'll look at a couple of examples to give us a sense of scale, starting with an example involving small objects that have been deliberately charged by rubbing or some other means so that they acquire charges of $\pm 10$ nC, a reasonable value for small objects that have been charged this way. How does the magnitude of the electric force compare to that of the weight force in such circumstances?

## EXAMPLE 20.5 Comparing electric and gravitational forces

A small plastic sphere is charged to $-10$ nC. It is held 1.0 cm above a small glass bead at rest on a table. The bead has a mass of 15 mg and a charge of $+10$ nC. Will the glass bead "leap up" to the plastic sphere?

**STRATEGIZE** We'll model the bead and the sphere as point charges, and measure distances from their centers. There is an attractive electric force from the bead that will tend to pull the sphere upward. If this force is greater than the downward weight force, then the bead will leap upward.

**PREPARE** We can start with a visual overview of the situation. FIGURE 20.20 establishes a $y$-axis, identifies the plastic sphere as $q_1$ and the glass bead as $q_2$, and shows a free-body diagram. The glass bead will rise if $F_{1\text{ on }2} > w$; if $F_{1\text{ on }2} < w$, the bead will remain at rest on the table, which then exerts a normal force $\vec{n}$ on the bead.

**SOLVE** Using the values provided, we have

$$F_{1\text{ on }2} = \frac{K|q_1||q_2|}{r^2} = 9.0 \times 10^{-3}\text{ N}$$

$$w = m_2 g = 1.5 \times 10^{-4}\text{ N}$$

$F_{1\text{ on }2}$ exceeds the bead's weight by a factor of 60, so the glass bead will leap upward.

**ASSESS** The values used in this example are realistic for spheres $\approx 2$ mm in diameter. These are small spheres that have been charged purposefully, given charges in the nC range. In such cases, for small separations, the electric force is considerable—much larger than the weight force.

FIGURE 20.20 A visual overview showing the charges and forces.

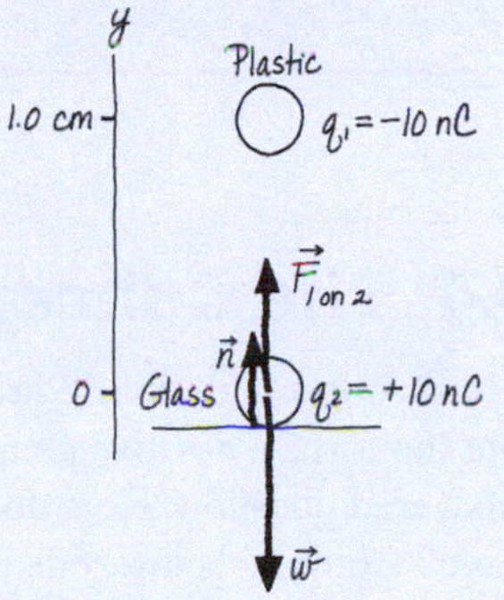


Some objects are deliberately charged, but others acquire charges as a matter of course; these charges tend to have smaller magnitudes. For example, bees develop an electric charge as they fly through the air. This causes the bee to pick up pollen (as we've seen) and can enhance the bee's electric sense (which we'll discuss later), but does this charge cause the bee itself to experience electric forces?

## EXAMPLE 20.6 What is the force between the bees? BIO

Two 0.10 g honeybees each acquire a charge of $+23$ pC ($1\text{ pC} = 1 \times 10^{-12}\text{ C}$) as they fly back to their hive. As they approach the hive entrance, they are 1.0 cm apart. Approximate the magnitude of the repulsive force between the two bees. How does this force compare with their weight?

**STRATEGIZE** The bees clearly aren't point charges, but modeling them as point charges will give us an approximate value for the electric force they will feel. We'll compute the repulsive force using Coulomb's law and then compare this to the weight force. You've certainly seen bees close together and they don't seem to experience a repulsive force, so we expect the weight force to be much greater than the electric force.

**PREPARE** The visual overview in FIGURE 20.21 is straightforward—two point charges separated by a distance $r = 1.0$ cm. The magnitudes of the forces on the bees are equal, so we compute only $F_{1\text{ on }2}$.

FIGURE 20.21 A visual overview showing a simplified model of the situation.

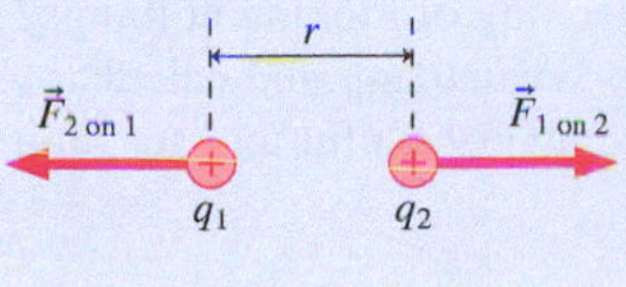


**SOLVE** The magnitude of the force is

$$F_{1\text{ on }2} = \frac{K|q_1||q_2|}{r^2} = \frac{(9.0 \times 10^9\text{ N}\cdot\text{m}^2/\text{C}^2)(23 \times 10^{-12}\text{ C})^2}{(0.010\text{ m})^2}$$
$$= 4.8 \times 10^{-8}\text{ N}$$

This is clearly a small force, but to put the size in perspective, we compare this force to the magnitude of the weight force acting on one bee:

$$w = mg = (0.10 \times 10^{-3}\text{ kg})(9.8\text{ m/s}^2) = 9.8 \times 10^{-4}\text{ N}$$

The weight force is greater than the electric force by a factor of 20,000, so the electric force is not likely to be noticeable to the bees.

**ASSESS** For objects that pick up charges naturally, the charges are much smaller—for the bees, in the pC range, not the nC range of the bead and sphere—so the forces are smaller as well. We don't expect to see significant forces between the bees. But the charge of the bee is large enough to attract a much lighter pollen grain from a much smaller distance. And, as we'll see, bees are very sensitive to nearby electric charges; it's likely that the charge on one bee, although not enough to push a nearby bee away, is large enough for a nearby bee to sense.

For the charged bead and sphere, the electric force was larger than the weight force; for the two bees, much less. On our macroscopic scale, electric forces dominate for small objects and deliberately charged objects. For much smaller charges, as on the bee, the electric force is much smaller as well. At an atomic level, the situation is more extreme. The opposite charges on water molecules create the attractive force of a hydrogen bond. The electric force on the molecules in this case is more than $10^{14}$ times larger than the weight force. At the atomic scale, the weight force is negligible; there's only one force that matters: the electric force between charged particles.

**STOP TO THINK 20.3** Charges 1 and 2 exert repulsive forces on each other. $q_1 = 4q_2$. Which statement is true?

A. $F_{1\text{ on }2} > F_{2\text{ on }1}$
B. $F_{1\text{ on }2} = F_{2\text{ on }1}$
C. $F_{1\text{ on }2} < F_{2\text{ on }1}$

## 20.4 The Concept of the Electric Field

**FIGURE 20.22** Visualizing the electric field.

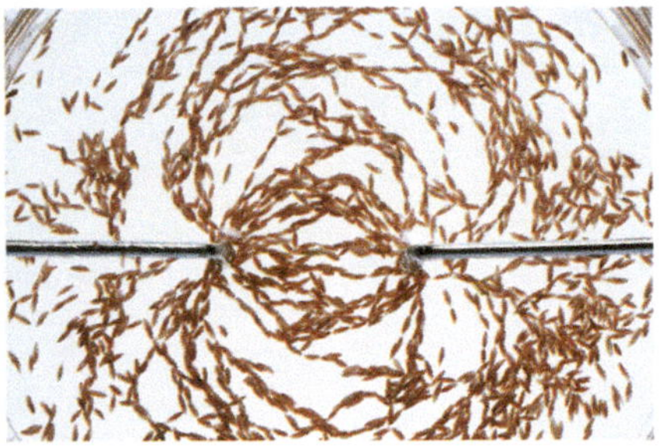

**Video** Figure 20.22

Coulomb's law is the basic law of electrostatics. We can use Coulomb's law to calculate the force a positive charge exerts on a nearby negative charge. But there is an unanswered question: How does the negative charge "know" that the positive charge is there? Coulomb's law tells us how to calculate the magnitude and direction of the force, but it doesn't tell us how the force is transmitted through empty space from one charge to the other. To answer this question, we will introduce the *field model,* first suggested in the early 19th century by Michael Faraday, a British investigator of electricity and magnetism.

FIGURE 20.22 shows a photograph of the surface of a shallow pan of oil with tiny grass seeds floating on it. When charged wires, one positive and one negative, touch the surface of the oil, the grass seeds line up to form a regular pattern. The pattern suggests that some kind of electric influence from the charges *fills the space* around the charges. Perhaps the grass seeds are reacting to this influence, creating the pattern that we see. This alteration of the space around the charges could be the *mechanism* by which the long-range Coulomb's law force is exerted.

**FIGURE 20.23** The force and field models for the interaction between two charges.

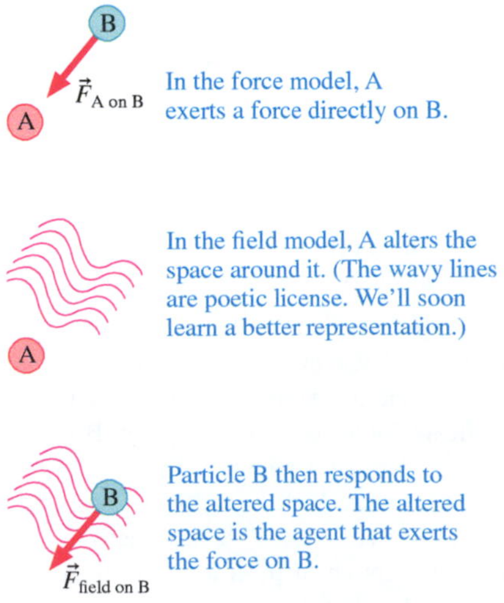


This is the essence of the field model. Consider the attractive force between a positive charge A and a negative charge B. FIGURE 20.23 shows the difference between the force model, which we have been using, and the field model. **In the field model, it is the alteration of space around charge A that is the agent that exerts a force on charge B.** This alteration of space is what we call a **field.** The charge makes an alteration *everywhere* in space. Other charges then respond to the alteration at their position.

The field model applies to many branches of physics. The space around a charge is altered to create the **electric field.** The alteration of the space around a mass is called the **gravitational field.** The alteration of the space around a magnet is called the **magnetic field,** which we will consider in Chapter 24.

### The Electric Field Is Real

The field model might seem arbitrary and abstract. We can already calculate forces between charges; why introduce another way of looking at things? We will find that the field model is a very useful tool for visualizing and calculating forces for complex arrangements of charges. It is also a very useful tool for understanding certain aspects of the natural world.

But the electric field is real; it's not just an abstraction. As we'll see, when your heart beats, a charge separation develops; the same is true of other muscles. This charge separation creates an electric field. You literally affect the space around you in a way that can be detected at a distance.

The platypus has an unusual bill that it uses to forage on the bottom of the creeks and ponds that it inhabits. The bill is very sensitive to vibrations produced by crustaceans and other prey, but it is also sensitive to the electric fields created by their heartbeats and muscle action. The platypus closes its eyes and ears as it forages, relying on its vibration sense and its electric sense to find its dinner. When a bee visits a flower, it transfers charge; this affects the electric field in the vicinity of the flower. The electric sense of a second bee allows it to determine, from a distance, that this flower was recently visited. A few mammals, some insects, and many fish have such an electric sense. To these creatures, the electric field is certainly not an abstraction; they actively sense the electric fields in their vicinity and use them to navigate, locate prey, and determine the best flowers to visit.

A platypus forages in the gravel at the bottom of a creek.

Of course, you use electric fields as well. Toward the end of this section, we'll explore electromagnetic waves, waves of electric and magnetic fields that carry energy and information through space. If you are reading this text on a device that has a wireless connection, these very words were transmitted to your device by the use of electric fields. Fields may seem abstract, but they are very, very real.

## The Field Model

We begin our investigation of electric fields by postulating a **field model** that describes how charges interact:

**Video** Electric Field

1. A group of charges, which we call the **source charges,** alter the space around them by creating an *electric field* $\vec{E}$.
2. If another charge is then placed in this electric field, it experiences a force $\vec{F}$ exerted *by the field.*

Suppose charge $q$ experiences an electric force $\vec{F}_{\text{on } q}$ due to other charges. The strength and direction of this force vary as $q$ is moved from point to point in space. This suggests that "something" is present at each point in space to cause the force that charge $q$ experiences. We define the electric field $\vec{E}$ at the point $(x, y, z)$ as

$$\vec{E} \text{ at } (x, y, z) = \frac{\vec{F}_{\text{on } q} \text{ at } (x, y, z)}{q} \tag{20.3}$$

Electric field at a point defined by the force on charge $q$

We're *defining* the electric field as a force-to-charge ratio; hence the units of the electric field are newtons per coulomb, or N/C. The magnitude $E$ of the electric field is called the **electric field strength.** Typical electric field strengths are given in TABLE 20.2. As the table shows, a very small field in a wire will keep charges in motion. For fields in air—the main situation we'll consider—typical electric fields are much larger than this. The electric field outdoors near the ground is usually about 100 N/C. You've never noticed this field before; it doesn't cause observable consequences except in special circumstances. A field of 1,000,000 N/C can cause a spark to jump; if the field around you approaches this value, you'll certainly notice it! The electric field inside the cell membrane is larger than this by a factor of 10, so the electrical nature of nerve and muscle cells is clearly quite important. And the electric field

**TABLE 20.2** Typical electric field strengths

| Field | Field strength (N/C) |
|---|---|
| Inside a current-carrying wire | $10^{-2}$ |
| Earth's field, near the earth's surface | $10^{2}$ |
| Near objects charged by rubbing | $10^{3}$ to $10^{6}$ |
| Needed to cause a spark in air | $10^{6}$ |
| Inside a cell membrane | $10^{7}$ |
| Inside an atom | $10^{11}$ |

inside an atom is 10,000 times larger still. At the atomic scale, electric forces are the only game in town.

You can think of using charge $q$ as a *probe* to determine whether an electric field is present at a point in space. If charge $q$ experiences an electric force at a point in space, as FIGURE 20.24a shows, then there is an electric field at that point causing the force. Further, we *define* the electric field at that point to be the vector given by Equation 20.3. FIGURE 20.24b shows the electric field at two points, but you can imagine "mapping out" the electric field by moving the charge $q$ throughout all space.

FIGURE 20.24 Charge $q$ is a probe of the electric field.

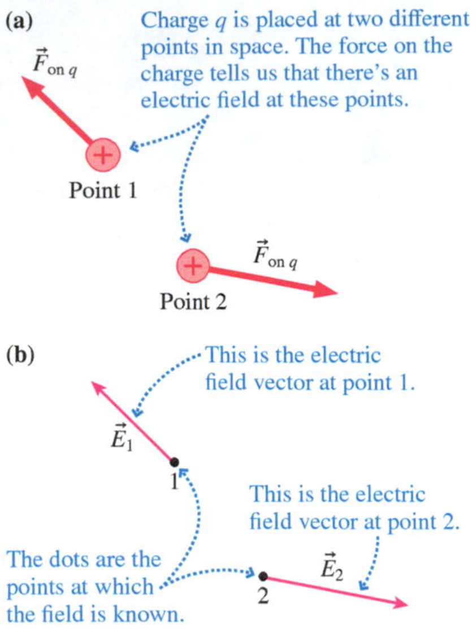


The basic idea of the field model is that **the field is the agent that exerts an electric force on a particle with charge *q*.** Notice three important things about the field:

1. The electric field, a vector, exists at every point in space. Electric field diagrams show a sample of the vectors, but there is an electric field vector at every point whether one is shown or not.
2. If the probe charge $q$ is positive, the electric field vector points in the same direction as the force on the charge; if negative, the electric field vector points opposite the force.
3. Because $q$ appears in Equation 20.3, you might think that the electric field depends on the magnitude of the charge used to probe the field. It doesn't. We know from Coulomb's law that the force $\vec{F}_{\text{on } q}$ is proportional to $q$. Thus the electric field defined in Equation 20.3 is *independent* of the charge $q$ that probes the field. The electric field depends on only the source charges that create the field.

## The Electric Field of a Point Charge

FIGURE 20.25 Charge $q'$ is used to probe the electric field of point charge $q$.

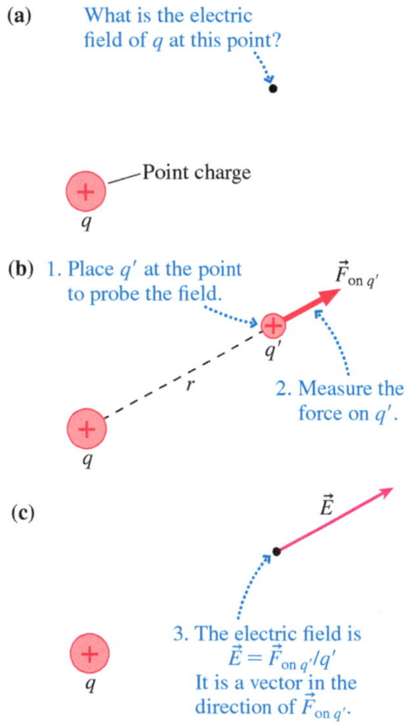


FIGURE 20.25a shows a point source charge $q$ that creates an electric field at all points in space. We can use a second charge, shown as $q'$ in FIGURE 20.25b, as a probe of the electric field created by charge $q$.

For the moment, assume both charges are positive. The force on $q'$, which is repulsive and points directly away from $q$, is given by Coulomb's law:

$$\vec{F}_{\text{on } q'} = \left(\frac{Kqq'}{r^2}, \text{ away from } q\right) \tag{20.4}$$

Equation 20.3 defines the electric field in terms of the force on the probe charge as $\vec{E} = \vec{F}_{\text{on } q'}/q'$, so for a positive charge $q$,

$$\vec{E} = \left(\frac{Kq}{r^2}, \text{ away from } q\right) \tag{20.5}$$

The electric field is shown in FIGURE 20.25c.

If $q$ is negative, the magnitude of the force on the probe charge is the same as in Equation 20.5, but the direction is toward $q$, so the general expression for the field is

$$\vec{E} = \left(\frac{K|q|}{r^2}, \begin{bmatrix} \text{away from } q \text{ if } q > 0 \\ \text{toward } q \text{ if } q < 0 \end{bmatrix}\right) \tag{20.6}$$

Electric field of point charge $q$ at a distance $r$ from the charge

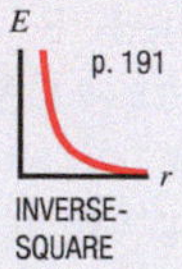


**NOTE** ► The expression for the electric field is similar to Coulomb's law. To distinguish the two, remember that Coulomb's law has the product of two charges in the numerator. It describes the force between *two* charges. The electric field has a single charge in the numerator. It is the field of a *single* charge. ◄

**EXAMPLE 20.7 Finding the electric field of a proton**

The electron in a hydrogen atom orbits the proton at a radius of 0.053 nm. What is the electric field due to the proton at the position of the electron?

**SOLVE** The proton's charge is $q = e$. At the distance of the electron, the magnitude of the field is

$$E = \frac{Ke}{r^2} = \frac{(9.0 \times 10^9 \text{ N} \cdot \text{m}^2/\text{C}^2)(1.60 \times 10^{-19} \text{ C})}{(5.3 \times 10^{-11} \text{ m})^2}$$
$$= 5.1 \times 10^{11} \text{ N/C}$$

Because the proton is positive, the electric field is directed away from the proton:

$$\vec{E} = (5.1 \times 10^{11} \text{ N/C, outward from the proton})$$

**ASSESS** This is a large field, but Table 20.2 shows that this is the correct magnitude for the field within an atom.

By drawing electric field vectors at a number of points around a positive point charge, we can construct an **electric field diagram** such as the one shown in FIGURE 20.26a. Notice that the field vectors all point straight away from charge $q$. We can draw a field diagram for a negative point charge in a similar fashion, as in FIGURE 20.26b. In this case, the field vectors point toward the charge, as this would be the direction of the force on a positive probe charge.

In the coming sections, as we use electric field diagrams, keep these ideas in mind:

- The diagram is just a representative sample of electric field vectors. The field exists at all the other points. A well-drawn diagram gives a good indication of what the field would be like at a neighboring point.
- The arrow indicates the direction and the strength of the electric field *at the point to which it is attached*—at the point where the *tail* of the vector is placed. The length of any vector is significant only relative to the lengths of other vectors.
- Although we have to draw a vector across the page, from one point to another, an electric field vector does not "stretch" from one point to another. Each vector represents the electric field at *one point* in space.

FIGURE 20.26 The electric field near a point charge.

(a) The electric field diagram of a positive point charge

(b) The electric field diagram of a negative point charge

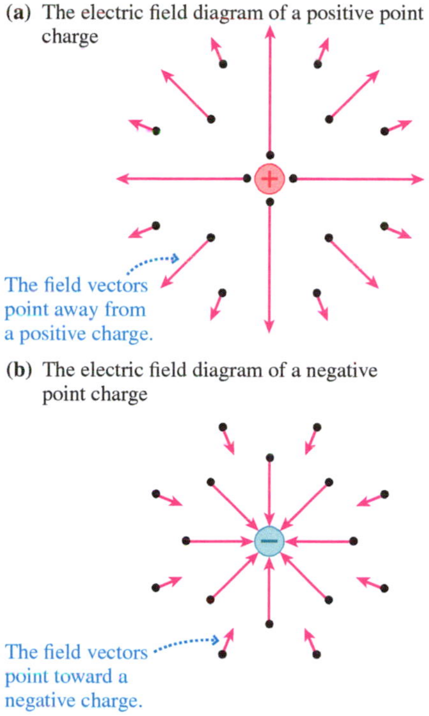


**STOP TO THINK 20.4** Rank in order, from largest to smallest, the electric field strengths $E_A$ to $E_D$ at points A to D.

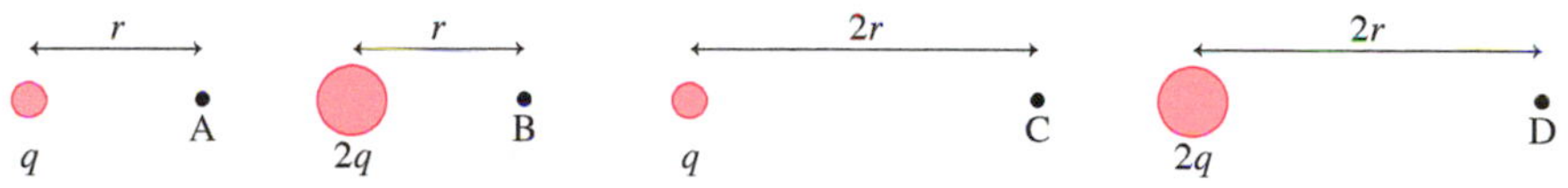


# 20.5 The Electric Field of Multiple Charges

At the start of the chapter, we noted that electrophoresis of DNA makes use of an electric field inside a gel to separate different fragments. To get good results, a uniform electric field is needed between the plates at the ends of the gel. How is such a uniform field made? In this section we'll look at the fields created by different arrangements of charges.

No matter the number of source charges, the electric field at a point in space can be found by looking at the force on a probe charge. Because the net force on the probe charge is the vector sum of the forces due to all of the individual charges, **the electric field due to multiple charges is the vector sum of the electric field due to each of the charges.**

## EXAMPLE 20.8 Finding the field near a dipole

A dipole consists of a positive and negative charge separated by 1.2 cm, as shown in FIGURE 20.27. What is the electric field strength along the line connecting the charges at a point 1.2 cm to the right of the positive charge?

FIGURE 20.27 Charges and distances for a dipole.

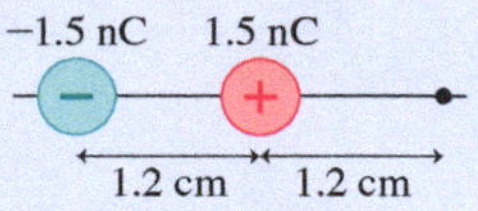


**STRATEGIZE** The dipole has no net charge, but it does have a net electric field. The electric field is the sum of the fields from the two charges.

**PREPARE** We define the $x$-axis to be along the line connecting the two charges, as in FIGURE 20.28. The point at which we calculate the field is 1.2 cm from the positive charge and 2.4 cm from the negative charge. Thus the electric field of the positive charge will be larger, as shown in Figure 20.28. The net electric field of the dipole is the vector sum of these two fields, so the electric field of the dipole at this point is in the positive $x$-direction.

FIGURE 20.28 Visual overview for finding the electric field.

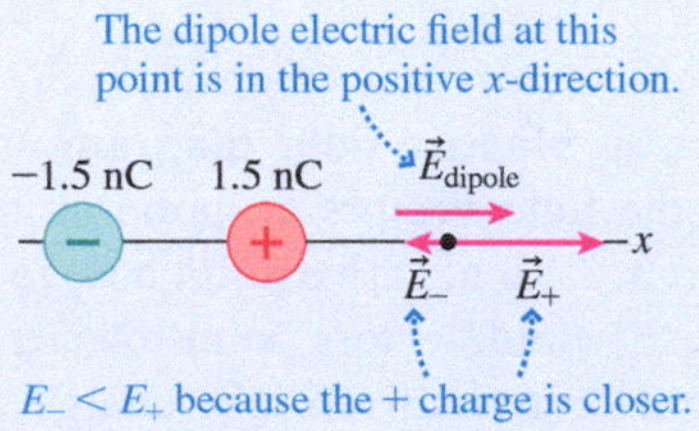


**SOLVE** The magnitudes of the fields of the two charges are given by Equation 20.6, so the magnitude of the dipole field is

$$E_{dipole} = E_+ - E_-$$

$$= \frac{\left(9.0 \times 10^9 \frac{N \cdot m^2}{C^2}\right)(1.5 \times 10^{-9}\ C)}{(0.012\ m)^2} - \frac{\left(9.0 \times 10^9 \frac{N \cdot m^2}{C^2}\right)(1.5 \times 10^{-9}\ C)}{(0.024\ m)^2}$$

$$= 7.0 \times 10^4\ N/C$$

**ASSESS** Table 20.2 lists the fields due to objects charged by rubbing as typically $10^3$ to $10^6$ N/C, and we've already seen that charges caused by rubbing are in the range of 1–10 nC. Our answer is in this range and thus is reasonable.

FIGURE 20.29 The electric field of a dipole.

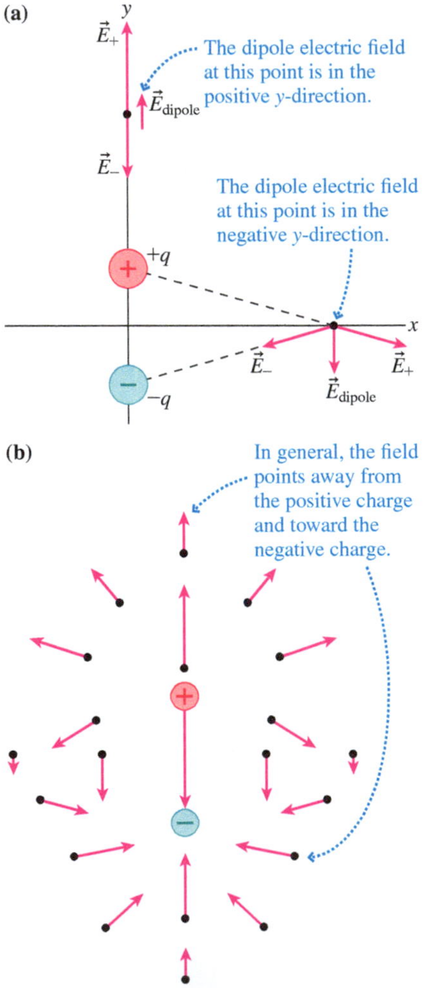


The electric dipole is an important charge distribution that we will see many times, so it's worth exploring the full field diagram. FIGURE 20.29 shows a dipole oriented along the $y$-axis. We can determine the field at any point by a vector addition of the fields of the two charges, as shown in Figure 20.29a. If we repeat this process at many points, we end up with the field diagram of Figure 20.29b. This is more complex than the field of a single charge, but it accurately shows how two charges alter the space around them.

## Uniform Electric Fields

FIGURE 20.30a shows another important practical situation, one we'll meet many times. Two conducting plates, called **electrodes,** are face-to-face with a narrow gap between them. One electrode has total charge $+Q$ and the other has total charge $-Q$. This arrangement of two electrodes, closely spaced and charged equally but oppositely, is called a **parallel-plate capacitor.** What is the electric field between the two plates? To keep things simple, we will focus on the field in the central region, far from the edges, as FIGURE 20.30b shows in a blown-up cross-section view.

At any point, the electric field is the vector sum of the fields from all of the positive charges and all of the negative charges on the plates. However, the field of a point charge decreases inversely with the square of its distance, so in practice only the nearby charges contribute to the field. As FIGURE 20.31a shows, the horizontal components of the individual fields cancel, while the vertical components add to give an electric field vector pointing from the positive plate toward the negative plate.

By mapping the electric field at many points, we find that the field inside a parallel-plate capacitor is the same—in both strength and direction—at every point. This is called a **uniform electric field.** FIGURE 20.31b shows that a uniform electric field is represented with parallel electric field vectors of equal length. A more detailed analysis finds that the electric field inside a parallel-plate capacitor is

$$\vec{E}_{\text{capacitor}} = \left(\frac{Q}{\epsilon_0 A}, \text{ from positive to negative}\right) \tag{20.7}$$

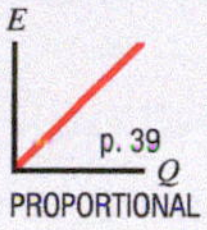


Electric field in a parallel-plate capacitor, plate area $A$ and charge $Q$

Equation 20.7 introduces a new constant $\epsilon_0$, pronounced "epsilon zero" or "epsilon naught," called the **permittivity constant.** Its value is related to the electrostatic constant as

$$\epsilon_0 = \frac{1}{4\pi K} = 8.85 \times 10^{-12}\ \text{C}^2/\text{N}\cdot\text{m}^2$$

There are a few things to note about the field in a parallel-plate capacitor:

- The field depends on the charge-to-area ratio $Q/A$, which is often called the *charge density*. If the charges are packed more closely, the field will be larger.
- Our analysis requires that the separation of the plates be small compared to their area. If this is true, the spacing between the plates does not affect the electric field, and this spacing does not appear in Equation 20.7.
- Although Figure 20.30 shows circular electrodes, the shape of the electrodes—circular or square or any other shape—is not relevant as long as the electrodes are very close together.

**NOTE** ▶ The charges on the plates are equal and opposite, $+Q$ and $-Q$, so the net charge is zero. The symbol $Q$ in Equation 20.7 is the *magnitude* of the charge on each plate. ◀

**FIGURE 20.30** A parallel-plate capacitor.

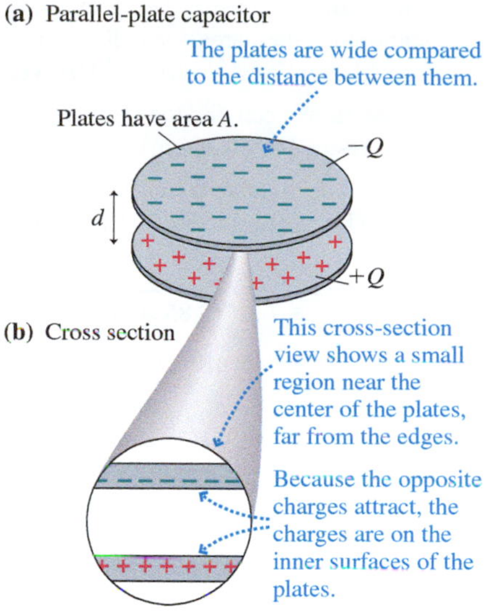


**FIGURE 20.31** The electric field inside a parallel-plate capacitor.

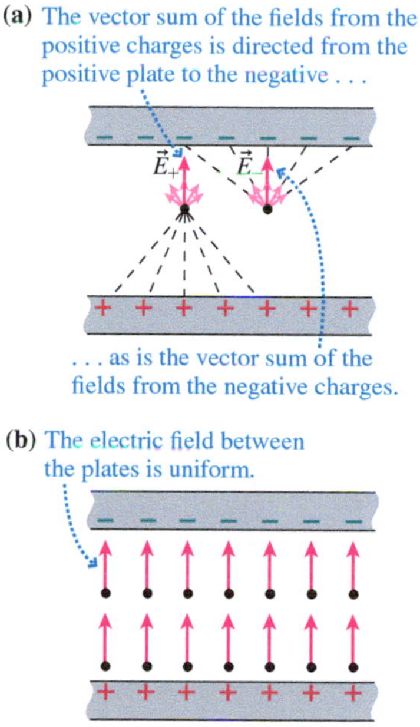


## EXAMPLE 20.9 Finding the field in an air cleaner

Long highway tunnels must have air cleaners to remove dust and soot coming from passing cars and trucks. In one type, known as an *electrostatic precipitator,* air passes between two oppositely charged metal plates, as in FIGURE 20.32. The large electric field between the plates ionizes dust and soot particles, which then feel a force due to the field. This force causes the charged particles to move toward and stick to one or the other plate, removing them from the air. A typical unit has dimensions and charges as shown in Figure 20.32. What is the electric field between the plates?

**FIGURE 20.32** An electrostatic precipitator.

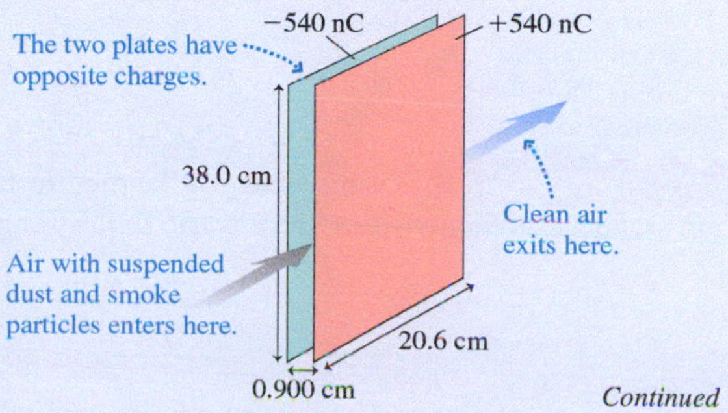


*Continued*

**STRATEGIZE** Because the spacing between the plates is much smaller than their area, we will treat this as a parallel-plate capacitor with a uniform electric field between the plates.

**PREPARE** We can find the field using Equation 20.7. The area of the plates is $A = (0.206\ \text{m})(0.380\ \text{m}) = 0.0783\ \text{m}^2$.

**SOLVE** The field direction is from the positive to the negative plate, which is to the left. The field strength between the plates is

$$E = \frac{Q}{\epsilon_0 A} = \frac{540 \times 10^{-9}\ \text{C}}{(8.85 \times 10^{-12}\ \text{C}^2/\text{N}\cdot\text{m}^2)(0.0783\ \text{m}^2)}$$
$$= 7.79 \times 10^5\ \text{N/C}$$

The question asks for the electric field, a vector, not just for the field strength. The electric field between the plates is

$$\vec{E} = (7.79 \times 10^5\ \text{N/C, to the left})$$

**ASSESS** Table 20.2 shows that a field of $10^6$ N/C can create a spark in air. The field we calculated between the plates is just a bit smaller than this, which makes sense. The field should be large, but not large enough to make a spark jump between the plates!

## Electric Field Lines

We can't see the electric field, so we use pictorial tools like electric field diagrams to help us visualize the electric field in a region of space. Another way to picture the field is to draw **electric field lines.** These are imaginary lines drawn through a region of space so that

- The tangent to a field line at any point is in the direction of the electric field $\vec{E}$ at that point, and
- The field lines are closer together where the electric field strength is greater.

FIGURE 20.33a shows the relationship between electric field lines and electric field vectors in one region of space. If we know what the field vectors look like, we can extrapolate to the field lines, as in FIGURES 20.33b and 20.33c for the electric field lines near a positive charge and between the plates of a capacitor.

KEY CONCEPT **FIGURE 20.33** Field vectors and field lines.

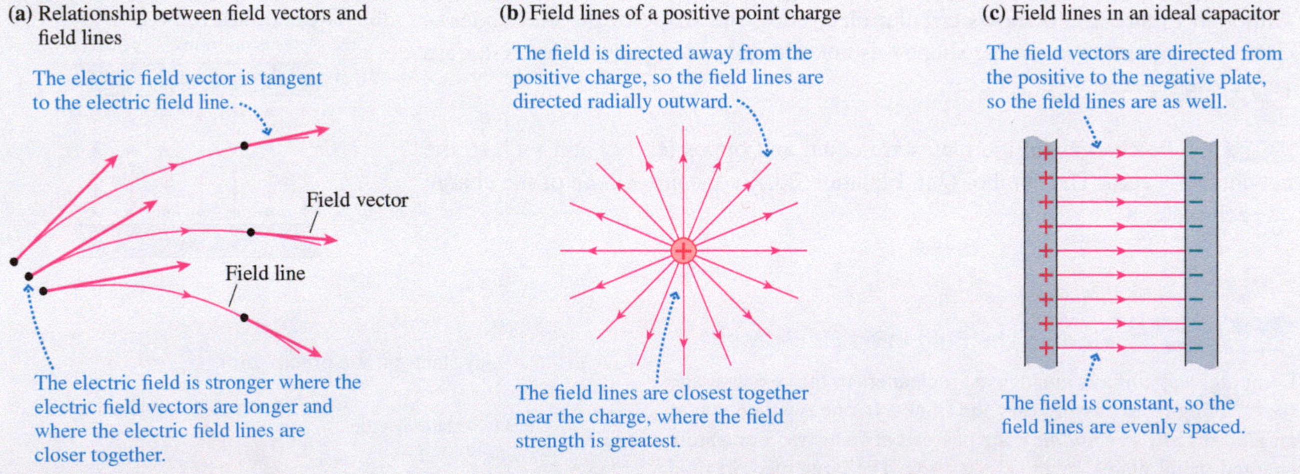

(a) Relationship between field vectors and field lines
The electric field vector is tangent to the electric field line.
Field vector
Field line
The electric field is stronger where the electric field vectors are longer and where the electric field lines are closer together.
(b) Field lines of a positive point charge
The field is directed away from the positive charge, so the field lines are directed radially outward.
The field lines are closest together near the charge, where the field strength is greatest.
(c) Field lines in an ideal capacitor
The field vectors are directed from the positive to the negative plate, so the field lines are as well.
The field is constant, so the field lines are evenly spaced.


**STOP TO THINK 20.5** Using what you learned in Figure 20.33, rank in order, from largest to smallest, noting any ties, the electric field strengths $E_1$ to $E_4$ at points 1 to 4.

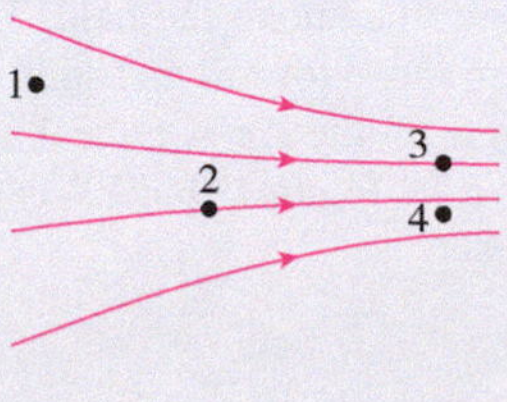

If we have an arrangement of charges, we can draw field lines as a guide to what the field looks like. As you generate a field line picture, there are two rules to keep in mind:

- Field lines cannot cross. The tangent to the field line is the electric field vector, which indicates the direction of the force on a positive charge. The force must be in a unique, well-defined direction, so two field lines cannot cross.
- The electric field is created by charges. Field lines start on a positive charge and end on a negative charge.

You can use the above information as the basis of a technique for sketching a field-line picture for an arrangement of charges.

- Draw field lines starting on positive charges and moving toward negative charges.
- Draw the field lines so that the field vectors are tangent to the field lines at each point.
- Make the lines close together where the field is strong, far apart where the field is weak.

For example, FIGURE 20.34 pictures the electric field of a dipole using electric field lines. You should compare this to Figure 20.29b, which illustrated the field with field vectors.

FIGURE 20.34 Electric field lines for a dipole.

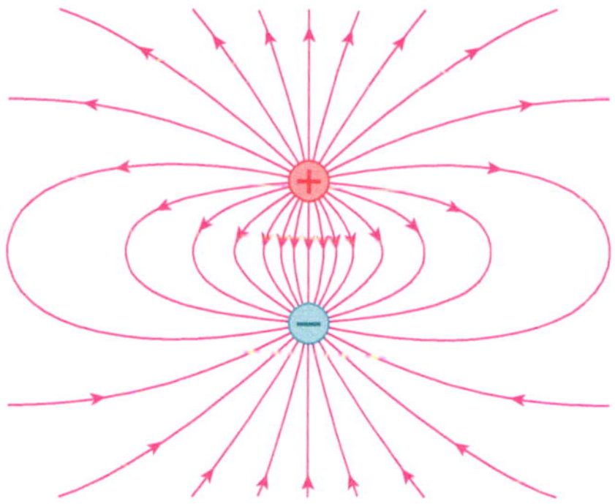

**CONCEPTUAL EXAMPLE 20.10** **Drawing field lines between a charged disk and a charged ring**

FIGURE 20.35 shows a disk with a negative charge in the center of a ring that has an equal amount of positive charge. Draw the field lines in the region between the disk and the ring.

***REASON*** We'll draw each field line by starting on a positive charge and moving in the direction of the force that a positive charge would feel—that is, in the direction of the field. The arrangement of charges has a symmetry that makes the drawing straightforward; the field lines go radially inward, each ending on a negative charge, as in FIGURE 20.36.

***ASSESS*** The field lines don't cross, and they are closer together near the disk, which makes sense. The charges on the disk are packed more closely than the charges on the ring, so we expect the field to be stronger near the disk. Our final picture seems reasonable.

FIGURE 20.35 Opposite charges on a disk and a ring.

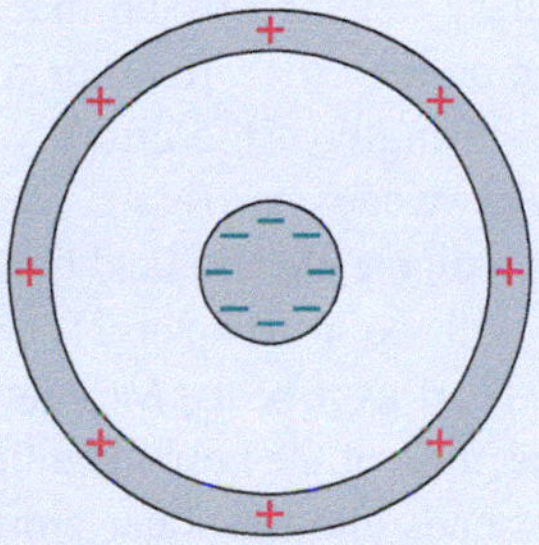

FIGURE 20.36 Field lines between the disk and the ring.

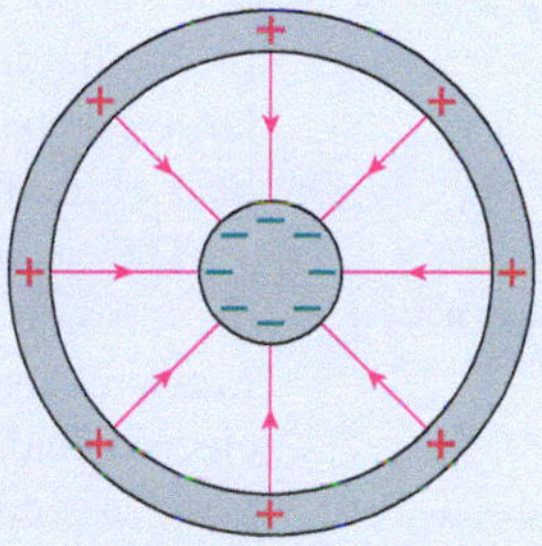

We can use this technique to get a qualitative picture of the electric field from various arrangements of charges. Before we consider more complex situations, it's worthwhile to sum up what we've learned about the electric field for two important cases, the field near a point charge and between the plates of a parallel-plate capacitor.

## SYNTHESIS 20.1 Two key electric fields

### The electric field due to a point charge

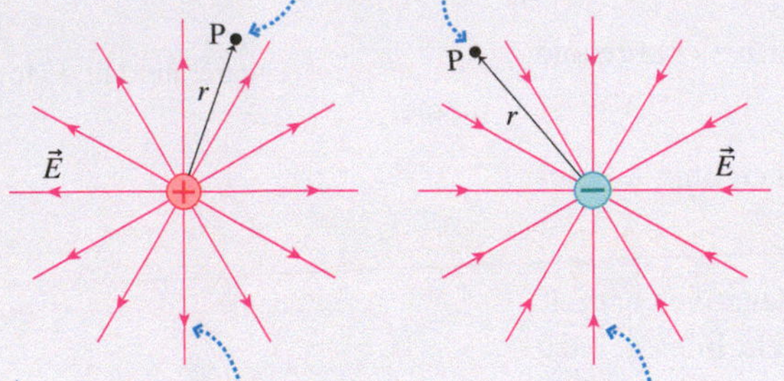


$$E = \frac{K|q|}{r^2}$$

Electrostatic constant → $K$; Magnitude of charge (C) → $|q|$; Magnitude of electric field (N/C) → $E$; Distance from charge to point P (m) → $r$

$$K = 9.0 \times 10^9\ \text{N}\cdot\text{m}^2/\text{C}^2$$

### The electric field inside a parallel-plate capacitor

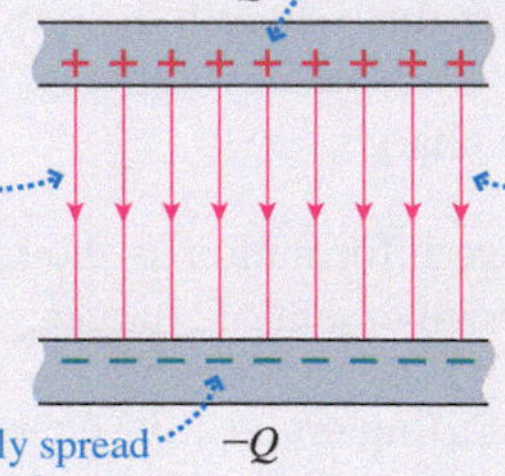


$$E = \frac{Q}{\epsilon_0 A}$$

Magnitude of electric field (N/C) → $E$; Magnitude of the charge on *each* plate (C) → $Q$; Permittivity constant → $\epsilon_0$; Area of each plate ($\text{m}^2$) → $A$

$$\epsilon_0 = \frac{1}{4\pi K} = 8.85 \times 10^{-12}\ \text{C}^2/\text{N}\cdot\text{m}^2$$

## The Electric Field of the Heart BIO

As we will see in detail in Chapter 23, a cell membrane is an insulator that encloses a conducting fluid and is surrounded by conducting fluid. While resting, the membrane is *polarized* with positive charges on the outside of the cell, negative charges on the inside. When a nerve or a muscle cell is stimulated, the polarity of the membrane switches; we say that the cell *depolarizes*. Later, when the charge balance is restored, we say that the cell *repolarizes*.

All nerve and muscle cells generate an electrical signal when depolarization occurs, but the largest electrical signal in the body comes from the heart. The rhythmic beating of the heart is produced by a highly coordinated wave of depolarization that sweeps across the tissue of the heart. As FIGURE 20.37a shows, the surface of the heart is positive on one side of the boundary between tissue that is depolarized and tissue that is not yet depolarized, negative on the other. In other words, the heart is a large electric dipole. The orientation and strength of the dipole change during each beat of the heart as the depolarization wave sweeps across it.

The electric dipole of the heart generates a dipole electric field that extends throughout the torso, as shown in FIGURE 20.37b. As we will see in Chapter 21, an *electrocardiogram* measures the changing electric field of the heart as it beats. Measurement of the heart's electric field can be used to diagnose the operation of the heart. The field from the beating heart of a creature that lives in the water extends into the water around it. A platypus, a shark, or any of the other predators with an electric sense can use this field to locate prey.

**FIGURE 20.37** BIO The beating heart generates a dipole electric field.

(a) The electric dipole of the heart

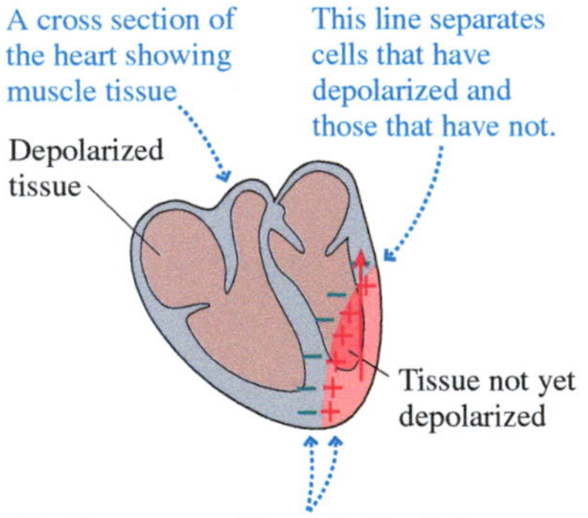


(b) The field of the heart in the body

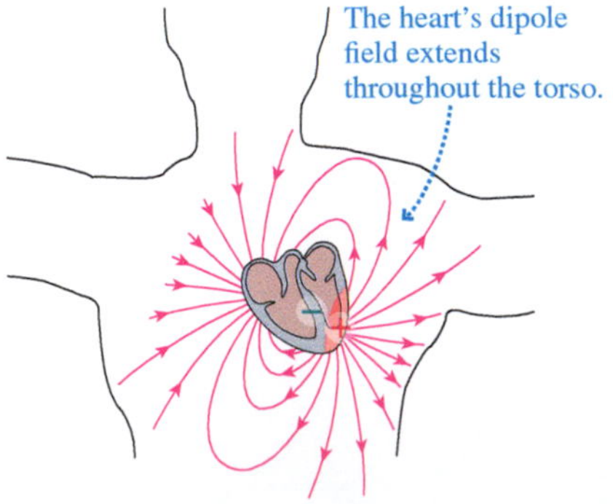


**STOP TO THINK 20.6** Which of the following is the correct representation of the electric field created by two positive charges?

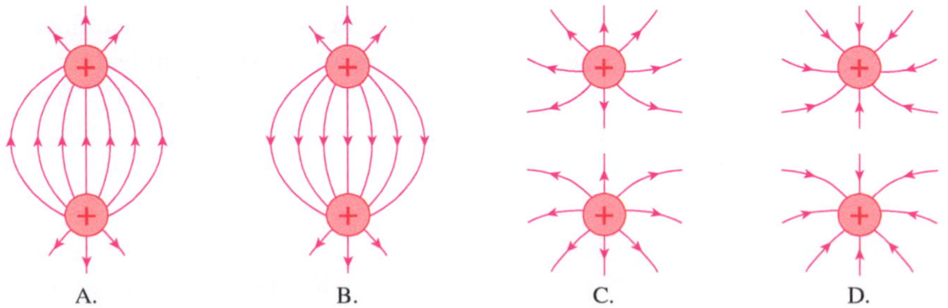

## 20.6 Conductors and Electric Fields

Consider a conductor in electrostatic equilibrium (recall that this means that none of the charges are moving). Suppose there were an electric field inside the conductor. Electric fields exert forces on charges, so an internal electric field would exert forces on the charges in the conductor. Because charges in a conductor are free to move, these forces would cause the charges to move. But that would violate the assumption that all the charges are at rest. Thus we're forced to conclude that **the electric field is zero at all points inside a conductor in electrostatic equilibrium.**

Because the electric field inside a conductor in electrostatic equilibrium is zero, any *excess* charge on the conductor must lie at its surface, as shown in FIGURE 20.38a. Any charge in the interior of the conductor would create an electric field there, in violation of our conclusion that the field inside is zero. Physically, excess charge ends up on the surface because the repulsive forces between like charges cause them to move as far apart as possible without leaving the conductor.

FIGURE 20.38b shows that **the electric field right at the surface of a charged conductor is perpendicular to the surface.** To see that this is so, suppose $\vec{E}$ had a component tangent to the surface. This component of $\vec{E}$ would exert a force on charges at the surface and cause them to move along the surface, thus violating the assumption that all charges are at rest. The only exterior electric field consistent with electrostatic equilibrium is one that is perpendicular to the surface.

FIGURE 20.38 The electric field inside and outside a charged conductor.

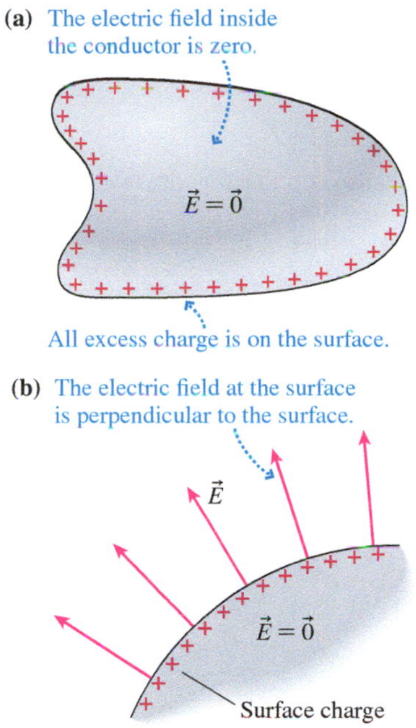


**CONCEPTUAL EXAMPLE 20.11 Drawing electric field lines for a charged sphere and plate**

FIGURE 20.39 shows a positively charged metal sphere above a conducting plate with a negative charge. Sketch the electric field lines.

FIGURE 20.39 The charged sphere and plate.

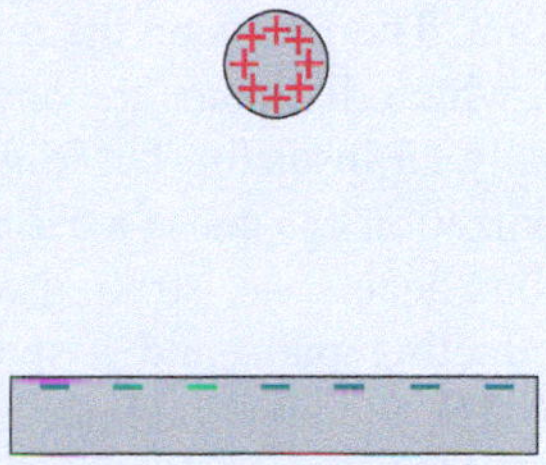

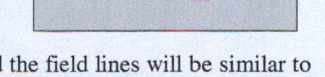

***REASON*** Field lines start on positive charges and end on negative charges. Thus we draw the field lines from the positive sphere to the negative plate, perpendicular to both surfaces, as shown in FIGURE 20.40. Near the sphere, the field is mostly determined by the presence of the sphere, and the field lines will be similar to those from a point charge. Near the plate, the field is mostly determined by the presence of the plate, and the field lines will be similar to those near one plate of a parallel-plate capacitor. As we draw the field lines from the sphere to the plate, we draw them to smoothly transition between the two cases.

FIGURE 20.40 Drawing field lines from sphere to plate.

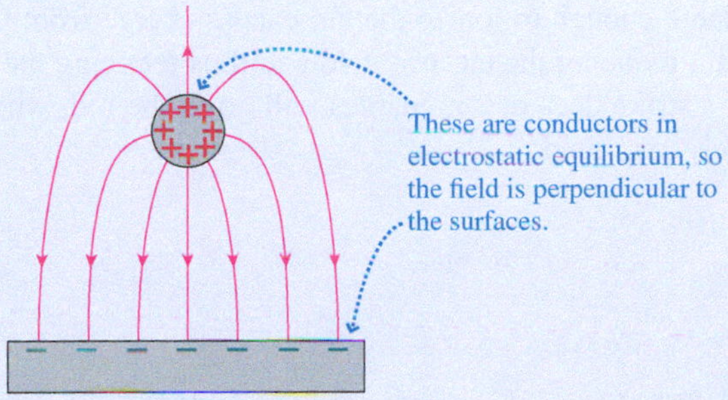

These are conductors in electrostatic equilibrium, so the field is perpendicular to the surfaces.


FIGURE 20.41 A region of space enclosed by conducting walls is screened from electric fields.

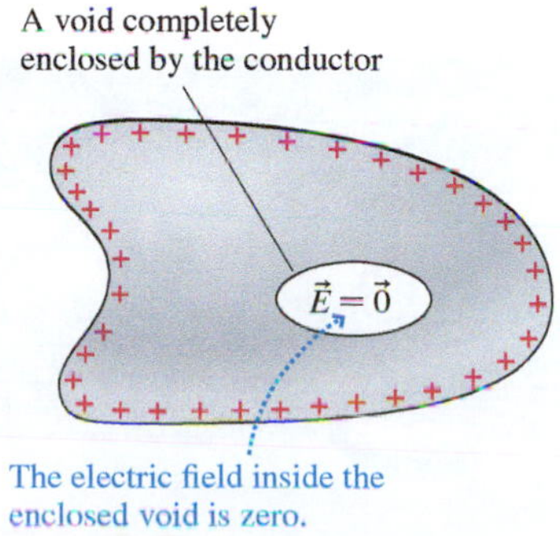


FIGURE 20.41 shows a practical use of these ideas. Here we see a charged conductor with a completely enclosed void. The excess charge on the conductor is at the surface and the electric field within the conductor is zero, so there's nothing that could create an electric field within the enclosure. We can conclude that **the electric field within a conducting enclosure is zero.**

A conducting box can be used to exclude electric fields from a region of space; this is called **screening.** Solid metal walls are ideal, but in practice wire screen or wire mesh provides sufficient screening for all but the most sensitive applications.

### CONCEPTUAL EXAMPLE 20.12 Analyzing static protection

Computer chips and other electronic components are very sensitive to electric charges and fields. Even a small static charge or field may damage them. Such components are shipped and stored in conducting bags. How do these bags protect the components stored inside?

***REASON*** Such a bag, when sealed, forms a conducting shell around its interior. All excess charge is on the surface of the bag, and the electric field inside is zero. A chip or component inside the bag is protected from damaging charges and fields.

**Video** Charged Conductor with Teardrop Shape

**FIGURE 20.42** The electric field is strongest at the pointed end.

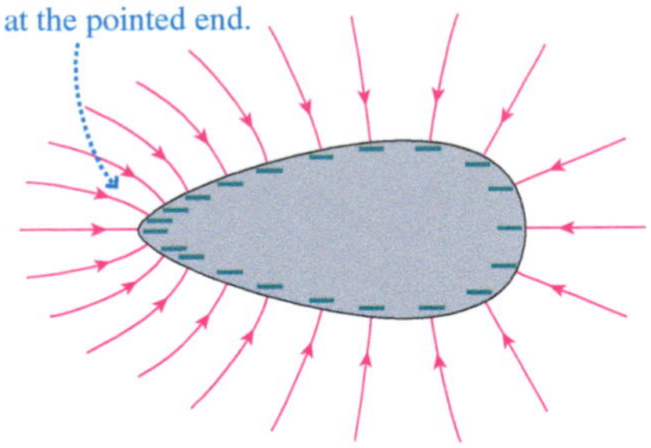


Although any excess charge on a conductor is found on the surface, it may not be uniformly distributed. **FIGURE 20.42** shows a charged conductor that is more pointed at one end than the other. The density of charge is highest—and thus the electric field is strongest—at the pointed end.

The sharper the point, the more closely packed the charges, and so the stronger the field. If there is a great deal of charge and a sharp point, the field near the point can be strong enough to tear electrons from air molecules, ionizing the air around the point. The forces on the resulting ions attract oppositely charged particles to the point, which dissipates the charge. But a more dramatic outcome is possible. Air is normally an insulator, but ionized air is a conductor. If the circumstances are right, the ionization can cause a cascade—electrons are torn from atoms, and the electrons and ions accelerate, striking other molecules and creating more ions. This creates a conducting path, and a spark can jump—or, in the atmosphere, lightning can strike. The electric field necessary for a spark to jump in air is typically between $1 \times 10^6$ N/C and $3 \times 10^6$ N/C; the exact value depends on air pressure, humidity, electrode shape, and other factors.

If you want to protect a building from a lightning strike, you want to defuse the situation before a strike occurs. Tall buildings may have pointed metal rods—lightning rods—on their high points. If charge begins to accumulate on the building, which means a lightning strike could be imminent, a large field develops at the tip of the rod. Once the field is large enough to ionize the air, excess charge from the building will dissipate into the air, reducing the electric field and thus reducing the probability of a lightning strike. If a strike does occur, charges will go to the rod, where a stout cable conducts them safely to ground.

The sharp point of a lightning rod reduces charge accumulation.

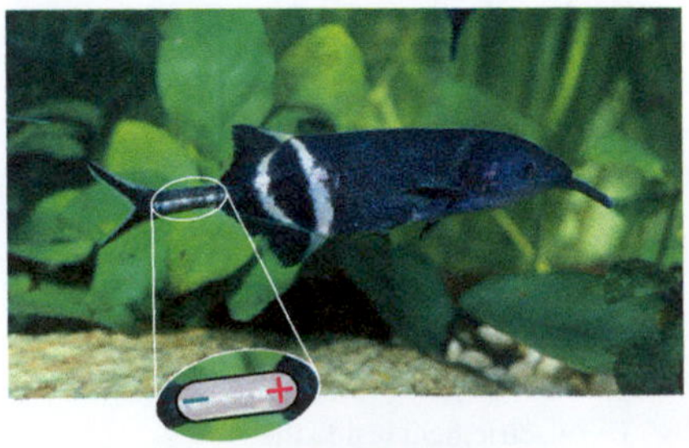

◀ **Electric fish** BIO The platypus can sense the electric fields of prey animals. This elephant nose fish goes a step further, generating its own electric field to explore the murky water where it lives, where vision is of little use. A stack of specially adapted cells called *electrocytes* in the tail create an electric dipole, which produces an electric field in the water around the fish. Sensors along the fish's body detect very small changes in this electric field due to the presence of nearby conductors—such as another fish. When an elephant nose explores a new environment, it often enters tail first, creating a probe field to check things out.

## 20.7 Forces and Torques in Electric Fields

The electric field was defined in terms of the force on a charge. In practice, we often want to turn the definition around to find the force exerted on a charge in a known electric field. If a charge $q$ is placed at a point in space where the electric field is $\vec{E}$, then according to Equation 20.3 the charge experiences an electric force

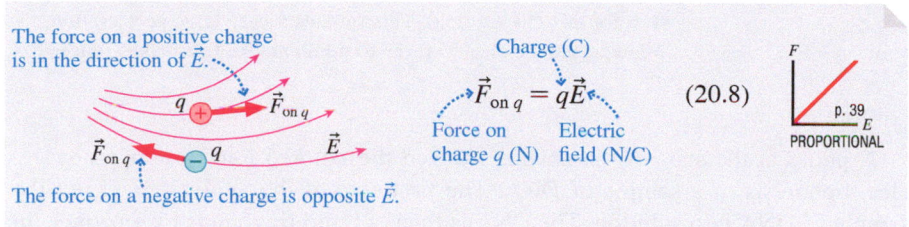

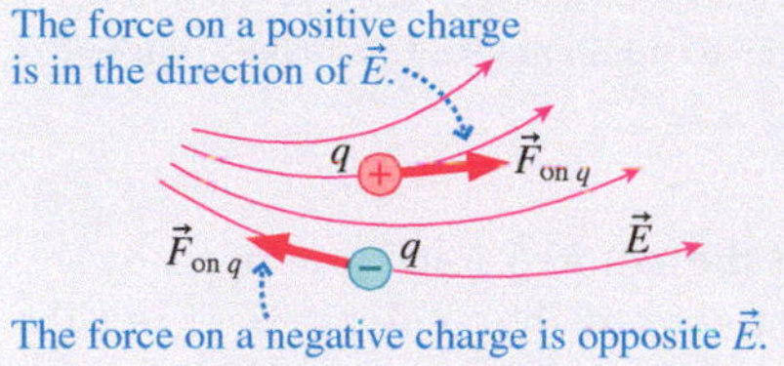


$$\vec{F}_{\text{on } q} = q\vec{E} \qquad (20.8)$$

Charge (C)

Force on charge $q$ (N)

Electric field (N/C)

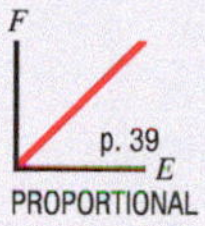


If $q$ is positive, the force on charge $q$ is in the direction of $\vec{E}$. The force on a negative charge is *opposite* the direction of $\vec{E}$.

Whether or not this force is important depends on the situation. When we looked at forces between charges, we saw that the electric force was small for ordinary objects, large for small objects that had been deliberately charged, and completely dominant at the atomic scale. We'll find a similar situation here.

### EXAMPLE 20.13 Comparing electric and gravitational forces

Earlier in the chapter, we saw that tumbling sand and dust grains can develop an electric charge. Typically, smaller grains develop a negative charge, while larger grains develop a positive charge. The smaller grains tend to be lofted higher by the wind, as we'd expect, so when a strong, steady wind blows across a sandy landscape, a charge separation develops. This produces an electric field near the ground, which can reach values as high as 150,000 N/C. At this field strength, how does the magnitude of the electric force compare to the weight force for a typical dust particle of mass $1.8 \times 10^{-12}$ kg and charge $1.2 \times 10^{-16}$ C?

**SOLVE** The magnitude of the electric force on the dust particle is

$$F = qE = (1.2 \times 10^{-16}\ \text{C})(150{,}000\ \text{N/C}) = 1.8 \times 10^{-11}\ \text{N}$$

Magnitude of the charge — Magnitude of the electric field

We'll compare this to the magnitude of the weight force. We haven't used this terminology before, but we can think of $g$ as the magnitude of the gravitational field of the earth, which makes clear the similarity between the electric force and the gravitational force. The magnitude of the weight force is given by $w = mg$; $g$ has units of m/s$^2$, but we can also write this as N/kg. The formula for the gravitational force, the weight force, then has exactly the same form as the formula for the electric force: It's a quantity that the field acts on (charge for the electric field, mass for the gravitational field) multiplied by the magnitude of the field:

$$w = mg = (1.8 \times 10^{-12}\ \text{kg})(9.8\ \text{N/kg}) = 1.8 \times 10^{-11}\ \text{N}$$

Magnitude of the mass — Magnitude of the gravitational field

For this situation, we see that the two forces have the same magnitude.

**ASSESS** The final result makes sense in terms of our earlier discussion. We noted that, in a sandstorm, the charges and fields can become large enough that the electric field could help lift particles from the ground, which intensifies the storm. For this to be true, the electric force needs to be of comparable magnitude to the weight force.

For dust particles, the forces were of comparable magnitude; for electrons and protons, the situation is very different; the electric force completely dominates the scene.

### EXAMPLE 20.14 Finding the force on an electron in the atmosphere

Under normal circumstances, the earth's electric field outdoors near ground level is uniform, about 100 N/C, directed down. What is the electric force on a free electron in the atmosphere? What acceleration does this force cause?

**PREPARE** The electric field is uniform, as shown in the field diagram of FIGURE 20.43. Whatever the position of the electron, it experiences the same field. Because the electron has a negative charge, the force on it is opposite the field—upward.

FIGURE 20.43 An electron in the earth's electric field.

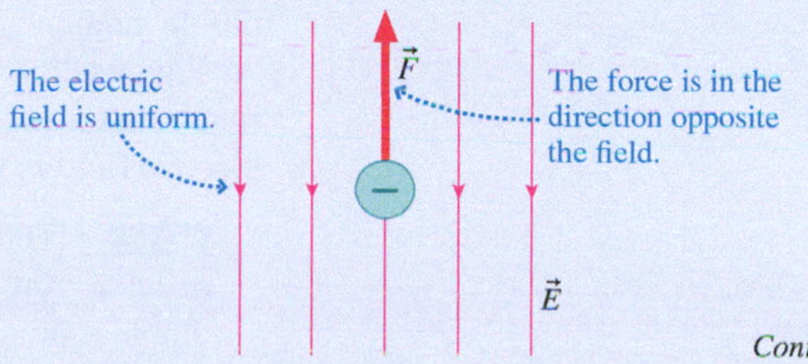


*Continued*

***SOLVE*** The magnitude of the force is given by Equation 20.8:

$$F = eE = (1.6 \times 10^{-19}\ \text{C})(100\ \text{N/C}) = 1.6 \times 10^{-17}\ \text{N}$$

This is much larger than the magnitude of the weight force, $w = mg = (9.1 \times 10^{-31}\ \text{kg})(9.8\ \text{m/s}^2) \sim 10^{-29}\ \text{N}$, so we can ignore the weight force in our calculation for the net force. Thus the net force on the electron is

$$\vec{F} = (1.6 \times 10^{-17}\ \text{N, upward})$$

The electron will accelerate upward, in the direction of the force. The magnitude of the acceleration is

$$a = \frac{F}{m} = \frac{1.6 \times 10^{-17}\ \text{N}}{9.1 \times 10^{-31}\ \text{kg}} = 1.8 \times 10^{13}\ \text{m/s}^2$$

***ASSESS*** The electric force completely dwarfs the weight force, even for this modest field. This implies a very large acceleration. We expected the electric force to dominate, so this makes sense.

FIGURE 20.44 BIO Electrophoresis of DNA samples.

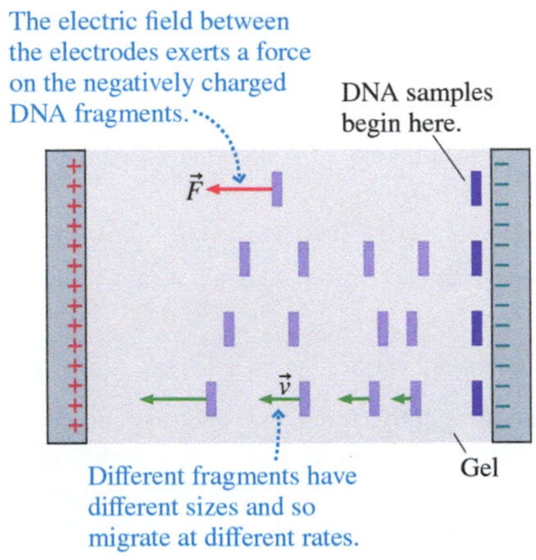


A photo at the start of the chapter showed the colored lines produced by gel electrophoresis of a sample of DNA. The first step of the analysis is to put the sample of DNA into solution. The DNA is then cut into fragments by enzymes. In solution, these fragments have a negative charge. Drops of solution containing the charged DNA fragments are placed in wells at one end of a container of gel. Electrodes at opposite ends of the gel create an electric field that exerts an electric force on the DNA fragments in the solution, as illustrated in FIGURE 20.44. The electric force makes the fragments move through the gel, but drag forces cause fragments of different sizes to migrate at different rates, with smaller fragments migrating faster than larger ones. After some time, the fragments sort themselves into distinct lines, creating a "genetic fingerprint." Two identical samples of DNA will produce the same set of fragments and thus the same pattern in the gel, but the odds are extremely small that two unrelated DNA samples would produce the same pattern.

If an electric dipole is placed in a uniform electric field, as shown in FIGURE 20.45a, the electric force on its negative charge is equal in magnitude but opposite in direction to the force on its positive charge. Thus an electric dipole in a uniform electric field experiences *no net force*. However, as you can see from FIGURE 20.45b, there is a net *torque* on the dipole that causes it to *rotate*.

FIGURE 20.45 Forces and torques on an electric dipole.

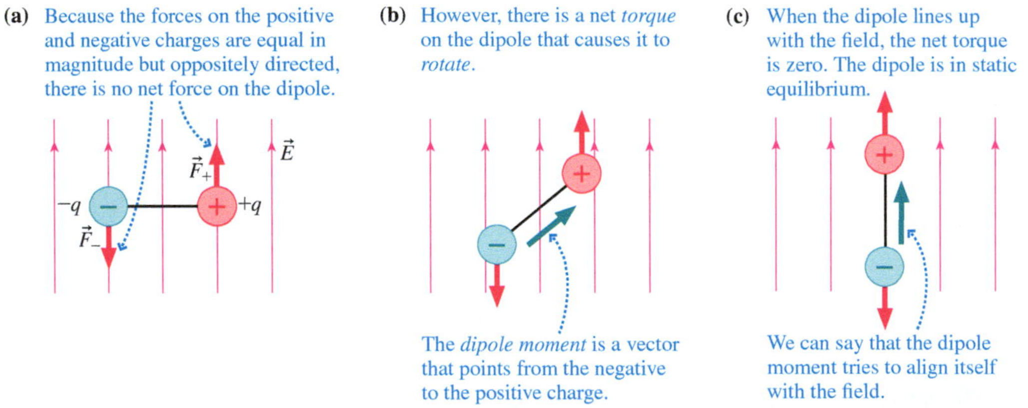

(a) Because the forces on the positive and negative charges are equal in magnitude but oppositely directed, there is no net force on the dipole.
$-q$
$+q$
$\vec{F}_+$
$\vec{F}_-$
$\vec{E}$
(b) However, there is a net *torque* on the dipole that causes it to *rotate*.
The *dipole moment* is a vector that points from the negative to the positive charge.
(c) When the dipole lines up with the field, the net torque is zero. The dipole is in static equilibrium.
We can say that the dipole moment tries to align itself with the field.


It is useful to define the **electric dipole moment,** which we often call simply the dipole moment, a vector that points from the negative to the positive charge of a dipole. As Figures 20.45b and 20.45c show, an electric dipole in a uniform electric field experiences a torque that causes it to rotate. **The equilibrium position of a dipole in an electric field is with the electric dipole moment aligned with the field.**

NOTE ▶ The convention for drawing the electric dipole moment is the opposite of what you may have learned in chemistry. We use this convention so that the dipole moment lines up with an applied electric field; we'll use the same convention for magnetic dipoles and magnetic fields. ◀

Earlier, we saw a photo of grass seeds lined up with the electric field from two charged electrodes. Now we can understand why the seeds line up as they do. First, the electric field polarizes the seeds, inducing opposite charges on their ends. The seeds are induced electric dipoles, with dipole moments along the axis of each seed. Torques on the dipole moments cause them to line up with the electric field, which reveals its structure.

**STOP TO THINK 20.7** Rank in order, from largest to smallest, the forces $F_A$ to $F_E$ a proton would experience if placed at points A to E in this parallel-plate capacitor.

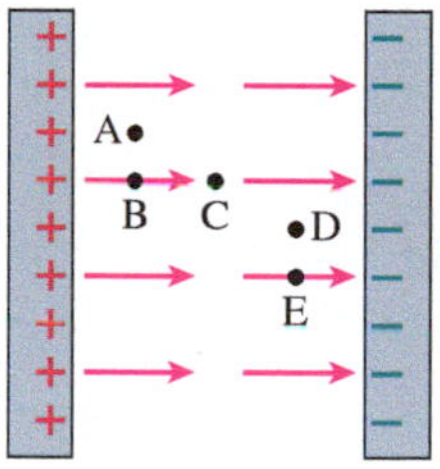


## INTEGRATED EXAMPLE 20.15 A cathode-ray tube

Older televisions and computer monitors, as well as some modern electronic instruments, use a *cathode-ray tube,* or *CRT,* to create an image on a screen. In a CRT, electrons are accelerated by an electric field inside an electron "gun," creating a beam of electrons all moving along in a straight line at the same high speed. A second electric field then steers these electrons to a particular point on a phosphor-coated glass screen, which causes the phosphor to glow brightly at that point. By rapidly varying the steering electric field and the intensity of the electron beam, the spot of electrons can be swept over the entire screen, resulting in the familiar glowing picture of a television.

**FIGURE 20.46** shows a simplified model of the internal structure of a CRT. Electrons—emitted from a hot filament—start with zero speed at the negative plate of a parallel-plate capacitor. The electric field inside this capacitor accelerates the electrons toward the positive plate, where they exit the capacitor with speed $v_1$ through a small hole. They then coast along at this speed until they enter the steering electric field of the deflector. This field causes them to follow a curved trajectory, exiting at an angle $\theta$ with respect to their original direction.

a. The CRT designer has specified that the electrons must leave the 4.0-cm-wide electron gun with a speed of $6.0 \times 10^7$ m/s. What electric field strength is needed inside the electron-gun capacitor?

b. The steering electric field has a constant strength of $1.5 \times 10^5$ N/C over the 5.0 cm length of the deflector. By what angle $\theta$ are the electrons deflected?

**STRATEGIZE** This problem has two separate parts, finding the field necessary to get the electron up to speed and then determining the deflection.

a. We know the speed the electron needs to gain; we will use this to find the necessary force. We will then find the field needed to create this force.

b. We know the speed of the electron, and we will find the electric force pushing it to the side. We will use these details to determine the resulting trajectory.

**PREPARE** We use a coordinate system in which the *x*-axis is horizontal and the *y*-axis vertical.

a. We can use constant-acceleration kinematics to find the electron's acceleration inside the electron gun. Newton's second law then gives the force on the electron, which we can relate to the electric field using Equation 20.3: $\vec{E} = \vec{F}_{\text{on } q}/q$.

b. Because the electric field is vertically down, the force on a negative electron is vertically up. An electron will accelerate vertically, but not horizontally, so the *x*-component of its velocity remains unchanged and equal to $v_1$ as it passes through the deflector. This is exactly analogous to the motion

**FIGURE 20.46** The electron gun and electron deflector of a CRT.

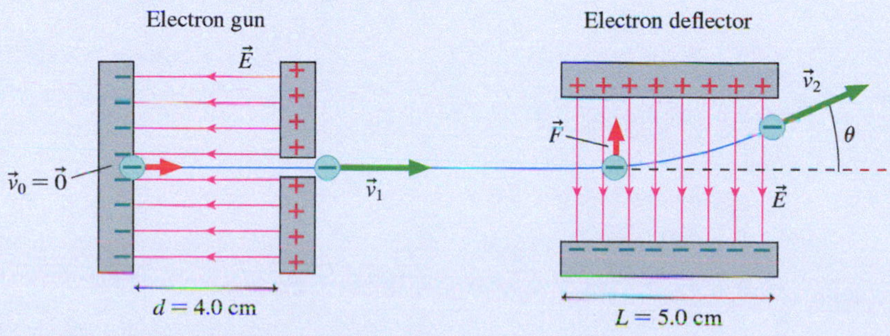


*Continued*

of a projectile, and the electrons follow a projectile-like parabolic trajectory. Just as with projectile motion, we can use the horizontal motion to find the time interval, then use the time interval to find the final velocity in the $y$-direction. As FIGURE 20.47 shows, the ratio of the electron's $y$- and $x$-components of velocity can be used to find $\theta$.

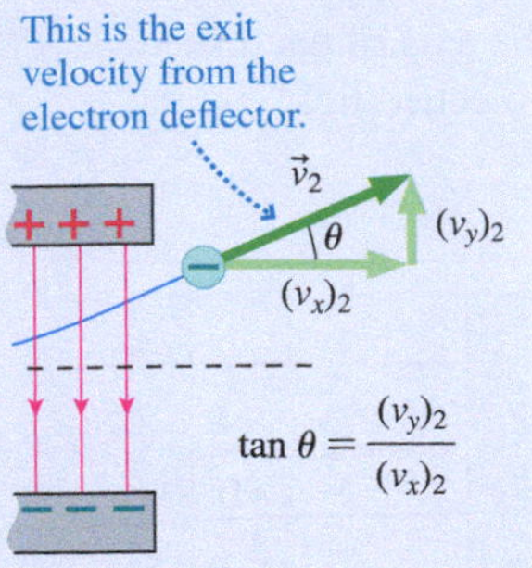


**FIGURE 20.47** The exit velocity of the electron.

**SOLVE**

a. One of the constant-acceleration kinematic equations from Chapter 2 was $(v_x)_1^2 = (v_x)_0^2 + 2a_x\,\Delta x$. Using $(v_x)_0 = 0$ and $\Delta x = d = 4.0$ cm, we find that an electron's acceleration inside the electron gun is

$$a_x = \frac{(v_x)_1^2}{2d} = \frac{(6.0 \times 10^7 \text{ m/s})^2}{2(0.040 \text{ m})} = 4.5 \times 10^{16} \text{ m/s}^2$$

Newton's second law tells us that the force causing this acceleration is

$$F_x = ma_x = (9.1 \times 10^{-31} \text{ kg})(4.5 \times 10^{16} \text{ m/s}^2)$$
$$= 4.1 \times 10^{-14} \text{ N}$$

Then, by Equation 20.3, the electric field is

$$E_x = \frac{F_x}{q} = \frac{4.1 \times 10^{-14} \text{ N}}{-1.6 \times 10^{-19} \text{ C}} = -2.6 \times 10^5 \text{ N/C}$$

This is a field in the negative $x$-direction, as we can see in Figure 20.46, with strength $2.6 \times 10^5$ N/C.

b. The $y$-component of the electron's acceleration in the deflector is

$$a_y = \frac{F_y}{m} = \frac{eE_y}{m} = \frac{(-1.6 \times 10^{-19} \text{ C})(-1.5 \times 10^5 \text{ N/C})}{9.1 \times 10^{-31} \text{ kg}}$$
$$= 2.6 \times 10^{16} \text{ m/s}^2$$

The negative electrons have an upward (positive) acceleration. This acceleration causes an electron to leave the deflector with a $y$-component of velocity

$$(v_y)_2 = a_y\,\Delta t$$

where $\Delta t$ is the time the electron spends in the deflector. Because the $x$-component of the velocity is constant, this time is simply

$$\Delta t = \frac{L}{(v_x)_1} = \frac{0.050 \text{ m}}{6.0 \times 10^7 \text{ m/s}} = 8.3 \times 10^{-10} \text{ s}$$

Thus

$$(v_y)_2 = a_y\,\Delta t = (2.6 \times 10^{16} \text{ m/s}^2)(8.3 \times 10^{-10} \text{ s})$$
$$= 2.2 \times 10^7 \text{ m/s}$$

Referring to Figure 20.47, and using $(v_x)_2 = (v_x)_1$ because there's no horizontal acceleration, we see that

$$\tan\theta = \frac{(v_y)_2}{(v_x)_2} = \frac{2.2 \times 10^7 \text{ m/s}}{6.0 \times 10^7 \text{ m/s}} = 0.37$$

so that

$$\theta = \tan^{-1}(0.37) = 20°$$

**ASSESS** A strong field accelerates the electrons in the $x$-direction, while a weaker one accelerates them in the $y$-direction. Thus it is reasonable that the ratio of the $y$- to the $x$-component of velocity is significantly less than 1.

The CRT shown in Figure 20.46 deflects electrons only vertically. A real CRT has a second electron deflector, rotated 90°, to provide a horizontal deflection. The two deflectors working together can scan the electron beam over all points on the screen.

# SUMMARY

**GOAL** To develop a basic understanding of electric phenomena in terms of charges, forces, and fields.

## GENERAL PRINCIPLES

### Charge

There are two kinds of charges, called **positive** and **negative.**

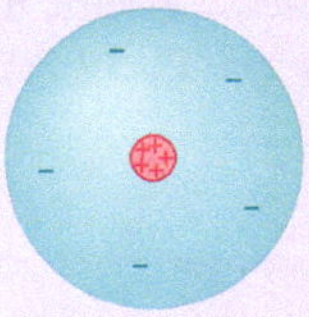

- Atoms consist of a nucleus containing positively charged protons surrounded by a cloud of negatively charged electrons.
- The **fundamental charge** $e$ is the magnitude of the charge on an electron or proton: $e = 1.60 \times 10^{-19}$ C.
- Matter with equal amounts of positive and negative charge is **neutral.**
- Charge is conserved; it can't be created or destroyed.

### Coulomb's Law

The forces between two charged particles $q_1$ and $q_2$ separated by distance $r$ are

$$F_{1\text{ on }2} = F_{2\text{ on }1} = \frac{K|q_1||q_2|}{r^2}$$

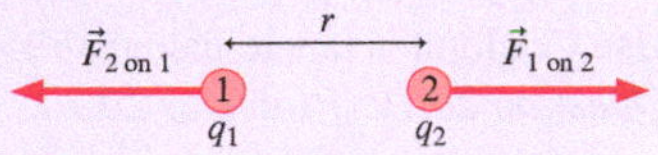


where $K = 8.99 \times 10^9\ \text{N}\cdot\text{m}^2/\text{C}^2 \approx 9.0 \times 10^9\ \text{N}\cdot\text{m}^2/\text{C}^2$ is the **electrostatic constant.** These forces are an action/reaction pair directed along the line joining the particles.

- The forces are repulsive for two like charges, attractive for two opposite charges.
- The net force on a charge is the vector sum of the forces from all other charges.
- The unit of charge is the coulomb (C).

## IMPORTANT CONCEPTS

### The Electric Field

Charges interact with each other via the electric field $\vec{E}$.

- Charge A alters the space around it by creating an electric field.

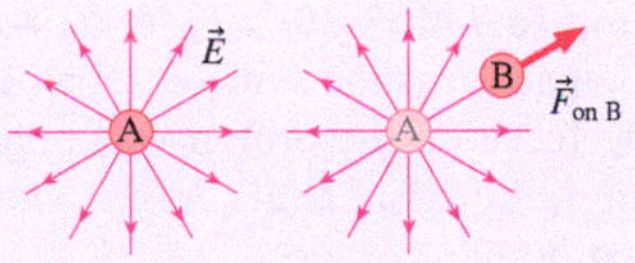


- The field is the agent that exerts a force on charge B. $\vec{F}_{\text{on B}} = q_B\vec{E}$
- An electric field is identified and measured in terms of the force on a probe charge $q$. The unit of the electric field is N/C. $\vec{E} = \dfrac{\vec{F}_{\text{on } q}}{q}$
- The electric field is a vector. The electric field from multiple charges is the vector sum of the fields from the individual charges. $\vec{E}_{\text{total}} = \vec{E}_1 + \vec{E}_2 + \cdots$

### Visualizing the electric field

The electric field exists at all points in space.

- An electric field vector shows the field at only one point, the point at the tail of the vector.
- A **field diagram** shows field vectors at several points.

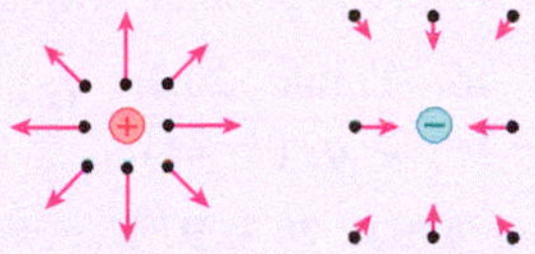

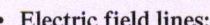

- **Electric field lines:**
  - are always parallel to the field vectors.
  - are close where the field is strong, far apart where the field is weak.
  - go from positive to negative charges.

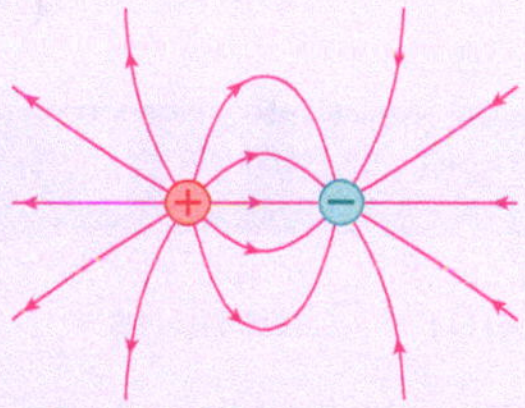

## APPLICATIONS

There are two types of material, **insulators** and **conductors.**

- Charge remains fixed on an insulator.
- Charge moves easily through conductors.
- Charge is transferred by contact between objects.

A **dipole** has no net charge, but has a field because the two charges are separated.

A dipole will rotate to align with an electric field.

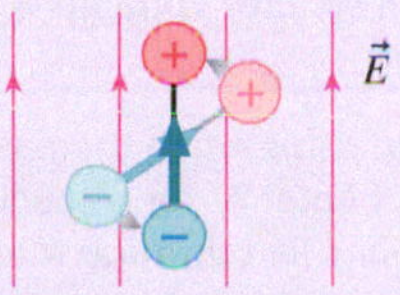


### Electric fields: important cases

The electric field of a **point charge** is

$$\vec{E} = \left(\frac{K|q|}{r^2}, \begin{bmatrix} \text{away from } q \text{ if } q > 0 \\ \text{toward } q \text{ if } q < 0 \end{bmatrix}\right)$$

The electric field inside a **parallel-plate capacitor** is uniform:

$$\vec{E} = \left(\frac{Q}{\epsilon_0 A}, \text{ from positive to negative}\right)$$

where $\epsilon_0 = 8.85 \times 10^{-12}\ \text{C}^2/\text{N}\cdot\text{m}^2$ is the **permittivity constant.**

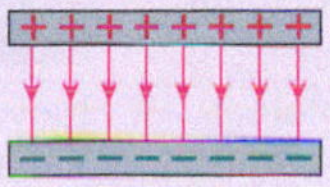

### Conductors in electric fields

- The electric field inside a conductor in **electrostatic equilibrium** is zero.
- Any excess charge is on the surface.

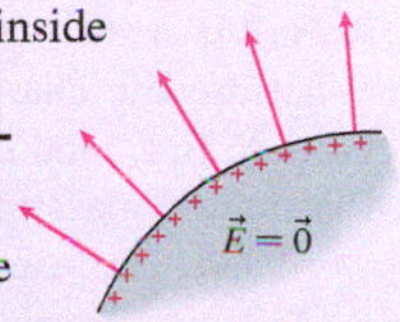


- The electric field is perpendicular to the surface.
- The density of charge and the electric field are highest near a pointed end.

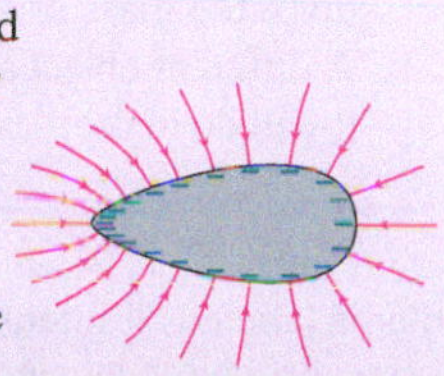

## Learning Objectives After studying this chapter, you should be able to:

- Use the charge model to explain basic electric phenomena. *Conceptual Questions 20.10, 20.26*
- Work with conductors and insulators, charging and discharging, and charge polarization. *Conceptual Questions 20.1, 20.2; Problems 20.4, 20.5, 20.6, 20.7, 20.8*
- Apply charge conservation and recognize the connection of charge to atoms and ions. *Conceptual Questions 20.12, 20.15; Problems 20.1, 20.3*
- Use Coulomb's law to calculate electric forces. *Conceptual Question 20.19; Problems 20.10, 20.11, 20.12, 20.14, 20.15*
- Understand and use the electric fields of point charges and dipoles. *Conceptual Questions 20.14, 20.23; Problems 20.20, 20.23, 20.24, 20.26, 20.29*
- Solve problems that involve parallel-plate capacitors and uniform electric fields. *Conceptual Questions 20.18, 20.22; Problems 20.31, 20.33, 20.34, 20.35, 20.36*
- Calculate electric forces and torques on charged particles and dipoles in electric fields. *Conceptual Questions 20.20, 20.33; Problems 20.39, 20.41, 20.42, 20.43, 20.47*

## STOP TO THINK ANSWERS

**Chapter Preview Stop to Think: A.** The $x$-component of the tension is directed to the left, so it is negative. This rules out answers B, D, and E. Also, the $x$-component of the tension is related to the cosine of the 30° angle, which, according to the question, is greater than the sine of 30°. Thus answer A, where the magnitude of the $x$-component is greater than that of the $y$-component, must be correct.

**Stop to Think 20.1: A.** The electroscope is originally given a positive charge. The charge spreads out, and the leaves repel each other. When a rod with a negative charge is brought near, some of the positive charge is attracted to the top of the electroscope, away from the leaves. There is less charge on the leaves, and so they move closer together.

**Stop to Think 20.2:** $q_E(+3e) > q_A(+1e) > q_D(0) > q_B(-1e) > q_C(-2e)$.

**Stop to Think 20.3: B.** The two forces are an action/reaction pair, opposite in direction but *equal* in magnitude.

**Stop to Think 20.4:** $E_B > E_A > E_D > E_C$. The field is proportional to the charge, and inversely proportional to the square of the distance.

**Stop to Think 20.5:** $E_3 = E_4 > E_2 > E_1$. The electric field is strongest where the electric field lines are closest together. At points 3 and 4 the spacing is the same, so the electric field strengths at these two points are equal.

**Stop to Think 20.6: C.** Electric field lines *start* on positive charges. Very near to each of the positive charges, the field lines should look like the field lines of a single positive charge.

**Stop to Think 20.7:** $F_A = F_B = F_C = F_D = F_E$. The field inside a capacitor is the same at all points. Because the field is uniform, the force on the proton will be the same at all points. The electric field exists at all points whether or not a vector is shown at that point.

Video Tutor Solution Chapter 20

# QUESTIONS

## Conceptual Questions

1. Four lightweight balls A, B, C, and D are suspended by threads. Ball A has been touched by a plastic rod that was rubbed with wool. When the balls are brought close together, without touching, the following observations are made:
   - Balls B, C, and D are attracted to ball A.
   - Balls B and D have no effect on each other.
   - Ball B is attracted to ball C.

   What are the charge states (positive, negative, or neutral) of balls A, B, C, and D? Explain.
2. Plastic and glass rods that have been charged by rubbing with wool and silk, respectively, hang by threads.
   a. An object repels the plastic rod. Can you predict what it will do to the glass rod? If so, what? If not, why not? Explain.
   b. A different object attracts the plastic rod. Can you predict what it will do to the glass rod? If so, what? If not, why not? Explain.
3. When you take clothes out of the dryer right after it stops, the clothes often stick to your hands and arms. Explain how the clothing acquires a charge, and why the items of clothing stick to you.
4. The positive charge in Figure Q20.4 is $+Q$. What is the negative charge if the electric field at the dot is zero?

FIGURE Q20.4

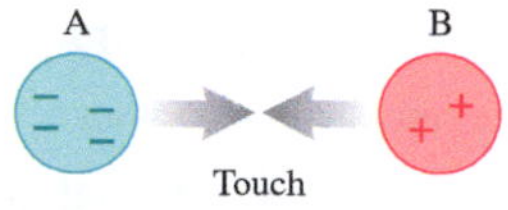


FIGURE Q20.5

5. As shown in Figure Q20.5, metal sphere A has 4 units of negative charge and metal sphere B has 2 units of positive charge. The two spheres are brought into contact. What is the final charge state of each sphere? Explain.

Problem difficulty is labeled as | (straightforward) to ||||| (challenging). Problems labeled INT integrate significant material from earlier chapters; Problems labeled BIO are of biological or medical interest.

The eText icon indicates when there is a video tutor solution available for the chapter or for a specific problem. To launch these videos, log into your eText through Mastering™ Physics or log into the Study Area.

6. Figure Q20.6 shows a positively charged rod held near, but not touching, a neutral metal sphere.
   a. Add plusses and minuses to the figure to show the charge distribution on the sphere.
   b. Does the sphere experience a net force? If so, in which direction? Explain.

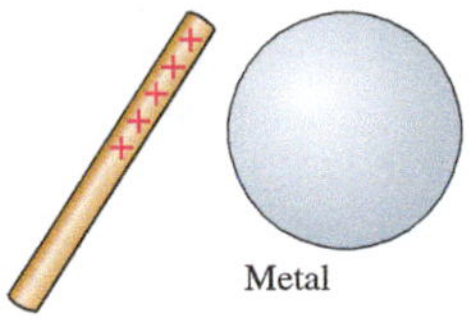


FIGURE Q20.6

7. A plastic balloon that has been rubbed with wool will stick to a wall.
   a. Can you conclude that the wall is charged? If not, why not? If so, where does the charge come from?
   b. Draw a charge diagram showing how the balloon is held to the wall.
8. You are given two metal spheres on portable insulating stands, a glass rod, and a piece of silk. Explain how to give the spheres *exactly* equal but opposite charges.
9. A metal rod A and a metal sphere B, on insulating stands, touch each other as shown in Figure Q20.9. They are originally neutral. A positively charged rod is brought near (but not touching) the far end of A. While the charged rod is still close, A and B are separated. The charged rod is then withdrawn. Is the sphere then positively charged, negatively charged, or neutral? Explain.

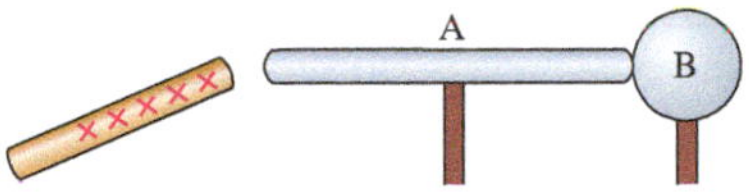


FIGURE Q20.9

10. Water that comes from the tap is a good conductor of electricity. In the chapter preview, you can see a photograph that shows a stream of water being attracted to a charged comb.
    a. Explain why the water stream is attracted to the comb.
    b. Water droplets that break off from the stream are strongly charged. If the comb has a positive charge, what is the charge of the droplets?
11. When you are painting a car with a sprayer, you get more even coverage (the paint droplets are evenly spread out) and less overspray (fewer droplets of paint that don't end up on the vehicle) if you give the droplets a strong electric charge.
    a. Why does charging the droplets give more even coverage?
    b. Why does charging the droplets help ensure that most of the paint ends up on the car?
12. BIO The relative proportions of cytosine-guanine and adenine-thymine bonds in a DNA sample can be estimated by measuring its "melting temperature," the temperature at which half of the DNA strands have pulled apart. Samples with a high percentage of cytosine-guanine pairs have a higher melting temperature than samples with a high percentage of adenine-thymine pairs. Explain why this is so, considering the nature of the bonds that hold the base pairs together (look back at Figure 20.14).
13. BIO A hummingbird gains a significant electric charge while flying and feeding. This has consequences: When a charged bird approaches a flower, the stamens of the flower bend toward the bird, even though the stamens are uncharged. Explain how this happens.

14. BIO A bumblebee can sense electric fields. The most likely mechanism underlying this sensitivity is the motion of hairs on the bee's body. The hairs bend in response to an electric field, and the bee senses this motion. Support for this mechanism as the source of the sensitivity is the fact that putting an electric charge on a bee (as occurs when a bee flies through the air) dramatically increases its electric field sensitivity. Explain why adding charge to a bee causes hairs on the bee's body to bend more in response to a field.
15. BIO Many drugs under development are delivered by nanoparticles in the bloodstream. To monitor changes in the nanoparticles, investigators can retrieve them from the blood by using a device with electrodes that apply oscillating electric fields. The nanoparticles, which are electrically conducting, are strongly attracted to the nearest electrode, while blood cells, which are poor conductors, experience only a weak force toward the nearest electrode, and suspended ions in the blood experience no net force. Explain why the conducting nanoparticles are strongly attracted to the electrodes while other components of the blood are not.
16. Each part of Figure Q20.16 shows two points near two charges. Compare the electric field strengths $E_1$ and $E_2$ at these two points. Is $E_1 > E_2$, $E_1 = E_2$, or $E_1 < E_2$?

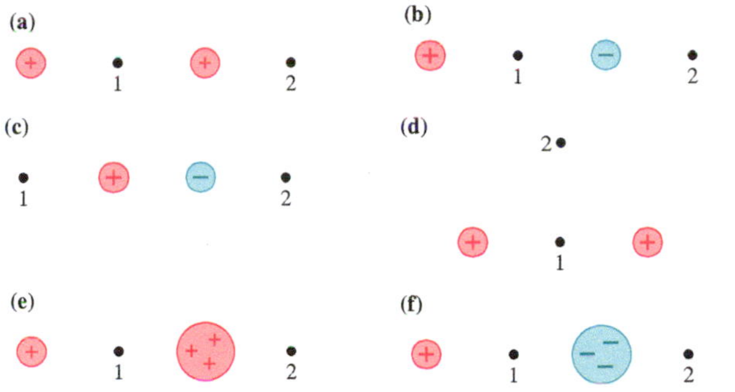


FIGURE Q20.16

17. BIO Iontophoresis is a noninvasive process that transports drugs through the skin without needles. In the photo, the red electrode is positive and the black electrode is negative. The electric field between the electrodes will drive the negatively charged molecules of an anesthetic through the skin. Should the drug be placed at the red or the black electrode? Explain.

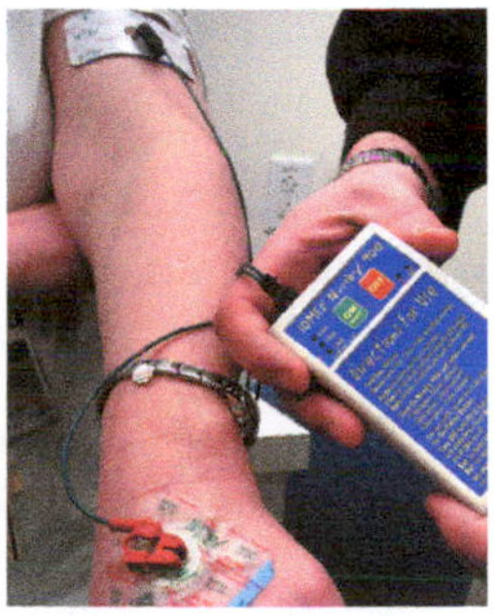

18. A positively charged particle is in the center of a parallel-plate capacitor that has charge $\pm Q$ on its plates. Suppose the distance between the plates is doubled, with the charged particle remaining in the center. Does the force on this particle increase, decrease, or stay the same? Explain.
19. Two charged particles are separated by 10 cm. Suppose the charge on each particle is doubled. By what factor does the electric force between the particles change?
20. A small positive charge $q$ experiences a force of magnitude $F_1$ when placed at point 1 in Figure Q20.20. In terms of $F_1$:
    a. What is the magnitude of the force on charge $q$ at point 3?
    b. What is the magnitude of the force on a charge $3q$ at point 1?
    c. What is the magnitude of the force on a charge $2q$ at point 2?
    d. What is the magnitude of the force on a charge $-2q$ at point 2?

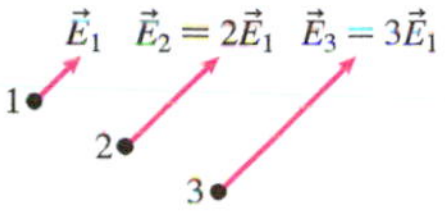


FIGURE Q20.20

21. A typical commercial airplane is struck by lightning about once per year. When this happens, the external metal skin of the airplane might be burned, but the people and equipment inside the aircraft experience no ill effects. Explain why this is so.

22. Microbes such as bacteria have small positive charges when BIO in solution. Public health agencies are exploring a new way to measure the presence of microbes in drinking water by using electric forces to concentrate the microbes. Water is sent between the two oppositely charged electrodes of a parallel-plate capacitor. Microbes in the water collect on one of the electrodes.
   a. On which electrode will the microbes collect?
   b. How could the microbes be easily removed from the electrodes for analysis?

23. a. Is there a point between a 10 nC charge and a 20 nC charge at which the electric field is zero? If so, which charge is this point closer to? If not, why not?
   b. Repeat part a for the case of a 10 nC charge and a −20 nC charge.

24. A Van de Graaff generator is meant to collect a large amount of charge on a spherical conductor. Explain why the spherical shape of the conductor is important for collecting as much charge as possible.

25. When the wind kicks up dust and sand, the dust grains are charged. The small grains tend to get a negative charge, and the large grains a positive charge. The small grains are lifted higher by the wind. What is the direction of the electric field produced by this charge separation?

## Multiple-Choice Questions

26. | Two lightweight, electrically neutral conducting balls hang from threads. Choose the diagram in Figure Q20.26 that shows how the balls hang after:
   a. Both are touched by a negatively charged rod.
   b. Ball 1 is touched by a negatively charged rod and ball 2 is touched by a positively charged rod.
   c. Both are touched by a negatively charged rod but ball 2 picks up more charge than ball 1.
   d. Only ball 1 is touched by a negatively charged rod.
   Note that parts a through d are independent; these are not actions taken in sequence.

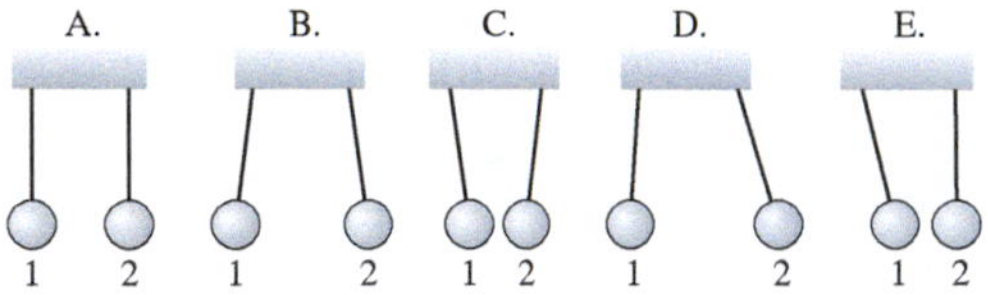


FIGURE Q20.26

27. | All the charges in Figure Q20.27 have the same magnitude. In which case does the electric field at the dot have the largest magnitude?

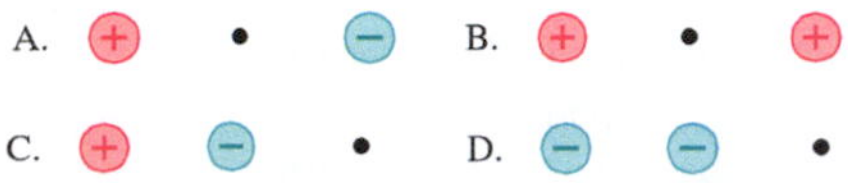


FIGURE Q20.27

28. | All the charges in Figure Q20.28 have the same magnitude. In which case does the electric field at the dot have the largest magnitude?

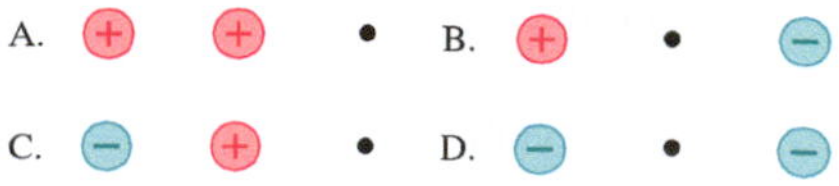


FIGURE Q20.28

29. | All the charges in Figure Q20.29 have the same magnitude. In which case does the electric field at the dot have the largest magnitude?

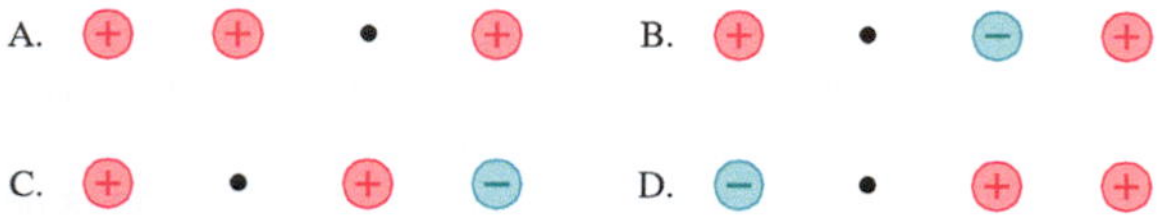


FIGURE Q20.29

30. | A glass bead charged to +3.5 nC exerts an $8.0 \times 10^{-4}$ N repulsive electric force on a plastic bead 2.9 cm away. What is the charge on the plastic bead?
   A. +2.1 nC B. +7.4 nC C. +21 nC D. +740 nC

31. | A +7.5 nC point charge and a −2.0 nC point charge are 3.0 cm apart. What is the electric field strength at the midpoint between the two charges?
   A. $3.3 \times 10^3$ N/C B. $5.7 \times 10^3$ N/C
   C. $2.2 \times 10^5$ N/C D. $3.8 \times 10^5$ N/C

32. || Three point charges are arranged as shown in Figure Q20.32. Which arrow best represents the direction of the electric field vector at the position of the dot?

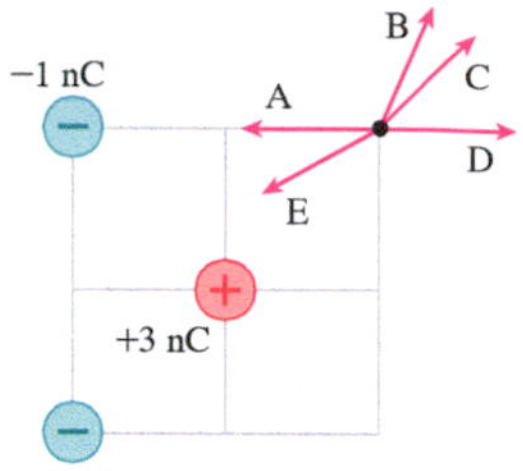


FIGURE Q20.32

33. | A positive charge is brought near to a dipole, as shown in Figure Q20.33. If the dipole is free to rotate, it

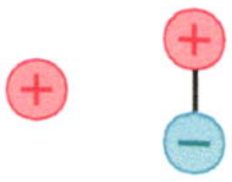

FIGURE Q20.33

   A. Begins to rotate in a clockwise direction.
   B. Begins to rotate in a counterclockwise direction.
   C. Remains stationary.

Questions 34 through 37 concern recent clinical results using electric fields to disrupt rapidly dividing tumor cells. When cells are dividing, some elements of the cell develop a small net charge, and others develop a dipole structure with a positive charge on one end and a negative charge on the other. When a uniform electric field is applied to the body, the field will not be uniform inside the body. The motion of charges and polarization effects, both of which vary strongly from place to place, create a field that is quite variable at the level of the cell. The field is strong in some places and weak in others. The varying field exerts forces and torques on the cell components. This can disrupt cell division, which can have an especially large impact on the rapidly dividing cells in a tumor. To illustrate this idea, we consider a cylindrical element with a dipole structure but no net charge, embedded in a dividing cell. The varying environment of the cell creates a local variation in the strength of the field. The field and the dipole element are shown in Figure Q20.34.

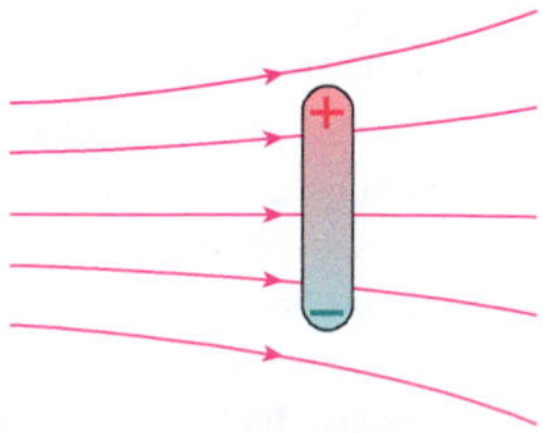

FIGURE Q20.34

34. | The strength of the field in Figure Q20.34 varies. The
BIO strength of the field is greatest
   A. On the left side of the figure.
   B. On the right side of the figure.
   C. On one of the field lines.
   D. In between the field lines.
35. || At the instant shown in Figure Q20.34, there is a net torque that will rotate the element
   A. Clockwise
   B. Counterclockwise
36. || The element will rotate into a position in which the net torque is zero. Once the element is in this position, the net force is directed
   A. To the right.   B. To the left.
37. || The net result of the rotation and the subsequent motion is that the element's dipole moment ____________, and the element moves toward the region of _______ field.
   A. lines up with the field, stronger
   B. lines up with the field, weaker
   C. lines up opposite the field, stronger
   D. lines up opposite the field, weaker

# PROBLEMS

### Section 20.1 Charges and Forces

### Section 20.2 Charges, Atoms, and Molecules

1. ||| A glass rod is charged to $+5.0$ nC by rubbing.
   a. Have electrons been removed from the rod or protons added? Explain.
   b. How many electrons have been removed or protons added?
2. | When a honeybee flies through the air, it develops a charge
BIO of $+17$ pC. How many electrons did it lose in the process of acquiring this charge?
3. ||| A plastic rod is charged to $-20$ nC by rubbing.
   a. Have electrons been added to the rod or protons removed? Explain.
   b. How many electrons have been added or protons removed?
4. | A housefly walking across a surface may develop a sig-
BIO nificant electric charge through a process similar to frictional charging. Suppose a fly picks up a charge of $+52$ pC. How many electrons does it lose to the surface it is walking across?
5. || A plastic rod that has been charged to $-15.0$ nC touches a metal sphere. Afterward, the rod's charge is $-10.0$ nC.
   a. What kind of charged particle was transferred between the rod and the sphere, and in which direction? That is, did it move from the rod to the sphere or from the sphere to the rod?
   b. How many charged particles were transferred?
6. || A glass rod that has been charged to $+12.0$ nC touches a metal sphere. Afterward, the rod's charge is $+8.0$ nC.
   a. What kind of charged particle was transferred between the rod and the sphere, and in which direction? That is, did it move from the rod to the sphere or from the sphere to the rod?
   b. How many charged particles were transferred?
7. ||| Two identical metal spheres A and B are in contact. Both are initially neutral. $1.0 \times 10^{12}$ electrons are added to sphere A, then the two spheres are separated. Afterward, what are the charge of A and the charge of B?
8. || Two identical metal spheres A and B are connected by a plastic rod. Both are initially neutral. $1.0 \times 10^{12}$ electrons are added to sphere A, then the connecting rod is removed. Afterward, what are the charge of A and the charge of B?
9. || If two identical conducting spheres are in contact, any excess charge will be evenly distributed between the two. Three identical metal spheres are labeled A, B, and C. Initially, A has charge $q$, B has charge $-q/2$, and C is uncharged. What is the final charge on each sphere if C is touched to B, removed, and then touched to A?

### Section 20.3 Coulomb's Law

10. || Falling raindrops frequently develop electric charges. Does
INT this create noticeable forces between the droplets? Suppose two 1.8 mg drops each have a charge of $+25$ pC; these are typical values. The centers of the droplets are at the same height and 0.40 cm apart. What is the approximate electric force between them? What horizontal acceleration does this force produce on the droplets?
11. || Two 1.0 kg masses are 1.0 m apart on a frictionless table.
INT Each has $+1.0\ \mu$C of charge.
   a. What is the magnitude of the electric force on one of the masses?
   b. What is the initial acceleration of each mass if they are released and allowed to move?
12. || Sodium chloride (NaCl) is an ionic compound. We can
INT effectively model it as sodium atoms with one electron removed next to chlorine atoms with one electron added. The attractive force between the positive and negative charges holds the crystal together. If we model the sodium and chlorine ions as point charges separated by 0.28 nm, the spacing in the crystal, what is the magnitude of the attractive force between two adjacent ions? How does this force compare to the weight of the sodium ion, which has a mass of 23 u?
13. ||| A small metal sphere has a mass of 0.15 g and a charge of $-23.0$ nC. It is 10.0 cm directly above an identical sphere that has the same charge. This lower sphere is fixed and cannot move. If the upper sphere is released, it will begin to fall. What is the magnitude of its initial acceleration?
14. ||| A small plastic sphere with a charge of $-5.0$ nC is near another small plastic sphere with a charge of $-12$ nC. If the spheres repel one another with a force of magnitude $8.2 \times 10^{-4}$ N, what is the distance between the spheres?

15. ||| A small metal bead, labeled A, has a charge of 25 nC. It is touched to metal bead B, initially neutral, so that the two beads share the 25 nC charge, but not necessarily equally. When the two beads are then placed 5.0 cm apart, the force between them is $5.4 \times 10^{-4}$ N. What are the charges $q_A$ and $q_B$ on the beads?

16. |||| A small glass bead has been charged to +20 nC. A tiny ball bearing 1.0 cm above the bead feels a 0.018 N downward electric force. What is the charge on the ball bearing?

17. | What are the magnitude and direction of the electric force on charge A in Figure P20.17?

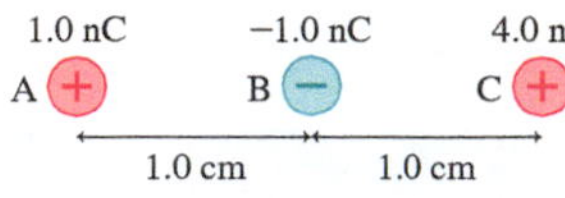


FIGURE P20.17

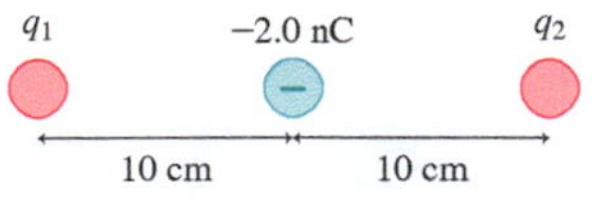


FIGURE P20.18

18. ||| In Figure P20.18, charge $q_2$ experiences no net electric force. What is $q_1$?

19. | Object A, which has been charged to +10 nC, is at the origin. Object B, which has been charged to −20 nC, is at $(x, y) = (0.0 \text{ cm}, 2.0 \text{ cm})$. What are the magnitude and direction of the electric force on each object?

## Section 20.4 The Concept of the Electric Field

20. || A platypus foraging for prey can detect an electric field as
BIO small as 0.002 N/C. To give an idea of the sensitivity of the platypus's electric sense, how far from a +10 nC point charge does the field have this magnitude?

21. || What magnitude charge creates a 1.0 N/C electric field at a point 1.0 m away?

22. || Honeybees accumulate charge as they fly, and they transfer
BIO charge to the flowers they visit. Honeybees are able to sense electric fields; tests show that they can detect a change in field as small as 0.77 N/C. Honeybees seem to use this sense to determine the charges on flowers in order to detect whether or not a flower has been recently visited, so they can plan their foraging accordingly. As a check on this idea, let's do a quick calculation using typical numbers for charges on flowers. If a bee is at a distance of 20 cm, can it detect the difference between flowers that have a +30 pC charge and a +40 pC charge?

23. || What are the strength and direction of the electric field 2.0 cm from a small glass bead that has been charged to +6.0 nC?

24. || A bumblebee can sense electric fields as the fields
BIO bend hairs on its body. Bumblebees have been conclusively shown to detect an electric field of 60 N/C. Could a bumblebee use this sense to detect the presence of another nearby bumblebee? Suppose a bumblebee has a charge of 24 pC. How far away could another bumblebee detect its presence?

25. | What are the strength and direction of the electric field 1.0 mm from (a) a proton and (b) an electron?

26. | A 30 nC charge experiences a 0.035 N electric force. What is the magnitude of electric field at the position of this charge?

27. | A −10 nC charge is located at the origin.
   a. What are the strengths of the electric fields at the positions $(x, y) = (0.0 \text{ cm}, 5.0 \text{ cm})$, $(-5.0 \text{ cm}, -5.0 \text{ cm})$, and $(-5.0 \text{ cm}, 5.0 \text{ cm})$?
   b. Draw a field diagram showing the electric field vectors at these points.

28. | A +10 nC charge is located at the origin.
   a. What are the strengths of the electric fields at the positions $(x, y) = (5.0 \text{ cm}, 0.0 \text{ cm})$, $(-5.0 \text{ cm}, 5.0 \text{ cm})$, and $(-5.0 \text{ cm}, -5.0 \text{ cm})$?
   b. Draw a field diagram showing the electric field vectors at these points.

29. ||| What are the strength and direction of the electric field at the position indicated by the dot in Figure P20.29? Specify the direction as an angle above or below horizontal.

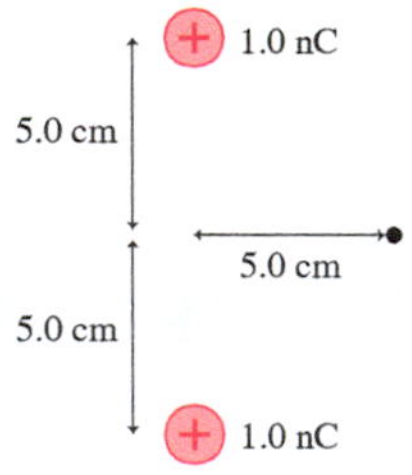


FIGURE P20.29

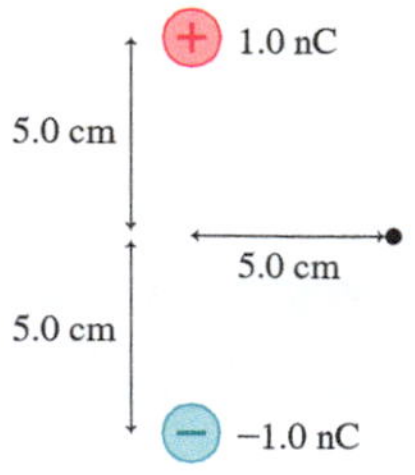


FIGURE P20.30

30. || What are the strength and direction of the electric field at the position indicated by the dot in Figure P20.30? Specify the direction as an angle above or below horizontal.

## Section 20.5 The Electric Field from Arrangements of Charges

31. || A parallel-plate capacitor is constructed of two square plates, size $L \times L$, separated by distance $d$. The plates are given charge $\pm Q$. Let's consider how the electric field changes if one of these variables is changed while the others are held constant. What is the ratio $E_f/E_i$ of the final electric field strength $E_f$ to the initial electric field strength $E_i$ if:
   a. $Q$ is doubled?
   b. $L$ is doubled?
   c. $d$ is doubled?

32. || Use what you know about drawing electric field lines to sketch the electric field of a *quadrupole,* two positive and two negative charges arranged as in Figure P20.32.

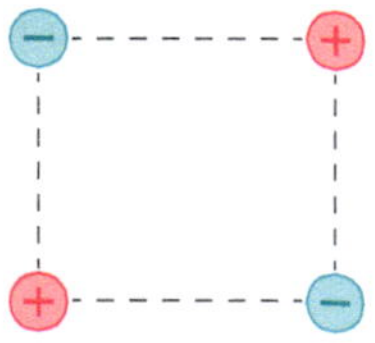

FIGURE P20.32

33. || A parallel-plate capacitor is formed from two 4.0 cm × 4.0 cm electrodes spaced 2.0 mm apart. The electric field strength inside the capacitor is $1.0 \times 10^6$ N/C. What is the charge (in nC) on each electrode?

34. | Two identical closely spaced circular disks form a parallel-plate capacitor. Transferring $1.5 \times 10^9$ electrons from one disk to the other causes the electric field strength between them to be $1.0 \times 10^5$ N/C. What are the diameters of the disks?

## Section 20.6 Conductors and Electric Fields

35. | A parallel-plate capacitor is constructed of two horizontal 12.0-cm-diameter circular plates. A 1.0 g plastic bead, with a charge of −6.0 nC, is suspended between the two plates by the force of the electric field between them.
   a. Which plate, the upper or the lower, is positively charged?
   b. What is the charge on the positive plate?

36. ||| Students in an introductory physics lab are performing an experiment with a parallel-plate capacitor made of two circular aluminum plates, each 20 cm in diameter, separated by 1.0 cm. How much charge can be added to each of the plates before a spark jumps between the two plates? For such flat electrodes, the field that causes a spark is at the high end of the range presented in the chapter; assume a value of $3 \times 10^6$ N/C.

37. ||| Storm clouds may build up large negative charges near their bottom edges. The earth is a good conductor, so the charge on the cloud attracts an equal and opposite charge on the earth under the cloud. The electric field strength near the earth depends on the shape of the earth's surface, as we can explain with a simple model. The top metal plate in Figure P20.37 has uniformly distributed negative charge. The bottom metal plate, which has a high point, has an equal and opposite charge that is free to move.
    a. Sketch the two plates and the region between them, showing the distribution of positive charge on the bottom plate.
    b. Complete your diagram by sketching electric field lines between the two plates. Be sure to note the direction of the field. Where is the field strongest?
    c. Explain why it is more dangerous to be on top of a hill or mountain during a lightning storm than on level ground.

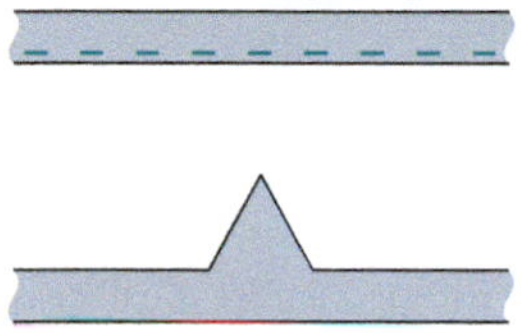

FIGURE P20.37

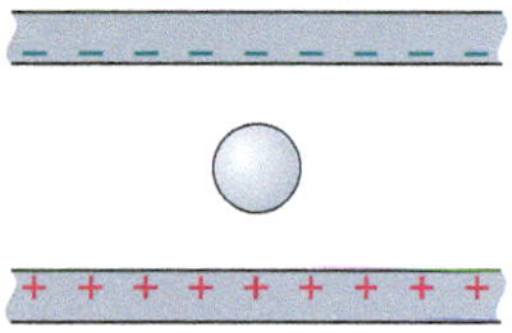

FIGURE P20.38

38. ||| A neutral conducting sphere is between two parallel charged plates, as shown in Figure P20.38. Sketch the electric field lines in the region between the plates. Be sure to include the effect of the conducting sphere.

### Section 20.7 Forces and Torques in Electric Fields

39. | What are the strength and direction of an electric field that INT will balance the weight of (a) a proton and (b) an electron?

40. || Raindrops acquire an electric charge as they fall. Suppose a INT 2.5-mm-diameter drop has a charge of +15 pC; these are both typical values. If the strength of the earth's electric field is 100 N/C, how does the magnitude of the electric force on the droplet compare to the weight force?

41. || A 0.10 g plastic bead is charged by the addition of $1.0 \times 10^{10}$ INT excess electrons. What electric field $\vec{E}$ (strength and direction) will cause the bead to hang suspended in the air?

42. || A housefly walking across a clean surface can accumulate a BIO INT significant positive or negative charge. In one experiment, the largest positive charge observed was +73 pC. A typical housefly has a mass of 12 mg. What magnitude and direction electric field would be necessary to "levitate" a housefly that has the maximum charge? Could such a field exist in air?

43. || One kind of e-book display consists of millions of very small INT spheres that float in a thin fluid layer between two conducting, transparent plates. Each sphere is black on one side and white on the other, and possesses an electric dipole moment directed from the white side to the black. When an electric field is applied between the plates, the spheres rotate so that their dipole moment lines up with the field. Depending on the field's direction, either the black or the white sides of the spheres can be made visible.

    The dipole moment of the spheres can be modeled as two opposite charges of magnitude $3.5 \times 10^{-15}$ C, separated by a distance of 100 $\mu$m. What is the maximum possible torque on a sphere if the electric field between the transparent plates is $4.0 \times 10^5$ N/C?

44. || Raindrops acquire an electric charge as they fall. Suppose INT a 2.0-mm-diameter drop has a charge of +12 pC; these are both very common values. In a thunderstorm, the electric field under a cloud can reach 15,000 N/C, directed upward. For a droplet exposed to this field, how do the magnitude and direction of the electric force compare to those of the weight force?

45. || A protein molecule in an electrophoresis gel has a negative BIO charge. The exact charge depends on the pH of the solution, but 30 excess electrons is typical. What is the magnitude of the electric force on a protein with this charge in a 1500 N/C electric field?

46. || Large electric fields in cell membranes cause ions to BIO move through the cell wall, as we will explore in Chapter 23. The field strength in a typical membrane is $1.0 \times 10^7$ N/C. What is the magnitude of the force on a calcium ion with charge $+e$?

47. ||| Molecules of carbon monoxide are permanent electric INT dipoles due to unequal sharing of electrons between the carbon and oxygen atoms. Figure P20.47 shows the distance and charges. Suppose a carbon monoxide molecule with a horizontal axis is in a vertical electric field of strength 15,000 N/C.

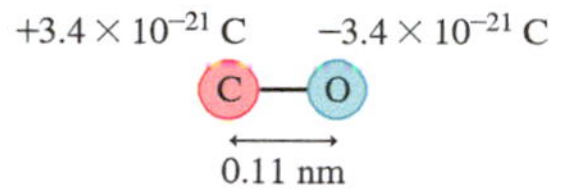


FIGURE P20.47

    a. What is the magnitude of the net force on the molecule?
    b. What is the magnitude of the torque on the molecule?

## General Problems

48. |||| A 2.0-mm-diameter copper ball is charged to +50 nC. What INT fraction of its electrons have been removed? The density of copper is 8900 kg/m$^3$.

49. || Two protons are 2.0 fm apart. (1 fm = 1 femtometer = INT $1 \times 10^{-15}$ m.)
    a. What is the magnitude of the electric force on one proton due to the other proton?
    b. What is the magnitude of the gravitational force on one proton due to the other proton?
    c. What is the ratio of the electric force to the gravitational force?

50. |||| The nucleus of a $^{125}$Xe atom (an isotope of the ele- INT ment xenon with mass 125 u) is 6.0 fm in diameter. It has 54 protons and charge $q = +54e$. (1 fm = 1 femtometer = $1 \times 10^{-15}$ m.)
    a. What is the electric force on a proton 2.0 fm from the surface of the nucleus?
    b. What is the proton's acceleration?

    **Hint:** Treat the spherical nucleus as a point charge.

51. ||| Two equally charged, 1.00 g spheres are placed with 2.00 cm INT between their centers. When released, each begins to accelerate at 225 m/s$^2$. What is the magnitude of the charge on each sphere?

52. |||| An electric dipole is formed from $\pm 1.0$ nC point charges spaced 2.0 mm apart. The dipole is centered at the origin, oriented along the $y$-axis. What is the electric field strength at the points (a) $(x, y) = (10$ mm, $0$ mm$)$ and (b) $(x, y) =$ $(0$ mm, $10$ mm$)$?

Watch Video Solution Problems 20.53 and 20.62

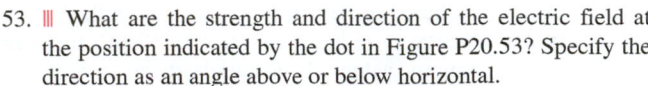

53. ||| What are the strength and direction of the electric field at the position indicated by the dot in Figure P20.53? Specify the direction as an angle above or below horizontal.

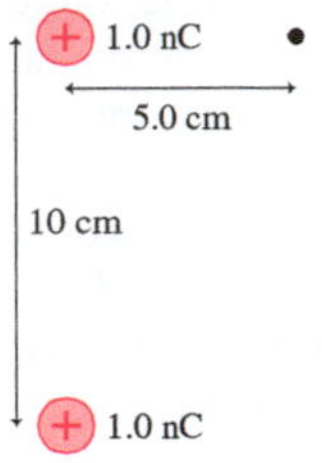


FIGURE P20.53

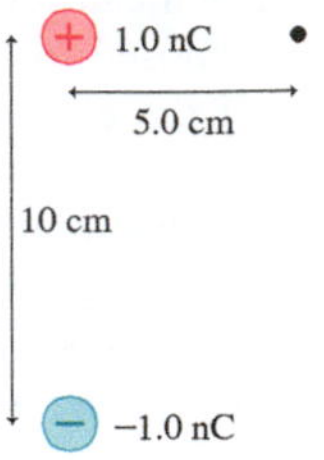


FIGURE P20.54

54. |||| What are the strength and direction of the electric field at the position indicated by the dot in Figure P20.54? Specify the direction as an angle above or below horizontal.

55. || What is the force on the 1.0 nC charge in Figure P20.55? Give your answer as a magnitude and a direction.

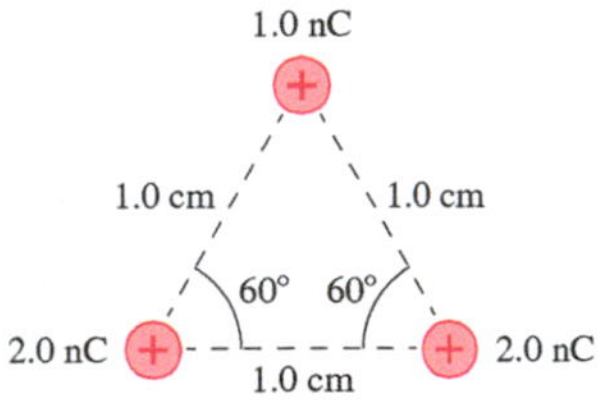


FIGURE P20.55

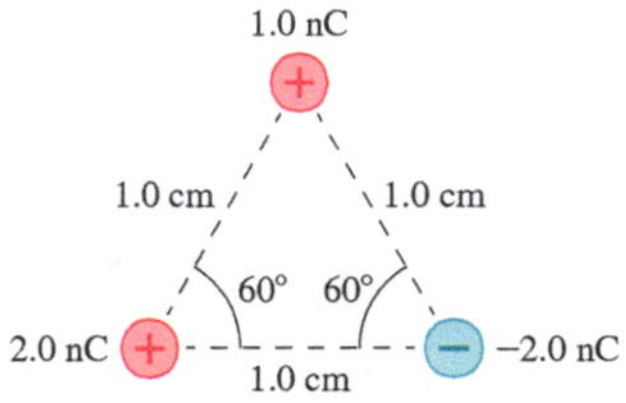


FIGURE P20.56

56. || What is the force on the 1.0 nC charge in Figure P20.56? Give your answer as a magnitude and a direction.

57. | What is the magnitude of the force on the 1.0 nC charge in the middle of Figure P20.57 due to the four other charges?

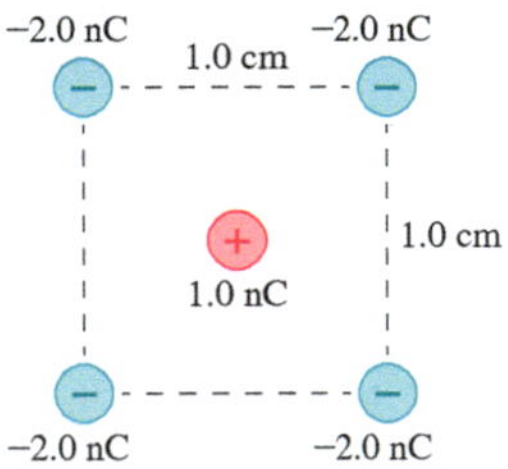


FIGURE P20.57

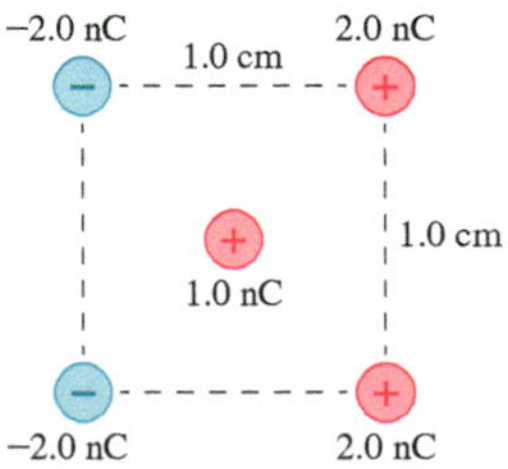


FIGURE P20.58

58. ||| What are the magnitude and direction of the force on the 1.0 nC charge in the middle of Figure P20.58 due to the four other charges?

59. || What are the magnitude and direction of the force on the 1.0 nC charge at the bottom of Figure P20.59?

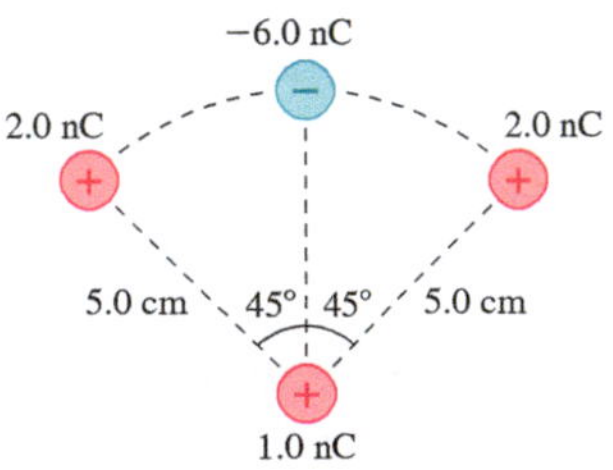


FIGURE P20.59

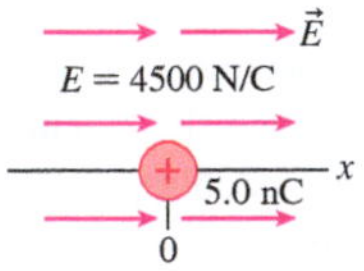


FIGURE P20.60

60. ||| As shown in Figure P20.60, a 5.0 nC charge sits at $x = 0$ in a uniform 4500 N/C electric field directed to the right. At what point along the $x$-axis would (a) a proton and (b) an electron experience no net force?

61. |||| Figure P20.61 shows four charges at the corners of a square of side $L$. What magnitude and sign of charge $Q$ will make the force on charge $q$ zero?

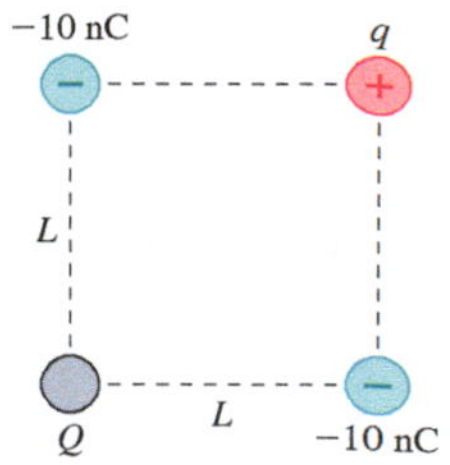


FIGURE P20.61

62. ||| In a simple model of the hydrogen INT atom, the electron moves in a circular orbit of radius 0.053 nm around a stationary proton. How many revolutions per second does the electron make?
**Hint:** What must be true for a force that causes circular motion?

63. ||| A 0.10 g honeybee acquires a charge of +23 pC while flying.
BIO
a. The electric field near the surface of the earth is typically 100 N/C, directed downward. What is the ratio of the electric force on the bee to the bee's weight?
b. What electric field strength and direction would allow the bee to hang suspended in the air?

64. ||| Two 2.0-cm-diameter disks face each other, 1.0 mm apart. INT They are charged to ±10 nC.
a. What is the electric field strength between the disks?
b. A proton is shot from the negative disk toward the positive disk. What launch speed must the proton have to just barely reach the positive disk?

65. |||| The electron gun in a television tube uses a uniform electric INT field to accelerate electrons from rest to $5.0 \times 10^7$ m/s in a distance of 1.2 cm. What is the electric field strength?

66. |||| A 0.020 g plastic bead hangs from INT a lightweight thread. Another bead is fixed in position beneath the point where the thread is tied. If both beads have charge $q$, the moveable bead swings out to the position shown in Figure P20.66. What is $q$?

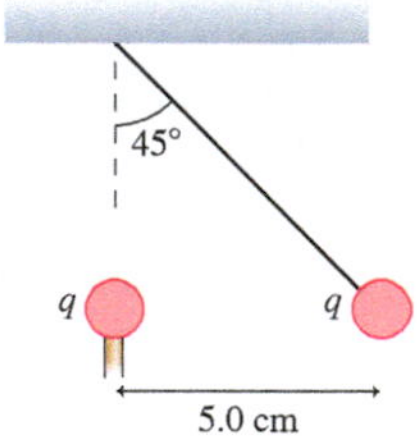


FIGURE P20.66

67. ||| A 4.0 mg bead with a charge of INT 2.5 nC rests on a table. A second bead, with a charge of − 5.6 nC, is directly above the first bead and is slowly lowered toward it. What is the closest the centers of the two beads can be brought together before the lower bead is lifted off the table?

68. ||||| Two 3.0 g spheres on 1.0-m-long threads repel each other INT after being equally charged, as shown in Figure P20.68. What is the charge $q$?

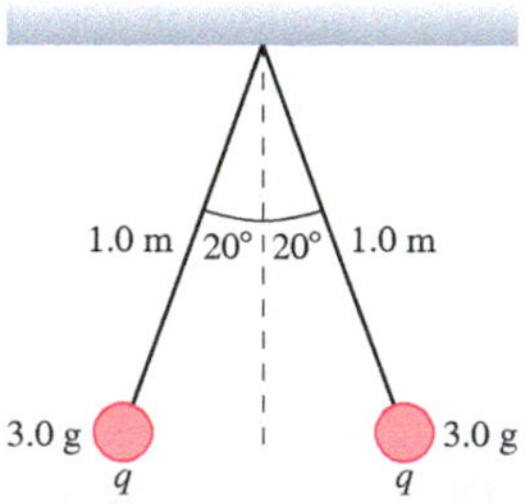


FIGURE P20.68

69. || A small charged bead has a mass of 1.0 g. It is held in a INT uniform electric field $\vec{E} = (200{,}000 \text{ N/C, up})$. When the bead is released, it accelerates upward with an acceleration of 20 m/s$^2$. What is the charge on the bead?

70. || A bead with a mass of 0.050 g and a charge of 15 nC is free INT to slide on a vertical rod. At the base of the rod is a fixed 10 nC charge. In equilibrium, at what height above the fixed charge does the bead rest?

71. |||| An electric field $\vec{E}$ = (100,000 N/C, right) causes the 5.0 g ball in Figure P20.71 to hang at a 20° angle. What is the charge on the ball?
INT

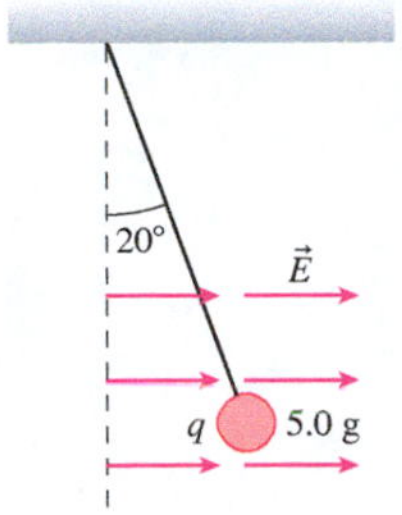

FIGURE P20.71

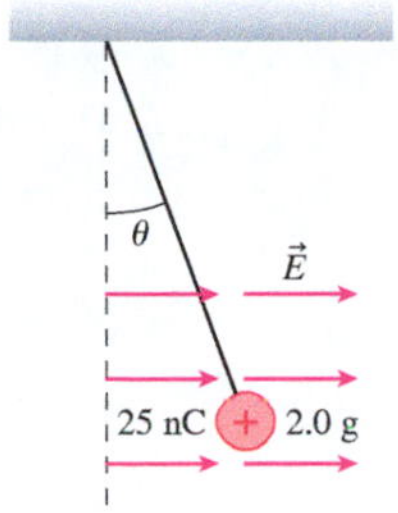

FIGURE P20.72

72. ||| An electric field $\vec{E}$ = (200,000 N/C, right) causes the 2.0 g ball in Figure P20.72 to hang at an angle. What is $\theta$?
INT

73. || A small bead with a positive charge $q$ is free to slide on a horizontal wire of length 4.0 cm. At the left end of the wire is a fixed charge $q$, and at the right end is a fixed charge $4q$. How far from the left end of the wire does the bead come to rest?
INT

74. ||| A parallel-plate capacitor consists of two plates, each with an area of 28 $cm^2$, separated by 3.0 mm. The charge on the capacitor is 8.3 nC. A proton is released from rest next to the positive plate. How long does it take for the proton to reach the negative plate?
INT

## MCAT-Style Passage Problems

### Flow Cytometry BIO

Flow cytometry, illustrated in Figure P20.75, is a technique used to sort cells by type. The cells are placed in a conducting saline solution which is then forced from a nozzle. The stream breaks up into small droplets, each containing one cell. A metal collar surrounds the stream right at the point where the droplets separate from the stream. Charging the collar polarizes the conducting liquid, causing the droplets to become charged as they break off from the stream. A laser beam probes the solution just upstream from the charging collar, looking for the presence of certain types of cells. All droplets containing one particular type of cell are given the same charge by the charging collar. Droplets with other desired types of cells receive a different charge, and droplets with no desired cell receive no charge. The charged droplets then pass between two parallel charged electrodes where they receive a horizontal force that directs them into different collection tubes, depending on their charge.

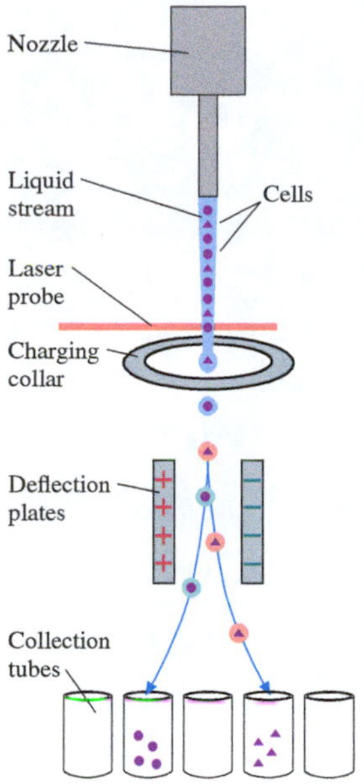

FIGURE P20.75

75. | If the charging collar has a positive charge, the net charge on a droplet separating from the stream will be
A. Positive. B. Negative.
C. Neutral. D. The charge will depend on the type of cell.

76. | Which of the following describes the charges on the droplets that end up in the five tubes, moving from left to right?
A. $+2q, +q, 0, -q, -2q$ B. $+q, +2q, 0, -2q, -q$
C. $-q, -2q, 0, +2q, +q$ D. $-2q, -q, 0, +q, +2q$

77. | Because the droplets are conductors, a droplet's positive and negative charges will separate while the droplet is in the region between the deflection plates. Suppose a neutral droplet passes between the plates. The droplet's dipole moment will point
A. Up. B. Down. C. Left. D. Right.

78. | Another way to sort the droplets would be to give each droplet the same charge, then vary the electric field between the deflection plates. For the apparatus as sketched, this technique will not work because
A. Several droplets are between the plates at one time, and they would all feel the same force.
B. The cells in the solution have net charges that would affect the droplet charge.
C. A droplet with a net charge would always experience a net force between the plates.
D. The droplets would all repel each other, and this force would dominate the deflecting force.

# 21 Electric Potential

A ray cruising over the ocean floor is able to detect weak electric potentials that signal the presence of creatures living below. What is the source of these electric potentials? And how is the ray able to sense them?

## LOOKING AHEAD

### Electric Potential

The *voltage* of a battery is the difference in **electric potential** between its two terminals.

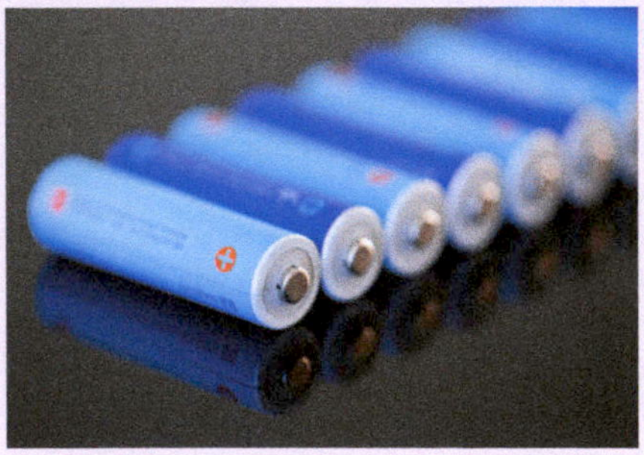

You'll learn how an electric potential is created when positive and negative charges are separated.

### Capacitors

The capacitors on this circuit board store charge and **electric potential energy.**

You'll learn how the energy stored in a capacitor depends on its charge.

### Potential and Field

There is an intimate connection between the electric potential and the electric field.

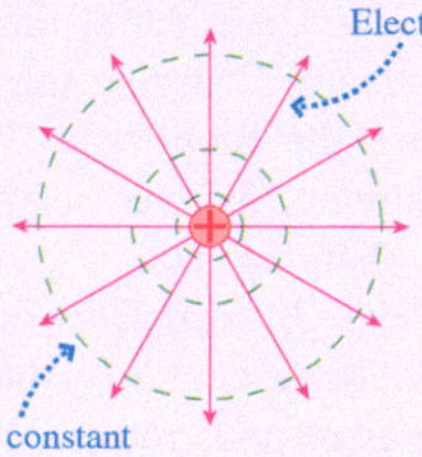


You'll learn how to move back and forth between field and potential representations.

**GOAL** **To calculate and use the electric potential and electric potential energy.**

## LOOKING BACK

### Work and Potential Energy

In Section 10.4 you learned that it is possible to store *potential energy* in a system of interacting objects. In this chapter, we'll learn about a new form of potential energy, electric potential energy.

This roller coaster is pulled to the top of the first hill by a chain. The tension in the chain does work on the coaster, increasing its gravitational potential energy.

**STOP TO THINK**

You lift a book at a constant speed. Which statement is true about the work $W$ done by your hand and the change in the book's gravitational potential energy $\Delta U_g$?

A. $W > \Delta U_g > 0$
B. $W < \Delta U_g < 0$
C. $W = \Delta U_g > 0$
D. $W = \Delta U_g < 0$

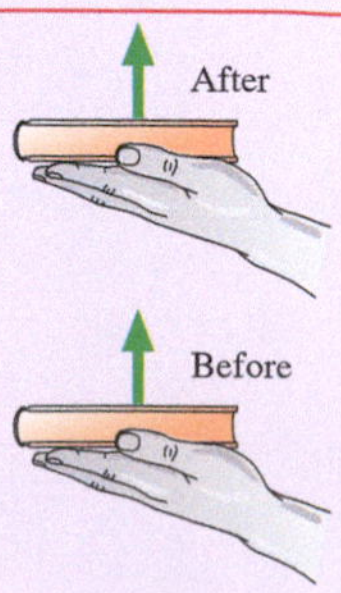

# 21.1 Electric Potential Energy and Electric Potential

You are already familiar with the concept of electric potential, although you don't know it. You've used the word "voltage" before, and voltage is, essentially, a measure of electric potential. In Chapter 20, you were introduced to a new force, the electric force; this new force has an associated form of potential energy. Once we've explored the properties of electric potential energy, we'll introduce the related concept of electric potential.

## Electric Potential Energy

Conservation of energy is a powerful tool for understanding the motion of mechanical systems. In ◀◀ SECTION 10.2 you learned that a system's energy can be changed by doing *work* on it; you will recall that work is done, and thus a system's energy is changed, when an external force acts on the system as the system undergoes a *displacement.*

To remind ourselves how conservation of energy works for mechanical systems, FIGURE 21.1a shows a hand lifting a book at a constant speed. The external force of the hand pushes on the book as the book rises, so the hand does work on the book, increasing its energy—in this case, its gravitational potential energy $U_g$. Mathematically, we can write this process as $\Delta U_g = W$: The work $W$ *changes* the book's potential energy.

FIGURE 21.1 The work done increases the potential energy for three systems.

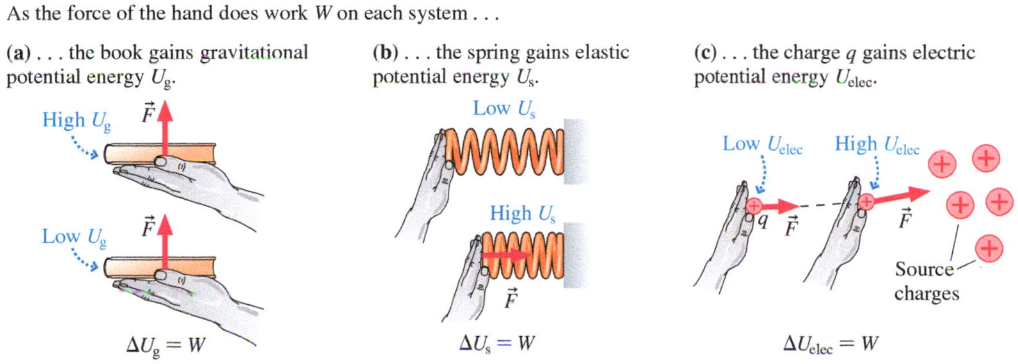


A similar process occurs when a spring is compressed, as in FIGURE 21.1b. Here, the hand pushes on the end of the spring as it compresses, doing work on the spring and increasing its elastic potential energy $U_s$, a process we can write as $\Delta U_s = W$.

Now let's apply these same ideas to the system of charged particles in FIGURE 21.1c. Here, several charges have been identified as *source charges* that are fixed and don't move. Because charge $q$ is repelled by the source charges, the hand has to *push* on charge $q$ in order to move it closer to the source charges. This means that the hand does work on $q$ as it moves it, transferring energy into the system of charges. Just as in the book and spring systems, this energy is in the form of potential energy—in this case, **electric potential energy** $U_{elec}$. Writing this process as $\Delta U_{elec} = W$, we see that **we can determine the electric potential energy of a charge when it's at a particular position by computing how much work it took to move the charge to that position.**

FIGURE 21.2 What is the electric potential energy near some source charges?

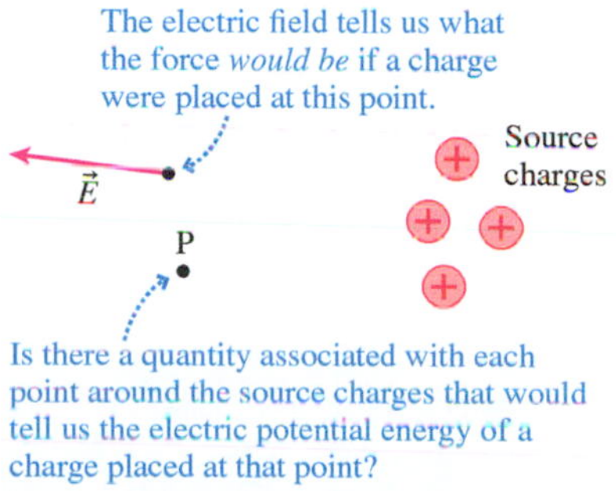


## Electric Potential

We introduced the concept of the *electric field* in Chapter 20. In the field model, the electric field is the agent by which charges exert a long-range force on another charge $q$. FIGURE 21.2 reminds us that the source charges first alter the space around them by creating an electric field $\vec{E}$ at every point in space. This electric field—not the source charges themselves—exerts a force $\vec{F}_{elec} = q\vec{E}$ on charge $q$. Also keep in mind that the electric field of the source charges is present throughout space whether or not charge $q$ is present to experience it. The electric field tells us what the force on the charge *would be* if the charge were placed there.

We can apply similar reasoning to electric potential energy. Consider again the source charges in Figure 21.2. If we place a charge $q$ at point P near the source charges, charge $q$ will have electric potential energy. If we place a different charge $q'$ at P, charge $q'$ will have a different electric potential energy. We will now define a quantity associated with point P that will tell us what the electric potential energy of $q$, $q'$, or any other charge would have at point P, without actually having to place the charge there.

FIGURE 21.3 Finding the electric potential energy and the electric potential.

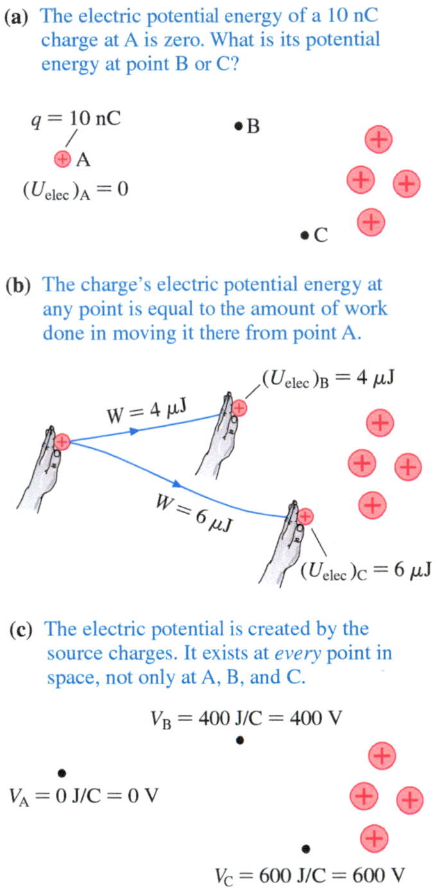

(a) The electric potential energy of a 10 nC charge at A is zero. What is its potential energy at point B or C?
q = 10 nC
A
B
C
(U_elec)_A = 0
(b) The charge's electric potential energy at any point is equal to the amount of work done in moving it there from point A.
(U_elec)_B = 4 μJ
W = 4 μJ
W = 6 μJ
(U_elec)_C = 6 μJ
(c) The electric potential is created by the source charges. It exists at every point in space, not only at A, B, and C.
V_B = 400 J/C = 400 V
V_A = 0 J/C = 0 V
V_C = 600 J/C = 600 V


First, we'll have to understand in a bit more detail how to find the electric potential energy of a charge $q$. Suppose, to be specific, we take $q = 10$ nC in FIGURE 21.3a and, for convenience, we let $(U_{\text{elec}})_A = 0$ J when the charge is at point A.

As we've already seen, to find charge $q$'s electric potential energy at any other point, such as point B or C, we need to find the amount of work it takes to move the charge from A to B, or A to C. In FIGURE 21.3b it takes the hand 4 $\mu$J of work to move the charge from A to B; thus its electric potential energy at B is $(U_{\text{elec}})_B = 4\ \mu\text{J}$. (Recall that the SI unit of energy is the joule, J.) Similarly, $q$'s electric potential energy at point C is $(U_{\text{elec}})_C = 6\ \mu\text{J}$ because it takes 6 $\mu$J of work to move it to point C.

What if we were to repeat this experiment with a different charge—say, $q = 20$ nC? According to Coulomb's law, the electric force on this charge due to the source charges will be twice that on the 10 nC charge. Consequently, the hand will have to push with twice as much force and thus do twice as much work in moving this charged particle from A to B. As a result, the 20 nC particle has $(U_{\text{elec}})_B = 8\ \mu\text{J}$ at B. A 5 nC particle would have $(U_{\text{elec}})_B = 2\ \mu\text{J}$ at B because the hand would have to work only half as hard to move it to B as it did to move the 10 nC particle to B. In general, **a charged particle's potential energy is proportional to its charge.**

Let's now look at the ratio of the electric potential energy to the electric charge at point B for each of these charges:

$$\frac{U_{(\text{elec})B}}{q} = \frac{2\ \mu\text{J}}{5\ \text{nC}} = \frac{4\ \mu\text{J}}{10\ \text{nC}} = \frac{8\ \mu\text{J}}{20\ \text{nC}} = 400\ \frac{\text{J}}{\text{C}}$$

$U/q$ for $q = 5$ nC; $U/q$ for $q = 10$ nC; $U/q$ for $q = 20$ nC; All three ratios are the *same*.

The ratio is the same for each charge, so the ratio is a property of point B, not of an individual charge. Thus we can write a simple expression for the electric potential energy that *any* charge $q$ *would have* if placed at point B:

$$(U_{\text{elec}})_B = \left(400\ \frac{\text{J}}{\text{C}}\right) q$$

This number is associated with point B. This part depends on the charge we place at B.

The number 400 J/C, which is associated with point B, tells us the *potential* for creating potential energy (there's a mouthful!) if a charge $q$ is placed at point B. This value is called the **electric potential,** and it is given the symbol $V$. At B, then, the electric potential is 400 J/C, as shown in FIGURE 21.3c. By similar reasoning, the electric potential at point C is 600 J/C.

A video to support a section's topic is embedded in the eText.

**Video** Electric Potential

This idea can be generalized. Any source charges create an electric potential at every point in the space around them. At a point where the potential is $V$, the electric potential energy of a charged particle $q$ is

$$U_{\text{elec}} = qV \qquad (21.1)$$

Relationship between electric potential and electric potential energy

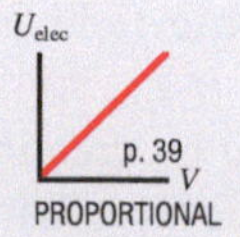

Notice the similarity to $\vec{F}_{elec} = q\vec{E}$. The electric potential, like the electric field, is created by the source charges and is present at all points in space. The electric potential is there whether or not charge $q$ is present to experience it. While the electric field tells us how the source charges would exert a *force* on $q$, the electric potential tells us how the source charges would provide $q$ with *potential energy*. Although we used the work done on a positive charge to justify Equation 21.1, it is also valid if $q$ is negative.

**NOTE** ▶ For potential energy, we found that we could choose a particular configuration of the system to have $U = 0$, the zero of potential energy. The same idea holds for electric potential. We can choose any point in space, wherever is convenient, to be $V = 0$. It will turn out that only *changes* in the electric potential are important, so the choice of a point to be $V = 0$ has no physical consequences. ◀

The unit of electric potential is the joule per coulomb, called the **volt** V:

$$1 \text{ volt} = 1 \text{ V} = 1 \text{ J/C}$$

Microvolts ($\mu$V), millivolts (mV), and kilovolts (kV) are commonly used units. TABLE 21.1 lists some typical electric potentials. We can now recognize that the electric potential in Figure 21.2—a potential due to the source charges—is 0 V at A and 400 V at B. This is shown in Figure 21.3c.

**NOTE** ▶ The symbol $V$ is widely used to represent the *volume* of an object, and now we're introducing the same symbol to mean *potential*. To make matters more confusing, V is the abbreviation for *volts*. In printed text, $V$ for potential is italicized while V for volts is not, but you can't make such a distinction in handwritten work. This is not a pleasant state of affairs, but these are the commonly accepted symbols. You must be especially alert to the *context* in which a symbol is used. ◀

**TABLE 21.1** Typical electric potentials

| Source of potential | Approximate potential |
|---|---|
| Brain activity at scalp (EEG) | 10–100 $\mu$V |
| Cells in human body | 100 mV |
| Battery | 1–10 V |
| Household electricity | 100 V |
| Static electricity | 10 kV |
| Transmission lines | 1000 kV |

**EXAMPLE 21.1 Finding the change in a charge's electric potential energy**

A 15 nC charged particle moves from point A, where the electric potential is 300 V, to point B, where the electric potential is −200 V. By how much does the electric potential change? By how much does the particle's electric potential energy change? How would your answers differ if the particle's charge were −15 nC?

***PREPARE*** The change in the electric potential $\Delta V$ is the potential at the final point B minus the potential at the initial point A. From Equation 21.1, we can find the change in the electric potential energy by noting that $\Delta U_{elec} = (U_{elec})_B - (U_{elec})_A = q(V_B - V_A) = q\,\Delta V$.

***SOLVE*** We have

$$\Delta V = V_B - V_A = (-200 \text{ V}) - (300 \text{ V}) = -500 \text{ V}$$

This change is *independent* of the charge $q$ because the electric potential is created by source charges.

The change in the particle's electric potential energy is

$$\Delta U_{elec} = q\,\Delta V = (15 \times 10^{-9} \text{ C})(-500 \text{ V}) = -7.5\ \mu\text{J}$$

A −15 nC charge would have $\Delta U_{elec}$ + 7.5 $\mu$J because $q$ changes sign while $\Delta V$ remains unchanged.

***ASSESS*** Because the electric potential at B is lower than that at A, the positive (+15 nC) charge will lose electric potential energy, while the negative (−15 nC) charge will gain energy.

**STOP TO THINK 21.1** A positively charged particle moves from point 1 to point 2. As it does, its electric potential energy

A. Increases.
B. Decreases.
C. Stays the same.

## 21.2 Sources of Electric Potential

When you scuff your feet on the carpet, you develop a charge, as we saw in ◀◀ SECTION 20.1, so there is an electric potential in your vicinity. This happens any time you separate charges; let's look at a specific example. Consider the uncharged

**Video** Sparks in the Air

capacitor shown in FIGURE 21.4a. There's no force on charge $q$, so no work is required to move it from A to B. Consequently, charge $q$'s electric potential energy remains *unchanged* as it is moved from A to B, so $(U_{elec})_B = (U_{elec})_A$. Then, because $U_{elec} = qV$, it must be the case that $V_B = V_A$. We say that the **potential difference** $\Delta V = V_B - V_A$ is *zero*.

Now consider what happens if electrons are transferred from the right side of the capacitor to the left, giving the left electrode charge $-Q$ and the right electrode charge $+Q$. The capacitor still has no net charge, but the charge has been *separated*. These separated charges exert a force $\vec{F}_{elec}$ on $q$, so, as FIGURE 21.4b shows, the hand must now do work on $q$ to move it from A to B, increasing its electric potential energy so that $(U_{elec})_B > (U_{elec})_A$. This means that the potential difference $\Delta V$ between A and B is no longer zero. What we've shown here is quite general: **A potential difference is created by *separating* positive charge from negative charge.**

FIGURE 21.4 Potential differences are created by charge separation.

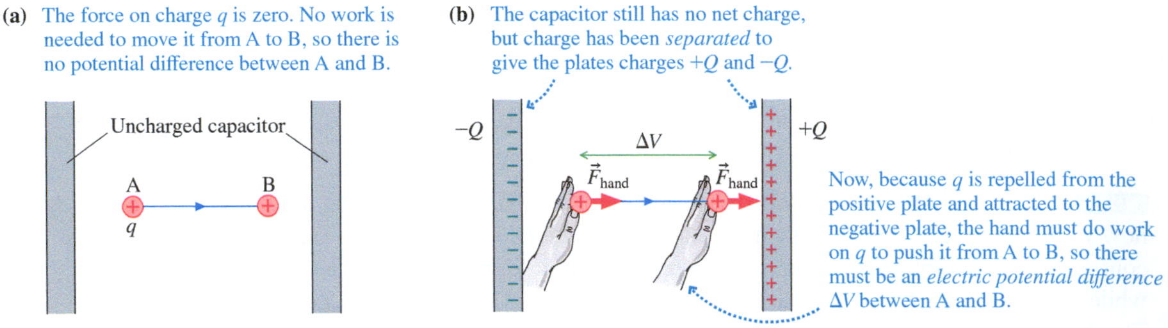

(a) The force on charge q is zero. No work is needed to move it from A to B, so there is no potential difference between A and B.
Uncharged capacitor
A
B
q
(b) The capacitor still has no net charge, but charge has been separated to give the plates charges +Q and −Q.
−Q
+Q
ΔV
F⃗hand
F⃗hand
Now, because q is repelled from the positive plate and attracted to the negative plate, the hand must do work on q to push it from A to B, so there must be an electric potential difference ΔV between A and B.


When you shuffle your feet across the carpet, you pick up charge from the floor, and there is a potential difference between your body and, say, a nearby doorknob that can be 10,000 V or more, enough to create a spark as the excess charge moves from higher to lower potential.

We saw in ◀◀ SECTION 20.1 that, during a sandstorm, sand and dust grains transfer charge when they collide. In the swirling ash cloud over a volcano, collisions between large and small particles transfer charges as well; the small particles acquire a negative charge and the large particles acquire a positive charge. Gases rushing upward loft the negatively charged small particles higher than the positively charged large particles. This creates a charge separation between the top and the bottom of the ash cloud. The resulting potential difference can lead to dramatic lightning strikes between the top and the bottom of the ash cloud. The same thing happens in storm clouds. Collisions between ice particles in the clouds transfer charge, and air currents separate particles. As this happens, the top of the cloud acquires a positive charge and the bottom of the cloud acquires a negative charge. This leads to lightning strikes inside the cloud, and, as the negative charges in the bottom of the cloud cause positive charges to accumulate in the ground below, lightning also flies between the cloud and the ground.

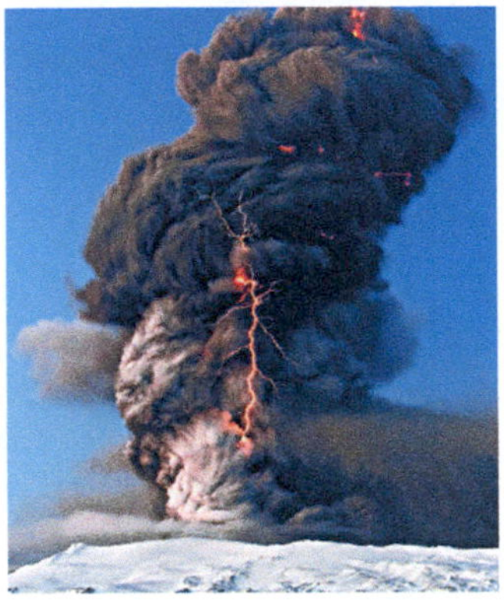
Lightning in the ash cloud over an erupting volcano.

Charge separation, and hence potential differences, can also be created by chemical processes. A **battery** uses chemical reactions to create an internal charge separation. This separation proceeds until a characteristic potential difference—about 1.5 V for a standard alkaline battery—appears between the two terminals of the battery. Different kinds of batteries maintain different potential differences between their terminals.

A battery maintains a fixed potential difference between its + and − terminals.

NOTE ▶ The potential difference between two points is often called the **voltage.** Thus we say that a battery's voltage is 1.5 V or 9 V, and we speak of the potential difference between a battery's terminals as the voltage "across" the battery. ◀

Chemical means of producing potential differences are crucial in biological systems. In nerve and muscle cells in your body, active processes create a potential difference across the cell membrane of about 70 mV, with the inside of the cell more negative than the outside. As illustrated in FIGURE 21.5, this *membrane potential* is caused by an imbalance of potassium ($K^+$) and sodium ($Na^+$) ions. The molar concentration of $K^+$ is higher inside the cell than outside, while the molar concentration of $Na^+$ is higher outside than inside. To keep the charge separated in the face of diffusion, which tends to equalize the ion concentrations, a *sodium-potassium exchange pump* continuously pumps sodium ions out of the cell and potassium ions into the cell. During one pumping cycle, three $Na^+$ are pushed out of the cell but only two $K^+$ are pushed in, giving a net transfer of one positive charge out of the cell. This continuous pumping leads to the charge separation that causes the membrane potential. We'll have a careful look at the electrical properties of nerve cells in Chapter 23.

FIGURE 21.5 BIO The membrane potential of a cell is due to a charge separation.

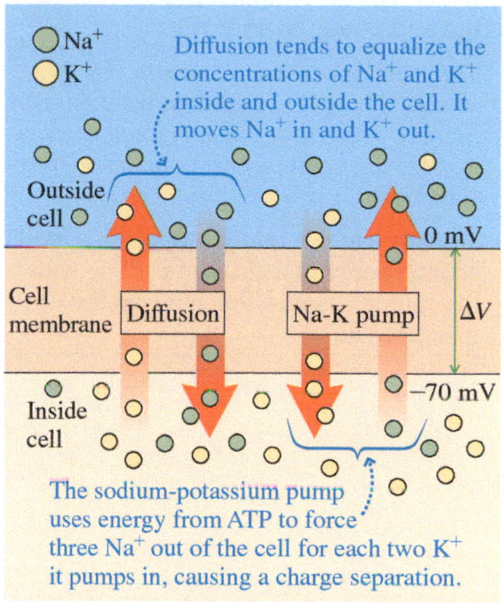


## Measuring Electric Potential

An electrocardiogram measures the potential difference between several locations on the body to diagnose possible heart problems; temperature is often measured using a *thermocouple,* a device that develops a potential difference proportional to temperature; your digital camera can sense when its battery is low by measuring the potential difference between its terminals.

Note that in all these applications, it's the potential *difference* between two points that's measured. The actual value of the potential at a given point depends on where we choose $V$ to be zero, but the difference in potential between two points is independent of this choice. Because of this, a **voltmeter,** the basic instrument for measuring potential differences, always has *two* inputs. Probes are connected from these inputs to the two points between which the potential difference is to be measured. We'll learn more in Chapter 23 about how voltmeters work.

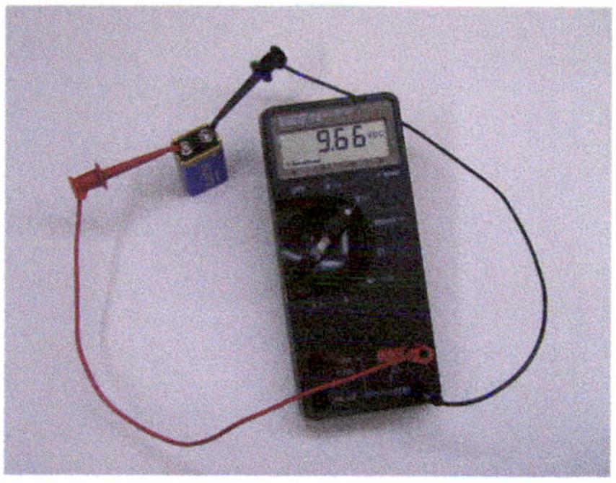


A voltmeter always uses two probes to measure a potential difference. Here, we see that the potential difference of a fresh 9 V battery is closer to 9.7 V.

As small as cells are, the membrane potential difference between the inside and outside of a cell can be measured by a (very small) probe connected to a voltmeter. FIGURE 21.6 is a micrograph of a nerve cell whose membrane potential is being measured. A very small glass pipette, filled with conductive fluid, is actually inserted through the cell's membrane. This pipette is one of the probes. The second probe need not be so small; it is simply immersed in the conducting fluid that surrounds the cell and can be quite far from the cell.

FIGURE 21.6 BIO Measuring the membrane potential.

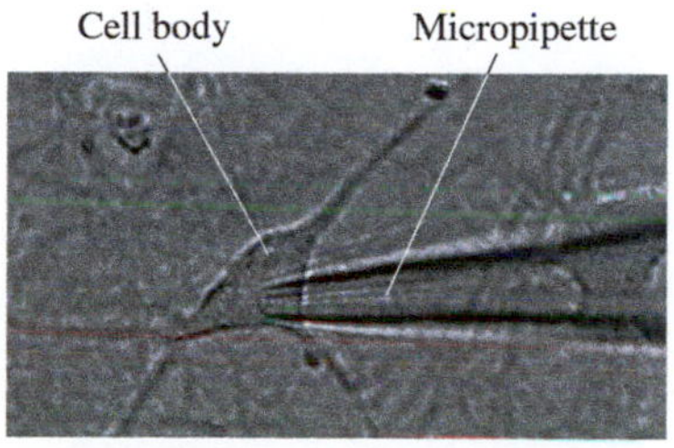


## Is High Voltage Dangerous?

The discussion of potential difference—voltage—might be at odds with your perceptions at this point. You've been told that you can acquire a potential difference of 10,000 V when you scuff your feet on the carpet. As you know, this potential difference can cause a spark to jump that is more annoying than painful. How, then, can the much lower 120 V of a household electric outlet produce a shock that is definitely painful and possibly fatal?

The reason you can feel a shock, and the reason electricity can be dangerous, is that nerve and muscle cells in your body have an electrical nature, as shown in Figure 21.5. You can stimulate a nerve or muscle cell by changing the cell potential, and this requires a transfer of charge. The quantity of charge transferred determines the physiological effect of a shock. The potential difference produced when you scuff your feet on the carpet is large, but the total amount of charge transferred when you touch a doorknob is quite small, in the nC range. This can stimulate the sensitive nerve cells in your fingertip (you'll feel it), but this is not enough charge to cause significant effects (it won't hurt you). The voltage of electric outlets is much smaller, but the charge transferred can be much greater. The much larger charge transfers (for a serious shock, tens or hundreds of mC per second) of household electricity can stimulate nerve and muscle cells throughout the body, possibly interfering with important functions. So, the 10,000 V potential of a power line—with the ability to transfer a lot of charge in a short time—is very dangerous. The 10,000 V potential of laundry fresh from the dryer—with much smaller quantities of charge that can be transferred—is not.

## 21.3 Electric Potential and Conservation of Energy

**TABLE 21.2** Distinguishing electric potential and potential energy

The *electric potential* is created by the source charges. The electric potential is present whether or not a charged particle is there to experience it. Potential is measured in J/C, or V.

The *electric potential energy* is the interaction energy of a charged particle with the source charges. Potential energy is measured in J.

The potential energy of a charged particle is determined by the electric potential: $U_{elec} = qV$. Although potential and potential energy are related and have similar names, they are not the same thing. TABLE 21.2 will help you distinguish between electric potential and electric potential energy.

As a roller coaster car moves from the top of the track to the bottom, its gravitational potential energy decreases. Because energy is conserved, this decrease in potential energy is accompanied by an *increase* in the car's kinetic energy $K$—the car speeds up. Similarly, when a particle with positive charge $q$ moves from a region of high electric potential to a region of low electric potential, its electric potential energy decreases and its kinetic energy increases. In ◀◀ SECTION 10.6, we introduced the before-and-after work-energy equation, a statement of the law of conservation of energy, which we used to analyze situations like a roller coaster car going down a hill. We can also apply conservation of energy to situations with moving electric charges. In such cases, we can generally ignore gravitational potential energy; the electric force is much stronger than the gravitational force. To treat cases of moving electric charges, we substitute electric potential energy for gravitational potential energy in our basic working equation. If we also rule out external forces that can do work, our statement of conservation of energy becomes

$$K_f + (U_{elec})_f = K_i + (U_{elec})_i$$

which we can write in terms of the electric potential $V$ as

$$K_f + qV_f = K_i + qV_i \tag{21.2}$$

Conservation of energy for a charged particle moving in an electric potential $V$

As usual, the subscripts i and f stand for the initial and final situations.

FIGURE 21.7 shows two positive charges moving through a region of changing electric potential. This potential has been created by source charges that aren't shown; our concern is only with the *effect* of this potential on the moving charge. Notice that we've used the before-and-after visual overview introduced earlier when we studied conservation of momentum and energy.

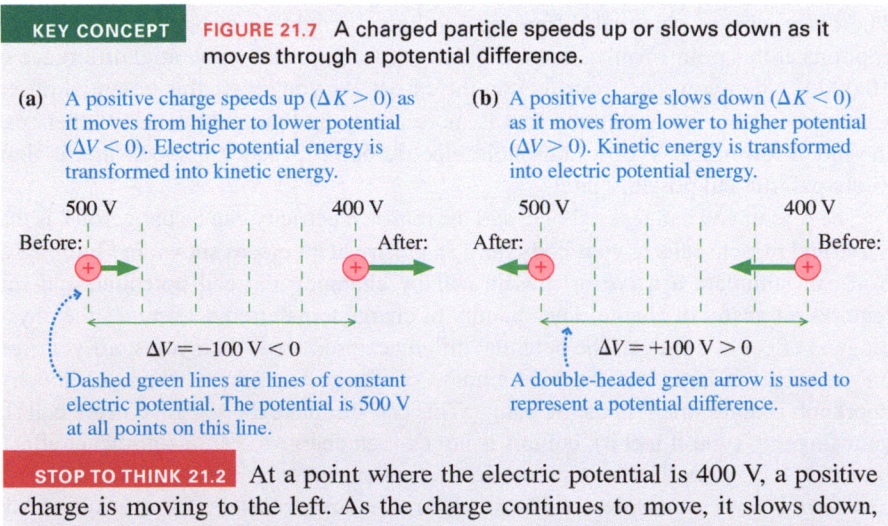

KEY CONCEPT **FIGURE 21.7** A charged particle speeds up or slows down as it moves through a potential difference.

(a) A positive charge speeds up ($\Delta K > 0$) as it moves from higher to lower potential ($\Delta V < 0$). Electric potential energy is transformed into kinetic energy.

(b) A positive charge slows down ($\Delta K < 0$) as it moves from lower to higher potential ($\Delta V > 0$). Kinetic energy is transformed into electric potential energy.

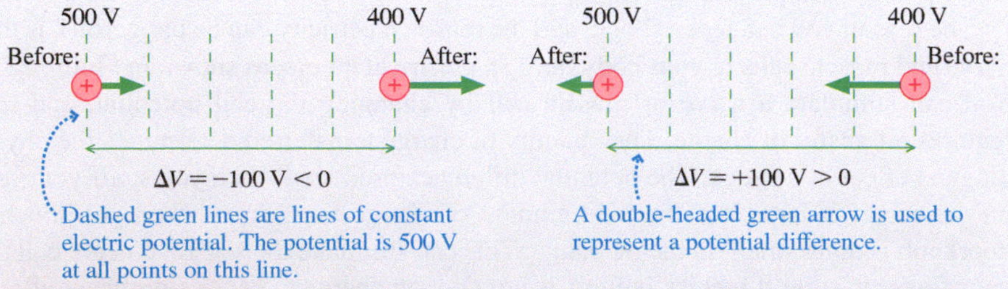


**STOP TO THINK 21.2** At a point where the electric potential is 400 V, a positive charge is moving to the left. As the charge continues to move, it slows down, stops, and then begins to move back to the right. Using what you learned in Figure 21.7, its speed as it passes its initial position is

A. Greater than its initial speed.
B. Less than its initial speed.
C. Equal to its initial speed.
D. There is not enough information to tell.

We can understand the motion of the charges if we rewrite Equation 21.2 as $K_f - K_i = -q(V_f - V_i)$, or

$$\Delta K = -q\,\Delta V \tag{21.3}$$

For the charge in Figure 21.7a, the change in potential—that is, the potential difference—as it moves from left to right is

$$\Delta V = V_f - V_i = 400\text{ V} - 500\text{ V} = -100\text{ V}$$

which is negative. Equation 21.3 then shows that $\Delta K$ is *positive,* indicating that the particle *speeds up* as it moves from higher to lower potential. Conversely, for the charge in Figure 21.7b the potential difference is +100 V; Equation 21.3 then shows that $\Delta K$ is negative, so the particle *slows down* in moving from lower to higher potential.

**NOTE** ► The situation is reversed for a negative charge. If $q < 0$, Equation 21.3 requires $K$ to increase as $V$ increases. A negative charge speeds up if it moves into a region of higher potential. ◄

**PROBLEM-SOLVING APPROACH 21.1**

## Conservation of energy in charge interactions

We use the principle of conservation of energy for electric charges in exactly the same way as we did for mechanical systems. The only difference is that we now consider electric potential energy instead of gravitational potential energy.

***STRATEGIZE*** Determine the object that is moving, how it is moving, and its charge. Think about the sign of the charge and the change in potential, and decide whether the object speeds up or slows down.

***PREPARE*** Draw a before-and-after visual overview. Define symbols that will be used in the problem, list known values, and identify what you're trying to find.

***SOLVE*** The mathematical representation is based on conservation of energy:

$$K_f + qV_f = K_i + qV_i$$

- Find the electric potential at both the initial and final positions. You may need to calculate it from a known expression for the potential, such as that of a point charge.
- $K_i$ and $K_f$ are the total kinetic energies of all moving particles.
- Some problems may need additional conservation laws, such as conservation of charge or conservation of momentum.

***ASSESS*** Check that your result has the correct units, is reasonable, and answers the question.

Exercise 17

**EXAMPLE 21.2** 

## Finding the final speed of a moving proton

A proton moves through an electric potential created by a number of source charges. Its speed is $2.5 \times 10^5$ m/s at a point where the potential is 1500 V. What will be the proton's speed a short time later when it reaches a point where the potential is −500 V?

***STRATEGIZE*** The proton has a positive charge, and it is moving from a higher potential to a lower potential. The situation is like that illustrated in Figure 21.7a; the proton will speed up as it moves.

***PREPARE*** FIGURE 21.8 is a visual overview of the situation showing the details of the initial and final states. We've drawn a longer velocity vector at the final position because we know that the proton speeds up. The mass of a proton is $m = 1.67 \times 10^{-27}$ kg.

FIGURE 21.8 A before-and-after visual overview for a proton through a potential difference.

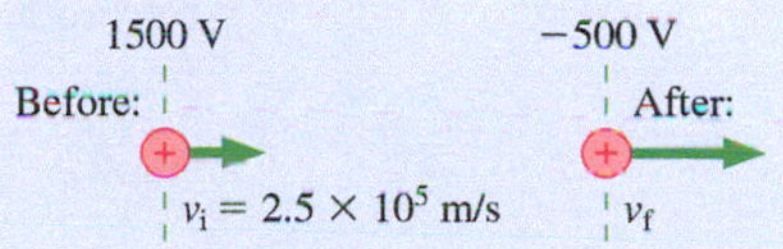


***SOLVE*** Conservation of energy gives

$$K_f + qV_f = K_i + qV_i$$

*Continued*

or

$$\frac{1}{2}mv_f^2 + qV_f = \frac{1}{2}mv_i^2 + qV_i$$

which we can write as

$$\frac{1}{2}mv_f^2 = \frac{1}{2}mv_i^2 + (qV_i - qV_f)$$

We can solve for $v_f^2$ by multiplying both sides by $2/m$ to get

$$v_f^2 = v_i^2 + \frac{2}{m}(qV_i - qV_f) = v_i^2 + \frac{2q}{m}(V_i - V_f)$$

or

$$v_f^2 = (2.5 \times 10^5\ \text{m/s})^2 + \frac{2(1.6 \times 10^{-19}\ \text{C})}{1.67 \times 10^{-27}\ \text{kg}}[1500\ \text{V} - (-500\ \text{V})]$$
$$= (4.46 \times 10^{11}\ \text{m/s})^2$$

Solving for the final speed gives

$$v_f = 6.7 \times 10^5\ \text{m/s}$$

**ASSESS** A positively charged particle speeds up as it moves from higher to lower potential, analogous to a particle speeding up as it slides down a hill from higher gravitational potential energy to lower gravitational potential energy. Our final result makes sense.

**FIGURE 21.9** A transformation of electric potential energy into thermal energy.

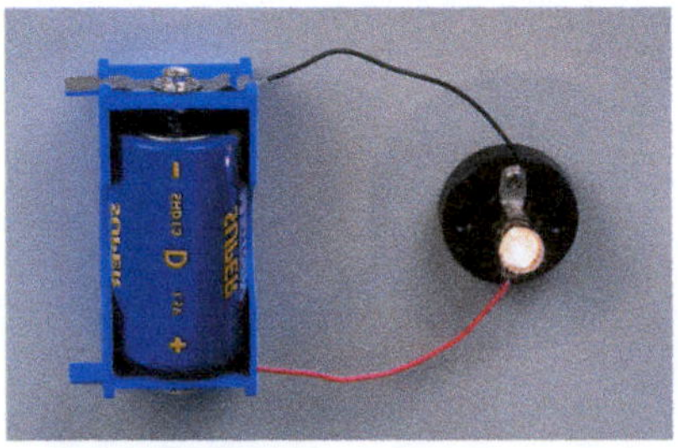

So far we've considered only the transformation of electric potential energy into kinetic energy as a charged particle moves from higher to lower electric potential. That is, we've studied the energy transformation $\Delta K = -q\,\Delta V$. But it's worth noting that electric potential energy can also be transformed into other kinds of energy; this is the basis of many applications of electricity. In FIGURE 21.9, for example, charges move in the wires from the high-potential terminal of the battery, through the lightbulb, and back to the low-potential terminal. In the bulb, their electric potential energy is transformed into thermal energy $E_{th}$, making the bulb hot enough to glow brightly. This energy transformation is $\Delta E_{th} = -q\,\Delta V$.

**EXAMPLE 21.3 Finding the power of a lightbulb**

Each second, 0.40 C of charge moves from the positive terminal of the 1.5 V battery in Figure 21.9, through the lightbulb, and to the negative terminal. What is the power of the bulb?

**STRATEGIZE** We will find the change in the potential energy of the charges in each second; this will equal the increase in thermal energy of the bulb. The bulb is hot, but it's temperature isn't changing, so this energy is transferred as heat and light to the environment.

**PREPARE** The potential difference that the charges move through is equal to the difference in potential between the positive and the negative terminals of the battery, so $\Delta V = -1.5$ V.

**SOLVE** The change in potential energy in 1 second is

$$\Delta U_{elec} = (-1.5\ \text{J/C})(0.40\ \text{C}) = -0.60\ \text{J}$$

The charges lose potential energy; this is the energy supplied to the bulb. The power of the bulb is thus

$$P = \frac{\Delta E}{\Delta t} = \frac{0.60\ \text{J}}{1\ \text{s}} = 0.60\ \text{J/s} = 0.60\ \text{W}$$

**ASSESS** This is a reasonable power for a small lightbulb.

## The Electron Volt

**Video** Energy Changes and Energy Units

The joule is a unit of appropriate size in mechanics and thermodynamics, where we deal with macroscopic objects, but it will be very useful to have an energy unit appropriate to atomic and nuclear events.

Suppose an electron accelerates through a potential difference $\Delta V = 1$ V. The electron might be accelerating from 0 V to 1 V, or from 1000 V to 1001 V. Regardless of the actual voltages, an electron, being negative, *speeds up* when it moves toward a higher potential, so that, according to Equation 21.3, the 1 V potential difference causes the electron, with $q = -e$, to gain kinetic energy

$$\Delta K = -q\,\Delta V = e\,\Delta V = (1.60 \times 10^{-19}\ \text{C})(1\ \text{V}) = 1.60 \times 10^{-19}\ \text{J}$$

We will define a new unit of energy, called the **electron volt,** as

$$1\ \text{electron volt} = 1\ \text{eV} = 1.60 \times 10^{-19}\ \text{J}$$

With this definition, the kinetic energy gained by the electron in our example is

$$\Delta K = 1\ \text{eV}$$

In other words, **1 electron volt is the kinetic energy gained by an electron (or proton) if it accelerates through a potential difference of 1 volt.**

**NOTE** ▶ The abbreviation eV uses a lowercase e but an uppercase V. Units of keV ($10^3$ eV), MeV ($10^6$ eV), and GeV ($10^9$ eV) are common. ◀

The electron volt can be a troublesome unit. One difficulty is its unusual name, which looks less like a unit than, say, "meter" or "second." A more significant difficulty is that the name suggests a relationship to volts. But *volts* are units of electric potential, whereas this new unit is a unit of energy! It is crucial to distinguish between the *potential V*, measured in volts, and an *energy* that can be measured either in joules or in electron volts. You can now use electron volts anywhere that you would previously have used joules. Doing so is no different from converting back and forth between units of centimeters and inches.

**NOTE** ▶ The joule remains the SI unit of energy. It will be useful to express energies in eV, but you *must* convert this energy to joules before doing most calculations. ◀

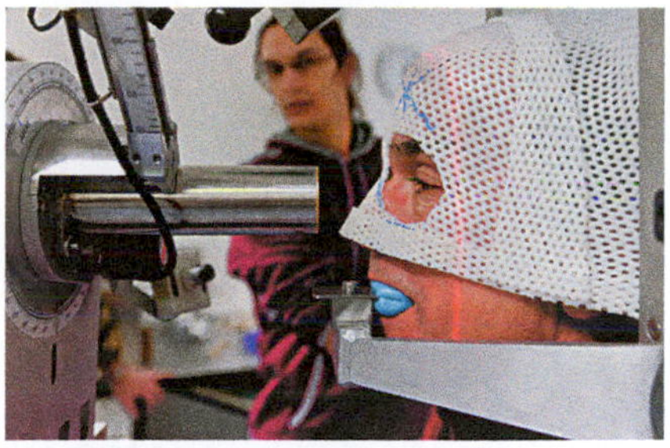

**Proton therapy** BIO Radiation can destroy tumors, but conventional radiation therapy also damages the surrounding tissue. A better choice may be a beam of protons, which can be directed precisely, concentrating the damage in the tumor. In a typical procedure for ocular tumors, the protons are accelerated through a potential difference of 60 MV, gaining a kinetic energy of 60 MeV.

**EXAMPLE 21.4 The speed of a proton**

Atomic particles are often characterized by their kinetic energy in MeV. What is the speed of an 8.7 MeV proton?

***SOLVE*** The kinetic energy of this particle is $8.7 \times 10^6$ eV. First, we convert the energy to joules:

$$K = 8.7 \times 10^6 \text{ eV} \times \frac{1.60 \times 10^{-19}\text{ J}}{1.0\text{ eV}} = 1.39 \times 10^{-12}\text{ J}$$

Now we can find the speed from

$$K = \frac{1}{2}mv^2$$

which gives

$$v = \sqrt{\frac{2K}{m}} = \sqrt{\frac{2(1.39 \times 10^{-12}\text{ J})}{1.67 \times 10^{-27}\text{ kg}}} = 4.1 \times 10^7 \text{ m/s}$$

Because the proton's charge and the electron's charge have the same magnitude, a general rule is that a proton or electron that accelerates (decelerates) through a potential difference of $V$ volts gains (loses) $V$ eV of kinetic energy. In Example 21.2, with a 2000 V potential difference, the proton gained 2000 eV of kinetic energy. In Example 21.4, the proton had to accelerate through a $8.7 \times 10^6$ V = 8.7 MV potential difference to acquire 8.7 MeV of kinetic energy.

**STOP TO THINK 21.3** A proton is released from rest at point Q, where the potential is 0 V. Afterward, the proton

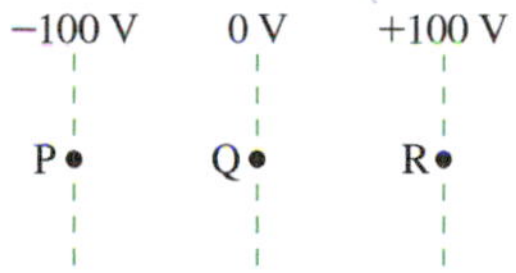


A. Remains at rest at Q.
B. Moves toward P with a steady speed.
C. Moves toward P with an increasing speed.
D. Moves toward R with a steady speed.
E. Moves toward R with an increasing speed.

## 21.4 Calculating the Electric Potential

Now that we understand what the electric potential is, let's calculate the electric potential for some important cases. We'll do so using Equation 21.1, the relationship between the potential energy of a charge $q$ at a point in space and the electric potential at that point. Rewriting Equation 21.1 slightly, we have

$$V = \frac{U_{\text{elec}}}{q} \qquad (21.4)$$

Our prescription for finding the potential at a certain point in space, then, is to first calculate the electric potential *energy* of a charge $q$ placed at that point. Then we can use Equation 21.4 to find the electric potential.

## The Electric Potential Inside a Parallel-Plate Capacitor

In Chapter 20 we learned that a *uniform* electric field can be created by placing equal but opposite charges on two parallel conducting plates—the **parallel-plate capacitor.** Thus finding the electric potential inside a parallel-plate capacitor is equivalent to finding the potential for the very important case of a uniform electric field.

FIGURE 21.10 Finding the potential of a parallel-plate capacitor.

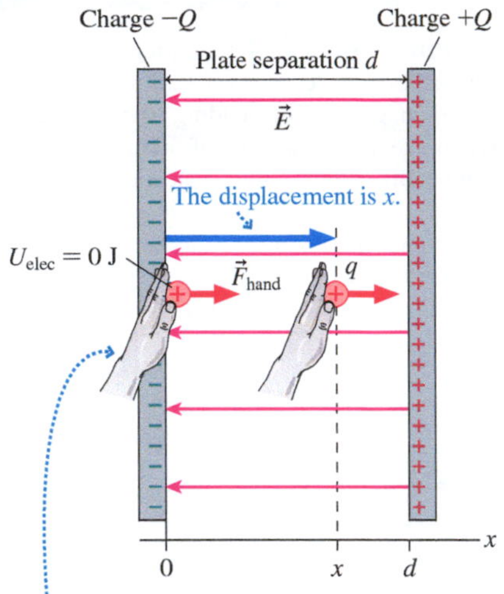


FIGURE 21.10 shows a cross-section view of a charged parallel-plate capacitor with separation $d$ between the plates. The charges $\pm Q$ on the plates are the source charges that create both the electric field and the electric potential in the space between the plates. As we found in ◀◀ SECTION 20.5, the electric field is $\vec{E} = (Q/\epsilon_0 A$, from positive to negative). We'll choose a coordinate system with $x = 0$ at the negative plate and $x = d$ at the positive plate.

We're free to choose the point of zero potential energy anywhere that's convenient, so we let $U_{elec} = 0$ when a mobile charge $q$ is at the negative plate. The charge's potential energy at any other position $x$ is then the amount of work an external force must do to move the charge at steady speed from the negative plate to that position. We'll represent the external force by a hand, although that's not really how charges get moved around.

The electric field in Figure 21.10 points to the left, so the force $\vec{F}_{elec} = q\vec{E}$ of the field on the charge is also to the left. To move the charge to the right at constant speed ($\vec{F}_{net} = \vec{0}$), the external force $\vec{F}_{hand}$ must push to the right with a force of the same magnitude: $F_{hand} = qE$. This force does work on the system as the charge is moved, changing the potential energy of the system.

Because the external force is constant and is parallel to the displacement, the work to move the charge to position $x$ is

$$W = \text{force} \times \text{displacement} = F_{hand}x = qEx$$

Consequently, the electric potential energy when charge $q$ is at position $x$ is

$$U_{elec} = W = qEx$$

As the final step, we can use Equation 21.4 to find that the electric potential of the parallel-plate capacitor at position $x$, measured from the positive plate, is

$$V = \frac{U_{elec}}{q} = Ex = \frac{Q}{\epsilon_0 A}x \tag{21.5}$$

where, in the last step, we wrote the electric field strength in terms of the amount of charge on the capacitor plates.

A first point to notice is that the electric potential increases linearly from the negative plate at $x = 0$, where $V_- = 0$, to the positive plate at $x = d$, where $V_+ = Ed$. The potential difference $\Delta V_C$ between the two capacitor plates is thus

$$\Delta V_C = V_+ - V_- = Ed \tag{21.6}$$

If capacitor voltage is fixed at some value $\Delta V_C$ by connecting its plates to a battery with a known voltage, the electric field strength inside the capacitor is determined from Equation 21.6 as

$$E = \frac{\Delta V_C}{d} \tag{21.7}$$

Electric field strength inside a parallel-plate capacitor

We can establish an electric field of known strength by applying a voltage across a capacitor whose spacing is known.

Notice that Equation 21.7 expresses the electric field as the ratio of a potential difference and a distance, which implies units of volts per meter, or V/m. This is equivalent to the units we have been using, N/C. Electric field strengths are usually expressed in V/m, but it may make sense to use N/C if you are interested in the forces on charges.

The parallel-plate capacitor may seem a very specialized case, but it is a model that can be applied to a wide variety of situations. Figure 21.5 shows a close-up view of a cell membrane. The charges inside and outside the membrane are separated by the fixed distance of the membrane thickness. We can model this as a parallel-plate capacitor, with the potential difference across the membrane maintained by the ion pumps.

**EXAMPLE 21.5 The electric field inside the cell membrane** BIO

A cell membrane is 7.0 nm thick, with a 70 mV potential difference across the membrane. What is the electric field strength inside the membrane?

**STRATEGIZE** We can model the cell membrane as a parallel-plate capacitor. The thickness of the membrane is the distance between the charge layers.

**PREPARE** We can use Equation 21.7 to calculate the electric field given the potential difference and the distance between the charges.

**SOLVE** With the numbers given, the field is

$$E = \frac{\Delta V_C}{d} = \frac{0.070 \text{ V}}{7.0 \times 10^{-9} \text{ m}} = 1.0 \times 10^7 \text{ V/m}$$

**ASSESS** Our result is consistent with the value we saw in Table 20.1, so we can have confidence in our results. This is a very large field—larger than that needed to cause a spark in air—so it's clear that the electrical nature of cells is strong indeed. We will explore the electrical nature of nerve cells in Chapter 23.

Returning to the electric potential, we can substitute Equation 21.7 for $E$ into Equation 21.5 for $V$. In terms of the capacitor voltage $\Delta V_C$, the electric potential at position $x$ inside the capacitor is

$$V = \frac{x}{d}\Delta V_C \qquad (21.8)$$

We can see that the potential increases linearly from $V = 0$ at $x = 0$ (the negative plate) to $V = \Delta V_C$ at $x = d$ (the positive plate).

Let's explore the electric potential inside the capacitor by looking at several different, but related, ways that the potential can be represented. In the table, a battery has established a 1.5 V potential difference across a parallel-plate capacitor with a 3 mm plate spacing.

**Video** Chapter 21 Electric Potential Table

**Graphical representations of the electric potential inside a capacitor**

A graph of potential versus $x$. You can see the potential increasing from 0 V at the negative plate to 1.5 V at the positive plate.

A three-dimensional view showing **equipotential surfaces.** These are mathematical surfaces, not physical surfaces, that have the same value of $V$ at every point. The equipotential surfaces of a capacitor are planes parallel to the capacitor plates. The capacitor plates are also equipotential surfaces.

A two-dimensional **equipotential map.** The green dashed lines represent slices through the equipotential surfaces, so $V$ has the same value everywhere along such a line. We call these lines of constant potential **equipotential lines** or simply **equipotentials.**

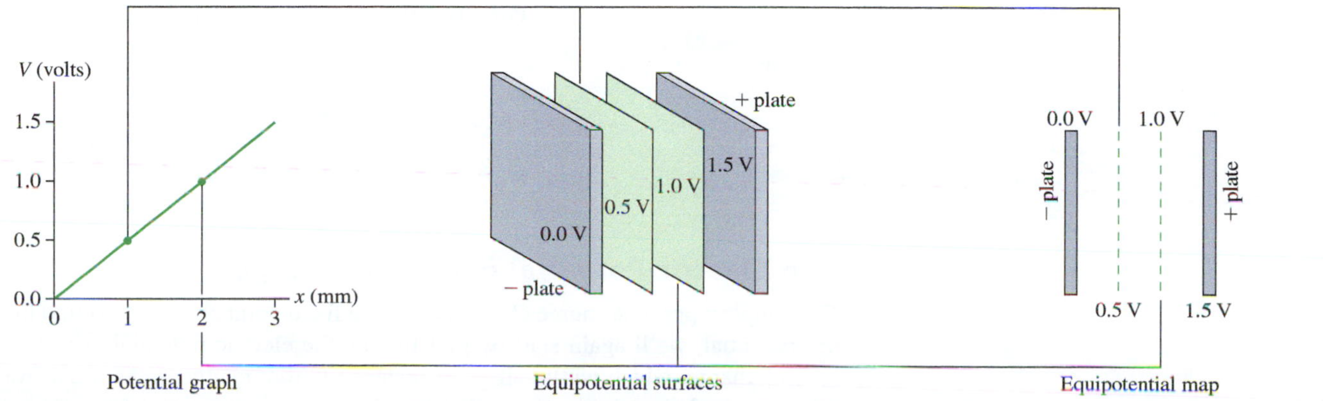

**NOTE** ▶ Equipotential lines are just the intersections of equipotential surfaces with the two-dimensional plane of the paper, and so are really just another way of representing equipotential surfaces. Because of this, we'll often use the terms "equipotential lines," "equipotential surfaces," and "equipotentials" interchangeably. ◀

### EXAMPLE 21.6 A proton in a capacitor

A parallel-plate capacitor is constructed of two disks spaced 2.00 mm apart. It is charged to a potential difference of 500 V. A proton is shot through a small hole in the negative plate with a speed of $2.0 \times 10^5$ m/s. What is the farthest distance from the negative plate that the proton reaches?

**STRATEGIZE** We'll treat this problem using conservation of energy. As the proton moves toward the higher potential, its potential energy will increase, so its kinetic energy will decrease. We are looking for the point at which its kinetic energy has gone to zero—this is as far as the proton will go.

**PREPARE** FIGURE 21.11 is a before-and-after visual overview of the proton in the capacitor. Equation 21.2 is an expression of conservation of energy for this situation. We set the starting point of the proton at the left plate, and we take the potential at this point to be zero, so $V_i = 0$. The potential between the plates of the capacitor is given by Equation 21.8. At the final position $x_f$ the kinetic energy is zero, so $K_f = 0$.

**SOLVE** We use the information we've assembled to work out the terms in Equation 21.2:

$$K_f + qV_f = K_i + qV_i$$

$$0 + e\Delta V_C \frac{x_f}{d} = \frac{1}{2}mv_i^2 + 0$$

FIGURE 21.11 A before-and-after visual overview of a proton moving in a capacitor.

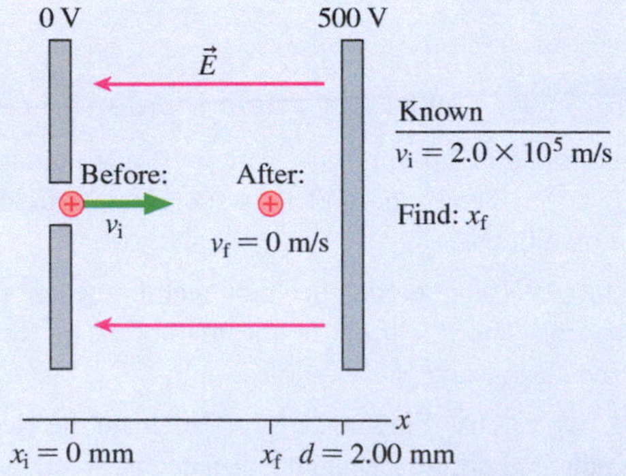


The final position is thus

$$x_f = \frac{mdv_i^2}{2e\Delta V_C} = 0.84 \text{ mm}$$

The proton travels 0.84 mm, a bit less than halfway across the region between the plates, before stopping and reversing course.

**ASSESS** Our final result is a position between the two plates, as it must be given the problem statement. This gives us confidence in our result.

Synthesis 21.1 brings together what we've learned about the electric field and electric potential of a parallel-plate capacitor.

### SYNTHESIS 21.1 The parallel-plate capacitor: Potential and electric field

There are several related expressions for the electric potential and field inside a parallel-plate capacitor.

**The variables used in describing the parallel-plate capacitor**

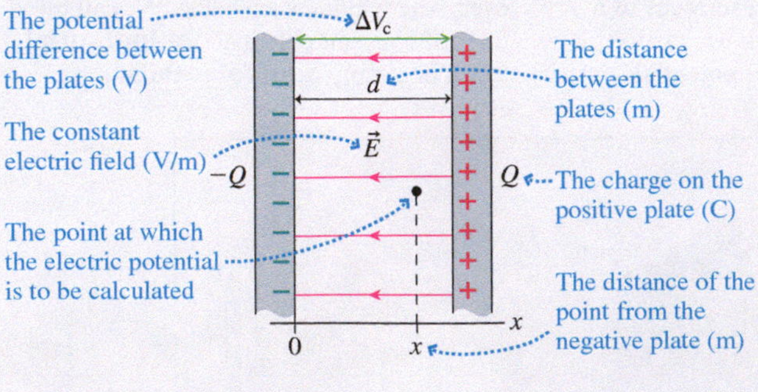


Three equivalent expressions for the electric potential are related by two expressions for the electric field inside the capacitor:

**Potential $V$** $\quad V = Ex \quad V = \frac{Q}{\epsilon_0 A}x \quad V = \frac{x}{d}\Delta V_C$

**Field $E$** $\quad E = \frac{Q}{\epsilon_0 A} \quad E = \frac{\Delta V_C}{d}$

## The Electric Potential of a Point Charge

The simplest possible source charge is a single fixed point charge $q$. To find the electric potential, we'll again start by first finding the electric potential energy when a second charge, which we'll call $q'$, is distance $r$ from charge $q$. As usual, we'll do this by calculating the work needed to bring $q'$ from a point where $U_{elec} = 0$ to distance $r$ from $q$. We're free to choose $U_{elec} = 0$ at any point that's convenient. Because the influence of a point charge goes to zero infinitely far from the charge, it is natural to choose $U_{elec} = 0$ (and hence $V = 0$) at a point that is infinitely distant from $q$.

We can't use the simple expression $W = Fd$ to find the work done in moving $q'$; this expression is valid for only a *constant* force $F$ and, as we know from Coulomb's law, the force on $q'$ gets larger and larger as it approaches $q$.

FIGURE 21.12 shows $q'$ at two different distances $r$ from the fixed charge $q$. When $q'$ is relatively far from $q$, the electric force on $q'$ is small. Not much external force is needed to push $q'$ closer to $q$ by a small displacement $d$, so the work done on $q'$ is small and the *change* $\Delta U_{elec}$ in the electric potential energy is small as well. This implies, as Figure 21.12 shows, that the graph of $U_{elec}$ versus $r$ is fairly flat when $q'$ is far from $q$.

However, the force on $q'$ is quite large when it gets near $q$, so the work required to move it through the same small displacement is much greater than before. The change in $U_{elec}$ is large, so the graph of $U_{elec}$ versus $r$ is steeper when $q'$ is near $q$.

The general shape of the graph of $U_{elec}$ must be as shown in Figure 21.12. When $q'$ is far from $q$, the potential energy is small. As $q'$ approaches $q$, the potential energy increases rapidly. An exact calculation finds the potential energy of two point charges to be

$$U_{elec} = K\frac{qq'}{r} = \frac{1}{4\pi\epsilon_0}\frac{qq'}{r} \tag{21.9}$$

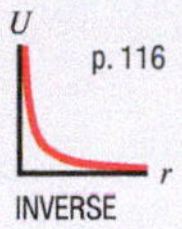


Electric potential energy of two charges $q$ and $q'$ separated by distance $r$

**NOTE** ▸ This expression is very similar to Coulomb's law. The difference is that the electric potential energy depends on the *inverse* of $r$—that is, on $1/r$—instead of the inverse-square dependence of Coulomb's law. Make sure you remember which is which! In this and in following equations we will generally give two different, equivalent, expressions, one with the electrostatic constant $K$, one with the permittivity constant $\epsilon_0$. In solving problems, you can use whichever expression is most convenient. ◂

Figure 21.12 was a graph of the potential energy for two like charges, where $qq'$ is positive. But Equation 21.9 is equally valid for *opposite* charges. In this case, the potential energy of the charges is *negative*. As FIGURE 21.13 shows, the potential energy of the two charges *decreases* as they get closer together. A particle speeds up as its potential energy decreases $(U \rightarrow K)$, so charge $q'$ accelerates toward the fixed charge $q$.

FIGURE 21.12 The electric potential energy of two point charges.

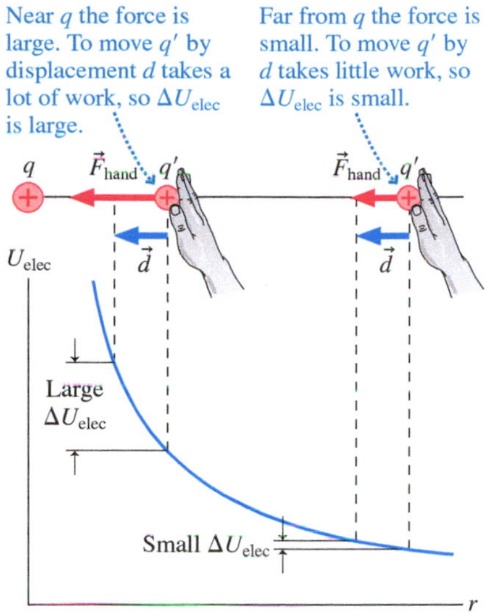


FIGURE 21.13 Potential-energy diagram for two opposite charges.

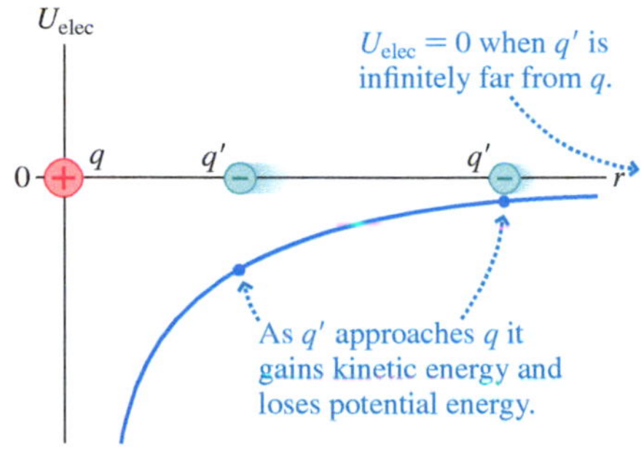


## EXAMPLE 21.7 Finding the escape velocity

An interaction between two elementary particles causes an electron and a positron (a positively charged electron) to be shot out back-to-back with equal speeds. What minimum speed must each particle have when they are 100 fm apart in order to end up far from each other? (Recall that 1 fm $= 10^{-15}$ m.)

**STRATEGIZE** As the particles move away from each other, the potential energy increases, so the kinetic energy decreases. The problem statement notes that the particles end up "far from each other." We can interpret this as meaning that they end up at a large enough distance, $r_f \simeq \infty$, such that $(U_{elec})_f = 0$. If the particles have the minimum speed to reach this point, they'll have zero kinetic energy when they get there: $K_f = 0$.

**PREPARE** FIGURE 21.14 shows the before-and-after visual overview. We need to interpret $U_{elec}$ as the potential energy of the electron + positron system. Similarly, $K$ is the total kinetic energy of the system. The electron and the positron, with equal masses and equal initial speeds, have equal initial kinetic energies.

FIGURE 21.14 The before-and-after visual overview of an electron and a positron flying apart.

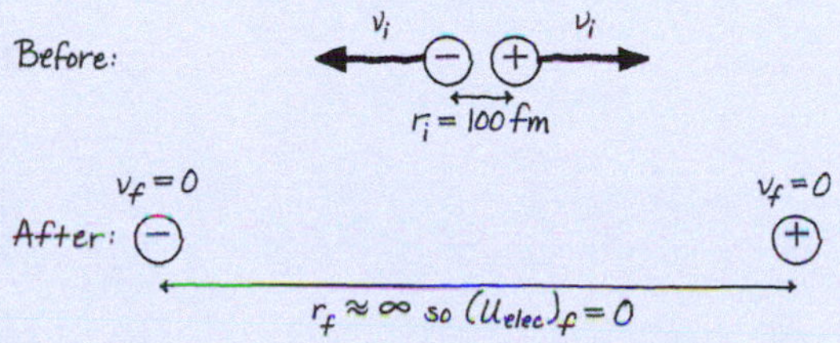


*Continued*

**SOLVE** The final kinetic energy is zero; so is the final potential energy. If the final energy is zero, the initial energy must be zero as well, so the expression for conservation of energy is

$$K_i + U_i = 0$$

If we write the initial speeds as $v_i$ and use the expression for the potential energy of the two charges in Equation 21.9, this becomes

$$\left(\frac{1}{2}mv_i^2 + \frac{1}{2}mv_i^2\right) + K\frac{qq'}{r_i} = 0$$

The two masses are the same—the mass of the electron. The charges are opposite; one charge is $+e$ and the other is $-e$. This gives

$$mv_i^2 + \frac{-Ke^2}{r_i} = 0$$

With $r_i = 100 \text{ fm} = 1.0 \times 10^{-13}$ m, we find

$$v_i = \sqrt{\frac{Ke^2}{mr_i}} = 5.0 \times 10^7 \text{ m/s}$$

**ASSESS** The particles are very close together; for comparison, the diameter of a typical atom is about $0.1 \text{ nm} = 1 \times 10^{-10}$ m. The potential energy at this small distance is negative with a very large magnitude. We expect that the particles would need to be moving quite rapidly to be able to separate.

Equation 21.9 gives the potential energy of a charge $q'$ when it is a distance $r$ from a point charge $q$. We know that the electric *potential* is related to the potential energy by $V = U_{elec}/q'$. Thus the electric potential of charge $q$ is

$$V = K\frac{q}{r} = \frac{1}{4\pi\epsilon_0}\frac{q}{r} \qquad (21.10)$$

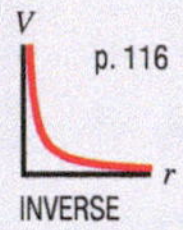

p. 116

Electric potential at distance $r$ from a point charge $q$

Notice that only the *source* charge $q$ appears in this expression. This is in line with our picture that a source charge *creates* the electric potential around it.

**EXAMPLE 21.8** **Calculating the potential of a point charge**

What is the electric potential 1.0 cm from a 1.0 nC charge? What is the potential difference between a point 1.0 cm away and a second point 3.0 cm away?

**PREPARE** We can use Equation 21.10 to find the potential at the two distances from the charge.

**SOLVE** The potential at $r = 1.0$ cm is

$$V_{1\text{ cm}} = K\frac{q}{r} = (9.0 \times 10^9 \text{ N} \cdot \text{m}^2/\text{C}^2)\left(\frac{1.0 \times 10^{-9} \text{ C}}{0.010 \text{ m}}\right) = 900 \text{ V}$$

We similarly calculate $V_{3\text{ cm}} = 300$ V. Thus the potential difference between these two points is $\Delta V = V_{1\text{ cm}} - V_{3\text{ cm}} = 600$ V.

**ASSESS** The final numbers are reasonable. We've noted that there can be a potential difference of 10,000 V between your finger and a doorknob when you've rubbed your feet on the carpet, when you acquire a charge of tens of nC. This is a smaller charge, so we expect a smaller potential.

FIGURE 21.15 shows three graphical representations of the electric potential of a point charge. These match the three representations of the electric potential inside a capacitor, and a comparison of the two is worthwhile. This figure assumes that $q$ is positive; think about how the representations would change if $q$ were negative.

FIGURE 21.15 Three graphical representations of the electric potential of a point charge.

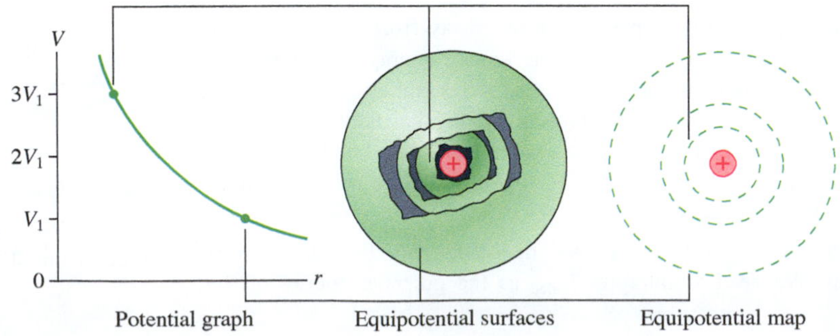

## The Electric Potential of a Charged Sphere

Equation 21.10 gives the electric potential of a point charge. It can be shown that the electric potential outside a charged sphere is the *same* as that of a point charge; that is,

$$V = K\frac{Q}{r} = \frac{1}{4\pi\epsilon_0}\frac{Q}{r} \qquad (21.11)$$

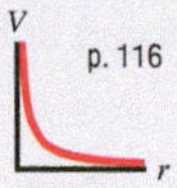

p. 116

Electric potential at a distance $r > R$ from the center of a sphere of radius $R$ and with charge $Q$

Suppose you connect a metal sphere to the positive terminal of a battery. This will give a positive charge to the sphere, which will be distributed across the surface. The potential at the surface will be the battery potential; if you connect the sphere to a 9 V battery, the potential at the surface will be 9 V. (In this case, we'd say that the sphere has been "charged to 9 V.") If we call the potential at the surface of the sphere $V_0$, then we can see from Equation 21.11 that

$$V_0 = V(\text{at } r = R) = \frac{KQ}{R} = \frac{Q}{4\pi\epsilon_0 R} \qquad (21.12)$$

Consequently, a sphere of radius R that is charged to potential $V_0$ has total charge

$$Q = \frac{RV_0}{K} = 4\pi\epsilon_0 RV_0 \qquad (21.13)$$

If we substitute this expression for $Q$ into Equation 21.11, we can write the potential outside a sphere that is charged to potential $V_0$ as

$$V = \frac{R}{r}V_0 \qquad (21.14)$$

Equation 21.14 tells us that the potential of a sphere is $V_0$ on the surface and decreases inversely with the distance from the center. Thus the potential at $r = 3R$ is $\frac{1}{3}V_0$.

**EXAMPLE 21.9** **Detecting charge on a sphere** BIO

We've seen that bees develop a positive charge as they fly through the air. When a bee lands on a flower, charge is transferred, and future bee visitors to the same flower can sense this charge and seek out a more promising bloom. Experimenters have done tests to determine the limits of bumblebee field sensitivity. In one test, bumblebees could detect, from a distance, the charge on a 3.0-cm-diameter sphere that had been charged to 30 V. What was the magnitude of the charge on the sphere?

**PREPARE** The sphere has radius $R = 0.015$ m; the sphere is charged to $V_0 = 30$ V.

**SOLVE** The charge is given by Equation 21.13:

$$Q = \frac{RV_0}{K} = 50 \text{ pC}$$

**ASSESS** This is comparable to the magnitude of the charge that a bee accumulates and that it might transfer to a flower, so our final result makes sense. A bumblebee with this level of sensitivity could, indeed, detect that another bee had visited a flower by sensing the transferred charge.

Synthesis 21.2 highlights the similarity of the potential due to a point charge and the potential outside a charged sphere.

## SYNTHESIS 21.2 Potential of a point charge and a charged sphere

The electric potential $V$ on or outside a charged sphere is the same as for a point charge: It depends only on the distance and the charge on the sphere.

**Point charge $q$**

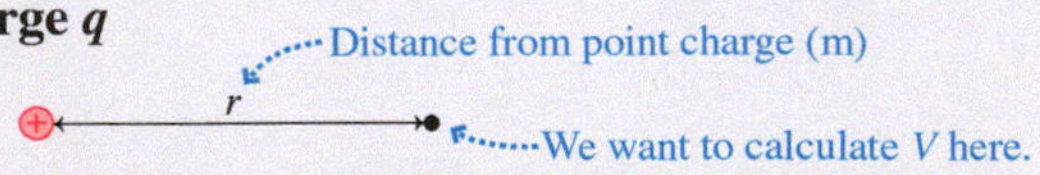


$$V = K\frac{q}{r} = \frac{1}{4\pi\epsilon_0}\frac{q}{r}$$

Electrostatic constant ($K$); Permittivity constant ($\epsilon_0$)

**Sphere with total charge $Q$**

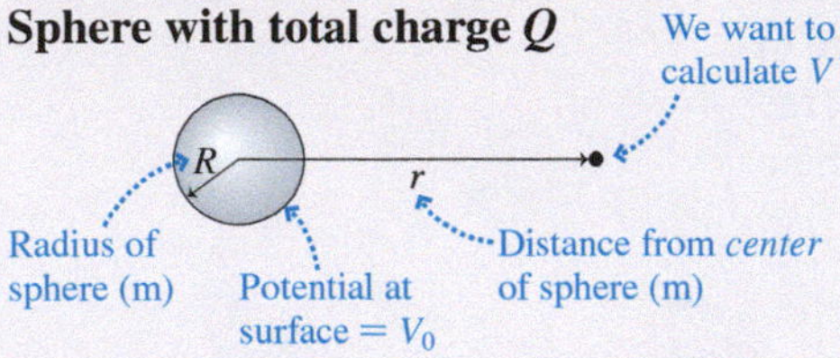


When the total charge $Q$ on the sphere is known:

$$V = K\frac{Q}{r} = \frac{1}{4\pi\epsilon_0}\frac{Q}{r}$$

When the potential $V_0$ at the surface of the sphere is known:

$$V = \frac{R}{r}V_0$$

These expressions are valid only *on* or *outside* the sphere, so $r \geq R$.

## Ionization Energy

We can model an atom as a sphere; this model is far from perfect, but it can give us some good insight. Suppose we want to separate an electron from an atom. Removing an electron from an atom leaves a positively charged ion behind, as shown in FIGURE 21.16. We can assume that the electron is at the surface of a sphere that is equal to the size of the atom. It requires energy to separate these two opposite charges.

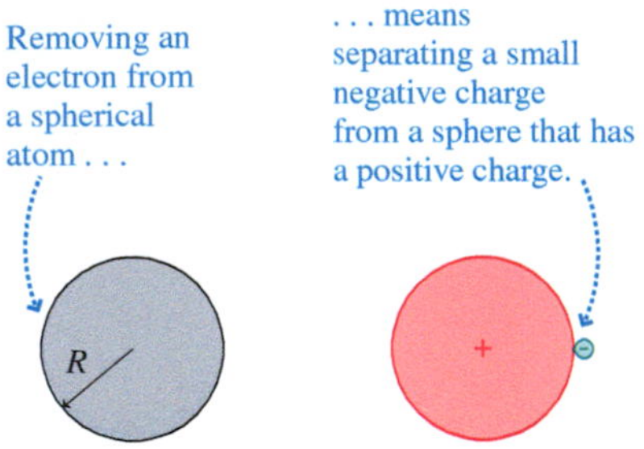


FIGURE 21.16 Ionizing an atom modeled as a sphere.

### EXAMPLE 21.10 Finding the ionization energy for an atom

The radius of a zinc atom is approximately 0.14 nm. What is the energy required to remove an electron from the atom—that is, the ionization energy?

**STRATEGIZE** We can model the atom as a sphere with radius 0.14 nm. Removing an electron means moving this negative charge away from a positively charged sphere, the ion left behind. We can find the difference in potential and use this to compute the change in potential energy; this is the energy necessary to remove the electron.

**PREPARE** The electron, with a negative charge, will have a negative potential energy at the surface of the positively charged sphere. Removing the electron means removing it to a great distance where $V = 0$.

**SOLVE** The electron begins at the surface and ends at a great distance. The initial potential at the surface of the sphere is $V_i = V_R$; the final potential is $V_f = 0$. The change in potential is thus

$$\Delta V = V_f - V_i = -V_R$$

$V_R$ is the potential at the surface of the sphere, which has the radius of the atom and a charge of $+e$:

$$V_R = K\frac{e}{R} = (9.0 \times 10^9\ \mathrm{N \cdot m^2/C^2})\frac{(1.6 \times 10^{-19}\ \mathrm{C})}{(0.14 \times 10^{-9}\ \mathrm{m})} = 10\ \mathrm{V}$$

This is a convenient time to use the energy unit eV. Moving the electron through a 10 V potential difference requires an energy of 10 eV; this is the ionization energy.

**ASSESS** Our final result, 10 eV, compares favorably with the measured value of 9.4 eV, which gives us confidence in our calculation and in the model underlying it.

The model we used to discuss ionization energy is simplistic, but it gives reasonable results, and it will let us make some qualitative deductions as well.

**CONCEPTUAL EXAMPLE 21.11 How much energy is needed to remove a second electron?**

It takes about 8 eV to remove one electron from a silicon atom. How much energy does it take to remove a second electron?

***REASON*** If we assume that the size of the atom stays the same (a reasonable assumption in this case), then the only difference is the charge of the ion that establishes the potential; it is $+2e$ instead of $+e$. Twice the potential means twice the energy, so it requires 16 eV to remove a second electron.

***ASSESS*** The second ionization energy for silicon (the energy required to remove a second electron) is, in fact, about 16 eV.

## The Electric Potential of Many Charges

Suppose there are many source charges $q_1, q_2, \ldots$. The electric potential $V$ at a point in space is the *sum* of the potentials due to each charge:

$$V = \sum_i K\frac{q_i}{r_i} = \sum_i \frac{1}{4\pi\epsilon_0}\frac{q_i}{r_i} \tag{21.15}$$

where $r_i$ is the distance from charge $q_i$ to the point in space where the potential is being calculated. Unlike the electric field of multiple charges, which required vector addition, the electric potential is a simple scalar sum. This makes finding the potential of many charges easier than finding the corresponding electric field. As an example, the equipotential map in FIGURE 21.17 shows that the potential of an electric dipole is the sum of the potentials of the positive and negative charges. We'll see later that the electric potential of the heart has the form of an electric dipole.

FIGURE 21.17 The electric potential of an electric dipole.

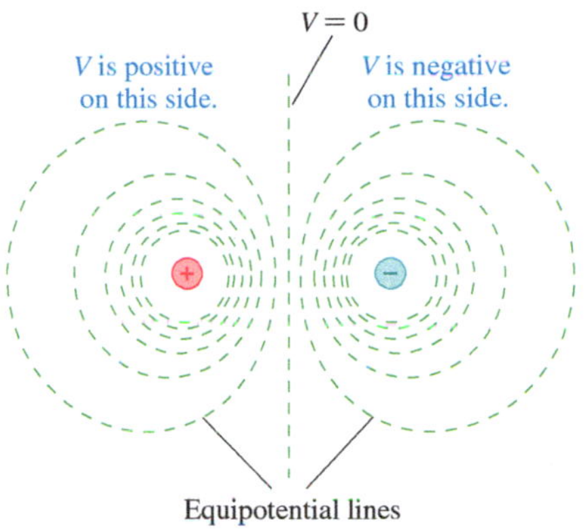


**EXAMPLE 21.12 Finding the potential of two charges**

What is the electric potential at the point indicated in FIGURE 21.18?

FIGURE 21.18 Finding the potential of two charges.

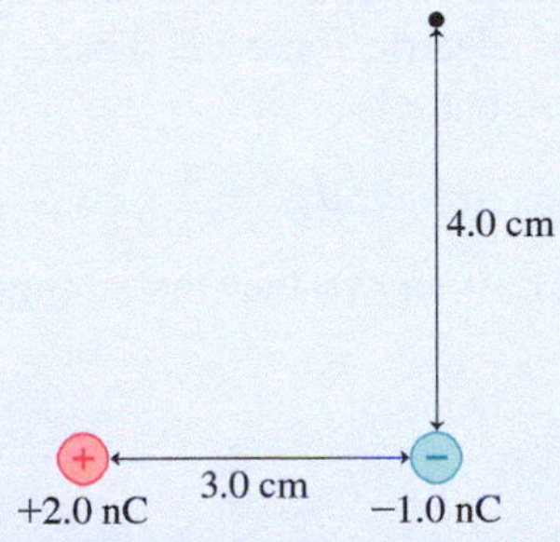


***STRATEGIZE*** The potential at the point is the sum of the potentials due to each charge. The potential is a scalar, so we don't need to use angles or find components.

***PREPARE*** The negative charge is 4.0 cm from the point at which we are calculating potentials. The distance to the positive charge is the length of the hypotenuse of a triangle with sides 3.0 cm and 4.0 cm, and so is 5.0 cm.

***SOLVE*** The potential at the indicated point is

$$V = \frac{Kq_1}{r_1} + \frac{Kq_2}{r_2}$$
$$= (9.0 \times 10^9\ \text{N}\cdot\text{m}^2/\text{C}^2)\left(\frac{2.0\times10^{-9}\ \text{C}}{0.050\ \text{m}} + \frac{-1.0\times10^{-9}\ \text{C}}{0.040\ \text{m}}\right)$$
$$= 140\ \text{V}$$

***ASSESS*** For charges and distances of these magnitudes, this is a typical magnitude for the potential, so our result seems reasonable.

**STOP TO THINK 21.4** Rank in order, from largest to smallest, the potential differences $\Delta V_{12}$, $\Delta V_{13}$, and $\Delta V_{23}$ between points 1 and 2, points 1 and 3, and points 2 and 3.

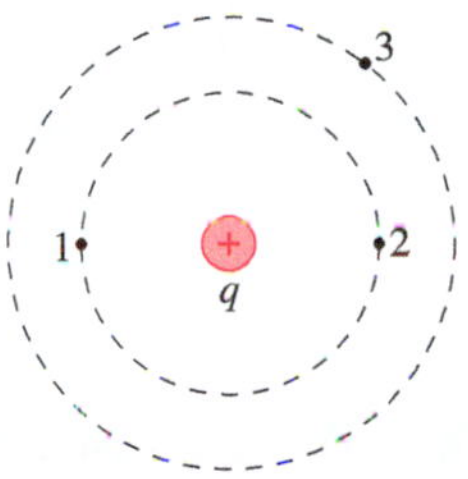


# 21.5 Connecting Potential and Field

In Chapter 20, we learned how source charges create an electric field around them; in this chapter, we found that source charges also create an electric potential everywhere in the space around them. But these two concepts of field and potential are clearly linked; we can calculate potential differences by considering the work done on a charge as it is pushed against the electric force due to the field.

FIGURE 21.19 The electric field is always perpendicular to an equipotential surface.

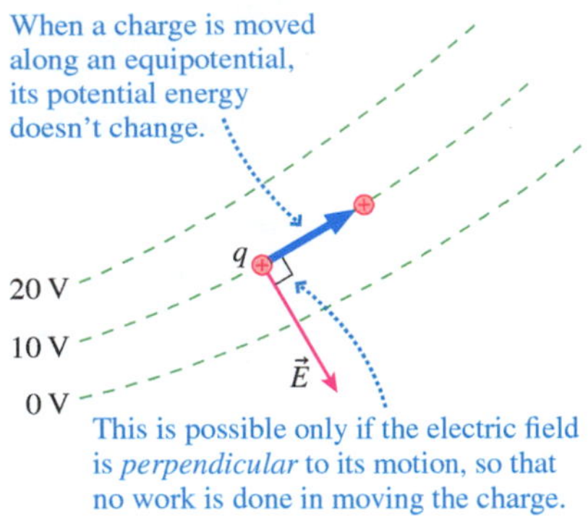


To make the connection between potential and field, FIGURE 21.19 shows an equipotential map of the electric potential due to some source charges (which aren't shown here). Suppose a charge $q$ moves a short distance along one of the equipotential surfaces. Because it moves along an equipotential, its potential and hence its potential energy are the *same* at the beginning and end of its displacement. This means that no work is done in moving the charge. As Figure 21.19 shows, the only way that no work can be done is if the electric field is *perpendicular* to the equipotential. (You should recall, from ◀◀ SECTION 10.2, that no work is done by a force perpendicular to a particle's displacement.) **The electric field at a point is perpendicular to the equipotential surface at that point.**

FIGURE 21.20 The electric field points "downhill."

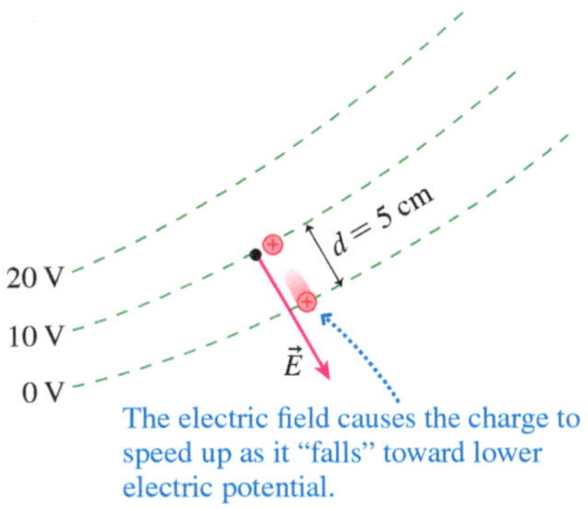


FIGURE 21.20 shows a positive charge released from rest starting at the 10 V equipotential. You learned in Section 21.3 that a positive charge speeds up as it moves from higher to lower potential, so this charge will speed up as it moves toward the lower 0 V equipotential. Because it is the electric field $\vec{E}$ in Figure 21.20 that pushes on the charge, causing it to speed up, $\vec{E}$ must point as shown, from higher potential (10 V) to lower potential (0 V). **The electric field points in the direction of** decreasing **potential.**

FIGURE 21.21 Finding the strength of the electric field.

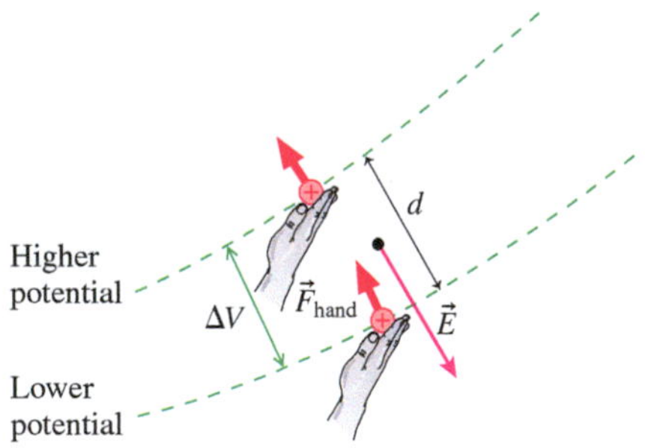


We can find an expression for the *magnitude* of $\vec{E}$ by considering the work required to move the charge, at constant speed, through a displacement that is directed *opposite* $\vec{E}$, as shown in FIGURE 21.21. The displacement is small enough that the electric field in this region can be considered as nearly constant. Conservation of energy requires that

$$W = \Delta U_{\text{elec}} = q\,\Delta V \tag{21.16}$$

Because the charge moves at a constant speed, the magnitude of the force of the hand $\vec{F}_{\text{hand}}$ is exactly equal to that of the electric force $q\vec{E}$. Thus, the work done on the charge in moving it through displacement $d$ is

$$W = F_{\text{hand}}\,d = qEd$$

Comparing this result with Equation 21.16 shows that the strength of the electric field is

$$E = \frac{\Delta V}{d} \tag{21.17}$$

Electric field strength in terms of the potential difference $\Delta V$ between two equipotential surfaces a distance $d$ apart

For example, in Figure 21.20 the magnitude of the potential difference is $\Delta V = (10\text{ V}) - (0\text{ V}) = 10\text{ V}$, while $d = 5\text{ cm} = 0.05\text{ m}$. Thus the electric field strength between these two equipotential surfaces is

$$E = \frac{\Delta V}{d} = \frac{10\text{ V}}{0.05\text{ m}} = 200\text{ V/m}$$

**NOTE** ▶ This expression for the electric field strength is similar to the result $E = \Delta V_C/d$ for the electric field strength inside a parallel-plate capacitor. A capacitor has a uniform field, the same at all points, so we can calculate the field using the full potential difference $\Delta V_C$ and the full spacing $d$. In contrast, Equation 21.17 applies only *locally,* at a point where the spacing between two nearby equipotential lines is $d$. ◀

Video What the Physics? Field of Light

FIGURE 21.22 summarizes what we've learned about the connection between potential and field.

FIGURE 21.22 The geometry of potential and field.

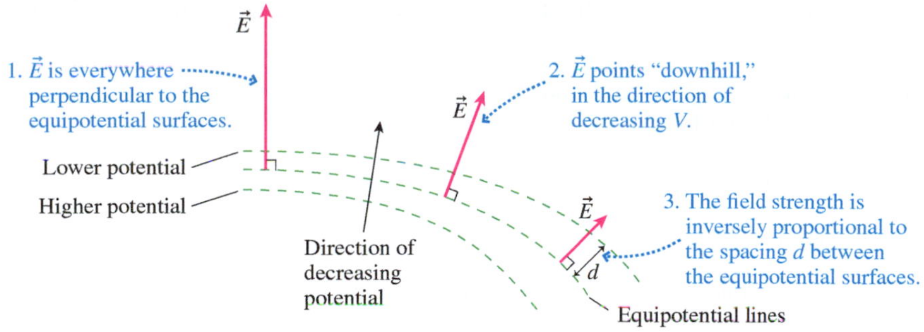


FIGURE 21.23 shows three important arrangements of charges that we've studied in this chapter and Chapter 20. Both electric field lines and equipotentials are shown, and you can see how the connections between field and potential summarized in Figure 21.22 apply in each case.

FIGURE 21.23 Electric field lines and equipotentials for three important cases.

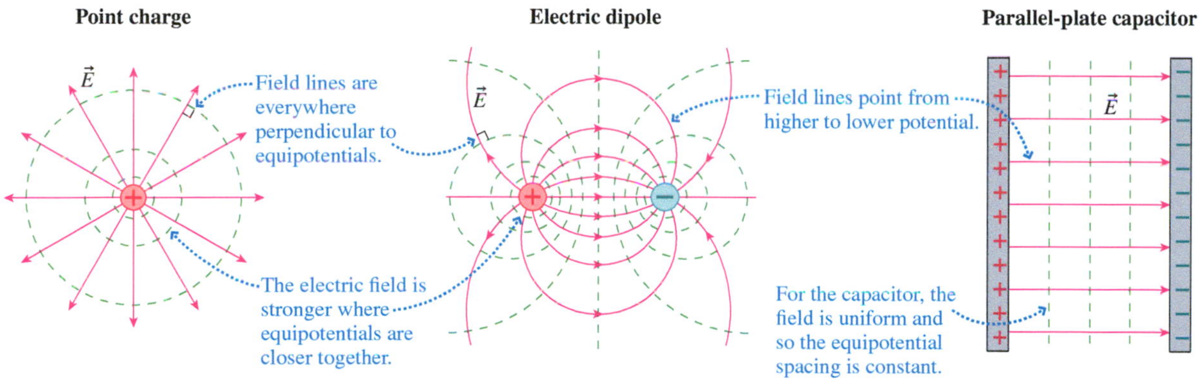


## EXAMPLE 21.13 Finding the electric field from equipotential lines

In FIGURE 21.24 a 1 cm × 1 cm grid is superimposed on an equipotential map of the potential. Estimate the strength and direction of the electric fields at points 1, 2, and 3. Show your results graphically by drawing the electric field vectors on the equipotential map.

FIGURE 21.24 Equipotential lines.

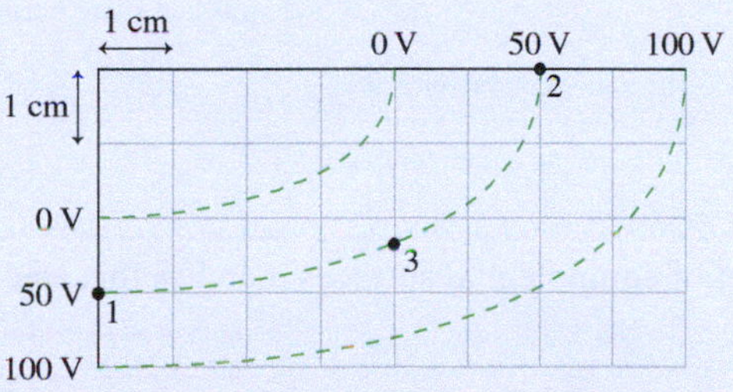


**STRATEGIZE** Some distant but unseen source charges have created electric fields and potentials. We do not need to see the source charges to determine the field; we will use the connections between field and potential.

**PREPARE** The electric field is perpendicular to the equipotential lines, points "downhill," and depends on the spacing between the equipotential lines. The potential is highest on the bottom and the right. Because $E = \Delta V/d$, the electric field is stronger where the equipotential lines are closer together and weaker where they are farther apart.

**SOLVE** FIGURE 21.25 shows how measurements of $d$ from the grid are combined with values of $\Delta V$ to determine $\vec{E}$. Point 3 requires an estimate of the spacing between the 0 V and the 100 V lines. Notice that we're using the 0 V and 100 V equipotential lines to determine $\vec{E}$ at a point on the 50 V equipotential.

FIGURE 21.25 The electric fields at points 1, 2, and 3.

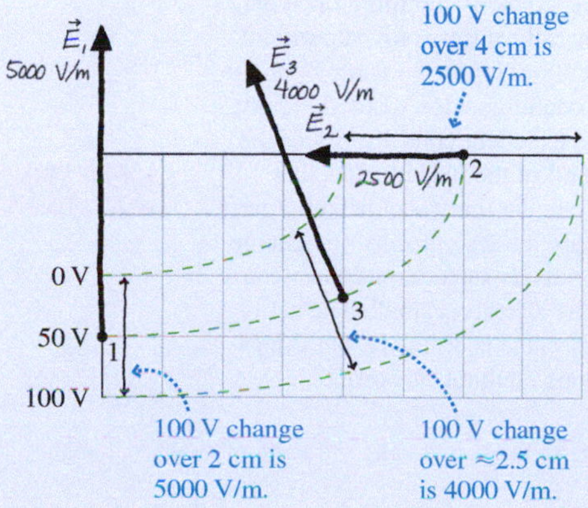


**ASSESS** A quick check of our work shows that it follows the rules we've learned: The electric field vectors point "downhill," toward lower potentials; the vectors are perpendicular to the equipotential lines; and the vectors are longer where the equipotentials are closer together.

## A Conductor in Electrostatic Equilibrium

In ◀◀ SECTION 20.6, you learned four important properties about conductors in electrostatic equilibrium:

1. Any excess charge is on the surface.
2. The electric field inside is zero.
3. The exterior electric field is perpendicular to the surface.
4. The field strength is largest at sharp corners.

Now we can add a fifth important property:

5. The entire conductor is at the same potential, and thus the surface is an equipotential surface.

FIGURE 21.26 All points inside a conductor in electrostatic equilibrium are at the same potential.

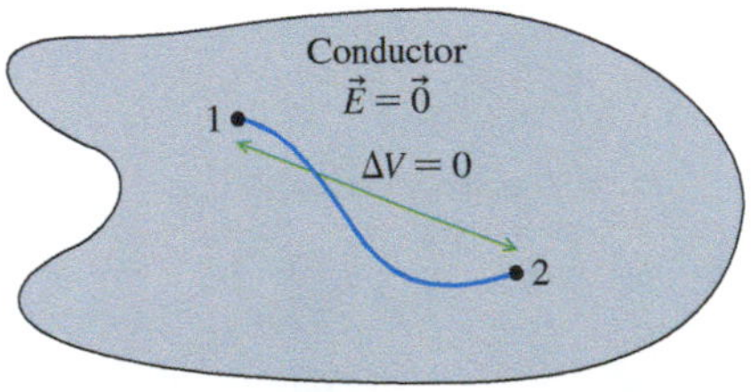


To see why this is so, FIGURE 21.26 shows two points inside a conductor connected by a line that remains entirely inside the conductor. We can find the potential difference $\Delta V = V_2 - V_1$ between these points by using an external force to push a charge along the line from 1 to 2 and calculating the work done. But because $\vec{E}=\vec{0}$, there is no force on the charge. The work is zero, and so $\Delta V = 0$. In other words, **any two points inside a conductor in electrostatic equilibrium are at the same potential.**

When a conductor is in electrostatic equilibrium, the *entire conductor* is at the same potential. If we charge a metal electrode, the entire electrode is at a single potential. The facts that the surface is an equipotential surface and that the exterior electric field is perpendicular to the surface can now be seen as a special case of our conclusion from the preceding section that electric fields are always perpendicular to equipotentials.

FIGURE 21.27 summarizes what we know about conductors in electrostatic equilibrium.

**Dazzled but not damaged** A typical jet airplane is struck by lightning once per year, but generally suffers only superficial damage. The lightning strike is rapid, so the equilibrium conditions don't exactly apply. However, the excellent conductivity of the aluminum shell of the plane means that potential differences between different parts of the plane are modest, most of the charge stays on the outside surface, and the electric field inside the aircraft is small. The action outside the airplane is dramatic, but things inside the plane continue as normal.

FIGURE 21.27 Electrical properties of a conductor in electrostatic equilibrium.

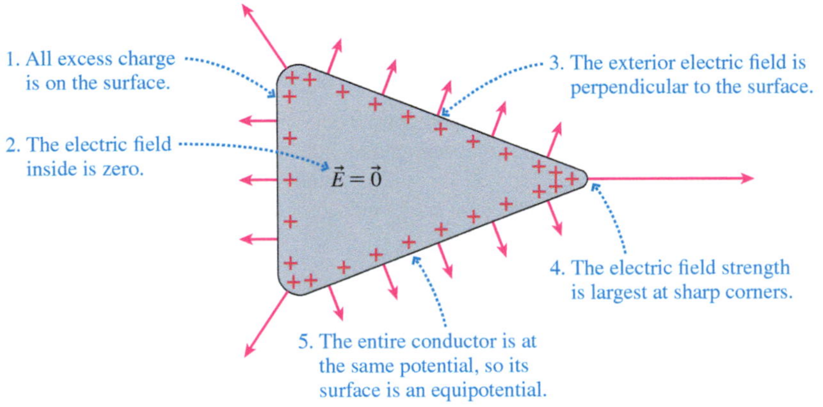


**STOP TO THINK 21.5** Which set of equipotential surfaces matches this electric field?

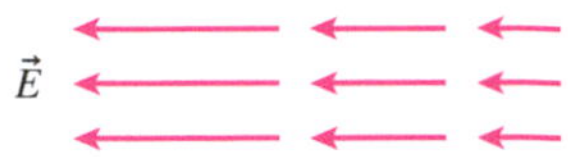


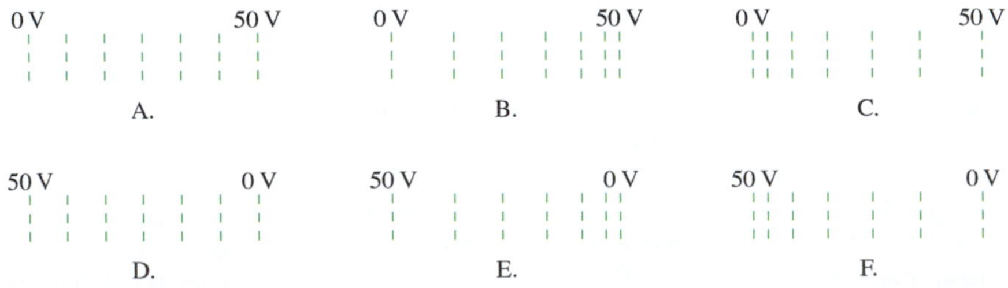

## 21.6 The Electrocardiogram BIO

As we saw in ◀◀ SECTION 20.5, the electrical activity of cardiac muscle cells makes the beating heart an electric dipole. A resting nerve or muscle cell is *polarized,* meaning that the outside is positive and the inside negative. Figure 21.5 showed this situation. Initially, all muscle cells in the heart are polarized. When triggered by an electrical impulse from the heart's sinoatrial node in the right atrium, heart cells begin to *depolarize,* moving ions through the cell wall until the outside becomes negative. This causes the muscle to contract. The depolarization of one cell triggers depolarization in an adjacent cell, which causes a "wave" of depolarization to spread across the tissues of the heart.

At any instant during this process, a boundary divides the negative charges of depolarized cells from the positive charges of cells that have not depolarized. As FIGURE 21.28a shows, this separation of charges creates an electric dipole and produces a dipole electric field and potential. FIGURE 21.28b shows the equipotential surfaces of the heart's dipole at one instant of time. These equipotential surfaces match those shown earlier in Figure 21.17.

FIGURE 21.28 BIO A contracting heart is an electric dipole.

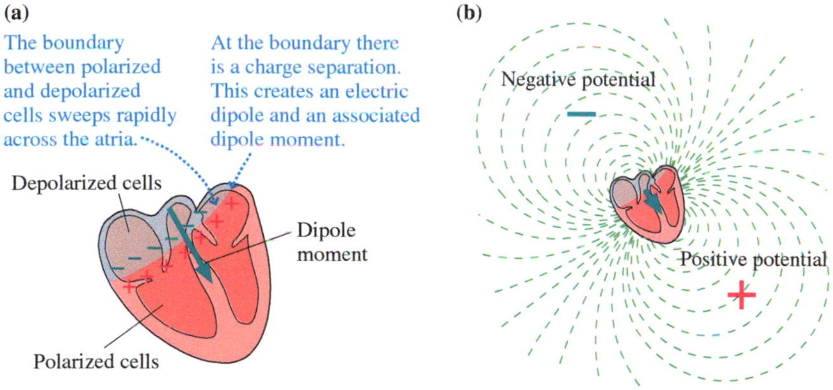


**NOTE** ▶ The convention is to draw diagrams of the heart as if you were facing the person whose heart is being drawn. The left side of the heart is thus on the right side of the diagram. ◀

A measurement of the electric potential of the heart is an invaluable diagnostic tool. Recall, however, that only potential *differences* are meaningful, so we need to measure the potential difference between two points on the torso. In practice, as FIGURE 21.29 shows, the potential difference is measured between several pairs of *electrodes* (often called *leads*). A chart of the potential differences is known as an **electrocardiogram,** abbreviated either ECG or, from its German origin, EKG. A common method of performing an EKG uses 12 leads and records 12 pairs of potential differences.

To understand the EKG, we need to consider how the field of a dipole relates to the potential measured between two electrodes at points. FIGURE 21.30 shows an electric dipole, the dipole moment vector, and the related equipotentials. Suppose we measure the potential difference between electrodes at points 1 and 2 shown on the diagram. The dipole moment vector is directed toward regions of higher potential, and, if the dipole moment is large, the potential difference should be large as well. If a line connecting the two electrodes is perpendicular to the dipole-moment vector, we expect zero potential difference; the largest potential difference between the two points will be seen if the line connecting the two points is parallel to the dipole-moment vector. The potential difference is thus related to the component vector of the dipole moment along the line that connects the two points. The component vector points toward the electrode at the higher potential, and the larger the magnitude of the component vector, the larger the potential difference.

FIGURE 21.29 BIO Measuring an EKG.

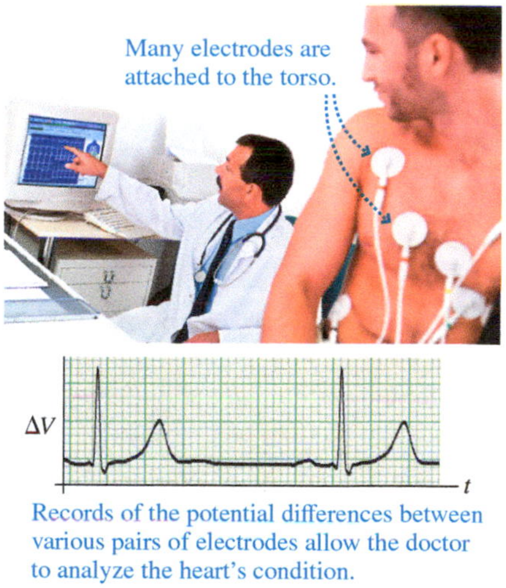

Many electrodes are attached to the torso.
ΔV
t
Records of the potential differences between various pairs of electrodes allow the doctor to analyze the heart's condition.

FIGURE 21.30 Determining the potential difference between two points due to a dipole potential.

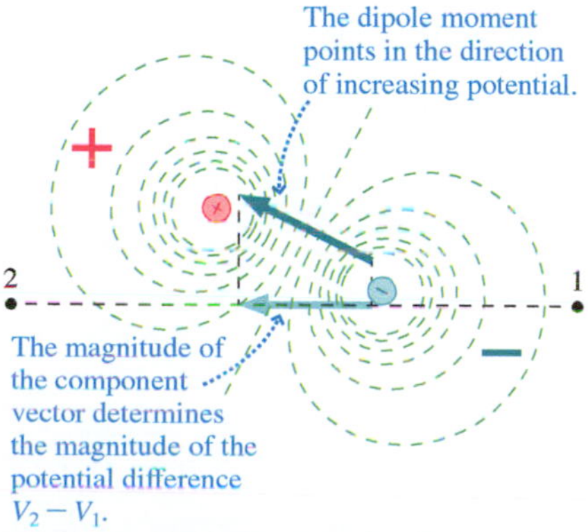

FIGURE 21.31 shows a simplified model of electrocardiogram measurement using only two electrodes, one on each arm. As the wave of depolarization moves across the heart muscle during each heartbeat, the dipole moment vector of the heart changes its magnitude and direction. As Figure 21.31 shows, both of these affect the potential difference between the electrodes, so each point on the EKG graph corresponds to a particular magnitude and orientation of the dipole moment.

FIGURE 21.31 BIO The potential difference between the electrodes changes as the heart beats.

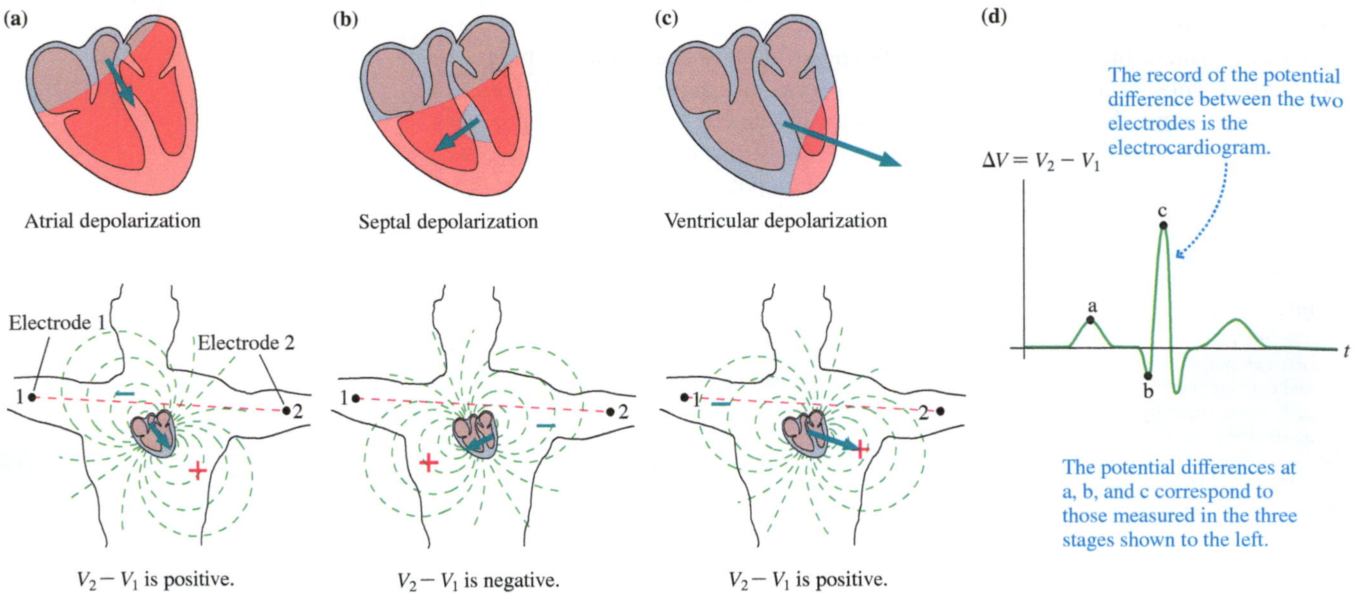


## CONCEPTUAL EXAMPLE 21.14 Determining the potential difference between EKG electrodes BIO

Technicians are recording an EKG for a patient using three electrodes as shown in FIGURE 21.32. At a particular instant, the dipole moment of the patient's heart is directed as shown. At this instant, which potential difference has the largest positive value: $V_3 - V_1$, $V_2 - V_1$, or $V_3 - V_2$?

FIGURE 21.32 The dipole moment of a patient's heart during an EKG.

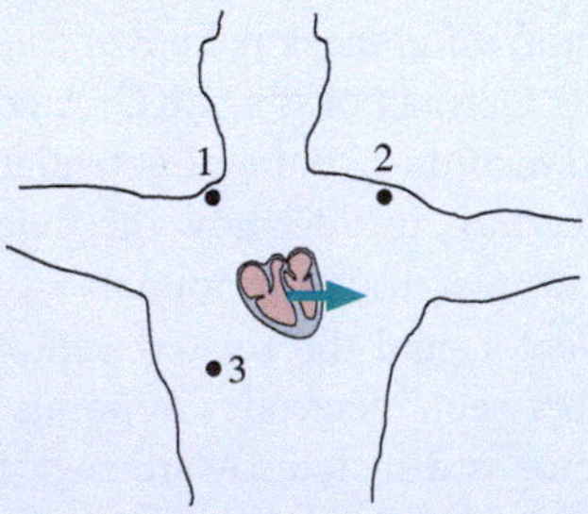


**REASON** The dipole moment is directed horizontally, along the line that connects electrodes 1 and 2, so the magnitude of the potential difference is largest between these two electrodes. The dipole moment vector points from electrode 1 to electrode 2, so $V_2 - V_1$ is positive. The potential difference $V_2 - V_1$ thus has the largest positive value.

**ASSESS** If we envision an overlay of the equipotential lines from the heart's dipole, we can see that this result makes sense.

The beating of the heart—and the action of other muscles in the body—creates electric fields and potentials. For animals such as humans that live surrounded by air, these potentials can be detected at the surface of the body. For animals that live in the water, the fields and potentials can be detected in the water at some distance from the animal. Many fish, and at least a few water-dwelling mammals, are able to detect the electric fields and potentials produced by the prey they seek. Sharks and rays have the greatest sensitivity. Sharks can detect electric fields as small as $1 \times 10^{-7}$ V/m, 1000 times smaller than the field from your heart. If you are swimming near a shark, the shark knows you are there. But the shark isn't really interested in you; it's looking for smaller, easier prey animals, which it can sense even when they are buried under the sand. The wide heads of hammerhead sharks and the wide bodies of rays, like the one in the photo that opened the chapter, enable them to scan a generous swath of ocean floor as they forage.

BIO A scalloped hammerhead shark uses its electric sense to find prey.

## 21.7 Capacitance and Capacitors

In Section 21.2 we found that potential differences are caused by the separation of charge. One common method of creating a charge separation, shown in FIGURE 21.33, is to move charge $Q$ from one initially uncharged conductor to a second initially uncharged conductor. This results in charge $+Q$ on one conductor and $-Q$ on the other. Two conductors with equal but opposite charge form a **capacitor.** The two conductors that make up a capacitor are its *electrodes,* or *plates.* We've already looked at some of the properties of parallel-plate capacitors, but capacitors can come in any shape. All that's needed is to have two electrodes separated by a region of space or an insulating material.

FIGURE 21.33 Charging a capacitor.

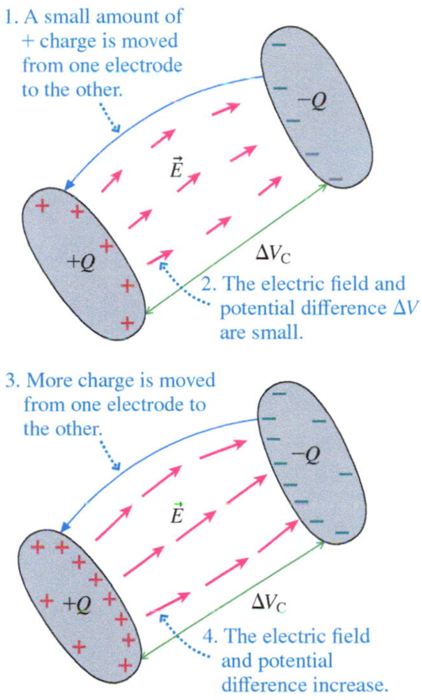


As Figure 21.33 shows, the electric field strength $E$ and the potential difference $\Delta V_C$ increase as the charge on each electrode increases. If we double the amount of charge on each electrode, the work required to move a charge from one electrode to the other doubles. This implies a doubling of the potential difference between the electrodes. Thus **the potential difference between the electrodes is directly proportional to their charge.**

Stated another way, **the charge of a capacitor is directly proportional to the potential difference between its electrodes.** A source of potential difference, such as a battery, is connected between the electrodes, causing a charge proportional to the potential difference to be moved from one electrode to the other. We can write the relationship between charge and potential difference as

$$Q = C\,\Delta V_C \qquad (21.18)$$

Charge on a capacitor with potential difference $\Delta V_C$

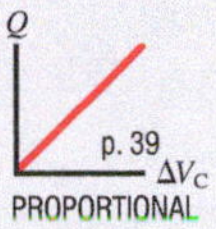


The constant of proportionality $C$ between $Q$ and $\Delta V_C$ is called the **capacitance** of the capacitor. Capacitance depends on the shape, size, and spacing of the two electrodes. A capacitor with a large capacitance holds more charge for a given potential difference than one with a small capacitance.

NOTE ▸ We will consider only situations where the charges on the electrodes are equal in magnitude but opposite in sign. When we say "A capacitor has charge $Q$," we mean that one electrode has charge $+Q$ and the other charge $-Q$. The potential difference between the electrodes is called the potential difference *of* the capacitor. ◂

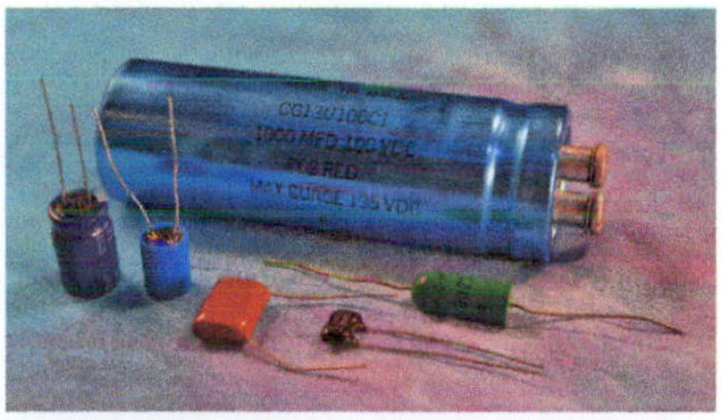

Capacitors are important elements in electric circuits. They come in a wide variety of sizes and shapes.

The SI unit of capacitance is the **farad.** One farad is defined as

$$1 \text{ farad} = 1 \text{ F} = 1 \text{ coulomb/volt} = 1 \text{ C/V}$$

The value of the capacitance, measured in farads, depends on the purpose of the capacitor. As you'll see in Chapter 23, capacitors are ubiquitous in electronic circuits used for timing, memory storage, and other functions. Such capacitors have values measured in microfarads ($\mu$F) or picofarads (pF). In recent years, devices with much larger capacitance have become available. "Supercapacitors" with values of many farads are used for energy storage in many applications.

## Charging a Capacitor

To "charge" a capacitor, we need to move charge from one electrode to the other. The simplest way to do this is to use a source of potential difference such as a battery, as shown in FIGURE 21.34. We learned earlier that a battery uses its internal chemistry to maintain a fixed potential difference between its terminals. If we connect a capacitor to a battery, charge flows from the negative electrode of the capacitor, through the battery, and onto the positive electrode. This flow of charge continues until the potential difference between the capacitor's electrodes is the same as the fixed potential difference of the battery. If the battery is then removed, the capacitor remains charged with a potential equal to that of the battery that charged it because there's no conducting path for charge on the positive electrode to move back to the negative electrode. Thus **a capacitor can be used to store charge.**

FIGURE 21.34 Charging a capacitor using a battery.

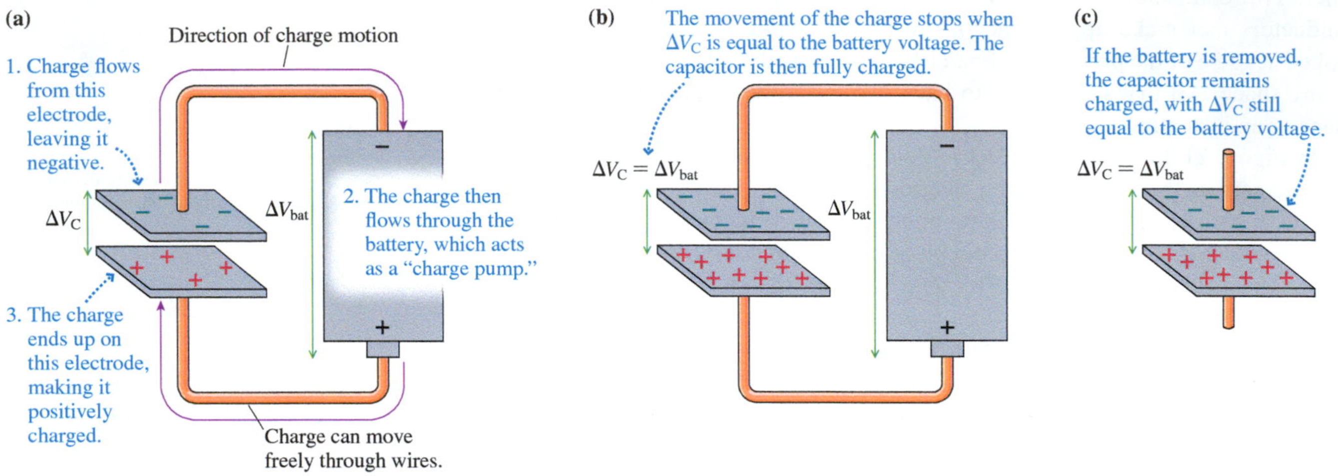


### EXAMPLE 21.15 Charging a capacitor

A 1.3$\mu$F capacitor is connected to a 1.5 V battery. What is the charge on the capacitor?

**PREPARE** Charge flows through the battery from one capacitor electrode to the other until the potential difference $\Delta V_C$ between the electrodes equals that of the battery, or 1.5 V.

**SOLVE** The charge on the capacitor is given by Equation 21.18:

$$Q = C\,\Delta V_C = (1.3 \times 10^{-6}\text{ F})(1.5\text{ V}) = 2.0 \times 10^{-6}\text{ C}$$

**ASSESS** The small charge is expected for the small capacitance and the small voltage.

## The Parallel-Plate Capacitor

As we've seen, the *parallel-plate capacitor* is important because it creates a uniform electric field between its flat electrodes. In ◀◀ SECTION 20.5, we found that the electric field of a parallel-plate capacitor is

$$\vec{E} = \left(\frac{Q}{\epsilon_0 A}, \text{ from positive to negative}\right)$$

where $A$ is the surface area of the electrodes and $Q$ is the charge on the capacitor. We can use this result to find the capacitance of a parallel-plate capacitor.

Earlier in this chapter, we found that the electric field strength of a parallel-plate capacitor is related to the potential difference $\Delta V_C$ and the plate spacing $d$ by

$$E = \frac{\Delta V_C}{d}$$

Combining these two results, we see that

$$\frac{Q}{\epsilon_0 A} = \frac{\Delta V_C}{d}$$

or, equivalently,

$$Q = \frac{\epsilon_0 A}{d} \Delta V_C \qquad (21.19)$$

If we compare Equation 21.19 to Equation 21.18, the definition of capacitance, we see that the capacitance of the parallel-plate capacitor is

$$C = \frac{\epsilon_0 A}{d} \qquad (21.20)$$

Capacitance of a parallel-plate capacitor with plate area $A$ and separation $d$

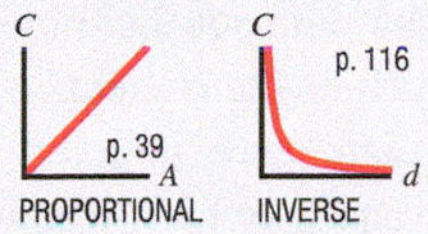


**NOTE** ▶ From Equation 21.20 you can see that the units of $\epsilon_0$ can be written as F/m. These units are useful when working with capacitors. ◀

## EXAMPLE 21.16 Making a parallel-plate capacitor

The capacitance of practical parallel-plate capacitors is quite small, in the pF range. Suppose you want to make a 1.0 μF capacitor using square plates separated by 1.0 mm. How large do the plates need to be?

***STRATEGIZE*** The capacitance depends on the area and the spacing. We know the capacitance and the spacing, so we can determine the area.

***PREPARE*** The area of the plates is the square of the edge length $L$: $A = L^2$.

***SOLVE*** Rearranging Equation 21.20, we solve for the area of the plates and thus the edge length:

$$L^2 = A = \frac{dC}{\epsilon_0}$$

$$L = \left(\frac{(0.0010\text{ m})(1.0 \times 10^{-6}\text{ F})}{(8.85 \times 10^{-12}\text{ F/m})}\right)^{1/2} = 11\text{ m}$$

***ASSESS*** To achieve this value of capacitance, you will need two conducting plates more than 35 feet on a side separated by 1/25 inch. This isn't really practical, but this result is expected. The problem statement noted that practical parallel-plate capacitors have a much smaller capacitance.

To obtain a reasonable value of capacitance we need a large electrode area and a small separation. In practice, this is accomplished by having large foil electrodes separated by a thin layer of insulator. The resulting "sandwich" is then rolled up, as in FIGURE 21.35. This construction makes it possible for large electrodes with a small separation to be squeezed into a small package. Even though the electrodes are no longer planes, the capacitance is reasonably well predicted from the parallel-plate-capacitor equation if $A$ is the area of the foils before they are folded and rolled. The insulating layer adds additional benefits, as the next section will show.

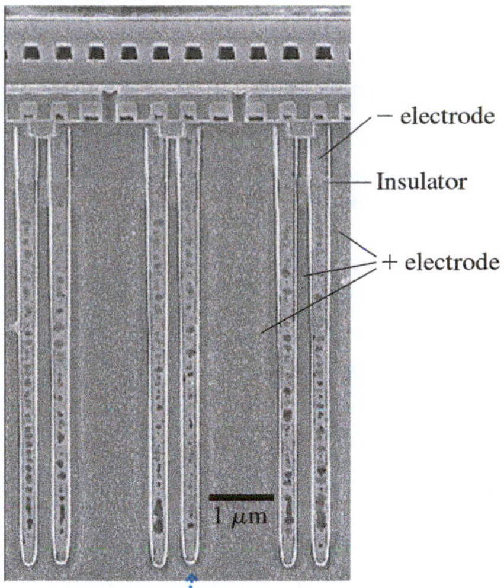


Each long structure is one capacitor.

**A capacity for memory** Your computer's random-access memory, or RAM, uses tiny capacitors to store the digital ones and zeroes that make up your data. A charged capacitor represents a one and an uncharged capacitor a zero. For a billion or more capacitors to fit on a single chip they must be very small. The micrograph is a cross section through the silicon wafer that makes up the memory chip. Each capacitor consists of a very long electrode separated by a thin insulating layer from the common electrode shared by all capacitors. Each capacitor's capacitance is only about $30 \times 10^{-15}$ F!

FIGURE 21.35 A capacitor disassembled to show its internal rolled-up structure.

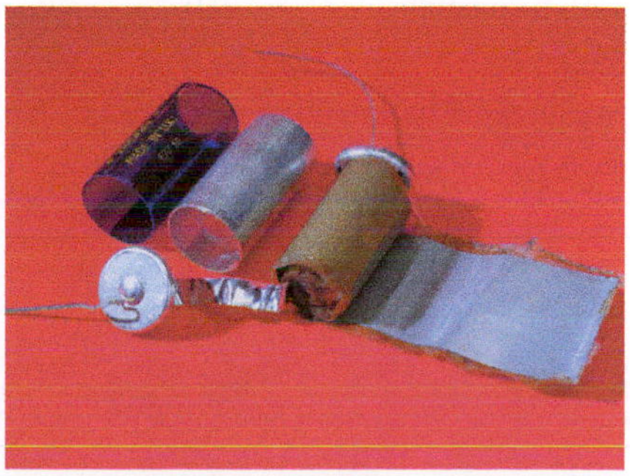

## Dielectrics and Capacitors

An insulator consists of vast numbers of atoms. When an insulator is placed in an electric field, each of its atoms polarizes. Recall, from ◀◀ SECTION 20.2, that *polarization* occurs when an atom's negative electron cloud and positive nucleus shift very slightly in opposite directions in response to an applied electric field. The net effect of all these tiny atomic polarizations is shown in FIGURE 21.36: An *induced* positive charge builds up on one surface of the insulator, and an induced negative charge on the other surface.

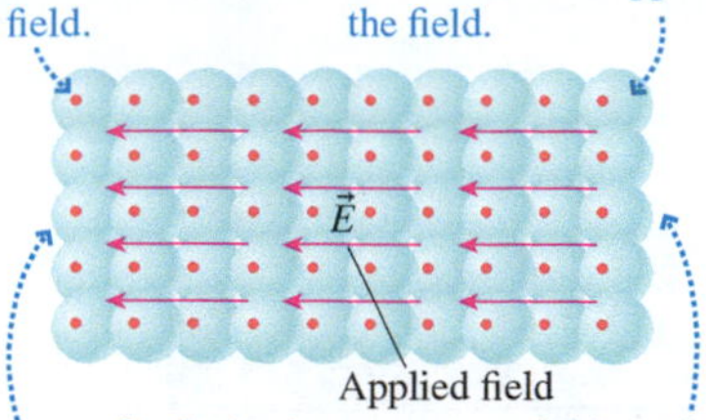


FIGURE 21.36 An insulator in an electric field becomes polarized.

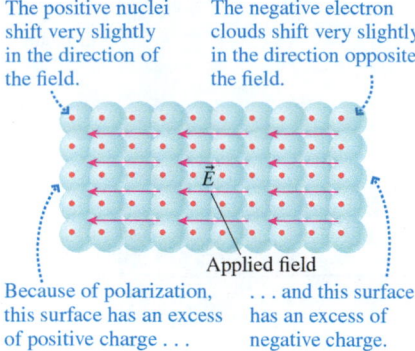

Notice that this distribution of charge—two equal but opposite layers—is identical to that of a parallel-plate capacitor, so this induced charge will create a uniform electric field, but one that is directed *opposite* to the applied electric field.

An insulator placed between the plates of a capacitor is called a **dielectric.** FIGURE 21.37a shows a dielectric in the uniform field of a parallel-plate capacitor with charge $Q$. The capacitor's electric field polarizes the dielectric; the polarized dielectric creates a field of its own directed opposite to the capacitor's field. As FIGURE 21.37b shows, these two fields add to give a net field in the same direction as the applied field, but *smaller.* Thus the electric field between the capacitor plates is smaller than it was without the dielectric.

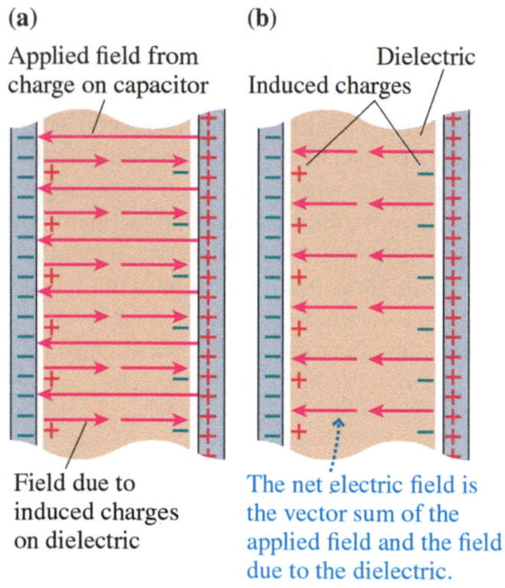


FIGURE 21.37 The electric field inside a dielectric.

When the dielectric is inserted, the capacitor's charge $Q$ doesn't change because its plates are isolated from each other. But the electric field between the plates decreases, which implies that the potential difference $\Delta V_C = Ed$ between the plates decreases as well. This implies that the capacitance $C = Q/\Delta V_C$ *increases.* **The presence of a dielectric results in an increased capacitance.**

Because some atoms are more easily polarized than others, the factor by which the capacitance is increased depends on the dielectric material. This factor is called the **dielectric constant** of the material, and it is given the symbol $\kappa$ (Greek letter kappa). If the capacitance without a dielectric present is $C_0$, then the capacitance with a dielectric present is

$$C = \kappa C_0 \tag{21.21}$$

Capacitance of a parallel-plate capacitor with a dielectric of dielectric constant $\kappa$

TABLE 21.3 lists the dielectric constants for a number of common substances. Note the high dielectric constant of water, which is of great importance in regulating the chemistry of biological processes. As we saw in ◀◀ SECTION 20.2, a molecule of water is a *permanent* electric dipole because the oxygen atom has a slight negative charge while the two hydrogen atoms each have a slight positive charge. Because the charge in a water molecule is *already* separated, the molecules' dipoles easily turn to line up with an applied electric field, leading to water's very high dielectric constant.

TABLE 21.3 Dielectric constants of some materials at 20°C

| Material | Dielectric constant $\kappa$ |
|---|---|
| Vacuum | 1 (exactly) |
| Air | 1.00054* |
| Teflon | 2.0 |
| Paper | 3.0 |
| Pyrex glass | 4.8 |
| Cell membrane | 9.0 |
| Ethanol | 24 |
| Water | 80 |
| Strontium titanate | 300 |

*Use 1.00 in all calculations.

### EXAMPLE 21.17 Finding the dielectric constant

A parallel-plate capacitor is charged using a 100 V battery; then the battery is removed. If a dielectric slab is slid between the plates, filling the space inside, the capacitor voltage drops to 30 V. What is the dielectric constant of the dielectric?

***STRATEGIZE*** After the capacitor is removed from the battery, the charge on the plates doesn't change. The field between the plates is reduced by the presence of the dielectric, so the voltage decreases. We will use the decrease in voltage to determine the dielectric constant.

***PREPARE*** The basic definition of capacitance, Equation 21.18, relates the charge to the voltage and the capacitance.

**SOLVE** The initial charge, $Q_1$, is the same as the charge $Q_2$ after the dielectric is inserted, so we have

$$Q_1 = C_1(\Delta V_C)_1 = Q_2 = C_2(\Delta V_C)_2$$

Inserting the dielectric increases the capacitance by a factor of $\kappa$, so that $C_2 = \kappa C_1$. Thus $C_1(\Delta V_C)_1 = \kappa C_1(\Delta V_C)_2$ or, canceling $C_1$, $(\Delta V_C)_1 = \kappa(\Delta V_C)_2$. The dielectric constant is then

$$\kappa = \frac{(\Delta V_C)_1}{(\Delta V_C)_2} = \frac{100\text{ V}}{30\text{ V}} = 3.3$$

**ASSESS** The voltage goes down by a factor of approximately 3, so we expect a dielectric constant of approximately 3, as we found.

As shown in Figure 21.35, a typical capacitor consists of two pieces of foil separated by a very thin layer of insulator. The insulator is generally selected to have a large dielectric constant, which increases the capacitance further.

**STOP TO THINK 21.6** If the potential difference across a capacitor is doubled, its capacitance

A. Doubles. B. Halves. C. Remains the same.

## 21.8 Energy and Capacitors

When you store charge in a capacitor, you also store energy. Suppose you connect a capacitor to a battery. Charge then flows from one plate, through the battery, and onto the other plate, leaving one plate with charge $-Q$ and the other with charge $+Q$. The battery must do *work* to transfer the charge; this work increases the electric potential energy of the charge on the capacitor. **A charged capacitor stores energy as electric potential energy.**

To find out how much energy is stored in a charged capacitor, recall that a charge $q$ moved through a potential difference $\Delta V$ gains potential energy $U = q\,\Delta V$. When a capacitor is charged, total charge $Q$ is moved from the negative plate to the positive plate. At first, when the capacitor is uncharged, the potential difference is $\Delta V = 0$. The last little bit of charge to be moved, as the capacitor reaches full charge, has to move through a potential difference $\Delta V \approx \Delta V_C$. *On average,* the potential difference across a capacitor while it's being charged is $\Delta V_{\text{average}} = \frac{1}{2}\Delta V_C$. Thus it seems plausible (and can be proved in a more advanced treatment) that the potential energy $U_C$ stored in a charged capacitor is

$$U_C = Q\,\Delta V_{\text{average}} = \frac{1}{2}Q\,\Delta V_C$$

We can use $Q = C\,\Delta V_C$ to write this in two different ways:

$$U_C = \frac{1}{2}\frac{Q^2}{C} = \frac{1}{2}C(\Delta V_C)^2 \qquad (21.22)$$

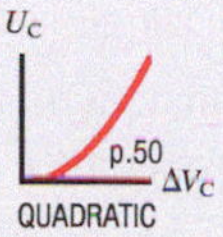


Electric potential energy of a capacitor with charge $Q$ and potential difference $\Delta V_C$

The potential energy stored in a capacitor depends on the *square* of the potential difference across it. This result is reminiscent of the potential energy $U_s = \frac{1}{2}k(\Delta x)^2$ stored in a spring, and a charged capacitor really is analogous to a stretched spring.

**Healthy plant, high capacitance** BIO Plants with fibrous roots have a very large total root surface area to maximize water and nutrient uptake. The surface area of a plant's roots is a good measure of the health of the plant; a healthy plant has a large root area. Rather than uproot the plant—which would damage it—an investigator can make an indirect, electrical measurement of the root area by measuring the capacitance of the plant. One "plate" is the soil, and the other is the inside of the plant. The membrane that separates the two has a reasonably constant thickness, so a larger capacitance implies a larger root area.

**Taking a picture in a flash** When you take a flash picture, the flash is fired using electric potential energy stored in a capacitor. Batteries are unable to deliver the required energy rapidly enough, but capacitors can discharge all their energy in only microseconds. A battery is used to slowly charge up the capacitor, which then rapidly discharges through the flash lamp.

A stretched spring holds energy until we release it; then that potential energy is transformed into kinetic energy. Likewise, a charged capacitor holds energy until we discharge it.

**EXAMPLE 21.18 Finding the power delivered by a capacitor**

A camera flash uses a capacitor to make a very bright—but short duration—pulse of light. A typical external flash unit has a 220 $\mu$F capacitor charged to 330 V. When the flash is triggered, the capacitor discharges in a time of 1.0 ms. What is the power delivered to the flash lamp?

**STRATEGIZE** We will find the energy stored in the capacitor and then use this energy and the discharge time to find the power.

**PREPARE** We know the voltage and the capacitance, so we can find the energy stored in the capacitor using $U_C = \frac{1}{2}C(\Delta V_C)^2$.

**SOLVE** The energy stored in the capacitor is

$$U_C = \frac{1}{2}(220 \times 10^{-6}\text{ F})(330\text{ V})^2 = 12\text{ J}$$

This is a good deal of energy, and given the short time, the power is large indeed:

$$P = \frac{\Delta E}{\Delta t} = \frac{12\text{ J}}{1.0 \times 10^{-3}\text{ s}} = 12{,}000\text{ W}$$

**ASSESS** This is a *lot* of power, but we know that the flash is very bright, so our result makes sense.

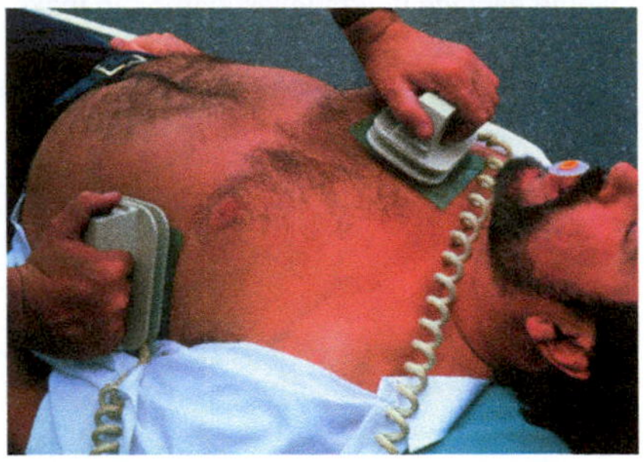

BIO A defibrillator, which can restore a heartbeat, discharges a capacitor through the patient's chest.

A heart attack or a serious injury can cause the heart to enter a state known as *fibrillation* in which the heart muscles twitch randomly and so cannot pump blood. A strong electric shock through the chest will stimulate all the muscles of the heart; this may interrupt the problematic rhythm and allow a normal pattern to resume. A great deal of charge needs to be transferred in a very short time. This is a perfect job for the capacitor in a *defibrillator.* The capacitor can store up to 360 J of energy, which is then released in a few milliseconds through electrodes pressed against a patient's chest.

The usefulness of the capacitors in the flash unit and the defibrillator stems from the fact that they can release their energy very quickly. A battery can store more *energy* per unit weight, so the energy to charge the capacitor in a camera flash comes from a battery. The capacitor can provide more *power* per unit weight, so it provides the dramatic surge in power needed for the pulse of the flash.

Batteries deliver their energy more slowly than capacitors, and they also take time to recharge. This is one of the limiting factors for an electric car—recharging the batteries takes a long time, longer than the time it takes to replenish a car's energy with gasoline. Capacitors can be charged much more quickly. Someday, capacitors will be able to match the energy storage of batteries, and electric cars with capacitor energy storage will be "refueled" in short order. These cars will also be capable of truly dramatic acceleration!

## The Energy in the Electric Field

FIGURE 21.38 A capacitor's energy is stored in the electric field.

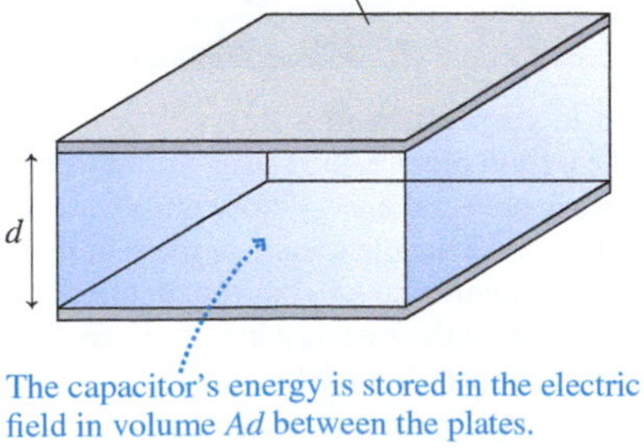


We can "see" the potential energy of a stretched spring in the tension of the coils. If a charged capacitor is analogous to a stretched spring, where is the stored energy? It's in the electric field!

FIGURE 21.38 shows a parallel-plate capacitor, filled with a dielectric with dielectric constant $\kappa$, in which the plates have area $A$ and are separated by distance $d$. Its capacitance is then $\kappa$ times that of a parallel-plate capacitor without a dielectric, or $C = \kappa\epsilon_0 A/d$. The potential difference across the capacitor is related to the electric field inside the capacitor by $\Delta V_C = Ed$. Substituting into Equation 21.22, we find that the energy stored in the capacitor is

$$U_C = \frac{1}{2}C(\Delta V_C)^2 = \frac{1}{2}\frac{\kappa\epsilon_0 A}{d}(Ed)^2 = \frac{1}{2}\kappa\epsilon_0(Ad)E^2 \qquad (21.23)$$

The quantity $Ad$ is the volume *inside* the capacitor, the region in which the capacitor's electric field exists. (Recall that an ideal capacitor has $\vec{E} = \vec{0}$ everywhere except between the plates.) Although we talk about "the energy stored in the capacitor," Equation 21.23 suggests that, strictly speaking, **the energy is stored in the capacitor's electric field.**

Because $Ad$ is the volume in which the energy is stored, we can define an **energy density** $u_E$ of the electric field:

$$u_E = \frac{\text{energy stored}}{\text{volume in which it is stored}} = \frac{U_C}{Ad} = \frac{1}{2}\kappa\epsilon_0 E^2 \qquad (21.24)$$

The energy density has units $J/m^3$. We've derived Equation 21.24 for a parallel-plate capacitor, but it turns out to be the correct expression for any electric field.

From this perspective, charging a capacitor stores energy in the capacitor's electric field as the field grows in strength. Later, when the capacitor is discharged, the energy is released as the field collapses.

We first introduced the electric field as a way to visualize how a long-range force operates. But if the field can store energy, **the field must be real,** not merely a pictorial device. We'll explore this idea further in Chapter 25, where we'll find that the energy transported by a light wave—the very real energy of warm sunshine—is the energy of electric and magnetic fields.

### EXAMPLE 21.19 Finding the energy density for a defibrillator

A portable defibrillator unit contains a 150 $\mu$F capacitor that is charged to 2000 V. The capacitor plates are separated by a 0.010-mm-thick dielectric with $\kappa = 300$.

a. What is the energy density in the electric field when the capacitor is charged?
b. How much total energy is stored in the capacitor?

**STRATEGIZE** We'll model the capacitor as a parallel-plate capacitor with a dielectric between the plates.

**PREPARE** To find the energy density, we need to know the electric field between the capacitor plates. Modeling it as a parallel-plate capacitor, we can find the field with Equation 21.7:

$$E = \frac{\Delta V_C}{d} = \frac{2000\text{ V}}{0.010 \times 10^{-3}\text{ m}} = 2.0 \times 10^8\text{ V/m}$$

**SOLVE**

a. With this electric field in hand, we use Equation 21.24 to find the energy density:

$$u_E = \frac{1}{2}\kappa\epsilon_0 E^2$$
$$= \frac{1}{2}(300)(8.85 \times 10^{-12}\text{ F/m})(2.0 \times 10^8\text{ V/m})^2$$
$$= 5.3 \times 10^7\text{ J/m}^3$$

b. The total energy stored is

$$U_C = \frac{1}{2}C(\Delta V_C)^2 = \frac{1}{2}(150\,\mu\text{F})(2000\text{ V})^2 = 300\text{ J}$$

**ASSESS** The electric field is quite large—larger than the field that would cause a spark in air—but a large field is needed to store a good deal of energy. The energy density is large, as we'd expect for a portable device, and the total energy stored is of the correct order, so our results are reasonable.

**STOP TO THINK 21.7** The plates of a parallel-plate capacitor are connected to a battery. if the distance between the plates is halved, the energy of the capacitor

A. Increases by a factor of 4.
B. Doubles.
C. Remains the same.
D. Is halved.
E. Decreases by a factor of 4.

## INTEGRATED EXAMPLE 21.20 Proton fusion in the sun

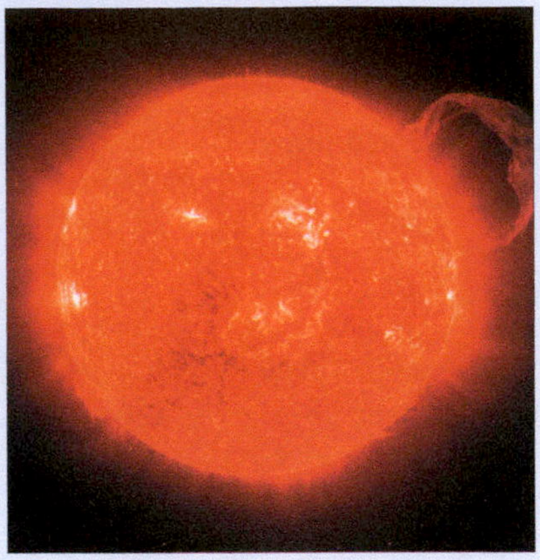

The sun's energy comes from nuclear reactions that fuse lighter nuclei into heavier ones, releasing energy in the process. The solar fusion process begins when two protons (the nuclei of hydrogen atoms) merge to produce a *deuterium* nucleus. Deuterium is the "heavy" isotope of hydrogen, with a nucleus consisting of a proton *and* a neutron. To become deuterium, one of the protons that fused has to turn into a neutron. The nuclear-physics process by which this occurs—and which releases the energy—will be studied in Chapter 30. Our interest for now lies not with the nuclear physics but with the conditions that allow fusion to occur.

Before two protons can fuse, they must come into contact. However, the energy required to bring two protons into contact is considerable because the electric potential energy of the two protons increases rapidly as they approach each other. Fusion occurs in the core of the sun because the ultra-high temperature there gives the protons the kinetic energy they need to come together.

a. A proton can be modeled as a charged sphere of diameter $d_p = 1.6 \times 10^{-15}$ m with total charge $e$. When two protons are in contact, what is the electric potential of one proton at the center of the other?
b. Two protons are approaching each other head-on, each with the same speed $v_0$. What value of $v_0$ is required for the protons to just come into contact with each other?
c. What does the temperature of the sun's core need to be so that the rms speed $v_{rms}$ of protons is equal to $v_0$?

**STRATEGIZE** We will treat this problem using conservation of energy, following the steps of Problem-Solving Approach 21.1 as the basis of our solution.

**PREPARE** FIGURE 21.39 shows a before-and-after visual overview. Both protons are initially moving with speeds $v_i = v_0$, so both contribute to the initial kinetic energy. We assume that they start out so far apart that $U_i \approx 0$. To "just touch" means that they've instantaneously come to rest ($K_f = 0$) at the point where the distance between their centers is equal to the diameter of a proton. We can use the potential of a charged sphere and the energy-conservation equation to find the speed $v_0$ required to achieve contact. Then we can use the results of Chapter 12 to find the temperature at which $v_0$ is the rms speed of the protons.

FIGURE 21.39 Visual overview of two protons coming into contact.

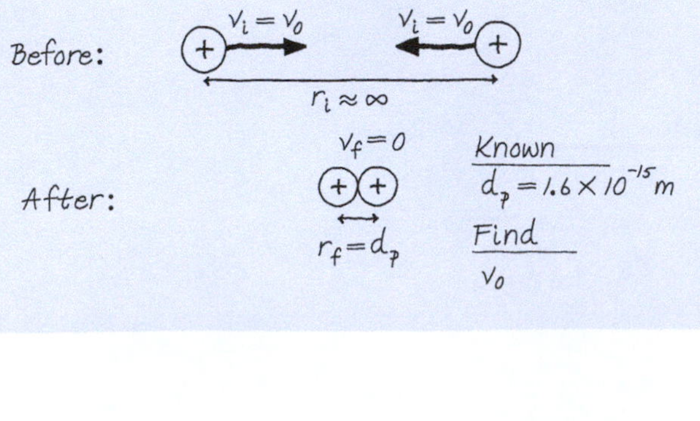


**SOLVE**

a. The electric potential at distance $r$ from a charged sphere was found to be $V = KQ/r$. When the protons are in contact, the distance between their centers is $r_f = d_p = 1.6 \times 10^{-15}$ m. Thus the potential of one proton, with $Q = e$, at the center of the other is

$$V = \frac{Ke}{r_f} = \frac{Ke}{d_p} = \frac{(9.0 \times 10^9 \text{ N} \cdot \text{m}^2/\text{C}^2)(1.6 \times 10^{-19} \text{ C})}{1.6 \times 10^{-15} \text{ m}}$$
$$= 9.0 \times 10^5 \text{ V}$$

b. The conservation of energy equation $K_f + qV_f = K_i + qV_i$ is

$$(0 + 0) + eV_f = \left(\frac{1}{2}mv_0^2 + \frac{1}{2}mv_0^2\right) + 0$$

where, as noted previously, both protons contribute to the initial kinetic energy, both end up at rest as the protons touch, and they started far enough apart that the initial potential energy (and potential) is effectively zero. When the protons meet, their potential energy is the charge of one proton ($e$) multiplied by the electric potential of the other—namely, the potential found in part a: $V_f = 9.0 \times 10^5$ V. Solving the energy equation for $v_0$, we get

$$v_0 = \sqrt{\frac{eV_f}{m}} = \sqrt{\frac{(1.6 \times 10^{-19} \text{ C})(9.0 \times 10^5 \text{ V})}{1.67 \times 10^{-27} \text{ kg}}}$$
$$= 9.29 \times 10^6 \text{ m/s}$$

c. In Section 12.2 we found that thetemperature of a gas is related to the average kinetic energy of the particles and thus to the rms speed of the particles by the equation

$$T = \frac{mv_{rms}^2}{3k_B}$$

It may seem strange to think of protons as a gas, but in the center of the sun, where all the atoms are ionized into nuclei and electrons, the protons are zooming around and do, indeed, act like a gas. For $v_{rms}$ of the protons to be equal to $v_0$ that we calculated in part b, the temperature would have to be

$$T = \frac{mv_0^2}{3k_B} = \frac{(1.67 \times 10^{-27} \text{ kg})(9.29 \times 10^6 \text{ m/s})^2}{3(1.38 \times 10^{-23} \text{ J/K})} = 3.5 \times 10^9 \text{ K}$$

**ASSESS** An extraordinarily high temperature—over 3 billion kelvin—is required to give an average solar proton a speed of $9.29 \times 10^6$ m/s. In fact, the core temperature of the sun is "only" about 14 million kelvin, a factor of $\approx 200$ less than we calculated. Protons can fuse at this lower temperature both because there are always a few protons moving much faster than average and because protons can reach each other even if their speeds are too low by the quantum-mechanical process of *tunneling*, which you'll learn about in Chapter 28. Still, because of the core's relatively "low" temperature, most protons bounce around in the sun for several billion years before fusing!

# SUMMARY

**GOAL** To calculate and use the electric potential and electric potential energy.

## GENERAL PRINCIPLES

### Electric Potential and Potential Energy

The **electric potential** $V$ is created by charges and exists at every point surrounding those charges.

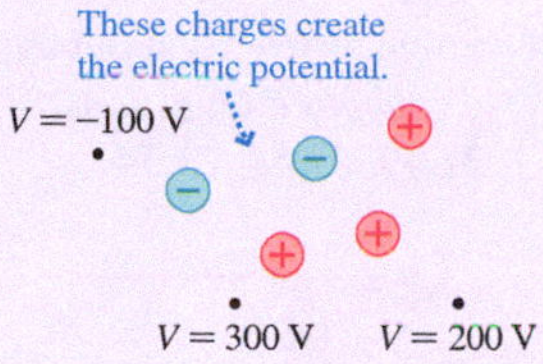


When a charge $q$ is brought near these charges, it acquires an **electric potential energy**

$$U_{\text{elec}} = qV$$

at a point where the other charges have created an electric potential $V$. Energy is conserved for a charged particle in an electric potential:

$$K_f + qV_f = K_i + qV_i$$

or

$$\Delta K = -q\,\Delta V$$

### Sources of Potential

Potential differences $\Delta V$ are created by a *separation of charge.* Two important sources of potential difference are

- A *battery,* which uses chemical means to separate charge and produce a potential difference.
- The opposite charges on the plates of a *capacitor,* which create a potential difference between the plates.

The electric potential of a point charge $q$ is $V = K\dfrac{q}{r}$

#### Connecting potential and field

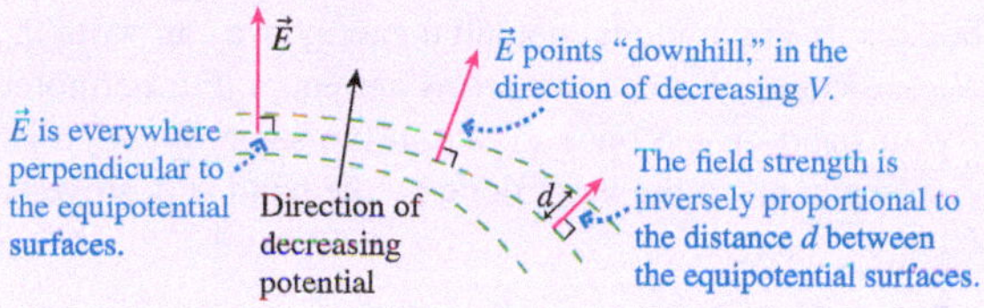


## IMPORTANT CONCEPTS

For a **conductor in electrostatic equilibrium**

- Any excess charge is on the surface.
- The electric field inside is zero.
- The exterior electric field is perpendicular to the surface.
- The field strength is largest at sharp corners.
- The entire conductor is at the same potential and so the surface is an equipotential.

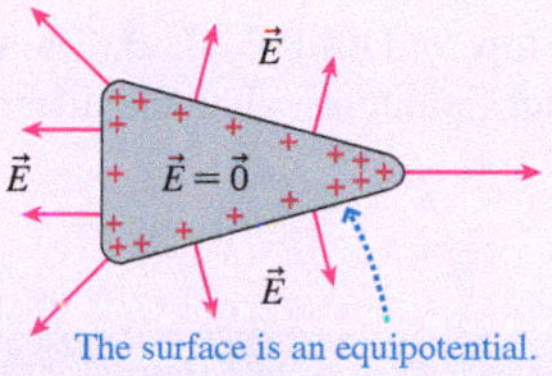


#### Graphical representations of the potential

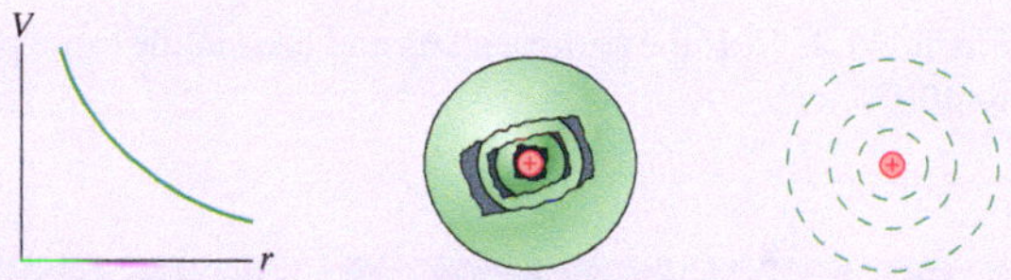


**Potential graph** **Equipotential surfaces** **Equipotential map**

## APPLICATIONS

### Capacitors and dielectrics

The charge $\pm Q$ on two conductors and the potential difference $\Delta V_C$ between them are proportional:

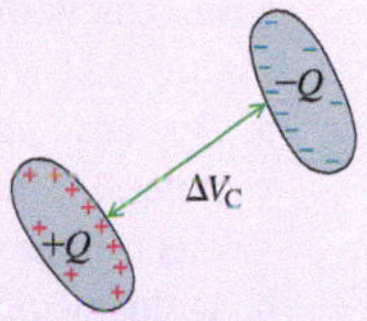


$$Q = C\,\Delta V_C$$

where $C$ is the **capacitance** of the two conductors.
A **parallel-plate capacitor** with plates of area $A$ and separation $d$ has a capacitance

$$C = \epsilon_0 \frac{A}{d}$$

When a **dielectric** is inserted between the plates of a capacitor, its capacitance is increased by a factor $\kappa$, the **dielectric constant** of the material.

The **energy stored in a capacitor** is $U_C = \frac{1}{2}C(\Delta V_C)^2$.

This energy is stored in the electric field, which has energy density

$$u_E = \frac{1}{2}\kappa\epsilon_0 E^2$$

### Parallel-plate capacitor

For a capacitor charged to $\Delta V_C$ the potential at distance $x$ from the negative plate is

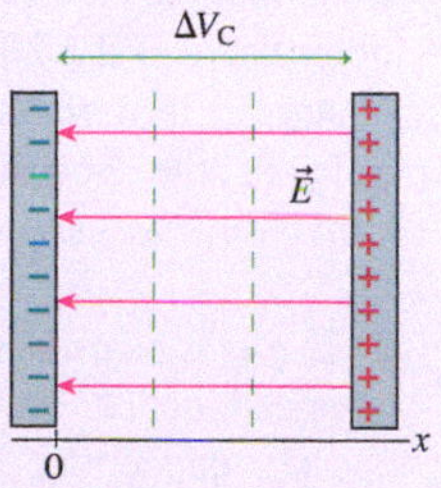


$$V = \frac{x}{d}\,\Delta V_C$$

The electric field inside is

$$E = \frac{\Delta V_C}{d}$$

### Units

- Electric potential: 1 V = 1 J/C
- Electric field: 1 V/m = 1 N/C
- Energy: 1 electron volt = 1 eV = $1.60 \times 10^{-19}$ J is the kinetic energy gained by an electron upon accelerating through a potential difference of 1 V.

## Learning Objectives After studying this chapter, you should be able to:

- Use electric potential energy and energy conservation to analyze the motion of charged particles. *Conceptual Questions 21.3, 21.5; Problems 21.6, 21.8, 21.10, 21.12, 21.14*
- Understand and use the concept of electric potential. *Conceptual Questions 21.1, 21.22; Problems 21.3, 21.7, 21.34, 21.35, 21.36*
- Work with and distinguish between volts and electron volts. *Conceptual Question 21.17; Problems 21.11, 21.13*
- Calculate the electric potential of important charge distributions. *Conceptual Question 21.11; Problems 21.15, 21.16, 21.19, 21.21, 21.23*
- Understand the relationship between the electric potential and the electric field. *Conceptual Questions 21.8, 21.13; Problems 21.26, 21.27, 21.28, 21.29, 21.32*
- Solve problems about capacitance, capacitors, and dielectrics. *Conceptual Questions 21.9, 21.16; Problems 21.37, 21.43, 21.49, 21.51, 21.54*

## STOP TO THINK ANSWERS

**Chapter Preview Stop to Think: C.** In general, conservation of energy dictates that the work done on a system increases its energy by the amount of work done. Because the only energy that changes for the book is its gravitational potential energy, we can write this as $W = \Delta U_g$, indicating that the two terms are equal. Furthermore, the force of your hand on the book points in the same direction as the book's displacement, so the work done by the hand is positive. Thus $W = \Delta U_g > 0$.

**Stop to Think 21.1: B.** If the charge were moved from 1 to 2 at a constant speed by a hand, the force exerted by the hand would need to be to the left, to oppose the rightward-directed electric force on the charge due to the source charges. Because the force due to the hand would be opposite the displacement, the hand would do *negative* work on the charge, decreasing its electric potential energy.

**Stop to Think 21.2: C.** If the particle starts and ends at the same point, where $V = 400$ V.

**Stop to Think 21.3: C.** The proton gains speed by losing potential energy. It loses potential energy by moving in the direction of decreasing electric potential.

**Stop to Think 21.4: $\Delta V_{13} = \Delta V_{23} > \Delta V_{12}$.** The potential depends on only the *distance* from the charge, not the direction. $\Delta V_{12} = 0$ because these points are at the same distance.

**Stop to Think 21.5: C.** $\vec{E}$ points "downhill," so $V$ must decrease from right to left. $E$ is larger on the left than on the right, so the equipotential lines must be closer together on the left.

**Stop to Think 21.6: C.** Capacitance is a property of the shape and position of the electrodes. It does not depend on the potential difference or charge.

**Stop to Think 21.7: B.** The energy is $\frac{1}{2}C(\Delta V_C)^2$. $\Delta V_C$ is constant, but $C$ doubles when the distance is halved.

Video Tutor Solution Chapter 21

# QUESTIONS

## Conceptual Questions

1. By moving a 10 nC charge from point A to point B, you determine that the electric potential at B is 150 V. What would be the potential at B if a 20 nC charge were moved from A to B?
2. Charge $q$ is fired through a small hole in the positive plate of a capacitor, as shown in Figure Q21.2.
   a. If $q$ is a positive charge, does it speed up or slow down inside the capacitor? Answer this question twice: (i) Using the concept of force. (ii) Using the concept of energy.
   b. Repeat part a if $q$ is a negative charge.

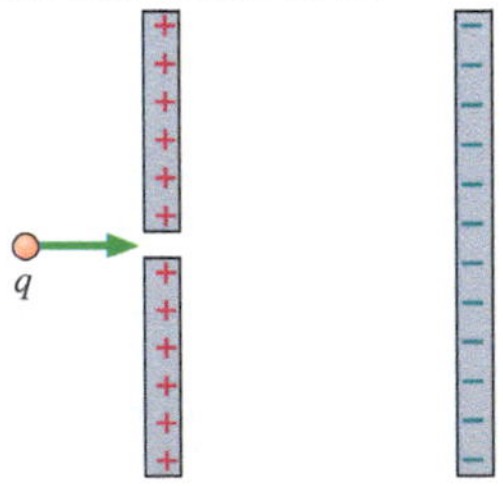

FIGURE Q21.2

3. An electron ($q = -e$) completes half of a circular orbit of radius $r$ around a nucleus with $Q = +3e$, as shown in Figure Q21.3.
   a. By how much does the electric potential energy change as the electron moves from i to f?
   b. Is the electron's speed at f greater than, less than, or equal to its speed at i?

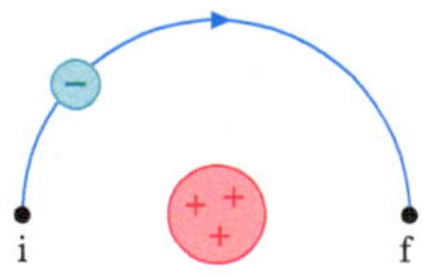

FIGURE Q21.3

4. A positively charged dust particle passes between the two oppositely charged plates of an electrostatic precipitator. There is an electric force directed toward the negative plate, but the drag force causes the particle to drift at a constant speed. Describe the energy transformation that takes place during this motion.

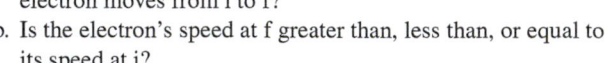

Problem difficulty is labeled as | (straightforward) to ||||| (challenging). Problems labeled INT integrate significant material from earlier chapters; Problems labeled BIO are of biological or medical interest.

The eText icon indicates when there is a video tutor solution available for the chapter or for a specific problem. To launch these videos, log into your eText through Mastering™ Physics or log into the Study Area.

5. An electron moves along the trajectory from i to f in Figure Q21.5.
   a. Does the electric potential energy increase, decrease, or stay the same? Explain.
   b. Is the electron's speed at f greater than, less than, or equal to its speed at i? Explain.

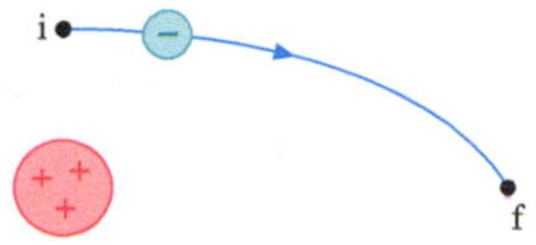


FIGURE Q21.5

6. As shown in Figure Q21.6, two protons are launched with the *same* speed from point 1 inside a parallel-plate capacitor. One proton moves along the path from 1 to 2, the other from 1 to 3. Points 2 and 3 are the *same* distance from the positive plate.
   a. Is $\Delta U_{1\to 2}$, the change in potential energy along the path $1 \to 2$, larger than, smaller than, or equal to $\Delta U_{1\to 3}$? Explain.
   b. Is the proton's speed $v_2$ at point 2 larger than, smaller than, or equal to the proton's speed $v_3$ at point 3? Explain.

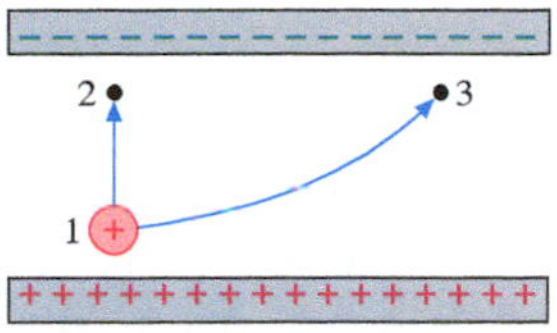


FIGURE Q21.6

7. Each part of Figure Q21.7 shows one or more point charges. The charges have equal magnitudes. If a positive charge is moved from position i to position f, does the electric potential energy increase, decrease, or stay the same? Explain.

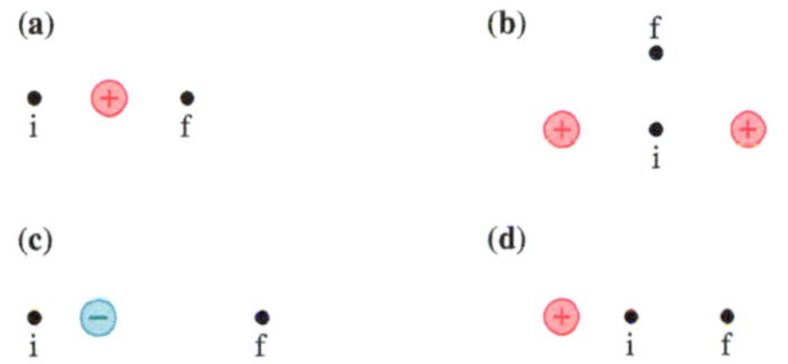


FIGURE Q21.7

8. Figure Q21.8 shows two points inside a capacitor. Let $V = 0$ V at the negative plate.
   a. What is the ratio $V_2/V_1$ of the electric potential at these two points? Explain.
   b. What is the ratio $E_2/E_1$ of the electric field strength at these two points? Explain.

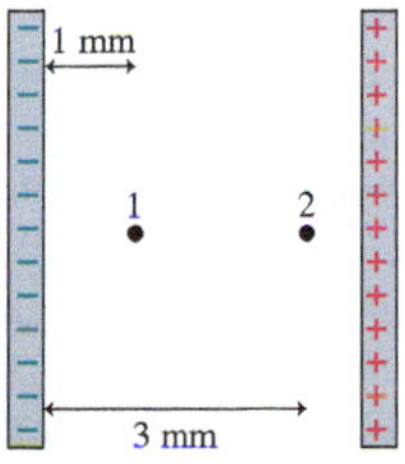


FIGURE Q21.8

9. A capacitor with plates separated by distance $d$ is charged to a potential difference $\Delta V_C$. All wires and batteries are disconnected, then the two plates are pulled apart (with insulated handles) to a new separation of distance $2d$.
   a. Does the capacitor charge $Q$ change as the separation increases? If so, by what factor? If not, why not?
   b. Does the electric field strength $E$ change as the separation increases? If so, by what factor? If not, why not?
   c. Does the potential difference $\Delta V_C$ change as the separation increases? If so, by what factor? If not, why not?
10. Figure Q21.10 shows two points near a positive point charge.
    a. What is the ratio $V_1/V_2$ of the electric potentials at these two points? Explain.
    b. What is the ratio $E_1/E_2$ of the electric field strengths at these two points? Explain.

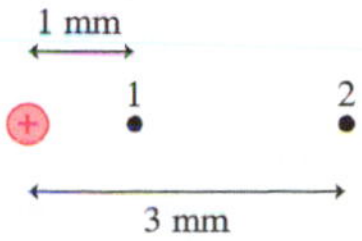


FIGURE Q21.10

11. Each part of Figure Q21.11 shows three points in the vicinity of two point charges. The charges have equal magnitudes. Rank in order, from largest to smallest, the potentials $V_1$, $V_2$, and $V_3$.

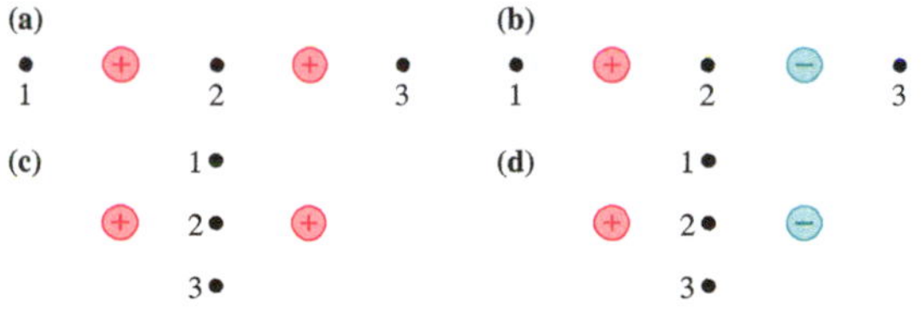


FIGURE Q21.11

12. Argon and potassium are neighbors in the periodic table but have very different sizes—a potassium atom is approximately 3 times the size of an argon atom. What does this difference imply about their ionization energies?
13. Rank in order, from largest to smallest, the electric field strengths $E_1$, $E_2$, $E_3$, and $E_4$ at the four labeled points in Figure Q21.13. Explain.

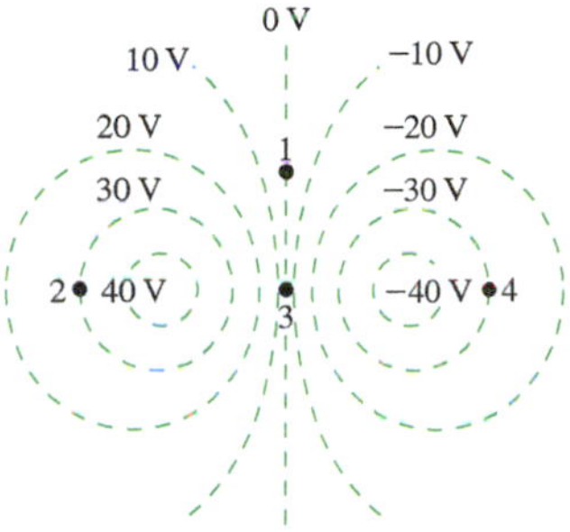


FIGURE Q21.13

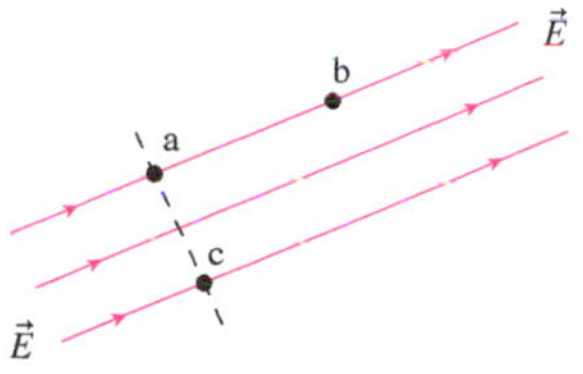


FIGURE Q21.14

14. Figure Q21.14 shows an electric field diagram. Rank in order, from highest to lowest, the electric potentials at points a, b, and c.
15. Rank in order, from largest to smallest, the energies $(U_C)_1$ to $(U_C)_4$ stored in each of the capacitors in Figure Q21.15. Explain.

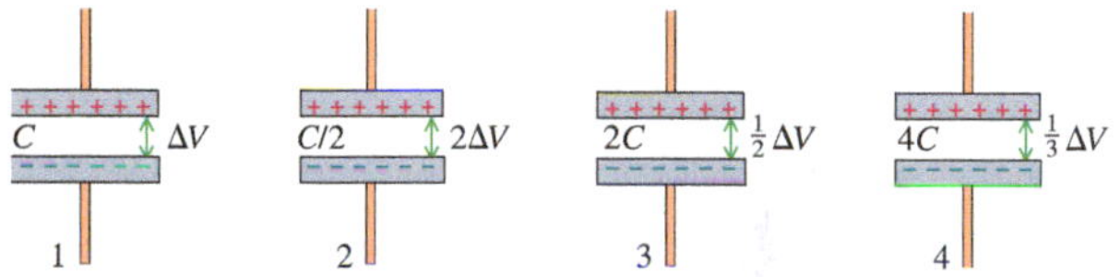


FIGURE Q21.15

16. A type of rain gauge consists of two separated vertical plates connected to a device that measures the capacitance of the two plates. How is this device able to detect the passage of raindrops?
17. A proton is launched from point 1 in Figure Q21.17 with an initial velocity of $3.9 \times 10^5$ m/s. By how much has its kinetic energy changed, in eV, by the time it passes through point 2?

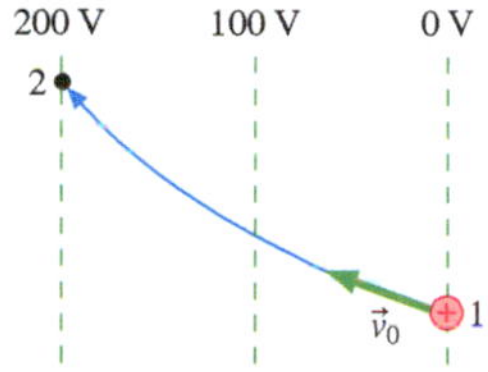


FIGURE Q21.17

## Multiple-Choice Questions

18. ‖ Under normal circumstances, a separation of charges between the earth and the atmosphere produces an electric field that points downward. If you charge a balloon on your hair, it acquires a negative charge. If you charge a helium balloon on your hair and then release it, as the balloon moves upward the electrical potential __________ and the balloon's electrical potential energy __________.
    A. increases, increases
    B. increases, decreases
    C. decreases, increases
    D. decreases, decreases

19. | A 1.0 nC positive point charge is located at point A in Figure Q21.19. The electric potential at point B is

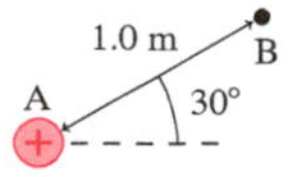

FIGURE Q21.19

A. 9.0 V B. 9.0 sin 30° V
C. 9.0 cos 30° V D. 9.0 tan 30° V

20. || In a fish tank, it is possible to have stray potentials from lights and other equipment. The fish in the tank position themselves to minimize the potential difference between any two points on their bodies. Suppose there is a dipole field in a fish tank resulting from the charges shown in the top view of the tank in Figure Q21.20. At which of the labeled positions and orientations are we most likely to find a fish?
BIO

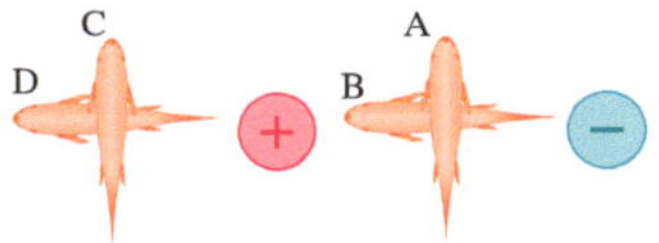

FIGURE Q21.20

21. || Figure Q21.21 shows two concentric circular electrodes, charged as noted. What can you conclude about the electric potential and the field strength at the two noted points between the two electrodes?

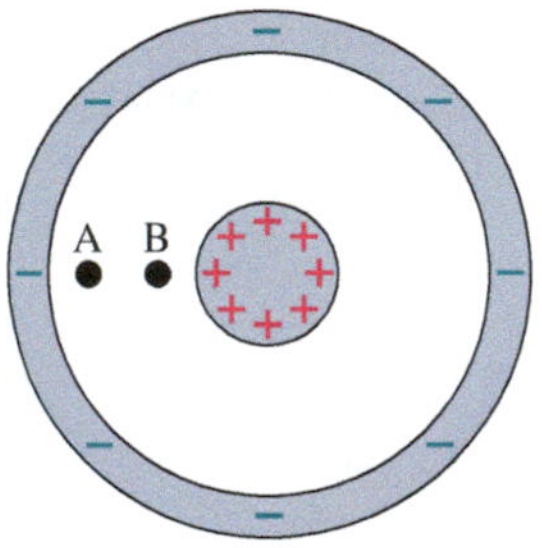

FIGURE Q21.21

A. The potential is greater at point A; the field strength is greater at point A.
B. The potential is greater at point A; the field strength is greater at point B.
C. The potential is greater at point B; the field strength is greater at point A.
D. The potential is greater at point B; the field strength is greater at point B.

Questions 22 through 26 refer to Figure Q21.22, which shows equipotential lines in a region of space. The equipotential lines are spaced by the same difference in potential, and several of the potentials are given.

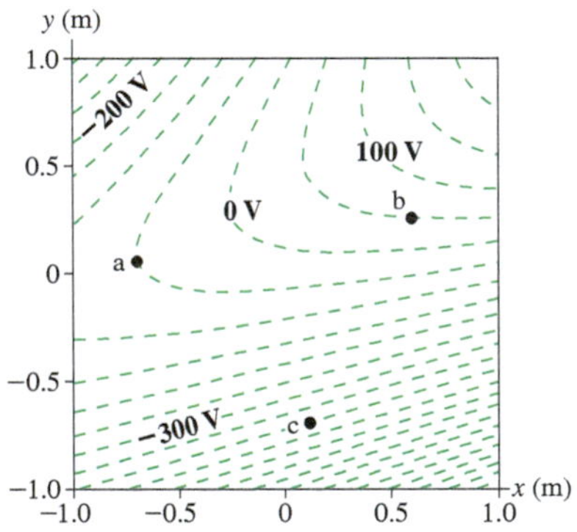

FIGURE Q21.22

22. | What is the potential at point c?
A. −400 V
B. −350 V
C. −100 V
D. 350 V
E. 400 V

23. | At which point, a, b, or c, is the magnitude of the electric field the greatest?

24. | What is the approximate magnitude of the electric field at point c?
A. 100 V/m B. 300 V/m C. 800 V/m
D. 1500 V/m E. 3000 V/m

25. | The direction of the electric field at point b is closest to which direction?
A. Right B. Up C. Left D. Down

26. || A +10 nC charge is moved from point c to point a. How much work is required in order to do this?
A. $3.5 \times 10^{-6}$ J B. $4.0 \times 10^{-6}$ J C. $3.5 \times 10^{-3}$ J
D. $4.0 \times 10^{-3}$ J E. 3.5 J

27. | A bug zapper consists of two metal plates connected to a high-voltage power supply. The voltage between the plates is set to give an electric field slightly less than $1 \times 10^6$ V/m. When a bug flies between the two plates, it increases the field enough to initiate a spark that incinerates the bug. If a bug zapper has a 4000 V power supply, what is the approximate separation between the plates?
A. 0.05 cm B. 0.5 cm C. 5 cm D. 50 cm

28. | An atom of helium and one of argon are singly ionized—one electron is removed from each. The two ions are then accelerated from rest by the electric field between two plates with a potential difference of 150 V. After accelerating from one plate to the other,
A. The helium ion has more kinetic energy.
B. The argon ion has more kinetic energy.
C. Both ions have the same kinetic energy.
D. There is not enough information to say which ion has more kinetic energy.

29. || The dipole moment of the heart is shown at a particular instant in Figure Q21.29. Which of the following potential differences will have the largest positive value?
BIO

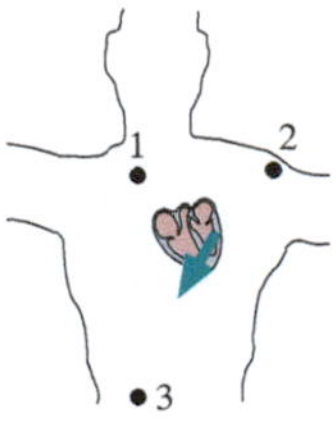

FIGURE Q21.29

A. $V_1 - V_2$ B. $V_1 - V_3$
C. $V_2 - V_1$ D. $V_3 - V_1$

Questions 30 through 32 concern Figure Q21.30, which shows the configuration of a gel electrophoresis experiment. Charged plates establish an electric field on a region of gel between the plates. A solution of DNA fragments, which acquire a negative charge in solution, is analyzed by using an electric field to drive the fragments through the gel. A drag force opposes the motion, causing the fragments to move at a constant speed. Larger fragments move more slowly, so the fragments separate by size to give a "fingerprint" of the sample.

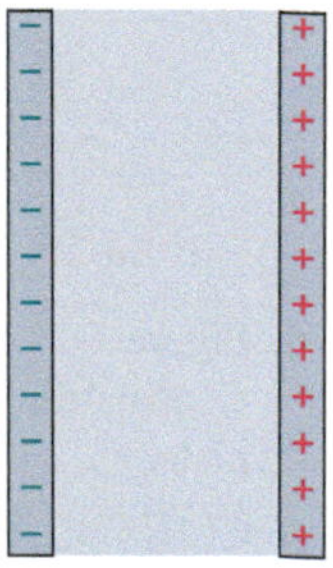
FIGURE Q21.30

30. | In the system shown in Figure Q21.30, the electric field is directed to the _____; the DNA fragments move to the _____.
A. right, right B. right, left
C. left, right D. left, left

31. || The fragments move in the direction of _________ electric potential and _________ electric potential energy.
A. higher, higher
B. higher, lower
C. lower, higher
D. lower, lower

32. || What is the energy transformation as the fragments move through the gel?
A. Electric potential energy to kinetic energy
B. Electric potential energy to thermal energy
C. Electric potential energy to kinetic energy and thermal energy
D. Kinetic energy to thermal energy

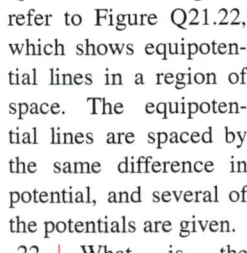

# PROBLEMS

## Section 21.1 Electric Potential Energy and Electric Potential

## Section 21.2 Sources of Electric Potential

1. ||| Moving a charge from point A, where the potential is 300 V, to point B, where the potential is 150 V, takes $4.5 \times 10^{-4}$ J of work. What is the value of the charge?
2. |||| The graph in Figure P21.2 shows the electric potential energy as a function of separation for two point charges. If one charge is $+0.44$ nC, what is the other charge?

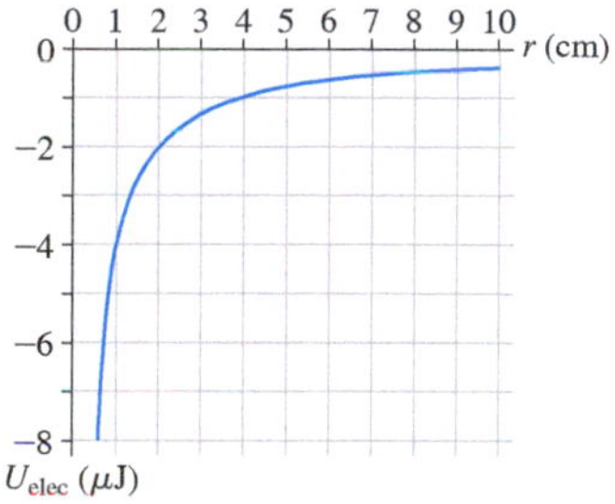


FIGURE P21.2

3. ||| It takes 3.0 μJ of work to move a 15 nC charge from point A to B. It takes −5.0 μJ of work to move the charge from C to B. What is the potential difference $V_C - V_A$?
4. || In 1 second, a battery charger moves 0.60 C of charge from the negative terminal to the positive terminal of a 1.5 V AA battery.
   a. How much work does the charger do?
   b. What is the power output of the charger in watts?
5. | A 20 nC charge is moved from a point where $V = 150$ V to a point where $V = -50$ V. How much work is done by the force that moves the charge?
6. | BIO In a typical mammalian cell, the net transport by the sodium-potassium exchange pump that maintains the 70 mV membrane potential is 500 singly charged ions per second. How much work does the pump do each second?
7. | At one point in space, the electric potential energy of a 15 nC charge is 45 μJ.
   a. What is the electric potential at this point?
   b. If a 25 nC charge were placed at this point, what would its electric potential energy be?

## Section 21.3 Electric Potential and Conservation of Energy

8. || BIO This scanning electron microscope image of a bacterium was produced using a beam of electrons accelerated through an 30 kV potential difference. What is the speed of the electrons?

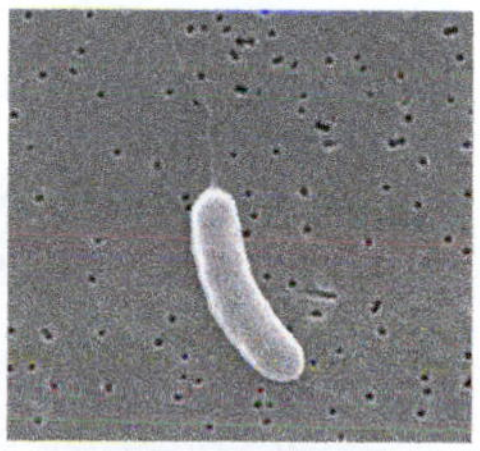

9. ||||| What potential difference is needed to accelerate a $He^+$ ion (charge $+e$, mass 4 u) from rest to a speed of $1.0 \times 10^6$ m/s?
10. || BIO A patient's tumor is being treated with proton-beam therapy. The protons are accelerated through a potential difference of 62 MV. What is the speed of the protons? (Note: The speed is high enough that, in principle, we should use a relativistic calculation—something you'll learn about in Chapter 27—but for this problem you should use the formulas you are already familiar with.)
11. || An electron with an initial speed of 500,000 m/s is brought to rest by an electric field.
    a. Did the electron move into a region of higher potential or lower potential?
    b. What was the potential difference that stopped the electron?
    c. What was the initial kinetic energy of the electron, in electron volts?
12. || BIO An electrophoresis gel rests between two parallel plates; the potential difference between the plates is 50 V. Each second, 50 mC of charge moves through the gel. What is the increase in thermal energy of the gel in 1.0 minute?
13. ||| A proton with an initial speed of 800,000 m/s is brought to rest by an electric field.
    a. Did the proton move into a region of higher potential or lower potential?
    b. What was the potential difference that stopped the proton?
    c. What was the initial kinetic energy of the proton, in electron volts?
14. || The Cassini spacecraft that visited Saturn and its moons did more than take pretty pictures; it also returned physical data. Saturn's moon Titan has a substantial atmosphere, and Cassini went close enough to detect atoms and molecules from Titan's atmosphere. It measured the atmospheric composition with a time-of-flight mass spectrometer. Ionized atoms and molecules were accelerated through a 30 kV potential difference and then sent through a chamber of length 18.8 cm. The time to travel the length of the chamber was measured, with heavier particles taking longer times. What is the travel time for a singly ionized molecule of water?

## Section 21.4 Calculating the Electric Potential

15. || The electric potential at a point that is halfway between two identical charged particles is 300 V. What is the potential at a point that is 25% of the way from one particle to the other?
16. || A 2.0 cm × 2.0 cm parallel-plate capacitor has a 2.0 mm spacing. The electric field strength inside the capacitor is $1.0 \times 10^5$ V/m.
    a. What is the potential difference across the capacitor?
    b. How much charge is on each plate?
17. ||| Two 2.00 cm × 2.00 cm plates that form a parallel-plate capacitor are charged to ± 0.708 nC. What are the electric field strength inside and the potential difference across the capacitor if the spacing between the plates is (a) 1.00 mm and (b) 2.00 mm?
18. | a. In Figure P21.18, which capacitor plate, left or right, is the positive plate?
    b. What is the electric field strength inside the capacitor?
    c. What is the potential energy of a proton at the midpoint of the capacitor?

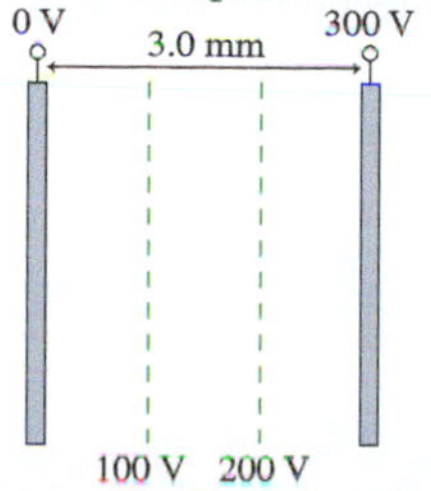


FIGURE P21.18

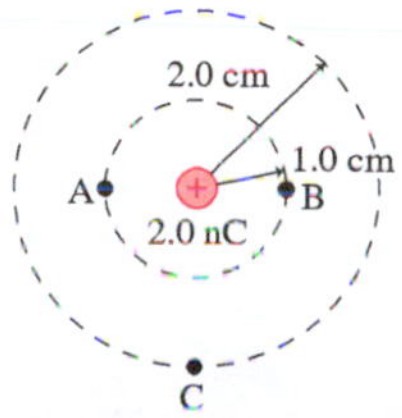


FIGURE P21.19

19. || What are the potential differences $\Delta V_{AB}$ and $\Delta V_{BC}$?

Watch Video Solution Problem 21.31

20. ‖ A flying hummingbird picks up charge as it moves through the air. This creates a potential near the bird. What is the "voltage of a hummingbird"? Assume that the bird acquires a charge of +200 pC, a typical value, and model the bird as a sphere of radius 3 cm.
BIO

21. ‖ What is the electric potential at the point indicated with the dot in Figure P21.21?

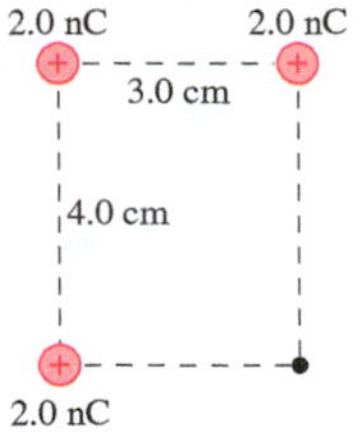


FIGURE P21.21

22. ‖ Raindrops acquire an electric charge as they fall. Suppose a 2.5-mm-diameter drop has a charge of +15 pC, fairly typical values. What is the potential at the surface of the raindrop?

23. ‖ a. What is the potential difference between the terminals of an ordinary AA or AAA battery? (If you're not sure, find one and look at the label.)
b. An AA battery is connected to a parallel-plate capacitor having 4.0-cm-diameter plates spaced 2 mm apart. How much charge does the battery move from one plate to the other?

24. ‖ The radius of a cadmium atom is approximately 0.16 nm. What is the ionization energy?

25. ‖ The radius of a calcium atom is approximately 0.23 nm. What is the ionization energy?

## Section 21.5 Connecting Potential and Field

26. ‖ Guiana dolphins are one of the few mammals able to detect electric fields. In a test of sensitivity, a dolphin was exposed to the variable electric field from a pair of charged electrodes. The magnitude of the electric field near the sensory organs was measured by detecting the potential difference between two measurement electrodes located 1.0 cm apart along the field lines. The dolphin could reliably detect a field that produced a potential difference of 0.50 mV between these two electrodes. What is the corresponding electric field strength?
BIO

27. ‖ a. In Figure P21.27, which point, A or B, has a higher electric potential?
b. What is the potential difference between A and B?

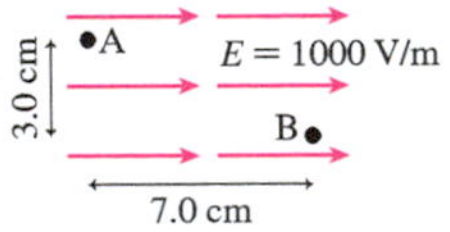


FIGURE P21.27

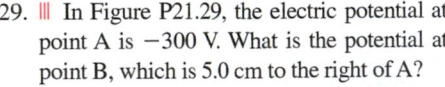

28. | Students in an introductory physics lab are producing a region of uniform electric field by applying a voltage to two 20-cm-diameter aluminum plates separated by 2.5 cm. They connect the two plates to the two terminals of a high-voltage power supply and then gradually turn up the voltage. In such a situation, sparks start to fly when the field exceeds $3 \times 10^6$ V/m. What is the highest voltage the students can use?

29. ||| In Figure P21.29, the electric potential at point A is −300 V. What is the potential at point B, which is 5.0 cm to the right of A?

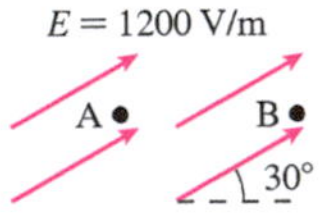


FIGURE P21.29

30. | Under typical atmospheric conditions, there is an electric field near the earth's surface, directed downward, with a magnitude of 100 V/m. If we say that the potential at the earth's surface is 0 V, what is the potential 1.0 km above the surface?

31. | What are the magnitude and direction of the electric field at the dot in Figure P21.31?

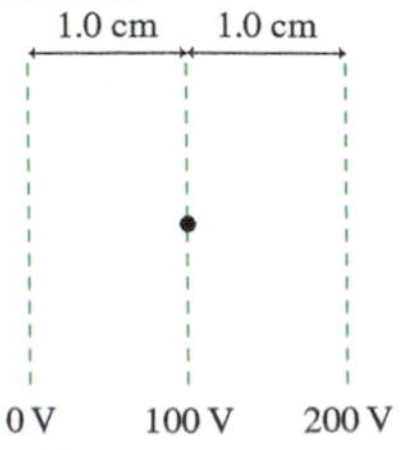


FIGURE P21.31

32. | What are the magnitude and direction of the electric field at the dot in Figure P21.32?

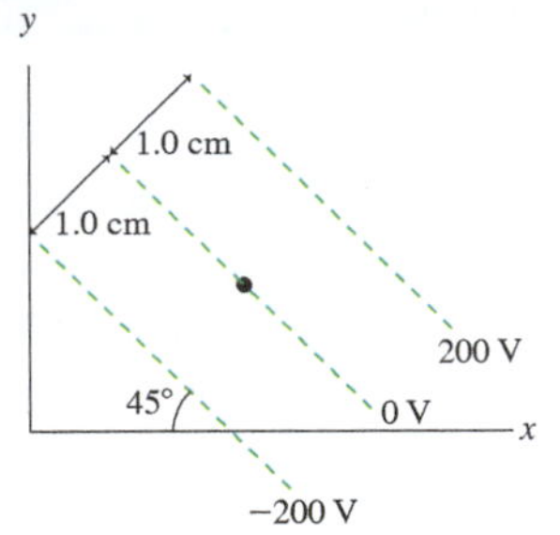


FIGURE P21.32

## Section 21.6 The Electrocardiogram

33. | One standard location for a pair of electrodes during an EKG is shown in Figure P21.33. The potential difference $\Delta V_{31} = V_3 - V_1$ is recorded. For each of the three instants a, b, and c during the heart's cycle shown in Figure 21.29, will $\Delta V_{31}$ be positive or negative? Explain.
BIO

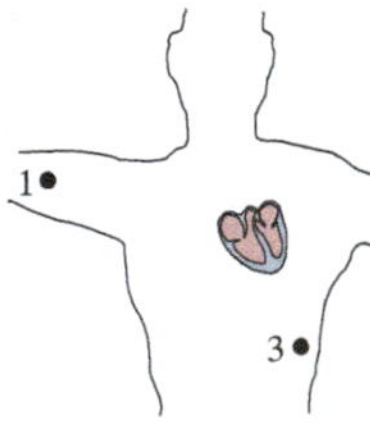


FIGURE P21.33

34. ‖ The eye has a modest dipole moment, with the cornea positive and the back of the eye negative. This dipole moment rotates with the eyeball. If electrodes are placed on either side of the eye, as in Figure P21.34a, the potential difference between the electrodes can be used to measure the orientation of the eye. A record of this potential is known as an electro-oculogram; it's similar in principle to the electrocardiogram.
BIO INT
a. Figure P21.34b is a top view showing the eyeball pointed straight ahead. For this case, is the potential difference $V_2 - V_1$ positive, negative, or zero?
b. Figure P21.34c is a top view showing the eyeball pointed to the right. For this case, is the potential difference $V_2 - V_1$ positive, negative, or zero?

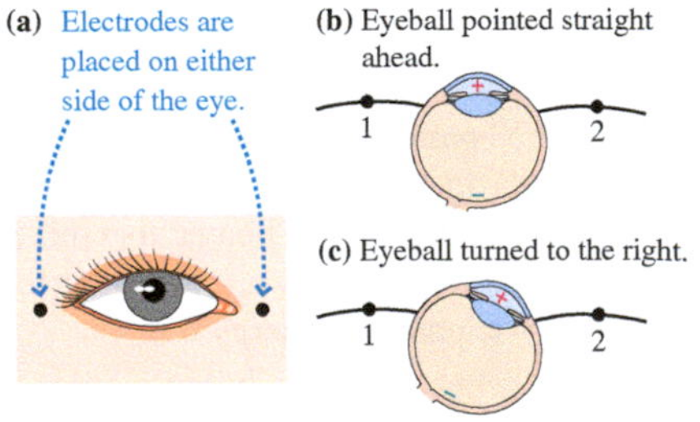


FIGURE P21.34

35. ||| In experimental tests, sharks have shown the ability to locate dipole electrodes (simulating the dipole fields of the heartbeats of prey animals) buried under the sand. In a test with young bonnethead sharks, sharks that detected the presence of a dipole usually swam toward the center of the dipole by following equipotential lines. Figure P21.35 shows a dipole electrode and three initial positions of a bonnethead shark. For each initial position, sketch the most likely path of the shark toward the center of the dipole.
BIO INT

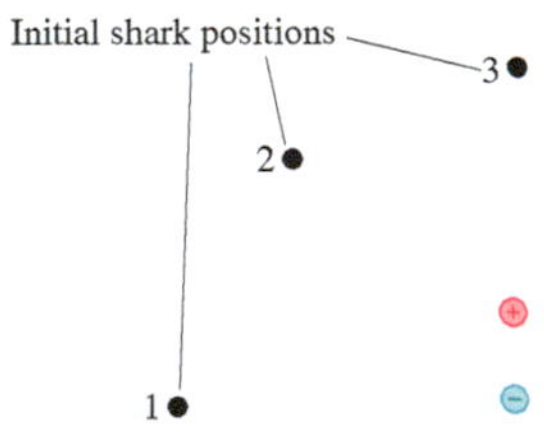


FIGURE P21.35

36. | Three electrodes, 1–3, are attached to a patient as shown in Figure P21.36. During ventricular depolarization (see Figure 21.31), across which pair of electrodes is the magnitude of the potential difference likely to be the smallest? Explain.
BIO

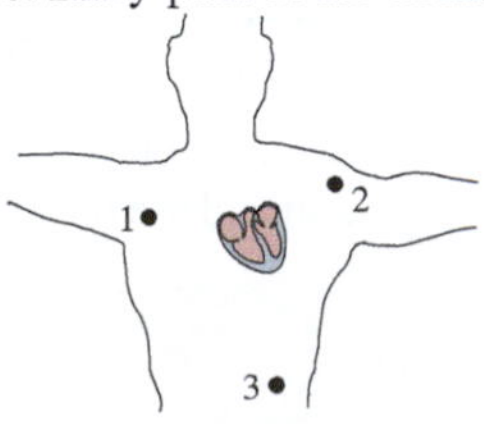


FIGURE P21.36

### Section 21.7 Capacitance and Capacitors

37. ||| Two 2.0 cm $\times$ 2.0 cm square aluminum electrodes, spaced 0.50 mm apart, are connected to a 100 V battery.
    a. What is the capacitance?
    b. What is the charge on the positive electrode?
38. || We've seen that bees develop a positive charge as they fly
BIO through the air. When a bee lands on a flower, charge is transferred, and an opposite charge is induced in the earth below the flower. The flower and the ground together make a capacitor; a typical value is 0.80 pF. If a flower is charged to 30 V relative to the ground, a bee can reliably detect the added charge and then avoids the flower in favor of flowers that have not been recently visited. Approximately how much charge must a bee transfer to the flower to create a 30 V potential difference?
39. ||| An uncharged capacitor is connected to the terminals of a 3.0 V battery, and 6.0 $\mu$C flows to the positive plate. The 3.0 V battery is then disconnected and replaced with a 5.0 V battery, with the positive and negative terminals connected in the same manner as before. How much additional charge flows to the positive plate?
40. || The earth is negatively charged, carrying 500,000 C of electric charge. This results in a 300 kV potential difference between the earth and the positively charged ionosphere. What is the capacitance of the earth–ionosphere system? If we assume that the bottom of the ionosphere is 60 km above the surface, what is the average electric field between the earth and the ionosphere?
41. ||| You need to construct a 100 pF capacitor for a science project. You plan to cut two $L \times L$ metal squares and place spacers between them. The thinnest spacers you have are 0.20 mm thick. What is the proper value of $L$?
42. || When a hummingbird visits a flower, its wings rub against
BIO the flower and leaves, and this can result in a noticeable charge on the bird. There is an opposite charge in the earth. We can consider the hummingbird and the earth to be the two electrodes of a capacitor. The capacitance for one species has been estimated to be 1.1 pF. If the bird accumulates a charge of +300 pC, what is the potential difference between the hummingbird and the earth?
43. | A switch that connects a battery to a 10 $\mu$F capacitor is closed. Several seconds later you find that the capacitor plates are charged to $\pm 30$ $\mu$C. What is the battery voltage?
44. || Investigators are exploring ways to treat milk for longer shelf
BIO life by using pulsed electric fields to destroy bacterial contamination. One system uses 8.0-cm-diameter circular plates separated by 0.95 cm. The space between the plates is filled with milk, which has a dielectric constant the same as that of water. The plates are briefly charged to 30,000 V. What is the capacitance of the system, and how much charge is on each plate when they are fully charged?
45. | Initially, the switch in Figure P21.45 is open and the capacitor is uncharged. How much charge flows through the switch after the switch is closed?

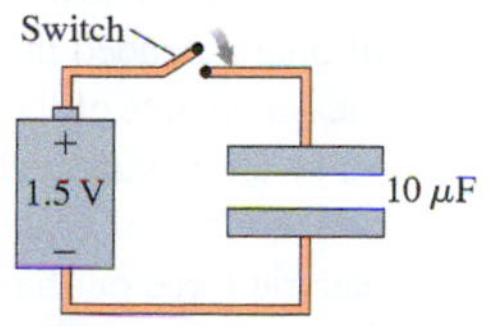


FIGURE P21.45

46. || A 1.2 nF parallel-plate capacitor has an air gap between its plates. Its capacitance increases by 3.0 nF when the gap is filled by a dielectric. What is the dielectric constant of that dielectric?
47. ||| A 25 pF parallel-plate capacitor with an air gap between the plates is connected to a 100 V battery. A Teflon slab is then inserted between the plates and completely fills the gap. What is the change in the charge on the positive plate when the Teflon is inserted?
48. || Nerve cells in your body can be electrically stimulated;
BIO a large enough change in a membrane potential triggers a
INT nerve impulse. Certain plants work the same way. A touch to *mimosa pudica*, the "sensitive plant," causes the leaflets to fold inward and droop. We can trigger this response electrically as well. In one experiment, investigators placed electrodes on the thick tissue at the base of a leaf. The electrodes were 3.5 mm apart. When the electrodes were connected to a 47 $\mu$F capacitor charged to 1.5 V, this stimulated a response from the plant.
    a. Eventually, all the charge on the capacitor was transferred to the plant. How much charge was transferred?
    b. What was the approximate electric field between the electrodes?
49. ||| A science-fair radio uses a homemade capacitor made of two 35 cm $\times$ 35 cm sheets of aluminum foil separated by a 0.25-mm-thick sheet of paper. What is its capacitance?
50. ||| A parallel-plate capacitor is connected to a battery and stores 4.4 nC of charge. Then, while the battery remains connected, a sheet of Teflon is inserted between the plates.
    a. Does the capacitor's charge increase or decrease?
    b. By how much does the charge change?
51. ||| A parallel-plate capacitor is charged by a 12.0 V battery, then the battery is removed.
    a. What is the potential difference between the plates after the battery is disconnected?
    b. What is the potential difference between the plates after a sheet of Teflon is inserted between them?

### Section 21.8 Energy and Capacitors

52. || The gecko in the photo
BIO is sticking upside down
INT to a smooth ceiling. The remarkable adhesion might be due to static electricity. Gecko feet are covered with microscopic hairs. When these hairs rub against a surface, charges separate, with the hair developing a positive charge and negative charge forming below the surface. There is an attractive force between the separated charges. This is an effective means of adhering to a surface, but it comes at a cost: Two planes of charge are like two charged plates of a capacitor, which takes energy to charge. Doubling the amount of charge on each surface increases the attractive force, but also increases the energy required to separate the charge. By what factor does this energy increase?

53. ||| To what potential should you charge a 1.0 $\mu$F capacitor to store 1.0 J of energy?
54. || To provide the pulse of energy needed for an intense bass, some car stereo systems add capacitors. One system uses a 2.0 F capacitor charged to 24 V, double the normal 12 V provided by the car's battery. How much energy does the capacitor store at 12 V? At 24 V?
55. | Capacitor 2 has half the capacitance and twice the potential difference as capacitor 1. What is the ratio $(U_C)_1/(U_C)_2$?
56. || The capacity of a battery to deliver charge, and thus power, decreases with temperature. The same is not true of capacitors. For sure starts in cold weather, a truck has a 500 F capacitor alongside a battery. The capacitor is charged to the full 13.8 V of the truck's battery. How much energy does the capacitor store? How does the energy density of the 9.0 kg capacitor system compare to the 130,000 J/kg of the truck's battery?

57. |||| 50 pJ of energy is stored in a 2.0 cm × 2.0 cm × 2.0 cm region of uniform electric field. What is the electric field strength?

58. |||| Two uncharged metal spheres, spaced 15.0 cm apart, have a capacitance of 24.0 pF. How much work would it take to move 12.0 nC of charge from one sphere to the other?

59. || A 2.0-cm-diameter parallel-plate capacitor with a spacing of 0.50 mm is charged to 200 V. What are (a) the total energy stored in the electric field and (b) the energy density?

## General Problems

60. ||| What is the change in electric potential energy of a 3.0 nC point charge when it is moved from point A to point B in Figure P21.60?

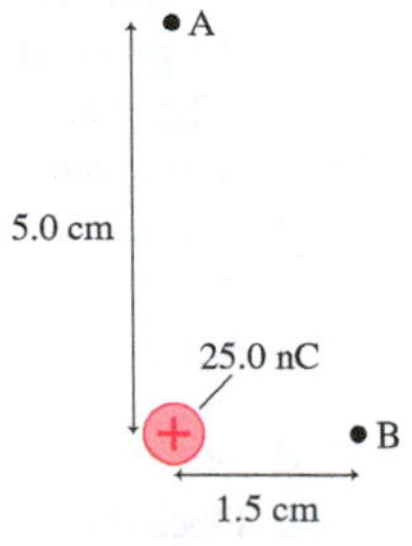


FIGURE P21.60

61. || A −50 nC charged particle is in a uniform electric field $\vec{E} = (10\text{ V/m, east})$. An external force moves the particle 1.0 m north, then 5.0 m east, then 2.0 m south, and finally 3.0 m west. The particle begins and ends its motion with zero velocity.
INT
   a. How much work is done on it by the external force?
   b. What is the potential difference between the particle's final and initial positions?

62. || The 4000 V equipotential surface is 10.0 cm farther from a positively charged particle than the 5000 V equipotential surface. What is the charge on the particle?

63. || What is the electric potential energy of the electron in Figure P21.63? The protons are fixed and can't move.

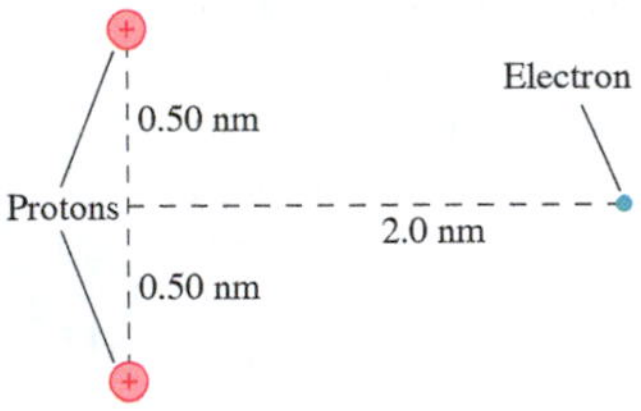


FIGURE P21.63

64. |||| Two point charges 2.0 cm apart have an electric potential energy −180 $\mu$J. The total charge is 30 nC. What are the two charges?

65. ||||| A +3.0 nC charge is at $x = 0$ cm and a −1.0 nC charge is at $x = 4$ cm. At what point or points on the $x$-axis is the electric potential zero?

66. ||||| A −3.0 nC charge is on the $x$-axis at $x = -9$ cm and a +4.0 nC charge is on the $x$-axis at $x = 16$ cm. At what point or points on the $y$-axis is the electric potential zero?

67. || A −2.0 nC charge and a +2.0 nC charge are located on the $x$-axis at $x = -1.0$ cm and $x = +1.0$ cm, respectively.
INT
   a. At what position or positions on the $x$-axis is the electric field zero?
   b. At what position or positions on the $x$-axis is the electric potential zero?
   c. Draw graphs of the electric field strength and the electric potential along the $x$-axis.

68. || Electric outlets have a voltage of approximately 120 V between the two parallel slots. Estimate the electric field strength between these two slots.

69. || A $Na^+$ ion moves from inside a cell, where the electric potential is −70 mV, to outside the cell, where the potential is 0 V. What is the change in the ion's electric potential energy as it moves from inside to outside the cell? Does its energy increase or decrease?
BIO

70. ||| Suppose that a molecular ion with charge −10$e$ is embedded within the 5.0-nm-thick cell membrane of a cell with membrane potential −70 mV. What is the electric force on the molecule?
BIO

71. | What is the electric potential at the point indicated with the dot in Figure P21.71?

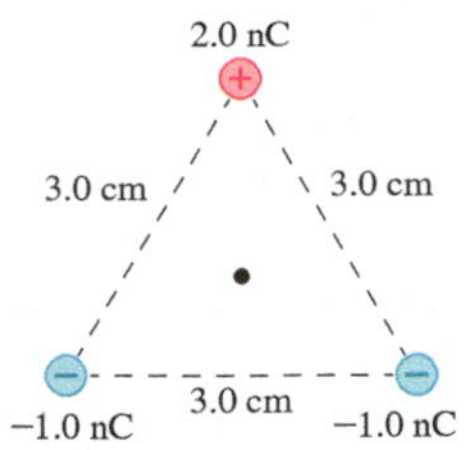


FIGURE P21.71

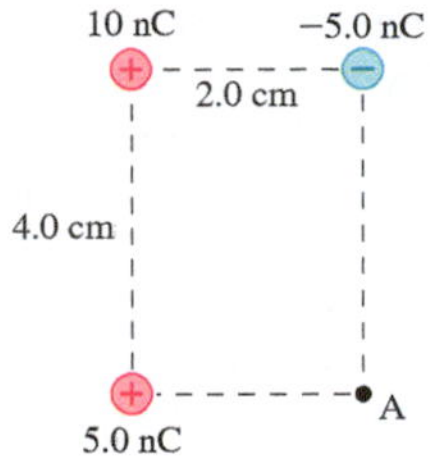


FIGURE P21.72

72. | a. What is the electric potential at point A in Figure P21.72?
   b. What is the potential energy of a proton at point A?

73. ||| A proton's speed as it passes point A is 50,000 m/s. It follows the trajectory shown in Figure P21.73. What is the proton's speed at point B?

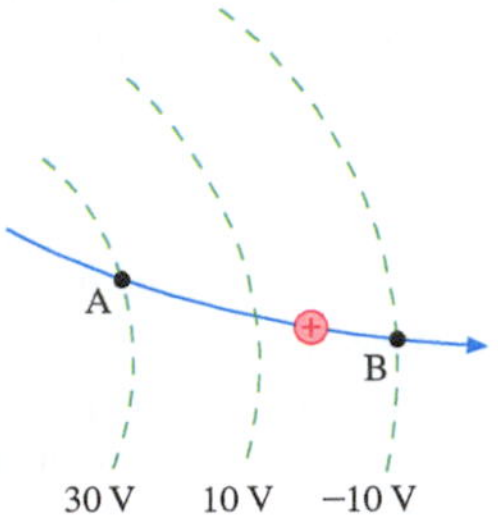


FIGURE P21.73

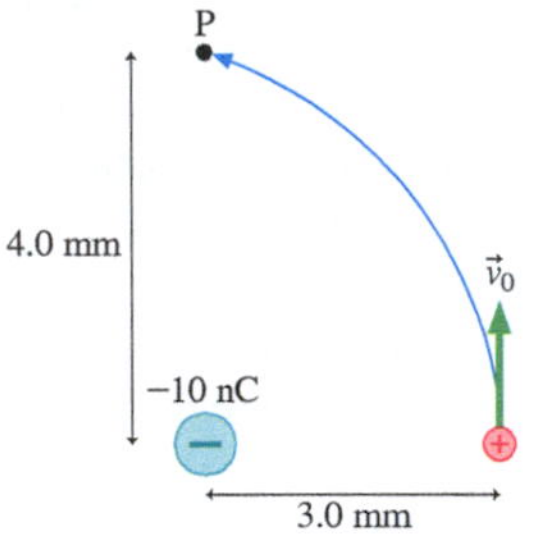


FIGURE P21.74

74. ||| A proton follows the path shown in Figure P21.74. Its initial speed is $v_0 = 1.9 \times 10^6$ m/s. What is the proton's speed as it passes through point P?

75. ||| A parallel-plate capacitor is charged to 5000 V. A proton is fired into the center of the capacitor at a speed of $3.0 \times 10^5$ m/s, as shown in Figure P21.75. The proton is deflected while inside the capacitor, and the plates are long enough that the proton will hit one of them before emerging from the far side of the capacitor. What is the impact speed of the proton?

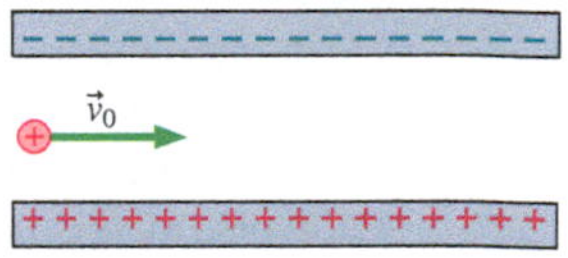


FIGURE P21.75

76. ||| A proton is released from rest at the positive plate of a parallel-plate capacitor. It crosses the capacitor and reaches the negative plate with a speed of 50,000 m/s. What will be the proton's final speed if the experiment is repeated with double the amount of charge on each capacitor plate?

77. || In the early 1900s, Robert Millikan used small charged droplets of oil, suspended in an electric field, to make the first quantitative measurements of the electron's charge. A 0.70-$\mu$m-diameter droplet of oil, having a charge of $+e$, is suspended in midair between two horizontal plates of a parallel-plate capacitor. The upward electric force on the droplet is exactly balanced by the downward force of gravity. The oil has a density of 860 kg/m$^3$, and the capacitor plates are 5.0 mm apart. What must the potential difference between the plates be to hold the droplet in equilibrium?
INT

78. |||| Two 2.0-cm-diameter disks spaced 2.0 mm apart form a parallel-plate capacitor. The electric field between the disks is $5.0 \times 10^5$ V/m.
   a. What is the voltage across the capacitor?
   b. How much charge is on each disk?
   c. An electron is launched from the negative plate. It strikes the positive plate at a speed of $2.0 \times 10^7$ m/s. What was the electron's speed as it left the negative plate?

79. || In *proton-beam therapy,* a high-energy beam of protons is fired at a tumor. The protons come to rest in the tumor, depositing their kinetic energy and breaking apart the tumor's DNA, thus killing its cells. For one patient, it is desired that 0.10 J of proton energy be deposited in a tumor. To create the proton beam, the protons are accelerated from rest through a 10 MV potential difference. What is the total charge of the protons that must be fired at the tumor to deposit the required energy?
BIO

80. ||| A 2.5-mm-diameter sphere is charged to −4.5 nC. An electron fired directly at the sphere from far away comes to within 0.30 mm of the surface of the target before being reflected.
   a. What was the electron's initial speed?
   b. At what distance from the surface of the sphere is the electron's speed half of its initial value?
   c. What is the acceleration of the electron at its turning point?

81. || A proton is fired from far away toward the nucleus of an iron atom. Iron is element number 26, and the diameter of the nucleus is 9.0 fm. (1 fm $= 10^{-15}$ m.) What initial speed does the proton need to just reach the surface of the nucleus? Assume the nucleus remains at rest.

82. ||| Determine the magnitude and direction of the electric field at points 1 and 2 in Figure P21.82.

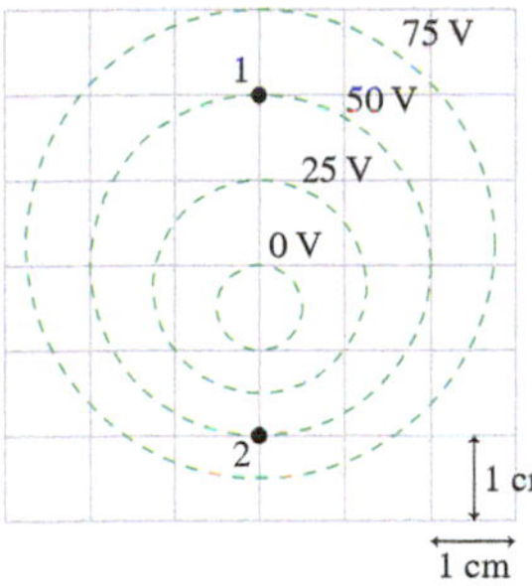


FIGURE P21.82

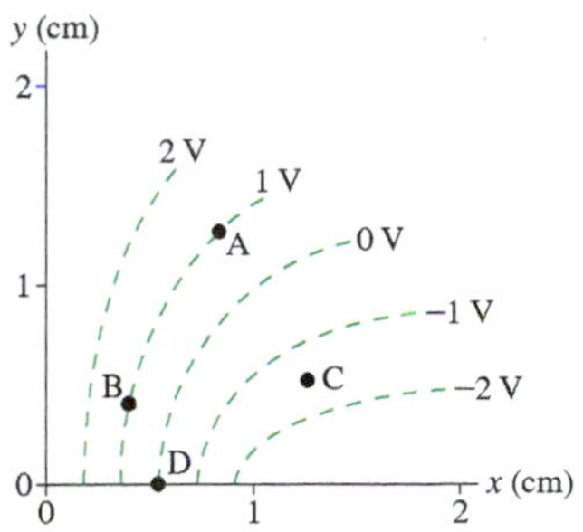


FIGURE P21.83

83. | Figure P21.83 shows a series of equipotential curves.
   a. Is the electric field strength at point A larger than, smaller than, or equal to the field strength at point B? Explain.
   b. Is the electric field strength at point C larger than, smaller than, or equal to the field strength at point D? Explain.
   c. Determine the electric field $\vec{E}$ at point D. Express your answer as a magnitude and direction.

84. ||| A capacitor consists of two 6.0-cm-diameter circular plates separated by 1.0 mm. The plates are charged to 150 V, then the battery is removed.
INT
   a. How much energy is stored in the capacitor?
   b. How much work must be done to pull the plates apart to where the distance between them is 2.0 mm?

85. ||| The dielectric in a capacitor serves two purposes. It increases the capacitance, compared to an otherwise identical capacitor with an air gap, and it increases the maximum potential difference the capacitor can support. If the electric field in a material is sufficiently strong, the material will suddenly become able to conduct, creating a spark. The critical field strength, at which breakdown occurs, is 3.0 MV/m for air, but 60 MV/m for Teflon.
   a. A parallel-plate capacitor consists of two square plates, 15 cm on a side, spaced 0.50 mm apart with only air between them. What is the maximum energy that can be stored by the capacitor?
   b. What is the maximum energy that can be stored if the plates are separated by a 0.50-mm-thick Teflon sheet?

86. | The highest magnetic fields in the world are generated when large arrays, or "banks," of capacitors are discharged through the copper coils of an electromagnet. At the National High Magnetic Field Laboratory, the total capacitance of the capacitor bank is 32 mF. These capacitors can be charged to 16 kV.
INT
   a. What is the energy stored in the capacitor bank when it is fully charged?
   b. When discharged, the entire energy from this bank flows through the magnet coil in 10 ms. What is the average power delivered to the coils during this time?

87. |||| The flash unit in a camera uses a special circuit to "step up" the 3.0 V from the batteries to 300 V, which charges a capacitor. The capacitor is then discharged through a flashlamp. The discharge takes 10 $\mu$s, and the average power dissipated in the flashlamp is $10^5$ W. What is the capacitance of the capacitor?
INT

## MCAT-Style Passage Problems

### A Lightning Strike

Storm clouds build up large negative charges, as described in the chapter. The charges dwell in *charge centers,* regions of concentrated charge. Suppose a cloud has −25 C in a 1.0-km-diameter spherical charge center located 10 km above the ground, as sketched in Figure P21.88. The negative charge center attracts a similar amount of positive charge that is spread on the ground below the cloud.

The charge center and the ground function as a charged capacitor, with a potential difference of approximately $4 \times 10^8$ V. The large electric field between these two "electrodes" may ionize the air, leading to a conducting path between the cloud and the ground. Charges will flow along this conducting path, causing a discharge of the capacitor—a lightning strike.

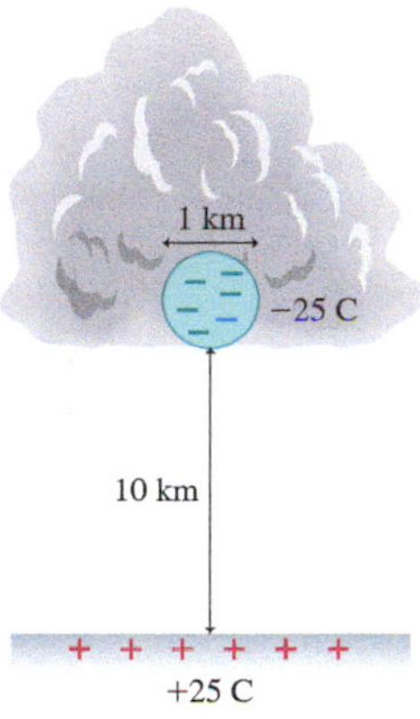


FIGURE P21.88

88. | What is the approximate magnitude of the electric field between the charge center and the ground?
   A. $4 \times 10^4$ V/m
   B. $4 \times 10^5$ V/m
   C. $4 \times 10^6$ V/m
   D. $4 \times 10^7$ V/m

89. | Which of the curves sketched in Figure P21.89 best approximates the shape of an equipotential drawn halfway between the charge center and the ground?

FIGURE P21.89

90. | What is the approximate capacitance of the charge center + ground system?
   A. $6 \times 10^{-8}$ F B. $2 \times 10^7$ F C. $4 \times 10^6$ F D. $8 \times 10^6$ F

91. | If 12.5 C of charge is transferred from the cloud to the ground in a lightning strike, what fraction of the stored energy is dissipated?
   A. 12% B. 25% C. 50% D. 75%

92. | If the cloud transfers all of its charge to the ground via several rapid lightning flashes lasting a total of 1 s, what is the average power?
   A. 1 GW B. 2 GW C. 5 GW D. 10 GW

# 22 Current and Resistance

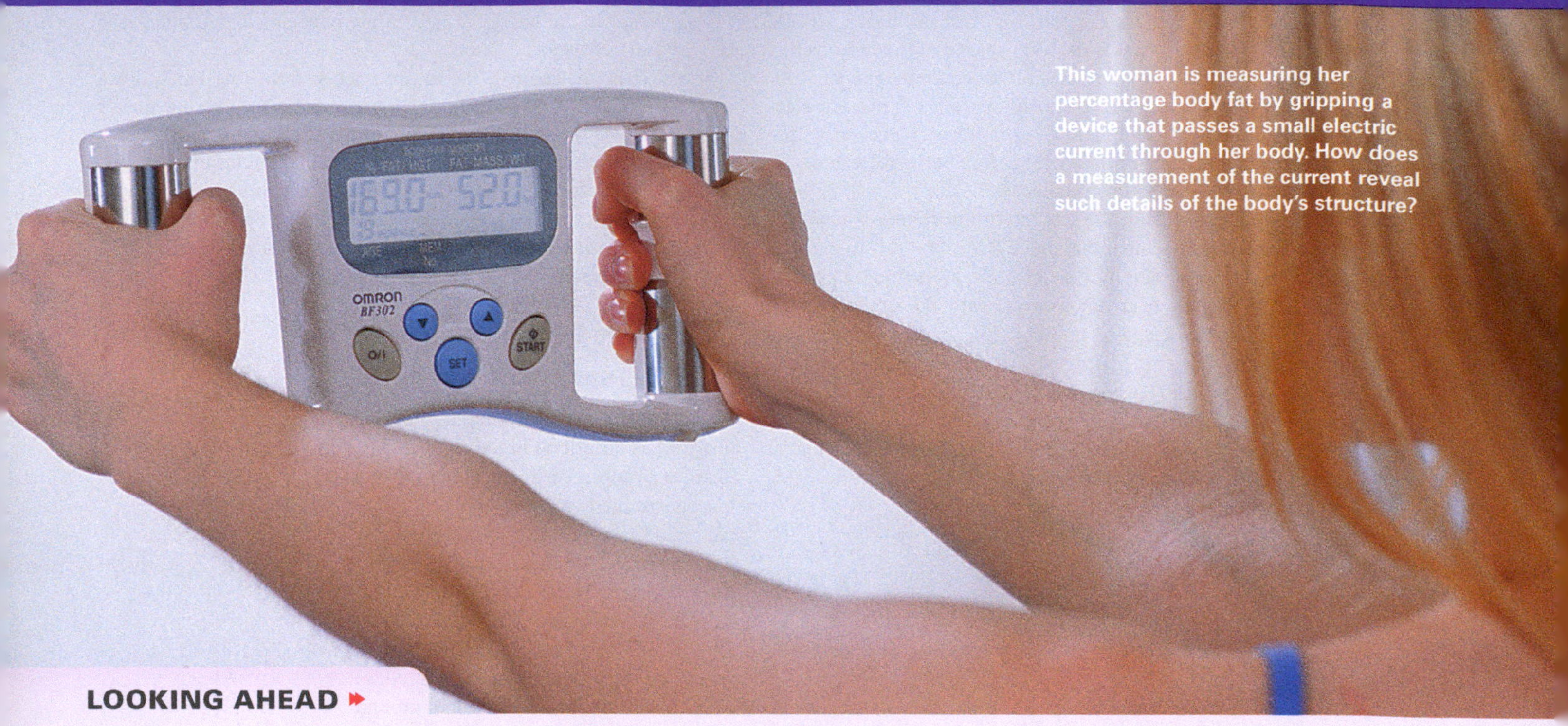


This woman is measuring her percentage body fat by gripping a device that passes a small electric current through her body. How does a measurement of the current reveal such details of the body's structure?

## LOOKING AHEAD ▸

### Current

The motion of charge through a conductor, like the wires connecting these lights, is called a **current.**

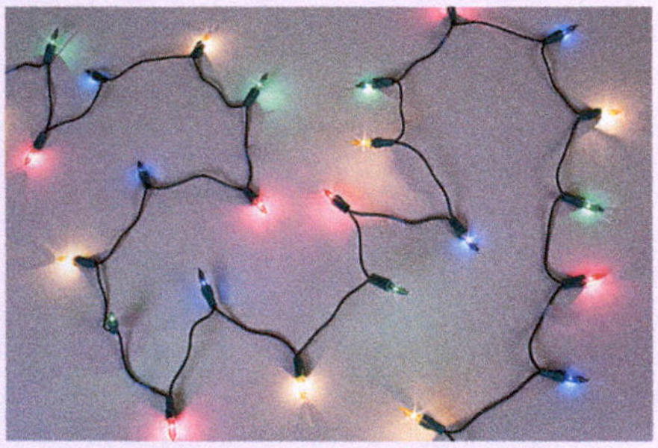

You'll learn why **conservation of current** ensures that each bulb is equally bright.

### Resistance and Ohm's Law

**Ohm's law** relates the current in this bulb to the battery's voltage and the bulb's **resistance** to the flow of charge.

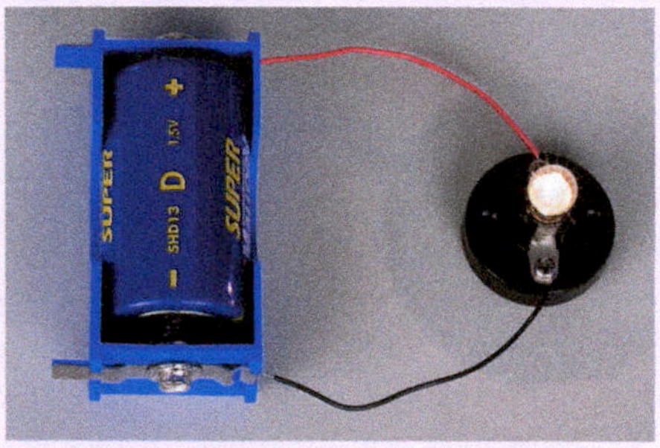


You'll learn how to relate the resistance of a wire to its size and composition.

### Electric Power

A hair dryer converts electric energy to thermal energy, leading to a blast of hot air.

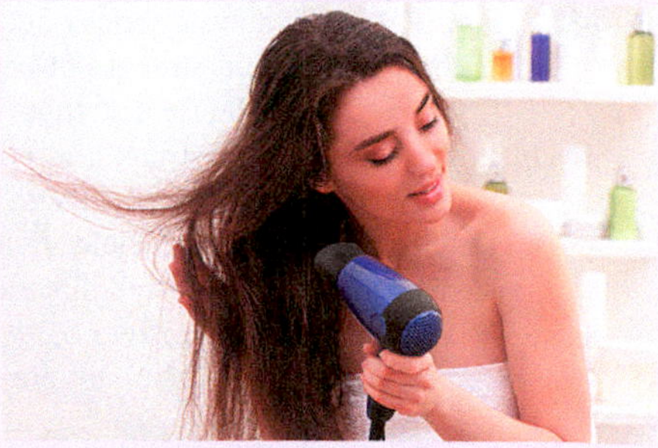

You'll learn the relationship of electric power to current, resistance, and voltage.

**GOAL** **To learn how and why charge moves through a conductor as a current.**

## LOOKING BACK ◂

### Electric Potential and Electric Field

In Section 21.5, you learned the connection between the electric potential and the electric field. In this chapter, you'll use this connection to understand why charges move in conductors.

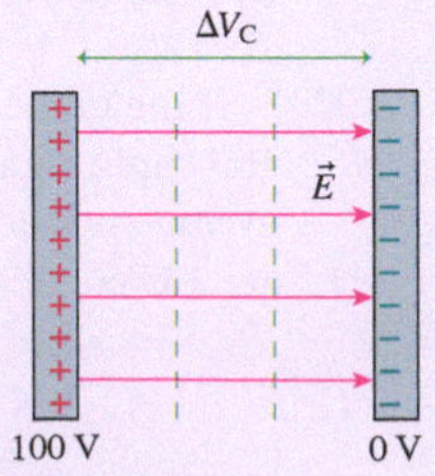


The electric field always points "downhill," from higher to lower potential, as in this parallel-plate capacitor.

**STOP TO THINK**

An electron is released from rest at the dot. The electron

A. Starts moving to the right.
B. Starts moving to the left.
C. Remains at rest.

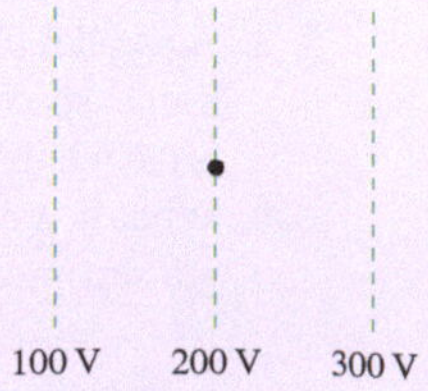

## 22.1 A Model of Current

FIGURE 22.1 A capacitor is discharged by a metal wire.

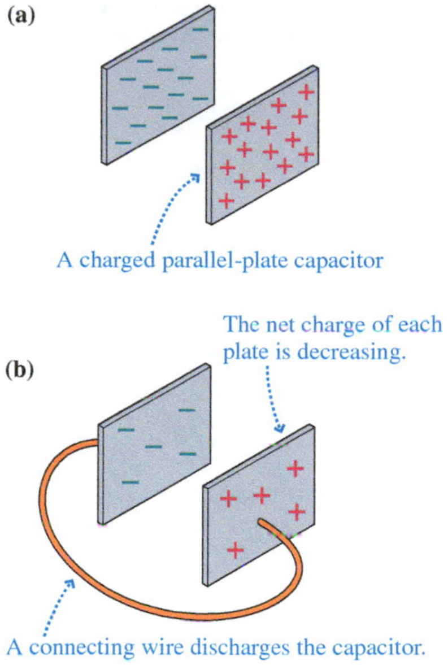


Let's start our exploration of current with a very simple experiment. FIGURE 22.1a shows a charged parallel-plate capacitor. If we connect the two capacitor plates to each other with a metal wire, as shown in FIGURE 22.1b, the plates quickly become neutral. We say that the capacitor has been *discharged.*

The wire is a *conductor,* a material through which charge easily moves. Apparently the excess charge on one capacitor plate is able to move through the wire to the other plate, neutralizing both plates. The motion of charges through a material is called a *current,* so the capacitor is discharged by a current in the connecting wire. Later in this chapter we will develop a quantitative expression for current, but for now the simple idea of current as charges in motion will suffice.

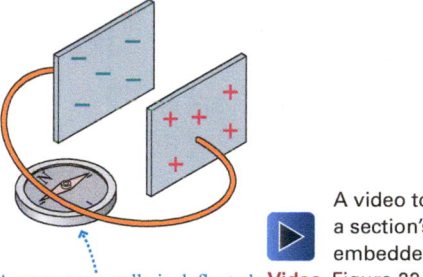

If we observe the capacitor discharge more closely, we see other effects. As FIGURE 22.2 shows, the connecting wire gets warmer. If the wire is very thin in places, such as the thin filament in a lightbulb, the wire gets hot enough to glow. The current-carrying wire also deflects a compass needle, a phenomenon we'll study in detail in Chapter 24. For now, we will use "makes the wire warmer" and "deflects a compass needle" as *indicators* that a current is present in a wire. In particular, we can use the brightness of a lightbulb to tell us the magnitude of the current; **more current means a brighter bulb.**

FIGURE 22.2 Indicators of a current.

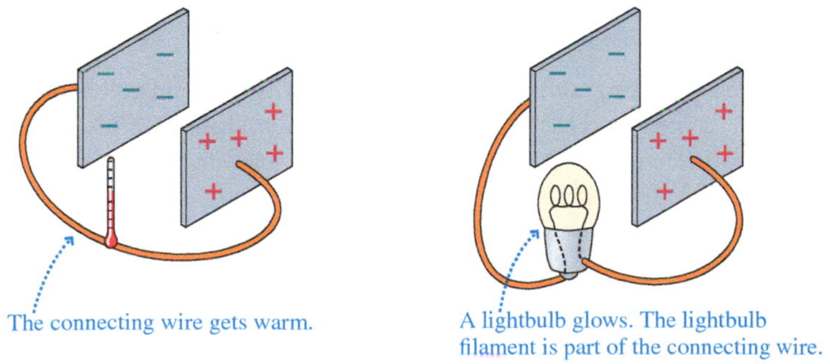


A video to support a section's topic is embedded in the eText.
**Video** Figure 22.2

The charges that move in a current are called *charge carriers.* In a metal, the charge carriers are electrons. As FIGURE 22.3 shows, it is the motion of the *conduction electrons,* which are free to move around, that forms a current—a flow of charge—in the metal. An *insulator* does not have such free charges and cannot carry a current. Although electrons are the charge carriers in metals, other materials may have different charge carriers. For instance, in ionic solutions such as seawater, blood, and intercellular fluids, the charge carriers are ions, both positive and negative.

FIGURE 22.3 Conduction electrons in a metal.

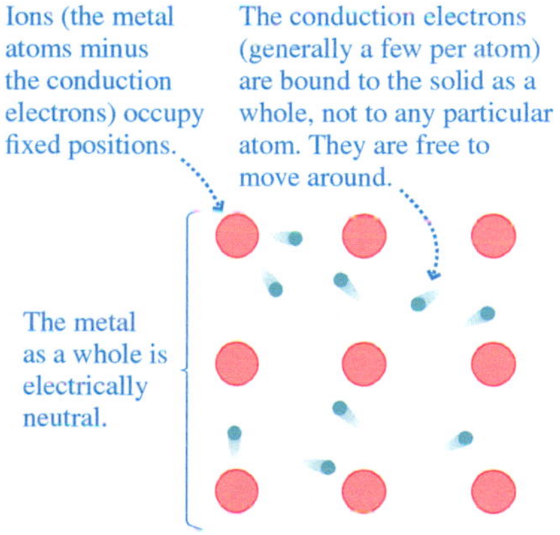

Ions (the metal atoms minus the conduction electrons) occupy fixed positions.
The conduction electrons (generally a few per atom) are bound to the solid as a whole, not to any particular atom. They are free to move around.
The metal as a whole is electrically neutral.

### Creating a Current

Suppose you want to slide a book across the table to your friend. You give it a quick push to start it moving, but it begins slowing down because of friction as soon as you take your hand off of it. The book's kinetic energy is transformed into thermal energy, leaving the book and the table slightly warmer. The only way to keep the book moving at a *constant* speed is to continue pushing it.

Something similar happens in a conductor. As we saw in ◀◀ SECTION 20.7, we can use an electric field to push on the electrons in a conductor. Suppose we take a piece of metal and apply an electric field, as in FIGURE 22.4 on the next page. The field exerts a force on the electrons, and they begin to accelerate. But the electrons aren't moving in a vacuum. Collisions between the electrons and the atoms of the metal slow them down, transforming the electrons' kinetic energy into the thermal

FIGURE 22.4 The motion of an electron in a conductor.

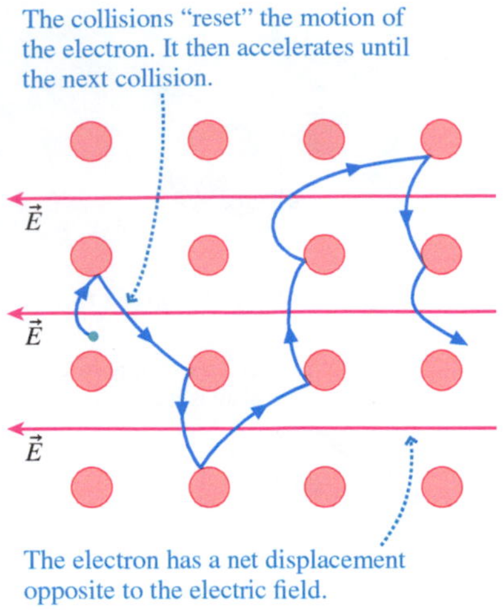


FIGURE 22.5 Creating a current in a wire.

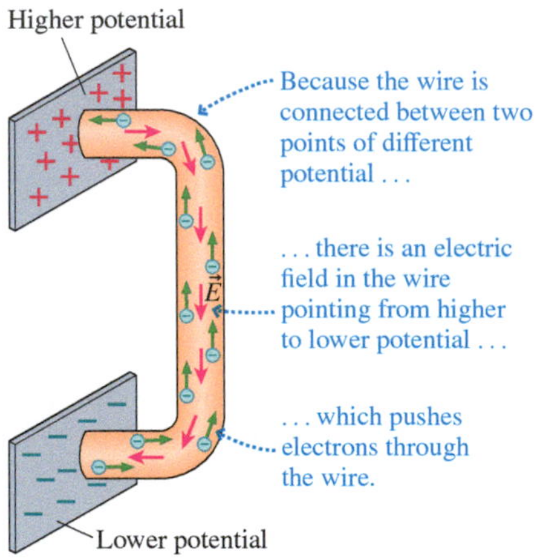


FIGURE 22.6 How does the current at A compare to the current at B?

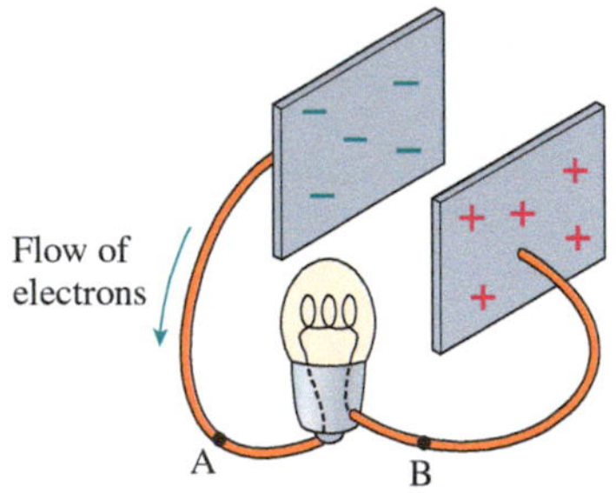


energy of the metal, which makes the metal warmer. (Recall that "makes the wire warmer" is one of our indicators of a current.) The motion of the electrons will cease *unless you continue pushing.* To keep the electrons moving, we must maintain an electric field. In a constant field, an electron's average motion will be opposite the field. We call this motion the electron's *drift velocity*. If the field goes to zero, so does the drift velocity.

How can you have an electric field inside a conductor? The important conclusion of ◀◀ SECTION 20.6 was that $\vec{E} = \vec{0}$ inside a conductor in electrostatic equilibrium. But a conductor with electrons moving through it is *not* in electrostatic equilibrium. The charges are in motion, so the field need not be zero.

FIGURE 22.5 shows how a wire connected between the plates of a charged capacitor causes it to discharge. (We've "unfolded" the plates to make the flow of charge easier to see.) The initial separation of charges creates a potential difference between the two plates. We saw in ◀◀ SECTION 21.5 that whenever there's a potential difference, an electric field points from higher potential toward lower potential. Connecting a wire between the plates establishes an electric field in the wire, and this electric field causes electrons to flow from the negative plate (which has an excess of electrons) toward the positive plate. **The potential difference creates the electric field that drives the current in the wire.**

As the current continues, and the charges flow, the plates discharge and the potential difference decreases. At some point, the plates will be completely discharged, meaning no more potential difference, no more field—and no more current. Finally, $\vec{E} = \vec{0}$ inside the conducting wire, and we have equilibrium.

## Conservation of Current

In FIGURE 22.6 a lightbulb has been added to the wire connecting two capacitor plates. The bulb glows while the current is discharging the capacitor. How does the current at point A compare to the current at point B? Are the currents at these points the same? Or is one larger than the other?

You might have predicted that the current at B is less than the current at A because the bulb, in order to glow, must use up some of the current. It's easy to test this prediction; for instance, we could compare the currents at A and B by comparing how far two compass needles at these positions are deflected. Any such test gives the same result: The current at point B is *exactly equal* to the current at point A. **The current leaving a lightbulb is exactly the same as the current entering the lightbulb.**

This is an important observation, one that demands an explanation. After all, "something" makes the bulb glow, so why don't we observe a decrease in the current? Electrons are charged particles. The lightbulb can't destroy electrons without violating both the law of conservation of mass and the law of conservation of charge. Thus the *number* of electrons is not changed by the lightbulb. Further, the lightbulb can't store electrons. Were it to do so, the bulb would become increasingly negatively charged until its repulsive force stopped the flow of new electrons and the bulb would go out. This doesn't happen. Every electron entering the lightbulb must be matched by an electron leaving the bulb, and thus the current at B is the same as at A.

Let's consider an analogy with water flowing through a pipe. Suppose we put a turbine in the middle of the pipe so that the flow of the water turns the turbine, as in FIGURE 22.7. Water flows *through* the turbine. It is not consumed by the turbine, and the number of gallons of water per minute leaving the pipe is exactly the same as the number entering. Nonetheless, the water must do work to turn the turbine, so there is an energy change as the water passes through.

Similarly, the lightbulb doesn't "use up" current, but, like the turbine, it *does* use energy. The energy is dissipated by atomic-level friction as the electrons move through the wire, making the wire hotter until, in the case of the lightbulb filament, it glows.

There are many other issues we'll need to examine, but we can draw a first important conclusion:

**Law of conservation of current** The current is the same at all points in a current-carrying wire.

**CONCEPTUAL EXAMPLE 22.1 Which bulb is brighter?**

The discharge of a capacitor lights two identical bulbs, as shown in FIGURE 22.8. Compare the brightness of the two bulbs.

***REASON*** Current is conserved, so any current that goes through bulb 1 must go through bulb 2 as well—the currents in the two bulbs are equal. We've noted that the brightness of a bulb is proportional to the current it carries. Identical bulbs carrying equal currents must have the same brightness.

***ASSESS*** This result makes sense in terms of what we've seen about the conservation of current. No charge is "used up" by either bulb.

FIGURE 22.8 Two bulbs lit by the current discharging a capacitor.

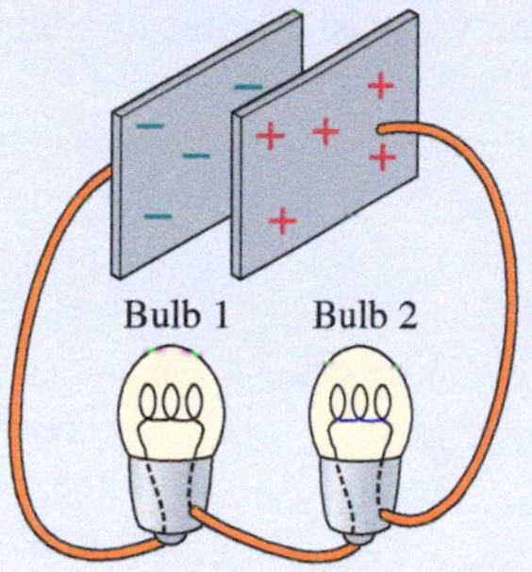


FIGURE 22.7 A water analogy for conservation of current.

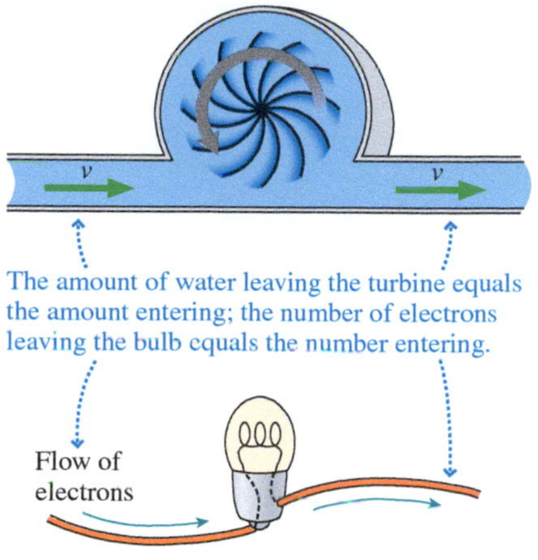


## 22.2 Defining and Describing Current

As we've seen, the current in a metal consists of electrons moving from low electric potential to high potential, in a direction opposite the electric field. But, as FIGURE 22.9 shows, a capacitor discharges in exactly the same way whether we consider the current to be negative charges moving opposite to the field or positive charges moving in the same direction as the field. In general, for *any* circuit, the current is in the same direction independent of which of these two perspectives we choose. We thus adopt the convention that **current is the flow of positive charge.** In Figure 22.9, then, the current is directed from the top plate to the bottom plate, whether it is produced by negative charges moving from bottom to top, or positive charges moving from top to bottom. From now on, we will think of a current as the flow of positive charge, even if the charges are actually electrons flowing in the opposite direction.

FIGURE 22.9 Negative and positive charges moving in opposite directions give the same current.

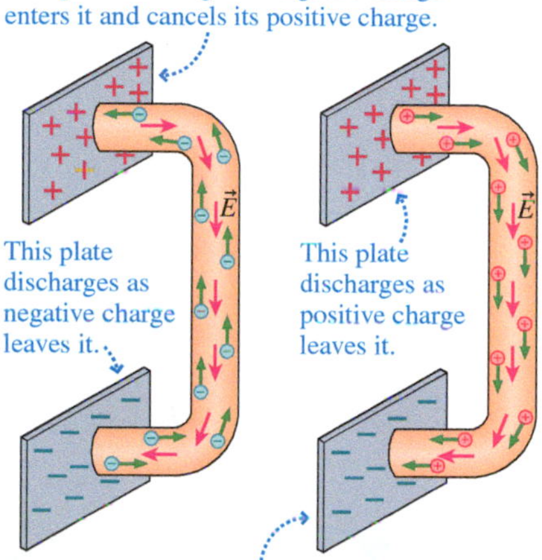

This plate discharges as negative charge enters it and cancels its positive charge.
This plate discharges as negative charge leaves it.
This plate discharges as positive charge leaves it.
This plate discharges as positive charge enters it and cancels its negative charge.

### Definition of Current

Because the coulomb is the SI unit of charge, and because currents are charges in motion, we define current as the *rate,* in coulombs per second, at which charge moves through a wire. FIGURE 22.10 shows a wire in which the electric field is $\vec{E}$. This electric field causes charges to move through the wire. Because we are considering current as the motion of positive charges, the motion is in the direction of the field.

The flow rate of water in a pipe measures the amount of water passing a cross-section area of the pipe per second. We use a similar convention for current. As illustrated in Figure 22.10, we can measure the amount of charge $\Delta q$ that passes through a cross section of the wire in a time interval $\Delta t$. We then define the current in the wire as

$$I = \frac{\Delta q}{\Delta t} \tag{22.1}$$

Definition of current

FIGURE 22.10 The current $I$.

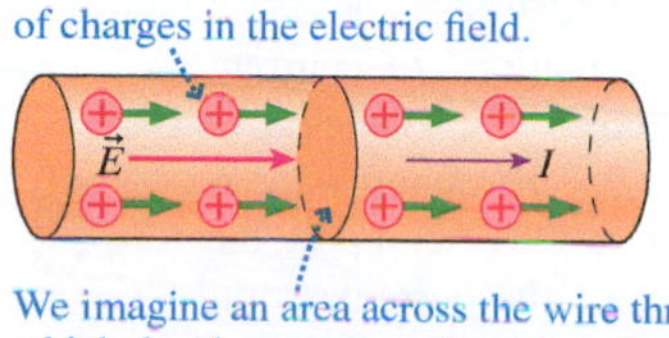

The current I is due to the motion of charges in the electric field.
We imagine an area across the wire through which the charges move. In a time Δt, charge Δq moves through this area.

With our positive-charge convention, the direction of the current in a wire is from higher potential to lower potential or, equivalently, in the direction of the electric field $\vec{E}$. The SI unit for current is based on the units for charge and time according to Equation 22.1. Current is measured in coulombs per second, which we define as the **ampere,** with the abbreviation A:

$$1 \text{ ampere} = 1 \text{ A} = 1 \text{ coulomb per second} = 1 \text{ C/s}$$

The *amp* is an informal shortening of ampere. Household currents are typically $\approx 1$ A. For example, the current through a 100 watt lightbulb is 0.83 A. The smaller currents in electronic devices are typically measured in milliamps ($1 \text{ mA} = 10^{-3} \text{ A}$) or microamps ($1 \mu\text{A} = 10^{-6} \text{ A}$).

For a *steady current,* which will be our primary focus, the total amount of charge that moves past a point in a wire carrying a current $I$ during the time interval $\Delta t$ is

$$q = I \Delta t \tag{22.2}$$

**EXAMPLE 22.2 Charge flow in a lightbulb**

A 100 W lightbulb carries a current of 0.83 A. How much charge flows through the bulb in 1 minute?

***STRATEGIZE*** Equation 22.2 gives the charge in terms of the current and the time interval.

***SOLVE*** According to Equation 22.2, the total charge passing through the bulb in 1 min = 60 s is

$$q = I \Delta t = (0.83 \text{ A})(60 \text{ s}) = 50 \text{ C}$$

***ASSESS*** The current corresponds to a flow of a bit less than 1 C per second, so our calculation seems reasonable, but the result is still somewhat surprising. That's a lot of charge! The enormous charge that flows through the bulb is a good check on the concept of conservation of current. If even a minuscule fraction of the charge stayed in the bulb, the bulb would become highly charged. For comparison, a Van de Graaff generator develops a potential of several hundred thousand volts due to an excess charge of just a few $\mu$C, a ten-millionth of the charge that flows through the bulb in 1 minute. Lightbulbs do not develop a noticeable charge, so the current into and out of the bulb must be exactly the same.

FIGURE 22.11 Kirchhoff's junction law.

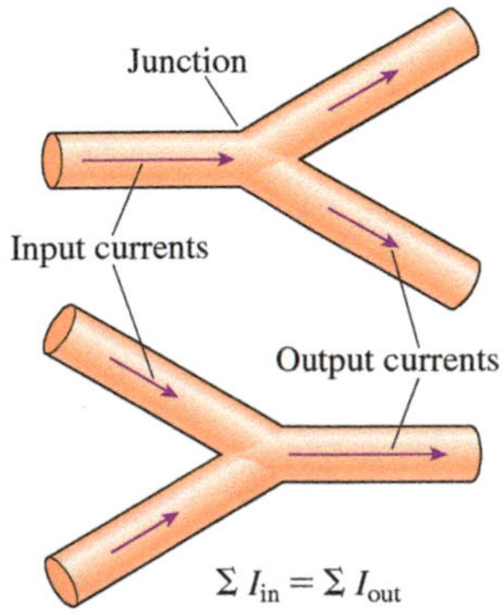


## Conservation of Current at a Junction

FIGURE 22.11 shows a wire splitting into two and two wires merging into one. A point where a wire branches is called a **junction.** The presence of a junction doesn't change the fact that current is conserved. We cannot create or destroy charges in the wire, and neither can we store them in the junction. The rate at which charges flow into one *or many* wires must be exactly balanced by the rate at which charges flow out of others. For a *junction,* the law of conservation of charge requires that

$$\Sigma I_{in} = \Sigma I_{out} \tag{22.3}$$

where, as usual, the $\Sigma$ symbol means "the sum of."

This basic conservation statement—that the sum of the currents into a junction equals the sum of the currents leaving—is called **Kirchhoff's junction law.** The junction law isn't a new law of physics; it is a consequence of the conservation of charge.

**EXAMPLE 22.3 Currents in a junction**

Four wires have currents as noted in FIGURE 22.12. What are the direction and the magnitude of the current in the fifth wire?

FIGURE 22.12 The junction of five wires.

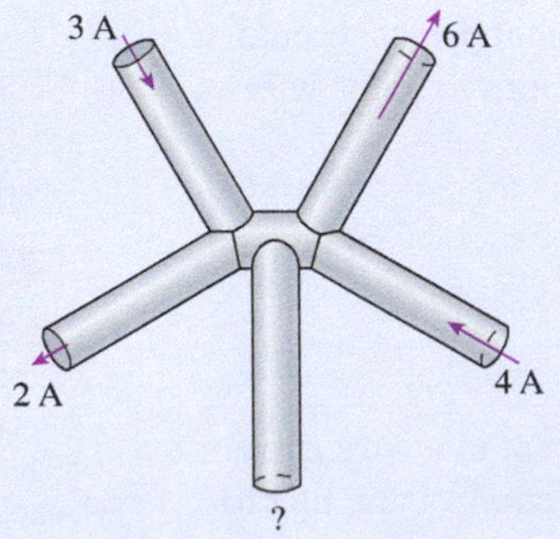


***STRATEGIZE*** This is a conservation of current problem for which we can use Kirchhoff's junction law. We compute the sum of the currents coming into the junction and the sum of the currents going out of the junction, and then compare these two sums. The unknown current is whatever is required to make the currents into and out of the junction "balance."

***SOLVE*** Two of the wires have currents into the junction:

$$\Sigma I_{in} = 3 \text{ A} + 4 \text{ A} = 7 \text{ A}$$

Two of the wires have currents out of the junction:

$$\Sigma I_{out} = 6 \text{ A} + 2 \text{ A} = 8 \text{ A}$$

To conserve current, the fifth wire must carry a current of 1 A into the junction.

***ASSESS*** If the unknown current is 1 A into the junction, a total of 8 A flows in—exactly what is needed to balance the current going out.

**STOP TO THINK 22.1** The discharge of a capacitor lights three bulbs. Comparing the current in bulbs 1 and 2, we can say that

A. The current in bulb 1 is greater than the current in bulb 2.
B. The current in bulb 1 is less than the current in bulb 2.
C. The current in bulb 1 is equal to the current in bulb 2.

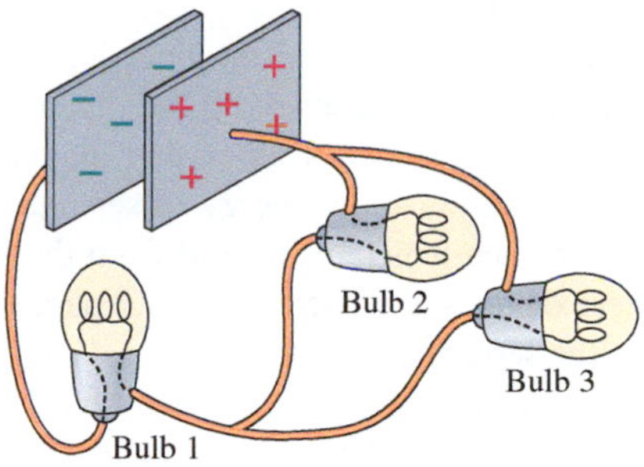


# 22.3 Batteries and emf

There are practical devices, such as a camera flash, that use the charge on a capacitor to create a current. But a camera flash gives a single, bright flash of light; the capacitor discharges and the current ceases. If you want a light to illuminate your way along a dark path, you need a *continuous* source of light such as a flashlight. Continuous light requires the current to be continuous as well.

FIGURE 22.13 shows a wire connecting the two terminals of a battery, much like the wire that connected the capacitor plates in Figure 22.1. Just like that wire, the wire connecting the battery terminals gets warm, deflects a compass needle, and makes a lightbulb inserted into it glow brightly, which indicates that charges are flowing through the wire from one battery terminal to the other. Evidently the current in a wire is the same whether it is supplied by a capacitor or a battery. Everything you've learned so far about current applies equally well to the current supplied by a battery, with one important difference—the duration of the current.

The wire connecting the battery terminals *continues* to deflect the compass needle and *continues* to light the lightbulb. A capacitor quickly runs out of excess charge, but a battery can keep the charges in motion.

How does a battery produce this sustained motion of charge? A real battery involves a series of chemical reactions, but FIGURE 22.14 shows a simple model of a battery that illustrates the motion of charges. The inner workings of a battery act like a *charge escalator* between the two terminals. Charges are removed from the low-potential negative terminal and "lifted" to the high-potential positive terminal. It is the charge escalator that sustains the current in the wire by providing a continuously renewed supply of charges at the positive terminal.

Once a charge reaches the positive terminal, it moves downhill through the wire—from high to low electric potential—until it reaches the negative terminal. The charge escalator then lifts the charge back to the positive terminal where it can start the loop all over again. This flow of charge in a continuous loop is what we call a **complete circuit.**

The charge escalator in the battery must be powered by some source of energy. It is lifting the charges "uphill" against an electric field. A battery consists of chemicals, called *electrolytes,* sandwiched between two electrodes made of different materials. The energy to move charges comes from chemical reactions between the electrolytes and the electrodes. These chemical reactions separate charge by moving positive ions to one electrode and negative ions to the other. In other words, chemical reactions, rather than a mechanical conveyor belt, transport charge from one electrode to the other.

As a battery creates a current in a circuit, the reactions that run the charge escalator deplete chemicals in the battery. A dead battery is one in which the supply of chemicals, and thus the supply of chemical energy, has been exhausted. You can "recharge" some types of batteries by forcing a current into the positive terminal, reversing the chemical reactions that move the charges, thus replenishing the chemicals and storing energy as chemical energy.

By separating charge, the charge escalator establishes the potential difference $\Delta V_{bat}$ between the terminals of the battery shown in Figure 22.14. The potential

FIGURE 22.13 There is a current in a wire connecting the terminals of a battery.

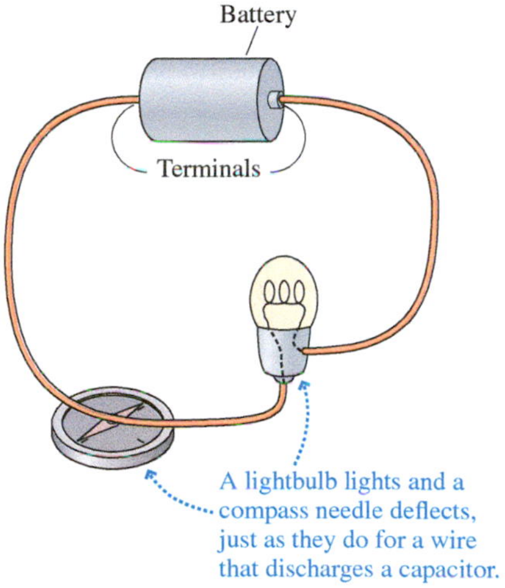

Battery
Terminals
A lightbulb lights and a compass needle deflects, just as they do for a wire that discharges a capacitor.

FIGURE 22.14 The charge escalator model of a battery.

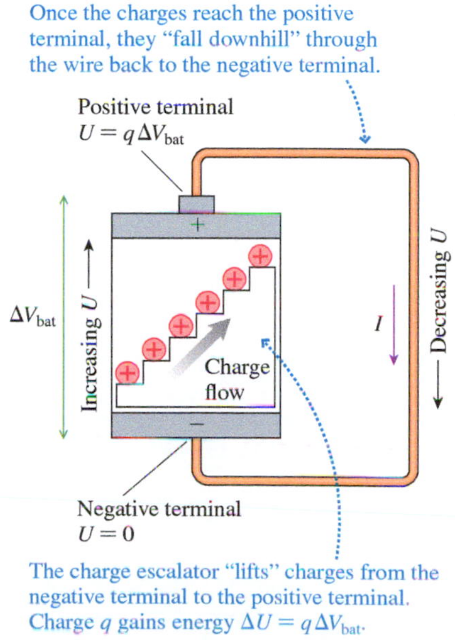

**A shocking predator?** BIO The torpedo ray captures and eats fish by paralyzing them with electricity. As we will see in Chapter 23, cells in the body use chemical energy to separate charge, just as in a battery. Special cells in the body of the ray called *electrocytes* produce an emf of a bit more than 0.10 V for a short time when stimulated. Such a small emf will not produce a large effect, but the torpedo ray has organs that contain clusters of hundreds of these electrocytes connected in a row. The total emf can be 50 V or more, enough to immobilize nearby prey.

difference established by a device, such as a battery, that can actively separate charge is called its **emf.** It is pronounced as the sequence of three letters "e-m-f." The symbol for emf is $\mathcal{E}$, a script E, and its units are volts.

There are devices other than batteries that can generate an emf. Rooftop solar panels use the sun's energy to separate charge and create an emf; the generator at a power plant generates its emf from the energy of steam or falling water. In contrast to these devices, a capacitor *stores* separated charges, but a capacitor has no means to create the separation. Hence a charged capacitor has a potential difference but not an emf.

The *rating* of a battery, such as 1.5 V, is the battery's emf. It is determined by the specific chemical reactions employed by the battery. An alkaline battery has an emf of 1.5 V; a rechargeable nickel-metal hydride (NiMH) battery has an emf of 1.2 V. Larger emfs are created by using several smaller "cells" in a row, much like going from the first to the fourth floor of a building by taking three separate escalators.

A battery with no current in it has a potential difference equal to its emf. With a current, the battery's potential difference is slightly less than the emf. We'll overlook this small difference and assume $\Delta V_{bat} = \mathcal{E}$.

**CONCEPTUAL EXAMPLE 22.4 Potential difference for batteries in series**

Three batteries are connected one after the other as shown in FIGURE 22.15; we say they are connected in *series*. What's the total potential difference?

**REASON** We can think of this as three charge escalators, one after the other. Each one lifts charges to a higher potential. Because each battery raises the potential by 1.5 V, the total potential difference of the three batteries in series is 4.5 V.

**ASSESS** Common AA and AAA batteries are 1.5 V batteries. Many consumer electronics, such as digital cameras, use two or four of these batteries. Wires inside the device connect the batteries in series to produce a total 3.0 V or 6.0 V potential difference.

FIGURE 22.15 Three batteries in series.

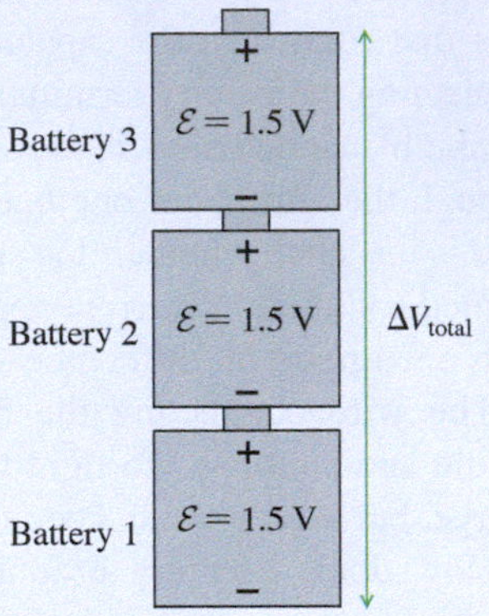


Because a battery maintains a constant potential difference $\Delta V_{bat} = \mathcal{E}$ across its terminals, as a charge $q$ moves from the negative terminal to the positive terminal, it gains electric potential energy $U = q\Delta V_{bat}$, as shown in Figure 22.14. Now we can understand the energy transformation in a battery: Chemical energy $E_{chem}$ is transformed into electric potential energy $U = q\Delta V_{bat}$ as charge moves through the battery. In a later section we'll examine the energy transformations that occur as this charge moves through a circuit element, such as a lightbulb, on its way back to the negative terminal.

**STOP TO THINK 22.2** A battery produces a current in a wire. As the current continues, which of the following quantities (perhaps more than one) decreases?

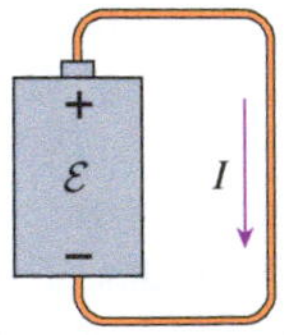


A. The positive charge in the battery
B. The emf of the battery
C. The chemical energy in the battery

## 22.4 Connecting Potential and Current

An important conclusion of the charge escalator model is that **a battery is a source of potential difference.** As FIGURE 22.16 shows, when charges flow through a wire that connects the battery terminals, this current is a *consequence* of the battery's potential difference.

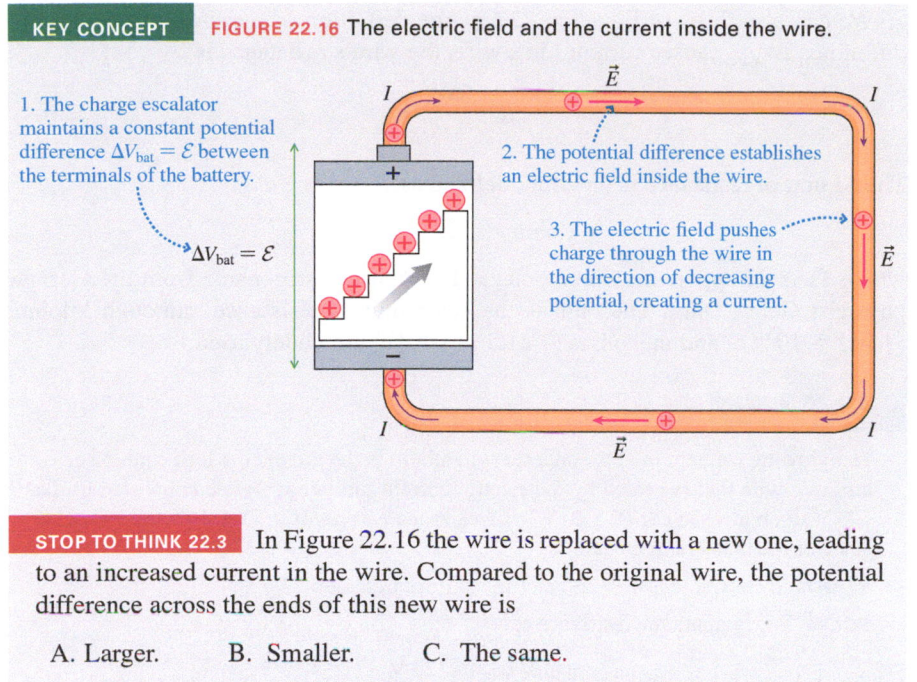

KEY CONCEPT **FIGURE 22.16** The electric field and the current inside the wire.

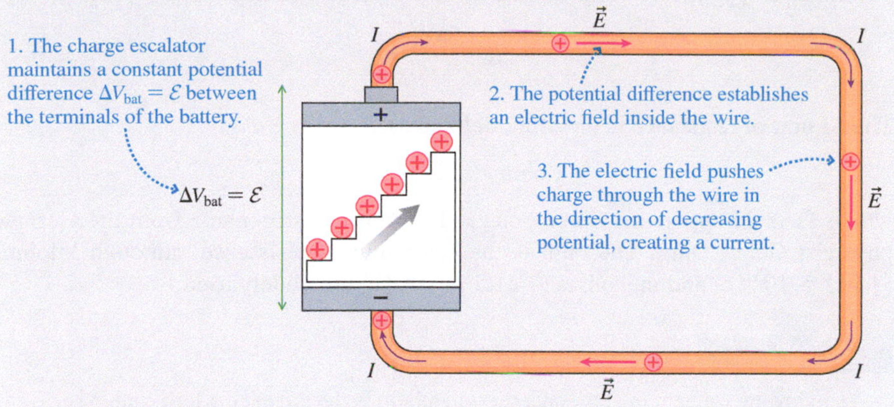


**STOP TO THINK 22.3** In Figure 22.16 the wire is replaced with a new one, leading to an increased current in the wire. Compared to the original wire, the potential difference across the ends of this new wire is

A. Larger. B. Smaller. C. The same.

Because the ends of the wire are connected to the terminals of the battery, the potential difference between the two ends of the wire is equal to the potential difference between the battery terminals:

$$\Delta V_{wire} = \Delta V_{bat} \tag{22.4}$$

This potential difference causes a current in the direction of decreasing potential.

## Resistance

Figure 22.16 explained the *direction* of the current in a wire connected between the terminals of a battery. **FIGURE 22.17** shows a series of experiments to determine what factors affect the current's *magnitude*. The experiments show that there are two factors that determine the current: the potential difference and the properties of the wire.

Figure 22.17a shows that adding a second battery in series increases the current, as you would expect. A larger potential difference creates a larger electric field that pushes charges through the wire faster. Careful measurements show that the current $I$ is proportional to $\Delta V_{wire}$.

Figures 22.17b and 22.17c illustrate the two properties of the wire that affect the current: the dimensions of the wire and the material of which it is made. Figure 22.17b shows that increasing the length of the wire decreases the current, while increasing the thickness of the wire increases the current. This seems reasonable because it should be harder to push charges through a long wire than a short one, and an electric field should be able to push more charges through a fat wire than a skinny one. Figure 22.17c shows that wires of different materials carry different currents—some materials are better conductors than others.

For any particular wire, we can define a quantity called the **resistance** that is a measure of how hard it is to push charges through the wire. We use the symbol $R$ for resistance. A large resistance implies that it is hard to move the charges through the wire; in a wire with small resistance, the charges move more easily. The current in the wire depends on both the potential difference $\Delta V_{wire}$ between the ends of the wire and the wire's resistance $R$:

$$I = \frac{\Delta V_{wire}}{R} \tag{22.5}$$

As we would expect, the smaller the resistance, the larger the current.

**FIGURE 22.17** Factors affecting the current in a wire.

(a) Changing potential

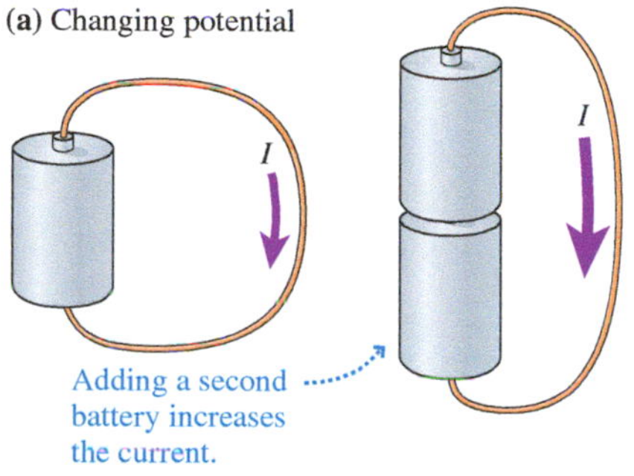


(b) Changing wire dimensions

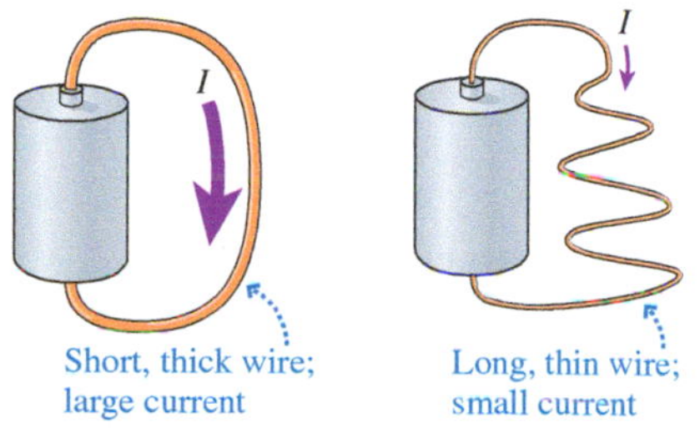


(c) Changing wire material

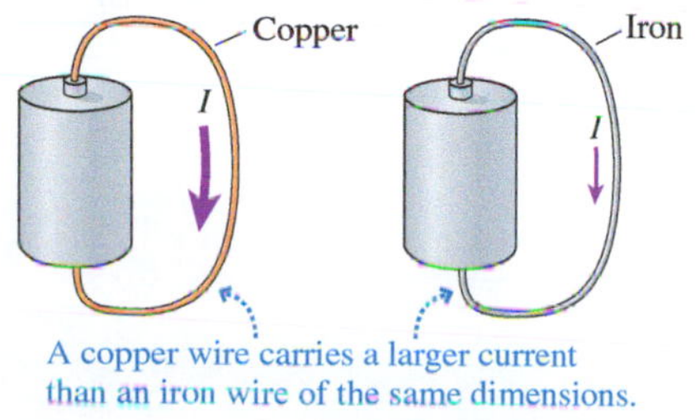

We can also think of Equation 22.5 as the definition of resistance. If a potential difference $\Delta V_{\text{wire}}$ causes current $I$ in a wire, the wire's resistance is

$$R = \frac{\Delta V_{\text{wire}}}{I} \tag{22.6}$$

The SI unit of resistance is the **ohm,** defined as

$$1 \text{ ohm} = 1\ \Omega = 1 \text{ V/A}$$

where $\Omega$ is an uppercase Greek omega. The unit takes its name from the German physicist Georg Ohm. The ohm is the basic unit of resistance, although kilohms ($1\ \text{k}\Omega = 10^3\ \Omega$) and megohms ($1\ \text{M}\Omega = 10^6\ \Omega$) are widely used.

**EXAMPLE 22.5 Resistance of a lightbulb**

The glowing element in an incandescent lightbulb is the *filament,* a long, thin piece of tungsten wire that is heated by the electric current through it. When connected to the 120 V of an electric outlet, a 60 W bulb carries a current of 0.50 A. What is the resistance of the filament in the lamp?

***STRATEGIZE*** The resistance is determined from Equation 22.6.

***SOLVE*** We compute the resistance as

$$R = \frac{\Delta V_{\text{wire}}}{I} = \frac{120 \text{ V}}{0.50 \text{ A}} = 240\ \Omega$$

***ASSESS*** As we will see, the resistance of the filament varies with temperature. This value holds for the lightbulb only when the bulb is glowing and the filament is hot.

## Resistivity

Figure 22.17c showed that the resistance of a wire depends on what it is made of. We define a quantity called **resistivity,** for which we use the symbol $\rho$ (lowercase Greek rho), to characterize the electrical properties of materials. Materials that are good conductors have low resistivities; materials that are poor conductors (and thus that are good insulators) have high resistivities. The resistivity $\rho$ has units of $\Omega \cdot \text{m}$. The resistivities of some common materials are listed in TABLE 22.1.

**TABLE 22.1** Resistivities of materials

| Material | Resistivity ($\Omega \cdot \text{m}$) |
|---|---|
| Copper | $1.7 \times 10^{-8}$ |
| Aluminum | $2.7 \times 10^{-8}$ |
| Tungsten (20°C) | $5.6 \times 10^{-8}$ |
| Tungsten (1500°C) | $5.0 \times 10^{-7}$ |
| Iron | $9.7 \times 10^{-8}$ |
| Nichrome | $1.5 \times 10^{-6}$ |
| Seawater | 0.22 |
| Blood (average) | 1.6 |
| Muscle | 13 |
| Fat | 25 |
| Pure water | $2.4 \times 10^{5}$ |
| Cell membrane | $3.6 \times 10^{7}$ |

Metals are generally very good conductors (and so have very low resistivities), but some metals are much better conductors than others. The resistivity of nichrome, an alloy used to make heating wires, is almost 100 times higher than that of copper. Pure water is a poor conductor, but the dissolved salts in seawater produce ions that can carry charge, so seawater is a moderately good conductor, with a resistivity one million times lower than that of pure water (but still ten million times higher than that of copper!). Glass is an excellent insulator with a resistivity in excess of $10^{14}\ \Omega \cdot \text{m}$, $10^{22}$ times that of copper.

The resistivity of a material depends on the temperature, as you can see from the two values for tungsten listed in Table 22.1. As the temperature increases, so do the thermal vibrations of the atoms. This makes them "bigger targets" for the moving electrons, causing collisions to be more frequent. Thus the resistivity of a metal increases with increasing temperature.

The resistance of a wire depends both on the resistivity of its material and on the dimensions of the wire. A wire made of a material of resistivity $\rho$, with length $L$ and cross-section area $A$, has resistance

$$R = \frac{\rho L}{A} \tag{22.7}$$

Resistance of a wire in terms of resistivity and dimensions

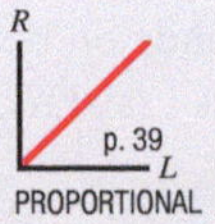


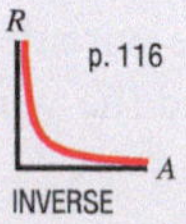

Resistance is a property of a *specific* wire or conductor because it depends on the conductor's length and diameter as well as on the resistivity of the material from which it is made.

**Video** Determining Resistance

**NOTE** ▸ It is important to distinguish between resistivity and resistance. *Resistivity* is a property of the *material,* not any particular piece of it. All copper wires (at the same temperature) have the same resistivity. *Resistance* characterizes a specific piece of the conductor that has a specific geometry. A short, thick copper wire has a smaller resistance than a long, thin copper wire. The relationship between resistivity and resistance is analogous to that between density and mass. ◂

### EXAMPLE 22.6 Finding the length of a lightbulb filament

We calculated in Example 22.5 that a 60 W lightbulb has a resistance of 240 Ω. At the operating temperature of the tungsten filament, its resistivity is approximately $5.0 \times 10^{-7}\ \Omega \cdot \text{m}$. If the wire used to make the filament is 0.040 mm in diameter (a typical value), how long must the filament be?

**STRATEGIZE** The resistance of a wire depends on its length, its cross-section area, and the material of which it is made.

**PREPARE** We need the cross-section area of the wire, which is $A = \pi r^2 = \pi(2.0 \times 10^{-5}\ \text{m})^2 = 1.26 \times 10^{-9}\ \text{m}^2$.

**SOLVE** Rearranging Equation 22.7 shows us that the filament must be of length

$$L = \frac{AR}{\rho} = \frac{(1.26 \times 10^{-9}\ \text{m}^2)(240\ \Omega)}{5.0 \times 10^{-7}\ \Omega \cdot \text{m}} = 0.60\ \text{m}$$

**ASSESS** This is quite long—nearly 2 feet. This result may seem surprising, but some reflection shows that it makes sense. The resistivity of tungsten is low, so the filament must be thin and long.

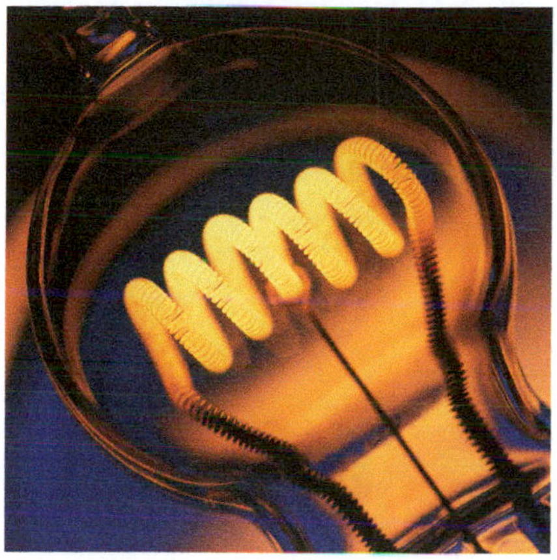

**Coils of coils** A close view of a lightbulb's filament shows that it is made of very thin wire that is coiled and then coiled again. The double-coil structure is necessary to fit the great length of the filament into the small space of the bulb's globe.

### EXAMPLE 22.7 Making a heater

An amateur astronomer uses a heater to warm her telescope eyepiece so moisture does not collect on it. The heater is a 20-cm-long, 0.50-mm-diameter nichrome wire that wraps around the eyepiece. When the wire is connected to a 1.5 V battery, what is the current in the wire?

**STRATEGIZE** The current in the wire depends on the emf of the battery and the resistance of the wire. The resistance of the wire depends on the resistivity of nichrome, given in Table 22.1, and the dimensions of the wire.

**PREPARE** Converted to meters, the relevant dimensions of the wire are $L = 0.20$ m and $r = 2.5 \times 10^{-4}$ m.

**SOLVE** The wire's resistance is

$$R = \frac{\rho L}{A} = \frac{\rho L}{\pi r^2} = \frac{(1.5 \times 10^{-6}\ \Omega \cdot \text{m})(0.20\ \text{m})}{\pi(2.5 \times 10^{-4}\ \text{m})^2} = 1.53\ \Omega$$

The wire is connected to the battery, so $\Delta V_{\text{wire}} = \Delta V_{\text{bat}} = 1.5$ V. The current in the wire is

$$I = \frac{\Delta V_{\text{wire}}}{R} = \frac{1.5\ \text{V}}{1.53\ \Omega} = 0.98\ \text{A}$$

**ASSESS** The emf of the battery is small, but so is the resistance of the wire, so this is a reasonable current, enough to warm the wire and the eyepiece.

## Electrical Measurements of Physical Properties

Because resistance depends sensitively on the properties of materials, a measurement of resistance can be a simple but effective probe of other quantities of interest. For example, the resistivity of water is strongly dependent on dissolved substances in the water, so one way to easily test water purity is by measuring its resistivity.

**CONCEPTUAL EXAMPLE 22.8** **Testing drinking water**

A house gets its drinking water from a well that has an intermittent problem with salinity. Before the water is pumped into the house, it passes between two electrodes in the circuit shown in FIGURE 22.18. The current passing through the water is measured with a meter. Which corresponds to increased salinity—an increased current or a decreased current?

***REASON*** Increased salinity causes the water's resistivity to decrease. This decrease causes a decrease in resistance between the electrodes. Current is inversely proportional to resistance, so this leads to an increase in current.

***ASSESS*** Increasing salinity means more ions in solution and thus more charge carriers, so an increase in current is expected. Electrical systems similar to this can therefore provide a quick check of water purity.

FIGURE 22.18 A water-testing circuit.

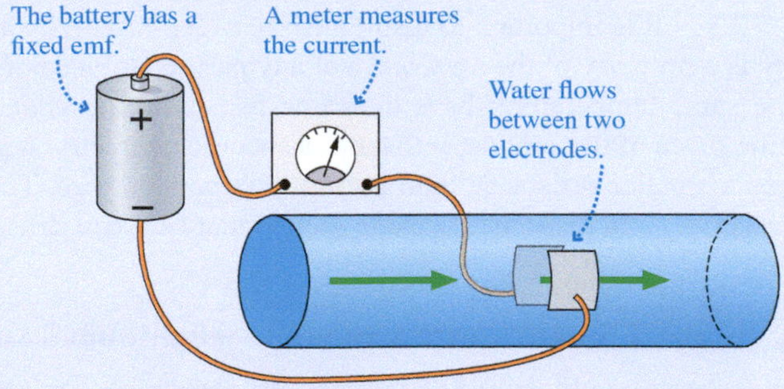


FIGURE 22.19 BIO An electrical impedance map showing the cross section of a healthy patient's torso.

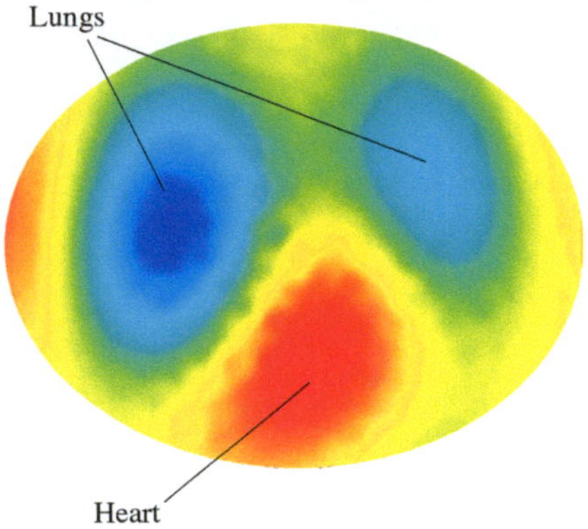


Different tissues in the body have different resistivities, as we saw in Table 22.1. For example, fat has a higher resistivity than muscle. Consequently, a routine test to estimate the percentage of fat in a person's body is based on a measurement of the body's resistance, as illustrated in the photo at the start of the chapter. A higher resistance of the body means a higher proportion of fat.

More careful measurements of resistance can provide more detailed diagnostic information. Passing a small, safe current between pairs of electrodes on opposite sides of a person's torso permits the resistance of the intervening tissue to be measured, a technique known as *electrical impedance tomography*. (Impedance is similar to resistance, but it also applies to AC circuits, which we will explore in Chapter 26.) FIGURE 22.19 shows an image of a patient's torso generated from measurements of resistance between many pairs of electrodes. The image shows the change in resistance between two subsequent measurements; decreasing resistance shows in red, increasing resistance in blue. This image was created during the resting phase of the heart, when blood was leaving the lungs and entering the heart. Blood is a better conductor than the tissues of the heart and lungs, so the motion of blood decreased the resistance of the heart and increased that of the lungs. This patient was healthy, but in a patient with circulatory problems any deviation from normal blood flow would lead to abnormal patterns of resistance that would be revealed in such an image.

**STOP TO THINK 22.4** A wire connected between the terminals of a battery carries a current. The wire is removed and stretched, decreasing its cross-section area and increasing its length. When the wire is reconnected to the battery, the new current is

A. Larger than the original current.
B. The same as the original current.
C. Smaller than the original current.

## 22.5 Ohm's Law and Resistor Circuits

The relationship between the potential difference across a conductor and the current passing through it that we saw in the preceding section was first deduced by Georg Ohm and is known as **Ohm's law:**

$$I = \frac{\Delta V}{R} \qquad (22.8)$$

Ohm's law for a conductor of resistance $R$

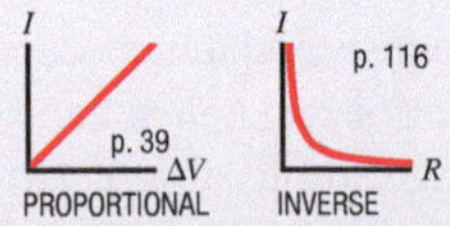


Video Basic Circuits

If we know that a wire of resistance $R$ carries a current $I$, we can compute the potential difference between the ends of the wire as $\Delta V = IR$.

Despite its name, Ohm's law is *not* a law of nature. It is limited to those materials whose resistance $R$ remains constant—or very nearly so—during use. Materials to which Ohm's law applies are called **ohmic.** FIGURE 22.20 shows that the current through an ohmic material is directly proportional to the potential difference; doubling the potential difference results in a doubling of the current. This is a linear relationship, and the resistance $R$ can be determined from the slope of the graph.

FIGURE 22.20 Current-versus-potential-difference for a conductor with resistance $R$.

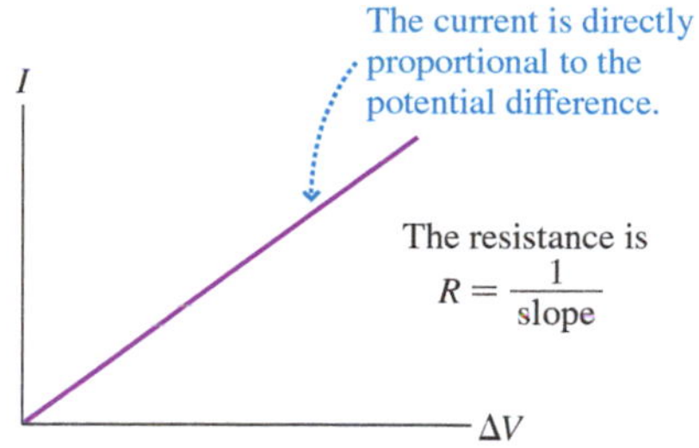


Other materials and devices are **nonohmic,** meaning that the current through the device is *not* directly proportional to the potential difference. Two important examples of nonohmic devices are batteries, where $\Delta V = \mathcal{E}$ is determined by chemical reactions, independent of $I$, and capacitors, where, as you'll learn in Chapter 26, the relationship between $I$ and $\Delta V$ is very different from that of a resistor.

**NOTE** ▶ Ohm's law applies only to resistive devices like wires or lightbulb filaments. It does *not* apply to nonohmic devices such as batteries or capacitors. ◀

## Resistors

The word "resistance" may have negative connotations—who needs something that slows charges and robs energy? In some cases resistance *is* undesirable. But in many other cases, circuit elements are designed to have a certain resistance for very practical reasons. We call these circuit elements **resistors.** There are a few basic types that will be very important as we start to look at electric circuits in detail.

### Examples of resistors

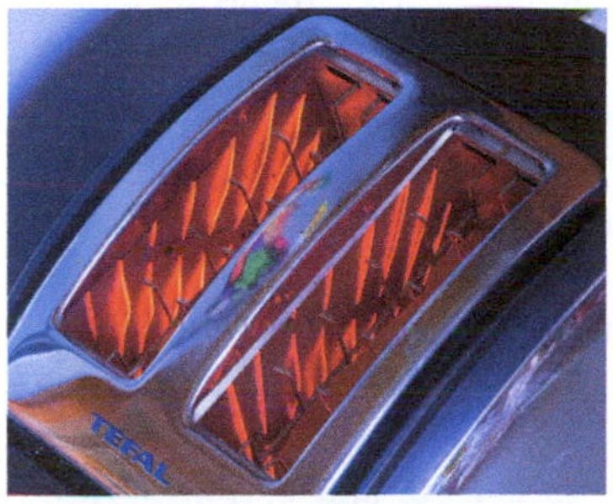

### Heating elements

As charges move through a resistive wire, their electric energy is transformed into thermal energy, heating the wire. Wires in a toaster, a stove burner, or the rear window defroster of a car are practical examples of this electric heating.

### Circuit elements

Inside many electronic devices is a circuit board with many small cylinders. These cylinders are resistors that help control currents and voltages in the circuit. The colored bands on the resistors indicate their resistance values.

### Sensor elements

A resistor whose resistance changes in response to changing circumstances can be used as a sensor. The resistance of this night-light sensor changes when daylight strikes it. A circuit detects this change and turns off the light during the day.

### CONCEPTUAL EXAMPLE 22.9 The changing current in a toaster

When you press the lever on a toaster, a switch connects the heating wires to 120 V. The wires are initially cool, but the current in the wires raises the temperature until they are hot enough to glow. As the wire heats up, how does the current in the toaster change?

***REASON*** As the wire heats up, its resistivity increases, as noted previously, so the resistance of the wires increases. Because the potential difference stays the same, an increasing resistance causes the current to decrease. The current through a toaster is largest when the toaster is first turned on.

***ASSESS*** This result makes sense. As the wire's temperature increases, the current decreases. This makes the system stable. If, instead, the current increased as the temperature increased, higher temperature could lead to more current, leading to even higher temperatures, and the toaster could overheat.

## Analyzing a Simple Circuit

FIGURE 22.21a shows the anatomy of a lightbulb. The key point is that a lightbulb, like a wire, has two "ends" and current passes *through* the bulb. Connections to the filament in the bulb are made at the tip and along the side of the metal cylinder. It is often useful to think of a lightbulb as a resistor that happens to give off light when a current is present. Now, let's look at a circuit using a battery, a lightbulb, and wires to make connections, as in FIGURE 22.21b. This is the basic circuit in a flashlight.

FIGURE 22.21 The basic circuit of a battery and a bulb.

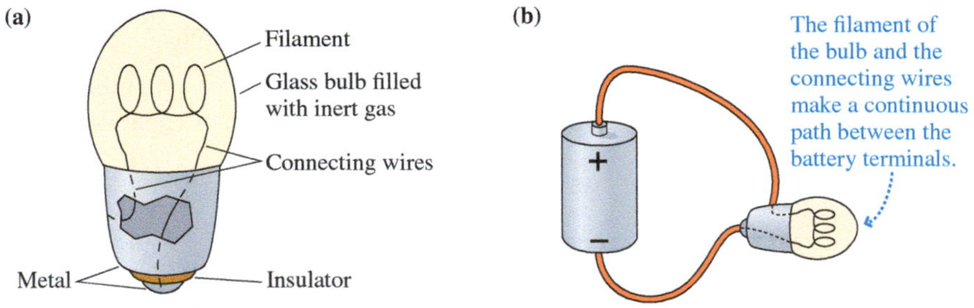


A typical flashlight bulb has a resistance of $\approx 3\ \Omega$, while the wires used to connect the bulb to the battery have a resistance of $\approx 0.01\ \Omega$. The resistance of the wires is so much less than that of the bulb that we can, with very little error, assume the resistance of the wires is *zero*. In this model of an **ideal wire,** with $R_{wire} = 0\ \Omega$, the potential difference $\Delta V$ between the ends of a connecting wire is zero, *even if there is a current in it.*

We know that, physically, the potential difference can't be zero. There must be an electric field in the wire for the charges to move, so it must have a potential difference. But in practice this potential difference is so small that we can assume it is zero.

FIGURE 22.22 shows how the ideal-wire model is used in the analysis of circuits. The resistor in Figure 22.22a is connected at each end to a wire, and current $I$ flows through all three. The current requires a potential difference $\Delta V_{resist} = IR_{resist}$ across the resistor, but there's no potential difference ($\Delta V_{wire} = 0$) for the ideal wires. Figure 22.22b shows this idea graphically by displaying the potential along the wire-resistor-wire combination. Current moves in the direction of decreasing potential, so there is a large *voltage drop*—a decrease in potential—across the resistor as we go from left to right, the direction of the current. The segments of the graph corresponding to the wires are horizontal because there's no voltage change along an ideal wire.

FIGURE 22.22 The potential along a wire-resistor-wire combination.

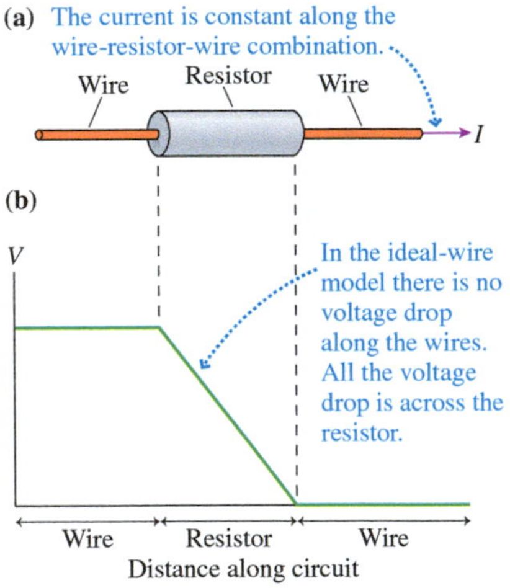


The linear variation in the potential across the resistor may remind you of the linear variation in potential between the plates of a parallel-plate capacitor. In a capacitor, this linear variation in potential corresponds to a uniform electric field; the same is true here. As we see in FIGURE 22.23, the electric field in a resistor carrying a current in a circuit is uniform; the strength of the electric field is

$$E = \frac{\Delta V}{L}$$

in analogy to Equation 21.6 for a parallel-plate capacitor. A larger potential difference corresponds to a larger field, as we would expect.

FIGURE 22.23 Electric field inside a resistor.

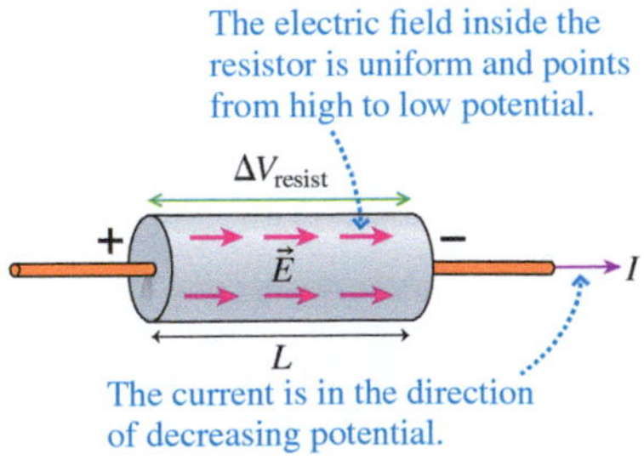

## EXAMPLE 22.10 Analyzing a single-resistor circuit

A 15 Ω resistor is connected to the terminals of a 1.5 V battery.

a. Draw a graph showing the potential as a function of distance traveled through the circuit, starting from $V = 0$ V at the negative terminal of the battery.
b. What is the current in the circuit?

**STRATEGIZE** The potential increases as we move through the battery, and drops as we move through the resistor. In the wires there is no voltage drop, so the voltage in the wires is a constant.

**PREPARE** To help us visualize the change in potential as charges move through the circuit, we begin with the sketch of the circuit in FIGURE 22.24. The zero point of potential is noted. We have drawn our sketch so that "up" corresponds to higher potential, which will help us make sense of the circuit. Charges are raised to higher potential in the battery, then travel "downhill" from the positive terminal through the resistor and back to the negative terminal. We assume ideal wires.

FIGURE 22.24 A single-resistor circuit.

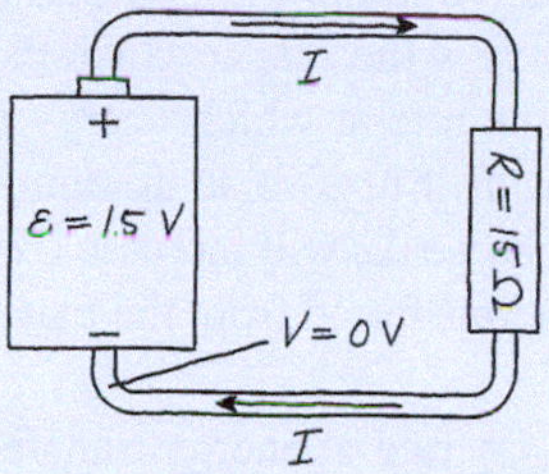


**SOLVE** a. FIGURE 22.25 is a graphical representation of the potential in the circuit. The distance $s$ is measured from the battery's negative terminal, where $V = 0$ V. As we move around the circuit to the starting point, the potential must return to its original value. Because the wires are ideal, there is no change in potential along the wires. This means that the potential difference across the resistor must be equal to the potential difference across the battery: $\Delta V_R = \mathcal{E} = 1.5$ V.

FIGURE 22.25 Potential-versus-position graph.

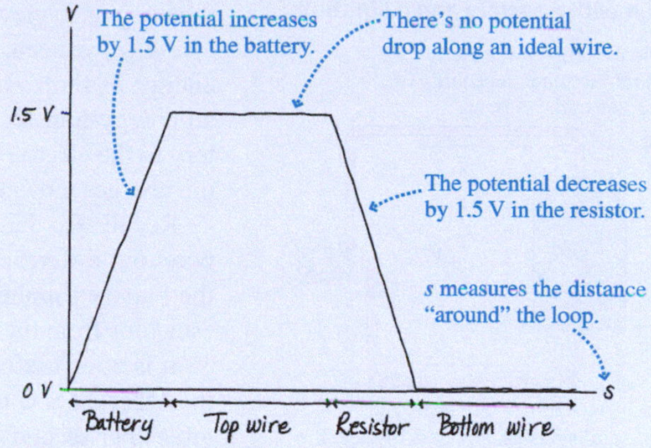


b. Now that we know the potential difference of the resistor, we can compute the current in the resistor by using Ohm's law:

$$I = \frac{\Delta V_R}{R} = \frac{1.5\ \text{V}}{15\ \Omega} = 0.10\ \text{A}$$

Because current is conserved, this is the current at any point in the circuit. In other words, the battery's charge escalator lifts charge at the rate 0.10 C/s, and charge flows through the wires and the resistor at the rate 0.10 C/s.

**ASSESS** This is a reasonable value of the current in a battery-powered circuit.

As we noted, there are many devices whose resistance varies as a function of a physical variable that it might be useful to measure, such as light intensity, temperature, or sound intensity. As the following example shows, we can use resistance measurements to monitor a physical variable.

## EXAMPLE 22.11 Using a thermistor

A *thermistor* is a device whose resistance varies with temperature in a well-defined way. A certain thermistor has a resistance of 2.8 kΩ at 20°C and 0.39 kΩ at 70°C. This thermistor is used in a water bath in a lab to monitor the temperature. The thermistor is connected in a circuit with a 1.5 V battery, and the current measured. What is the change in current in the circuit as the temperature rises from 20°C to 70°C?

**STRATEGIZE** We can use Ohm's law to find the currents at the two temperatures.

**SOLVE** The currents are

$$I(20°\text{C}) = \frac{\Delta V}{R} = \frac{1.5\ \text{V}}{2.8 \times 10^3\ \Omega} = 0.54\ \text{mA}$$

$$I(70°\text{C}) = \frac{\Delta V}{R} = \frac{1.5\ \text{V}}{0.39 \times 10^3\ \Omega} = 3.8\ \text{mA}$$

The change in current is thus 3.3 mA.

**ASSESS** A modest change in temperature leads to a large change in current, which is reasonable—this is a device intended to provide a sensitive indication of a temperature change.

**STOP TO THINK 22.5** Two identical batteries are connected in series in a circuit with a single resistor. $V = 0$ V at the negative terminal of the lower battery. Rank in order, from highest to lowest, the potentials $V_A$ to $V_E$ at the labeled points, noting any ties. Assume the wires are ideal.

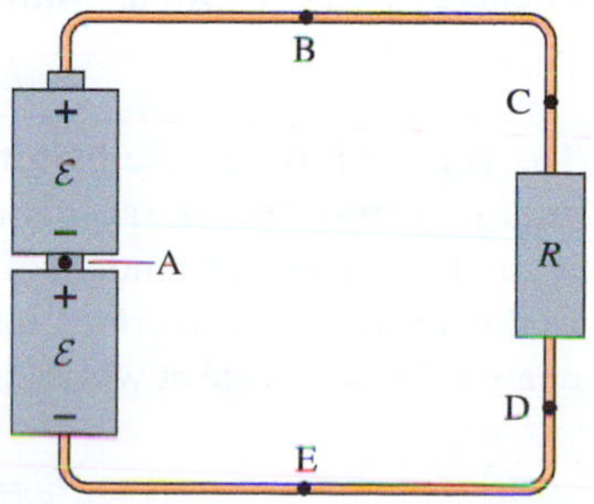

## 22.6 Energy and Power

FIGURE 22.26 Energy transformations in a circuit with a battery and a lightbulb.

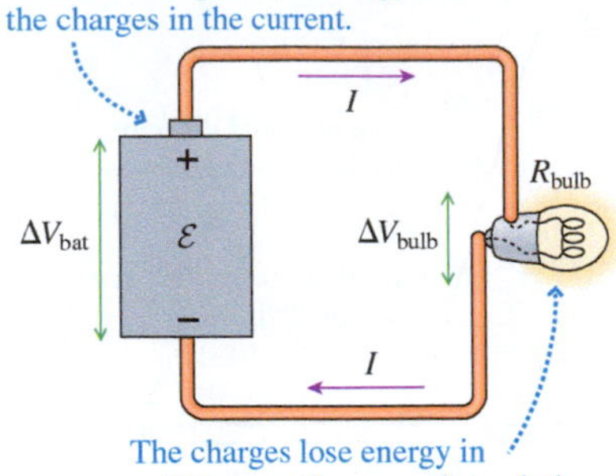


When you flip the switch on a flashlight, a battery is connected to a lightbulb, which then begins to glow. The bulb is radiating energy. Where does this energy come from?

As we've seen, a battery not only supplies a potential difference but also supplies energy, as shown in the battery and bulb circuit of FIGURE 22.26. The charge escalator is an energy-transfer process, transferring the chemical energy $E_{chem}$ stored in the battery to the electric potential energy $U$ of the charges. That energy is then dissipated as the charges move through the lightbulb, keeping the filament warm and glowing.

Recall also that charge $q$ gains potential energy $\Delta U = q\,\Delta V$ as it moves through a potential difference $\Delta V$. Because the potential difference of a battery is $\Delta V_{bat} = \mathcal{E}$, the battery supplies energy $\Delta U = q\mathcal{E}$ to charge $q$ as it lifts the charge up the charge escalator from the negative to the positive terminal.

It is often useful to know the *rate* at which the battery supplies energy. You learned in ◀◀ SECTION 10.10 that the rate at which energy is transformed is *power,* measured in joules per second or *watts.* Suppose an amount of charge $\Delta q$ moves through the battery in a time $\Delta t$. The charge $\Delta q$ will increase its potential energy by $\Delta U = (\Delta q)\mathcal{E}$. The *rate* at which energy is transferred from the battery to the moving charges is

$$P_{bat} = \text{rate of energy transfer} = \frac{\Delta U}{\Delta t} = \frac{\Delta q}{\Delta t}\mathcal{E} \tag{22.9}$$

But $\Delta q/\Delta t$, the rate at which charge moves through the battery, is the current $I$. Hence the power supplied by a battery or any source of emf is

$$P_{emf} = I\mathcal{E} \tag{22.10}$$

Power delivered by a source of emf

$I\mathcal{E}$ has units of J/s, or W.

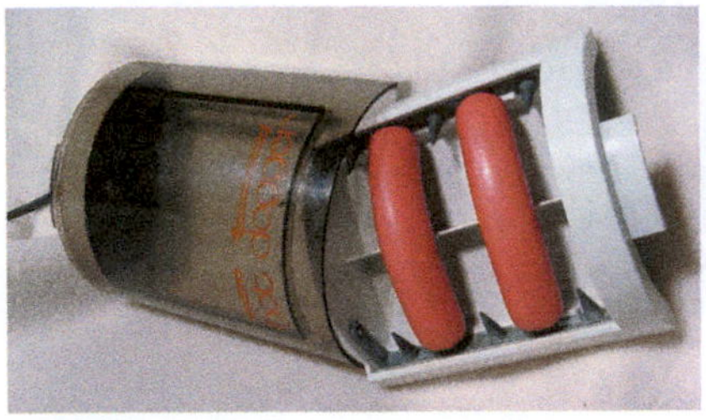

**Hot dog resistors** Before microwave ovens were common, there were devices that used a decidedly lower-tech approach to cook hot dogs. Prongs connected the hot dog to the 120 V of household electricity, making it the resistor in a circuit. The current through the hot dog dissipated energy as thermal energy, cooking the hot dog in about 2 minutes.

**EXAMPLE 22.12 Power delivered by a car battery**

A car battery has $\mathcal{E} = 12$ V. When the car's starter motor is running, the battery current is 320 A. What power does the battery supply?

***STRATEGIZE*** The power is the product of the emf of the battery and the current.

***SOLVE*** The battery supplied by the battery is

$$P_{bat} = I\mathcal{E} = (320\text{ A})(12\text{ V}) = 3.8\text{ kW}$$

***ASSESS*** This is a lot of power (about 5 hp), but this amount makes sense because turning over a car's engine is hard work. Car batteries are designed to reliably provide such intense bursts of power for starting the engine.

FIGURE 22.27 The power from the battery is dissipated in the resistor.

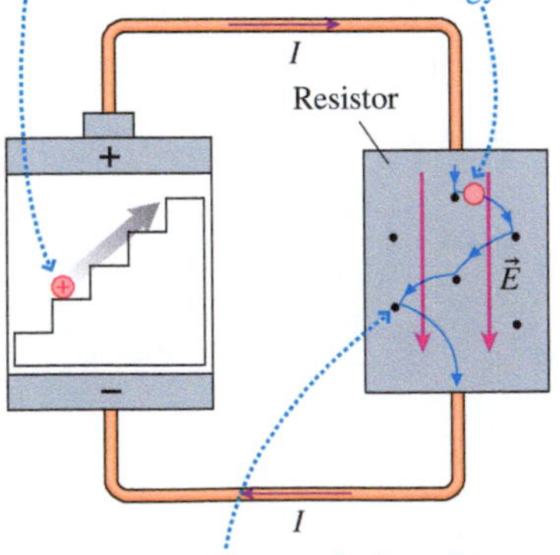


Suppose we consider a circuit consisting of a battery and a single resistor. $P_{bat} = I\mathcal{E}$ is the energy transferred per second from the battery's store of chemicals to the moving charges that make up the current. FIGURE 22.27 shows the entire sequence of energy transformations, which looks like

$$E_{chem} \rightarrow U \rightarrow K \rightarrow E_{th}$$

The net result is that **the battery's chemical energy is transferred to the thermal energy of the resistor,** raising its temperature.

In the resistor, the amount of charge $\Delta q$ loses potential energy $\Delta U = (\Delta q)(\Delta V_R)$ as this energy is transformed into kinetic energy and then into the resistor's thermal energy. Thus the rate at which energy is transferred from the current to the resistor is

$$P_R = \frac{\Delta U}{\Delta t} = \frac{\Delta q}{\Delta t}\Delta V_R = I\,\Delta V_R \tag{22.11}$$

We say that this power—so many joules per second—is *dissipated* by the resistor as charge flows through it.

Our analysis of the single-resistor circuit in Example 22.10 found that $\Delta V_R = \mathcal{E}$. That is, the potential difference across the resistor is exactly the emf supplied by the battery. Because the current is the same in the battery and the resistor, a comparison of Equations 22.10 and 22.11 shows that

$$P_R = P_{bat} \qquad (22.12)$$

The power dissipated in the resistor is exactly equal to the power supplied by the battery. The *rate* at which the battery supplies energy is exactly equal to the *rate* at which the resistor dissipates energy. This is, of course, exactly what we would have expected from energy conservation.

Most household appliances, such as a 100 W lightbulb or a 1500 W hair dryer, have a power rating. These appliances are intended for use at a standard household voltage of 120 V, and their rating is the power they will dissipate if operated with a potential difference of 120 V. Their power consumption will differ from the rating if they are operated at any other potential difference—for instance, if you use a lightbulb with a dimmer switch.

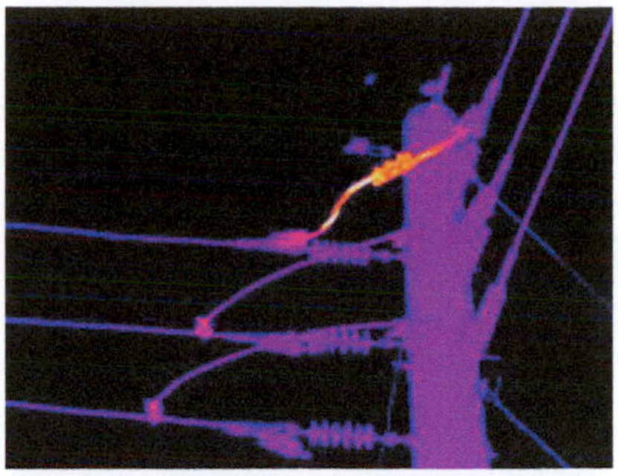

**Hot wire** This thermal camera image of power lines shows that the lines are warm—as we'd expect, given the large currents that they carry. Spots where corrosion has thinned the wires get especially warm, making such images helpful for monitoring the condition of power lines.

**EXAMPLE 22.13 Finding the current in a lightbulb**

How much current is "drawn" by a 75 W lightbulb connected to a 120 V outlet?

***STRATEGIZE*** Equation 22.10 relates the current to the power dissipated in the bulb and the outlet's emf.

***PREPARE*** We can model the lightbulb as a resistor.

***SOLVE*** Because the lightbulb is operating as intended, it will dissipate 75 W of power. We can rearrange Equation 22.11 to find

$$I = \frac{P_R}{\Delta V_R} = \frac{75\text{ W}}{120\text{ V}} = 0.63\ A$$

***ASSESS*** We've said that we expect currents on the order of 1 A for lightbulbs and other household items, so our result seems reasonable.

A resistor obeys Ohm's law: $I = \Delta V_R / R$. This gives us two alternative ways of writing the power dissipated by a resistor. We can either substitute $IR$ for $\Delta V_R$ or substitute $\Delta V_R / R$ for $I$. Thus

$$P_R = I\,\Delta V_R = I^2 R = \frac{(\Delta V_R)^2}{R} \qquad (22.13)$$

Power dissipated by resistance $R$ with current $I$ and potential difference $\Delta V_R$

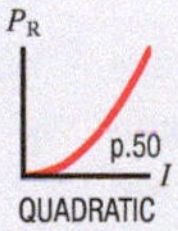


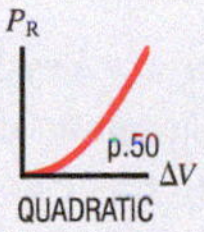


**Video** Practical Electricity Applications

It is worth writing the different forms of this equation to illustrate that the power varies as the square of both the current and the potential difference.

**EXAMPLE 22.14 Finding the power of a dim bulb**

How much power is dissipated by a 60 W (120 V) lightbulb when operated, using a dimmer switch, at 100 V?

***STRATEGIZE*** We will find the resistance of the bulb from the 60 W it dissipates at 120 V. We will then use this resistance to find the power at 100 V.

***PREPARE*** The 60 W rating is for operation at 120 V. We assume that the resistance doesn't change if the bulb is run at a lower power—not quite right, but a reasonable approximation for this case in which the voltage is only slightly different from the rated value.

***SOLVE*** The lightbulb dissipates 60 W at $\Delta V_R = 120$ V. Thus the filament's resistance is

$$R = \frac{(\Delta V_R)^2}{P_R} = \frac{(120\text{ V})^2}{60\text{ W}} = 240\ \Omega$$

The power dissipation when operated at $\Delta V_R = 100$ V is

$$P_R = \frac{(\Delta V_R)^2}{R} = \frac{(100\text{ V})^2}{240\ \Omega} = 42\text{ W}$$

***ASSESS*** Reducing the voltage by 17% leads to a 30% reduction of the power. This makes sense; the power is proportional to the square of the voltage, so we expect a proportionally larger change in power.

**EXAMPLE 22.15** **Determining the voltage of a stereo**

Most stereo speakers are designed to have a resistance of 8.0 Ω. If an 8.0 Ω speaker is connected to a stereo amplifier with a rating of 100 W, what is the maximum possible potential difference the amplifier can apply to the speakers?

**STRATEGIZE** The rating of an amplifier is the *maximum* power it can deliver. Most of the time it delivers far less, but the maximum might be needed for brief, intense sounds. The maximum potential difference will occur when the amplifier is providing the maximum power, so we will make our computation with this figure.

**PREPARE** We can model the speaker as a resistor.

**SOLVE** The maximum potential difference occurs when the power is at a maximum. At the maximum power of 100 W,

$$P_R = 100\text{ W} = \frac{(\Delta V_R)^2}{R} = \frac{(\Delta V_R)^2}{8.0\ \Omega}$$

$$\Delta V_R = \sqrt{(8.0\ \Omega)(100\text{ W})} = 28\text{ V}$$

This is the maximum potential difference the amplifier might provide.

**ASSESS** As a check on our result, we note that the resistance of the speaker is less than that of a lightbulb, so a smaller potential difference can provide 100 W of power.

**STOP TO THINK 22.6** Rank in order, from largest to smallest, the powers $P_A$ to $P_D$ dissipated in resistors A to D.

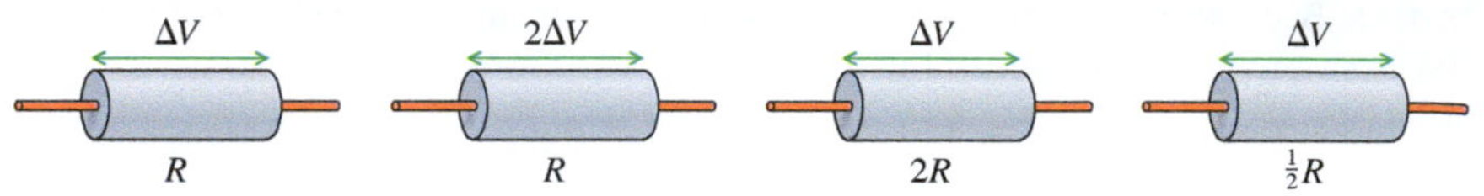


**INTEGRATED EXAMPLE 22.16** **Electrical measurements of body composition** BIO

The woman in the photo at the start of the chapter is gripping a device that passes a small current through her body. How does this permit a determination of body fat?

The exact details of how the device works are beyond the scope of this chapter, but the basic principle is quite straightforward: The device applies a small potential difference and measures the resulting current. Comparing multiple measurements allows the device to determine the resistance of one part of the body, the upper arm. The resistance of the upper arm depends sensitively on the percentage of body fat in the upper arm, and the percentage of body fat in the upper arm is a good predictor of the percentage of fat in the body overall. Let's make a simple model of the upper arm to show how the resistance of the upper arm varies with the percentage of body fat.

The model of a person's upper arm in FIGURE 22.28 ignores the nonconductive elements (such as the skin and the mineralized portion of the bone) and groups the conductive elements into two distinct sections—muscle and fat—that form two parallel segments. The resistivity of each tissue type is shown. This simple model isn't a good description of the actual structure of the arm, but it predicts the electrical character quite well.

FIGURE 22.28 A simple model of the resistance of the upper arm.

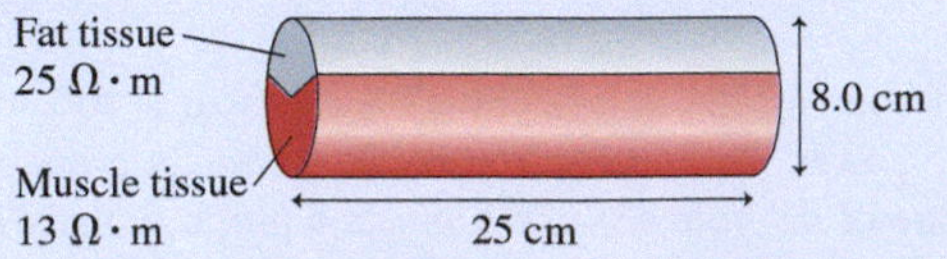


a. An experimental subject's upper arm, with the dimensions shown in the figure, is 40% fat and 60% muscle. A potential difference of 0.60 V is applied between the elbow and the shoulder. What current is measured?

b. A 0.60 V potential difference applied to the upper arm of a second subject with an arm of similar dimensions gives a current of 0.87 mA. What are the percentages of muscle and fat in this person's upper arm?

**STRATEGIZE** FIGURE 22.29 shows how we can model the upper arm as two resistors that are connected together at the ends. We will use the ideal-wire model in which there's no "loss" of potential along the wires. Consequently, the potential difference across each of the two segments is the full 0.60 V of the battery. The current "splits" at the junction between the two resistors, and conservation of current tells us that

$$I_{total} = I_{muscle} + I_{fat}$$

The resistance of each segment depends on its resistivity (given in Figure 22.28) and on its length and cross-section area.

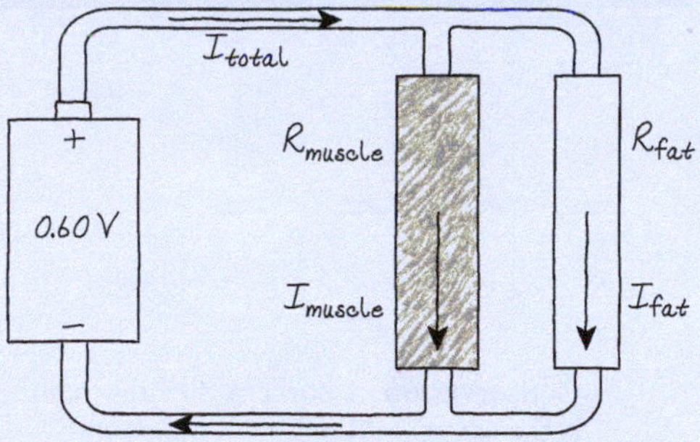


FIGURE 22.29 Current through the tissues of the upper arm.

**PREPARE** The cross-section area of the whole arm is $A = \pi r^2 = \pi(0.040\text{ m})^2 = 0.00503\text{ m}^2$; the area of each segment is this number multiplied by the appropriate fraction.

**SOLVE** a. An object's resistance is related to its geometry and the resistivity of the material by $R = \rho L/A$. Thus the resistances of the muscle (60% of the area) and fat (40% of the area) segments are

$$R_{\text{muscle}} = \frac{\rho_{\text{muscle}} L}{A_{\text{muscle}}} = \frac{(13\ \Omega\cdot\text{m})(0.25\text{ m})}{(0.60)(0.00503\text{ m}^2)} = 1080\ \Omega$$

$$R_{\text{fat}} = \frac{\rho_{\text{fat}} L}{A_{\text{fat}}} = \frac{(25\ \Omega\cdot\text{m})(0.25\text{ m})}{(0.40)(0.00503\text{ m}^2)} = 3110\ \Omega$$

The potential difference across each segment is 0.60 V. We can then use Ohm's law, $I = \Delta V/R$, to find that the current in each segment is

$$I_{\text{muscle}} = \frac{0.60\text{ V}}{1080\ \Omega} = 0.556\text{ mA}$$

$$I_{\text{fat}} = \frac{0.60\text{ V}}{3110\ \Omega} = 0.193\text{ mA}$$

The conservation of current equation then gives the total current as the sum of these two values:

$$I_{\text{total}} = 0.556\text{ mA} + 0.193\text{ mA} = 0.75\text{ mA}$$

b. If we know the current, we can determine the amount of muscle and fat. Let the fraction of muscle tissue be $x$; the fraction of fat tissue is then $(1 - x)$. We repeat the steps of the earlier calculation with these expressions in place:

$$R_{\text{muscle}} = \frac{\rho_{\text{muscle}} L}{A} = \frac{(13\ \Omega\cdot\text{m})(0.25\text{ m})}{(x)(0.00503\text{ m}^2)} = \frac{646\ \Omega}{x}$$

$$R_{\text{fat}} = \frac{\rho_{\text{fat}} L}{A} = \frac{(25\ \Omega\cdot\text{m})(0.25\text{ m})}{(1-x)(0.00503\text{ m}^2)} = \frac{1240\ \Omega}{1-x}$$

In terms of these values, the current in each segment is

$$I_{\text{muscle}} = \frac{0.60\text{ V}}{646\ \Omega}x = 0.929(x)\text{ mA}$$

$$I_{\text{fat}} = \frac{0.60\text{ V}}{1240\ \Omega}(1-x) = 0.484(1-x)\text{ mA}$$

The sum of these currents is the total current:

$$I_{\text{total}} = 0.87\text{ mA} = 0.929(x)\text{ mA} + 0.484(1-x)\text{ mA}$$

Rearranging the terms on the right side gives

$$0.87\text{ mA} = (0.484 + 0.445x)\text{ mA}$$

Finally, we can solve for $x$:

$$x = 0.87$$

This person therefore has 87% muscle and 13% fatty tissue in the upper arm.

**ASSESS** A good check on our work is that the total current we find in part a is small—important for safety—and reasonably close to the value given in part b; the arms are the same size, and the variation in body fat between individuals isn't all that large, so we expect the numbers to be similar. The current given in part b is greater than we found in part a. Muscle has a lower resistance than fat, so the subject of part b must have a higher percentage of muscle—exactly what we found.

# SUMMARY

**GOAL** To learn how and why charge moves through a conductor as a current.

## GENERAL PRINCIPLES

### Batteries, Resistors, and Current

The **current** is defined to be the motion of positive charges:

$$I = \frac{\Delta q}{\Delta t}$$

The battery transforms its chemical energy into electric potential energy of the charges passing through it, raising their electric potential. The potential difference of the battery is its **emf** $\mathcal{E}$.

A **battery** is a source of potential difference. Chemical processes in the battery separate charges. We use a **charge escalator** model to show the lifting of charges to higher potential.

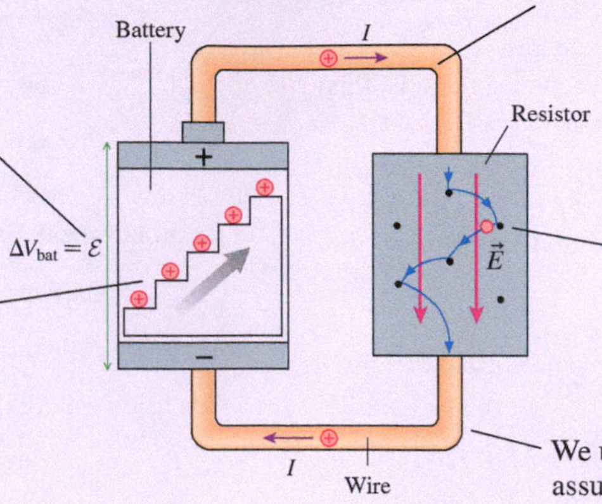


**Conservation of current** dictates that the current is the same at all points in the circuit.

The battery creates an electric field in the circuit that causes charges to move. Positive charges move in the direction of the electric field, which is the direction of decreasing potential.

Random collisions of charges with atoms impede the flow of the charges and are the source of **resistance.** The collisions increase the thermal energy of the **resistor.**

We use the **ideal-wire model** in which we assume that there is no resistance in the wires.

## IMPORTANT CONCEPTS

### Resistance, resistivity, and Ohm's law

The **resistivity** $\rho$ is a property of a particular material.

- Good conductors, such as copper, have low resistivity.
- Poor conductors, such as pure water, have high resistivity.

The **resistance** is a property of a particular wire or conductor. The resistance of a wire depends on its resistivity and dimensions:

$$R = \frac{\rho L}{A}$$

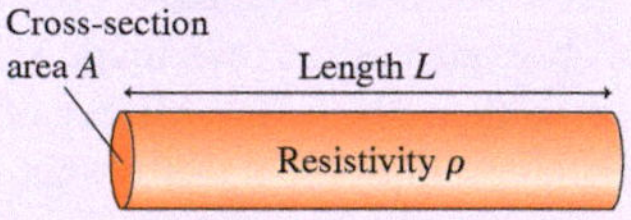


**Ohm's law** describes the relationship between the potential difference across and the current in a resistor:

$$I = \frac{\Delta V}{R}$$

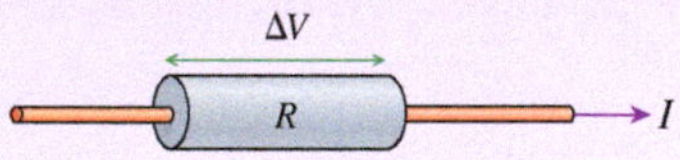


### Energy and power

The energy used by a circuit is supplied by the emf of the battery through a series of energy transformations:

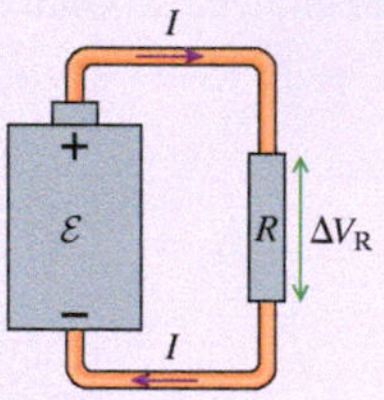


$$E_{chem} \rightarrow U_{elec} \rightarrow K \rightarrow E_{th}$$

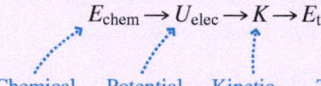

Chemical energy in the battery | Potential energy of separated charges | Kinetic energy of moving charges | Thermal energy of atoms in the resistor

The battery *supplies* power at the rate

$$P_{emf} = I\mathcal{E}$$

The resistor *dissipates* power at the rate

$$P_R = I\,\Delta V_R = I^2R = \frac{(\Delta V_R)^2}{R}$$

## APPLICATIONS

### Conducting materials

When a potential difference is applied to a wire, if the relationship between potential difference and current is linear, the material is **ohmic.**

The resistance is

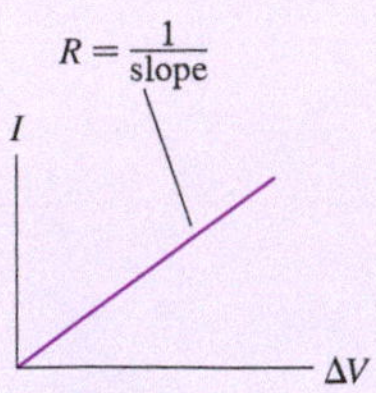


**Resistors** are made of ohmic materials and have a well-defined value of resistance:

$$R = \frac{\Delta V}{I}$$

### Batteries in series

Batteries connected one after the other are in *series.* The total potential difference is the sum of the potential differences of each battery.

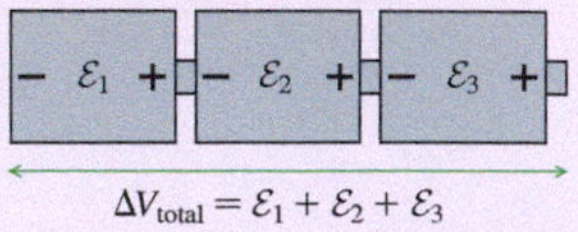

## Learning Objectives After studying this chapter, you should be able to:

- Use the current model to explain the motion of charged particles through a conductor. *Conceptual Questions 22.3, 22.4; Problems 22.8, 22.10*
- Apply the law of conservation of current. *Conceptual Questions 22.5, 22.13; Problems 22.3, 22.4, 22.6, 22.9, 22.12*
- Understand and use the emf of a battery. *Conceptual Questions 22.8, 22.14; Problems 22.13, 22.14, 22.15, 22.16, 22.17*
- Find the resistance of a wire from the resistivity of the material. *Conceptual Questions 22.10, 22.11; Problems 22.19, 22.20, 22.23, 22.25, 22.28*
- Use Ohm's law to connect potential, current, and resistance. *Conceptual Questions 22.15, 22.16; Problems 22.18, 22.22, 22.30, 22.31, 22.32*
- Calculate the power delivered by an emf and dissipated by a resistor. *Conceptual Questions 22.20, 22.21; Problems 22.36, 22.37, 22.38, 22.40, 22.41*

## STOP TO THINK ANSWERS

**Chapter Preview Stop to Think: A.** The electric field points from high to low potential—here, from right to left. The electron, being negatively charged, feels a force opposite the field, or to the right. It will start moving in that direction.

**Stop to Think 22.1: A.** From Kirchhoff's junction law the current through bulb 1 is the sum of the currents through bulbs 2 and 3. Bulb 1 carries a larger current than bulb 2, so it will be brighter.

**Stop to Think 22.2: C.** Charge flows out of one terminal of the battery but back into the other; the amount of charge in the battery does not change. The emf is determined by the chemical reactions in the battery and is constant. But the chemical energy in the battery steadily decreases as the battery converts it to the potential energy of charges.

**Stop to Think 22.3: C.** Because they are connected end-to-end, the potential difference across the wire is equal to that across the battery. But the battery supplies a *constant* potential difference, its emf, independent of the current it provides.

**Stop to Think 22.4: C.** Stretching the wire decreases the area and increases the length. Both of these changes increase the resistance of the wire. When the wire is reconnected to the battery, the resistance is greater but the potential difference is the same as in the original case, so the current will be smaller.

**Stop to Think 22.5: $V_B = V_C > V_A > V_D = V_E$.** There's no potential difference along ideal wires, so $V_B = V_C$ and $V_D = V_E$. Potential increases in going from the − to the + terminal of a battery, so $V_A > V_E$ and $V_B > V_A$. These imply $V_C > V_D$, which was expected because potential decreases as current passes through a resistor.

**Stop to Think 22.6: $P_B > P_D > P_A > P_C$.** The power dissipated by a resistor is $P_R = (\Delta V_R)^2/R$. Increasing $R$ decreases $P_R$; increasing $\Delta V_R$ increases $P_R$. But changing the potential has a larger effect because $P_R$ depends on the square of $\Delta V_R$.

Video Tutor Solution Chapter 22

# QUESTIONS

## Conceptual Questions

1. What is the current in the wire in Figure Q22.1?

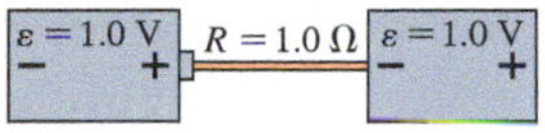


FIGURE Q22.1

2. Will the bulb in Figure Q22.2 light? Explain.

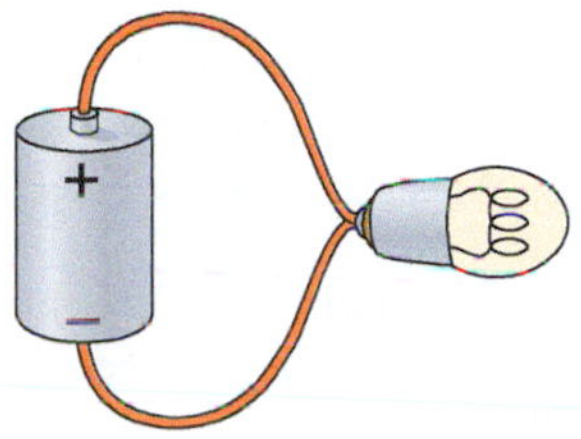


FIGURE Q22.2

3. A current flows in a circuit that consists of a lightbulb and a battery, as in Figure 22.21b of this chapter. How do the electrons in the bulb's filament, far from the battery and its emf, "know" that they should flow?
4. A lightbulb is connected to a battery by two copper wires of equal lengths but different thicknesses. A thick wire connects one side of the lightbulb to the positive terminal of the battery and a thin wire connects the other side of the bulb to the negative terminal.
   a. Which wire carries a greater current? Or is the current the same in both? Explain.
   b. If the two wires are switched, will the bulb get brighter, dimmer, or stay the same? Explain.

Problem difficulty is labeled as | (straightforward) to ||||| (challenging). Problems labeled INT integrate significant material from earlier chapters; Problems labeled BIO are of biological or medical interest.

The eText icon indicates when there is a video tutor solution available for the chapter or for a specific problem. To launch these videos, log into your eText through Mastering™ Physics or log into the Study Area.

5. All wires in Figure Q22.5 are made of the same material and have the same diameter. Rank in order, from largest to smallest, the currents $I_1$ to $I_4$. Explain.

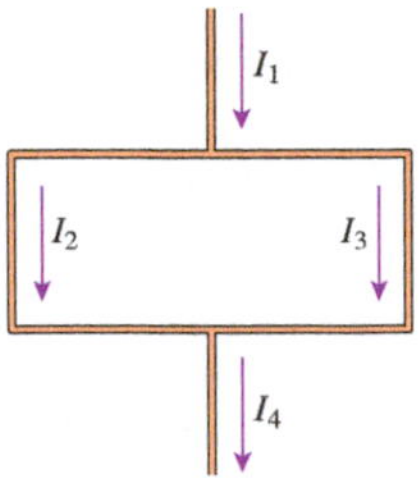

FIGURE Q22.5

6. A wire carries a 4 A current. What is the current in a second wire that delivers twice as much charge in half the time?
7. An electrolyte contains equal numbers of positive and negative ions in solution. When an electric field is applied, the positive ions move in the direction of the field and the negative ions move opposite the field. If the negative ions move faster than the positive ions, is the net current in the direction of the field, is it opposite the field, or do the two ion currents cancel to give no net current? Explain.
8. Cells in the nervous system have a potential difference of 70 mV across the cell membrane separating the interior of the cell from the extracellular fluid. This potential difference is maintained by ion pumps that move charged ions across the membrane. Is this an emf? BIO
9. a. Which direction—clockwise or counterclockwise—does an electron travel through the wire in Figure Q22.9? Explain.
   b. Does an electron's electric potential energy increase, decrease, or stay the same as it moves through the wire? Explain.

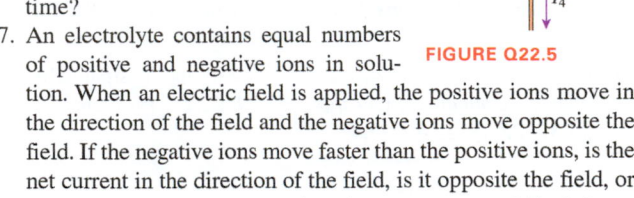

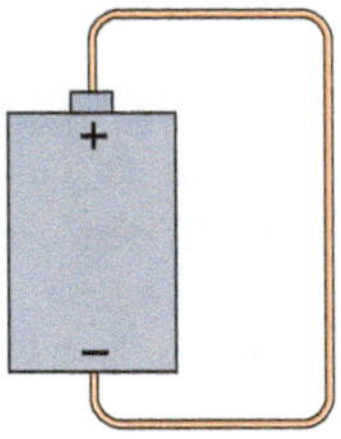

FIGURE Q22.9

   c. If you answered "decrease" in part b, where does the energy go? If you answered "increase" in part b, where does the energy come from?
   d. Which way—up or down—does an electron move through the *battery?* Explain.
   e. Does an electron's electric potential energy increase, decrease, or stay the same as it moves through the battery? Explain.
   f. If you answered "decrease" in part e, where does the energy go? If you answered "increase" in part e, where does the energy come from?
10. The wires in Figure Q22.10 are all made of the same material; the length and radius of each wire are noted. Rank in order, from largest to smallest, the resistances $R_1$ to $R_5$ of these wires. Explain.

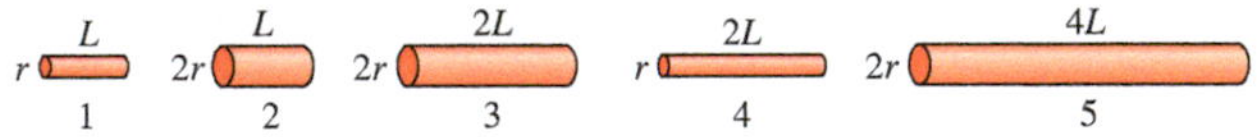

FIGURE Q22.10

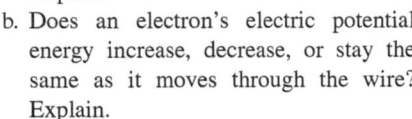

11. The wires in Figure Q22.10 all have the same resistance; the length and radius of each wire are noted. Rank in order, from largest to smallest, the resistivities $\rho_1$ to $\rho_5$ of these wires.
12. The two circuits in Figure Q22.12 use identical batteries and wires made of the same material and of equal diameters. Rank in order, from largest to smallest, the currents $I_1$,$I_2$,$I_3$, and $I_4$ at points 1 to 4.

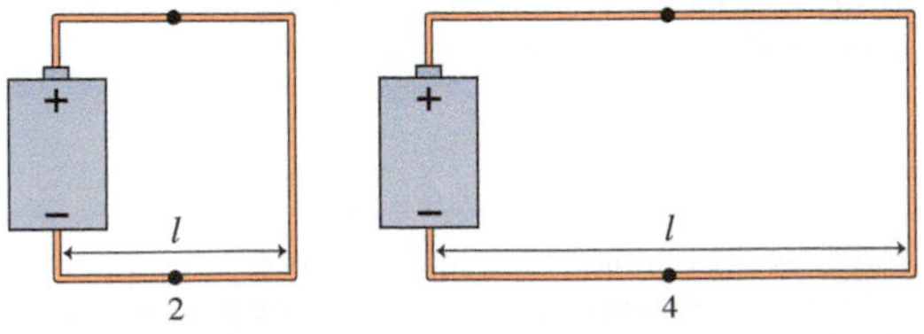

FIGURE Q22.12

13. The two circuits in Figure Q22.13 use identical batteries and wires made of the same material and of equal diameters. Rank in order, from largest to smallest, the currents $I_1$ to $I_7$ at points 1 to 7. Explain.

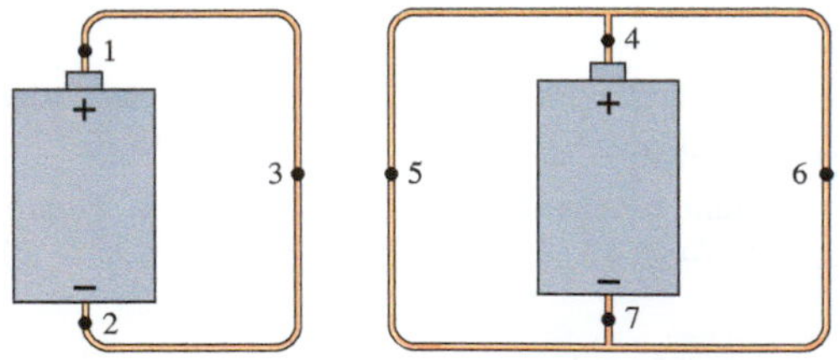

FIGURE Q22.13

14. Which, if any, of these statements are true? (More than one may be true.) Explain your choice or choices.
   a. A battery supplies energy to a circuit.
   b. A battery is a source of potential difference. The potential difference between the terminals of the battery is always the same.
   c. A battery is a source of current. The current leaving the battery is always the same.
15. Rank in order, from largest to smallest, the currents $I_1$ to $I_4$ through the four resistors in Figure Q22.15. Explain.

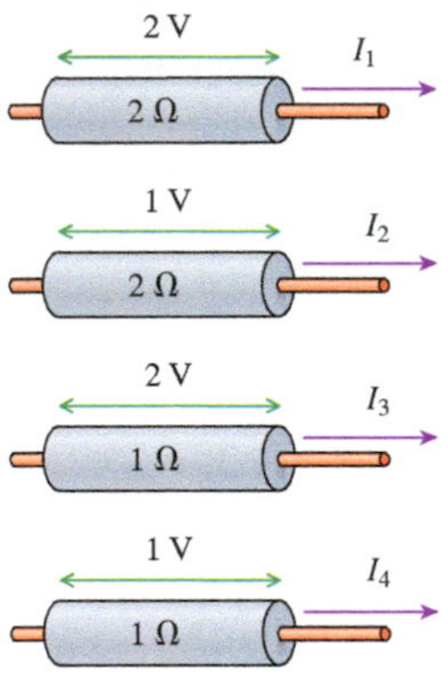

FIGURE Q22.15

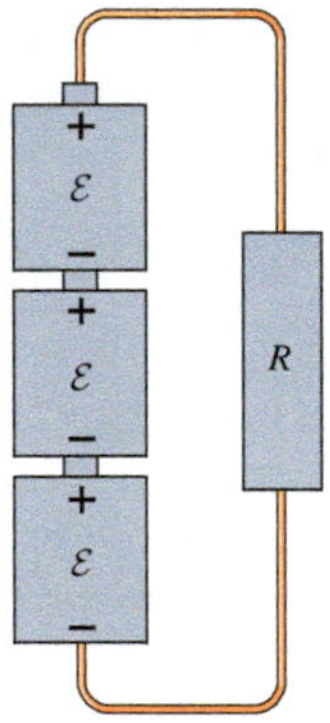

FIGURE Q22.16

16. The circuit in Figure Q22.16 has three batteries of emf $\mathcal{E}$ in series. Assuming the wires are ideal, sketch a graph of the potential as a function of distance traveled around the circuit, starting from $V = 0$ V at the negative terminal of the bottom battery. Note all important points on your graph.
17. A battery and a resistor are wired into a circuit. The resistor dissipates 0.50 W. Now two batteries, each identical to the original one, are connected in series with the resistor. What power does it dissipate?
18. A battery and a resistor are wired into a circuit. The resistor dissipates 4.0 W. Now the resistor is replaced with one that has twice the original resistance. What power does the new resistor dissipate?
19. Over time, atoms "boil off" the hot filament in an incandescent bulb and the filament becomes thinner. How does this affect the brightness of the lightbulb?
20. A 100 W lightbulb and a 60 W lightbulb each operate at a voltage of 120 V. Which bulb carries more current?
21. A 100 W lightbulb is brighter than a 60 W lightbulb when both operate at the same voltage of 120 V. If, instead, they were both operated at the same current of 0.5 A, which would be brighter? Explain.

## Multiple-Choice Questions

22. | Lightbulbs are typically rated by their power dissipation when operated at a given voltage. Which of the following lightbulbs has the largest current through it when operated at the voltage for which it's rated?
A. 0.8 W, 1.5 V B. 6 W, 3 V
C. 4 W, 4.5 V D. 8 W, 6 V

23. || Lightbulbs are typically rated by their power dissipation when operated at a given voltage. Which of the following lightbulbs has the largest resistance when operated at the voltage for which it's rated?
A. 0.8 W, 1.5 V B. 6 W, 3 V
C. 4 W, 4.5 V D. 8 W, 6 V

24. | A copper wire is stretched so that its length increases and its diameter decreases. As a result,
A. The wire's resistance decreases, but its resistivity stays the same.
B. The wire's resistivity decreases, but its resistance stays the same.
C. The wire's resistance increases, but its resistivity stays the same.
D. The wire's resistivity increases, but its resistance stays the same.

25. | The potential difference across a length of wire is increased. Which of the following does *not* increase as well?
A. The electric field in the wire
B. The power dissipated in the wire
C. The resistance of the wire
D. The current in the wire

26. | A resistor dissipates 0.25 W when current of 20 mA passes through it. How much current would be needed for the resistor to dissipate 0.50 W?
A. 28 mA B. 40 mA C. 56 mA D. 80 mA

27. | A resistor connected to a 3.0 V battery dissipates 1.0 W. If the battery is replaced by a 6.0 V battery, the power dissipated by the resistor will be
A. 1.0 W B. 2.0 W
C. 3.0 W D. 4.0 W

28. | If a 1.5 V battery stores 5.0 kJ of energy (a reasonable value for an inexpensive C cell), for how many minutes could it sustain a current of 1.2 A?
A. 2.7
B. 6.9
C. 9.0
D. 46

29. | Figure Q22.29 shows a side view of a wire of varying circular cross section. Rank in order the currents flowing in the three sections.

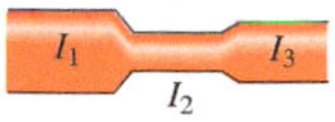

FIGURE Q22.29

A. $I_1 > I_2 > I_3$
B. $I_2 > I_3 > I_1$
C. $I_1 = I_2 = I_3$
D. $I_1 > I_3 > I_2$

30. ||| A person gains weight by adding fat—and therefore adding
BIO girth—to his body and his limbs, with the amount of muscle remaining constant. How will this affect the electrical resistance of his limbs?
A. The resistance will increase.
B. The resistance will stay the same.
C. The resistance will decrease.

31. | The resistance of an 8.3-m-long, 0.50-mm-diameter wire is 4.1 Ω. From what material is this wire likely made?
A. Copper B. Aluminum C. Iron D. Nichrome

# PROBLEMS

### Section 22.1 A Model of Current

### Section 22.2 Defining and Describing Current

1. ||| The current in an electric hair dryer is 10 A. How much charge and how many electrons flow through the hair dryer in 5.0 min?
2. | Electroplating uses electrolysis to coat one metal with another. In a copper-plating bath, copper ions with a charge of $+2e$ move through the electrolyte from the copper anode to the cathode—the metal object to be plated. If the current through the system is 1.2 A, how many copper ions reach the cathode each second?
3. | Three wires meet at a junction. Wire 1 has a current of 0.40 A into the junction. The current of wire 2 is 0.65 A out of the junction. (a) How many electrons per second move past a point in wire 3? (b) In which direction do the electrons move—into or out of the junction?
4. | When a nerve cell depolarizes, charge is transferred across
BIO the cell membrane, changing the potential difference. For a typical nerve cell, 9.0 pC of charge flows in a time of 0.50 ms. What is the average current?
5. | A wire carries a 15 $\mu$A current. How many electrons pass a given point on the wire in 1.0 s?
6. || In a typical lightning strike, 2.5 C flows from cloud to ground in 0.20 ms. What is the current during the strike?
7. | Electroconvulsive therapy is a last-line treatment for certain
BIO mental disorders. In this treatment, an electric current is passed directly through the brain, inducing seizures. The total charge that passes through the brain is called the *dose*. In a typical session, a dose of 0.093 C is applied, using a pulse of current that lasts 1.0 ms. What is the current during this pulse?
8. || In an ionic solution, $5.0 \times 10^{15}$ positive ions with charge $+2e$ pass to the right each second while $6.0 \times 10^{15}$ negative ions with charge $-e$ pass to the left. What are the magnitude and direction of current in the solution?
9. || In proton beam therapy, a beam of high-energy protons is
BIO used to deliver radiation to a tumor, killing its cancerous cells. In one session, the radiologist calls for a dose of $4.9 \times 10^8$ protons to be delivered at a beam current of 120 nA. How long should the beam be turned on to deliver this dose?
10. ||| When a current passes through the body, it is mainly carried
BIO by positively charged sodium ions ($Na^+$) with charge $+e$ and negatively charged chloride ions ($Cl^-$) with charge $-e$. In a given electric field, these two kinds of ions do not move at the same speed; the chloride ions move 50% faster than the sodium ions. If a current of 150 $\mu$A is passed through the leg of a patient, how many chloride ions pass a cross section of the leg per second?

Watch Video Solution Problems 22.15 and 22.21

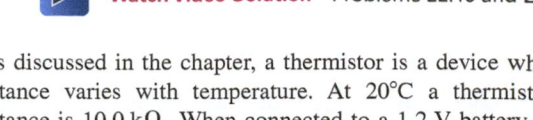

11. | | What are the values of currents $I_B$ and $I_C$ in Figure P22.11? The directions of the currents are as noted.

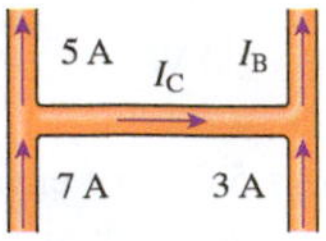


FIGURE P22.11

12. | The currents through several segments of a wire object are shown in Figure P22.12. What are the magnitudes and directions of the currents $I_B$ and $I_C$ in segments B and C?

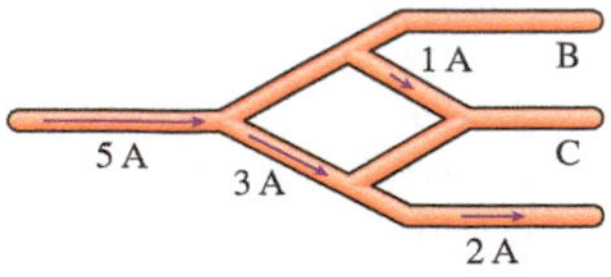


FIGURE P22.12

### Section 22.3 Batteries and emf

13. | How much electric potential energy does 1.0 $\mu$C of charge gain as it moves from the negative terminal to the positive terminal of a 1.5 V battery?

14. | What is the emf of a battery that increases the electric potential energy of 0.050 C of charge by 0.60 J as it moves it from the negative to the positive terminal?

15. | | A 9.0 V battery supplies a 2.5 mA current to a circuit for 5.0 h.
   a. How much charge has been transferred from the negative to the positive terminal?
   b. How much electric potential energy has been gained by the charges that passed through the battery?

16. | A laptop battery has an emf of 10.8 V. The laptop uses 0.70 A while running.
   a. How much charge moves through the battery each second?
   b. By how much does the electric potential energy of this charge increase as it moves through the battery?

17. | BIO An electric catfish can generate a significant potential difference using stacks of special cells called *electrocytes.* Each electrocyte develops a potential difference of 110 mV. How many cells must be connected in series to give the 350 V a large catfish can produce?

### Section 22.4 Connecting Potential and Current

18. | A wire with resistance $R$ is connected to the terminals of a 6.0 V battery. What is the potential difference $\Delta V_{ends}$ between the ends of the wire and the current $I$ through it if the wire has the following resistances? (a) 1.0 $\Omega$ (b) 2.0 $\Omega$ (c) 3.0 $\Omega$.

19. | Wires 1 and 2 are made of the same metal. Wire 2 has twice the length and twice the diameter of wire 1. What are the ratios (a) $\rho_2/\rho_1$ of the resistivities and (b) $R_2/R_1$ of the resistances of the two wires?

20. | BIO Nerve impulses are carried along *axons,* the elongated fibers that transmit neural signals. We can model an axon as a tube with an inner diameter of 10 $\mu$m. The tube wall is insulating, but the fluid inside it has a resistivity of 0.50 $\Omega \cdot$ m. What is the resistance of a 1-mm-long axon?

21. | | | BIO Resistivity measurements on the leaves of corn plants are a good way to assess stress and overall health. The leaf of a corn plant has a resistance of 2.0 M$\Omega$ measured between two electrodes placed 20 cm apart along the leaf. The leaf has a width of 2.5 cm and is 0.20 mm thick. What is the resistivity of the leaf tissue? Is this greater than or less than the resistivity of muscle tissue in the human body?

22. | | | As discussed in the chapter, a thermistor is a device whose resistance varies with temperature. At 20°C a thermistor's resistance is 10.0 k$\Omega$. When connected to a 1.2 V battery, the current through the thermistor increases by 30 $\mu$A as the temperature increases to 30°C. What is the thermistor's resistance at this higher temperature?

23. | | | A motorcyclist is making an electric vest that, when connected to the motorcycle's 12 V battery, will warm her on cold rides. She is using 0.25-mm-diameter copper wire, and she wants a current of 4.0 A in the wire. What length wire must she use?

24. | | BIO The femoral artery is the large artery that carries blood to the leg. A person's femoral artery has an inner diameter of 1.0 cm. What is the resistance of a 20-cm-long column of blood in this artery?

25. | | A 3.0 V potential difference is applied between the ends of a 0.80-mm-diameter, 50-cm-long nichrome wire. What is the current in the wire?

26. | | BIO Impedance plethysmography is a technique that can be used to detect thrombosis—the presence of clots in a blood vessel—by measuring the electrical resistance of a limb, such as the calf of the leg. In a typical clinical setting, a current of 200 $\mu$A is passed through the leg from the upper thigh to the foot. The voltage is measured at two points along the calf separated by 13 cm.
   a. If the voltage measured is 17 mV, what is the resistance of the calf between the electrodes?
   b. If the average calf diameter between the electrodes is 12 cm, what is the average resistivity of this part of the leg?

27. | | | | BIO The relatively high resistivity of dry skin, about $1 \times 10^6\ \Omega \cdot$ m, can safely limit the flow of current into deeper tissues of the body. Suppose an electrical worker places his palm on an instrument whose metal case is accidentally connected to a high voltage. The skin of the palm is about 1.5 mm thick. Estimate the area of skin on the worker's palm that would contact a flat panel, then calculate the approximate resistance of the skin of the palm.

28. | | A rear window defroster consists of a long, flat wire bonded to the inside surface of the window. When current passes through the wire, it heats up and melts ice and snow on the window. For one window the wire has a total length of 12.2 m, a width of 1.8 mm, and a thickness of 0.11 mm. The wire is connected to the car's 12.0 V battery and draws 7.5 A. What is the resistivity of the wire material?

### Section 22.5 Ohm's Law and Resistor Circuits

29. | Figure P22.29 shows the current-versus-potential-difference graph for a resistor.
   a. What is the resistance of this resistor?
   b. Suppose the length of the resistor is doubled while keeping its cross section the same. (This requires doubling the amount of material the resistor is made of.) Copy the figure and add to it the current-versus-potential-difference graph for the longer resistor.

FIGURE P22.29

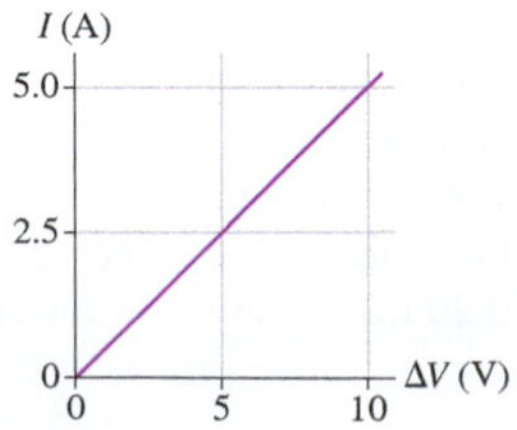

30. | Figure P22.30 is a current-versus-potential-difference graph for a cylinder. What is the cylinder's resistance?

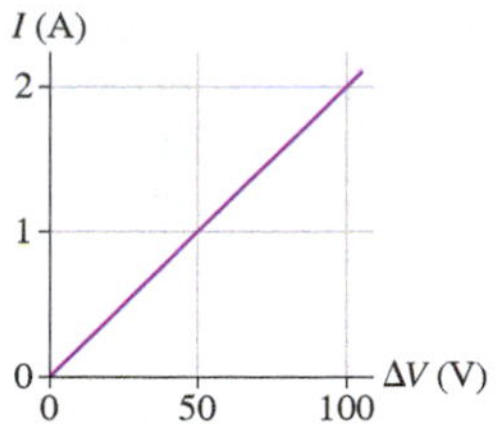


FIGURE P22.30

31. | In Example 22.6 the length of a 60 W, 240 Ω lightbulb filament was calculated to be 60 cm.
   a. If the potential difference across the filament is 120 V, what is the strength of the electric field inside the filament?
   b. Suppose the length of the bulb's filament were doubled without changing its diameter or the potential difference across it. What would the electric field strength be in this case?
   c. Remembering that the current in the filament is proportional to the electric field, what is the current in the filament following the doubling of its length?
   d. What is the resistance of the filament following the doubling of its length?
32. | The electric field inside a 30-cm-long copper wire is 0.010 V/m. What is the potential difference between the ends of the wire?
33. || A copper wire is 1.0 mm in diameter and carries a current of 20 A. What is the electric field strength inside this wire?
34. ||| Two identical lightbulbs are connected in series to a single 9.0 V battery.
   a. Sketch the circuit.
   b. Sketch a graph showing the potential as a function of distance through the circuit, starting with $V = 0$ V at the negative terminal of the battery.

### Section 22.6 Energy and Power

35. | On a sunny day, a rooftop solar panel delivers 60 W of power to the house at an emf of 17 V. How much current flows through the panel?
36. | a. What is the resistance of a 1500 W (120 V) hair dryer?
   b. What is the current in the hair dryer when it is used?
37. || A 1.5 V D-size battery stores 54 kJ of energy. What steady current could this battery provide for 20 h before being depleted?
38. || A 70 W electric blanket runs at 18 V.
   a. What is the resistance of the wire in the blanket?
   b. How much current does the wire carry?
39. | A 60-cm-long heating wire is connected to a 120 V outlet. If the wire dissipates 45 W, what are (a) the current in and (b) the resistance of the wire?
40. ||| BIO An electric eel develops a potential difference of 450 V, driving a current of 0.80 A for a 1.0 ms pulse. For this pulse, find (a) the power, (b) the total energy, and (c) the total charge that flows.
41. ||| INT When running on its 11.4 V battery, a laptop computer uses 8.3 W. The computer can run on battery power for 9.0 h before the battery is depleted.
   a. What is the current delivered by the battery to the computer?
   b. How much energy, in joules, is this battery capable of supplying?
   c. How high off the ground could a 75 kg person be raised using the energy from this battery?

## General Problems

42. | A 3.0 V battery powers a flashlight bulb that has a resistance of 6.0 Ω. How much charge moves through the battery in 10 min?
43. | A heating element in a toaster dissipates 900 W when run at 120 V. How much charge passes through the heating element in 1 minute?
44. ||| INT The 120 V electric heater in a coffee maker has a resistance of 15 Ω. How long will it take for this heater to raise 5 cups (1100 g) of water from 20°C to the ideal brewing temperature of 90°C?
45. |||| INT For a science experiment you need to electroplate a 100-nm-thick zinc coating onto both sides of a very thin, 2.0 cm × 2.0 cm copper sheet. You know that the charge carriers in the ionic solution are divalent (charge $2e$) zinc ions. The density of zinc is 7140 kg/m$^3$. If the electroplating apparatus operates at 1.0 mA, how long will it take the zinc to reach the desired thickness?
46. ||| INT The hot dog cooker described in the chapter heats hot dogs by connecting them to 120 V household electricity. A typical hot dog has a mass of 60 g and a resistance of 150 Ω. How long will it take for the cooker to raise the temperature of the hot dog from 20°C to 80°C? The specific heat of a hot dog is approximately 2500 J/kg · K.
47. |||| INT Air isn't a perfect electric insulator, but it has a very high resistivity. Dry air has a resistivity of approximately $3 \times 10^{13}$ Ω · m. A capacitor has square plates 10 cm on a side separated by 1.2 mm of dry air. If the capacitor is charged to 250 V, what fraction of the charge will flow across the air gap in 1 minute? Make the approximation that the potential difference doesn't change as the charge flows.
48. || BIO The biochemistry that takes place inside cells depends on various elements, such as sodium, potassium, and calcium, that are dissolved in water as ions. These ions enter cells through narrow pores in the cell membrane known as *ion channels*. Each ion channel, which is formed from a specialized protein molecule, is selective for one type of ion. Measurements with microelectrodes have shown that a 0.30-nm-diameter potassium ion ($K^+$) channel carries a current of 1.8 pA. How many potassium ions pass through if the ion channel opens for 1.0 ms?
49. || BIO High-resolution measurements have shown that an ion channel (see Problem 48) is a 0.30-nm-diameter cylinder with length of 5.0 nm. The intracellular fluid filling the ion channel has resistivity 0.60 Ω · m. What is the resistance of the ion channel?
50. | BIO When an ion channel opens in a cell wall (see Problem 48), monovalent (charge $e$) ions flow through the channel at a rate of $1.0 \times 10^7$ ions/s.
   a. What is the current through the channel?
   b. The potential difference across the ion channel is 70 mV. What is the power dissipation in the channel?
51. ||| The total charge a battery can supply is rated in mA · h, the product of the current (in mA) and the time (in h) that the battery can provide this current. A battery rated at 1000 mA · h can supply a current of 1000 mA for 1.0 h, 500 mA current for 2.0 h, and so on. A typical AA rechargeable battery has a voltage of 1.2 V and a rating of 1800 mA · h. For how long could this battery drive current through a long, thin wire of resistance 22 Ω?
52. | Laptop batteries are rated in W · h, the product of the power (in W) that the battery can provide and the time (in h) that it can provide this power. For instance, a 50 W · h battery can provide 50 W for 1.0 h, 25 W for 2.0 h, and so on. A 10.8 V laptop battery is rated at 76 W · h. What is the total charge the battery can provide to the laptop before the battery is depleted?

Watch Video Solution Problem 22.61

53. || The heating element of a simple heater consists of a 2.0-m-long, 0.60-mm-diameter nichrome wire. When plugged into a 120 V outlet, the heater draws 8.0 A of current when hot.
   a. What is the wire's resistance when it is hot?
   b. Use your answer to part a to calculate the resistivity of nichrome in this situation. Why is it not the same as the value of $\rho$ given for nichrome in Table 22.1?

54. ||| Variations in the resistivity of blood can give valuable clues
BIO to changes in the blood's viscosity and other properties. The resistivity is measured by applying a small potential difference and measuring the current. Suppose a medical device attaches electrodes into a 1.5-mm-diameter vein at two points 5.0 cm apart. What is the blood resistivity if a 9.0 V potential difference causes a 230 $\mu$A current through the blood in the vein?

55. |||| A 40 W (120 V) lightbulb has a tungsten filament of diameter 0.040 mm. The filament's operating temperature is 1500°C.
   a. How long is the filament?
   b. What is the resistance of the filament at 20°C?

56. ||| The resistivity of blood is related to its *hematocrit,* the
BIO volume fraction of red blood cells in the blood. A commonly used equation relating the hematocrit $h$ to the blood resistivity $\rho$ (in $\Omega \cdot$ m) is $\rho = 1.32/(1 - h) - 0.79$. In one experiment, blood filled a graduated cylinder with an inner diameter of 0.90 cm. The resistance of the blood between the 1.0 cm and 2.0 cm marks of the cylinder was measured to be 234 $\Omega$. What was the hematocrit for this blood?

57. ||| When the starter motor on a car is engaged, there is a 300 A current in the wires between the battery and the motor. Suppose the wires are made of copper and have a total length of 1.0 m. What minimum diameter can the wires have if the voltage drop along the wires is to be less than 0.50 V?

58. |||| One way to sterilize food products such as meats and poultry
BIO is to irradiate them with a beam of high-energy electrons. In one
INT food plant, the electrons in the beam have an energy of 10 MeV and the beam current is 0.30 mA. 7000 J of electron energy must be deposited in a 1 kg package of meat to properly sterilize it. How long should the beam be turned on to deposit this energy?

59. | The two segments of the wire in Figure P22.59 have equal diameters and equal lengths but different resistivities $\rho_1$ and $\rho_2$. Current $I$ passes through this wire. If the resistivities have the ratio $\rho_2/\rho_1 = 2$, what is the ratio $\Delta V_1/\Delta V_2$ of the potential differences across the two segments of the wire?

FIGURE P22.59

60. ||| The 400 V battery of a Tesla Model S electric car stores $3.0 \times 10^8$ J of energy. At 65 mph, the car can drive 250 mi before the battery is depleted. At this speed, what is the current delivered by the battery?

61. || A wire is 2.3 m long and has a diameter of 0.38 mm. When connected to a 1.2 V battery, there is a current of 0.61 A. What material is the wire likely made of?

62. | The filament of a 100 W (120 V) lightbulb is a tungsten wire 0.035 mm in diameter. At the filament's operating temperature, the resistivity is $5.0 \times 10^{-7}$ $\Omega \cdot$ m. How long is the filament?

63. ||| You've made the finals of the Science Olympics! As one of
INT your tasks, you're given 1.0 g of copper and asked to make a wire, using all the metal, with a resistance of 1.0 $\Omega$. Copper has a density of 8900 kg/m$^3$. What length and diameter will you choose for your wire?

64. ||| In proton beam therapy, a beam of high-energy protons is
BIO used to kill cancerous cells in a tumor. In one system, the beam, which consists of protons with an energy of $2.8 \times 10^{-11}$ J, has a current of 80 nA. The protons in the beam mostly come to rest within the tumor. The radiologist has ordered a total dose corresponding to $3.6 \times 10^{-3}$ J of energy to be deposited in the tumor.
   a. How many protons strike the tumor each second?
   b. How long should the beam be on in order to deliver the required dose?

65. || An immersion heater used to boil water for a single cup of
INT tea plugs into a 120 V outlet and is rated at 300 W.
   a. What is the resistance of the heater?
   b. Suppose your super-size, super-insulated tea mug contains 400 g of water at a temperature of 18°C. How long will this heater take to bring the water to a boil? You can ignore the energy needed to raise the temperature of the mug and the heater itself.

66. |||| The graph in Figure P22.66 shows the current through a 1.0 $\Omega$ resistor as a function of time.
   a. How much charge flowed through the resistor during the 10 s interval shown?
   b. What was the total energy dissipated by the resistor during this time?

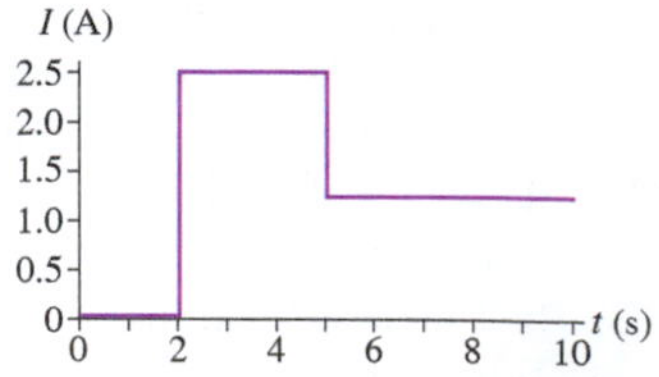

FIGURE P22.66

67. ||| It's possible to estimate the percentage of fat in the body by
BIO measuring the resistance of the upper leg rather than the upper arm; the calculation is similar. A person's leg measures 40 cm between the knee and the hip, with an average leg diameter (ignoring bone and other poorly conducting tissue) of 12 cm. A potential difference of 0.75 V causes a current of 1.6 mA. What are the fractions of (a) muscle and (b) fat in the leg?

68. | If you touch the two terminals of a power supply with your
BIO two fingertips on opposite hands, the potential difference will produce a current through your torso. The maximum safe current is approximately 5 mA.
   a. If your hands are completely dry, the resistance of your body from fingertip to fingertip is approximately 500 k$\Omega$. If you accidentally touch both terminals of your 120 V household electricity supply with dry fingers, will you receive a dangerous shock?
   b. If your hands are moist, your resistance drops to approximately 1 k$\Omega$. If you accidentally touch both terminals of your 120 V household supply with moist fingers, will you receive a dangerous shock?

69. | The average resistivity of the human body (apart from sur
BIO face resistance of the skin) is about 5.0 $\Omega \cdot$ m. The conducting path between the right and left hands can be approximated as a cylinder 1.6 m long and 0.10 m in diameter. The skin resistance can be made negligible by soaking the hands in salt water.
   a. What is the resistance between the hands if the skin resistance is negligible?
   b. If skin resistance is negligible, what potential difference between the hands is needed for a lethal shock current of 100 mA? Your result shows that even small potential differences can produce dangerous currents when skin is damp.

## MCAT-Style Passage Problems

### Lightbulb Failure

You've probably observed that the most common time for an incandescent lightbulb to fail is the moment when it is turned on. Let's look at the properties of the bulb's filament to see why this happens.

The current in the tungsten filament of a lightbulb heats the filament until it glows. The filament is so hot that some of the atoms on its surface fly off and end up sticking on a cooler part of the bulb. Thus the filament gets progressively thinner as the bulb ages. There will certainly be one spot on the filament that is a bit thinner than elsewhere. This thin segment will have a higher resistance than the surrounding filament. More power will be dissipated at this spot, so it won't only be a thin spot, it also will be a hot spot.

Now, let's look at the resistance of the filament. The graph in Figure P22.70 shows data for the current in a lightbulb as a function of the potential difference across it. The graph is not linear, so the filament is not an ohmic material with a constant resistance. However, we can define the resistance at any particular potential difference $\Delta V$ to be $R = \Delta V/I$. This ratio, and hence the resistance, increases with $\Delta V$ and thus with temperature.

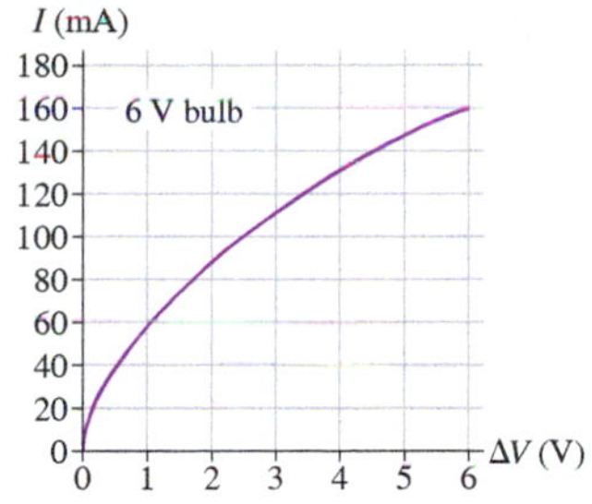


FIGURE P22.70

When the bulb is turned on, the filament is cold and its resistance is much lower than during normal, high-temperature operation. The low resistance causes a surge of higher-than-normal current lasting a fraction of a second until the filament heats up. Because power dissipation is $I^2R$, the power dissipated during this first fraction of a second is much larger than the bulb's rated power. This current surge concentrates the power dissipation at the high-resistance thin spot, perhaps melting it and breaking the filament.

70. | For the bulb in Figure P22.70, what is the approximate resistance of the bulb at a potential difference of 6.0 V?
    A. 7.0 Ω
    B. 17 Ω
    C. 27 Ω
    D. 37 Ω
71. | As the bulb ages, the resistance of the filament
    A. Increases.
    B. Decreases.
    C. Stays the same.
72. | Which of the curves in Figure P22.72 best represents the expected variation in current as a function of time in the short time interval immediately after the bulb is turned on?

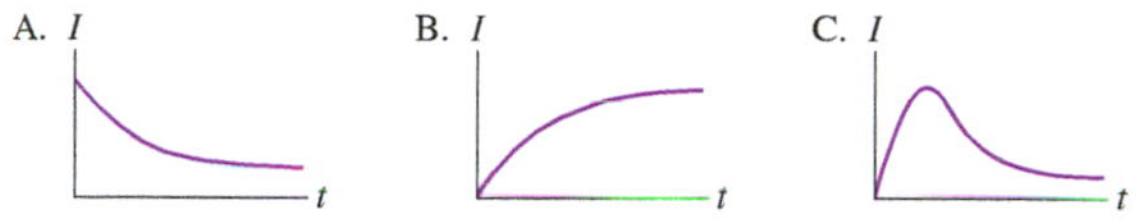


FIGURE P22.72

73. | There are devices to put in a light socket that control the current through a lightbulb, thereby increasing its lifetime. Which of the following strategies would increase the lifetime of a bulb without making it dimmer?
    A. Reducing the average current through the bulb
    B. Limiting the maximum current through the bulb
    C. Increasing the average current through the bulb
    D. Limiting the minimum current through the bulb

# 23 Circuits

The electric eel isn't really an eel; it's a fish. But it is electric, producing pulses of up to 600 V that it uses to stun prey. How does the fish produce such a large potential difference?

## LOOKING AHEAD ➧

### Analyzing Circuits

Practical circuits consist of many elements—resistors, batteries, capacitors—connected together.

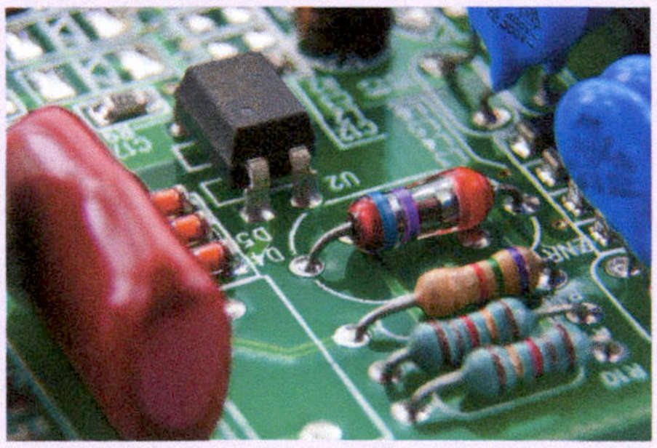

You'll learn how to analyze complex circuits by breaking them into simpler pieces.

### Series and Parallel Circuits

There are two basic ways to connect resistors together and capacitors together: **series circuits** and **parallel circuits.**

You'll learn why holiday lights are wired in series but headlights are in parallel.

### Electricity in the Body

Your nervous system works by transmitting electrical signals along *axons*, the long nerve fibers shown here.

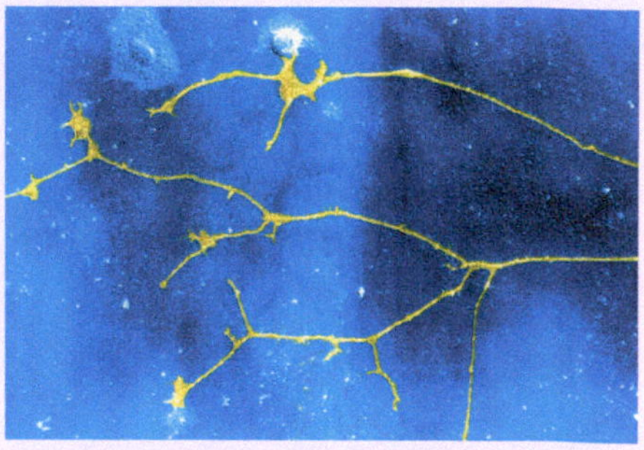

You'll learn how to understand nerve impulses in terms of the resistance, capacitance, and electric potential of nerve cells.

**GOAL** **To understand the fundamental physical principles that govern electric circuits.**

## LOOKING BACK ➧

### Ohm's Law

In Section 22.5 you learned Ohm's law, the relationship between the current through a resistor and the potential difference across it.

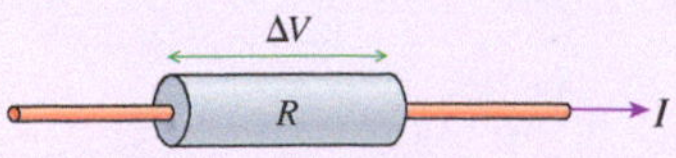


$$I = \frac{\Delta V}{R}$$

In this chapter, you'll use Ohm's law when analyzing more complex circuits consisting of multiple resistors and batteries.

**STOP TO THINK**

Rank in order, from smallest to largest, the resistances $R_1$ to $R_4$ of the four resistors.

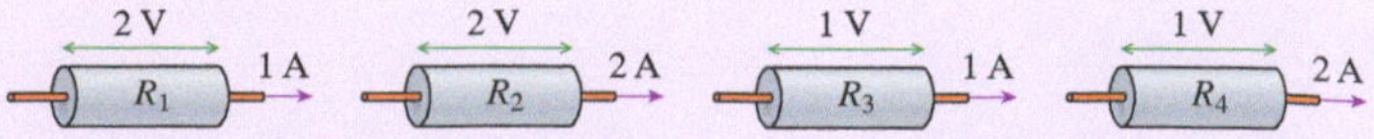

## 23.1 Circuit Elements and Diagrams

In Chapter 22 we analyzed a very simple circuit, a resistor connected to a battery. In this chapter, we will explore more complex circuits involving more and different elements. As was the case with other topics in this text, we will learn a good deal by making appropriate drawings. To do so, we need a system for representing circuits symbolically in a manner that highlights their essential features.

FIGURE 23.1 shows an electric circuit in which a resistor and a capacitor are connected by wires to a battery. To understand the operation of this circuit, we do not need to know whether the wires are bent or straight, or whether the battery is to the right or to the left of the resistor. The literal picture of Figure 23.1 provides many irrelevant details. It is customary when describing or analyzing circuits to use a more abstract picture called a **circuit diagram.** A circuit diagram is a *logical* picture of what is connected to what. The actual circuit, once it is built, may *look* quite different from the circuit diagram, but it will have the same logic and connections.

In a circuit diagram we replace pictures of the circuit elements with symbols. FIGURE 23.2 shows the basic symbols that we will need.

FIGURE 23.1 An electric circuit.

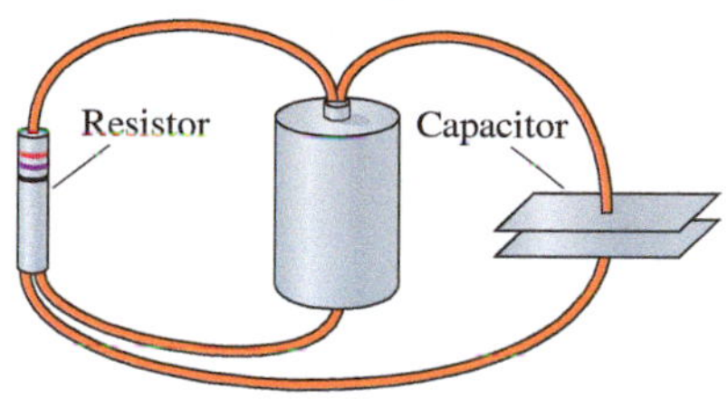


FIGURE 23.2 A library of basic symbols used for electric circuit drawings.

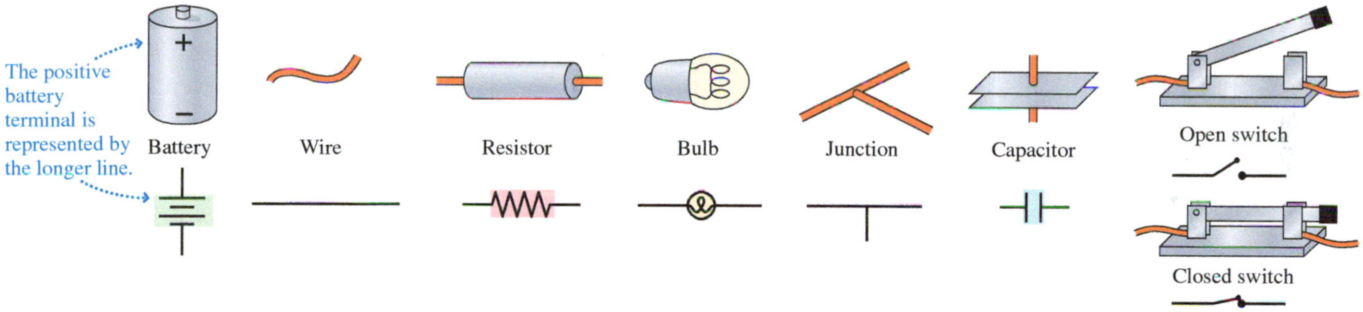


FIGURE 23.3 is a circuit diagram of the circuit shown in Figure 23.1. Notice how circuit elements are labeled. The battery's emf $\mathcal{E}$ is shown beside the battery, and the resistance $R$ of the resistor and capacitance $C$ of the capacitor are written beside them. We would include numerical values for $\mathcal{E}$, $R$, and $C$ if they are known. The wires, which in reality may bend and curve, are shown as straight-line connections between the circuit elements. The positive potential of the battery is toward the top of the diagram; in general, we try to put higher potentials toward the top. You should get into the habit of drawing your own circuit diagrams in a similar fashion.

FIGURE 23.3 A circuit diagram for the circuit of Figure 23.1.

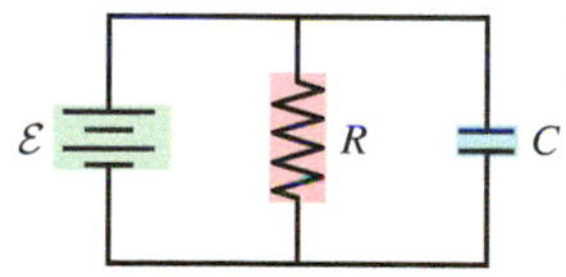


**STOP TO THINK 23.1** Which of these diagrams represent the same circuit?

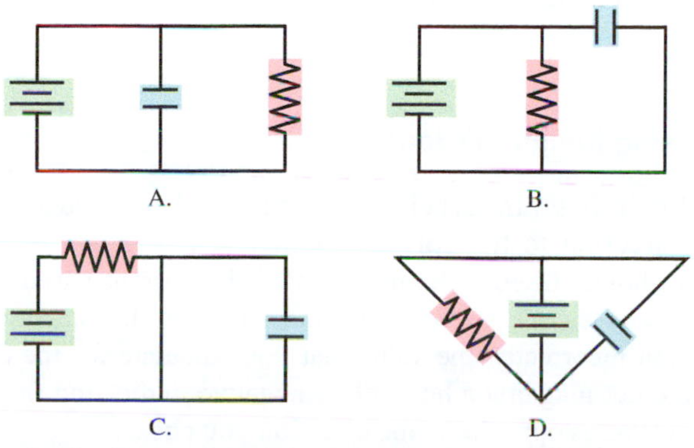

## 23.2 Kirchhoff's Laws

Once we have a diagram for a circuit, we can analyze it. Our tools and techniques for analyzing circuits will be based on the physical principles of potential differences and currents.

You learned in ◀◀ SECTION 22.2 that, as a result of charge and current conservation, the total current into a junction must equal the total current leaving the junction, as in FIGURE 23.4. This result was called *Kirchhoff's junction law,* which we wrote as

$$\sum I_{in} = \sum I_{out} \tag{23.1}$$

Kirchhoff's junction law

FIGURE 23.4 Kirchhoff's junction law.

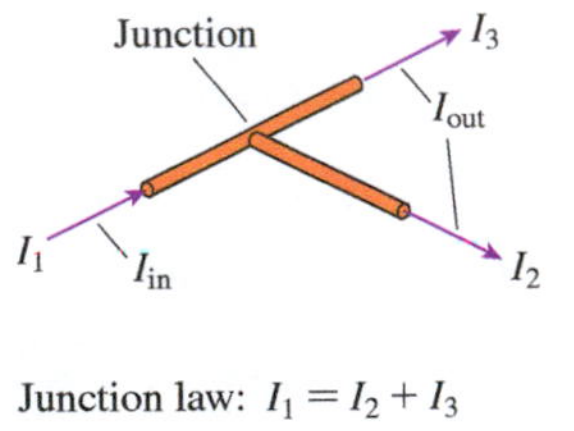


Kirchhoff's junction law isn't a new law of nature. It's an application of a law we already know: the conservation of charge. We can also apply the law of conservation of energy to circuits. When we learned about gravitational potential energy in Chapter 10, we saw that the gravitational potential energy of an object depends only on its position, not on the path it took to get to that position. The same is true of electric potential energy, as you learned in Chapter 21 and as we discussed in ◀◀ SECTION 22.5. If a charged particle moves around a closed loop and returns to its starting point, there is no net change in its electric potential energy: $\Delta U_{elec} = 0$. Because $V = U_{elec}/q$, **the net change in the electric potential around any loop or closed path must be zero** as well.

FIGURE 23.5a shows a circuit consisting of a battery and two resistors. If we start at point a in the lower left corner, at the negative terminal of the battery, and plot the potential around the loop, we get the graph shown in FIGURE 23.5b. The potential increases as we move "uphill" through the battery, then decreases in two "downhill" steps, one for each resistor. Ultimately, the potential ends up where it started, as it must. This is a general principle that we can apply to any circuit, as shown in FIGURE 23.5c. If we add all of the potential differences around the loop formed by the circuit, the sum must be zero. This result is known as **Kirchhoff's loop law:**

$$\Delta V_{loop} = \sum_i \Delta V_i = 0 \tag{23.2}$$

Kirchhoff's loop law

FIGURE 23.5 Kirchhoff's loop law.

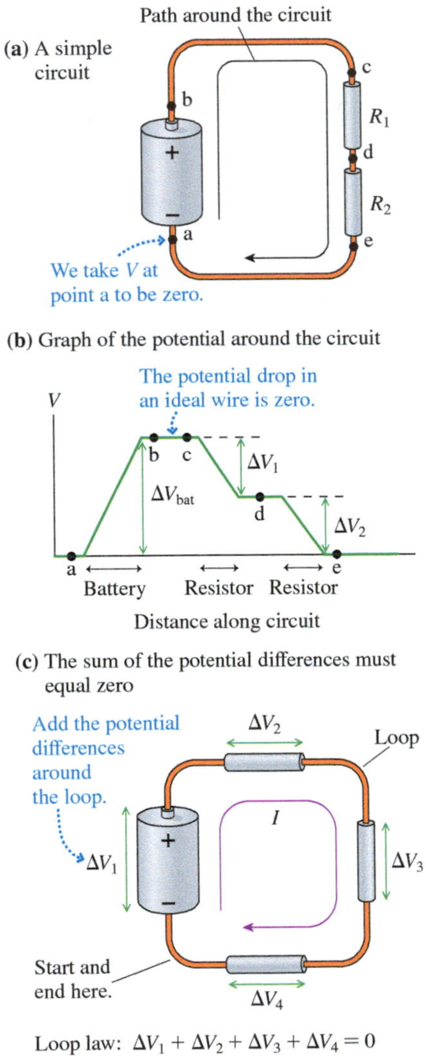


In Equation 23.2, $\Delta V_i$ is the potential difference across the $i$th component in the loop.

Kirchhoff's loop law can be true only if at least one of the potential differences $\Delta V_i$ is negative. To apply the loop law, we need to explicitly identify which potential differences are positive and which are negative.

### TACTICS BOX 23.1 Using Kirchhoff's loop law

1. **Draw a circuit diagram.** Label all known and unknown quantities.
2. **Assign a direction to the current.** Draw and label a current arrow $I$ to show your choice. Choose the direction of the current based on how the batteries or sources of emf "want" the current to go. If you choose the current direction incorrectly, the value that you calculate for the current will have the correct magnitude but will be negative, indicating that the actual current direction is opposite to the direction you chose.

❸ **"Travel" around the loop.** Start at any point in the circuit, then go all the way around the loop in the direction you assigned to the current in step 2. As you go through each circuit element, $\Delta V$ is interpreted to mean $\Delta V = V_{\text{downstream}} - V_{\text{upstream}}$.

- For a battery with current in the negative-to-positive direction:

$$\Delta V_{\text{bat}} = +\mathcal{E}$$

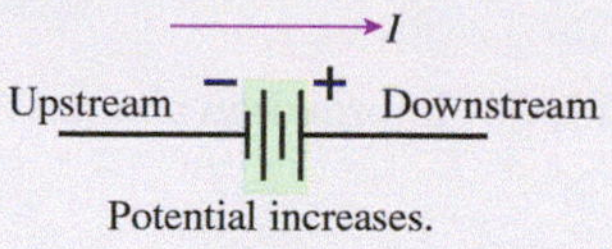


- For a battery with current in the positive-to-negative direction:

$$\Delta V_{\text{bat}} = -\mathcal{E}$$

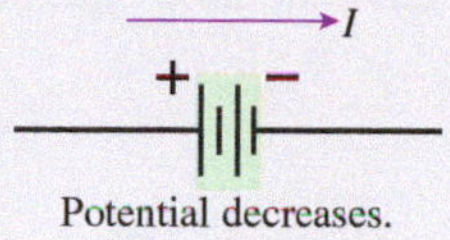


- For a resistor: $\Delta V_R = -IR$

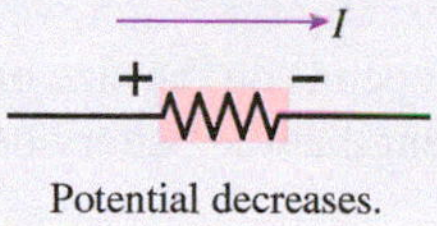


❹ **Apply the loop law:** $\sum \Delta V_i = 0$

Exercises 8, 9

We usually think of the current in a battery as flowing in the negative-to-positive direction, as it certainly does in a simple circuit with one battery and one resistor. But in circuits that have more than one battery, the current can go through a battery in the "wrong," positive-to-negative direction when it is forced to do so by other, higher-voltage batteries.

Although $\Delta V_{\text{bat}}$ can be positive or negative for a battery, $\Delta V_R$ for a resistor is always negative because the potential in a resistor *decreases* along the direction of the current—charge flows "downhill," as we saw in ◀◀ SECTION 22.4. Because the potential across a resistor always decreases, we often speak of the *voltage drop* across the resistor.

**NOTE** ▸ The equation for $\Delta V_R$ in Tactics Box 23.1 seems to be the opposite of Ohm's law, but Ohm's law was concerned only with the *magnitude* of the potential difference. Kirchhoff's law requires us to recognize that the electric potential inside a resistor *decreases* in the direction of the current. ◂

The most basic electric circuit is a single resistor connected to the two terminals of a battery, as in FIGURE 23.6. We considered this circuit in ◀◀ SECTION 22.5, but let's now apply Kirchhoff's laws to its analysis.

FIGURE 23.6 The basic circuit of a resistor connected to a battery.

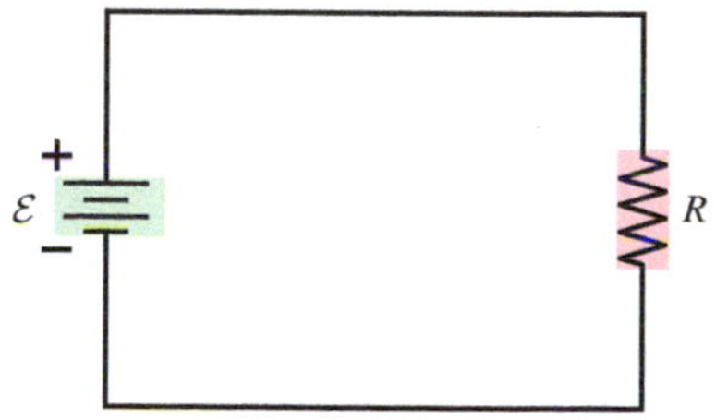


This basic circuit has no junctions, so the current is the same in all parts of the circuit. Kirchhoff's junction law is not needed. Kirchhoff's loop law is the tool we need to analyze this circuit.

FIGURE 23.7 shows the first three steps of Tactics Box 23.1. Notice that we're assuming the ideal-wire model in which there are no potential differences along the connecting wire. The fourth step is to apply Kirchhoff's loop law, $\sum \Delta V_i = 0$:

$$\Delta V_{\text{loop}} = \sum_i \Delta V_i = \Delta V_{\text{bat}} + \Delta V_R = 0 \tag{23.3}$$

FIGURE 23.7 Analysis of the basic circuit using Kirchhoff's loop law.

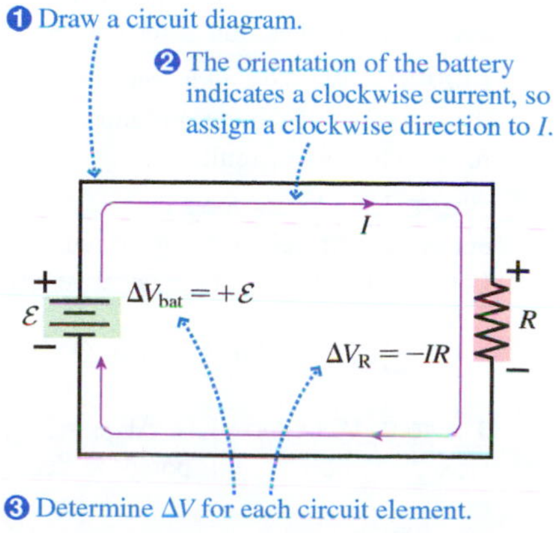


Let's look at each of the two terms in Equation 23.3:

1. The potential *increases* as we travel through the battery on our clockwise journey around the loop, as we see in the conventions in Tactics Box 23.1. We enter the negative terminal and, farther downstream, exit the positive terminal after having gained potential $\mathcal{E}$. Thus

$$\Delta V_{\text{bat}} = +\mathcal{E}$$

2. The *magnitude* of the potential difference across the resistor is $\Delta V = IR$, but Ohm's law does not tell us whether this should be positive or negative—and the difference is crucial. The potential of a resistor *decreases* in the direction of the current, which we've indicated with the + and − signs in Figure 23.7. Thus

$$\Delta V_R = -IR$$

With this information about $\Delta V_{bat}$ and $\Delta V_R$, the loop equation becomes

$$\mathcal{E} - IR = 0 \tag{23.4}$$

We can solve the loop equation to find that the current in the circuit is

$$I = \frac{\mathcal{E}}{R} \tag{23.5}$$

This is exactly the result we saw in Chapter 22. Notice again that the current in the circuit depends on the size of the resistance. The emf of a battery is a fixed quantity; the current that the battery delivers depends jointly on the emf and the resistance.

## EXAMPLE 23.1 Analyzing a circuit with two batteries

What is the current in the circuit of FIGURE 23.8? What is the potential difference across each resistor?

FIGURE 23.8 The circuit with two batteries.

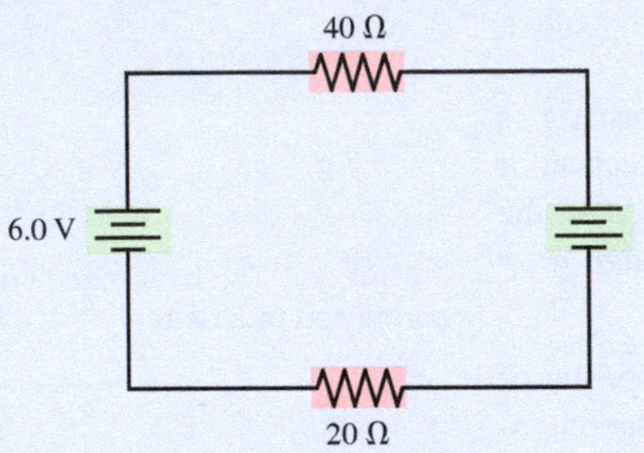


**STRATEGIZE** We will first use Kirchhoff's loop law, outlined in Tactics Box 23.1, to find the current in the circuit. Then we will use Ohm's law to find the potential differences across the resistors.

Which direction should we assign to the current in this circuit with its *two* batteries? We expect that the current will go in the direction that the larger emf—the 9.0 V battery—"wants" it to go—that is, counterclockwise. Remember, though, that if we incorrectly choose the current direction to be clockwise, we will still get the correct magnitude for the current, but it will be *negative*—telling us that the actual direction of the current is counterclockwise.

**PREPARE** We have redrawn the circuit in FIGURE 23.9, showing the direction of the current and the direction of the potential difference for each circuit element.

FIGURE 23.9 Analyzing the circuit.

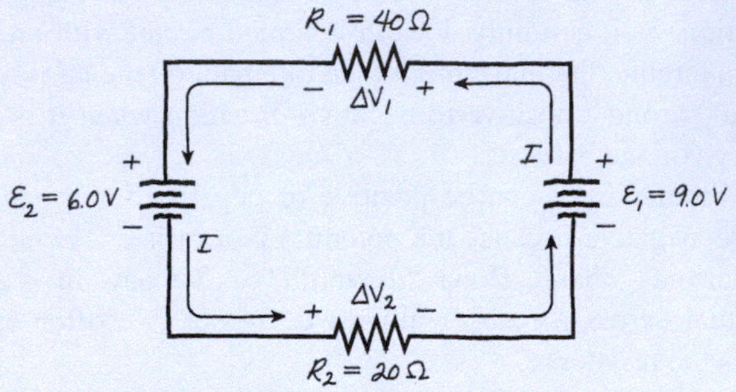


**SOLVE** Kirchhoff's loop law tells us to add the potential differences as we travel around the circuit in the direction of the current. Let's do this starting at the negative terminal of the 9.0 V battery:

$$\sum_i \Delta V_i = +9.0\text{ V} - I(40\ \Omega) - 6.0\text{ V} - I(20\ \Omega) = 0$$

The 6.0 V battery has $\Delta V_{bat} = -\mathcal{E}$, in accord with Tactics Box 23.1, because the potential decreases as we travel through this battery in the positive-to-negative direction. We can solve this equation for the current:

$$I = \frac{3.0\text{ V}}{60\ \Omega} = 0.050\text{ A} = 50\text{ mA}$$

Now that the current is known, we can use Ohm's law, $\Delta V = IR$, to find the magnitude of the potential difference across each resistor. For the 40 Ω resistor,

$$\Delta V_1 = (0.050\text{ A})(40\ \Omega) = 2.0\text{ V}$$

and for the 20 Ω resistor,

$$\Delta V_2 = (0.050\text{ A})(20\ \Omega) = 1.0\text{ V}$$

**ASSESS** The Assess step will be very important in circuit problems. There are generally other ways that you can analyze a circuit to check your work. In this case, you can do a final application of the loop law. If we start at the lower right-hand corner of the circuit and travel counterclockwise around the loop, the potential increases by 9.0 V in the first battery, then decreases by 2.0 V in the first resistor, decreases by 6.0 V in the second battery, and decreases by 1.0 V in the second resistor. The total decrease is 9.0 V, so the charge returns to its starting potential, a good check on our calculations.

**STOP TO THINK 23.2** What is the potential difference across resistor $R$?

A. −3.0 V B. −4.0 V
C. −5.0 V D. −6.0 V
E. −10 V

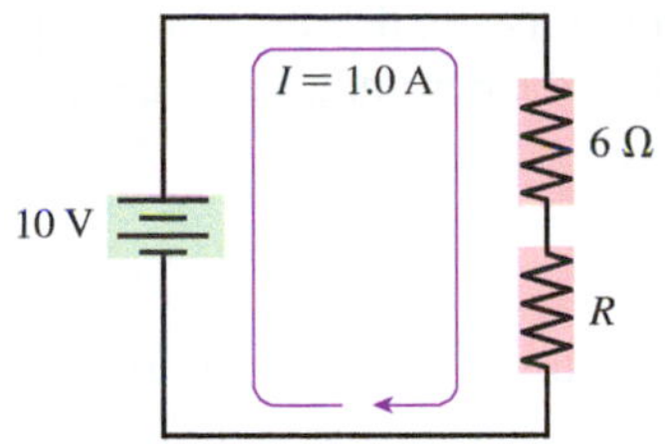


## 23.3 Series and Parallel Circuits

Example 23.1 involved a circuit with multiple elements—two batteries and two resistors. As we introduce more circuit elements, we have possibilities for different types of connections. Suppose you use a single battery to light two lightbulbs. There are two possible ways that you can connect the circuit, as shown in FIGURE 23.10. These *series* and *parallel* circuits have very different properties. We will consider these two cases in turn.

We say two bulbs or resistors are connected in **series** if they are connected directly to each other with no junction in between. All series circuits share certain characteristics.

FIGURE 23.10 Series and parallel circuits.

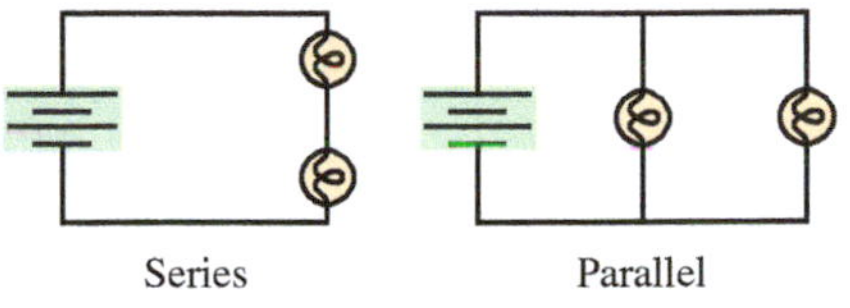


**CONCEPTUAL EXAMPLE 23.2 Determining the brightness of bulbs in series**

FIGURE 23.11 shows two identical lightbulbs connected in series. Which bulb is brighter: A or B? Or are they equally bright?

FIGURE 23.11 Two bulbs in series.

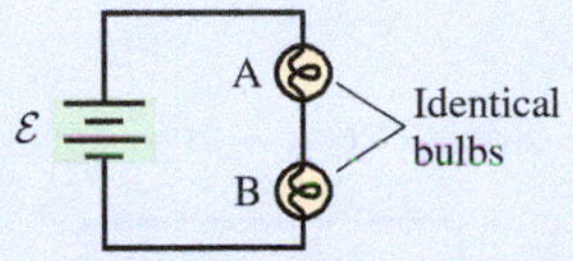


***REASON*** Current is conserved, and there are no junctions in the circuit. Thus, as FIGURE 23.12 shows, the current is the same at all points.

FIGURE 23.12 The current in the series circuit.

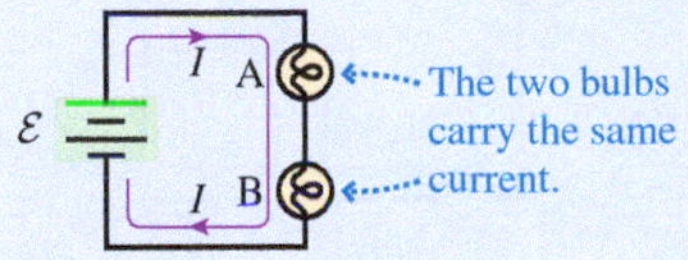


We learned in ◀◀ SECTION 22.6 that the power dissipated by a resistor is $P = I^2R$. If the two bulbs are identical (i.e., the same resistance) and have the same current through them, the power dissipated by each bulb is the same. This means that the brightness of the bulbs must be the same. The voltage across each of the bulbs will be the same as well because $\Delta V = IR$.

***ASSESS*** It's perhaps tempting to think that bulb A will be brighter than bulb B, thinking that something is "used up" before the current gets to bulb B. It is true that *energy* is being transformed in each bulb, but current must be conserved and so both bulbs dissipate energy at the same rate. We can extend this logic to a special case: If one bulb burns out, so that current can no longer pass through it, the second bulb will go dark as well. If one bulb can no longer carry a current, neither can the other.

### Series Resistors

FIGURE 23.13a shows two resistors in series connected to a battery. Because there are no junctions, the current $I$ must be the same in both resistors.

We can use Kirchhoff's loop law to look at the potential differences. Starting at the battery's negative terminal and following the current clockwise around the circuit, we find

$$\sum_i \Delta V_i = \mathcal{E} + \Delta V_1 + \Delta V_2 = 0 \qquad (23.6)$$

The voltage drops across the two resistors, in the direction of the current, are $\Delta V_1 = -IR_1$ and $\Delta V_2 = -IR_2$, so we can use Equation 23.6 to find the current in the circuit:

$$\mathcal{E} = -\Delta V_1 - \Delta V_2 = IR_1 + IR_2$$

$$I = \frac{\mathcal{E}}{R_1 + R_2} \qquad (23.7)$$

Suppose, as in FIGURE 23.13b, we replace the two resistors with a single *equivalent* resistor having the value $R_{eq} = R_1 + R_2$. The total potential difference across this

FIGURE 23.13 Replacing two series resistors with an equivalent resistor.

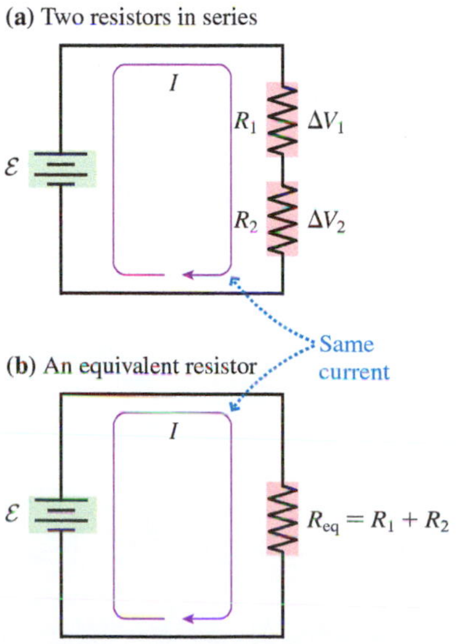

resistor is still $\mathcal{E}$ because the potential difference is established by the battery. Further, the current in this single-resistor circuit is

$$I = \frac{\mathcal{E}}{R_{eq}} = \frac{\mathcal{E}}{R_1 + R_2}$$

which is the same as in the original two-resistor circuit. In other words, this single resistor is equivalent to the two series resistors in the sense that the circuit's current and potential difference are the same in both cases. Nothing anywhere else in the circuit would differ if we took out resistors $R_1$ and $R_2$ and replaced them with resistor $R_{eq}$.

We can extend this analysis to a case with more resistors. If we have $N$ resistors in series, their **equivalent resistance** is the sum of their $N$ individual resistances:

$$R_{eq} = R_1 + R_2 + \cdots + R_N \qquad (23.8)$$

Equivalent resistance of $N$ series resistors

A video to support a section's topic is embedded in the eText.
**Video** Circuits: Warm-Up Exercise

The current and the power output of the battery will be unchanged if the $N$ series resistors are replaced by the single resistor $R_{eq}$.

## EXAMPLE 23.3 Analyzing a series resistor circuit

What is the current in the circuit of FIGURE 23.14?

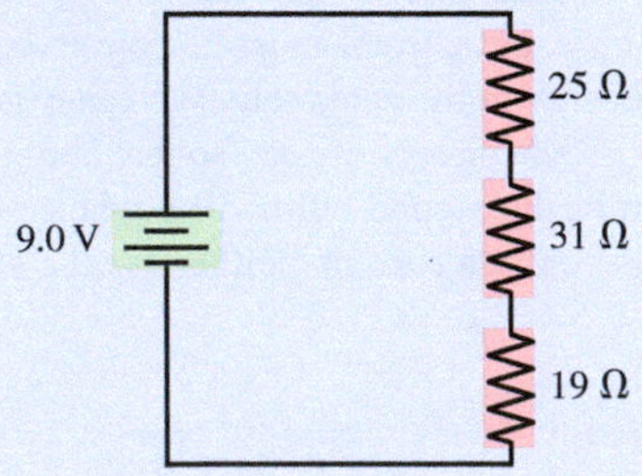


FIGURE 23.14 A series resistor circuit.

**STRATEGIZE** The three resistors are in series, so we can replace them with a single equivalent resistor, then calculate the current using Ohm's law.

**PREPARE** The circuit with its equivalent resistor is shown in FIGURE 23.15.

**SOLVE** The equivalent resistance is calculated using Equation 23.8:

$$R_{eq} = 25\ \Omega + 31\ \Omega + 19\ \Omega = 75\ \Omega$$

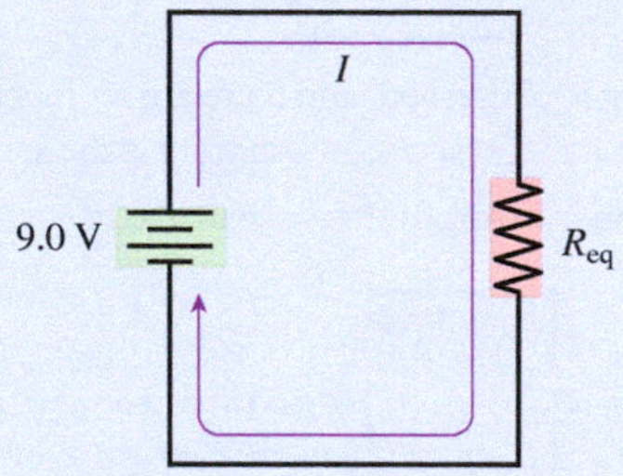


FIGURE 23.15 Analyzing a circuit with series resistors.

The current in the equivalent circuit of Figure 23.15 is

$$I = \frac{\mathcal{E}}{R_{eq}} = \frac{9.0\ \text{V}}{75\ \Omega} = 0.12\ \text{A}$$

This is also the current in the original circuit.

**ASSESS** The current in the circuit is the same whether there are three resistors or a single equivalent resistor. The equivalent resistance is the sum of the individual resistance values, and so it is always greater than any of the individual values. This is a good check on your work.

## EXAMPLE 23.4 Finding the potential difference of a string of minilights

A string of Christmas-tree minilights consists of 50 bulbs wired in series. What is the potential difference across each bulb when the string is plugged into a 120 V outlet?

**STRATEGIZE** The 50 bulbs are in series, so the current in the circuit, and hence through each bulb, will be the same if we replace the 50 bulbs with their equivalent resistance. Once we know the current, we will find the potential difference across one bulb using Ohm's law.

**PREPARE** FIGURE 23.16 shows the circuit, which has 50 bulbs in series. We assume that each bulb has resistance $R$.

FIGURE 23.16 50 bulbs connected in series.

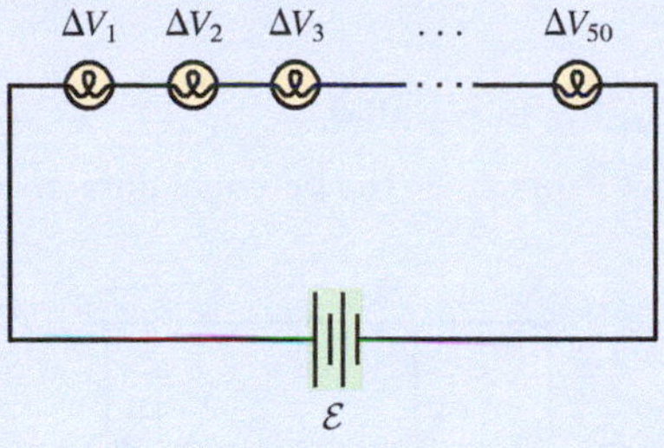


**SOLVE** The equivalent resistance of the 50 bulbs, each with resistance $R$, is

$$R_{eq} = R_1 + R_2 + \cdots + R_{50} = R + R + \cdots + R = 50R$$

Applying Ohm's law to a circuit with only this single equivalent resistor $R_{eq}$, we find that the current is

$$I = \frac{\mathcal{E}}{R_{eq}} = \frac{\mathcal{E}}{50R}$$

This current $I$ through the equivalent resistor is the same as the current through each series resistor $R$. Thus we apply Ohm's law to find the potential drop across one bulb of resistance $R$:

$$\Delta V = IR = \left(\frac{\mathcal{E}}{50R}\right)R = \frac{\mathcal{E}}{50} = \frac{120\text{ V}}{50} = 2.4\text{ V}$$

**ASSESS** This result seems reasonable. The potential difference is "shared" by the bulbs in the circuit. Since the potential difference is shared among 50 bulbs, the potential difference across each bulb will be quite small.

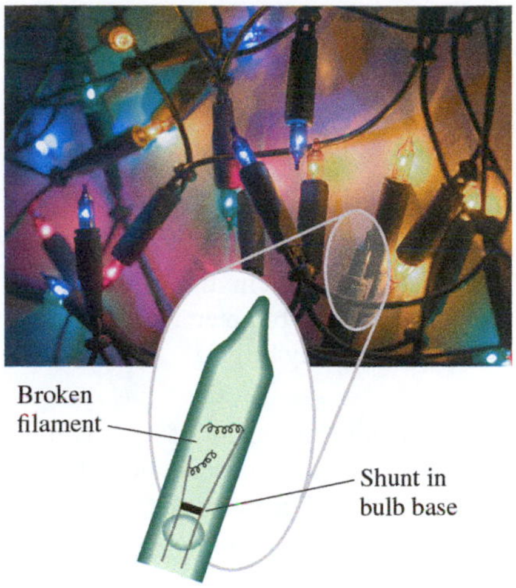


**A seasonal series circuit puzzle**
Christmas-tree minilights are connected in series. This is easy to verify: When you remove one bulb from a string of lights, the circuit is not complete, and the entire string of lights goes out. But when one bulb *burns* out, meaning its filament has broken, the string of lights stays lit. How is this possible? The secret is a *shunt* in the base of the bulb. Initially, the shunt is a good insulator. But if the filament breaks, the shunt is activated, and its resistance drops. The shunt can now carry the current, so the other bulbs will stay lit.

Minilights are wired in series because the bulbs can be inexpensive low-voltage bulbs. But there is a drawback that is true of all series circuits: If one bulb is removed, there is no longer a complete circuit, and there will be no current. Indeed, if you remove a bulb from a string of minilights, the entire string will go dark.

Let's use our knowledge of series circuits to look at another lightbulb puzzle. FIGURE 23.17 shows two different circuits, one with one battery and one lightbulb and a second with one battery and two lightbulbs. All of the batteries and bulbs are identical. You now know that B and C, which are connected in series, are equally bright, but how does the brightness of B compare to that of A?

Suppose the resistance of each identical lightbulb is $R$. In the first circuit, the battery drives current $I_A = \mathcal{E}/R$ through bulb A. In the second circuit, bulbs B and C are in series, with an equivalent resistance $R_{eq} = R_A + R_B = 2R$, but the battery has the same emf $\mathcal{E}$. Thus the current through bulbs B and C is $I_{B,C} = \mathcal{E}/R_{eq} = \mathcal{E}/2R = \frac{1}{2}I_A$. Bulb B has only half the current of bulb A, so B is dimmer.

Many people predict that A and B should be equally bright. It's the same battery, so shouldn't it provide the same current to both circuits? No—recall that **a battery is a source of potential difference, *not* a source of current.** In other words, the battery's emf is the same no matter how the battery is used. When you buy a 1.5 V battery you're buying a device that provides a specified amount of potential difference, not a specified amount of current. The battery does provide the current to the circuit, but the *amount* of current depends on the resistance. Your 1.5 V battery causes 1 A to pass through a 1.5 Ω resistor but only 0.1 A to pass through a 15 Ω resistor.

This is a critical idea for understanding circuits. A battery provides a fixed emf (potential difference). It does *not* provide a fixed and unvarying current. **The amount of current depends jointly on the battery's emf *and* the resistance of the circuit attached to the battery.**

FIGURE 23.17 How does the brightness of bulb B compare to that of bulb A?

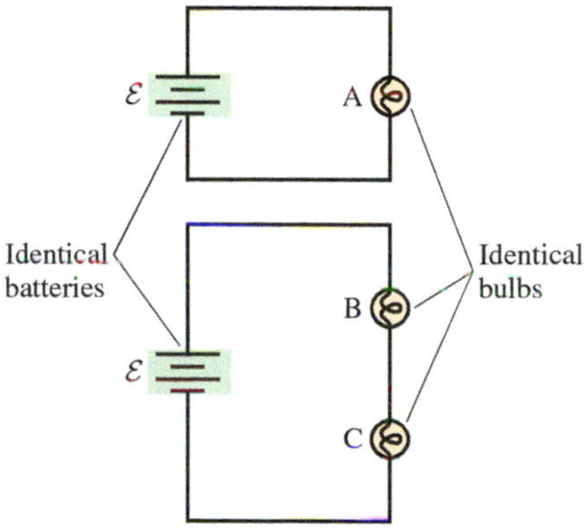


Video Figure 23.17

## Parallel Resistors

In the next example, we consider the second way of connecting two bulbs in a circuit. The two bulbs in FIGURE 23.18 are connected at *both* ends. We say that they are connected in **parallel.**

### CONCEPTUAL EXAMPLE 23.5 Comparing the brightness of bulbs in parallel

Which lightbulb in the circuit of Figure 23.18 is brighter: A or B? Or are they equally bright?

FIGURE 23.18 Two bulbs in parallel.

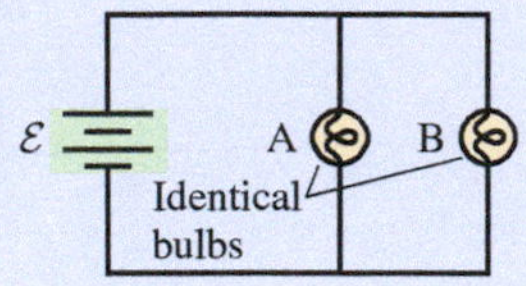


FIGURE 23.19 The potential differences of the bulbs.

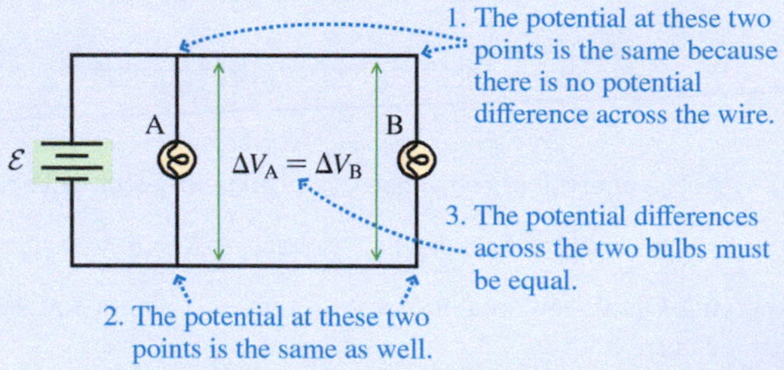


***REASON*** Both ends of the two lightbulbs are connected together by wires. Because there's no potential difference along ideal wires, the potential at the top of bulb A must be the same as the potential at the top of bulb B. Similarly, the potentials at the bottoms of the bulbs must be the same. This means that the potential *difference* $\Delta V$ across the two bulbs must be the same, as we see in FIGURE 23.19. Because the bulbs are identical (i.e., equal resistances), the currents $I = \Delta V/R$ through the two bulbs are equal and thus the bulbs are equally bright.

***ASSESS*** One might think that A would be brighter than B because current takes the "shortest route." But current is determined by potential difference, and two bulbs connected in parallel have the same potential difference.

**Video** Analyzing More Complex Circuits

FIGURE 23.20 Replacing two parallel resistors with an equivalent resistor.

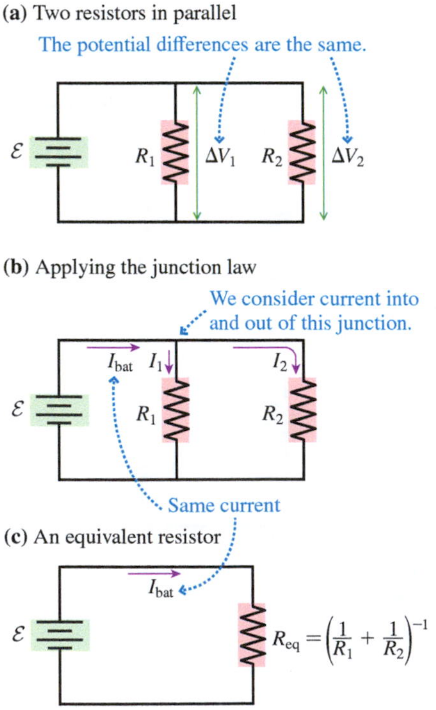


Let's look at parallel circuits in more detail. The circuit of FIGURE 23.20a has a battery and two resistors connected in parallel. If we assume ideal wires, the potential differences across the two resistors are equal. In fact, the potential difference across each resistor is equal to the emf of the battery, because both resistors are connected directly to the battery with ideal wires; that is, $\Delta V_1 = \Delta V_2 = \mathcal{E}$.

Now we apply Kirchhoff's junction law. The current $I_{bat}$ from the battery splits into currents $I_1$ and $I_2$ at the top junction noted in FIGURE 23.20b. According to the junction law,

$$I_{bat} = I_1 + I_2 \tag{23.9}$$

We can apply Ohm's law to each resistor to find that the battery current is

$$I_{bat} = \frac{\Delta V_1}{R_1} + \frac{\Delta V_2}{R_2} = \frac{\mathcal{E}}{R_1} + \frac{\mathcal{E}}{R_2} = \mathcal{E}\left(\frac{1}{R_1} + \frac{1}{R_2}\right) \tag{23.10}$$

Can we replace a group of parallel resistors with a single equivalent resistor as we did for series resistors? To be equivalent, the potential difference across the equivalent resistor must be $\Delta V = \mathcal{E}$, the same as for the two resistors it replaces. Further, the current through the equivalent resistor must be the same as it was through the parallel resistors, so that $I = I_{bat}$. A resistor with this current and potential difference must have resistance

$$R_{eq} = \frac{\Delta V}{I} = \frac{\mathcal{E}}{I_{bat}} = \left(\frac{1}{R_1} + \frac{1}{R_2}\right)^{-1} \tag{23.11}$$

where we used Equation 23.10 for $I_{bat}$. This is the *equivalent resistance,* so a single resistor $R_{eq}$ acts exactly the same as the two resistors $R_1$ and $R_2$ as shown in FIGURE 23.20c.

We can extend this analysis to the case of $N$ resistors in parallel. For this circuit, the equivalent resistance is the inverse of the sum of the inverses of the $N$ individual resistances:

$$R_{eq} = \left(\frac{1}{R_1} + \frac{1}{R_2} + \cdots + \frac{1}{R_N}\right)^{-1} \tag{23.12}$$

Equivalent resistance of $N$ parallel resistors

The current and the power output of the battery will be unchanged if the $N$ parallel resistors are replaced by the single resistor $R_{eq}$.

**NOTE** ▶ When you use Equation 23.12, don't forget to take the inverse of the sum that you compute. ◀

In Figure 23.20 each of the resistors is subject to the full potential difference of the battery. If one resistor were removed, the conditions of the second resistor would not change. This is an important property of parallel circuits.

▶ **Parallel circuits for safety** You have certainly seen cars with only one headlight lit. This tells us that automobile headlights are connected in parallel: The currents in the two bulbs are independent, so the loss of one bulb doesn't affect the other. The parallel wiring is very important so that the failure of one headlight will not leave the car without illumination.

Now, let's look at a final lightbulb puzzle. FIGURE 23.21 shows two different circuits: one with one battery and one lightbulb and a second with one battery and two lightbulbs. As before, the batteries and the bulbs are identical. You know that B and C, which are connected in parallel, are equally bright, but how does the brightness of B compare to that of A?

Each of the bulbs A, B, and C is connected to the same potential difference, that of the battery, so they each have the *same* brightness. Though all of the bulbs have the same brightness, there is a difference between the circuits. In the second circuit, the battery must power two lightbulbs, and so it must provide twice as much current. Recall that the battery is a source of fixed potential difference; the current depends on the circuit that is connected to the battery. Adding a second lightbulb doesn't change the potential difference, but it does increase the current from the battery.

FIGURE 23.21 How does the brightness of bulb B compare to that of bulb A?

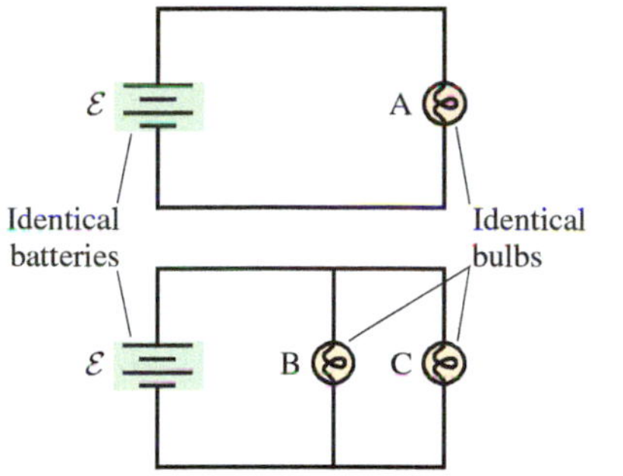


## EXAMPLE 23.6 Current in a parallel resistor circuit

The three resistors of FIGURE 23.22 are connected to a 12 V battery. What current is provided by the battery?

FIGURE 23.22 A parallel resistor circuit.

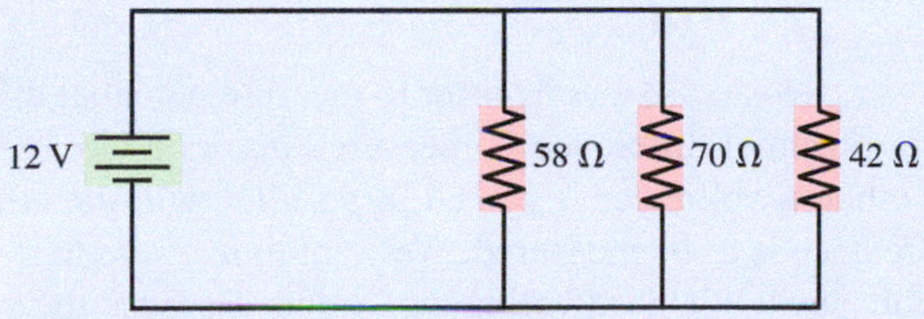


***STRATEGIZE*** The three resistors are in parallel, so we can reduce them to a single equivalent resistor. Then we will use Ohm's law to find the current.

***PREPARE*** The equivalent resistor corresponding to the three original resistors is shown in FIGURE 23.23.

FIGURE 23.23 Analyzing a circuit with parallel resistors.

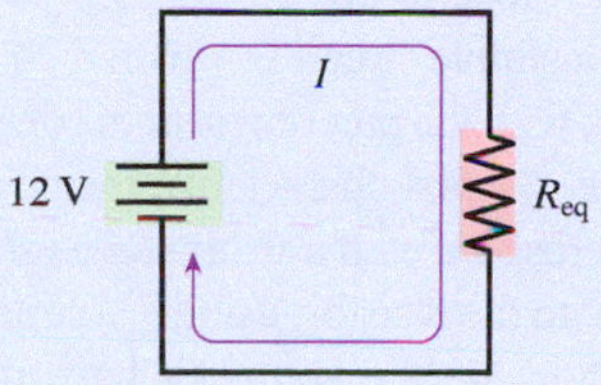


***SOLVE*** We use Equation 23.12 to calculate the equivalent resistance:

$$R_{eq} = \left(\frac{1}{58\ \Omega} + \frac{1}{70\ \Omega} + \frac{1}{42\ \Omega}\right)^{-1} = 18.1\ \Omega$$

Once we know the equivalent resistance, we use Ohm's law to calculate the current leaving the battery:

$$I = \frac{\mathcal{E}}{R_{eq}} = \frac{12\ \text{V}}{18.1\ \Omega} = 0.66\ \text{A}$$

Because the battery can't tell the difference between the original three resistors and this single equivalent resistor, the battery in Figure 23.22 provides a current of 0.66 A to the circuit.

***ASSESS*** As we'll see, the equivalent resistance of a group of parallel resistors is less than the resistance of any of the resistors in the group. 18 Ω is less than any of the individual values, a good check on our work.

The value of the total resistance in this example may seem surprising. The equivalent of a parallel combination of 58 Ω, 70 Ω, and 42 Ω is only 18 Ω. Shouldn't more resistors imply more resistance? The answer is yes for resistors in series, but not for resistors in parallel. Even though a resistor is an obstacle to the flow of charge, parallel resistors provide more pathways for charge to get through. Consequently, **the equivalent resistance of several resistors in parallel is always *less* than any single resistor in the group.** As an analogy, think about driving in heavy traffic. If there is an alternate route or an extra lane for cars to travel, more cars will be able to "flow."

**STOP TO THINK 23.3** Rank in order, from brightest to dimmest, the identical bulbs A to D.

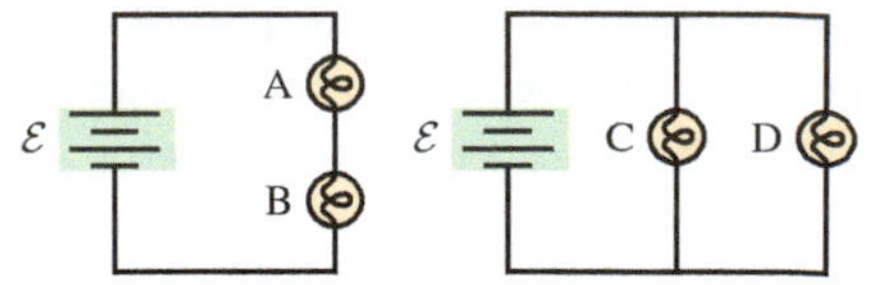


## 23.4 Measuring Voltage and Current

When you use a meter to measure the voltage or the current in a circuit, how do you connect the meter? The connection depends on the quantity you wish to measure.

A device that measures the current in a circuit element is called an **ammeter.** Because charge flows *through* circuit elements, an ammeter must be placed *in series* with the circuit element whose current is to be measured so that all the current that passes through the circuit element also passes through the meter.

FIGURE 23.24a shows a simple one-resistor circuit with a fixed emf $\mathcal{E} = 1.5$ V and an unknown resistance $R$. To determine the resistance, we must know the current in the circuit, which we measure using an ammeter. We insert the ammeter in the circuit as shown in FIGURE 23.24b. We have to *break the connection* between the battery and the resistor in order to insert the ammeter. The resistor and the ammeter now have the same current because they are in series, so the reading of the ammeter is the current through the resistor.

FIGURE 23.24 An ammeter measures the current in a circuit.

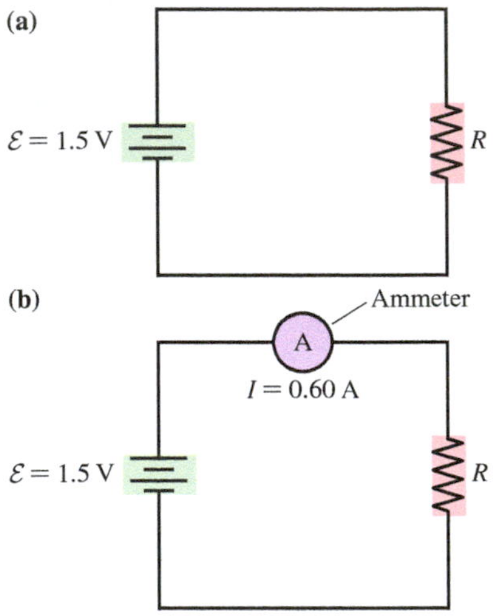


Because the ammeter is in series with resistor $R$, the total resistance seen by the battery is $R_{eq} = R + R_{meter}$. In order to *measure* the current without *changing* the current, the ammeter's resistance must be much less than $R$. Thus **the resistance of an ideal ammeter is zero.** Real ammeters come quite close to this ideal.

The ammeter in Figure 23.24b reads 0.60 A, meaning that the current in the ammeter—and in the resistor—is $I = 0.60$ A. If the ammeter is ideal, which we will assume, then there is no potential difference across the ammeter ($\Delta V = IR = 0$ if $R = 0\ \Omega$) and thus the potential difference across the resistor is $\Delta V = \mathcal{E}$. The resistance can then be calculated as

$$R = \frac{\mathcal{E}}{I} = \frac{1.5\ \text{V}}{0.60\ \text{A}} = 2.5\ \Omega$$

As we saw in ◀◀ SECTION 21.2, we can use a **voltmeter** to measure potential differences in a circuit. Because a potential difference is measured *across* a circuit element, from one side to the other, a voltmeter is placed in *parallel* with the circuit element whose potential difference is to be measured. We want to *measure* the voltage without *changing* the voltage—without affecting the circuit. Because the voltmeter is in parallel with the resistor, the voltmeter's resistance must be very large so that it draws very little current. **An ideal voltmeter has infinite resistance.** Real voltmeters come quite close to this ideal.

FIGURE 23.25 A voltmeter measures the potential difference across a circuit element.

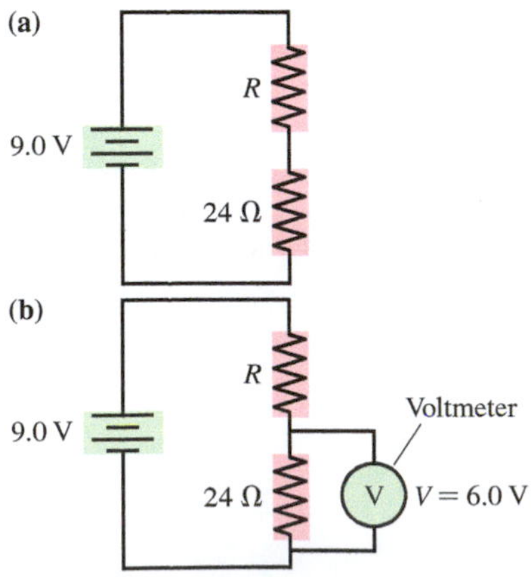


FIGURE 23.25a shows a simple circuit in which a 24 Ω resistor is connected in series with an unknown resistance, with the pair of resistors connected to a 9.0 V battery. To determine the unknown resistance, we first characterize the circuit by measuring the potential difference across the known resistor with a voltmeter as shown in FIGURE 23.25b. The voltmeter is connected in parallel with the resistor; using a voltmeter does *not* require that we break the connections. The resistor and the voltmeter have the same potential difference because they are in parallel, so the reading of the voltmeter is the voltage across the resistor.

The voltmeter in Figure 23.25b tells us that the potential difference across the 24 Ω resistor is 6.0 V, so the current through the resistor is

$$I = \frac{\Delta V}{R} = \frac{6.0\ \text{V}}{24\ \Omega} = 0.25\ \text{A} \tag{23.13}$$

The two resistors are in series, so this is also the current in unknown resistor $R$. We can use Kirchhoff's loop law and the voltmeter reading to find the potential difference across the unknown resistor:

$$\sum_i \Delta V_i = 9.0\text{ V} + \Delta V_R - 6.0\text{ V} = 0 \tag{23.14}$$

from which we find $\Delta V_R = -3.0$ V. We can now use $\Delta V_R = -IR$ to calculate

$$R = \frac{-\Delta V_R}{I} = -\frac{(-3.0\text{ V})}{0.25\text{ A}} = 12\ \Omega \tag{23.15}$$

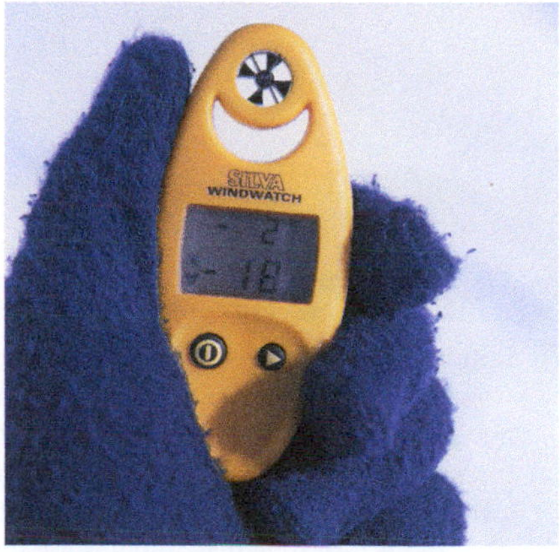

**A circuit for all seasons** This device displays wind speed and temperature, but these are computed from basic measurements of voltage and current. The wind turns a propeller attached to a generator; a rapid spin means a high voltage. A circuit in the device contains a *thermistor,* whose resistance varies with temperature; low temperatures mean high resistance and thus a small current.

**STOP TO THINK 23.4** Which is the right way to connect the meters to measure the potential difference across and the current through the resistor?

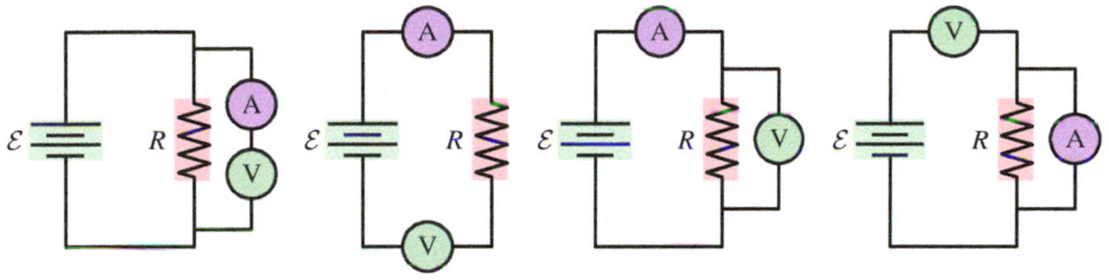


## 23.5 More Complex Circuits

In this section, we will consider circuits that involve both series and parallel resistors. Combinations of resistors can often be reduced to a single equivalent resistance through a step-by-step application of the series and parallel rules.

**EXAMPLE 23.7 Combining resistors**

What is the equivalent resistance of the group of resistors shown in FIGURE 23.26?

***STRATEGIZE*** We will analyze this circuit by looking for pairs of resistors that are in series or parallel, and then replacing them with their equivalent resistors. We will continue this process until the entire network of resistors is reduced to a single equivalent resistor.

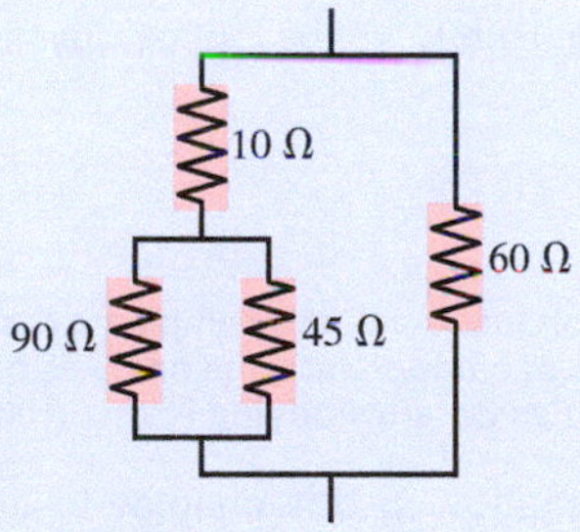


FIGURE 23.26 A resistor circuit.

***SOLVE*** The process of simplifying the circuit is shown in FIGURE 23.27. Note that the 10 Ω and 60 Ω resistors are *not* in parallel. They are connected at their top ends but not at their bottom ends. Resistors must be connected at *both* ends to be in parallel. Similarly, the 10 Ω and 45 Ω resistors are *not* in series because of the junction between them.

***ASSESS*** The last step in the process is to reduce a combination of parallel resistors. The resistance of parallel resistors is always less than the smallest of the individual resistance values, so our final result must be less than 40 Ω. This is a good check on the result.

FIGURE 23.27 A combination of resistors is reduced to a single equivalent resistor.

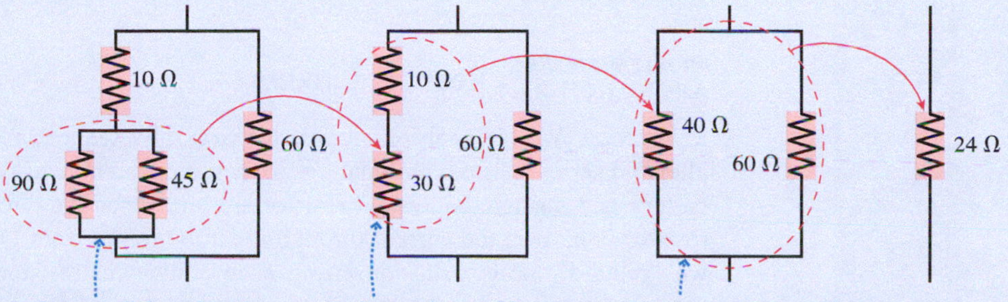


Reduce parallel combination: $R_{eq} = \left(\frac{1}{90\ \Omega} + \frac{1}{45\ \Omega}\right)^{-1} = 30\ \Omega$

Reduce series combination: $R_{eq} = 30\ \Omega + 10\ \Omega = 40\ \Omega$

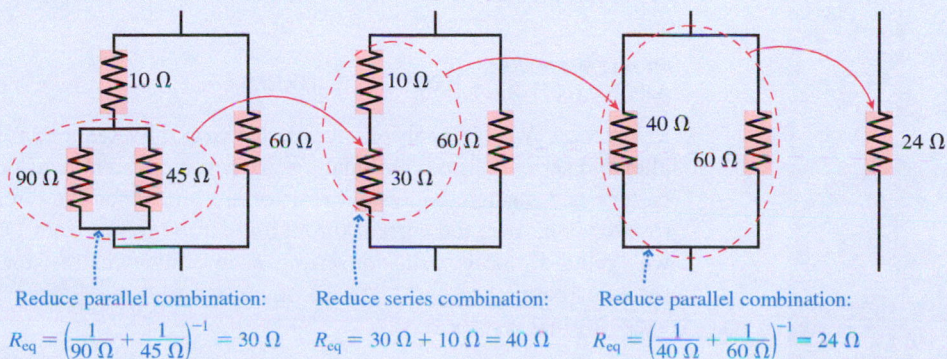

Reduce parallel combination: $R_{eq} = \left(\frac{1}{40\ \Omega} + \frac{1}{60\ \Omega}\right)^{-1} = 24\ \Omega$

Two special cases (worth remembering for reducing circuits) are the equivalent resistances of two identical resistors $R_1 = R_2 = R$ in series and in parallel:

Two identical resistors in series: $R_{eq} = 2R$

Two identical resistors in parallel: $R_{eq} = \dfrac{R}{2}$

## EXAMPLE 23.8 How does the brightness of bulbs change when a switch is closed?

Initially the switch in FIGURE 23.28 is open so that no current can flow through it. Bulbs A and B are equally bright, and bulb C is not glowing. What happens to the brightness of A and B when the switch is closed, connecting the wires on each side of the switch? And how does the brightness of C then compare to that of A and B? Assume that all bulbs are identical.

FIGURE 23.28 A lightbulb circuit.

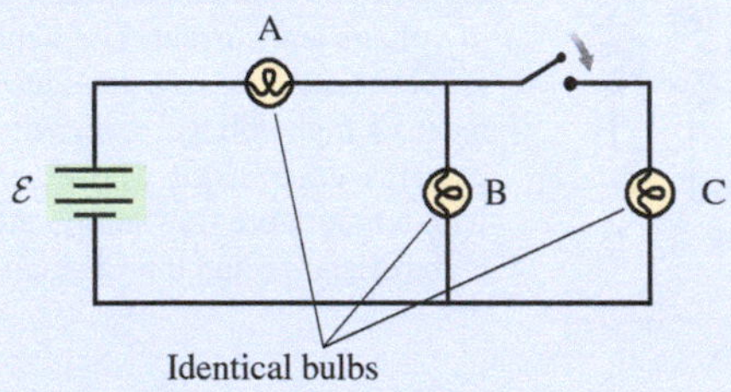


***STRATEGIZE*** We will reduce series and parallel combinations of bulbs to their equivalent resistors.

***PREPARE*** We assume the bulbs are identical, with resistance $R$. We can then use the equivalent-resistors rules for identical resistors.

***SOLVE*** Initially, before the switch is closed, bulbs A and B are in series; bulb C is not part of the circuit. A and B are identical resistors in series, so their equivalent resistance is $2R$ and the current from the battery is

$$I_{before} = \frac{\mathcal{E}}{R_{eq}} = \frac{\mathcal{E}}{2R} = \frac{1}{2}\frac{\mathcal{E}}{R}$$

This is the initial current in bulbs A and B, so they are equally bright.

Closing the switch places bulbs B and C in parallel with each other. The equivalent resistance of the two identical resistors in parallel is $R_{B,C} = R/2$. This equivalent resistance of B and C is in series with bulb A; hence the total resistance of the circuit is $R_{eq} = R + \frac{1}{2}R = \frac{3}{2}R$, and the current leaving the battery is

$$I_{after} = \frac{\mathcal{E}}{R_{eq}} = \frac{\mathcal{E}}{3R/2} = \frac{2}{3}\frac{\mathcal{E}}{R} > I_{before}$$

Closing the switch *decreases* the total circuit resistance and thus *increases* the current leaving the battery.

All the current from the battery passes through bulb A, so A *increases* in brightness when the switch is closed. The current $I_{after}$ then splits at the junction. Bulbs B and C have equal resistance, so the current divides equally. The current in B is $\frac{1}{3}(\mathcal{E}/R)$, which is *less* than $I_{before}$. Thus B *decreases* in brightness when the switch is closed. With the switch closed, bulbs B and C are in parallel, so bulb C has the same brightness as bulb B.

***ASSESS*** Our final results make sense. Initially, bulbs A and B are in series, and all of the current that goes through bulb A also goes through bulb B. But when we add bulb C, the current has another option—it can go through bulb C. This will increase the total current, and all that current must go through bulb A, so we expect a brighter bulb A. But now the current through bulb A can go through both bulbs B and C. The current splits, so we'd expect that bulb B will be dimmer than before.

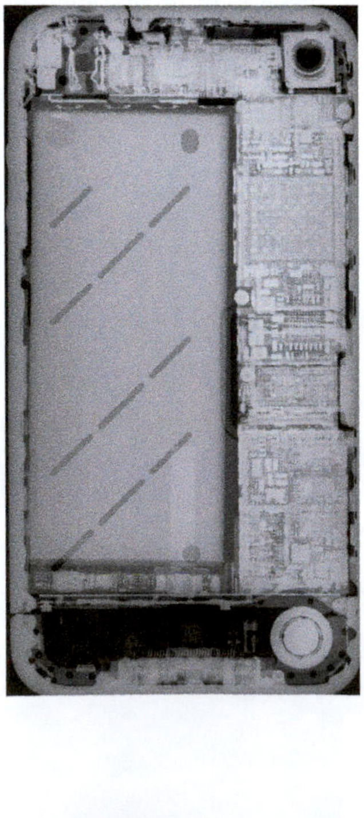

◀ This x-ray image of a cell phone shows the complex circuitry inside. Though there are thousands of components, the analysis of such a circuit starts with the same basic rules we are studying in this chapter.

We can use the information in this chapter to analyze more complex but more realistic circuits. This will give us a chance to bring together the many ideas of this chapter and to see how they are used in practice. The techniques that we use for this analysis are quite general.

## PROBLEM-SOLVING APPROACH 23.1 Resistor Circuits

***STRATEGIZE*** We can analyze any resistor circuit by sequentially reducing parallel and series resistor combinations to their equivalent resistors until only the battery and a single equivalent resistor are left. Once the circuit's equivalent resistor is known, the current through it can be found from Ohm's law. Then we "rebuild" the circuit, converting each equivalent resistor to its original series or parallel resistors. As we do so, we find the current and voltage for each original resistor.

***PREPARE*** Draw a circuit diagram. Label all known and unknown quantities.

***SOLVE*** Base your mathematical analysis on Kirchhoff's laws and on the rules for series and parallel resistors:

- Step by step, reduce the circuit to the smallest possible number of equivalent resistors.
- Determine the current through and potential difference across the equivalent resistors.
- Rebuild the circuit, using the facts that the current is the same through resistors in series and the potential difference is the same across parallel resistors.

***ASSESS*** Use two important checks as you rebuild the circuit.

- Verify that the sum of the potential differences across series resistors matches $\Delta V$ for the equivalent resistor.
- Verify that the sum of the currents through parallel resistors matches $I$ for the equivalent resistor.

Exercise 23

## EXAMPLE 23.9 Analyzing a complex circuit

Find the current through and the potential difference across each of the four resistors in the circuit shown in FIGURE 23.29.

FIGURE 23.29 A multiple-resistor circuit.

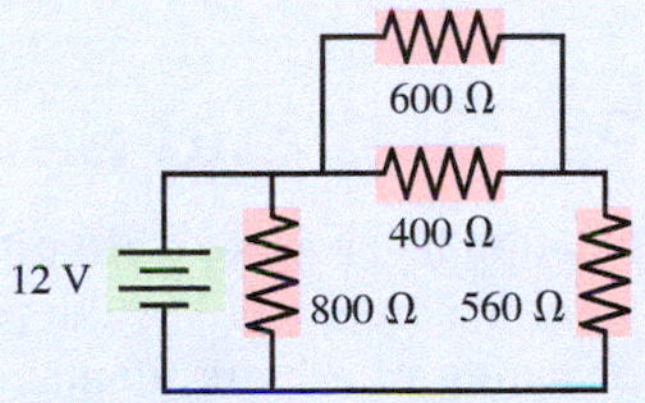


***STRATEGIZE*** We will analyze this complicated multi-resistor circuit by using the ideas of Problem-Solving Approach 23.1.

***PREPARE*** FIGURE 23.30 shows the circuit diagram. We'll keep redrawing the diagram as we analyze the circuit.

***SOLVE*** First, we break down the circuit, step-by-step, into one with a single resistor. Figure 23.30a does this in three steps, using the rules for series and parallel resistors. The final battery-and-resistor circuit is one that is easy to analyze. The potential difference across the 400 Ω equivalent resistor is $\Delta V_{400} = \Delta V_{\text{bat}} = \mathcal{E} = 12$ V. The current is

$$I = \frac{\mathcal{E}}{R} = \frac{12\text{ V}}{400\ \Omega} = 0.030\text{ A} = 30\text{ mA}$$

FIGURE 23.30 The step-by-step circuit analysis.

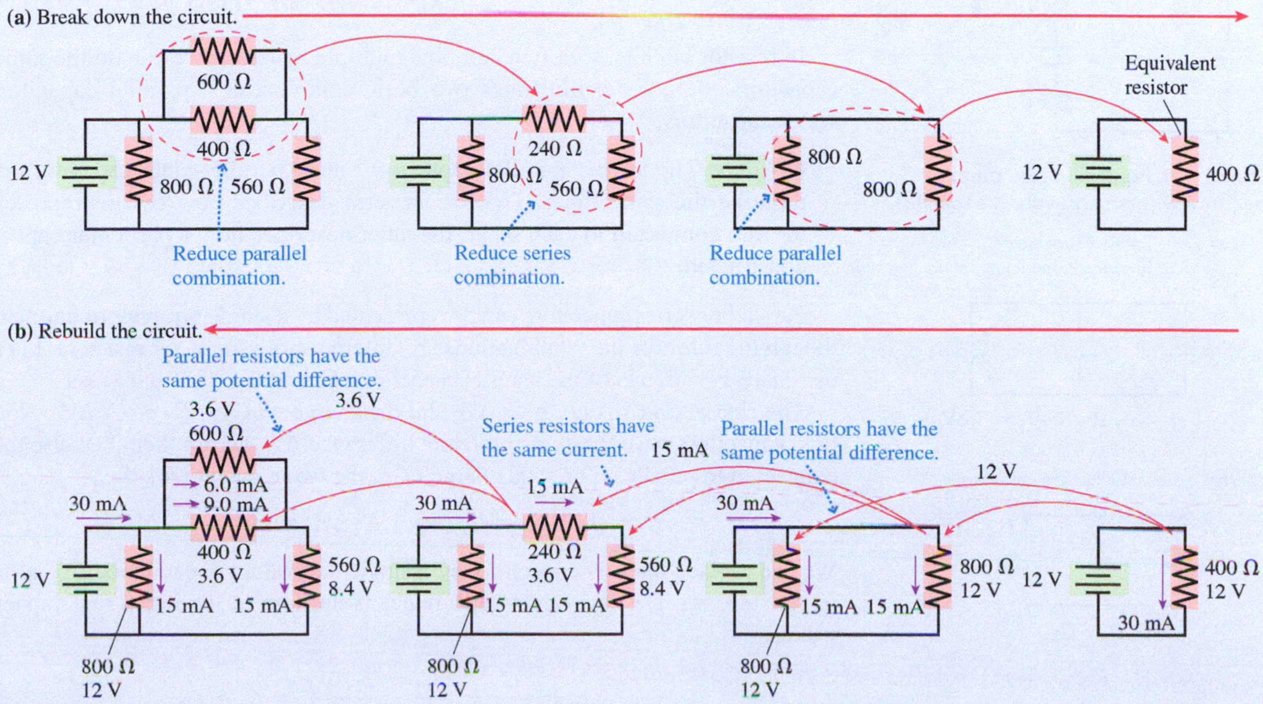


*Continued*

Second, we rebuild the circuit, step-by-step, finding the currents and potential differences at each step. Figure 23.30b repeats the steps of Figure 23.30a exactly, but in reverse order. The 400 Ω resistor came from two 800 Ω resistors in parallel. Because $\Delta V_{400} = 12$ V, it must be true that each $\Delta V_{800} = 12$ V. The current through each 800 Ω resistor is then $I = \Delta V/R = 15$ mA. A check on our work is to note that 15 mA + 15 mA = 30 mA.

The rightmost 800 Ω resistor was formed by combining 240 Ω and 560 Ω in series. Because $I_{800} = 15$ mA, it must be true that $I_{240} = I_{560} = 15$ mA. The potential difference across each is $\Delta V = IR$, so $\Delta V_{240} = 3.6$ V and $\Delta V_{560} = 8.4$ V. Here the check on our work is to note that 3.6 V + 8.4 V = 12 V = $\Delta V_{800}$, so the potential differences add as they should.

Finally, the 240 Ω resistor came from 600 Ω and 400 Ω in parallel, so they each have the same 3.6 V potential difference as their 240 Ω equivalent. The currents are $I_{600} = 6.0$ mA and $I_{400} = 9.0$ mA. Note that 6.0 mA + 9.0 mA = 15 mA, which is a third check on our work. We now know all currents and potential differences.

**ASSESS** We *checked our work* at each step of the rebuilding process by verifying that currents summed properly at junctions and that potential differences summed properly along a series of resistances. This "check as you go" procedure is extremely important. It provides you, the problem solver, with a built-in error finder that will immediately inform you if you've made a mistake.

**STOP TO THINK 23.5** Rank in order, from brightest to dimmest, the identical bulbs A to D.

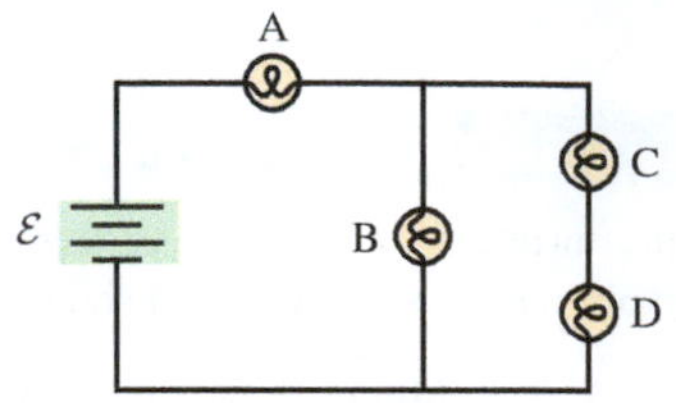


FIGURE 23.31 Simple capacitor circuit.

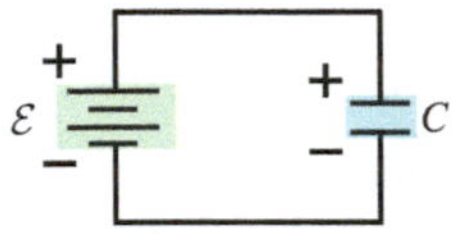


FIGURE 23.32 Parallel and series capacitors.

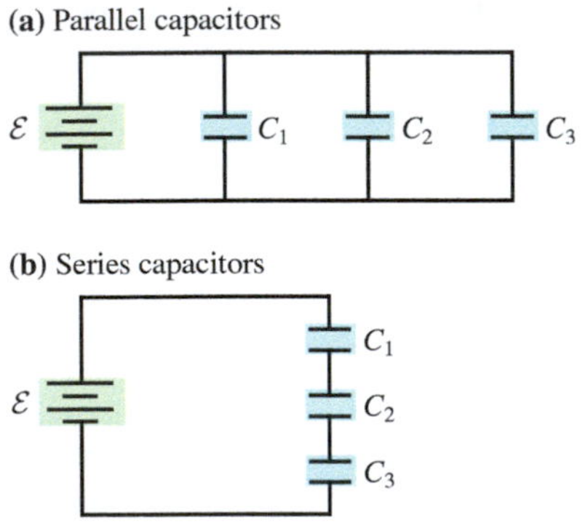


FIGURE 23.33 Replacing two parallel capacitors with an equivalent capacitor.

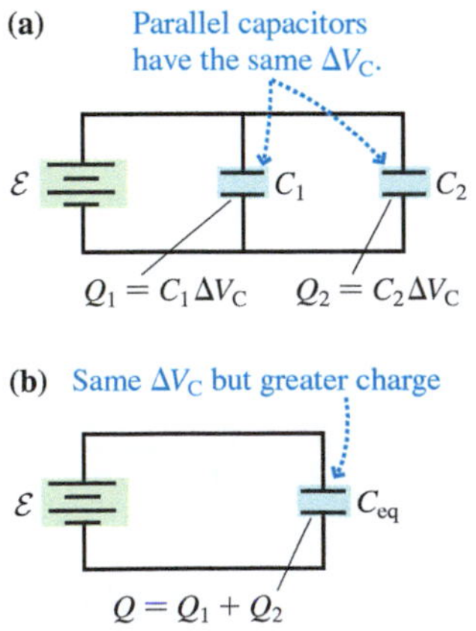


## 23.6 Capacitors in Parallel and Series

Two conductors separated by an insulating layer make a circuit element called a *capacitor*, a device we have considered in some detail in the past few chapters. FIGURE 23.31 shows a basic circuit consisting of a battery and a capacitor. When we connect the capacitor to the battery, charge will flow to the capacitor, increasing its potential difference until $\Delta V_C = \mathcal{E}$. Once the capacitor is fully charged, there will be no further current. We saw in ◀◀ SECTION 21.7 that the magnitude of the charge on each plate of the capacitor at this point will be $Q = C\,\Delta V_C = C\mathcal{E}$.

In resistor circuits, we often combine multiple resistors; we can do the same with capacitors. FIGURE 23.32 illustrates two basic combinations: parallel capacitors and series capacitors.

**NOTE** ▶ The terms "parallel capacitors" and "parallel-plate capacitor" do not describe the same thing. The former term describes how two or more capacitors are connected to each other; the latter describes how a particular capacitor is constructed. ◀

Parallel or series capacitors can be represented by a single **equivalent capacitance,** though the rules for the combinations are different from those for resistors. Let's start our analysis with two of the parallel capacitors, $C_1$ and $C_2$, of FIGURE 23.33a.

The charge on $C_1$ is $Q_1 = C_1\,\Delta V_C$ and the charge on $C_2$ is $Q_2 = C_2\,\Delta V_C$. Note that both capacitors have the *same* potential difference $\Delta V_C$ across them because they are connected in parallel. The total charge $Q$ on the two capacitors is then

$$Q = Q_1 + Q_2 = C_1\,\Delta V_C + C_2\,\Delta V_C = (C_1 + C_2)\Delta V_C$$

We can replace the two capacitors by a single equivalent capacitance $C_{eq}$, as shown in FIGURE 23.33b. The equivalent capacitance is the sum of the individual capacitance values:

$$C_{eq} = \frac{Q}{\Delta V_C} = \frac{(C_1 + C_2)\Delta V_C}{\Delta V_C} = C_1 + C_2 \qquad (23.16)$$

We can easily extend this analysis to more than two capacitors. If $N$ capacitors are in parallel, their equivalent capacitance is the sum of the individual capacitances:

$$C_{eq} = C_1 + C_2 + C_3 + \cdots + C_N \tag{23.17}$$

Equivalent capacitance of $N$ parallel capacitors

Neither the battery nor any other part of a circuit can tell if the parallel capacitors are replaced by a single capacitor having capacitance $C_{eq}$.

**NOTE** ▸ Adding another capacitor in parallel adds more capacitance. The formula for *parallel* capacitors is thus similar to the formula for *series* resistors. ◂

Now let's look at two capacitors connected in series, as shown in the circuit of FIGURE 23.34a. The center section, consisting of the bottom plate of $C_1$, the top plate of $C_2$, and the connecting wire, is electrically isolated so that the net charge on this section is fixed; in practice, this net charge is *zero*. This means that the charge on the bottom plate of $C_1$ and the charge on the top plate of $C_2$ are equal in magnitude but opposite in sign.

Suppose that the charge on capacitor $C_1$ is $Q$, which means that the charge on its top plate is $Q$ and the charge on its bottom plate is $-Q$. Because the center section has no net charge, the charge on the top plate of $C_2$ must therefore be $+Q$, implying that the charge of $C_2$ is also $Q$, the same as $C_1$. This shows that **two capacitors in series have the same charge.**

Using this fact that both capacitors have the same charge $Q$, we can calculate the potential differences across the two capacitors to be $\Delta V_1 = Q/C_1$ and $\Delta V_2 = Q/C_2$. The total potential difference across both capacitors is $\Delta V_C = \Delta V_1 + \Delta V_2$. Suppose, as in FIGURE 23.34b, we replaced the two capacitors with a single capacitor having charge $Q$ and potential difference $\Delta V_C = \Delta V_1 + \Delta V_2$. This capacitor is equivalent to the original two because the battery has to establish the same potential difference and move the same amount of charge in either case.

The inverse of the capacitance of this equivalent capacitor is

$$\frac{1}{C_{eq}} = \frac{\Delta V_C}{Q} = \frac{\Delta V_1 + \Delta V_2}{Q} = \frac{\Delta V_1}{Q} + \frac{\Delta V_2}{Q} = \frac{1}{C_1} + \frac{1}{C_2} \tag{23.18}$$

**FIGURE 23.34** Replacing two series capacitors with an equivalent capacitor.

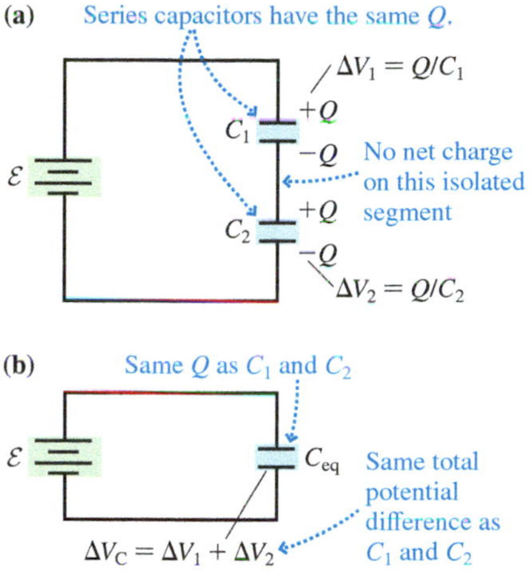


We can easily extend this analysis to more than two capacitors. If $N$ capacitors are in series, their equivalent capacitance is the inverse of the sum of the inverses of the individual capacitances:

$$C_{eq} = \left(\frac{1}{C_1} + \frac{1}{C_2} + \frac{1}{C_3} + \cdots + \frac{1}{C_N}\right)^{-1} \tag{23.19}$$

Equivalent capacitance of $N$ series capacitors

**NOTE** ▸ Adding capacitors in *series* reduces the total capacitance, just like adding resistors in *parallel*. ◂

## EXAMPLE 23.10 Analyzing a capacitor circuit

a. Find the equivalent capacitance of the combination of capacitors in the circuit of FIGURE 23.35.

b. What charge flows through the battery as the capacitors are being charged?

**FIGURE 23.35** A capacitor circuit.

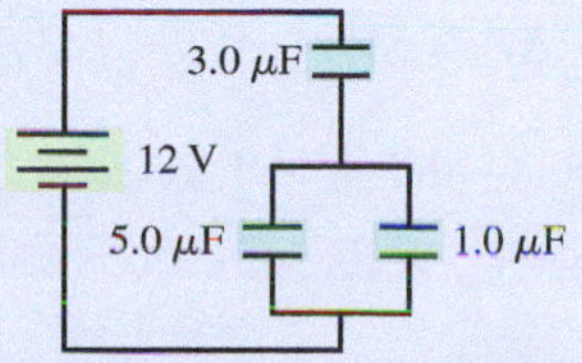


***STRATEGIZE*** We will use the relationships for parallel and series capacitors to reduce the capacitors to a single equivalent capacitance, much as we did for resistor circuits. We can then compute the charge through the battery using this value of capacitance.

*Continued*

**SOLVE**

a. FIGURE 23.36 shows how we find the equivalent capacitance by reducing parallel and series combinations.

b. The battery sees a capacitance of 2.0 $\mu$F. To establish a potential difference of 12 V, the charge that must flow is

$$Q = C_{eq}\,\Delta V_C = (2.0 \times 10^{-6}\text{ F})(12\text{ V}) = 2.4 \times 10^{-5}\text{ C}$$

**ASSESS** We check our work by noting that the equivalent capacitance of the parallel combination is *greater* than either of its two capacitances, and the equivalent capacitance of the series capacitance is *less* than either of its two capacitances.

FIGURE 23.36 Analyzing a capacitor circuit.

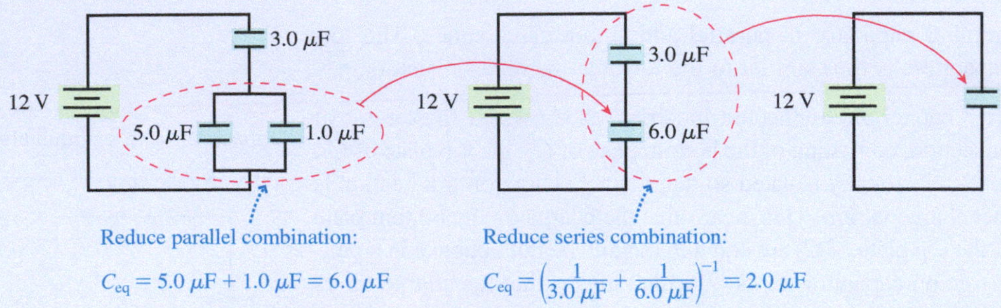


**STOP TO THINK 23.6** Rank in order, from largest to smallest, the equivalent capacitance $(C_{eq})_A$ to $(C_{eq})_C$ of circuits A to C.

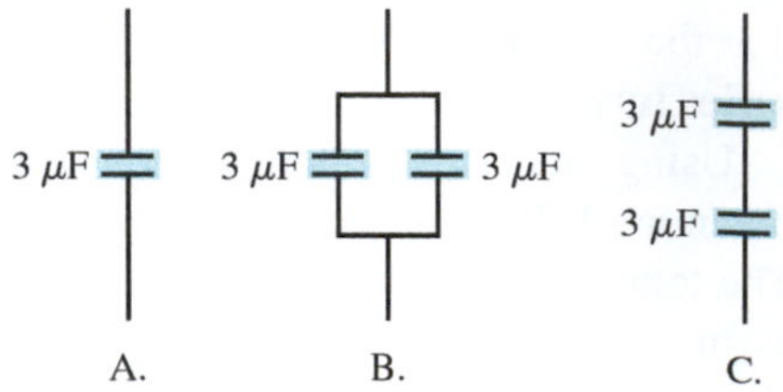


## 23.7 *RC* Circuits

The resistor circuits we have seen have a steady current. If we add a capacitor to a resistor circuit, we can make a circuit in which the current varies with time. Circuits that contain resistors and capacitors are known as ***RC* circuits.** They are very common in electronic equipment. A simple example of an *RC* circuit is the flashing bike light in the photograph. As we will see, the values of the resistance and capacitance in an *RC* circuit determine the *time* it takes the capacitor to charge or discharge. In the case of the bike light, this time determines the time between flashes. A large capacitance causes a slow cycle of on-off-on; a smaller capacitance means a more rapid flicker.

The rear flasher on a bike blinks on and off. The timing is controlled by an *RC* circuit.

### Discharging a Capacitor

FIGURE 23.37a shows an *RC* circuit consisting of a charged capacitor, an open switch, and a resistor. The capacitor has initial charge $Q_0$ and potential difference $(\Delta V_C)_0 = Q_0/C$. There is no current, so the potential difference across the resistor is zero. Then, at $t = 0$, the switch closes and the capacitor begins to discharge through the resistor.

FIGURE 23.37b shows the circuit *immediately* after the switch closes. The capacitor voltage is still $(\Delta V_C)_0$ because the capacitor hasn't yet had time to lose any charge, but now there's a current $I_0$ in the circuit that's starting to discharge the capacitor. Applying Kirchhoff's loop law, going around the loop clockwise, we find

$$\sum_i \Delta V_i = \Delta V_C + \Delta V_R = (\Delta V_C)_0 - I_0 R = 0$$

Thus the *initial* current—the initial rate at which the capacitor begins to discharge—is

$$I_0 = \frac{(\Delta V_C)_0}{R} \qquad (23.20)$$

**FIGURE 23.37** Discharging an *RC* circuit.

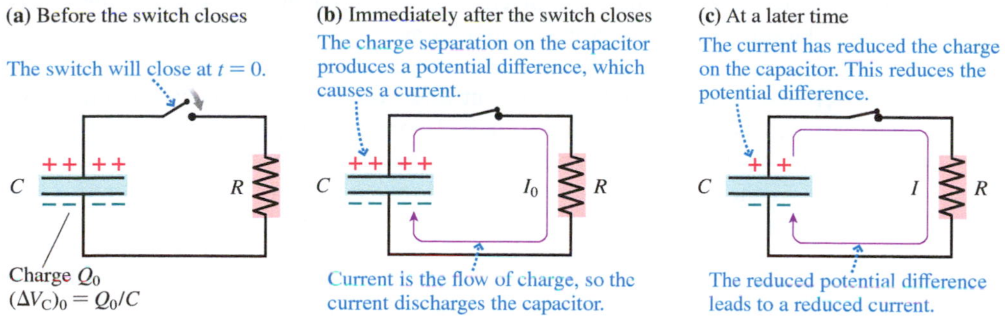


As time goes by, the current continues and the charge on the capacitor decreases. **FIGURE 23.37c** shows the circuit some time after the switch is closed; as we can see, both the charge on the capacitor (and thus the potential difference) and the current in the circuit have decreased. When the capacitor voltage has decreased to $\Delta V_C$, the current has decreased to

$$I = \frac{\Delta V_C}{R} \tag{23.21}$$

As time increases further, the charge continues to decrease, as do both $\Delta V_C$ and $I$. Because the current is the rate at which charge leaves the capacitor, the charge leaves more and more slowly as time goes on, leading to a "flattening out" of the charge-versus-time curve. Eventually, the capacitor is completely discharged, and $\Delta V_C$ and $I$ are zero.

If we use a voltmeter and an ammeter to measure the capacitor voltage and the current in the circuit of Figure 23.37 as a function of time, we find the variation shown in the graphs of **FIGURE 23.38.** At $t = 0$, when the switch closes, the potential difference across the capacitor is $(\Delta V_C)_0$ and the current suddenly jumps to $I_0$. The current and the capacitor voltage then "decay" to zero, but *not* linearly.

The graphs in Figure 23.38 have the same shape as the graph for the decay of the amplitude of a damped simple harmonic oscillator we saw in Chapter 14. Both the current and the voltage are *exponential decays* given by the equations

**FIGURE 23.38** Current and capacitor voltage in an *RC* discharge circuit.

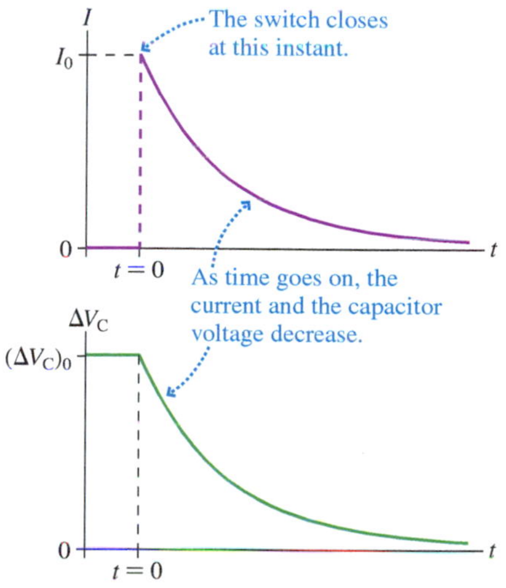


$$I = I_0 e^{-t/RC}$$
$$\Delta V_C = (\Delta V_C)_0 e^{-t/RC} \tag{23.22}$$

Current and voltage during a capacitor discharge

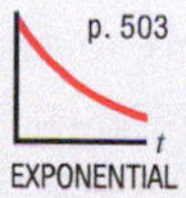


In ◀◀ SECTION 14.6, we saw that we could characterize exponential decay by a **time constant** $\tau$. The time constant is really a *characteristic time* for the circuit. A long time constant implies a slow decay; a short time constant, a rapid decay. The time constant for the decay of current and voltage in an *RC* circuit is

$$\tau = RC \tag{23.23}$$

If you work with the units, you can show that the product of ohms and farads is seconds, so the quantity *RC* really is a time. In terms of this time constant, the current and voltage equations are

$$I = I_0 e^{-t/\tau}$$
$$\Delta V_C = (\Delta V_C)_0 e^{-t/\tau} \tag{23.24}$$

The current and voltage in the circuit do not drop to zero after one time constant; that's not what the time constant means. Instead, each increase in time by one time constant causes the voltage and current to decrease by a factor of $e^{-1} = 0.37$, as we see in **FIGURE 23.39.**

**Video** Discharge Speed for Series and Parallel Capacitors

**FIGURE 23.39** The meaning of the time constant in an *RC* circuit.

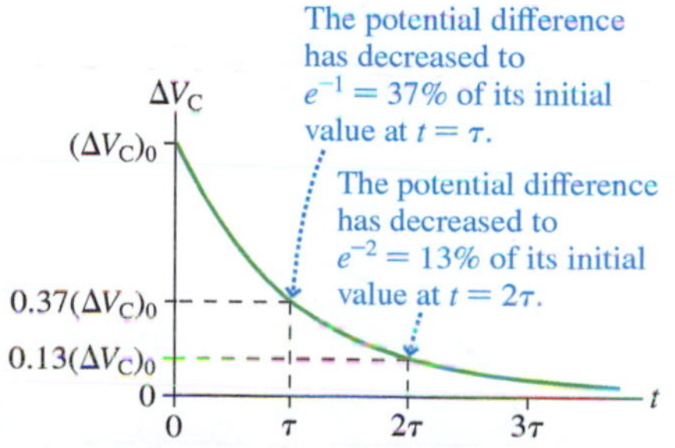

We can understand why the time constant has the form $\tau = RC$. A large value of resistance opposes the flow of charge, so increasing $R$ increases the decay time. A larger capacitance stores more charge, so increasing $C$ also increases the decay time.

After one time constant, the current and voltage in a capacitor circuit have decreased to 37% of their initial values. When is the capacitor fully discharged? There is no exact time that we can specify because $\Delta V_C$ approaches zero gradually. But after $5\tau$ the voltage and current have decayed to less than 1% of their initial values. For most purposes, we can say that the capacitor is discharged at this time.

### EXAMPLE 23.11 Finding the current in an *RC* circuit

The switch in the circuit of FIGURE 23.40 has been in position a for a long time, so the capacitor is fully charged. The switch is changed to position b at $t = 0$. What is the current in the circuit immediately after the switch is closed? What is the current in the circuit 25 $\mu$s later?

FIGURE 23.40 The *RC* circuit.

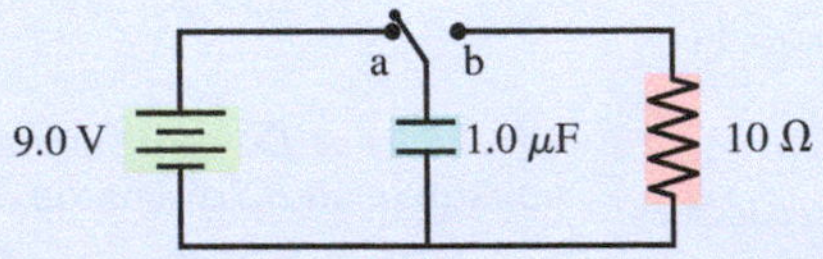


**STRATEGIZE** When the switch is moved to position b, the battery no longer plays any role in the circuit; the current then falls exponentially with a time constant determined by the resistor and capacitor.

**PREPARE** The time constant for the decay is given by Equation 23.23:

$$\tau = (10\ \Omega)(1.0 \times 10^{-6}\ \text{F}) = 1.0 \times 10^{-5}\ \text{s} = 10\ \mu\text{s}$$

**SOLVE** The capacitor is connected across the battery terminals, so initially it is charged to $(\Delta V_C)_0 = 9.0$ V. When the switch is moved to position b, the initial current is given by Equation 23.20:

$$I_0 = \frac{(\Delta V_C)_0}{R} = \frac{9.0\ \text{V}}{10\ \Omega} = 0.90\ \text{A}$$

As charge flows, the capacitor discharges. The current in the circuit as a function of time is given by Equation 23.22. 25 $\mu$s after the switch is moved to position b, the current is

$$I = I_0 e^{-t/\tau} = (0.90\ \text{A})e^{-(25\ \mu\text{s})/(10\ \mu\text{s})} = 0.074\ \text{A}$$

**ASSESS** This result makes sense. 25 $\mu$s after the switch has closed is 2.5 time constants, so we expect the current to decrease to a small fraction of the initial current. Notice that we left times in units of $\mu$s; this is one of the rare cases where we needn't convert to SI units. Because the exponent is $-t/\tau$, which involves a ratio of two times, we need only be certain that both $t$ and $\tau$ are in the same units.

FIGURE 23.41 A circuit for charging a capacitor.

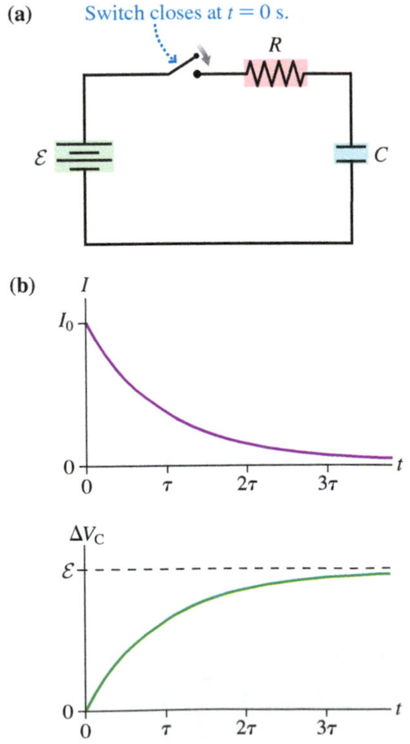


## Charging a Capacitor

FIGURE 23.41a shows a circuit that charges a capacitor. After the switch is closed, the potential difference of the battery causes a current in the circuit, and the capacitor begins to charge. As the capacitor charges, it develops a potential difference that opposes the current, so the current decreases. As the current decreases, so does the rate of charging of the capacitor. The capacitor charges until $\Delta V_C = \mathcal{E}$, when the charging current ceases.

If we measure the current in the circuit and the potential difference across the capacitor as a function of time, we find that they vary according to the graphs in FIGURE 23.41b. The characteristic time for this charging circuit is the same as for the discharge, the time constant $\tau = RC$.

When the switch is first closed, the potential difference across the uncharged capacitor is zero, so the initial current is

$$I_0 = \frac{\mathcal{E}}{R}$$

The equations that describe the current and the capacitor voltage as a function of time are

$$I = I_0 e^{-t/RC}$$
$$\Delta V_C = \mathcal{E}(1 - e^{-t/RC}) \qquad (23.25)$$

Current and voltage while charging a capacitor

The time constant $\tau$ in an *RC* circuit can be used to control the behavior of a circuit. For example, a bike flasher uses an *RC* circuit that alternately charges and discharges, over and over, as a switch opens and closes. A separate circuit turns the light on when the capacitor voltage exceeds some threshold voltage and turns the light off when the capacitor voltage goes below this threshold. The time constant of the *RC* circuit determines how long the capacitor voltage stays above the threshold and thus sets the length of the flashes. More complex *RC* circuits provide timing in computers and other digital electronics. As we will see in the next section, we can also use *RC* circuits to model the transmission of nerve impulses, and the time constant will be a key factor in determining the speed at which signals can be propagated in the nervous system.

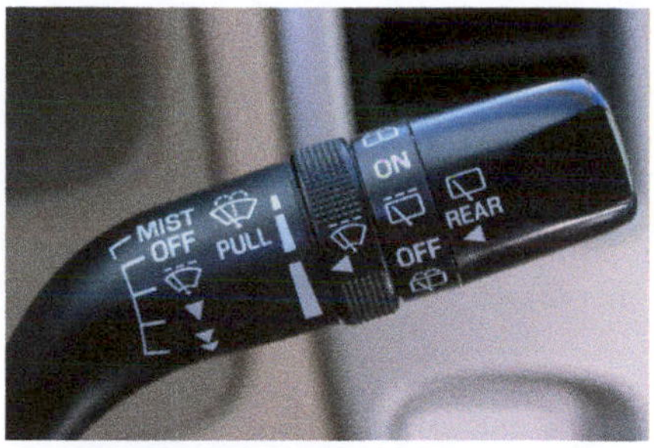


**A rainy-day *RC* circuit** When you adjust the dial to control the delay of the intermittent windshield wipers in your car, you are adjusting a variable resistor in an *RC* circuit that triggers the wipers. Increasing the resistance increases the time constant and thus produces a longer delay between swipes of the blades. A light mist calls for a long time constant and thus a large resistance.

**STOP TO THINK 23.7** The time constant for the discharge of this capacitor is

A. 5 s
B. 4 s
C. 2 s
D. 1 s

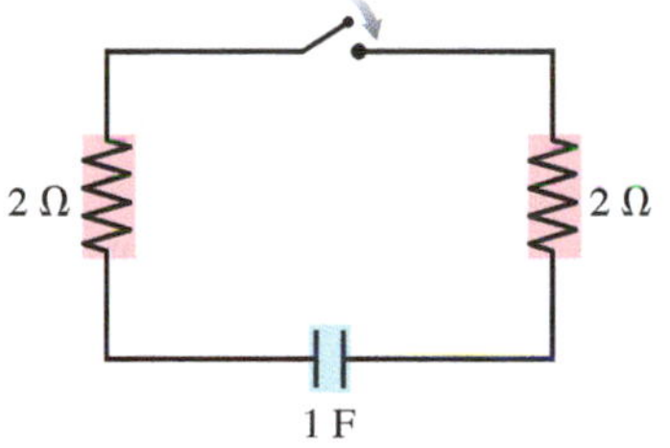


## 23.8 Electricity in the Nervous System BIO

In the late 1700s, the Italian scientist Galvani discovered that animal tissue has an electrical nature. He found that a frog's leg would twitch when stimulated with electricity, even when no longer attached to the frog. Further investigations by Galvani and others revealed that electrical signals can animate muscle cells, and that a small potential applied to the *axon* of a nerve cell can produce a signal that propagates down its length.

Our goal in this section will be to understand the nature of electrical signals in the nervous system. When your brain orders your hand to move, how does the signal get from your brain to your hand? Answering this question will use our knowledge of fields, potential, resistance, capacitance, and circuits, all of the knowledge and techniques that we have learned so far in Part VI.

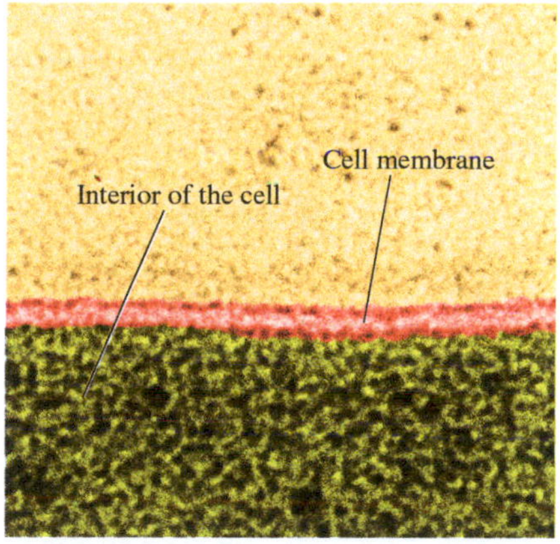

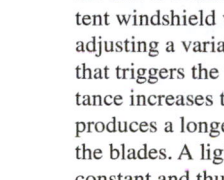

A close-up view of the cell membrane, the insulating layer that divides the conducting fluids inside and outside a cell.

### The Electrical Nature of Nerve Cells

We start our analysis with a very simple *model* of a nerve cell that allows us to describe its electrical properties. The model begins with a *cell membrane,* an insulating layer of lipids approximately 7 nm thick that separates regions of conducting fluid inside and outside the cell.

As we saw in ◀◀ SECTION 21.2, the cell membrane is not a passive structure. It has channels and pumps that transport ions between the inside and the outside of the cell. Ions, rather than electrons, are the charge carriers of the cell. In our simple model we will consider the transport of only two positive ions, sodium ($Na^+$) and potassium ($K^+$), though other ions are also important to cell function. Ions can slowly diffuse through the cell membrane. In addition, sodium and potassium ions are transported via the following structures:

- *Sodium-potassium exchange pumps*. These pump $Na^+$ ions out of the cell and $K^+$ ions in. In the cell's resting state, the concentration of sodium ions outside the cell is about ten times the concentration on the inside. Potassium ions are more concentrated on the inside.
- *Sodium and potassium channels*. These channels in the cell membrane are usually closed. When they are open, ions move in the direction of lower concentration. Thus $Na^+$ ions flow into the cell and $K^+$ ions flow out.

Our simple model, illustrated in FIGURE 23.42, ignores many of the features of real cells, but it allows us to accurately describe the reaction of nerve cells to a stimulus and the conduction of electrical signals.

FIGURE 23.42 A simple model of a nerve cell.

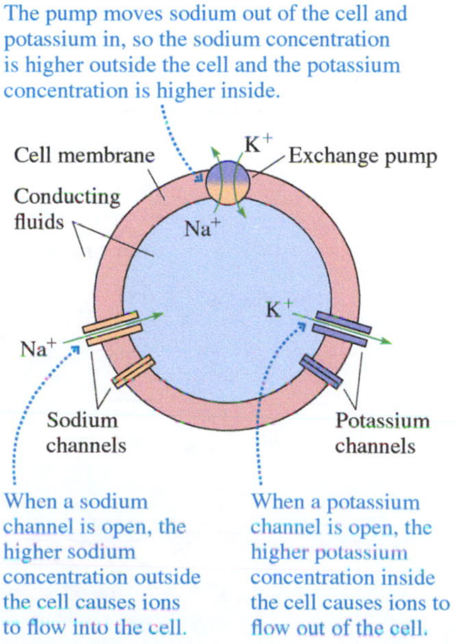

The ion exchange pumps act much like the charge escalator of a battery, using chemical energy to separate charge by transporting ions. The transport and subsequent diffusion of charged ions lead to a separation in charge across the cell membrane. Consequently, **a living cell generates an emf.** This emf takes energy to create and maintain. The ion pumps that produce the emf of nerve cells account for 25–40% of the energy usage of the brain.

The charge separation produces an electric field inside the cell membrane and results in a potential difference between the inside and the outside of the cell, as shown in FIGURE 23.43. The potential inside a nerve cell is typically 70 mV less than that outside the cell. This is called the cell's *resting potential.* Because this potential difference is produced by a charge separation across the membrane, we say that the membrane is *polarized.* And, because the potential difference is entirely across the membrane, we may call this potential difference the *membrane potential.*

FIGURE 23.43 The resting potential of a nerve cell.

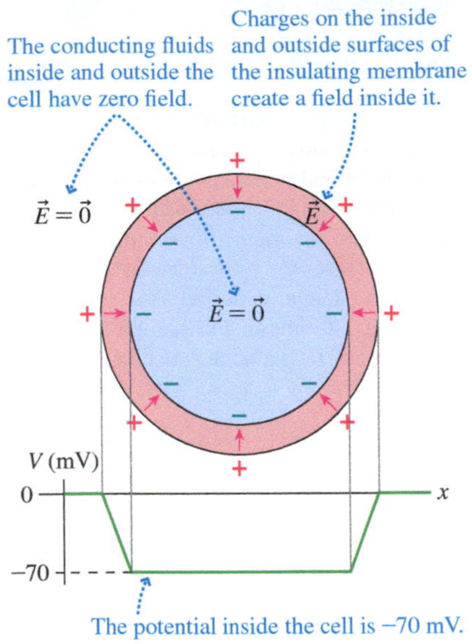


### EXAMPLE 23.12 Finding the electric field in a cell membrane BIO

The thickness of a typical nerve cell membrane is 7.0 nm. What is the electric field inside the membrane of a resting nerve cell?

**STRATEGIZE** The inner and outer surfaces of the membrane are equipotentials. We learned in Chapter 21 that the electric field is perpendicular to the equipotentials and is related to the potential difference by $E = \Delta V/d$.

**PREPARE** The magnitude $\Delta V$ of the potential difference between the inside and the outside of the cell is 70 mV.

**SOLVE** The electric field strength is

$$E = \frac{\Delta V}{d} = \frac{70 \times 10^{-3}\ \text{V}}{7.0 \times 10^{-9}\ \text{m}} = 1.0 \times 10^{7}\ \text{V/m}$$

The field points from positive to negative, so the electric field is

$$\vec{E} = (1.0 \times 10^{7}\ \text{V/m, inward})$$

**ASSESS** This is a very large electric field; in air it would be large enough to cause a spark! But we expect the fields to be large to explain the cell's strong electrical character.

### EXAMPLE 23.13 Finding the resistance of a cell membrane BIO

Charges can move across the cell membrane, so it is not a perfect insulator; the cell membrane has a certain resistance. The resistivity of the cell membrane was given in Chapter 22 as $3.6 \times 10^{7}\ \Omega \cdot \text{m}$. What is the resistance of the 7.0-nm-thick membrane of a spherical cell with diameter 0.050 mm?

**STRATEGIZE** The membrane potential will cause charges to move *through* the membrane. As we learned in ◀◀ SECTION 22.4, an object's resistance depends on its resistivity, length, and cross-section area. What this means for a cell membrane is noted in FIGURE 23.44.

**PREPARE** The area of the membrane is the surface area of a sphere, $4\pi r^2$.

**SOLVE** We can calculate the resistance using the equation for the resistance of a conductor of length $L$ and cross-section area $A$ from Chapter 22:

$$R_{\text{membrane}} = \frac{\rho L}{A} = \frac{(3.6 \times 10^{7}\ \Omega \cdot \text{m})(7.0 \times 10^{-9}\ \text{m})}{4\pi(2.5 \times 10^{-5}\ \text{m})^2}$$
$$= 3.2 \times 10^{7}\ \Omega = 32\ \text{M}\Omega$$

**ASSESS** The resistance is quite high; the membrane is a good insulator, as we noted.

FIGURE 23.44 The cell membrane can be modeled as a resistor.

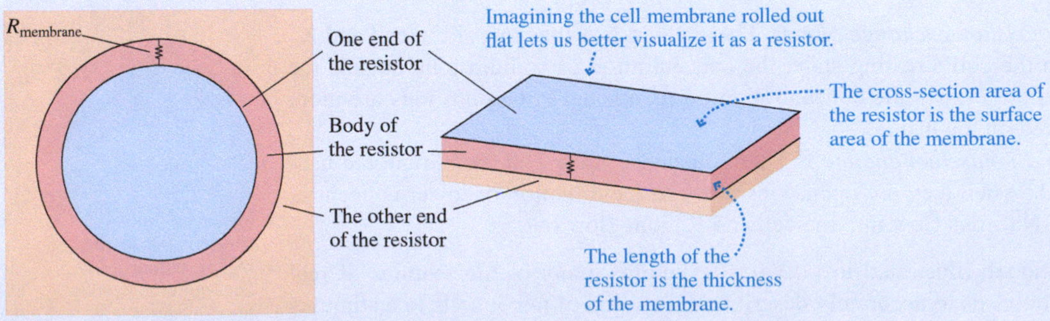

We can associate a resistance with the cell membrane, but we can associate other electrical quantities as well. The fluids inside and outside of the membrane are good conductors; they are separated by the membrane, which is not. Charges therefore accumulate on the inside and outside surfaces of the membrane. A cell thus looks like two charged conductors separated by an insulator—a capacitor.

**EXAMPLE 23.14 Finding the capacitance of a cell membrane** BIO

What is the capacitance of the membrane of the spherical cell specified in Example 23.13? The dielectric constant of a cell membrane is approximately 9.0.

**STRATEGIZE** If we imagine opening up a cell membrane and flattening it out, we would get something that looks like a parallel-plate capacitor with the plates separated by a dielectric, as illustrated in FIGURE 23.45.

**PREPARE** The relevant dimensions are the same as those in Example 23.13.

**SOLVE** The capacitance of the membrane is that of a parallel-plate capacitor filled with a dielectric, so its capacitance is $\kappa$ times that of a parallel-plate capacitor without a dielectric, or $C_{\text{membrane}} = \kappa\epsilon_0 A/d$. Inserting the dimensions from Example 23.13, we find

$$C_{\text{membrane}} = \frac{\kappa\epsilon_0 A}{d} = \frac{9.0(8.85\times10^{-12}\ \text{C}^2/\text{N}\cdot\text{m}^2)\ 4\pi(2.5\times10^{-5}\ \text{m})^2}{7.0\times10^{-9}\ \text{m}}$$
$$= 8.9\times10^{-11}\ \text{F}$$

**ASSESS** Though the cell is small, the cell membrane has a reasonably large capacitance of $\approx 90$ pF. This makes sense because the membrane is quite thin.

FIGURE 23.45 The cell membrane can also be modeled as a capacitor.

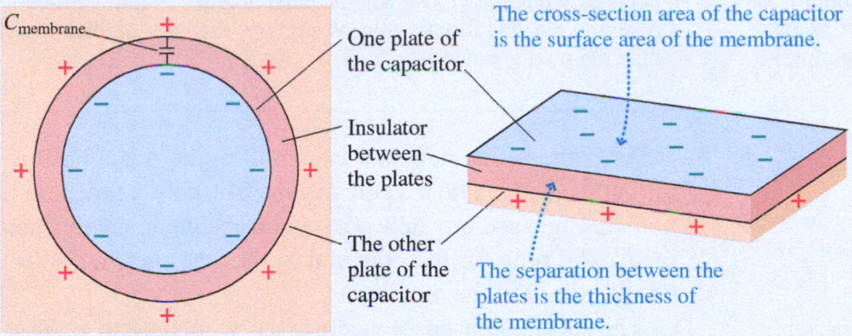


Because the cell membrane has both resistance and capacitance, it can be modeled as an $RC$ circuit, as shown in FIGURE 23.46. The membrane, like any $RC$ circuit, has a time constant. The previous examples calculated the resistance and capacitance of the 7.0-nm-thick membrane of a 0.050-mm-diameter cell. We can use these numbers to compute the membrane's time constant:

$$\tau = RC = (3.2\times10^7\ \Omega)(8.9\times10^{-11}\ \text{F}) = 2.8\times10^{-3}\ \text{s} \approx 3\ \text{ms}$$

Indeed, if we raise the membrane potential of a real nerve cell by 10 mV (large enough to easily measure but not enough to trigger a response in the cell), the potential will decay back to its resting value with a time constant of a few ms.

But the real action happens when some stimulus *is* large enough to trigger a response in the cell. In this case, ion channels open and the potential changes in much less time than the cell's time constant, as we will see next.

FIGURE 23.46 The cell membrane can be modeled as an $RC$ circuit.

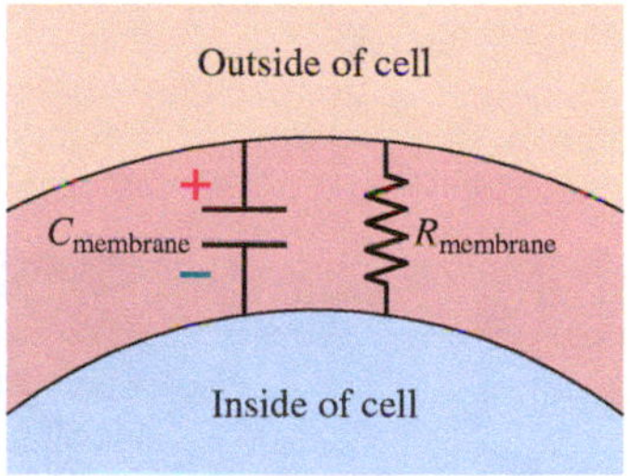


## The Action Potential

As we've seen, the potential in the interior of a cell is approximately 70 mV less than outside the cell. If we take the conducting fluid outside the cell to be at 0 V, then the interior of the cell will be at −70 mV. Now suppose a nerve cell is sitting quietly at its resting potential, so that the membrane potential is −70 mV. However, this potential can change drastically in response to a *stimulus*. *Neurons*—nerve cells—can be stimulated by neurotransmitter chemicals released at synapse junctions. A neuron can also be electrically stimulated by a changing potential, which is why Galvani saw the frog's leg jump. Whatever the stimulus, the result is a rapid change called an *action potential*—the "firing" of a nerve cell. There are three phases in the action potential, as outlined in the table on the next page.

### The action potential

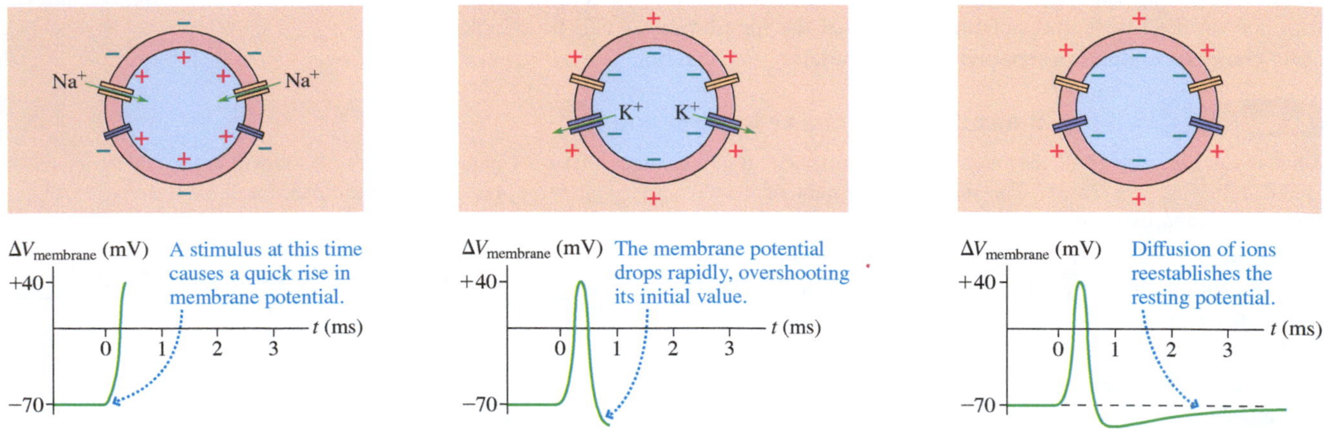


**Depolarization**

A cell *depolarizes* when a stimulus causes the opening of the sodium channels. The concentration of sodium ions is much higher outside the cell, so positive sodium ions flow into the cell, rapidly raising its potential to 40 mV, at which point the sodium channels close.

**Repolarization**

The cell *repolarizes* as the potassium channels open. The higher potassium concentration inside the cell drives these ions out of the cell. The potassium channels close when the membrane potential reaches about −80 mV, slightly *less* than the resting potential.

**Reestablishing resting potential**

The reestablishment of the resting potential after the sodium and potassium channels close is a relatively slow process controlled by the motion of ions across the membrane.

**Video** What the Physics? Sensitive Plants

After the action potential is complete, there is a brief resting period, after which the cell is ready to be triggered again. The action potential is driven by ionic conduction through sodium and potassium channels, so the potential changes are quite rapid. The time for the potential to rise and then to fall is much less than the 3 ms time constant of the membrane.

This discussion has concerned nerve cells, but muscle cells undergo a similar cycle of depolarization and repolarization. The resulting potential changes are responsible for the signal that is measured by an electrocardiogram, which we learned about in ◀◀ SECTION 21.6. The potential differences in the human body are small because the changes in potential are small. But some fish have electric organs in which the action potentials of thousands of specially adapted cells are added in series, leading to very large potential differences—hundreds of volts in the case of the electric eel.

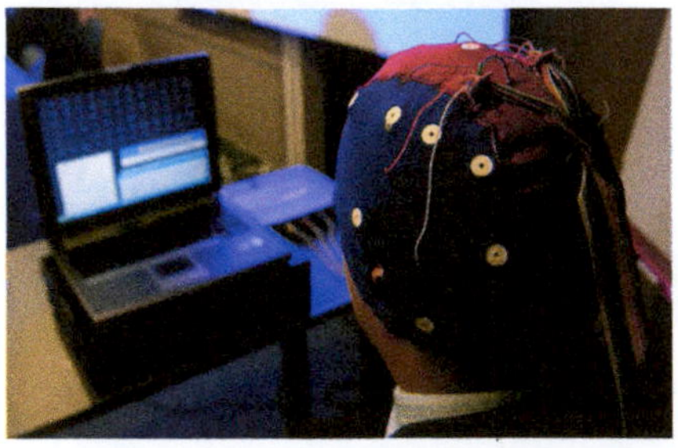

◀ **Touchless typing** Different thought processes lead to different patterns of action potentials among the many neurons of the brain. The electrical activity of the cells and the motion of ions through the conducting fluid surrounding them lead to measurable differences in potential between points on the scalp. You can't use these potential differences to "read someone's mind," but it is possible to program a computer to recognize patterns and perform actions when they are detected. This man is using his thoughts—and the resulting pattern of electric potentials—to select and enter letters.

**EXAMPLE 23.15 Counting ions moving through a channel** BIO

Investigators can measure the ion flow through a single ion channel with the *patch clamp* technique, as illustrated in FIGURE 23.47. A micropipette, a glass tube $\approx 1\ \mu\text{m}$ in diameter, makes a seal on a patch of cell membrane that includes one sodium channel. This tube is filled with a conducting saltwater solution, and a very sensitive ammeter measures the current as sodium ions flow into the cell. A sodium channel passes an average current of 4.0 pA during the 0.40 ms that the channel is open during an action potential. How many sodium ions pass through the channel?

FIGURE 23.47 Measuring the current in a single sodium channel.

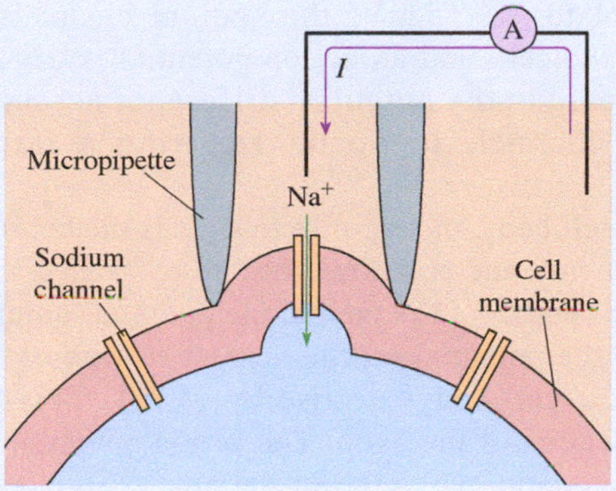


**STRATEGIZE** The current is a measure of how much charge flows through the ion channel per second; we can thus use the values of the current and the time to find this charge. The number of ions can then be found from this total charge and the charge per ion.

**PREPARE** Each $Na^+$ ion has a charge $q = +e$.

**SOLVE** In ◀◀ SECTION 22.2, we saw that the charge delivered by a steady current in time $\Delta t$ is $Q = I\,\Delta t$. The amount of charge flowing through the channel in $\Delta t = 4.0 \times 10^{-4}$ s is

$$Q = I\,\Delta t = (4.0 \times 10^{-12}\text{ A})(4.0 \times 10^{-4}\text{ s}) = 1.6 \times 10^{-15}\text{ C}$$

This charge is due to $N$ ions, each with $q = e$, so the number of ions is

$$N = \frac{Q}{e} = \frac{1.6 \times 10^{-15}\text{ C}}{1.6 \times 10^{-19}\text{ C}} = 10{,}000$$

**ASSESS** The number of ions flowing through one channel is not large, but a cell has a great many channels. The patch clamp technique and other similar procedures have allowed investigators to elucidate the details of the response of the cell membrane to a stimulus.

## The Propagation of Nerve Impulses

Let's return to the question posed at the start of the section: How is a signal transmitted from the brain to a muscle in the hand? The primary cells of the nervous system responsible for signal transmission are known as **neurons.** The transmission of a signal to a muscle is the function of a *motor neuron,* whose structure is sketched in FIGURE 23.48. The transmission of signals takes place along the *axon* of the neuron, a long fiber—up to 1 m in length—that connects the cell body to a muscle fiber. This particular neuron has a myelin sheath around the axon, though not all neurons do.

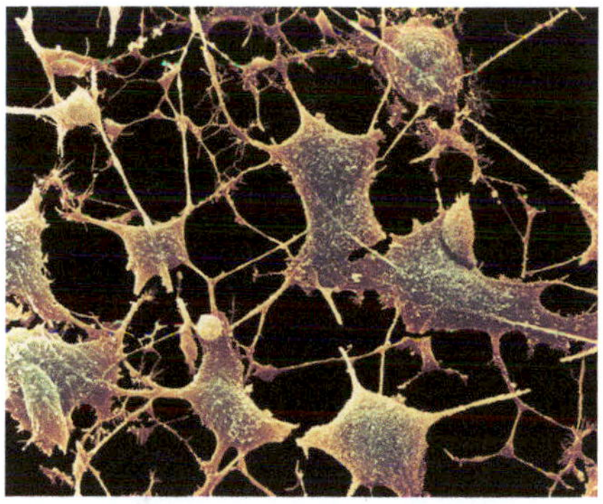

The axons connecting these nerve cells are clearly visible in the micrograph.

How is a signal transmitted along an axon? The axon is long enough that different points on its membrane may have different potentials. When one point on the axon's membrane is stimulated, the membrane will depolarize at this point. The resulting action potential may trigger depolarization in adjacent parts of the membrane. Stimulating the axon's membrane at one point can trigger a *wave* of action potential—a nerve impulse—that travels along the axon. When this impulse reaches a muscle cell, the muscle cell depolarizes and produces a mechanical response.

FIGURE 23.48 A motor neuron.

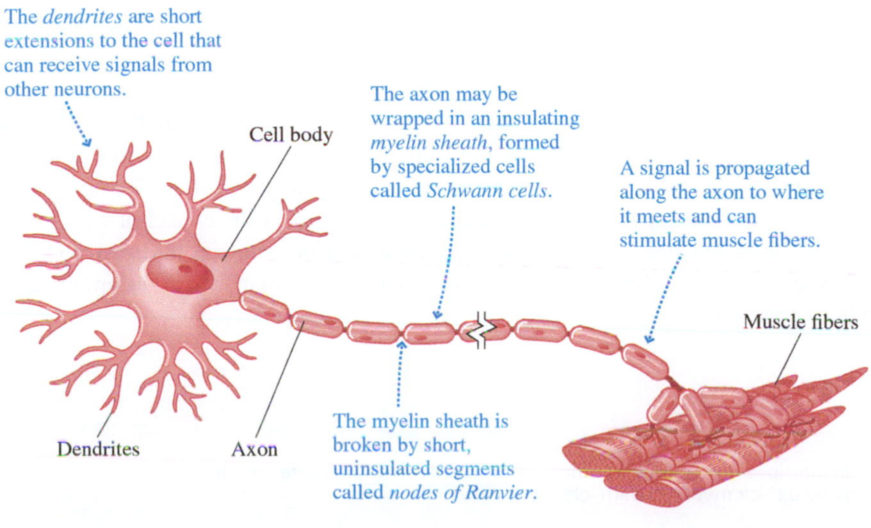

The *dendrites* are short extensions to the cell that can receive signals from other neurons.
Cell body
The axon may be wrapped in an insulating *myelin sheath,* formed by specialized cells called *Schwann cells.*
A signal is propagated along the axon to where it meets and can stimulate muscle fibers.
Muscle fibers
Dendrites
Axon
The myelin sheath is broken by short, uninsulated segments called *nodes of Ranvier.*

FIGURE 23.49 Propagation of a nerve impulse.

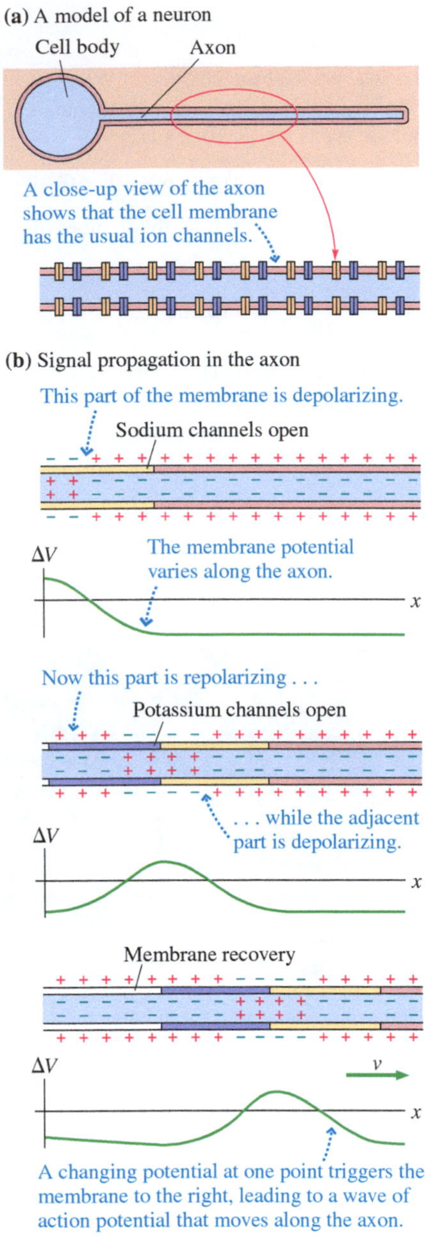


Let's look at this process in more detail. We will start with a simple model of an axon with no myelin sheath in FIGURE 23.49a. The sodium channels are normally closed, but if the potential at some point is raised by $\approx 15$ mV, from the resting potential of $-70$ mV to $\approx -55$ mV, the sodium channels suddenly open, sodium ions rush into the cell, and an action potential is triggered. This is the key idea: **A small increase in the potential difference across the membrane causes the sodium channels to open, triggering a large action-potential response.**

This process begins at the cell body, in response to signals the neuron receives at its dendrites. If the cell's membrane potential goes up by $\approx 15$ mV, an action potential is initiated in the cell body. As the membrane potential quickly rises to a peak of $+40$ mV, it causes the potential on the nearest section of the axon—where the axon attaches to the cell body—to rise by 15 mV. This triggers an action potential in this first section of the axon. The action potential in the first section of the axon triggers an action potential in the next section of the axon, which triggers an action potential in the next section, and so on down the axon until reaching the end.

As FIGURE 23.49b shows, this process causes a wave of action potential to propagate down the axon. The signal moves relatively slowly. At each point on the membrane, channels must open and ions must diffuse through, which takes time. On a typical axon with no myelin sheath, the action potential propagates at a speed of about 1 m/s. If all nerve signals traveled at this speed, a signal telling your hand to move would take about 1 s to travel from your brain to your hand. Clearly, at least some neurons in the nervous system must transmit signals at a higher speed than this!

One way to make the signals travel more quickly is to increase an axon's diameter. The giant axon in the squid triggers a rapid escape response when the squid is threatened. This axon may have a diameter of 1 mm, a thousand times that of a typical axon, providing for the necessary rapid signal transmission. But your nervous system consists of 300 billion neurons, and they can't all be 1 mm in diameter—there simply isn't enough space in your body. In your nervous system, higher neuron signal speed is achieved in a totally different manner.

## Increasing Nerve-Impulse Speed by Insulation

The axons of motor neurons and most other neurons in your body can transmit signals at very high speeds because they are insulated with a myelin sheath. Look back at the structure of a motor neuron in Figure 23.48. Schwann cells wrap the axon with myelin, insulating it electrically and chemically, with breaks at the nodes of Ranvier. The ion channels are concentrated in these nodes because this is the only place where the extracellular fluid is in contact with the cell membrane. In an insulated axon, a signal propagates by jumping from one node to the next. This process is called *saltatory conduction,* from the Latin *saltare,* "to leap."

FIGURE 23.50a shows an electrical model for saltatory conduction in which we model one segment of the axon, between successive nodes, as an *RC* circuit. As shown in FIGURE 23.50b, when the potential at a node increases by $\approx 15$ mV an action potential is triggered, which we model as closing a battery's switch. The emf of this battery charges the capacitance $C$ through resistance $R$. As the capacitor charges, its potential soon reaches 15 mV, triggering the action potential in the next node. As shown in FIGURE 23.51, this process continues, with the depolarization "jumping" from node to node down the axon.

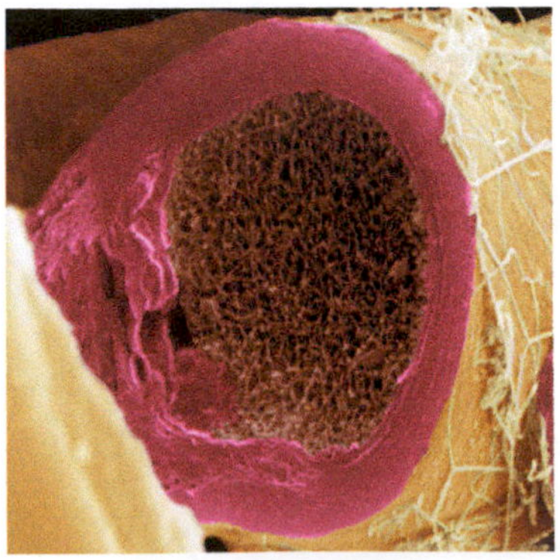

◀ The interior of this axon is insulated from the surrounding intercellular fluids by a thick myelin sheath, clearly visible in the cross-section view.

KEY CONCEPT **FIGURE 23.50** A circuit model of nerve-impulse propagation along myelinated axons.

(a) An electrical model of a myelinated axon (b) Impulse propagation in the myelinated axon

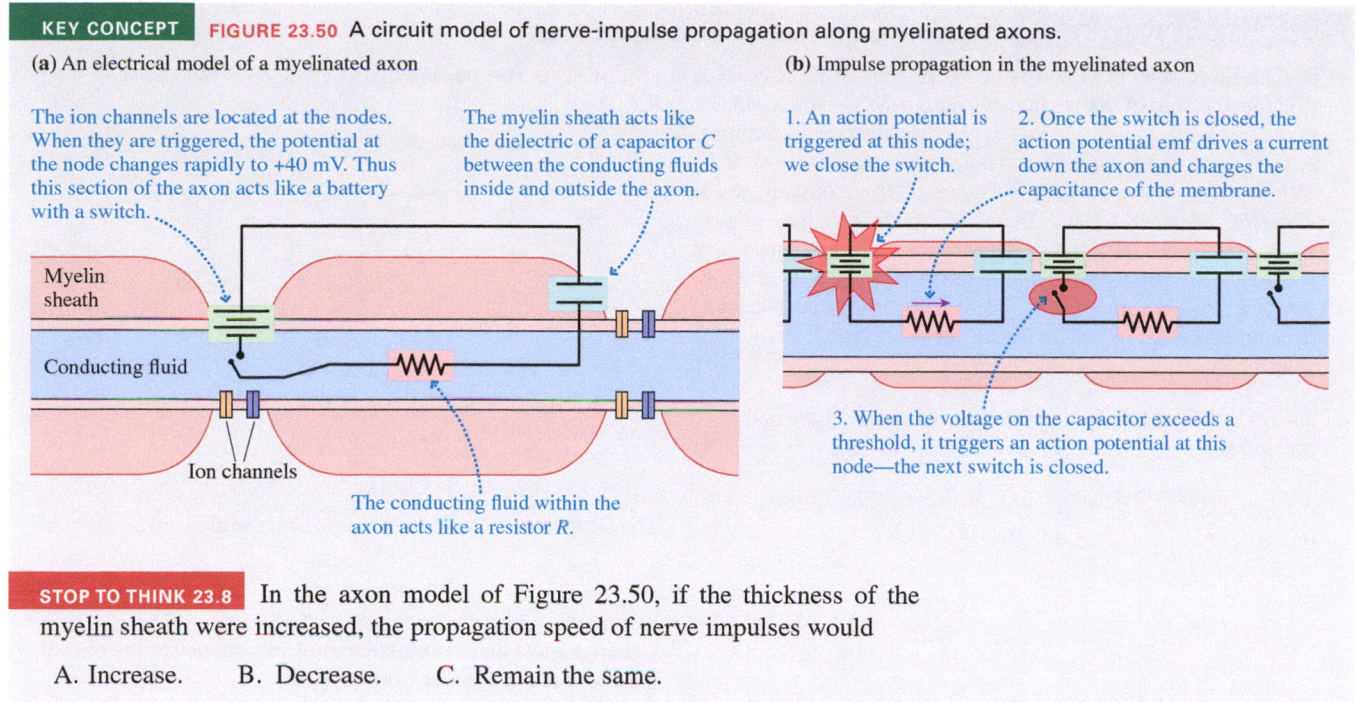

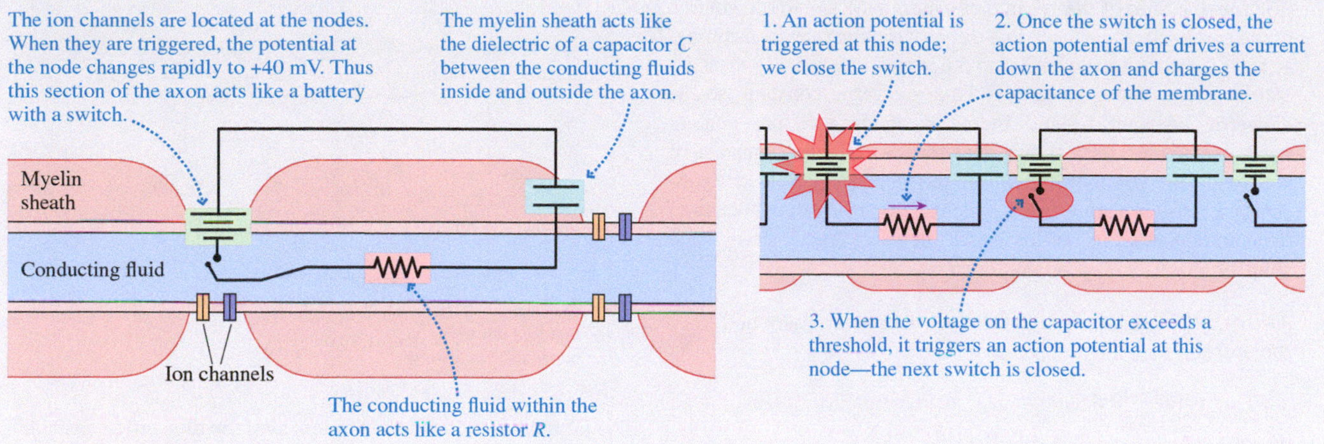


STOP TO THINK 23.8 In the axon model of Figure 23.50, if the thickness of the myelin sheath were increased, the propagation speed of nerve impulses would

A. Increase. B. Decrease. C. Remain the same.

How rapidly does an impulse move down a myelinated axon? The critical time for propagation is the time constant $\tau = RC$ for charging the capacitance of the segments of the axon.

The resistance of an axon between one node and the next is $\approx 25\ \text{M}\Omega$. The myelin insulation increases the separation between the inner conducting fluid and the outer conducting fluid. Because the capacitance of a capacitor depends inversely on the electrode spacing $d$, the myelin reduces the capacitance of the membrane from the $\approx 90$ pF we calculated earlier to $\approx 1.6$ pF per segment. With these values, the time constant for charging the capacitor in one segment is

$$\tau = R_{\text{axon}} C_{\text{membrane}} = (25 \times 10^6\ \Omega)(1.6 \times 10^{-12}\ \text{F}) = 40\ \mu\text{s}$$

We've modeled the axon as a series of such segments, and the time constant is a good estimate of how much time it takes for a signal to jump from one node to the next. Because the nodes of Ranvier are spaced about 1 mm apart, the speed at which the nerve impulse travels down the axon is approximately

$$v = \frac{L_{\text{node}}}{\tau} = \frac{1.0 \times 10^{-3}\ \text{m}}{40 \times 10^{-6}\ \text{s}} = 25\ \text{m/s}$$

Although our model of nerve-impulse propagation is very simple, this predicted speed is just about right for saltatory conduction of signals in myelinated axons. This speed is 25 times faster than that in unmyelinated axons; at this speed, your brain can send a signal to your hand in $\approx \frac{1}{25}$ s.

Your electrical nature might not be as apparent as that of the electric eel, but the operation of your nervous system is inherently electrical. When you decide to move your hand, the signal from your brain travels to your hand in a process that is governed by the electrical nature of the cells in your body.

**FIGURE 23.51** Nerve-impulse propagation along a myelinated axon.

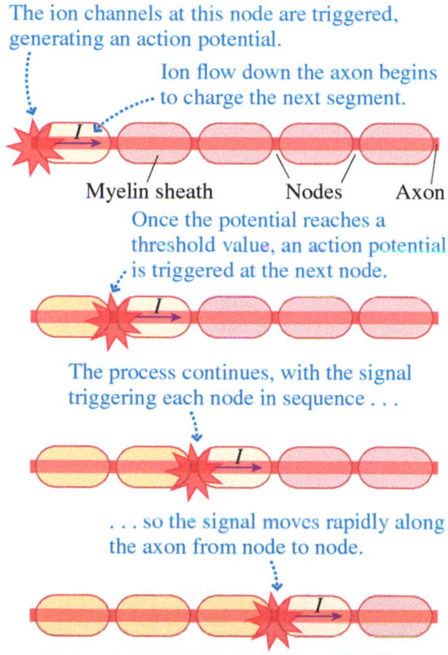

The ion channels at this node are triggered, generating an action potential.
Ion flow down the axon begins to charge the next segment.
Myelin sheath
Nodes
Axon
Once the potential reaches a threshold value, an action potential is triggered at the next node.
The process continues, with the signal triggering each node in sequence . . .
. . . so the signal moves rapidly along the axon from node to node.

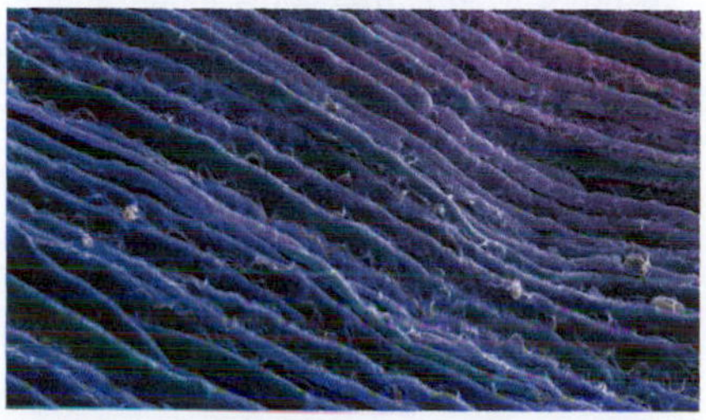

▶ Myelinated axons in the spinal cord carry electrical signals between the brain and body. You can see the nodes of Ranvier on some of the axons.

## INTEGRATED EXAMPLE 23.16 Measuring the moisture content of soil

The moisture content of soil is given in terms of its *volumetric fraction,* the ratio of the volume of water in the soil to the volume of the soil itself. Measuring this directly is quite time-consuming, so soil scientists are eager to find other means to reliably measure soil moisture. Water has a very large dielectric constant, so the dielectric constant of soil increases as its moisture content increases. FIGURE 23.52 shows data for the dielectric constant of soil versus the volumetric fraction; the increase in dielectric constant with soil moisture is quite clear. This strong dependence of dielectric constant on soil moisture allows a very sensitive—and simple—electrical test of soil moisture.

FIGURE 23.52 Variation of the dielectric constant with soil moisture.

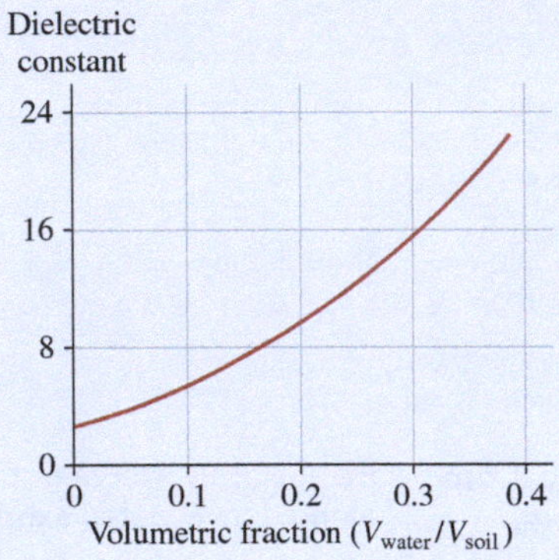


A soil moisture meter has a probe with two separated electrodes. When the probe is inserted into the soil, the electrodes form a capacitor whose capacitance depends on the dielectric constant of the soil between them. A circuit charges the capacitor probe to 3.0 V, then discharges it through a resistor. The decay time depends on the capacitance—and thus the soil moisture—so a measurement of the time for the capacitor to discharge allows a determination of the amount of moisture in the soil.

In air, the probe's capacitor takes 15 $\mu$s to discharge from 3.0 V to 1.0 V. In one particular test, when the probe was inserted into the ground, this discharge required 150 $\mu$s. What was the approximate volumetric fraction of water for the soil in this test?

***STRATEGIZE*** FIGURE 23.53 is a sketch of the measurement circuit of the soil moisture meter. The capacitor is first charged, then connected across a resistor to form an *RC* circuit. The decay of the capacitor voltage is governed by the time constant for the circuit. The time constant depends on the resistance and on the electrode capacitance, which depends on the dielectric constant of the soil between the electrodes.

***PREPARE*** The capacitance of the probe in air is that of a parallel-plate capacitor: $C_{air} = \epsilon_0 A/d$. The capacitance of the probe in soil differs only by the additional factor of the dielectric constant of the medium between the plates—in this case, the soil:

$$C_{soil} = \kappa_{soil}\left(\frac{\epsilon_0 A}{d}\right) = \kappa_{soil} C_{air}$$

FIGURE 23.53 The measurement circuit of the soil moisture meter.

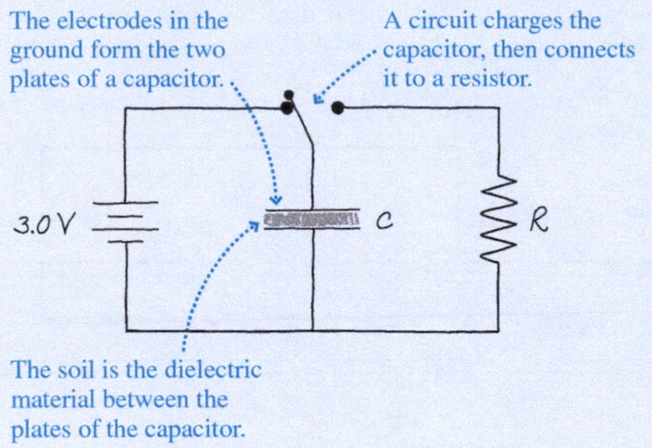

The electrodes in the ground form the two plates of a capacitor.
A circuit charges the capacitor, then connects it to a resistor.
3.0 V
C
R
The soil is the dielectric material between the plates of the capacitor.

The ratio of the capacitance values gives the dielectric constant:

$$\kappa_{soil} = \frac{C_{soil}}{C_{air}}$$

Once we know the dielectric constant, we can determine the volumetric fraction of water from the graph.

***SOLVE*** We are given the times for the decay in air and in soil, but not the capacitance or the resistance of the probe. That's not a problem, though; we don't actually need the capacitance, only the *ratio* of the capacitances in air and in soil. Equation 23.22 gives the voltage decay of an *RC* circuit: $\Delta V_C = (\Delta V_C)_0\, e^{-t/RC}$. In air, the decay is

$$1.0\text{ V} = (3.0\text{ V})e^{-(15\times10^{-6}\text{ s})/RC_{air}}$$

In soil, the decay is

$$1.0\text{ V} = (3.0\text{ V})e^{-(150\times10^{-6}\text{ s})/RC_{soil}}$$

Because the starting and ending points for the decay are the same, the exponents of the two expressions must be equal:

$$\frac{15\times10^{-6}\text{ s}}{RC_{air}} = \frac{150\times10^{-6}\text{ s}}{RC_{soil}}$$

We can solve this for the ratio of the capacitances in soil and air:

$$\frac{C_{soil}}{C_{air}} = \frac{150\times10^{-6}\text{ s}}{15\times10^{-6}\text{ s}} = 10$$

We saw previously that this ratio is the dielectric constant, so $\kappa_{soil} = 10$. We then use the graph of Figure 23.52 to determine that this dielectric constant corresponds to a volumetric water fraction of approximately 0.20.

***ASSESS*** The decay times in air and in soil differ by a factor of 10, so the capacitance in the soil is much larger than that in air. This implies a large dielectric constant, meaning that there is a lot of water in the soil. A volumetric fraction of 0.20 means that 20% of the soil's volume is water (that is, 1.0 cm$^3$ of soil contains 0.20 cm$^3$ of water)—which is quite a bit, so our result seems reasonable.

# SUMMARY

**GOAL** To understand the fundamental physical principles that govern electric circuits.

## GENERAL PRINCIPLES

### Kirchhoff's loop law

For a closed loop:

- Assign a direction to the current.
- Add potential differences around the loop:

$$\sum_i \Delta V_i = 0$$

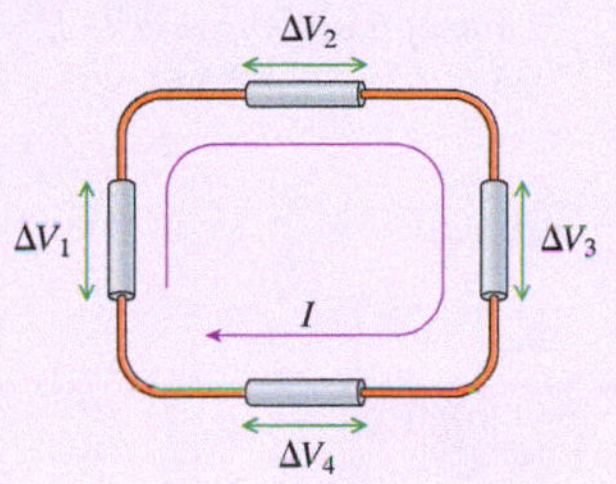


### Kirchhoff's junction law

For a junction:

$$\sum I_{in} = \sum I_{out}$$

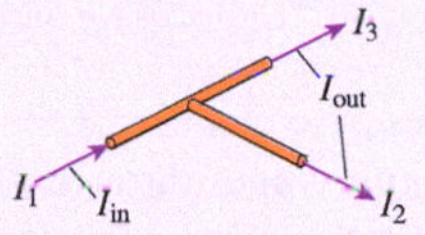


### Analyzing Circuits

***STRATEGIZE*** Sequentially reduce parallel and series resistor combinations to their equivalent resistors.

***PREPARE*** Draw a circuit diagram.

***SOLVE*** *Break the circuit down:*

- Reduce the circuit to the smallest possible number of equivalent resistors.
- Find the current and potential difference.

*Rebuild the circuit:*

- Find the current and potential difference for each resistor.

***ASSESS*** Verify that

- The sum of the potential differences across series resistors equals the potential difference across the equivalent resistor.
- The sum of the currents through parallel resistors equals the current through the equivalent resistor.

## IMPORTANT CONCEPTS

### Series elements

A series connection has no junction. The current in each element is the same.

Resistors in series can be reduced to an equivalent resistance:

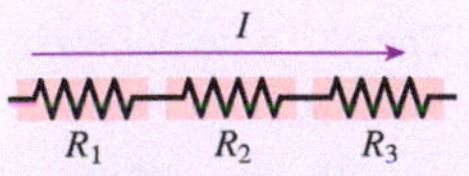


$$R_{eq} = R_1 + R_2 + R_3 + \cdots$$

Capacitors in series can be reduced to an equivalent capacitance:

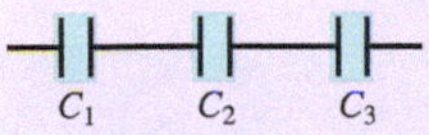


$$C_{eq} = \left(\frac{1}{C_1} + \frac{1}{C_2} + \frac{1}{C_3} + \cdots\right)^{-1}$$

### Parallel elements

Elements connected in parallel are connected by wires at both ends. The potential difference across each element is the same.

Resistors in parallel can be reduced to an equivalent resistance:

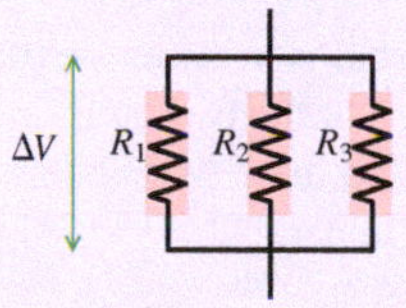


$$R_{eq} = \left(\frac{1}{R_1} + \frac{1}{R_2} + \frac{1}{R_3} + \cdots\right)^{-1}$$

Capacitors in parallel can be reduced to an equivalent capacitance:

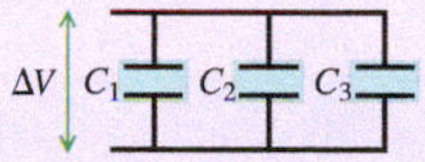


$$C_{eq} = C_1 + C_2 + C_3 + \cdots$$

## APPLICATIONS

### *RC* circuits

The discharge of a capacitor through a resistor is an exponential decay:

$$\Delta V_C = (\Delta V_C)_0 e^{-t/RC}$$

$$I = I_0 e^{-t/RC}$$

The **time constant** for the decay is

$$\tau = RC$$

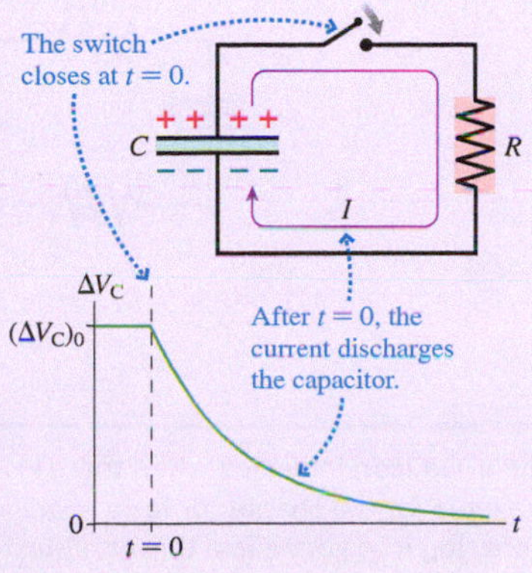


### Electricity in the nervous system

Cells in the nervous system maintain a negative potential inside the cell membrane. When triggered, the membrane depolarizes and generates an *action potential.*

An action potential travels as a wave along the axon of a neuron. More rapid saltatory conduction can be achieved by insulating the axon with myelin, causing the action potential to jump from node to node.

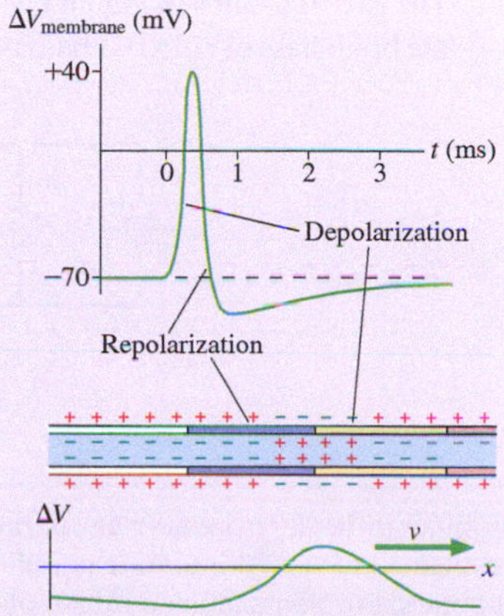

## Learning Objectives After studying this chapter, you should be able to:

- Use Kirchhoff's and Ohm's laws to analyze circuits. *Conceptual Questions 23.18, 23.19; Problems 23.9, 23.20, 23.26, 23.27, 23.31*
- Calculate the equivalent resistance of resistors in series and in parallel. *Conceptual Questions 23.12, 23.13; Problems 23.10, 23.11, 23.19, 23.20, 23.21*
- Calculate the equivalent capacitance of capacitors in series and in parallel. *Conceptual Question 23.26; Problems 23.36, 23.37, 23.40, 23.41, 23.42*
- Analyze the growth and decay of current and potential in an *RC* circuit. *Conceptual Questions 23.27, 23.30; Problems 23.44, 23.45, 23.46, 23.48, 23.49*
- Apply a model of signal propagation in the nervous system. *Conceptual Questions 23.32, 23.33; Problems 23.50, 23.51, 23.52, 23.53, 23.55*

## STOP TO THINK ANSWERS

**Chapter Preview Stop to Think: $R_4 < R_3 = R_2 < R_1$.** Ohm's law can be used to solve for $R$: $R = \Delta V/I$. Thus we have $R_1 = (2\text{ V})/(1\text{ A}) = 2\ \Omega$, $R_2 = (2\text{ V})/(2\text{ A}) = 1\ \Omega$, $R_3 = (1\text{ V})/(1\text{ A}) = 1\ \Omega$, and $R_4 = (1\text{ V})/(2\text{ A}) = \frac{1}{2}\ \Omega$.

**Stop to Think 23.1: A, B, and D.** These three are the same circuit because the logic of the connections is the same. In each case, there is a junction that connects one side of each circuit element and a second junction that connects the other side. In C, the functioning of the circuit is changed by the extra wire connecting the two sides of the capacitor.

**Stop to Think 23.2: B.** The potential difference in crossing the battery is its emf, or 10 V. The potential drop across the 6 Ω resistor is $\Delta V = -IR = -6.0$ V. To make the sum of the potential differences zero, as Kirchhoff's loop law requires, the potential difference across resistor $R$ must be $-4.0$ V.

**Stop to Think 23.3: C = D > A = B.** The two bulbs in series are of equal brightness, as are the two bulbs in parallel. But the two bulbs in series have a larger resistance than a single bulb, so there will be less current through the bulbs in series than the bulbs in parallel.

**Stop to Think 23.4: C.** The voltmeter must be connected in parallel with the resistor, and the ammeter in series.

**Stop to Think 23.5: A > B > C = D.** All the current from the battery goes through A, so it is brightest. The current divides at the junction, but not equally. Because B is in parallel with C + D, but has half the resistance of the two bulbs together, twice as much current travels through B as through C + D. So B is dimmer than A but brighter than C and D. C and D are equally bright because of conservation of current.

**Stop to Think 23.6: $(C_{eq})_B > (C_{eq})_A > (C_{eq})_C$.** Two capacitors in parallel have a larger capacitance than either alone; two capacitors in series have a smaller capacitance than either alone.

**Stop to Think 23.7: B.** The two 2 Ω resistors are in series and equivalent to a 4 Ω resistor. Thus $\tau = RC = 4$ s.

**Stop to Think 23.8: A.** A thicker sheath would space the "plates" of the capacitor—the inner and outer surfaces of the sheath—farther apart, thereby reducing the capacitance $C$. The $RC$ time constant $\tau$ would thus decrease, increasing the speed $v = L_{node}/\tau$.

Video Tutor Solution Chapter 23

# QUESTIONS

## Conceptual Questions

1. The two lightbulbs in Figure Q23.1 are glowing. What happens to the brightness of bulb B if bulb A is removed from the circuit?

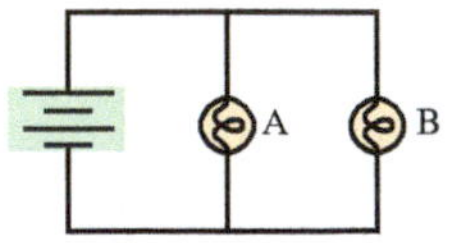


FIGURE Q23.1

2. A battery, a lightbulb, and two resistors are in the circuit shown in Figure Q23.2. What happens to the brightness of the bulb if the two resistors are exchanged?

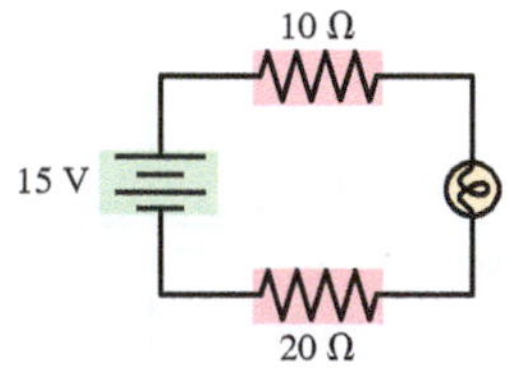


FIGURE Q23.2

Problem difficulty is labeled as | (straightforward) to ||||| (challenging). Problems labeled INT integrate significant material from earlier chapters; Problems labeled BIO are of biological or medical interest.

The eText icon indicates when there is a video tutor solution available for the chapter or for a specific problem. To launch these videos, log into your eText through Mastering™ Physics or log into the Study Area.

3. Current $I_{in}$ flows into three resistors connected together one after the other as shown in Figure Q23.3. The accompanying graph shows the value of the potential as a function of position.
   a. Is $I_{out}$ greater than, less than, or equal to $I_{in}$? Explain.
   b. Rank in order, from largest to smallest, the three resistances $R_1$, $R_2$, and $R_3$. Explain.

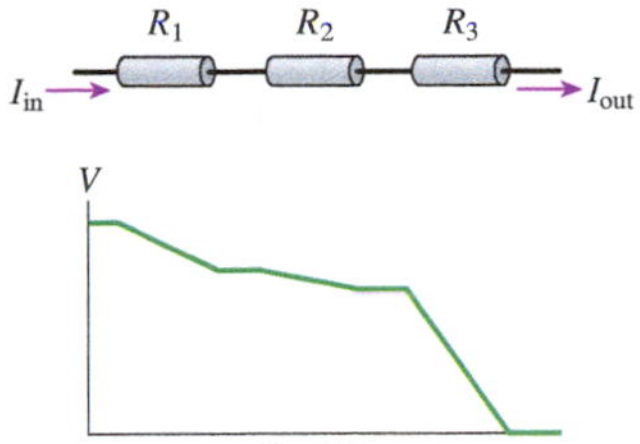


FIGURE Q23.3

4. The circuit in Figure Q23.4 has two resistors, with $R_1 > R_2$. Which resistor dissipates the larger amount of power? Explain.

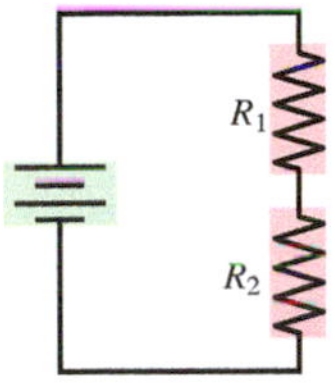


FIGURE Q23.4

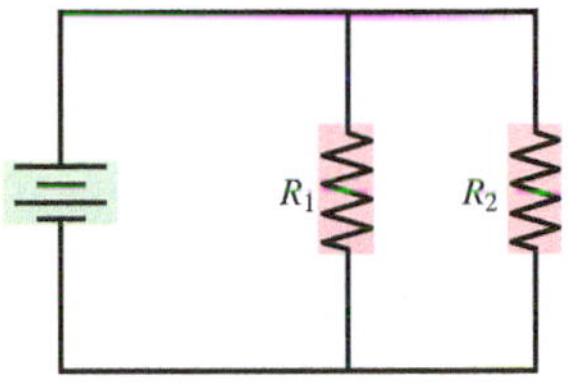


FIGURE Q23.5

5. The circuit in Figure Q23.5 has a battery and two resistors, with $R_1 > R_2$. Which resistor dissipates the larger amount of power? Explain.
6. If the resistors and batteries in Figures Q23.4 and Q23.5 are all the same, which of the two circuits dissipates more total power? Explain.
7. The currents through three resistors in a circuit are shown. Which of the following statements is true? Explain.
   A. $I_2 = I_1 + I_3$ B. $I_1 = I_2 + I_3$
   C. $I_3 = I_1 + I_2$ D. None of these is correct.

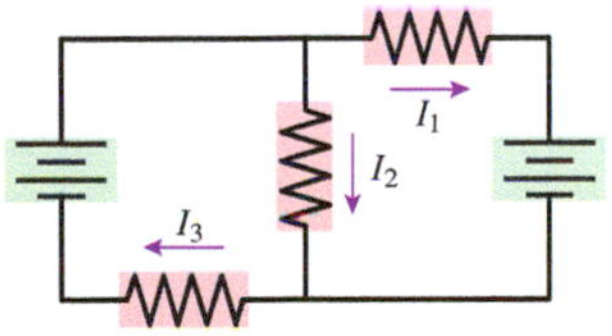


FIGURE Q23.7

8. In the circuit shown in Figure Q23.8, bulbs A and B are glowing. Then the switch is closed. What happens to each bulb? Does it get brighter, stay the same, get dimmer, or go out? Explain.

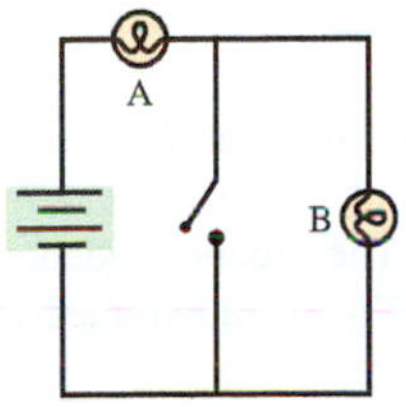


FIGURE Q23.8

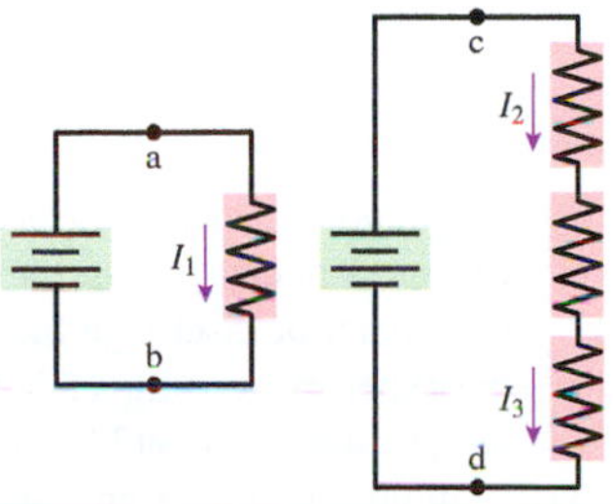


FIGURE Q23.9

9. Figure Q23.9 shows two circuits. The two batteries are identical and the four resistors all have exactly the same resistance.
   a. Is $\Delta V_{ab}$ larger than, smaller than, or equal to $\Delta V_{cd}$? Explain.
   b. Rank in order, from largest to smallest, the currents $I_1$, $I_2$, and $I_3$. Explain.

10. Figure Q23.10 shows two circuits. The two batteries are identical and the four resistors all have exactly the same resistance.
    a. Compare $\Delta V_{ab}$, $\Delta V_{cd}$, and $\Delta V_{ef}$. Are they all the same? If not, rank them in order from largest to smallest. Explain.
    b. Rank in order, from largest to smallest, the five currents $I_1$ to $I_5$. Explain.

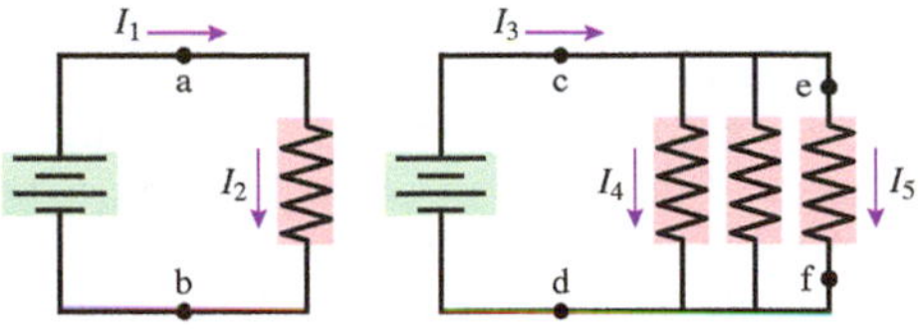


FIGURE Q23.10

11. a. In Figure Q23.11, what fraction of current $I$ goes through the 3 Ω resistor?
    b. If the 9 Ω resistor is replaced with a larger resistor, will the fraction of current going through the 3 Ω resistor increase, decrease, or stay the same?

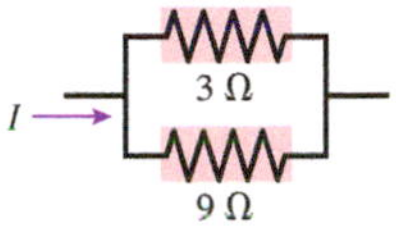


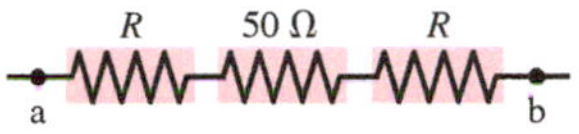


FIGURE Q23.11 FIGURE Q23.12

12. Two of the three resistors in Figure Q23.12 are unknown but equal. Is the total resistance between points a and b less than, greater than, or equal to 50 Ω? Explain.
13. Two of the three resistors in Figure Q23.13 are unknown but equal. Is the total resistance between points a and b less than, greater than, or equal to 200 Ω? Explain.

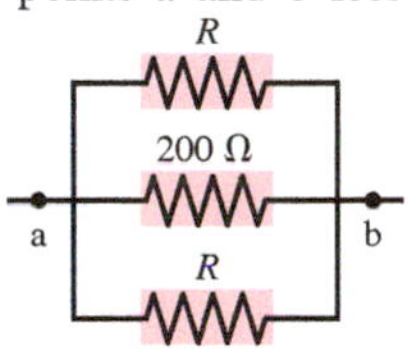


FIGURE Q23.13

14. Rank in order, from largest to smallest, the currents $I_1$, $I_2$, and $I_3$ in the circuit diagram in Figure Q23.14.

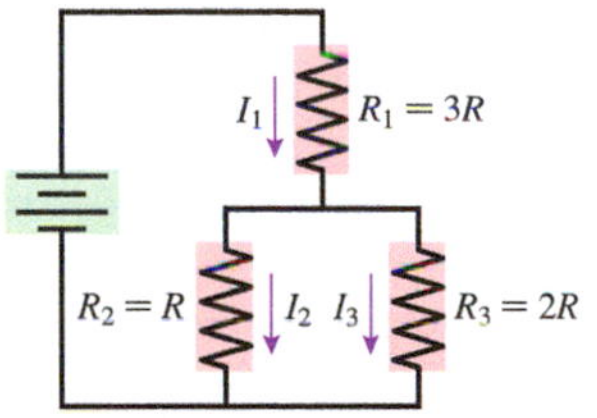


FIGURE Q23.14

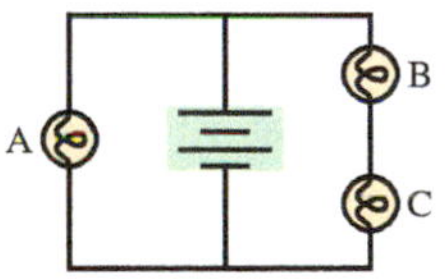


FIGURE Q23.15

15. The three bulbs in Figure Q23.15 are identical. Rank the bulbs from brightest to dimmest. Explain.
16. The four bulbs in Figure Q23.16 are identical. Rank the bulbs from brightest to dimmest. Explain.

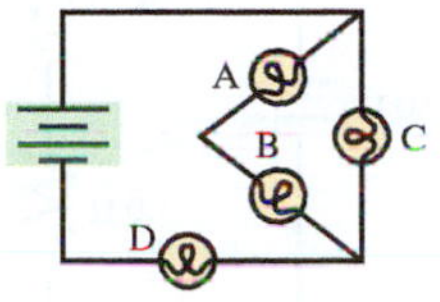


FIGURE Q23.16

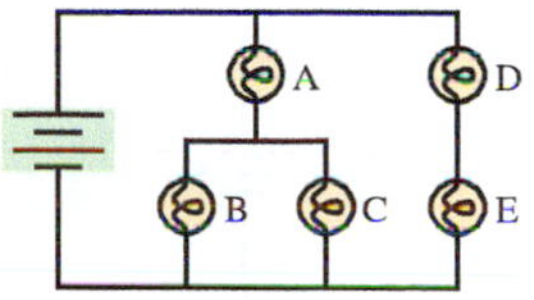


FIGURE Q23.17

17. Figure Q23.17 shows five identical bulbs connected to a battery. All the bulbs are glowing. Rank the bulbs from brightest to dimmest. Explain.

18. a. The three bulbs in Figure Q23.18 are identical. Rank the bulbs from brightest to dimmest. Explain.
    b. Suppose a wire is connected between points 1 and 2. What happens to each bulb? Does it get brighter, stay the same, get dimmer, or go out? Explain.

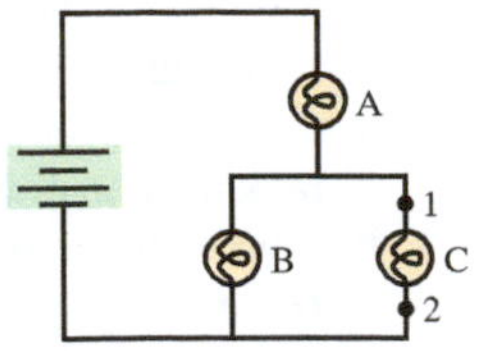

FIGURE Q23.18

19. The three bulbs in Figure Q23.18 are identical. When bulb C is removed from the circuit, what happens to the brightness of bulb A? Of bulb B? Explain.
20. What should the value of $R$ be in the circuit of Figure Q23.20 such that the power dissipated in $R$ is twice that dissipated in the 100 Ω resistor?

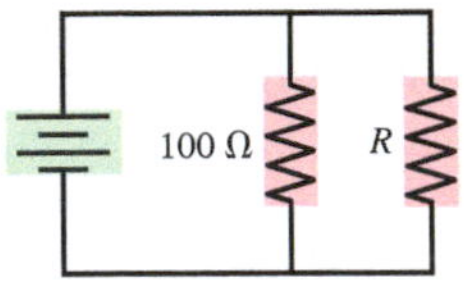

FIGURE Q23.20

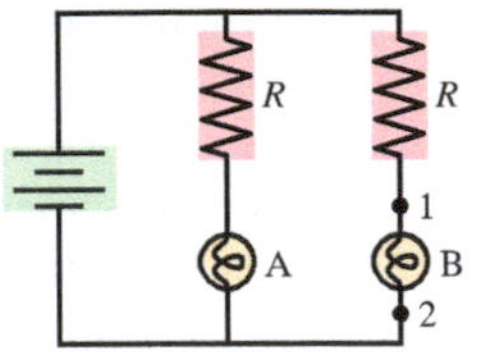

FIGURE Q23.21

21. Initially, bulbs A and B in Figure Q23.21 are both glowing. Bulb B is then removed from its socket. Does removing bulb B cause the potential difference $\Delta V_{12}$ between points 1 and 2 to increase, decrease, stay the same, or become zero? Explain.
22. a. Consider the points a and b in Figure Q23.22. Is the potential difference $\Delta V_{ab}$ between points a and b zero? If so, why? If not, which point is more positive?
    b. If a wire is connected between points a and b, does it carry a current? If so, in which direction—to the right or to the left? Explain.

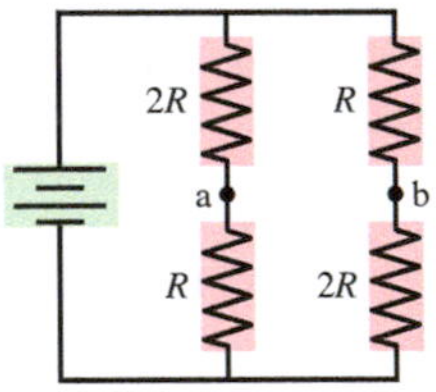

FIGURE Q23.22

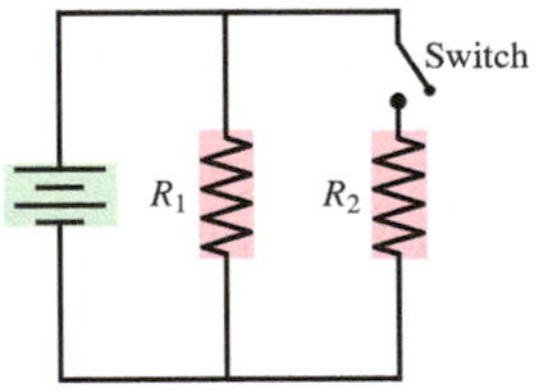

FIGURE Q23.23

23. When the switch in Figure Q23.23 is closed,
    a. Does the current through the battery increase, decrease, or stay the same? Explain.
    b. Does the current through $R_1$ increase, decrease, or stay the same? Explain.
24. A voltmeter is (incorrectly) inserted into a circuit as shown in Figure Q23.24.
    a. What is the current in the circuit?
    b. What does the voltmeter read?
    c. How would you change the circuit to correctly connect the voltmeter to measure the potential difference across the resistor?

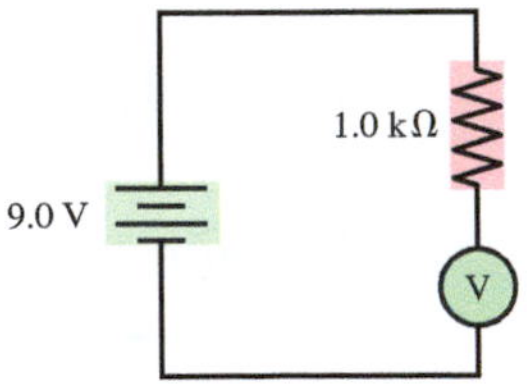

FIGURE Q23.24

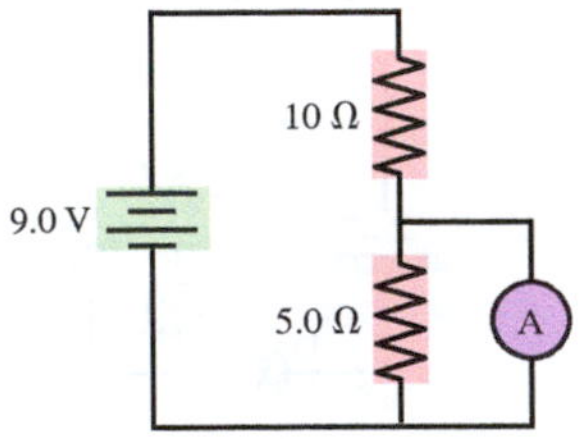

FIGURE Q23.25

25. An ammeter is (incorrectly) inserted into a circuit as shown in Figure Q23.25.
    a. What is the current through the 5.0 Ω resistor?
    b. How would you change the circuit to correctly connect the ammeter to measure the current through the 5.0 Ω resistor?
26. Rank in order, from largest to smallest, the equivalent capacitances $(C_{eq})_1$ to $(C_{eq})_4$ of the four groups of capacitors shown in Figure Q23.26.

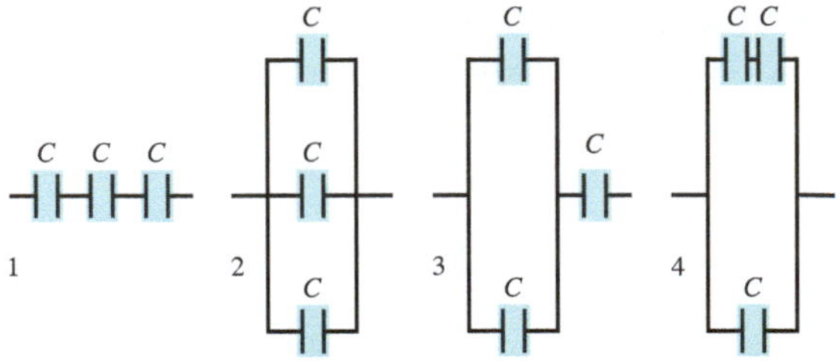

FIGURE Q23.26

27. Figure Q23.27 shows a circuit consisting of a battery, a switch, two identical lightbulbs, and a capacitor that is initially uncharged.
    a. *Immediately* after the switch is closed, are either or both bulbs glowing? Explain.
    b. If both bulbs are glowing, which is brighter? Or are they equally bright? Explain.
    c. For any bulb (A or B or both) that lights up immediately after the switch is closed, does its brightness increase with time, decrease with time, or remain unchanged? Explain.

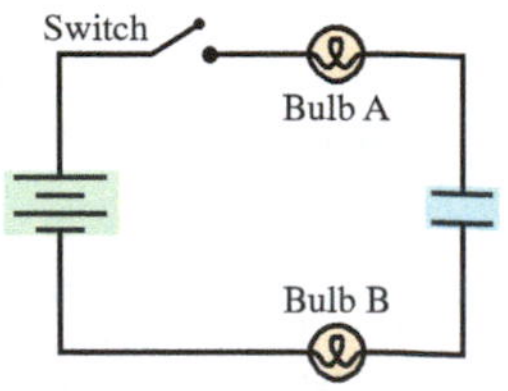

FIGURE Q23.27

28. Figure Q23.28 shows the voltage as a function of time across a capacitor as it is discharged (separately) through three different resistors. Rank in order, from largest to smallest, the values of the resistances $R_1$ to $R_3$.

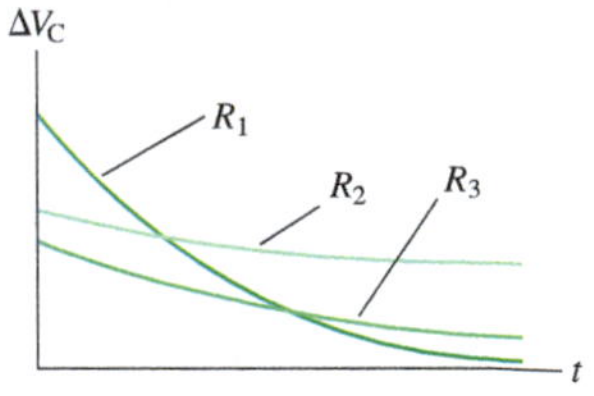

FIGURE Q23.28

29. A charged capacitor could be connected to two identical resistors in either of the two ways shown in Figure Q23.29. Which configuration will discharge the capacitor in the shortest time once the switch is closed? Explain.

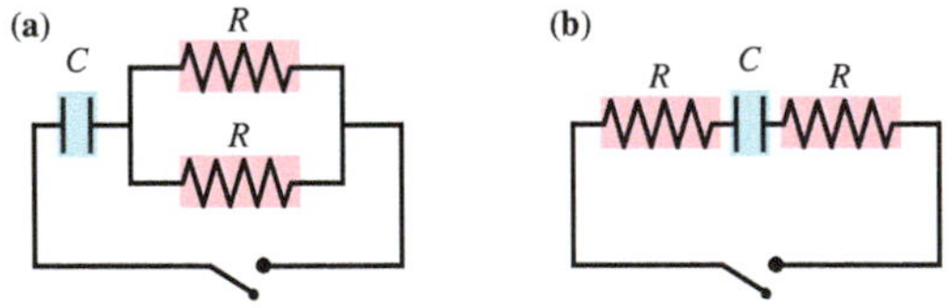

FIGURE Q23.29

30. Two students measure the time constant of an $RC$ circuit. The first student charges the capacitor using a 12 V battery, then lets the capacitor discharge the resistor. The second student repeats the experiment using a 5 V battery. Which student measures the longer time constant? Explain.
31. BIO A device to make an electrical measurement of skin moisture has electrodes that form two plates of a capacitor; the skin is the dielectric between the plates. Adding moisture to the skin means adding water, which has a large dielectric constant. If a circuit repeatedly charges and discharges the capacitor to determine the capacitance, how will an increase in skin moisture affect the charging and discharging time? Explain.

32. Consider the model of nerve conduction in myelinated axons BIO presented in the chapter. Suppose the distance between the nodes of Ranvier was halved for a particular axon.
   a. How would this affect the resistance and the capacitance of one segment of the axon?
   b. How would this affect the time constant for the charging of one segment?
   c. How would this affect the signal propagation speed for the axon?

33. Adding a myelin sheath to an axon results in faster signal prop- BIO agation. It also means that less energy is required for a signal to propagate down the axon. Explain why this is so.

## Multiple-Choice Questions

34. | What is the current in the circuit of Figure Q23.34?
   A. 1.0 A  B. 1.7 A
   C. 2.5 A  D. 4.2 A

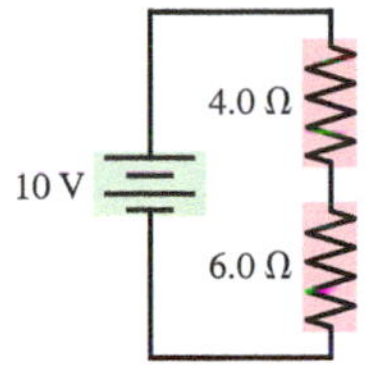

FIGURE Q23.34

35. | Which resistor in Figure Q23.34 dissipates the most power?
   A. The 4.0 Ω resistor.
   B. The 6.0 Ω resistor.
   C. Both dissipate the same power.

36. || Normally, household lightbulbs are connected in parallel to a power supply. Suppose a 40 W and a 60 W lightbulb are, instead, connected in series, as shown in Figure Q23.36. Which bulb is brighter?
   A. The 60 W bulb.
   B. The 40 W bulb.
   C. The bulbs are equally bright.

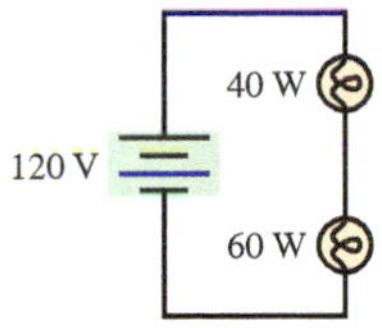

FIGURE Q23.36

37. ||| A metal wire of resistance $R$ is cut into two pieces of equal length. The two pieces are connected together side by side. What is the resistance of the two connected wires?
   A. $R/4$  B. $R/2$  C. $R$  D. $2R$  E. $4R$

38. | In Figure Q23.38, the current $I_{bat}$ in the battery is equal to
   A. $I_1/2$  B. $2I_1/3$  C. $I_1$  D. $3I_1/2$  E. $2I_1$

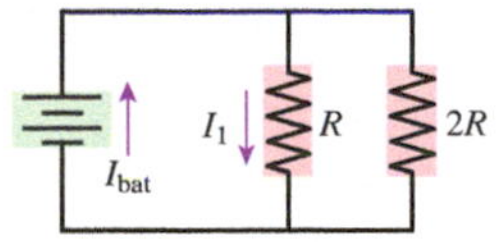

FIGURE Q23.38

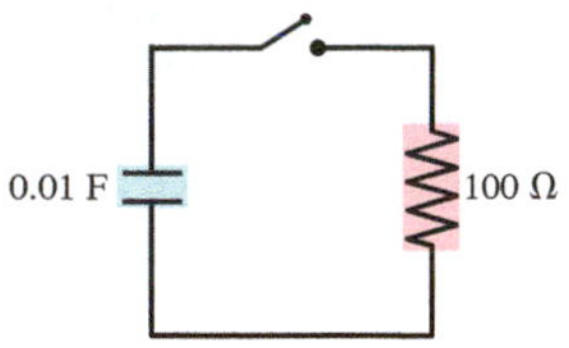

FIGURE Q23.39

39. || With the switch open, the capacitor in Figure Q23.39 has an initial charge of 1.0 C. The switch is then closed. 2.0 s later, the charge on the capacitor is
   A. 1.0 C  B. 0.5 C  C. $(1/e)(1.0\text{ C}) = 0.37\text{ C}$
   D. $(1/e^2)(1.0\text{ C}) = 0.14\text{ C}$

40. | If a cell's membrane thickness doubles but the cell stays the BIO same size, how do the resistance and the capacitance of the cell membrane change?
   A. The resistance and the capacitance increase.
   B. The resistance increases, the capacitance decreases.
   C. The resistance decreases, the capacitance increases.
   D. The resistance and the capacitance decrease.

41. ||| If a cell's diameter is reduced by 50% without changing the BIO membrane thickness, how do the resistance and capacitance of the cell membrane change?
   A. The resistance and the capacitance increase.
   B. The resistance increases, the capacitance decreases.
   C. The resistance decreases, the capacitance increases.
   D. The resistance and the capacitance decrease.

42. || Which of the following would increase the speed of a nerve BIO impulse along an axon?
   A. Decreasing the spacing between the nodes of the axon
   B. Decreasing the resistance between nodes
   C. Decreasing the thickness of the myelin sheath
   D. Increasing the resting potential of the axon

# PROBLEMS

### Section 23.1 Circuit Elements and Diagrams

1. || Draw a circuit diagram for the circuit of Figure P23.1.

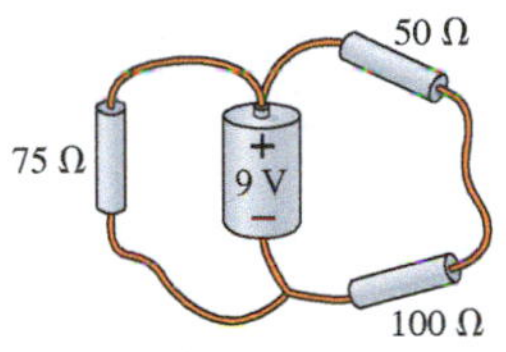

FIGURE P23.1

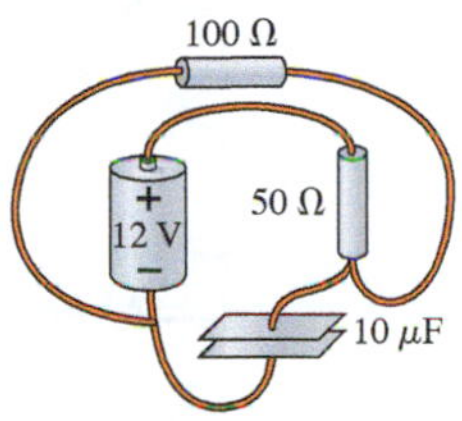

FIGURE P23.2

2. || Draw a circuit diagram for the circuit of Figure P23.2.
3. || Draw a circuit diagram for the circuit of Figure P23.3.

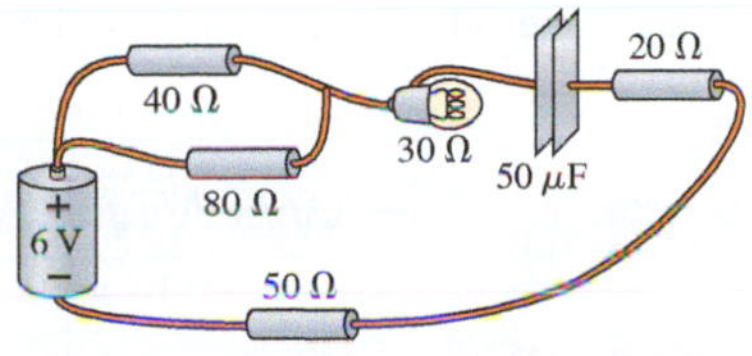

FIGURE P23.3

### Section 23.2 Kirchhoff's Laws

4. || In Figure P23.4, what is the current in the wire above the junction? Does charge flow toward or away from the junction?

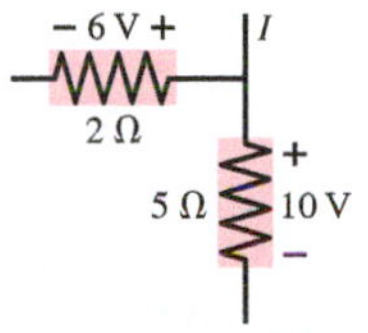

FIGURE P23.4

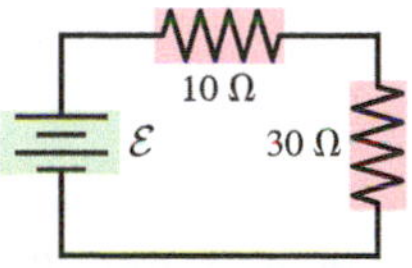

FIGURE P23.5

5. | The current through the 30 Ω resistor in Figure P23.5 is measured to be 0.25 A. What is the emf ℰ of the battery?

6. | a. What is the potential difference across each resistor in Figure P23.6?
   b. Draw a graph of the potential as a function of the distance traveled through the circuit, traveling clockwise from $V = 0$ V at the lower left corner. See Figure P23.9 for an example of such a graph.

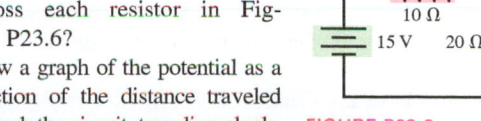

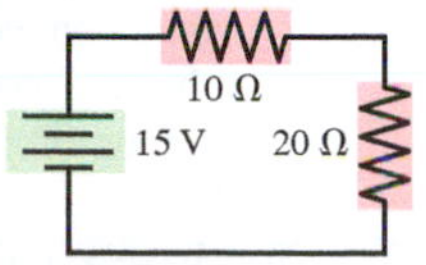

FIGURE P23.6

Watch Video Solution Problem 23.7

7. | a. What are the magnitude and direction of the current in the 18 Ω resistor in Figure P23.7?
   b. Draw a graph of the potential as a function of the distance traveled through the circuit, traveling clockwise from $V = 0$ V at the lower left corner. See Figure P23.9 for an example of such a graph.

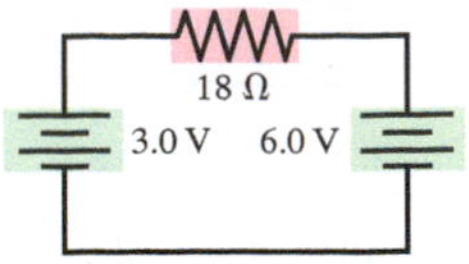


FIGURE P23.7

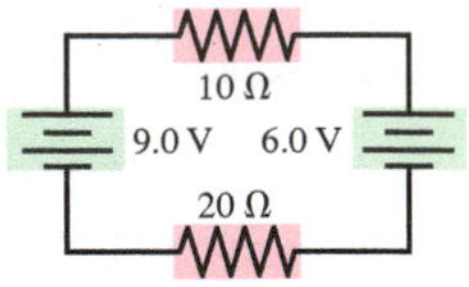


FIGURE P23.8

8. ||| a. What are the magnitude and direction of the current in the 20 Ω resistor in Figure P23.8?
   b. Draw a graph of the potential as a function of the distance traveled through the circuit, traveling clockwise from the lower left corner, where $V = 0$ V. See Figure P23.9 for an example of such a graph.

9. | The current in a circuit with only one battery is 2.0 A. Figure P23.9 shows how the potential changes when going around the circuit in the clockwise direction, starting from the lower left corner. Draw the circuit diagram.

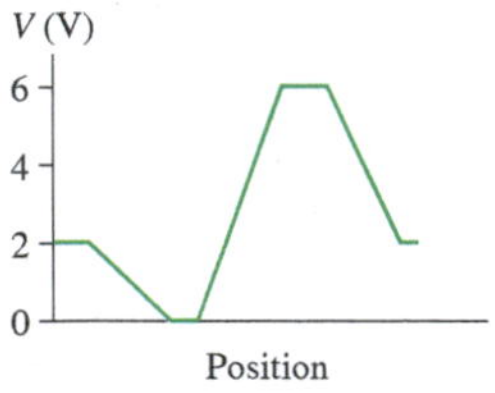


FIGURE P23.9

## Section 23.3 Series and Parallel Circuits

10. | What is the equivalent resistance of each group of resistors shown in Figure P23.10?

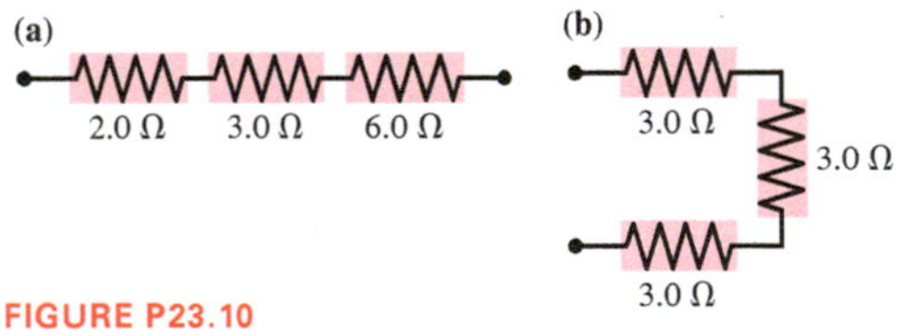


FIGURE P23.10

11. | What is the equivalent resistance of each group of resistors shown in Figure P23.11?

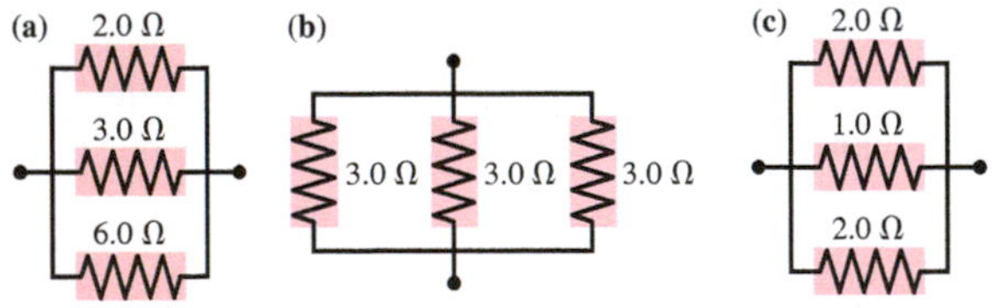


FIGURE P23.11

12. | Three resistors in parallel have an equivalent resistance of 5.0 Ω. Two of the resistors have resistances of 10 Ω and 30 Ω. What is the resistance of the third resistor?

13. | Three identical resistors have an equivalent resistance of 30 Ω when connected in parallel. What is their equivalent resistance when connected in series?

14. | You have a collection of 1.0 kΩ resistors. How can you connect four of them to produce an equivalent resistance of 0.25 kΩ?

15. | You have a collection of six 1.0 kΩ resistors. What is the smallest resistance you can make by combining them?

16. || You have six 1.0 kΩ resistors. How can you connect them to produce a total equivalent resistance of 1.5 kΩ?

17. |||| When two resistors are wired in series with a 12 V battery, the current through the battery is 0.30 A. When they are wired in parallel with the same battery, the current is 1.6 A. What are the values of the two resistors?

18. ||| What is the value of resistor $R$ in Figure P23.18?

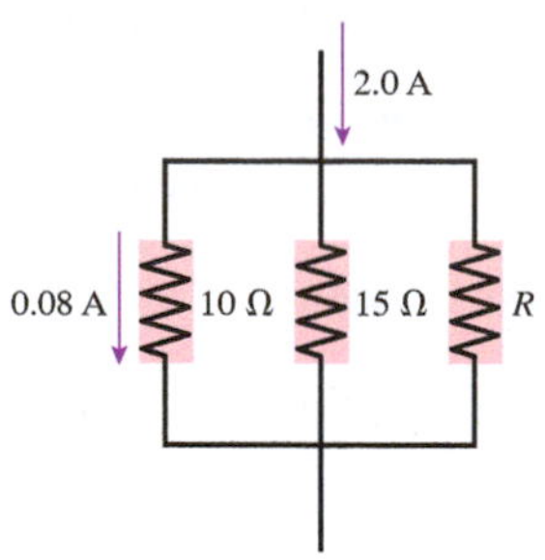


FIGURE P23.18

## Section 23.4 Measuring Voltage and Current

## Section 23.5 More Complex Circuits

19. || What is the equivalent resistance between points a and b in Figure P23.19?

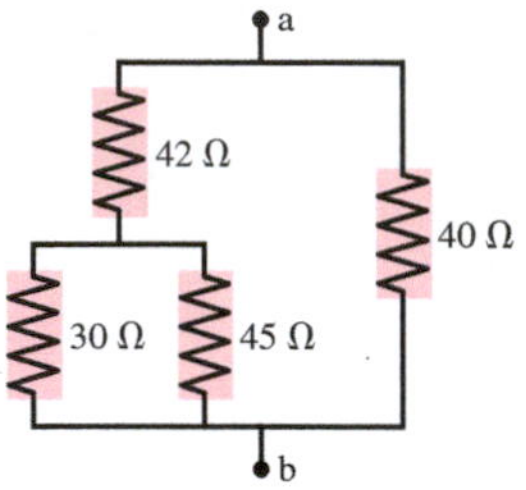


FIGURE P23.19

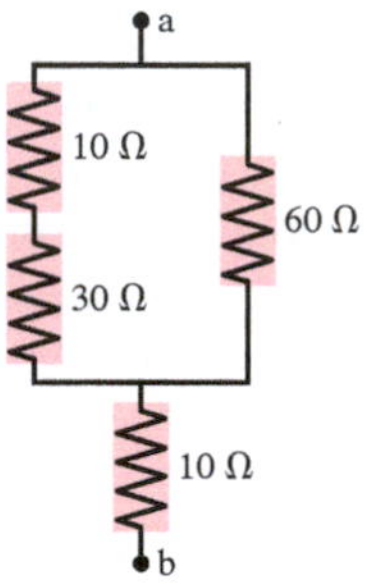


FIGURE P23.20

20. | What is the equivalent resistance between points a and b in Figure P23.20?

21. || You have three 1.0 kΩ resistors. What are the values of all the equivalent resistances that can be formed using all three of these resistors?

22. || Two batteries supply current to the circuit in Figure P23.22. The figure shows the potential difference across two of the resistors and the value of the third resistor. What current is supplied by the batteries?

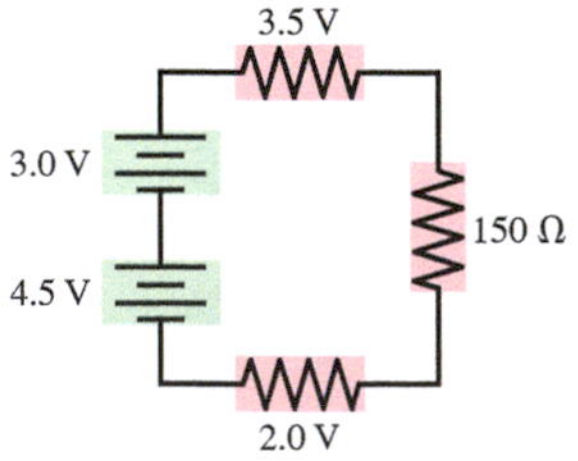


FIGURE P23.22

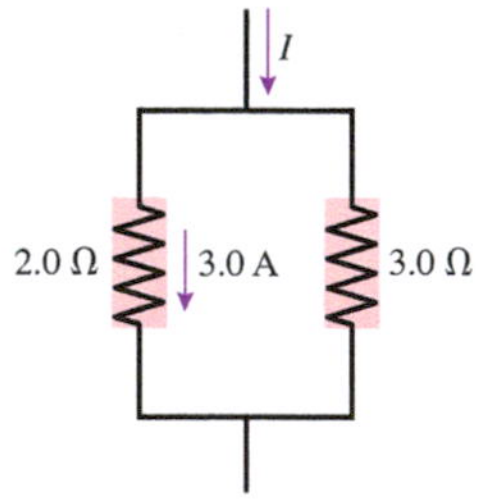


FIGURE P23.23

23. | Part of a circuit is shown in Figure P23.23.
   a. What is the current through the 3.0 Ω resistor?
   b. What is the value of the current $I$?

24. | What is the value of resistor $R$ in Figure P23.24?

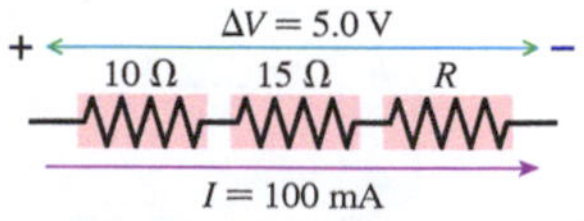


FIGURE P23.24

25. || What are the resistances $R$ and the emf of the battery in Figure P23.25?

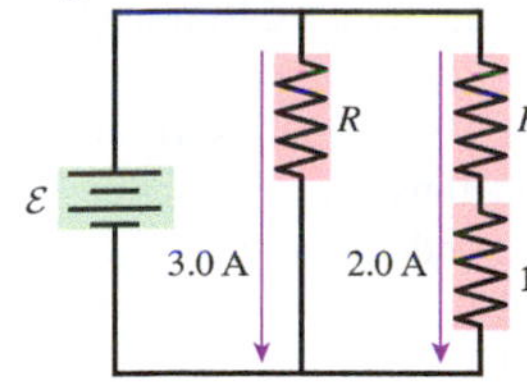


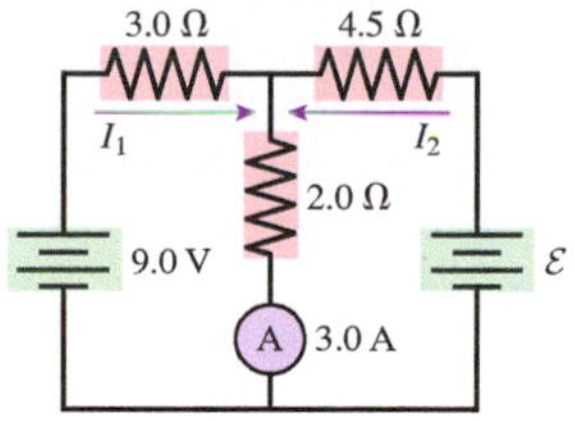


FIGURE P23.25 FIGURE P23.26

26. || The ammeter in Figure P23.26 reads 3.0 A. Find $I_1$, $I_2$, and $\mathcal{E}$.

27. || Find the current through and the potential difference across each resistor in Figure P23.27.

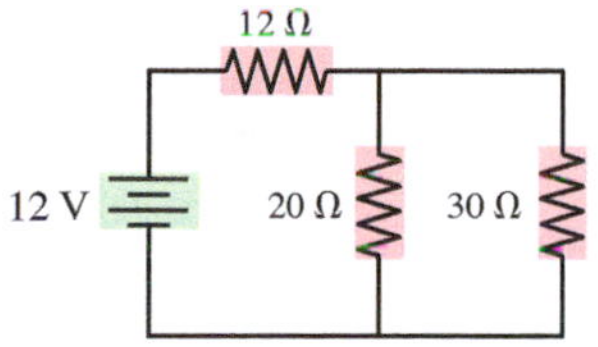


FIGURE P23.27

28. ||| Find the current through and the potential difference across each resistor in Figure P23.28.

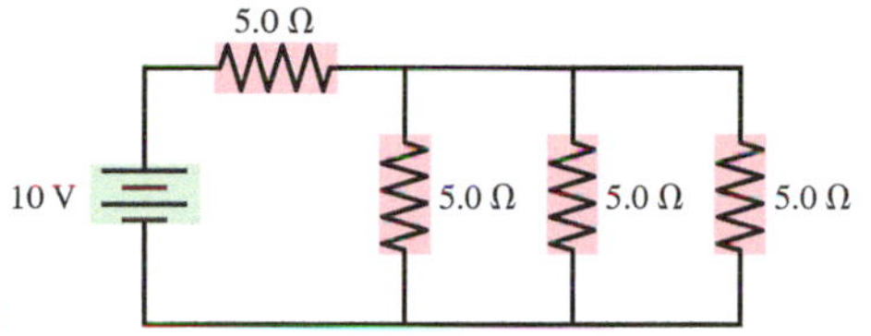


FIGURE P23.28

29. ||| Find the current through and the potential difference across each resistor in Figure P23.29.

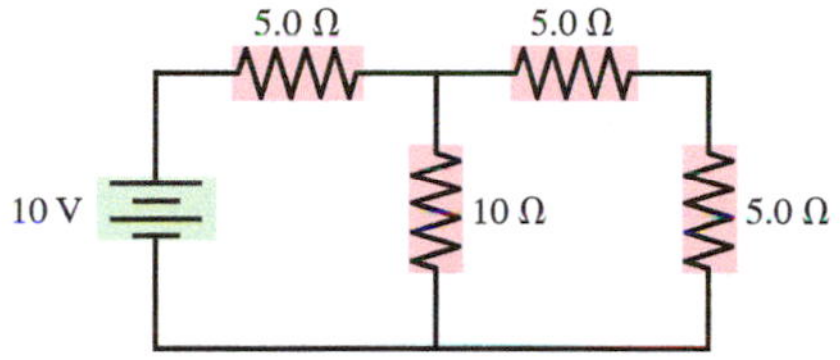


FIGURE P23.29

30. ||| What is the current through the battery in Figure P23.30?

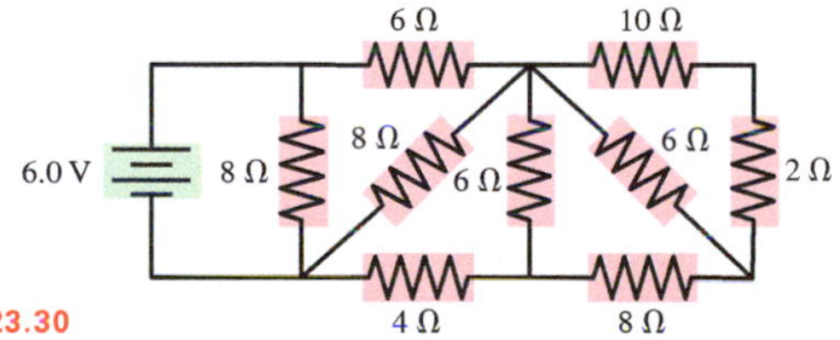


FIGURE P23.30

31. || For the circuit shown in Figure P23.31, find the current through and the potential difference across each resistor. Place your results in a table for ease of reading.

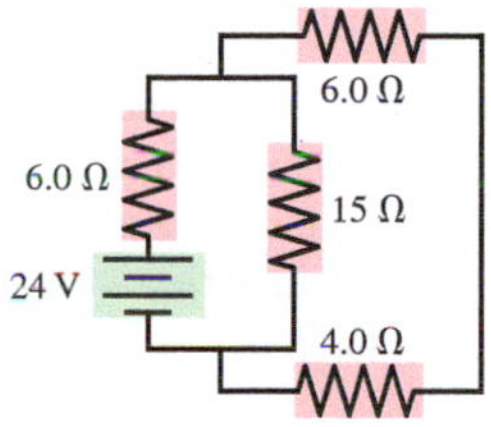


FIGURE P23.31

32. |||| Consider the potential differences between pairs of points in Figure P23.32. What are the magnitudes of the potential differences $\Delta V_{14}$, $\Delta V_{24}$, and $\Delta V_{34}$?

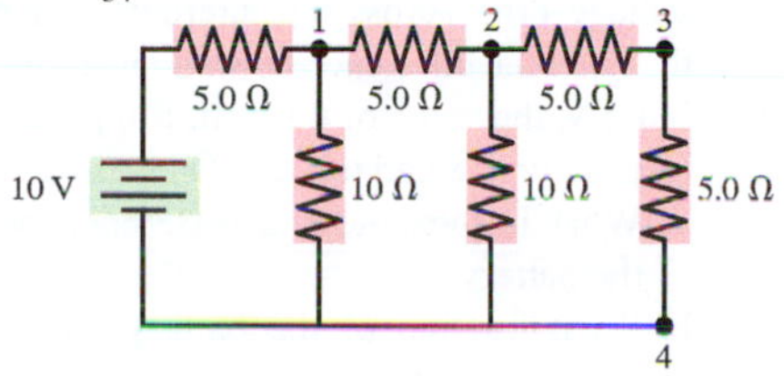


FIGURE P23.32

33. || For the circuit shown in Figure P23.33, find the current through and the potential difference across each resistor. Place your results in a table for ease of reading.

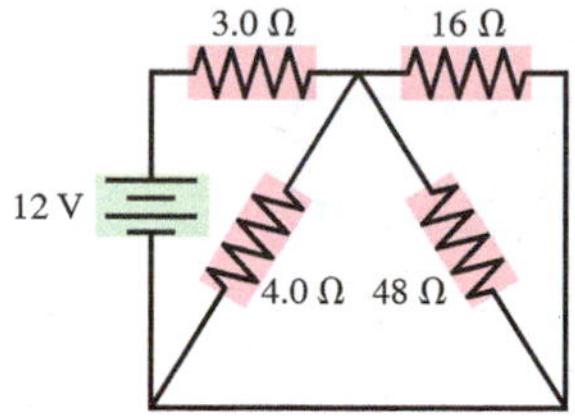


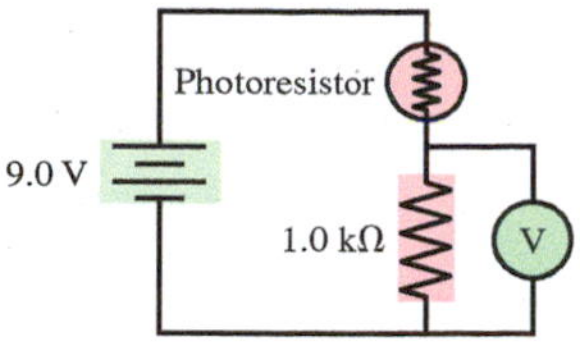


FIGURE P23.33 FIGURE P23.34

34. || A photoresistor, whose resistance decreases with light intensity, is connected in the circuit of Figure P23.34. On a sunny day, the photoresistor has a resistance of 0.56 kΩ. On a cloudy day, the resistance rises to 4.0 kΩ. At night, the resistance is 20 kΩ.
   a. What does the voltmeter read for each of these conditions?
   b. Does the voltmeter reading increase or decrease as the light intensity increases?

35. ||| The two unknown resistors in Figure P23.35 have the same resistance $R$. When the switch is closed, the current through the battery increases by 50%. What is $R$?

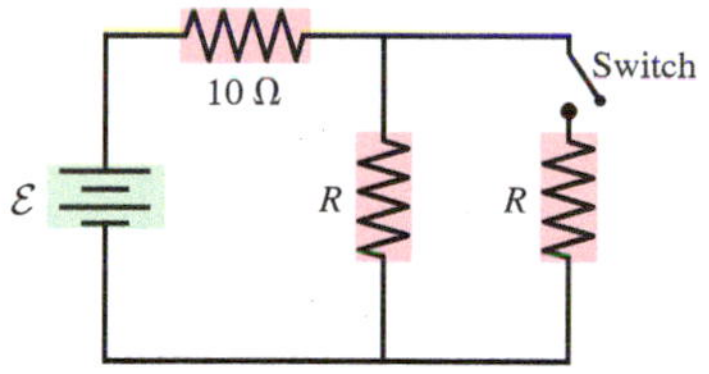


FIGURE P23.35

## Section 23.6 Capacitors in Parallel and Series

36. | A 6.0 $\mu$F capacitor, a 10 $\mu$F capacitor, and a 16 $\mu$F capacitor are connected in parallel. What is their equivalent capacitance?

37. | A 6.0 $\mu$F capacitor, a 10 $\mu$F capacitor, and a 16 $\mu$F capacitor are connected in series. What is their equivalent capacitance?

38. | You need a capacitance of 50 $\mu$F, but you don't happen to have a 50 $\mu$F capacitor. You do have a 30 $\mu$F capacitor. What additional capacitor do you need to produce a total capacitance of 50 $\mu$F? Should you join the two capacitors in parallel or in series?

39. | You need a capacitance of 50 $\mu$F, but you don't happen to have a 50 $\mu$F capacitor. You do have a 75 $\mu$F capacitor. What additional capacitor do you need to produce a total capacitance of 50 $\mu$F? Should you join the two capacitors in parallel or in series?

40. || What is the equivalent capacitance of the three capacitors in Figure P23.40?

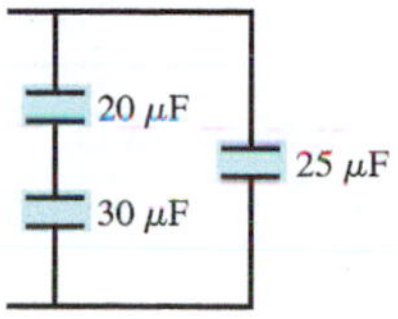


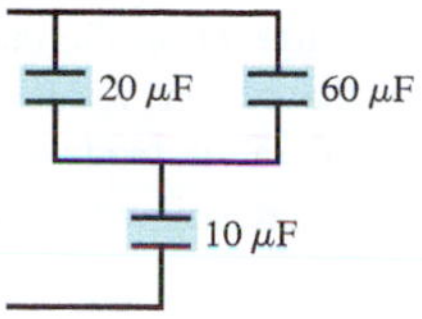


FIGURE P23.40 FIGURE P23.41

41. | What is the equivalent capacitance of the three capacitors in Figure P23.41?

Watch Video Solution Problems 23.55 and 23.59

42. ||| For the circuit of Figure P23.42,
   a. What is the equivalent capacitance?
   b. How much charge flows through the battery as the capacitors are being charged?

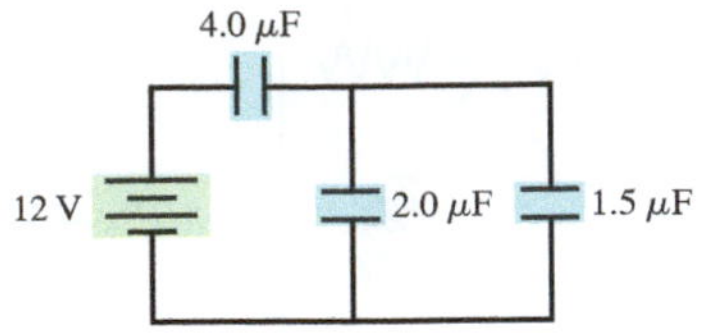


FIGURE P23.42

43. ||| For the circuit of Figure P23.43,
   a. What is the equivalent capacitance?
   b. What is the charge on each capacitor?

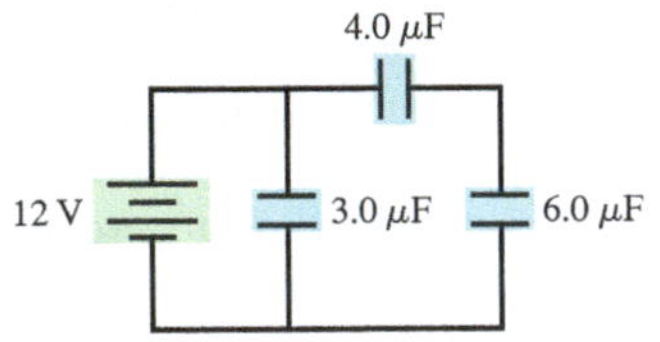


FIGURE P23.43

### Section 23.7 *RC* Circuits

44. || What is the time constant for the discharge of the capacitor in Figure P23.44?

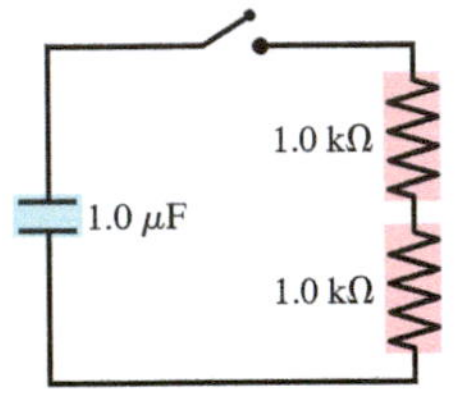


FIGURE P23.44

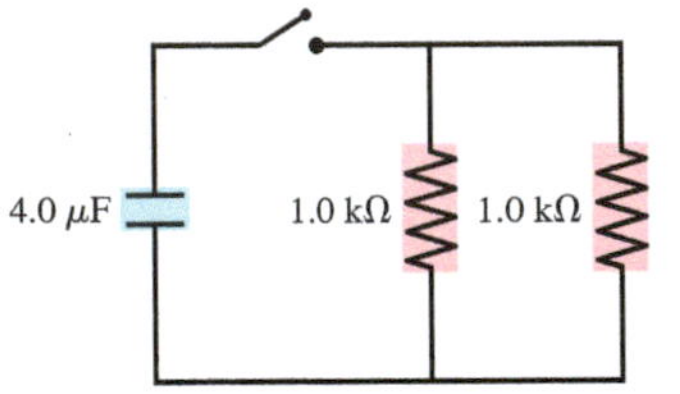


FIGURE P23.45

45. || What is the time constant for the discharge of the capacitor in Figure P23.45?

46. ||| With the switch open, the potential difference across the capacitor in Figure P23.44 is 10.0 V. After the switch is closed, how long will it take for the potential difference across the capacitor to decrease to 5.0 V?

47. |||| A 10 $\mu$F capacitor initially charged to 20 $\mu$C is discharged through a 1.0 k$\Omega$ resistor. How long does it take to reduce the capacitor's charge to 10 $\mu$C?

48. | A capacitor charging circuit consists of a battery, an uncharged 20 $\mu$F capacitor, and a 4.0 k$\Omega$ resistor. At $t = 0$ s, the switch is closed; 0.15 s later, the current is 0.46 mA. What is the battery's emf?

49. || The switch in Figure P23.49 has been in position a for a long time. It is changed to position b at $t = 0$ s. What are the charge $Q$ on the capacitor and the current $I$ through the resistor (a) immediately after the switch is changed? (b) At $t = 50$ $\mu$s? (c) At $t = 200$ $\mu$s?

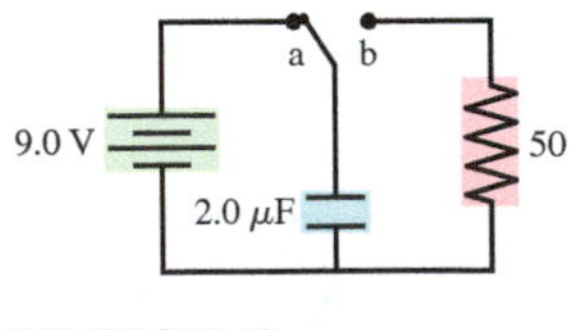


FIGURE P23.49

### Section 23.8 Electricity in the Nervous System

50. | A 9.0-nm-thick cell membrane undergoes an action potential that follows the curve in the table on page 822. What is the strength of the electric field inside the membrane just before the action potential and at the peak of the depolarization?
BIO

51. ||| A cell membrane has a resistance and a capacitance and thus a characteristic time constant. What is the time constant of a 9.0-nm-thick membrane surrounding a 0.040-mm-diameter spherical cell?
BIO

52. | Changing the thickness of the myelin sheath surrounding an axon changes its capacitance and thus the conduction speed. A myelinated nerve fiber has a conduction speed of 55 m/s. If the spacing between nodes is 1.0 mm and the resistance of segments between nodes is 25 M$\Omega$, what is the capacitance of each segment?
BIO

53. ||| A particular myelinated axon has nodes spaced 0.80 mm apart. The resistance between nodes is 20 M$\Omega$; the capacitance of each insulated segment is 1.2 pF. What is the conduction speed of a nerve impulse along this axon?
BIO

54. | To measure signal propagation in a nerve in the arm, the nerve is triggered near the armpit. The peak of the action potential is measured at the elbow and then, 4.0 ms later, 24 cm away from the elbow at the wrist.
BIO
   a. What is the speed of propagation along this nerve?
   b. A determination of the speed made by measuring the time between the application of a stimulus at the armpit and the peak of an action potential at the elbow or the wrist would be inaccurate. Explain the problem with this approach, and why the noted technique is preferable.

55. || A myelinated axon conducts nerve impulses at a speed of 40 m/s. What is the signal speed if the thickness of the myelin sheath is halved but no other changes are made to the axon?
BIO

## General Problems

56. || How much power is dissipated by each resistor in Figure P23.56?
INT

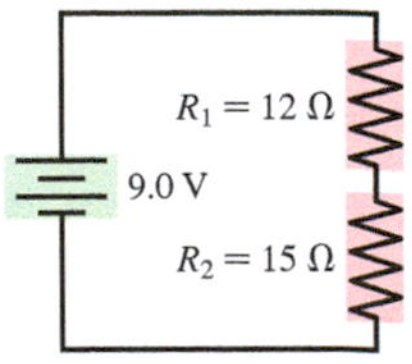


FIGURE P23.56

57. |||| Two 75 W (120 V) lightbulbs are wired in series, then the combination is connected to a 120 V supply. How much power is dissipated by each bulb?
INT

58. ||| A *voltage divider* circuit uses two resistors in series to produce a potential difference that is less than that of a circuit's emf. Figure P23.58 shows the circuit. What should the resistance $R$ be such that the potential difference across $R$ is one-third that of the battery?

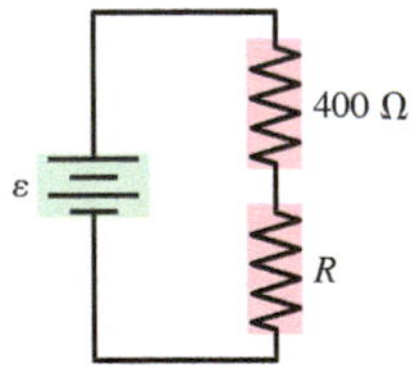


FIGURE P23.58

59. |||| A real battery is not just an emf. We can model a real 1.5 V battery as a 1.5 V emf in series with a resistor known as the "internal resistance," as shown in Figure P23.59. A typical battery has 1.0 $\Omega$ internal resistance due to imperfections that limit current through the battery. When there's no current through the battery, and thus no voltage drop across the internal resistance, the potential difference between its terminals is 1.5 V, the value of the emf. Suppose the terminals of this battery are connected to a 2.0 $\Omega$ resistor.
INT

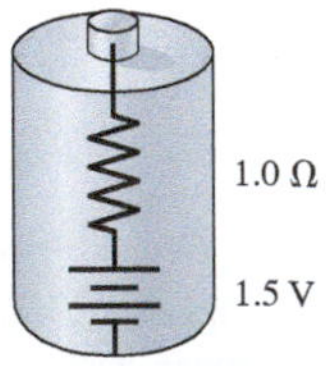


FIGURE P23.59

   a. What is the potential difference between the terminals of the battery?
   b. What fraction of the battery's power is dissipated by the internal resistance?

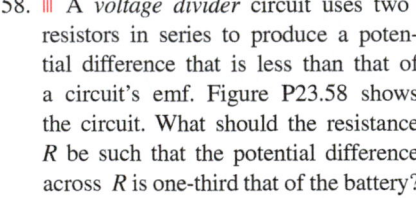

60. ||| If the battery in Figure P23.60 were ideal, lightbulb A would not dim when the switch is closed. However, real batteries have a small *internal resistance*, which we can model as the 0.3 Ω resistor shown inside the battery. In this case, the brightness of bulb A changes when the switch is closed.

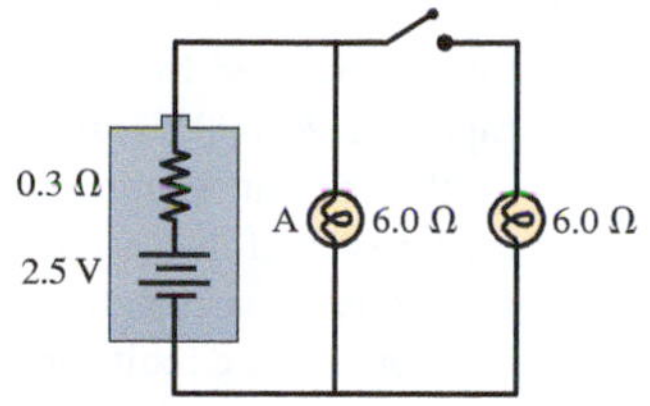


FIGURE P23.60

a. How much power does bulb A dissipate when the switch is open?
b. How much power does bulb A dissipate when the switch is closed?

61. |||| Batteries are recharged by connecting them to a power supply INT (i.e., another battery) of greater emf in such a way that the current flows *into* the positive terminal of the battery being recharged, as was shown in Example 23.1. This reverse current through the battery replenishes its chemicals. The current is kept fairly low so as not to overheat the battery being recharged by dissipating energy in its internal resistance.
a. Suppose the real battery of Figure P23.59 is rechargeable. What emf power supply should be used for a 0.75 A recharging current?
b. If this power supply charges the battery for 10 minutes, how much energy goes into the battery? How much is dissipated as thermal energy in the internal resistance?

62. ||||| Three 10 Ω resistors are in the circuit shown in Figure P23.62. The left end of the circuit is at a potential of 10 V, while the right end is at 0 V.
a. What are the electric potentials at points 2 and 3 of the circuit?
b. From the potential differences across each resistor, what are the currents in each resistor, including their directions (left or right)?
c. What are the currents, including directions, in each wire?
d. What is the equivalent resistance of this circuit measured between points 1 and 4?

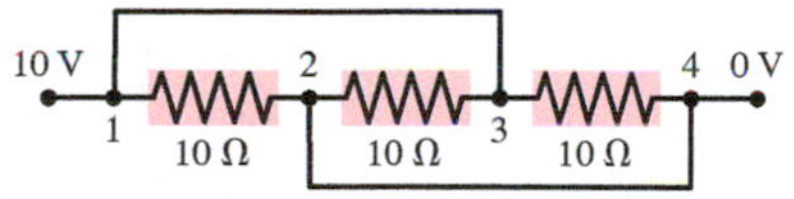


FIGURE P23.62

63. || The 10 Ω resistor in Figure P23.63 is dissipating 40 W of INT power. How much power are the other two resistors dissipating?

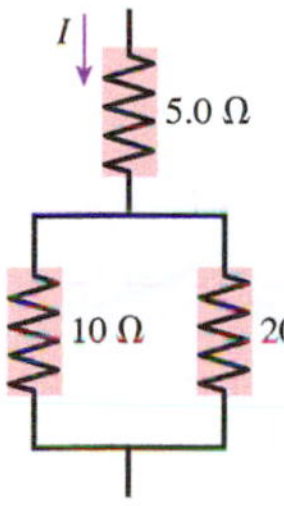


FIGURE P23.63

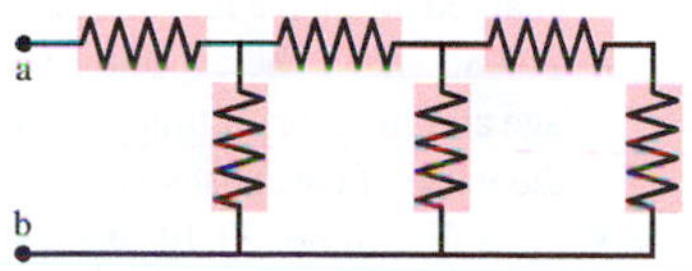


FIGURE P23.64

64. ||| The resistors in Figure P23.64 are all 1.0 Ω. What is the equivalent resistance between points a and b?

65. | What is the equivalent resistance between points a and b in Figure P23.65?

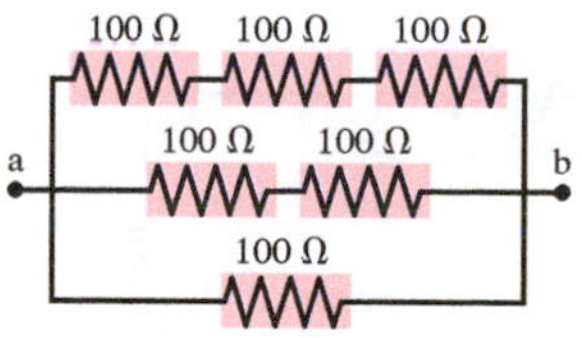


FIGURE P23.65

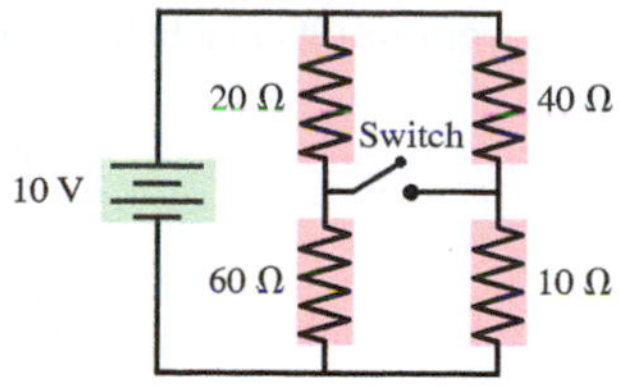


FIGURE P23.66

66. ||| What is the current through the battery in Figure P23.66 when the switch is (a) open and (b) closed?

67. || What is the ratio $P_{parallel}/P_{series}$ of the total power dissipated by INT two identical resistors connected in parallel to a battery to the total power when they are connected in series to the same battery?

68. ||| You have a device that needs a voltage reference of 3.0 V, but you have only a 9.0 V battery. Fortunately, you also have several 10 kΩ resistors. Show how you can use the resistors and the battery to make a circuit that provides a potential difference of 3.0 V.

69. | There is a current of 0.25 A in the circuit of Figure P23.69.
INT a. What is the direction of the current? Explain.
b. What is the value of the resistance $R$?
c. What is the power dissipated by $R$?
d. Make a graph of potential versus position, starting from $V = 0$ V in the lower left corner and proceeding clockwise. See Figure P23.9 for an example.

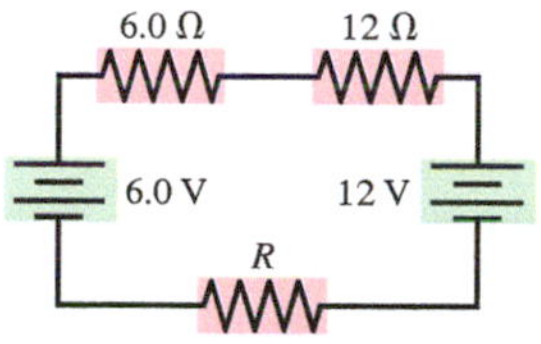


FIGURE P23.69

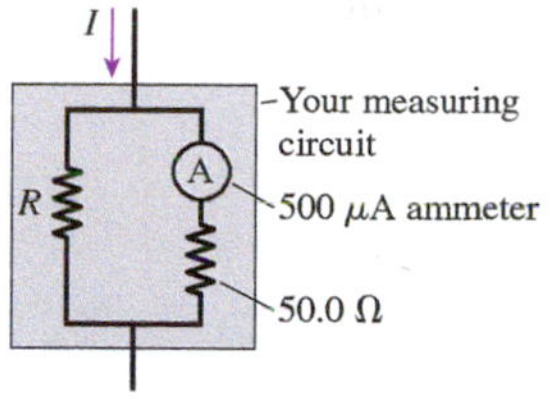


FIGURE P23.70

70. || A circuit you're building needs an ammeter that goes from 0 mA to a full-scale reading of 50.0 mA. Unfortunately, the only ammeter in the storeroom goes from 0 μA to a full-scale reading of only 500 μA. Fortunately, you can make this ammeter work by putting it in a measuring circuit, as shown in Figure P23.70. This lets a certain fraction of the current pass through the meter; knowing this value, you can deduce the total current. Assume that the ammeter is ideal.
a. What value of $R$ must you use so that the meter will go to full scale when the current $I$ is 50.0 mA?
**Hint:** When $I = 50.0$ mA, the ammeter should be reading its maximum value.
b. What is the equivalent resistance of your measuring circuit?

71. || A circuit you're building needs a voltmeter that goes from 0 V to a full-scale reading of 5.0 V. Unfortunately, the only meter in the storeroom is an *ammeter* that goes from 0 μA to a full-scale reading of 500 μA. It is possible to use this meter to measure voltages by putting it in a measuring circuit as shown in Figure P23.71. What value of $R$ must you use so that the meter will go to full scale when the potential difference $\Delta V$ is 5.0 V? Assume that the ammeter is ideal.

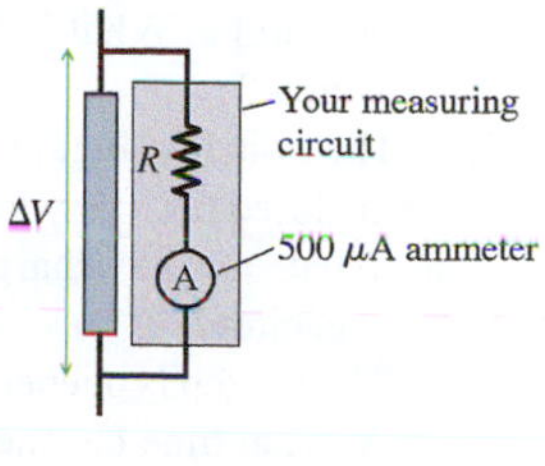


FIGURE P23.71

Watch Video Solution Problem 23.77

72. || For the circuit shown in Figure P23.72, find the current through and the potential difference across each resistor. Place your results in a table for ease of reading.

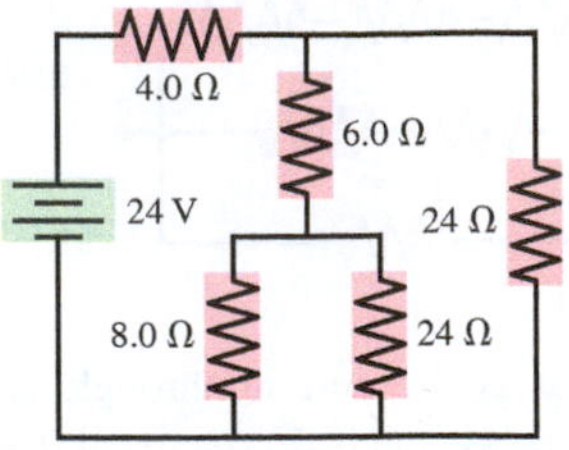


FIGURE P23.72

73. || You have three 12 $\mu$F capacitors. Draw diagrams showing how you could arrange all three so that their equivalent capacitance is (a) 4.0 $\mu$F, (b) 8.0 $\mu$F, (c) 18 $\mu$F, and (d) 36 $\mu$F.

74. || Initially, the switch in Figure P23.74 is in position a and capacitors $C_2$ and $C_3$ are uncharged. Then the switch is flipped to position b. Afterward, what are the charge on and the potential difference across each capacitor?

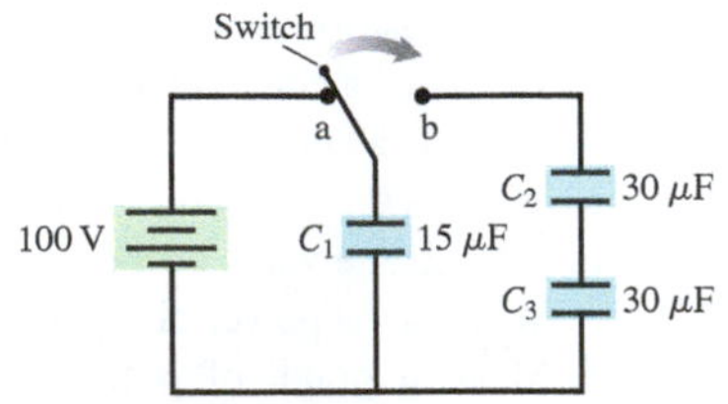


FIGURE P23.74

75. |||| A 10 V battery is wired in series with an uncharged capacitor, a resistor, and a switch to make an $RC$ circuit. At $t = 0$ s the switch is closed. Figure P23.75 shows the current through the resistor.

a. What is the value of the resistor?
b. What is the value of the capacitor?

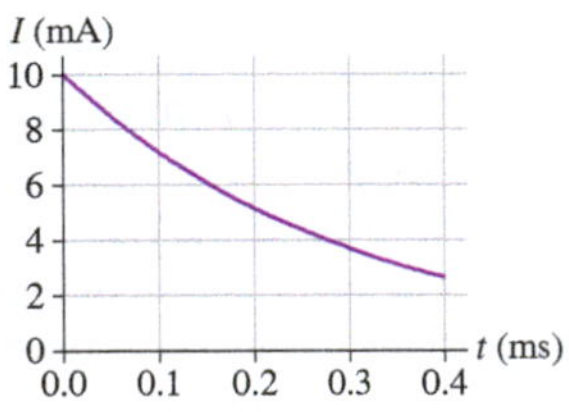


FIGURE P23.75

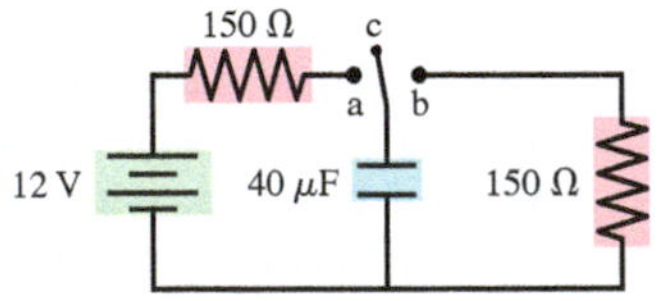


FIGURE P23.76

76. ||| The capacitor in Figure P23.76 is initially uncharged and the switch, in position c, is not connected to either side of the circuit. The switch is now flipped to position a for 10 ms, then to position b for 10 ms, and then brought back to position c. What is the final potential difference across the capacitor?

77. |||| The switch in Figure P23.77 has been closed for a very long time.

a. What is the charge on the capacitor?
b. The switch is opened at $t = 0$ s. At what time has the charge on the capacitor decreased to 10% of its initial value?

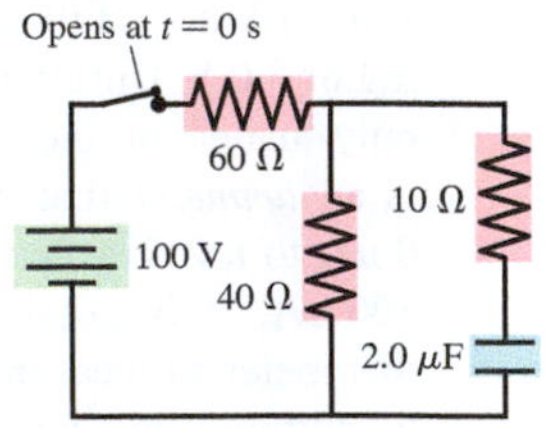


FIGURE P23.77

78. ||| The switch in Figure 23.77 has been open for a very long time. The switch is closed at $t = 0$ s.

a. At $t = 0$ s, what are the currents in the three resistors?
b. After a very long time, what are the currents in the resistors?

79. ||| As Figure P23.79 shows, a single cell in a sample of tissue can be modeled by a circuit consisting of two resistors and a capacitor. $R_I$ is the resistance of the fluid in the cell's interior, $R_E$ the resistance of the fluid surrounding the cell, and $C$ the capacitance of the cell membrane.
BIO

a. In one experiment, a 100 nA current $I$ has been passing through the circuit for a long time. What is the potential difference across the capacitor?
b. The current is suddenly turned off. How long does it take for the potential difference between points a and b to reach one-half of its initial value?

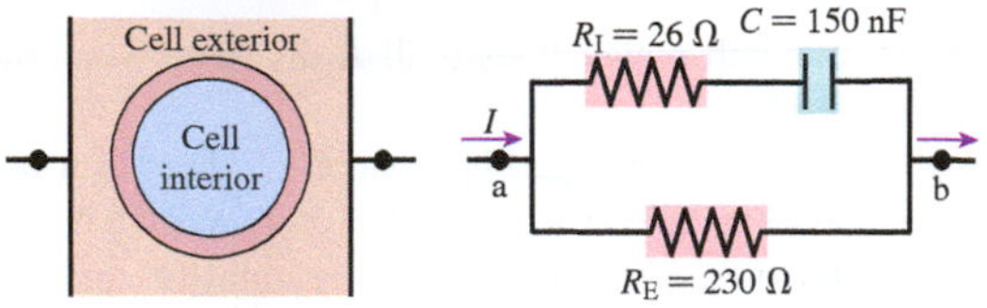


FIGURE P23.79

80. |||| A 0.15 F capacitor is charged to 25 V. It is then discharged through a 1.2 kΩ resistor.
INT

a. What is the power dissipated by the resistor just when the discharge is started?
b. What is the total energy dissipated by the resistor during the entire discharge interval?

81. ||||| Intermittent windshield wipers use a variable resistor in an $RC$ circuit to set the delay between successive passes of the wipers. A typical circuit is shown in Figure P23.81. When the switch closes, the capacitor (initially uncharged) begins to charge and the potential at point b begins to increase. A sensor measures the potential difference between points a and b, triggering a pass of the wipers when $V_b = V_a$. (Another part of the circuit, not shown, discharges the capacitor at this time so that the cycle can start again.)

a. What value of the variable resistor will give 12 seconds from the start of a cycle to a pass of the wipers?
b. To decrease the time, should the variable resistance be increased or decreased?

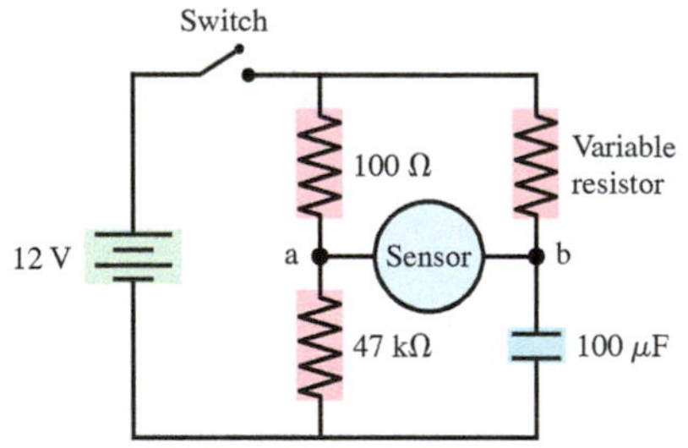


FIGURE P23.81

82. ||| A 50 $\mu$F capacitor that had been charged to 30 V is discharged through a resistor. Figure P23.82 shows the capacitor voltage as a function of time. What is the value of the resistance?

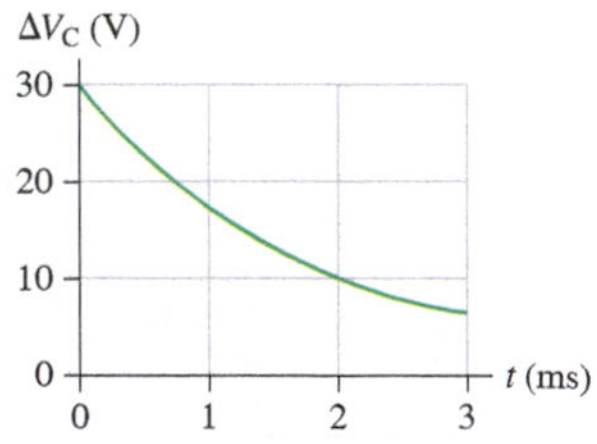


FIGURE P23.82

83. ||| In Example 23.14 we estimated the capacitance of the cell membrane to be 89 pF, and in Example 23.15 we found that approximately 10,000 $Na^+$ ions flow through an ion channel when it opens. Based on this information and what you learned in this chapter about the action potential, estimate the total number of sodium channels in the membrane of a nerve cell.
BIO

84. |||| The giant axon of a squid is 0.5 mm in diameter, 10 cm long, and not myelinated. Unmyelinated cell membranes behave as capacitors with 1 $\mu$F of capacitance per square centimeter of membrane area. When the axon is charged to the $-70$ mV resting potential, what is the energy stored in this capacitance?
BIO INT

85. || A cell has a 7.0-nm-thick membrane with a total membrane area of $6.0 \times 10^{-9}$ m$^2$.
BIO
a. We can model the cell as a capacitor, as we have seen. What is the magnitude of the charge on each "plate" when the membrane is at its resting potential of $-70$ mV?
b. How many sodium ions does this charge correspond to?

## MCAT-Style Passage Problems

### The Defibrillator BIO

A defibrillator is designed to pass a large current through a patient's torso in order to stop dangerous heart rhythms. Its key part is a capacitor that is charged to a high voltage. The patient's torso plays the role of a resistor in an *RC* circuit. When a switch is closed, the capacitor discharges through the patient's torso. A jolt from a defibrillator is intended to be intense and rapid; the maximum current is very large, so the capacitor discharges quickly. This rapid pulse depolarizes the heart, stopping all electrical activity. This allows the heart's internal nerve circuitry to reestablish a healthy rhythm.

A typical defibrillator has a 32 $\mu$F capacitor charged to 5000 V. The electrodes connected to the patient are coated with a conducting gel that reduces the resistance of the skin to where the effective resistance of the patient's torso is 100 $\Omega$.

86. | Which pair of graphs in Figure P23.86 best represents the capacitor voltage and the current through the torso as a function of time after the switch is closed?

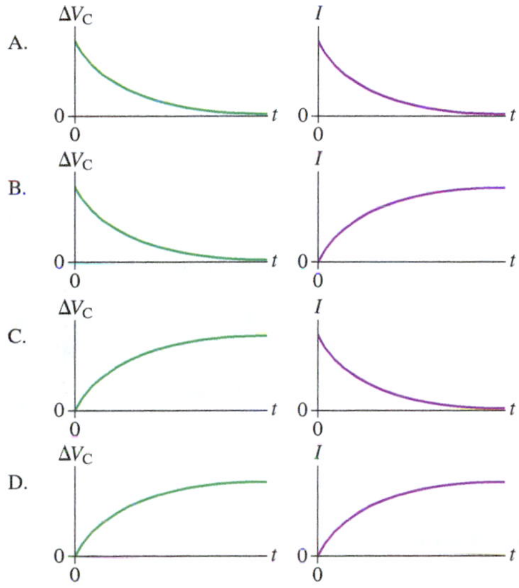

FIGURE P23.86

87. | For the values noted in the passage above, what is the time constant for the discharge of the capacitor?
A. 3.2 $\mu$s B. 160 $\mu$s C. 3.2 ms D. 160 ms

88. | If a patient receives a series of jolts, the resistance of the torso may increase. How does such a change affect the initial current and the time constant of subsequent jolts?
A. The initial current and the time constant both increase.
B. The initial current decreases, the time constant increases.
C. The initial current increases, the time constant decreases.
D. The initial current and the time constant both decrease.

89. | In some cases, the defibrillator may be charged to a lower voltage. How will this affect the time constant of the discharge?
A. The time constant will increase.
B. The time constant will not change.
C. The time constant will decrease.

### Electric Fish BIO INT

The voltage produced by a single nerve or muscle cell is quite small, but there are many species of fish that use multiple action potentials in series to produce significant voltages. The electric organs in these fish are composed of specialized disk-shaped cells called *electrocytes.* The cell at rest has the usual potential difference between the inside and the outside, but the net potential difference *across* the cell is zero. An electrocyte is connected to nerve fibers that initially trigger a depolarization in one side of the cell but not the other. For the very short time of this depolarization, there is a net potential difference across the cell, as shown in Figure P23.90. Stacks of these cells connected in series can produce a large total voltage. Each stack can produce a small current; for more total current, more stacks are needed, connected in parallel.

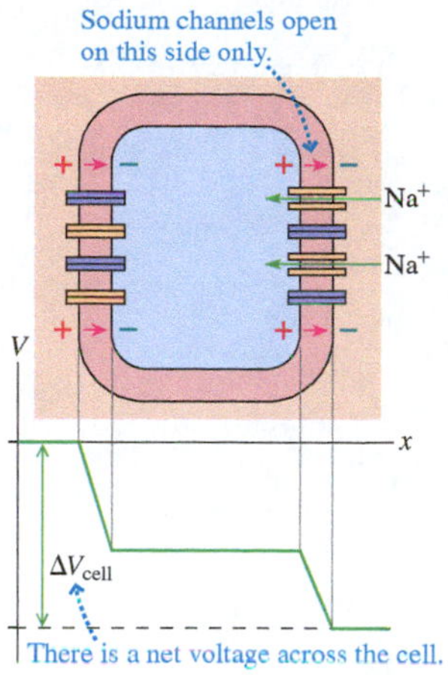

FIGURE P23.90

90. | In an electric eel, each electrocyte can develop a voltage of 150 mV for a short time. For a total voltage of 450 V, how many electrocytes must be connected in series?
A. 300 B. 450 C. 1500 D. 3000

91. | An electric eel produces a pulse of current of 0.80 A at a voltage of 500 V. For the short time of the pulse, what is the instantaneous power?
A. 400 W B. 500 W C. 625 W D. 800 W

92. | Electric eels live in fresh water. The torpedo ray is an electric fish that lives in salt water. The electrocytes in the ray are grouped differently than in the eel; each stack of electrocytes has fewer cells, but there are more stacks in parallel. Assuming that the electric eel and the torpedo ray are adapted to their environments, we can assume that
A. In the lower resistivity of salt water, it is adaptive for an electric fish to provide a higher output current at a lower output voltage.
B. In the lower resistivity of salt water, it is adaptive for an electric fish to provide a lower output current at a higher output voltage.
C. In the higher resistivity of salt water, it is adaptive for an electric fish to provide a higher output current at a lower output voltage.
D. In the higher resistivity of salt water, it is adaptive for an electric fish to provide a lower output current at a higher output voltage.

93. | The electric catfish is another electric fish that produces a voltage pulse by means of stacks of electrocytes. As the fish grows in length, the magnitude of the voltage pulse the fish produces grows as well. The best explanation for this change is that, as the fish grows,
A. The voltage produced by each electrocyte increases.
B. More electrocytes are added to each stack.
C. More stacks of electrocytes are added in parallel to the existing stacks.
D. The thickness of the electrocytes increases.

# 24 Magnetic Fields and Forces

This detailed image of the skeletal system of a dolphin wasn't made with x rays; it was made with magnetism. How is this done?

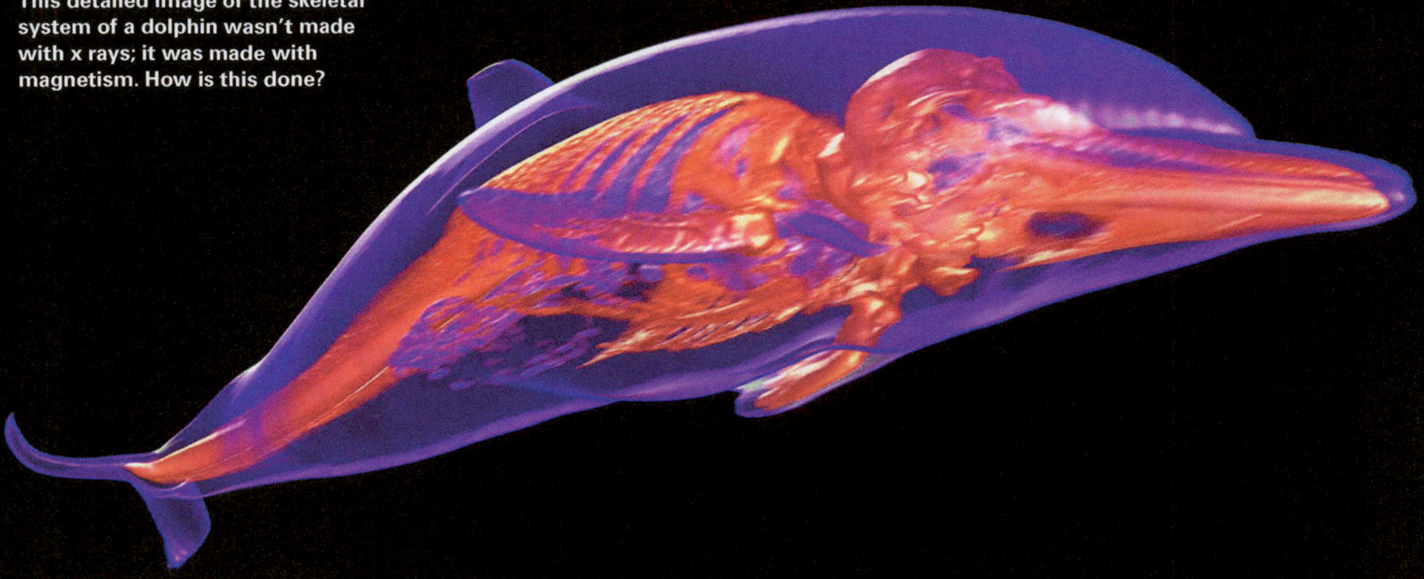

## LOOKING AHEAD ➧

### Magnetic Fields

A compass is a small magnet that will rotate to line up with a magnetic field.

You'll learn how to use compasses and other tools to map magnetic fields.

### Sources of the Field

Magnets produce a magnetic field; so do current-carrying wires, loops, and coils.

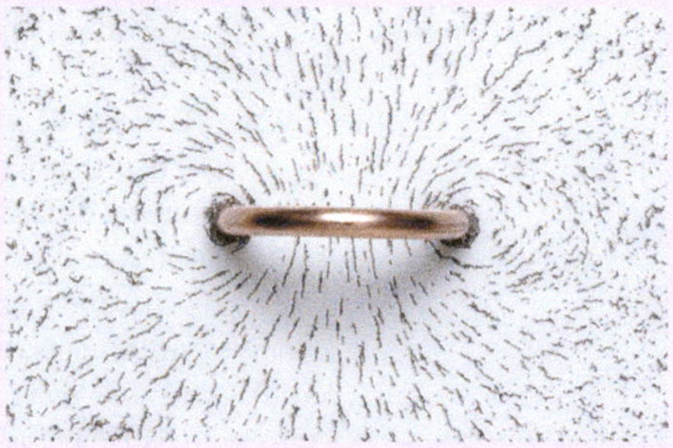

You'll learn to describe the magnetic fields created by currents. These iron filings show the magnetic field shape for this current-carrying wire loop.

### Effects of the Field

Magnetic fields exert forces on moving charged particles and electric currents.

You'll see how the motion of charged particles in the earth's magnetic field gives rise to the aurora.

**GOAL** To learn about magnetic fields and how magnetic fields exert forces on currents and moving charges.

## LOOKING BACK ➧

### Electric Fields

In Chapter 20, we described electric interactions between charged objects in terms of the field model.

You learned how to draw and interpret the electric field of a dipole. In this chapter, you'll see how a magnetic dipole creates a magnetic field with a similar structure.

**STOP TO THINK**

An electric dipole in a uniform electric field experiences no net force, but it does experience a net torque. The rotation of this dipole will be

A. Clockwise.
B. Counterclockwise.

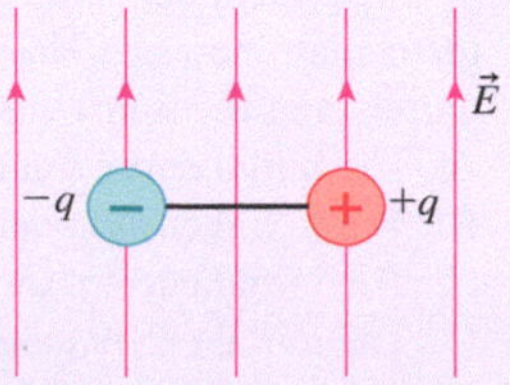

## 24.1 Magnetism

We began our investigation of electricity in Chapter 20 by looking at the results of simple experiments with charged rods. In the following table, we'll investigate how magnets interact with compasses, other magnets, and various types of materials.

### Exploring magnetism

**Experiment 1**

If a bar-shaped magnet is taped to a piece of cork and allowed to float in a dish of water, it turns to align itself in an approximate north-south direction. The end of a magnet that points north is called the **north pole**. The other end is the **south pole**. In this text, we represent north poles as red and south poles as white.

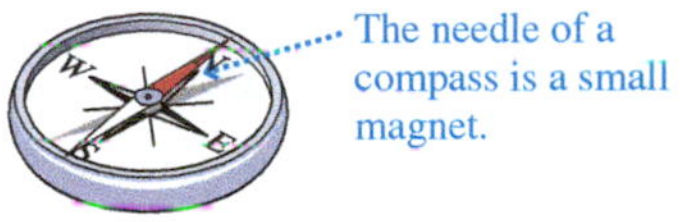

The needle of a compass is a small magnet.


A magnet that is free to pivot like this is called a **compass**. In addition to pointing north, a compass will pivot to line up with a nearby magnet.

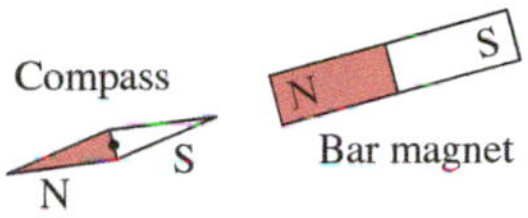


**Experiment 2**

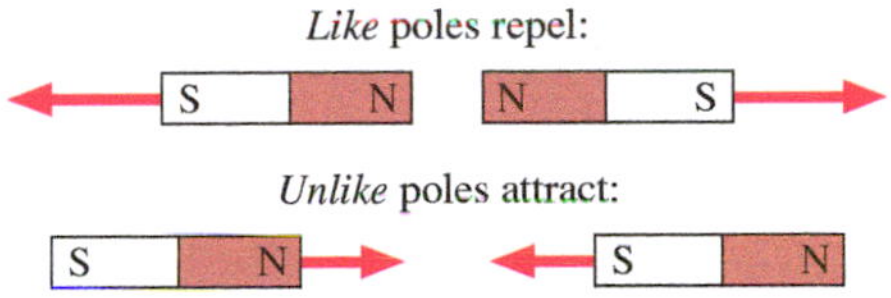


If the north pole of one magnet is brought near the north pole of another magnet, they repel each other. Two south poles also repel each other, but the north pole of one magnet exerts an attractive force on the south pole of another magnet.

**Experiment 3**

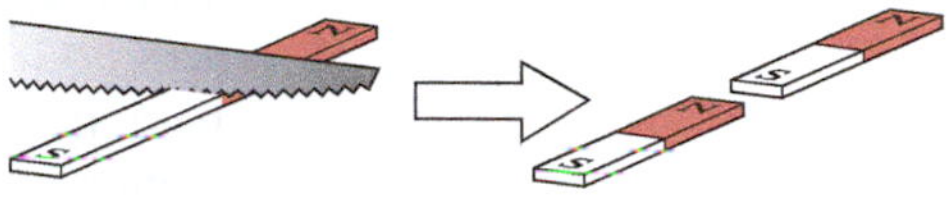

Cutting a bar magnet in half produces two weaker but still complete magnets, each with a north pole and a south pole.

**Experiment 4**

Magnets can pick up some objects, such as paper clips, but not all. If an object is attracted to one pole of a magnet, it is also attracted to the other pole. Most materials, including copper, aluminum, glass, and plastic, experience no force from a magnet.

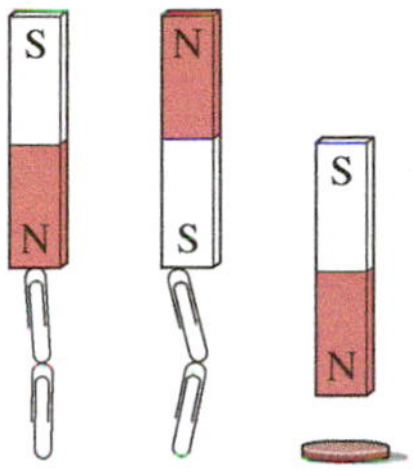


**Experiment 5**

When a magnet is brought near an electroscope, the leaves of the electroscope remain undeflected. If a charged rod is brought near a magnet, there is a small polarization force like the ones we studied in Chapter 21, as there would be on any metal bar, but there is no other effect.

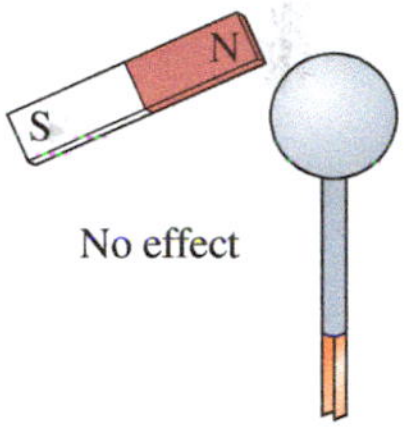


What do these experiments tell us?

- Experiment 5 reveals that magnetism is not the same as electricity. **Magnetic poles and electric charges share some similar behavior, but they are not the same.**
- Experiment 2 shows that magnetism is a long-range force. Magnets need not touch each other to exert a force on each other.
- Experiments 1 and 3 show that magnets have two types of poles, north and south, and thus are **magnetic dipoles.** Cutting a magnet in half yields two weaker but still complete magnets, each with a north pole and a south pole. The basic unit of magnetism is thus a magnetic dipole.
- Experiment 1 shows how the poles of a bar magnet can be identified by using it as a compass. Then, as Experiment 2 shows, the known poles of this bar magnet can be used to identify the poles of other magnets: A pole that repels a known south pole and attracts a known north pole must be a south magnetic pole.
- Experiment 4 reveals that only certain materials, called **magnetic materials,** are attracted to a magnet. The most common magnetic material is iron. Magnetic materials are attracted to both poles of a magnet.

**STOP TO THINK 24.1** Does the compass needle rotate?

A. Yes, clockwise.
B. Yes, counterclockwise.
C. No, not at all.

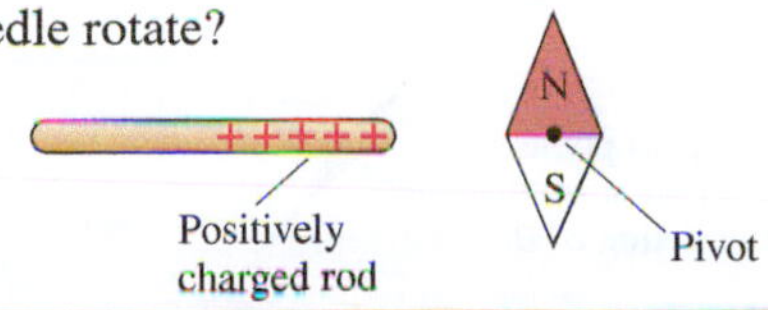

## 24.2 The Magnetic Field

When we studied the *electric* force between two charges in ◂◂ SECTION 20.4, we developed a new way to think about forces between charges—the *field model.* In this view, the space around a charge is not empty: The charge alters the space around it by creating an *electric field.* A second charge brought into this electric field then feels a force due to the *field.*

The concept of a field can also be used to describe the force that turns a compass to line up with a magnet: **Every magnet sets up a *magnetic* field in the space around it.** If another magnet—such as a compass needle—is then brought into this field, the second magnet will feel the effects of the *field* of the first magnet. In this section, we'll see how to define the magnetic field, and then we'll study what the magnetic field looks like for some common shapes and arrangements of magnets.

### Measuring the Magnetic Field

What does the direction a compass needle points tell us about the magnetic field at the position of the compass? Recall how an *electric* dipole behaves when placed in an electric field, as shown in FIGURE 24.1a. In ◂◂ SECTION 20.7, you learned that an electric dipole experiences a *torque* when placed in an electric field, a torque that tends to align the axis of the dipole with the field. This means that the *direction* of the electric field is the same as the direction of the aligned dipole's axis. The torque on the dipole is greater when the electric field is stronger; hence, the *magnitude* of the field, which we also call the *strength* of the field, is proportional to the torque on the dipole.

The magnetic dipole of a compass needle behaves very similarly when it is in a magnetic field. The magnetic field exerts a torque on the compass needle, causing the needle to rotate until it points in the direction of the field, as shown in FIGURE 24.1b.

FIGURE 24.1 Dipoles in electric and magnetic fields.

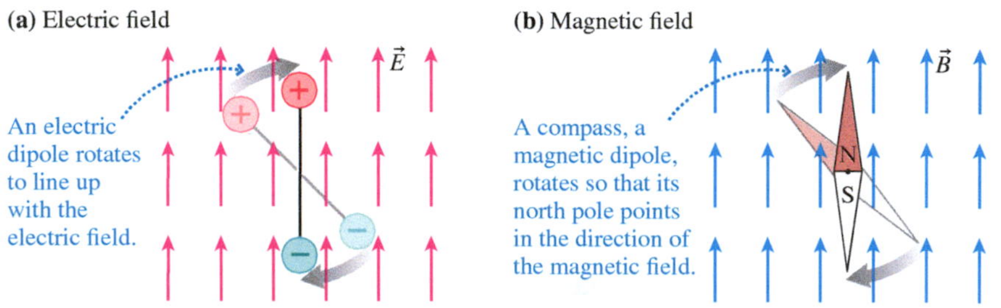


Because the magnetic field has both a direction and a magnitude, we represent it using a *vector*. We will use the symbol $\vec{B}$ to represent the magnetic field and $B$ to represent the magnitude or strength of the field. FIGURE 24.2 shows how to use a compass to determine the magnitude and direction of the magnetic field. The direction of the magnetic field is the direction that the north pole of a compass needle points; the magnitude of the magnetic field is proportional to the torque felt by the compass needle as it turns to line up with the field direction.

FIGURE 24.2 Determining the direction and magnitude of a magnetic field.

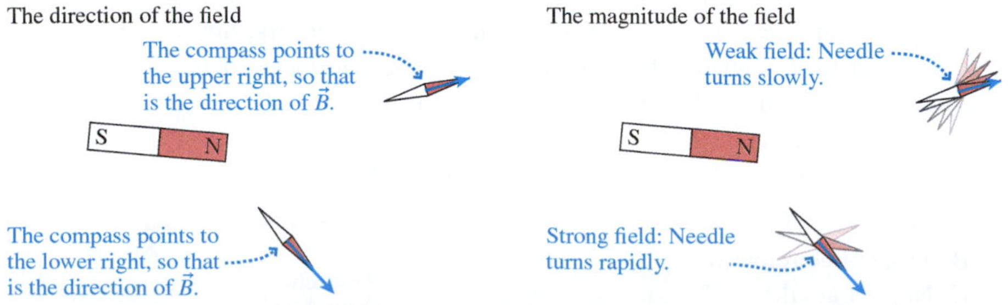

We can produce a "picture" of the magnetic field by using *iron filings*—very small elongated grains of iron. If there are enough grains, iron filings can give a very detailed representation of the magnetic field, as shown in FIGURE 24.3. The compasses that we use to determine field direction show us that **the magnetic field of a magnet points *away* from the north pole and *toward* the south pole.**

FIGURE 24.3 Revealing the magnetic field of a bar magnet using iron filings.

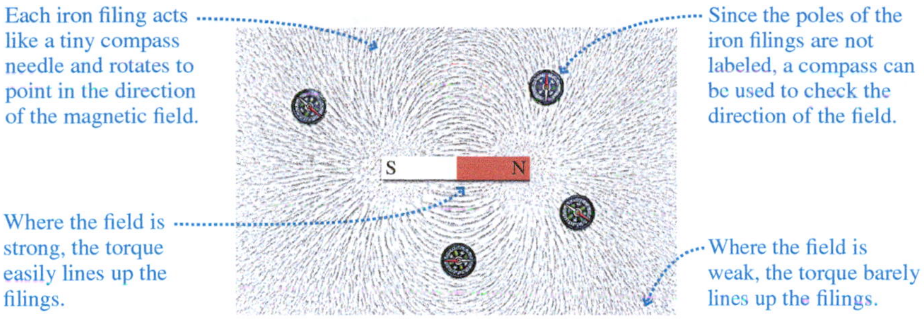

Each iron filing acts like a tiny compass needle and rotates to point in the direction of the magnetic field.
Since the poles of the iron filings are not labeled, a compass can be used to check the direction of the field.
S
N
Where the field is strong, the torque easily lines up the filings.
Where the field is weak, the torque barely lines up the filings.

## Magnetic Field Vectors and Field Lines

We can draw the field of a magnet such as the one shown in Figure 24.3 in either of two ways. When we want to represent the magnetic field at one particular point, the **magnetic field vector** representation is especially useful. But if we want an overall representation of the field, **magnetic field lines** are often simpler to use. These two representations are similar to the *electric field* vectors and lines used in Chapter 20, and we'll use similar rules to draw them.

As shown in FIGURE 24.4, we can imagine placing a number of compasses near the magnet to measure the direction and magnitude of the magnetic field. To represent the field at the location of one of the compasses, we then draw a vector with its *tail* at that location. Figure 24.4 shows how to choose the direction and magnitude of this vector. Although we've drawn magnetic field vectors at only a few points around the magnet, it's important to remember that the magnetic field exists at *every* point around the magnet.

FIGURE 24.4 Mapping out the magnetic field of a bar magnet using compasses.

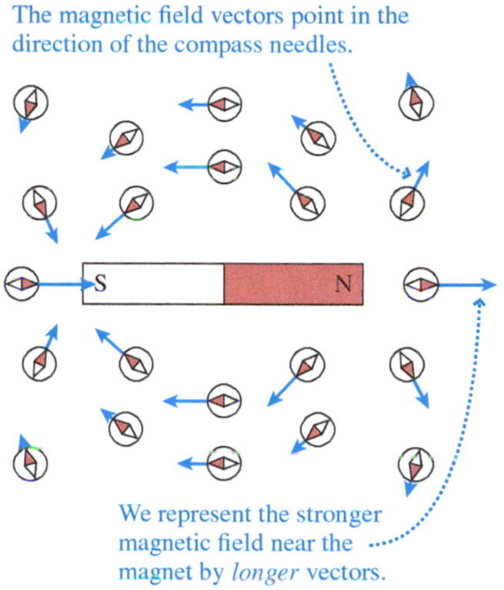

The magnetic field vectors point in the direction of the compass needles.
S
N
We represent the stronger magnetic field near the magnet by *longer* vectors.

We can also represent the magnetic field using magnetic field lines. The rules for drawing these lines are similar to those for drawing the electric field lines of Chapter 20. Electric field lines begin on positive charges and end on negative charges; magnetic field lines go from a north magnetic pole to a south magnetic pole. The direction and the spacing of the field lines show the direction and the strength of the field, as illustrated in FIGURE 24.5.

FIGURE 24.5 Drawing the magnetic field lines of a bar magnet.

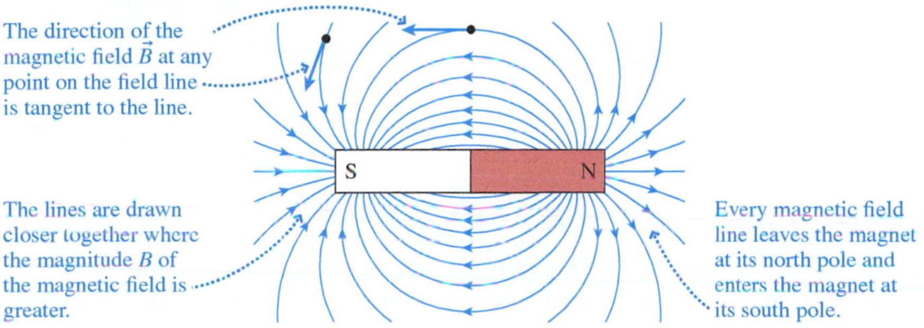


Now that we know how to think about magnetic fields, let's look at magnetic fields from magnets with different arrangements. We'll use the iron filing method to show the lines from real magnets, along with a drawing of the field lines.

**An atlas of magnetic fields produced by magnets**

| A single bar magnet | A single bar magnet (closeup) | Two bar magnets, unlike poles facing | Two bar magnets, like poles facing |
|---|---|---|---|

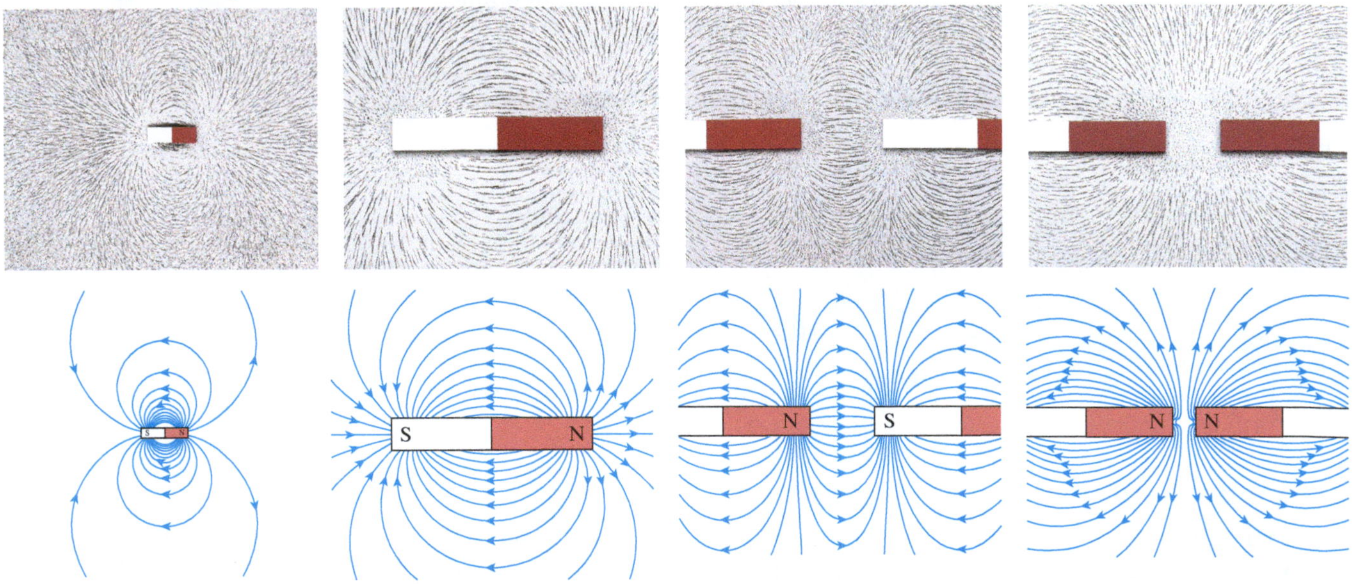


The magnetic field lines start on the north pole and end on the south pole.

As you move away from the magnet, the field lines are farther apart, indicating a weaker field.

Closer to the magnet, we can more clearly see how the field lines always start on north poles and end on south poles.

With more than one magnet, the field lines still start on a north pole and end on a south pole. But they can start on the north pole of one magnet, and end on the south pole of another.

With two like poles placed nearby, the field lines starting on the north poles curve sharply toward their south poles in order to avoid the north pole of the other magnet.

You are probably most familiar with bar magnets that have a north pole on one end and a south pole on the other, but magnets can have more than one pair of north-south poles, and the poles need not be at the ends of the magnet. Flexible refrigerator magnets have an unusual arrangement of long, striped poles, as shown in FIGURE 24.6. Most of the field exits the plain side of the magnet, so this side sticks better to your refrigerator than the label side does.

FIGURE 24.6 The magnetic field of a refrigerator magnet.

(a)

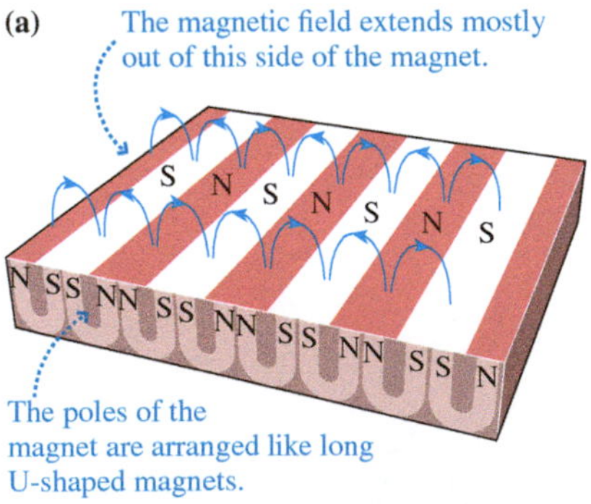

The magnetic field extends mostly out of this side of the magnet.
S N S N S N S
N S S N N S S N N S S N N S S N
The poles of the magnet are arranged like long U-shaped magnets.

(b)

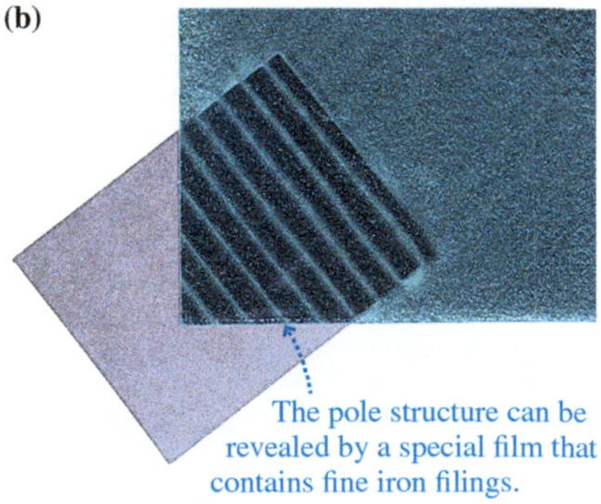

The pole structure can be revealed by a special film that contains fine iron filings.

## The Magnetic Field of the Earth

As we've seen, a bar magnet that is free to pivot—a compass—always swings so that its north pole points geographically north. But we've also seen that if a magnet is brought near a compass, the compass swings so that its north pole faces the south pole of the magnet. These observations can be reconciled if the earth itself is a large magnet, as shown in FIGURE 24.7a, where **the south pole of the earth's magnet is located near—but not exactly at the same place as—the north geographic pole of the earth.** The north pole of a compass needle placed at the equator will point toward the south pole of the earth's magnet—that is, to the north.

The earth's magnetic field has components both parallel to the ground (horizontal) and perpendicular to the ground (vertical). An ordinary north-pointing compass responds to only the horizontal component of the field, but a compass free to pivot vertically will tilt downward as well. FIGURE 24.7b shows that, near the equator, the earth's magnetic field is nearly parallel to the ground. The angle from the horizontal, called the **dip angle,** is quite small. Near the poles, the field is more nearly vertical; near the north pole, the field points down more than it points north. Sea turtles seem to use the dip angle of the earth's field to determine their latitude.

FIGURE 24.7 The earth's magnetic field.

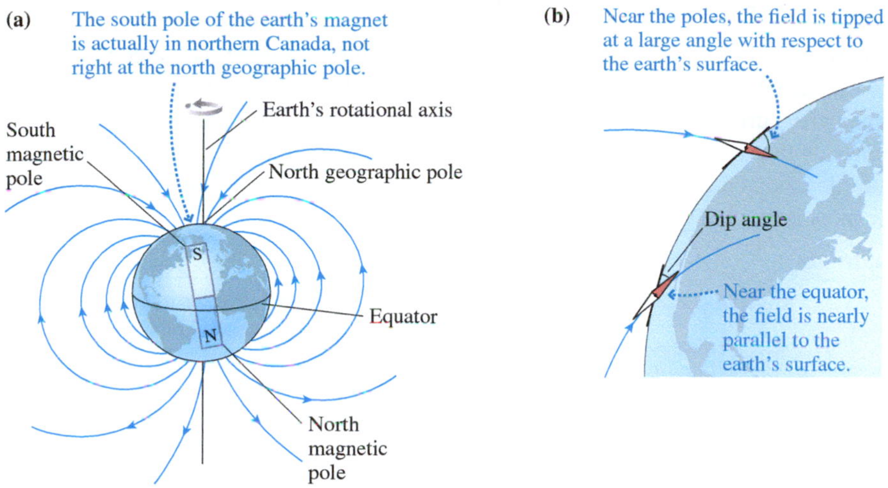


Other animals also use the earth's magnetic field to navigate. You might imagine that animals would use an internal compass to determine orientation, to know which way is north, but the most unequivocal example of a creature that navigates using the earth's field is one that uses the field to tell up from down, not north from south. Magnetotactic bacteria have strongly magnetized bits of iron in their bodies, as you can see in FIGURE 24.8. Their bodies rotate to line up parallel to the earth's field, just like a compass. These bacteria prefer the low-oxygen zone near the bottom of bodies of water. If the water is disturbed and they are displaced upward, they travel in the direction of the earth's field, following the vertical component of the field downward.

FIGURE 24.8 BIO The internal compass of a magnetotactic bacterium.

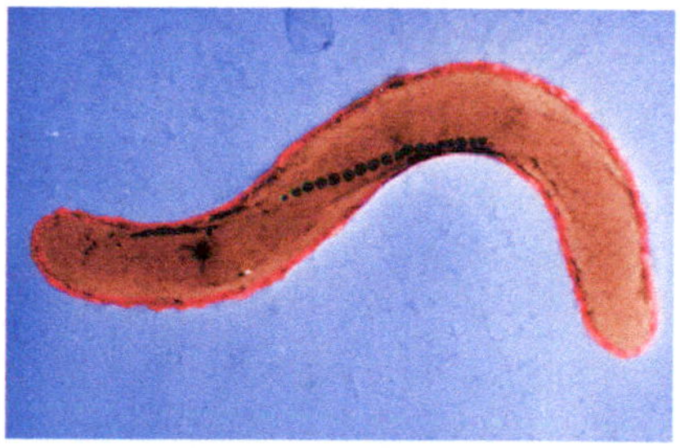

## CONCEPTUAL EXAMPLE 24.1 Homing pigeon navigation BIO

There is evidence that pigeons use the earth's magnetic field to navigate. Although the exact mechanism is not understood, researchers suspect that magnetic sensors in the bird's head are responsible for detecting the earth's field. To investigate this question, researchers attached small magnets to the backs of pigeons and measured their ability to find their way home. FIGURE 24.9 shows a pigeon, with a magnet on its back, flying north from its release point in New York State. In order to make the bird think that it's flying south, should the pole P of the magnet be a north or a south pole?

FIGURE 24.9 Is the pole P a north or a south pole?

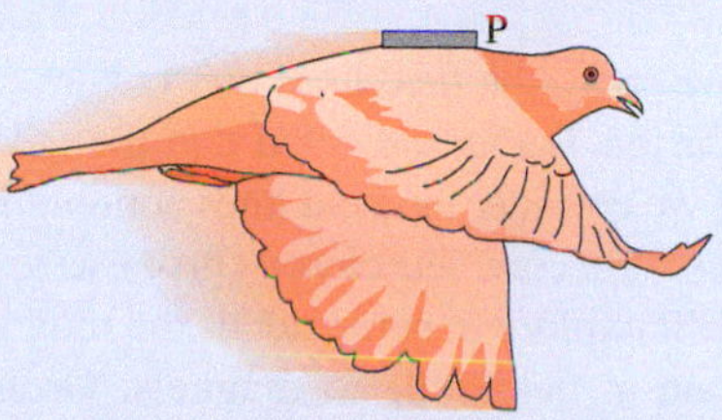


***REASON*** Figure 24.7a shows that the earth's field at northern latitudes points to the north and down; FIGURE 24.10 shows the earth's field $\vec{B}_{\text{earth}}$ pointing in this direction. For the pigeon to think it's flying south, the magnetic field at its head would need to be flipped north to south, as shown by vector $\vec{B}_{\text{south}}$. The magnet should thus create a field that adds to $\vec{B}_{\text{earth}}$ to give $\vec{B}_{\text{south}}$. As Figure 24.10 shows, this field is directed to the south, or, in Figure 24.9, to the left, *into* the magnet. This shows that pole P is a south pole.

FIGURE 24.10 Magnetic fields for the pigeon.

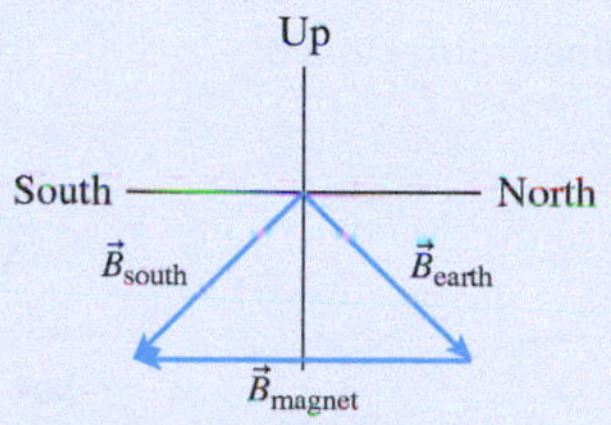


***ASSESS*** When the pigeon is flying north, it is flying toward the *south* pole of the earth's magnet. It makes sense, then, that a stronger south pole *behind* it would turn its magnetic orientation around.

**STOP TO THINK 24.2** A compass is placed next to a bar magnet as shown. Which figure shows the correct alignment of the compass?

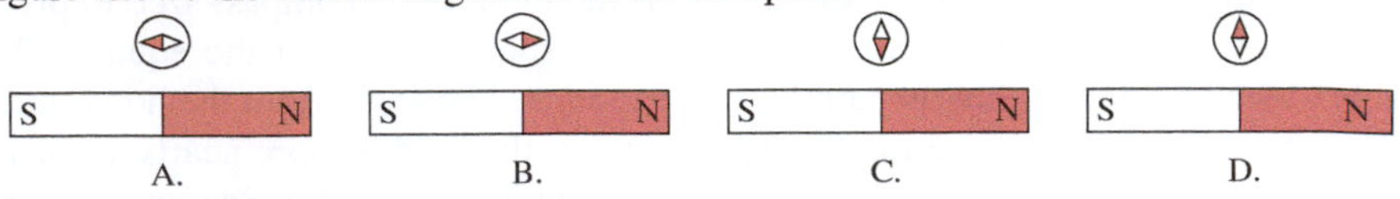


## 24.3 Electric Currents Also Create Magnetic Fields

You expect a compass to react to the presence of a bar magnet. But a compass will also deflect if you place the compass near a wire and pass a current through the wire. When the current stops, the compass goes back to its original orientation. This shows us that an *electric* current produces a *magnetic* field. The shape of the field lines depends on the shape of the current-carrying wire. The following table shows the most important configurations we'll study.

**An atlas of magnetic fields produced by currents**

**Current in a long, straight wire**

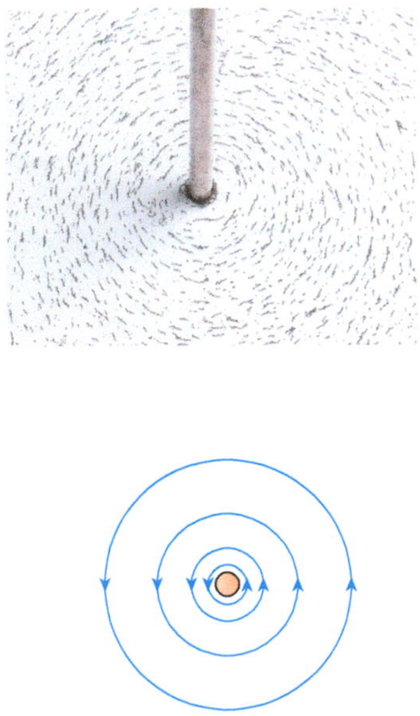

The magnetic field lines form *circles* around the wire. The iron filings are less affected by the field as the distance from the wire increases, indicating that the field is getting weaker as the distance from the wire increases.

**A circular loop of current**

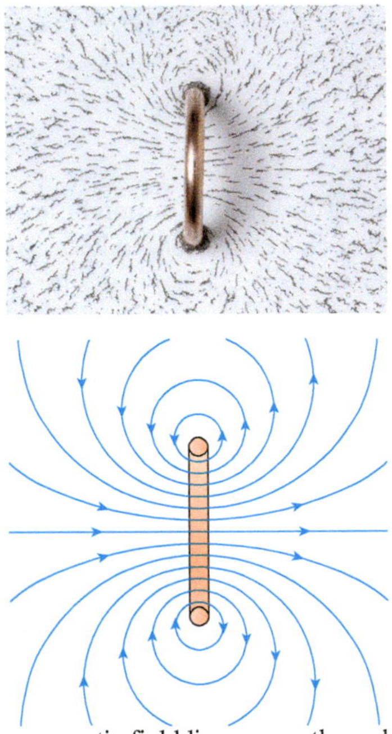

The magnetic field lines curve through the center of the loop, around the outside, and back through the loop's center, forming complete closed curves. The field lines far from the loop look like the field lines far from a bar magnet.

**Current in a solenoid**

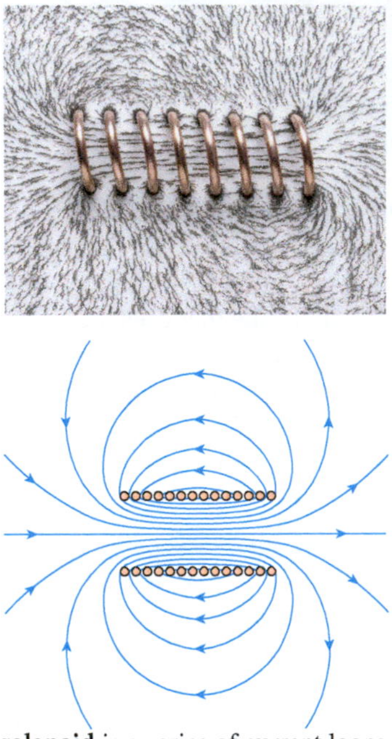

A **solenoid** is a series of current loops placed along a common axis. The field outside is very weak compared to the field inside. Inside the solenoid, the magnetic field lines are reasonably evenly spaced; the field inside is nearly uniform.

FIGURE 24.11 Field lines form closed curves for magnets, too.

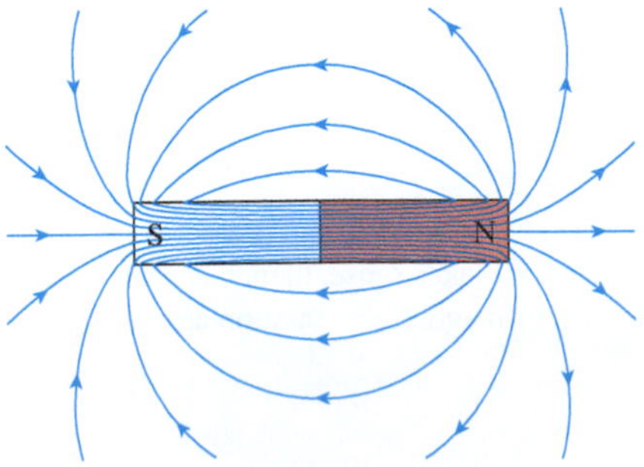


As the pictures in the atlas above show, the field lines due to currents have no start or end: They form complete closed curves. Earlier we noted that the field lines of magnets start and end on magnetic poles. If we consider the field lines continuing *inside* a magnet, however, we find that these lines also form complete closed curves, as shown in FIGURE 24.11.

In this and the next section, we'll explore in some detail the magnetic fields created by currents; later, we'll look again at fields due to magnets. Ordinary magnets are often called **permanent magnets** to distinguish their unchanging magnetism from that caused by currents that can be switched on and off. We look at magnetism in this order because magnetism from currents is easier to understand, but keep in mind that currents and magnets are both equally important sources of magnetic fields.

## The Magnetic Field of a Straight, Current-Carrying Wire

We have seen in the pictures in the atlas that iron filings line up in *circles* around a straight, current-carrying wire. As FIGURE 24.12 shows, we also can use our basic instrument, the compass, to determine the direction of the magnetic field.

FIGURE 24.12 How compasses respond to a current-carrying wire.

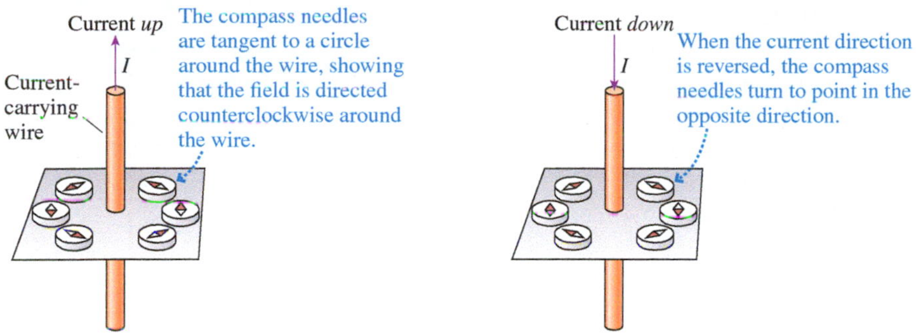


To help remember in which direction compasses will point, we use the *right-hand rule* shown in Tactics Box 24.1. We'll use this same rule later to find the direction of the magnetic field due to several other shapes of current-carrying wire, so we'll call this rule the **right-hand rule for fields.**

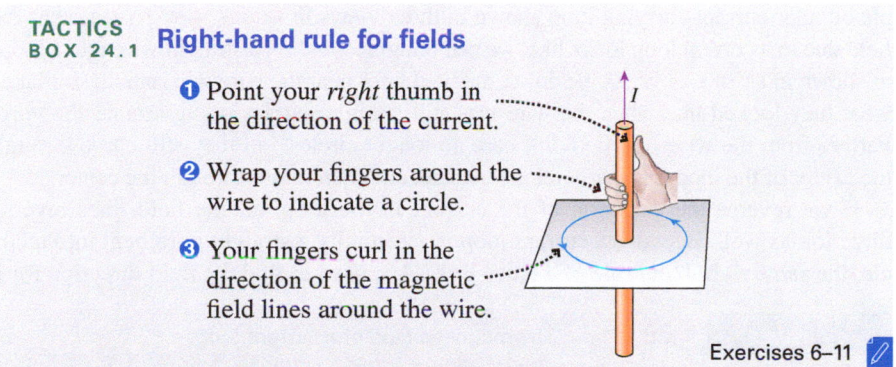

TACTICS BOX 24.1 **Right-hand rule for fields**

1. Point your *right* thumb in the direction of the current.
2. Wrap your fingers around the wire to indicate a circle.
3. Your fingers curl in the direction of the magnetic field lines around the wire.

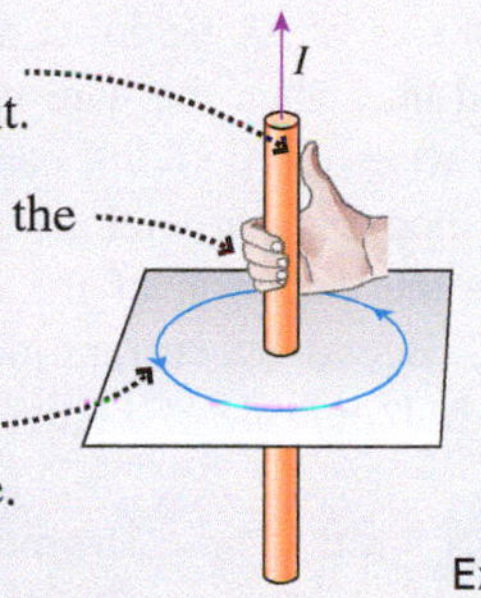


Exercises 6–11

Magnetism often requires three-dimensional figures of the sort shown in Tactics Box 24.1. But since two-dimensional figures are easier to draw, we will make as much use of them as we can. Consequently, we will often need to indicate field vectors or currents that are perpendicular to the page. FIGURE 24.13a shows the notation we will use. FIGURE 24.13b demonstrates this notation by showing the compasses around a current that is directed into the page. To use the right-hand rule with this drawing, point your right thumb into the page. Your fingers will curl clockwise, giving the direction in which the north poles of the compass needles point.

FIGURE 24.13 The notation for vectors and currents that are perpendicular to the page.

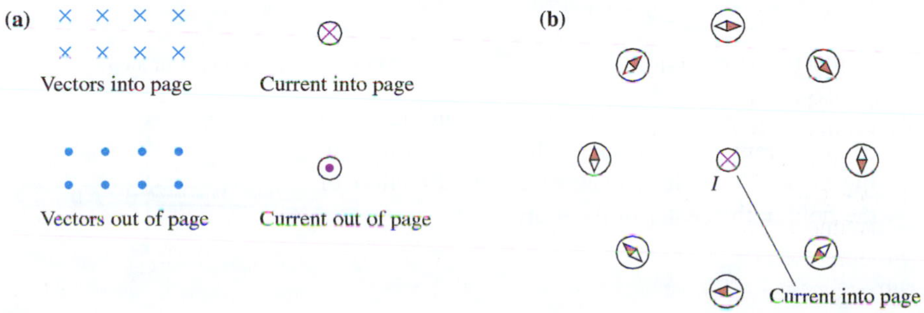

**CONCEPTUAL EXAMPLE 24.2** **Drawing the magnetic field of a current-carrying wire**

Sketch the magnetic field of a long, current-carrying wire, with the current going into the paper. Draw both magnetic field line and magnetic field vector representations.

***REASON*** From the iron filings pictures in the atlas, we have seen that the field lines form circles around the wire, and the magnetic field becomes weaker as the distance from the wire is increased. FIGURE 24.14 shows how we construct both field line and field vector representations of such a field.

FIGURE 24.14 Drawing the magnetic field of a long, straight, current-carrying wire.

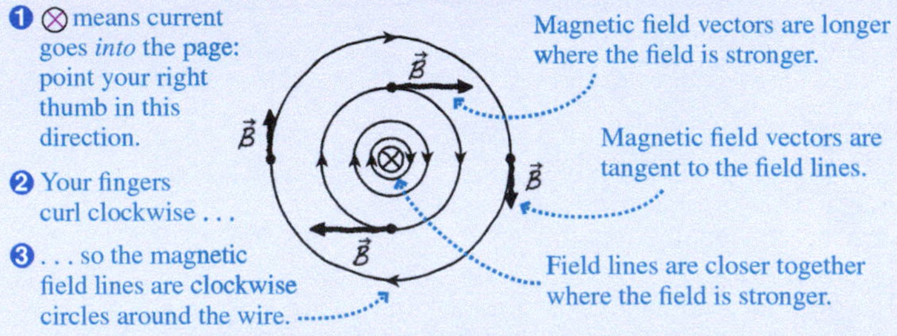


***ASSESS*** Figure 24.14 illustrates the key features of the field. The direction of the field vectors and field lines matches what we saw in Figure 24.13, and the field strength drops off with distance, as we learned in the atlas figure. We don't expect you to draw such a figure, but it's worth looking at the full 3-D picture of the field in FIGURE 24.15. This conveys the idea that the field lines exist in every plane along the length of the wire.

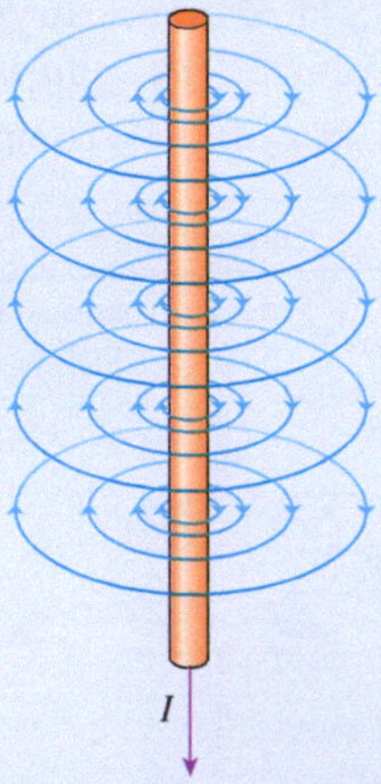


FIGURE 24.15 Field lines exist everywhere along the wire.

## The Magnetic Field of a Current Loop

We can extend our understanding of the field from a long, straight, current-carrying wire to the fields due to other shapes of current-carrying wires. Let's start with the simple circular current-carrying loop shown in three views in FIGURE 24.16. To see what the field due to a current loop looks like, we can imagine bending a straight wire into a loop, as shown in FIGURE 24.17. As we do so, the field lines near the wire will remain similar to what they looked like when the wire was still straight: circles going around the wire. Farther from the wires the field lines are no longer circles, but they still curve through the center of the loop, back around the outside, and then return through the center.

If we reverse the direction of the current in the loop, all the field lines reverse direction as well. Because a current loop is essentially a straight wire bent into a circle, the same right-hand rule of Tactics Box 24.1, used to find the field direction for a

FIGURE 24.16 Three views of a current loop.

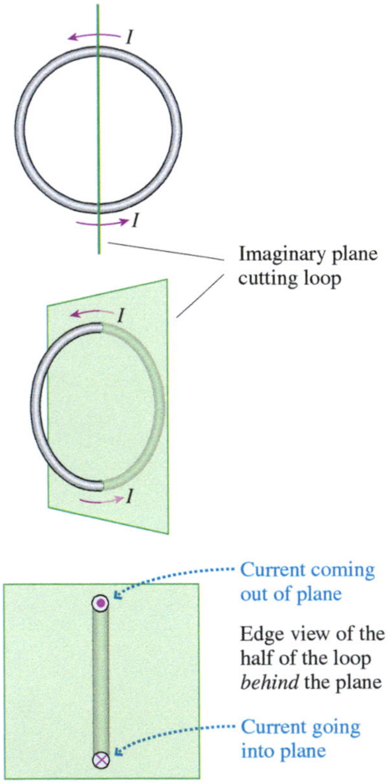

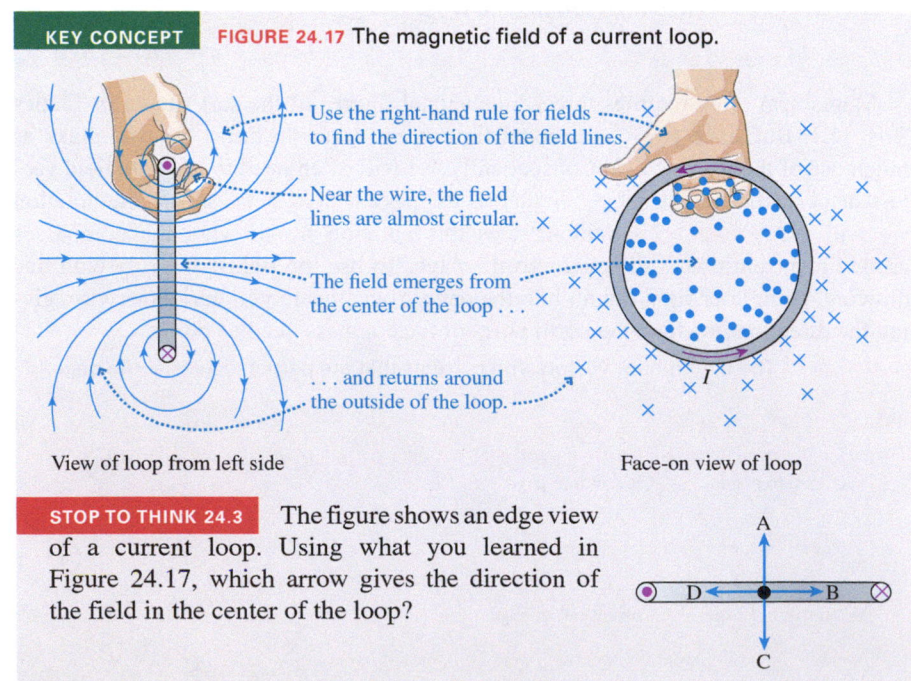

**KEY CONCEPT** FIGURE 24.17 The magnetic field of a current loop.

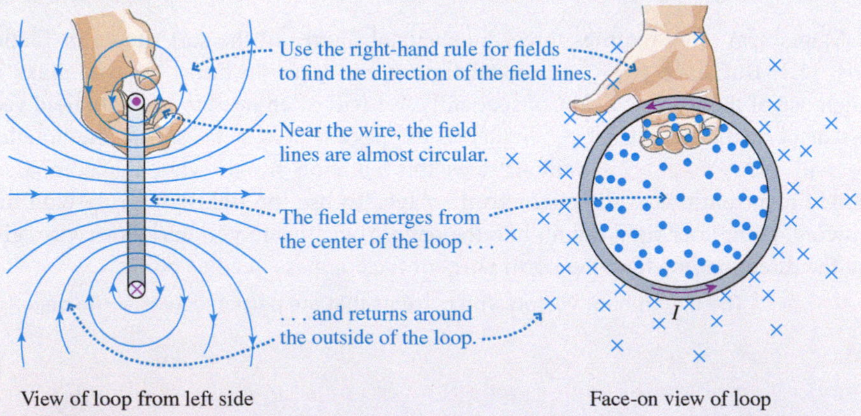


**STOP TO THINK 24.3** The figure shows an edge view of a current loop. Using what you learned in Figure 24.17, which arrow gives the direction of the field in the center of the loop?

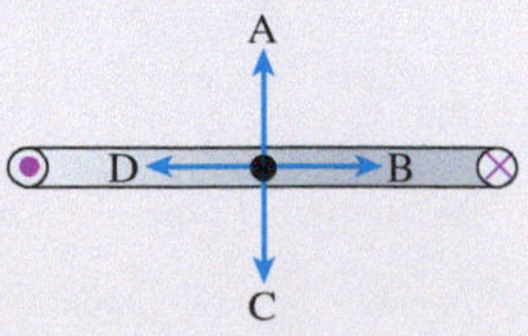


A video to support a section's topic is embedded in the eText.

**Video** Figure 24.17

long, straight wire, can also be used to find the field direction for a current loop. As shown in Figure 24.17, you again point your thumb in the direction of the current in the loop and let your fingers curl through the center of the loop. Your fingers are then pointing in the direction in which $\vec{B}$ passes through the *center* of the loop.

### The Magnetic Field of a Solenoid

There are many applications of magnetism, such as the MRI system used to make the image at the beginning of this chapter, for which we would like to generate a **uniform magnetic field,** a field that has the same magnitude and the same direction at every point within some region of space. As we've seen, a quite uniform magnetic field can be generated with a solenoid. A solenoid, as shown in FIGURE 24.18, is a long coil of wire that has the same current $I$ passing through each loop in the coil. Solenoids may have hundreds or thousands of loops, often called *turns,* sometimes wrapped in several layers.

The iron filing picture in the atlas on page 844 shows us that **the field within the solenoid is strong, mainly parallel to the axis, and quite uniform, whereas the field outside the solenoid is very weak.** FIGURE 24.19 reviews these ideas and shows why the field inside is much stronger than the field outside. The field direction inside the solenoid can be determined by using the right-hand rule for any of the loops that form it.

FIGURE 24.18 A solenoid.

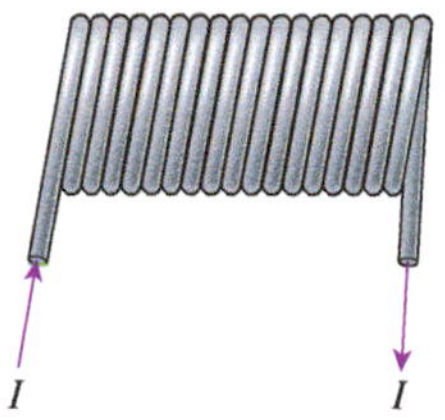


FIGURE 24.19 The field of a solenoid.

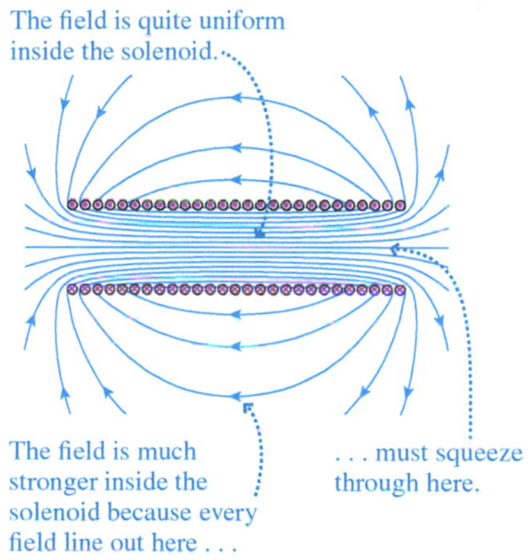


**STOP TO THINK 24.4** A compass is placed above a long wire. When a large current is turned on in the direction shown, in which direction will the compass point?

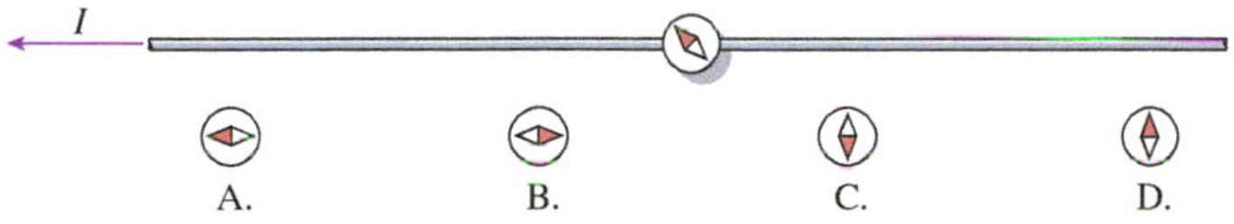


## 24.4 Calculating the Magnetic Field Due to a Current

We've now seen the appearance of the magnetic field for several shapes of wires. Next, we'll learn how to calculate magnetic fields. In SI units, magnetic field strengths are measured in **tesla,** abbreviated as T. One tesla is quite a large field, so most of the field strengths you will work with are much less than 1 T, as you can see from the field strengths listed in TABLE 24.1, so we often express field strengths in $\mu$T ($10^{-6}$ T) or mT ($10^{-3}$ T). The earth's magnetic field strength varies from place to place, but 50 $\mu$T is a good average value that you can use for solving problems.

TABLE 24.1 Typical magnetic field strengths

| Field source and location | Field strength (T) |
|---|---|
| 10 cm from a wire with 1 A current | $2\times10^{-6}$ |
| Surface of the earth | $5\times10^{-5}$ |
| 1 cm from a wire with 10 A current | $2\times10^{-4}$ |
| Refrigerator magnet | $5\times10^{-3}$ |
| 100-turn coil, 1 cm diameter, with 1 A current | $1\times10^{-2}$ |
| Surface of the sun, in a sunspot | $1\times10^{-1}$ |
| Near a rare-earth magnet | 1 |
| MRI solenoid | 1 |
| World's strongest magnet | 45 |

### Wires

Now, let's look at the expression for the field around a long, straight, current-carrying wire. We've seen already that the magnetic field lines form circles around the wire and that the field gets weaker as the distance from the wire increases. Not surprisingly, the magnitude of the field also depends on the *current* through the wire, increasing in proportion to the current. Putting these experimental observations together, we find that the magnitude of the magnetic field (in T) a distance $r$ from the wire carrying current $I$ is given by

$$B = \frac{\mu_0 I}{2\pi r} \tag{24.1}$$

Magnetic field due to a long, straight, current-carrying wire

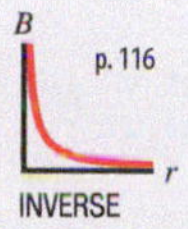

Equation 24.1 is only exact for an infinitely long wire, but it is quite accurate if the length of the wire is much greater than the distance $r$ from the wire.

The constant $\mu_0$, which relates the strength of the magnetic field to the currents that produce it, is called the **permeability constant.** Its role in magnetic field expressions is similar to the role of the permittivity constant in electric field expressions. Its value is

$$\mu_0 = 1.26 \times 10^{-6}\,\mathrm{T \cdot m/A}$$

According to Equation 24.1, the magnetic field strength near a current-carrying wire decreases with distance; the strength of the field is inversely proportional to the distance from the wire. FIGURE 24.20 illustrates this relationship.

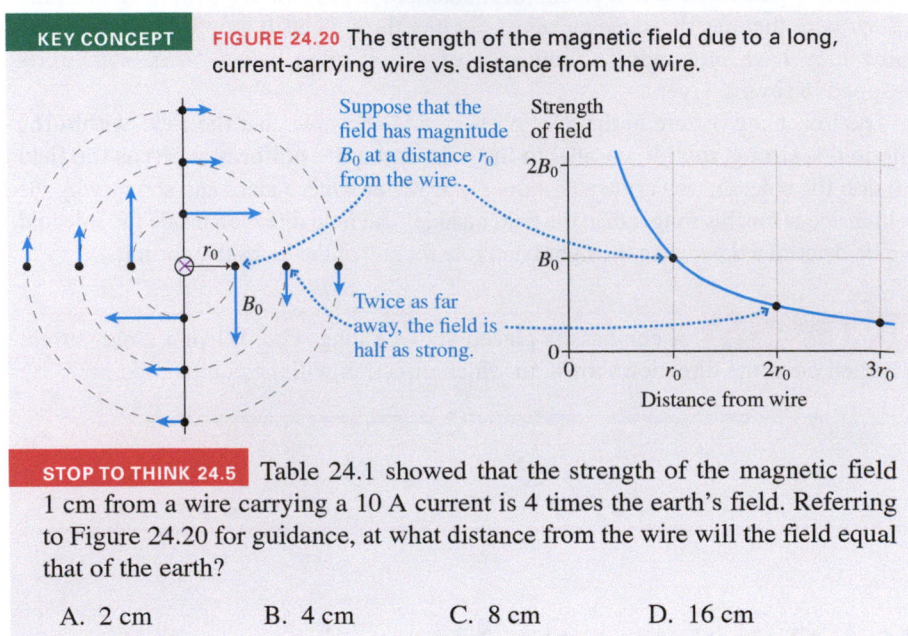

KEY CONCEPT FIGURE 24.20 The strength of the magnetic field due to a long, current-carrying wire vs. distance from the wire.

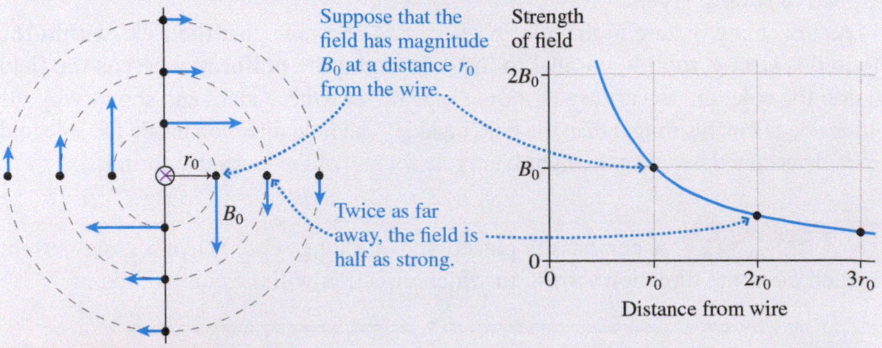


**STOP TO THINK 24.5** Table 24.1 showed that the strength of the magnetic field 1 cm from a wire carrying a 10 A current is 4 times the earth's field. Referring to Figure 24.20 for guidance, at what distance from the wire will the field equal that of the earth?

A. 2 cm B. 4 cm C. 8 cm D. 16 cm

FIGURE 24.21 Adding magnetic fields due to more than one source.

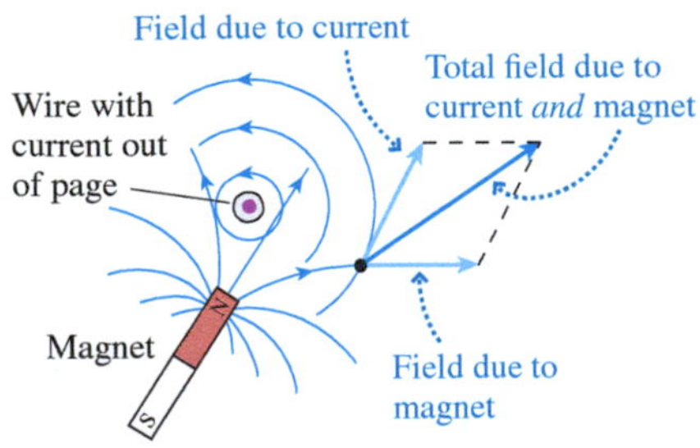


## Magnetic Fields from More Than One Source

When two or more sources of magnetic fields are brought near each other, how do we find the total magnetic field at any particular point in space? For electric fields, we used the principle of superposition: The total electric field at any point is the *vector* sum of the individual fields at that point. The same principle holds for magnetic fields as well. FIGURE 24.21 illustrates the principle of superposition applied to *magnetic* fields.

### EXAMPLE 24.3 Finding the magnetic field of two parallel wires

Two long, straight wires lie parallel to each other, as shown in FIGURE 24.22. They each carry a current of 5.0 A, but in opposite directions. What is the magnetic field at point P?

FIGURE 24.22 Two parallel, current-carrying wires.

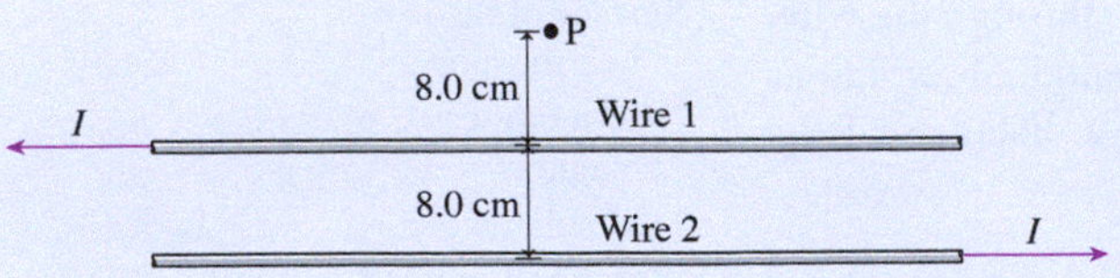


***STRATEGIZE*** The magnetic field at point P is the vector sum of the fields at P due to each of the wires. Because the magnetic field due to a wire forms circles around the wire, the fields due to wires 1 and 2 point either out of or into the plane of the page. Vectors are much easier to visualize when they are in the plane of the page, so we redraw the two wires as seen from their *ends,* so that their magnetic fields lie in this plane.

***PREPARE*** The wires as redrawn this way are shown in FIGURE 24.23; we have also added a coordinate system. In this view, the right-hand rule for fields tells us that the magnetic field at P from wire 1 points to the right and that from wire 2 to the left. Because wire 2 is twice as far from P as

FIGURE 24.23 View of the wires from their right ends.

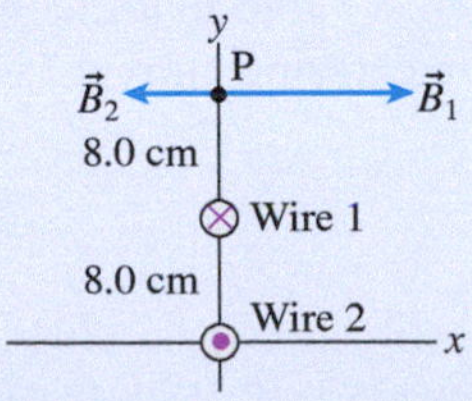

wire 1, we've drawn its field $\vec{B}_2$ half as long as the field $\vec{B}_1$ from wire 1.

**SOLVE** From Figure 24.23, we can see that the total field $\vec{B}$—the vector sum of the two fields $\vec{B}_1$ and $\vec{B}_2$—points to the right. To find the magnitude of $\vec{B}$, we need the magnitudes of $\vec{B}_1$ and $\vec{B}_2$.

We use Equation 24.1 to find these magnitudes. From Figure 24.23, $r$ for wire 1 is 8.0 cm (or 0.080 m), while $r$ for wire 2 is 16 cm.

We then have

$$B_1 = \frac{\mu_0 I}{2\pi r} = \frac{(1.26 \times 10^{-6}\ \text{T}\cdot\text{m/A})(5.0\ \text{A})}{2\pi(0.080\ \text{m})} = 1.253 \times 10^{-5}\ \text{T}$$

and

$$B_2 = \frac{\mu_0 I}{2\pi r} = \frac{(1.26 \times 10^{-6}\ \text{T}\cdot\text{m/A})(5.0\ \text{A})}{2\pi(0.16\ \text{m})} = 0.627 \times 10^{-5}\ \text{T}$$

Because $\vec{B}_2$ points to the left, its $x$-component is negative. Thus, the $x$-component of the total field $\vec{B}$ is

$$\begin{aligned} B_x &= (B_1)_x + (B_2)_x \\ &= (1.253 \times 10^{-5}\ \text{T}) - (0.627 \times 10^{-5}\ \text{T}) = 0.63 \times 10^{-5}\ \text{T} \\ &= 6.3\ \mu\text{T} \end{aligned}$$

In terms of the original view of the problem in Figure 24.22, we can write

$$\vec{B} = (6.3\ \mu\text{T, into the page})$$

**ASSESS** $B_x$ is positive, which tells us that the total field points to the right, in the same direction as the field due to wire 1. This makes sense because P is closer to wire 1 than to wire 2. Table 24.1 gives the field 10 cm from a wire carrying 1 A as 2 $\mu$T. The field in this problem should be about the same size. The distances are about the same; the currents are larger, but there are two opposing fields. So, the magnitude of our answer makes sense.

## Current Loops

The magnetic field due to a current loop is more complex than that of a straight wire, as we can see from FIGURE 24.24, but there is a simple expression for the field at the *center* of the loop. Because the loop can be thought of as a wire bent into a circle, the expression for the strength of the field is similar to that of a wire. The magnitude of the field at the center is

FIGURE 24.24 The magnetic field at the center of a current loop.

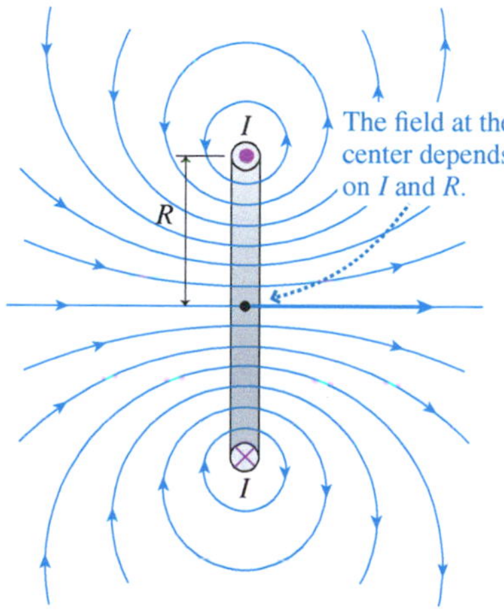


$$B = \frac{\mu_0 I}{2R} \tag{24.2}$$

Magnetic field at the center of a current loop of radius $R$

where $I$ is the current in the loop.

If $N$ loops of wire carrying the same current $I$ are all tightly wound into a single flat coil, then the magnitude of the field at the center is just $N$ times bigger (since we're superimposing $N$ individual current loops):

$$B = \frac{\mu_0 N I}{2R} \tag{24.3}$$

Magnetic field at the center of a thin coil with $N$ turns

► **The magnetocardiogram** BIO When the heart muscle contracts, action potentials create a dipole electric field that can be measured to create an electrocardiogram. These action potentials also cause charges to circulate around the heart, creating a current loop. This current loop creates a small, but measurable, magnetic field. A record of the heart's magnetic field, a *magnetocardiogram,* can provide useful information about the heart in cases where an electrocardiogram is not possible. This image shows the magnetic field of a fetal heartbeat measured at the surface of the mother's abdomen. Here, blue represents a field pointing into the body and red a field pointing out of the body. This is just the field expected from a current loop whose plane lies along the black line between the two colored areas.

**EXAMPLE 24.4** **Canceling the earth's magnetic field** BIO

Green turtles are thought to navigate by using the dip angle of the earth's magnetic field. To test this hypothesis, green turtle hatchlings were placed in a 72-cm-diameter tank with a 60-turn coil of wire wrapped around the outside. A current in the coil created a magnetic field at the center of the tank that exactly canceled the vertical component of the earth's 50 $\mu$T field. At the location of the test, the earth's field was directed 60° below the horizontal. What was the current in the coil?

**STRATEGIZE** FIGURE 24.25 shows the earth's magnetic field passing downward through the coil. To cancel the vertical component of this field, the current in the coil must generate an upward field of equal magnitude. We can use the right-hand rule (see Figure 24.17) to find that the current must circulate around the coil as shown. Viewed from above, the current will be counterclockwise.

FIGURE 24.25 The coil field needed to cancel the vertical component of the earth's field.

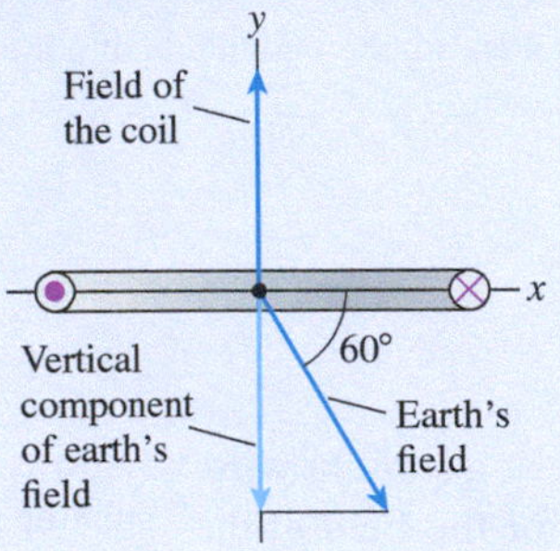


**SOLVE** The vertical component of the earth's field is

$$(B_{\text{earth}})_y = -(50 \times 10^{-6}\ \text{T})\sin(60°) = -4.33 \times 10^{-5}\ \text{T}$$

The field of the coil, given by Equation 24.3, must have the same magnitude at the center. The $2R$ in the equation is just the diameter of the coil, 72 cm or 0.72 m. Thus

$$B_{\text{coil}} = \frac{\mu_0 NI}{2R} = 4.33 \times 10^{-5}\ \text{T}$$

$$I = \frac{(4.33 \times 10^{-5}\ \text{T})(2R)}{\mu_0 N}$$

$$= \frac{(4.33 \times 10^{-5}\ \text{T})(0.72\ \text{m})}{(1.26 \times 10^{-6}\ \text{T}\cdot\text{m/A})(60)} = 0.41\ \text{A}$$

As noted, this current is counterclockwise as viewed from above.

**ASSESS** Equation 24.3 shows that the field in the center of a coil is proportional to the number of turns in the coil, proportional to the current, and inversely proportional to the radius of the coil. Table 24.1 gives 0.01 T for the field in the center of a 100-turn coil that is 1 cm in diameter and carries a current of 1.0 A. The coil in this problem is nearly 100 times larger, carries about half as much current, and has about half as many turns. We'd therefore predict a field that is less than the Table 24.1 value by a factor of a few hundred, which it is. A rough estimate of the answer agrees with our result, so it seems reasonable.

## Solenoids

FIGURE 24.26 The magnetic field inside a solenoid.

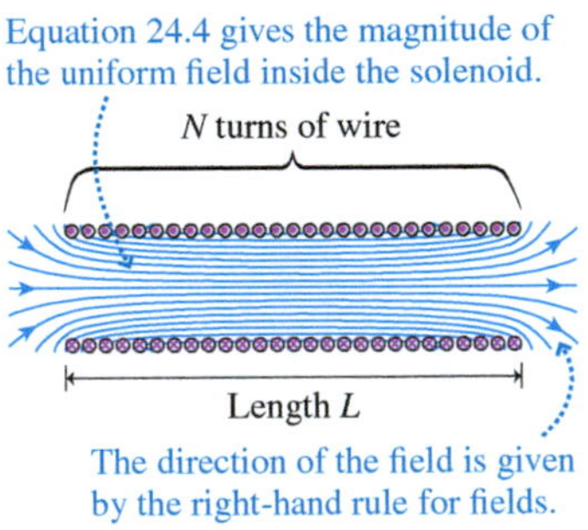


As we've seen, the field inside a solenoid is quite uniform, while the field outside is very small; the greater a solenoid's length in comparison to its diameter, the better these statements hold. Measurements that need a uniform magnetic field are often conducted inside a solenoid, which can be built quite large. The cylinder that surrounds a patient undergoing magnetic resonance imaging (MRI), such as the one shown at the top of the next page, contains a large solenoid. Consider a solenoid of length $L$ with $N$ turns of wire, as in FIGURE 24.26. We expect that the more turns we can pack into a solenoid of a given length—that is, the greater the ratio $N/L$—the stronger the field inside will be. We further expect that the strength of the field will be proportional to the current $I$ in the turns. Somewhat surprisingly, the field inside a solenoid does *not* depend on its radius. For this reason, the radius $R$ doesn't appear in the equation for the field inside a solenoid:

$$B = \frac{\mu_0 NI}{L} \tag{24.4}$$

Magnetic field inside a solenoid of length $L$ with $N$ turns

**EXAMPLE 24.5** **Generating an MRI magnetic field** BIO

A typical MRI solenoid has a length of about 1 m and a diameter of about 1 m. A typical field inside such a solenoid is about 1 T. How many turns of wire must the solenoid have to produce this field if the largest current the wire can carry is 100 A?

**STRATEGIZE** This solenoid is not very long compared to its diameter, so using Equation 24.4 will give only an approximate result. This is acceptable, since we have only rough estimates of the field $B$ and the length $L$.

**PREPARE** Equation 24.4 gives the magnetic field $B$ of a solenoid in terms of the current $I$, the number of turns $N$, and the length $L$. Here, however, we want to find the number of turns in terms of the other variables. We need $B = 1$ T, $I = 100$ A, and $L = 1$ m.

**SOLVE** We can solve Equation 24.4 for $N$ to get

$$N = \frac{LB}{\mu_0 I} = \frac{(1\ \text{m})(1\ \text{T})}{(1.26 \times 10^{-6}\ \text{T}\cdot\text{m/A})(100\ \text{A})} = 8000\ \text{turns}$$

to one significant figure.

**ASSESS** The number of turns required is quite large, but the field is quite large, so this makes sense.

In the above example, the diameter of the solenoid was about 1 m. The length of wire in each turn is thus $\pi \times 1$ m, or about 3 m. The total length of wire is then about 8000 turns $\times$ 3 m/turn = 24,000 m = 24 km $\approx$ 15 miles! If this magnet used ordinary copper wire, the total resistance $R$ would be about 35 $\Omega$. The power dissipated by a resistor is equal to $I^2R$, so the total power would be about $(100\text{ A})^2(35\ \Omega) = 350{,}000$ W, a huge and impractical value. MRI magnets must use *superconducting* wire, which when cooled to near absolute zero has *zero* resistance. This allows 24 km of wire to carry 100 A with no power dissipation at all.

At this point, it's worthwhile to bring together the relationships for magnetic fields from currents to help you see the connections between the different situations and remember the meanings of the different variables.

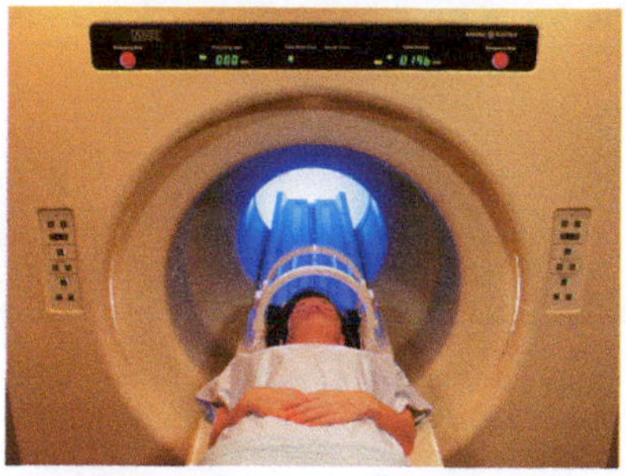
A patient's head in the solenoid of an MRI scanner.

**SYNTHESIS 24.1 Fields from currents**

An electric current produces a magnetic field. The geometry of the field depends on the geometry of the conductor.

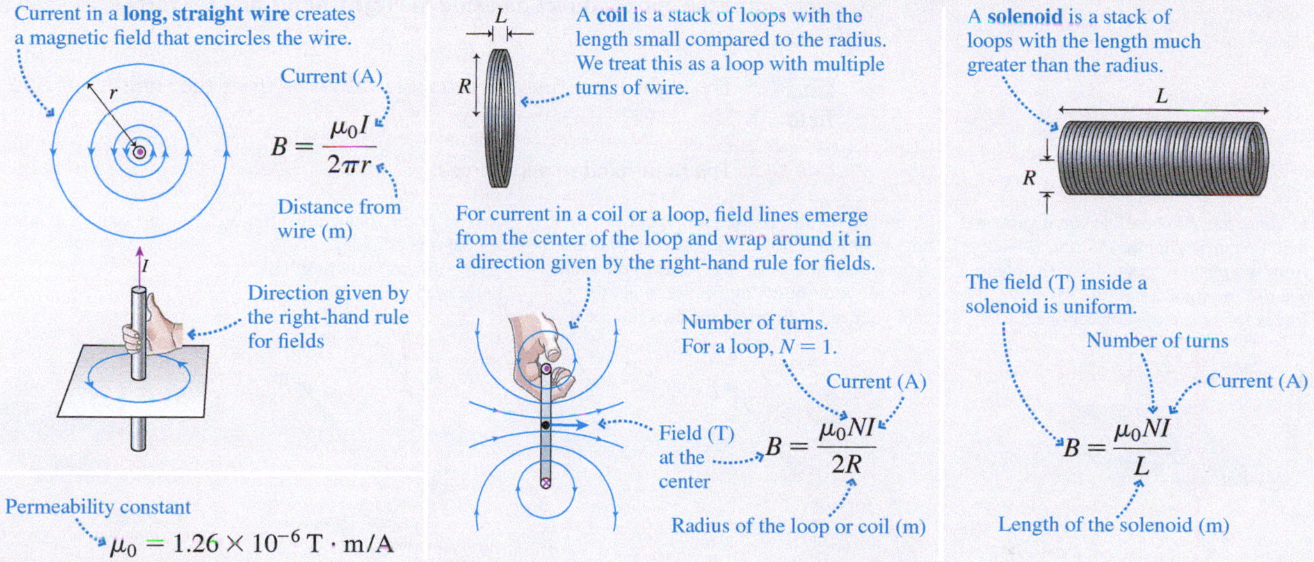


**STOP TO THINK 24.6** An investigator has made a solenoid by wrapping 100 turns of wire on a tube that is 10 cm long and 2 cm in diameter. The power supply is providing as much current as it can, but a stronger field is needed, so the solenoid must be rewrapped. Which of the following will result in a stronger field?

A. Wrapping 100 turns of wire on a tube that is 20 cm long and 2 cm in diameter.
B. Wrapping 100 turns of wire on a tube that is 10 cm long and 1 cm in diameter.
C. Wrapping 100 turns of wire on a tube that is 5 cm long and 2 cm in diameter.

## 24.5 Magnetic Fields Exert Forces on Moving Charges

It's time to switch our attention from what magnetic fields look like and how they are created to what they actually *do*. We've already seen that magnetic fields exert forces and torques on magnets, such as the torque that causes a compass needle to line up with the field. Now we'll see that magnetic fields also exert forces on moving charged particles and, later on, on electric currents in wires. The experimental results in the following table illustrate the nature of the magnetic force on a moving charged particle.

**The force on a charged particle moving in a magnetic field**

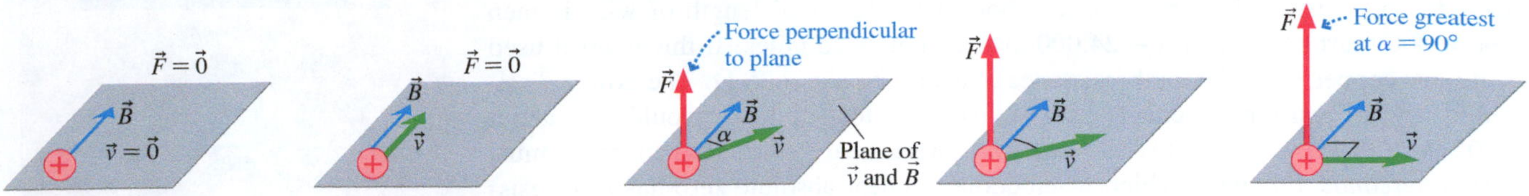


There is no magnetic force on a charged particle at rest.

There is no magnetic force on a charged particle moving *parallel* to a magnetic field.

As the angle $\alpha$ between the velocity and the magnetic field increases, the magnetic force also increases. The force is greatest when the angle is 90°. The magnetic force is always perpendicular to the plane containing $\vec{v}$ and $\vec{B}$.

As the experiments show, the magnetic force is quite different from the electric force. First, there is no magnetic force if the charged particle is at rest or if it's moving parallel to the magnetic field. Second, the force is always *perpendicular* to the plane containing $\vec{v}$ and $\vec{B}$. Because there are two directions perpendicular to a plane, we determine the correct direction using the **right-hand rule for forces,** as shown in FIGURE 24.27.

NOTE ▶ The right-hand rule for forces is different from the right-hand rule for fields. ◀

FIGURE 24.27 The right-hand rule for forces.

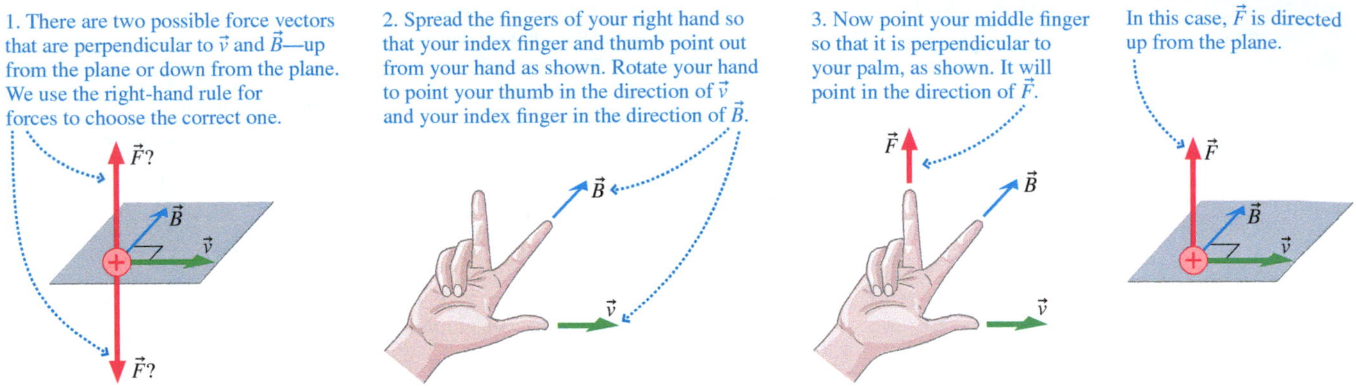
1. There are two possible force vectors that are perpendicular to $\vec{v}$ and $\vec{B}$—up from the plane or down from the plane. We use the right-hand rule for forces to choose the correct one.
$\vec{F}$?
$\vec{B}$
$\vec{v}$
$\vec{F}$?
2. Spread the fingers of your right hand so that your index finger and thumb point out from your hand as shown. Rotate your hand to point your thumb in the direction of $\vec{v}$ and your index finger in the direction of $\vec{B}$.
$\vec{B}$
$\vec{v}$
3. Now point your middle finger so that it is perpendicular to your palm, as shown. It will point in the direction of $\vec{F}$.
$\vec{F}$
$\vec{B}$
$\vec{v}$
In this case, $\vec{F}$ is directed up from the plane.
$\vec{F}$
$\vec{B}$
$\vec{v}$

NOTE ▶ The right-hand rule for forces gives the direction of the force on a *positive* charge. For a negative charge, the force is in the opposite direction. ◀

We can organize all of the experimental information about the magnetic force on a moving charged particle into a single equation. The magnitude and direction of the force $\vec{F}$ on a charged particle moving in a magnetic field are given by

The *direction* of the force

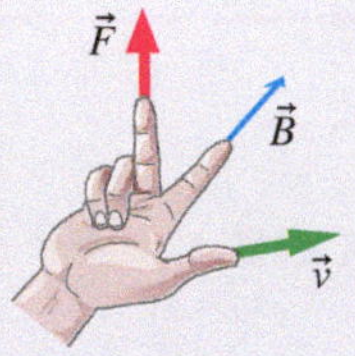


The *magnitude* of the force

Magnitude of charge (C) · Magnetic field (T)

$$F = |q| vB \sin\alpha \tag{24.5}$$

Speed of particle (m/s) · Angle between $\vec{v}$ and $\vec{B}$

The velocity and the magnetic field are perpendicular in many practical situations. In this case $\alpha$ is 90°, and the magnitude of the magnetic field simplifies to

$$F = |q| vB \tag{24.6}$$

with the direction of the field again determined by the right-hand rule for forces. The following Tactics Box summarizes and shows how to use the above information.

Video Magnet and Electron Beam

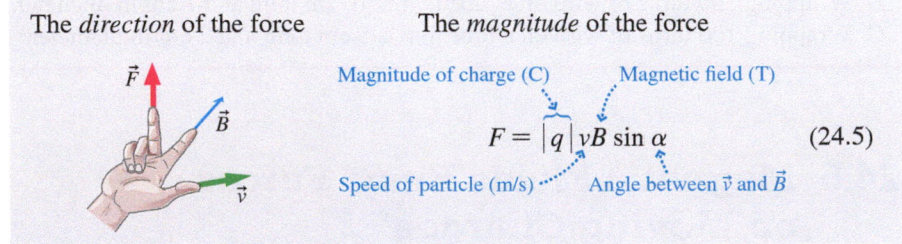

**TACTICS BOX 24.2 Determining the magnetic force on a moving charged particle**

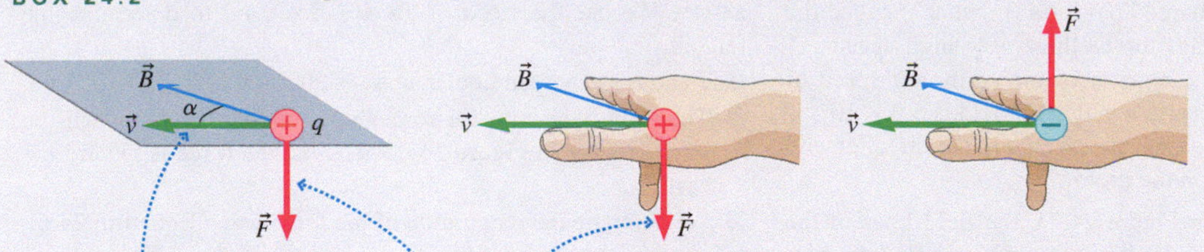

$$F = |q|vB\sin\alpha$$

$$F = |q|vB$$

❶ Note the direction of $\vec{v}$ and $\vec{B}$, and find the angle $\alpha$ between them.

❷ The force is perpendicular to the plane containing $\vec{v}$ and $\vec{B}$. The direction of $\vec{F}$ is given by the right-hand rule.

❸ For a negative charge, the force is in the direction opposite to that predicted by the right-hand rule.

❹ The magnitude of the force is given by Equation 24.5 or Equation 24.6.

Exercises 17, 18

**CONCEPTUAL EXAMPLE 24.6 Earth's magnetic field and a digital compass**

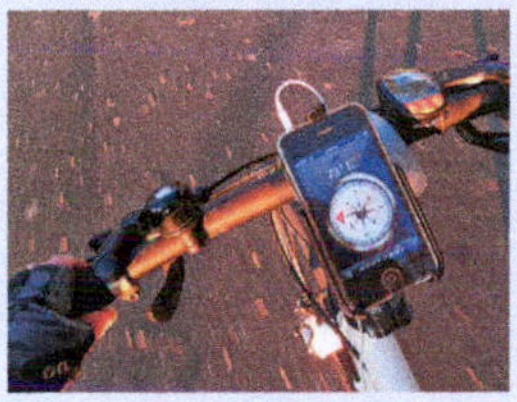

Your smartphone probably has a built-in digital compass. There's no actual pivoting needle, of course. Instead, a sensor in your phone detects the earth's magnetic field by the force it exerts on electrons moving in a conductor. FIGURE 24.28 shows one of these electrons moving to the right when the earth's field points in the direction shown. What is the direction of the magnetic force?

FIGURE 24.28 An electron moving in the earth's magnetic field.

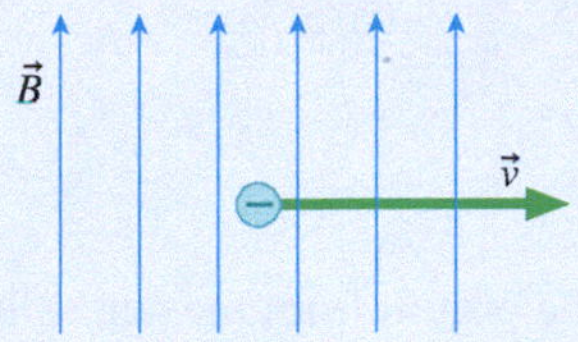

***REASON*** FIGURE 24.29 shows how the right-hand rule for forces is applied to this situation:

- Point your right thumb in the direction of the electron's velocity and your index finger in the direction of the magnetic field.
- Bend your middle finger to be perpendicular to your index finger. Your middle finger, which now points out of the page, is the direction of the force on a positive charge. But the electron is negative, so the force on the electron is *into* the page.

FIGURE 24.29 Using the right-hand rule.

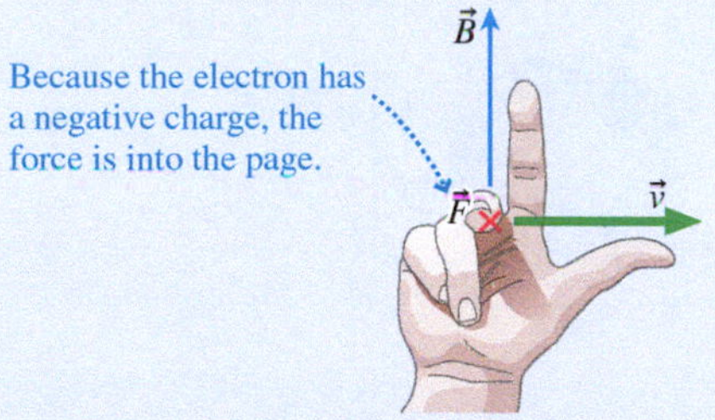

Because the electron has a negative charge, the force is into the page.


***ASSESS*** The force is perpendicular to both the velocity and the magnetic field, as it must be. The force on an electron is into the page; the force on a proton would be out of the page.

**CONCEPTUAL EXAMPLE 24.7 Determining the force on a charged particle moving near a current-carrying wire**

A proton is moving to the right above a horizontal wire that carries a current to the right. What is the direction of the magnetic force on the proton?

***REASON*** The current in the wire creates a magnetic field; this magnetic field exerts a force on the moving proton. We follow three steps to solve the problem:

1. Sketch the situation, as shown in FIGURE 24.30a.
2. Determine the direction of the field at the position of the proton due to the current in the wire (FIGURE 24.30b).
3. Determine the direction of the force that this field exerts on the proton (FIGURE 24.30c). In this case, the force is down.

***ASSESS*** As it should be, the magnetic force is perpendicular to both $\vec{v}$ and $\vec{B}$. This example also gives insight into how two parallel, current-carrying wires exert forces on each other, a topic we'll study in a later section. The proton moving to the right, which is essentially a small current in the same direction as the current in the wire, is *attracted* toward the wire.

FIGURE 24.30 Determining the direction of the force.

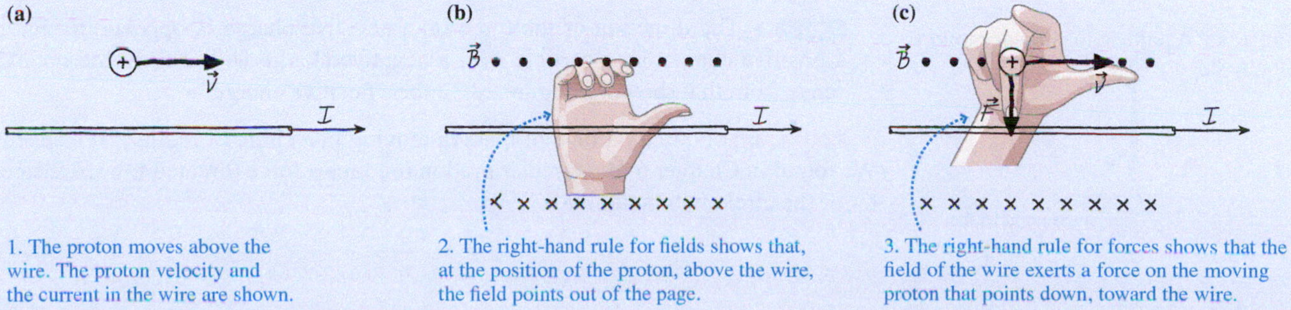

(a)
1. The proton moves above the wire. The proton velocity and the current in the wire are shown.
(b)
2. The right-hand rule for fields shows that, at the position of the proton, above the wire, the field points out of the page.
(c)
3. The right-hand rule for forces shows that the field of the wire exerts a force on the moving proton that points down, toward the wire.

**EXAMPLE 24.8** **Finding the force on a charged particle in the earth's magnetic field**

The sun emits streams of charged particles (in what is called the *solar wind*) that move toward the earth at very high speeds. A proton is moving toward the equator of the earth at a speed of 500 km/s. At this point, the earth's magnetic field is $5.0 \times 10^{-5}$ T directed parallel to the earth's surface. What are the direction and the magnitude of the force on the proton?

**STRATEGIZE** As we saw in Figure 24.7a, the field lines of the earth go from the earth's south pole to the earth's north pole; near the equator they are parallel to the earth's surface. Using this magnetic field direction and the direction of the proton's motion, we will use the steps of Tactics Box 24.2 to find the force on the proton.

**PREPARE** FIGURE 24.31 shows a picture of the proton entering the earth's magnetic field. We need to convert the proton's velocity to m/s:

$$500\ \text{km/s} = 5.0 \times 10^2\ \text{km/s} \times \frac{1 \times 10^3\ \text{m}}{1\ \text{km}} = 5.0 \times 10^5\ \text{m/s}$$

**SOLVE** We use the steps of Tactics Box 24.2 to determine the force.

❶ $\vec{v}$ and $\vec{B}$ are perpendicular, so $\alpha = 90°$.

❷ The right-hand rule for forces tells us that the force will be into the page in Figure 24.31. That is, the force is toward the east.

❸ We compute the magnitude of the force using Equation 24.6:

$$F = |q|vB = (1.6 \times 10^{-19}\ \text{C})(5.0 \times 10^5\ \text{m/s})(5.0 \times 10^{-5}\ \text{T})$$
$$= 4.0 \times 10^{-18}\ \text{N}$$

**ASSESS** This is a small force, but the proton has an extremely small mass of $1.67 \times 10^{-27}$ kg. Consequently, this force produces a very large acceleration: approximately 200 million times the acceleration due to gravity!

FIGURE 24.31 A proton in the magnetic field of the earth.

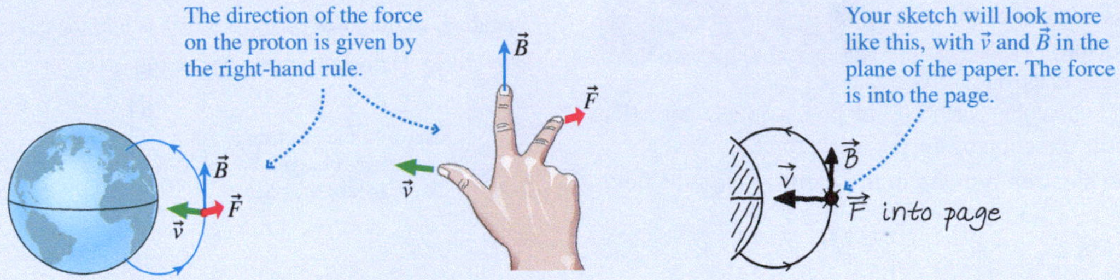


FIGURE 24.32 A charged particle moving perpendicular to a uniform magnetic field.

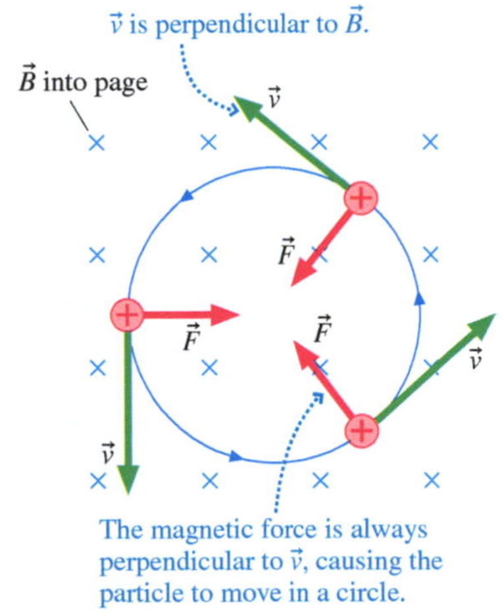


Example 24.8 looked at the force on a proton from the sun as it reaches the magnetic field of the earth. How does this force affect the *motion* of the proton? Let's consider the general question of how charged particles move in magnetic fields.

## Paths of Charged Particles in Magnetic Fields

We know that the magnetic force on a moving charged particle is always perpendicular to its velocity. Suppose a positively charged particle is moving perpendicular to a uniform magnetic field $\vec{B}$, as shown in FIGURE 24.32. In ◂◂ SECTION 6.3, we looked at the motion of objects subject to a force that was always perpendicular to the velocity. The result was *circular motion at a constant speed*. For a ball moving in a circle at the end of a string, the tension force is always perpendicular to $\vec{v}$. For a satellite moving in a circular orbit, the gravitational force is always perpendicular to $\vec{v}$. Here, for a charged particle moving in a magnetic field, the magnetic force is always perpendicular to $\vec{v}$ and so causes the particle to move in a circle, as Figure 24.32 shows. Thus, **a particle moving perpendicular to a uniform magnetic field undergoes uniform circular motion at constant speed.**

NOTE ▸ The direction of the force on a negative charge is opposite to that on a positive charge, so a particle with a negative charge will orbit in the opposite sense from that shown in Figure 24.32 for a positive charge. ◂

FIGURE 24.33 A particle in circular motion.

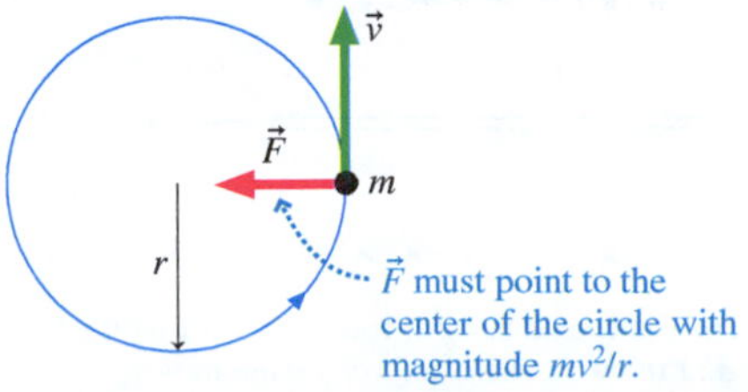


FIGURE 24.33 shows a particle of mass $m$ moving in a circle of radius $r$ at a speed $v$. We found in Chapter 6 that circular motion requires a force directed toward the center of the circle with magnitude

$$F = \frac{mv^2}{r} \tag{24.7}$$

For a charged particle moving in a magnetic field, this force is provided by the magnetic force. In Figure 24.32 we assumed that the velocity was perpendicular to the magnetic field, so the magnitude of the force on the charged particle due to the magnetic field is given by Equation 24.6. This is the force that produces the circular motion, so we can equate it to the force in Equation 24.7:

$$F = |q|vB = \frac{mv^2}{r}$$

Solving for $r$, we find that the radius of the circular orbit for a charged particle moving in a magnetic field is given by

$$r = \frac{mv}{|q|B} \tag{24.8}$$

Particle mass (kg) · Particle velocity (m/s) · Particle charge (C) · Magnetic field strength (T)

A particle moving *perpendicular* to a magnetic field moves in a circle. In the table at the start of this section, we saw that a particle moving *parallel* to a magnetic field experiences no magnetic force, and so continues in a straight line. A more general situation in which a charged particle's velocity $\vec{v}$ is neither parallel to nor perpendicular to the field $\vec{B}$ is shown in FIGURE 24.34. The net result is a circular motion due to the perpendicular component of the velocity coupled with a constant velocity parallel to the field: The charged particle spirals around the magnetic field lines in a helical trajectory.

As we discussed in Example 24.8, high-energy particles stream out from the sun in the solar wind. Some of these charged particles become trapped in the earth's magnetic field. As FIGURE 24.35 shows, the particles spiral in helical trajectories along the earth's magnetic field lines. Some of these particles enter the atmosphere near the north and south poles, ionizing gas and creating the ghostly glow of the **aurora**.

FIGURE 24.34 A charged particle in a magnetic field follows a helical trajectory.

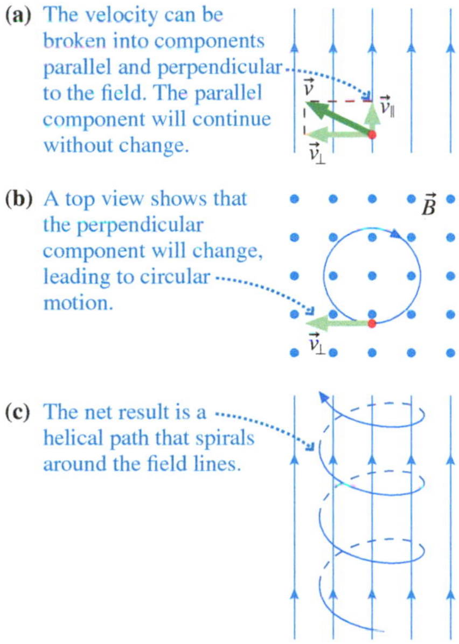

(a) The velocity can be broken into components parallel and perpendicular to the field. The parallel component will continue without change.
$\vec{v}$
$\vec{v}_\parallel$
$\vec{v}_\perp$
(b) A top view shows that the perpendicular component will change, leading to circular motion.
$\vec{B}$
$\vec{v}_\perp$
(c) The net result is a helical path that spirals around the field lines.


FIGURE 24.35 Charged particles in the earth's magnetic field create the aurora.

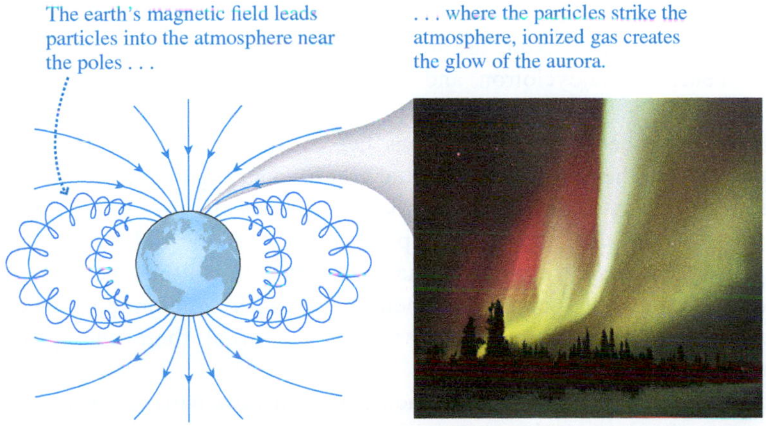


## EXAMPLE 24.9 Force on a charged particle in the earth's magnetic field, revisited

In Example 24.8, we considered a proton in the solar wind moving toward the equator of the earth, where the earth's magnetic field is $5.0 \times 10^{-5}$ T, at a speed of 500 km/s ($5.0 \times 10^5$ m/s). We now know that the proton will move in a circular orbit around the earth's field lines. What are the radius and the period of the orbit?

***STRATEGIZE*** We will find the radius of the proton's orbit using Equation 24.8. Its period can be found from this radius and the proton's known speed.

***PREPARE*** We begin with a sketch of the situation, noting the proton's orbit, as shown in FIGURE 24.36 on the next page.

*Continued*

FIGURE 24.36 A proton orbits the earth's magnetic field lines.

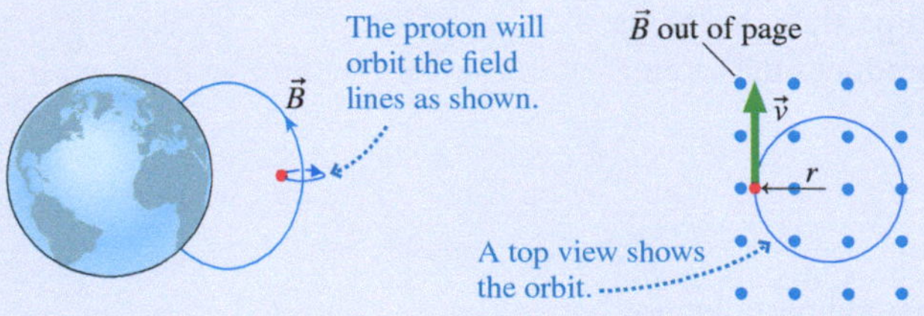


**SOLVE** Before we use any numbers, we will do some work with symbols. The period $T$ for one orbit is the distance of one orbit (the circumference $2\pi r$) divided by the speed:

$$T = \frac{2\pi r}{v}$$

We then substitute Equation 24.8 for the radius of the orbit to get

$$T = \frac{2\pi}{v} r = \frac{2\pi}{\cancel{v}}\left(\frac{m\cancel{v}}{qB}\right) = \frac{2\pi m}{qB}$$

The speed $v$ cancels, and so doesn't appear in the final expression. All protons in the earth's magnetic field orbit with the same period, regardless of their speed; a higher speed just means a larger circle, completed in the same time. Using values for mass, charge, and field, we compute the radius and the period of the orbit:

$$r = \frac{(1.67 \times 10^{-27}\ \text{kg})(5.0 \times 10^5\ \text{m/s})}{(1.60 \times 10^{-19}\ \text{C})(5.0 \times 10^{-5}\ \text{T})} = 100\ \text{m}$$

$$T = \frac{2\pi(1.67 \times 10^{-27}\ \text{kg})}{(1.60 \times 10^{-19}\ \text{C})(5.0 \times 10^{-5}\ \text{T})} = 0.0013\ \text{s}$$

**ASSESS** We can do a quick check on our math. We've found the radius of the orbit and the period, so we can compute the speed:

$$v = \frac{2\pi r}{T} = \frac{2\pi(100\ \text{m})}{0.0013\ \text{s}} = 5 \times 10^5\ \text{m/s}$$

This is the speed that we were given in the problem statement, which is a good check on our work.

FIGURE 24.37 The cyclotron.

(a) Interior view of a cyclotron

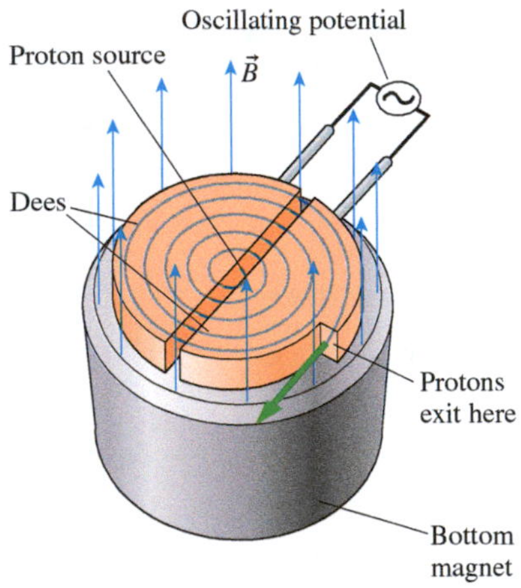


(b) A cyclotron used for the production of medical isotopes

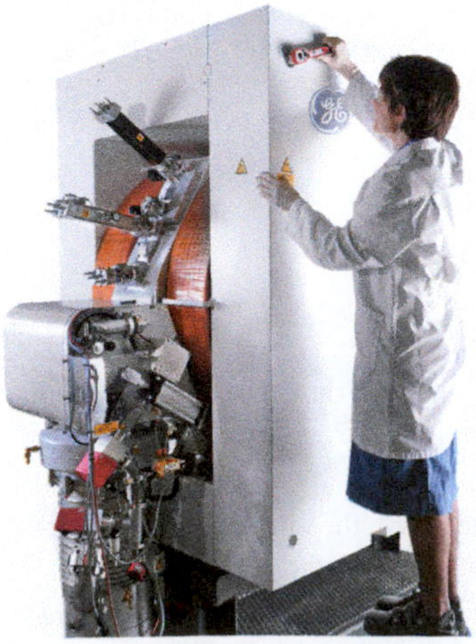

## The Cyclotron

The medical imaging technique of *positron-emission tomography* (*PET*), which you'll learn about in Chapter 30, is used to make images of the internal *biological activity* of the body. PET requires the fluorine isotope $^{18}$F, which, because of its short half-life of only 110 minutes, must be created at or near the medical facility in which it's required. This is done using a **cyclotron** to fire energetic protons at $^{18}$O atoms in water, replacing a neutron with a proton to create $^{18}$F. A cyclotron, shown in FIGURE 24.37a, consists of an evacuated chamber within a large, uniform magnetic field. Inside the chamber are two hollow conductors called "dees," separated by a small gap. Protons are injected into the magnetic field from a source near the center of the cyclotron, and they begin to move in a circular orbit in the magnetic field.

The key to accelerating the protons lies in the oscillating electric potential applied between the dees. This potential creates a strong electric field in the gap between the dees, which accelerates the protons passing through the gap. Soon, these protons, traveling in a circular path in the magnetic field, reach the opposite gap. By this time, the potential has changed sign, so the electric field again accelerates the protons, increasing their speed and kinetic energy. This process repeats each time the protons cross a gap, so that their energy increases continuously.

According to Equation 24.8, the radius of a charged particle's orbit is proportional to its speed, so as the protons speed up, the radius of their orbit increases. When the protons reach the outer edge of the magnet, where their speed and energy are the highest, they exit the cyclotron as a continuous high-energy beam that can be directed to an $^{18}$O target.

Example 24.9 showed that the period of a charged particle moving in a uniform magnetic field is *independent* of the particle's speed and orbit radius. For a cyclotron, this means that the period with which the electric field must be switched is the same for protons moving slowly near the cyclotron's center as for those moving much faster near its outer edge. This fact greatly simplifies the instrument's design.

**EXAMPLE 24.10 A medical cyclotron BIO**

It takes a proton with a kinetic energy of 11 MeV to efficiently change $^{18}O$ nuclei into $^{18}F$. If the magnetic field inside the cyclotron is 1.2 T, what is the radius of the protons' orbit just before they exit the cyclotron?

**STRATEGIZE** Equation 24.8 relates the radius of a charged particle's orbit to its speed. We will find the speed of the protons from their kinetic energy.

**PREPARE** We need to convert the protons' energy from MeV to J. An 11 MeV proton's kinetic energy in J is

$$K = (11 \times 10^6\ \text{eV}) \times \frac{1.60 \times 10^{-19}\ \text{J}}{1\ \text{eV}} = 1.76 \times 10^{-12}\ \text{J}$$

**SOLVE** The kinetic energy of a particle is $K = \frac{1}{2}mv^2$, so the proton's speed is

$$v = \sqrt{\frac{2K}{m}} = \sqrt{\frac{2(1.76 \times 10^{-12}\ \text{J})}{1.67 \times 10^{-27}\ \text{kg}}} = 4.59 \times 10^7\ \text{m/s}$$

We then find the radius of the orbit corresponding to this speed from Equation 24.8:

$$r = \frac{mv}{|q|B} = \frac{(1.67 \times 10^{-27}\ \text{kg})(4.59 \times 10^7\ \text{m/s})}{(1.60 \times 10^{-19}\ \text{C})(1.2\ \text{T})} = 0.40\ \text{m}$$

**ASSESS** The woman next to the cyclotron in FIGURE 24.37b is roughly 2 m tall, so a 0.80 m diameter for a proton's orbit as it exits the cyclotron seems reasonable.

## Electromagnetic Flowmeters

Blood contains many kinds of ions, such as $Na^+$ and $Cl^-$. When blood flows through a vessel, these ions move with the blood. An applied magnetic field will produce a force on these moving charges. We can use this principle to make a completely noninvasive device for measuring the blood flow in an artery: an *electromagnetic flowmeter.*

**Video** Electromagnetic Flowmeter

A flowmeter probe clamped to an artery has two active elements: magnets that apply a strong field across the artery and electrodes that contact the artery on opposite sides, as shown in FIGURE 24.38. The blood flowing in an artery carries a mix of positive and negative ions. Because these ions are in motion, the magnetic field exerts a force on them that produces a measurable voltage. We know from Equation 24.5 that the faster the blood's ions are moving, the greater the forces separating the positive and negative ions. The greater the forces, the greater the degree of separation and the higher the voltage. The measured voltage is therefore directly proportional to the velocity of the blood.

**Go with the flow** Many scientists and resource managers rely on accurate measurements of stream flows. The easiest way to get a quick measurement of the speed of a river or creek is to use an electromagnetic flowmeter similar to the one used for measuring flow in blood vessels. Water flows between the poles of a strong magnet. Two electrodes measure the resulting potential difference, which is proportional to the flow speed.

FIGURE 24.38 The operation of an electromagnetic flowmeter.

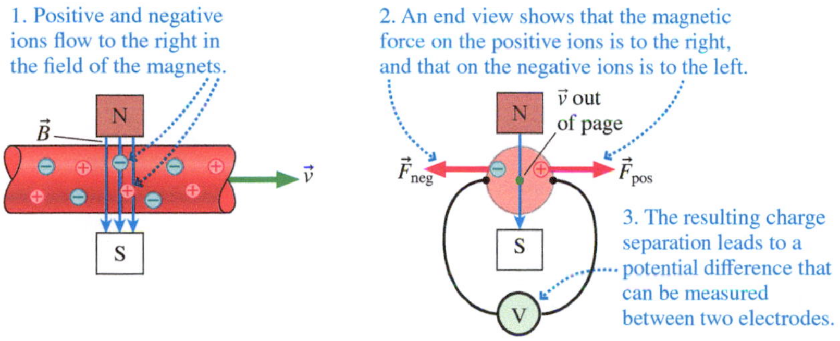


**STOP TO THINK 24.7** These charged particles are traveling in circular orbits with velocities and field directions as noted. Which particles have a negative charge?

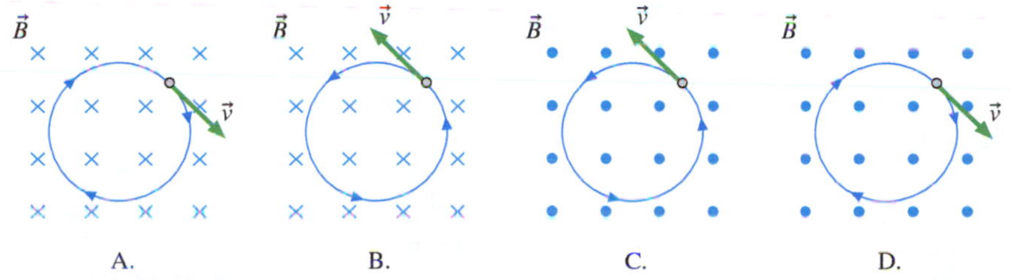

## 24.6 Magnetic Fields Exert Forces on Currents

We have seen that a magnetic field exerts a force on a moving charge. Because the current in a wire consists of charges moving through the wire, we expect that a magnetic field will exert a force on a current-carrying wire as well. This force is responsible for the operation of loudspeakers, electric motors, and many other devices.

### The Form of the Magnetic Force on a Current

In the table at the start of Section 24.5, we saw that a magnetic field exerts no force on a charged particle moving parallel to a magnetic field. If a current-carrying wire is *parallel* to a magnetic field, we also find that the force on it is zero.

However, there is a force on a current-carrying wire that is *perpendicular* to a magnetic field, as shown in FIGURE 24.39a.

FIGURE 24.39 Magnetic force on a current-carrying wire.

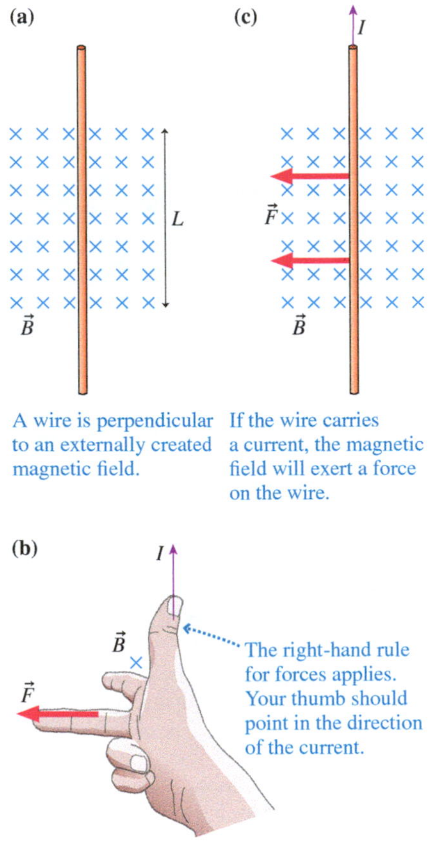


**NOTE** ▶ The magnetic field shown in Figure 24.39 is an *external* field, created by a permanent magnet or by other currents; it is *not* the field of the current $I$ in the wire. ◀

The direction of the force on the current is found by considering the force on each charge in the current. We model current as the flow of positive charge, so **the right-hand rule for forces applies to currents in the same way it does for moving charges.** With your fingers aligned as usual, point your right thumb in the direction of the current (the direction of the motion of positive charges) and your index finger in the direction of $\vec{B}$. Your middle finger is then pointing in the direction of the force $\vec{F}$ on the wire, as in FIGURE 24.39b. Consequently, the entire length of wire that is within the magnetic field experiences a force perpendicular to both the current direction and the field direction, as shown in FIGURE 24.39c.

If the length $L$ of the part of the wire that is in the magnetic field, the current $I$, or the magnetic field $B$ is increased, then the magnitude of the force on the wire will also increase. We can show that the magnetic force on a current-carrying wire is given by

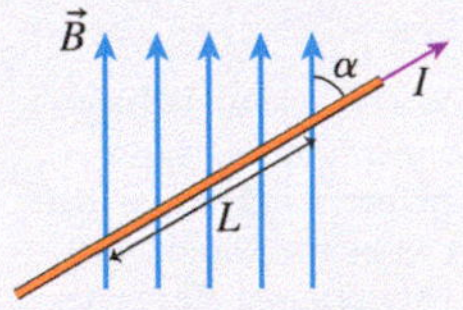

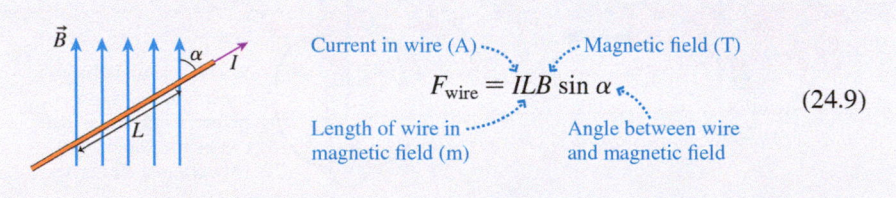

$$F_{\text{wire}} = ILB \sin \alpha \qquad (24.9)$$

Current in wire (A); Magnetic field (T); Length of wire in magnetic field (m); Angle between wire and magnetic field

Video What the Physics? Dancing Filament

In many practical cases the wire will be perpendicular to the field, so that $\alpha = 90°$. In this case,

$$F_{\text{wire}} = ILB \qquad (24.10)$$

If we rewrite Equation 24.10 as $B = F_{\text{wire}}/IL$, we can see that the unit for magnetic field, the tesla, can be defined in terms of other units:

$$1\ \text{T} = 1\ \frac{\text{N}}{\text{A}\cdot\text{m}}$$

Video Current-Carrying Wire in Magnetic Field

**EXAMPLE 24.11 Finding the magnetic force on a current-carrying wire** BIO

Scientists have studied the aerodynamic forces generated by a flying insect by tethering it to the center of a taut, horizontal wire. As the insect "flies" in a wind tunnel, the insect exerts forces on the wire that a researcher can detect by measuring the wire's deflection. To calibrate the wire, a known force must be applied to it. This is done by passing a current through the wire while it is positioned in a magnetic field that is perpendicular to the wire. The researcher requires the force on the wire to point upward with a magnitude of $4.2 \times 10^{-4}$ N. The length of the wire that is in the field is 50 mm, and the wire carries a current of 30 mA from left to right. What direction and magnitude of magnetic field will provide the necessary force?

***STRATEGIZE*** Because the magnetic field is perpendicular to the wire, we will use Equation 24.10 to calculate the force. The right-hand rule for forces will be used to find the direction of the field.

***PREPARE*** A sketch of the situation is shown in FIGURE 24.40a, in which the magnetic force on the wire is shown pointing up, as required. We have assumed that the current is directed to the right.

FIGURE 24.40 A current-carrying wire in a magnetic field.

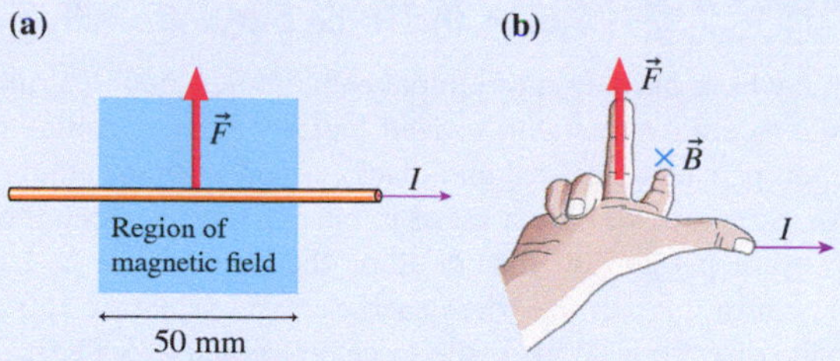


***SOLVE*** As shown in FIGURE 24.40b, the right-hand rule for forces shows that the direction of the magnetic field is into the page. Solving Equation 24.10 for the magnetic field gives

$$B = \frac{F_{\text{wire}}}{IL} = \frac{4.2 \times 10^{-4}\ \text{N}}{(0.030\ \text{A})(0.050\ \text{m})} = 0.28\ \text{T}$$

***ASSESS*** Table 24.1 shows that 0.28 T is a reasonable field for a laboratory magnet.

## Forces Between Currents

Because a current produces a magnetic field, and a magnetic field exerts a force on a current, it follows that two current-carrying wires will exert forces on each other. To see how, FIGURE 24.41a shows a wire carrying current $I_1$. From the right-hand rule for fields, we know that this current creates a magnetic field $\vec{B}_1$ pointing out of the page to the left of the wire and into the page to its right. If, as in FIGURE 24.41b, a second wire with current $I_2$ is placed next to the first wire, the current in the second wire will feel a magnetic force due to $\vec{B}_1$. Using the right-hand rule for forces, we can see that when $I_2$ is in the same direction as $I_1$ the force on the second wire is to the left; that is, the second wire is *attracted* to the first wire.

We can repeat this process to find the force on the first wire due to the magnetic field of the current $I_2$ in the second wire; you should be able to show that this force points to the right, so the first wire is also attracted to the second wire. Note that the forces on the two wires form a Newton's third law action/reaction pair. Thus, not only are they directed oppositely, as we've just seen, but they must have the *same* magnitude.

The attractive forces between two wires when the currents are in the same direction are shown in FIGURE 24.42a. By a similar argument, we can show that the forces between two wires are *repulsive* when the currents are in opposite directions, as shown in FIGURE 24.42b.

FIGURE 24.41 How one current exerts a force on another current.

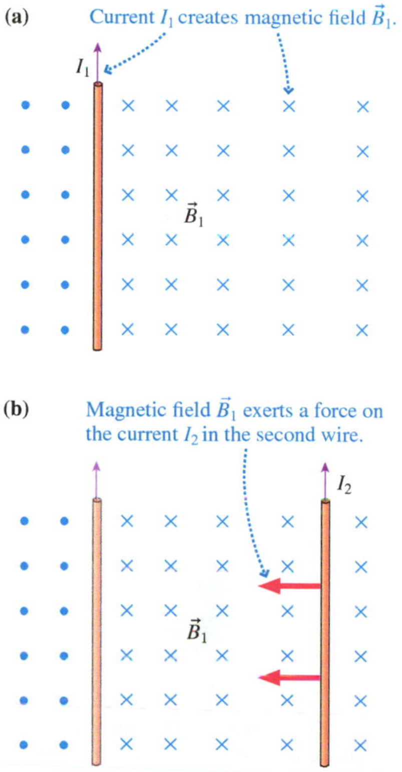


FIGURE 24.42 Forces between currents.

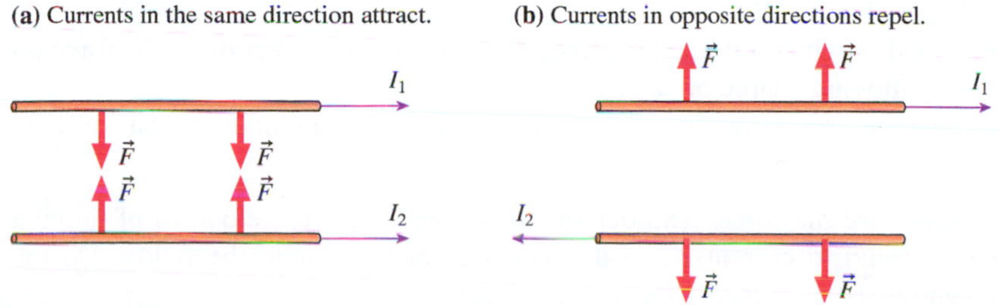

The following example shows how to calculate the *magnitude* of the force between two current-carrying wires.

### EXAMPLE 24.12 Finding the force between wires in jumper cables

You may have used a set of jumper cables connected to a running vehicle to start a car with a dead battery. Jumper cables are a matched pair of wires, red and black, joined together along their length. Suppose we have a set of jumper cables in which the two wires are separated by 1.2 cm along their 3.7 m (12 ft) length. While starting a car, the wires each carry a current of 150 A, in opposite directions. What is the force between the two wires?

**STRATEGIZE** According to our earlier discussion, the magnitudes of the forces on the wires are equal. Thus we only need to find the force on one of the wires—say, the red one. Finding this force is a two-step process. First, we will find the magnetic field at the position of the red wire due to the current in the black wire. Then we will find the force on the current in the red wire due to this magnetic field.

**PREPARE** We start by sketching the situation, noting distances and currents, as shown in FIGURE 24.43.

FIGURE 24.43 Jumper cables carrying opposite currents.

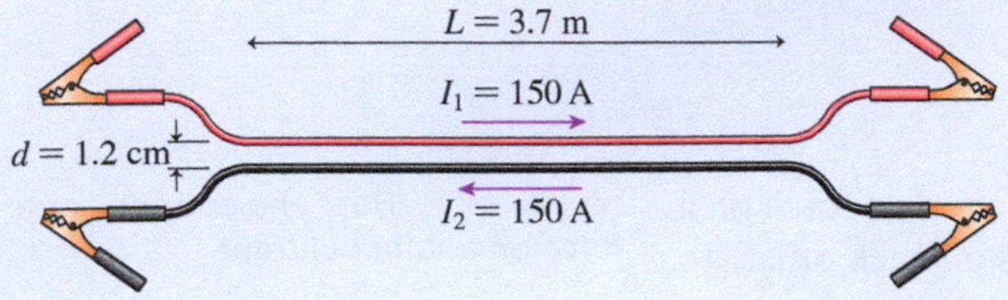


**SOLVE** The magnetic field at the position of the red wire, due to the current in the black wire, is

$$B = \frac{\mu_0 I}{2\pi d} = \frac{(1.26 \times 10^{-6}\,\text{T}\cdot\text{m/A})(150\,\text{A})}{2\pi(0.012\,\text{m})} = 2.51 \times 10^{-3}\,\text{T}$$

According to the right-hand rule for fields, this magnetic field is directed into the page. The magnitude of the force on the red wire is then

$$F_{\text{wire}} = ILB = (150\,\text{A})(3.7\,\text{m})(2.51 \times 10^{-3}\,\text{T}) = 1.4\,\text{N}$$

The direction of the force can be found using the right-hand rule for forces. The magnetic field at the position of the red wire is into the page, while the current is to the right. This means that the force on the red wire is in the plane of the page, directed *away* from the black wire. Thus the force between the two wires is repulsive, as we expect when their currents are directed oppositely.

**ASSESS** These wires are long, close together, and carry very large currents. But the force between them is quite small—much less than the weight of the wires. In practice, the forces between currents are not an important consideration unless there are many coils of wire, which leads to a large total force, such as in an MRI solenoid.

### PROBLEM-SOLVING APPROACH 24.1 Magnetic force problems

Magnetic fields exert forces on both moving charges and current-carrying wires, but in both cases the force can be found using the same approach.

**STRATEGIZE** There are two key factors to identify in magnetic force problems:

- The source of the magnetic field
- The charges or currents that feel a force due to this magnetic field

**PREPARE** In magnetic force problems, the force, the field, and the wire or the direction of the moving charge can all point in different directions, so it is especially important to prepare a careful sketch of the situation.

**SOLVE** First, determine the magnitude and direction of the magnetic field at the position of the charges or currents that are of interest.

- In some problems the magnetic field is given.
- If the field is due to a current, use the right-hand rule for fields to determine the field direction and the appropriate equation to determine its magnitude.

Next, determine the force this field produces. Working with the charges or currents you identified previously,

- Use the right-hand rule for forces to determine the direction of the force on any moving charge or current.
- Use the appropriate equation to determine the magnitude of the force on any moving charge or current.

**ASSESS** Are the forces you determine perpendicular to velocities of moving charges and to currents? Are the forces perpendicular to the fields? Do the magnitudes of the forces seem reasonable?

Exercise 28

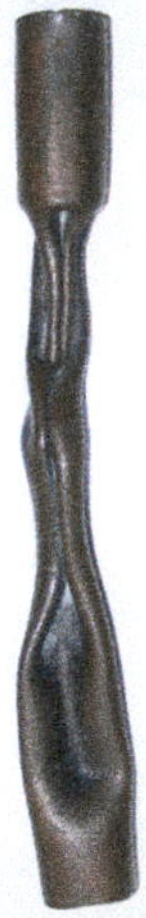

**Crushed by currents** The forces between currents are normally quite small, but the enormous currents in a lightning strike can produce remarkable forces. This hollow rod carried tens of thousands of amps following a lightning strike. The currents in all parts of the rod were parallel, and the size of the current led to attractive forces strong enough to actually crush the rod.

## Forces Between Current Loops

Now let's consider the forces between two current loops. Doing so will allow us to begin to make connections with some of the basic phenomena of magnetism that we saw earlier in the chapter.

We've seen that there is an attractive force between two parallel wires that carry currents in the same direction. If these two wires are bent into loops, as in FIGURE 24.44, then the force between the two loops will also be attractive. The forces will be repulsive if the currents are in opposite directions.

Early in the chapter, we examined the fields from various permanent magnets and current arrangements. FIGURE 24.45a reminds us that a bar magnet is a magnetic dipole, with a north and a south pole. Field lines come out of its north pole, loop back around, and go into the south pole. FIGURE 24.45b shows that the field of a current loop is very similar to that of a bar magnet. This leads us to the conclusion that **a current loop, like a bar magnet, is a magnetic dipole,** with a north and a south pole, as indicated in Figure 24.45b. We can use this conclusion to understand the forces between current loops. In FIGURE 24.46, we show how the poles of each current loop can be represented by a magnet. This model helps us understand the forces between current loops; in the next section, we'll use this model to understand torques on current loops as well.

FIGURE 24.44 Forces between parallel current loops.

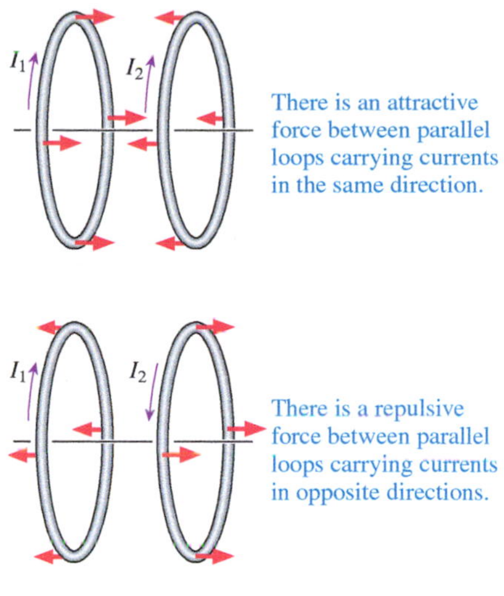

$I_1$
$I_2$
There is an attractive force between parallel loops carrying currents in the same direction.
$I_1$
$I_2$
There is a repulsive force between parallel loops carrying currents in opposite directions.


FIGURE 24.45 We can picture the loop as a small bar magnet.

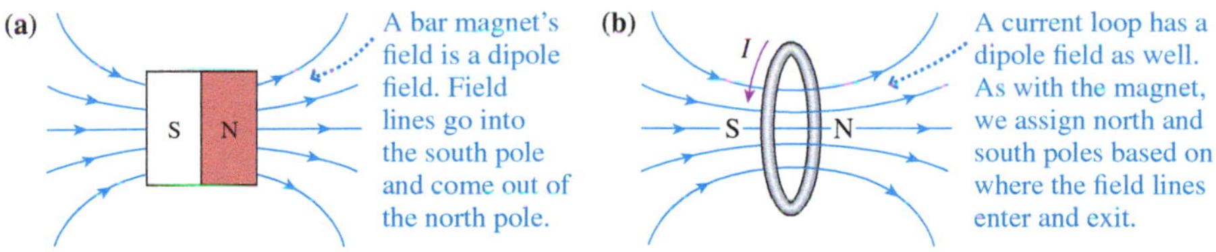

(a)
S
N
A bar magnet's field is a dipole field. Field lines go into the south pole and come out of the north pole.
(b)
$I$
S
N
A current loop has a dipole field as well. As with the magnet, we assign north and south poles based on where the field lines enter and exit.


FIGURE 24.46 Forces between current loops can be understood in terms of their magnetic poles.

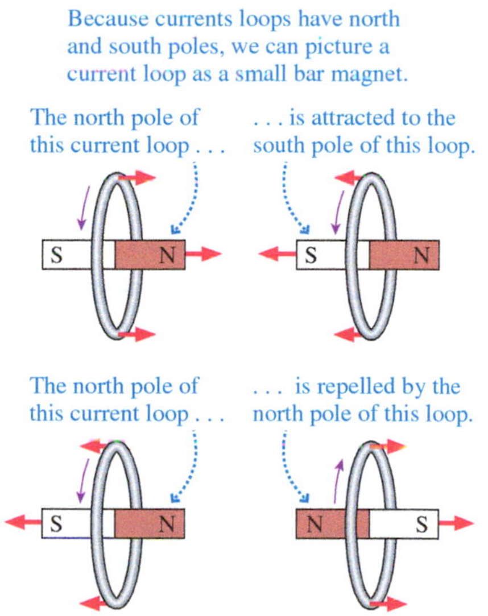


**STOP TO THINK 24.8** Four wires carry currents in the directions shown. A uniform magnetic field is directed into the paper as shown. Which wire experiences a force to the left?

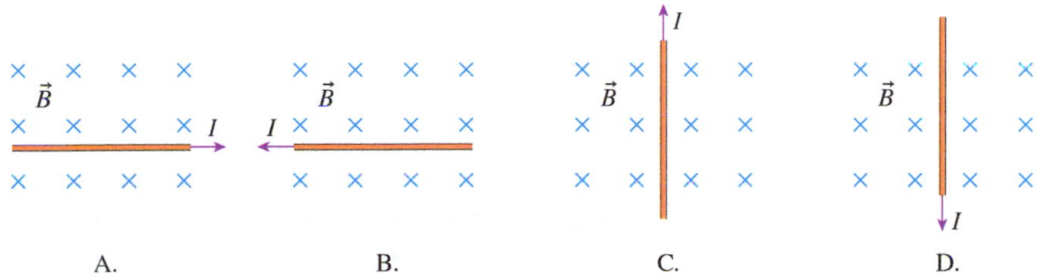


# 24.7 Magnetic Fields Exert Torques on Dipoles

At the beginning of this chapter, we found that a compass needle is actually a small magnet that turns to align itself with a magnetic field. This implies that a magnetic dipole—a magnet or a current loop—experiences a *torque* when placed in a magnetic field. In this section, we'll use what we've learned about magnetic forces to understand the torque on a magnetic dipole. We will consider only the case of a current loop, but the results will be equally applicable to permanent magnets.

## A Current Loop in a Uniform Field

FIGURE 24.47 shows a current loop—a magnetic dipole—in a uniform magnetic field $\vec{B}$. The current in each of the four sides of the loop experiences a magnetic force due to $\vec{B}$. Because the field is uniform, the forces on opposite sides of the loop are of

FIGURE 24.47 A loop in a uniform magnetic field experiences a torque.

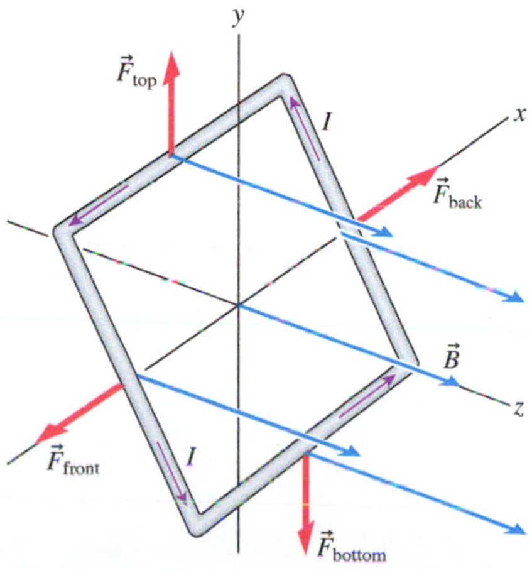

equal magnitude. The direction of each force is determined by the right-hand rule for forces. You can see that the forces $\vec{F}_{front}$ and $\vec{F}_{back}$ produce no net force or torque. The forces $\vec{F}_{top}$ and $\vec{F}_{bottom}$ also give no net force, but they will rotate the loop by exerting a torque on it.

Although we've shown a current loop, the conclusion is true for any magnetic dipole: **In a uniform field, a dipole experiences a torque but no net force.** For example, a compass needle is not attracted to the earth's poles; it merely feels a torque that lines it up with the earth's magnetic field.

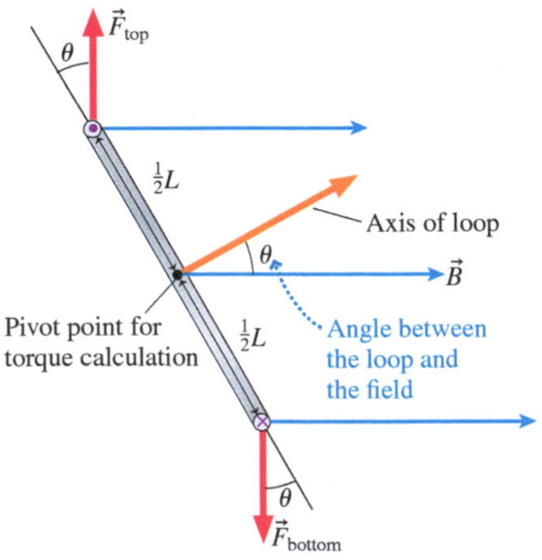

FIGURE 24.48 Calculating the torque on a current loop in a uniform magnetic field.

We can calculate the torque by looking at a side view of the current loop of Figure 24.47. This is shown in FIGURE 24.48, where you can see that the forces $F_{top}$ and $F_{bottom}$ act to rotate the loop clockwise. The angle $\theta$ between the axis of the loop and the field will be important. We've drawn a vector from the center of the loop that represents the axis of the loop; this vector will be used to define the angle $\theta$.

In ◀◀ SECTION 7.3, we calculated the torque due to a force $F$ as $rF \sin\theta$, where $r$ is the distance from the pivot to the point where the force acts, and $\theta$ is the angle between the force vector and a line drawn from the pivot to that point. In Figure 24.48, the net torque is then

$$\begin{aligned}\tau = \tau_{top} + \tau_{bottom} &= \left(\tfrac{1}{2}L\right)F_{top}\sin\theta + \left(\tfrac{1}{2}L\right)F_{bottom}\sin\theta \\ &= \left(\tfrac{1}{2}L\right)(ILB)\sin\theta + \left(\tfrac{1}{2}L\right)(ILB)\sin\theta \\ &= (IL^2)B\sin\theta\end{aligned}$$

$L^2$ is the area $A$ of the square loop. Using this, we can generalize the result to any loop of area $A$:

$$\tau = (IA)B\sin\theta = \mu B\sin\theta \tag{24.11}$$

There are two things to note about this torque:

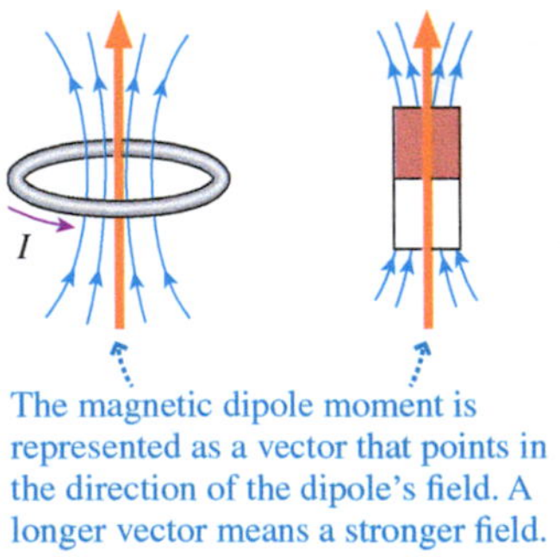

$I$
The magnetic dipole moment is represented as a vector that points in the direction of the dipole's field. A longer vector means a stronger field.

FIGURE 24.49 The dipole moment vector.

1. **The torque depends on properties of the current loop:** its area $A$ and the current $I$. The quantity $\mu = IA$, known as the **magnetic dipole moment** (or simply *magnetic moment*), is a measure of how much torque a dipole will feel in a magnetic field. A permanent magnet can be assigned a magnetic dipole moment as well; in this case, there is no clear meaning to $I$ and $A$, and we'll simply use the symbol $\mu$ for the magnetic moment. We'll represent the magnetic moment as an arrow pointing in the direction of the dipole's magnetic field, as in FIGURE 24.49. This figure shows the magnetic dipole moment for a current loop and a bar magnet. You can now see that the vector defining the axis of the loop in Figure 24.48 is simply the magnetic dipole moment.
2. **The torque depends on the angle between the magnetic dipole moment and the magnetic field.** The torque is maximum when $\theta$ is 90°, when the magnetic moment is perpendicular to the field. The torque is zero when $\theta$ is 0° (or 180°), when the magnetic moment is parallel to the field. As we see in FIGURE 24.50, a magnetic dipole free to rotate in a field will do so until $\theta$ is zero, at which point it will be stable. **A magnetic dipole will rotate to line up with a magnetic field** just as an electric dipole will rotate to line up with an electric field.

FIGURE 24.50 Torque on a dipole in an externally created magnetic field.

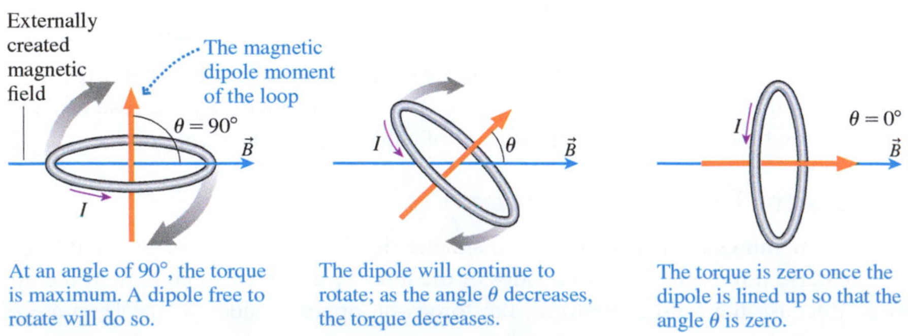

A compass needle, which is a dipole, also rotates until its north pole is in the direction of the magnetic field, as we noted at the start of the chapter.

**CONCEPTUAL EXAMPLE 24.13** **Does the loop rotate?**

Two nearby current loops are oriented as shown in FIGURE 24.51. Loop 1 is fixed in place; loop 2 is free to rotate. Will it do so?

FIGURE 24.51 Will loop 2 rotate?

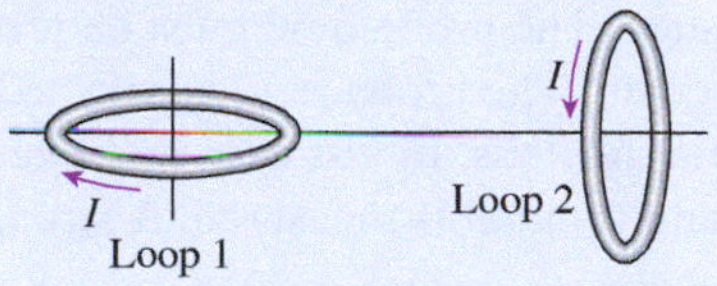


**REASON** The current in loop 1 generates a magnetic field. As FIGURE 24.52 shows, the field of loop 1 is upward as it passes loop 2. Because the field is perpendicular to the magnetic moment of loop 2, the field exerts a torque on loop 2 and causes loop 2 to rotate until its magnetic moment lines up with the field from loop 1.

**ASSESS** These two loops align so that their magnetic moments point in opposite directions. We can think of this in terms of their poles: When aligned this way, the north pole of loop 1 is closest to the south pole of loop 2. This makes sense because these opposite poles attract each other.

FIGURE 24.52 How the field of loop 1 affects loop 2.

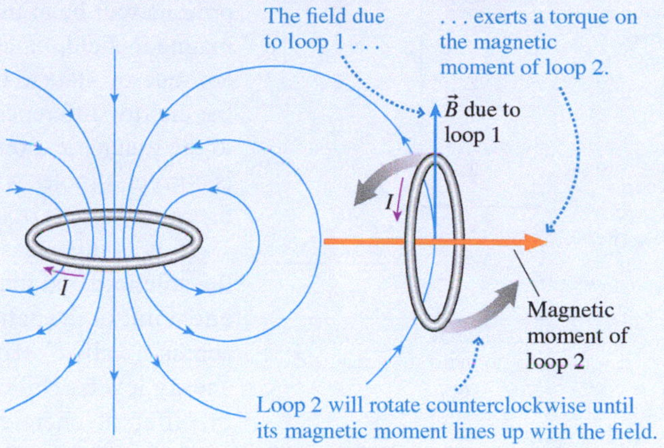


For any dipole in a field, there are actually two angles for which the torque is zero, $\theta = 0°$ and $\theta = 180°$, but there is a difference between these two cases. The $\theta = 0°$ case is *stable:* Once the dipole is in this configuration, it will stay there. The $\theta = 180°$ case is *unstable.* There is no torque if the alignment is perfect, but the slightest rotation will result in a torque that rotates the dipole until it reaches $\theta = 0°$.

We can make a gravitational analogy with this situation in FIGURE 24.53. For an upside-down pendulum, there will be no torque if the mass is directly above the pivot point. But, as we saw in ◀◀ SECTION 7.4, this is a position of unstable equilibrium. If displaced even slightly, the mass will rotate until it is below the pivot point. This is the point of lowest potential energy.

We can see, by analogy with the upside-down pendulum, that the unstable alignment of a magnetic dipole has a higher energy. Given a chance, the magnet will rotate "downhill" to the position of lower energy and stable equilibrium. This difference in energy is the key to understanding how the magnetic properties of atoms can be used to image tissues in the body in MRI.

FIGURE 24.53 Going from unstable to stable equilibrium.

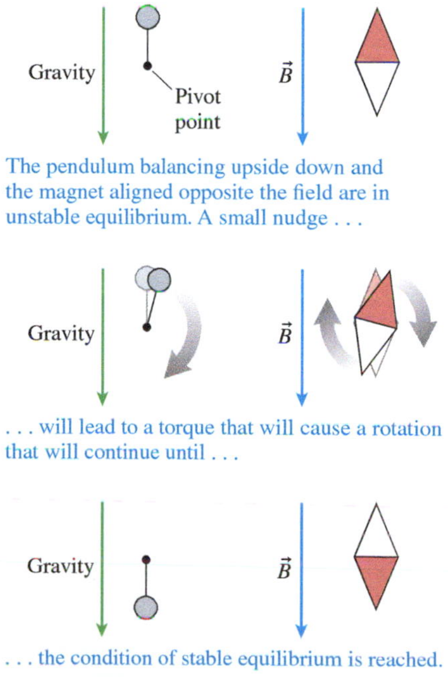

Gravity
Pivot point
$\vec{B}$
The pendulum balancing upside down and the magnet aligned opposite the field are in unstable equilibrium. A small nudge . . .
Gravity
$\vec{B}$
. . . will lead to a torque that will cause a rotation that will continue until . . .
Gravity
$\vec{B}$
. . . the condition of stable equilibrium is reached.


## Magnetic Resonance Imaging (MRI)

Magnetic resonance imaging is a modern diagnostic tool that provides detailed images of tissues and structures in the body with no radiation exposure. The key to this imaging technique is the magnetic nature of atoms. **The nuclei of individual atoms have magnetic moments and behave like magnetic dipoles.** Atoms of different elements have different magnetic moments; therefore, a magnetic field exerts different torques on different kinds of atoms.

A person receiving an MRI scan is placed in a large solenoid that has a strong magnetic field along its length. Think about one hydrogen atom in this person's body. The nucleus of the hydrogen atom is a single proton. The proton has a magnetic moment, but the proton is a bit different from a simple bar magnet: It is subject to the rules of quantum mechanics, which we will learn about in Chapter 28. A bar magnet can have any angle with the field, but the proton can only line up either *with*

FIGURE 24.54 The energy difference between the two possible orientations of a proton's magnetic moment during MRI.

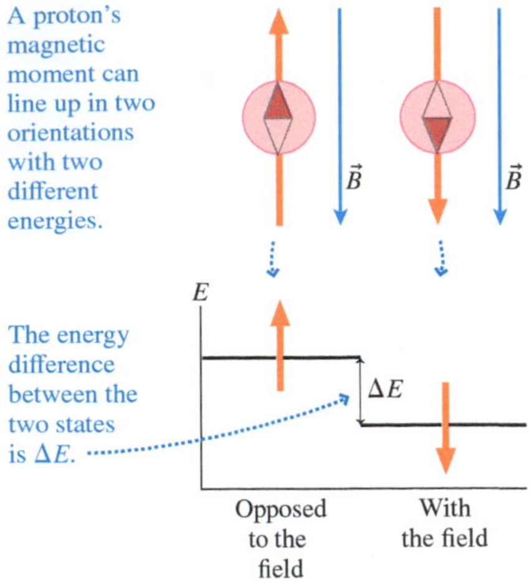


FIGURE 24.55 BIO Cross-sectional image of the brain from an MRI scan.

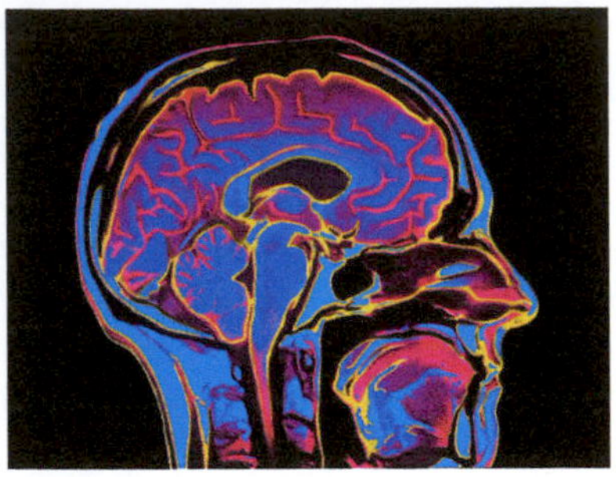

*the field* (the low-energy state) or *opposed to the field* (the high-energy state), as we see in FIGURE 24.54.

The energy difference $\Delta E$ between these two orientations of the proton depends on two key parameters: the magnetic moment of the proton and the strength of the magnetic field.

A tissue sample will have many hydrogen atoms with many protons. Some of the protons will be in the high-energy state and some in the low-energy state. A second magnetic field, called a *probe field,* can be applied to "flip" the dipoles from the low-energy state to the high-energy state. The probe field must be precisely tuned to the energy difference $\Delta E$ for this to occur. The probe field is selected to correspond to the magnetic moment of a particular nucleus, in this case hydrogen. If the tuning is correct, dipoles will change state and a signal is measured. A strong signal means that many atoms of this kind are present.

How is this measurement turned into an image? The magnetic field strengths of the solenoid and the probe field are varied so that the correct tuning occurs at only one point in the patient's body. The position of the point of correct tuning is swept across a "slice" of tissue. Combining atoms into molecules slightly changes the energy levels. Different tissues in the body have different concentrations of atoms in different chemical states, so the strength of the signal at a point will vary depending on the nature of the tissue at that point. As the point of correct tuning is moved, the intensity of the signal is measured at each point; a record of the intensity versus position gives an image of the structure of the interior of the body, as in FIGURE 24.55.

## Electric Motors

The torque on a current loop in a magnetic field is the basis for how an electric motor works. The *armature* of a motor is a loop of wire wound on an axle that is free to rotate. This loop is in a strong magnetic field. A current in the loop causes it to feel a torque due to this field, as FIGURE 24.56 shows. The loop will rotate to align itself with the external field. If the current were steady, the armature would simply rotate until its magnetic moment was in its stable position. To keep the motor turning, a device called a *commutator* reverses the current direction in the loop every 180°. As the loop reaches its stable configuration, the direction of current in the loop switches, putting the loop back into the unstable configuration, so it will keep rotating to line up in the other direction. This process continues: Each time the loop nears a stable point, the current switches. The loop will keep rotating as long as the current continues.

FIGURE 24.56 The operation of a simple motor depends on the torque on a current loop in a magnetic field.

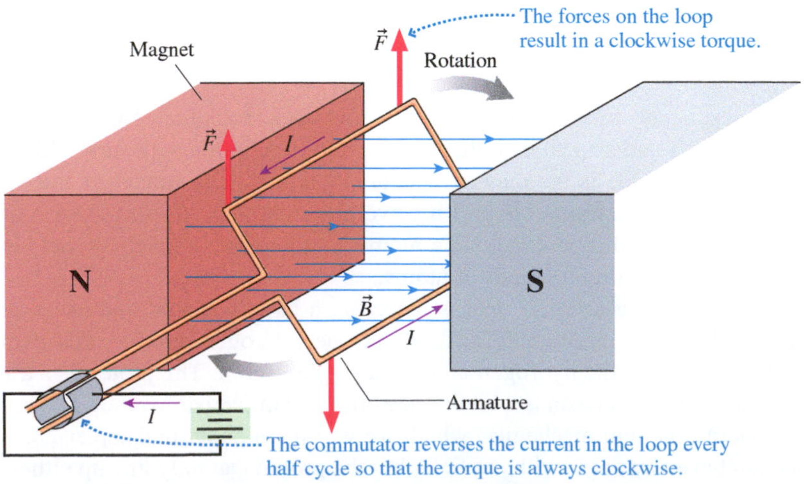

**STOP TO THINK 24.9** Which way will this current loop rotate?

A. Clockwise.
B. Counterclockwise.
C. The loop will not rotate.

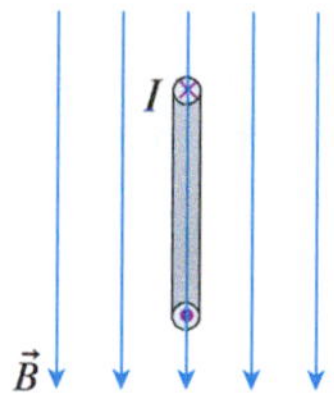


FIGURE 24.57 Magnetic moment of the electron.

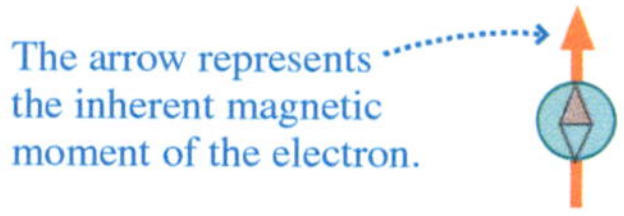


## 24.8 Magnets and Magnetic Materials

We started the chapter by looking at permanent magnets. We know that permanent magnets produce a magnetic field, but what is the source of this field? There are no electric currents in these magnets. Why can you make a magnet out of certain materials but not others? Why does a magnet stick to the refrigerator? The goal of this section is to answer these questions by developing an atomic-level view of the magnetic properties of matter.

### Ferromagnetism

Iron, nickel, and cobalt are elements that have very strong magnetic behavior: A chunk of iron (or steel, which is mostly iron) will stick to a magnet, and the chunk can be magnetized so that it is itself a magnet. Other metals—such as aluminum and copper—do not exhibit this property. We call materials that are strongly attracted to magnets and that can be magnetized **ferromagnetic** (from the Latin for iron, *ferrum*).

The key to understanding magnetism at the atomic level is that electrons, just like protons and nuclei, have an *inherent magnetic moment,* as we see in FIGURE 24.57. Magnetism, at an atomic level, is due to the inherent magnetic moment of electrons.

If the magnetic moments of all the electrons in an atom pointed in the same direction, the atom would have a very strong magnetic moment. But this doesn't happen. In atoms that have many electrons, the electrons usually occur in pairs that have magnetic moments in opposite directions, as we'll see in Chapter 29. Only the electrons that are unpaired are able to give the atom a net magnetic moment.

Even so, atoms with magnetic moments don't necessarily form a solid with magnetic properties. For most elements whose atoms have magnetic moments, the magnetic moments of the atoms are randomly arranged when the atoms join together to form a solid. As FIGURE 24.58 shows, this random arrangement produces a solid whose net magnetic moment is very close to zero. Ferromagnetic materials have atoms with net magnetic moments that tend to line up and reinforce each other, as in FIGURE 24.59. This alignment of moments occurs in only a few elements and alloys, and this is why a small piece of iron has such a strong overall magnetic moment. Such a piece has a north and a south magnetic pole, generates a magnetic field, and aligns parallel to an external magnetic field. In other words, it is a magnet—and a very strong one at that.

In a large sample of iron, the magnetic moments will be lined up in local regions called **domains,** each looking like the ordered situation of Figure 24.59, but there will be no long-range ordering. Inside a domain, the atomic magnetic moments will be aligned, but the magnetic moments of individual domains will be randomly oriented. The individual domains are quite small—on the order of 0.1 mm or so—so a piece of iron the size of a nail has thousands of domains. The random orientation of the magnetic moments of the domains, as shown in FIGURE 24.60, means that there is no overall magnetic moment. So, how can you magnetize a nail?

FIGURE 24.58 The random magnetic moments of the atoms in a typical solid.

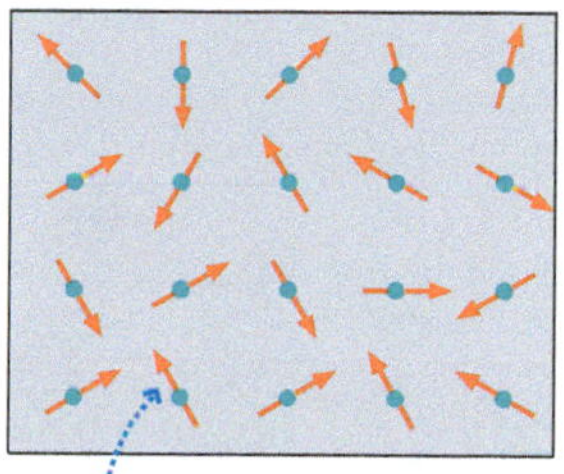


FIGURE 24.59 In a ferromagnetic solid, the atomic magnetic moments align.

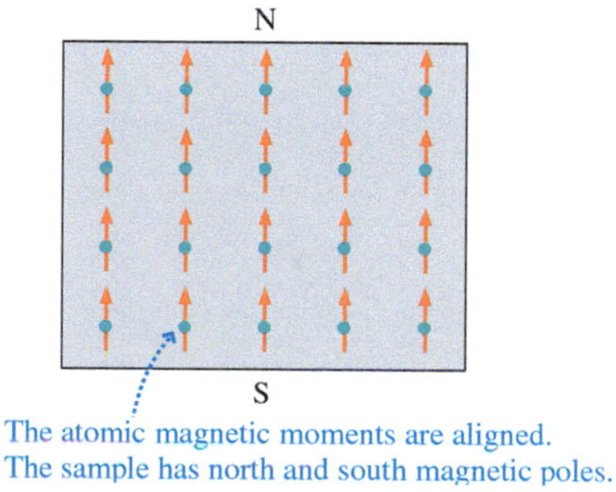


FIGURE 24.60 Magnetic domains in a ferromagnetic material.

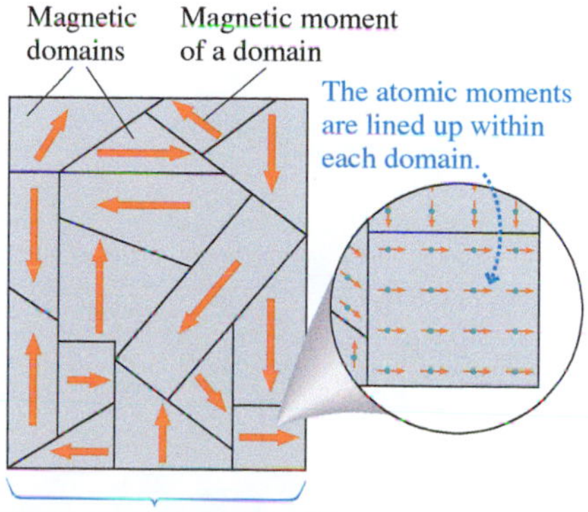

Magnetic domains
Magnetic moment of a domain
The atomic moments are lined up within each domain.
The magnetic moments of the domains tend to cancel one another. The sample as a whole possesses no net magnetic moment.

FIGURE 24.61 Inducing a magnetic moment in a piece of iron.

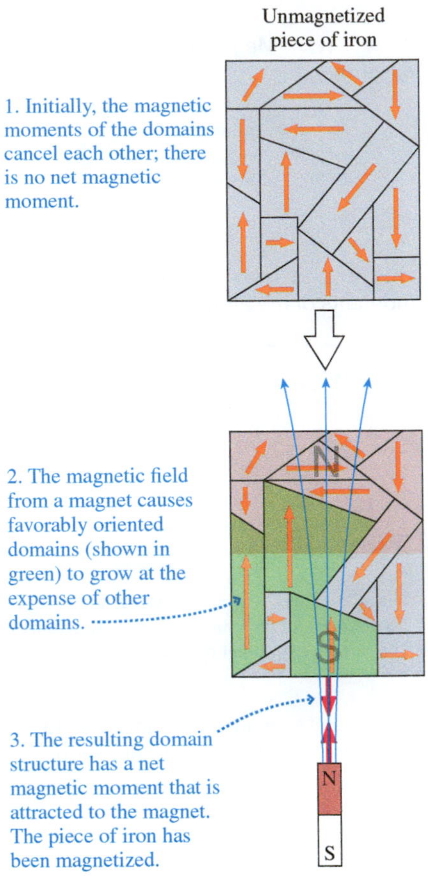

Unmagnetized piece of iron
1. Initially, the magnetic moments of the domains cancel each other; there is no net magnetic moment.
2. The magnetic field from a magnet causes favorably oriented domains (shown in green) to grow at the expense of other domains.
3. The resulting domain structure has a net magnetic moment that is attracted to the magnet. The piece of iron has been magnetized.
N
S
N
S


## Induced Magnetic Moments

When you bring a magnet near a piece of iron, as in FIGURE 24.61, the magnetic field of the magnet penetrates the iron and creates torques on the atomic magnetic moments. The atoms will stay organized in domains, but domains aligned with the external field become larger at the expense of domains opposed to the field. After this shift in domain boundaries, the magnetic moments of the domains no longer cancel out. The iron will have a net magnetic moment that is aligned with the external field. The iron will have developed an *induced magnetic moment.*

**NOTE** ► Inducing a magnetic moment with a magnetic field is analogous to inducing an electric dipole with an electric field, which we saw in Chapter 21. ◄

Looking at the pole structure of the induced magnetic moment in the iron, we can see that the iron will now be attracted to the magnet. The fact that a magnet attracts and picks up ferromagnetic objects was one of the basic observations about magnetism with which we started the chapter. Now we have an explanation of how this works, based on three facts:

1. Electrons are microscopic magnets due to their inherent magnetic moment.
2. In a ferromagnetic material, these atomic magnetic moments are aligned. Regions of aligned moments form magnetic domains.
3. The individual domains shift in response to an external magnetic field to produce an induced magnetic moment for the entire object. This induced magnetic moment will be attracted by a magnet that produced the orientation.

When a piece of iron is near a magnet, the iron becomes a magnet as well. But when the applied field is taken away, the domain structure will (generally) return to where it began: The induced magnetic moment will disappear. In the presence of a *very* strong field, however, a piece of iron can undergo more significant changes to its domain structure, and some domains may permanently change orientation. When the field is removed, the iron may retain some of this magnetic character: The iron will have become permanently magnetized. But pure iron is a rather poor permanent magnetic material; it is very easy to disrupt the ordering of the domains that has created the magnetic moment of the iron. For instance, if you heat (or even just drop!) a piece of magnetized iron, the resulting random atomic motions tend to destroy the alignment of the domains, destroying the magnetic character in the process.

Alloys of ferromagnetic materials often possess more robust magnetic characters. Alloys of iron and other ferromagnetic elements with rare-earth elements can make permanent magnets of incredible strength.

### CONCEPTUAL EXAMPLE 24.14 Sticking things to the refrigerator

Everyone has used a magnet to stick papers to the fridge. Why does the magnet stick to the fridge through a layer of paper?

***REASON*** When you bring a magnet near the steel door of the refrigerator, the magnetic field of the magnet induces a magnetic moment in the steel. The direction of this induced moment will be such that it is attracted to the magnet—thus the magnet will stick to the fridge. Magnetic fields go through nonmagnetic materials such as paper, so the magnet will hold the paper to the fridge.

***ASSESS*** This result helps makes sense of another observation you've no doubt made: You can't stick a thick stack of papers to the fridge. This is because the magnetic field of the magnet decreases rapidly with distance. Because the field is weaker, the induced magnetic moment is smaller.

Induced magnetic moments are used to store information on computer hard disk drives. As shown in FIGURE 24.62, a hard drive consists of a rapidly rotating disk with a thin magnetic coating on its surface. Information for computers is stored as digital zeros and ones, and on the disk these are stored as tiny magnetic domains, each less than 100 nm long! The direction of the domains can be changed by the write head—a tiny switchable magnet that skims over the surface of the disk. The magnetic field of the write head changes the orientation of the domains, encoding information on the disk. The information can then be retrieved by the read head—a small probe that is sensitive to the magnetic fields of the tiny domains.

**FIGURE 24.62** Computer hard disks store information using magnetic fields.

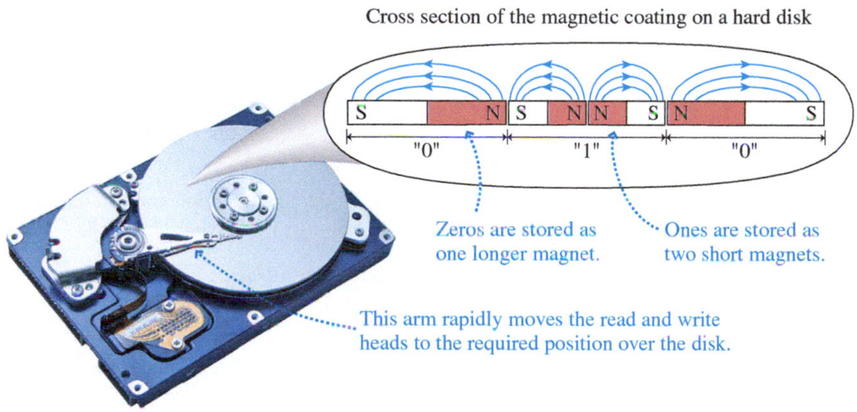


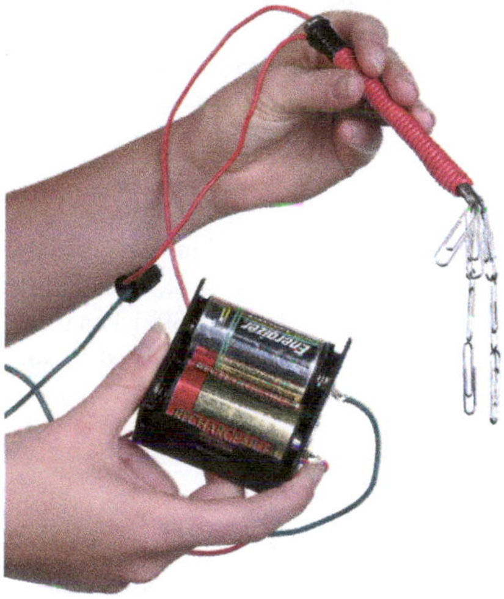

An electric current in the wire produces a magnetic field that magnetizes the nail around which the wire is wound.

## Electromagnets

The magnetic domains in a ferromagnetic material have a strong tendency to line up with an applied magnetic field. This means that it is possible to use a piece of iron or other ferromagnetic material to increase the strength of the field from a current-carrying wire. For example, suppose a solenoid is wound around a piece of iron. When current is passed through the wire, the solenoid's magnetic field lines up the domains in the iron, thus magnetizing it. The resulting **electromagnet** may produce a field that is hundreds of times stronger than the field due to the solenoid itself.

In the past few sections, we've begun to see examples of the deep connection between electricity and magnetism. This connection was one of the most important scientific discoveries of the 1800s and is something we will explore in more detail in the next chapter.

**Video** Magnetic Materials
**Video** Magnetic Fields and Current

**STOP TO THINK 24.10** A chain of paper clips is hung from a permanent magnet. Which diagram shows the correct induced pole structure of the paper clips?

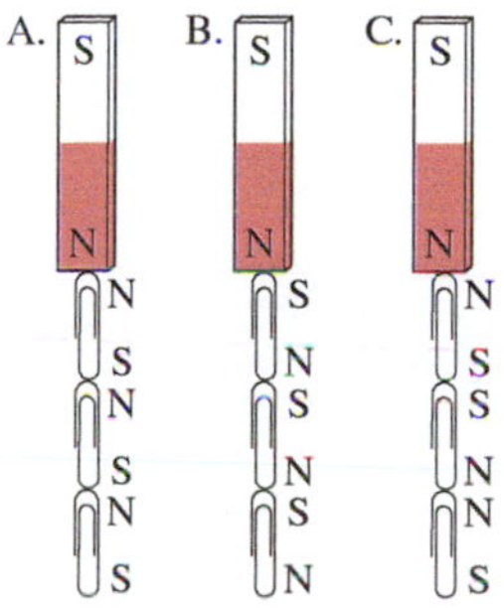

## INTEGRATED EXAMPLE 24.15 Rotating magnetic bacteria BIO

As we have seen, magnetotactic bacteria contain magnetic material in their bodies that allows them to use the earth's magnetic field to navigate. Biologists have directly observed the effect of this internal magnet by suddenly reversing the direction of the magnetic field in which the bacteria are swimming; because the bacteria are small magnetic dipoles, they rotate to line up with the new magnetic field. FIGURE 24.63 shows the path of one magnetotactic bacterium as it swims; we can clearly observe the reversal of its direction when the field is switched.

A bacterium's dipole moment can be estimated by observing the time it takes it to reorient when the field is reversed. As the magnetic torque acts to turn the bacterium's dipole to line up with the field, a *drag torque* due to the surrounding water acts to resist the rotation. If we approximate the bacterium as a sphere of radius *R*, the drag torque is

$$\tau_{drag} = 8\pi\eta R^3\omega$$

where $\omega$ is the bacterium's angular speed and $\eta$ is the viscosity of water. In one experiment, a bacterium with a diameter of 5.0 $\mu$m took 1.4 s to rotate by 180° when the direction of the $5.0 \times 10^{-5}$ T magnetic field was reversed. Estimate the magnetic dipole moment of this bacterium.

**STRATEGIZE** In Chapter 5, we found that small organisms moving in water always move at their terminal speed, the speed at which the external force acting on them equals the drag force. Similarly, we expect that our magnetic bacterium will always rotate at its terminal *angular* speed $\omega_{term}$, determined by the condition that the external magnetic *torque* $\tau_{magnetic}$ is equal to the drag torque $\tau_{drag}$. We will thus set these two expressions equal and solve for the magnetic moment.

The magnetic torque is given by Equation 24.11, where $\mu$ is the bacterium's magnetic moment that we wish to find. As the bacterium rotates, the term $\sin\theta$ in Equation 24.11 changes, which complicates the mathematics. Because we're only asked for an *estimate,* we'll simplify things by assuming that the torque always has its maximum value—that is, that $\sin\theta = 1$.

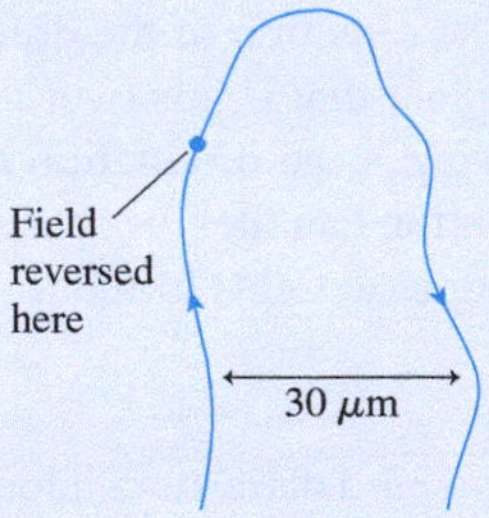


FIGURE 24.63 A bacterium reverses its direction when the magnetic field is switched from up to down.

**PREPARE** The bacterium takes 1.4 s to rotate through half a circle, so its period of rotation *T* is 2.8 s. Its angular velocity is then $\omega_{term} = 2\pi/T = 2\pi/(2.8\text{ s}) = 2.2$ rad/s. From Table 5.3, the viscosity of water at 20°C is $1.0 \times 10^{-3}$ Pa · s.

**SOLVE** At terminal angular speed, the magnetic and drag torques are equal, so

$$\mu B \sin\theta = 8\pi\eta R^3\omega_{term}$$

If we set $\sin\theta = 1$, we solve for the magnetic moment $\mu$ as

$$\mu = \frac{8\pi\eta R^3\omega_{term}}{B\cdot 1}$$

$$= \frac{8\pi(1.0\times10^{-3}\text{ Pa}\cdot\text{s})(2.5\times10^{-6}\text{ m})^3(2.2\text{ rad/s})}{5.0\times10^{-5}\text{ T}}$$

$$= 1.7\times10^{-14}\text{ A}\cdot\text{m}^2$$

**ASSESS** This very small magnetic moment is difficult to assess but it is in line with results from experiments. The magnetic material found in magnetotactic bacteria is magnetite, or $Fe_3O_4$. The magnetic moment we found in this example corresponds to a sphere of magnetite with a diameter of about 1 $\mu$m. This seems plausible for the magnetic material inside a bacterium.

# SUMMARY

**GOAL** To learn about magnetic fields and how magnetic fields exert forces on currents and moving charges.

## GENERAL PRINCIPLES

### Sources of Magnetism

Magnetic fields can be created by either:

- Electric currents or
- Permanent magnets

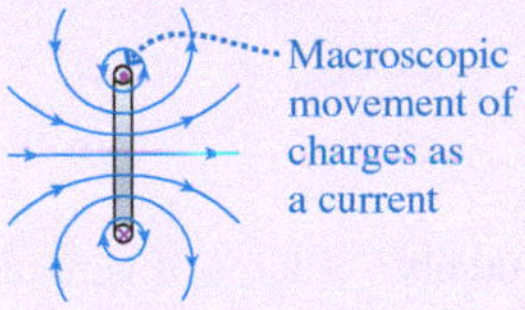


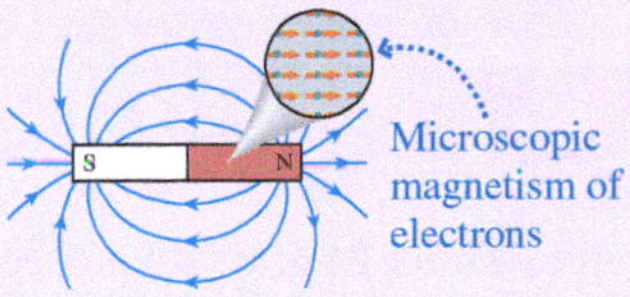


The most basic unit of magnetism is the **magnetic dipole,** which consists of a north and a south pole.

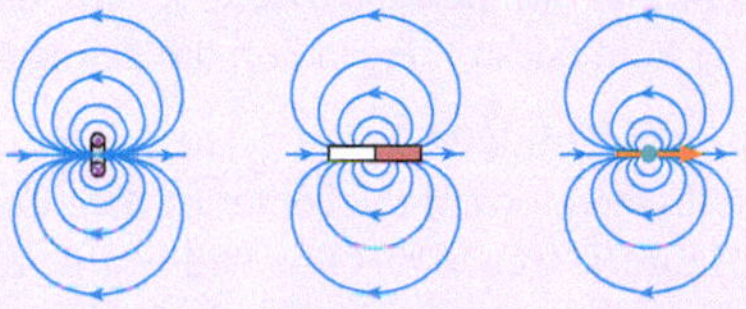

Three basic kinds of dipoles are: Current loop, Permanent magnet, Atomic magnet

### Consequences of Magnetism

Magnetic fields exert long-range forces on magnetic materials and on moving charges or currents.

- Unlike poles of magnets attract each other; like poles repel each other.

- A magnetic field exerts a force on a moving charged particle.

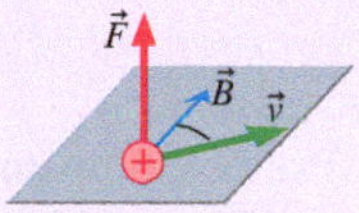


- Parallel wires with currents in the same direction attract each other; when the currents are in opposite directions, the wires repel each other.

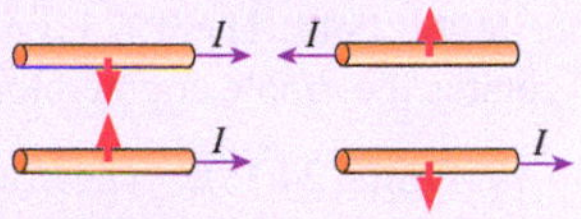


Magnetic fields exert torques on magnetic dipoles, aligning their axes with the field.

## IMPORTANT CONCEPTS

### Magnetic Fields

The **direction of the magnetic field**

- is the direction in which the north pole of a compass needle points.
- due to a current can be found from the **right-hand rule for fields.**

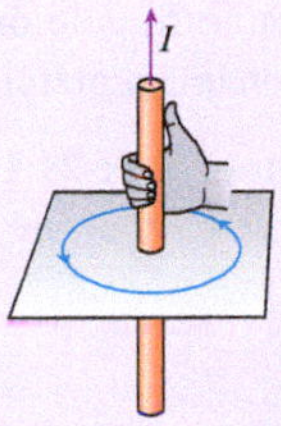


The **strength of the magnetic field** is

- proportional to the torque on a compass needle when turned slightly from the field direction.
- measured in tesla (T).

### Magnetic Forces and Torques

The magnitude of the magnetic force on a *moving* charge depends on its charge $q$, its speed $v$, and the angle $\alpha$ between the velocity and the field:

$$F = |q|vB \sin \alpha$$

The direction of this force on a positive charge is given by the **right-hand rule for forces.**

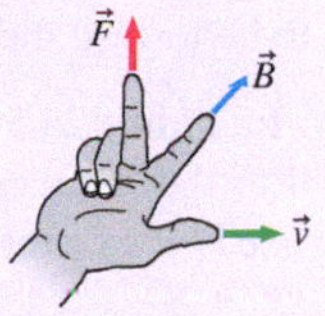


The magnitude of the force on a *current-carrying wire* perpendicular to the magnetic field depends on the current and the length of the wire: $F = ILB$.

The torque on a *current loop* in a magnetic field depends on the current, the loop's area, and how the loop is oriented in the field: $\tau = (IA)B \sin \theta$.

## APPLICATIONS

### Fields due to common currents

Long, straight wire

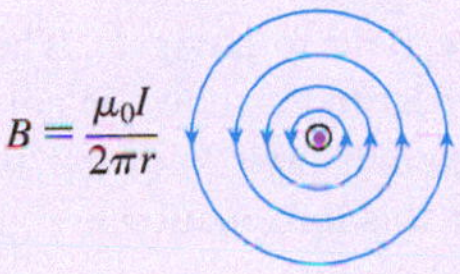

$$B = \frac{\mu_0 I}{2\pi r}$$

Current loop

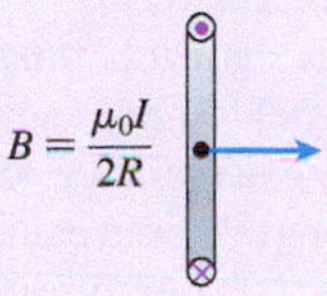

$$B = \frac{\mu_0 I}{2R}$$

Solenoid

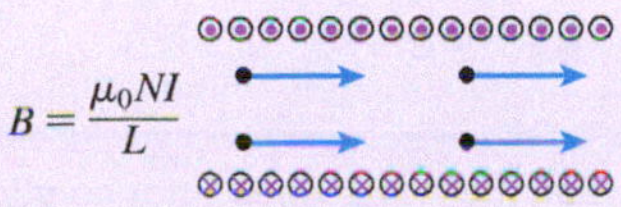

$$B = \frac{\mu_0 N I}{L}$$

### Charged-particle motion

There is no force if $\vec{v}$ is parallel to $\vec{B}$.

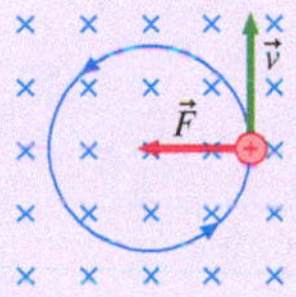


If $\vec{v}$ is perpendicular to $\vec{B}$, the particle undergoes uniform circular motion with radius $r = mv/|q|B$.

### Stability of magnetic dipoles

A magnetic dipole is stable (in a lower energy state) when aligned with the external magnetic field. It is unstable (in a higher energy state) when aligned opposite to the field.

The probe field of an MRI scanner measures the flipping of magnetic dipoles between these two orientations.

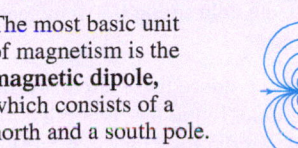

## Learning Objectives After studying this chapter, you should be able to:

- Recognize basic magnetic phenomena. *Conceptual Questions 24.2, 24.6*
- Calculate the magnetic fields of important current distributions. *Conceptual Questions 24.10, 24.14; Problems 24.1, 24.3, 24.4, 24.13, 24.16*
- Use the right-hand rules for magnetic fields and forces. *Conceptual Questions 24.8, 24.15; Problems 24.5, 24.6, 24.7, 24.22, 24.33*
- Determine the magnetic force on a moving charged particle. *Conceptual Questions 24.21, 24.22; Problems 24.23, 24.24, 24.25, 24.28, 24.30*
- Calculate magnetic forces and torques on currents. *Conceptual Questions 24.1, 24.27; Problems 24.34, 24.35, 24.36, 24.38, 24.39*
- Understand the atomic basis of magnetism. *Conceptual Question 24.28; Problems 24.42, 24.43*

## STOP TO THINK ANSWERS

**Chapter Preview Stop to Think: B.** The positive charge feels a force in the direction of the field, the negative charge feels a force opposite the field. There is no net force, but there will be a torque that rotates the dipole counterclockwise.

**Stop to Think 24.1: C.** The compass needle will not rotate, since there is no force between the stationary charges on the rod and the magnetic poles of the compass needle.

**Stop to Think 24.2: A.** The compass needle will rotate to line up with the field of the magnet, which goes from the north to the south pole.

**Stop to Think 24.3: A.** Using the right-hand rule for currents on any part of the loop shows that the field at the center of the loop is directed toward the top of the page.

**Stop to Think 24.4: D.** The compass needle will rotate to line up with the field circling the wire. The right-hand rule for fields shows this to be toward the top of the paper in the figure.

**Stop to Think 24.5: B.** The strength of the field is inversely proportional to the distance from the wire. If the distance from the wire is increased by a factor of 4, the field strength decreases by a factor of 4.

**Stop to Think 24.6: C.** If you keep the number of turns of wire the same and decrease the overall length, the field strength will increase. The field doesn't depend on the diameter.

**Stop to Think 24.7: A, C.** The force to produce these circular orbits is directed toward the center of the circle. Using the right-hand rule for forces, we see that this will be true for the situations in A and C if the particles are negatively charged.

**Stop to Think 24.8: C.** The right-hand rule for forces gives the direction of the force. With the field into the paper, the force is to the left if the current is toward the top of the paper.

**Stop to Think 24.9: B.** Looking at the forces on the top and the bottom of the loop, we can see that the loop will rotate counterclockwise. Alternatively, we can look at the dipole structure of the loop: With a north pole on the left and a south pole on the right, the loop will rotate counterclockwise.

**Stop to Think 24.10: B.** All of the induced dipoles will be aligned with the field of the bar magnet.

Video Tutor Solution Chapter 24

# QUESTIONS

## Conceptual Questions

1. In Figure Q24.1, suppose the magnet on the right is fixed in place and the magnet on the left is free to pivot about its center. Will the magnet on the left start to rotate? If so, will it initially rotate clockwise or counterclockwise?

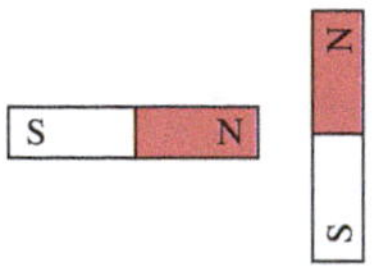


FIGURE Q24.1

2. Can two magnetic field lines cross? If so, give an example. If not, explain why not.
3. When you are in the southern hemisphere, does a compass point north or south?
4. If you were standing directly at the earth's north magnetic pole, in what direction would a compass point if it were free to swivel in any direction? Explain.
5. BIO If you took a sample of magnetotactic bacteria from the northern hemisphere to the southern hemisphere, would you expect them to survive? Explain.
6. BIO Green turtles use the earth's magnetic field to navigate. They seem to use the field to tell them their latitude—how far north or south of the equator they are. Explain how knowing the direction of the earth's field could give this information.

Problem difficulty is labeled as | (straightforward) to ||||| (challenging). Problems labeled INT integrate significant material from earlier chapters; Problems labeled BIO are of biological or medical interest.

The eText icon indicates when there is a video tutor solution available for the chapter or for a specific problem. To launch these videos, log into your eText through Mastering™ Physics or log into the Study Area.

7. A compass is placed at point P in Figure Q24.7. In which direction, shown by compasses A–D, will the compass point? Explain.

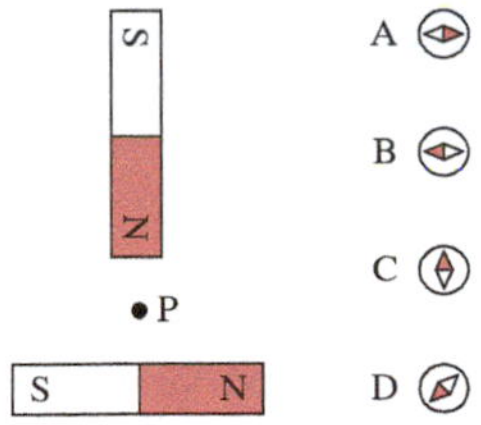


FIGURE Q24.7

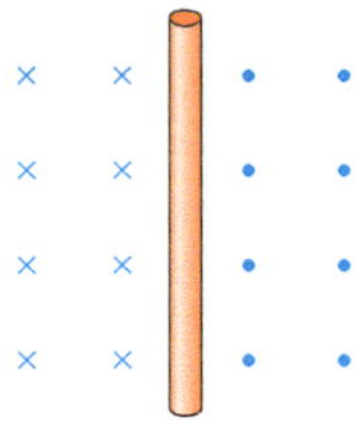

FIGURE Q24.8

8. What is the current direction in the wire of Figure Q24.8?
9. What is the current direction in the wire of Figure Q24.9? Explain.

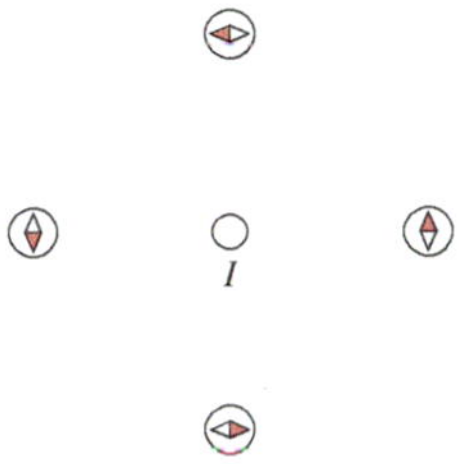


FIGURE Q24.9

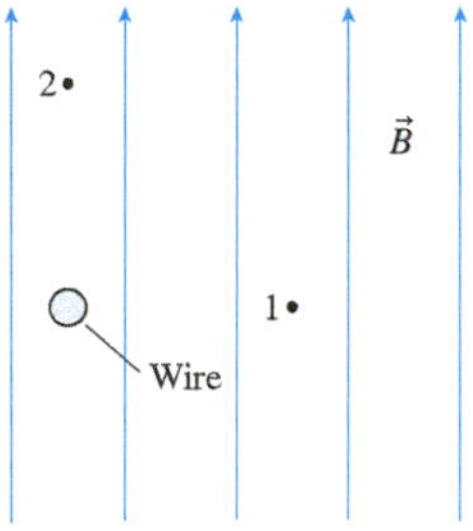


FIGURE Q24.10

10. As shown in Figure Q24.10, a uniform magnetic field points upward, in the plane of the paper. Then the current is turned on in a long wire perpendicular to the paper. The magnetic field at point 1 is then found to be zero. Draw the magnetic field vector at point 2 when the current is on.
11. Two concentric current loops lie in the same plane. The outer loop has twice the diameter of the inner loop. The inner loop carries a 1.0 A current in a clockwise direction. What current, and which direction, should the outer loop carry such that the magnetic field in the center of the loops is zero?
12. An electron is moving in a circular orbit in a uniform magnetic field. Is the kinetic energy of the electron changing? Explain.
13. Figure Q24.13 shows a solenoid as seen in cross section. Compasses are placed at points 1, 2, and 3. In which direction will each compass point when there is a large current in the direction shown? Explain.

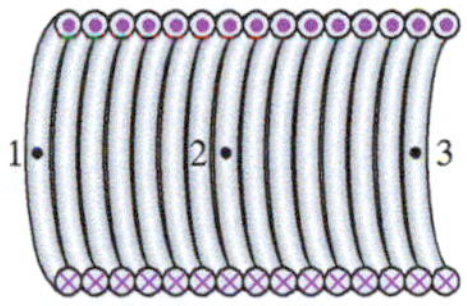


FIGURE Q24.13

14. One long solenoid is placed inside another solenoid. Both solenoids have the same length and the same number of turns of wire, but the outer solenoid has twice the diameter of the inner solenoid. Each solenoid carries the same current, but the two currents are in opposite directions, as shown in Figure Q24.14. What is the magnetic field at the center of the inner solenoid? Explain.

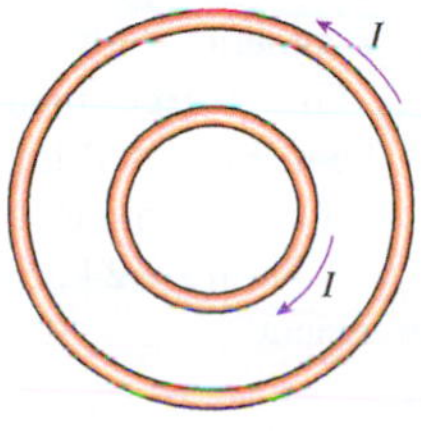


FIGURE Q24.14

15. What is the *initial* direction of deflection for the charged particles entering the magnetic fields shown in Figure Q24.15?

FIGURE Q24.15

16. Describe the force on the charged particles after they enter the magnetic fields shown in Figure Q24.16.

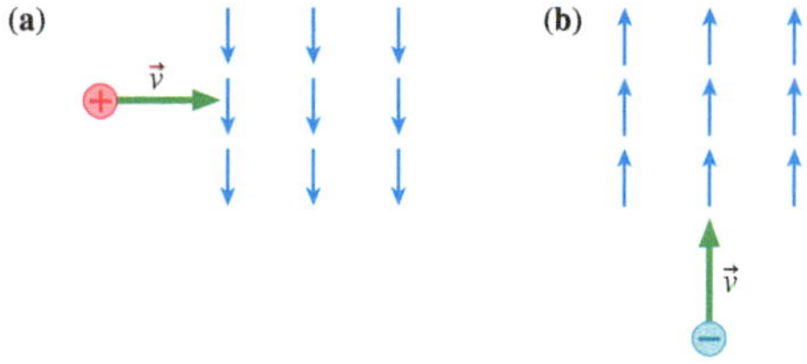


FIGURE Q24.16

17. Which, if any, of the three scenarios A–C for a moving charged particle shown in Figure Q24.17 is physically possible? Explain.

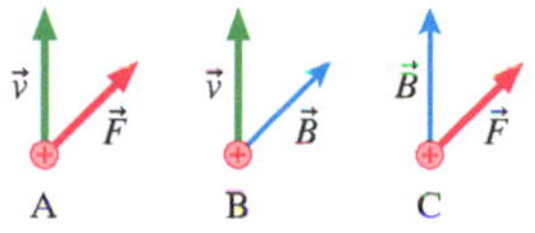


FIGURE Q24.17

18. A positive charge is moving in the direction shown in Figure Q24.18. The magnetic force on the charge is out of the page. Which of the magnetic field directions A–D could result in this force? There may be more than one correct answer.

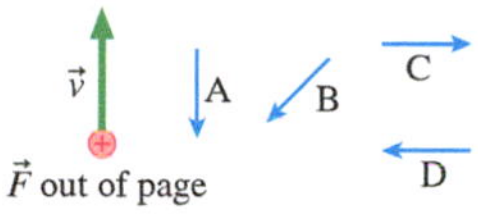


FIGURE Q24.18

19. An electron is moving near a long, current-carrying wire, as shown in Figure Q24.19. What is the direction of the magnetic force on the electron?

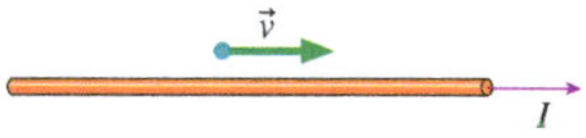


FIGURE Q24.19

20. Two positive charges are moving in a uniform magnetic field with velocities as shown in Figure Q24.20. The magnetic force on each charge is also shown. In which direction does the magnetic field point?

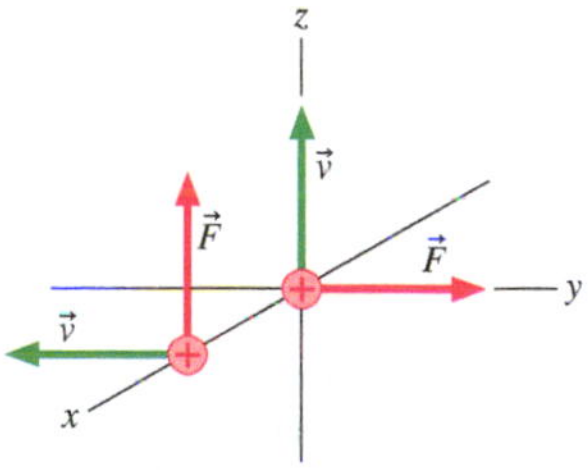


FIGURE Q24.20

21. Three charged particles move in a magnetic field as shown in Figure Q24.21. All the particles have the same mass and the same magnitude of charge. Which particle is moving the fastest? Which particles have positive charges, and which have negative charges?

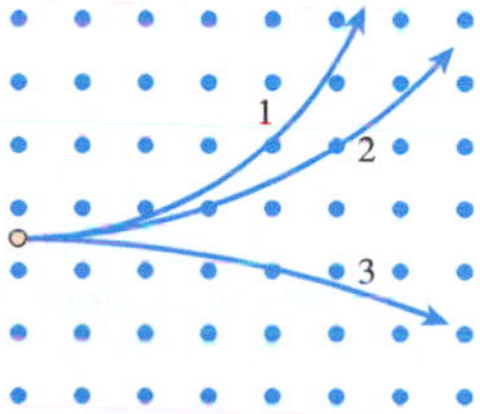


FIGURE Q24.21

22. An electron and a proton are moving in circular orbits in the earth's magnetic field, high above the earth's atmosphere. The two particles move at the same speed. Which particle takes more time to complete one orbit? Explain.
23. A proton moves in a region of uniform magnetic field, as shown in Figure Q24.23. The velocity at one instant is shown. Will the subsequent motion be a clockwise or counterclockwise orbit?

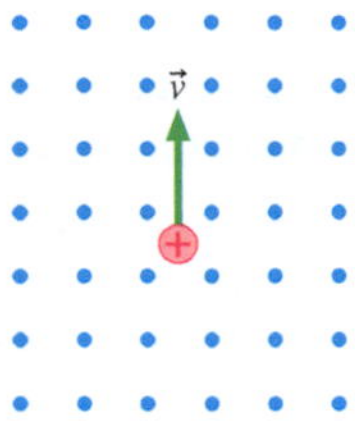

FIGURE Q24.23

24. A proton is moving near a long, current-carrying wire. When the proton is at the point shown in Figure Q24.24, in which direction is the force on it?

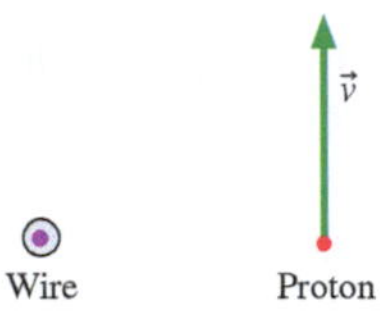

FIGURE Q24.24

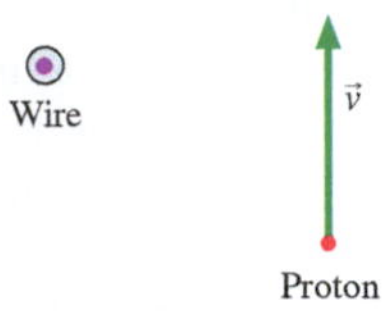

FIGURE Q24.25

25. A proton is moving near a long, current-carrying wire. When the proton is at the point shown in Figure Q24.25, in which direction is the force on it?
26. A long wire and a square loop lie in the plane of the paper. Both carry a current in the direction shown in Figure Q24.26. In which direction is the net force on the loop? Explain.

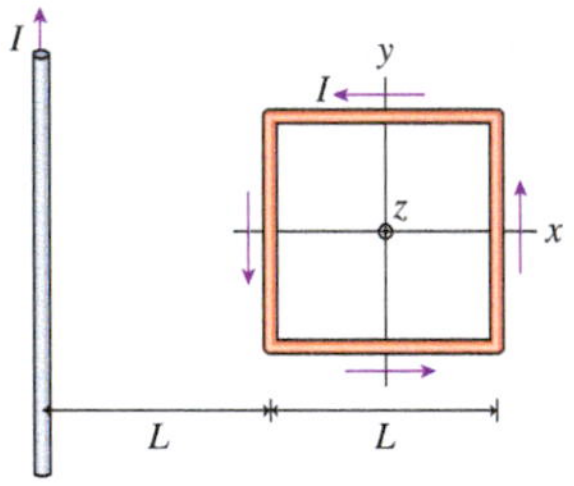

FIGURE Q24.26

27. A solenoid carries a current that produces a field inside it. A wire carrying a current lies inside the solenoid, at the center, carrying a current along the solenoid's axis. Is there a force on this wire due to the field of the solenoid? Explain.
28. Archaeologists can use instruments that measure small variations in magnetic field to locate buried walls made of fired brick, . When fired, the magnetic moments in the clay become randomly aligned; as the clay cools, the magnetic moments line up with the earth's field and retain this alignment even if the bricks are subsequently moved. Explain how this leads to a measurable magnetic field variation over a buried wall.

## Multiple-Choice Questions

29. || Two magnets with unlabeled poles are arranged as shown in Figure Q24.29. A compass points in the direction shown. What are the two poles 1 and 2?
A. 1 = N, 2 = S
B. 1 = N, 2 = N
C. 1 = S, 2 = S
D. 1 = S, 2 = N

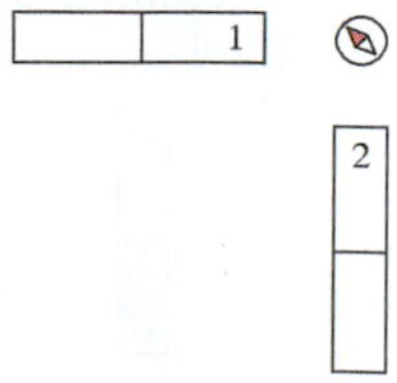

FIGURE Q24.29

30. || If a compass is placed above a current-carrying wire, as in Figure Q24.30, the needle will line up with the field of the wire. Which of the views shows the correct orientation of the needle for the noted current direction?

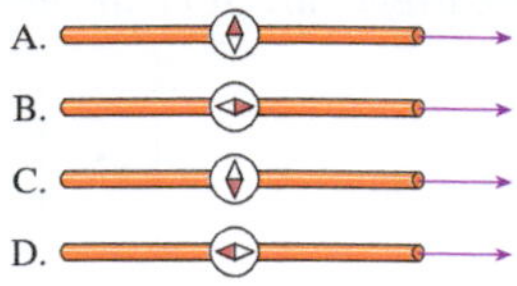

FIGURE Q24.30

31. | Two wires carry equal and opposite currents, as shown in Figure Q24.31. At a point directly between the two wires, the field is

FIGURE Q24.31

A. Directed up, toward the top of the page.
B. Directed down, toward the bottom of the page.
C. Directed to the left.
D. Directed to the right.
E. Zero.
32. | Figure Q24.32 shows four particles moving to the right as they enter a region of uniform magnetic field, directed into the paper as noted. All particles move at the same speed and have the same charge. Which particle has the largest mass?

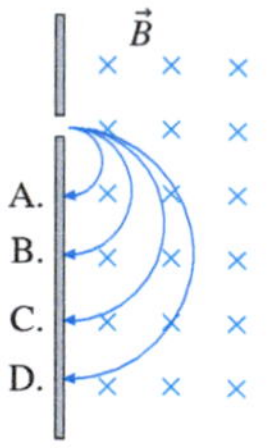

FIGURE Q24.32

33. || A charged particle moves in a circular path in a uniform magnetic field. Which of the following would increase the period of its motion?
A. Increasing its mass
B. Increasing its charge
C. Increasing the field strength
D. Increasing its speed
34. | If all of the particles shown in Figure Q24.34 are electrons, what is the direction of the magnetic field that produced the indicated deflection?

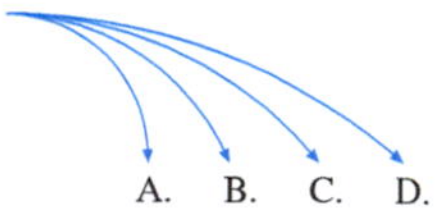

FIGURE Q24.34

A. Up (toward the top of the page).
B. Down (toward the bottom of the page).
C. Out of the plane of the paper.
D. Into the plane of the paper.
35. | If two compasses are brought near enough to each other, the magnetic fields of the compasses themselves will be larger than the field of the earth, and the needles will line up with each other. Which of the arrangements of two compasses shown in Figure Q24.35 is a possible stable arrangement?

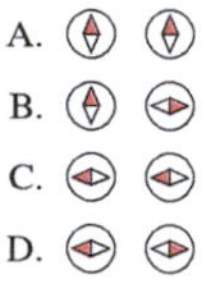

FIGURE Q24.35

# PROBLEMS

### Section 24.1 Magnetism

### Section 24.2 The Magnetic Field

### Section 24.3 Electric Currents Also Create Magnetic Fields

### Section 24.4 Calculating the Magnetic Field Due to a Current

1. | Table 24.1 notes that the magnetic field 10 cm from a wire carrying a 1 A current is 2 $\mu$T. What is the field 1 cm from the wire?

2. || An investigator places a sample 1.0 cm from a wire carrying a large current; the strength of the magnetic field has a particular value at this point. Later, she must move the sample to a 5.0 cm distance, but she would like to keep the field the same. By what factor must she increase the current?

3. || The magnetic field at the center of a 1.0-cm-diameter loop is 2.5 mT.
   a. What is the current in the loop?
   b. A long, straight wire carries the same current you found in part a. At what distance from the wire is the magnetic field 2.5 mT?

4. || For a particular scientific experiment, it is important to be completely isolated from any magnetic field, including the earth's field. The earth's field is approximately 50 $\mu$T, but at any particular location it may be a bit more or less than this. A 1.00-m-diameter current loop with 200 turns of wire is adjusted to carry a current of 0.215 A; at this current, the coil's field at the center is exactly equal to the earth's field in magnitude but opposite in direction, so that the total field at the center of the coil is zero. What is the strength of the earth's magnetic field at this location?

5. || What should the current $I$ be in the lower wire of Figure P24.5 such that the magnetic field at point P is zero?

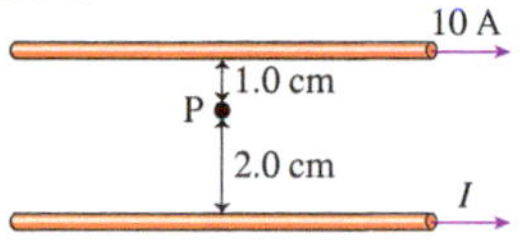

FIGURE P24.5

6. | What are the magnetic field strength and direction at points 1, 2, and 3 in Figure P24.6?

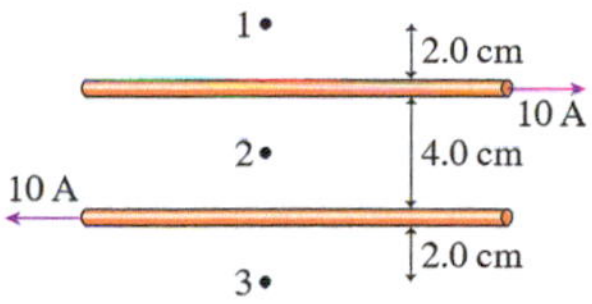

FIGURE P24.6

7. ||| Two long wires carry 10 A currents in the directions shown in Figure P24.7. One wire is 10 cm above the other. What are the direction and magnitude of the magnetic field at a point halfway between them?

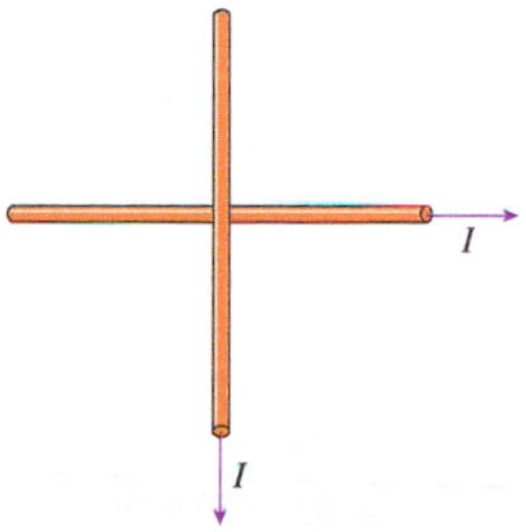

FIGURE P24.7

8. | Although the evidence is weak, there has been concern in BIO recent years over possible health effects from the magnetic fields generated by transmission lines. A typical high-voltage transmission line is 20 m off the ground and carries a current of 200 A. Estimate the magnetic field strength on the ground underneath such a line. What percentage of the earth's magnetic field does this represent?

9. | Some consumer groups urge pregnant women not to use BIO electric blankets, in case there is a health risk from the magnetic fields from the approximately 1 A current in the heater wires.
   a. Estimate, stating any assumptions you make, the magnetic field strength a fetus might experience. What percentage of the earth's magnetic field is this?
   b. It is becoming standard practice to make electric blankets with minimal external magnetic field. Each wire is paired with another wire that carries current in the opposite direction. How does this reduce the external magnetic field?

10. ||| A long wire carrying a 5.0 A current perpendicular to the $xy$-plane intersects the $x$-axis at $x = -2.0$ cm. A second, parallel wire carrying a 3.0 A current intersects the $x$-axis at $x = +2.0$ cm. At what point or points on the $x$-axis is the magnetic field zero if (a) the two currents are in the same direction and (b) the two currents are in opposite directions?

11. | A biophysics experiment uses a very sensitive magnetic field BIO probe to determine the current associated with a nerve impulse traveling along an axon. If the peak field strength 1.0 mm from an axon is 8.0 pT, what is the peak current carried by the axon?

12. ||| The wire in Figure P24.12 carries 10 times the current of the loop; the direction of the current in the wire is shown. The magnetic field in the center of the loop is zero. If the radius $R$ of the loop is 1.3 cm,
    a. What is the direction of the current in the loop?
    b. What is the distance $d$ from the wire to the center of the loop?

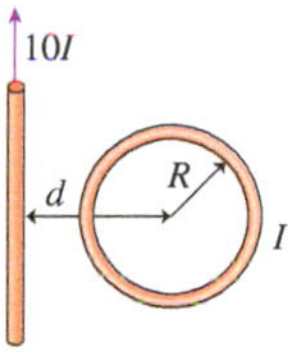

FIGURE P24.12

13. || A solenoid used to produce magnetic fields for research purposes is 2.0 m long, with an inner radius of 30 cm and 1000 turns of wire. When running, the solenoid produces a field of 1.0 T in the center. Given this, how large a current does it carry?

14. | Two concentric current loops lie in the same plane. The smaller loop has a radius of 3.0 cm and a current of 12 A. The bigger loop has a current of 20 A. The magnetic field at the center of the loops is found to be zero. What is the radius of the bigger loop?

15. | The magnetic field of the brain has been measured to be BIO approximately $3.0 \times 10^{-12}$ T. Although the currents that cause this field are quite complicated, we can get a rough estimate of their size by modeling them as a single circular current loop 16 cm (the width of a typical head) in diameter. What current is needed to produce such a field at the center of the loop?

16. || A researcher would like to perform an experiment in zero magnetic field, which means that the field of the earth must be canceled. Suppose the experiment is done inside a solenoid of diameter 1.0 m, length 4.0 m, with a total of 5000 turns of wire. The solenoid is oriented to produce a field that opposes and exactly cancels the 52 $\mu$T local value of the earth's field. What current is needed in the solenoid's wire?

17. ||| What is the magnetic field at the center of the loop in Figure P24.17?

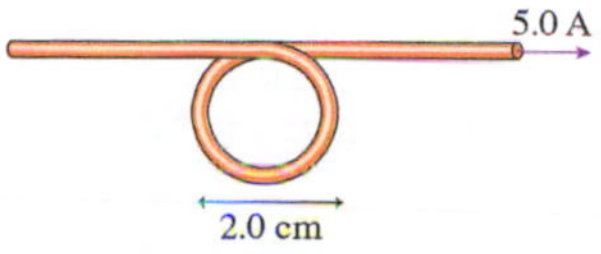

FIGURE P24.17

Watch Video Solution Problems 24.23 and 24.29

18. | Experimental tests have shown that hammerhead sharks BIO can detect magnetic fields. In one such test, 100 turns of wire were wrapped around a 7.0-m-diameter cylindrical shark tank. A magnetic field was created inside the tank when this coil of wire carried a current of 1.5 A. Sharks trained by getting a food reward when the field was present would later unambiguously respond when the field was turned on.
   a. What was the magnetic field strength in the center of the tank due to the current in the coil?
   b. Is the strength of the coil's field at the center of the tank larger or smaller than that of the earth?

19. | We have seen that the heart produces a magnetic field BIO that can be used to diagnose problems with the heart. The magnetic field of the heart is a dipole field produced by a loop current in the outer layers of the heart. Suppose that the field at the center of the heart is 90 pT (a pT is $10^{-12}$ T) and that the heart has a diameter of approximately 12 cm. What current circulates around the heart to produce this field?

20. ||| Young domestic chickens have BIO the ability to orient themselves in the earth's magnetic field. Researchers used a set of two coils to adjust the magnetic field in the chicks' pen. Figure P24.20 shows the two coils, whose centers coincide, seen edge-on. The axis of coil 1 is parallel to the ground and points to the north; the axis of coil 2 is oriented vertically. Each coil has 43 turns and a radius of 1.0 m. At the location of the experiment, the earth's field had a magnitude of $5.6 \times 10^{-5}$ T and pointed to the north, tilted up from the horizontal by 61°. If the researchers wished to exactly cancel the earth's field, what currents would they need to apply in coil 1 and in coil 2?

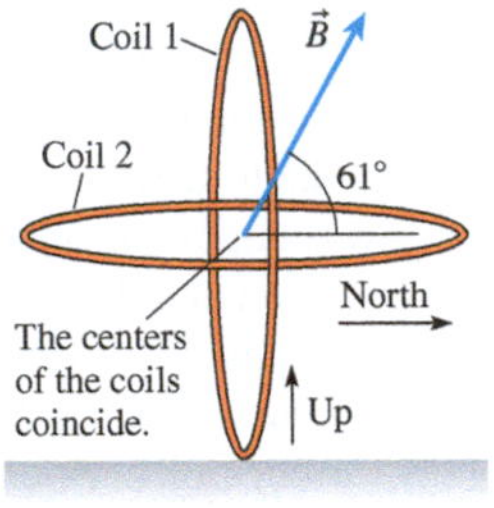


FIGURE P24.20

21. |||| In the Bohr model of the hydrogen atom, the electron moves in a circular orbit of radius $5.3 \times 10^{-11}$ m with speed $2.2 \times 10^6$ m/s. According to this model, what is the magnetic field at the center of a hydrogen atom due to the motion of the electron?
**Hint:** Determine the *average* current of the orbiting electron.

## Section 24.5 Magnetic Fields Exert Forces on Moving Charges

22. | A proton moves with a speed of $1.0 \times 10^7$ m/s in the directions shown in Figure P24.22. A 0.50 T magnetic field points in the positive $x$-direction. For each, what is the magnetic force on the proton? Give your answers as a magnitude and a direction.

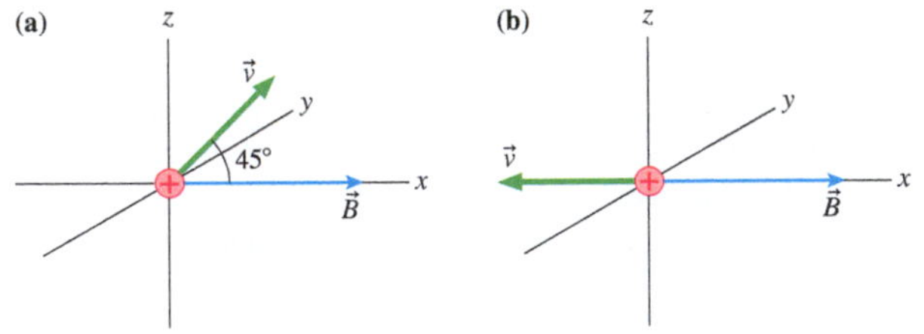


FIGURE P24.22

23. || An electron moves with a speed of $1.0 \times 10^7$ m/s in the directions shown in Figure P24.23. A 0.50 T magnetic field points in the positive $x$-direction. For each, what is the magnetic force on the electron? Give your answers as a magnitude and a direction.

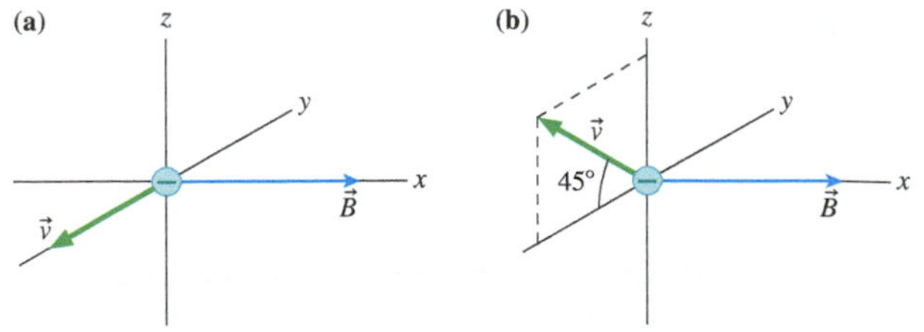


FIGURE P24.23

24. | An electromagnetic flowmeter applies a magnetic field of BIO 0.20 T to blood flowing through a coronary artery at a speed of 15 cm/s. What force is felt by a chlorine ion with a single negative charge?

25. | The aurora is caused when electrons and protons, moving in the earth's magnetic field of $\approx 5.0 \times 10^{-5}$ T, collide with molecules of the atmosphere and cause them to glow. What is the radius of the circular orbit for
   a. An electron with speed $1.0 \times 10^6$ m/s?
   b. A proton with speed $5.0 \times 10^4$ m/s?

26. ||| Problem 24.25 describes two particles that orbit the earth's magnetic field lines. What is the *frequency* of the circular orbit for
   a. An electron with speed $1.0 \times 10^6$ m/s?
   b. A proton with speed $5.0 \times 10^4$ m/s?

27. | In a simplified model of the hydrogen atom, its electron moves in a circular orbit with a radius of $5.3 \times 10^{-10}$ m at a frequency of $6.6 \times 10^{15}$ Hz. What magnetic field would be required to cause an electron to undergo this same motion?

28. ||| Medical cyclotrons need efficient sources of protons to inject BIO into their center. In one kind of ion source, hydrogen atoms (i.e., protons with one orbiting electron) are fed into a chamber where there is a strong magnetic field. Electrons in this chamber are trapped in tight orbits, which greatly increases the chance that they will collide with a hydrogen atom and ionize it. One such source uses a magnetic field of 50 mT, and the electrons' kinetic energy is 2.0 eV. If the electrons travel in a plane perpendicular to the field, what is the radius of their orbits?

29. |||| The microwaves in a microwave oven are produced in a special tube called a *magnetron.* The electrons orbit in a magnetic field at a frequency of 2.4 GHz, and as they do so they emit 2.4 GHz electromagnetic waves. What is the strength of the magnetic field?

30. ||| A cyclotron is used to produce a beam of high-energy deu- BIO INT terons that then collide with a target to produce radioactive isotopes for a medical procedure. Deuterons are nuclei of deuterium, an isotope of hydrogen, consisting of one neutron and one proton, with total mass $3.34 \times 10^{-27}$ kg. The deuterons exit the cyclotron with a kinetic energy of 5.00 MeV.
   a. What is the speed of the deuterons when they exit?
   b. If the magnetic field inside the cyclotron is 1.25 T, what is the diameter of the deuterons' largest orbit, just before they exit?
   c. If the beam current is 400 $\mu$A, how many deuterons strike the target each second?

31. ||| A medical cyclotron used in the production of medical isotopes accelerates protons to 6.5 MeV. The magnetic field in the cyclotron is 1.2 T.
BIO
   a. What is the diameter of the largest orbit, just before the protons exit the cyclotron?
   b. A proton exits the cyclotron 1.0 ms after starting its spiral trajectory in the center of the cyclotron. How many orbits does the proton complete during this 1.0 ms?

32. |||| A *mass spectrometer* uses a magnetic field to determine the masses of ions. A beam of ions, each with a mass 62 times that of a proton, enters a 0.30 T magnetic field, as shown in Figure P24.32. If the speed of the ions is $1.4 \times 10^5$ m/s, at what angle $\theta$ do the ions leave the field?

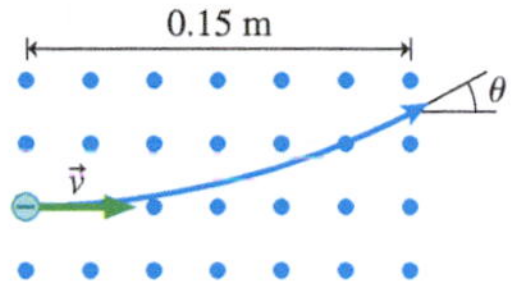

FIGURE P24.32

## Section 24.6 Magnetic Fields Exert Forces on Currents

33. | What magnetic field strength and direction will levitate the 2.0 g wire in Figure P24.33?

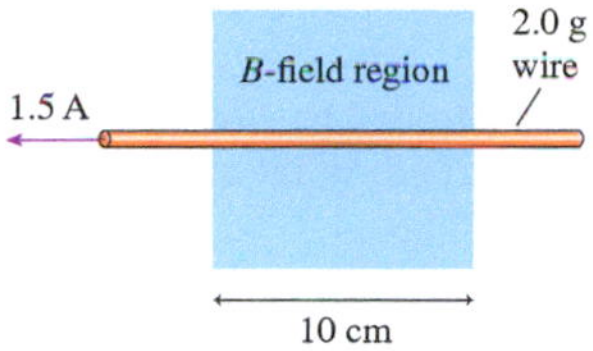

FIGURE P24.33

34. | What is the net force (magnitude and direction) on each wire in Figure P24.34?

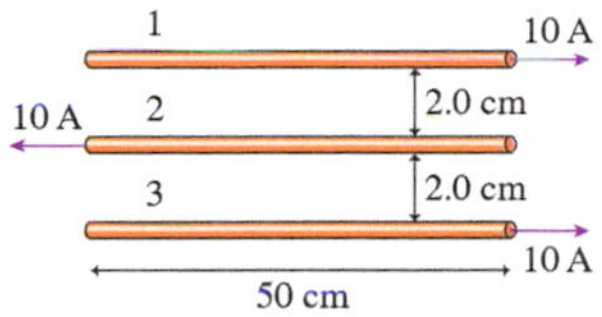

FIGURE P24.34

35. | The unit of current, the ampere, is defined in terms of the force between currents. If two 1.0-meter-long sections of very long wires a distance 1.0 m apart each carry a current of 1.0 A, what is the force between them? (If the force between two actual wires has this value, the current is defined to be exactly 1 A.)

36. | A uniform 2.5 T magnetic field points to the right. A 3.0-m-long wire, carrying 15 A, is placed at an angle of 30° to the field, as shown in Figure P24.36. What is the force (magnitude and direction) on the wire?

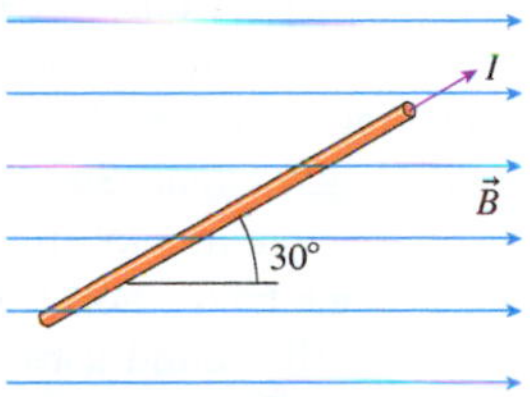

FIGURE P24.36

37. ||| The right half of the square loop of wire shown in Figure P24.37 is in a 0.85 T magnetic field directed into the page. The current in the loop is 1.3 A in a clockwise direction. What is the magnitude of the force on the loop, and in which direction does it act?

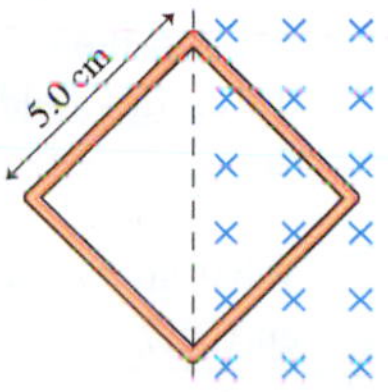

FIGURE P24.37

## Section 24.7 Magnetic Fields Exert Torques on Dipoles

38. || A current loop in a motor has an area of 0.85 cm². It carries a 240 mA current in a uniform field of 0.62 T. What is the magnitude of the maximum torque on the current loop?

39. || A square current loop 5.0 cm on each side carries a 500 mA current. The loop is in a 1.2 T uniform magnetic field. The axis of the loop, perpendicular to the plane of the loop, is 30° away from the field direction. What is the magnitude of the torque on the current loop?

40. || People have proposed driving motors with the earth's magnetic field. This is possible in principle, but the small field means that unrealistically large currents are needed to produce noticeable torques. Suppose a 20-cm-diameter loop of wire is oriented for maximum torque in the earth's field. What current would it need to carry in order to experience a very modest $1.0 \times 10^{-3}$ N · m torque?

41. | As we've seen, protons have an inherent magnetic moment, which has been measured to be $1.4 \times 10^{-26}$ A · m². What would be the torque on a proton whose moment was oriented at 90° to a 1.2 T magnetic field?

## Section 24.8 Magnets and Magnetic Materials

42. | All ferromagnetic materials have a *Curie temperature,* a temperature above which they will cease to be magnetic. Explain in some detail why you might expect this to be so.

43. | A solenoid is near a piece of iron, as shown in Figure P24.43. When a current is present in the solenoid, a magnetic field is created. This magnetic field will magnetize the iron, and there will be a net force between the solenoid and the iron.
   a. Make a sketch showing the direction of the magnetic field from the solenoid. On your sketch, label the induced north magnetic pole and the induced south magnetic pole in the iron.
   b. Will the force on the iron be attractive or repulsive?
   c. Suppose this force moves the iron. Which way will the iron move?

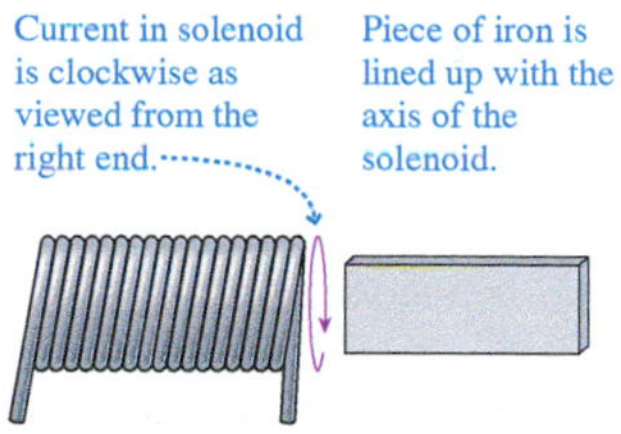

FIGURE P24.43

# General Problems

44. | The right edge of the circuit in Figure P24.44 extends into a 50 mT uniform magnetic field. What are the magnitude and direction of the net force on the circuit?
INT

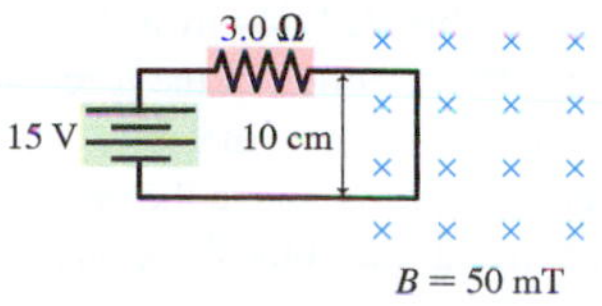

FIGURE P24.44

Watch Video Solution Problems 24.47 and 24.53

45. || INT The two 10-cm-long parallel wires in Figure P24.45 are separated by 5.0 mm. For what value of the resistor $R$ will the force between the two wires be $5.4 \times 10^{-5}$ N?

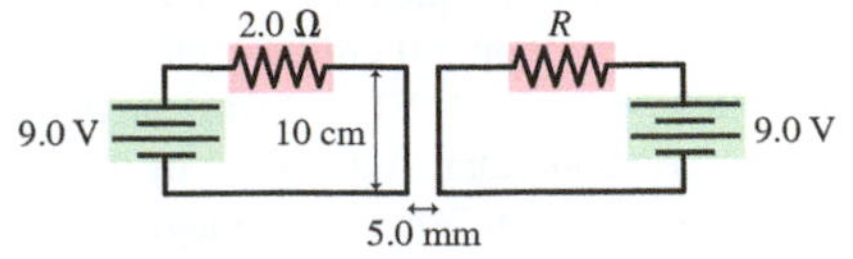


FIGURE P24.45

46. || INT An electron travels with speed $1.0 \times 10^7$ m/s between the two parallel charged plates shown in Figure P24.46. The plates are separated by 1.0 cm and are charged by a 200 V battery. What magnetic field strength and direction will allow the electron to pass between the plates without being deflected?

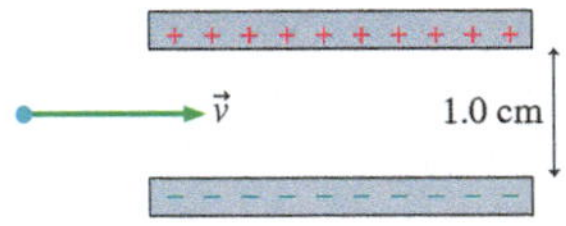


FIGURE P24.46

47. || INT The two springs in Figure P24.47 each have a spring constant of 10 N/m. They are stretched by 1.0 cm when a current passes through the wire. How big is the current?

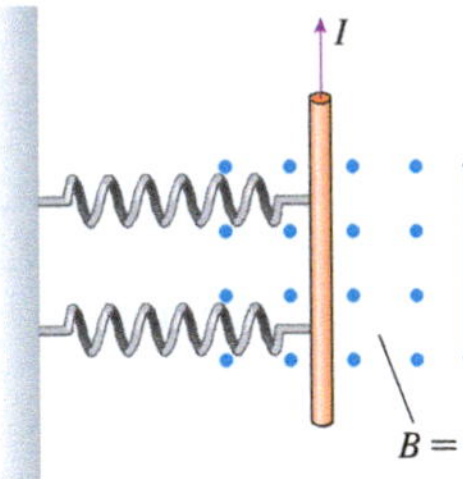


FIGURE P24.47

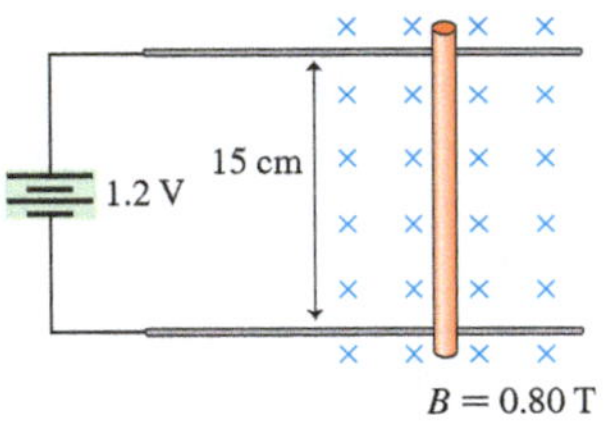


FIGURE P24.48

48. || INT A device called a *railgun* uses the magnetic force on currents to launch projectiles at very high speeds. An idealized model of a railgun is illustrated in Figure P24.48. A 1.2 V power supply is connected to two conducting rails. A segment of copper wire, in a region of uniform magnetic field, slides freely on the rails. The wire has a 0.85 mΩ resistance and a mass of 5.0 g. Ignore the resistance of the rails. When the power supply is switched on,
   a. What is the current?
   b. What are the magnitude and direction of the force on the wire?
   c. What will be the wire's speed after it has slid a distance of 1.0 mm?

49. ||| Irrigation channels that require regular flow monitoring are often equipped with electromagnetic flowmeters in which the magnetic field is produced by horizontal coils embedded in the bottom of the channel. A particular coil has 100 turns and a diameter of 6.0 m. When it's time for a measurement, a 5.0 A current is turned on. The large diameter of the coil means that the field in the water flowing directly above the center of the coil is approximately equal to the field in the center of the coil.
   a. What is the magnitude of the field at the center of the coil?
   b. If the field is directed downward and the water is flowing east, what is the direction of the force on a positive ion in the water above the center of the coil?
   c. If the water is flowing above the center of the coil at 1.5 m/s, what is the magnitude of the force on an ion with a charge $+e$?

50. | BIO INT Typical blood velocities in the coronary arteries range from 10 to 30 cm/s. An electromagnetic flowmeter applies a magnetic field of 0.25 T to a coronary artery with a blood velocity of 15 cm/s. As we saw in Figure 24.38, this field exerts a force on ions in the blood, which will separate. The ions will separate until they make an electric field that exactly balances the magnetic force. This electric field produces a voltage that can be measured.
   a. What force is felt by a singly ionized (positive) sodium ion?
   b. Charges in the blood will separate until they produce an electric field that cancels this magnetic force. What will be the resulting electric field?
   c. What voltage will this electric field produce across an artery with a diameter of 3.0 mm?

51. |||| INT The coil of the electric motor shown in Figure P24.51 has 100 turns and an area of 0.015 m². Its rotation axis is attached to a pulley with a radius of 3.0 cm. At the moment shown, the plane of the coil is parallel to the 0.29 T magnetic field. What current in the coil will hold the 1.0 kg weight in equilibrium?

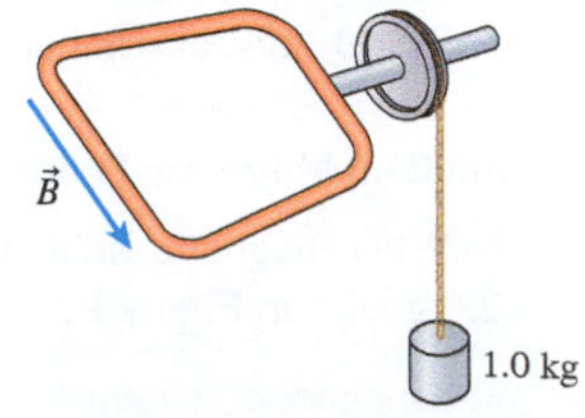


FIGURE P24.51

52. ||| INT A loudspeaker creates sound by pushing air back and forth with a paper cone that is driven by a magnetic force on a wire coil at the base of the cone. Figure P24.52 shows the details. The cone is attached to a coil of wire that sits in the gap between the poles of a circular magnet. The 0.18 T magnetic field, which points radially outward from N to S, exerts a force on a current in the wire, moving the cone. The coil of wire that sits in this gap has a diameter of 5.0 cm, contains 20 turns of wire, and has a resistance of 8.0 Ω. The speaker is connected to an amplifier whose instantaneous output voltage of 6.0 V creates a clockwise current in the coil as seen from above. What is the magnetic force on the coil at this instant?

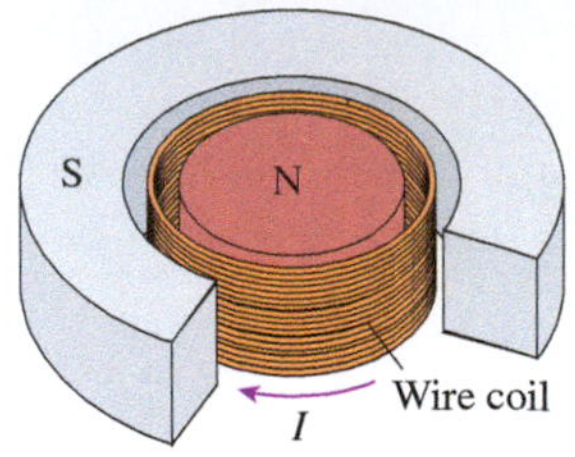


FIGURE P24.52

53. ||| BIO Bats are capable of navigating using the earth's field—a plus for an animal that may fly great distances from its roost at night. If, while sleeping during the day, bats are exposed to a field of a similar magnitude but different direction than the earth's field, they are more likely to lose their way during their next lengthy night flight. Suppose you are a researcher doing such an experiment in a location where the earth's field is 50 μT at a 60° angle below horizontal. You make a 50-cm-diameter, 100-turn coil around a roosting box; the sleeping bats are at the center of the coil. You wish to pass a current through the coil to produce a field that, when combined with the earth's field, creates a net field with the same strength and dip angle (60° below horizontal) as the earth's field but with a horizontal component that points south rather than north. What are the proper orientation of the coil and the necessary current?

54. ||| At the equator, the earth's field is essentially horizontal; near
BIO the north pole, it is nearly vertical. In between, the angle varies. As you move farther north, the dip angle, the angle of the earth's field below horizontal, steadily increases. Green turtles seem to use this dip angle to determine their latitude. Suppose you are a researcher wanting to test this idea. You have gathered green turtle hatchlings from a beach where the magnetic field strength is 50 $\mu$T and the dip angle is 56°. You then put the turtles in a 1.2-m-diameter circular tank and monitor the direction in which they swim as you vary the magnetic field in the tank. You change the field by passing a current through a 100-turn horizontal coil wrapped around the tank. This creates a field that adds to that of the earth. What current should you pass through the coil, and in what direction, to produce a net field in the center of the tank that has a dip angle of 62°?

55. || Internal components of cathode-ray-tube televisions and computer monitors can become magnetized; the resulting magnetic field can deflect the electron beam and distort the colors on the screen. Demagnetization can be accomplished with a coil of wire whose current switches direction rapidly and gradually decreases in amplitude. Explain what effect this will have on the magnetic moments of the magnetic materials in the device, and how this might eliminate any magnetic ordering.

56. ||| A 1.0-m-long, 1.0-mm-diameter copper wire carries a cur-
INT rent of 50.0 A to the east. Suppose we create a magnetic field that produces an upward force on the wire exactly equal in magnitude to the wire's weight, causing the wire to "levitate." What are the field's direction and magnitude?

57. || An insulated copper wire is wrapped around an iron nail. The resulting coil of wire consists of 240 turns of wire that cover 1.8 cm of the nail, as shown in Figure P24.57. A current of 0.60 A passes through the wire. If the ferromagnetic properties of the nail increase the field by a factor of 100, what is the magnetic field strength inside the nail?

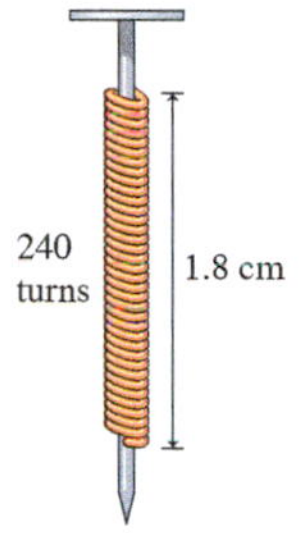


FIGURE P24.57

58. ||| A long, straight wire with a linear mass density of 50 g/m is
INT suspended by threads, as shown in Figure P24.58. There is a uniform magnetic field pointing vertically downward. A 10 A current in the wire experiences a horizontal magnetic force that deflects it to an equilibrium angle of 10°. What is the strength of the magnetic field $\vec{B}$?

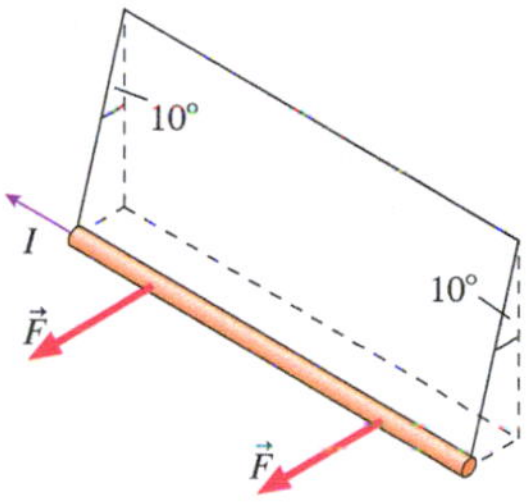


FIGURE P24.58

## MCAT-Style Passage Problems

### The Velocity Selector INT

In experiments where all the charged particles in a beam are required to have the same velocity (for example, when entering a mass spectrometer), scientists use a *velocity selector*. A velocity selector has a region of uniform electric and magnetic fields that are perpendicular to each other and perpendicular to the motion of the charged particles. Both the electric and magnetic fields exert a force on the charged particles. If a particle has precisely the right velocity, the two forces exactly cancel and the particle is not deflected. Equating the forces due to the electric field and the magnetic field gives the following equation:

$$qE = qvB$$

Solving for the velocity, we get:

$$v = \frac{E}{B}$$

A particle moving at this velocity will pass through the region of uniform fields with no deflection, as shown in Figure P24.59. For higher or lower velocities than this, the particles will feel a net force and will be deflected. A slit at the end of the region allows only the particles with the correct velocity to pass.

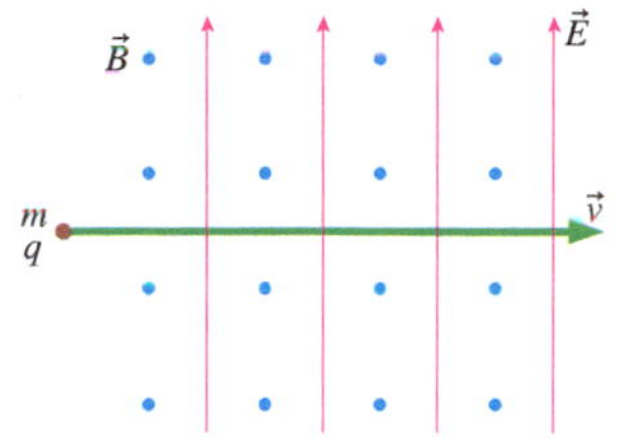


FIGURE P24.59

59. | Assuming the particle in Figure P24.59 is positively charged, what are the directions of the forces due to the electric field and to the magnetic field?
   A. The force due to the electric field is directed up (toward the top of the page); the force due to the magnetic field is directed down (toward the bottom of the page).
   B. The force due to the electric field is directed down (toward the bottom of the page); the force due to the magnetic field is directed up (toward the top of the page).
   C. The force due to the electric field is directed out of the plane of the paper; the force due to the magnetic field is directed into the plane of the paper.
   D. The force due to the electric field is directed into the plane of the paper; the force due to the magnetic field is directed out of the plane of the paper.

60. | How does the kinetic energy of the particle in Figure P24.59 change as it traverses the velocity selector?
   A. The kinetic energy increases.
   B. The kinetic energy does not change.
   C. The kinetic energy decreases.

61. | Suppose a particle with twice the velocity of the particle in Figure P24.59 enters the velocity selector. The path of this particle will curve
   A. Upward (toward the top of the page).
   B. Downward (toward the bottom of the page).
   C. Out of the plane of the paper.
   D. Into the plane of the paper.

62. | Next, a particle with the same mass and velocity as the particle in Figure P24.59 enters the velocity selector. This particle has a charge of $2q$—twice the charge of the particle in Figure P24.59. In this case, we can say that
   A. The force of the electric field on the particle is greater than the force of the magnetic field.
   B. The force of the magnetic field on the particle is greater than the force of the electric field.
   C. The forces of the electric and magnetic fields on the particle are still equal.

### Ocean Potentials INT

The ocean is salty because it contains many dissolved ions. As these charged particles move with the water in strong ocean currents, they feel a force from the earth's magnetic field. Positive and negative charges are separated until an electric field develops that balances this magnetic force. This field produces measurable potential differences that can be monitored by ocean researchers.

The Gulf Stream moves northward off the east coast of the United States at a speed of up to 3.5 m/s. Assume that the current flows at this maximum speed and that the earth's field is 50 $\mu$T tipped 60° below horizontal.

63. | What is the direction of the magnetic force on a singly ionized negative chlorine ion moving in this ocean current?
A. East B. West C. Up D. Down
64. | What is the magnitude of the force on this ion?
A. $2.8 \times 10^{-23}$ N B. $2.4 \times 10^{-23}$ N
C. $1.6 \times 10^{-23}$ N D. $1.4 \times 10^{-23}$ N
65. | What magnitude electric field is necessary to exactly balance this magnetic force?
A. $1.8 \times 10^{-4}$ N/C B. $1.5 \times 10^{-4}$ N/C
C. $1.0 \times 10^{-4}$ N/C D. $0.9 \times 10^{-4}$ N/C
66. | The electric field produces a potential difference. If you place one electrode 10 m below the surface of the water, you will measure the greatest potential difference if you place the second electrode
A. At the surface.
B. At a depth of 20 m.
C. At the same depth 10 m to the north.
D. At the same depth 10 m to the east.

### The Mass Spectrometer

If you have a sample of unknown composition, a first step at analysis might be a determination of the masses of the atoms and molecules in the sample. A *mass spectrometer* to make such an analysis can take various forms, but for many years the best technique was to determine the masses of ionized atoms and molecules in a sample by observing their circular paths in a uniform magnetic field, as illustrated in Figure P24.67. A sample to be analyzed is vaporized, then singly ionized. The ions are accelerated through an electric field, and ions of a known speed selected. These ions travel into a region of uniform magnetic field, where they follow circular paths. An exit slit allows ions that have followed a particular path to be counted by a detector, producing a record of the masses of the particles in the sample.

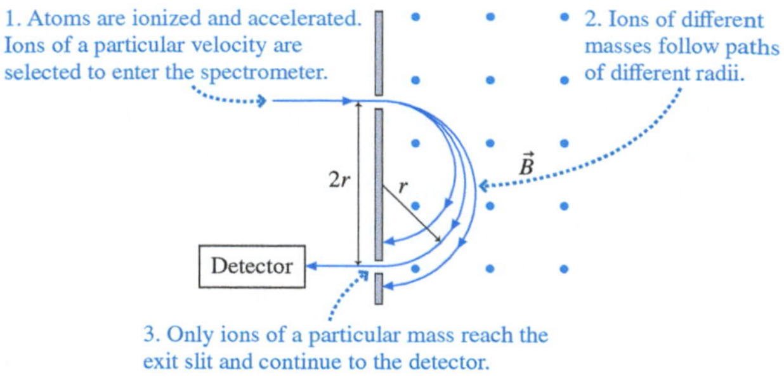


FIGURE P24.67

67. | In the spectrometer shown in Figure P24.67, do the ions have positive or negative charge?
A. Positive
B. Negative
68. || The moving ions can be thought of as a current loop, and it will produce its own magnetic field. The direction of this field at the center of the particles' circular orbit is
A. In the same direction as the spectrometer's magnetic field.
B. Opposite the direction of the spectrometer's magnetic field.
69. || Why is it important that the ions have a known speed?
A. The radius of the orbit depends on the mass, the charge, and the speed. If the charge and the speed are the same, the orbit depends on only the mass.
B. The orbit must be circular, and this is the case for only a certain range of speeds.
C. If the ions are moving too fast, the magnetic field will not be able to bend their path to the detector.
D. The ions are all accelerated by the same electric field, and so will all have the same speed anyway.
70. || A mass spectrometer similar to the one in Figure P24.67 is designed to analyze biological samples. Molecules in the sample are singly ionized, then they enter a 0.80 T uniform magnetic field at a speed of $2.3 \times 10^5$ m/s. If a molecule has a mass 85 times the mass of the proton, what will be the approximate distance between the points where the ion enters and exits the magnetic field?
A. 25 cm
B. 50 cm
C. 75 cm
D. 100 cm

# 25 EM Induction and EM Waves

The photo of the flower on the left shows how it appears to our eyes, in visible light. But there's more to the story! The false-color view of the flower on the right shows its appearance in the ultraviolet, beyond the range of human vision, revealing pigments we can't see. Whose eyes are these pigments intended for?

## LOOKING AHEAD

### Magnetism and Electricity

The turning windmill blades rotate a wire coil in a *magnetic* field, producing an *electric* current.

You'll continue to explore the deep connections between magnetism and electricity.

### Induction

A physician programs a pacemaker by using a rapidly changing magnetic field to **induce** a voltage in the implanted device.

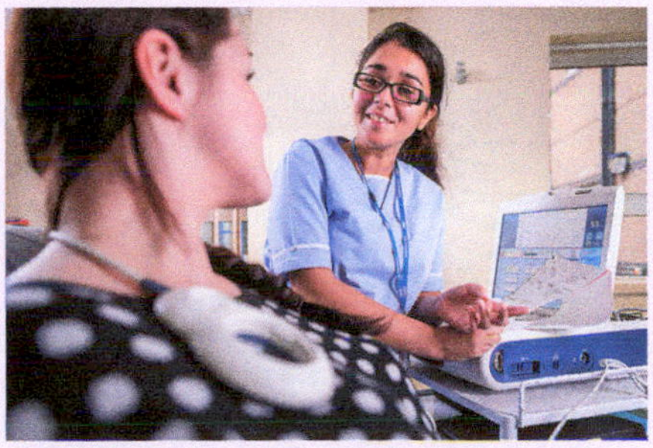

You'll learn to analyze **electromagnetic induction** qualitatively and quantitatively.

### Electromagnetic (EM) Waves

The vertical antennas on this cell phone tower emit **electromagnetic waves,** waves of electric and magnetic fields.

You'll learn the properties of different electromagnetic waves, from radio waves to light waves.

**GOAL** To understand the nature of electromagnetic induction and electromagnetic waves.

## LOOKING BACK

### Traveling Waves

In Chapter 15 you learned the properties of traveling waves. For sinusoidal waves, the wave speed is the product of the wave's frequency and wavelength.

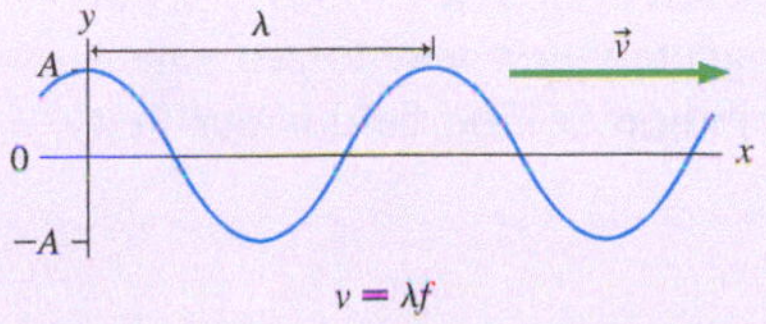


In this chapter, you'll see how the properties of traveling waves are used to describe electromagnetic waves.

**STOP TO THINK**

A microwave oven uses 2.4 GHz electromagnetic waves. A cell phone uses electromagnetic waves at a slightly lower 1.9 GHz frequency. What can you say about the wavelengths of the two?

A. The waves from the oven have a longer wavelength.
B. The waves from the phone have a longer wavelength.
C. The waves from the oven and the phone have the same wavelength.

## 25.1 Induced Currents

In Chapter 24, we learned that a current can create a magnetic field. As soon as this discovery was widely known, investigators began considering a related question: Can a magnetic field create a current?

One of the early investigators was Michael Faraday, who was experimenting with two coils of wire wrapped around an iron ring, as shown in FIGURE 25.1, when he made a remarkable discovery. He had hoped that the magnetic field generated by a current in the coil on the left would create a magnetic field in the iron, and that the magnetic field in the iron might then somehow produce a current in the circuit on the right.

This technique failed to generate a steady current, but Faraday noticed that the needle of the current meter jumped ever so slightly at the instant when he closed the switch in the circuit on the left. After the switch was closed, the needle immediately returned to zero. Faraday's observation suggested to him that a current was generated only if the magnetic field was *changing* as it passed through the coil. Faraday set out to test this hypothesis through a series of experiments, shown in the following table.

FIGURE 25.1 Faraday's discovery of electromagnetic induction.

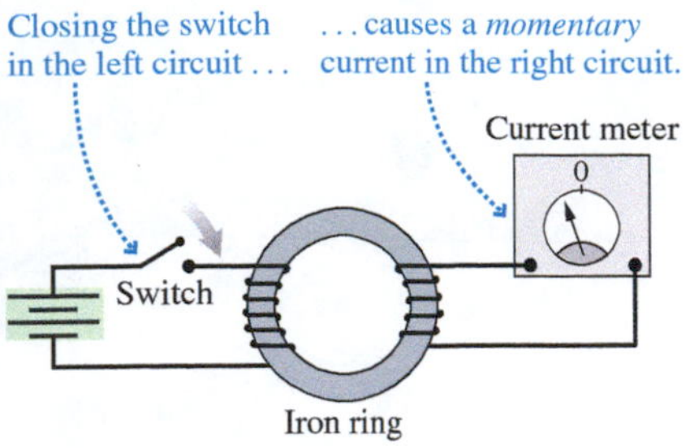


**Faraday investigates electromagnetic induction**

Faraday placed one coil directly above the other, without the iron ring. There was no current in the lower coil while the switch was held in the open or closed position, but a momentary current appeared whenever the switch was opened or closed.

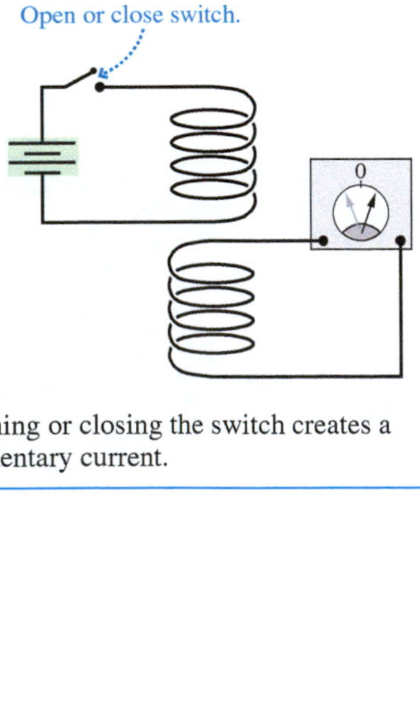


Opening or closing the switch creates a momentary current.

Faraday pushed a bar magnet into a coil of wire. This action caused a momentary deflection of the needle in the current meter, although *holding* the magnet inside the coil had no effect. A quick withdrawal of the magnet deflected the needle in the other direction.

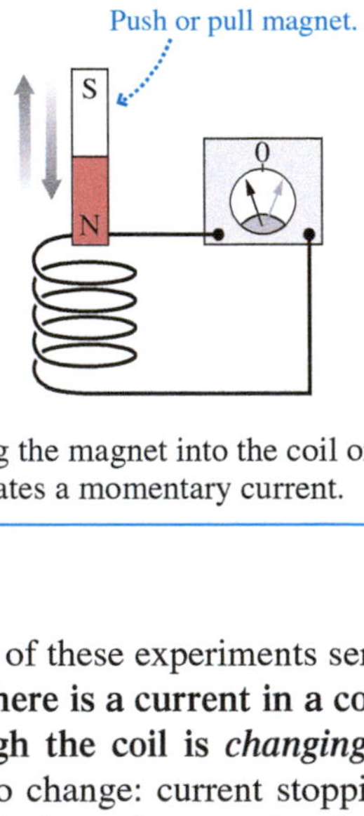


Pushing the magnet into the coil or pulling it out creates a momentary current.

Must the magnet move? Faraday created a momentary current by rapidly pulling a coil of wire out of a magnetic field, although there was no current if the coil was stationary in the magnetic field. Pushing the coil *into* the magnet caused the needle to deflect in the opposite direction.

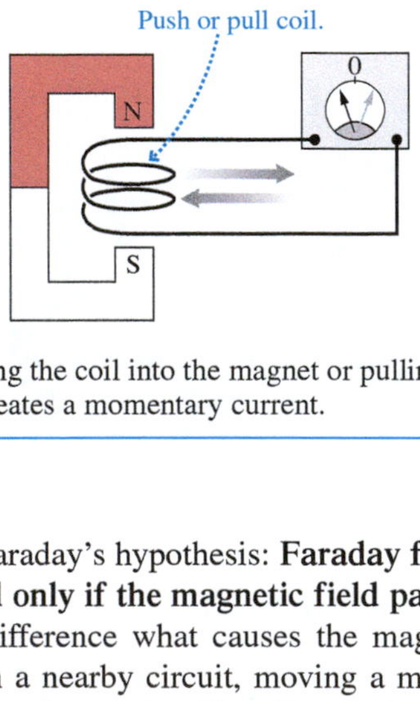


Pushing the coil into the magnet or pulling it out creates a momentary current.

All of these experiments served to bolster Faraday's hypothesis: **Faraday found that there is a current in a coil of wire if and only if the magnetic field passing through the coil is *changing*.** It makes no difference what causes the magnetic field to change: current stopping or starting in a nearby circuit, moving a magnet through the coil, or moving the coil into and out of a magnet. The effect is the same in all cases. There is no current if the field through the coil is not changing, so it's not the magnetic field itself that is responsible for the current but, instead, it is the *changing of the magnetic field.*

The current in a circuit due to a changing magnetic field is called an **induced current.** Opening the switch or moving the magnet *induces* a current in a nearby circuit. An induced current is not caused by a battery; it is a completely new way to generate a current. The creation of an electric current by a changing magnetic field is our first example of **electromagnetic (EM) induction.**

## 25.2 Motional emf

A satellite tethered to the space shuttle.

In 1996, astronauts on the space shuttle deployed a satellite at the end of a 20-km-long conducting tether. A potential difference of up to 3500 V developed between the shuttle and the satellite as the wire between the two swept through the earth's magnetic field. Why would the motion of a wire in a magnetic field produce such a large voltage? Let's explore the mechanism behind this *motional emf.*

To begin, consider a conductor of length $l$ that moves with velocity $\vec{v}$ through a uniform magnetic field $\vec{B}$, as shown in FIGURE 25.2. The charge carriers inside the conductor—assumed to be positive, as in our definition of current—also move with velocity $\vec{v}$, so they each experience a magnetic force. For simplicity, we will assume that $\vec{v}$ is perpendicular to $\vec{B}$, in which case the magnitude of the force is $F_B = qvB$. This force causes the charge carriers to move. For the geometry of Figure 25.2, the right-hand rule tells us that the positive charges move toward the top of the moving conductor, leaving an excess of negative charge at the bottom.

FIGURE 25.2 The magnetic force on the charge carriers in a moving conductor creates an electric field inside the conductor.

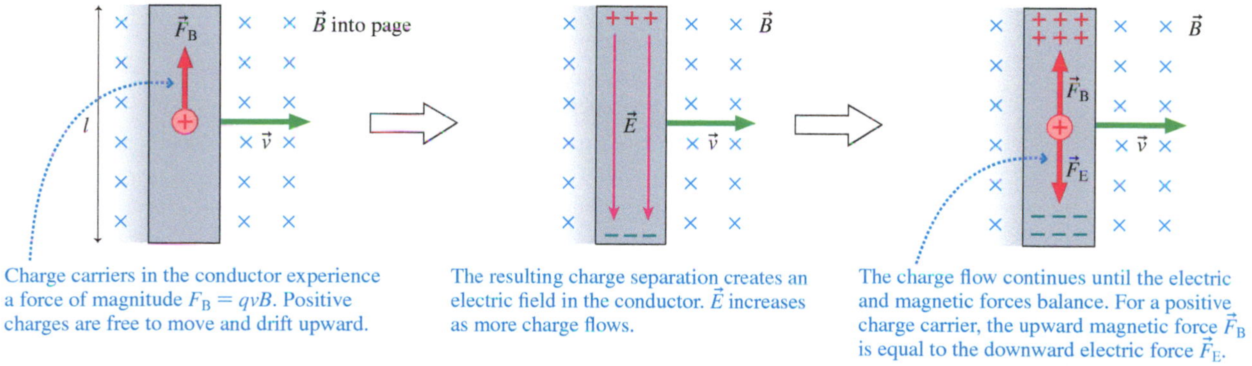


This motion of the charge carriers cannot continue forever. The separation of the charge carriers creates an electric field. The resulting electric force *opposes* the separation of charge, so the charge separation continues only until the electric force has grown to exactly balance the magnetic force:

$$F_E = qE = F_B = qvB$$

When this balance occurs, the charge carriers experience no net force and thus undergo no further motion. The electric field strength at equilibrium is

$$E = vB \tag{25.1}$$

Thus, **the magnetic force on the charge carriers in a moving conductor creates an electric field $E = vB$ inside the conductor.**

The electric field, in turn, creates an electric potential difference between the two ends of the moving conductor. We found in ◀◀ SECTION 21.5 that the potential difference between two points separated by distance $l$ parallel to an electric field $E$ is $\Delta V = El$. Thus the motion of the wire through a magnetic field *induces* a potential difference

$$\Delta V = vlB \tag{25.2}$$

between the ends of the conductor. The potential difference depends on the strength of the magnetic field and on the wire's speed through the field. This is similar to the action of the electromagnetic flowmeter that we saw in Chapter 24.

There's an important analogy between this potential difference and the potential difference of a battery. FIGURE 25.3a reminds you that a battery uses a nonelectric force—which we called the charge escalator—to separate positive and negative

FIGURE 25.3 Two different ways to generate an emf.

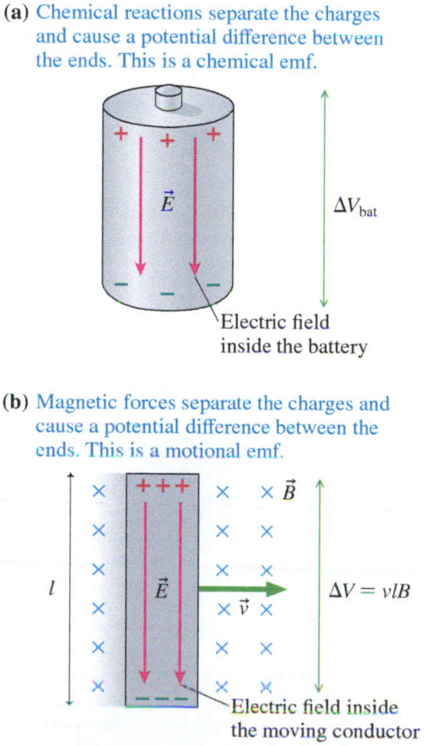

charges. We refer to a battery, where the charges are separated by chemical forces, as a source of *chemical emf*. By contrast, the moving conductor of FIGURE 25.3b separates charges using *magnetic* forces. The charge separation creates a potential difference—an emf—due to the conductor's *motion*. We can thus define the **motional emf** of a conductor of length $l$ moving with velocity $\vec{v}$ perpendicular to a magnetic field $\vec{B}$ to be

$$\mathcal{E} = vlB \qquad (25.3)$$

emf due to motion of conductor (V); Speed of conductor perpendicular to magnetic field (m/s); Length of conductor (m); Magnetic field strength (T)

**EXAMPLE 25.1 Finding the motional emf for an airplane**

A Boeing 747 aircraft with a wingspan of 65 m is cruising at 260 m/s over northern Canada, where the magnetic field of the earth (magnitude $5.0 \times 10^{-5}$ T) is directed straight down. What is the potential difference between the tips of the wings?

***STRATEGIZE*** The wing is a conductor moving through a magnetic field, so there will be a motional emf.

***PREPARE*** We can visualize a top view of this situation exactly as in Figure 25.3b, with the wing as the moving conductor.

***SOLVE*** The magnetic field is perpendicular to the velocity, so we compute the potential difference using Equation 25.3:

$$\Delta V = vlB = (260 \text{ m/s})(65 \text{ m})(5.0 \times 10^{-5} \text{ T}) = 0.85 \text{ V}$$

***ASSESS*** The earth's magnetic field is small, so the motional emf will be small as well unless the speed and the length are quite large. The tethered satellite generated a much higher voltage due to its much greater speed and the great length of the tether, the moving conductor.

◀ **A head for magnetism?** BIO Hammerhead sharks seem to navigate using the earth's magnetic field. It's likely that they detect the *magnetic* field using their keen *electric* sense, detecting the earth's magnetic field by sensing the motional emf as they move through the water. The width of their oddly shaped heads is an asset because the magnitude of the potential difference ($\mathcal{E}$) is proportional to the length of the moving conductor ($l$).

## Induced Current in a Circuit

The moving conductor of Figure 25.3 developed an emf, but it couldn't sustain a current because the charges had nowhere to go. We can change this by including the moving conductor in a circuit.

FIGURE 25.4 shows a length of wire with resistance $R$ sliding with speed $v$ along a fixed U-shaped conducting rail. (For simplicity, we will assume that the rail has zero resistance.) The wire and the rail together form a closed conducting loop—a circuit.

Suppose a magnetic field $\vec{B}$ is perpendicular to the plane of the circuit. Charges in the moving wire will be pushed to the ends of the wire by the magnetic force, just as they were in Figure 25.3, but now the charges can continue to flow around the circuit. The moving wire acts like the battery in a circuit.

**The current in the circuit is an induced current,** due to magnetic forces on moving charges. The total resistance of the circuit is just the resistance $R$ of the moving wire, so the induced current is given by Ohm's law:

$$I = \frac{\mathcal{E}}{R} = \frac{vlB}{R} \qquad (25.4)$$

FIGURE 25.4 A current is induced in the circuit as the wire moves through a magnetic field.

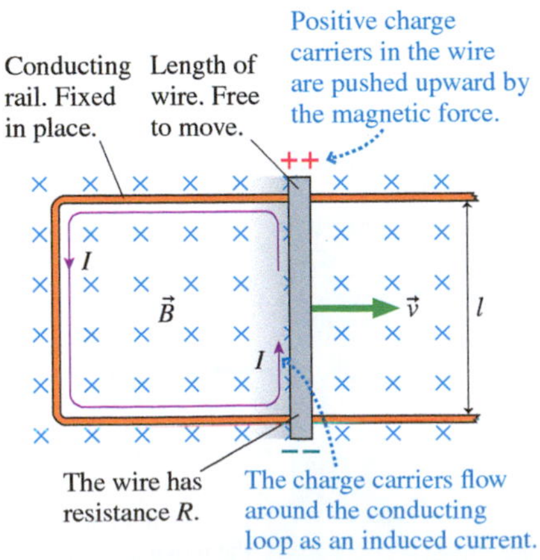

We've assumed that the wire is moving along the rail at constant speed. But we must apply a continuous pulling force $\vec{F}_{pull}$ to make this happen; FIGURE 25.5 shows why. The moving wire, which now carries induced current $I$, is in a magnetic field. You learned in ◀◀ SECTION 24.6 that a magnetic field exerts a force on a current-carrying wire. According to the right-hand rule, the magnetic force $\vec{F}_{mag}$ on the moving wire points to the left. This "magnetic drag" will cause the wire to slow down and stop *unless* we exert an equal but opposite pulling force $\vec{F}_{pull}$ to keep the wire moving.

**NOTE** ▶ Think about this carefully. As the wire moves to the right, the magnetic force $\vec{F}_B$ pushes the charge carriers *parallel* to the wire. Their motion, as they continue around the circuit, is the induced current $I$. Now, because we have a current, a second magnetic force $\vec{F}_{mag}$ enters the picture. This force on the current is *perpendicular* to the wire and acts to slow the wire's motion. ◀

FIGURE 25.5 A pulling force is needed to move the wire to the right.

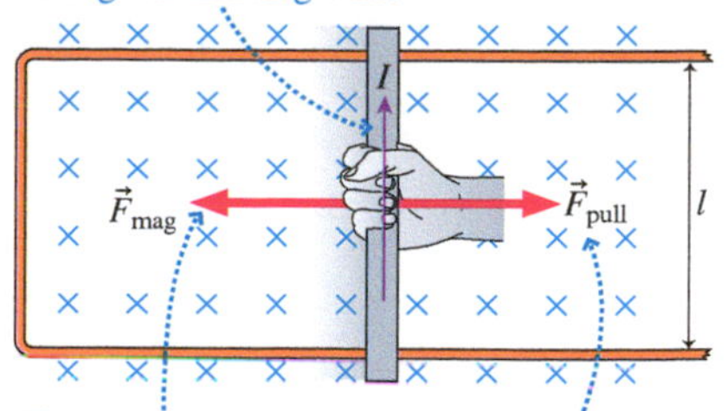


The magnitude of the magnetic force on a current-carrying wire was found in Chapter 24 to be $F_{mag} = IlB$. Using that result, along with Equation 25.4 for the induced current, we find that the force required to pull the wire with a constant speed $v$ is

$$F_{pull} = F_{mag} = IlB = \left(\frac{vlB}{R}\right)lB = \frac{vl^2B^2}{R} \tag{25.5}$$

## Energy Considerations

FIGURE 25.6 is another look at the wire moving on a conducting rail. Because a force is needed to pull the wire through the magnetic field at a constant speed, we must do work to keep the wire moving. You learned in Chapter 10 that the power exerted by a force pushing or pulling an object with velocity $v$ is $P = Fv$, so the power provided to the circuit by the force pulling on the wire is

$$P_{input} = F_{pull}v = \frac{v^2l^2B^2}{R} \tag{25.6}$$

This is the rate at which energy is added to the circuit by the pulling force.

But the circuit dissipates energy in the resistance of the circuit. You learned in Chapter 22 that the power dissipated by current $I$ as it passes through resistance $R$ is $P = I^2R$. Equation 25.4 for the induced current $I$ gives us the power dissipated by the circuit of Figure 25.6:

$$P_{dissipated} = I^2R = \frac{v^2l^2B^2}{R} \tag{25.7}$$

Equations 25.6 and 25.7 have identical results. This makes sense: The rate at which work is done on the circuit is exactly balanced by the rate at which energy is dissipated. The fact that our final result is consistent with energy conservation is a good check on our work.

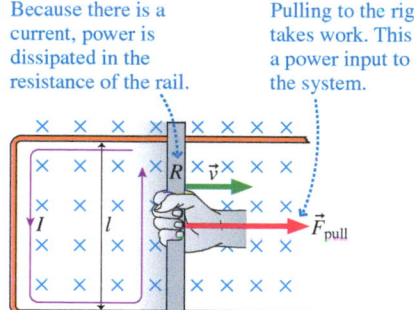

FIGURE 25.6 Power into and out of an induced-current circuit.

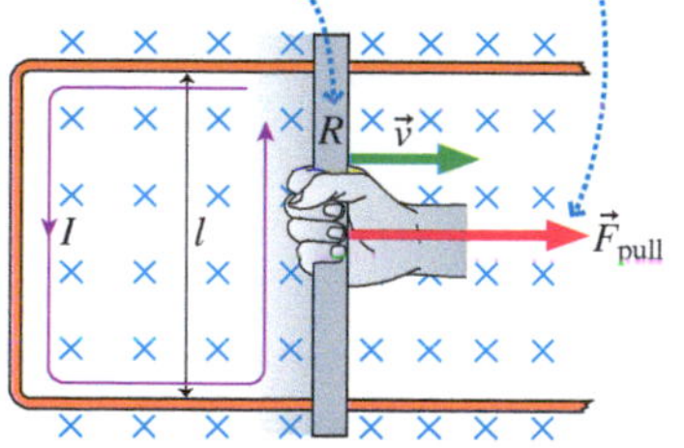


## Generators

A device that converts mechanical energy to electric energy is called a **generator.** The example of Figure 25.6 is a simple generator, but it is not very practical. Rather than move a straight wire, it's more practical to rotate a coil of wire, as in FIGURE 25.7 on the next page. The coil is rotated by some external torque, due perhaps to a steam turbine or windmill blades. As the coil rotates, the left edge always moves upward through the magnetic field while the right edge always moves downward. The motion of the wires through the magnetic field induces a current to flow, as noted in the figure. The induced current is removed from the rotating loop by *brushes* that press up against rotating *slip rings*. The circuit is completed as shown in the figure.

FIGURE 25.7 A generator using a rotating loop of wire.

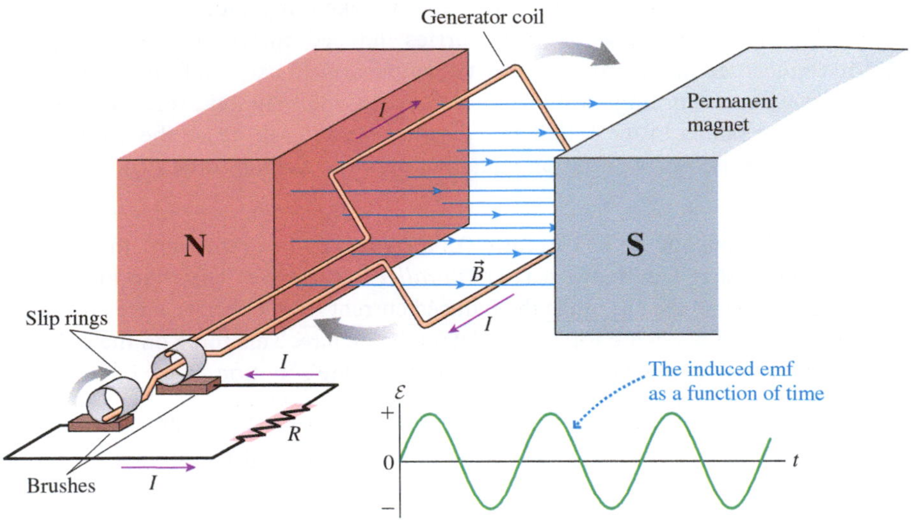


As the coil in the generator of Figure 25.7 rotates, the sense of the emf changes, giving a sinusoidal variation of emf as a function of time. Electric power plants use generators of this sort, so the electricity in your house has a varying voltage. The alternating sign of the voltage produces an *alternating current,* so we call such electricity *AC*. We'll have more to say about this type of electricity in Chapter 26.

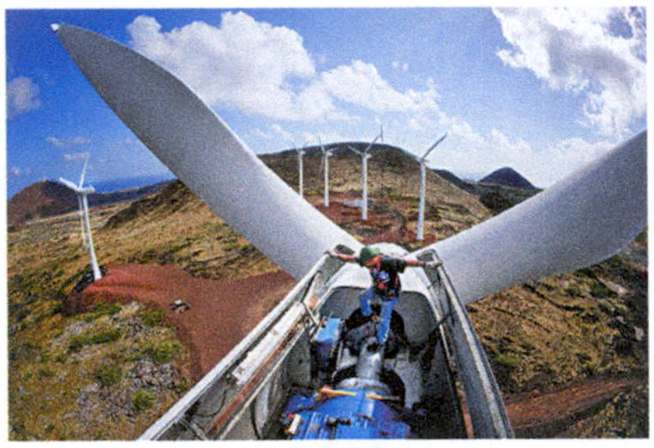

◄ **Turning wind into electric power** The rotating blades of a wind turbine turn coils of wire in a magnetic field, generating a large emf of around 1000 V. When the blades are running at full power, some 3000 A flow through these coils, which leads to an enormous magnetic drag force on them. The force to overcome this drag comes from the wind itself. The kinetic energy of the wind is transformed into electric energy from the generator.

**STOP TO THINK 25.1** A square conductor moves through a uniform magnetic field directed out of the page. Which of the figures shows the correct charge distribution on the conductor?

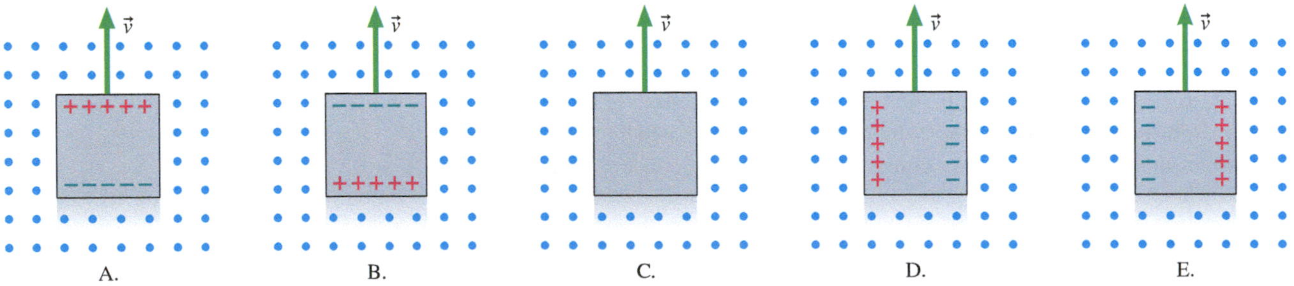


## 25.3 Magnetic Flux and Lenz's Law

We've begun our exploration of electromagnetic induction by analyzing a circuit in which one wire moves through a magnetic field. You might be wondering what this has to do with Faraday's discovery. Faraday found that a current is induced when the amount of magnetic field passing through a coil or a loop of wire changes. But that's exactly what happens as the slide wire moves down the rail in Figure 25.4! As the circuit expands, more magnetic field passes through the larger loop. It's time to define more clearly what we mean by "the amount of field passing through a loop."

FIGURE 25.8 The amount of air flowing through a loop depends on the angle of the loop.

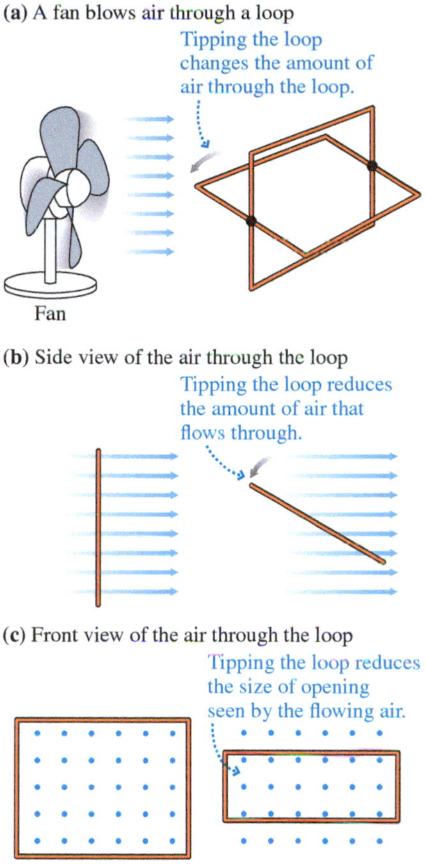


## Magnetic Flux

Imagine holding a rectangular loop of wire in front of a fan, as shown in FIGURE 25.8a. The arrows represent the flow of the air. If you want to get the most air through the loop, you know that you should hold the loop perpendicular to the direction of the flow. If you tip the loop from this position, less air will pass through the loop. FIGURE 25.8b is a side view that makes this reduction clear—fewer arrows pass through the tipped loop. Yet another way to visualize this situation is FIGURE 25.8c, which shows a front view with the air coming toward you; the dots represent the front of the arrows. From this point of view it's clear why the flow is smaller: The tipped loop presents a smaller area to the moving air. We say that the *effective area* of the loop has been reduced.

We can apply this idea to a magnetic field passing through a loop. FIGURE 25.9a shows a side view of a loop in a uniform magnetic field. To have the most field vectors going through the loop, we need to turn the loop to be perpendicular to the magnetic field vectors, just as we did for airflow. We define the *axis* of the loop to be a line through the center of the loop that is perpendicular to the plane of the loop. We see that the largest number of field vectors go through the loop when its axis is lined up with the field. Tipping the loop by an angle $\theta$ reduces the number of vectors passing through the loop, just as for the air from the fan.

FIGURE 25.9b is a front view of the loop with the dimensions noted. When the loop is tipped by an angle $\theta$, fewer field vectors pass through the loop because the effective area is smaller. We can define the effective area as

$$A_{\text{eff}} = ab\cos\theta = A\cos\theta \tag{25.8}$$

FIGURE 25.9 The amount of magnetic field passing through a loop depends on the angle of the loop.

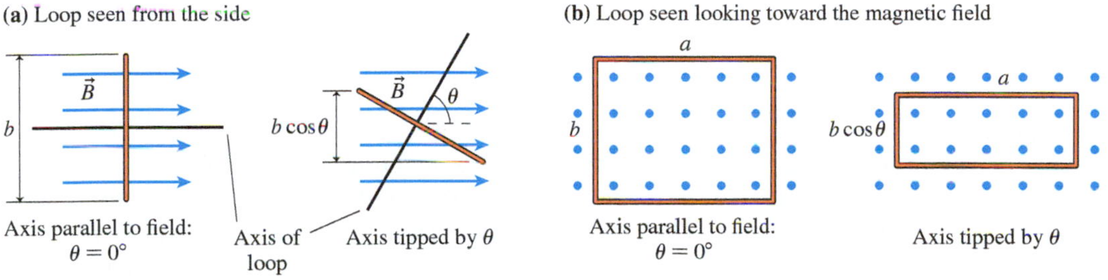


Ultimately, the amount of field that "goes through" the loop depends on two things: the strength of the field and the effective area of the loop. With this in mind, let's define the **magnetic flux** $\Phi$ as

$$\Phi = A_{\text{eff}}B = AB\cos\theta \tag{25.9}$$

Magnetic flux through area $A$ at angle $\theta$ to field $B$

FIGURE 25.10 Definition of magnetic flux.

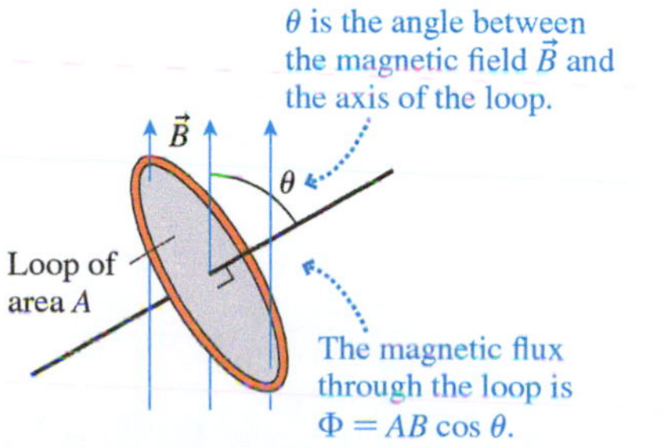


The magnetic flux measures the amount of magnetic field passing through a loop of area $A$ if the loop is tilted at angle $\theta$ from the field. The SI unit of magnetic flux is the **weber.** From Equation 25.9 you can see that

$$1 \text{ weber} = 1 \text{ Wb} = 1 \text{ T}\cdot\text{m}^2$$

The relationship of Equation 25.9 is illustrated in FIGURE 25.10.

## EXAMPLE 25.2 Finding the flux of the earth's magnetic field through a vertical loop

At a particular location, the earth's magnetic field is 50 $\mu$T tipped at an angle of 60° below horizontal. A 10-cm-diameter circular loop of wire sits flat on a table. What is the magnetic flux through the loop?

**STRATEGIZE** The flux through the loop is given by Equation 25.9. We will need the angle $\theta$ and the area $A$.

FIGURE 25.11 Finding the flux of the earth's field through a loop.

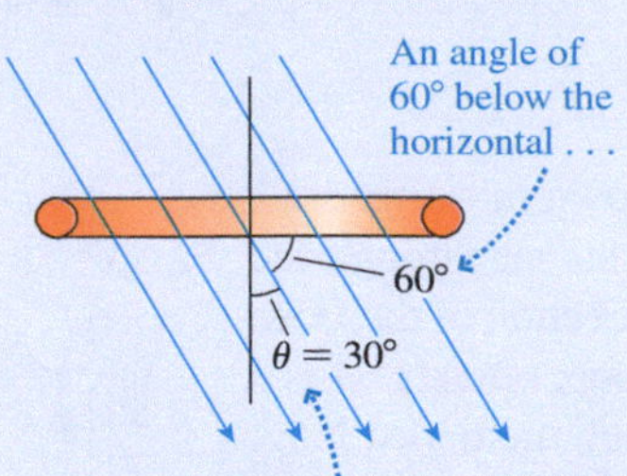


**PREPARE** FIGURE 25.11 shows the loop and the field of the earth. The field is tipped by 60°, so the angle of the field with respect to the axis of the loop is $\theta = 30°$. The radius of the loop is 5.0 cm, so the area of the loop is $A = \pi r^2 = \pi(0.050\text{ m})^2 = 0.0079\text{ m}^2$.

**SOLVE** The flux through the loop is then

$$\Phi = AB\cos\theta = (0.0079\text{ m}^2)(50 \times 10^{-6}\text{ T})\cos 30°$$
$$= 3.4 \times 10^{-7}\text{ Wb}$$

**ASSESS** It's a small loop and a small field, so a very small flux seems reasonable.

## Lenz's Law

Some of the induction experiments from earlier in the chapter could be explained in terms of motional emf, but others had no motion. What they all have in common, though, is that one way or another the magnetic flux through the coil or loop *changes*. We can summarize all of the discoveries as follows: **Current is induced in a loop of wire when the magnetic flux through the loop changes.**

For example, a momentary current is induced in the loop of FIGURE 25.12 as the bar magnet is pushed toward the loop because the magnetic field through the loop, and hence the flux, increases. Pulling the magnet away from the loop, which decreases the flux, causes the current meter to deflect in the opposite direction. How can we predict the *direction* of the current in the loop?

FIGURE 25.12 Pushing a bar magnet toward the loop induces a current in the loop.

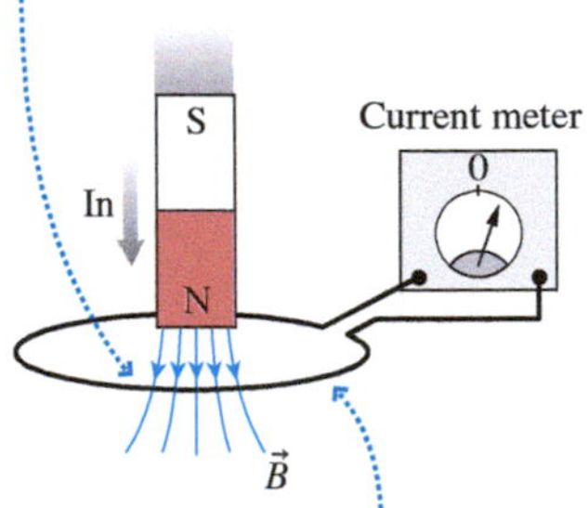


The German physicist Heinrich Lenz began to study electromagnetic induction after learning of Faraday's discovery. Lenz developed a rule for determining the direction of the induced current. We now call his rule **Lenz's law,** and it can be stated as follows:

> **Lenz's law** There is an induced current in a closed, conducting loop if and only if the magnetic flux through the loop is changing. The direction of the induced current is such that the induced magnetic field opposes the *change* in the flux.

Lenz's law is rather subtle, and it takes some practice to see how to apply it.

**NOTE** ► One difficulty with Lenz's law is the term "flux," from a Latin root meaning "flow." In everyday language, the word "flux" may imply that something is changing. Think of the phrase "The situation is in flux." In physics, "flux" simply means "passes through." A steady magnetic field through a loop creates a steady, *un*changing magnetic flux. ◄

Lenz's law tells us to look for situations where the flux is *changing*. We'll see three ways in which this comes about:

- The magnetic field through the loop changes (increases or decreases).
- The loop changes in area or angle.
- The loop moves into or out of a magnetic field.

We can understand Lenz's law this way: If the flux through a loop changes, a current is induced in a loop. That current generates *its own* magnetic field $\vec{B}_{\text{induced}}$. **It is this induced field that opposes the flux change.** Let's look at an example to clarify what we mean by this statement.

FIGURE 25.13 shows a magnet above a coil of wire. The field of the magnet creates a downward flux through the loop. The three parts of the figure show how the induced current in the loop develops in response to the motion of the magnet and thus the changing flux in the coil.

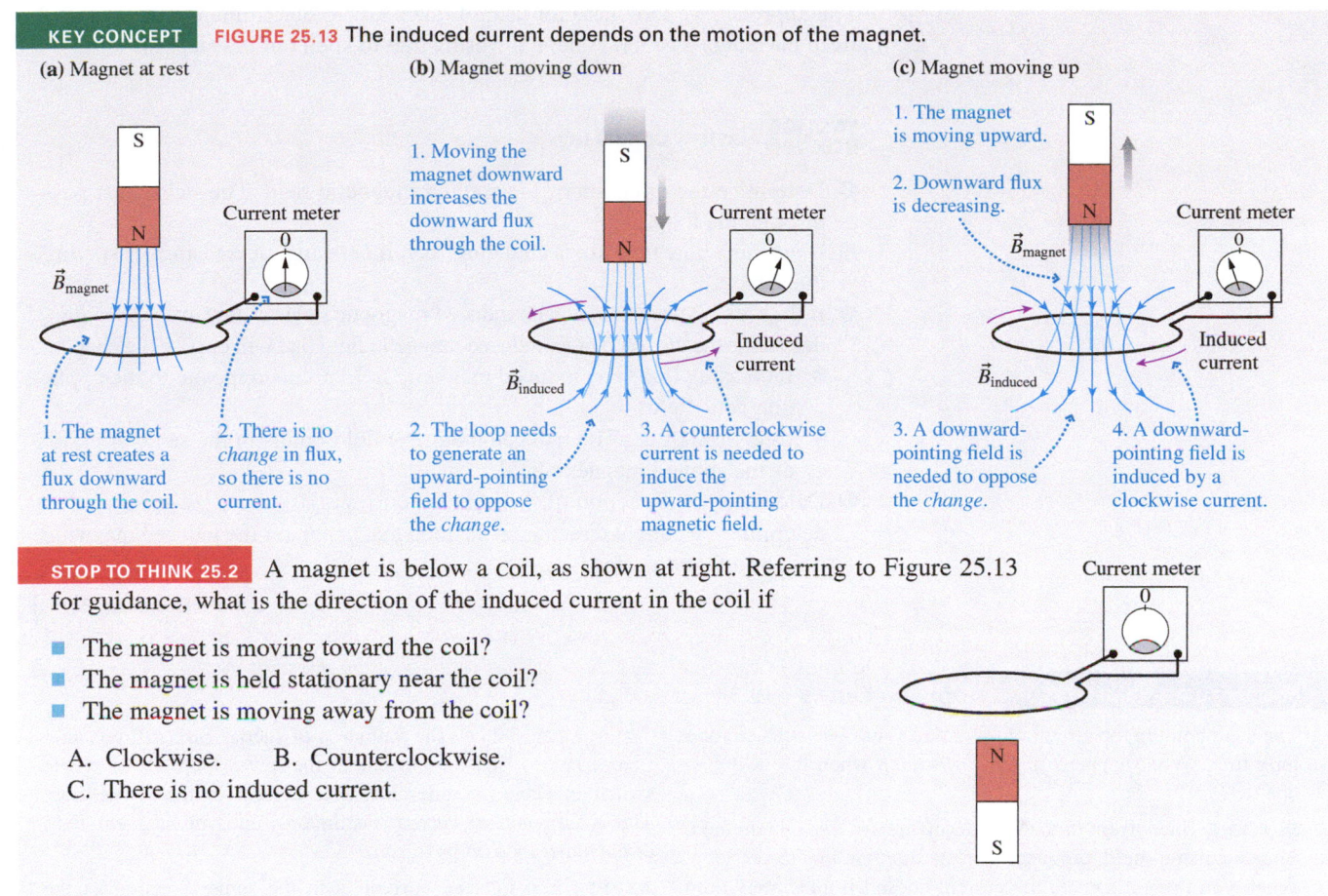

**KEY CONCEPT** **FIGURE 25.13** The induced current depends on the motion of the magnet.

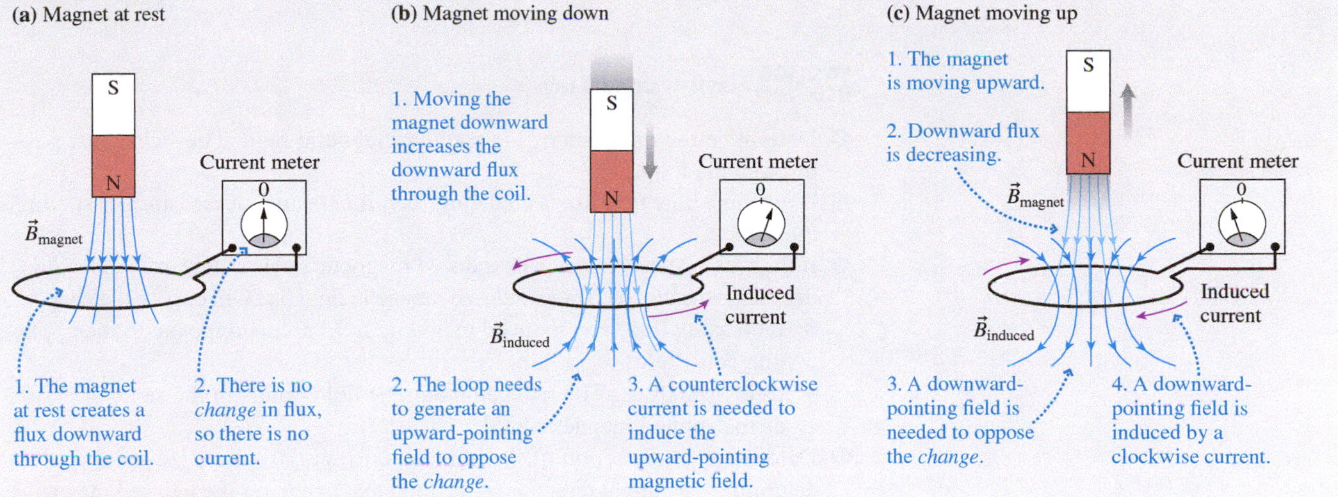


**STOP TO THINK 25.2** A magnet is below a coil, as shown at right. Referring to Figure 25.13 for guidance, what is the direction of the induced current in the coil if

- The magnet is moving toward the coil?
- The magnet is held stationary near the coil?
- The magnet is moving away from the coil?

A. Clockwise. B. Counterclockwise.
C. There is no induced current.

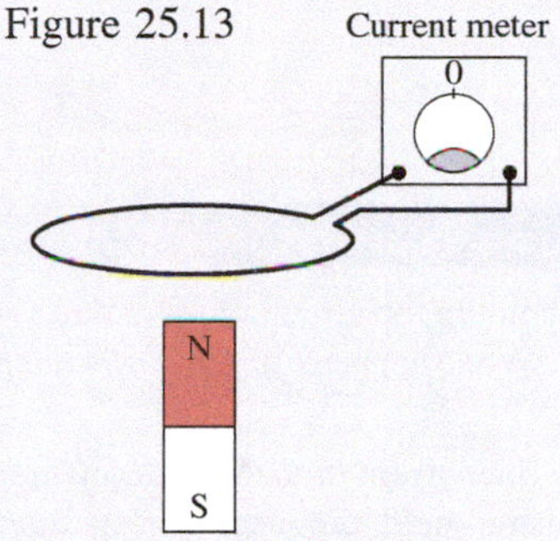


- In part a, the magnet is not moving. There is a flux through the loop, but because there is no *change* in the flux, no current is induced in the loop.
- In part b, the magnet is moving toward the loop, so the downward magnetic flux through the loop increases. According to Lenz's law, the loop will generate a field that opposes this change. To oppose an *increase in the downward flux,* the loop itself needs to generate an *upward-pointing magnetic field*. The induced magnetic field at the center of the loop will point upward if the current is counterclockwise, according to the right-hand rule you learned in Chapter 24. Thus pushing the north end of a bar magnet toward the loop induces a counterclockwise current around the loop. This induced current ceases as soon as the magnet stops moving.
- Now suppose the bar magnet is pulled back away from the loop, as in part c. There is a downward magnetic flux through the loop, but the flux decreases as the magnet moves away. According to Lenz's law, the induced magnetic field of the loop will oppose this decrease. To oppose a *decrease in the downward flux,* the loop itself needs to generate a *downward-pointing magnetic field*. The induced current is clockwise, opposite the induced current of part b.

The magnetic field of the bar magnet is pointing downward in each part of the figure, but the induced magnetic field can be zero, directed upward, or directed downward depending on the motion of the magnet. It is not the *flux* due to the magnet that the induced current opposes, but the *change in the flux*. This is a subtle but critical distinction. When the field of the magnet points downward and is increasing, the induced current *opposes the increase* by generating an upward field. When the field of the magnet points downward but is decreasing, the induced current *opposes the decrease* by generating a downward field.

The approach we took here for using Lenz's law to determine the direction of an induced current is a general one; it's worthwhile to spell out the steps.

### TACTICS BOX 25.1 Using Lenz's law

1. Determine the direction of the applied magnetic field. The field must pass through the loop.
2. Determine how the flux is changing. Is it increasing, decreasing, or staying the same?
3. If the flux is steady, there is no induced magnetic field. If the flux is changing, determine the direction of an induced magnetic field that will oppose the change:
   - Increasing flux: The induced magnetic field points opposite to the applied magnetic field.
   - Decreasing flux: The induced magnetic field points in the same direction as the applied magnetic field.
4. Determine the direction of the induced current. Use the right-hand rule to determine the current direction in the loop that generates the induced magnetic field you found in step 3.

Exercises 8–11

### CONCEPTUAL EXAMPLE 25.3 Applying Lenz's law 1

The switch in the top circuit of FIGURE 25.14 has been closed for a long time. What happens in the lower loop when the switch is opened?

***REASON*** The current in the upper loop creates a magnetic field. This magnetic field produces a flux through the lower loop. When you open a switch, the current doesn't immediately drop to zero; it falls off over a short time. As the current changes in the upper loop, the flux in the lower loop changes.

FIGURE 25.15 shows the four steps of Tactics Box 25.1 on using Lenz's law to find the current in the lower loop. Opening the switch induces a counterclockwise current in the lower loop. This is a momentary current, lasting only until the magnetic field of the upper loop drops to zero.

***ASSESS*** The induced current is in the same direction as the original current. This makes sense, because the induced current is opposing the change, a decrease in the current.

FIGURE 25.14 Circuits for Example 25.3.

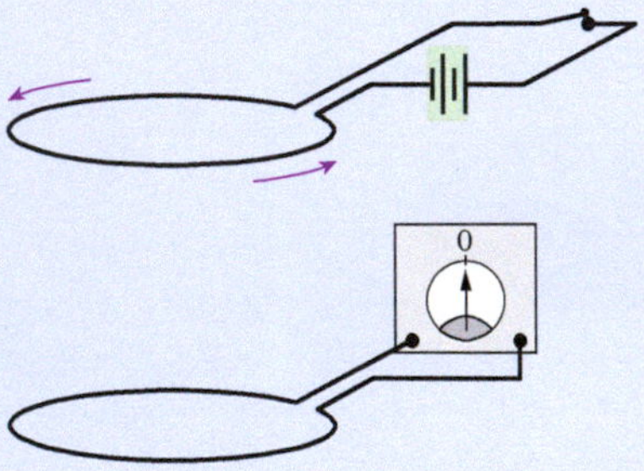


FIGURE 25.15 Finding the induced current.

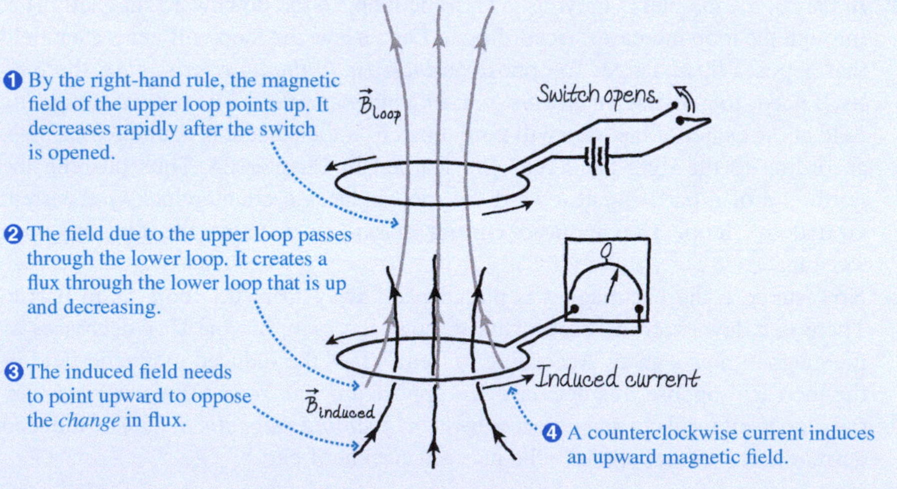


### CONCEPTUAL EXAMPLE 25.4 Applying Lenz's law 2

A loop is moved toward a current-carrying wire as shown in FIGURE 25.16. As the wire is moving, is there a clockwise current around the loop, a counterclockwise current, or no current?

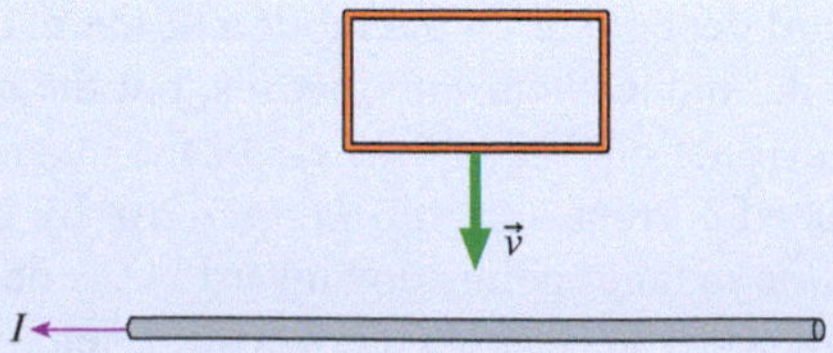


FIGURE 25.16 The moving loop.

**REASON** FIGURE 25.17 shows that the magnetic field above the wire points into the page. We learned in Chapter 24 that the magnetic field of a straight, current-carrying wire is proportional to $1/r$, where $r$ is the distance away from the wire, so the field is stronger closer to the wire.

As the loop moves toward the wire, the flux through the loop increases. To oppose the *change* in the flux—the increase into the page—the magnetic field of the induced current must point out of the page. Thus, according to the right-hand rule, a counterclockwise current is induced, as shown in Figure 25.17.

**ASSESS** The loop moves into a region of stronger field. To oppose the increasing flux, the induced field should be opposite the existing field, so our answer makes sense.

FIGURE 25.17 The motion of the loop changes the flux through the loop and induces a current.

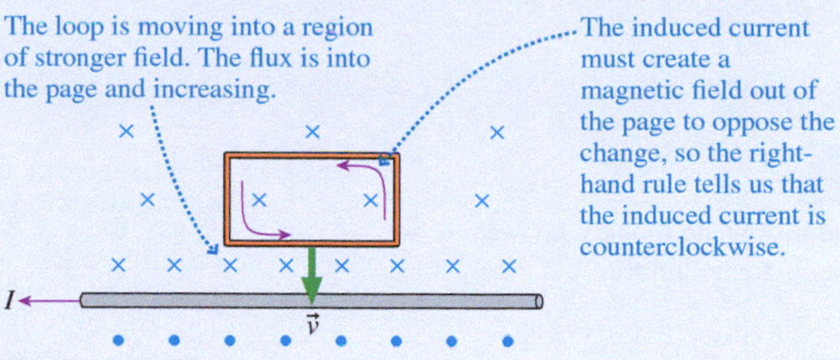

The loop is moving into a region of stronger field. The flux is into the page and increasing.
The induced current must create a magnetic field out of the page to oppose the change, so the right-hand rule tells us that the induced current is counterclockwise.
I
$\vec{v}$


**STOP TO THINK 25.3** As a coil moves to the right at constant speed, it passes over the north pole of a magnet and then moves beyond it. Which graph best represents the current in the loop for the time of the motion? A counterclockwise current as viewed from above the loop is a positive current, clockwise is a negative current.

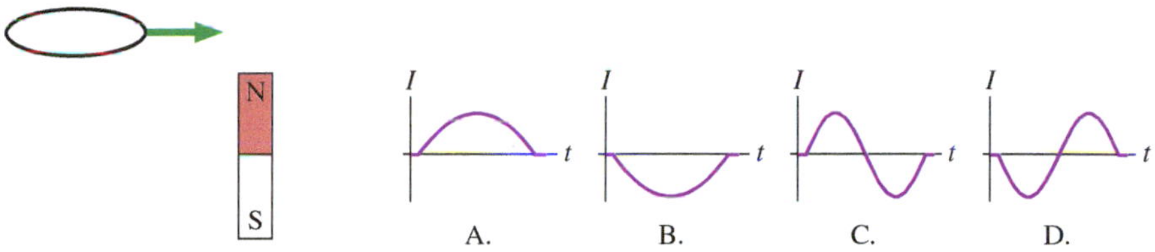


## 25.4 Faraday's Law

A current is induced when the magnetic flux through a conducting loop changes. Lenz's law allows us to find the direction of the induced current. To put electromagnetic induction to practical use, we also need to know the *size* of the induced current.

In the preceding examples, we found that a change in the flux caused a current to flow in a loop of wire. But we also know that a current doesn't just flow by itself—there must be an emf in the circuit to keep the current flowing. Evidently, **a changing flux in a circuit causes an emf,** which we call an **induced emf $\mathcal{E}$.** If this emf is induced in a complete circuit having resistance $R$, a current

$$I_{\text{induced}} = \frac{\mathcal{E}}{R} \tag{25.10}$$

is established in the wire as a *consequence* of the induced emf. The direction of the current is given by Lenz's law. The last piece of information we need is the size of the induced emf $\mathcal{E}$.

The research of Faraday and others led to the discovery of the basic law of electromagnetic induction, which we now call **Faraday's law.**

**Faraday's law** An emf $\mathcal{E}$ is induced in a conducting loop if the magnetic flux through the loop changes. If the flux changes by $\Delta\Phi$ during time interval $\Delta t$, the magnitude of the emf is

$$\mathcal{E} = \left|\frac{\Delta\Phi}{\Delta t}\right| \tag{25.11}$$

and the direction of the emf is such as to drive an induced current in the direction given by Lenz's law.

In other words, the magnitude of the induced emf is the *rate of change* of the magnetic flux through the loop.

**Inducing music** As you can see in the photo, the pickups on an electric guitar have small disk-shaped magnets that magnetize the nearby steel strings. What you can't see are the small coils below each magnet. As the magnetized strings vibrate, the flux they create in the coils changes rapidly, inducing a changing emf in the coils. When amplified, this emf can drive a loudspeaker, which converts the strings' motion to sound.

A video to support a section's topic is embedded in the eText.
**Video** Faraday's Law

A coil of wire consisting of $N$ turns in a changing magnetic field acts like $N$ batteries in series, so the induced emf of the entire coil is

$$\mathcal{E}_{\text{coil}} = N\left|\frac{\Delta\Phi_{\text{per turn}}}{\Delta t}\right| \tag{25.12}$$

**Video** Making Music with Magnetism
**Video** Eddy Currents

**PROBLEM-SOLVING APPROACH 25.1** **Electromagnetic induction**

Faraday's law allows us to find the *magnitude* of induced emfs and currents; Lenz's law allows us to determine the *direction*.

***STRATEGIZE*** There will be an induced emf in a loop or coil when the magnetic flux through the loop *changes*. This can occur if the magnetic field strength changes, the loop rotates, the loop moves into or out of the field, or the area of the loop changes.

***PREPARE*** Make simplifying assumptions about wires and magnetic fields. Draw a picture or a circuit diagram. Use Lenz's law to determine the direction of the induced current.

***SOLVE*** The mathematical representation is based on Faraday's law

$$\mathcal{E} = \left|\frac{\Delta\Phi}{\Delta t}\right|$$

For an $N$-turn coil, multiply by $N$. The size of the induced current is $I = \mathcal{E}/R$.

***ASSESS*** Check that your result has the correct units, is reasonable, and answers the question.

Exercise 15

Let's return to the situation of Figure 25.4, where a wire moves through a magnetic field by sliding on a U-shaped conducting rail. We looked at this problem as an example of motional emf; now, let's look at it using Faraday's law.

**EXAMPLE 25.5** **Finding the emf using Faraday's law**

**FIGURE 25.18** shows a wire of resistance $R$ sliding on a U-shaped conducting rail. Assume that the conducting rail is an ideal wire. Use Faraday's law and the steps of Problem-Solving Approach 25.1 to derive an expression for the current in the wire.

**FIGURE 25.18** A wire sliding on a rail.

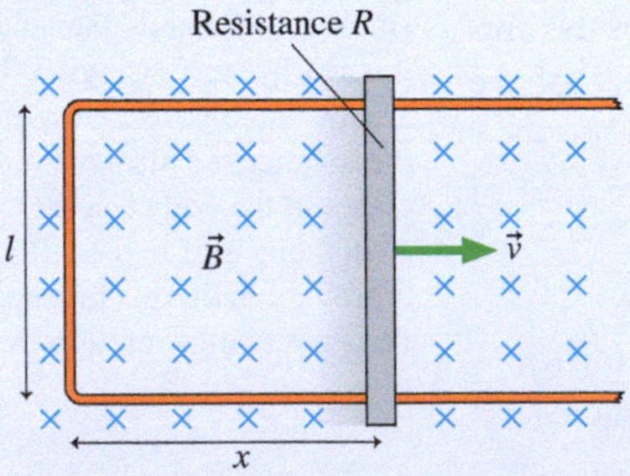


***STRATEGIZE*** Even though the magnetic field is constant, the flux is changing in this loop because its *area* is changing. To apply Faraday's law, we'll need to find the rate at which this area is increasing.

***PREPARE*** **FIGURE 25.19** shows the current loop formed by the wire and the rail. The flux is into the page and is increasing because

**FIGURE 25.19** Induced current in the sliding wire.

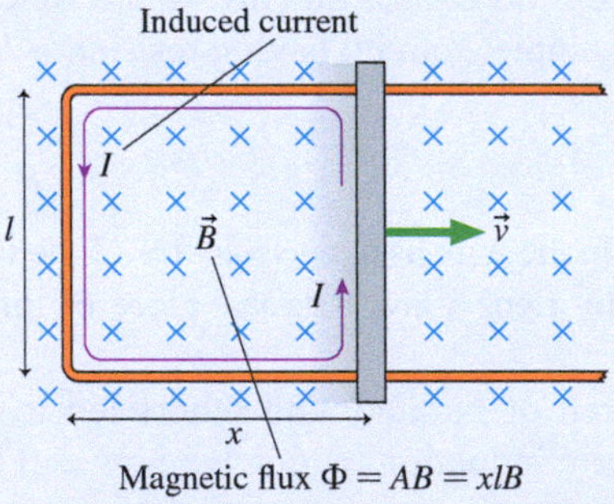


the loop's area is increasing. According to Lenz's law, the induced current must be counterclockwise so as to oppose the change, because the induced magnetic field must be out of the page.

***SOLVE*** The magnetic field $\vec{B}$ is perpendicular to the plane of the loop, so $\theta = 0°$ and the magnetic flux is $\Phi = AB$, where $A$ is the area of the loop. If the sliding wire is distance $x$ from the end, as in Figure 25.19, the area of the loop is $A = xl$ and the flux at that instant of time is

$$\Phi = AB = xlB$$

The flux through the loop increases as the wire moves and $x$ increases. This flux change induces an emf, according to Faraday's law, so we write

$$\mathcal{E} = \left|\frac{\Delta\Phi}{\Delta t}\right| = \left|\frac{\Delta(AB)}{\Delta t}\right| = \left|\frac{\Delta(xlB)}{\Delta t}\right|$$

The only quantity in the final ratio that is changing is the position $x$, so we can write

$$\mathcal{E} = \left|\frac{\Delta(xlB)}{\Delta t}\right| = lB\left|\frac{\Delta x}{\Delta t}\right|$$

But $|\Delta x/\Delta t|$ is the wire's speed $v$, so the induced emf is

$$\mathcal{E} = vlB$$

The wire and the loop have a total resistance $R$; thus the magnitude of the induced current is

$$I = \frac{\mathcal{E}}{R} = \frac{vlB}{R}$$

**ASSESS** This is exactly the same result we found in Section 25.2, where we analyzed this situation by considering the force on moving charge carriers. This is a good check on our work and a nice connection between the ideas of motional emf and Faraday's law.

## EXAMPLE 25.6 Finding the induced current in a bracelet during an MRI scan BIO

A patient having an MRI scan has neglected to remove a copper bracelet. The bracelet is 6.0 cm in diameter and has a resistance of 0.010 Ω. The magnetic field in the MRI solenoid is directed along the person's body from head to foot; her bracelet is perpendicular to $\vec{B}$. As a scan is taken, the magnetic field in the solenoid decreases from 1.00 T to 0.40 T in 1.2 s. What are the magnitude and direction of the current induced in the bracelet?

**STRATEGIZE** Here, the loop area is fixed but the magnetic field is changing, which leads to a changing flux and an emf. Once we know $\mathcal{E}$, we can use Ohm's law to find $I$.

**PREPARE** We follow the steps in Problem-Solving Approach 25.1, beginning with a sketch of the situation. FIGURE 25.20 shows the bracelet and the applied field looking down along the patient's body. The field is directed down through the loop; as the applied field decreases, the flux into the loop decreases. To oppose the decreasing flux, as required by Lenz's law, the field from the induced current must be in the direction of the applied field. Thus, from the right-hand rule, the induced current in the bracelet must be clockwise.

FIGURE 25.20 A circular conducting loop in a decreasing magnetic field.

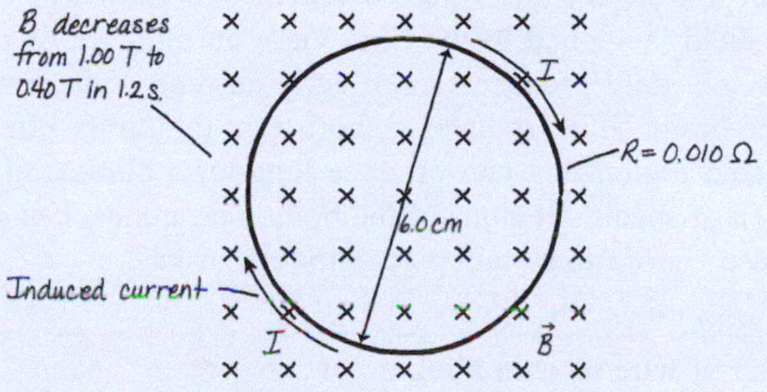


**SOLVE** The magnetic field is perpendicular to the plane of the loop; hence $\theta = 0°$ and the magnetic flux is $\Phi = AB = \pi r^2 B$. The area of the loop doesn't change with time, but $B$ does, so $\Delta\Phi = \Delta(AB) = A\Delta B = \pi r^2\,\Delta B$. The change in the magnetic field during $\Delta t = 1.2$ s is $\Delta B = 0.40\text{ T} - 1.00\text{ T} = -0.60$ T. According to Faraday's law, the magnitude of the induced emf is

$$\mathcal{E} = \left|\frac{\Delta\Phi}{\Delta t}\right| = \pi r^2\left|\frac{\Delta B}{\Delta t}\right| = \pi(0.030\text{ m})^2\left|\frac{-0.60\text{ T}}{1.2\text{ s}}\right|$$
$$= \pi(0.030\text{ m})^2(0.50\text{ T/s}) = 0.0014\text{ V}$$

The current induced by this emf is

$$I = \frac{\mathcal{E}}{R} = \frac{0.0014\text{ V}}{0.010\ \Omega} = 0.14\text{ A}$$

The decreasing magnetic field causes a 0.14 A clockwise current during the 1.2 s that the field is decreasing.

**ASSESS** The emf is quite small, but, because the resistance of the metal bracelet is also very small, the current is respectable. Electromagnetic induction produces currents large enough for practical applications, so this result seems plausible. The induced current could easily distort the readings of the MRI machine, and a larger current could cause enough heating to be potentially dangerous, so operators are careful to have patients remove all metal before an MRI scan.

As these two examples show, there are two fundamentally different ways to change the magnetic flux through a conducting loop:

- The loop can move or expand or rotate, creating a motional emf.
- The magnetic field can change.

Faraday's law tells us that the induced emf is simply the rate of change of the magnetic flux through the loop, *regardless* of what causes the flux to change.

FIGURE 25.21 Eddy currents.

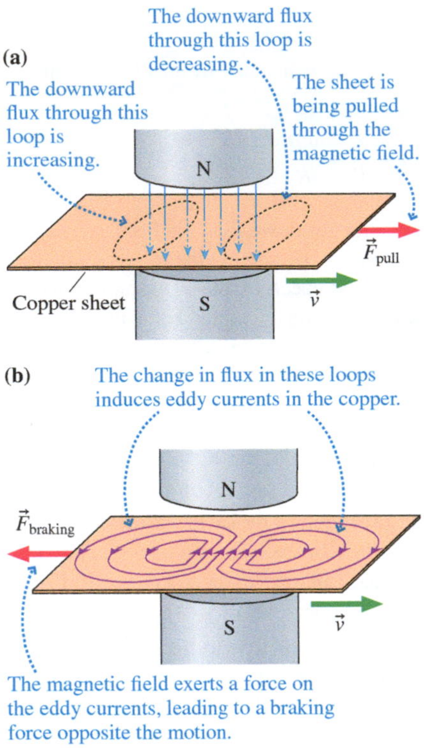


Video What the Physics? Magnetic Braking

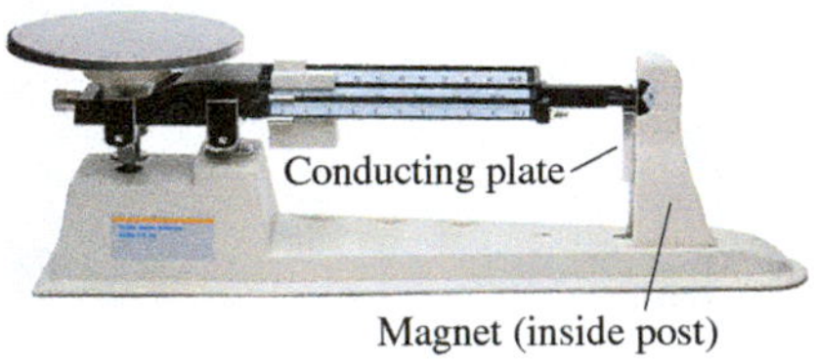


## Eddy Currents

Here is a remarkable physics demonstration that you can try: Take a sheet of copper and place it between the pole tips of a strong magnet, as shown in FIGURE 25.21a. Now, pull the copper sheet out of the magnet as fast as you can. Copper is not a magnetic material and thus is not attracted to the magnet, but, surprisingly, it takes a significant effort to pull the metal through the magnetic field.

Let's analyze this situation to discover the origin of the force. Figure 25.21a shows two "loops" lying entirely inside the metal sheet. The loop on the right is leaving the magnetic field, and the flux through it is decreasing. According to Faraday's law, the flux change will induce a current to flow around this loop, just as in a loop of wire, even though this current does not have a wire to define its path. As a consequence, a clockwise—as given by Lenz's law—"whirlpool" of current begins to circulate in the metal, as shown in FIGURE 25.21b. Similarly, the loop on the left is entering the field, and the flux through it is increasing. Lenz's law requires this whirlpool of current to circulate the opposite way. These spread-out whirlpools of induced current in a solid conductor are called **eddy currents.**

Figure 25.21b shows the direction of the eddy currents. Notice that both whirlpools are moving in the same direction as they pass through the center of the magnet. The magnet's field exerts a force on this current. By the right-hand rule, this force is to the left, opposite to the direction of the pull, and thus it acts as a *braking* force. Because of the braking force, **an external force is required to pull a metal through a magnetic field.** If the pulling force ceases, the magnetic braking force quickly causes the metal to decelerate until it stops. No matter which way the metal is moved, the magnetic forces on the eddy currents act to oppose the motion of the metal. *Magnetic braking* uses the braking force associated with eddy currents to slow trains and transit-system vehicles.

Magnetic braking is also used in many laboratory balances, such as the familiar triple-beam balance shown, to damp out oscillations that occur as the weights are moved. At the end of the beam, a conducting plate passes through a magnetic field. As the beam moves up and down, the magnetic force—which is always directed opposite to the motion—serves to quickly slow the beam. This method is superior to damping based on friction because static friction can cause the beam to stop at a point away from its true equilibrium position. Magnetic damping, however, gets smaller as the beam slows down; when the beam is at rest, there is no magnetic force to disturb equilibrium.

Eddy currents can also be induced by changing fields; this has practical applications as well. In a technique known as *transcranial magnetic stimulation* (TMS), a large oscillating magnetic field is applied to the head via a current-carrying coil. FIGURE 25.22 illustrates how this field produces small eddy currents that stimulate neurons in the tissue of the brain. This produces a short-term inhibitory effect on the neurons in the stimulated region that can produce long-term clinical effects. And by inhibiting the action of specific regions of the brain, researchers can determine the importance of these regions to certain perceptions or tasks.

FIGURE 25.22 BIO Transcranial magnetic stimulation.

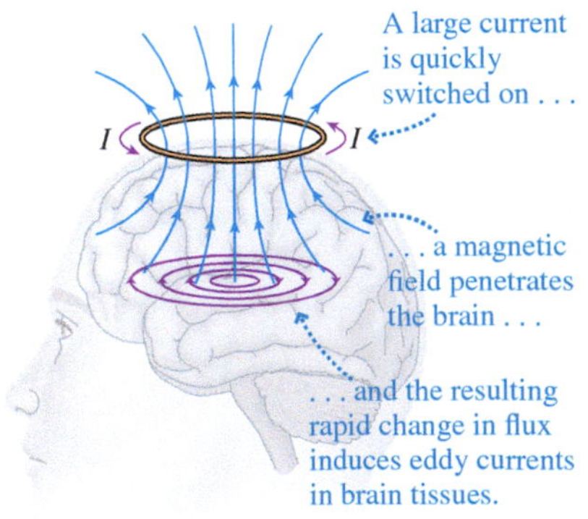


**STOP TO THINK 25.4** A loop of wire rests in a region of uniform magnetic field. The magnitude of the field is increasing, thus inducing a current in the loop. Which of the following changes to the situation would make the induced current larger? Choose all that apply.

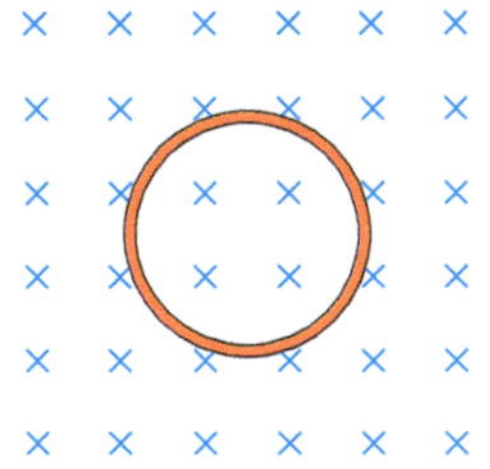

A. Increase the rate at which the field is increasing.
B. Replace the loop with one of the same resistance but larger diameter.
C. Replace the loop with one of the same resistance but smaller diameter.
D. Orient the loop parallel to the magnetic field.
E. Replace the loop with one of lower resistance.

## 25.5 Electromagnetic Waves

We will start this section with the puzzle shown in FIGURE 25.23. A long, tightly wound solenoid of radius $r_1$ passes through the center of a conducting loop having a larger radius $r_2$. The solenoid carries a current and generates a magnetic field. What happens to the loop if the solenoid current changes?

You learned in Chapter 24 that the magnetic field is strong inside a long solenoid but essentially zero outside. Even so, changing the field inside the solenoid causes the flux through the loop to change, so our theory predicts an induced current in the loop. But the loop is completely outside the solenoid, where the magnetic field is zero. How can the charge carriers in the conducting loop possibly "know" that the magnetic field inside the solenoid is changing?

FIGURE 25.23 Is there a current in the loop when the solenoid current changes?

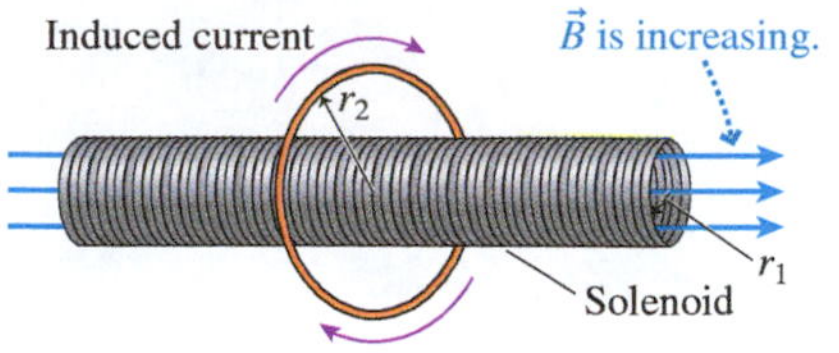


### Induced Fields

To answer this question, we will first consider a related question: When a changing flux through a loop induces a current, what actually *causes* the current? What *force* pushes the charges around the loop against the resistive forces of the metal? When we considered currents in Chapter 22, it was an *electric field* that moved charges through a conductor. Somehow, changing a magnetic field must create an electric field.

In fact, a changing magnetic field *does* cause what we call an **induced electric field.** FIGURE 25.24a shows a conducting loop in an increasing magnetic field. According to Lenz's law, there is an induced current in the counterclockwise direction. Something has to act on the charge carriers to make them move, so we can infer that the current is produced by an induced electric field tangent to the loop at all points.

But the induced electric field exists whether there is a conducting loop or not. The space in which the magnetic field is changing is filled with the pinwheel pattern of induced electric fields shown in FIGURE 25.24b.

FIGURE 25.24 An induced electric field creates a current in the loop.

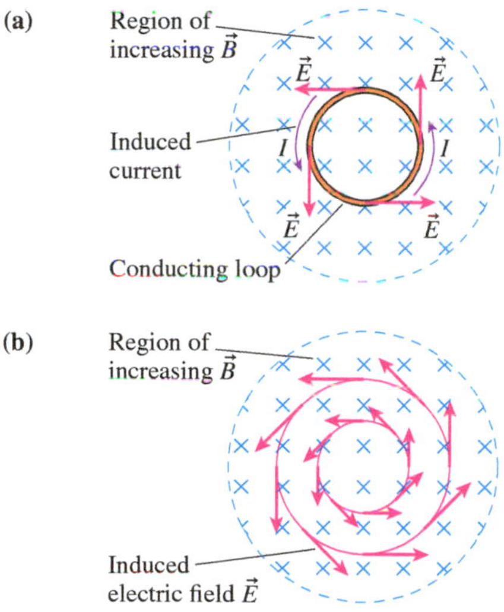


A changing magnetic field induces an electric field. What about a changing *electric* field? Early investigators looking for connections between electricity and magnetism wondered about this and hypothesized that **a changing electric field creates an induced magnetic field.**

This hypothesis leads to a surprising conclusion: If a changing magnetic field can induce an electric field in the absence of any charges, and if a changing electric field can induce a magnetic field in the absence of any currents, then it should be possible to establish self-sustaining electric and magnetic fields *independent of any charges or currents*. A changing electric field $\vec{E}$ creates a magnetic field $\vec{B}$, which then changes in just the right way to recreate the electric field, which then changes in just the right way to again recreate the magnetic field, with the fields continuously recreated through electromagnetic induction. In fact, electric and magnetic fields *can* sustain themselves, free from any charges or currents, if they take the form of an **electromagnetic wave.**

### Properties of Electromagnetic Waves

To sustain itself and travel through space, an electromagnetic wave must have a very specific geometry, shown in FIGURE 25.25. You can see that **an electromagnetic wave is a transverse wave.** $\vec{E}$ and $\vec{B}$ are perpendicular to each other as well as perpendicular to the direction of travel.

A mathematical analysis shows that such a wave travels with speed

$$v_{em} = \frac{1}{\sqrt{\epsilon_0 \mu_0}} \quad (25.13)$$

where $\epsilon_0$ and $\mu_0$ are the permittivity and permeability constants from our expressions for electric and magnetic fields. If you insert the values we've seen for these constants, you find $v_{em} = 3.00 \times 10^8$ m/s. This is a value you have seen before—it is the speed of light!

FIGURE 25.25 A sinusoidal electromagnetic wave.

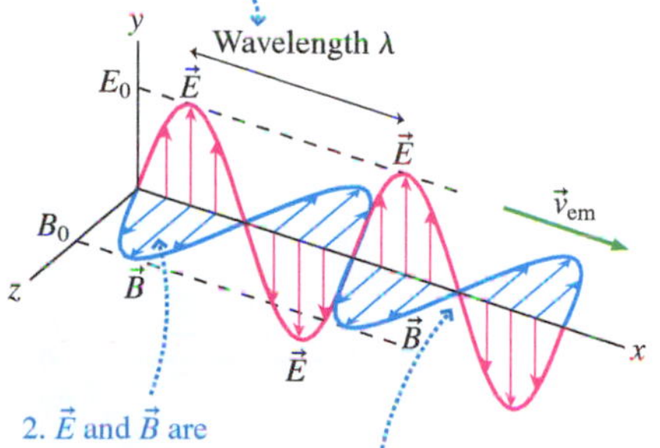

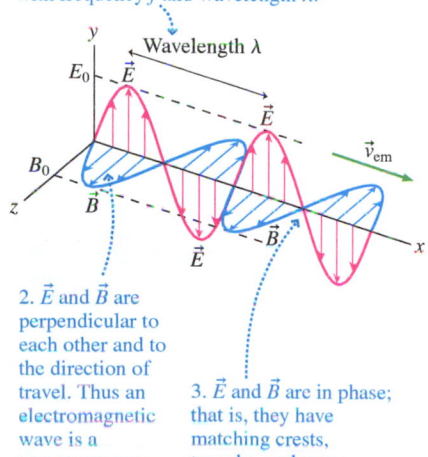

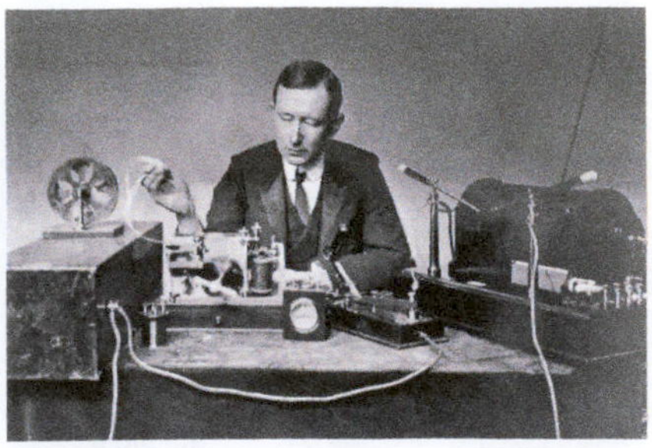

**The grandfather of the cell phone** Not long after Maxwell determined the properties of electromagnetic waves, experimenters used these waves to carry signals across a laboratory. By the late 1800s, Guglielmo Marconi and other investigators had developed the technology for reliably producing and detecting these electromagnetic waves. Marconi was able to send signals with no wires at the speed of light to others at great distances away—something we have come to take for granted.

In a vacuum, all electromagnetic waves must travel at the same speed, the speed of light $c$. James Clerk Maxwell, the first to make this analysis, made a bold leap and concluded that **light is an electromagnetic wave.** We studied the wave properties of light in Part V, but at that time we didn't discuss just what is "waving." Now we know— light is a wave of electric and magnetic fields.

The amplitudes of the electric and magnetic fields vary in tandem; at every point on the wave, the electric and magnetic field strengths are related by

$$\frac{E}{B} = c \tag{25.14}$$

Figure 25.25 shows the values of the electric and magnetic fields at points along a single line, the $x$-axis.

NOTE ▸ An $\vec{E}$ vector pointing in the $y$-direction says that at that point on the $x$-axis, where the vector's tail is, the electric field points in the $y$-direction and has a certain strength. Nothing is "reaching" to a point in space above the $x$-axis. ◂

Electromagnetic waves are oscillations of the electric and magnetic fields, but they are still waves, and all the general principles we have learned about waves apply. In ◂◂ SECTION 15.3, we learned that we could characterize sinusoidal waves by their speed, wavelength, and frequency, with these variables related by the fundamental relationship $v = \lambda f$. All electromagnetic waves move at the speed of light, $v_{em} = c$, so for sinusoidal electromagnetic waves this relationship becomes

$$c = \lambda f \tag{25.15}$$

The spectrum of electromagnetic waves ranges from waves of long wavelength and (relatively) low frequency (radio waves and microwaves) to waves of short wavelength and high frequency (visible light, ultraviolet, and x rays). We'll have more to say about the electromagnetic spectrum later in the chapter.

Suppose the electromagnetic wave of Figure 25.25 is a plane wave traveling in the $x$-direction. Recall from Chapter 15 that the displacement of a plane wave is the same at *all points* in any plane perpendicular to the direction of motion. In this case, the fields are the same in any $yz$-plane. If you were standing on the $x$-axis as the wave moves toward you, the electric and magnetic fields would vary as in the series of pictures of FIGURE 25.26. The $\vec{E}$ and $\vec{B}$ fields at each point in the $yz$-plane oscillate in time, but they are always synchronized with all the other points in the plane. As the plane wave passed you, you would see a uniform oscillation of the $\vec{E}$ and $\vec{B}$ fields of the wave.

FIGURE 25.26 The fields of an electromagnetic plane wave moving toward you, shown every one-eighth period for half a cycle.

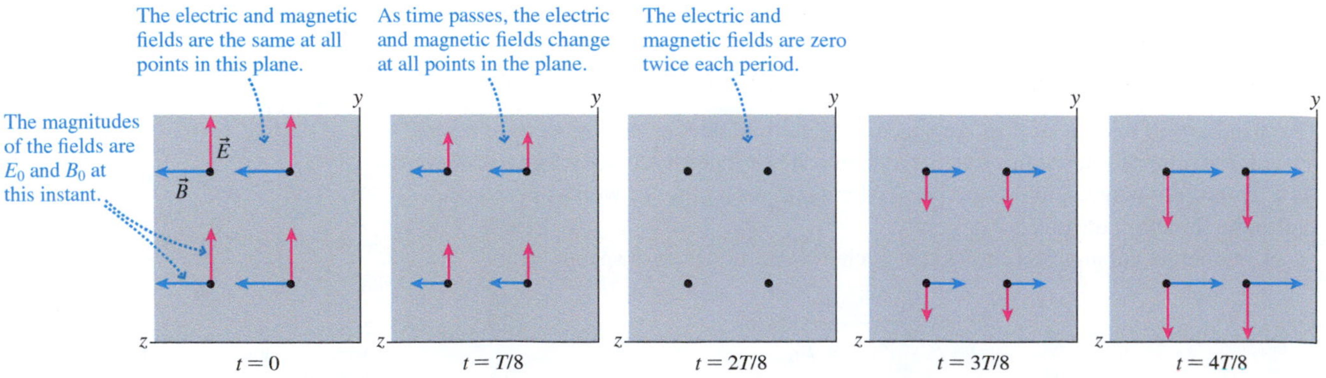


We can adapt our equation for traveling waves from Chapter 15 to electromagnetic waves. If a plane electromagnetic wave moves in the $x$-direction with the electric field along the $y$-axis, then the magnetic field is along the $z$-axis. The equations for the electric and magnetic fields of a wave with wavelength $\lambda$ and period $T$ are

$$E_y = E_0 \sin\left(2\pi\left(\frac{x}{\lambda} - \frac{t}{T}\right)\right) \qquad B_z = B_0 \sin\left(2\pi\left(\frac{x}{\lambda} - \frac{t}{T}\right)\right) \quad (25.16)$$

$E_0$ and $B_0$ are the amplitudes of the oscillating fields. The amplitudes of the fields must match the relationship between the fields given in Equation 25.14, so they must be related like this:

$$\frac{E_0}{B_0} = c \quad (25.17)$$

Relationship between field amplitudes for an electromagnetic wave

Since the value of $c$ is so large, the electric field amplitude in N/C is much larger than the magnetic field amplitude in T.

FIGURE 25.27 The polarization of an electromagnetic wave.

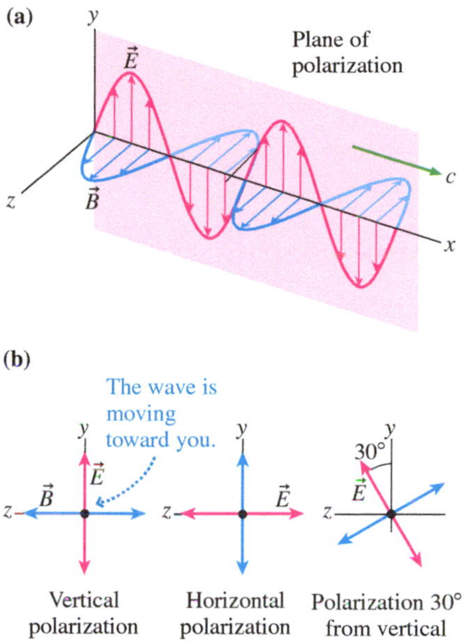


## Polarization

As we've seen, the electric field vectors of an electromagnetic wave lie in a plane perpendicular to the direction of propagation. The plane containing the electric field vectors is called the **plane of polarization.** FIGURE 25.27a shows a wave traveling along the $x$-axis; the plane of polarization is the $xy$-plane. If the wave were moving toward you, it would appear as in the first diagram in FIGURE 25.27b. This particular wave is *vertically polarized* ($\vec{E}$ oscillating along the $y$-axis). The second diagram of Figure 25.27b shows a wave that is *horizontally polarized* ($\vec{E}$ oscillating along the $z$-axis). The plane of polarization needn't be horizontal or vertical; it can have any orientation, as shown in the third diagram.

NOTE ▶ This use of the term "polarization" is completely independent of the idea of *charge polarization* that you learned about in Chapter 20. ◀

Not all electromagnetic waves are polarized. Consider sunlight or light from an incandescent lightbulb. Each atom in the sun's hot atmosphere or in the bulb's filament emits light independently of all the other atoms. Although the wave from each individual emitter is polarized, it is polarized in a random direction with respect to the waves from all its neighbors. The resulting wave, a superposition of waves with electric fields in all possible directions, is *unpolarized*: At each point in space, the direction of the electric field changes very rapidly in a random way.

**Video** Parallel-Wire Polarizer for Microwaves

## Energy of Electromagnetic Waves

All waves transfer energy. Ocean waves erode beaches, sound waves set your eardrum to vibrating, and light from the sun warms the earth. In all of these cases, the waves carry energy from the point where they are emitted to another point where their energy is transferred to an object. In water waves, the wave energy is the kinetic and gravitational potential energy of water; for electromagnetic waves, the wave energy is the energy of the electric and magnetic fields. The energy of the wave depends on the amplitudes of these fields; if the electric and magnetic fields have greater amplitudes, an electromagnetic wave will carry more energy.

In ◀◀ SECTION 15.5, we defined the *intensity* of a wave (measured in W/m$^2$) to be $I = P/A$, where $P$ is the power (energy transferred per second) of a wave that impinges on area $A$. The intensity of an electromagnetic wave can be written in terms of the amplitudes of the oscillating electric and magnetic fields:

$$I = \frac{P}{A} = \frac{1}{2} c\epsilon_0 E_0^2 = \frac{1}{2}\frac{c}{\mu_0} B_0^2 \quad (25.18)$$

Intensity of an electromagnetic wave with field amplitudes $E_0$ and $B_0$

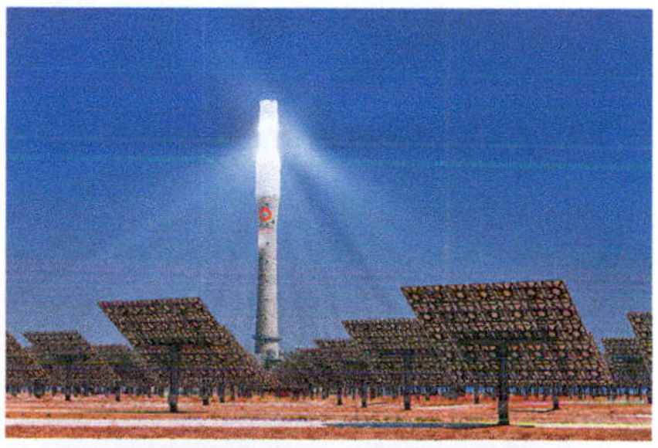

**Solar power** The energy from the sun is carried through space by electromagnetic waves—that is, by electric and magnetic fields. This power plant in Spain uses 2650 mirrors across 480 acres to concentrate the sun's power at a tower, where molten salts are liquefied at 565°C. The energy of this molten salt can be used to power electric generators up to 15 hours later.

The intensity of a plane wave, such as that of a laser beam, does not change with distance. But, as we saw in Section 15.5, the intensity of a spherical wave, spreading out from a point, must decrease with the square of the distance to conserve energy. If a source with power $P_{source}$ emits waves *uniformly* in all directions, the wave intensity at distance $r$ is

$$I = \frac{P_{source}}{4\pi r^2} \tag{25.19}$$

The intensities of the electromagnetic waves from antennas, cell phones, and other "point sources" are reasonably well described by Equation 25.19.

It's worthwhile to pull together all of the details that we've seen for electromagnetic waves. Seeing the "big picture" will help us as we start to solve electromagnetic wave problems.

## SYNTHESIS 25.1 Electromagnetic waves

An electromagnetic wave is a transverse wave of oscillating electric and magnetic fields.

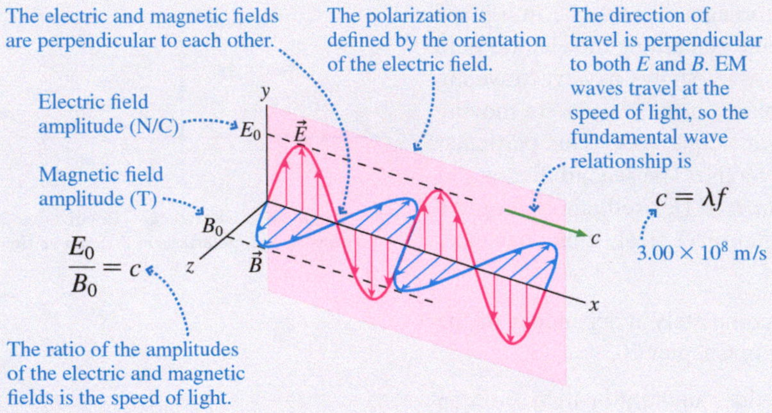


The intensity of the EM wave depends on the field amplitudes:

$$I = \frac{1}{2}c\epsilon_0 E_0^2 = \frac{1}{2}\frac{c}{\mu_0}B_0^2$$

Intensity (W/m$^2$): $I$

Permittivity constant, $8.85 \times 10^{-12}\ \mathrm{C^2/N \cdot m^2}$

Permeability constant, $1.26 \times 10^{-6}\ \mathrm{T \cdot m/A}$

The intensity of an EM point source decreases with distance from the source:

$$I = \frac{P_{source}}{4\pi r^2}$$

$P_{source}$: Source power (W)

$r$: Distance from the source (m)

## EXAMPLE 25.7 Electric and magnetic fields of a cell phone

A digital cell phone emits 0.60 W of 1.9 GHz radio waves. What are the amplitudes of the electric and magnetic fields at a distance of 10 cm?

***STRATEGIZE*** We will solve this problem using details from Synthesis 25.1. We can approximate the cell phone as a point source, so we can use the second intensity equation to find the intensity at 10 cm. Once we know the intensity, we can use the first intensity equation to compute the field amplitudes.

***SOLVE*** The intensity at a distance of 10 cm is

$$I = \frac{P_{source}}{4\pi r^2} = \frac{0.60\ \mathrm{W}}{4\pi(0.10\ \mathrm{m})^2} = 4.8\ \mathrm{W/m^2}$$

We can rearrange the first intensity equation to solve for the amplitude of the electric field:

$$E_0 = \sqrt{\frac{2I}{c\epsilon_0}} = \sqrt{\frac{2(4.8\ \mathrm{W/m^2})}{(3.0\times10^8\ \mathrm{m/s})(8.85\times10^{-12}\ \mathrm{C^2/N\cdot m^2})}}$$
$$= 60\ \mathrm{V/m}$$

We can then use the relationship between field amplitudes to find the amplitude of the magnetic field:

$$B_0 = \frac{E_0}{c} = 2.0 \times 10^{-7}\ \mathrm{T}$$

***ASSESS*** The electric field amplitude is reasonably small. For comparison, the typical electric field due to atmospheric electricity is 100 V/m; the field near a charged Van de Graaff generator can be 1000 times larger than this. The scale of the result thus seems reasonable; we know that the electric fields near a cell phone's antenna aren't large enough to produce significant effects. The magnetic field is smaller yet, only 1/250th of the earth's field, which, as you know, is quite weak. This makes sense as well; you haven't noticed magnetic effects while making a phone call!

## Polarizers and Changing Polarization

If we have an unpolarized source of light, we can transform it into polarized light by sending it through a *polarizing filter*. A typical polarizing filter is a plastic sheet containing long organic molecules called polymers, as shown in

FIGURE 25.28. The molecules are aligned to form a grid, like the metal bars in a barbecue grill, then treated so they conduct electrons along their length.

As a light wave travels through a polarizing filter, the component of the electric field oscillating parallel to the polymer grid drives the electrons up and down the molecules. The electrons absorb energy from the light wave, so the parallel component of $\vec{E}$ is absorbed in the filter. Thus the light wave emerging from a polarizing filter is polarized perpendicular to the polymer grid. We call the direction of the transmitted polarization the *axis* of the polarizer.

Suppose a *polarized* light wave with electric field amplitude $E_{\text{incident}}$ approaches a polarizing filter with a vertical axis (that is, the filter transmits only vertically polarized light). What is the intensity of the light that passes through the filter? FIGURE 25.29 shows that the oscillating electric field of the polarized light can be decomposed into horizontal and vertical components. The vertical component will pass; the horizontal component will be blocked. As we see in Figure 25.29, the magnitude of the electric field of the light transmitted by the filter is

$$E_{\text{transmitted}} = E_{\text{incident}} \cos\theta \tag{25.20}$$

Because the intensity depends on the square of the electric field amplitude, the transmitted intensity is related to the incident intensity by what is known as **Malus's law:**

$$I_{\text{transmitted}} = I_{\text{incident}}(\cos\theta)^2 \tag{25.21}$$

Malus's law for transmission of polarized light by a polarizing filter

FIGURE 25.30a shows how Malus's law can be demonstrated with two polarizing filters. The first, called the *polarizer,* is used to produce polarized light of intensity $I_0$. The second, called the *analyzer,* is rotated by angle $\theta$ relative to the polarizer. As the photographs of FIGURE 25.30b show, the transmission of the analyzer is (ideally) 100% when $\theta = 0°$ and steadily decreases to zero when $\theta = 90°$. Two polarizing filters with perpendicular axes, called *crossed polarizers,* block all the light.

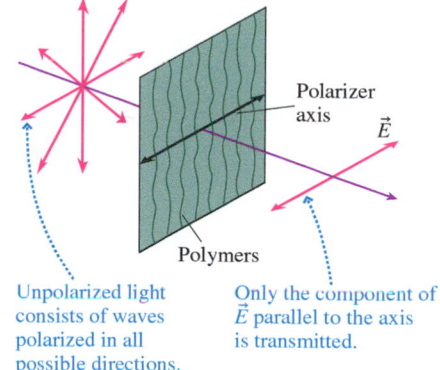

FIGURE 25.28 A polarizing filter.

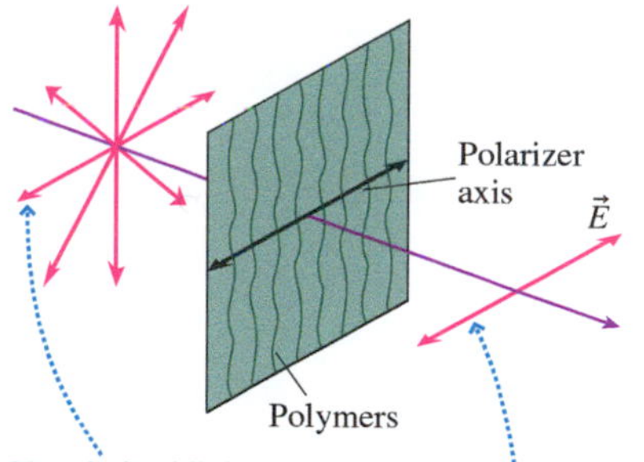


FIGURE 25.29 Components of the electric field parallel and perpendicular to a polarizer's axis.

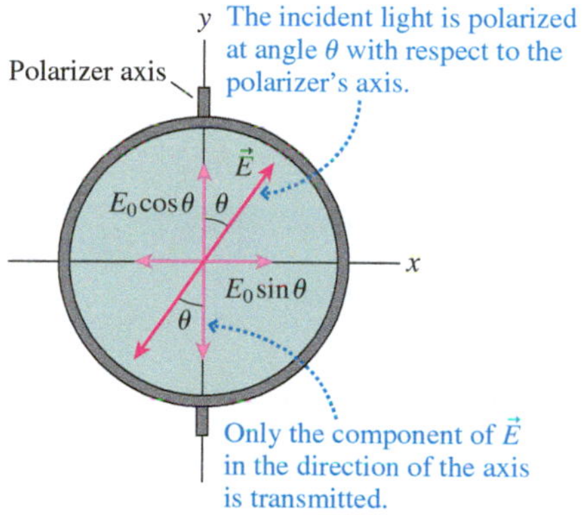

y
The incident light is polarized at angle θ with respect to the polarizer's axis.
Polarizer axis
E
E0cosθ
θ
x
E0sinθ
θ
Only the component of E in the direction of the axis is transmitted.

FIGURE 25.30 The intensity of the transmitted light depends on the angle between the polarizing filters.

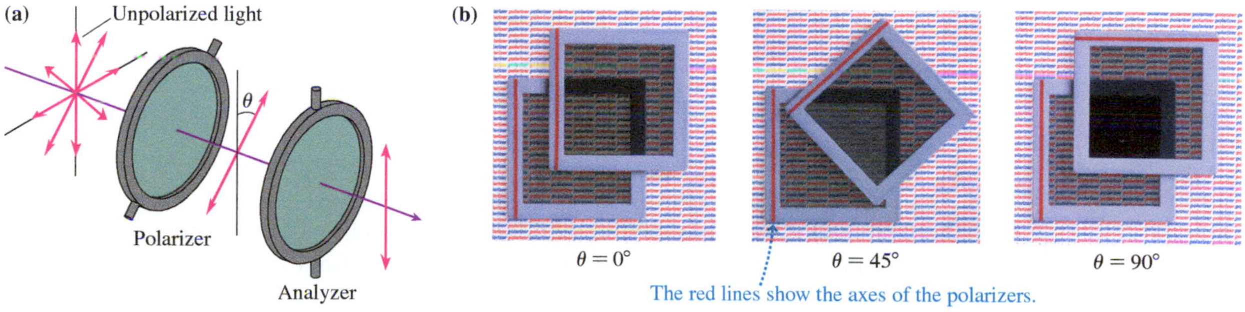


Video Figure 25.30

Suppose you place a transparent object between two crossed polarizers. Normally, no light would make it through the analyzer, and the object would appear black. But if the object is able to *change* the angle of polarization of the light, some of the light emerging from the object will be able to pass through the analyzer. This can be a valuable analytical technique. FIGURE 25.31 shows a micrograph of a very thin section of molar teeth as it appears when viewed between crossed polarizers. Different minerals and different materials in the teeth affect the polarization of the light in different ways, giving an image that clearly highlights the different tissues in the teeth.

When light passes through a glucose solution, the plane of polarization is rotated by an amount that is proportional to the concentration of glucose. Higher concentrations of glucose mean larger angles of rotation. In principle, this could be used as a test of blood glucose—a test that doesn't involve poking a fingertip.

FIGURE 25.31 BIO Polarized light micrograph of a thin section of molar teeth.

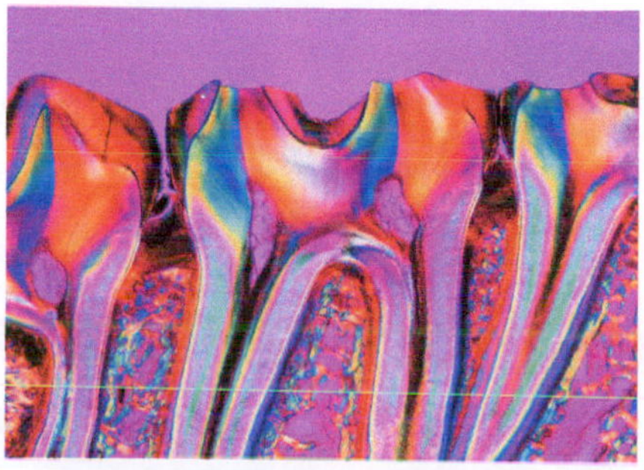

**EXAMPLE 25.8** **Using polarization to test for blood glucose** BIO

Doctors have long sought a noninvasive technique for patients, such as diabetics, who need to monitor their blood glucose level several times a day. One possibility is to measure the change in the polarization of light transmitted through the clear aqueous humor in the front of the eye, where the glucose level is known to be almost the same as that in the blood. At the expected glucose level, the change in polarization is quite small, but measurable, as this example—using realistic numbers—shows.

A laser is passed through a polarizer and emerges with intensity 55 W/m$^2$. In a first measurement, the polarized laser light is passed through an analyzer, which is adjusted to an angle of 90°, giving zero transmitted intensity. Then, in a second measurement, the polarized laser light is passed through the aqueous humor in a patient's eye before reaching the analyzer, which is still set at 90°; in this case, the rotation of the plane of polarization by the glucose allows some light to pass, giving a transmitted intensity of $8.7 \times 10^{-7}$ W/m$^2$. What is the angle of rotation of the polarization by the glucose?

**STRATEGIZE** We will use Malus's law (Equation 25.21) to relate the intensity transmitted through the analyzer to the intensity incident on the analyzer. For the initial measurement, $\theta = 90°$ exactly because the analyzer is adjusted so that no light passes. In the second test, some light is transmitted because the glucose rotates the polarization so that the angle is smaller than 90°. The difference between the two angles is the rotation of the plane of polarization.

**PREPARE** We can use Equation 25.21 to find the rotation angle for the second test. The incident intensity is the 55 W/m$^2$ of the polarized laser light and the transmitted intensity is $8.7 \times 10^{-7}$ W/m$^2$.

**SOLVE** From Equation 25.21, the angle of rotation is

$$\cos\theta = \sqrt{\frac{I_{\text{transmitted}}}{I_{\text{incident}}}} = \sqrt{\frac{8.7 \times 10^{-7}\ \text{W/m}^2}{55\ \text{W/m}^2}} = 1.26 \times 10^{-4}$$

The polarization angle is thus

$$\theta = \cos^{-1}(1.26 \times 10^{-4}) = 89.9928°$$

We've kept extra significant figures because we are interested in the small difference between this angle and 90°. The angle $90° - 89.9928° = 0.0072°$ is the angle by which the glucose has rotated the plane of polarization.

**ASSESS** The plane of polarization rotates by a very small angle, but we expect this from the problem statement. The rotation, though small, can be readily measured with modern optical equipment.

**Making a beeline** BIO At an angle of 90° to the sun, the sky's light is strongly polarized. In the left photo, this polarization causes different transmission through polarizers with polarization axes given by the red arrows. Opposite to the sun, the sky's light is unpolarized, causing equal transmission for all of the polarizers in the right photo. In the photo of the sky at 90° to the sun, the arrows in the dark polarizers define a line that points toward the sun; you can tell where the sun is though it does not appear in the photo. Honeybees and other insects—unlike you—have eyes that are sensitive to polarization. Bees can reliably navigate in dense forest cover; as long as they can see a small patch of blue sky, they can determine the position of the sun.

In polarizing sunglasses, the polarization axis is vertical (when the glasses are in the normal orientation) so that the glasses transmit only vertically polarized light. *Glare*—the reflection of the sun and the skylight from lakes and other horizontal surfaces—has a strong horizontal polarization. This light is almost completely blocked, so the sunglasses "cut glare" without affecting the main scene.

**STOP TO THINK 25.5** Unpolarized light of equal intensity is incident on four pairs of polarizing filters. Rank in order, from largest to smallest, the intensities $I_A$ to $I_D$ transmitted through the second polarizer of each pair.

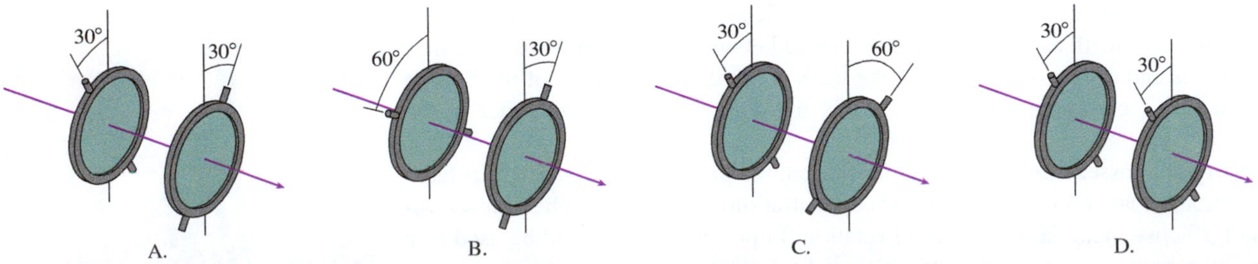

# 25.6 The Photon Model of Electromagnetic Waves

FIGURE 25.32 shows three images made with a camera in which the film has been replaced by a special high-sensitivity detector. A correct exposure, at the bottom, shows a perfectly normal image of a woman. But with very faint illumination (top), the picture is *not* just a dim version of the properly exposed image. Instead, it is a collection of dots. A few points on the detector have registered the presence of light, but most have not. As the illumination increases, the density of these dots increases until the dots form a full picture.

FIGURE 25.32 Images made with an increasing level of light intensity.

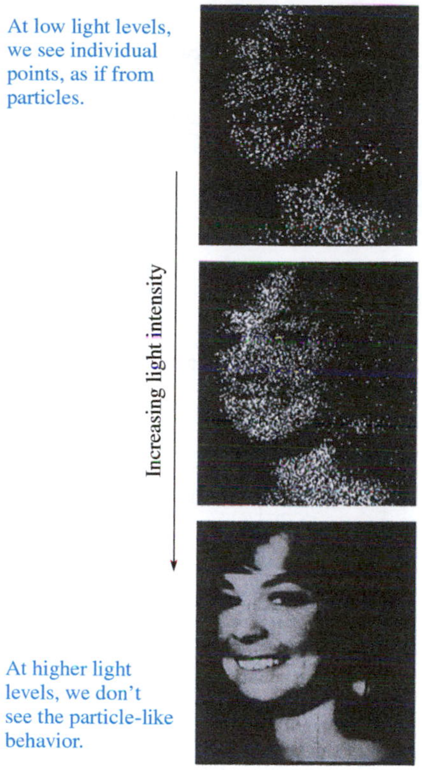

At low light levels, we see individual points, as if from particles.
Increasing light intensity
At higher light levels, we don't see the particle-like behavior.

This is not what we might expect. If light is a wave, reducing its intensity should cause the picture to grow dimmer and dimmer until it disappears, but the entire picture would remain present. Instead, the top image in Figure 25.32 looks as if someone randomly threw "pieces" of light at the detector, causing full exposure at some points but no exposure at others.

If we did not know that light is a wave, we would interpret the results of this experiment as evidence that light is a stream of some type of particle-like object. If these particles arrive frequently enough, they overwhelm the detector and it senses a steady "river" instead of the individual particles in the stream. Only at very low intensities do we become aware of the individual particles.

As we will see in Chapter 28, many experiments convincingly lead to the surprising result that **electromagnetic waves have a particle-like nature.** These particle-like components of electromagnetic waves are called **photons.**

The **photon model** of electromagnetic waves consists of three basic postulates:

1. Electromagnetic waves consist of discrete, massless units called photons. A photon travels in vacuum at the speed of light, $3.00 \times 10^8$ m/s.
2. Each photon has energy
$$E_{\text{photon}} = hf \tag{25.22}$$
where $f$ is the frequency of the wave and $h$ is a *universal constant* called **Planck's constant.** The value of Planck's constant is
$$h = 6.63 \times 10^{-34}\ \text{J} \cdot \text{s}$$
In other words, the electromagnetic waves come in discrete "chunks" of energy $hf$. The higher the frequency, the more energetic the chunks.
3. The superposition of a sufficiently large number of photons has the characteristics of a continuous electromagnetic wave.

### EXAMPLE 25.9 Finding the energy of a photon of visible light

550 nm is the approximate average wavelength of visible light.

a. What is the energy of a photon with a wavelength of 550 nm?
b. A 40 W incandescent lightbulb emits about 1 J of visible light energy every second. Estimate the number of visible light photons emitted per second.

**STRATEGIZE** The energy and frequency of a photon are related by Equation 25.22. Once we know the photon energy, we can find the number of photons it would take to have a total energy of 1 J.

**SOLVE** a. The frequency of the photon is
$$f = \frac{c}{\lambda} = \frac{3.00 \times 10^8\ \text{m/s}}{550 \times 10^{-9}\ \text{m}} = 5.45 \times 10^{14}\ \text{Hz}$$
Equation 25.22 gives us the energy of this photon:
$$E_{\text{photon}} = hf = (6.63 \times 10^{-34}\ \text{J} \cdot \text{s})(5.45 \times 10^{14}\ \text{Hz})$$
$$= 3.61 \times 10^{-19}\ \text{J}$$
This is an extremely small energy! In fact, photon energies are so small that they are usually measured in electron volts (eV) rather than joules. Recall that $1\ \text{eV} = 1.60 \times 10^{-19}$ J. With this, we find that the photon energy is
$$E_{\text{photon}} = 3.61 \times 10^{-19}\ \text{J} \times \frac{1\ \text{eV}}{1.60 \times 10^{-19}\ \text{J}} = 2.3\ \text{eV}$$

b. The photons emitted by a lightbulb span a range of energies, because the light spans a range of wavelengths, but the *average* photon energy corresponds to a wavelength near 550 nm. Thus we can estimate the number of photons in 1 J of light as
$$N \approx \frac{1\ \text{J}}{3.61 \times 10^{-19}\ \text{J/photon}} \approx 3 \times 10^{18}\ \text{photons}$$
A typical lightbulb emits about $3 \times 10^{18}$ photons every second.

**ASSESS** The number of photons emitted per second is staggeringly large. It's not surprising that in our everyday life we sense only the river and not the individual particles within the flow.

**TABLE 25.1** Energies of some atomic and molecular processes

| Process | Energy |
|---|---|
| Breaking a hydrogen bond between two water molecules | 0.24 eV |
| Energy released in metabolizing one molecule of ATP | 0.32 eV |
| Breaking the bond between atoms in a water molecule | 4.7 eV |
| Ionizing a hydrogen atom | 13.6 eV |

As we saw, a single photon of light at a wavelength of 550 nm has an energy of 2.3 eV. Let's look at just what 2.3 eV "buys" in interactions with atoms and molecules. **TABLE 25.1** lists some energies required for typical atomic and molecular processes. These values show that 2.3 eV is a significant amount of energy on an atomic scale. It is certainly enough to cause a molecular transformation (as it does in the sensory system of your eye), and photons with a bit more energy (higher frequency and thus shorter wavelength) can break a covalent bond. The photon model of light will be essential as we explore the interaction of electromagnetic waves with matter.

**STOP TO THINK 25.6** Two FM radio stations emit radio waves at frequencies of 90.5 MHz and 107.9 MHz. Each station emits the same total power. If you think of the radio waves as photons, which station emits the larger number of photons per second?

A. The 90.5 MHz station.
B. The 107.9 MHz station.
C. Both stations emit the same number of photons per second.

**FIGURE 25.33** The electromagnetic spectrum.

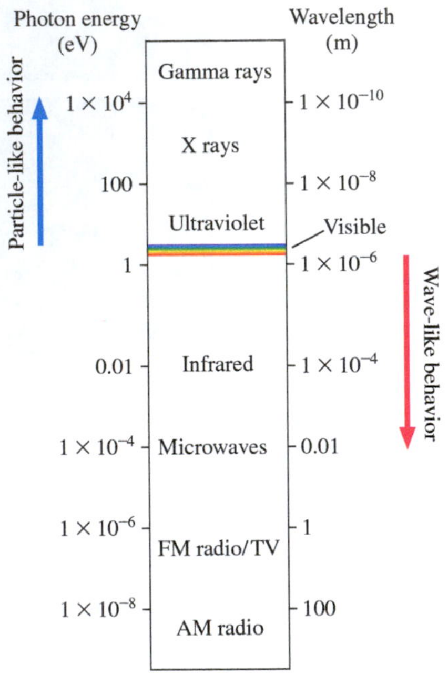


## 25.7 The Electromagnetic Spectrum

We have now seen two very different ways to look at electromagnetic waves: as oscillating waves of the electric and magnetic fields, and as particle-like units of the electromagnetic field called photons. This dual nature of electromagnetic waves is something we will discuss at length in Chapter 28. For now, we will note that each view is appropriate in certain circumstances. For example, we speak of radio *waves* but of x *rays*. The "ray" terminology tells us that x rays are generally better described as photons than as waves.

**FIGURE 25.33** shows the *electromagnetic spectrum* with photon energy (in eV) and wavelength (in m) scales. As you can see, electromagnetic waves span an extraordinarily wide range of wavelengths and energies. Radio waves have wavelengths of many meters but very low photon energies—only a few billionths of an eV. Because the photon energies are so small, radio waves are well described by Maxwell's theory of electromagnetic waves. At the other end of the spectrum, x rays and gamma rays have very short wavelengths and very high photon energies—large enough to ionize atoms and break molecular bonds. Consequently, x rays and gamma rays, although they do have wave-like characteristics, are best described as photons. Visible light is in the middle.

### Radio Waves and Microwaves

An electromagnetic wave is self-sustaining, needing no charges or currents to keep it propagating through space. However, charges and currents are needed at the *source* of an electromagnetic wave. Radio waves and microwaves are generally produced by the motion of charged particles in an antenna.

**FIGURE 25.34** The electric field of an oscillating dipole.

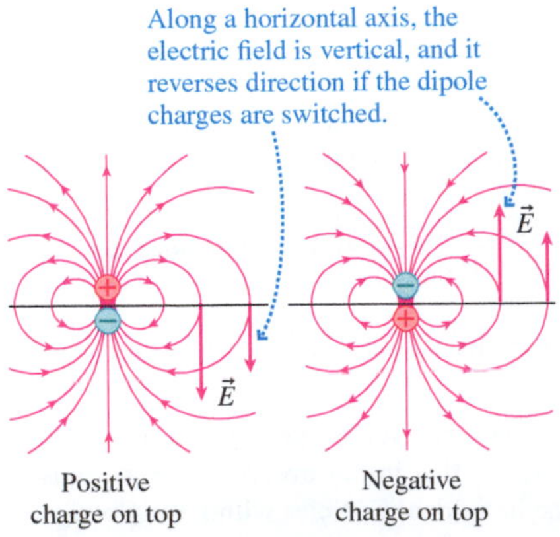


**FIGURE 25.34** reminds you what the electric field of an electric dipole looks like. If the dipole is vertical, the electric field $\vec{E}$ at points along the horizontal axis in the figure is also vertical. Reversing the dipole, by switching the charges, reverses $\vec{E}$. If the charges were to *oscillate* back and forth, switching position at frequency $f$, then $\vec{E}$ would oscillate in a vertical plane. The changing $\vec{E}$ would then create an induced magnetic field $\vec{B}$, which could then create an $\vec{E}$, which could then create a $\vec{B}$, . . . , and a vertically polarized electromagnetic wave at frequency $f$ would radiate out into space.

This is exactly what an **antenna** does. **FIGURE 25.35** shows two metal wires attached to the terminals of an oscillating voltage source. The figure shows an instant when the top wire is negative and the bottom is positive, but these will reverse in half a cycle. The wire is basically an oscillating dipole, and it creates an oscillating electric field. The oscillating $\vec{E}$ induces an oscillating $\vec{B}$, and they take off as an electromagnetic wave at speed $v_{em} = c$. The wave does need oscillating charges as a *wave source*, but once created it is self-sustaining and independent of the source.

Radio waves are *detected* by antennas as well. The electric field of the radio wave drives a current up and down a conductor, producing a potential difference that can be amplified. For best reception, the antenna length should be about $\frac{1}{4}$ of a wavelength. A typical cell phone works at 1.9 GHz, with wavelength $\lambda = c/f = 16$ cm. Thus a cell phone antenna should be about 4 cm long, or about $1\frac{1}{2}$ inches; it is generally hidden inside the phone itself.

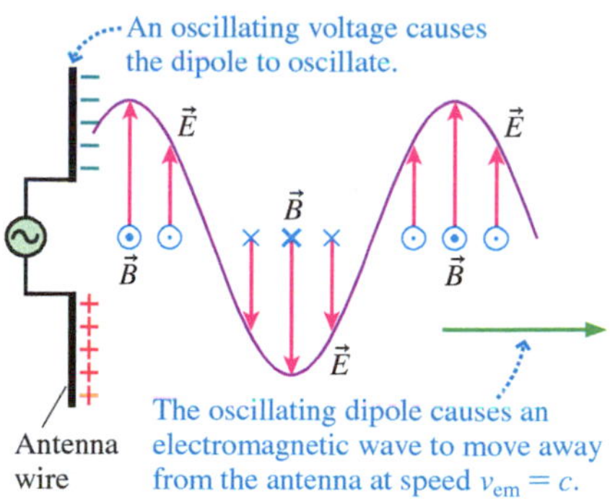


**FIGURE 25.35** An antenna generates a self-sustaining electromagnetic wave.

**CONCEPTUAL EXAMPLE 25.10 Wildlife tracking** BIO

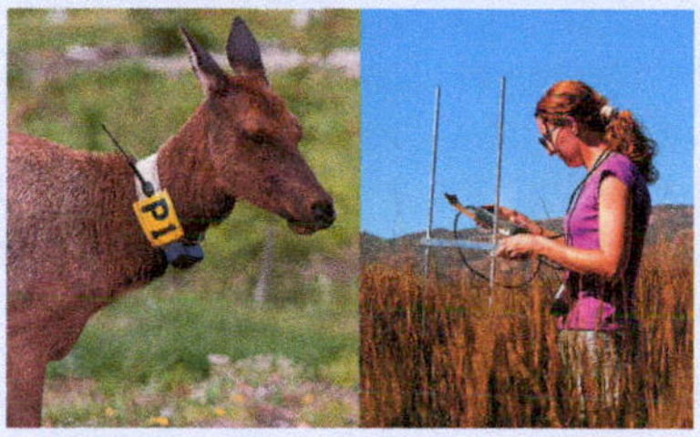

The elk shown in the left photo wears a radio collar with a vertical broadcast antenna. A wildlife biologist can search for the signal from the elk's collar by using a receiving antenna. How should the receiving antenna be oriented?

***REASON*** We have seen that an antenna generates a wave that is polarized in the same direction as the antenna itself. As the photo on the right shows, the receiving antenna should be oriented in the direction of the wave's polarization—here, vertically—so that the wave's electric field can drive charges up and down the antenna, creating a measurable potential difference. If the receiving antenna were horizontal, charges would be driven only along the tiny diameter of the antenna wire, creating no potential difference along its length.

***ASSESS*** In reality, tracked animals are always moving, and their antennas may change orientation. To obtain the largest signal, the biologist must adjust the receiving antenna to match.

In materials with no free charges, the electric fields of radio waves and microwaves can still interact with matter by exerting a torque on molecules, such as water, that have a permanent electric dipole moment, as shown in **FIGURE 25.36**. The molecules acquire kinetic energy from the wave, then their collisions with other molecules transform that energy into thermal energy, increasing the temperature.

This is how a microwave oven heats food. Water molecules, with their large dipole moment, rotate in response to the electric field of the microwaves, then transfer this energy to the food via molecular collisions.

**FIGURE 25.36** A radio wave interacts with matter.

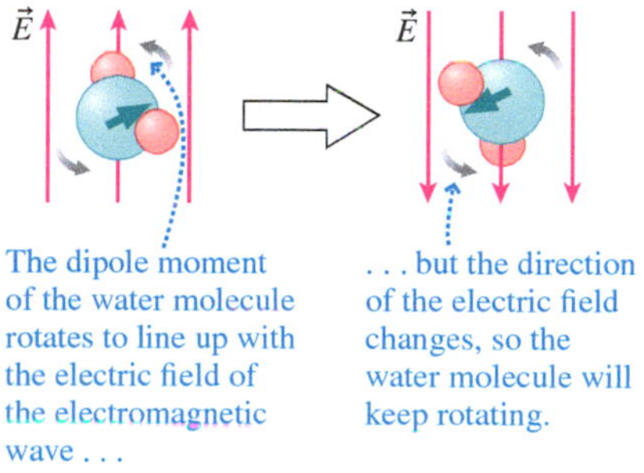


## Infrared, Visible Light, and Ultraviolet

Radio waves can be produced by oscillating charges in an antenna. At the higher frequencies of infrared, visible light, and ultraviolet, the "antennas" are individual atoms. This portion of the electromagnetic spectrum is *atomic radiation.*

Nearly all the atomic radiation in our environment is *thermal radiation* due to the thermal motion of the atoms in an object. As we saw in Chapter 12, thermal radiation—a form of heat transfer—is described by Stefan's law: If heat energy $Q$ is radiated in a time interval $\Delta t$ by an object with surface area $A$ and absolute temperature $T$, the *rate* of heat transfer $Q/\Delta t$ (joules per second) is

$$\frac{Q}{\Delta t} = e\sigma A T^4 \tag{25.23}$$

The constant $e$ in this equation is the object's emissivity, a measure of its effectiveness at emitting electromagnetic waves, and $\sigma$ is the Stefan-Boltzmann constant, $\sigma = 5.67 \times 10^{-8}\ \text{W/(m}^2 \cdot \text{K}^4)$.

In Chapter 12 we considered the amount of energy radiated and its dependence on temperature. The filament of an incandescent bulb glows simply because it is hot. If you increase the current through a lightbulb filament, the filament temperature increases and so does the total energy emitted by the bulb, in accordance with Stefan's law. The three pictures in **FIGURE 25.37** show a glowing lightbulb with the

**Video** Microwaves

**FIGURE 25.37** The brightness of the bulb varies with the temperature of the filament.

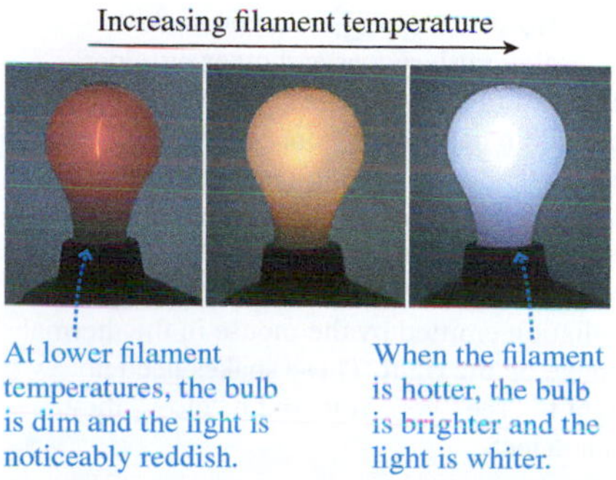

Increasing filament temperature
At lower filament temperatures, the bulb is dim and the light is noticeably reddish.
When the filament is hotter, the bulb is brighter and the light is whiter.

filament at successively higher temperatures. We can clearly see an increase in brightness in the sequence of three photographs.

But it's not just the brightness that varies. The *color* of the emitted radiation changes as well. At low temperatures, the light from the bulb is quite red. Looking at the change in color as the temperature of the bulb rises in Figure 25.37, we see that **the spectrum of thermal radiation changes with temperature.**

If we measure the intensity of thermal radiation as a function of wavelength for an object at three temperatures, 3500 K, 4500 K, and 5500 K, the data appear as in FIGURE 25.38. Notice two important features:

- Increasing the temperature increases the intensity at all wavelengths. **Making the object hotter causes it to emit more radiation across the entire spectrum.**
- Increasing the temperature causes the peak intensity to shift to a shorter wavelength. **The higher the temperature, the shorter the wavelength of the peak of the spectrum.**

**FIGURE 25.38** A thermal emission spectrum depends on the temperature.

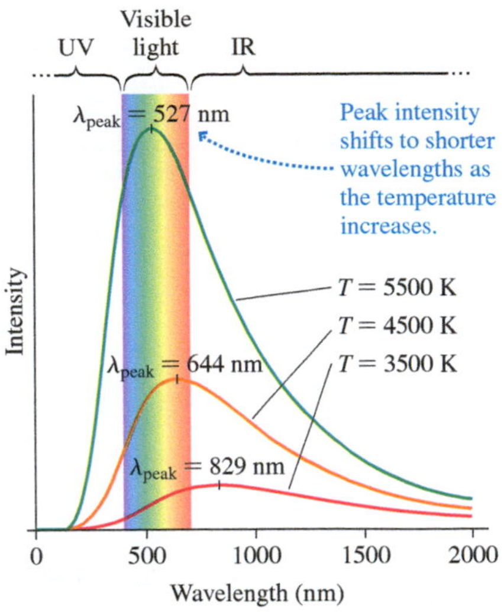


It is this variation of the peak wavelength that causes the change in color of the glowing filament in Figure 25.37. The temperature dependence of the peak wavelength of thermal radiation is known as *Wien's law*:

$$\lambda_{peak}\,(\text{in nm}) = \frac{2.9 \times 10^6\ \text{nm}\cdot\text{K}}{T\,(\text{in K})} \qquad (25.24)$$

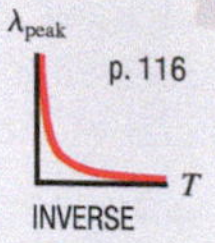


Wien's law for the peak wavelength of a thermal emission spectrum

**EXAMPLE 25.11 Finding peak wavelengths of radiating objects**

What are the wavelengths of peak intensity and the corresponding spectral regions for radiating objects at (a) the normal human body temperature of 37°C, (b) the temperature of the filament in an incandescent lamp, 1500°C, and (c) the temperature of the surface of the sun, 5800 K?

***STRATEGIZE*** All of the objects emit thermal radiation, so the peak wavelengths are given by Equation 25.24.

***PREPARE*** We need to convert temperatures to kelvin. The temperature of the human body is $T = 37 + 273 = 310$ K, and the filament temperature is $T = 1500 + 273 = 1773$ K.

***SOLVE*** We use Equation 25.24 to find the wavelengths of peak intensity:

$$\text{a. } \lambda_{peak}(\text{body}) = \frac{2.9 \times 10^6\ \text{nm}\cdot\text{K}}{310\ \text{K}} = 9.4 \times 10^3\ \text{nm} = 9.4\ \mu\text{m}$$

$$\text{b. } \lambda_{peak}(\text{filament}) = \frac{2.9 \times 10^6\ \text{nm}\cdot\text{K}}{1773\ \text{K}} = 1600\ \text{nm}$$

$$\text{c. } \lambda_{peak}(\text{sun}) = \frac{2.9 \times 10^6\ \text{nm}\cdot\text{K}}{5800\ \text{K}} = 500\ \text{nm}$$

The peak of the emission curve at body temperature is far into the infrared region of the spectrum, well below the range of sensitivity of human vision. You don't see someone "glow," although people do indeed emit significant energy in the form of electromagnetic waves, as we saw in Chapter 12. The sun's emission peaks right in the middle of the visible spectrum, which seems reasonable. Interestingly, most of the energy radiated by an incandescent bulb is *not* visible light. The tail of the emission curve extends into the visible region, but the peak of the emission curve—and most of the emitted energy—is in the infrared region of the spectrum. A 100 W bulb emits only a few watts of visible light. This inefficiency of incandescent lamps is why they are rapidly being replaced by high-efficiency fluorescent and LED lamps.

***ASSESS*** The temperature spans a wide range of kelvin temperatures; it makes sense that the wavelengths span an equally wide range.

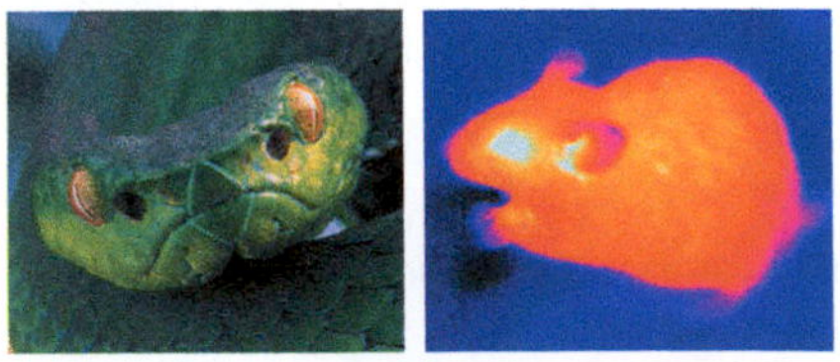

**It's the pits . . .** BIO Certain snakes—including rattlesnakes and other pit vipers—can hunt in total darkness. The viper in the left photo has pits in front of its eyes. These are a second set of vision organs; they have sensitive tissue at the bottom that allows them to detect the thermal radiation emitted by warm-blooded prey, such as the thermal radiation emitted by the mouse in the thermal image on the right. These snakes need no light to "see" you. You emit a "glow" they can detect.

Infrared radiation, with its relatively long wavelength and low photon energy, produces effects in tissue similar to those of microwaves—heating—but the penetration is much less than for microwaves. Infrared is absorbed mostly by the top layer of your skin and simply warms you up, as you know from sitting in the sun or under a heat lamp. The wave picture is generally most appropriate for infrared.

In contrast, ultraviolet photons have enough energy to interact with molecules in entirely different ways, ionizing molecules and breaking molecular bonds. The cells in skin are altered by ultraviolet radiation, causing sun tanning and sun burning. DNA molecules can be permanently damaged by ultraviolet radiation. There is a sharp threshold for such damage at 290 nm (corresponding to 4.3 eV photon energy). At longer wavelengths, damage to cells is slight; at shorter wavelengths, it can be extensive. The interactions of ultraviolet radiation with matter are best understood from the photon perspective, with the absorption of each photon being associated with a particular molecular event.

Visible light is at a transition point in the electromagnetic spectrum. Your studies of wave optics in Chapter 17 showed you that light has a wave nature. At the same time, the energy of photons of visible light is large enough to cause molecular transitions—which is how your eye detects light. When we work with visible light, we will often move back and forth between the wave and photon models.

**EXAMPLE 25.12 Seeing (infra)red BIO**

Normally, wavelengths longer than about 750 nm—in the infrared—are invisible to the human eye. However, recent experiments have shown that infrared light is visible under certain conditions but, surprisingly, often appears to be green or yellow. The effect occurs when *two* infrared photons simultaneously excite the same photoreceptor in the eye. The photoreceptor behaves as if it had absorbed a *single* photon with an energy equal to the sum of the two infrared photon energies. If, when observing infrared light, a person perceives 500 nm green light, what is the wavelength of the infrared light?

**STRATEGIZE** We will start by finding the energy of a photon of green light. The energy of each infrared photon is then half this value. From this infrared photon energy we can find the infrared wavelength.

**PREPARE** The frequency of an electromagnetic wave is related to its photon energy by Equation 25.22; the frequency and wavelength are related by $\lambda = c/f$.

**SOLVE** The energy of the perceived green photon is

$$E_{\text{green}} = hf = \frac{hc}{\lambda} = \frac{(6.63\times10^{-34}\ \text{J}\cdot\text{s})(3.00\times10^{8}\ \text{m/s})}{500\times10^{-9}\ \text{m}}$$
$$= 3.98\times10^{-19}\ \text{J}$$

The energy of each infrared photon is then

$$E_{\text{IR}} = \frac{3.98\times10^{-19}\ \text{J}}{2} = 1.99\times10^{-19}\ \text{J}$$

from which the infrared wavelength is

$$\lambda_{\text{IR}} = \frac{hc}{E_{\text{IR}}} = \frac{(6.63\times10^{-34}\ \text{J}\cdot\text{s})(3.00\times10^{8}\ \text{m/s})}{1.99\times10^{-19}\ \text{J}} = 1000\ \text{nm}$$

**ASSESS** This wavelength is well into the infrared region and would not normally be visible. We see that the infrared wavelength is twice that of the perceived green light, which makes sense—photon energy and wavelength are inversely related, so the infrared photons, with half the energy, will have twice the wavelength.

## Color Vision

The cones, the color-sensitive cells in the retina of the eye, each contain one of three slightly different forms of a light-sensitive photopigment. A single photon of light can trigger a reaction in a photopigment molecule, which ultimately leads to a signal being produced by a cell in the retina. The energy of the photon must be matched to the energy of a molecular transition for the photon energy to be absorbed. Each photopigment has a range of photon energies to which it is sensitive. Our color vision is a result of the differential response of three types of cones that contain three slightly different pigments, shown in FIGURE 25.39.

Humans have three types of cone cells in the eye, mice have two, and chickens four—giving a chicken keener color vision than a human. The three color photopigments that bees possess give them excellent color vision, but a bee's color sense is different from a human's. The peak sensitivities of a bee's photopigments are in the yellow, blue, and ultraviolet regions of the spectrum. A bee can't see the red of a rose, but it is quite sensitive to ultraviolet wavelengths well beyond the range of human vision. The flower in the right-hand photo at the start of the chapter looks pretty to us, but its coloration is really intended for other eyes. The ring of ultraviolet-absorbing pigments near the center of the flower, which is invisible to humans, helps bees zero in on the pollen.

**FIGURE 25.39** The sensitivity of different cone cells in the human eye.

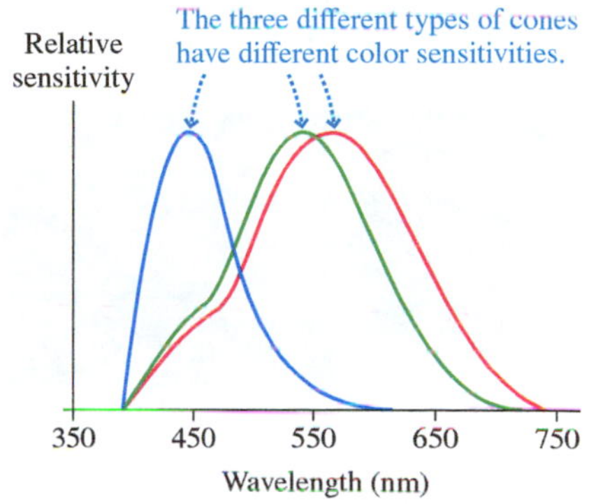

## X Rays and Gamma Rays

FIGURE 25.40 A simple x-ray tube.

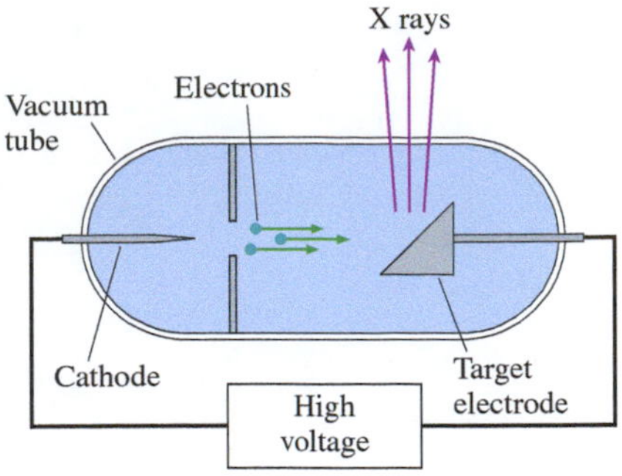


At the highest energies of the electromagnetic spectrum we find x rays and gamma rays. There is no sharp dividing line between these two regions of the spectrum; the difference is the source of radiation. High-energy photons emitted by electrons are called x rays. If the source is a nuclear process, we call them gamma rays.

We will look at the emission of x rays in atomic processes and gamma rays in nuclear processes in Part VII. For now, we will focus on the "artificial" production of x rays in an x-ray tube, such as the one shown in FIGURE 25.40. Electrons are emitted from a cathode and accelerated to a kinetic energy of several thousand eV by the electric field between two electrodes connected to a high-voltage power supply. The electrons make a sudden stop when they hit a metal target electrode. The rapid deceleration of an electron can cause the emission of a single photon with a significant fraction of the electron's kinetic energy. These photons, with energies well in excess of 1000 eV, are x rays. The x rays pass through a window in the tube and then may be used to produce an image or treat a disease.

### EXAMPLE 25.13 Determining x-ray energies in a medical x-ray tube

An x-ray tube used for medical work has an accelerating voltage of 30 kV. What is the maximum energy of an x-ray photon that can be produced in this tube? What is the wavelength of this x ray?

**STRATEGIZE** An electron accelerated through a potential difference of 30 kV acquires a kinetic energy of 30 keV. When this electron hits the metal target and stops, energy may be converted to an x ray. The maximum energy that could be converted is 30 keV, so this is the maximum possible energy of an x-ray photon from the tube. Once we know the photon's energy *E*, its frequency is *E/h*, from which we can find its wavelength.

**SOLVE** We find the maximum x-ray energy in joules as

$$E = 30 \times 10^3 \text{ eV} \times \frac{1.60 \times 10^{-19}\text{ J}}{1 \text{ eV}} = 4.8 \times 10^{-15}\text{ J}$$

For electromagnetic waves, $c = \lambda f$, so we can calculate

$$\lambda = \frac{c}{f} = \frac{c}{E/h} = \frac{hc}{E} = \frac{(6.63 \times 10^{-34}\text{ J}\cdot\text{s})(3.00 \times 10^8\text{ m/s})}{4.8 \times 10^{-15}\text{ J}}$$
$$= 4.1 \times 10^{-11}\text{ m} = 0.041\text{ nm}$$

**ASSESS** This is a very short wavelength, comparable to the spacing between atoms in a solid.

X rays and gamma rays (and the short-wavelength part of the ultraviolet spectrum) are **ionizing radiation;** the individual photons have sufficient energy to ionize atoms. When such radiation strikes tissue, the resulting ionization can produce cellular damage. When people speak of "radiation" they often mean "ionizing radiation."

At several points in this chapter we have hinted at places where a full understanding of the phenomena requires some new physics. We have used the photon model of electromagnetic waves, and we have mentioned that nuclear processes can give rise to gamma rays. There are other questions that we did not raise, such as why the electromagnetic spectrum of a hot object has the shape that it does. These puzzles began to arise in the late 1800s and early 1900s, and it soon became clear that the physics of Newton and Maxwell was not sufficient to fully describe the nature of matter and energy. Some new rules, some new models, were needed. We will return to these puzzles as we begin to explore the exciting notions of quantum physics in Part VII.

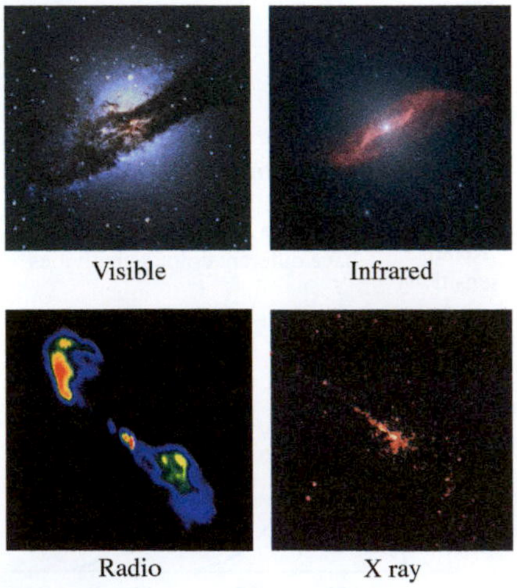


◀ **Seeing the universe in a different light** These four images of the Centaurus A galaxy have the same magnification and orientation, but they are records of different types of electromagnetic waves. (All but the visible-light image are false-color images.) The visible-light image shows a dark dust lane cutting across the galaxy. In the infrared, this dust lane glows quite brightly—telling us that the dust particles are hot. The radio and x-ray images show jets of matter streaming out of the galaxy's center, hinting at the presence of a massive black hole. Views of the cosmos beyond the visible range are important tools of modern astronomy.

**STOP TO THINK 25.7** A group of four stars, all the same size, have the four different surface temperatures given below. Which of these stars emits the most red light?

A. 3000 K B. 4000 K C. 5000 K D. 6000 K

## INTEGRATED EXAMPLE 25.14 Space circuits

The very upper part of the atmosphere, where the space shuttle orbits, is called the *ionosphere*. The few atoms and molecules that remain at this altitude are mostly ionized by intense ultraviolet radiation from the sun. The thin gas of the ionosphere thus consists largely of positive ions and negative electrons, so it can carry an electric current. This was crucial to the operation of the tethered satellite system that was tested on the space shuttle in the 1990s. As described in the chapter, a probe was deployed on a conducting wire that tethered the probe at a great distance from the craft. As the space shuttle orbited, the wire moved through the earth's magnetic field, creating a potential difference between the ends of the wire. Charge flowing through the ionosphere back to the shuttle created a complete circuit.

In the final test of the tethered satellite system, a potential difference of 3500 V was generated across 20 km of cable as the shuttle orbited at 7800 m/s.

a. To produce the noted potential difference, what was the component of the magnetic field perpendicular to the wire?
b. The 3500 V potential created a current of 480 mA in the ionosphere. What was the total resistance of the circuit thus formed?
c. How much power was dissipated in this circuit?
d. What is the drag force on the wire due to its motion in the earth's magnetic field?
e. The ionization of the upper atmosphere is due to solar radiation at wavelengths of 95 nm and shorter. In what part of the spectrum is this radiation? What is the lowest-energy photon, in eV, that contributes to the ionization?

**STRATEGIZE** The motion of the wire connecting the tethered satellite to the space shuttle leads to a motional emf that drives a current through this wire and back through the ionosphere.

**PREPARE** FIGURE 25.41 shows how we can model this process as an electric circuit. We know the voltage and the current in this circuit, so we can find the resistance and the power. Because the wire carries a current, the earth's field will exert a force on it, which is the drag force we are asked to find.

FIGURE 25.41 The tethered satellite circuit.

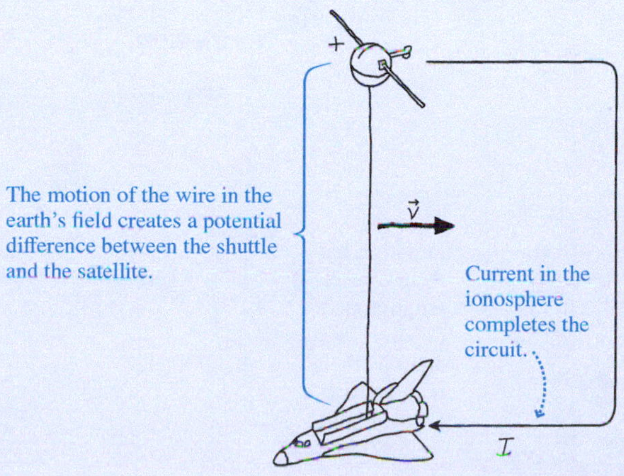


**SOLVE** a. We know the magnitude of the velocity and the length of the wire, so we can use the equation for the motional emf, Equation 25.3, to find the magnitude of the component of the field perpendicular to the wire:

$$B = \frac{\mathcal{E}}{vl} = \frac{3500\text{ V}}{(7800\text{ m/s})(20 \times 10^3\text{ m})} = 2.24 \times 10^{-5}\text{ T}$$

To two significant figures, this is $B = 22\ \mu\text{T}$.

b. The 3500 V potential difference produced a current of 480 mA. From Ohm's law, the resistance of the circuit was thus

$$R = \frac{\Delta V}{I} = \frac{3500\text{ V}}{480 \times 10^{-3}\text{ A}} = 7300\ \Omega$$

c. We know the voltage and the current, so we can compute the power:

$$P = I\,\Delta V = (0.48\text{ A})(3500\text{ V}) = 1700\text{ W}$$

d. The component of the magnetic field perpendicular to the current-carrying wire exerts a drag force on the wire. We learned in Chapter 24 that the force on a current-carrying wire is $F = IlB$. Thus

$$F = IlB = (0.48\text{ A})(20 \times 10^3\text{ m})(2.24 \times 10^{-5}\text{ T}) = 0.22\text{ N}$$

e. Radiation with wavelengths of 95 nm and shorter is in the ultraviolet region of the spectrum. The lowest-energy photon in this region has the lowest frequency and thus the longest wavelength—namely, 95 nm. The frequency is

$$f = \frac{c}{\lambda} = \frac{3.0 \times 10^8\text{ m/s}}{95 \times 10^{-9}\text{ m}} = 3.2 \times 10^{15}\text{ Hz}$$

The photon energy is then given by Equation 25.22:

$$E_{\text{photon}} = hf = (6.63 \times 10^{-34}\text{ J}\cdot\text{s})(3.2 \times 10^{15}\text{ Hz}) = 2.1 \times 10^{-18}\text{ J}$$

Converting to eV, we find

$$E_{\text{photon}} = 2.1 \times 10^{-18}\text{ J} \times \frac{1\text{ eV}}{1.6 \times 10^{-19}\text{ J}} = 13\text{ eV}$$

**ASSESS** We have many good chances to check our work to verify that it makes sense. First, the field component that we calculate is about half the value we typically use for the earth's field, which seems reasonable—we'd be suspicious if the field we calculated was more than the earth's field.

The product of the drag force and the speed is the power dissipated by the drag force:

$$P = Fv = (0.22\text{ N})(7800\text{ m/s}) = 1700\text{ W}$$

This is exactly what we found for the electric power dissipated in the circuit, a good check on our work. The two values must be equal, as they are.

A final check on our work is the value we calculate for the photon energy. Table 25.1 shows that it takes about 13 eV to ionize a hydrogen atom. Photons with wavelengths shorter than 95 nm are able to ionize hydrogen atoms, so it seems likely they would also ionize the nitrogen and oxygen molecules of the upper atmosphere.

# SUMMARY

**GOAL** To understand the nature of electromagnetic induction and electromagnetic waves.

## GENERAL PRINCIPLES

### Electromagnetic Induction

The **magnetic flux** measures the amount of magnetic field passing through a surface:

$$\Phi = AB\cos\theta$$

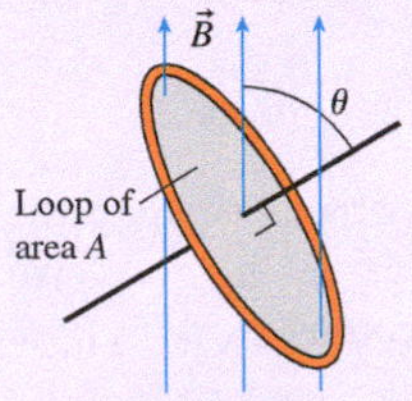


**Lenz's law** specifies that there is an induced current in a closed conducting loop if the magnetic flux through the loop is changing. The direction of the induced current is such that the induced magnetic field opposes the *change* in flux.

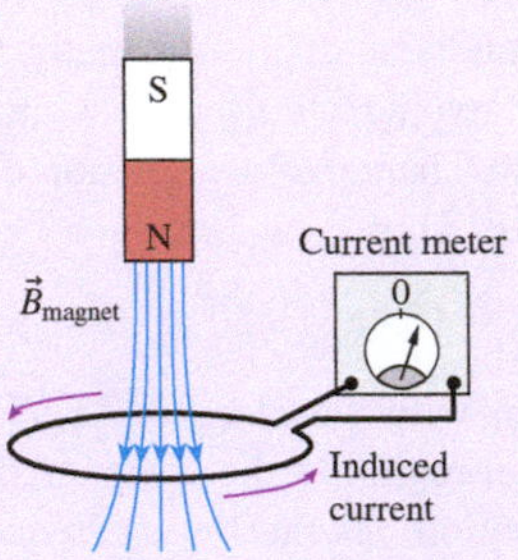


**Faraday's law** specifies the magnitude of the induced emf in a closed loop:

$$\mathcal{E} = \left|\frac{\Delta\Phi}{\Delta t}\right|$$

Multiply by $N$ for an $N$-turn coil.

The size of the induced current is

$$I = \frac{\mathcal{E}}{R}$$

### Electromagnetic Waves

An electromagnetic wave is a self-sustaining oscillation of electric and magnetic fields.

- The wave is a transverse wave with $\vec{E}$, $\vec{B}$, and $\vec{v}$ mutually perpendicular.
- The wave propagates with speed

$$v_{em} = c = \frac{1}{\sqrt{\epsilon_0\mu_0}} = 3.00 \times 10^8 \text{ m/s}$$

- The wavelength, frequency, and speed are related by

$$c = \lambda f$$

- The amplitudes of the fields are related by

$$\frac{E_0}{B_0} = c$$

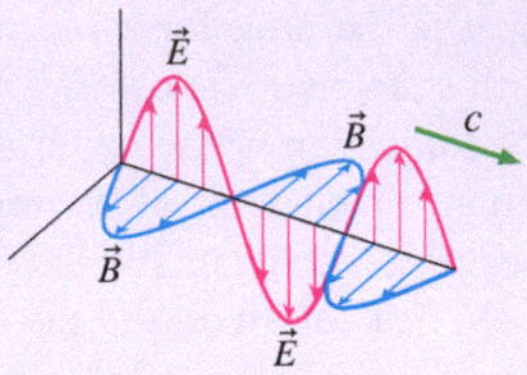


## IMPORTANT CONCEPTS

### Motional emf

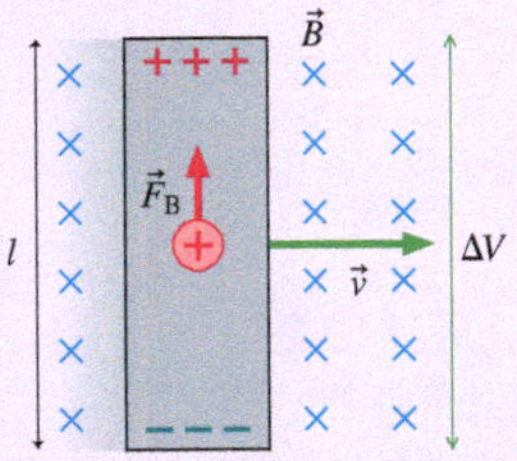


The motion of a conductor through a magnetic field produces a force on the charges. The separation of charges leads to an emf:

$$\mathcal{E} = vlB$$

### The photon model

Electromagnetic waves appear to be made of discrete units called photons. The energy of a photon of frequency $f$ is

$$E = hf$$

This photon view becomes increasingly important as the photon energy increases.

### The electromagnetic spectrum

Electromagnetic waves come in a wide range of wavelengths and photon energies.

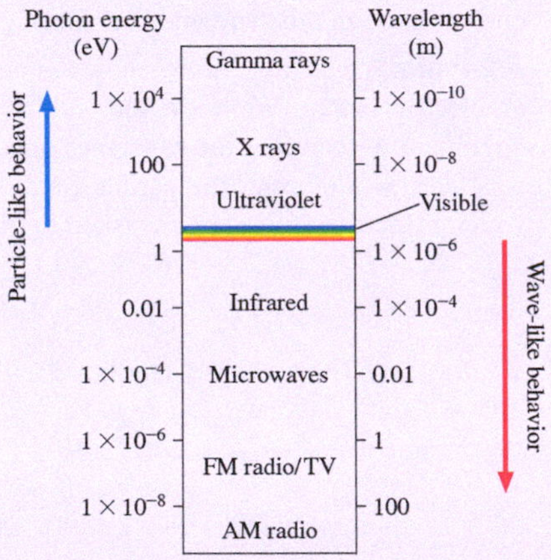


## APPLICATIONS

A changing flux in a solid conductor creates **eddy currents.**

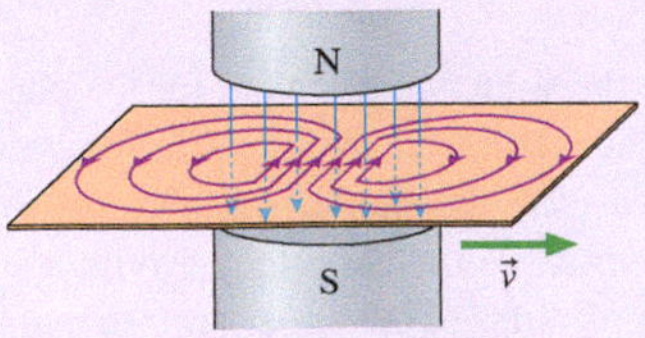


The plane of the electric field of an electromagnetic wave defines its **polarization.** The intensity of polarized light transmitted through a polarizing filter is given by **Malus's law:**

$$I = I_0\cos^2\theta$$

where $\theta$ is the angle between the electric field and the polarizer axis.

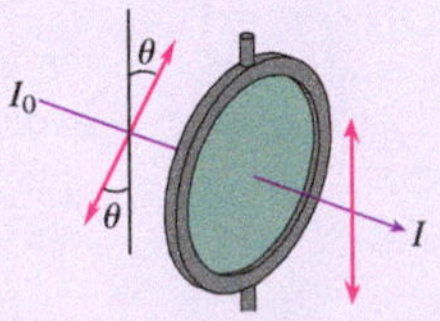


Thermal radiation has a peak wavelength that depends on an object's temperature according to **Wien's law:**

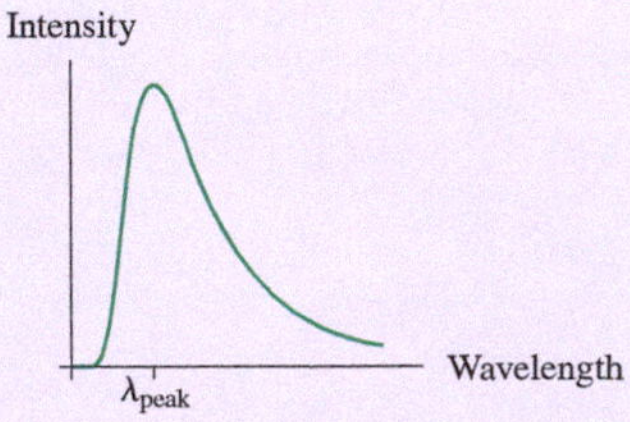


$$\lambda_{peak} \text{ (in nm)} = \frac{2.9 \times 10^6 \text{ nm} \cdot \text{K}}{T}$$

## Learning Objectives After studying this chapter, you should be able to:

- Calculate the motional emf of a conductor moving in a magnetic field. *Conceptual Questions 25.2, 25.18; Problems 25.1, 25.2, 25.3, 25.4, 25.8*
- Calculate the magnetic flux through a loop. *Conceptual Questions 25.4, 25.5; Problems 25.9, 25.10, 25.11, 25.12*
- Use Lenz's law to find the direction of an induced current. *Conceptual Questions 25.3, 25.9; Problems 25.13, 25.14*
- Use Faraday's law to calculate the induced emf and induced current of a changing magnetic flux. *Conceptual Questions 25.1, 25.8; Problems 25.15, 25.17, 25.20, 25.22, 25.24*
- Understand electromagnetic waves and their properties. *Conceptual Questions 25.31, 25.33; Problems 25.25, 25.26, 25.27, 25.29, 25.33*
- Work with polarization and polarizers. *Conceptual Questions 25.19, 25.21; Problems 25.35, 25.37, 25.38, 25.39*
- Apply the wave and photon models of light to the electromagnetic spectrum. *Conceptual Questions 25.22, 25.23; Problems 25.43, 25.46, 25.47, 25.50, 25.53*

### STOP TO THINK ANSWERS

**Chapter Preview Stop to Think: B.** We can rewrite the relationship given in the Looking Back box as $\lambda = c/f$. Lower frequencies correspond to longer wavelengths, so the 1.9 GHz cell phone waves have a longer wavelength than the 2.4 GHz oven waves.

**Stop to Think 25.1: E.** According to the right-hand rule, the magnetic force on a positive charge carrier is to the right.

**Stop To Think 25.2: A, C, B.** The magnetic field is directed upward out of the north pole of the magnet, so:

- As the magnet moves toward the coil, the upward flux through the loop increases. Opposing this change requires an induced downward field, meaning a clockwise current, choice A.
- The magnet isn't moving, so there is no change in the flux and no induced current, choice C.
- As the magnet moves away from the coil, the upward flux through the loop decreases. Opposing this change requires an induced upward field, meaning a counterclockwise current, choice B.

**Stop to Think 25.3: D.** The field of the bar magnet emerges from the north pole and points upward. As the coil moves toward the pole, the flux through it is upward and increasing. To oppose the increase, the induced field must point downward. This requires a clockwise (negative) current. As the coil moves away from the pole, the upward flux is decreasing. To oppose the decrease, the induced field must point upward. This requires a counterclockwise (positive) current.

**Stop to Think 25.4: A, B, E.** The induced emf is equal to the rate of change of the flux; the current is proportional to this emf and inversely proportional to the resistance. Increasing the rate of change of the field will increase the rate of change of the flux. If the loop is larger, the flux will be greater, and therefore the rate of change of the flux will also be greater. And a given emf will induce a larger current in the loop if the resistance is decreased.

**Stop to Think 25.5: $I_D > I_A > I_B = I_C$.** The intensity depends upon $\cos^2\theta$, where $\theta$ is the angle *between* the axes of the two filters. The filters in D have $\theta = 0°$, so all light is transmitted. The two filters in both B and C are crossed ($\theta = 90°$) and transmit no light at all.

**Stop to Think 25.6: A.** The photon energy is proportional to the frequency. The photons of the 90.5 MHz station each have lower energy, so more photons must be emitted per second.

**Stop to Think 25.7: D.** A hotter object emits more radiation across the *entire* spectrum than a cooler object. The 6000 K star has its maximum intensity in the blue region of the spectrum, but it still emits more red radiation than the somewhat cooler stars.

Video Tutor Solution Chapter 25

# QUESTIONS

## Conceptual Questions

1. The magnetic flux through a loop of wire is zero. Can there be an induced current in the loop at this instant? Explain.

2. An inventor designs a model plane that uses the emf that develops across its metal wings as it flies through the earth's magnetic field to power the plane's propeller. Will this plane be able to power itself this way? Explain.

Problem difficulty is labeled as | (straightforward) to ||||| (challenging). Problems labeled INT integrate significant material from earlier chapters; Problems labeled BIO are of biological or medical interest.

The eText icon indicates when there is a video tutor solution available for the chapter or for a specific problem. To launch these videos, log into your eText through Mastering™ Physics or log into the Study Area.

3. Parts a through f of Figure Q25.3 show one or more metal wires sliding on fixed metal rails in a magnetic field. For each, determine if the induced current is clockwise, counterclockwise, or zero.

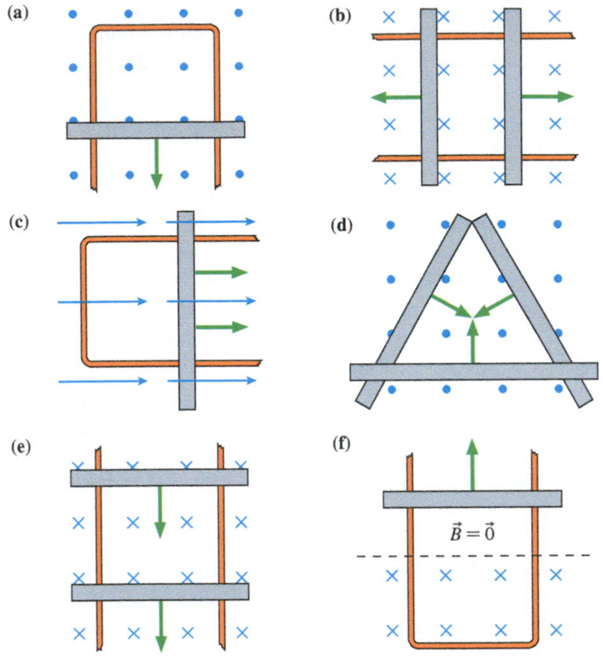


FIGURE Q25.3

4. Figure Q25.4 shows four different loops in a magnetic field. The numbers indicate the lengths of the sides and the strength of the field. Rank in order the magnetic fluxes $\Phi_1$ through $\Phi_4$, from the largest to the smallest. Some may be equal. Explain.

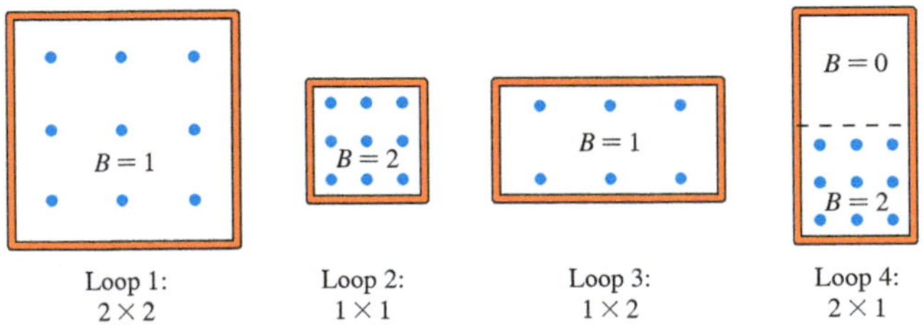


FIGURE Q25.4

5. Figure Q25.5 shows four different circular loops that are perpendicular to the page. The radius of loops 3 and 4 is twice that of loops 1 and 2. The magnetic field is the same for each. Rank in order the magnetic fluxes $\Phi_1$ through $\Phi_4$, from the largest to the smallest. Some may be equal. Explain.

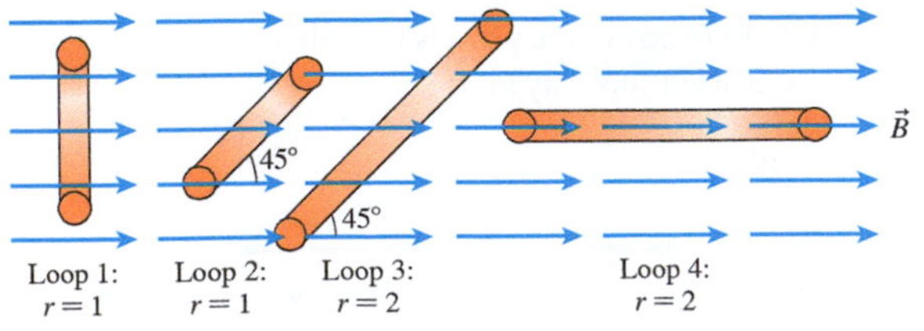


FIGURE Q25.5

6. A circular loop rotates at constant speed about an axle through the center of the loop. Figure Q25.6 shows an edge view and defines the angle $\phi$, which increases from 0° to 360° as the loop rotates.
   a. At what angle or angles is the magnetic flux a maximum?
   b. At what angle or angles is the magnetic flux a minimum?
   c. At what angle or angles is the magnetic flux *changing* most rapidly?

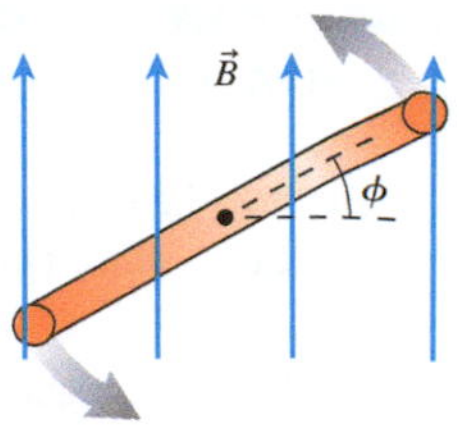


FIGURE Q25.6

7. The power lines that run through your neighborhood carry *alternating currents* that reverse direction 120 times per second. As the current changes, so does the magnetic field around a line. Suppose you wanted to put a loop of wire up near the power line to extract power by "tapping" the magnetic field. Sketch a picture of how you would orient the coil of wire next to a power line to develop the maximum emf in the coil. (Note that this is dangerous and illegal, and not something you should try.)

8. The magnetic flux passing through a coil of wire varies as shown in Figure Q25.8. During which time interval(s) will an induced current be present in the coil? Explain.

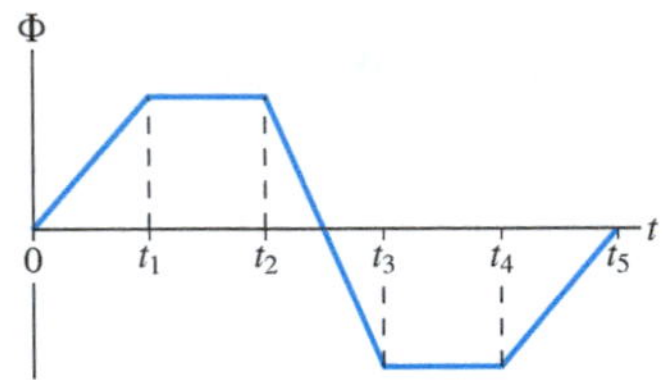


FIGURE Q25.8

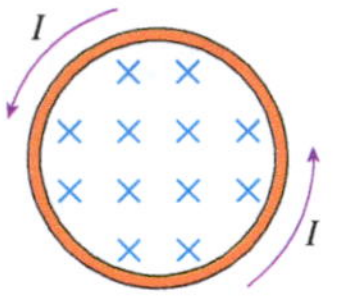


FIGURE Q25.9

9. There is a counterclockwise induced current in the conducting loop shown in Figure Q25.9. Is the magnetic field inside the loop increasing in strength, decreasing in strength, or steady?

10. Many roller coasters use magnetic brakes to slow the train at the end of its run. A conducting plate on the train passes by powerful magnets fixed on the track.
    a. Explain the principle behind this magnetic braking.
    b. The system needs an ordinary friction-based brake to bring the train to a full stop. Explain why the magnetic brake is not very efficient when the train is moving slowly.

11. The conducting loop in Figure Q25.11 is moving into the region between the magnetic poles shown.
    a. Is the induced current (viewed from above) clockwise or counterclockwise?
    b. Is there an attractive magnetic force that tends to pull the loop in, like a magnet pulls on a paper clip? Or do you need to push the loop in against a repulsive force?

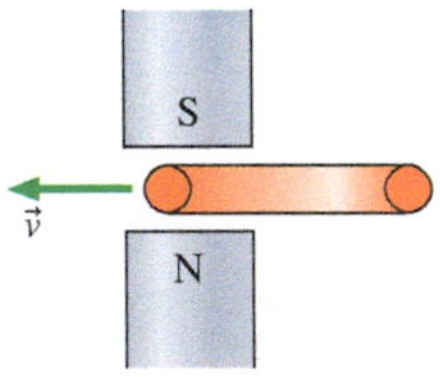


FIGURE Q25.11

12. Figure Q25.12 shows two concentric, conducting loops. We will define a counterclockwise current (viewed from above) to be positive, a clockwise current to be negative. The graph shows the current in the outer loop as a function of time. Sketch a graph that shows the induced current in the inner loop. Explain.

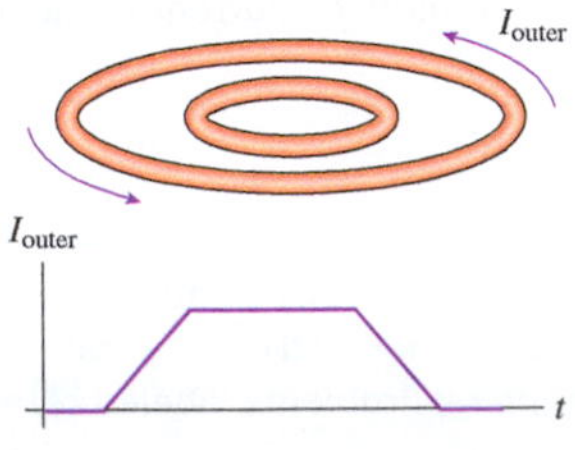


FIGURE Q25.12

13. Figure Q25.13 shows conducting loops next to each other. We will define a counterclockwise current (viewed from above) to be positive, a clockwise current to be negative. The graph shows the current in the left loop as a function of time. Sketch a graph that shows the induced current in the right loop. Explain.

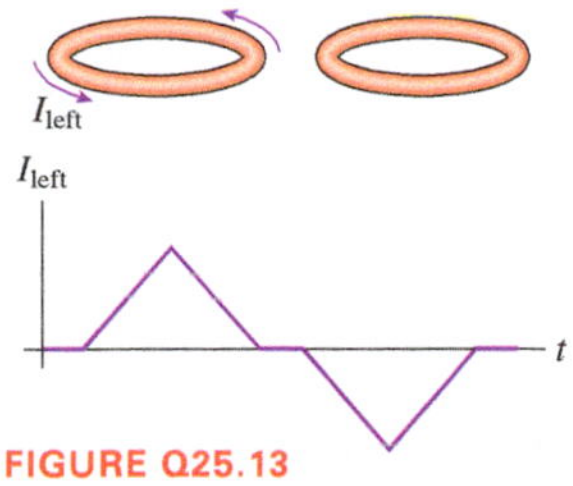

FIGURE Q25.13

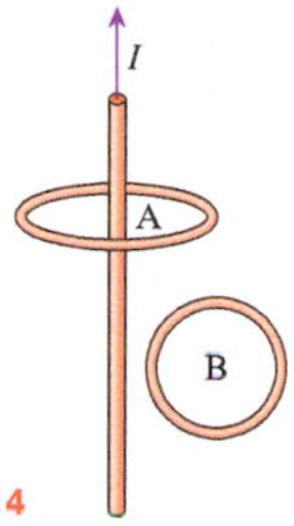

FIGURE Q25.14

14. The current in the wire shown in Figure Q25.14 is increasing. What is the direction of the induced current in each loop?
15. A loop of wire is horizontal. A bar magnet is pushed toward the loop from below, along the axis of the loop, as shown in Figure Q25.15.
   a. In what direction is the current in the loop? Explain.
   b. Is there a magnetic force on the loop? If so, in which direction? Explain.

   **Hint:** Recall that a current loop is a magnetic dipole.
   c. Is there a magnetic force on the magnet? If so, in which direction?

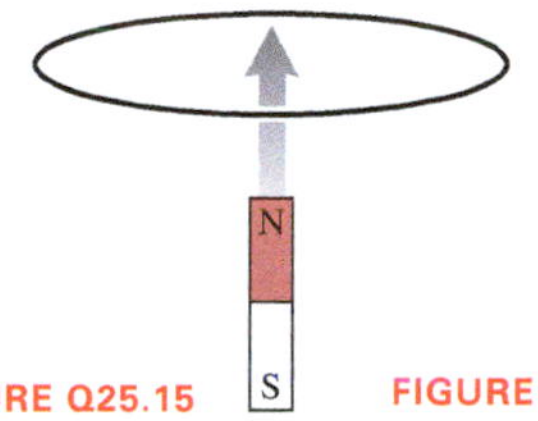

FIGURE Q25.15

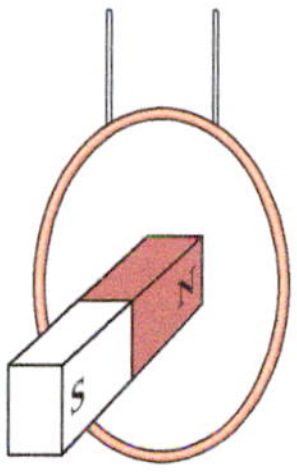

FIGURE Q25.16

16. A copper loop hangs from two strings, as shown in Figure Q25.16. In which direction (toward or away from the magnet) does the loop swing if the magnet
   a. Is stationary?
   b. Is moving toward the loop?
   c. Is moving away from the loop?
17. A conducting loop around a region of strong magnetic field contains two lightbulbs, as shown in Figure Q25.17. The wires connecting the bulbs are ideal. The magnetic field is increasing rapidly.
   a. Do the bulbs glow? Why or why not?
   b. If they glow, which bulb is brighter? Or are they equally bright? Explain.

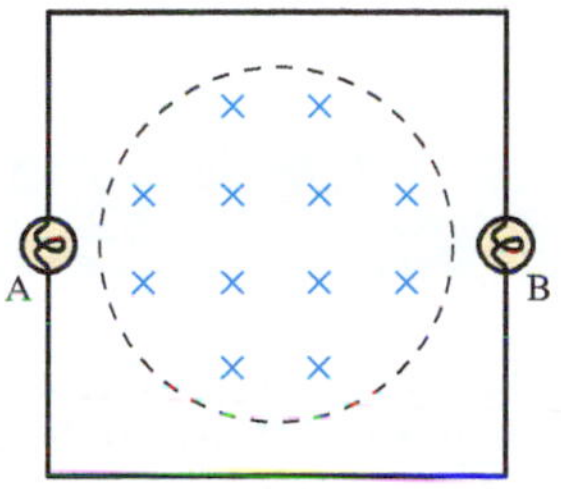

FIGURE Q25.17

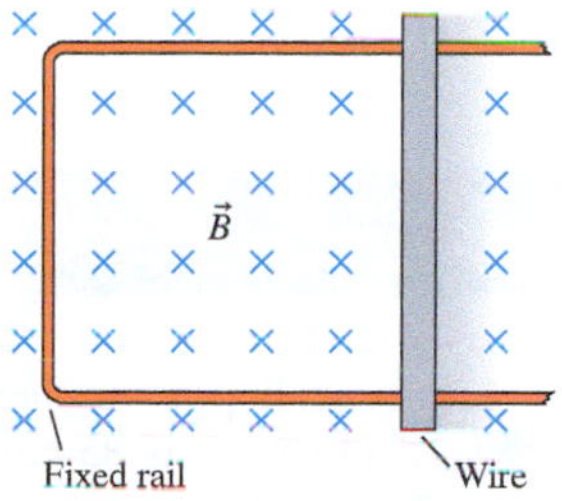

FIGURE Q25.18

18. A metal wire is resting on a U-shaped conducting rail, as shown in Figure Q25.18. The rail is fixed in position, but the wire is free to move.
   a. If the magnetic field is increasing in strength, which way does the wire move?
   b. If the magnetic field is decreasing in strength, which way does the wire move?
19. Although sunlight is unpolarized, the light that reflects from smooth surfaces may be partially polarized in the direction parallel to the plane of the reflecting surface. How should the axis of the polarizers in sunglasses be oriented—vertically or horizontally—to reduce the glare from a horizontal surface such as a road or a lake?
20. Can sound waves be polarized? How about traveling waves on a string? Explain.
21. The wavelengths of AM radio waves are so long—more than 200 m—that it would be impractical to make a $\frac{1}{4}$ wavelength antenna to detect their electric field. Instead, AM radios detect the *magnetic* field of the wave using a coil of wire. The changing flux of the wave's magnetic field induces an emf in the coil that is detected and amplified by the receiver. If you are standing to the west of an AM station that is broadcasting a vertically polarized wave, in which direction should the axis of the coil be oriented in order to best detect the wave?
22. Three laser beams have wavelengths $\lambda_1 = 400$ nm, $\lambda_2 = 600$ nm, and $\lambda_3 = 800$ nm. The power of each laser beam is 1 W.
   a. Rank in order, from largest to smallest, the photon energies $E_1$, $E_2$, and $E_3$ in these three laser beams. Explain.
   b. Rank in order, from largest to smallest, the number of photons per second $N_1$, $N_2$, and $N_3$ delivered by the three laser beams. Explain.
23. The intensity of a beam of light is increased but the light's frequency is unchanged. As a result, which of the following (perhaps more than one) are true? Explain.
   A. The photons travel faster.
   B. Each photon has more energy.
   C. The photons are larger.
   D. There are more photons per second.
24. A vertically polarized light wave passes through the five polarizers in Figure Q25.24. Rank in order, from largest to smallest, the transmitted intensities $I_a$ to $I_e$.

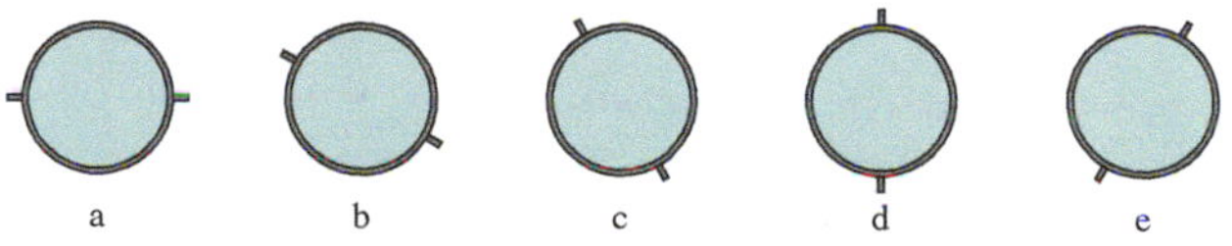

FIGURE Q25.24

25. Arc welding uses electric current to make an extremely hot electric arc that can melt metal. The arc emits ultraviolet light that can cause sunburn and eye damage if a welder is not wearing protective gear. Why does the arc give off ultraviolet light?

## Multiple-Choice Questions

26. | A circular loop of wire has an area of 0.30 m$^2$. It is tilted by 45° with respect to a uniform 0.40 T magnetic field. What is the magnetic flux through the loop?
   A. 0.085 T · m$^2$
   B. 0.12 T · m$^2$
   C. 0.38 T · m$^2$
   D. 0.75 T · m$^2$
   E. 1.3 T · m$^2$

27. | In Figure Q25.27, a square loop is rotating in the plane of the page around an axis through its center. A uniform magnetic field is directed into the page. What is the direction of the induced current in the loop?
   A. Clockwise.
   B. Counterclockwise.
   C. There is no induced current.

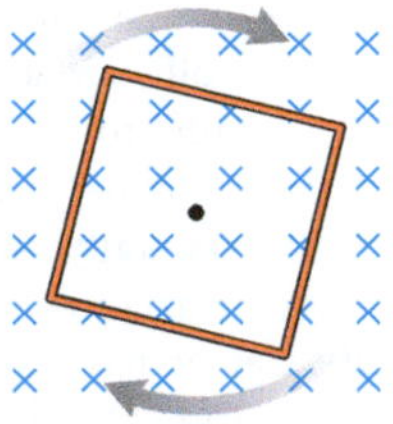

FIGURE Q25.27

28. | A diamond-shaped loop of wire is pulled at a constant velocity through a region where the magnetic field is directed into the paper in the left half and is zero in the right half, as shown in Figure Q25.28. As the loop moves from left to right, which graph best represents the induced current in the loop as a function of time? Let a clockwise current be positive and a counterclockwise current be negative.

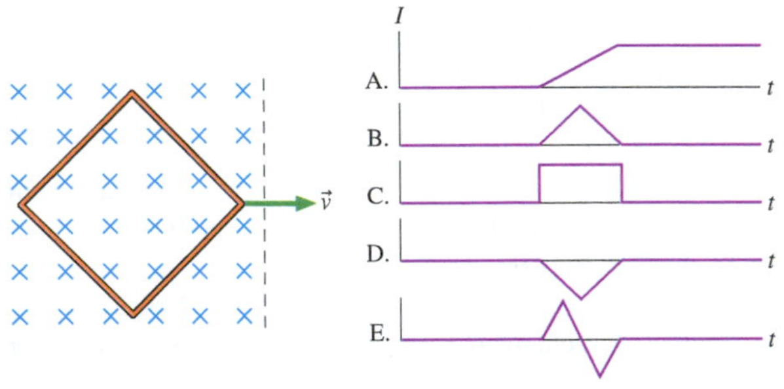


FIGURE Q25.28

29. || Figure Q25.29 shows a triangular loop of wire in a uniform magnetic field. If the field strength changes from 0.30 to 0.10 T in 50 ms, what is the induced emf in the loop?
   A. 0.08 V   B. 0.12 V
   C. 0.16 V   D. 0.24 V
   E. 0.36 V

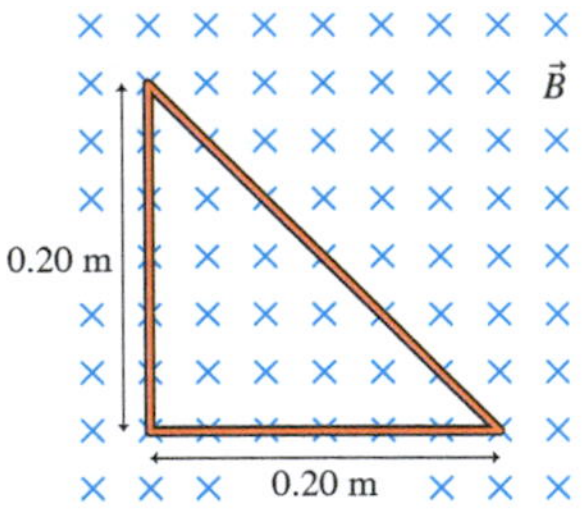


FIGURE Q25.29

30. || A device called a *flip coil* can be used to measure the earth's magnetic field. The coil has 100 turns and an area of 0.010 $m^2$. It is oriented with its plane perpendicular to the earth's magnetic field, then flipped 180° so the field goes through the coil in the opposite direction. The earth's magnetic field is 0.050 mT, and the coil flips over in 0.50 s. What is the average emf induced in the coil during the flip?
   A. 0.050 mV   B. 0.10 mV   C. 0.20 mV   D. 1.0 mV

31. | The electromagnetic waves that carry FM radio range in frequency from 87.9 MHz to 107.9 MHz. What is the range of wavelengths of these radio waves?
   A. 500–750 nm   B. 0.87–91.08 m
   C. 2.78–3.41 m   D. 278–341 m
   E. 234–410 km

32. || The beam from a laser is focused with a lens, reducing the area of the beam by a factor of 2. By what factor does the amplitude of the electric field increase?
   A. The amplitude does not change.
   B. The amplitude increases by a factor of $\sqrt{2}$.
   C. The amplitude increases by a factor of 2.
   D. The amplitude increases by a factor of $2\sqrt{2}$.
   E. The amplitude increases by a factor of 4.

33. | A spacecraft in orbit around the moon measures its altitude by reflecting a pulsed 10 MHz radio signal from the surface. If the spacecraft is 10 km high, what is the time between the emission of the pulse and the detection of the echo?
   A. 33 ns   B. 67 ns
   C. 33 $\mu$s   D. 67 $\mu$s

34. | A 6.0 mW vertically polarized laser beam passes through a polarizing filter whose axis is 75° from vertical. What is the laser-beam power after passing through the filter?
   A. 0.40 mW   B. 1.0 mW
   C. 1.6 mW   D. 5.6 mW

35. ||| By using antenna rods of several different lengths, the antenna in Figure Q25.35 is designed to pick up signals across a broad frequency range. The highest frequency it can pick up is 32 MHz. What is approximately the lowest frequency it can efficiently detect?

FIGURE Q25.35

   A. 120 MHz
   B. 80 MHz
   C. 50 MHz
   D. 25 MHz
   E. 12 MHz

36. || How many photons are emitted during 5.0 s of operation of a red laser pointer? The device outputs 2.8 mW at a 635 nm wavelength.
   A. $4.5 \times 10^{10}$
   B. $4.5 \times 10^{11}$
   C. $4.5 \times 10^{15}$
   D. $4.5 \times 10^{16}$

# PROBLEMS

## Section 25.1 Induced Currents

## Section 25.2 Motional emf

1. ||| A potential difference of 53 mV is developed across the ends of a 12.0-cm-long wire as it moves through a 0.27 T uniform magnetic field at a speed of 5.0 m/s. The magnetic field is perpendicular to the axis of the wire. What is the angle between the magnetic field and the wire's velocity?

2. || A scalloped hammerhead shark swims at a steady speed of
BIO 1.5 m/s with its 85-cm-wide head perpendicular to the earth's 50 $\mu$T magnetic field. What is the magnitude of the emf induced between the two sides of the shark's head?

3. || A 10-cm-long wire is pulled along a U-shaped conducting rail in a perpendicular magnetic field. The total resistance of the wire and rail is 0.20 Ω. Pulling the wire with a force of 1.0 N causes 4.0 W of power to be dissipated in the circuit.
   a. What is the speed of the wire when pulled with a force of 1.0 N?
   b. What is the strength of the magnetic field?

4. | Figure P25.4 shows a 15-cm-long metal rod pulled along two frictionless, conducting rails at a constant speed of 3.5 m/s. The rails have negligible resistance, but the rod has a resistance of 0.65 Ω.
   a. What is the current induced in the rod?
   b. What force is required to keep the rod moving at a constant speed?

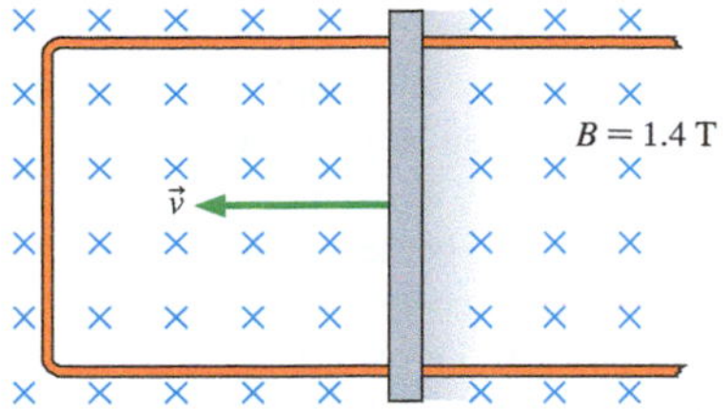

FIGURE P25.4

5. ||| A 50 g horizontal metal bar, 12 cm long, is free to slide up and down between two tall, vertical metal rods that are 12 cm apart. A 0.060 T magnetic field is directed perpendicular to the plane of the rods. The bar is raised to near the top of the rods, and a 1.0 Ω resistor is connected across the two rods at the top. Then the bar is dropped. What is the terminal speed at which the bar falls? Assume the bar remains horizontal and in contact with the rods at all times.

6. || In the rainy season, the Amazon flows fast and runs deep. In one location, the river is 23 m deep and moves at a speed of 4.0 m/s toward the east. The earth's 50 $\mu$T magnetic field is parallel to the ground and directed northward. If the bottom of the river is at 0 V, what is the potential (magnitude and sign) at the surface?

7. || A delivery truck with 2.8-m-high aluminum sides is driving west at 75 km/h in a region where the earth's magnetic field is $\vec{B} = (5.0 \times 10^{-5}\text{ T, north})$.
   a. What is the potential difference between the top and the bottom of the truck's side panels?
   b. Will the tops of the panels be positive or negative relative to the bottoms?

8. || Microphones are used to convert sound vibrations to electrical signals. In one kind of microphone, sound waves move a coil of wire up and down in a magnetic field, as shown in Figure P25.8. The coil shown has a diameter of 1.2 cm and has 100 turns; the 0.010 T magnetic field points radially out from the north pole to the south. If the microphone's emf at some instant is 10 mV, what is the speed of the coil?

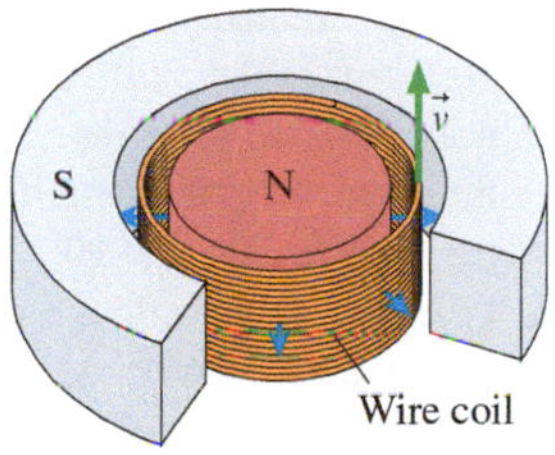

FIGURE P25.8

## Section 25.3 Magnetic Flux and Lenz's Law

9. | Figure P25.9 is an edge-on view of a 10-cm-diameter circular loop rotating in a uniform 0.050 T magnetic field. What is the magnetic flux through the loop when $\theta$ is 0°, 30°, 60°, and 90°?

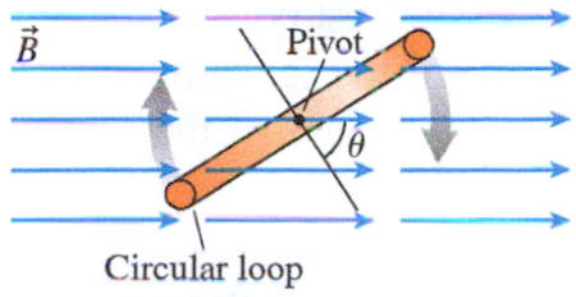

FIGURE P25.9

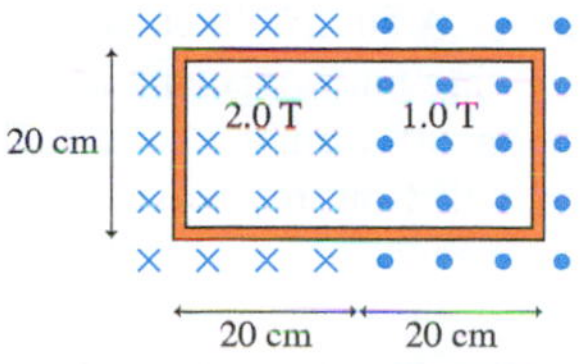

FIGURE P25.10

10. || What is the magnetic flux through the loop shown in Figure P25.10?

11. ||| The 2.0-cm-diameter solenoid in Figure P25.11 passes through the center of a 6.0-cm-diameter loop. The magnetic field inside the solenoid is 0.20 T. What is the magnetic flux through the loop (a) when it is perpendicular to the solenoid and (b) when it is tilted at a 60° angle?

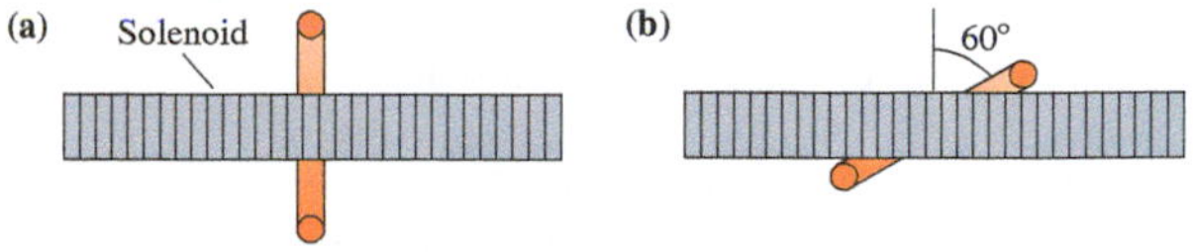

FIGURE P25.11

12. ||| You are given 13.0 m of thin wire. You form the wire into a circular coil with 50 turns. If this coil is placed with its axis parallel to a 0.12 T magnetic field, what is the flux through the coil?

13. || The metal equilateral triangle in Figure P25.13, 20 cm on each side, is halfway into a 0.10 T magnetic field.
   a. What is the magnetic flux through the triangle?
   b. If the magnetic field strength decreases, what is the direction of the induced current in the triangle?

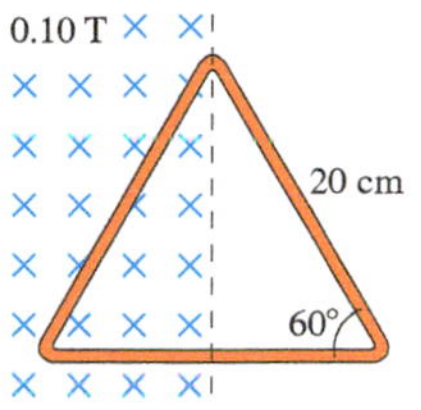

FIGURE P25.13

## Section 25.4 Faraday's Law

14. || A magnet and a coil are oriented as shown in Figure P25.14. The magnet is moved rapidly into the coil, held stationary in the coil for a short time, and then rapidly pulled back out of the coil. Sketch a graph showing the reading of the ammeter as a function of time. The ammeter registers a positive value when current goes into the "+" terminal.

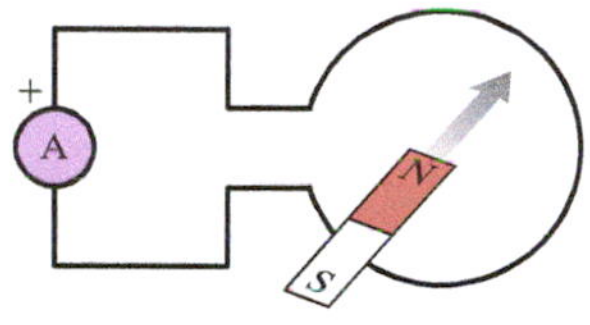

FIGURE P25.14

15. ||| A 1000-turn coil of wire 2.0 cm in diameter is in a magnetic field that drops from 0.10 T to 0 T in 10 ms. The axis of the coil is parallel to the field. What is the emf of the coil?

16. | Figure P25.16 shows a 100-turn coil of wire of radius 12 cm in a 0.15 T magnetic field. The coil is rotated 90° in 0.30 s, ending up parallel to the field. What is the average emf induced in the coil as it rotates?

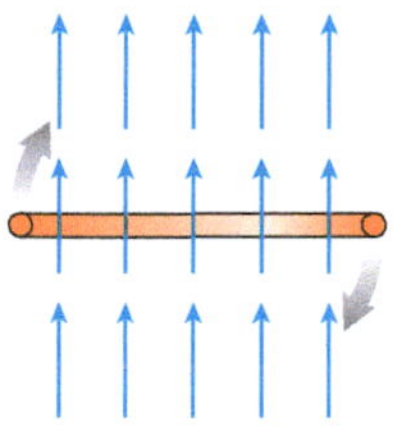
FIGURE P25.16

17. | Figure P25.17 shows a 10-cm-diameter loop in three different magnetic fields. The loop's resistance is 0.10 Ω. For each case, determine the induced emf, the induced current, and the direction of the current.

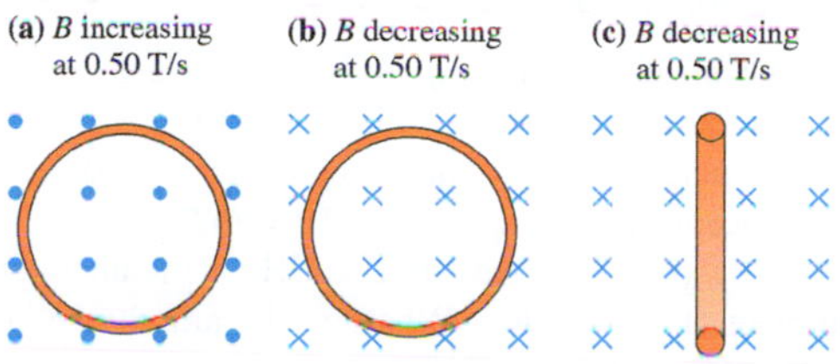

FIGURE P25.17

Watch Video Solution Problem 25.19

18. A wire with a resistance of 3.2 Ω is bent into the two-loop shape shown in Figure P25.18. The length of the sides of the large square is 9.0 cm; those of the smaller square are half that. If the magnetic field shown is increasing at a rate of 0.075 T/s, what is the current in the wire, and in which direction does it flow in the larger loop?

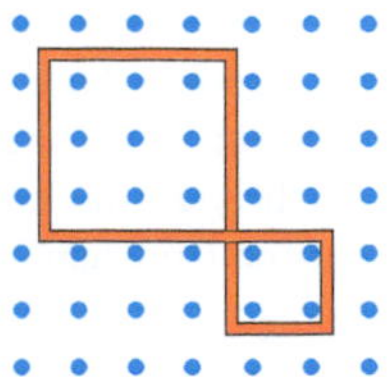
FIGURE P25.18

19. BIO Patients undergoing an MRI occasionally report seeing flashes of light. Some practitioners assume that this results from electric stimulation of the eye by the emf induced by the rapidly changing fields of an MRI solenoid. We can do a quick calculation to see if this is a reasonable assumption. The human eyeball has a diameter of approximately 25 mm. Rapid changes in current in an MRI solenoid can produce rapid changes in field, with $\Delta B/\Delta t$ as large as 50 T/s. What emf would this induce in a loop circling the eyeball? How does this compare to the 15 mV necessary to trigger an action potential?

20. The loop in Figure P25.20 has an induced current as shown. The loop has a resistance of 0.10 Ω. Is the magnetic field strength increasing or decreasing? What is the rate of change of the field, $\Delta B/\Delta t$?

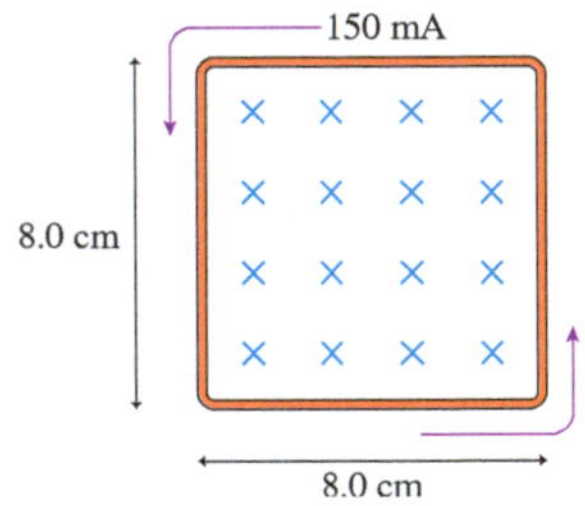

FIGURE P25.20

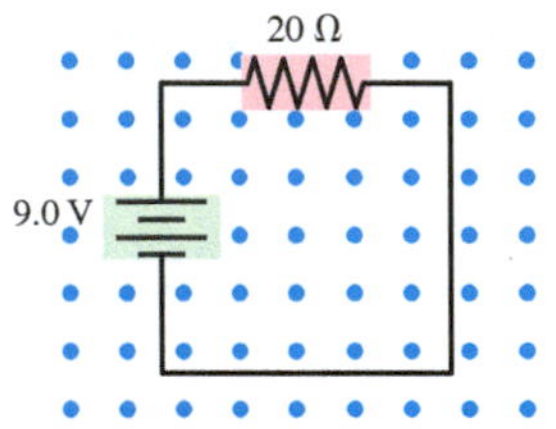

FIGURE P25.21

21. The circuit of Figure P25.21 is a square 20 cm on a side. The magnetic field increases steadily from 0 T to 0.50 T in 10 ms. What is the current in the resistor during this time?

22. INT A solenoid passes through the center of a wire loop, as shown in Figure 25.23 in the chapter. The solenoid has 1200 turns, a diameter of 2.0 cm, and is 7.5 cm long. The resistance of the loop is 0.032 Ω. If the current in the solenoid is increased by 1.3 A in 50 ms, what is the induced current in the loop?

23. A 20-cm-circumference loop of wire has a resistance of 0.12 Ω. The loop is placed between the poles of an electromagnet, and a field of 0.55 T is switched on in a time of 15 ms. What is the induced current in the loop?

24. The magnetic field at the earth's surface can vary in response to solar activity. During one intense solar storm, the vertical component of the magnetic field changed by 2.8 $\mu$T per minute, causing voltage spikes in large loops of the power grid that knocked out power in parts of Canada. What emf is induced in a square 100 km on a side by this rate of change of field?

### Section 25.5 Electromagnetic Waves

25. What is the electric field amplitude of an electromagnetic wave whose magnetic field amplitude is 2.0 mT?

26. What is the magnetic field amplitude of an electromagnetic wave whose electric field amplitude is 10 V/m?

27. A microwave oven operates at 2.4 GHz with an intensity inside the oven of 2500 W/m$^2$. What are the amplitudes of the oscillating electric and magnetic fields?

28. The maximum allowed leakage of microwave radiation from a microwave oven is 5.0 mW/cm$^2$. If microwave radiation outside an oven has the maximum value, what is the amplitude of the oscillating electric field?

29. A typical helium-neon laser found in supermarket checkout scanners emits 633-nm-wavelength light in a 1.0-mm-diameter beam with a power of 1.0 mW. What are the amplitudes of the oscillating electric and magnetic fields in the laser beam?

30. BIO Biologists often study the patterns of migratory birds by using radar (1–10 GHz electromagnetic waves) to track their flight. To check whether radar waves influence the birds' flight, researchers tracked the birds visually, both with the radar on and with it off. The 9 GHz radar waves had an intensity of 400 W/m$^2$ at 250 m. What was the amplitude of the electric field at this distance? (The experiments showed that the radar did not affect the birds.)

31. At what distance from a 10 mW point source of electromagnetic waves is the electric field amplitude 0.010 V/m?

32. The intensity of sunlight at the earth's surface is approximately 1000 W/m$^2$. What is the electric field amplitude of sunlight?

33. A radio antenna broadcasts a 1.0 MHz radio wave with 25 kW of power. Assume that the radiation is emitted uniformly in all directions.
    a. What is the wave's intensity 30 km from the antenna?
    b. What is the electric field amplitude at this distance?

34. BIO Biologists use *optical tweezers* to manipulate micron-sized objects using a beam of light. In this technique, a laser beam is focused to a very small-diameter spot. Because small particles are attracted to regions of high light intensity, the focused beam can be used to "grab" onto particles and manipulate them for various experiments. In one experiment, a 10 mW laser beam is focused to a spot that has a diameter of 0.62 $\mu$m.
    a. What is the intensity of the light in this spot?
    b. What is the amplitude of the electric field?

35. The intensity of a polarized electromagnetic wave is 10 W/m$^2$. What will be the intensity after passing through a polarizing filter whose axis makes the following angles with the plane of polarization? (a) $\theta = 0°$, (b) $\theta = 30°$, (c) $\theta = 45°$, (d) $\theta = 60°$, (e) $\theta = 90°$.

36. A low-power college radio station broadcasts 10 W of electromagnetic waves. At what distance from the antenna is the electric field amplitude $2.0 \times 10^{-3}$ V/m, the lower limit at which good reception is possible?

37. Only 25% of the intensity of a polarized light wave passes through a polarizing filter. What is the angle between the electric field and the axis of the filter?

38. BIO The minimum change in light intensity that is detectable by the human eye is about 1%. Light is sent through a pair of polarizers whose axes are at an angle of 45° to each other. By what angle should the second polarizer be turned so that the change in the intensity of the light that exists is barely perceptible?

39. A 50 mW laser beam is polarized horizontally. It then passes through two polarizers. The axis of the first polarizer is oriented at 30° from the horizontal, and that of the second is oriented at 60° from the horizontal. What is the power of the transmitted beam?

### Section 25.6 The Photon Model of Electromagnetic Waves

40. What is the energy (in eV) of a photon of visible light that has a wavelength of 500 nm?

41. | What is the energy (in eV) of an x-ray photon that has a wavelength of 1.0 nm?

42. | What is the wavelength of a photon whose energy is twice that of a photon with a 600 nm wavelength?

43. || One recent study has shown that x rays with a wavelength of BIO 0.0050 nm can produce mutations in human cells.
   a. Calculate the energy in eV of a photon of radiation with this wavelength.
   b. Assuming that the bond energy holding together a water molecule is typical, use Table 25.1 to estimate how many molecular bonds could be broken with this energy.

44. | Rod cells in the retina of the eye detect light using a pho- BIO topigment called rhodopsin. 1.8 eV is the lowest photon energy that can trigger a response in rhodopsin. What is the maximum wavelength of electromagnetic radiation that can cause a transition? In what part of the spectrum is this?

45. || The thermal emission of the human body has maximum BIO intensity at a wavelength of approximately 9.5 $\mu$m. What photon energy corresponds to this wavelength?

46. || The absorption of light BIO by chlorophyll, the pigment in plants that is responsible for converting light to useful energy, is shown in Figure P25.46. Two sharp peaks are visible in the violet and red parts of the spectrum (very little light is absorbed in the green part of the spectrum, which is why leaves appear green). What are the energies in eV of the photons that are absorbed at the two peaks?

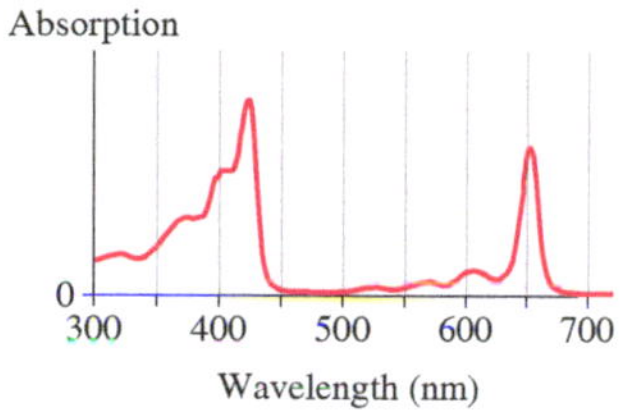

FIGURE P25.46

47. || The intensity of electromagnetic radiation from the sun reaching the earth's upper atmosphere is 1.37 kW/m$^2$. Assuming an average wavelength of 680 nm for this radiation, find the number of photons per second that strike a 1.00 m$^2$ solar panel directly facing the sun on an orbiting satellite.

48. || A 193-nm-wavelength UV laser for eye surgery emits a BIO 0.500 mJ pulse. How many photons does the light pulse contain?

49. ||| The human eye can barely detect a star whose intensity at BIO the earth's surface is $1.6 \times 10^{-11}$ W/m$^2$. If the dark-adapted eye has a pupil diameter of 7.0 mm, how many photons per second enter the eye from the star? Assume the starlight has a wavelength of 550 nm.

50. || A particular species of copepod, a small marine crustacean, BIO emits a flash of light consisting of $1.4 \times 10^{10}$ photons at a wavelength of 490 nm. The flash lasts for 2.4 s. What is the power of the flash?

## Section 25.7 The Electromagnetic Spectrum

51. || The spectrum of a glowing filament has its peak at a wavelength of 1200 nm. What is the temperature of the filament, in °C?

52. | Our sun's 5800 K surface temperature gives a peak wavelength in the middle of the visible spectrum. What is the minimum surface temperature for a star whose emission peaks at some wavelength less than 400 nm—that is, in the ultraviolet?

53. | The hottest ordinary star in our galaxy has a surface temperature of 53,000 K. What is the peak wavelength of its thermal radiation?

54. || The star Sirius is much hotter than the sun, with a peak wavelength of 290 nm compared to the sun's 500 nm. It is also larger, with a diameter 1.7 times that of the sun. By what factor does the energy emitted by Sirius exceed that of the sun?

55. | The photon energies used in different types of medical x-ray BIO imaging vary widely, depending upon the application. Single dental x rays use photons with energies of about 25 keV. The photon energy used for x-ray microtomography, a process that allows repeated imaging in single planes at varying depths within the sample, is 2.5 times greater. What are the wavelengths of the x rays used for these two purposes?

56. | A python can detect thermal radiation from objects that dif- INT fer in temperature from their environment as long as the received intensity of thermal radiation is greater than 0.60 W/m$^2$. Your body emits a good deal of thermal radiation; a typical human body has a surface area of 1.8 m$^2$, a surface temperature of 30°C, and an emissivity $e = 0.97$ at infrared wavelengths. As we've seen, the intensity of a source of radiation decreases with the distance from the source. If you are outside on a cool, day, what is the maximum distance from which a python could detect your presence?

57. | If astronomers look toward any point in outer space, they see radiation that matches the emission spectrum of an object at 2.7 K, a remnant of the Big Bang. What is the peak wavelength of this radiation? What part of the electromagnetic spectrum is it in?

## General Problems

58. ||| A square loop of wire, with sides of length 12 cm, is bent by 90° along a line halfway between two of its opposite sides, as shown in Figure P25.58. A uniform magnetic field of 0.85 T is then applied at an angle of 45° from both faces of the bent loop. What is the magnetic flux through the loop?

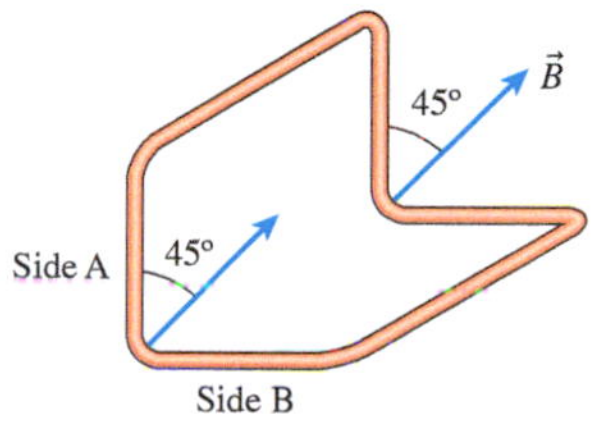

FIGURE P25.58

59. |||| People immersed in strong unchanging magnetic fields occa- BIO sionally report sensing a metallic taste. Some investigators suspect that motion in the constant field could produce a changing flux and a resulting emf that could stimulate nerves in the tongue. We can make a simple model to see if this is reasonable by imagining a somewhat extreme case. Suppose a patient having an MRI is immersed in a 3.0 T field along the axis of his body. He then quickly tips his head to the side, toward his right shoulder, tipping his head by 30° in the rather short time of 0.15 s. Estimate the area of the tongue; then calculate the emf that could be induced in a loop around the outside of the tongue by this motion of the head. How does this emf compare to the approximately 15 mV necessary to trigger an action potential? Does it seem reasonable to suppose that an induced emf is responsible for the noted effect?

60. || Some electronic devices such as phones and watches can be charged wirelessly. In the charger, there is a coil that generates a rapidly changing magnetic field. In the device, there is a second, 100-turn rectangular coil with side lengths of 3.0 cm and 4.0 cm. If the maximum rate of change of the magnetic field in this second coil is 53 T/s, what is the emf induced in this coil?

Watch Video Solution Problems 25.65 and 25.67

61. |||| Currents induced by rapid field changes in an MRI solenoid can, in some cases, heat tissues in the body, but under normal circumstances the heating is small. We can do a quick estimate to show this. Consider the "loop" of muscle tissue shown in Figure P25.61. This might be muscle circling the bone of your arm or leg. Muscle tissue is not a great conductor, but current will pass through muscle and so we can consider this a conducting loop with a rather high resistance. Suppose the magnetic field along the axis of the loop drops from 1.6 T to 0 T in 0.30 s, as it might in an MRI solenoid.
BIO INT

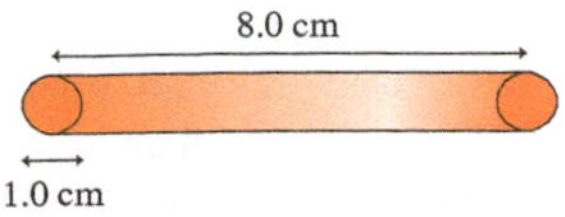

FIGURE P25.61

a. How much energy is dissipated in the loop?
b. By how much will the temperature of the tissue increase? Assume that muscle tissue has resistivity 13 $\Omega \cdot m$, density $1.1 \times 10^3$ kg/m$^3$, and specific heat 3600 J/kg $\cdot$ K.

62. || The loop in Figure P25.62 is being pushed into the 0.20 T magnetic field at 50 m/s. The resistance of the loop is 0.10 $\Omega$. What are the direction and magnitude of the current in the loop?

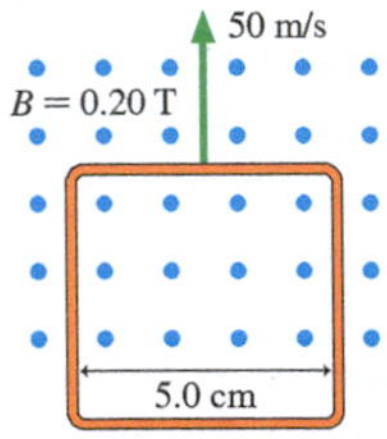

FIGURE P25.62

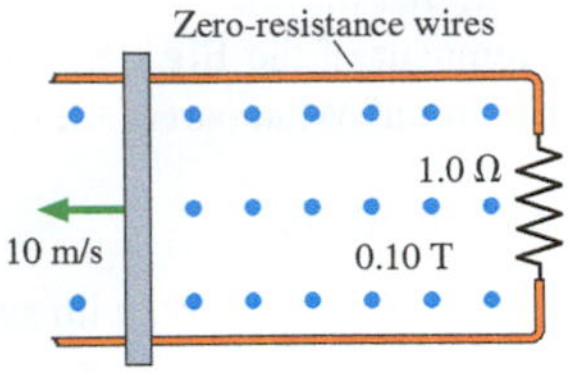

FIGURE P25.63

63. ||| A 20-cm-long, zero-resistance wire is pulled outward, on zero-resistance rails, at a steady speed of 10 m/s in a 0.10 T magnetic field. (See Figure P25.63.) On the opposite side, a 1.0 $\Omega$ carbon resistor completes the circuit by connecting the two rails. The mass of the resistor is 50 mg.
INT
a. What is the induced current in the circuit?
b. How much force is needed to pull the wire at this speed?
c. How much does the temperature of the carbon increase if the wire is pulled for 10 s? The specific heat of carbon is 710 J/kg $\cdot$ K. Neglect thermal energy transfer out of the resistor.

64. |||| A TMS (transcranial magnetic stimulation) device creates very rapidly changing magnetic fields. The field near a typical pulsed-field machine rises from 0 T to 2.5 T in 200 $\mu$s. Suppose a technician holds his hand near the device so that the axis of his 2.0-cm-diameter wedding band is parallel to the field.
BIO INT
a. What emf is induced in the ring as the field changes?
b. If the band is made of a gold alloy with resistivity $6.2 \times 10^{-8}$ $\Omega \cdot m$ and has a cross-section area of 4.0 mm$^2$, what is the induced current?

65. || The 10-cm-wide, zero-resistance wire shown in Figure P25.65 is pushed toward the 2.0 $\Omega$ resistor at a steady speed of 0.50 m/s. The magnetic field strength is 0.50 T.

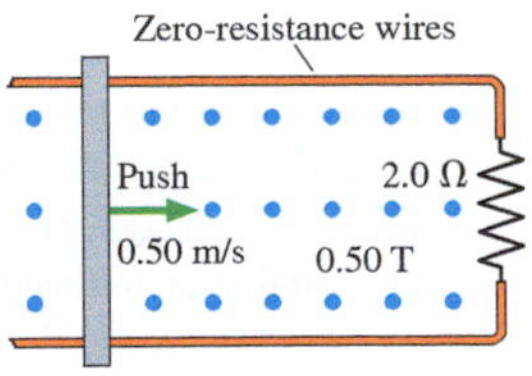

FIGURE P25.65

a. What is the magnitude of the pushing force?
b. How much power does the pushing force supply to the wire?
c. What are the direction and magnitude of the induced current?
d. How much power is dissipated in the resistor?

66. |||| Experiments to study vision often need to track the movements of a subject's eye. One way of doing so is to have the subject sit in a magnetic field while wearing special contact lenses that have a coil of very fine wire circling the edge. A current is induced in the coil each time the subject rotates his eye. Consider an experiment in which a 20-turn, 6.0-mm-diameter coil of wire circles the subject's cornea while a 1.0 T magnetic field is directed as shown in Figure P25.66. The subject begins by looking straight ahead. What emf is induced in the coil if the subject shifts his gaze by 5.0° in 0.20 s?
BIO

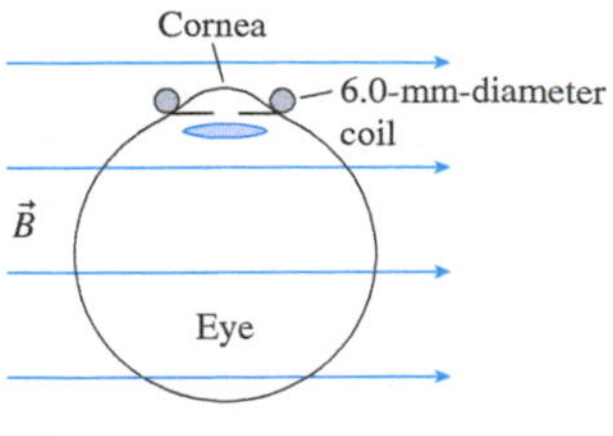

FIGURE P25.66

67. || A LASIK vision correction system uses a laser that emits 10-ns-long pulses of light, each with 2.5 mJ of energy. The laser is focused to a 0.85-mm-diameter circle. (a) What is the average power of each laser pulse? (b) What is the electric field strength of the laser light at the focus point?
BIO INT

68. |||| Satellite TV is broadcast from satellites that orbit 37,000 km above the earth's surface. Their antennas broadcast a 15 kW microwave signal that covers most of North America, an area of about $2.5 \times 10^7$ km$^2$.
a. What is the total power that strikes a 46-cm-diameter ground-based dish antenna?
b. The dish antenna focuses the incoming wave to a 1.0 cm$^2$ area. What is the amplitude of the electric field at this focus?

69. || A new cordless phone emits 4.0 mW at 5.8 GHz. The manufacturer claims that the phone has a range of 100 feet. If we assume that the wave spreads out evenly with no obstructions, what is the electric field strength at the base unit 100 feet from the phone?

70. || In reading the instruction manual that came with your garage-door opener, you see that the transmitter unit in your car produces a 250 mW signal and that the receiver unit is supposed to respond to a radio wave of the correct frequency if the electric field amplitude exceeds 0.10 V/m. You wonder if this is really true. To find out, you put fresh batteries in the transmitter and start walking away from your garage while opening and closing the door. Your garage door finally fails to respond when you're 42 m away. Are the manufacturer's claims true?

71. ||| Unpolarized light passes through a vertical polarizing filter, emerging with an intensity $I_0$. The light then passes through a horizontal filter, which blocks all of the light; the intensity transmitted through the pair of filters is zero. Suppose a third polarizer with axis 45° from vertical is inserted between the first two. What is the transmitted intensity now?

72. ||| A light-emitting diode (LED) connected to a 3.0 V power supply emits 440 nm blue light. The current in the LED is 10 mA, and the LED is 60% efficient at converting electric power input into light power output. How many photons per second does the LED emit?
INT

73. | A 1000 kHz AM radio station broadcasts with a power of 20 kW. How many photons does the transmitting antenna emit each second?

74. |||| The human body has a surface area of approximately 1.8 m$^2$, a surface temperature of approximately 30°C, and a typical emissivity at infrared wavelengths of $e = 0.97$. If we make the approximation that all photons are emitted at the wavelength of peak intensity, how many photons per second does the body emit?
BIO

## MCAT-Style Passage Problems

### Electromagnetic Wave Penetration BIO

Radio waves and microwaves are used in therapy to provide "deep heating" of tissue because the waves penetrate beneath the surface of the body and deposit energy. We define the *penetration depth* as the depth at which the wave intensity has decreased to 37% of its value at the surface. The penetration depth is 15 cm for 27 MHz radio waves. For radio frequencies such as this, the penetration depth is proportional to $\sqrt{\lambda}$, the square root of the wavelength.

75. | What is the wavelength of 27 MHz radio waves?
    A. 11 m B. 9.0 m C. 0.011 m D. 0.009 m
76. | If the frequency of the radio waves is increased, the depth of penetration
    A. Increases. B. Does not change. C. Decreases.
77. | For 27 MHz radio waves, the wave intensity has been reduced by a factor of 3 at a depth of approximately 15 cm. At this point in the tissue, the electric field amplitude has decreased by a factor of
    A. 9 B. $3\sqrt{3}$ C. 3 D. $\sqrt{3}$

### The Metal Detector

Metal detectors use induced currents to sense the presence of any metal—not just magnetic materials such as iron. A metal detector, shown in Figure P25.78, consists of two coils: a transmitter coil and a receiver coil. A high-frequency oscillating current in the transmitter coil generates an oscillating magnetic field along the axis and a changing flux through the receiver coil. Consequently, there is an oscillating induced current in the receiver coil.

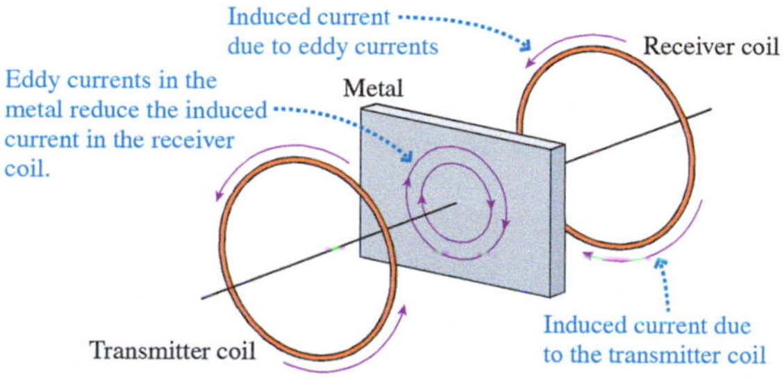


FIGURE P25.78

If a piece of metal is placed between the transmitter and the receiver, the oscillating magnetic field in the metal induces eddy currents in a plane parallel to the transmitter and receiver coils. The receiver coil then responds to the superposition of the transmitter's magnetic field and the magnetic field of the eddy currents. Because the eddy currents attempt to prevent the flux from changing, in accordance with Lenz's law, the net field at the receiver decreases when a piece of metal is inserted between the coils. Electronic circuits detect the current decrease in the receiver coil and set off an alarm.

78. | The metal detector will not detect insulators because
    A. Insulators block magnetic fields.
    B. No eddy current can be produced in an insulator.
    C. No emf can be produced in an insulator.
    D. An insulator will increase the field at the receiver.
79. | A metal detector can detect the presence of metal screws used to repair a broken bone inside the body. This tells us that
    A. The screws are made of magnetic materials.
    B. The tissues of the body are conducting.
    C. The magnetic fields of the device can penetrate the tissues of the body.
    D. The screws must be perfectly aligned with the axis of the device.
80. | Which of the following changes would *not* produce a larger eddy current in the metal?
    A. Increasing the frequency of the oscillating current in the transmitter coil
    B. Increasing the magnitude of the oscillating current in the transmitter coil
    C. Increasing the resistivity of the metal
    D. Decreasing the distance between the metal and the transmitter

# 26 AC Electricity

Transmission lines carry alternating current at voltages that can exceed 500,000 V. Why are such high voltages used? And how can birds perch safely on these high-voltage wires?

## LOOKING AHEAD

### Household Electricity

Electricity in your home is **alternating current,** abbreviated as **AC.** The current reverses direction 120 times per second.

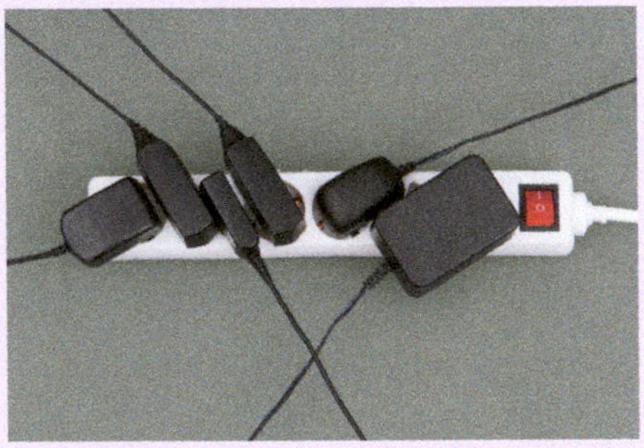

AC allows the use of transformers, such as the ones shown, to convert voltages for electronic devices. You'll learn the details of how this is done.

### Electrical Safety

Nerve and muscle cells respond to electrical signals, so exposure can be dangerous. Insulating gloves protect this lineman.

You'll learn a model for electrical conduction in the body that allows you to determine the risks from different electrical exposures.

### AC Circuits

Tuning a radio means adjusting the resonance frequency of an **oscillation circuit** to match a station's broadcast frequency.

You'll learn how these oscillation circuits are used in radios, televisions, cell phones, and other devices.

**GOAL** To understand and apply basic principles of AC electricity, electrical safety, and household electricity.

## LOOKING BACK

### DC Circuits

In Chapter 23, you learned to calculate the voltage and current in DC circuits with resistors.

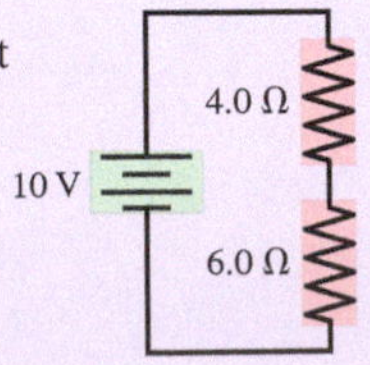


In this chapter, you'll extend the analysis to AC circuits that also include capacitors and **inductors**.

**STOP TO THINK**

Circuit 1 and circuit 2 at right each has a combination of resistors connected between points a and b. Which circuit has the greater equivalent resistance between points a and b?

A. Circuit 1
B. Circuit 2
C. Both circuits have the same resistance.

Circuit 1

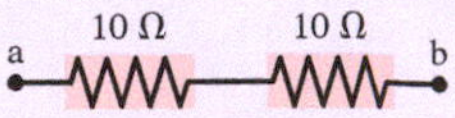


Circuit 2

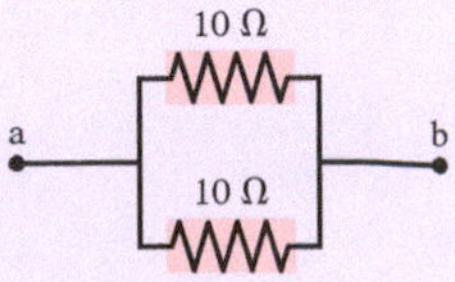

## 26.1 Alternating Current

A battery creates a constant emf. In a battery-powered flashlight, the bulb carries a constant current and glows with a steady light. The electricity distributed to homes in your neighborhood is different. The picture on the right is a long-exposure photo of a string of LED minilights swung through the air. Each bulb appears as a series of dashes because each bulb in the string flashes on and off 60 times each second. This isn't a special property of the bulbs, but of the electricity that runs them. Household electricity does not have a constant emf; it has a sinusoidal variation that causes the light output of the bulbs to vary. The bulbs light when the emf is positive but not when the emf is negative. The resulting flicker is too rapid to notice under normal circumstances.

LED minilights flash on and off 60 times a second.

In ◀◀ SECTION 25.2 we saw that an electrical generator—whether powered by steam, water, or wind—works by rotating a coil of wire in a magnetic field. The steady rotation of the coil causes the emf and the induced current in the coil to oscillate sinusoidally, alternately positive and then negative. This oscillation forces the charges to flow first in one direction and then, a half cycle later, in the other—an **alternating current,** abbreviated as AC. (If the emf is constant and the current is always in the same direction, we call the electricity *direct current,* abbreviated as DC.) The electricity from power outlets in your house is *AC electricity,* with an emf oscillating at a frequency of 60 Hz. Audio, radio, television, computer, and telecommunication equipment also make extensive use of AC circuits, with frequencies ranging from approximately $10^2$ Hz in audio circuits to approximately $10^9$ Hz in cell phones.

The instantaneous emf of an AC voltage source, shown graphically in FIGURE 26.1, can be written as

$$\mathcal{E} = \mathcal{E}_0 \cos(2\pi f t) = \mathcal{E}_0 \cos\left(\frac{2\pi t}{T}\right) \qquad (26.1)$$

emf of an AC voltage source

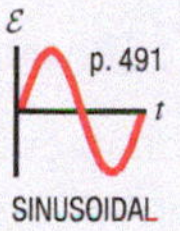


where $\mathcal{E}_0$ is the peak, or maximum, emf (recall that the units of emf are volts); $T$ is the period of oscillation (in s); and $f = 1/T$ is the oscillation frequency (in cycles per second, or Hz).

FIGURE 26.1 The emf of an AC voltage source.

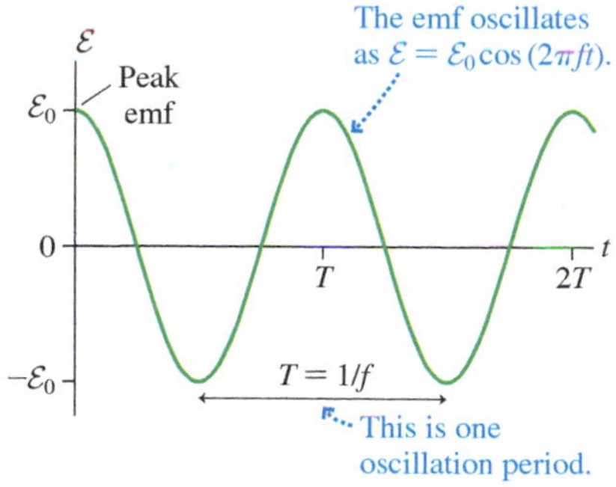


### Resistor Circuits

In Chapter 23 you learned to analyze a circuit in terms of the current $I$ and potential difference $\Delta V$. Now, because the current and voltage are changing with time, we will use a lowercase $i$ to represent the *instantaneous* current through a circuit element, the value of the current at one particular instant of time. Similarly, we will use a lowercase $v$ for the circuit element's instantaneous voltage.

FIGURE 26.2 shows the instantaneous current $i_R$ through a resistor $R$. The potential difference across the resistor, which we call the *resistor voltage* $v_R$, is given by Ohm's law:

$$v_R = i_R R \qquad (26.2)$$

FIGURE 26.2 The instantaneous current through a resistor.

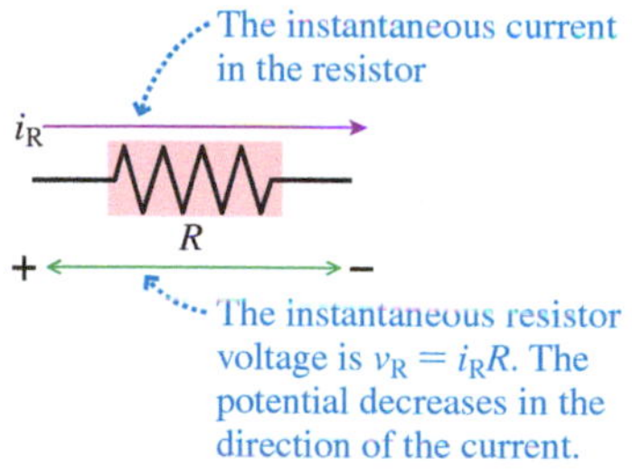


FIGURE 26.3 shows a resistor $R$ connected across an AC emf $\mathcal{E}$. The circuit symbol for an AC generator is . We can analyze this circuit in exactly the same way we analyzed a DC resistor circuit. Kirchhoff's loop law says that the sum of all the potential differences around a closed path is zero, so we can write

$$\sum \Delta V = \Delta V_{source} + \Delta V_R = \mathcal{E} - v_R = 0 \qquad (26.3)$$

The minus sign appears, just as it did in the equation for a DC circuit, because the potential *decreases* when we travel through a resistor in the direction of the current. Thus we find from the loop law that $v_R = \mathcal{E} = \mathcal{E}_0 \cos(2\pi f t)$. This isn't surprising because the resistor is connected directly across the terminals of the emf.

FIGURE 26.3 An AC resistor circuit.

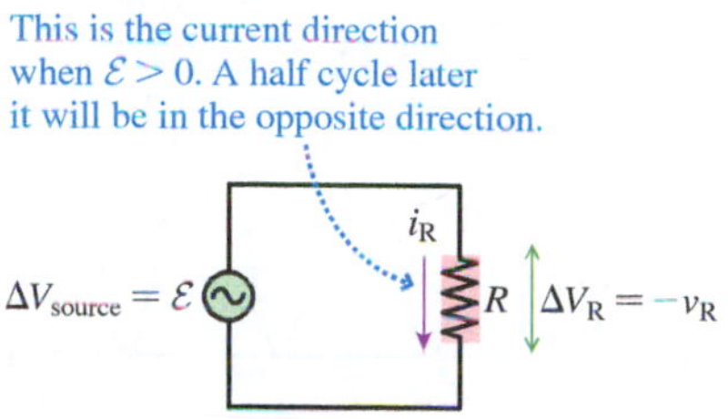

Because the resistor voltage is a sinusoidal voltage at frequency $f$, it is useful to write

$$v_R = V_R \cos(2\pi f t) \tag{26.4}$$

In this equation $V_R$ is the peak or maximum voltage, the amplitude of the sinusoidally varying voltage. (You can see that $V_R = \mathcal{E}_0$ in the single-resistor circuit of Figure 26.3.) Thus the current through the resistor is

$$i_R = \frac{v_R}{R} = \frac{V_R \cos(2\pi f t)}{R} = I_R \cos(2\pi f t) \tag{26.5}$$

where $I_R = V_R / R$ is the peak current.

**NOTE** ▸ It is important to understand the distinction between instantaneous and peak quantities. The instantaneous current $i_R$, for example, is a quantity that is always changing with time according to Equation 26.5. However, the peak current $I_R$ is a fixed number, the maximum value that the instantaneous current reaches. The instantaneous current oscillates between $+I_R$ and $-I_R$. ◂

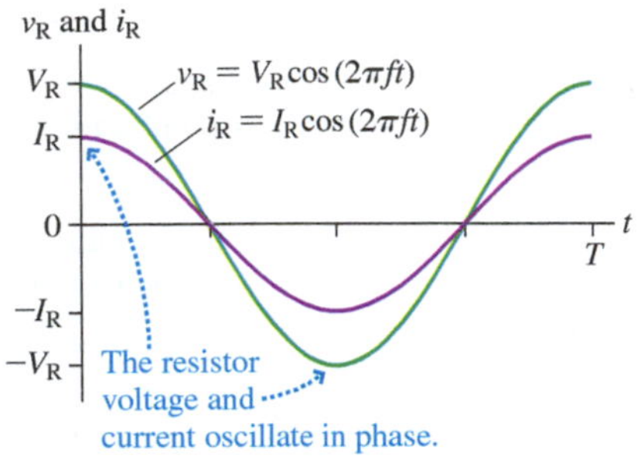


FIGURE 26.4 Graph of the current through and voltage across a resistor.

The resistor's instantaneous current and voltage both oscillate as $\cos(2\pi f t)$. FIGURE 26.4 shows the voltage and the current simultaneously on the same graph. The fact that the peak current $I_R$ is drawn as being less than $V_R$ has no significance. Current and voltage are measured in different units, so in a graph like this you can't compare the value of one to the value of the other. Showing the two different quantities on a single graph—a tactic that can be misleading if you're not careful—simply illustrates that they oscillate *in phase:* **The current is at its maximum value when the voltage is at its maximum, and the current is at its minimum value when the voltage is at its minimum.**

## AC Power in Resistors

In Chapter 23 you learned that the power dissipated by a resistor is $P = I\,\Delta V_R = I^2R$. In an AC circuit, the resistor current $i_R$ and voltage $v_R$ are constantly changing, as we saw in Figure 26.4, so the instantaneous power loss $p = i_R v_R = i_R^2 R$ (note the lowercase $p$) is constantly changing as well. We can use Equations 26.4 and 26.5 to write this instantaneous power as

$$p = i_R^2 R = [\,I_R \cos(2\pi f t)\,]^2 R = I_R^2 R[\,\cos(2\pi f t)\,]^2 \tag{26.6}$$

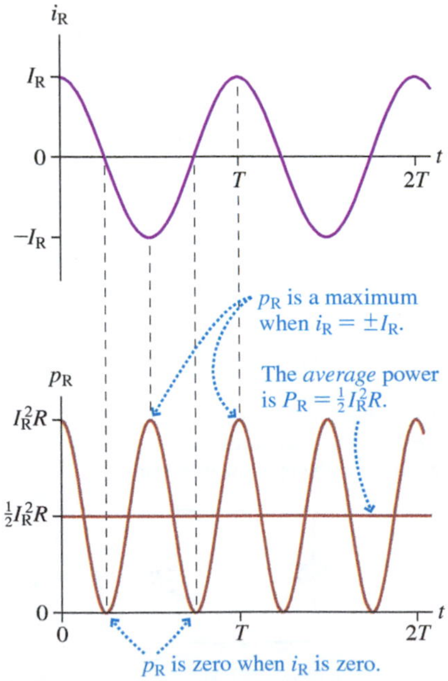


FIGURE 26.5 The instantaneous power loss in a resistor.

FIGURE 26.5 shows the instantaneous power graphically. You can see that, because the cosine is squared, the power oscillates *twice* during every cycle of the emf: The energy dissipation peaks both when $i_R = I_R$ and when $i_R = -I_R$. The energy dissipation doesn't depend on the current's direction through the resistor.

The current in an incandescent lightbulb reverses direction 120 times per second (twice per cycle), so the power reaches a maximum 120 times a second. But the hot filament of the bulb glows steadily, so it makes more sense to pay attention to the *average power* than the instantaneous power. Figure 26.5 shows that the **average power** $P_R$ is related to the peak power as follows:

$$P_R = \frac{1}{2} I_R^2 R \tag{26.7}$$

We could do a similar analysis for voltage to show that the power is $P_R = \frac{1}{2}{V_R}^2/R$. Recall that, in a DC circuit, the power is $P_R = I^2R = V^2/R$. We can write more useful expressions for power that mirror the DC expressions by defining a **root-mean-square current** and a **root-mean-square voltage** as follows. (We use the subscript "rms" to signify root-mean-square current and voltage.)

$$I_{rms} = \frac{I_R}{\sqrt{2}} \qquad V_{rms} = \frac{V_R}{\sqrt{2}} \tag{26.8}$$

Root-mean-square current and root-mean-square voltage

An rms value is a single number that serves as the measure of the size of a varying quantity. We saw this idea in Chapter 12, where we introduced the rms speed of molecules in a gas. If we rewrite the expressions for the average power dissipated by a resistor in an AC circuit in terms of the rms expressions, we obtain

$$P_R = (I_{rms})^2 R = \frac{(V_{rms})^2}{R} = I_{rms} V_{rms} \tag{26.9}$$

Average power loss in a resistor

The "120 V" on this lightbulb is its operating rms voltage. The "100 W" is its average power dissipation at this voltage.

These expressions are the same as the DC expressions, with $I_{rms}$ replacing $I$ and $V_{rms}$ replacing $V$. **As long as you work with rms voltages and currents, all the expressions you learned for DC power carry over to AC power.** The average power loss for a resistor in an AC circuit with $I_{rms} = 1$ A is the same as in a DC circuit with $I = 1$ A. In what follows, we'll use rms values as a general rule; if you are given a single voltage for an AC power supply, you can assume that this is the rms voltage. For instance, we've noted that household lamps and appliances in the United States operate at the 120 V present at wall outlets. This voltage is the rms value $\mathcal{E}_{rms}$; the peak voltage is higher by a factor of $\sqrt{2}$, so $\mathcal{E}_0 = 170$ V.

**EXAMPLE 26.1 The resistance and current of a toaster**

The hot wire in a toaster dissipates 580 W when plugged into a 120 V outlet.

a. What is the wire's resistance?
b. What are the rms and peak currents through the wire?

**STRATEGIZE** The 120 V outlet voltage is an rms value. We will solve Equation 26.9 for $R$ and $I_{rms}$, then use Equation 26.8 to find the peak current.

**PREPARE** In Equation 26.9, we have $V_{rms} = 120$ V and $P_R = 580$ W.

**SOLVE**

a. We rearrange Equation 26.9 to find the resistance from the rms voltage and the average power:

$$R = \frac{(V_{rms})^2}{P_R} = \frac{(120\text{ V})^2}{580\text{ W}} = 25\ \Omega$$

b. A second rearrangement of Equation 26.9 allows us to find the current in terms of the power and the resistance, both of which are known:

$$I_{rms} = \sqrt{\frac{P_R}{R}} = \sqrt{\frac{580\text{ W}}{25\ \Omega}} = 4.8\text{ A}$$

From Equation 26.8, the peak current is

$$I_R = \sqrt{2}\, I_{rms} = \sqrt{2}\,(4.8\text{ A}) = 6.8\text{ A}$$

**ASSESS** We can do a quick check on our work by calculating the power for the rated voltage and computed current:

$$P_R = I_{rms} V_{rms} = (4.8\text{ A})(120\text{ V}) = 580\text{ W}$$

This agrees with the value in the problem statement, giving us confidence in our solution.

**STOP TO THINK 26.1** An AC current with a peak value of 1.0 A passes through bulb A. A DC current of 1.0 A passes through an identical bulb B. Which bulb is brighter?

A. Bulb A
B. Bulb B
C. Both bulbs are equally bright.

## 26.2 AC Electricity and Transformers

Your cell phone runs at about 3.5 V. To charge it up from a wall outlet at $V_{rms} = 120$ V requires the use of a **transformer**, a device that takes an AC voltage as an input and produces either a higher or lower AC voltage as its output. As we'll see, the operation of a transformer is based on the emf produced by changing magnetic fields, so the input must be AC electricity.

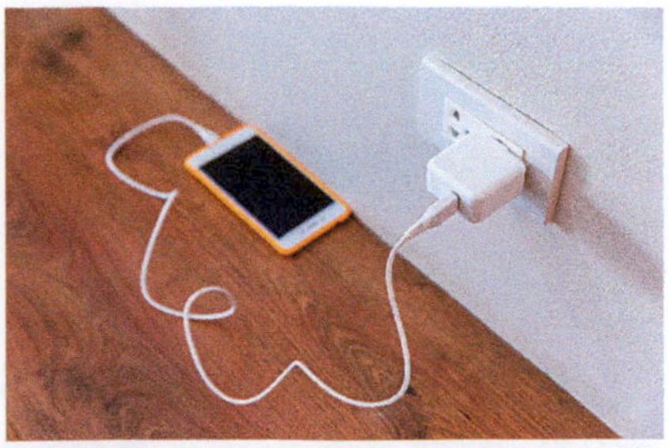

Your cell phone charger incorporates a transformer that provides the necessary voltage reduction.

## Transformer Operation

FIGURE 26.6a shows a simplified version of a transformer, consisting of two coils of wire wrapped on a single iron core. The left coil is called the **primary coil** (or simply the *primary*). It has $N_1$ turns of wire connected to an AC voltage source of amplitude $V_1$, which creates an AC current in the coil. This current in turn creates an oscillating magnetic field that magnetizes the iron of the core to produce a much stronger magnetic field—and a correspondingly large magnetic flux—through the primary coil.

The magnetic field lines created by the primary coil follow the iron core, as shown in Figure 26.6a, so nearly all of the field lines from the primary coil also go through the right coil of wire, which has $N_2$ turns and is called the **secondary coil** (or the *secondary*). Because the magnetic field in the iron is oscillating, the flux through the secondary is changing with time; this induces an emf, an AC voltage of amplitude $V_2$, in the secondary coil. To complete the picture, this emf is connected to a resistor $R$, which we call the *load*. The emf of the secondary causes a current in this resistor, which dissipates power in it.

FIGURE 26.6 A transformer and its circuit symbol.

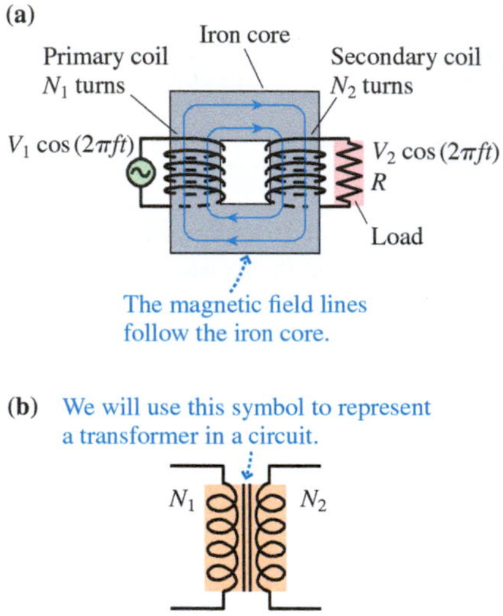


FIGURE 26.6b shows the symbol we'll use to represent a transformer in a circuit, with $N_1$ and $N_2$ specifying the numbers of turns of wire in the primary and secondary coils.

How is the peak voltage $V_2$ at the secondary related to the voltage $V_1$ at the primary? Suppose at some instant in time the primary voltage is $v_1$. We've seen that the primary voltage creates a flux $\Phi$ through the primary coil; according to Faraday's law (Equation 25.11), the emf across the coil—that is, $v_1$—is related to the rate of change of the flux by

$$v_1 = N_1 \frac{\Delta\Phi}{\Delta t} \tag{26.10}$$

In an **ideal transformer,** *all* of the flux is "guided" by the iron core through the secondary coil. Consequently, the rate at which the flux changes through the secondary coil is also $\Delta\Phi/\Delta t$. This changing flux induces an emf across the secondary coil given by

$$v_2 = N_2 \frac{\Delta\Phi}{\Delta t} \tag{26.11}$$

Because $\Delta\Phi/\Delta t$ is the same in Equations 26.10 and 26.11, we can write

$$\frac{v_1}{v_2} = \frac{N_1}{N_2} \tag{26.12}$$

Equation 26.12 gives the ratio of the instantaneous voltages. But the instantaneous values of an AC voltage are proportional to both the peak and rms voltages, so we can also write

$$V_2 = \frac{N_2}{N_1} V_1 \quad \text{and} \quad (V_2)_{\text{rms}} = \frac{N_2}{N_1} (V_1)_{\text{rms}} \tag{26.13}$$

Transformer voltages for primary and secondary coils with $N_1$ and $N_2$ turns

Depending on the ratio $N_2/N_1$, the voltage $V_2$ across the load can be transformed to a higher or a lower voltage than $V_1$. A *step-up transformer,* with $N_2 > N_1$, increases the voltage, while a *step-down transformer,* with $N_2 < N_1$, lowers the voltage.

Equation 26.13 relates the voltage at a transformer's secondary to the voltage at its primary. We can also relate the *currents* in the secondary and primary by considering energy conservation. If a transformer is connected to a load that draws an rms

A video to support a section's topic is embedded in the eText.

**Video** The Ring Shooter

current $(I_2)_{rms}$ from the secondary, the average power supplied to the load by the transformer, given by Equation 26.9, is

$$P_2 = (V_2)_{rms}(I_2)_{rms}$$

The primary coil draws a current $(I_1)_{rms}$ from the voltage source to which it's connected. This source provides power

$$P_1 = (V_1)_{rms}(I_1)_{rms}$$

to the transformer.

We'll assume that our ideal transformer has no loss of electric energy, so $P_1 = P_2$, or $(V_1)_{rms}(I_1)_{rms} = (V_2)_{rms}(I_2)_{rms}$. We can solve this for the current in the secondary:

$$(I_2)_{rms} = \frac{(V_1)_{rms}}{(V_2)_{rms}}(I_1)_{rms} = \frac{N_1}{N_2}(I_1)_{rms}$$

A similar expression holds for the peak current as well. We can therefore write expressions for the current in the coils of a transformer that mirror the expressions for the voltage in Equation 26.13:

$$I_2 = \frac{N_1}{N_2}I_1 \quad \text{and} \quad (I_2)_{rms} = \frac{N_1}{N_2}(I_1)_{rms} \tag{26.14}$$

Transformer currents for primary and secondary coils with $N_1$ and $N_2$ turns

Comparing Equations 26.13 and 26.14, you can see that **a step-up transformer raises voltage but lowers current; a step-down transformer lowers voltage but raises current.** This must be the case in order to conserve energy.

NOTE ▸ We've made some assumptions about the ideal transformer in completing this derivation. Real transformers come quite close to this ideal, so you can use the previous equations for computations on real transformers. ◂

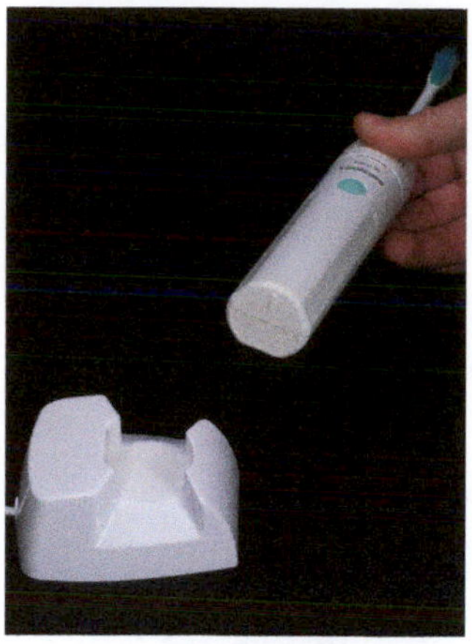

**Getting a charge** There is no direct electrical contact between the primary and secondary coils in a transformer; the energy is carried from one coil to the other by the magnetic field. This makes it possible to charge devices that are completely sealed, with no external electrical contacts, such as the electric toothbrush shown. A primary coil in the base creates an alternating magnetic field that induces an alternating emf in a secondary coil in the brush's handle. This emf, after conversion to DC, is used to charge the toothbrush's battery.

Video Transformers

## EXAMPLE 26.2 Analyzing a step-down transformer

A book light has a 1.4 W, 4.8 V bulb that is powered by a transformer connected to a 120 V electric outlet. The secondary coil of the transformer has 20 turns of wire. How many turns does the primary coil have? What is the current in the primary coil?

**STRATEGIZE** We know the voltages of the primary and the secondary, so we can compute the turns in the primary coil using Equation 26.13. We know the voltage and the power of the bulb, so we can find the current in the bulb. This is the current in the secondary, which we can use in Equation 26.14 to find the current in the primary, the current provided by the outlet.

**PREPARE** FIGURE 26.7 is a sketch of the circuit: this is the basic transformer circuit of Figure 26.6a, with the bulb as the load. All of the circuit quantities given in the problem statement are included in the sketch.

FIGURE 26.7 A transformer provides power to a book light.

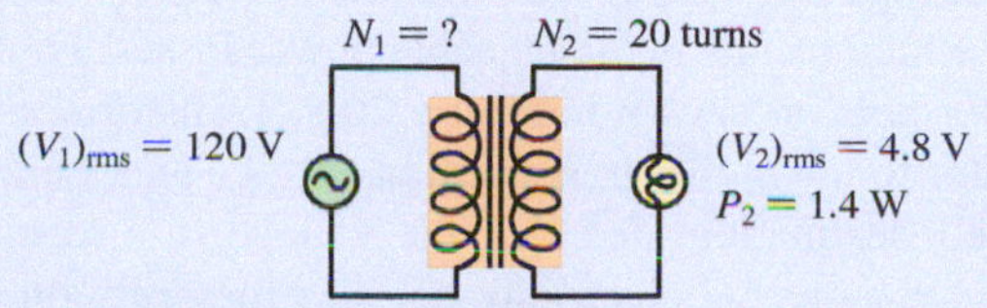


**SOLVE** The bulb is rated at 4.8 V; this is the rms voltage at the secondary, so $(V_2)_{rms} = 4.8$ V. The power outlet has the usual $(V_1)_{rms} = 120$ V, so we can rearrange Equation 26.13 to find

$$N_1 = N_2\frac{(V_1)_{rms}}{(V_2)_{rms}} = (20 \text{ turns})\left(\frac{120 \text{ V}}{4.8 \text{ V}}\right) = 500 \text{ turns}$$

The bulb connected to the secondary dissipates 1.4 W at 4.8 V; this is an rms voltage, so the rms current in the secondary is

$$(I_2)_{rms} = \frac{P_2}{(V_2)_{rms}} = \frac{1.4 \text{ W}}{4.8 \text{ V}} = 0.29 \text{ A}$$

We can then rearrange Equation 26.14 to find the rms current in the primary:

$$(I_1)_{rms} = (I_2)_{rms}\frac{N_2}{N_1} = (0.29 \text{ A})\frac{20}{500} = 0.012 \text{ A}$$

**ASSESS** We can check our results by looking at the power supplied by the wall outlet. This is $P_1 = (120 \text{ V})(0.012 \text{ A}) = 1.4$ W, the same as the power dissipated by the bulb, as must be the case because we've assumed the transformer is ideal.

## Power Transmission

Long-distance electrical transmission lines run at very high voltages—up to 1,000,000 volts! FIGURE 26.8 outlines the steps in transmitting electricity from a power plant to a city; the voltages shown are typical values. A generator in the power plant produces electricity at 25 kV, which is stepped up to 500 kV for transmission, then stepped down in two stages to give you 120 V at the outlet. There are three stages of transformers involved. Why go to all this trouble to transmit electric power at such high voltages? In a word: efficiency, as we show in the next example.

FIGURE 26.8 Electricity transmission.

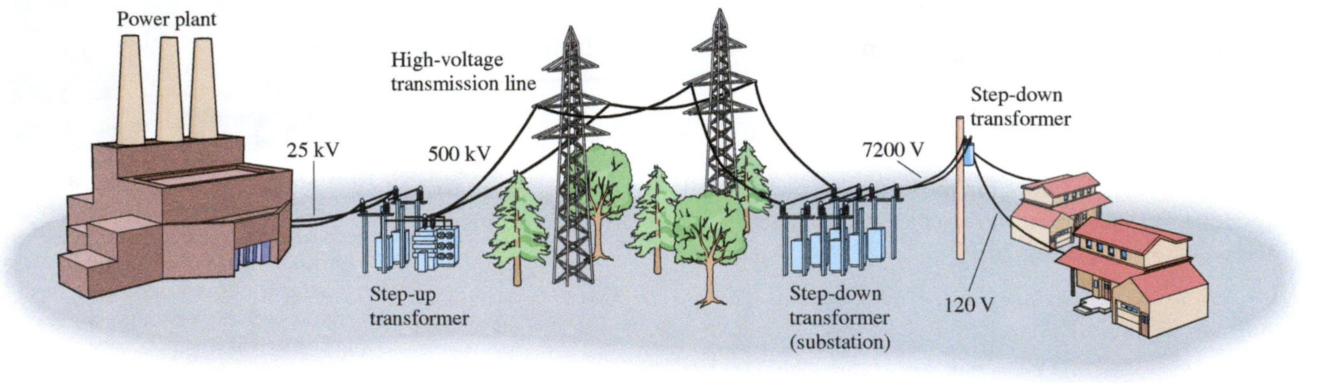


### EXAMPLE 26.3 Using high voltages to efficiently transmit electric power

To provide power to a small city, a power plant generates 40 MW of AC electricity. The power plant is 50 km from the city (a typical distance), and the 100 km of wire used in the transmission line (to the city and back) has a resistance of 7.0 Ω.

a. To provide 40 MW of power at the generator voltage of 25,000 V, what current is required?
b. What is the power dissipated in the resistance of the transmission line for this current?
c. To provide 40 MW of power at 500,000 V, what current is required?
d. What is the power dissipated in the resistance of the transmission line for this higher voltage?

**STRATEGIZE** We will treat the city—and the wires that transmit power to it—as a load. We will find the current required to provide the 40 MW of power, and from this current the power dissipated by the wire, from Equation 26.9.

**PREPARE** The voltages are rms values, so we use $V_{\text{rms}} = 25{,}000$ V for part a and $V_{\text{rms}} = 500{,}000$ V for part c.

**SOLVE**

a. To provide 40 MW at the generator voltage of 25,000 V, the current is

$$I_{\text{rms}} = \frac{40 \times 10^6\ \text{W}}{25 \times 10^3\ \text{V}} = 1600\ \text{A}$$

b. Passing this current through the transmission lines will result in power dissipation in the 7.0 Ω resistance of the wires. We don't know the voltage drop across the wires, but we do know the current and resistance, so we can compute

$$P_{\text{dissipated in wires}} = (I_{\text{rms}})^2 R = (1600\ \text{A})^2\,(7.0\ \Omega) = 18\ \text{MW}$$

This is nearly half the power generated, clearly an unacceptable loss.

c. Increasing the transmission voltage to 500 kV reduces the necessary current:

$$I_{\text{rms}} = \frac{40 \times 10^6\ \text{W}}{500 \times 10^3\ \text{V}} = 80\ \text{A}$$

This is a remarkably small current to supply a city. If you use several high-power appliances at one time, you could easily use this much current in your house. But the necessary current for the city can be so small because the voltage is so large.

d. The relatively small current means that the power dissipated in the resistance of the wires will be small as well:

$$P_{\text{dissipated in wires}} = (I_{\text{rms}})^2 R = (80\ \text{A})^2\,(7.0\ \Omega) = 0.045\ \text{MW}$$

This is only about 0.1% of the power generated, which is quite reasonable.

**ASSESS** The final result—the power dissipated in the wires is dramatically reduced for an increased transmission voltage—is just what the example was designed to illustrate.

Transmitting electricity at high voltages means that the current in the transmission lines can be decreased, which results in manageable power losses. High-voltage transmission requires transformers, as shown in Figure 26.8. Transformers increase the voltage from the generator to the higher transmission voltage and then, once the transmission lines reach their destination, decrease the voltage to a safe value for household use. Transformers require AC electricity because they rely on changing flux for their operation. This is the main reason we use AC power even though (as you'll see in the next section) it is slightly more dangerous than DC.

**STOP TO THINK 26.2** Each of the transformers shown here has its primary connected to a 120 V AC power supply and its secondary connected to a 1000 Ω load resistor. Rank the transformers, from highest to lowest, by the rms voltage they create across the resistor.

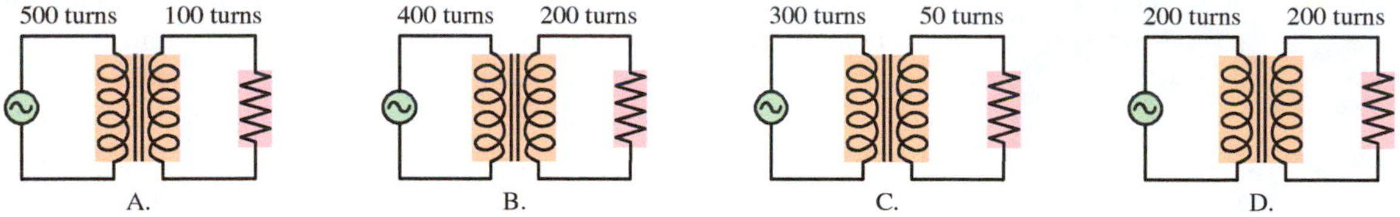


## 26.3 Household Electricity

The electricity in your home can be understood using the techniques of circuit analysis we've developed, but we need to add one more concept. So far we've dealt with potential *differences*. Although we are free to choose the zero point of potential anywhere that is convenient, our analysis of circuits has not suggested any need to establish a zero point. Potential differences are all we have needed.

Difficulties can begin to arise, however, if you want to connect two different circuits together. Perhaps you would like to connect your game system to your TV, or connect your computer monitor to the computer itself. Incompatibilities can arise unless all the circuits to be connected have a common reference point for the potential. This is the reason for having an electric *ground*.

### Electric Outlets Are Grounded Parallel Circuits

You learned previously that the earth itself is a conductor. Suppose we have two circuits. If we connect one point of each circuit to the earth by an ideal wire, then both circuits have a common reference point. A circuit connected to the earth in this way is said to be **grounded.** In practice, we also agree to call the potential of the earth $V_{earth} = 0$ V. **FIGURE 26.9** shows a circuit with a ground connection. Under normal circumstances, the ground connection does not carry any current because it is not part of a complete circuit. In this case, it does not alter the behavior of the circuit.

**FIGURE 26.10** shows a circuit diagram for the outlets in your house. The 120 V electric supply is provided by the power company. It is transmitted to outlets throughout your house by wires in the walls. One terminal of the electric supply is grounded; we call this the **neutral** side. The other side is a 120 V AC voltage; we call this the **hot** side. Each electric outlet has two slots, one connected to the hot side and one connected to the neutral side. When you insert a plug into an electric outlet, the prongs of the plug connect to the two terminals of the electric supply. The device you've plugged in completes a circuit between the two terminals, the potential difference across the device leads to a current, and the device turns on.

The multiple outlets in a room or area of your house are connected in parallel so that each works when the others are not being used. Because the outlets on a single circuit are in parallel, when you plug in another device, the total current in the circuit

**FIGURE 26.9** A grounded circuit.

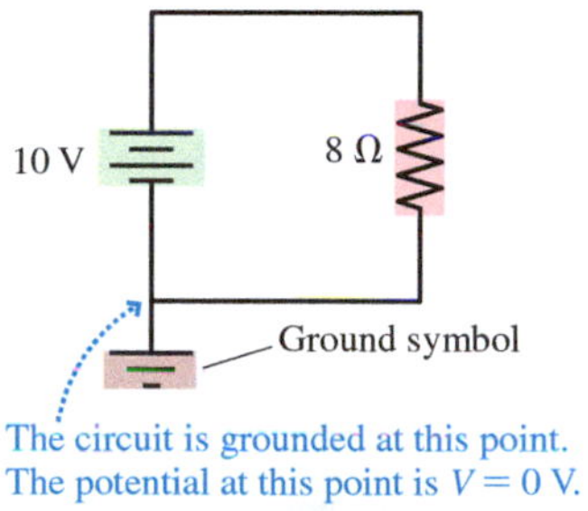

Household circuits are literally "grounded"—a metal rod is driven into the ground and the neutral side of the circuit is connected to it.

**FIGURE 26.10** Multiple outlets on one circuit.

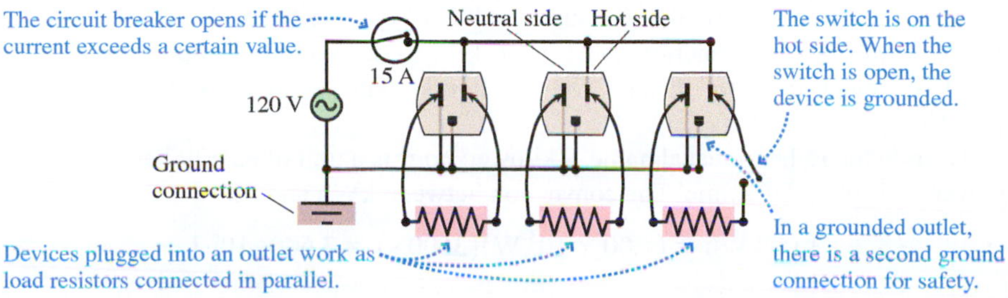

The circuit breaker opens if the current exceeds a certain value.
Neutral side
Hot side
15 A
120 V
The switch is on the hot side. When the switch is open, the device is grounded.
Ground connection
Devices plugged into an outlet work as load resistors connected in parallel.
In a grounded outlet, there is a second ground connection for safety.

**Video** Figure 26.10

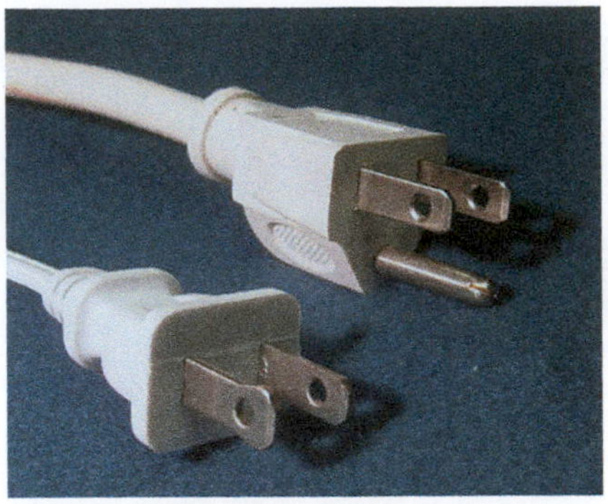

A two-prong "polarized" plug has one large prong and one small one. When plugged into a standard outlet, the large prong is grounded. A three-prong plug has a round pin that makes a second ground connection.

increases. This can create a problem; although the wires in your walls are good conductors, they aren't ideal. The wires have a small resistance and heat up when carrying a current. If there is too much current, the wires could get hot enough to cause a fire.

The circuits in your house are protected with *circuit breakers* to limit the current in each circuit. A circuit breaker consists of a switch and an ammeter that measures the current in the circuit. If the ammeter measures too much current (typically $I_{\text{rms}} \geq 15\text{ A}$), it sends a signal to open the switch to disconnect the circuit. You have probably had the experience of having a circuit breaker "trip" if you have too many things plugged in. Before you reset the breaker, some devices need to be turned off, unplugged, or moved to a different circuit.

Grounding of household circuits provides an important reference potential, but the main reason for grounding is safety. The two slots in a standard outlet are different sizes; the neutral slot is a bit larger. Most electric devices are fitted with plugs that can be inserted into an outlet in only one orientation. A lamp will almost certainly have this sort of plug. When you turn the lamp off, the switch disconnects the hot wire, not the neutral wire. The lamp is then grounded, and thus safe, when it is switched off.

The round hole in a standard electric outlet is a second ground connection. If a device has a metal case, the case will likely be connected to the ground. If a wire comes loose inside the device and contacts the metal case, a person touching the case could get a shock. But if the case is grounded, its potential is always 0 V and it is always safe to touch. In addition, a hot wire touching the grounded case would be a short circuit, causing a sudden very large current that would trip the circuit breaker, disconnecting the hot wire and preventing any danger.

**EXAMPLE 26.4 Will the circuit breaker open?**

A circuit in a student's room has a 15 A circuit breaker. One evening, she plugs in a computer (240 W), a lamp (with two 60 W bulbs), and a space heater (1200 W). Will this be enough to trip the circuit breaker?

**STRATEGIZE** The circuit breaker will trip if the current through it exceeds 15 A. We will find the current through the circuit by adding the currents through each appliance.

**PREPARE** We start by sketching the circuit, as in FIGURE 26.11. Because the three devices are in the same circuit, they are connected in parallel. We can model each of them as a resistor.

**SOLVE** The current in the circuit is the sum of the currents in the individual devices:

$$(I_{\text{total}})_{\text{rms}} = (I_{\text{computer}})_{\text{rms}} + (I_{\text{lamp}})_{\text{rms}} + (I_{\text{heater}})_{\text{rms}}$$

Equation 26.9 gives the power as the rms current times the rms voltage, so the current in each device is the power divided by the rms voltage:

FIGURE 26.11 The circuit with the circuit breaker.

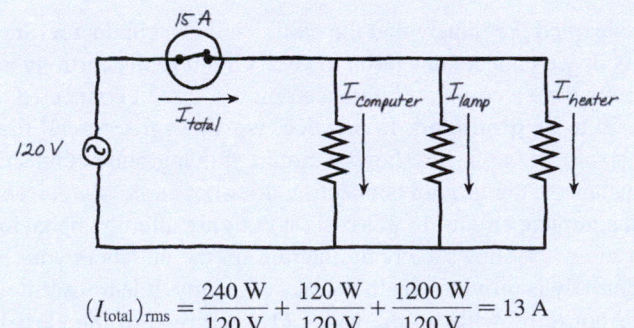

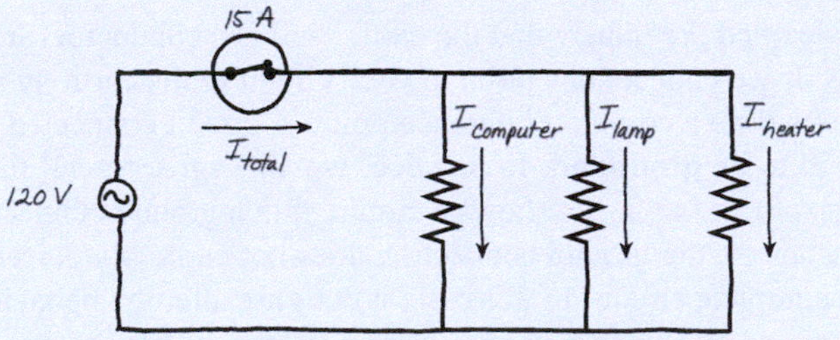


$$(I_{\text{total}})_{\text{rms}} = \frac{240\text{ W}}{120\text{ V}} + \frac{120\text{ W}}{120\text{ V}} + \frac{1200\text{ W}}{120\text{ V}} = 13\text{ A}$$

This is almost but not quite enough to trip the circuit breaker.

**ASSESS** Generally all of the outlets in one room (and perhaps the lights as well) are on the same circuit. You have quite possibly used electric devices with this much total power in one room without problems, so this result seems reasonable.

## Kilowatt Hours

Your home's electric meter records the kilowatt hours of electricity you use.

The product of watts and seconds is joules, the SI unit of energy. However, your local electric company prefers to use a different unit, called *kilowatt hours,* to measure the energy you use each month.

A device in your home that consumes $P$ kW of electricity for $\Delta t$ hours has used $P\Delta t$ kilowatt hours of energy, abbreviated kWh. For example, suppose you run a 1500 W electric water heater for 10 hours. The energy used in kWh is $(1.5\text{ kW})(10\text{ h}) = 15\text{ kWh}$.

Despite the rather unusual name, a kilowatt hour is a unit of energy because it is a power multiplied by a time. The conversion between kWh and J is

$$1.00\text{ kWh} = (1.00 \times 10^3\text{ W})(3600\text{ s}) = 3.60 \times 10^6\text{ J}$$

Your monthly electric bill specifies the number of kilowatt hours you used last month. This is the amount of energy that the electric company delivered to you that you transformed into light and thermal energy inside your home.

**EXAMPLE 26.5 Computing the cost of electric energy**

A typical electric space heater draws an rms current of 12.5 A on its highest setting. If electricity costs 12¢ per kilowatt hour (an approximate national average), how much does it cost to run the heater for 2 hours?

***STRATEGIZE*** We can find the energy in kWh as the product of the heater's power in kW and the time it runs in hours.

***SOLVE*** The power dissipated by the heater is

$$P = V_{rms} I_{rms} = (120\text{ V})(12.5\text{ A}) = 1500\text{ W} = 1.5\text{ kW}$$

In 2 hours, the energy used is $(1.5\text{ kW})(2.0\text{ h}) = 3.0\text{ kWh}$. At 12¢ per kWh, the cost is 36¢.

**STOP TO THINK 26.3** Rank the following four devices by the energy they use for the time period given, from highest to lowest:

A. Electric blanket (60 W) used for 8 h
B. Flat-screen television (120 W) for 3 h
C. Whole-house fan (250 W) for 8 h
D. Air conditioner (1000 W) for 3 h

## 26.4 Biological Effects and Electrical Safety BIO

You can handle an ordinary 9 V battery without the slightest danger, but the 120 volts from an electric outlet can lead to a nasty shock. Yet the girl in FIGURE 26.12 is safely touching a Van de Graaff generator at a potential of 400,000 V. What makes electricity either safe or dangerous?

**The relative safety of electric sources isn't governed by their voltage but by the current they can push through your body.** Current is what produces physiological effects and damage because it mimics nerve impulses and causes muscles to involuntarily contract.

Higher voltages are generally more dangerous than lower voltages because they tend to produce larger currents, but the amount of current also depends on resistance and on the ability of the voltage source to deliver current. The Van de Graaff generator in Figure 26.12 is at a high potential with respect to the ground, but the girl is standing on an insulating platform. The high resistance of the platform means that very little current is passing through her to the ground; she won't feel a thing. Even if she touches a grounded object, the current will be modest—the total charge on the generator is quite small, so a dangerous current simply isn't possible. You can get a much worse shock from a 120 V household circuit because it is capable of providing a much larger current.

TABLE 26.1 lists approximate values of current that produce different physiological effects. Currents through the chest cavity are particularly dangerous because they can interfere with respiration and the proper rhythm of the heart. It requires less AC current than DC current to produce physiological effects. An AC current larger than 100 mA can induce fibrillation of the heart, in which it beats in a rapid, chaotic, uncontrolled fashion.

To calculate likely currents through the body, we model the body as several connected resistors, as shown in FIGURE 26.13. (These are average values; there is significant individual variation.) Because the interior of the body has a high saltwater content, its resistance is fairly low. But current must pass through the skin before getting inside the

FIGURE 26.12 High voltages are not necessarily dangerous.

TABLE 26.1 Physiological effects of currents passing through the body

| Physiological effect | AC current (rms) (mA) | DC current (mA) |
|---|---|---|
| Threshold of sensation | 1 | 3 |
| Paralysis of respiratory muscles | 15 | 60 |
| Heart fibrillation, likely fatal | > 100 | > 500 |

FIGURE 26.13 Resistance model of the body.

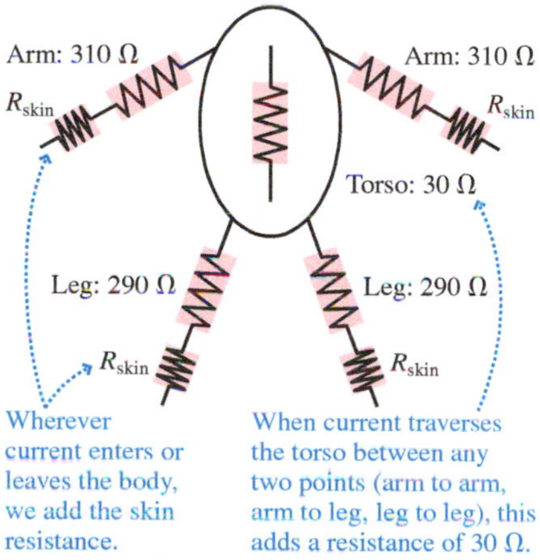

body, and the skin generally has a fairly high resistance. If you touch a wire with the dry skin of a finger, the skin's resistance might be greater than 1 MΩ. Moist skin and larger contact areas can reduce the skin's resistance to less than 10 kΩ.

## EXAMPLE 26.6 Is the worker in danger? BIO

A worker in a plant grabs a bare wire that he does not know is connected to a 480 V (rms) AC supply. His other hand is holding a grounded metal railing. The skin resistance of each of his hands, in full contact with a conductor, is 2200 Ω. He will receive a shock. Is it large enough to be dangerous?

***STRATEGIZE*** We will use the resistance model of the body shown in Figure 26.13. Here, the current path goes through the skin of one hand, up one arm, across the torso, down the other arm, and through the skin of the other hand.

***PREPARE*** We can draw a circuit model for this situation as in FIGURE 26.14a; the worker's body completes a circuit between two points at a potential difference of 480 V. The current will depend on this potential difference and the resistance of his body, including the resistance of the skin.

***SOLVE*** The individual resistances that determine the current are shown in FIGURE 26.14b. The equivalent resistance of the series combination is 5050 Ω, so the AC current through his body is

$$I_{rms} = \frac{\Delta V_{rms}}{R_{eq}} = \frac{480\text{ V}}{5050\ \Omega} = 95\text{ mA}$$

FIGURE 26.14 A circuit model for the worker.

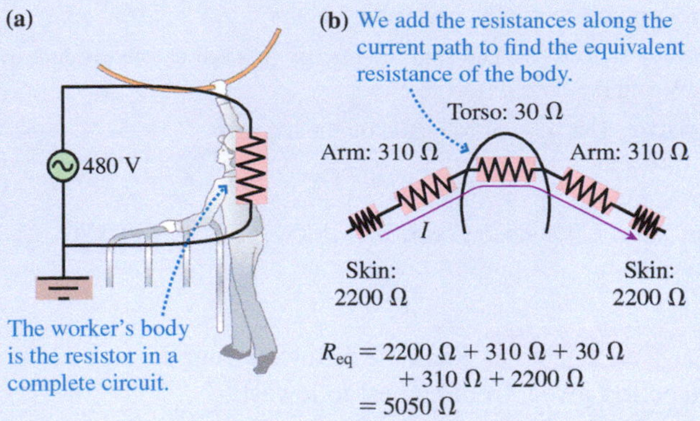


From Table 26.1 we see that this is a very dangerous, possibly fatal, current.

***ASSESS*** The voltage is high and the resistance relatively low, so it's no surprise to find a dangerous level of current.

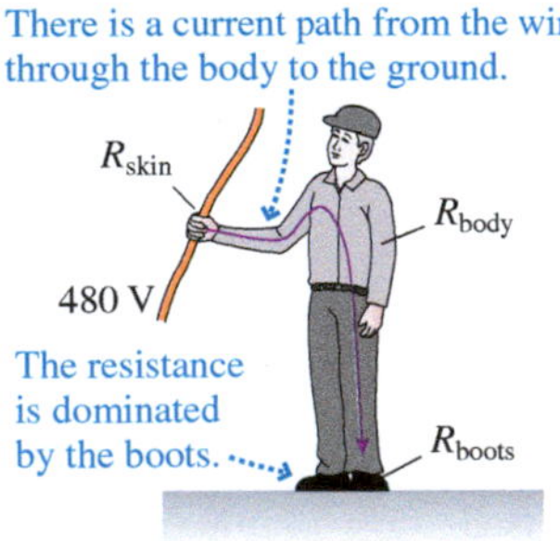


FIGURE 26.15 Wearing electrically insulating boots increases resistance.

Because of the danger of electric shock, workers who might accidentally contact electric lines wear protective clothing. The resistance of the protective clothing is much greater than that of the body or the skin. Boots, in particular, are made with soles that have very high electrical resistance, because workers are usually standing on the ground. Suppose a worker wearing protective boots touches a live wire at 480 V, as shown in FIGURE 26.15. There is a 480 V potential difference between his hand and the ground, so there will be a current through his body. It will travel through his skin, then his body, then his boots. The total resistance is the sum of these series resistances. For typical boots, the electrical resistance of the soles is at least 10 MΩ; this is so much greater than the resistance of the skin and the resistance of the body that the total resistance can be approximated as being equal to the resistance of the soles of the boots. If we assume a resistance of 10 MΩ for the current path, then the current that passes through the worker's body is only

$$I_{rms} = \frac{\Delta V_{rms}}{R_{eq}} \approx \frac{\Delta V_{rms}}{R_{boots}} = \frac{480\text{ V}}{10\text{ M}\Omega} = 48\ \mu\text{A}$$

In this case, the worker will be fine. Because the current is much less than the threshold for sensation, he won't even feel a shock!

Now, think about the birds sitting on the wire at the start of the chapter. How are they able to perch on the high-voltage wire? Think about the examples we've seen; we always consider the *potential difference* between two parts of the body. The wire is at an elevated potential with respect to the ground, so each foot is at a high potential, but there is only a very small potential *difference* between the feet due to the potential decrease along the wire. For the high-voltage power line considered earlier in the chapter, the potential difference between the two points along the wire where a bird's feet touch is less than 1 mV, not something the bird will notice! If a bird touches a wire and a grounded pole, or two neighboring wires, however, the result can be very different. This is a problem for birds with large wingspans that can establish a connection between widely separated conductors at very different voltages.

**Video** Electrical Safety

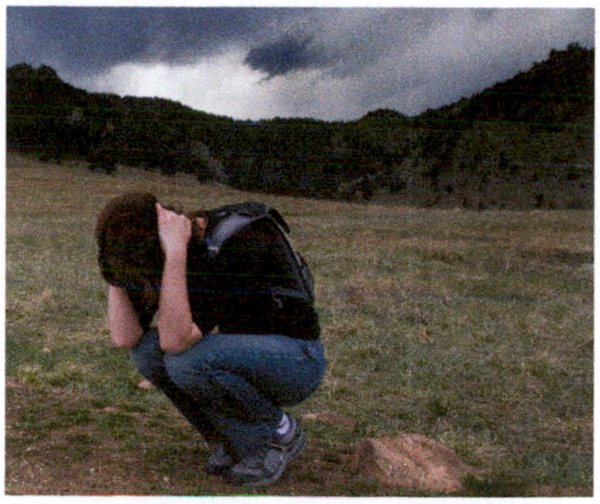

▶ **The lightning crouch** Most lightning injuries come from the resulting *ground current,* not the lightning itself. If you are standing near a point where lightning strikes the ground, there will be potential difference $\Delta V = Ed$ between your feet, which can cause a dangerous current up one leg and down the other. If you are caught outdoors in a lightning storm, you can minimize the potential difference $\Delta V$ by minimizing $d$, the distance between your feet. Experts advise you to assume the lightning crouch, shown in the photo, with your feet close together and your heels off the ground.

## GFI Circuits

If you are standing in good electrical contact with the ground, you are grounded. If you then accidentally touch a hot wire with your hand, a dangerous current could pass through you to ground. In kitchens and bathrooms, where grounding on damp floors is a good possibility, building codes require *ground fault interrupter* outlets, abbreviated GFI. A GFI outlet, as shown in FIGURE 26.16, has two buttons, one red and one black, for testing and resetting the circuitry. Some devices, such as hair dryers, are generally constructed with a GFI in the power cord.

GFI outlets have a built-in sensing circuit that compares the currents in the hot and neutral wires of the outlet. In normal operation, all the current coming in through the hot wire passes through the device and then back out through the neutral wire, so the currents in the hot and neutral wires should always be equal. If the current in the hot wire does *not* equal the current in the neutral wire, some current from the hot wire is finding an alternative path to ground—perhaps through a person. This is a *ground fault,* and the GFI disconnects the circuit. GFIs are set to trip at current differences of about 5 mA—large enough to feel, but not large enough to be dangerous.

FIGURE 26.16 A GFI outlet.

**STOP TO THINK 26.4** Suppose a current enters a person's body at one point and exits at another. Using the model of resistance of the body presented in Figure 26.13, rank the following current paths in terms of resistance, from highest to lowest. Assume that the skin resistance is zero.

A. Right hand to left hand
B. Right hand to right foot
C. Right hand to left foot
D. Right foot to left foot

**Video** What the Physics? Don't Try This at Home

# 26.5 Capacitor Circuits

In ◀ SECTION 23.7 we analyzed the one-time charging or discharging of a capacitor in an $RC$ circuit. In this chapter we will look at capacitors in circuits with an AC source of emf that repeatedly charges and discharges the capacitor.

FIGURE 26.17a shows a current $i_C$ charging a capacitor with capacitance $C$. The instantaneous capacitor voltage is $v_C = q/C$, where $\pm q$ is the charge on the two capacitor plates at this instant of time. FIGURE 26.17b illustrates the basic capacitor circuit, in which capacitance $C$ is connected across an AC source of emf $\mathcal{E}$. The capacitor is in parallel with the source, so the capacitor voltage equals the emf: $v_C = \mathcal{E} = \mathcal{E}_0 \cos(2\pi f t)$. It is useful to write

$$v_C = V_C \cos(2\pi f t) \qquad (26.15)$$

where $V_C$ is the peak or maximum voltage across the capacitor. $V_C = \mathcal{E}_0$ in this single-capacitor circuit.

Charge flows to and from the capacitor plates but not through the gap between the plates. But the charges $\pm q$ on the opposite plates are always of equal magnitude, so the currents into and out of the capacitor must be equal. We can find the current $i_C$ by considering how the charge $q$ on the capacitor varies with time. In FIGURE 26.18a on the next page we have plotted the oscillating voltage $v_C$ across the capacitor.

FIGURE 26.17 An AC capacitor circuit.

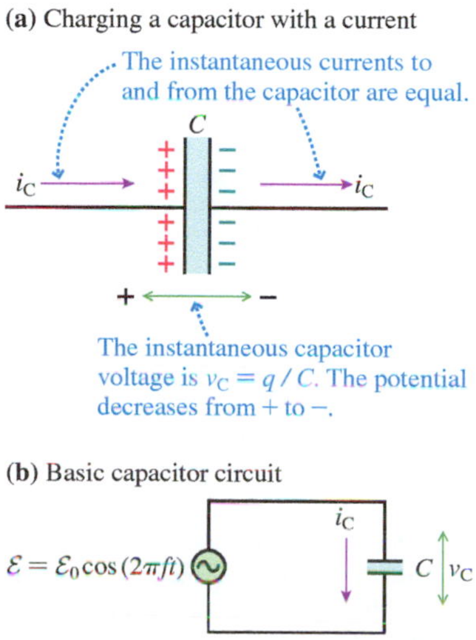

FIGURE 26.18 Voltage, charge, and current graphs for a capacitor in an AC circuit.

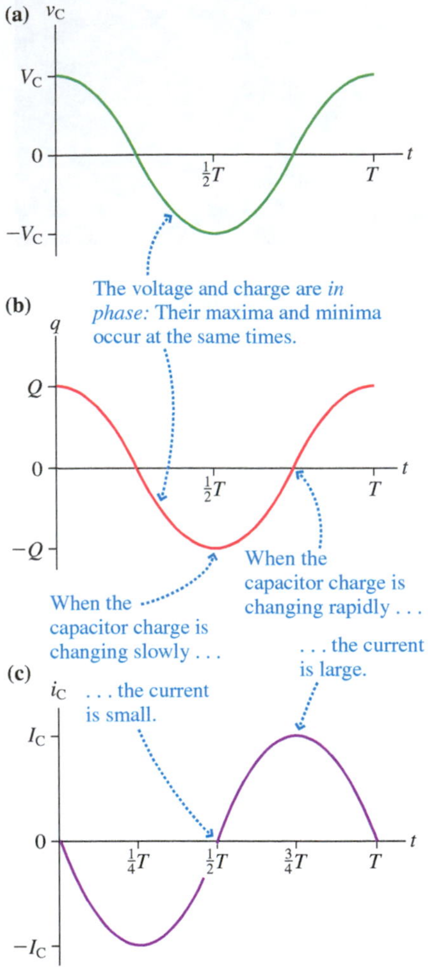


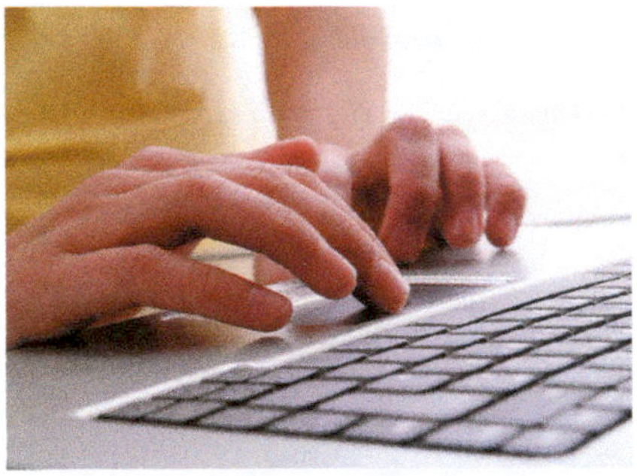

**Detecting digits** Under the surface of a trackpad or a touchscreen is an array of tiny capacitors. When you touch the pad or screen, your finger's high dielectric constant (it's largely water) changes the capacitance of nearby capacitors, altering the current in the capacitor circuits and thereby revealing the location of your finger. The eraser end of a pencil, with its small dielectric constant, won't work—try it!

Because the charge and the capacitor voltage are directly proportional, with $q = Cv_C$, the graph of $q$, shown in FIGURE 26.18b, looks like the graph of $v_C$. We say that the charge is *in phase* with the voltage.

In Chapter 22, we defined current to be $\Delta q/\Delta t$. The capacitor current, $i_C$, will thus be related to the charge on the capacitor by

$$i_C = \frac{\Delta q}{\Delta t}$$

Thus the current through the capacitor is equal to the *rate* $\Delta q/\Delta t$ at which the capacitor's charge is changing. From Figure 26.18b you can see that this rate of change is zero at $t = 0, \frac{T}{2}$, and $T$; it is large and negative at $t = \frac{T}{4}$, and is large and positive at $t = \frac{3T}{4}$. The current through the capacitor, which is proportional to these rates of change, is plotted in FIGURE 26.18c.

You can see that a capacitor's voltage and current are *not* in phase, as they are for a resistor: The maxima of the current do not occur at the same time as the maxima of the voltage, and the current minima do not occur at the same time as the voltage minima. More precisely, if we compare Figures 26.18a and 26.18c, we see that the current peaks at $\frac{3}{4}T$, one-quarter period *before* the voltage peaks. We say that **the AC current through a capacitor *leads* the capacitor voltage.**

## Capacitive Reactance

The phase of the current is not the whole story. We would also like to know the current's peak value, which we can also determine from $i_C = \Delta q/\Delta t$. Because $q = Cv_C$, this can be written as

$$i_C = C\,\frac{\Delta v_C}{\Delta t} \qquad (26.16)$$

Let's now reason by analogy. In ◀◀ SECTION 14.3 the position of a simple harmonic oscillator was given by $x = A\cos(2\pi ft)$. Further, the oscillator's velocity $v = \Delta x/\Delta t$ was found to be $v = -v_{max}\sin(2\pi ft)$, with a maximum, or peak, value $v_{max} = 2\pi fA$. Here we have an oscillating voltage $v_C = V_C\cos(2\pi ft)$, analogous to position, and we need to find the quantity $\Delta v_C/\Delta t$, analogous to velocity. Thus $\Delta v_C/\Delta t$ must have a maximum, or peak, value $(\Delta v_C/\Delta t)_{max} = 2\pi fV_C$.

The peak capacitor current $I_C$ occurs when $\Delta v_C/\Delta t$ is maximum, so the peak current is

$$I_C = C(2\pi fV_C) = (2\pi fC)V_C \qquad (26.17)$$

For a resistor, the peak current and voltage are related through Ohm's law:

$$I_R = \frac{V_R}{R}$$

We can write Equation 26.17 in a form similar to Ohm's law if we define the **capacitive reactance** $X_C$ to be

$$X_C = \frac{1}{2\pi fC} \qquad (26.18)$$

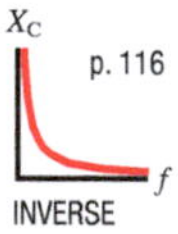


With this definition of capacitive reactance, Equation 26.17 becomes

$$I_C = \frac{V_C}{X_C} \quad \text{or} \quad V_C = I_C X_C \qquad (26.19)$$

Peak current through or voltage across a capacitor

The units of reactance, like those of resistance, are ohms.

NOTE ▶ Reactance relates the *peak* voltage $V_C$ and current $I_C$. It does *not* relate the *instantaneous* capacitor voltage and current because they are out of phase; that is, $v_C \neq i_C X_C$. ◀

A capacitor's reactance $X_C$ depends inversely on the frequency. The reactance becomes very large at low frequencies (i.e., the capacitor is a large impediment to current). The reactance decreases as the frequency increases until, at very high frequencies, $X_C \approx 0$ and the capacitor begins to act like an ideal wire.

**EXAMPLE 26.7 Finding the capacitive reactance**

What is the capacitive reactance of a 0.100 $\mu$F capacitor at a 100 Hz audio frequency and at a 100 MHz FM-radio frequency?

***STRATEGIZE*** The capacitive reactance is given by Equation 26.18.

***SOLVE*** At 100 Hz,

$$X_C(\text{at } 100 \text{ Hz}) = \frac{1}{2\pi f C} = \frac{1}{2\pi(100 \text{ Hz})(1.00 \times 10^{-7} \text{ F})} = 15{,}900\ \Omega$$

Increasing the frequency by a factor of $10^6$ decreases $X_C$ by a factor of $10^6$, giving

$$X_C(\text{at } 100 \text{ MHz}) = 0.0159\ \Omega$$

***ASSESS*** A capacitor with a substantial reactance at audio frequencies has virtually no reactance at FM radio frequencies.

**EXAMPLE 26.8 Finding a capacitor's current**

A 10 $\mu$F capacitor is connected to a 1000 Hz oscillator with a peak emf of 5.0 V. What is the peak current through the capacitor?

***STRATEGIZE*** The current is found from Equation 26.19. We'll also need the capacitive reactance from Equation 26.18.

***PREPARE*** The circuit diagram is as in Figure 26.17b. This is a simple one-capacitor circuit.

***SOLVE*** The capacitive reactance at $f = 1000$ Hz is

$$X_C = \frac{1}{2\pi f C} = \frac{1}{2\pi(1000 \text{ Hz})(10 \times 10^{-6} \text{ F})} = 16\ \Omega$$

The peak voltage across the capacitor is $V_C = \mathcal{E}_0 = 5.0$ V; hence the peak current is

$$I_C = \frac{V_C}{X_C} = \frac{5.0 \text{ V}}{16\ \Omega} = 0.31 \text{ A}$$

***ASSESS*** Using reactance and Equation 26.19 is just like using resistance and Ohm's law, but don't forget that it applies to only the *peak* current and voltage, not the instantaneous values. Further, reactance, unlike resistance, depends on the frequency of the signal.

**STOP TO THINK 26.5** A capacitor is attached to an AC voltage source. Which change will result in a doubling of the current?

A. Halving the voltage and doubling the frequency
B. Doubling the frequency
C. Halving the frequency
D. Doubling the voltage and halving the frequency

## 26.6 Inductors and Inductor Circuits

**FIGURE 26.19** A changing current through a solenoid induces an emf across the solenoid.

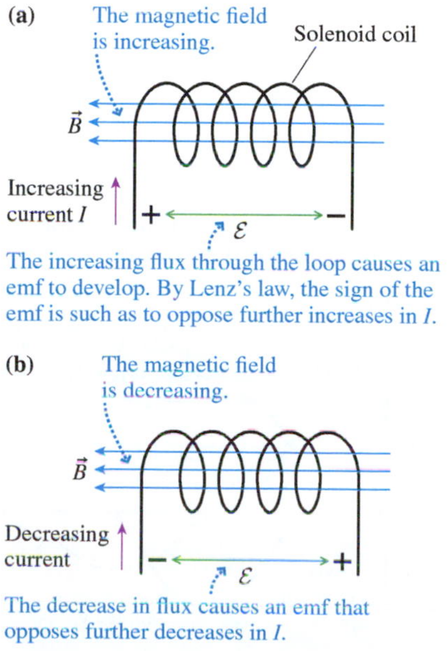


FIGURE 26.19 shows a length of wire formed into a coil, making a solenoid. In Chapter 24 you learned that current in a solenoid creates a magnetic field inside the solenoid. If this current is *increasing*, as shown in Figure 26.19a, then the magnetic field—and thus the flux—inside the coil increases as well. According to Faraday's law, this changing flux causes an emf—a potential difference—to develop across the coil. The *direction* of the emf can be inferred from Lenz's law: Its direction will *oppose* the increase in the flux; that is, it will oppose the increase in the current. The emf will have the opposite sign if the current through the coil is decreasing, as shown in Figure 26.19b.

Coils of this kind, called **inductors,** are widely used in AC circuits. The circuit symbol for an inductor is . There are two primary things to remember about an inductor. First, **an inductor develops a potential difference across it if**

**the current through it is *changing*.** Second, because the direction of this potential difference opposes the change in the current, **an inductor resists changes in the current through it.**

**Video** The Inductor

**Is anybody up there?** Detectors beneath the pavement can sense the presence of cars waiting at intersections and on stretches of road before traffic lights. A slot is cut in the pavement and a coil of wire is sealed into the slot; this coil of wire is the inductor in a detection circuit. The presence of the steel in a car above the coil greatly increases the coil's inductance, signaling that a car is present.

## Inductance

By Faraday's law, the voltage developed across an inductor is proportional to the rate at which the flux through the coil changes. In addition, because the flux is proportional to the coil's current, the instantaneous inductor voltage $v_L$ must be proportional to $\Delta i_L/\Delta t$, the rate at which the current through the inductor changes. Thus we can write

$$v_L = L\frac{\Delta i_L}{\Delta t} \qquad (26.20)$$

The constant of proportionality $L$ is called the **inductance** of the inductor. A coil with many turns has a higher inductance than a similarly sized coil with fewer turns. Inductors often have an iron core inside their coil windings to increase their inductance. The magnetic field from the current magnetizes the iron core, which greatly increases the overall field through the windings. This gives a larger change in flux through the windings and hence a larger induced emf. Equation 26.20 shows that this implies a larger value of $L$.

From Equation 26.20 we see that inductance has units of V · s/A. It's convenient to define an SI unit of inductance called the **henry,** in honor of Joseph Henry, an early investigator of magnetism. We have

$$1 \text{ henry} = 1 \text{ H} = 1 \text{ V}\cdot\text{s/A}$$

Practical inductances are usually in the range of millihenries (mH) or microhenries ($\mu$H).

FIGURE 26.20 An AC inductor circuit.

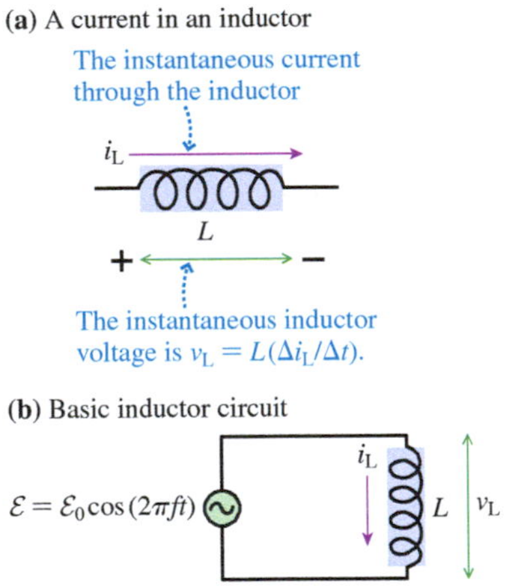


## Inductor Circuits

FIGURE 26.20a shows the instantaneous current $i_L$ through an inductor. If the current is changing, the instantaneous inductor voltage is given by Equation 26.20.

FIGURE 26.20b, where inductance $L$ is connected across an AC source of emf $\mathcal{E}$, is the simplest inductor circuit. The inductor is in parallel with the source, so the inductor voltage equals the emf: $v_L = \mathcal{E} = \mathcal{E}_0\cos(2\pi ft)$. We can write

$$v_L = V_L\cos(2\pi ft) \qquad (26.21)$$

where $V_L$ is the peak or maximum voltage across the inductor. You can see that $V_L = \mathcal{E}_0$ in this single-inductor circuit.

We can find the inductor current $i_L$ by considering again Equation 26.20, which tells us that the inductor voltage is high when the current is changing rapidly (i.e., $\Delta i_L/\Delta t$ is large) and is low when the current is changing slowly (i.e., $\Delta i_L/\Delta t$ is small). From this, we can graphically find the current, as shown in FIGURE 26.21.

Just as for a capacitor, the current and voltage are not in phase. There is again a phase difference of one-quarter cycle, but for an inductor the current peaks one-quarter period *after* the voltage peaks. **The AC current through an inductor *lags* the inductor voltage.**

FIGURE 26.21 Voltage and current graphs for an inductor in an AC circuit.

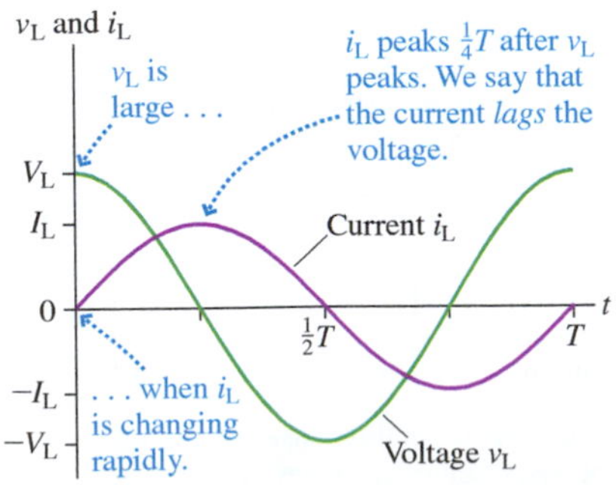


## Inductive Reactance

To find the relationship between the peak values of the inductor's current and voltage, we can use Equation 26.20 and, as we did for the capacitor, the analogy with simple harmonic motion. For the oscillating capacitor voltage, with peak value $(v_C)_{max} = V_C$, the maximum value of $\Delta v_C/\Delta t$ was $(\Delta v_C/\Delta t)_{max} = 2\pi f V_C$. If an oscillating inductor current has peak value $(i_L)_{max} = I_L$, then, by exactly the same reasoning, the maximum value of $\Delta i_L/\Delta t$ is $(\Delta i_L/\Delta t)_{max} = 2\pi f I_L$. If we use this result in Equation 26.20, we see that the maximum or peak value of the inductor voltage is

$$V_L = L(2\pi f I_L) = (2\pi f L)I_L \qquad (26.22)$$

We can write Equation 26.22 in a form reminiscent of Ohm's law:

$$I_L = \frac{V_L}{X_L} \quad \text{or} \quad V_L = I_L X_L \tag{26.23}$$

Peak current through or voltage across an inductor

where the **inductive reactance,** analogous to the capacitive reactance, is defined as

$$X_L = 2\pi f L \tag{26.24}$$

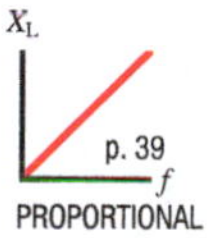


The inductive reactance increases linearly as the frequency increases. This makes sense. Faraday's law tells us that the induced voltage across a coil increases as the rate of change of $B$ increases, and $B$ is directly proportional to the inductor current.

Now that we've seen how resistors, capacitors, and inductors behave in circuits, it's worthwhile to collect the details to compare the similarities and differences.

**Clean power** The digital circuits inside computers generate high-frequency AC signals that can "leak" through the power supply and propagate through a household's electricity supply. To help prevent this, there are one or more inductors—in this case, a copper coil with an iron core—on the board inside a computer that connect it to your household electricity. At high frequencies, the inductive reactance is high enough to significantly reduce the unwanted transmissions.

## SYNTHESIS 26.1 AC circuit elements

The graphs of current and voltage are both sinusoidal, but their phases are different for the three circuit elements.

**Resistor** The current and voltage are *in phase:* they peak or are zero at the same time.

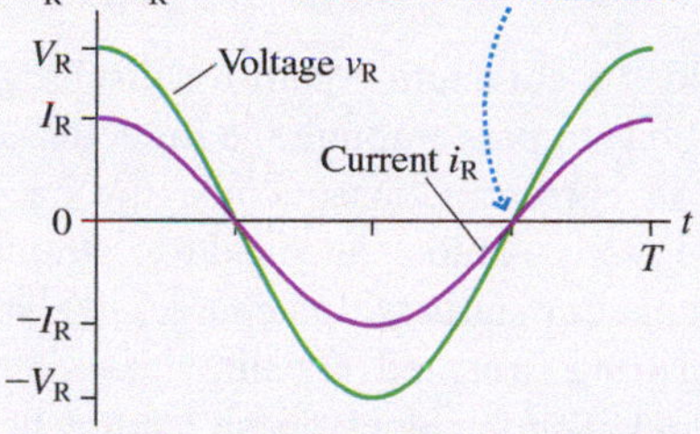

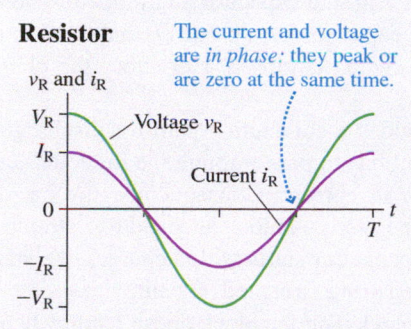

**Capacitor** The current *leads* the voltage: the current peaks before the voltage does.

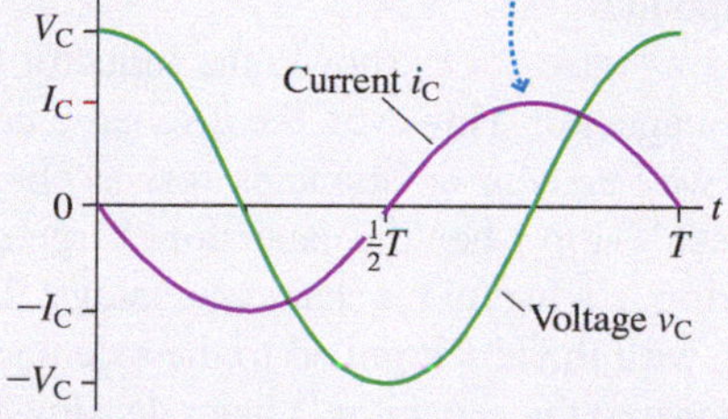


**Inductor** The current *lags* the voltage: the current peaks after the voltage does.

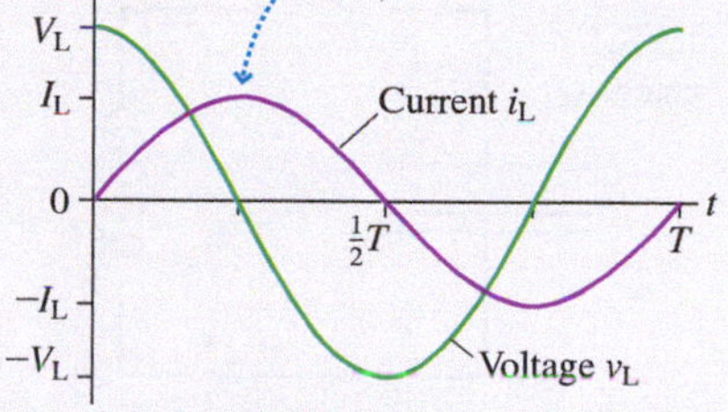


The *peak* values of current and voltage are related by the resistance or the reactance:

$V_R = I_R R$ ← Resistance (Ω)

$V_C = I_C X_C$ ← Capacitive reactance (Ω)

$V_L = I_L X_L$ ← Inductive reactance (Ω)

Resistance is constant, but capacitive and inductive reactance vary with frequency:

Does not depend on frequency → $R = \text{constant}$

Inversely proportional to frequency → $X_C = \dfrac{1}{2\pi f C}$ (Capacitance (F); Frequency (Hz))

Proportional to frequency → $X_L = 2\pi f L$ (Inductance (H); Frequency (Hz))

## EXAMPLE 26.9 Finding the current and voltage of a radio's inductor

A 0.25 $\mu$H inductor is used in an FM radio circuit that oscillates at 100 MHz. The current through the inductor reaches a peak value of 2.0 mA at $t = 5.0$ ns. What is the peak inductor voltage, and when, closest to $t = 5.0$ ns, does it occur?

**STRATEGIZE** Synthesis 26.1 shows that the inductor current lags the voltage; the voltage reaches its peak value one-quarter period before the current reaches its peak.

**PREPARE** The inductor is in a circuit like that in Figure 26.20b.

**SOLVE** The inductive reactance at $f = 100$ MHz $= 1.0 \times 10^8$ Hz is

$$X_L = 2\pi f L = 2\pi(1.0 \times 10^8\ \text{Hz})(0.25 \times 10^{-6}\ \text{H}) = 157\ \Omega$$

Thus the peak voltage is $V_L = I_L X_L = (0.0020\ \text{A})(157\ \Omega) = 0.31$ V. The voltage peak occurs one-quarter period before the current peaks, and we know that the current peaks at $t = 5.0$ ns. The period of a 100 MHz oscillation is 10 ns, so the voltage peaks at

$$t = 5.0\ \text{ns} - \frac{10\ \text{ns}}{4} = 2.5\ \text{ns}$$

**ASSESS** One period of a 100 MHz signal is 10 ns, so 2.5 ns—a quarter of a period—sounds reasonable for the lag between the current and voltage.

**STOP TO THINK 26.6** An inductor is attached to an AC voltage source. Which change will result in a halving of the current?

A. Halving the voltage and doubling the frequency
B. Doubling the frequency
C. Halving the frequency
D. Doubling the voltage and halving the frequency

## 26.7 Oscillation Circuits

Each radio station in your city broadcasts at its own frequency, but you can tune a radio to pick up one station and no other. This is done using an *oscillation circuit,* a circuit that is designed to have a particular frequency at which it "wants" to oscillate. Tuning your radio means adjusting the frequency of the oscillation circuit to equal that of the station you want to listen to.

### *LC* Circuits

You learned in Chapter 23 that the voltage across a charged capacitor decays exponentially if the capacitor is connected to a resistor to form an *RC* circuit. Something very different occurs if the resistor is replaced with an *inductor.* Instead of decaying to zero, the capacitor voltage now undergoes sinusoidal *oscillations.*

FIGURE 26.22 An *LC* circuit.

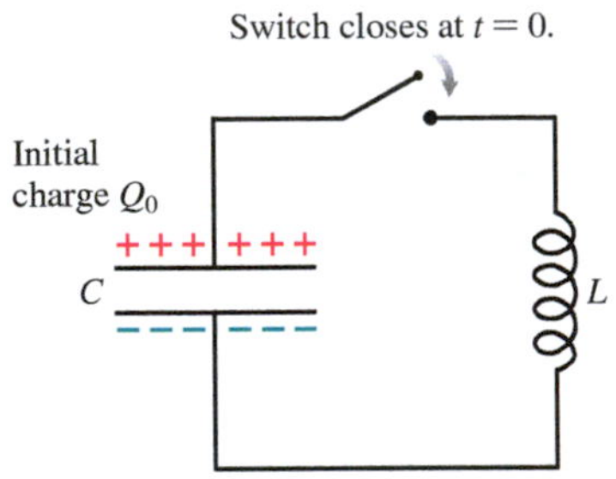


To understand how this occurs, let's start with the capacitor and inductor shown in the ***LC* circuit** of FIGURE 26.22. Initially, the capacitor has charge $Q_0$ and there is no current in the inductor. Then, at $t = 0$, the switch is closed. How does the circuit respond?

As FIGURE 26.23 shows, the inductor provides a conducting path for discharging the capacitor. However, the discharge current has to pass through the inductor, and, as we've seen, an inductor resists changes in current. Consequently, the current doesn't stop when the capacitor charge reaches zero. A block attached to a stretched spring is a useful mechanical analogy. The capacitor starts with a charge, like starting with the block pulled to the side and the spring stretched. Closing the switch to discharge the capacitor is like releasing the block. But the block doesn't stop when it reaches the origin—it keeps going until the spring is fully compressed. Likewise, the current continues until it has recharged the capacitor with the opposite polarization. This process repeats over and over, charging the capacitor first one way, then the other. The charge and current *oscillate.*

Recall that the oscillation frequency of a mass $m$ on a spring with spring constant $k$ is

$$f = \frac{1}{2\pi}\sqrt{\frac{k}{m}}$$

The oscillation frequency thus depends only on the two basic parameters $k$ and $m$ of the system and not on the amplitude of the oscillation. Similarly, the frequency of an *LC* oscillator is determined solely by the values of its inductance and capacitance. We would expect larger values of $L$ and $C$ to cause an oscillation with a lower frequency, because a larger inductance means the current changes more slowly and a larger capacitance takes longer to discharge. A detailed analysis shows that the frequency has a form reminiscent of that for a mass and spring:

$$f = \frac{1}{2\pi\sqrt{LC}} \quad (26.25)$$

Frequency of an *LC* oscillator

KEY CONCEPT **FIGURE 26.23** The capacitor charge oscillates much like a block attached to a spring.

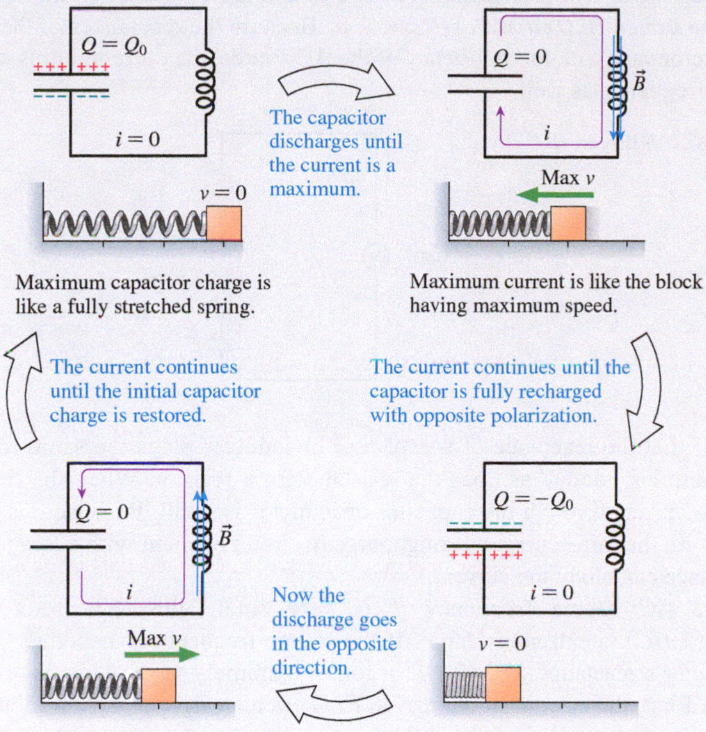


**STOP TO THINK 26.7** Consider the oscillation circuit shown in Figure 26.23. During the portion of the cycle when the capacitor's top plate is positive and the amount of charge is increasing, the current through the inductor is

A. Clockwise and increasing.
B. Clockwise and decreasing.
C. Zero.
D. Counterclockwise and increasing.
E. Counterclockwise and decreasing.

**EXAMPLE 26.10 Determining the frequency of an *LC* oscillator**

An *LC* circuit consists of a 10 $\mu$H inductor and a 500 pF capacitor. What is the oscillator's frequency?

**STRATEGIZE** Equation 26.25 relates an LC circuit's frequency to its capacitance and inductance.

**SOLVE** From Equation 26.25 we have

$$f = \frac{1}{2\pi\sqrt{LC}} = \frac{1}{2\pi\sqrt{(10\times10^{-6}\,\text{H})(500\times10^{-12}\,\text{F})}}$$

which gives $f = 2.3\times10^6$ Hz $= 2.3$ MHz.

**ASSESS** The frequencies of *LC* oscillators are generally *much* higher than those of mechanical oscillators. Frequencies in the MHz or even GHz ($10^9$ Hz) range are typical.

## *RLC* Circuits

Once the oscillations of the ideal *LC* circuit of Figure 26.23 are started, they will continue forever. All real circuits, however, have some resistance, which dissipates energy, causing the amplitudes of the voltages and currents to decay.

We can model this situation by adding a series resistor *R* to our *LC* circuit, as shown in FIGURE 26.24. The circuit still oscillates, but the peak values of the voltage and current decrease with time as the current through the resistor transforms the electric and magnetic energy of the circuit into thermal energy. Because the power loss is proportional to *R*, a larger resistance causes the oscillations to decay more

**FIGURE 26.24** An *RLC* circuit.

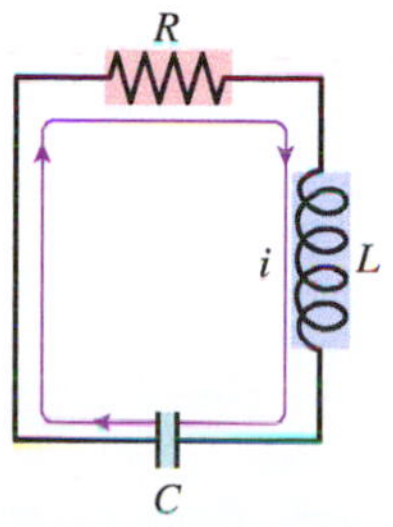

FIGURE 26.25 An *RLC* circuit exhibits damped oscillations.

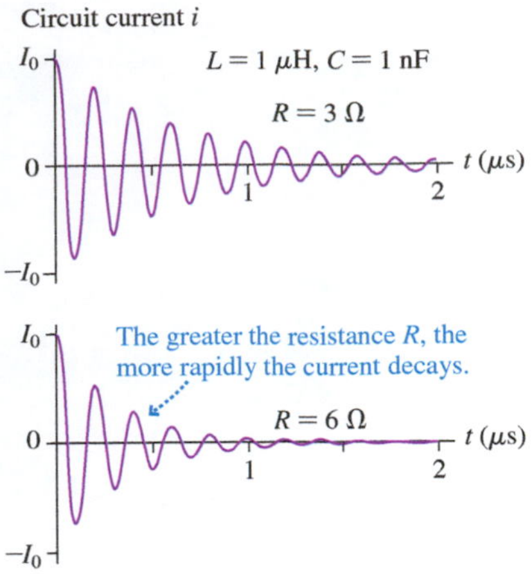

**Nuclear magnetic resonance,** or NMR, is an important analytic technique in chemistry and biology. In a large magnetic field, the magnetic moments of atomic nuclei rotate at tens to hundreds of MHz. This motion generates a changing magnetic field that induces an AC emf in a coil placed around the sample. Each kind of nucleus rotates at a characteristic frequency. Adding a capacitor to the coil creates an *RLC* resonance circuit that responds strongly to only one frequency—and hence to one kind of nucleus.

rapidly, as shown in FIGURE 26.25. This behavior is analogous to the damped harmonic oscillator discussed in ◀◀ SECTION 14.6.

To keep the circuit oscillating, we need to add an AC source to the circuit. This makes the **driven *RLC* circuit** of FIGURE 26.26. Because the reactances of the capacitor and inductor vary with the frequency of the AC source, the current in this circuit varies with frequency as well.

FIGURE 26.26 A driven *RLC* circuit.

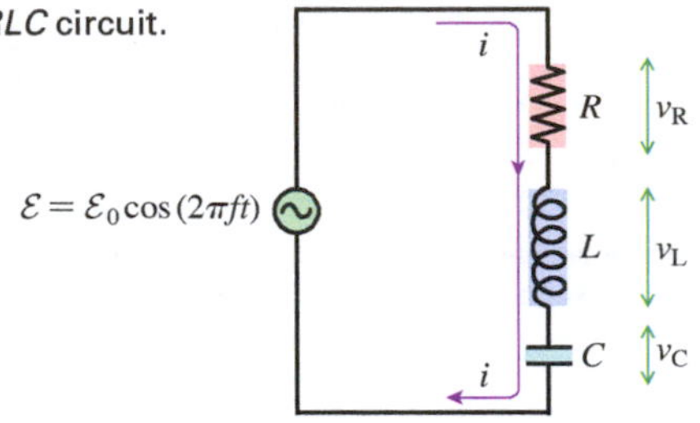


Recall that the reactance of a capacitor or inductor plays the same role for the peak quantities $I$ and $V$ as does the resistance of a resistor. When the reactance is large, the current through the capacitor or inductor is small. Because this is a series circuit, with the same current throughout, any circuit element with a large resistance or reactance can block the current.

If the AC source frequency $f$ is very small, the capacitor's reactance $X_C = 1/(2\pi f C)$ is extremely large. If the source frequency $f$ becomes very large, the inductor's reactance $X_L = 2\pi f L$ becomes extremely large. This has two consequences. First, the current in the driven *RLC* circuit will approach zero at very low and very high frequencies. Second, there must be some intermediate frequency, where neither $X_C$ nor $X_L$ is too large, at which the circuit current $I$ is a maximum. The frequency $f_0$ at which the current is at its maximum value is called the **resonance frequency.**

Resonance occurs at the frequency at which the capacitive reactance equals the inductive reactance. If the reactances are equal, the capacitor voltage and the inductor voltage are the same. But these two voltages are out of phase with the current, with the current *leading* the capacitor voltage and *lagging* the inductor voltage. At resonance, the capacitor and inductor voltages have equal magnitudes but are exactly out of phase with each other. When we add all the voltages around the loop in Kirchhoff's loop law, the capacitor and inductor voltages cancel and the current is then limited only by the resistance $R$. Thus the condition for resonance is $X_C = X_L$, or

$$\frac{1}{2\pi f_0 C} = 2\pi f_0 L$$

which gives

$$f_0 = \frac{1}{2\pi\sqrt{LC}} \tag{26.26}$$

This is the same frequency at which the circuit would oscillate if it had no resistor. At this frequency, maximum current is determined by the magnitude of the emf and the resistance:

$$I_{max} = \frac{\mathcal{E}_0}{R} \tag{26.27}$$

The driven *RLC* circuit is directly analogous to the driven, damped oscillator that you studied in ◀◀ SECTION 14.7. A mechanical oscillator exhibits resonance by having a large-amplitude response when the driving frequency matches the system's natural frequency. Equation 26.26 is the natural frequency of the driven *RLC* circuit, the frequency at which the current "wants" to oscillate. The circuit has a large current response when the oscillating emf matches this frequency.

FIGURE 26.27 shows the peak current $I$ of a driven $RLC$ circuit as the emf frequency $f$ is varied. Notice how the current increases until reaching a maximum at frequency $f_0$, then decreases. This is the hallmark of resonance.

As $R$ decreases, causing the damping to decrease, the maximum current becomes larger and the curve in Figure 26.27 becomes narrower. You saw exactly the same behavior for a driven mechanical oscillator. The emf frequency must be very close to $f_0$ in order for a lightly damped system to respond, but the response at resonance is very large—exactly what is needed for a tuning circuit.

It is possible to derive an expression for the current graphs shown in Figure 26.27. The peak current in the $RLC$ circuit is

$$I = \frac{\mathcal{E}_0}{\sqrt{R^2 + (X_L - X_C)^2}} = \frac{\mathcal{E}_0}{\sqrt{R^2 + (2\pi f L - 1/2\pi f C)^2}} \qquad (26.28)$$

Peak current in an $RLC$ circuit

Note that the denominator is smallest, and hence the current is largest, when $X_L = X_C$; that is, at resonance. The three peak voltages, if you need them, are then found from $V_R = IR$, $V_L = IX_L$, and $V_C = IX_C$.

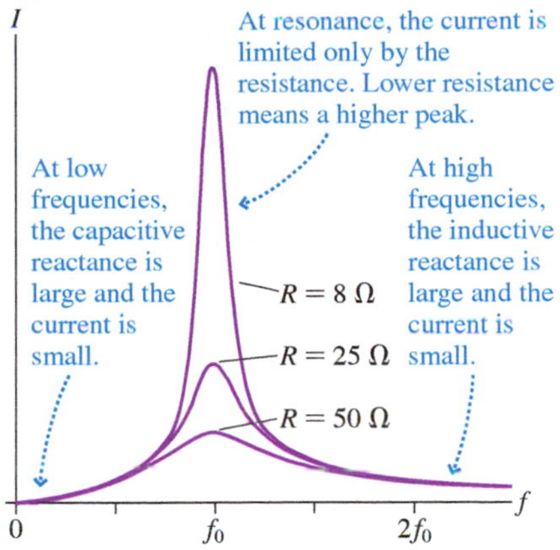


FIGURE 26.27 A graph of the current $I$ versus emf frequency for a series $RLC$ circuit.

## EXAMPLE 26.11 Designing a radio receiver

An AM radio antenna picks up a 1000 kHz signal with a peak voltage of 5.0 mV. The tuning circuit consists of a 60 $\mu$H inductor in series with a variable capacitor. The inductor coil has a resistance of 0.25 Ω, and the resistance of the rest of the circuit is negligible.

a. To what value should the capacitor be tuned to listen to this radio station?
b. What is the peak current through the circuit at resonance?
c. A stronger station at 1050 kHz produces a 10 mV antenna signal. What is the current at this frequency when the radio is tuned to 1000 kHz?

**STRATEGIZE** The inductor's 0.25 Ω resistance can be modeled as a resistance in series with the inductance, so we have a series RLC circuit. To have the maximum response to the desired 1000 kHz radio frequency, the capacitor should be adjusted so that the resonance frequency of the circuit is also 1000 kHz. Then the peak current is given by Equation 26.27. We will find the current through the circuit for the 1050 kHz station using Equation 26.28.

**PREPARE** The antenna signal at $f = 1000\text{ kHz} = 10^6\text{ Hz}$ is the emf $\mathcal{E}_0$. The circuit looks like that in Figure 26.26.

**SOLVE**

a. Because $f_0 = 1/2\pi\sqrt{LC}$, the appropriate capacitance is

$$C = \frac{1}{L(2\pi f_0)^2} = \frac{1}{(60 \times 10^{-6}\text{ H})(2\pi \times 10^6\text{ Hz})^2}$$
$$= 4.22 \times 10^{-10}\text{ F} \approx 420\text{ pF}$$

b. The maximum current is

$$I_{max} = \frac{\mathcal{E}_0}{R} = \frac{5.0 \times 10^{-3}\text{ V}}{0.25\ \Omega} = 0.020\text{ A} = 20\text{ mA}$$

c. The 1050 kHz signal is "off resonance," so the reactances $X_L$ and $X_C$ are not equal: $X_L = 2\pi f L = 396\ \Omega$ and $X_C = 1/2\pi f C = 359\ \Omega$ at $f = 1050$ kHz. The peak voltage of this signal is $\mathcal{E}_0 = 10$ mV. With these values, Equation 26.28 for the peak current is

$$I = \frac{\mathcal{E}_0}{\sqrt{R^2 + (X_L - X_C)^2}} = 0.27\text{ mA}$$

**ASSESS** These are realistic values for the input stage of an AM radio. You can see that the signal from the 1050 kHz station is strongly suppressed when the radio is tuned to 1000 kHz. The resonant circuit has a large response to the selected station, but not nearby stations.

## INTEGRATED EXAMPLE 26.12 The ground fault interrupter

As we've seen, a GFI disconnects a household circuit when the currents in the hot and neutral wires are unequal. Let's look inside a GFI outlet to see how this is done.

When you plug in an appliance such as a hair dryer, its heating coils—the load resistance $R$—complete the circuit between the hot and neutral wires. At any instant the AC currents $i$ in the hot and neutral wires are equal and opposite, as FIGURE 26.28 shows.

FIGURE 26.28 Currents in the wires in an electric outlet.

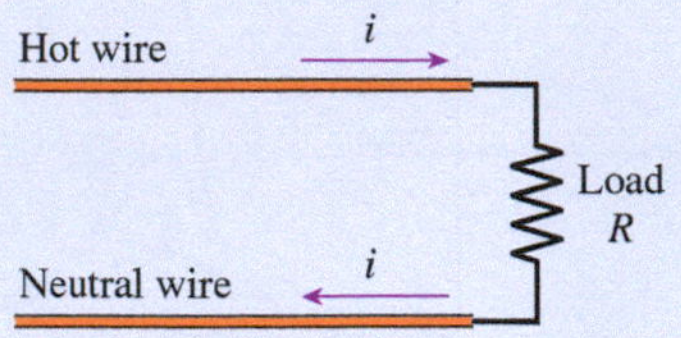


If someone accidentally touches the hot wire—a potentially dangerous situation—some of the current in the hot wire is diverted to a path through the person's body. The currents in the hot and neutral wires are no longer equal. This difference is sensed by the GFI, and the outlet is switched off.

To check for any difference in current, inside the GFI the hot and neutral wires are threaded through an iron ring, as FIGURE 26.29 shows. A coil of fine wire wraps around the ring; this coil is connected to a circuit that detects any current in the coil.

FIGURE 26.29 The working elements of a GFI outlet.

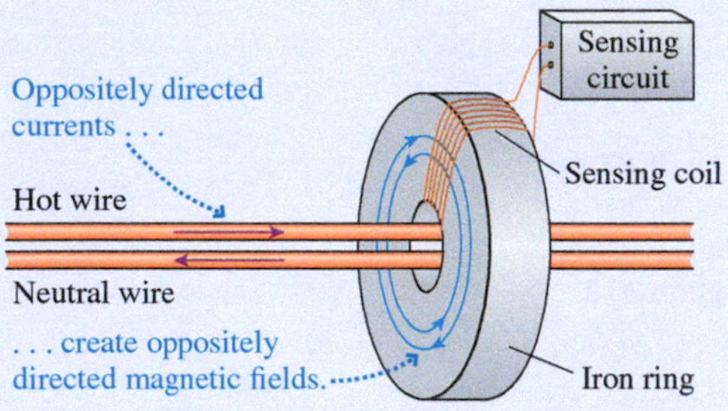


If the currents in the hot and neutral wires are equal and opposite, the magnetic fields of the two wires are also equal and opposite, so there is no net magnetic field in the iron ring and thus no flux through the sensing coil. But if the two currents aren't equal, there are both a net field and a net flux. Because the current is AC, the flux through the sensing coil is changing rapidly; according to Faraday's law, this changing flux induces a current in the coil. The sensing circuit detects this current and opens a small circuit breaker to turn the outlet off within a few milliseconds, in time to prevent any injury.

a. A GFI will break the circuit if the difference in current between the hot and neutral wires is 5.0 mA. Suppose that, at some instant, there is an excess current of 5.0 mA to the right in the hot wire of Figure 26.29. What are the direction and the magnitude of the magnetic field in the iron ring due to the current in the wire? The ring has an average diameter of 0.80 cm.

b. Consider a worst-case scenario: A person sitting in a bathtub reaches out of the tub and accidentally touches a hot wire with a wet, soapy hand that has minimal skin resistance. What would be the peak current through the person's body? Would this be dangerous? Would a GFI disconnect the circuit?

**STRATEGIZE** Part a is about the magnetic fields of the wires. The currents in the hot and neutral wires create a field around the wires, as we saw in Chapter 24. We can use Equation 24.1 to find the magnitude of the magnetic field. Part b is an electrical safety question; we will need to find the resistance of the current path through the body to find the current.

**PREPARE** For part a, we need Equation 24.1 for the magnetic field of a long wire, which is $B = \mu_0 I/2\pi r$. The iron ring has an average diameter of 0.80 cm, so we need to find the field at a distance $r = 0.40$ cm from the wire. Equal currents produce no net field, so we only need to consider the field due to the excess current in the hot wire, so that $I = 5.0$ mA. For part b, we note that the person in the tub is sitting in conducting water that is well grounded. If this person touches a hot wire, there is a current path from the hot wire to ground, as FIGURE 26.30a shows. Resistance values for elements of the body are as shown in Figure 26.13. Because of the negligible skin resistance of a wet, soapy hand, the equivalent resistance of the body is the sum of the resistances of the arm and the torso: $R_{eq} = 340\ \Omega$.

Given that the neutral wire of the electricity supply is also grounded, there is a complete circuit through the person, as shown in FIGURE 26.30b.

FIGURE 26.30 The resistance of the path through the body and the resulting circuit.

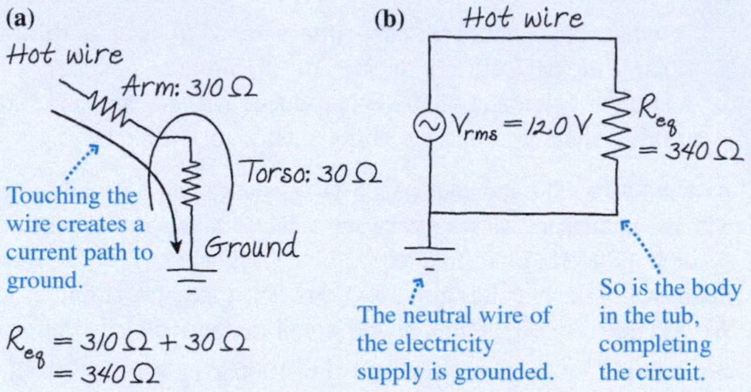


**SOLVE**

a. The magnetic field due to the currents is

$$B = \frac{\mu_0 I}{2\pi r} = \frac{(1.26 \times 10^{-6}\ \text{T}\cdot\text{m/A})(5.0 \times 10^{-3}\ \text{A})}{2\pi(0.0040\ \text{m})}$$
$$= 2.5 \times 10^{-7}\ \text{T} = 0.25\ \mu\text{T}$$

From Tactics Box 24.1, we can use the right-hand rule for fields to find that the field is clockwise through the iron ring in the view of Figure 26.29.

b. The peak current in the circuit of Figure 26.30b, occurring at the peak of the sinusoidal household AC voltage ($\mathcal{E}_0 = 170$ V), is

$$I_R = \frac{\mathcal{E}_0}{R_{eq}} = \frac{170\ \text{V}}{340\ \Omega} = 0.50\ \text{A}$$

Table 26.1 shows that this is a very dangerous level of current, likely to be fatal. Fortunately, it's also well above the threshold for detection by the GFI, which will quickly disconnect the circuit.

**ASSESS** The worst-case scenario is one that you could imagine happening if a person in a bathtub touches a faulty radio or picks up a faulty hair dryer. It's a situation you would expect to be dangerous, in which we'd expect a GFI to come into play.

# SUMMARY

**GOAL** To understand and apply basic principles of AC electricity, electrical safety, and household electricity.

## IMPORTANT CONCEPTS

### AC circuits

**AC circuits** are driven by an emf that oscillates with frequency $f$.

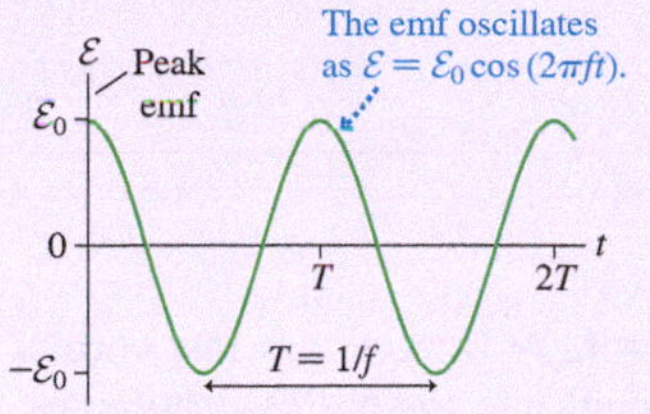


- Peak values of voltages and currents are denoted by capital letters: $I$, $V$.
- Instantaneous values of voltages and currents are denoted by lowercase letters: $i$, $v$.

### Circuit elements used in AC circuits

| | Resistor | Capacitor | Inductor |
|---|---|---|---|
| **Symbol** |  | 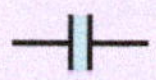 |  |
| **Reactance** | Resistance $R$ is constant | $X_C = 1/(2\pi f C)$ | $X_L = 2\pi f L$ |
| **$I$ and $V$** | $V_R = I_R R$ | $V_C = I_C X_C$ | $V_L = I_L X_L$ |
| **Graph** | 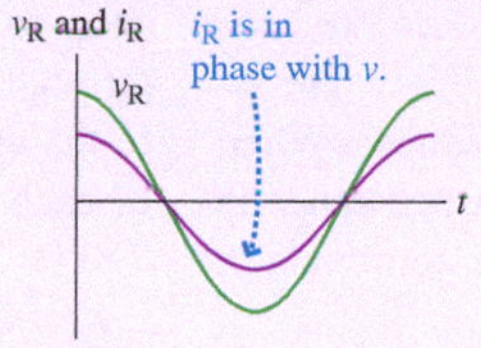 | 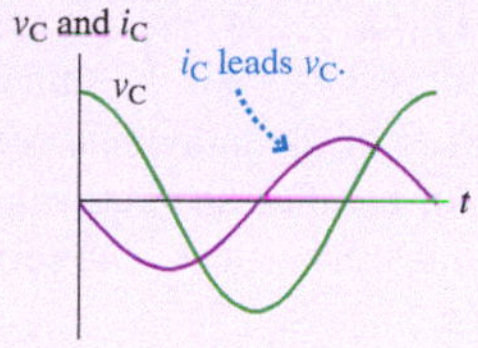 | 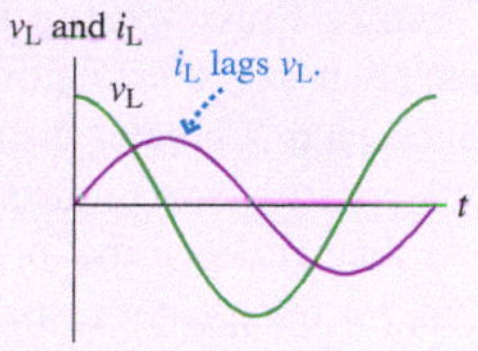 |



### Power in AC resistor circuits

The average power dissipated by a resistor is

$$P_R = (I_{rms})^2 R = \frac{(V_{rms})^2}{R} = I_{rms} V_{rms}$$

where $I_{rms} = I_R/\sqrt{2}$ and $V_{rms} = V_R/\sqrt{2}$ are the root-mean-square (rms) voltage and current.

### Electrical safety and biological effects

Currents passing through the body can produce dangerous effects. Currents larger than 15 mA (AC) and 60 mA (DC) are potentially fatal.

The body can be modeled as a network of resistors. If the body forms a circuit between two different voltages, the current is given by Ohm's law: $I = \Delta V/R$.

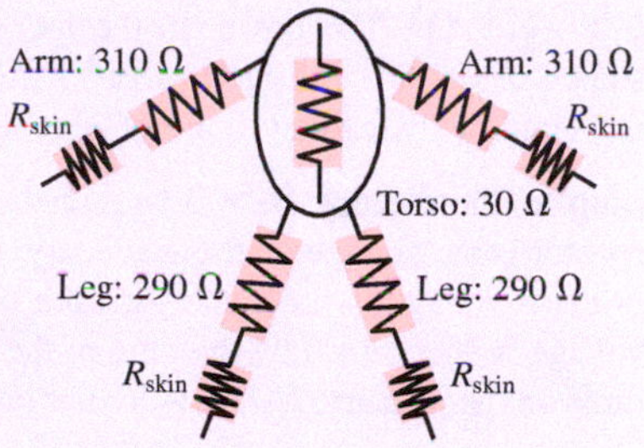


## APPLICATIONS

### Transformers

**Transformers** are used to increase or decrease an AC voltage. The rms voltage at the secondary is related to the rms voltage at the primary by

$$(V_2)_{rms} = \frac{N_2}{N_1}(V_1)_{rms}$$

and the currents are related by

$$(I_2)_{rms} = \frac{N_1}{N_2}(I_1)_{rms}$$

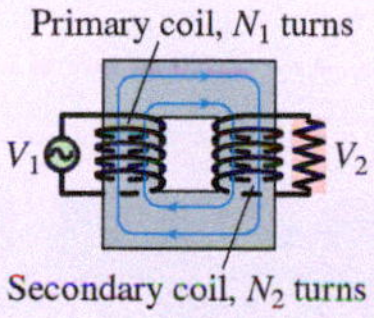


### *LC* and *RLC* circuits

In an $LC$ circuit, the current and voltages oscillate with frequency

$$f = \frac{1}{2\pi\sqrt{LC}}$$

In the $RLC$ circuit, the oscillations decay as energy is dissipated in the resistor.

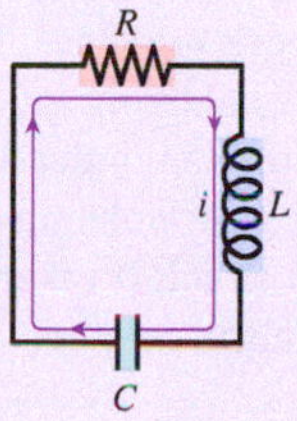


### The driven *RLC* circuit

If an AC source of amplitude $\mathcal{E}_0$ is placed in an $RLC$ circuit, then the voltages and currents oscillate continuously.

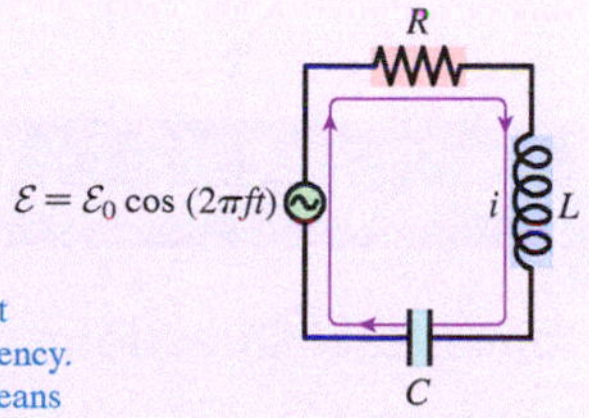


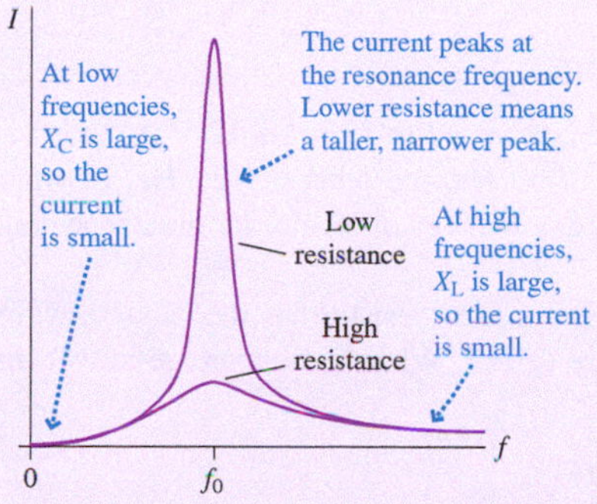


- The **resonance frequency** is $f_0 = \dfrac{1}{2\pi\sqrt{LC}}$.
- The maximum value of the current is $I_{max} = \mathcal{E}_0/R$.
- The peak current at any frequency $f$ is given by

$$I = \frac{\mathcal{E}_0}{\sqrt{R^2 + (X_L - X_C)^2}}$$

## Learning Objectives

After studying this chapter, you should be able to:

- Analyze AC circuits consisting of resistors, capacitors, and inductors. *Conceptual Questions 26.13, 26.18; Problems 26.6, 26.32, 26.35, 26.39, 26.40*
- Understand what a transformer does. *Conceptual Questions 26.5, 26.7; Problems 26.10, 26.11, 26.12, 26.13, 26.15*
- Recognize how electricity gets to and is used in the home. *Problems 26.17, 26.18, 26.19, 26.20, 26.23*
- Understand the safety implications of using and working with electricity. *Conceptual Questions 26.9, 26.11; Problems 26.24, 26.25, 26.27, 26.28, 26.30*
- Analyze oscillation circuits and resonance. *Conceptual Questions 26.21, 26.22; Problems 26.44, 26.46, 26.47, 26.50, 26.51*

## STOP TO THINK ANSWERS

**Chapter Preview Stop to Think: A.** When two resistors are connected in series, the equivalent resistance is the sum of the two resistances and is greater than either of the individual resistances. When two resistors are connected in parallel, the equivalent resistance is less than either of the individual resistances. The series combination thus has the greater resistance.

**Stop to Think 26.1: B.** The power in the AC circuit is proportional to the square of the rms current, or to $(I_{rms})^2 = (I_R/\sqrt{2})^2 = \frac{1}{2}I_R^2 = \frac{1}{2}(1\text{ A})^2$. The power in the DC circuit is proportional to the square of the DC current, or to $I_R^2 = (1\text{ A})^2$. Thus the power in the DC circuit is twice that in the AC circuit, so bulb B is brighter.

**Stop to Think 26.2: D > B > A > C.** The rms voltage on the load resistor is the voltage on the secondary, given by Equation 26.13. The primary voltage is the same for each transformer, so the secondary voltage is determined by the ratio of the turns on the secondary to the turns on the primary, $N_2/N_1$. A higher ratio means a higher voltage.

**Stop to Think 26.3: D > C > A > B.** The energy the devices use, in kWh, is the power in kW multiplied by the time used in hours: A. 0.48 kWh, B. 0.36 kWh, C. 2.0 kWh, D. 3.0 kWh.

**Stop to Think 26.4: A > B = C > D.** Any way that current goes through the body, it goes through the torso; this contributes 30 Ω to the current path in every case. Figure 26.13 shows the resistance of each arm to be 310 Ω and that of each leg to be slightly lower, 290 Ω. The highest resistance is thus a path through two arms, the lowest is a path through two legs, with a path through one arm and one leg in the middle.

**Stop to Think 26.5: B.** The current is $I_C = V_C/X_C = 2\pi f C V_C$. Thus $I_C$ is proportional to both $f$ and $V_C$. Doubling the frequency while keeping $V_C$ constant will double the current.

**Stop to Think 26.6: B.** The current is $I_L = V_L/X_L = V_L/2\pi f L$. Thus $I_L$ is *inversely* proportional to $f$. Doubling the frequency while keeping $V_L$ constant will halve the current.

**Stop to Think 26.7: E.** The charge on the capacitor is increasing, so the current must be counterclockwise. As the capacitor charges up, the capacitor voltage opposes the current, so the current is decreasing. This corresponds to the time between the two figures on the left side of Figure 26.23.

Video Tutor Solution Chapter 26

# QUESTIONS

### Conceptual Questions

1. Identical resistors are connected to separate 12 V AC sources. One source operates at 60 Hz, the other at 120 Hz. In which circuit, if either, does the resistor dissipate the greater average power?
2. The AC voltage across a 100 Ω resistor has the "square wave" shape shown in Figure Q26.2. What is the average power dissipated in the resistor?

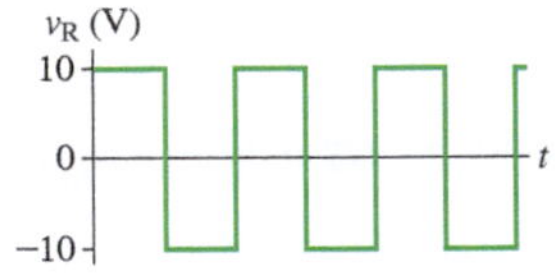


FIGURE Q26.2

3. Most battery-powered devices won't work if you put the battery in backward. But for a device that you plug in, you can often reverse the orientation of the plug with no problem. Explain the difference.
4. If an incandescent lightbulb is connected to a 120 V, 60 Hz electric outlet, how many times a second does the bulb reach peak brightness? (This flicker is too fast for you to see.)
5. A soldering gun contains a transformer that lowers the 120 V from an outlet to a few volts. It's possible to have a current of 150 A in the tip of the soldering iron even though the outlet has a circuit breaker that permits no more than 15 A. Explain how this is possible.

Problem difficulty is labeled as | (straightforward) to ||||| (challenging). Problems labeled INT integrate significant material from earlier chapters; Problems labeled BIO are of biological or medical interest.

The eText icon indicates when there is a video tutor solution available for the chapter or for a specific problem. To launch these videos, log into your eText through Mastering™ Physics or log into the Study Area.

6. A 12 V DC power supply is connected to the primary coil of a transformer. The primary coil has 100 turns and the secondary coil has 200. What is the rms voltage across the secondary?
7. Figure Q26.7 shows three wires wrapped around an iron core. The figure shows the number of turns and the direction of each of the windings. At one particular instant, $V_A - V_B = 20$ V. At that same instant, what is $V_C - V_D$?

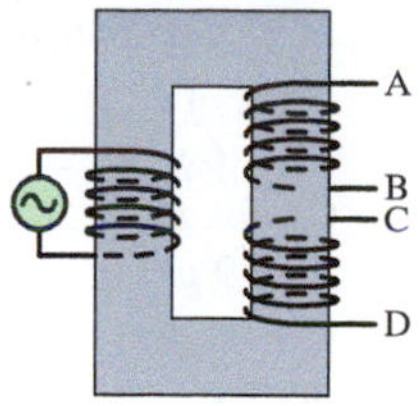

FIGURE Q26.7

8. Women usually have higher resistance of their arms and legs than men. Why might you expect to see this variation in resistance? BIO
9. If you work out enough to visibly increase the diameter of your biceps, will this increase or decrease your susceptibility to electric shock? Explain. BIO
10. If you stuck a paperclip in each of the two slots of a GFI-protected outlet, then grabbed one paperclip with your right hand and the other paperclip with your left hand, would the GFI protect you? Explain.
11. New homes are required to have GFI-protected outlets in bathrooms, kitchens, and any outdoor locations. Why is GFI protection required in these locations but not, say, in bedrooms? BIO
12. The peak current through a resistor is 2.0 A. What is the peak current if
    a. The resistance $R$ is doubled?
    b. The peak emf $\mathcal{E}_0$ is doubled?
    c. The frequency $f$ is doubled?
13. The peak current through a capacitor is 2.0 A. What is the peak current if
    a. The peak emf $\mathcal{E}_0$ is doubled?
    b. The capacitance $C$ is doubled?
    c. The frequency $f$ is doubled?
14. An AC voltage source is connected in turn to a resistor, a capacitor, and an inductor. Graphs A, B, and C in Figure Q26.14 show the resulting currents. Which graph corresponds to the resistor, which to the capacitor, and which to the inductor? Explain.

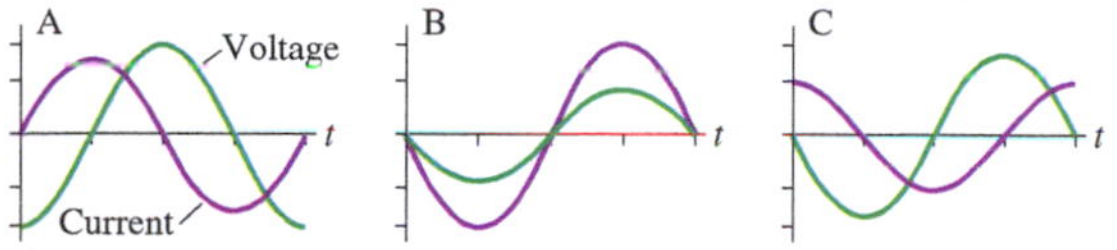

FIGURE Q26.14

15. The ignition system in a car uses a transformer whose primary has 200 turns and carries 10 A; its secondary, with 20,000 turns, is connected to the spark plug. An electronic switch suddenly stops the current in the primary, creating a brief, 40,000 V pulse in the secondary that makes a spark in the plug, which ignites the fuel mixture. Why does the voltage pulse occur only as the primary current is stopped? Why is it important that the primary current be stopped suddenly?
16. An inductive loop buried in a roadway detects the presence of cars above it, as described in the chapter. If the loop is connected to an AC supply, will the current increase or decrease when a car drives above the loop?
17. Figure Q26.17 shows two inductors and the potential difference across them at time $t = 0$ s.
    a. Can you tell which of these inductors has the larger current flowing through it at $t = 0$ s? If so, which one? If not, why not?
    b. Can you tell through which inductor the current is changing more rapidly at $t = 0$ s? If so, which one? If not, why not?

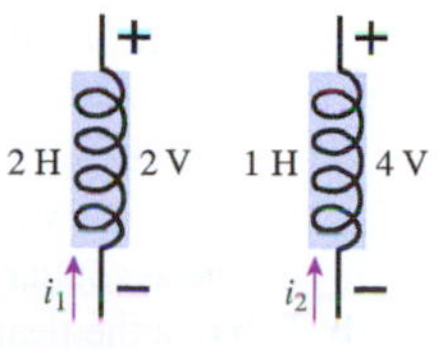

FIGURE Q26.17

18. The peak current passing through an inductor is 2.0 A. What is the peak current if
    a. The peak emf $\mathcal{E}_0$ is doubled?
    b. The inductance $L$ is doubled?
    c. The frequency $f$ is doubled?
19. Consider the four circuits in Figure Q26.19. Rank in order, from largest to smallest, the inductive reactances $(X_L)_1$ to $(X_L)_4$. Explain.

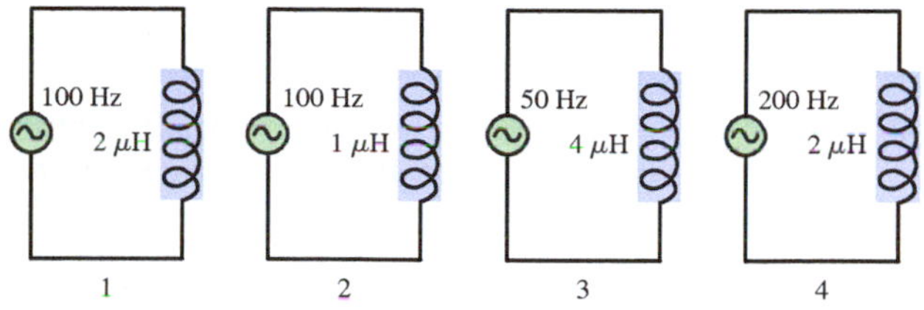

FIGURE Q26.19

20. Figure 26.23 in the chapter illustrates a block-and-spring analogy to an $LC$ circuit. What mechanical property of the block-and-spring system corresponds to the inductance $L$ of the $LC$ circuit? What mechanical property corresponds to the capacitance $C$?
21. The resonance frequency of a driven $RLC$ circuit is 1000 Hz. What is the resonance frequency if
    a. The resistance $R$ is doubled?
    b. The inductance $L$ is doubled?
    c. The capacitance $C$ is doubled?
    d. The peak emf $\mathcal{E}_0$ is doubled?
    e. The emf frequency $f$ is doubled?
22. Consider the four circuits in Figure Q26.22. They all have the same resonance frequency $f_0$ and are driven by the same emf. Rank in order, from largest to smallest, the maximum currents $(I_{max})_1$ to $(I_{max})_4$. Explain.

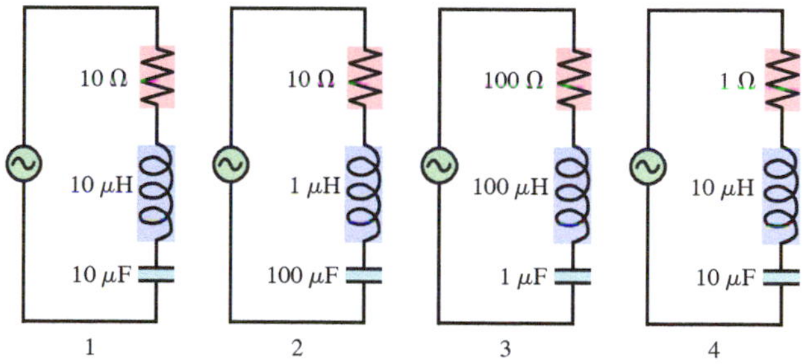

FIGURE Q26.22

## Multiple-Choice Questions

23. | A transformer has 1000 turns in the primary coil and 100 turns in the secondary coil. If the primary coil is connected to a 120 V outlet and draws 0.050 A, what are the voltage and current of the secondary coil?
    A. 1200 V, 0.0050 A
    B. 1200 V, 0.50 A
    C. 12 V, 0.0050 A
    D. 12 V, 0.50 A
24. | An inductor is connected to an AC generator. As the generator's frequency is increased, the current in the inductor
    A. Increases.
    B. Decreases.
    C. Does not change.

Watch Video Solution Problems 26.5 and 26.11

25. | A capacitor is connected to an AC generator. As the generator's frequency is increased, the current in the capacitor
    A. Increases.
    B. Decreases.
    C. Does not change.
26. | An AC source is connected to a series combination of a lightbulb and a variable inductor. If the inductance is increased, the bulb's brightness
    A. Increases.
    B. Decreases.
    C. Does not change.
27. || An AC source is connected to a series combination of a lightbulb and a variable capacitor. If the capacitance is increased, the bulb's brightness
    A. Increases.
    B. Decreases.
    C. Does not change.
28. | The circuit shown in Figure Q26.28 has a resonance frequency of 15 kHz. What is the value of $L$?
    A. 1.6 $\mu$H B. 2.4 $\mu$H
    C. 5.2 $\mu$H D. 18 $\mu$H
    E. 59 $\mu$H

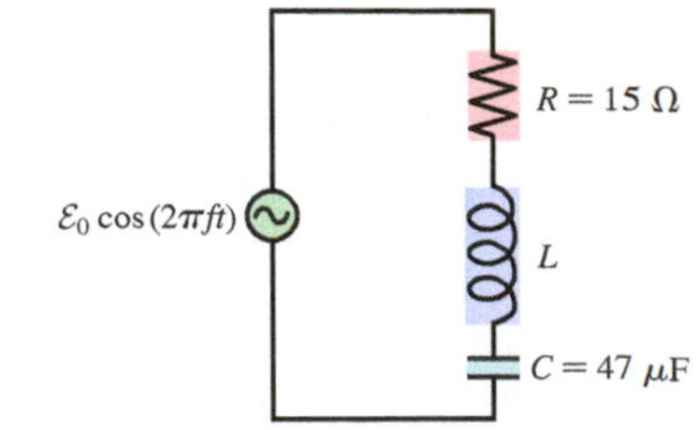


FIGURE Q26.28

29. ||| At resonance, a driven $RLC$ circuit has $V_C = 5.0$ V, $V_R = 8.0$ V, and $V_L = 5.0$ V. What is the peak voltage across the entire circuit?
    A. 18 V B. 10 V C. 8.0 V D. 5.0 V E. 3.0 V
30. ||| A driven $RLC$ circuit has $V_C = 5.0$ V, $V_R = 7.0$ V, and $V_L = 9.0$ V. The driving frequency is
    A. Higher than the resonance frequency.
    B. Equal to the resonance frequency.
    C. Lower than the resonance frequency.

# PROBLEMS

## Section 26.1 Alternating Current

1. | A 200 $\Omega$ resistor is connected to an AC source with $\mathcal{E}_0 = 10$ V. What is the peak current through the resistor if the emf frequency is (a) 100 Hz? (b) 100 kHz?
2. | Figure P26.2 shows voltage and current graphs for a resistor.
    a. What is the value of the resistance $R$?
    b. What is the emf frequency $f$?

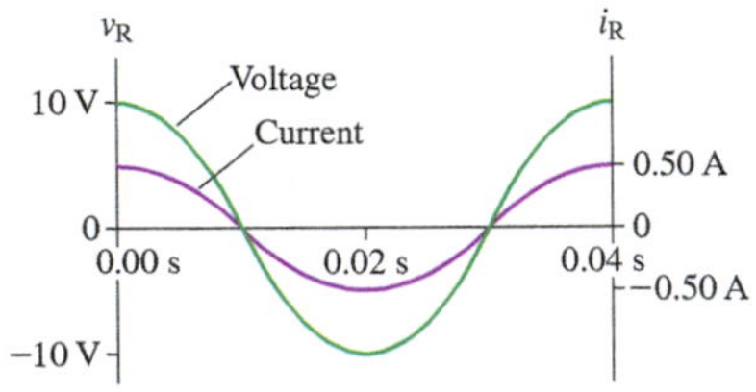


FIGURE P26.2

3. || A resistor dissipates 2.00 W when the rms voltage of the emf is 10.0 V. At what rms voltage will the resistor dissipate 10.0 W?
4. | The heating element of a hair dryer dissipates 1500 W when connected to a 120 V outlet. What is the resistance?
5. || A toaster oven is rated at 1600 W for operation at 120 V, 60 Hz.
    a. What is the resistance of the oven heater element?
    b. What is the peak current through it?
    c. What is the peak power dissipated by the oven?
6. ||| The instantaneous power dissipated by a 20 $\Omega$ resistor connected to an AC source is shown in Figure P26.6.
    a. What is the rms voltage across the resistor?
    b. What is the peak current through the resistor?
    c. What is the frequency of the AC source?

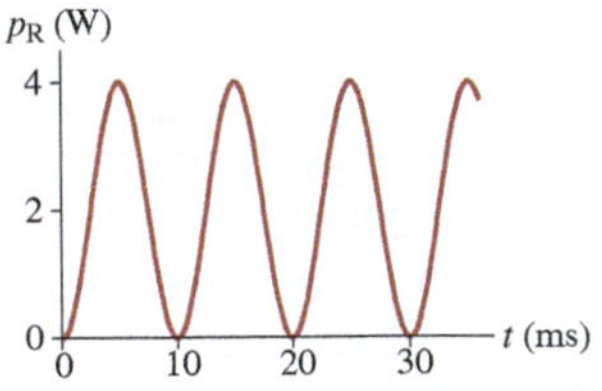


FIGURE P26.6

7. || A generator produces 40 MW of power and sends it to town at an rms voltage of 75 kV. What is the rms current in the transmission lines?
8. | BIO Soles of boots that are designed to protect workers from electric shock are rated to pass a maximum rms current of 1.0 mA when connected across an 18,000 V AC source. What is the minimum allowed resistance of the sole?

## Section 26.2 AC Electricity and Transformers

9. | The primary coil of a transformer is connected to a 120 V wall outlet. The secondary coil is connected to a lamp that dissipates 60 W. What is the rms current in the primary coil?
10. || A microscope illuminator uses a transformer to step down the 120 V AC of the wall outlet to power a 12.0 V, 50 W microscope bulb.
    a. What is the resistance of the bulb filament?
    b. What is the rms current in the bulb filament?
    c. What is the rms current in the primary coil?
11. || A power pack charging a cell phone battery has an output of 0.40 A at 5.2 V (both rms). What is the rms current at the 120 V wall outlet where the power pack is plugged in?
12. || A neon sign transformer has a 450 W AC output with an rms voltage of 15 kV when connected to a normal household outlet. There are 500 turns of wire in the primary coil.
    a. How many turns of wire does the secondary coil have?
    b. When the transformer is running at full power, what is the current in the secondary coil? The current in the primary coil?
13. || The "power cube" transformer for a portable CD player has an output of 4.5 V and 600 mA (both rms) when plugged into a 120 V outlet.
    a. If the primary coil has 400 turns of wire, how many turns are on the secondary coil?
    b. What is the peak current in the primary coil?

14. | Electricity is distributed to neighborhoods at a relatively high AC voltage, often 7200 V. Transformers mounted on utility poles then transform this high voltage down to the 120 V used in homes. A typical transformer of this kind can handle as much as 15 kW of electric power flowing through it from its primary to its secondary. What is the primary current at this maximum power?

15. | A generator produces 250 kW of electric power at 7.2 kV. The current is transmitted to a remote village through wires with a total resistance of 15 Ω.
    a. What is the power loss due to resistance in the wires?
    b. What is the power loss if the voltage is increased to 30 kV?

### Section 26.3 Household Electricity

16. || In an old house, the wires leading to a 120 V outlet have a total resistance of 0.45 Ω. When you plug in a hair dryer, it draws a 12 A current.
    a. How much does the outlet voltage decrease due to the voltage drop across the wires?
    b. What is the power dissipated as heat in the wires?
17. || A typical American family uses 1000 kWh of electricity a month. What is the average rms current in the 120 V power line to a typical house?
18. || If you use an extension cord, current travels from the 120 V outlet, along one wire inside the cord, through the appliance you've plugged into the cord, and back to the outlet through a second wire in the cord. The total resistance of the two wires in a light-duty extension cord is 0.40 Ω; the current through such a cord should be limited to 13 A. If a 13 A shop vacuum is powered using this cord,
    a. What is the voltage drop across the vacuum?
    b. How much power does the vacuum use?
    c. How much power is dissipated in the cord?
19. || INT The following appliances are connected to a single 120 V, 15 A circuit in a kitchen: a 330 W blender, a 1000 W coffeepot, a 150 W coffee grinder, and a 750 W microwave oven. If these are all turned on at the same time, will they trip the circuit breaker?
20. || In Fort Collins, Colorado, the sun shines for an average of 1850 hours per year. A homeowner installs a set of solar panels that provide 4.0 kW of electric power when the sun shines. If the local utility charges \$0.13/kWh, how much will she save each year on electricity because of her solar panels?
21. || A 60 W (120 V) night light is turned on for an average of 12 h a day year round. What is the annual cost of electricity at a billing rate of \$0.10/kWh?
22. || Electricity prices vary widely from state to state. In Connecticut, for instance, electricity costs \$0.209/kWh, while in Illinois the cost is \$0.125/kWh. On average, residents of both states use about 720 kWh of electricity per month. How much more per year does a Connecticuter (look it up!) pay for electricity than an Illinoisan?
23. ||| INT The manufacturer of an electric table saw claims that it has a 3.0 horsepower motor. 1 horsepower is approximately 750 W. It is designed to be used on a normal 120 V outlet with a 15 A circuit breaker. Is this claim reasonable? Explain.

### Section 26.4 Biological Effects and Electrical Safety

24. || BIO John is changing a lightbulb in a lamp. It's a warm summer evening, and the resistance of his damp skin is only 4000 Ω. While one hand is holding the grounded metal frame of the lamp, the other hand accidentally touches the hot electrode in the base of the socket. What is the current through his torso?
25. || BIO In some countries AC outlets near bathtubs are restricted to a maximum of 25 V to minimize the chance of dangerous shocks while bathing. A man is in the tub; the lower end of his torso is well grounded, and the skin resistance of his wet, soapy hands is negligible. He reaches out and accidentally touches a live electric wire. What voltage on the wire would produce a dangerous 100 mA current?
26. || BIO If you touch the terminal of a battery, the small area of contact means that the skin resistance will be relatively large; 50 kΩ is a reasonable value. What current will pass through your body if you touch the two terminals of a 9.0 V battery with your two hands? Will you feel it? Will it be dangerous?
27. || BIO A person standing barefoot on the ground 20 m from the point of a lightning strike experiences an instantaneous potential difference of 300 V between his feet. If we assume a skin resistance of 1.0 kΩ, how much current goes up one leg and back down the other?
28. ||| BIO Occupational safety experts have developed an alternative criterion for electrical safety. They have found that shocks lasting less than 3 s will be nonlethal if the product of the voltage drop across the body, the current through the body, and the time ($\leq 3.0$ s) that the current flows does not exceed $13.5\ \text{V}\cdot\text{A}\cdot\text{s} = 13.5\ \text{J}$. Suppose that one hand of a potential victim is grounded and the other hand touches a voltage source; suppose further that his skin resistance is negligible—a worst-case scenario. Using the criterion above, what is the lowest voltage that will not be lethal for a shock that lasts 1.0 s?
29. ||| BIO A fisherman has netted a torpedo ray. As he picks it up, this electric fish creates a short-duration 50 V potential difference between his hands. His hands are wet with salt water, and so his skin resistance is a very low 100 Ω. What current passes through his body? Will he feel this DC pulse?

Problems 30 and 31 concern a high-voltage transmission line. Such lines are made of bare wire; they are not insulated. Assume that the wire is 100 km long, has a resistance of 7.0 Ω, and carries 200 A.

30. ||| BIO A bird is perched on the wire with its feet 2.0 cm apart. What is the potential difference between its feet?
31. ||| BIO INT Would it be possible for a person to safely hang from this wire? Assume that the hands are 15 cm apart, and assume a skin resistance of 2200 Ω.

### Section 26.5 Capacitor Circuits

32. || A 0.30 $\mu$F capacitor is connected across an AC generator that produces a peak voltage of 10.0 V. What is the peak current through the capacitor if the emf frequency is (a) 100 Hz? (b) 100 kHz?
33. || A 20 $\mu$F capacitor is connected across an AC generator that produces a peak voltage of 6.0 V. The peak current is 0.20 A. What is the oscillation frequency in Hz?
34. | The peak current through a capacitor is 10.0 mA. What is the current if
    a. The emf frequency is doubled?
    b. The emf peak voltage is doubled (at the original frequency)?
    c. The frequency is halved and, at the same time, the emf is doubled?

35. ||| Figure P26.35 shows voltage and current graphs for a capacitor.
   a. What is the frequency of the voltage across the capacitor?
   b. What is the value of the capacitance?

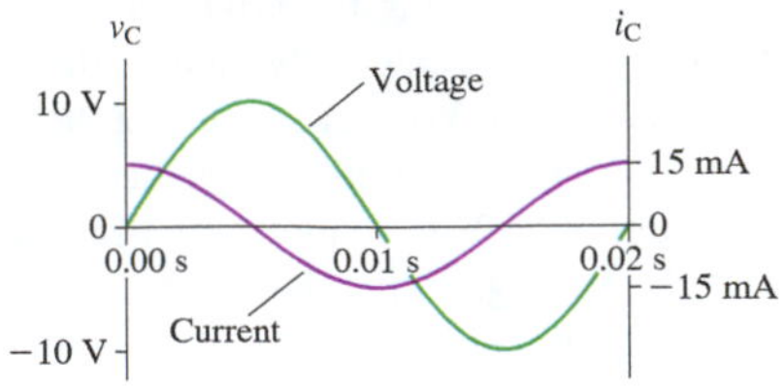


FIGURE P26.35

36. || A capacitor is connected across an AC source. At one instant in time, the current through the capacitor is 23 mA when the voltage across the capacitor is changing at a rate of $7.2 \times 10^4$ V/s. What is the value of the capacitance?

37. | The peak current through a capacitor is 8.0 mA when connected to an AC source with a peak voltage of 1.0 V. What is the capacitive reactance of the capacitor?

### Section 26.6 Inductors and Inductor Circuits

38. || Magnetic resonance imaging instruments use very large BIO magnets that consist of many turns of superconducting wire. A typical such magnet has an inductance of 40 H. When the magnet is initially powered up, the current through it must be increased slowly so as not to "quench" the wires out of their superconducting state. One such magnet is specified to have its current increased from 0 A to 150 A over 200 min. What constant voltage needs to be applied to yield this rate?

39. || A 20 mH inductor is connected across an AC generator that produces a peak voltage of 10.0 V. What is the peak current through the inductor if the emf frequency is (a) 100 Hz? (b) 100 kHz?

40. ||| Figure P26.40 shows voltage and current graphs for an inductor.
   a. What is the frequency of the voltage across the inductor?
   b. What is the value of the inductance?

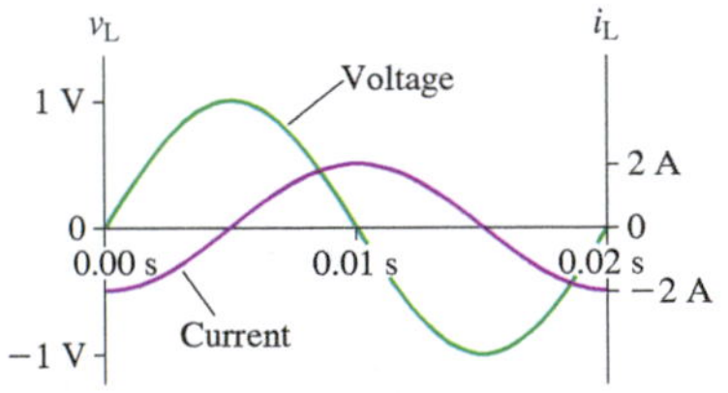


FIGURE P26.40

41. | A 500 $\mu$H inductor is connected across an AC generator that produces a peak voltage of 5.0 V.
   a. At what frequency $f$ is the peak current 50 mA?
   b. What is the instantaneous value of the emf at the instant when $i_L = I_L$?

42. ||| An inductor is connected to a 15 kHz oscillator that produces an rms voltage of 6.0 V. The peak current is 65 mA. What is the value of the inductance $L$?

43. | The peak current through an inductor is 12.5 mA when connected to an AC source with a peak voltage of 1.0 V. What is the inductive reactance of the inductor?

### Section 26.7 Oscillation Circuits

44. || A 2.0 mH inductor is connected in parallel with a variable capacitor. The capacitor can be varied from 100 pF to 200 pF. What is the range of oscillation frequencies for this circuit?

45. || An FM radio station broadcasts at a frequency of 100 MHz. What inductance should be paired with a 10 pF capacitor to build a receiver circuit for this station?

46. || The inductor in the $RLC$ tuning circuit of an AM radio has a value of 350 mH. What should be the value of the variable capacitor in the circuit to tune the radio to 740 kHz?

47. || At what frequency $f$ do a 1.0 $\mu$F capacitor and a 1.0 $\mu$H inductor have the same reactance? What is the value of the reactance at this frequency?

48. || What capacitor in series with a 100 $\Omega$ resistor and a 20 mH inductor will give a resonance frequency of 1000 Hz?

49. || What inductor in series with a 100 $\Omega$ resistor and a 2.5 $\mu$F capacitor will give a resonance frequency of 1000 Hz?

50. | A series $RLC$ circuit has a 200 kHz resonance frequency. What is the resonance frequency if the capacitor value is doubled and the inductor value is halved?

51. ||| An $RLC$ circuit with a 10 $\mu$F capacitor is connected to a variable-frequency power supply with an rms output voltage of 6.0 V. The rms current in the circuit as a function of the driving frequency appears as in Figure P26.51. What are the values of the resistor and the inductor?

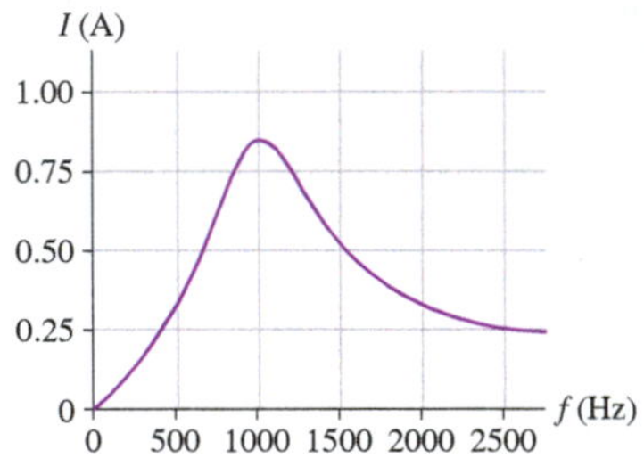


FIGURE P26.51

52. |||| In a home stereo system, low sound frequencies are handled by large "woofer" speakers, and high frequencies by smaller "tweeter" speakers. For the best sound reproduction, low-frequency currents from the amplifier should not reach the tweeter. One way to do this is to place a capacitor in series with the 8.0 $\Omega$ resistance of the tweeter; one then has an $RLC$ circuit with no inductor $L$ (that is, an $RLC$ circuit with $L = 0$). What value of $C$ should be chosen so that the current through the tweeter at 200 Hz is half its value at very high frequencies?

53. ||||| A series $RLC$ circuit consists of a 280 $\Omega$ resistor, a 25 $\mu$H inductor, and an 18 $\mu$F capacitor. What is the rms current if the emf is supplied by a standard 120 V, 60 Hz wall outlet?

54. |||| An $RLC$ circuit has a 24 $\Omega$ resistor and a 58 $\mu$H inductor; when connected to an AC source, its resonance frequency is 320 kHz. At what frequency, greater than the resonance frequency, will the current in the circuit be one-half its value at resonance?

## General Problems

55. || Electric outlets in England are 230 V. Alice brings her electric kettle from England, where it draws 13 A, and wants to use it in the United States. She uses a step-up transformer to increase the 120 V outlet voltage to 230 V, then plugs her kettle into the secondary. What is the current in the primary in the few seconds before the 15 A circuit breaker trips?

56. ‖ The charger for your cell phone contains a small transformer. While charging, it provides 5.0 W to your phone at 5.0 V. Assuming an ideal transformer, how much current does this transformer draw from the 120 V wall socket?

57. ‖ BIO INT The girl in Figure 26.12 of the chapter has her hand on the sphere of a Van de Graaff generator that is at a potential of 400,000 V. She is standing on an insulating platform, so no current flows through her. But what happens if she touches something that is grounded? Is she still safe? Figure P26.57 shows the equivalent circuit. $C_{VDG} = 20$ pF represents the capacitance of the Van de Graaff sphere, $C_{girl} = 100$ pF is the capacitance of the girl's body, and $R = 5$ kΩ is the resistance of her body, from one hand to the other. The switch is closed when she touches ground.

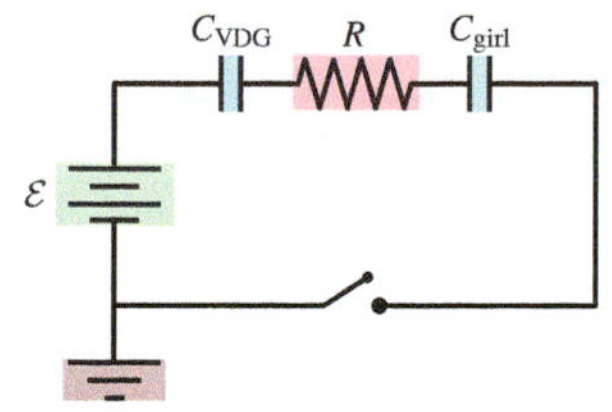

FIGURE P26.57

   a. What is the initial current at the instant the switch is closed?
   b. What is the time constant for the current to decay?
   c. Occupational safety experts have found that a shock is safe if the product of the voltage, the current, and the time the current is delivered is less than 13.5 V · A · s = 13.5 J (see Problem 26.28). Estimate this product. Is this shock safe?

58. ‖ The voltage across a 60 μF capacitor is described by the equation $v_C = (18\text{ V})\cos(200t)$, where $t$ is in seconds.
   a. What is the voltage across the capacitor at $t = 0.010$ s?
   b. What is the capacitive reactance?
   c. What is the peak current?

59. ‖ The voltage across a 75 μH inductor is described by the equation $v_L = (25\text{ V})\cos(60t)$, where $t$ is in seconds.
   a. What is the voltage across the inductor at $t = 0.10$ s?
   b. What is the inductive reactance?
   c. What is the peak current?

60. ||| An electronics hobbyist is building a radio set to receive the AM band, with frequencies from 520 kHz to 1700 kHz. The antenna, which also serves as the inductor in an $LC$ circuit, has an inductance of 230 μH. She needs to add a variable capacitor whose capacitance she can adjust to tune the radio. What is the minimum capacitance the capacitor must have? The maximum value?

61. ‖ For the circuit of Figure P26.61,
   a. What is the resonance frequency?
   b. At resonance, what is the peak current through the circuit?

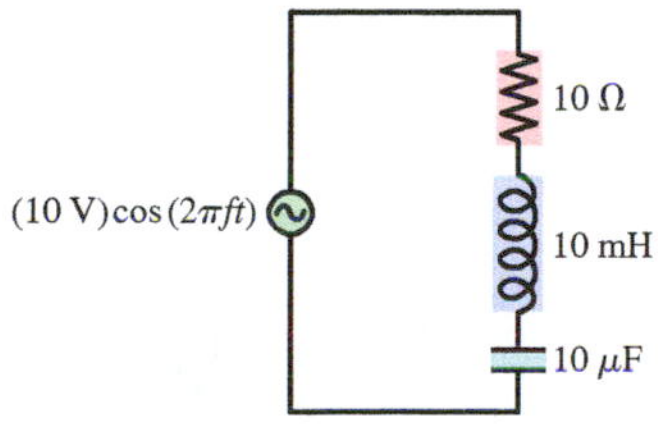

FIGURE P26.61

62. | For the circuit of Figure P26.62,
   a. What is the resonance frequency?
   b. At resonance, what is the peak current through the circuit?

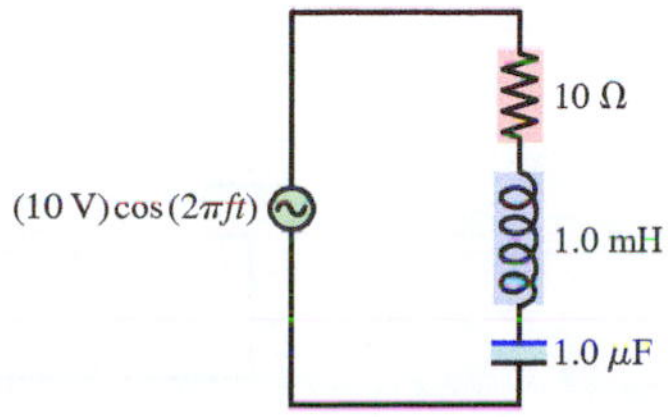

FIGURE P26.62

63. ‖ An $RLC$ circuit consists of a 48 Ω resistor, a 200 μF capacitor, and an inductor. The rms current is 2.5 A when the circuit is connected to a 120 V, 60 Hz outlet. What is the inductance?

## MCAT-Style Passage Problems

### Cell Membrane Capacitance BIO

The capacitance of biological membranes is about 1.0 μF per cm$^2$ of membrane area, so investigators can determine the surface area of a cell membrane by using intracellular electrodes to measure the membrane's capacitive reactance. An investigator applies a 1.0 μA peak current at 40 kHz to a cell and measures the peak out-of-phase voltage—that is, the component of the voltage due to the capacitive reactance of the cell membrane—to be 0.16 V.

64. ‖ If the frequency is doubled to 80 kHz and the current is kept the same, what will be the peak out-of-phase voltage?
   A. 0.32 V  B. 0.16 V
   C. 0.080 V  D. 0.040 V

65. ‖ What is the approximate capacitance of the cell membrane?
   A. $20 \times 10^{-11}$ F
   B. $10 \times 10^{-11}$ F
   C. $5.0 \times 10^{-11}$ F
   D. $2.5 \times 10^{-11}$ F

66. ‖ If the capacitance of a cell membrane is measured to be $6.0 \times 10^{-11}$ F, what is the area?
   A. $6.0 \times 10^{-13}$ m$^2$
   B. $6.0 \times 10^{-11}$ m$^2$
   C. $6.0 \times 10^{-9}$ m$^2$
   D. $6.0 \times 10^{-7}$ m$^2$

67. ‖ If the investigator applies a 1.0 μA peak current at 40 kHz to a cell with twice the membrane area of the cell noted in the passage, what will be the peak out-of-phase voltage?
   A. 0.32 V  B. 0.16 V
   C. 0.080 V  D. 0.040 V

### Halogen Bulbs

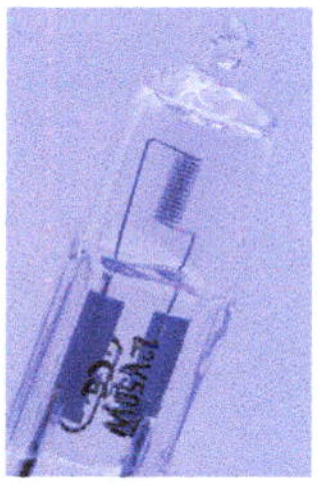

Halogen bulbs have some differences from standard incandescent lightbulbs. They are generally smaller, the filament runs at a higher temperature, and they have a quartz (rather than glass) envelope. They may also operate at lower voltage. Consider a 12 V, 50 W halogen bulb for use in a desk lamp. The lamp plugs into a 120 V, 60 Hz outlet, and it has a transformer in its base.

68. | The 12 V rating of the bulb refers to the rms voltage. What is the peak voltage across the bulb?
   A. 8.5 V  B. 12 V
   C. 17 V  D. 24 V

69. | Suppose the transformer in the base of the lamp has 500 turns of wire on its primary coil. How many turns are on the secondary coil?
   A. 50  B. 160
   C. 500  D. 5000

70. | How much current is drawn by the lamp at the outlet? That is, what is the rms current in the primary?
   A. 0.42 A  B. 1.3 A
   C. 4.2 A  D. 13 A

71. | What will be the voltage across the bulb if the lamp's power cord is accidentally plugged into a 240 V, 60 Hz outlet?
   A. 2 V  B. 24 V
   C. 36 V  D. 48 V

# PART VI SUMMARY Electricity and Magnetism

**KNOWLEDGE STRUCTURE VI**

**BASIC GOALS** How do charged particles interact? How do electric circuits work? What are the properties and characteristics of electric and magnetic fields?

**CROSS-CUTTING CONCEPTS** The **field model** explains how charged particles experience forces from other charges, and how magnets and moving charges experience forces from other magnets or currents. Electric fields are created by charges and changing magnetic fields; magnetic fields are created by magnets, currents, and changing electric fields.

**GENERAL PRINCIPLES**

The force between charges is given by **Coulomb's law:**

$$F_{1\text{ on }2} = F_{2\text{ on }1} = \frac{K|q_1||q_2|}{r^2}$$

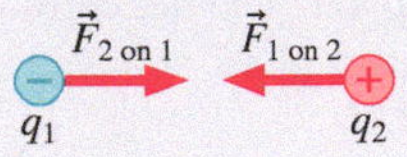


Opposite charges attract, like charges repel.

The force on a charge in an electric field is

$$\vec{F} = q\vec{E}$$

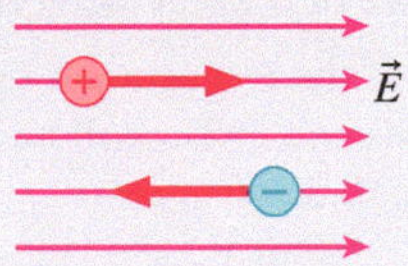


The force on a moving charge in a magnetic field is

$$F = |q|\, vB \sin\alpha$$

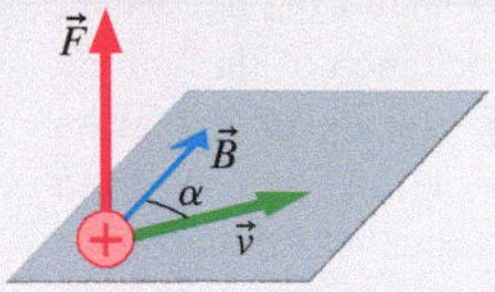


The force is perpendicular to $\vec{B}$ and $\vec{v}$

The induced emf due to a changing flux (**Faraday's law**) is

$$\mathcal{E} = \left|\frac{\Delta\Phi}{\Delta t}\right|$$

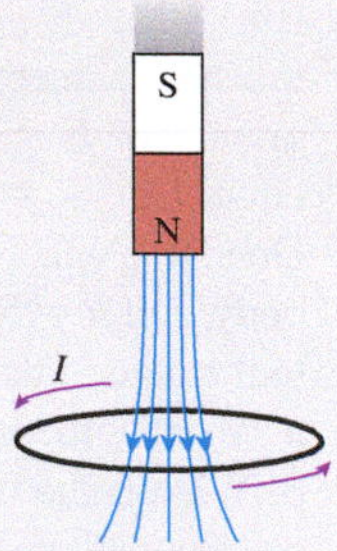


## Electric fields and forces

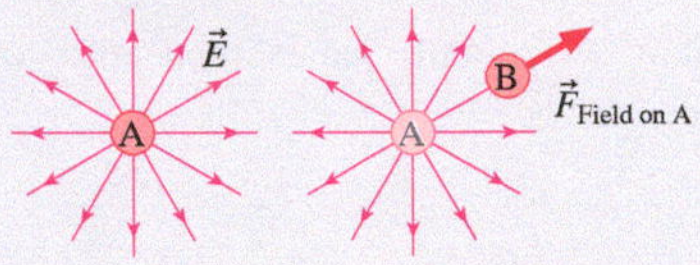


Charge A *creates* an electric field; charge B feels a force due to this field.

Electric fields also exert torques on dipoles.

## Electric potential

The interaction of charged particles can also be described in terms of an electric potential $V$.

- Only potential differences $\Delta V$ are important.
- If the potential of a particle of charge $q$ changes by $\Delta V$, its potential energy changes by $\Delta U = q\,\Delta V$.
- When two equipotential surfaces with potential difference $\Delta V$ are separated by distance $d$, the electric field strength is $E = \Delta V/d$.

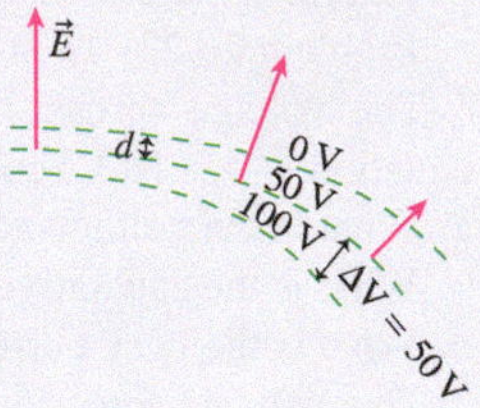


## Magnetic fields

Magnetic fields are created by either:

- Permanent magnets
- Electric currents

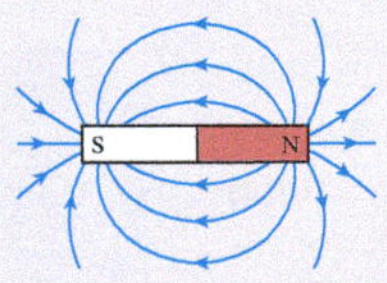


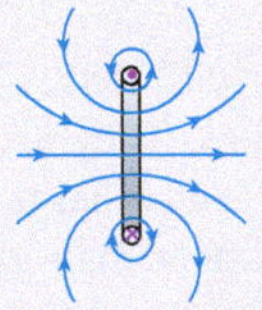

Magnetic fields exert forces on:

- Moving charges
- Current-carrying wires

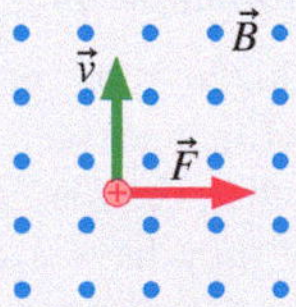


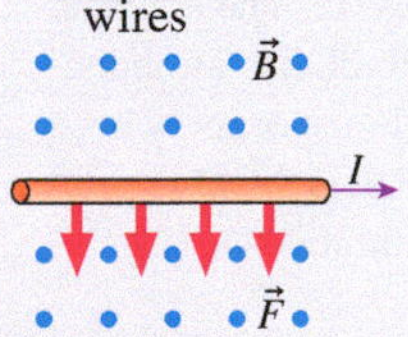


Magnetic fields exert torques on magnetic dipoles such as compass needles.

## Electromagnetic waves

An electromagnetic wave is a self-sustaining oscillation of electric and magnetic fields. In vacuum, all electromagnetic waves travel at the same speed $c$.

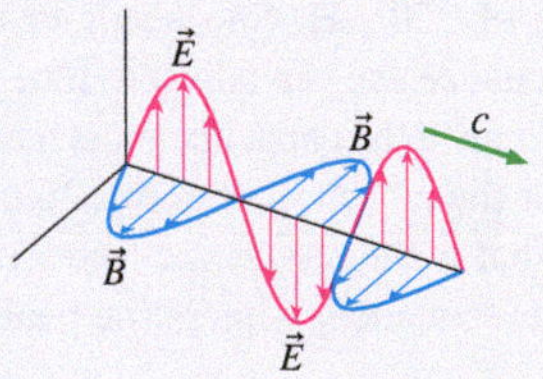


## Current and circuits

Potential differences $\Delta V$ drive current in circuits. The **current** $I$ is defined to be the motion of positive charges.

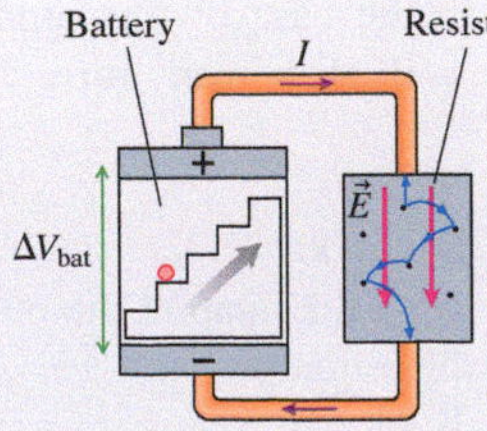


According to the **charge escalator model,** a battery lifts charges from low to high potential.

The current through a resistor is $I = \Delta V_R / R$. This is **Ohm's law.**

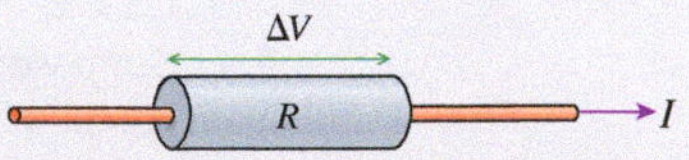


A resistor transforms electrical energy into thermal energy. It dissipates this power at a rate

$$P_R = I\Delta V_R = I^2 R = \frac{(\Delta V_R)^2}{R}$$

Circuits obey Kirchhoff's loop law...

$$\mathcal{E}_1 - IR - \mathcal{E}_2 = 0\ldots$$

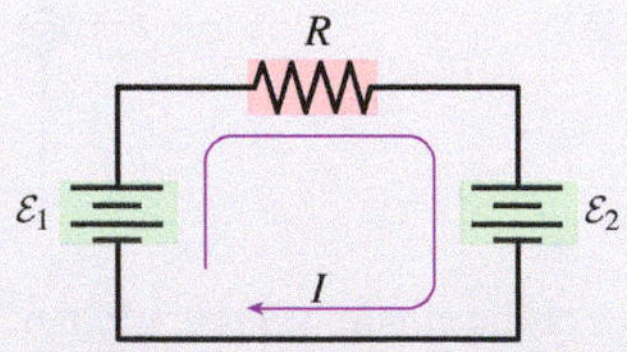


... and Kirchhoff's junction law: $I_{in} = I_1 = I_{out} = I_2 + I_3$.

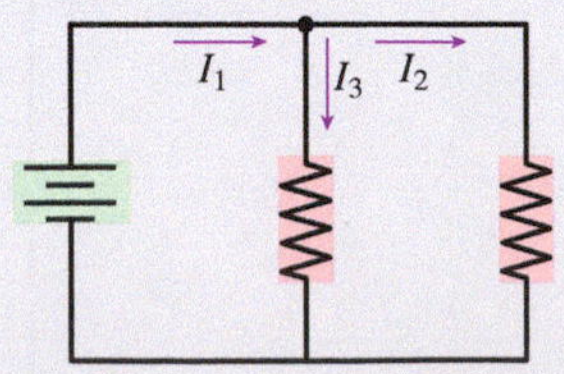

## The Greenhouse Effect and Global Warming

Electromagnetic waves are real, and we depend on them for our very existence; energy carried by electromagnetic waves from the sun provides the basis for all life on earth. Because of the sun's high surface temperature, it emits most of its thermal radiation in the visible portion of the electromagnetic spectrum. As the figure below shows, the earth's atmosphere is transparent to the visible and near-infrared radiation, so most of this energy travels through the atmosphere and warms the earth's surface.

Although seasons come and go, *on average* the earth's climate is very steady. To maintain this stability, the earth must radiate thermal energy—electromagnetic waves—back into space at exactly the same average rate that it receives energy from the sun. Because the earth is much cooler than the sun, its thermal radiation is long-wavelength infrared radiation that we cannot see. A straightforward calculation using Stefan's law finds that the average temperature of the earth should be $-18°C$, or 0°F, for the incoming and outgoing radiation to be in balance.

This result is clearly not correct; at this temperature, the entire earth would be covered in snow and ice. The measured global average temperature is actually a balmier 15°C, or 59°F. The straightforward calculation fails because it neglects to consider the earth's atmosphere. At visible wavelengths, as the figure shows, the atmosphere has a wide "window" of transparency, but this is not true at the infrared wavelengths of the earth's thermal radiation. The atmosphere lets in the visible radiation from the sun, but the outgoing thermal radiation from the earth sees a much smaller "window." Most of this radiation is absorbed in the atmosphere.

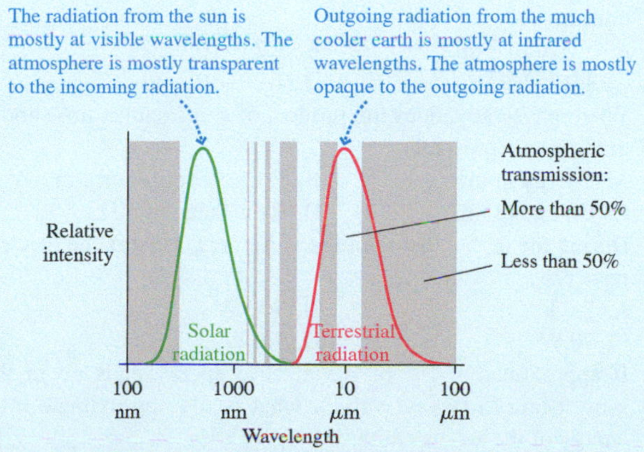


Thermal radiation curves for the sun and the earth. The shaded bands show regions for which the atmosphere is transparent (no shading) or opaque (shaded) to electromagnetic radiation.

Because it's easier for visible radiant energy to get in than for infrared to get out, the earth is warmer than it would be without the atmosphere. The additional warming of the earth's surface because of the atmosphere is called the **greenhouse effect.** The greenhouse effect is a natural part of the earth's physics; it has nothing to do with human activities, although it's doubtful any advanced life forms would have evolved without it.

The atmospheric gases most responsible for the greenhouse effect are carbon dioxide and water vapor, both strong absorbers of infrared radiation. These **greenhouse gases** are of concern today because humans, through the burning of fossil fuels (oil, coal, and natural gas), are rapidly increasing the amount of carbon dioxide in the atmosphere. Preserved air samples show that carbon dioxide made up 0.027% of the atmosphere before the industrial revolution. In the last 150 years, human activities have increased the amount of carbon dioxide by nearly 50%, to about 0.040%. By 2050, the carbon dioxide concentration will likely increase to 0.054%, double the pre-industrial value, unless the use of fossil fuels is substantially reduced.

Carbon dioxide is a powerful absorber of infrared radiation. And good absorbers are also good emitters. The carbon dioxide in the atmosphere radiates energy back to the surface of the earth, warming it. Increasing the concentration of carbon dioxide in the atmosphere means more radiation; this increases the average surface temperature of the earth. The net result is **global warming.**

There is strong evidence that the earth has warmed nearly 1°C in the last 100 years because of increased greenhouse gases. What happens next? Climate scientists, using sophisticated models of the earth's atmosphere and oceans, calculate that a doubling of the carbon dioxide concentration will likely increase the earth's average temperature by an additional 2°C ($\approx 3°F$) to 6°C ($\approx 9°F$) There is some uncertainty in these calculations; the earth is a large and complex system. Perhaps the earth will get cloudier as the temperature increases, moderating the increase. Or perhaps the arctic ice cap will melt, making the earth less reflective and leading to an even more dramatic temperature increase.

But the basic physics that leads to the greenhouse effect, and to global warming, is quite straightforward. Carbon dioxide in the atmosphere keeps the earth warm; more carbon dioxide will make it warmer. How much warmer? That's an important question, one that many scientists around the world are attempting to answer with ongoing research. But large or small, change *is* coming. Global warming is one of the most serious challenges facing scientists, engineers, and all citizens in the 21st century.

# PART VI PROBLEMS

*The following questions are related to the passage "The Greenhouse Effect and Global Warming" on the previous page.*

1. The intensity of sunlight at the top of the earth's atmosphere is approximately 1400 $W/m^2$. Mars is about 1.5 times as far from the sun as the earth. What is the approximate intensity of sunlight at the top of Mars's atmosphere?
   A. 930 $W/m^2$
   B. 620 $W/m^2$
   C. 410 $W/m^2$
   D. 280 $W/m^2$
2. Averaged over day, night, seasons, and weather conditions, a square meter of the earth's surface receives an average of 240 W of radiant energy from the sun. The average power radiated back to space is
   A. Less than 240 W.
   B. More than 240 W.
   C. Approximately 240 W.
3. The thermal radiation from the earth's surface peaks at a wavelength of approximately 10 $\mu$m. If the surface of the earth warms, this peak will
   A. Shift to a longer wavelength.
   B. Stay the same.
   C. Shift to a shorter wavelength.
4. The thermal radiation from the earth's surface peaks at a wavelength of approximately 10 $\mu$m. What is the energy of a photon at this wavelength?
   A. 2.4 eV B. 1.2 eV
   C. 0.24 eV D. 0.12 eV
5. Electromagnetic waves in certain wavelength ranges interact with water molecules because the molecules have a large electric dipole moment. The electric field of the wave
   A. Exerts a net force on the water molecules.
   B. Exerts a net torque on the water molecules.
   C. Exerts a net force and a net torque on the water molecules.

*The following passages and associated questions are based on the material of Part VI.*

### Taking an X Ray BIO

X rays are a very penetrating form of electromagnetic radiation. X rays pass through the soft tissue of the body but are largely stopped by bones and other more dense tissues. This makes x rays very useful for medical and dental purposes, as you know.

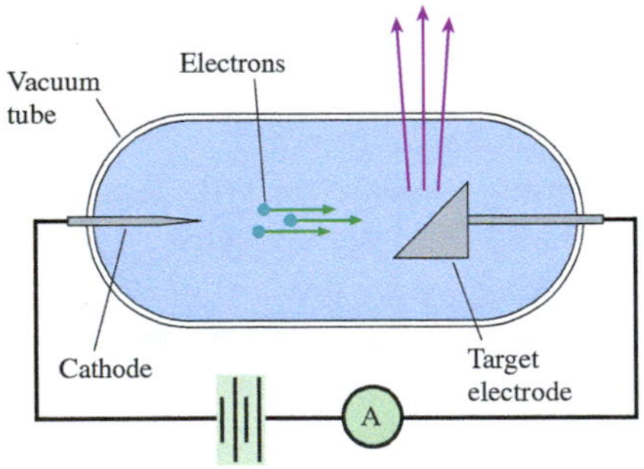


FIGURE VI.1

A schematic view of an x-ray tube and a driver circuit is given in Figure VI.1. A filament warms the cathode, freeing electrons. These electrons are accelerated by the electric field established by a high-voltage power supply connected between the cathode and a metal target. The electrons accelerate in the direction of the target. The rapid deceleration when they strike the target generates x rays. Each electron will emit one or more x rays as it comes to rest.

An x-ray image is essentially a shadow; x rays darken the film where they pass, but the film stays unexposed, and thus light, where bones or dense tissues block x rays. An x-ray technician adjusts the quality of an image by adjusting the energy and the intensity of the x-ray beam. This is done by adjusting two parameters: the accelerating voltage and the current through the tube. The accelerating voltage determines the energy of the x-ray photons, which can't be greater than the energy of the electrons. The current through the tube determines the number of electrons per second and thus the number of photons emitted. In clinical practice, the exposure is characterized by two values: "kVp" and "mAs." kVp is the peak voltage in kV. The value mAs is the product of the current (in mA) and the time (in s) to give a reading in mA · s. This is a measure of the total number of electrons that hit the target and thus the number of x rays emitted.

Typical values for a dental x ray are a kVp of 70 (meaning a peak voltage of 70 kV) and mAs of 7.5 (which comes from a current of 10 mA for 0.75 s, for a total of 7.5 mAs). Assume these values in all of the problems that follow.

6. In Figure VI.1, what is the direction of the electric field in the region between the cathode and the target electrode?
   A. To the left
   B. To the right
   C. Toward the top of the page
   D. Toward the bottom of the page
7. If the distance between the cathode and the target electrode is approximately 1.0 cm, what will be the maximum acceleration of the free electrons? Assume that the electric field is uniform.
   A. $1.2 \times 10^{18}$ m/s$^2$ B. $1.2 \times 10^{16}$ m/s$^2$
   C. $1.2 \times 10^{15}$ m/s$^2$ D. $1.2 \times 10^{12}$ m/s$^2$
8. What, physically, does the product of a current (in mA) and a time (in s) represent?
   A. Energy in mJ B. Potential difference in mV
   C. Charge in mC D. Resistance in m$\Omega$
9. During the 0.75 s that the tube is running, what is the electric power?
   A. 7.0 kW B. 700 W
   C. 70 W D. 7.0 W
10. If approximately 1% of the electric energy ends up in the x-ray beam (a typical value), what is the approximate total energy of the x rays emitted?
    A. 500 J B. 50 J
    C. 5 J D. 0.5 J
11. What is the maximum energy of the emitted x-ray photons?
    A. $70 \times 10^3$ J B. $1.1 \times 10^{-11}$ J
    C. $1.1 \times 10^{-14}$ J D. $1.6 \times 10^{-18}$ J

## Electric Cars

In recent years, practical hybrid cars have hit the road—cars in which the gasoline engine runs a generator that charges batteries that run an electric motor. These cars offer increased efficiency, but significantly greater efficiency could be provided by a purely electric car run by batteries that you charge by plugging into an electric outlet in your house.

But there's a practical problem with such vehicles: the time necessary to recharge the batteries. If you refuel your car with gas at the pump, you add 130 MJ of energy per gallon. If you add 20 gallons, you add a total of 2.6 GJ in about 5 minutes. That's a lot of energy in a short time; the electric system of your house simply can't provide power at this rate.

There's another snag as well. Suppose there were electric filling stations that could provide very high currents to recharge your electric car. Conventional batteries can't recharge very quickly; it would still take longer for a recharge than to refill with gas.

One possible solution is to use capacitors instead of batteries to store energy. Capacitors can be charged much more quickly, and as an added benefit, they can provide energy at a much greater rate—allowing for peppier acceleration. Today's capacitors can't store enough energy to be practical, but future generations will.

12. A typical home's electric system can provide 100 A at a voltage of 220 V. If you had a charger that ran at this full power, approximately how long would it take to charge a battery with the equivalent of the energy in one gallon of gas?
    A. 100 min  B. 50 min
    C. 20 min  D. 5 min
13. The Tesla Model S, a production electric car, has a 400 V battery system that can provide a power of 270 kW. At this peak power, what is the current supplied by the batteries?
    A. 75 kA  B. 1900 A
    C. 680 A  D. 75 A
14. To charge the batteries in a Tesla Model S, a transformer is used to step up the voltage of the household supply. If you step a 220 V, 100 A system up to 400 V, what is the maximum current you can draw at this voltage?
    A. 180 A  B. 100 A
    C. 55 A  D. 45 A
15. One design challenge for a capacitor-powered electric car is that the voltage would change with time as the capacitors discharged. If the capacitors in a car were discharged to half their initial voltage, what fraction of energy would still be left?
    A. 75%  B. 67%
    C. 50%  D. 25%

## Wireless Power Transmission

Your laptop has wireless communications connectivity, and you might even have a wireless keyboard or mouse. But there's one wire you haven't been able to get rid of yet—the power cord.

Researchers are working on ways to circumvent the need for a direct electrical connection for power, and they are experiencing some success. Recently, investigators were able to use current flowing through a primary coil to power a 60 W lightbulb connected to a secondary coil 2.0 m away, with approximately 15% efficiency. The coils were large and the efficiency low, but it's a start.

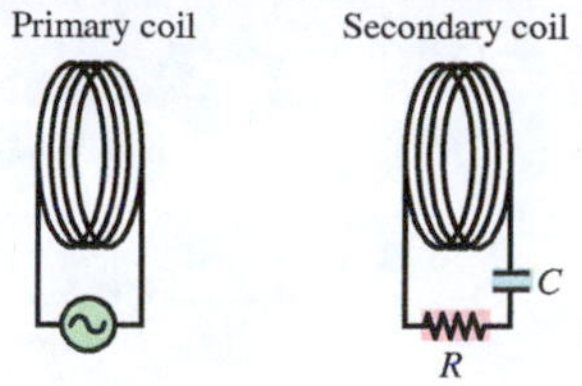

FIGURE VI.2

The wireless power transfer system is outlined in Figure VI.2. An AC supply generates a current through the primary coil, creating a varying magnetic field. This field induces a current in the secondary coil, which is connected to a resistance (the lightbulb) and a capacitor that sets the resonance frequency of the secondary circuit to match the frequency of the primary circuit.

16. At a particular moment, the current in the primary coil is clockwise, as viewed from the secondary coil. At the center of the secondary coil, the field from the primary coil is
    A. To the right.
    B. To the left.
    C. Zero.
17. At a particular moment, the magnetic field from the primary coil points to the right and is increasing in strength. The field due to the induced current in the secondary coil is
    A. To the right.
    B. To the left.
    C. Zero.
18. The power supply drives the primary coil at 9.9 MHz. If this frequency is doubled, how must the capacitor in the secondary circuit be changed?
    A. Increase by a factor of 2
    B. Increase by a factor of $\sqrt{2}$
    C. Decrease by a factor of 2
    D. Decrease by a factor of 4
19. What are the rms and peak currents for a 60 W bulb? (The rms voltage is the usual 120 V.)
    A. 0.71 A, 0.71 A  B. 0.71 A, 0.50 A
    C. 0.50 A, 0.71 A  D. 0.50 A, 0.50 A

## Additional Integrated Problems

20. A 20 Ω resistor is connected across a 120 V source. The resistor is then lowered into an insulated beaker, containing 1.0 L of water at 20° C, for 60 s. What is the final temperature of the water?
21. As shown in Figure VI.3, a square loop of wire, with a mass of 200 g, is free to pivot about a horizontal axis through one of its sides. A 0.50 T horizontal magnetic field is directed as shown. What current $I$ in the loop, and in what direction, is needed to hold the loop steady in a horizontal plane?

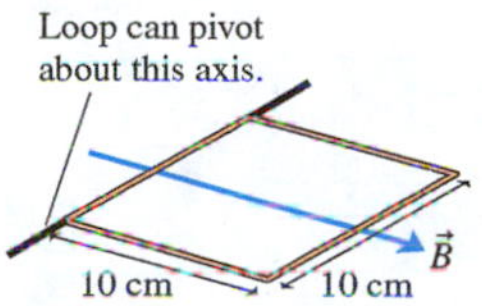

FIGURE VI.3

PART

# VII Modern Physics

Sunlight doesn't penetrate this cave, so creatures must provide their own light. These aptly named glowworms are *bioluminescent;* chemical reactions in the cells in their bodies turn chemical energy into light energy, resulting in an eerie greenish glow. Understanding the processes that lead to this cool light means understanding the interaction of light and atoms, which we'll explore in the coming chapters.

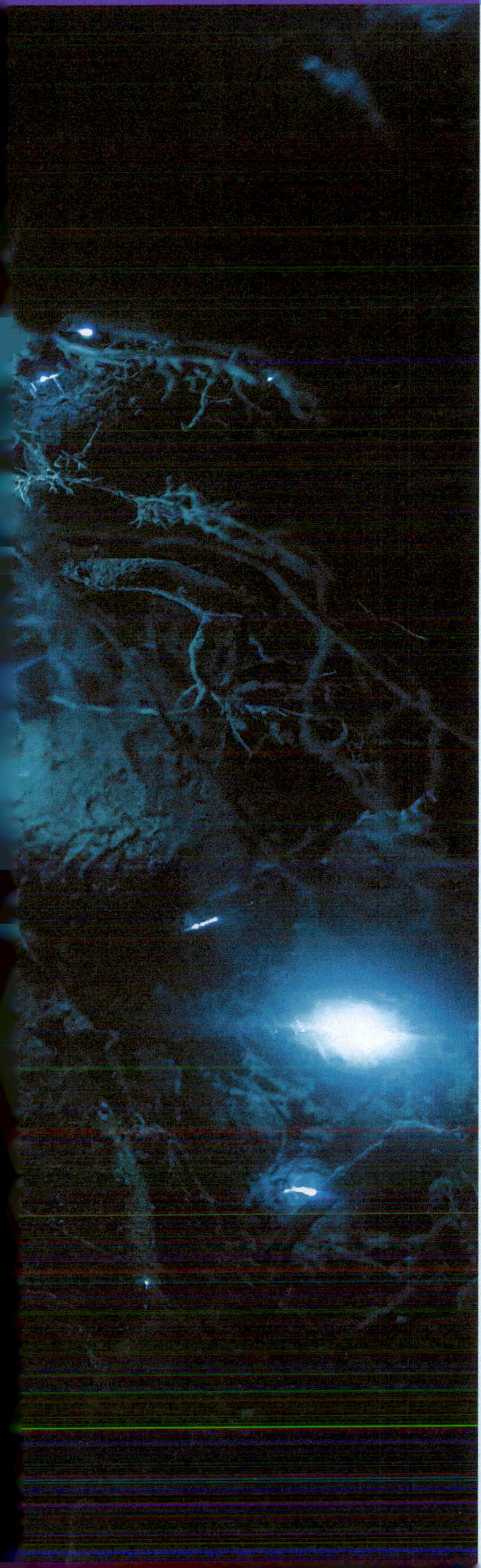

OVERVIEW

## New Ways of Looking at the World

Newton's mechanics and Maxwell's electromagnetism are remarkable theories that explain a wide range of physical phenomena, as we have seen in the past 26 chapters—but our story doesn't stop there. In the early 20th century, a series of discoveries profoundly altered our understanding of the universe at the most fundamental level, forcing scientists to reconsider the very nature of space and time and to develop new models of light and matter.

### Relativity

The idea of measuring distance with a meter stick and time with a clock or stopwatch seems self-evident. But Albert Einstein, as a young, unknown scientist, realized that Maxwell's theory of electromagnetism could be consistent only if an additional rather odd assumption was made: that the speed of light is the same for all observers, no matter how they might be moving with respect to each other or to the source of the light. This assumption changes the way that we think about space and time. When we study Einstein's theory of *relativity,* you will see how different observers can disagree about lengths and time intervals. We need to go beyond stopwatches and meter sticks. Time can pass at different rates for different observers; time is, as you will see, relative. Our exploration will end with the most famous equation in physics, Einstein's $E = mc^2$. Matter can be converted to energy, and energy to matter.

### Quantum Physics

We've seen that light, a wave, sometimes acts like a particle, a photon. We'll now find that particles such as electrons or atoms sometimes behave like waves. All of the characteristics of waves, such as diffraction and interference, will also apply to particles. This odd notion—that there's no clear distinction between particles and waves—is the core of a new model of light and matter called *quantum physics.* The wave nature of particles will lead to the *quantization* of energy. A particle confined in a box—or an electron in an atom—can have only certain energies. This idea will be a fruitful one for us, allowing us to understand the spectra of gases and phenomena such as bioluminescence.

### Atoms, Nuclei, and Particles

We have frequently used the atomic model, explaining the properties of matter by considering the behavior of the atoms that comprise it. But when you get right down to it, what *is* an atom? And what's inside an atom? As you know, an atom has a tiny core called a *nucleus.* We'll look at what goes on inside the nucleus. One remarkable discovery will be that the nuclei of certain atoms spontaneously decay, turning the atom from one element to another. This phenomenon of *radioactivity* will give us a window into the nature of atoms and the particles that comprise them.

Once we know that the nucleus of an atom is composed of protons and neutrons, there's a natural next question to ask: What's inside a proton or neutron? There is an answer to this question, an answer we will learn in the final chapter of this text.

# 27 Relativity

Researchers studying African wild dogs fit them with collars that use the global positioning system (GPS) to pinpoint the locations of these wide-ranging predators. (The collars also protect the dogs from snares and thus reduce mortality.) How does maintaining the accuracy of the GPS require the use of relativity?

## LOOKING AHEAD ➧

### Simultaneity

The lightning strikes are simultaneous to you, but to someone who is moving relative to you they occur at different times.

You'll learn how to compute the order in which two events occur according to observers moving relative to each other.

### Time and Space

Time itself runs faster on the surface of the earth than on GPS satellites that are in rapid motion relative to the earth.

You'll learn how moving clocks run slower and moving objects are shorter than when they are at rest.

### Mass and Energy

The sun's energy comes from converting 4 billion kilograms of matter into energy every second.

You'll learn how Einstein's famous equation $E = mc^2$ shows that mass and energy are equivalent.

**GOAL** **To understand how Einstein's theory of relativity changes our concepts of time and space.**

## LOOKING BACK ➧

### Relative Motion

In Section 3.8 you learned how to find the velocity of a ball relative to Ana given its velocity relative to Carlos.

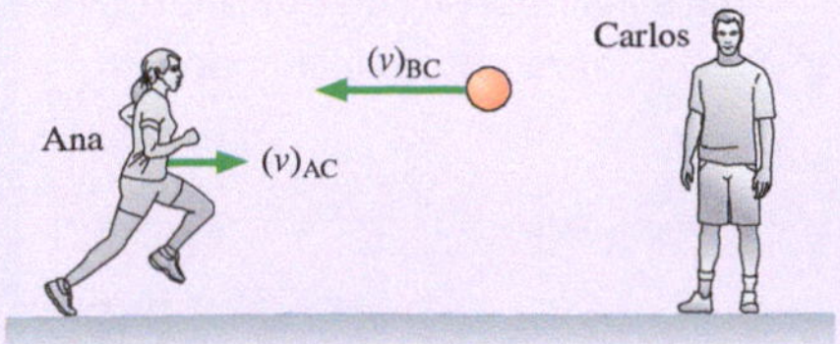


In this chapter, you'll see how our commonsense ideas about relative motion break down when one or more of the velocities approach the speed of light.

**STOP TO THINK**

The car is moving at 10 m/s relative to Bill. How fast does Amy see the car as moving?

A. 5 m/s B. 10 m/s C. 15 m/s D. 20 m/s

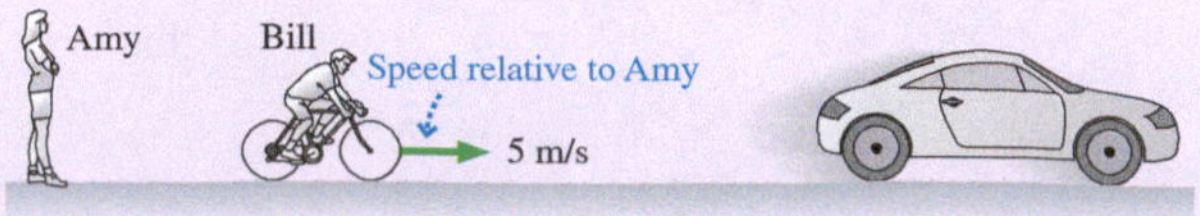

## 27.1 Relativity: What's It All About?

Albert Einstein (1879–1955) was one of the most influential thinkers in history.

What do you think of when you hear the phrase "theory of relativity"? A white-haired Einstein? $E = mc^2$? Black holes? Time travel? There is, without doubt, a certain mystique associated with relativity, an aura of the strange and exotic. The good news is that understanding the ideas of relativity is well within your grasp. Einstein's *special theory of relativity,* the portion of relativity we'll study, is not mathematically difficult at all. The challenge is conceptual because relativity questions deeply held assumptions about the nature of space and time. In fact, that's what relativity is all about—space and time.

In Newtonian mechanics, space and time are absolute quantities; the length of a meter stick and the time between ticks of a clock are the same to any observer, whether moving or not. But relativity challenges these commonsense notions. As we'll see, ground-based observers measure the length of a fast-moving rocket to be *shorter,* and a clock on the rocket to run *slower,* compared to when the rocket is at rest.

Because relativity challenges our basic notions about space and time, to make progress we will have to exercise the utmost care with regard to logic and precision. We will need to state very precisely just how it is that we know things about the physical world and then ruthlessly follow the logical consequences. The challenge is to stay on this path and not to let our prior assumptions—assumptions that are deeply ingrained in all of us—lead us astray.

### What's Special, and What's Relative, About Special Relativity?

A bird flies through the air. How fast is it moving? This is a question that doesn't have a unique answer. The bird might be moving at 10 m/s relative to the air, but only 6 m/s relative to the ground if it is flying into a 4 m/s headwind. To answer the question "How fast is the bird moving?" you need to first answer another question: "Relative to what?" We explored this idea of relative motion back in ◀◀ SECTION 3.8. This notion of relativity will become even more important in this chapter. We'll find that seemingly straightforward questions don't have straightforward answers. If you ask, "What time is it?" there is, surprisingly, not a unique answer. Time is, as we'll see, also relative.

Einstein's first paper on relativity dealt exclusively with objects that move relative to each other with constant velocity. Motion at constant velocity is a "special case" of motion—namely, motion for which the acceleration is zero. Hence Einstein's first theory of relativity has come to be known as *special relativity*. It is "special" in the sense of being a restricted, particular case of a more general theory that came later, not special in the everyday sense meaning distinctive or exceptional.

## 27.2 Galilean Relativity

You may have had the experience of sitting on a train and looking up to see another train moving slowly past. It can be hard to tell if the other train is moving past your stationary train, or if you're moving in the opposite direction past a stationary train. Only the *relative* velocity between the trains has meaning.

Galileo was the first to understand how the laws of physics depended on the relative motion between different observers. A firm grasp of *Galilean relativity* is necessary to appreciate and understand what is new in Einstein's theory. We'll begin by reviewing and extending the ideas of relative motion.

### Reference Frames

Suppose you're passing me as we both drive in the same direction along a freeway. My car's speedometer reads 55 mph while your speedometer shows 60 mph. Is 60 mph your "true" speed? That is certainly your speed relative to someone standing beside the road, but your speed relative to me is only 5 mph. Your speed is 120 mph relative to a driver approaching from the other direction at 60 mph.

A moving object does not have a "true" speed or velocity. The very definition of velocity, $v = \Delta x/\Delta t$, assumes the existence of a coordinate system in which, during some time interval $\Delta t$, the displacement $\Delta x$ is measured. The best we can manage is to specify an object's velocity relative to, or with respect to, the coordinate system in which it is measured.

Let's define a **reference frame** to be a coordinate system in which experimenters equipped with meter sticks, stopwatches, and any other needed equipment make position and time measurements on moving objects. Three ideas are implicit in our definition of a reference frame:

- A reference frame extends infinitely far in all directions.
- The experimenters are at rest in the reference frame.
- The number of experimenters and the quality of their equipment are sufficient to measure positions and velocities to any level of accuracy needed.

FIGURE 27.1 The standard reference frames $S$ and $S'$.

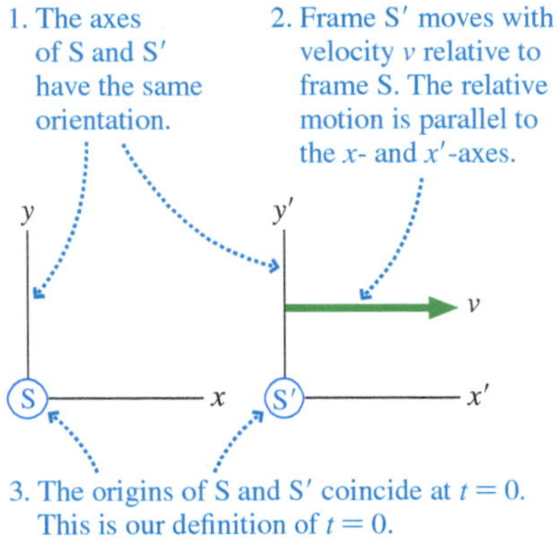


The first two points are especially important. It is often convenient to say "the laboratory reference frame" or "the reference frame of the rocket." These are shorthand expressions for "a reference frame, infinite in all directions, in which the laboratory (or the rocket) and a set of experimenters happen to be at rest."

NOTE ▸ A reference frame is not the same thing as a "point of view." That is, each experimenter does not have his or her own private reference frame. **All experimenters at rest relative to each other share the same reference frame.** ◂

FIGURE 27.1 shows how we represent two reference frames, S and S′, that are in relative motion. The coordinate axes in S are $x$, $y$, $z$ and those in S′ are $x'$, $y'$, $z'$. Reference frame S′ moves with velocity $v$ relative to S or, equivalently, S moves with velocity $-v$ relative to S′. There's no implication that either reference frame is "at rest." Notice that the zero of time, when experimenters start their stopwatches, is the instant when the origins of S and S′ coincide.

## Inertial Reference Frames

FIGURE 27.2 Two reference frames.

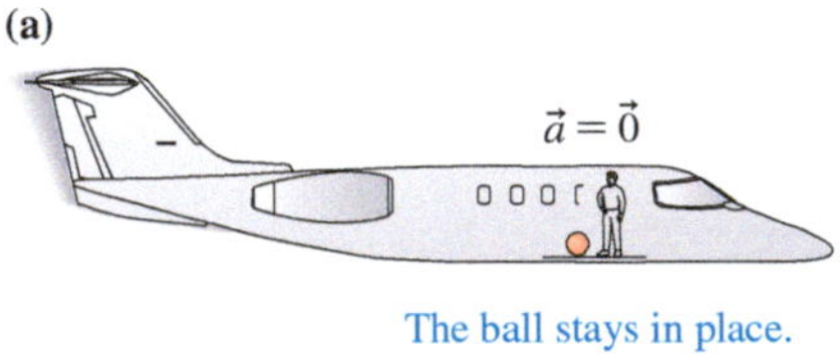


A ball with no horizontal forces stays at rest in an airplane cruising at constant velocity. The airplane is an inertial reference frame.

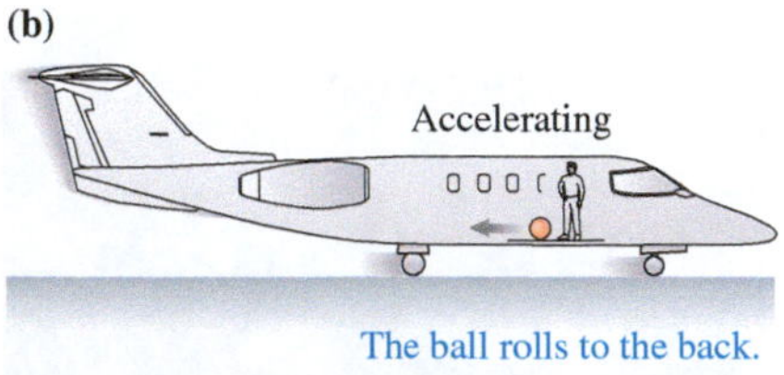


The ball rolls to the back of the plane during takeoff. An accelerating plane is not an inertial reference frame.

Certain reference frames are especially simple. FIGURE 27.2a shows a student cruising at constant velocity in an airplane. If he places a ball on the floor, it stays there. There are no horizontal forces, and the ball remains at rest relative to the airplane. That is, $\vec{a} = \vec{0}$ in the airplane's coordinate system when $\vec{F}_{net} = \vec{0}$, so Newton's first law is satisfied. Similarly, if the student drops the ball, it falls straight down—relative to the student—with an acceleration of magnitude $g$, satisfying Newton's second law.

We define an **inertial reference frame** as one in which Newton's first law is valid. That is, an inertial reference frame is one in which an isolated particle, one on which there are no forces, either remains at rest or moves in a straight line at constant speed, as measured by experimenters at rest in that frame.

Not all reference frames are inertial. The student in FIGURE 27.2b conducts the same experiment during takeoff. He carefully places the ball on the floor just as the airplane starts to accelerate down the runway. You can imagine what happens. The ball rolls to the back of the plane as the passengers are being pressed back into their seats. If the student measures the ball's motion using a meter stick attached to the plane, he will find that the ball accelerates *in the plane's reference frame.* Yet he would be unable to identify any force on the ball that would act to accelerate it toward the back of the plane. This violates Newton's first law, so the plane is *not* an inertial reference frame during takeoff. **In general, accelerating reference frames are not inertial reference frames.**

NOTE ▸ An inertial reference frame is an idealization. A true inertial reference frame would need to be floating in deep space, far from any gravitational influence. In practice, an earthbound laboratory is a good approximation of an inertial reference frame because the accelerations associated with the earth's rotation and motion around the sun are too small to influence most experiments. ◂

These ideas are in accord with your everyday experience. If you're in a jet flying smoothly at 600 mph—an inertial reference frame—Newton's laws are valid: You can toss and catch a ball, or pour a cup of coffee, exactly as you would on the ground. But if the plane were diving, or shaking from turbulence, simple "experiments" like these would fail. A ball thrown straight up would land far from your hand, and the stream of coffee would bend and turn on its way to missing the cup. These apparently simple observations can be stated as the *Galilean principle of relativity:*

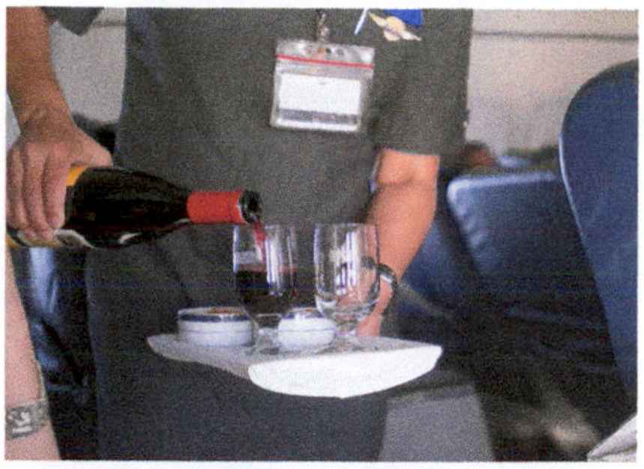

This flight attendant pours wine on a smoothly flying airplane moving at 600 mph just as easily as she does at the terminal. These ideas were first discussed by Galileo in 1632 in the context of pouring water while on a moving ship.

**Galilean principle of relativity** Newton's laws of motion are valid in all inertial reference frames.

In our study of relativity, we will restrict our attention to inertial reference frames. This implies that the relative velocity $v$ between two reference frames is constant. Any reference frame that moves at constant velocity with respect to an inertial reference frame is itself an inertial reference frame. Conversely, a reference frame that accelerates with respect to an inertial reference frame is not an inertial reference frame. Although special relativity can be used for accelerating reference frames, we will confine ourselves here to the simple case of inertial reference frames moving with respect to each other at constant velocity.

**STOP TO THINK 27.1** Which of these is an inertial reference frame (or a very good approximation)?

A. Your bedroom
B. A car rolling down a steep hill
C. A train coasting along a level track
D. A rocket being launched
E. A roller coaster going over the top of a hill
F. A sky diver falling at terminal speed

## The Galilean Velocity Transformation

Suppose Sue is standing beside a highway as Jim drives by at 50 mph, as shown in FIGURE 27.3. Let S be Sue's reference frame—a reference frame attached to the ground—and let S′ be the reference frame moving with Jim, attached to his car. We see that the velocity of reference frame S′ relative to S is $v = 50$ mph.

Now suppose a motorcyclist blasts down the highway, traveling in the same direction as Jim. Sue measures the motorcycle's velocity to be $u = 75$ mph. What is the cycle's velocity $u'$ measured relative to Jim? We can answer this on the basis of common sense. If you're driving at 50 mph, and someone passes you going 75 mph, then his speed *relative to you* is 25 mph. This is the *difference* between his speed relative to the ground and your speed relative to the ground. Thus Jim measures the motorcycle's velocity to be $u' = 25$ mph.

FIGURE 27.3 A motorcycle's velocity as seen by Sue and by Jim.

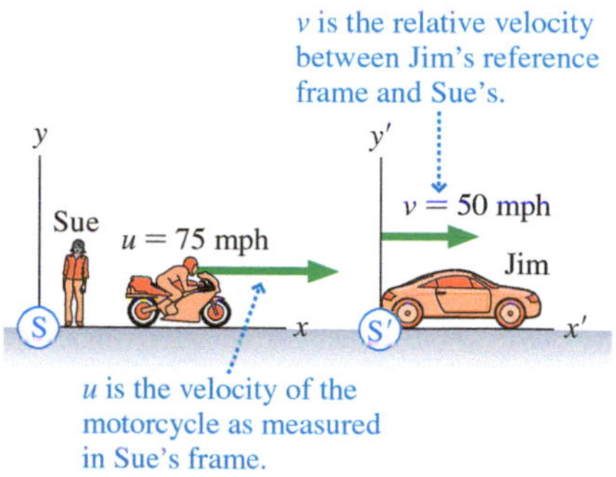

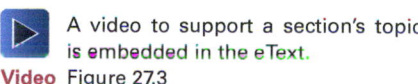

A video to support a section's topic is embedded in the eText.
Video Figure 27.3

**NOTE** ▶ In this chapter, we will use $v$ to represent the velocity of one reference frame relative to another. We will use $u$ and $u'$ to represent the velocities of objects with respect to reference frames S and S′. In addition, we will assume that all motion is parallel to the $x$-axis. ◀

We can state this idea as a general rule. An object's velocity measured in frame S is related to its velocity measured in frame S′ by

$$u' = u - v \quad \text{and} \quad u = u' + v \tag{27.1}$$

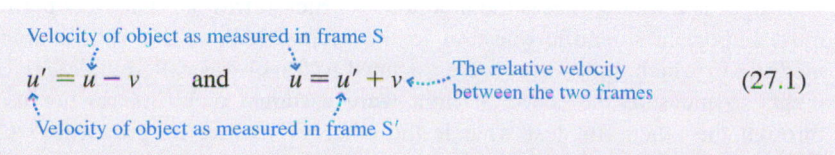

Velocity of object as measured in frame S
Velocity of object as measured in frame S′
The relative velocity between the two frames

Equations 27.1 are the **Galilean velocity transformations.** If you know the velocity of a particle as measured by the experimenters in one inertial reference frame, you can use Equations 27.1 to find the velocity that would be measured by experimenters in any other inertial reference frame.

**EXAMPLE 27.1 Finding the speed of sound**

An airplane is flying at speed 200 m/s with respect to the ground. Sound wave 1 is approaching the plane from the front, while sound wave 2 is catching up from behind. Both waves travel at 340 m/s relative to the ground. What is the velocity of each wave relative to the plane?

**STRATEGIZE** This is a relative motion problem, similar to problems we saw in Chapter 3. We'll treat it using the concept of frames of reference and the velocity transformations given in Equations 27.1.

**PREPARE** Assume that the earth (frame S) and the airplane (frame S′) are inertial reference frames. Frame S′, in which the airplane is at rest, moves with velocity $v = 200$ m/s relative to frame S. FIGURE 27.4 shows the airplane and the sound waves.

FIGURE 27.4 Experimenters in the plane measure different speeds for the sound waves than do experimenters on the ground.

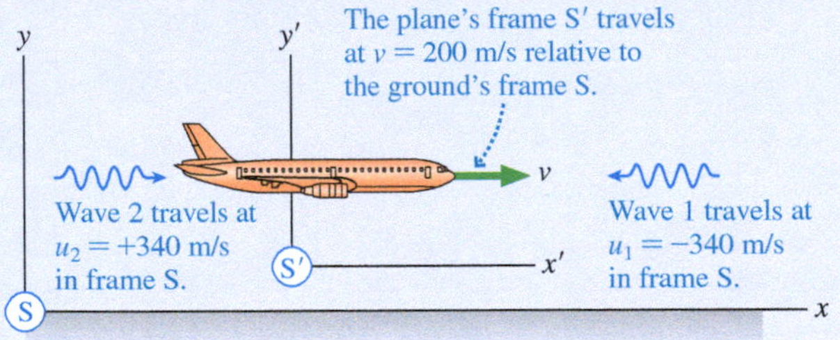


**SOLVE** The speed of a mechanical wave, such as a sound wave or a wave on a string, is its speed *relative to its medium.* Thus the *speed of sound* is the speed of a sound wave through a reference frame in which the air is at rest. This is reference frame S, where wave 1 travels with velocity $u_1 = -340$ m/s and wave 2 travels with velocity $u_2 = +340$ m/s. Notice that the Galilean transformation uses *velocities,* with appropriate signs, not just speeds.

The airplane travels to the right with reference frame S′ at velocity $v$. We can use the Galilean transformation of velocity to find the velocities of the two sound waves in frame S′:

$$u_1' = u_1 - v = -340 \text{ m/s} - 200 \text{ m/s} = -540 \text{ m/s}$$

$$u_2' = u_2 - v = 340 \text{ m/s} - 200 \text{ m/s} = 140 \text{ m/s}$$

Thus wave 1 approaches the plane with a *speed* of 540 m/s, while wave 2 approaches with a speed of 140 m/s.

**ASSESS** This isn't surprising. If you're driving at 50 mph, a car coming the other way at 55 mph is approaching you at 105 mph. A car coming up behind you at 55 mph seems to be gaining on you at the rate of only 5 mph. Mechanical wave speeds behave the same. Notice that a mechanical wave would appear to be stationary to a person moving at the wave speed. To a surfer, the crest of the ocean wave remains at rest under his or her feet.

**STOP TO THINK 27.2** Ocean waves are approaching the beach at 10 m/s. A boat heading out to sea travels at 6 m/s. How fast are the waves moving in the boat's reference frame?

A. 16 m/s B. 10 m/s C. 6 m/s D. 4 m/s

## 27.3 Einstein's Principle of Relativity

The 19th century was an era of optics and electromagnetism. Thomas Young demonstrated in 1801 that light is a wave, and by midcentury scientists had devised techniques for measuring the speed of light. Faraday discovered electromagnetic induction in 1831, setting in motion a series of events leading to Maxwell's conclusion, in 1864, that light is an electromagnetic wave.

If light is a wave, what is the medium in which it travels? This was perhaps the most important scientific question in the second half of the 19th century. The medium in which light waves were assumed to travel was called the *ether.* Experiments to measure the speed of light were assumed to be measuring its speed through the ether. But just what is the ether? What are its properties? Can we collect a jar full of ether to study? Despite the significance of these questions, experimental efforts to detect the ether or measure its properties kept coming up empty-handed.

Maxwell's theory of electromagnetism didn't help the situation. The crowning success of Maxwell's theory was his prediction that light waves travel with speed

$$c = \frac{1}{\sqrt{\epsilon_0 \mu_0}} = 3.00 \times 10^8 \text{ m/s}$$

This is a very specific prediction with no wiggle room. The difficulty with such a specific prediction was the implication that Maxwell's laws of electromagnetism, enshrined in a set of four equations, are valid *only* in the reference frame of the ether. After all, as **FIGURE 27.5** shows, the light speed should certainly be faster or slower than $c$ in a reference frame moving through the ether, just as the sound speed is different to someone moving through the air.

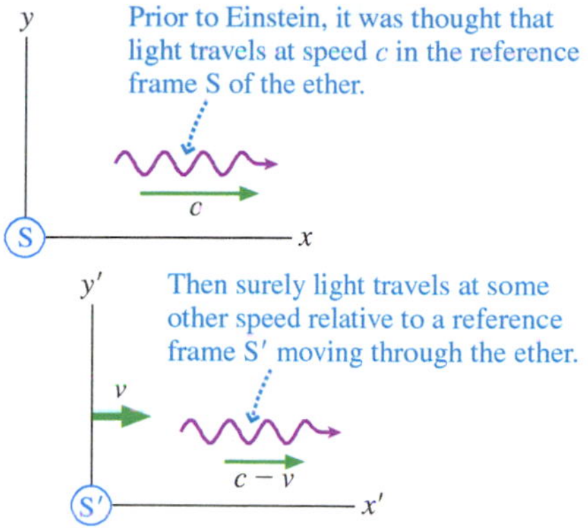


**FIGURE 27.5** It seems as if the speed of light should differ from $c$ in a reference frame moving through the ether.

As the 19th century closed, it appeared that Maxwell's theory did not obey the classical principle of relativity. There was just one reference frame, the reference frame of the ether, in which the laws of electromagnetism seemed to be true. And to make matters worse, the fact that no one had been able to detect the ether meant that no one could identify the one reference frame in which Maxwell's equations "worked."

It was in this muddled state of affairs that a young Albert Einstein made his mark on the world. Even as a teenager, Einstein had wondered how a light wave would look to someone "surfing" the wave, traveling alongside the wave at the wave speed. You can do that with a water wave or a sound wave, but light waves seemed to present a logical difficulty. An electromagnetic wave sustains itself by virtue of the fact that a changing magnetic field induces an electric field and a changing electric field induces a magnetic field. But to someone moving with the wave, *the fields would not change.* How could there be an electromagnetic wave under these circumstances?

In certain rivers, tides send waves upriver that can be surfed for miles. From the reference frame of these surfers, the waves are standing still. If you could move along with a light wave, would the electric and magnetic fields appear motionless?

Several years of thinking about the connection between electromagnetism and reference frames led Einstein to the conclusion that *all* the laws of physics, not just the laws of mechanics, should obey the principle of relativity. In other words, the principle of relativity is a fundamental statement about the nature of the physical universe. The Galilean principle of relativity stated only that Newton's laws hold in any inertial reference frame. Einstein was able to state a much more general principle:

**Principle of relativity** All the laws of physics are the same in all inertial reference frames.

All of the results of Einstein's theory of relativity flow from this one simple statement.

## The Constancy of the Speed of Light

If Maxwell's equations of electromagnetism are laws of physics, and there's every reason to think they are, then, according to the principle of relativity, Maxwell's equations must be true in *every* inertial reference frame. On the surface this seems to be an innocuous statement, equivalent to saying that the law of conservation of momentum is true in every inertial reference frame. But follow the logic:

1. Maxwell's equations are true in all inertial reference frames.
2. Maxwell's equations predict that electromagnetic waves, including light, travel at speed $c = 3.00 \times 10^8$ m/s.
3. Therefore, **light travels at speed $c$ in all inertial reference frames.**

FIGURE 27.6 Light travels at speed $c$ in all inertial reference frames, regardless of how the reference frames are moving with respect to the light source.

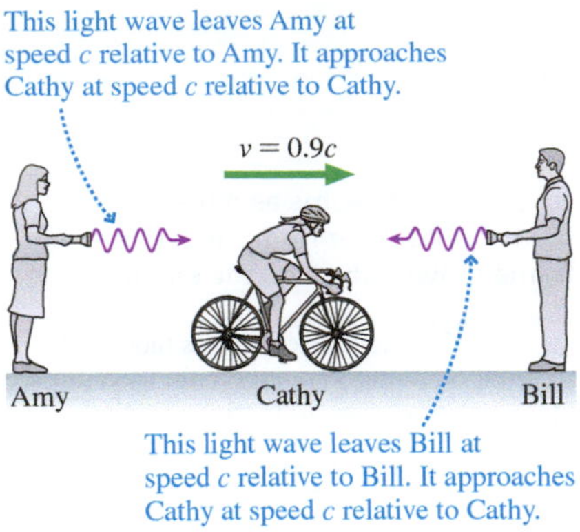


FIGURE 27.7 Experiments find that the photons travel through the laboratory with speed $c$, not the speed $1.99975c$ that you might expect.

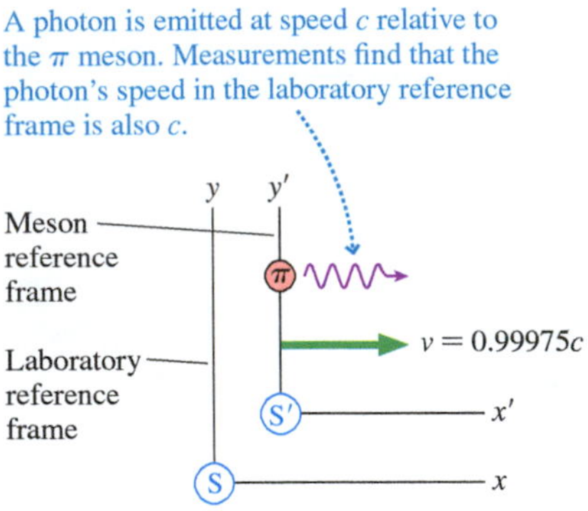


FIGURE 27.6 shows the implications of this conclusion. *All* experimenters, regardless of how they move with respect to each other, find that *all* light waves, regardless of the source, travel in their reference frame with the *same* speed $c$. If Cathy's velocity toward Bill and away from Amy is $v = 0.9c$, Cathy finds, by making measurements in her reference frame, that the light from Bill approaches her at speed $c$, not at $c + v = 1.9c$. And the light from Amy, which left Amy at speed $c$, catches up from behind at speed $c$ *relative to Cathy*, not the $c - v = 0.1c$ you would have expected.

Although this prediction goes against common sense, the experimental evidence for it is strong. Laboratory experiments are difficult because even the highest laboratory speed is insignificant in comparison to $c$. In the 1930s, however, the physicists R. J. Kennedy and E. M. Thorndike realized that they could use the earth itself as a laboratory. The earth's speed as it circles the sun is about 30,000 m/s. The velocity of the earth in January differs by 60,000 m/s from its velocity in July, when the earth is moving in the opposite direction. Kennedy and Thorndike were able to use a very sensitive and stable interferometer to show that the numerical values of the speed of light in January and July differ by less than 2 m/s, well within experimental error.

More recent experiments have used unstable elementary particles, called $\pi$ mesons, that decay into high-energy photons, or particles of light. The $\pi$ mesons, created in a particle accelerator, move through the laboratory at 99.975% the speed of light, or $v = 0.99975c$, as they emit photons at speed $c$ in the $\pi$ meson's reference frame. As FIGURE 27.7 shows, you would expect the photons to travel through the laboratory with speed $c + v = 1.99975c$. Instead, the measured speed of the photons in the laboratory was, within experimental error, $3.00 \times 10^8$ m/s.

In summary, *every* experiment designed to compare the speed of light in different reference frames has found that light travels at $3.00 \times 10^8$ m/s in every inertial reference frame, regardless of how the reference frames are moving with respect to each other.

## How Can This Be?

You're in good company if you find this impossible to believe. Suppose I shot a ball forward at 50 m/s while driving past you at 30 m/s. You would certainly see the ball traveling at 80 m/s relative to you and the ground. What we're saying with regard to light is equivalent to saying that the ball travels at 50 m/s relative to my car and *at the same time* travels at 50 m/s relative to the ground, even though the car is moving across the ground at 30 m/s. It seems logically impossible.

You might think that this is merely a matter of semantics. If we can just get our definitions and use of words straight, then the mystery and confusion will disappear. Or perhaps the difficulty is a confusion between what we "see" versus what "really happens." In other words, a better analysis, one that focuses on what really happens, would find that light "really" travels at different speeds in different reference frames.

Alas, what "really happens" is that light travels at $3.00 \times 10^8$ m/s in every inertial reference frame, regardless of how the reference frames are moving with respect to each other. It's not a trick. There remains only one way to escape the logical contradictions.

FIGURE 27.8 shows how the speed of a light ray would be measured by Laura, in stationary frame S, and by Dan and Eric, in frame S′, which is moving to the right at

FIGURE 27.8 Measuring the speed of light in two different reference frames.

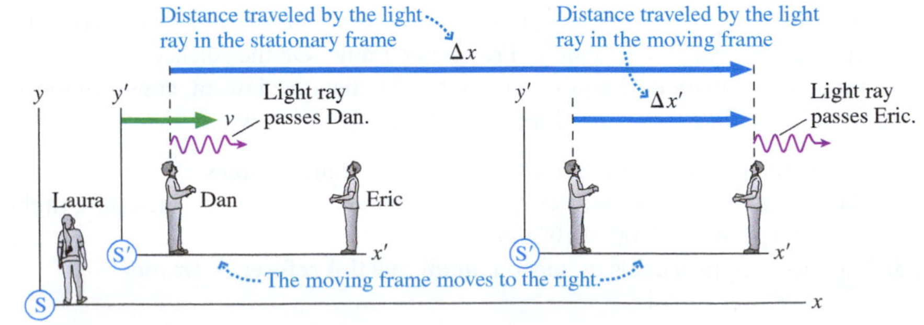

relative velocity $v$. As the ray moves from Dan to Eric, they measure it as having traveled a distance $\Delta x'$, the distance between them in their frame. To Laura, however, the ray will have traveled the longer distance $\Delta x$, simply because Eric is moving to the right so, as seen by Laura, the ray has to travel farther before it reaches him.

Recall that the definition of velocity is $\Delta x/\Delta t$, the ratio of the distance traveled to the time interval in which the travel occurs. Laura measures a *different* value for the distance traveled by the light ray than do Dan and Eric, but according to the principle of relativity, both sets of observers must measure the *same* value for the speed of light $c$. The only way this is possible is that the time $\Delta t$ that, according to Laura, it takes light to make the trip is *not the same* as the time $\Delta t'$ that Dan and Eric measure.

We've assumed, since the beginning of this text, that time is simply time. It flows along like a river, and all experimenters in all reference frames simply use it. But this is evidently not the case. Something must be wrong with assumptions that we've made about the nature of time. The principle of relativity has painted us into a corner, and our only way out is to reexamine our understanding of time.

## 27.4 Events and Measurements

To question some of our most basic assumptions about space and time requires extreme care. We need to be certain that no assumptions slip into our analysis unnoticed. Our goal is to describe the motion of a particle in a clear and precise way, making the barest minimum of assumptions.

### Events

The fundamental element of relativity is called an **event.** An event is a physical activity that takes place at a definite point in space and at a definite instant of time. A firecracker exploding is an event. A collision between two particles is an event. A light wave hitting a detector is an event.

Events can be observed and measured by experimenters in different reference frames. An exploding firecracker is as clear to you as you drive by in your car as it is to me standing on the street corner. We can quantify where and when an event occurs with four numbers: the coordinates $(x, y, z)$ and the instant of time $t$. These four numbers, illustrated in FIGURE 27.9, are called the **spacetime coordinates** of the event.

FIGURE 27.9 The location and time of an event are described by its spacetime coordinates.

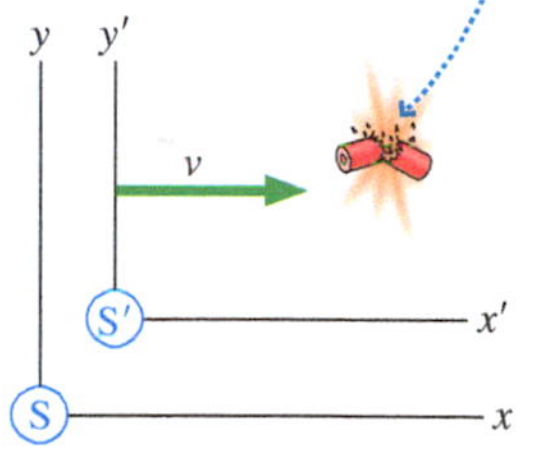


The spatial coordinates of an event measured in reference frames S and S′ may differ. But it now appears that the instant of time recorded in S and S′ may also differ. Thus the spacetime coordinates of an event measured by experimenters in frame S are $(x, y, z, t)$, and the spacetime coordinates of the *same event* measured by experimenters in frame S′ are $(x', y', z', t')$.

### Measurements

Events are what "really happen," but how do we learn about an event? That is, how do the experimenters in a reference frame determine the spacetime coordinates of an event? This is a problem of *measurement.*

We defined a reference frame to be a coordinate system in which experimenters can make position and time measurements. That's a good start, but now we need to be more precise as to *how* the measurements are made. Imagine that a reference frame is filled with a cubic lattice of meter sticks, as shown in FIGURE 27.10. At every intersection is a clock, and all the clocks in a reference frame are *synchronized.* We'll return in a moment to consider how to synchronize the clocks, but assume for the moment it can be done.

FIGURE 27.10 The spacetime coordinates of an event are measured by a lattice of meter sticks and clocks.

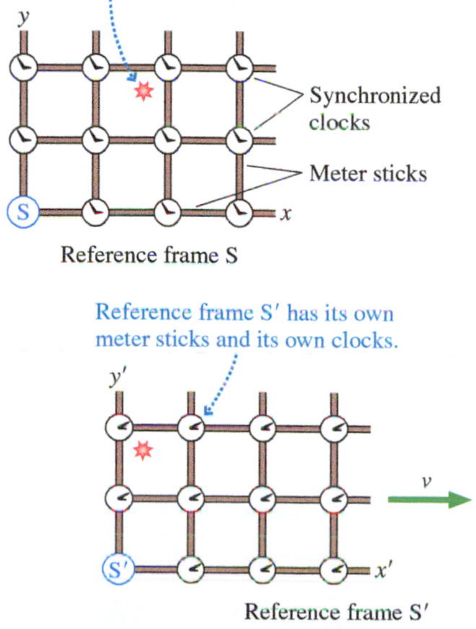


Now, with our meter sticks and clocks in place, we can use a two-part measurement scheme:

- The $(x, y, z)$ coordinates of an event are determined by the intersection of meter sticks closest to the event.
- The event's time $t$ is the time displayed on the clock nearest the event.

Several important issues need to be noted:

- The clocks and meter sticks in each reference frame are imaginary, so they have no difficulty passing through each other.
- Measurements of position and time made in one reference frame must use only the clocks and meter sticks in that reference frame.
- There's nothing special about the sticks being 1 m long and the clocks 1 m apart. The lattice spacing can be altered to achieve whatever level of measurement accuracy is desired.
- We'll assume that the experimenters in each reference frame have assistants sitting beside every clock to record the position and time of nearby events.
- Perhaps most important, $t$ is the time at which the event *actually happens,* not the time at which an experimenter sees the event or at which information about the event reaches an experimenter.
- All experimenters in one reference frame agree on the spacetime coordinates of an event. In other words, **an event has a unique set of spacetime coordinates in each reference frame.**

**STOP TO THINK 27.3** A carpenter is working on a house two blocks away. You notice a slight delay between seeing the carpenter's hammer hit the nail and hearing the blow. At what time does the event "hammer hits nail" occur?

A. At the instant you hear the blow
B. At the instant you see the hammer hit
C. Very slightly before you see the hammer hit
D. Very slightly after you see the hammer hit

## Clock Synchronization

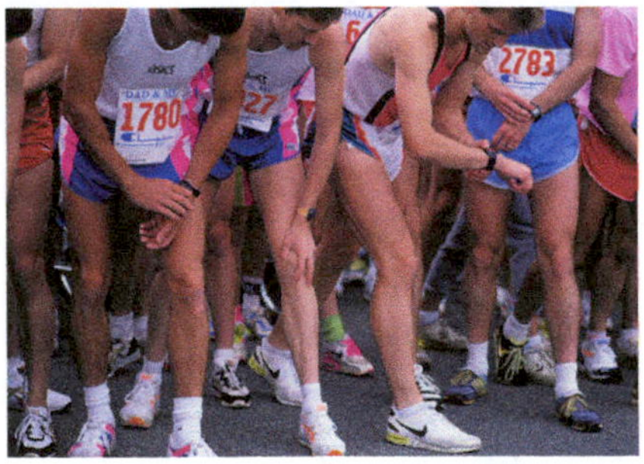

It's easy to synchronize clocks that are all in one place, but synchronizing distant clocks takes some care.

It's important that all the clocks in a reference frame be **synchronized,** meaning that all clocks in the reference frame have the same reading at any one instant of time. We would not be able to use a sequence of events to track the motion of a particle if the clocks differed in their readings. Thus we need a method of synchronization. One idea that comes to mind is to designate the clock at the origin as the *master clock.* We could then carry this clock around to every clock in the lattice, adjust that clock to match the master clock, and finally return the master clock to the origin.

This would be a perfectly good method of clock synchronization in Newtonian mechanics, where time flows along smoothly, the same for everyone. But we've been driven to reexamine the nature of time by the possibility that time is different in reference frames moving relative to each other. Because the master clock would *move,* we cannot assume that the master clock keeps time in the same way as the stationary clocks.

We need a synchronization method that does not require moving the clocks. Fortunately, such a method is easy to devise. Each clock is resting at the intersection of meter sticks, so by looking at the meter sticks, the assistant knows, or can calculate, exactly how far each clock is from the origin. Once the distance is known, the assistant can calculate exactly how long a light wave will take to travel from the origin to each clock. For example, light will take 1.00 $\mu$s to travel to a clock 300 m from the origin.

**NOTE** ▶ It's handy for many relativity problems to know that the speed of light is $c = 300\ \text{m}/\mu\text{s}$. ◀

To synchronize the clocks, the assistants begin by setting each clock to display the light travel time from the origin, but they don't start the clocks. Next, as

FIGURE 27.11 shows, a light flashes at the origin and, simultaneously, the clock at the origin starts running from $t = 0$ s. The light wave spreads out in all directions at speed $c$. A photodetector on each clock recognizes the arrival of the light wave and, without delay, starts the clock. The clock had been preset with the light travel time, so each clock as it starts reads exactly the same as the clock at the origin. Thus all the clocks will be synchronized after the light wave has passed by.

FIGURE 27.11 Synchronizing the clocks.

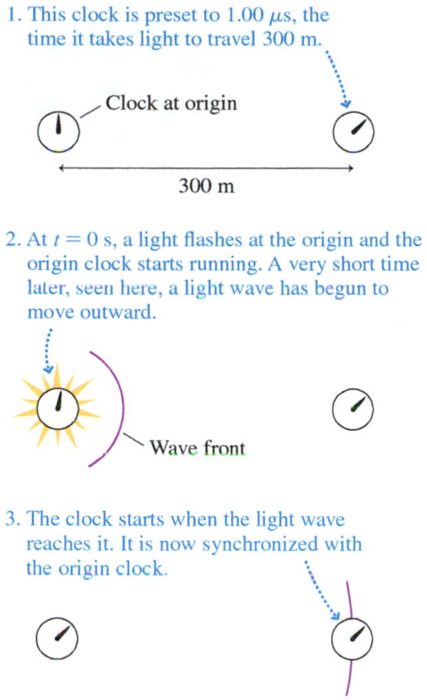


## Events and Observations

We noted previously that $t$ is the time the event *actually happens.* This is an important point, one that bears further discussion. Light waves take time to travel. Messages, whether they're transmitted by light pulses, telephone, or courier on horseback, take time to be delivered. An experimenter *observes* an event, such as an exploding firecracker, only *at a later time* when light waves reach his or her eyes. But our interest is in the event itself, not the experimenter's observation of the event. The time at which the experimenter sees the event or receives information about the event is not when the event actually occurred.

Suppose at $t = 0$ s a firecracker explodes at $x = 300$ m. The flash of light from the firecracker will reach an experimenter at the origin at $t_1 = 1.0\ \mu$s. The sound of the explosion will reach the experimenter at the origin at $t_2 = 0.88$ s. Neither of these is the time $t_{\text{event}}$ of the explosion, although the experimenter can work backward from these times, using known wave speeds, to determine $t_{\text{event}}$. In this example, the spacetime coordinates of the event—the explosion—are (300 m, 0 m, 0 m, 0 s).

**EXAMPLE 27.2 Finding the time of an event**

Experimenter A in reference frame S stands at the origin looking in the positive $x$-direction. Experimenter B stands at $x = 900$ m looking in the negative $x$-direction. A firecracker explodes somewhere between them. Experimenter B sees the light flash at $t = 3.00\ \mu$s. Experimenter A sees the light flash at $t = 4.00\ \mu$s. What are the spacetime coordinates of the explosion?

***STRATEGIZE*** The explosion of the firecracker is an event—it happens at a particular place at a particular instant of time. The two experimenters are in the same reference frame, so they will agree on the spacetime coordinates of the event. They detect the light flash at different times because of the travel time of the light. We will use the details of their observations to determine the actual time and location of the event.

***PREPARE*** FIGURE 27.12 shows the two experimenters and the explosion at unknown position $x$.

FIGURE 27.12 The light wave reaches the experimenters at different times. Neither of these is the time at which the event actually happened.

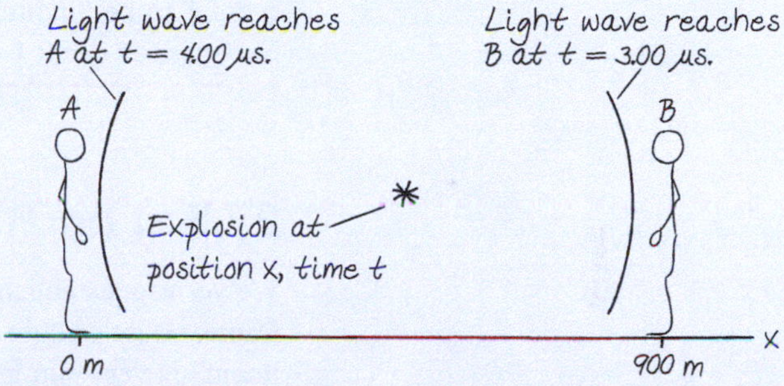


***SOLVE*** The two experimenters observe light flashes at two different instants, but there's only one event. Light travels at 300 m/$\mu$s, so the additional 1.00 $\mu$s needed for the light to reach experimenter A implies that distance $(x - 0\text{ m})$ from $x$ to A is 300 m longer than distance $(900\text{ m} - x)$ from B to $x$; that is,

$$(x - 0\text{ m}) = (900\text{ m} - x) + 300\text{ m}$$

This is easily solved to give $x = 600$ m as the position coordinate of the explosion. The light takes 1.00 $\mu$s to travel 300 m to experimenter B and 2.00 $\mu$s to travel 600 m to experimenter A. The light is received at 3.00 $\mu$s and 4.00 $\mu$s, respectively; hence it was emitted by the explosion at $t = 2.00\ \mu$s. The spacetime coordinates of the explosion are (600 m, 0 m, 0 m, 2.00 $\mu$s).

***ASSESS*** Although the experimenters *see* the explosion at different times, they agree that the explosion actually *happened* at $t = 2.00\ \mu$s.

## Simultaneity

Two events 1 and 2 that take place at different positions $x_1$ and $x_2$ but at the *same time* $t_1 = t_2$, as measured in some reference frame, are said to be **simultaneous** in that reference frame. Simultaneity is determined by when the events actually happen, not when they are seen or observed. In general, simultaneous events are not *seen* at the same time because of the difference in light travel times from the events to an experimenter.

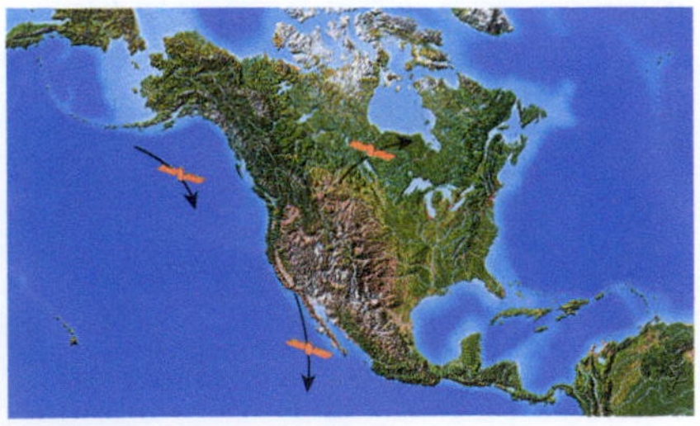

**Timing is everything** The global positioning system (GPS) is a network of 24 satellites that orbit the earth. The orbits of the satellites are precisely known, and the satellites emit synchronized signals. A GPS receiver can use the known positions of the satellites and the relative travel times of the signals to pinpoint its position with remarkable accuracy. The signals travel at the speed of light, moving 300 m in 1 $\mu$s, so determining a location to within a few meters clearly requires extremely precise synchronization.

**EXAMPLE 27.3 Using relative timing to determine location**

An experimenter stands along a line between two firecrackers that are 1500 m apart. The two firecrackers explode simultaneously. She sees the flash of the light from the firecracker to her left 1.0 $\mu$s before she sees the flash from the firecracker to her right. Where is the experimenter standing?

**STRATEGIZE** The experimenter and the two firecrackers are in the same reference frame. In this frame, the two explosions happen simultaneously. We can use the relative timing of when the light reaches the experimenter—when she sees the explosions—to determine her position.

**PREPARE** If she is a distance $x$ from the left firecracker, she is a distance 1500 m $- x$ from the right firecracker. Light from both explosions travels toward her at 300 m/$\mu$s.

**SOLVE** The light from the right firecracker takes 1.0 $\mu$s longer to reach her, so it must travel 300 m farther than the light from the left firecracker. This means that

$$x + 300\text{ m} = 1500\text{ m} - x$$

This gives her distance from the left firecracker as $x = 600$ m.

**ASSESS** The experimenter is clearly closer to the left firecracker, as we found. The difference in the distances is 300 m, so the 1.0 $\mu$s time difference makes sense.

In this example, the relative timing of the signals from the two simultaneous explosions lets us determine a person's location. This is similar, in principle, to the use of GPS signals for determining location.

**STOP TO THINK 27.4** A tree and a pole are 3000 m apart. Each is suddenly hit by a bolt of lightning. Mark, who is standing at rest midway between the two, sees the two lightning bolts at the same instant of time. Nancy is at rest under the tree. Define event 1 to be "lightning strikes tree" and event 2 to be "lightning strikes pole." For Nancy, does event 1 occur before, after, or at the same time as event 2?

## 27.5 The Relativity of Simultaneity

We've now established a means for measuring the time of an event in a reference frame, so let's begin to investigate the nature of time. The following "thought experiment" is very similar to one suggested by Einstein.

FIGURE 27.13 shows a long railroad car traveling to the right with a velocity $v$ that may be an appreciable fraction of the speed of light. A firecracker is tied to each end of the car, just above the ground. Each firecracker is powerful enough that, when it explodes, it will make a burn mark on the ground at the position of the explosion.

**FIGURE 27.13** A railroad car traveling to the right with velocity $v$.

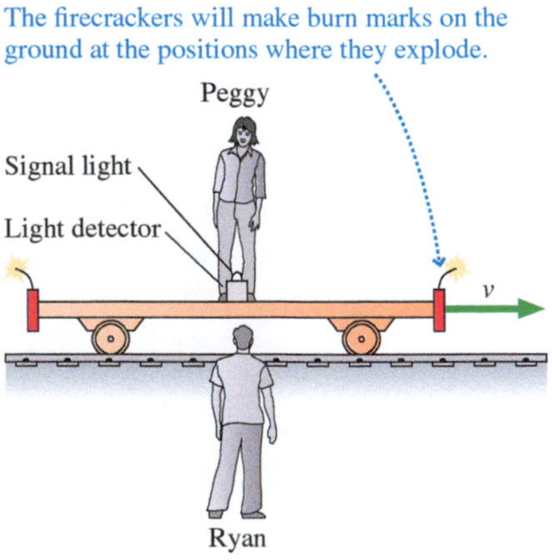


Ryan is standing on the ground, watching the railroad car go by. Peggy is standing in the exact center of the car with a special box at her feet. This box has two light detectors, one facing each way, and a signal light on top. The box works as follows:

- If a flash of light is received at the detector facing right, as seen by Ryan, before a flash is received at the left detector, then the light on top of the box will turn green.
- If a flash of light is received at the left detector before a flash is received at the right detector, or if two flashes arrive simultaneously, the light on top will turn red.

The firecrackers explode as the railroad car passes Ryan, and he sees the two light flashes from the explosions simultaneously. He then measures the distances to the two burn marks and finds that he was standing exactly halfway between the marks. Because light travels equal distances in equal times, Ryan concludes that the two explosions were simultaneous in his reference frame, the reference frame of the ground. Further, because he was midway between the two ends of the car, he was directly opposite Peggy when the explosions occurred.

FIGURE 27.14 Exploding firecrackers seen in two different reference frames.

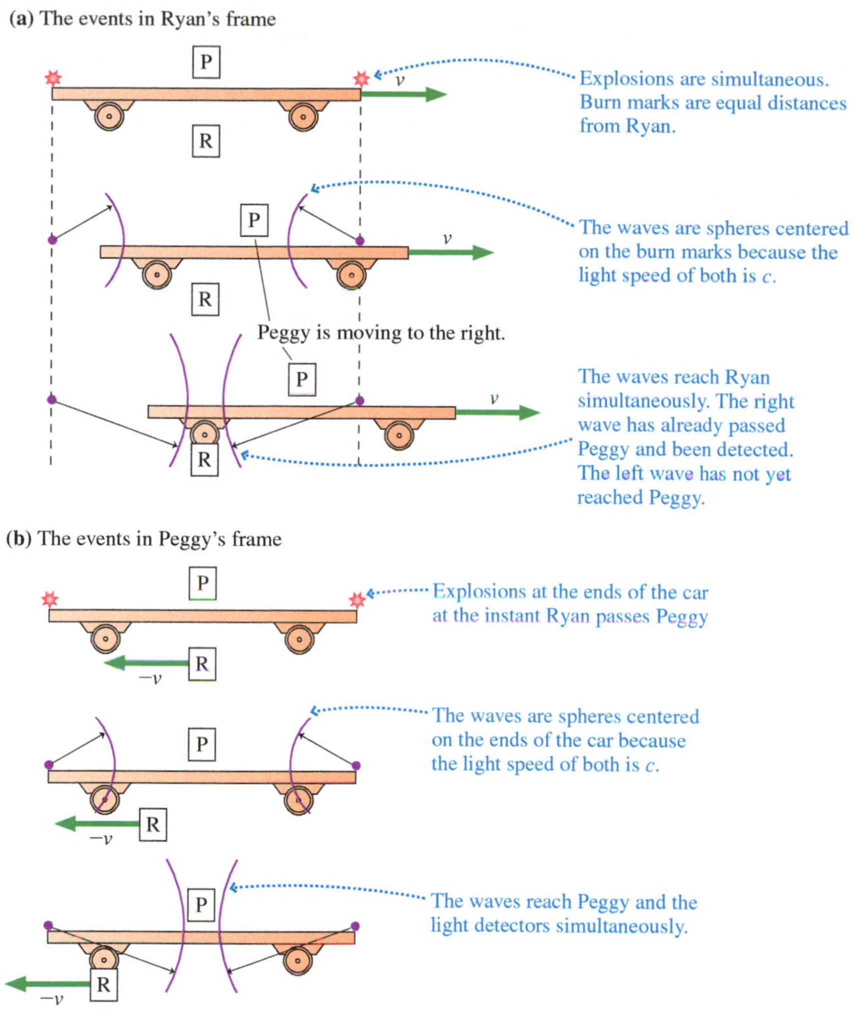


FIGURE 27.14a shows the sequence of events in Ryan's reference frame. Light travels at speed $c$ in all inertial reference frames, so, although the firecrackers were moving, the light waves are spheres centered on the burn marks. Ryan determines that the light wave coming from the right reaches Peggy and the box before the light wave coming from the left. Thus, according to Ryan, the signal light on top of the box turns green.

How do things look in Peggy's reference frame, a reference frame moving to the right at velocity $v$ relative to the ground? As FIGURE 27.14b shows, Peggy sees Ryan moving to the left with speed $v$. Light travels at speed $c$ in all inertial reference frames, so the light waves are spheres centered on the ends of the car. If the explosions are simultaneous, as Ryan has determined, the two light waves reach her and the box simultaneously. Thus, according to Peggy, the signal light on top of the box turns red!

Now the light on top must be either green or red. *It can't be both!* Later, after the railroad car has stopped, Ryan and Peggy can place the box in front of them. It has either a red light or a green light. Ryan can't see one color while Peggy sees the other. Hence we have a paradox. It's impossible for Peggy and Ryan both to be right. But who is wrong, and why?

What do we know with absolute certainty?

- Ryan detected the flashes simultaneously.
- Ryan was halfway between the firecrackers when they exploded.
- The light from the two explosions traveled toward Ryan at equal speeds.

The conclusion that the explosions were simultaneous in Ryan's reference frame is unassailable. The light is green.

Peggy, however, made an assumption. It's a perfectly ordinary assumption, one that seems sufficiently obvious that you probably didn't notice, but an assumption nonetheless. Peggy assumed that the explosions were simultaneous.

Didn't Ryan find them to be simultaneous? Indeed, he did. Suppose we call Ryan's reference frame S, the explosion on the right event R, and the explosion on the left event L. Ryan found that $t_R = t_L$. But Peggy has to use a different set of clocks, the clocks in her reference frame S′, to measure the times $t'_R$ and $t'_L$ at which the explosions occurred. The fact that $t_R = t_L$ in frame S does *not* allow us to conclude that $t'_R = t'_L$ in frame S′.

In fact, the right firecracker must explode *before* the left firecracker in frame S′. Figure 27.14b, with its assumption about simultaneity, was incorrect. FIGURE 27.15 shows the situation in Peggy's reference frame with the right firecracker exploding first. Now the wave from the right reaches Peggy and the box first, as Ryan had concluded, and the light on top turns green.

FIGURE 27.15 The real sequence of events in Peggy's reference frame.

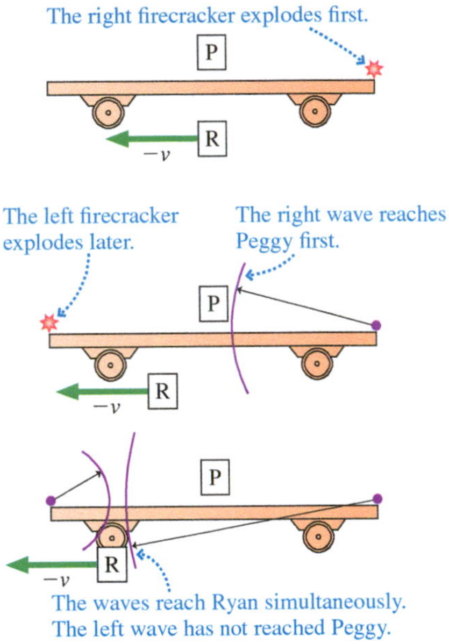

The right firecracker explodes first.
P
R
−v
The left firecracker explodes later.
The right wave reaches Peggy first.
P
R
−v
P
R
−v
The waves reach Ryan simultaneously. The left wave has not reached Peggy.

One of the most disconcerting conclusions of relativity is that **two events occurring simultaneously in reference frame S are *not* simultaneous in any reference frame S′ that is moving relative to S.** This is called the **relativity of simultaneity.**

The two firecrackers *really* explode at the same instant of time in Ryan's reference frame. And the right firecracker *really* explodes first in Peggy's reference frame. It's not a matter of when they see the flashes. Our conclusion refers to the times at which the explosions actually occur.

The paradox of Peggy and Ryan contains the essence of relativity, and it's worth careful thought. First, review the logic until you're certain that there *is* a paradox, a logical impossibility. Then convince yourself that the only way to resolve the paradox is to abandon the assumption that the explosions are simultaneous in Peggy's reference frame. If you understand the paradox and its resolution, you've made a big step toward understanding what relativity is all about.

**STOP TO THINK 27.5** A tree and a pole are 3000 m apart. Each is suddenly hit by a bolt of lightning. Mark, who is standing at rest midway between the two, sees the two lightning bolts at the same instant of time. Nancy is flying her rocket at $v = 0.5c$ in the direction from the tree toward the pole. The lightning hits the tree just as she passes by it. Define event 1 to be "lightning strikes tree" and event 2 to be "lightning strikes pole." For Nancy, does event 1 occur before, after, or at the same time as event 2?

## 27.6 Time Dilation

The principle of relativity has driven us to the logical conclusion that the time at which an event occurs may not be the same for two reference frames moving relative to each other. Our analysis thus far has been mostly qualitative. It's time to start developing some quantitative tools that will allow us to compare measurements in one reference frame to measurements in another reference frame.

FIGURE 27.16a shows a special clock called a **light clock.** The light clock is a box of height $h$ with a light source at the bottom and a mirror at the top. The light source emits a very short pulse of light that travels to the mirror and reflects back to a light detector beside the source. The clock advances one "tick" each time the detector receives a light pulse, and it immediately, with no delay, causes the light source to emit the next light pulse.

FIGURE 27.16 The ticking of a light clock can be measured by experimenters in two different reference frames.

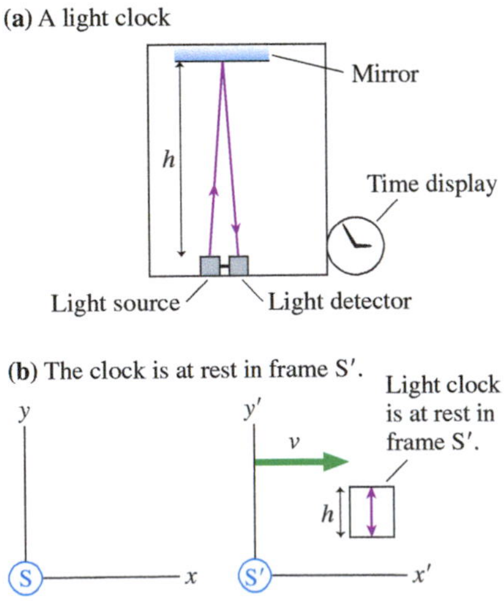


Our goal is to compare two measurements of the interval between two ticks of the clock: one taken by an experimenter standing next to the clock and the other by an experimenter moving with respect to the clock. To be specific, FIGURE 27.16b shows the clock at rest in reference frame S′. We call this the **rest frame** of the clock. Reference frame S′ moves to the right with velocity $v$ relative to reference frame S.

Relativity requires us to measure *events,* so let's define event 1 to be the emission of a light pulse and event 2 to be the detection of that light pulse. Experimenters in both reference frames are able to measure where and when these events occur *in their frame.* In frame S, the time interval $\Delta t = t_2 - t_1$ is one tick of the clock. Similarly, one tick in frame S′ is $\Delta t' = t_2' - t_1'$.

It's simple to calculate the tick interval $\Delta t'$ observed in frame S′, the rest frame of the clock, because the light simply goes straight up and back down. The total distance traveled by the light is $2h$, so the time of one tick is

$$\Delta t' = \frac{2h}{c} \tag{27.2}$$

**FIGURE 27.17** shows the light clock as seen in frame S. As seen in S, the clock is moving to the right at speed $v$. Thus the mirror has moved a distance $\frac{1}{2}v(\Delta t)$ during the time $\frac{1}{2}(\Delta t)$ in which the light pulse moves from the source to the mirror. To move from the source to the mirror, as seen from frame S, the light must move along the *diagonal path* shown. That is, the light must travel *farther* from source to mirror than it did in the rest frame of the clock.

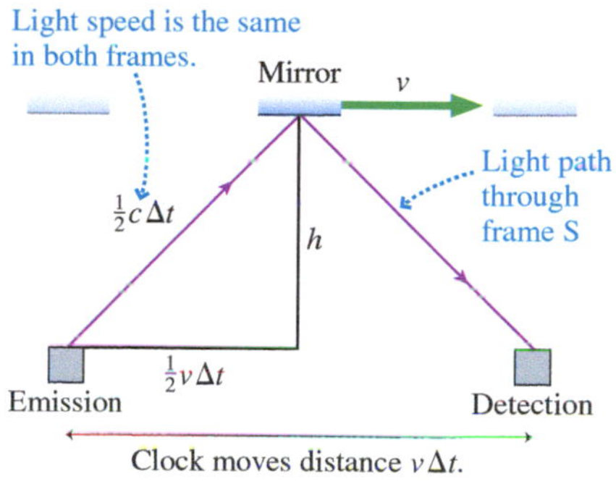


**FIGURE 27.17** A light clock analysis in which the speed of light is the same in all reference frames.

The length of this diagonal is easy to calculate because, according to special relativity, the speed of light is equal to $c$ in all inertial frames. Thus the diagonal length is simply

$$\text{distance} = \text{speed} \times \text{time} = c\left(\frac{1}{2}\Delta t\right) = \frac{1}{2}c\,\Delta t$$

We can then apply the Pythagorean theorem to the right triangle in Figure 27.17 to find that

$$h^2 + \left(\frac{1}{2}v\,\Delta t\right)^2 = \left(\frac{1}{2}c\,\Delta t\right)^2 \tag{27.3}$$

We can solve for $\Delta t$ by first rewriting Equation 27.3 as

$$h^2 = \left(\frac{1}{2}c\,\Delta t\right)^2 - \left(\frac{1}{2}v\,\Delta t\right)^2 = \left[\left(\frac{1}{2}c\right)^2 - \left(\frac{1}{2}v\right)^2\right]\Delta t^2 = \left(\frac{1}{2}\right)^2(c^2 - v^2)\,\Delta t^2$$

so that

$$\Delta t^2 = \frac{h^2}{\left(\frac{1}{2}\right)^2(c^2 - v^2)} = \frac{(2h)^2}{c^2 - v^2} = \frac{(2h/c)^2}{1 - v^2/c^2}$$

from which we find

$$\Delta t = \frac{2h/c}{\sqrt{1 - v^2/c^2}} = \frac{\Delta t'}{\sqrt{1 - v^2/c^2}} \tag{27.4}$$

where we have used Equation 27.2 to write $2h/c$ as $\Delta t'$. The time interval between two ticks in frame S is *not* the same as in frame S′.

It's useful to define $\beta = v/c$, the speed as a fraction of the speed of light. For example, a reference frame moving with $v = 2.4 \times 10^8$ m/s has $\beta = 0.80$. In terms of $\beta$, Equation 27.4 is

$$\Delta t = \frac{\Delta t'}{\sqrt{1 - \beta^2}} \tag{27.5}$$

If reference frame S′ is at rest relative to frame S, then $\beta = 0$ and $\Delta t = \Delta t'$. In other words, experimenters in both reference frames measure time to be the same. But two experimenters moving relative to each other will measure *different* time intervals between the same two events. We're unaware of these differences in our everyday lives because our typical speeds are so small compared to $c$. But the differences are easily measured in a laboratory, and they do affect the precise timekeeping needed to make accurate location measurements with a GPS receiver.

**Moving fast means running slow** GPS satellites orbit at a speed of about 14,000 km/h. This is fast enough to make the moving satellites' clocks run slow by about 7 ms a day, an amount that would soon cause unacceptable errors. Therefore, a relativistic correction is implemented before the satellites are launched. The clocks are designed to run slightly fast, by a factor of 1.000000000447. Once in orbit, the clocks slow down to precisely match an earthbound clock.

## Proper Time

Frame S′ has one important distinction. It is the *one and only* inertial reference frame in which the clock is at rest. Consequently, it is the one and only inertial reference frame in which the times of both events—the emission of the light and the detection of the light—are measured at the *same* position. You can see that the light pulse in Figure 27.16a, the rest frame of the clock, starts and ends at the same position, while in Figure 27.17 the emission and detection take place at *different* positions in frame S.

The time interval between two events that occur at the *same position* is called the **proper time** $\Delta\tau$. Only one inertial reference frame measures the proper time, and it can do so with a single clock that is present at both events. Experimenters in an inertial reference frame moving with speed $v = \beta c$ relative to the proper-time frame must use two clocks to measure the time interval: one at the position of the first event, the other at the position of the second event. We can rewrite Equation 27.5, where the time interval $\Delta t'$ is the proper time $\Delta\tau$, to find that the time interval in any other reference frame is

$$\Delta t = \frac{\Delta\tau}{\sqrt{1-\beta^2}} \geq \Delta\tau \qquad (27.6)$$

Time dilation in terms of proper time $\Delta\tau$ (where $\beta = v/c$)

Because $\beta = v/c$, and $v$ is always less than $c$, $\beta$ is always less than 1. This means that the factor $1/\sqrt{1-\beta^2}$ appearing in Equation 27.6 is always *equal to* (when $v = 0$) *or greater than 1*. Thus $\Delta t \geq \Delta\tau$. Recalling that $\Delta t$ is the time between clock ticks in a frame such as S in which the clock is moving, while the proper time $\Delta\tau$ is the time between ticks in a frame at which the clock is at rest, we can interpret Equation 27.6 as saying that **the time interval between two ticks is the shortest in the reference frame in which the clock is at rest.** The time interval between two ticks is longer when it is measured in any reference frame in which the clock is moving. Because a longer tick interval implies a clock that runs more slowly, we can also say that a **moving clock runs slowly compared to an identical clock at rest.** This "stretching out" of the time interval implied by Equation 27.6 is called **time dilation.** FIGURE 27.18 illustrates the sense in which a moving clock runs slow.

KEY CONCEPT

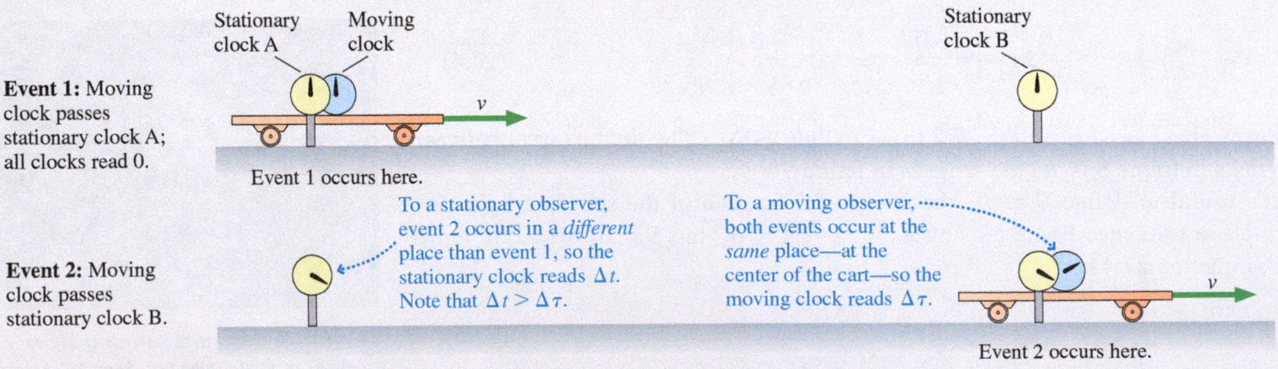

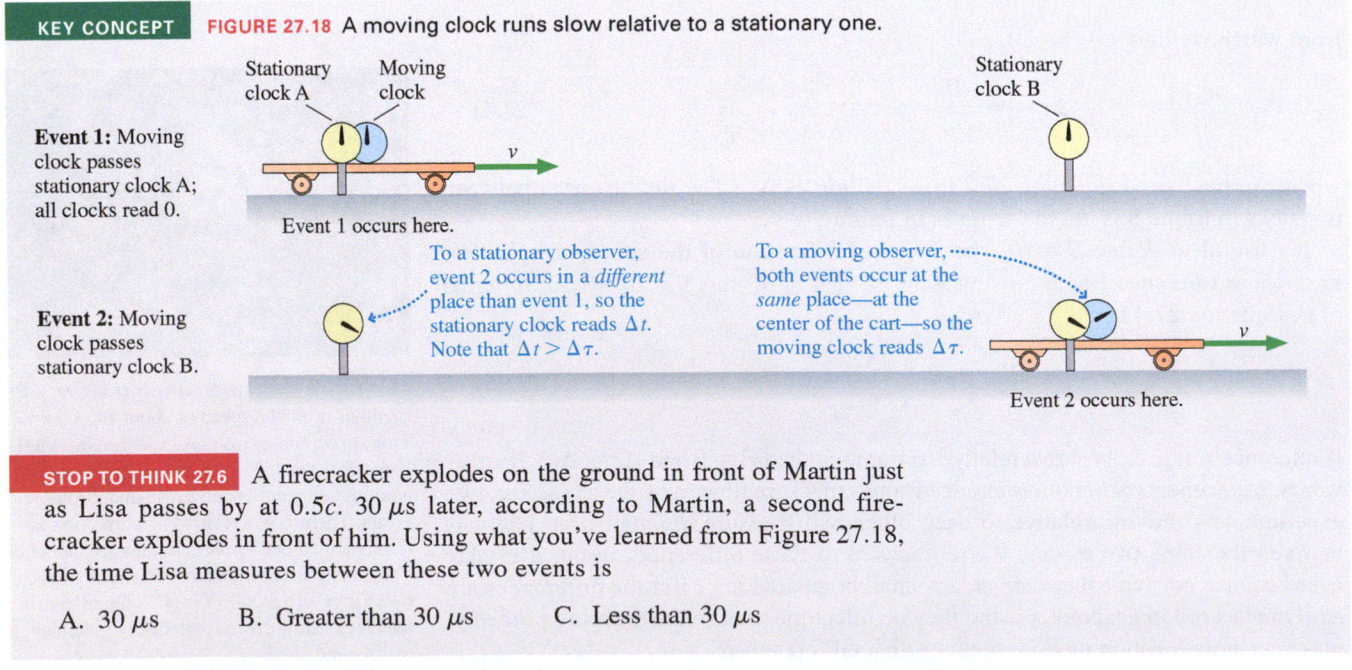

FIGURE 27.18 A moving clock runs slow relative to a stationary one.

**STOP TO THINK 27.6** A firecracker explodes on the ground in front of Martin just as Lisa passes by at $0.5c$. 30 $\mu$s later, according to Martin, a second firecracker explodes in front of him. Using what you've learned from Figure 27.18, the time Lisa measures between these two events is

A. 30 $\mu$s B. Greater than 30 $\mu$s C. Less than 30 $\mu$s

Equation 27.6 was derived using a light clock because the operation of a light clock is clear and easy to analyze. But the conclusion is really about time itself. Any clock, regardless of how it operates, behaves the same. For example, suppose you and a light clock are traveling in a very fast spaceship. The light clock happens to tick at the same rate as your heart beats—say, 60 times a minute, or once a second. Because the light clock is at rest in your frame, it measures the proper time between two successive beats of your heart; that is, $\Delta\tau = 1$ s. But to an experimenter stationed on the ground, watching you pass by at an enormous speed, the time interval $\Delta t = \Delta\tau/\sqrt{1-\beta^2}$ between two ticks of the clock—and hence two beats of your heart—would be *longer.* If, for instance, $\Delta t = 2$ s, the ground-based experimenter would conclude that your heart is beating only 30 times per minute. To the experimenter, *all* processes on your spaceship, including all your biological processes, would appear to run slowly.

**EXAMPLE 27.4 Journey time from the sun to Saturn**

Saturn is $1.43 \times 10^{12}$ m from the sun. A rocket travels along a line from the sun to Saturn at a constant speed of exactly $0.9c$ relative to the solar system. How long does the journey take as measured by an experimenter on earth? As measured by an astronaut on the rocket?

***STRATEGIZE*** These questions are about time intervals in different frames of reference. We know that time runs more slowly for a moving clock, so we expect that the astronaut will measure a shorter time interval.

***PREPARE*** Let the solar system be in reference frame S and the rocket be in reference frame S′ that travels with velocity $v = 0.9c$ relative to S. Relativity problems must be stated in terms of *events.* Let event 1 be "the rocket and the sun coincide" (the experimenter on earth says that the rocket passes the sun; the astronaut on the rocket says that the sun passes the rocket) and event 2 be "the rocket and Saturn coincide."

FIGURE 27.19 shows the two events as seen from the two reference frames. Notice that the two events occur at the *same position* in S′, the position of the rocket.

**FIGURE 27.19** Visual overview of the trip as seen in frames $S$ and $S'$.

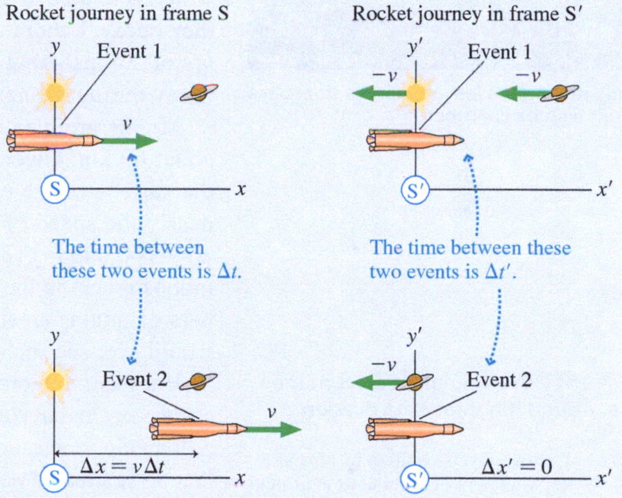


***SOLVE*** The time interval measured in the solar system reference frame, which includes the earth, is simply

$$\Delta t = \frac{\Delta x}{v} = \frac{1.43 \times 10^{12}\text{ m}}{0.9 \times (3.00 \times 10^8\text{ m/s})} = 5300\text{ s}$$

Relativity hasn't abandoned the basic definition $v = \Delta x/\Delta t$, although we do have to be sure that $\Delta x$ and $\Delta t$ are measured in just one reference frame and refer to the same two events.

How are things in the rocket's reference frame? The two events occur at the *same position* in S′. Thus the time measured by the astronaut is the *proper time* $\Delta\tau$ between the two events. We can then use Equation 27.6 with $\beta = 0.9$ to find

$$\Delta\tau = \sqrt{1-\beta^2}\,\Delta t = \sqrt{1-0.9^2}\,(5300\text{ s}) = 2310\text{ s}$$

***ASSESS*** The time interval measured between these two events by the astronaut is less than half the time interval measured by the experimenter on earth. We expected a shorter time interval for the astronaut, so we have confidence in our result.

Earlier in the chapter, we looked at examples in which measured times differed because of the travel time of light. In the previous example, the time interval between *seeing* the events from earth, which would have to allow for the travel time for light, would be something other than 5300 s. But the travel time of light wasn't a factor in our calculations; $\Delta t$ was the time interval between the rocket passing the sun and then subsequently passing Saturn, as measured by synchronized clocks at the Sun and at Saturn. The measured time intervals in the two frames are different because time passes differently in the two reference frames that are moving relative to each other. And recall that time dilation is not about clocks; it's about time. If you were on the moving rocket, it would seem like less time had passed between the two events—because less time *did* pass for you.

**STOP TO THINK 27.7** Molly flies her rocket past Nick at constant velocity $v$. Molly and Nick both measure the time it takes the rocket, from nose to tail, to pass Nick. Which of the following is true?

A. Both Molly and Nick measure the same amount of time.
B. Molly measures a shorter time interval than Nick.
C. Nick measures a shorter time interval than Molly.

## The Evidence from Cosmic Rays

Cosmic ray muons leave tracks as they pass through a spark chamber.

Is there any evidence for the crazy idea that clocks moving relative to each other tell time differently? Indeed, there's plenty.

Some of the most compelling evidence comes from *cosmic rays,* high-energy particles traveling through space that strike the earth's atmosphere. Most of the particles are protons, with which you are familiar. But when these protons hit air molecules, the extreme energy of the collisions can lead to the creation of exotic particles that aren't part of your daily life. A common product of such collisions is the *muon,* which is related to the electron but has 200 times as much mass. Muons are unstable; they decay. Laboratory studies have shown that stationary muons decay with a *half-life* of 1.5 ms—that is, half the muons decay within 1.5 ms, half of those remaining decay during the next 1.5 ms, and so on. These decays can be used as a clock.

Muons are created, in abundance, at the top of the atmosphere, at a height of about 60 km. Given their short lifetime, you might predict that they couldn't reach the surface of the earth. The muons travel down through the atmosphere at very nearly the speed of light. The time needed to reach the ground, assuming $v \approx c$, is $\Delta t \approx (60{,}000 \text{ m})/(3 \times 10^8 \text{ m/s}) = 200\ \mu\text{s}$. This is 133 half-lives, so the fraction of muons reaching the ground should be $\approx (1/2)^{133} = 10^{-40}$. Even with the vast numbers of muons created in cosmic rays, if this was all there was to the story, no one would ever see such a muon reach the surface of the earth.

FIGURE 27.20 We wouldn't detect muons at the ground if not for time dilation.

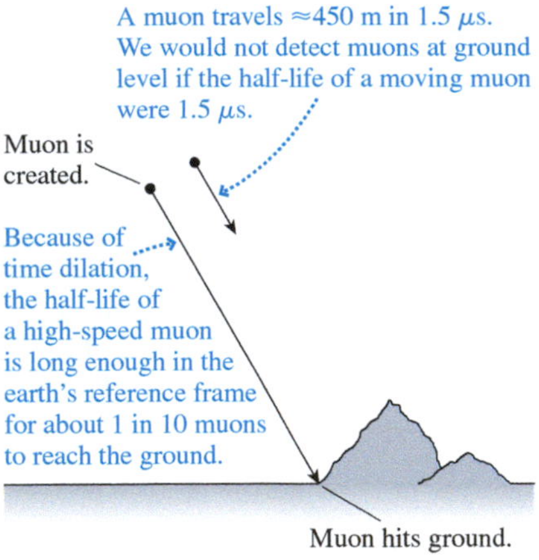


But there is more to the story. Muons *do* reach the surface of the earth; dozens of cosmic-ray muons pass through your body each second. This is a dramatic illustration—and a dramatic vindication—of the concept of time dilation. In FIGURE 27.20, the two events "muon is created" and "muon hits ground" take place at two different places in the earth's reference frame. However, these two events occur at the *same position* in the muon's reference frame. (The muon is like the rocket in Example 27.4.) Thus the muon's internal clock measures the proper time. The time-dilated interval $\Delta t = 200\ \mu\text{s}$ in the earth's reference frame corresponds to a time of only $\Delta t' \approx 5\ \mu\text{s}$ in the muon's reference frame. That is, in the muon's reference frame it takes only 5 $\mu$s from its creation at the top of the atmosphere until the ground runs into it. This is 3.3 half-lives, so the fraction of muons reaching the ground is $(1/2)^{3.3} = 0.1$, or 1 out of 10. With this time dilation, muons will, indeed, reach the ground, as observed.

## The Twin Paradox

The most well-known relativity paradox is the twin paradox. George and Helen are twins. On their 25th birthday, Helen departs on a starship voyage to a distant star. Let's imagine, to be specific, that her starship accelerates almost instantly to a speed of $0.95c$ and that she travels to a star that is 9.5 light years from earth. Upon arriving, she discovers that the planets circling the star are inhabited by fierce aliens, so she immediately turns around and heads home at $0.95c$.

A **light year,** abbreviated ly, is the distance that light travels in one year. A light year is vastly larger than the diameter of the solar system. The distance between two neighboring stars is typically a few light years. For our purpose, we can write the speed of light as $c = 1$ ly/year. That is, light travels 1 light year per year.

This value for $c$ allows us to determine how long, according to George and his fellow earthlings, it takes Helen to travel out and back. Her total distance is 19 ly and, due to her rapid acceleration and rapid turnaround, she travels essentially the entire distance at speed $v = 0.95c = 0.95$ ly/year. Thus the time she's away, as measured by George, is

$$\Delta t_G = \frac{19 \text{ ly}}{0.95 \text{ ly/year}} = 20 \text{ years}$$

George will be 45 years old when his sister Helen returns.

While she's away, George takes a physics class and studies Einstein's theory of relativity. He realizes that time dilation will make Helen's clocks run more slowly than his clocks, which are at rest relative to him. Her heart—a clock—will beat fewer times and the minute hand on her watch will go around fewer times. In other words, she's aging more slowly than he is. Although she is his twin, she will be younger than he is when she returns.

Calculating Helen's age is not hard. We simply have to identify Helen's clock, because it's always with Helen as she travels, as the clock that measures proper time $\Delta\tau$. From Equation 27.6,

$$\Delta t_H = \Delta\tau = \sqrt{1-\beta^2}\,\Delta t_G = \sqrt{1-0.95^2}\,(20 \text{ years}) = 6.25 \text{ years}$$

George will have just celebrated his 45th birthday as he welcomes home his 31-year-and-3-month-old twin sister.

This may be unsettling, because it violates our commonsense notion of time, but it's not a paradox. There's no logical inconsistency in this outcome. So why is it called "the twin paradox"?

Helen, knowing that she had quite of bit of time to kill on her journey, brought along several physics books to read. As she learns about relativity, she begins to think about George and her friends back on earth. Relative to her, they are all moving away at $0.95c$. Later they'll come rushing toward her at $0.95c$. Time dilation will cause their clocks to run more slowly than her clocks, which are at rest relative to her. In other words, as FIGURE 27.21 shows, Helen concludes that people on earth are aging more slowly than she is. Alas, she will be much older than they when she returns.

Finally, the big day arrives. Helen lands back on earth and steps out of the starship. George is expecting Helen to be younger than he is. Helen is expecting George to be younger than she is.

Here's the paradox! It's logically impossible for each to be younger than the other at the time when they are reunited. Where, then, is the flaw in our reasoning? It seems to be a symmetrical situation—Helen moves relative to George and George moves relative to Helen—but symmetrical reasoning has led to a conundrum.

But are the situations really symmetrical? George goes about his business day after day without noticing anything unusual. Helen, however, experiences three distinct periods during which the starship engines fire, she's crushed into her seat, and free dust particles that had been floating inside the starship are no longer, in the starship's reference frame, at rest or traveling in a straight line at constant speed. In other words, George spends the entire time in an inertial reference frame, *but Helen does not.* The situation is *not* symmetrical.

**The principle of relativity applies *only* to inertial reference frames.** Our discussion of time dilation was for inertial reference frames. Thus George's analysis and calculations are correct. Helen's analysis and calculations are *not* correct because she was trying to apply an inertial reference frame result to a noninertial reference frame.

Helen is younger than George when she returns. This is strange, but not a paradox. It is a consequence of the fact that time flows differently in two reference frames moving relative to each other.

This odd result, that the traveling twin is younger than the twin who stayed behind, is experimentally testable. An experiment in 1971 sent an atomic clock around the world on a jet plane while an identical clock remained in the laboratory. This experiment was difficult because the traveling clock's speed was so small compared to $c$, but measuring the small differences between the time intervals was just barely within the capabilities of atomic clocks. The scientists found that, upon its return, the airborne clock was 60 ns behind the stay-at-home clock, exactly as predicted by relativity.

Alpha Centauri (arrow) is one of the closest stars to the sun, at a distance of 4.3 ly. If you traveled there and back at $0.99c$, your earthbound friends would be 8.6 years older, while you would have aged only 1.2 years.

FIGURE 27.21 The twin paradox.

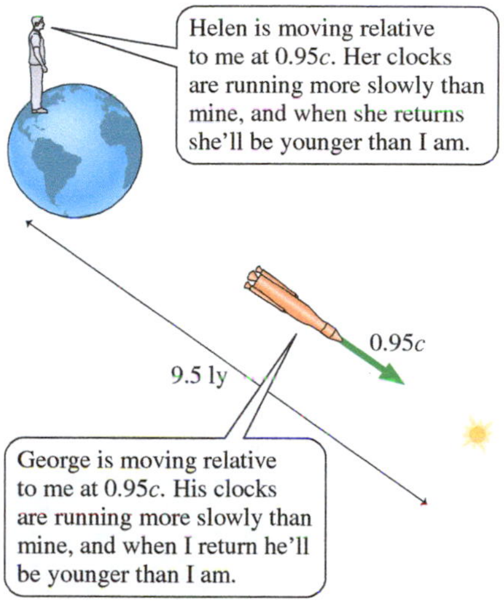

FIGURE 27.22 The length of a train car as measured by Ryan and by Peggy.

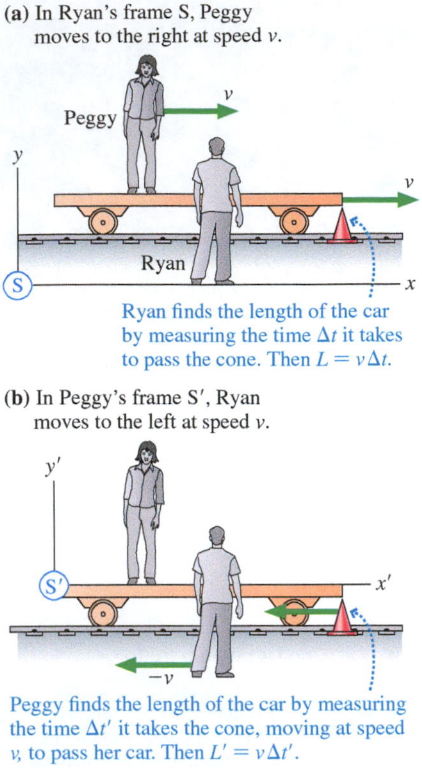

**Two perspectives on a relativistic trip** The Stanford Linear Accelerator is a 3.2-km-long electron accelerator that accelerates electrons to a speed of $0.99999999995c$. From our perspective, the electrons take a time $\Delta t = (3200\text{ m})/c = 11\ \mu\text{s}$ to make the trip. However, we see their "clocks" run slowly, ticking off only 110 ps during the trip. What do the electrons see? In their reference frame the end of the accelerator is coming toward them at $0.99999999995c$, but its length is contracted to only 3.3 cm. Thus it arrives in a time $(3.3\text{ cm})/c = 110$ ps. The same result, but from a different perspective.

## 27.7 Length Contraction

We've seen that relativity requires us to rethink our idea of time. Now let's turn our attention to the concepts of space and distance. Consider again Peggy on her train car, which is reference frame S′, moving past Ryan, who is at rest in frame S, at relative speed $v$. Ryan wants to measure the length $L$ of Peggy's car as it moves past him. As shown in FIGURE 27.22a, he can do so by measuring the time $\Delta t$ that it takes the car to pass the fixed cone; he then calculates that $L = v\,\Delta t$.

FIGURE 27.22b shows the situation in Peggy's reference frame S′, where the car is at rest. Peggy wants to measure the length $L'$ of her car; we'll soon see that $L'$ need not be the same as $L$. Peggy can measure $L'$ by finding the time $\Delta t'$ that the cone, moving at speed $v$, takes to move from one end of the car to the other. In this way, she finds that $L' = v\,\Delta t'$.

Speed $v$ is the relative speed between S and S′ and is the same for both Ryan and Peggy. From Ryan's and Peggy's measurements of the car's length we can then write

$$v = \frac{L}{\Delta t} = \frac{L'}{\Delta t'} \tag{27.7}$$

The time interval $\Delta t$ measured in Ryan's frame S is the proper time $\Delta\tau$ because the two events that define the time intervals—the front end and back end of the car passing the cone—occur at the same position (the cone) in Ryan's frame. We can use the time-dilation result, Equation 27.6, to relate $\Delta\tau$ measured by Ryan to $\Delta t'$ measured by Peggy. Equation 27.7 then becomes

$$\frac{L}{\Delta\tau} = \frac{L'}{\Delta t'} = \frac{L'}{\Delta\tau/\sqrt{1-\beta^2}} \tag{27.8}$$

The $\Delta\tau$ cancels, so the car's length $L$ in Ryan's frame is related to its length $L'$ in Peggy's frame by

$$L = (\sqrt{1-\beta^2})L' \tag{27.9}$$

Surprisingly, we find that **the length of the car in Ryan's frame is different from its length in Peggy's frame.**

Peggy's frame S′, in which the car's length is $L'$, has one important distinction. It is the *one and only* inertial reference frame in which the car is at rest. Experimenters in frame S′ can take all the time they need to measure $L'$ because the car isn't moving. The length of an object measured in the reference frame in which the object is at rest is called the **proper length** $\ell$. When you measure the length of an everyday object, such as a curtain rod or tabletop, it is usually at rest in your reference frame, so everyday length measurements are of proper length.

We can use the proper length $\ell$ to write Equation 27.9 as

$$L = (\sqrt{1-\beta^2})\ell \le \ell \tag{27.10}$$

Length contraction in terms of proper length $\ell$

Because $\beta \ge 0$, the factor $\sqrt{1-\beta^2}$ is less than or equal to 1. This means that $L \le \ell$. Because $\ell$, the proper length, is measured in a reference frame in which the object is at rest while $L$ is measured in a frame in which the object is moving, we see that **the length of an object is greatest in the reference frame in which the object is at rest.** This "shrinking" of the length of an object or the distance between two objects, as measured by an experimenter moving with respect to the object(s), is called **length contraction.**

NOTE ► A moving object's length is contracted only in the direction in which it's moving (its length along the $x$-axis in Figure 27.22). The object's length in the $y$- and $z$-directions doesn't change. ◄

**EXAMPLE 27.5** **Length contraction of a ladder**

Dan holds a 5.0-m-long ladder parallel to the ground. He then gets up to a good sprint, eventually reaching 98% of the speed of light. How long is the ladder according to Dan, once he is running, and according to Carmen, who is standing on the ground as Dan goes by?

**STRATEGIZE** This is a question about length contraction. The ladder is moving relative to Carmen and so will appear shorter to her.

**PREPARE** Let reference frame S′ be attached to Dan. The ladder is at rest in this reference frame, so Dan measures the proper length of the ladder: $\ell = 5.0$ m. Dan's frame S′ moves relative to Carmen's frame S with velocity $v = 0.98c$.

**SOLVE** We can find the length of the ladder in Carmen's frame from Equation 27.10. We have

$$L = (\sqrt{1-\beta^2})\,\ell = \sqrt{1-0.98^2}\,(5.0\text{ m}) = 1.0\text{ m}$$

**ASSESS** The length of the moving ladder as measured by Carmen is only one-fifth its length as measured by Dan. These lengths are different because *space is different* in two reference frames moving relative to each other.

The conclusion that space is different in reference frames moving relative to each other is a direct consequence of the fact that time is different. Experimenters in both reference frames agree on the relative velocity $v$, leading to Equation 27.7: $v = L/\Delta t = L'/\Delta t'$. Because of time dilation, Ryan (who measures proper time) finds that $\Delta t < \Delta t'$. Thus $L$ *has* to be less than $L'$. That is the only way Ryan and Peggy can reconcile their measurements.

## The Binomial Approximation

A useful mathematical tool is the **binomial approximation.** Suppose we need to evaluate the quantity $(1+x)^n$. If $x$ is much less than 1, it turns out that an excellent approximation is

$$(1+x)^n \approx 1 + nx \qquad \text{if} \qquad x \ll 1 \tag{27.11}$$

You can try this on your calculator. Suppose you need to calculate $1.01^2$. Comparing this expression with Equation 27.11, we see that $x = 0.01$ and $n = 2$. Equation 27.11 then tells us that

$$1.01^2 \approx 1 + 2 \times 0.01 = 1.02$$

The exact result, using a calculator, is 1.0201. The approximate answer is good to about 99.99%! The smaller the value of $x$, the better the approximation.

The binomial approximation is very useful when we need to calculate a relativistic expression for a speed much less than $c$, so that $v \ll c$. Because $\beta = v/c$, a reference frame moving with $v^2/c^2 \ll 1$ has $\beta^2 \ll 1$. In these cases, we can write

$$\begin{aligned} \sqrt{1-\beta^2} &= (1 - v^2/c^2)^{1/2} \approx 1 - \frac{1}{2}\frac{v^2}{c^2} \\ \frac{1}{\sqrt{1-\beta^2}} &= (1 - v^2/c^2)^{-1/2} \approx 1 + \frac{1}{2}\frac{v^2}{c^2} \end{aligned} \tag{27.12}$$

The following example illustrates the use of the binomial approximation.

**EXAMPLE 27.6** **The shrinking school bus**

An 8.0-m-long school bus drives past at 30 m/s. By how much is its length contracted?

**STRATEGIZE** This is another length contraction problem, but with a difference: The speed is much, much less than the speed of light. We will find a difference in length, but it will be quite small.

**PREPARE** The school bus is at rest in an inertial reference frame S′ moving at velocity $v = 30$ m/s relative to the ground frame S. The given length, 8.0 m, is the proper length $\ell$ in frame S′.

**SOLVE** In frame S, the school bus is length-contracted to

$$L = (\sqrt{1-\beta^2})\,\ell$$

The bus's speed $v$ is much less than $c$, so we can use the binomial approximation to write

$$L \approx \left(1 - \frac{1}{2}\frac{v^2}{c^2}\right)\ell = \ell - \frac{1}{2}\frac{v^2}{c^2}\ell$$

*Continued*

The *amount* of the length contraction is

$$\ell - L = \frac{1}{2}\frac{v^2}{c^2}\ell = \frac{1}{2}\left(\frac{30\text{ m/s}}{3.0\times 10^8\text{ m/s}}\right)^2(8.0\text{ m})$$
$$= 4.0\times 10^{-14}\text{ m} = 40\text{ fm}$$

where 1 fm = 1 femtometer = $10^{-15}$ m.

**ASSESS** The amount the bus "shrinks" is only slightly larger than the diameter of the nucleus of an atom. It's no wonder that we're not aware of length contraction in our everyday lives. If you had tried to calculate this number exactly, your calculator would have shown $\ell - L = 0$. The difficulty is that the difference between $\ell$ and $L$ shows up only in the 14th decimal place. A scientific calculator determines numbers to 10 or 12 decimal places, but that isn't sufficient to show the difference. The binomial approximation provides an invaluable tool for finding the very tiny difference between two numbers that are nearly identical.

**STOP TO THINK 27.8** Peggy, standing on a moving railroad car, passes Ryan at velocity $v$. Peggy and Ryan both measure the length of the car, from one end to the other. The length Peggy measures is _______ the length Ryan measures.

A. Longer than B. The same as C. Shorter than

## 27.8 Velocities of Objects in Special Relativity

In Section 27.2 we discussed Galilean relativity, which is applicable to objects that are moving at speeds much less than the speed of light. We found that if the velocity of an object is $u$ in reference frame S, then its velocity measured in frame S′, moving at velocity $v$ relative to frame S, is $u' = u - v$.

But we soon learned that this expression is invalid for objects moving at an appreciable fraction of the speed of light. In particular, light itself moves at speed $c$ as measured by *all* observers, independent of their relative velocities. The Galilean transformation of velocity needs to be modified for objects moving at relativistic speeds.

Einstein's relativity includes a velocity-addition expression valid for *any* velocities. An object's velocity measured in frame S is related to its velocity measured in frame S′ by the Lorentz velocity transformation:

Velocity of object as measured in frame S

$$u' = \frac{u - v}{1 - uv/c^2} \quad \text{or} \quad u = \frac{u' + v}{1 + u'v/c^2} \tag{27.13}$$

The relative velocity between the two frames

Velocity of object as measured in frame S′

These equations were discovered by Dutch physicist H. A. Lorentz a few years before Einstein published his theory of relativity, but Lorentz didn't completely understand their implications for space and time. Notice that the denominator is $\approx 1$ if either $u$ or $v$ is much less than $c$. In other words, these equations agree with the Galilean velocity transformation when velocities are nonrelativistic (i.e., $\ll c$), but they differ as velocities approach the speed of light.

**EXAMPLE 27.7 A speeding bullet**

A rocket flies past the earth at precisely $0.9c$. As it goes by, the rocket fires a bullet in the forward direction at precisely $0.95c$ with respect to the rocket. What is the bullet's speed with respect to the earth?

**STRATEGIZE** We will compute the speed in the earth's frame using the velocity transformation of Equation 27.13.

**PREPARE** The rocket and the earth are inertial reference frames. Let the earth be frame S and the rocket be frame S′. The velocity of frame S′ relative to frame S is $v = 0.9c$. The bullet's velocity in frame S′ is $u' = 0.95c$.

**SOLVE** We can use the Lorentz velocity transformation to find

$$u = \frac{u' + v}{1 + u'v/c^2} = \frac{0.95c + 0.90c}{1 + (0.95c)(0.90c)/c^2} = 0.997c$$

The bullet's speed with respect to the earth is 99.7% of the speed of light.

**NOTE** ► Many relativistic calculations are much easier when velocities are specified as a fraction of $c$. ◄

**ASSESS** The Galilean transformation of velocity would give $u = 1.85c$. Now, despite the very high speed of the rocket and of the bullet with respect to the rocket, the bullet's speed with respect to the earth remains less than $c$ as it must, giving us confidence in the final result.

Suppose the rocket in Example 27.7 fired a laser beam in the forward direction as it traveled past the earth at velocity $v$. The laser beam would travel away from the rocket at speed $u' = c$ in the rocket's reference frame S′. What is the laser beam's speed in the earth's frame S? According to the Lorentz velocity transformation, it must be

$$u = \frac{u' + v}{1 + u'v/c^2} = \frac{c + v}{1 + cv/c^2} = \frac{c + v}{1 + v/c} = \frac{c + v}{(c + v)/c} = c$$

Light travels at speed $c$ in both frame S and frame S′. This important consequence of the principle of relativity is "built into" the Lorentz velocity transformation.

**STOP TO THINK 27.9** Sam flies past earth at $0.75c$. As he goes by, he fires a bullet forward at $0.75c$. Suzy, on the earth, measures the bullet's speed to be

A. $1.5c$ B. $c$ C. Between $0.75c$ and $c$
D. $0.75c$ E. Less than $0.75c$

## 27.9 Relativistic Momentum

In Newtonian mechanics, the total momentum of a system is a conserved quantity. Further, the law of conservation of momentum, $P_f = P_i$, is true in all inertial reference frames *if* the particle velocities in different reference frames are related by the Galilean velocity transformation.

In relativity, however, we know that particle velocities in different frames are related by the Lorentz velocity transformation, which differ dramatically from the Galilean transformation as particle speeds approach $c$. If we use the Lorentz transformation, it's not difficult to show that a particle's Newtonian momentum $p = mu$ is *not* conserved in a frame moving relative to a frame in which momentum *is* conserved.

Yet momentum conservation is such a central and important feature of mechanics that it seems likely to hold in relativity as well. Indeed, a relativistic analysis of particle collisions shows that momentum conservation does hold, provided that we redefine the momentum of a particle as

$$p = \frac{mu}{\sqrt{1 - u^2/c^2}} \tag{27.14}$$

You can see that Equation 27.14 reduces to the classical expression $p = mu$ when the particle's speed $u \ll c$.

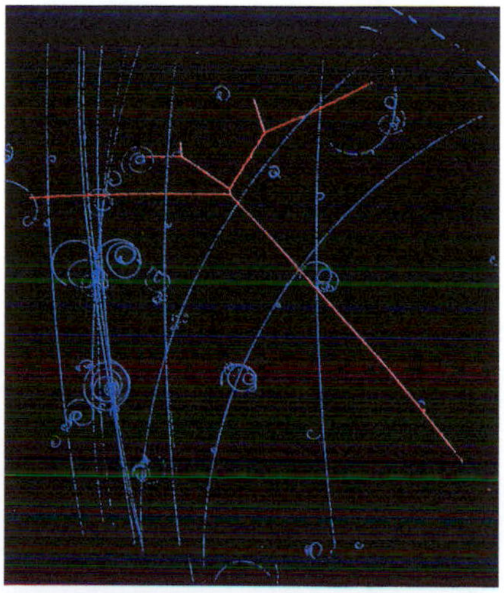

In this photograph, the track (shown in red) of a high-energy proton enters from the lower right. The proton then collides with other protons, sending them in all directions, where further collisions occur. Momentum is conserved, but at these high speeds the relativistic expression for momentum must be used.

To simplify our notation, let's define the quantity

$$\gamma = \frac{1}{\sqrt{1 - u^2/c^2}} \tag{27.15}$$

With this definition of $\gamma$, the momentum of a particle is

$$p = \gamma mu \tag{27.16}$$

Relativistic momentum for a particle with mass $m$ and speed $u$

### EXAMPLE 27.8 Momentum of a subatomic particle

Electrons in a particle accelerator reach a speed of $0.999c$ relative to the laboratory. One collision of an electron with a target produces a muon that moves forward with a speed of $0.950c$ relative to the laboratory. The muon mass is $1.90 \times 10^{-28}$ kg. What is the muon's momentum in the laboratory frame and in the frame of the electron beam?

**STRATEGIZE** The speeds are high enough that we'll need to use the relativistic formula to compute the momentum. We'll also need to determine the velocity of the muon in the electron beam frame.

**PREPARE** Let the laboratory be reference frame S. The reference frame S′ of the electron beam (i.e., a reference frame in which the electrons are at rest) moves in the direction of the electrons at $v = 0.999c$. The muon velocity in frame S is $u = 0.95c$.

**SOLVE** $\gamma$ for the muon in the laboratory reference frame is

$$\gamma = \frac{1}{\sqrt{1 - u^2/c^2}} = \frac{1}{\sqrt{1 - 0.95^2}} = 3.203$$

Thus the muon's momentum in the laboratory is

$$\begin{aligned} p &= \gamma mu \\ &= (3.203)(1.90 \times 10^{-28}\ \text{kg})(0.95 \times 3.00 \times 10^8\ \text{m/s}) \\ &= 1.73 \times 10^{-19}\ \text{kg} \cdot \text{m/s} \end{aligned}$$

The momentum is a factor of 3.2 larger than the Newtonian momentum $mu$. To find the momentum in the electron-beam reference frame, we must first use the velocity transformation equation to find the muon's velocity in frame S′:

$$u' = \frac{u - v}{1 - uv/c^2} = \frac{0.95c - 0.999c}{1 - (0.95c)(0.999c)/c^2} = -0.9617$$

In the laboratory frame, the faster electrons are overtaking the slower muon. Hence the muon's velocity in the electron-beam frame is negative. $\gamma'$ for the muon in frame S′ is

$$\gamma' = \frac{1}{\sqrt{1 - u'^2/c^2}} = \frac{1}{\sqrt{1 - 0.9617^2}} = 3.648$$

The muon's momentum in the electron-beam reference frame is

$$\begin{aligned} p' &= \gamma' mu' \\ &= (3.648)(1.90 \times 10^{-28}\ \text{kg})(-0.9617 \times 3.00 \times 10^8\ \text{m/s}) \\ &= -2.00 \times 10^{-19}\ \text{kg} \cdot \text{m/s} \end{aligned}$$

**ASSESS** From the laboratory perspective, the muon moves only slightly slower than the electron beam. But it turns out that the muon moves faster with respect to the electrons, although in the opposite direction, than it does with respect to the laboratory.

FIGURE 27.23 The speed of a particle cannot reach the speed of light.

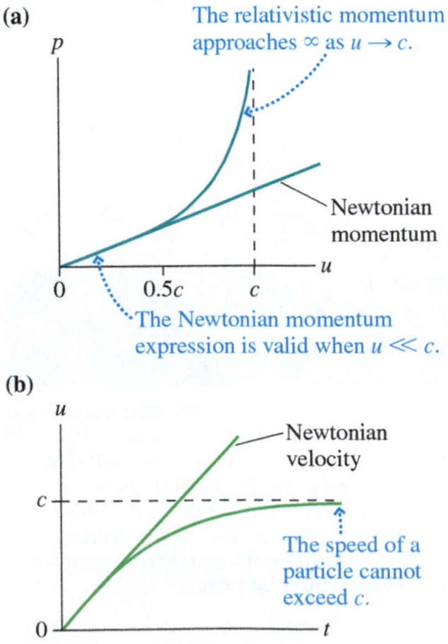


## The Cosmic Speed Limit

FIGURE 27.23a is a graph of momentum versus velocity. For a Newtonian particle, with $p = mu$, the momentum is directly proportional to the velocity. The relativistic expression for momentum agrees with the Newtonian value if $u \ll c$, but $p$ approaches $\infty$ as $u \to c$.

The implications of this graph become clear when we relate momentum to force. Consider a particle subjected to a constant force, such as a rocket that never runs out of fuel. From the impulse-momentum theorem we have $\Delta p = F\,\Delta t$, or $p = mu = Ft$ if the rocket starts from rest at $t = 0$. If Newtonian physics were correct, a particle would go faster and faster as its velocity $u = p/m = (F/m)t$ increased without limit. But the relativistic result, shown in FIGURE 27.23b, is that the particle's velocity approaches the speed of light ($u \to c$) as $p$ approaches $\infty$. Relativity gives a very different outcome than Newtonian mechanics.

The speed $c$ is a "cosmic speed limit" for material particles. A force cannot accelerate a particle to a speed higher than $c$ because the particle's momentum becomes infinitely large as the speed approaches $c$. The amount of effort required for each additional increment of velocity becomes larger and larger until no amount of effort can raise the velocity any higher.

Actually, at a more fundamental level, $c$ is a speed limit for *any* kind of **causal influence.** If you throw a rock and break a window, your throw is the *cause* of the breaking window and the rock is the causal influence. A causal influence can be any kind of particle, wave, or information that travels from A to B and allows A to be the cause of B.

For two unrelated events—a firecracker explodes in Tokyo and a balloon bursts in Paris—the relativity of simultaneity tells us that in one reference frame the firecracker may explode before the balloon bursts, but in some other reference frame the balloon may burst first.

However, for two causally related events—A causes B—it would be nonsense for an experimenter in any reference frame to find that B occurs before A. No experimenter in any reference frame, no matter how it is moving, will find that you are born before your mother is born.

But according to relativity, a causal influence traveling faster than light could result in B causing A, a logical absurdity. Thus **no causal influence of any kind—a particle, wave, or other influence—can travel faster than $c$.**

The existence of a cosmic speed limit is one of the most interesting consequences of the theory of relativity. This speed limit is a well-established law of nature. No matter how clever, scientists and engineers can't create a spaceship that can travel at the speed of light or faster, any more than they can create a spaceship that accelerates with no energy input.

**STOP TO THINK 27.10** Lauren is flying in her spaceship at $0.3c$. If she doubles her speed to $0.6c$, her momentum

A. Doubles. B. More than doubles. C. Less than doubles.

## 27.10 Relativistic Energy

Energy is our final topic in this chapter on relativity. Space, time, velocity, and momentum are changed by relativity, so it seems inevitable that we'll need a new view of energy. Indeed, one of the most profound results of relativity, and perhaps the one with the most far-reaching consequences, was Einstein's discovery of the fundamental relationship between energy and mass.

Consider an object of mass $m$ moving at speed $u$. Einstein found that the **total energy** of such an object is

$$E = \frac{mc^2}{\sqrt{1 - u^2/c^2}} = \gamma mc^2 \qquad (27.17)$$

Total energy of an object of mass $m$ moving at speed $u$

where $\gamma = 1/\sqrt{1 - u^2/c^2}$ was defined in Equation 27.15.

To understand this expression, let's start by examining its behavior for objects traveling at speeds much less than the speed of light. In this case we can use the binomial approximation you learned in Section 27.7 to write

$$\gamma = \frac{1}{\sqrt{1 - u^2/c^2}} \approx 1 + \frac{1}{2}\frac{u^2}{c^2}$$

For low speeds $u$, then, the object's total energy is

$$E \approx mc^2 + \frac{1}{2}mu^2 \qquad (27.18)$$

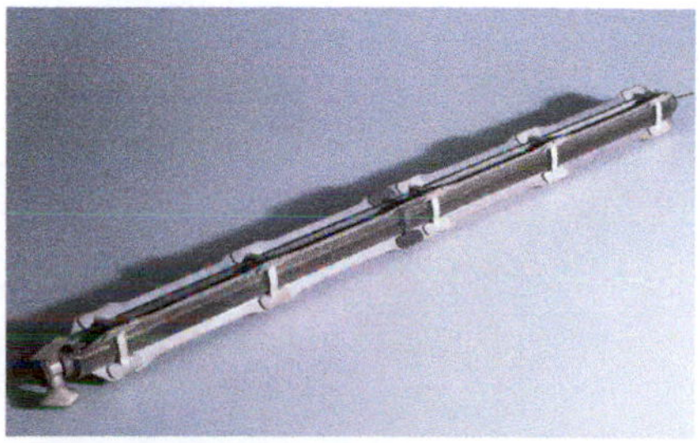

This fuel rod for a nuclear power reactor contains about 5 kg of uranium. Its usable energy content, which comes from the conversion of a small fraction of the uranium's mass to energy, is equivalent to that of about 10 million kg of coal.

The second term in this expression is the familiar Newtonian kinetic energy $K = \frac{1}{2}mu^2$ you studied in ◀◀ SECTION 10.3, written here in terms of velocity $u$ rather than $v$. But there is an additional term in the total energy, the **rest energy** given by

$$E_0 = mc^2 \tag{27.19}$$

When a particle is at rest, with $u = 0$, it still has energy $E_0$. Indeed, because $c$ is so large, the rest energy can be enormous. Equation 27.19 is, of course, Einstein's famous $E = mc^2$, perhaps the most famous equation in all of physics. It tells us that there is a fundamental equivalence between mass and energy, an idea we'll explore later in this section.

**Video** Mass-Energy Equivalence

**EXAMPLE 27.9** **The rest energy of an apple**

What is the rest energy of a 200 g apple?

***SOLVE*** From Equation 27.19 we have

$$E_0 = mc^2 = (0.20\text{ kg})(3.0 \times 10^8\text{ m/s})^2 = 1.8 \times 10^{16}\text{ J}$$

***ASSESS*** This is an enormous energy, enough to power a medium-sized city for about a year.

Equation 27.18 suggests that the total energy of an object is the sum of a rest energy, which is a new idea, and the familiar kinetic energy. But Equation 27.18 is valid only when the object's speed is low compared to $c$. For higher speeds, we need to use the full energy expression, Equation 27.17. We can use Equation 27.17 to find a relativistic expression for the kinetic energy $K$ by subtracting the rest energy $E_0$ from the total energy. Doing so gives

$$K = E - E_0 = \gamma mc^2 - mc^2 = (\gamma - 1)mc^2 = (\gamma - 1)E_0 \tag{27.20}$$

Thus we can write the total energy of an object of mass $m$ as

$$E = \underbrace{mc^2}_{\text{Rest energy } E_0} + \underbrace{(\gamma - 1)mc^2}_{\text{Kinetic energy } K} \tag{27.21}$$

For speeds less than 10% of the speed of light, Equation 27.20 will give a value nearly equal to that predicted by the Newtonian kinetic energy formula. For speeds greater than this, we'll need to compute kinetic energy using the relativistic formula.

**EXAMPLE 27.10** **Relativistic electrons**

Electrons in an electron microscope are accelerated to $2.1 \times 10^8\text{m/s}$.

a. Use the Newtonian formula to calculate the kinetic energy. By what factor does this understate the actual kinetic energy?
b. What is the total energy of the electrons?

***STRATEGIZE*** The electron is moving at more than 2/3 the speed of light, so we expect a significant difference between the Newtonian and relativistic results.

***PREPARE*** For this motion,

$$\gamma = \frac{1}{\sqrt{1 - u^2/c^2}} = 1.40$$

The rest energy is

$$E_0 = mc^2 = 8.2 \times 10^{-14}\text{ J}$$

***SOLVE***

a. The Newtonian formula gives a kinetic energy of

$$K = \frac{1}{2}mv^2 = \frac{1}{2}(9.1 \times 10^{-31})(2.1 \times 10^8\text{ m/s})^2 = 2.0 \times 10^{-14}\text{ J}$$

Equation 27.20 gives the correct, higher value:

$$K = (\gamma - 1)\,mc^2 = (0.40)(8.2 \times 10^{-14}\text{ J}) = 3.3 \times 10^{-14}\text{ J}$$

This is larger than the Newtonian value by a factor of 1.6.

b. The total energy is the sum of the kinetic energy and the rest energy:

$$E = E_0 + K = 11.5 \times 10^{-14}\text{ J}$$

***ASSESS*** There is a significant difference between the Newtonian and relativistic results, as we expected. The electrons are clearly moving at relativistic speeds, so we expect that the kinetic energy will be a significant fraction of the rest energy, as we found.

The result of this example shows that, for a device that you are aware of—and may have used—speeds are high enough that relativity is a factor. Relativity may seem esoteric, but its results have implications in areas that are familiar to you.

**STOP TO THINK 27.11** An electron moves through the lab at a speed such that $\gamma = 1.5$. The electron's kinetic energy is

A. Greater than its rest energy.
B. Equal to its rest energy.
C. Less than its rest energy.

## The Equivalence of Mass and Energy

Now we're ready to explore the significance of Einstein's famous equation $E = mc^2$. FIGURE 27.24 shows an experiment that has been done countless times at particle accelerators around the world. An electron that has been accelerated to $u \approx c$ is aimed at a target material. When a high-energy electron collides with an atom in the target, it can easily knock one of the electrons out of the atom. Thus we would expect to see two electrons leaving the target: the incident electron and the ejected electron. Instead, *four* particles emerge from the target: three electrons and a positron. A *positron,* or positive electron, is the antimatter version of an electron, identical to an electron in all respects other than having charge $q = +e$. In particular, a positron has the same mass $m_e$ as an electron.

FIGURE 27.24 An inelastic collision between electrons can create an electron-positron pair.

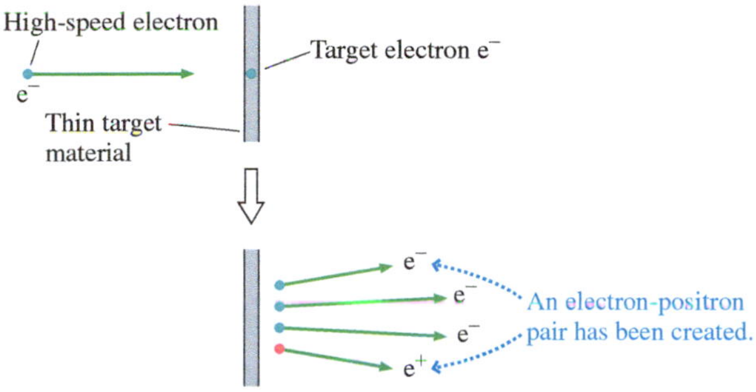

The tracks of elementary particles in a bubble chamber show the creation of two electron-positron pairs. The negative electron and positive positron curve in opposite directions in the magnetic field.

In chemical-reaction notation, the collision is

$$e^-(\text{fast}) + e^-(\text{at rest}) \rightarrow e^- + e^- + e^- + e^+$$

An electron and a positron have been created, apparently out of nothing. Mass $2m_e$ before the collision has become mass $4m_e$ after the collision. (Notice that charge has been conserved in this collision.)

Although the mass has increased, it wasn't created "out of nothing." If you measured the energies before and after the collision, you would find that the kinetic energy before the collision was *greater* than the kinetic energy after. In fact, the decrease in kinetic energy is exactly equal to the rest energy of the two particles that have been created: $\Delta K = 2m_e c^2$. The new particles have been created *out of energy*!

Not only can particles be created from energy, particles can return to energy. FIGURE 27.25 shows an electron colliding with a positron, its antimatter partner. When a particle and its antiparticle meet, they *annihilate* each other. The mass disappears, and the energy equivalent of the mass is transformed into two high-energy photons. Photons have no mass and represent pure energy. Positron-electron annihilation is also the basis of the medical procedure known as a positron-emission tomography, or PET scans. We'll study this important diagnostic tool in detail in Chapter 30.

FIGURE 27.25 The annihilation of an electron-positron pair.

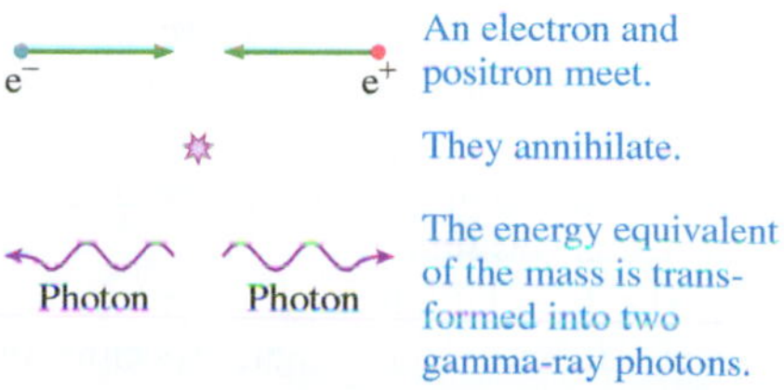

## Conservation of Energy

The creation and annihilation of particles with mass, processes strictly forbidden in Newtonian mechanics, are vivid proof that neither mass nor the Newtonian definition of energy is conserved. Even so, the *total* energy—the kinetic energy *and* the energy equivalent of mass—remains a conserved quantity.

**Law of conservation of total energy** The energy $E = \sum E_i$ of an isolated system is conserved, where $E_i = \gamma_i m_i c^2$ is the total energy of particle $i$.

Mass and energy are not the same thing, but as the last few examples have shown, they are *equivalent* in the sense that mass can be transformed into energy and energy can be transformed into mass as long as the total energy is conserved.

Probably the most well-known application of the conservation of total energy is nuclear fission. The uranium isotope $^{236}U$, containing 236 protons and neutrons, does not exist in nature. It can be created when a $^{235}U$ nucleus absorbs a neutron, increasing its atomic mass from 235 to 236. The $^{236}U$ nucleus quickly fragments into two smaller nuclei and several extra neutrons, a process known as **nuclear fission.** The nucleus can fragment in several ways, but one is

$$n + {}^{235}U \rightarrow {}^{236}U \rightarrow {}^{144}Ba + {}^{89}Kr + 3n \qquad (27.22)$$

Ba and Kr are the atomic symbols for barium and krypton; n is the symbol for a neutron.

TABLE 27.1 Mass before and after fission of $^{235}U$

| Initial nucleus | Initial mass (u) | Final nucleus | Final mass (u) |
|---|---|---|---|
| $^{235}U$ | 235.0439 | $^{144}Ba$ | 143.9229 |
| n | 1.0087 | $^{89}Kr$ | 88.9176 |
| | | 3n | 3.0260 |
| Total | 236.0526 | | 235.8665 |

This reaction seems like an ordinary chemical reaction—until you check the masses. The masses of atomic isotopes are known with great precision from many decades of measurement in instruments called mass spectrometers. As shown in TABLE 27.1, if you add up the masses on both sides, you find that the mass of the products is 0.186 u less than the mass of the initial neutron and $^{235}U$, where $1\text{ u} = 1.66 \times 10^{-27}$ kg is the atomic mass unit. Converting to kilograms gives us the mass loss of $3.09 \times 10^{-28}$ kg.

Mass has been lost, but the energy equivalent of the mass has not. As FIGURE 27.26 shows, the mass has been converted to kinetic energy, causing the two product nuclei and three neutrons to be ejected at very high speeds. The kinetic energy is easily calculated: $\Delta K = m_{\text{lost}}c^2 = 2.8 \times 10^{-11}$ J.

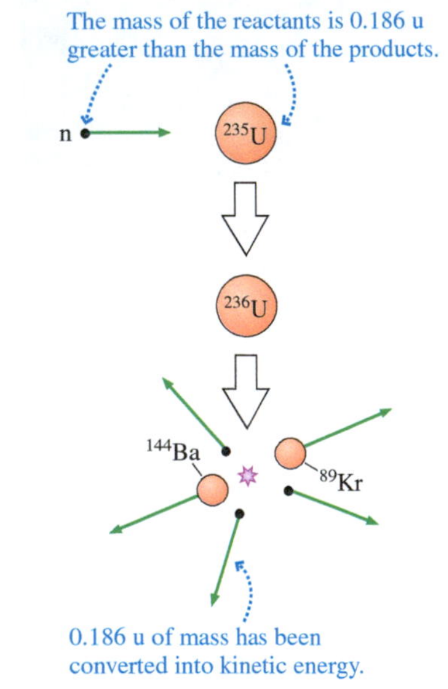


FIGURE 27.26 In nuclear fission, the energy equivalent of lost mass is converted into kinetic energy.

This is a very tiny amount of energy, but it is the energy released from *one* fission. The number of nuclei in a macroscopic sample of uranium is on the order of $N_A$, Avogadro's number. Hence the energy available if *all* the nuclei fission is enormous. This energy, of course, is the basis for both nuclear power reactors and nuclear weapons.

We started this chapter with an expectation that relativity would challenge our basic notions of space and time. We end by finding that relativity changes our understanding of mass and energy. Most remarkable of all is that each and every one of these new ideas flows from one simple statement: The laws of physics are the same in all inertial reference frames.

**INTEGRATED EXAMPLE 27.11** **The global positioning system**

The wild dog in the photo that opens this chapter is being tracked using the global positioning system (GPS), a system of 24 satellites in circular orbits high above the earth. Each satellite, with an orbital speed of 3900 m/s, carries an atomic clock whose time is accurate to $\pm 1$ ns per day. Every 30 s, each satellite sends out a radio signal giving its precise location in space and the exact time the signal was sent.

The dog's GPS receiver records the time at which the signal is received. Because the signal specifies the time at which it was sent, the receiver can easily calculate how long it took the signal to reach the dog. Then, because radio waves are electromagnetic waves traveling at the speed of light, the precise distance to the satellite can be calculated.

The signal from one satellite actually locates the dog's position only along a large sphere centered on the satellite. To pinpoint the location exactly requires the signals from four or more satellites. In this problem, we'll ignore these complicating effects.

a. As the satellite passes overhead, suppose two lights flash: one at the front of the satellite (the side toward which the satellite is moving) and one at the rear (the side opposite the direction of motion). In the reference frame of the satellite, the two flashes are simultaneous. Are they simultaneous to an experimenter on the earth? If not, which flash occurs first?
b. In one day, how much time does the clock running on the satellite gain or lose compared to an identical clock on earth?
c. If the clock error in part b were not properly taken into account, by how much would the dog's position on earth be in error after one day?

**STRATEGIZE** The relative timing of the two flashes is a question of simultaneity, similar to an example earlier in the chapter. We know that time dilation will cause the moving clock to tick more slowly, so it will lose time compared to an earth clock.

**PREPARE** Consider an astronaut standing on the satellite halfway between the lights. Because the two flashes are simultaneous in the reference frame of the satellite, the light from these flashes will reach the astronaut at the same time. An experimenter on the earth will also see these flashes reaching the astronaut at the same time, but will not agree that the flashes occurred simultaneously. We can use these observations to decide which flash occurred first to an earthbound experimenter.

In part b, because of time dilation the clock on the moving satellite runs slow compared to one on the earth, so this clock will *lose* time.

**SOLVE**

a. FIGURE 27.27 shows the satellite (here represented by a rod) as seen by an experimenter on earth. When the light waves from the two flashes meet at the satellite's center, the one from the rear of the satellite will have traveled farther than the one from the front. This is because the flash coming from the rear needs to "catch up" with the center of the moving satellite. Since both waves travel at speed $c$, but the wave from the rear travels farther, the rear light must have flashed *earlier* according to an earthbound experimenter.
b. The clock on the satellite measures proper time $\Delta\tau$ because this clock is at rest in the reference frame of the satellite. We want to know how much time the satellite's clock measures during an interval $\Delta t = 24\text{ h} = 86{,}400\text{ s}$ measured on the earth. We can rearrange Equation 27.6 as

$$\Delta\tau = \Delta t\sqrt{1-\beta^2}$$

FIGURE 27.27 The satellite as seen by an observer on earth.

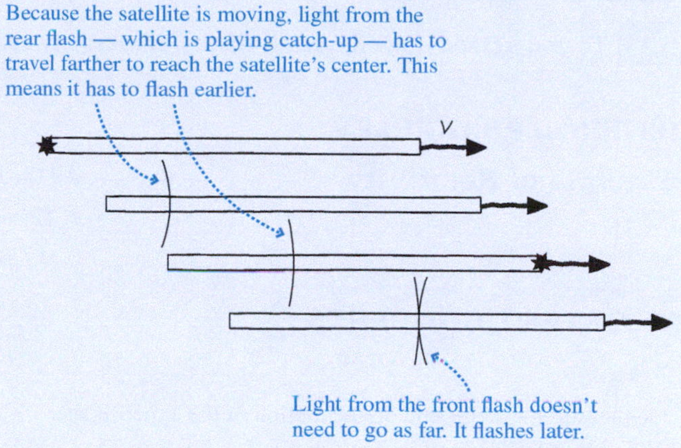

Because the satellite is moving, light from the rear flash — which is playing catch-up — has to travel farther to reach the satellite's center. This means it has to flash earlier.
v
Light from the front flash doesn't need to go as far. It flashes later.

Even for a fast-moving satellite, $\beta = v/c$ is so small that the term $\sqrt{1-\beta^2}$ will be exactly 1 on most calculators. We must therefore use the binomial approximation, Equation 27.12, to write

$$\Delta\tau = \Delta t\sqrt{1-\beta^2} \approx \Delta t\left(1 - \frac{1}{2}\frac{v^2}{c^2}\right) = \Delta t - \frac{1}{2}\frac{v^2}{c^2}\Delta t$$

The *difference* between the satellite and earth clocks is

$$\Delta t - \Delta\tau = \frac{1}{2}\frac{v^2}{c^2}\Delta t = \frac{1}{2}\left(\frac{3900\text{ m/s}}{3.0\times 10^8\text{ m/s}}\right)^2(86{,}400\text{ s}) = 7.3\ \mu\text{s}$$

Because this result is positive, $\Delta\tau$ is less than $\Delta t$: The moving clock *loses* 7.3 $\mu$s per day compared to a clock on earth.

c. The radio signal from the satellite travels at speed $c$. In 7.3 $\mu$s, the signal travels a distance

$$\Delta x = c\,\Delta t = (3.0\times 10^8\text{ m/s})(7.3\times 10^{-6}\text{ s}) = 2200\text{ m}$$

If the satellite clocks were not corrected for this relativistic effect, all GPS receivers on earth would miscalculate their positions by 2.2 km after just one day.

**ASSESS** Even for a very fast-moving object like a satellite, the corrections due to relativity are small. But these small corrections can have large effects on high-precision measurements.

# SUMMARY

**GOAL** To understand how Einstein's theory of relativity changes our concepts of space and time.

## GENERAL PRINCIPLES

### Principle of Relativity

**All the laws of physics are the same in all inertial reference frames.**

- The speed of light $c$ is the same in all inertial reference frames.
- No particle or causal influence can travel at a speed greater than $c$.

## IMPORTANT CONCEPTS

### Time

Time measurements depend on the motion of the experimenter relative to the events.

**Proper time** $\Delta\tau$ is the time interval between two events measured in a reference frame in which the events occur at the same position. The time interval $\Delta t$ between the same two events, in a frame moving with relative velocity $v$, is

$$\Delta t = \Delta\tau/\sqrt{1-\beta^2} \geq \Delta\tau$$

where $\beta = v/c$. This is called **time dilation.**

### Space

Spatial measurements depend on the motion of the experimenter relative to the events.

**Proper length** $\ell$ is the length of an object measured in a reference frame in which the object is at rest. The length $L$ in a frame in which the object moves with velocity $v$ is

$$L = (\sqrt{1-\beta^2})\,\ell \leq \ell$$

This is called **length contraction.**

### Momentum

The law of conservation of momentum is valid in all inertial reference frames if the momentum of a particle with velocity $u$ is $p = \gamma m u$, where

$$\gamma = 1/\sqrt{1-u^2/c^2}$$

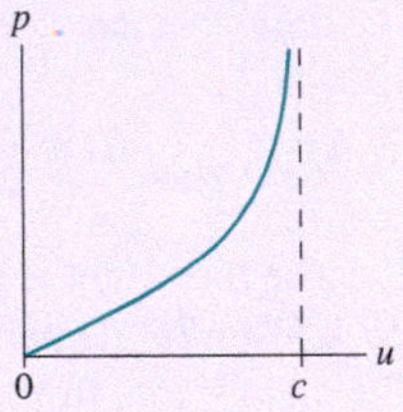


The momentum approaches $\infty$ as $u \to c$.

### Energy

The **total energy** of a particle is $E = \gamma mc^2$. This can be written as

$$E = \underbrace{mc^2}_{\text{Rest energy } E_0} + \underbrace{(\gamma - 1)mc^2}_{\text{Kinetic energy } K}$$

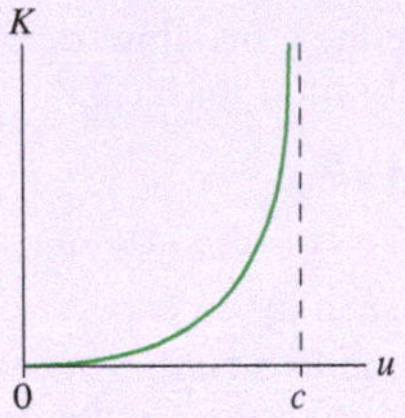


$K$ approaches $\infty$ as $u \to c$.

The total energy of an isolated system is conserved.

### Simultaneity

Events that are simultaneous in reference frame S are not simultaneous in frame S′ moving relative to S.

### Mass-energy equivalence

Mass $m$ can be transformed into energy $E = mc^2$.

Energy can be transformed into mass $m = E/c^2$.

## APPLICATIONS

An **event** happens at a specific place in space and time. Spacetime coordinates are $(x, t)$ in frame S and $(x', t')$ in frame S′.

A **reference frame** is a coordinate system with meter sticks and clocks for measuring events. Experimenters at rest relative to each other share the same reference frame.

If an object has velocity $u$ in frame S and $u'$ in frame S′, the two velocities are related by the **Lorentz velocity transformation:**

$$u' = \frac{u - v}{1 - uv/c^2} \qquad u = \frac{u' + v}{1 + u'v/c^2}$$

where $v$ is the relative velocity between the two frames.

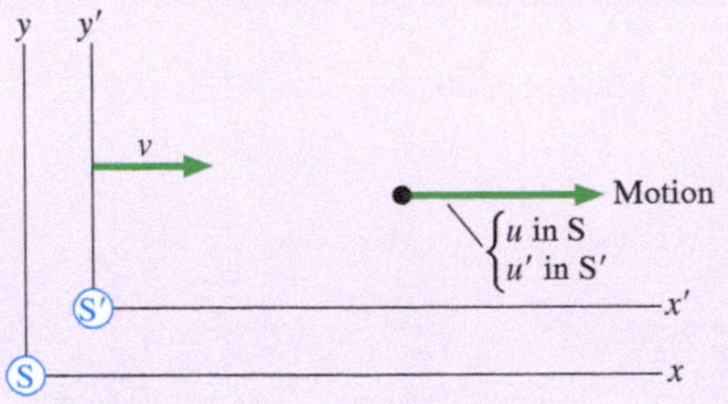

## Learning Objectives After studying this chapter, you should be able to:

- Understand reference frames and the principle of relativity. *Conceptual Questions 27.5, 27.6; Problems 27.5, 27.6, 27.7, 27.8, 27.10*
- Use events and simultaneity. *Conceptual Questions 27.9, 27.12; Problems 27.13, 27.14, 27.15, 27.16, 27.17*
- Calculate proper time and time dilation. *Conceptual Questions 27.13, 27.14; Problems 27.18, 27.19, 27.21, 27.22, 27.24*
- Calculate proper length and length contraction. *Conceptual Questions 27.15, 27.16; Problems 27.26, 27.28, 27.29, 27.30, 27.33*
- Use the Lorentz velocity transformations. *Conceptual Question 27.18; Problems 27.35, 27.36, 27.37, 27.38*
- Solve problems using relativistic momentum and energy. *Conceptual Questions 27.19, 27.20; Problems 27.39, 27.40, 27.43, 27.48, 27.52*

## STOP TO THINK ANSWERS

**Chapter Preview Stop to Think: C.** The relative velocity $(v)_{CA}$ of the car (C) relative to Amy (A) is $(v)_{CA} = (v)_{CB} + (v)_{BA} =$ 10 m/s + 5 m/s = 15 m/s.

**Stop to Think 27.1: A, C, and F.** These move at constant velocity, or very nearly so. The others are accelerating.

**Stop to Think 27.2: A.** $u' = u - v = -10$ m/s $- 6$ m/s $= -16$ m/s. The *speed* is 16 m/s.

**Stop to Think 27.3: C.** Even the light has a slight travel time. The event is the hammer hitting the nail, not your seeing the hammer hit the nail.

**Stop to Think 27.4: At the same time.** Mark is halfway between the tree and the pole, so the fact that he *sees* the lightning bolts at the same time means they *happened* at the same time. It's true that Nancy *sees* event 1 before event 2, but the events actually occurred before she sees them. Mark and Nancy share a reference frame, because they are at rest relative to each other, and all experimenters in a reference frame, after correcting for any signal delays, *agree* on the spacetime coordinates of an event.

**Stop to Think 27.5: After.** This is the same as the case of Peggy and Ryan. In Mark's reference frame, as in Ryan's, the events are simultaneous. Nancy *sees* event 1 first, but the time when an event is seen is not when the event actually happens. Because all experimenters in a reference frame agree on the spacetime coordinates of an event, Nancy's position in her reference frame cannot affect the order of the events. If Nancy had been passing Mark at the instant the lightning strikes occurred in Mark's frame, then Nancy would be equivalent to Peggy. Event 2, like the firecracker at the front of Peggy's railroad car, occurs first in Nancy's reference frame.

**Stop to Think 27.6: A.** The two events—firecracker 1 explodes and firecracker 2 explodes—occur in the *same* place, in front of Martin. Thus the proper time is the time measured in Martin's frame, or 30 $\mu$s. The time interval measured by Lisa must be *longer* than the proper time.

**Stop to Think 27.7: C.** Nick measures proper time because Nick's clock is present at both the "nose passes Nick" event and the "tail passes Nick" event. Proper time is the smallest measured time interval between two events.

**Stop to Think 27.8: A.** The car is at rest in Peggy's frame, so she measures the proper length $\ell$. The length measured in any other frame is shorter than the proper length, so Peggy measures the car to be longer than does Ryan.

**Stop to Think 27.9: C.** The bullet must obviously travel faster than Sam, eliminating answers D and E. And we know that no object can travel at or faster than the speed of light, eliminating A and B. The bullet travels faster than Sam, or faster than $0.75c$, but slower than $c$.

**Stop to Think 27.10: B.** Inspection of the graph in Figure 27.23a shows that the graph for relativistic momentum rapidly rises above the linear graph for Newtonian momentum. Thus if Lauren's speed doubles, her momentum must more than double.

**Stop to Think 27.11: C.** The kinetic energy is $(\gamma - 1)mc^2 = 0.5mc^2$, which is less than the rest energy $mc^2$.

Video Tutor Solution Chapter 27

# QUESTIONS

## Conceptual Questions

1. You are in an airplane cruising smoothly at 600 mph. What experiment, if any, could you do that would demonstrate that you are moving, while those on the ground are at rest?

2. Frame S′ moves relative to frame S as shown in Figure Q27.2.
   a. A ball is at rest in frame S′. What are the speed and direction of the ball in frame S?
   b. A ball is at rest in frame S. What are the speed and direction of the ball in frame S′?

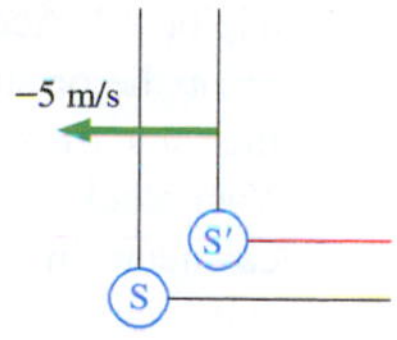

FIGURE Q27.2

Problem difficulty is labeled as | (straightforward) to ||||| (challenging). Problems labeled INT integrate significant material from earlier chapters; Problems labeled BIO are of biological or medical interest.

The eText icon indicates when there is a video tutor solution available for the chapter or for a specific problem. To launch these videos, log into your eText through Mastering[TM] Physics or log into the Study Area.

3. a. Two balls move as shown in Figure Q27.3. What are the speed and direction of each ball in a reference frame that moves with ball 1?
   b. What are the speed and direction of each ball in a reference frame that moves with ball 2?

FIGURE Q27.3

4. A lighthouse beacon alerts ships to the danger of a rocky coastline.
   a. According to the lighthouse keeper, with what speed does the light leave the lighthouse?
   b. A boat is approaching the coastline at speed $0.5c$. According to the captain, with what speed is the light from the beacon approaching her boat?
5. As a rocket passes the earth at $0.75c$, it fires a laser perpendicular to its direction of travel as shown in Figure Q27.5.
   a. What is the speed of the laser beam relative to the rocket?
   b. What is the speed of the laser beam relative to the earth?

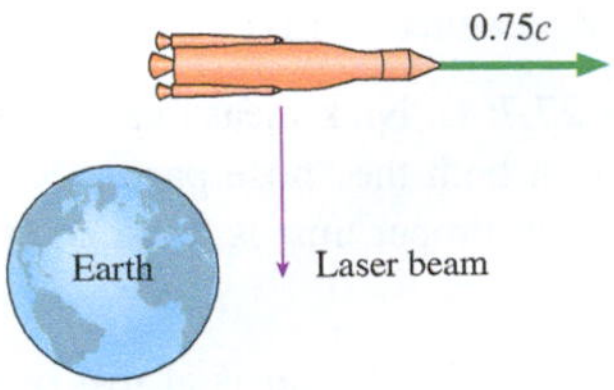


FIGURE Q27.5

6. You are sending a message to residents of another star system using a laser beam, with information encoded in flashes of light. You realize that you sent a sequence meaning "prepare for an invasion" when you meant to send "prepare for a visit." You dispatch a rocket to intercept the signal and block it before it reaches the recipients. Is there any hope for this mission? Explain.
7. At the instant that a clock standing next to you reads $t = 1.0\ \mu s$, you look at a second clock, 300 m away, and see that it reads $t = 0\ \mu s$. Are the two clocks synchronized? If not, which one is ahead?
8. Firecracker 1 is 300 m from you. Firecracker 2 is 600 m from you in the same direction. You see both explode at the same time. Define event 1 to be "firecracker 1 explodes" and event 2 to be "firecracker 2 explodes." Does event 1 occur before, after, or at the same time as event 2? Explain.
9. Firecrackers 1 and 2 are 600 m apart. You are standing exactly halfway between them. Your lab partner is 300 m on the other side of firecracker 1. You see two flashes of light, from the two explosions, at exactly the same instant of time. Define event 1 to be "firecracker 1 explodes" and event 2 to be "firecracker 2 explodes." According to your lab partner, based on measurements he or she makes, does event 1 occur before, after, or at the same time as event 2? Explain.
10. Your clocks and calendars are synchronized with the clocks and calendars in a star system exactly 10 ly from earth that is at rest relative to the earth. You receive a TV transmission from the star system that shows a date and time display. The date it shows is June 17, 2050. When you glance over at your own wall calendar, what date does it show?
11. Two trees are 600 m apart. You are standing exactly halfway between them and your lab partner is at the base of tree 1. Lightning strikes both trees.
   a. Your lab partner, based on measurements he makes, determines that the two lightning strikes were simultaneous. What did you see? Did you see the lightning hit tree 1 first, hit tree 2 first, or hit them both at the same instant of time? Explain.
   b. Lightning strikes the trees again. This time your lab partner sees both flashes of light at the same instant of time. What did you see? Did you see the lightning hit tree 1 first, hit tree 2 first, or hit them both at the same instant of time? Explain.
   c. In the scenario of part b, were the lightning strikes simultaneous? Explain.
12. Figure Q27.12 shows Peggy standing at the center of her railroad car as it passes Ryan on the ground. Firecrackers attached to the ends of the car explode. A short time later, the flashes from the two explosions arrive at Peggy at the same time.
   a. Were the explosions simultaneous in Peggy's reference frame? If not, which exploded first? Explain.
   b. Were the explosions simultaneous in Ryan's reference frame. If not, which exploded first? Explain.

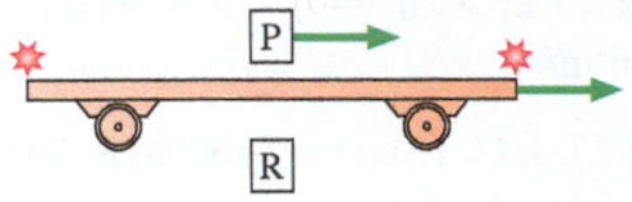


FIGURE Q27.12

13. In Figure Q27.13, clocks $C_1$ and $C_2$ in frame S are synchronized. Clock $C'$ moves at speed $v$ relative to frame S. Clocks $C'$ and $C_1$ read exactly the same as $C'$ goes past. As $C'$ passes $C_2$, is the time shown on $C'$ earlier, later, or the same as the time shown on $C_2$? Explain.

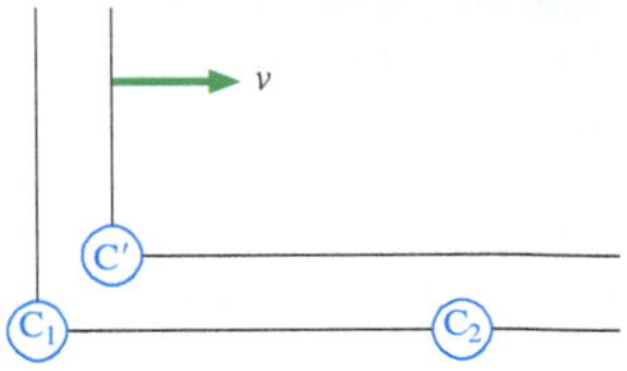


FIGURE Q27.13

14. A moving muon lives longer, from the observer's point of view, than a muon at rest. Is there a frame of reference in which a muon's lifetime is *shorter* than its lifetime in a frame in which the muon is at rest?
15. Your spaceship zips past a spherical asteroid at $0.80c$. What is the shape of the asteroid from your point of view?
16. A 100-m-long train is heading for an 80-m-long tunnel. If the train moves sufficiently fast, is it possible, according to experimenters on the ground, for the entire train to be inside the tunnel at one instant of time? Explain.
17. Dan picks up a 15-m-long pole and begins running very fast, holding the pole horizontally and pointing in the direction he's running. He heads toward a barn that is 12 m long and has open doors at each end. Dan runs so fast that, to Farmer Brown standing by his barn, the ladder is only 5 m long. As soon as the pole is completely inside the barn, Farmer Brown closes both doors so that Dan and the pole are inside with both doors shut. Then, just before Dan reaches the far door, Farmer Brown opens both doors and Dan emerges, still moving at high speed. According to Dan, however, the barn is contracted to only 4 m and the pole has its full 15 m length. Farmer Brown sees the pole completely inside the barn with both doors closed. What does Dan see happening?
18. The rocket speeds shown in Figure Q27.18 are relative to the earth. Is the speed of A relative to B greater than, less than, or equal to $0.8c$?

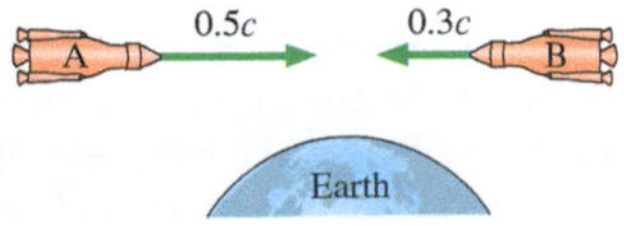


FIGURE Q27.18

19. Can a particle of mass $m$ have total energy less than $mc^2$? Explain.
20. In your chemistry classes, you have probably learned that, in a chemical reaction, the mass of the products is equal to the mass of the reactants. That is, the mass of the substances produced in a chemical reaction is equal to the mass of the substances consumed in the reaction. Is this absolutely true, or is there actually a small difference? Explain.

## Multiple-Choice Questions

21. | Lee and Leigh are twins. At their first birthday party, Lee is placed on a spaceship that travels away from the earth and back at a steady 0.866$c$. The spaceship eventually returns, landing in the swimming pool at Leigh's eleventh birthday party. When Lee emerges from the ship, it is discovered that
    A. He is still only 1 year old. B. He is 6 years old.
    C. He is also 11 years old. D. He is 21 years old.
22. || A space cowboy wants to eject from his spacecraft 100,000 km after passing a space buoy, as seen by spectators at rest with respect to the buoy. To do this, the cowboy sets a timer on his craft that will start as he passes the buoy. He plans to cruise by the buoy at 0.300$c$. How much time should he allow between passing the buoy and ejecting?
    A. 1.01 s B. 1.06 s C. 1.11 s
    D. 1.33 s E. 1.58 s
23. | In an intergalactic competition, spaceship pilots compete to see who can cover the distance between two asteroids in the shortest time. The judges are at rest with respect to the two asteroids. From the judges' point of view, a pilot has covered the 3-million-km course in 20 seconds. From the pilots' point of view,
    A. The course is longer than 3 million km, the time longer than 20 seconds.
    B. The course is longer than 3 million km, the time shorter than 20 seconds.
    C. The course is shorter than 3 million km, the time longer than 20 seconds.
    D. The course is shorter than 3 million km, the time shorter than 20 seconds.
24. | Energy in the sun is produced by the fusion of four hydrogen atoms into a helium atom. The process involves several steps, but the net reaction is simply $4\,{}^1\mathrm{H} \rightarrow {}^4\mathrm{He}$ + energy. Given this, you can say that
    A. One helium atom has more mass than four hydrogen atoms.
    B. One helium atom has less mass than four hydrogen atoms.
    C. One helium atom has the same mass as four hydrogen atoms.
25. || A particle moving at speed 0.40$c$ has momentum $p_0$. The speed of the particle is increased to 0.80$c$. Its momentum is now
    A. Less than $2p_0$ B. Exactly $2p_0$ C. Greater than $2p_0$
26. | A particle moving at speed 0.40$c$ has kinetic energy $K_0$. The speed of the particle is increased to 0.80$c$. The kinetic energy is now
    A. Less than $4K_0$ B. Exactly $4K_0$ C. Greater than $4K_0$

# PROBLEMS

### Section 27.2 Galilean Relativity

1. ||| A sprinter crosses the finish line of a race. The roar of the crowd in front approaches her at a speed of 355 m/s. The roar from the crowd behind her approaches at 335 m/s. What are the speed of sound and the speed of the sprinter?
2. | A baseball pitcher can throw a ball with a speed of 40 m/s. He is in the back of a pickup truck that is driving away from you. He throws the ball in your direction, and it floats toward you at a lazy 10 m/s. What is the speed of the truck?
3. | A boy on a skateboard coasts along at 5 m/s. He has a ball that he can throw at a speed of 10 m/s. What is the ball's speed relative to the ground if he throws the ball (a) forward or (b) backward?
4. || A boat takes 3.0 hours to travel 30 km down a river, then 5.0 hours to return. How fast is the river flowing?
5. || When the moving sidewalk at the airport is broken, as it often seems to be, it takes you 50 s to walk from your gate to baggage claim. When it is working and you stand on the moving sidewalk the entire way, without walking, it takes 75 s to travel the same distance. How long will it take you to travel from the gate to baggage claim if you walk while riding on the moving sidewalk?
6. || Two children, one at the front of a parade float and one at the back, are playing catch with a ball. Each sees the other as tossing the ball at 4.0 m/s. The float moves down the road at 2.0 m/s. At what speed do the two children toss the ball from the point of view of a spectator watching the float move by?

### Section 27.3 Einstein's Principle of Relativity

7. || An out-of-control alien spacecraft is diving into a star at a speed of $1.0 \times 10^8$ m/s. At what speed, relative to the spacecraft, is the starlight approaching?
8. | A spaceship measures bright flashes of light from a distant star. The spacecraft now heads toward the star at 0.90$c$. From the spacecraft's point of view, at what speed do the pulses approach?
9. || A starship blasts past the earth at $2.0 \times 10^8$ m/s. Just after passing the earth, the starship fires a laser beam out its back. With what speed does the laser beam approach the earth?
10. || You are flying at 0.99$c$ with respect to Kara. At the exact instant you pass Kara, she fires a very short laser pulse in the same direction you're heading.
    a. After 1.0 s has elapsed on Kara's watch, what does Kara say the distance is between you and the laser pulse?
    b. After 1.0 s has elapsed on your watch, what do you say the distance is between you and the laser pulse?

### Section 27.4 Events and Measurements

### Section 27.5 The Relativity of Simultaneity

11. || Bjorn is standing at $x = 600$ m. Firecracker 1 explodes at the origin and firecracker 2 explodes at $x = 900$ m. The flashes from both explosions reach Bjorn's eye at $t = 3.0\ \mu$s. At what time did each firecracker explode?

Watch Video Solution Problems 27.17 and 27.25

12. |||| Bianca is standing at $x = 600$ m. Firecracker 1, at the origin, and firecracker 2, at $x = 900$ m, explode simultaneously. The flash from firecracker 1 reaches Bianca's eye at $t = 3.0\ \mu s$. At what time does she see the flash from firecracker 2?

13. || You are standing at $x = 9.0$ km. Lightning bolt 1 strikes at $x = 0$ km and lightning bolt 2 strikes at $x = 12.0$ km. Both flashes reach your eye at the same time. Your assistant is standing at $x = 3.0$ km. Does your assistant see the flashes at the same time? If not, which does she see first and what is the time difference between the two?

14. || A light flashes at position $x = 0$ m. One microsecond later, a light flashes at position $x = 1000$ m. In a second reference frame, moving along the $x$-axis at speed $v$, the two flashes are simultaneous. Is this second frame moving to the right or to the left relative to the original frame?

15. ||| Jose is looking to the east. Lightning bolt 1 strikes a tree 300 m from him. Lightning bolt 2 strikes a barn 900 m from him in the same direction. Jose sees the tree strike 1.0 $\mu$s before he sees the barn strike. According to Jose, were the lightning strikes simultaneous? If not, which occurred first and what was the time difference between the two?

16. || Your 1000-m-long starship has warning lights at each end that, to you, flash simultaneously every minute. You are moving directly away from the planet Zerkon at $0.70c$. To a Zerkonian, do the lights flash simultaneously? If not, which flashes first—the light at the front of your ship or the trailing one?

17. || There is a lightbulb exactly halfway between the front and rear of a long hallway in your spaceship. Your ship is traveling at $0.5c$ relative to the earth. The bulb is suddenly turned on. Let event 1 be when the light from the bulb first strikes the front end of the hall, and event 2 be when it strikes the rear end. Which event occurs first, or are they simultaneous, in the frame of (a) the spaceship and (b) a person standing on the earth?

### Section 27.6 Time Dilation

18. || The fastest spaceship ever launched was the Helios 2 spacecraft. At its closest approach to the sun, it was moving at 69,000 m/s relative to the sun. If it maintained this speed for one year, how much time would it lose compared to an observer at rest relative to the sun?

19. ||| A cosmic ray travels 60 km through the earth's atmosphere in 400 $\mu$s, as measured by experimenters on the ground. How long does the journey take according to the cosmic ray?

20. || INT You leave on a mission to a distant star. Your craft accelerates at $0.95g$ for 30 days. From your point of view, you spend the next 1000 days moving at this speed. From the point of view of an earth observer, how long is this period of steady travel?

21. || At what speed relative to a laboratory does a clock tick at half the rate of an identical clock at rest in the laboratory? Give your answer as a fraction of $c$.

22. || INT The muon is a heavier relative of the electron; it is unstable, as we've seen. The tauon is an even heavier relative of the muon and the electron, with a half-life of only $2.9 \times 10^{-13}$ s. A tauon is moving through a detector at $0.999c$. If the tauon lives for one half-life, how far will it travel through the detector before decaying?

23. ||| A starship voyages to a distant planet 10 ly away. The explorers stay 1 yr, return at the same speed, and arrive back on earth 26 yr after they left. Assume that the time needed to accelerate and decelerate is negligible.
    a. What is the speed of the starship?
    b. How much time has elapsed on the astronauts' chronometers?

24. || An astronaut travels to a star system 4.5 ly away at a speed of $0.90c$. Assume that the time needed to accelerate and decelerate is negligible.
    a. How long does the journey take according to Mission Control on earth?
    b. How long does the journey take according to the astronaut?
    c. How much time elapses between the launch and the arrival of the first radio message from the astronaut saying that she has arrived?

25. || A subatomic particle moves through the laboratory at $0.90c$. Laboratory experimenters measure its lifetime, from creation to annihilation, to be 2.3 ps $(1\text{ ps} = 1\text{ picosecond} = 10^{-12}\text{ s})$. According to the particle, how long did it live?

### Section 27.7 Length Contraction

26. | At what speed, as a fraction of $c$, will a moving rod have a length 60% that of an identical rod at rest?

27. | Jill claims that her new rocket is 100 m long. As she flies past your house, you measure the rocket's length and find that it is only 80 m. Should Jill be cited for exceeding the $0.5c$ speed limit?

28. | A high-energy proton leaves the sun at $0.430c$. In the frame of the proton, what is the distance to the earth?

29. || A muon travels 60 km through the atmosphere at a speed of $0.9997c$. According to the muon, how thick is the atmosphere?

30. || INT Electrons in an x-ray tube are accelerated to $0.65c$. The beam then travels 1.0 cm to a target. What is this distance in the frame of the electrons?

31. || The Stanford Linear Accelerator (SLAC) accelerates electrons to $v = 0.99999997c$ in a 3.2-km-long tube. If they travel the length of the tube at full speed (they don't, because they are accelerating), how long is the tube in the electrons' reference frame?

32. | Our Milky Way galaxy is 100,000 ly in diameter. A spaceship crossing the galaxy measures the galaxy's diameter to be a mere 1.0 ly.
    a. What is the speed of the spaceship relative to the galaxy?
    b. How long is the crossing time as measured in the galaxy's reference frame?

33. | The X-15 rocket-powered plane holds the record for the fastest speed ever attained by a manned aircraft, at 2020 m/s. At this speed, by how much is the 15.5-m-long aircraft length contracted?
    **Hint:** Use the binomial approximation.

34. ||| You're standing on an asteroid when you see your best friend rocketing by in her new spaceship. As she goes by, you notice that the front and rear of her ship coincide exactly with the 400-m-diameter of another nearby asteroid that is stationary with respect to you. However, you happen to know that your friend's spaceship measured 500 m long in the showroom. What is your friend's speed relative to you?

### Section 27.8 Velocities of Objects in Special Relativity

35. |||| A rocket cruising past earth at $0.800c$ shoots a bullet out the back door, opposite the rocket's motion, at $0.900c$ relative to the rocket. What is the bullet's speed relative to the earth?
36. || A high-energy proton is ejected from the sun at $0.300c$; it is gaining on a proton ejected at $0.250c$. According to the slower proton, with what speed is the faster proton gaining on it?
37. ||| A base on Planet X fires a missile toward an oncoming space fighter. The missile's speed according to the base is $0.85c$. The space fighter measures the missile's speed as $0.96c$. How fast is the space fighter traveling relative to Planet X?
38. || Two rockets approach each other. Each is traveling at $0.75c$ in the earth's reference frame. What is the speed of one rocket relative to the other?

### Section 27.9 Relativistic Momentum

39. | A proton is accelerated to $0.999c$.
    a. What is the proton's momentum?
    b. By what factor does the proton's momentum exceed its Newtonian momentum?
40. ||| A 1.0 g particle has momentum 400,000 kg · m/s. What is the particle's speed?
41. | At what speed is a particle's momentum twice its Newtonian value?
42. || What is the speed of a particle whose momentum is $mc$?

### Section 27.10 Relativistic Energy

43. || What are the kinetic energy, the rest energy, and the total energy of a 1.0 g particle with a speed of $0.80c$?
44. || $^{98}_{43}$Tc decays by emitting a 140 keV gamma ray—a high-energy photon. The mass of the nucleus is approximately 98 u. By what fraction does the mass of the nucleus decrease when the gamma ray is emitted?
45. | A quarter-pound hamburger with all the fixings has a mass of 200 g. The food energy of the hamburger (480 food calories) is 2 MJ.
    a. What is the energy equivalent of the mass of the hamburger?
    b. By what factor does the energy equivalent exceed the food energy?
46. || INT An astronaut in a spacesuit on a spacewalk is a completely sealed system, with no matter lost or gained. In principle, the energy the astronaut uses will result in a decrease in mass. Suppose on a gentle spacewalk his metabolic power plus the power of the suit is 250 W. What is the change in mass, in kg, on a 1 hour spacewalk?
47. || How fast must an electron move so that its total energy is 10% more than its rest mass energy?
48. || INT Relativity might seem far removed from daily life, but even some older technologies had to take it into account. The electrons that draw the picture on the screen in an old-fashioned television set are accelerated through a potential difference of 17 kV.
    a. Calculate the speed of the electrons using the Newtonian formula for kinetic energy.
    b. Calculate the speed of the electrons using the relativistic formula for kinetic energy.
    c. By what factor does the Newtonian result exceed the relativistic result?
49. | At what speed is a particle's kinetic energy twice its rest energy?
50. || INT The electrons that are used in an electron microscope are accelerated through a potential difference of 70 kV.
    a. Calculate the speed of the electrons using the Newtonian formula for kinetic energy.
    b. Calculate the speed of the electrons using the relativistic formula for kinetic energy.
    c. By what factor does the Newtonian result exceed the relativistic result?
51. || Under the right circumstances, a photon of high enough energy can give rise to an electron-positron pair. What minimum energy photon is required?
52. || INT An electron and a positron meet and completely annihilate, creating two photons, each with the same energy. What is the wavelength of each of the photons?
53. || The chemical energy of gasoline is 46 MJ/kg. If gasoline's mass could be completely converted into energy, what mass of gasoline would be needed to equal the chemical energy content of 1.0 kg of gasoline?
54. |||| INT A standard nuclear power plant generates 3.0 GW of thermal power from the fission of $^{235}$U. Experiments show that, on average, 0.19 u of mass is lost in each fission of a $^{235}$U nucleus. How many kilograms of $^{235}$U undergo fission each year in this power plant?

## General Problems

55. ||| A firecracker explodes at $x = 0$ m, $t = 0\ \mu$s. A second explodes at $x = 300$ m, $t = 2.0\ \mu$s. What is the proper time between these events?
56. ||| A muon is created 60 km above the surface of the earth. In the earth's frame, the muon is traveling directly downward at $0.9999c$.
    a. In the earth's frame of reference, how much time elapses between the muon's creation and the time it reaches the surface?
    b. Use time dilation to determine, in the muon's frame, how much time elapses between the muon's creation and the time it reaches the surface.
    c. In the muon's frame of reference, length contraction reduces the distance between the point at which it is created and the surface of the earth. What is this distance?
    d. The ratio of the distance you calculated in part c and the time you calculated in part b is a speed. Calculate this speed; what do you notice?
57. || You and Maria each own identical spaceships. As you fly past Maria, you measure her ship to be 90 m long and your own ship to be 100 m long.
    a. How long does Maria measure your ship to be?
    b. How fast is Maria moving relative to you?
58. ||| A very fast-moving train car passes you, moving to the right at $0.50c$. You measure its length to be 12 m. Your friend David flies past you to the right at a speed relative to you of $0.80c$. How long does David measure the train car to be?
59. ||| A spaceship heads directly toward an asteroid at a relative speed of $0.75c$. When it is $3.0 \times 10^8$ m from the asteroid, as measured in the asteroid's frame, the rocket fires a laser pulse at the asteroid. According to the rocket captain, what is the time interval between when the pulse hits the asteroid and when the rocket passes the asteroid, barely missing it?

Watch Video Solution Problems 27.69 and 27.71

60. ||| Because of the earth's rotation, a person living on top of a mountain moves at a faster speed than someone at sea level. The mountain dweller's clocks thus run slowly compared to those at sea level. If the average life span of a hermit is 80 years, on average how much longer would a hermit dwelling on the top of a 3000-m-high mountain live compared to a sea-level hermit?
INT

61. |||| Two spaceships are at the sun, and they set their clocks to zero. They then travel in opposite directions at speeds of $0.80c$ and $0.90c$ relative to the sun. When they are $5.0 \times 10^9$ m apart, as measured in the sun's reference frame, what does each ship's clock read?

62. ||| You are on a spacecraft traveling away from the earth at $0.50c$. A beacon on earth flashes exactly once per second.
    a. From the point of view of a person on earth, what is the time between one flash reaching your ship and the next?
    b. From your point of view, what is the time between one flash reaching your ship and the next?

63. || A spaceship flies past an experimenter who measures its length to be one-half the length he had measured when the spaceship was at rest. An astronaut aboard the spaceship notes that his clock ticks at 1-second intervals. What is the time between ticks as measured by the experimenter?

64. ||| Marissa's spaceship approaches Joseph's at a speed of $0.99c$. As Marissa passes Joseph, they synchronize their clocks to both read $t = 0$ s. When Marissa's clock reads 100 s, she sends a light signal back to Joseph. According to his clock, when does he receive this signal?

65. ||| At a speed of $0.90c$, a spaceship travels to a star that is 9.0 ly distant.
    a. According to a scientist on earth, how long does the trip take?
    b. According to a scientist on the spaceship, how long does the trip take?
    c. According to the scientist on the spaceship, what is the distance traveled during the trip?
    d. At what speed do observers on the spaceship see the star approaching them?

66. || In an attempt to reduce the extraordinarily long travel times for voyaging to distant stars, some people have suggested traveling at close to the speed of light. Suppose you wish to visit the red giant star Betelgeuse, which is 430 ly away, and that you want your 20,000 kg rocket to move so fast that you age only 20 years during the round trip.
INT
    a. How fast must the rocket travel relative to earth?
    b. How much energy is needed to accelerate the rocket to this speed?
    c. How many times larger is this energy than the total energy used by the United States in a typical year, roughly $1.0 \times 10^{20}$ J?

67. |||| A rocket traveling at $0.500c$ sets out for the nearest star, Alpha Centauri, which is 4.25 ly away from earth. It will return to earth immediately after reaching Alpha Centauri. What distance will the rocket travel and how long will the journey last according to (a) stay-at-home earthlings and (b) the rocket crew?

68. || A distant quasar is found to be moving away from the earth at $0.80c$. A galaxy closer to the earth and along the same line of sight is moving away from us at $0.20c$. What is the recessional speed of the quasar as measured by astronomers in the other galaxy?

69. |||| A space beacon on Planet Karma emits a pulse of light every second. Spaceman Trevor flies directly toward the beacon at a speed of $0.95c$. According to Trevor, what is the time between one pulse reaching his ship and the next?
    **Hint:** First find the time between pulses reaching the ship in Planet Karma's reference frame.

70. ||| Two rockets, A and B, approach the earth from opposite directions at speed $0.800c$. The length of each rocket measured in its rest frame is 100 m. What is the length of rocket A as measured by the crew of rocket B?

71. ||| The highest-energy cosmic ray ever detected had an energy of about $3.0 \times 10^{20}$ eV. Assume that this cosmic ray was a proton.
    a. What was the proton's speed as a fraction of $c$?
    b. If this proton started at the same time and place as a photon traveling at the speed of light, how far behind the photon would it be after traveling for 1 ly?

72. ||| What is the speed of an electron after being accelerated from rest through a $20 \times 10^6$ V potential difference?
INT

73. ||| What is the speed of a proton after being accelerated from rest through a $50 \times 10^6$ V potential difference?
INT

74. || The half-life of a muon at rest is 1.5 $\mu$s. Muons that have been accelerated to a very high speed and are then held in a circular storage ring have a half-life of 7.5 $\mu$s.
    a. What is the speed of the muons in the storage ring?
    b. What is the total energy of a muon in the storage ring? The mass of a muon is 207 times the mass of an electron.

75. || What is the total energy, in MeV, of
INT
    a. A proton traveling at 99.0% of the speed of light?
    b. An electron traveling at 99.0% of the speed of light?

76. || The factor $\gamma$ appears in many relativistic expressions. A value $\gamma = 1.01$ implies that relativity changes the Newtonian values by approximately 1% and that relativistic effects can no longer be ignored. At what kinetic energy, in MeV, is $\gamma = 1.01$ for (a) an electron, and (b) a proton?

77. | The sun radiates energy at the rate $3.8 \times 10^{26}$ W. The source of this energy is fusion, a nuclear reaction in which mass is transformed into energy. The mass of the sun is $2.0 \times 10^{30}$ kg.
INT
    a. How much mass does the sun lose each year?
    b. What percentage is this of the sun's total mass?
    c. Estimate the lifetime of the sun.

78. || The radioactive element radium (Ra) decays by a process known as *alpha decay,* in which the nucleus emits a helium nucleus. (These high-speed helium nuclei were named alpha particles when radioactivity was first discovered, long before the identity of the particles was established.) The reaction is ${}^{226}\text{Ra} \rightarrow {}^{222}\text{Rn} + {}^{4}\text{He}$, where Rn is the element radon. The accurately measured atomic masses of the three atoms are 226.025, 222.017, and 4.003. How much energy is released in each decay? (The energy released in radioactive decay is what makes nuclear waste "hot.")

79. | The nuclear reaction that powers the sun is the fusion of four protons into a helium nucleus. The process involves several steps, but the net reaction is simply $4\text{p} \rightarrow {}^{4}\text{He} + \text{energy}$. The mass of a helium nucleus is known to be $6.64 \times 10^{-27}$ kg.
   a. How much energy is released in each fusion?
   b. What fraction of the initial rest mass energy is this energy?

80. || When antimatter (which we'll learn more about in Chapter 30)
INT interacts with an equal mass of ordinary matter, both matter and antimatter are converted completely into energy in the form of photons. In an antimatter-fueled spaceship, a staple of science fiction, the newly created photons are shot from the back of the ship, propelling it forward. Suppose such a ship has a mass of $2.0 \times 10^{6}$ kg, and carries a mass of fuel equal to 1% of its mass, or $1.0 \times 10^{4}$ kg of matter and an equal mass of antimatter.

   What is the final speed of the ship, assuming it starts from rest, if all energy released in the matter-antimatter annihilation is transformed into the kinetic energy of the ship?

## MCAT-Style Passage Problems

### Pion Therapy BIO

Subatomic particles called *pions* are created when protons, accelerated to speeds very near $c$ in a particle accelerator, smash into the nucleus of a target atom. Charged pions are unstable particles that decay into muons with a half-life of $1.8 \times 10^{-8}$ s. Pions have been investigated for use in cancer treatment because they pass through tissue doing minimal damage until they decay, releasing significant energy at that point. The speed of the pions can be adjusted so that the most likely place for the decay is in a tumor.

Suppose pions are created in an accelerator, then directed into a medical bay 30 m away. The pions travel at the very high speed of $0.99995c$. Without time dilation, half of the pions would have decayed after traveling only 5.4 m, not far enough to make it to the medical bay. Time dilation allows them to survive long enough to reach the medical bay, enter tissue, slow down, and then decay where they are needed, in a tumor.

81. | What is the half-life of a pion in the reference frame of the patient undergoing pion therapy?
   A. $1.8 \times 10^{-10}$ s
   B. $1.8 \times 10^{-8}$ s
   C. $1.8 \times 10^{-7}$ s
   D. $1.8 \times 10^{-6}$ s

82. | According to the pion, what is the distance it travels from the accelerator to the medical bay?
   A. 0.30 m
   B. 3.0 m
   C. 30 m
   D. 3000 m

83. | The proton collision that creates the pion also creates a gamma-ray photon traveling in the same direction as the pion. The photon will get to the medical bay first because it is moving faster. What is the speed of the photon in the pion's reference frame?
   A. $0.00005c$
   B. $0.5c$
   C. $0.99995c$
   D. $c$

84. | If the pion slows down to $0.99990c$, about what percentage of its kinetic energy is lost?
   A. 0.03%
   B. 0.3%
   C. 3%
   D. 30%

# 28 Quantum Physics

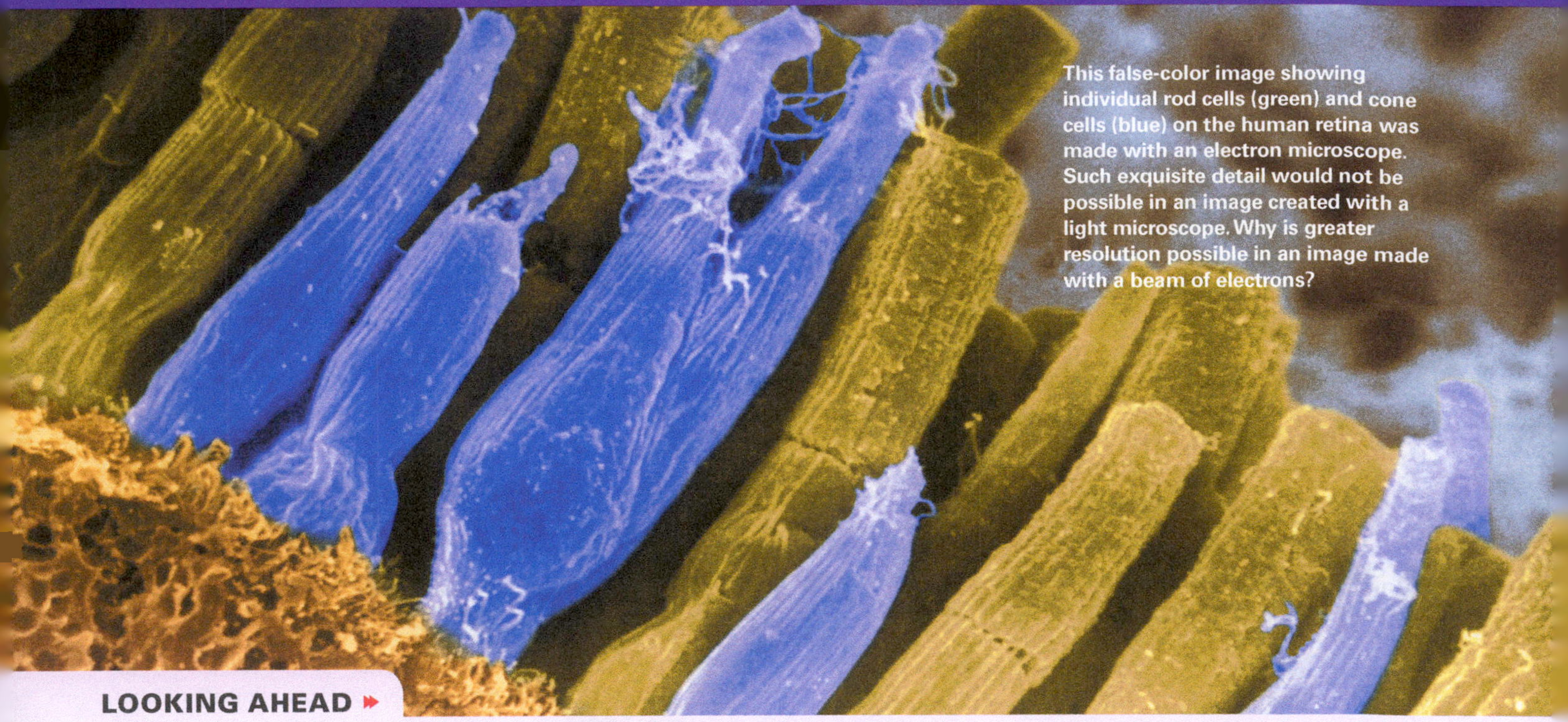

This false-color image showing individual rod cells (green) and cone cells (blue) on the human retina was made with an electron microscope. Such exquisite detail would not be possible in an image created with a light microscope. Why is greater resolution possible in an image made with a beam of electrons?

## LOOKING AHEAD ➤

### Waves Behave Like Particles

An interference pattern made with very low-intensity light shows that the light hits the screen in discrete "chunks" called **photons.**

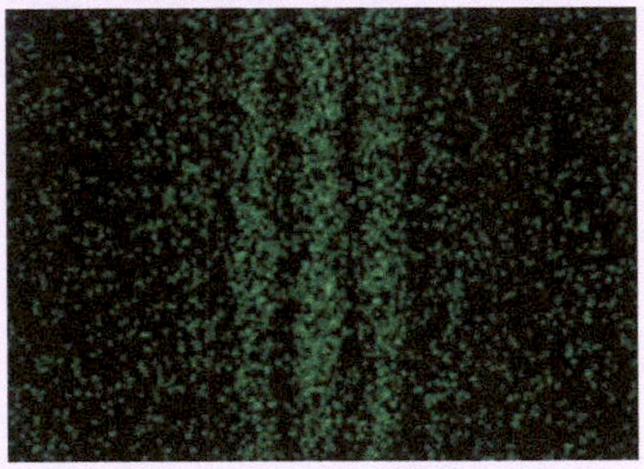

You'll learn how light sometimes behaves like a wave and sometimes like a particle.

### Particles Behave Like Waves

This image of electrons diffracting from an aluminum target shows that, surprisingly, particles have a wave nature.

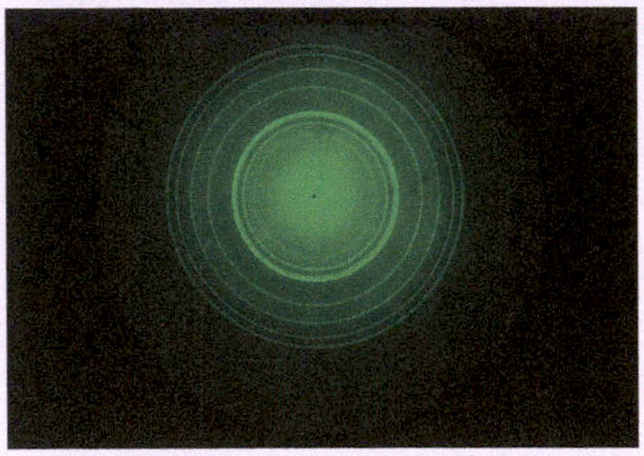

You'll learn that the wavelength of a particle is directly related to its momentum.

### Quantization of Energy

These discrete colors are emitted by helium atoms as their electrons "jump" between quantized **energy levels.**

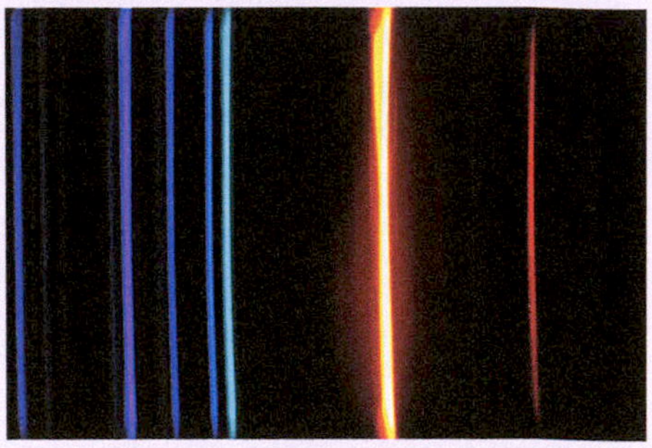

You'll learn how to calculate photon energies and wavelengths as quantum systems emit or absorb photons.

**GOAL** To understand the quantization of energy for light and matter.

## LOOKING BACK ◀

### Double-Slit Interference

In Section 17.2 you learned that light waves spreading out from two narrow slits interfere, producing distinct fringes on a screen. This interference is clear evidence of the wave nature of light.

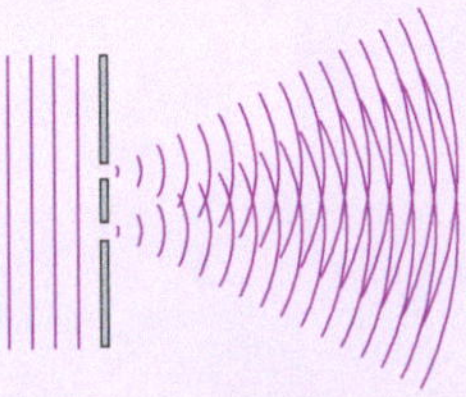

In this chapter, the double-slit interference experiment will reveal striking new properties of light and a surprising wave-like behavior of electrons.

**STOP TO THINK**

A laser illuminates two slits, leading to the interference pattern shown below. After the right-hand slit is covered up, what will the pattern look like?

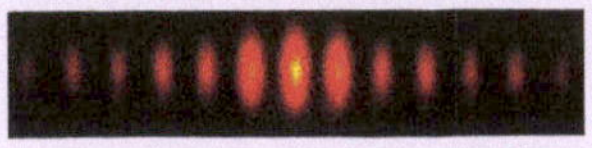

A.
B.
C.
D.

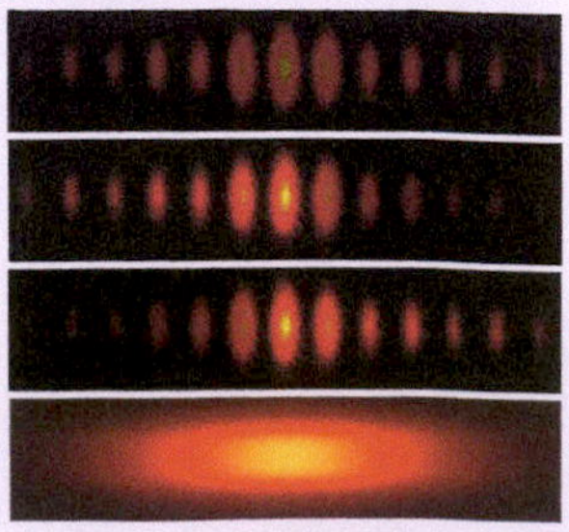

Except for relativity, everything we have studied until this point in the text was known by 1900. Newtonian mechanics, thermodynamics, and the theory of electromagnetism form what we call *classical physics*. It is an impressive body of knowledge with immense explanatory power and a vast number of applications.

But a spate of discoveries right around 1900 showed that classical physics, though remarkable, was incomplete. Investigations into the nature of light and matter led to many astonishing discoveries that classical physics simply could not explain. Sometimes, as you will see, light refuses to act like a wave and seems more like a collection of particles. Other experiments found that electrons sometimes behave like waves. These discoveries eventually led to a radical new theory of light and matter called *quantum physics*.

This chapter will introduce you to this strange but wonderful quantum world. We will take a more historical approach than in previous chapters. As we introduce new ideas, we will describe in some detail the key experiments and the evolution of theories to explain them.

## 28.1 X Rays and X-Ray Diffraction

The rules of quantum physics apply at the scale of atoms and electrons. Experiments to elucidate the nature of the atom and the physics of atomic particles produced results that defied explanation with classical theories. Investigators saw things no one had ever seen before, phenomena that needed new principles and theories to explain them.

In 1895, the German physicist Wilhelm Röntgen was studying how electrons travel through a vacuum. He sealed an electron-producing cathode and a metal target electrode into a vacuum tube. A high voltage pulled electrons from the cathode and accelerated them to very high speed before they struck the target electrode. One day, by chance, Röntgen left a sealed envelope containing film near the vacuum tube. He was later surprised to discover that the film had been exposed even though it had never been removed from the envelope. Some sort of penetrating radiation from the tube had exposed the film.

Röntgen had no idea what was coming from the tube, so he called them x rays, using the algebraic symbol x meaning "unknown." X rays were unlike anything, particle or wave, ever discovered before. Röntgen was not successful at reflecting the rays or at focusing them with a lens. He showed that they travel in straight lines, like particles, but they also pass right through most solid materials with very little absorption, something no known particle could do. The experiments of Röntgen and others led scientists to conclude that these mysterious rays were electromagnetic waves with very short wavelengths, as we learned in ◀◀ SECTION 25.7. These short-wavelength waves were produced in Röntgen's apparatus by the collision of fast electrons with a metal target. X rays are still produced this way, as shown in the illustration of the operation of a modern x-ray tube in FIGURE 28.1.

FIGURE 28.1 The operation of a modern x-ray tube.

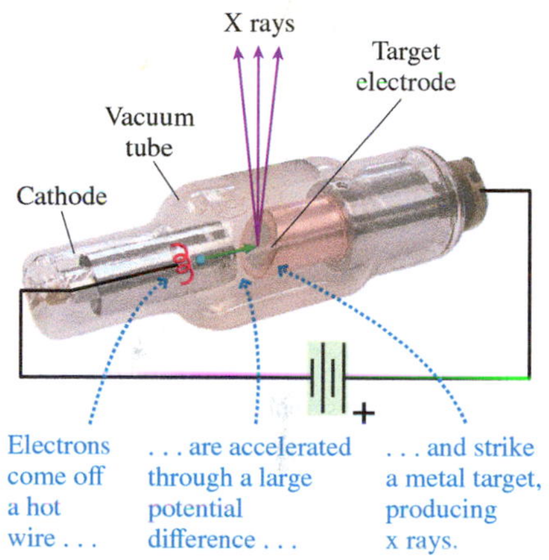


### X-Ray Images BIO

X rays are penetrating, and Röntgen immediately realized that x rays could be used to create an image of the interior of the body. One of Röntgen's first images showed the bones in his wife's hand, dramatically demonstrating the medical potential of these newly discovered rays. Substances with high atomic numbers, such as lead or the minerals in bone, are effective at stopping the rays; materials with low atomic numbers, such as the water and organic compounds of soft tissues in the body, diminish them only slightly. As illustrated in FIGURE 28.2, an x-ray image is essentially a shadow of the bones and dense components of the body; where these tissues stop the x rays, the film is not exposed. The basic procedure for producing an x-ray image on film is little changed from Röntgen's day.

FIGURE 28.2 Creating an x-ray image.

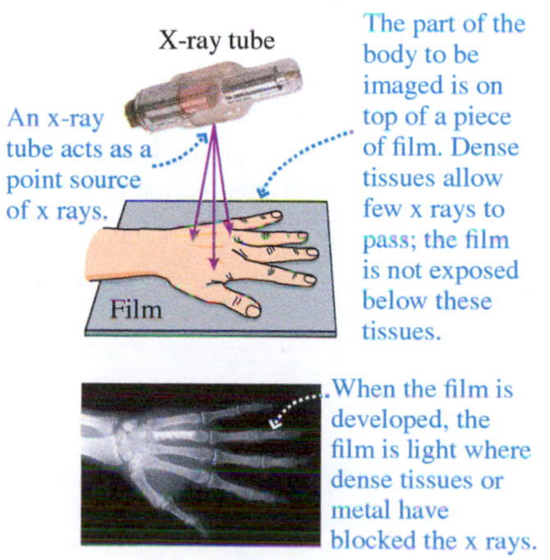

X-ray tube
An x-ray tube acts as a point source of x rays.
The part of the body to be imaged is on top of a piece of film. Dense tissues allow few x rays to pass; the film is not exposed below these tissues.
Film
When the film is developed, the film is light where dense tissues or metal have blocked the x rays.

This use of x rays was of tremendous practical importance, but more important to the development of our story is the use of x rays to probe the structure of matter at an atomic scale.

## X-Ray Diffraction

At about the same time scientists were first concluding that x rays were very-short-wavelength electromagnetic waves, researchers were also deducing that the size of an atom is $\approx 0.1$ nm, and it was suggested that solids might consist of atoms arranged in a regular crystalline *lattice*. In 1912, the German scientist Max von Laue noted that x rays passing through a crystal ought to undergo diffraction from the "three-dimensional grating" of the crystal in much the same way that visible light diffracts from a diffraction grating. Such x-ray diffraction by crystals was soon confirmed experimentally, and measurements confirmed that x rays are indeed electromagnetic waves with wavelengths in the range 0.01 nm to 10 nm—a much shorter wavelength than visible light.

FIGURE 28.3 X rays incident on a simple cubic lattice crystal.

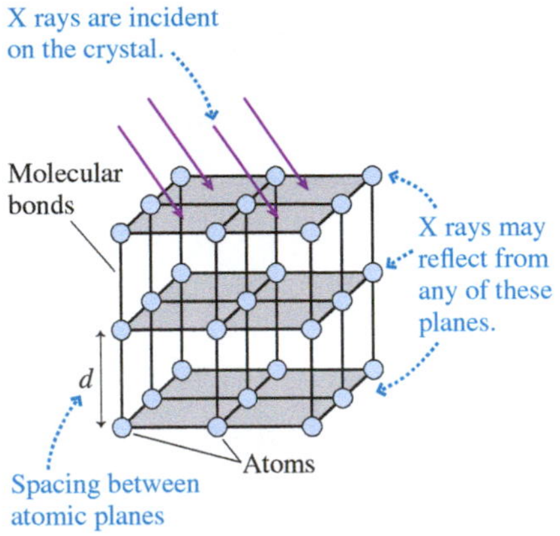


To understand x-ray diffraction, we begin by looking at the arrangement of atoms in a solid. FIGURE 28.3 shows x rays striking a crystal with a *simple cubic lattice.* This is a very straightforward arrangement, with the atoms in planes with spacing $d$ between them.

FIGURE 28.4a shows a side view of the x rays striking the crystal, with the x rays incident at angle $\theta$. Most of the x rays are transmitted through the plane, but a small fraction of the wave is reflected, much like the weak reflection of light from a sheet of glass. The reflected wave obeys the law of reflection—the angle of reflection equals the angle of incidence—and the figure has been drawn accordingly.

FIGURE 28.4 X-ray reflections from parallel atomic planes.

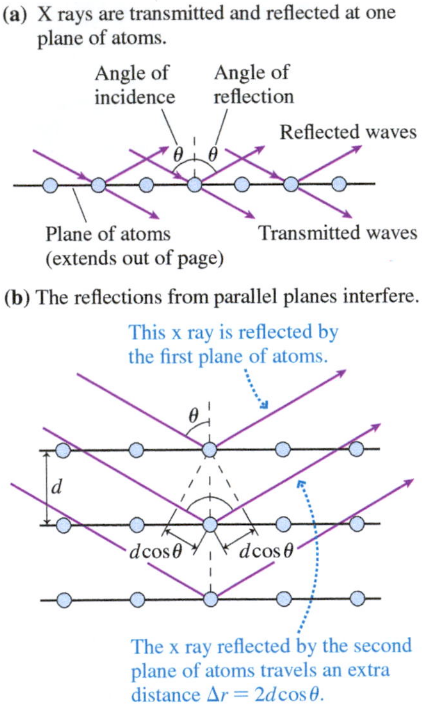


As we saw in Figure 28.3, a solid has not one single plane of atoms but many parallel planes. As x rays pass through a solid, a small fraction of the wave reflects from each of the parallel planes of atoms shown in FIGURE 28.4b. The net reflection from the solid is the *superposition* of the waves reflected by each atomic plane. For most angles of incidence, the reflected waves are out of phase and their superposition is very nearly zero. However, as in the thin-film interference we studied in ◀◀ SECTION 17.4, there are a few specific angles of incidence for which the reflected waves are in phase. For these angles of incidence, the reflected waves interfere constructively to produce a strong reflection. This strong x-ray reflection at a few specific angles of incidence is called **x-ray diffraction.**

You can see from Figure 28.4b that the wave reflecting from any particular plane travels an extra distance $\Delta r = 2d\cos\theta$ before combining with the reflection from the plane immediately above it, where $d$ is the spacing between the atomic planes. If $\Delta r$ is a whole number of wavelengths, then these two waves will be in phase when they recombine. But if the reflections from two neighboring planes are in phase, then *all* the reflections from *all* the planes are in phase and will interfere constructively to produce a strong reflection. Consequently, x rays will reflect from the crystal when the angle of incidence $\theta_m$ satisfies the **Bragg condition:**

$$\Delta r = 2d\cos\theta_m = m\lambda \qquad m = 1, 2, 3, \ldots \tag{28.1}$$

The Bragg condition for constructive interference of x rays reflected from a solid

NOTE ▶ This formula is similar to that for constructive interference for light passed through a grating that we saw in Chapter 17. In both cases, we get constructive interference at only a few well-defined angles. ◀

**EXAMPLE 28.1 Analyzing x-ray diffraction**

X rays with a wavelength of 0.105 nm are diffracted by a crystal with a simple cubic lattice. Diffraction maxima are observed at angles 31.6° and 55.4° and at no angles between these two. What is the spacing between the atomic planes that cause this diffraction?

**STRATEGIZE** The angles must satisfy the Bragg condition. We don't know the values of $m$, but we know that they are two consecutive integers. We'll assume that the two maxima occur at $m$ and $m + 1$.

**PREPARE** In Equation 28.1 $\theta_m$ *decreases* as $m$ increases, so 31.6° corresponds to the larger value of $m$. We will assume that 55.4° corresponds to $m$ and 31.6° to $m + 1$.

**SOLVE** The values of $d$ and $\lambda$ are the same for both diffractions, so we can use the Bragg condition to find

$$\frac{m+1}{m} = \frac{\cos 31.6°}{\cos 55.4°} = 1.50 = \frac{3}{2}$$

Thus 55.4° is the second-order diffraction and 31.6° is the third-order diffraction. With this information we can use the Bragg condition with the second-order diffraction data to find

$$d = \frac{2\lambda}{2\cos\theta_2} = \frac{0.105\text{ nm}}{\cos 55.4°} = 0.185\text{ nm}$$

**ASSESS** We learned above that the size of atoms is $\approx$0.1 nm, so this is a reasonable value for the atomic spacing in a crystal.

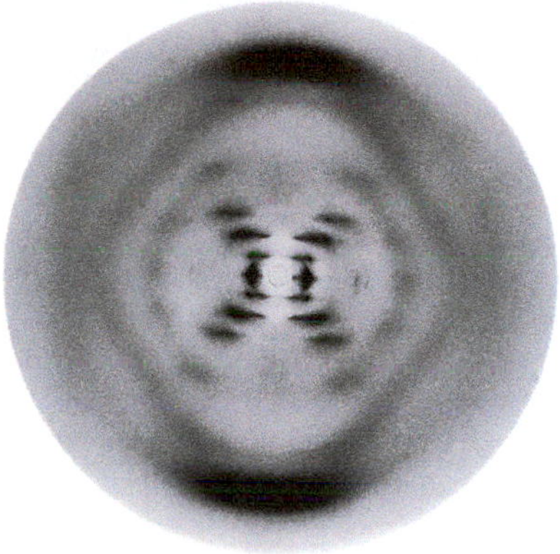

**X marks the spot** BIO Rosalind Franklin, an English chemist and x-ray crystallographer, obtained this x-ray diffraction pattern for DNA in 1953. The cross of dark bands in the center of the diffraction pattern reveals something about the arrangement of atoms in the DNA molecule—that the molecule has the structure of a helix. This x-ray diffraction image was a key piece of information in the effort to unravel the structure of the DNA molecule.

Example 28.1 shows that an x-ray diffraction pattern reveals details of the crystal that produced it. The structure of the crystal was quite simple, so the example was straightforward. More complex crystals produce correspondingly complex patterns that can help reveal the structure of the crystals that produced them. As investigators developed theories of atoms and atomic structure, x rays were an invaluable tool—as they still are. X-ray diffraction is still widely used to decipher the three-dimensional structure of biological molecules such as proteins.

**STOP TO THINK 28.1** The first-order diffraction of x rays from two crystals with simple cubic structure is measured. The first-order diffraction from crystal A occurs at an angle of 20°. The first-order diffraction of the same x rays from crystal B occurs at 30°. Which crystal has the larger atomic spacing?

## 28.2 The Photoelectric Effect

In ◀◀ SECTION 25.6, we introduced the idea that light can be thought of as *photons*, packets with a particular amount of energy. This is an idea that you have likely heard before, but when it was first introduced, it was truly revolutionary. For such an odd idea to find broad acceptance, compelling experimental evidence was needed. This evidence was provided by studies of the *photoelectric effect*, which we will explore in detail in this section to recognize the rationale for and the impact of what was a startling new concept.

The first hints about the photon nature of light came in the late 1800s with the discovery that a negatively charged electroscope could be discharged by shining ultraviolet light on it. The English physicist J. J. Thomson found that the ultraviolet light was causing the electroscope to emit electrons, as illustrated in FIGURE 28.5. The emission of electrons from a substance due to light striking its surface came to be called the **photoelectric effect**. This seemingly minor discovery became a pivotal event that opened the door to the new ideas we discuss in this chapter.

**FIGURE 28.5** Ultraviolet light discharges an electroscope.

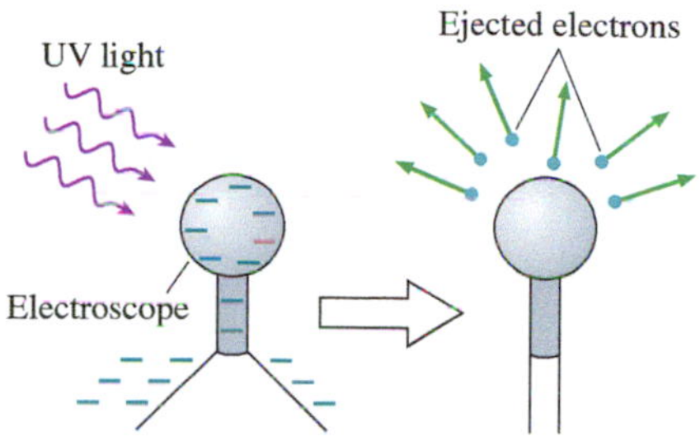


Ultraviolet light discharges a negatively charged electroscope by causing it to emit electrons.

## Characteristics of the Photoelectric Effect

FIGURE 28.6 An experimental device to study the photoelectric effect.

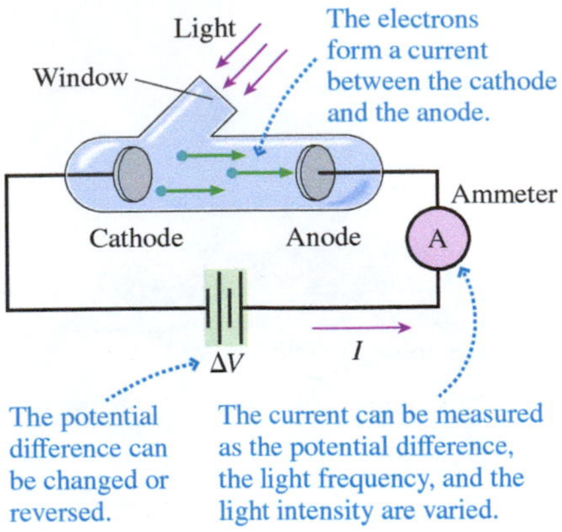


A video to support a section's topic is embedded in the eText.
**Video** Figure 28.6

FIGURE 28.6 shows an evacuated glass tube with two facing electrodes and a window. When ultraviolet light shines on the cathode, a steady counterclockwise current (clockwise flow of electrons) passes through the ammeter. There are no junctions in this circuit, so the current must be the same all the way around the loop. The current in the space between the cathode and the anode consists of electrons moving freely through space (i.e., not inside a wire) at the *same rate* as the current in the wire. There is no current if the electrodes are in the dark, so electrons don't spontaneously leap off the cathode. Instead, the light causes electrons to be ejected from the cathode at a steady rate.

The battery in Figure 28.6 establishes an adjustable potential difference $\Delta V$ between the two electrodes. With it, we can study how the current $I$ varies as the potential difference and the light's wavelength and intensity are changed. Doing so reveals the following characteristics of the photoelectric effect:

- The current $I$ is directly proportional to the light intensity. If the light intensity is doubled, the current also doubles.
- The current appears without delay when the light is applied.
- Electrons are emitted *only* if the light frequency $f$ exceeds a **threshold frequency** $f_0$. This is shown in the graph of FIGURE 28.7a.
- The value of the threshold frequency $f_0$ depends on the type of metal from which the cathode is made.
- If the potential difference $\Delta V$ is more than about 1 V positive (anode positive with respect to the cathode), the current changes very little as $\Delta V$ is increased. If $\Delta V$ is made negative (anode negative with respect to the cathode), by reversing the battery, the current decreases until at some voltage $\Delta V = -V_{stop}$ the current reaches zero. The value of $V_{stop}$ is called the **stopping potential.** This behavior is shown in FIGURE 28.7b.
- The value of $V_{stop}$ is the same for both weak light and intense light. A more intense light causes a larger current, but in both cases the current ceases when $\Delta V = -V_{stop}$.

FIGURE 28.7 The photoelectric current depends on the light frequency $f$ and the battery potential difference $\Delta V$.

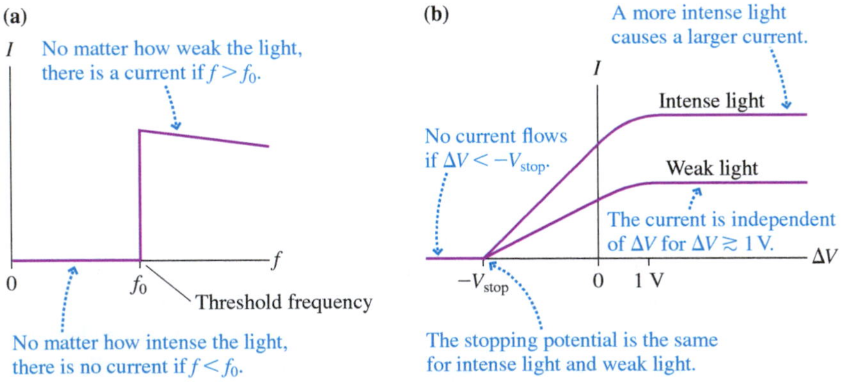


**Video** The Photoelectric Effect

**NOTE** ► We're defining $V_{stop}$ to be a *positive* number. The potential difference that stops the electrons is $\Delta V = -V_{stop}$, with an explicit minus sign. ◄

## Understanding the Photoelectric Effect

You learned in Chapter 22 that electrons are the charge carriers in a metal and they move around freely inside like a sea of negatively charged particles. The electrons are bound inside the metal and do not spontaneously spill out of an electrode at room temperature.

A useful analogy, shown in FIGURE 28.8, is the water in a swimming pool. Water molecules do not spontaneously leap out of the pool if the water is calm. To remove a water molecule, you must do *work* on it to lift it upward, against the force of gravity, to the edge of the pool. A minimum energy is needed to extract a water molecule—namely, the energy needed to lift a molecule that is right at the surface. Removing a water molecule that is deeper requires more than the minimum energy.

Similarly, a *minimum* energy is needed to free an electron from a metal. To extract an electron, you need to increase its energy until its speed is fast enough to escape. The minimum energy $E_0$ needed to free an electron is called the **work function** of the metal. Some electrons, like deeper water molecules, may require more energy than $E_0$ to escape, but all will require *at least* $E_0$. TABLE 28.1 lists the work functions in eV of some elements. (Recall that the conversion to joules is $1\text{ eV} = 1.60 \times 10^{-19}$ J.)

Now, let's return to the photoelectric-effect experiment of Figure 28.6. When ultraviolet light shines on the cathode, electrons leave with some kinetic energy. An electron with energy $E_{\text{elec}}$ inside the metal loses energy $\Delta E$ as it escapes, so it emerges as an electron with kinetic energy $K = E_{\text{elec}} - \Delta E$. The work function energy $E_0$ is the *minimum* energy needed to remove an electron, so the *maximum* possible kinetic energy of an ejected electron is

$$K_{\text{max}} = E_{\text{elec}} - E_0$$

The electrons, after leaving the cathode, move out in all directions. FIGURE 28.9 shows what happens as the potential difference $\Delta V$ between the cathode and the anode is varied:

- If the potential difference between the cathode and the anode is $\Delta V = 0$, there will be no electric field between the plates. Some electrons will reach the anode, creating a measurable current, but many do not.
- If the anode is positive, it attracts *all* of the electrons to the anode. A further increase in $\Delta V$ does not cause any more electrons to reach the anode and thus does not cause a further increase in the current $I$. This is why the curves in Figure 28.7b become horizontal for $\Delta V$ more than about 1 V positive.
- If the anode is negative, it repels the electrons. However, an electron leaving the cathode with sufficient kinetic energy can still reach the anode, just as a ball hits the ceiling if you toss it upward with sufficient kinetic energy. A slightly negative anode voltage turns back only the slowest electrons. The current steadily decreases as the anode voltage becomes increasingly negative until, as the left side of Figure 28.7b shows, at the stopping potential, *all* electrons are turned back and the current ceases.

We can use conservation of energy to relate the maximum kinetic energy to the stopping potential. When $\Delta V$ is negative, as in the bottom panel of Figure 28.9, electrons are "going uphill," converting kinetic energy to potential energy as they slow down. That is, $\Delta U = -e\,\Delta V = -\Delta K$, where we've used $q = -e$ for electrons and $\Delta K$ is negative because the electrons are losing kinetic energy. When $\Delta V = -V_{\text{stop}}$, where the current ceases, the very fastest electrons, with $K_{\text{max}}$, are being turned back *just* as they reach the anode. They're converting 100% of their kinetic energy into potential energy, so $\Delta K = -K_{\text{max}}$. Thus $e\,V_{\text{stop}} = K_{\text{max}}$, or

$$V_{\text{stop}} = \frac{K_{\text{max}}}{e} \qquad (28.2)$$

In other words, **measuring the stopping potential tells us the maximum kinetic energy of the electrons.**

## Einstein's Explanation

When light shines on the cathode in a photoelectric-effect experiment, why do electrons leave the metal at all? Early investigators suggested explanations based on classical physics. It was known that a heated electrode spontaneously emits electrons, so it was natural to suggest that the light falling on the cathode simply heated it, causing it to emit electrons. But this explanation is not consistent with experiment. For light to

FIGURE 28.8 A swimming pool analogy of electrons in a metal.

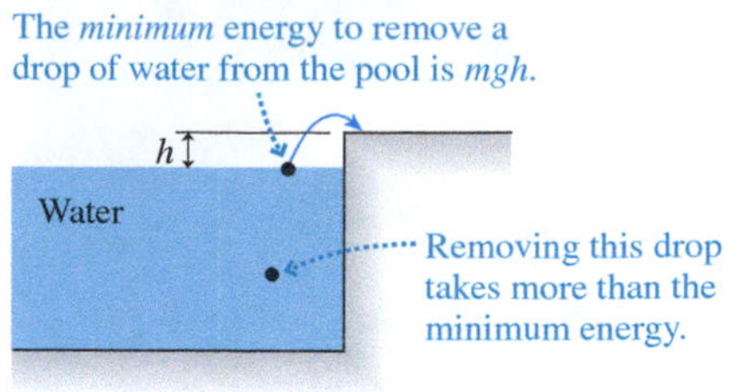


TABLE 28.1 The work functions for some metals

| Element | $E_0$ (eV) |
|---|---|
| Potassium | 2.30 |
| Sodium | 2.75 |
| Aluminum | 4.28 |
| Tungsten | 4.55 |
| Copper | 4.65 |
| Iron | 4.70 |
| Gold | 5.10 |

FIGURE 28.9 The effect of different voltages between the anode and cathode.

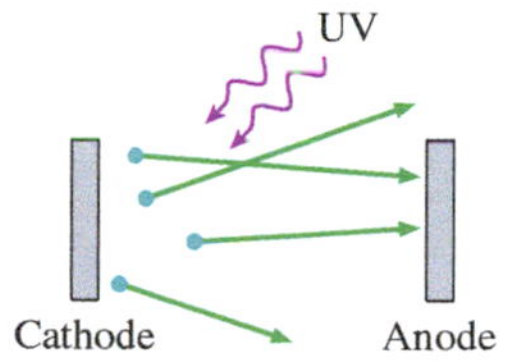


$\Delta V = 0$: The electrons leave the cathode in all directions. Only some reach the anode.

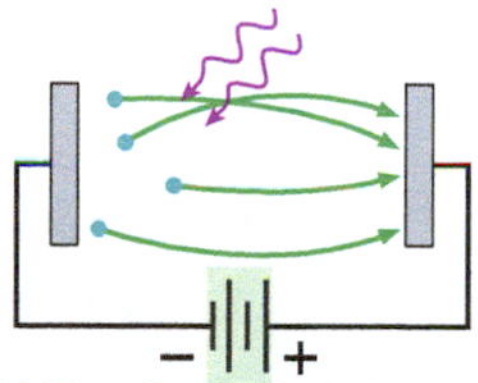


$\Delta V > 0$: Making the anode positive creates an electric field that pushes all the electrons to the anode.

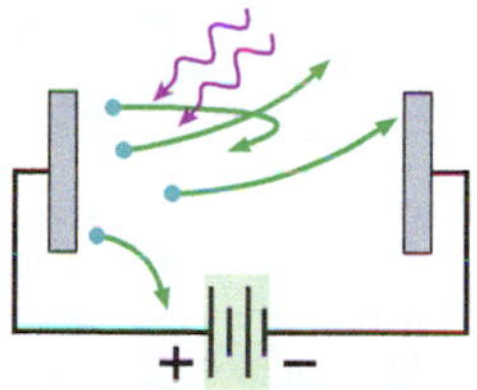


$\Delta V < 0$: Making the anode negative repels the electrons. Only the very fastest make it to the anode.

**Einstein's "Miracle Year"** Albert Einstein was a little-known young man of 26 in 1905. In that single year, Einstein published three papers on three different topics, each of which would revolutionize physics. One was his initial paper on the theory of relativity. A second paper used statistical mechanics to explain a phenomenon called *Brownian motion,* the random motion of small particles suspended in water. It is Einstein's third paper of 1905, on the nature of light, in which we are most interested in this chapter.

heat the metal would take an appreciable time, but, as we've seen, the electrons are emitted without delay when the light is applied.

The heating hypothesis also fails to explain the threshold frequency. If a weak intensity at a frequency just slightly above the threshold can generate a current, then certainly a strong intensity at a frequency just slightly below the threshold should be able to do so—it will heat the metal even more. There is no reason that a slight change in frequency should matter. Yet the experimental evidence shows a sharp frequency threshold, as we've seen.

A new physical theory was needed to fully explain the photoelectric-effect data. The currently accepted solution came in a 1905 paper by Albert Einstein in which he offered an exceedingly simple but amazingly bold idea that explained all of the noted features of the data.

Einstein's paper extended the work of the German physicist Max Planck, who had found that he could explain the form of the spectrum of a glowing, incandescent object that we saw in ◀◀ SECTION 25.7 only if he assumed that the oscillating atoms inside the heated solid vibrated in a particular way. The energy of an atom vibrating with frequency $f$ had to be one of the specific energies $E = 0, hf, 2hf, 3hf, \ldots,$ where $h$ is a constant. That is, the vibration energies are **quantized.** The constant $h$, now called **Planck's constant,** is

$$h = 6.63 \times 10^{-34}\ \text{J}\cdot\text{s} = 4.14 \times 10^{-15}\ \text{eV}\cdot\text{s}$$

The first value, with SI units, is the proper one for most calculations, but you will find the second to be useful when energies are expressed in eV.

Einstein was the first to take Planck's idea seriously. Einstein went even further and suggested that **electromagnetic radiation itself is quantized!** That is, light is not really a continuous wave but, instead, arrives in small packets or bundles of energy. Einstein called each packet of energy a **light quantum,** and he postulated that the energy of one light quantum is directly proportional to the frequency of the light. That is, each quantum of light, which is now known as a **photon,** has energy

**Video** What the Physics? Photonography

$$E = hf \qquad (28.3)$$

The energy of a photon, a quantum of light, of frequency $f$

where $h$ is Planck's constant. Higher-frequency light is composed of higher-energy photons—it is composed of bundles of greater energy. This seemingly simple assumption was completely at odds with the classical understanding of light as a wave, but it allowed Einstein to explain all of the properties of the photoelectric effect.

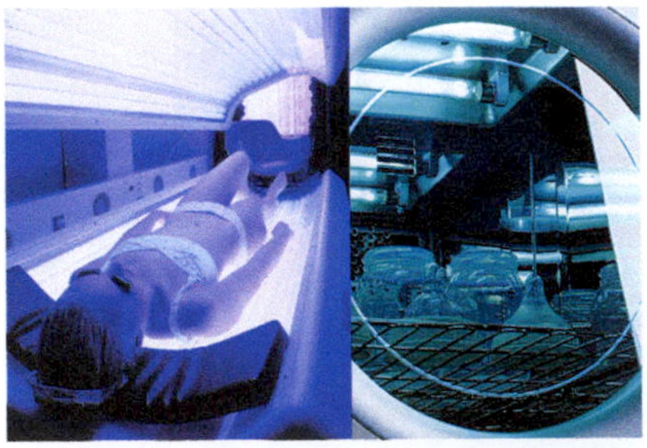

**Not all ultraviolet is created equal** BIO
In Example 28.2, you see that ultraviolet light sources with small differences in wavelength can have very different biological effects. Tanning beds emit nearly all of their energy at wavelengths greater than 315 nm. This light stimulates cells to produce melanin—resulting in a tan—but produces little short-term cell damage. Germicidal lamps use ultraviolet peaked at 254 nm, which will damage and even kill cells. Exposure to such a source will result in very painful sunburn.

### EXAMPLE 28.2 Finding the energy of ultraviolet photons

Ultraviolet light at 290 nm does 250 times as much cellular damage as an equal intensity of ultraviolet at 310 nm; there is a clear threshold for damage at about 300 nm. What is the energy, in eV, of photons with a wavelength of 300 nm?

***PREPARE*** The energy of a photon is related to its frequency by $E = hf$.

***SOLVE*** The frequency at wavelength 300 nm is

$$f = \frac{c}{\lambda} = \frac{3.00 \times 10^{8}\ \text{m/s}}{300 \times 10^{-9}\ \text{m}} = 1.00 \times 10^{15}\ \text{Hz}$$

We can now use Equation 28.3 to calculate the energy, using the value of $h$ in eV · s:

$$E = hf = (4.14 \times 10^{-15}\ \text{eV}\cdot\text{s})(1.00 \times 10^{15}\ \text{Hz}) = 4.14\ \text{eV}$$

***ASSESS*** This number seems reasonable. We saw in Chapter 25 that splitting a bond in a water molecule requires an energy of 4.7 eV. We'd expect photons with energies in this range to be able to damage the complex organic molecules in a cell. As the problem notes, there is a sharp threshold for this damage. For energies larger than about 4.1 eV, photons can disrupt the genetic material of cells. Lower energies have little effect.

## Einstein's Postulates and the Photoelectric Effect

The idea that light is quantized is now widely understood and accepted. But at the time of Einstein's paper, it was a truly revolutionary idea. Though we have used the photon model before, it is worthwhile to look at the theoretical underpinnings in more detail. In his 1905 paper, Einstein framed three postulates about light quanta and their interaction with matter:

1. Light of frequency $f$ consists of discrete quanta, each of energy $E = hf$. Each photon travels at the speed of light $c$.
2. Light quanta are emitted or absorbed on an all-or-nothing basis. A substance can emit 1 or 2 or 3 quanta, but not 1.5. Similarly, an electron in a metal cannot absorb half a quantum but only an integer number.
3. A light quantum, when absorbed by a metal, delivers its entire energy to *one* electron.

**NOTE** ▶ These three postulates—that light comes in chunks, that the chunks cannot be divided, and that the energy of one chunk is delivered to one electron—are crucial for understanding the new ideas that will lead to quantum physics. ◀

Let's look at how Einstein's postulates apply to the photoelectric effect. We now think of the light shining on the metal as a torrent of photons, each of energy $hf$. Each photon is absorbed by *one* electron, giving that electron an energy $E_{elec} = hf$. This leads us to several interesting conclusions:

- An electron that has just absorbed a quantum of light energy has $E_{elec} = hf$. FIGURE 28.10 shows that this electron can escape from the metal if its energy exceeds the work function $E_0$, or if

$$E_{elec} = hf \geq E_0 \quad (28.4)$$

In other words, there is a *threshold frequency*

$$f_0 = \frac{E_0}{h} \quad (28.5)$$

for the ejection of electrons. If $f$ is less than $f_0$, even by just a small amount, none of the electrons will have sufficient energy to escape no matter how intense the light. But even very weak light with $f \geq f_0$ will give a few electrons sufficient energy to escape **because each photon delivers all of its energy to one electron.** This threshold behavior is exactly what the data show.

- A more intense light delivers a larger number of photons to the surface. These eject a larger number of electrons and cause a larger current, exactly as observed.
- There is a distribution of kinetic energies, because different electrons require different amounts of energy to escape, but the *maximum* kinetic energy is

$$K_{max} = E_{elec} - E_0 = hf - E_0 \quad (28.6)$$

As we noted in Equation 28.2, the stopping potential $V_{stop}$ is a measure of $K_{max}$. Einstein's theory predicts that the stopping potential is related to the light frequency by

$$V_{stop} = \frac{K_{max}}{e} = \frac{hf - E_0}{e} \quad (28.7)$$

According to Equation 28.7, the stopping potential does *not* depend on the intensity of the light. Both weak light and intense light will have the same stopping potential. This agrees with the data.

- If each photon transfers its energy $hf$ to just one electron, that electron immediately has enough energy to escape. The current should begin instantly, with no delay, exactly as experiments had found.

FIGURE 28.10 The ejection of an electron.

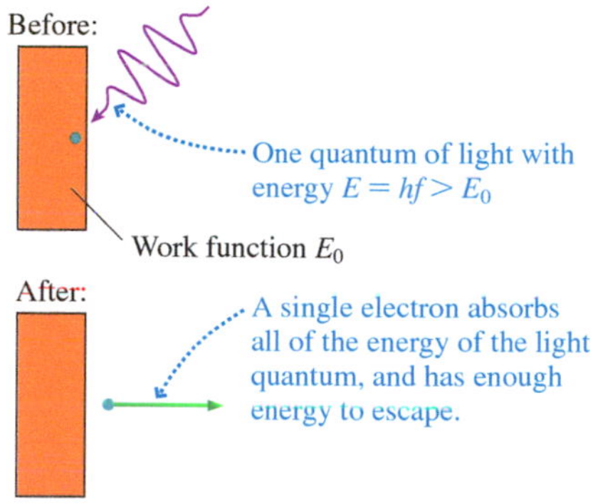

**Seeing the world in a different light** BIO
Plants use photosynthesis to convert the energy of light to chemical energy. Photons of visible light have sufficient energy to trigger the necessary molecular transitions, but infrared photons do not, so visible photons are absorbed while infrared photons are reflected. In this infrared photo, infrared photons strongly reflected from the trees' leaves make the trees appear a ghostly white.

Ultimately, Einstein's postulates are able to explain all of the observed features of the data for the photoelectric effect, though they require us to think of light in a very different way.

FIGURE 28.11 A pebble transfers energy to the water.

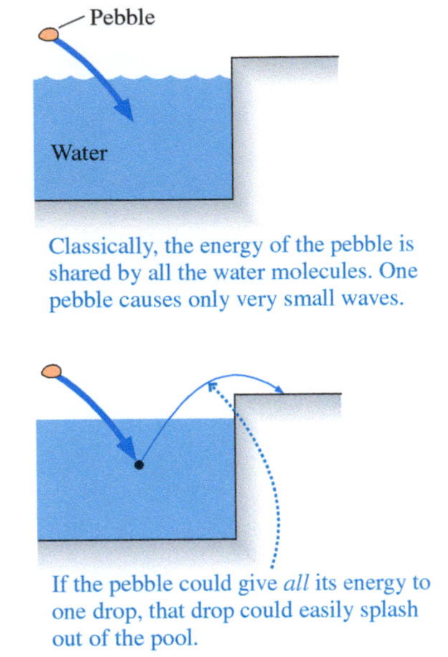

Pebble
Water
Classically, the energy of the pebble is shared by all the water molecules. One pebble causes only very small waves.
If the pebble could give *all* its energy to one drop, that drop could easily splash out of the pool.


Let's use the swimming pool analogy again to help us visualize the photon model. FIGURE 28.11 shows a pebble being thrown into the pool. The pebble increases the energy of the water, but the increase is shared among all the molecules in the pool. The increase in the water's energy is barely enough to make ripples, not nearly enough to splash water out of the pool. But suppose *all* the pebble's energy could go to *one drop* of water that didn't have to share it. That one drop of water would easily have enough energy to leap out of the pool. Einstein's hypothesis that a light quantum transfers all its energy to one electron is equivalent to the pebble transferring all its energy to one drop of water.

Einstein was awarded the Nobel Prize in 1921 not for his theory of relativity, as many would suppose, but for his explanation of the photoelectric effect. Einstein showed convincingly that energy is quantized and that light, even though it exhibits wave-like interference, comes in the particle-like packets of energy we now call photons. This was the first big step in the development of the theory of quantum physics.

### EXAMPLE 28.3 Finding the photoelectric threshold frequency

What are the threshold frequencies and wavelengths for electron emission from sodium and from aluminum?

**STRATEGIZE** Table 28.1 gives values for the work functions of different metals in eV. With these values, we can use Equation 28.5 to find the threshold frequencies. To save unit conversion, we use the value of $h$ with units of eV · s. Once we have calculated the frequencies, we can find the wavelengths.

**PREPARE** From Table 28.1, the work function for sodium is $E_0 = 2.75$ eV and that for aluminum is $E_0 = 4.28$ eV.

**SOLVE** With these values, we calculate the threshold frequencies as

$$f_0 = \frac{E_0}{h} = \begin{cases} 6.64 \times 10^{14}\text{ Hz} & \text{sodium} \\ 10.34 \times 10^{14}\text{ Hz} & \text{aluminum} \end{cases}$$

We use these frequencies to calculate the wavelengths with $\lambda = c/f$, giving

$$\lambda = \begin{cases} 452\text{ nm} & \text{sodium} \\ 290\text{ nm} & \text{aluminum} \end{cases}$$

**ASSESS** Sodium has a much lower work function, so we expect it to have a much lower threshold frequency and thus a much longer threshold wavelength; our results make sense. The photoelectric effect can be observed with sodium for $\lambda < 452$ nm. This includes blue and violet visible light but not red, orange, yellow, or green. Aluminum, with a larger work function, needs ultraviolet wavelengths, $\lambda < 290$ nm.

### EXAMPLE 28.4 Determining the maximum electron speed

What are the maximum electron speed and the stopping potential if sodium is illuminated with light of wavelength 300 nm?

**STRATEGIZE** The photons of the incoming light have a certain energy. Each photon will give its energy to one electron. Some of the energy goes to overcoming the work function; the balance goes to kinetic energy of the electron.

**PREPARE** The frequency of the incoming light is $f = c/\lambda = 1.00 \times 10^{15}$ Hz. The energy of the photon, in eV, is thus

$$hf = (4.14 \times 10^{-15}\text{ eV} \cdot \text{s})(1.00 \times 10^{15}\text{ Hz}) = 4.14\text{ eV}$$

The work function of sodium is $E_0 = 2.75$ eV.

**SOLVE** The maximum kinetic energy of the electron is given by Equation 28.6; this is just the difference between the photon energy and the work function:

$$K_{max} = hf - E_0 = 4.14\text{ eV} - 2.75\text{ eV} = 1.39\text{ eV}$$
$$= 2.22 \times 10^{-19}\text{ J}$$

Because $K = \frac{1}{2}mv^2$, where $m$ is the electron's mass, not the mass of the sodium atom, the maximum speed of an electron leaving the cathode is

$$v_{max} = \sqrt{\frac{2K_{max}}{m}} = 6.99 \times 10^5\text{ m/s}$$

Note that $K_{max}$ must be in J, the SI unit of energy, in order to calculate a speed in m/s.

Now that we know the maximum kinetic energy of the electrons, we can use Equation 28.7 to calculate the stopping potential:

$$V_{stop} = \frac{K_{max}}{e} = 1.39 \text{ V}$$

An anode voltage of $-1.39$ V will be just sufficient to stop the fastest electrons and thus reduce the current to zero.

**ASSESS** The stopping potential has the *same numerical value* as $K_{max}$ expressed in eV, which makes sense. An electron with a kinetic energy of 1.39 eV can go "uphill" against a potential difference of 1.39 V, but no more.

**STOP TO THINK 28.2** The work functions of metals A, B, and C are 3.0 eV, 4.0 eV, and 5.0 eV, respectively. UV light shines on all three metals, causing electrons to be emitted. Rank in order, from largest to smallest, the stopping voltages for A, B, and C.

## 28.3 Photons

We've now seen compelling evidence for the photon nature of light, but this leaves an important question: Just what *are* photons? To begin our explanation, let's return to the experiment that showed most dramatically the wave nature of light—the double-slit interference experiment. We will make a change, though: We will dramatically lower the light intensity by inserting filters between the light source and the slits. The fringes will be too dim to see with the naked eye, so we will use a detector that can build up an image over time. (This is the same sort of detector we imagined using for the extremely low-light photograph in ◀◀ SECTION 25.6.)

FIGURE 28.12 shows the outcome of such an experiment at four different times. At early times, very little light has reached the detector, and there are no discernible fringes. Instead, the detector shows dots; it is registering the arrival of particle-like objects.

As the detector builds up the image for a longer time, we see that the positions of the dots are not entirely random. They are grouped into bands at *exactly* the positions where we expect to see bright constructive-interference fringes. As the detector continues to gather light, the light and dark fringes become quite distinct. After a long time, the individual dots overlap and the image looks exactly like those we saw in Chapter 17.

The dots of light on the screen, which we'll attribute to the arrival of individual photons, are particle-like, but the overall picture clearly does not mesh with the classical idea of a particle. A classical particle, when faced with a double-slit apparatus, would go through one slit or the other. If light consisted of classical particles, we would see two bright areas on the screen, corresponding to light that has gone through one or the other slit. Instead, we see particle-like dots forming wave-like interference fringes.

This experiment was performed with a light level so low that only one photon at a time passed through the apparatus. If particle-like photons arrive at the detector in a banded pattern as a consequence of wave-like interference, as Figure 28.12 shows, but if only one photon at a time is passing through the experiment, what is it interfering with? The only possible answer is that the photon is somehow interfering *with itself.* Nothing else is present. But if each photon interferes with itself, rather than with other photons, then each photon, despite the fact that it is a particle-like object, must somehow go through *both* slits! This is something only a wave could do.

This all seems pretty crazy. But crazy or not, this is the way light behaves in real experiments. **Sometimes it exhibits particle-like behavior and sometimes it exhibits wave-like behavior.** The thing we call *light* is stranger and more complex than it first appeared, and there is no way to reconcile these seemingly contradictory behaviors. We have to accept nature as it is, rather than hoping that nature will conform to our expectations. Furthermore, as we will see, this half-wave/half-particle behavior is not restricted to light.

FIGURE 28.12 A double-slit experiment performed with light of very low intensity.

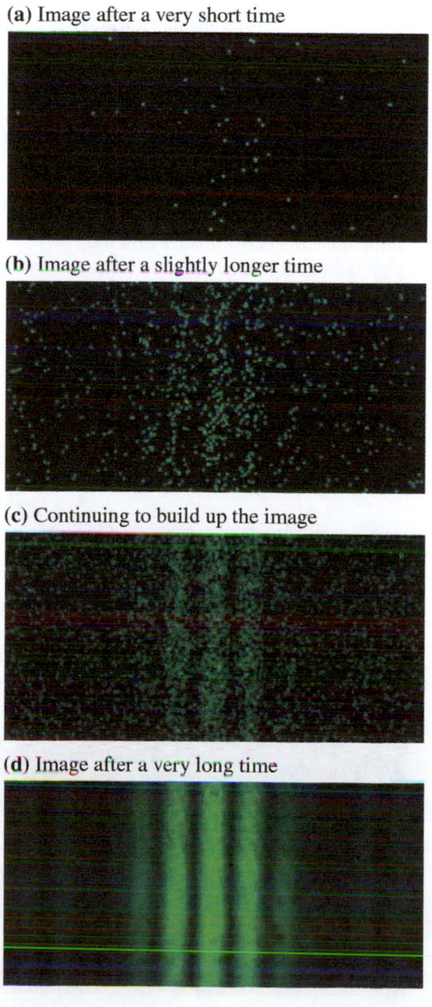

## The Photon Rate

The photon nature of light isn't apparent in most cases. Most light sources with which you are familiar emit such vast numbers of photons that you are aware of only their wave-like superposition, just as you notice only the roar of a heavy rain on your roof and not the individual raindrops. Only at extremely low intensities does the light begin to appear as a stream of individual photons, like the random patter of raindrops when it is barely sprinkling.

### EXAMPLE 28.5 How many photons per second does a laser emit?

The 1.0 mW light beam from a laser pointer ($\lambda = 670$ nm) shines on a screen. How many photons strike the screen each second?

**STRATEGIZE** We can use the power of the beam to find the energy that reaches the screen each second. The energy arrives in chunks of a certain size—the energy of an individual photon. The ratio of these two values is the number of photons.

**PREPARE** The power of the beam is 1.0 mW, or $1.0 \times 10^{-3}$ J/s, so $1.0 \times 10^{-3}$ J reaches the screen each second. This energy arrives as individual photons. The frequency of the photons is $f = c/\lambda = 4.48 \times 10^{14}$ Hz, so the energy of an individual photon is

$$E = hf = (6.63 \times 10^{-34}\ \text{J} \cdot \text{s})(4.48 \times 10^{14}\ \text{Hz}) = 2.97 \times 10^{-19}\ \text{J}$$

**SOLVE** The number of photons reaching the screen each second is the total energy reaching the screen each second divided by the energy of an individual photon:

$$\frac{1.0 \times 10^{-3}\ \text{J/s}}{2.97 \times 10^{-19}\ \text{J/photon}} = 3.4 \times 10^{15}\ \text{photons per second}$$

**ASSESS** Each photon carries a small amount of energy, so there must be a huge number of photons per second to produce even this modest power. Our answer makes sense.

### CONCEPTUAL EXAMPLE 28.6 Comparing photon rates

A red laser pointer and a green laser pointer have the same power. Which one emits a larger number of photons per second?

**REASON** Red light has a longer wavelength and thus a lower frequency than green light, so the energy of a photon of red light is less than the energy of a photon of green light. The two pointers emit the same amount of light energy per second. Because the red laser emits light in smaller "chunks" of energy, it must emit more chunks per second to have the same power. The red laser emits more photons each second.

**ASSESS** This result can seem counterintuitive if you haven't thought hard about the implications of the photon model. Light of different wavelengths is made of photons of different energies, so these two lasers with different wavelengths—though they have the same power—must emit photons at different rates.

**FIGURE 28.13** The operation of a solar cell.

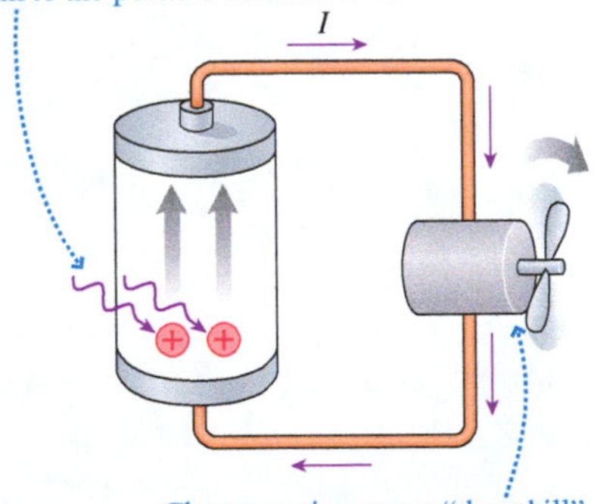

Photons with energy greater than the threshold give their energy to charge carriers, increasing their potential energy and lifting them to the positive terminal of the solar cell.
I
Charge carriers move "downhill" through the circuit. Their energy can be used to run useful devices.

## Detecting Photons

Early light detectors, which used the photoelectric effect directly, consisted of a polished metal plate in a vacuum tube. When light fell on the plate, an electron current was generated that could trigger an action, such as sounding an alarm, or could provide a measurement of the light intensity.

Modern devices work on similar principles. In a *solar cell,* incoming photons give their energy to charge carriers, lifting them into higher-energy states. Recall the charge escalator model of a battery in ◀◀SECTION 22.3. The solar cell works much like a battery, but the energy to lift charges to a higher potential comes from photons, not chemical reactions, as shown in FIGURE 28.13. The photon energy must exceed some minimum value to cause this transition, so solar cells have a threshold frequency, just like a device that uses the photoelectric effect directly. For a silicon-based solar cell, the most common type, the energy threshold is about 1.1 eV, corresponding to a wavelength of about 1200 nm, just beyond the range of the visible-light spectrum, in the infrared.

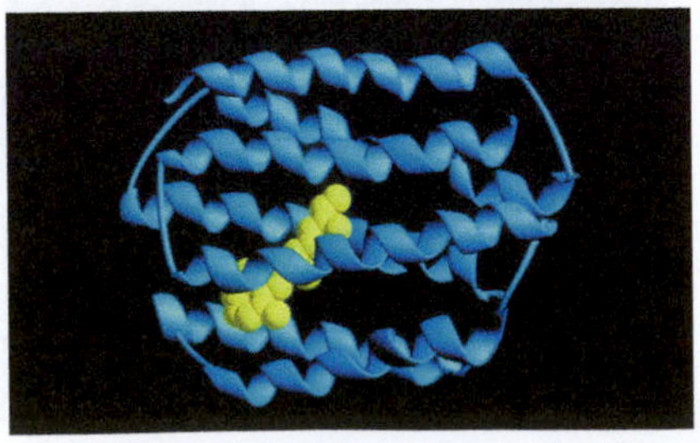

◀ **Seeing photons** BIO The basis of vision is the detection of single photons by specially adapted molecules in the rod and cone cells of the eye. This image shows a molecule of *rhodopsin* (blue) with a molecule called *retinal* (yellow) nested inside. A single photon of the right energy triggers a transition of the retinal molecule, changing its shape so that it no longer fits inside the rhodopsin "cage." The rhodopsin then changes shape to eject the retinal, and this motion leads to an electrical signal in a nerve fiber.

**EXAMPLE 28.7 Finding the current from a solar cell**

1.0 W of monochromatic light of wavelength 550 nm illuminates a silicon solar cell, driving a current in a circuit. What is the maximum possible current this light could produce?

**STRATEGIZE** The wavelength is shorter than the 1200 nm threshold wavelength noted for a silicon solar cell, so the photons will have sufficient energy to cause charge carriers to flow. Each photon of the incident light will give its energy to a single charge carrier. The maximum number of charge carriers that can possibly flow in each second is thus equal to the number of photons that arrive each second.

**PREPARE** The power of the light is $P = 1.0\text{ W} = 1.0\text{ J/s}$. The frequency of the light is $f = c/\lambda = 5.45 \times 10^{14}$ Hz, so the energy of individual photons is $E = hf = 3.61 \times 10^{-19}$ J.

**SOLVE** The number of photons arriving per second is $(1.0\text{ J/s})/(3.61 \times 10^{-19}\text{ J/photon}) = 2.77 \times 10^{18}$. Each photon can set at most one charge carrier into motion, so the maximum current is $2.77 \times 10^{18}$ electrons/s. The current in amps—coulombs per second—is the electron flow rate multiplied by the charge per electron:

$$I_{max} = (2.77 \times 10^{18}\text{ electrons/s})(1.6 \times 10^{-19}\text{ C/electron}) = 0.44\text{ C/s} = 0.44\text{ A}$$

**ASSESS** The key concept underlying the solution is that one photon gives its energy to a single charge carrier. We've calculated the current if all photons give their energy to charge carriers. The current in a real solar cell will be less than this because some photons will be reflected or otherwise "lost" and will not transfer their energy to charge carriers.

The *charge-coupled device* (CCD) or *complementary metal oxide semiconductor* (CMOS) detector in a digital camera consists of millions of *pixels,* each a microscopic silicon-based photodetector. Each photon hitting a pixel (if its frequency exceeds the threshold frequency) liberates one electron. These electrons are stored inside the pixel, and the total accumulated charge is directly proportional to the light intensity—the number of photons—hitting the pixel. After the exposure, the charge in each pixel is read and the value stored in memory; the pixel is then reset to be ready for the next picture.

**STOP TO THINK 28.3** The intensity of a beam of light is increased but the light's frequency is unchanged. Which one (or perhaps more than one) of the following is true?

A. The photons travel faster.
B. Each photon has more energy.
C. There are more photons per second.

## 28.4 Matter Waves

Prince Louis-Victor de Broglie was a French graduate student in 1924. It had been 19 years since Einstein had shaken the world of physics by introducing photons and thus blurring the distinction between a particle and a wave. As de Broglie thought about these issues, it seemed that nature should have some kind of symmetry. If light waves could have a particle-like nature, why shouldn't material particles have some kind of wave-like nature? In other words, could **matter waves** exist?

With no experimental evidence to go on, de Broglie reasoned by analogy with Einstein's equation $E = hf$ for the photon and with some of the ideas of his theory of relativity. De Broglie determined that *if* a material particle of momentum $p = mv$ has a wave-like nature, its wavelength must be given by

$$\lambda = \frac{h}{p} = \frac{h}{mv} \tag{28.8}$$

De Broglie wavelength for a moving particle

where $h$ is Planck's constant. This wavelength is called the **de Broglie wavelength.**

**EXAMPLE 28.8** **Calculating the de Broglie wavelength of an electron**

What is the de Broglie wavelength of an electron with a kinetic energy of 1.0 eV?

**STRATEGIZE** The wavelength depends on the electron's speed and mass. We can find the speed from the kinetic energy, and then use this to determine the wavelength.

**PREPARE** An electron with kinetic energy $K = \frac{1}{2}mv^2 = 1.0\text{ eV} = 1.6 \times 10^{-19}\text{ J}$ has speed

$$v = \sqrt{\frac{2K}{m}} = 5.9 \times 10^5\text{ m/s}$$

Although fast by macroscopic standards, the electron gains this speed by accelerating through a potential difference of a mere 1 V.

**SOLVE** The de Broglie wavelength is

$$\lambda = \frac{h}{mv} = 1.2 \times 10^{-9}\text{ m} = 1.2\text{ nm}$$

**ASSESS** The electron's wavelength is small, but it is similar to the wavelengths of x rays and larger than the approximately 0.1 nm spacing of atoms in a crystal. We can observe x-ray diffraction, so if an electron has a wave nature, it should be easily observable.

**FIGURE 28.14** A double-slit interference pattern created with electrons.

What would it mean for matter—an electron or a proton or a baseball—to have a wavelength? Would it obey the principle of superposition? Would it exhibit diffraction and interference? Surprisingly, **matter exhibits all of the properties that we associate with waves.** For example, FIGURE 28.14 shows the intensity pattern recorded after 50 keV electrons passed through two narrow slits separated by 1.0 $\mu$m. The pattern is clearly a double-slit interference pattern, and the spacing of the fringes is exactly as the theory of Chapter 17 would predict for a wavelength given by de Broglie's formula. **The electrons are behaving like waves!**

But if matter waves are real, why don't we see baseballs and other macroscopic objects exhibiting wave-like behavior? The key is the wavelength. We found in Chapter 17 that diffraction, interference, and other wave-like phenomena are observed when the wavelength is comparable to or larger than the size of an opening a wave must pass through. As Example 28.8 just showed, a typical electron wavelength is somewhat larger than the spacing between atoms in a crystal, so we expect to see wave-like behavior as electrons pass through matter or through microscopic slits. But the de Broglie wavelength is inversely proportional to an object's mass, so the wavelengths of macroscopic objects are millions or billions of times smaller than the wavelengths of electrons—vastly smaller than the size of any openings these objects might pass through. The wave nature of macroscopic objects is unimportant and undetectable because their wavelengths are so incredibly small, as the following example shows.

**EXAMPLE 28.9** **Calculating the de Broglie wavelength of a smoke particle**

One of the smallest macroscopic particles we could imagine using for an experiment would be a very small smoke or soot particle. These are $\approx 1\ \mu$m in diameter, too small to see with the naked eye and just barely at the limits of resolution of a visible-light microscope. A particle this size has mass $m \approx 10^{-18}$ kg. Estimate the de Broglie wavelength for a 1-$\mu$m-diameter particle moving at the very slow speed of 1 mm/s.

**SOLVE** The particle's momentum is $p = mv \approx 10^{-21}\text{ kg}\cdot\text{m/s}$. The de Broglie wavelength of a particle with this momentum is

$$\lambda = \frac{h}{p} \approx 7 \times 10^{-13}\text{ m}$$

**ASSESS** The wavelength is much, much smaller than the particle itself—much smaller than an individual atom! We don't expect to see this particle exhibiting wave-like behavior.

The example shows that a very small particle moving at a very slow speed has a wavelength that is too small to be of consequence. For larger objects moving at higher speeds, the wavelength is even smaller. A pitched baseball will have a wavelength of about $10^{-34}$ m, so a batter cannot use the wave nature of the ball as an excuse for not getting a hit. With such unimaginably small wavelengths, it is little wonder that we do not see macroscopic objects exhibiting wave-like behavior.

## The Interference and Diffraction of Matter

Though de Broglie made his hypothesis in the absence of experimental data, experimental evidence was soon forthcoming. FIGURES 28.15a and b show diffraction patterns produced by x rays and electrons passing through an aluminum-foil target. The primary observation to make from Figure 28.15 is that **electrons diffract and interfere exactly like x rays.**

FIGURE 28.15 Diffraction patterns produced by x rays, electrons, and neutrons.

(a) Diffraction pattern produced by x rays passing through aluminum.

(b) Diffraction pattern produced by electrons passing through aluminum.

(c) Diffraction pattern produced by neutrons passing through a sodium chloride crystal.

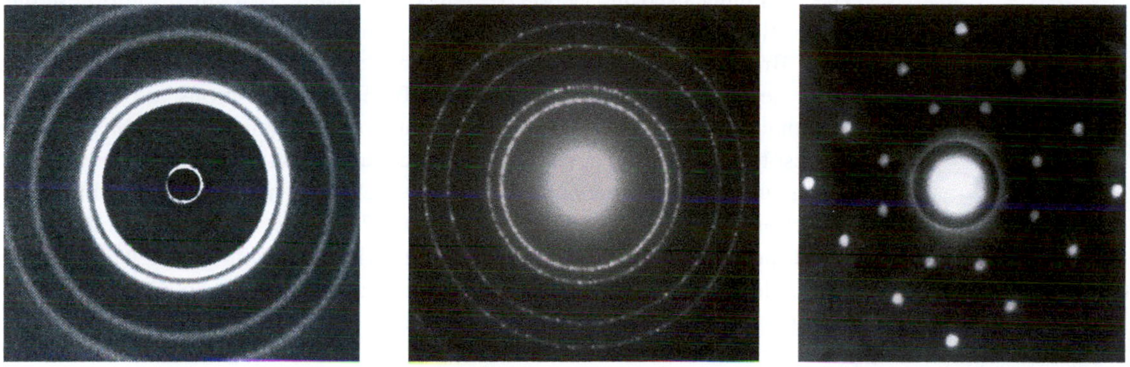

Later experiments demonstrated that de Broglie's hypothesis applies to other material particles as well. Neutrons have a much larger mass than electrons, which tends to decrease their de Broglie wavelength, but it is possible to generate very slow neutrons. The much slower speed compensates for the heavier mass, so neutron wavelengths can be made comparable to electron wavelengths. FIGURE 28.15c shows a neutron diffraction pattern. The pattern appears different than the x-ray and electron patterns because the regular pattern of the sodium chloride crystal leads to well-defined diffraction maxima. The appearance of these maxima is a clear indication that a neutron, too, is a matter wave. In recent years it has become possible to observe the interference and diffraction of atoms and even large molecules!

## The Electron Microscope

As you learned in ◀◀ SECTION 19.7, the wave nature of light limits the ultimate resolution of an optical microscope—the smallest resolvable separation between two objects—to about half a wavelength of light. For visible light, the smallest feature that can be resolved, even with perfect lenses, is about 200 or 250 nm. But the picture of the retina at the start of the chapter can show details much finer than this because it wasn't made with light—it was made with a beam of electrons.

The electron microscope, invented in the 1930s, works much like a light microscope. In the absence of electric or magnetic fields, electrons travel through a vacuum in straight lines much like light rays. Electron trajectories can be bent with electric or magnetic fields. A coil of wire carrying a current can produce a magnetic field that bends parallel electron trajectories so that they all cross at a single point; we call this an *electron lens.* An electron lens focuses electrons in the same way a glass lens bends and focuses light rays.

FIGURE 28.16 shows how a *transmission electron microscope* (*TEM*) works. This is purely classical physics; the electrons experience electric and magnetic forces, and they follow trajectories given by Newton's second law. Our ability to control electron trajectories allows electron microscopes to have magnifications far exceeding those of light microscopes. But just as in a light microscope, the resolution is ultimately limited by wave effects. Electrons are not classical point particles; they have wave-like properties and a de Broglie wavelength $\lambda = h/p$.

FIGURE 28.16 The electron microscope.

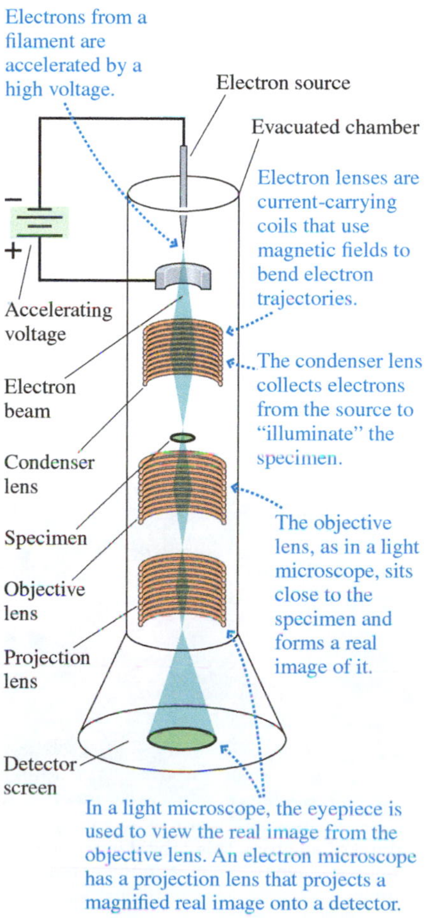

Electrons from a filament are accelerated by a high voltage.
Electron source
Evacuated chamber
Electron lenses are current-carrying coils that use magnetic fields to bend electron trajectories.
−
+
Accelerating voltage
Electron beam
The condenser lens collects electrons from the source to "illuminate" the specimen.
Condenser lens
Specimen
The objective lens, as in a light microscope, sits close to the specimen and forms a real image of it.
Objective lens
Projection lens
Detector screen
In a light microscope, the eyepiece is used to view the real image from the objective lens. An electron microscope has a projection lens that projects a magnified real image onto a detector.

**CONCEPTUAL EXAMPLE 28.10 Which wavelength is shorter?**

An electron is accelerated through a potential difference $\Delta V$. A second electron is accelerated through a potential difference that is twice as large. Which electron has a shorter de Broglie wavelength?

**REASON** The wavelength is inversely proportional to the speed. The electron that is accelerated through the larger potential difference will be moving faster and so will have a shorter de Broglie wavelength.

**ASSESS** Creating an electron micrograph requires high-speed electrons. Higher accelerating voltages mean higher speeds and shorter wavelengths, which would—in principle—allow for better resolution.

**FIGURE 28.17** BIO TEM image of a pigment molecule from a crustacean shell.

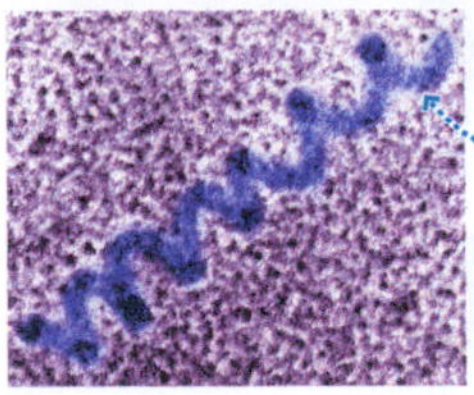

The great resolving power of the electron microscope allows the imaging of incredibly fine detail, in this case the actual structure of a molecule.
20 nm
The total image size is 1/10 the size of the smallest feature that can be resolved by a light microscope. A light microscope could not detect this individual molecule, let alone show its structure.

The reasoning used in Section 19.7 to determine maximum resolution applies equally well to electrons. Thus the resolving power is, at best, about half the electrons' de Broglie wavelength. For a 100 kV accelerating voltage, which is fairly typical, the de Broglie wavelength is $\lambda \approx 0.004$ nm (the electrons are moving fast enough that the momentum has to be calculated using relativity) and thus the theoretical resolving power of an electron microscope is about 0.002 nm.

In practice, the resolving powers of the best electron microscopes are limited by imperfections in the electron lenses to about 0.07 nm, just sufficient to resolve individual atoms with diameters of about 0.1 nm. This resolving power is about 3000 times smaller than can be achieved with light microscopes, as noted in FIGURE 28.17. Good light microscopes function at their theoretical limit, but there's still room to improve electron microscopes if a clever scientist or engineer can make a better electron lens.

**STOP TO THINK 28.4** A beam of electrons, a beam of protons, and a beam of oxygen atoms each pass at the same speed through a 1-$\mu$m-wide slit. Which will produce the widest central maximum on a detector behind the slit?

A. The beam of electrons.
B. The beam of protons.
C. The beam of oxygen atoms.
D. All three patterns will be the same.
E. None of the beams will produce a diffraction pattern.

## 28.5 Energy Is Quantized

De Broglie hypothesized that material particles have wave-like properties, and you've now seen experimental evidence that this must be true. Not only is this bizarre, but the implications are profound.

You learned in ◀◀ SECTION 16.3 that the waves on a string fixed at both ends form standing waves. Wave reflections from both ends create waves traveling in both directions, and the superposition of two oppositely directed waves produces a standing wave. Could we do something like this with particles? Is there such a thing as a "standing matter wave"? In fact, you are probably already familiar with standing matter waves—the atomic electron orbitals that you learned about in chemistry.

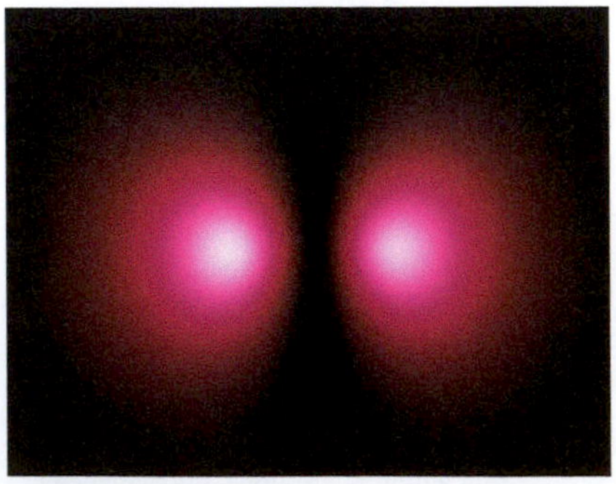
This computer simulation shows the $p$ orbital of an atom. This orbital is an electron standing wave with a clear node at the center.

We'll have more to say about these orbitals in Chapter 29. For now, we'll start our discussion of standing matter waves with a simpler physical system called a "particle in a box." For simplicity, we'll consider one-dimensional motion, a particle that moves back and forth along the $x$-axis. The "box" is defined by two fixed ends, and the particle bounces back and forth between these boundaries, as in FIGURE 28.18. We'll assume that collisions with the ends of the box are perfectly elastic, with no loss of kinetic energy.

Figure 28.18a shows a classical particle, such as a ball or a dust particle, in the box. This particle simply bounces back and forth at constant speed. But if particles have wave-like properties, perhaps we should consider a *wave* reflecting back and forth from the ends of the box. The reflections will create the standing wave shown in Figure 28.18b. This standing wave is analogous to the standing wave on a string that is tied at both ends.

What can we say about the properties of this standing matter wave? We can use what we know about matter waves and standing waves to make some deductions.

For waves on a string, we saw that there were only certain possible modes. The same will be true for the particle in a box. In Chapter 16, we found that the wavelength of a standing wave is related to the length $L$ of the string by

$$\lambda_n = \frac{2L}{n} \qquad n = 1, 2, 3, 4, \ldots \tag{28.9}$$

The wavelength of the particle in a box will follow the same formula, but the wave describing the particle must also satisfy the de Broglie condition $\lambda = h/p$. Equating these two expressions for the wavelength gives

$$\frac{h}{p} = \frac{2L}{n} \tag{28.10}$$

Solving Equation 28.10 for the particle's momentum $p$, we find

$$p_n = n\left(\frac{h}{2L}\right) \qquad n = 1, 2, 3, 4, \ldots \tag{28.11}$$

This is a remarkable result; it is telling us that the momentum of the particle can have only certain values, the ones given by the equation. Other values simply aren't possible. The energy of the particle is related to its momentum by

$$E = \frac{1}{2}mv^2 = \frac{p^2}{2m} \tag{28.12}$$

If we use Equation 28.11 for the momentum, we find that the particle's energy is also restricted to a specific set of values given by

$$E_n = \frac{1}{2m}\left(\frac{hn}{2L}\right)^2$$

or

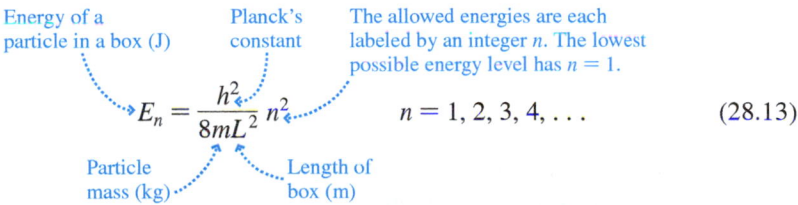

$$E_n = \frac{h^2}{8mL^2}n^2 \qquad n = 1, 2, 3, 4, \ldots \tag{28.13}$$

Energy of a particle in a box (J) → $E_n$
Planck's constant → $h$
The allowed energies are each labeled by an integer $n$. The lowest possible energy level has $n = 1$.
Particle mass (kg) → $m$
Length of box (m) → $L$

This conclusion is one of the most profound discoveries of physics. Because of the wave nature of matter, **a confined particle can have only certain energies.** This result—that a confined particle can have only discrete values of energy—is called the **quantization** of energy. More informally, we say that energy is *quantized.* The number $n$ is called the **quantum number,** and each value of $n$ characterizes one **energy level** of the particle in the box.

The lowest possible energy the particle in the box can have is

$$E_1 = \frac{h^2}{8mL^2} \tag{28.14}$$

In terms of this lowest possible energy, all other possible energies are

$$E_n = n^2E_1 \tag{28.15}$$

This quantization is in stark contrast to the behavior of classical objects. It would be as if a baseball pitcher could throw a baseball at only 10 m/s, or 20 m/s, or 30 m/s, and so on, but at no speed in between. Baseball speeds aren't quantized, but the energy levels of a confined electron are—a result that has far-reaching implications.

**FIGURE 28.18** A particle of mass $m$ confined in a box of length $L$.

(a) A classical particle of mass $m$ bounces back and forth between two boundaries.

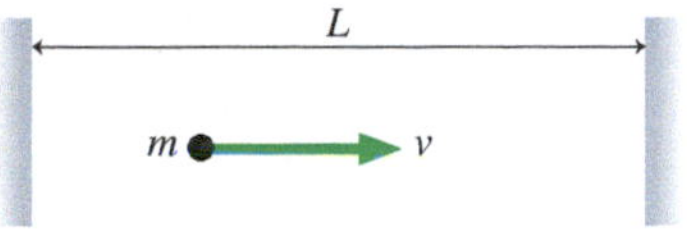


(b) Matter waves moving in opposite directions create standing waves.

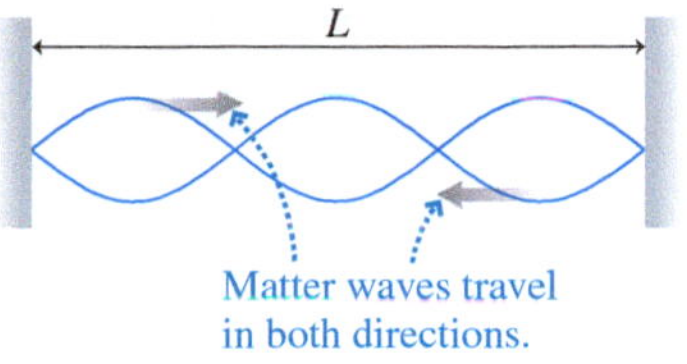


**Video** Quantized Energy

### EXAMPLE 28.11 Finding the allowed energies of a confined electron

An electron is confined to a region of space of length 0.19 nm—comparable in size to an atom. What are the first three allowed energies of the electron?

**STRATEGIZE** We'll model this system as a particle in a box, with a box of length 0.19 nm.

**PREPARE** The possible energies are given by Equation 28.13.

**SOLVE** The mass of an electron is $m = 9.11 \times 10^{-31}$ kg. Thus the first allowed energy is

$$E_1 = \frac{h^2}{8mL^2} = 1.7 \times 10^{-18}\ \text{J} = 10\ \text{eV}$$

This is the lowest allowed energy. The next two allowed energies are

$$E_2 = 2^2 E_1 = 40\ \text{eV}$$

$$E_3 = 3^2 E_1 = 90\ \text{eV}$$

**ASSESS** These energies are significant; $E_1$ is larger than the work function of any metal in Table 28.1. Confining an electron to a region the size of an atom limits its energy to states separated by differences in energy that are significant for describing the behavior of an electron. Clearly, our treatment of electrons in atoms must be a quantum treatment, which is as we expect.

The energies allowed by Equation 28.13 are inversely proportional to both $m$ and $L^2$. Both $m$ and $L$ have to be exceedingly small before energy quantization has any significant effect. If we treat a baseball along a line between the pitcher and the batter as a particle in a box, the lowest energy state is $1 \times 10^{-69}$ J! This unimaginably small number tells us that we needn't worry about quantum mechanics affecting the outcome of a baseball game.

### EXAMPLE 28.12 The energy levels of a virus

The treatment of electrons in atoms must be a quantum treatment, but classical physics still works for baseballs. Where is the dividing line? Suppose we consider a spherical virus, with a diameter of 30 nm, constrained to exist in a long, narrow cell of length 1.0 $\mu$m. If we treat the virus as a particle in a box, what is the lowest energy level? Is a quantum treatment necessary for the motion of the virus?

**STRATEGIZE** We are treating the virus as a particle in a box; the length of the box is the length of the cell.

**PREPARE** The density of a virus is very close to that of water, so the mass is

$$m = \rho \frac{4}{3}\pi r^3 = (1000\ \text{kg/m}^3)\frac{4}{3}\pi(15 \times 10^{-9}\ \text{m})^3 = 1.4 \times 10^{-20}\ \text{kg}$$

**SOLVE** Equation 28.14 gives the lowest energy level as

$$E_1 = \frac{h^2}{8mL^2} = \frac{(6.63 \times 10^{-34}\ \text{J} \cdot \text{s})^2}{8(1.4 \times 10^{-20}\ \text{kg})(1.0 \times 10^{-6}\ \text{m})^2}$$

$$= 3.9 \times 10^{-36}\ \text{J} = 2.4 \times 10^{-17}\ \text{eV}$$

This is such an incredibly small amount of energy that there is no hope of distinguishing between energies of $E_1$ or $4E_1$ or $9E_1$. In principle, the energy is quantized, but the allowed energies are so closely spaced that they will seem to be perfectly continuous. There is no need to use quantum physics to describe the motion of the virus.

**ASSESS** This result seems reasonable. A virus is small, but at a few million atomic mass units, a typical virus is much larger than electrons and other particles that must be treated with quantum physics.

Erwin Schrödinger, one of the early architects of quantum mechanics.

An atom is certainly more complicated than a simple one-dimensional box, but an electron is "confined" within an atom. Thus the electron orbits must, in some sense, be standing waves, and **the energy of the electrons in an atom must be quantized.** This has important implications for the physics of atomic systems, as we'll see in the next section.

**STOP TO THINK 28.5** A particle in a box, with the standing matter wave shown, has an energy of 8.0 eV. What is the lowest energy that this particle can have?

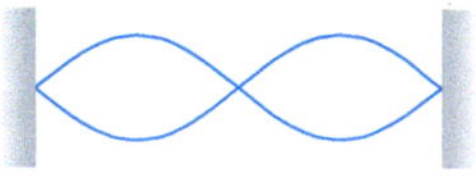

A. 1.0 eV B. 2.0 eV C. 4.0 eV D. 8.0 eV

## 28.6 Energy Levels and Quantum Jumps

Einstein and de Broglie introduced revolutionary new ideas—a blurring of the distinction between waves and particles, and the quantization of energy—but the first to develop a full-blown theory of quantum physics, in 1925, was the Austrian physicist Erwin Schrödinger. Schrödinger's theory is now called *quantum mechanics*. It describes how to calculate the quantized energy levels of systems from a particle in a box to electrons in atoms. Quantum mechanics also describes another important piece of the puzzle: How does a quantized system gain or lose energy?

## Energy-Level Diagrams

The full theory of quantum mechanics shows that, just as for a particle in a box, the energy of a real physical system, such as an atom, is quantized: Only certain energies are allowed while all other energies are forbidden.

An **energy-level diagram** is a useful visual representation of the quantized energies. As an example, FIGURE 28.19a is the energy-level diagram for an electron in a 0.19-nm-long box. We computed these energies in Example 28.11. In an energy-level diagram, the vertical axis represents energy, but the horizontal axis is not a scale. Think of it as a ladder in which the energies are the rungs of the ladder. The lowest rung, with energy $E_1$, is called the **ground state.** Higher rungs, called **excited states,** are labeled by their quantum numbers, $n = 2, 3, 4, \ldots$. Whether it is a particle in a box, an atom, or the nucleus of an atom, quantum physics requires the system to be on one of the rungs of the ladder.

If a quantum system changes from one state to another, its energy changes. One thing that has not changed in quantum physics is the conservation of energy: If a system drops from a higher energy level to a lower, the excess energy $\Delta E_{\text{system}}$ must go somewhere. In the systems we will consider, this energy generally ends up in the form of an emitted photon. As FIGURE 28.19b shows, a quantum system in energy level $E_i$ that "jumps down" to energy level $E_f$ loses an energy $\Delta E_{\text{system}} = |E_f - E_i|$. This jump corresponds to the emission of a photon of frequency

$$f_{\text{photon}} = \frac{\Delta E_{\text{system}}}{h} \tag{28.16}$$

Conversely, if the system absorbs a photon, it can "jump up" to a higher energy level, as shown in FIGURE 28.19c. In this case, the frequency of the absorbed photon must follow Equation 28.16 as well. Such jumps are called **transitions,** or **quantum jumps.**

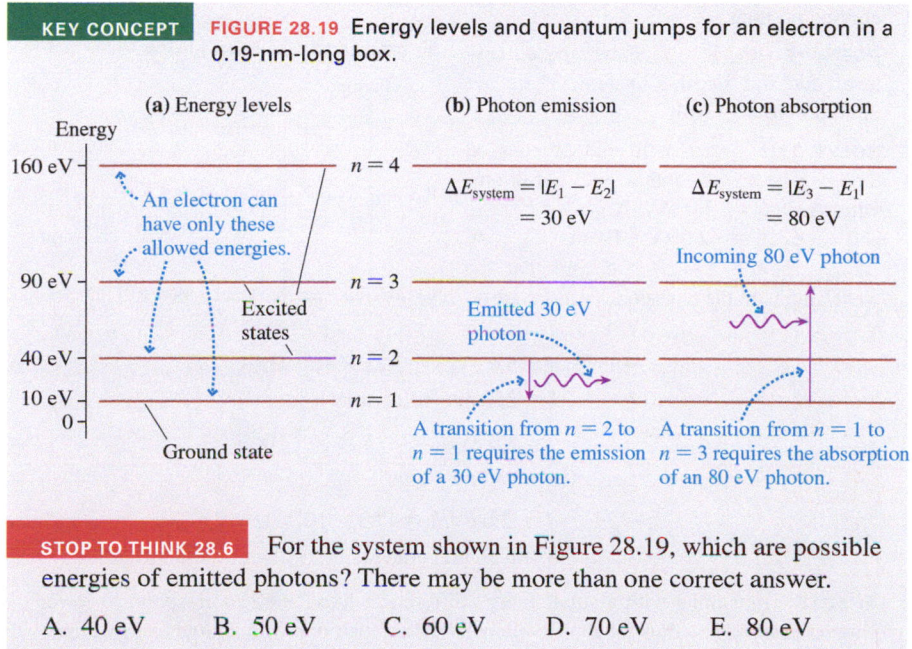

KEY CONCEPT **FIGURE 28.19** Energy levels and quantum jumps for an electron in a 0.19-nm-long box.

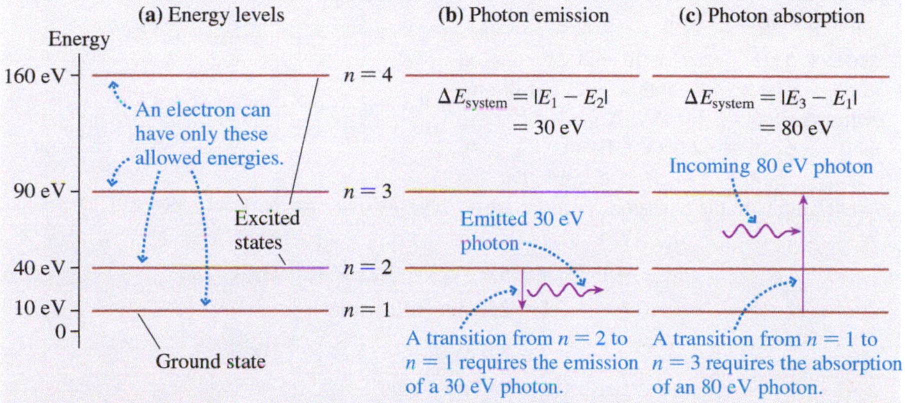


**STOP TO THINK 28.6** For the system shown in Figure 28.19, which are possible energies of emitted photons? There may be more than one correct answer.

A. 40 eV B. 50 eV C. 60 eV D. 70 eV E. 80 eV

Notice that Equation 28.16 links Schrödinger's quantum theory to Einstein's earlier idea about the quantization of light energy. According to Einstein, a photon of frequency $f$ has energy $E_{\text{photon}} = hf$. If a system jumps from an initial state with energy $E_i$ to a final state with *lower* energy $E_f$, energy will be conserved if the system emits a photon with $E_{\text{photon}} = \Delta E_{\text{system}}$. The photon must have exactly the frequency given by Equation 28.16 if it is to carry away exactly the right amount of energy. As we'll see in the next chapter, these photons form the *emission spectrum* of the quantum system.

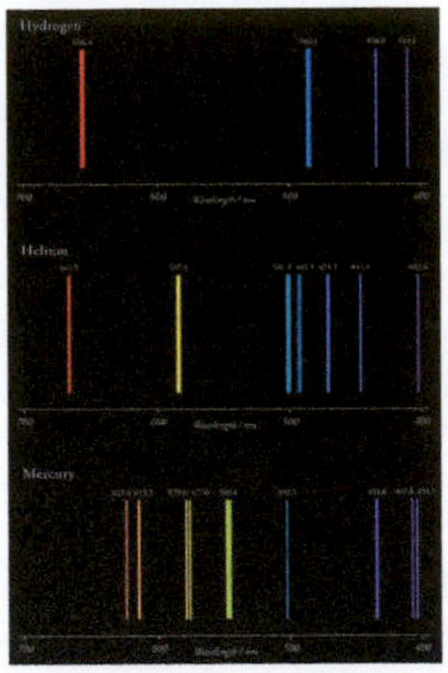

As we see in this diagram, the spectrum of each gas shows a discrete set of wavelengths, corresponding to the energies of possible transitions. Each gas has a different set of energy states, which leads to different possible transitions for each gas, and thus different spectra.

Similarly, a system can conserve energy while jumping to a higher-energy state, for which additional energy is needed, by absorbing a photon of frequency $f_{\text{photon}} = \Delta E_{\text{system}}/h$. The photon will not be absorbed unless it has exactly this frequency. The frequencies absorbed in these upward transitions form the system's *absorption spectrum.*

Let's summarize what quantum physics has to say about the properties of atomic-level systems:

- **The energies are quantized.** Only certain energies are allowed; all others are forbidden. This is a consequence of the wave-like properties of matter.
- **The ground state is stable.** Quantum systems seek the lowest possible energy state. A particle in an excited state, if left alone, will jump to lower and lower energy states until it reaches the ground state. Once in its ground state, there are no lower energy states to which a particle can jump.
- **Quantum systems emit and absorb a *discrete spectrum* of light.** Only those photons whose frequencies match the energy *intervals* between the allowed energy levels can be emitted or absorbed. Photons of other frequencies cannot be emitted or absorbed without violating energy conservation.

We'll use these ideas in the next two chapters to understand the properties of atoms and nuclei.

## EXAMPLE 28.13 Determining an emission spectrum from quantum states

An electron in a quantum system has allowed energies $E_1 = 1.0$ eV, $E_2 = 4.0$ eV, and $E_3 = 6.0$ eV. What wavelengths are observed in the emission spectrum of this system?

**STRATEGIZE** Photons are emitted when the system undergoes a quantum jump from a higher energy level to a lower energy level. We will determine what jumps are possible and then determine the energy differences; this will tell us the possible energies of the emitted photons.

**PREPARE** FIGURE 28.20 shows the energy-level diagram for this system. There are three possible transitions.

FIGURE 28.20 The system's energy-level diagram and quantum jumps.

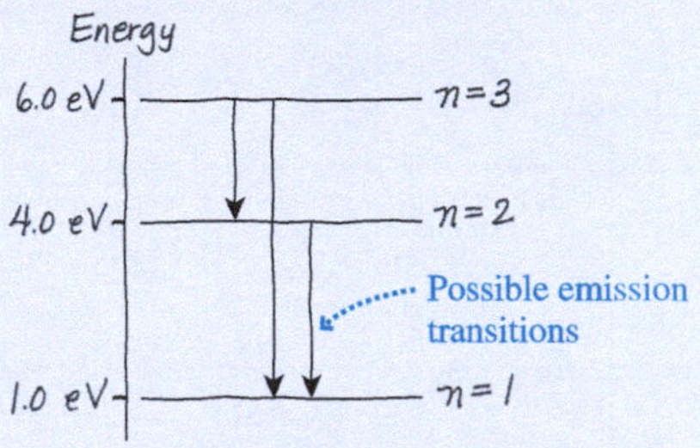


**SOLVE** This system will emit photons on the $3 \to 1$, $2 \to 1$, and $3 \to 2$ transitions, with $\Delta E_{3\to1} = 5.0$ eV, $\Delta E_{2\to1} = 3.0$ eV, and $\Delta E_{3\to2} = 2.0$ eV. From $f_{\text{photon}} = \Delta E_{\text{system}}/h$ and $\lambda = c/f$, we find that the wavelengths in the emission spectrum are

$$3 \to 1 \qquad f = 5.0\ \text{eV}/h = 1.21 \times 10^{15}\ \text{Hz}$$
$$\lambda = 250\ \text{nm (ultraviolet)}$$
$$2 \to 1 \qquad f = 3.0\ \text{eV}/h = 7.25 \times 10^{14}\ \text{Hz}$$
$$\lambda = 410\ \text{nm (blue)}$$
$$3 \to 2 \qquad f = 2.0\ \text{eV}/h = 4.83 \times 10^{14}\ \text{Hz}$$
$$\lambda = 620\ \text{nm (orange)}$$

**ASSESS** Transitions with a small energy difference, like $3 \to 2$, correspond to lower photon energies and thus longer wavelengths than transitions with a large energy difference like $3 \to 1$, as we would expect.

**STOP TO THINK 28.7** A photon with a wavelength of 410 nm has energy $E_{\text{photon}} = 3.0$ eV. Do you expect to see a spectral line with $\lambda = 410$ nm in the emission spectrum of the system represented by this energy-level diagram?

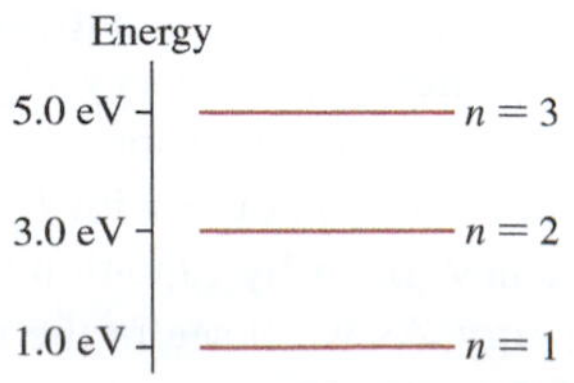

## 28.7 The Uncertainty Principle

One of the strangest aspects of the quantum view of the world is an inherent limitation on our knowledge: **For a particle such as an electron, if you know where it is, you can't know exactly how fast it is moving.** This very counterintuitive notion is a result of the wave nature of matter and is worth a bit of explanation.

FIGURE 28.21 shows an experiment in which electrons moving along the $y$-axis pass through a slit of width $a$. Because of the wave nature of electrons, the slit causes them to spread out and produce a diffraction pattern.

But we can think of the experiment in a different way—as making a measurement of the position of the electrons. As an electron goes through the slit, we know something about its horizontal position. Our knowledge isn't perfect; we just know it is somewhere within the slit. We can establish an *uncertainty,* a limit on our knowledge. The uncertainty in the horizontal position is $\Delta x = a$, the width of the slit.

But, after passing through the slit, the electrons' waves spread out as they produce the diffraction pattern, causing the electrons to strike the screen over a range of positions. For this to happen, the electrons must have acquired a *horizontal* component of velocity $v_x$ that varies from electron to electron. By sending the electrons through a slit—by trying to pin down their horizontal position—we've created an uncertainty in their horizontal velocity. Gaining knowledge of the *position* of the electrons has introduced uncertainty into our knowledge of the *velocity* of the electrons.

FIGURE 28.21 An experiment to illustrate the uncertainty principle.

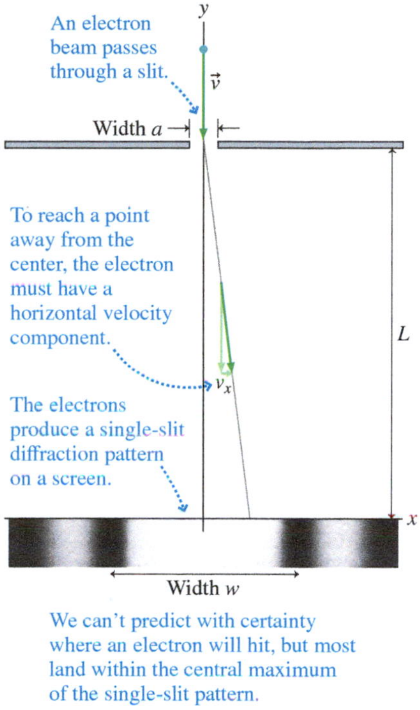


**CONCEPTUAL EXAMPLE 28.14 Changing the uncertainty**

Suppose we narrow the slit in the experiment we just discussed, allowing us to determine the electron's horizontal position more precisely. How does this affect the diffraction pattern? How does this change in the diffraction pattern affect the uncertainty in the velocity?

**REASON** We learned in Chapter 17 that the width of the central maximum of the single-slit diffraction pattern is $w = 2\lambda L/a$. Making the slit narrower—decreasing the value of $a$—increases the value of $w$, making the central fringe wider. If the fringe is wider, the spread of horizontal velocities must be greater, so there is a greater uncertainty in the horizontal velocity.

**ASSESS** Improving our knowledge of the position decreases our knowledge of the velocity.

We've made this argument by considering a particular experiment, but it is an example of a general principle. In 1927, the German physicist Werner Heisenberg proved that, for any particle, the product of the uncertainty $\Delta x$ in its position and the uncertainty $\Delta p_x$ in its $x$-momentum has a lower limit. This is the **Heisenberg uncertainty principle:**

$$\Delta x \, \Delta p_x \geq \frac{h}{4\pi} \qquad (28.17)$$

Heisenberg uncertainty principle for position and momentum

A decreased uncertainty in position—knowing more precisely where a particle is—comes at the expense of an increased uncertainty in velocity and thus in momentum. But the relationship also goes the other way: Knowing a particle's velocity or momentum more precisely requires an increase in the uncertainty about its position.

NOTE ▸ In this text we will use the following convention for uncertainties: If the uncertainty calculated from Heisenberg's uncertainty principle is $\Delta x$, this means that the possible values of $x$ are in the range of $\pm \Delta x/2$ centered on the average value of $x$. ◂

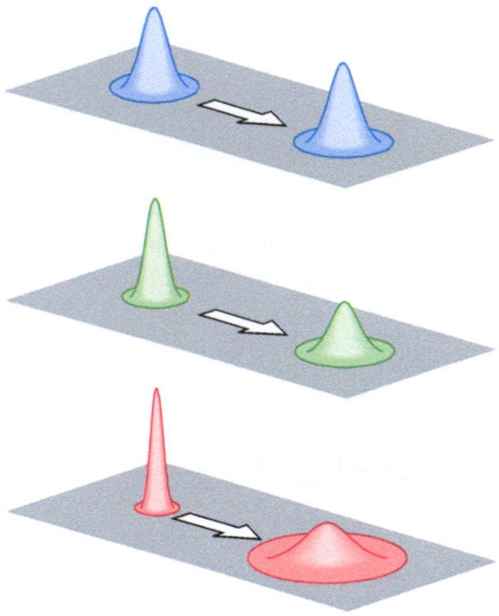

◀ **If I know where you are, I don't know where you're going** In quantum physics, we represent particles by a *wave function* that describes their wave nature. This series of diagrams shows simulations of the evolution of three traveling wave functions. The top diagram shows the broad wave function of a particle whose position is not precisely defined. The uncertainty in momentum (and thus velocity) is small, so the wave function doesn't spread out much as it travels. The lower graphs show particles with more sharply peaked wave functions, implying less uncertainty in their initial positions. A reduced uncertainty in position means a larger uncertainty in velocity, so the wave functions spread out more quickly.

Uncertainties are associated with all experimental measurements, but better procedures and techniques can reduce those uncertainties. Classical physics places no limits on how small the uncertainties can be. A classical particle at any instant of time has an exact position $x$ and an exact momentum $p_x$, and with sufficient care we can measure both $x$ and $p_x$ with such precision that we can make the product $\Delta x\ \Delta p_x$ as small as we like. There are no inherent limits to our knowledge.

In the quantum world, it's not so simple. No matter how clever you are, and no matter how good your experiment, you *cannot* measure both $x$ and $p_x$ simultaneously with arbitrarily good precision. Any measurements you make are limited by the condition that $\Delta x\ \Delta p_x \geq h/4\pi$. **The position and the momentum of a particle are *inherently* uncertain.**

Why? Because of the wave-like nature of matter! The "particle" is spread out in space, so there simply is not a precise value of its position $x$. Our belief that position and momentum have precise values is tied to our classical concept of a particle. As we revise our ideas of what atomic particles are like, we must also revise our ideas about position and momentum.

Let's revisit particles in a one-dimensional "box," now looking at uncertainties.

### EXAMPLE 28.15 Determining uncertainties

What constraints does the uncertainty principle place on our knowledge of the world? To get a sense of scale, we'll look at the uncertainty in speed for a confined electron and for a confined dust particle.

a. What range of velocities might an electron have if confined to a 0.30-nm-wide region, about the size of an atom?

b. A 1.0-$\mu$m-diameter dust particle ($m \approx 10^{-15}$ kg) is confined within a 5-$\mu$m-long box. Can we know with certainty if the particle is at rest? If not, within what range is its velocity likely to be found?

**STRATEGIZE** Localizing a particle means specifying its position with some accuracy—so there must be an uncertainty in the velocity. We can estimate the uncertainty by using Heisenberg's uncertainty principle.

**PREPARE** For part a, we aren't given the exact position of the particle, only that it is within a 0.30-nm-wide region. This means that we have specified the electron's position within the range $\Delta x = 3.0 \times 10^{-10}$ m. For part b, we know that the particle is somewhere in the box, so we have specified its position within the range $\Delta x = L = 5\ \mu$m.

**SOLVE**

a. The uncertainty principle specifies that the least possible uncertainty in the momentum is

$$\Delta p_x = \frac{h}{4\pi\ \Delta x}$$

The uncertainty in the velocity is thus approximately

$$\Delta v_x = \frac{\Delta p_x}{m} \approx \frac{h}{4\pi m\ \Delta x} \approx 2 \times 10^5 \text{ m/s}$$

Because the *average* velocity is zero (the particle is equally likely to be moving right or left), the best we can do is to say that the electron's velocity is somewhere in the interval $-1 \times 10^5$ m/s $\leq v_x \leq 1 \times 10^5$ m/s. **It is simply not possible to specify the electron's velocity more precisely than this.**

b. With a finite $\Delta x$, the uncertainty $\Delta p_x$ *cannot* be zero. **We cannot know with certainty if the particle is at rest inside the box.** No matter how hard we try to bring the particle to rest, the uncertainty in our knowledge of the particle's momentum will be approximately $\Delta p_x \approx h/(4\pi\ \Delta x) = h/(4\pi L)$. Consequently, the range of possible velocities is

$$\Delta v_x = \frac{\Delta p_x}{m} \approx \frac{h}{4\pi m L} \approx 1.0 \times 10^{-14} \text{ m/s}$$

This range of possible velocities will be centered on $v_x = 0$ m/s if we have done our best to have the particle be at rest. Therefore all we can know with certainty is that the particle's velocity is somewhere within the interval $-5 \times 10^{-15}$ m/s $\leq v_x \leq 5 \times 10^{-15}$ m/s.

**ASSESS** Our uncertainty about the electron's velocity is enormous. For an electron confined to a region of this size, the best we can do is to state that its speed is less than two hundred thousand miles per hour! The uncertainty principle clearly sets real, practical limits on our ability to describe electrons. The situation for the dust particle is different. We can't say for certain that the particle is absolutely at rest. But knowing that its speed is less than $5 \times 10^{-15}$ m/s means that the particle is at rest for all practical purposes. At this speed, the dust particle would require nearly 6 hours to travel the width of one atom! Again we see that the quantum view has profound implications at the atomic scale but need not affect the way we think of macroscopic objects.

**STOP TO THINK 28.8** The speeds of an electron and a proton have been measured to the same uncertainty. Which one has a larger uncertainty in position?

A. The proton, because it's more massive.
B. The electron, because it's less massive.
C. The uncertainty in position is the same, because the uncertainty in velocity is the same.

## 28.8 Applications and Implications of Quantum Theory

Quantum theory seems bizarre to those of us living at a scale where the rather different rules of classical physics apply. In this section we consider some of the implications of quantum theory and some applications that confirm these unusual notions.

### Tunneling and the Scanning Tunneling Microscope

The fact that particles have a wave nature allows for imaging at remarkably small scales—the scale of single atoms! The *scanning tunneling microscope* doesn't work like other microscopes you have seen, but instead builds an image of a solid surface by scanning a probe near the surface.

FIGURE 28.22 shows the tip of a very, very thin metal needle, called the *probe tip,* positioned above the surface of a solid sample. The space between the tip and the surface is about 0.5 nm, only a few atomic diameters. Electrons in the sample are attracted to the positive probe tip, but no current should flow, according to classical physics, because the electrons cannot cross the gap between the sample and the probe; it is an incomplete circuit. As we found with the photoelectric effect, the electrons are bound inside the sample and not free to leave.

However, electrons are not classical particles. The electron has a wave nature, and waves don't have sharp edges. One startling prediction of Schrödinger's quantum mechanics is that the electrons' wave functions extend very slightly beyond the edge of the sample. When the probe tip comes close enough to the surface, close enough to poke into an electron's wave function, an electron that had been in the sample might suddenly find itself in the probe tip. In other words, a quantum electron *can* cross the gap between the sample and the probe tip, thus causing a current to flow in the circuit. This process is called **tunneling** because it is rather like tunneling through an uncrossable mountain barrier to get to the other side. Tunneling is completely forbidden by the laws of classical physics, so the fact that it occurs is a testament to the reality of quantum ideas.

The probability that an electron will tunnel across the gap is very sensitive to the size of the gap, which makes the **scanning tunneling microscope,** or STM, possible. When the probe tip passes over an atom or over an atomic-level bump on the surface, the gap narrows and the current increases as more electrons are able to tunnel across. Similarly, the tunneling probability decreases when the probe tip passes across an atomic-size valley, and the current falls. The current-versus-position data are used to construct an image of the surface.

The STM was the first technology that allowed imaging of individual atoms, and it was one of a handful of inventions in the 1980s that jump-started the current interest in nanotechnology. STM images offer a remarkable view of the world at an atomic scale. The STM image of FIGURE 28.23a clearly shows the hexagonal arrangement of the individual atoms on the surface of graphite. The image of a DNA molecule in FIGURE 28.23b shows the actual twists of the helical structure. Current research efforts aim to develop methods for sequencing DNA with scanning tunneling microscopes and other nanoprobes—to directly "read" a single strand of DNA!

FIGURE 28.22 The scanning tunneling microscope.

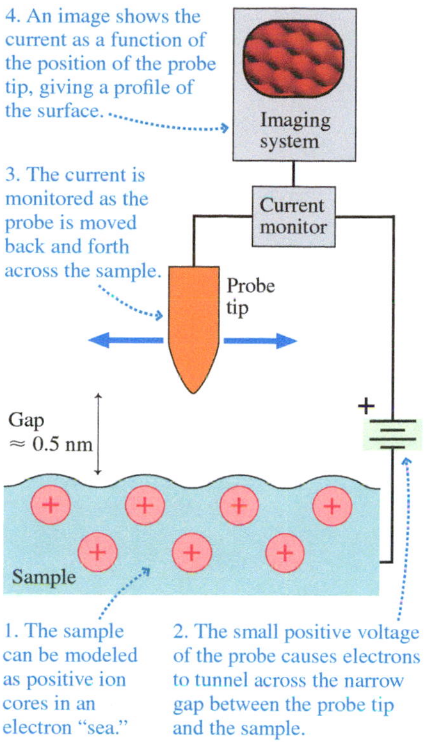


FIGURE 28.23 STM images.

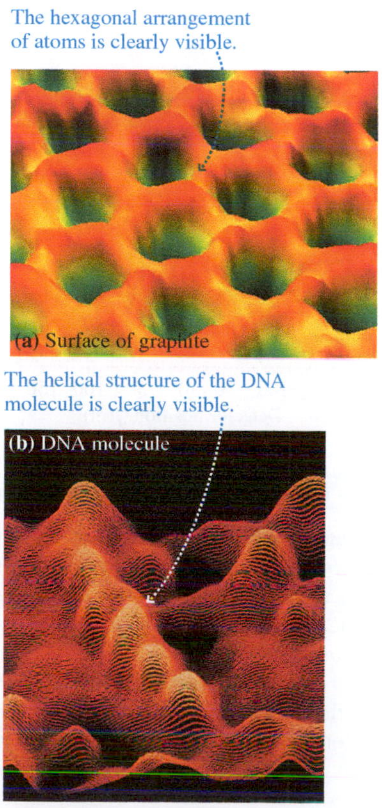

The hexagonal arrangement of atoms is clearly visible.
(a) Surface of graphite
The helical structure of the DNA molecule is clearly visible.
(b) DNA molecule

## Wave–Particle Duality

One common theme that has run through this chapter is the idea that, in quantum theory, things we think of as being waves have a particle nature, while things we think of as being particles have a wave nature. What is the true nature of light, or an electron? Are they particles or waves?

The various objects of classical physics are *either* particles *or* waves. There's no middle ground. Planets and baseballs are particles or collections of particles, while sound and light are clearly waves. Particles follow trajectories given by Newton's laws; waves obey the principle of superposition and exhibit interference. This wave–particle dichotomy seemed obvious until physicists encountered irrefutable evidence that light sometimes acts like a particle and, even stranger, that matter sometimes acts like a wave.

You might at first think that light and matter are *both* a wave *and* a particle, but that idea doesn't quite work. The basic definitions of particleness and waviness are mutually exclusive. Two sound waves can pass through each other and can overlap to produce a larger-amplitude sound wave; two baseballs can't. It is more reasonable to conclude that light and matter are *neither* a wave *nor* a particle. At the microscopic scale of atoms and their constituents—a physical scale not directly accessible to our five senses—the classical concepts of particles and waves turn out to be simply too limited to explain the subtleties of nature.

Although matter and light have both wave-like aspects and particle-like aspects, they show us only one face at a time. If we arrange an experiment to measure a wave-like property, such as interference, we find photons and electrons acting like waves, not particles. An experiment to look for particles will find photons and electrons acting like particles, not waves. These two aspects of light and matter are *complementary* to each other, like a two-piece jigsaw puzzle. Neither the wave nor the particle model alone provides an adequate picture of light or matter, but taken together they provide us with a basis for understanding these elusive but most fundamental constituents of nature. This two-sided point of view is called *wave–particle duality.*

For over two hundred years, scientists and nonscientists alike felt that the clockwork universe of Newtonian physics was a fundamental description of reality. But wave–particle duality, along with Einstein's relativity, undermines the basic assumptions of the Newtonian worldview. The certainty and predictability of classical physics have given way to a new understanding of the universe in which chance and uncertainty play key roles—the universe of quantum physics.

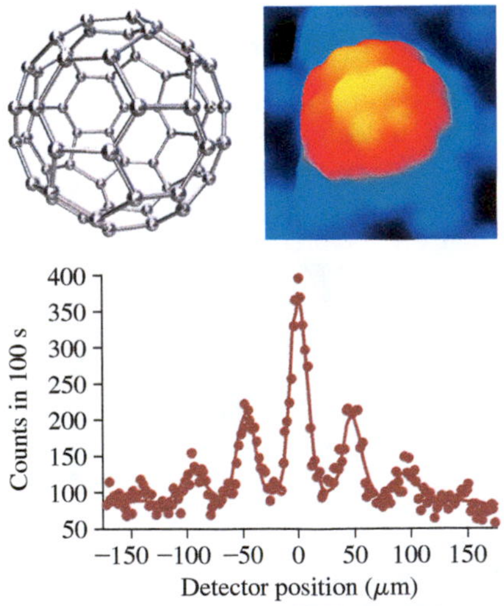


◀ **The dual nature of a buckyball** Treating atomic-level structures involves frequent shifts between particle and wave views. 60 carbon atoms can create the molecule diagrammed in the top-left image, known as $C_{60}$, or *buckminsterfullerene.* The scanning electron microscope image of a $C_{60}$ molecule shown in the top-right image is a particle-like view of the molecule with individual carbon atoms clearly visible. The $C_{60}$ molecule, though we can make a picture of it—showing the atoms that make it up—also has a wave nature. A beam of $C_{60}$ sent through a grating will produce a diffraction pattern similar to those you studied for light waves, as the bottom graph shows.

### INTEGRATED EXAMPLE 28.16 Magnetic resonance imaging BIO

In Chapter 24, we learned that the magnetism of permanent magnets arises because the inherent magnetic moment of electrons causes them to act like little compass needles. Protons also have an inherent magnetic moment, and this is the basis for *magnetic resonance imaging* (MRI) in medicine.

Although a compass needle would prefer to align with a magnetic field, the needle can point in *any* direction. This isn't the case for the magnetic moment of a proton. Quantum physics tells us that the proton's energy must be quantized. There are only two possible energy levels—and thus two possible orientations—for protons in a magnetic field:

$E_1 = -\mu B$ magnetic moment aligned with the field

$E_2 = +\mu B$ magnetic moment aligned opposite the field

where $\mu = 1.41 \times 10^{-26}$ J/T is the known value of the proton's magnetic moment. FIGURE 28.24 shows the two possible energy states. The magnetic moment, like a compass needle, "wants" to align with the field, so that is the lower-energy state.

FIGURE 28.24 Energy levels for a proton in a magnetic field.

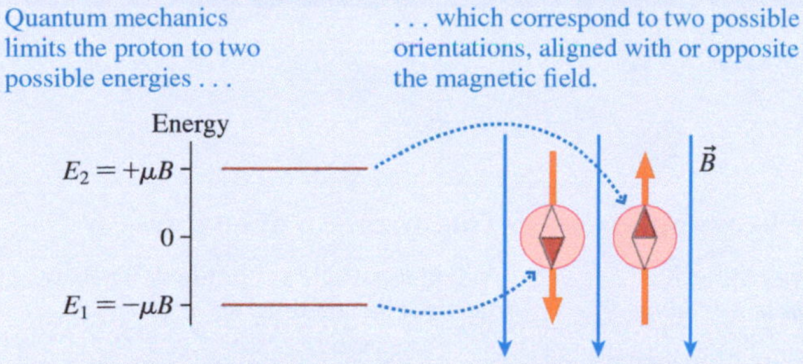


Human tissue is mostly water. Each water molecule has two hydrogen atoms whose nuclei are single protons. In a magnetic field, the protons go into one or the other quantum state. A photon of just the right energy can "flip" the orientation of a proton's magnetic moment by causing a quantum jump from one state to the other. The energy difference between the states is small, so the relatively low-frequency photons are in the radio portion of the electromagnetic spectrum. These photons are provided by a *probe coil* that emits radio waves. When the probe is tuned to just the right frequency, the waves are *in resonance* with the energy levels of the protons, thus giving us the name magnetic *resonance* imaging.

The rate of absorption of these low-energy photons is proportional to the density of hydrogen atoms. Hydrogen density varies with tissue type, so an MRI image—showing different tissues—is formed by measuring the variation across the body of the rate at which photons cause quantum jumps between the two proton energy levels. A figure showing the absorption rate versus position in the body is an image of a "slice" through a patient's body, as in FIGURE 28.25.

FIGURE 28.25 An MRI image shows the cross section of a patient's head.

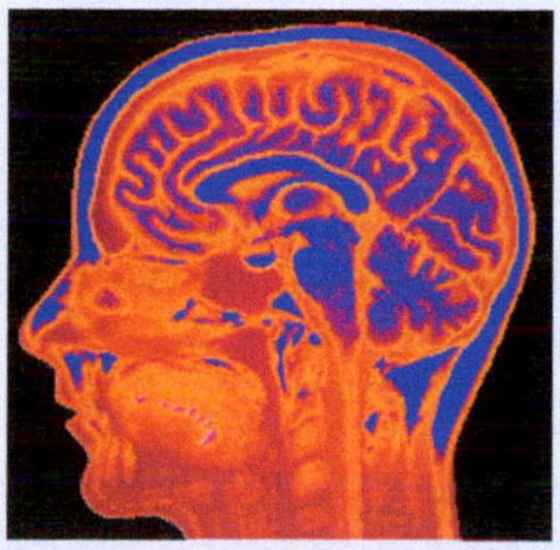

a. An MRI patient is placed inside a solenoid that creates a strong magnetic field. If the field strength is 2.00 T, to what frequency must the probe coil be set? What is the wavelength of the photons produced?
b. In a uniform magnetic field, all protons in the body would absorb photons of the same frequency. To form an image of the body, the magnetic field is designed to vary from point to point in a known way. Because the field is different at each point in the body, each point has a unique frequency of photons that will be absorbed. The actual procedure is complex, but consider a simple model in which the field strength varies only along the axis of the patient's body, which we will call the $x$-axis. In particular, suppose that the magnetic field strength in tesla is given by $B = 2.00 + 1.60x$, where $x$, measured from a known reference point, is in meters. The probe coil is first tuned to the resonance frequency at the reference point. As the frequency is increased, a strong signal is observed at a frequency 4.7 MHz above the starting frequency. What is the location in the body, relative to the reference point, of the tissue creating this strong signal?

**STRATEGIZE** If a photon's energy equals the energy difference between the high- and low-energy proton states, the photon will be able to cause the proton to undergo a quantum jump to the higher state—the photon will be absorbed. We can determine the energy difference between the two proton states, and thus we can determine the photon frequency necessary to cause a quantum jump. If the magnetic field varies, so will the energy difference. Photons of a particular frequency will be able to cause transitions only at locations where the field has a certain value.

**PREPARE** The photon energy $E_{\text{photon}}$ must be equal to the energy difference between the two states: $\Delta E_{\text{system}} = 2\mu B$. If we know the field, we can determine the correct photon frequency; if we know the photon frequency, we can use this to determine the value of the field.

**SOLVE**

a. At 2.00 T, the energy difference between the two proton states is

$$\Delta E_{\text{system}} = E_2 - E_1 = 2\mu B = 2\,(1.41 \times 10^{-26}\ \text{J/T})(2.00\ \text{T})$$
$$= 5.64 \times 10^{-26}\ \text{J}$$

This is a very low energy—only $3.5 \times 10^{-7}\ \text{eV}$. A photon will be absorbed if $E_{\text{photon}} = hf = E_{\text{system}}$. Thus the photon frequency must be

$$f = \frac{\Delta E_{\text{system}}}{h} = \frac{5.64 \times 10^{-26}\ \text{J}}{6.63 \times 10^{-34}\ \text{J} \cdot \text{s}}$$
$$= 85.1 \times 10^6\ \text{Hz} = 85.1\ \text{MHz}$$

This corresponds to a wavelength of $\lambda = c/f = 3.53$ m.

b. The magnetic field at the reference point ($x = 0$ m) is 2.00 T, so the probe frequency at this point is the 85.1 MHz we found in part a. The strong signal is 4.7 MHz above this, or 89.8 MHz. We can solve $E_{\text{photon}} = hf = E_{\text{system}} = 2\mu B$ to find the magnetic field at the point creating this strong signal:

$$B = \frac{hf}{2\mu} = 2.11\ \text{T}$$

We can then use the field-versus-distance formula given in the problem to find the position of this signal:

$$B = 2.11\ \text{T} = 2.00 + 1.60x$$
$$x = 0.069\ \text{m}$$

Thus there is a high density of protons 6.9 cm from the reference point.

**ASSESS** The frequency of the probe coil is in the radio portion of the electromagnetic spectrum, as we expected. The strong signal of part b is at a higher frequency, so this corresponds to a higher field and a positive value of $x$, as we found. The frequency is only slightly different from the original frequency, so we expect the point to be close to the reference position, as we found.

This is a simplified model of MRI, but the key features are present: A magnetic field that varies with position creates different energy levels for protons at different positions in the body, then tuned radio-wave photons measure the proton density at these different positions by causing and detecting quantum jumps between the two proton energy levels.

# SUMMARY

**GOAL** To understand the quantization of energy for light and matter.

## GENERAL PRINCIPLES

### Light has particle-like properties

- The energy of a light wave comes in discrete packets (light quanta) we call **photons.**
- For light of frequency $f$, the energy of each photon is $E = hf$, where $h = 6.63 \times 10^{-34}\ \text{J} \cdot \text{s}$ is **Planck's constant**.
- When light strikes a metal surface, all of the energy of a single photon is given to a single electron.

### Matter has wave-like properties

- The **de Broglie wavelength** of a particle of mass $m$ is $\lambda = h/mv$.
- The wave-like nature of matter is seen in the interference patterns of electrons, protons, and other particles.

### Quantization of energy

When a particle is confined, it sets up a de Broglie standing wave.

The fact that standing waves can have only certain allowed wavelengths leads to the conclusion that a confined particle can have only certain allowed energies.

### Wave–particle duality

- Experiments designed to measure wave properties will show the wave nature of light and matter.
- Experiments designed to measure particle properties will show the particle nature of light and matter.

### Heisenberg uncertainty principle

A particle with wave-like characteristics does not have a precise value of position $x$ or a precise value of momentum $p_x$. Both are uncertain. The position uncertainty $\Delta x$ and momentum uncertainty $\Delta p_x$ are related by

$$\Delta x\, \Delta p_x \geq \frac{h}{4\pi}$$

The more you pin down the value of one, the less precisely the other can be known.

## IMPORTANT CONCEPTS

### Photoelectric effect

Light with frequency $f$ can eject electrons from a metal only if $f \geq f_0 = E_0/h$, where $E_0$ is the metal's **work function.** Electrons will be ejected even if the intensity of the light is very small.

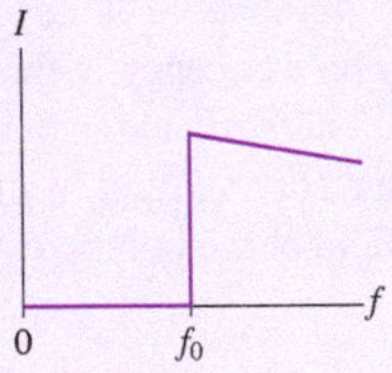


The **stopping potential** that stops even the fastest electrons is

$$V_{\text{stop}} = \frac{K_{\max}}{e} = \frac{hf - E_0}{e}$$

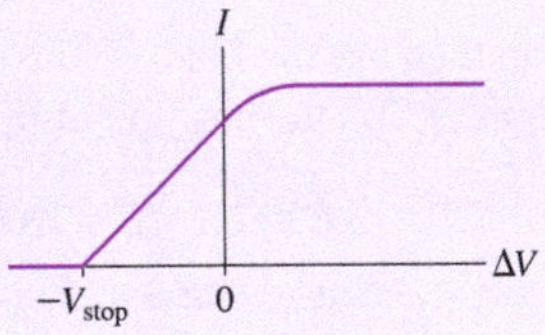


The details of the photoelectric effect could not be explained with classical physics. New models were needed.

### X-ray diffraction

X rays with wavelength $\lambda$ undergo strong reflections from atomic planes spaced by $d$ when the angle of incidence satisfies the **Bragg condition:**

$$2d \cos\theta = m\lambda$$
$$m = 1, 2, 3, \ldots$$

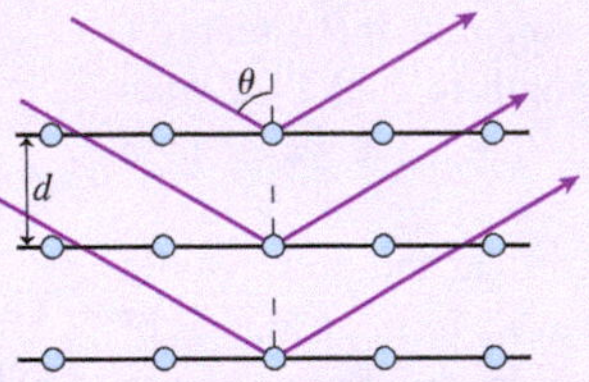


### Energy levels and quantum jumps

The localization of electrons leads to quantized energy levels. An electron can exist only in certain energy states. An electron can jump to a higher level if a photon is absorbed, or to a lower level if a photon is emitted. The energy difference between the levels equals the photon energy.

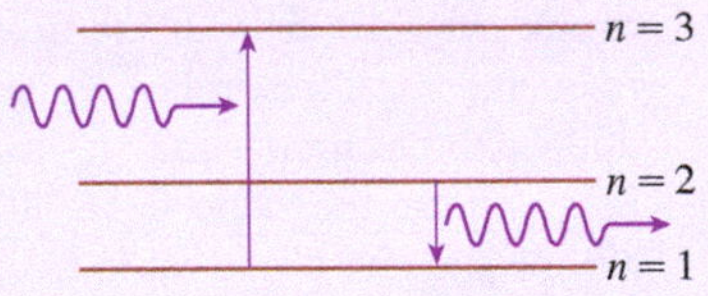


## APPLICATIONS

The wave nature of light limits the resolution of a light microscope. A more detailed image may be made with an **electron microscope** because of the very small de Broglie wavelength of fast electrons.

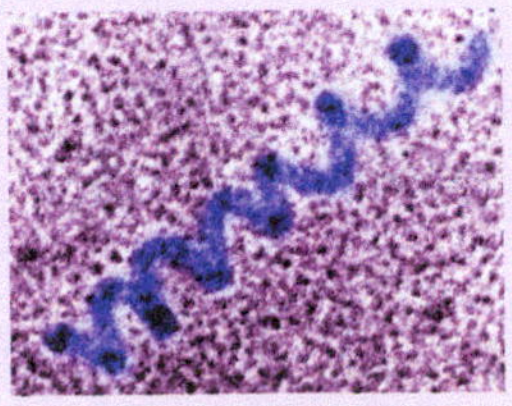

The wave nature of electrons allows them to **tunnel** across an insulating gap to the tip of a **scanning tunneling microscope,** revealing details of the atoms on a surface.

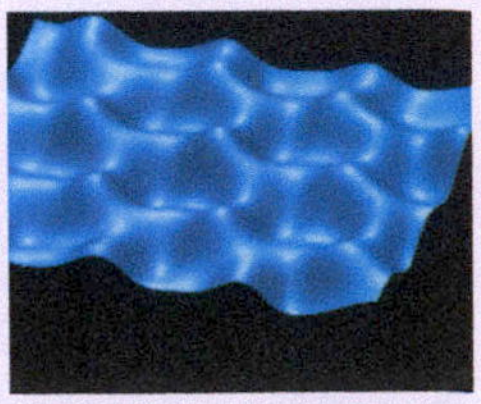

## Learning Objectives After studying this chapter, you should be able to:

- Solve problems about x-ray diffraction. *Conceptual Questions 28.1, 28.2; Problems 28.1, 28.2, 28.3, 28.4, 28.5*
- Understand the significance of the photoelectric effect and use Einstein's postulates. *Conceptual Questions 28.5, 28.6; Problems 28.8, 28.10, 28.12, 28.14, 28.15*
- Apply the photon model of light. *Conceptual Questions 28.11, 28.12; Problems 28.18, 28.19, 28.22, 28.24, 28.26*
- Recognize the experimental evidence for the wave nature of matter, and calculate the de Broglie wavelength. *Conceptual Questions 28.15, 28.16; Problems 28.33, 28.34, 28.35, 28.37, 28.38*
- Work with quantized energy levels and quantum jumps. *Conceptual Questions 28.21, 28.25; Problems 28.43, 28.44, 28.46, 28.47, 28.49*
- Apply the uncertainty principle. *Conceptual Question 28.27; Problems 28.53, 28.54, 28.55, 28.56, 28.57*

## STOP TO THINK ANSWERS

**Chapter Preview Stop to Think: D.** The fringes result from the interference of the waves spreading out from each of the two slits. With only one slit there is no interference, and only a broad, spread-out pattern is observed.

**Stop to Think 28.1: B.** The Bragg condition $2d\cos\theta_1 = \lambda$ tells us that larger values of $d$ go with larger values of $\theta_1$.

**Stop to Think 28.2: $V_A > V_B > V_C$.** For a given wavelength of light, electrons are ejected faster from metals with smaller work functions because it takes less energy to remove an electron. Faster electrons need a larger negative voltage to stop them.

**Stop to Think 28.3: C.** Photons always travel at $c$, and a photon's energy depends on only the light's frequency, not its intensity. Greater intensity means more energy each second, which means more photons.

**Stop to Think 28.4: A.** The widest diffraction pattern occurs for the largest wavelength. The de Broglie wavelength is inversely proportional to the particle's mass, and so will be largest for the least massive particle.

**Stop to Think 28.5: B.** The quantum number $n$ is related to the wavelength as $\lambda_n = 2L/n$. The wave shown in the figure has a wavelength equal to $L$, so for this wave $n = 2$. The lowest energy is the ground state with $n = 1$. According to Equation 28.15, the $n = 1$ state has 1/4 the energy of the $n = 2$ state, or 2.0 eV.

**Stop to Think 28.6: B, D, and E.** The emitted photons' energies must be the difference in energies between two allowed energy levels. Thus $E_3 - E_2 = 50$ eV, $E_4 - E_3 = 70$ eV, and $E_3 - E_1 = 80$ eV are possible photon energies. 40 eV and 60 eV do not correspond to the difference between any two levels.

**Stop to Think 28.7: No.** The energy of an emitted photon is the energy *difference* between two allowed energies. The three possible quantum jumps have energy differences of 2.0 eV, 2.0 eV, and 4.0 eV.

**Stop to Think 28.8: B.** Because $\Delta p_x = m\,\Delta v_x$, the uncertainty in position is $\Delta x = \dfrac{h}{\Delta p_x} = \dfrac{h}{m\,\Delta v_x}$. A more massive particle has a smaller position uncertainty.

Video Tutor Solution Chapter 28

# QUESTIONS

## Conceptual Questions

1. If the photon energy of an x-ray beam is increased, how does this affect the angle for a particular order or constructive interference from a crystal?
2. X rays of increasing wavelength are reflected from a crystal. Is there a threshold wavelength above which no constructive interference is observed?
3. Explain the reasoning by which we claim that the stopping potential $V_{stop}$ measures the maximum kinetic energy of the electrons in a photoelectric-effect experiment.
4. How does Einstein's explanation account for each of these characteristics of the photoelectric effect?
   a. The photoelectric current is zero for frequencies below some threshold.
   b. The photoelectric current increases with increasing light intensity.
   c. The photoelectric current is independent of $\Delta V$ for $\Delta V \gtrsim 1$ V.
   d. The photoelectric current decreases slowly as $\Delta V$ becomes more negative.
   e. The stopping potential is independent of the light intensity.

   Which of these *cannot* be explained by classical physics? Explain.

Problem difficulty is labeled as | (straightforward) to ||||| (challenging). Problems labeled INT integrate significant material from earlier chapters; Problems labeled BIO are of biological or medical interest.

The eText icon indicates when there is a video tutor solution available for the chapter or for a specific problem. To launch these videos, log into your eText through Mastering™ Physics or log into the Study Area.

5. A current is detected in a photoelectric-effect experiment when the cathode is illuminated with green light. Will a current necessarily be detected if the cathode is illuminated with blue light? With red light?
6. Figure Q28.6 shows the typical photoelectric behavior of a metal as the anode-cathode potential difference $\Delta V$ is varied.
   a. Why do the curves become horizontal for $\Delta V \gtrsim 1$ V? Shouldn't the current increase as the potential difference increases? Explain.
   b. Why doesn't the current immediately drop to zero for $\Delta V < 0$ V? Shouldn't $\Delta V < 0$ V prevent the electrons from reaching the anode? Explain.
   c. The current is zero for $\Delta V < -2.0$ V. Where do the electrons go? Are no electrons emitted if $\Delta V < -2.0$ V? Or if they are, why is there no current? Explain.

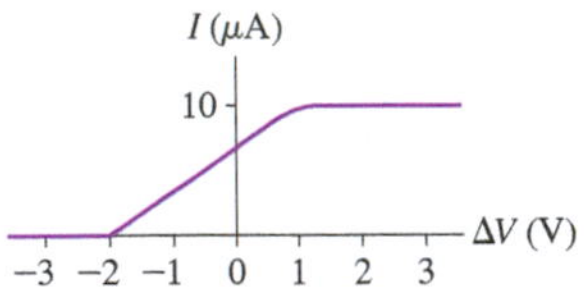

FIGURE Q28.6

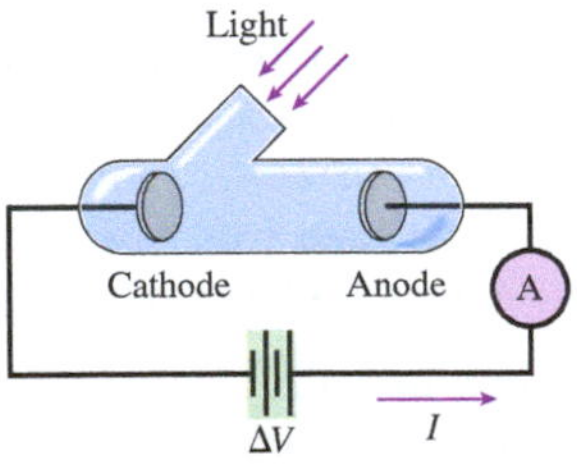

FIGURE Q28.7

7. In the photoelectric-effect experiment, as illustrated by Figure Q28.7, a current is measured while light is shining on the cathode. But this does not appear to be a complete circuit, so how can there be a current? Explain.
8. Metal surfaces on spacecraft in bright sunlight develop a net electric charge. Do they develop a negative or a positive charge? Explain.
9. Metal 1 has a larger work function than metal 2. Both are illuminated with the same short-wavelength ultraviolet light. Do electrons from metal 1 have a higher speed, a lower speed, or the same speed as electrons from metal 2? Explain.
10. A gold cathode is illuminated with light of wavelength 250 nm. It is found that the current is zero when $\Delta V = 1.0$ V. Would the current change if
    a. The light intensity is doubled?
    b. The anode-cathode potential difference is increased to $\Delta V = 5.5$ V?
11. Three laser beams have wavelengths $\lambda_1 = 400$ nm, $\lambda_2 = 600$ nm, and $\lambda_3 = 800$ nm. The power of each laser beam is 1 W.
    a. Rank in order, from largest to smallest, the photon energies $E_1$, $E_2$, and $E_3$ in these three laser beams. Explain.
    b. Rank in order, from largest to smallest, the number of photons per second $N_1$, $N_2$, and $N_3$ delivered by the three laser beams. Explain.
12. A beam of white light has the same intensity at all visible-light wavelengths. Does the beam contain more red-light photons or more blue-light photons?
13. An investigator is measuring the current in a photoelectric effect experiment. The cathode is illuminated by light of a single wavelength. What happens to the current if the wavelength of the light is reduced by a factor of two while keeping the intensity constant?
14. A photon striking a solar cell can set one electron in motion in a circuit. For a particular solar cell, the energy delivered to a circuit by each electron is 0.5 eV. Will the energy-conversion efficiency of the solar cell be greater for red light or for blue light?
15. For a given accelerating voltage, an electron microscope has a certain theoretical resolution. Suppose that instead of a beam of electrons you used a beam of protons, with the same accelerating voltage. How would this affect the theoretical resolution?
16. When you cool a gas, how does this affect the de Broglie wavelength of the gas atoms?
17. An electron and a proton are accelerated from rest through potential differences of the same magnitude. Afterward, which particle has the larger de Broglie wavelength? Explain.
18. A neutron is shot straight up with an initial speed of 100 m/s. As it rises, does its de Broglie wavelength increase, decrease, or not change? Explain.
19. Electrons are accelerated from rest through an 8000 V potential difference. By what factor would their de Broglie wavelength increase if they were instead accelerated through a 2000 V potential?
20. Can an electron with a de Broglie wavelength of 2 $\mu$m pass through a slit that is 1 $\mu$m wide? Explain.
21. a. For the allowed energies of a particle in a box to be large, should the box be very big or very small? Explain.
    b. Which is likely to have larger values for the allowed energies: an atom in a molecule, an electron in an atom, or a proton in a nucleus? Explain.
22. Figure Q28.22 shows the standing de Broglie wave of a particle in a box.
    a. What is the quantum number?
    b. Can you determine from this picture whether the "classical" particle is moving to the right or to the left? If so, which is it? If not, why not?

FIGURE Q28.22

23. A particle in a box of length $L_a$ has $E_1 = 2$ eV. The same particle in a box of length $L_b$ has $E_2 = 50$ eV. What is the ratio $L_a/L_b$?
24. Imagine that the horizontal box of Figure 28.18 is instead oriented vertically. Also imagine the box to be on a neutron star where the gravitational field is so strong that the particle in the box slows significantly, nearly stopping, before it hits the top of the box. Make a *qualitative* sketch of the $n = 3$ de Broglie standing wave of a particle in this box.
    **Hint:** The nodes are *not* uniformly spaced.
25. Figure Q28.25 shows a standing de Broglie wave.
    a. Does this standing wave represent a particle that travels back and forth between the boundaries with a constant speed or a changing speed? Explain.
    b. If the speed is changing, at which end is the particle moving faster and at which end is it moving slower?

FIGURE Q28.25

26. BIO The molecules in the rods and cones in the eye are tuned to absorb photons of particular energies. The retinal molecule, like many molecules, is a long chain. Electrons can freely move along one stretch of the chain but are reflected at the ends, thus behaving like a particle in a one-dimensional box. The absorption of a photon lifts an electron from the ground state into the first excited state. Do the molecules in a red cone (which are tuned to absorb red light) or the molecules in a blue cone (tuned to absorb blue light) have a longer "box"?

27. Science fiction movies often use devices that transport people and objects rapidly from one position to another. To "beam" people in this fashion means taking them apart atom by atom, carefully measuring each position, and then sending the atoms in a beam to the desired final location where they reassemble. How do the principles of quantum mechanics pose problems for this futuristic means of transportation?

## Multiple-Choice Questions

28. | A light sensor is based on a photodiode that requires a minimum photon energy of 1.7 eV to create mobile electrons. What is the longest wavelength of electromagnetic radiation that the sensor can detect?
    A. 500 nm  B. 730 nm
    C. 1200 nm  D. 2000 nm
29. | In a photoelectric-effect experiment, the frequency of the light is increased while the intensity is held constant. As a result,
    A. There are more electrons.B.
    B. The electrons are faster.
    C. Both A and B.
    D. Neither A nor B.
30. | In a photoelectric-effect experiment, the intensity of the light is increased while the frequency, which is above the threshold frequency, is held constant. As a result,
    A. There are more electrons.  B. The electrons are faster.
    C. Both A and B.  D. Neither A nor B.
31. | In the photoelectric effect, electrons are never emitted from a metal if the frequency of the incoming light is below a certain threshold value. This is because
    A. Photons of lower-frequency light don't have enough energy to eject an electron.
    B. The electric field of low-frequency light does not vibrate the electrons rapidly enough to eject them.
    C. The number of photons in low-frequency light is too small to eject electrons.
    D. Low-frequency light does not penetrate far enough into the metal to eject electrons.
32. || Visible light has a wavelength of about 500 nm. A typical radio wave has a wavelength of about 1.0 m. How many photons of the radio wave are needed to equal the energy of one photon of visible light?
    A. 2,000  B. 20,000
    C. 200,000  D. 2,000,000
33. | Two radio stations have the same power output from their antennas. One broadcasts AM at a frequency of 1000 kHz and one broadcasts FM at a frequency of 100 MHz. Which statement is true?
    A. The FM station emits more photons per second.
    B. The AM station emits more photons per second.
    C. The two stations emit the same number of photons per second.
34. | Light consisting of 2.7 eV photons is incident on a piece of potassium, which has a work function of 2.3 eV. What is the maximum kinetic energy of the ejected electrons?
    A. 2.3 eV  B. 2.7 eV
    C. 5.0 eV  D. 0.4 eV

Questions 35 and 36 concern the photoelectric effect in a cosmic context. Ultraviolet radiation striking dust grains composed of the moon's dominant rock type, feldspar (with work function 4.5 eV), leaves them with a small electric charge. Smaller grains levitate due to the resulting electrostatic repulsion, giving the surface a diffuse dust halo.

35. | What sign of charge do the dust grains have?
    A. Negative
    B. Positive
    C. Some grains are negative, some are positive.
36. || 200 nm ultraviolet strikes a dust grain and ejects an electron. What is the electron's kinetic energy?
    A. 1.7 eV  B. 4.5 eV
    C. 6.2 eV  D. 10.7 eV
37. | You shoot a beam of electrons through a double slit to make an interference pattern. After noting the properties of the pattern, you then double the speed of the electrons. What effect would this have?
    A. The fringes would get closer together.
    B. The fringes would get farther apart.
    C. The positions of the fringes would not change.
38. | Photon P in Figure Q28.38 moves an electron from energy level $n = 1$ to energy level $n = 3$. The electron jumps down to $n = 2$, emitting photon Q, and then jumps down to $n = 1$, emitting photon R. The spacing between energy levels is drawn to scale. What is the correct relationship among the wavelengths of the photons?

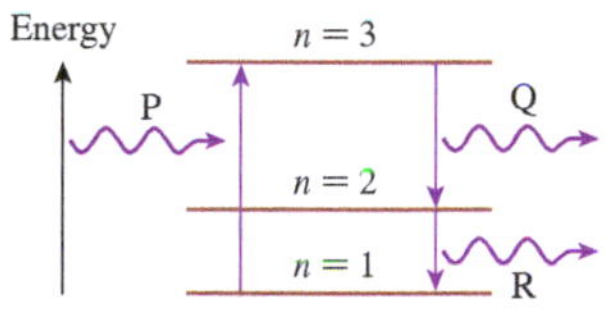

FIGURE Q28.38

    A. $\lambda_P < \lambda_Q < \lambda_R$  B. $\lambda_R < \lambda_P < \lambda_Q$
    C. $\lambda_Q < \lambda_P < \lambda_R$  D. $\lambda_R < \lambda_Q < \lambda_P$

# PROBLEMS

### Section 28.1 X Rays and X-Ray Diffraction

1. | X rays diffract from a crystal in which the spacing between atomic planes is 0.175 nm. The second-order diffraction occurs at 45.0°. What is the angle of the first-order diffraction?
2. || A crystal sample of bacteriorhodpsin, a light-sensitive pro-
BIO tein found in halobacteria that responds to light energy, has crystal planes separated by 0.20 nm. If a beam of 11 keV x rays illuminates a sample, what angles will give diffraction maxima?
3. | X rays with a wavelength of 0.20 nm undergo first-order diffraction from a crystal at a 54° angle of incidence. At what angle does first-order diffraction occur for x rays with a wavelength of 0.15 nm?
4. || The spacing between atomic planes in a crystal is 0.110 nm. If 12.0 keV x rays are diffracted by this crystal, what are the angles of (a) first-order and (b) second-order diffraction?
5. || X rays with a wavelength of 0.085 nm diffract from a crystal in which the spacing between atomic planes is 0.18 nm. How many diffraction orders are observed?

## Section 28.2 The Photoelectric Effect

6. | Which metals in Table 28.1 exhibit the photoelectric effect for (a) light with $\lambda = 400$ nm and (b) light with $\lambda = 250$ nm?
7. | Electrons are emitted when a metal is illuminated by light with a wavelength less than 388 nm but for no greater wavelength. What is the metal's work function?
8. || Electrons in a photoelectric-effect experiment emerge from a copper surface with a maximum kinetic energy of 1.10 eV. What is the wavelength of the light?
9. || A metal surface is illuminated by light with a wavelength of 350 nm. The maximum kinetic energy of the emitted electrons is found to be 1.50 eV. What is the maximum electron kinetic energy if the same metal is illuminated by light with a wavelength of 250 nm?
10. ||| Light with a wavelength of 375 nm illuminates a metal cathode. The maximum kinetic energy of the emitted electrons is 0.76 eV. What is the longest wavelength of light that will cause electrons to be emitted from this cathode?
11. || You need to design a photodetector that can respond to the entire range of visible light. What is the maximum possible work function of the cathode?
12. || A photoelectric-effect experiment finds a stopping potential of 1.93 V when light of 200 nm wavelength is used to illuminate the cathode.
    a. From what metal is the cathode made?
    b. What is the stopping potential if the intensity of the light is doubled?
13. || Zinc has a work function of 4.3 eV.
    a. What is the longest wavelength of light that will release an electron from a zinc surface?
    b. A 4.7 eV photon strikes the surface and an electron is emitted. What is the maximum possible speed of the electron?
14. || Image intensifiers used in night-vision devices create a bright image from dim light by letting the light first fall on a *photocathode*. Electrons emitted by the photoelectric effect are accelerated and then strike a phosphorescent screen, causing it to glow more brightly than the original scene. Recent devices are sensitive to wavelengths as long as 900 nm, in the infrared.

    a. If the threshold wavelength is 900 nm, what is the work function of the photocathode?
    b. If light of wavelength 700 nm strikes such a photocathode, what will be the maximum kinetic energy, in eV, of the emitted electrons?
15. || Light with a wavelength of 350 nm shines on a metal surface, which emits electrons. The stopping potential is measured to be 1.25 V.
    a. What is the maximum speed of emitted electrons?
    b. Calculate the work function and identify the metal.
16. || A nickel crystal's work function is measured to be 5.22 eV at 25°C. As the temperature increases by 300°C, the work function drops by 50 meV. By how much does this shift the threshold wavelength for photoelectric emission?
17. || Potassium and gold cathodes are used in a photoelectric-effect experiment. For each cathode, find:
    a. The threshold frequency
    b. The threshold wavelength
    c. The maximum electron ejection speed if the light has a wavelength of 220 nm
    d. The stopping potential if the wavelength is 220 nm

## Section 28.3 Photons

18. | BIO When an ultraviolet photon is absorbed by a molecule of DNA, the photon's energy can be converted into vibrational energy of the molecular bonds. Excessive vibration damages the molecule by causing the bonds to break. Ultraviolet light of wavelength less than 290 nm causes significant damage to DNA; ultraviolet light of longer wavelength causes minimal damage. What is the threshold photon energy, in eV, for DNA damage?
19. | BIO Your eyes have three different types of cones with maximum absorption at 437 nm, 533 nm, and 564 nm. What photon energies correspond to these wavelengths?
20. || BIO A firefly glows by the direct conversion of chemical energy to light. The light emitted by a firefly has peak intensity at a wavelength of 550 nm.

    a. What is the minimum chemical energy, in eV, required to generate each photon?
    b. One molecule of ATP provides 0.30 eV of energy when it is metabolized in a cell. What is the minimum number of ATP molecules that must be consumed in the reactions that lead to the emission of one photon of 550 nm light?
21. | What is the wavelength, in nm, of a photon with energy (a) 0.30 eV, (b) 3.0 eV, and (c) 30 eV? For each, is this wavelength visible light, ultraviolet, or infrared?
22. || BIO INT A study of photosynthesis in phytoplankton in the open ocean used short pulses of laser light to trigger photosynthetic reactions. The investigator's system used 0.10 mW pulses of 640 nm laser light of length 200 ps. How many photons were contained in each pulse?
23. || Station KAIM in Hawaii broadcasts on the AM dial at 870 kHz, with a maximum power of 50,000 W. At maximum power, how many photons does the transmitting antenna emit each second?
24. || BIO INT The wavelengths of light emitted by a firefly span the visible spectrum but have maximum intensity near 550 nm. A typical flash lasts for 100 ms and has a power of 1.2 mW. If we assume that all of the light is emitted at the peak-intensity wavelength of 550 nm, how many photons are emitted in one flash?
25. ||| BIO At 510 nm, the wavelength of maximum sensitivity of the human eye, the dark-adapted eye can sense a 100-ms-long flash of light of total energy $4.0 \times 10^{-17}$ J. (Weaker flashes of light may be detected, but not reliably.) If 60% of the incident light is lost to reflection and absorption by tissues of the eye, how many photons reach the retina from this flash?
26. || BIO INT One molecule of ATP provides 0.30 eV when it is used to power cellular processes. Photosynthesis in a typical plant requires 8 photons at 550 nm to produce 1 molecule of ATP. What is the overall efficiency of this process?
27. | 550 nm is the average wavelength of visible light.
    a. What is the energy of a photon with a wavelength of 550 nm?
    b. A typical incandescent lightbulb emits about 1 J of visible light energy every second. Estimate the number of visible photons emitted per second.

28. || *Dinoflagellates* are single-cell creatures that float in the world's oceans; many types are bioluminescent. When disturbed by motion in the water, a typical bioluminescent dinoflagellate emits 100,000,000 photons in a 0.10-s-long flash of light of wavelength 460 nm. What is the power of the flash in watts?
BIO

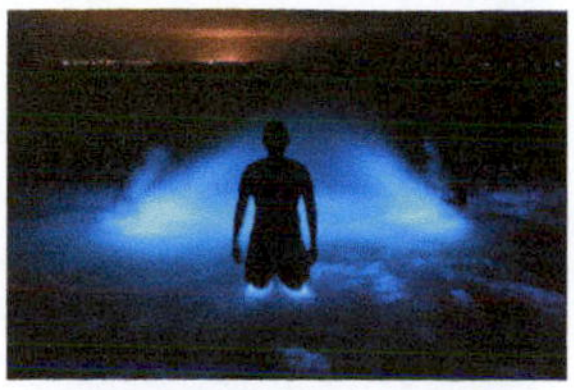

29. || A circuit employs a silicon solar cell to detect flashes of light lasting 0.25 s. The smallest current the circuit can detect reliably is 0.42 $\mu$A. Assuming that all photons reaching the solar cell give their energy to a charge carrier, what is the minimum power of a flash of light of wavelength 550 nm that can be detected?

30. || In a laser range-finding experiment, a pulse of laser light is fired toward an array of reflecting mirrors left on the moon by Apollo astronauts. By measuring the time it takes for the pulse to travel to the moon, reflect off the mirrors, and return to earth, scientists can calculate the distance to the moon to within a few centimeters. A single mirror receives 0.38 W of power during a 100-ps-long pulse of 532-nm-wavelength laser light. How many photons are in the pulse?

31. ||| Exposure to a sufficient quantity of ultraviolet light will redden the skin, producing *erythema*—a sunburn. The amount of exposure necessary to produce this reddening depends on the wavelength. For a 1.0 cm$^2$ patch of skin, 3.7 mJ of ultraviolet light at a wavelength of 254 nm will produce reddening; at 300 nm wavelength, 13 mJ are required.
BIO
   a. What is the photon energy corresponding to each of these wavelengths?
   b. How many total photons does each of these exposures correspond to?
   c. Explain why there is a difference in the number of photons needed to provoke a response in the two cases.

32. ||| A silicon solar cell behaves like a battery with a 0.50 V terminal voltage. Suppose that 1.0 W of light of wavelength 600 nm falls on a solar cell and that 50% of the photons give their energy to charge carriers, creating a current. What is the solar cell's efficiency—that is, what percentage of the energy incident on the cell is converted to electric energy?
INT

### Section 28.4 Matter Waves

33. | a. What is the de Broglie wavelength of a 200 g baseball with a speed of 30 m/s?
   b. What is the speed of a 200 g baseball with a de Broglie wavelength of 0.20 nm?

34. | a. What is the speed of an electron with a de Broglie wavelength of 0.20 nm?
   b. What is the speed of a proton with a de Broglie wavelength of 0.20 nm?

35. || What is the kinetic energy, in eV, of an electron with a de Broglie wavelength of 1.0 nm?

36. || A paramecium is covered with motile hairs called cilia that propel it at a speed of 1 mm/s. If the paramecium has a volume of $2 \times 10^{-13}$ m$^3$ and a density equal to that of water, what is its de Broglie wavelength when in motion? What fraction of the paramecium's 150 $\mu$m length does this wavelength represent?
BIO INT

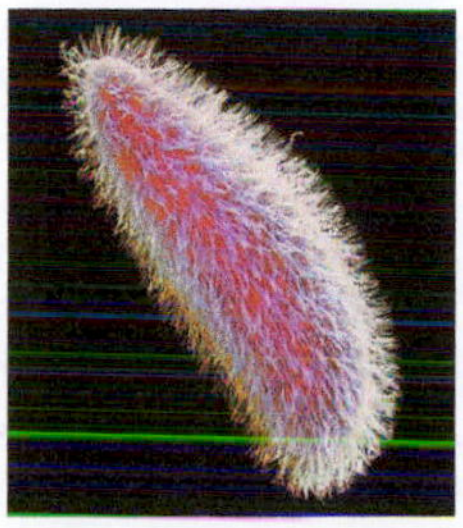

37. || The diameter of an atomic nucleus is about 10 fm ($1 \text{ fm} = 10^{-15}$ m). What is the kinetic energy, in MeV, of a proton with a de Broglie wavelength of 10 fm?

38. ||| At very low temperature, the de Broglie wavelengths of the atoms in a gas can become larger than the distance between atoms, which leads to new states of matter. In an early study, a gas of $^{23}$Na was cooled to 200 nK. What is the de Broglie wavelength of the sodium atoms at this temperature? To get a sense of scale, at everyday temperatures and pressures, the distance between air molecules is about 3 nm. How does the de Broglie wavelength compare to this distance? (The gas in the study was much more diffuse than this, but the atoms were close enough for some overlap.)
INT

39. || Through what potential difference must an electron be accelerated from rest to have a de Broglie wavelength of 500 nm?

40. ||| If a diffuse gas of atoms is trapped, cooled, and compressed, the de Broglie wavelengths of some atoms can become larger than the distance between atoms. The overlap between atoms leads to new states of matter. In one experiment, $^{87}$Rb atoms were cooled to 20 nK. What is their de Broglie wavelength at this temperature? How does this compare to the 0.47 nm diameter of the atoms?
INT

41. || What is the de Broglie wavelength of a red blood cell with a mass of $1.00 \times 10^{-11}$ g that is moving with a speed of 0.400 cm/s? Do we need to be concerned with the wave nature of the blood cells when we describe the flow of blood in the body?
BIO

### Section 28.5 Energy Is Quantized

42. || What is the length of a box in which the minimum energy of an electron is $1.5 \times 10^{-18}$ J?

43. ||| What is the length of a one-dimensional box in which an electron in the $n = 1$ state has the same energy as a photon with a wavelength of 600 nm?

44. ||| An electron confined in a one-dimensional box is observed, at different times, to have energies of 12 eV, 27 eV, and 48 eV. What is the length of the box?

45. | The nucleus of a typical atom is 5.0 fm ($1 \text{ fm} = 10^{-15}$ m) in diameter. A very simple model of the nucleus is a one-dimensional box in which protons are confined. Estimate the energy of a proton in the nucleus by finding the first three allowed energies of a proton in a 5.0-fm-long box.

46. || Investigators have created structures consisting of linear chains of ionized atoms on a smooth surface. Electrons are restricted to travel along the chain. The energy levels of the electrons match the results of the particle-in-a-box model. For a 5.0-nm-long chain, what are the energies (in eV) of the first three states?

### Section 28.6 Energy Levels and Quantum Jumps

47. || Figure P28.47 is an energy-level diagram for a quantum system. What wavelengths appear in the system's emission spectrum?

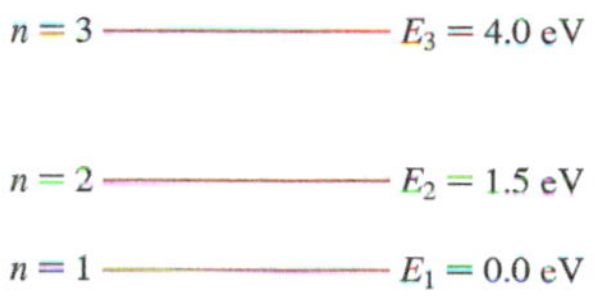


FIGURE P28.47

48. || The allowed energies of a quantum system are 1.0 eV, 2.0 eV, 4.0 eV, and 7.0 eV. What wavelengths appear in the system's emission spectrum?

Watch Video Solution Problems 28.49 and 28.65

49. || A quantum system has three energy levels, so three wavelengths appear in its emission spectrum. The shortest observed wavelength is 248 nm; light with a 414 nm wavelength is also observed. What is the third wavelength?

50. ||| The allowed energies of a quantum system are 0.0 eV, 4.0 eV, and 6.0 eV.
   a. Draw the system's energy-level diagram. Label each level with the energy and the quantum number.
   b. What wavelengths appear in the system's emission spectrum?

51. ||| The allowed energies of a quantum system are 0.0 eV, 1.5 eV, 3.0 eV, and 6.0 eV. How many different wavelengths appear in the emission spectrum?

52. ||| The color of dyes results from the preferential absorp-tion of certain wavelengths of light. Certain dye molecules consist of symmetric pairs of rings joined at the center by a chain of carbon atoms, as shown in Figure P28.52. Electrons of the bonds along the chain of carbon atoms are shared among the atoms in the chain, but are repelled by the nitrogen-containing rings at the end of the chain. These electrons are thus free to move along the chain but not beyond its ends. They look very much like a particle in a one-dimensional box. For the molecule shown, the effective length of the "box" is 0.85 nm. Assuming that the electrons start in the lowest energy state, what are the three longest wavelengths this molecule will absorb?

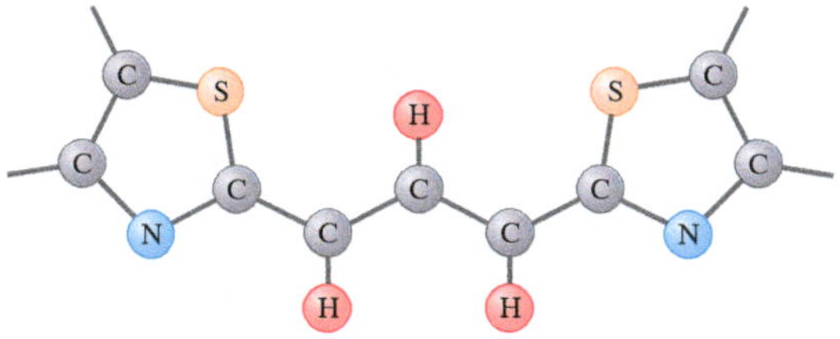


FIGURE P28.52

### Section 28.7 The Uncertainty Principle

53. || The speed of an electron is known to be between $3.0 \times 10^6$ m/s and $3.2 \times 10^6$ m/s. Estimate the smallest possible uncertainty in its position.

54. || What is the smallest box in which you can confine an electron if you want to know for certain that the uncertainty in the electron's speed is no more than 20 m/s?

55. ||| BIO INT A spherical virus has a diameter of 50 nm. It is contained inside a long, narrow cell of length $1 \times 10^{-4}$ m. What uncertainty does this imply for the velocity of the virus along the length of the cell? Assume the virus has a density equal to that of water.

56. ||| As computer structures get smaller and smaller, quantum rules start to create difficulties. Suppose electrons move through a channel in a microprocessor. If we know that an electron is somewhere along the 50 nm length of the channel, what is $\Delta v_x$? If we treat the electron as a classical particle moving at a speed at the outer edge of the uncertainty range, how long would it take to traverse the channel?

57. ||| A proton is confined within an atomic nucleus of diameter 4 fm ($1 \text{ fm} = 10^{-15}$ m). Estimate the uncertainty in speed for a proton in the nucleus.

## General Problems

58. || X rays with a wavelength of 0.0700 nm diffract from a crystal. Two adjacent angles of x-ray diffraction are 45.6° and 21.0°. What is the distance in nm between the atomic planes responsible for the diffraction?

59. |||| In a photoelectric-effect experiment, the maximum kinetic energy of electrons is 2.8 eV. When the wavelength of the light is increased by 50%, the maximum energy decreases to 1.1 eV. What are (a) the work function of the cathode and (b) the initial wavelength?

60. || Light of constant intensity but varying wavelength was used to illuminate the cathode in a photoelectric-effect experiment. The graph of Figure P28.60 shows how the stopping potential depended on the frequency of the light. What is the work function, in eV, of the cathode?

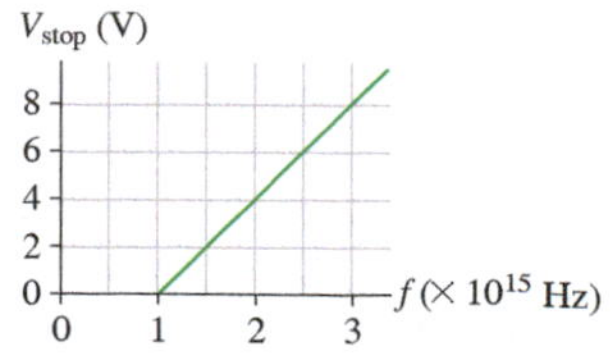


FIGURE P28.60

61. || BIO Suppose you need to image the structure of a virus with a diameter of 50 nm. For a sharp image, the wavelength of the probing wave must be 5.0 nm or less. We have seen that, for imaging such small objects, this short wavelength is obtained by using an electron beam in an electron microscope. Why don't we simply use short-wavelength electromagnetic waves? There's a problem with this approach: As the wavelength gets shorter, the energy of a photon of light gets greater and could damage or destroy the object being studied. Let's compare the energy of a photon and an electron that can provide the same resolution.
   a. For light of wavelength 5.0 nm, what is the energy (in eV) of a single photon? In what part of the electromagnetic spectrum is this?
   b. For an electron with a de Broglie wavelength of 5.0 nm, what is the kinetic energy (in eV)?

62. | Gamma rays are photons with very high energy.
   a. What is the wavelength of a gamma-ray photon with energy 625 keV?
   b. How many visible-light photons with a wavelength of 500 nm would you need to match the energy of this one gamma-ray photon?

63. || BIO INT A typical incandescent lightbulb emits approximately $3 \times 10^{18}$ visible-light photons per second. Your eye, when it is fully dark adapted, can barely see the light from an incandescent lightbulb 10 km away. How many photons per second are incident at the image point on your retina? The diameter of a dark-adapted pupil is 6 mm.

64. || BIO The intensity of sunlight hitting the surface of the earth on a cloudy day is about 0.50 kW/m$^2$. Assuming your pupil can close down to a diameter of 2.0 mm and that the average wavelength of visible light is 550 nm, how many photons per second of visible light enter your eye if you look up at the sky on a cloudy day?

65. ||| INT A red LED (light emitting diode) is connected to a battery; it carries a current. As electrons move through the diode, they jump between states, emitting photons in the process. Assume that each electron that travels through the diode causes the emission of a single 630 nm photon. What current is necessary to produce 5.0 mW of emitted light?

66. ||| INT Electrons with a speed of $2.0 \times 10^6$ m/s pass through a double-slit apparatus. Interference fringes are detected with a fringe spacing of 1.5 mm.
   a. What will the fringe spacing be if the electrons are replaced by neutrons with the same speed?
   b. What speed must neutrons have to produce interference fringes with a fringe spacing of 1.5 mm?

67. | The electron interference pattern of Figure 28.14 was made INT by shooting electrons with 50 keV of kinetic energy through two slits spaced 1.0 $\mu$m apart. The fringes were recorded on a detector 1.0 m behind the slits.
   a. What was the speed of the electrons? (The speed is large enough to justify using relativity, but for simplicity do this as a nonrelativistic calculation.)
   b. Figure 28.14 is greatly magnified. What was the actual spacing on the detector between adjacent bright fringes?
68. |||| It is stated in the text that special relativity must be used to INT calculate the de Broglie wavelength of electrons in an electron microscope. Let us discover how much of an effect relativity has. Consider an electron accelerated through a potential difference of $1.00 \times 10^5$ V.
   a. Using the Newtonian (nonrelativistic) expressions for kinetic energy and momentum, what is the electron's de Broglie wavelength?
   b. The de Broglie wavelength is $\lambda = h/p$, but the momentum of a relativistic particle is not $mv$. Using the relativistic expressions for kinetic energy and momentum, what is the electron's de Broglie wavelength?
69. || An electron confined to a one-dimensional box of length 0.70 nm jumps from the $n = 2$ level to the ground state. What is the wavelength (in nm) of the emitted photon?
70. || What is the length of a box in which the difference between an electron's first and second allowed energies is $1.0 \times 10^{-19}$ J?
71. || Two adjacent allowed energies of an electron in a one-dimensional box are 2.0 eV and 4.5 eV. What is the length of the box?
72. ||| An electron confined to a box has an energy of 1.28 eV. Another electron confined to an identical box has an energy of 2.88 eV. What is the smallest possible length for those boxes?
73. ||| It can be shown that the allowed energies of a particle of mass $m$ in a two-dimensional square box of sided $L$ are

$$E_{nl} = \frac{h^2}{8mL^2}(n^2 + l^2)$$

The energy depends on two quantum numbers, $n$ and $l$, both of which must have an integer value 1, 2, 3, . . . .
   a. What is the minimum energy for a particle in a two-dimensional square box of side $L$?
   b. What are the five lowest allowed energies? Give your values as multiples of $E_{min}$.
74. ||| An electron confined in a one-dimensional box emits a 200 nm photon in a quantum jump from $n = 2$ to $n = 1$. What is the length of the box?
75. ||| A proton confined in a one-dimensional box emits a 2.0 MeV gamma-ray photon in a quantum jump from $n = 2$ to $n = 1$. What is the length of the box?
76. ||| As an electron in a one-dimensional box of length 0.600 nm jumps between two energy levels, a photon of energy 8.36 eV is emitted. What are the quantum numbers of the two levels?
77. || Magnetic resonance is used in imaging; it is also a useful tool for analyzing chemical samples. Magnets for magnetic resonance experiments are often characterized by the proton resonance frequency they create. What is the field strength of an 800 MHz magnet?
78. ||| The electron has a magnetic moment, so you can do mag- INT netic resonance measurements on substances with unpaired electron spins. The electron has a magnetic moment $\mu = 9.3 \times 10^{-24}$ J/T. A sample is placed in a solenoid of length 15 cm with 1200 turns of wire carrying a current of 3.5 A. A probe coil provides radio waves to "flip" the spins. What is the necessary frequency for the probe coil?

## MCAT-Style Passage Problems

### Compton Scattering

Further support for the photon model of electromagnetic waves comes from *Compton scattering*, in which x rays scatter from electrons, changing direction and frequency in the process. Classical electromagnetic wave theory cannot explain the change in frequency of the x rays on scattering, but the photon model can.

Suppose an x-ray photon is moving to the right. It has a collision with a slow-moving electron, as in Figure P28.79. The photon transfers energy and momentum to the electron, which recoils at a high speed. The x-ray photon loses energy, and the photon energy formula $E = hf$ tells us that its frequency must decrease. The collision looks very much like the collision between two particles.

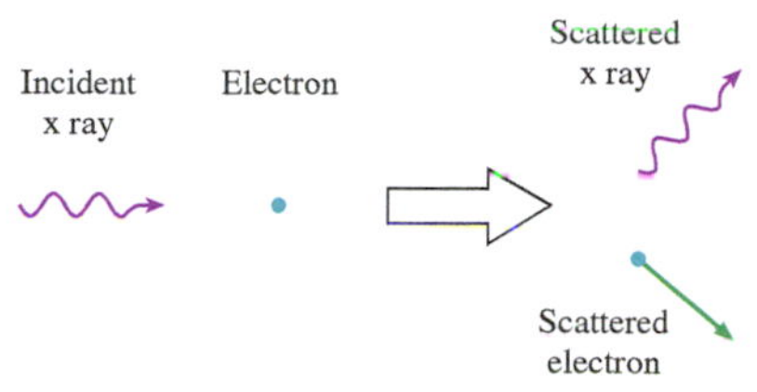


FIGURE P28.79

79. | When the x-ray photon scatters from the electron,
   A. Its speed increases.
   B. Its speed decreases.
   C. Its speed stays the same.
80. | When the x-ray photon scatters from the electron,
   A. Its wavelength increases.
   B. Its wavelength decreases.
   C. Its wavelength stays the same.
81. | When the electron is struck by the x-ray photon,
   A. Its de Broglie wavelength increases.
   B. Its de Broglie wavelength decreases.
   C. Its de Broglie wavelength stays the same.
82. | X-ray diffraction can also change the direction of a beam of x rays. Which statement offers the best comparison between Compton scattering and x-ray diffraction?
   A. X-ray diffraction changes the wavelength of x rays; Compton scattering does not.
   B. Compton scattering changes the speed of x rays; x-ray diffraction does not.
   C. X-ray diffraction relies on the particle nature of the x rays; Compton scattering relies on the wave nature.
   D. X-ray diffraction relies on the wave nature of the x rays; Compton scattering relies on the particle nature.

### Electron Diffraction

A series of experiments by Clinton Davisson and Lester Germer in the 1920s gave a clear indication of the wave nature of matter. The investigators scattered a relatively low energy electron beam from a nickel crystal. They found very strong reflections at certain angles that varied with the energy of the electron beam. The strong reflections were analogous to those observed in x-ray diffraction. The angles at which the intensity of the reflected beam peaks agreed with the Bragg condition if the electrons were assumed to have a wavelength given by the de Broglie formula. This was conclusive experimental proof of the wave nature of the electron.

Davisson and Germer used an electron beam that was directed perpendicular to the surface, as shown in Figure P28.83. They observed a particularly strong reflection, corresponding to $m = 1$ in the Bragg condition, at $\phi = 50°$. At this angle, the spacing between the scattering planes was $d = 0.091$ nm.

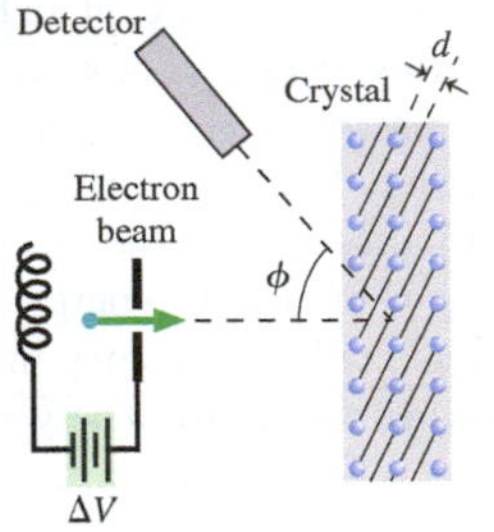


FIGURE P28.83 Electron beam scattered from a crystal.

83. | The angle $\phi$ in the figure is not the angle $\theta_1$ in the Bragg condition. For $\phi = 50°$, what is $\theta_1$?
   A. 25° B. 50° C. 65° D. 140°
84. ||| What is the de Broglie wavelength of electrons in the beam?
   A. 0.077 nm B. 0.120 nm C. 0.154 nm D. 0.165 nm
85. || What is the accelerating voltage?
   A. 45 V B. 55 V
   C. 65 V D. 75 V
86. | If the accelerating voltage is increased, what happens to the de Broglie wavelength of the electrons?
   A. It increases.
   B. It does not change.
   C. It decreases.
87. | If an investigator wanted to reproduce the results with a beam of protons, to get a strong $m = 1$ reflection at $\phi = 50°$ would require
   A. Higher speed particles than for the electron beam.
   B. Particles moving at the same speed as for the electron beam.
   C. Lower speed particles than for the electron beam.

## Collisional Excitation

Electrically excited mercury atoms have particularly strong emission of 4.9 eV photons, corresponding to a transition of one of the atom's electrons from a higher energy state to a lower. Mercury vapor absorbs light of this wavelength as well; the energy of the photon moves an electron from the lower state to the higher.

It's also possible to produce an excitation through other means. If an electron with a kinetic energy of 4.9 eV strikes a mercury atom, it can transfer energy, moving an electron in the mercury atom from the lower level to the higher. The electron loses kinetic energy in the process; this is an inelastic collision. Electrons with kinetic energies lower than this transition energy undergo elastic collisions, leaving their kinetic energy unchanged.

This was the idea behind the Franck-Hertz experiment, a classic experiment of early-20th-century physics. The basic setup is illustrated in Figure P28.88a. A tube is filled with mercury vapor. A heated electrode emits slow-moving electrons, and a variable power supply provides a voltage to accelerate the electrons toward a second electrode.

The current varies as the voltage is changed, as shown in Figure P28.88b. As the voltage is increased, this initially leads to an increased current. At a certain point, the electrons have enough energy to excite the mercury atoms, and the collisions lead to a loss in the energy of the moving electrons and a reduction in current, a clear demonstration of the existence of quantized energy levels in the mercury atom.

In modern versions of this experiment performed in student laboratories, the onset of inelastic collisions that excite the mercury atoms leads to a visible glow in the tube. This is illustrated in Figure P28.88a. The position of the glowing gas shows the location in the tube where the accelerating electrons reach the proper energy.

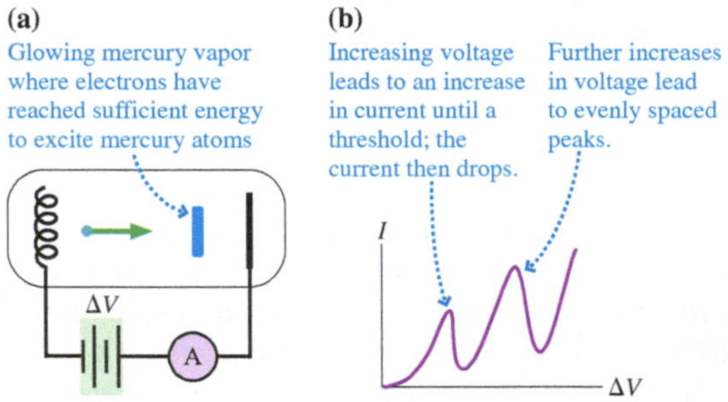


FIGURE P28.88 The setup of the Franck-Hertz experiment and typical data.

88. | If the temperature of the heated electrode is increased, more electrons are emitted. What change does this produce in the graph in Figure P28.88b?
   A. The peaks get taller.
   B. The peaks get shorter.
   C. The peaks occur at higher voltages.
   D. The peaks occur at lower voltages.
89. | What is the approximate wavelength of the 4.9 eV photons?
   A. 250 nm B. 300 nm
   C. 350 nm D. 400 nm
   E. 450 nm
90. || Approximately how fast must an electron move to excite a mercury atom in a collision?
   A. $9.3 \times 10^5$ m/s B. $1.3 \times 10^6$ m/s
   C. $1.9 \times 10^6$ m/s D. $2.6 \times 10^6$ m/s
91. || If the voltage is increased past the first threshold, the current drops and a visibly glowing region appears in the tube, as in Figure P28.88a. If the voltage is increased further, this glowing region will
   A. Move to the right, toward the positive electrode.
   B. Move to the left, toward the negative electrode.
   C. Stay at the same position.
92. || If the electrons that lose energy to a mercury atom continue to accelerate, they can acquire enough kinetic energy to excite another atom, leading to a drop in current. At what voltage does the drop in current begin?
   A. 4.9 V B. 7.4 V
   C. 9.8 V D. 14.7 V

# 29 Atoms and Molecules

A protein from this species of jellyfish glows green when illuminated with short-wavelength light. How does this shift in wavelength occur, and how is this molecule used in biological research?

## LOOKING AHEAD ▸

### Atomic Models

When light from a glowing gas is dispersed into colors, each element is seen to emit a unique set of wavelengths.

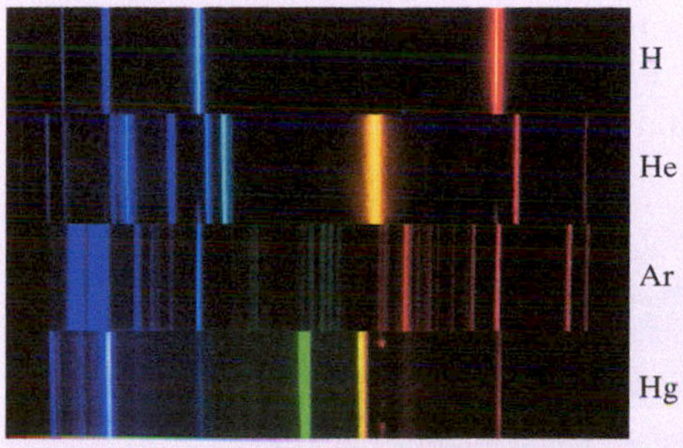


You'll learn how these **spectral lines** are important clues into the nature of the atom.

### Molecules

Specialized molecules in certain microscopic plankton can emit light. These glowing waves are caused by such *bioluminescent* creatures.

You'll learn why, unlike atoms, the light given off by molecules contains a broad *band* of wavelengths.

### Lasers

The intense beams of light at this laser light show result from the **stimulated emission** of photons by atoms.

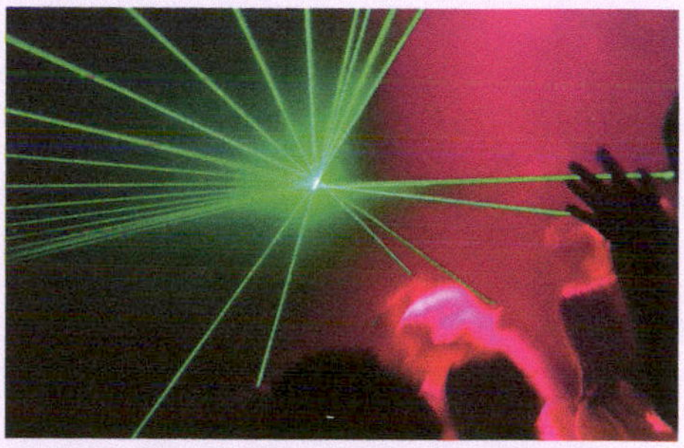

You'll learn that when atoms give off light, they can stimulate other atoms to emit light of exactly the same wavelength and phase.

**GOAL** **To use quantum physics to understand the properties of atoms, molecules, and their spectra.**

## LOOKING BACK ◂

### Energy-Level Diagrams

In Sections 28.5 and 28.6 you learned that the energies of atomic-sized systems are *quantized*. The allowed energies of such a system can be represented in an energy-level diagram.

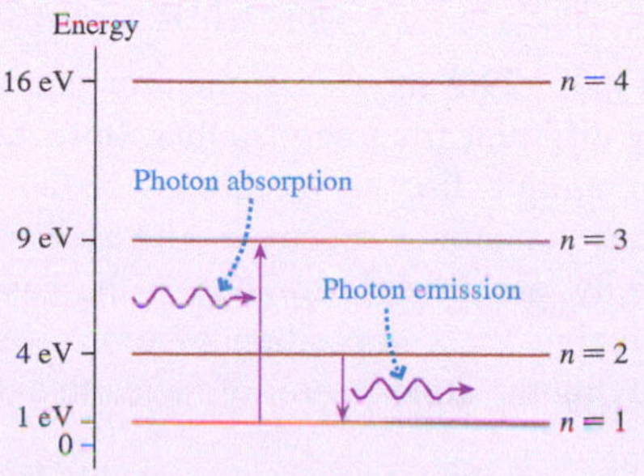


In this chapter, you'll use the ideas of quantization to understand in detail the energy levels and spectra of atoms and molecules.

**STOP TO THINK**

After the $n = 1$ to $n = 3$ absorption shown in the energy-level diagram to the left, which is a possible energy of an emitted photon?

A. 1 eV
B. 4 eV
C. 5 eV
D. 9 eV

## 29.1 Spectroscopy

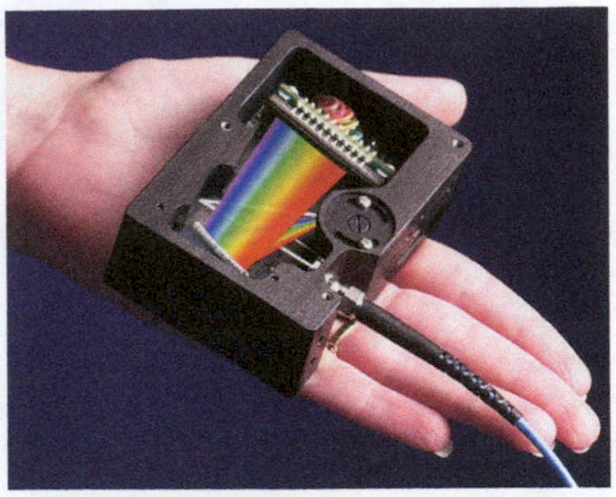

Some modern spectrometers are small enough to hold in your hand. (The rainbow has been added to show the paths that different colors take.)

FIGURE 29.1 A diffraction spectrometer.

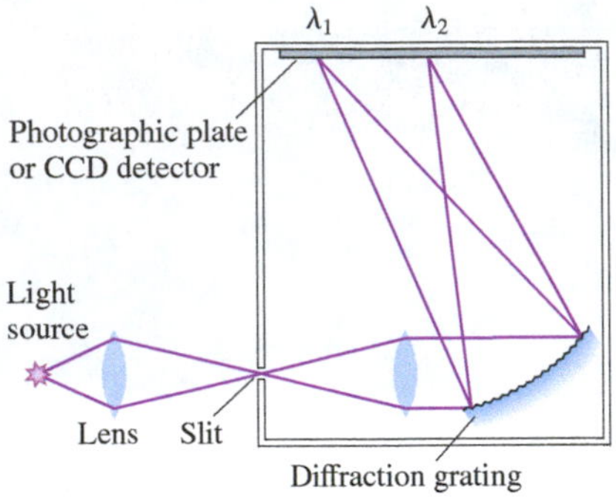


In Chapter 28 we took a historical approach, considering pivotal experiments and the developing understanding of quantum physics. We'll take a similar approach in this chapter as we explore the development of models of the atom.

Our story starts with the development of tools for studying light and the use of light as a tool for understanding matter. The interference and diffraction of light were well understood by the end of the 19th century, and the knowledge was used to design practical tools for measuring wavelengths with great accuracy. The primary instrument for measuring the wavelengths of light is a **spectrometer,** such as the one shown in FIGURE 29.1. The heart of a spectrometer is a diffraction grating that causes different wavelengths of light to diffract at different angles.

Each wavelength is focused to a different position on the photographic plate or, more likely today, a CCD detector like the one in your digital camera. The distinctive pattern of wavelengths emitted by a source of light and recorded on the detector is called the **spectrum** of the light.

As you learned in ◀◀ SECTION 25.7, hot, self-luminous objects, such as the sun or the filament of an incandescent lightbulb, emit a **continuous spectrum** in which a rainbow is formed by light being emitted at every possible wavelength. As you learned, this spectrum depends on only the object's temperature and thus contains no information about the atoms that the object is made of.

To learn about the light emitted and absorbed by individual atoms, we need to investigate them in the form of a low-pressure gas, where the atoms are far apart and isolated from one another. If a high voltage is applied to two electrodes sealed in a glass tube filled with a low-pressure gas, the gas begins to glow. In contrast to a hot solid object like a lightbulb filament, the light emitted by such a *gas discharge tube* contains only certain discrete, individual wavelengths. Such a spectrum is called a **discrete spectrum.**

FIGURE 29.2 shows examples of discrete spectra as they appear on the detector of a spectrometer. Each bright line in a discrete spectrum, called a **spectral line,** represents *one* specific wavelength present in the light emitted by the source. A discrete spectrum is sometimes called a *line spectrum* because of its appearance on the detector. The familiar neon sign is actually a gas discharge tube that contains neon. Such a sign has a reddish-orange color because, as Figure 29.2 shows, nearly all of the wavelengths emitted by neon atoms fall within the wavelength range 600–700 nm that we perceive as orange and red.

FIGURE 29.2 Examples of spectra in the visible wavelength range 400–700 nm.

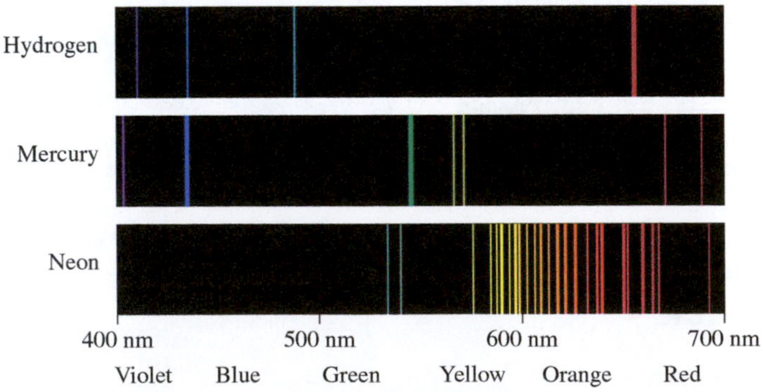


A video to support a section's topic is embedded in the eText.
Video Figure 29.2

Figure 29.2 shows that the *atomic spectra* of hydrogen, mercury, and neon look very different from one another. In fact, every element in the periodic table has its own, unique discrete spectrum. The fact that each element emits a unique spectrum means that atomic spectra can be used as "fingerprints" to identify elements. Consequently, atomic spectroscopy is the basis of many contemporary technologies for analyzing the composition of unknown materials, monitoring air pollutants, and studying the atmospheres of the earth and other planets.

Gases can also *absorb* light. In an absorption experiment, as shown in FIGURE 29.3a, a white-light source emits a continuous spectrum that, in the absence of a gas, uniformly illuminates the detector. When a sample of gas is placed in the light's path, any wavelengths absorbed by the gas are missing and the detector is dark at that wavelength.

Gases not only emit discrete wavelengths, but also absorb discrete wavelengths. But there is an important difference between the emission spectrum and the absorption spectrum of a gas: **Every wavelength that is absorbed by the gas is also emitted, but *not* every emitted wavelength is absorbed.** The wavelengths in the absorption spectrum appear as a *subset* of the wavelengths in the emission spectrum. As an example, FIGURE 29.3b shows both the absorption and the emission spectra of sodium atoms. All of the absorption wavelengths are prominent in the emission spectrum, but there are many emission lines for which no absorption occurs.

What causes atoms to emit or absorb light? Why a discrete spectrum? Why are some wavelengths emitted but not absorbed? Why is the spectrum emitted by each element different? These questions arose from experiments and explorations in the late 19th and early 20th centuries, but the physics of the time was incapable of providing answers.

In the search for explanations, the first step was to look for patterns in the data. While the spectra of other atoms have dozens or even hundreds of wavelengths, the visible spectrum of hydrogen, between 400 nm and 700 nm, consists of a mere four spectral lines (see Figure 29.2 and TABLE 29.1). If any spectrum could be understood, it should be that of the first element in the periodic table. The breakthrough came in 1885, not by an established and recognized scientist but by a Swiss school teacher, Johann Balmer. Balmer showed that the wavelengths in the hydrogen spectrum could be represented by the simple formula

$$\lambda = \frac{91.1\text{ nm}}{\left(\dfrac{1}{2^2} - \dfrac{1}{n^2}\right)} \qquad n = 3, 4, 5, \ldots \tag{29.1}$$

Later experimental evidence, as ultraviolet and infrared spectroscopy developed, showed that Balmer's result could be generalized to

$$\lambda = \frac{91.1\text{ nm}}{\left(\dfrac{1}{m^2} - \dfrac{1}{n^2}\right)} \tag{29.2}$$

$m$ can be 1, 2, 3, . . .
$n$ can be any integer greater than $m$.

We now refer to Equation 29.2 as the **Balmer formula,** although Balmer himself suggested only the original version in which $m = 2$. Other than at the very highest levels of resolution, the Balmer formula accurately describes *every* wavelength in the emission spectrum of hydrogen.

The Balmer formula is what we call *empirical knowledge*. It is an accurate mathematical representation found through experimental evidence, but it does not rest on any physical principles or physical laws. Yet the formula was so simple that it must, everyone agreed, have a simple explanation. It would take 30 years to find it.

**STOP TO THINK 29.1** The black lines show the emission or absorption lines observed in two spectra of the same element. Which one is an emission spectrum and which is an absorption spectrum?

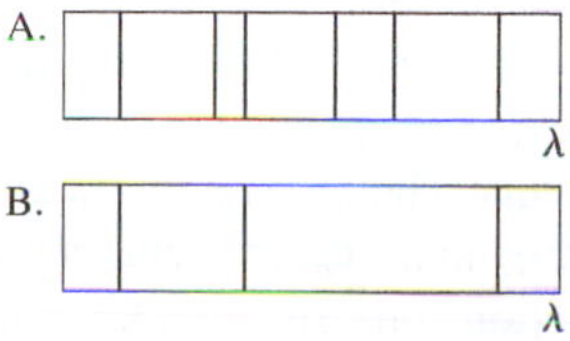


FIGURE 29.3 Measuring an absorption spectrum.

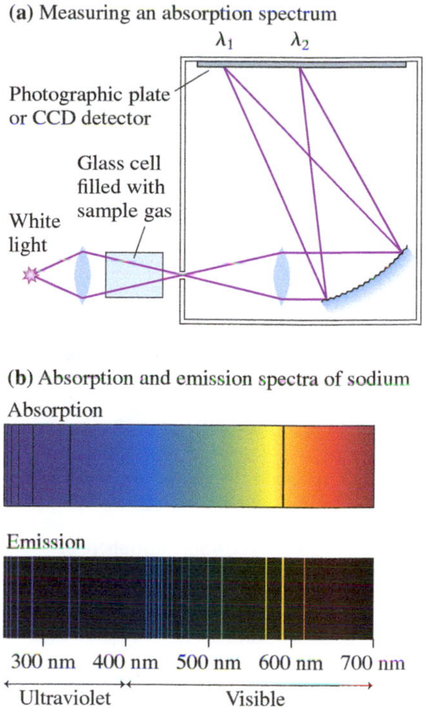


TABLE 29.1 Wavelengths of visible lines in the hydrogen spectrum

| |
|---|
| 656 nm |
| 486 nm |
| 434 nm |
| 410 nm |

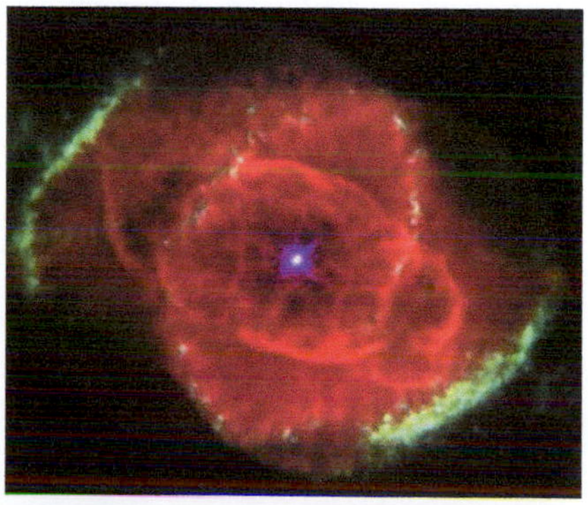

**Astronomical colors** The red color of this nebula is due to the emission of light from hydrogen atoms. The atoms are excited by intense ultraviolet light from the star in the center. They then emit red light, with $\lambda = 656$ nm, as predicted by the Balmer formula with $m = 2$ and $n = 3$.

## 29.2 Atoms

It was the ancient Greeks who first had the idea of atoms as indivisible units of matter, but experimental evidence for atoms didn't appear until Dalton, Avogadro, and others began to formulate the laws of chemistry in the early 19th century. The existence of atoms with diameters of approximately $10^{-10}$ m was widely accepted by 1890, but it was still unknown if atoms were indivisible little spheres or if they had some kind of internal structure.

The 1897 discovery of the electron by J. J. Thomson had two important implications. First, atoms are not indivisible; they are built of smaller pieces. The electron was the first *subatomic* particle to be discovered. And second, the constituents of the atom are *charged particles*. Hence it seems plausible that the atom must be held together by electric forces.

FIGURE 29.4 Thomson's raisin-cake model of the atom.

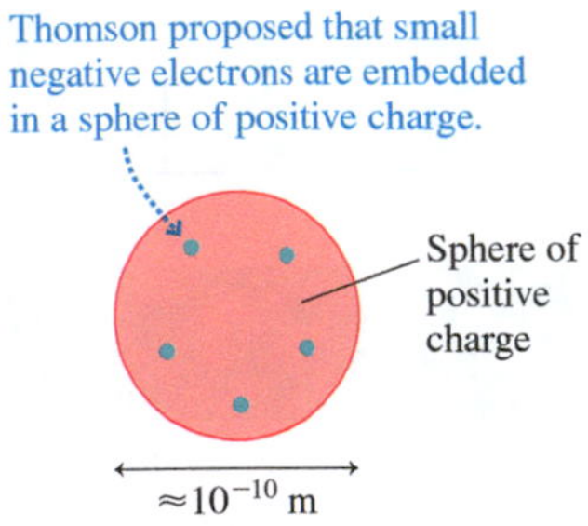


Within a few years, measurements of the electron's mass $m_e$ revealed that the electron is much less massive than even the smallest atom. Because the electrons are very small and light compared to the whole atom, it seemed reasonable to think that the positively charged part (protons were not yet known) had most of the mass and would take up most of the space. Thomson suggested that the atom consists of a spherical "cloud" of positive charge, roughly $10^{-10}$ m in diameter, in which the smaller negative electrons are embedded. The positive charge exactly balances the negative, so the atom as a whole has no net charge. This model of the atom has been called the "plum-pudding model" or the "raisin-cake model" for reasons that should be clear from the picture of FIGURE 29.4. However, Thomson's model of the atom did not stand the test of time.

Almost simultaneously with Thomson's discovery of the electron, the French physicist Henri Becquerel announced his discovery that some new form of "rays" were emitted by crystals of uranium. One of Thomson's former students, Ernest Rutherford, began a study of these new rays and discovered that a uranium crystal actually emits two *different* rays. **Beta rays** were eventually found to be high-speed electrons emitted by the uranium. **Alpha rays** (or alpha particles, as we now call them) consist of helium nuclei, with mass $m = 6.64 \times 10^{-27}$ kg, emitted at high speed from the sample.

**EXAMPLE 29.1 Finding the speed of an alpha particle**

The very high speeds of alpha particles make them suitable for experiments that probe the nature of matter. A nucleus ejects an alpha particle with a kinetic energy of 8.3 MeV, a typical energy. How fast is the alpha particle moving?

**STRATEGIZE** The kinetic energy is related to the speed. We'll assume that we can ignore relativity and use the classical formula for kinetic energy, $K = \frac{1}{2}mv^2$.

**PREPARE** We need to convert the kinetic energy from eV to J:

$$K = 8.3 \times 10^6 \text{ eV} \times \frac{1.60 \times 10^{-19} \text{ J}}{1.00 \text{ eV}} = 1.3 \times 10^{-12} \text{ J}$$

**SOLVE** Now, using the alpha-particle mass $m = 6.64 \times 10^{-27}$ kg given previously, we can find the speed:

$$K = \frac{1}{2}mv^2 = 1.3 \times 10^{-12} \text{ J}$$

$$v = \sqrt{\frac{2K}{m}} = 2.0 \times 10^7 \text{ m/s}$$

**ASSESS** This is quite fast, about 7% of the speed of light. We were told to expect a high speed, so this result makes sense. The speed isn't so large that we need to use a relativistic calculation, though; our calculation can stand.

### The First Nuclear Physics Experiment

Rutherford soon realized that he could use these high-speed alpha particles as projectiles to probe inside other atoms. In 1909, Rutherford and his students set up the experiment shown in FIGURE 29.5 to shoot alpha particles through very thin metal foils. The alpha particle is charged, and it experiences electric forces from the positive and negative charges of the atoms as it passes through the foil. According to Thomson's raisin-cake model of the atom, the forces exerted on the alpha particle by the positive atomic charges should roughly cancel the forces from the negative electrons, causing the alpha particles to experience only slight deflections. Indeed, this was the experimenters' initial observation.

At Rutherford's suggestion, his students then set up the apparatus to see if any alpha particles were deflected at *large* angles. It took only a few days to find the answer. Not only were alpha particles deflected at large angles, but a very few were reflected almost straight backward toward the source!

How can we understand this result? FIGURE 29.6a shows that an alpha particle passing through a Thomson atom would experience only a small deflection. But if an atom has a small positive core, such as the one in FIGURE 29.6b, a few of the alpha particles can come very close to the core. Because the electric force varies with the inverse square of the distance, the very large force of this very close approach can cause a large-angle scattering or even a backward deflection of the alpha particle.

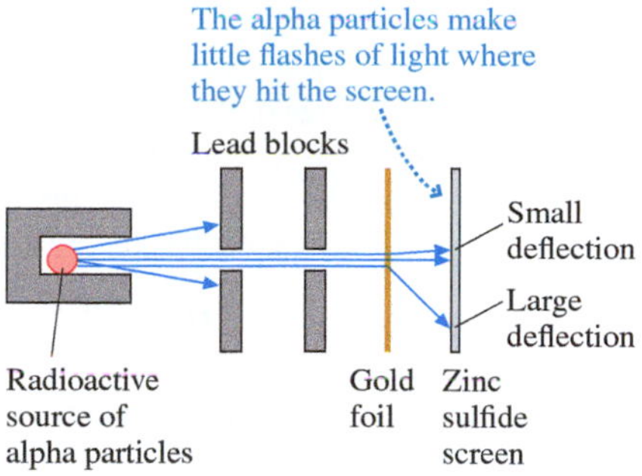

FIGURE 29.5 Rutherford's experiment to shoot high-speed alpha particles through a thin gold foil.

FIGURE 29.6 Alpha particles interact differently with a concentrated positive nucleus than they would with the spread-out charges in Thomson's model.

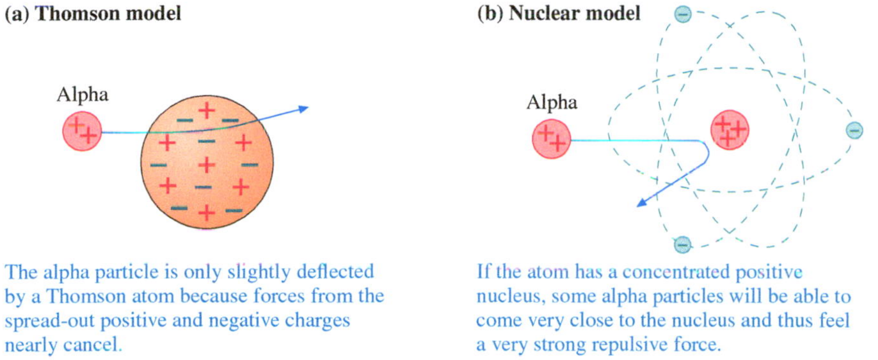

(a) Thomson model
Alpha
The alpha particle is only slightly deflected by a Thomson atom because forces from the spread-out positive and negative charges nearly cancel.
(b) Nuclear model
Alpha
If the atom has a concentrated positive nucleus, some alpha particles will be able to come very close to the nucleus and thus feel a very strong repulsive force.

The discovery of large-angle scattering of alpha particles led Rutherford to envision an atom in which negative electrons orbit a small, massive, positive **nucleus,** rather like a miniature solar system. This is the **nuclear model of the atom.** Further experiments showed that the diameter of the atomic nucleus is $\approx 1 \times 10^{-14}$ m $= 10$ fm ($1$ fm $= 1$ femtometer $= 10^{-15}$ m), a mere 0.01% the diameter of the atom itself. Thus nearly all of the atom is empty space—the void!

**EXAMPLE 29.2** **Going for the gold**

An 8.3 MeV alpha particle is shot directly toward the nucleus of a gold atom (atomic number 79). What is the distance of closest approach of the alpha particle to the nucleus?

***STRATEGIZE*** We can treat this problem using conservation of energy; it's similar to problems we saw in Chapter 21. As the positively charged alpha particle moves closer to the nucleus, the potential energy increases and the kinetic energy decreases. At its closest approach, the alpha particle will be at rest; the kinetic energy has been completely transformed into potential energy. We can ignore the interaction of the alpha particle with the gold atom's electrons; they are spread out over a distance that is very large compared to the size of the nucleus, and their mass is much less than that of the alpha particle. The particle will easily push the electrons aside with little change in its velocity.

***PREPARE*** The gold nucleus has about 50 times the mass of the alpha particle, so we can assume that, in this interaction, the gold nucleus does not move; we can consider the motion of the alpha particle in the potential of the stationary gold nucleus. Remember that the potential near a sphere of charge is the same as near a point charge. FIGURE 29.7 is a before-and-after visual overview of the situation. In the "before" picture, the alpha particle is very far from the gold nucleus, with a large kinetic energy. We assume that the distance is very large, so the potential energy can be assumed to be zero. In the "after" picture, the alpha particle is at rest at its distance of closest approach. The kinetic energy has gone to zero, with a corresponding increase in potential energy.

FIGURE 29.7 A before-and-after visual overview of an alpha particle colliding with a nucleus.

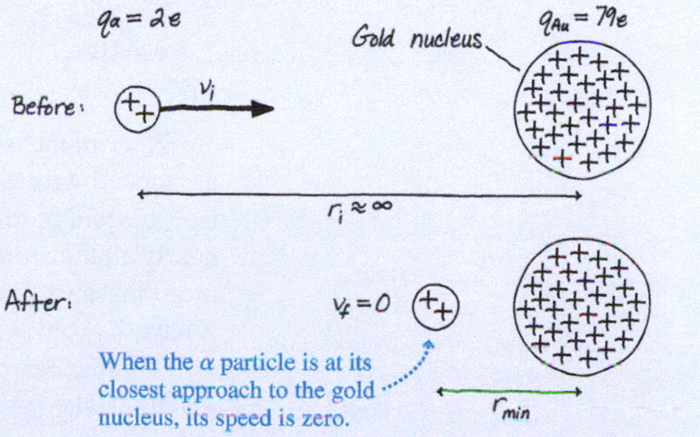


*Continued*

**SOLVE** Recall, from Chapter 21, that the electric potential near a sphere of charge is

$$V = K\frac{Q}{r}$$

where $Q$ is the charge of the sphere, and $r$ is the distance from the center. The sphere is the gold nucleus, so $Q = q_{Au}$; at the distance of closest approach, $r = r_{min}$. At this point, the alpha particle, with charge $q_\alpha$, has a potential energy

$$U = K\frac{q_\alpha q_{Au}}{r_{min}}$$

The conservation of energy statement $K_f + U_f = K_i + U_i$ is

$$0 + K\frac{q_\alpha q_{Au}}{r_{min}} = \frac{1}{2}mv_i^{\,2} + 0$$

where $q_\alpha$ is the alpha-particle charge and we've treated the gold nucleus as a point charge $q_{Au}$. The solution for $r_{min}$ is

$$r_{min} = K\frac{2q_\alpha q_{Au}}{mv_i^{\,2}}$$

The mass of the alpha particle is $m = 6.64 \times 10^{-27}$ kg and its charge is $q_\alpha = 2e = 3.20 \times 10^{-19}$ C. Gold has atomic number 79, so $q_{Au} = 79e = 1.26 \times 10^{-17}$ C. In Example 29.1 we found that an 8.3 MeV alpha particle has speed $v = 2.0 \times 10^7$ m/s. With this information, we can calculate

$$r_{min} = 2.7 \times 10^{-14}\ \text{m}$$

**ASSESS** This distance is much less than the size of the atom, on the scale of the size of the nucleus. The particle will get close enough to the nucleus to experience a very large force, as we expected.

## Using the Nuclear Model

The nuclear model of the atom makes it easy to picture atoms and understand such processes as ionization. For example, the **atomic number** of an element, its position in the periodic table of the elements, is the number of orbiting electrons (of a neutral atom) and the number of units of positive charge in the nucleus. The atomic number is represented by Z. Hydrogen, with $Z = 1$, has one electron orbiting a nucleus with charge $+1e$. Helium, with $Z = 2$, has two orbiting electrons and a nucleus with charge $+2e$. Because the orbiting electrons are very light, an x-ray photon or a rapidly moving particle, such as another electron, can knock one of the electrons away, creating a positive *ion.* Removing one electron makes a singly charged ion, with $q_{ion} = +e$. Removing two electrons creates a doubly charged ion, with $q_{ion} = +2e$. This is shown for lithium $(Z = 3)$ in FIGURE 29.8.

FIGURE 29.8 Different ionization stages of the lithium atom $(Z = 3)$.

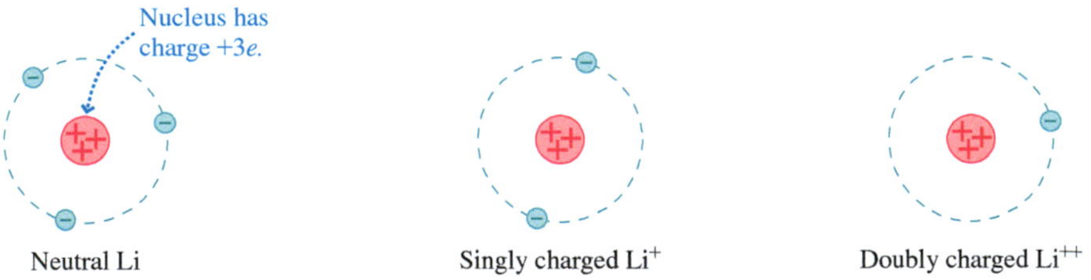


Experiments soon led to the recognition that the positive charge of the nucleus is associated with a positive subatomic particle called the **proton.** The proton's charge is $+e$, equal in magnitude but opposite in sign to the electron's charge. Further, with nearly all the atomic mass associated with the nucleus, the proton is about 1800 times more massive than the electron: $m_p = 1.67 \times 10^{-27}$ kg. Atoms with atomic number $Z$ have $Z$ protons in the nucleus, giving the nucleus charge $+Ze$.

This nuclear model of the atom allows us to understand why, during chemical reactions and when an object is charged by rubbing, electrons are easily transferred but protons are not. The protons are tightly bound in the nucleus, shielded by all the electrons, but outer electrons are easily stripped away. Rutherford's nuclear model has explanatory power that was lacking in Thomson's model.

But there was a problem. Helium, with atomic number 2, has twice as many electrons and protons as hydrogen. Lithium, $Z = 3$, has three electrons and protons. If a nucleus contains $Z$ protons to balance the $Z$ orbiting electrons, and if nearly all the atomic mass is contained in the nucleus, then helium should be twice as massive as hydrogen and lithium three times as massive. But it was known from chemistry measurements that helium is *four times* as massive as hydrogen and lithium is *seven times* as massive.

This difficulty was not resolved until the discovery, in 1932, of a third subatomic particle. This particle has essentially the same mass as a proton but *no* electric charge. It is called the **neutron.** Neutrons reside in the nucleus, with the protons, where they contribute to the mass of the atom but not to its charge. As you'll see in Chapter 30, neutrons help provide the "glue" that holds the nucleus together.

We now know that a nucleus contains $Z$ protons plus $N$ neutrons, as shown in FIGURE 29.9, giving the atom a **mass number** $A = Z + N$. The mass number, which is a dimensionless integer, is *not* the same thing as the atomic mass $m$. But because the proton and neutron masses are both $\approx 1$ u, where

$$1\text{ u} = 1 \text{ atomic mass unit} = 1.66 \times 10^{-27}\text{ kg}$$

the mass number $A$ is *approximately* the mass in atomic mass units.

There is a *range* of neutron numbers that happily form a nucleus with $Z$ protons, creating a series of nuclei with the same $Z$-value (i.e., they are all the same chemical element) but different masses. Such a series of nuclei are called **isotopes.** The notation used to label isotopes is $^{A}Z$, where the mass number $A$ is given as a *leading* superscript. For example, the most common isotope of neon has $Z = 10$ protons and $N = 10$ neutrons. Thus it has mass number $A = 20$ and is labeled $^{20}$Ne. The neon isotope $^{22}$Ne has $Z = 10$ protons (that's what makes it neon) and $N = 12$ neutrons. Helium has the two isotopes shown in FIGURE 29.10. $^{3}$He is rare, but it can be isolated and has important uses in scientific research.

FIGURE 29.9 The nucleus of an atom contains protons and neutrons.

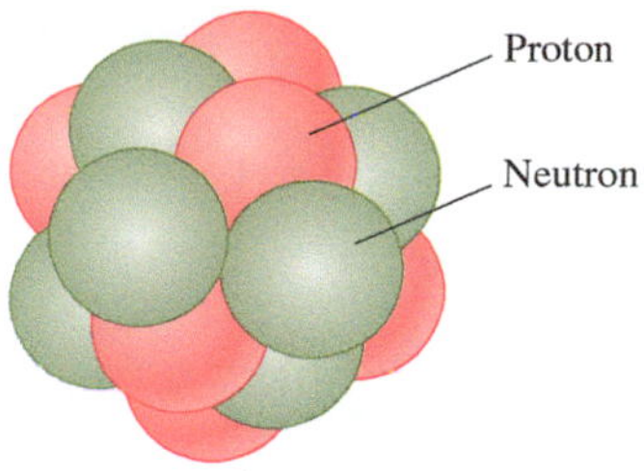


FIGURE 29.10 The two isotopes of helium.

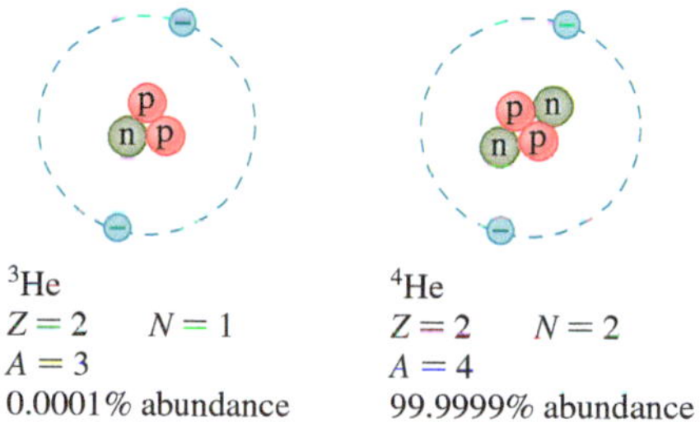


**STOP TO THINK 29.2** Carbon is the sixth element in the periodic table. How many protons and how many neutrons are there in a nucleus of the isotope $^{14}$C?

## 29.3 Bohr's Model of Atomic Quantization

Rutherford's nuclear model was an important step toward understanding atoms, but it had a serious shortcoming. According to Maxwell's theory of electricity and magnetism, the electrons orbiting the nucleus should act as small antennas and radiate electromagnetic waves. As FIGURE 29.11 shows, the energy carried off by the waves would cause the electrons to quickly—in less than a microsecond—spiral into the nucleus! In other words, classical Newtonian mechanics and electromagnetism predict that an atom with electrons orbiting a nucleus would be highly unstable and would immediately self-destruct. This clearly does not happen.

FIGURE 29.11 The fate of a Rutherford atom.

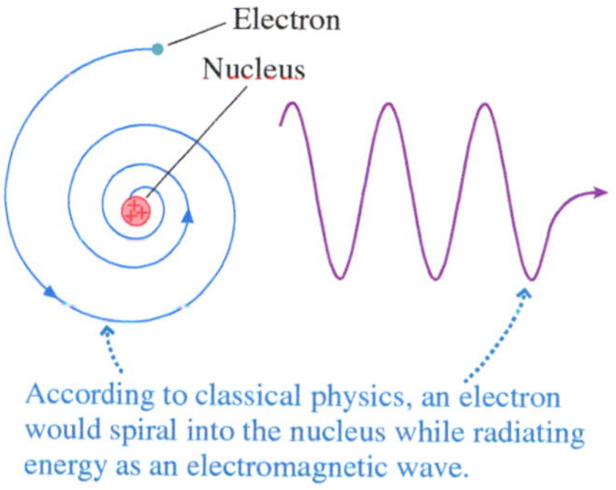

Electron
Nucleus
According to classical physics, an electron would spiral into the nucleus while radiating energy as an electromagnetic wave.

The experimental efforts of the late 19th and early 20th centuries had been impressive, and there could be no doubt about the existence of electrons, about the small positive nucleus, and about the unique discrete spectrum emitted by each atom. But the theoretical framework for understanding such observations had lagged behind.

A missing piece of the puzzle, although not recognized as such for a few years, was Einstein's 1905 introduction of light quanta. If light comes in discrete packets of energy, which we now call photons, and if atoms emit and absorb light, what does that imply about the structure of the atoms? This was the question posed by the Danish physicist Niels Bohr.

Bohr wanted to understand how a solar-system–like atom could be stable and not radiate away all its energy. He soon recognized that Einstein's light quanta had profound implications about the structure of atoms, and in 1913 Bohr proposed a radically new model of the atom in which he added quantization to Rutherford's nuclear atom. The basic assumptions of the **Bohr model of the atom** are shown on the next page.

**KEY CONCEPT** **Understanding Bohr's model**

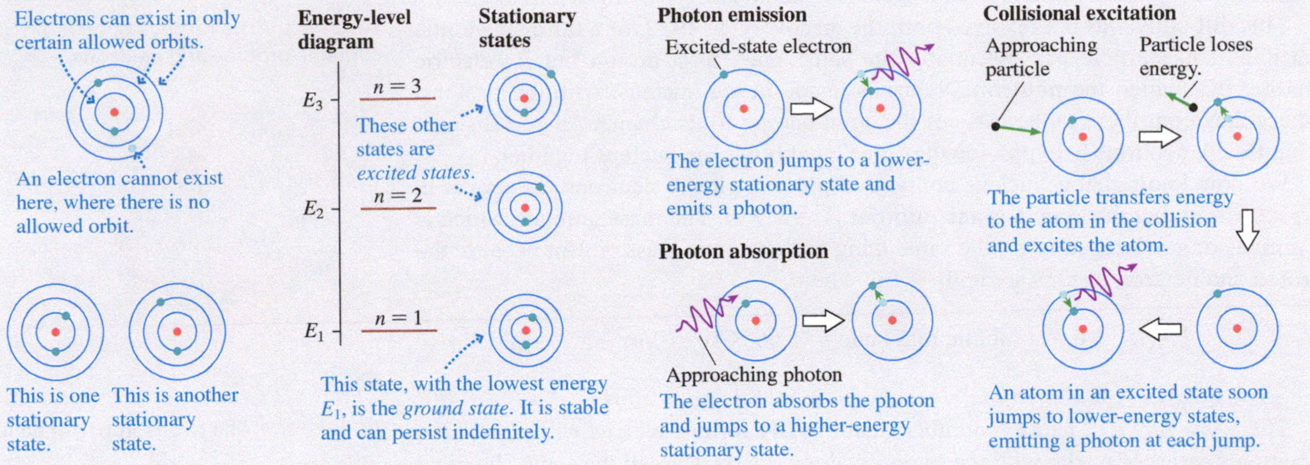


The electrons in an atom can exist in only certain *allowed orbits*. A particular arrangement of electrons in these orbits is called a **stationary state.**

Each stationary state has a discrete, well-defined energy $E_n$, as the energy-level diagram shows. That is, atomic energies are *quantized*. The stationary states are labeled by the *quantum number n* in order of increasing energy: $E_1 < E_2 < E_3 < \cdots$.

An atom can undergo a *transition,* or *quantum jump,* from one stationary state to another by emitting or absorbing a photon whose energy is exactly equal to the energy difference between the two stationary states.

Atoms can also move from a lower-energy state to a higher-energy state by absorbing energy in a collision with an electron or other atom in a process called **collisional excitation.** The excited atoms soon jump down to lower states, eventually ending in the stable ground state.

**STOP TO THINK 29.3** 3.0 eV photons excite atoms in a gas from their ground state to the $n = 3$ excited state. The atoms are then observed emitting 2.0 eV photons. Using what you learned in the table above, what other emitted photon energies would be observed?

A. 3.0 eV and 5.0 eV B. 3.0 eV and 1.0 eV C. 1.0 eV and 4.0 eV
D. 3.0 eV only E. 1.0 eV only

The implications of Bohr's model are profound. In particular:

- **Matter is stable.** Once an atom is in its ground state, there are no states of any lower energy to which it can jump. It can remain in the ground state forever.
- **Atoms emit and absorb a *discrete spectrum.*** When an atom jumps from an initial state with energy $E_i$ to a final state with energy $E_f$, conservation of energy requires that it emit or absorb a photon with energy $E_{photon} = \Delta E_{atom} = |E_f - E_i|$. Because $E_{photon} = hf$, this photon must have frequency $f_{photon} = \Delta E_{atom}/h$. Photons of other frequencies cannot be emitted or absorbed without violating energy conservation.
- **Emission spectra can be produced by collisions.** Energy from collisions can kick an atom up to an excited state. The atom then emits photons in a discrete emission spectrum as it jumps down to lower-energy states.
- **Absorption wavelengths are a subset of the wavelengths in the emission spectrum.** Recall that all the lines seen in an absorption spectrum are also seen in emission, but many emission lines are *not* seen in absorption. According to Bohr's model, **most atoms, most of the time, are in their lowest energy state,** the $n = 1$ ground state. Thus the absorption spectrum consists of *only* those transitions such as $1 \rightarrow 2$, $1 \rightarrow 3$, . . . in which the atom jumps from $n = 1$ to a higher value of $n$ by absorbing a photon. Transitions such as $2 \rightarrow 3$ are not observed

because there are essentially no atoms in $n = 2$ at any instant of time to do the absorbing. However, atoms that have been excited to the $n = 3$ state by collisions can emit photons corresponding to transitions $3 \rightarrow 1$ *and* $3 \rightarrow 2$. Thus the wavelength corresponding to $\Delta E_{atom} = E_3 - E_1$ is seen in both emission and absorption, but photons with $\Delta E_{atom} = E_3 - E_2$ occur in emission only.

- **Each element in the periodic table has a unique spectrum.** The energies of the stationary states are just the energies of the orbiting electrons. Different elements, with different numbers of electrons, have different stable orbits and thus different stationary states. States with different energies will emit and absorb photons of different wavelengths.

**EXAMPLE 29.3 Wavelengths in emission and absorption spectra**

An atom has stationary states $E_1 = 0.0$ eV, $E_2 = 2.0$ eV, and $E_3 = 5.0$ eV. What wavelengths are observed in the absorption spectrum and in the emission spectrum of this atom?

**STRATEGIZE** Photons are emitted when an atom undergoes a quantum jump from a higher energy level to a lower energy level. Photons are absorbed in a quantum jump from a lower energy level to a higher energy level. However, most of the atoms are in the $n = 1$ ground state, so the only quantum jumps in the absorption spectrum start from the $n = 1$ state.

**PREPARE** FIGURE 29.12 shows the energy-level diagram for the atom, with possible transitions noted.

FIGURE 29.12 The atom's energy-level diagram.

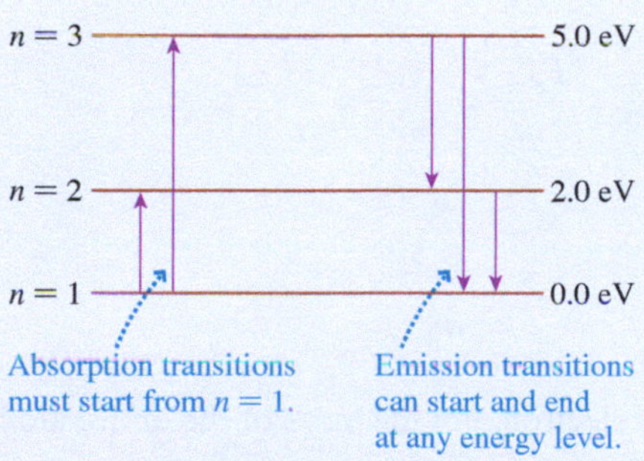


**SOLVE** This atom absorbs photons on the $1 \rightarrow 2$ and $1 \rightarrow 3$ transitions, with $\Delta E_{1\rightarrow 2} = 2.0$ eV and $\Delta E_{1\rightarrow 3} = 5.0$ eV. From $f_{photon} = \Delta E_{atom}/h$ and $\lambda = c/f$, we find that the wavelengths in the absorption spectrum are

$$1 \rightarrow 3 \quad f_{photon} = 5.0\text{ eV}/h = 1.2 \times 10^{15}\text{ Hz}$$
$$\lambda = 250\text{ nm (ultraviolet)}$$
$$1 \rightarrow 2 \quad f_{photon} = 2.0\text{ eV}/h = 4.8 \times 10^{14}\text{ Hz}$$
$$\lambda = 620\text{ nm (orange)}$$

The emission spectrum also has the 620 nm and 250 nm wavelengths due to the $2 \rightarrow 1$ and $3 \rightarrow 1$ quantum jumps. In addition, the emission spectrum contains the $3 \rightarrow 2$ quantum jump with $\Delta E_{3\rightarrow 2} = 3.0$ eV that is *not* seen in absorption because there are too few atoms in the $n = 2$ state to absorb. A similar calculation finds $f_{photon} = 7.3 \times 10^{14}$ Hz and $\lambda = c/f = 410$ nm. Thus the emission wavelengths are

$$2 \rightarrow 1 \quad \lambda = 620\text{ nm (orange)}$$
$$3 \rightarrow 2 \quad \lambda = 410\text{ nm (blue)}$$
$$3 \rightarrow 1 \quad \lambda = 250\text{ nm (ultraviolet)}$$

**STOP TO THINK 29.4** A photon with a wavelength of 410 nm has energy $E_{photon} = 3.0$ eV. Do you expect to see a spectral line with $\lambda = 410$ nm in the emission spectrum of the atom represented by this energy-level diagram? If so, what transition or transitions will emit it? Do you expect to see a spectral line with $\lambda = 410$ nm in the absorption spectrum? If so, what transition or transitions will absorb it?

## 29.4 The Bohr Hydrogen Atom

Bohr's hypothesis was a bold new idea, yet there was still one enormous stumbling block: What *are* the stationary states of an atom? Everything in Bohr's model hinges on the existence of these stationary states, of there being only certain electron orbits that are allowed. But nothing in classical physics provides any basis for such orbits. And Bohr's model describes only the *consequences* of having stationary states, not how to find them. If such states really exist, we will have to go beyond classical physics to find them.

NOTE ▶ Bohr's analysis of the hydrogen atom is sometimes called the *Bohr atom.* It's important not to confuse this analysis, which applies only to hydrogen, with the more general postulates of the *Bohr model of the atom.* Those postulates, which we looked at in the previous section, apply to any atom. To make the distinction clear, we'll call this analysis of hydrogen the *Bohr hydrogen atom.* This analysis follows a different approach than Bohr used. We'll make use of de Broglie matter waves, which weren't proposed until after Bohr's work. But the end point is the same, and we'll be in a better position to understand the work that came after Bohr. ◀

## The Stationary States of the Hydrogen Atom

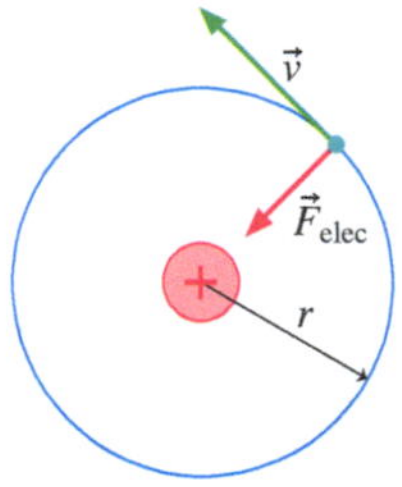


FIGURE 29.13 A Rutherford hydrogen atom. The size of the nucleus is greatly exaggerated.

FIGURE 29.13 shows a Rutherford hydrogen atom, with a single electron orbiting a proton. We will assume a circular orbit of radius $r$ and speed $v$. We will also assume the proton remains stationary while the electron revolves around it, a reasonable assumption because the proton is roughly 1800 times as massive as the electron.

The electron, as we are coming to understand it, has both particle-like and wave-like properties. First, let's treat the electron as a charged particle. The proton exerts a Coulomb electric force on the electron:

$$\vec{F}_{elec} = \left(\frac{Ke^2}{r^2}, \text{ toward center}\right) \tag{29.3}$$

This force gives the electron an acceleration $\vec{a}_{elec} = \vec{F}_{elec}/m$ that also points to the center. This is a centripetal acceleration, causing the particle to move in its circular orbit. The centripetal acceleration of a particle moving in a circle of radius $r$ at speed $v$ is $v^2/r$, so that

$$a_{elec} = \frac{F_{elec}}{m} = \frac{Ke^2}{mr^2} = \frac{v^2}{r}$$

Rearranging, we find

$$v^2 = \frac{Ke^2}{mr} \tag{29.4}$$

NOTE ▶ $m$ is the mass of the electron, *not* the mass of the entire atom. ◀

Equation 29.4 is a *constraint* on the motion. The speed $v$ and radius $r$ must satisfy Equation 29.4 if the electron is to move in a circular orbit. This constraint is not unique to atoms; we earlier found a similar relationship between $v$ and $r$ for orbiting satellites.

Now let's treat the electron as a de Broglie wave. In ◀◀ SECTION 28.5 we found that a particle confined to a one-dimensional box sets up a standing wave as it reflects back and forth. A standing wave, you will recall, consists of two traveling waves moving in opposite directions. When the round-trip distance in the box is equal to an integer number of wavelengths ($2L = n\lambda$), the two oppositely traveling waves interfere constructively to set up the standing wave.

FIGURE 29.14 An $n = 10$ electron standing wave around the orbit's circumference.

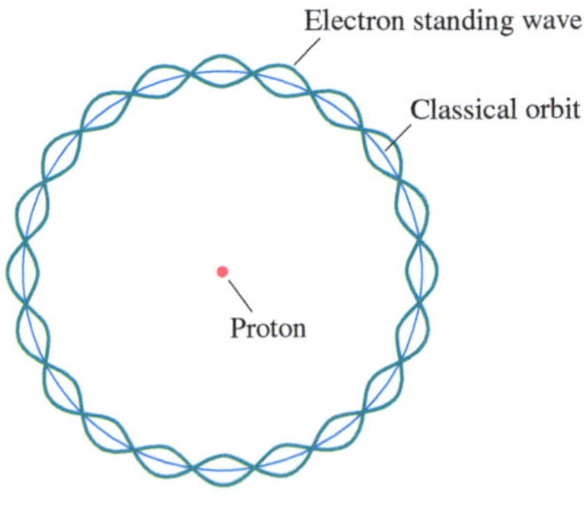


Suppose that, instead of traveling back and forth along a line, our wave-like particle travels around the circumference of a circle. The particle will set up a standing wave, just like the particle in the box, if there are waves traveling in both directions and if the round-trip distance is an integer number of wavelengths. This is the idea we want to carry over from the particle in a box. As an example, FIGURE 29.14 shows a standing wave around a circle with $n = 10$ wavelengths.

The mathematical condition for a circular standing wave is found by replacing the round-trip distance $2L$ in a box with the round-trip distance $2\pi r$ on a circle. Thus a circular standing wave will occur when

$$2\pi r = n\lambda \qquad n = 1, 2, 3, \ldots \tag{29.5}$$

But the de Broglie wavelength for a particle *has* to be $\lambda = h/p = h/mv$. Thus the standing-wave condition for a de Broglie wave is

$$2\pi r = n\frac{h}{mv}$$

This condition is true only if the electron's speed is

$$v = \frac{nh}{2\pi mr} \qquad n = 1, 2, 3, \ldots \tag{29.6}$$

In other words, the electron cannot have just any speed, only the discrete values given by Equation 29.6.

The quantity $h/2\pi$ occurs so often in quantum physics that it is customary to give it a special name. We define the quantity $\hbar$, pronounced "h bar," as

$$\hbar = \frac{h}{2\pi} = 1.05 \times 10^{-34}\ \text{J} \cdot \text{s} = 6.58 \times 10^{-16}\ \text{eV} \cdot \text{s}$$

With this definition, we can write Equation 29.6 as

$$v = \frac{n\hbar}{mr} \qquad n = 1, 2, 3, \ldots \tag{29.7}$$

This, like Equation 29.4, is another relationship between $v$ and $r$. This is the constraint that arises from treating the electron as a wave.

Now if the electron can act as both a particle *and* a wave, then both the Equation 29.4 *and* Equation 29.7 constraints have to be obeyed. That is, $v^2$ as given by the Equation 29.4 particle constraint has to equal $v^2$ of the Equation 29.7 wave constraint. Equating these gives

$$v^2 = \frac{Ke^2}{mr} = \frac{n^2\hbar^2}{m^2r^2}$$

We can solve this equation to find that the radius $r$ is

$$r_n = n^2\frac{\hbar^2}{Kme^2} \qquad n = 1, 2, 3, \ldots \tag{29.8}$$

where we have added a subscript $n$ to the radius $r$ to indicate that it depends on the integer $n$.

The right-hand side of Equation 29.8, except for the $n^2$, is just a collection of constants. Let's group them all together and define the **Bohr radius** $a_B$ to be

$$a_B = \text{Bohr radius} = \frac{\hbar^2}{Kme^2} = 5.29 \times 10^{-11}\ \text{m} = 0.0529\ \text{nm}$$

With this definition, Equation 29.8 for the radius of the electron's orbit becomes

$$r_n = n^2 a_B \qquad n = 1, 2, 3, \ldots \tag{29.9}$$

Allowed radii of the Bohr hydrogen atom

For example, $r_1 = 0.053$ nm, $r_2 = 0.212$ nm, and $r_3 = 0.476$ nm.

These are the stationary states we were looking for. **A hydrogen atom can exist *only* if the radius of the electron's orbit is one of the values given by Equation 29.9.** Intermediate values of the radius, such as $r = 0.100$ nm, cannot exist because the electron cannot set up a standing wave around the circumference. The possible orbits are *quantized,* and integer $n$ is the quantum number.

## Hydrogen Atom Energy Levels

The energy of a hydrogen atom is the kinetic energy of its electron plus the potential energy of the electron's interaction with the proton. This is

$$E = K + U = \frac{1}{2}mv^2 + \frac{Kq_{\text{elec}}q_{\text{proton}}}{r} = \frac{1}{2}mv^2 - \frac{Ke^2}{r} \qquad (29.10)$$

where we used $q_{\text{elec}} = -e$ and $q_{\text{proton}} = +e$.

Now we can proceed to find the energies of the stationary states. Knowing the possible radii, we can return to Equation 29.7 and find the possible electron speeds:

$$v_n = \frac{n\hbar}{mr_n} = \frac{1}{n}\frac{\hbar}{ma_B} = \frac{v_1}{n} \qquad n = 1, 2, 3, \ldots \qquad (29.11)$$

where $v_1 = \hbar/ma_B = 2.18 \times 10^6$ m/s is the electron's speed in the $n = 1$ orbit. The speed decreases as $n$ increases.

Finally, we can determine the energies of the stationary states by using Equations 29.9 and 29.11 for $r$ and $v$ in Equation 29.10 for the energy. After some work, the result simplifies to

$$E_n = \frac{1}{2}m{v_n}^2 - \frac{Ke^2}{r_n} = -\frac{1}{n^2}\left(\frac{Ke^2}{2a_B}\right) \qquad (29.12)$$

Let's define

$$E_1 = \frac{Ke^2}{2a_B} = 13.60 \text{ eV}$$

We can then write the energy levels of the stationary states of the hydrogen atom as

$$E_n = -\frac{E_1}{n^2} = -\frac{13.60 \text{ eV}}{n^2} \qquad n = 1, 2, 3, \ldots \qquad (29.13)$$

Allowed energies of the Bohr hydrogen atom

This has been a lot of math, so we need to see where we are and what we have learned. TABLE 29.2 shows values of $r_n$, $v_n$, and $E_n$ for quantum numbers $n = 1$ to 4. We do indeed seem to have found stationary states of the hydrogen atom. Each state, characterized by its quantum number $n$, has a unique radius, speed, and energy. These are displayed graphically in FIGURE 29.15, in which the orbits are drawn to scale. Notice how the atom's diameter increases very rapidly as $n$ increases. At the same time, the electron's speed decreases.

TABLE 29.2 Radii, speeds, and energies for the first four states of the Bohr hydrogen atom

| $n$ | $r_n$ (nm) | $v_n$ (m/s) | $E_n$ (eV) |
|---|---|---|---|
| 1 | 0.053 | $2.18 \times 10^6$ | −13.60 |
| 2 | 0.212 | $1.09 \times 10^6$ | −3.40 |
| 3 | 0.476 | $0.73 \times 10^6$ | −1.51 |
| 4 | 0.847 | $0.55 \times 10^6$ | −0.85 |

FIGURE 29.15 The first four stationary states, or allowed orbits, of the Bohr hydrogen atom drawn to scale.

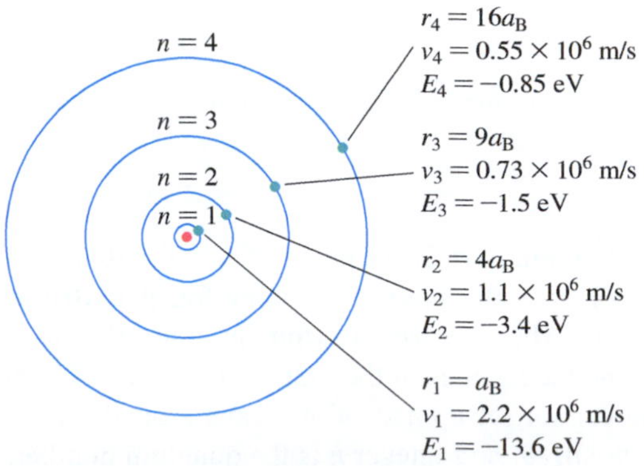

**EXAMPLE 29.4 Possible electron speeds in a hydrogen atom**

Can an electron in a hydrogen atom have a speed of $3.60 \times 10^5$ m/s? If so, what are its energy and the radius of its orbit? What about a speed of $3.65 \times 10^5$ m/s?

**PREPARE** To be in a stationary state, the electron must have speed $v_n = v_1/n$, with $n = v_1/v$ an integer. Only if $v_1/v$ is an integer is $v$ an allowed electron speed.

**SOLVE** A speed of $3.60 \times 10^5$ m/s would require quantum number

$$n = \frac{v_1}{v} = \frac{2.19 \times 10^6 \text{ m/s}}{3.60 \times 10^5 \text{ m/s}} = 6.08$$

This is not an integer, so the electron *cannot* have this speed. But if $v = 3.65 \times 10^5$ m/s, then

$$n = \frac{2.19 \times 10^6 \text{ m/s}}{3.65 \times 10^5 \text{ m/s}} = 6$$

This is the speed of an electron in the $n = 6$ excited state. An electron in this state has energy

$$E_6 = -\frac{13.60 \text{ eV}}{6^2} = -0.378 \text{ eV}$$

and the radius of its orbit is

$$r_6 = 6^2 a_B = 6^2 (0.0529 \text{ nm}) = 1.90 \text{ nm}$$

It is important to understand why the energies of the stationary states are negative. An electron and a proton bound into an atom have *less* energy than they do when they're separated. We know this because we would have to do work (i.e., *add* energy) to pull the electron and proton apart.

When the electron and proton are completely separated $(r \rightarrow \infty)$ and at rest $(v = 0)$, their potential energy $U = Kq_1q_2/r$ and kinetic energy $K = mv^2/2$ are zero. As the electron moves closer to the proton to form a hydrogen atom, its potential energy *decreases,* becoming negative, while the kinetic energy of the orbiting electron increases. Equation 29.12, however, shows that the potential energy decreases faster than the kinetic energy increases, leading to an overall negative energy for the atom.

**STOP TO THINK 29.5** What is the quantum number of this hydrogen atom?

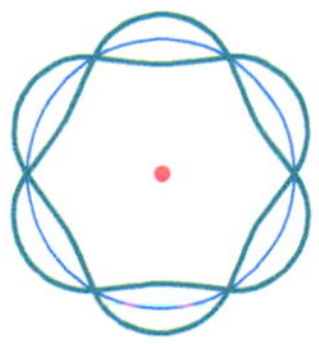

## The Hydrogen Spectrum

The most important experimental evidence that we have about the hydrogen atom is its spectrum, so the primary test of the Bohr hydrogen atom is whether it correctly predicts the spectrum. FIGURE 29.16 is an energy-level diagram for the hydrogen atom. As we noted in ◀◀ SECTION 28.6, the energies are like the rungs of a ladder. The lowest rung is the ground state, with $E_1 = -13.60$ eV. The top rung, with $E = 0$ eV, corresponds to a hydrogen ion in the limit $n \rightarrow \infty$. This top rung is called the **ionization limit.** In principle there are an infinite number of rungs, but only the lowest few are shown. The energy levels with higher values of $n$ are all crowded together just below the ionization limit at $n = \infty$.

The figure shows a $1 \rightarrow 4$ transition in which a photon is absorbed and a $4 \rightarrow 2$ transition in which a photon is emitted. For two quantum states $m$ and $n$, where $n > m$ and $E_n$ is the higher-energy state, an atom can *emit* a photon in an $n \rightarrow m$ transition or *absorb* a photon in an $m \rightarrow n$ transition.

According to Bohr's model of atomic quantization, the frequency of the photon emitted in an $n \rightarrow m$ transition is

$$f_{\text{photon}} = \frac{\Delta E_{\text{atom}}}{h} = \frac{E_n - E_m}{h} \tag{29.14}$$

We can use Equation 29.13 for the energies $E_n$ and $E_m$ to predict that the emitted photon has frequency

$$f_{\text{photon}} = \frac{1}{h}\left(\frac{-13.60 \text{ eV}}{n^2} - \frac{-13.60 \text{ eV}}{m^2}\right) = \frac{13.60 \text{ eV}}{h}\left(\frac{1}{m^2} - \frac{1}{n^2}\right)$$

The frequency is a positive number because $m < n$ and thus $1/m^2 > 1/n^2$.

**FIGURE 29.16** The energy-level diagram of the hydrogen atom.

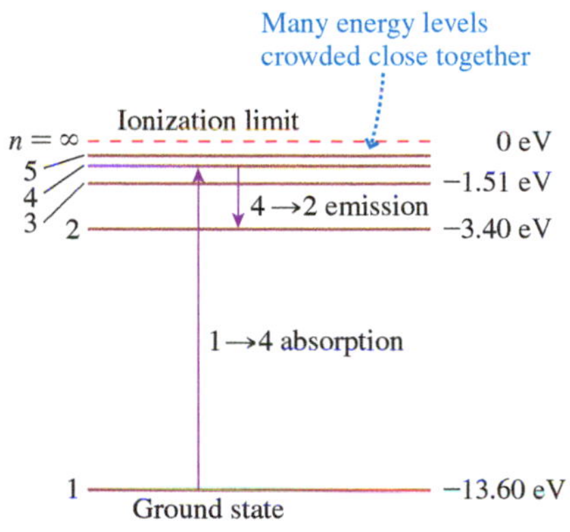

We are more interested in wavelength than frequency because wavelengths are the quantity measured by experiment. The wavelength of the photon emitted in an $n \to m$ quantum jump is

$$\lambda_{n\to m} = \frac{c}{f_{\text{photon}}} = \frac{\lambda_0}{\left(\dfrac{1}{m^2} - \dfrac{1}{n^2}\right)} \qquad \begin{array}{l} m = 1, 2, 3, \ldots \\ n = m+1, m+2, \ldots \end{array} \tag{29.15}$$

with $\lambda_0 = 91.1$ nm. This should look familiar. It is the Balmer formula, Equation 29.2.

Unlike previous atomic models, **the Bohr hydrogen atom correctly predicts the discrete spectrum of the hydrogen atom.** FIGURE 29.17 shows two series of transitions that give rise to wavelengths in the spectrum. The *Balmer series,* consisting of transitions ending on the $m = 2$ state, gives visible wavelengths, and this is the series that Balmer initially analyzed. The *Lyman series,* ending on the $m = 1$ ground state, is in the ultraviolet region of the spectrum and was not measured until later. These series, as well as others in the infrared, are observed in a discharge tube where collisions with electrons excite the atoms upward from the ground state to state $n$. They then decay downward by emitting photons. Only the Lyman series is observed in the absorption spectrum because, as noted previously, essentially all the atoms in a quiescent gas are in the ground state.

FIGURE 29.17 Transitions producing the Lyman series and the Balmer series of lines in the hydrogen spectrum.

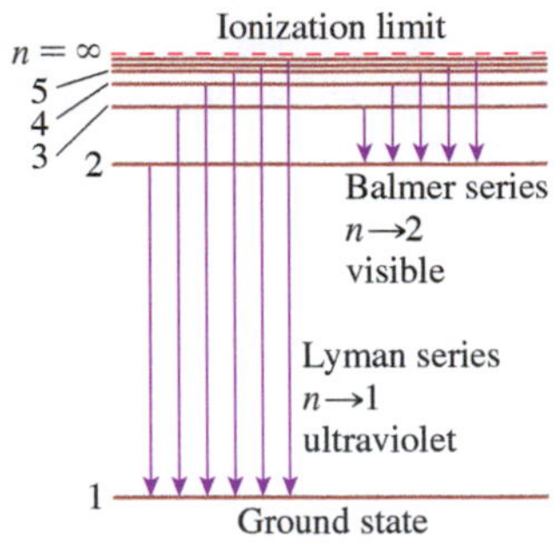


**EXAMPLE 29.5 Wavelengths in galactic hydrogen absorption**

Whenever astronomers look at distant galaxies, they find that the light has been strongly absorbed at the wavelength of the $1 \to 2$ transition in the Lyman series of hydrogen. This absorption tells us that interstellar space is filled with vast clouds of hydrogen left over from the Big Bang. What is the wavelength of the $1 \to 2$ absorption in hydrogen?

**SOLVE** Equation 29.15 predicts the *absorption* spectrum of hydrogen if we let $m = 1$. The absorption seen by astronomers is from the ground state of hydrogen $(m = 1)$ to its first excited state $(n = 2)$. The wavelength is

$$\lambda_{1\to 2} = \frac{91.1\text{ nm}}{\left(\dfrac{1}{1^2} - \dfrac{1}{2^2}\right)} = 121\text{ nm}$$

**ASSESS** This wavelength is far into the ultraviolet. Ground-based astronomy cannot observe this region of the spectrum because the wavelengths are strongly absorbed by the atmosphere, but with space-based telescopes, first widely used in the 1970s, astronomers see 121 nm absorption in nearly every direction they look.

## 29.5 The Quantum-Mechanical Hydrogen Atom

By introducing stationary states, together with Einstein's ideas about light quanta, Bohr was able to provide the first solid understanding of discrete spectra and, in particular, to predict the Balmer formula for the wavelengths in the hydrogen spectrum. But the analysis of the previous section can't explain the spectra of any other atom. The idea of stationary states has validity, but there must be more to the story.

In 1925, Erwin Schrödinger introduced his general theory of *quantum* mechanics. This theory is capable of calculating the allowed energy levels (i.e., the stationary states) of any system.

The Bohr hydrogen atom was characterized by a single quantum number $n$. In contrast, Schrödinger's quantum-mechanical analysis of the hydrogen atom found that it must be described by *four* quantum numbers. Energy has an associated quantum number, and so does angular momentum, a quantity that you learned about in ◀◀ SECTION 9.7. In addition, there are two other quantum numbers related to the orientation of the orbit and the electron's spin. The full set of quantum numbers and the related quantization conditions are as follows:

1. Schrödinger found that the *energy* of the hydrogen atom is given by the same expression found by Bohr, or

$$E_n = -\frac{13.60\text{ eV}}{n^2} \qquad n = 1, 2, 3, \ldots \tag{29.16}$$

The integer $n$ is called the **principal quantum number.**

2. The angular momentum $L$ of the electron's orbit must be one of the values

$$L = \sqrt{l(l+1)}\,\hbar \qquad l = 0, 1, 2, 3, \ldots, n-1 \qquad (29.17)$$

The integer $l$ is called the **orbital quantum number.**

3. The plane of the electron's orbit can be tilted, but only at certain discrete angles. Each allowed angle is characterized by a quantum number $m$, which must be one of the values

$$m = -l, -l+1, \ldots, 0, \ldots, l-1, l \qquad (29.18)$$

The integer $m$ is called the **magnetic quantum number** because it becomes important when the atom is placed in a magnetic field.

4. The electron's *spin*—discussed later in this section—can point only up or down. These two orientations are described by the **spin quantum number** $m_s$, which must be one of the values

$$m_s = -\frac{1}{2} \text{ or } +\frac{1}{2} \qquad (29.19)$$

In other words, each stationary state of the hydrogen atom is identified by a quartet of quantum numbers $(n, l, m, m_s)$, and each quantum number is associated with a physical property of the atom.

**NOTE** ▶ For a hydrogen atom, the energy of a stationary state depends on only the principal quantum number $n$, not on $l$, $m$, or $m_s$. ◀

### EXAMPLE 29.6 Listing quantum numbers

List all possible states of a hydrogen atom that have energy $E = -3.40$ eV.

***SOLVE*** Energy depends on only the principal quantum number $n$. From Equation 29.16, states with $E = -3.40$ eV have

$$n = \sqrt{\frac{-13.60 \text{ eV}}{-3.40 \text{ eV}}} = 2$$

An atom with principal quantum number $n = 2$ can have either $l = 0$ or $l = 1$, but $l \geq 2$ is ruled out. If $l = 0$, the only possible value for the magnetic quantum number $m$ is $m = 0$. If $l = 1$, then the atom can have $m = -1$, $m = 0$, or $m = +1$. For each of these, the spin quantum number can be $m_s = +\frac{1}{2}$ or $m_s = -\frac{1}{2}$. Thus the possible sets of quantum numbers are

| $n$ | $l$ | $m$ | $m_s$ | $n$ | $l$ | $m$ | $m_s$ |
|---|---|---|---|---|---|---|---|
| 2 | 0 | 0 | $+\frac{1}{2}$ | 2 | 0 | 0 | $-\frac{1}{2}$ |
| 2 | 1 | 1 | $+\frac{1}{2}$ | 2 | 1 | 1 | $-\frac{1}{2}$ |
| 2 | 1 | 0 | $+\frac{1}{2}$ | 2 | 1 | 0 | $-\frac{1}{2}$ |
| 2 | 1 | $-1$ | $+\frac{1}{2}$ | 2 | 1 | $-1$ | $-\frac{1}{2}$ |

These eight states all have the same energy.

## Energy and Angular Momentum Are Quantized

The energy of the hydrogen atom depends on only the principal quantum number $n$. For other atoms, however, the allowed energies depend on both $n$ and $l$. Consequently, it is useful to label the stationary states of an atom by their values of $n$ and $l$. The lowercase letters shown in TABLE 29.3 are customarily used to represent the various values of quantum number $l$. Using these symbols, we call the ground state of the hydrogen atom, with $n = 1$ and $l = 0$, the 1*s* state; the 3*d* state has $n = 3$, $l = 2$.

FIGURE 29.18 is an energy-level diagram for the hydrogen atom in which the rows are labeled by $n$ and the columns by $l$. The left column contains all of the $l = 0$ (or *s*) states, the next column is the $l = 1$ (or *p*) states, and so on.

As Equation 29.17 shows, the orbital quantum number $l$ of an allowed state must be less than that state's principal quantum number $n$. For the ground state, with $n = 1$, only $l = 0$ is possible, so that the only $n = 1$ state is the 1*s* state. When $n = 2$, $l$ can be 0 or 1, leading to both a 2*s* and a 2*p* state. For $n = 3$, there are 3*s*, 3*p*, and 3*d* states; and so on. Figure 29.18 shows only the first few energy levels for each value of $l$, but there really are an infinite number of levels, as $n \to \infty$, crowding together beneath $E = 0$. The dotted line at $E = 0$ is the atom's *ionization limit,* the energy of a hydrogen atom in which the electron has been moved infinitely far away to form an $H^+$ ion.

**TABLE 29.3** Symbols used to represent quantum number $l$

| $l$ | Symbol | $l$ | Symbol |
|---|---|---|---|
| 0 | *s* | 2 | *d* |
| 1 | *p* | 3 | *f* |

**FIGURE 29.18** The energy-level diagram for the hydrogen atom.

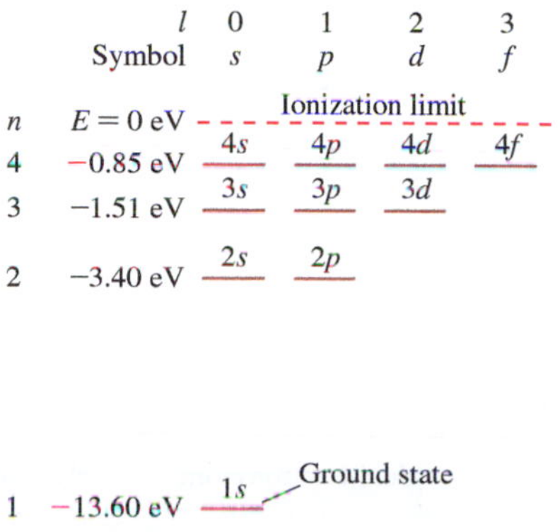

Classically, the angular momentum $L$ of an orbiting electron can have any value. Not so in quantum mechanics. Equation 29.17 tells us that **the electron's orbital angular momentum is quantized.** The magnitude of the orbital angular momentum must be one of the discrete values

$$L = \sqrt{l(l+1)}\,\hbar = 0, \sqrt{2}\,\hbar, \sqrt{6}\,\hbar, \sqrt{12}\,\hbar, \ldots$$

where $l$ is an integer.

A particularly interesting prediction is that the ground state of hydrogen, with $l = 0$, has *no* angular momentum. A classical particle cannot orbit unless it has angular momentum, but apparently a quantum particle does not have this requirement.

**STOP TO THINK 29.6** What are the quantum numbers $n$ and $l$ for a hydrogen atom with $E = -(13.60/9)$ eV and $L = \sqrt{2}\,\hbar$?

## The Electron Spin

You learned in ◀◀ SECTION 24.8 that an electron has an inherent magnetic dipole moment—it acts as a tiny bar magnet with a north and a south pole. In association with its magnetic moment, an electron also has an intrinsic *angular momentum* called the *electron spin.* In the early years of quantum mechanics, it was thought that the electron was a very tiny ball of negative charge spinning on its axis, which would give the electron both a magnetic dipole moment and spin angular momentum. However, a spinning ball of charge would violate the laws of relativity and other physical laws. As far as we know today, the electron is truly a point particle that happens to have an intrinsic magnetic dipole moment and angular momentum.

The two possible spin quantum numbers $m_s = \pm\frac{1}{2}$ mean that the electron's intrinsic magnetic dipole points in the $+z$-direction or the $-z$-direction. These two orientations are called *spin up* and *spin down.* It is convenient to picture a little vector that can be drawn ↑ for a spin-up state and ↓ for a spin-down state. We will use this notation in the next section.

# 29.6 Multi-electron Atoms

One of the first big successes of Schrödinger's quantum mechanics was an ability to calculate the stationary states and energy levels of *multi-electron atoms,* atoms in which $Z$ electrons orbit a nucleus with $Z$ protons. As we've seen, a major difference between multi-electron atoms and the simple one-electron hydrogen is that the energy of an electron in a multi-electron atom depends on both quantum numbers $n$ *and* $l$. Whereas the $2s$ and $2p$ states in hydrogen have the same energy, their energies are different in a multi-electron atom. The difference arises from the electron-electron interactions that do not exist in a single-electron hydrogen atom.

FIGURE 29.19 shows an energy-level diagram for the electrons in a multi-electron atom. (Compare this to the hydrogen energy-level diagram in Figure 29.18.) For comparison, the hydrogen-atom energies are shown on the right edge of the figure. Two features of this diagram are of particular interest:

- For each $n$, the energy increases as $l$ increases until the maximum-$l$ state has an energy very nearly that of the same $n$ in hydrogen. States with small values of $l$ are significantly lower in energy than the corresponding state in hydrogen.
- As the energy increases, states with different $n$ begin to alternate in energy. For example, the $3s$ and $3p$ states have lower energy than a $4s$ state, but the energy of an electron in a $3d$ state is slightly higher. This will have important implications for the structure of the periodic table of the elements.

FIGURE 29.19 An energy-level diagram for electrons in a multi-electron atom.

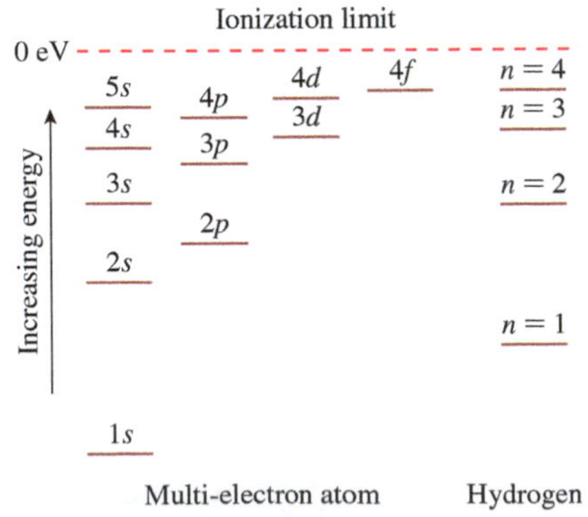

## The Pauli Exclusion Principle

By definition, the ground state of a quantum system is the state of lowest energy. What is the ground state of an atom that has $Z$ electrons and $Z$ protons? Because the $1s$ state is the lowest energy state, it seems that the ground state should be one in which all $Z$ electrons are in the $1s$ state. However, this idea is not consistent with the experimental evidence.

In 1925, the Austrian physicist Wolfgang Pauli hypothesized that no two electrons in a quantum system can be in the same quantum state. That is, **no two electrons can have exactly the same set of quantum numbers ($n, l, m, m_s$).** If one electron is present in a state, it *excludes* all others. This statement is called the **Pauli exclusion principle.** It turns out to be an extremely profound statement about the nature of matter.

The exclusion principle is not applicable to hydrogen, where there is only a single electron, but in helium, with $Z = 2$ electrons, we must make sure that the two electrons are in different quantum states. This is not difficult. For a $1s$ state, with $l = 0$, the only possible value of the magnetic quantum number is $m = 0$. But there are *two* possible values of $m_s$—namely, $-\frac{1}{2}$ and $+\frac{1}{2}$. If a first electron is in the spin-down $1s$ state $(1, 0, 0, -\frac{1}{2})$, a second $1s$ electron can still be added to the atom as long as it is in the spin-up state $(1, 0, 0, +\frac{1}{2})$. This is shown schematically in FIGURE 29.20, where the dots represent electrons on the rungs of the "energy ladder" and the arrows represent spin down or spin up.

FIGURE 29.20 The ground state of helium.

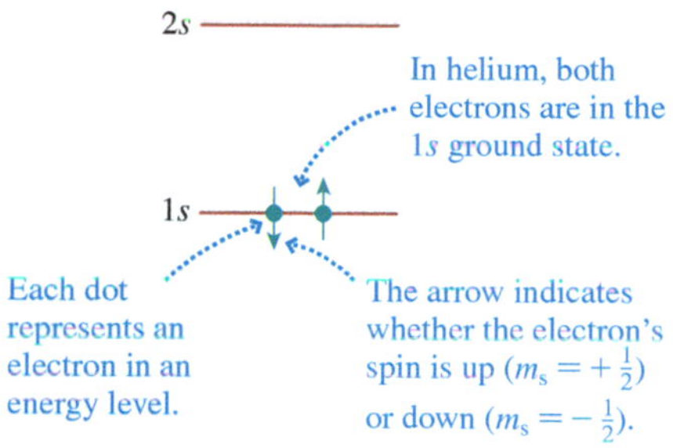


The Pauli exclusion principle does not prevent both electrons of helium from being in the $1s$ state as long as they have opposite values of $m_s$, so we predict this to be the ground state. A list of an atom's occupied energy levels is called its **electron configuration.** The electron configuration of the helium ground state is written $1s^2$, where the superscript 2 indicates two electrons in the $1s$ energy level.

The states $(1, 0, 0, -\frac{1}{2})$ and $(1, 0, 0, +\frac{1}{2})$ are the only two states with $n = 1$. The ground state of helium has one electron in each of these states, so all the possible $n = 1$ states are filled. Consequently, the electron configuration $1s^2$ is called a **closed shell.**

FIGURE 29.21 The ground state of lithium.

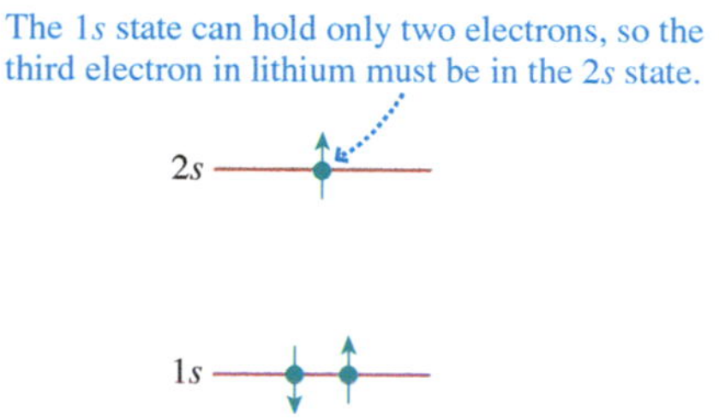


The next element, lithium, has $Z = 3$ electrons. The first two electrons can go into $1s$ states, with opposite values of $m_s$, but what about the third electron? The $1s^2$ shell is closed, and there are no additional quantum states having $n = 1$. The only option for the third electron is the next energy state, $n = 2$. Figure 29.19 showed that for a multi-electron atom, the next level above the $1s$ level is the $2s$ state, so lithium's third ground-state electron will be $2s$. FIGURE 29.21 shows the electron configuration with the $2s$ electron being spin up, but it could equally well be spin down. The electron configuration for the lithium ground state is written $1s^2\,2s$. This indicates two $1s$ electrons and a single $2s$ electron.

## The Periodic Table of the Elements

The Russian chemist Dmitri Mendeléev was the first to propose, in 1867, a *periodic* arrangement of the elements based on the regular recurrence of chemical properties. He did so by explicitly pointing out "gaps" where, according to his hypothesis, undiscovered elements should exist. He could then predict the expected properties of the missing elements. The subsequent discovery of these elements verified Mendeléev's organizational scheme, which came to be known as the *periodic table of the elements.*

One of the great triumphs of the quantum-mechanical theory of multi-electron atoms is that it explains the structure of the periodic table. We can understand this structure by looking at the energy-level diagram of FIGURE 29.22, which is an expanded version of the energy-level diagram of Figure 29.19. Just as for helium and lithium, atoms with larger values of $Z$ are constructed by placing $Z$ electrons into the lowest energy levels that are consistent with the Pauli exclusion principle.

FIGURE 29.22 An energy-level diagram showing how many electrons can occupy each subshell.

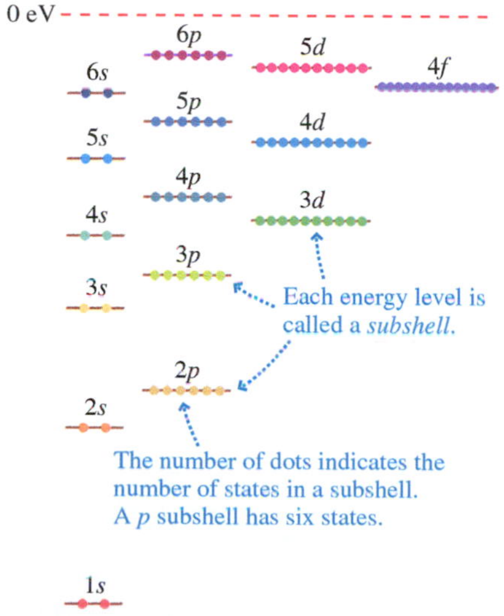


The $s$ states of helium and lithium can each hold two electrons—one spin up and the other spin down—but the higher-angular-momentum states that will become

TABLE 29.4 Number of states in each subshell of an atom

| Subshell | $l$ | Number of states |
|---|---|---|
| $s$ | 0 | 2 |
| $p$ | 1 | 6 |
| $d$ | 2 | 10 |
| $f$ | 3 | 14 |

filled for higher-$Z$ atoms can hold more than two electrons. For each value $l$ of the orbital quantum number, there are $2l + 1$ possible values of the magnetic quantum number $m$ and, for each of these, two possible values of the spin quantum number $m_s$. Consequently, each energy *level* in Figure 29.22 is actually $2(2l + 1)$ different *states* that, taken together, are called a *subshell.* TABLE 29.4 lists the number of states in each subshell. Each state in a subshell is represented in Figure 29.22 by a colored dot. The dots' colors correspond to the periodic table in FIGURE 29.23, which is color coded to show which subshells are being filled as $Z$ increases.

FIGURE 29.23 The periodic table of the elements. The elements are color coded to the states in the energy-level diagram of Figure 29.22.

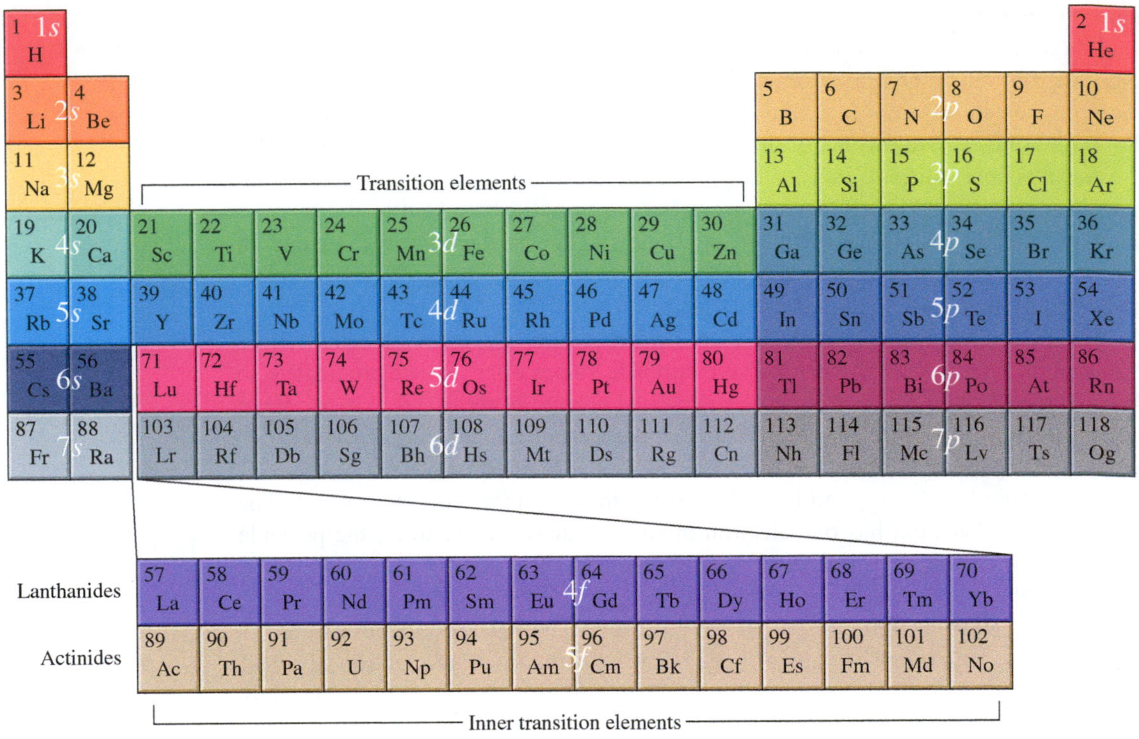


We can now use Figure 29.22 to construct the periodic table in Figure 29.23. We've already seen that lithium has two electrons in the $1s$ state and one electron in the $2s$ state. Four-electron beryllium ($Z = 4$) comes next. We see that there is still an empty state in the $2s$ subshell for this fourth electron to occupy, so beryllium closes the $2s$ subshell and has electron configuration $1s^2\,2s^2$.

As $Z$ increases further, the next six electrons can each occupy states in the $2p$ subshell. These are the elements boron (B) through neon (Ne), completing the second row of the periodic table. Neon, which completes the $2p$ subshell, has ground-state configuration $1s^2\,2s^2\,2p^6$.

The $3s$ subshell is the next to be filled, leading to the elements sodium and magnesium. Filling the $3p$ subshell gives aluminum through argon, completing the third row of the table.

The fourth row is where the periodic table begins to get complicated. You might expect that once the $3p$ subshell in argon was filled, the $3d$ subshell would start to fill, starting with potassium. But if you look back at the energy-level diagram of Figure 29.22, you can see that the $3d$ state is slightly *higher* in energy than the $4s$ state. Because the ground state is the *lowest energy state* consistent with the Pauli exclusion principle, potassium finds it more favorable to fill a $4s$ state than to fill a $3d$ state. After the $4s$ state is filled (at calcium), the 10 *transition elements* from scandium (Sc) through zinc (Zn) fill the 10 states of the $3d$ subshell.

The same pattern applies to the fifth row, where the 5*s*, 4*d*, and 5*p* subshells fill in succession. In the sixth row, however, after the initial 6*s* states are filled, the 4*f* subshell has the lowest energy, so it begins to fill *before* the 5*d* states. The elements corresponding to the 4*f* subshell, lanthanum through ytterbium, are known as the lanthanides, and they are traditionally drawn as a row separated from the rest of the table. The seventh row follows this same pattern.

Thus the entire periodic table can be built up using our knowledge of the energy-level diagram of a multi-electron atom along with the Pauli exclusion principle.

**EXAMPLE 29.7** **The ground state of arsenic**

Predict the ground-state electron configuration of arsenic.

**SOLVE** The periodic table shows that arsenic (As) has $Z = 33$, so we must identify the states of 33 electrons. Arsenic is in the fourth row, following the first group of transition elements. Argon ($Z = 18$) filled the 3*p* subshell, then calcium ($Z = 20$) filled the 4*s* subshell. The next 10 elements, through zinc ($Z = 30$), filled the 3*d* subshell. The 4*p* subshell starts filling with gallium ($Z = 31$), and arsenic is the third element in this group, so it will have three 4*p* electrons. Thus the ground-state configuration of arsenic is

$$1s^2 2s^2 2p^6 3s^2 3p^6 4s^2 3d^{10} 4p^3$$

**STOP TO THINK 29.7** Which element has the ground-state electron configuration $1s^2 2s^2 2p^6 3s^2 3p^3$?

A. P B. Al C. B D. Ge

## 29.7 Excited States and Spectra

The periodic table organizes information about the *ground states* of the elements. These states are chemically most important because most atoms spend most of the time in their ground states. All the chemical ideas of valence, bonding, reactivity, and so on are consequences of these ground-state atomic structures. But the periodic table does not tell us anything about the excited states of atoms. It is the excited states that hold the key to understanding atomic spectra, and that is the topic to which we turn next.

Sodium ($Z = 11$) is a multi-electron atom that we will use to illustrate excited states. The ground-state electron configuration of sodium is $1s^2 2s^2 2p^6 3s$. The first 10 electrons completely fill the 1*s*, 2*s*, and 2*p* subshells, creating a *neon core* whose electrons are tightly bound together. The 3*s* electron, however, is a *valence electron* that can be easily excited to higher energy levels. If this electron were excited to the 3*p* state, for instance, then we would write the electron configuration as $1s^2 2s^2 2p^6 3p$.

The excited states of sodium are produced by raising the valence electron to a higher energy level. The electrons in the neon core are unchanged. FIGURE 29.24 is an energy-level diagram showing the ground state and some of the excited states of sodium. The 1*s*, 2*s*, and 2*p* states of the neon core are not shown on the diagram. These states are filled and unchanging, so only the states available to the valence electron are shown. Notice that the zero of energy has been shifted to the ground state. As we have discovered before, the zero of energy can be located where it is most convenient. With this choice, the excited-state energies tell us how far each state is above the ground state. The ionization limit now occurs at the value of the atom's ionization energy, which is 5.14 eV for sodium.

Left to itself, an atom will be in its lowest-energy ground state. How does an atom get into an excited state? The process of getting it there is called **excitation,** and there are two basic mechanisms: absorption and collision. We'll begin by looking at excitation by absorption.

FIGURE 29.24 The 3*s* ground state of the sodium atom and some of the excited states.

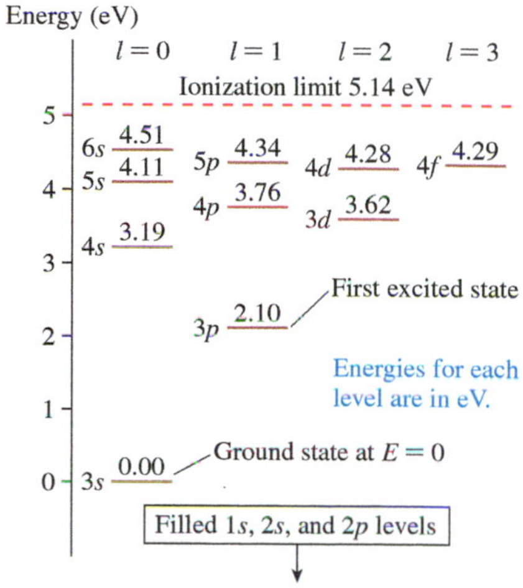

## Excitation by Absorption

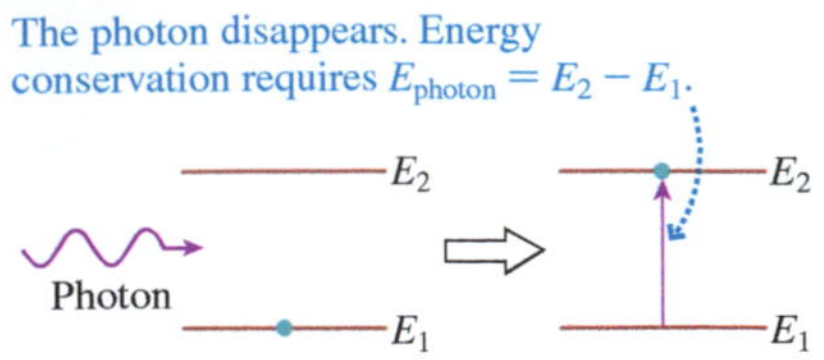


FIGURE 29.25 Excitation by photon absorption.

One of the postulates of the basic Bohr model is that an atom can jump from one stationary state, of energy $E_1$, to a higher-energy state $E_2$ by absorbing a photon of frequency $f_{photon} = \Delta E_{atom}/h$. This process is shown in FIGURE 29.25. Because we are interested in spectra, it is more useful to write this in terms of the wavelength:

$$\lambda = \frac{c}{f_{photon}} = \frac{hc}{\Delta E_{atom}}$$

Energy transitions are generally expressed in eV and wavelengths in nm. If we express $h$ in eV · s and c in nm/s, we get a version of the above equation that allows us to quickly find wavelengths in nm given a transition energy in eV:

$$\lambda(\text{in nm}) = \frac{1240 \text{ eV} \cdot \text{nm}}{\Delta E_{atom}(\text{in eV})} \tag{29.20}$$

A quantum-mechanical analysis of how the electrons in an atom interact with a light wave shows that transitions must also satisfy the following **selection rule:** Transitions (either absorption or emission) from a state with orbital quantum number $l$ can occur to only another state whose orbital quantum number differs from the original state by $\pm 1$, or

$$\Delta l = l_2 - l_1 = \pm 1 \tag{29.21}$$

Selection rule for emission and absorption

**EXAMPLE 29.8 Analyzing absorption in sodium**

What are the two longest wavelengths in the absorption spectrum of sodium? What are the transitions?

***STRATEGIZE*** Figure 29.24 shows the ground state and excited states of the sodium atom. We will look for the lowest-energy transitions that satisfy the rules given.

***PREPARE*** An absorption transition starts from the ground state; the only possible transitions are those for which $l$ changes by $\pm 1$.

***SOLVE*** As Figure 29.24 shows, the sodium ground state is $3s$. Starting from an $s$ state ($l = 0$), the selection rule permits quantum jumps only to $p$ states ($l = 1$). The lowest excited state is the $3p$ state and $3s \rightarrow 3p$ is an allowed transition ($\Delta l = 1$), so this will be the longest wavelength. You can see from the data in Figure 29.24 that $\Delta E_{atom} = 2.10 \text{ eV} - 0.00 \text{ eV} = 2.10 \text{ eV}$ for this transition. The corresponding wavelength is

$$\lambda = \frac{1240 \text{ eV} \cdot \text{nm}}{2.10 \text{ eV}} = 590 \text{ nm}$$

(Because of rounding, the calculation gives $\lambda = 590$ nm. The experimental value is actually 589 nm.)

The next excited state is $4s$, but a $3s \rightarrow 4s$ transition is not allowed by the selection rule. The next allowed transition is $3s \rightarrow 4p$, with $\Delta E_{atom} = 3.76$ eV. The wavelength of this transition is

$$\lambda = \frac{1240 \text{ eV} \cdot \text{nm}}{3.76 \text{ eV}} = 330 \text{ nm}$$

***ASSESS*** If you look at the sodium spectrum shown earlier in Figure 29.3b, you will see that 589 nm and 330 nm are, indeed, the two longest wavelengths in the absorption spectrum.

## Collisional Excitation

An electron traveling with a speed of $1.0 \times 10^6$ m/s has a kinetic energy of 2.85 eV. If this electron collides with a ground-state sodium atom, a portion of its energy can be used to excite the atom to a higher-energy state, such as its $3p$ state. This process is called **collisional excitation** of the atom.

Collisional excitation differs from excitation by absorption in one fundamental way. In absorption, the photon disappears. Consequently, *all* of the photon's energy must be transferred to the atom. Conservation of energy then requires $E_{photon} = \Delta E_{atom}$. In contrast, the electron is still present after collisional excitation and can still have some kinetic energy. That is, the electron does *not* have to transfer its entire energy to the atom. If the electron has an incident kinetic energy of 2.85 eV, it could transfer 2.10 eV to the sodium atom, thereby exciting it to the $3p$ state, and still depart the collision with a speed of $5.1 \times 10^5$ m/s and 0.75 eV of kinetic energy.

To excite the atom, the incident energy of the electron (or any other matter particle) merely has to *exceed* $\Delta E_{atom}$; that is, $E_{particle} \geq \Delta E_{atom}$. There's a threshold energy for exciting the atom, but no upper limit. It is all a matter of energy conservation. FIGURE 29.26 shows the idea graphically.

Collisional excitation by electrons is the predominant method of excitation in electrical discharges such as fluorescent lights, street lights, and neon signs. A gas is sealed in a tube at reduced pressure ($\approx 1$ mm Hg), then a fairly high voltage ($\approx 1000$ V) between electrodes at the ends of the tube causes the gas to ionize, creating a current in which both ions and electrons are charge carriers. The electrons accelerate in the electric field, gaining several eV of kinetic energy, then transfer some of this energy to the gas atoms upon collision.

**NOTE** ▶ In contrast to photon absorption, there are no selection rules for collisional excitation. Any state can be excited if the colliding particle has sufficient energy. ◀

FIGURE 29.26 Excitation by electron collision.

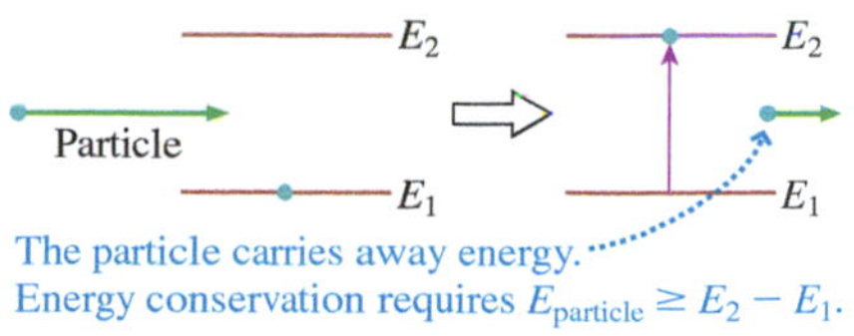


### CONCEPTUAL EXAMPLE 29.9 Possible excitation of hydrogen?

Can an electron with a kinetic energy of 11.4 eV cause a hydrogen atom to emit the prominent red spectral line ($\lambda = 656$ nm, $E_{photon} = 1.89$ eV) in the Balmer series?

***REASON*** The electron must have sufficient energy to excite the upper state of the transition. The electron's energy of 11.4 eV is significantly greater than the 1.89 eV energy of a photon with wavelength 656 nm, but don't confuse the energy of the photon with the energy of the excitation. The red spectral line in the Balmer series is emitted in an $n = 3 \rightarrow 2$ quantum jump with $\Delta E_{atom} = 1.89$ eV, but to cause this emission, the electron must excite an atom from its *ground state,* with $n = 1$, up to the $n = 3$ level. From Figure 29.16, the necessary excitation energy is

$$\Delta E_{atom} = E_3 - E_1 = (-1.51 \text{ eV}) - (-13.60 \text{ eV}) = 12.09 \text{ eV}$$

The electron does *not* have sufficient energy to excite the atom to the state from which the emission would occur.

***ASSESS*** As our discussion of absorption spectra showed, almost all excitations of atoms begin from the ground state. Quantum jumps down in energy, however, can begin and end at any two states allowed by selection rules.

## Emission Spectra

Understanding emission hinges on the three ideas shown in FIGURE 29.27. Once we have determined the energy levels of an atom, from quantum mechanics, we can immediately predict its emission spectrum.

As an example, FIGURE 29.28a shows some of the transitions and wavelengths observed in the emission spectrum of sodium. This diagram makes the point that each wavelength represents a quantum jump between two well-defined energy levels. Notice that the selection rule $\Delta l = \pm 1$ is obeyed in the sodium spectrum. The $5p$ levels can undergo quantum jumps to $3s$, $4s$, or $3d$ but *not* to $3p$ or $4p$.

FIGURE 29.28b shows the emission spectrum of sodium as it would be recorded in a spectrometer. (Many of the lines seen in this spectrum start from higher excited

FIGURE 29.27 Generation of an emission spectrum.

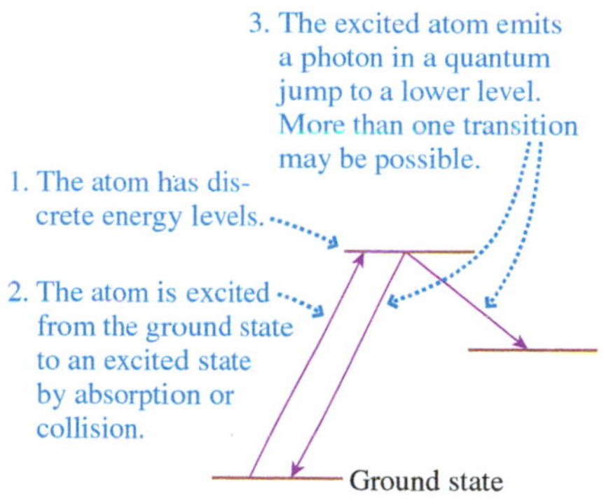

3. The excited atom emits a photon in a quantum jump to a lower level. More than one transition may be possible.
1. The atom has discrete energy levels.
2. The atom is excited from the ground state to an excited state by absorption or collision.
Ground state

FIGURE 29.28 The emission spectrum of sodium.

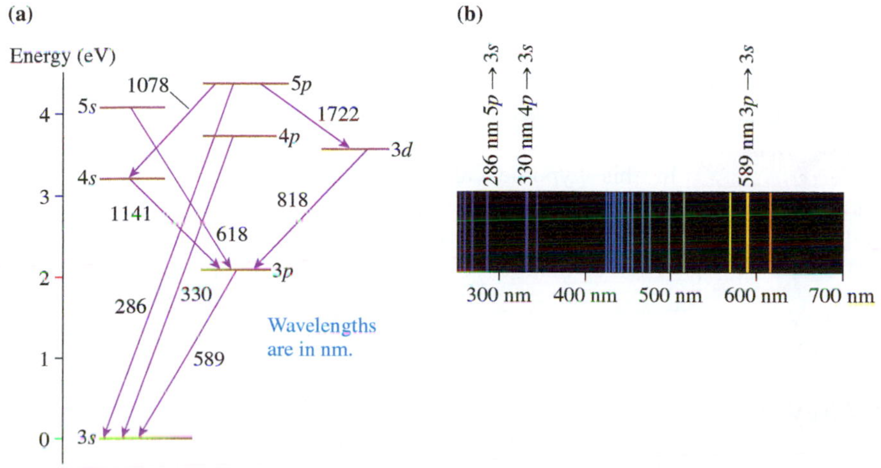

states that are not seen in the rather limited energy-level diagram of Figure 29.28a.) By comparing the spectrum to the energy-level diagram, you can recognize that the spectral lines at 589 nm, 330 nm, and 286 nm form a *series* of lines due to possible $np \rightarrow 3s$ transitions. They are the dominant features in the sodium spectrum.

The most obvious visual feature of sodium emission is its bright yellow color, produced by the 589 nm photons emitted in the $3p \rightarrow 3s$ transition. This is the basis of the *flame test* used in chemistry to test for sodium: A sample is held in a Bunsen burner, and a bright yellow glow indicates the presence of sodium. The 589 nm emission is also prominent in the pinkish-yellow glow of the common sodium-vapor street lights. These operate by creating an electrical discharge in sodium vapor. Most sodium-vapor lights use high-pressure lamps to increase their light output. The high pressure, however, causes the formation of $Na_2$ molecules, and these molecules emit the pinkish portion of the light.

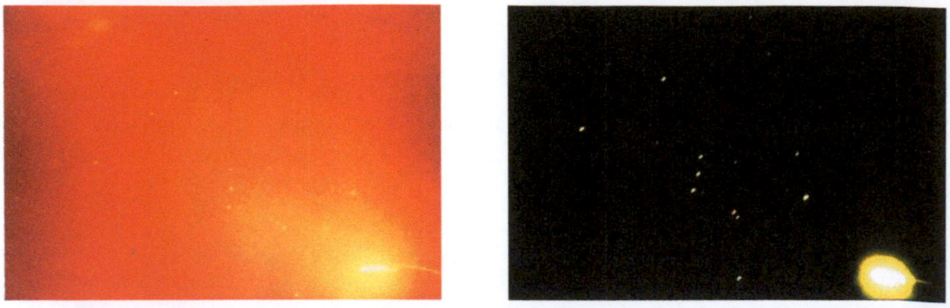

▲ **Seeing the light** For those of us who live in cities, the beauty of the night sky is obscured by the light from streetlights and other sources. Some cities close to astronomical observatories use low-pressure sodium lights, and these emit most of their light at the distinctively yellow 589 nm. This is not a wavelength of key importance to astronomers, so it does not affect their work. There are benefits for amateur astronomers as well. The very specific 589 nm emission from sodium can be removed with an interference filter that blocks this wavelength but passes others. Adding such a filter to a telescope gives a clear view of the night sky. The photos show the sky near a sodium streetlight without (left) and with (right) such a filter.

## X Rays

FIGURE 29.29 The generation of x rays from copper atoms.

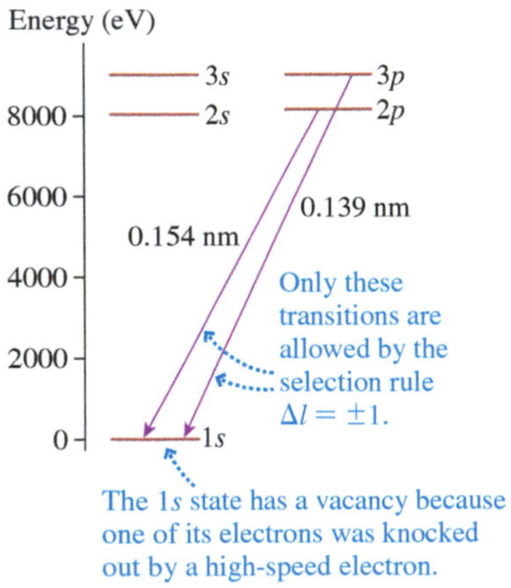


Chapter 28 noted that x rays are produced when very-high-speed electrons, accelerated with potential differences of many thousands of volts, crash into metal targets. Rather than exciting the atom's valence electrons, such as happens in a gas discharge tube, these high-speed projectiles are capable of knocking inner-shell electrons out of the target atoms, producing an *inner-shell vacancy.* As FIGURE 29.29 shows for copper atoms, this vacancy is filled when an electron from a higher shell undergoes a quantum jump into the vacancy, emitting a photon in the process.

In heavy elements, such as copper or iron, the energy difference between the inner and outer shells is very large—typically 10 keV. Consequently, the photon has energy $E_{photon} \approx 10$ keV and wavelength $\lambda \approx 0.1$ nm. These high-energy photons are the x rays discovered by Röntgen. X-ray photons are about 10,000 times more energetic than visible-light photons, and the wavelengths are about 10,000 times smaller. Even so, the underlying physics is the same: A photon is emitted when an electron in an atom undergoes a quantum jump.

**STOP TO THINK 29.8** In this hypothetical atom, what is the photon energy $E_{photon}$ of the longest-wavelength photons emitted by atoms in the $5p$ state?

A. 1.0 eV
B. 2.0 eV
C. 3.0 eV
D. 4.0 eV

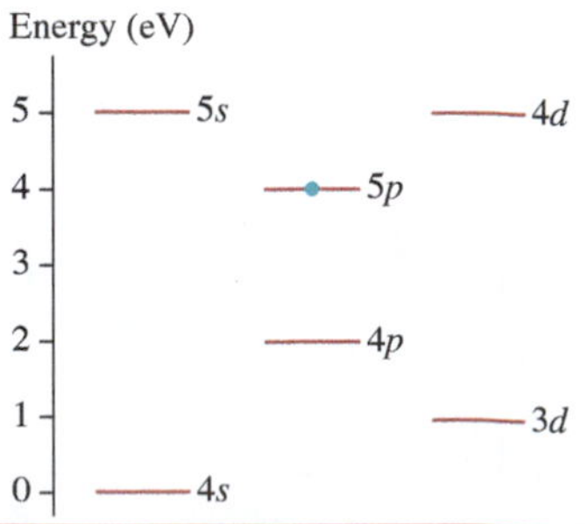

# 29.8 Molecules

Quantum mechanics applies to molecules just as it does to atoms, but molecules are more complex because they have internal modes of storing energy. In particular, molecules can *rotate* about their center of mass, and the atoms can *vibrate* back and forth as if the molecular bonds holding them together were little springs. For the most part, we'll overlook this internal motion and focus our attention on the electrons and the electron energy levels.

A quantum-mechanical analysis of molecules finds the following general results:

- The energy is quantized. Electrons can exist in only certain allowed energy levels.
- The number of energy levels is so extraordinarily high, and the energy levels are jammed so close together, that for all practical purposes the energy levels group into *bands* of allowed energy.
- In thermal equilibrium, nearly all molecules are in the very lowest energy levels.

FIGURE 29.30 shows a generic molecular energy-level diagram for a medium-size molecule. Whereas an atom has a well-defined ground state, a molecule has a broad band of lower energy levels. Similarly, a single excited state, such as the 2*s* state of the hydrogen atom, has been replaced by a band of excited energy levels. Despite the vast number of allowed energy levels, nearly all molecules spend nearly all their time in the very lowest energy levels.

FIGURE 29.30 The molecular energy-level diagram for a medium-size molecule.

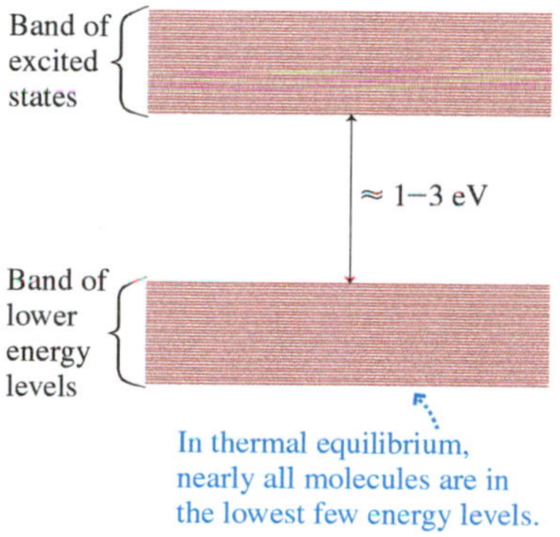


FIGURE 29.31 uses the energy-level diagram to explain two important phenomena of molecular spectroscopy: absorption and fluorescence. Whereas the absorption spectrum of an atom consists of discrete spectral lines, a molecule has a continuous *absorption band.* The absorption of light at a higher frequency (shorter wavelength) followed by the emission of light at a lower frequency (longer wavelength) is called **fluorescence.** Fluorescence occurs in molecules, but not atoms, because molecules can transform some of the absorbed energy into the vibrational energy of the atoms and thus increase the thermal energy of the molecules.

FIGURE 29.31 Molecular absorption and fluorescence.

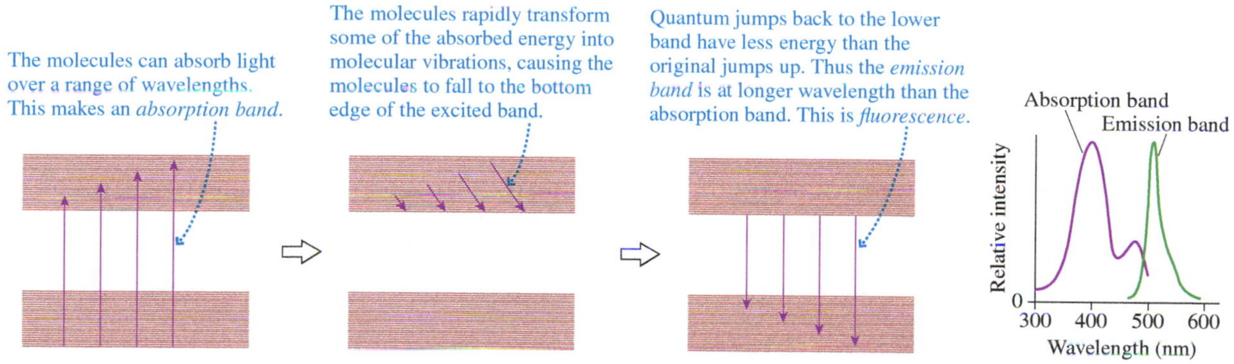

The molecules can absorb light over a range of wavelengths. This makes an *absorption band.*
The molecules rapidly transform some of the absorbed energy into molecular vibrations, causing the molecules to fall to the bottom edge of the excited band.
Quantum jumps back to the lower band have less energy than the original jumps up. Thus the *emission band* is at longer wavelength than the absorption band. This is *fluorescence.*
Absorption band
Emission band
Relative intensity
0
300
400
500
600
Wavelength (nm)

▶ **Whiter than white** Laundry detergents often contain fluorescent dyes that absorb invisible ultraviolet light and then fluoresce in a broad band of visible wavelengths; they give off white light. The white glow of the laundered shirt is evident in this photo of a person working under ultraviolet light. But this effect is useful even under normal circumstances. Sunlight contains lots of UV, so outdoors your white clothes not only reflect the white visible sunlight, but fluorescence produces even more white light—your clothes appear "whiter than white."

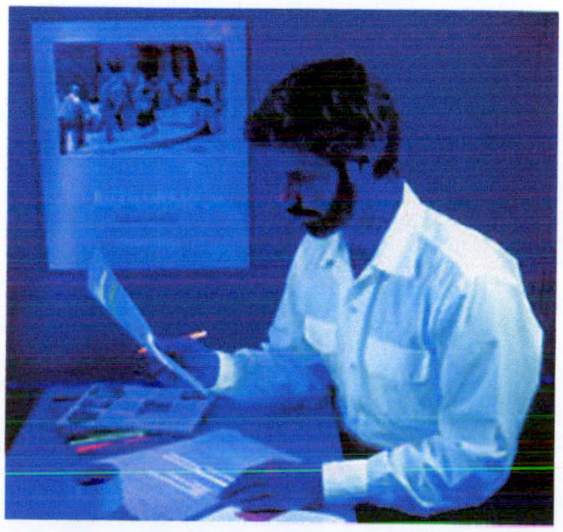

Many biological molecules are fluorescent. A good example is chlorophyll, the green pigment in plants that allows photosynthesis to happen. When illuminated by blue or ultraviolet light, chlorophyll fluoresces an unexpected dark red color. This property of chlorophyll is used by marine biologists to measure the concentration of phytoplankton—microscopic plants—in seawater.

**Video** Fluorescence

FIGURE 29.32 BIO Green fluorescent protein shows the locations in mosquito larvae at which a particular gene is being expressed.

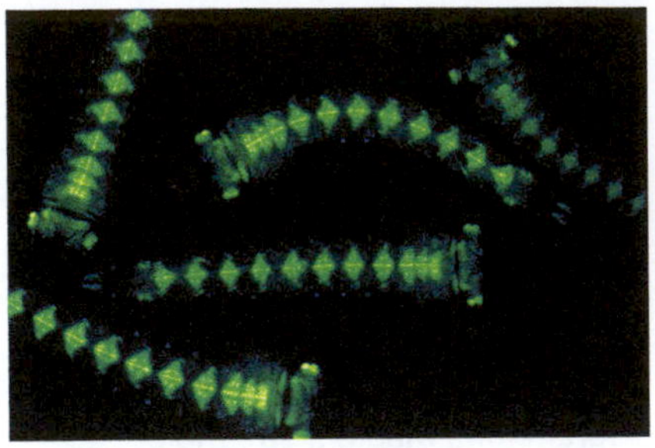

Recently, scientists isolated a fluorescent protein from a species of jellyfish. This protein fluoresces green when illuminated with ultraviolet light, so it's been dubbed GFP, for *green fluorescent protein.* GFP has become an important tool in genetics because it can be used to identify when particular genes are being expressed in living cells. This is done by fusing the jellyfish GFP gene to the gene being studied. When the gene is active, the cell manufactures GFP in addition to the usual protein coded by the gene. If the cells are observed under ultraviolet light, a bright green glow from the GFP indicates that the gene is turned on. The cells are dark where the gene is not active. FIGURE 29.32 shows an example.

The color of GFP may remind you of *bioluminescence*—the summer flashes of fireflies or the green glow of various deep-sea fish—and, indeed, there is a connection. Bioluminescence is actually a form of *chemiluminescence,* the production of light in chemical reactions. Some chemical reactions create reactant molecules in an excited state. These molecules then emit light as they jump to lower energy levels, just as if they had first absorbed shorter-wavelength light. The emitted light has exactly the same spectrum as fluorescence, but the method by which the molecules are excited is different. Light sticks, which come in many colors, are an example of chemiluminescence.

Bioluminescence is just chemiluminescence in a biological organism. Some biochemical reaction—often a catalyzed reaction in an organism—produces a molecule in an excited state, and light is emitted as the molecule jumps to a lower state. In fireflies, the reaction involves an enzyme with the intriguing name *luciferase.* In the jellyfish *Aequoria victoria,* seen in the photo at the start of the chapter, the source of GFP, bioluminescent reactions actually create blue light. However, the blue light is absorbed by the green fluorescent protein—just as predicted by the absorption curve in Figure 29.31—and re-emitted as green fluorescence, giving the jellyfish an eerie green glow when seen in the dark ocean.

## 29.9 Stimulated Emission and Lasers

FIGURE 29.33 Three types of radiative transitions.

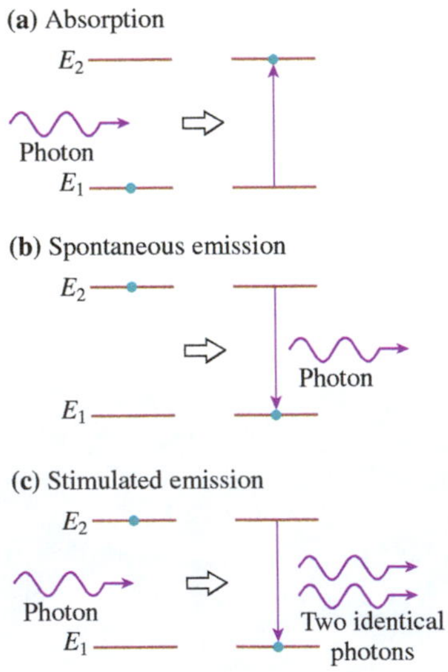


We have seen that an atom can jump from a lower-energy level $E_1$ to a higher-energy level $E_2$ by absorbing a photon. FIGURE 29.33a illustrates the process. Once in level 2, as shown in FIGURE 29.33b, the atom emits a photon of the same energy as it jumps back to level 1. Because this transition occurs spontaneously, it is called **spontaneous emission.**

In 1917, Einstein proposed a second mechanism by which an atom in state 2 can make a transition to state 1. The left part of FIGURE 29.33c shows a photon approaching an atom in its excited state 2. According to Einstein, if the energy of the photon is exactly equal to the energy difference $E_2 - E_1$ between the two states, this incoming photon can *induce* the atom to make the $2 \rightarrow 1$ transition, emitting a photon in the process. This process is called **stimulated emission.**

The incident photon is *not* absorbed in the process, so now there are *two* photons. And, interestingly, the emitted photon is *identical* to the incident photon. This means that as the two photons leave the atom they have exactly the same frequency and wavelength, are traveling in exactly the same direction, and are exactly in phase with each other. In other words, **stimulated emission produces a second photon that is an exact clone of the first.**

### Lasers

The word **laser** is an acronym for the phrase *light amplification by the stimulated emission of radiation*. But what *is* a laser? Basically it is a device that produces a beam of highly *coherent* and essentially monochromatic (single-color) light as a result of stimulated emission. **Coherent** light is light in which all the electromagnetic waves have the same phase, direction, and amplitude. It is the coherence of a laser beam that allows it to be very tightly focused or to be rapidly modulated for communications.

A spectacular laser light show depends on three key properties of coherent laser light: It can be very intense, its color is extremely pure, and the laser beam is narrow with little divergence.

Let's take a brief look at how a laser works. FIGURE 29.34 represents a system of atoms that have a lower energy level $E_1$ and a higher energy level $E_2$. Suppose that there are $N_1$ atoms in level 1 and $N_2$ atoms in level 2. Left to themselves, all the atoms would soon end up in level 1 because of the spontaneous emission $2 \rightarrow 1$. To prevent this, we can imagine that some type of excitation mechanism, perhaps an electrical discharge, continuously produces new excited atoms in level 2.

Let a photon of frequency $f = (E_2 - E_1)/h$ be incident on this group of atoms. Because it has the correct frequency, it could be absorbed by one of the atoms in level 1. Another possibility is that it could cause stimulated emission from one of the level 2 atoms. Ordinarily $N_2 \ll N_1$, so absorption events far outnumber stimulated emission events. Even if a few photons were generated by stimulated emission, they would quickly be absorbed by the vastly larger group of atoms in level 1.

**FIGURE 29.34** Energy levels 1 and 2, with populations $N_1$ and $N_2$.

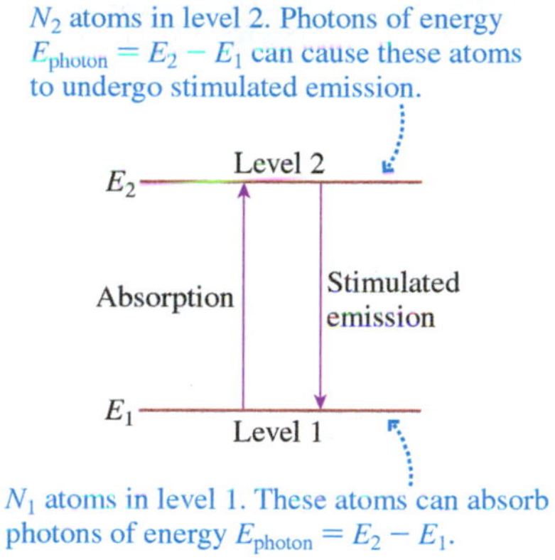


But what if we could somehow arrange to place *every* atom in level 2, making $N_1 = 0$? Then the incident photon, upon encountering its first atom, will cause stimulated emission. Where there was initially one photon of frequency $f$, now there are two. These will strike two additional excited-state atoms, again causing stimulated emission. Then there will be four photons. As FIGURE 29.35 shows, there will be a *chain reaction* of stimulated emission until all $N_2$ atoms emit a photon of frequency $f$.

In stimulated emission, each emitted photon is *identical* to the incident photon. The chain reaction of Figure 29.35 will lead not just to $N_2$ photons of frequency $f$, but to $N_2$ *identical* photons, all traveling together in the same direction with the same phase. If $N_2$ is a large number, as would be the case in any practical device, the one initial photon will have been *amplified* into a gigantic, coherent pulse of light!

**FIGURE 29.35** Stimulated emission creates a chain reaction of photon production in a population of excited atoms.

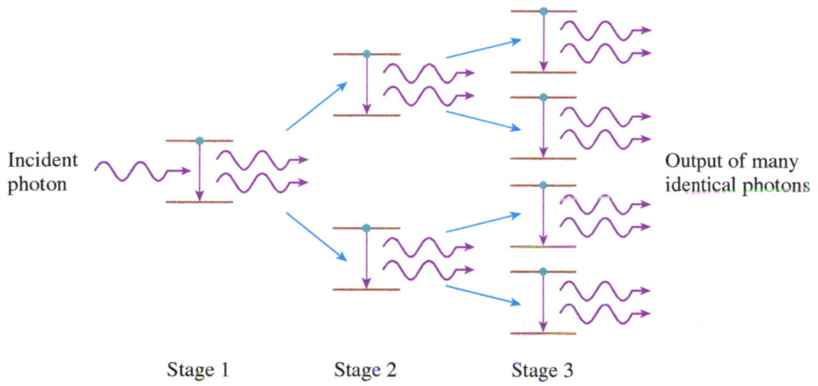


Although the chain reaction of Figure 29.35 illustrates the idea most clearly, it is not necessary for every atom to be in level 2 for amplification to occur. All that is needed is to have $N_2 > N_1$ so that stimulated emission exceeds absorption. Such a situation is called a **population inversion.** The stimulated emission is sustained by placing the *lasing medium*—the sample of atoms that emits the light—in an **optical cavity** consisting of two facing mirrors. As FIGURE 29.36 shows, the photons interact repeatedly with the atoms in the medium as they bounce back and forth. This repeated interaction is necessary for the light intensity to build up to a high level. If one of the mirrors is partially transmitting, some of the light emerges as the *laser beam.*

**FIGURE 29.36** Lasing takes place in an optical cavity.

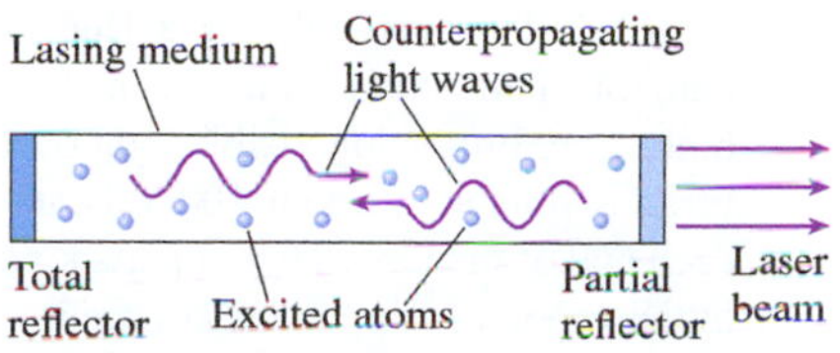


## Lasers in Medicine

The invention of lasers was followed almost immediately by medical applications. Even a small-power laser beam can produce a significant amount of very localized *heating* if focused with a lens. More powerful lasers can easily *cut* through tissue by literally vaporizing it, replacing a stainless steel scalpel with a beam of light. Not only can laser surgery be very precise, but it generally has less blood loss than conventional surgery because the heat of the laser seals the blood vessels and capillaries.

One common medical use of lasers is to remove plaque from artery walls, thus reducing the risk of stroke or heart attack. In this procedure, an optical fiber is threaded through arteries to reach the site. A powerful laser beam is then fired through the fiber to carefully vaporize the plaque. Laser beams traveling through optical fibers are also used to kill cancer cells in *photodynamic therapy*. In this case, light-sensitive chemicals are injected into the bloodstream and are preferentially taken up by cancer cells. The optical fiber is positioned next to the tumor and illuminates it with just the right wavelength to activate the light-sensitive chemicals and kill the cells. These procedures are minimally invasive, and they can reach areas of the body not readily accessible by conventional surgery.

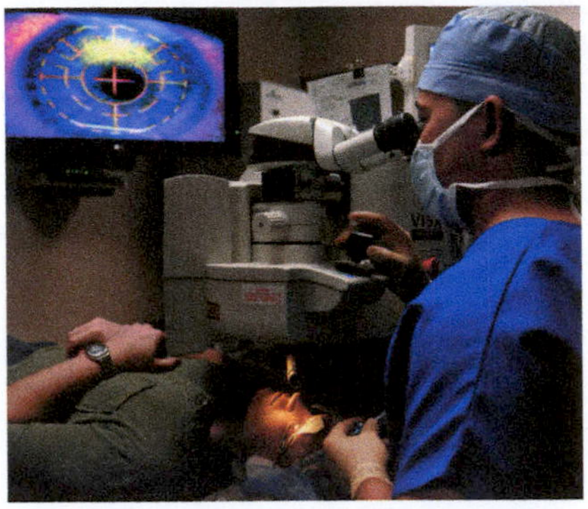

◀ **Laser vision** BIO Laser-based LASIK surgery can correct for vision defects, such as near- or farsightedness, that result from an incorrect refractive power of the eye. You learned in Chapter 19 that the majority of your eye's focusing power occurs at the surface of the cornea. In LASIK, a special knife first cuts a small, thin flap in the cornea, and this flap is folded out of the way. A computer-controlled ultraviolet laser very carefully vaporizes the underlying corneal tissue to give it the desired shape; then the flap is folded back into place. The procedure takes only a few minutes and requires only a few numbing drops in the eye.

## INTEGRATED EXAMPLE 29.10 Compact fluorescent lighting

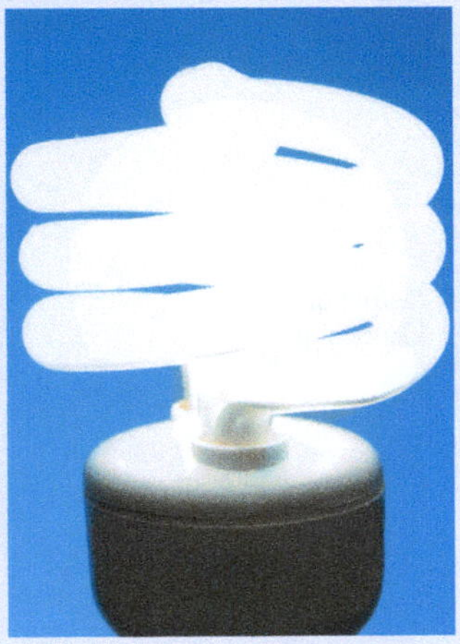

You learned in Chapters 22 and 25 how an ordinary incandescent bulb works: Current passes through a filament, heating it until it glows white hot. But such bulbs are very inefficient, giving off only a few watts of visible light for every 100 W of electric power supplied to the bulb. Most of the power is, instead, converted into thermal energy.

This is why, in recent years, incandescent lighting has given way to other types of lighting. The compact fluorescent lamp has become a symbol of energy efficiency. These bulbs are about four times more efficient than incandescent bulbs at transforming electric energy into visible light.

Inside the glass tube of a fluorescent bulb is a very small amount of mercury, which is in the form of a vapor when the bulb is on. Producing visible light occurs by a three-step process. First, a voltage of about 100 V is applied between electrodes at each end of the tube. This imparts kinetic energy to free electrons in the vapor, causing them to slam into mercury atoms and excite them by collisional excitation. Second, the excited atoms then jump to lower-energy states, emitting UV photons. Finally, the UV photons strike a *phosphor* that coats the inside of the tube, causing it to fluoresce with visible light. This is the light you see.

a. A mercury atom, after being collisionally excited by an electron, emits a photon with a wavelength of 185 nm in a quantum jump back to the ground state. If the electron starts from rest, what minimum distance must it travel to gain enough kinetic energy to cause this excitation? The 60-cm-long tube has 120 V applied between its ends.

b. After being collisionally excited, atoms sometimes emit two photons by jumping first from a high energy level to an intermediate level, giving off one photon, and then from this intermediate level to the ground state, giving off a second photon. An atom is excited to a state that is 7.79 eV above the ground state. It emits a 254-nm-wavelength photon and then a second photon. What is the wavelength of the second photon?

c. The energy-level diagram of the molecules in the phosphor is shown in FIGURE 29.37. After excitation by UV photons, what range of wavelengths can the phosphor emit by fluorescence?

FIGURE 29.37 Energy-level diagram of the phosphor molecules.

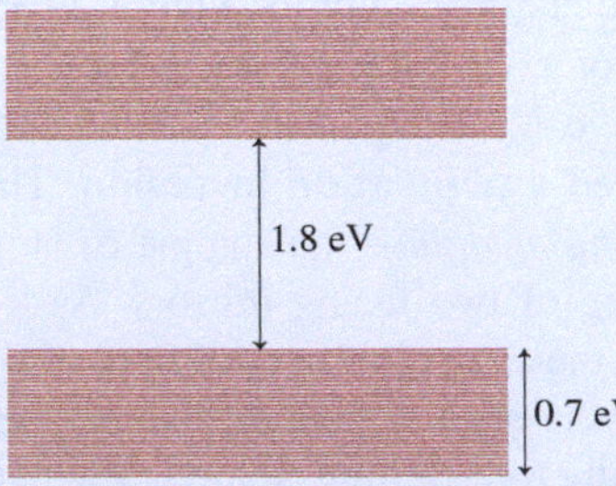


***STRATEGIZE*** There are three parts to the problem. Each part is concerned with the energy change as electrons move between different states. The initial excitation is produced by the kinetic energy of a moving electron; the electron must have an energy greater than or equal to the energy difference between the two states. When the atom drops into a lower energy state, the difference in energy is converted into photons. If we know the energy of the transition and the energy of one photon, we can find the energy of the second. Finally, we consider the energy-level diagram of the phosphor; the possible differences in energy of electrons moving between states correspond to the energies of emitted photons.

**PREPARE** An electron collisionally excites the mercury atom from its ground state to an excited state, increasing the atom's energy by $\Delta E_{\text{atom}}$. Then the atom decays back to the ground state, giving off a 185-nm-wavelength photon. In order for collisional excitation to work, the kinetic energy of the incident electron must equal or exceed $\Delta E_{\text{atom}}$; that is, it must equal or exceed the energy of a 185-nm-wavelength photon. The free electrons gain kinetic energy by accelerating through the potential difference inside the tube. We can use conservation of energy to find the distance an electron must travel to gain $\Delta E_{\text{atom}}$, the minimum kinetic energy needed to cause an excitation.

In part b, the excited energy of the atom, 7.79 eV, is converted into photon energies. If we can find the energy of the first 254-nm-wavelength photon, the remaining energy must be that of the second photon, from which we can find its wavelength.

An inspection of Figure 29.31 shows that quantum jumps during fluorescence all begin at the bottom of the band of excited states but can end anywhere in the lower energy band. This range of energies will give us the range of wavelengths of the emitted photons.

**SOLVE**

a. The minimum kinetic energy of the electron equals the energy of the 185-nm-wavelength photon that is subsequently emitted. This energy is

$$K_{\text{min}} = \Delta E_{\text{min}} = E_{\text{photon}} = \frac{hc}{\lambda} = \frac{1240\ \text{eV}\cdot\text{nm}}{185\ \text{nm}} = 6.7\ \text{eV}$$

where we have used the value of $hc$ from Equation 29.20.

Recall that 1 eV is the kinetic energy gained by an electron as it accelerates through a 1 V potential difference. Here, the electron must gain a kinetic energy of 6.7 eV, so it must accelerate through a potential difference of 6.7 V. The fluorescent tube has a total potential drop of 120 V in 60 cm, or 2.0 V/cm. Thus to accelerate through a 6.7 V potential difference, the electron must travel a distance

$$\Delta x = \frac{6.7\ \text{V}}{2.0\ \text{V/cm}} = 3.4\ \text{cm}$$

b. The first emitted photon has energy

$$E_{\text{photon}} = \frac{hc}{\lambda} = \frac{1240\ \text{eV}\cdot\text{nm}}{254\ \text{nm}} = 4.88\ \text{eV}$$

The energy remaining to the second photon is then $7.79\ \text{eV} - 4.88\ \text{eV} = 2.91\ \text{eV}$; the wavelength of this photon is

$$\lambda = \frac{hc}{E_{\text{photon}}} = \frac{1240\ \text{eV}\cdot\text{nm}}{2.91\ \text{eV}} = 426\ \text{nm}$$

c. The energy of the photon emitted during a quantum jump from the bottom of the upper energy band to the top of the lower energy band is 1.8 eV, corresponding to a wavelength of

$$\lambda = \frac{hc}{E_{\text{photon}}} = \frac{1240\ \text{eV}\cdot\text{nm}}{1.8\ \text{eV}} = 690\ \text{nm}$$

The photon energy for a jump to the bottom of the lower energy band is $1.8\ \text{eV} + 0.7\ \text{eV} = 2.5\ \text{eV}$, with a wavelength of

$$\lambda = \frac{hc}{E_{\text{photon}}} = \frac{1240\ \text{eV}\cdot\text{nm}}{2.5\ \text{eV}} = 500\ \text{nm}$$

Thus this phosphor, after absorbing UV photons, emits visible light with a wavelength range of 500–690 nm.

**ASSESS** For part a, it seems reasonable that the electron travels only a small fraction of the tube length before gaining enough energy to collisionally excite an atom. The bulb would be very inefficient if electrons had to travel the full length before colliding with a mercury atom. The photon wavelengths in parts b and c also seem reasonable. In particular, the range of wavelengths emitted by the phosphor is a little more than half the visible spectrum, from green through red but missing blue and violet. Compact fluorescent tubes have three different phosphors, each with a somewhat different range of emission wavelengths, to give the full spectrum of white light. Slightly altering the balance between these phosphors distinguishes a "warm white" bulb from a "cool white" bulb.

# SUMMARY

**GOAL** To use quantum physics to understand the properties of atoms, molecules, and their spectra.

## IMPORTANT CONCEPTS

### The Structure of an Atom

An atom consists of a very small, positively charged nucleus, surrounded by orbiting electrons.

- The number of protons is the atom's **atomic number Z.**
- The **atomic mass number** *A* is the number of protons + the number of neutrons.

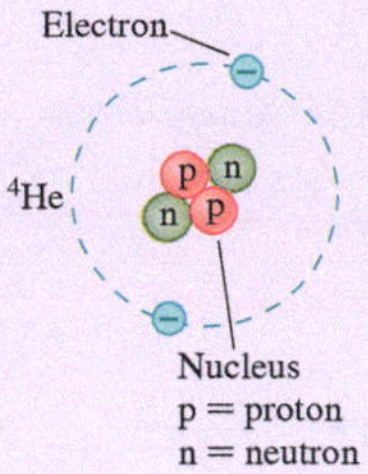


### The Bohr Atom

In Bohr's model,

- The atom can exist in only certain **stationary states.** These states correspond to different electron orbits. Each state is numbered by **quantum number** $n = 1, 2, 3, \ldots$.
- Each state has a discrete, well-defined energy $E_n$.
- The atom can change its energy by undergoing a **quantum jump** between two states by emitting or absorbing a photon of energy $E_{\text{photon}} = \Delta E_{\text{atom}} = |E_f - E_i|$.

### The Hydrogen Atom

In **Bohr's model of the hydrogen atom** the stationary states are found by requiring an integer number of de Broglie wavelengths to fit around the circumference of the electron's orbit: $2\pi r = n\lambda$. The integer *n* is the *principal quantum number.*

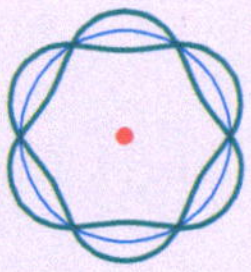

This leads to energy quantization with

$$E_n = -\frac{13.60 \text{ eV}}{n^2}$$

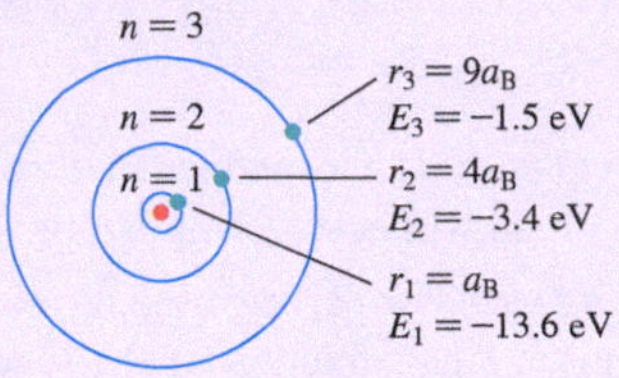


and orbit radii $r_n = n^2 a_B$, where $a_B = 0.053$ nm is the **Bohr radius.**

The wavelengths of light in the hydrogen atom spectrum are given by the **Balmer formula:**

$$\lambda_{n \to m} = \frac{91.1 \text{ nm}}{\left(\frac{1}{m^2} - \frac{1}{n^2}\right)} \qquad \begin{aligned} &m = 1, 2, 3, \ldots \\ &n = m+1, m+2, \ldots \end{aligned}$$

**Beyond the Bohr model,** quantum mechanics adds other quantized parameters, each with its own quantum number:

- The *orbital angular momentum,* quantum number *l*:

$$L = \sqrt{l(l+1)}\,\hbar \qquad l = 0, 1, 2, 3, \ldots, n-1$$

- The *angle of the electron's orbit,* quantum number *m*:

$$m = -l, -l+1, \ldots, 0, \ldots, l-1, l$$

- The *direction of the electron* **spin,** quantum number $m_s$:

$$m_s = -\tfrac{1}{2} \text{ or } +\tfrac{1}{2}$$

The energy of a hydrogen atom depends only on *n*:

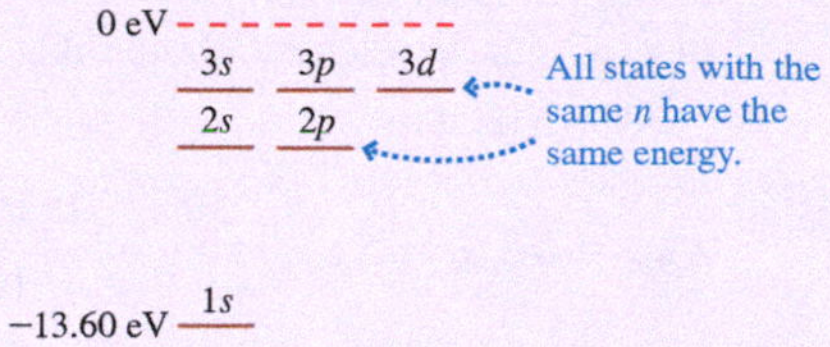


### Multi-electron atoms

Each electron is described by the same quantum numbers $(n, l, m, m_s)$ used for the hydrogen atom, but the energy now depends on *l* as well as *n*.

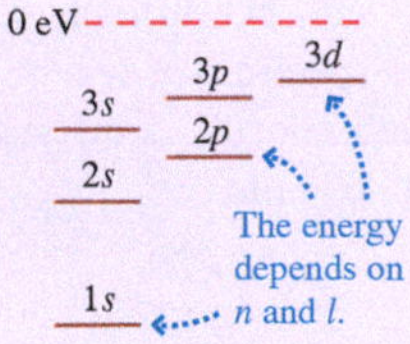


The **Pauli exclusion principle** states that no more than one electron can occupy each quantum state.

### Molecules

In molecules, the states are spaced very closely into **bands** of states. Because electrons can be excited to and from many states, the spectra of molecules are broad, not discrete.

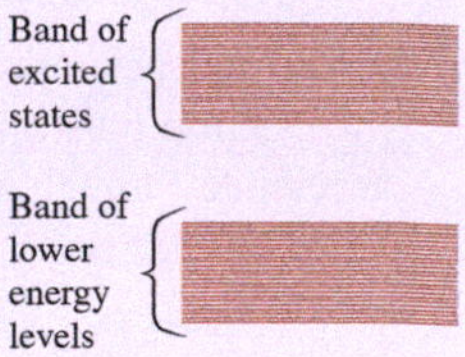


## APPLICATIONS

**Atomic emission spectra** are generated by excitation followed by a photon-emitting quantum jump.

- **Excitation** occurs by absorption of a photon or by collision.
- A quantum jump can occur only if $\Delta l = \pm 1$.
- Quantized energies give rise to a **discrete spectrum.**

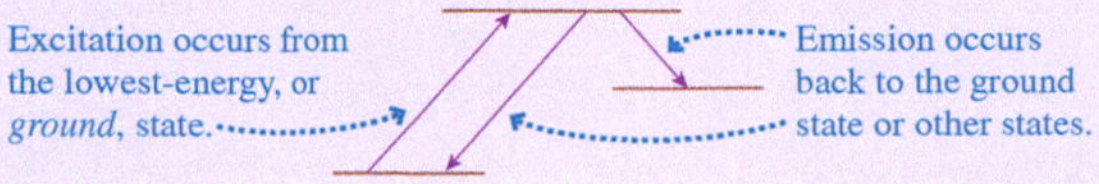


### Lasers

A photon with energy $E_{\text{photon}} = E_2 - E_1$ can induce **stimulated emission** of a second photon identical to the first. These photons can then induce more atoms to emit photons. If more atoms are in state 2 than in state 1, this process can rapidly build up an intense beam of identical photons. This is the principle behind the laser.

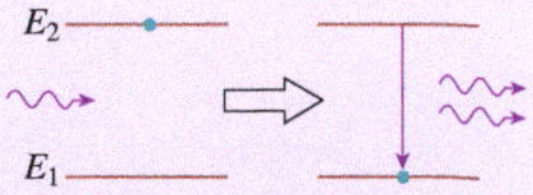

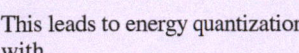

## Learning Objectives After studying this chapter, you should be able to:

- Work with and distinguish between emission and absorption spectra. *Conceptual Question 29.2; Problems 29.1, 29.2, 29.3, 29.4, 29.5*
- Understand and apply the nuclear model of the atom. *Conceptual Question 29.4; Problems 29.6, 29.7, 29.8, 29.10, 29.12*
- Understand and apply Bohr's model of atomic quantization. *Conceptual Questions 29.1, 29.5; Problems 29.14, 29.15, 29.16, 29.17*
- Calculate energy levels and spectra of the hydrogen atom. *Conceptual Questions 29.7, 29.8; Problems 29.22, 29.23, 29.26, 29.27, 29.30*
- Apply quantum ideas to multi-electron atoms and their spectra. *Conceptual Questions 29.17, 29.22; Problems 29.31, 29.36, 29.39, 29.42, 29.43*
- Understand how lasers work. *Conceptual Questions 29.24, 29.25; Problems 29.47, 29.48, 29.49, 29.50, 29.51*

## STOP TO THINK ANSWERS

**Chapter Preview Stop to Think: C.** Once the system is in the $n = 3$ state, it can emit a photon as it jumps directly to the $n = 1$ ground state, or two photons as it jumps first to the $n = 2$ state and then to the $n = 1$ state. Energy conservation requires that the energy of the emitted photon equal the *difference* between the energies of the two states. These differences are $E_3 - E_1 = 8$ eV, $E_3 - E_2 = 5$ eV, and $E_2 - E_1 = 3$ eV.

**Stop to Think 29.1: A is emission, B is absorption.** All wavelengths in the absorption spectrum are seen in the emission spectrum, but not all wavelengths in the emission spectrum are seen in the absorption spectrum.

**Stop to Think 29.2: 6 protons and 8 neutrons.** The number of protons is the atomic number, which is 6. That leaves $14 - 6 = 8$ neutrons.

**Stop to Think 29.3: B.** Because the atoms absorb 3.0 eV photons, the $n = 3$ state must be 3.0 eV above the ground state. The excited atom can then jump directly to the ground state, emitting a 3.0 eV photon in the process. But the atom could also undergo a $3 \rightarrow 2$ transition followed by a $2 \rightarrow 1$ transition. Suppose it is the $2 \rightarrow 1$ transition that emits the 2.0 eV photon. Then the $n = 2$ state must have an energy of 2.0 eV, and so the $3 \rightarrow 2$ transition must emit a 1.0 eV photon. (The same result is found if it is the $3 \rightarrow 2$ transition that emits the 2.0 eV photon.)

**Stop to Think 29.4: In emission from the $n = 3$ to $n = 2$ transition, but not in absorption.** The photon energy has to match the energy *difference* between two energy levels. Absorption is from the ground state, at $E_1 = 0$ eV. There's no energy level at 3 eV to which the atom could jump.

**Stop to Think 29.5: $n = 3$.** Each antinode is half a wavelength, so this standing wave has three full wavelengths in one circumference.

**Stop to Think 29.6: $n = 3, l = 1$, or a $3p$ state.**

**Stop to Think 29.7: A.** An inspection of the periodic table in Figure 29.23 shows that the element that has three of the possible six $3p$ states filled is phosphorus (P).

**Stop to Think 29.8: C.** Emission is a quantum jump to a lower-energy state. The $5p \rightarrow 4p$ transition is not allowed because $\Delta l = 0$ violates the selection rule. The lowest-energy allowed transition is $5p \rightarrow 3d$, with $E_{photon} = \Delta E_{atom} = 3.0$ eV.

Video Tutor Solution Chapter 29

# QUESTIONS

## Conceptual Questions

1. A neon discharge emits a bright reddish-orange spectrum. But a glass tube filled with neon is completely transparent. Why doesn't the neon in the tube absorb orange and red wavelengths?
2. The two spectra shown in Figure Q29.2 belong to the same element, a fictional Element X. Explain why they are different.

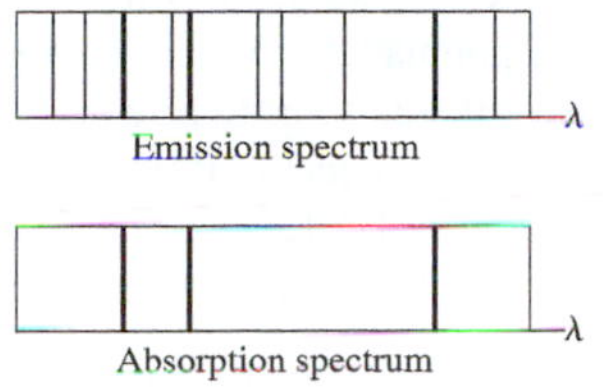


FIGURE Q29.2

3. Is a spectral line with wavelength 656 nm seen in the absorption spectrum of hydrogen atoms? Why or why not?
4. Can nuclei of the same element have different values of $Z$? Of $N$? Of $A$? Can nuclei of different elements have the same values of $Z$? Of $N$? Of $A$?
5. An atom has four energy levels. How many lines are in its emission spectrum? In its absorption spectrum?
6. An atom has exactly two lines in its absorption spectrum. How many energy levels does it have? How many lines are in its emission spectrum?
7. The $n = 3$ state of hydrogen has $E_3 = -1.51$ eV.
   a. Why is the energy negative?
   b. What is the physical significance of the specific number 1.51 eV?

Problem difficulty is labeled as | (straightforward) to ||||| (challenging). Problems labeled INT integrate significant material from earlier chapters; Problems labeled BIO are of biological or medical interest.

The eText icon indicates when there is a video tutor solution available for the chapter or for a specific problem. To launch these videos, log into your eText through Mastering™ Physics or log into the Study Area.

8. For a hydrogen atom, list all possible states $(n, l, m, m_s)$ that have $E = -1.51$ eV.
9. What are the $n$ and $l$ values of the following states of a hydrogen atom: (a) $4d$, (b) $5f$, (c) $6s$?
10. How would you label the hydrogen-atom states with the following $(n, l, m)$ quantum numbers: (a) $(4, 3, 0)$, (b) $(3, 2, 1)$, (c) $(3, 2, -1)$?
11. A hydrogen atom is in a state with principal quantum number $n = 5$. What possible values of the orbital quantum number $l$ could this atom have?
12. In a multi-electron atom, does a $5s$ electron have higher or lower energy than a $4f$ electron? Explain.
13. Do the following electron configurations represent a possible state of an element? If so, (i) identify the element and (ii) determine if this is the ground state or an excited state. If not, why not?
    a. $1s^2 2s^2 2p^6 3s^2$
    b. $1s^2 2s^2 2p^7 3s$
    c. $1s^2 2s^2 2p^4 3s^2 3p^2$
14. Why is the section of the periodic table labeled as "transition elements" exactly 10 elements wide in all rows?
15. The lanthanides are a series of elements in the periodic table, from $Z = 57$ to $Z = 70$, that appear in a row set below the main table. Explain why there are exactly 14 elements in this row.
16. Suppose that electrons had three possible values of the spin quantum number. What would be the electron configuration of an element with $Z = 10$?
17. In recent years, the periodic table has been expanded through element 118, which has a filled $7p$ subshell. If element 119 is synthesized, the extra electron is likely to go into the $8s$ subshell. Where would this element go in the periodic table—which element would be directly above it in the table?
18. An electron is in an $f$ state. Can it undergo a quantum jump to an $s$ state? A $p$ state? A $d$ state? Explain.
19. Figure Q29.19 shows the energy-level diagram of Element X.
    a. What is the ionization energy of Element X?
    b. An atom in the ground state absorbs a photon, then emits a photon with a wavelength of 1240 nm. What conclusion can you draw about the energy of the photon that was absorbed?
    c. An atom in the ground state has a collision with an electron, then emits a photon with a wavelength of 1240 nm. What conclusion can you draw about the initial kinetic energy of the electron?

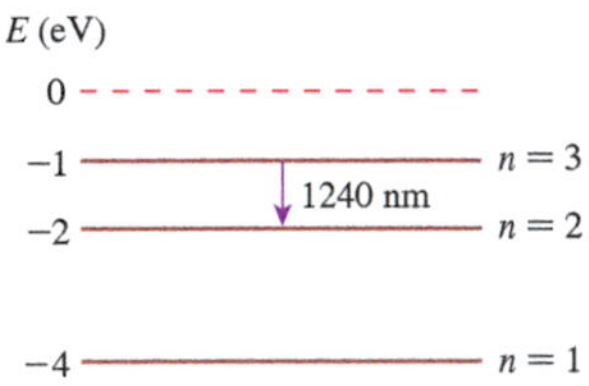


FIGURE Q29.19

20. a. Which states of a hydrogen atom can be excited by a collision with an electron with kinetic energy $K = 12.5$ eV? Explain.
    b. After the collision the atom is not in its ground state. What happens to the electron? (i) It bounces off with $K > 12.5$ eV, (ii) It bounces off with $K = 12.5$ eV, (iii) It bounces off with $K < 12.5$ eV, (iv) It is absorbed by the atom. Explain your choice.
    c. After the collision, the atom emits a photon. List all the possible $n \rightarrow m$ transitions that might occur as a result of this collision.
21. What *is* an atom's ionization energy? In other words, if you know the ionization energy of an atom, what is it that you know about the atom?
22. Figure Q29.22 shows the energy levels of a hypothetical atom.
    a. What *minimum* kinetic energy (in eV) must an electron have to collisionally excite this atom and cause the emission of a 620 nm photon? Explain.
    b. Can an electron with $K = 6$ eV cause the emission of 620 nm light from this atom? If so, what is the final kinetic energy of the electron? If not, why not?
    c. Can a 6 eV photon cause the emission of 620 nm light from this atom? Why or why not?
    d. Can a 7 eV photon cause the emission of 620 nm light from this atom? Why or why not?

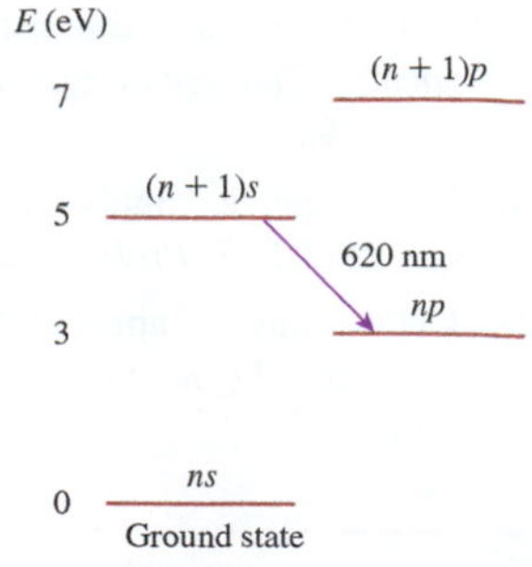


FIGURE Q29.22

23. Seven possible transitions are identified on the energy-level diagram in Figure Q29.23. For each, is this an allowed transition? If allowed, is it an emission or an absorption transition, and is the photon infrared, visible, or ultraviolet? If not allowed, why not?

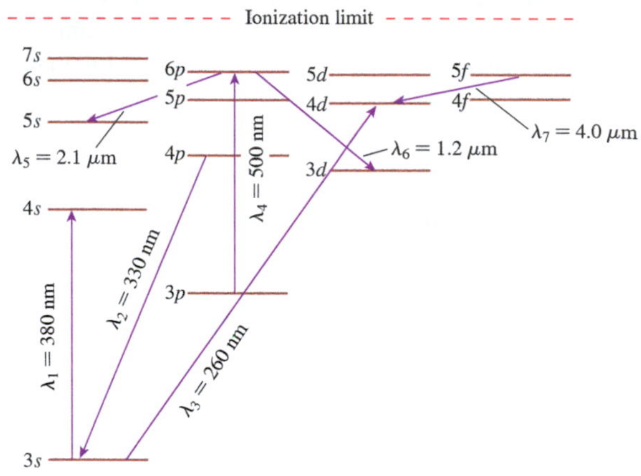


FIGURE Q29.23

24. A 2.0 eV photon is incident on an atom in the $p$ state, as shown in the energy-level diagram in Figure Q29.24. Does the atom undergo an absorption transition, a stimulated emission transition, or neither? Explain.

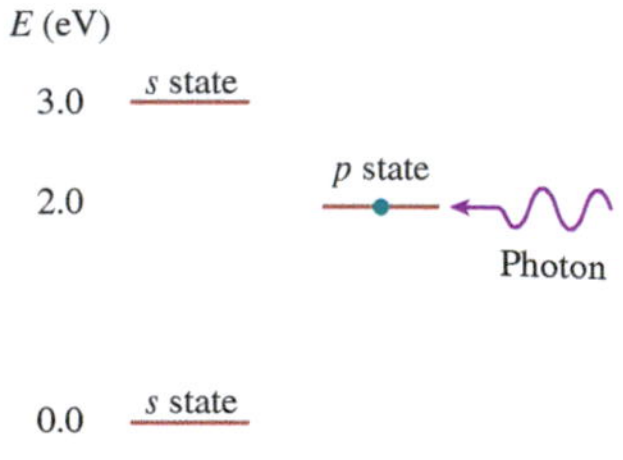


FIGURE Q29.24

25. A glass tube contains $2 \times 10^{11}$ atoms, some of which are in the ground state and some of which are excited. Figure Q29.25 shows the populations for the atoms' three energy levels. Is it possible for these atoms to be a laser? If so, on which transition would laser action occur? If not, why not?

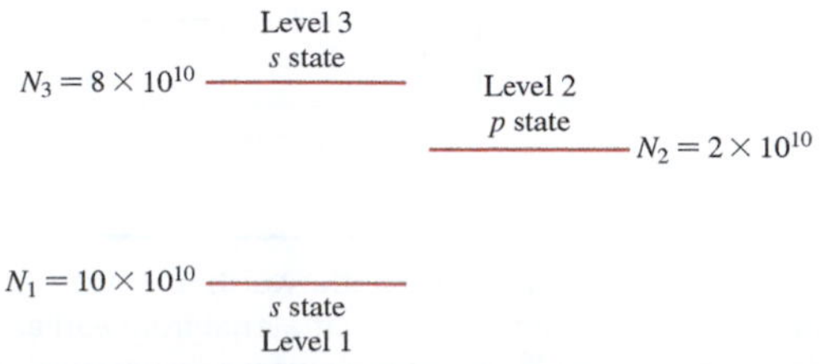


FIGURE Q29.25

26. We usually associate fluorescence with ultraviolet light—objects glow under "black light." But it's possible to excite fluorescence with visible light as well. The light from a green laser pointer causes certain dyes to emit red light. Could the laser pointer cause a dye to emit blue light? Explain.

## Multiple-Choice Questions

27. | An electron collides with an atom in its ground state. The atom then emits a photon of energy $E_{photon}$. In this process the *change* $\Delta E_{elec}$ in the electron's energy is
    A. Greater than $E_{photon}$.
    B. Greater than or equal to $E_{photon}$.
    C. Equal to $E_{photon}$.
    D. Less than or equal to $E_{photon}$.
    E. Less than $E_{photon}$.
28. || How many states are in the $l = 4$ subshell?
    A. 8 B. 9 C. 16 D. 18 E. 22
29. | What is the ground-state electron configuration of calcium ($Z = 20$)?
    A. $1s^2 2s^2 2p^6 3s^2 3p^8$
    B. $1s^2 2s^2 2p^6 3s^2 3p^6 4s^1 4p^1$
    C. $1s^2 2s^2 2p^6 3s^2 3p^6 4s^2$
    D. $1s^2 2s^2 2p^6 3s^2 3p^6 4p^2$
30. | An atom emits a photon with a wavelength of 275 nm. By how much does the atom's energy change?
    A. 0.72 eV
    B. 1.06 eV
    C. 2.29 eV
    D. 3.06 eV
    E. 4.51 eV
31. || The energy of a hydrogen atom is $-3.40$ eV. What is the electron's kinetic energy?
    A. 1.70 eV
    B. 2.62 eV
    C. 3.40 eV
    D. 5.73 eV
    E. 6.80 eV
32. | A "soft x-ray" photon with an energy of 41.8 eV is absorbed by a hydrogen atom in its ground state, knocking the atom's electron out. What is the speed of the electron as it leaves the atom?
    A. $1.84 \times 10^5$ m/s
    B. $3.08 \times 10^5$ m/s
    C. $8.16 \times 10^5$ m/s
    D. $3.15 \times 10^6$ m/s
    E. $3.83 \times 10^6$ m/s

# PROBLEMS

### Section 29.1 Spectroscopy

1. | Figure 29.2 and Table 29.1 showed the wavelengths of the first four lines in the visible spectrum of hydrogen.
   a. Determine the Balmer formula $n$ and $m$ values for these wavelengths.
   b. Predict the wavelength of the fifth line in the spectrum.
2. | The wavelengths in the hydrogen spectrum with $m = 1$ form a series of spectral lines called the Lyman series. Calculate the wavelengths of the first four members of the series.
3. | The Paschen series is analogous to the Balmer series, but with $m = 3$. Calculate the wavelengths of the first three members in the Paschen series. What part(s) of the electromagnetic spectrum are these in?
4. || The Brackett series in the hydrogen spectrum corresponds to transitions that have a final state of $m = 4$. What are the wavelengths of the first three lines in this series? What part of the electromagnetic spectrum are these lines in?
5. || The Pfund series in the hydrogen spectrum corresponds to transitions that have a final state of $m = 5$. What are the wavelengths of the first three lines in this series? What part of the electromagnetic spectrum are these lines in?

### Section 29.2 Atoms

6. | How many electrons, protons, and neutrons are contained in the following atoms or ions: (a) $^{6}Li$, (b) $^{13}C^{+}$, and (c) $^{18}O^{++}$?
7. | How many electrons, protons, and neutrons are contained in the following atoms or ions: (a) $^{9}Be^{+}$, (b) $^{12}C$, and (c) $^{15}N^{+++}$?
8. | Write the symbol for an atom or ion with
   a. Four electrons, four protons, and five neutrons.
   b. Six electrons, seven protons, and eight neutrons.
9. | Write the symbol for an atom or ion with
   a. Three electrons, three protons, and five neutrons.
   b. Five electrons, six protons, and eight neutrons.
10. || Consider the lead isotope $^{207}Pb$.
    INT a. How many electrons, protons, and neutrons are in a neutral $^{207}Pb$ atom?
    b. The lead nucleus has a diameter of 14.2 fm. What is the density of matter in a lead nucleus?
    c. The density of lead is 11,400 kg/m$^3$. How many times the density of lead is your answer to part b?
11. | Consider the gold isotope $^{197}Au$.
    INT a. How many electrons, protons, and neutrons are in a neutral $^{197}Au$ atom?
    b. The gold nucleus has a diameter of 14.0 fm. What is the density of matter in a gold nucleus?
    c. The density of gold is 19,300 kg/m$^3$. How many times the density of gold is your answer to part b?
12. || In a student lab experiment, 5.3 MeV alpha particles from the decay of $^{210}Po$ are directed at a piece of thin platinum foil. If an alpha particle is directed straight toward the nucleus of a platinum atom, what is the distance of closest approach? How does this compare to the approximately 7 fm radius of the nucleus?
13. |||| A 20 MeV alpha particle is
    INT fired toward a $^{238}U$ nucleus. It follows the path shown in Figure P29.13. What is the alpha particle's speed when it is closest to the nucleus, 20 fm from its center? Assume that the nucleus doesn't move.

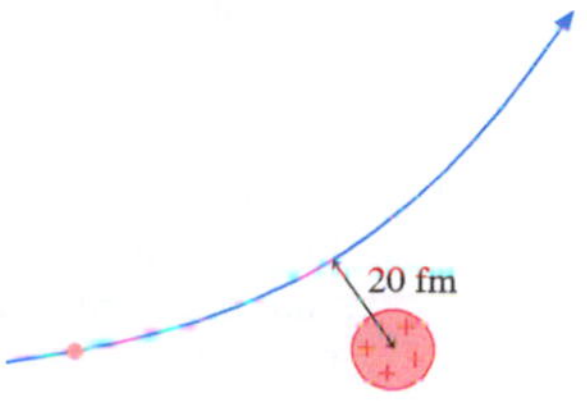

FIGURE P29.13

### Section 29.3 Bohr's Model of Atomic Quantization

14. | Figure P29.14 is an energy-level diagram for a simple atom. What wavelengths appear in the atom's (a) emission spectrum and (b) absorption spectrum?

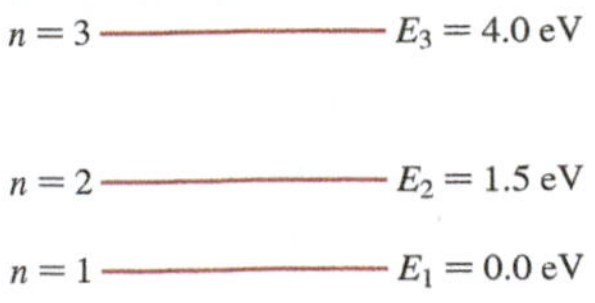


FIGURE P29.14

15. || An electron with 2.0 eV of kinetic energy collides with an atom whose energy-level diagram is shown in Figure P29.14. The electron kicks the atom into an excited state. What is the electron's kinetic energy after the collision?
16. | The allowed energies of a simple atom are 0.0 eV, 4.0 eV, and 6.0 eV.
    a. Draw the atom's energy-level diagram. Label each level with the energy and the principal quantum number.
    b. What wavelengths appear in the atom's emission spectrum?
    c. What wavelengths appear in the atom's absorption spectrum?
17. ||| The allowed energies of a simple atom are 0.0 eV, 4.0 eV, and 6.0 eV. An electron traveling at a speed of $1.6 \times 10^6$ m/s collisionally excites the atom. What are the minimum and maximum speeds the electron could have after the collision?

### Section 29.4 The Bohr Hydrogen Atom

18. || A researcher observes hydrogen emitting photons of energy 1.89 eV. What are the quantum numbers of the two states involved in the transition that emits these photons?
19. | A hydrogen atom is in the $n = 3$ state. In the Bohr model, how many electron wavelengths fit around this orbit?
20. || A hydrogen atom is in its $n = 1$ state. In the Bohr model, what is the ratio of its kinetic energy to its potential energy?
21. || Infrared light with a wavelength of 1870 nm is emitted from hydrogen. What are the quantum numbers of the two states involved in the transition that emits this light?
22. ||| A hydrogen atom is excited from its ground state to the $n = 4$ state. The atom subsequently emits three photons, one of which has a wavelength of 656 nm. What are the wavelengths of the other two photons?
23. | a. Calculate the de Broglie wavelength of the electron in the $n = 1$, 2, and 3 states of the hydrogen atom. Use the information in Table 29.2.
    b. Show numerically that the circumference of the orbit for each of these stationary states is exactly equal to $n$ de Broglie wavelengths.
    c. Sketch the de Broglie standing wave for the $n = 3$ orbit.
24. || If the outermost electron in an atom is excited to a very high energy state, its orbit is far beyond that of the other electrons. To a good approximation, we can think of the electron as orbiting a compact core with a charge equal to the charge of a single proton. The outer electron in such a *Rydberg atom* thus has energy levels corresponding to those of hydrogen.
    a. What is the radius of the $n = 100$ state of the Bohr hydrogen atom?
    b. Sodium is a common element for such studies. How does the radius you calculated in part a compare to the approximately 0.2 nm radius of a typical sodium atom?
25. | Determine all possible wavelengths of photons that can be emitted from the $n = 4$ state of a hydrogen atom.

### Section 29.5 The Quantum-Mechanical Hydrogen Atom

26. | List the quantum numbers of (a) all possible $3p$ states and (b) all possible $3d$ states.
27. || When all quantum numbers are considered, how many different quantum states are there for a hydrogen atom with $n = 1$? With $n = 2$? With $n = 3$? List the quantum numbers of each state.
28. | What is the angular momentum of a hydrogen atom in (a) a $4p$ state and (b) a $5f$ state? Give your answers as a multiple of $\hbar$.
29. || The energy of a hydrogen atom is 12.09 eV above its ground-state energy. As a multiple of $\hbar$, what is the largest angular momentum that this atom could have?
30. || A hydrogen atom is in the $5p$ state. Determine (a) its energy, (b) its angular momentum, (c) its quantum number $l$, and (d) the possible values of its magnetic quantum number $m$.

### Section 29.6 Multi-electron Atoms

31. | Predict the ground-state electron configurations of Mg, Sr, and Ba.
32. | If elements beyond $Z = 120$ are ever synthesized, electrons in these heavy atoms will begin filling a $g$ subshell, corresponding to $l = 4$. How many states will be in a $g$ subshell?
33. || Predict the ground-state electron configurations of Si, Ge, and Pb.
34. | Identify the element for each of these electron configurations. Then determine whether this configuration is the ground state or an excited state.
    a. $1s^2 2s^2 2p^5$
    b. $1s^2 2s^2 2p^6 3s^2 3p^6 3d^{10} 4s^2 4p$
35. || a. With what element is the $3s$ subshell first completely filled?
    b. With what element is the $5d$ subshell first half filled?
36. | Identify the element for each of these electron configurations. Then determine whether this configuration is the ground state or an excited state.
    a. $1s^2 2s^2 2p^6 3s^2 3p^6 4s^2 3d^9$
    b. $1s^2 2s^2 2p^6 3s^2 3p^6 4s^2 3d^{10} 4p^6 5s^2 4d^{10} 5p^6 6s^2 4f^{14} 5d^7$
37. | Explain what is wrong with these electron configurations:
    a. $1s^2 2s^2 2p^8 3s^2 3p^4$
    b. $1s^2 2s^3 2p^4$

### Section 29.7 Excited States and Spectra

38. | In a hydrogen atom, the small magnetic moment of the proton interacts with the magnetic moment of the electron. This results in a 5.87 $\mu$eV energy difference between the $m_s = +\frac{1}{2}$ and $m_s = -\frac{1}{2}$ states. What wavelength photon is emitted in a transition between these two states?
39. | An electron with a speed of $5.00 \times 10^6$ m/s collides with an atom. The collision excites the atom from its ground state (0 eV) to a state with an energy of 3.80 eV. What is the speed of the electron after the collision?
40. | Hydrogen gas absorbs light of wavelength 103 nm. Afterward, what wavelengths are seen in the emission spectrum?
41. || What is the longest wavelength of light that can excite the $4s$ state of sodium?
42. || An electron with a kinetic energy of 3.90 eV collides with a sodium atom. What possible wavelengths of light are subsequently emitted?

43. | a. Is a $4p \rightarrow 4s$ transition allowed in sodium? If so, what is its wavelength? If not, why not?
b. Is a $3d \rightarrow 4s$ transition allowed in sodium? If so, what is its wavelength? If not, why not?

44. ||| INT A dye molecule has electrons that are free to travel up and down a chain of atoms, giving electron energy states that are given by the particle-in-a-box model, with $E_n = (0.30\text{ eV})n^2$.
a. Sketch the energy levels that correspond to $n = 1$ to $n = 4$.
b. What are the three longest wavelengths of visible light that the molecule will absorb?

### Section 29.8 Molecules

45. || Figure P29.45 shows a molecular energy-level diagram. What are the longest and shortest wavelengths in (a) the molecule's absorption spectrum and (b) the molecule's fluorescence spectrum?

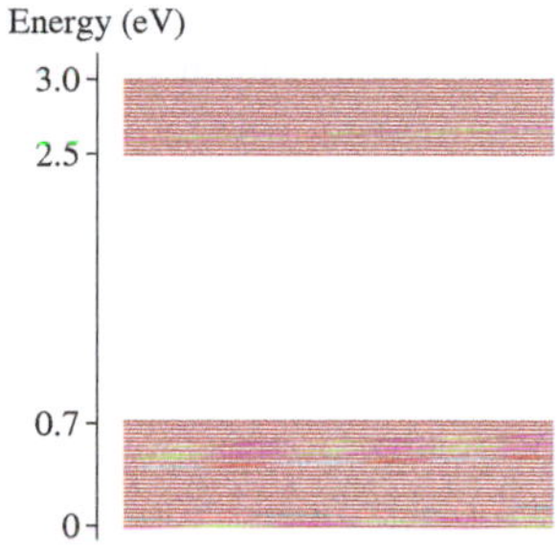

FIGURE P29.45

46. | The molecule whose energy-level diagram is shown in Figure P29.45 is illuminated by 2.7 eV photons. What is the longest wavelength of light that the molecule can emit?

### Section 29.9 Stimulated Emission and Lasers

47. || A 1.00 mW helium-neon laser emits a visible laser beam with a wavelength of 633 nm. How many photons are emitted per second?

48. | A 1000 W carbon dioxide laser emits an infrared laser beam with a wavelength of 10.6 $\mu$m. How many photons are emitted per second?

49. || BIO INT In LASIK surgery, a laser is used to reshape the cornea of the eye to improve vision. The laser produces extremely short pulses of light, each containing 1.0 mJ of energy.
a. In each pulse there are $9.7 \times 10^{14}$ photons. What is the wavelength of the laser?
b. Each pulse lasts only 20 ns. What is the average power delivered to the eye during a pulse?

50. || BIO Port-wine birthmarks, which are caused by malformed capillaries close to the skin, can be removed with laser pulses. Laser light of the right color will pass through the skin with little absorption but will be strongly absorbed by oxyhemoglobin in the blood of these capillaries, which destroys them without damaging adjacent tissue. A typical system uses a series of 6900 W pulses of 585 nm laser light. Each pulse lasts 0.45 ms. How many photons are in each pulse?

51. || A ruby laser emits an intense pulse of light that lasts a mere 10 ns. The light has a wavelength of 690 nm, and each pulse has an energy of 500 mJ.
a. How many photons are emitted in each pulse?
b. What is the *rate* of photon emission, in photons per second, during the 10 ns that the laser is "on"?

## General Problems

52. ||| A 2.55 eV photon is emitted from a hydrogen atom. What are the Balmer formula $n$ and $m$ values corresponding to this emission?

53. | Two of the wavelengths emitted by a hydrogen atom are 102.6 nm and 1876 nm.
a. What are the Balmer formula $n$ and $m$ values for each of these wavelengths?
b. For each of these wavelengths, is the light infrared, visible, or ultraviolet?

54. |||| INT In Example 29.2 it was assumed that the initially stationary gold nucleus would remain motionless during a head-on collision with an 8.3 MeV alpha particle. What is the actual recoil speed of the gold nucleus after that elastic collision? Assume that the mass of a gold nucleus is exactly 50 times the mass of an alpha particle.
**Hint:** Review the discussion of perfectly elastic collisions in Chapter 10 .

55. ||| The diameter of an atom is $1.2 \times 10^{-10}$ m and the diameter of its nucleus is $1.0 \times 10^{-14}$ m. What percent of the atom's volume is occupied by mass and what percent is empty space?

56. | INT If the nucleus is a few fm in diameter, the distance between the centers of two protons must be $\approx 2$ fm.
a. Calculate the repulsive electric force between two protons that are 2.0 fm apart.
b. Calculate the attractive gravitational force between two protons that are 2.0 fm apart. Could gravity be the force that holds the nucleus together?

57. || The absorption spectrum of an atom consists of the wavelengths 200 nm, 300 nm, and 500 nm.
a. Draw the atom's energy-level diagram.
b. What wavelengths are seen in the atom's emission spectrum?

58. || INT The oxygen nucleus $^{16}O$ has a radius of 3.0 fm.
a. With what speed must a proton be fired toward an oxygen nucleus to have a turning point 1.0 fm from the surface? Assume that the nucleus is heavy enough to remain stationary during the collision.
b. What is the proton's kinetic energy in MeV?

59. || The first three energy levels of the fictitious element X are shown in Figure P29.59.
a. What wavelengths are observed in the absorption spectrum of element X? Give your answers in nm.
b. State whether each of your wavelengths in part a corresponds to ultraviolet, visible, or infrared light.
c. An electron with a speed of $1.4 \times 10^6$ m/s collides with an atom of element X. Shortly afterward, the atom emits a 1240 nm photon. What was the electron's speed after the collision? Assume that, because the atom is so much more massive than the electron, the recoil of the atom is negligible.
**Hint:** The energy of the photon is *not* the energy transferred to the atom in the collision.

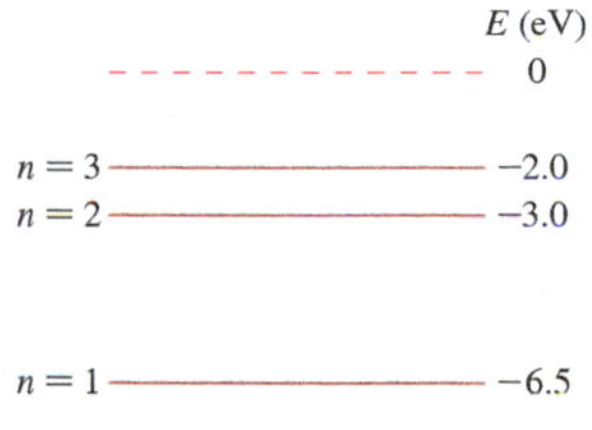

FIGURE P29.59

60. ||| A simple atom has only two absorption lines, at 250 nm and 600 nm. What is the wavelength of the one line in the emission spectrum that does not appear in the absorption spectrum?

61. | A hydrogen atom in the ground state absorbs a 12.75 eV photon. Immediately after the absorption, the atom undergoes a quantum jump to the next-lowest energy level. What is the wavelength of the photon emitted in this quantum jump?

62. || What is the energy of a Bohr hydrogen atom with a 5.18 nm diameter?

63. ||| A beam of electrons is incident on a gas of hydrogen atoms.
INT a. What minimum speed must the electrons have to cause the emission of 656 nm light from the $3 \rightarrow 2$ transition of hydrogen?
b. Through what potential difference must the electrons be accelerated to have this speed?

64. ||| Two hydrogen atoms collide head-on. The collision brings
INT both atoms to a halt. Immediately after the collision, both atoms emit a 121.6 nm photon. What was the speed of each atom just before the collision?

65. ||| A hydrogen atom in its fourth excited state emits a photon with a wavelength of 1282 nm. What is the atom's maximum possible orbital angular momentum after the emission? Give your answer as a multiple of $\hbar$.

66. ||| Germicidal lamps are used to sterilize tools in biological
BIO and medical facilities. One type of germicidal lamp is a low-
INT pressure mercury discharge tube, similar to a fluorescent lightbulb, that has been optimized to emit ultraviolet light with a wavelength of 254 nm. A 15-mm-diameter, 25-cm-long tube emits 4.5 W of ultraviolet light. The mercury vapor pressure inside the tube is 1.0 Pa at the operating temperature of 40°C.
a. On average, how many ultraviolet photons does each mercury atom emit per second?
b. The lifetime of the excited state in mercury is a rather long 120 ns. On average, what fraction of its time does a mercury atom spend in the excited state?

67. | Figure P29.67 shows the first few energy levels of the lithium atom. Make a table showing all the allowed transitions in the emission spectrum. For each transition, indicate
a. The wavelength, in nm.
b. Whether the transition is in the infrared, the visible, or the ultraviolet spectral region.
c. Whether or not the transition would be observed in the lithium absorption spectrum.

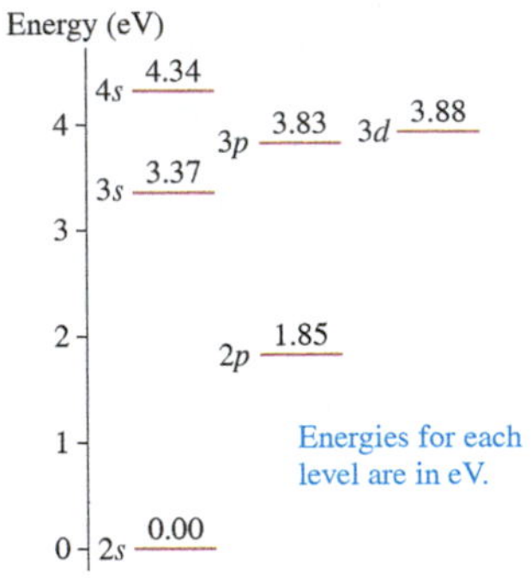


FIGURE P29.67

68. || Figure P29.68 shows a few energy levels of the mercury atom. One valence electron is always in the 6s state; the other electron changes states. Make a table showing all of the allowed transitions in the emission spectrum. For each transition, indicate
a. The wavelength, in nm.
b. Whether the transition is in the infrared, the visible, or the ultraviolet spectral region.
c. Whether or not the transition would be observed in the mercury absorption spectrum.

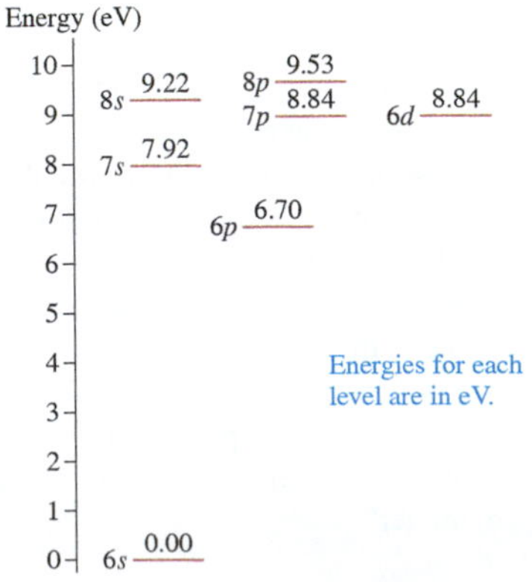


FIGURE P29.68

69. | Fluorescence microscopy, discussed in Section 29.8, is an
BIO important tool in modern cell biology. A variation on this tech-
INT nique depends on a phenomenon known as two-photon excitation. If two photons are absorbed simultaneously (i.e., within about $10^{-16}$ s), their energies can add. A molecule that is normally excited by a 350 nm photon can be excited by two photons each having half as much energy. For this process to be useful, photons must illuminate the sample at the very high rate of at least $10^{29}$ photons/m$^2 \cdot$ s. This is achieved by focusing a laser beam to a small spot and by concentrating the power of the laser into very short ($10^{-13}$ s) pulses that are fired $10^8$ times each second. Suppose a biologist wants to use two-photon excitation to excite a molecular species that would be excited by 500 nm light in normal one-photon fluorescence microscopy. What minimum intensity (W/m$^2$) must the laser beam have during each pulse?

## MCAT-Style Passage Problems

### Light-Emitting Diodes

Light-emitting diodes, known by the acronym LED, produce the familiar green and red indicator lights used in a wide variety of consumer electronics. LEDs are *semiconductor devices* in which the electrons can exist only in certain energy levels. Much like molecules, the energy levels are packed together close enough to form what appears to be a continuous band of possible energies. Energy supplied to an LED in a circuit excites electrons from a *valence band* into a *conduction band*. An electron can emit a photon by undergoing a quantum jump from a state in the conduction band into an empty state in the valence band, as shown in Figure P29.70.

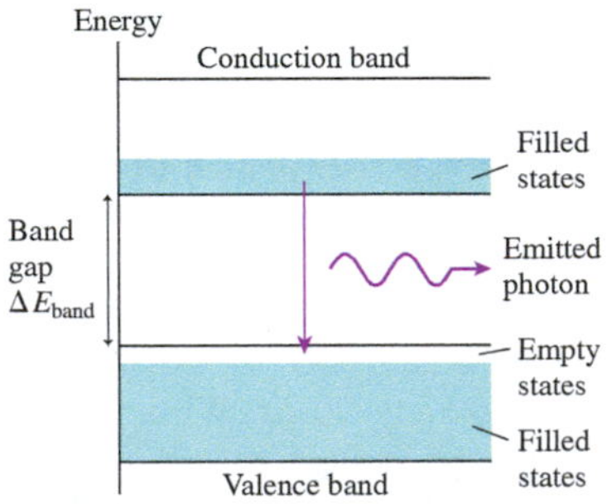


FIGURE P29.70 Energy-level diagram of an LED.

The size of the band gap $\Delta E_{band}$ determines the possible energies—and thus the wavelengths—of the emitted photons. Most LEDs emit a narrow range of wavelengths and thus have a distinct color. This makes them well-suited for traffic lights and other applications where a certain color is desired, but it makes them less desirable for general illumination. One way to make a "white" LED is to combine a blue LED with a substance that fluoresces yellow when illuminated with the blue light. The combination of the two colors makes light that appears reasonably white.

70. | An LED emits green light. Increasing the size of the band gap could change the color of the emitted light to
A. Red
B. Orange
C. Yellow
D. Blue

71. | Suppose the LED band gap is 2.5 eV, which corresponds to a wavelength of 500 nm. Consider the possible electron transitions in Figure P29.70. 500 nm is the
A. Maximum wavelength of the LED.
B. Average wavelength of the LED.
C. Minimum wavelength of the LED.

72. | The same kind of semiconducting material used to make an LED can also be used to convert absorbed light into electric energy, essentially operating as an LED in reverse. In this case, the absorption of a photon causes an electron transition from a filled state in the valence band to an unfilled state in the conduction band. If $\Delta E_{band} = 1.4$ eV, what is the minimum wavelength of electromagnetic radiation that could lead to electric energy output?
A. 140 nm B. 890 nm
C. 1400 nm D. 8900 nm

73. | The efficiency of a light source is the percentage of its energy input that gets radiated as visible light. If some of the blue light in an LED is used to cause a fluorescent material to glow,
A. The overall efficiency of the LED is increased.
B. The overall efficiency of the LED does not change.
C. The overall efficiency of the LED decreases.

### The Spectrum of Singly Ionized Helium

Singly ionized helium has a single orbiting electron, so the mathematics of the Bohr hydrogen atom will apply, with one important difference: The charge of the nucleus is twice that of the single proton at the center of a hydrogen atom. This changes the energy levels; the magnitude of each energy is greater than the corresponding Bohr level by a factor of $2^2 = 4$:

$$E_n = -\frac{1}{n^2}\left(\frac{K(2e)^2}{2a_B}\right) = \frac{4(-13.6\text{ eV})}{n^2} = \frac{-54.4\text{ eV}}{n^2}$$

The Balmer and Lyman series of spectral lines in hydrogen have analogs in singly ionized helium, but at shorter wavelengths; the photons corresponding to these transitions are beyond the visible-light spectrum. The transitions that end on the $n = 4$ state produce a set of spectral lines called the Pickering series. The visible-light lines in this series were first seen in the light from certain hot stars, but some of the lines overlap the hydrogen Balmer series lines, so these lines were initially missed. This led to an initial mischaracterization of the source of the lines.

74. | What energy is required to remove the remaining electron from singly ionized helium?
A. 108.8 eV B. 54.4 eV
C. 27.2 eV D. 13.6 eV

75. || What is, approximately, the longest wavelength that will be absorbed by ionized helium?
A. 30 nm B. 60 nm
C. 90 nm D. 120 nm

76. || Consider the transitions in singly ionized helium that end on $n = 3$. What is the longest wavelength emitted in this series of transitions?
A. 470 nm B. 320 nm
C. 270 nm D. 250 nm

77. ||| The longest wavelength in the hydrogen Balmer series corresponds to a transition from $n = 3$ to $n = 2$. What transition in the ionized helium Pickering series will overlap this Balmer line?
A. $n = 5$ to $n = 4$ B. $n = 6$ to $n = 4$
C. $n = 7$ to $n = 4$ D. $n = 8$ to $n = 4$

78. || The Paschen series of wavelengths in the hydrogen spectrum correspond to transitions that end on $m = 3$. There is a series of wavelengths in the ionized helium spectrum that partially overlap this series, as the Pickering series partially overlaps the Balmer series. What is the end state for these transitions?
A. $m = 5$ B. $m = 6$
C. $m = 8$ D. $m = 16$

### Muonic Hydrogen

A muon is a short-lived particle we first met in Chapter 27. Muons are created by cosmic rays; they can also be created by particle accelerators. The muon is similar to an electron but has a larger mass: $m_\mu \approx 200m_e$. During its brief lifetime, a muon can combine with a proton to create a system that is similar to atomic hydrogen called a *muonic hydrogen atom*. The larger mass of the muon makes some of the assumptions of the Bohr hydrogen atom treatment less accurate, but using the mathematics of the Bohr hydrogen atom to analyze this system will give approximate results that allow us to understand how the changing mass affects the properties of the system.

79. || How does the radius of a muonic hydrogen atom compare to that of a regular hydrogen atom?
A. It is approximately 200 times larger.
B. It is approximately the same size.
C. It is approximately 200 times smaller.

80. ||| How does the energy required to ionize a muonic hydrogen atom compare to that required to ionize a regular hydrogen atom?
A. It is greater.
B. It is approximately the same.
C. It is less.

81. ||| Consider the spectral lines of a muonic hydrogen atom. If we consider the series of lines analogous to the Balmer series in hydrogen, how do the wavelengths compare to the wavelengths of the Balmer series?
A. They are longer.
B. They are approximately the same.
C. They are shorter.

82. ||| The larger mass of the muon complicates an accurate mathematical treatment similar to that of the Bohr hydrogen atom because
A. The de Broglie wavelength of the muon is shorter than that of the electron.
B. The relatively small difference in mass between the muon and the proton means that we can't ignore the motion of the proton.
C. The short lifetime of the muon means that it is less likely to form a stationary state.

# 30 Nuclear Physics

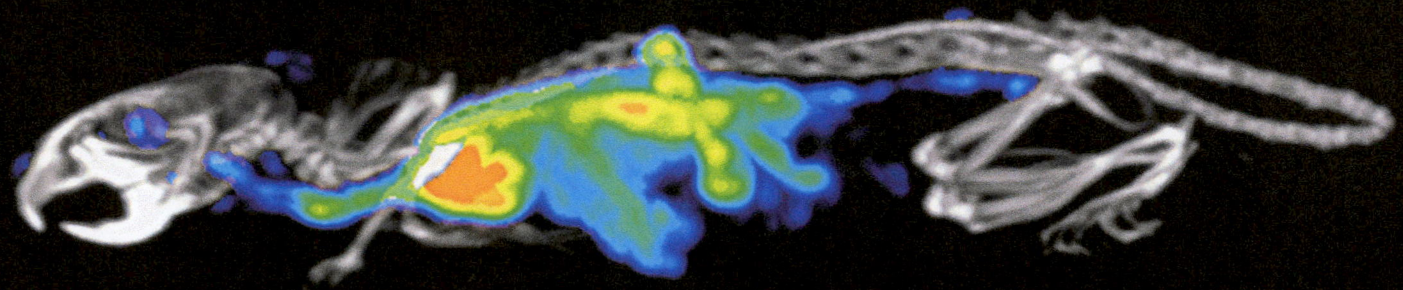

The colored patches superimposed on the image of the mouse's skeleton shows the presence of radioactive nuclei introduced into the mouse's body, giving a clear picture of the tissues where these particles are accumulating. How—and why—does an investigator create an image that shows the location of particular nuclei in the body?

## LOOKING AHEAD

### Nuclei and Isotopes

The ratio of hydrogen **isotopes** in rainwater varies with geographical location, which allows biologists to track the migration of birds.

You'll learn about nuclear structure. The number of protons determines the element; the number of neutrons, the isotope.

### Radioactivity and Radiation

The **radioactive** nuclei in this tank are unstable. They decay, emitting high-energy particles—**radiation**—that ionize the water.

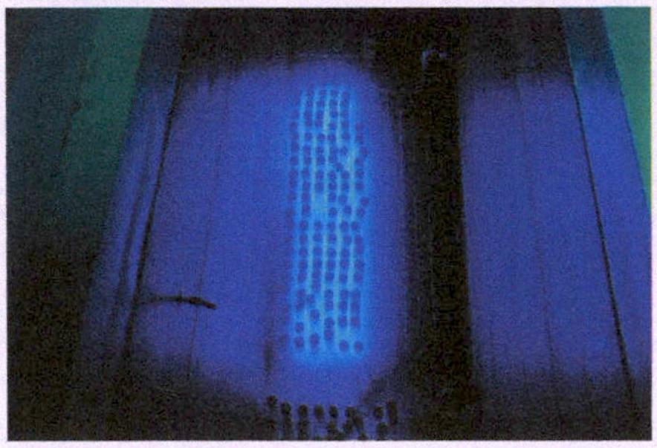

You'll learn about different nuclear decay modes (alpha, beta, and gamma) and the resulting radiation for each.

### Decay and Half-Life

Measurements of carbon isotopes in these cave drawings show that they are 30,000 years old.

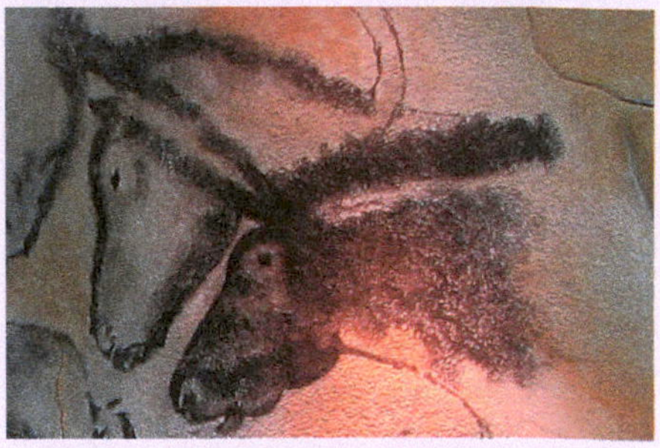

In any sample of $^{14}C$, half the nuclei decay every 5700 years. You'll see how to use this half-life to calculate an object's age.

**GOAL** To understand the physics of the nucleus and some of the applications of nuclear physics.

## LOOKING BACK

### Energy levels in atoms

In Chapter 29, you learned how the periodic table of the elements is based on the energy levels of multi-electron atoms.

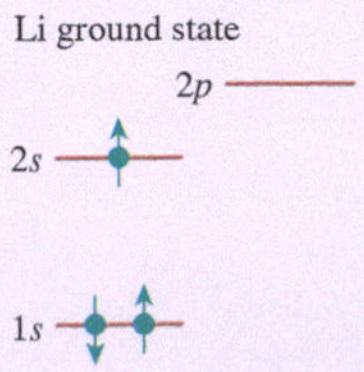


The protons and neutrons in nuclei also have energy levels. Understanding these energy levels will allow you to understand nuclear decay modes.

**STOP TO THINK**

This energy-level diagram represents a neutral atom with four electrons. What element is this? And is this the ground state of the atom or an excited state?

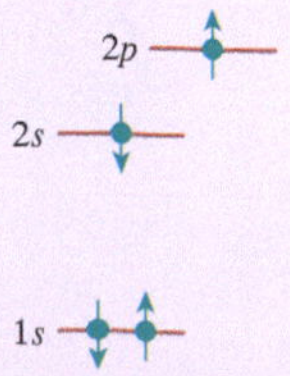


A. Lithium, ground state
B. Lithium, excited state
C. Beryllium, ground state
D. Beryllium, excited state
E. Boron, ground state
F. Boron, excited state

# 30.1 Nuclear Structure

For 29 chapters, we've made frequent references to properties of atoms that are due to the electrons that surround the nucleus. In this final chapter it's time to dig deeper, to talk about the nucleus itself. In particular:

- What is nuclear matter? What are its properties?
- What holds the nucleus together? Why doesn't the electrostatic force blow it apart?
- What is the connection between the nucleus and radioactivity?

These are questions of **nuclear physics.**

As you learned in Section 29.2, an atom consists of a "cloud" of electrons orbiting a very small nucleus. You also learned that the nucleus is composed of two types of particles: positively charged protons and uncharged neutrons. Together, protons and neutrons are referred to as **nucleons.**

The nucleus is a tiny speck in the center of a vastly larger atom. As FIGURE 30.1 shows, the nuclear diameter of roughly $10^{-14}$ m is only about 1/10,000 the size of the atom, defined by the diameter of its electron cloud. What we call *matter* is overwhelmingly empty space!

FIGURE 30.1 The nucleus is a tiny speck within an atom.

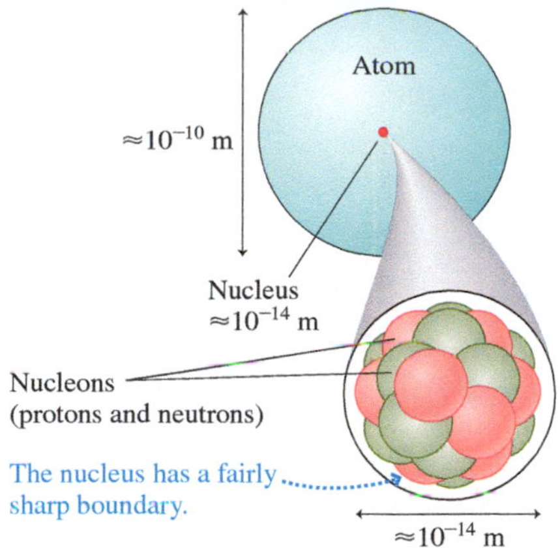


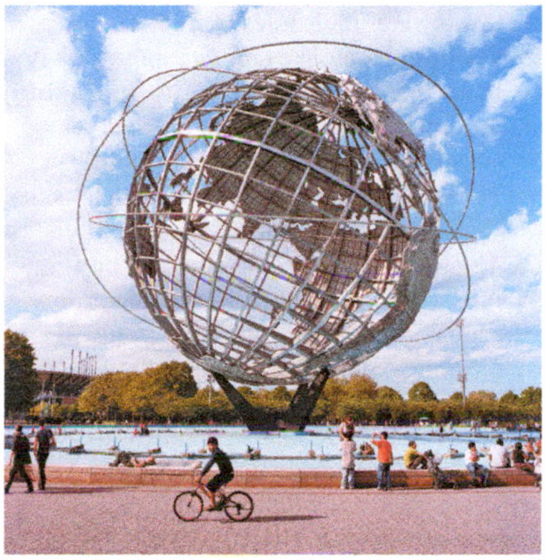

If an atom were the size of the Unisphere in New York City, the nucleus would be only the size of a pea.

The number of protons $Z$ in a nucleus is the element's atomic number. Which element (hydrogen, carbon, gold, etc.) a particular nucleus corresponds to is determined by the number of protons in the nucleus, not by the number of orbiting electrons. Electrons are easily added and removed to form negative and positive ions, but doing so doesn't change the element. The mass number $A$ is the total number of nucleons—the total number of protons and neutrons—in the nucleus. Thus we have $A = Z + N$, where $N$ is the neutron number.

**NOTE** ▶ The mass number, which is dimensionless, is *not* the same thing as the atomic mass $m$. We'll look at actual atomic masses later. ◀

Protons and neutrons are virtually identical other than the fact that the proton has one unit of the fundamental charge $e$, whereas the neutron is electrically neutral. The neutron is slightly more massive than the proton, but the difference is very small. TABLE 30.1 lists the basic properties of protons and neutrons.

TABLE 30.1 Protons and neutrons

| | Proton | Neutron |
|---|---|---|
| Number | $Z$ | $N$ |
| Charge $q$ | $+e$ | 0 |
| Mass, in u | 1.00728 | 1.00866 |

A video to support a section's topic is embedded in the eText.

**Video** Nuclear Physics Notation

## Isotopes

As we learned in Chapter 29, not all atoms of the same element (and thus the same $Z$) have the same mass. There is a *range* of neutron numbers that happily form a nucleus with $Z$ protons, creating a series of nuclei having the same $Z$-value (i.e., they are all the same chemical element) but different $A$-values. Each $A$-value in a series of nuclei with the same $Z$-value is called an *isotope.* Isotopes for some of the elements are given in a table in Appendix C.

The notation used to label isotopes uses the mass number $A$ as a *leading* superscript, as shown in FIGURE 30.2. Hence ordinary carbon, which has six protons and six neutrons in the nucleus (and thus has $A = 12$), is written $^{12}$C and pronounced "carbon twelve." The radioactive form of carbon used in carbon-dating archaeological artifacts is $^{14}$C. It has six protons, making it carbon, and eight neutrons, for a total of 14 nucleons. The isotope $^{2}$H is a hydrogen atom in which the nucleus is not simply a proton but a proton and a neutron. Although the isotope is a form of hydrogen, it is called **deuterium.** Sometimes, for clarity, we will find it useful to include the atomic number as a leading *subscript.* Ordinary carbon is then written as $^{12}_{6}$C; deuterium as $^{2}_{1}$H.

The chemical behavior of an atom is largely determined by the orbiting electrons. Different isotopes of the same element have very similar *chemical* properties. $^{14}$C will form the same chemical compounds as $^{12}$C and will generally be treated the same by the body, a fact that permits the use of $^{14}$C to determine the age of a sample. But the *nuclear* properties of these two isotopes are quite different, as we will see.

FIGURE 30.2 Three isotopes of carbon.

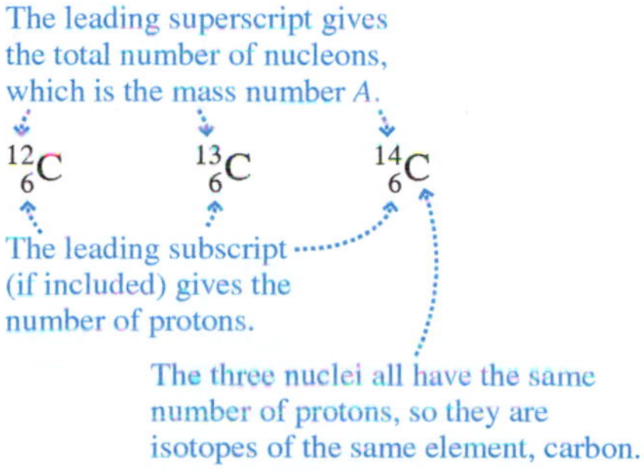

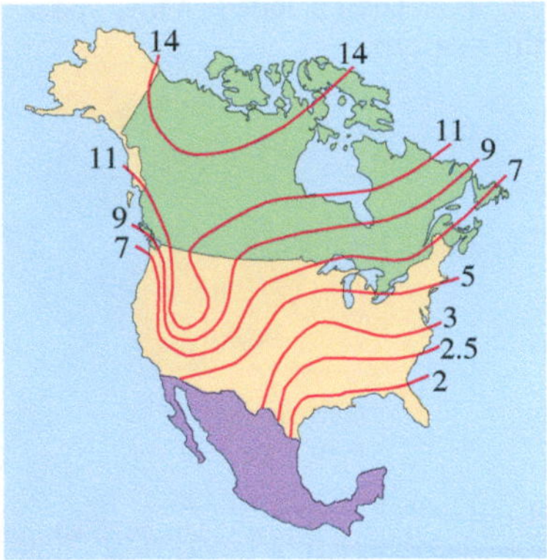


**Tracking migrating birds** The natural abundance of deuterium in fresh water is about 0.016%. Rainwater, however, has a lower natural abundance of this isotope. This is because water molecules that contain a heavy deuterium atom evaporate more slowly than those that contain ordinary hydrogen. In colder climates, this difference in evaporation rates is more pronounced, so rainfall (and hence surface water) in colder regions has a lower deuterium content than rainfall in warmer areas, as the map shows. The numbers are the percent deficit of deuterium in the local rainfall compared to a standard water sample. Biologists have used this fact to help track the migratory patterns of birds. A bird's claws and feathers contain water whose isotopic abundance of deuterium provides a clue to the bird's geographical origin.

Most elements have multiple naturally occurring isotopes. For each element, the fraction of naturally occurring nuclei represented by one particular isotope is called the **natural abundance** of that isotope. For instance, oxygen has two primary isotopes, $^{16}O$ and $^{18}O$. The data in Appendix C show that the natural abundance of $^{16}O$ is 99.76%, meaning that 9976 out of every 10,000 naturally occurring oxygen atoms are the isotope $^{16}O$. Most of the remaining 0.24% of naturally occurring oxygen is the isotope $^{18}O$, which has two extra neutrons. These two isotopes of oxygen are chemically equivalent, but the mass difference can lead to subtle differences in macroscopic behaviors. Atmospheric water vapor is always slightly deficient in water molecules that contain $^{18}O$ compared to water in the ocean because the lighter molecules that contain $^{16}O$ evaporate more readily.

More than 3000 isotopes are known. The majority of these are **radioactive,** meaning that the nucleus is not stable but, after some period of time, will *decay* by either fragmenting into smaller nuclei or emitting some kind of subatomic particle to reach a more stable state. Many of these radioactive isotopes are created by nuclear reactions in the laboratory and have only a fleeting existence. Only 253 isotopes are **stable** (i.e., nonradioactive) and occur in nature. In addition, there are a handful of radioactive isotopes with such long decay times, measured in billions of years, that they also occur naturally.

## Atomic Mass

You learned in ◂◂ SECTION 12.1 that atomic masses are specified in terms of the *atomic mass unit* u, defined such that the atomic mass of the isotope $^{12}C$ is exactly 12 u. The conversion to SI units is

$$1\ \text{u} = 1.6605 \times 10^{-27}\ \text{kg}$$

Alternatively, as we saw in ◂◂ SECTION 27.10, we can use Einstein's $E_0 = mc^2$ to express masses in terms of their energy equivalent. The energy equivalent of 1 u of mass is

$$\begin{aligned} E_0 &= (1.6605 \times 10^{-27}\ \text{kg})(2.9979 \times 10^8\ \text{m/s})^2 \\ &= 1.4924 \times 10^{-10}\ \text{J} = 931.49\ \text{MeV} \end{aligned} \tag{30.1}$$

A mass of 1 u has an energy equivalent of 931.49 MeV, so we can use the following equation to find the energy equivalent of any atom or particle whose mass is given in atomic mass units:

$$E_0\ (\text{in MeV}) = m\,(\text{in u}) \times (931.49\ \text{MeV/u}) \tag{30.2}$$

By noting that Einstein's formula implies $m = E_0/c^2$, we can also write 1 u in the following form:

$$1\ \text{u} = \frac{E_0}{c^2} = 931.49\left(\frac{\text{MeV}}{c^2}\right) \tag{30.3}$$

It may seem unusual, but the units MeV/$c^2$ are units of mass. This will be a useful unit for us when we need to compute energy equivalents. The energy equivalent of mass 1 MeV/$c^2$ is simply 1 MeV.

TABLE 30.2 shows some important atomic mass values. Notice that the mass of a hydrogen atom is equal to the sum of the masses of a proton and an electron. But a quick calculation shows that the mass of a helium atom (2 protons, 2 neutrons, and 2 electrons) is 0.03038 u *less* than the sum of the masses of its constituents. The difference is due to the *binding energy* of the nucleus, a topic we'll look at in Section 30.2.

TABLE 30.2 Some atomic masses

| Particle | Symbol | Mass (u) | Mass (MeV/$c^2$) |
|---|---|---|---|
| Electron | e | 0.000549 | 0.51 |
| Proton | p | 1.007276 | 938.27 |
| Neutron | n | 1.008665 | 939.56 |
| Hydrogen | $^1H$ | 1.007825 | 938.78 |
| Helium | $^4He$ | 4.002602 | 3728.38 |

NOTE ▸ The atomic masses of the neutron and the proton are both ≈1 u. In earlier chapters, we often used the approximation that the atomic mass in u is equal to the mass number $A$. This approximation is sufficient in many contexts, such as when we calculated the speeds of gas molecules in Chapter 12. But many nuclear calculations involve the small difference between two masses that are almost the same. The two masses must be calculated or specified to four or five significant figures if their difference is to be meaningful. In calculations of nuclear energies, you should use more significant figures than usual, and you should start with the accurate values for nuclear masses given in Table 30.2 or Appendix C. ◂

The *chemical* atomic mass shown on the periodic table of the elements is the *weighted average* of the atomic masses of all naturally occurring isotopes. For example, chlorine has two stable isotopes: $^{35}$Cl, with atomic mass $m = 34.97$ u, has an abundance of 75.8% and $^{37}$Cl, at 36.97 u, has an abundance of 24.2%. The average, weighted by abundance, is $(0.758 \times 34.97\text{ u}) + (0.242 \times 36.97\text{ u}) = 35.45$ u. This is the value shown on the periodic table and is the correct value for most chemical calculations, but it is not the mass of any particular isotope of chlorine.

**NOTE** ▸ Nuclear physics calculations involve the masses of specific isotopes. The mass values for specific isotopes are given in Appendix C; these are the values you'll need for calculations in this chapter. Don't use the chemical atomic masses given in the periodic table! ◂

**STOP TO THINK 30.1** Three electrons orbit a neutral $^6$Li atom. How many electrons orbit a neutral $^7$Li atom?

## 30.2 Nuclear Stability

Because nuclei are characterized by two independent numbers, $N$ and $Z$, it is useful to show the known nuclei on a plot of neutron number $N$ versus proton number $Z$. **FIGURE 30.3** shows such a plot. Stable nuclei are represented by blue diamonds and unstable, radioactive nuclei by red dots.

**FIGURE 30.3** Stable and unstable nuclei shown on a plot of neutron number $N$ versus proton number $Z$.

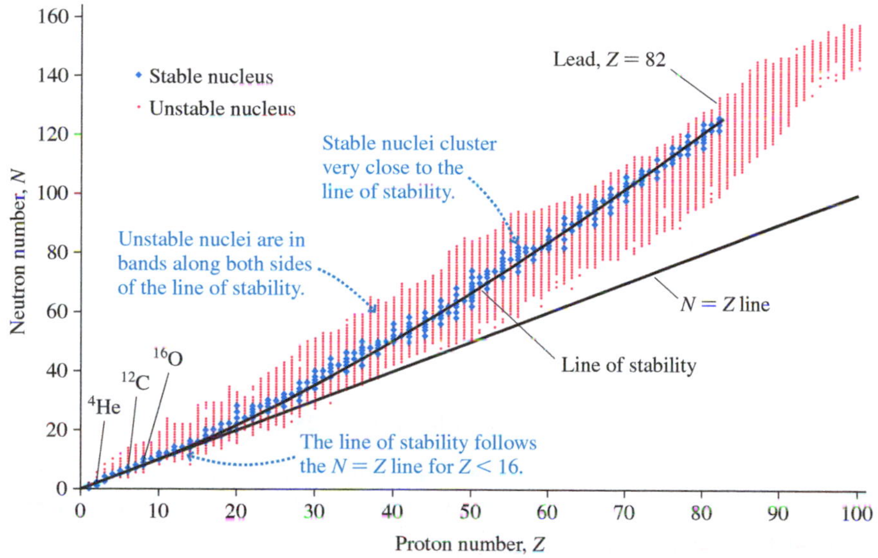


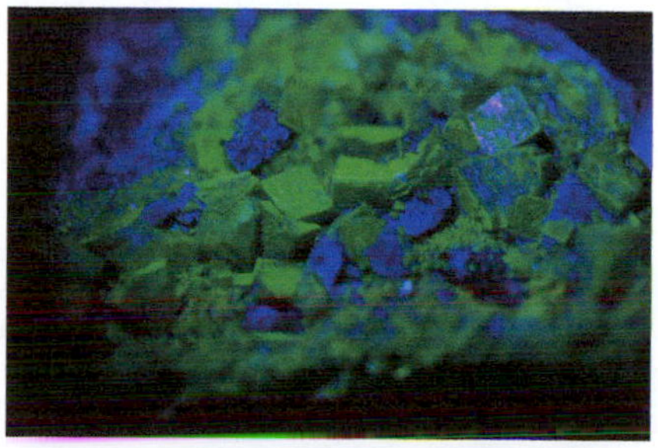

**Unstable but ubiquitous uranium** All the isotopes of uranium are unstable, but one is very long-lived. Half the $^{238}$U that was present at the formation of the earth is still around—and it is found all around you at a low concentration in nearly all of the rocks and soil on the earth's surface. Much of the radiation that you are exposed to comes from this naturally occurring and widely distributed unstable element.

We can make several observations from this graph:

- The stable nuclei cluster very close to the curve called the **line of stability.**
- There are no stable nuclei with $Z > 82$ (lead). Heavier elements (up to $Z = 92$ (uranium)) are found in nature, but they are radioactive. Even heavier elements can be produced in the laboratory.

- Unstable nuclei are in bands along both sides of the line of stability.
- The lightest elements, with $Z < 16$, are stable when $N \approx Z$. The familiar isotopes $^{4}He$, $^{12}C$, and $^{16}O$ all have equal numbers of protons and neutrons.
- As Z increases, the number of neutrons needed for stability grows increasingly larger than the number of protons. The $N/Z$ ratio is $\approx 1.2$ at $Z = 40$ but has grown to $\approx 1.5$ at $Z = 80$.

These observations—especially the fact that $N \approx Z$ for small Z but $N > Z$ for large Z—will be explained by the model of the nucleus that we'll explore in Section 30.3.

## Binding Energy

FIGURE 30.4 The nuclear binding energy.

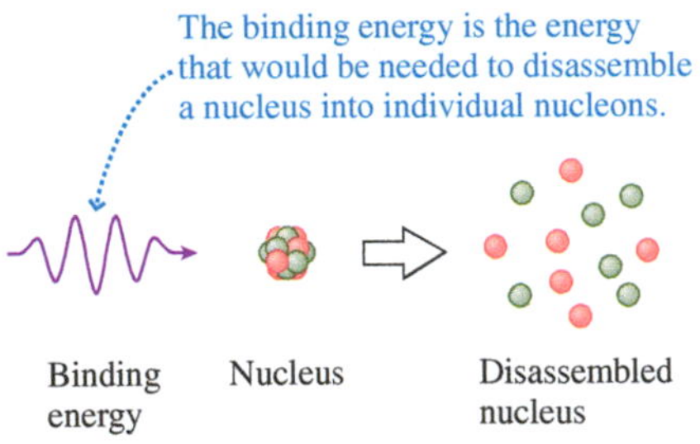


A nucleus is a *bound system.* That is, you would need to supply energy to disperse the nucleons by breaking the nuclear bonds between them. FIGURE 30.4 shows this idea schematically.

You learned a similar idea in atomic physics. The energy levels of the hydrogen atom are negative numbers because the bound system has less energy than a free proton and electron. The energy you must supply to an atom to remove an electron is called the *ionization energy.*

The energy you would need to supply to a nucleus to disassemble it into individual protons and neutrons is called the **binding energy.** Whereas ionization energies of atoms are only a few eV, the binding energies of nuclei are tens or hundreds of MeV.

We can use conservation of energy to learn something surprising about the mass of a nucleus compared to the total mass of the individual nucleons that make it up. Using Einstein's formula $E_0 = mc^2$, we find that the conservation of energy equation that corresponds to the situation in Figure 30.4 is

$$B + m_{\text{nucleus}}c^2 = Zm_{\text{p}}c^2 + Nm_{\text{n}}c^2$$

where $B$ is the binding energy, $Zm_{\text{p}}$ is the mass of the Z protons, and $Nm_{\text{n}}$ is the mass of the $N$ neutrons. We can rearrange this equation as $((Zm_{\text{p}} + Nm_{\text{n}}) - m_{\text{nucleus}})c^2 = B$. Because $B$ is positive, we see that the mass of the nucleus is less than the mass of its constituents. In nuclear physics, mass is *not* conserved—only total energy.

FIGURE 30.5 Calculating the binding energy of the helium nucleus.

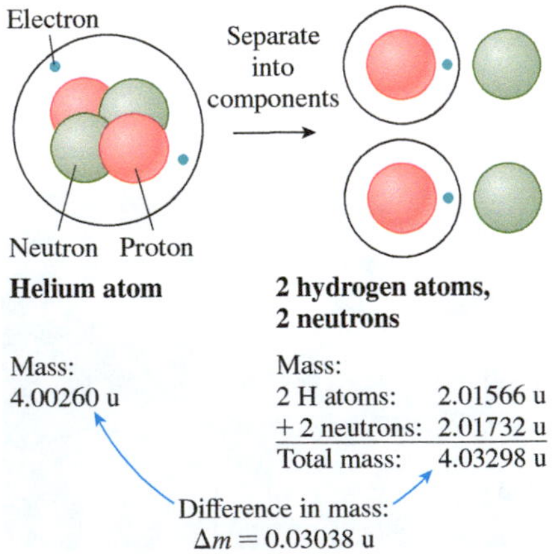


We can illustrate these ideas by considering a helium atom that consists of the two protons and two neutrons in its nucleus plus its two electrons. Suppose we break this helium atom into its constituent parts, as shown in FIGURE 30.5. Here we take the basic components of a helium atom to be two neutrons plus two *hydrogen atoms;* by taking hydrogen atoms (a proton plus an electron) instead of just protons as a basic component, we automatically take into account the mass of the electrons that were in the original helium atom.

Figure 30.5 shows that the difference in mass between the helium atom and its components is $\Delta m = 0.03038$ u. From our previous discussion, this mass difference arises from the binding energy $B$ that was put into the system to separate the tightly bound nucleons. We can use Equation 30.2, which gives the energy equivalent of this mass difference, to find the binding energy:

$$B = (0.03038\ \text{u})(931.49\ \text{MeV/u}) = 28.30\ \text{MeV}$$

This discussion shows that, in general, we can compute the nuclear binding energy by considering the mass difference between the atom and its separated components, $Z$ hydrogen atoms and $N$ neutrons:

$$B = (Zm_{\text{H}} + Nm_{\text{n}} - m_{\text{atom}}) \times (931.49\ \text{MeV/u}) \qquad (30.4)$$

Nuclear binding energy for an atom of mass $m_{\text{atom}}$ with $Z$ protons and $N$ neutrons

EXAMPLE 30.1 **Finding the binding energy of iron**

What is the nuclear binding energy of $^{56}Fe$ to the nearest MeV?

**STRATEGIZE** We will use Equation 30.4 to find the binding energy.

**PREPARE** Appendix C gives the atomic mass of $^{56}Fe$ as 55.934940 u. Iron has atomic number 26, so an atom of $^{56}Fe$ could be separated into 26 hydrogen atoms and 30 neutrons. The mass of the separated components is more than that of the iron nucleus; the difference gives us the binding energy.

**SOLVE** We solve for the binding energy using Equation 30.4. The masses of the hydrogen atom and the neutron are given in Table 30.2. We find

$$B = (26(1.007825\text{ u}) + 30(1.008665\text{ u}) - 55.934940\text{ u})(931.49\text{ MeV/u})$$
$$= (0.52846\text{ u})(931.49\text{ MeV/u}) = 492.26\text{ MeV} \approx 492\text{ MeV}$$

**ASSESS** The difference in mass between the nucleus and its components is a small fraction of the mass of the nucleus, so we must use several significant figures in our mass values. The mass difference is small—about half that of a proton—but the energy equivalent, the binding energy, is enormous.

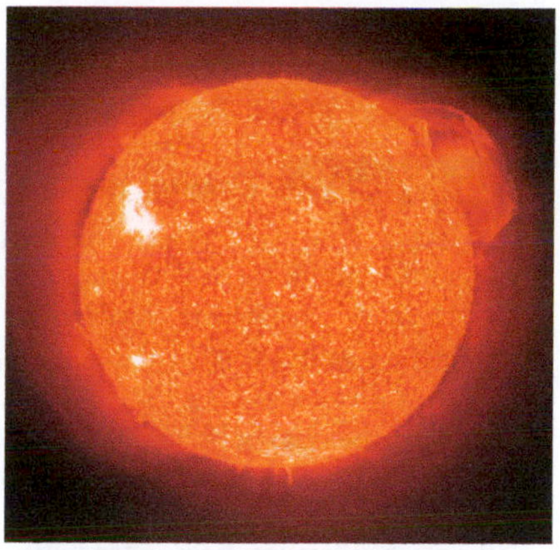

**A nuclear fusion weight-loss plan** The sun's energy comes from reactions that combine four hydrogen atoms to create a single atom of helium—a process called **nuclear fusion.** The mass of a helium atom is less than that of four hydrogen atoms, so energy is released in this process. As the fusion reactions continue, the mass of the sun decreases—by 130 trillion tons per year! That's a lot of mass, but given the sun's enormous size, this change will amount to only a few hundredths of a percent of the sun's mass over its 10-billion-year lifetime.

How much energy is 492 MeV? To make a comparison with another energy value we have seen, the binding energy of a single iron nucleus is equivalent to the energy released in the metabolism of nearly *2 billion* molecules of ATP! The energy scale of nuclear processes is clearly quite different from that of chemical processes.

As $A$ increases, the nuclear binding energy increases, simply because there are more nuclear bonds. A more useful measure for comparing one nucleus to another is the quantity $B/A$, called the *binding energy per nucleon.* Iron, with $B = 492$ MeV and $A = 56$, has 8.79 MeV per nucleon. This is the amount of energy, on average, you would need to supply in order to remove *one* nucleon from the nucleus. Nuclei with larger values of $B/A$ are more tightly held together than nuclei with smaller values of $B/A$.

FIGURE 30.6 is a graph of the binding energy per nucleon versus mass number $A$. The line connecting the points is often called the **curve of binding energy.**

For small values of $A$, adding more nucleons increases the binding energy per nucleon. If two light nuclei can be joined together to make a single, larger nucleus, the final nucleus will have a higher binding energy per nucleon. Because the final nucleus is more tightly bound, energy will be released in this *nuclear fusion* process. Nuclear fusion of hydrogen to helium is the basic reaction that powers the sun.

FIGURE 30.6 The curve of binding energy for the stable nuclei.

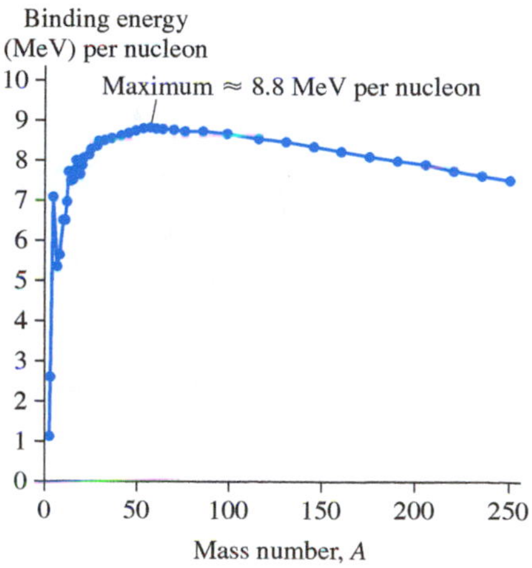


The curve of binding energy has a broad maximum at $A \approx 60$. Nuclei with $A > 60$ become more stable if their mass *decreases* because for these nuclei *subtracting* nucleons increases the binding energy per nucleon. *Alpha decay,* one of the three basic types of radioactive decay that we'll examine later in this chapter, occurs when a heavy nucleus becomes more stable by ejecting a group of four nucleons in order to decrease its mass, releasing energy in the process. The decrease in binding energy per nucleon as mass increases helps explain why there are no stable nuclei beyond $Z = 82$.

A few very heavy nuclei, especially some isotopes of uranium and plutonium, are so unstable that they can be induced to fragment into two lighter nuclei in the process known as *nuclear fission.* For example, the collision of a slow-moving neutron with a $^{235}U$ nucleus can cause the reaction

$$n + {}^{235}U \rightarrow {}^{236}U \rightarrow {}^{90}Sr + {}^{144}Xe + 2n \qquad (30.5)$$

The $^{235}U$ nucleus absorbs the neutron to become $^{236}U$, but $^{236}U$ is so unstable that it immediately fragments—in this case into a $^{90}Sr$ nucleus, a $^{144}Xe$ nucleus, and two neutrons. The less massive $^{90}Sr$ and $^{144}Xe$ nuclei are more tightly bound than the original $^{235}U$ nucleus, so a great deal of energy is released in this reaction. As we've seen, nuclei with lower values of $Z$ have relatively smaller numbers of neutrons, meaning there will be neutrons "left over" after the reaction. Equation 30.5 shows some free neutrons among the reaction products, but the two nuclear fragments have "extra" neutrons as well—they have too many neutrons and will be unstable. This is generally true for the products of a fission reaction. The fact that the waste products of nuclear fission are radioactive has important consequences for the use of nuclear fission as a source of energy.

**EXAMPLE 30.2 Finding the energy released in a nuclear reaction**

How much energy is released in the nuclear reaction in Equation 30.5? If 1.0 kg of $^{235}U$ were to undergo this reaction, how much energy would be released?

**STRATEGIZE** We will assume that the kinetic energies of the original neutron and the $^{235}U$ atom are negligible. Then we will compare the mass of the particles on the right-hand side of Equation 30.5 with the mass of the particles on the left-hand side. According to Einstein's equivalence between mass and energy, the mass difference is converted into energy.

**PREPARE** We can find the masses of the isotopes and the neutron from Appendix C.

**SOLVE** The mass of the particles on the left-hand side is

$$m_L = 1.0087\text{ u} + 235.0439\text{ u} = 236.0526\text{ u}$$

The mass of the particles on the right-hand side is

$$m_R = 89.9077\text{ u} + 143.9389\text{ u} + 2(1.0087\text{ u}) = 235.8640\text{ u}$$

The difference in the masses—the mass that is lost in the reaction—is then

$$\Delta m = 236.0526\text{ u} - 235.8640\text{ u} = 0.1886\text{ u}$$

The mass that is lost is converted into energy according to Equation 30.2:

$$\Delta E = (0.1886\text{ u})(931.49\text{ MeV/u}) = 176\text{ MeV}$$

This energy appears as kinetic energy of the Sr and Xe nuclei and of the neutrons.

To find the energy released by 1.0 kg of $^{235}U$, we need to know how many atoms are in 1.0 kg of $^{235}U$. We can do so by first finding the mass, in kg, of a single $^{235}U$ atom, which is

$$m_U = (235\text{ u})(1.66 \times 10^{-27}\text{ kg/u}) = 3.90 \times 10^{-25}\text{ kg}$$

Then the number of atoms in 1.0 kg is

$$n = \frac{1.0\text{ kg}}{3.90 \times 10^{-25}\text{ kg}} = 2.56 \times 10^{24}$$

The energy released by this many nuclei undergoing the reaction of Equation 30.5 is then

$$E = n\Delta E = (2.56 \times 10^{24})(176 \times 10^6\text{ eV})(1.60 \times 10^{-19}\text{ J/eV}) = 7.2 \times 10^{13}\text{ J}$$

**ASSESS** From Figure 30.6, we can see that in going from a nucleus with $A \approx 240$ to nuclei with $A \approx 120$, the binding energy changes by roughly 1 MeV per nucleon. Given the 240 or so nucleons involved, an energy change of 176 MeV seems plausible. The energy released by 1.0 kg of $^{235}U$ is extremely large. You may have heard of the "kiloton," a unit of energy equal to $4.2 \times 10^{12}$ J, used to express the energy of atomic weapons. In these units, the energy released is 17 kilotons, a typical yield for an atomic bomb.

◄ **A formidable chain reaction** As we've seen, if the nucleus of certain isotopes of uranium or plutonium is struck by a neutron, the nucleus can split into two more tightly bound fragments, releasing tremendous energy. Extra neutrons are left over after the split, each of which can cause the fission of another nucleus—releasing more neutrons, which can produce further reactions. The net result is a nuclear fission **chain reaction.** A controlled reaction is the energy source of a nuclear power plant. An uncontrolled reaction is responsible for the terrible destructive power of a nuclear explosion, shown here in an above-ground test in Nevada in 1957.

**STOP TO THINK 30.2** $^{238}U$ is long-lived but ultimately unstable; it will eventually spontaneously break into two fragments, a $^{4}He$ nucleus and a $^{234}Th$ nucleus, in a process called alpha decay, which we'll learn about in the next section. A great deal of energy is released in the process. What must be true about the masses of the nuclei involved?

A. $m_U > m_{Th} + m_{He}$
B. $m_U = m_{Th} + m_{He}$
C. $m_U < m_{Th} + m_{He}$

## 30.3 Forces and Energy in the Nucleus

The nucleus of the atom is made up of protons, which are positively charged, and neutrons, which have no charge. Why doesn't the repulsive electrostatic force between the protons simply cause the nucleus to fly apart? There must be another force at work within the nucleus that holds the nucleons together. This force is called the **strong nuclear force,** or just the *strong force*.

The strong force has four important properties:

1. It is an *attractive* force between any two nucleons.
2. It does not act on electrons.
3. It is a *short-range* force, acting only over nuclear distances. The strong force is negligible until the nucleons are very close together.
4. Over the range where it acts, it is *stronger* than the electrostatic force that tries to push two protons apart.

FIGURE 30.7 summarizes the properties of the strong force. Many decades of research have shown that the strong force between two nucleons is independent of whether they are protons or neutrons. Charge is the basis for electromagnetic interactions, but it is of no relevance to the strong force. Protons and neutrons are identical as far as nuclear forces are concerned.

A given proton in the nucleus feels a repulsive electrostatic force from all the other protons because the electrostatic force is a long-range force. However, because of the short range of the strong force, a proton feels an attractive force from only the very few other protons with which it is in close contact. A nucleus with too many protons will be unstable because the repulsive electrostatic forces will overcome the attractive strong forces. Because neutrons participate in the strong force but exert no repulsive forces, **the neutrons provide the extra "glue" that holds the nucleus together.** In small nuclei, one neutron per proton is sufficient for stability. Hence small nuclei have $N \approx Z$. But as the nucleus grows, the repulsive force increases faster than the binding energy. More neutrons are needed for stability, so heavy nuclei have $N > Z$.

In ◀◀ SECTION 28.5 you learned that confining particles to a region of space results in the quantization of energy: Confined particles occupy *discrete* energy levels. In Chapter 29 you saw the consequences of this quantization for electrons bound in atoms: Atoms are built up by the addition of electrons to successively higher energy levels. The protons and neutrons bound in the nucleus are also confined to a small region of space, so these nucleons also occupy discrete energy levels. Thus, just as with electrons in atoms, nuclear states are built up by the addition of nucleons to successively higher energy levels. Neutrons and protons have spin and follow the Pauli principle. Each energy level can hold only a certain number of spin-up particles and spin-down particles, depending on the quantum numbers; additional nucleons must go into higher energy levels.

Although there is a great deal of similarity between the descriptions of nucleon energy levels in the nucleus and electron energy levels in the atom, there is one important difference: the energy scale. You learned in Section 28.5 that the energy-level spacing for confined particles increases rapidly as the size of the containing region decreases—the energy-level spacing for a small "box" is much greater than for a large one. Thus electron energy levels, whose "box" is the size of an atom, are typically separated by a few eV, but proton and neutron energy levels, determined by the much smaller size of the nucleus, are separated by a few MeV—a million times as much.

## Low-*Z* Nuclei

As our first example of an energy-level description of the nucleus, we'll consider the energy levels of low-$Z$ nuclei ($Z < 8$). Because these nuclei have so few protons, we can neglect the electrostatic potential energy due to proton-proton repulsion and consider only the much larger nuclear potential energy. In that case, the energy levels of the protons and neutrons are essentially identical.

FIGURE 30.8 shows the three lowest allowed energy levels for protons and neutrons, and the maximum number of protons and neutrons the Pauli principle allows in each. Energy values vary from nucleus to nucleus, but the spacing between these levels is several MeV.

FIGURE 30.7 Forces between pairs of particles in the nucleus.

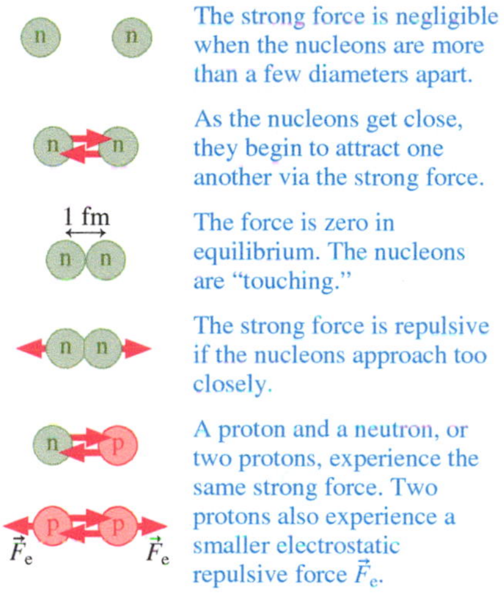


FIGURE 30.8 The three lowest energy levels of neutrons and protons in a low-$Z$ nucleus.

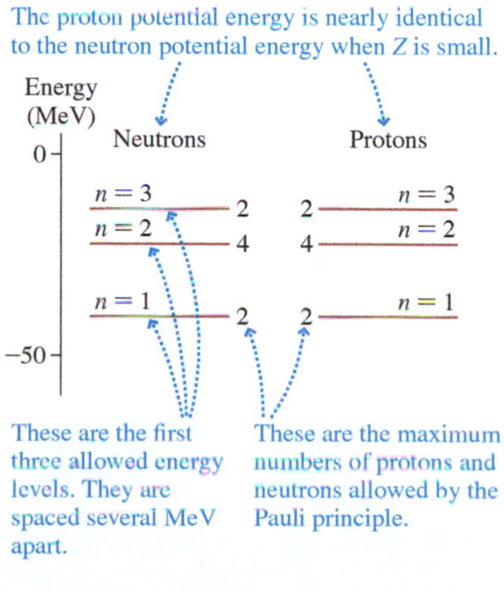

The proton potential energy is nearly identical to the neutron potential energy when Z is small.
Energy (MeV)
0
−50
Neutrons
Protons
n = 3
n = 2
n = 1
2
4
2
These are the first three allowed energy levels. They are spaced several MeV apart.
These are the maximum numbers of protons and neutrons allowed by the Pauli principle.

Suppose we look at a series of nuclei, all with $A = 12$ but with different numbers of protons and neutrons: $^{12}B$, $^{12}C$, $^{12}N$. All have 12 nucleons, but only $^{12}C$ is a stable isotope; the other two are not. Why do we see this difference in stability?

FIGURE 30.9 Nuclear energy-level diagrams of $^{12}C$, $^{12}B$, and $^{12}N$.

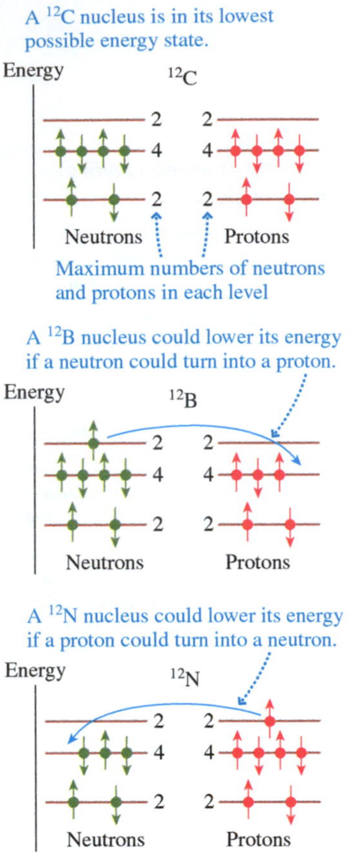


FIGURE 30.9 shows the energy-level diagrams of $^{12}C$, $^{12}B$, and $^{12}N$. Look first at $^{12}C$, a nucleus with six protons and six neutrons. You can see that exactly six protons are allowed in the $n = 1$ and $n = 2$ proton energy levels. The same is true for the six neutrons. No other arrangement of the nucleons would lower the total energy, so this nucleus is stable.

$^{12}B$ has five protons and seven neutrons. The sixth neutron fills the $n = 2$ neutron energy level, so the seventh neutron has to go into the $n = 3$ energy level. (This is just like the third electron in Li having to go into the $n = 2$ energy level because the first two electrons have filled the $n = 1$ energy level.) The $n = 2$ proton energy level has one vacancy because there are only five protons.

In atoms, electrons in higher energy levels move to lower energy levels by emitting a photon as the electron undergoes a quantum jump. That can't happen here because the higher-energy nucleon in $^{12}B$ is a neutron whereas the vacant lower energy level is that of a proton. But an analogous process could occur *if* a neutron could somehow turn into a proton, allowing it to move to a lower energy level. $^{12}N$ is just the opposite, with the seventh proton in the $n = 3$ energy level. If a proton could somehow turn into a neutron, it could move to a lower energy level, lowering the energy of the nucleus.

You can see from the diagrams that the $^{12}B$ and $^{12}N$ nuclei have significantly more energy—by several MeV—than $^{12}C$. If a neutron could turn into a proton, and vice versa, these nuclei could move to a lower-energy state—that of $^{12}C$. In fact, that's exactly what happens! Both $^{12}B$ and $^{12}N$ decay into the more stable $^{12}C$ in a process known as *beta decay*. We'll explore the details of how this happens in Section 30.4.

## High-*Z* Nuclei

We can use the energy levels of protons and neutrons in the nucleus to give a qualitative explanation for one more observation. FIGURE 30.10 shows the neutron and proton energy levels of a high-*Z* nucleus. In a nucleus with many protons, the increasing electrostatic potential energy raises the proton energy levels but not the neutron energy levels. Protons and neutrons now have a different set of energy levels.

FIGURE 30.10 The proton energy levels are displaced upward in a high-*Z* nucleus because of the electric potential energy.

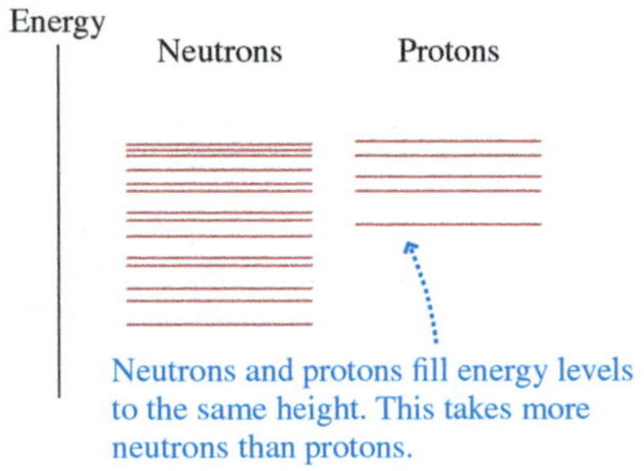


As a nucleus is "built" by the addition of protons and neutrons, the proton energy levels and the neutron energy levels must fill to just about the same height. If there were neutrons in energy levels above vacant proton levels, the nucleus would lower its energy by changing neutrons into protons. Similarly, a proton would turn into a neutron if there were a vacant neutron energy level beneath a filled proton level. **The net result is that the filled levels for protons and neutrons are at just about the same height.**

Because the neutron energy levels start at a lower energy, *more neutron states* are available than proton states. Consequently, a high-*Z* nucleus will have more neutrons than protons. This conclusion is consistent with our observation in Figure 30.3 that $N > Z$ for heavy nuclei.

**STOP TO THINK 30.3** Based on the model of nuclear energy levels and transitions you have seen, would you expect $^{13}C$ to be stable?

# 30.4 Radiation and Radioactivity

When a nucleus is unstable, it will decay. But early investigators didn't observe the decay directly; instead, they observed the high-energy particles or rays emitted in the decay of unstable nuclei. By measuring the properties of the emitted radiation, they identified three different types of decay, which they called alpha, beta, and gamma. We now define *radioactivity,* or *radioactive decay,* as the spontaneous emission of particles or high-energy photons from unstable nuclei as they decay from higher-energy to lower-energy states. In this section, we will explore the decay mechanisms that result in the three different types of radiation, considering the changes in the nuclei that produce these decays.

**Video** Radioactive Decay

## Alpha Decay

Many large nuclei are unstable; it is energetically favorable for them to spontaneously break apart into smaller fragments. A combination of two neutrons and two protons, a $^4$He nucleus, is an especially stable nuclear combination. When a large nucleus spontaneously decays by breaking into two smaller fragments, one of these fragments is almost always a helium nucleus—an **alpha particle,** symbolized by $\alpha$.

An unstable nucleus that ejects an alpha particle loses two protons and two neutrons, so $Z$ decreases by 2 and $A = Z + N$ decreases by 4. Thus this decay can be written as

$$^{A}_{Z}\text{X} \rightarrow {}^{A-4}_{Z-2}\text{Y} + \alpha + \text{energy} \qquad (30.6)$$

Alpha decay of a nucleus

FIGURE 30.11 Alpha decay.

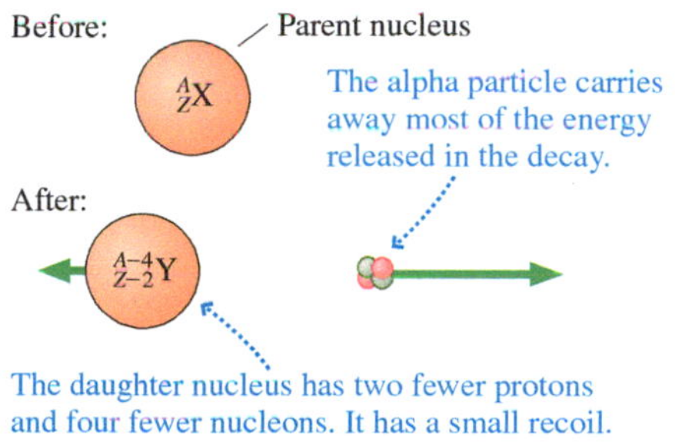


FIGURE 30.11 shows the alpha-decay process. The original nucleus X is called the **parent nucleus,** and the decay-product nucleus Y is the **daughter nucleus.**

Energy conservation tells us that an alpha decay can occur only when the mass of the parent nucleus is greater than the mass of the daughter nucleus plus the mass of the alpha particle. This requirement is often met for heavy, high-$Z$ nuclei beyond the maximum of the curve of binding energy of Figure 30.6. It is energetically favorable for these nuclei to eject an alpha particle because the daughter nucleus is more tightly bound than the parent nucleus.

The daughter nucleus, which is much more massive than an alpha particle, undergoes only a slight recoil, as we see in Figure 30.11. Consequently, **the energy released in an alpha decay ends up mostly as the kinetic energy of the alpha particle.** This energy is approximately equal to the mass energy difference between the initial and final states:

$$K_\alpha \approx \Delta E = (m_\text{X} - m_\text{Y} - m_\text{He})c^2 \qquad (30.7)$$

### EXAMPLE 30.3 Analyzing alpha decay in a smoke detector

An isotope of americium, $^{241}$Am, is part of the sensing circuit in most smoke detectors. $^{241}$Am decays by emitting an alpha particle. These alpha particles ionize the surrounding air, creating a current between two charged plates. Smoke particles entering the region between the plates interrupt the current, which triggers the alarm. When the $^{241}$Am decays, what is the daughter nucleus?

***STRATEGIZE*** Equation 30.6 shows that alpha decay causes the atomic number to decrease by 2 and the atomic mass number to decrease by 4. We need to find an isotope that has these properties.

***SOLVE*** Let's write an equation for the decay showing the alpha particle as a helium nucleus, including the atomic weight superscript and the atomic number subscript for each element. There is no change in the total number of neutrons or protons, so the subscripts and superscripts must "balance" in the reaction:

$$^{241}_{95}\text{Am} \rightarrow {}^{237}_{93}\text{Y} + {}^{4}_{2}\text{He} + \text{energy}$$

(Here "Y" stands for the unknown daughter nucleus, not the element Yttrium!) A quick glance at the periodic table reveals the unknown element in this equation, the daughter nucleus, to be an isotope of neptunium, $^{237}_{93}$Np.

***ASSESS*** Balancing the two sides of the above reaction is similar to balancing the equation for a chemical reaction.

**EXAMPLE 30.4 Finding the energy of an emitted alpha particle**

The uranium isotope $^{238}U$ undergoes alpha decay to an isotope of thorium, $^{234}Th$. What is the kinetic energy, in MeV, of the alpha particle?

**STRATEGIZE** The decay products have less mass than the initial nucleus. This difference in mass is released as energy, most of which goes to the kinetic energy of the alpha particle. Because the energy of the alpha particle is only approximately equal to the reaction energy, we needn't use the full accuracy that the values in Appendix C provide.

**PREPARE** To 4 decimal places, the atomic mass of $^{238}U$ is 238.0508 u, that of $^{234}Th$ is 234.0436 u, and that of $^{4}He$—the alpha particle—is 4.0026 u.

**SOLVE** We can calculate the kinetic energy of the alpha particle using Equation 30.7:

$$K_\alpha = (238.0508\text{ u} - 234.0436\text{ u} - 4.0026\text{ u})c^2$$
$$= (0.0046\text{ u})c^2$$

If we convert from u to MeV/$c^2$, using the conversion factor 1 u = 931.49 MeV/$c^2$, the $c^2$ cancels, and we end up with

$$K_\alpha = \left(0.0046\text{ u} \times \frac{931.49\text{ MeV}/c^2}{1\text{ u}}\right)c^2 = 4.3\text{ MeV}$$

**ASSESS** This is a typical alpha-particle energy, corresponding to a speed of about 5% of the speed of light. Notice that with a careful use of conversion factors we never had to evaluate $c^2$.

## Beta Decay

In beta decay, a nucleus decays by emitting an electron, $e^-$. A typical example of beta decay occurs in the carbon isotope $^{14}C$, which undergoes the beta-decay process

$$^{14}C \rightarrow {}^{14}N + e^-$$

Carbon has $Z = 6$ and nitrogen has $Z = 7$. Because $Z$ increases by 1 but $A$ doesn't change, it appears that a neutron within the nucleus has changed itself into a proton by emitting an electron. That is, the basic decay process appears to be

$$n \rightarrow p + e^- \qquad (30.8)$$

FIGURE 30.12 Beta decay.

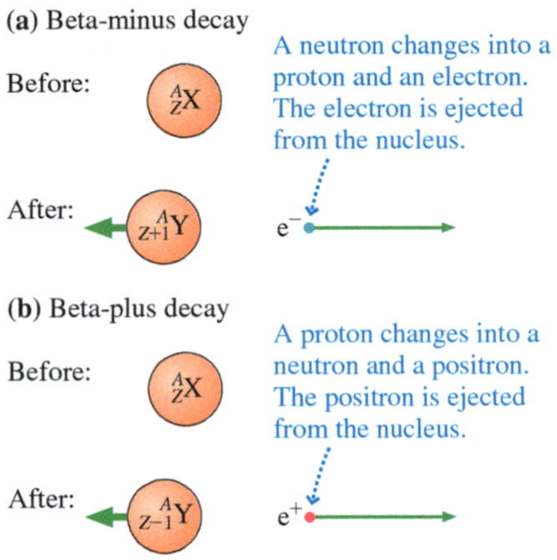


As FIGURE 30.12a shows, in this process the electron is ejected from the nucleus but the proton is not. Because a negatively charged electron is ejected, we call the process *beta-minus decay*. In beta-minus decay, $Z$ increases by 1 while $A$ remains the same, so we can write this decay as

$$^{A}_{Z}X \rightarrow {}^{A}_{Z+1}Y + e^- + \text{energy} \qquad (30.9)$$

Beta-minus decay of a nucleus

Do neutrons *really* turn into protons? It turns out that a free neutron—one not bound in a nucleus—is *not* a stable particle. It decays into a proton and an electron, with a half-life of approximately 10 minutes. This decay conserves energy because $m_n > m_p + m_e$. Furthermore, it conserves charge.

Whether a neutron *within* a nucleus can decay depends not only on the masses of the neutron and proton but also on the masses of the parent and daughter nuclei, because energy has to be conserved for the entire nuclear system. **Beta decay occurs only if $m_X > m_Y$.** $^{14}C$ can undergo beta decay to $^{14}N$ because $m(^{14}C) > m(^{14}N)$. But $m(^{12}C) < m(^{12}N)$, so $^{12}C$ is stable and its neutrons will not decay.

A few nuclei undergo a slightly different form of beta decay by emitting a *positron*. A positron, for which we use the symbol $e^+$, is identical to an electron except that it has a positive charge. As we saw in ◀◀ SECTION 27.10, the positron is the *antiparticle* of the electron. We call this emission of a positron *beta-plus decay*.

**NOTE** ▶ We use the terms "beta-minus decay" and "beta-plus decay" when it's important to distinguish between these two forms of decay. But when it's not important to make this distinction, we call the general process simply "beta decay." ◀

Inside a nucleus undergoing beta-plus decay, a proton changes into a neutron and a positron:

$$p^+ \rightarrow n + e^+ \qquad (30.10)$$

The full decay process, shown in FIGURE 30.12b, is

$$^{A}_{Z}\text{X} \rightarrow {}^{\;\;A}_{Z-1}\text{Y} + \text{e}^{+} + \text{energy} \tag{30.11}$$

Beta-plus decay of a nucleus

Beta-plus decay does *not* happen for a free proton because $m_p < m_n$. It *can* happen within a nucleus as long as energy is conserved for the entire nuclear system, but it is far less common than beta-minus decay.

In our earlier discussion of Section 30.3 we noted that the $^{12}$B and $^{12}$N nuclei could reach a lower-energy state if a proton could change into a neutron, and vice versa. Now we see that such changes can occur if the energy conditions are favorable. And, indeed, $^{12}$B undergoes beta-minus decay to $^{12}$C, while $^{12}$N undergoes beta-plus decay to $^{12}$C.

In general, beta decay is a process in which nuclei with too many neutrons or too many protons move closer to the line of stability that was shown in Figure 30.3. FIGURE 30.13 illustrates this for a series of nuclei with $A = 135$. Nuclei with $Z < 56$ (barium) have too many neutrons, so they undergo beta-minus ($\beta^-$) decay, changing a neutron into a proton. Nuclei with $Z > 56$ have too many protons, so they undergo beta-plus ($\beta^+$) decay, changing a proton into a neutron. Note that in each beta decay, the daughter nucleus has a lower mass than its parent, as conservation of energy requires. The nucleus $^{135}_{56}$Ba has the lowest mass in this series and so is stable.

FIGURE 30.13 Nuclei move toward the stability line by undergoing beta decay.

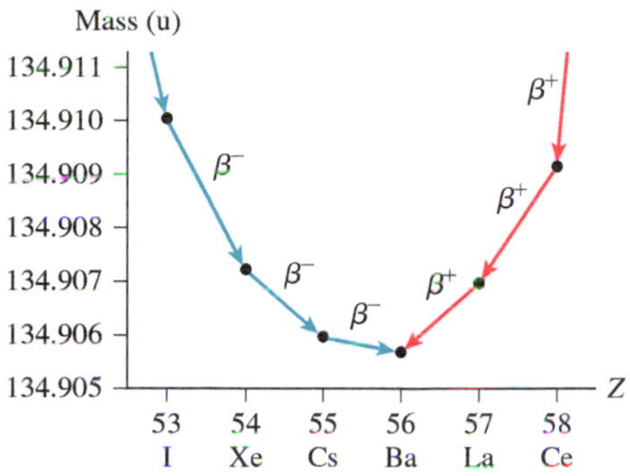


**NOTE** ► The electron emitted in beta decay has nothing to do with the atom's valence electrons. The beta particle is created in the nucleus and ejected directly from the nucleus when a neutron is transformed into a proton and an electron. ◄

**EXAMPLE 30.5 Analyzing beta decay in the human body** BIO

Your body contains several radioactive isotopes. Approximately 20% of the radiation dose you receive each year comes from the radioactive decay of these atoms. Most of this dose comes from one potassium isotope, $^{40}$K, which most commonly decays by beta-minus emission. What is the daughter nucleus?

***STRATEGIZE*** In beta-minus decay, Equation 30.9 shows that $Z$ increases by 1 while $A$ remains unchanged.

***SOLVE*** Rewriting Equation 30.9 as

$$^{40}_{19}\text{K} \rightarrow {}^{40}_{20}\text{Y} + \text{e}^{-} + \text{energy}$$

we see that the daughter nucleus must be the calcium isotope $^{40}_{20}$Ca.

## Gamma Decay

Gamma decay is similar to quantum processes you saw in earlier chapters. In Chapter 28, you learned that an atomic system can emit a photon with $E_{\text{photon}} = \Delta E_{\text{atom}}$ when an electron undergoes a quantum jump from an excited energy level to a lower energy level. Nuclei are no different. A proton or a neutron in an excited nuclear state, such as the one shown in FIGURE 30.14, can undergo a quantum jump to a lower-energy state by emitting a high-energy photon. This is the gamma-decay process.

The spacing between atomic energy levels is only a few eV. Nuclear energy levels, by contrast, are on the order of 1 MeV apart, meaning gamma-ray photons will have energies $E_{\text{gamma}} \approx 1$ MeV. Photons with this much energy are quite penetrating and deposit an extremely large amount of energy at the point where they are finally absorbed.

Nuclei left to themselves are usually in their ground states and thus cannot emit gamma-ray photons. However, alpha and beta decay often leave the daughter nucleus in an excited nuclear state, so gamma emission is often found to accompany alpha and beta emission.

FIGURE 30.14 Gamma decay.

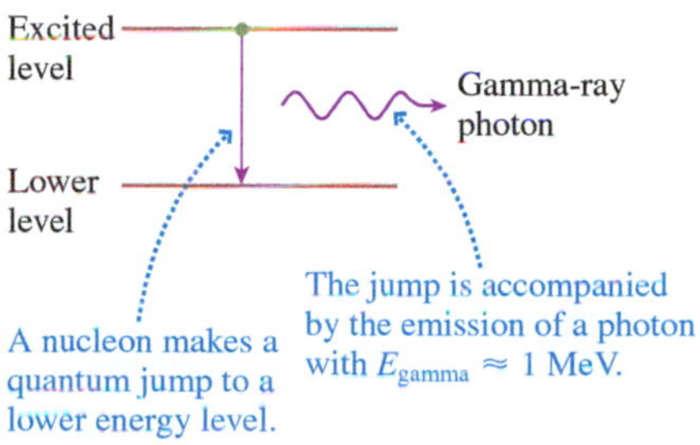

Let's look at an example. One of the most important isotopes for medical imaging is $^{99}$Tc, an isotope of the element technetium. An excited state of $^{99}$Tc is produced in the beta decay of the molybdenum isotope $^{99}$Mo:

$$^{99}\text{Mo} \rightarrow {^{99}\text{Tc}^*} + e^- + \text{energy}$$

The asterisk signifies that the technetium nucleus is in an excited state. The excited nucleus then makes a transition to a lower-energy state via the emission of a 140 keV gamma ray:

$$^{99}\text{Tc}^* \rightarrow {^{99}\text{Tc}} + \gamma$$

The full decay process is shown in FIGURE 30.15. The final state of the technetium nucleus is much more stable than the excited state.

It's useful to collect information about the different possible decays that we've seen, so that we can compare and contrast the different decay modes.

FIGURE 30.15 $^{99}$Tc$^*$, a gamma emitter, is produced in the beta decay of $^{99}$Mo.

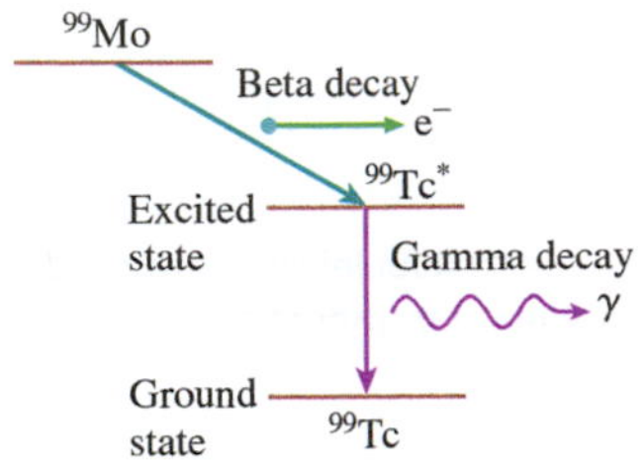


## SYNTHESIS 30.1 Nuclear decay modes

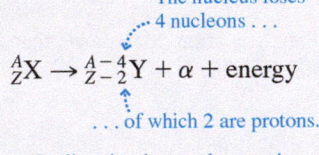

### Alpha decay

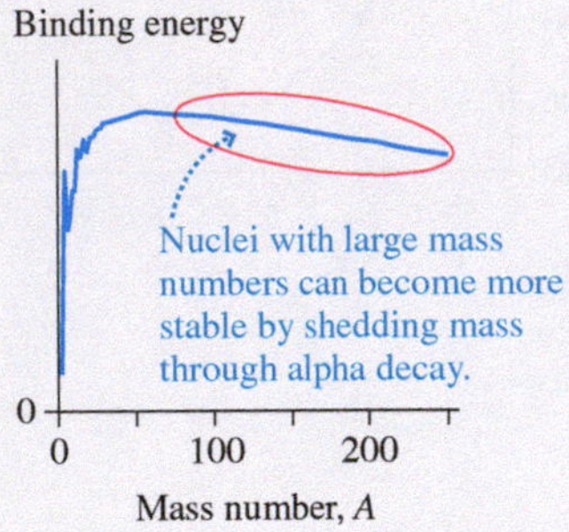


The nucleus loses 4 nucleons . . .

$$^A_Z\text{X} \rightarrow {^{A-4}_{Z-2}\text{Y}} + \alpha + \text{energy}$$

. . . of which 2 are protons.

Radium is a heavy element that is produced by decay of uranium.

$$^{226}_{88}\text{Ra} \rightarrow {^{222}_{86}\text{Rn}} + \alpha + \text{energy}$$

### Beta decay

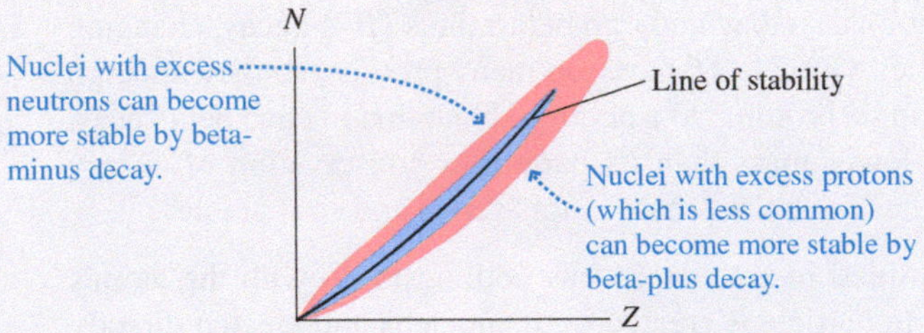

N
Nuclei with excess neutrons can become more stable by beta-minus decay.
Line of stability
Nuclei with excess protons (which is less common) can become more stable by beta-plus decay.
Z

*Beta-minus decay*

A neutron turns into a proton, so:

The number of nucleons stays the same . . .

$$^A_Z\text{X} \rightarrow {_{Z+1}^{\ \ \ A}\text{Y}} + e^- + \text{energy}$$

. . . and the nucleus gains a proton.

This chlorine isotope is present at low levels in the environment.

$$^{36}_{17}\text{Cl} \rightarrow {^{36}_{18}\text{Ar}} + e^- + \text{energy}$$

*Beta-plus decay*

A proton turns into a neutron, so:

The number of nucleons stays the same . . .

$$^A_Z\text{X} \rightarrow {_{Z-1}^{\ \ \ A}\text{Y}} + e^+ + \text{energy}$$

. . . and the nucleus loses a proton.

This carbon isotope is produced in cyclotrons.

$$^{11}_{6}\text{C} \rightarrow {^{11}_{5}\text{B}} + e^+ + \text{energy}$$

### Gamma decay

Nuclei left in an excited state by alpha or beta decay can drop into a lower energy state through gamma decay.

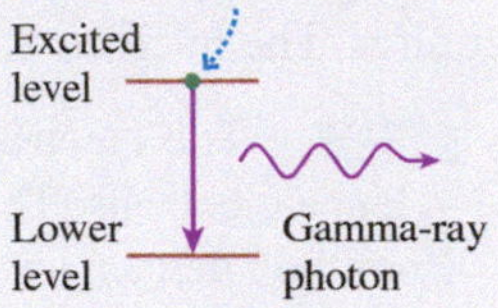


No change in the number of neutrons or protons means no change of element or isotope.

$$^A_Z\text{X}^* \rightarrow {^A_Z\text{X}} + \gamma$$

This excited form of nickel is produced in the beta-minus decay of $^{60}$Co.

$$^{60}_{28}\text{Ni}^* \rightarrow {^{60}_{28}\text{Ni}} + \gamma$$

### EXAMPLE 30.6 What type of decay?

Phosphorus has one stable isotope, $^{31}$P. The isotope $^{32}$P is a neutron-rich radioactive isotope that is used in nuclear medicine. What is the likely daughter nucleus of $^{32}$P decay?

**STRATEGIZE** Synthesis 30.1 contains details of when different decay modes are expected. $A = 32$ isn't an especially large mass number, so alpha decay is unlikely. But $^{32}$P is neutron rich; it has one more neutron than the only stable phosphorus isotope, so it is likely to undergo beta-minus decay.

**PREPARE** Phosphorus has atomic number 15.

**SOLVE** When $^{32}_{15}$P undergoes beta-minus decay, the daughter nucleus has the same number of nucleons and one more proton. Following the details in Synthesis 30.1, we can write the decay as

$$^{32}_{15}\text{P} \rightarrow {^{32}_{16}\text{S}} + e^-$$

The daughter nucleus of this decay is $^{32}$S.

**ASSESS** $^{32}_{16}$S is a stable isotope of sulfur, so our solution has a radioactive isotope decaying to a more stable state. This gives us confidence in our result.

## Decay Series

A radioactive nucleus decays into a daughter nucleus. In many cases, the daughter nucleus is also radioactive and decays to produce its own daughter nucleus. The process continues until reaching a daughter nucleus that is stable. The sequence of isotopes, starting with the original unstable isotope and ending with the stable isotope, is called a **decay series.**

The elements with $Z > 82$ present in the earth's crust are part of the decay series of a few long-lived isotopes of uranium and thorium. As an example, FIGURE 30.16 shows the decay series of $^{235}$U, an isotope of uranium with a 700-million-year half-life. There are many unstable nuclei in the decay series. Ultimately, all $^{235}$U nuclei end as the $^{207}$Pb isotope of lead, a stable nucleus.

Notice that some nuclei can decay by either alpha *or* beta decay. Thus there are a variety of paths that a decay can follow, but they all end at the same point.

FIGURE 30.16 $^{235}$U decay series.

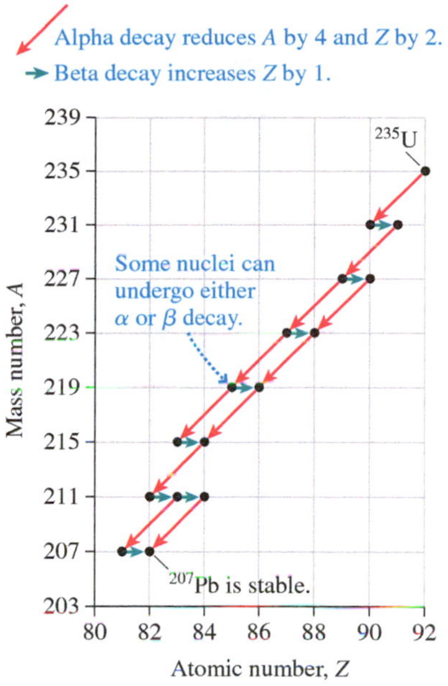


## Nuclear Radiation Is a Form of Ionizing Radiation BIO

When the sun shines on your skin, it warms it; the low-energy photons are absorbed, and their energy is converted to thermal energy. The energies of the alpha and beta particles and the gamma-ray photons of nuclear decay have much higher energies, typically in the range 0.1–10 MeV, roughly a million times larger. These energies are much higher than the ionization energies of atoms and molecules, which, as we've seen, are a few eV. Rather than simply being absorbed and increasing an object's thermal energy, nuclear radiation *ionizes* matter and *breaks* molecular bonds. FIGURE 30.17 shows this process on the atomic scale. As a high-energy alpha particle passes an atom, there is an attractive electric force between the positively charged alpha particle and the negatively charged electrons of the atom. This force can be large enough that an electron is "torn" from the atom as the alpha particle passes by, leaving a positively charged ion in its wake. Because the ionization energy of an atom is $\approx$ 10 eV, the alpha particle loses roughly 10 eV of kinetic energy each time it ionizes an atom.

FIGURE 30.17 An alpha particle can ionize an atom by pulling off one of its electrons.

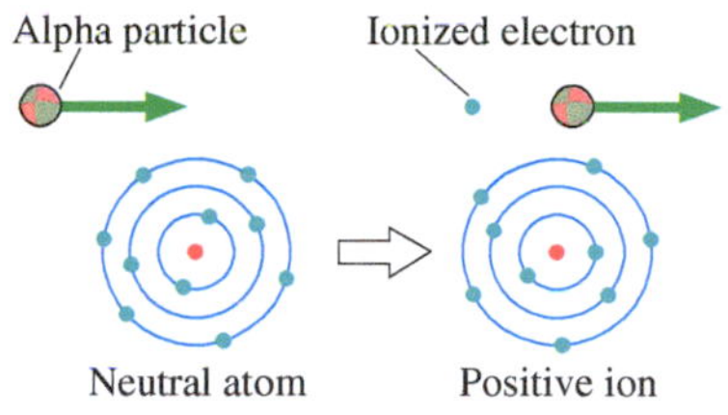


The mass of an alpha particle is much greater than that of an electron, so the path of an alpha particle is barely disturbed in the ionization process. The alpha particle continues to move in a straight-line path, ionizing many atoms as it goes. An alpha particle with an energy of 1 MeV, losing about 10 eV per ionization event, can ionize $\approx$ 100,000 atoms or molecules before finally stopping. FIGURE 30.18 shows the *ionization trail* left by a high-energy alpha particle as it moves through matter.

FIGURE 30.18 A high-energy charged particle leaves a trail of ionization along its path.

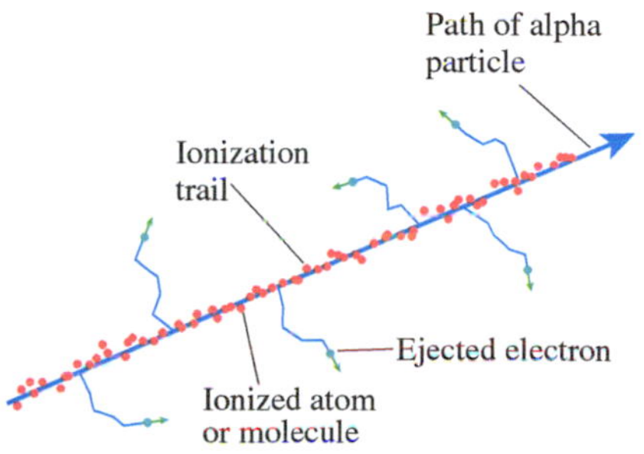


High-energy beta rays leave an ionization trail similar to that of alpha particles, although they travel farther between ionization events. The interaction of gamma rays with matter is quite different. Because gamma rays—high-energy photons—are not charged, they do not directly ionize matter through which they travel. Instead, a gamma ray can pass quite far through matter before it eventually interacts with an electron, giving it most or all of its energy. This high-energy electron then travels through the material, ionizing atoms in its wake, just as a beta ray does.

As alpha or beta particles travel along their ionization trail, they lose energy in each ionization event. Eventually all their initial kinetic energy is lost and they come to rest. Because alpha particles travel a very short distance between ionizations, they lose energy rapidly and can travel only a short distance in matter; alpha particles can be stopped by a sheet of paper. Beta particles travel somewhat farther in matter than alpha particles, with a few millimeters of aluminum or a centimeter or so of water needed to stop them. Because they are uncharged, gamma rays have the largest range, penetrating 10–20 cm in water before giving up their energy to electrons.

When ionizing radiation enters the human body, it causes damage in two ways. First, the ions drive chemical reactions that wouldn't otherwise occur. These reactions may damage the machinery of cells. Very large doses of ionizing radiation upset the delicate ionic balance that drives cellular transport and can rapidly lead to cell death. For this reason, large doses of penetrating gamma rays are sometimes used to sterilize medical equipment inside and out.

**Video** What the Physics? Cloud Chamber

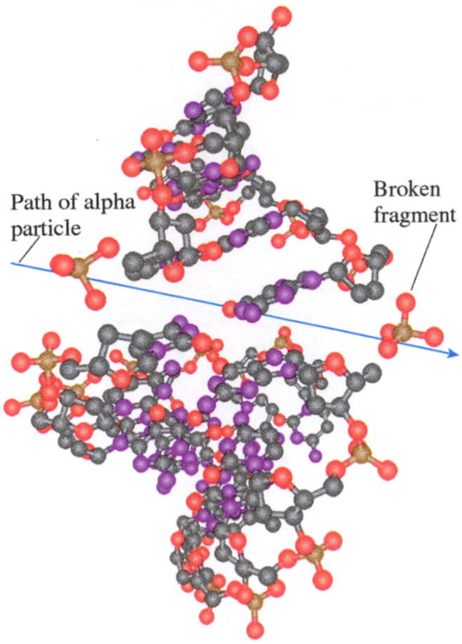


**FIGURE 30.19** Ionizing radiation can severely damage DNA molecules.

Second, ionizing radiation can damage DNA molecules by ionizing them and breaking bonds. FIGURE 30.19 shows how this might occur: A single high-energy alpha particle breaks several bonds in a DNA molecule, cleaving it in two as shown. If the damage is this extensive, cellular repair mechanisms will not be able to cope, and the DNA will be permanently damaged, possibly creating a mutation or a tumor. Tissues that have rapidly proliferating cells, such as bone marrow, are quite sensitive to ionizing radiation. Those that have less-active cell reproduction, such as the nervous system, are much less sensitive.

**CONCEPTUAL EXAMPLE 30.7 Sterilizing food by irradiation BIO**

Radiation is used in commercial facilities to sterilize food. One important use of this process is to kill insect pests in bulk grains. Which kind of radiation—alpha, beta, or gamma—is most effective for this process?

***REASON*** Alpha particles can be stopped by a sheet of paper, so they would not be able to penetrate the grain beyond a thin surface layer. Beta rays are more penetrating, but still cannot penetrate beyond a few centimeters. Gamma rays, however, are highly penetrating, and so would be able to deposit their energy throughout an entire bulk sample.

***ASSESS*** In fact, electrons are typically used to sterilize thin food products like meat, while gamma rays (from $^{60}$Co) are used for bulk products like wheat or vegetables on pallets.

**NOTE** ▸ Ionizing radiation causes damage to materials and tissues, but **objects irradiated with alpha, beta, or gamma radiation do not become radioactive.** Ionization drives chemical processes involving the electrons. An object could become radioactive only if its nuclei were somehow changed, and that does not happen. ◂

**STOP TO THINK 30.4** The cobalt isotope $^{60}\text{Co}\,(Z = 27)$ decays to the nickel isotope $^{60}\text{Ni}\,(Z = 28)$. The decay process is

A. Alpha decay.
B. Beta-minus decay.
C. Beta-plus decay.
D. Gamma decay.

## 30.5 Nuclear Decay and Half-Lives

The decay of nuclei is different from other types of decay you are familiar with. A tree branch that falls to the forest floor decays. It darkens, becomes soft, and crumbles. You might be able to tell, just by looking at it, about how long it had been decaying. Nuclear decay is different. The nucleus doesn't "age" in any sense. Instead, a nucleus has a certain probability that, within the next second, it will spontaneously turn into a different nucleus and, in the process, eject an alpha or beta particle or a gamma ray.

We can use an analogy here: If you toss a coin, it always has a 50% probability of showing tails, no matter what previous tosses might have been. You might "expect" heads if you've tossed 10 tails in a row, but the 11th toss still has a 50% chance of coming up tails. Likewise with nuclei. If a nucleus doesn't decay in this second, it is no more or less likely to decay in the next. The nucleus remains just as it was, without any change, until the decay finally occurs.

In fact, the mathematics of radioactive decay is the same as that of tossing coins. Suppose you have a large number $N_0$ of coins. You toss them all and then keep those that come up heads while setting aside those that come up tails. Probability dictates that about half the coins will show tails and be set aside. Now you repeat the process over and over. With each subsequent toss, about half the coins are set aside—they "decay."

**NOTE** ► The coins you put aside represent the daughter nuclei to which the original nuclei decay. It's important to remember that decaying nuclei don't vanish—they simply change into different nuclei. ◄

After the first toss, the number of coins you have left is about $(1/2)N_0$ because you set aside about half the coins. After the second toss, when you set aside about half of that half, the number of remaining coins is about $(1/2) \times (1/2)N_0$, or $(1/2)^2 N_0 = N_0/4$. Half of these coins will be set aside in the third toss, leaving you with $(1/2)^3 N_0$, or 1/8 of what you started with. After $m$ tosses—assuming you started with a very large number of coins—the number of coins left is $N = (1/2)^m N_0$.

Something similar happens with radioactive nuclei. If you start with $N_0$ unstable nuclei, after an interval of time we call one *half-life,* you'll have $N = (1/2)N_0$ nuclei remaining. The **half-life** $t_{1/2}$ is the average time required for one-half the nuclei to decay. This process continues, with one-half the remaining nuclei decaying in each successive half-life. The number of nuclei $N$ remaining at time $t$ is

$$N = N_0 \left(\frac{1}{2}\right)^{t/t_{1/2}} \qquad (30.12)$$

Number of atoms remaining after time $t$ → $N$

Number of atoms at the start, $t = 0$ → $N_0$

The units for $t$ and $t_{1/2}$ must be the same.

Thus $N = N_0/2$ at $t = t_{1/2}$, $N = N_0/4$ at $t = 2t_{1/2}$, $N = N_0/8$ at $t = 3t_{1/2}$, and so on, with the ratio $t/t_{1/2}$ playing the role of the "number of tosses." **No matter how many nuclei there are at any point in time, the number decays by half during the next half-life.**

**NOTE** ► Each isotope that is unstable and decays has a characteristic half-life, which can range from a fraction of a second to billions of years. Appendix C provides nuclear data and half-lives for the isotopes referred to in this text. ◄

The nature of the decay process is shown graphically in FIGURE 30.20.

KEY CONCEPT **FIGURE 30.20** The decay of a sample of radioactive nuclei.

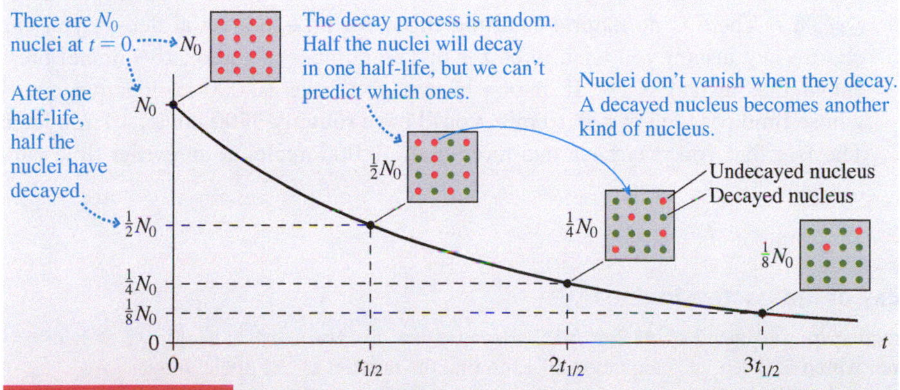


Video Figure 30.20

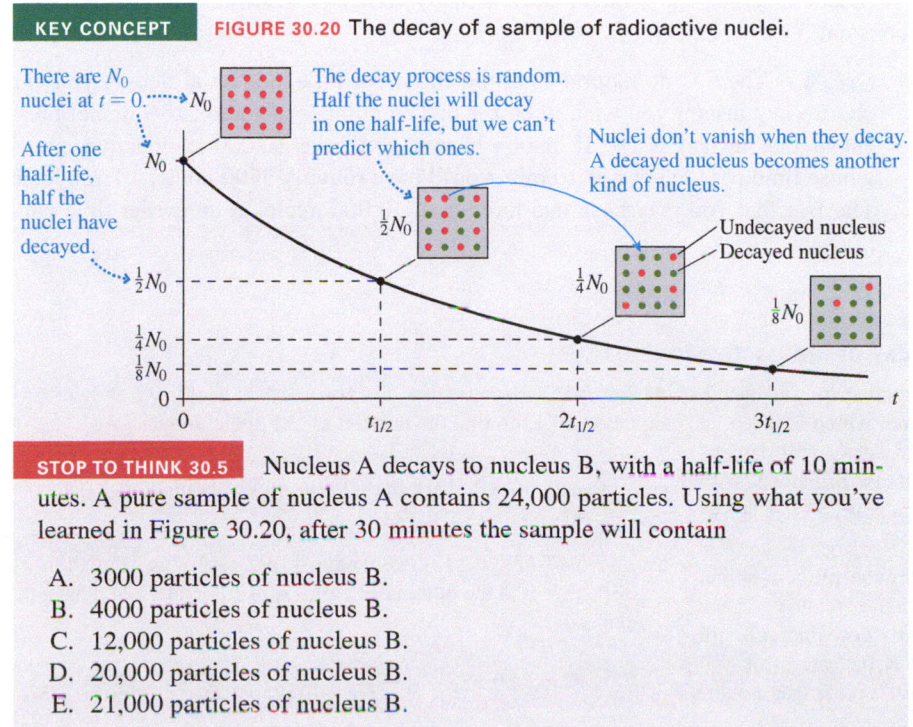

**STOP TO THINK 30.5** Nucleus A decays to nucleus B, with a half-life of 10 minutes. A pure sample of nucleus A contains 24,000 particles. Using what you've learned in Figure 30.20, after 30 minutes the sample will contain

A. 3000 particles of nucleus B.
B. 4000 particles of nucleus B.
C. 12,000 particles of nucleus B.
D. 20,000 particles of nucleus B.
E. 21,000 particles of nucleus B.

The graph in Figure 30.20 has a form we have seen before: exponential decay. The exponential decay of the number of nuclei is analogous to the exponential decay of the capacitor voltage in an *RC* circuit or the amplitude of a damped harmonic oscillator. We can write Equation 30.12 in the form of an exponential decay in terms of a *time constant* $\tau$ that is related to the half-life:

$$N = N_0 e^{-t/\tau}$$
$$\tau = \frac{t_{1/2}}{\ln 2} = (1.44)t_{1/2} \qquad (30.13)$$

The time constant $\tau$ is proportional to the half-life $t_{1/2}$.

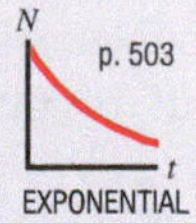


We can demonstrate the relationship between the half-life and the time constant because we know, by definition, that $N = N_0/2$ at $t = t_{1/2}$. Thus, according to Equation 30.13,

$$\frac{N_0}{2} = N_0 e^{-t_{1/2}/\tau}$$

The $N_0$ cancels, and we can then take the natural logarithm of both sides to find

$$\ln\left(\frac{1}{2}\right) = -\ln 2 = -\frac{t_{1/2}}{\tau}$$

We can rearrange this equation in two different ways, finding the time constant in terms of the half-life, or the half-life in terms of the time constant:

$$\tau = \frac{t_{1/2}}{\ln 2} = 1.44 t_{1/2}$$
$$t_{1/2} = \tau \ln 2 = 0.693\tau \qquad (30.14)$$

FIGURE 30.21 The number of radioactive atoms decreases exponentially with time.

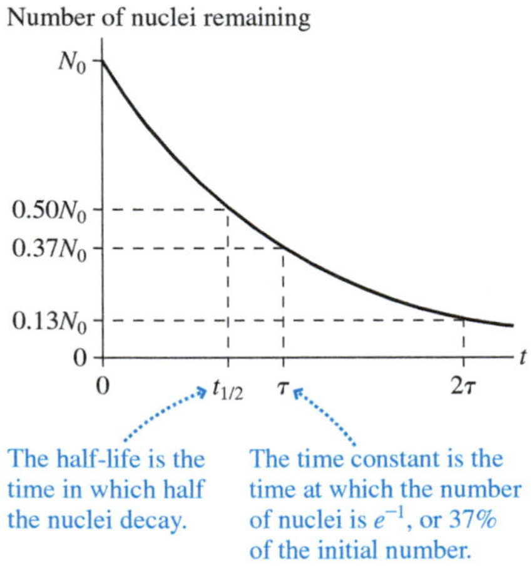


Whether we use Equation 30.12 and the half-life or Equation 30.13 and the time constant is a matter of convenience; both equations describe the same decay.

**FIGURE 30.21** is a graphical representation of Equation 30.13; it is the same graph as that of Figure 30.20, simply written in a different mathematical form. The number of radioactive nuclei decreases from $N_0$ at $t = 0$ to $N = N_0 e^{-1} = 0.37N_0$ at time $t = \tau$. In practical terms, the number decreases by roughly two-thirds during one time constant.

**NOTE** ▸ There is no natural "starting time" for an exponential decay; you can choose any instant you wish to be $t = 0$. The number of radioactive nuclei present at that instant is $N_0$. If at one instant you have 10,000 radioactive nuclei whose time constant is $\tau = 10$ min, you'll have roughly 3700 nuclei 10 min later. The fact that you may have had more than 10,000 nuclei at an earlier time isn't relevant. ◂

## EXAMPLE 30.8 Determining the decay of radioactive iodine BIO

Patients with Graves disease have an overactive thyroid gland. A common treatment uses radioactive iodine, which is taken up by the thyroid. The radiation emitted in its decay will damage the tissues of the gland. A single pill is produced with $4.0 \times 10^{14}$ atoms of the isotope $^{131}I$, which has a half-life of 8.0 days.

a. How many atoms remain 24 hours after the pill's creation, when the pill is delivered to a hospital?

b. Although the iodine in the pill is constantly decaying, it is still usable as long as it contains at least $1.1 \times 10^{14}$ atoms of $^{131}I$. What is the maximum delay before the pill is no longer usable?

***STRATEGIZE*** Because we are given the half-life, we will use Equation 30.12 to find the number of $^{131}I$ nuclei left.

***PREPARE*** The starting number of atoms is $N_0 = 4.0 \times 10^{14}$. For convenience, we will work in the time units of hours.

***SOLVE***

a. The half-life is $t_{1/2} = 8.0 \text{ days} = 192 \text{ h}$. Using Equation 30.12, we find the number of atoms remaining after 24 h have elapsed:

$$N = (4.0 \times 10^{14})\left(\frac{1}{2}\right)^{24/192} = 3.7 \times 10^{14} \text{ atoms}$$

b. The time after which $1.1 \times 10^{14}$ atoms remain is given by

$$1.1 \times 10^{14} = (4.0 \times 10^{14})\left(\frac{1}{2}\right)^{t/192}$$

To solve for $t$, we write this as

$$\frac{1.1 \times 10^{14}}{4.0 \times 10^{14}} = \left(\frac{1}{2}\right)^{t/192}$$

or

$$0.275 = \left(\frac{1}{2}\right)^{t/192}$$

Now, we take the natural logarithm of both sides:

$$\ln(0.275) = \ln\left(\left(\frac{1}{2}\right)^{t/192}\right)$$

We can solve for $t$ by using the fact that $\ln(a^x) = x\ln(a)$. This allows us to "pull out" the $t/192$ exponent to find

$$\ln(0.275) = \left(\frac{t}{192}\right)\ln\left(\frac{1}{2}\right)$$

Solving for $t$, we find that the pill ceases to be useful after

$$t = 192\frac{\ln(0.275)}{\ln(1/2)} = 360 \text{ h} = 15 \text{ days}$$

**ASSESS** The weakest usable concentration of iodine is approximately one-fourth of the initial concentration. This means that the decay time should be approximately equal to two half-lives, which is what we found.

## Activity

The **activity** $R$ of a radioactive sample is the number of decays that occur per second. Each decay corresponds to an alpha, beta, or gamma emission, so the activity is a measure of how much radiation is being given off. A detailed treatment of the mathematics of decay shows that the activity of a sample of $N$ nuclei having time constant $\tau$ (and half-life $t_{1/2}$) is

$$R = \frac{N}{\tau} = \frac{0.693N}{t_{1/2}} \tag{30.15}$$

A sample with $N = 1.0 \times 10^{10}$ nuclei decaying with time constant $\tau = 100$ s would, at that instant, have activity $R = 1.0 \times 10^8$ decays/s—in each second, $1.0 \times 10^8$ atoms would decay. Note that at later times $R$ would be smaller because the number of nuclei $N$ would be smaller.

We see from Equation 30.15 that **activity is inversely proportional to the half-life.** If two samples have the same number of nuclei, the sample with the shorter half-life has the larger activity. We can combine Equation 30.15 with Equations 30.12 and 30.13 to obtain expressions for the variation of activity with time:

$$R = \frac{N}{\tau} = \frac{N_0}{\tau}\left(\frac{1}{2}\right)^{t/t_{1/2}} = \frac{N_0}{\tau}e^{-t/\tau}$$

$N_0/\tau$ is the initial activity $R_0$, so we find the following expression for the decay of activity:

Activity after time $t$
Activity at the start, $t = 0$

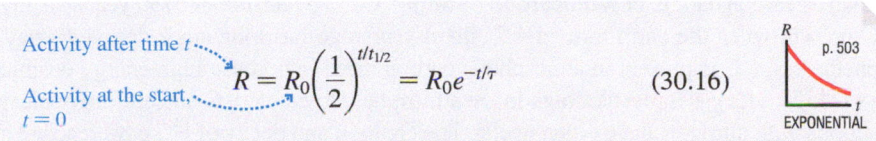

$$R = R_0\left(\frac{1}{2}\right)^{t/t_{1/2}} = R_0e^{-t/\tau} \tag{30.16}$$

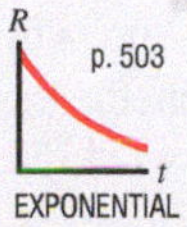


This equation has the same form as that for the decay of the sample. The activity of a sample decreases exponentially along with the number of remaining nuclei.

The SI unit of activity is the **becquerel,** defined as

$$1 \text{ becquerel} = 1 \text{ Bq} = 1 \text{ decay/s or } 1 \text{ s}^{-1}$$

An older unit of activity, but one that continues in widespread use, is the **curie.** The conversion factor is

$$1 \text{ curie} = 1 \text{ Ci} = 3.7 \times 10^{10} \text{ Bq}$$

1 Ci is a substantial amount of radiation. The radioactive samples used in laboratory experiments are typically $\approx 1\ \mu$Ci or, equivalently, $\approx 40{,}000$ Bq. These samples can be

**Powered by decay** This sphere is made of an oxide of $^{238}$Pu, a relatively short-lived ($t_{1/2} = 88$ yr) isotope of plutonium that undergoes alpha decay. The short half-life means that this sphere has a very high activity, so high that the alpha particles heat the sphere enough to make it glow. Radioactive spheres like this one are used to power spacecraft on voyages far from the sun, by using the heat generated by the radioactive decay to produce electric energy. Even after 20 years, *thermoelectric generators* of this kind still produce 75% of their original power. For many years, plutonium "batteries" were used to power heart pacemakers as well.

handled with only minor precautions. Larger sources of activity require thick shielding and other special precautions to prevent exposure to a high level of radiation.

**CONCEPTUAL EXAMPLE 30.9 Relative activities of isotopes in the body**

$^{40}$K ($t_{1/2} = 1.3 \times 10^9$ yr) and $^{14}$C ($t_{1/2} = 5.7 \times 10^3$ yr) are two radioactive isotopes found in measurable quantities in your body. Suppose you have 1 mole of each. Which is more radioactive—that is, which has a greater activity?

***REASON*** Equation 30.15 shows that the activity of a sample is proportional to the number of atoms and inversely proportional to the half-life. Because both samples have the same number of atoms, the sample of $^{14}$C, with its much shorter half-life, has a much greater activity.

**EXAMPLE 30.10 Determining the decay of activity BIO**

A $^{60}$Co (half-life 5.3 yr) source used to provide gamma rays to irradiate tumors has an activity of 0.43 Ci.

a. How many $^{60}$Co atoms are in the source?
b. What will be the activity of the source after 10 yr?

***STRATEGIZE*** The activity of the source depends on the half-life and the number of atoms. If we know the activity and the half-life, we can compute the number of atoms. The number of atoms will undergo exponential decay, and so will the activity.

***SOLVE***

a. Equation 30.15 relates the activity and the number of atoms. Rewriting the equation, we can relate the initial activity and the initial number of atoms as $N_0 = t_{1/2}R_0/0.693$. To use this equation, we need numbers in SI units. In Bq, the initial activity of the source is

$$R_0 = (0.43\text{ Ci})\left(\frac{3.7 \times 10^{10}\text{ Bq}}{1\text{ Ci}}\right) = 1.59 \times 10^{10}\text{ Bq}$$

The half-life $t_{1/2}$ in s is

$$t_{1/2} = (5.3\text{ yr})\left(\frac{3.16 \times 10^7\text{ s}}{1\text{ yr}}\right) = 1.67 \times 10^8\text{ s}$$

Thus the initial number of $^{60}$Co atoms in the source is

$$N_0 = \frac{t_{1/2}R_0}{0.693} = \frac{(1.67 \times 10^8\text{ s})(1.59 \times 10^{10}\text{ Bq})}{0.693}$$
$$= 3.8 \times 10^{18}\text{ atoms}$$

b. The variation of activity with time is given by Equation 30.16. After 10 yr, the activity is

$$R = R_0\left(\frac{1}{2}\right)^{t/t_{1/2}} = (1.59 \times 10^{10}\text{ Bq})\left(\frac{1}{2}\right)^{10/5.3}$$
$$= 4.3 \times 10^9\text{ Bq} = 0.12\text{ Ci}$$

***ASSESS*** For part a, the initial activity of 0.43 Ci is roughly $10^{10}$ Bq, or $10^{10}$ decays/s. For this to continue for 10 years or so, roughly $10^{18}$ atoms seems plausible. For part b, 10 years is about two half-lives, so we would expect the activity to fall by about a factor of 4. Our answer is close to this.

## Radioactive Dating

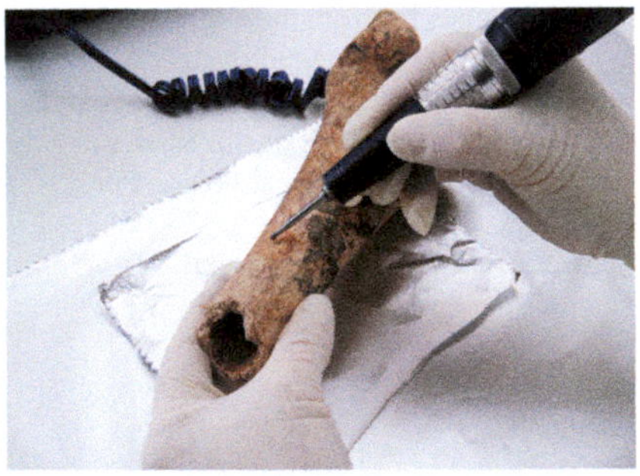

A researcher is extracting a small sample of an ancient bone. By measuring the ratio of carbon isotopes present in the sample she will determine the age of the bone.

Many geological and archaeological samples can be dated by measuring the decays of naturally occurring radioactive isotopes.

The most well-known dating technique uses the radioactive carbon isotope $^{14}$C and is known as carbon dating, or **radiocarbon dating.** $^{14}$C has a half-life of 5730 years, so any $^{14}$C present when the earth formed 4.5 billion years ago has long since decayed away. Nonetheless, $^{14}$C is present in atmospheric carbon dioxide because high-energy cosmic rays collide with gas molecules high in the atmosphere to create $^{14}$C nuclei from nuclear reactions with nitrogen and oxygen nuclei. The creation and decay of $^{14}$C have reached a steady state in which the $^{14}$C/$^{12}$C ratio is relatively stable at $1.3 \times 10^{-12}$.

All living organisms constantly exchange carbon dioxide with the environment, so the $^{14}$C/$^{12}$C ratio in living organisms is also $1.3 \times 10^{-12}$. As soon as an organism dies, the $^{14}$C in its tissue begins to decay and no new $^{14}$C is added. As time goes on, the $^{14}$C decays at a well-known rate. Thus, a measurement of the activity of an ancient organic sample permits a determination of the age. "New" samples have a higher fraction of $^{14}$C than "old" samples.

The first step in radiocarbon dating is to extract and purify carbon from the sample. The carbon is then placed in a shielded chamber and its activity measured. This activity is then compared to the activity of an identical modern sample. Equation 30.16 relates the activity of a sample at a time $t$ to its initial activity. If we assume that the original activity $R_0$ was the same as the activity of the modern sample, we can determine the time since the decay began—and thus the age of the sample.

**EXAMPLE 30.11 Carbon-dating a tooth** BIO

A rear molar from a mammoth skeleton is dated using a measurement of its $^{14}C$ content. Carbon from the tooth is chemically extracted and formed into benzene. The benzene sample is placed in a shielded chamber. Decays from the sample come at an average rate of 11.5 counts per minute. A modern benzene sample of the exact same size gives 54.9 counts per minute. What is the age of the skeleton?

***STRATEGIZE*** We can assume that, thousands of years ago, the sample had an initial activity of 54.9 counts per minute—identical to the activity of a modern sample. The present activity is lower due to the decay of the $^{14}C$ since the death of the mammoth.

***PREPARE*** Equation 30.16 gives the decrease of the activity as a function of time as $R = R_0(1/2)^{t/t_{1/2}}$. The current activity is $R = 11.5$ counts per minute, and we assume that the initial activity was $R_0 = 54.9$. $t$ is the time since the mammoth died—the age of the skeleton.

***SOLVE*** We solve for $t$ by rearranging terms and computing a natural logarithm, as in

$$\frac{R}{R_0} = \left(\frac{1}{2}\right)^{t/t_{1/2}}$$

$$\ln\left(\frac{R}{R_0}\right) = \left(\frac{t}{t_{1/2}}\right)\ln\left(\frac{1}{2}\right)$$

We then solve for the time $t$:

$$t = \frac{t_{1/2}}{\ln(1/2)}\ln\left(\frac{R}{R_0}\right) = \frac{5730\text{ yr}}{\ln(1/2)}\ln\left(\frac{11.5}{54.9}\right) = 12{,}900\text{ yr}$$

***ASSESS*** The final time is in years, the same unit we used for the half-life. This is a realistic example of how such radiocarbon dating is done; the numbers and details used in this example come from an actual experimental measurement. The age of the sample places it at the end of the last ice age, when mammoths last roamed the earth, so our result seems reasonable.

**CONCEPTUAL EXAMPLE 30.12 Source contamination**

One possible problem with carbon dating is contamination with modern carbon sources. Suppose an archaeologist has unearthed and carbon-dated a fragment of wood that has absorbed carbon of recent vintage from organic molecules in groundwater. Does he underestimate or overestimate the age of the wood?

***REASON*** Because the wood has absorbed modern carbon, it will have more $^{14}C$ than it would had it decayed undisturbed. The present activity is higher than it would otherwise be. This will lead to an underestimate of the age of the wood.

**Responsible dating** BIO Measuring the activity of the carbon in a sample to determine the fraction of $^{14}C$ can require a significant amount of organic material—perhaps 25 g. For an artifact of great historical importance, such as this parchment fragment from the Dead Sea scrolls, this would be unacceptable. Instead, dates are obtained by using a mass spectrometer to directly measure the ratio of carbon isotopes, from which the age can be determined. This can be done with excellent accuracy on as little as 0.1 g of material.

Carbon dating can be used to date skeletons, wood, paper, fur, food material, and anything else made of organic matter. It is quite accurate for ages to about 15,000 years, about three half-lives. Items are dated to about 50,000 years with a fair degree of reliability.

Isotopes with longer half-lives are used to date geological samples. Potassium-argon dating, using $^{40}K$ with a half-life of 1.25 billion years, is especially useful for dating rocks of volcanic origin.

**STOP TO THINK 30.6** A sample of 1000 radioactive atoms has a 10 minute half-life. How old is the sample when 750 atoms have decayed?

A. 10 minutes B. 15 minutes C. 20 minutes D. 30 minutes

# 30.6 Medical Applications of Nuclear Physics BIO

Nuclear physics has brought both peril and promise to society. Radioactivity can cause tumors. At the same time, radiation can be used to diagnose and cure some cancers. This section is a brief survey of medical applications of nuclear physics.

## Radiation Dose

Nuclear radiation disrupts a cell's machinery by altering and damaging biological molecules, as we saw in Section 30.4. The biological effects of radiation depend on two factors. The first is the physical factor of how much energy is absorbed by the body. The second is the biological factor of how tissue reacts to different forms of radiation.

Ionizing radiation damages cells of the body, but it also damages bacteria and other pathogens. This gamma source is used for sterilizing medical equipment. The blue glow is due to the ionization of the air around the source.

Suppose a beta particle travels through tissue, losing kinetic energy as it ionizes atoms it passes. The energy lost by the beta particle is a good measure of the number of ions produced and thus the amount of damage done. In a certain volume of tissue, more ionization means more damage. For this reason, we define the radiation **dose** as the energy from ionizing radiation absorbed by 1 kg of tissue. The SI unit for the dose is the **gray,** abbreviated Gy. The Gy is defined as

$$1 \text{ Gy} = 1.00 \text{ J/kg of absorbed energy}$$

The number of Gy depends only on the energy absorbed, not on the type of radiation or on what the absorbing material is. Another common unit for dose is the *rad;* 1 rad = 0.01 Gy.

A 1 Gy dose of gamma rays and a 1 Gy dose of alpha particles have different biological consequences. To account for such differences, the **relative biological effectiveness** (RBE) is defined as the biological effect of a given dose relative to the biological effect of an equal dose of x rays. TABLE 30.3 lists the relative biological effectiveness of different forms of radiation. Higher values correspond to larger biological effects.

TABLE 30.3 Relative biological effectiveness of radiation

| Radiation type | RBE |
|---|---|
| X rays | 1 |
| Gamma rays | 1 |
| Beta particles | 1 |
| Protons | 5 |
| Neutrons | 5–20 |
| Alpha particles | 20 |

The radiation **dose equivalent** is the product of the energy dose in Gy and the relative biological effectiveness. Dose equivalent is measured in **sieverts,** abbreviated Sv. To be precise,

$$\text{dose equivalent in Sv} = \text{dose in Gy} \times \text{RBE}$$

One Sv of radiation produces the same biological damage regardless of the type of radiation. Another common unit of dose equivalent (also called biologically equivalent dose) is the *rem;* 1 rem = 0.01 Sv.

NOTE ▸ In practice, the term "dose" is often used for both dose and dose equivalent. Use the units as a guide. If the unit is Sv or rem, it is a dose equivalent; if Gy or rad, a dose. ◂

**EXAMPLE 30.13 Finding energy deposited in radiation exposure** BIO

A 75 kg patient is given a bone scan. A phosphorus compound containing the gamma-emitter $^{99}$Tc is injected into the patient. It is taken up by the bones, and the emitted gamma rays are measured. The procedure exposes the patient to 3.6 mSv (360 mrem) of radiation. What is the total energy deposited in the patient's body, in J and in eV?

**STRATEGIZE** The exposure is given in Sv, so it is a dose equivalent, a combination of deposited energy and biological effectiveness. The RBE for gamma rays is 1. Gamma rays are penetrating, and the source is distributed throughout the body, so this is a whole-body exposure. Each kg of the patient's body will receive approximately the same energy.

**PREPARE** The dose in Gy is the dose equivalent in Sv divided by the RBE. In this case, because RBE = 1, the dose in Gy is numerically equal to the equivalent dose in Sv. The dose is thus 3.6 mGy $= 3.6 \times 10^{-3}$ J/kg.

**SOLVE** The radiation energy absorbed in the patient's body is

$$\text{absorbed energy} = (3.6 \times 10^{-3} \text{ J/kg})(75 \text{ kg}) = 0.27 \text{ J}$$

In eV, this is

$$\text{absorbed energy} = (0.27 \text{ J})(1 \text{ eV}/1.6 \times 10^{-19} \text{ J}) = 1.7 \times 10^{18} \text{ eV}$$

**ASSESS** The total energy deposited, 0.27 J, is quite small; there will be negligible heating of tissue. But radiation produces its effects in other ways, as we have seen. Because it takes only $\approx 10$ eV to ionize an atom, this dose is enough energy to ionize over $10^{17}$ atoms, meaning it can cause significant disruption to the cells of the body.

FIGURE 30.22 BIO Natural sources of radiation dose.

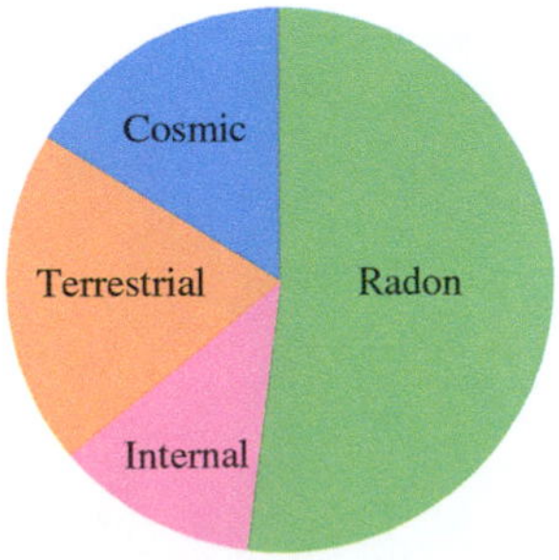


## Sources of Radiation and Its Effects

The question inevitably arises: What is a safe dose of radiation? Unfortunately, there is no simple or clear definition of a safe dose. A prudent policy is to avoid unnecessary exposure and to weigh the significance of an exposure in relation to the natural background. We are all exposed to continuous radiation from a variety of natural sources, as FIGURE 30.22 shows. The largest contribution to the background radiation is due to the radon isotope $^{222}$Rn. Radon is a decay product of the uranium that is always present in soil. Radon is an inert gas and so it easily seeps into basements, where its daughter isotopes attach to dust particles and can be inhaled into the lungs. Terrestrial radiation is mostly in the form of gamma rays from uranium and its decay products in the soil and in certain building materials. Cosmic rays are high-energy ionizing particles that stream from the sun and more distant astronomical sources. Internal radiation originates mostly from radioactive $^{40}K$, present at the 0.012% level in the potassium we ingest in many foods; about 4000 $^{40}K$ nuclei decay in your body every second.

This natural background dose averages about 3 mSv (300 mrem) per year, although there are wide regional variations depending on elevation and soil type. In high-altitude Colorado, where the thinner atmosphere absorbs fewer cosmic rays, the annual cosmic-ray dose is nearly twice the dose at sea level. The Colorado Plateau area also has higher levels of uranium in the soil, and the terrestrial contribution to the background there is nearly five times higher than it is along the Atlantic coast.

TABLE 30.4 lists the expected exposure from several different sources that are not of natural origin. A dental x ray subjects a person to approximately 1% of the yearly natural background that he or she would normally receive and so is likely not a cause for significant worry. Mammograms involve a much larger dose, concentrated in a small region of the body. A nuclear medicine procedure, such as a PET scan (which is discussed in a later section), may involve an exposure that is larger than the typical yearly background dose. This significant dose must be weighed against the medical benefits of the procedure.

At what dose is there a clear, elevated risk? It is difficult to separate out the effects of artificial radiation from natural sources, as well as other potential causes of cancer, such as chemicals and smoking. But large-scale studies suggest that a slightly increased chance of cancer developing over one's lifetime begins at an annual dose of 100 mSv. As Table 30.4 also shows, doses above 1000 mSv received in a short period—such as a nuclear accident—are extremely dangerous, and those above 5000 mSv are likely to be fatal.

**TABLE 30.4** Radiation exposure and effects

| Radiation Limit or Source | Exposure (mSv) |
|---|---|
| Fatal dose* | 10,000 |
| Dose resulting in 50% fatalities* | 5000 |
| Onset of radiation poisoning symptoms* | 500 |
| Astronauts on the International Space Station | 150/year |
| Allowable dose for radiation workers | 50/year |
| Whole-body x ray scan | 10 |
| PET scan | 7 |
| Mammogram | 0.70 |
| Chest x ray | 0.30 |
| Transatlantic airplane flight | 0.050 |
| Dental x ray | 0.030 |

*If received in a short period

## Nuclear Medicine

The tissues in the body most susceptible to radiation are those that are rapidly proliferating—including tumors. The goal in *radiation therapy* is to apply a large enough dose of radiation to destroy or shrink a tumor while producing minimal damage to surrounding healthy tissue.

In FIGURE 30.23a, a patient with a brain tumor is fitted with a metal *collimator* that absorbs gamma rays except for those traveling along desired paths. The collimator is fashioned so that gamma rays from an external source will be concentrated on the tumor, as shown in FIGURE 30.23b. Because the rapidly dividing cells of a tumor are much more sensitive to radiation than the tissues of the brain, and because surgical options carry risks of significant complications, radiation is a common means of treating tumors of the brain.

Other tumors are treated by surgically implanting radioactive "seeds" within the tumor. One common type of seed contains $^{125}I$, which undergoes a nuclear decay followed by emission of a 27.5 keV photon. The relatively low-energy photon has a very short range so that it will damage only tissue close to the seed.

Some tissues in the body will preferentially take up certain isotopes, allowing for treatment by isotope ingestion. A common treatment for hyperthyroidism, in which the thyroid gland is overactive, is to damage the gland with the isotope $^{131}I$, a beta-emitter with a half-life of 8.0 days. A patient is given a tablet containing $^{131}I$. The iodine in the blood is taken up and retained by the thyroid gland, resulting in a reduction of the gland's activity with minimal disruption of surrounding tissue.

FIGURE 30.23 BIO The use of gamma rays to treat a tumor in the brain.

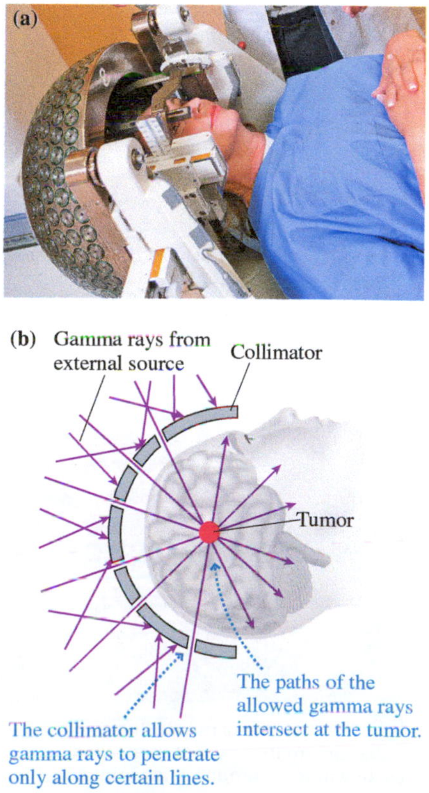

**EXAMPLE 30.14 Using radiation to treat disease** BIO

Patients who have prostate cancer can be treated by implanting radioactive metal "seeds" into the gland, as shown in FIGURE 30.24. For one patient's treatment, the seeds contain the samarium isotope $^{153}$Sm, which decays with a half-life of 1.9 days by emitting a 0.81 MeV beta particle. On average, 50% of the decay energy is absorbed by the 24 g prostate gland. If the seeds have an activity of 100 MBq when first implanted, what dose equivalent in Sv does the gland receive during the first day?

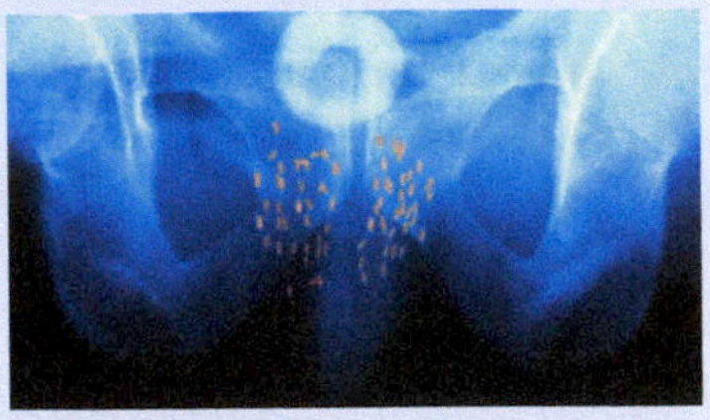

FIGURE 30.24 Radioactive seeds are implanted in the prostate gland to shrink a tumor.

**STRATEGIZE** To find the dose equivalent, we will need to know how many nuclei have decayed and how much energy each nucleus deposits in the body. We can find the initial number of nuclei from the activity, as we did in Example 30.10. Then, using the known half-life, we can find the number that remain after one day. The difference between these two numbers is the number of decays in the first day.

**PREPARE** We need to convert the half-life from days to seconds:

$$t_{1/2} = (1.9\text{ days})\left(\frac{8.64 \times 10^4\text{ s}}{1\text{ day}}\right) = 1.64 \times 10^5\text{ s}$$

We also need the absorbed energy in joules. Each atom that decays emits a 0.81 MeV beta particle, and half of this energy is deposited in the gland. In joules, this is

$$\text{energy deposited per decay} = (0.5)(0.81\text{ MeV})(1.60 \times 10^{-13}\text{ J/MeV}) = 6.48 \times 10^{-14}\text{ J}$$

**SOLVE** The initial number of radioactive nuclei is found from Equation 30.15:

$$N_0 = \frac{t_{1/2}R_0}{0.693} = \frac{(1.64 \times 10^5\text{ s})(100 \times 10^6\text{ decays/s})}{0.693} = 2.37 \times 10^{13}$$

After 1.0 day, the number of radioactive nuclei in the sample has decreased to

$$N = N_0\left(\frac{1}{2}\right)^{t/t_{1/2}} = (2.37 \times 10^{13})\left(\frac{1}{2}\right)^{1.0\text{ day}/1.9\text{ days}} = 1.65 \times 10^{13}$$

The number of decays during the first day is the difference between these numbers:

$$N_0 - N = (2.37 \times 10^{13}) - (1.65 \times 10^{13}) = 7.2 \times 10^{12}\text{ decays}$$

The energy deposited in the gland by these decays is

$$\text{total energy deposited} = (6.48 \times 10^{-14}\text{ J/decay})(7.2 \times 10^{12}\text{ decays}) = 0.467\text{ J}$$

The radiation dose to the 0.024 kg gland is thus

$$\text{dose} = \frac{0.467\text{ J}}{0.024\text{ kg}} = 19\text{ Gy}$$

Beta particles have RBE = 1, so the dose equivalent in Sv is the same:

$$\text{dose equivalent} = 19\text{ Sv}$$

**ASSESS** This is a very large dose equivalent, but the goal is to destroy the tumor, so this is a reasonable result.

In this example, the dose received by the prostate gland was 19 Sv. From Table 30.4, this dose appears to be well above a fatal dose. But the values in Table 30.4 are for a *full-body* dose, with all organs of the body exposed. The body can tolerate high doses that are localized to small regions, a fact that makes nuclear medicine possible.

## Nuclear Imaging

X rays from an external source may be used to make an image of the body, as described in Chapter 28. *Nuclear imaging* uses an internal source—radiation from isotopes in the body—to produce an image of tissues in the body. The image of the mouse that opened the chapter is an example of nuclear imaging.

There is a key difference between x rays and nuclear imaging procedures. **An x ray is an image of anatomical structure;** it is excellent for identifying structural problems like broken bones. **Nuclear imaging creates an image of the biological activity of tissues in the body.** For example, nuclear imaging can detect reduced metabolic activity of brain tissue after a stroke.

Let's look at an example that illustrates the difference between a conventional x-ray image or CAT scan and an image made with a nuclear imaging technique. Suppose a doctor suspects a patient has cancerous tissue in the bones. An x ray does not show anything out of the ordinary; the tumors may be too small or may appear similar to normal bone. The doctor then orders a scan with a **gamma camera,** a device that can measure and produce an image from gamma rays emitted within the body.

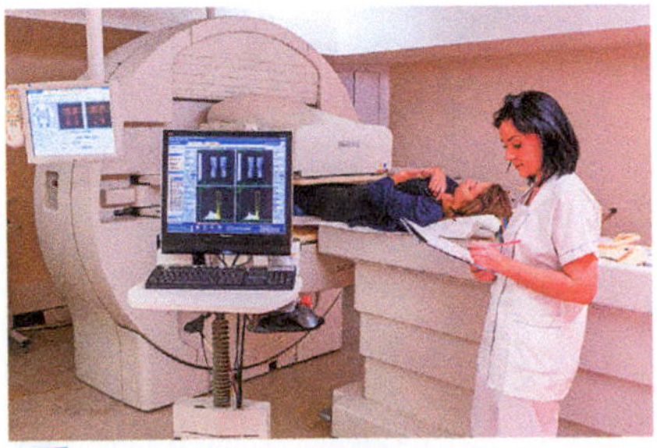

BIO The detector is measuring gamma radiation emitted by isotopes taken up by tissues in the woman's knee joints.

The patient is given a dose of a phosphorus compound labeled with the gamma-emitter $^{99}Tc$. This compound is taken up and retained in bone tissue where active growth is occurring. The $^{99}Tc$ will be concentrated in the bones where there has been recent injury or inflammation—or where a tumor is growing. The patient is then scanned with a gamma camera. FIGURE 30.25a shows how the gamma camera can pinpoint the location of the gamma-emitting isotopes in the body and produce an image that reveals their location and intensity. A typical image is shown in FIGURE 30.25b. The bright spots show high concentrations of $^{99}Tc$, revealing areas of tumor growth. The tumors may be too small to show up on an x ray, but their activity is easily detected with the gamma camera. With such early detection, the patient's chance of a cure is greatly improved. Combining such images with a traditional x-ray or CT scan, as in the photo at the start of the chapter, can provide even more information.

FIGURE 30.25 BIO The operation of a gamma camera.

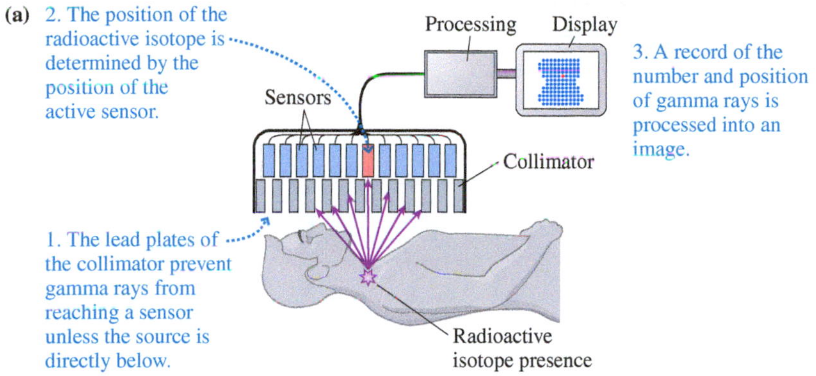


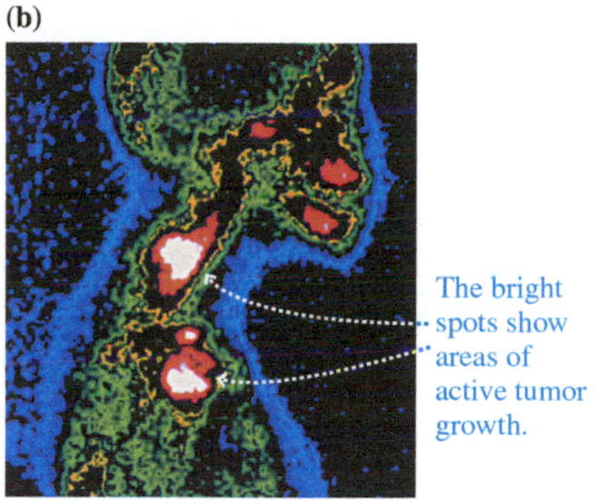

(b)
The bright spots show areas of active tumor growth.


## CONCEPTUAL EXAMPLE 30.15 Using radiation to diagnose disease BIO

A patient suspected of having kidney disease is injected with a solution containing molecules that are taken up by healthy kidney tissue. The molecules have been "tagged" with radioactive $^{99}Tc$. A gamma camera scan of the patient's abdomen gives the image in FIGURE 30.26. In this image, blue corresponds to the areas of highest activity. Which of the patient's kidneys has reduced function?

FIGURE 30.26 A gamma scan of a patient's kidneys.

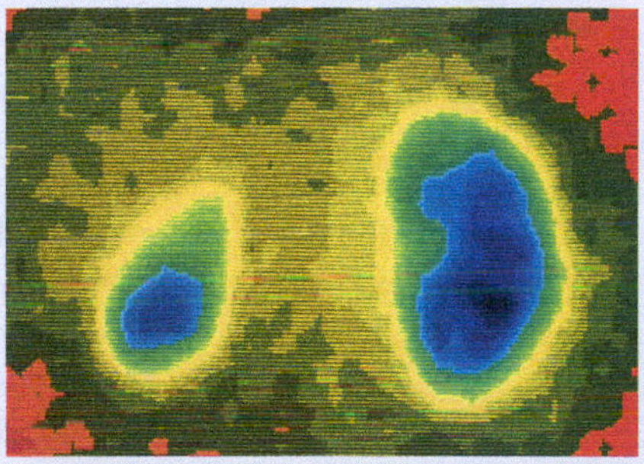

**REASON** Healthy tissue should show up in blue on the scan because healthy tissue will absorb molecules with the $^{99}Tc$ attached and will thus emit gamma rays. The kidney imaged on the right shows normal activity throughout; the kidney imaged on the left appears smaller, so it has a smaller volume of healthy tissue. The patient is ill; the problem is with the kidney imaged on the left.

**ASSESS** Depending on the isotope and how it is taken up by the body, either healthy tissue or damaged tissue could show up on a gamma camera scan.

**Video** Positron Emission Tomography

## Positron-Emission Tomography

We have seen that a small number of radioactive isotopes decay by the emission of a positron. Such isotopes can be used for an imaging technique known as *positron-emission tomography*, or *PET*. PET is particularly important for imaging the brain.

The imaging process relies on the mass–energy conversion resulting from the combination of an electron and a positron. Suppose an electron and a positron are at rest, or nearly so; the combined momentum is nearly zero. The electron and positron have opposite charges and so will attract each other. When they meet, as we saw in ◀◀ SECTION 27.10, they completely annihilate—but energy and momentum are still conserved. The conservation of energy means that the annihilation will produce one or more high-energy photons—gamma rays. We learned in Chapter 27 that photons have momentum, so the annihilation can't produce a single photon because that photon would leave the scene of the annihilation with momentum. Instead the most likely result is a pair of photons directed exactly opposite each other, as shown in FIGURE 30.27a.

FIGURE 30.27 Positron-emission tomography.

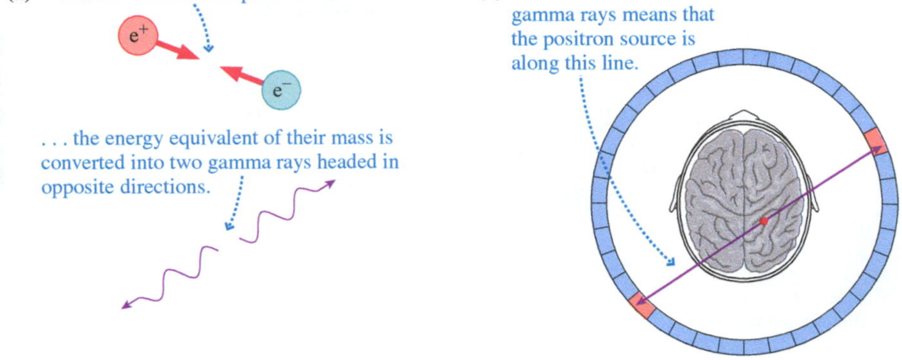


Most PET scans use the fluorine isotope $^{18}F$, which emits a positron as it undergoes beta-plus decay to $^{18}O$ with a half-life of 110 minutes. $^{18}F$ is used to create an analog of glucose called fluorine-18 fluorodeoxyglucose (F-18 FDG). This compound is taken up by tissues in the brain. Areas that are more active are using more glucose, so the F-18 FDG is concentrated in active brain regions. When a fluorine atom in the F-18 FDG decays, the emitted positron immediately collides with a regular electron. The two annihilate to produce two gamma rays that travel out of the brain in opposite directions, as shown in Figure 30.27a.

FIGURE 30.27b shows a patient's head surrounded by a ring of gamma-ray detectors. Because the gamma rays from the positron's annihilation are emitted back to back, simultaneous detection of two gamma rays on opposite sides of the subject indicates that the annihilation occurred somewhere along the line between those detectors. Recording many such pairs of gamma rays shows with great accuracy where the decays are occurring. A full scan will show more activity in regions of the brain where metabolic activity is enhanced, less activity in regions where metabolic activity is depressed. An analysis of these scans can provide a conclusive diagnosis of stroke, injury, or Alzheimer's disease.

**STOP TO THINK 30.7** A patient ingests a radioactive isotope to treat a tumor. The isotope provides a dose of 0.10 Gy. Which type of radiation will give the highest dose equivalent in Sv?

A. Alpha particles B. Beta particles C. Gamma rays

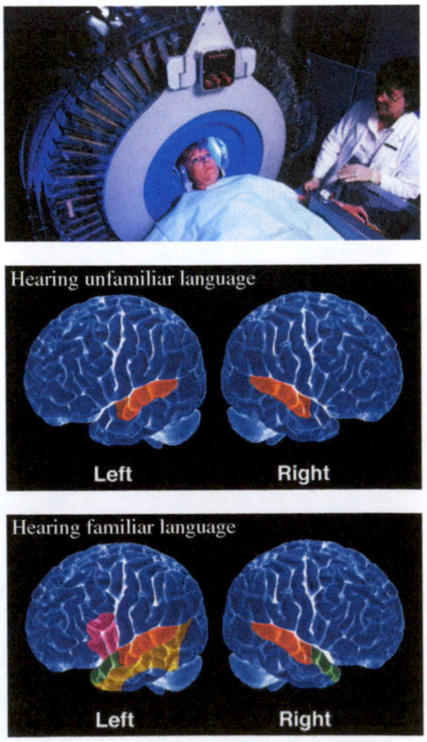


◀ **This is your brain on PET** BIO The woman in the top photo is undergoing a PET scan not to diagnose disease but to probe the workings of the brain. While undergoing a PET scan, a subject is asked to perform different mental tasks. The lower panels show functional images from a PET scan superimposed on anatomical images from a CAT scan. The subject first listened to speech in an unfamiliar language; the active areas of the brain were those responsible for hearing. Next, she listened to speech in a familiar language, resulting in activity in the parts of the brain responsible for speech and comprehension.

# 30.7 The Ultimate Building Blocks of Matter

As we've seen, modeling the nucleus as being made of protons and neutrons allows a description of all of the elements in the periodic table in terms of just three basic particles—protons, neutrons, and electrons. But are protons and neutrons *really* basic building blocks? Molecules are made of atoms. Atoms are made of a cloud of electrons surrounding a positively charged nucleus. The nucleus is composed of protons and neutrons. Where does this process end? Are electrons, protons, and neutrons the basic building blocks of matter, or are they made of still smaller subunits?

This question takes us into the domain of what is known as **particle physics**—the branch of physics that deals with the basic constituents of matter and the relationships among them. Particle physics starts with the constituents of the atom, the proton, neutron, and electron, but there are many other particles below the scale of the atom. We call these particles **subatomic particles.**

## Antiparticles

We've described the positron as the *antiparticle* to the electron. In what sense is a positron an *anti*electron? As we've seen, when a positron and an electron meet, they annihilate each other, turning the energy equivalent of their masses into the pure energy of two photons. Mass disappears and light appears in one of the most spectacular confirmations of Einstein's relativity.

Every subatomic particle that has been discovered has an antiparticle twin that has the same mass and the same spin but opposite charge. In addition to positrons, there are antiprotons (with $q = -e$), antineutrons (also neutral, but not the same as regular neutrons), and antimatter versions of all the various subatomic particles we will see. The notation to represent an antiparticle is a bar over the top of the symbol. A proton is represented as p, an antiproton as $\bar{\text{p}}$.

Antiparticles provide interesting opportunities for creating "exotic" subatomic particles. When a particle meets its associated antiparticle, the two annihilate, leaving nothing but their energy behind. This energy must go somewhere. Sometimes it is emitted as gamma-ray photons, but this energy can also be used to create other particles.

The major tool for creating and studying subatomic particles is the *particle collider.* These machines use electric and magnetic fields to accelerate particles and their antiparticles, such as e and $\bar{\text{e}}$, or p and $\bar{\text{p}}$, to speeds very close to the speed of light. These particles then collide head-on. As they collide and annihilate, their mass-energy and kinetic energy combine to produce exotic particles that are not part of ordinary matter. These particles come in a dizzying variety—pions, kaons, lambda particles, sigma particles, and dozens of others—each with its own antiparticle. Most live no more than a trillionth of a second before decaying into other particles.

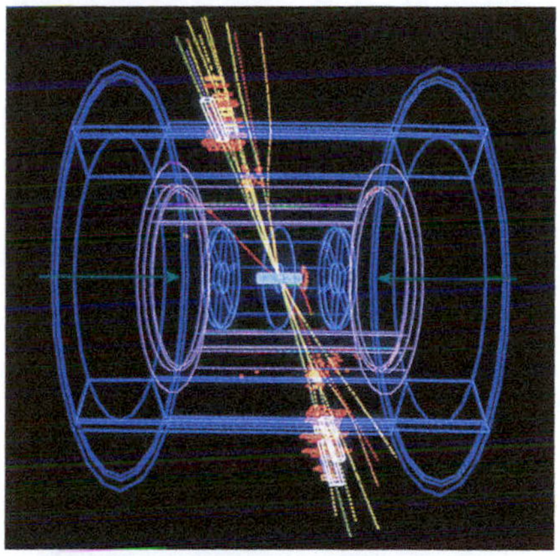

**Subatomic crash tests** The above picture is the record of a collision between an electron and a positron (paths represented by the green arrows) brought to high speeds in a collider. The particles annihilated in the center of a detector that measured the paths of the particles produced in the collision. In this case, the annihilation of the electron and positron created a particle known as a Z boson. The Z boson then quickly decayed into the two jets of particles seen coming out of the detector.

**EXAMPLE 30.16 Determining a possible outcome of a proton-antiproton collision**

When a proton and an antiproton annihilate, the resulting energy can be used to create new particles. One possibility is the creation of electrically neutral particles called *neutral pions.* A neutral pion has a rest mass of 135 MeV/$c^2$. How many neutral pions could be produced in the annihilation of a proton and an antiproton? Assume the proton and antiproton are moving very slowly as they collide.

**STRATEGIZE** Because the proton and antiproton are moving slowly, with essentially no kinetic energy, the total energy available for creating new particles is the energy equivalent of the masses of the proton and the antiproton.

**PREPARE** The mass of a proton is given in Table 30.2 as 938 MeV/$c^2$. The mass of an antiproton is the same.

**SOLVE** The total energy from the annihilation of a proton and an antiproton is the energy equivalent of their masses:

$$E = (m_{\text{proton}} + m_{\text{antiproton}})c^2 = (938\text{ MeV}/c^2 + 938\text{ MeV}/c^2)c^2 = 1876\text{ MeV}$$

*Continued*

It takes 135 MeV to create a neutral pion. The ratio

$$\frac{\text{energy available}}{\text{energy required to create a pion}} = \frac{1876\text{ MeV}}{135\text{ MeV}} = 13.9$$

tells us that we have enough energy to produce 13 neutral pions from this process, but not quite enough to produce 14.

***ASSESS*** Because the mass of a pion is much less than that of a proton or an antiproton, the annihilation of a proton and antiproton can produce many more pions than the number of particles at the start. Though the production of 13 neutral pions is a possible outcome of a proton-antiproton interaction, it is not a likely one. In addition to the conservation of energy, there are many other physical laws that determine what types of particles, and in what quantities, are likely to be produced.

**A big detector for a small particle** The rubber raft in the photo is floating inside a particle detector designed to measure neutrinos. Neutrinos are so weakly interacting that a neutrino produced in a nuclear reaction in the center of the sun will likely pass through the entire mass of the sun and escape. Of course, the neutrino's weakly interacting nature also means that it is likely to pass right through a detector. The Super Kamiokande experiment in Japan monitors interactions in an enormous volume of water in order to spot a very small number of neutrino interactions.

## Neutrinos

The most abundant particle in the universe is not the electron, proton, or neutron; it is a particle you may have never heard of—the **neutrino,** a neutral, nearly massless particle that interacts only weakly with matter.

The neutrino is represented by the symbol $\nu$, a lowercase Greek nu. There are three types of neutrinos. The neutrino involved in beta decay is the *electron neutrino* $\nu_e$; it shows up in processes involving electrons and positrons. The electron neutrino of course has an antiparticle, the antineutrino $\bar{\nu}_e$. The full descriptions of beta-minus and beta-plus decays, including the neutrinos, are

$$n \rightarrow p^+ + e^- + \bar{\nu}_e$$

$$p^+ \rightarrow n + e^+ + \nu_e$$

**NOTE** ▸ If we are concerned with the accurate balance of energy and momentum in a beta decay, we must include the neutrino or the antineutrino, but we can generally ignore the presence of this weakly interacting particle for the problems we solve in this chapter. ◂

We have seen that the sun is powered by a nuclear fusion process in which hydrogen is converted into helium. In this process, prodigious numbers of neutrinos are created. The number of these neutrinos passing through your body each second is staggering—over 600 trillion. But perhaps even more amazing is that over your entire lifetime, only one or two of these neutrinos will interact with the nuclei of your body; the rest just pass through without leaving any trace of their passage. Alpha particles can be stopped by a sheet of paper, beta particles by a few millimeters of aluminum. But it would take a piece of lead 1 light-year thick to stop a neutrino!

It was initially thought that the neutrino, like the photon, was a *massless* particle. However, experiments in the last few years have shown that the neutrino mass, while tiny, is not zero. The best current evidence suggests a mass less than one-millionth the mass of an electron. Experiments now under way will determine a more precise value. Because the neutrino is so abundant in the universe, this small mass may be of great cosmological significance.

## Quarks

The process of beta decay, in which a neutron can change into a proton, and vice versa, gives a hint that the neutron and the proton are *not* fundamental units but are made of smaller subunits.

There is another reason to imagine such subunits: the existence of dozens of subatomic particles—muons, pions, kaons, omega particles, and so on. Just as the periodic table explains the many different atomic elements in terms of three basic particles, perhaps it is possible to do something similar for this "subatomic zoo."

We now understand protons and neutrons to be composed of smaller charged particles whimsically named **quarks.** The quarks that form protons and neutrons are called **up quarks** and **down quarks,** symbolized as u and d, respectively. The nature of these quarks and the composition of the neutron and the proton are shown in FIGURE 30.28.

**NOTE** ► It seems surprising that the charges of quarks are *fractions* of $e$. Don't charges have to be integer multiples of $e$? It's true that atoms, molecules, and all macroscopic matter must have $q = Ne$ because these entities are constructed from electrons and protons. But no law of nature prevents other types of matter from having other amounts of charge. ◄

A neutron and a proton differ by one quark. Beta decay can now be understood as a process in which a down quark changes to an up quark, or vice versa. Beta-minus decay of a neutron can be written as

$$\mathrm{d} \rightarrow \mathrm{u} + \mathrm{e}^- + \overline{\nu}_\mathrm{e}$$

The existence of quarks thus provides an explanation of how a neutron can turn into a proton.

**FIGURE 30.28** The quark content of the proton and neutron.

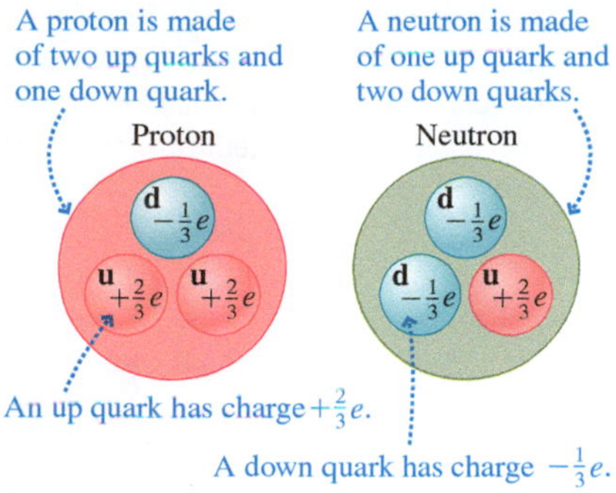


**CONCEPTUAL EXAMPLE 30.17 Quarks and beta-plus decay**

What is the quark description of beta-plus decay?

***REASON*** In beta-plus decay, a proton turns into a neutron, with the emission of a positron and an electron neutrino. To turn a proton into a neutron requires the conversion of an up quark into a down quark; the total reaction is thus

$$\mathrm{u} \rightarrow \mathrm{d} + \mathrm{e}^+ + \nu_\mathrm{e}$$

## Fundamental Particles

Our current understanding of the truly *fundamental* particles—the ones that cannot be broken down into smaller subunits—is that they come in two basic types: **leptons** (particles like the electron and the neutrino) and quarks (which combine to form particles like the proton and the neutron). The leptons and quarks are listed in TABLE 30.5. A few points are worthy of note:

- Each particle has an associated antiparticle.
- There are three *families* of leptons. The first is the electron and its associated neutrino, and their antiparticles. The other families are based on the muon and the tau, heavier siblings to the electron. Only the electron and positron are stable.
- There are also three families of quarks. The first is the up-down family that makes all "normal" matter. The other families are pairs of heavier quarks that form more exotic particles.

**TABLE 30.5** Leptons and quarks

| Leptons | | Antileptons | |
|---|---|---|---|
| Electron | $e^-$ | Positron | $e^+$ |
| Electron neutrino | $\nu_e$ | Electron antineutrino | $\overline{\nu}_e$ |
| Muon | $\mu^-$ | Antimuon | $\mu^+$ |
| Muon neutrino | $\nu_\mu$ | Muon antineutrino | $\overline{\nu}_\mu$ |
| Tau | $\tau^-$ | Antitau | $\tau^+$ |
| Tau neutrino | $\nu_\tau$ | Tau antineutrino | $\overline{\nu}_\tau$ |
| **Quarks** | | **Antiquarks** | |
| Up | u | Antiup | $\overline{\mathrm{u}}$ |
| Down | d | Antidown | $\overline{\mathrm{d}}$ |
| Strange | s | Antistrange | $\overline{\mathrm{s}}$ |
| Charm | c | Anticharm | $\overline{\mathrm{c}}$ |
| Bottom | b | Antibottom | $\overline{\mathrm{b}}$ |
| Top | t | Antitop | $\overline{\mathrm{t}}$ |

As far as we know, this is where the trail ends. Matter is made of molecules; molecules of atoms; atoms of protons, neutrons, and electrons; protons and neutrons of quarks. Quarks and electrons seem to be truly fundamental. But scientists of the early 20th century thought they were at a stopping point as well—they thought that they knew all of the physics that there was to know. As we've seen over the past few chapters, this was far from true. New tools such as the next generation of particle colliders will certainly provide new discoveries and new surprises.

The early chapters of this text, in which you learned about forces and motion, had very obvious applications to things in your daily life. But in these past few chapters we see that even modern discoveries—discoveries such as antimatter, which may seem like science fiction—can be put to very practical use. As we come to the close of this text, we hope that you have gained an appreciation not only for what physics tells us about the world, but also for the wide range of problems it can be used to solve.

## INTEGRATED EXAMPLE 30.18 Čerenkov radiation

We know that nothing can travel faster than $c$, the speed of light in a vacuum. But we also know that light itself goes slower as it travels through a medium. Consequently particles moving at high speeds through a medium can be traveling faster than a light wave in the medium.

FIGURE 30.29 is a photo of the core of a nuclear reactor. The core is immersed in water, which carries away the thermal energy produced in the reactor. As high-energy electrons emerge with a speed very close to $c$ from nuclear reactions in the core, they move through the water faster than the speed of light in water.

FIGURE 30.29 Čerenkov light illuminates the water surrounding a nuclear reactor core.

Recall that a shock wave—a sonic boom—is produced when an airplane moves faster than the speed of sound. An electron moving faster than the speed of light in water makes an electromagnetic shock wave—a light pulse analogous to a sonic boom. This particular type of electromagnetic radiation is known as **Čerenkov radiation,** or Čerenkov light, and it is responsible for the blue glow around the reactor core in Figure 30.29.

One source of high-speed electrons is the beta-minus decay of $^{133}$Xe, a radioactive isotope of xenon that is produced in the fission of uranium and accumulates in the reactor core.

a. What is the daughter nucleus of this decay?

b. Assume that all of the energy released in the decay goes to kinetic energy of the emitted electron. What is the electron's kinetic energy?

c. Use the equations of special relativity to determine the electron's speed.

d. Would the emitted electron be moving at a speed high enough to cause Čerenkov light in water?

e. Based on the color of the Čerenkov light you can see in Figure 30.29, which of the following describes the spectrum of Čerenkov light?

- The intensity of Čerenkov light is uniform at all frequencies.
- The intensity of Čerenkov light is proportional to frequency.
- The intensity of Čerenkov light is proportional to wavelength.

Explain.

***STRATEGIZE*** $^{133}$Xe undergoes beta decay, so the mass of this nucleus must be greater than that of the daughter nucleus. The "lost" mass is converted to energy, which we assume goes to the kinetic energy of the electron, so the kinetic energy of the electron will be $K = \Delta m \cdot c^2$. Once we know the electron's kinetic energy, we can determine the speed to see if it exceeds the speed of light in water.

***PREPARE*** The speed of light in a medium is given by $v = c/n$, where $n$ is the index of refraction. Water has $n = 1.33$.

***SOLVE***

a. Equation 30.9 tells us that beta-minus decay increases $Z$ by 1 while leaving $A$ unchanged. Xe has $Z = 54$; $Z = 55$ is Cs (cesium), so the daughter nucleus, still with $A = 133$, is $^{133}$Cs.

b. We use the data in Appendix C to find the mass difference between the parent and daughter nuclei:

$$\Delta m = m(^{133}\text{Xe}) - m(^{133}\text{Cs})$$
$$= 132.905906 \text{ u} - 132.905436 \text{ u} = 0.00047 \text{ u}$$

We assume that the energy corresponding to this mass difference is the kinetic energy of the emitted electron, so the electron's kinetic energy is

$$K = \Delta m \cdot c^2 = (0.00047 \text{ u})(931.49 \text{ MeV}/c^2)c^2 = 0.44 \text{ MeV}$$

c. The kinetic energy of the electron is large enough that we'll need to consider relativity—a classical treatment won't be sufficient. Equation 27.20 gives the relationship between an object's kinetic energy and its rest energy $E_0$ as $K = (\gamma - 1)E_0$. We can rearrange this to give $\gamma$ in terms of a ratio of two energies that we know: the electron's rest energy, 0.51 MeV, and the electron's kinetic energy, 0.44 MeV. Because it's a ratio, we need not convert units:

$$\gamma = 1 + \frac{K}{E_0} = 1 + \frac{0.44 \text{ MeV}}{0.51 \text{ MeV}} = 1.9$$

We can now use the definition of $\gamma$ to solve for the electron's speed:

$$\gamma = \frac{1}{\sqrt{1 - (v/c)^2}}$$
$$v = c\sqrt{1 - 1/\gamma^2} = 2.5 \times 10^8 \text{ m/s}$$

d. Čerenkov light will be emitted if the speed of the emitted electron is greater than the speed of a light wave in the water. The speed of light in water is

$$v = \frac{3.00 \times 10^8 \text{ m/s}}{1.33} = 2.3 \times 10^8 \text{ m/s}$$

The electron is moving faster than this, and so it will emit Čerenkov light.

e. The photo reveals that Čerenkov light appears blue, so more high-frequency blue light is emitted than lowfrequency red light. The intensity is greater at higher frequencies, so the intensity is proportional to frequency—at least for visible light. The index of refraction decreases for very high frequencies, returning to $n = 1$ for x rays. Light speed at these very high frequencies is no longer slower than the particle speed, so Čerenkov light "cuts off" at very high frequencies.

***ASSESS*** The energy of the beta particle is reasonably typical for particles emitted by nuclear decays, and we know that Čerenkov light is observed around reactor cores, so it's reasonable to expect the beta particle to be moving fast enough to emit Čerenkov light.

# SUMMARY

**GOAL** To understand the physics of the nucleus and some of the applications of nuclear physics.

## GENERAL PRINCIPLES

### The Nucleus

The nucleus is a small, dense, positive core at the center of an atom.

$Z$ protons, charge $+e$, spin $\frac{1}{2}$

$N$ neutrons, charge 0, spin $\frac{1}{2}$

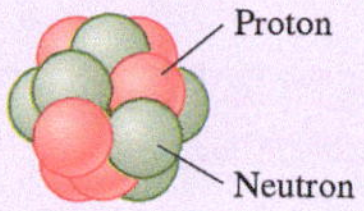


The **mass number** is

$$A = Z + N$$

**Isotopes** of an element have the same value of $Z$ but different values of $N$.

The strong force holds nuclei together:

- It acts between any two nucleons.
- It is short range.

Adding neutrons to a nucleus allows the strong force to overcome the repulsive Coulomb force between protons.

The **binding energy** $B$ of a nucleus depends on the mass difference between an atom and its constituents:

$$B = (Zm_H + Nm_n - m_{atom}) \times (931.49 \text{ MeV/u})$$

### Nuclear Stability

Most nuclei are not **stable.** Unstable nuclei undergo **radioactive decay.** Stable nuclei cluster along the **line of stability** in a plot of the isotopes.

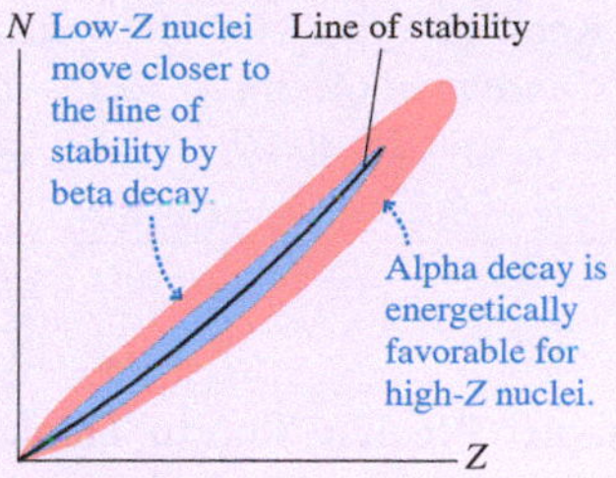


Mechanisms by which unstable nuclei decay:

| Decay | Particle |
|---|---|
| alpha | $^4$He nucleus |
| beta-minus | $e^-$ |
| beta-plus | $e^+$ |
| gamma | photon |

Alpha and beta decays change the nucleus; the daughter nucleus is a different element.

Alpha decay:

$$^A_Z\text{X} \rightarrow {}^{A-4}_{Z-2}\text{Y} + \alpha + \text{energy}$$

Beta-minus decay:

$$^A_Z\text{X} \rightarrow {}_{Z+1}^{\;\;A}\text{Y} + \beta + \text{energy}$$

## IMPORTANT CONCEPTS

### Energy levels

Nucleons fill nuclear energy levels, similar to filling electron energy levels in atoms. Nucleons can often jump to lower energy levels by emitting beta particles or gamma photons.

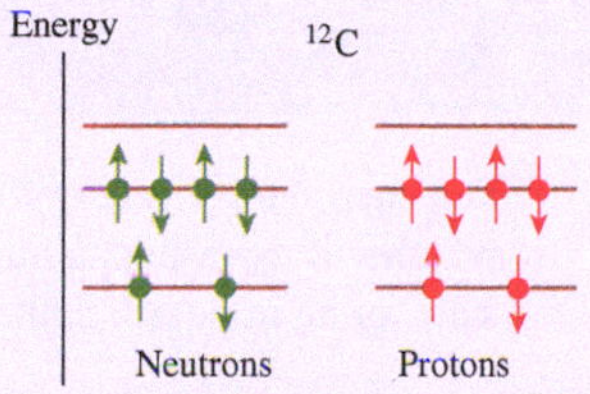


### Quarks

Nucleons (and other particles) are made of quarks. Quarks and leptons are fundamental particles.

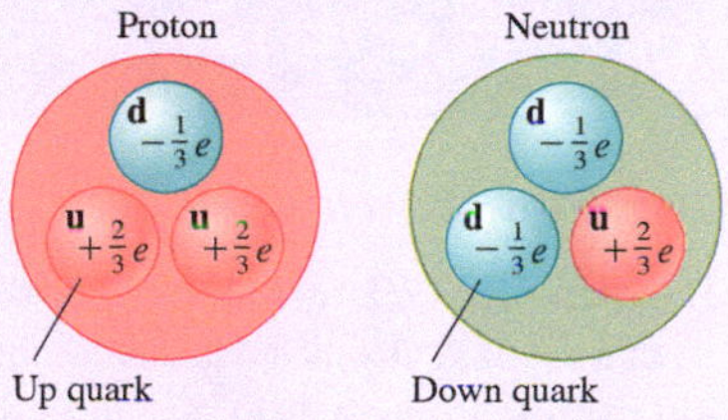


## APPLICATIONS

### Radioactive decay

The number of undecayed nuclei decreases exponentially with time $t$:

$$N = N_0 e^{-t/\tau}$$

$$N = N_0 \left(\frac{1}{2}\right)^{t/t_{1/2}}$$

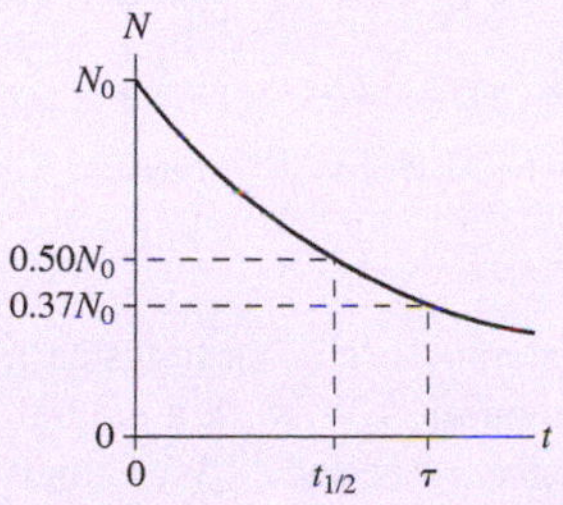


The **half-life**

$$t_{1/2} = \tau \ln 2 = 0.693\tau$$

is the time in which half of any sample decays.

### Measuring radiation

The **activity** of a radioactive sample is the number of decays per second. Activity is related to the half-life as

$$R = \frac{0.693N}{t_{1/2}} = \frac{N}{\tau}$$

The radiation **dose** is measured in grays, where

$$1 \text{ Gy} = 1.00 \text{ J/kg of absorbed energy}$$

The **relative biological effectiveness** (RBE) is the biological effect of a dose relative to the biological effects of x rays. The **dose equivalent** is measured in sieverts, where

$$\text{dose equivalent in Sv} = \text{dose in Gy} \times \text{RBE}$$

## Learning Objectives After studying this chapter, you should be able to:

- Interpret nuclear structure and binding energy. *Conceptual Questions 30.5, 30.6; Problems 30.1, 30.5, 30.11, 30.16, 30.17*
- Work with nuclear energy levels. *Conceptual Question 30.18; Problems 30.18, 30.19, 30.20, 30.21*
- Know the types of nuclear radiation and their properties. *Conceptual Questions 30.15, 30.17; Problems 30.22, 30.24, 30.27, 30.28, 30.30*
- Calculate activity and half-lives for nuclear decay. *Conceptual Questions 30.12, 30.13; Problems 30.31, 30.32, 30.38, 30.41, 30.42*
- Understand radiation dose and the applications of nuclear physics to medicine. *Conceptual Questions 30.20, 30.26; Problems 30.45, 30.46, 30.50, 30.52, 30.56*
- Recognize and distinguish between the fundamental subatomic particles. *Conceptual Questions 30.27, 30.28; Problems 30.57, 30.58*

### STOP TO THINK ANSWERS

**Chapter Preview Stop to Think: D.** There are four electrons, so this is atomic number 4—beryllium. There is one electron in the 2*s* level and one in the 2*p* level. The 2*s* level could hold two electrons, so this is an excited state; if the electron drops from 2*p* to 2*s*, the energy is decreased.

**Stop to Think 30.1: Three.** Different isotopes of an element have different numbers of neutrons but the same number of protons. The number of electrons in a neutral atom matches the number of protons.

**Stop to Think 30.2: A.** Energy is released in the decay process, so the mass must decrease.

**Stop to Think 30.3: Yes.** $^{12}C$ has filled levels of protons and neutrons; the neutron we add to make $^{13}C$ will be in a higher energy level, but there is no "hole" in a lower level for it to move to, so we expect this nucleus to be stable.

**Stop to Think 30.4: B.** An increase of $Z$ with no change in $A$ occurs when a neutron changes to a proton and an electron, ejecting the electron.

**Stop to Think 30.5: E.** 30 minutes is $3t_{1/2}$ so the number of nuclei A decreases by a factor of $2^3 = 8$ to 3000. This means that 21,000 A nuclei have decayed, and each became nucleus B.

**Stop to Think 30.6: C.** One-quarter of the atoms are left. This is one-half of one-half, or $(1/2)^2$, so two half-lives, 20 minutes, have elapsed.

**Stop to Think 30.7: A.** Dose equivalent is the product of dose in Gy (the same for each) and RBE (highest for alpha particles).

Video Tutor Solution Chapter 30

# QUESTIONS

## Conceptual Questions

1. Atom A has a larger atomic mass than atom B. Does this mean that atom A also has a larger atomic number? Explain.
2. Given that $m_H = 1.007825$ u, is the mass of a hydrogen atom $^1H$ greater than, less than, or equal to 1/12 the mass of a $^{12}C$ atom? Explain.
3. Nucleus A has 5 protons and 6 neutrons, while nucleus B has 6 protons and 5 neutrons. Are these nuclei isotopes of the same element, or are they different elements?
4. Rounding slightly, the nucleus $^3He$ has a binding energy of 2.5 MeV/nucleon and the nucleus $^6Li$ has a binding energy of 5 MeV/nucleon.
   a. What is the binding energy of $^3He$?
   b. What is the binding energy of $^6Li$?
   c. Is it energetically possible for two $^3He$ nuclei to join or fuse together into a $^6Li$ nucleus? Explain.
   d. Is it energetically possible for a $^6Li$ nucleus to split or fission into two $^3He$ nuclei? Explain.
5. For each nuclear energy-level diagram in Figure Q30.5, state whether it represents a nuclear ground state, an excited nuclear state, or an impossible nucleus.

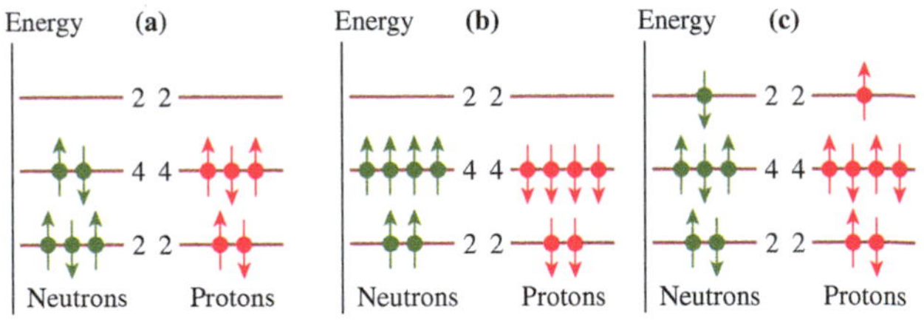


FIGURE Q30.5

6. Naturally occurring tungsten metal is composed of three stable isotopes: $^{182}_{74}W$, with a natural abundance of 25%; $^{183}_{74}W$, abundance 14%; and $^{184}_{74}W$, abundance 31%. Tungsten also has the highest melting point of any element at 3695 K. If a sample of tungsten were held at its melting point for 10 h, would you expect the natural abundances of its isotopes to change? If so, which would increase the most? Which would decrease the most?

Problem difficulty is labeled as | (straightforward) to |||| (challenging). Problems labeled INT integrate significant material from earlier chapters; Problems labeled BIO are of biological or medical interest.

The eText icon indicates when there is a video tutor solution available for the chapter or for a specific problem. To launch these videos, log into your eText through Mastering™ Physics or log into the Study Area.

7. Figure Q30.7 shows how the number of nuclei of one particular isotope varies with time. What is the half-life of the nucleus?

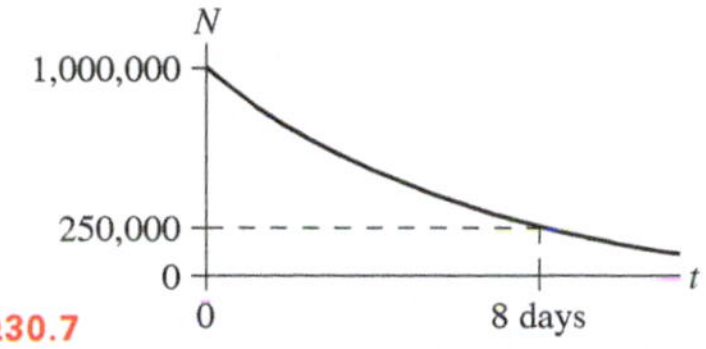


FIGURE Q30.7

8. A radioactive sample has a half-life of 3 days. Its current activity is 100 $\mu$Ci. What was its activity 6 days ago?
9. Nucleus A decays into the stable nucleus B with a half-life of 10 s. At $t = 0$ s there are 1000 A nuclei and no B nuclei. At what time will there be 750 B nuclei?
10. A radioactive sample's half-life is 1.0 min, so each nucleus in the sample has a 50% chance of undergoing a decay sometime between $t = 0$ and $t = 1$ min. One particular nucleus has not decayed at $t = 15$ min. What is the probability this nucleus will decay between $t = 15$ and $t = 16$ min?
11. The material that formed the earth was created in a supernova explosion approximately 6 billion years ago. The two most common isotopes of uranium, $^{238}$U and $^{235}$U, were created in roughly equal amounts. Today, the earth contains more than 100 times as much $^{238}$U as $^{235}$U. Which isotope has a longer half-life?
12. BIO $^{99}$Mo beta decays with a half-life of 66 h to $^{99}$Tc$^*$, an isotope that is widely used in medical imaging. The $^{99}$Mo itself is created in a nuclear reactor by bombarding $^{235}$U nuclei with neutrons. One particular $^{99}$Mo was created in this way 66 h ago, while another was created just now. Which nucleus is more likely to decay in the next 66 h?
13. Oil and coal generally contain no measurable $^{14}$C. What does this tell us about how long they have been buried?
14. Radiocarbon dating assumes that the abundance of $^{14}$C in the environment has been constant. Suppose $^{14}$C was less abundant 10,000 years ago than it is today. Would this cause a lab using radiocarbon dating to overestimate or underestimate the age of a 10,000-year-old artifact? (In fact, the abundance of $^{14}$C in the environment does vary slightly with time. But the issue has been well studied, and the ages of artifacts are adjusted to compensate for this variation.)
15. Identify the unknown X in the following decays:
    a. $^{222}_{86}\text{Rn} \rightarrow {}^{218}_{84}\text{Po} + \text{X}$  b. $^{228}_{88}\text{Ra} \rightarrow {}^{228}_{89}\text{Ac} + \text{X}$
    c. $^{140}_{54}\text{Xe} \rightarrow {}^{140}_{55}\text{Cs} + \text{X}$  d. $^{64}_{29}\text{Cu} \rightarrow {}^{64}_{28}\text{Ni} + \text{X}$
16. Are the following decays possible? If not, why not?
    a. $^{232}_{90}\text{Th} \rightarrow {}^{236}_{92}\text{U} + \alpha$  b. $^{238}_{94}\text{Pu} \rightarrow {}^{236}_{92}\text{U} + \alpha$  c. $^{33}_{15}\text{P} \rightarrow {}^{32}_{16}\text{S} + e^-$
17. What kind of decay, if any, would you expect for the nuclei with the energy-level diagrams shown in Figure Q30.17?

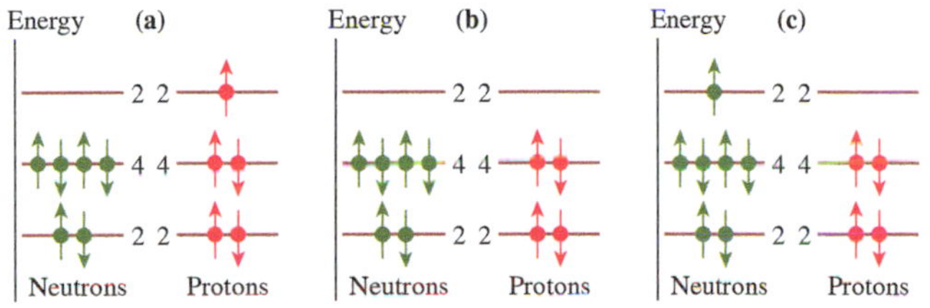


FIGURE Q30.17

18. The nuclei of $^4$He and $^{16}$O are very stable and are often referred to as "doubly magic" nuclei. Use what you know about energy levels to explain what is special about these particular nuclei.
19. BIO A and B are fresh apples. Apple A is strongly irradiated by nuclear radiation for 1 hour. Apple B is not irradiated. Afterward, in what ways are apples A and B different?
20. BIO A patient's tumor is irradiated with gamma rays from an external source. Afterward, is his body radioactive? Explain.
21. BIO It's possible that a bone tumor will not show up on an x-ray image but will show up in a gamma scan. Explain why this is so.
22. BIO Four radiation doses are as follows: Dose A is 0.10 Gy with an RBE of 1, dose B is 0.20 Gy with an RBE of 1, dose C is 0.10 Gy with an RBE of 2, and dose D is 0.20 Gy with an RBE of 2.
    a. Rank in order, from largest to smallest, the amount of energy delivered by these four doses.
    b. Rank in order, from largest to smallest, the biological damage caused by these four doses.
23. BIO Two different sources of radiation give the same dose equivalent in Sv. Does this mean that the radiation from each source has the same RBE? Explain.
24. BIO Some types of MRI can produce images of resolution and detail similar to PET. Though the images are similar, MRI is generally preferred over PET for studies of brain function involving healthy subjects. Why?
25. BIO Sulfur colloid particles tagged with $^{99}$Tc are taken up and retained by cells in the liver and spleen. A patient is suspected of having a liver tumor that would destroy these cells. Explain how a gamma camera scan could be used to confirm or rule out the existence of a tumor.
26. BIO Alpha emitters that are breathed into the lungs are especially dangerous. Give two reasons why this is so.

The following two questions concern an uncommon nuclear decay mode known as *electron capture*. Certain nuclei that are proton-rich but energetically prohibited from undergoing beta-plus decay can capture an electron from the 1*s* shell, which then combines with a proton to make a neutron. The basic reaction is

$$p + e^- \rightarrow n + \nu_e$$

27. Give a description of the electron capture process in terms of quarks.
28. Electron capture is usually followed by the emission of an x ray. Why?

## Multiple-Choice Questions

29. | BIO $^{74}$As is a beta-plus emitter used for locating tumors with PET. What is the daughter nucleus?
    A. $^{73}_{33}$As  B. $^{74}_{32}$Ge
    C. $^{74}_{34}$Se  D. $^{75}_{33}$As
30. || When uranium fissions, the fission products are radioactive because the nuclei are neutron-rich. What is the most likely decay mode for these nuclei?
    A. Alpha decay  B. Beta-minus decay
    C. Beta-plus decay  D. Gamma decay
31. | A certain watch's luminous glow is due to zinc sulfide paint that is energized by beta particles given off by *tritium*, the radioactive hydrogen isotope $^3$H, which has a half-life of 12.3 years. This glow has about 1/10 of its initial brightness. How many years old is the watch?

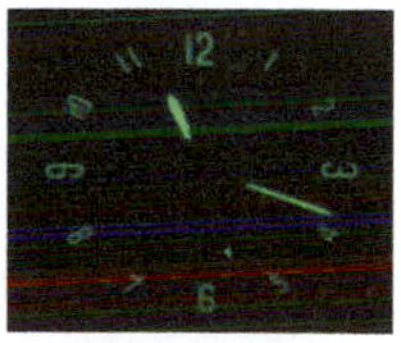

    A. 20 yr  B. 30 yr  C. 40 yr  D. 50 yr
32. | What is the unknown isotope in the following fission reaction: $n + {}^{235}\text{U} \rightarrow {}^{131}\text{I} + ? + 3n$
    A. $^{86}$Rb  B. $^{102}$Rb  C. $^{89}$Y  D. $^{102}$Y
33. | An investigator has 0.010 $\mu$g samples of two isotopes of strontium, $^{89}$Sr ($t_{1/2} = 51$ days) and $^{90}$Sr ($t_{1/2} = 28$ years). The samples contain approximately the same number of atoms. What can you say about the activity of the two samples?
    A. The $^{89}$Sr sample has a higher activity.
    B. The $^{90}$Sr sample has a higher activity.
    C. The two samples have about the same activity.

Watch Video Solution Problems 30.13 and 30.25

34. ‖ The uranium in the earth's crust is 0.7% $^{235}$U and 99.3% $^{238}$U. Two billion years ago, $^{235}$U comprised approximately 3% of the uranium in the earth's crust. This tells you something about the relative half-lives of the two isotopes. Suppose you have a sample of $^{235}$U and a sample of $^{238}$U, each with exactly the same number of atoms.
   A. The sample of $^{235}$U has a higher activity.
   B. The sample of $^{238}$U has a higher activity.
   C. The two samples have the same activity.
35. | Suppose you have a 1 g sample of $^{226}$Ra, half-life 1600 years. How long will it be until only 0.1 g of radium is left?
   A. 1600 yr B. 3200 yr C. 5300 yr D. 16,000 yr
36. | A sample of $^{131}$I, half-life 8.0 days, is registering 100 decays per second. How long will be it before the sample registers only 1 decay per second?
   A. 8 days B. 53 days C. 80 days D. 800 days
37. | The complete expression for the decay of the radioactive hydrogen isotope *tritium* may be written as $^3\text{H} \rightarrow {}^3\text{He} + \text{X} + \text{Y}$. The symbols X and Y represent
   A. $\text{X} = \text{e}^+, \text{Y} = \bar{\nu}_\text{e}$ B. $\text{X} = \text{e}^-, \text{Y} = \nu_\text{e}$
   C. $\text{X} = \text{e}^+, \text{Y} = \nu_\text{e}$ D. $\text{X} = \text{e}^-, \text{Y} = \bar{\nu}_\text{e}$
38. | The quark compositions of the proton and neutron are, respectively, uud and udd, where u is an up quark (charge $+\frac{2}{3}e$) and d is a down quark (charge $-\frac{1}{3}e$). There are also antiup $\bar{\text{u}}$ (charge $-\frac{2}{3}e$) and antidown $\bar{\text{d}}$ (charge $+\frac{1}{3}e$) quarks. The combination of a quark and an antiquark is called a *meson*. The mesons known as *pions* have the composition $\pi^+ = \text{u}\bar{\text{d}}$ and $\pi^- = \bar{\text{u}}\text{d}$. Suppose a proton collides with an antineutron. During such collisions, the various quarks and antiquarks annihilate whenever possible. When the remaining quarks combine to form a single particle, it is a
   A. Proton B. Neutron C. $\pi^+$ D. $\pi^-$

# PROBLEMS

## Section 30.1 Nuclear Structure

1. | How many protons and how many neutrons are in (a) $^3$H, (b) $^{40}$Ar, (c) $^{40}$Ca, and (d) $^{239}$Pu?
2. | How many protons and how many neutrons are in (a) $^3$He, (b) $^{20}$Ne, (c) $^{60}$Co, and (d) $^{226}$Ra?
3. ‖ Use the data in Appendix C to calculate the chemical atomic mass of lithium, to two decimal places.
4. | Use the data in Appendix C to calculate the chemical atomic mass of neon, to two decimal places.
5. | What are the energy equivalents (in MeV) of the mass of (a) an electron, (b) a proton, (c) a hydrogen atom, and (d) a helium atom?
6. ‖ INT In *heavy water*, each of the hydrogen atoms is replaced by a deuterium atom. If the density of ordinary water at a given temperature is 1.000 g/cm$^3$, what is the density of heavy water? Assume that the average distance between the molecules is the same in both cases.

## Section 30.2 Nuclear Stability

7. | Calculate (in MeV) the total binding energy and the binding energy per nucleon (a) for $^3$H and (b) for $^3$He.
8. | Calculate (in MeV) the total binding energy and the binding energy per nucleon (a) for $^{40}$Ar and (b) for $^{40}$K.
9. | Calculate (in MeV) the binding energy per nucleon for $^3$He and $^4$He. Which is more tightly bound?
10. ‖ Calculate the binding energy per nucleon for $^{62}$Ni, which has the highest value of any isotope.
11. | Calculate (in MeV) the binding energy per nucleon for (a) $^{14}$N, (b) $^{56}$Fe, and (c) $^{207}$Pb.
12. ‖ When a nucleus of $^{235}$U undergoes fission, it breaks into two smaller, more tightly bound fragments. Calculate the binding energy per nucleon for $^{235}$U and for the fission product $^{137}$Cs.
13. ‖ When a nucleus of $^{240}$Pu undergoes fission, it breaks into two smaller, more tightly bound fragments. Calculate the binding energy per nucleon for $^{240}$Pu and for the fission product $^{133}$Xe.
14. ‖ a. Compute the binding energy of the reactants and of the products in the nuclear fusion reaction

$$^2\text{H} + {}^6\text{Li} \rightarrow {}^4\text{He} + {}^4\text{He}$$

   b. What is the change in binding energy per nucleon in this reaction?
15. ‖ a. Compute the binding energy of the reactants and of the products in the nuclear fusion reaction

$$^3\text{H} + {}^3\text{He} \rightarrow {}^2\text{H} + {}^4\text{He}$$

   b. What is the change in binding energy per nucleon in this reaction?
16. ‖| Compute the energy released in the fission reaction

$$\text{n} + {}^{239}\text{Pu} \rightarrow {}^{137}\text{Xe} + {}^{100}\text{Zr} + 3\text{n}$$

17. ‖| The sun's energy comes from a reaction that combines four hydrogen atoms to create a helium atom plus two positrons:

$$4{}^1\text{H} \rightarrow {}^4\text{He} + 2e^+$$

How much energy is released in this process?

## Section 30.3 Forces and Energy in the Nucleus

18. ‖ Draw an energy-level diagram, similar to Figure 30.9, for the protons and neutrons in $^{11}$Be. Do you expect this nucleus to be stable?
19. ‖ Draw energy-level diagrams, similar to Figure 30.9, for all $A = 10$ nuclei listed in Appendix C. Show all the occupied neutron and proton levels. Given the diagrams, are any of the nuclei obviously unstable?
20. ‖ Draw energy-level diagrams, similar to Figure 30.9, for all $A = 14$ nuclei listed in Appendix C. Show all the occupied neutron and proton levels. Given the diagrams, are any of the nuclei obviously unstable?
21. ‖ INT The strong force between nucleons has a magnitude of approximately $2.5 \times 10^4$ N for two nucleons whose centers are 1.0 fm apart. What is the electrostatic force between two protons this distance apart? Compare the electrostatic force to the strong force.

## Section 30.4 Radiation and Radioactivity

22. | BIO $^{15}$O and $^{131}$I are isotopes used in medical imaging. $^{15}$O is a beta-plus emitter, $^{131}$I a beta-minus emitter. What are the daughter nuclei of the two decays?
23. | Spacecraft have been powered with energy from the alpha decay of $^{238}$Pu. What is the daughter nucleus?
24. | Identify the unknown isotope X in the following decays.
   a. $^{234}\text{U} \rightarrow \text{X} + \alpha$ b. $^{32}\text{P} \rightarrow \text{X} + \text{e}^-$
   c. $\text{X} \rightarrow {}^{30}\text{Si} + \text{e}^+$ d. $^{24}\text{Mg} \rightarrow \text{X} + \gamma$
25. | Identify the unknown isotope X in the following decays.
   a. $\text{X} \rightarrow {}^{224}\text{Ra} + \alpha$ b. $\text{X} \rightarrow {}^{207}\text{Pb} + \text{e}^-$
   c. $^7\text{Be} + \text{e}^- \rightarrow \text{X}$ d. $\text{X} \rightarrow {}^{60}\text{Ni} + \gamma$

26. ‖ The New Horizons spacecraft, launched in 2006, spent 9.5 years on its journey to Pluto. It derives its electric power from the heat generated by 10 kg of $^{238}$Pu, which has an activity of $6.3 \times 10^{15}$ Bq. Each decay emits an alpha particle with an energy of 5.6 MeV. What is the total thermal power generated by this plutonium source?

27. ||| *Neutron probes* are used in agronomy to measure the moisture content of soil. A pellet of $^{241}$Am emits alpha particles that cause a beryllium disk to emit neutrons. These neutrons move out into the soil where they are reflected back into the probe by the hydrogen nuclei in water. The neutron count is thus indicative of the moisture content near the probe. What is the energy of the alpha particle emitted by the $^{241}$Am?

28. | Medical gamma imaging is generally done with the tech-
BIO netium isotope $^{99}$Tc$^*$, which decays by emitting a gamma-ray photon with energy 140 keV. What is the mass loss of the nucleus, in u, upon emission of this gamma ray?

29. ‖ Cobalt has one stable isotope, $^{59}$Co. What are the likely decay modes and daughter nuclei for (a) $^{56}$Co and (b) $^{62}$Co?

30. ‖ Manganese has one stable isotope, $^{55}$Mn. What are the likely decay modes and daughter nuclei for (a) $^{51}$Mn and (b) $^{59}$Mn?

### Section 30.5 Nuclear Decay and Half-Lives

31. | The radioactive hydrogen isotope $^{3}$H is called tritium. It decays by beta-minus decay with a half-life of 12.3 years.
    a. What is the daughter nucleus of tritium?
    b. A watch uses the decay of tritium to energize its glowing dial. What fraction of the tritium remains 20 years after the watch was created?

32. | The barium isotope $^{133}$Ba has a half-life of 10.5 years. A sample begins with $1.0 \times 10^{10}$ $^{133}$Ba atoms. How many are left after (a) 2 years, (b) 20 years, and (c) 200 years?

33. | The cadmium isotope $^{109}$Cd has a half-life of 462 days. A sample begins with $1.0 \times 10^{12}$ $^{109}$Cd atoms. How many are left after (a) 50 days, (b) 500 days, and (c) 5000 days?

34. |||| Uranium has two very long-lived isotopes, $^{235}$U and $^{238}$U. What were the percent abundances of these two isotopes when the earth was formed $4.54 \times 10^{9}$ y ago? (The relevant data are given in Appendix C.)

35. ‖ The Chernobyl reactor accident in what is now Ukraine was the
BIO worst nuclear disaster of all time. Fission products from the reactor core spread over a wide area. The primary radiation exposure to people in western Europe was due to the short-lived (half-life 8.0 days) isotope $^{131}$I, which fell across the landscape and was ingested by grazing cows that concentrated the isotope in their milk. Farmers couldn't sell the contaminated milk, so many opted to use the milk to make cheese, aging it until the radioactivity decayed to acceptable levels. How much time must elapse for the activity of a block of cheese containing $^{131}$I to drop to 1.0% of its initial value?

36. |||| What is the age in years of a bone in which the $^{14}$C/$^{12}$C ratio is measured to be $1.65 \times 10^{-13}$?

37. ‖ $^{85}$Sr is a short-lived (half-life 65 days) isotope used in bone
BIO scans. A typical patient receives a dose of $^{85}$Sr with an activity of 0.10 mCi. If all of the $^{85}$Sr is retained by the body, what will be its activity in the patient's body after one year has passed?

38. ‖ The smallest $^{14}$C/$^{12}$C ratio that can be reliably measured is about $3.0 \times 10^{-15}$, setting a limit on the oldest carbon specimens that can be dated. How old would a sample with this carbon ratio be?

39. ||| What is the activity, in Bq and Ci, of 1.0 g of $^{226}$Ra? Marie
INT Curie was the discoverer of radium; can you see where the unit of activity named after her came from?

40. ‖ Many medical PET scans use the isotope $^{18}$F, which has a half-life
BIO of 1.8 h. A sample prepared at 10:00 A.M. has an activity of 20 mCi. What is the activity at 1:00 P.M., when the patient is injected?

41. ||| An investigator collects a sample of a radioactive isotope with an activity of 370,000 Bq. 48 hours later, the activity is 120,000 Bq. What is the half-life of the sample?

42. | $^{235}$U decays to $^{207}$Pb via the decay series shown in Figure 30.16. The first decay in the chain, that of $^{235}$U, has a half-life of $7.0 \times 10^{8}$ years. The subsequent decays are much more rapid, so we can take this as the half-life for the complete decay of $^{235}$U to $^{207}$Pb. Certain minerals exclude lead but not uranium from their crystal structure, so when the minerals form they have no lead, only uranium. As time goes on, the uranium decays to lead, so measuring the ratio of lead atoms to uranium atoms allows investigators to determine the ages of the minerals. If a sample of a mineral contains 3 atoms of $^{207}$Pb for every 1 atom of $^{235}$U, how many years ago was it formed?

43. ‖ A sample of $1.0 \times 10^{10}$ atoms that decay by alpha emission has a half-life of 100 min. How many alpha particles are emitted between $t = 50$ min and $t = 200$ min?

44. ‖ The technetium isotope $^{99}$Tc
BIO is useful in medical imaging, but its short 6.0 h half-life means that shipping it from a source won't work; it must be created where it will be used. Hospitals extract the $^{99}$Tc daughter product from the decay of the molybdenum isotope $^{99}$Mo. The $^{99}$Mo source has a half-life of 2.7 days, so it must be replaced weekly. If a hospital receives a 20 GBq $^{99}$Mo source, what is its activity one week later?

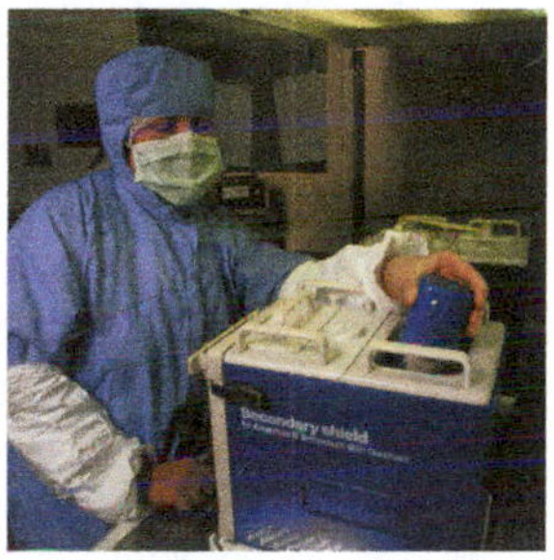

### Section 30.6 Medical Applications of Nuclear Physics

45. | A passenger on an airplane flying across the Atlantic receives
BIO an extra radiation dose of about 5 $\mu$Sv per hour from cosmic rays. How many hours of flying would it take in one year for a person to double his or her yearly radiation dose? Assume there are no other significant radiation sources besides natural background.

46. ‖ A 50 kg nuclear plant worker is exposed to 20 mJ of neutron
BIO radiation with an RBE of 10. What is the dose in mSv?

47. ‖ The decay chain of uranium includes radon, a noble gas.
BIO When uranium in the soil decays to radon, it may seep into houses; this can be a significant source of radiation exposure. Most of the exposure comes from the decay products of radon, but some comes from alpha decay of the radon itself. If radon in the air in your home is at the maximum permissible level, the gas in your lungs will have an activity of about 0.22 Bq. Each decay generates an alpha particle with 5.5 MeV of energy, and essentially all that energy is deposited in lung tissue. Over the course of 1 year, what will be the dose equivalent in Sv to the approximately 0.90 kg mass of your lungs?

48. ‖ A gamma scan showing the
BIO active volume of a patient's lungs can be created by having a patient breathe the radioactive isotope $^{133}$Xe, which undergoes beta-minus decay with a subsequent gamma emission from the daughter nucleus. A typical procedure gives a dose of 3.0 mSv to the lungs. How much energy is deposited in the 1.2 kg mass of a patient's lungs?

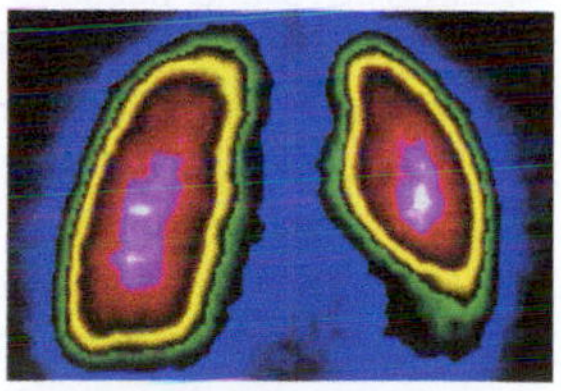

49. ‖ How many Gy of gamma-ray photons cause the same
BIO biological damage as 0.30 Gy of alpha radiation?

50. | 1.5 Gy of gamma radiation are directed into a 150 g tumor.
BIO How much energy does the tumor absorb?

51. |||| Certain cancers of the liver can be treated by injecting micro-
BIO scopic glass spheres containing radioactive $^{90}$Y into the blood vessels that supply the tumor. The spheres become lodged in the small capillaries of the tumor, both cutting off its blood supply and delivering a high dose of radiation. $^{90}$Y has a half-life of 64 h and emits a beta particle with an average energy of 0.89 MeV. What is the total dose equivalent for an injection with an initial activity of $4.0 \times 10^7$ Bq if all the energy is deposited in a 40 g tumor?

52. ||| $^{131}$I undergoes beta-minus
BIO decay with a subsequent gamma emission from the daughter nucleus. Iodine in the body is almost entirely taken up by the thyroid gland, so a gamma scan using this isotope will show a bright area corresponding to the thyroid gland with the surrounding tissue appearing dark. Because the isotope is concentrated in the gland, so is the radiation dose, most of which results from the beta emission. In a typical procedure, a patient receives 0.050 mCi of $^{131}$I. Assume that all of the iodine is absorbed by the 0.15 kg thyroid gland. Each $^{131}$I decay produces a 0.97 MeV beta particle. Assume that half the energy of each beta particle is deposited in the gland. What dose equivalent in Sv will the gland receive in the first hour?

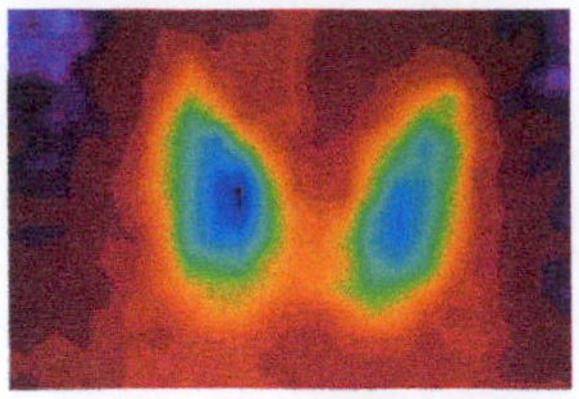

53. ||| $^{90}$Sr decays with the emission of a 2.8 MeV beta particle.
BIO Strontium is chemically similar to calcium and is taken up by bone. A 75 kg person exposed to waste from a nuclear accident absorbs $^{90}$Sr with an activity of 370,000 Bq. Assume that all of this $^{90}$Sr ends up in the skeleton. The skeleton forms 17% of the person's body mass. If 50% of the decay energy is absorbed by the skeleton, what dose equivalent in Sv will be received by the person's skeleton in the first month?

54. ||| A patient receives a gamma scan of his liver. He ingests
BIO 3.7 MBq of $^{198}$Au, which decays with a 2.7 day half-life by emitting a 1.4 MeV beta particle. Medical tests show that 60% of this isotope is absorbed and retained by the liver. If all of the radioactive decay energy is deposited in the liver, what is the total dose equivalent received by the patient's 1.5 kg liver once all of the atoms have decayed?

55. ||| A 75 kg patient swallows a 30 $\mu$Ci beta emitter with a half-
BIO life of 5.0 days, and the radioactive nuclei are quickly distributed throughout his body. The beta particles are emitted with an average energy of 0.35 MeV, 90% of which is absorbed by the body. What dose equivalent does the patient receive in the first week?

### Section 30.7 The Ultimate Building Blocks of Matter

56. || What are the minimum energies of the two oppositely
BIO directed gamma rays in a PET procedure?

57. |||| Positive and negative pions, denoted $\pi^+$ and $\pi^-$, are antiparticles of each other. Each has a rest mass of 140 MeV/$c^2$. Suppose a collision between an electron and positron, each with kinetic energy $K$, produces a $\pi^+$, $\pi^-$ pair. What is the smallest possible value for $K$?

58. || In a particular beta-minus decay of a free neutron (that is, one not part of an atomic nucleus), the emitted electron has exactly the same kinetic energy as the emitted electron antineutrino. What is the value, in MeV, of that kinetic energy? Assume that the recoiling proton has negligible kinetic energy.

## General Problems

59. ||| Use the graph of binding energy of Figure 30.6 to estimate the total energy released if a nucleus with mass number 240 fissions into two nuclei with mass number 120.

60. |||| As we've seen, the sun's energy comes from fusion reactions that combine four hydrogen atoms to produce a single helium atom. Even in the sun's core, where these reactions proceed most rapidly, the reaction rate is very slow, with only about $6.7 \times 10^{13}$ reactions per second occurring in 1 cubic meter of the core. How much power is produced by these fusion reactions per cubic meter? Compare this number with the 300 W/m$^3$ metabolic power produced by a resting reptile.

61. ||| You are assisting in an anthropology lab over the summer
BIO by carrying out $^{14}$C dating. A graduate student found a bone he believes to be 20,000 years old. You extract the carbon from the bone and prepare an equal-mass sample of carbon from modern organic material. To determine the activity of a sample with the accuracy your supervisor demands, you need to measure the time it takes for 10,000 decays to occur.
   a. The activity of the modern sample is 1.06 Bq. How long does that measurement take?
   b. It turns out that the graduate student's estimate of the bone's age was accurate. How long does it take to measure the activity of the ancient carbon?

62. ||| Wood was excavated from Stonehenge and found, using $^{14}$C
INT dating, to be 4500 years old. What is the $^{14}$C activity in 1.00 g of carbon taken from this wood?

63. |||| Figure P30.63 shows the decay scheme for $^{137}$Cs, which has two possible beta decay modes. The first, labeled $\beta_1$, is a decay directly to the ground state of $^{137}$Ba. The second beta decay ($\beta_2$) is to an excited state $^{137}$Ba$^*$. This excited state subsequently undergoes gamma decay to the ground state. In beta decay, the maximum possible energy of the emitted beta particle is equal to the difference in energy between the initial and final states of the nucleus. The maximum energy of $\beta_2$ has been measured to be 0.512 MeV.
   a. What is the maximum energy of the beta particle $\beta_1$?
   b. What is the energy of the gamma ray?

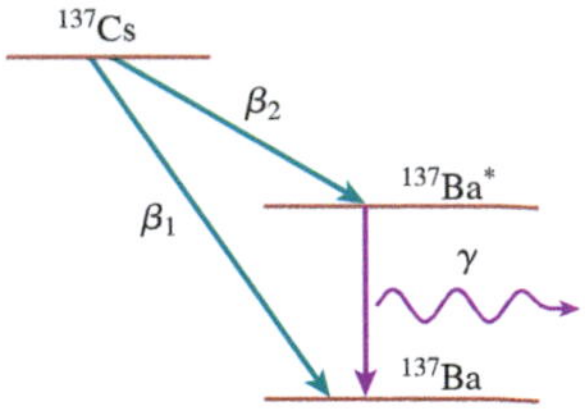


FIGURE P30.63

64. |||| A sample of wood from an archaeological excavation is
BIO dated by using a mass spectrometer to measure the fraction of $^{14}$C atoms. Suppose 100 atoms of $^{14}$C are found for every $1.0 \times 10^{15}$ atoms of $^{12}$C in the sample. What is the wood's age?

65. |||| In Example 30.11, carbon from an ancient tooth was
BIO extracted and formed into a sample of benzene. Using the information from Example 30.11 and its preceding discussion, what is the mass of the carbon, both $^{12}$C and $^{14}$C, in the sample tested in Example 30.11?

66. ||||| 7200 years ago, the ratio of $^{14}$C to $^{12}$C in the atmosphere was 10% higher than it is today. A piece of wood is known to be 7200 years old. If its age were determined using radiocarbon dating, how old would it appear to be? (The variation in atmospheric $^{14}$C with time is well understood and is corrected for in actual radiocarbon dating.)

67\. |||| The technique known as potassium-argon dating is used to date volcanic rock and ash, and thus establish dates for nearby fossils, like this 1.8-million-year-old hominid skull. The potassium isotope $^{40}K$ decays with a 1.28-billion-year half-life and is naturally present at very low levels. The most common decay mode is beta-minus decay into the stable isotope $^{40}Ca$, but 10.9% of decays result in the stable isotope $^{40}Ar$. The high temperatures in volcanoes drive argon out of solidifying rock and ash, so there is no argon in newly formed material. After formation, argon produced in the decay of $^{40}K$ is trapped, so $^{40}Ar$ builds up steadily over time. Accurate dating is possible by measuring the ratio of the number of atoms of $^{40}Ar$ and $^{40}K$. 1.8 million years after its formation,

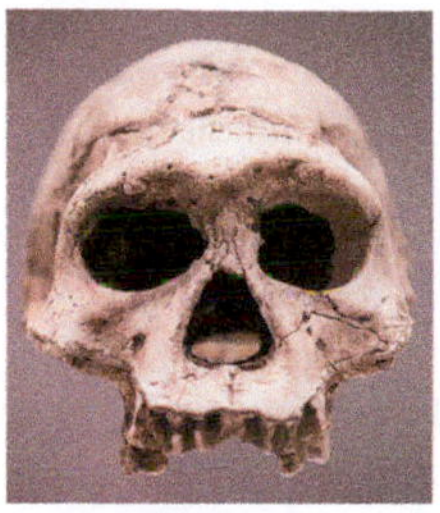

a. What fraction of the $^{40}K$ initially present in a sample has decayed?

b. What is the $^{40}Ar/^{40}K$ ratio of the sample?

68\. |||| BIO Corals take up certain elements from seawater, including uranium but not thorium. After the corals die, the uranium isotopes slowly decay into thorium isotopes. A measurement of the relative fraction of certain isotopes therefore provides a determination of the coral's age. A complicating factor is that the thorium isotopes decay as well. One scheme uses the alpha decay of $^{234}U$ to $^{230}Th$. After a long time, the two species reach an equilibrium in which the number of $^{234}U$ decays per second (each producing an atom of $^{230}Th$) is exactly equal to the number of $^{230}Th$ decays per second. What is the relative concentration of the two isotopes—the ratio of $^{234}U$ to $^{230}Th$—when this equilibrium is reached?

69\. ||| All the very heavy atoms found in the earth were created long ago by nuclear fusion reactions in a supernova, an exploding star. The debris spewed out by the supernova later coalesced to form the sun and the planets of our solar system. Nuclear physics suggests that the uranium isotopes $^{235}U$ ($t_{1/2} = 7.04 \times 10^8$ yr) and $^{238}U$ ($t_{1/2} = 4.47 \times 10^9$ yr) should have been created in roughly equal amounts. Today, 99.28% of uranium is $^{238}U$ and 0.72% is $^{235}U$. How long ago did the supernova occur?

70\. |||| BIO About 12% of your body mass is carbon; some of this is radioactive $^{14}C$, a beta-emitter. If you absorb 100% of the 49 keV energy of each $^{14}C$ decay, what dose equivalent in Sv do you receive each year from the $^{14}C$ in your body?

71\. |||| BIO Gamma rays may be used to kill pathogens in ground beef. One irradiation facility uses a $^{60}C$ source that has an activity of $1.0 \times 10^6$ Ci. $^{60}C$ undergoes beta decay and then gives off two gamma rays, at 1.17 and 1.33 MeV; typically 30% of this gamma-ray energy is absorbed by the meat. The dose required to kill all pathogens present in the beef is 4000 Gy. How many kilograms of meat per hour can be processed in this facility?

72\. ||| BIO A 70 kg human body typically contains 140 g of potassium. Potassium has a chemical atomic mass of 39.1 u and has three naturally occurring isotopes. One of those isotopes, $^{40}K$, is radioactive with a half-life of 1.3 billion years and a natural abundance of 0.012%. Each $^{40}K$ decay deposits, on average, 1.0 MeV of energy into the body. What yearly dose in Gy does the typical person receive from the decay of $^{40}K$ in the body?

73\. || BIO A chest x ray uses 10 keV photons. A 60 kg person receives a 30 mrem dose from one x ray that exposes 25% of the patient's body. How many x-ray photons are absorbed in the patient's body?

## MCAT-Style Passage Problems

### Plutonium-Powered Exploration BIO

The Curiosity rover sent to explore the surface of Mars has an electric generator powered by heat from the radioactive decay of $^{238}Pu$, a plutonium isotope that decays by alpha emission with a half-life of 88 years. At the start of the mission, the generator contained $9.6 \times 10^{24}$ nuclei of $^{238}Pu$.

74\. | What is the daughter nucleus of the decay?
A. $^{238}Am$ B. $^{238}Pu$ C. $^{238}Np$
D. $^{236}Th$ E. $^{234}U$

75\. || What was the approximate activity of the plutonium source at the start of the mission?
A. $2 \times 10^{21}$ Bq B. $2 \times 10^{19}$ Bq C. $2 \times 10^{17}$ Bq
D. $2 \times 10^{15}$ Bq E. $2 \times 10^{13}$ Bq

76\. || The generator initially provided 125 W of power. If you assume that the power of the generator is proportional to the activity of the plutonium, by approximately what percent did the power output decrease over the first two years of the rover's mission?
A. 1.5% B. 2.0% C. 2.5% D. 3.0% E. 3.5%

### Nuclear Fission

The uranium isotope $^{235}U$ can *fission*—break into two smaller-mass components and free neutrons—if it is struck by a free neutron. A typical reaction is

$$^{1}_{0}n + ^{235}_{92}U \rightarrow ^{141}_{56}Ba + ^{92}_{36}Kr + 3^{1}_{0}n$$

As you can see, the subscripts (the number of protons) and the superscripts (the number of nucleons) "balance" before and after the fission event; there is no change in the number of protons or neutrons. Significant energy is released in this reaction. If a fission event happens in a large chunk of $^{235}U$, the neutrons released may induce the fission of other $^{235}U$ atoms, resulting in a chain reaction. This is how a nuclear reactor works.

The number of neutrons required to create a stable nucleus increases with atomic number. When the heavy $^{235}U$ nucleus fissions, the lighter reaction products are thus neutron rich and are likely unstable. Many of the short-lived radioactive nuclei used in medicine are produced in fission reactions in nuclear reactors.

77\. | What statement can be made about the masses of atoms in the above reaction?
A. $m(^{235}_{92}U) > m(^{141}_{56}Ba) + m(^{92}_{36}Kr) + 2m(^{1}_{0}n)$
B. $m(^{235}_{92}U) < m(^{141}_{56}Ba) + m(^{92}_{36}Kr) + 2m(^{1}_{0}n)$
C. $m(^{235}_{92}U) = m(^{141}_{56}Ba) + m(^{92}_{36}Kr) + 2m(^{1}_{0}n)$
D. $m(^{235}_{92}U) = m(^{141}_{56}Ba) + m(^{92}_{36}Kr) + 3m(^{1}_{0}n)$

78\. | Because the decay products in the above fission reaction are neutron rich, they will likely decay by what process?
A. Alpha decay B. Beta decay C. Gamma decay

79\. || $^{235}U$ is radioactive, with a long half-life of 704 million years. The decay products of a $^{235}U$ fission reaction typically have half-lives of a few minutes. This means that the decay products of a fission reaction have
A. Much higher activity than the original uranium.
B. Much lower activity than the original uranium.
C. The same activity as the original uranium.

80\. | If a $^{238}_{92}U$ nucleus is struck by a neutron, it may absorb the neutron. The resulting nucleus then rapidly undergoes beta-minus decay. The daughter nucleus of that decay is
A. $^{239}_{91}Pa$ B. $^{239}_{92}U$ C. $^{239}_{93}Np$ D. $^{239}_{94}Pu$

KNOWLEDGE STRUCTURE VII

**BASIC GOALS**

What are the properties and characteristics of space and time?
What do we know about the nature of light and atoms?
How are atomic and nuclear phenomena explained by energy levels, wave functions, and photons?

**CROSS-CUTTING CONCEPTS**

**The nature of matter**

Matter is composed of atoms. An atom consists of a nucleus—a small, dense, positively charged core—surrounded by negatively charged electrons.

The nucleus is composed of protons, which have a positive charge, and neutrons, which have no charge.

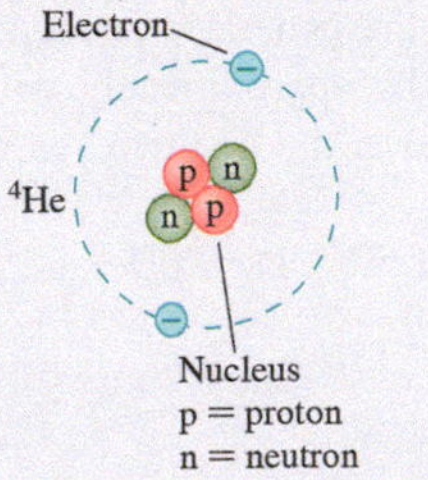


**Quantum states of electrons, neutrons, and protons**

An atom's electrons can exist in only certain quantum states.

- Only one electron can occupy each state.
- States are filled from lower energies to higher.
- The properties of the states and the order in which they are filled determine the structure of the periodic table.
- Electrons can move between states if energy is absorbed or emitted.

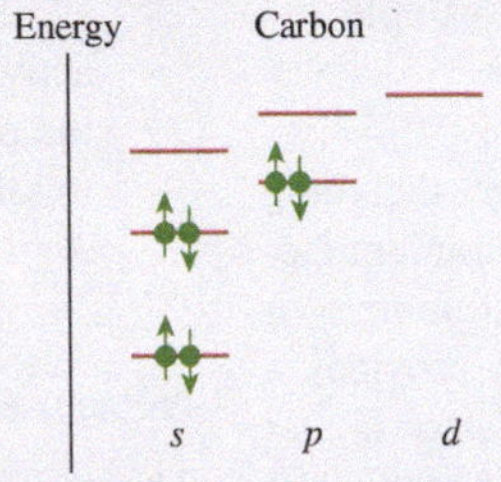


The neutrons and protons in the nucleus occupy energy levels analogous to those of the electrons.

- There is one set of states for the neutrons, one for the protons.
- States are filled from lower energies to higher.
- If the protons and neutrons can be rearranged to form a lower energy state, the nucleus will be unstable.

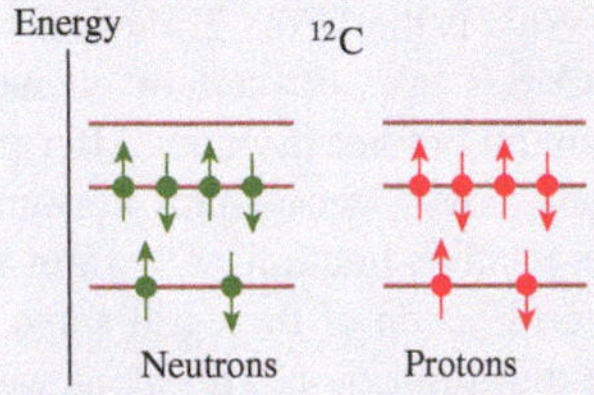


**GENERAL PRINCIPLES**

| | |
|---|---|
| **Principle of relativity** | All the laws of physics are the same in all inertial reference frames. |
| **Quantization of energy** | Particles of matter and photons of light have only certain allowed energies. |
| **Uncertainty principle** | $\Delta x\, \Delta p \geq h/4\pi$ |
| **Pauli exclusion principle** | No more than one electron, neutron, or proton can occupy the same quantum state. |

## Relativity

- The speed of light $c$ is the same in all inertial reference frames.
- No particle or causal influence can travel faster than $c$.
- Length contraction: The length of an object in a reference frame in which the object moves with speed $v$ is
$$L = \sqrt{1-\beta^2}\,\ell \leq \ell$$
where $\ell$ is the proper length.
- Time dilation: The proper time interval $\Delta\tau$ between two events is measured in a reference frame in which the two events occur at the same position. The time interval $\Delta t$ in a frame moving with relative speed $v$ is
$$\Delta t = \Delta\tau/\sqrt{1-\beta^2} \geq \Delta\tau$$
- Particles have energy even when at rest. Mass can be transformed into energy and vice versa: $E_0 = mc^2$.

## Quantum physics

- Matter has wave-like properties. A particle has a de Broglie wavelength
$$\lambda = \frac{h}{mv}$$
- Light has particle-like properties. A photon of light of frequency $f$ has energy
$$E_{\text{photon}} = hf = \frac{hc}{\lambda}$$
- The wave nature of matter leads to quantized energy levels in atoms and nuclei. A transition between quantized energy levels involves the emission or absorption of a photon.

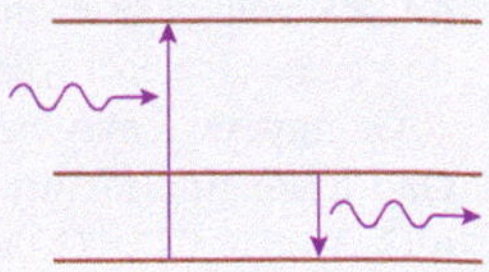

## Properties of atoms

- Quantized energy levels depend on quantum numbers $n$ and $l$.
- An atom can jump from one state to another by emitting or absorbing a photon of energy $E_{\text{photon}} = \Delta E_{\text{atom}}$.
- The ground-state electron configuration is the lowest-energy configuration consistent with the Pauli principle.

## Properties of nuclei

- The nucleus is held together by the strong force, an attractive short-range force between any two nucleons.

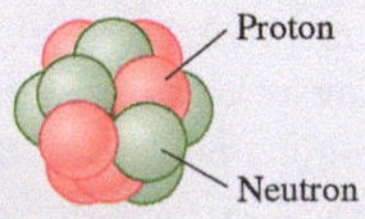


- Unstable nuclei decay by alpha, beta, or gamma decay. The number of undecayed nuclei decreases exponentially with time:
$$N = N_0\left(\frac{1}{2}\right)^{t/t_{1/2}}$$

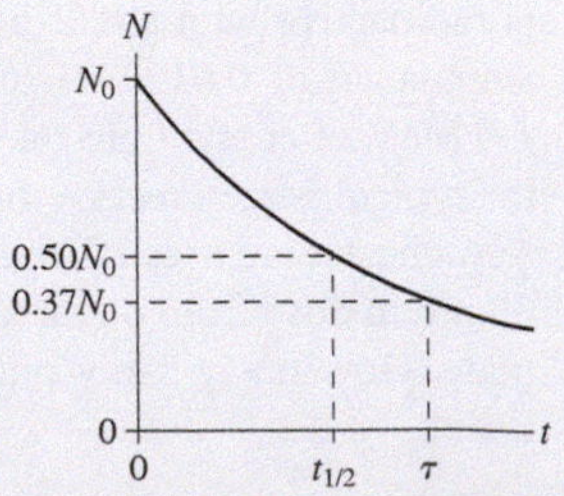

## The Physics of Very Cold Atoms

Modern physics is a study of extremes. Relativity deals with the physics of objects traveling at near-light speeds. Quantum mechanics is about the physics of matter and energy at very small scales. Nuclear physics involves energies that dwarf anything dreamed of in previous centuries.

Some of the most remarkable discoveries of recent years are at another extreme—very low temperatures, mere billionths of a degree above absolute zero. Let's look at how such temperatures are achieved and some new physics that emerges.

You learned in Part III that the temperature of a gas depends on the speeds of the atoms in the gas. Suppose we start with atoms at or above room temperature. Cooling the gas means slowing the atoms down. How can we drastically reduce their speeds, bringing them nearly to a halt? The trick is to slow them, thus cooling them, using the interactions between light and atoms that we explored in Chapters 28 and 29.

Photons have momentum, and that momentum is transferred to an atom when a photon is absorbed. Part a of the figure shows an atom moving "upstream" against a laser beam tuned to an atomic transition. Photon absorptions transfer momentum, slowing the atom. Subsequent photon emissions give the atom a "kick," but in random directions, so on average the emissions won't speed up the atom in the same way that the absorptions slow it. A beam of atoms moving "upstream" against a correctly tuned laser beam is slowed down—the "hot" beam of atoms is cooled.

Once a laser cools the atoms, a different configuration of laser beams can trap them. Part b of the figure shows six overlapped laser beams, each tuned slightly *below* the frequency of an atomic absorption line. If an atom tries to leave the overlap region, it will be moving "upstream" against one of the laser beams. The atom will see that laser beam Doppler-shifted to a higher frequency, matching the transition frequency. The atom will then absorb photons from this laser beam, and the resulting kick will nudge it back into the overlap region. The atoms are trapped in what is known as *optical molasses* or, more generally, an *atom trap*. More effective traps can be made by adding magnetic fields. The final cooling of the atoms is by evaporation—letting the more energetic atoms leave the trap.

Ultimately, these techniques produce a diffuse gas of atoms moving at only $\approx 1$ mm/s. This corresponds to a nearly unbelievable temperature of just a few nanokelvin—billionths of a degree above absolute zero. This is colder than outer space. The coldest spot in the universe is inside an atom trap in a physics lab.

Once the atoms are cooled, some very remarkable things happen. As we saw in Part VII, all particles, including atoms, have a wave nature. As the atoms slow, their wavelengths increase. In a correctly prepared gas at a low enough temperature, the de Broglie wavelength of an individual atom is larger than the spacing between atoms, and the wave functions of multiple atoms overlap. When this happens, some atoms undergo *Bose-Einstein condensation,* coalescing into one "super atom," with thousands of atoms occupying the same quantum state. The resulting Bose-Einstein condensate illustrates the counterintuitive nature of the quantum world. The atoms in the condensate—that is, their wave functions—are all in the same place at the same time! This is a new and truly bizarre state of matter.

Are there applications for Bose-Einstein condensation? Current talk of atom lasers and other futuristic concepts aside, no one really knows, just as the early architects of quantum mechanics didn't know that their theory would be used to design the chips that power personal computers.

At the start of the 20th century, there was a worry that everything in physics had been discovered. There is no such worry at the start of the 21st century, which promises to be full of wonderful discoveries and remarkable applications. What do you imagine the final chapter of a physics textbook will look like 100 years from now?

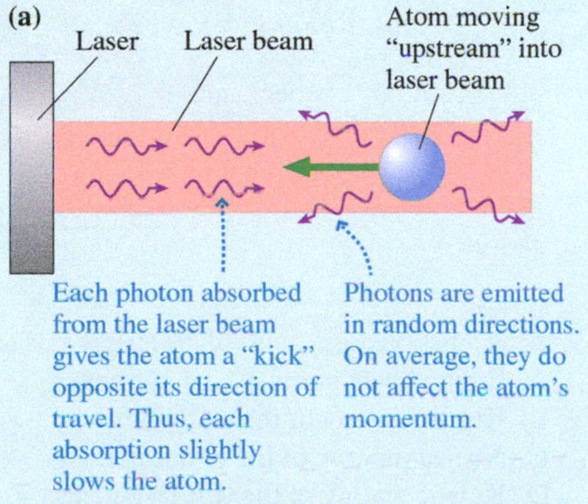


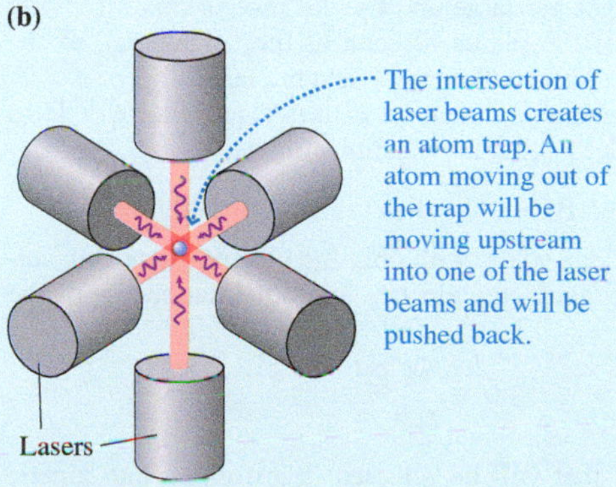


Laser cooling and trapping.

*The following questions are related to the passage "The Physics of Very Cold Atoms" on the preceding page.*

1. Why is it useful to create an assembly of very slow-moving cold atoms?
   A. The atoms can be more easily observed at slow speeds.
   B. Lowering the temperature this way permits isotopes that normally decay in very short times to persist long enough to be studied.
   C. At low speeds the quantum nature of the atoms becomes more apparent, and new forms of matter emerge.
   D. At low speeds the quantum nature of the atoms becomes less important, and they appear more like classical particles.
2. The momentum of a photon is given by $p = h/\lambda$. Suppose an atom emits a photon. Which of the following photons will give the atom the biggest "kick"—the highest recoil speed?
   A. An infrared photon  B. A red-light photon
   C. A blue-light photon  D. An ultraviolet photon
3. When an atom moves "upstream" against the photons in a laser beam, the energy of the photons appears to be _______ if the atom were at rest.
   A. Greater than  B. Less than  C. The same as
4. A gas of cold atoms strongly absorbs light of a specific wavelength. Warming the gas causes the absorption to decrease. Which of the following is the best explanation for this reduction?
   A. Warming the gas changes the atomic energy levels.
   B. Warming the gas causes the atoms to move at higher speeds, so the atoms "see" the photons at larger Doppler shifts.
   C. Warming the gas causes more collisions between the atoms, which affects the absorption of photons.
   D. Warming the gas makes it more opaque to the photons, so fewer enter the gas.
5. A gas of cold atoms starts at a temperature of 100 nK. The average speed of the atoms is then reduced by half. What is the new temperature?
   A. 71 nK  B. 50 nK  C. 37 nK  D. 25 nK
6. Rubidium is often used for the type of experiments noted in the passage. At a speed of 1.0 mm/s, what is the approximate de Broglie wavelength of an atom of $^{87}$Rb?
   A. 5 nm  B. 50 nm  C. 500 nm  D. 5000 nm
7. A gas of rubidium atoms and a gas of sodium atoms have been cooled to the same very low temperature. What can we say about the de Broglie wavelengths of typical atoms in the two gases?
   A. The sodium atoms have the longer wavelength.
   B. The wavelengths are the same.
   C. The rubidium atoms have the longer wavelength.

*The following passages and associated questions are based on the material of Part VII.*

### Splitting the Atom

"Splitting" an atom in the process of nuclear fission releases a great deal of energy. If all the atoms in 1 kg of $^{235}$U undergo nuclear fission, $8.0 \times 10^{13}$ J will be released, equal to the energy from burning $2.3 \times 10^{6}$ kg of coal. What is the source of this energy? Surprisingly, the energy from this nuclear disintegration ultimately comes from the electric potential of the positive charges that make up the nucleus.

The protons in a nucleus exert repulsive forces on each other, but this force is less than the short-range attractive nuclear force. If a nucleus breaks into two smaller nuclei, the nuclear force will hold each of the fragments together, but it won't bind the two positively charged fragments to each other. This is illustrated in Figure VII.1. The two fragments feel a strong repulsive electrostatic force. The charges are large and the distance is small (roughly equal to the sum of the radius of each of the fragments), so the force—and thus the potential energy—is quite large.

In a fission reaction, a neutron causes a nucleus of $^{235}$U to split into two smaller nuclei; a typical reaction is

$$n + {}^{235}\text{U} \rightarrow {}^{87}\text{Br} + {}^{147}\text{La} + \text{neutrons} + \approx 200\ \text{MeV}$$

Right after the nucleus splits, with only the electric force now acting on the two fragments, the electrostatic potential energy of the two fragments is

$$U = \frac{kq_1q_2}{r_1 + r_2}$$

This is the energy that will be released, transformed into kinetic energy, when the fragments fly apart. If we use reasonable estimates for the radii of the two fragments, we compute a value for the energy that is close to the experimentally observed value of 200 MeV for the energy released in the fission reaction. The energy released in this *nuclear* reaction is actually *electric* potential energy.

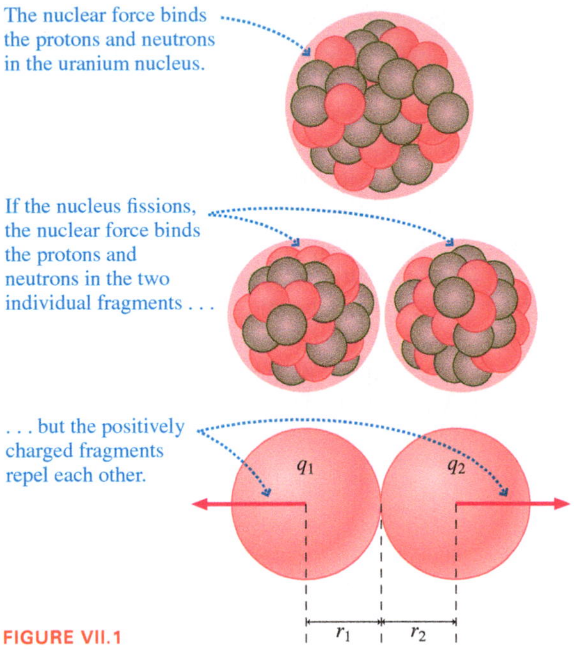

The nuclear force binds the protons and neutrons in the uranium nucleus.
If the nucleus fissions, the nuclear force binds the protons and neutrons in the two individual fragments . . .
. . . but the positively charged fragments repel each other.
$q_1$
$q_2$
$r_1$
$r_2$


FIGURE VII.1

8. How many neutrons are "left over" in the noted fission reaction?
   A. 1  B. 2  C. 3  D. 4
9. After a fission event, most of the energy released is in the form of
   A. Emitted beta particles and gamma rays.
   B. Kinetic energy of the emitted neutrons.
   C. Nuclear energy of the two fragments.
   D. Kinetic energy of the two fragments.
10. Suppose the original nucleus is at rest in the fission reaction described previously. If we neglect the momentum of the neutrons, after the two fragments fly apart,
    A. The Br nucleus has more momentum.
    B. The La nucleus has more momentum.
    C. The momentum of the Br nucleus equals that of the La nucleus.

11. Suppose the original nucleus is at rest in the fission reaction noted above. If we neglect the kinetic energy of the neutrons, after the two fragments fly apart,
    A. The Br nucleus has more kinetic energy.
    B. The La nucleus has more kinetic energy.
    C. The kinetic energy of the Br nucleus equals that of the La nucleus.
12. 200 MeV is a typical energy released in a fission reaction. To get a sense for the scale of the energy, if we were to use this energy to create electron-positron pairs, approximately how many pairs could we create?
    A. 50 B. 100 C. 200 D. 400
13. The two fragments of a fission reaction are isotopes that are neutron-rich; each has more neutrons than the stable isotopes for their nuclear species. They will quickly decay to more stable isotopes. What is the most likely decay mode?
    A. Alpha decay B. Beta decay C. Gamma decay

### Detecting and Deciphering Radiation

A researcher has placed a sample of radioactive material in an enclosure and blocked all emissions except those that travel in a particular direction, creating a beam of radiation. The beam then passes through a uniform magnetic field, as shown in Figure VII.2, before reaching a bank of detectors. Only three of the detectors record significant signals, so the researcher deduces that the particles coming from the source have taken three different paths, illustrated in the figure, and concludes that the sample is emitting three different kinds of radiation. Assume that the emitted particles move with similar speeds.

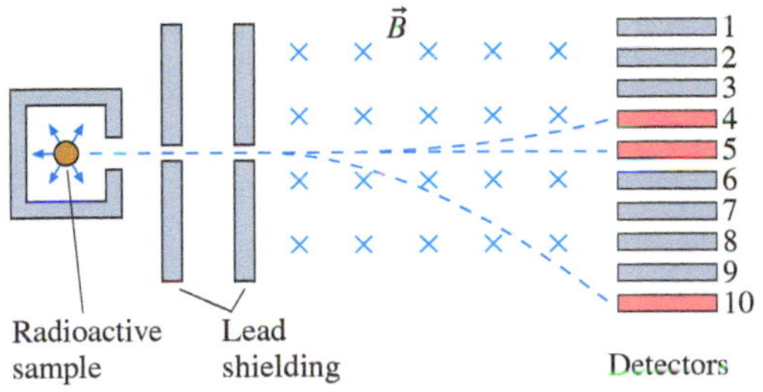


FIGURE VII.2

14. What type of radiation is detected by detector 4?
    A. Alpha B. Beta-minus C. Beta-plus D. Gamma
15. What type of radiation is detected by detector 5?
    A. Alpha B. Beta-minus C. Beta-plus D. Gamma
16. What type of radiation is detected by detector 10?
    A. Alpha B. Beta-minus C. Beta-plus D. Gamma

### Additional Integrated Problems

17. The glow-in-the-dark dials on some watches and some key-chain lights shine with energy provided by the decay of radioactive tritium, $^3_1$H. Tri- tium is a radioactive isotope of hydrogen with a half-life of 12 years. Each decay emits an electron with an energy of 19 keV. A typical new watch has tritium with a total activity of 15 MBq.

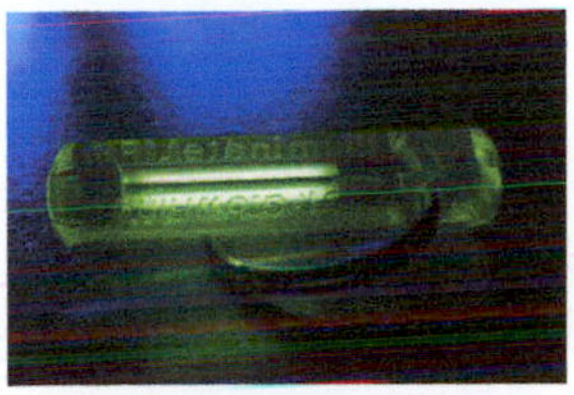

A keychain light powered by the decay of tritium.

    a. What is the speed of the emitted electron? (This speed is high enough that you'll need to do a relativistic calculation.)
    b. What is the power, in watts, provided by the radioactive decay process?
    c. What will be the activity of the tritium in a watch after 5 years, assuming none escapes?
18. BIO An x-ray tube is powered by a high-voltage supply that delivers 700 W to the tube. The tube converts 1% of this power into x rays of wavelength 0.030 nm.
    a. Approximately how many x-ray photons are emitted per second?
    b. If a 75 kg technician is accidentally exposed to the full power of the x-ray beam for 1.0 s, what dose equivalent in Sv does he receive? Assume that the x-ray energy is distributed over the body, and that 80% of the energy is absorbed.
19. Many speculative plans for spaceships capable of interstellar travel have been developed over the years. Nearly all are powered by the fusion of light nuclei, one of a very few power sources capable of providing the incredibly large energies required. A typical design for a fusion-powered craft has a $1.7 \times 10^6$ kg ship brought up to a speed of $0.12c$ using the energy from the fusion of $^2_1$H and $^3_2$He. Each fusion reaction produces a daughter nucleus and one free proton with a combined kinetic energy of 18 MeV; these high-speed particles are directed backward to create thrust.
    a. What is the kinetic energy of the ship at the noted top speed? For the purposes of this problem you can do a nonrelativistic calculation.
    b. If we assume that 50% of the energy of the fusion reactions goes into the kinetic energy of the ship (a very generous assumption), how many fusion reactions are required to get the ship up to speed?
    c. How many kilograms of $^2_1$H and of $^3_2$He are required to produce the required number of reactions?
20. A muon is a lepton that is a higher-mass (rest mass 105 MeV/$c^2$) sibling to the electron. Muons are produced in the upper atmosphere when incoming cosmic rays collide with the nuclei of gas molecules. As the muons travel toward the surface of the earth, they lose energy. A muon that travels from the upper atmosphere to the surface of the earth typically begins with kinetic energy 6.0 GeV and reaches the surface of the earth with kinetic energy 4.0 GeV. The energy decreases by one-third of its initial value. By what fraction does the speed of the muon decrease?
21. A muon is a lepton that is a higher-mass (rest mass 105 MeV/$c^2$) sibling to the electron. Muons are produced in the upper atmosphere when incoming cosmic rays collide with the nuclei of gas molecules. The muon half-life is 1.5 $\mu$s, but atmospheric muons typically live much longer than this because of time dilation, as we saw in Chapter 27. Suppose 100,000 muons are created 120 km above the surface of the earth, each with kinetic energy 10 GeV. Assume that the muons don't lose energy but move at a constant velocity directed straight down toward the surface of the earth. How many muons survive to reach the surface?

APPENDIX A

# Mathematics Review

## Algebra

**Using exponents:**

$a^{-x} = \dfrac{1}{a^x}$ $\quad a^x a^y = a^{(x+y)}$ $\quad \dfrac{a^x}{a^y} = a^{(x-y)}$ $\quad (a^x)^y = a^{xy}$

$a^0 = 1$ $\quad a^1 = a$ $\quad a^{1/n} = \sqrt[n]{a}$

**Fractions:**

$\left(\dfrac{a}{b}\right)\left(\dfrac{c}{d}\right) = \dfrac{ac}{bd}$ $\quad \dfrac{a/b}{c/d} = \dfrac{ad}{bc}$ $\quad \dfrac{1}{1/a} = a$

**Logarithms:**

Natural (base $e$) logarithms: If $a = e^x$, then $\ln(a) = x$ $\quad \ln(e^x) = x$ $\quad e^{\ln(x)} = x$

Base-10 logarithms: If $a = 10^x$, then $\log_{10}(a) = x$ $\quad \log_{10}(10^x) = x$ $\quad 10^{\log_{10}(x)} = x$

The following rules hold for both natural and base-10 algorithms:

$\ln(ab) = \ln(a) + \ln(b)$ $\quad \ln\left(\dfrac{a}{b}\right) = \ln(a) - \ln(b)$ $\quad \ln(a^n) = n\ln(a)$

The expression $\ln(a + b)$ cannot be simplified.

**Linear equations:**

The graph of the equation $y = ax + b$ is a straight line. $a$ is the slope of the graph. $b$ is the $y$-intercept.

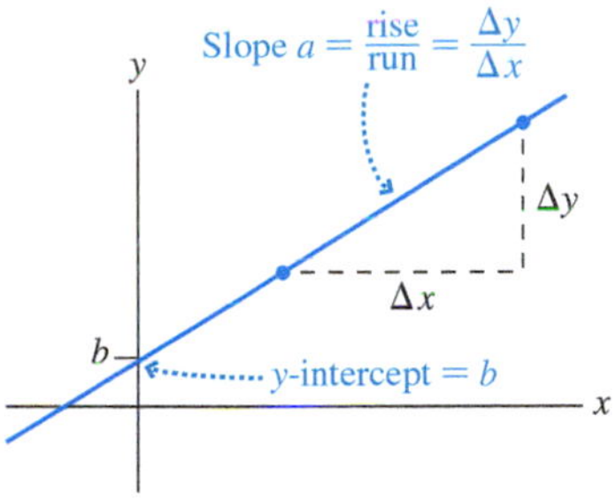


**Proportionality:**

To say that $y$ is proportional to $x$, written $y \propto x$, means that $y = ax$, where $a$ is a constant. Proportionality is a special case of linearity. A graph of a proportional relationship is a straight line that passes through the origin. If $y \propto x$, then

$$\frac{y_1}{y_2} = \frac{x_1}{x_2}$$

**Quadratic equation:**

The quadratic equation $ax^2 + bx + c = 0$ has the two solutions $x = \dfrac{-b \pm \sqrt{b^2 - 4ac}}{2a}$.

## Geometry and Trigonometry

**Area and volume:**

Rectangle

$A = ab$

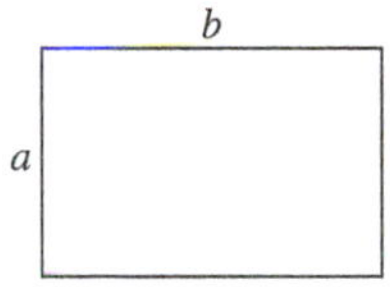


Rectangular box

$V = abc$

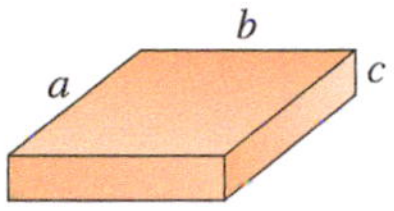


Triangle

$A = \frac{1}{2}ab$

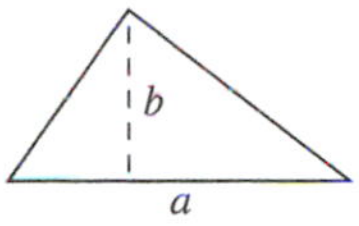


Right circular cylinder

$V = \pi r^2 l$

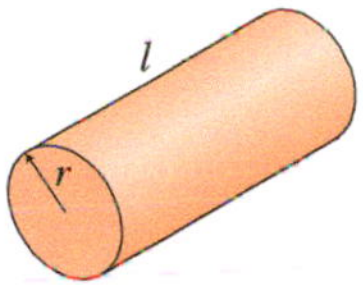


Circle

$C = 2\pi r$

$A = \pi r^2$

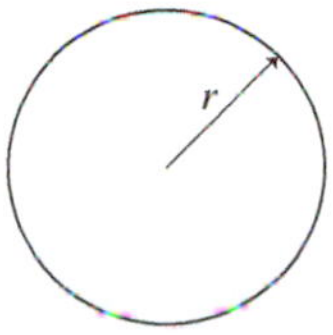


Sphere

$A = 4\pi r^2$

$V = \frac{4}{3}\pi r^3$

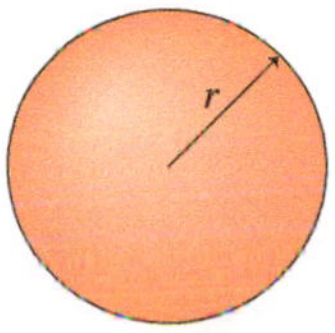

**Arc length and angle:** The angle $\theta$ in radians is defined as $\theta = s/r$.
The arc length that spans angle $\theta$ is $s = r\theta$.
$2\pi$ rad $= 360°$

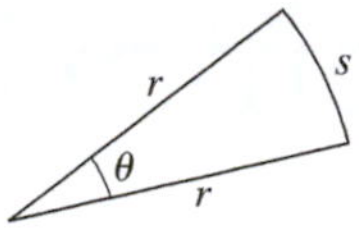


**Right triangle:** Pythagorean theorem $c = \sqrt{a^2 + b^2}$ or $a^2 + b^2 = c^2$

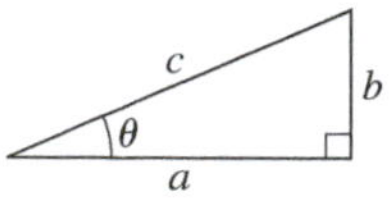


$$\sin\theta = \frac{b}{c} = \frac{\text{far side}}{\text{hypotenuse}} \qquad \theta = \sin^{-1}\left(\frac{b}{c}\right)$$

$$\cos\theta = \frac{a}{c} = \frac{\text{adjacent side}}{\text{hypotenuse}} \qquad \theta = \cos^{-1}\left(\frac{a}{c}\right)$$

$$\tan\theta = \frac{b}{a} = \frac{\text{far side}}{\text{adjacent side}} \qquad \theta = \tan^{-1}\left(\frac{b}{a}\right)$$

In general, if it is known that the sine of an angle $\theta$ is $x$, so $x = \sin\theta$, then we can find $\theta$ by taking the *inverse sine* of $x$, denoted $\sin^{-1}x$. Thus $\theta = \sin^{-1}x$. Similar relationships apply for cosines and tangents.

**General triangle:** $\alpha + \beta + \gamma = 180° = \pi$ rad

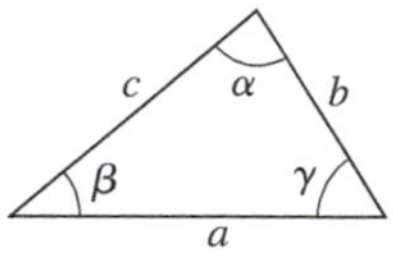


**Identities:**

$$\tan\alpha = \frac{\sin\alpha}{\cos\alpha} \qquad \sin^2\alpha + \cos^2\alpha = 1$$

$$\sin(-\alpha) = -\sin\alpha \qquad \cos(-\alpha) = \cos\alpha$$

$$\sin(2\alpha) = 2\sin\alpha\cos\alpha \qquad \cos(2\alpha) = \cos^2\alpha - \sin^2\alpha$$

## Expansions and Approximations

**Binomial approximation:** $(1 + x)^n \approx 1 + nx$ if $x \ll 1$

**Small-angle approximation:** If $\alpha \ll 1$ rad, then $\sin\alpha \approx \tan\alpha \approx \alpha$ and $\cos\alpha \approx 1$.

The small-angle approximation is excellent for $\alpha < 5°$ ($\approx 0.1$ rad) and generally acceptable up to $\alpha \approx 10°$.

# Periodic Table of Elements

Key: Atomic number — 27; Symbol — Co; Atomic mass — 58.9

| Period | | | | | | | | | | | | | | | | | | |
|---|---|---|---|---|---|---|---|---|---|---|---|---|---|---|---|---|---|---|
| 1 | 1 H 1.0 | | | | | | | | | | | | | | | | | 2 He 4.0 |
| 2 | 3 Li 6.9 | 4 Be 9.0 | | | | | | | | | | | 5 B 10.8 | 6 C 12.0 | 7 N 14.0 | 8 O 16.0 | 9 F 19.0 | 10 Ne 20.2 |
| 3 | 11 Na 23.0 | 12 Mg 24.3 | Transition elements | | | | | | | | | | 13 Al 27.0 | 14 Si 28.1 | 15 P 31.0 | 16 S 32.1 | 17 Cl 35.5 | 18 Ar 39.9 |
| 4 | 19 K 39.1 | 20 Ca 40.1 | 21 Sc 45.0 | 22 Ti 47.9 | 23 V 50.9 | 24 Cr 52.0 | 25 Mn 54.9 | 26 Fe 55.8 | 27 Co 58.9 | 28 Ni 58.7 | 29 Cu 63.5 | 30 Zn 65.4 | 31 Ga 69.7 | 32 Ge 72.6 | 33 As 74.9 | 34 Se 79.0 | 35 Br 79.9 | 36 Kr 83.8 |
| 5 | 37 Rb 85.5 | 38 Sr 87.6 | 39 Y 88.9 | 40 Zr 91.2 | 41 Nb 92.9 | 42 Mo 95.9 | 43 Tc [98] | 44 Ru 101.1 | 45 Rh 102.9 | 46 Pd 106.4 | 47 Ag 107.9 | 48 Cd 112.4 | 49 In 114.8 | 50 Sn 118.7 | 51 Sb 121.8 | 52 Te 127.6 | 53 I 126.9 | 54 Xe 131.3 |
| 6 | 55 Cs 132.9 | 56 Ba 137.3 | 71 Lu 175.0 | 72 Hf 178.5 | 73 Ta 180.9 | 74 W 183.9 | 75 Re 186.2 | 76 Os 190.2 | 77 Ir 192.2 | 78 Pt 195.1 | 79 Au 197.0 | 80 Hg 200.6 | 81 Tl 204.4 | 82 Pb 207.2 | 83 Bi 209.0 | 84 Po [209] | 85 At [210] | 86 Rn [222] |
| 7 | 87 Fr [223] | 88 Ra [226] | 103 Lr [262] | 104 Rf [267] | 105 Db [268] | 106 Sg [271] | 107 Bh [274] | 108 Hs [277] | 109 Mt [278] | 110 Ds [281] | 111 Rg [281] | 112 Cn [285] | 113 Nh [286] | 114 Fl [289] | 115 Mc [289] | 116 Lv [293] | 117 Ts [294] | 118 Og [294] |

| | | | | | | | | | | | | | | |
|---|---|---|---|---|---|---|---|---|---|---|---|---|---|---|
| Lanthanides 6 | 57 La 138.9 | 58 Ce 140.1 | 59 Pr 140.9 | 60 Nd 144.2 | 61 Pm 144.9 | 62 Sm 150.4 | 63 Eu 152.0 | 64 Gd 157.3 | 65 Tb 158.9 | 66 Dy 162.5 | 67 Ho 164.9 | 68 Er 167.3 | 69 Tm 168.9 | 70 Yb 173.0 |
| Actinides 7 | 89 Ac [227] | 90 Th 232.0 | 91 Pa 231.0 | 92 U 238.0 | 93 Np [237] | 94 Pu [244] | 95 Am [243] | 96 Cm [247] | 97 Bk [247] | 98 Cf [251] | 99 Es [252] | 100 Fm [257] | 101 Md [258] | 102 No [259] |

Inner transition elements

An atomic mass in brackets is that of the longest-lived isotope of an element with no stable isotopes.

APPENDIX C

# Atomic and Nuclear Data

| Atomic Number ($Z$) | Element | Symbol | Mass Number ($A$) | Atomic Mass (u) | Percent Abundance | Decay Mode | Half-Life $t_{1/2}$ |
|---|---|---|---|---|---|---|---|
| 0 | (Neutron) | n | 1 | 1.008 665 | | $\beta^-$ | 10.4 min |
| 1 | Hydrogen | H | 1 | 1.007 825 | 99.985 | stable | |
| | Deuterium | D | 2 | 2.014 102 | 0.015 | stable | |
| | Tritium | T | 3 | 3.016 049 | | $\beta^-$ | 12.33 yr |
| 2 | Helium | He | 3 | 3.016 029 | 0.000 1 | stable | |
| | | | 4 | 4.002 602 | 99.999 9 | stable | |
| | | | 6 | 6.018 886 | | $\beta^-$ | 0.81 s |
| 3 | Lithium | Li | 6 | 6.015 121 | 7.50 | stable | |
| | | | 7 | 7.016 003 | 92.50 | stable | |
| | | | 8 | 8.022 486 | | $\beta^-$ | 0.84 s |
| 4 | Beryllium | Be | 9 | 9.012 174 | 100 | stable | |
| | | | 10 | 10.013 534 | | $\beta^-$ | $1.5 \times 10^6$ yr |
| 5 | Boron | B | 10 | 10.012 936 | 19.90 | stable | |
| | | | 11 | 11.009 305 | 80.10 | stable | |
| | | | 12 | 12.014 352 | | $\beta^-$ | 0.020 2 s |
| 6 | Carbon | C | 10 | 10.016 854 | | $\beta^+$ | 19.3 s |
| | | | 11 | 11.011 433 | | $\beta^+$ | 20.4 min |
| | | | 12 | 12.000 000 | 98.90 | stable | |
| | | | 13 | 13.003 355 | 1.10 | stable | |
| | | | 14 | 14.003 242 | | $\beta^-$ | 5 730 yr |
| | | | 15 | 15.010 599 | | $\beta^-$ | 2.45 s |
| 7 | Nitrogen | N | 12 | 12.018 613 | | $\beta^+$ | 0.011 0 s |
| | | | 13 | 13.005 738 | | $\beta^+$ | 9.96 min |
| | | | 14 | 14.003 074 | 99.63 | stable | |
| | | | 15 | 15.000 108 | 0.37 | stable | |
| | | | 16 | 16.006 100 | | $\beta^-$ | 7.13 s |
| | | | 17 | 17.008 450 | | $\beta^-$ | 4.17 s |
| 8 | Oxygen | O | 15 | 15.003 065 | | $\beta^+$ | 122 s |
| | | | 16 | 15.994 915 | 99.76 | stable | |
| | | | 17 | 16.999 132 | 0.04 | stable | |
| | | | 18 | 17.999 160 | 0.20 | stable | |
| | | | 19 | 19.003 577 | | $\beta^-$ | 26.9 s |
| 9 | Fluorine | F | 18 | 18.000 937 | | $\beta^+$ | 109.8 min |
| | | | 19 | 18.998 404 | 100 | stable | |
| | | | 20 | 19.999 982 | | $\beta^-$ | 11.0 s |
| 10 | Neon | Ne | 19 | 19.001 880 | | $\beta^+$ | 17.2 s |
| | | | 20 | 19.992 435 | 90.48 | stable | |
| | | | 21 | 20.993 841 | 0.27 | stable | |
| | | | 22 | 21.991 383 | 9.25 | stable | |
| 17 | Chlorine | Cl | 35 | 34.968 853 | 75.77 | stable | |
| | | | 36 | 35.968 307 | | $\beta^-$ | $3.0 \times 10^5$ yr |
| | | | 37 | 36.965 903 | 24.23 | stable | |

| Atomic Number ($Z$) | Element | Symbol | Mass Number ($A$) | Atomic Mass (u) | Percent Abundance | Decay Mode | Half-Life $t_{1/2}$ |
|---|---|---|---|---|---|---|---|
| 18 | Argon | Ar | 36 | 35.967 547 | 0.34 | stable | |
| | | | 38 | 37.962 732 | 0.06 | stable | |
| | | | 39 | 38.964 314 | | $\beta^-$ | 269 yr |
| | | | 40 | 39.962 384 | 99.60 | stable | |
| | | | 42 | 41.963 049 | | $\beta^-$ | 33 yr |
| 19 | Potassium | K | 39 | 38.963 708 | 93.26 | stable | |
| | | | 40 | 39.964 000 | 0.01 | $\beta^-$ | $1.28 \times 10^9$ yr |
| | | | 41 | 40.961 827 | 6.73 | stable | |
| 26 | Iron | Fe | 54 | 54.939 613 | 5.9 | stable | |
| | | | 56 | 55.934 940 | 91.72 | stable | |
| | | | 57 | 56.935 396 | 2.1 | stable | |
| | | | 58 | 57.933 278 | 0.28 | stable | |
| | | | 60 | 59.934 072 | | $\beta^-$ | $1.5 \times 10^6$ yr |
| 27 | Cobalt | Co | 59 | 58.933 198 | 100 | stable | |
| | | | 60 | 59.933 820 | | $\beta^-$ | 5.27 yr |
| 38 | Strontium | Sr | 84 | 83.913 425 | 0.56% | stable | |
| | | | 86 | 85.909 262 | 9.86% | stable | |
| | | | 87 | 86.908 879 | 7.00% | stable | |
| | | | 88 | 87.905 614 | 82.58% | stable | |
| | | | 89 | 88.907 450 | | $\beta^-$ | 50.53 days |
| | | | 90 | 89.907 738 | | $\beta^-$ | 27.78 yr |
| 53 | Iodine | I | 127 | 126.904 474 | 100 | stable | |
| | | | 129 | 128.904 984 | | $\beta^-$ | $1.6 \times 10^7$ yr |
| | | | 131 | 130.906 124 | | $\beta^-$ | 8 days |
| 54 | Xenon | Xe | 128 | 127.903 531 | 1.9 | stable | |
| | | | 129 | 128.904 779 | 26.4 | stable | |
| | | | 130 | 129.903 509 | 4.1 | stable | |
| | | | 131 | 130.905 069 | 21.2 | stable | |
| | | | 132 | 131.904 141 | 26.9 | stable | |
| | | | 133 | 132.905 906 | | $\beta^-$ | 5.4 days |
| | | | 134 | 133.905 394 | 10.4 | stable | |
| | | | 136 | 135.907 215 | 8.9 | stable | |
| 55 | Cesium | Cs | 133 | 132.905 436 | 100 | stable | |
| | | | 137 | 136.907 078 | | $\beta^-$ | 30 yr |
| 82 | Lead | Pb | 204 | 203.973 020 | 1.4 | stable | |
| | | | 206 | 205.974 440 | 24.1 | stable | |
| | | | 207 | 206.975 871 | 22.1 | stable | |
| | | | 208 | 207.976 627 | 52.4 | stable | |
| | | | 210 | 209.984 163 | | $\alpha, \beta^-$ | 22.3 yr |
| | | | 211 | 210.988 734 | | $\beta^-$ | 36.1 min |
| 83 | Bismuth | Bi | 209 | 208.980 374 | 100 | stable | |
| | | | 211 | 210.987 254 | | $\alpha$ | 2.14 min |
| | | | 215 | 215.001 836 | | $\beta^-$ | 7.4 min |
| 86 | Radon | Rn | 219 | 219.009 477 | | $\alpha$ | 3.96 s |
| | | | 220 | 220.011 369 | | $\alpha$ | 55.6 s |
| | | | 222 | 222.017 571 | | $\alpha, \beta^-$ | 3.823 days |

| Atomic Number ($Z$) | Element | Symbol | Mass Number ($A$) | Atomic Mass (u) | Percent Abundance | Decay Mode | Half-Life $t_{1/2}$ |
|---|---|---|---|---|---|---|---|
| 88 | Radium | Ra | 223 | 223.018 499 | | $\alpha$ | 11.43 days |
| | | | 224 | 224.020 187 | | $\alpha$ | 3.66 days |
| | | | 226 | 226.025 402 | | $\alpha$ | 1 600 yr |
| | | | 228 | 228.031 064 | | $\beta^-$ | 5.75 yr |
| 90 | Thorium | Th | 227 | 227.027 701 | | $\alpha$ | 18.72 days |
| | | | 228 | 228.028 716 | | $\alpha$ | 1.913 yr |
| | | | 229 | 229.031 757 | | $\alpha$ | 7 340 yr |
| | | | 230 | 230.033 127 | | $\alpha$ | $7.54 \times 10^4$ yr |
| | | | 231 | 231.036 299 | | $\alpha, \beta^-$ | 25.52 h |
| | | | 232 | 232.038 051 | 100 | $\alpha$ | $1.40 \times 10^{10}$ yr |
| | | | 234 | 234.043 593 | | $\beta^-$ | 24.1 days |
| 92 | Uranium | U | 233 | 233.039 630 | | $\alpha$ | $1.59 \times 10^5$ yr |
| | | | 234 | 234.040 946 | | $\alpha$ | $2.45 \times 10^5$ yr |
| | | | 235 | 235.043 924 | 0.72 | $\alpha$ | $7.04 \times 10^8$ yr |
| | | | 236 | 236.045 562 | | $\alpha$ | $2.34 \times 10^7$ yr |
| | | | 238 | 238.050 784 | 99.28 | $\alpha$ | $4.47 \times 10^9$ yr |
| 93 | Neptunium | Np | 237 | 237.048 168 | | $\alpha$ | $2.14 \times 10^6$ yr |
| | | | 238 | 238.050 946 | | $\beta^-$ | 2.12 days |
| | | | 239 | 239.052 939 | | $\beta^-$ | 2.36 days |
| 94 | Plutonium | Pu | 238 | 238.049 555 | | $\alpha$ | 87.7 yr |
| | | | 239 | 239.052 157 | | $\alpha$ | $2.412 \times 10^4$ yr |
| | | | 240 | 240.053 808 | | $\alpha$ | 6560 yr |
| | | | 242 | 242.058 737 | | $\alpha$ | $3.73 \times 10^6$ yr |
| | | | 244 | 244.064 200 | | $\alpha$ | $8.1 \times 10^7$ yr |
| 95 | Americium | Am | 241 | 241.056 823 | | $\alpha$ | 432.21 yr |
| | | | 243 | 243.061 375 | | $\alpha$ | 7 370 yr |

# Answers

## Chapter 1

### Answers to odd-numbered multiple-choice questions

19. A
21. C
23. B
25. C
27. B
29. A

### Answers to odd-numbered problems

1.

3.

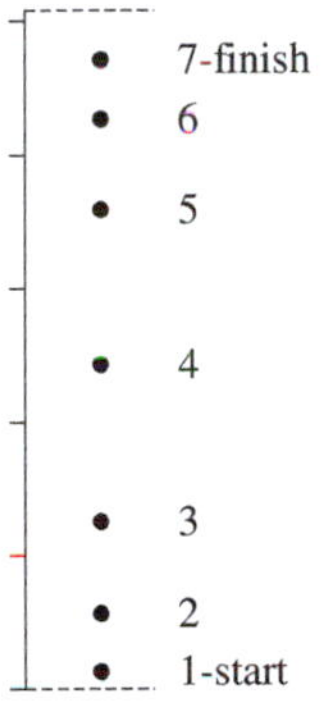


5. a. 2 mi b. 2 mi
7. −22 m
9. 800 m
11. Bike, boy, cat, toy car
13. −1.0 m/s
15. a. Second 100 m b. 9.88 m/s
17. a. 0.20 m b. 20 m/s c. 27 m/s
19. a. 3 b. 3 c. 3 d. 2
21. a. 846 b. 7.9 c. 5.77 d. 13.1
23. $8.8480 \times 10^3$ m
25. We get $6 \times 10^{-9}$ m/s; your answer should be close.
27. 487 m
29. $1.16 \times 10^3$ ft
31. 71 m, 45° south of west
33. 38 km
35. 6.1 m
37. 177 m
39. 150 cm
41. a. 4.6 m b. 6.1 m
43.

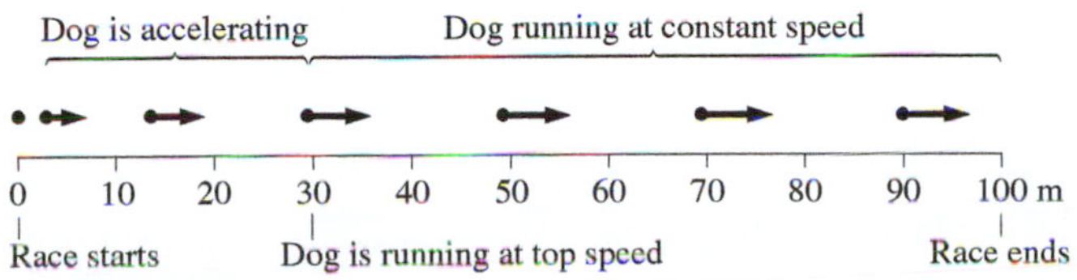


45.

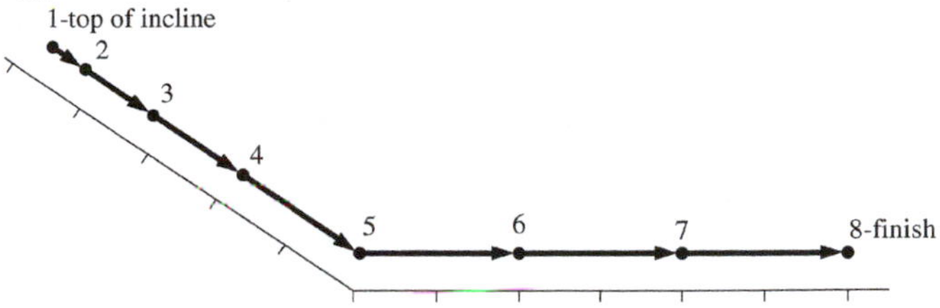


47.

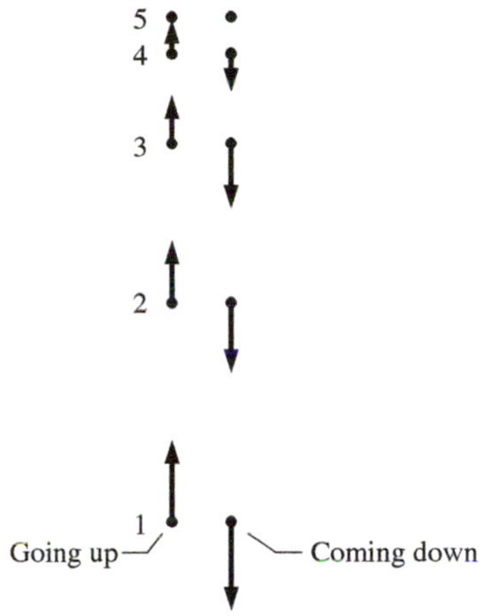


49.

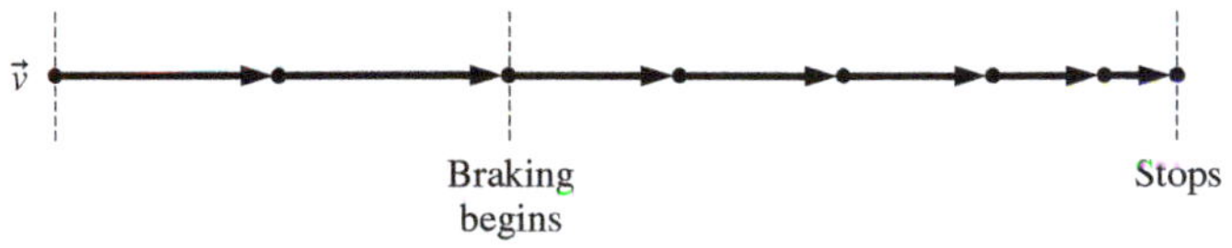


55. $2.5 \times 10^9$ s
57. 69 mph
59. 9.3% longer
61. $3.0 \times 10^4$ m/s
63. 26 mph
65. a. 25 km b. 0 km c. 28 m/s
67. a. 22 mph
69. a. AB and CD b. All segments c. AB and CD
71. a. 21° b. 990m c. 4.1 m/s
73. a. 220 m b. 220 m c. 260 s
75. a. 1.41 km b. 1.06 m/s
77. A

## Chapter 2

### Answers to odd-numbered multiple-choice questions

15. C
17. C
19. D
21. B
23. A
25. C
27. B

## Answers to odd-numbered problems

1. b.

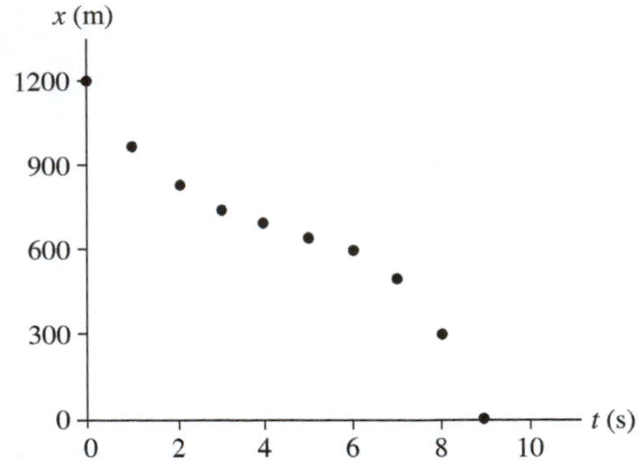


3. a. 8 s b.

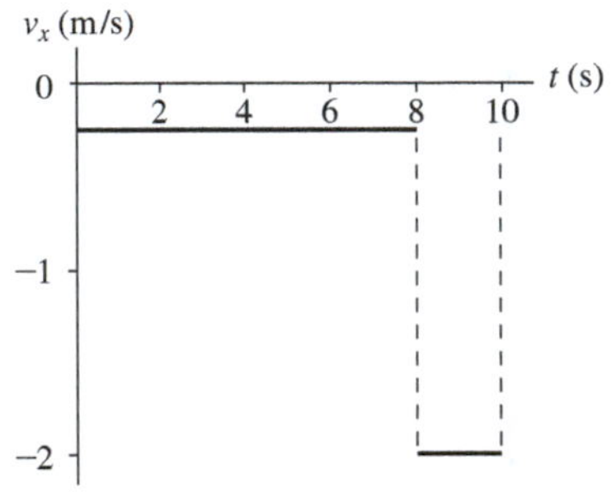


5. a.

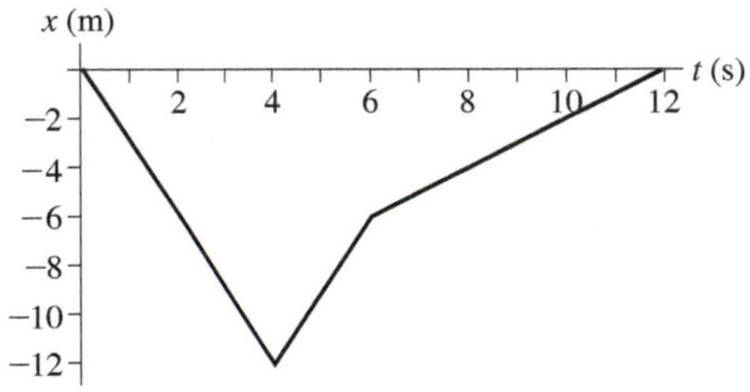


b. 0 m

7.

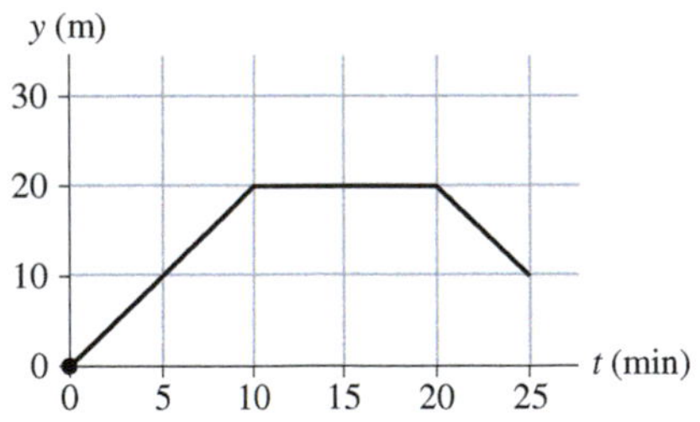


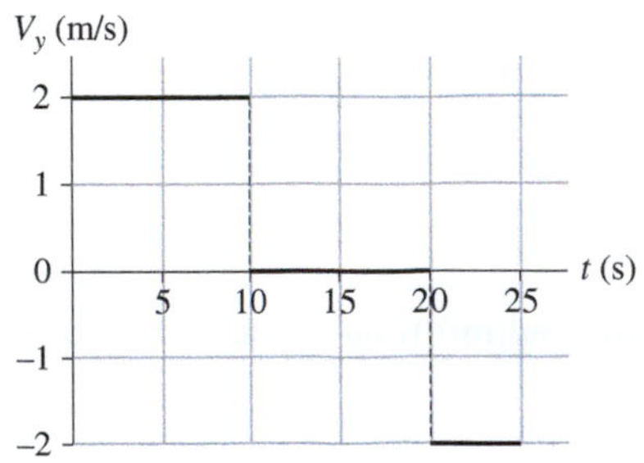


9. 0.43 s
11. a. Beth b. 20 min
13. 4.3 min
15. 12 m/s
17. a.

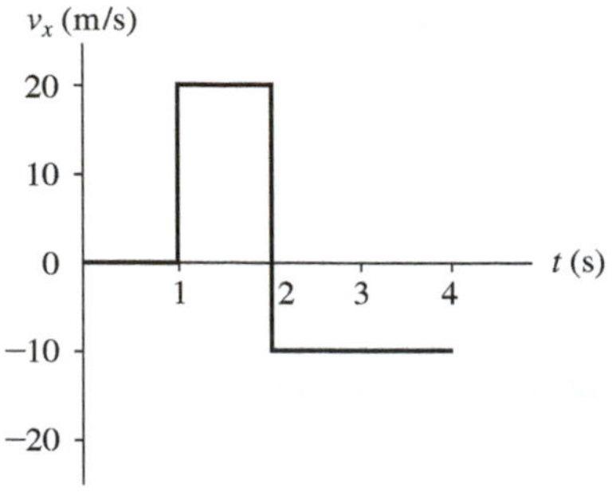


b. Yes, at $t = 2$ s

19. a. 26 m, 28 m, 26 m b. Yes, at $t = 3$ s
21.

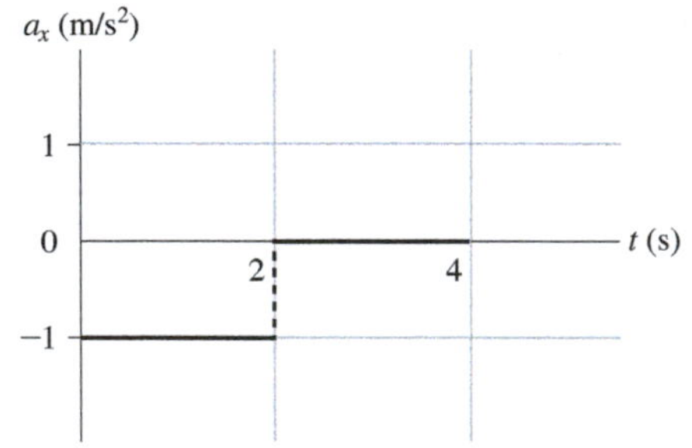


23.

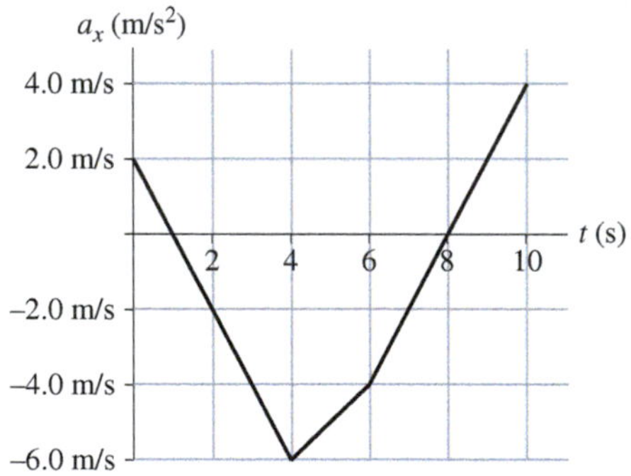


25. 6.1 m/s$^2$, 2.5 m/s$^2$, 1.5 m/s$^2$
27. Trout
29. a. 2.7 m/s$^2$ b. 0.27$g$ c. 134 m, 440 ft
31. a. 15 m/s b. 0.22 m
33. 2.8 m/s$^2$
35. 37 m/s$^2$
37. Yes
39. a. 5 m b. 22 m/s
41. a. 2.8 s b. 31 m
43. 10.0 s
45. 0.18 s
47. 52 m
49. 1.1 s
51. 4.6 m/s
53. 7.7 m/s
55. a. 3.0 s b. 15 m/s c. −31 m/s, −35 m/s
57. 57 mph

59. b.

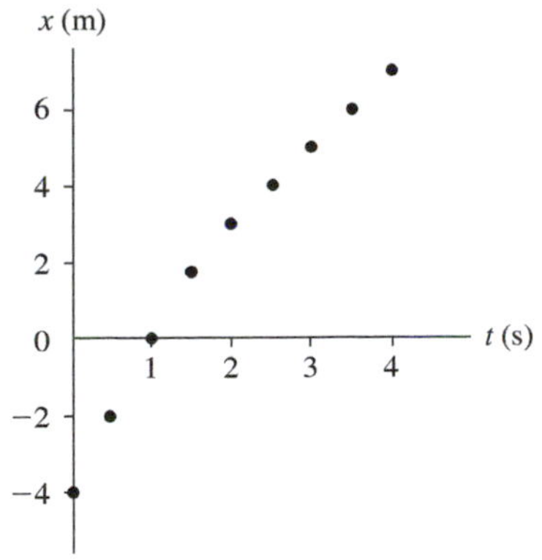


c. 4 m d. 4 m e. 4 m/s f. 2 m/s g. $-2$ m/s$^2$

61. a. 2.3 m/s$^2$, 0.23$g$ b. 35 s c. 4.2 km
63. a. 16 m/s$^2 = 1.6g$ b. 5 m/s$^2 = 0.5g$ c. 2 cm
65. a. $\approx$1.0 cm b. 35 m/s$^2$ c. 0.84 m/s
67. a. 1000 m/s$^2$ b. 1.0 m/s c. 5.1 cm
69. 3.2 s
71. a. 1.7 s b. $-9.9$ m/s
73. 110 m
75. a. 4.1 s b. Equal speeds
77. Man: 10.2 s; horse: 10.7 s; man wins
79. Honda wins by 1.0 s
81. 5.5 m/s$^2$
83. B

## Chapter 3

### Answers to odd-numbered multiple-choice questions

21. C
23. a. B b. E c. C d. D e. B f. A
25. A
27. C
29. C
31. E

### Answers to odd-numbered problems

5. 8.0 m
7. 87 m/s
9. a. $d_x = 71$ m, $d_y = -71$ m b. $v_x = 280$ m/s, $v_y = 100$ m/s
   c. $a_x = 0.0$ m/s$^2$, $a_y = -5.0$ m/s$^2$
11. a. 45 m/s, 63° b. 6,3 m/s$^2$, $-72°$
13. 7.1 km, 7.5 km
15. 25 m
17. 530 m
19. 1.3 s
25. Vector C, to the right
27. a.

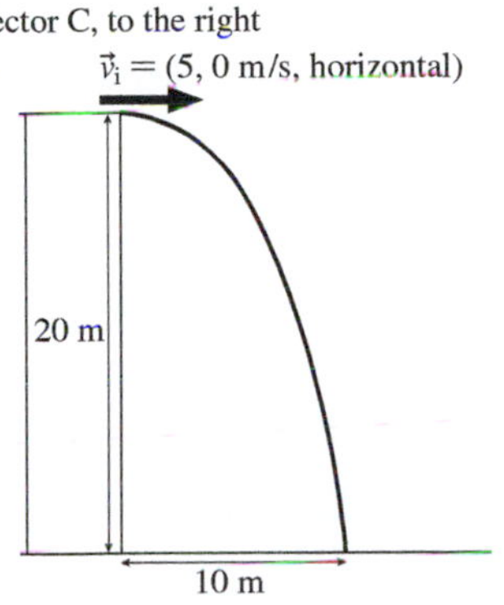


d. 10 m

29. 1.1 m
31. a. 10 m/s b. 50°
33. a. 12 m/s b. 12 m/s
35. 6.8 m
37. 10 m/s$^2$, 1.0$g$
39. $2.7 \times 10^{-3}$ m/s$^2$
41. a. 4.0 m/s$^2$ b. 32 m/s$^2$
43. 27 m
45. Ball 1: 1.5 m/s; ball 2: 15 m/s
47. Ball 1: 15 m/s; ball 2: 5 m/s
49. 2.0 km/h
51. a. (8, 7) c. $\sqrt{113}$, 41°
53. a. $(-2, 2)$ b. 2.8, 135°
55. $2.4 \times 10^3$ ft
57. a. 26 m/s b. 70°
59. a. 63 m b. 7.1 s
61. a. $(v_x)_0 = 2.0$ m/s $(v_y)_0 = 4.0$ m/s
   $(v_x)_1 = 2.0$ m/s $(v_y)_0 = 2.0$ m/s
   $(v_x)_2 = 2.0$ m/s $(v_y)_0 = 0.0$ m/s
   $(v_x)_3 = 2.0$ m/s $(v_y)_0 = -2.0$ m/s

   b. 2.0 m/s$^2$ c. 63°
63. a. 6.0 m/s b. Yes
65. 6.0 m
67. 0.82 m
69. Yes, by 1.0 m
71. 16° from the vertical
73. 6.1 m
75. 41° south of west
77. a. 7.2° south of east b. 2.5 h
79. 30 s
81. 5.0 m/s
83. a. 2.9 m/s$^2$, 0.30$g$ b. 57 mph
85. B
87. C

## Chapter 4

### Answers to odd-numbered multiple-choice questions

21. C
23. D
25. D
27. C
29. C
31. B

### Answers to odd-numbered problems

1. First is rear-end; second is head-on
5.

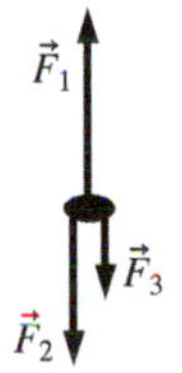


7. Gravity by ground
9. Weight, normal force by ground, kinetic friction force by ground
11. Weight, normal force by slope, kinetic friction force by slope
13. $m_1 = 0.080$ kg and $m_3 = 0.50$ kg
15. 3.0 m/s$^2$
17. a. 16 m/s$^2$ b. 4.0 m/s$^2$ c. 8.0 m/s$^2$ d. 32 m/s$^2$
19. 0.71$g$
21. 0.25 kg
23. a. No b. 10 m/s$^2$ to the left
25. 6.6 N
27. $1.1 \times 10^3$ N

33.

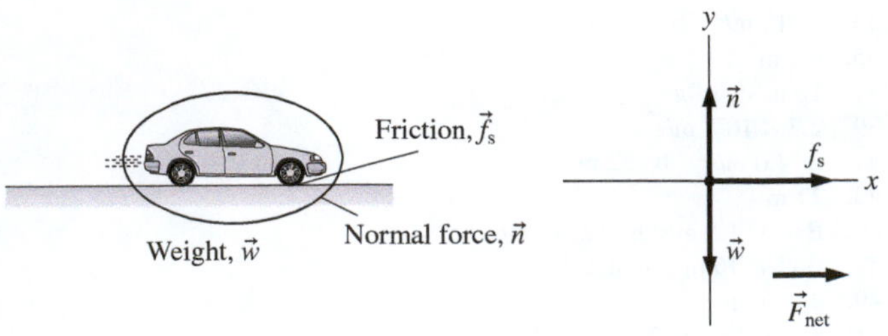


35.

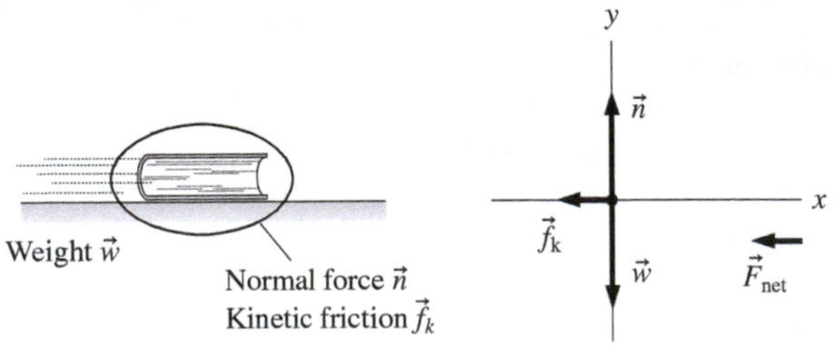


37.

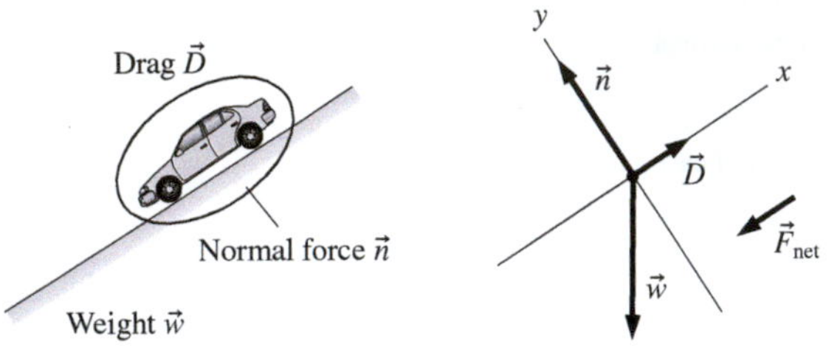


39.

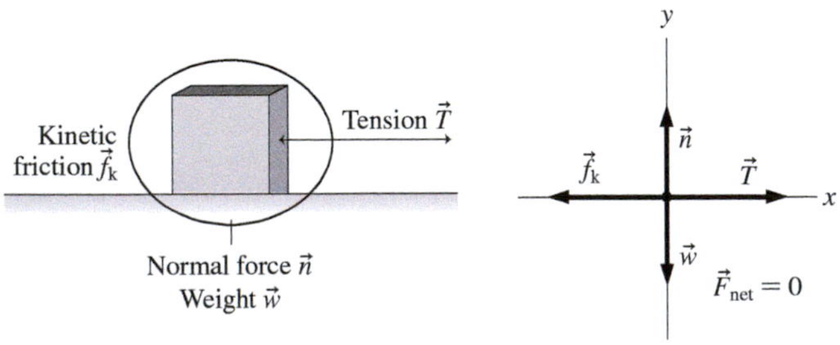


41.

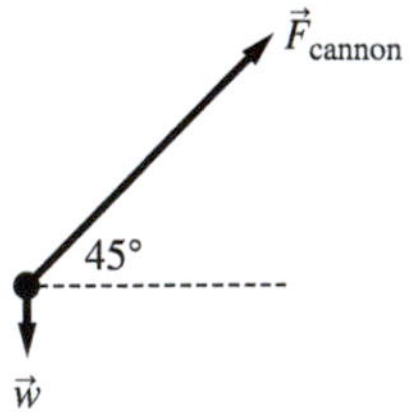


43.

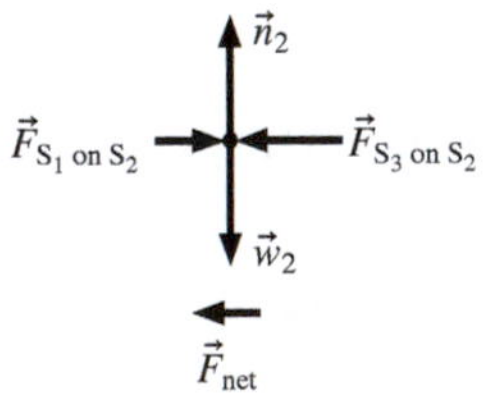


45. Normal force of road on car and of car on road; friction force of road on car and of car on road

47. **Motion diagram**

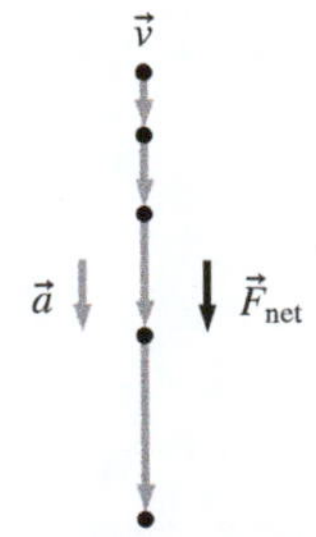


49. **Motion diagram**

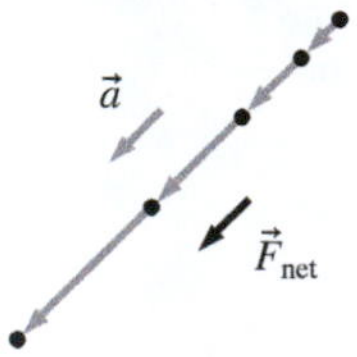


51.

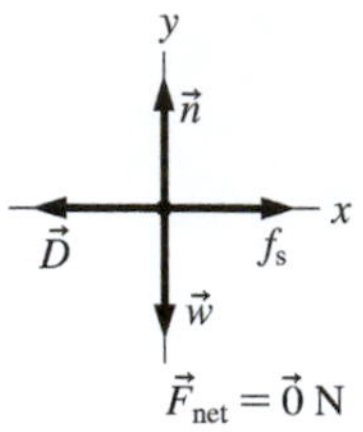


53.

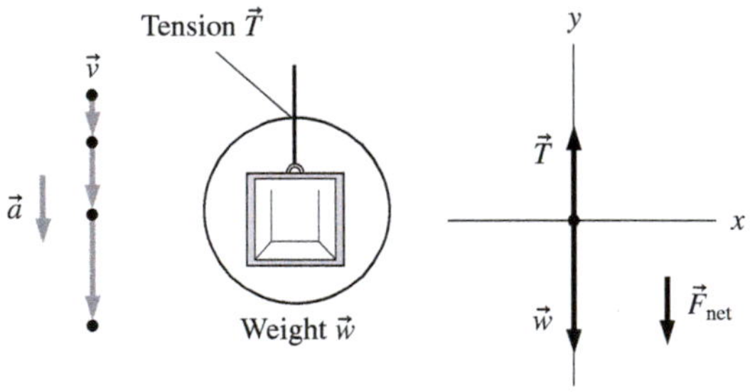


55.

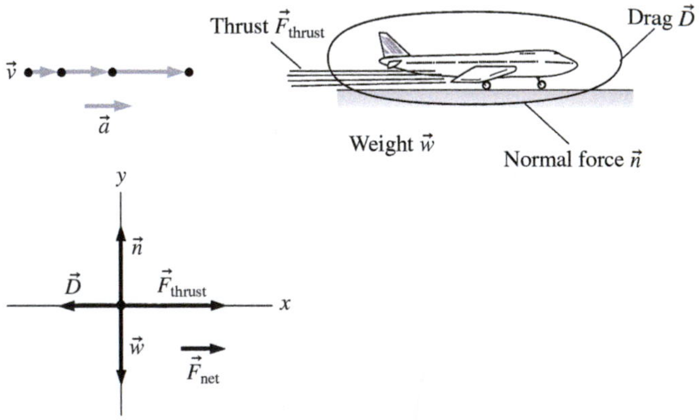

57.

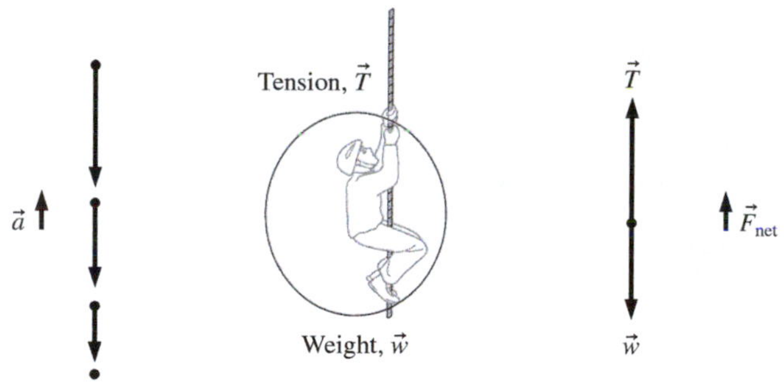


59.

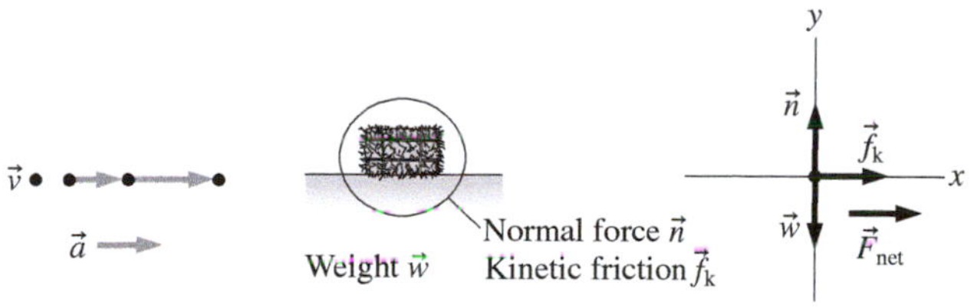


61.

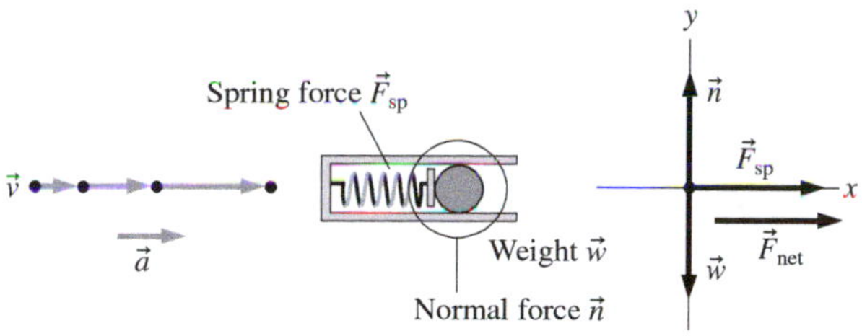


63.

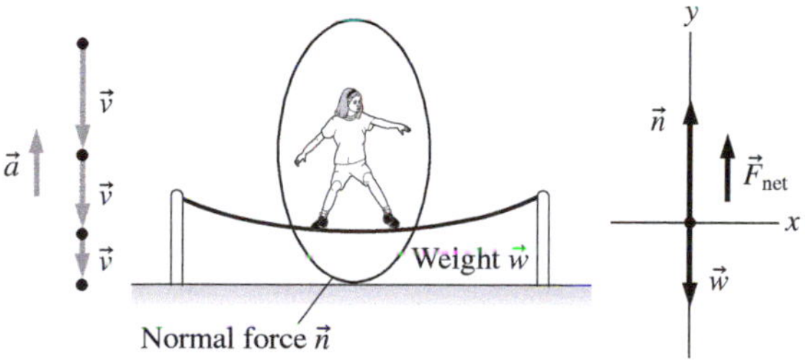


65.

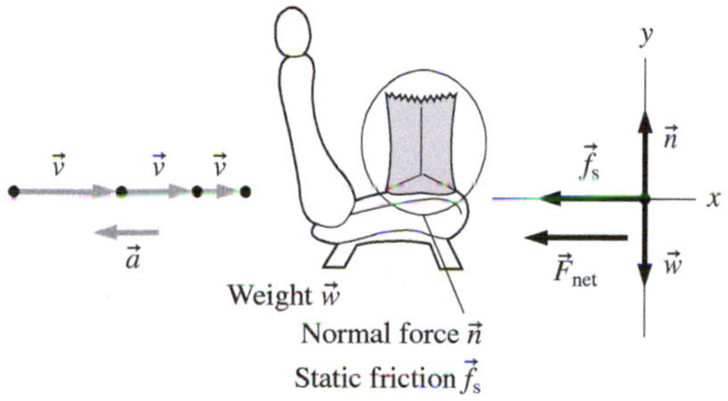


67. a. Ground on greyhound b. 320 N c. 80 m
69. b. 160 N c. 1/5
71. 0.084 N
73. Greater on the way up
75. C
77. C

# Chapter 5

## Answers to odd-numbered multiple-choice questions

23. B
25. D
27. a. B b. D
29. C
31. C
33. D

## Answers to odd-numbered problems

1. $T_1 = 87$ N, $T_2 = 50$ N
3. $T_l = 3.6 \times 10^3$ N, $T_r = 3.9 \times 10^3$ N
5. 49 N
7. $T_l = 4.1 \times 10^2$ N, $T_r = 5.0 \times 10^2$ N
9. 2.4 m/s
11. $a_x = 1.0$ m/s$^2$, $a_y = 0.0$ m/s$^2$
13. $1.2 \times 10^2$ N
15. 310 N
17. $3.8 \times 10^2$ N
19. a. 590 N b. 740 N c. 590 N
21. a. 780 N b. 1100 N
23. 1000 N, 740 N, 590 N
25. b. 180 N
27. 0.25
29. 140 m
31. a.

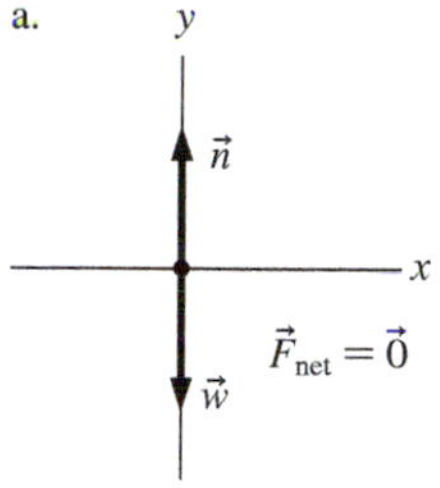


b.

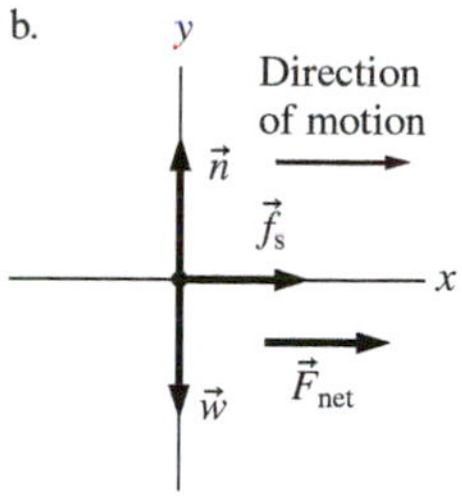


33. $\approx 500$ s, $\approx 3000$ m
35. $2.4 \times 10^4$ N
37. $4.6 \times 10^2$ s $\approx 7.7$ min
39. 7.4 m/s$^2$
41. a. 3000 N b. 3000 N
43. a. 6.0 N b. 10 N
45. $3.9 \times 10^2$ N
47. a. 1.0 N b. 50 N
49. a. 530 N b. 5300 N
51. $F_{net}(1\text{ s}) = 8$ N, $F_{net}(4\text{ s}) = 0$ N, $F_{net}(7\text{ s}) = -12$ N
53. a. 490 N b. 740 N
55. a. 490 N b. 240 N
57. 3200 N, 5 times actual weight
59. a. 6800 N b. $1.4 \times 10^6$ N c. 11 times, 2300 times
61. a. 5.2 m/s$^2$ b. 1000 kg
63. 160 N
65. 60 m
67. $\approx 36$ mph
69. a. $-10.2$ m/s$^2$ b. 35 m/s
71. a. 11 kg b. 57 kg
73. 0.98 kg
75. Stay at rest
77. $T_1 = 17$ N, $T_2 = 27$ N
79. 14 N
81. a. 220 N b. 440 N
83. Down at 0.93 m/s$^2$
85. C
87. B

## Chapter 6

### Answers to odd-numbered multiple-choice questions

21. D
23. E
25. B
27. A
29. A
31. D

### Answers to odd-numbered problems

1. 3.9 m/s
3. a. 0.56 rev/s b. 1.8 s
5. a. 2.3 Hz b. 31 m/s c. $4.5 \times 10^2$ m/s$^2$
7. a. $3.0 \times 10^4$ m/s b. $6.0 \times 10^{-3}$ m/s$^2$
9. a. 5.1$g$ b. 57%
11. 0.56 Hz
13. $T_3 > T_1 = T_4 > T_2$
15. 1.8
17. 9400 N, toward center, static friction
19. a. 1700 m/s$^2$ b. 240 N
21. 13 m/s
23. 270 N
25. $4.7 \times 10^2$ N
27. 20 m/s
29. 78%, or just over 3/4
31. a. $1.8 \times 10^4$ m/s$^2$ b. $4.4 \times 10^3$ m/s$^2$
33. 99 min
35. 1/2
37. $6.0 \times 10^{-4}$
39. 3.9 m/s$^2$
41. a. $3.53 \times 10^{22}$ N b. $1.99 \times 10^{29}$ N c. 0.564%
43. a. 3.77 m/s$^2$ b. 25.9 m/s$^2$
45. 4.0
47. 280 y
49. 92 min, 7700 m/s
51. 0.039 au
53. 11 h
57. $9.5 \times 10^3$ N
59. 12%
61. 1.1
63. 7.4 m/s
65. No
67. 5.4 m/s
69. 22 m/s
71. 2400 m
73. a. $3.0 \times 10^{24}$ kg b. 0.89 m/s$^2$
75. (12 cm, 0 cm)
77. $6.5 \times 10^{23}$ kg
79. 0.48 m/s
81. C
83. A

## Chapter 7

### Answers to odd-numbered multiple-choice questions

19. C
21. D
23. A
25. A
27. D
29. D

### Answers to odd-numbered problems

1. a. $\theta = \pi/2$ b. $\theta = 0$ c. $\theta = 4\pi/3$
3. $1.7 \times 10^{-3}$ rad/s
5. a. 1.3 rad, 72° b. $3.9 \times 10^{-5}$ rad/s
7. 3.0 rad
9. a. 12.5 rad b. −2.5 rad/s
11. a. 0.105 rad/s b. 0.00105 m/s
13. 11 m/s
15. 960 rev
17. a. 160 rad/s$^2$ b. 50 rev
19. $\tau_1 < \tau_2 = \tau_3 < \tau_4$
21. −0.20 N · m
23. 28.3 N
25. 7.5 N · m
27. 5.5 N · m
29. 2.5 N · m
31. −4.9 N · m
33. a. 34 N · m b. 24 N · m
35. $x = 0.0333$ m, $y = 0.0333$ m
37. 2.93 mm
39. $7.2 \times 10^{-7}$ kg · m$^2$
41. 4.5 cm
43. 1.8 kg
45. $6.0 \times 10^{-3}$ N · m
47. 8.0 N · m
49. 0.047 N · m
51. 17 rad/s$^2$
53. 0.11 N · m
55. 0.50 s
57. a. 14 rad/s b. 11 m/s c. 7.9 m/s
59. 2.0 m
61. b.

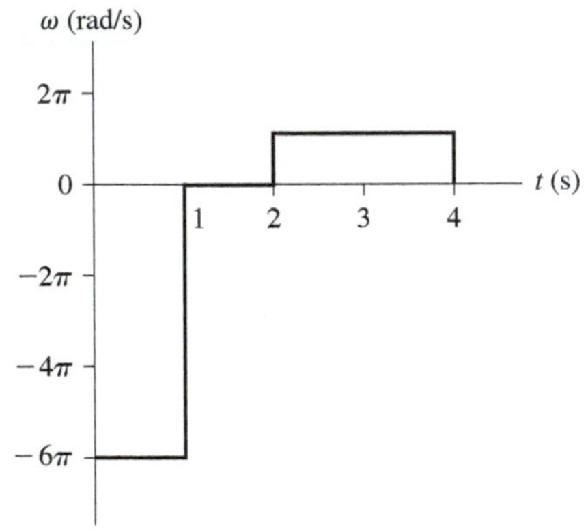


63. 55 rotations
65. −0.94 N · m
67. 7.5 cm
69. a. 1.4 kg · m$^2$ b. Increase
71. 3.5 rad/s$^2$
73. 180 s
75. −0.28 N · m
77. 2.2 N
79. B
81. B
83. B

## Chapter 8

### Answers to odd-numbered multiple-choice questions

19. A
21. B
23. D
25. B
27. B

### Answers to odd-numbered problems

1. Right 470 N; left 160 N
3. 15 cm
5. 140 N
7. 160 N
9. 1.4 m
11. 590 N
13. 98 N
15. 16°
17. 50°
19. 0.93 m
21. 0.96 N
23. 0.30 N/m
25. a. 390 N/m b. 18 cm
27. a. 0.13 m b. 0.08 m
29. 98 N/m
31. a. 2 mm b. 0.25 mm
33. 2.0 kg
35. 16 cm
37. 0.0078%
39. 1.0 mm
41. a. 3.8 mN b. 9 mm
43. 3.3%
45. 78 N
47. 1100 N
49. a. 340 N b. 8.6
51. $F_1 = 750$ N, $F_2 = 1000$ N
53. 350 N
55. 1.0 m
57. a. 49 N b. 1500 N/m c. 3.4 cm
59. a. 1.0 cm b. 4.5 N
61. a. 2300 N b. 1.0 m
63. 0.25 = 25%
65. a. $4.8 \times 10^{-2}$ m$^2$ b. 70,000 kg c. 0.14
67. a. 206 N b. 84 N
69. B
71. C

## Part I Problems

### Answers to odd-numbered problems

1. A
3. A
5. C
7. B
9. D
11. A
13. B
15. D
17. B
19. A
21. b. 2.8 N

## Chapter 9

### Answers to odd-numbered multiple-choice questions

23. C
25. A
27. A
29. C

### Answers to odd-numbered problems

1. 15 m/s
3. 2.6 kg · m/s
5. 60 N
7. 1500 N
9. 0.50 s
11. 960 N
13. −110 N
15. a. −19 kN b. −280 kN
17. a. 5.08 kg · m/s b. 3400 N
19. 0.205 m/s
21. 0.31 m/s
23. 1.4 m/s
25. 0.47 m/s
27. 4.8 m/s
29. 5.5 cm/s
31. 2.3 m/s
33. 1.0 kg
35. 1.7 m/s at 45° north of east
37. (−2 kg · m/s, 4 kg · m/s)
39. 14 m/s at 45° north of east
41. 0.025 kg · m$^2$/s, into the page
43. 1.3 rev/s
45. $2.8 \times 10^{34}$ kg · m$^2$/s
47. 0.20 s
49. 930 N
51. 3.8 m/s
53. 12 kg · m/s, left; 15° above the horizontal
55. a. 0.59 m/s d. Yes
57. 1.1 m/s
59. The quarterback
61. 6.3 m
63. 1.7 m/s, 55° below the $x$-axis
65. a. 8.0 m/s b. 100$g$
67. 20 m/s downward
69. 2.0 m/s
71. $1.5 \times 10^7$ m/s, forward
73. 0.85 m/s, 72° below the $x$-axis
75. 400 rpm
77. 0
79. 18 rev/s, clockwise
81. B
83. C

## Chapter 10

### Answers to odd-numbered multiple-choice questions

25. C
27. C
29. C
31. C

### Answers to odd-numbered problems

1. a. −30 J b. 30 J
3. Rope 1: 0.919 kJ; rope 2: 0.579 kJ
5. a. 0 J b. −43 J c. 43 J
7. −220 J
9. 110 km/h
11. a. 14 m/s b. Factor of 4
13. 4.0 m/s
15. 0.86 kJ
17. 0.049 J
19. 1.9 kg · m$^2$
21. 6.2 m
23. 14 kJ
25. 10 J
27. 9.7 J
29. 16.5 kJ
31. 6.2 kJ
33. 110 N

35. a. 5.1 m b. 10 m/s c. 22 m/s
37. 4.5 m/s
39. a. Yes b. 14 m/s
41. 3.1 kJ
43. 26 m
45. 17 kJ
47. 24 m/s
49. a. Kinetic energy to gravitational potential energy b. 1.0 m c. 4.4 m/s
51. a. Right b. 17.3 m/s c. $x = 1.0$ m and 6.0 m
53. 0.08 nm
55. 0.86 m/s and 2.9 m/s
57. 1/2
59. a. $1.8 \times 10^2$ J b. 59 W
61. a. $9.5 \times 10^5$ J b. $2.0 \times 10^4$ W
63. 45 kW
65. 34 min
67. a. 500 N b. $1.7 \times 10^4$ W
69. $2.0 \times 10^4$ W
71. 91 kJ
73. a. 20 J b. 0.74 J c. 13 J d. 6.3 J
75. a. 2.0 mJ b. 1/2
77. 4.2 m/s
79. 5.8 m
81. a. 0.34 m
83. a. $\sqrt{\dfrac{(m+M)\,kd^2}{m^2}}$ b. $2.0 \times 10^2$ m/s c. 99.8%
85. a. 1.7 m/s b. 2/3
87. a. 2.6 m/s b. 33 cm
89. 1.3 m/s
91. D
93. B
95. B
97. B
99. B

## Chapter 11

### Answers to odd-numbered multiple-choice questions

33. D
35. C
37. D

### Answers to odd-numbered problems

1. 1.7 MJ
3. 3.3%
5. 8 LEDs, 8.0 W
7. 2060 Cal
9. 370,000 J = 89,000 cal = 89 Cal
11. 490 Cal
13. ≈ 1.4 km
15. 2.0 h
17. 710 m
19. 1800 reps
21. a. 196 J b. 15,680 J c. 0.0094 donuts
23. 1200 km
25. $2.9 \times 10^{-2}$ g
27. −290°C, −452°F
29. 1.3
31. 700 J from the system
33. 200 J
35. 10 J, increases
37. 0.40
39. 5000 J
41. 67%
43. 47%, 35%
45. 1.5
47. a. 200 J b. 250 J
49. 7.3
51. 6.3 kW
53. a. Refrigerator c b. Refrigerator b
55. 1 slice
57. a. 670 Cal b. 1200 W
59. 830 W
61. a. 80 W b. 5 people
63. a. 950 Cal b. 46 W, 54 W less
65. 230 K
67. a. 110 m b. 32%
69. 92
71. a. 86 MJ b. 2.8 $mi^2$
73. A
75. A
77. D
79. A

## Part II Problems

### Answers to odd-numbered problems

1. A
3. B
5. D
7. C
9. A
11. C
13. D
15. B
17. C
19. B
21. D
23. 0.22 m/s to the west
25. a. 3.0 m/s b. 120 N c. 60 J

## Chapter 12

### Answers to odd-numbered multiple-choice questions

31. D
33. B
35. A
37. B

### Answers to odd-numbered problems

1. Carbon
3. $3.5 \times 10^{24}$
5. 0.024 $m^3$
7. 313°C
9. −9.3°C
11. 11°C
13. 1.1 N
15. 47.9 psi
17. 350 m/s
19. $8.9 \times 10^8$ molecules
21. 700 L
23. 6.0°C
25. $5.5 \times 10^3$ $m^3$
27. 1.6 atm
29. a. 9500 kPa
31. a. 98 $cm^3$
33. a. Isochoric b. 910 K, 300 K
35. a. Isobaric b. 120°C c. 0.0094 mol
37. a. 2.5 $cm^3$ b. Water to the bubble
39. 35 psi

41. 32 psi
43. 4.7 m
45. 0.5 cm
47. 0.83%
49. 13 kJ
51. a. 36°C b. 3000 J
53. 2300 W
55. 71 min
57. 12 W, 250 Cal/d
59. 14°F
61. 76°C
63. 28°C
65. 0.218 kg
67. 610 g
69. a. 91 J b. 140°C
71. 9.0°C
73. 830 W
75. 6.0 W
77. $4.7 \times 10^{-4}$ m$^2$
79. 300 W
81. 8 ms
83. No
85. 60 J
87. a. 80 J b. 1.3 W
89. 0.12°C
91. $2.4 \times 10^6$ kg/min
93. 0.54 kg/h
95. 3.8 kg/h
97. a. 83 J/(kg · K) b. $2.0 \times 10^5$ J/kg
99. 60 J
101. a. 810 J b. −490 J c. 0 J
103. a. 320 J b. 0.13
105. a. 84 W b. 180 W c. Radiation d. Chilly
107. A
109. B
111. B

## Chapter 13

### Answers to odd-numbered multiple-choice questions

33. A
35. D
37. C
39. A
41. B
43. D

### Answers to odd-numbered problems

1. 1200 kg/m$^3$
3. 14.1 kg
5. 11 kg/m$^3$
7. 1100 atm
9. 3.8 kPa
11. 0.10 atm
13. 3.2 km
15. 88,000 Pa
17. a. $1.1 \times 10^5$ Pa b. 4400 Pa for both A–B and A–C
19. 3.68 mm
21. 0.82 m
23. a. 3.0 cm b. 3.2 cm
25. 1.9 N
27. 1065 kg/m$^3$
29. 63 kg
31. 0.96 N
33. 3.2 m/s
35. a. 1.0 m/s, 16 m/s b. $3.1 \times 10^{-4}$ m$^3$/s
37. 110 kPa
39. 3.0 m/s
41. $1.8 \times 10^5$ Pa
43. 260 Pa
45. 3200 Pa
47. 3.1 kPa
49. a. $2.8 \times 10^{-4}$ m/s b. $3.5 \times 10^2$ Pa
51. 17 Pa
53. $4.8 \times 10^{23}$ atoms
55. 27 cm
57. 750 kg/m$^3$
59. 44 N
61. a. 0.38 N b. 20 m/s
63. $2 \times 10^{-3}$ m/s
65. a. $p_2 = p_{atmos}$ b. 4.6 m
67. 4.1 cm
69. $8.2 \times 10^5$ Pa
71. $8.8 \times 10^{-5}$ m$^3$/s
73. B
75. C

## Part III Problems

### Answers to odd-numbered problems

1. A
3. B
5. D
7. B
9. C
11. A
13. C
15. D
17. A
19. 300 N
21. a. 0.071 mol b. $6.7 \times 10^{-4}$ m$^3$ c. 0.019 mol

## Chapter 14

### Answers to odd-numbered multiple-choice questions

23. a. B b. C c. C
25. a. B b. B c. B d. E e. E
27. B
29. B

### Answers to odd-numbered problems

1. 2.3 ms
3. 0.80 s, 1.3 Hz
5. 40 N
7. a. 13 cm b. 9.0 cm
9. a. 20 cm b. 0.25 Hz
11. $2.0 \times 10^{-2}$ m
13. a. 0.2 m/s$^2$ b. 1/50
15. a. 0.67 m/s b. 5.2 m/s$^2$
17. 1.6 mm
19. a. $U = 1/4E$ $K = 3/4E$ b. $A/\sqrt{2}$
21. a. 2.83 s b. 1.41 s c. 2.00 s d. 1.41 s
23. a. 2.0 cm b. 0.63 s c. 5.0 N/m d. 20 cm/s e. $1.0 \times 10^{-3}$ J f. 15 cm/s
25. a. 0.50 s b. 5.5 cm c. 70 cm/s d. 0.049 J
27. a. 4.00 s b. 5.66 s c. 2.83 s d. 4.00 s
29. 35.7 cm
31. 3.67 m/s$^2$
33. $8.7 \times 10^{-2}$ kg · m$^2$
35. 1.9 s
37. a. 1.24 s b. 97 steps/min

39. 10.0 s
41.

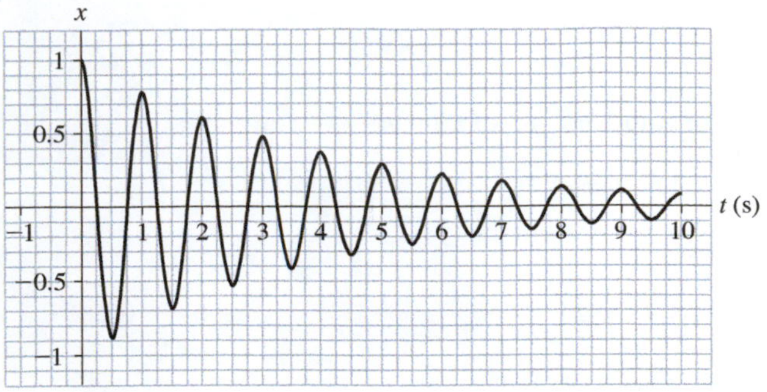


43. a. 5.0 kHz b. $4.8 \times 10^{-3}$ s
45. 2 Hz, 0.25 s
47. 2.8 s
49. 250 N/m
51. 1.1 m/s
53. a. 0.169 kg b. 0.565 m/s
55. 2.8 Hz
57. 5.86 $m/s^2$
59. a. 0.25 Hz, 3.0 s b. 6.0s, 1.5 s c. 9/4
61. a. 6.40 cm b. 28.3 cm/s
63. 1.8 Hz
65. 11 cm. 1.7 s
67. a. 7.9 m/s b. $8.0 \times 10^2$ N
69. a. 1.02 m/s b. 3.2 times
71. a. 376 s b. 1.06%
73. 73 cm
75. 240 oscillations
77. D
79. A
81. A
83. C

## Chapter 15

### Answers to odd-numbered multiple-choice questions

21. D
23. D
25. D

### Answers to odd-numbered problems

1. 280 m/s
3. 0.076 s
5. Neon
7. 6.0 s
9.

History graph at $x = 6$ m

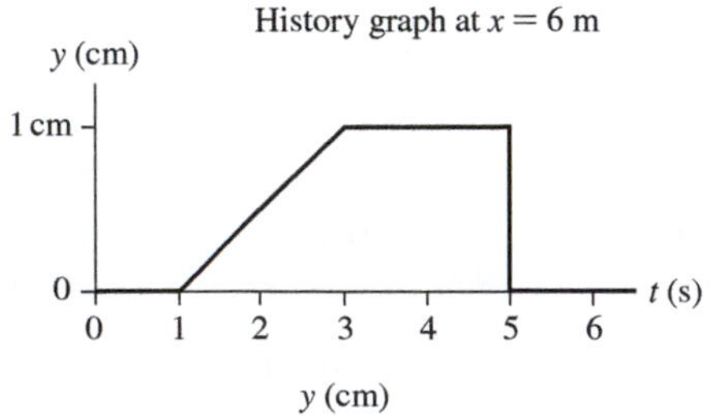


11. 1.0 cm/s
13. y (cm)

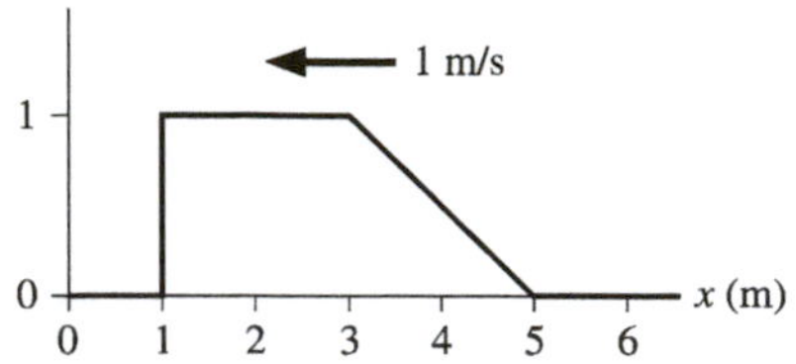


Snapshot graph at $t = 0$ s

15. 10 m/s
17. a. 8.6 mm b. 15 ms
19. a., b.

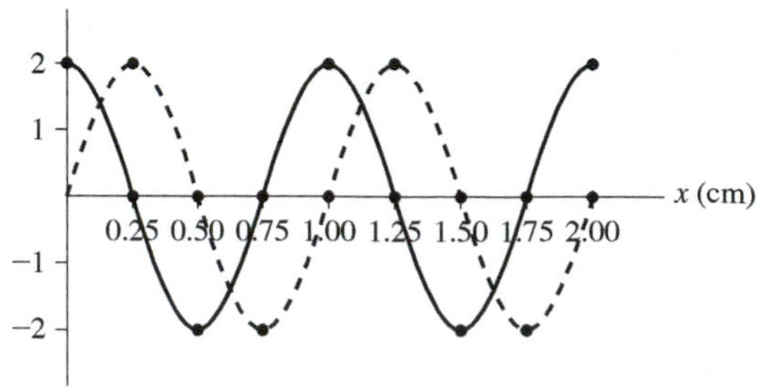


c. 2.0 m/s
21. 6.0 cm, 1.7 Hz, 1.2 m
23. 8.0 ms
25. 15 cm
27. a. $6.7 \times 10^{14}$ Hz b. $4.6 \times 10^{14}$ Hz
29. 0.96 s
31. 55 pJ
33. 600 kJ
35. 11 W
37. a. $6.7 \times 10^1$ W b. $8.5 \times 10^{10}$ $W/m^2$
39. $2.8 \times 10^{24}$ $W/m^2$
41. 25 $pW/m^2$, 14 dB
43. a. 3.2 $mW/m^2$ b. 95 dB
45. 1.4 $mW/m^2$
47. 103 dB
49. 49 m
51. a. 650 Hz b. 560 Hz
53. 16 m/s away
55. a. 431 Hz b. 429 Hz
57. 1.3 kHz
59. 589 m/s
61. 1.0 m/s
63. a. 29 s b. Longitudinal c. 6.4 km d. 5.1 cm
65. 11 m/s
67. a. 9 ms b. 0.47 Hz
69. $5.8 \times 10^3$ m/s, 2230 Hz
71. $y(x, t) = (5.0\text{ cm}) \cos\left[2\pi\left(\frac{x}{50\text{ cm}} + \frac{t}{0.125\text{ s}}\right)\right]$
73. $3.2 \times 10^{-6}$
75. 0.78°C
77. 94 dB
79. 4.7 m
81. 620 Hz, 580 Hz
83. D
85. B

## Chapter 16

### Answers to odd-numbered multiple-choice questions

19. A
21. A
23. A
25. D

### Answers to odd-numbered problems

1.

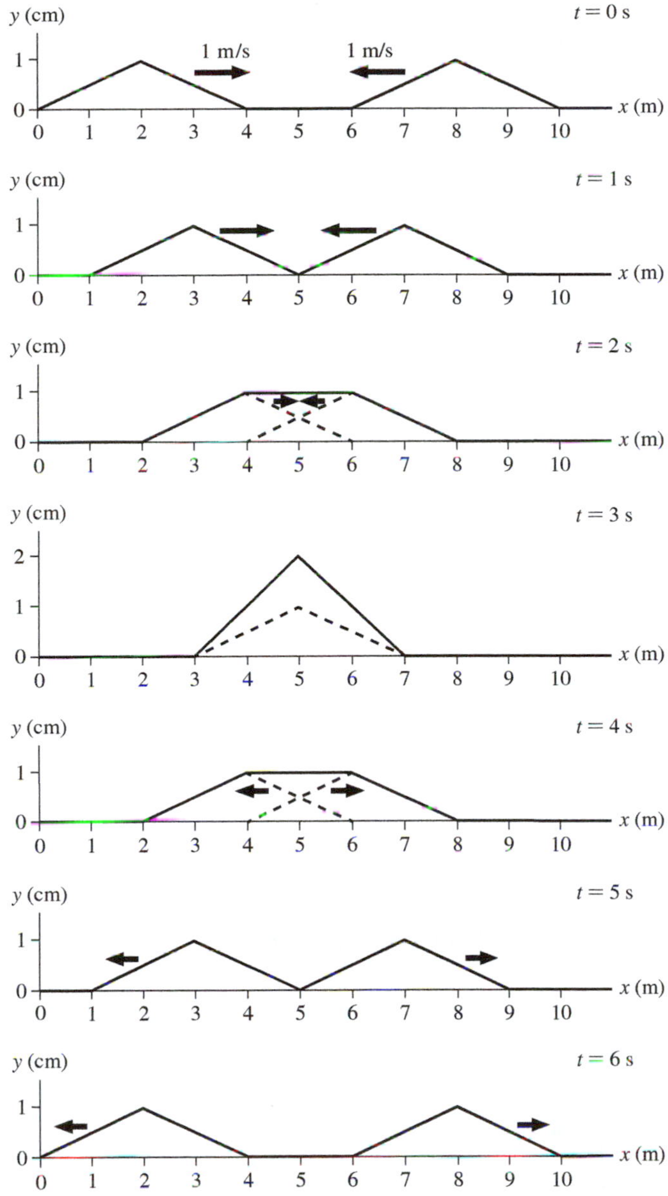


3.

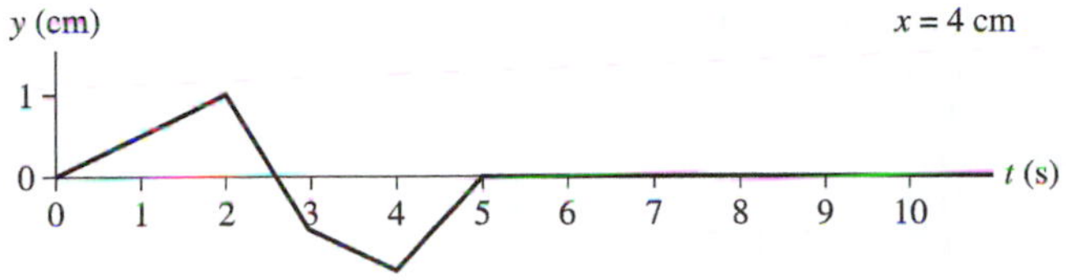


5.

| $t$(s) | $y$(cm) |
|---|---|
| 0 | 0 |
| 1 | 0 |
| 2 | 0 |
| 3 | 0 |
| 4 | 0 |
| 5 | 0 |
| 6 | 0 |

7. 5.0 s
9. 60 Hz
11. 53 m/s
13. a. 4.80 m, 2.40m, 1.60 m b. 75.0 Hz
15. a. 700 Ha b. 56 N
17. 2180 N
19. 13 m, 5.9 s
21. a. $2.6 \times 10^2$ Hz b. 1.3 kHz
23. 8.8 Hz
25. 928 Hz
27. a. 5.67 Hz b. 22.7 Hz c. The frequency will decrease.
29. 44 cm
31. 6700 Hz
33. 290 Hz, 0.30 m
35. 580 Hz, 4900 Hz
37. a. 20 cm b. 40 cm
39. 515 Hz
41. 1.0 m, 3.0 m, 5.0 m
43. Perfectly destructive
45. 438 Hz
47. 203 Hz
49. 527 Hz
51. Yes
53. 160 Hz, 320 Hz, 480 Hz, . . .
55. $2.4 \times 10^{-6}$ N
57. 81 Hz
59. 408 Hz
61. 1210 Hz
63. 1400 Hz
65. 2.98 m, 5.62 m, 17.88 m
67. 2 Hz
69. 1750 Hz
71. 0.20 m/s
73. B
75. B

## Part IV Problems

### Answers to odd-numbered problems

1. $v_{light} > v_{earthquake} > v_{sound} > v_{tsunami}$
3. D
5. B
7. B
9. C
11. B
13. C
15. C
17. C
19. C
21. 500

## Chapter 17

### Answers to odd-numbered multiple-choice questions

23. B
25. C
27. B
29. D
31. C

### Answers to odd-numbered problems

1. $1.5 \times 10^{-11}$ s
3. 0.40 ns
5. 459 nm
7. 1.1°, 2.4 cm
9. 0.050 rad
11. 1.2 mm
13. 500 nm
15. 746 nm
17. 3.2°, 6.3°
19. 17° to 32°
21. a. 1.3m b. 7
23. 3, 15°, 31°, 50°
25. 200 nm
27. 121 nm
29. 906 nm
31. 410 nm, 690 nm, purple
33. 0.200 mm
35. 4.0 mm
37. 610 nm
39. 4.9 mm
41. 78 cm
43. 0.01525 rad, 0.87°
45. 6.4 cm
47. 500 nm
49. 14.5 cm
51. 1.0 nm
53. 670 lines/mm
55. 16°
57. 477 nm
59. $\approx 1.4$
61. 0.10 mm
63. a. Double b. 0.15 mm
65. 0.1 mm
67. a. Diffraction b. 0.044° c. 4.6 mm d. 1.5 m
69. D
71. A

## Chapter 18

### Answers to odd-numbered multiple-choice questions

21. D
23. B
25. C
27. B
29. A

### Answers to odd-numbered problems

1. 11 ft
3. 8.00 cm
5. 433 cm
7. 8.7 times
9. 1.0 m
11. 180°
13. 1.4
15. 35°
17. 4.0 m
19. 1.3 cm
21. 1.30
23. 61.3°
25. 3.2 cm
27. 23 cm
29. a. Closer b. Farther
31. $s' = -15$ cm, upright and virtual
33. $s' = -6.7$ cm, upright and virtual
35. $s' = -6$ cm, upright and virtual
37. $s' = -40$ cm, upright and virtual
39. $s' = 30$ cm, inverted and real
41. Concave, $f = 2.4$ cm
43. $f = -36$ cm
45. $s' = -15$ cm, $h' = 1.5$ cm
47. $s' = 50$ cm, $h' = 0.67$ cm
49. $s' = -20$ cm, $h' = 0.33$ cm
51. $s' = -16$ cm, $h' = 1.1$ cm
53. $s' = 56$ cm, $h' = 3.7$ cm
55. 70 cm
57. 18 cm
59. 1.1 cm
61. 0.20 m/s
63. 1.2 m
65. 82.8°
67. 4.7 m
69. 26°
71. 42°
73. 110 cm
75. Raised by 0.10 mm
77. $s' = -40$ cm, $h' = 2.5$ cm
79. $s' = 160$ cm, $h' = 0.50$ cm; $s' = 40$ cm, $h' = 8.0$ cm
81. 67 cm
83. A
85. B

## Chapter 19

### Answers to odd-numbered multiple-choice questions

19. C
21. C
23. C
25. B
27. B

### Answers to odd-numbered problems

1. 4.4 cm
3. 0.29 m
5. 0.58 cm away from the detector
7. 17 mm
9. $-1.0$ cm
11. 11 cm
13. 1.2 cm
15. 4.0 D
17. 2.0 D
19. $-0.33$ D
21. $-4.3$ D
23. Warren > Carol > Bernie = Maria > Janet
25. 2.6 cm
27. 9 D
29. 2.1 m
31. 5.0 cm
33. a. 1.20 cm b. 1.67 cm
35. 11 mm
37. 2.5 cm
39. 4.5 cm in front of second lens, $h' = -1.0$ cm, opposite that of the object
41. 12

43. 3.6°
45. a. 14° b. 1.67 cm
47. 0.28 mm, violet
49. a. 0.068 radian b. 0.17 radian c. Yes
51. a. 380 nm b. 270 nm
53. 4.9 mm
55. 3.1
57. 0.70 cm
59. b. $h'' = 2.0$ cm, $s'' = -20$ cm
61. 5.0 mm
63. 2.0 cm, 98 cm
65. 122.45 cm, 0.45 cm away from the objective
67. 6.7 cm
69. 2.0 m
71. $2.8 \times 10^{-7}$ rad
73. 15 km
75. No
77. A
79. B

## Part V Problems

### Answers to odd-numbered problems

1. C
3. A
5. A
7. B
9. B
11. C
13. D
15. C
17. Aberrations are reduced.
19. 7.2 mm, 0.069 W/m$^2$

## Chapter 20

### Answers to odd-numbered multiple-choice questions

27. A
29. D
31. D
33. A
35. A
37. C

### Answers to odd-numbered problems

1. a. Electrons have been removed. b. $3.1 \times 10^{10}$
3. a. Electrons have been added. b. $1.25 \times 10^{11}$
5. a. Electrons have been added to the sphere. b. $3.1 \times 10^{10}$
7. −80 nC
9. A = $3q/8$, B = $-q/4$, C = $3q/8$
11. a. $9.0 \times 10^{-3}$ N b. $9.0 \times 10^{-3}$ m/s$^2$
13. 6.6 m/s$^2$
15. $q_A = 10$ nC or 15 nC; $q_B = 10$ nC or 15 nC
17. $9.0 \times 10^{-5}$ N in the $-x$-direction
19. $\vec{F}_{\text{B on A}} = (4.5 \times 10^{-3}$ N, $+y$-direction$)$,
$\vec{F}_{\text{A on B}} = (4.5 \times 10^{-3}$ N, $-y$-direction$)$
21. 0.11 nC
23. $1.4 \times 10^5$ N/C, away from the bead
25. a. $1.4 \times 10^{-3}$ N/C, away from the proton
b. $1.4 \times 10^{-3}$ N/C, toward the electron
27. a. $3.6 \times 10^4$ N/C, $1.8 \times 10^4$ N/C, $1.8 \times 10^4$ N/C
b.

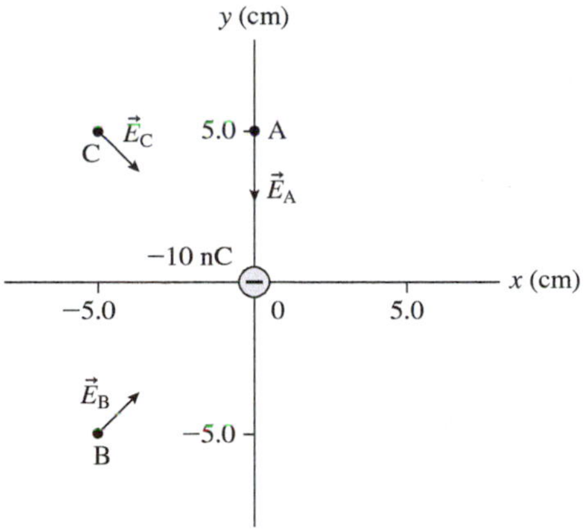


29. 2500 N/C, horizontal along $+x$-axis
31. a. The ratio is doubled. b. The ratio increases by a factor of 4.
c. 1
33. ±14 nC
35. a. Upper b. $1.6 \times 10^7$ C
39. a. $1.0 \times 10^{-7}$ N/C b. $5.6 \times 10^{-11}$ N/C
41. $6.1 \times 10^5$ N/C, downward
43. $1.4 \times 10^{-13}$ N·m
45. $17.2 \times 10^{-15}$ N
47. a. 0 b. $5.6 \times 10^{-27}$ N·m
49. a. 57.6 N b. $4.7 \times 10^{-35}$ N c. $1.2 \times 10^{36}$
51. 100 nC
53. 4000 N/C, 9.3°above the horizontal
55. $3.1 \times 10^{-4}$ N, upward
57. 0 N
59. $1.1 \times 10^{-5}$ N, upward
61. 28 nC
63. a. $2.3 \times 10^{-6}$ b. $4.3 \times 10^7$ N/C
65. $5.9 \times 10^5$ N/C
67. 5.7 cm
69. 150 nC
71. 180 nC
73. 1.3 cm
75. B
77. D

## Chapter 21

### Answers to odd-numbered multiple-choice questions

19. A
21. D
23. C
25. D
27. B
29. A
31. B

### Answers to odd-numbered problems

1. −3.0 μC
3. 530 V
5. −4.0 μJ
7. a. 3000 V b. 75 μJ
9. $-2.1 \times 10^4$ V
11. a. Lower b. −0.71 V c. 0.71 eV
13. a. Higher b. 3300 V c. 3300 eV
15. 400 V
17. a. $E = 200$ kV/m, $\Delta V_C = 200$ V b. $E = 200$ kV/m, $\Delta V_C = 400$ V
19. 0 V, −900 V
21. 1400 V
23. a. 1.5 V b. 8.3 pC

25. $1.4 \times 10^{-18}$ J
27. a. A b. 70 V
29. −352 V
31. 10 kV/m, left
33. a. Positive b. Negative c. Positive
37. a. 7.1 pF b. 0.71 nC
39. 4.0 $\mu$C
41. 4.8 cm
43. 3.0 V
45. 15 $\mu$C
47. 2.5 nC
49. 13 nF
51. a. 12.0 V b. 24 V
53. 1400 V
55. 1/2
57. 1200 V/m
59. a. $1.1 \times 10^{-7}$ J b. 0.71 J/m$^3$
61. a. $1.1 \times 10^{-6}$ J b. −20 V
63. $-2.2 \times 10^{-19}$ J
65. $x = 6.0$ cm, $x = 3.0$ cm
67. a. $+\infty, -\infty$ b. $x = \pm\infty$
69. 0.070 eV
71. 0.0 V
73. $1.0 \times 10^5$ m/s
75. $7.5 \times 10^5$ m/s
77. 47 V
79. 10 nC
81. $4.0 \times 10^7$ m/s
83. a. Smaller b. Smaller c. $\vec{E} = 5.5$ V/cm, in the $+x$-direction
85. a. $5.5 \times 10^{-4}$ J b. 0.36 J
87. 22 $\mu$F
89. A
91. D

## Chapter 22

### Answers to odd-numbered multiple-choice questions

23. C
25. C
27. D
29. D
31. C

### Answers to odd-numbered problems

1. 3000 C, $1.9 \times 10^{22}$ m/s
3. a. $1.6 \times 10^{18}$ electrons/s b. Into
5. $9.4 \times 10^{13}$ electrons
7. 93 A
9. $6.5 \times 10^{-4}$ s
11. $I_B = 5$ A, $I_C = -2$ A
13. 1.5 $\mu$J
15. a. 45 C b. $4.1 \times 10^2$ J
17. 3200
19. a. 1.0 b. 0.50
21. 50 $\Omega \cdot$ m
23. 8.7 m
25. 2.0 A
27. $1 \times 10^5$ $\Omega$
29. a. 2.0 $\Omega$
b.

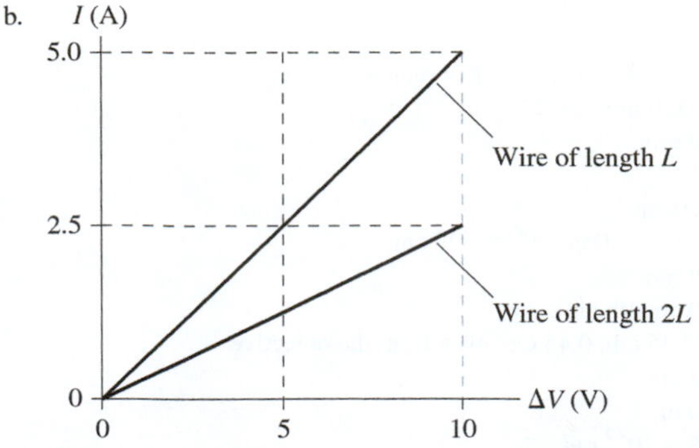


31. a. 200 V/m b. 100 V/m c. 0.25 A d. 480 $\Omega$
33. 0.43 V/m
35. 3.5 A
37. 0.50 A
39. a. 0.38 A b. 320 $\Omega$
41. a. 0.73 A b. 270 kJ c. $3.7 \times 10^2$ m
43. 450 C
45. 28 min
47. 20%
49. 42 G$\Omega$
51. 33 h
53. a. 15 $\Omega$ b. $2.1 \times 10^{-6}$ $\Omega \cdot$ m
55. a. 090 m b. 40 $\Omega$
57. 3.6 mm
59. 1/2
61. Iron
63. 2.6 m, 0.24 mm
65. a. 48 $\Omega$ b. 7.6 min
67. a. 25/26 b. 1/26
69. a. 1.0 $\Omega$ b. 100 V
71. A
73. B

## Chapter 23

### Answers to odd-numbered multiple-choice questions

35. B
37. A
39. D
41. B

### Answers to odd-numbered problems

1.

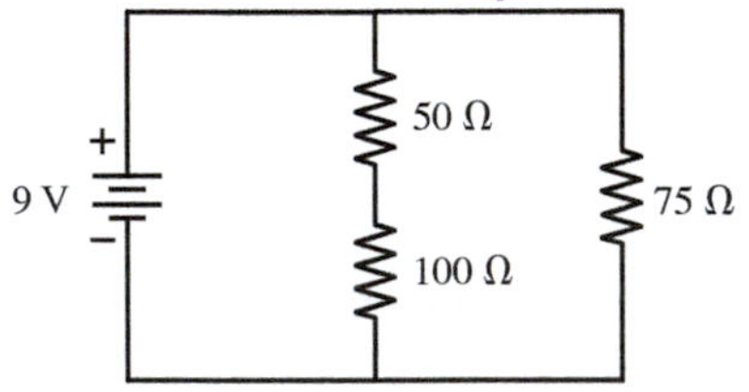

3.

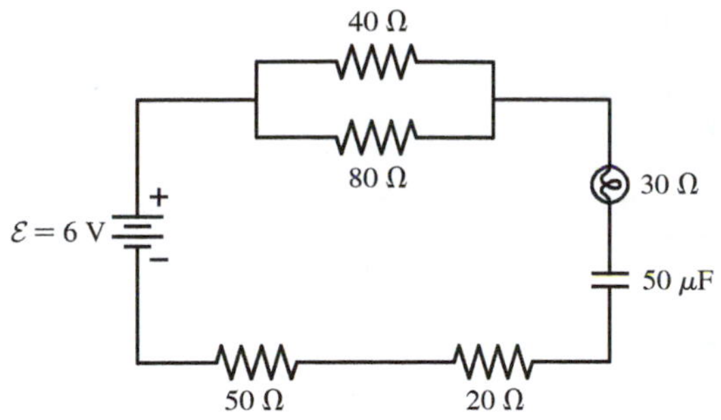


5. 10 V
7. a. 0.17 A, to the left
9.

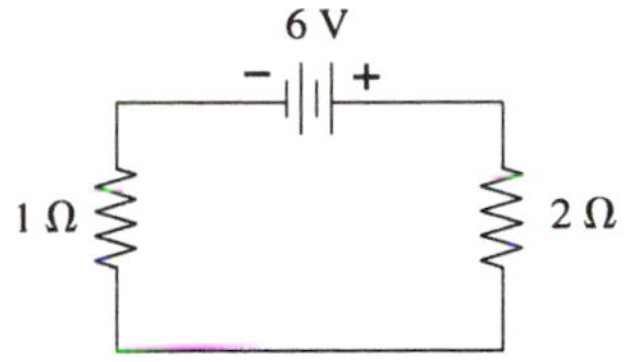


11. a. 1.0 Ω b. 1.0 Ω c. 0.50 Ω
13. 270 Ω
15. 170 Ω
17. 10 Ω, 30 Ω
19. 24 Ω
21. 0.33 $k\Omega$, 3.0 $k\Omega$, 0.67 $k\Omega$, 1.5 $k\Omega$,
23. a. 2.0 A b. 5.0 A
25. 20 Ω, 60 V
27. $I_1 =$ 0.50 A, $\Delta V_1 = 6.0$ V
$I_2 =$ 0.30 A, $\Delta V_2 = 6.0$ V
$I_3 =$ 0.20 A, $\Delta V_3 = 6.0$ V

29.

| ***R*** | ***I* (A)** | **Δ*V*(V)** |
|---|---|---|
| $R_1$ | 1.0 | 5.0 V |
| $R_2$ | 0.50 | 5.0 V |
| $R_3$ | 0.50 | 2.5 V |
| $R_4$ | 0.50 | 2.5 V |

31.

| ***R*** | ***I* (A)** | **Δ*V*(V)** |
|---|---|---|
| 6.0 Ω (left) | 2.0 | 12 |
| 15 Ω | 0.8 | 12 |
| 6.0 Ω (top) | 1.2 | 7.2 |
| 4.0 Ω | 1.2 | 4.8 |

33.

| ***R*** | ***I* (A)** | **Δ*V*(V)** |
|---|---|---|
| 3 Ω | 2.0 | 6.0 |
| 4 Ω | 1.5 | 6.0 |
| 48 Ω | 0.13 | 6.0 |
| 16 Ω | 0.38 | 6.0 |

35. 20 Ω
37. 3.0 μF
39. 150 μF in series
41. 8.9 μF
43. a. 5.4 μF
b. 4.0 μF: 28.8 μC
6.0 μF: 28.8 μC
3.0 μF: 36 μC
45. 2.0 ms
47. 6.9 ms
49. a. 18 μC, 180 mA b. 11 μC, 110 mA c. 2.4 μC, 24 mA
51. 2.9 ms
53. 33 m/s
55. 20 m/s
57. 19 W
59. a. 1.0 V b. 1/3
61. a. 2.3 V b. 1.0 kJ, 340 J
63. $P_5 = 45$ W, $P_{20} = 20$ W
65. 55 Ω
67. 4
69. a. Counterclockwise b. 6.0 Ω c. 0.38 W
71. 10 Ω
73.

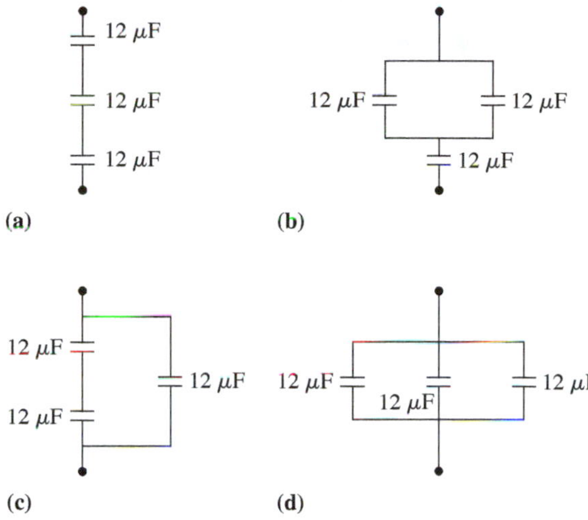


75. a. 1.0 kΩ b. $3.0 \times 10^{-7}$ F
77. a. 18 μF b. 0.23 ms
79. a. $2.3 \times 10^{-5}$ C b. 27 μs
81. a. 19 kΩ b. Increased
83. 6100
85. a. 4.8 pC b. $3.0 \times 10^{7}$
87. C
89. B
91. A
93. B

## Chapter 24

### Answers to odd-numbered multiple-choice questions

29. D
31. A
33. A
35. C

### Answers to odd-numbered problems

1. 20 μT
3. a. 20 A b. 1.6 mm
5. 20 A
7. $5.7 \times 10^{-5}$ T, 45° clockwise from the $+x$-axis
9. a. 2 μT, 4% b. The two fields nearly cancel.
11. 40 nA
13. 1.6 kA
15. $3.8 \times 10^{-7}$ A
17. $4.1 \times 10^{-4}$ T
19. 8.6 μA
21. 13 T
23. a. ($8.0 \times 10^{-13}$ N, $-z$-direction) b. ($8.0 \times 10^{-13}$ N, $-z$-direction)
b. ($8.0 \times 10^{-13}$ N, 45° clockwise from $-z$-axis in $yz$ plane)
25. a. 11 cm b. 10 m
27. $2.4 \times 10^{5}$ T

29. 0.086 T
31. a. 0.61 m b. 18,000 orbits
33. 0.13 T, out of the page
35. $2.0 \times 10^{-7}$ N
37. 0.078 N to the right
39. $7.5 \times 10^{-4}$ N·m
41. $1.7 \times 10^{-26}$ N·m
43. a.

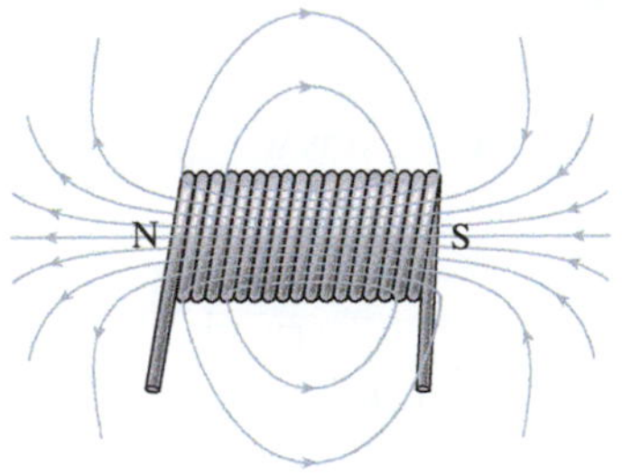


b. Attractive c. To the left, toward the solenoid
45. 3.0 Ω
47. 2.0 A
49. a. 0.11 mT b. North c. $2.5 \times 10^{-23}$ N
51. 0.68 A
53. 0.20 A, axis of coil north-south
55. Rapid switching doesn't allow the moments to align.
57. 1.0 T
59. A
61. B
63. A
65. B
67. A
69. A

## Chapter 25

### Answers to odd-numbered multiple-choice questions

27. C
29. A
31. C
33. D
35. E

### Answers to odd-numbered problems

1. 19°
3. a. 4.0 m/s b. 2.2 T
5. 9500 m/s
7. a. 2.9 mV b. Negative
9. For $\theta = 0°$, $3.9 \times 10^{-4}$ Wb; for $\theta = 30°$, $3.4 \times 10^{-4}$ Wb; for $\theta = 36°$, $2.0 \times 10^{-4}$ Wb; for $\theta = 90°$, 0 Wb
11. a. $6.3 \times 10^{-5}$ Wb b. $6.3 \times 10^{-5}$ Wb
13. a. $8.7 \times 10^{-4}$ Wb b. Clockwise
15. 3.1 V
17. a. 3.9 mV, 39 mA, clockwise b. 3.9 mv, 39 mA, clockwise c. 0 V, 0 A
19. 25 mV
21. 0.55 A
23. 0.97 A
25. $6.0 \times 10^5$ V/m
27. $E_0 = 1.4 \times 10^3$ V/m, $B_0 = 4.6 \times 10^{-6}$ T
29. $E_0 = 90$ V/m, $B_0 = 3.3 \times 10^{-6}$ T
31. 77 m
33. a. $2.2 \times 10^{-6}$ W/m$^2$ b. 0.041 V/m
35. a. 10 W/m$^2$ b. 7.5 W/m$^2$ c. 5.0 W/m$^2$ d. 2.5 W/m$^2$ e. 0 W/m$^2$
37. 60°
39. 28 mW
41. 1.2 keV
43. a. 250 keV b. $1.8 \times 10^6$ bonds
45. $2.1 \times 10^{-20}$ J
47. $4.68 \times 10^{21}$ photons
49. 1700 photons/s
51. 2100°C
53. 55 nm
55. $\lambda_{dental} = 50$ pm; $\lambda_{microtomography} = 20$ pm
57. 1.1 mm
61. a. $5.2 \times 10^{-9}$ J b. $6.6 \times 10^{-11}$ K
63. a. 0.20 A b. $4.0 \times 10^{-3}$ N c. 11°C
65. a. $6.3 \times 10^{-4}$ N b. $3.1 \times 10^{-4}$ W c. $1.3 \times 10^{-2}$ A, counterclockwise d. $3.1 \times 10^{-4}$ W
67. a. 250 kW b. $1.8 \times 10^7$ V/m
69. 0.016 V/m
71. $I_0/4$
73. $3.0 \times 10^{31}$
75. A
77. D
79. C

## Chapter 26

### Answers to odd-numbered multiple-choice questions

23. D
25. A
27. A
29. C

### Answers to odd-numbered problems

1. a. 50 mA b. 50 mA
3. 22.4 V
5. a. 9.0 Ω b. 19 A c. 3200 W
7. 530 A
9. 0.50 A
11. 17 mA
13. a. 15 turns b. 32 mA
15. a. 18 kW b. 1.0 kW
17. 12 A
19. Yes
21. $26.30
23. No
25. 34 V
27. 190 mA
29. 59 mA, yes
31. Yes
33. 270 Hz
35. a. 50 Hz b. 4.8 μF
37. 130 Ω
39. a. 0.80 A b. 0.80 mA
41. a. $3.2 \times 10^4$ Hz b. 0 V
43. 80 Ω
45. 0.25 μH
47. 1.0 Ω
49. 10 mH
51. 7.1 Ω, 2.5 mH
53. 0.38 A
55. 25 A
57. a. 80 A b. 83 ns c. Yes
59. a. 24 V b. 4.5 mΩ c. 5600 A
61. a. 500 Hz b. 1.0 A
63. 35 mH
65. D
67. C
69. A
71. B

## Part VI Problems

### Answers to odd-numbered problems

1. B
3. C
5. B
7. A
9. B
11. C
13. C
15. D
17. B
19. C
21. 20 A, clockwise

## Chapter 27

### Answers to odd-numbered multiple-choice questions

21. B
23. D
25. C

### Answers to odd-numbered problems

1. $v_{sound} = 345$ m/s, $v_{sprinter} = 10$ m/s
3. a. 15 m/s b. 5 m/s
5. 30 s
7. $3.0 \times 10^8$ m/s
9. $3.0 \times 10^8$ m/s
11. $t_1 = 1.0\ \mu s$, $t_2 = 2.0\ \mu s$
13. She sees flash 2 40 $\mu$s after flash 1.
15. Barn 1.0 $\mu$s before the tree.
17. a. Simultaneous b. Event 2 occurs first.
19. 350 $\mu$s
21. $0.87c$
23. a. $0.80c$ b. 15 yr
25. 1.0 ps
27. Yes
29. 1.5 km
31. 0.78 m
33. 0.35 nm
35. $0.375c$
37. $0.60c$
39. a. $1.12 \times 10^{-17}$ kg · m/s b. 22.4
41. $0.87c$
43. $6.0 \times 10^{13}$ J, $9.0 \times 10^{13}$ J, $1.5 \times 10^{14}$ J
45. a. $1.8 \times 10^{16}$ J b. $9.0 \times 10^9$
47. $0.42c$
49. $0.94c$
51. 1.02 meV
53. $5.1 \times 10^{-10}$ kg
55. 1.7 $\mu$s
57. a. 90 m b. $0.44c$
59. 0.50 s
61. 5.9 s, 4.3 s
63. 2 s
65. a. 10 yr b. 4.4 yr c. 3.9 ly d. $0.90c$
67. a. 170 yr b. 14.7 yr
69. 0.16 s
71. a. $(1-4/9 \times 10^{-24})c$ b. 46 nm
73. $0.31c$
75. a. 6660 MeV b. 3.63 MeV
77. a. $1.3 \times 10^{17}$ kg b. $6.7 \times 10^{-12}$% c. $1.5 \times 10^{13}$ yr
79. a. $4 \times 10^{-12}$ J b. 0.6%
81. D
83. D

## Chapter 28

### Answers to odd-numbered multiple-choice questions

29. B
31. A
33. B
35. B
37. A

### Answers to odd-numbered problems

1. 69.3°
3. 64°
5. 4
7. 3.20 eV
9. 3.47 eV
11. 1.8 eV
13. a. 290 nm b. $3.7 \times 10^5$ m/s
15. a. $6.63 \times 10^5$ m/s b. 2.30 eV, potassium
17.

| Metal | $E_0$ (eV) | $f_0$ (Hz) | $\lambda_0$ (nm) | $v_{max}$ (m/s) | $V_{stop}$ (V) |
|---|---|---|---|---|---|
| Potassium | 2.30 | $5.56 \times 10^{14}$ | 540 | $10.8 \times 10^5$ | 3.35 |
| Gold | 5.10 | $1.23 \times 10^{15}$ | 244 | $4.40 \times 10^5$ | 0.55 |

19. 2.84 eV, 2.33 eV, 2.20 eV
21. a. 4140 nm, infrared b. 410 nm, visible b. 41 nm, ultraviolet
23. $8.7 \times 10^{31}$
25. 41
27. a. 2.3 eV b. $3 \times 10^{18}$
29. $9.5 \times 10^{-7}$ W
31. a. $7.83 \times 10^{-12}$ J, $6.63 \times 10^{-19}$ J b. $4.7 \times 10^{15}$ photons
33. a. $1.1 \times 10^{-34}$ m b. $1.7 \times 10^{-23}$ m/s
35. 1.5 eV
37. 8.2 MeV
39. 6.0 $\mu$V
41. No
43. 0.43 nm
45. $E_1 = 1.3 \times 10^{-12}$ J, $E_2 = 5.3 \times 10^{-12}$ J, $E_3 = 1.2 \times 10^{-11}$ J
47. $\lambda =$ 830 nm for the $2 \rightarrow 1$ transition
    $\lambda =$ 500 nm for the $3 \rightarrow 2$ transition
    $\lambda =$ 310 nm for the $3 \rightarrow 1$ transition
49. 618 nm
51. 3.0 eV, 4.5 eV, 6.0 eV
53. $5.8 \times 10^{-10}$ m
55. $8.1 \times 10^{-12}$
57. 0 m/s to $1 \times 10^7$ m/s
59. a. 2.3 eV b. 240 nm
61. a. 250 eV, x-ray region of spectrum b. 0.060 eV
63. $7.4 \times 10^4\ s^{-1}$
65. 2.3 mA
67. a. $1.32 \times 10^8$ m/s b. $5.51 \times 10^{-6}$ m
69. 540 nm
71. 0.87 nm
73. a. $\dfrac{h^2}{4mL^2}$ b. $E_{min}, \frac{5}{2}E_{min}, 4E_{min}, 5E_{min}, \frac{13}{2}E_{min}$
75. 18 fm
77. 19 T
79. C
81. B
83. A
85. B
87. C
89. A
91. B

## Chapter 29

### Answers to odd-numbered multiple-choice questions

27. B
29. C
31. C

### Answers to odd-numbered problems

1. a.

| Wavelength (nm) | $m$ | $n$ |
|---|---|---|
| 656.6 | 2 | 3 |
| 486.3 | 2 | 4 |
| 434.2 | 2 | 5 |
| 410.3 | 2 | 6 |

b. 397 nm

3. 1870 nm, 1280 nm, 1090 nm, infrared
5. 7.45 $\mu$m, 4.65 $\mu$m, 3.74 $\mu$m, infrared
7. a. 3 electrons, 4 protons, 5 neutrons
   b. 6 electrons, 6 protons, 6 neutrons
   c. 4 electrons, 7 protons, 8 neutrons
9. a. $^{8}$Li b. $^{14}$C$^{+}$
11. a. 79 electrons, 79 protons, 118 neutrons b. $2.29 \times 10^{17}$ kg/m$^3$
    c. $1.19 \times 10^{13}$
13. $1.8 \times 10^7$ m/s
15. a. Yes b. 0.50 eV
17. $6.7 \times 10^5$ m/s, $1.1 \times 10^6$ m/s
19. 3
21. $4 \rightarrow 3$
23. a. 0.332 nm, 0.665 nm, 0.997 nm
    b. 0.332 nm $= (2\pi(0.053 \text{ nm}))/1$, 0.665 nm $= (2\pi(0.232 \text{ nm}))/2$, 0.997 nm $= (2\pi(0.476 \text{ nm}))/3$
    c.

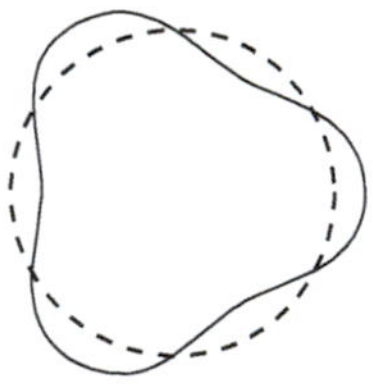

25. 97.2 nm, 486, 1870 nm
27. $n = 1$: $\left(1, 0, 0, \pm\frac{1}{2}\right)$ $n = 2$: $\left(2, 0, 0, \pm\frac{1}{2}\right)$, $\left(2, 1, -1, \pm\frac{1}{2}\right)$, $\left(2, 1, 0, \pm\frac{1}{2}\right)$, $\left(2, 1, 1, \pm\frac{1}{2}\right)$, $\left(2, 1, -1, \pm\frac{1}{2}\right)$,
    $n = 3$: $\left(3, 0, 0, \pm\frac{1}{2}\right)$, $\left(3, 1, -1, \pm\frac{1}{2}\right)$, $\left(3, 1, 0, \pm\frac{1}{2}\right)$, $\left(3, 1, 1, \pm\frac{1}{2}\right)$, $\left(3, 2, -2, \pm\frac{1}{2}\right)$, $\left(3, 2, -1, \pm\frac{1}{2}\right)$, $\left(3, 2, 0, \pm\frac{1}{2}\right)$, $\left(3, 2, 1, \pm\frac{1}{2}\right)$, $\left(3, 2, 2, \pm\frac{1}{2}\right)$,
29. $3\hbar$
31. $1s^2 2s^2 2p^6 3s^2 3p^6 4s^2 3d^{10} 4p^6 5s^2 4d^{10} 5p^6 6s^2$
33. $1s^2 2s^2 2p^6 3s^2 3p^2$, $1s^2 2s^2 2p^6 3s^2 3p^6 4s^2 3d^{10} 4p^2$, $1s^2 2s^2 2p^6 3s^2 3p^6 4s^2 3d^{10} 4p^6 5s^2 4d^{10} 5p^6 6s^2 4f^{14} 5d^{10} 6p^2$
35. a. Magnesium b. Molybdenum
37. a. Can't have 8 electrons in a $p$ state
    b. Can't have 3 electrons in an $s$ state
39. $4.86 \times 10^6$ m/s
41. 330 nm
43. a. Yes, 2180 nm b. No, violates $\Delta t = 1$
45. 690 nm
47. $3.18 \times 10^{15}$ s$^{-1}$
49. a. 192 nm b. 50 kW
51. a. $1.7 \times 10^{18}$ b. $1.7 \times 10^{26}$ photons/s
53. a. 102.6 nm: $n = 3, m - 1$; 1876 nm: $n = 4, m - 3$
    b. Ultraviolet, infrared
55. 0.000000000058%, 99.999999999942%
57. b. $\lambda_{41} = 200$ nm $\lambda_{31} = 300$ nm $\lambda_{21} = 500$ nm
    $\lambda_{42} = 333$ nm $\lambda_{43} = 600$ nm $\lambda_{32} = 750$ nm
59. a.

| Transition | $E_f$ (eV) | $E_i$ (eV) | $\Delta E$ (eV) | $\lambda$ (nm) |
|---|---|---|---|---|
| $1 \rightarrow 2$ | −3.0 | −6.5 | 3.5 | 360 |
| $1 \rightarrow 3$ | −2.0 | −6.5 | 4.5 | 280 |

b. Both waves are ultraviolet. c. $6.2 \times 10^5$ m/s

61. 1870 nm
63. a. $2.06 \times 10^6$ m/s b. 12.1 V
65. $\sqrt{6}\hbar$
67.

| Transition | a. Wavelength | b. Type | c. Absorption |
|---|---|---|---|
| $2p \rightarrow 2s$ | 670 nm | VIS | Yes |
| $3s \rightarrow 2p$ | 816 nm | IR | No |
| $3p \rightarrow 2s$ | 324 nm | UV | Yes |
| $3p \rightarrow 3s$ | 2696 nm | IR | No |
| $3d \rightarrow 2p$ | 611 nm | VIS | No |
| $3d \rightarrow 3p$ | 25 $\mu$m | IR | No |
| $4s \rightarrow 2p$ | 498 nm | VIS | No |
| $4s \rightarrow 3p$ | 2430 nm | IR | No |

69. $2.0 \times 10^{10}$ W/m$^2$
71. A
73. C
75. A
77. B
79. C
81. C

## Chapter 30

### Answers to odd-numbered multiple-choice questions

29. B
31. C
33. A
35. C
37. D

### Answers to odd-numbered problems

1. a. 1 proton, 2 neutrons b. 18 protons, 22 neutrons
   c. 20 protons, 20 neutrons d. 94 protons, 145 neutrons
3. 6.94 u
5. a. 0.511 MeV b. 938.27 Me V c. 938.78 MeV d. 3728.4 MeV
7. a. Total 8.48 MeV, per nucleon 2.38 MeV
   b. Total 7.72 MeV, per nucleon 2.57 MeV
9. $^{3}$He: 2.57 MeV, $^{4}$He: 7.07 MeV; $^{4}$He is more tightly bound.
11. 1. 7.58 MeV b. 8.79 MeV c. 7.87 MeV
13. $^{240}$Pu: 7.56 MeV, $^{133}$Xe: 8.41 MeV
15. a $^{3}$H: 8.482 MeV, $^{3}$He: 7.718 MeV; $^{2}$H: 2.224 MeV, $^{4}$He: 28.297 MeV
    b. 2.389 MeV
17. 25.7 MeV
19. None are obviously unstable.
21. 230 N
23. $^{234}$U
25. a. $^{228}$Th b. $^{207}$Tl c. $^{7}$Li d. $^{60}$Ni
27. 5.64 MeV
29. a. Beta-plus b. Beta-minus
31. a. $^{3}$He b. 32%

33. a. $9.3 \times 10^{11}$ b. $4.7 \times 10^{11}$ c. $5.5 \times 10^{8}$
35. 53 d
37. 2.0 $\mu$Ci
39. $3.7 \times 10^{10}$ Bq, 1.0 Ci
41. 30 h
43. $4.6 \times 10^{9}$
45. 600 h
47. 0.14 mSv
49. 600 Gy
51. 47 Gy
53. 17 mSv
55. 0.29 mSv
57. 139 MeV
59. 200 MeV
61. a. 157 min b. 29.5 h
63. a. 664 keV b. 0.152 MeV
65. $^{12}C = 3.66 \times 10^{-3}$ kg, $^{14}C = 5.55 \times 10^{-15}$ kg
67. a. 0.10% b. 0.00098
69. $5.9 \times 10^{9}$ yr
71. 4,000 kg/hr
73. $2.8 \times 10^{12}$
75. D
77. A
79. C

## Part VII Problems

### Answers to odd-numbered problems

1. C
3. A
5. D
7. A
9. D
11. A
13. B
15. D
17. a. $8.0 \times 10^{7}$ m/s b. $4.6 \times 10^{-8}$ W c. 11 MBq
19. a. $1.1 \times 10^{21}$ J b. $7.7 \times 10^{32}$ reactions
    c. $2.6 \times 10^{6}$ kg of $^{2}_{1}H$, $3.8 \times 10^{6}$ kg of $^{3}_{2}H$
21. 15,000

# Credits

**FM** **P. iii top** Sally Knight/Pearson Education; **p. iii middle** John Eisele/Colorado State University; **bottom** Stuart Field; **p. xii** Ethan Daniels/Shutterstock; **p. xx** Jim Zipp/Science Source; **p. xxi left** Gabriel12/Shutterstock; **right** Kate Grishakova/Shutterstock; **p. xxii left** Arno Vlooswijk/TService/Science Source; **right** Andy Harmer/Alamy Stock Photo; **p.xxiii** Juniors Bildarchiv GmbH/Alamy Stock Photo; **p. xxiv** CBCK Christine/Alamy Stock Photo; **p. xxv** Lisa Werner/Alamy Stock Photo.

**PART I OPENER** **P. 2** Interfoto/Zoology/Alamy Stock Photo.

**CHAPTER 1** **P. 4 top** Jim Zipp/Science Source; **bottom left** Faltermaier/Vario images, Inh. Susanne Baumgarten e.K./Alamy Stock Photo; **bottom right** Johnny Greig/Science Source; **p. 5 top left** Theo Allofs/Corbis/Getty Imagers; **top right** Thomas Peter/Reuters; **bottom left** Richard Megna/Fundamental Photographs, NYC.; **bottom right** Steve Allen/Getty Images; **p. 8 bottom left** Chris Carroll/Corbis/Getty Images; **p. 14 top** Stuart Field; **center** Stuart Field; **bottom** Clive Rose/Getty Images; **p. 16** NASA Earth Observatory; **p. 18 top** Niels Van Kampenhout/Alamy Stock Photo; **center** Vikram Hoshing/Ephotocorp/Alamy Stock Photo; **p. 19 top left** Darren Baker/Corbis/Getty Images; **p. 19 top center** Michael Burrell/Alamy Stock Photo; **top right** C Squared Studios/Photodisc/Getty images; **p. 19 middle** Khaled Desouki/AFP/Getty Images; **bottom** Balfour Studios/Alamy Stock Photo; **p. 20** Alan Schein Photography/Corbis/Getty Images; **p. 23** Mehau Kulyk/Science Source; **p. 28 Problem 22** Jupiterimages/Stockbyte/Getty Images; **p. 29 Problems 41–42** Philip Dalton/Alamy Stock Photo; **p. 30 Problem 68** Sebastian Kaulitzki/Alamy Stock Photo.

**CHAPTER 2** **P. 32 top** Eduard KyslynskyyShutterstock; **p. 32 middle, left to right** Jhorrocks/E+/Getty Images; Steve Bloom Images/Alamy Stock Photo; EpicStockMedia/Shutterstock; **p. 41** Milos Muller/Shutterstock; **p. 44** Texas Transportation Institute; **p. 47** NASA; **p. 50** Vincent Munier/Nature Picture Library; **p. 52** Kuritafsheen/RooM the Agency/Alamy Stock Photo; **p. 55** NASA; **p. 56** Mike Brinson/The Image Bank/Getty Images; **p. 58** Steve Bloom Images/Alamy Stock Photo; **p. 62 Question 3** Jim Barber/Shutterstock; **p. 66 Problem 28** Krzysztof Odziomek/Shutterstock; **p. 67 Problem 40** Courtesy Hövding; **p. 68 Problem 45** Richard Megna/Fundamental Photographs, NYC; **p. 69 Problem 66** EcoPrint/Shutterstock; **Problem 68** Mirko Graul/Shutterstock.

**CHAPTER 3** **P. 71 top** Martin Harvey/Alamy Stock Photo; **p. 71 middle, left to right** Erik Isakson/Tetra Images/Alamy Stock Photo; Kyle Krause/Photolibrary/Getty Images; Kristy-Anne Glubish/Design Pics Inc./Alamy Stock Photo; **p. 71 bottom right** PCN Chrome/PCN Photography/Alamy Stock Photo; **p. 74** Tom Uhlman/Alamy Stock Photo; **p. 80** Scott Markewitz/The Image Bank/Getty Images; **p. 81** ML Harris/The Image Bank/Getty Images; **p. 83** Ivan Chudakov/Shutterstock; **p. 85** Richard Megna/Fundamental Photographs, NYC.; **p. 89** Gustoimages/Science Source; **p. 90** Don Henry/Alamy Stock Photo; **p. 92** Phil Noble/Reuters; **p. 93** David Madison/Allsport Concepts/Getty Images; **p. 95** Jerry Young/Dorling Kindersley, Ltd.; **p. 98 Question 7** Andy Lyons/Getty Images; **Question 14** Jupiterimages/Stockbyte/Getty Images; **p. 100 Problem 15** Don Mason/Corbis/Getty Images; **Problem 18** Mike Dobel/Alamy Stock Photo; **p. 101 Problem 42** frogmo9/Fotolia; **p. 103 Problem 70** Kai Pfaffenbach/Reuters.

**CHAPTER 4** **P. 105 top** MarclSchauer/Shutterstock; **p. 105 left to right** Wittek, R./Arco Images GmbH/Alamy Stock Photo; Bernd Friedel/123RF; PCN Photography/Alamy Stock Photo; **bottom** Dlewis33/iStock/Getty Images; **p. 106 left** NASA; **p. 107 top, left to right** National Highway Traffic Safety Administration; **p. 107 left, top to bottom** Tony Freeman/PhotoEdit Inc.; Dmitry Kalinovsky/Shutterstock; Comstock/Stockbyte/Getty Images; Dorling Kindersley, Ltd.; StudioOneNine/Shutterstock; Creative Digital Vision/Pearson Education; **p. 110** Sportpoint/Shutterstock; **p. 113** D.P. Wilson/FLPA/Science Source; **p. 117** Richard Megna/Fundamental Photographs, NYC.; **p. 118** Stan Honda/AFP/Getty Images; **p. 121** Roger Bacon/REUTERS/Alamy Stock Photo; **p. 123** Joppo/Shutterstock; **p. 125** Robert salthouse/Alamy Stock Photo; **p. 128 Question 7** Richard Megna/Fundamental Photographs, NYC.; **Question 19** Rein Tafenau/Alamy Stock Photo; **p. 129 Problem 3** Longview/The Image Bank/Getty Images.

**CHAPTER 5** **P. 134 top** Nick Garbutt/Nature Picture Library/Alamy Stock Photo; **p. 134 middle, left to right** Digital Vision/Getty Images; Eric Schrader/Pearson Education, Inc.; Radius Images/Alamy Stock Photo; **p. 135** Cesar Rangel/AFP/Getty Images; **p. 141** NASA; **p. 142** Erin Patrice O'Brien/The Image Bank/Taxi/Getty Images; **p. 144 top** NASA; **bottom** NASA; **p. 148 top** Nulinukas/Shutterstock; **bottom** Roman Krochuk/Shutterstock; **p. 151 top** Eric Schrader/Pearson Education, Inc.; **bottom** Marc Xavier/Fotolia; **p. 168 Problem 2** Dumrath/F1online digitale Bildagentur GmbH/Alamy Stock Photo; **p. 169 Problem 5** Martin M Rotker/Science Source/Getty Images; **Problem 9** Sport In Pictures/Alamy Stock Photo; **p. 170 Problem 20** J. D. Pooley/AP Images; **p. 173 Problem 73** Richard Megna/Fundamental Photographs, NYC.; **Problem 74** Martyn Hayhow/AFP/Getty Images.

**CHAPTER 6** **P. 175 top** Gabriel12/Shutterstock; **middle, left to right** Blue Images/Corbis/Getty Images; Matthew Jacques/Shutterstock; NASA Earth Observing System; **bottom right** Tom Carter/Alamy Stock Photo; **p. 176** Kristy-Anne Glubish/Design Pics Inc./Alamy Stock Photo; **p. 177** Val lawless/Shutterstock; **p. 178** Mark A Leman/The Image Bank/Getty Images; **p. 182 top** Mark Scheuern/Alamy Stock Photo; **bottom** Robert Laberge/Getty Images; **p. 183 top** Matt Perrin Sport/Alamy Stock Photo; **bottom** Richard Megna/Fundamental Photographs, NYC.; **p. 184** FPG/Hulton Archive/Getty Images; **p. 185** Susumu Nishinaga/Science Source; **p. 186 top** Hubble Heritage Team/NASA; **middle** Wacky Racing/Picturesbyrob/Alamy Stock Photo; **p. 187** Richard Megna/Fundamental Photographs, NYC.; **p. 189 top** NASA; **bottom** M.G.M /Album/M.G.M/Album/Newscom; **p. 192** Ralph White/NASA; **p. 193** NASA; **p. 195** Jet Propulsion Laboratory/NASA; **p. 198 Question 3** Mike Fuchslocher / Alamy Stock Photo; **p. 199 Question 15** Jupiterimages/Stockbyte/Getty Images; **p. 200 Problem 8** Rafa Irusta/Shutterstock; **Problem 10** Images-USA/Alamy Stock Photo; **p. 201 Problem 14** Todd Korol/Reuters; **Problem 18** Exra Shaw/Getty Images; **Problem 25** Barry Lewis/Alamy Stock Photo; **p. 203 Problem 63** Frilet Patrick/Hemis.fr/Hemis/Alamy Stock Photo; **Problem 64** Dwphotos/Shutterstock; **p. 204 Problem 79** NASA.

**CHAPTER 7** **P. 205 top** Anthony Collins / Alamy Stock Photo; **p. 205 middle, left to right** Hadrian/Shutterstock; Dmbaker/123RF; Oleg Mikhaylov/Shutterstock; **bottom right** Graficart.net/Alamy Stock Photo; **p. 206** BMJ/Shutterstock; **p. 211** Dennis D. Potokar/Science Source; **p. 216 left top** PhotoAbility/Alamy Stock Photo; **left bottom** Chris Mole/Shutterstock; **p. 216** Stock Image/Pixland/Alamy Stock Photo; **p. 218** Sputnik/Alamy Stock Photo; **p. 219** Image 100/Alamy Stock Photo; **p. 222** Franck Fife/Afp Photo/Newscom; **p. 224** Stuart Field; **p. 226** Tiffany Brown/AP Images; **p. 228** Cordier Sylvain/Hemis.fr/Alamy Stock Photo; **p. 231 top** Richard Megna/Fundamental Photographs, NYC.; **bottom** John Mead/Science Source; **p. 232** Danilo Calilung/Alamy Stock Photo; **p. 236 Question 4** Kari Marttila/Alamy Stock Photo; **Question 12** Richard Megna/Fundamental Photographs, NYC.; **Question 14** Alex Milan Tracy/Sipa USA/Newscom; **Question 15** Dennis Frates/Alamy Stock Photo; **p. 237 Questions 27–29** Atsushi Tomura/Aflo Sport/Newscom; **p. 238 Problem 14** SCDNR's Southeastern Regional Taxonomic Center (SERTC).; **p. 240 Problem 52** ZUMA Press, Inc./Alamy Stock Photo; **p. 241 Problem 60** Harry Todd/Fox Photos/Hulton Archive/Getty Images; **p. 242 Problem 72** Lucian Coman/Shutterstock; **Problem 78** Images by J.Edwards/D. Whitaker/M.Laskwaski/A.Acosta; **p. 243 Problem 81** John Terence Turner/Alamy Stock Photo.

**CHAPTER 8** **P. 244 top** Eugene Mynzul/Shutterstock; **p. 244 middle, left to right** Larin Andrey Vladimirovich/Shutterstock; Christopher Cassidy/Alamy Stock Photo; Helen Birkin/Shutterstock; **bottom right** Mr Doomits/Shutterstock; **p. 247** James Kay/SCPhotos/Alamy Stock Photo; **p. 251 top** Stuart Field; **bottom** Richard Megna/Fundamental Photographs, NYC.; **p. 252** Ted Kinsman/Science Source; **p. 256** Javier Larrea/Age fotostock/Getty images; **p. 257** Harry Taylor/Dorling Kindersley, Ltd.; **p. 258 Figure 8.25 left** Arco/G. Lacz/Arco Images GmbH/Alamy Stock Photo; **right** Courtesy of Ellen Meiselman/Flickr; **Figure 8.26 (a)** John Eisele/Colorado State University;**p. 261** Adam Pearlstein/Pearson Education, Inc.; **p. 264 Question 4** Uz Foto /Shutterstock; **p. 265 Question 5, left** Fizkes/Shutterstock; **Question 5, right** Fuse/Corbis/Getty Images; **Question 13** Mint Images/REX/Shutterstock; **Question 14, left and right** John Eisele/Colorado State University/Colorado State University Photography; **p. 267 Problem 16** Richie Chan/Shutterstock; **p. 269 Problem 45** Ratmaner/Shutterstock; **Problem 46** Ivan Smuk/123RF; **Problem 50** Chris Curtis/Shutterstock; **p. 270 Problem 62** Kai Pfaffenbach/Reuters;

**PART I SUMMARY** **P. 273** Bill Schoening, Vanessa Harvey, REU Program/NOA/AURA/NSF; **p. 274 left** Owen Humphreys/AP Images; **right** SPL/Science Source.

**PART II OPENER** **P. 276** Jim Zipp/Science Source.

**CHAPTER 9** **P. 278 top** James Hager/Robertharding/Alamy Stock Photo; **middle, left to right** Pete Fontaine/WireImage/Getty Images; INTERFOTO/Alamy Stock Photo; Dave Fleetham/Design Pics/Perspectives/Getty Images; **bottom** iStock/Getty Images; **p. 279** Brian Jones; **p. 281** Freebilly/Shuterstock; **p. 284** Petr Zurek/Alamy Stock Photo; **p. 288** Richard Megna/Fundamental Photographs, NYC; **p. 292** Georgette Douwma/Nature Picture Library; **p. 295** Charlie Edwards/Photodisc/Getty images; **p. 297** Emery Wells/Image Bank Film Signature/Getty Images; **p. 298 top** NASA; **bottom** Ivan Kuzmin/Alamy Stock Photo; **p. 299 left** Stockfolio/Alamy Stock Photo; **right** Kostas Tsironis/AP Images; **p. 302 Question 12** NASA; **Question 18** Vespasian/Alamy Stock Photo; **p. 305 Problem 44** Macor/123RF; **p. 306 Problem 52** Anthony Pierce/Alamy Stock Photo.

**CHAPTER 10** **P. 309 top** Kate Grishakova/Shutterstock; **middle, left to right** Muratart/Shutterstock; Chris Rout/Alamy Stock Photo; Kali9/iStock/Getty Images; **p. 310 left to right** Doug Martin/Science Source; David Kleyn/Alamy Stock Photo; Erik Isakson/Tetra Images/Alamy Stock Photo; **p. 311 top, left to right** Danita Delimont/Gallo Images/Getty Images; Nicholas Toh/Shutterstock; Photo Researchers, Inc/Alamy Stock Photo; **p. 311 left, top to bottom** Stuart Franklin/Getty Images; Spencer Grant/AGE Fotostock; Scott Sroka/National Geographic/Getty Images; David Gray/Reuters; **p. 312 left to right** Kimimasa Mayama/REUTERS/Alamy Stock Photo; Gary Morrison/The Image Bank/Getty Images; Gart Buss/The Image Bank/Getty Images; **p. 315 top left** Al Grillo/AP Images; **bottom right** Danny Beier/Panther Media/AGE Fotostock; **p. 317** InsectWorld/Shutterstock; **p. 320** NASA; **p. 321** lassedesignen/Shutterstock; **p. 322** Meredith Davenport/REUTERS/Alamy Stock Photo; **p. 325 top right** Brian Jones; **middle left** Chris Salvo/The Image Bank/Getty Images; **middle right** Chip Forelli/The Image Bank/Getty Images; **p. 328** High Speed Nature/Alamy Stock Photo; **p. 340** Ted Kinsman/Science Source; **p. 341 top** Neil Setchfield/Planet Images/Getty Images; **bottom** David Madison/The Image Bank/Getty Images; **p. 344** Charles D. Winters/Science Source; **p. 348 Problem 18** Tom Heaton/REX/Shutterstock; **p. 349 Problem 40** Paul Miguel/Alamy Stock Photo; **p. 351 Problem 70** Kristel Segeren/Shutterstock; **p. 353 Problem 88** Paulo Oliveira/Alamy Stock Photo.

**CHAPTER 11** **P. 354 top** Mitchell Funk/The Image Bank/Getty Images; **middle, left to right** Dirima/Shutterstock; Tony McConnell/Science Source; Robas/Vetta/Getty Images; **p. 355 top, left to right** Brian Jones; Willard Clay/The Image Bank/Getty Images; David Woodfall/The Image Bank/Getty Images; **bottom, left to right** Brian Jones; Steve Gorton/Dorling Kindersley, Ltd.; Steve Allen/Getty Images; **p. 356** Eye35.pix/Alamy Stock Photo; **p. 357** Christina Richards/Shutterstock; **p. 359** Pearson Education, Inc.; **p. 360** Science Photo Library/Alamy Stock Photo; **p. 361 top** Asgeir Helgestad/Nature Picture Library; **bottom** Jurgen Freund/Nature Picture Library; **p. 363** MaximImages archive/Alamy Stock Photo; **p. 364** CJM Photography/Alamy Stock Photo; **p. 365 top** Richard Megna/Fundamental Photographs, NYC. ; **bottom** W. Phillips, NIST; **p. 368** Richard Megna/Fundamental Photographs, NYC.; **p. 370 left to right** Chris Ladd/The Image Bank/Getty Images; Tony McConnell/Science Source; Brian Jones; **p. 373** Wagan Corporation; **p. 378** Basement Stock/Alamy Stock Photo; **p. 380** Andy Sacks/The Image Bank/Getty Images; **p. 382** Eye35.pix/Alamy Stock Photo; **p. 383 Question 5** Tom Reichner/Shutterstock; **p. 384 Question 20** Richard Megna/Fundamental Photographs, NYC.; **p. 386 Problems 24–26** Dennis Jacobsen/Shutterstock; **Problem 41** David Nunuk/All Canada Photos/Alamy Stock Photo; **p. 387 Problem 54** Brian Jones; **p. 388 Problem 63** Anthony Kaminju/Reuters; **p. 389 Problem 72** Martin Harvey/Science Source;

**PART II SUMMARY** **P. 391 left** Scott Camazine/Science Source; **right** National Optical Astronomy Observatories/Science Source; **p. 392** Rex/Shutterstock; **p. 393** NASA Earth Observing System.

**PART III OPENER** **P. 394–395** Vladsilver/Shutterstock.

**CHAPTER 12** **P. 396 top** Arno Vlooswijk/TService/Science Source; **middle, left to right** Nui7711/Fotolia; Steve Hamblin/Alamy Stock Photo; NHPA/PhotoShot/Superstock; **bottom** B.A.E. Inc./Alamy Stock Photo; **p. 397 top** Wolfelarry/Fotolia; **middle** Coprid/Fotolia; **bottom** Image Source/IS020/Alamy Stock Photo; **p. 398** Richard Megna/Fundamental Photographs, NYC.; **p. 400** STS-129 Crew, Expedition 21 Crew/NASA; **p. 401** JPL/NASA; **p. 403** Tom McHugh/Science Source; **p. 404** Adam36/Fotolia; **p. 411** U of Illinois At Urbana-Champaign, Dept. of Atmospheric Sciences; **p. 412** Richard Choy/Photolibrary/Getty Images; **p. 414** TongRo Images Inc/Getty Images; **p. 416 top** Kenneth B. Storey/Carleton University; **bottom** Pete Orelup/Flickr RF/Getty Images; **p. 418 top, left to right** Alessandro Trovati/AP Images; Stephen Oliver/Alamy Stock Photo; Tom Brakefield/Photodisc/Getty Images; **middle** Pinkomelet/Shutterstock; **p. 423 left to right** Cn Boon/Alamy Stock Photo; Gary S. Settles & Jason Listak/Science Source; Pascal Goetgheluck/Science Source; Brian McEntire/Shutterstock; **p. 424 top** Gillian Merritt/Alamy Stock Photo; **bottom** Kwest/Shutterstock; **p. 425 top** Tony McConnell/Science Source; **bottom** 2000 John Hopkins University/Applied Physics Laboratory LLC; **p. 426** Ted Kinsman/Science Source; **p. 429 top** Robert Hamilton/Alamy Stock Photo; **middle** Power and Syred/Science Source; **p. 430** Brian Jones; **p. 433 Question 12** Eric Schrader/Pearson Education, Inc.; **p. 434 Question 27** Matthew Johnston/Alamy Stock Photo; **p. 437 Question 55** Brian Jones; **p. 438 Problem 75** B. Mauck et al., "Thermal windows on the trunk of hauled-out seals: hot spots for thermoregulatory evaporation?" J Exp Biol 206, 1727-1738, doi: 10.1242/?jeb.00348; **Problem 81** Eye of Science/Science Source; **p. 439 Problem 95** Four Oaks/Shutterstock.

**CHAPTER 13** **P. 441 top** David Tipling Photo Library/Alamy Stock Photo; **middle, left to right** Hank Shiffman/Shutterstock; Attila Volgyi/Xinhua/Alamy Stock Photo; Andrey Popov/Shutterstock; **bottom** Christopher Gardiner/Shutterstock; **p. 445** Woods Hole Oceanographic Institution/Oceanus Magazine; **p. 447** NOAA; **p. 451** Harald Sund/The Image Bank/Getty Images; **p. 453 top** Altrendo Images/Getty Images; **bottom** Vladimir Wrangel/Shutterstock; **p. 455** Alix/Science Source; **p. 456** Don Farrall/Photodisc/Getty Images; **p. 457** Culture-images GmbH/Alamy; **p. 459** Ed Endicott / Alamy Stock Photo; **p. 460** Zephyr/Science Source; **p. 463** Monkey Business Images/Shutterstock; **p. 465 top** Martin Harvey/Alamy Stock Photo; **bottom** DuCane Medical Imaging Ltd./Science Source; **p. 466** David M. Phillips/Science Source; **p. 468 top** Rusla Ruseyn/Shutterstock; **middle** Vincent Hazat/PhotoAlto/Alamy Stock Photo; **p. 472 Question 21** Vicspacewalker/Shutterstock; **Question 24** Suzanne Oliver; **p. 474 Problem 12** Reinhard Dirscherl/Alamy Stock Photo; **p. 475 Problem 22** Denis-Huot/Nature Picture Library/Alamy Stock Photo;

**PART III SUMMARY** **P. 480** Egill Bjarnason/Alamy Stock Photo.

**PART IV OPENER** **P. 482–483** D-Robert Franz/Image State/Alamy Stock Photo.

**CHAPTER 14** **P. 484 top** Tom McHugh/Science Source; **middle, left to right** With Love Photographs/Flickr RF/Getty Images; Richard Megna/Fundamental Photographs, NYC.; Wan Xiang/Newscom; **p. 486 top** Oscar Burriel/Science Source; **bottom** CORBIS RF/AGE Fotostock; **p. 490** Randall Fung/Corbis/Getty Images; **p. 493** Serge Kozak/AGE Fotostock; **p. 494** Barry Bland/Alamy Stock Photo; **p. 496** NASA; **p. 497** H. Craighead/Cornell Univ.; **p. 498** Aris Messinis/AFP/Getty Images; **p. 499** Tim Wright/Corbis Documentary/Getty Images; **p. 502** Drubig-photo/Fotolia; **p. 503** Steve P./Alamy Stock Photo; **p. 504** Laszlo Podor/Alamy Stock Photo; **p. 505** Tom Williams/Roll Call Photos/Newscom; **p. 507** Sirtravelalot/Shutterstock; **p. 511 Question 16** Image Werks Co.,Ltd./Alamy Stock Photo; **Question 19** Leo Mason/Popperfoto/Getty Images; **p. 514 Problem 37** Josep Pena/Fotolia; **p. 515 Problem 46** John Henshall/Alamy Stock Photo; **p. 516 Problem 67** MJTH/Shutterstock; **p. 517 Problem 81** Ted Horowitz/Corbis/Getty Images.

**CHAPTER 15** **P. 518 top** Andy Harmer/Alamy Stock Photo; **middle, left to right** Stuart Field; Richard Megna/Fundamental Photographs, NYC; Aldo Liverani/Andia/Alamy Stock Photo; **bottom** Brian Jones; **p. 519 top** Olga Selyutina/Shutterstock; **bottom** Eddy Lemaistre/Corbis Sport/Getty Images; **p. 521 top** Aleksandr Belugin/Alamy Stock Photo; **bottom** James Gerhold/Photolibrary/Getty Images; **p. 523** Yevgeniy11/Shutterstock; **p. 530** B. Benoit/Science Source; **p. 533** BMJ/Shutterstock; **p. 534** Mark Carwardine/Photolibrary/Getty Images; **p. 535** Steve Gschmeissner/Science Source; **p. 536** NASA; **p. 538 top** NASA; **bottom** NOAA/Science Source; **p. 539 left** NASA; **right** Robert Harding/Robert Harding World Imagery/Alamy Stock Photo; **p. 543 Question 12** David M Schrader/Fotolia; **Question 19** John Roman/Shutterstock; **p. 545 Problem 28** Monitor National Marine Sanctuary/NOAA; **p. 546 Problem 38** Sihasakprachum/Shutterstock.

**CHAPTER 16** **P. 549 top** Martin Philbey/Redferns/Getty Images; **middle, left to right** Amana images inc./Alamy Stock Photo; Richard Megna/Fundamental Photographs, NYC.; Wayhome studio/Shutterstock; **p. 550** Brian Jones; **p. 551** Pearson Education, Inc.; **p. 554** Brian Jones; **p. 556** Brian Jones; **p. 557 top** Stuart Field; **bottom** Photodisc/Getty Images; **p. 558** Field, Stuart; **p. 560** Dave Maiullo as seen in "THAT PHYSICS SHOW" Off-Broadway www.thatphysicsshow.com; **p. 562** Courtesy Brookhaven National Laboratory; **p. 565** ImageState/Alamy Stock Photo; **p. 569** Davidf/iStock/Getty Images; **p. 571** Utekhina Anna/Shutterstock; **p. 573 Question 1** Dale W. Brown/Shutterstock; **p. 574 Question 10** Jim West/Alamy Stock Photo; **p. 576 Problem 22** Allstar Picture Library/Alamy Stock Photo; **Problem 26** Joe Tucciarone/Science Source; **Problem 28** PCJones/Alamy Stock Photo; **p. 578 Problem 59** ldutko/Shutterstock;

**PART IV SUMMARY** **P. 581** Courtesy D.Vatvani, Deltares, the Netherlands. ; **p. 583** Zoonar GmbH/Alamy Stock Photo.

**PART V OPENER** **P. 584–585** Larry Lynch Nature Photography.

**CHAPTER 17** **P. 586 top** Mbolina/Shutterstock; **middle, left to right** Vasilius/Shutterstock; Ervin Monn/Shutterstock; GIPhotoStock/Science Source; **p. 587 middle** Sigitas Baltramaitis/Alamy Stock Photo; **p. 587 bottom, left and right** Richard Megna/Fundamental Photographs, NYC.; **p. 595 bottom left** Adam Pearlstein/Pearson Education, Inc.; **bottom right** Ed Marshall/Alamy Stock Photo; **p. 597** Adam Pearlstein/Pearson Education, Inc.; **p. 598** Ainsley E Seago/Parrish Brady/Jean-Pol Vigneron/Tom D Schultz/Royal Society Publishing; **p. 599** Stuart Field; **p. 601** Pearson Education, Inc.; **p. 602 top** NOAA Office of Ocean Exploration and Research, Bioluminescence 2009 Expedition; **bottom** Terry Oakley/Alamy Stock Photo; **p. 603 top to bottom** imagebroker/Alamy Stock Photo; Richard Megna/Fundamental Photographs, NYC.; John Eisele/Colorado State University/Colorado State University Photography; Ted Kinsman/ Science Source; Marc Anderson/Alamy Stock Photo; **p. 604** Adam Pearlstein/Pearson Education, Inc.; **p. 605** FotoRequest/Shutterstock; **p. 606** Nir Darom/Shutterstock; **p. 608 top** Ken Kay/Fundamental Photographs, NYC.; **bottom** Adam Pearlstein/Pearson Education, Inc.; **p. 610** Adam Pearlstein/Pearson Education, Inc.; **p. 611** NASA Earth Observing System; **p. 614 Question 18** Leonid S. Shtandel/Shutterstock; **p. 615 Problem 6** Sam72/Shutterstock; **p. 616 Problem 20** Yoshi0511/Shutterstock; **Problem 22** Steve Gschmeissner/Science Source; **Problem 28** Yusran Abdul Rahman/Shutterstock; **p. 617 Problem 32** Bill Gozansky/Alamy Stock Photo; **p. 619 Problem 68** Tom Zagwodzki/Goddard Space Flight Center/NASA; **Problem 69** P. Vukusic, Univ. of Exeter.

**CHAPTER 18** **P. 620 top** Michael Kappeler/DDP/Getty Images; **middle, left to right** Kjekol/Fotolia; Dan Suzio/Science Source; Wavebreakmedia/Shutterstock; **bottom** Richard Megna/Fundamental Photographs, NYC.; **p. 622 top, left to right** Lvaschenko Roman/Shutterstock; Dynamic Graphics Group/IT Stock Free/Alamy Stock Photo; Mike Stone/Alamy Stock Photo; Paul Rapson/Alamy Stock Photo; **bottom** Stuart Field; **p. 623** Stuart Field; **p. 626** Tyler Olson/Shutterstock; **p. 627** Pearson Education, Inc.; **p. 628** Brian Jones; **p. 629** Ken Kay/Fundamental Photographs, NYC.; **p. 630** Pete Atkinson/Stone/Getty Images; **p. 631 top** Samrith Na Lumpoon/Shutterstock; **bottom** Pearson Education, Inc.; **p. 633** Richard Megna/Fundamental Photographs, NYC.; **p. 635** Pearson Education, Inc.; **p. 637** Getty Images; **p. 638** Pixelklex/Shutterstock; **p. 641** Solvin Zankl/Alamy Stock Photo; **p. 642** Alper Çugun; **p. 643** Paul Silverman/Fundamental Photographs, NYC.; **p. 647** Chrispo/Shutterstock; **p. 650 Questions 1 and 3** NASA; **p. 651 Question 4** Saga 1966/Shutterstock; **Question 7** National Gallery, London/Art Resource, NY; **Question 8** Liz Van Steenburgh/Shutterstock; **Question 9** AKG Images; **Question 15** Natalia Harper/Alamy Stock Photo; **Question 16** Brian Jones; **p. 652 Problem 2** Brian Jones; **p. 653 Problem 22** Arsgera/Shutterstock; **Problem 26** Dubois Ch/Andia/Alamy Stock Photo; **p. 654 Problem 29** Vstock/Alamy Stock Photo; **Problem 38** Jones, Brian; **p. 655 Problem 57** Nano Calvo/Vwpics/Visual&Written SL/Alamy Stock Photo; **p. 656 Problem 83** David Hancock/Alamy Stock Photo.

**CHAPTER 19** **P. 657 top** Juniors Bildarchiv GmbH/Alamy Stock Photo; **middle, left to right** Jupiter Images/Alamy Stock Photo; John Eisele/Colorado State University/Colorado State University Photography; Biophoto Associates/Science Source; **p. 658** Doug Perrine/Nature Picture Library; **p. 660 top to bottom** Alexander Raths/Fotolia; NASA; Tetra Images/Alamy Stock Photo; **p. 664** Chris Ryan/OJO Images Ltd/Alamy Stock Photo; **p. 667** Shackleton/Panther Media GmbH/Alamy Stock Photo; **p. 668 left and right** Stuart Field; **p. 670 left and right** Stuart Field; **p. 671** Tetra Images/Alamy Stock Photo; **p. 672** Biophoto Associates/Science Source; **p. 673** NASA; **p. 674 Moon** Brian Jones; **Epigraph** *Philosophical Transactions* (No. 80, February 19, 1671/72); **p. 676** Eric Schrader/Pearson Education, Inc.; **p. 678** NASA; **p. 679 top** Granger, CE; **(a)–(c)** Spencer Bliven; **p. 680 left** James Cavallini/Science Source; **right** Steve Gschmeissner/Science Photo Library/Alamy Stock Photo; **p. 682** Daniel Hebert/Shutterstock; **p. 683 center** Spencer Bliven; **p. 684 Question 1** Stuart Field; **Question 3** Detail Heritage/Alamy Stock Photo; **p. 685 Question 17** Sam Ogden/Science Source; **p. 687 Problem 30** Valentin Valkov/Shutterstock; **p. 688 Problem 53** Alfred Pasieka/Science Source; **p. 689 Problem 71** NASA;

**PART V SUMMARY** **P. 690** Spencer Bliven; **p. 691 left and right** Rob Meyer/Meyer Instruments.

**PART VI OPENER** **P. 694–695** Mark Conlin/Alamy Stock Photo.

**CHAPTER 20** **P. 696 top** CBCK-Christine/Alamy Stock Photo; **middle, left to right** Charles D- Winters/Science Source; yevgeniy11/Fotolia; Scott Sinklier/Alamy Stock Photo; **p. 697** Pearson Education, Inc.; **p. 699** Ahmad A Atwah/Shutterstock; **p. 700** Vilevi/YAY Media AS/Alamy Stock Photo; **p. 702** Steve Hopkin/The Image Bank/Getty Images; **p. 706** Science Pictures Limited/Science Source; **p. 710** Richard Megna/Fundamental Photographs, NYC.; **p. 711** Thierry Grun/Alamy Stock Photo; **p. 720 top** Brian Jones; **middle** Uhryn Larysa/Shutterstock; **bottom** ARCO/Reinhard, H/Arco Images/AGE Fotostock; **p. 727 Question 13** Ondrej Prosicky/Shutterstock; **Question 17** David Lassman/The Image Works; **p. 728 Question 24** Adam Pearlstein/Pearson Education, Inc.

**CHAPTER 21** **P. 734 top** Nick Everett/Alamy Stock Photo; **middle left** Steheap/Fotolia; **middle right** Kokhanchikov/Shutterstock; **bottom** Marianthi Lainas/Alamy Stock Photo; **p. 738 top** Erlend Haarberg/naturepl.com/Nature Picture Library/Alamy Stock Photo; **bottom** Focal point/Shutterstock; **p. 739 middle** Stuart Field; **bottom** Department of

Neuroscience at Scripps Florida; **p. 742** Pearson Education; **p. 743** Geilert/Agencja Fotograficzna Caro/Alamy Stock Photo; **p. 749** Roomanald/Shutterstock; **p. 754** David Morrison/123RF; **p. 755** Stocklite/Shutterstock; **p. 757** Matt9122/Shutterstock; Alicia Armentrout/Pearson Education, Inc.; **p. 759** Andrew Lambert Photography/Science Source; **p. 761 top** Swapan/Shutterstock; **bottom** Dimitri Otis/The Image Bank/Getty Images; **p. 762** Adam Hart-Davis/Science Source; **p. 764** NASA Earth Observing System; **p. 769 Problem 8** Janice Haney Carr/Centers for Disease Control and Prevention; **Problem 14** JPL-Caltech/Space Science Institute/NASA; **p. 771 Problem 52** Gearsolid/iStock/Getty Images.

**CHAPTER 22** **P. 774 top** May/Science Source; **middle, left to right** Pearson Education, Inc.; Pearson Education, Inc.; Africa Studio/Shutterstock; **p. 780** Peter Scoones/Nature Picture Library; **p. 783** Dennis O'Clair/The Image Bank/Getty Images; **p. 784** Dr. Jennifer Mueller.; **p. 785 left to right** Pixel shepherd/Alamy Stock Photo; Igor S./Shutterstock; Brian Jones; **p. 788** Brian Jones; **p. 789** Osmose Utilities Services, Inc; **p. 796 Problem 17** Juniors Bildarchiv/R312/GmbH/Alamy Stock Photo.

**CHAPTER 23** **P. 800 top** Mark Newman/Lonely Planet Images/Getty Images; **middle, left to right** Pavlo Sachek/Fotolia; Patrick Batchelder/Alamy Stock Photo; CNRI/Science Source; **p. 807** Dynamic Graphics Group/IT Stock Free/Alamy; **p. 809** Abelstock/Alamy Stock Photo; **p. 811** Simon Fraser/Science Source; **p. 812** Kc_film/Alamy Stock Photo; **p. 816** Pearson Education, Inc.; **p. 819 top** Eric Schrader/Pearson Education, Inc.; **middle** Photo by: Biophoto Associates/Science Source; Colorization by: Mary Martin; **p. 822** Stephane DeSakutin/AFP/Getty Images; **p. 823** SPL/Science Source; **p. 824** STEVE GSCHMEISSNER/SCIENCE PHOTO LIBRARY, photoquest/Science Photo Library/Newscom; **p. 825** STEVE GSCHMEISSNER/Science Photo Library/Alamy.

**CHAPTER 24** **P. 838 top** Thierry Berrod/Mona Lisa Production/Science Source; **middle, left to right** Daniel Dempster/Alamy Stock photo; John Eisele/Colorado State University; Romko_chuk/iStockphoto/Getty Images; **p. 841** John Eisele/Colorado State University; **p. 842 all** John Eisele/Colorado State University; **p. 843** Atsushi Arakaki et al., "Formation of magnetite by bacteria and its application" J R Soc Interface. 2008 September 6; 5(26): 977–999. doi: 10.1098/rsif.2008.0170; **p. 844 all** John Eisele/Colorado State University; **p. 849** Kathleen Gustafson; **p. 851** Charles Thatcher/The Image Bank/Getty Images; **p. 853** Stacy Walsh Rosenstock/Alamy Stock Photo; **p. 855** Chris Madeley/Science Source; **p. 856** GE Healthcare; **p. 857** Hach Company; **p. 860** Brian Jones; **p. 864** UHB Trust/The Image Bank/Getty Images; **p. 867 left** Vitaly Korovin/Shutterstock; **right** John Eisele/Colorado State University.

**CHAPTER 25** **P. 879 top left and right** Ted Kinsman/Science Source; **middle, left to right** Yegor Korzh/Shutterstock; ABERRATION FILMS LTD/SCIENCE SOURCE; Jakelv7500/Shutterstock; **p. 881** Martin Marietta Corporation/Science Source; **p. 882** wildestanimal/Shutterstock; **p. 884** Lance Cheung/U.S. Air Force; **p. 889** Steve Tulley/Alamy Stock Photo; **p. 892** Martin Shields/Alamy Stock Photo; **p. 894** SSPL/The Image Works; **p. 895** Steve Morgan/Alamy Stock Photo; **p. 897 middle** Richard Megna/Fundamental Photographs, NYC.; **bottom** Innerspace Imaging/Science Source; **p. 898** Brian Jones; **p. 899** Pearson Education, Inc.; **p. 901 middle left** Jim West/Alamy Stock Photo; **middle right** GFC Collection/Alamy Stock Photo; **p. 901 bottom, left to right** Brian Jones; **p. 902 left** Ashok Captain/ephotocorp/Alamy Stock Photo; **right** Edwrd Kinsman/Science Source; **p. 904 clockwise from upper left** National Hurricane Center/NOAA; NASA; Jet Propulsion Laboratory/NASA; Image courtesy of NRAO/AUI; **p. 909 Question 25** SSPL/The Image Works; **p. 910 Question 35** Courtesy of Cushcraft Amateur Antennas.

**CHAPTER 26** **P. 916 top** Duncan Usher/Alamy Stock Photo; **middle, left to right** Biehler Michael/Shutterstock; Tom Carter/Alamy Stock Photo; Eric Audras/ONOKY/Getty Images; **p. 917** Pearson Education; **p. 919** Seanami/iStockphoto/Getty Images; **p. 920** 2p2play/Shutterstock; **p. 921** Pearson Education; **p. 923** Photosthai/Shutterstock; **p. 924 top** Brian Jones; **bottom** 1125089601/Shutterstock; **p. 925** Paul A. Souders/Corbis/VCG/Getty Images; **p. 927 top** Pearson Education; **bottom** Steve Wisbauer/Stockbyte/Getty Images; **p. 928** AllOver images/KHG/Alamy Stock Photo; **p. 930** Eric Schrader/Pearson Education; **p. 931** Galyna Andrushko/Shutterstock; **p. 934** David Burton/Alamy Stock Photo; **p. 941 Problem 14** Bob Betenia/Alamy Stock Photo; **p. 943 Problem 68** Steve Sant/Alamy Stock Photo.

**PART VII OPENER** **PP. 948–949** Moritz Wolf/ImageBROKER/Alamy Stock Photo.

**CHAPTER 27** **P. 950 top** Conservation Photojournalism/Shutterstock; **middle, left to right** MVH/E+/Getty Images; NASA; Sadeq/Fotolia; **p. 951 top** Topham/The Image Works; **bottom** Kaz Chiba/The Image Bank/Getty Images; **p. 953** EStock Photo/Alamy Stock Photo ; **p. 955** Hemis/Cintract Romain/Alamy Stock Photo; **p. 958** Jim Corwin/Alamy Stock Photo; **p. 960** Arid Ocean/Shutterstock; **p. 963** U.S. Department of Defense Visual Information Center; **p. 966** Science Source; **p. 967** Eckhard Slawik/Science Source; **p. 968** Stanford Linear Accelerator Center/Science Source; **p. 971** Lawrence Berkeley Laboratory/SPL/Science Source; **p. 973** SSPL/The Image Works; **p. 975** LBNL/Science Source; **p. 982 Problem 13** Dynamic Graphics/Age Fotostock; **p. 984 Problem 68** John Chumack/Science Source.

**CHAPTER 28** **P. 986 top** Science Source; **middle, left to right** T. L. Dimitrova, A. Weis American Journal of Physics 76, 137 (2008); doi: 10.1119/1.2815364 ; Science Source; Ted Kinsman/Science Source; **bottom** GIPhotoStock/Science Source; **p. 987 top** ORAU; **bottom** Neil Borden/Science Source; **p. 989** Raymond Gosling/King's College London; **p. 992 top** Bettmann/Getty images; **bottom left** Travelib/Alamy Stock Photo; **bottom right** Nor Gal/Shutterstock; **p. 993** RUZvOLD/Shutterstock; **p. 995** T. L. Dimitrova, A. Weis American Journal of Physics 76, 137 (2008); doi: 10.1119/1.2815364; **p. 996** Kenneth Eward/BioGrafx/Science Source; **p. 998** Pearson Education; **p. 999 (a) and (b)** Educational Development Center, Inc.; **(c)** Oak Ridge National Laboratory/Courtesy AIP ESVA/Science Source; **p. 1000 top** Science Source; **bottom** Brian Jones; **p. 1002** Bettmann/Getty Images; **p. 1004** SPL/Science Source; **p. 1007 top** Colin Cuthbert/Science Source; **middle** IBM Research/Science Source; **bottom** Lawrence Livermore National Laboratory/Science Source; **p. 1008 left** Ogwen/Shutterstock; **right** Dr. Tomáš Samuely/Institute of Physics/Pavol Jozef Šafárik University, Košice; **p. 1009** Mehau Kulyk/Science Source; **p. 1010 left** Science Source; **right** IBM Research/Science Source; **p. 1014 Problem 14** Martin Dohrn/Nature Picture Library/Alamy Stock Photo; **Problem 20** Darwin Dale/Science Source; **p. 1015 Problem 28** Solent News/Splash News/Newscom; **Problem 36** SPL/Science Source.

**CHAPTER 29** **P. 1019 top** Lisa Werner/Alamy Stock Photo; **middle, left to right** Stuart Field; Doug Perrine/Nature Picture Library/Alamy Stock Photo; G.K./Shutterstock; **p. 1020** Image courtesy of Ocean Optics; **p. 1021** NASA; **p. 1040 left and right** Lowell Observatory; **p. 1041** Richard Megna/Fundamental Photographs, NYC.; **p. 1042** Sinclair Stammers/Science Source; **p. 1043** Frankwalker.de/Fotolia; **p. 1044 top** U.S. Air Force photo/Tech. Sgt. Larry A. Simmons; **middle** Pawel Gaul/iStockphoto/Getty Images.

**CHAPTER 30** **P. 1054 top** Oivashchenko; **middle, left to right** Blickwinkel/McPHOTO/NBT/Alamy Stock Photo; Earl Roberge/Science Source; HTO/Pearson Education; **p. 1055** Ball and Albanese/Alamy Stock Photo; **p. 1057** Roberto de Gugliemo/Science Source; **p. 1059** NASA; **p. 1060** U.S. Department of Energy; **p. 1063** Chimpinski/Shutterstock; **p. 1071** Photo courtesy Los Alamos National Laboratory; **p. 1072** James King-Holmes/Science Source; **p. 1073** Bert de Ruiter/Alamy Stock Photo; **p. 1074** Hank Morgan/Science Source; **p. 1075** Stockbroker/MBI/Alamy Stock Photo; **p. 1076 top** Chauvaint Chapon/Phanie/SuperStock; **bottom** Philippe Voisin/Phanie/Alamy Stock Photo; **p. 1077 top** CNRI/Science Source; **bottom** CNRI/Science Source; **p. 1078 top** Hank Morgan/Science Source; **middle** Zephyr/Science Source; **bottom** Zephyr/Science Source; **p. 1079** CERN/Science Source; **p. 1080** Kamioka Observatory; **p. 1082** U. S. Department of Energy; **p. 1085 Question 31** Brian Jones; **p. 1087 Problem 44** David Parker/Science Source; **Problem 48** Jean-Perrin/CNRI/Science Source; **p. 1088 Problem 52** Science Source; **p. 1089 Problem 67** John R. Foster/Science Source;

**PART VII SUMMARY** **P. 1093** Bgran, Used under a Creative Commons License, http://creativecommons.org/licenses/by/2.0/deed.en.

# Index

For users of the two-volume edition, pages 1–583 are in Volume 1 and pages 584–1093 are in Volume 2.


## A

Aberrations, 677–678
Absolute pressure, 403–404
Absolute temperature scale, 365
Absolute zero, 365
Absorption band, 1041
Absorption spectra, 1004, 1025–1027
  collisional excitation, 1026–1027
  excitation by, 1038
  gases, 1021
  hydrogen, 1031–1032
  spectroscopy, 1021
AC circuits, 916–947
  AC sources, 917–919
  capacitor circuits, 927–929
  inductor circuits, 929–932
  *LC* circuits, 932–933
  power in, 918–919
  *RC* circuits, 816–819
  resistor circuits, 918–919
  *RLC* circuits, 933–935
  transformers and, 919–923
Acceleration, 43–50, 83–85, 90–92
  angular, 212–214
  centripetal, 91–92, 176
  in circular motion, 90–92, 176
  constant, 47–50, 53–55, 115–116
  forces and, 115–118
  free-fall, 55–60, 86–87
  linear (one-dimensional) motion, 43–50, 55–60
  problem-solving strategy for, 53–55
  in projectile motion, 86–87
  rotational motion and, 211–215
  sign determinations, 46–47
  in simple harmonic motion, 489–490
  tangential, 214–215
  two-dimensional motion and, 83–85
  units of, 44, 57
Acceleration vectors, 46–47, 83–85, 86–87, 90–91
Acceleration-versus-time graphs, 45–46, 489–491, 493
Accommodation (in vision), 662–663
Achromatic doublet, 678
Action potential, 821–823
Action/reaction pairs of forces, 122–123, 157–163. *See also* Newton's third law of motion
  interacting objects, 122–123, 157–159
  rope and pulley systems, 159–163
Activation energy, 337
Activity, of radioactive sample, 1071–1072
Adiabatic processes (adiabats), 411–412
Agents (of force), 107
Air
  atmospheric pressure, 447–448
  density of, 442–443
  drag force (resistance) in, 152
Air conditioners, 372
Air resistance, 151. *See also* Drag force
Alpha decay, 1059, 1063–1064, 1066
Alpha particles (or rays), 1022–1024, 1063
Alternating current (AC), 884, 916–947. *See also* AC circuits
Ammeters, 810
Ampere (A), 778
Amplitude, 487, 496–497, 502–503, 525–528
  displacement from equilibrium, 487, 525–528
  linear restoring force, 487
  oscillations and, 487, 502–503
  simple harmonic motion from, 496–497
Angle of incidence, 624, 627
Angle of reflection, 624
Angle of refraction, 627
Angular acceleration, 212–214
Angular displacement, 207–209, 315
Angular magnification, 668
Angular momentum, 295–298
  conservation of, 296
  electron spin, 1034
  moment of inertia and, 295–298
  orbital, 1033–1034
  quantization of, 1033–1034
  total, 296
Angular position, 206, 209
Angular position-versus-time graphs, 209–210
Angular resolution, 679
Angular size, 668
Angular speed, 208, 210
Angular velocity, 207–209, 211
Angular velocity-versus-time graphs, 209–210
Antennas, 900–901
Antimatter, 975, 1079–1080
Antinodes, 552–553, 559–560
Antiparticles, 1079–1080
Antireflection coatings, 599, 601
Apparent size, 668
Apparent weight, 142–144, 184–186
Arc length, 206
Archimedes' principle, 449–450
Astigmatism, 667
Atmospheric pressure, 447–448
Atomic data tables, A-4–A-6
Atomic mass number (A), 397, 1025, 1055, 1056–1057
Atomic mass unit (u), 397, 1056–1057
Atomic models, 110–111, 397–405
  conduction, 423
  ideal gases, 399–405, 442
  liquids, 397, 442
  matter, 397–405
  molecular bonds, 110
  normal force, 110–111
  solids, 397
  tension force, 110
Atomic number, 1024–1025
Atomic physics, 1024–1037
  emission spectra, 1020, 1025–1027, 1039–1040
  excited states, 1003, 1037–1040
  hydrogen atom, 1027–1032
  lasers, 1042–1044
  multi-electron atoms, 1034–1037
  nuclear model for, 1024–1025
  periodic table of the elements, 1035–1037
  stimulated emission, 1042–1044
Atomic spectra, 1020–1021
Atoms, 703–705, 1019–1040, 1079–1081
  Bohr model of, 1025–1032
  electric charge and, 703–705
  electron configurations of, 703, 1035
  nuclear model of, 1022–1025. *See also* Nucleus
  particle physics, 1079–1081
  plum-pudding (raisin-cake) model of, 1022
  Schrödinger's quantum-mechanical model of, 1032–1033
  shell electron configurations, 1035–1037
  spectroscopy for, 1020–1021
Aurora, 855
Avogadro's number ($N_A$), 398
Axon, 823–825

## B

Babinet's principle, 608
Balance, 250–251
Balmer formula, 1021
Balmer series, 1032
Barometers, 447–448
Base of support, 250
Batteries, 738, 762, 768, 788–789
  charge escalator model, 778–779
  charging capacitors with, 768
  current in, 802–804
  electric potential and, 738, 768
  emf of, 778–779, 788–789
  energy transformation in, 788–789
  potential difference of, 780–781
  recharging, 762
Beats, 569–570
Becquerel (Bq), 1071
Bernoulli effect, 458
Bernoulli's equation, 459–460
Beta decay, 1064–1065, 1066
Beta particles, 1065
Big Bang theory, 538
Binding energy, 1056, 1058–1060
Binomial approximation, 969
Biologically equivalent dose, radiation, 1073–1074
Bioluminescence, 1042
Blood pressure, 463–465
Blue shift, 538
Bohr models
  atomic quantization, 1025–1027
  hydrogen atom, 1027–1032
Bohr radius, 1029
Boiling point, 416
Boltzmann's constant, 399
Bond energy, 336
Bond length, 335
Boundary conditions (standing waves), 553–555
Bonds, hydrogen, 704–705
Brachiation, 502
Bragg condition, 988–989
Bright fringes, 592–593
Brownian motion, 426–427, 992
Buoyancy, 448–454
Buoyant force, 448–449

## C

Calories (cal), 359
Calorimetry, 418–420
  problem-solving strategy for, 419
Cameras, 658–660
Capacitance, 757–761
  electric potential and, 757–761
  equivalent, 814–816
Capacitive reactance, 928–929
Capacitors, 757–763, 814–819, 927–929
  AC circuits, 927–929, 932–933
  charging, 757–758, 818–819
  dielectrics and, 760–761
  discharging, 775, 816–818
  electric fields inside, 762–763
  electric potential in, 757–763
  energy storage in, 761–763
  in parallel and series, 814–816
  time constant for, 817–819
Carbon dating, 1072–1073
Cartesian coordinates, 74

Cathode-ray tubes (CRT), 723–724
Causal influences, 973
Cell membrane, 819–820
Cells, 738–739, 819–825
Celsius scale, 365
Center of gravity, 219–223, 247–249
  finding, 220–222
  static equilibrium and, 247–249
  torque and, 219–223, 247–249
Central maxima, 591, 609
Centrifugal force, 184, 186–187
Centripetal acceleration, 91–92, 176
Čerenkov radiation, 1082
Chain reactions, 1060
Charge carriers, 775
Charge diagrams, 698–699, 701
Charge distribution, 881–883. *See also* Electric fields
  electric dipoles, 722–723
Charge escalator, 779–780
Charge models, 699–701, 779–780
Charge polarization, 702
Charge-coupled device (CCD), 660
Charged particles. *See* Electrons; Protons
Charges. *See* Electric charges
Charging, 697–699, 703, 712–720
  atomic view of, 703
  capacitors, 757–758
  process of, 697–699
Chemical energy, 311, 335–339
  activation energy, 337
  catalysts, 338
  conservation of, 338–339
  endothermic reactions, 338
  exothermic reactions, 338
  molecular bonds and, 335–337
  reaction coordinates, 337
  reaction energy diagrams for, 337–338
  reaction rates, 338
Chemiluminescence, 1042
Chlorophyll, 677
Chromatic aberration, 677–678
Circuit breakers, 924
Circuit diagrams, 801
Circuit elements, 785, 801, 931. *See also* Capacitors; Inductors; Resistors
Circuits. *See* Electric circuits
Circular motion, 5, 90–92, 175–204, 206–211. *See also* Uniform circular motion
  acceleration and velocity vectors for, 90–92, 176
  angular position, 206
  angular velocity, 207–209
  apparent weight in, 184–186
  centrifugal force, 184, 186–187
  direction of, 206
  frequency (revolutions) of, 177–178
  linear motion quantity relations, 212–213
  orbital, 187–189, 193–196
  period of, 177–178
  problem-solving strategy for, 180
  simple harmonic motion and, 491–494
  speed of, 177–178, 187
  weightlessness and, 189
Circular orbits. *See* Orbital motion
Circular waves, 531
Circular-aperture diffraction, 609–610
Circulatory system, fluid flow in, 463–468
Clock synchronization, 957, 958–959
Coefficient of kinetic friction, 147
Coefficient of linear expansion, 413
Coefficient of performance (COP), 372–373
Coefficient of rolling friction, 147–148
Coefficient of static friction, 147
Coefficient of volume expansion, 412
Coherent light, 1042–1043
Cold reservoir, 368
Collisional excitation, 1025–1027, 1038–1039
Collisions, 279–295
  conservation of energy for, 340
  conservation of momentum for, 287–292, 340
  elastic, 279–287, 340–341
  energy in, 339–341
  inelastic, 292–295, 339
  mass-energy equivalence and, 975
  molecular particles, 302, 404
  pressure in gases, 302, 404
  relativistic energy and, 975
  two-dimensional, 294–295
Color, 602–605, 674–677, 904
Combination motion, 211. *See also* Rolling motion
Compasses, 839, 840–841
Complete circuit, 779
Component vectors, 75, 135
Compression, 252–258
  adiabatic, 411–412
  elastic potential energy and, 323–324
  equilibrium and, 252–258
  of gases, 407–409, 411–412, 442
  Hooke's law for, 253–254
  incompressibility of fluids, 442, 444, 455
  isobaric (constant pressure), 407–408
  isothermal (constant temperature), 408–409
  restoring forces and, 255, 257
  sound waves and, 529, 558–559
  Young's modulus for, 255–256
Condensation point, 416
Conduction, 423–424
Conduction electrons, 775
Conductors, 700–701, 704
  charge polarization, 702
  charging process, 697–699, 704
  discharging, 700
  in electric fields, 719–720
  in electrostatic equilibrium, 701, 754
  insulators and, 700–702, 704
  isolated, 701
Conservation of angular momentum, 296
Conservation of charge, 699, 703–704, 802
Conservation of current, 776–777
Conservation of energy, 313–314, 327–331. *See also* First Law of Thermodynamics
  in charge interactions, 740–743
  for chemical energy, 338–339
  in collisions, 340
  energy conservation and, 378
  in fluid flow, 459–460
  in isolated systems, 313, 327–328, 331
  Kirchhoff's loop law, 802–804
  law of, 313, 327, 338
  problem-solving strategy for, 328, 741
  quantum physics and, 991, 1003–1004
  relativity and, 976
  work-energy equation for, 313, 327
Conservation of momentum, 287–292, 296, 971
Constant force, 115–117, 314–316
Constant-pressure (isobaric) processes, 407–408, 421–422
Constant-temperature (isothermal) processes, 408–409
Constant-volume processes, 406–407, 421–422
Constructive interference, 550, 553, 565–569
Contact forces, 107
Continuity, equation of, 455–456
Continuous spectrum, 1020
Convection, 424
Convection cells, 391
Converging lenses, 633–638
Converging (concave) mirrors, 639–642
Conversion factor, 17
Coordinate, 9
Coordinate systems, 8–9
  Cartesian, 74
  motion and, 8–9, 74–78
  with tilted axes, 79
Cosmic ray muons, 966
Cosmic speed limit, 972–973
Coulomb (C), 703, 777
Coulomb's law, 705–710, 747
Critical angle, 250, 629–630
Critical speed, 186
Curie (Ci), 1071
Current. *See* Electric current; Induced electric current
Current loops, 844
  as magnetic dipole, 849
  magnetic fields of, 844, 846–847
  magnetic field in center of, 849–850
  magnetic flux in, 884–889
  magnetic force between, 861
  torque and forces on, 861–863
Current-carrying wire, 844
  length of, 847–848
  magnetic fields of, 844–846
  magnetic force on and between, 858–860
  transformer coils, 920–921
Curves, circular motion on, 182–183
Cyclotrons, 856–857

## D

Damped oscillations, 502–504
Dark fringes, 593
Dark matter, 273
Daughter nucleus, 1063
DC circuits, 917, 918–919
de Broglie wavelength, 997–998
Decay. *See* Nuclear decay; Radioactive decay
Decibels (dB), 534–535
Decomposed (resolved) vector, 75–76
Defibrillators, 762
Degrees vs. radians, 206
Density, 151–152, 442–443, 450–454
  average, 450–451
  Bernoulli equation and, 459–460
  buoyancy and, 450–454
  drag and, 151–152
  energy, 763
  fluids, 450–454, 459–460
  linear, 521
  mass, 442–443
  units of, 152
Depolarization, 718, 755–756, 822
  cellular, 718, 822
Descriptive models, 7
Destructive interference, 551, 553, 566–569
Deuterium, 1055
Diatomic gas, 398, 422
Dielectric constant, 760
Dielectrics, 760–761
Diffraction, 587, 590, 594–599, 605–610
  circular-aperture, 609–610
  double-slit, 590–593, 675, 995
  Huygens' principle, 605–606
  of light waves, 590
  of matter, 999
  order of, 595
  ray model for, 610
  single-slit, 605–609
  spectroscopy for, 596–597
  x-ray, 988–989
Diffraction fringes, 605, 607–608
Diffraction grating, 594–599
Diffuse reflection, 622, 625
Diffusion, 426–430
  cell membrane, 819–820
  Fick's law, 428–429
  passive, 428
  random walk, 426–427
  root-mean-square distance, 427
  thermal properties of, 426–429
Diffusion constant, 427
Diopter (D), 660
Dip angle, 843
Dipoles. *See* Electric dipoles; Magnetic dipoles
Direct current (DC), 917, 918–919
Discharging electric charge, 700
Discontinuity, 554
Discrete spectra, 1020–1021
Dispersion of light, 674–677
Displacement, 10, 19–20, 40, 72–73
  amplitude of, 487, 525–528
  angular, 207–209, 315
  from equilibrium, 323–324, 487–489
  net, 72–73
  of waves, 525–528

position change, 10
velocity, 12, 40
work and, 314–318
Displacement vectors, 19–20, 72
Disturbance of the medium, 519
Diverging lenses, 633, 638–639
Diverging (convex) mirrors, 639–640, 642–643
Doppler effect, 536–540
Doppler ultrasound, 539
Dose, radiation, 1073–1074
Double-slit interference, 590–599, 675
analysis of, 591–594
diffraction gratings and, 594–599
photon behavior and, 995
Young's, 590–591
Drag coefficients, 152
Drag force, 112, 151–157, 502
in air (resistance), 151, 154–155
damped oscillations, 502
Reynolds number (Re) for, 151–153, 154–156
Stoke's law, 154
terminal speed, 153–154
units for, 152
viscosity and, 151–152
in water, 151, 155–156
Drift velocity, 776
Driven oscillations, 504–506
Driven *RLC* circuit, 934
Driving frequency, 504–505
Dynamic equilibrium, 135, 137–138
Dynamics, 106. *See also* Fluid dynamics; Thermodynamics
friction force and, 182
maximum walking speed, 183–184
moments of inertia, 223–227
Newton's second law and, 138–141, 178–181, 227–230
problem-solving strategies for, 139, 180
rotational motion, 215–230
torque, 215–223
uniform circular motion, 178–184

## E

Echolocation, 529, 538
Eddy currents, 892
Efficiency, 356–358, 370–373
coefficient of performance (COP), 372–373
of energy transformations, 356–358
entropy and, 376–377
of heat engines, 370–371
of heat pumps, 372–373
of human body, 360–362
problem-solving strategy for, 357
second law of thermodynamics and, 370–371, 372
Elastic collisions, 279–287, 340–341
Elastic limit, 256
Elastic potential energy, 310, 323–324, 331
Elastic region, 256
Elasticity, 252–258. *See also* Collisions
biological materials, 257–258
Hooke's law and, 253–254
solid materials, 255–258
springs, 252–255
tensile strength, 256–258
Young's modulus for, 255–256, 257
Electric charges, 697–705, 712–720. *See also* Charging; Electrons; Point charges
atomic view of, 703–705
charge models, 699–701, 779–780
capacitor energy storage and, 761–763
conductors, 700–702, 704, 719–720
conservation of, 699, 703–704, 802
dipoles, 704, 714
electric fields and, 710–723
electric potential and, 735–743, 749–750
electrostatic equilibrium, 701
energy conservation in, 740–743
insulators, 700–701, 704
polarization, 702, 718, 755–757
problem-solving approach to, 741
separation of, 738–739
transferring, 703–705
visualizing, 698–699
Electric circuits, 800–837, 917–919. *See also* AC circuits; Resistor circuits
alternating current (AC), 917–923, 927–935
capacitors, 814–819, 927–929
complex, 811–814
direct current, 917, 918–919
household, 923–925, 927
induced current in, 880, 882–883
inductors, 929–932
Kirchhoff's laws, 802–805
*LC*, 932–933
oscillation, 932–935
parallel, 805, 808–814, 815–816
problem-solving strategy for, 812–813
*RC*, 816–819
*RLC*, 933–935
series, 805–807, 811–814, 814–816
single-resistor, 786–787
symbols for, 801
voltage and current measurements, 810–811
Electric current, 774–799, 844–851, 880–884, 916–947
alternating (AC), 884, 916–947
batteries, 779–780, 788–789
charge carriers, 774
conservation of, 776–777, 778–779
creating, 775–776
direction of, 775–776, 802–804
energy transfer and, 788–790
ground fault interrupter (GFI) for, 927
induced, 880, 882–892
at a junction, 778–779
*LC* circuit capacitors, 932–933
magnetic fields and, 844–851, 858–861
magnetic forces on, 858–861
measuring, 810–811
model of, 775–777
peak, 918, 928, 935
potential difference and, 776, 780–787
power and, 788–790
resistance and, 774–799
root-mean-square, 918–919
safety and, 925–927
Electric dipole moment, 722
Electric dipoles, 704
electric charge and, 704
electric field of, 714
electric potential of, 751
induced, 704, 722–723, 866–867
magnetic dipole vs., 840
torque on, 722–723, 840
Electric field diagrams, 713
Electric field lines, 716–718
Electric field strength, 711–712
Electric field vectors, 712–713, 714–718
Electric fields, 710–723
calculating, 711–713
in conductors, 719–720, 754
Coulomb's law and, 710
dipoles and. *See* Electric dipoles
electric potential and, 751–754
electrostatic equilibrium and, 754
from equipotential surfaces, 752–753
field model, 710, 711–712
forces and torques in, 721–723
induced, 893
in parallel-plate capacitor, 714–716, 718
of point charges, 712–713, 718
potential energy storage in, 762–763
uniform, 714–716
visualizing, 710
Electric forces, 113, 696–733. *See also* Electric fields
charges and, 697–705, 712–720
Coulomb's law, 705–710
electric dipoles and, 722–723
field model for, 711–712
problem-solving approach for, 707
torque and, 721–723
Electric generators, 883–884
Electric motors, 864
Electric potential, 734–773
calculating, 743–751
capacitors and, 757–763
of charged sphere, 749–750
current and, 776, 780–787
of dipoles, 751
electric field and, 751–754
finding, 736–737
graphical representations of, 745–746, 748
in parallel-plate capacitor, 744–746, 758–759
of point charges, 746–748, 750
sources of, 737–739
units of, 737, 742–734
Electric potential energy, 735–737, 740–743, 762–763
conservation of, 735, 740–743
in electric field, 735–737, 762–763
electric potential vs., 740–743
Electric power, 788–790
current and, 788–790
transmission of, 922
Electrical impedance tomography, 784
Electricity. *See also* Electric circuits; Electric current; Electric fields
atoms and, 703–705
biological effects of, 925–927
charge model, 699–701
conservation of charge, 699, 703–704
Coulomb's law, 705–710, 747
household, 923–925
insulators, 700–701, 704, 775, 824–825
in nervous system, 819–825
power transmission of, 922
safety of, 925–927
Electrocardiograms, 718, 755–757
Electrodes, 714–715, 757
Electrolytes, 779
Electromagnetic flowmeters, 857
Electromagnetic (EM) induction, 879–915, 929–932. *See also* Induced electric current
AC circuits, 929–932
eddy currents, 892
Faraday's law, 889–892, 929, 930
induced currents, 800, 802–804
*LC* circuits, 932–933
Lenz's law, 886–889
magnetic flux, 884–892
motional emf, 880–884
problem-solving strategy for, 890
in solenoids, 893–895, 929–930
in wire loop, 886–889
Electromagnetic radiation. *See* Radiation
Electromagnetic spectrum, 530, 894, 900–904
Electromagnetic (EM) waves, 519, 523, 530–531, 588, 879–915
antennas, 900–901
energy of, 895–896
greenhouse effect and, 945
induced fields, 893
infrared radiation, 903
light, 523, 530, 588, 894, 896–898
Maxwell's theory of, 894
microwaves, 531, 900–901
nodes, 558
photon model of, 899–904, 993–995
polarization of, 895, 896–898
properties of, 893–895, 896
radio waves, 900–901
speed of light, 588, 893–894
standing, 558
thermal radiation and, 901–902, 945
traveling. *See* Traveling waves
visible light, 900, 903–904
wave model, 519, 899–900
wave-particle duality, 899–900
Electromagnets, 867
Electromotive force. *See* emf
Electron configuration, 1035
Electron microscope, 681, 999–1000
Electron spin, 1033, 1034
Electron volt (eV), 742–743

Electrons, 703–705
alpha particles (rays), 1022–1024
beta particles (rays), 1022, 1065
closed shell, 1035
conduction, 704, 775
current and, 775–776
de Broglie wavelength, 997–998
diffraction pattern, 999
drift velocity, 776
electric charge and, 703–705, 1022
excited states, 1037–1040
ground state and, 1035
Heisenberg uncertainty principle, 1005–1007
magnetic dipole moment, 1034
quantization and, 1000–1002
photoelectric effect, 989, 990–991, 993–995
sea of, 704
shell configurations, 1035–1037
tunneling, 1007
valence, 704, 1037
wavelength of, 997–1000
wave-like properties of, 999–1002
Electrostatic constant, 706
Electrostatic equilibrium, 701, 754
Elements, periodic table of, 1035–1037, A-3
emf, 778–779, 788–789, 881–882, 889
of a battery, 778–779, 788–789
Faraday's law for, 889–891
of generators, 883–884
induced, 889
induced current and, 882–884
motional, 881–884
peak, 917
Emission spectra, 1003, 1020–1021, 1025–1027
collisional excitation, 1026–1027
excitation and, 1039–1040
hydrogen, 1031–1032
spectroscopy, 1020–1021
Emissivity, 425
Empirical formulas, 413
Endothermic reactions, 338
Energy, 309–353, 354–389. *See also* Conservation of energy; Energy transformations; Kinetic energy; Potential energy; Thermal energy; Work
binding, 1058–1060
capacitor storage, 761–763
chemical, 311, 335–339
in collisions, 339–341
conservation of, 378, 740–743
electric potential and, 740–743
of electromagnetic waves, 895–896
entropy and, 375–380
in human body (system), 358–363
hydrogen atoms, 1033–1034
ionization, 750–751
mechanical, 312–313, 340
models of, 310–314
of motional emf, 883
nuclear, 311, 1060–1062
power and, 341–343
problem-solving strategy for, 328
quantization of, 992, 1000–1004, 1033–1034
relativistic, 973–976
rest, 974
self-organizing systems, 391
simple harmonic motion and, 494–499
spontaneous reactions, 379–380
spreading of, 375
stored, 310, 363
systems of, 310–311
thermal, 311, 325–326
thermodynamics, 367–377
total, 310–311, 313, 331, 973–974, 1058
Energy density, 763
Energy diagrams, 331–338
Energy-level diagrams, 1002–1004
Energy levels, 1001–1004, 1025–1027
hydrogen atoms, 1030–1031
quantization and, 1001–1004, 1025–1027
quantum jumps, 1002–1004
Energy-mass equivalence, 975
Energy reservoir, 368
Energy transfer, 312–313, 366–367. *See also* Energy transformations
calorimetry, 418–420
efficiency of, 370–371, 376–377
entropy and, 375–380
heat as, 312, 366–367
heat engines, 369–372
thermal equilibrium, 366
by waves, 519, 532–533, 535
work as, 312–313, 369–372
Energy-transfer diagrams, 368–369
Energy transformations, 311–312, 355–363
efficiency of, 356–358, 360–362
energy input for, 358–359
energy output from, 359–360
in the human body, 358–363
locomotion analysis, 363
metabolic energy use, 358–360
power from, 341–343
representation of, 311–312
English units, 17. *See also* Units of measurement
Entropy, 375–380, 391
Equation of continuity, 455–456
Equations of motion, 34–35, 38, 47–50
Equilibrium, 135–138, 244–271. *See also* Newton's second law
center of gravity, 247–249
displacement from, 323–324, 487–489
dynamic, 135, 137–138
electrostatic, 701, 754
elasticity and, 252–258
energy diagrams for positions of, 334–335
hydrostatic, 445
oscillation and, 485–486
phase, 416
pivot point, 245–247, 249
problem-solving strategies for, 135
restoring forces, 252–257, 485, 487–489
second law of thermodynamics, 374–377
stable, 250–251, 334–335
stability and balance, 250–251
static, 135, 136–137, 245–249
thermal, 366–367
unstable, 250–251, 334
Equipotential lines (map), 745–746, 748
Equipotential surfaces, 745, 748, 752–753
Equivalent capacitance, 814–816
Equivalent resistance, 805–812
Escape velocity, 747–748
Estimation of measurements, 18–19
Evaporation, 417–418
Events (relativity), 957, 959, 964–965
Exact numbers, 15
Excitation, 1037–1044
absorption, 1038
collisional, 1038–1039
emission spectra, 1039–1040
molecules, 1041–1042
stimulated emission, 1042–1044
x-ray, 1040
Excited states, 1003
Exclusion principle, 1035
Exothermic reactions, 338
Expansion, 407–409, 411–413
adiabatic, 411–412
constant-pressure (isobaric) process, 407–408
constant-temperature (isothermal) process, 408–409
of gases, 407–409, 411–412
thermal, 412–413
Explanatory models, 7
Explosions, 291–292
Exponential decay, 502–503, 817, 1070
External forces, 288, 314
Eyepiece, 671

## F

Fahrenheit scale, 365
Far point, 662
Farad (F), 757–758
Faraday's law, 889–892, 929, 930
Ferromagnetism, 865
Fiber optics, 630–631
Fick's law, 428–429
Field diagrams, electric, 713
Fields, 710. *See also* Electric fields; Gravitational fields; Magnetic fields
Filters
colored, 676–677
polarizing, 896–898
First law of thermodynamics, 367–369. *See also* Conservation of energy
Fission, 1059–1060
Flame test, 1040
Floatation, 450–454. *See also* Buoyancy
Fluid dynamics, 454–463
Bernoulli effect, 458
Bernoulli's equation, 459–460
equation of continuity, 455–456
lift, 458
Poiseuille's equation, 461–463
pressure forces and, 458–463
viscosity and, 455, 461–463
Fluid element, 457
Fluid flow, 455–468
Bernoulli effect, 458
in circulatory system, 463–468
equation of continuity, 455–456
laminar, 455
pressure and, 455–468
pressure gradient, 458
streamlines, 456
turbulent, 455
viscous, 461–463
volume flow rate, 456
Fluid statics, 442–457
Fluids, 151–152, 441–477
Archimedes' principle, 449–450
Bernoulli's equation, 459–460
buoyancy, 448–454
density of, 151–152, 442–443, 450–454, 459–460
drag through, 151–152
incompressibility of, 442, 444, 455
laws of biology, 479
in motion, 454–463
pressure, 443–448, 458
viscosity of, 151–152, 455, 461–463
Fluorescence, 1040–1042
Fluorescent lighting, 1044–1045
Focal point and length, 633, 660–661, 669
Focus
cameras, 658–659
human eye, 661–662
refractive power, 660–661
Force vectors, 107, 108
Force-versus-time graph, 279–280
Forces, 94, 105–133. *See also* Electric forces; Magnetic forces; Pressure
acceleration and, 115–118
action/reaction pairs, 122–123, 157–163
atomic model for, 110–111
buoyant, 448–449
centrifugal force, 184, 186–187
combining, 109
constant, 115–117, 314–316
contact, 107
drag, 112, 151–157, 502
elasticity and, 252–257, 258–261
electric, 113
external, 288, 314
free-body diagrams for, 120–122
friction, 111–112, 124, 146–151, 157
gravitational, 190–193
in the human body, 258–261
identifying, 113–115
impulsive, 279
interacting objects, 122–123
internal, 288, 314
long-range, 107
magnetic, 113
magnitude of, 119
mass and, 116–118
net, 109, 118

no work from, 317
normal, 110–111, 144–146, 157
polarization, 702
propulsion, 124
restoring, 252–257, 485, 487–489
spring, 110, 252–254
strong nuclear, 1060–1062
tension, 110, 159–163
thrust, 112–113, 124
vs. torque, 216
units of, 119
weight, 109, 141–144, 157
work and, 314–318
Formants, 564–565
Free-body diagrams, 120–122
Free fall, 55–60, 86–87
acceleration, 55–60, 86
gravity effects on, 55–57, 86, 188–189, 192–193
inclined plane acceleration and, 80
magnitude of, 56–57
orbital motion, 188–189
pendulum SHM and, 486
planetary gravity and, 192–193
projectile motion, 86–87, 188–189
weight and, 141
Frequency, 177–178, 485
angular speed, 208–209
beats, 569–570
circular motion, 177–178
driving, 504–505
electromagnetic waves, 530–531
fundamental, 556–557
hearing (sound) range, 529
natural, 504
oscillations, 485, 490
period and, 485, 496
resonance, 506, 934
revolutions, 177
for simple harmonic motion, 496–497, 500, 502
sinusoidal waves, 526–528
threshold, 990, 994
units of, 177, 485
Frequency shifts, 538–539
Frequency spectrum, 564
Freezing point, 416
Friction, 106, 111–112, 124, 146–151, 157
causes of, 151
charging by, 698–701
coefficients of, 147–148
conservation of energy and, 327–331
kinetic, 111, 147, 149, 157
model of, 148–150
rolling, 148–149
sliding/slipping and, 146–151
static, 112, 124, 146–147, 149, 157
thermal energy and, 325–326
uniform circular motion and, 183
Fringes, 591–593
bright, 592–593, 609
central maxima and, 591, 609
circular-aperture diffraction, 609
dark, 593, 605, 607–608
diffraction, 605, 607–608
double-slit interference, 591–593
interference, 591–593, 598, 602
secondary maxima and, 605, 607–608
single-slit diffraction, 605, 607–608
thin-film interference, 602
Fringe spacing, 592
Fundamental charge, 703–704
Fundamental frequency, 556–557
Fundamental particles, 1081
Fusion, 416–417, 764, 1059

## G

Galilean relativity, 951–954
principle of, 953
reference frames, 951–953
transformation of velocity, 953–954
Gamma decay, 1065–1066
Gamma rays, 904, 1075
Gas constant, 405
Gases, 399–412. *See also* Fluids; Ideal gases
absorption spectra, 1021, 1026
atomic model of, 399–405, 442
atmospheric pressure, 447–448
compression of, 407–409, 411–412
condensation point, 416
density of, 442–443
diatomic, 398
expansion of, 407–409, 411–412
fluid properties of, 442–448
heat transfer in, 422–426
monatomic, 398
phase changes, 415–416
pressure in, 402–405, 407–408, 443–444
specific heats of, 421–422
speed of sound in, 522–523
Gauge pressure, 403–404, 444
Generators, 803–804
Gibbs free energy, 379
Global positioning system (GPS), 960, 976–977
Global warming, 945
Graphical representations, 33–37, 213–214, 489–494, 524–529
Gravitational constant, 190
Gravitational field, 710
Gravitational force, 190–193, 721
Gravitational potential energy, 310, 322–323, 331
Gravitational torque, 219–223
Gravity, 55–57, 187–196. *See also* Center of gravity; Free-fall
acceleration (free-fall) due to, 55–57, 86, 188–189, 192
Newton's law of, 190–193
orbital motion and, 188–189, 193–196
planetary, 192–193
projectile motion, 86
weightlessness, 188
Gray (Gy), 1074
Greenhouse effect, 945
Ground fault interrupters (GFI), 927, 936
Ground state, 1003, 1026, 1034, 1035
Grounded electrical circuits, 923–924

## H

Half-lives, 966, 1068–1073
Harmonics, 556–557, 564–565
Hearing, 529, 561–562, 564
frequency range of, 529
frequency spectrum, 564
resonance in, 506, 561–562
sound intensity level, 533–536
Heart, electric fields of, 718
Heat, 312, 364–367. *See also* Heat transfer; Thermal energy; Thermal properties of matter; Thermodynamics
atomic model of, 366–367
energy transfer process, 312, 366–367
*pV* diagrams for, 406
specific, 414, 421–422
temperature and, 364–367
thermal equilibrium, 366–367
work-energy equation with, 367
Heat engines, 369–372
efficiency of, 370–371
operation of, 369–370
second law of thermodynamics for, 370–371
thermal energy transfer in, 370–371
Heat of fusion, 416–417
Heat of transformation, 416–417
Heat of vaporization, 416–417
Heat pumps, 372–374
Heat transfer, 367–374, 422–426
conduction, 423–424
convection, 424
first law of thermodynamics and, 367–369
heat engine process, 369–372
heat pump process, 372–374
radiation, 425–426
Heating elements, 785. *See also* Resistors
Heisenberg uncertainty principle, 1005–1007
Henry (H), 930
Hertz (Hz), 485
History graphs, 524–525
Hooke's law, 253–254, 256
Hot reservoir, 368
Household electricity, 923–925, 927
Huygens' principle, 605–606
Hydrogen
absorption spectra, 1031–1032
angular momentum of, 1033–1034
Bohr atomic model, 1027–1032
electron spin, 1033, 1034
emission spectra, 1031–1032
energy levels, 1030–1031
quantization, 1033–1034
quantum-mechanical model, 1032–1034
Schrödinger's quantum-mechanical atom, 1032–1033
spectrum, 1031–1032
stationary states of, 1028–1031
Hydrogen bonds, 704–705
Hydrostatic equilibrium, 445
Hydrostatic pressure, 444–447
Hyperopia, 664–665

## I

Ideal-gas law, 404–405
Ideal-gas models, 399–405
Ideal-gas processes, 406–412. *See also* Compression
adiabatic, 411–412
constant pressure (isobaric), 407–408
constant temperature (isothermal), 408–409
constant volume, 406–407
*pV* diagrams for, 406
thermodynamics of, 409–411
work in, 409–411
Ideal-wire model, 786–787
Image distance, 625–626, 637–639
Image formation, 625–626, 631–643
magnification and, 636–637
by plane mirrors, 625–626
ray tracing, 633–639
real images, 634–636
by refraction, 631–632
by spherical mirrors, 639–643
by thin lenses, 633–639
thin-lens equation for, 644–647
virtual, 637–638
Image plane, 635
Impulse, 279–288
conservation of momentum and, 287–288
force-versus-time graph, 279–280
impulse approximation, 286–287
impulse-momentum theorem, 280–284
problem-solving strategies for, 285–287
Impulsive force, 279
Inclined planes, 79–82, 145–146
Incompressibility of fluids, 442, 444, 455
Index of refraction, 588–589, 627–629
Induced dipoles
electric, 704, 722–723
magnetic moments, 866–867
Induced electric current, 880
eddy currents, 892
magnetic flux and, 884–892
motional emf and, 882–884
in MRI scans, 891
Induced electric fields, 893–895, 929–932
Induced emf, 889
Induced magnetic fields, 893–895
Inductance, 930
Induction. *See* Electromagnetic induction
Inductive reactance, 930–931
Inductors, 929–931
Inelastic collisions, 292–293, 339
Inertia, 117, 151
angular momentum and, 295–298
drag and, 151
moment of, 223–227, 295–298

Inertial reference frames, 952–953
Infrared radiation, 901, 903
Inner-shell vacancy, 1040
Instantaneous velocity, 41–43
Insulators, 700–702, 704, 775, 782–783
  conductors and, 700–701, 775, 782–783
  electric dipoles and, 704
  in nerve conduction, 824–825
  resistivity, 782–783
Intensity of waves, 532–533, 535
  electromagnetic, 896
  light, 897–898
  photoelectric effect and, 990–995
  sound, 534–535
Interacting objects, 122–123. *See also* Newton's third law of motion
  action/reaction pairs, 122–123, 157–160
  Newton's laws applied to, 157–160
  objects in contrast, 157–158
  rope and pulley systems, 161–164
Interference, 550–551, 565–569, 590–605. *See also* Diffraction
  along a line, 565–567
  color from, 602–605
  constructive and destructive, 550–551, 553, 565–569
  diffraction grating and, 594–599
  double-slit, 590–593, 675
  identifying, 568
  of light waves, 590–605, 999
  of matter, 999
  from multiple slits, 594–595
  of sound waves, 565–569
  of spherical waves, 567–569
  superposition as, 550–551
  thin-film, 599–605
Interference fringes, 591–593, 598, 602. *See also* Fringes
Internal forces, 288, 314
Inversely proportional relationships, 116
Inverse-square relationships, 190–191
Inverted image, 635
Ion cores, 704
Ionization, 703
Ionization energy, 750–751
Ionization limit, 1031, 1033
Ionizing radiation, 904, 1067–1068
Ions, 703
Iridescent colors, 604–605
Irreversible processes, 374, 375
Isobaric (constant-pressure) processes, 407–408
Isolated system, 289, 313, 327–328, 375–376
Isotherm, 408
Isothermal (constant-temperature) processes, 408–409
Isotopes, 1025, 1055–1056

## J

Joule (J), 314, 743
Junctions, 778–779

## K

Kelvin scale, 365
Kilowatt hours (kWh), 924–925
Kinematic equations of motion, 34–35, 38, 42–43, 56–57, 88
Kinematics, 33–37
  angular acceleration, 212–214
  circular motion, 176–178, 186–189
  constant acceleration, 47–50
  free-fall, 55–60, 187–189
  instantaneous velocity, 41–43
  one-dimensional motion, 33–38, 41–43, 55–60
  projectile motion, 85–90, 186–189
  rotational motion, 195, 211–223
  simple harmonic motion, 489–494
  two-dimensional motion, 70–104, 294–295
  uniform motion, 37–40
Kinetic energy, 310, 318–321. *See also* Collisions; Conservation of energy
  average, 400
  charged particles and, 740–743, 775, 991
  in collisions, 339–340, 775, 975
  conservation of, 327–331
  electric potential energy and, 740–743
  locomotion analysis for, 363
  in photoelectric effect, 991
  relativistic, 973–976
  rotational, 318, 320–321
  simple harmonic motion, 494–497
  total, 320
  transformation of, 363
  translational, 318–319
  work and, 318–321
Kinetic friction, 111, 147, 149, 157
Kirchhoff's laws, 778–779, 802–805
  junction law, 778–779, 802
  loop law, 802–804

## L

Laminar flow, 455
Laser beams, 558, 622–623, 630–631, 1043–1044
Laser cavity, 558
Lasers, 558, 691, 1042–1044
Launch angle, 86
Law of reflection, 624
*LC* circuits, 932–933
Length, measurement of, 14
Length contraction, 968–970
Lens plane, 633
Lenses, 633–639, 644–647
  aberrations, 677–678
  camera, 658–659
  combinations of, 661, 670
  converging, 633–638
  diverging, 633, 638–639
  of eyes, 660–667
  focal point and length, 633, 660, 661
  image formation with, 633–639
  magnification, 636–637
  magnifiers, 667–670
  microscope, 670–672, 679–681, 999–1000
  optical instruments using, 658–674
  ray tracing, 633–639
  refractive power of, 660–661
  resolution and, 678–681
  sign conventions for, 638, 645
  telescope, 673–674
  thin, 633–639
  thin-lens equation using, 644–647
Lenz's law, 886–889
Leptons, 1081
Lever arm, 216
Lifetime of damped oscillations, 503
Lift, 458
Light, 587. *See also* Color; Electromagnetic waves; Light rays; Light waves; Ray optics; Wave optics
  coherent, 1042–1043
  diffuse reflection, 623, 625
  dispersion, 674–677
  electromagnetic waves and, 523, 588, 894–898, 899–900, 904
  filtering, 676–677
  intensity of, 897–898, 899–900
  particle-like behavior of, 995
  photon model of, 587, 899–904, 993–995
  polarization of, 895, 896–898
  quanta of, 587, 992. *See also* Photons
  quantum theory of. *See* Quantum mechanics; Quantum physics
  ray model of, 587, 610, 621–624
  reflection, 624–626, 629–630
  refraction, 588–589, 627–632
  scattering, 623
  sources of, 621–622
  spectrum of, 588, 675, 899–904, 1020–1021
  speed of, 523, 588, 893–894, 955–957, 972
  wave model, 587, 605–606
  wavelengths of, 589, 593, 678, 824, 894, 900
Light clock, 962
Light quanta, 992. *See also* Photons
Light rays, 587, 620–656. *See also* Ray optics
  image formation with, 625–626, 631–643
  ray diagrams, 622
  ray tracing, 633–639
Light waves, 523, 530. *See also* Wave optics
  Doppler effect, 538
  electromagnetic spectrum, 530
  interference, 590–605
  phase changes of, 599–600
  propagation of, 587–588
  red/blue shift effect, 538
  reflected, 538, 599–605
  speed of, 523
  thin films and, 599–605
Light year (ly), 966
Like charges, 698
Line of action, 216
Line of stability, 1057
Linear density, 521
Linear motion, 32–70, 207, 212–213. *See also* Drag force; One-dimensional motion
Linear restoring force, 487–489, 499
Liquids, 397. *See also* Fluids
  atomic model of, 442
  boiling point, 416
  buoyancy, 448–454
  density of, 442–443, 452
  evaporation, 417–418
  fluid properties of, 442–448
  fusion, 416–417
  phase changes, 397, 415–416
  pressure, 443–448, 458–463
  vaporization, 416–417
  viscosity of, 455, 461–463
Locomotion, 363, 500–502
Logarithmic scale, 534
Longitudinal waves, 520
Long-range forces, 107
Lorentz velocity transformations, 970–971
Lyman series, 1032

## M

Macroscopic objects, 110, 375
Magnetic dipole moments, 862–864
Magnetic dipoles, 839, 840, 861–865
  current loops, 861–863
  electric dipole vs., 840
  induced, 866–867
  torque on, 840, 861–865
Magnetic domains, 866–867
Magnetic field lines, 841–842
Magnetic field vectors, 841–842
Magnetic fields, 710, 838–878, 880, 893–895. *See also* Magnetic forces; Magnets
  calculating/measuring, 840–841, 847–851
  charged particles, 851–857
  of current loops, 844, 846–847, 849–850
  of earth, 842–843, 855–856, 886
  electric currents and, 844–851, 880–884
  electromagnetic induction, 880–884
  induced, 893–895
  magnet representation of, 841–842
  Maxwell's theory of, 894
  paths of charged particles in, 854–856
  right-hand rule for, 845, 846
  of solenoids, 844, 847, 850–851, 893–895
  of straight current-carrying wire, 844, 845–846, 847–848
  superposition for, 848–849
  torque on dipoles, 840, 861–865
  uniform, 847, 861–863
Magnetic flux, 884–892
  in closed loop wires, 884–889
  eddy currents, 892
  Faraday's law, 889–892
  induced emf, 889–890
  Lenz's law, 886–889
Magnetic forces, 113, 851–878
  on current-carrying wires, 858–860
  between current loops, 861

on moving charges, 851–857
problem-solving strategy for, 860
right-hand rule for, 852–853, 858–859
torques on dipoles, 861–865
Magnetic materials, 839, 865–867
Magnetic moments, 862–864, 1034
Magnetic poles, 842, 863
Magnetic quantum number, 1033
Magnetic resonance imaging (MRI), 850–851
induced magnetic current, 891
magnetic moments, 863–864, 1008–1009
energy quantization of, 1008–1009
Magnetism, 839–851
atomic structure and, 865–866
biological effects of, 849, 850–851, 856–857
electromagnetism, 867
ferromagnetism, 865
induced current, 879–915
induced magnetic moments, 866–867
Magneto cardiogram, 849
Magnets, 839–844, 865–867
atomic view of, 865–866
magnetic field representation, 841–842
permanent, 844–845, 866–867
Magnification, 636–637, 667–674
angular and apparent size, 668
magnifying glass, 667–670
microscope, 670–672
telescope, 673–674
Magnifiers, 667–670
Magnitude, 19, 34, 56–57, 119
electric field strength, 711–712
Malus's law, 897
Mass
atomic, 397
force and, 116–118
inertial, 117, 225
measuring, 141–142
molar, 398
vs. weight, 141–142
Mass density, 442–443
Mass-energy equivalence, 975
Mass number (A), 397, 1025, 1055, 1056–1057
Massless approximations, 160, 253
Mathematical relationships, 38–39
Matter
antimatter, 975, 1079–1080
atomic models, 397–405, 442
elasticity of, 252–257
evaporation, 417–418
gases, 399–412
heat of transformation, 416–417
ideal-gas processes, 406–412
moles and molar mass, 398
phases of, 397, 414–418
thermal properties of, 396–440
Matter waves, 519, 997–1000
de Broglie wavelength, 997–998
electron microscope, 999–1000
interference and diffraction, 999
quantization of, 1000–1002
wave model, 519
Maxwell's theory of electromagnetism, 894, 954–955
Mechanical advantage, 258–259
Mechanical energy, 312–313, 340
elastic collisions and, 340
work as, 312–313
Mechanical waves, 519
Mechanics, 106
Medium, 519
of mechanical waves, 519
wave speed and, 521–523
Melting point, 416
Membrane potential, 739, 820
Metabolic energy use, 358–360
Microscopes, 670–672, 999–1000. *See also* Magnifiers
electron, 681, 999–1000
lenses, 670–671
magnification of, 670–672
resolution of, 679–681
scanning confocal, 691
scanning tunneling (STM), 1007
transmission electron (TEM), 999–1000
Microwaves, 531, 900–901
Miles per hour, 13
Mirrors, 625–626, 639–643
converging (concave), 639–642
diverging (convex), 639–640, 642–643
image formation with, 639–643
plane, 625–626, 640
ray tracing method, 639–643
reflecting telescopes, 674
reflection and, 625–626
sign conventions for, 645
spherical, 639–643
thin-lens equation using, 644–647
Mode number, 554, 558
Models, 7–8. *See also* Wave models
electric field, 710, 711–712
nuclear (Rutherford), 1022–1025
photon (EM waves), 587, 899–904, 993–995
ray (light), 587, 610, 621–624
wave, 519–520, 587, 605–606
Modes of standing waves, 554–556, 559–561
Modulation, of sound, 569–570
Molar mass, 398
Molar specific heat, 421–422
Mole (mol), 398
Molecular bonds, 110, 335–339
atomic models, 110
bond energy, 336
bond length, 335
chemical energy and, 337–339
energy diagrams for, 335–336
hydrogen, 704–705
Molecular energy levels, 1041–1042
Molecular mass unit (u), 397
Molecular speeds, 400–402
Molecules, 1041–1042
Moment arm (lever arm), 216
Moment of inertia, 223–227, 295–298
Momentum, 278–308. *See also* Angular momentum
change in, 281–284
collisions, 279–295
conservation of, 287–292, 971
cosmic speed limit, 972–973
explosions, 291–292
impulse and, 279–288
particle models of, 282–283
problem-solving strategies for, 285–287, 290
relativistic, 971–973
system definition for, 288–291
total, 284, 288–289
Momentum vectors, 280–282
Monatomic gases, 398
Moon
gravity on, 192–193
orbital motion of, 188
Motion, 5–6. *See also* Dynamics; Kinematics; Newton's laws; Rotational motion; Simple harmonic motion
causes of, 106–108. *See also* Forces
circular, 90–92
with constant acceleration (free fall), 47–50, 55–60
with constant velocity, 79
equations of, 34–35, 38, 47–50
graphical representation of, 33–37, 39, 41, 46, 49
one-dimensional (linear), 8, 32–70, 207
orbital, 187–190
particle model, 7–8
problem-solving strategies for, 51–55
projectile, 85–90
relative, 93–95
rolling, 231–233
strength vs. range of, 259–261
two-dimensional (planar), 70–104
uniform (constant velocity), 11–12, 37–40
units of, 14, 16–18
Motion diagrams, 5–6, 9–10, 75–79
frames, 5
operational definitions, 6
time progression, 9–10
vectors used on, 75–79
Motional emf, 881–884
MRI. *See* Magnetic resonance imaging (MRI) scans
Multi-electron atoms, 1034–1037
Muons, 966
Musical instruments, 557–558, 562–563, 564
Myopia, 664–665


## N


Natural abundance, of isotopes, 1056
Natural frequency, 504
Near point, 662–663
Nervous system, 819–825
Net force, 109, 118, 178–180, 183
Net torque, 218–219, 245–247
Net (total) work, 317
Neurons, 819–825
action potential, 821–823
nerve impulses, 823–825
Neutral buoyancy, 451
Neutral objects, charging of, 699, 702
Neutrinos, 1080
Neutron number *(N)*, 1025, 1057–1058
Neutrons, 703, 1024–1025, 1055, 1060–1062
beta decay, 1064–1065, 1066
diffraction pattern, 999
quarks, 1080–1081
range of, 1055
strong nuclear force of, 1060–1062
Newton (N), 119
Newton's first law of motion, 106–107
inertial reference frames, 952–953
Newton's law of gravity, 190–193
Newton's second law of motion, 117–119, 135, 178–181, 185–186
apparent weight and, 185–186
component forms of, 135–141
dynamics and, 138–141, 178–181
equilibrium and, 135–138
net force and, 178–180
problem-solving with, 135, 139, 227
rotational dynamics and, 223–224, 227–230
uniform circular motion and, 178–181
Newton's third law of motion, 122–125, 157–158, 267–292
action/reaction pairs of forces, 122–123, 157–158
conservation of momentum and, 267–292
propulsion and, 124
Nodes, 552–553, 558, 559–560
Nonequilibrium thermodynamics, 391
Nonohmic materials, 785
Normal forces, 110–111, 144–146, 157
Nuclear binding energy, 1056, 1058–1060
Nuclear data tables, A-4–A-6
Nuclear decay, 1062–1073
alpha, 1063–1064, 1066
beta, 1064–1065, 1066
decay series, 1066–1067
gamma, 1065–1066
half-lives, 1066–1073
time constant, 1070–1071
Nuclear energy, 311
Nuclear fission, 976, 1059–1060
Nuclear force, strong, 1060–1062
Nuclear fusion, 1059
Nuclear imaging and medicine, 1075–1077
Nuclear model (Rutherford), 1022–1025
Nuclear physics, 1054–1093
atomic structures, 1022–1025, 1055–1062
decay mechanisms, 1062–1073
medical applications, 1073–1078
nuclear stability, 1057–1060
nuclear structure, 1055–1057
quarks, 1080–1081
radioactivity, 1056, 1062–1068
strong and weak forces, 1060–1062
subatomic particles, 1022, 1079–1080
Nuclear stability, 1057–1060
Nucleons, 1055, 1060–1062
Nucleotides, 705

Nucleus, 703, 1023–1025, 1055–1062, 1063–1064. *See also* Nuclear decay
alpha decay, 1063–1064
daughter, 1063
high-Z, 1062
low-Z, 1061–1062
parent, 1063
quark model, 1080–1081
Rutherford's model, 1023–1025
strong nuclear force, 1060–1062
Numerical aperture (NA), 680

## O

Object distance, 625–626
Object plane, 635
Objects, *See* Interacting objects
Objective lens, 671
Ocean waves, 581
Ohm (Ω), 782
Ohmic materials, 785
Ohm's law, 784–787, 803
One-dimensional motion, 8, 32–70
acceleration, 43–50
circular motion quantity relations, 212–213
coordinate systems, 8–10
displacement, 10–11
free fall, 55–60
position-versus-time graphs, 33–37, 41-42
problem-solving strategy for, 51–55
rotational dynamics quantity relations, 225
uniform, 37–40
velocity-versus-time graphs, 36–37, 40, 42, 45–46
Operational definitions, 6
Opposite charges, 698. *See also* Electric dipoles
Optical axis, 631–632
Optical cavity, 1043
Optical instruments, 657–689
aberrations, 677–678
cameras, 658–660
magnifiers, 667–670
microscopes, 670–672, 679–681, 691
resolution of, 677–681
spectrometers, 1020
telescopes, 673–674
Optics, 586–620. *See also* Lenses; Ray optics; Wave optics
Orbital angular momentum, 1033–1034
Orbital motion, 187–189, 193–196
dark matter and, 273
free fall in, 188–189
gravity and, 193–196
of the moon, 189
planetary, 193–195
projectiles and, 188–189
weightlessness in, 189
Orbital quantum number, 1033
Order of diffraction, 595
Ordered systems, 391. *See also* Entropy
Order-of-magnitude estimate (~), 18
Origin, in coordinate systems, 8–9
Oscillating system, 7
Oscillation circuits, 932–935
Oscillations, 484–517. *See also* Simple harmonic motion; Waves
amplitude, 487, 502–503
damped, 502–504
driven, 504–507
equilibrium and, 485–486
frequency of, 485, 490, 496–497, 500, 501, 504–505, 917
pendulums, 486, 488–489, 499–502
period of, 485, 491, 496
resonance and, 504–507
spring systems, 486, 487–488, 494–499
Output power, 341–342

## P

Parabolic trajectory, 188
Parallel circuits, 805, 808–810, 811–816, 923–924
capacitors, 814–816
complex, 811–814
electrical outlets, 923–924
resistors, 808–810, 811–814
Parallelogram rule, 73
Parallel-plate capacitors, 714–716, 718, 744–746, 758–759
Parent nucleus, 1063
"Particle in a box" systems, 1000–1002
Particle model of motion, 7–8
Particle physics, 1079–1081
Particles (elementary), 7. *See also* Electrons; Neutrons; Protons
alpha, 1022–1024
antiparticles, 1079–1080
beta, 1022
energy quantization, 1000–1002
fundamental, 1081
matter waves of, 997–1000
neutrinos, 1080
quarks, 1080–1081
subatomic, 1022–1025, 1079–1081
wave-particle duality, 995, 1000–1002
Pascal (Pa), 152, 403, 443
Pascal's principle, 446
Passive diffusion, 428
Path-length difference, 566
Pauli exclusion principle, 1035
Pendulum systems, 486, 488–489, 499–502
frequency of, 500, 501
locomotion and, 500–502
oscillation of, 486, 488
restoring forces, 488–489
simple harmonic motion, 499–502
Perfectly elastic collisions, 340
Perfectly inelastic collisions, 293, 339
Period, 177–178, 485, 496
angular speed, 208–209
of circular motion, 177–178
frequency and,
of oscillation, 485, 496
of sinusoidal waves, 491
Periodic motion, 177
Periodic table of elements, 1035–1037, A-3
Permanent magnets, 844–845, 866–867
Permeability constant, 848
Permittivity constant, 715
PET scan, 1078
Phase, of waves, 566, 599
Phase changes, 397, 414–418
calorimetry, 418–420
evaporation, 417–418
fusion, 416–417
heat of transformation, 416–417
light waves, 599–600
temperature and, 415–416
vaporization, 416–417
Phase equilibrium, 416
Photodynamic therapy, 1044
Photoelectric effect, 989–995
characteristics of, 990
classical interpretation of, 990–991
Einstein's explanation of, 991–992
Einstein's postulates for light quanta, 993–994
electrons and, 989, 990–991
threshold frequency, 994
Photon model of EM waves, 587, 899–900, 993–995
Photon rate, 996
Photons, 587, 899, 992–997
absorption, 1025–1027, 1031–1032
detection of, 996–997
in electromagnetic waves, 899
emission, 996, 1025–1027, 1031–1032
energy of, 899, 992, 993
frequency of, 1025–1027
in quantum theory, 587, 992–995
Physical pendulums, 500–502
Pictorial representations, 51–52
Pigments, 677
Pinhole camera, 658
Pitch, 529, 563–565
Pivot point, 215–216
finding, 246–247
mechanical advantage of, 258–259
moment arm, 216
radial line, 215
rotational motion and, 215–216, 220
static equilibrium and, 245–247
Planck's constant, 899, 992
Plane mirrors, 625–626
Plane of polarization, 895
Plane waves, 531, 895–896
Planetary gravity, 192–193
Planetary motion, 193–196. *See also* Orbital motion
Pliant materials, 255
"Plum-pudding" model, 1022
Point charges, 706
Coulomb's law for, 706, 747
electric fields, 712–713, 718
electric potential of, 746–748, 750, 751
graphic representation of, 748
multiple, 751
potential energy of, 746–748
Point sources of light, 605
Poiseuille's equation, 461–463
Polarization
cellular, 718, 822
charge, 702, 718, 755–757
electromagnetic (EM) waves, 895, 896–898
of light, 896–898
Malus's law for, 897
potential difference and, 755–757
Polarization force, 702
Polarizing filters, 896–898
Poles, magnetic, 839
Population inversion, 1043
Position, 8–11
motion and, 8–11, 33–37, 40
sign conventions for, 33
uncertainty in, 1005–1007
Position-versus-time graphs, 33–37, 41-42, 489–491, 493
one-dimensional motion, 33–37, 41-42
simple harmonic motion, 489–491
two-dimensional motion, 87–90
Positron, 975, 1064
Positron-emission tomography (PET), 1078
Potential difference, 738–741, 776, 780–787
across batteries, 738, 758, 780–781
across biological systems, 738–739
across capacitors, 757–761
current and, 776, 780–787
across inductors, 929
in lightening strikes, 927
measurement of, 739
resistance and, 781–782
across resistors, 785–787
Potential energy, 310, 321–324
conservation of, 327–331
elastic (spring), 310, 323–324
electric. *See* Electric potential; Electric potential energy
gravitational, 310, 322–323
simple harmonic motion, 494–495
Potential-energy curve (PE), 332–333
Potential graphs, 745, 748
Power, 341–343, 788–790
average, 918–919
electric, 788–790
energy transformation for, 341–343, 788–790
output, 341–342
refractive, 660–661
resistors, 918–919
specific, 342–343
transmission, 922
units of, 341, 788
wave intensity and, 532–533, 535
Prefixes used in science, 17
Presbyopia, 663–664
Pressure, 402–405, 407–408
absolute, 403
atmospheric, 447–448
Bernoulli effect, 458

blood, 463–465
in circulatory system, 463–468
constant-pressure (isobaric) processes, 407–408, 421–422
differences in system components, 465–468
in fluid, 443–448, 458
fluid flow and, 455–468
in gases, 402–405, 407–408, 443–444
gauge, 403–404
hydrostatic, 444–447
in liquids, 444–448
measurement of, 403–404, 443–444, 447–448, 458
molecular collisions and, 402, 404
molar specific heat, 421–422
units of, 403, 443, 448
Pressure drop, 462–467
Pressure gauge, 447
Pressure gradient, 458
Pressure-volume *(pV)* diagrams, 406
Primary coil, 920–921
Principle of relativity, 953, 954–957
Einstein's, 954 –957
Galilean, 953
Principle of superposition, 550–551. *See also* Superposition
Principle quantum number, 1032
Prisms, 674–675
Problem-solving, overview, 51–55
Projectile motion, 5, 85–90, 186–189
acceleration and velocity vectors for, 86–87
apparent weight and, 186–187
free fall and, 188–189
launch angle, 86
orbital motion and, 187–189
problem-solving strategy for, 87–89
range of projectile, 89–90
Proper length, 968
Proper time, 964–965
Proportional relationships, 39
Proportionality constant, 39
Propulsion, 124, 292
Protons, 703, 764, 1024–1025, 1055, 1060–1062
beta decay, 1064–1065, 1066
electric charge and, 703, 743
fusion, 764
quarks, 1080–1081
strong nuclear force, 1060–1062
Pulleys, 160–163
*pV* (pressure-volume) diagrams, 406

## Q

Quadratic drag, 152
Quadratic relationships, 50
Quanta of light, 587
Quantization, 992. *See also* Quantum jumps; Quantum mechanics
of angular momentum, 1033–1034
atomic, Bohr's models and, 1025–1027
of energy, 992, 1000–1004, 1025–1027, 1030–1031, 1033–1034
of hydrogen spectrum, 1031–1032
of matter waves, 1000–1002
photoelectric effect and, 992
Schrödinger's quantum theory, 1002–1004
Quantum jumps, 1002–1004, 1025–1027
Quantum mechanics, 1002–1004, 1032–1034. *See also* Atomic physics; Quantum physics
absorption spectra, 1004
atomic structure and, 1032–1034
emission spectrum, 1003, 1039–1040
energy-level diagrams for, 1002–1004
excited states, 1003, 1037–1040
ground state, 1003
hydrogen atom, 1032–1034
lasers and stimulated emission, 1042–1044
molecules, 1041–1042
multi-electron atoms, 1034–1037
"particle in a box" systems, 1000–1002
Pauli exclusion principle, 1035
Quantum numbers, 1001, 1025–1027, 1032–1033, 1035–1037
Quantum physics, 986–1018. *See also* Atomic physics; Quantization; Quantum mechanics
antimatter, 975, 1079–1080
Bohr atom, 1025–1032
de Broglie wavelength, 997–998
energy quantization, 992, 1000–1004
energy-level diagrams, 1003–1004
Heisenberg uncertainty principle, 1005–1007
matter waves, 997–1000
photoelectric effect, 989–995
photons, 992–997
tunneling, 1007
wave-particle duality, 995, 1000–1002, 1008
x rays and x-ray diffraction, 987–989
Quarks, 1080–1081

## R

Radial line, 215
Radian (rad), 206
Radiation, 425–426, 1062–1073
alpha, 1063–1064, 1066
atomic, 901
beta, 1064–1065, 1066
Čerenkov, 1082
decay series, 1066–1067
electromagnetic, 901–903
gamma, 904, 1065–1066
global warming from, 945
half-lives, 1066–1073
heat transfer, 425–426
infrared, 901, 902
ionizing, 904, 1067–1068
medical applications, 1073–1078
microwave, 531, 900–901
nuclear decay, 1062–1073
sources and effects of, 1074–1075
spectra, 900–903
thermal, 425–426, 901–902
ultraviolet, 904
x rays, 904
Radiation dose, 1073–1075
Radio waves, 900–901
Radioactive decay
activity of samples, 1071–1072
alpha, 1063–1064, 1066
beta, 1064–1065, 1066
decay series, 1066–1067
gamma, 1065–1066
half-lives, 1066–1073
muons and, 966
pi ($\pi$) mesons, 956
Radioactivity, 1056, 1062–1068
Radiocarbon dating, 1072–1073
Rainbows, 676
"Raisin-cake" model, 1022
Ramps, motion on, 79–82, 145–146
Random walk, 426–427
Range, of projectiles, 89–90
Rarefaction, 529, 558–559
Ratio reasoning, 39
Ray diagrams, 622
Ray model of light, 587, 610, 621–624
Ray optics, 587, 620–656. *See also* Lenses; Reflection; Refraction
dispersion, 674–677
mirrors, 625–626, 639–643
resolution, 677–681
thin lenses, 633–639
Ray tracing, 633–643
lenses, 633–639
mirrors, 639–643
Rayleigh's criterion, 679–680
*RC* circuits, 816–819
Reaction coordinates, 337
Reaction energy diagrams, 337–338
Real images, 634–636
Red shift, 538
Reference frames, 951–953, 962–963
Reflecting telescope, 674
Reflection (light), 624–626
diffuse, 523, 626
fiber optics, 630–631
law of, 624
mirrors, 626–627, 639–643, 674
specular, 624
thin-film interference, 599–605
total internal, 629–631
Reflection (mechanical waves), 553–554
at a discontinuity, 554
frequency shift, 538–539
of sound waves, 538–539, 553–554
of standing waves on a string, 553–554
Reflection grating, 597–598
Reflective objects, 622
Refracting telescope, 674
Refraction, 588–589, 627–632
dispersion, 675–676
fiber optics, 630–631
image formation by, 631–632
index of, 588–589, 627–629
rainbows, 676
Snell's law of, 627
total internal reflection and, 629–631
Refractive power, 660–661
Refrigerators, 372–374
Relative biological effectiveness (RBE), 1074
Relative motion, 93–95
Doppler effect, 537–538
Galilean relativity, 951–954
Relative velocity, 93–94
Relativistic energy, 973–976
Relativistic momentum, 971–973
Relativity, 950–985
clock synchronization, 957, 958–959
conservation of energy and, 976
cosmic speed limit, 972–973
Einstein's theory of, 951, 954–957
energy and, 973–976
events in, 957, 959
Galilean, 951–954
length contraction, 968–970
Lorentz velocity transformations, 970–971
measurements and, 957–960
momentum and, 971–973
reference frames, 951–953, 970–971
simultaneity in, 959–962
space-time coordinates, 957
special, 951, 970–971
simultaneity, 959–962
time dilation, 962–967
velocity transformation, 953–954
rem, 1074
Repolarization (cellular), 718, 822
Resistance, 774–799, 802–812
current and, 774–799
equivalent, 805–812
measurement of, 783–784, 787
potential difference and, 781–782
Resistivity, 782–783
Resistors, 784–787, 917–919
AC power in, 918–919
combined, 811–814
ideal-wire model, 786–787
Ohm's law for, 784–787, 803
parallel, 808–810, 811–814
potential difference, 785–787
problem-solving strategy for, 812–813
series, 805–807, 811–814
Resolution, 677–681
angular, 679
microscopes, 679–681
optical instruments, 677–681
wave nature of light, 678–679
Resolved vector, 75
Resolving power (RP), 680
Resonance, 504–506, 556
in AC circuits, 934–935
driven oscillations and, 504–506
in hearing, 506, 561–562
standing-wave, 556, 561–562

Resonance frequency, 505–506, 934
Resonant modes, 556
Response curve, 505
Rest energy, 974
Rest frame, 962
Resting potential, 820, 822
Restoring forces, 252–257, 485, 487–489
  compressing materials, 255, 257
  elasticity and, 252–257
  equilibrium and, 252–257, 485, 487–489
  Hooke's law for, 252–254, 256
  linear, 487–489, 499
  oscillation and, 485, 487–489
  pendulums, 486, 488–489, 499
  springs, 487–488
  stretching materials, 255–256
Resultant vector, 73
Reversible process, 374
Revolution (rev), 177, 206
Reynolds number (Re), 151–153, 154–156, 157, 455. *See also* Drag force
Right-hand rules
  magnetic fields, 845, 846
  magnetic forces, 852–853, 858–859
Rigid bodies, rotational motion of, 195, 211–215
Rigid materials, 255
*RLC* circuits, 933–935
Rocket propulsion, 124, 292
Rolling constraint, 231
Rolling friction, 148–149
Rolling motion, 211–215, 231–233, 320–321
Root-mean-square current, 918–919
Root-mean-square distance, 427
Root-mean-square speed, 400–401
Root-mean-square voltage, 918–919
Rope and pulley systems
  constraints due to, 229–230
  massless approximation, 230
  tension, 159–163
Rotational dynamics, 223–230
  linear motion quantity relations, 225
  moment of inertia and, 223–227
  Newton's second law for, 223–224, 227–230
  problem-solving strategy for, 227
Rotational kinetic energy, 318, 320–321
Rotational motion, 5, 195, 206, 211–215
  angular acceleration, 212–214, 228–229
  fixed-axis, 227–230
  graphing of, 213–214
  gravity effects on, 195
  moment of inertia and, 223–227
  Newton's second law for, 224–225, 227–230
  rigid body, 195, 211–215
  rolling (combination motion), 211–215, 231–233
  tangential acceleration, 214–215, 229–230
  torque, 215–219

## S

Saltatory conduction, 824
Satellites, 194
Scalar quantities, 19
Scanning confocal microscope, 691
Scanning tunneling microscope (STM), 1007
Scattering, of light, 623
Schrödinger's quantum theory, 1002–1004, 1032–1033. *See also* Quantum mechanics
Scientific notation, 14, 15–16
Screening, 720
Sea of electrons, 704
Second law of thermodynamics, 370–371, 372, 374–377
Secondary coil, 920–921
Secondary maxima, 605, 607–608
Seivert (Sv), 1074
Selection rules, 1038
Self-luminous objects, 622
Self-organizing systems, 391
Sensor elements, 785. *See also* Resistors
Series circuits, 805–807, 811–816
  capacitors, 814–816
  complex, 811–814
  resistors, 805–807, 811–814
Shadows, 623–624
Shell electron configurations, 1035–1037
Shock waves, 536–537, 539–540
SI (metric) units, 17. *See also* Units of measurement
  of acceleration, 44, 57
  of angles, 206
  of angular acceleration, 212
  of angular momentum, 296
  of angular speed, 208–209
  of angular velocity, 207
  of capacitance, 757–758
  of charge, 703
  of conduction, 423–424
  converting to English units, 17–18
  of current, 777–778
  of density, 152, 442
  of electric potential, 737, 742–744
  of energy, 319, 359, 742–734
  of energy transitions, 1038
  of force, 119
  of frequency, 177, 485
  impulse, 279
  of inductance, 930
  of intensity, 532
  of kinetic energy, 319
  of magnetic field, 847
  of magnetic flux, 885
  of mass density, 442
  of momentum, 281
  of power, 341, 788
  of pressure, 403, 443, 448
  of radiation dose, 1074
  of radioactive sample activity, 1071
  of resistance, 782
  of speed, 13
  of stress, 255
  of temperature, 365
  of torque, 216
  of velocity, 13, 34
  of viscosity, 152, 461
  of volume, 398
  of work, 314
Significant figures, 14–15
Simple harmonic motion, 486–499
  amplitude of, 496–497
  circular motion and, 491–494
  energy in, 494–499
  frequency of, 496–497, 500, 501
  kinematics of, 489–494
  linear restoring force, 487–489
  oscillation in, 486
  pendulum systems, 486
  sinusoidal waves, 486, 487, 489–493
  spring systems, 486, 494–499
Simultaneity, 959–962
Single-slit diffraction, 605–609
Sinusoidal functions and relationships, 486, 489–491
Sinusoidal waves, 486, 487, 489–493, 525–528. *See also* Simple harmonic motion
Slipping/sliding motion, 146–151, 164
Slopes
  motion on, 79–82, 145–146
  position-versus-time graphs, 35–38
  velocity-versus-time graphs, 41, 45
Small-angle approximation, 488
Snapshot graphs, 524–528
Snell's law, 627
Snell's window, 630
Solar cells, 996
Solenoids, 844, 929
  induced electric fields, 893–895, 929–930
  length of, 850–851
  magnetic fields of, 844, 847, 850–851, 893
  in magnetic resonance imaging, 850–851
Solids, 397
  atomic model of, 397
  phase changes, 415–416
  specific heat of, 414
Sonic boom, 539
Sound. *See also* Sound waves
  beats, 569–570
  decibel scale (dB), 534–535
  formants, 564–565
  frequency spectrum, 564
  intensity level (loudness, dB scale), 533–536, 562
  musical instruments, 557–558, 562–563, 564
  pitch, 529, 563–565
  resonance, 506, 561–562
  speech and hearing, 506, 561–562, 563–565
  timbre, 564
  tone quality, 557, 564
  vowels, 564–565
Sound waves, 521, 528–530, 532–540
  compression, 529, 558–559
  Doppler effect, 536–540
  in gases, 522–523
  harmonics of, 556–557, 564–565
  hearing and, 529, 561–562, 564
  interference among, 565–569
  from moving source, 537–539
  fundamental frequency, 556–557
  medium properties of, 521–523
  modes, 559–561
  modulation, 569–570
  musical instruments, 557–558, 562–563
  nodes, 559–560
  principle of superposition, 550–551
  problem-solving strategy for, 561
  rarefactions, 529, 558–559
  reflection of, 538–539, 553–554, 559
  speed of, 521–523
  standing, 558–563
Source charges, 711
Spacetime coordinates, 957
Special relativity, 951, 970–971
Specific heat, 414, 421–422, 479
Specific metabolic rate, 479
Specific power, 342–343
Spectra. *See also* Electromagnetic spectrum
  absorption, 1004, 1021, 1025–1027, 1031–1032, 1038
  atomic, 1020–1021
  Balmer formula, 1021
  Balmer series, 1032
  Bohr's model of atom and, 1025–1027
  continuous, 1020
  discrete, 1020–1021
  electromagnetic, 894, 899–904
  emission, 1003, 1020–1021, 1025–1027, 1031–1032, 1039–1040
  excited states and, 1037–1040
  of hydrogen atom, 1031–1032
  Lyman series, 1032
  quantization and, 1003–1004
  of visible light, 675, 900, 1020
Spectral line, 1020
Spectrometer, 1020
Spectrophotometer, 597
Spectroscopy, 596–597, 1020–1021
Specular reflection, 624
Speech, 563–565
Speed, 11–13, 34–35
  angular, 208–209, 210
  of circular motion, 177–178, 186
  curve modifications for, 182–183
  equation of continuity, 455–456
  of fluid flow, 456–463
  of light, 523, 588, 955–957, 972
  maximum walking, 183
  Poiseuille's equation for, 461–463
  root-mean-square, 400–401
  of sound, 521–523, 537–538, 954
  of uniform motion, 11–12, 182–184
  volume flow rate, 456
  wave, 519–520, 521–523, 527–528
Spheres, potential energy of charged, 749–750
Spherical mirrors, 639–643
Spherical waves, 531–533, 535
Spin angular momentum, 1034
Spin quantum number, 1033, 1034, 1036

Spontaneous emission, 1042
Spontaneous reactions, 379–380
Spring constant, 252–253
Spring force, 110, 252–254
Springs, 110, 252–254
elasticity of, 252–254
frequency of, 496–497
Hooke's law for, 253–254
massless, 254
motion of mass on, 494–495
oscillation, 486, 487–488
potential energy of, 323–324
restoring force, 252–254
simple harmonic motion of, 486, 494–499
Square-root relationships, 401
Stability
balance and, 250–251
nuclear, 1057–1060
Stable equilibrium, 250–251, 334–335
Standard atmosphere (atm), 403
Standing waves, 549–583
antinodes, 552–553, 559–560
creating, 554–556
electromagnetic, 558
interference of, 550–551, 553, 565–569
modes, 554–556, 559–561
nodes, 552–553, 558, 559–560
phases of, 555
problem-solving strategy for, 561
sound, 558–563
on a string, 520–521
transverse, 558
wavelength, 554–555
Static equilibrium, 135, 136–137, 245–249
center of gravity for, 247–249
problem-solving strategies for, 135, 247
pivot point for, 245–247, 249
torque of, 245–249
Static friction, 112, 124, 146–147, 149, 157
Stationary states, 1026–1027, 1028–1031
Stefan-Boltzmann constant, 901
Stefan's law, 425, 901
Stimulated emission, 1042
Stoke's law, 154
Stopping potential, 990
Straight-line motion, 5, 8–11, 38. *See also* One-dimensional motion
Strain, 255–256
Streamlines, 457
Strength, range of motion vs., 259–261
Stress, 255–258
Stretching, 255–258
elastic limit, 256–257
elastic potential energy and, 323–324
tensile strength, 256–258
Young's modulus for, 255–256, 257
Strong nuclear force, 1060–1062
Subatomic particles, 1022–1025, 1079–1081
Subshells, 1035–1037
Superposition, 549–583
beats, 569–570
electromagnetic waves, 558
interference and, 550–551, 565–569
magnetic fields, 848–849
principle of, 550–551
Supersonic speeds, 539
Synchronization, 957, 958–959
Systeme Internationale d'Unitès. *See* SI (metric) units

## T

Tangential acceleration, 214–215, 229–230
Telescopes, 673–674
Temperature, 364–367
absolute zero, 365
atomic view of, 364
Boltzmann's constant for, 399
constant-temperature (isothermal) processes, 408–409
heat of transformation and, 416–417
molecular speeds and, 400–402
phase changes and, 415–416
pressure and, 405
specific heat, 414
thermal energy and, 364
Temperature scales, 365
Tensile strength, 256–258
Tensile stress, 255
Tension force, 110
atomic model, 110
rope and pulley systems, 159–163
Terminal speed, 153–154
Tesla (T), 847
Thermal conductivity, 423–424
Thermal energy, 311, 325–326, 364–367. *See also* Conservation of energy; Heat engines; Thermodynamics
atomic view of, 364
calorimetry, 418–420
charged particles and, 742
current and, 775–776
efficiency of, 326
electric potential energy and, 742
entropy and, 375–380
heat transfer of, 366–367
friction and, 325–326
macroscopic interactions, 375
temperature and, 364
work from, 325–326
Thermal equilibrium, 366–367
Thermal expansion, 412–413
Thermal properties of matter, 396–440
diffusion, 426–429
evaporation, 417–418
gases, 399–412
heat of transformation, 416–417
heat transfer, 422–426
phase changes, 397, 414–418
problem-solving strategy for, 419
specific heat, 414–418, 421–422
thermal conductivity, 423–424
Thermal radiation, 425–426, 901–902
Thermistors, 787
Thermodynamics, 367–377. *See also* Heat engines
entropy and, 375–380, 391
first law of, 367–369
heat engines, 369–372
heat pumps, 372–374
ideal gas processes and,
second law of, 370–371, 372, 374–377
Thermometers, 365
Thin lenses, 633–639. *See also* Lenses; Magnification
converging, 633–638
diverging, 633, 638–638
image distance, 637–639
Thin-film interference, 599–605
Thin-lens equation, 644–647
Thomson's "plum-pudding/raisin-cake" model, 1022
Threshold frequency, 990, 994
Thrust, 112–113, 124
Tilted axes coordinates, 79
Time, 8, 9–10, 957–965
dilation, 962–967
events and, 957–962, 964–965
proper, 964–965
simultaneity, 959–962
Time constant, 502–503, 817–819, 1070–1071
Time intervals, 10–11
Tip-to-tail rule, 73
Tone quality, 557
Torque, 215–219. *See also* Rotational motion
center of gravity and, 247–249
current loops and, 861–863
electric dipoles and, 722–723
in electric fields, 721–723
gravitational, 219–223
in the human body, 259–261
magnetic dipoles and, 840, 861–865
net, 218–219, 245–247
pivot point, 215–216, 220, 245–247, 249, 258–261
static equilibrium and, 245–249
Total energy, 310–311, 313, 331, 973–974
Total internal reflection, 629–631
Total momentum, 284, 288–289
Trajectory, 5. *See also* Orbital motion
Transcranial magnetic stimulation (TMS), 892
Transformers, 919–923
Transitions. *See* Quantum jumps
Translational kinetic energy, 318–319
Translational motion, 211
Transmission electron microscope (TEM), 999–1000
Transmission (diffraction) grating, 594–596
Transverse waves, 520, 558, 893
Traveling waves, 519, 520–524. *See also* Electromagnetic waves; Sound Waves
amplitude of, 525–528
circular, 531
displacement of, 525–528
Doppler effect, 536–540
electromagnetic, 523, 530–531
energy transfer by, 519, 532–533
frequency of, 526–528
graphs of, 524–529
intensity of, 532–535
light. *See* Light waves
microwaves, 531
plane, 531
shock, 537–537, 539–540
sinusoidal, 525–528
sound, 521, 529–530, 532–540
speed, 519–520, 521–523
spherical, 531–533
on a string, 520–521
wave models, 519–520
wavelength of, 525–528
Tsunamis, 581
Tube analogy, sound waves, 558–562
Tunneling, 1007
Turbulent flow, 455, 466
Turning point, 489
Twin paradox, 966–967
Two-dimensional motion, 70–104, 294–295
circular, 90–92
collisions, 294–295
on inclined planes, 79–82
projectile, 85–90
relative, 93–95

## U

Ultrasound, 529–530, 439
Ultraviolet radiation, 904
Uncertainty principle, 1005–1007
Uniform circular motion, 90–91, 176–184
angular displacement and velocity, 207–209
angular position, 206
centripetal acceleration in, 176
curve modifications for, 182–183
dynamics of, 178–184
frequency and speed, 177–178
friction force and, 183
maximum walking speed, 183
period of, 177–178
simple harmonic motion and, 492–494
velocity in, 176
Uniform electric fields, 714–716
Uniform magnetic fields, 847, 861–863
Uniform motion, 11–12, 37–40
Units of measurement, 14, 16–18, 44, 57. *See also* SI (metric) units
conversions, 17–18, 448
prefixes, 17
Unstable equilibrium, 250–251, 334
Unstable nuclei, 1057–1060
Upright image, 637

## V

Vacuum, 403
Valence electrons, 704
Van de Graaff generator, 925
Vaporization, heat of, 416–417
Vector quantities, 19
Vector sum, 20, 118

Vectors, 19–23, 70–104
acceleration, 46–47, 83–85, 86–87, 90–91
addition of, 20, 72–73, 77
components of, 75–79
coordinate systems for, 74
decomposed (resolved), 75–76
displacement, 19–20, 72
electric field, 712–713, 714–718
force, 107, 108
magnetic field, 841–842
magnitude of, 19, 72, 73
momentum, 280–282
multiplication of, 73, 78
resultant, 73
subtraction of, 73–74, 78
trigonometry and, 20–22
two-dimensional motion, 70–104
velocity, 22–23, 34, 43, 72, 82–83, 85, 86–87
Velocity, 11–13, 34–37, 41, 90–92
angular, 207–209, 211
average, 12–13, 41
calculating graphically, 35–37, 41–43
in circular motion, 90–92, 176, 188–189, 207–209
drift, 776
escape, 747–748
Galilean transformation of, 953–954
instantaneous, 13, 41–43
Lorentz transformation of, 970–971
in projectile motion, 86–87
relative, 93–94
sign conventions for, 34
uniform circular motion and, 176
Velocity vectors, 22–23, 34, 43, 72, 82–83, 85, 86–87, 90–91
Velocity-versus-time graphs, 36–37, 40, 42, 45–46, 209–210, 489–491, 493
Venturi tube, 458
Virtual images, 625, 637–638
Viscosity, 151–152, 455, 461–463
drag and, 151–152
Poiseuille's equation and, 461–463
Reynolds number (Re), 455
units of, 152
Visible light, 675, 899, 900, 903–904, 1020. *See also* Light
Visible spectrum, 588, 900
Vision, 602–605, 622–623, 660–668. *See also* Lenses; Reflection; Refraction
accommodation, 662–663
angular and apparent size, 668
color sensitivity, 603–605, 904
focusing, 662–663
human eye functions, 660
lenses for correction of, 660–661, 664–666
in low-light conditions, 602
surgical correction of, 666–667
thin-film interference and, 602–605
Visual overview, 52–53
Volt (V), 737
Voltage, 738. *See also* Potential difference
AC circuits, 917–923
capacitors, 927–929
danger of high, 739
inductors, 929–932
peak, 927, 930–931
resistors, 810–811, 917–919
root-mean-square, 918–919
transformers, 919–923
Voltage drop, 786, 803
Voltmeters, 739, 810
Volume, 398–399, 406–407
constant processes, 406–407, 421–422
conversion factors, 398–399
fluid element, 457
macroscopic system, 398
molar specific heat, 421–422
Volume flow rate, 456
Vowel sounds, 564–565


## W


Walking gait, 183
Water
hydrogen bonds in, 704–705
phase changes of, 415–416
specific heat of, 414
waves (tsunami), 581
Watt (W), 341
Wave equations, 527
Wave fronts, 531
Wave models, 519–520, 587, 605–606
Wave optics, 586–619
circular-aperture diffraction, 609–610
diffraction, 587, 594–599, 605–610
double-slit interference, 590–599, 675, 995
Huygens' principle, 605–606
photon model, 587
ray model, 587
reflection grating, 597–598
single-slit diffraction, 605–609
spectroscopy, 596–567, 1020–1021
thin-film interference, 599–605
transmission (diffraction) grating, 594–596
wave model, 587, 605–606
Wave pulses, 520–521
Wave speed, 519–520, 521–523
Wavelength, 525–528
de Broglie, 997–998
dispersion and, 675
electromagnetic waves, 894, 900
light waves, 589, 593, 675, 678
matter waves, 997–1000
standing waves, 554–555
traveling waves, 525–528
Wave-particle duality, 899–900, 995, 1000–1002, 1008
Waves 518–548. *See also* Amplitude; Electromagnetic waves; Light waves; Matter Waves; Simple harmonic motion; Sound Waves; Standing waves; Traveling waves
antinodes, 552–553, 559–560
circular, 531
energy transfer by, 532–533
frequency shift, 538–539
graphical representation of, 486, 489–494
history graphs, 524–525
Huygens' principle, 605–606
intensity of, 532–535, 896, 990–912. *See also* Intensity of waves
interference, 550–551, 553, 565–569
longitudinal, 520
mechanical, 519
nodes, 552–553, 558, 559–560
phases of, 566
plane, 531
sinusoidal, 486, 489–491, 525–528
snapshot graphs, 524–528
speed of, 519–520, 521–523, 527–528
spherical, 531–533
superposition of, 549–583
transverse, 520
Waves on a string, 520–521
creating, 554–556
discontinuity, 554
electromagnetic, 558–559
fundamental frequency, 556–557
higher harmonics, 556–557
musical instruments, 557–558
reflection of, 553–554
standing, 553–559
traveling, 520–521
Weak nuclear force, 1060–1062
Weber (Wb), 885
Weight, 109, 141–144, 157
apparent, 142–144, 184–186
circular motion and, 184–186
free fall and force of, 141
gravity and, 191–193
measurement of, 141–142
vs. mass, 141–142
Weightlessness, 144, 189
Wein's law, 902
Wires. *See also* Current-carrying wires; Current loops; Solenoids
magnetic fields produced by, 844–847, 847–848
magnetic forces between currents, 859–861
magnetic forces on currents, 858–859
Work, 312–313, 314–318. *See also* Conservation of energy; Thermal energy
angular displacement and, 315
calculating, 316
conservation of energy and, 313, 327–331
displacement and, 314–318
electric potential and, 735
energy transfer process, 312–313
in first law of thermodynamics, 367–369
forces and, 314–318
ideal gas processes, 409–411
kinetic energy and, 318–321
potential energy and, 321–324
power and, 341–343
thermal energy and, 325–326
total (net), 317
Work-energy equation, 313, 326, 327, 331, 367
Work function, 991


## X


X rays, 904, 987–989
biological effects of, 1073–1074
Bragg condition for, 988–989
creating an image, 987–988
diffraction pattern, 999
excitation by, 1040
X-ray diffraction, 988–989


## Y


Young's double-slit interference experiment, 590–593, 675, 995
Young's modulus, 255–256, 257


## Z


Zero, absolute (temperature), 365
Zero of electric potential energy, 737
Zero vector, 73